American College of Veterinary Internal Medicine Forum 2010

Anaheim, California, USA
9-12 June 2010

Editors:

Dianne McFarlane **Jeff Toll**

ISBN: 978-1-61738-830-9

Printed from e-media with permission by:

Curran Associates, Inc.
57 Morehouse Lane
Red Hook, NY 12571

Some format issues inherent in the e-media version may also appear in this print version.

Printed by Curran Associates, Inc. (2010)

For permission requests, please contact the American College of Veterinary Internal Medicine at the address below.

American College of Veterinary Internal Medicine
1997 Wadsworth Boulevard Suite A
Lakewood CO 80214-5293

Phone: (800) 245-9081
Fax: (303) 231-0880

acvim@acvim.org

Additional copies of this publication are available from:

Curran Associates, Inc.
57 Morehouse Lane
Red Hook, NY 12571 USA
Phone: 845-758-0400
Fax: 845-758-2634
Email: curran@proceedings.com
Web: www.proceedings.com

ACVIM SPECIALTY SYMPOSIUM (PRE-FORUM) PROCEEDINGS

Cardiology

Large Animal Internal Medicine - Equine

Large Animal Internal Medicine - Food Animal

VETERINARY SCIENTIFIC PROGRAM PROCEEDINGS

AAVN / ACVN

ACVCP / AAVPT

ACVIM Business Management

ACVIM Cardiology

Session# **Title / Author** **Page#**

ACVIM Equine

ACVIM Food Animal Internal Medicine

ACVIM Neurology

ACVIM Oncology

ACVIM One Medicine

ACVIM Small Animal Internal Medicine

ASCVP

AVHTM

CGS

ISCAID

Liver Study Group

SCE

SVN/U

VCRS (SA & Equine)

VECCS

VUS

TECHNICIAN PROGRAM PROCEEDINGS

WELCOME TO THE 2010 ACVIM FORUM IN ANAHEIM, CA

Welcome to Anaheim! I am pleased and honored that you have chosen to attend the 2010 ACVIM Forum, the premier veterinary internal medicine continuing education meeting! Much of the credit for this distinction goes to the ACVIM Forum Program Committee as well as the many specialists who deliver state-of-the-art clinical science and veterinary internal medicine advances to our attendees.

This year's agenda is full, with a core program of approximately 280 scientific lectures, more than 350 research abstracts, and a technician lecture series. All sessions of the core program are open to every attendee. Thursday through Saturday you'll find a variety of special interest group (SIG) sessions (workshops, case discussions, and labs) that address diverse topics in veterinary internal medicine. Many of these sessions are available in the evenings. Veterinary technicians will be pleased to see a scientific program and workshops designed specifically for them. Veterinary technicians also are welcome to attend other scientific sessions.

Make plans to attend the Opening Night Reception, sponsored by Hill's Pet Nutrition and the ACVIM on Thursday evening, and the 2010 Bayer Dance Party on Friday evening. On Friday morning, all attendees are invited to the Keynote Address: Extra-Ordinary Veterinarian: Dr. Paul Pion, DACVIM (Nominated by ACVIM Diplomates to inspire attendees with the story of his unusual career path). After the Keynote address, please stay for the Small Animal or Large Animal Consensus Statement DRAFT session.

Research posters are on display throughout the meeting. Content may be discussed with the authors on Friday evening during the Wine and Cheese Reception sponsored by Nestlé Purina.

There is an abundance of quality session topics on the daily scientific sessions grid. However, if you are unable attend a session of interest the Proceedings will help to fill in the gaps.

Make sure to leave plenty of time to meet with our many exhibitors in the Exhibit Hall. You will be surprised by the number and diversity of companies that come to bring you the very latest in veterinary technology, pharmaceutical advances, superior nutritional products, educational materials, and career opportunities.

The ACVIM is indebted to the 2010 ACVIM Forum Program Committee (chaired by Dr. Marlene Hauck) for the incredible effort it takes to put together a program of this magnitude. We also want to reiterate our thanks to our many speakers, facilitators, and moderators who are willing to share their knowledge, talent, and time. We are very grateful to our sponsors for their recognition, presence, and support of our efforts to provide an unequaled educational experience for all attendees. The ACVIM Forum continues to grow substantially in both size and quality due to a wide range of sponsor support.

The dedicated ACVIM staff members have handled all of the organizational details needed to bring this meeting to you, and are available at the registration desk to help you with any special needs or requests. Please take a few minutes to personally thank them for a job well done.

The members of the ACVIM Board of Regents realize that you have choices regarding where to spend your continuing education dollars, and we appreciate the fact that you have chosen to join us in Anaheim. We hope that you have an outstanding educational experience, and that your attendance at the 2010 ACVIM Forum will facilitate your continued commitment to the highest quality of service in veterinary practice, academia, or industry. Mark you calendars now to join us for the 2011 ACVIM Forum, June 15-18, in Denver, Colorado.

Wishing you a stimulating and productive meeting,

Dr. Richard LeCouteur, BVSc, PhD, DACVIM (Neurology), DECVN
Chair, ACVIM Board of Regents

LETTER FROM THE EDITORS

We are pleased to be able to offer the Proceedings in CD-ROM format and online. You may purchase additional copies by contacting the ACVIM office. The unit cost is $49.00 (US) for a CD-Rom, which includes domestic shipping (additional charges apply with International shipping including Canada). Payment must accompany your order. To order by mail, send your check or money order (US funds only) to the ACVIM, 1997 Wadsworth Blvd., Suite A, Lakewood, CO 80214-5293. To place your order using Visa, MasterCard or American Express, call the office at 800-245-9081, 303-231-9933, or e-mail ACVIM@ACVIM.org.

Format

The manuscripts are arranged according to specialty group, then by program number. This organizational format was developed in response to overwhelming requests by prior attendees. The sections designated in the Table of Contents correspond to the specialty headings in the grid. All programs for the Liver Study Group are listed together, by program number; all programs for the ACVIM Large Animal Internal Medicine Group are listed together, by program number; etc. The research abstracts, author index, and key word index are also listed in the Table of Contents with their corresponding page numbers.

We acknowledge that not all manuscripts presented for publication conform to the guidelines distributed (e.g. outlines and paper length). We are thankful that the majority of speakers were able to keep their paper within the length and format specified.

Acknowledgements

These Proceedings represent a collaborative effort by a number of people. We greatly appreciate the time and commitment of the speakers that make these Proceedings possible.

Dianne McFarlane
Diplomate ACVIM (Large Animal Internal Medicine)
Co-Editor

Jeff Toll
Diplomate ACVIM (Small Animal Internal Medicine)
Co-Editor

Jess Green
ACVIM Forum Event Planner & Publications Manager

THANK YOU TO THE 2010 ACVIM FORUM SPONSORS

DIAMOND SPONSORS > $150,000

PLATINUM SPONSORS $75,000 - $149,999

Hill's™

GOLD SPONSORS $50,000 - $74,999

Bayer HealthCare Animal Health

INFINITI MEDICAL™ Advanced Technology for Vets and Pets

Nestlé PURINA

Pfizer Animal Health

SILVER SPONSORS $25,000 - $49,999

ROYAL CANIN VETERINARY DIET

BRONZE SPONSORS $10,000 - $24,999

Elanco

Intervet Schering-Plough Animal Health

MERIAL

VIN

GENERAL PROGRAM SUPPORT < $10,000

ALOKA

AnimalScan

Boehringer Ingelheim

BrightHeart Veterinary Centers

CareCredit

Core Ultrasound

IDEXX Laboratories

Karl Storz Veterinary Endoscopy

Live Oak Bank

Novartis

Sound-Eklin

SoundVision

Universal Ultrasound

Vetel Diagnostics

VetImaging

Virbac Animal Health

MEDIA SPONSORS & IN-KIND PARTNERS

International Veterinary Information Systems (IVIS)

Journal of Veterinary Internal Medicine (JVIM)

The Horse

Veterinary Cancer Society (VCS)

SPEAKER DISCLOSURE INDEX
VETERINARY PROGRAM

Session	Speaker	Disclosure
PFC01	Kathryn Montgomery Meurs	Co-investigator on a patent held by Ohio State Univeristy and Washington State University for genetic testing for Maine Coons and Boxers
PFC01	D. David Sisson	Nothing Disclosed
PFC01	William P. Thomas	Nothing Disclosed
PFC02	Robert L. Hamlin	Consultant to interpret ECGs, help design studies, evaluate data
PFC03	Robert L. Hamlin	Consultant to interpret ECGs, help design studies, evaluate data
PFC04	Matthew W. Miller	Nothing Disclosed
PFC05	Ronald G. Grifka	Nothing Disclosed
PFC06	Ronald G. Grifka	Nothing Disclosed
PFC06	Daniel F. Hogan	Nothing Disclosed
PFC06	Matthew W. Miller	Nothing Disclosed
PFC06	E. Christopher Orton	Nothing Disclosed
PFC06	D. David Sisson	Nothing Disclosed
PFC06	Anthony H. Tobias	Nothing Disclosed
PFC07	William P. Thomas	Nothing Disclosed
PFL01	Noah D. Cohen	Nothing Disclosed
PFL01	Kenneth W. Hinchcliff	Co-Editor, *Journal of Veterinary Internal Medicine*
PFL02	Noah D. Cohen	Nothing Disclosed
PFL02	Kenneth W. Hinchcliff	Co-Editor, *Journal of Veterinary Internal Medicine*
PFL03	Jennifer M. MacLeay	Nothing Disclosed
PFL04	Robert P. Magnus	Nothing Disclosed
PFL05	Benjamin R. Buchanan	Nothing Disclosed
PFL05	Robert P. Magnus	Nothing Disclosed
PFL05	Harold C. McKenzie	Nothing Disclosed
PFL05	Edward D. Voss	Nothing Disclosed
PFL05	Dana N. Zimmel	Nothing Disclosed
PFL06	Caryn E. Plummer	Nothing Disclosed
PFL07	Wendy E. Vaala	Employed by Intervet Schering Plough
PFL08	Sanja Modric	Nothing Disclosed
PFL08	Abigail Salyers	Nothing Disclosed
PFL09	Gilles Fecteau	Nothing Disclosed
PFL09	Paul H. Walz	Nothing Disclosed
PFL10	Robert P. Magnus	Nothing Disclosed
PFN01	Peter J. Dickinson	Nothing Disclosed
PFN02	Douglas C. Wallace	Nothing Disclosed
PFN03	Douglas C. Wallace	Nothing Disclosed
PFN05	Paul A. Cuddon	Nothing Disclosed
PFN06	Curtis Wells Dewey	Nothing Disclosed
PFO01	Melissa C. Paoloni	Nothing Disclosed
PFO03	Craig A. Clifford	Advisory board member
PFO03	Kevin August Hahn	Employee of Hill's Pet Nutrition, Inc.
PFO03	Douglas H. Thamm	Nothing Disclosed
PFO04	Anthony Rusk	Nothing Disclosed
PFO05	Kristy Richards	Speaker's Bureau, Cephalon; Consultant for Merck
PFO06	Kristy Richards	Speaker's Bureau, Cephalon; Consultant for Merck
PFO07	Philip J. Bergman	Consultant for Merial, Pfizer Animal Health, Boerhinger, Elanco
PFO08	Daniel L. Gustafson	Nothing Disclosed
PFS01	Clarence Rawlings	Nothing Disclosed
PFS02	Nancy D. Kay	Nothing Disclosed
PFS03	George E. Moore	Nothing Disclosed
PFS05	Clarence Rawlings	Nothing Disclosed
PFS06	Nancy D. Kay	Nothing Disclosed
PFS07	Katrin Hartmann	Nothing Disclosed
PFS08	John Tait	Nothing Disclosed
PFS10	Demetrius Bagley	Consultant for Bard Urological, Cook Urological, Boston Scientific, Olympus Gyrus ACMI
PFS11	Lisa M. Freeman	Nothing Disclosed

PFS12a	Brett A. Sponseller	Nothing Disclosed
PFS12b	Richard E. Goldstein	Nothing Disclosed
PFS13	Helio Autran de Morais	Nothing Disclosed
PFS14	Demetrius Bagley	Consultant for Bard Urological, Cook Urological, Boston Scientific, Olympus Gyrus ACMI
PFS15	Carolyn J. Henry	Nothing Disclosed
PFS16a	Urs Giger	Nothing Disclosed
PFS16b	Leslie Lyons	Nothing Disclosed
PFS17	Ashley Hill	Nothing Disclosed
PFS18	Allyson C. Berent	Advisor for Infiniti Medical, LLC
PFS19	Jonathan F. Bach	Nothing Disclosed
PFS20a	Jeffrey A Brockman	Employeed by Hill's Pet Nutrition, Inc.
PFS20b	Samer Al-Murrani	Nothing Disclosed
PFS21	Paul S. Morley	Nothing Disclosed
PFS22	Allyson C. Berent	Advisor for Infiniti Medical, LLC
PFS23	Shannon P Holmes	Nothing Disclosed
PFS24a	Melanie Craven	Nothing Disclosed
PFS24b	Stijn Johannes Maria Niessen	Nothing Disclosed
PFS25	George E. Moore	Nothing Disclosed
1	E. Christopher Orton	Nothing Disclosed
2	Daniel F. Hogan	Nothing Disclosed
3	Kathryn Montgomery Meurs	Co-investigator on a patent held by Ohio State Univeristy and Washington State University for genetic testing for Maine Coons and Boxers.
4	Michael Podell	Nothing Disclosed
5	Colette D. Williams	Nothing Disclosed
6	Adam Mamelak	Nothing Disclosed
9a	Marlene L. Hauck	Nothing Disclosed
9b	Erika L. Krick	Nothing Disclosed
10	Amy E. DeClue	Consultant for company that offers diagnostic test.
11	Michael R. Lappin	Nothing Disclosed
12	Remo Lobetti	Nothing Disclosed
13	Robert T. O'Brien	Nothing Disclosed
14	Anne Barger	Nothing Disclosed
15a	Jerome Elie Leemans	Nothing Disclosed
15b	Claire Rebecca Sharp	Nothing Disclosed
16	Shelly Lu	Nothing Disclosed
17	Sharon A. Center	Nothing Disclosed
18	Philipp Mayhew	Nothing Disclosed
19	Jennifer Buur	Nothing Disclosed
20	Tomas Martin-Jimenez	Nothing Disclosed
21	Cynthia Cole	Employee of NAH
22	Katherine Cole MacGillivray	Nothing Disclosed
23	Carter E. Judy	Nothing Disclosed
24	Stefania Bucca	Nothing Disclosed
25	David E. Anderson	Nothing Disclosed
26	Simon Francis Peek	Nothing Disclosed
27	Stacey Renee Byers	Nothing Disclosed
28	Sarel R. Van Amstel	Nothing Disclosed
29	Jan Shearer	Nothing Disclosed
30	André Desrochers	Nothing Disclosed
31	Stephanie J. Valberg	Drs. Valberg, Mickelson and McCue own the license for PSSM.
32	Monica R. Aleman	Nothing Disclosed
33	Dennis O'Brien	Nothing Disclosed
34	Andrea Fischer	Nothing Disclosed
35	Susan E. Lana	Nothing Disclosed
36	Steven E. Suter	Nothing Disclosed
37	Steven E. Suter	Nothing Disclosed
38	Katrina Larson Mealey	Washington State University holds a patent for MDR1 genetic testing.
39	Amy E. DeClue	Nothing Disclosed
40	Richard E. Goldstein	Nothing Disclosed
41	Katrin Hartmann	Nothing Disclosed

42a	Leah Ann Cohn	Nothing Disclosed
42b	Sanjay Kapil	Research funding from Intervet/Schering Plough Animal Health
43	Joerg M. Steiner	Director of the GI Lab at Texas A&M; Consultant for Idexx Laboratories
44	Jan S. Suchodolski	Nothing Disclosed
45	Karin Allenspach	Funding by Probiotics Ltd UK, CEVA Sante Animale, FR, AB Science, FR
46	Keith P. Richter	Nothing Disclosed
47	Kurt Allan Grimm	Nothing Disclosed
48	Butch KuKanich	Nothing Disclosed
49	Dawn M. Boothe	Nothing Disclosed
50	Tamara Grubb	Speaker for Pfizer Animal Health and Abbott Animal Health
51	Urs Giger	Scientific advisor to DMS, DiaMed, Alvedia, and IDEXX
52	Stephanie A. Smith	Nothing Disclosed
53	Beth Davidow	Nothing Disclosed
54	Anne S. Hale	Minority owner and current CEO of the world's largest veterinary blood bank. Patent holder for a gel tube assay developed for point of care crossmatch evaluation.
55	John S. Adams	Nothing Disclosed
56	Dennis J. Chew	Nothing Disclosed
57	Gregory A. Brent	Nothing Disclosed
58	Mark E. Peterson	Nothing Disclosed
59	James A. Roth	Research support, consulting income, speaker honorarium, or sponsorship of the OIE Veterinary Biologics Training Program from: Bayer Animal Health, Boehringer Ingelheim Vetmedica, Inc, Fort Dodge Animal Health, Immunobiosciences, Merial Ltd., Schering Plough/Intervet, and Transboundary Animal Biologics, Inc.
60	James A. Roth	Research support, consulting income, speaker honorarium, or sponsorship of the OIE Veterinary Biologics Training Program from: Bayer Animal Health, Boehringer Ingelheim Vetmedica, Inc, Fort Dodge Animal Health, Immunobiosciences, Merial Ltd., Schering Plough/Intervet, and Transboundary Animal Biologics, Inc.
61	James A. Roth	Research support, consulting income, speaker honorarium, or sponsorship of the OIE Veterinary Biologics Training Program from: Bayer Animal Health, Boehringer Ingelheim Vetmedica, Inc, Fort Dodge Animal Health, Immunobiosciences, Merial Ltd., Schering Plough/Intervet, and Transboundary Animal Biologics, Inc.
62	James A. Roth	Research support, consulting income, speaker honorarium, or sponsorship of the OIE Veterinary Biologics Training Program from: Bayer Animal Health, Boehringer Ingelheim Vetmedica, Inc, Fort Dodge Animal Health, Immunobiosciences, Merial Ltd., Schering Plough/Intervet, and Transboundary Animal Biologics, Inc.
63	Simon Francis Peek	Nothing Disclosed
64	David E. Anderson	Nothing Disclosed
65	David E. Anderson	Nothing Disclosed
66	André Desrochers	Nothing Disclosed
67a	Suzanne M. Cunningham	Nothing Disclosed
67b	Sophia Gry Moesgaard	Nothing Disclosed
68a	Torkel Falk	Nothing Disclosed
68b	Scott D. Forney	Nothing Disclosed
69	Emmanuel Mignot	Nothing Disclosed
70	Emmanuel Mignot	Nothing Disclosed
71a	Rowan J. Milner	Nothing Disclosed
71b	Jeffrey Carl Phillips	Advisor to Biothermatics
72a	Laura D. Garrett	Nothing Disclosed
72b	Douglas H. Thamm	Nothing Disclosed
73	Victor Test	Research support, Advisory Boards, Speakers Bureau, Actelion Pharmaceuticals; CME Speakers Bureau and Speakers Bureau, Gilead Pharmceuticals; Speakers Bureau United Therapeutics; Research support, Lung Rx
74	Victor Test	Research support, Advisory Boards, Speakers Bureau, Actelion Pharmaceuticals; CME Speakers Bureau and Speakers Bureau, Gilead Pharmceuticals; Speakers Bureau United Therapeutics; Research support, Lung Rx
75a	Craig G. Ruaux	Consultant for Gastrointestinal Laboratory at Texas A&M University
75b	Sara L. Ford	Nothing Disclosed
76a	Lynda D. Melendez	Employed by Hill's Pet Nutrition, Inc.
76b	Steven C. Zicker	Employed by Hill's Pet Nutrition, Inc.
77	Jeanne A. Barsanti	Nothing Disclosed
78	Dawn M. Boothe	Nothing Disclosed

79	Caroline Mansfield	Nothing Disclosed
80	Katrina Larson Mealey	Washington State University has submitted a patent for the ABCB4 technology.
81	Marilyn E. Dunn	Nothing Disclosed
82	Andrew J. Mackin	Nothing Disclosed
83	Rebecka Shoshana Hess	Nothing Disclosed
84	Linda Mary Fleeman	Nothing Disclosed
85	Demosthenes Pappagianis	Nothing Disclosed
86	Jane E. Sykes	Nothing Disclosed
87	Connie Gebhart	Research Support: Boehringer Ingelheim Vetmedica, Novartis Animal Health, Elanco Animal Health
88	Allison Jean Stewart	Nothing Disclosed
89a	Jenifer R. Gold	Nothing Disclosed
89b	Harold C. Schott	Nothing Disclosed
90a	Kate E. Borer	BBSRC CASE studentship for PhD; Collaborator for Waltham Centre for Pet Nutrition, Funding by Waltham Centre for Pet Nutrition
90b	Veronique Anne Lacombe	Nothing Disclosed
91	Abigail Salyers	Nothing Disclosed
92	Abigail Salyers	Nothing Disclosed
93	Thomas Ryan	Nothing Disclosed
94	Thomas Ryan	Nothing Disclosed
95	John D. Bonagura	Nothing Disclosed
96	D. David Sisson	Nothing Disclosed
97	Lisa M. Freeman	Nothing Disclosed
98	Dan G. Ohad	Nothing Disclosed
99	Henry W. Green	Nothing Disclosed
100a	Etienne Coté	Nothing Disclosed
100b	Robert Prosek	Nothing Disclosed
101a	David S. Bruyette	Nothing Disclosed
101b	Sarah C. Charney	Nothing Disclosed
102a	Sarah Anne Moore	Nothing Disclosed
102b	Roger M. Clemmons	Nothing Disclosed
103a	Roger M. Clemmons	Nothing Disclosed
103b	Joan R. Coates	Patent and Inter-Institutional agreement in relation to Prediction and Diagnosis of Canine Degenerative Myelopathy
104a	Renee Barber	Nothing Disclosed
104b	Ronaldo Casimiro da Costa	Nothing Disclosed
105	James Bradner	Nothing Disclosed
106	James Bradner	Nothing Disclosed
107a	Kerry Christine Rissetto	Nothing Disclosed
107b	Erin Malone	Nothing Disclosed
108a	Jeffrey N. Bryan	Nothing Disclosed
108b	Douglas H. Thamm	Nothing Disclosed
109	Chand Khanna	Nothing Disclosed
110	Melissa C. Paoloni	Nothing Disclosed
111	Douglas H. Thamm	Previous sponsored research agreement with AB Science ; Previous consulting and paid lecture honoraria from Pfizer Animal Health
112	Cheryl A. London	Nothing Disclosed
113a	Tracy L. Hill	Nothing Disclosed
113b	Tracy L. Hill	Nothing Disclosed
114a	Barrak M. Pressler	Nothing Disclosed
114b	Eran Dvir	Nothing Disclosed
115a	Jill D. Brunker	Nothing Disclosed
115b	Michael R. Lappin	Nothing Disclosed
116a	Stephanie A. Smith	Nothing Disclosed
116b	Sharon A. Center	Nothing Disclosed
117	Catharine Scott-Moncrieff	Nothing Disclosed
118	Catharine Scott-Moncrieff	Nothing Disclosed
119	Catharine Scott-Moncrieff	Nothing Disclosed
120	Catharine Scott-Moncrieff	Nothing Disclosed
121	Mary B Nabity	Nothing Disclosed
122	Mary B Nabity	Nothing Disclosed

123	John A. Christian	Nothing Disclosed
124	John A. Christian	Nothing Disclosed
125	Karl E. Jandrey	Nothing Disclosed
126	Karl E. Jandrey	Nothing Disclosed
127	Karl E. Jandrey	Nothing Disclosed
128	Karl E. Jandrey	Nothing Disclosed
129	Bruce Klein	Nothing Disclosed
130	Amy M. Grooters	Nothing Disclosed
131	Joseph Wheat	Owner, President and Director of MiraVista diagnostics and MiraBella Technologies
132	Alfred M. Legendre	Nothing Disclosed
133	Richard Miller	Nothing Disclosed
134	Richard Miller	Nothing Disclosed
135	Harold C. Schott	Nothing Disclosed
136	K. Gary Magdesian	Nothing Disclosed
137	Raymond J. Geor	Nothing Disclosed
138a	Jessica K Suagee	Nothing Disclosed
138b	Megan Leigh Shepherd	Nothing Disclosed
139a	Shannon E. Pratt-Phillips	Nothing Disclosed
139b	Nicholas Frank	Consultant for Boehringer Ingelheim; Research funding and consultant for Lloyd, Inc.
140	David A. Dzanis	Consultant for approximately 65 pet food, animal feed, supplement and ingredient manufacturers
141a	Stacey Renee Byers	Nothing Disclosed
141b	Erica C. McKenzie	Nothing Disclosed
142a	Maria E. Prado	Nothing Disclosed
142b	Tim Potter	Nothing Disclosed
143a	Jean A. Hall	Nothing Disclosed
143b	Peter D. Constable	Consultant for Bayer and Boehringer Ingelheim
144a	Maisie E. Dawes	Nothing Disclosed
144b	Alexandre Secorum Borges	Nothing Disclosed
145	Daniela Bedenice	Nothing Disclosed
146	Daniela Bedenice	Nothing Disclosed
147	Tamara Gull	Nothing Disclosed
148a	Ana Paula M. Pacheco	Nothing Disclosed
148b	Judith Wagner	Nothing Disclosed
149a	O. Lynne Nelson	Nothing Disclosed
149b	Sonja Fonfara	Nothing Disclosed
150a	Ingrid Ljungvall	Nothing Disclosed
150b	Margaret M. Sleeper	Nothing Disclosed
151a	Gerhard Wess	Nothing Disclosed
151b	Roberto A. Santilli	Nothing Disclosed
152a	Sonya Gale Gordon	Consultant for Boehringer Ingelheim
152b	Lori Drourr	Nothing Disclosed
153	Sophy Jesty	Nothing Disclosed
154	M. Kimberly McGurrin	Co-applicant on patent with Rhythm Technologies a subsidiary of Cardiac Output Technologies
155	James W. Buchanan	Nothing Disclosed
156	Brian A. Scansen	Nothing Disclosed
157a	Karen R. Munana	Nothing Disclosed
157b	Ines Carrera	Nothing Disclosed
158a	Andrea Fischer	Nothing Disclosed
158b	Steven De Decker	Nothing Disclosed
159a	Karen R. Munana	Nothing Disclosed
159b	Jacqueline Johnson	Nothing Disclosed
160a	Stephen M. Reed	Nothing Disclosed
161	Sofia Cerda-Gonzalez	Nothing Disclosed
162	J. Andy Shores	Nothing Disclosed
163	Thomas Goedde	Nothing Disclosed
163	Kaspar Matiasek	Nothing Disclosed
164	Patrick James Kenny	Nothing Disclosed
165	Sarah Boston	Nothing Disclosed
165	J. Paul Woods	Nothing Disclosed

166	Sarah Boston	Nothing Disclosed
166	J. Paul Woods	Nothing Disclosed
167	Sarah Boston	Nothing Disclosed
167	J. Paul Woods	Nothing Disclosed
168	Sarah Boston	Nothing Disclosed
168	J. Paul Woods	Nothing Disclosed
169	N. Edward Robinson	Nothing Disclosed
170	J. Scott Weese	Nothing Disclosed
171	George E. Lees	Nothing Disclosed
172	Adam J. Moeser	Nothing Disclosed
173	Linda A. Ross	Nothing Disclosed
174	Gregory F. Grauer	Nothing Disclosed
175	Hervé Lefebvre	Research funding from Royal Canin
176a	Natalie Clare Finch	PhD funding by Royal Canin and Waltham
176b	Joseph William Bartges	Nothing Disclosed
177	Christopher Mariani	Nothing Disclosed
178	Christopher Mariani	Nothing Disclosed
179	Christopher Mariani	Nothing Disclosed
180	Christopher Mariani	Nothing Disclosed
181	Robin W. Allison	The clinical pathology laboratory at OSU is owned and operated by Antech Diagnostics
182	Robin W. Allison	The clinical pathology laboratory at OSU is owned and operated by Antech Diagnostics
183	Amy Leone MacNeill	Nothing Disclosed
184	Amy Leone MacNeill	Nothing Disclosed
185	Justine A. Lee	Nothing Disclosed
186	Justine A. Lee	Nothing Disclosed
187	Yvette S. Nout	Nothing Disclosed
188	Yvette S. Nout	Nothing Disclosed
189	Steven C. Zicker	Employee of Hill's Pet Nutrition, Inc.
190	Claudia Ann Kirk	Nothing Disclosed
191	Linda Mary Fleeman	Nothing Disclosed
192	David A. Dzanis	Consultant for American Pet Products Association and approximately 65 pet food, animal feed, supplement, and ingredient manufacturers
193a	Morgan Keith Chaffin	Nothing Disclosed
193b	Katharina L. Lohmann	Nothing Disclosed
194a	Chris Sanchez	Nothing Disclosed
194b	Renaud Leguillette	Nothing Disclosed
195a	Jonathan H. Foreman	Nothing Disclosed
195b	Stephanie A. Smith	Nothing Disclosed
196a	Stefania Bucca	Nothing Disclosed
196b	Carrie Jacqueline Finno	Nothing Disclosed
197	Alison Clode	Nothing Disclosed
198	Alison Clode	Nothing Disclosed
199	Caryn E. Plummer	Nothing Disclosed
200	Caryn E. Plummer	Nothing Disclosed
201	Hans Coetzee	Nothing Disclosed
202	Hans Coetzee	Nothing Disclosed
203	Suzanne Millman	Nothing Disclosed
204	Suzanne Millman	Nothing Disclosed
205	Paul J. Plummer	Nothing Disclosed
206	Paul J. Plummer	Nothing Disclosed
207	Christopher Chase	Consultant for Novartis Animal Health; Intervet-Schering Plough;Pfizer Animal Health; Boehringer Ingeheim Vetmedica; Merial
208	Christopher Chase	Consultant for Novartis Animal Health; Intervet-Schering Plough;Pfizer Animal Health; Boehringer Ingeheim Vetmedica; Merial
209	Robin Brogdon	Nothing Disclosed
209	Karen E Felsted	Nothing Disclosed
210	Robin Brogdon	Nothing Disclosed
210	Karen E Felsted	Nothing Disclosed
211	Robin Brogdon	Nothing Disclosed
211	Karen E Felsted	Nothing Disclosed
212	Robin Brogdon	Nothing Disclosed

212	Karen E Felsted	Nothing Disclosed
213	Adrian Boswood	Research funding from IDEXX Ltd.
214	Andrew W. Beardow	Nothing Disclosed
215	Stephen J. Ettinger	Nothing Disclosed
216	Robert Prosek	Nothing Disclosed
217	Karin Hultin Jaderlund	Nothing Disclosed
218	Karen M. Vernau	Nothing Disclosed
219	Kaspar Matiasek	Nothing Disclosed
220	G. Diane Shelton	Nothing Disclosed
221	Carlos H de M. Souza	Nothing Disclosed
222	Susan M. LaRue	Nothing Disclosed
223	Susan M. LaRue	Nothing Disclosed
224	Douglas H. Thamm	Nothing Disclosed
225	Carol R. Reinero	Nothing Disclosed
225	Kim A. Selting	Nothing Disclosed
226	Carol R. Reinero	Nothing Disclosed
226	Kim A. Selting	Nothing Disclosed
227	Ned F. Kuehn	Nothing Disclosed
228	Ned F. Kuehn	Nothing Disclosed
229	Patty A. Lathan	Nothing Disclosed
230	Audrey Karen Cook	Consultant for Dechra PLC
231	Harriet Merlin Syme	Nothing Disclosed
232	Patricia Ann Schenck	Nothing Disclosed
233	Michael D. Willard	Nothing Disclosed
234	Michael D. Willard	Nothing Disclosed
235	Michael D. Willard	Nothing Disclosed
236	Michael D. Willard	Nothing Disclosed
237	Tobias Schwarz	Nothing Disclosed
237	Susanne Steiger	Nothing Disclosed
238	Tobias Schwarz	Nothing Disclosed
238	Susanne Steiger	Nothing Disclosed
239	Sally Christine Perea	Co-owner of Davis Veterinary Medical Consulting, Inc. (DVMC). Employed by Natura Pet Products, Inc.
240	Sally Christine Perea	Co-owner of Davis Veterinary Medical Consulting, Inc. (DVMC). Employed by Natura Pet Products, Inc.
241	Lauren A. Trepanier	Nothing Disclosed
242	Lauren A. Trepanier	Nothing Disclosed
243	Stephanie A. Smith	Nothing Disclosed
244	Andrew J. Mackin	Nothing Disclosed
245a	Jody P. Lulich	Nothing Disclosed
245b	Claudia Ann Kirk	Nothing Disclosed
246a	Dennis J. Chew	Nothing Disclosed
246b	Joseph William Bartges	Nothing Disclosed
247a	Larry D. Cowgill	Nothing Disclosed
247b	David J. Polzin	Nothing Disclosed
248a	Catherine E. Langston	Nothing Disclosed
248b	Mark J. Acierno	Nothing Disclosed
249	Matthew A. Wright	Nothing Disclosed
250	Allison Zwingenberger	Nothing Disclosed
251	Rachel E. Pollard	Nothing Disclosed
252	Anne Barger	Nothing Disclosed
253	Paul S. Morley	Nothing Disclosed
254	M. Kimberly McGurrin	Nothing Disclosed
255	Carter E. Judy	Nothing Disclosed
256	Stephanie J. Valberg	Drs. Valberg, Mickelson and McCue own the license for PSSM
257	Sophy Jesty	Nothing Disclosed
258	Sophy Jesty	Nothing Disclosed
259	Nicholas Frank	Consultant for Boehringer Ingelheim; Research funding and consultant for Lloyd, Inc.
260	Nicholas Frank	Consultant for Boehringer Ingelheim; Research funding and consultant for Lloyd, Inc.
261	Sheila M. McGuirk	Nothing Disclosed
262	Hans Coetzee	Nothing Disclosed

263	Christopher Chase	Consultant for Novartis Animal Health; Intervet-Schering Plough;Pfizer Animal Health; Boehringer Ingeheim Vetmedica; Merial
264	Fiona Pauline Maunsell	Nothing Disclosed
265	Peter D. Constable	Consultant for Boehringer, Bayer and Fort Dodge
265	Allen J. Roussel	Nothing Disclosed
266	Peter D. Constable	Consultant for Boehringer and Bayer
266	Allen J. Roussel	Nothing Disclosed
267	Jim Reynolds	Nothing Disclosed
268	Jim Reynolds	Nothing Disclosed
269	Robin Brogdon	Nothing Disclosed
269	Karen E Felsted	Nothing Disclosed
270	Robin Brogdon	Nothing Disclosed
270	Karen E Felsted	Nothing Disclosed
271	Robin Brogdon	Nothing Disclosed
271	Karen E Felsted	Nothing Disclosed
272	Robin Brogdon	Nothing Disclosed
272	Karen E Felsted	Nothing Disclosed

SPEAKER DISCLOSURE INDEX
TECHNICIAN PROGRAM

Session	Speaker	Disclosure
PFT01a	Cindy Meeker	Nothing Disclosed
PFT01b	Kristine M. Obssuth	Nothing Disclosed
PFT01c	Tamberlyn D. Moyers	Nothing Disclosed
PFT02a	Amber Lee Caress	Nothing Disclosed
PFT02b	Amanda Jo Callens	Nothing Disclosed
PFT02c	Dee DiPiero	Nothing Disclosed
PFT03a	Gina Galyon	Nothing Disclosed
PFT03b	Laura Batoon	Nothing Disclosed
PFT03c	Rachel V. Poulin	Nothing Disclosed
PFT04a	Bobbie Werbe	Nothing Disclosed
PFT04b	Jessica Marie Weber	Nothing Disclosed
T05	Pamela Harrigan	Nothing Disclosed
T06	H. Edward Durham	Nothing Disclosed
T07	Pamela Harrigan	Nothing Disclosed
T08	Romain Pariaut	Nothing Disclosed
T09	Romain Pariaut	Nothing Disclosed
T10	Kimberly A. Baldwin	Nothing Disclosed
T11	Kim Zwerenz Miks	Nothing Disclosed
T12	Colleen Suzanne Ruderman	Nothing Disclosed
T13	Meri Frances Hall	Nothing Disclosed
T14	Sean Owens	Nothing Disclosed
T15	John E. Madigan	Nothing Disclosed
T16	Sarah le Jeune	Nothing Disclosed
T17	M. Kimberly McGurrin	Co-applicant on patent with Rhythm Technologies a subsidiary of Cardiac Output Technologies
T18	Laurel E. Williams	Nothing Disclosed
T19	Anjanette M. Price	Nothing Disclosed
T20	Laura D. Garrett	Nothing Disclosed
T21	Tara Marmulak	Nothing Disclosed
T22	Liz Kelley	Nothing Disclosed
T23	Ann Wortinger	Nothing Disclosed
T24	Kara Burns	Employeed by Hill's Pet Nutrition, Inc.
T25	Laurel E. Williams	Nothing Disclosed
T26	Theresa E. Pancotto	Nothing Disclosed
T27	Lauren A. Trepanier	Nothing Disclosed
T28	Carissa D. Sparks	Nothing Disclosed
T29	Christopher Mariani	Nothing Disclosed
T30	Yvette S. Nout	Nothing Disclosed
T31	Karen R. Munana	Nothing Disclosed
T32	Regina Fairman	Nothing Disclosed
T33	Kathi L. Smith	Nothing Disclosed
T34	Regina Fairman	Nothing Disclosed
T35	Justine A. Lee	Nothing Disclosed
T36	Kristin Miguel	Nothing Disclosed
T37	Birgit Puschner	Nothing Disclosed
T38	Susan L. Cox	Nothing Disclosed
T38	Marika Pappagianis	Nothing Disclosed
T39	K. Gary Magdesian	Nothing Disclosed
T40	Danielle Teresa Russ	Nothing Disclosed
T41	Danielle Teresa Russ	Nothing Disclosed
T42	Katy W. Waddell	Nothing Disclosed
T43	Katy W. Waddell	Nothing Disclosed
T44	Chad M. Johannes	Employed by Pfizer Animal Health
T45	Enrico Pierluigi Spugnini	Owner, Biopulse S.r.l.
T46	Kara Burns	Employed by Pfizer Animal Health
T47	Katrina Larson Mealey	Washington State University holds a patent on MDR1 genetic testing
T48	Lisa M. Freeman	Nothing Disclosed

T49	Stacey B. Leach	Nothing Disclosed
T50	Anne Myers	Nothing Disclosed
T51	June Boon	Employed to teach instructional seminars with Sound Technologies

KEYNOTE SESSION

2010 ACVIM Award Winners & Honorees

2010 ACVIM Forum Sponsor Recognition

KEYNOTE ADDRESS

Nominated by ACVIM Diplomates, Dr. Paul Pion – one of the founders of the Veterinary Information Network (VIN) – will share his life story, how he was influenced by those around him to make choices beyond the usual and customary, and challenge us to reflect on the choices we make that shape our lives.

UNUSUAL CASE PRESENTATIONS IN VETERINARY CARDIOLOGY

William P. Thomas, DVM, DACVIM
Davis, CA

INTRODUCTION

Veterinary cardiologists encounter animals with a variety of cardiac abnormalities, the majority of which are common and familiar to an experienced clinician. However, some of the most interesting cases presented to a specialist involve uncommon conditions or unusual presentations of familiar ones. Throughout a year, a cardiologist can expect to encounter patients with unusual conditions that require an ability to interpret findings based on experience and an understanding of cardiac anatomy and physiology. In this interactive session, a series of clinical vignettes will be presented, demonstrating unusual conditions or unusual findings or complications of more common conditions. The audience will be encouraged to participate in the brief discussions of each patient, especially those presented as "unknowns". This is a visual presentation of unusual findings which requires audience presence and participation to be successful. Cases will be presented from the following categories:

Echocardiographic Abnormalities

These cases demonstrate unusual, unique abnormalities identified during echocardiographic imaging, such as foreign bodies or unusual anatomic defects.

Cardiac Pacing Anomalies

Complications during or following pacemaker implantation are reportedly common. The cases presented will demonstrate uncommon or unanticipated problems during or following transvenous pacemaker implantation.

Cath Lab Experiences

Several trans-vascular procedures have become common in veterinary cardiology (balloon dilatation, pacemaker implantation, PDA closure). These cases will demonstrate some unusual complications or outcomes in patients during or following cardiac catheterization and interventions.

Surgical Cases

Although cardiac surgery is performed infrequently on veterinary patients, some interesting and unexpected complications or outcomes may occur. These cases will demonstrate unusual results or complications of attempts at cardiac surgery.

REFERENCES

[1.] Wess G, et al. **J Vet Intern Med** 2006;20:877. [2.] Oyama MA, et al. **J Vet Intern Med** 2001;15:229 [3.] Sereda NC, et al. **J Vet Cardiol** 2009;11:53. [4.] Campbell FE, et al. **J Vet Intern Med** 2006;20:83. [5.] Kienle RD & Thomas WP. In: Nyland T & Mattoon J, eds. **Small Animal Diagnostic Ultrasound. 2nd Ed** (Philadelphia, WB Saunders) 2001:198. [6.] Thomas WP & Sisson DD. In Fox PR, et al, eds. **Textbook of Canine and Feline Cardiology, 2nd Ed** (Philadelphia, WB Saunders) 1999;173. [7.] Fossum TW. In: Fossum TW, et al, eds. **Small Animal Surgery, 3rd Ed** (St. Louis, Mosby) 2007;775.

HIGHLIGHTS FROM HAVEMEYER: OPHTHALMOLOGY

Caryn E. Plummer, DVM, DACVO
Gainesville, FL

The first IEOC Symposium was held April 3-4, 2009 in Lexington, Kentucky. The Dorothy Havemeyer Foundation of New York, NY supported this workshop. The IEOC is a group of clinicians and scientists dedicated to the study of ocular disease of horses and the goal of this symposium is to share clinical and basic research on equine ophthalmology and to facilitate the development of multi-centered collaborative research. There were 50 attendees from around the world (i.e., from USA, Belgium, Italy, Germany, Spain, and the UK). The format included "state of the art" lectures, free paper presentations, and round-table discussions on common equine ocular diseases including adnexal neoplasia, infectious keratitis, non-ulcerative keratitis, glaucoma, vision, cataract surgery, recurrent uveitis, and retinal disease.

This first IEOC symposium met its goals of gathering a group of leading equine ophthalmology clinicians and researchers to indentify the challenges of the field. To facilitate collaboration, notes from round-table discussions, including the ideas and plans that were discussed were complied and distributed to the attendees. Development of an IEOC membership organization and website was discussed and supported by the group in an effort to further advance the science of equine ophthalmology. If you are interested in joining the IEOC or learning more about equine ophthalmology and the collaborative clinical trials currently underway, please visit www.equineophtho.com for more information.

While the presentations and discussion sessions touched briefly on most aspects of ophthalmology that are frequently seen and addressed by practicing equine generalists and veterinary ophthalmologists, more detailed discussion centered upon approaches to the catastrophic cornea and the latest surgical approaches to inflammatory corneal disease, the latest approaches to periocular neoplasia and the current state of research into the pathogenesis and therapeutic approaches to equine recurrent uveitis.

Clinicians at the University of Florida continue their work on different surgical techniques to stabilize compromised, infected and damaged corneas focusing on ways to reduce scar formation, improve visual function and decrease healing times. In an attempt to minimize local scarring and to avoid some of the potential complications of full-thickness penetrating keratoplasties, partial thickness excisional procedures have been developed, including the posterior lamellar keratoplasty and the deep lamellar endothelial keratoplasty. Keratoplasty procedures that address only the diseased portion of the cornea, leaving the healthy portion of the host cornea intact, are now the preferred method of surgical therapy for focal corneal disease. In addition to advances in corneal transplantation, UF faculty have pioneered the use of amniotic membrane transplantation for corneal disease. It has anti-inflammatory properties, is very strong and has been shown to decrease corneal scarring. It allows for a much more cosmetic and functional outcome than the traditional conjunctival graft. Ongoing studies are examining ways to ease handling and storage of this tissue and are investigating its utility as a substrate for the expansion of corneal epithelial cells for transplantation.

Clinicians at the University of Missouri are developing protocols for and currently have clinical trials underway that employ photodynamic therapy for the treatment of periocular squamous cell carcinoma. Since there is so little skin available for blepharoplastic procedures in the horse and so many periocular tumors are presented for evaluation and treatment at advanced stages of disease, traditional surgical and chemotherapeutic approaches to SCC are often unsuccessful. Photodynamic therapy uses the effects of light on a phototoxic substance injected directly into the tumor. Preliminary results are encouraging.

North Carolina State University continues to lead the field in the investigation of ERU, looking into both the pathogenesis and treatment thereof. ERU is being increasingly recognized as a spontaneous model of human immune-mediated uveitis, so studies of this disease phenomenon have the potential to benefit not just the veterinary community, but also potentially human beings. FDA approval of a surgically-implanted sustained-release suprachoroidal cyclosporine implant are pending. Studies to date have had very promising results in cases that have not yet had extensive damage or long-standing disease. Recent studies aim to evaluate the utility of other immune-modulating compounds such as sirolimus.

REFERENCES

1. Gilger BC, Brooks DE, **Equine Vet J** 2009; 41: 606. 2. Gilger BC. **www.equineophtho.com** 2009.

HIGHLIGHTS FROM THE 22ND CONFERENCE OF THE WORLD ASSOCIATION FOR THE ADVANCEMENT OF VETERINARY PARASITOLOGY 2009 – CALGARY, CANADA

Wendy Vaala, VMD, dipACVIM
Alma, Wisconsin

The WAAVP Symposium was held in Calgary, Canada, August 8 – 13, 2009. The theme of the symposium was: One World, One Health: Parasites in a Changing Landscape. There were more than 550 abstracts and posters presented at the conference. Less than 35 presentations were focused on equine parasitology. This sobering statistic highlights the desperate need for more clinical trials and basic research in equine parasitology if we, as a profession, are to generate the data and diagnostic tools necessary to describe and understand the changes occurring in parasite transmission, pathology and drug resistance and to begin to develop sustainable parasite control programs for horses.

Selected Abstracts

The following abstracts are those that focused on some aspect of clinical or diagnostic parasitology in the horse and / or donkey. The abstracts and scientific program are available on the following website: www.waavp2009.com.

1. Fatal *Halicephalobus gingivalis* (Nematoda) Brain Infection in an Icelandic Horse
Eydal M, Bambir SH et al. Instit for Experimental Pathology, Univ of Iceland, Reykjavik, Iceland

There are less than 50 published cases of *Halicephalobus gingivalis* infection in horses worldwide. The first case of fatal *H gingivalis* infection in an Icelandic stallion is described. A few months following an accident that resulted in mouth wounds, the stallion developed severe neurological signs that necessitated euthanasia. Histopathology revealed severe multifocal malacia associated with numerous intralesional nematodes. Mouth wounds were the suspected route of entry for the parasites.

2. The Grazing Horse Gets the Worm: the Use of Coprological and Coproantigen ELISA Analyses for Diagnosing *Anoplocephala perfoliata* Infection in Southern Alberta Horses
Skotarek SL, Colwell DD et al. Univ of Calgary, Airdrie, AB, Canada

The prevalence and distribution of *A perfoliata* in western Canada was described using fecal examination and a coproantigen ELISA. Over an 18-month period (winter 2006 – summer 2007) fecal samples were collected from privately owned horses in realistic management settings and were evaluated using standard centrifugation flotation. A subset of samples was further evaluated using a coproantigen ELISA. Samples collected at an abattoir served as the gold standard for evaluating diagnostic techniques. There was a positive correlation between the presence of tapeworm antigen and infection intensity using the coproantigen ELISA. The ELISA detected more infections than fecal analysis alone. The prevalence of *A perfoliata* varied among seasons and ranges from 18% in winter to 8% in summer based on fecal analysis. Overall tapeworm prevalence ranged from 25 – 62%.

3. Inaccurate Diagnosis Between *Anoplocephala perfoliata* and *A. Magna* in Naturally Infected Horses Meana A, Bohorquez A et al. Dept of Animal Health, Madrid, Spain

In Spain the overall tapeworm prevalence is 30% (*A perfoliata*, 40%, *A magna*, 20%). The probability of *A magna* infections is higher in younger animals and data suggests the absence of *A magna* infections in immune competent adult horses. *A magna* resides in the jejunum and ileum and is easier to detect by fecal analysis than *A perfoliata*. The role of *A perfoliata* in colic in horses continues to be debated in part due to the unreliability of the currently available antemortem diagnostic assays.

The use of antibody titers using ELISA tests to identify *A perfoliata* and *A magna* infected horses was questioned due to the assay's moderate sensitivity and low specificity to differentiate between the two tapeworm species. The nested PCR assay, developed to detect and amplify *A perfoliata* DNA within fecal samples, was recommended as a valuable diagnostic tool.

4. Parasitological and Serological Diagnosis of *Trypanosoma evansi* in Experimetnally Inoculated Donkeys Hilali MA, Cairo University, Cairo, Egypt

T evansi is an important parasite in horses, donkeys and camels in Africa. Seronegative donkeys were used in an experimental infection model to study the prepatent and patent periods of T evansi and to compare to different diagnostic tests: parasitological evaluation of a blood sample and the Card agglutination test (CATT). The prepatent period varied from 1 – 4 days post infection (PI). The patent period ranged from 7 – 48 days PI. An increase in *T evansi* antibodies was detected as early as day 4 post infection. The CATT was easier to perform and more reliable than direct parasitological examination of blood samples.

5. Donkey Parasites in the UK; Infection Levels, Treatment Intervals and Anthelmintic Use Burden F, Trawford, A. The Donkey Sanctuary, Sidmouth UK

Donkeys are often perceived as important parasite reservoirs, often blamed for being asymptomatic carriers of lungworm and shedders of drug resistant cyathostomes. Coprological techniques were used to determine the levels of parasite infection in donkeys new to The Donkey Sanctuary over a 4-year period (2004-08). 735 new additions were studied. 73% of those donkeys had an identifiable strongyle infection, 4% had liver flukes, 3% had tapeworms and 4% had lungworm infection. According to owners, 13% of the new additions had never been dewormed. Of those that had received anthelmintic treatments, ivermectin was the most frequently used dewormer (41% of donkeys) followed by benzimidazoles (21%). The level of lungworm infection was lower than that usually quoted for donkeys.

6. *Gastrophilus Nasalis* (Diptera:Ostridae): A Major Cause of Rectal Prolapse in Working Donkeys in Ethiopia Getachew M et al. Faculty of Vet Med, Debre Zeit, Ethiopia

Gastrophilus larvae are obligate parasites in the gastrointestinal tract of equids. This retrospective study revealed that Gastrophilus nasalis was respsonsible for 83.7% of the rectal prolapses in working donkeys in Ethiopia. The larvae re-attach temporarily to the rectal mucosa after passing through the digestive tract. This may stimulate an increased frequency of defecation, irritation, inflammation and intense tenesmus, leading to rectal prolapse. Cases of rectal prolapse were more common during the wet, cool season compared to the dry, warm season. A significant decrease in rectal prolapse was noted after a strategic deworming program was initiated.

7. Structure of Strongylid Communities of Wild and Domestic Equids in Ukraine and Modern Methods of Parasite Control Kuzmina T et al. Schmalhausen Institute of Zoology NAS of Ukraine, Kyiv, Ukraine

The purpose of the study was to examine the structure of strongylid communities of 5 equid species and analyze the current methods of parasite control in Ukraine. 156 animals (84 horses, 30 donkeys, 22 Przewalski's horses, 11 zebras, 9 Turkmenian kulans) were involved in the study. Animals were treated with Univerm® (0.2% avermectin) and fecal analysis performed 24, 36, 48 and 60 hours post-treatment. The number of different strongylid species were identified in all hosts with the following results: horses – 26 species, Przewalski's horses – 31 species, donkeys – 24 species, zebras – 17 species, kulans – 24 species. The shape of the prevalence frequency distribution of strongylid species was bimodal in horses and donkeys and multimodal in Przewalski's horses, zebras and kulans. Frequent anthelmintic treatments lead to strongylid community structure destruction and cause the development of resistance in nematodes. It is recommended that an integrated control program should be a prospective approach for long-term parasite control in various types of horse-keeping systems in Ukraine.

8. Seasonal Variation in Development and Survival of *Parascaris equorum* Eggs in Pasture or on Gravel Surface Lindgren K, et al. Swedish Institute of Agriculture and Environmental Engineering, Uppsala, Sweden

A popular theory suggests a major source of *P equorum* infection in foals from overwintered eggs deposited by the previous generation of foals. To study the transmission in different environments standardized fecal pats with a known number of P equorum eggs were deposited in plots with either short grass or gravel in August 2008. A second fecal pat deposition was conducted in October 2008. Triplicate samples from the pats were analyzed for the number of surviving eggs and their developmental stages at 1, 2, 4, 8, 16, 32, 52 and 84 weeks after deposition. Air and ground temperatures and relative humidity were constantly recorded near the fecal pats. Eight weeks after summer deposition the egg counts in the fecal pats decreased by approximately 50% on gravel and 90% on grass, which was partly due to the fact that fecal pats disappeared quicker on grass. Eight weeks after autumn deposition the number of eggs had decreased by approximately 10% in the fecal pats on both gravel and grass. The difference between FECs between seasons was related to the higher activity of dung degrading organisms in fecal pats deposited in the summer. There were 10-200 times more eggs in the soil samples from the grass surface than from the gravel.

9. Ivermectin Resistance in Cyathostomes in Four Donkey Herds at the Donkey Sanctuary, UK Trawford A, Burden F. The Donkey Sanctuary, Sidmouth, UK

The Donkey Sanctuary has 2500 donkeys residing on its nine farms in the south west of the UK. In 2005 reduced egg reappearance periods (ERP) were noted? for moxidectin in the donkey herds. Since 2005 the FECs in the herds were monitored on a four weekly basis and subsequently identified a reduction in the ERP for ivermectin. To identify ivermectin efficacy in four geographically separate locations (A, B, C, and D) FECR tests were performed 14 days after deworming with an ivermectin paste formulation (Eraquell™). In the four different locations the FECR tests showed mean reductions of 91% (range 0 – 100%), 97% (range 50 – 100%), 84% (range 0 – 100%), and 94% (range 33 – 100%). The number of donkeys that had positive fecals post-dosing were 25%, 10%, 39% and 17% in locations A, B, C, and D respectively. A lack of efficacy of ivermectin was demonstrated based on the reduced ERP and FECRT results and was attributed to the presence of ivermectin resistant cyathostomes.

10. A Larval Migration Assay for Assessing Macrocylic lactone Sensitivity in Equine Cyathostomin Populations McArthur C, Robinson A et al. Moredun Research Institute, Penicuik, near Edinburgh, UK

Traditional *in vivo* methods for detecting resistance (e.g., FECRT) lack sensitivity and standardization. A more promising technique for diagnosing macrocyclic lactone (ML) resistance is an *in vitro* larval migration assay (LMA) for cyathostomins. The "cidal" activity of ML anthelmintics involves paralysis of the parasite. The LMA measures the effect of a range of drug concentrations on the motility of infective third stage cyathostome larvae. Cyathostomin populations of varying ML sensitivity were subjected to ivermectin-LMA and dose-response curves and larval migration inhibition 50 (LMI50) values generated. Two populations of ML-resistant cyathostomins exhibited LMI50 values of 2.41 and 2.86 μg/ml which were higher than the LMI50 values derived from two ML-naïve populations (0.15 and 0.26 μg/ml).

11. The Effect of Sampling Handling and Storage on the Accuracy and Repeatability of Equine Strongyle Fecal Egg Counts Andersen U, Pilegaard K, et al. University of Copenhagen, Taastrup, DE

This study examined the effect of a combination of storage time and storage temperature as well as airtight vs uncovered storage on the number of strongyle eggs (FEC performed using the Stoll technique) and their degree of embryonation. Feces were stored airtight for a total of 120 hours at -18C, 4C, 20C and 37C. During the first 12 hours temperature had no influence on the FEC.

Storage at 4C did not reduce the FEC for the entire 120 hours of storage. Significant reductions were found after 24 hours at -18C and 37C. Availability of air had no significant influence on FEC or degree of embryonation after 24 hrs of storage at room temperature (11.8 – 16.9C).

12. Determination of Ivermectin Strongyle Efficacy and Egg Reappearance Period (ERP) on Danish Horse Farms Larsen M, et al. University of Copenhagen, Taastrup, DE

In Denmark anthelmintic usage has been restricted by veterinary prescription only for a decade. Efficacy of ivermectin (0.2 mg/kg) was evaluated on a total of 196 horses from farms using selective therapy. FECRTs, using the McMAster method, were performed 14 days post treatment. 79 of these horses were infected with *Parascaris equorum*. 117 horses were infected with strongyles only. Overall efficacy of ivermectin was 96.9% against *P equorum* and 100% against strongyles.

13. Prevalence of *Strongylus vulgaris* and Pyrantel Resistant Cyathostomins on Danish Horse Farms Using Selective Therapy Nielsen M et al. University of Copenhagen, Taastrup, DE

Fecal samples were collected from 1644 horses on 64 farms. Horses with FECs of 200 epg or higher were treated with pyrantel embonate paste (19 mg/kg). Larval cultures were performed on all pre-treatment fecals for identification of *Strongylus vulgaris*. Pyrantel resistance (FECRT < 88%) was detected on 19 farms (30%); suspect resistance (FECRT 88 – 92%) on 9 farms (14%) and the remainder of the farms (56%) showed no evidence of resistance (FECRT > 92%). *S vulgaris* was detected in 78 (5%) of individual horses and on 31 (48%) of farms. *S vulgaris* was not associated with pyrantel efficacy; however low pretreatment FECs were associated with its presence.

14. Cyathostome Nematofes (Nematoda-cyathostominae) of Horses: Ecology of Larvae on Grass *Brachiaria humidicola* in the Seasona in Seropedica, RJ, Brazil Rodrigues LA et al. Universidade Federal Rural RJ, IV DPA, RJ-Brazile, Rio de Janeiro, Brazil

The objective was to examine how climatic conditions influence the development and survival of pre-parasitic stages of cyathostomes by examining the ecology of cyathostome larvae in Brachiaria humidicola grass in different seasons in a region of Rio de Janeiro. Feces of naturally infected horses were deposited on grass at the beginning of each season. Weekly FECs were performed on samples collected at 8 am and 5 pm using the Baermann technique. Temperature and precipitation were recorded at the time of sample collection. In the dry, mild seasons (winter and spring) there was higher survival and transmission of L3 to the grass. During rainy months (summer and autumn) there was a lower precalence of L3 in the samples of feces and grass.

Amino-Acetonitrile Derivatives (AADs)

A new class of anthelmintics, **Amino-Acetonitrile Derivatives** was introduced at this symposium. Monepantel (ZOLVIX®), the first member of this new class of synthetic chemicals, has been licensed by Novartis in New Zealand as an anthelmintic for sheep. These sulfur-containing compounds are transformed into sulfoxides that are readily metabolized to sulfones. Comparison of blood concentrations of the sulfone metabolite after intravenous and oral application of monepantel revealed good oral absorption. Within a few hours after oral application, the sulfone metabolite reaches about twice the concentration of the parent drug. This phenomenon is attributed to a strong hepatic first pass effect. Since the metabolite has similar anthelmintic potency compared to the parent drug, the *in vivo* efficacy of monepantel is attributed to the sum of the parent drug and the sulfone metabolite. Monepantel sulfone has a high volume of distribution.

AADs act via the nicotinic acetylcholine receptors of the nematode-specific DEG-3 subfamily. The drug paralyzes the worm by a attacking a previously undiscovered receptor (H_{CO} – MPTL-1) present only in nematodes. Monepantel has demonstrated efficacy against a variety of pathogenic nematode species, both larval and adult stages, at an oral dose of 2.5 mg/kg in sheep. The same dose was sufficient to cure animals infected with resistant or multi-drug resistant nematode isolates, including Haemonchus contortus. The drug also demonstrates good tolerability and low toxicity to mammals. In sheep, the margin of safety is at least 10X the maximum recommended dose. Field studies, conducted at multiple sites, including Scotland, England, France, Switzerland, Australia and New Zealand, confirmed that ZOLVIX was highly effective against field infections of *Haemonchus*, *Teladorsagia*, *Trichostrongylus*, *Cooperia*, *Nematodirus*, *Chabertia* and *Oesophagostomum*.

Additional Comments

There were numerous presentations examining humoral and cellular immune responses stimulated by various parasites in their respective hosts. Investigators from around the world reported on the incidence of drug resistant parasites in ruminants, horses, companion animals, swine, and fish. All drug classes were discussed. New research focuses on identifying the specific mutations that confer resistance among different parasite species. In the future, new drug classes may be discovered not by random experimentation, but by designing a drug with a targeted mode of action. Several sessions discussed the efficacy non-pharmacological parasite control strategies that included improving the nutritional status of the host, the feeding of various plants or shrubs suspected to have anti-parasitic properties. New diagnostic assays were presented and included the use of nested PCR testing in ruminants and other mammals to identify specific parasite species including nematodes, cestodes, and coccidia. The larval migration inhibition assay was discussed as a promising tool to use for the detection of ML-resistance among parasites.

ISSUES IMPACTING THE USE OF ANTIMICROBIAL DRUGS IN FOOD-PRODUCING ANIMALS: CENTER FOR VETERINARY MEDICINE'S PERSPECTIVE

Sanja Modric, D.V.M., Ph.D., Center for Veterinary Medicine, Food and Drug Administration

(Specialty Panel on Antimicrobial Resistance Considerations in Food Animal Internal Medicine)

Major regulatory issues impacting the use of antimicrobial drugs in food-producing animals include the extralabel use of certain drugs, judicious use, veterinary oversight, and non-therapeutic uses. All these issues are recognized in their role for minimizing the risk of emergence and spread of antimicrobial resistance, and are currently being evaluated by the Food and Drug Administration's Center for Veterinary Medicine.

A – Extralabel Use

Extra-label drug use (ELU) refers to the use of an approved drug in a manner that is not in accordance with the approved label directions. The Animal Medicinal Drug Use Clarification Act (AMDUCA) of 1994 provided for legal ELU of approved animal and human drugs in animals (the ELU regulations were implemented in 1996). AMDUCA allows veterinarians to prescribe extra-label uses of certain approved animal drugs and approved human drugs for animals under specified conditions. By statue, FDA also has an authority to issue a prohibition order if evidence shows that extralabel use of a drug in food-producing animals has caused, or is likely to cause, a public health risk (21 CFR 530 – Extralabel Drug Use in Animals).

On July 3, 2008, FDA issued an order prohibiting the extralabel use of cephalosporin antimicrobial drugs in food-producing animals, including, but not limited to, cattle, swine, chickens, and turkeys. The intention of the order was to help further protect consumers against antimicrobial-resistant strains of zoonotic foodborne bacterial pathogens. The prohibition order was based on the emergence of cephalosporin-resistant zoonotic foodborne pathogens, increased resistance to ceftiofur in food animal isolates, and increased resistance to ceftiofur and ceftriaxone in human isolates. The order, which was to take effect on November 30, 2008, would have changed 21 CFR 530.41 to list cephalosporins as prohibited from extralabel use in food-producing animals. The agency received many substantive comments on the order of prohibition, the majority of which fell into two major categories: (1) the order was too broad; (2) it did not meet a legal burden of proof. In order for FDA to fully consider the comments, the agency decided to revoke the order on November 25, 2008. FDA recognizes the importance of veterinary access to effective therapies and is aware of the ELU as a critical tool. Therefore, it carefully considers any decision to exercise its authority to limit extralabel use and does so only when it believes it is necessary to protect the public health. FDA has completed its review of the public comments and intends to issue a revised order prohibiting certain extralabel uses of cephalosporin drugs in food-producing animals. Such revised order will be issued following the procedures in 21 CFR 530.25 that provide for a public comment period prior to implementing the order.

B – Judicious Use

As antimicrobial resistance continues to be identified as a significant public health concern within the public health community, among consumers, and in Congress, one of the often discussed public health strategies focuses on the judicious use of antimicrobial drugs. CVM has decided to take a more proactive approach in this discussion by engaging its stakeholders in the dialogue on the judicious use of antimicrobials, and by trying to seek a consensus on a way forward. CVM's current position is that the continued availability of effective antimicrobial drugs for use in animals must be defended, as they are important for treatment, control, and prevention of disease in animals. Judicious use can help curb the emergence of resistance through reducing unnecessary or inappropriate use. CVM believes there are two key factors that are important for assuring judicious use of medically-important drugs: (a) focusing the use on therapeutic purposes; and (b) veterinary involvement/consultation. Therapeutic purposes include uses that are necessary for ensuring the health and welfare of animals to treat, control, and prevent specifically identified diseases. CVM believes that such uses of medically-important drugs are consistent with "judicious use" principles.

C – Non-Therapeutic Uses

Unlike the therapeutic uses, production (non-therapeutic or subtherapeutic) uses of antimicrobial drugs are not associated with any identified diseases. As such, CVM does not consider non-therapeutic uses of antimicrobial drugs consistent with their judicious use.

D – Veterinary Oversight

Veterinary oversight is required for the use of certain FDA-approved antimicrobials, or antimicrobials used in an extralabel manner. Veterinarians play a critical role in the diagnosis of disease and in the decision-making process related to instituting measures to treat, control, and prevent diseases. Such involvement is valuable for helping to ensure that these drugs are used as judiciously as possible in order to minimize antimicrobial resistance development.

Prior to 1996, there were only two options for regulating the distribution of animal drugs: OTC or prescription-use. In 1996, Congress passed the Animal Drug Availability Act (ADAA) to facilitate the approval and marketing of new animal drugs and medicated feeds. As part of the ADAA, Congress determined that certain new animal drugs should be approved for use in animal feed but only if these medicated feeds were administered under a veterinarian's order and professional supervision. Currently, most feed- and water-use antimicrobials are available over-the-counter (OTC), since most were approved over 30 years ago. Two more recent feed-based products were approved as "veterinary feed directives" (tilmicosin, florfenicol), and no new injectable antimicrobials have been approved as OTC in over 25 years. CVM is exploring approaches for phasing in greater veterinary oversight of previously approved antimicrobial drugs, including those intended for use in feed and water.

In conclusion, there are ongoing public health concerns about the development of antimicrobial resistance and the resulting failure of antimicrobial therapies in humans. In particular, there are concerns raised regarding the use of antimicrobials of human importance in food-producing animals for non-therapeutic purposes. CVM, through its role in promoting public health and protecting animal health, works to ensure that needed drugs are available for treating, controlling, and preventing disease.

COLLABORATION BETWEEN PRIVATE PRACTICE & ACADEMIA

C.W. Dewey, DVM, MS, DACVIM (Neurology), DACVS
Ithaca, NY

INTRODUCTION

For decades, the question over which is "better" for specialists-private practice or academia, has been raised between individuals as well as within individuals. The answer to that question is, of course, dependent on the goals and aspirations of a specific specialist. In the "old days", private specialty practices were few and far between, and university hospitals tended to have the best equipment and the most complete complement of specialists in a single location. During the last 20 years, the landscape of specialized veterinary medicine has changed dramatically. Large, fully-staffed and fully-equipped private referral hospitals now abound and most veterinary specialists seek careers in the private sector following residency training. In addition, many university faculty members have left academic positions for private referral practice. Coincident with the seemingly explosive growth of private referral practices, many university hospitals have experienced either limited growth or steady losses of both faculty and revenue. The advantages of private practice life in comparison with academia have historically included dramatically higher salaries, more freedom to choose a location in which to live, more flexible working hours, limited to no pressure to publish and less bureaucracy. Advantages to an academic career compared to private referral practice include opportunities for research and scholarly pursuits, more teaching opportunities, and job security (assuming that tenure is secured); in addition, although there may be less choice in terms of flexibility of working hours, academia tends to be, in my experience, less hectic than private practice life. Within the last 2 years, the landscape of specialty veterinary medicine has once again changed. And it was not a good change for academia or the private sector. Since the recession has unfolded, many specialists in private practice have experienced salary cuts, and some have had to look for work elsewhere. The days of escalating salary offers from specialty referral centers competing for the same prospective employees have ended. Hiring freezes, salary freezes, faculty position losses, as well as discontinuation of internship and residency positions are some of the recent fallout effects of the recession at many veterinary colleges across the country. Although the financial status of both private and academic referral medicine will likely improve, it is unlikely that things will ever be the same. In some ways, this "leveling effect" of our new economic environment may have some positive value. It is my hope that our collective need to do more with less will be an incentive for increased collaboration between private practice and academia. Related to this hope is that the previously rigid lines of division between what is considered "academic" and what is "private" will become increasingly blurred. This discussion will cover some ideas concerning collaboration between our two worlds in relation to staff/faculty, house officer training, research and publication, and student teaching. Since none of the ideas in this discussion are absolutes, and much of what I will discuss reflects my personal observations and opinions, I invite other viewpoints and ideas from the members of our college during and after the presentation.

STAFF/FACULTY

There are distinct advantages for staff members of a private referral hospital and faculty members at a university hospital to collaborate with each other. Although academic salaries have improved substantially in recent years, there remains a fairly wide gap between academia and the private sector in this regard. I believe that this gap is becoming narrower in our post-recession economy, but it will not likely disappear. It has become commonplace for faculty members to consult at private referral centers. In some instances, the academic specialist provides a service that is not currently available in the private practice; in other cases, the academic specialist is complementary to one or more specialists in the same discipline currently employed at the private practice. In either case, the consulting specialist may provide new and useful clinical information and expertise to the private referral center. In addition, a liaison is formed between the private practice and the university practice through the consultant that can eventually lead to much more than a fee-for-service arrangement. Most universities allow faculty members a certain number of days per month for consultation at other hospitals. Though not specifically stated in many cases, there sometimes exists a tone of disapproval from administrators regarding consulting activity by clinical faculty members. For multiple reasons, I feel that such consulting activity should be encouraged rather than discouraged. There is nothing wrong with making extra money, unless doing so detracts from the individual's responsibilities as a faculty member. This may seem to some to be a bold statement, but it is bold only in the context of the indefensible misconception that academic specialists should not perform outside consultation and make money doing so. In my opinion and experience, such consulting activity actually contributes to the productivity of an academic clinician. The ability to complement one's income by such consultative activity is likely to help close the salary gap between academic and private referral practice and, by doing so, help retain specialists in academia. Consulting at a private referral hospital is likely to enhance the faculty member's reputation as a specialist and provide increased opportunities for clinical research. It has been my experience that consulting tends to increase caseload at both the private practice and the academic institution. More pet owners become aware of the services available and will seek out that service, and often seek out the specific individual who provides that service. The allure of a specialist in private practice offering consultative service to an academic clinical service is not as obvious. Certainly, the financial rewards of such activity will not be a compelling factor. However, there are a substantial number of private practice specialists who actively pursue clinical research. These individuals are not doing this for money, but financial rewards are likely to be obtained from this activity. I have noticed that the thrill of discovery and advancement of our profession is alive and well in the private specialty referral community. I have met numerous private practice specialists who have the drive to investigate new treatment options for their patients, but lack the infrastructure and support required to conduct clinical research projects. This is compounded by the increase in private practice residency programs we have experienced over the years. Since many residency programs require that the resident complete a research project, this is sometimes a difficult task to accomplish in a private practice setting. Consulting at an academic institution can also take the form of teaching. There are some situations in which the academic faculty in a certain discipline may be less experienced than the collaborating private practice specialist. In this scenario, the private practice specialist may fulfill a mentoring role for new faculty members, in some cases performing complicated procedures with those faculty members- simultaneously providing expertise and teaching. Such individuals may be offered adjunct professor titles as an acknowledgement of their contributions to the college.

HOUSE OFFICER TRAINING

For the ACVIM (Neurology) College, the percentage of private practice residency programs is approaching the 50% mark. For some other specialty colleges, the majority of residency programs are in the private sector. There are a number of challenges faced by both private referral centers and academia with regard to residency training programs. These challenges include funding shortages, inadequate caseload, and lack of availability of specialists in other disciplines to provide necessary training requirements. It is very likely that such challenges can, in most cases, be mitigated or eliminated by combining forces. The opportunities for sharing resources between academic and private institutions are numerous and can be as simple as out-rotations (e.g., a private practice resident spending 2 wks with a pathologist) or as involved as "hybrid" residency programs (i.e., the resident spends part of the program in a referral practice, part in a university residency). With the increasing availability of distance learning technology, having group rounds sessions with residents and/or interns at two or more locations simultaneously via the internet is something worth considering. In particular, such an integrated program involving pathologists with expertise in neuropathology may help solve the difficulty facing many residents in fulfilling neuropathology training requirements. The concept of a "hybrid" residency training program is not entirely new, and I feel that the potential advantages of such programs far outweigh the disadvantages. In some cases, funding for residency programs is not available. Splitting the cost of such a program between a private practice and academic institution may allow for an individual to pursue residency training who otherwise would not have had the opportunity to do so. Such a program would benefit both institutions and expose the resident to the private practice and academia models of residency training. In addition to receiving training from a variety of skilled specialists, the resident would serve as a conduit for ideas between the two participating institutions. Such "hybrid" programs would be best suited for referral centers that are not too widely separated geographically. If the institutions involved were already engaged in collaboration/consultative activity among faculty and staff members, this would enhance the facilitation of the residency program.

RESEARCH AND PUBLICATION

For most academic specialists, research and publication is a facet of our career that invokes mixed emotions. Though most of us in academia actually enjoy pursuing this activity, it is not without its frustrations. The "publish or perish" environment of academia necessarily removes some of the joy from these endeavors and replaces it with anxiety. One issue that often arises is the problem of garnering enough cases for conducting clinical trials. For some disorders, it is even difficult to find case records for meaningful retrospective studies. As mentioned previously, many private referral practitioners are actively involved in clinical research or would very much like to be. When I was in private practice, I enjoyed conducting clinical research and publishing papers. I did this because I liked to do it, not because I had to do it. I still perform clinical research and publish in the literature because I really like to do so. My assessment of private practice specialists' frustrations as they related to research and publishing may not apply to everyone in practice with research and publishing proclivities, but I think they may be familiar to many. Some of my colleagues in referral practice have had similar frustrations. These stumbling blocks include difficulty finding reference material, time constraints (no real "off-clinic" time in most private referral settings), lack of statistical support and difficulty procuring funding for research. Part of the time constraint issue in private practice is due to high caseload; though not an absolute, caseload tends to be higher in private referral practice compared to academic referral practice. As with the scenario for residency training, the challenges we face collectively in research and publication can be drastically mitigated by combining forces. Clinical trials can be completed much more rapidly when cases from 2 or more institutions are used instead of just one, journal articles and textbook chapters can be collected at university libraries and shared with private practice collaborators, research grants can be co-authored and submitted using the university infrastructure, and assistance with study design and statistical analysis can be provided by the university site, if needed. The value of publication to an academician is measurable in the sense that a group of people will count all that academician's papers following a certain time period and decide if the quality and quantity are sufficient. The real value of publication in the veterinary literature is difficult to enumerate. These publications advance our profession and/or educate the veterinary community. We find new and better ways to treat our patients, and this provides those involved in such advancements a sense of accomplishment and pride. In addition to the personal satisfaction of discovering and reporting new ideas and treatments in our profession, such activities enhance our reputations in our profession as well as that of our respective institutions. I firmly believe that publications have market value, though this may be difficult or impossible to quantify. I regularly get referrals from veterinarians because of papers I have written, and I also regularly get clients who have found my publications mentioned on the internet and request referral from their family veterinarian. Both clients and veterinarians assume that you have something a little extra to offer their pet if you have published something meaningful regarding that pet's disorder. And it is, I think, a valid assumption. There is nothing wrong with reaping benefits from research and publication that has helped out patients and their owners and advanced our profession. Like most beneficial things, it would be best if we shared in the rewards of our collective hard work.

STUDENT TEACHING

This is a tricky one. There is no doubt in my mind that veterinary students would benefit immensely if they were regularly rotated through private referral practices. In addition to experiencing a high case volume and gaining different perspectives from many specialists, such students would be exposed to practice management concepts that they may not have been exposed to during their formal education. The problems with the concept of regular student rotations in a private referral hospital arise because private referral hospitals do not function like university hospitals and university hospitals do not operate like private referral hospitals. The infrastructure and scheduling of university hospitals is centered on student teaching; this is as it should be. Students are integrated into the day-to-day hospital operations in university hospitals, so that the lack of students on a busy clinical rotation can be a difficult situation to manage. In contrast, private referral practices do not have veterinary students assisting with hospital operations, and introducing students to a private practice setting may interfere with the normal pattern of clinic flow. A large group of veterinary students spending rotations at a private referral hospital is likely to be overwhelming for that hospital. Also, the lack of these students from the university hospital may adversely affect the smooth functioning of clinical rotations at the university. Perhaps a reasonable solution would be for small groups of students (2 or 3) that have expressed an interest in a certain discipline and have completed a university clinical rotation in that discipline, be allowed to rotate through a clinical rotation in a private referral practice in the same specialty discipline. In some university veterinary hospitals, there are specialties that are not represented among the faculty. In these cases, there exists an opportunity for neighboring private referral specialists in those missing disciplines to positively impact the clinical education of veterinary students. Such teaching opportunities may take the form of student externships as already described and/or didactic teaching to the students either directly or remotely (i.e., via the internet).

`For some of the concepts discussed in the above paragraphs, an obvious question is raised: where will funding come from? Who will pay for "hybrid" residency programs, collaborative research efforts, coordinated student externships, etc? There are many potential sources for such funding, including the institutions sponsoring such programs. These sources include private industry, specialty foundations, research foundations and private donors. In addition to the advantages of private referral practices and university hospitals sharing expenses for such endeavors, there are advantages to sharing efforts to procure outside funding for our common goals.

COMPARATIVE ONCOLOGY TRIALS CONSORTIUM (COTC) UPDATES

Melissa C. Paoloni, DVM, DACVIM (Oncology)
Bethesda, MD

INTRODUCTION

The COTC represents a large-scale collaborative effort of the NCI and extramural veterinary academic oncology programs. Through the COTC, the rapid evaluation of early cancer drugs, biologics, medical devices and imaging strategies using biologically intensive trials in companion animals (dogs) with cancer inform their development for human cancer patients. Currently there are 19 members of the COTC, active in trial accrual and/or biological endpoint evaluation (COTC PD Core).

COTC studies are designed to address or inform drug development via the following parameters:

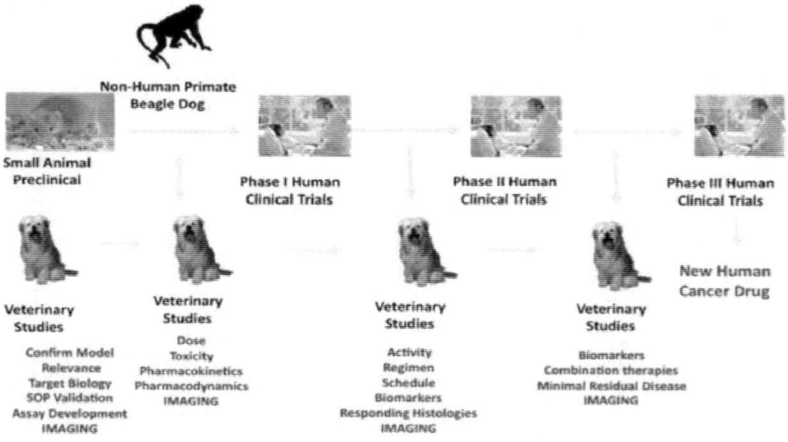

COTC DATA SAFETY MONITORING BOARD

This year has seen the establishment of the COTC DSMB (Data Safety Monitoring Board). The board is a group of individuals with pertinent experience that reviews on a regular basis accumulating data from ongoing COTC trials. The DSMB will advise the COP regarding the continuing safety of current participants and those yet to be recruited, as well as the continuing validity and scientific merit of COTC trials. This advancement serves the interests of our patients, sponsors and allows us to manage comparative oncology trials with the utmost safety and compliance. It comes out of our interest in defining FDA guidelines for reporting in Comparative Oncology trials.

COTC TRIAL UPDATES

- **COTC003/8**: Evaluation of the mTOR inhibitor Rapamycin in Dogs with Osteosarcoma
- **COTC010:** Evaluation of NHS-IL12-IL2 and. NHS-IL12 fusion proteins in tumor bearing dogs
- **COTC013:** Evaluation of Orally Administered mTOR inhibitor Rapamycin in Tumor Bearing Dogs
- **COTC007:** Preclinical Comparison of Three Indenoisoquinolines Candidates in Tumor-Bearing Dogs: A Pilot Study of Topotecan in Dogs with Lymphoma

REFERENCES

[1] COP website: http://ccr.nci.nih.gov/resources/cop

ADVANCES IN LYMPHOMA RESEARCH & TREATMENT: 2010

Kristy L. Richards, PhD, MD
Chapel Hill, NC

INTRODUCTION

In 2009, there were ~66,000 new cases of non-Hodgkin lymphoma (NHL) and ~8500 new cases of Hodgkin lymphoma (HL) in this country[1]. The incidence of NHL has nearly doubled since 1975, while the incidence of HL has remained constant[2,3]. Despite the rise in NHL cases, the mortality from both NHL and HL has been dropping over the past 10-15 years[3], likely due in large part to improved treatments. During the past year, there have been several new advances in the treatment of lymphoma and improved classification of lymphoma subtypes, which likely will continue to positively impact mortality rates in these disorders.

NEW DRUGS

FDA approvals

Of the 26 newly approved drugs by the Food and Drug Administration (FDA) in 2009, five of them were oncology drugs[4]. Of these five, three were drugs approved for B- and T-cell malignancies. The first, pralatrexate, is an antimetabolite similar to methotrexate, and is the first and only drug approved for peripheral T-cell lymphoma[5]. The second, ofatumumab, is a fully humanized anti-CD20 monoclonal antibody approved for relapsed or refractory chronic lymphocytic leukemia (CLL)[6]. The third, romidepsin, is a histone deacetylase inhibitor that is indicated for cutaneous T-cell lymphoma that has received at least one prior therapy[5,7,8].

Investigational agents

Many additional agents, both chemical and biologic, are in the developmental pipeline[5,9,10]. Some of these appear quite promising in early phase trials. New agents are generally targeted therapies, and can broadly divided into two categories, monoclonal antibodies and small molecule inhibitors. Monoclonal antibody development, spurred on by the spectacularly successful anti-CD20 antibody, rituximab, has expanded to include a number of targets in both B- and T-cell lymphomas. New targets include CD22, CD23, CD30, CD40, and CD80. In addition, antibody conjugates, where a toxin is covalently attached, are increasingly being tested. Hybrid, bispecific antibodies that recruit T-cells are also showing promise in early trials.

Small molecules are being developed that target a number of pathways/processes important in lymphocyte survival. These include proteasomes, the mTOR pathway, the BCL2 anti-apoptotic pathway, cyclin dependent kinases, and histone deacetylation. Immunomodulatory agents (-imids) are also showing promise, although their mechanism of action is less well defined. Combinations of these agents, both with each other, and with traditional cytotoxic chemotherapy are being explored and have the potential to rapidly increase the therapeutic armamentarium in both HL and NHL.

NEW TREATMENT REGIMENS

Bendamustine plus rituximab for low-grade NHL

Bendamustine is a relatively new drug in the U.S., but has been in use in Eastern Europe for many years. The FDA approved bendamustine for CLL and relapsed indolent non-Hodgkin lymphoma in 2008[11]. At the American Society of Hematology (ASH) meeting in December, 2009, results from a phase III trial were presented comparing rituximab plus bendamustine (BR) with rituximab, cyclophosphamide, doxorubicin, vincristine, and prednisone (R-CHOP) in the treatment of front-line indolent and mantle cell lymphomas[12]. Results of this study revealed not only that BR was less toxic than R-CHOP, including no hair loss, but also that BR had improved progression free survival compared with R-CHOP. Among 513 patients, overall response rates were 94% in each treatment arm, but a higher complete remission rate (40% vs. 31%) was seen in patients treated with BR. Progression-free survival was significantly longer for BR versus R-CHOP (55 months vs. 35 months; *P*=0.0002). Outcomes were better for BR-treated patients in all lymphoma subgroups studied, but overall survival was similar in both treatment groups (median follow-up, 32 months). Notably, this regimen had activity in mantle cell lymphoma, providing a needed therapeutic option for this difficult-to-treat malignancy. Results of this trial make BR a reasonable choice for first-line treatment of indolent as well as mantle cell lymphomas.

Cardioprotective alternative to R-CHOP

Another trial presented at the ASH annual meeting addressed alternatives to standard anthracyline containing regimens. In this trial, presented by Moccia et al., 81 patients with diffuse large B-cell lymphoma (DLBCL) who had cardiac contraindications to anthracycline therapy (i.e., R-CHOP) were treated with an alternative regimen substituting etoposide for doxorubicin (R-CEOP, rather than R-CHOP)[13]. When compared with historical matched controls treated with R-CHOP, patients who received R-CEOP showed similar five-year time-to-progression rates (62% and 57%, respectively; median follow-up, 28 months) indicating similar efficacy of these regimens. Five-year overall survival rates were significantly worse for patients treated with R-CEOP (49% vs. 64%; *P*=0.02); this was likely secondary to underlying comorbidities of these patients. Toxicities were comparable in patients treated with R-CEOP and those who received R-CHOP. R-CEOP is therefore a viable alternative to R-CHOP, and is a potentially curative option for patients with DLBCL who have contraindications to anthracycline-based chemotherapy.

Reduced intensity regimens for Hodgkin lymphoma

Given the high success rates in treating Hodgkin lymphoma, attempts are now being made to minimize treatment toxicity while maintaining the excellent cure rates that have been achieved. One recent study gives an excellent example of this strategy. Engert et al. recently presented results from a study in 1370 patients with early stage (I or II) favorable HL. Favorable disease lacks the following risk factors: >2 involved nodal areas, large mediastinal mass, extranodal disease, or elevated sedimentation rate. Patients were randomized into one of four treatment groups: 2 vs. 4 cycles of ABVD (doxorubicin, bleomycin, vinblastine, and dacarbazine), combined with 20 vs. 30 cGy of involved-field radiotherapy (IFRT) for consolidation. Complete remissions were achieved in 97% to 99% of patients across all groups; after a median follow-up of 79 to 91 months, no difference in 5-year overall survival, progression-free survival, or freedom from treatment failure were observed (all treatments were associated with rates >90% for all measures). Toxicities were increased with 4 cycles than 2 cycles of ABVD and with 30 Gy than 20 Gy IFRT. These results demonstrate that treatment with 2 cycles of ABVD and 20 Gy IFRT is adequate for the treatment of early stage HL, and should be the new standard of care.

NEW WHO CLASSIFICATION

The classification of lymphomas has evolved over the last fifty years from a purely morphological based system, to a system more reflective of disease biology. This is due to improved understanding of the normal lymphocyte biology, as well as the discovery of cell surface markers able to differentiate subpopulations of lymphocytes. Pathogenetic insights allow a more rational classification of the complex set of disease entities that comprise NHLs and HLs. Infectious agents as well as characteristic chromosomal translocations have been identified that aid in disease classification. To incorporate these scientific advances, and improve pathologic characterization of these diseases, as a route to more systematic treatments, the World Health Organization (WHO) classification of lymphoid malignancies was updated in late 2008.

Features of this new classification system include new categories defining gray-zone lymphomas that show morphologic and immunophenotypic features of two different entities, identification of disease entities by their site of occurrence, age of the patient, or other clinical factors, and recognition of small clonal lymphoid populations. Attempts were made to align diagnostic categories with clinical entities having distinctly different outcomes, although this remains a work in progress for some of the most common lymphomas, such as follicular lymphoma, DLBCL, and peripheral T-cell lymphoma.

New gray-zone lymphomas defined by the 2008 WHO criteria include an intermediate provisional entity between primary mediastinal large B-cell lymphoma (PMBL), a subtype of DLBCL, and the mediastinal nodular sclerosis subtype of classical Hodgkin lymphoma (NSCHL). These overlap cases tend to occur in young men, and to be more aggressive than either PMBL or NSHCL. Another newly defined provisional borderline category is an intermediate malignancy with features of both DLBCL and Burkitt lymphoma. These malignancies are often high-grade, and are often "double-hit" lymphomas, carrying translocations of both 8;14 (IgH-cMyc) and 14;18 (IgH-BCL2). They have an aggressive course and poor prognosis.

Examples of new disease entities defined by clinical features such as site of occurrence or age of the patient include EBV^+ DLBCL of the elderly, an entity thought to be caused by decreased immune surveillance in older people, and several cutaneous T- and B-cell lymphomas that are now recognized as distinctive by their anatomic location.

Clonal, potentially premalignant populations are now recognized to occur, detectable due to more sensitive diagnostic capability as well as better understanding of disease origin and progression. An example of this is monoclonal B lymphocytosis which occurs in 3-5% of healthy adults. Circulating B-cells have the immunophenotype of CLL, and can cause an elevated white blood count that is below the level needed to diagnose CLL. This condition can progress to CLL, but often is characterized by chromosomal abnormalities associated with good-prognosis CLL, so often remains indolent.

SUMMARY

HL and NHL continue to be significant clinical problems in oncology, but progress in understanding disease biology leading to better classifications and better treatments are cause for hope. Further work is needed, but current improvements in treatment and future agents in development are encouraging. Future progress will rely on continuing scientific and clinical advances, and will reamin a model for rational progress in disease classification and treatment leading to improved outcomes for patients.

REFERENCES

[1.] Jemal A, Siegel R, Ward E, Hao Y, Xu J, Thun MJ. Cancer statistics, 2009. **CA Cancer J Clin**. 2009;59:225-249. [2.] Hayat MJ, Howlader N, Reichman ME, Edwards BK. Cancer statistics, trends, and multiple primary cancer analyses from the Surveillance, Epidemiology, and End Results (SEER) Program. Oncologist. 2007;12:20-37. [3.] Horner MJ RL, Krapcho M, Neyman N, Aminou R, Howlader N, Altekruse SF, Feuer EJ, Huang L, Mariotto A, Miller BA, Lewis DR, Eisner MP, Stinchcomb DG, Edwards BK (eds). SEER Cancer Statistics Review, 1975-2006. November 2008 SEER data submission (ed 2009): National Cancer Institute; 2009. [4.] Hughes B. 2009 FDA drug approvals. **Nat Rev Drug Discov**;9:89-92. [5.] Cheson BD. Novel therapies for peripheral T-cell non-Hodgkin's lymphomas. **Curr Opin Hematol**. 2009;16:299-305. [6.] Osterborg A. Ofatumumab, a human anti-CD20 monoclonal antibody. **Expert Opin Biol Ther**;10:439-449. [7.] Campas-Moya C. Romidepsin for the treatment of cutaneous T-cell lymphoma. **Drugs Today** (Barc). 2009;45:787-795. [8.] Piekarz RL, Frye R, Turner M, et al. Phase II multi-institutional trial of the histone deacetylase inhibitor romidepsin as monotherapy for patients with cutaneous T-cell lymphoma. **J Clin Oncol**. 2009;27:5410-5417. [9.] Costello R, Sanchez C, Le Treut T, Rihet P, Imbert J, Sebahoun G. Peripheral T-cell lymphoma gene expression profiling and potential therapeutic exploitations. **Br J Haematol.** 2009. [10.] Tay K, Dunleavy K, Wilson WH. Novel agents for B-cell non-Hodgkin lymphoma: Science and the promise. Blood Rev. [11.] Dennie TW, Kolesar JM. Bendamustine for the treatment of chronic lymphocytic leukemia and rituximab-refractory, indolent B-cell non-Hodgkin lymphoma. **Clin Ther**. 2009;31 Pt 2:2290-2311. [12.] Rummel MJ, Niederle N, Maschmeyer G, et al. Bendamustine Plus Rituximab Is Superior in Respect of Progression Free Survival and CR Rate When Compared to CHOP Plus Rituximab as First-Line Treatment of Patients with Advanced Follicular, Indolent, and Mantle Cell Lymphomas: Final Results of a Randomized Phase III Study of the StiL (Study Group Indolent Lymphomas, Germany). **American Society of Hematology**. New Orleans, LA; 2009. [13.] Moccia AA, Schaff K, Hoskins P, et al. R-CHOP with Etoposide Substituted for Doxorubicin (R-CEOP): Excellent Outcome in Diffuse Large B Cell Lymphoma for Patients with a Contraindication to Anthracyclines. **American Society of Hematology**. New Orleans, LA; 2009.

OF MICE & MEN (AND DOGS!) – VACCINES FOR CANINE MALIGNANT MELANOMA

Philip J. Bergman DVM, MS, PhD, Diplomate ACVIM, Oncology
Armonk, NY

INTRODUCTION

Canine malignant melanoma (CMM) of the oral cavity, nail bed, foot pad and mucocutaneous junction is a spontaneously occurring, highly aggressive and frequently metastatic neoplasm. CMM is a relatively common diagnosis representing ~ 4% of all canine tumors and it is the most common oral tumor in the dog. CMM and advanced human melanoma (HM) are diseases that are initially treated with aggressive local therapies including surgery and/or fractionated radiation therapy; however, systemic metastatic disease is a common sequela. Based on these similarities, CMM appears to be a good clinical model for evaluating new treatments for advanced HM.

Canine patients with advanced disease (WHO stage II, III or IV) have a reported median survival time of 1-5 months with standardized therapies. A combination of hypo-fractionated radiation therapy and chemotherapy have a reported median survival time of one year in stage I oral CMM. Human patients with deep AJCC stage II or stage III disease (locally advanced or regional lymph node involvement) have at least a 50% chance of recurrence after surgical resection; patients with stage IV melanoma (distant metastases) have a median survival of less than ten months and most of these patients eventually die of melanoma. Standard systemic therapy is dacarbazine chemotherapy in HM, and carboplatin chemotherapy in CMM. Unfortunately, response rates to chemotherapy in humans or dogs with advanced melanoma range from 8-28% with little evidence that treatment improves survival. It is easily evident that new approaches to this disease are desperately needed and multiple methodologies have been reported to date.

Active immunotherapy in the form of vaccines represents one potential therapeutic strategy for melanoma. The advent of DNA vaccination circumvents some of the previously encountered hurdles in vaccine development. DNA is relatively inexpensive and simple to purify in large quantity. The antigen of interest is cloned into a bacterial expression plasmid with a constitutively active promoter. The plasmid is introduced into the skin or muscle with an intradermal or intramuscular injection. Once in the skin or muscle, professional antigen presenting cells, particularly dendritic cells, are able to present the transcribed and translated antigen in the proper context of major histocompatibility complex and costimulatory molecules. The bacterial and plasmid DNA itself contains immunostimulatory sequences that may act as a potent immunological adjuvant in the immune response. In clinical trials for infectious disease, DNA immunization has been shown to be safe and effective in inducing immune responses to malaria and human immunodeficiency virus. Although DNA vaccines have induced immune responses to viral proteins, vaccinating against tissue specific self-proteins on cancer cells is clearly a more difficult problem. One way to induce immunity against a tissue specific differentiation antigen on cancer cells is to vaccinate with xenogeneic antigen or DNA that is homologous to the cancer antigen. It has been shown that vaccination of mice with DNA encoding cancer differentiation antigens is ineffective when self-DNA is used, but tumor immunity can be induced by orthologous DNA from another species.

We have chosen to target defined melanoma differentiation antigens of the tyrosinase family. Tyrosinase is a melanosomal glycoprotein, essential in melanin synthesis. The full length human tyrosinase gene was shown to consist of five exons and was localized to chromosome 11q14-q21. Immunization with xenogeneic human DNA encoding tyrosinase family proteins induced antibodies and cytotoxic T cells against syngeneic B16 melanoma cells in C57BL/6 mice, but immunization with mouse tyrosinase-related DNA did not induce detectable immunity. In particular, xenogeneic DNA vaccination induced tumor protection from syngeneic melanoma challenge and autoimmune hypopigmentation. Thus, xenogeneic DNA vaccination could break tolerance against a self tumor differentiation antigen, inducing antibody, T-cell and anti-tumor responses.

EXPERIMENTAL DESIGN

From April 2000 to June 2007, approximately 500 dogs with previously histologically confirmed spontaneous malignant melanoma were treated with xenogeneic DNA vaccinations. Pre-trial evaluation included complete physical examination, a complete blood count and platelet count, serum chemistry profile, urinalysis, lactate dehydrogenase, anti-nuclear antibody, and three-dimensional measurements of the primary tumor if present (or maximal tumor size from medical records if patient was treated prior to pre-trial considerations). For evaluation of metastatic disease, 3-view radiographs of the thorax were obtained and regional lymph nodes were evaluated with fine needle aspiration/cytology and/or biopsy/histopathology. All dogs were clinically staged according to the WHO staging system of stage I (tumor < 2 cm diameter), II (tumors 2-4 cm. diameter, negative nodes), stage III (tumor > 4 cm. and/or positive nodes) or stage IV (distant metastatic disease). The numbers of previous treatments with surgery, radiation and/or chemotherapy were recorded. Dogs with WHO stage II, III or IV histologically confirmed malignant melanoma were allowed entrance onto the various studies due to the lack of effective available systemic treatments. Due to a strong safety profile, dogs with stage I melanoma were allowed inclusion from 2005 on. Additional entry criteria included: an estimated life expectancy of 6 weeks or more, free of clinically detectable brain metastases, no previous therapy (surgery, radiation and/or chemotherapy) for at least 3 weeks and no serious intercurrent medical illnesses. Written consent for entry onto this trial was obtained from each dog's owner prior to entry onto the study; this consent included request for necropsy upon death due to any reason. These studies were performed under Animal Medical Center IRB approval.

Cohorts of three dogs each received increasing doses of xenogeneic plasmid DNA encoding either human tyrosinase (huTyr; 100, 500 or 1500mcg), murine GP75 (muGP75; 100, 500 or 1500mcg), murine tyrosinase (muTyr; 5 dogs at 100mcg or 500mcg each), or muTyr ± HuGM-CSF (9 dogs at 50mcg muTyr, 3 dogs each at 100, 400 or 800mcg HuGM-CSF, or 3 dogs each at 50mcg muTyr with 100, 400 or 800mcg HuGM-CSF) intramuscularly biweekly for a total of 4 vaccinations in the left caudal thigh with the Biojector 2000 jet delivery device with #3 (intramuscular) Bioject syringes. The Biojector2000 is a carbon dioxide powered jet delivery device which is FDA approved for administration of intramuscular injections and has been used in DNA vaccine clinical investigations. Subjective pain level responses and post-vaccinal presence of a wheal or other reaction were assessed and recorded by the veterinarian administering the DNA vaccination. The dogs did not receive any concomitant treatments during the course of vaccination.

RESULTS

The signalments of dogs on this study have been similar to those in previously reported CMM studies. No toxicity was seen in any dogs receiving the aforementioned vaccines with the exception of minimal to mild pain responses at vaccination, one muGP75 dog experienced mild aural depigmentation, and one muTyr dog has experienced moderate foot pad vitiligo. Dogs with stage II-III loco-regionally controlled CMM across the xenogeneic vaccine studies have a Kaplan-Meier (KM) median survival time (MST) of > 2 years (median not yet reached). The KM MST for all stage II-IV dogs treated with huTyr, muGP75 and muTyr are 389, 153 and 224 days, respectively. The KM MST for stage II-IV dogs treated with 50mcg MuTyr, 100/400/800mcg HuGM-CSF or combination MuTyr/HuGM-CSF are 242, 148 and > 900 (median not reached, 6/9 dogs still alive) days, respectively. For dogs on the Phase Ib MuTyr/HuGM-CSF/Combination trial, significant differences in MST were noted across pre-vaccination stage (stage IV MST = 99 days, stage III = 553 days and stage II > 401 days, P < .001). The results from dogs vaccinated with huTyr were published in 2003 (Bergman et al, Clin Cancer Res 2003).

DEVELOPMENT OF SPECIFIC ANTI-TYROSINASE HUMORAL IMMUNE RESPONSES

We have recently investigated the humoral responses of dogs receiving HuTyr as a potential explanation for the long-term survivals seen in some of the dogs on this study. Utilizing standard ELISA with mammalian expressed purified human tyrosinase protein as the target of interest (kind gift of C Andreoni & JC Audonnet, Merial, Inc.), we have found 3/9 dogs with 2-5 fold post-vaccinal humoral responses compared to pre-immune sera. We have confirmed these findings utilizing a flow-cytometric-based assay of pre- and post-vaccinal sera in permeabilized human SK-MEL melanoma cells expressing endogenous human tyrosinase. Interestingly, the three dogs with post-vaccinal anti-HuTyr humoral responses are dogs with unexpected long-term tumor control (Liao et al, Cancer Immunity, 2006). Co-Investigators have also determined that normal dogs receiving the HuTyr-based melanoma vaccine develop Ag-specific IFN-γ T cells (Goubier et al, Vaccine, 2008).

CONCLUSIONS

The results of these trials demonstrate that xenogeneic DNA vaccination in CMM is: 1) safe, 2) develops specific anti-tyrosinase humoral and cell-mediated immune responses, 3) potentially therapeutic with particularly exciting results in stage II/III local-regional controlled disease and dogs receiving MuTyr/HuGM-CSF combination, and 4) an attractive candidate for further evaluation in an adjuvant, minimal residual disease Phase II setting for CMM. A safety and efficacy USDA licensure multi-institutional trial investigating HuTyr in dogs with locally controlled stage II/III oral melanoma was initiated in April, 2006. Human trials of xenogeneic tyrosinase DNA vaccination have recently initiated. In March 2007 and December 2009, we received conditional followed by full licensure (respectively) from the USDA for the canine melanoma vaccine. This represents the first US-government approved vaccine for the treatment of cancer across species.

In summary, CMM is a more clinically faithful therapeutic model for HM when compared to more traditional mouse systems as both human and canine diseases are chemoresistant, radioresistant, share similar metastatic phenotypes/site selectivity, and occur spontaneously in an outbred, immuno-competent scenario. In addition, this work also shows that veterinary cancer centers and human cancer centers can work productively together to benefit veterinary and human patients afflicted with cancer.

MINIMALLY INVASIVE SURGERY

Clarence A. Rawlings, DVM, Ph.D, DACVS
Athens, GA

INTRODUCTION

Rigid endoscopy to examine and obtain biopsies of the nose, urogenital system, ear, and abdomen is standard of practice for most internists. These applications should also be practiced in the high quality general small animal practice and by general surgeons who are attempting to provide good diagnostic examinations in these systems. Minimally invasive surgery as a treatment is the next step to improved patient care. Interventional treatments in the nose, ears, and urogenital system have been developed and are applicable to many patients. The addition of laparoscopic-assisted and thoracoscopic-assisted treatments both reduce surgical invasiveness and trauma, as well as being procedures within the surgical abilities expected in sub-specialists. This presentation on minimally invasive surgery will stress the indications/contraindications for procedures, some technical differences for these procedures, and my philosophy of introducing endoscopy into endoscopy.

CYSTOSCOPY

Rigid cystoscopy should be a standard tool for any clinician attempting complete evaluation and treatment of the urinary system. When learned in a very consistent manner, cystoscopy is our basic introduction to rigid endoscopy and should be performed early in any internal medicine and surgery residency. Cystoscopy is frequently the next step after a thorough history, physical examination, CBC and chemistry profile, and urinalysis. Many patients should also have non-contrast radiographs and ultrasonography performed before cystoscopy. Specific indications for cystoscopy include chronic/recurrent urinary tract disease, hematuria, dysuria, pollakiuria, trauma, calculi, incontinence, and imaging findings from radiographs and ultrasonography. Although the 2.7 cystoscope is the first purchased and standard cystoscope, the referral specialist frequently needs an integrated 1.9 cystoscopy for dogs less than 6 kg and cats, and a 2.5/2.8 flexible fibroscope for the male urethra. A larger cystoscope (3.5 or 4.0) is needed to thoroughly examine the bladder of dogs over 20 kg. This can also be done with the extended length cystoscope used for transcervical catheterization. Transurethral cystoscopy in females is performed in dorsal recumbency and the animals perineum at the table's end. Although the vestibule, urethra, and bladder are examined on the way in, the exit route provides the best examination, digital images, movies, and biopsies. At least 8 biopsies of masses are usually obtained by a flexible biopsy cup forceps passed through the operating channel. When possible, better biopsies can be obtained from the vestibule, and distal urethra and vagina using a 3 mm oval biopsy cup forceps passed beside the cystoscope. The vagina should always be examined.

Cystoscopy has become the **treatment tool for calculi removal, lithotripsy, ectopic ureters, polypectomy, injection augmentation for spay incontinence, and transitional cell cancinoma**

Calculi removal using a 3 or 4-wire basket passed through the cystoscope's operating channel is preferred when urethral and cystic calculi are small enough to be pulled through the urethra. My current guidelines are calculi with a diameter less than 2.5 to 3 times the diameter of the appropriate cystoscope that can be passed through the urethra. During cystoscopic examination of the urethra and bladder of male cats following perineal urethrostomy, calculi have been removed using a 1.9 cystoscope and basket retrieval instruments. After examining the lower urinary tract and flushing the bladder, the basket is passed through the operating channel of preferably a 2.7 cystoscope. The bladder should be only mildly distended and the lavage flow kept low. The goal is to concentrate calculi and reduce the swirling effect when calculi are subjected to high fluid flows. Having the patient in dorsal recumbency with the body tilted with the head up can also move the calculi toward the outflow tract. External abdominal palpation of the bladder can be helpful. The basket is opened in the area of the calculi and gradually closed around the calculi. The procedure is repeated until remaining calculi are removed. Vigorous flushing may also be used to flush the smallest diameter calculi. Leaving the cystoscopic sheath in the urethra with the cranial end in the outflow tract, while squeezing on the bladder, can provide a conduit for small calculi to be flushed from the bladder. Completeness of calculi removal is examined by cystoscopy and a lateral view radiograph.

Most calculi that are too large to be removed by transurethral cystoscopy can be removed by either **laparoscopic-assisted cystoscopy or keyhole transvesicular cystourethroscopy**. Advantages of these techniques are limited peritoneal contamination with urine, improved visibility of the entire lower urinary tract, and more complete removal of calculi. Laparoscopic-assisted cystoscopy is definitely preferred when other procedures such as liver biopsies are combined or when a prior abdominal surgery has been performed. With the bladder secured to the abdominal wall, a small cystotomy is performed and the cystoscope placed through it. A 2.7 mm cystoscope is used, except in small dogs and cats where a 1.9 mm cystoscope is preferred. Calculi are removed with a variety of instruments depending on their size and number. Various alligator forceps, 5 mm Babcock forceps, and arthroscopic grasping forceps can be passed parallel to the cystoscope to grasp and retrieve calculi. Another calculi removal technique is to use a wire basket retrieval instrument. The entire assembly of cystoscope and forceps or basket retrieval device are removed with the calculi, and then the cystoscope and retrieval devices are replaced into the bladder to retrieve the next calculi (below). Following removal of calculi from the bladder and outflow tract, the urethra is examined with the rigid cystoscope in the female and a 2.5/2.8 flexible fiber optic urethroscope in the male dog. It is common to watch a urethral catheter pass around irregularly shaped calculi without feeling resistance to the catheter passage. Urethral strictures just proximal to the os penis from prior calculi obstruction and trauma are common and can justify a scrotal urethrostomy. Urethral calculi are removed with either retrograde flushing or a basket retrieval device passed beside the urethroscope. Again, cystoscopy and a radiograph are used to verify calculi removal.

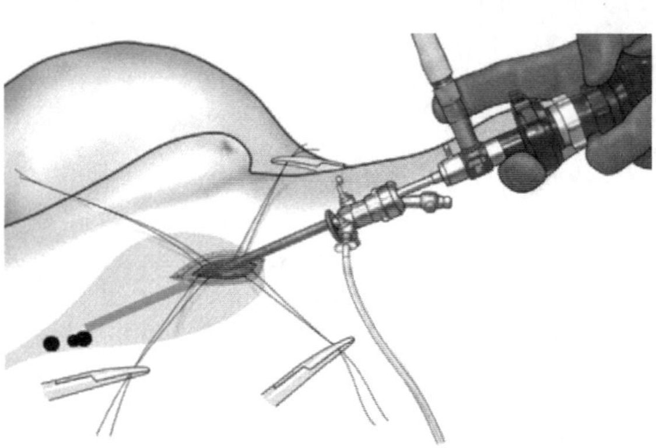

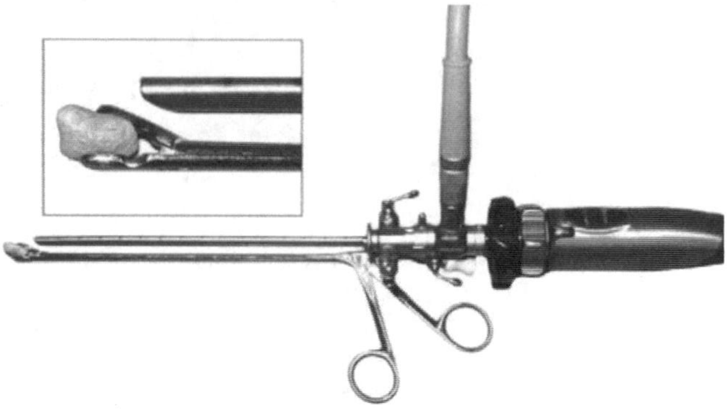

Rarely is it necessary to resort to traditional laparotomy and cystotomy to remove cystic calculi. Typical laparotomy and cystotomy indications are in patients with multiple abdominal surgical indications or very large calculi. Transurethral lithotripsy is also an acceptable calculi treatment, but its wide spread applications is reduced by the contraindications of multiple calculi and large calculi requiring prolonged and potentially traumatic cystoscopy.

Transurethral cystoscopy been used to treat other lesions, such as included strictures, transitional carcinoma, inflammatory polyps, and persisting cystotomy closure suture which may have been related to recurrent urinary tract signs and infections. **Ectopic ureters** are best treated by cutting the membrane between the urethra and intramural ectopic ureter with a diode laser (below). The same laser has also been used to resect transitional cell carcinomas, polyps, urethral strictures, and narrow vestibular-vaginal junctions. **Injection augmentation** as an interventional treatment of sphincter deficit incontinence has gained popularity. It suffers the same disadvantages of colposuspension in that both require supplemental therapy and persistent monitoring as their effects are transient. Injection augmentation by a single cystoscopist is much easier and better when the cystoscope sheath is designed for submucosal injection (below).

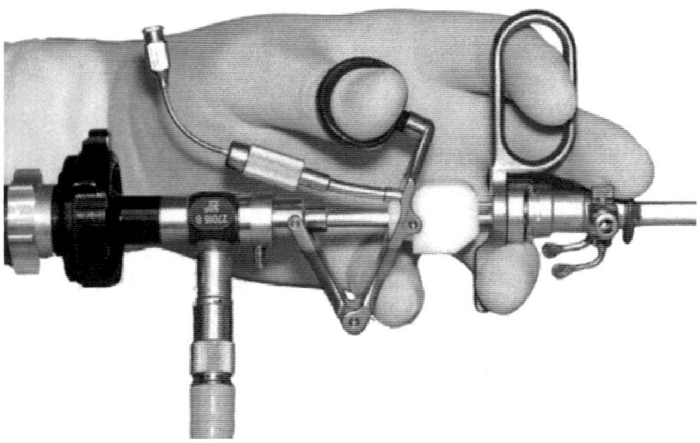

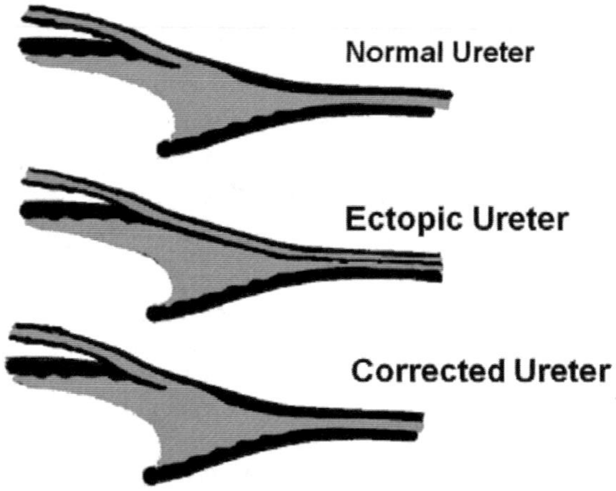

RHINOSCOPY

Rigid endoscopy of the nose has advantages over flexible endoscopy with the ability to flush out hemorrhage and nasal discharge, producing a better optical space. Rigid endoscopy can be used to examine the full length of the nose, avoiding the need for retrograde flexible endoscopy of the nasal pharynx. The primary disadvantage of rigid endoscopy is the difficulty of examining the frontal sinuses, unless there is extensive oteolysis. Small trephine holes can be used to gain rigid endoscopy access to the sinuses. I prefer to have the patient in ventral recumbency, no packing in the pharynx, a slightly tipping down of the nose, and an occlusive endotracheal tube cuff to prevent aspiration. The most useful endoscope is the 2.7 mm, 18 cm long endoscope with a 3.5 mm OD arthroscopy examination sheath. Cats and small dogs are examined with the 1.9 mm cystoscope or an arthroscope. The preferred biopsy device is a 3 mm oval biopsy forceps, which procures large samples for imprint cytology and formalin fixed pathology. This combination of endoscope and biopsy forceps requires some practice, as they are passed parallel to each other during sampling; the tips of the forceps may not always remain in sight when learning the technique (below). After the initial examination and focused biopsies, manipulation and lavage should be vigorous. Many of the best biopsies are manipulated free and flushed into the pharynx and mouth.

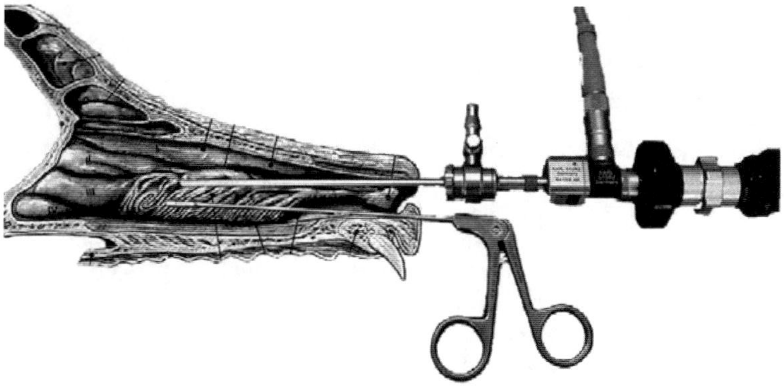

Interventional treatments have included foreign body removal, treatment of choanal stricture, and debulking of nasal cancer. Advanced imaging (CT and MRI) are superior to radiographs for determining the site and extent of nasal disease. Foreign bodies and cancer can be specifically identified and localized. This anatomic knowledge speeds and simplifies endoscopic examination. Patients with nasal cancer may undergo trans nares curettage at the time of diagnostic rhinoscopy. Surgical resection of nasal tumors may prolong survival time and improves quality of life for cancer patients. Debulking provides large, diagnostic samples for histopathologic examination and imprint cytology. Patients quickly over-come clinical signs of nasal obstruction, and repeated curettage in 4-6 months can prolong survival and improve quality of life. We also recommend radiation therapy, but many clients opt for the convenience and lower cost of curettage alone.

VIDEO-OTOSCOPY

Video-otoscopy is frequently relegated to the dermatologist. This is understandable as many ear conditions are a sentinel of systemic dermatologic conditions requiring a more holistic approach. However, ear conditions are common, can produce other signs such as neurologic and respiratory signs, and are frequently seen in patients referred for other problems. Skull radiographs are helpful, but limited in their diagnostic sensitivity for middle ear disease. Computed tomography (CT) and magnetic resonance imagining (MRI} are useful for examining the tympanic bulla. Undiagnosed and incompletely managed middle ear disease is a common cause of persistent otitis externa. In addition, failure to rid the ear of exudate makes it nearly impossible to appropriately medicate the external ear. Otoscope and small endoscopes (1.9 and 2.7 mm) with a cystoscope sheath are used while infusing fluid in anesthetized patients.

Interventional otoscopic treatments are used to manage **middle ear disease in dogs and inflammatory polyps in cats**. Chronic external ear disease with middle ear signs in dogs are common referrals to surgeons. In many dogs, otoscopy can be used to thoroughly clean the bulla via the ear canal. **Aggressive and persistent treatment can markedly reduce the requirement for end-stage ear surgery**. A key indication for salvage ear surgery is the failure to maintain an acceptable external canal. Inflammatory polyps in cats appear to start in the cranial lateral compartment, but the caudal medial compartment frequently contains mucous and debris. The partition between these two compartments is perforated and both compartments are cleaned. Treatment works in most cats, but a few may have recurrence.

LAPAROSCOPY

Laparoscopy is many clinicians' perception of rigid endoscopy and minimally invasive surgery. The minimally invasive approach for diagnostic studies is useful to convince clients to proceed with our frequent recommendations for biopsies of liver, abdominal masses, kidneys, pancreas, and full thickness biopsy selected from several areas of the intestine. Our most common instrumentation is 5 mm except for a 10 mm Babcock or Duvall forceps for grasping the antrum during gastropexy and 3 mm equipment for small dogs and cats. In these smaller patients a 2.7 endoscope with it 30° viewing angle works well. We strongly prefer to place the initial (scope) trocar on the ventral midline using either a Hasson technique or a 12 fr catheter to insufflate the abdomen. After the other trocars are placed, the patient can be rotated to achieve best expose the surgical site. A tilt table is highly desirable. The figures below demonstrate the midline trocars and how gravity is the best retractor when the patient is rotated.

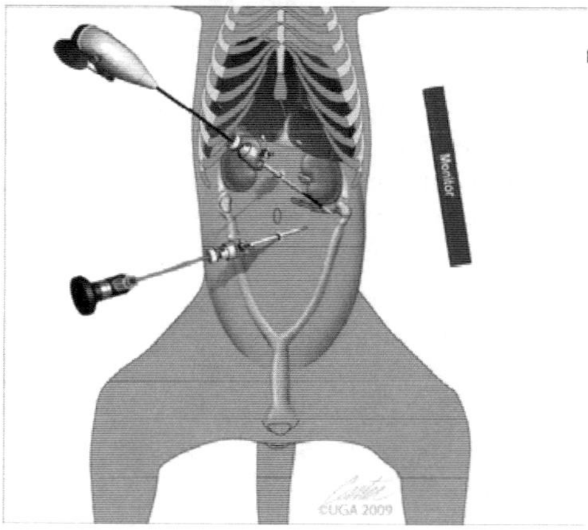

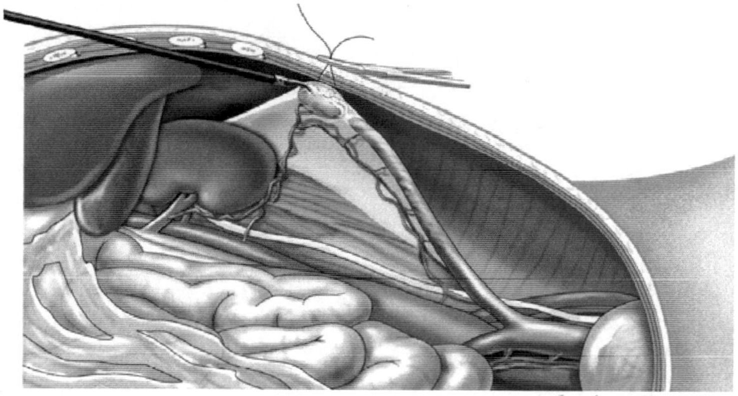

©UGA 2009

The most common laparoscopic diagnostic procedures are for the liver and abdominal masses/cancer/staging. Three to four biopsies should be taken from the liver, selecting the various types of tissue for samples. When cholecentesis is desired, a 1 ¼ inch needle is passed through the quadrate lobe into the gall bladder with the insufflation reduced. The places the abdominal wall close to the gall bladder and the quadrate lobe seals the bladder perforation. Abdominal masses can both be sampled for biopsies as well as to access whether resection is likely. If the mass can be potentially surgically treated, laparoscopy is converted to laparotomy for definitive surgery. Patients with a good quality of liver and a laparoscopic terminal diagnosis can be awakened with only the stress of 2 small trocar incisions. In selected cases, laparoscopic assisted resection of intestinal and splenic masses can be performed.

This century has seen a major expansion of laparoscopic treatments in small animal practice, probably because of the application of laparoscopic-assisted treatments. Briefly, after exploration of the abdomen, the surgical target is grasped and taken outside the abdomen through an enlarged trocar site, where traditional surgery is done. This simplifies endoscopic surgery so that thousands of veterinarians have been trained and many are successfully performing laparoscopic-assisted surgeries. Laparoscopic-assisted **cryptorchidectomy** (below) is an excellent example. **Other laparoscopic-assisted procedures include gastropexy, colopexy, cystopexy, feeding tube placement, cystoscopic calculi removal, cystoscopic polyp resection, ovariohysterectomy, full-thickness intestinal biopsy, gastrointestinal foreign body removal, resection of intestinal tumors, and complete enteroscopy with a flexible scope.**

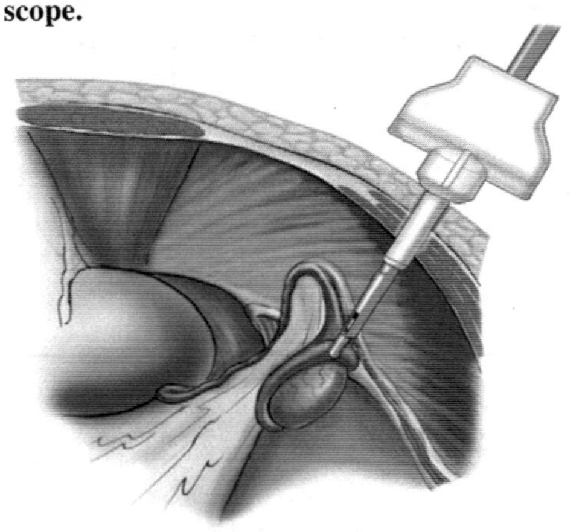

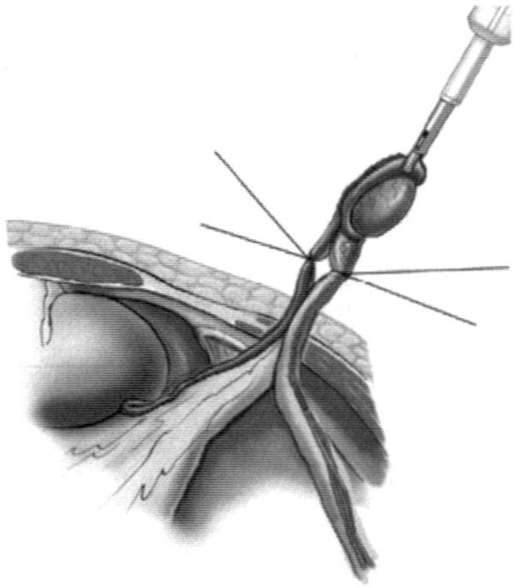

The first laparoscopic-assisted treatment was laparoscopic-assisted gastropexy. It was developed out of frustrations of developing involved treatments that only sub-specialists with expensive tools could consistently perform. These have limited impact on improved patient care. Laparoscopic-assisted procedures are easily learned and should be an important skill development stage as endoscopic surgeons advance to more difficult procedures. For a procedure to impact veterinary care more practices need endoscopic instrumentation and veterinary staff need to be trained. Good progress has been made on both and a focus needs to be **educating pet owners to the advantages of minimally invasive surgery**. To aid these efforts, an endoscopy website has been developed at the University of Georgia (**www.vet.uga.edu/mis**). Practices are encouraged to share this web site with their clients and use it as a link on their website. The third edition of Small Animal Endoscopy (Editors: Tams and Rawlings) will also include a DVD with downloaded client education brochures that practices can add their practice's identification and distribute.

Minimally invasive surgery to spay dogs and cats, as well as **cryptorchidectomy** has been promoted by many general practitioners. Few minimally invasive procedures are as rewarding as removing abdominal testes when compared to traditional laparotomy. The abdomen is easily explored with the patient in Trendelenburg position and with minimal trauma to abdominal organs. The testicle is grasped with 5 mm Babcock forceps and the trocars' opening is enlarged as the testicle is withdrawn to outside the patient. It is ligated and amputated. If both testicles are retained, the more difficult is removed first. Since the trocar site had to be lengthened to remove the testicle, a moist gauze is placed around the trocar and forceps to obstruct gas escaping from abdomen while locating the remaining retained testicle. **Ovariohysterectomy** can be done laparoscopically, but is most efficiently performed by laparoscopic-assistance. Most **ovariectomy** procedures are laparoscopic. There are multiple techniques that are used, but all avoid breaking the suspensory ligaments, probably one of the reasons that laparoscopic spays are less painful after surgery than traditional spays. **Residual ovaries** are best removed laparoscopically. They are easier to identify and remove than through a midline laparotomy. Energy devices are essential for efficiently and effectively completing laparoscopic spays.

Over 1,000 dogs have been operated and hundreds of veterinarians trained to perform **laparoscopic-assisted gastropexies**. Complications following gastropexy have been minimal. Anecdotal experience is that dogs with laparoscopic gastropexies have not developed gastric dilation-volvulus, although some have bloated without twisting. We recommend an ultrasound examination at 6 to 12 months after the gastropexy to insure that a gastropex is present. A trocar for insufflation and the laparoscope is placed on the midline, just caudal to the umbilicus. The second trocar is used for a 10 mm Babcock or Duvall forceps and is placed lateral to the right edge of the rectus abdominus muscle and 3-4 cm caudal to the last rib. The antrum is lifted to the trocar site, which is lengthened sufficiently to perform an incisional gastropexy. The incisional gastropexy is done with a 5 cm long antral incision midway between the greater and lesser curvatures. The seromuscular layer of the antrum is sutured to the transverses abdominus in a fashion typical of an incisional gastropexy. Some bloated dogs without a gastropexy have also been treated laparoscopically. When laparoscopically operating a bloated dog, the surgeon should have permission and be willing to make the decision to convert to a traditional laparotomy.

Enterostomy tubes are indicated when the stomach or antrum is not functional, or when pancreatitis is present. Prior to tube placement, biopsies may be taken from the pancreas and liver. An enterostomy feeding tube can be placed into the descending duodenum with the intestine approached in similar fashion as for gastropexy. The duodenum is apposed to the abdominal wall about the enterostomy site using at least 4 interrupted sutures through the abdominal wall and into the duodenum. A purse-string is placed in the antimesenteric side of the duodenum and an 8 Fr catheter passed throughout the purse-string into the small intestine. The duodenum should be examined with a laparoscope to insure that the tube is advanced aborally. A **gastrostomy tube** can also be done on the left side.

The most frequent laparoscopic-assisted intestinal surgery is to obtain multiple full thickness biopsies. This typically is after failure to obtain a rational histologic diagnosis from biopsies obtained by gastro-duodenoscopy and in patients requiring other biopsies. It is common for medical patients to have sufficient abnormal examinations and laboratory tests to justify tissue from the liver, pancreas, and masses in addition to intestinal biopsies. These biopsies are taken prior to the intestinal biopsies. Again, two trocars are required and a third is needed if "running" the small intestine. The intestine is exteriorized sufficiently to limit contamination from the biopsy site. **Foreign bodies** have been removed from both the stomach and small intestines. Removal of gastric foreign bodies is preferably done by gastroscopy; thus avoiding incisions. Some objects such as golf balls can be difficult to remove via the esophagus. Removal of intestinal foreign bodies are typically focal and in patients with limited disease. Three trocars are typically required and there are advantages to extending the midline trocar site for exteriorization of the intestine containing foreign body. If the bowel appears to be questionable, conversion to laparotomy should be considered. **Large and small intestinal tumors and cecal-colic inversions** have been removed with laparoscopic-assistance. It is critical that these suspected cancer patients have biopsies taken from regional lymph nodes and that target organs for metastasis such as the liver be examined thoroughly. For all intestinal surgeries, the greater omentum is used to patch the enterotomy site. When medical management and a purse-string about the anus have failed to correct recurrent rectal prolapse, **colopexy** has been used to prevent recurrence. A similar technique has been used for **cystopexy** in cases of retroflexed bladder into a perineal hernia. Some have proposed cystopexies for urinary incontinence in spayed female dogs.

THORASCOPY

Compared to thoracotomy, thoracoscopy is much less painful and provides superior examination of the thorax. Indications include **undiagnosed pleural effusion, pericardial effusion, undiagnosed pulmonary disease, mediastinal mass, lymphadenopathy, chylothroax, persistent right aortic arch, patent ductus arteriosus, spontaneous pneumothorax, and pulmonary masses**. Due to the unexpected nature of some of these conditions, conversion to a thoracotomy is a frequent consideration. Thoracoscopy should be done by those experienced and competent in thoracotomy surgeries and who have the capability for patient care following thoracic surgery. Thorascopic-assisted surgery can simplify lung resections while still reducing the trauma of a thoracotomy. Anesthesia can be challenging in cardiopulmonary compromised patients. Endobronchial blockers are used for most thoracoscopies, to provide single lung ventilation; primarily for lung resections, as both lungs are ventilated. .

Examination and biopsies of pleural tissue, mediastinal tissue, and lung are excellent diagnostic procedures. Pericardial resection and thoracic duct ligature are our approach to chylothorax. Many clinicians have resected portions of the pericardium. Our success in treating chylothorax appears to be improved when thorascopically obstructing the thoracic duct, which is much easier to see and expose that when done by thoracotomy. All thoracoscopy patients are managed like thoracotomy patients following surgery. This includes local and systemic analgesic treatment, chest tubes, intranasal oxygen, etc.

OTHER RIGID ENDOSCOPY APPLICATIONS

The surgeon comfortable with rigid endoscopy can frequently apply scopes in traditional surgeries needing improved lighting and visualization. Examples include exploration of fistula, a distended biliary system, cystic and tubular structures, and the inside of the kidney, ureter, bladder, and urethra. The diagnosis and treatment of disease of some of these areas has occurred only when endoscopes were used.

ADVANCED ENDOSCOPIC SURGERY

This is an exciting phase for minimally invasive surgery as a large number of innovative clinicians are continually developing and improving techniques. This is great, but can lead to a limited amount of "evidenced-based medicine" to guide clinical decisions. Care must be taken to develop indications/contraindications for procedures , rather than doing a procedure because it can be done. A critical component of **expanding minimally invasive surgery is the individual's development of endoscopic psychomotor skills**. At Georgia, models to develop psychomotor skills are used prior to animal endoscopy. This includes the **model for gastroenteroscopy, cystoscopy, and laparoscopy** developed in collaboration with Pacific Research Laboratory (Sawbones). Endoscopic surgery can require ligature placement and suture apposition of tissue as a primary means of treatment and to manage complications. Psychomotor excercises in the **SAGES "Fundamentals of Laparoscopic Surgery"** include peg transfer, pattern cutting, ligation of tubular structure with ligature loop and suturing with extra- and intracorporeal techniques. Mastery of these skills aid tremendously to prepare surgeons for the more involved minimally invasive surgeries.

More involved surgeries include **cholecystectomy, adrenalectomy of adrenal tumors, thoracic duct ligature, persistent right aortic arch, patent ductus arteriosus, and intracorporeal intestinal resections and anastomoses.** These procedures are currently being performed in clinical medicine and the list is expanding. The same considerations for intervention for minimally invasive surgery for these conditions are similar to those for traditional surgery.

An exciting combination of flexible endoscopy using natural orificies (**NOTES**) is currently being evaluated for its clinical application in small animal practice. Examples include spays and gastropexies performed via gastroscopy and vaginoscopy using a flexible scope and entries through these viscus organs.

COMPLICATIONS AND CONVERSIONS

Although complications during endoscopic procedures should be rare when properly performed, all client consents should include discussion that complications can occur and that we are prepared to correct them. This includes a signed consent for conversion, ".....Has permission to convert to a traditional (open) procedure if deemed necessary during endoscopic surgery." The operator and team should be prepared to deal with complications.

CONCLUSIONS

Even though minimally invasive should produce less postoperative stress, good practice requires analgesia and critical monitoring after endoscopy. Endoscopy and minimally invasive surgery should be an appropriate extension of a clinician's area of expertise, not just a neat new tool. Regardless of whether a generalist or specialist is performing minimally invasive surgery, there should be appropriate patient selection for endoscopy, good equipment, trained support team, and competent operators.

REFERENCES

Adams L, et al. **J Am Vet Med Assoc**, 2008; 232(7):1026. Bevan JM, et al: **JAVMA** 2009; 234: 1286. Cole TL, et al:. **J Am Vet Med Assoc** 2002; 220: 1483. Fossum TW, et al**: J Vet Intern Med** 2004;18:307. Lulich JP, et al: **JAVMA** 2009; 234: 1279. Miller JM, et al: **J Am Anim Hosp Assoc** 2006; 42: 160. Miller N, et al: **J Am Vet Med Assoc**. 2004;24: 875. Palmeiro BS: **JAVMA** 2004;225:548. Radlinsky MG, et al. **Vet Surg** 2002;31:138. Rawlings CA, **et al**: **J Am Vet Med Assoc.** 2003; 222: 759. Rawlings CA, et al: **Am J Vet Res**. 2001; 62: 871. Rawlings CA, et al: **Am J Vet Res.** 2002;63: 1313. Rawlings CA, et al: **Am J Vet Res** 2002;63: 1226. Rawlings CA, et al: **J Am Vet Med Assoc** 2002;221:1576. Rawlings CA, et al: **J Am Vet Med Assoc**. 2003;222:759. Rawlings CA, et al**: J Am Vet Med Assoc**. 2003;223: 317.

TAKING THE PROBLEM OUT OF THE PROBLEM CLIENT: THE ART & SCIENCE OF CLIENT COMMUNICATION

Nancy Kay, DVM, DACVIM
Rohnert Park, CA

INTRODUCTION

When seeking the fortitude necessary to work with challenging or problem clients, it helps to think back to the reasons why we all chose to do this job in the first place. It's a no-brainer that a big part of why we all chose to become veterinarians is because we love animals (I don't think many of us signed on so we could become rich and famous!). Perhaps the second most common reason we chose this profession is that we appreciate and enjoy the human-animal connection, and our profession is a direct way to interact with and positively influence that bond. No matter how kooky or eccentric or problematic the client might be, remind yourself that most of the time, underlying it all, is a loving human-animal relationship.

THE SCIENCE OF CLIENT COMMUNICATION

There has been a plethora of research on physician-client relationships, [1,2] and some recent work pertaining to veterinarian-client interactions [3,4,5,6,7,8]. There are two basic models of physician-client interactions that have been described, both of which readily apply to veterinarian-client interactions as well. They are referred to as:

Paternalistic Care

The physician assumes the role of patient guardian, and dominates the encounter. The patient's voice is diminished. There is very little two-way communication and the content of the conversation is purely medical. Discussion of options is minimized and the doctor makes strong recommendations. This approach would sound something like, "Take this three times a day and call me next week if things aren't any better."

Relationship Centered Care

The physician acts as an advisor or counselor, and there is negotiation and shared decision-making. Options are discussed, the patient's perspective is important, empathy is provided, and the conversation includes discussion of non-medical topics (inquiries about, family, work, mental well being). This might sound like, "We have a few different options to consider and they are as follows……" This is the model of care that most positively impacts the patient's physical and emotional well being. Additionally, human studies have documented that improved two-way communication directly correlates with:

1. Improved emotional health for both client and physician
2. Improved compliance
3. Improved pain control
4. Improved headache control
5. Improved control of hypertension
6. Improved regulation of diabetes mellitus
7. Fewer malpractice suits

Most of us learned about client communication by the seat of our pants- kind of like having to stick ones finger into the fire and get burned repeatedly in order to figure out a more successful approach! The Argus Institute [9] has raised the instructional bar by providing in depth client communication training to students at Colorado State University. Gradually, more and more veterinary schools are including formal communication training in their curricula. Such instruction is a must when one considers the following trends:

1. Clients are better educated, in large part, because of the Internet.
2. Given changes in accessible technology, there are far more options to discuss with clients.
3. Clients have greater expectations of us, and far more questions.
4. They have access to resources that teach them how to be savvy medical advocates [10] and know it is reasonable to expect relationship centered care.
5. Clients are becoming more litigious and less forgiving of our blunders, especially when communication has been inadequate.
6. Our responsibilities have expanded to include the mental health and well being of our clients.

One can pay a hefty price for inadequate communication- the number one complaint lodged by dissatisfied clients to veterinary regulatory bodies. In fact, a single unhappy client, on average, reports their dissatisfaction to 26 other people. On the other hand, effective communication has many paybacks including: increased job satisfaction, decreased staff turnover, improved patient outcomes, increased client loyalty, financial success, and a decrease in the number of problem clients.

THE ART OF CLIENT COMMUNICATION

When I present information on the Art of Client Communication, I'd like to issue the following disclaimer. This is not information that is backed up by science. Rather, this information represents my impressions gleaned from many years of practicing client communication, and observing behavior modeled by other veterinarians who are happy in their professional lives and have a plethora of satisfied clients.

Empathy

Consider the following question. Which one of the reasons below is the LEAST likely cause of your client's bad behavior?

1. He is scared.
2. She is angry.
3. He is worried.
4. She is in a state of denial.
5. He is physically or emotionally not well.
6. She simply wants to make your life miserable.

I assume (and hope) that you have selected answer number 6. Yet, in the heat of the moment while confronted with a problem client, don't we often reflexively assume that answer number 6 is the reason they are misbehaving? When one recognizes that a client's behavior is a result of overwhelming worry or fear, it is much easier to respond in a supportive, constructive fashion. It helps to differentiate between the client's intentions and their actions. Usually, it is their actions that evoke negative responses. Most clients, however, even those who seem misguided, have genuinely good intentions. By focusing on their intentions and not their actions, it will be much easier to deliver forth patience and compassion.

A by-product of concentrating on intentions is the ability to provide **EMPATHY** to your clients. Here are three descriptions of what empathy means:

1. Empathy is a respectful understanding of what others are experiencing.
2. Empathy occurs only when we have successfully shed all preconceived ideas and judgments.
3. In spite of all similarities, every living situation has, like a newborn child, a new face that has never been before and will never come again. It demands of you a reaction that cannot be prepared beforehand. It demands nothing of what is past. It demands presence, responsibility- it demands you!

Empathy is the single most important ingredient for successful client communication- it is powerful and magical stuff. A wonderful description of the use of empathy can be found in *Nonviolent Communication.* [11] Empathic communication replaces withdrawing, critical judgments, defensiveness, and attacking behaviors. Empathy lets the client to know that you care and understand. It creates a positive and trusting foundation for future interactions and communications. It allows you to:

1. Observe your client without judgment.
2. Learn what your client is feeling.
3. Learn what your client is needing.
4. Hear what your client is requesting.

It's easy to recognize how empathetic communication benefits the clients. But, how does it benefit the veterinarian? The biggest payoff for the veterinarian is that problem clients will become far less problematic! Empathy allows one to "get into the client's head" and what one learns is often downright interesting. It allows one to turn thoughts of "Oh no!" into thoughts of, "How interesting!"

EUTHANASIA

One could spend an entire presentation or more on the euthanasia process. For the sake of time, end-of-life communication will be discussed only briefly here. The manner in which the veterinarian communicates with a client during the period surrounding the euthanasia process has the potential to alleviate or aggravate grief. The most important points for ensuring client satisfaction with the euthanasia process are as follows:

1. The client encounters compassionate and caring attitudes.
2. The client feels that the decision to euthanize was theirs rather than someone else's. Folks who feel that the decision-making was taken out of their hands often have great difficulty making progress through the grief process.
3. The client has been made aware of all choices surrounding the euthanasia and care of remains processes.
4. The client has the option to be present.
5. If performed within the veterinary hospital, a private place is provided for the euthanasia process.
6. The client is informed and prepared for what will happen when their pet is euthanized.
7. Employees are trained to attend to the client's emotional needs.
8. There is appropriate follow-up such as a sympathy card, memorial donation, telephone call, and referral to a pet loss support group or appropriate reading material.

In many situations, the grief process is an ante mortem as well as a post mortem process. It may begin just as soon as a client gets the notion that their pet's illness is serious and might be terminal. Not everyone passes through every stage of grief (denial, anger, bargaining, depression, acceptance), and these stages can occur in any order. Folks can skip from one stage to another, and then back again depending on what suits them emotionally. For example, the client who initially seemed grateful may appear angry a week or two later.

Guilt is not listed as an official stage of grief, but as common as it is amongst our clients, it probably should be given the status of an "honorary" stage of the grief process in veterinary medicine. One study found that 50% of clients reported feeling guilty about their decision to euthanize. Many clients seemingly get "stuck" in guilt. Unlike the five traditional stages of grief, guilt seems to serve no constructive purpose whatsoever. Consider the following strategies to escort a client away from guilt and back towards the healthier stages of grieving.

1. Acknowledge the client's feelings. For example, "It sounds like you feel that you let Roxy down somehow." This will hopefully open up the door to conversation that will then allow for some positive coaching.
2. Don't try to talk the client out of their feelings. Unfortunately, this just doesn't work although, it's so automatic to want to say this, that it sure would be nice if it were a useful counseling strategy. In other words, avoid saying things like, "You shouldn't feel guilty."
3. Try redirecting their focus to their good intentions. Remind them of the wonderful care they've provided over the years. If appropriate, reassure them that, had they brought their pet in earlier, it wouldn't have made a bit of difference. Acknowledge their deep love and concern for their pet.
4. However possible, it is most important that the guilty feeling client not feel judged by any member of the veterinary staff. This can be enormously difficult to do, especially if one feels that the client failed to do something that could have made a difference. In that situation, it's best to fake it. Remember, guilt serves no positive purpose.

VETERINARY ETIQUETTE 101

The following are some basic "Veterinary Etiquette 101" communication recommendations. They should ideally apply to every face-to-face veterinarian-client interaction.

1. Make eye contact.
2. Greet the client solidly- don't ignore the human attached to the other end of the leash.
3. Address the client by name. Use first or last names depending on comfort level. It doesn't matter which name is used- just use it!
4. Be aware of the patient's name, age, breed, and gender.
5. Find something special about the pet to comment on. Sometimes, finding something positive to say is a challenge, but it's kind of a fun challenge. For example, the cat that requires general anesthesia just to be touched- one might comment on how beautiful her eyes are. Remember, that animal in front of you is incredibly special to that person in front of you and it pays to get on board with that.

Just as one learns to do a physical examination or perform an ultrasound evaluation in a set pattern of steps, one should establish a regular routine of healthy habits pertaining to client communication during office visits. This author's ritualistic sequence (Ten Healthy Habits for Improved Client Communication) is as follows:

Step number 1: Summon forth the grown up in myself.
Step number 2: Be prepared to spend the time.
Step number 3; Prepare mentally to be empathetic.
Step number 4: Greet the client solidly.
Step number 5: Check in on how the client is feeling.
Step number 6: Be a superb listener- wonderfully empathetic.
Step number 7: Provide lots of information.
Step number 8: Provide and discuss all reasonable options.
Step number 9: Invite questions.
Step number 10: Repeat the check-in on how the client is feeling

REFERENCES

[1.] Kurtz SM, et al. Teaching and Learning Communication Skills in Medicine; **Radcliffe Medical Press** 2005 [2.] Silverman J, et al. Skills for Communicating with Patients; **Radcliffe Publishing** 2008 [3.] Lewis RE et al. **JAVMA** 2003; 222:1690 [4.] Shaw JR, et al. **JAVMA** 2004; 225:222 [5.] Shaw, JR, et al. **JAVMA** 2004; 224:676 [6.] Kogan, LR, et al. **JAVMA** 2004; 224:504 [7.] Shaw JR, et al. **JAVMA** 2006; 228:714 [8.] Cornell KK, et al. **Vet Clinics of North America**, 2007; 37 [9.] Argus Institute: Colorado State University: **www.argusinstitute.colostate.edu** [10.] Kay ND, Speaking for Spot: Be the Advocate Your Dog Needs to Live a Happy, Healthy, Longer Life; **Trafalgar Square Books** 2008 [11.] Rosenberg MB. Nonviolent Communication; **Puddledancer Press** 2003

DISEASE REPORT & SURVEILLANCE – THE ROLE OF SMALL ANIMALS

George E. Moore DVM, MS, PhD, DACVPM, DACVIM
West Lafayette, IN

INTRODUCTION

Infectious diseases have been an impetus for disease surveillance in people and animals, although primarily due to concern for zoonoses, i.e. transmission of disease from vertebrate animals to man. Disease surveillance is currently conducted for selected diseases of people and of animals by organizations at many different levels, from global down to local.[1] Information for diseases/conditions that are not infectious, e.g. toxicologic, environmental, cancer, birth defects, etc, is also captured by many (human) hospital-based systems in the US but is much less commonly collected for animals.

BIG PLAYERS – PEOPLE AND LARGE ANIMALS

The World Health Organization (WHO), Geneva, Switzerland, is the major disease surveillance organization internationally. The WHO is concerned about rabies, a companion animal zoonotic disease with more than 50,000 deaths reported annually around the world – often in children. Many infectious diseases reportable to WHO are arthropod-borne diseases and can affect companion animals. Global animal disease surveillance is conducted by the Office International des Epizooties (OIE), now termed the World Organization for Animal Health, Paris, France. Although its list of diseases notifiable to the OIE includes diseases of companion animals, e.g. rabies, leptospirosis, tularemia, and anthrax, the disease lists are commonly grouped by large animal species. The OIE is buoyed by the economic value of animals, and the negative impact of animal diseases, within society.

Human disease surveillance for nationally notifiable infectious diseases in the US is spearheaded by the Centers for Disease Control and Prevention (CDC), Atlanta, GA, under the federal Department of Health and Human Services. A list of diseases which should be nationally notifiable is determined by representatives from the Council of State and Territorial Epidemiologists (CSTE), with input from CDC. The CDC operates several disease surveillance systems, many with a focus on specific diseases or etiologies, e.g. West Nile virus, foodborne and waterborne diseases. These systems only collect and report data for human cases of disease, although numbers of animal (including cat and dog) cases of rabies are collected from states as part of rabies surveillance.

The U.S. Department of Agriculture (USDA) Animal and Plant Health Inspection Service (APHIS) has national responsibility for animal health issues, coordinating various aspects of surveillance through the overarching National Animal Health Surveillance System (NAHSS). The goal of the NAHSS is to provide greater protection from endemic, emerging, and foreign animal diseases that could affect U.S. livestock, poultry, and wildlife populations. One component of the multi-faceted NAHSS is the National Animal Health Reporting System (NAHRS). This system was designed for surveillance of confirmed OIE reportable diseases in specific commercial livestock, poultry, and aquaculture species in the United States.

Thus the impetus and the funding for disease surveillance has been the health of the national populous or the national economy.

IN OUR TOWN, COUNTY, OR STATE

The United States has approximately 60 state (or tribal or territorial) health departments and more than 3,000 health departments below state level. The diseases considered notifiable and the requirements for reporting them vary by state. Surveillance may be conducted by state departments of health (who may have a veterinarian on staff) and by local public health offices or officials. This division of responsibility and authority between state and local health departments also varies substantially by state. In some states, rabies is the only zoonotic disease required to be reported.[2] Veterinarians, when on these staffs, play an vital role in informing officials about the importance of animal diseases – including those of companion animals.

Animal disease reporting is usually received by state agencies such as departments of agriculture or boards of animal health. Due to the rising population of companion animals, these agencies increasingly have an individual responsible for disease issues in small animals. Their regulatory concerns are often focused on communicable disease threats from intrastate and interstate movement of animals. These agencies, however, may not have the resources to conduct animal disease prevention and control activities at a local level. The limited resources of these agencies have also created a reliance on passive surveillance for diseases.

AT HOSPITALS

Due to federal and state mandated reporting, human hospitals and clinics have been involved in disease surveillance for many years. Likewise numerous registries and databases have been developed – usually each one independent of the others. For many years, the only system which routinely collected medical information on companion animal disease was the Veterinary Medical Data Base (VMDB) (http://www.vmdb.org), currently housed in Urbana, Illinois. The VMDB collects standardized information on all cases seen at Colleges/Schools of Veterinary Medicine in North America. Data submission however is voluntary and many, but not all, universities participate. Another limitation of the VMDB is the lack of impetus for timely reporting, and institutions lag behind at different rates. Thus the VMDB would not provide timely information on disease surveillance, but could (and has) provide(d) valuable historical trends. A strength of the database is the source of the information – namely, board-certified specialists at university teaching hospitals, thus providing user confidence in the recorded diagnoses. Because these institutions typically operate as referral hospitals, the ability to generalize data results to a larger population is a reasonable concern for some diseases.

DISCONNECTED SYSTEMS

As already described, in the US the reporting requirements for zoonotic diseases can vary widely from state to state in terms of the specific diseases to be reported and the level or agency to receive the report. The intersection of farm animal, companion animal, wildlife, and human health policy are often quite distant. As various agencies pursue their mandates, there are no comprehensive laws or approaches to monitor, report on, or manage companion animal zoonoses.[3] Although not focusing on diseases of companion animals, a National Academy of Sciences report has proposed a coordinated mechanism for improving collaboration and cooperation among local, state, and federal agencies, and the private sector health community.[4] It remains to be determined who will take the lead on this proposal, or how it will receive funding and support for coordinating activities.

IMPORTANT ATTRIBUTES OF REPORTING, SURVEILLANCE, AND INFORMATION SYSTEMS

CDC has published Guidelines for Evaluating Surveillance Systems to promote the best use of public health resources.[5] This guidance provided 9 system attributes which can be assessed as part of an overall system evaluation, and such attributes should also be considered for any system used in companion animal disease surveillance.

Surveillance systems should be as simple as possible while still capable of meeting the system's objectives. This simplicity should be evident from the user interface for data entry through the analysis phase and the generation of reports. Simplicity can also affect the system's acceptability by participants and end-users. Reduced acceptability can affect reporting rates, timeliness of reporting, and completeness of reports. A flexible surveillance system can adapt to changing needs with little additional time, personnel, or allocated funds. This flexibility is important when considering the addition of new or additional diseases, new data streams, or new information technology. System stability impacts its availability and reliably for timely data collection, management, and analysis. Unexpected outages, downtimes, or lack of dedicated resources can dramatically reduce the utility and effectiveness of disease surveillance. Timeliness is important even for routine operations. Since the goal of surveillance is to provide information leading to disease prevention and/or control, this information must be disseminated in time to 'make a difference', e.g. curtail an outbreak. Hand entry of data delays timeliness but may be necessary in some systems. Electronic data collection and data interchange can improve timeliness, but steps may still be required to validate entered or captured data. Thus, data quality is critical to the success of any surveillance system. Data must be complete yet also standardized in entry, e.g. abbreviations, for easy retrieval and analysis. Data that are complete and valid are also necessary to determine the representativeness of the collected information relative to actual events. Systems often collect information from a limited (sample) population with inferences then made to a larger target population. Assuming the aforementioned system attributes are satisfactory, a surveillance system must be capable of identifying an appropriate proportion of the health events of interest. The sensitivity of a system however is often influenced by factors apart from the technical structure of the system. Identification of diseases can be affected by the actual rate of occurrence of the disease, the likelihood of a patient entering the health care system utilizing surveillance, and the skill and testing methodologies of the medical professionals potentially reporting cases. Some systems seek to improve timeliness and sensitivity by collecting information without waiting for confirmation of the suspected diagnosis. A response however is ideally founded on a high proportion of reported cases that actually have the disease under surveillance – the predictive value positive (PVP). A system with low PVP for a health event can result in incorrect communication pertaining to the event, and even misdirected resources. Because the PVP is an indicator of erroneous, or false-positive, reports it may also point toward larger scale problems in data validity, and ultimately in acceptability of the system.

NEW DATA SOURCES FOR COMPANION ANIMAL DISEASE SURVEILLANCE

Changes in veterinary practices and practice management are providing valuable new resources that increase our capabilities in companion animal disease surveillance. Some privately- and corporately-owned practices in the United States however now have tens, or even hundreds, of hospital locations and provide primary care to thousands, or even millions, of patients annually. Although such large numbers of locations and patients are extremely valuable for surveillance, medical information from different locations must be linked, in compatible language, and accessible (via networks and/or permissions).

Many companion animal practitioners rely on commercial diagnostic laboratories for prompt pick-up, delivery, and testing of diagnostic samples and specimens from their patients. Although patient information (e.g. signalment, history, and clinical signs) associated with the sample or specimen is usually limited, the typically large geographic area of service and number of submissions makes veterinary laboratory databases a rich source of information for surveillance. Veterinary laboratory databases have already been used to investigate the epidemiology of infectious diseases using serologic and microbiologic test results.

CONCLUSION

Nationally directed support, coordination, and funding for companion animal disease reporting and surveillance has been, and most likely will continue to be, negatively impacted by the lack of a direct economic or health impact on people from these diseases. Visionary and non-parochial leaders in practice, academia, industry, and government are needed to pull together resources for improved companion animal disease reporting and surveillance at multiple levels and from different sources.

REFERENCES

[1]Moore GE, et al. **Vet Clin North Am Small Anim Pract** 2009;39(2):225. [2]Wohl JS, et al. **J Am Vet Med Assoc** 2007;230(4):494. [3]Reaser JK, et al. **Zoonoses Public Health** 2008;55(8-10):385. [4]National Research Council. **National Academies Press**, 2005. [5]CDC. **MMWR Recomm Rep** 2001;50(RR-13):1.

GREAT EXPECTATIONS WHAT YOUR CLIENT EXPECTS OF YOU IN 2010

Nancy Kay, DVM, DACVIM
Rohnert Park, CA

INTRODUCTION

Unlike human medical doctors who need to think only in terms of their "patient" (and in school, they need only learn the medical nuances of one, rather than multiple species), we are doubly blessed with both "patients" and "clients." What we do with our patients is usually the simple part. What we do with our clients is far more complex, often requiring significant expenditures of our time and creative energy. When you think about it, our roles as veterinary health professionals extend far beyond preserving and restoring the health of our clients' pets. On a daily basis, any one of us might be expected to assume the role of social worker, financial planner, decision-making advisor, and even psychotherapist. It's no wonder we're constantly running behind schedule!

Our clients' expectations are evolving at a rapid pace, seemingly in association with their ready access to lots of information via the Internet. Additionally we have entered the age of the "empowered patient" in human medicine and the "empowered client" in our profession.

HOW CLIENT EXPECTATIONS ARE CHANGING

Relationship Centered Care

Our clients are better informed than ever before and know that it is reasonable to expect to be involved in every aspect of their pet's health care. Gone are the days of simply following doc's orders. They are looking for relationship centered care defined by collaborative communication with their veterinarians. [(1)] They seek active exchange of information, consideration of their personal circumstances, and empathic feedback. Paternalistic care (a style of communication that diminishes the voice of the client) is falling by the wayside.

Discussion About Their Internet Research

It's a fact of life- our clients are going to avail themselves of the Internet to research their pet's diagnosis and symptoms. It's also a fact of life that, those who are brave enough, will want to discuss what they've learned with us. My recommendation is, "If you can't fight 'em, join 'em!" A key to making these Internet discussions as palatable as possible is to steer our clients in the direction of Websites that provide valuable information. I encourage you to teach your clients how to find instructive, accurate information while avoiding "online junk food." In addition to supplying your own Website recommendations, provide your clients with the following guidelines:

-Veterinary college Web sites invariably provide reliable information. Search for them by entering "veterinary college" or "veterinary school" after the name of the disease or symptom being researched.

-Web addresses ending in ".org," ".edu," and ".gov," represent nonprofit organizations, educational institutions, and governmental agencies, respectively. They will likely be sources of objective and accurate information.

-If researching a breed-specific disease, pay a visit to the site hosted by that specific breed's national organization.

-Avoid business-sponsored Web sites that stand to make money if one believes and act open what they profess (especially if it involves purchasing something).

- Be ever so wary of anecdotal information. It's perfectly okay to indulge in remarkable tales (how global warming is the cause of hip dsyplasia), but one should view that which is read as fiction rather than fact. As fascinating as these *National Enquirer* type stories may be, they are not worthy of discussion during an office visit (unless some comic relief is needed!).

-Check out disease-specific online forums such as those sponsored by Yahoo (http://groups.yahoo.com). Not only do they provide a wealth of educational information, members can be a wonderful source of emotional support- always a good thing for those of us who share their homes and hearts with an animal. Look for a group that focuses on a specific disease (kidney failure, diabetes, etc), has lots of members, and has been around for several years. For example, an excellent Yahoo group is AddisonDogs (addresses issues pertaining to dogs with Addison's Disease) has 3,509 members (as of February, 2010) and has been up and running for eight years. Large groups such as this typically have multiple moderators who screen participants, screen comments to keep things on topic, present more than one point of view (always a good thing), and provide greater round-the-clock availability for advice and support. The moderators present cited references from well-respected veterinary journals that allow them to provide evidence based recommendations. Such groups should have a homepage that explains the focus of the group and provides the number of members and posts per month (the more the better). They may have public archives of previous posts that can provide a wealth of information.

Internet Access To You

Some clients prefer to communicate via email rather than telephone. This may help eliminate the game of phone tag we often play with our clients. The disadvantage is that can be harder to impose limits on the amount of time spent discoursing with our clients.

Access To The Entire Hospital and Being Present for Procedures

These days many clients express reluctance to relinquish their pets to a nurse who will take them to "the back" of the hospital. Needless to say, such requests often require some compromise. We are all well versed in the potential pitfalls and disadvantages of allowing clients into the "bowels of the hospital". However, there are some significant potential advantages. The client's medical understanding might be enhanced if they get to see what you see (radiograph, real time ultrasound image, etc.). The patient may be calmer and happier when their favorite human is present (of course, we've all seen cases where this has backfired). Additionally, when clients actually sees the bustling staff and fancy equipment that we all take for granted, they may have greater perceived value when it comes time to pay their bill.

Discussion And Open Mindedness About Vaccinations

Fewer clients behave like "Stepford Wives" by simply handing their pet to receive the myriad of vaccinations listed on the vaccine reminder card. They know that the "once a year habit" is no longer the norm. They expect discussion about the pros and cons of vaccinating, vaccine selection based on disease exposure, and separation of vaccinations (giving only one vaccination at a time). Additionally, more and more clients these days prefer to forego ongoing adult vaccinations altogether. If you haven't already done so, I encourage you to familiarize yourself with the vaccination guidelines set forth by the American Animal Hospital Association.[(2)]

Nutrition/Diet Information And Advice

Unfortunately, most of us received meager training on the topic of nutrition. Yet, more than ever before, our clients are demanding information about what their cats and dogs should be eating. Books that provide recipes for homemade diets are selling like hotcakes (no pun intended)[(3)] and the raw food trend is escalating. While it is easy for everyone to agree that a diet should be "balanced" and that certain diseases should be treated with specific dietary modifications, there is little evidence based medicine to support much of anything else we might recommend or discourage.

Access To Progressive Veterinary Medicine

These days, clients are more informed than ever about all the "bells and whistles" in terms of what veterinary medicine has to offer. They have access to resources that teach them how to become savvy medical advocates for their four-legged family members.[(4)] They expect discussion about second opinions, referral to specialists, holistic care, twenty-four hour care, and access to advanced diagnostic technology and therapeutics (joint replacement, advanced imaging, rehabilitation therapy, cancer therapy aggressive pain management, stem cell therapy, etc.).

More "Bark For Their Buck"

Unless things have changed drastically between the due date for these proceedings (February, 2010) and now, we are still going to be faced with more clients than ever before who need to creatively stretch their veterinary health care dollars. Here are some ways they expect to accomplish this:

-Filling their prescriptions less expensively. The AVMA mandates, "Veterinarians should honor client requests to prescribe rather than dispense a drug (AVMA Principles of Veterinary Medical Ethics). The client has the option of filling a prescription at any pharmacy."[(5)]
-Written cost estimates before services are provided.
-Discussion of all options, regardless of cost.
-Information about payment plans.
-Information about insurance providers.

Discussion About All End-Of-Life Options Including Hospice Care

Some clients wish to discuss the option of euthanasia outside of the veterinary hospital setting. If we cannot accommodate this request ourselves, it is important to be able to refer them to other veterinarians who can accommodate this request. Additionally, it is ideal if we can provide them with resources for emotional support (both prior to and following the loss of their pet). Some people are opposed to humane euthanasia for moral, philosophical, emotional, or religious reasons. For years, such clients have created a conundrum for us, pitting our professional obligation to ease patient suffering against the client's convictions. The answer to this moral/ethical dilemma comes in the form of hospice care- an alternative that emphasizes patient comfort and emotional support for both the patient and the family. The hospice movement is rapidly growing.[(5)] Some clients choose such palliative/supportive care until it is clear that euthanasia is indicated. For others, hospice care is provided until death occurs naturally. It behooves us to either provide hospice care or be able to refer to a veterinarian who does.

REFERENCES

[1.] Shaw JR, et al. **JAVMA** 2004; 224:676 [2.] American Animal Hospital Association Canine Vaccine Guidelines. 2006; **www.aahanet.org/PublicDocuments/VaccineGuidelines06Revised.pdf** [3.] Pitcairn RH, et al, Dr. Pitcairn's **New Complete Guide to Natural Health for Dogs and Cats**; Rodale Books 2005 [4.] Kay ND, **Speaking for Spot**: Be the Advocate Your Dog Needs to Live a Happy, Healthy, Longer Life; Trafalgar Square Books 2008 [5.] **AVMA** Policy Internet Pharmacies: www.avma.org/issues/policy/internet_pharmacies.asp [6.] **JAVMA** News; Pet Hospice Movement Gaining Momentum 2009; www.avma.org/onlnews/javma/apr09/090415o.asp [7.] PDR for Nonprescription Drugs, Dietary Supplements, and Herbs; PDR Network, LLC; 2010

INFLUENZA VIRUS INFECTIONS IN CATS – DO WE HAVE TO WORRY?

Katrin Hartmann, Prof., Dr. med. vet., Dr. habil., DECVIM
Munich, Germany

INTRODUCTION

Influenza viruses are negative-sense, single-stranded, segmented RNA viruses belonging to the *Orthomyxoviridae* that infect mammals and birds. Influenza viruses are enveloped and relatively susceptible to environmental influences. Survival is longer in colder temperatures and moist conditions. Common disinfectants, warm temperatures, and low humidity are effective inactivations. In the absence of organic material, most common disinfectants inactivate the virus within 10 minutes. Influenza virus primarily spreads through the transfer of virus-containing respiratory secretions *via* aerosolized particles from an infected to susceptible individuals. Fomite transmission through contact with contaminated surfaces or common use of items also can occur.

Aquatic birds are generally asymptomatic reservoirs for the viruses. Almost every winter, two genera (usually types A and B) produce episodic outbreaks of an acute self-limiting febrile illness in people. Types B and C are mainly human pathogens and are only rarely seen in animals. Type A viruses have the widest host range and greatest mutability and can rapidly evolve into new genetic subtypes. Cats and dogs living in close proximity to humans may occasionally become infected during human influenza type A outbreaks. Furthermore, cats or dogs that have close contact to poultry, swine, or horses, or that eat uncooked meat byproducts may become infected. Subtypes of influenza A viruses are specified by the antigenic nature of 2 surface glycoproteins (hemagglutinin [H1-H16] and neuraminidase [N1-N9]). Influenza viruses continuously evolve through point mutations in their nucleic acid which leads to a gradual "drift" in their genome encoding surface proteins. Mutations in strains with hemagglutinins of subtype H5 or H7 can sometimes result in highly pathogenic avian influenza virus (HPAIV) strains that may cause significant mortality. Furthermore, the segmented genome of influenza viruses allows reassortment during coinfection within a single host species. This drives evolution of influenza viruses in avian species but also in mammals.

In experimental studies cats are susceptible to different influenza A viruses, including human (H3N2, H2N2), avian (H7N3), and seal (H7N7) influenza virus isolates. In some of the experiments, infection could also be transmitted to uninfected cage mates. After infection, cats developed antibodies, sometimes even shed virus, but never became sick. These earlier studies suggested that cats might become a host for influenza viruses in nature.[1] Surveys in healthy field cats in the 1970s found antibodies against influenza viruses of subtype H3N2. No antibodies against the subtypes H1, H2, H4 through H8 were detected. As human epidemics with influenza H3N2 viruses occurred in the countries of investigation, it might be that the cats had been exposed to this particular virus. Because of the close association of pets with humans, concern has been raised, although unnecessarily, that cats and dogs may be important in the spread or maintenance of influenza infection in people. Influenza infection that jumps between different species is often a self-limiting spread and does not transmit between in-contact animals. No evidence suggests that the virus replicates sufficiently to spread from infected pets back to humans. Prevention of human infection involves minimizing direct contact with animal reservoirs such as swine and poultry.

The first instance of clinically apparent influenza virus infection associated with sustained transmission in canine populations occurred in the United States in Florida. In 1999, a strain of H3N8 equine influenza virus that was circulating in horses adaptively evolved *via* genetic mutation-induced alterations in the HA protein and readily infected racing greyhound dogs in the United States. Isolation of the virus was first made from dogs affected in a respiratory disease outbreak in Florida in 2004 and homologous isolates were found in greyhounds and other breeds in this and other states.[2] H3N8 virus isolates from the Florida outbreak were determined to be a host-adapted canine influenza virus; it readily replicated in dogs and was shown to have spread in a sporadic fashion throughout the United States. There is no evidence that the H3N8 canine influenza virus can spread from dogs to people, and there is no evidence that the canine H3N8 virus infects other animals, such as cats.

SWINE INFLUENZA (H1N1) VIRUS INFECTIONS IN CATS

During the recent pandemic outbreak of swine origin influenza virus A H1N1 infection in human beings in 2009, cases of H1N1 infection in cats have been reported. Prior to reports in cats, the virus had been only detected in human beings, pigs, birds, and ferrets. The first reports of cats with H1N1 occured in separate incidents in Iowa and Utah, United States, where two 13-year old cats developed signs of a respiratory infection after family members in the household were ill.[3] These cats had antibodies against H1N1. The cats and their owners have recovered from their illnesses and there is no evidence that the cats passed the virus to any people. Polymerase chain reaction (PCR) results, targeting the H1N1 virus, were negative in these 2 cats; however, this may relate to the low rate and transient nature of shedding in heterologous hosts.

AVIAN INFLUENZA (H5N1) VIRUS INFECTIONS IN CATS

All influenza H and N subtypes can be found in different combinations in different aquatic birds which represent the natural reservoir of all influenza viruses. Generally, influenza virus infections in aquatic birds are of low pathogenicity and are not associated with clinical signs. Avian influenza virus of subtype H5 and H7 are exceptional in that they can mutate enabling systemic replication of the virus in the infected avian host. Infections with these highly pathogenic avian influenza viruses (HPAIV) are associated with various symptoms in water birds and severe clinical signs in domestic poultry. Some of these viruses also acquire a broadened host spectrum.

Pandemic HPAIV of subtype H5N1 originated in China in 1996 and spread until 2003 to other countries in Southeast Asia. The viruses, continuously shaped by several reassortment events and by antigenic drift, established endemic infections in this area, driven by subclinically infected duck populations. Apart from causing severe clinical signs in gallinaceous bird populations, HPAIV

H5N1 also exhibited a potential to infect mammals including humans. Next to human beings, naturally infected mammal species included members of the order *Carnivora* (*Canidae*, *Felidae*, *Mustelidae*, and *Viveridae*).

Several outbreaks of infection with HPAIV H5N1 have been reported in felids under natural conditions. The first outbreak was seen in 2003, when 2 tigers and 2 leopards suffering from high fever and respiratory distress died in a zoo in Suphanburi, Thailand.[4] Shortly after, a clouded leopard died in a zoo in Chonburi, Thailand, from infection with influenza H5N1. One month later, a tiger at the same zoo was found to be infected but recovered from the disease.[5] During an outbreak in a tiger zoo in Sriracha, Thailand, a total of 147 tigers died or were euthanized.[6] First evidence that domestic cats are at risk was in 2004, when 3 domestic cats from a household in Thailand, in which 14 cats had died, were tested positive for influenza H5N1.[5] The virus was also detected in a domestic cat in Thailand that had died showing high fever, dyspnea, convulsions, and ataxia.[7] Experimental infections using HPAIV H5N1 confirmed that cats can develop severe clinical signs after infection.[8,9] The first cases of HPAIV H5N1 infection in domestic cats in Europe were detected during the outbreak of avian influenza on the German Isle of Rügen in February 2006, during which 3 free-roaming cats found dead were harboring the virus.[10] At approximately the same time, 3 cats that did not show clinical signs tested positive by PCR for influenza H5N1 in an animal shelter in Graz, Austria, after an infected swan had been brought to the shelter.[11]

Epidemiology

Infections of felids with HPAIV H5N1 have been documented in areas with HPAIV H5N1 infections in poultry or wild birds in the surrounding area.[4,6,11] Phylogenetic analyses have shown that viral isolates from cats and tigers are highly similar to the avian virus circulating in poultry at the same time, indicating that no genetic reassortment events with mammalian influenza viruses had occurred. Several point mutations have been identified that are associated with higher virulence in mammals; however, none of them seems to be essential for the infection in felids.[4,12,13,14] Infections in felids were reported from 8 countries in Asia (China, Thailand, Vietnam, Indonesia, Iraq) and Europe (Austria, Germany). Presence of antibodies and clinical evidence of HPAIV H5N1 infections in different species of wild cats (e.g., Asiatic Golden Cat, Clouded Leopard) has been reported from Cambodia.[15] Despite the described sporadic HPAIV H5N1 cases in felids, the prevalence among pet cats in European countries seens to be low. A study conducted in Germany included 171 cats with outdoor access in areas in which infected birds had been found at the time of investigation. Neither virus excretion nor antibodies were detected among the cats.[16] In Italy, another epidemiologic study included cats in the city of Milan.[17] In this study, there was also no evidence of antibodies against HPAIV H5N1. As no outbreaks of avian influenza had occurred in the sampling area at that time, the likelihood of finding antibody-positive cats was most probably low. As the tests performed in these studies are able to detect antibodies against all influenza A viruses, the investigations also confirmed that cats are not prone to infections with the seasonal influenza viruses circulating in humans. In contrast to these investigations, 7% of field cats (8/111) tested in an unpublished study of the National Institute of Animal Health in Bangkok and 20% of 500 cats tested in Indonesia had antibodies to influenza A H5N1. It is possible that the virus strains circulating in Asia have a higher virulence for cats compared with the European virus strains, and thus, more cats become infected in Asia. Within HPAIV H5N1, at least 2 genetically and antigenically distinct lineages (clades 1 and 2) exist with different geographic distributions in Asia. Clade 1 was isolated mainly in Vietnam and Thailand, whereas clade 2 was found mainly in China and Indonesia. From there, the Qinghai-like sublineage spread to the Middle East, Europe, and Africa, and separated into 3 different subclusters. All Asian cases reported in felids were caused by infections with clade 1 viruses. The outbreak in domestic cats in Iraq and the fatal cases in Germany were caused by Qinghai-like clade 2 viruses. Therefore, cats seem to be susceptible to different circulating H5N1 virus strains.[14,18] Whether these differences contribute to the different antibody prevalences reported from Asia and Europe remains to be investigated. It seems more likely that the difference in prevalence among cats is caused by higher risk of exposure to the virus in Asia. Compared with the situation in Asia, only a relatively small number of wild birds and poultry farms were affected in Europe during the last years. Additionally, cats included in the European studies were pet cats, which were fed by their owners and thus, would have been most probably only been exposed during bird hunting.

Pathogenesis

Infection of felids occurs mostly through direct contact with infected birds, particularly through ingestion of infected poultry. Both inhalation and ingestion seem to be potential routes of virus entry.[8,18,19] The virus spreads to the lower respiratory tract, where it can cause severe pneumonia.[4,7,10] Immunhistologic investigations show that influenza H5N1 in cats in the lungs predominantly attaches to type II pneumocytes and alveolar macrophages.[20] The pattern of influenza virus H5N1 attachment to the lower respiratory tract in cats is similar to that in human tissue and is consistent with the respective pathology and immunohistochemistry results of experimental H5N1 infection in cats.[20] The predominant involvement of the lower respiratory tract and the inability of the virus to attach to cells of the upper respiratory tract may be a reason why cats excrete virus at relatively low concentrations.[8,20]

Infected cats develop clinical signs, including a significant rise in body temperature from day one post infection onwards, decreased activity, protrusion of the nictitating membrane, conjunctivitis, and laboured breathing by day 2 post infection. Unlike other influenza viruses that usually show a restricted replication in the respiratory tract in mammals, HPAIV H5N1 infection in cats leads to viremia and systemic infection causing severe necrosis and inflammation in many organs with a fatal disease outcome. This has also been shown after experimental intratracheal and oculo-nasopharyngeal infection.[8,9,19]

Viral excretion occurs in diseased animals *via* the respiratory and digestive tract, demonstrated through viral detection and viral isolation in pharyngeal, nasal, and rectal swabs as well as urine samples. Viral excretion starts between day 1 and 3 post infection and lasts up to seven days post infection.[8,19] Feeding of raw infected chicken also indicates that ingestion of infected birds can be a route of transmission to cats under natural conditions.[8] Both the respiratory and gastrointestinal routes of infection may also cause horizontal transmission which has been demonstrated after experimental infections[8] and was also assumed under natural conditions in the Sriracha tiger zoo outbreak in Thailand.[6]

Clinical Signs

Infection with HPAIV H5N1 also can result in severe clinical signs but also in subclinical infection. Three cats housed in an animal shelter in Austria excreted virus after contact with an infected swan, and two cats developed antibodies to influenza H5N1 virus, but none of them developed clinical signs.[11] Experimentally, the development of clinical signs proved to be dose-dependent. Infection with a high viral dose results in viral shedding and in fatal disease. When cats are exposed to a moderate dose, they do not develop clinical signs, but develop antibodies and shed small amounts of virus. Cats exposed to a low dose of virus do not become infected.[9]

The incubation period varies depending on the route of infection, and is shorter after direct experimental infection through respiratory and oral routes (1 to 2 days) than after cat-to-cat transmission (5 days).[19,21] Clinical signs observed most commonly are caused by disorders of the respiratory tract and include fever, lethargy, labored breathing, conjunctivitis, and protrusion of the third eyelid.[4,6,7,8] Neurologic signs result from cerebral and cerebellar congestion and nonsuppurative meningoencephalitis accompanied by vasculitis.[6,7] Diarrhea, as described in affected poultry and also in human beings infected with HPAIV H5N1, has not been observed in affected cats. Laboratory abnormalities described in tigers include severe leukopenia and thrombocytopenia and increased activities of the liver enzymes alanine aminotransferase and aspartate aminotransferase.[6] Markedly increased liver enzyme activities also have been found in domestic cats.[10] Sudden death may occur in infected cats as soon as two days after the onset of clinical signs.[7,9]

Diagnosis

Due to the systemic replication of the virus in cats, detection is not only possible in pharyngeal, nasal, and rectal swabs but also in fecal and urine samples, organ tissues, and in pleural fluid.[7,18,19] Pharyngeal swabs are convenient to obtain and can also detect subclinically infected cats.[9,11] The swabs should be placed in a sterile tube with a few drops of saline. Influenza viral RNA can be identified by real-time reverse transcription PCR with primers specific for the matrix gene. Subsequently, a subtype-specific real-time reverse transcription PCR can be used to identify the involved hemagglutinin and neuraminidase genes.[4,10] Viral isolation can be performed by inoculation of swab sample suspensions into embryonated hen eggs or in cell cultures and be identified subsequently by real-time reverse transcription PCR or hemagglutination inhibition assays.[7,19,22] Immunohistochemistry can be used to detect influenza virus antigen in affected organs.[7,19] Diagnostic in-practice tests designed to specifically screen cats for infection are not commercially available. All human tests investigated in cats lacked the sensitivity to produce reliable results in samples derived from experimentally infected cats.[23] Antibodies can be detected in serum samples within 2 weeks post infection.[9] Hemagglutination inhibition assays and a commercially available, species-independent competitive ELISA based on the nucleoprotein are more sensitive than virus neutralization assays.[9,11]

Treatment

The prognosis for infected cats in the field is unclear. Subclinical infections are possible, but all domestic cats that developed clinical signs to date have died or had to be euthanized. Therefore, the value of symptomatic treatment remains unknown. Not much is known about the efficacy of antiviral treatment in cats infected with HPAIV H5N1. The neuraminidase inhibitor oseltamivir has shown good antiviral activity against HPAIV H5N1 *in vitro*[25] as well as in experimentally infected mice and ferrets[26,27] and is recommended for treatment and prophylaxis of HPAIV H5N1 infection in people. However, treatment with oseltamivir (75 mg/60 kg q12h) was unsuccessful in tigers during the outbreak in the Sriracha tiger zoo in 2004. It failed to protect animals from infection and disease. This might be due to differences between humans and felids in drug pharmacokinetics and host metabolisms. Sequence analysis of the virus obtained from treated tigers excluded the possibility that drug-resistant mutants had emerged.[12]

Prevention

Although the transmissions of HPAIV H5N1 in cats in Asia and Europe did not result in sustained transmission chains among cats, the number of countries reporting these infections demonstrate that the transmissions do not represent exceptional single events but rather an expected event if the exposure or the infection pressure in the environment is high enough. The role of cats in the adaptation of HPAIV H5N1 to mammals remains to be elucidated and therefore, vaccination regimens have been discussed. So far, however, no licensed influenza vaccine for cats is available. However, contact of cats with infected birds must be avoided. It is advised to keep cats indoors in areas where HPAIV outbreaks occur. Cats should not be fed uncooked poultry meat.

There have been no reports of human beings contracting the infection from diseased dogs or cats. However, preventive measures must be taken to minimize the risk of transmission to human beings when cats are suspected to be infected with HPAIV H5N1. If animals are suspected to be infected or infection is confirmed, they should be isolated and contacts should be restricted to a minimum. Personnel handling infected animals should wear protective clothing (gloves, face masks, etc.).[28,29] Trans-species transmission of HPAIV H5N1 infections between cats and dogs seems to be low. At least under experimental conditions, when cats and dogs had close contact and shared water and food bowls, no transmission from cats to dogs occurred.[30]

REFERENCES

[1]Hinshaw VS, et al. **Infect Immun** 1981;34:354. [2]Dubovi EJ, Njaa LB. **Vet Clin Small Anim** 2009;38:827. [3]www.promedmail.org;archive nr. 20091105.3816. [4]Keawcharoen J, et al. **Emerg Infect Dis** 2004;10:2189. [5]Enserink M, Kaiser J. **Science** 2004;305:1385. [6]Thanawongnuwech R, et al. **Emerg Infect Dis** 2005;11:699. [7]Songserm T, et al. **Emerg Infect Dis** 2006;12:681. [8]Kuiken T, et al. **Science** 2004;306:241. [9]Vahlenkamp TW, et al. **J Gen Virol** 2008;89:968. [10]Klopfleisch R, et al. **Vet Pathol** 2007;44:261. [11]Leschnik M, et al. **Emerg Infect Dis** 2007;13:243. [12]Amonsin A, et al. **Virology** 2006;344:480. [13]Amonsin A, et al. **Arch Virol** 2007;152:1925. [14]Weber S, et al. **J Gen Virol** 2007;88:554. [15]Desvaux S, et al. **Emerg Infect Dis** 2009;15:475. [16]Marschall J, et al. **J Feline Med Surg** 2008;10:355. [17]Paltrinieri S, et al. **Emerg Infect Dis** 2007;13:662. [18]Yingst SL, et al. **Emerg Infect Dis** 2006;12:1295. [19]Rimmelzwaan GF, et al. **Am J Pathol** 2006;168:176. [20]Van Riel D, et al. **Science** 2006;312:399. [21]Kuiken T, **Nature** 2006;440:741. [22]Songserm T, et al. **Emerg Infect Dis** 2006;12:1744. [23]Marschall J, et al. **Transbound Emerg Dis** 2008;55:315. [24]Karaca K, et al. **Clin Diagn Lab Immunol** 2005;12:1340. [25]Hurt AC, al. **Antiviral Res** 2007;73:228. [26]Leneva IA, et al. **Antiviral Res** 2000;48:101. [27]Govorkova EA, et al. **Antimicrob Agents Chemother** 2007;51:1414. [28]Thiry E, et al. **Vet Microbiol** 2007;122:25. [29]Thiry E, et al. **Feline Med Surg** 2009;11:615. [30]Giese M, et al. **Emerg Infect Dis** 2008;14:308.

MANAGING WEIGHT LOSS IN CHRONIC DISEASE

Lisa M. Freeman, DVM, PhD, DACVN
North Grafton, MA

Cachexia, a loss of lean body mass, has been described for over 2000 years. It is estimated to be present in more than 5 million people in the United States. Cachexia occurs in a variety of chronic diseases in people, including cancer, AIDS, heart failure, chronic obstructive pulmonary disease, renal failure, and rheumatoid arthritis. The syndrome of cachexia also appears to be common in companion animals with chronic diseases, such as heart failure, renal failure, hepatic failure, and cancer. However, this is an area that has only recently begun to be studied so much additional information is needed.

The weight loss that occurs in cachexia is unlike that seen in a healthy dog, cat, or person that loses weight. In a healthy animal that is receiving insufficient calories to meet requirements, metabolic adaptations allow fat to be used as the primary fuel source, thus preserving lean body mass. Conversely, the primary fuel source in animal with chronic disease is amino acids from muscle so that these animals quickly catabolize muscle and lean body mass. Thus, the distinguishing feature of cachexia is a loss of lean body mass which has direct and deleterious effects on strength, immune function, and survival. In people, cachexia is an independent predictor of survival.

One aspect of cachexia that needs further study in both people and companion animals is its definition. There is currently debate on whether it is a loss of body weight or lean body mass that defines cachexia. Although the main underlying issue is a loss of lean body mass, this is difficult to detect clinically, especially in the early stages. In addition, there are certain types of cachexia (eg, rheumatoid cachexia, cardiac cachexia with fluid accumulation) in which the weight loss is masked by accumulation of fat or water. Therefore, waiting until weight loss occurs often prevents an early diagnosis and misses the underlying issue of inflammation and lean body mass losses. Cachexia is often mistakenly viewed as an end-stage syndrome manifested by an emaciated dog or cat. In fact, cachexia is a slowly progressive process of lean body mass loss that can be very subtle initially and can even occur in an obese animal. Recognizing the process of cachexia at an early stage provides improved opportunities to manage it effectively.

One of the forms of cachexia that has been studied in dogs and cats is cardiac cachexia, the loss of lean body mass that occurs in animals with heart failure. In one study of dogs with dilated cardiomyopathy (DCM), over 50% of patients had some degree of cachexia (Freeman et al, 1998). The loss of lean body mass in cardiac cachexia is a multifactorial process caused by the anorexia, increased energy requirements, and metabolic alterations (Freeman and Roubenoff, 1994). The anorexia may be secondary to the fatigue or dyspnea or may be due to medication toxicity or an unpalatable diet (Freeman and Roubenoff, 1994). Anorexia is present in 34-84% of dogs and cats with heart disease (Mallery et al, 1999; Freeman et al, 2003; Torin et al, 2007). However, it is the increased production of inflammatory cytokines, such as tumor necrosis factor (TNF) and interleukin-1 (IL-1), that are the primary mediators of cachexia (Freeman and Roubenoff, 1994; Freeman et al, 1998; Meurs et al, 2002). These inflammatory cytokines are known to directly cause anorexia, to increase energy requirements, and to increase the catabolism of lean body mass. Of particular pertinence to heart disease, TNF and IL-1 also cause cardiac myocyte hypertrophy and fibrosis and have negative inotropic effects.

The deleterious effects of cachexia have been emerging and recent studies have emphasized the role of body weight and body composition in heart failure. While obesity is a risk factor for development of heart disease in people, obesity may actually be associated with a protective effect once heart failure is present - this is known as the obesity paradox. A recent large meta-analysis on body condition in people with heart failure concluded that obesity and overweight were associated with lower all-cause and cardiovascular mortality and that underweight patients consistently had a higher risk of death. Given the adverse effects associated with cachexia, the association between obesity and improved survival in heart failure appears to be due to a *lack* of cachexia, rather than to the obesity per se. This is due to the increased reserve of lean body mass in overweight and obese people. The obesity paradox also has been demonstrated in dogs and cats with heart failure (Slupe et al, 2008). These data emphasize the importance of avoiding weight (and muscle) loss in the patient with heart failure.

Cardiac cachexia is typically recognized only after congestive heart failure (CHF) has developed. Loss of lean body mass is most readily evident in the epaxial, gluteal, scapular, or temporal muscles. Nutritional considerations for cardiac cachexia should include management of anorexia, if present and nutritional modulation of cytokine production also may be helpful. One method of decreasing the production and effects of cytokines is with n-3 polyunsaturated fatty acid supplementation. Supplementation of fish oil, which is high in n-3 fatty acids can decrease cytokine production in dogs with CHF and improve cachexia (Freeman et al, 1998). In some, but not all dogs with CHF-induced anorexia, fish oil supplementation can improve food intake (Freeman et al, 1998). In addition, reduction of cytokines has been correlated with survival in dogs with CHF (Freeman et al, 1998).

In people, cancer is one of the most common diseases in which cachexia is present. In people with cancer, it is estimated that over 50% lose weight unintentionally, although the prevalence depends on the type of cancer. In dogs, only 4% had a low body condition score ($\leq$3/9) at the time of diagnosis and 55% had a body condition score $\geq$6/9 (Michel et al, 2004). However, 35% of dogs had mild to severe muscle wasting. In addition, 69% of dogs for which pre-diagnosis body weights were available had experienced some weight loss (31% had <5% weight loss, 14% had lost 5-10%, and 23% had a weight loss of >10%; (Michel et al, 2004). A recent study in cats with cancer showed muscle loss in 91% of patients (Baez et al, 2007). In addition, cats that were below optimal body condition had a significantly shorter survival time compared to those with a body condition score of five or greater. Therefore, specific attention to muscle mass, as well as changes in body weight, are important in order to detect cancer cachexia.

For animals with chronic diseases in which weight loss is a component (eg, renal, heart or hepatic failure; cancer; respiratory disease), a variety of treatments can be employed. However, in many cases, practical methods to help the owner in managing their animals' appetite are critical to success. This is particularly important because anorexia is one of the most common contributing causes to an owner's decision to euthanize their pet. Dietary modification, assisted feeding, or feeding strategies often are beneficial in improving food intake and quality of life for these patients.

Cachexia plays an important and deleterious role in the pathogenesis of many chronic diseases and a better understanding of this syndrome is critical to optimize patient care. There is now a great deal of interest in studying the causes of this syndrome and in developing drugs and other treatments to specifically target cachexia. These are currently in various stages of development but have a great deal of promise for reducing the morbidity and mortality of many common diseases.

REFERENCES

[1.] Baez JL, Michel KE, Sorenmo K, Shofer FS. A prospective investigation of the prevalence and prognostic significance of weight loss and changes in body condition in feline cancer patients. [2.] Freeman LM, Roubenoff R. The nutrition implications of cardiac cachexia. **Nutr Rev**, 52: 340, 1994. [3.] Freeman LM, Rush JE, Kehayias JJ, Ross JN, Meydani SN, Brown DJ, Dolnikowski GG, Marmor BN, White ME, Dinarello CA, Roubenoff R. Nutritional alterations and the effect of fish oil supplementation in dogs with heart failure. **J Vet Int Med**, 12:440-448, 1998. [4.] Freeman LM, Rush JE, Cahalane AK, Kaplan PM, Markwell PJ. Dietary patterns in dogs with cardiac disease. **J Am Vet Med Assoc** 2003; 223: 1301-1305. [5.] Freeman LM. The pathophysiology of cardiac cachexia. **Current Opinion in Supportive and Pallative Care** 2009; 3: 276-281 [6.] Mallery KF, Freeman LM, Harpster NK, Rush JE. Factors contributing to the euthanasia decision in dogs with congestive heart failure. **J Am Vet Med Assoc**, 214:1201-1204, 1999. [7.] Meurs KM, Fox PR, Miller MW, Kapadia S, Mann DL. Plasma concentrations of tumor necrosis factor-alpha in cats with congestive heart failure. **Am J Vet Res** 2002; 63:640-642. [8.] Michel KE, Sorenmo K, Shofer FS. Evaluation of body condition score and weight loss in dogs presented to a veterinary oncology service. **J Vet Intern Med** 2004; 18: 692-695. [9.] Slupe JL, Freeman LM, Rush JE. The relationship between body weight, body condition, and survival in dogs with heart failure. **J Vet Intern Med** 2008; 22: 561-565.[10.] Torin DS, Freeman LM, Rush JE. Dietary patterns of cats with cardiac disease. **J Am Vet Med Assoc** 2007; 230: 862-867.

Nutrition, cachexia, cancer, heart disease, renal disease

APPARENT REVERSE ZOONOTIC TRANSMISSION OF PANDEMIC H1N1 INFLUENZA VIRUS INFECTION IN DOMESTIC CATS: CLINICAL DIAGNOSES AND DISEASE

BA Sponseller[1,2], JD Trujillo[1,2], A Jergens[1], C Prior[3], A West[3], E Strait[1], K Harmon[1], S Ramamoorthy[1], T Pearce[2] H Bender[1], K Waller[1], K Miles[1], L Koster[3], M Jenkins-Moore[3], M Killian[3], S Swenson[3], K-J Yoon[1], P Nara[1,2]. [1]College of Veterinary Medicine and [2]Center for Advanced Host Defenses, Immunobiotics, and Translational Comparative Medicine, Iowa State University, Ames, IA; [3]Park City Animal Clinic, Park City, UT; [4]USDA National Veterinary Services Laboratories, Ames, IA.

Influenza viruses are important pathogens in humans and animals that are typically host specific yet have aquatic birds as the primary reservoir. Interspecies transmission of influenza virus is a health concern due to 1) the potential of emergence of a novel strains (i.e., reassortment) and reverse transmission and 2) host adaptation and novel endemic influenza in human and other animal species. Although this virus appears to be human host adapted, occasional transmission of pH1N1 influenza virus from humans to animals is occurring. Recent examples of typical and atypical species transmissions amplify this concern and include infection of swine with human H1N2 virus, a reassortment event between human and swine influenza viruses and the emergence of canine influenza through host adaptation of equine H3N8 virus. Here we report the first two documented feline cases of pH1N1 influenza virus infection, clinical manifestations of disease, methods of antemortem diagnosis and the impact of this discovery. **Case A:** The 13 year-old cat was presented for respiratory disease, lethargy and inappetance following influenza-like illness (ILI) in 2 of 3 family members. Clinical examination revealed that the cat had bilateral wheezes, was afebrile and clinically dehydrated. Radiographs revealed a bilateral caudo-dorsal alveolar radiographic pattern. Moderate lymphopenia and modest hemoconcentration were evident. BAL cytology supported a viral pneumonia. Molecular testing comprising two separate real time PCR assays (influenza A screening and differential matrix gene assay for pH1N1) were positive for RNA from the BAL at presentation. Presumptive positive results were confirmed by the USDA-NVSL by real time PCR assays and virus isolation (A/Feline/IA/NVSL026991/2009). **Case B:** The 14 year-old cat was presented with inappetance, dyspnea and tachypnea following ILI and confirmed pH1N1 infection in 1 of 2 family members. Moderate lymphopenia developed over 3 days. Radiographs revealed a non-specific hilar and caudo-dorsal bilateral alveolar pattern. Influenza A/B rapid antigen test (OSOM) was positive at presentation. Serum collected 6 days later had a specific hemagglutinin inhibition titer of 1:160 for pH1N1. Infection with pH1N1 should be considered in cats presenting with lower airway disease, lymphopenia and a history of influenza-like illness in the household. These atypical index cases at the start of a new human pandemic emphasize the role of human/animal contact in interspecies transmission. This discovery reinforces the need for the one health initiative. Additional information regarding intra/interspecies transmission may be presented as this sero-molecular epidemiological study is ongoing.

FELINE PRIMARY HYPEROXALURIA – GENETICS, BIOCHEMICAL, CLINICAL AND HISTOPATHOLOGIC MANIFESTATIONS RE. Goldstein, AM. Chandler, DF. Paul and RM Peters. College of Veterinary Medicine, Cornell University, Ithaca, NY.

Primary hyperoxaluria (PH) is a rare, potentially fatal inherited disease in humans resulting from mutations in the alanine:glyoxylate aminotransferase (AGT) gene (PH1) or in the glyoxylate reductase (GRHPR) gene (PH2). An autosomal recessive disease in cats, mimicking PH2, has been described. We identified 3 kittens from a commercial cattery with renal lesions consistent with feline PH on necropsy. The goals of this study were to identify the genetic cause of PH in these cats and to characterize the disease in cats through the establishing a research colony of cats with PH to aid in the study of feline as well as human conditions.

The genetic component of the study was accomplished utilizing DNA extracted from one blood and 2 formalin-fixed kidney tissue samples from the 3 affected kittens, a blood sample from the affected cats' sire, (suspected carrier) and blood samples from 2 healthy cats. The exons and flanking intronic regions of the feline GRHPR gene were amplified and sequenced. A point mutation in a hypothesized splice site was found: G to A at the acceptor site of intron 4. Affected cats were homozygous for the mutation, the carrier was heterozygous and it was not present in the healthy cats' DNA. An abnormal and shortened transcript was confirmed at the RNA level, purified from formalin fixed tissues.

The population of the cattery was screened for the PH mutation and carrier cats were identified for the establishment of the feline PH colony. To date 14 kittens have been closely evaluated from 6 different litters. Nine affected, 2 carriers, and 2 healthy have been included based on genetic testing at 6 weeks of age. Cats were evaluated by serial physical examinations, blood and urine chemistries, as well as hepatic GRHPR activity and histopathology in some cats. No abnormalities have been noted in carrier or healthy cats. The affected cats exhibit normal physical examination and blood chemistry findings when first tested at approximately 8 weeks of age. Azotemia developed in 8 affected cats to date at the median age of 84 days (range 79-109). Clinical signs mainly including anorexia, weight loss and weakness occurred in 7 affected cats to date. Euthanasia has been performed because of increasing morbidity on 7 cats median age 94 days (range 81-116). GRHPR activity was markedly decreased when assayed from liver in 2 affected cats when compared to healthy and carrier littermates. Common necropsy findings on 7 affected cats to date included varying degrees of lymphoplasmacytic tubulointerstitial nephritis with fibrosis, parenchymal collapse and tubular degeneration, regeneration and dilation. All cats had extensive intratubular crystals compatible with calcium oxalate. At the conclusion of this study we will have characterized the genetics, biochemical and clinical aspects of feline PH and will have established an ideal model for the study of PH and other oxalate disorders in cats and humans.

ONCOLOGY UPDATE FOR THE INTERNIST

Carolyn J. Henry, DVM, MS, DACVIM (Oncology)
Columbia, MO

INTRODUCTION

Some recent advances in veterinary oncology may have direct impact on the way that internists will diagnose, treat, and monitor therapy of their cancer-bearing patients. Given the dynamic nature of the field of veterinary oncology and the fact that many new findings are not yet published or are published in the human, not veterinary, medical literature, it can be difficult to stay current if one is not privy to list serve discussions or specialty meeting abstracts. This talk is intended to provide internists with a brief summary of some newer findings relative to the diagnosis and treatment of the more common cancers encountered in small animal internal medicine practices. Further updates will be provided as available.

CANINE MAST CELL TUMOR

The most relevant recent change in our approach to treatment of canine mast cell tumors (MCT) has been the introduction of receptor tyrosine kinase (RTK) inhibitors to the veterinary market. The only RTK inhibitor currently FDA approved for veterinary use is toceranib phosphate, or Palladia.[1]This drug inhibits the receptor tyrosine kinase, Kit, which is mutated in up to 30% of canine MCT, and causes an antiproliferative effect.[1] Palladia also targets VEGFR-2 on endothelial cells and PDGFR-B on pericytes to provide an antiangiogenic effect. Palladia is an oral drug approved for use in dogs with Grade II or III recurrent MCT with or without lymph node metastasis. The drug's approval was based on results of a prospective, randomized double-blinded study in which dogs received active drug or placebo for 6 weeks, followed by an unblinded phase of the study in which all dogs received active drug.[2] The overall response rate (complete and partial remissions) was 42.8% and the most common side effects were gastrointestinal and managed by a drug break, followed by resumption at a lower dosage. Subsequent experience with the drug in the clinical setting post-FDA approval has been similar. Palladia is initially given every other day, usually at night. If gastrointestinal side effects are noted, the drug is stopped until signs resolve, after which treatment is resumed, but on a Monday-Wednesday-Friday schedule, rather than every other day. Neutropenia is another potential side effect; thus, regular monitoring with complete blood counts is essential. The other RTK inhibitor developed for veterinary use is masitanib (Kinavet, AB Science). This latter drug is approved for use in Europe, but not yet FDA approved for use in the US. Kinavet targets KIT, PDGFRα and PDGFRβ, and FGFR-3. In a prospective clinical trial, dogs with measurable Grade II or III MCT without metastasis that were randomized to receive Kinavet as first-line therapy had a significantly longer time to tumor progression (TTP = 253 days) than dogs receiving placebo, regardless of Kit mutation status.[3] Observed side effects in clinical trial? Clinicians wishing to use Kinavet prior to FDA approval may obtain the drug via personal import from AB Science SA. The request must be on clinic letterhead and it is necessary to establish an account with AB Science SA. Each patient request requires a separate letter. The drug is shipped as a 3-month supply free of charge, but the requesting clinic must pay $140 for shipping. Both Palladia and Kinavet offer new options for a disease that is commonplace in veterinary medicine and for which we went decades without having good treatment options. However, neither drug is without the potential for significant side effects. Internists are urged to educate referring veterinarians and clients about these side effects and to require close monitoring of any patient for whom they dispense these drugs.

FELINE ORAL SQUAMOUS CELL CARCINOMA

Feline oral squamous cell carcinoma (FOSCC) is perhaps one of the most frustrating oncologic diagnoses clinicians must face in feline practice. Despite advances in medical, surgical, and radiation oncology, the one-year survival rate has averaged only 10% for cats with oral SCC. However, in 2008, an abstract was submitted for the combined meeting of the European Society of Veterinary Oncologists and the Veterinary Cancer Society, that described a protocol combining bleomycin (20 IU/m^2 SC on days 1-3, then weekly for a maximum of 20 doses) and methotrexate (15 to 20 mg/m^2 SC weekly on a 4 weeks on/4 weeks off cycle) for treatment of FOSCC.[4]The abstract reported that seven of eight cats (87.5%) with SCC treated with the protocol responded, including five complete remissions and two partial remissions. This study was retrospective in nature and included too few cats to use this as a basis for case management recommendations. However, given the dramatic improvement in response rate over that reported with other treatments, along with the reported lack of serious side effects, further evaluation of this protocol certainly seems warranted. A modification of the protocol is suggested in order to minimize the likelihood of adverse effects (specific adverse effects?). Based on the experience of the abstract author and others, we currently use a protocol in which the bleomycin dosage is reduced to 15 IU/m^2 SC weekly for a maximum of 20 doses and the methotrexate is administered at a dosage of 15 mg/m^2 SC weekly, titrating towards 20 mg/m^2 or the highest dose that does not cause anorexia. Methotrexate is administered for 4 weeks on and 4 weeks off while the cat is receiving bleomycin. Case outcome is currently being assessed and will be updated accordingly.

Bisphosphonates offer an additional option in the case management of FOSCC. Although results of prospective studies are pending, the bisphosphonate, pamidronate, has been used to slow bone loss in cats affected by invasive oral squamous cell carcinoma. The dosage most often utilized is 1 to 1.5 mg/kg intravenously over two hours in 50 ml 0.9% NaCl every 21 to 28 days. Renal function must be monitored closely in patients undergoing pamidronate therapy[5]

CANCER SCREENING TESTS

Within the past five to ten years, there has been increased interest in developing screening tests with the potential for application to veterinary oncology. Some tests are specific for certain tumor types, such as the V-BTA test for bladder cancer and the

PetScreen test for lymphoma, while others are more general screens for malignancy such as measurement of thymidine kinase or lactate dehydrogenase. [6-9] Table 1 summarizes some of the currently available cancer screening tests that will be discussed.

TABLE 1: Currently Available Veterinary Cancer Screening Tests

Test	Company	Sample	Website	Phone	Test Type
Canine lymphoma blood test	PetScreen, LTD	Red top tube	vets@pet-screen.com	516 874 3701	Send out
Liason Thymidine Kinase Assay	DiaSorin	Serum; EDTA plasma	www.diasorin.com	(800) 328-1482	In-House
PARR (lymphoma)	Colorado State University	various	www.dlab.colostate.edu/security2/test_price.cfm	970-297-1281	Send out
V-BTA	Polymedco	Free catch urine	vetinfo@polymedco.com	1-800-431-2123	In-house

CANINE ORAL MELANOMA

Canine oral malignant melanoma is recognized as the most common oral malignancy in dogs. With a metastatic rate in excess of 80%, even when local control can be achieved with surgery or radiation, systemic disease is likely. Medical therapy has been attempted with various agents including carboplatin, cisplatin, and piroxicam, but response rates are 28% or less. The USDA conditional approval of a xenogeneic plasmid DNA encoding human tyrosinase for use against canine oral melanoma in 2007 provided a new option by which to address micrometastatic disease.[10] The vaccine is intended for use in dogs with oral melanoma that is locally controlled via surgery or radiation therapy. Vical Incorporated announced in January 2010 that the USDA had granted the company's licensee Merial Ltd full licensure for the vaccine to be marketed under the name, Oncept™. Dogs are vaccinated via intramuscular injection every 2 weeks for 4 doses, and then administered a booster vaccine every 6 months. In their press release, Vical Incorporated stated that dogs vaccinated with Oncept™ after surgery had significantly better survival times than control (unvaccinated) dogs ($p<0.0001$). [http://www.globenewswire.com/newsroom/news.html?d=181546] They were unable to report median survival time (MST) for vaccinated dogs at the time of study publication, as more than 50% of the treated dogs were still alive. The vaccine may prove to be a useful adjunct to local therapy of canine malignant melanoma and has not been associated with adverse effects in dogs treated to date.

CANINE THYROID CARCINOMA

While the use of radioactive iodine to treat functional thyroid carcinomas in dogs is not novel, the use of this treatment modality for dogs lacking clinically overt hyperthyroidism and without elevation of T4 levels has only recently gained acceptance amongst oncologists. In their 2006 manuscript, Turrel et al, reported results of I-131 therapy for 39 dogs with thyroid carcinoma.[11] Twenty-one of the dogs were hyperthyroid, 16 were euthyroid and two were hypothyroid. Serum T4 concentration prior to I-131 treatment did not affect survival time. Dogs that had metastasis had shorter survival times (366 days) than dogs without metastatic disease (839 days). A 2005 Australian study reported similar findings for 43 dogs undergoing I-131 treatment, with or without surgery.[12] Dogs in the radioiodide alone group had a MST of 30 months compared to 34 months for dogs in the surgery/radioiodide treatment group. These results compare favorably to the 3-month MST reported for untreated dogs. On the basis of these newer reports and similar anecdotal reports elsewhere, many oncologists now recommend thyroid scintigraphy as a routine part of the diagnostic evaluation for dogs with thyroid carcinoma, even if they have normal or low T4 levels.[13] If this approach is recommended, it is important to remember that the iodine-containing contrast agents used with computed tomography (CT) interfere with nuclear scintigraphic imaging of thyroid carcinomas. Therefore, one should plan the order of diagnostic tests such that thyroid scintigraphy precedes CT imaging whenever possible. If the CT is performed prior to thyroid scintigraphy, it should be done without use of iodinated contrast agents. We generally recommend CT imaging of the neck even when scintigraphy shows marked thyroid uptake. The reason for this suggestion is that the combined imaging modalities are needed to fully appreciate the extent of disease and to determine the likelihood of success with I-131, surgery, external beam radiation therapy, or a combination thereof. In addition, CT imaging can be used to monitor response to therapy. Sites offering canine I-131 therapy are listed in Table 2 below.[14]

TABLE 2: US Veterinary Centers Known to Offer I-131 for Dogs		
Site	**Web address**	**Phone**
Arizona Veterinary Specialists Gilbert, AZ	www.azvs.com	**480-635-1110**
Colorado State University Fort Collins, CO	www.csuanimalcancercenter.org	**970-297-4195**
Southeast Veterinary Oncology Orange Park, FL	www.petcancercare.com	**904-278-3870**
University of Missouri Columbia, MO	www.cvm.missouri.edu/oncology/index.html	**573-882-7821**
University of Tennessee Knoxville, TN	www.vet.utk.edu	**865-974-8387**
Veterinary Specialty Center Buffalo Grove, IL	www.vetspecialty.com	**847-459-7535**

REFERENCES

[1]London CA, et al. Topics in **Comp Anim Med** 2009;24:106; [2]London C et al. **Clin Cancer Res** 2009; 15:3856; [3]Hahn KA, et al**. J Vet Intern Med** 2008;22:1301; [4]Lowe R. Proceedings of the Combined ESVONC and VCS meeting; Copenhagen, Denmark, 2008: 7 (abstract); [5]de Lorimier LP, et al. In: Henry and Higginbotham: Cancer Management in Small Animal Practice, 2010:187; [6]von Euller, et al. **Int J Oncology** 2008;34:1505; [7]Ratcliffe L, et al**. J Vet Comparative Oncol** 2008;7:92; [8]Lana S, et al. **J Vet Intern Med**, 2008; 20:329 ;[9]Henry CJ, et al. **Am J Vet Res**, 2003:64(8):1017 [10]Bergman PJ, et al. Topics in Comp Anim Med 2009;24:130; [11]Turrel JM, et al. JAVMA 2006;229:542; [12]Worth AJ, et al. **Australian Vet J** 2005;83:208; [13] Bryan JN, et al. In: Henry and Higginbotham: Cancer Management in Small Animal Practice, 2010:362; [14]Lattimer JC, et al. In: Henry and Higginbotham: Cancer Management in Small Animal Practice, 2010:151

CLINICAL, BIOCHEMICAL, AND MOLECULAR GENETIC CHARACTERIZATION OF PORPHYRIA IN CATS. U Giger[1], S Clavero[2], DF Bishop[2], Haskins ME[1], RJ Desnick[2]. [1]Section of Medical Genetics, School of Veterinary Medicine, University of Pennsylvania, Philadelphia, PA, and [2]Department of Genetics and Genomic Sciences, Mount Sinai School of Medicine, New York, NY.

Porphyrias are a group of inborn errors resulting from the accumulation of porphyrins due to deficient activities of specific enzymes in the heme biosynthetic pathway. They have been well characterized in humans and also reported in cats, dogs, cows, sheep, pigs, chickens, and squirrels. In humans, they are clinically classified as either erythroid with cutaneous involvement or hepatic with acute neurovisceral attacks. Here we describe the clinical, biochemical, and molecular genetic studies of one cat with congenital erythropoietic porphyria (CEP), an autosomal recessive disorder, and five unrelated lines of cats that appeared clinically to have CEP, but were ultimately identified to have acute intermittent porphyria (AIP), an autosomal dominant disorder in humans.

The porphyric cats had erythrodontia (brown discolored that fluoresced pink under UV light), porphyrinuria, and a compensated hemolytic disorder but no evidence of acute life-threatening neurovisceral attacks or cutaneous lesions (with near normal life-expectancies). Urinary uroporphyrin I and coproporphyrin I concentrations were increased in the porphyric cats. However, cats with AIP had high urinary 5-aminolevulinic acid and porphobilinogen concentrations, while the cat with CEP had normal or low levels of these porphyrin precursors. Porphyrin analysis of the discolored teeth from the cats with AIP revealed markedly elevated uroporphyrin I and coproporphyrin I isomers, thereby explaining their CEP-like phenotype. Cats with AIP had ≤50% of normal hydroxymethylbilane synthase (HMBS) activity in erythrocytes and tissues, while the HMBS activity was slightly increased in the cat with CEP. In contrast, the erythrocyte and/or hepatic uroporphyrinogen III-synthase (UROS) activity was markedly decreased in the cat with CEP and normal in the cats with AIP.

Sequencing the feline HMBS gene revealed different mutations in each of the five lines of cats with AIP including one duplication, two deletions and two missense mutations. Interestingly, four cats were heterozygous for the HMBS mutations, while one was homozygous for a mild missense mutation. Mutagenesis and prokaryotic expression studies confirmed reduced residual HMBS activities and family studies further confirmed the modes of inheritance. While all cats with AIP had a normal UROS gene sequence, the cat with CEP was homozygous for a missense mutation in UROS resulting in drastically reduced enzyme activity and stability.

Thus, cats with discolored teeth and normal or mild hemolysis may have either deficient HMBS or UROS activity representing animal models of human AIP or CEP, resp. The biochemical and molecular characterization facilitates screening of affected cats. Porphyric cats may also permit further characterization of the pathogenesis of porphyria as well as the evaluation of therapeutic strategies for patients with AIP and CEP including gene, transplantation, and stem cell therapies.

CAT GENOMICS: DNA ARRAYS AND THE VETERINARIAN'S ROLE IN CAT DISEASES.

LA Lyons. School of Veterinary Medicine, University of California – Davis, Davis, CA.

The NIH – National Human Genome Research Institute (NHGRI) has previously supported a low coverage (2X) and now an in depth coverage (~10X) genomic sequencing effort for the domestic cat. Both the Washington University at St. Louis Sequencing Center and the Broad Institute of MIT, in collaboration with AgenCourt, have performed the cat genome sequencing and genome assembly. These genome projects have been augmented by a private sequencing effort conducted by Hills Pet Food, Inc. Combined, the domestic cat should have one of the most in depth mammalian genomic sequences available for companion animals.

The base genomic sequence was developed from an inbred Abyssinian cat. In addition, DNA pools of 4 cats from each of several breeds and random bred populations were also sequenced. This additional and important effort assists with the identification of genetic variation between the difference cat breeds, identifying single nucleotide polymorphisms (SNPs) and DNA insertions /deletions (DIPs) that are important to the genetic make-up of each breed and each cat. Hundreds of thousands of SNPs have been identified and will be used to develop a DNA array, which will be an imperative resource to decipher the genetic component of cat diseases. Funding donated by Hills Pet Food, Inc. to the Morris Animal Foundation is supporting the construction of a 74K array. The cat DNA arrays or "chips" will not only easily identify genetic regions for simple genetic traits, such as those caused by a single dominant or recessive gene, but more importantly, the arrays should be able to decipher the genetic contribution of complex diseases, such as asthma, diabetes, hyperthyroidism, feline urinary syndromes, obesity, and lymphomas. Most of these traits have some environmental component, such as nutrition or allergen exposure, the DNA arrays will help to find the genes that contribute to disease onset and progression.

Veterinarians are imperative to the future of cat genetics. The DNA arrays allow the development of case – control studies, where 30 – 100 cases and matched controls for a given disease are assayed on the arrays. The cases and controls required are dependent on the level of relative risk of the disease or trait in the population under study. Because the arrays are most important for complex diseases, the correct clinical diagnosis of disease is absolutely vital for a successful study. The analysis of arrays is a high cost endeavor, improper samples can lead to a huge waste of time, resources and funds. This research report will present the status of the cat genome sequencing effort, the production of the cat DNA array, review proper study design and the viral role of veterinary diagnostics for the future of cat genetics and disease eradication.

VETERINARY ENDOUROLOGY: REVERSE TRANSLATIONAL MEDICINE FROM HUMAN TO VETERINARY MEDICINE

Allyson C. Berent, DVM, DACVIM
New York, NY

The written definition of **"endourology"** /en·do·urol·o·gy/ (-ūr-ol′ah-je) is "the branch of urologic surgery concerned with closed procedures for visualizing or manipulating the urinary tract." This is typically reserved for dealing with conditions in the renal pelvis, ureters, bladder and urethra. In human medicine endourology has tremendously advanced over the past 20-30 years, and in veterinary medicine similar progress is following suite.

Training in various subspecialties in human medicine has traditionally relied heavily upon using animals as a research model. There are thousands of examples of this in all specialties. Veterinary endourology has progressed dramatically over the past 5-10 years. Interestingly, many of the procedures we are currently performing in veterinary medicine were adapted from current procedures being performed in human medicine for similar conditions. This we termed "reverse translational medicine:" using a human model for naturally occurring animals disease. The collaboration and adaptations made in the various instrumentation, imaging, procedures, devices and techniques has made the subspecialty in veterinary urology advance dramatically. The collaboration and experience between both veterinary and human urology is quite fascinating and this lecture will compare the similarities, differences, and adaptations in the process of training and witnessing translational/reverse translational medicine between both physicians and veterinarians.

This talk is an overview of the process many of us who have trained in endourology have gone through, focusing on the differences between human and veterinary patients with very similar diseases but very different challenges. These challenges have resulted in adaptation, creativity and manipulation of the currently performed techniques compared to our human counterparts. The subsequent lecture will discuss the details of the more commonly performed endourologic procedures actively being performed in veterinary patients, while this lecture will focus on how we got there.

Similarities

When comparing disease processes in human and veterinary patients, they are very similar. The most common dilemma in both human and veterinary urology is dealing with recurrent urolithiasis. In humans nearly 10% of all men and 5% of all women in America will develop a kidney stone in their lifetime. In humans, kidney and ureteral stones are the most significant clinical problem encountered urologically, where in veterinary medicine bladder and urethral stones are seemingly more prevalent. Calcium oxalate stones are the most common stone type seen in both human and veterinary patients. In humans calcium oxalate nephrolithiasis account for 70-80% of stone type, which is similar in our feline patients, but canine nephroliths are estimated to be about 60% calcium oxalate and 40% struvite. Feline patients have been documented to have >95% of ureteroliths being calcium oxlate in nature. Typically urate and cystine stones are seen with genetic defects. In humans uric acid stones are as common as struvite stones, where this is not the case in veterinary medicine. Because humans will not tolerate a chronic urinary tract infection, infection induced struvite stones are not nearly as common in developed countries, as they are seen commonly in our canine patients. Stone recurrence is also very common in both human and veterinary patients, with calcium oxalate stones reported to recur in similar numbers (40-60% or more).

Both human and veterinary stones respond similarly to lithotripsy (both laser and shockwave), though the decision on technique chosen for the treatment of various patients is different, due to patient size, sex and species. ESWL is very effective and commonly performed in humans for most stone types in the upper tract (excluding cystine), and this is similar in canine patients. Due to the small diameter of the feline ureter and the fragility of the feline kidney, ESWL is not commonly recommended in cats.

In humans, ureteral stenting is commonly performed for ureteral decompression, passive dilation for the ease of ureteroscopy and ESWL, and the prevention of stricture formation after ureteral surgery or ureteroscopy. This has been performed for similar reasons in veterinary medicine as well. It has been documented that after about 2 weeks with an indwelling double pigtail ureteral stent, the ureter will passively dilate, resulting in improved urine drainage and the passage of a larger ureteral stent, when necessary. This phenomenon has also been documented in canine and feline patients in the authors practice in as little as 2 days, improving our success with ureteral stenting. Other similarities include some similar congenital anomalies like circumcaval ureters resulting in ureteral strictures and congenital ureteral stenosis, both of which are rare.

Transitional cell carcinoma (TCC) is the most common neoplasia of the urinary tract in veterinary and human medicine. The difference is that human TCC is typically of a non-invasive form, which can be endoscopically treated with laser resection and intravesicular chemotherapy with great success. In veterinary medicine 95-99% of the time TCC is highly invasive to the muscular layer of the urinary tract, making debulking surgery, laser ablation and local infusions minimally effective. Systemic therapy is often required to slow tumor progression, while local intervention like stenting or laser ablation is typically necessary if an obstruction pursues.

Differences

There are many more differences than similarities that have been discovered during the development of urinary interventions in small animal veterinary patients. These differences have resulted in manipulation and adaptation of currently performed procedures in human medicine to make them as effective as possible considering our limitations. Most of these differences have made the procedures in veterinary patients more challenging than in human medicine, but there are a few benefits we have discovered as well. The 3 main differences are the fact that 1) all veterinary patients need to be placed under general anesthesia for endourological procedures, where in humans conscious sedation or epidural analgesia is commonly performed for minor procedures. This typically adds a level of complexity to each procedure when compared to human urology; 2) most veterinary patients present with urologic disease at a much later stage than our human counterparts. When we see a feline patient with a ureteral obstruction they are typically azotemic, sometimes severely, with a "big kidney little kidney syndrome," suggesting that one ureter was already obstructed and left untreated, and the current obstructed kidney is resulting in a life threatening state. In humans, unless the ureteral obstruction is resulting in a severe pyelonephritis, they are rarely end-stage. Humans will have a stone discovered much earlier in the course of the disease, as ureteral and renal colic/pain, hematuria, or malaise results in earlier detection, and treatment is typically immediate. This makes the prognosis more precarious in veterinary patients, particularly when traditional surgery is performed, as the outcomes are met with a high morbidity/mortality; 3) the anatomy of each part of the urinary tract (kidney, renal pelvis, ureter, bladder and urethra) is subtly different, and in veterinary patients there are many more challenges in gaining access for various treatments.

The most common differences are anatomy related. The human **kidney** has long, large calyces that often entrap nephroliths, making access and retrieval difficult at times. Dogs and cats have very short calyces, so most nephroliths are contained within the renal pelvis, rather than the renal calyx. This makes finding a stone and fragmentation using ESWL and percutaneous nephrolithotomy (PCNL) easier in veterinary patients than human patients. Though access into the kidney for a PCNL in veterinary patients is much more difficult due to the mobility of the kidney. For ESWL the issue arises in veterinary patients during stone passage since the ureteral diameter is much smaller in dogs and cats when compared to humans. This makes fragment passage after ESWL more difficult, often requiring concurrent ureteral stent placement for dilation and protection. Nephrostomy tubes have been placed in both dog and cat patients, but are always considered temporary, being placed for stabilization prior to a definitive procedure. In humans nephrostomy tubes are often considered a long-term treatment option as tube management can be accomplished on an outpatient basis. This is not the case in our patients and the risk of infection, dislodgement and leakage is too high to consider this a long-term option. In humans the renal pelvis extends beyond the renal parenchyma to the ureteropelvic junction (UPJ), where is dogs and cats this extension is much less dramatic. In cats the UPJ is hidden within the parenchyma of the kidney, making a true pyelotomy or laparoscopic pyelotomy for stone retrieval very difficult to successfully perform and close without leakage post-operatively.

The canine and feline **ureter** is much smaller than the human ureter. Typically, an adult human ureter is approximately 4.0-5.0 mm in diameter or bigger and a pediatric ureter is 2-3 mm in diameter. The feline ureter is normally 0.3-0.4 mm in diameter and the canine ureter is 1.0-2.5 mm in diameter. This makes intervention much more difficult in small animals than humans. Currently, the smallest ureteroscope made is 2.7 mm at the tip and 3.1 mm along the shaft, clearly showing that this is typically too large for successful ureteroscopy in small animals. We have been able to gently pass a ureteroscope up a canine ureter in an animal >15-20 kg in size, and this is still considered technically challenging. Because of the narrow ureter, devices made for human urology are typically too wide for our veterinary patients. This makes us rely on pediatric products, which are often too large or too long. Canine and feline ureters also have a "J shape" loop commonly seen as the ureter enters the urinary bladder at the UVJ. This makes ureteral guidewire access slightly more challenging. With proper techniques this can be straightened out to successfully pass the ureteral stent. The feline UVJ is also more distal than the human or the dog, situated in the proximal urethra, rather than the bladder neck. It is also extremely small and subtle in appearance. Knowing this makes UVJ wire access easier, needing smaller cystoscopes and guidewires. One main benefit we have discovered in our dog and cat patients, when compared to humans, is that ureteral stents are much better tolerated. In humans ureteral stents cause some degree of dysuria in approximately 80% of patients. They are also recommended to be removed within 3-6 months, and maximally 12 months, due to the risk of stent encrustation and diffuse ureteral obstruction. In our experience, this condition has not been documented in any dog or cat patient (over 125 stents placed to date), and the longest indwelling stent is in place for over 4.5 years without any concern of encrustation or obstruction. Dysuria is seen in <5% of veterinary patients and is usually short-lived (3-5 days) and resolves.

The canine and feline **bladder** is considered peritoneal, as compared to retroperitoneal in humans. This makes the common procedure of suprapubic percutaneous cystolithotomy (PCCL) more difficult in veterinary patients when a large stone burden is present. Male dogs have a long curved urethra, unlike humans. This makes cystourethroscopy more difficult, requiring a small diameter flexible scope. Due to the narrow, long urethra in male dogs, and the fact that most dogs with bladder stones are small breed male dogs, transurethral retrieval is difficult. Dogs also typically have a large stone burden, as stones will accumulate and grow in size prior to the owner noticing significant signs warranting evaluation. Stones will likely lie in the bladder for longer periods of time compared to a human due to the fact that dogs are quadrapeds, and gravity is not used in their favor. Male cats have a long urethra and a reasonable size urethral lumen, except at the penile urethra, which is the size limiting factor in permitting cystoscopy. Most cat patients can not accommodate an instrument bigger than 1.5 mm (4.5-5 french). Dogs and cats also do not have a clear urethral sphincter, like humans. They have more of a sphincter mechanism, so the control of continence is more reliant on the entire urethra. This allows for a better outcome in urethral stenting, when necessary for chronic strictures or obstructive neoplasia.

Adaptations

The success in veterinary endourology is in the manipulations and creativity people have used to account for the significant differences described. This started with creating smaller rigid cystoscopes nearly 15-20 years ago and progressed to flexible cystoscopes, nephroscopes, ureteroscopes, and now semi-rigid male cat cystoscopes. In addition to this endoscopic progress, devices (i.e. feline ureteral stents) and techniques (i.e. percutaneous cystolithotomy, ureteral stenting, subcutaneous ureteral bypass, etc) have also been created and manipulated for our veterinary patients. This progress was made possible by collaboration with human urologists, medical device companies and persistence.

For nearly all of our procedures we will use general anesthesia. For procedures in the kidney some examples of our adaptations include: using smaller cystoscopes and laser lithotripsy for PCNL rather than larger nephroscopes and ultrasonic lithotripsy; performing a ureteral bypass, resulting in an indwelling nephrostomy tube drainage, rather than an external nephrostomy tube; when a nephrostomy tube is necessary, a nephropexy will typically be performed surgically in feline patients to prevent kidney mobility and the risk of inadvertent removal (this is typically unnecessary in dogs); the placement of a double pigtail ureteral stent prior to ESWL to prevent a ureteral obstruction for large nephroliths (termed steinstrasse); etc. For <u>ureteral</u> interventions some examples include: using surgical and endoscopic/fluoroscopic assistance for the placement of feline ureteral stents; using ureteral stents as a long-term management for ureteral obstructions (stones, tumors, or strictures), since they are well tolerated, do not encrust, and have been shown to decrease the morbidity and mortality rates compared to traditional surgery; creating smaller stents for feline patients with an improved design when compared to the currently available human pediatric stents; addressing a ureteral obstruction as early as possible in the course of the disease, prior to the development of severe azotemia; using smaller coronary balloons for ureteral stricture dilation and treatment, etc. For <u>bladder and urethral</u> interventions we have manipulated the PCCL procedure to aid in closure of the port site, and creating a minimally invasive procedure for patients of any size, sex, breed, stone type or stone number; using a urethral stent for a malignant urethral obstruction, while concurrently treating with systemic chemotherapy to address the local disease, understanding the invasive nature in our patients; creating a cystoscope for male cats that has a working channel for urethral and ureteral intervention; etc.

Conclusion

In conclusion, endourologic procedures are vast in both human and veterinary medicine. By progressively manipulating current devices and techniques from our M.D. counterparts, the veterinary community has progressed rapidly and successfully in providing a high level of minimally invasive veterinary care, particularly pertaining to the urinary system. Understanding the similarities and differences has allowed veterinarians to advance dramatically.

REFERENCES

Available upon request.

VENTILATOR VIGNETTES

Jonathan Bach, DVM, DACVIM, DACVECC
Madison, WI

BACKGROUND AND INDICATIONS FOR MECHANICAL VENTILATION

During The 1952 polio epidemic in Copenhagen, tracheostomy and intermittent positive pressure ventilation (IPPV) were routinely utilized. Principles of basic airway management were employed and mortality fell from 80 to 25%. The Salk & Sabin vaccines eradicated polio, but IPPV had become an established standard. Further expansion subsequently occurred in the operating theater. Surgeons and anesthetists alike recognized the benefits and other machines were developed.[1] Continued expansion in surgery, anesthesia, recovery and intensive care furthered the use and application of mechanical ventilation (MV). A ventilator performs some or all of the work of breathing. Patients requiring MV can be categorized into one of three groups: 1) Hypoxemia, 2) Hypoventilation, and 3) Respiratory muscle fatigue. These distinctions are easy to remember when one recalls the two major functions of the respiratory system: oxygenation of the blood, and removal of waste (carbon dioxide) or ventilation.

Hypoxemia

Hypoxemia is defined as an arterial partial pressure of oxygen (PaO_2) less than 80 mmHg (approximately 94% SaO_2%), and severe hypoxemia is less than 60 mmHg (approximately 90% SaO_2%). Hemoglobin bound oxygen is reflected by the SaO_2%. Hemoglobin bound and dissolved oxygen have approximate correlations, and the hemoblobin oxygen dissociation curve can shift with acidosis, temperature, and red blood cell 2,3 DPG content. Dissolved oxygen (PaO_2) is only assessed with an arterial blood gas and is considered the gold standard. While saturation can be measured with pulse oximetry (SpO_2%); most blood gas analyzers estimate SaO_2% from the PaO_2.

Hypoxemia may result from several underlying thoracic problems including parenchymal diseases (eg: pneumonia, neoplasia, contusions, inflammation), pleural space diseases (eg: effusion, neoplasia), vascular pathology (eg: PTE, right to left shunt), and disease of the airways (eg: bronchitis).

Patients with hypoxemia will have an increased respiratory rate and effort, which can be mild; or they may have severe distress upon presentation with open mouthed breathing, cyanosis, orthopnea or paradoxical respiration. Many times they are tachycardic as well. Carefully observation of a patients breathing pattern, once they have acclimated to their surroundings, can heighten the clinician's index of suspicion. Patients with pulmonary parenchymal disease often have a restrictive pattern of breathing which is shallow, rapid and require greater effort (i.e. – increased work of breathing (↑WOB)).

Detection of cyanosis is an extremely insensitive method to detect hypoxemia. Cyanosis is only apparent when hypoxemia is severe (PaO2 < ~44 mmHg). Further cyanosis requires that adequate hemoglobin be present; hence animals with marked anemia may not develop cyanosis. Whenever possible, pulmonary function should be assessed with an arterial blood gas and/or pulse oximetry. Blood gas availability has increased; however increased technical skill is required. Further, obtaining an arterial blood gas in a patient with respiratory distress can be very stressful for the patient. Pulse oximetry can be readily performed at the bedside. Many scenarios can falsely lower the SpO_2% including motion, pigment, hypothermia, and vasoconstriction. Severe hypoxemia is defined as a PaO_2 < 60 mmHg (SpO_2 ~90%). Values < 80 mmHg (SpO_2 ~94%) are an indication for supplemental oxygen. If the PaO_2 remains < 60 mmHg (SpO_2 ~90%) despite supplemental oxygen, intubation and mechanical ventilation (MV) should strongly be considered. An intubated patient can easily be administered 100% oxygen.

Hypoventilation

Hypoventilation has several causes. These include severe parenchymal disease, significant pleural space lesions, and upper airway obstruction. Patients presenting with lesions affecting these respective components of the respiratory system will have readily apparent respiratory distress or audible upper airway noises. Hypoventilation from neuromuscular disease is many times more insidious and requires a keen index of suspicion. Differentials include centrally mediated hypoventilation involving the brain (eg: respiratory depressants & anesthetics, brain lesions such as trauma or neoplasia), spinal cord (eg: IVDD, trauma, FCE, inflammation), and lower motor neuron diseases (eg: myasthenia gravis, tick paralysis, botulism, polyradiculoneuritis, neuromuscular blockers).

Hypoventilation results in carbon dioxide (CO_2) retention and elevation of arterial and venous CO_2 partial pressures. Normal PCO_2 is approximately between 30-40 mmHg. Values greater than 45 mmHg are cause for concern, and values greater than 60 mmHg indicate hypoventilation that warrants intubation and mechanical ventilation. Unfortunately, PCO_2 can only be measure with blood gas analysis. But thankfully a venous sample is reliable in the absence of marked hypovolemia. Hypoxemia that associates hypoventilation can easily be overcome with supplemental oxygen. Only monitoring an SpO_2% will likely result in hypoventilation going undiagnosed.

Respiratory Muscle Fatigue

Respiratory muscle fatigue is a concept, rather than a well defined objective entity. The normal respiratory system has several auxiliary muscles (scalenes, sternomastoids, and abdominal muscles for forced expiration) that are engaged in cases of respiratory distress. It is activation of these muscles that is the earliest indication of abnormal respiratory system loading. Long before arterial blood gases or pulse oximetry indicate severe hypoxemia (PaO_2 < 60 mmHg), assessment of these respiratory changes provide clues towards diagnosing the degree of respiratory distress and discomfort. Although it is unfortunate that an "objective" definition for respiratory fatigue is not available, assessment of such condition can be rapidly made by quick, noninvasive methods including observing and examining the patient. Orthopnea serves to decrease resistance of the respiratory system. It involves holding the head & neck in an extended position. Opening of the mouth greatly increases orophayngeal diameter, maximizing air flow. The elbows are abducted to aid is chest wall expansion. And often patients sit which maximizes the mechanical advantage of the diaphragm fibers on their length tension curve. Patients may have paradoxical breathing in which the abdomen increases in size during inspiration and decreases during expiration. These forceful movements of the diaphragm are no longer counter acted by the chest wall muscles resulting in chest wall collapse during inspiration. A patient's normal work of breathing (WOB) is approximately 2% of cardiac output, while those with respiratory distress can consume as much as 25-30% of cardiac output.

The entire toll of respiratory muscle fatigue is indication that the respiratory system is failing and the patient requires support. When observing this condition and merely having the thought that this patient may be fatiguing, is a good indication that they need assistance.

PATIENT VENTILATION

Goals and Terminology

When one follows four simple goals, prescribing and managing a patient who is undergoing MV becomes much less daunting. The four goals include: 1) stabilize ventilation and target $PaCO_2$ approximately between 35 & 50 mmHg. Patients with parenchymal disease can be markedly tachypneic and hyperventilate decreasing their $PaCO_2$ quite low; this does not require intervention unless the patient has concurrent brain injury. End tidal CO_2 ($ETCO_2$)can be used as a surrogate for $PaCO_2$. If the $ETCO_2$ is more than 5 mmHg higher than $PaCO_2$, this indicates increased alveolar dead space ventilation. 2) Stabilize oxygenation such that PaO_2 achieves 80 to 120 mmHg with the least aggressive ventilator settings. 3) Achieve patient comfort and patient/ventilator synchrony. When patient/ventilator asynchronies occur, one is often tempted to increase the use of sedative / anesthetic drugs. While this may be effective, higher dosages of medications also bring greater chances of medication side effects (e.g. hypotension from propofol or opiates, respiratory depression from propofol or barbiturates, hepaotoxicity from benzodiazepines, etc). Instead the clinician should target their efforts to adjusting the ventilator such that it more closely matches the patient. First evaluate the system for possible problems such as kink, occlusion, or dislodgement of the respiratory circuit. Perhaps the patient has developed a pneumothorax, become hypoxemic, hypercapnic or hyperthermic; is the patient truly inadequately sedate (e.g. jaw tone, chewing, palpebral reflex). If the circuit and patient are not the cause of asynchrony, consider adjustments include decreasing the trigger sensitivity, adjusting the tidal volume, or inspiratory time. 4) The last goal of MV is to avoid complications, outlined further below.

Ventilator terminology

The breath controlled or limited by volume or pressure. During volume controlled you set rate (f, frequency) and tidal volume (TV), and the ventilator delivers the determined volume. TV = flow * time. TV will remain constant, but airway pressures will vary. Alveolar over-distension is a concern. During pressure controlled ventilation you set inspiratory pressure (Pisp) and inspiratory time (Ti). The ventilator delivers a flow rate up to the pressure limit, for the desired time. TV will vary from breath to breath. Decreases in compliance, or increased resistance will decrease the TV and minute ventilation (MV). Positive end expiratory pressure (PEEP) is intended to prevent airway collapse and atlectrauma associated w/ cyclic opening/closing of airways. PEEP improves gas exchange through four mechanisms: increased functional residual capacity (FRC), stabilizes alveoli after recruitment, redistributes extravascular lung water, and improves V/Q matching. How much PEEP is highly debated. Commonly applied levels in people are between 8 and 20 cmH_20. PEEP has adverse cardiovascular affects including reduced CO via restricting right ventricular filling and precluding thoracic venous inflow. Inspired oxygenation (FiO_2) concentrations can vary from 21% to 100%. Typically patients with parenchymal disease will require higher FiO_2; many times they will also require higher respiratory rate (f, frequency) and more PEEP as well.

After designing the basic starting breath (volume or pressure controlled, TV or Pisp respectively, f, FiO_2, and PEEP), one must consider the various modes, or *how* and *when* the ventilator will deliver breaths.

- IMV: intermittent mandatory ventilation. Think anesthesia ventilators. Two breath types can occur: ventilator controlled breaths, and unsupported spontaneous breaths.
- AC: assist controlled ventilation. Assists every breath initiated by a patient. However if no patient efforts are detected, delivers a mandatory breath according to the desired rate (f).
- SIMV: synchronized intermittent mandatory ventilation – Similar to IMV, however there is a trigger window (typically last 25% of expiratory time) where the patient can initiate a breath and the ventilator will synchronize with this effort to deliver a mechanically ventilated breath. If no patient efforts are detected, the ventilator delivers a breath according to the set rate (f). Can be a useful weaning mode as the f is decreased, the clinician can assess the patient's capabilities.
- CPAP/PS: pressure support – patient controls all aspects of breath including initiation, tidal volume, termination of the breath, and respiratory rate while the ventilator delivers positive airway pressure according to the clinicians desired level. Can be utilized during weaning. Also, PS can be provided in SIMV mode to breaths falling outside the trigger window.

Patient monitoring

Typical patient monitoring will include most of the following: continuously monitored EKG, SpO_2, $ETCO_2$, BP, and temperature in most patients undergoing MV. Additionally, several intermittently monitored parameters includes central venous pressure, urine production, and arterial blood gas (ABG). Hourly values should be recorded for most parameters, and the ventilator setting should be recorded hourly. Notation of ventilator settings and parameter readings should be made with each blood gas analysis; this is very helpful when comparing the trends of blood gas results and making adjustments to ventilator settings. Diligent nursing care should be a component of a patient undergoing mechanical ventilation including but not limited to eye lube with artificial tears and oral rinses with chlorhexidine solution every 4 hours. A patient's weight should be shifted adjusting their weight predominantly from one side to the other. Passive range of motion can also be considered. Patients should have diligent hygiene of IV and urinary catheters as well as ET tubes.

Complications

Numerous complications can occur during MV. Ventilator Associated Lung Injury (VALI) is used to describe changes in clinical patients. The term Ventilator Induced Lung Injury (VILI) is used in experimental models. Studies reveal that larger tidal volumes result in rapid pulmonary changes that mimic ARDS. Injurious ventilator settings results in diffuse alveolar edema, local production of inflammatory mediators with subsequent systemic leakage. The propensity to injury is partly related to the inhomogeneity in distensibility of the injured lung. Open areas tend to be hyperinflated, and injured areas underinflated. Barotraum can result from excessive airway pressures that lead to increased alveolar shear forces which contributes to epithelial disruption and loss of barrier function of the alveolar epithelium. Decreases in TV in people have had encouraging results and are being adapted to additional populations and utilized in the veterinary setting also. Ventilator Associated Pneumonia can result as the normal host defense mechanisms are compromised and atelectasis predisposes to impaired bacterial clearance from alveoli. There also be by increased populations of gram negative bacteria in oropharynx; some studies have linked oropharyngeal colonization rates with H2 blocker use. Further the oropharynx may be colonized by nosocomial organisms that may have a multidrug resistance patterns. The diagnosis of VAP is based upon identification of new exudate with the airways, or newly identified pulmonary infiltrates with concurrent deterioration of oxygenation. Due to the challenge of transportation for imaging, bedside endobronchoscopy is appealing; appropriate samples can be obtained for cytology, gram stain, and culture, guiding appropriate therapy. Recommendations to prevent VAP in people include elevation of the head 10-15 degrees, suction of subglottic secretions, protecting the airway (eg - minimizing ET tube withdrawal, adequate cuff inflation), preventing regurgitation by assessing gastric residual volumes, and selective gastrointestinal decontamination although this treatment is highly debated. Oxygen Toxicity can result from prolonged high inspired oxygen tensions which lead to the development of reactive oxygen species and free radicals. These substances can further damage cellular structures (proteins, enzymes, cell wall) and lead to generation of inflammatory mediators and further lung inflammation. Recommendations are to decrease the FiO2 to 0.6 or less within 12-18 hours, if possible. Currently there is no specific treatment, and the clinician should presume a degree of oxygen toxicity is present if these recommendations cannot be achieved. Also, Altered Hemodynamics can result from decreased venous return and decreased stroke volume which contributes to decreases in cardiac output. The effects of PPV are more severe during the inspiratory phase of respiration and with concurrent hypvolemia. Tachycardia and systemic hypotension may result, furthering organ hypoperfusion contributing to dysfunction. Efforts must be made to correct hypovolemia and limit airway pressures (ie, lower tidal volume), shorten the inspired time, reduce PEEP.

Veterinary Experience

Undoubtedly patients with hypoventilation who do not have pulmonary parenchymal disease have the best prognosis. Reported survival rates are from 39-71%.[4, 8] Dogs with cervical disk disease have better prognosis than hypoventilation that is due to LMN disease. Those with only parenchymal disease have survival rates of 20-22%. Patients suffering from both parenchymal disease and hypoventilation only had survival rates of 8%.[8] Cats are reported to have slightly lower weaning and survival rates (15-21%) than dogs.[5,8] This may be due to size limitations and the challenges of delivering an accurate tidal volume to smaller animals where the respiratory circuit accounts for a greater proportion of resistance. Smaller canine patients also have lower survival rates.[7,13] Patients with contusions have reported survival rates of 30%.[13] As critical care ventilators advance and offer additional options such as dedicated pediactric or neonatal settings, and various modes, the apparent size factor may be overcome in the future.

REFERENCES

[1]Rosengart MR. **Surg Clin N Am** 2006;86:1305. [2]Haitsma **J. Crit Care Clin** 2007;23:117. [3]King LG: **Textbook of Respiratory disease in dogs & cats.** W.B. Saunders Co, Philadelphia; 2004, pp 217. [4]Beal MW, et al. **JAVMA** 2001;218(10):1598. [5]Lee J, et al. **JAVMA** 2005;226(6):924. [6]Kube S, et al. **JAAHA** 2003;39:513. [7]King LG, et al. **JAVMA** 1994;204(7):1045. [8]Hopper K, et al. **JAVMA** 2007;230(1):64. [9]Drellich S. **VCNA** 2002;32:1087. [10]Hopper K, et al. **Compendium** 2005, April:256. [11]Hopper K, Clare M. **Compendium** 2005, April:195. [12]Vassilev E, McMichael M. **JVECC** 2004;14:15. [13]Campbell VL, King LG. **JAVMA** 2000;217(10):1505

AGE RELATED GENE EXPRESSION CHANGES IN CATS. J. Brockman, S. Al-Murrani, S. Yu Hills Pet Nutrition Center, Topeka, KS.

The purpose of this study was to compare the gene expression profiles of healthy adult cats (age 1-7 years) to the gene expression profiles of healthy senior cats (age > 7 years) to better understand the molecular mechanisms and biological processes involved in feline aging.

Gene expression profiling was performed on whole blood RNA from 48 adult cats and 54 senior cats using the Affymetrix platform with a custom feline GeneChip representing approximately 22,000 genes. Differentially expressed genes (>1.25 fold, $q < 0.003$) were analyzed for enrichment of biological pathways and functions using the GeneGo MetaCore 5.3 software. Relative cortisol levels were determined using Mass spectrometry.

These analyses indicate that age related gene expression differences in cat are consistent with hallmarks of the reactive oxygen species (ROS) theory on aging. Age related deficits in biological processes were observed including mitochondrial function, metabolism, immune function, oxidative stress, cell cycle, DNA repair, and proteostasis. One of the key mechanistic observations that impacts most of these biological processes is an up-regulation of the glucocorticoid receptor (GCR) in aged cats and a corresponding increase in it's ligand, cortisol (1.67 fold $p<0.0001$).

Nutritional intervention into aging and age related disease may be most effective during the adult stage of life. It is becoming clear that the accumulation from damage induced by ROS is central to the aging process. This damage can be mitigated through nutritional intervention by maintaining balanced metabolism (decreasing excess ROS production) and proper antioxidant support (neutralizing ROS before damage occurs).

GENOMIC & PROTEOMIC ANALYSES OF WHOLE BLOOD & SERUM SAMPLES IN ARTHRITIC & NON-ARTHRITIC CATS.

Xiangming Gao, Junyu Li, Sukhaswami Malladi, Kevin Hahn and Samer Al-Murrani
Hill's Pet Nutrition, Topeka, KS.

Introduction

Feline arthritis remains a poorly defined and diagnosed disease. There are many interconnected molecular mechanisms at play and, therefore, this report attempts to identify and discuss the major molecular changes that are observed in the blood/serum of arthritic cats and how they relate to and corroborate each other. Whole blood and serum samples from arthritic and non-arthritic cats were used for GeneChip-based and Mass Spectrometry (MS)-based analyses. The large scale study of gene/protein expression in relation to arthritis in cats has never been attempted before due to the lack of appropriate samples and reagents. The proteomic analysis was conducted on serum from arthritic and non-arthritic cats with the hope that this approach would offer a window to assess molecular changes linked to the disease.

The findings reported here have consequences for how the disease is diagnosed and eventually treated. These findings are not the product of a single experiment or a single platform but reflect the outcome of several experiments using multiple technical platforms that all converge to give a holistic picture of the disease in the blood.

Study

Experimental Design & Statistical Analysis

Whole blood and serum samples from 44 arthritic and 36 non-arthritic cats were used in this study. Whole blood samples were used for GeneChip analysis using our custom feline gene chip designed by Affymetrix. Data analysis was performed using Partek® GS software for Gene Expression data. The Robust Multichip Average algorithm was used for background adjustment, normalization, and probe-level summarization of the raw data. ANOVA analysis was performed to find significant differentially expressed genes between any two groups with a minimal False Discovery Rate control of 0.1 and a fold change equal to or greater than 1.3 in each direction.

Serum samples were equalized for the proteomic analysis using Bio-Rad's ProteoMinor beads and labeled fluorescently with Cy 3 or Cy 5. The protein mixtures were separated using Differential In-Gel Electrophoresis (DIGE). Differentially expressed protein spots were identified using SameSpot software from Nonlinear Dynamics. Differentially expressed proteins ($P < 0.05$ and fold change $>= 1.3$) were cut and tryptic digests were performed in gel and identified by LC/MS/MS using an LTQ-Orbitrap Mass Spectrometer. Protein identification was performed using BSI Peaks combined with Xtandem and Sequest search engines. A confidence score of 0.99 was set for positive identification.

Results and Discussion

It is widely accepted that cell-mediated immune functions decline with age [1]. These age associated immunological changes render the individual more susceptible to infections, cancer and autoimmune diseases [1]. Thymic involution has been implicated as the fundamental cause of the degradation of the immune system. Thymic involution is accompanied by a pronounced decrease in the generation of T-cells and therefore, the replacement of naïve T-Cells. With increasing age the proportion of naïve T-Cells decreases and that of memory T-Cells increases [2]. It has also been reported that many alterations in immune cell function that are found in aging individuals are also seen in autoimmune diseases such as rheumatoid arthritis and systemic lupus erythematosus [1]. Our studies revealed a major theme that involves the down-regulation of key genes involved in orchestrating the immune response. All cell surface proteins involved in T-Cell signaling and most of those involved in B-Cell signaling were down-regulated in cats with arthritis. This involved the T-Cell receptor itself as well as its co-receptor CD3 and other associated molecules. Additionally, CD4 and CD8 were both down-regulated indicating a dampening of cytotoxic as well as humoral immunity. Furthermore, some B-Cell signaling molecules were also down-regulated, including CD74, CD79a and b and CD84. Moreover, our proteomic and genomic studies both showed a reduction in the expression of the IL-2 receptor in arthritic cats compared to non-arthritic controls, which is a well documented consequence of aging in other species. The significant increase in the magnitude of this particular response in arthritic cats only lends further support to the notion that arthritic cats seem to show evidence consistent with accelerated aging based on the types of genes that are changed. The overall reduction in immune activity is not uncommon in aging cats (most of the arthritic cats are older cats), however, the arthritic cats seemed to have a more pronounced effect compared to their non-arthritic controls. Historically, such observations have been documented in cases of autoimmune disease in other species [1]. The down-regulation of those key immune system molecules can be counter-productive in two ways: by reducing the ability of the cells of the immune system to be properly activated in case of a challenge and also by reducing the response of these less active cells to apoptosis which is a major mechanism by which the system eliminates non-functional cells.

In the immune system, apoptosis plays a crucial role in selection of the T- Cell repertoire in the thymus, deletion of self-reactive T- and B-Cells, regulation of immunological memory, deletion of effector T-Cells following an effective immune response, and in cytotoxicity of target cells by CD8+ T-Cells and NK Cells [3]. Two major pathways lead to apoptosis: the intrinsic cell death pathway controlled by Bcl-2 family members and the extrinsic cell death pathway controlled by death receptor signalling. These two pathways work together to regulate T lymphocyte development and function [3]. Our data suggests that key pro-apoptotic genes from both of the major apoptotic pathways are down-regulated in arthritic animals compared to age matched controls. Furthermore, key anti-apoptotic genes are up-regulated. The overall net effect is the dampening of apoptosis as a regulatory mechanism in arthritic cats.

Growing evidence suggests a role for polyamines in apoptosis. In various cellular models, polyamine depletion following treatment with polyamine biosynthesis inhibitors appears to inhibit mitochondrial and death receptor pathways of apoptosis by affecting key proteins. These studies indicate that inhibition of polyamine biosynthesis may prevent or reduce the apoptotic response triggered by a variety of stimuli in non-tumoral cells, such as cardiac cells, stem cells, chondrocytes, macrophages and intestinal epithelial cells [4].

Our data indicates that key proteins involved in polyamine synthesis are down-regulated in arthritic compared to non-arthritic cats.

Furthermore, in a recent study by Teixeira, et. al. [5], large-scale gene expression profiling was performed using peripheral blood mononuclear cells from rheumatoid arthritis patients compared to controls. The report suggested that 238 genes were found to be down-regulated and 101 genes up-regulated in RA patients compared to controls (at FDR cut-off of <5% ($q < 0.05$)). A review of the genes that are up-regulated and down-regulated in this report revealed very good agreement with the cat data reported here.

In addition to the GeneChip analysis we also conducted DIGE and MS-based analysis of the sera associated with the arthritic and non-arthritic cats used in this study. The data from this analysis revealed that several of the "usual suspect" pathways associated with the development of rheumatoid arthritis are found in these samples. This corroborates the radiographic evidence of arthritis in the cats enrolled in this study

The cytoskeleton is a three-dimensional network comprising actin microfilaments, tubulin microtubules and intermediate filaments. Intermediate filaments, mainly keratin and Vimentin, are cross-linked by special binding proteins, such as Plectin, Desmoplakin, Envoplakin, Periplakin, Epiplakin, and Plakophilins [6-8]. The major cytoskeletal networks are disorganized or often absent in arthritis [9]. The data presented here are consistent with these findings with the major intermediate filaments (Keratin 1 and Vimentin) being down-regulated as well as their binding proteins including Envoplakin, plectin and Plakophilins. Vimentin is an intermediate filament which is abundantly expressed in synovial fibroblasts [10] and plays a role in the stability of the cytoplasmic architecture and in inflammation [11, 12]. In Arthritis, Vimentin can be citrullinated [13]. The down regulation of Vimentin could be a result of citrullination and caspase-3 cleavage to form different fragments [14, 15]. Antibodies directed against citrullinated Vimentin are among the most specific serological markers for the diagnosis of rheumatoid arthritis [16] with sensitivity and specificity of 69.5% and 90.8% respectively in ELISA based experiment [17].

Components of the Complement system are involved in the initiation of rheumatoid arthritis. Depletion of serum C3 levels by cobra venom factor delayed the onset of arthritis until C3 levels returned to normal [19]. Complement inhibition in experimental models points to a novel therapeutic approach to the treatment of arthritis [20]. Other damage in arthritis could come from the late components of the Complement system. These late components C5b-9 are assembled to form Membrane Attack Complex (MAC) or Terminal Complement Complex [21]. Once MAC inserts into the cell membrane, the water balance across the membrane is disrupted, the complex also allows lethal amounts of calcium to enter the cell [22]. Clusterin is a ubiquitous glycoprotein with multiple biological functions and constitutively expressed in most tissues. Clusterin interacts with lipids, complement membrane attack complex, TGFR, and immunoglobulin and its expression is related to apoptotic phenomena, tissue injury, or autoimmune damage [23]. Clusterin regulates plasma complement via direct interaction with C9 to prevent formation of MAC pores thus inhibiting complement-mediated cytolysis [24]. Individuals lacking sufficient amounts of clusterin systemically would likely have poor control of antibody mediated inflammation [25, 26].

Summary

Several lines of evidence suggest that feline arthritis exhibits many molecular hallmarks associated with rheumatoid arthritis in other species as well as other changes that are associated with accelerated aging and immune system dysregulation. Although this is not proof positive that feline arthritis is a form of RA or SLE the findings do suggest that the feline disease is different than what is known to occur in osteoarthritis in dogs and humans and maybe more complex than is currently accepted. Furthermore, our proteomic studies confirm findings that are consistent with the presence of arthritis (as is known to occur in other species) in the serum samples used in this study. The findings from the proteomic studies also confirm the findings from the gene expression studies. The data overall indicates that feline arthritis is a complex disease that is manifested far beyond the synovium with clear ramifications for future intervention strategies.

REFERENCES

[1]Fulop, T., Jr., et al., Arthritis Res Ther, 2003. **5**(6): p. 290-302;[2]Hasler, P. and M. Zouali, Cell Immunol, 2005. **233**(2): p. 102-8;[3]Gupta, S., et al., Immun Ageing, 2005. **2**: p. 12; [4]Flamigni, F., et al., Amino Acids, 2007. **33**(2): p. 197-202;[5]Teixeira, V.H., et al., PLoS ONE, 2009, **4**(8): e6803; [6]Leung, C.L., K.J. Green, and R.K. Liem, Trends Cell Biol, 2002. **12**(1): p. 37-45; [7]Jang, S.I., et al., J Cell Sci, 2005. **118**(Pt 4): p. 781-93; [8]Fontao, L., et al., Mol Biol Cell, 2003. **14**(5): p. 1978-92; [9]Blain, E.J.Exp Pathol, 2009. **90**(1): p. 1-15;[10]Osung, O.A., M. Chandra, and E.J. Holborow, Ann Rheum Dis, 1982. **41**(1): p. 74-7; [11]Fuchs, E. and K. Weber, Annu Rev Biochem, 1994. **63**: p. 345-82; [12]Mor-Vaknin, N., et al., Nat Cell Biol, 2003. **5**(1): p. 59-63; [13]Sghiri, R., et al., Rheumatol Int, 2008. **29**(1): p. 59-62.; [14]Tilleman, K., et al., Rheumatology (Oxford), 2008. **47**(5): p. 597-604; [15]Byun, Y., et al., Cell Death Differ, 2001. **8**(5): p. 443-50; [16]Dejaco, C., et al., Scand J Rheumatol, 2009. **38**(1): p. 66-7; [17]Dejaco, C., et al., Arthritis Res Ther, 2006. **8**(4): p. R119; [18]Kinloch, A., et al., Arthritis Res Ther, 2005. **7**(6): p. R1421-9; [19]Morgan, K., et al., Arthritis Rheum, 1981. **24**(11): p. 1356-62; [20]Okroj, M., et al., Ann Med, 2007. **39**(7): p. 517-30; [21]Corvetta, A., et al., Clin Exp Rheumatol, 1992. **10**(5): p. 433-8; [22]Podack, E.R. and H.J. Muller-Eberhard, J Biol Chem, 1979. **254**(19): p. 9808-14; [23]Jones, S.E. and C. Jomary, Int J Biochem Cell Biol, 2002. **34**(5): p. 427-31; [24]Tschopp, J., et al., J Immunol, 1993. **151**(4): p. 2159-65; [25]Devauchelle, V., et al., J Immunol, 2006. **177**(9): p. 6471-9; [26]Newkirk, M.M., et al., J Rheumatol, 1999. **26**(3): p. 597-603.

VETERINARY ENDOUROLOGY: FLUOROSCOPY & ENDOSCOPY FOR NOVEL THERAPIES

Allyson C. Berent, DVM, DACVIM
New York, NY

INTRODUCTION

Interventional Endoscopy (IE) uses contemporary imaging modalities, such as fluoroscopy and endoscopy, to perform diagnostic and therapeutic procedures in various body parts. A majority of IE procedures currently performed in veterinary medicine pertain to the urinary tract, and this subspecialty has been termed "endourology." The written definition of **"endourology"** /en·do·urol·o·gy/ (-ūr-olʹah-je) is "the branch of urologic surgery concerned with closed procedures for visualizing or manipulating the urinary tract." This is typically reserved for dealing with conditions in the renal pelvis, ureters, bladder and urethra. In human medicine endourology has tremendously advanced over the past 20-30 years, and in veterinary medicine similar progress is only just beginning.

This talk is an overview of some of the more commonly performed endourologic procedures actively being performed in veterinary patients including: percutaneous nephrolithotomy procedures (PCNL), nephrostomy tube placement, ureteral stenting for various causes in both canine and feline patients, cystoscopic-guided laser ablation of ectopic ureters, laser and shockwave lithotripsy for upper and lower tract stone disease, and urethral stenting for urothelial tumors and strictures. We will also discuss some promising future endourological applications currently under investigation.

The relatively common incidence of upper and lower urinary tract obstructions (stones, strictures, tumors, etc), troublesome uroliths, and ureteral ectopia, combined with the invasiveness and morbidity associated with many of the traditional surgical techniques, or the lack of therapeutic options at times, makes the use of minimally invasive endourologic procedures incredibly appealing. For decades, such procedures have been considered the standard-of-care in human medicine. Over the past 5 years the reality of such possibilities in small animal veterinary medicine has also been proven, thanks to the mentorship of our M.D colleauges. The goal of this lecture is not just to show what is possible in veterinary medicine, but geared toward showing how to perform some of these procedures with the proper equipment and techniques. It is clear that many of these procedures hold a place in the practice of veterinary urology and understanding the technique is key.

EQUIPMENT

Various flexible and rigid endoscopes are used for traditional endourologic procedures. Rigid cystoscopy is commonly performed in female animals for urethral, bladder, and ureteral access. Diameters range from 1.9 to 7.5 mm. Flexible ureteroscopes are used for urethral and bladder access in male dogs (2.7 mm) and for ureteral access in animals of acceptable sizes (>15-20 kg). Rigid nephroscopes are used (2.8-7.3 mm diameter) for percutaneous renal procedures. Different types of intracorporeal lithotripters and lasers are available for these procedures such as ultrasonic, pneumatic, electrohydraulic, and the holmium:YAG laser. The diode laser is used for endoscopic tissue resection, resulting in very effective tissue cutting and tissue coagulation, limiting bleeding. Extracorporeal shock wave lithotripsy (ESWL) also has great application for small to moderate size nephroliths and ureteroliths.

Stents are commonly used in the urinary tract to open an inappropriately closed system, typically the ureter or the urethra. A urethane soft double pigtail ureteral catheter is the "classic" indwelling ureteral stent. A NITINOL (**Ni**ckel and **Ti**tanium alloy from the **N**aval **O**rdnance **L**aboratory) self expanding metallic stent (SEMS) is classically used for urethral obstructions. These procedures will be discussed in great detail during this lecture.

For many of the more commonly performed IR procedures, a traditional fluoroscopy unit is sufficient. A C-arm fluoroscopy unit has the advantage of mobility of the image intensifier, permitting various tangential views without moving the patient. Ultrasonography is useful for percutaneous needle access into the urinary bladder or renal pelvis for the placement of percutaneous cystostomy or nephrostomy tubes. Various guidewires of an assortment of sizes, shapes and stiffness', as well as catheters, are needed for each procedure (see below).

PROCEDURES

Kidneys and Ureter

Percutaneous Nephrolithotomy (PCNL)

Nephrolithiasis or proximal ureteral obstructions secondary to ureteroliths can result in progressive renal insufficiency, intractable pyelonephritis, ureteral colic, and hydronephrosis. If the stone is small enough it may pass, however others require surgery to relieve the obstruction or avoid permanent nephron damage. Nephrotomies, pyelotomies or ureterotomies can be prolonged, invasive, and complicated surgeries, potentially resulting in significant morbidity and progressive decline in the GFR. In people, percutaneous nephrolithotomy (PCNL) is considered the standard-of-care for nephroliths too large to be treated with extracorporeal shockwave lithotripsy (ESWL) or retrograde ureteroscopy with laser lithotripsy. This procedure has recently been performed successfully in veterinary patients of various sizes (as small as 3.1 kg). This procedure aims to minimize morbidity and preserve as much renal function as possible. With the advent of powerful and safe extracorporeal shockwave lithotrites (ESWL), the success rate for the breakage and passage of large nephroliths, typically considered too big for ESWL, has been shown successful, making the need for PCNL less common.

Extracorporeal shock-wave lithotripsy is a minimally invasive alternative for the removal of upper tract calculi in the renal pelvis, or ureters. ESWL uses external shockwaves that is passes through a water medium directed under fluoroscopic guidance in 2 planes. The stone is shocked anywhere from 1000-3500 times at different energy levels to allow for implosion and powdering of a stone. The debris is then left to pass down the ureter into the urinary bladder over a 1-2 week period. It has been seen to take as long as 3 months to full stone clearance. This procedure can be performed safely for nephroliths smaller than 10 mm, and ureteroliths smaller than 5 mm. In larger stone burdens an indwelling double pigtail ureteral stent is placed prior to ESWL to aide in stone debris passage via passage ureteral dilation.

Percutaneous Nephrostomy Tube Placement

Ureteral obstructions secondary to ureteroliths or malignancy can result in severe hydronephrosis and/or life-threatening azotemia when present bilaterally or in animals with concurrent renal insufficiency. Some patients can be managed with supportive care until a ureterolith passes, others may require surgery to avoid permanent damage and/or hemodialysis to stabilize the patient prior to a prolonged anesthesia. Ureterotomies can be relatively prolonged and complicated surgeries in these often debilitated patients. One possibility is to place a nephrostomy tube percutaneously using ultrasound and fluoroscopic guidance to quickly relieve the obstruction and determine whether adequate renal function remains before prolonged anesthesia for ureteral surgery is performed. We use a locking-loop pigtail catheter. We typically use a 5 french catheter (over a 0.018" guidewire) in cats and a 6 french catheter (over a 0.035" guidewire) in dogs.

Ureteral Stenting

Ureteral stenting is performed for a variety of disorders to divert urine from the renal pelvis into the urinary bladder. This technique can be useful in patients with ureteral obstruction due various causes: ureterolithiasis, ureteral or trigonal obstructive neoplasia, ureteral stenosis/strictures, ureteral tears, ureteral spasm, etc. In addition, the presence of the ureteral stent results in passive ureteral dilation (over 2 days to 2 weeks) which may permit the passage of previously obstructive ureteroliths and urine. This is currently the most common procedure performed in feline patients with obstructive ureterolithiasis in the author's practice. Ureteral stenting is also ideal in patients with nephroliths or ureteroliths that are undergoing ESWL, to aid in fragment passage following treatment. Various stenting materials, shapes and sizes have been prototyped for feline and canine ureteral obstructions and after prospective evaluation better materials and sizes are currently available for our small animal veterinary patients. The author recommends a 2.5 french multifenestrated stent for cats and a 3.7 french multifenestrated stent for small dogs and a 4.7 french multifenestrated stent for large dogs. Animals with trigonal and UVJ induced malignant obstructions are typically stented with a 6 french double pigtail ureteral stent.

Cystoscopic-Guided Ectopic Ureter Laser Ablation

Ectopic ureters are a common congenital anatomic deformity in dogs with the ureteral orifice being positioned distal to the bladder trigone within the ureter, vagina, vestibule or uterus. Over 99% of dogs with ectopic ureters transverse intramurally and are candidates for this minimally invasive procedure. Endoscopic repair of ectopic ureters has been performed in over 50 dogs in the author's practice. This is done with the use of fluoroscopy, cystoscopy and a diode or holmium:YAG laser. This procedure is performed on an out-patient basis at the time of cystoscopic diagnosis, avoiding the need for more than one anesthetic procedure for fixation. This procedure can be successfully done in both male and female dogs and female cats. Overall, surgical fixation of ectopic ureters reports results of continued incontinence with concurrent medical intervention anywhere from 40-71% of cases. This is likely due to concurrent sphincter mechanism incompetence of the urethra (USMI), short urethral syndrome, or an intrapelvic bladder. Thus far, in the author's experience, continence has been maintained in nearly 80% of patients either alone (50%) or with (30%) concurrent treatments (phenylpropanolamine, collagen injections or the placement of a hydraulic occluder).

Ureteroscopy for Idiopathic Renal Hematuria

Idiopathic renal hematuria is a rare condition in which a focal area of bleeding in the upper urinary tract results in long term hematuria, iron deficient anemia (chronically) and the potential for clot formation, or calculi due to blood clots, resulting in ureteral colic or signs of lower urinary tract disease. In people, the presence of a hemangioma, venous rupture, angioma, or vascular malformations have been visualized ureteroscopically in the renal pelvis, which is typically cauterized or lasered through the working channel of a ureteroscope. This is possible in dogs over 15-20 kg.

Urinary Bladder and Urethra

Laser lithotripsy is an innovative technique involving the intracorporeal fragmentation of uroliths.. Small diameter fibers (200, 365, 550 microns) are guided through the working channel of small diameter flexible or rigid cystoscopes/ureteroscopes. The laser energy is focused on the urolith surface, directed via cystoscopy. As the fiber tip is advanced less than 5 mm from the calculus, the vapor bubble comes in contact with and impacts the stone. The closer the fiber tip is to the target, the larger the effect. The stone is fragmented until the pieces are small enough to be removed normograde through the urethral orifice, either via voiding urohydropropulsion or with the assistance of a stone basket. This process is useful for renal, ureteral, cystic and urethral calculi. All stone types are able to be fragmented using laser lithotripsy. The authors do not typically recommend this procedure on male dogs <7 kg or male dogs with excessive numbers of bladder stones. This is very reasonable on female dogs, female cats and male dogs with urethral stones. There are various other endourologic procedure for the removal of urinary bladder stones including: cystoscopic stone basketing and percutaneous cystolithotomy (PCCL). These procedure will be briefly discussed in this lecture.

Urethral Stenting for Obstruction

Obstructions of the urethra can cause severe discomfort, dysuria and life-threatening azotemia. Greater than 80% of animals with transitional cell carcinoma (TCC) or prostatic carcinoma experience dysuria, and approximately 10-30% develop complete urinary tract obstruction. Chemotherapy and radiation therapy has been successful in slowing tumor growth but complete cure is uncommon and when signs of obstruction occur, more aggressive therapy is indicated. Various invasive surgical procedures have been attempted though most are associated with an undesirable outcome. Placement of self-expanding metallic stents using fluoroscopic guidance through a transurethral approach can be a fast, reliable, and safe alternative to establish urethral patency in both males and females dogs and cats, with a > 85% good to excellent palliative outcome. Urethral stenting is also be useful in patients with benign urethral strictures, extraluminal compressions, or reflex dysynnergia, particularly when traditional therapies have failed or when surgery is refused or impossible. Nearly all animals that died after stent placement were due to reasons other than urinary obstruction, most of which being distant metastatic disease.

Antegrade Urethral Catheterization

Urethral catheterization is typically a fairly simple and routinely performed procedure in veterinary patients, most often due to urethral obstructions or to monitor urine output. Occasionally, standard retrograde catheterization can be difficult in very small (female) patients, female patients with obstructive tumors, or patients with urethral tears. Antegrade urethral catheterization performed under direct fluoroscopic visualization can be performed rapidly, easily, and safely in patients in whom attempts at routine retrograde catheterization have failed. This procedure will be discussed.

Conclusion

In conclusion, endourologic procedures are vast in both human and veterinary medicine. By progressively manipulating current devices and techniques from our M.D. counterparts, the veterinary community has progressed rapidly and successfully in providing a high level of minimally invasive veterinary care, particularly pertaining to the urinary system.

REFERENCES

[1]Barth A, et al. **J Am Vet Med Assoc** 2005; 226(1). [2] Weisse C, et al. **CVT XIV** (in press), 2006. [3] Berent,. A. Proceedings, **ACVS**. Washington DC, 2006. [3.] Fries CL et al. **Vet Surg** 1991; 20(2):104. [3] Knapp DW, et al. **Urol Oncol** 2000; 5:47. [4] Liptak JM, et al. **Vet Surg** 2004; 33:505. [5] Mishina M, et al. **J Vet. Med Sci** 1997; 59(4): 297. [6] Norris AM, et al. **J Vet Intern Med** 1992; 16:145. [7]Stiffler KS, et al. **J Am Vet Med Assoc** 2003; 223(3):325. [8] Stone EA, et al. **Vet Surg** 1988; 17:147. [9] Tawfiek ER, et al. **J of Urol** 1998; 160: 700. [10] Kyles A,et al. **J Am Vet Med Assoc** 2005; 226(6): 937-944. [11] Mayhew P, et al. **J Am Vet Med Assoc** 2006; 229(3): 389-393. [12] McLaughlin M, et al. **Clin Tech Small Anim Pract** 2000; 15(1): 17-24. [13] Denstedt, JD, et al. **J of Endourology** 1995; 9, 255-258. [14] Wollin T, et al. **J Clin Laser Med Surg** 1998; 16(1): 13-20. [15]Weisse C, et al. **J Am Vet Med Assoc** 2006; 229(2): 226-234. [16] Barth A, et al. **J AmVet Med Assoc** 2005; 226(1): 73-76. [17]Ell C, et al. **Gut** 1988; 29:746-751. [18] Howard RD, et al. **J Am Vet Med Assoc** 1998; 212:1600-1603. [19]Kopecky KK, et al. **Investig Radiol** 1990; 25:627-630. [20] Halland SK, et al. **J Am Vet Med Assoc** 2002; 220: 1831-1834. [21] Adams LG et al. **Vet Clin North America Small Anim Pract** 1999; 29:293-302. [22]Conigliaro R, et al. **J Clin Gastroenterol** 2006; 40(3):213-9. [23] Wamsteker EJ. **Curr Opin Gastroenterol** 2006; 22(3):300-4. [24] Streeter RN, et al. **J Am Vet Med Assoc** 2001; 219(5): 640-643. [25]Block G, et al. **J Am Vet Med Assoc**. 1996; 208(4): 531-536. [26]Matsuoka K, et al. **Urology** 1995; 45(6): 947-952. [27]Bagley DH. **Current Opin Urology** 2002; 12: 277-280. [28]Lulich JP,et al. **J Am Vet Med Assoc** 2001; 218(10):1583-1586. [29]Bartges JW, et al. **Vet Clin North Am Small Anim Pract** 2004; 34(4): 969-987. [30]Lane IF. **Vet Clin North Am Small Anim Pract** 2004; 34(4): 1011-1025.

MRI FOR INTERNISTS

Shannon P. Holmes, DVM, MSc, Dip. ACVR
Athens, GA

Due to its higher sensitivity in delineating differences between various tissues compared to radiographs, ultrasound and computed tomography (CT), **magnetic resonance imaging (MRI)** is considered the optimal imaging modality for many disease processes. Neurologists and orthopedists currently use MRI as a mainstay in routine diagnostic imaging. Several properties of MRI give it a distinct advantage over other imaging modalties. It provides high resolution multiplanar images of anatomic areas; CT has this capability, but only through image reconstruction from transverse plane images. The ability of an imaging modality to resolve differences in adjacent tissues of the body is known as contrast resolution. MRI soft tissue contrast resolution is superior to CT in that it is capable of distinguishing effacing soft tissues because of differences in the proton chemistry and relaxation. CT is governed by x-ray attenuation, and is therefore limited to 5 tissue types (metal, mineral, soft tissue/fluid, fat, gas).

MR images represent pixel maps of hydrogen proton relaxation characteristics in the three dimensional body of a human, dog or cat. Contrast in MR images is produced by manipulating proton relaxation characteristics. Alteration of MR imaging parameters creates different intrinsic contrasts of tissue, which are defined by the unique chemical nature of each tissue. Definable MR contrast includes T1 and T2 relaxation, gradient echo imaging (GRE) [or T2-star (T2*)], proton density (PD), short tau inversion recovery (STIR), fluid attenuation inversion recovery (FLAIR) and diffusion (DWI). In addition, the availability of multiple intravenous MRI contrast agents differs from the other modalities. These are typically gadolinium- and iron-based contrast agents, with the former being the most widely used in clinical studies.

MR BASICS & INTERPRETATION

The physics of MRI are complex, especially in comparison to other imaging modalities. MR examinations differ from CT examinations, because the latter is limited to pre- and post-contrast exams. A typical MR examination will have multiple different imaging sequences; the selection of imaging sequences is based upon optimal tissue characterization and pathologic processes in the examined anatomic region. For example, the imaging sequence protocol used for a MR scan of the brain would be different than that used in an abdominal study. Intravenous contrast enhancement is often a component of a MR study, but may not contribute as much information about tissue composition as the pre-contrast images. T2-weighted images have been the foundation of clinical imaging, and appear in virtually every MR examination regardless of the body region examined. T1-weigthed images provide less tissue characterization, and are predominantly included for pre- and post-contrast studies using gadolinium contrast. Depending on the area examined, different imaging sequences (T2*, PD, STIR, FLAIR and DWI) will be included. It is beyond the scope of these notes explore interpretation of all MR sequences, but examples will be provided in reference to their clinical utility. In the table below, typical tissue characteristics are provided for commonly used imaging sequences to aid in MR interpretation. Small changes in imaging parameters can produce marked changes in the appearance of the resultant MR image. Without a strong understanding of MR physics, image characteristics and cross-sectional anatomy, interpreting MR studies can be challenging and confusing.

Pulse Sequence	**Use in Veterinary MRI**	**Signal Intensity (SI)**		
		Bright	**Intermediate**	**Dark**
T1	Primarily with gadolinium contrast agent (CA)	Fat Slow flowing blood Paramagnetic CAs Methemoglobin	Muscle Brain Fluid/Cysts	Bone/Calcification Fast flowing blood Tendons
T2	Ubiquitous - very sensitive for pathologic processes	Fluid (CSF, synovial) *Edema/Inflammation* *Some tumors*	Muscle Brain (Gray-White distinction)	Bone/Calcification Hemosiderin T2 CAs
PD	Neurologic & Orthopedic imaging	Superior gray-white matter distinction. Hyperintense lesions in tendons, ligaments & muscles are consistent with fiber pattern disruption and edema.		
FLAIR	Neurologic imaging ↳ Suppression of low protein or cellular fluids (like CSF)	T1- or T2-weighted, similar image SIs to above. ***Specifically used to suppress signal from CSF*** URINE		
STIR	Oncologic imaging ↳ Increased conspicuity of neoplastic lesions	Can be T1- or T2-weighted, similar image SIs to above. ***FAT*** *is most commonly suppressed tissue, but any tissue can be* *T2-weighting is most common (bright pathology on dark background).*		
GRE **T2***	Fast imaging ↳ Ideal for abdominal scans Hemorrhage detection	Can be T1-, PD, or T2 weighted. Fluid & flowing blood commonly bright. Hemorrhage produces "blooming" signal voids.		
DWI **(w ADC map)**	Neurologic → stroke detection Oncologic → malignant vs. benign	SI on DW images and on the ADC maps distinguished free from restricted diffusion, like in CSF and intracellular fluid, respectively.		

MR APPLICATIONS FOR INTERNAL MEDICINE

HEAD & NECK

The head and neck region has extremely complex anatomy with many different soft tissue types confined to a small cross-sectional area. MRI has emerged as the preferred imaging study in human medicine for many extracranial head and neck disease processes that in the past required CT. The thyroid glands, parathyroid tissue, oropharynx, nasopharynx, retropharynx, tongue, larynx, salivary glands, cervical soft tissues, and paranasal sinuses are all well delineated on MR images. Each tissue possesses unique normal signal intensity characteristics that allow their distinction from adjacent structures. Knowledge of normal relative signal intensity of a structure makes identification of abnormal tissue easier in this area, but anatomic symmetry of the skull and neck also aid in interpretation of abnormalities.

The head and neck have been extensively studied by veterinary neurology groups; the remaining anatomy in these regions is less studied. There are 2 reports on nasal MRI in dogs. Extensive osteolysis and a mass effect are strongly associated with nasal neoplasms. Both CT and MR can be used to detect an extracranial neoplasm; however, MR more accurately delineates the extent of invasion via signal intensity variations. MR can be a better method than CT for concurrent evaluations of nasal passage, orbital, and brain involvement. Nasal tumor tissue is easily discriminated from sinus fluid, which should translate into better radiation treatment planning. Garosi *et al.* demonstrated the utility of MRI for vestibular disease, defining peripheral processes like otitis media, otitis interna, and middle ear neoplasia, as well as identifying patients with no macroscopic (therefore presumed microscopic) abnormalities to explain vestibular signs. The normal appearance of the thyroid glands has been described. Clinically, we have found MR useful in determining the resectability of thyroid carcinomas, since information can be obtained regarding tumor extent and vascular invasion. Tumor tissue can also be distinguished from surrounding tissue reaction, which is commonly seen at the margins of these tumors. Small tonsillar tumors have been detected, which were not identified with CT. Foreign body detection using MRI is excellent; large scan regions are possible with exceptional soft tissue detail. MR commonly detects and localizes the foreign body/bodies and documents the number and communication of sinus tracts. These findings can be critical for complete resection and cure.

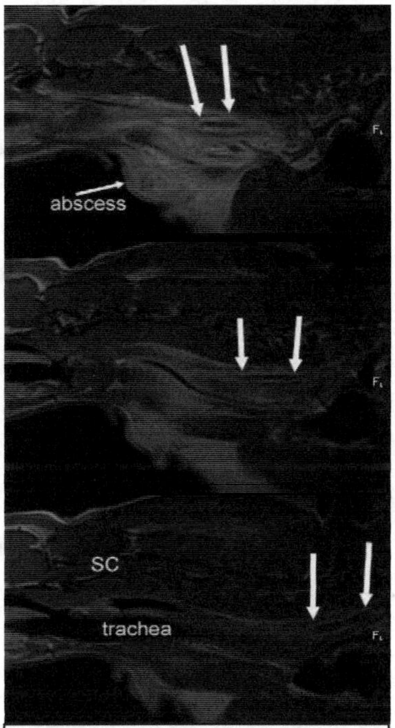

T1W Post-Gadolinium (in sagittal plane: 7cm long stick foreign body – not identified on 2 prior ultrasound examinations and during 4 surgeries

THORAX, MEDIASTINUM & PLEURA

Cardiac and respiratory gating, as well as fast scanning techniques, have made MRI studies of the chest possible and because of its superior soft tissue contrast resolution, the imaging procedure of choice. In humans, MRI has proven extremely useful in diagnosing medistinal and hilar masses, and detecting lymph node enlargement. It has the unique ability to distinguish vessels from soft tissue masses without the use of intravenous contrast agents, and without the streak artifacts that can be encountered with CT images.

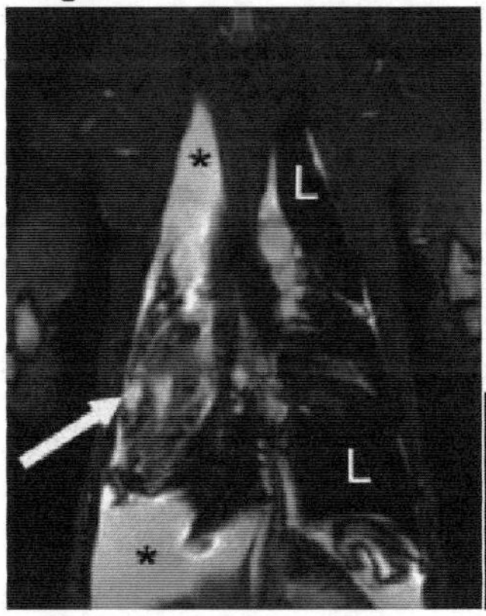

There have been no veterinary reports of MR imaging of this body region. Examples of this anatomic region are limited to clinical cases. Pulmonary masses have been detected within collapsed lung, where the mass was masked on CT images. In a preliminary comparison of cardiac hemagiosarcomas examined using CT and MR, location and extent of invasion was better visualized on MR. Small ectopic parathyroid tissue has been successfully detected using MR imaging, where the technetium-99m sestamibi nuclear study was negative. The abnormal tissue in this case was localized to cranial mediastinal lymph nodes that were judged to be normal on contrast enhanced CT. These cases, although anecdotal, demonstrate the precision of MR in finding the proverbial "needle in a haystack."

Thoracic MR – STIR image in the dorsal plane. Aerated lung is seen as black (labeled "L"). Fluid within the pleural space is bright/white (labeled "*") – which is bilateral but worse in the right thorax. The right caudal lung is collapsed and possesses intermediate signal intensity. A mass is seen within the collapsed lung (labeled with white arrow). The different proton composition and density distinguished the mass from collapsed lung parenchyma. Enlarged tracheobronchial lymph nodes were also identified (images not included).

ABDOMEN & PELVIS

In veterinary medicine, the standard imaging modality used to examine the abdomen is ultrasound. This modality is widely available, economical, and non-invasive. However, there are recognized limitations, such as the superimposition of gastrointestinal gas, dependence on operator skill, and insufficient ultrasound beam penetration in large patients. Ultrasound is used less extensively in human medicine and has largely been replaced by cross-sectional imaging. The intrapelvic region is challenging to image without rectal ultrasound. This anatomic region is ideally suited to MR examination, due to improved delineation of complex soft tissue anatomy and the absence of bone artifact commonly encountered with CT.

Special imaging sequences and fast scanning techniques have facilitated development of robust abdominal MR protocols in human medicine. These protocols are applicable to veterinary medicine, as similar disease processes occur across species.

Respiratory motion is a major limitation in abdominal MR imaging, which can be mitigated with breath-hold and respiratory gating techniques. Rapid acquisition mechanisms additionally reduce motion artifact, and allow cross-sectional abdominal imaging that is similar or superior to CT imaging. The detection of hepatic metastases is probably the best example of the sensitivity of CT versus MR. CT misses between 20 to 30% of liver metastases, which is an integral part of cancer staging. When tri-phasic enhanced CT and gadolinium or superparamagnetic iron oxide (SPIO) enhanced MR examinations are compared in the detection sensitivity of liver metastases, both types of enhanced MR studies are superior to enhanced CT exams. MRI is an excellent screening study for the liver, as well as other organs, because of its sensitivity in detecting soft tissue lesions. The reported advantages of abdominal MR imaging is the distinction of primary or metastatic masses from cysts, hematomas, abscesses, high resolution imaging of parenchymal organs, delineation of tumor extent, presence of vascular invasion, retroperitoneal lesions and detection of lymphadenopathy.

There are relatively limited veterinary reports on the application of abdominal MR in veterinary patients. Clifford et al. (2004) used the dynamic contrast enhanced MRI to correlate lesions detected on MR images with hepatic and splenic histology/cytology. Lesion characterization was made in conjunction with lesion appearance on T2-weighted images, but enhancement patterns tended to better distinguish between benign and malignant lesions. Of the 35 focal hepatic or splenic lesions, the appearance on MR images classified a lesion as malignant or benign with an overall accuracy 94% (33 of 35 lesions). The overall sensitivity was 100%, where as the overall specificity was 90%, representing unparalleled diagnostic accuracy compared to other imaging modalities.

MR angiography has been studied in the detection and depiction of portosystemic vascular anomalies (PSVA). Sequin *et al. (1999)*, Bruehschwein *et al.* (2010) and Holmes *et al.*(in preparation) used MR portography, and reported a sensitivity and specificity approached 100% with experienced reviewers; this is similar to reports on CT angiography for PSVA but avoids ionizing radiation. In addition to being superior to ultrasound and nuclear scintigraphy, cross-sectional images and 3-D reconstructions provide surgeons a virtual "roadmap" of anomalous vessels, which has translated into reduced intra-operative times in potentially metabolically compromised patients.

MR Portogram (3-D reconstruction):
4 minute acquisition

Two vascular malformations are identified. The patient has a duplex caudal vena cava, which bifurcates cranial to the left renal vein (labeled with thick white arrow). The second is a single extrahepatic portosystemic vascular anomaly. The vessels originates as a splenic vein, dissects into the retroperitoneal space and around the left kidney and the inserts on the left portion of the duplexed caudal vena cava. An initial exploratory laparotomy did not identify the anomaly. Identification of its retroperitoneal location on the MR images facilitated localization of the anomalous vessel and surgical attenuation.

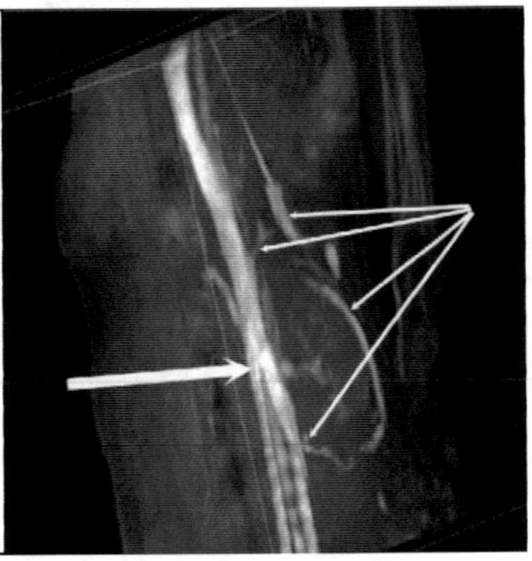

Clinically, numerous other applications have been explored at institutes where MR is available. It has proven useful in delineating adrenal tumor type through chemical shift imaging, and in identifying vascular invasion. Small tumors, such as insulinomas and gastrinomas, can be detected on MR. Neoplastic tissue possesses a different signal intensity than the normal pancreatic parenchyma, allowing discrimination of pathologically altered tissue. Metastatic lymph nodes are more accurately predicted on MR images, because of detectable alterations in the signal intensity, rather than only relying on their size and contrast enhancement pattern.

The cost of MR studies and the need for general anesthesia are commonly cited as a reason for its limited usage. The reported sensitivity of initial veterinary abdominal studies, as well as similarity to human disease processes, provides a cost-benefit to veterinary patients, which includes superior diagnostic and therapeutic plans. This is especially beneficial in chronic and complex disease processes, where other imaging modalities may fail to adequately define and diagnose pathologic changes. As more expensive treatment options are made available to veterinary clients, the need for accurate diagnoses and delineation of disease extent become paramount.

CONCLUSION

The use of MRI is no longer limited to veterinary neurologists. Through techniques developed in human medicine, virtually any region of the body of a dog or cat is amenable to MR examination. Although more costly, the superior soft tissue contrast of MRI delineates disease processes with high sensitivity and specificity. These diseases may be less well resolved or go undetected on other imaging modalities. Many internal medicine specialist and general practitioners encounter complicated cases in which a global picture of a body region or many body regions is needed to completely understand the disease process. MR has the ability to characterize both large and small areas, and offer more definitive diagnoses.

REFERENCES

[1.] Drees R, *et al.* **J Small Anim Pract.** 2009;50(7):334-40 [2.] Agthe P, *et al.* **J Small Anim Pract**. 2009; 50(12): 641-8. [3.] Miles MS, *et al.* **JAVMA** 2008;232(12): 1844-9. [4.] Garosi LS, *et al.* **JAVMA** 2001;218(3): 385-91. [5.] Taeymans O, *et al.* **Vet Radiol Ultrasound**. 2008;49(3): 238-42. [6.] Hatabu H, *et al.* **Radiol Clin North Am**. 2000 May;38(3):593-620. [7.] Martin DR, *et al.* **Radiol Clin North Am.** 2005 Sep;43(5):861-8 [8.] Hekimoglu K, *et al.* **Eur J Radiol**. 2009. [9.] EPub. Hosch WP, *et al.* **AJR Am J Roentgenol**. 2006; 186(6): 1502-11. [10.] Clifford CA, *et al.* **J Vet Intern Med.** 2004 May-Jun;18(3):330-8. [11.] Seguin B, *et al.* **Vet Radiol Ultrasound** 1999;40:251–258. [12.] Bruehschwein A, *et al.* **Vet Radiol Ultrasound** (2010) [13.] Epub. Holmes *et al.* **in preparation**. [14.] Halefoglu AM, *et al.* **Acta Radiol**. 2009 Nov;50(9):1071-9.

GENOME-WIDE ANALYSIS OF GRANULOMATOUS COLITIS IN THE BOXER DOG.

M Craven[1], GM Acland[1], JG Mezey[1], AR Boyko[1], W Wang[1], K Meurs[2], SP McDonough[1], KW Simpson[1]. 1. Cornell University, Ithaca, NY. 2. Department of Veterinary Clinical Sciences, Washington State University, Pullman, WA, USA.

Granulomatous colitis (GC), or histiocytic ulcerative colitis (HUC) is a breed-specific inflammatory bowel disease (IBD) of young Boxer dogs. Culture-independent molecular analysis has recently uncovered a correlation between GC and mucosally invasive *E.coli*. Eradication of intramucosal *E. coli* using antimicrobials (typically enrofloxacin) is associated with clinical and histologic disease remission. The breed specificity and disease characteristics suggest a heritable immunologic genetic susceptibility in the Boxer dog, which we sought to investigate using a genome-wide association scan (GWAS).

The GC phenotype was confirmed in 18 Boxer dogs (10F, 8M) by presence on histology of PAS+ macrophages. Mucosally invasive *E.coli* were detected by FISH analysis in all dogs. The control group consisted of 11 Boxer dogs (4M, 7F), >8 years old, with no history of gastrointestinal disease. Genomic DNA was harvested from blood samples collected by venipuncture into sodium EDTA, and a GWAS was performed using the Canine Affymetrix v2.0 single nucleotide polymorphism (SNP) array. An analysis pipeline was established to interpret genotypes using algorithms developed specifically for multiple locus GWA analysis, enabling statistical comparisons of case and control populations (Fisher).

Genotyping revealed GC-associated SNPs (rs8955971, p=4.3e-06 and rs24450648, p=9.12e-05) in the gene encoding neutrophil cytosolic factor 2 (NCF2, chromosome 7: 19,875,317-19,904,798). The product of NCF2 is a cytosolic subunit of the multi-protein complex NADPH oxidase, found in phagocytes. The NADPH complex is an essential component of bacterial killing, required for production of the 'respiratory burst' by delivery of reactive oxygen species to the phagosome. Dysfunction of the NADPH complex results in an ineffective respiratory burst and inability to eliminate intracellular pathogens, particularly catalase producing bacteria and fungi. Mutations in NCF2, as well as 3 other subunits (NCF1, CYBBa, CYBBb), are known to cause chronic granulomatous disease (CGD) in people, a disease complex comprising immunodeficiency disorders and predisposition to chronic infections. Patients with CGD may develop colitis characterized by 'exuberant, pigment laden macrophages' (Schäppi, Pediatr Gastroenterol Nutr '03;36) and are frequently maintained on chronic antibiotic therapy, though the specific bacteria involved are undefined. A recent GWAS in human IBD has also identified Crohn's disease-associated SNPs in the gene encoding an additional NADPH oxidase complex subunit, NCF4.

In summary, GWA analysis has identified NCF2 as a candidate gene for GCB, which guides further disease mapping and phenotypic characterization by neutrophil function testing. The known causative NCF2 mutations in human CGD add weight to this finding, and our results suggest a potential role for *E. coli* in the pathogenesis of CGD colitis in people.

NON-VIRAL GENE THERAPY FOR CANINE DIABETES MELLITUS: CANINE MUSCLE MADE TO PRODUCE CANINE INSULIN. SJM Niessen[1,3], M Fernandez-Fuente[1,2], RJ Piercy[1,2], BC Catchpole[1], A Holder[1], A Mahmoud[3], A Aldibbiat[3], AE Brown[3], SC Campbell[3], DB Church[1], JAM Shaw[3]. 1. Royal Veterinary College, Hatfield, Herts, UK. 2. Dubowitz Neuromuscular Centre, London, Greater London, UK. 3. Newcastle Medical School, Tyne and Wear, Newcastle, UK.

Development of alternative therapeutic options for canine diabetes mellitus (DM) is attractive in view of its increasing prevalence, its potential as model for human latent auto-immune diabetes of the adult, the significant impact of traditional treatment methods on pet owner lifestyle, as well as owners' reluctance to perform injections and associated costs, leading to a proportion of diabetic pets being euthanased instead of being treated. Muscle targeted gene therapy (MTGT) using insulin genes has the potential to provide an inexpensive, low maintenance alternative or adjunctive treatment method for canine DM, as has been suggested for human DM. MTGT is based on the injection of protein-coding genes into skeletal muscle, exploiting muscle's intrinsic ability to read such genes and subsequently produce the corresponding protein, which is then secreted into the general circulation by virtue of the extensive vascularisation of peripheral muscles.

A canine muscle cell line was established through primary culture of a biopsy of the biceps femoris of a young dog, as well as through transdifferentiation of canine fibroblasts after infection with a myo-differentiation gene containing adenovirus-vector (pAdTrack-CMV-IRES-GFP-MyoD). The generated canine muscle cells were subsequently transfected with non-viral plasmids (pVR1012) containing mutant furin-cleavable canine insulin gene (cppI4). cppI4 was created by modification of the native canine preproinsulin gene in order to produce proinsulin which can be cleaved by furin, an enzyme present in muscle, resulting in processing of proinsulin into mature insulin and C-peptide, a process normally exclusive to the pancreatic beta-cell. Insulin and desmin immunostaining were used to prove insulin production by canine muscle cells; canine insulin ELISA to prove insulin secretion into the medium.

A canine myoblast culture was successfully established through both methods, as proven by positive desmin-immunostaining. The cells tolerated transfection with cppI4-containing plasmid and 48 hours (hrs) later double immunostaining for insulin and desmin proved present. Canine insulin ELISA assessment of medium of a cppI4-transfected canine myo- and fibroblast-mixture proved presence of canine insulin 24 and 48 hrs post-transfection (24hrs: $10x10^3$+/-$0.7x10^3$pmol/l; 48hrs: $18x10^3$+/-$4x10^3$pmol/l, mean +/- standard deviation).

The current study provides proof of principle that canine muscle cells can be induced to produce and secrete mature canine insulin upon transfection with non-viral plasmid DNA containing a mutant canine preproinsulin-gene which produces furin-cleavable canine proinsulin. This technology could be developed to provide an alternative canine DM treatment option, potentially using additional oral drug- or glucose-responsive plasmid elements or to provide a constant source for background insulin, as well as C-peptide.

PITFALLS OF PROSPECTIVE STUDIES

George E. Moore DVM, MS, PhD, DACVPM, DACVIM
West Lafayette, IN

INTRODUCTION

Prospective studies are typically much stronger methodologically than retrospective studies, but they are also much more expensive in terms of time and labor. For these and other reasons, many (too many?) retrospective studies have been published in the veterinary literature. While retrospective studies are convenient, as you work with the data 'on hand', prospective studies should have extensive detailed planning in order to minimize the pitfalls associated with the different phases of each particular study. Randomized clinical trials (the methodologically strongest prospective study) as well as cohort studies are very susceptible to bias or systematic error at these different phases. Avoiding pitfalls often means eliminating, or minimizing to the degree possible, various kinds of biases.

CHECKLISTS – A GREAT START !!

Fortunately, expert panels in the last decade have published recommendations/checklists to strengthen reporting of randomized trials and observational studies. By post hoc evaluation of published studies, these experts determined weaknesses and inconsistencies in reporting and thus developed a checklist of essential items that should be included in the final study report submitted for peer-review and publication. These recommendations for human studies have been published as the CONSORT Statement (*Con*solidated *S*tandards *o*f *R*eporting *T*rials)[1] and the STROBE Statement (*S*trengthening *t*he *R*eporting of *Ob*servational Studies in *E*pidemiology)[2].

The advantage of the checklists within these Statements is, if a component of a study is to be reported at its conclusion, then it should be incorporated at the planning phase. Thinking about it at the end of the study (after 'the horse is out of the barn') is too late to prevent or even remove errors that occurred during the study.

Evaluation of the use of these guidelines for clinical trials in small animals[3] and in livestock[4], with suggested modifications, has been recently published in the Journal of Veterinary Internal Medicine.

DID YOU PLAN FOR PROBLEMS?

Perhaps the checklists should have been printed on paper wider than 8.5 inches. For good study planning, add two columns to the right of each specific topic recommendation entitled: 'What could go wrong in this phase?' and 'what is our fix or contingency plan?'. Many weaknesses of prospective studies are preventable, and preventions are not necessarily costly. Proper planning does not prevent all problems but it certainly can minimize their impact.

GETTING ENOUGH PATIENTS

Appropriate planning includes sample size calculation – often a requirement of grant proposals. While sample size calculations are indeed one of the most important parts of a proposal, they are also one of the most difficult. Although inclined to minimize sample size (by over-estimating differences) in order to reduce overall study costs and possibly length, investigators should 'round up' the numbers needed as much as possible as other factors will later act to reduce the number of patients contributing analyzable data. Anticipate drop-outs. Considering larger numbers needed for a study also compels the investigator to consider *a priori* how to maximize patient recruitment and retention.

CHASING TOO MANY RABBITS

As data collection is increasingly easy due to advances in information technology, it is customary to collect too much data – that is, data on a very large number of variables. In fact, you may have more variables than you have patients/subjects in the study. The investigator should not lose sight of the original primary objective, nor ignore the importance of confounding variables, but inclusion of too many variables in the analysis greatly reduces the power to find significant differences when controlling for many variables.

CONCLUSION

Using checklists from the aforementioned references, cited below, will greatly help a clinician-investigator diminish the problems that may arise during or at the end of a prospective study. Although one individual will need to serve as primary or lead investigator, getting input from and brainstorming with other clinicians/researchers while planning (a team approach) is a valuable and under-utilized resource.

REFERENCES

[1]Altman DG, et al. **Ann Intern Med** 2001;134(8):663. [2]Vandenbroucke JP, et al. **Ann Intern Med** 2007;147(8):W163. [3]Sargeant JM, et al. **J Vet Intern Med** 2010;24:44. [4]O'Connor AM, et al. **J Vet Intern Med** 2010;24:57.

KEYWORDS: prospective studies, cohort, clinical trials

DIETARY CARBOHYDRATES: IMPLICATIONS FOR LAMINITIS & EQUINE METABOLIC SYNDROME

Raymond J. Geor, BVSc, PhD, DACVIM
East Lansing, MI

INTRODUCTION

Carbohydrates (CHO) constitute the primary source of dietary energy for horses and ponies, with forages, cereal grains, and grain by-products the primary sources. As a nonruminant herbivore, horses are well adapted to the digestion (fermentation) of fibrous CHO. Feral equids spend much of the day browsing fiber-rich grasses, shrubs and sedges, the composition of which change slowly during transition from season to season. In contrast, domesticated horses are often provided low forage diets that emphasize bolus feedings of high-starch cereal grains. Additionally, 'improved' pastures contain forage species with far higher sugar content (including fructans) when compared to pastures of evolutionary grasslands. Both circumstances are thought to increase risk for development of gastrointestinal disturbances and laminitis. Indeed, the bolus administration of large doses of starch or oligofructose (a synthetic fructan) is a reliable means for the experimental induction of laminitis, wherein rapid fermentation of the administered CHO triggers systemic inflammatory events that culminate in laminar epithelial dysfunction and failure.[1-4] Similar events are thought to contribute to laminitis that develops when horses and ponies are exposed to pasture forage rich in sugars and/or fructans (*pasture-associated laminitis*). However, a key feature of pasture-associated laminitis is that in any population, certain individuals are susceptible to recurrent episodes of laminitis whilst others on the same pasture are apparently resistant to the condition, suggesting that additional factors contribute to laminitis risk. In this context, there is gathering evidence that susceptibility to laminitis, especially the pasture-associated form, is linked to an insulin-resistant phenotype, commonly termed *equine metabolic syndrome* (EMS).[5,6]

This presentation will briefly review terminology used to describe dietary CHO, and then discuss current concepts linking dietary (especially forage) CHO with laminitis as well as the implications for dietary management of equids considered high-risk for laminitis.

DIETARY CARBOHYDRATES: TERMINOLOGY AND MEASUREMENT

Plant CHO can be divided into three groups: simple sugars, storage forms such as starch and fructan, and structural polysaccharides such as cellulose and hemicellulose. From the perspective of equine digestive physiology, Hoffman et al.[7] proposed a three-way division of dietary CHO based on primary site of digestion and speed of fermentation:

1) Hydrolyzable carbohydrates (CHO-H) – can be digested in the small intestine but also are subject to rapid fermentation;
2) Rapidly fermented carbohydrates (CHO-F_R) – readily available for microbial fermentation in the large intestine; and
3) Slowly fermentable carbohydrates (CHO-F_S).

The hydrolyzable fraction includes hexoses, disaccharides, some oligosaccharides, and the non-resistant starches. The primary product of digestion of these compounds is glucose, and thus energy yield is relatively high. The CHO-F_R fraction includes pectins, fructans, and starches that are resistant to small intestinal digestion. These compounds are fermented to lactate and/or propionate. The CHO-F_S fraction includes cellulose and hemicelluloses, with acetate the primary product of fermentation in the large intestine. This system recognizes that *both* CHO-H and CHO-F_R are subject to rapid bacterial fermentation in the hindgut and therefore are risk factors for digestive and metabolic disturbances. It also allows better evaluation of the energy value of the CHO portion of the feed as it separates the components that are absorbed as glucose from those that are fermented to volatile fatty acids.[7]

A number of collective terms are used describe the CHO fractions in feeds and forages.[8] The *water soluble carbohydrate (WSC)* fraction includes simple (e.g. glucose, sucrose, fructose) and more complex (oligosaccharides and fructans) sugars. Some commercial laboratories measure WSC but report it as "sugar", while others use the term sugar to describe the free sugar fraction only. *Ethanol soluble carbohydrates (ESC)* is a relatively new term that has been introduced in an attempt to better characterize the component of WSC that is digestible by mammalian enzymes, i.e. simple sugars. The difference between the WSC and ESC fractions has been used to approximate the amount of fructan in a particular feedstuff. However, this estimate may be inaccurate, in that some oligosaccharides (including some fructans) may be measured in the ESC fraction. The *non-structural carbohydrates (NSC)* fraction comprises mono- and disaccharides, oligosaccharides (including fructan), fructan polysaccharides, and starch. There is still some debate over the optimal way to measure the NSC content of feeds and some confusion due to different methods being used in different publications. The most common system, developed by Van Soest, estimates NSC *'by difference'* where: **NSC = 100 - (crude protein % + NDF % + moisture % + fat % + ash %)**. This estimate was taken to represent the combined sugar, starch and fructan content of the feed. However, it also includes pectins, gums and mucilages that, unlike starch, sugar, and fructan, are not subject to very rapid fermentation and therefore do not induce the marked changes in lactic acid concentrations and pH within the hindgut that can occur with excess intake of sugars, starch, or fructan. Therefore, the NSC 'by difference' fraction is now referred to as *nonfiber carbohydrate or NFC.* The quantitative difference between measured NSC and NFC is small for some feeds (e.g. cereal grains) but can be quite large for other feeds (e.g. feeds with substantial pectin such as sugar beet pulp). Few commercial feed analysis laboratories completely fractionate the carbohydrates that make up the NSC, but in most feeds the amount of NSC can be approximated by summing the amount of starch and WSC.[8] This determination probably more accurately reflects the potential for a feed to be rapidly fermented to produce lactic acid when compared to the estimate of NSC content calculated 'by difference'.

FORAGE CARBOHYDRATES AND LAMINITIS

In broad terms, the ingestion of pasture forage may trigger laminitis via induction of digestive and/or metabolic disturbances. At certain times of the year, pasture forage is rich in NSC, including simple sugars, starches, and fructans. Rapid fermentation of these carbohydrates in the hindgut (cecum and colon) may trigger intestinal disturbances, triggering a chain of events that culminates in laminitis. Additionally, intake of feeds rich in these carbohydrate fractions may exacerbate hyperinsulinemia in predisposed animals, with the high circulating insulin also contributing to development of laminitis.[9] Much of our current knowledge of the pathogenesis of pasture-associated laminitis has been extrapolated from experimental models in which the condition is induced by the administration of large doses (~17 g/kg bwt) of starch or oligofructose (7-12 g/kg bwt). In both circumstances, rapid fermentation of starch or fructan initiates changes in hindgut bacterial flora with proliferation of Gram positive organisms, especially lactic acid-producing lactobacilli and streptococci, leading to a decrease in the intraluminal pH and an increase in intestinal permeability.[2,10] Systemically, there is evidence of an inflammatory response with increases in circulating mRNA for interleukin (IL)-1β, IL-6, IL-8, and IL-10 and evidence of laminar inflammation prior to the onset of lameness.[10] The relevance of these experimental models to the understanding of naturally-occurring laminitis has been questioned because of their extreme nature (i.e., large dose of CHO administered as a bolus). Nonetheless, it is likely that similar, albeit less severe, intestinal and systemic events are involved in the development of laminitis in horses and ponies consuming pasture forages rich in sugars, starches, and/or fructans (or over-consuming cereal grain-based feeds).

At certain times of the year, the quantity of pasture NSC ingested by grazing equids may approach or exceed the amount of starch or fructan known to induce laminitis when administered as a single dose.[11] A large number of environmental factors influence the accumulation of starches and fructans in pasture plants; these include the intensity and duration of sunlight, temperature (ambient and soil), soil fertility, water availability, and nitrogen status. Studies in several Northern European countries have shown fructan content of perennial ryegrass to vary between <100 g and >400 g/kg DM depending on the season and growing conditions.[11] In general, pasture NSC is highest in spring, lowest in mid-summer, and intermediate in the fall. For example, in tall fescue/Kentucky bluegrass pastures in the US, NSC content is highest in April and May (>15-20% DM, i.e. 150-200 g/kg), intermediate in the fall, and lowest in mid-winter and summer (<5-7% DM). However, there also can be marked daily fluctuations in NSC content that coincide with patterns of energy storage (photosynthetic activity) and utilization. Thus, pasture NSC tends to rise during the morning, reaching maxima in the afternoon and then declining overnight. Therefore, horses grazing in the afternoon, when compared to nighttime or the morning, may ingest between two and four times as much NSC. Stress conditions that restrict plant growth (and therefore energy demands) result in accumulation of NSC. These stress conditions include low temperatures, killing frosts, applications of non-lethal herbicides, and low soil fertility.[11]

For horses with 24-hour access to pasture, daily forage intake likely ranges between 2% to 3% of bodyweight (as DM), or 10 to 15 kg DM intake for a 500-kg horse. Thus, NSC intake would range between 0.75 and 1.5 kg DM/day and 2.25 and 4.5 kg DM/day for, respectively, pastures with NSC content of 100 g/kg DM and 300 g/kg DM. The higher end of forage NSC intake approaches the amount of starch or fructan known to induce digestive disturbances and laminitis, albeit consumed over a 12 to 17 hour period rather than as a single bolus. However, it is possible that the dosage of NSC (e.g. as fructan) required to trigger digestive and metabolic disturbances in susceptible animals is considerably lower than that needed to reliably induce disease in healthy experimental animals. It is also possible that laminitis-prone animals ingest larger quantities of forage as a percentage of bodyweight.

Role of Insulin Resistance and Hyperinsulinemia

A second potential mechanism linking the consumption of forage NSC to laminitis is exacerbation of insulin resistance (IR) and hyperinsulinemia in predisposed animals.[5,12] Studies of healthy grazing horses have shown a strong relationship between pasture NSC content and circulating insulin concentrations,[13] with exacerbation of hyperinsulinemia in laminitis-prone ponies when they are grazing spring pasture.[6,12] It is possible that episodes of hyperinsulinemia induce laminitis in grazing horses or ponies, supported by results of a recent landmark study in which Obel Grade 2 laminitis was experimentally induced in healthy, non-obese ponies by infusing exogenous insulin for up to 72 hours, to achieve serum insulin concentration of approximately 1000 mU/L.[14] It is also worth noting that endotoxemia secondary to disturbances in the hindgut microenvironment could exacerbate pre-existing IR and hyperinsulinemia in association with increased NSC intake from pasture forage. In healthy horses, IV administration of lipopolysaccharide (20 ng/kg bwt) decreased insulin sensitivity and increased the pancreatic insulin response to a glucose load.[15]

Insulin resistance also may lower the threshold for disease in the face of other conditions that trigger laminitis, e.g. hindgut disturbances associated with rapid fermentation of starch or fructan ingested in pasture forage that invoke a systemic inflammatory response and/or alter digital vascular hemodynamics. Insulin is a vasoregulatory hormone, invoking vasodilatation through pathways similar to those of insulin-mediated glucose metabolism.[16] In insulin resistant states, insulin's ability to counteract endothelin-1 associated vasoconstriction may be compromised due to decreased nitric oxide synthesis, while compensatory hyperinsulinemia might stimulate increased endothelin-1 production.[16] Insulin also modulates inflammatory responses, and there is evidence from studies in human patients and experimental animals to suggest that the microvascular dysfunction (e.g. platelet and leukocyte adhesion; leukocyte emigration) associated with sepsis or other inflammatory stimuli is exacerbated in insulin-resistant states.[17] Similarly, in horses and ponies with an insulin resistant phenotype (i.e. EMS), a pro-inflammatory state could amplify impairments to lamellar and/or digital vascular function associated with carbohydrate overload from pasture, thereby lowering the threshold for laminitis.

COUNTERING LAMINITIS RISK IN PREDISPOSED ANIMALS

Strategies for avoidance of laminitis in high-risk animals focus on two areas: 1) mitigation of metabolic and endocrine predispositions (IR, obesity, PPID) in high-risk horses and ponies, and 2) feeding management that limits intake of NSC.[18] Horses and ponies with a history of laminitis or physical characteristics suggestive of EMS or PPID should be carefully evaluated, including assessment of body condition and the presence of abnormal fat deposits, and screening tests for IR. Horses with IR and/or hyperinsulinemia require interventions to improve insulin sensitivity, including strategies for induction of weight loss (restriction in dietary energy intake, increased physical activity) and possibly use of pharmacologic agents (e.g. levothyroxine sodium) that increase insulin sensitivity and/or promote weight loss. A more detailed discussion on these aspects is presented elsewhere in these proceedings (see paper by Dr. Nicholas Frank). As pasture-associated laminitis occurs at pasture, the most obvious way to avoid the condition is to prevent access to pasture and to feed forage alternatives that are low in the carbohydrates known to be involved in triggering the disease. However, complete elimination of pasture access is not always necessary and many horses or ponies that have had one or more episodes of pasture laminitis can return to grazing activity providing there has been successful implementation of countermeasures to obesity and IR. Decisions regarding whether and to what extent affected animals can be allowed access to pasture must be made on a case-by-case basis, but in general –

- The horse or pony should be held off pasture until there has been complete resolution of the acute laminitis episode. If there is no evidence of EMS or PPID, a gradual re-introduction to pasture may be considered. Start with 1 to 2 hours of grazing, once or twice per day, or turnout for longer periods if the horse is fitted with a grazing muzzle.
- Obese, IR horses should be held off pasture for a longer period (e.g. 2-3 months), allowing time for implementation of management changes (i.e. dietary restriction, increased physical activity) that result in improved insulin sensitivity. *Even then, it is advisable to severely restrict or avoid any grazing during periods in which the pasture forage NSC content is likely to be high, e.g. spring and early summer; after summer and fall rains that cause the grass to turn green; pastures that have been frosted or drought-stressed (both can result in fructan accumulation).*
- Some IR horses and ponies with history of repeated episodes of laminitis require permanent housing in a dry lot because they appear to be susceptible to further episodes of laminitis in the face of even small variations in forage quality and quantity.

Feeds and Forages for Insulin Resistant Horses

The primary goal of feeding management of insulin-resistant equids is avoidance of feedstuffs rich in NSC that may increase risk of laminitis, either by exacerbation of hyperinsulinemia or via disturbances to the hindgut environment. General feeding rules include:

- Removal of cereal grains and grain-based feeds (e.g. sweet feed) from the diet.
- A diet based on grass hay (or hay substitute) with low (<10% dry matter) NSC content, with a ration balancer supplement provided to meet vitamin and mineral requirements.
- Feeding for maintenance of bodyweight and body condition score.

The soaking of hay in water (for 30-60 min) prior to feeding has been recommended as a strategy to reduce WSC content. However, recent studies have suggested that the extent of WSC loss during soaking is variable and unpredictable.[19] Accordingly, with respect to the selection and feeding of forage for insulin-resistant, laminitis-prone equids, hay soaking should be used as an adjunct rather than primary management strategy. Ideally, the results of proximate nutrient analysis, including direct measurement of starch and WSC, should be reviewed before selection of the hay. Forage-only diets may not provide adequate protein, minerals or vitamins. Therefore, this author recommends supplementing the forage diet with a low calorie commercial ration balancer product that contains sources of high-quality protein and a mixture of vitamins and minerals to balance the low vitamin E, copper, zinc, selenium and other minerals typically found in mature grass hays. These products are often designed to be fed in small quantities (e.g. 0.5-1.0 kg/day).

Not all insulin-resistant horses are obese and, in some animals, a hay-only ration will not meet energy needs – especially if the horse is thin and some weight gain is desired. In these circumstances, additional energy-providing feeds may be needed, e.g. sugar beet pulp, vegetable oils, stabilized rice bran. Many feed companies now offer products with fairly low NSC content (<20% DM) when compared to traditional sweet feeds or cereal grains, and these products are often marketed for use in horses and ponies at high risk for laminitis. Digestible fibers (beet pulp and soya hulls) and vegetable oils are included in place of starch-rich ingredients. In theory, these lower-NSC feeds mitigate post-feeding glycemic and insulinemic responses and carry lower risk for disturbances in hindgut function associated with rapid fermentation of starch and/or WSC. To the author's knowledge, there have been no published controlled studies in target populations and therefore data are not available to support or refute efficacy claims. Nonetheless, clinical experience has indicated that these products can be useful in the management of insulin-resistant horses.

REFERENCES

[1]Garner H, et al. **Am J Vet Res** 1975; 34:441. [2]Bailey S, et al. **Vet J** 2004: 167:129. [3]Longland A, et al. **Proc 16th ENPS** 1999:258. [4]van Eps A, Pollitt C. **Equine Vet J** 2006; 38:203. [5]Treiber K, et al. **JAVMA** 2006; 228:1538. [6]Carter R, et al. **Equine Vet J** 2009; 41:171. [7]Hoffman R, et al. **J Anim Sci** 2001; 79:500. [8]Harris P, Geor R. Vet Clin North Am Equine Pract 2009; 25:23. [9]Geor R. **Vet Clin North Am Equine Pract** 2009; 25:39. [10]Belknap JK, et al. **Equine Vet J** 2007; 39:42. [11]Longland A, Byrd B. **J Nutr** 2006; 136:2099S. [12]Treiber K, et al. **JVIM** 2008; 22:735. [13]Byrd BM, et al. **J Anim Sci** 2006; 84:330. [14]Asplin KE, et al. Vet J 2007; 174:530. [15]Toth F, et al. **Am J Vet Res** 2008; 69:82. [16]Kim JK, et al. **Circulation** 2006; 113:1888. [17]Singer G, Granger DN. **Microcirc** 2007; 14:375. [18]Geor R, Harris P. **Vet Clin North Am Equine Pract** 2009; 25:51. [19]Longland A, et al. **J Eq Vet Sci** 2009; 29:383.

THE GLYCEMIC / INSULINEMIC RESPONSE OF HORSES TO 3 DIETS

Jessica K. Suagee, M.S., Benjamin A. Corl, Ph.D., Candace D. Flinn, Ray J. Geor, BVSC, Ph.D., DACVIM
Blacksburg, VA

INTRODUCTION

One of the main functions of insulin is to promote efficient glucose utilization through facilitating glucose uptake into muscle, adipose, and liver tissue. The responsiveness and sensitivity of these tissues to insulin stimulation varies depending on the physiological state of the animal. Reduced insulin sensitivity, or insulin resistance, is a physiological state in which normal concentrations of insulin produce a less than normal biological response.[1] Thus, an insulin-resistant animal will require a greater circulating concentration of insulin (hyperinsulinemia) in order to control blood glucose following ingestion of a meal. They may also require greater insulin concentrations in order to maintain blood glucose, even in the unfed state. Horses considered insulin-resistant often demonstrate fasting hyperinsulinemia.[2]

Obesity (body condition score of 8 or 9, on a scale of 1-9) and the feeding of high glycemic diets are two conditions often associated with insulin resistance in horses.[3-6] A diet is considered high glycemic if it has significant non-structural carbohydrate (NSC) content, with typical ranges of 40-60% NSC. The NSC component of a diet includes simple sugars and starches that are digested and absorbed in the small intestine, thus directly contributing to postprandial blood glucose elevation. Dietary inclusion of starch has a dose-dependent effect on postprandial insulin concentrations in horses; horses fed 0.3 g starch/kg BW had a peak plasma insulin concentration of 43 μIU/mL compared to 202 μIU/mL for horses fed a meal of 2.0 g starch/kg BW. Further, the duration of increased plasma insulin concentration is longer in the horse fed the latter diet.[7] Therefore, a horse being fed a high NSC diet would have a much greater daily exposure to insulin than a horse being fed a low NSC diet. In other species, hyperinsulinemia has been shown to alter insulin sensitivity. Mice with transgenic hyperinsulinemia become insulin-resistant, while isolated rat adipocytes cultured with high concentrations of insulin develop reduced cell responsiveness.[8,9] Potential explanations include either down-regulation of or an increase in cellular internalization of the insulin receptor, as well as reduced activity of the proteins in the intracellular insulin signaling cascade.[10] In horses, repeated exposure of peripheral tissues to large amounts of insulin could alter the responsiveness of the insulin-reactive tissues, ultimately resulting in insulin resistance.

Insulin sensitivity may be reduced in horses fed high glycemic diets for a period of weeks to months; however, the fact that most of these studies have utilized horses across a range of BCS makes the influence of obesity unclear. The interaction between obesity and diet was initially identified by Hoffman et al.[5] In that study, horses with a BCS of 7-8 exhibited lower insulin sensitivity than horses of moderate BCS (5-6), regardless of diet. However, irrespective of body condition, horses that were fed the high glycemic diet had lower insulin sensitivity. These interactions have also been investigated following the induction of weight gain. A 16 wk overfeeding period resulted in increased fasting insulin concentrations and decreased insulin sensitivity.[4] The diet used for weight gain was highly glycemic and therefore possibly also influenced insulin sensitivity. Quinn et al. induced weight gain through overfeeding of either a high or low glycemic diet and found that horses maintained insulin sensitivity even after an increase of BCS from 4 to 7 when fed the low glycemic diet, compared to reduced insulin sensitivity in the horses fed the high glycemic diet.[6] In a cross-sectional study of horses maintained on grass pasture, Vick et al. reported a correlation coefficient of 0.53 and 0.57 between BCS (range 4-9) and plasma insulin concentration and insulin sensitivity, respectively.[11] Although the NSC content of the pasture was not reported, it was likely of lower glycemic value than concentrate diets used in the other studies, suggesting an independent relationship between reduced insulin sensitivity and BCS. Several potential mechanisms could explain the relationship between obesity and insulin resistance, including increased inflammation and altered lipid metabolism.[12] The goals of the present study were to evaluate the effect of high glycemic diets on insulin sensitivity and insulin dynamics, independent of the influence of obesity or weight gain.

MATERIALS AND METHODS

Eighteen mature Thoroughbred mares were used for this study. Horses were assigned to one of three treatment diets (VT10-9.7% NSC, VT20-21.4% NSC, and VT60-58.9% NSC) such that treatment groups were balanced for age and BCS. At all times throughout the study, horses were fed 60% of DE from hay and 40% DE from concentrate. Total concentrate ration was divided in half and fed twice daily, at 0800 and 1400 hrs, while 40% of the total hay ration was offered in the AM and 60% offered in the PM. Hay was group-fed and calculated to include a 10% refusal rate. Horses were fed VT10 for 30 d while adapting to treatment conditions. Horses were switched to their treatment diets on d 1 and fed for 90 days. On d 0 and d 91, insulin sensitivity was tested using an intravenous glucose tolerance test that utilized an intravenous infusion of dextrose (150 mg/kg). For this test, 10 mL blood samples were collected at -30, -15, -1, 2.5, 5, 7.5, 10, 15, 20, 25, 30, 45, 60, 75, 90, 120, and 180 min post dextrose administration. On d 1, 15, 30, and 90, blood samples were collected at -15, -1, 15, 30, 45, 60, 75, 90, 105, 120, 150, 180, 210, and 240 min relative to the 0800 feeding. For these tests, horses were fasted overnight. Plasma was harvested from all blood samples and used for analysis of glucose and insulin concentrations.

RESULTS

Initial and final bodyweight ($P = 0.826$) and BCS ($P = 0.767$) were not different for any treatment group. Insulin sensitivity, as assessed through peak plasma insulin concentrations and area under the curve during the intravenous glucose tolerance test, was not influenced by 90 d of feeding for any treatment ($P = 0.682$, $P = 0.305$, respectively). Peak plasma glucose concentrations following ingestion of a meal were different across diets ($P = 0.003$), whereby VT10, VT20, and VT60 increased glucose to a peak concentration of 117.2 ± 6.7, 136.5 ± 6.7, and 158.5 ± 6.9 mg/dL, respectively. There was no interaction between diet and day of study ($P = 0.334$). Similarly, peak plasma insulin concentrations were different across diets ($P < 0.001$), but within diet, did not change across days of the study ($P = 0.999$). Peak plasma insulin concentrations (mean, lower 95% confidence interval, upper 95% confidence interval) in horses on d 1 of the study were 33.9 (19.5, 58.8), 101.5 (58.5, 176.0), and 153.2 (83.5, 281.1) while on d 90 were 40.2 (23.2, 69.7), 141.1 (78.4, 254.2), and 185.2 (100.9, 339.9) for VT10, VT20, and VT60 fed horses, respectively.

DISCUSSION

Diets fed during this experiment provided 0.3, 0.5, or 1.14 g NSC/kg BW per meal for VT10, VT20, and VT60 fed horses, respectively. Thus, the observed peak postprandial insulin concentrations were similar to those previously reported by Vervuert et al.[7] The NSC content of the VT60 diet was similar to diets that resulted in reduced insulin sensitivity in previous studies.[6] Thus, the lack of change in insulin sensitivity following 90 d of feeding could be due to our exclusion of obese animals or a feeding period of insufficient duration. Obesity in horses has been associated with increased white blood cell expression of inflammatory proteins[11] and increased plasma lipids,[13] both of which have been linked in other species with altered insulin function. Several mechanisms could account for the interactions between diet, obesity, and insulin sensitivity. Results of this study suggest that in horses of moderate body condition, 90 d of feeding a high glycemic diet will not alter whole-body insulin dynamics.

REFERENCES

[1] Kahn CR. **Metabolism.** 1978; 12:1893-1902. [2] Treiber KH, et al. **J Nutr.** 2006;136:2094S-2098S. [3] Henneke DR, et al. **Equine Vet J.** 1983;15:371-372. [4] Carter RA, et al. **Am J Vet Res.** 2009;70:1250-1258. [5] Hoffman RM, et al. **J Anim Sci.** 2003;81:2333-2342. [6] Quinn RW, et al. **J Equine Vet Sci.** 2008;28:728-735. [7] Vervuert I, et al. **Vet J.** 2009;182:68-72. [8] Marban SL, et al. **Lessons in Animal Diabetes VI 6th ed.** 1996. [9] Marshall S, et al. **J Clin Invest.** 1980;66:763-772. [10] Shanik MH, et al. **Diabetes Care.** 2008;31:S262-S268. [11] Vick MM, et al. **J Anim Sci.** 2007;85:1144-1155. [12] Chen H. **Pharmacol Res.** 2006;53:469-477. [13] Frank N, et al. **J Am Vet Med Assoc.** 2006;228:1383-1390.

INSULIN & GLUCOSE LEVELS OF HORSES FED GRASS HAY DIETS

Megan L Shepherd, DVM, William S Swecker Jr, DVM, PhD, DACVN
Blacksburg, VA

INTRODUCTION

Laminitis is a costly and debilitating inflammatory condition of the equine hoof. In 2000, 13% of horse owners/operations reported having horses with laminitis, and approximately 5% of affected horses either died or were euthanized.[1] Risk factors for laminitis include obesity (BCS 7-9/9), hyperinsulinemia, and consumption of a diet with high concentrations of non-structural carbohydrates (NSC). In 2007, Thatcher et al.[2] evaluated a population of 300 adult horses randomly selected from the Virginia-Maryland Regional College's hospital client database and reported that 51% of horses were overweight or obese (BCS 6.5-9/9). In the same population, Geor et. al[3] reported a positive association between plasma insulin concentrations and body condition score. The definition of hyperinsulinemia in horses depends upon the fasted or fed state; >20μIU/L after a 6-hour fast[4] to >40μIU/L in non-fasted forage-fed horses.[3,5-7] Several studies have highlighted the association between consumption of increased NSC concentrations and hyperinsulinemia or reduced insulin sensitivity.[8-10]

Pasture forages are a risk factor for laminitis, as 45.6% laminitis cases are pasture-induced.[1] Forages include both grasses and legumes, which can be further divided into cool-season vs warm-season grasses and legumes. These plant types differ in photosynthetic pathways by which they store excess energy as non-structural carbohydrates (NSC), specifically starch polysaccharides and fructan oligo/polysaccharides. Fructans accumulate in the leaf and stem vacuoles of cool season grasses, while starch accumulates in the leaf chloroplast of legumes and warm-season grasses.[11] Fructan storage is not self-limiting, and cool season grasses have the ability to accumulate high concentrations of fructans (279g fructan/kg forage DM).[11,12] In a Virginia study, average water-soluble carbohydrate (WSC: mono/disaccharides + fructan oligo/polysaccharides) concentrations were recorded as high as 256 g fructan/kg forage DM during the month of April in orchardgrass.[13] Fructans are classified as fibers because the *beta*-glycosidic bonds that link the monosaccharide units cannot be degraded by mammalian enzymes; they are fermented in the equine cecum and large colon by microbes.[14,15] Conversely, non-fiber carbohydrates (mono/disaccharides and starch polysaccharides) are digested by mammalian enzymes and subsequently absorbed in the small intestine. Non-fiber carbohydrates directly influence plasma glucose and insulin concentrations. Forages contain high concentrations of insoluble or slowly-fermentable fibers (i.e. cellulose). However, concentrations of non-fiber carbohydrates and rapidly fermentable fibers (i.e. fructans) in forages vary based on stage of plant maturity, climate, and forage species. The variation in forage carbohydrates has a metabolic consequence for horses, as laminitis can be induced by hyperinsulinemia[16] and enteral administration of fructans.[17,18] Obesity, insulin dynamics, and forage carbohydrates share an intimate relationship with laminitis; however, the exact links between these factors and laminitis are still being explored. Veterinarians currently recommend feeding hays with <10% NSC on a dry matter basis (DMB) for obese and/or hyperinsulinemic horses to minimize the risk of laminitis.[4] However, the effect of variable NSC concentrations in cool-seaon grass hay on insulin and glucose concentrations in horses has not been evaluated.

ANIMALS & STUDY DESIGN

The objective of this study was to evaluate insulin and glucose concentrations along with physical parameters in overweight Arabian geldings fed cool-season grass hays with with either high or moderate NSC (WSC + starch) concentrations. We hypothesized that geldings fed a high NSC hay would have higher fasted serum insulin and plasma glucose concentrations than horses fed a moderate NSC hay.

Eight adult overweight (average BCS 7/9) Arabian geldings were fed each of two orchardgrass hays, high NSC (18% DMB) and moderate NSC (12% DMB), in a cross-over design during two 28-day periods. Geldings were fed to maintain their body weight as recorded at the beginning of the study[19]. Total daily hay was divided into two equal portions and fed at 0800 and 1600 hours. The hay ration was supplemented with a vitamin-mineral supplement to meet requirements for adult maintenance. Water was provided free choice via automatic waterers throughout the study.

Geldings were weighed and condition scored, and venous blood samples were collected from the jugular vein on days 0, 7, 14, 21, and 28 of each period. Sampling was performed prior to the morning feeding. The geldings were weighed on a scale on days 0 and 28 of each period. Attitude, appetite, and digital pulses were monitored daily.

RESULTS

Effects of hay, period, day, and day*hay on glucose concentrations were not detected ($p>0.05$). Insulin was influenced by hay ($p=0.001$), day ($p=0.03$), and day*hay ($p=0.04$). Insulin concentrations were higher ($p=0.0007$) on day 7 in the high NSC group (15.6μIU/ml) than the moderate NSC group (9.5μIU/ml). Digital pulses remained unremarkable through the study, and lameness was not detected in the geldings.

DISCUSSION

Geldings fed high NSC hay had higher average insulin concentrations than when fed the moderate NSC hay, yet a within-day effect was only detected on day 7. The increase in insulin concentrations in the geldings in this study did not exceed10μIU/ml, an increase associated with an exacerbation of insulin resistance.[6] Hyperinsulinemia (>20μIU/ml)[4] did not develop in either group of overweight geldings. This study demonstrates that overweight Arabian geldings fed either a high (18% DMB) or moderate (12% DMB) NSC hay do not develop hyperinsulinemia or clinical evidence of laminitis.

REFERENCES

[1.] USDA-NAHMS 2000. [2.] Thatcher CD, et al. **J Anim Physiol Anim Nutr** 2008;92:222. [3.] Geor R, et al. **J Vet Intern Med** 2007;21:601. [4.] Frank N. **J Equine Vet Sci** 2009;29:259. [5.] Frank N. AAEP 2006;51. [6.] Frank N. **Compend Eq** 2008;348. [7.] Bailey S, et al. **J Am Vet Med Assoc** 2007;231:1365. [8.] Hoffman RM, et al. **J Anim Sci** 2003;81:2333. [9.] Treiber K, et al. **J Anim Sci** 2005;83:2357. [10.] Treiber K, et al. **J Anim Physiol Anim Nutr** 2005;89:427. [11.] Longland A, Byrd B. **J Nutr** 2006;136:2099S. [12.] Chatterton N, et al. **J Plant Physiol** 1989;134:169. [13.] Shepherd M, et al. AAVN Nutrition and Research Symposium 2009;7. [14.] Niness K. **J Nutr** 1999;129:1402S. [15.] James S, et al. **Intern Med J** 2003;33:291. [16.] Asplin K, et al. **Vet J** 2007;174:530. [17.] van Eps A, Pollitt C. **Equine Vet J** 2006;38:203. [18.] Kalck KA, et al. **Am J Vet Res** 2009;70:624. [19.] NRC 2007.

ASSESSING INSULIN SENSITIVITY IN CLINICAL & RESEARCH SETTINGS

S Pratt-Phillips, Ph.D.
Raleigh, NC

INTRODUCTION

Insulin sensitivity is described as the ability of insulin to exert its physiologic effect of stimulating the movement of glucose from the bloodstream into tissues such as adipose and skeletal muscle. In contrast, insulin resistance is the ineffectiveness of insulin to exert its effect on these tissues. The link between insulin resistance and several disease states in horses necessitates accurate and relatively simple techniques for the quantification of insulin sensitivity, both in clinical situations and research studies. As such, quantification of insulin sensitivity should directly assess the movement of glucose in the presence of insulin; however, several tests commonly used in horses are imprecise measures of insulin sensitivity and may lead to misinterpretation. This brief review aims to address several methods of assessing insulin sensitivity in horses in both clinical and research settings. More thorough reviews are available elsewhere.[1-5]

BASAL MEASUREMENTS

In clinical settings, tests should be simple, minimally invasive, and cost-effective. Thus, the most common tests to address insulin and glucose dynamics involve a single blood sample to determine basal glucose and insulin concentrations in serum or plasma. In humans, the HOMA-IR (homeostatic model of insulin resistance; *glucose mmol/L * insulin μU/mL /22.5*), QUICKI (quantitative insulin sensitivity check index; *1 / (log(insulin μU/mL) + log(glucose mg/dL)),* and the glucose-to-insulin ratio (G/I) are commonly used to estimate insulin resistance (HOMA-IR) and sensitivity (QUICKI and G/I), and are derived from insulin and glucose concentrations in a fasting blood sample. While these measures are simply estimates, they have been shown to be correlated to more quantitative measures of insulin sensitivity such as the euglycemic-hyperinsulinemic clamp (EHC) and minimal model analysis (described below).[6,7] These indices are less frequently used in horses; in fact, basal plasma insulin concentrations are most commonly used to identify problems with insulin dynamics in horses in clinical settings. Proxy measurements (RISQI, the reciprocal of the square root of insulin and MIRG, the modified insulin to glucose ratio)[8] have been developed from basal insulin and glucose concentrations, are correlated to values of insulin sensitivity derived from the minimal model (described below), and have been successfully used to identify at-risk groups of horses[9] (though hyperinsulinemia alone has also been used to identify such groups).[10] Hyperinsulinemia can indicate impaired insulin action, as more insulin is required to manage glucose, while hyperglycemia can suggest uncontrolled glucose due to insulin resistance or reduced β-cell function. There is some disagreement as to what degree of hyperinsulinemia indicates a problem with insulin dynamics, but typically values greater than 20 μU/mL require further investigation. In true cases of diabetes, however, insulin values may appear normal or low and may therefore be misleading. The conditions of blood sampling are important when using basal indices, as insulin and insulin:glucose ratios can be highly variable over several days.[11] Care should be taken to ensure that horses have not had any concentrate or (ideally) any high non-structural carbohydrate forage for approximately 4 hours prior to testing. Basal measurements are useful for clinical screening of horses for insulin resistance due to ease of sampling, though further testing with more quantitative tests should be considered in some situations. Basal measurements may also be useful for large-scale epidemiological-type studies in which more complex procedures are not feasible.

TOLERANCE TESTS

Tolerance testing assesses the ability to deal with a bolus of glucose or insulin[5] and has been used extensively in equine research; [12-14] though glucose intolerance may suggest insulin resistance, tolerance tests do not specifically quantify insulin sensitivity. This is because the glucose clearance rate does not distinguish between hepatic glucose production and the exogenous glucose, nor does it accommodate for differences in insulin secretion or glucose-stimulated glucose uptake. Another potential criticism of such tests is the possibility of urinary glucose spilling due to hyperglycemia.[15] Data from an oral glucose tolerance test (OGTT; typically 1g/kg BW) may further be flawed because of differences in gastric emptying and glucose absorption. An intravenous glucose tolerance test (IVGTT; typically 100-500mg/kg BW) avoids such differences but still does not specifically quantify insulin sensitivity. An intravenous insulin tolerance test (IVITT; also called an insulin tolerance test or insulin sensitivity test) determines the effect of a bolus of insulin on blood glucose concentrations and therefore can quantify insulin sensitivity; however, risk of hypoglycemia makes the test impractical in many situations. Furthermore, the natural response of the body to maintain euglycemia through the action of counter-regulatory hormones may discredit the results. Data generated from tolerance tests include the areas under (or above, in the case of the IVITT) the curve for glucose and insulin, as well as glucose and insulin ratios. Several human studies have applied equations to results of glucose tolerance tests to derive estimates of insulin sensitivity, some with successful correlation with those derived from the EHC method.[16,17] Minimal modeling of data generated from an IVGTT can generate more accurate insulin sensitivity estimates (described below).

CGIT

The combined intravenous glucose and insulin test was designed to generate both glucose tolerance data (from the infusion of glucose) and insulin sensitivity data (through the infusion of exogenous insulin) in a simple test that could be practical for clinical settings.[18] The test infuses glucose (150mg/kg BW) then immediately insulin (0.1U/kg BW), followed by 14 blood samples over 150 min; the test may be shortened to 60 min in some clinical situations. Because glucose is infused prior to the insulin, any resulting hypoglycemia due to the insulin is less severe (though may still cause adverse reactions in the horse). In this test, insulin resistance is suspected when glucose concentrations remain higher than baseline for more than 45 min.[19] The test is repeatable with a CV of approximately 15%.[18]

MINIMAL MODEL

The minimal model analysis is a mathematical program applied to glucose and insulin values obtained during either an IVGTT (also called the frequently sampled intravenous glucose tolerance test, FSIGTT) or more commonly, an insulin-modified IVGTT[19]. The latter procedure involves the infusion of glucose, followed at 20 min by a bolus of insulin. Recently, optimal doses of both glucose and insulin have been established for horses, to both minimize urinary glucose spill and improve accuracy.[15] It is now recommended to use 100mg/kg BW of dextrose and 20 mU of insulin. The mathematical model derives estimates of insulin sensitivity (termed SI), glucose effectiveness (Sg; the ability of glucose to promote its own uptake), acute insulin response (AIRg; the acute insulin response to glucose infusion) and the disposition index (DI; the β-cell responsiveness, which is a product of AIRg and SI).[21] The minimal model method is advantageous over other tests because of this additional information. The minimal model has been used to estimate insulin sensitivity in several equine studies.[21-23] While believed to be more sensitive than the EHC method described below,[24] minimal modeling of an IVGTT has been reported to have a repeatability of approximately 24%.[25]

EUGLYCEMIC-HYPERINSULINEMIC CLAMP

The euglycemic-hyperinsulinemic clamp method is often considered the "gold standard" method of insulin sensitivity quantification in human research[26]. In this test, insulin is infused at a constant rate (typically 3mU/kg BW/min) to achieve hyperinsulinemia. At the same time glucose is infused at a variable rate to maintain euglycemia (5 mM or baseline values), with blood samples collected every 5 minutes to monitor blood glucose concentrations. Glucose concentrations are "clamped," and steady-state is typically achieved after 60-90 min. The glucose infusion rate required to maintain euglycemia is an estimate of insulin sensitivity, as a horse that is highly sensitive to insulin will require more glucose to maintain euglycemia, while a horse that is insulin-resistant will require a low glucose infusion rate. In general, the average glucose infusion rate (GIR) during the last hour of a 3-hr clamp is reported, though "steady state" must be achieved (where the GIR does not fluctuate more than 10% and euglycemia is maintained). Authors may also report the insulin-stimulated glucose uptake (M), which accounts for a space correction for any non-metabolic movement in the glucose space, or M/I which is the glucose uptake per unit of insulin (a specific estimate of tissue sensitivity to insulin). An alternative 2-hr clamp procedure involves the infusion of a priming dose of insulin prior to the constant rate infusion to achieve hyperinsulinemia and steady state more quickly.[27] Despite generating an accurate measure of glucose disposal per unit of insulin, the EHC method is criticized because of its complexity and duration (2-3 hr) and because it does not reflect the physiologic nature of glucose and insulin. Further, unless the methodology is consistent study-to-study, it is difficult to compare results. Nonetheless, this test has been used successfully in equine research [28-30] and is repeatable with a CV of 14%.[25] This test also provides the researcher with a constant level of hyperinsulinemia, in which tissue samples may be obtained to examine the molecular effects of insulin.

SUMMARY

In clinical settings, accurate quantification of insulin sensitivity (or perhaps more importantly, insulin resistance) is an essential screening tool for identifying horses that may be at risk for development of diseases such as laminitis. In such settings, simple and inexpensive tests are warranted and most clinicians prefer to test the basal insulin and glucose concentrations. While not a specific measure of insulin sensitivity, hyperinsulinemia along with other risk factors (obesity, hyperleptinemia[10]) may suggest insulin resistance. Because of the variability in basal insulin concentrations, blood samples collected on sequential days may strengthen the diagnosis. Further confirmation may be achieved with the CGIT or other more specific tests.

For research investigating disease states and treatments, investigators require a test that is practical, repeatable, and accurate. Thus, in laboratory settings, the minimal modeling of an insulin-modified IVGTT (using the "new" doses described above) or the "gold standard" EHC are ideal to quantify insulin sensitivity, depending on the desired data (Sg or AIRg vs. insulin-stimulated tissue samples). In epidemiological settings, however, these tests are not practical and single blood samples may have to suffice.

REFERENCES

[1]Firshman A, et. al. **Equine Vet J** 2007; 39: 567. [2]Matsuda M. **Nutr Metab Cardiovasc Dis** 2009; (Epub). [3]Monzillo L, et al. Nutr Rev 2003;61: 397. [4]Muniyappa R, et. al**. Am J Physiol Endocrinol Metab** 2008; 294: E15. [5]Radziuk J **J Clin Endocrinol Metab** 2000; 85: 4426. [6]Katz A, et. al. **J Clin Endocrinol Metab** 2000; 85: 2402. [7]Matthews D, et. al. **Diabetologia** 1985; 28:412. [8]Treiber K, et. al. **Am J Vet Res** 2005; 66: 2114. [9]Treiber K, et. al. **J Am Vet Med Assoc** 2006; 228: 1538. [10]Carter R, et. al. **Equine Vet J** 2009; 41: 171. [11]Pratt S, et. al. **J Eq Vet Sci** 2009; 29: 507.[12]Coffman J, et. al. **Can J Comp Med** 1983; 47: 347. [13]Freestone J, et. al. **Equine Vet J Suppl** 1992; 11: 13. [14]Garcia M, et. al. **Am J Vet Res** 1986; 47: 570. [15]Toth F, et al. **Equine Vet J** 2009; 41: 844. [16]Matsuda M, et. al. **Diabetes Care** 1999; 22: 1462. [17]Soonthornpun S, et. al. **J Clin Endocrin Metab** 2003; 88: 1019. [18]Eiler H, et. al. **Am J Vet Res** 2005; 66: 1598. [19]Frank N, et. al. **J Am Vet Med Assoc** 2006; 228: 1383. [20]Bergman, R et. al. **Am J Physiol** 1979; 236: E667. [21]Hoffman R, et. al. **J Anim Sci** 2003; 81: 2333. [22]Frank N, et. al. **Am J Vet Res** 2008 69: 76. [23]Kalck K, et. al. **Am J Vet Res** 2009; 70: 624. [24]Kronfeld D, et. al. **J Am Vet Med Assoc** 2005; 226:712. [25]Pratt S, et al **J Vet Intern Med** 2005; 19: 883. [26]De Fronzo R, et, al. **Am J Phys** 1979; 237: E214. [27]Pratt S, et. al. **J Appl Physiol** 2007; 103: 1063. [28]Firshman A, et. al. **Am J Vet Res** 2008; 69: 818. [29]Pratt S, et. al. **Equine Vet J Suppl** 2006; 39: 579. [30]Tiley H, et. al. **Am J Vet Res** 2008; 69: 51.

MANAGEMENT OF THE INSULIN-RESISTANT HORSE

Nicholas Frank, DVM, PhD, DACVIM
Knoxville, TN

INTRODUCTION

The management plan for an insulin-resistant horse should include: 1) immediate actions to prevent exacerbation of insulin resistance (IR), 2) a weight reduction program if the horse is obese, and 3) long-term control of IR through diet and exercise interventions.

MANAGEMENT PLAN

1. Preventing exacerbation of insulin resistance

Insulin sensitivity reflects the interaction between the physiology of the horse and the diet consumed. The most important components of the diet are the water-soluble carbohydrate content of the pasture and the amount of grass available for consumption. Carbohydrate intake cannot be regulated when horses are grazing, so the first step in any management program is to remove the insulin-resistant horse from pasture. An insulin-resistant horse should be moved to a dry lot or small grass paddock (size of two tennis courts) and ideally provided with a companion from the herd to reduce stress. These options are preferable to stall confinement because results of a recent study performed by our research group indicate that insulin sensitivity decreases in horses when they are brought in from pasture and confined to stalls.[1]

2. Weight reduction program (obese horses)

Diet – Obesity and IR are associated in equids,[2, 3] so obese horses must be placed on a weight reduction program to improve insulin sensitivity.[4] Horse owners should be educated about the dangers of obesity in horses because obesity-associated IR is an important risk factor for pasture-associated laminitis.[5, 6] Obese horses are placed on a weight reduction program that reduces caloric intake while increasing energy expenditure through exercise. In cases of overfeeding, weight loss can sometimes be achieved by simply eliminating all concentrates other than a vitamin/mineral/protein ration balancer and providing hay to meet the horse's digestible energy for maintenance (DE_m). Amounts provided depend upon the caloric content of the hay, so it is preferable to analyze samples and obtain the digestible energy per lb dry matter content. Hay samples can be sent to the Equi-Analytical, Inc. laboratory for analysis. The DE_m for a horse can be calculated using the formula: DE_m (MCal; multiply by 1,000 for kcal) = 1.4 + (0.03 × bwt in kg).[7] A more straightforward approach is to feed hay in amounts equivalent to 1.5% of current body weight (e.g. 18 lbs for a 1,200 lb horse) initially and then lower this amount to 1.5% of ideal body weight (e.g. 15 lbs for a 1,000 lb horse) after 2 weeks. If the horse does not begin to lose weight after an additional 4 weeks, the amount of hay fed should be lowered to 1.0% of ideal body weight (10 lb for a 1,000 lb horse), which is the minimum requirement for forage. Energy intake cannot be regulated if horses are allowed free access to pasture, so obese horses must be housed in dirt paddocks at this time. The underlying metabolic status of the horse often determines the success of the weight loss program, with obese ponies being the most difficult animals to manage.[4]

Exercise – Increased physical activity promotes weight loss by increasing energy expenditure, so horses that are free of laminitis should be exercised as frequently as possible. It is recommended that obese horses be exercised under saddle (or on a longe line) 4 to 7 days a week for a minimum of 30 minutes at a trot or canter, excluding the time required for warm up and cool down.

Levothyroxine sodium – Drug therapy should be reserved for specific situations, and veterinarians have a responsibility to discourage horse owners from administering drugs to their horse as a substitute for diet and management changes. The two indications for levothyroxine treatment are 1) short-term (3 to 6 months) treatment to accelerate weight loss and improve insulin sensitivity while diet and exercise interventions are taking effect, and 2) treatment of refractory patients. The Thyro L® product manufactured by Lloyd, Inc. has been evaluated in several studies conducted by the author's research group.[1, 8-10] Administration of levothyroxine at higher dosages has been associated with weight loss and increased insulin sensitivity,[8-10] and pretreatment with levothyroxine for 14 days prevented horses from developing insulin resistance after endotoxin infusion.[1] Levothyroxine has been administered at an approximate dosage of 0.1 mg/kg, which has been rounded to 48 mg/day for horses weighing 450 to 525 kg. It is assumed that levothyroxine induces weight loss by raising circulating thyroxine concentrations and stimulating basal metabolic rate. Serum thyroxine concentrations usually fall between 40 and 80 ng/mL (reference range: 10 to 40 ng/mL) in treated horses, but clinical signs of hyperthyroidism are not observed. Horses should always be placed on a weight reduction diet when levothyroxine is administered. In a recent study, obese horses with Equine Metabolic Syndrome (EMS) were treated with levothyroxine without calorie restriction to determine whether the drug was effective in chronically affected animals. Interestingly, a higher dosage of 72 mg/day (approximately 0.15 mg/kg) was required to induce significant weight loss and improvement in insulin responses in these severely affected horses. Current recommendations are therefore to initiate treatment at 48 mg/day (with restricted caloric intake) and then increase to 72 mg/day after 3 months if the horse remains obese. Most horses are treated for 6 months and then taken off the drug by administering 24 mg/day for 2 weeks and then 12 mg/day for 2 weeks.

3. Long-term management of insulin resistance

Hay and ration balancer diet – Most insulin-resistant horses can be maintained on a simple diet of hay plus ration balancer. Hay should be analyzed to determine the non-structural carbohydrate (NSC) content of the hay, which is calculated by adding water-soluble carbohydrate (WSC) and starch percentages together (dry matter basis). Nonstructural carbohydrates are found within the plant cell, whereas structural carbohydrates such as cellulose make up the cell wall. Simple sugars are examples of non-structural carbohydrates that raise blood glucose concentrations and therefore contribute to the postprandial glycemic response. Starches also contribute to the glycemic response by yielding monosaccharides that are absorbed within the small intestine. The NSC value estimates the amount of carbohydrate that is likely to challenge the glucose/insulin metabolism of the horse. Some veterinarians and nutritionists prefer to calculate NSC by calculating the sum of the ethanol-soluble carbohydrate (ESC) and starch percentages because this excludes long-chain fructans that are unlikely to contribute to the glycemic response. Since low-NSC hay often has a lower DE content, more of it can be fed to provide the same of amount of digestible energy, and this reduces boredom.

The importance of selecting hay with low NSC content depends upon the severity of IR and history of laminitis. If obesity and IR have been induced by overfeeding grain, then insulin sensitivity is likely to improve with weight loss. In this situation, hay analysis may be unnecessary once the problem of obesity has been addressed. In contrast, a horse with EMS that has suffered from repeated laminitis episodes requires intensive management, and hay should always be analyzed before purchase. Hay containing < 10% NSC is preferred, but < 12% hay may be acceptable if soaked in cold water for 60 minutes. Grass or alfalfa hay is appropriate, provided that the NSC content has been measured. Clients should weigh their hay to ensure that correct amounts are fed.

Low-NSC pelleted feeds – Insulin-resistant horses with leaner body condition scores require additional calories, but care must be taken to provide energy without exacerbating IR. A commercial pelleted feed with low NSC content should be selected; there are many of these feeds available. Low-NSC pelleted feeds contain sugar beet pulp, which is primarily digested within the large intestine and has a minimal effect on the glycemic response. Molasses-free sugar beet pulp can also be fed alone as a less expensive option, but each batch should be analyzed to ensure that the sugar content is low. Beet pulp should be soaked before feeding to reduce the likelihood of esophageal obstruction. Horses with marked IR should be reassessed 2 weeks after starting a new diet by rechecking blood glucose and insulin concentrations. Minor changes in insulin concentrations can be attributed to temporal variability, but an increase of greater than 10 μU/mL signals that the feed is exacerbating IR. Some low-NSC feeds are less palatable, so it may take up to 7 days for a horse to adapt to a new feed. It is advisable to lower glycemic responses by providing smaller meals more frequently (3 to 4 times daily) and feeding hay before pellets.

Metformin hydrochloride – Glucophage® (Bristol-Myers Squibb) and generic forms of metformin have been used for decades in human medicine. Metformin is a biguanide drug that is administered to control hyperglycemia and increase tissue insulin sensitivity in humans with diabetes mellitus. Only a small number of studies have been performed to date in horses. Durham et al.[11] reported that resting insulin concentrations and proxy measures of insulin sensitivity improved in horses and ponies with presumed IR when they were treated with metformin at a dosage of 15 mg/kg PO q12h. Administration of metformin at this dosage was associated with positive clinical outcomes, but Hustace et al.[12] subsequently reported that the oral bioavailability of 3 grams metformin was 7.1 ± 1.5% in fasted horses and 3.9 ± 1.0% in fed animals. Our current recommendation for metformin is therefore 30 mg/kg PO q12h, but this may be revised as new information becomes available.

Pergolide mesylate – This drug is a treatment for pituitary pars intermedia dysfunction (PPID; also called equine Cushing's disease) but is mentioned here because this endocrine disorder can exacerbate IR. In the author's experience, horses with pre-existing IR (as part of EMS) become more insulin resistant when PPID develops as a concurrent problem. Pergolide is prescribed on a total dose basis using a starting dosage of 1.0 mg/day for horses and ponies with PPID.

Supplements – Many supplements are currently marketed for the insulin-resistant horse, but there is little scientific information to support their use. Based upon studies performed in humans, magnesium and chromium are most likely to be beneficial, but more research is required. An association exists between whole-body magnesium deficits and IR in humans, and chromium is a co-factor for the tyrosine kinase insulin receptor pathway. Horse owners must be cautioned that supplements (and drugs) are not substitutes for management changes. Unfortunately, owners worry about the psychological welfare of their horse when it is taken off pasture, so supplements are sometimes perceived as attractive alternatives.

Return to pasture – This is one of the most difficult questions to address because insulin-resistant horses are at risk for developing pasture-associated laminitis. Before making this decision, the clinician should first review the history and response to management practices. If the horse suffered from obesity-associated IR caused by overfeeding and responded well to management changes, then it can be transitioned back to pasture grazing once a lean body condition has been attained and glucose and insulin concentrations return to reference range. This return to pasture should always be gradual, and access must remain limited because affected animals are predisposed to obesity. Grazing muzzles, strip grazing, or restricted turnout are options in these cases. In contrast, markedly insulin-resistant horses that respond slowly to management changes or require medical treatment should only be allowed to graze for very short periods (< 1 hour, 2-3 times daily) or must be kept off pasture altogether. These horses should remain in a dry lot or small grass paddock.

REFERENCES

[1]Tóth F, et al. **Am J Vet Res** 2010;71:60. [2]Vick MM, et al. **J Anim Sci** 2007;85:1144. [3]Carter RA, et al. **Am J Vet Res** 2009;70:1250. [4]Van Weyenberg S, et al. **J Anim Physiol Anim Nutr (Berl)** 2008;92:538. [5]Treiber KH, et al. **J Am Vet Med Assoc** 2006;228:1538. [6]Carter RA, et al. **Equine Vet J** 2009;41:171. [7]Nutrient Requirements of Horses 5th Ed. **NRC** 1989:2.[8]Frank N, et al. **Am J Vet Res** 2008;69:68. [9]Frank N, et al. **Am J Vet Res** 2008;69:76. [10]Sommardahl CS, et al. **Am J Vet Res** 2005;66:1025. [11]Durham AE, et al. **Equine Vet J** 2008;40:493. [12]Hustace JL, et al. **Am J Vet Res** 2009;70:665.

CARBOHYDRATE-RELATED CLAIMS FOR HORSE FEEDS: REGULATORY CONSIDERATIONS

David A. Dzanis, DVM, PhD, DACVN
Santa Clarita, CA

INTRODUCTION

The science of nutrition and the law governing the regulation of foods and feeds are rarely in complete harmony. Scientists and regulators approach their tasks in a similar manner, in that both evaluate a given situation based on a set of established (but not dogmatic) principles and use rational processes in reaching a conclusion. However, since these two groups approach a situation from different viewpoints, the proper course of action needed to resolve an issue as determined by a scientist may not coincide with that of a regulator. The regulation of carbohydrate-related claims on horse feed labels is a good example of this disparity.

US REGULATION OF HORSE FEEDS

The Center for Veterinary Medicine within the US Food & Drug Administration (FDA) has primary jurisdiction over all horse feeds in interstate commerce (including imports).[1] The Federal Food, Drug, & Cosmetic Act of 1938 (FFDCA) grants FDA federal authority to regulate complete feeds, concentrates, treats, supplements, and the ingredients intended to be incorporated into these products. The determination of what constitutes interstate commerce of a product considers the origins of its ingredients as well as the location of product manufacture and the point of sale or receipt of the product. Most commercial feeds are distributed in multiple states, so FDA has federal jurisdiction. Even if manufactured and sold in a single state, most feeds are comprised of at least some ingredients obtained from sources outside of the state in which they are manufactured, so they would fall under federal authority as well. However, horse feeds that are wholly sourced, produced, and distributed within a state's borders normally would not be subject to FDA oversight.

Even if the distribution of a feed is limited to intrastate commerce, the feed most often is still subject to regulation. Most state governments also exercise authority over animal feeds that are distributed within their respective jurisdictions. This constitutes another layer of regulatory oversight that is more extensive than is required for most human food items. The state's laws and regulations are enforced by the state feed control official, typically through an office in that state's department of agriculture but occasionally through other agencies (e.g., state chemist's office). Although feeds such as hays and unmixed whole or physically altered seeds may be exempt from labeling requirements under the state's commercial feed law, the state can still take action against a feed found to be adulterated.

Compliance with 50 different sets of state regulations can prove difficult for a manufacturer of a feed product in national distribution. The Association of American Feed Control Officials (AAFCO) was formed in 1909 with the mission to foster uniformity in state regulation of animal feeds.[2] It is neither a government body empowered to act under authority of law, nor a trade association whose goal (as ascribed by its critics) is to mitigate the impact of regulation on the industry. Rather, it is a private body whose members are wholly comprised of federal, state, and foreign government officials. It strives to achieve its mission through publication of the AAFCO Model Bill and Regulations, establishment of definitions for feed ingredients, and setting of other standards via policy or guidance that represent a consensus among regulators as to what constitutes the appropriate regulation of animal feed. None of what AAFCO publishes has any direct power of law. However, states are encouraged to adopt the AAFCO Models into their own laws and regulations, at which time they become enforceable by that state. While not all states follow AAFCO Models, enough do such that it has nation-wide impact.

The federal and state governments' mandate is to enforce the laws relating to horse feed manufacture and distribution. FFDCA and equivalent state laws prohibit the distribution of feeds that are either adulterated or misbranded.[1] The term "adulterated" may refer to the presence of a chemical, microbiologic, or physical contaminant, including any ingredient that is not a GRAS ("generally recognized as safe") substance, an approved food additive, or a substance otherwise sanctioned (e.g., via the AAFCO Feed Ingredient Definition process) for use in horse feeds. Failure for a product to meet stated nutrient guarantees or to conform to ingredient or nutritional representations is also a form of adulteration.[2] The term "misbranded" relates to false or misleading claims or labeling that otherwise is not in compliance with federal or state regulations.

Most state laws (particularly those that follow AAFCO Models) mandate a form of periodic product registration and/or company licensure as a condition of distributing in that state.[2] Typically, the process requires submission of product labels for review by the feed control official. Any product label found to be misbranded may be denied sale in that state. States may also collect samples of products for laboratory analysis to evaluate nutrient content and look for evidence of contamination. Depending on the egregiousness of any violations found, the product may be seized by the regulator, or the company may be notified and allowed time to remedy the violation. While these actions would appear to affect only that jurisdiction, in practice, it is infeasible for horse feed companies to distribute products in different states with different labels. Hence, labeling found to be objectionable by just one state may result in revision of labeling for product sold in all 50 states.

THE CARBOHYDRATE CONUNDRUM

Section 5(a)(3) of the AAFCO Model Bill mandates the feed label to bear a guaranteed analysis as "required to advise the user of the composition of the feed or to support claims made in the labeling." [2] The Model Regulations specify the minimum required guarantees for all horse feeds, including crude protein, crude fat, crude fiber, calcium, phosphorus, copper, selenium, zinc, and vitamin A (if added). Label claims for other nutrients, such as iron or vitamin E, mandates additional label guarantees under the Model Bill. Therefore, under normal circumstances, a claim related to carbohydrate content of the feed would require a guarantee for carbohydrates to appear on the label.

However, other guidance within AAFCO confuses this matter. Under AAFCO's Statements for Uniform Interpretation and Policy, Policy Statement #1 reads as follows:

> ***Nitrogen-Free Extract and Carbohydrate Guarantees*** *– Nitrogen-free extract and carbohydrate guarantees are no longer considered as necessary and meaningful for purchaser information, therefore, their future use is discouraged (Combination of four resolutions adopted prior to 1923, Amended 1963).*[2]

As a result of this policy, the majority of states do not allow carbohydrate guarantees on horse feed labels. However, since under the Model Bill a nutrient-related label claim requires a supporting guarantee, this creates a conundrum for both industry and regulators. Because carbohydrates cannot be guaranteed, carbohydrate-related claims therefore are also prohibited. AAFCO recognized the current policy hinders the ability for labels to bear potentially meaningful nutrition information, so a working group was established to review and offer recommendations to remedy the situation.

AAFCO ACTIONS

The Carbohydrate Working Group, comprised of state feed control officials, as well as representatives of the horse feed industry and the American College of Veterinary Nutrition, reviewed Policy Statement #1 and recommended that it NOT be deleted or amended. In this case, the intended purpose of guarantees was considered to be to provide information to help the consumer make meaningful judgments regarding carbohydrate content as they relate to the fractions that may affect glucose metabolism in the horse. As a term encompassing a broad range of compounds, "carbohydrate" was considered too vague to allow for meaningful guarantees. However, it was felt that specified carbohydrate fractions were amenable to definition and quantification. To that end, the Group defined the fractions of nutritional interest as follows:

> ***Dietary Starch (nutrient)*** *– An alpha-linked-glucose carbohydrate of or derived from plants, animals and/or microbes from which glucose is released through the hydrolytic actions of purified alpha-amylases and amyloglucosidases that are specifically active only on the alpha-(1,4) and alpha-(1,6) linkages in samples that have been gelatinized in heated, mildly acidic buffer. Its concentration in feed is determined by enzymatically converting the alpha-linked-glucose carbohydrate to glucose and then measuring the liberated glucose. This definition encompasses plant starch, glycogen, maltooligosaccharides, and maltose/isomaltose.*
>
> ***Sugars (nutrient)*** *– The sum of all free disaccharides and monosaccharides such as sucrose, lactose, glucose, fructose and galactose or others digestible by enzymes found in an animal's digestive tract.*
>
> ***Fructans (nutrient)*** *– Polysaccharides and oligosaccharides in which fructose is the major constituent and glucose is the minor constituent. Glucose content is 33% or less.*

These definitions were recommended to the AAFCO Ingredient Definitions Committee. Subsequently, they have been adopted by AAFCO and now appear in the Official Publication.[2] There are still issues with laboratory methods to directly measure these feed components so that feed control officials have means to verify guarantees (especially for fructans), but consensus on methodology is being resolved by the AAFCO Laboratory Methods and Services Committee.

With definitions for the fractions of interest in place, the next charge of the Carbohydrate Working Group was to determine how these substances could be claimed on the label. Much effort was spent attempting to define what constitutes "low" and "less" starch, sugars, and fructans in horse feeds, but a consensus could not be reached. While nutritionists participated in the discussion, other practical considerations regarding the feasibility of regulation took precedence, so that in the end, the Group recommended a very broadly-applicable regulation, as follows:

> ***Proposed amendment (addition) to AAFCO Model Regulation 4: Expression of guarantees***
> *(i) Guarantees for dietary starch, sugars, and fructans for commercial feeds, other than customer-formula feed, pet food, and specialty pet food products.*
>
> *(1) A commercial feed which bears on its labeling a claim in any manner for levels of "dietary starch", "sugars," "fructans," or words of similar designation, shall include on the label:*
>
> *(a) Guarantees for maximum percentage of dietary starch and maximum percentage sugars, in the guaranteed analysis section immediately following the crude fiber guarantee.*
>
> *(b) A maximum percentage guarantee for fructans immediately following sugars, if the feed contains forage products.*
>
> *(c) Feeding directions shall indicate the proper use of the feed product and a recommendation to consult with a veterinarian or nutritionist for a recommended diet.*

The recommendations of the Carbohydrate Working Group for the proposed amendment were accepted by the AAFCO Feed Labeling Committee at its meeting in January, 2010. The proposal must now be reviewed by the AAFCO Model Bill and Regulations Committee to ensure that the language is consistent and does not create a conflict with existing Model Regulations. The AAFCO

Board of Directors, and eventually the entire membership of AAFCO, must vote favorably on the proposal before it is enacted and appears in the Official Publication.

THE VETERINARIAN'S ROLE

Assuming the above-mentioned proposal is eventually enacted into the AAFCO Model Regulations and subsequently incorporated into state regulations, it will help horse owners in some regards but may potentially confuse them in others. The regulation will require labeling to include recommendations to consult with a veterinarian or nutritionist on use of the product. Therefore, veterinarians may be asked for advice regarding the appropriateness of a horse feed for an individual animal and how to feed it to achieve desired results.

Under the regulation, a claim of "low carbs" will not be permitted on horse feed labels. However, claims related to a specified fraction or combination of fractions, i.e., dietary starch, sugars, and/or fructans, will be allowed by regulators. A claim for a single fraction, e.g., "low dietary starch," will require guarantees for both dietary starch and sugars, as well as for fructans if the feed contains forage products (e.g., alfalfa meal, other hay-based products).

Of most importance to note is the fact that as written, the regulation does not set any standards or limits as to actual carbohydrate fraction content. Theoretically, a horse feed label may bear a claim for "low sugars" when in fact the actual content of sugars and starch are well above any physiologic benefit to the horse. It is presumed that competition between companies will hinder misleading claims, so that the marketplace will weed out the ineffective products. However, there is no regulatory recourse against manufacturers who do make claims in bad faith, so it is likely that products with misleading labeling will continue to be available on the market. Therefore, veterinarians are advised to regard any carbohydrate-related label claim on a horse feed label with due scrutiny. Before offering any advice to horse owners, the veterinarian must consider the guarantees, the feeding directions, and the carbohydrate fraction content of all other components of the ration in determining whether the product in question is of potential benefit to an individual horse.

CONCLUSION

The intent of animal feed labeling regulations is in part to help safeguard animal health by providing the consumer with information necessary to make informed purchasing decisions. While a panel of academic nutritionists may have reached a different conclusion, the proposed changes to the AAFCO Model Regulations to allow for guarantees for specified carbohydrate fractions should be of benefit to horse owners and veterinarians who know how to use the guarantees in formulating for a horse that may benefit from a "low carb" ration. However, due caution is prudent in interpretation of any label claims regarding carbohydrate content.

REFERENCES

[1]Federal Food, Drug and Cosmetic Act. [Online]. Available: http://www.fda.gov/RegulatoryInformation/Legislation/FederalFoodDrugandCosmeticActFDCAct/default.htm. Accessed: February 1, 2010. [2] Official Publication. **Association of American Feed Control Officials**. Oxford, IN, 2010.

COMPARATIVE ASPECTS OF GLUCOSE METABOLISM & REGULATION IN COMPANION ANIMALS

Steve Zicker DVM, Ph.D. DACVIM, DACVN
Topeka, KS

INTRODUCTION

Glucose metabolic regulation in mammals has progressed from elucidation of catabolic and anabolic pathways in single tissues to complex interrelationships between and within organ systems. Initial conceptualization of pathways being "on" or "off" has progressed to mathematical models of flux under different physiologic or nutritional conditions. Despite extensive knowledge, there is still much to learn about intricacies and relationships regarding regulation of this homeostatic system within, and between species. Glucose metabolic regulation will be reviewed with differences between dogs, cats and horses highlighted. In addition, relationships between foods relatively high or low in simple carbohydrates will be addressed.

GLUCOSE HOMEOSTASIS-OVERVIEW

Dogs, cats and horses absorb sugars as monosaccharides in the small intestine after digestion of polysaccharides in foods. Horses also ferment structural and non-structural carbohydrates in their colon and cecum to volatile fatty acids (VFA) for absorption. Absorbed monosaccharides, or VFA are then carried via portal circulation to the liver, or beyond, to appropriate metabolic pathways which transform these molecules via catabolic or anabolic pathways to produce energy or macromolecules.

Whole-body glucose homeostasis requires integrated control of absolute, as well as differential, responses to changes in concentrations over time, by way of multiple inter-organ relationships. Brain tissue which is pre-eminent in organ priority of glucose usage, requires steady-state concentrations above approximately 5 mM. Brain tissue cannot manufacture glucose and it is therefore imperative for survival that whole body inter-organ homeostatic interrelationships maintain a constant supply of glucose. Conversely, if hypoglycemia must be avoided, attenuation of hyperglycemia is penultimate in importance to mitigate complications of diabetes.

CLASSIC UNDERSTANDING OF GLUCOSE TRANSPORTERS IN HOMEOSTASIS

Glucose is highly hydrophilic and requires specific transport molecules to cross hydrophobic cell membranes. Intestinal glucose absorption first requires crossing of apical brush border membranes. Glucose must first be transported against a concentration gradient across the apical cell membrane by active transport mechanisms. An active sodium gradient is required to drive sodium dependent glucose co-transport proteins (SGLT1).[1]

Glucose transporter proteins (GLUT or SLC2A) utilize facilitated diffusion of glucose down a concentration gradient. Thirteen different GLUT receptors have been identified and classified into 3 classes, I, II and III, based on kinetic properties and substrate specificity (**Table 1**). Class I is comprised of GLUT1-4 while GLUT5 is a member of class II. GLUT1 and 3 are high-affinity transporters (K_m ~ 1 mM), while GLUT4 is medium (K_m ~ 2-5 mM) affinity compared to GLUT2 (K_m ~ 15-20 mM). GLUT4 is dependent on insulin to be translocated to cell surfaces. GLUT2 has the highest Km and is the last transporter to bind glucose. The differential Km values, and tissue distribution, allows low Km transporter tissues (GLUT 1&3) first priority of glucose utilization. However, in terms of homeostatic regulation GLUT2 and GLUT4 expressing tissues are of more interest especially when considering glucose sensing functions associated with GLUT2.[2]

Table 1- Physicochemical characteristics of glucose transporters in mammals

TYPE	AFFINITY (Km)/ CAPACITY	SUBSTRATE	LOCATION	TISSUE
GLUT1/ SLC2A1	High/Low	Glucose	Cell surface	Most tissues, endothelium of blood brain barrier, placenta, fetal,
GLUT2/SLC2A2	Low/High	Glucose, fructose, galactose	Cell surface or subcellular	Liver (sinusoidal membrane), β-cells of pancreas, basolateral and apical membrane of small intestine and proximal tubule
GLUT3/SLC2A3	High	Glucose	Cell surface	Neurons, placenta
GLUT4/SLC2A4	Medium	Glucose	Subcellular (requires insulin to translocate)	Adipose (brown and white), striated and cardiac muscle
GLUT5/ SLC2A5	Medium	Fructose	Cell surface	Jejunum (apical membrane)

GLUT-2 IN THE SMALL INTESTINE

Classic mechanisms described glucose entry into enterocytes solely via apical SGLT-1 active transport followed by basolateral GLUT-2 facilitated transport into blood. Evidence for GLUT-2 transporter translocation to apical membranes exists.[3] Apical GLUT2 may be responsible for up to 75% of glucose absorption in response to high glucose containing meals when concentrations are high within the lumen. Translocation of GLUT2 occurs quickly to apical membranes and is mediated by calcium signals whereas removal from apical membranes is mediated by increased insulin in response to elevated blood glucose.

Both SGLT1 and GLUT2 are responsive to dietary influences in rats.[4] Increased dietary carbohydrate enhanced expression of intestinal transporter proteins, SGLT1, GLUT2 and GLUT5, which increased hexose flux across the small intestine. Streptozotocin induced diabetes increased activities of these same three glucose transporters while insulin administration reversed these changes in rats.[5] Diabetes resistance induced by high fructose or saturated fat diets results in permanent imbedding of GLUT2 in apical membranes which may further contribute to lack of glycemic control.[6]

INCRETINS

Plasma insulin is significantly greater following oral administration of glucose compared to intravenous administration.[7] Gut associated potentiation of insulin release has been attributed to humoral or neural factors termed incretins. Two polypeptides in particular, glucose-dependent insulinotropic polypeptide (GIP) and glucagon-like peptide-1 (GLP-1), are predominately responsible for this effect by stimulating glucose-dependent insulin secretion via second messengers in pancreatic beta-islet cells.[8]

GLUT-2 IMPORTANCE IN PANCREAS AND LIVER

GLUT2 is present in pancreatic beta-cells, intestine, hepatocytes and hypothalamic neurons of most species which allows it to act as a functional glucose sensor. As glucose in blood increases above normal physiologic concentrations, where GLUT1 is 100% bound, GLUT2 transporters increase binding and transport of glucose into these specific tissues. Once inside these tissues a variety of responses are initiated, "glucose sensing", that act to return blood glucose to normal physiologic levels.

GLUCOKINASE AND HEXOKINASE

Glucose within cells is metabolized by hexokinases (HK) which phosphorylate glucose to glucose-6-phosphate (G6P). HK I-III have higher affinity for glucose than HK IV, glucokinase (GK). HK I-II are near V_{max} at normal concentrations of blood glucose and display typical Michaelis-Menten kinetics. All important to the action of GK is its inherent sigmoidal kinetic constant which features an $S_{0.5}$ of approximately 8 mM (144 mg/dL) and a lower end inflection of approximately 4 mM (72 mg/dL) for glucose. In addition, GK is resistant to G6P inhibition characteristic of other HK. Not all cells express GK and HK to the same extent which helps to traffic metabolic signals in the body.

Normal pancreatic beta islet cells do not express HK I-II to a significant degree but do have GK.[9] As blood glucose increases, transport into beta cells increases via GLUT2, and GK activity is increased because of increased substrate concentration. Subsequent phosphorylation to G6P increases its concentration in beta cells which in turn directly induces GK mRNA expression as well as preproinsulin production and release of insulin.[10] Insulinoma cells express HK I which may contribute to continual or uncontrolled release of insulin at lower concentrations of glucose in blood.[11]

Liver cells predominately express GK, of which the majority is found bound to a nuclear protein, glucokinase regulatory protein (GKRP). GKRP inhibits release and activation of GK to its active location in cytosol by way of fructose-6-phosphate to fructose-1-phosphate ratios. GKRP is not present in pancreatic beta cells which allows for balancing of metabolic responsibilities between these tissues. Additionally, liver requires insulin to express increased GK as compared to pancreas where G6P directly induces GK production.[12]

INSULIN

Insulin is released by pancreatic beta-cells in response to increased concentrations of glucose in blood. In muscle and adipose, insulin stimulates translocation of GLUT4 to the cell surface and allows for facilitated absorption of glucose. This may increase whole body glucose disposal by up to 10 fold. In liver, insulin acts to decrease gluconeogenesis by decreasing expression of two key enzymes, phoshoenolpyruvate carboxykinase (PEPCK) and glucose-6-phosphatase (G-6-Pase). Insulin also inhibits fructose 1,6-bisphosphatase via the fructose 2,6 bisphosphate allosteric system. In intestine, insulin acts to reverse GLUT2 translocation to apical membranes thus decreasing glucose absorption from gut lumen. These are the major feedback mechanisms in place which are responsible for return of glucose concentration to normal in most mammals

TISSUE DIFFERENCES

Intestine

Cats, dogs and horses digest and absorb sugars from starch and simple carbohydrates.[13] SGLT1 is present in horses in highest concentration in duodenum with decreasing activity in posterior small intestine.[14] SGLT1 has been indirectly demonstrated in dogs by inhibition of active transport with phloridzin which completely eliminates glucose transport.[15] This implies that apical facilitated transport of sugar is not manifested in dogs as in rats and humans. Characterization of dog and cat intestinal sugar transport protein expression of GLUT transporters is incomplete, however horses do express GLUT5 in small intestine.[16]

Functional Intestinal Glucose Transport

In contrast to omnivores, cats do not appear to regulate their intestinal sugar transport activities in response to diet but do show temporal changes from weaning to adult.[17,18] Intestinal amino acid transport proteins of cats do increase in response to increasing dietary protein.[17] SGLT1 in horses is adaptive and will increase in activity in response to increased dietary hydrolysable carbohydrate.[19] Exercise in dogs increases both facilitated, and active, glucose transport.

Pancreas

GLUT-2 mRNA has been reported in beta islet cells of the feline pancreas.[20] Specifics of GLUT receptors in the pancreas of dogs and horses are incomplete. Glucokinase has been detected via PCR in dogs and cats.

Functional Pancreatic Tests

Infusion of various secretagogues, such as glucose, stimulate insulin release from the pancreas. In horses, butyrate, a VFA, functions as a secretagogue of insulin in addition to glucose.[21] Arginine stimulates insulin release in cats and dogs.[22,23] Prolonged infusion of glucose with hyperglycemia in cats results in beta-cell toxicity.[24]

Liver

Liver prefers carbon skeletons of amino acids for fuel under fed conditions. GK, as well as GKRP, is absent in cats but present in dogs. Comparison of liver glucose metabolizing enzymes revealed activities of hexokinase, fructokinase, pyruvate kinase, glucose-6-phosphate dehydrogenase, fructose-1,6-bisphosphatase and glucose-6-phosphatase were higher in cats compared to dogs.[25]

WHOLE BODY GLUCOSE DISPOSAL

Whole body glucose disposal, utilizing a euglycemic hyperinsulinemic clamp, assumes that a constant infusion of insulin will completely inhibit hepatic gluconeogenesis. The amount of glucose infused which maintains euglycemia at steady state insulin infusion is the whole body glucose disposal rate. Hyperglycemic clamp models are similar but rely on endogenous insulin release in response to infused glucose to maintain concentrations above normal physiologic values. Species differences are difficult to compare as protocols vary with respect to amount of insulin or glucose infused. Nonetheless, the following range of values have been reported (μmol/kg/min): Cat (15.9-1500);[26,27] Dog (79-113);[27,28] Pony (7.3);[29] Standardbred (35-39);[30] and Warmblood (14).[29]

ENDOGENOUS GLUCOSE PRODUCTION

Cats, dogs and horses have approximately equal rates of hepatic endogenous glucose production (EGP, **Table 2**).[28,31-32] Obesity increases hepatic glucose production in dogs whereas it is decreased in obese cats. Insulin infusion in obese dogs does not inhibit EGP but completely inhibits EGP in lean dogs. Exercise in horses stimulates EGP, probably via increased Cori cycle flux.

Table 2-Endogenous Glucose Production (mg/kg-min) in Dogs, Cats, and Horses

Species	Basal-lean	Basal-obese	Clamp-lean	Clamp-obese	Exercise
Cat	2.08	1.58			
Dog	2.1	3.3	-0.3	2.7	
Horse	1.5				6.9

INFLUENCE OF DIET

Dietary influences on glucose metabolism are complex and yet to be completely evaluated. Obesity in cats results in decreases in: peripheral GLUT4 expression, endogenous glucose production and insulin-sensitivity.[31,33-34] Cats fed high protein foods had lower liver and muscle glycogen stores compared to cats fed high carbohydrate foods but maintained blood glucose better.[35] PEPCK and gluconeogenesis are decreased in cats fed high carbohydrate foods compared to high protein food when measured during the fed state. Gluconeogenic capacity is adaptive to fed and fasted states when cats are fed high carbohydrate, but not high protein foods, as manifested by suppression of PEPCK and gluconeogenesis in the fed state of the high carbohydrate fed cats.[35]

Dogs fed a high fat diet developed hepatic insulin resistance with increased endogenous glucose production. As in cats, peripheral GLUT4 expression is decreased in obese dogs which results in decreased insulin sensitivity.[36] High-protein foods fed to dogs revealed no effect on fasting blood glucose compared to when high carbohydrate foods were fed.[37] Glucose utilization rate was proportional to glucose ingested and mitochondrial PEPCK was increased in high-protein fed dogs.

Horses also show insulin resistance in response to increased body weight above normal.[38] Increased levels of digestible carbohydrates have been implicated in development of insulin resistance and laminitis in horses.[39] As such, low glycemic feeds have been recommended for management of this syndrome.[40]

SUMMARY

Regulation of glucose within specific concentrations is a well choreographed and complex process requiring participation of multiple organ systems. The above review is generic to models in use which are mainly of rat origin. However, recent elucidation of the equine and feline genomes should rapidly yield specific information regarding glucose homeostasis specific to each species.

REFERENCES

[1]Uldry M, et.al. **Pflugers Arch** 2004;447(5):480, [2]Thorens B. **Am J Physiol (Gastrointest Liver Physiol)** 1996;270(33):G541, [3]Kellet GL, et.al. **Annu Rev Nutr** 2008;28:35, [4]Cheesman CI, et.al. **J Physiol Lond** 1991;437:563, [5]Burant CF. **J Clin Invest** 1994;93:578, [6]Tobin V, et.al. **Diabetes** 2008;57:555, [7]Ishida T, et.al. **J Clin Invest** 1983;72(2):590, [8]Drucker DJ. **Cell Metabolism** 2006;3:53, [9]Quintens R, et.al. **Biochem Soc Trans** 2008;36(Pt 3):300, [10]Matchinsky RM. **Diabetes** 2002;51(Suppl 3):S394, [11]Jackson TC, et.al. **J Vet Intern Med** 2009;23(2):383, [12]Nussey SS, et.al. Endocrinology, an integrated approach. BIOS Scientific Publishers, Limited, Oxford 2001:23-70, [13]de-Oliveira LD, et.al. **J Anim Sci** 2008;86:2237-2246, [14]Dyer J, et.al. **Equine Vet J** 2002;34(4):349, [15]Pencek RR, et.al. **J Nutr** 2002;132:1929, [16]Merediz EF, et.al. **Equine Vet J**. 2004;36(6):532, [17]Buddington RK et.al. **Am J Physiol Regul Integr Comp Physiol** 1991;261:793, [18]Buddington RK, et.al. **Am J Physiol** 1992;263(5 Pt 1):G605, [19]Dyer J, et.al. **Pflugers Arch** 2009:58(2):419, [20]Zini E, et.al. **Vet J**. 2008 in press, [21]Holley DC, et.al. **J Anim Sci** 1979;49:1021, [22]Kawai K, et.al. **Diabetes** 1987;36(3):256, [23]Kitamura T, et.al. **J Vet Intern Med** 1999;13:549, [24]Zini E et.al. **Diabetologia** 2009;52:336, [25]Tanaka A, et.al. **Vet Res Commun** 2005;29(6):477, [26]Petrus DJ, et.al. **Res Vet Sci** 1998;65(2):179, [27]Hasegawa S, et.al. **J Vet Med Sci** 1992:54;583, [28]Kim SP et.al. **Diabetes** 2003:52;2453, [29]Rijnen KE et.al. **Am J Vet Res** 2003:64;1260, [30]Pratt SE et.al. **J Vet Intern Med** 2005:19;883, [30]Kley SK, et.al. **Am J Physiol Regul Integr Comp Physiol** 2009:296;R936, [32]Geor RJ, et.al. **J Appli Physiol** 2000:88;1765, [33]Brennan CL, et.al. **Dom Anim Endocrin** 2004:26;291, [34]Hoenig M, et.al. **Dom Anim Endocrin** 2006:31;373, [35]Kettelhut IC, et.al. **Am J Physiol Regula Integr Comp Physiol** 1980:239;437, [36]Gayet C, et.al. **Br J Nutr** 2007;98:497, [37]Belo PS, et.al. **J Nutr** 1976;106:1465, [38]Carter RA, et.al. **Am J Vet Res** 2009:70;1250-1258, [39]Kronfeld DS, et.al. **J Nutr** 2006:136;2090S, [40]Harris P, et.al. **Vet Clin North Am Equine Pract** 2009:25(1);23.

SWEET SUCCESS: NUTRITIONAL MANAGEMENT OF THE CANINE DIABETIC

Linda M. Fleeman BVSc PhD MACVSc
Melbourne, Australia

WHAT IS CANINE DIABETES MELLITUS?

Diabetes mellitus is caused by absolute or relative deficiency of insulin. This results in altered carbohydrate, fat, and protein metabolism, which manifests as hyperglycemia, hyperlipidemia, polyuria, lethargy, weight loss, polyphagia, poor coat, and reduced immunity. The majority of diabetic dogs likely have absolute insulin deficiency due to type 1 diabetes mellitus caused by immune destruction of pancreatic beta cells[1-4]. In approximately 30%, loss of beta cells is due to extensive pancreatic damage from chronic pancreatitis[1,5]. Canine diabetes also occurs in association with corticosteroid therapy, hyperadrenocorticism[6], and progesterone-induced acromegaly[7]. In entire bitches, a form analogous to human gestational diabetes can occur during diestrus[8] or pregnancy[9].

Most diabetic dogs are over 5 years of age with the highest prevalence occurring in the 8-12 year age range[10,11,12]. Entire females are at increased risk[12], especially if also overweight[13]. This is probably because of an additive effect of diestrus-associated and obesity-associated insulin resistance. Although mixed breed dogs are more commonly affected, some pure breeds are over-represented. In the USA, breeds at increased risk for developing diabetes include the Miniature Schnauzer, Bichon Frise, Miniature Poodle, and Samoyed[10]. Breed predisposition for developing diabetes varies with geographic region. In the UK, the Samoyed, Tibetan Terrier, Cairn Terrier, Miniature Schnauzer, Yorkshire Terrier, Border Terrier, and Labrador Retriever breeds are over-represented[3], while in Sweden the most commonly affected breeds are Australian Terriers, Samoyeds, Swedish Elkhounds, and Swedish Lapphunds[12].

Chronic pancreatitis in diabetic dogs can have important clinical implications. In addition to the possibility of episodes of acute pancreatitis, progressive destruction of both endocrine and exocrine tissue will result in loss of insulin-secreting beta cells, glucagon-secreting alpha cells, and exocrine acinar cells. Glucagon has an important role in the counter-regulatory response to hypoglycemia and is therefore crucial in protecting diabetic dogs from the life-threatening consequences of insulin over-dose. Impaired glucagon counter-regulatory response to insulin-induced hypoglycemia has been identified in diabetic dogs and was associated with episodes of clinical hypoglycemia[14]. A likely mechanism is loss of pancreatic alpha cells due to chronic pancreatitis.

Major improvements in the level of care and therapy available for diabetic dogs have been achieved over recent decades and treated diabetic dogs now have a similar chance of survival to that of non-diabetic dogs of the same age[10,15]. Administration of exogenous insulin is the mainstay of therapy for all diabetic dogs and remains the most reliable and effective means of managing clinical signs. The primary goals of therapy for diabetic dogs are 1) resolution of all clinical signs and 2) avoidance of insulin-induced hypoglycemia.

THE ROLE OF NUTRITION IN THE MANAGEMENT OF CANINE DIABETES MELLITUS

Feeding consistent meals at fixed times each day is crucial to successful management of diabetes in dogs. Commercial dog foods usually result in postprandial elevation of plasma glucose concentration from 1 to 6 hours following consumption by dogs[16] and meals should be timed so that maximal exogenous insulin activity occurs during the postprandial period. Thus, dogs should be fed within 1 hour of administration of lente insulin[17] (Vetsulin or Caninsulin, Intervet) or neutral protamine Hagedorn (NPH) insulin[18] (Humulin N, Eli Lilly), or within 6 hours of protamine zinc insulin (PZI)[19]. When a twice-daily insulin dosing regimen is used, a feasible compromise is to feed the dog immediately following the insulin injection. Ideally, every meal should contain the same ingredients and calorie content. The majority of diabetic dogs will readily consume meals twice-daily following the insulin injections if the meals are highly palatable and contain half their daily caloric requirement. They are more likely to readily accept a diet that has a formulation similar to the diet they were consuming before diagnosis of diabetes. A practical approach is to obtain a diet history at the time of diagnosis and then negotiate an agreement with the owner on a set meal that will be fed to the diabetic dog every 12 hours. In the majority of cases, the most appropriate meal comprises complete and balanced commercial dog food formulated for adult maintenance with <10% of the calories from treats or other foods. For finicky eaters, the meal should be fed at the time of insulin administration and remain available until the expected end of the period of maximal exogenous insulin activity. If treats are fed, they should also be consumed by the dog during the expected period of maximal exogenous insulin activity, and all treats containing high sugar or fat should be avoided.

Dietary manipulation that reduces the risk of insulin-induced hypoglycemia affords important clinical benefit for diabetic dogs. Severe hypoglycemia resulting from insulin overdose can cause irreversible brain damage and death, and avoidance of insulin-induced hypoglycemia is one of the primary aims of therapy in diabetic dogs. Owner compliance for feeding consistent meals at fixed times each day should be strongly encouraged because the risk of hypoglycemia may be increased when diabetic dogs are fed at irregular intervals. Mild clinical signs of hypoglycemia include weakness, restlessness, pacing, anorexia, and diarrhea. In severe cases, these can progress to ataxia, blindness, seizures, and coma[20]. If mild signs of hypoglycemia are observed, the owner should feed a meal of the dog's usual food. Hand-feeding might be necessary to encourage the dog to eat. If the dog is unwilling or unable to eat, syrup containing a high glucose concentration can be administered orally. Suitable syrups are marketed for use by human diabetics and should be kept in reserve by all owners of diabetic dogs. When the dog recovers, a meal of the dog's usual food should be fed immediately, and then the owner should contact their veterinarian before the next insulin injection is due, at which point reduction of the insulin dose is usually recommended. As a general rule, half the usual dose of insulin should be administered whenever a diabetic dog does not eat the accompanying meal. Owners should seek prompt veterinary advice whenever a diabetic dog shows inappetence or anorexia because there is an increased risk of hypoglycemia if insulin is administered when the dog does not eat. Additionally, any concurrent illness in diabetic patients that causes inappetence or anorexia can be rapidly complicated by dehydration, depression, and ketosis. The majority of diabetic dogs that present with diabetic ketoacidosis have at least one concurrent disease, with acute pancreatitis the most common diagnosis[21].

WHAT IS THE RECOMMENDED DIET FOR DOGS WITH DIABETES MELLITUS?

The major determinant of the postprandial glycemic response in dogs is the total carbohydrate content of the meal[22] so particular care should be applied to ensure consistent source and content of dietary carbohydrate. As many of the clinical signs of diabetes mellitus are attributable to hyperglycemia, it is rational to limit the glycemic load following each meal. Therefore, a moderately carbohydrate-restricted diet (for example carbohydrate <30% metabolizable energy (ME)) is recommended for diabetic dogs. The postprandial glycemic response to dietary carbohydrate is also potentially influenced by the type of carbohydrate and by the way it has been processed. For commercial dog foods that are processed using standard methods and contain similar total dietary carbohydrate content, the major potential influence on the postprandial glycemic response is the dietary carbohydrate source. Low glycemic index carbohydrate sources are likely preferable and recommended carbohydrate sources for diabetic dogs include sorghum and barley, while rice is unlikely to be an optimal carbohydrate source[23]. However, caution is required when extrapolating the results of dietary carbohydrate studies in non-diabetic dogs to clinical recommendations for diabetic dogs. This is because all diabetic dogs require exogenous insulin therapy, which has an overwhelming effect on carbohydrate metabolism and the postprandial glycemic response.

Several studies indicate that high-fiber diets, compared with low-fiber diets, might be associated with improved glycemic control, possibly because low-fiber formulations often contain increased digestible dietary carbohydrate content. However, there has been no clear demonstration of clinical benefit for diabetic dogs of feeding a high-fiber diet compared with feeding a typical adult maintenance diet with moderate-fiber content (30-40 g/1000 kcal)[24]. Based on the evidence of a recent randomized clinical trial, routine recommendation of high-fiber, moderate-carbohydrate, moderate-fat diets is not indicated for diabetic dogs managed with lente insulin (Vetsulin or Caninsulin, Intervet)[25]. The diets typical of those traditionally recommended for management of canine diabetes did not reduce exogenous insulin requirement or improve glycemic control when compared with a commercial, canine adult maintenance diet with moderate-fiber, low-carbohydrate, and high-fat content. Importantly, the traditionally recommended high-fiber, moderate-fat diet caused significant weight loss, and thus is not suitable for diabetic dogs with thin body condition[25]. For diabetic dogs managed with NPH insulin (Humulin N, Eli Lilly), there is some evidence that the traditionally recommended high-fiber diet can actually cause greater postprandial hyperglycemia than other diets[18]. In addition, increased dietary fiber content in excess of that routinely recommended for healthy dogs might be associated with other unwanted side effects in diabetic dogs, such as inappetence, diarrhea, and flatulence.

Dietary fat restriction (for example fat <30% ME) should be considered for diabetic dogs with concurrent chronic pancreatitis or persistent hypertriglyceridemia. Hypertriglyceridemia has been proposed as a possible inciting cause of canine pancreatitis and is commonly seen in diabetic dogs, indicating that the diabetic state might also be a risk factor for pancreatitis. However, dogs with diabetes secondary to chronic pancreatitis likely also have substantial loss of exocrine pancreatic function and so will be at increased risk for loss of body weight when fed a restricted fat diet. A more targeted approach is to only recommend dietary fat restriction for diabetic dogs if there is persistent hypertriglyceridemia once good glycemic control has been established. It is recommended that clinical response then be evaluated by monitoring body weight and serum triglyceride concentration.

The optimal dietary protein for diabetic dogs has not been determined but it is rational that recommendations would be no different than for non-diabetic dogs. As there is frequently a requirement for both carbohydrate- and fat-restriction, dietary protein will often provide a substantial source of calories for diabetic dogs (for example protein >30% ME).

MONITORING AND MANAGEMENT OF CLINICAL NUTRITION PROGRAMS IN DIABETIC DOGS

As one of the primary goals of treatment of canine diabetes is resolution of all clinical signs, it is essential that these be carefully monitored. One of the key clinical signs of untreated diabetes mellitus is loss of body weight and condition despite polyphagia. It is therefore important to record both body weight and body condition at each reassessment. With institution of appropriate medical and nutritional therapy, weight loss is usually arrested before optimal glycemic control is achieved. If body weight loss continues despite adequate glycemic control, the possibility of concurrent disease should be considered and serum trypsin-like immunoreactivity (cTLI) concentration assayed to evaluate exocrine pancreatic function.

Dogs with poorly controlled diabetes have a decreased ability to metabolize the nutrients absorbed from their gastrointestinal tract and lose glucose and amino acids in their urine, so require more calories for maintenance than healthy dogs. Conversely, weight gain is often a feature of chronic insulin over-dose in diabetic dogs. Although a proportion of diabetic dogs will be overweight at the time of diagnosis and will ultimately benefit from managed weight loss, it is advisable to first establish that pathological weight loss is arrested in response to treatment and that the dog can maintain body condition before beginning a program of calorie restriction.

Glycemic monitoring plus observation of clinical signs is used to evaluate response to the insulin and dietary regimen[26]. Exogenous insulin therapy has an overwhelming effect on carbohydrate metabolism and the postprandial glycemic response, and consequently good glycemic control is primarily achieved by regular reappraisal and adjustment of the insulin dose (Figure 1). Compliance of both the owner and the patient with the feeding recommendations must also be routinely evaluated to ensure appropriate timing and consistency of the meals. An approach that is often successful following each reappraisal is to recommend the 'dose' of food along with the dose of insulin to educate the owner of the importance of carefully weighing or measuring the food for their dog in the same manner that they carefully prepare the dose of insulin.

Fasting serum triglyceride concentrations can be monitored to identify persistent hypertriglyceridemia and to monitor the response to feeding a fat-restricted diet (Figure 2). Exogenous insulin therapy will result in resolution of hypertriglyceridemia in some diabetic dogs, while others require dietary fat restriction in addition to insulin therapy. Dietary fat restriction <30% ME should be recommended for all diabetic dogs with fasting serum triglyceride concentration >500 mg/dL (>5.5 mmol/L). For diabetic dogs with good glycemic control, dietary fat restriction <30% ME is recommended if fasting serum triglyceride concentration is >400 mg/dL (>4.4 mmol/L.) It is expected that fasting triglyceride levels will decrease in response to dietary fat restriction. Therefore if fasting serum triglyceride concentration is >400 mg/dL (>4.4 mmol/L) when the dog is being fed a diet with <30% ME fat, further restriction of dietary fat to <20% ME is recommended. An alternative or adjunct to monitoring fasting serum triglyceride concentrations is to monitor postprandial triglyceride concentrations in diabetic dogs. Fasting triglyceride concentration might not accurately predict the postprandial peak concentration in dogs[27], which could be an important limitation when monitoring response to dietary fat restriction. As serial postprandial blood sampling is routine in diabetic dogs when assessing glycemia, it is convenient to also evaluate triglycerides at the same time using point-of-care meters such as the CardioChek PA test system.

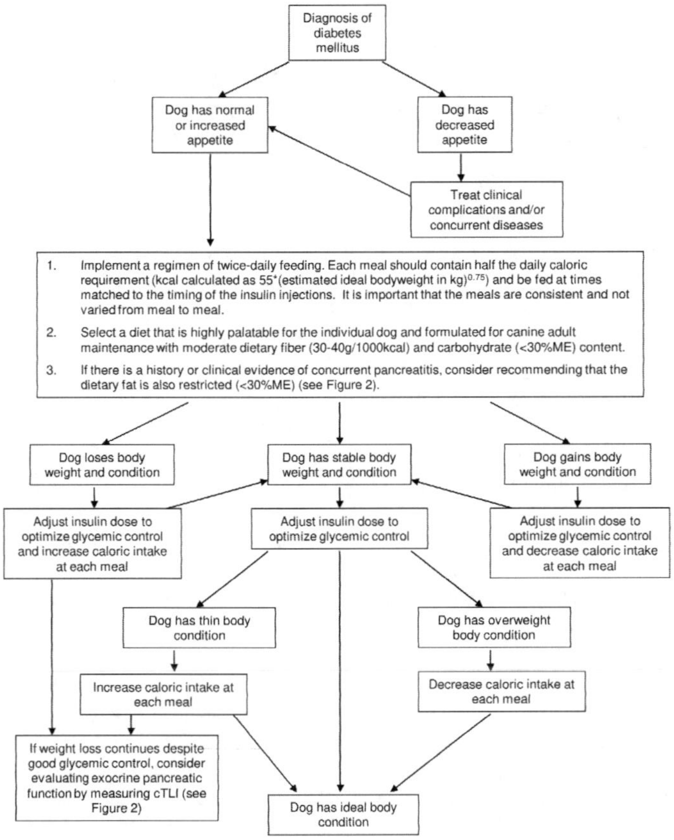

Figure 1: Algorithm for adjustment of insulin dose and caloric intake in diabetic dogs.

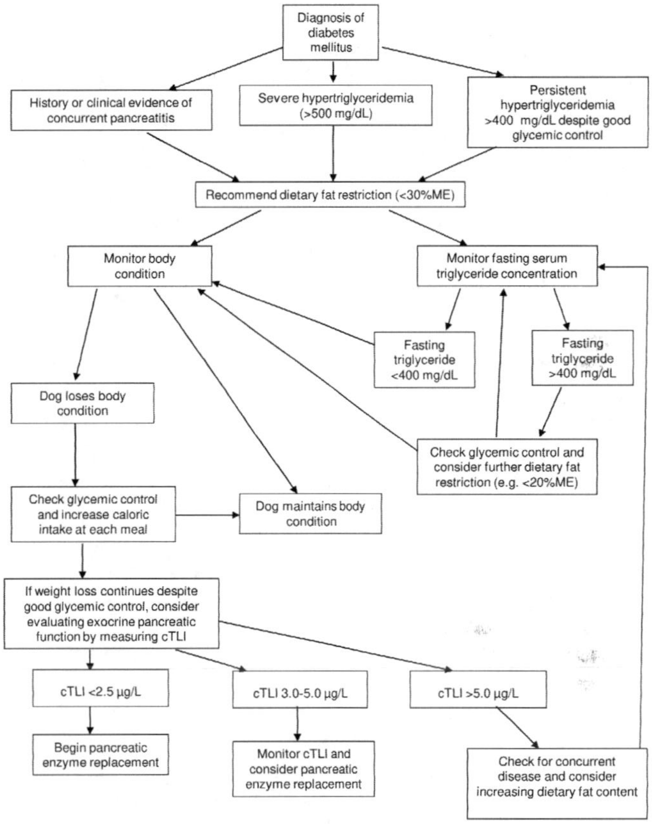

Figure 2: Algorithm for monitoring fasting triglyceride and adjustment of dietary fat intake in diabetic dogs.

REFERENCES

[1]Rand JS, et. al. **J Nutr** 2004;134(8S):2072S. [2]Catchpole B, et. al. **Diabetologia** 2005;48(10):1948. [3]Catchpole B, et. al. **J Small Anim Pract** 2008;49:4. [4]Davison LJ. **ACVIM Proceedings** 2009. [5]Alejandro R, et. al. **J Am Vet Med Assoc** 1988;193:1050. [6]Peterson ME. **Res Vet Sci** 1984;36:177. [7]Eigenmann JE, et. al. **Acta Endocrinologica** 1983;104(2):167. [8]Scaramal JD, et. al. **Medicina** 1997;57:169. [9]Fall T, et. al. **J Vet Intern Med** 2008;22:1296. [10]Guptill L, et. al. **Vet J** 2003;165(3):240. [11]Davison LJ, et. al. **Vet Record** 2005;156:467. [12]Fall T, et. al. **J Vet Intern Med** 2007;21:1209. [13]Klinkenberg H, et. al. **J Nutr** 2006;136:1985S. [14]Duesberg C, et. al. **J Vet Intern Med** 1995;9:181. [15]Graham PA & Nash AS. **J Vet Intern Med** 1997;11:142. [16]Elliott KF, et. al. **Res Vet Sci** (accepted, under review). [17]Fleeman LM & Rand JS. **Vet Record** 2009;164:232. [18]Palm CA, et. al. **J Vet Intern Med** 2009;23(1):50. [19]Stenner VJ, et. al. **J Vet Intern Med** 2004;18(3):444. [20]Whitley NT, et. al. **J Am Vet Med Assoc** 1997;211(3):326. [21]Hess RS, et. al. **J Am Vet Med Assoc** 2000;217:1166. [22]Nguyen P, et. al. **J Nutr** 1998;128(12 Suppl):2654S. [23]Sunvold GD & Bouchard GF. In: Recent advances in canine and feline nutrition Vol II. Orange Frazer Press; 1998;123. [24]Fleeman LM & Rand JS. The Royal Canin Canine Nutrition Encyclopedia. 2006;192. [25]Fleeman LM, et. al. **J Small Anim Pract** 2009;50(11):604. [26]Fleeman LM & Rand JS. **Vet Clin North Am Small Anim Pract** 2001;31(5):855. [27]Elliott KF, et. al. **Am J Vet Res** (accepted).

CARBOHYDRATE-RELATED CLAIMS FOR PET FOODS: REGULATORY CONSIDERATIONS

David A. Dzanis, DVM, PhD, DACVN
Santa Clarita, CA

INTRODUCTION

The science of nutrition and the law governing the regulation of foods and feeds are rarely in complete harmony. Scientists and regulators approach their tasks in a similar manner, in that both evaluate a given situation based on a set of established (but not dogmatic) principles, and use rational processes in reaching a conclusion. However, since these two groups approach a situation from different viewpoints, the proper course of action needed to resolve an issue as determined by a scientist may not coincide with that of a regulator. The regulation of carbohydrate-related claims on pet food labels is a good example of this disparity.

US REGULATION OF PET FOODS

The Center for Veterinary Medicine within the US Food & Drug Administration (FDA) has primary jurisdiction over all animal feeds in interstate commerce (including imports).[1] The Federal Food, Drug, and Cosmetic Act of 1938 (FFDCA) grants FDA federal authority to regulate all pet foods, including complete and balanced pet foods, treats and snacks, supplements, edible chews (e.g., rawhides, pig ears), and the ingredients intended to be incorporated into these products. While the US Department of Agriculture has oversight of meat and poultry products intended for human consumption, these same products intended for animal consumption fall within FDA's authority. The determination of what constitutes interstate commerce of a product considers the origins of its ingredients as well as the location of product manufacture and the point of sale or receipt of the product. Most commercial pet foods are distributed in multiple states, so FDA has federal jurisdiction. Even if manufactured and sold in a single state, most products are comprised of at least some ingredients obtained from sources outside of the state in which they are manufactured, so they would fall under federal authority as well. However, pet foods that are wholly sourced, produced and distributed within a state's borders normally would not be subject to FDA oversight.

Even if the distribution of a pet food is limited to intrastate commerce, the pet food most often is still subject to regulation. Most state governments also exercise authority over pet feeds that are distributed within their respective jurisdictions. This constitutes another layer of regulatory oversight that is more extensive than is required for most human food items. The state's laws and regulations are enforced by the state feed control official, typically through an office in that state's department of agriculture but occasionally through other agencies (e.g., state chemist's office).

Compliance with 50 different sets of state regulations can prove difficult for a manufacturer of a pet food product in national distribution. The Association of American Feed Control Officials (AAFCO) was formed in 1909 with the mission to foster uniformity in state regulation of animal feeds.[2] It is neither a government body empowered to act under authority of law, nor a trade association whose goal (as ascribed by its critics) is to mitigate the impact of regulation on the industry. Rather, it is a private body whose members are wholly comprised of federal, state and foreign government officials. It strives to achieve its mission through publication of the AAFCO Model Bill and Regulations (including separate Model Regulations for Pet Food and Specialty Pet Food), establishment of definitions for feed and pet food ingredients and setting of other standards via policy or guidance that represent a consensus among regulators as to what constitutes the appropriate regulation of animal feed. None of what AAFCO publishes has any direct power of law. However, states are encouraged to adopt the AAFCO Models into their own laws and regulations, at which time they become enforceable by that state. While not all states follow AAFCO Models, enough do wherein it has nation-wide impact.

The federal and state governments' mandate is to enforce the laws relating to pet food manufacture and distribution. FFDCA and equivalent state laws prohibit the distribution of foods that are either adulterated or misbranded.[1] The term "adulterated" may refer to the presence of a chemical, microbiologic or physical contaminant, including the inclusion of any ingredient that is not a GRAS ("generally recognized as safe") substance, an approved food additive, or a substance otherwise sanctioned (e.g., via the AAFCO Feed Ingredient Definition process) for use in pet foods. Failure for a product to meet stated nutrient guarantees or conform to ingredient or nutritional representations is also a form of adulteration.[2] The term "misbranded" relates to false or misleading claims or labeling that otherwise are not in compliance with federal or state regulations.

Most state laws (particularly those that follow AAFCO Models) mandate a form of periodic product registration and/or company licensure as a condition of distributing in that state.[2] Typically, the process requires submission of product labels for review by the feed control official. Any labeling found to be misbranded may be denied for sale in that state. States may also collect samples of products for laboratory analysis to evaluate nutrient content and look for evidence of contamination. Depending on the egregiousness of any violations found, the product may be seized by the regulator, or the company may be notified and allowed time to remedy the violation. While these actions would appear to affect only that jurisdiction, in practice, it is infeasible for pet food companies to distribute products in different states with different labels. Hence, labeling found to be objectionable by just one state may result in revision of labeling for product sold in all 50 states.

THE CARBOHYDRATE CONUNDRUM

Section 5(a)(3) of the AAFCO Model Bill mandates the feed or pet food label to bear a guaranteed analysis as "required to advise the user of the composition of the feed or to support claims made in the labeling."[2] The Model Regulations for Pet Food and Specialty Pet Food specify the minimum required guarantees for all pet foods, including crude protein, crude fat, crude fiber, and moisture. Label claims for other nutrients, such as calcium or vitamin E, mandate additional label guarantees under the Model Bill. Therefore, under normal circumstances, a claim related to carbohydrate content of the pet food would require a guarantee for carbohydrates to appear on the label.

However, other guidance within AAFCO confuses this matter. Under AAFCO's Statements for Uniform Interpretation and Policy, Policy Statement #1 reads as follows:

> ***Nitrogen-Free Extract and Carbohydrate Guarantees*** – *Nitrogen-free extract and carbohydrate guarantees are no longer considered as necessary and meaningful for purchaser information, therefore, their future use is discouraged (Combination of four resolutions adopted prior to 1923, Amended 1963).*[2]

As a result of this policy, the majority of states do not allow carbohydrate guarantees on pet food labels. However, since under the Model Bill a nutrient-related label claim requires a supporting guarantee, this creates a conundrum for both industry and regulators. Because carbohydrates cannot be guaranteed, carbohydrate-related claims therefore are also prohibited. AAFCO recognized the current policy hinders the ability for labels to bear potentially meaningful nutrition information, so established a working group to review and offer recommendations to remedy the situation. Although this action was primarily driven by concerns with claims for horse feeds, pet food also was to be considered in the group's deliberations.

AAFCO ACTIONS

The Carbohydrate Working Group, comprised of state feed control officials as well as representatives of the pet food and horse feed industries and the American College of Veterinary Nutrition (ACVN), reviewed Policy Statement #1 and recommended that it NOT be deleted or amended. In this case, the intended purpose of guarantees was considered to be to provide information to help the consumer make meaningful judgments regarding carbohydrate content as they relate to fractions that may affect glucose metabolism in dogs and cats. As a term encompassing a broad range of compounds, "carbohydrate" was considered too vague to allow for meaningful guarantees. However, it was felt that specified carbohydrate fractions were amenable to definition and quantification. To that end, the Group defined the fractions of nutritional interest, as follows:

> ***Dietary Starch (nutrient)*** – *An alpha-linked glucose carbohydrate of or derived from plants, animals and/or microbes from which glucose is released through the hydrolytic actions of purified alpha-amylases and amyloglucosidases that are specifically active only on the alpha-(1,4) and alpha-(1,6) linkages in samples that have been gelatinized in heated, mildly acidic buffer. Its concentration in feed is determined by enzymatically converting the alpha-linked-glucose carbohydrate to glucose and then measuring the liberated glucose. This definition encompasses plant starch, glycogen, maltooligosaccharides and maltose/isomaltose.*
>
> ***Sugars (nutrient)*** – *The sum of all free disaccharides and monosaccharides such as: sucrose, lactose, glucose, fructose and galactose or others digestible by enzymes found in an animal's digestive tract.*

The Group also defined "fructans" to address its presence in horse feeds. These definitions were recommended to the AAFCO Ingredient Definitions Committee. Subsequently, they have been adopted by AAFCO and now appear in the Official Publication.[2] There are still issues with laboratory methods needed to directly measure these feed components so that feed control officials have means to verify guarantees, especially for fructans, but consensus on methodology is being resolved by the AAFCO Laboratory Methods and Services Committee.

With definitions for the fractions of interest in place, the next charge of the Carbohydrate Working Group was to determine how these substances could be claimed on the label. Much effort was spent on attempting to define what constitutes "low" and/or "less" starch and sugars in pet foods. To effectively compare carbohydrate content between products of widely variable moisture and calorie contents, a straight percentage of starch and sugars would be meaningless. Although ACVN expressed concern regarding the true usefulness of such claims at all on pet foods labels, it proposed maximum limits for pet foods based on caloric content, as follows:

ACVN proposed amendment of AAFCO Model Regulations for Pet Food and Specialty Pet Food, November 2009

1. Renumber Regulation PF4(a)(1)E and F to PF4(a)(1)G and H, respectively, and insert a new Regulation PF4(a)(1)E and F

E. Maximum percentage of dietary starch, if required by Regulation PF10;

F. Maximum percentage of sugars, if required by Regulation PF10;

2. Insert into Regulation PF10 Descriptive terms:

(c) Carbohydrate related terms

(1) "Low" Dietary Starch and Sugars

A dog or cat food product that is labeled as complete and balanced as provided in Regulation PF7(c)(1) and which bears on its label a claim of "low dietary starch, "low sugars" or words of similar designation, shall:

- A. *Contain no more than 55 g/1000 kcal ME of total dietary starch and sugars, determined by dividing the total of the percent maximum guarantees for dietary starch and sugars by the calorie content in kcal ME/kg, then multiplying the result by 10,000.*
- B. *Include on the label:*
    - i. *Maximum percent guarantees for both dietary starch and sugars in the guaranteed analysis section immediately following the crude fiber guarantee in addition to the other mandatory guaranteed analysis information; and*

ii. A calorie content statement in accordance with the format provided in Regulation PF9.

(2) "Less" or "Reduced" Dietary Starch and Sugars

A. A dog or cat food product that is labeled as complete and balanced as provided in Regulation PF7(c)(1) and which bears on its label a claim of "less dietary starch", "reduced sugars," or words of similar designation, shall include on the label:

i. The name of the product of comparison and the percentage of reduction in dietary starch and sugars (expressed on an equal calorie basis) explicitly stated and juxtaposed with the largest or most prominent use of the claim on each panel of the label on which the term appears: and

ii. The comparative statement printed in type of the same color and style and not less than one-half the size used in the claim; and

iii. Maximum percent guarantees for both dietary starch and sugars in the guaranteed analysis section immediately following the crude fiber guarantee in addition to the other mandatory guaranteed analysis information; and

iv. A calorie content statement in accordance with the format provided in Regulation PF9.

B. A comparison between products in different categories of moisture content (i.e., less than 20%, 20% or more but less than 65%, 65% or more) is misleading.

The Carbohydrate Working Group could not reach a consensus on the proposed amendment. Presently, the matter is being moved from the Carbohydrate Working Group to the AAFCO Pet Food Committee for further deliberation. Whether it will be accepted in present form, be radically revised, or dropped altogether is not known at this juncture.

THE VETERINARIAN'S ROLE

Notwithstanding regulatory efforts to curtail "low carb" claims on pet food labels, the marketplace is replete with products bearing claims of this nature. Without concurrent label guarantees, such claims are of minimal usefulness and may mislead the consumer. Some companies have attempted to support their claims with nitrogen-free extract guarantees, but have been forced to remove the guarantees (and claims) to comply with state feed control official enforcement actions.

Unless and until dietary starch and sugars guarantees are allowed on pet food labels, carbohydrate content can be calculated by difference from the other guarantees [i.e., NFE% = 100 – (crude protein% + crude fat% + crude fiber% + moisture% + ash%)]. This method has its limits, but allows for rough comparisons between products.

Although "low carb" pet foods are popular among some pet owners, there is considerable concern regarding the usefulness of "low" or "less" carbohydrate-related claims for pet foods. While restriction in non-structural carbohydrate intake may be beneficial to small populations of dogs or cats, for the majority of animals such claims would only serve to connote a health benefit that does not exist. For animals in which such a diet may help, scientific data are lacking to accurately specify a "low" level in a pet food at which there would be clear physiological benefit to these animals. Also, "less" claims may invite unfair comparisons that have no meaningful physiologic bases. Veterinarians are in the best position to convey these limitations to clients who ask about these types of pet foods.

CONCLUSION

The intent of animal feed label regulations is in part to help safeguard animal health by providing the consumer with information necessary to make informed purchasing decisions. Although the true benefits to the animal are arguable, the changes to the AAFCO Model Regulations for Pet Food and Specialty Pet Food as proposed by ACVN would help establish a level playing field among pet food companies. All claims would have to meet the same standards, so consumers would be able to reasonably compare products and make better informed purchasing decisions.

REFERENCES

[1]Federal Food, Drug and Cosmetic Act. [Online]. Available: http://www.fda.gov/RegulatoryInformation/Legislation/FederalFoodDrugandCosmeticActFDCAct/default.htm. Accessed: February 1, 2010. [2] Official Publication. **Association of American Feed Control Officials**. Oxford, IN, 2010.

DEMYSTIFINGY DRUG LABELS: IDENTIFYING KEY COMPONENTS

Jennifer L. Buur, DVM, PhD, DACVCP
Pomona, CA

INTRODUCTION

Rational use of pharmacological agents, like other forms of evidenced based medicine, requires knowledge of the pharmacokinetic and pharmacodynamic properties of each substance being considered. There are thousands of formulations approved for use in veterinary medicine and even more in human medicine. Given that more formulations are being approved every year and the ability to use drugs in an off label manner[1], it becomes impossible for even the most astute and conscientious clinician to keep up with all the current information. Thankfully, there are many great resources to help medical professionals make rational decisions that improve patient care.

One of the most important and underutilized resources is the drug label or package insert that accompanies every drug approved for use by the United State Food and Drug Administration (US-FDA). The drug label provides a complete synopsis of benefits and risks associated with the use of the medication. This session will provide a systematic overview on each section found on a drug label (focus on package inserts) and provide evidence on how this information can be useful in a clinical setting.

PURPOSE & FUNCTION OF A DRUG LABEL

The Food Drug and Cosmetic Act requires the development of a product label in order to ensure that the product is used appropriately[2]. It is intended as the primary means of communication between pharmaceutical sponsor, US-FDA, and clinician. A drug label is defined as any written, graphic, or printed matter accompanying the drug. This includes labels on the drug bottle, package inserts, wrappers, and so on. The purpose of a drug label is to provide information on the risks and benefits of each formulation so that health professionals can make informed and rational decisions that result in better patient care. This includes safety information for the patient, the administrator of the drug, the consumer of animal products (if food animal species), and the environment.

PARTS OF A DRUG LABEL

All drug labels contain the same basic information. And while there are few differences found between companion and food animal species, most drug labels follow a systematic approach starting with a description of the product and ending with the manufacturer's information. The sections can be divided roughly into identification, safety, and efficacy[3].

IDENTIFICATION

Description

The description of the product includes the brand name, the generic drug name, the formulation, and the marketing designation of prescription (Rx only), over the counter (OTC), or veterinary feed directive (VFD). Additionally, chemical structures, chemical names of active and inactive ingredients, drug classification, and the physical description of the formulation may be included in this section. The purpose of this section is for proper identification of the formulation. This section has practical application in confirming dosing regimens with clients and in identification after possible overdose exposure.

How Supplied

This is a description of the possible formulations, numbers of tablets in containers, and dosing strengths which is useful when you are considering use of this formulation in patients that are of a different size or during extralabel use.

NADA/ANADA/FDA Approval Statement

These statements allow a practitioner to easily access Freedom of Information Summaries online at www.FDA.gov. This number is also important in reporting of adverse effects.

Storage Information

This section provides information needed to preserve the safety and efficacy of the product including light sensitivity and stability after reconstitution.

Manufacturer Information

This provides contact information for the manufacturer as well as patent information.

SAFETY

Contraindications

This is the most severe of the safety statements. These adverse effects can result in mortality, life threatening morbidity requiring the expertise of a medical professional, or significant reproductive loss.

Warnings

Warnings include serious adverse effects, safety hazards, and potential treatments in case adverse events occur. These risks are considered reasonable in light of the benefits of the formulations and are based on reasonable evidence. Also included in this section are any Human Warnings including Food Residue Warnings and information for physicians in the case of accidental human exposure.

Precautions

This section includes information on appropriate patient selection, laboratory testing for screening purposes, drug-drug interactions, and anything else not covered in Contraindication or Warning sections. Taken in combination with Contraindications and Warnings, these three section provide a concise summary of possible risk associated with use of the drug.

Adverse Reactions

This section is based on field evidence (approval and post approval studies) and reports events that could be associated with the drug use in predictable or idiosyncratic ways. The most important aspect in this section is the toll free number provided for technical assistance or reporting of adverse events.

Animal Safety

This reports the results of the Target Animal Safety studies completed prior to approval. This information can be helpful when contemplating using the formulation off label for unapproved conditions that may require higher doses.

EFFICACY

Indications & Use

This is a list of the approved species, diseases, and/or conditions for the formulation.

Dosage & Administration

This lists the approved dose or dose range for each species and condition.

Information for Owner or Person Treating the Animal

This information is considered to be what is important for owners to know in order to maximize the safety and efficacy of the product and should be communicated to the owner or person in charge of treatment. This may be in the form of a client specific insert. This section may not be found in all drug labels.

Clinical Pharmacology

This section includes both the pharmacokinetic and pharmacodynamic data available for the formulation. Summary data for both properties are included for all target species and disease processes identified in the Indications and Use section. Drugs approved for human use will have data for approved populations (adults, pediatrics, etcetera). Additional information regarding efficacy can be found in this section including microbiology (MIC), general drug class effects, possible interactions, and subpopulation information.

Effectiveness

This is a summary of the efficacy studies required for approval. Full details of these studies are provided in the Freedom of Information Summaries found online.

CONCLUSION

Rational drug use requires a veterinarian to understand the risks and benefits associated with the use of drugs for a given species with a given disease process. By law, veterinarians are required to use products labeled for that specific prior to going off label. Even when using drugs in an approved off label manner, it is essential to have a good understanding of their pharmacokinetic and pharmacodynamic properties. Product labels provide this information to practicing veterinarians and other health care professionals. They are a vital, but underutilized tool in providing high quality patient care.

REFERENCES

[1] **Animal Medicinal Drug Use Clarification Act**. United States Food and Drug Administration. Title 21 Code of Federal Regulations, part 530, 1994. [2] **Federal Food, Drug, and Cosmetic Act.** 1982. [3] ***Veterinary Pharmacology and Therapeutics 9th Edition***. Ames, Iowa: Blackwell Publishing, 2009 .

PHARMACOKINETICS & THE DRUG LABEL: WHAT CAN YOU USE?

Tomás Martín-Jiménez, DVM, PhD, DACVCP, DECVPT.
Knoxville, TN

INTRODUCTION

Clinical pharmacokinetics is concerned with the rational, safe, and effective use of drugs. It studies the factors that determine the time course of the plasma concentration of a drug and its variability. The ideal goal is to tailor the drug and dosing regimen to the unique characteristics of each patient[1]. Pharmacokinetic information contained in the drug package insert can be an important tool when there is a need to adapt the dosage regimen to particular clinical scenarios (e.g. renal impairment, change in bacterial susceptibility, etc). Pharmacokinetic information is most useful when the range of doses (or concentrations) in which the drug exhibits linear pharmacokinetics is known. Linear pharmacokinetics implies direct proportionality between dose and exposure. For example, under linear kinetics, a 20 mg/kg dose provides two-fold the exposure (but not necessarily the pharmacological effect) of a 10 mg/kg dose. Some drugs exhibit linear pharmacokinetics within a wide range of doses and concentrations, while others do not. Therefore, when examining the pharmacokinetic information in the drug insert we can start by determining the PK linearity and range. Under nonlinear pharmacokinetics, there is no proportionality between dose and exposure, and therefore dose adjustment becomes a much more difficult task. The principles discussed by the author are based on the assumption of linear PK (also called first-order PK), which is by far the most common scenario.

Clinical pharmacodynamics is concerned with the rational, safe and effective use of drugs. It studies the relationship between drug exposure and pharmacological or toxic effect. Pharmacodynamic studies often allow defining exposure targets that are optimally related to the desired level of pharmacological effect (e.g. a range of AUC values that optimizes the probability of bacterial eradication during an infection)[2].

Important therapeutic questions that may arise during the course of drug administration include:

What should be the dose interval? Is compliance likely to be a problem? Are there any potential interaction with the patient's other drugs? What if the "normal average" dosing regimen is not appropriate for this patient? Should liver or renal impairment be taken into account? How do I tailor the dose to the changing bacterial susceptibility in my patient? How should I interpret TDM findings?

When both pharmacokinetic and pharmacodynamic information are available, clinicians are better equipped to provide rational answers to these and other questions as well as to calculate adjusted dosage regimens. The objectives of this lecture are to review the significance of the pharmacokinetic and pharmacodynamic information included in drug package inserts, as well as discussing some practical applications.

RELEVANT PHARMACOKINETIC INFORMATION IN THE DRUG PACKAGE INSERT

Fraction of Drug Unbound in Plasma (f_u)

Free drug in plasma can diffuse to tissues and act on receptors to produce an effect. While drug displacement from plasma protein binding sites by a competing drug is almost always unimportant, protein binding may be significant in the interpretation of plasma concentrations of drugs (e.g. when interpreting the results of therapeutic drug monitoring for displaced drugs or drugs with nonlinear plasma protein binding). Protein binding is also important with antimicrobials. When adjusting the dosage regimen of antimicrobials based on PKPD indices of efficacy (e.g. AUC/MIC or T>MIC), the exposure parameters (Cp, or AUC) should be expressed in terms of unbound drug (e.g. $Cp_u = Cp \times f_u$). Expressed in this way, PKPD indices of efficacy can be extrapolated across species (e.g. from mice to dogs).

Area under the Plasma Concentration-Time Curve (AUC) and Other Measures of Exposure

The AUC represents the total area contained between the concentration-time profile and the X and Y axis of a pharmacokinetic curve. For a single dose, the AUC represents the complete exposure of an animal to a drug, i.e. the exposure from time 0 to infinity. As such, AUC after single dose is usually represented as $AUC_{0\text{-}inf}$. When multiple doses are administered at regular dose intervals (τ) (with τ < 3-4 half-lives), the plasma concentration of the drug accumulates until a plateau is reached. At plateau or steady-state, the peak and trough concentrations remain unchanged with each dose[3]. Assuming time- and dose-independent PK, the AUC over a dose interval at plateau ($AUC_{0\text{-}\tau}$) equals the $AUC_{0\text{-}inf}$ following a single dose. In the case of antimicrobials administered by multiple doses, PK-PD relationships are based on the unbound $AUC_{0\text{-}24}$, or the area corresponding to a 24-hr dosing interval at steady-state[4] (or the corresponding sum of areas if the drug is administered at shorter intervals).

Other measures of exposure that can be relevant for dosage adjustment include C_{max}, or peak plasma concentration after single or multiple doses; Cp_{ss}, or concentration at steady-state after IV constant-rate infusion, $Cpavg_{ss}$, or average concentration at steady state after multiple doses with constant dose interval; and C_{min}, or trough concentration at steady-state after multiple doses. These measures of exposure are generally related to the volume of distribution and/or systemic clearance of a drug, and therefore are amenable to become targets for drug adjustment, as long as their values are associated to a given probability of clinical outcome.

Under certain circumstances, these measures of exposure may need to be adjusted by the fraction of free drug reaching the target tissue. In those cases, information on drug tissue distribution may allow the corresponding adjustment to be made.

Systemic Availability (F)

When a drug is administered by an extra-vascular route, a fraction of the administered dose may not reach the systemic circulation. This may be due to poor absorption, drug inactivation in the stomach, or enzymatic first-pass effect (in the intestinal wall or the liver). The bioavailability of a drug is formally defined as the rate and extent of absorption. Although the rate of absorption can be important when quick onset of pharmacological effects is needed, it is not usually relevant for long-term drug administration. The extent of absorption, known as systemic availability, represents the fraction of the total dose that reaches the systemic circulation after extravascular administration (usually PO, SQ, or IM) and may have a value between 0 and 1. Absolute systemic availability is determined by comparison between the dose-adjusted AUCs after IV and extravascular administrations, while relative systemic availability refers to the same calculation but for two extravascular administrations (e.g. PO vs IM):

$$F = \frac{AUC_{PO}}{AUC_{IV}} \times \frac{Dose_{IV}}{Dose_{PO}}$$

Food may affect absorption and first-pass metabolism, typically in opposite ways. In monogastric species, food usually decreases the oral availability of drugs that are not very lipid soluble, while it tends to increase the oral availability of drugs that undergo high first-pass metabolism. Likewise, food components may specifically bind to certain drugs (e.g. calcium and tetracyclines) altering the absorption profile of the latter. While food effects are typically managed by appropriately timing the dose administration relative to the feeding process, dose adjustments can be made by applying an appropriate correcting factor to the value of F.

Systemic Clearance (Cl)

In first order pharmacokinetic systems, the clearance of a drug from plasma is a proportionality constant that relates the rate of elimination (mass/time) to the total plasma concentration (mass/volume). Systemic clearance is defined as the volume of plasma cleared of drug per unit time and is expressed in units of flow (volume/time). *Clearance is the single most important* pharmacokinetic parameter, since it determines the maintenance dose-rate (dose/unit time) required to maintain a given level of drug exposure. For example, for a drug administered intravascularly, the AUC is solely dependent on the dose and the drug systemic clearance. For a drug administered extravascularly, the factors affecting exposure are the dose, F, and Cl.

$$AUC = \frac{F \times Dose}{Cl};$$

Therefore, to calculate the dose for a given target AUC and F, we only need to know the value of clearance:

$$Dose = \frac{Cl \times AUC}{F}$$

When the overall elimination profile of a drug obeys linear kinetics, systemic clearance is considered constant and independent of the dose over the clinically useful range of plasma concentrations. If the eliminating organ is significantly impaired or a concurrently administered drug affects the clearance of the original drug, the dose or dose interval may need to be changed in proportion to the change in clearance.

Different animals treated with similar doses will exhibit different AUCs on account of the biological variability in clearance and systemic availability. Therefore, for a given value of F, it will be difficult to predict the individual AUC of drugs with very variable clearance[5]. Although a complete discussion on pharmacokinetic variability is beyond the scope of this lecture, it is important to recognize its influence on drug exposure predictions and dose calculations. As a general rule (assuming normal distribution), the AUC of drugs with coefficients of variation (standard deviation/average value) in clearance above 50% may be hard to predict in the individual patient. This may be of particular importance with critical patients, antimicrobials, and narrow therapeutic index drugs.

Other measures of exposure are linearly related to the systemic clearance as follows:

$$Cp_{ss} = \frac{R_0}{Cl}$$

Therefore, for a given target Cp_{ss}

$$R_0 = Cl \times Cp_{ss}$$

Where R_0 is the zero-order rate of IV infusion.

When multiple doses are administered at constant dose intervals the average concentration at steady state ($Cpavg_{ss}$) is a function of the dose rate, F, and clearance in the same way as described above.

Volume of Distribution (Vd)

This is the second most important pharmacokinetic parameter for dose determination and adjustment. The volume of distribution is a function of the lipid solubility and plasma and tissue protein binding of the drug. It represents the volume into which a drug appears to be distributed with a concentration similar to that of plasma. This parameter is useful for determining the loading dose, i.e. the dose required to achieve the target plasma concentration (e.g. $Cp_{avg_{ss}}$) in as little time as possible. This calculation may be important for the initial treatment of drugs with long half-lives (e.g. phenobarbital), since 4 half-lives are needed to reach concentrations in plasma above 90% of the target average concentration at steady state. For drugs that can be approximated with monocompartmental kinetic equations, the loading dose can be computed as follows:

$$LD = \frac{Vd \times target\ Cp_{avg-ss}}{F}$$

Elimination Half-Life (t½)

The terminal elimination half-life is the time that it takes for the concentration of the drug in plasma at any time to decrease by one-half. When the amounts of drug in plasma and tissue are in equilibrium, the half-life may also represent the time that it takes for the total amount of drug in the body to decrease by one-half. This parameter provides an indication of:

- The time-course of drug elimination (by 4 half-lives, the plasma concentration of the drug will have decreased by about 93%).
- The time-course of drug accumulation (when the dose interval is short enough relative to the half-life, it takes 4 half-lives to reach about 93% of the steady-state concentration).
- The required dose interval, in particular for drugs with pharmacodynamic effects paralleling plasma concentrations. The choice of dose interval is also affected by the therapeutic index of the drug and the expected level of compliance. Half-lives that allow for once or twice-a-day administration are usually associated to increased compliance, in comparison to more frequent administration. Notice that since the half-life is a hybrid parameter of Vd and Cl:

$$t½ = \frac{0.693 \times Vd}{Cl}$$

If Cl decreases (e.g. because of renal dysfunction), it may be possible to increase the dose interval (e.g. from 8 to 12 hr) increasing the chances for adequate compliance.

PKPD-BASED DOSAGE REGIMEN DESIGN FOR ANTIBACTERIAL DRUGS

Outcome of infections in animal models has been shown to correlate well with one of three PKPD parameters, namely, the ratio of peak plasma concentration of the drug to the minimum inhibitory concentration (Cmax/MIC), the ratio of the 24-h area under the curve to the minimum inhibitory concentration (AUC/MIC), or the fraction of the dose interval during which the concentration of the drug in plasma remains above the minimum inhibitory concentration (T>MIC)[6,7].

Pharmacokinetic (Cl, f_u, C_{max}, PK linearity range, etc) and bacterial susceptibility information (MIC_{50}, MIC_{90}) included in the package insert can be integrated with published information on well established PKPD targets and the bacterial susceptibility profile of a specific case in order to tailor the dosage regimen of antibiotics to the observed level of bacterial susceptibility.

Several examples will be presented during the lecture illustrating the use of pharmacokinetic information from the package insert to make rational decisions about dosage regimens.

REFERENCES

[1]Rodman JH, et al. **Applied Pharmacokinetics and Pharmacodynamics**, 4th ed. 2006; Burton ME, Shaw LM, Schentag JJ, Evans WE: Lippincott Williams & Wilkins, pp. 40-59. [2]Martin-Jimenez T. **Farmacologia y Terapeutica Veterinaria** 1st ed. 2002; Botana LM, Landoni F, Martin-Jimenez T: McGraw-Hill Interamericana, pp. 58-77. [3]Winter ME. Basic Clinical Pharmacokinetics, 4th ed. 2004; Winter ME: Lippincott Williams & Wilkins, pp.2-111. [4]Mouton JW, et al. **J Antimicrob Chemother** 2005;55:601. [5]Martin-Jimenez T, et al. **Comparative Pharmacokinetics. Principes, Techniques and Applications**, 1st ed. 1999; Riviere JE: Iowa State University Press, pp. 259-282. [6]Andes D, et al. **Antibiotic Optimization. Concepts and Strategies in Clinical Practice**, 1st ed. 2005; Owens R, Ambrose PG, Nightingale CH: Taylor and Francis, pp. 65-88. [7]Martinez M, et al. **Antimicrobial Therapy in Veterinary Medicine**, 4th ed. 2006; Giguere S, Prescott JF, Baggot JD, Walker RD, Dowling PM: Blackwell Publishing; pp. 81-106.

SAFETY & EFFICACY: WHAT A DRUG LABEL DOES & DOES NOT TELL YOU

Cynthia Cole DVM, PhD, DACVCP
Greensboro, NC

INTRODUCTION

The information that must be provided on the labels of medications that have been approved by the Food and Drug Administration – Center for Veterinary Medicine (FDA-CVM) is strictly regulated. The authority of the agency to regulate label content is provided for in the Federal Food, Drug and Cosmetic Act (FFDCA) and the Code of Federal Regulations. [1,2] The regulations are very specific on what information must be provided on the label and what is not allowed. In addition, there are some differences between products that require a prescription (RX) and those available over the counter (OTC), those that are administered orally versus parenterally, and all of these products and Type B and C Medicated Feeds. The label, the package insert, and the Freedom of Information (FOI) Summary provide a plethora of information that is often not utilized by clinicians. It is important to understand what information is provided on medication labels and how to interpret it.

What is on the label?

All FDA approved products are required to have specific information on their label. That information includes: the product name, established name, active ingredient, name and address of manufacturer, packer, or distributor, ingredients, dosage recommendation, indications for safe use, any warnings or cautions, side effects, manufacturers control number, and country of origin. Very small containers may be exempt, but the required information must still be provided on a package insert. The package insert will generally contain even more information than the label including information for the owner or person treating the animal, brief summaries of the efficacy and safety studies, and pertinent clinical pharmacology data. Finally the FOI Summary contains abstracts of the efficacy and safety studies which were used to demonstrate that the product was both safe and efficacious when used per the label directions.

WARNINGS AND CAUTIONS

Warning Section

User Safety

When a drug poses a hazard to human health by any means of exposure the label will indicate how that risk can be prevented or minimized. In addition treatment recommendations and specific first aid measures will be included if appropriate. The Material Safety Data Sheet (MSDS) will generally contain more detailed safety information and a source for that sheet such as a phone number and or website would be included again if appropriate.

Human Food Safety

When a drug is intended to be used in food producing animals the label will indicate the time interval between the last use of the drug and when the animal can be slaughtered or its products can be used for food (i.e., withdrawal time). If a withdrawal time has not been established in a particular production class this will also be indicated.

Animal Safety

Serious adverse reactions that may occur with the use of the drug will be listed.

Environmental Safety

When the use or disposal of a drug can pose a risk to the environment the label will indicate how to decrease or prevent environmental exposure as well as the proper disposal of the container and any unused product.

Hazard Warning Levels

There are several types of hazard warnings that may appear on the label. The level of severity of the warning increases from Precaution to Warning to Contraindication. Precautions topics would include recommended screening or diagnostics tests, information about drug interactions, use in specific subgroups (e.g., pediatric, geriatric, or specific disease states), etc. In some cases Precautions are general for an entire drug class. For example, a Precaution may indicate, that as a class, cyclooxygenase inhibitory NSAIDs may be associated with gastrointestinal and renal toxicity. A Warning may address user safety, human food safety, environmental safety or serious adverse reactions and potential safety hazards to the animal receiving the drug. A Contraindication statement clearly stipulates that a drug should never be used for a particular indication because the risk of its use in the animal outweighs the benefits. In terms of animal safety, the difference between a Warning and a Contraindication is that the former indicates that serious adverse events may occur, while the latter indicates that they are very likely to occur. In addition, it is not necessary to prove that an adverse event is caused by the drug in order for a Warning to be required, but generally there should be strong evidence that the drug caused the reaction for a Contraindication to be required. Warning and Contraindications may also be required to address issues surrounding extra-label use or accidental exposures. Finally, if the adverse events associated with use of the product may lead to death or serious injury a "boxed warning" may be required. These are often called "black box warnings" and are meant to draw the attention of the user/prescriber to the information on the label.

Cautions should not be confused with hazard warnings. Cautions are general broad statements by the FDA. For example, a caution for all prescription drugs indicates that federal law restricts their use by or on the order of a licensed veterinarian. Other cautions prohibit the extra-label use of certain drugs, such as clenbuterol, chloramphenicol, and a number of others, in food producing animals

INDICATIONS

Indications are also referred to as product claims or uses and they fall into several categories. Treatment is appropriate when a disease or condition can be diagnosed as present and the application of the drug results in restoration of health or recovery from disease. This is often used for antibiotics and parasiticides. Treatment and control are terms used for anthelmintic products. Treatment indicates that there is a 90% or greater kill of adult and/or larval stages of susceptible parasites. Post treatment, biologically active metabolites may remain in the host that may be effective in controlling the development of new infections by killing infective larval stages for some time post treatment. Prevention is an appropriate label when the drug is given to animals at risk of developing a disease and keeps them free of that disease. For example, Drug X is indicated for the prevention of heartworm disease due to *Dirofilaria immitis* in dogs. Other indications include Control, Management, and numerous Production Indications. The latter indication often refers to specific species and production classes, for example replacement dairy heifers, lactating dairy cows, dry dairy cows, and dairy calves.

CLINICAL PHARMACOLOGY

On some package inserts, pharmacokinetic information on the drug product may be provided if it is important for the safe and effective use of the drug. For antimicrobials this section will provide information regarding susceptibility of target animal pathogens. It is required that the sponsor makes clear what information is based on in vivo versus in vitro study results. Clinicians may find the specific MIC_{50} and MIC_{90} parameters and breakpoint information particularly useful.

FOI SUMMARIES

FOI Summaries are generated by the sponsor, but the material is reviewed and approved by the FDA-CVM. In the efficacy section a clinician will find information on how the dose of the drug was determined, as well as the studies that provided "substantial evidence" of efficacy. For field trials this will include the name and location of all the investigators, the number of animals, the study design, parameters evaluated, and results. In the safety section the design and results of all of the safety studies, such as target animal safety, are discussed. Typically for therapeutic agents a tiered dosing study at 1X, 3X, and 5X the label dose will be conducted over an extended period of time. The effects of even higher doses administered for shorter periods of time may also reported. Although the administration of these higher than label doses is not recommended, the clinician can glean important trends from the safety studies that may not be apparent on the label. It should be remembered, however, that everything found on post mortem examination during these safety studies, which is abnormal, is reported, regardless of whether it is likely to be drug related or not.

CONCLUSIONS

Medication labels, package inserts and FOI Summaries can provide the clinicians with valuable information that can help them use the product safely and effectively. Particularly for new products, clinicians may find a thorough review of these materials to be well worth their time.

REFERENCES

[1] 21 CFR 201.105 (http://ecfr.gpoaccess.gov). [2] http://www.fda.gov/RegulatoryInformation/Legislation/FederalFoodDrugandCosmeticActFDCAct/default.htm

DETECTING & TREATING SEVERE ACUTE PAIN MANAGEMENT: WHAT WORKS?

Kurt A. Grimm, DVM, MS, PhD, Diplomate of the American College of Veterinary Anesthesiologists, Diplomate of the American College of Veterinary Clinical Pharmacology
Conifer, CO

Patient stress is probably a contributing factor in some cases of adverse patient outcome. Stress during induction of anesthesia can increase circulating catecholamine concentration predisposing the heart to arrhythmias. Additionally, stress or anxiety can lead to increased doses of anesthetic agents resulting in excessive anesthetic depth once the patient is anesthetized. Premedication with a tranquilizer or sedative will help reduce anxiety and stress during the perioperative period.

Use of analgesics prior to surgery (preemptive analgesia) may also be beneficial. Opioids are commonly incorporated into premedication protocols to facilitate sedation and analgesia. When opioids are used, anesthetic drug associated respiratory depression may be enhanced, but adequate patient monitoring will facilitate early detection of significant respiratory depression and allow appropriate management. Non-steroidal anti-inflammatory drugs are commonly being administered during the perioperative period(1). Debate exists as to when to administer the drugs (preoperatively vs. postoperatively). Since intraoperative hypotension is always possible, the author prefers to administer NSAIDs during or after surgery once adequate perfusion pressures have been assured during anesthesia. While data exists to support the safe use of preoperative NSAIDs in healthy patients undergoing relatively short periods of anesthesia, if the approach of reducing risk during anesthesia is taken, it is rational to wait until surgery is nearing completion before administering NSAIDs.

Pain Management

Acute clinical pain typically arises from soft tissue trauma or inflammation, with the most common example being postoperative surgical pain. Though it does not serve a protective function in the sense that physiologic pain does, acute pain does play a biologically adaptive role by facilitating tissue repair and healing. This is achieved by hyper-sensitizing the injured area (primary hyperalgesia) as well as the surrounding tissues (secondary hyperalgesia) to all types of stimuli, such that contact with any external stimulus is avoided and the reparative process can proceed undisturbed. This realization is not, however, a license to allow patients to suffer needlessly in the postoperative period or upon presentation in the emergency room. By having an appreciation of the underlying functional basis of such pain the practitioner is able to initiate appropriate pain management strategies while taking steps to optimize wound healing.

Pain Physiology

An important conceptual breakthrough in understanding pain physiology is the recognition that pain following most types of noxious stimulation is usually protective and quite distinct from pain resulting from overt damage to tissues or nerves. It plays an integral adaptive role as part of the body's normal defense mechanisms, warning of contact with potentially damaging environmental insults and initiating behavioral and reflex avoidance strategies. It is also often referred to as nociceptive pain because it is only elicited when intense noxious stimuli threaten to injure tissue. It is characterized by a high stimulus threshold, is well localized and transient, and demonstrates a stimulus-response relationship similar to the other somatosensations. This protective mechanism is facilitated by a highly specialized network of nociceptors and primary sensory neurons which encode the intensity, duration and quality of noxious stimuli and, by virtue of their topographically organized projections to the spinal cord, its location.

Nociception

The physiologic component of pain is termed nociception, which is comprised of the processes of transduction, transmission and modulation of neural signals generated in response to an external noxious stimulus. It is a physiologic process that, when carried to completion, results in the conscious perception of pain. In its simplest form the pain pathway can be considered as a three neuron chain, with the first order neuron originating in the periphery and projecting to the spinal cord, the second order neuron ascending the spinal cord, and the third order neuron projecting to the cerebral cortex. On a more complex level, the pathway involves a network of branches and communications with other sensory neurons and descending inhibitory neurons from the midbrain that modulate afferent transmission of painful stimuli.

Descending Modulation of Pain

The descending modulatory system has been described as having four tiers. The final, and perhaps most important, site involved in the descending modulation of nociceptive information is at the level of the spinal cord. Just as dorsal horn processing is vital to the integration of ascending noxious input, its role in anti-nociception is equally crucial. Dense concentrations of GABA, glycine, serotonin, norepinephrine and the endogenous opioid peptides (enkephalins, endorphins and dynorphins) have been identified in dorsal horn neurons, and all produce inhibitory effects on nociceptive transmission. Specifically, the spinal opioid system fine-tunes descending control mechanisms by acting both presynaptically, as well as postsynaptically. Communication among dorsal horn neurons involves complex interactions, and it is now apparent that a single neuron may be influenced by many neurotransmitters, that each neurotransmitter may have numerous actions in a given region, and that multiple neurotransmitters may exist within a single neuron. Simply stated on a more global level, nociceptive processing is a three-neuron chain with dual input at each level. Discriminative and affective aspects of pain are transmitted in related, and yet distinct ascending pathways, with modifications made by both segmental and descending modulatory systems.

Critical Anthropomorphic Evaluation of Pain

Animal behaviour differs between species, between individuals of the same species, and even within the same individual over time and with different environmental influences(2). Creating a rapid and simple method for evaluation of pain perception is unlikely. However many clinicians advocate the use of pain scoring to measure treatment effectiveness to prevent the unnecessary suffering of feline patients. The most useful approach to clinical treatment incorporates some of our own human experiences and feelings (anthropomorphic) but modifying those expected feelings based on knowledge of feline behavior and physiology. While not scientifically rigorous, this critical anthropomorphic approach incorporates our knowledge of pain in humans with our understanding of how we expect animals to react to noxious stimulation to clinically guide our treatment of pain.

Management of Surgical Pain

Most clinical pain syndromes associated with surgery are complex and often involve more than one type of pain. It can be very difficult to predict the mechanisms mediating pain associated with multiple types of tissue and neuronal damage. Acute and chronic pain states may occur simultaneously. An animal with osteosarcoma may present with classic symptoms of chronic inflammatory pain and hypersensitivity, while surgery to amputate the affected limb will generate pain sensation typical of acute tissue injury. Amputation following a traumatic event may also initiate neuropathic pain associated with large nerve transection. It should not be surprising that a single drug administered at a "standard" dose would not be an effective strategy for managing all types of pain associated with surgical treatment. The clinical objective following surgery should be to minimize debilitating pathologic pain while maintaining the protective, adaptive aspects associated with physiologic pain.

One of the strategies used to achieve this objective is the concept of preemptive analgesia. The plasticity of the nervous system in response to noxious input has been well established. Initiating treatment prior to, or as early as possible after, acute insult is believed to inhibit peripheral and central sensitization processes. A second strategy involves combining drugs and techniques to achieve beneficial additive or synergistic analgesic effects (multimodal or balanced analgesia), and a third is to incorporate regional nerve blocks whenever possible prior to surgical insult of tissue. With this approach, lower anesthetic doses can be used thereby reducing potential undesirable side effects during the perioperative period.

Analgesic Agents

Opioids are a diverse group of natural and synthetic drugs used extensively in the management of post-operative pain in man and animals. Use in the surgical patient is becoming more widespread with the development of novel delivery systems such as the transdermal patch. The traditional view that a given drug always behaves as either an agonist or an antagonist at a particular receptor is a gross over-simplification, and recent studies have demonstrated that a number of variables appear to contribute to the effectiveness of various opioids in the clinical setting. Dosage, species, and stimulus intensity, character and duration can all alter the overall analgesic effect of an opioid. Opioids dampen peripheral and central afferent nociceptive transmission and thus, are extremely effective in treating acute inflammatory pain associated with recent surgical trauma. They are not, however, equally efficacious in managing all types of pain. Neuropathic pain syndromes are often characterized by a poor or short-lived response to opioid therapy.

Local anesthetics act either by blocking sodium channels that prevent nerve impulse transmission and nociceptor excitation, or by inhibiting modulatory nociceptive processing when administered centrally. In addition to their well-known topical, local and regional effects, recent studies have documented the efficacy of low dose intravenous lidocaine infusions in the management of hyperalgesia and neuropathic pain states induced by trauma or surgical procedure.

Non-steroidal anti-inflammatory drugs continue to be the mainstay of chronic pain management in both human and veterinary patients. Recent development of more selective agents has generated considerable interest in their use for intra- and postoperative pain as well. Traditionally, it has been believed that the analgesic effects of NSAIDs are related to their ability to inhibit cyclooxygenase activity and prevent prostaglandin synthesis and peripheral nociceptor sensitization. However, there is considerable evidence that at least some NSAIDs have a central spinal site of action, and may act synergistically with other analgesic compounds(3).

Alpha 2 adrenergic agonists bind to alpha 2 receptors located in the dorsal horn of the spinal cord, modulating the release of substance P, calcitonin gene-related peptide and various other neurotransmitters involved in rostral transmission of nociceptive information. Opioids likely exert their analgesic action through similar modulatory pathways and co-administration may result in additive or synergistic drug interactions. In humans alpha 2 agonists are often used as "rescue therapy" when opioid tolerance has developed. Use in surgical patients should be reserved for patients that have been stabilized with normal cardiopulmonary function.

Nonpharmacologic Therapy

In addition to drug intervention, there are a number of nonpharmacologic methods being employed to supplement analgesia. The importance of good nursing care in animals recovering from surgery cannot be overemphasized. Techniques such as physiotherapy, chiropractic or massage therapy may help to alleviate pain following trauma, and may facilitate an earlier return of function.

REFERENCES

[1] Lascelles BD, **Vet Surg.** 1998 Nov-Dec;27(6):568-82. [2] Gross DR **J Am Vet Med Assoc**. 2003 Jun 1;222(11):1505-10. [3] Ko JC, **J Am Anim Hosp Assoc.** 2009 Jan-Feb;45(1):19-23.

MANAGING SEVERE CHRONIC PAIN: MAINTAINING QUALITY OF LIFE IN DOGS

Butch KuKanich, DVM, PhD, DACVCP
Manhattan, KS

INTRODUCTION

Dogs are living much longer due to better health care, preventative medicine, and client care. The extended life of dogs has also seen increases in the incidence of chronic painful conditions such as osteoarthritis, degenerative joint disease, neurologic pain and degeneration and cancer associated pain. Of dogs presenting for necropsy 23% of dogs over 1.5 years of age died of cancer (Bronson, 1982). Therefore there is a need for safe and effective medications for chronic pain in dogs.

PHARMACEUTICAL MANAGEMENT

The nonsteroidal anti-inflammatory drugs (NSAIDs) are widely used in veterinary medicine for their analgesic effect. These are the only FDA approved analgesic drugs for oral administration to dogs. The NSAIDs probably provide the most consistent effects, but their effects are limited to mild and moderate pain. Adverse effects of NSAIDs are severe enough to warrant discontinuation in up to 5% of dogs treated and can include gastrointestinal, renal, and hepatic adverse effects. In rare and severe cases NSAIDs can cause death in dogs. In the authors opinion, opioids produce more adverse effects than NSAIDs, such as panting, sedation, vomiting, anorexia, but the adverse effects are less severe than those produced by the NSAIDs. Many of the opioids have poor oral bioavailability, due to first pass (presystemic) metabolism in the intestines and the liver and therefore have little to no efficacy in dogs. Morphine (including extended release forms) (KuKanich, et al, 2005a; Aragon, et al, 2009), methadone (KuKanich, et al, 2005b), and oxycodone () have such poor oral bioavailability their clinical use in dogs is not recommended.

Codeine

Codeine has been recommended for use in dogs at a dose of 1.1-2.2 mg/kg PO q 6-12 hours. The oral bioavailability of codeine in dogs is low, ~ 4%, compared to ~ 60% in humans. Approximately 10% of codeine in metabolized to morphine in humans, but morphine was not detected in Greyhound dogs after oral administration of codeine 2 mg/kg PO. However large amounts of a metabolite, codeine-6-glucuronide, were formed in dogs which may provide an analgesic effect (KuKanich, 2010). The concentrations of acetaminophen were low and rapidly eliminated after administration of the oral combination tablet. The efficacy of oral codeine in dogs has not been demonstrated, but the high concentrations of a potential active metabolite is suggestive it may provide some effect.

Hydrocodone

Recent studies of hydrocodone have been conducted in Greyhound dogs. Hydrocodone concentrations after oral administration of a tablet containing 10 mg hydrocodone with 325 mg acetaminophen at a targeted dose of 0.5 mg/kg hydrocodone resulted in hydrocodone plasma concentrations ~ 50% of those achieved in humans administered 10 mg. However hydrocodone was partially metabolized to hydromorphone (an active metabolite) which persisted for at least 8 hours after oral hydrocodone administration. Clinical efficacy studies have not been conducted in dogs. Hydrocodone tablets are available in combination with acetaminophen in addition to hydrocodone with homatropine, an anticholinergic. Hydrocodone tablets with acetaminophen should NEVER BE ADMINISTERED TO CATS. Hydrocodone has a high abuse potential and may be best not to stock this drug in your hospital due to liability and crime risks. Hydrocodone is inexpensive and can be scripted to a local pharmacy.

Tramadol

Tramadol is a central acting analgesic with multiple mechanisms of action. Tramadol and its active metabolite (o-desmethyltramdol or M1) are serotonin and norepinephrine reuptake inhibitors which may contribute to its analgesic effects by stimulating descending pain inhibitory pathways. Tramadol's active metabolite also appears to act through the mu opiate receptors to elicit analgesic effects. Although no efficacy studies are available in dogs, studies in humans demonstrate the metabolite is necessary for tramadol to be effective as an analgesic. Studies in dogs and cats have identified the active metabolite, although the formation appears variable. Selective serotonin reuptake inhibitors (SSRI's), such as fluoxetine and paroxetine, inhibit the metabolism of tramadol to the active metabolite decreasing its efficacy and increasing the risk of toxicity (serotonin syndrome). Other drug-drug interactions due to serotonin toxicity with tramadol include tricyclic antidepressants (clomipramine, amitriptyline) and serotonin norepinephrine reuptake inhibitors (venlafaxine, duloxetine). Ondansetron (Zofran) and other serotonin antagonists (5HT3) may decrease the efficacy of tramadol, but the results have been variable. Tramadol is inexpensive and currently not a scheduled drug in the USA. Tramadol may lower the seizure threshold, therefore should be used cautiously in animals prone to seizures such as epileptics. Current dosing recommendations for dogs are 5 mg/kg q 6-8 hours and up to 10 mg/kg PO q 6-8 hours (KuKanich, et al, 2004; KuKanich & Papich, *in press*) in non-responsive dogs and 1-2 mg/kg q 12 hours for cats. The efficacy of these dosages has not been evaluated. It is important to remember that maximum analgesic effects may not occur immediately and may be delayed up to 14 days for chronic pain conditions such as cancer and degenerative joint disease. Tramadol can be administered with NSAIDs, gabapentin, and amantadine.

Gabapentin

Gabapentin in a γ-amino-butyric acid (GABA) analogue which may increase brain GABA concentrations in humans, but the precise mechanism of action has not been determined. Gabapentin does not appear to act directly on GABA receptors, but may increase GABA concentration by decreasing GABA metabolism, increase non-synaptic GABA release, or decrease reuptake. Gabapentin may also decrease the release of glutamate, a stimulatory neurotransmitter, in the spinal cord which may contribute to its analgesic effects. Gabapentin inhibits the voltage gated calcium channels in the spinal cord which may also contribute to its analgesic effects. Gabapentin is labeled for use in humans as an anticonvulsant and for the management of neuropathic pain. Studies in humans have demonstrated variable efficacy of gabapentin for chronic and cancer pain. Gabapentin exhibits additive to synergistic effects when combined with tramadol. Recent findings in a small clinical trial in humans indicated gabapentin was effective in decreasing acute postoperative pain and analgesic consumption although other studies found no benefit compared to a placebo. In dogs approximately 40% of gabapentin is metabolized (hepatic) with the remainder eliminated by renal mechanisms. No studies have evaluated the efficacy of gabapentin in dogs or cats for treating acute or chronic pain, however pharmacokinetic studies suggest dosages of 10-20 mg/kg PO q 8 h) in dogs (KuKanich & Cohen, 2009). Adverse effects in humans include sedation, ataxia, fatigue, tremors, and rebound seizures upon discontinuation of therapy. Gabapentin can be combined with NSAIDs, tramadol, and amantadine for the management of pain.

Amantadine

Amantadine was originally introduced as an antiviral drug. However, it may be effective in treating chronic pain and as an adjunct in acute pain due to its activity as an n-methyl-d-aspartate (NMDA) receptor antagonist. Another effect of amantadine is as a dopamine agonist. As an NMDA antagonist, amantadine's primary action is to decrease central sensitization (windup pain) and does not typically provide analgesia as a stand-alone treatment. Therefore amantadine is often used in combination with analgesics such as tramadol, NSAIDs, or opioids. Some clinical trials in humans have demonstrated efficacy in chronic pain conditions, however little information is available on appropriate dosages, efficacy, and adverse effects in animals. A recent study in dogs (Lascelles, et al 2008) that were poorly responsive to NSAIDs alone co-administration of amantadine (3-5 mg/kg PO q 24h) resulted in significantly improved activity by day 42. Common adverse effects in humans (up to 33% of patients) include nausea, anorexia, insomnia, and difficulty concentrating. In humans, adverse effects may increase when amantadine is administered concurrently with antihistamines, anticholinergics, and psychotropic drugs. Tolerance may develop with long-term use of amantadine. The current dose recommendations are 3-5 mg/kg PO q 24h and are based anecdotally.

Tricyclic antidepressants (clomipramine, amitriptyline)

Tricyclic antidepressants (TCAs) have been used in the management of chronic pain in humans due to their activity as serotonin and norepinephrine reuptake inhibitors. Serotonin and norepinephrine act to modulate pain by enhancing the descending inhibitory pathways. In humans, adverse effects are typically dose limiting, primarily due to their anticholinergic effects. TCAs should not be combined with tramadol as serotonin syndrome (toxicity) may occur with effects such as tremors, rigidity, coma, incoordination, agitation, arrhythmias, hyper- and hypotension, and dyspnea among others. The efficacy of TCAs for the control of pain in dogs and cats have not been extensively evaluated.

MULTIMODAL ANALGESIA

Combining drugs with different mechanisms of action (i.e. opioid, NSAID, GABA-ergic) may increase the overall effectiveness of pain management. In humans, tramadol combined with NSAIDS or gabapentin produces better pain control than any of the drugs by themselves. Gabapentin combined with NSAIDs provides better analgesic than either drug alone. Although little data is available on drug combinations in animals, many believe it is an effective management strategy.

REFERENCES

[1.] Bronson, **AJVR** 1982:43:2057-9; [2.] KuKanich, **JVPT**, 2010:33:15-21; [3.] KuKanich et al, **JVPT**, 2005a:28:371-376; [4.] Aragon, et al, 2009**: JVPT**:32:129-136; [5.] KuKanich, et al, 2005b, **JVPT**:28:461-466; [6.] KuKanich & Papich, 2004, **JVPT**:27:239-246; [7.] KuKanich & Papich, **AJVR**, *in press*; [8.] KuKanich & Cohen, 2009:**Vet J: EPUB**; [9.] Lascelles, et al, 2008: **JVIM**:22:53-59.

TREATMENT OF CHRONIC PAIN IN CATS: WHAT ARE THE OPTIONS

Dawn Merton Boothe, DVM, PhD, DACVIM, DACVCP
Auburn, AL

The cat as a species represents a therapeutic challenge when trying to use NSAIDs safey, including the newer drugs. Nonsteroidal anti-inflammatory drugs block the first step of prostaglandin synthesis by binding to and inhibiting cyclooxygenase This action is both dose and drug dependent. The pharmacologic effects of this class of drugs and the relative dose associated with the effect include (antithrombotic, especially aspirin) <<< antipyresis < analgesia < control of inflammation. The role of PGs in normal physiology might best be understood by considering them as protective in nature. Their formation is mediated by one of at least two isoforms of cyclooxygenases. An inducible form (COX-2_ is measured following endotoxin stimulation (induction) of PG in macrophages. The constitutive form is characterized by baseline activity; designated COX-1, it is generally measured as thromboxane B2 synthesis from platelets. These or similar tests are used to generate a COX1:COX2 ratio which indicates the concentration of the NSAID that is needed to to inhibit 50% of the activity of either enzyme. A ratio > 1 indicates COX 2 is more easily inhibited than COX-1, often interepreted as a safer drug. However, these ratios must be studied in the species of interest and even then should serve only for screening purposes. Currently, one veterinary drug is approved for use for cats in the US (compared to 5 for dogs), although several others have been used effectively. What is not clear is which of these drugs prefer COX-2 to COX-1 in cats; to date, it appears that a ratio in cats has been reported only for carprofen (similar to dogs). NSAIDS with data addressing use specifically in cats include piroxicam, ketoprofen and carprofen. Tepoxalin (Zubrin®), the newest NSAID approved for use in dogs, is a "dual inhibitor", targeting both COX and LOX, thus targeting prostaglandins and leukotrienes.

Cox 1 versus Cox 2: Physiology and Pathology. In general, inhibition of COX-1 is responsible for efficacy whereas inhibition of COX-2 is responsible for side effects. However, strict adherence to this simplistic approach will lead to therapeutic failure and increased morbidity with NSAID use. *Central Nervous System*: In addition to pain, COX-2 has an important role in the pathogenesis of Alzheimer's Disease (AZD), and acute brain injury. *Pain:* PGs have been implicated in causing increased pain perception (allodonia) in damaged compared to normal tissues. Induced COX-2 PGE as been associated with hyperalgesia (exaggerated response to pain) in either the spinal cord (primary hyperalgesia) or at nociceptors in peripheral tissues (secondary hyperalgesia and central sensitization, manifested as a change in excitability threshold. Through COX-2, prostaglandins influence other CNS neurotransmitters (eg, glycine inhibition or glutamate stimulation) or other receptors (eg, NMDA). PGs also appear to mediate the anorexia and lethargy associated with chronic pain. *Gastrointestinal tract:* Both COX-1 and COX-2 are constitutively expressed in the GI tract. Although COX-1 provides the major role in the protection of the GI tract, induction of COX-2 is important to healing of GI damage. *Healing:* The application of PGE-2 facilitates bone healing in experimental animals; 4 weeks of ibuprofen (16%) or rofecoxib (Vioxx®; 87%) compared to placebo (0%) led to mal-union in rats with experimentally-induced fractures. *Cardiovascular Disease:* Platelets contain thromboxane synthetase, which catalyzes the formation of thromboxane from arachidonic acid. Thrombosis reflects platelet aggregation and vasoconstriction. The formation of a thrombus is kept "in check" by the presence of prostacyclin synthetase in vascular endothelial cells. This enzyme catalyzes metabolism of arachidonic acid to prostacyclin (PGI2), a vasodilatory and platelet inhibiting prostaglandin endproduct. However, whereas TXA2 is associated with COX-2, prostacyclin synthetase is associated with COX-1. The preferential inhibition of COX-2 may allow thrombus formation to go unchecked, increasing the risk of thromboembolic disorders. Thus, care must be taken when using these drugs in patients with hypertrophic cardiomyopathy. *Kidney:* In the kidney; both COX-1 and 2 are constitutively expressed. Both are formed in the macula densa of humans and animals, but COX-2 may have a more important role than COX-1. In (nonhuman) animals, inhibition of COX-2 causes sodium and potassium retention in salt depleted, but not normal, animals. However, in humans, COX-2 appears to influence renal vasculature and podocytes. *Lungs*: The role of PGs in the lungs is less than that of leukotrienes; however, inflammatory diseases, such as asthma, are associated with smooth muscle proliferation which is inhibited by COX-2. Thus, like the GI tract, COX-2 appears to have a protective role in the diseased lung. Note that with older NSAIDs, inhibition of COX resulted in shunting of AA to LOX and increased production of leukotrienes. While this is not longer an issue in human medicine, the issue may remain in cats until the role of COX inhibition is clarified for each of the drugs. As such, NSAIDs should be used only with caution in cats with asthma. An exception might be tepoxalin. *Cancer:* COX-2 increases markedly in a variety of soft-tissue tumors in humans and in transitional cell carcinoma in dogs (Knapp 2004). These studies suggest that benefits of NSAIDs in cancer may reflect inhibited COX-2. Depending on the model, inhibition of COX-2 associated with tumors reduces cell proliferation, increases apoptosis, and reduces metastasis. Inhibitors may also enhance anti-tumor effects of radiation, although toxicity is increased. However, at least for NSAIDs, GI toxicity is also increased when combined with antimetabolite anticancer drugs, presumably reflecting a combined toxic effect on the GI tract.

Pharmacokinetic Data for Selected NSAIDs Adult Cats

	Route	Dose	Cmax (1)	Half-life
		mg/kg	mcg/ml	hr
Aspirin	PO	2.5		27-45
n=5	IV	20	118	22+3
		44		37.6
Carprofen	IV	4	24 ± 9.7	20.1±16.6
n=5 R (-)				21.3±9.1
S (+)				14.6+5.8
Deracoxib	PO	1	0.28 + 1.3	8
Flunixin (n=4)	IV	2	12.5/0.11	6.6 ± 2.4
Ketoprofen				1.5
Meloxicam	SC	0.3	1.5+0.9	15.1+5
Piroxicam (n=8)	PO	0.3	0.52	13
Tepoxalin	PO	10	2.3±1.8	4.7±0.8
metabolite			1.8± 1.2	3.5±0.4

Disposition. Poor metabolizers have been described for some drugs in humans (eg. Asian descent) and dogs. It is tempting to speculate that breed differences will also emerge in cats of Asian descent. Adding to the complex impact of disposition on NSAID safety is the impact of enantiomers. Enantiomers are molecules that are mirror images of one another (R- or S+). They occur for many drugs which contain a chiral carbon around which its bonds rotate. Although chemically similar, the body handles each enantiomer differently in regards to disposition (see carprofen in table). Further, the pharmacodynamic effects of enantiomers are likely to be different. In essence, the body treats each enantiomer as a different drug. Most NSAIDs contain a chiral carbon, and are sold as racemic mixtures (50:50) of the R(-) or S(+) enantiomers. As such, extrapolation among species, genders, breeds, ages, etc becomes increasingly complex.

Safety:*<u>Gastrointestinal</u>* damage is the most common and serious side effect of the NSAIDs among species. Cats are likely to be more sensitive to the GI side effects of NSAIDs compared to people. Because COX-2 is important to the healing ulcer, any NSAID should be discontinued at the first indication of GI upset. GI ulceration should be anticipated in an any animal receiving an NSAID. Treatment for GI toxicity should protect the damaged mucosa, and if necessary, control gastric acid secretion. Because (intestinal) ulceration is difficult without acid, the single most important treatment may be antisecretory drugs. In the face of severe ulceration, a proton pump blocker such as omeprazole may be most effective. However, its inhibitory effects on drug metabolizing should be avoided in patients receiving NSAIDs. Accordingly, famotidine (also can be given once daily and is minimally involved with drug interactions) can be used, especially from a preventative standpoint. As an inhibitor of hepatic drug metabolizing enzymes, omeprazole might be reserved for treatment of ulceration, when the NSAID is to be discontinued. The benefits of sucralfate include binding to and thus protecting damaged mucosa, as well as increased PGs synthesis, angiogenesis, and sulfhydryl (oxygen radical scavengers) production. However, sucralfate must bind to damaged tissue to be effective and thus is helpful in treating, but not preventing, ulceration. *<u>Hemostasis:</u>* All traditional NSAIDs are able to impair platelet activity due to impaired prostaglandin (thromboxane) synthesis. At pharmacologic doses, aspirin selectively and irreversibly acetylates a serine residue of a platelet cyclooxgenase and, accordingly, will always have very low Cox 1: Cox2 ratios. However, newer NSAIDs inhibit prostacyclin, which acts to impair platelet aggregation, while minimally affecting thromboxane synthetase. As a result, thrombosis can occur relatively unchecked, predisposing patients to thrombosis (eg, Viox®). Unfortunately, it is unclear which of these drugs target predominantly Cox-2 in cats, decreasing confidence in use. <u>Renal:</u> The cat may be predisposed to nephrotoxicity (compared to the dog or human), perhaps in part because of differences in renal function. In the kidney, vasodilatory PGs are protective, assuring that medullary vasodilation and urinary output continue during states of renal arterial vasoconstriction. The loss of this protective effect becomes important in patients with compromised renal function. Newer NSAIDs do not appear any less likely to be associated with this effect than the traditional NSAIDs in humans. In cats, a review of adverse drug events at the FDA's CVM website reveals that the proportion of adverse events of cats that reflect the kidneys (eg, 45% of meloxicam and 24% of carpfofen) is greater than that in dogs (9%), and is greater for meloxicam compared to carprofen. Unfortunately, this is the only comparative data available for NSAIDs and it has been removed from the website.

FDA Adverse events for Cats *(Dog)

	Meloxicam		Carprofen	
Sign	(n=497)	%	(n=489)	%
Creatinine	230	46	109	22
Anorexia	226	45	138	28
BUN	224	45 *(9)	118	24 *(9)
Depression	200	40	153	31
Vomiting	166	33	231	47
Renal Failure	116	23	28	6
Death	67	13	45	9
Azotemia	59	12		0
Death, Euth	54	11		0

Drugs: Meloxicam, like piroxicam, is a member of the oxicam group of NSAIDs The COX2:COX1 ratio for meloxicam, unlike that for piroxicam, favors selective COX2 inhibition in humans, suggesting that it has a wider margin of safety than most other. The drug is more potent (although not necessarily more efficacious) than aspirin, indomethacin, and piroxicam; hence, its dose is smaller. The disposition of meloxicam has been studied in cats. Meloxicam is among the NSAIDs characterized by a shorter half-life in cats compared to dogs, and is one of the few NSAIDS that appear to be well tolerated in cats. It's safe use in Canada for several years predated its approval for use in cats in the US. Meloxicam is the only NSAID approved for use in cats in the United States. However, despite its apparent safety compared to other NSAIDs, the therapeutic margin of meloxicam is relatively narrow. Cats do not tolerate doses greater than or equal to 0.3 mg/kg gastric ulceration and death has occurred at 3X to 6X the normal for 10 days. Several studies support the efficacy of meloxicam in cats. In one study, the optimal dose of meloxicam to prevent endotoxin-induced fever in cats was 0.3 mg/kg, a dose also noted for its ability to cause toxicity. Meloxicam (0.2 mg/kg SC), carprofen (4 mg/kg SC), tolfenamic acid (4 mg/kg SC) and ketoprofen (2 mg/kg SC) did not differ in their ability to control post-operative pain following OHE in cats in one study. Meloxicam has recently been approved in Australia for chronic use at a dose of 0.05 mg/kg daily in cats. Genuw (J Fel Med Surg 2008) recently reported the safe used of meloxicam in adult (aged) cats using a case-control design (n=46; treated cats, age 12.9$\pm$4.2 yr) in Australia. Meloxicam was administered at 0.1 mg/cat (0.01-0.03 mg/kg (0.1 mg/cat) with food once daily following a 0.1 mg/kg once daily for 4 days oral loading dose. Mean duration of treatment was 5.8 months. However, there are some issues with the study. Since it was case controlled, no placebo of blinding took place and efficacy comparisons were based on comparison at baseline. Thus, it is difficult to assess efficacy. Illnesses in both the treatment and control group were diverse, with 25% of each group suffering from some disease. Ten pairs of cats were matched by disease. Cats with renal disease were included if disease was stable. One cat in each group died due to chronic renal insufficiency. Creatinine was measured only in the first 10 pairs of cats and only until 1 month of therapy. No statistical differences were found in creatinine between the treatment and control group (data not provided) at 1 month; however, the power of the study was not reported. One cat in each group died from chronic renal insufficiency; creatinine increased in cats with pre-existing disease treated with meloxicam numerically more than it did in control cats with the exception of 6 months (data not provided to evaluate variability). In addition, 4 cats in the meloxicam group and one in the control group developed vomiting. Caution is suggested when interpreting the results of this study as evidence of safety of meloxicam in regards to renal dysfunction associated with long term therapy. Caution is thus encouraged when using any NSAID, and perhaps meloxicam, in cats with pre-existing renal disease; this may include that population of cats whose serum creatinines are "high" normal. **Ketoprofen.** Although not firmly established, the efficacy of ketoprofen has also been attributed to its ability to inhibit some lipooxygenases and thus formation of leukotrienes. Ketoprofen is not approved for use in small animals in the United States but is approved for both dogs and cats in Europe. The half-life of the drug is short in cats (1.5 hrs) compared to dogs. Ketoprofen has been used as an analgesic in cats, particularly in Canada (1 mg/kg every 24 hours for 7 to 10 days). The antipyretic effect of ketoprofen (2 mg/kg subcutaneously followed by 1 mg/kg once daily orally) in febrile cats was rapid, being evident in 4 hours with temperatures normalized at that time.. Temperatures did not change in the antibiotic-only treated cats. The use of ketoprofen as an analgesic is variable . In cats subjected to ovariohysterectomy, ketoprofen (2 mg/kg subcutaneously) compared favorably with buprenorphine (0.006 mg/kg or 6 mcg/kg intramuscularly) and mepiridine as gas anesthesia was discontinued. Response better for both drugs compared with the control at both 4 and 8 hours but still present for buprenorphine only compared with control at 18 hours. **Carprofen** is approved for use in cats in select countries outside the United States. The disposition of the drug has been studied in cats, including enantiomers at the dose associated with control of inflammation (4 mg/kg). A smaller clearance for carprofen in cats results in a elimination half-life that is at least twice as long in cats compared to dogs. The S isomer is cleared almost 3 times as rapidly as the R isomer resulting in a shorter half-life for S compared to R in the cat. The relative COX2 selectivity of carprofen that occurs in dogs has not been well documented in cats, although Brideau's data dose suggest relative selectivity similar to that in dogs. Clinically, however, this may not be true. The gastrointestinal effects of single dose carprofen (4 mg/kg IV) or aspirin (20 mg/kg IV) were studied in cats (n=5) using endoscopy and a randomized crossover design. Lesions in the stomach and duodenum 8 h postinjection were limited to minor pinpoint erosion in one cat. Clinical laboratory tests were not affected by either drug. However, duodenal perforation has been reported in cats using oral administration of carprofen (2.2 mg/kg twice daily for 7 days) following ovariohysterectomy. The ulceration was no doubt exacerbated by flunixin meglumine and dexamethasone treatment prior to referral. As a postoperative analgesic in cats, carprofen compares favorably with pethidine (meperidine), providing equal but longer analgesia (at least 24 hours) when administered a t 4 mg/kg subcutaneously postoperatively (Balmer et al., 1998). In a clinical trial of cats undergoing OHE, no difference was found in the analgesic effects of carprofen, ketoprofen, meloxicam or tolfenamic acid. based on the visual analogue scale and a nociceptive threshold at the incision site. With 9/10 cats responding, all responses were described as good, although none prevented wound tenderness. **Tepoxalin** is among the dual acting NSAIDs. It is a potent anti-inflammatory and analgesic pyrazole derivative that also inhibits production of IL-1, suppreses NFkB activation and dependent gene expression (Keier 2004). Tepoxalin also has been studied in cats. Whereas tepoxalin was well tolerated in cats when administered at 100 mg/kg once daily for 3 consecutive days, saturation kinetics occur when 60 mg/kg was administered as two doses four hours apart. Signs suggestive of CNS ADE occurred (drunken-like state), a response not recorded in any other species. Tepoxalin is a potent antipyretic agent in cats at doses between 5 and 10 mg/kg and provides analgesia at least equivalent to butorphanol at 10 mg/kg for onychectomy (Personal Communication, Gerryll Hall, Technical Service Veterinarian, Schering Plough, April 2004).

Other drugs: Tramadol (Ultram) is a synthetic analogue of codeine currently marketed as a racemic (1:1) mixture of ± enantiomers. Tramadol appears to have multiple mechanisms of analgesia, with interaction among the pathways perhaps contributing to its efficacy. Opioid analgesia reflects agonistic interaction with mu receptors; additionally, tramadol enhances spinal pain inhibitory pathways through inhibition of neuronal re-uptake of serotonin (5-HT) and noradrenaline (NA), and release of 5-HT. As such, tramadol might be indicated for a broad array of conditions associated with pain, including chronic pain: studies suggest that the analgesia associated with 5-HT1A receptor agonists increases with chronic or repeat administration. Both tramadol stereoisomers are associated with analgesia, although receptor interactions appear to differ between the isomers. Tramadol is de-ethylated to an active metabolite, O-desmethyltramadol (ODT) which may be responsible for the majority of analgesic activity. The disposition of tramadol and ODT after either IV (2 mg/kg) or oral (5 mg/kg) administration of tramadol has been described in cats).The AUC of active metabolites is equal to or surpasses that of the AUC of the parent compound. The elimination half-life of both parent and ODT is approximately 2 fold higher in cats. Cats tolerated tramadol by either route, although they exhibited euphoria for several hours. A 5 mg/kg oral dose achieves for tramadol and well exceeds for ODT the minimum effective analgesic concentrations suggested in humans. It is not clear if a similar level of analgesia would be expected in cats, but based on the higher concentrations of ODT in cats, a dose of 2 mg/kg twice daily is a reasonable starting dose. Steagall studied the antinociceptive effects of tramadol (1 mg/kg SC) with or without acepromazine (0.1 mg/kg) in cats (n=8) subjected to thermal stimulation. Animals had only a limited response to tramadol, although the effect was increased when tramadol was combined with acepromazine. Care must be taken to NOT prescribe UltramPlus which contains acetaminophen. **Gabapentin** was initially developed as human anti-epileptic drugs. It is structurally related to GABA; gabapentin is a molecule of GABA covalently bound to a lipophilic cyclohexane ring. Although their mechanism of action was intended to be a GABA agonist, pharmacologically, they do not bind to an portion of the GABA receptor, nor do the appear to interfere with degradation or other aspects of GABA receptor activity. Rather, their mechanism of analgesic action appears to occur through binding to CaVa2-d proteins on voltage gated calcium channels. Decreased calcium influx prevents release of neurotransmitters otherwise stimulated by a variety of chemical signals. Gabapentin appears to be a safe anticonvulsant in cats and accordingly should be considered for long term control of pain .

Selected References: Pypendop BH, Ilkiw JE: Pharmacokinetics of tramadol, and its metabolite *O*-desmethyl-tramadol, in cats, *J Vet Pharmacol Therap* 31:52-59, 2007. doi: 10.1111/j.1365 2885.2007.00921. Steagall PV, Taylor PM, Brondani JT, et al: Antinociceptive effects of tramadol and acepromazine in cats, *J Feline Med Surg* Feb;10(1):24-31, 2008. Slingsby LS, Waterman-Pearson AE: Comparison of pethidine, buprenorphine and ketoprofen for postoperative analgesia after ovariohysterectomy in the cat, *Vet Rec* 143:185-189, 1998. Taylor PM, Delatour P, Landoni FM, et al: Pharmacodynamics and enantioselective pharmacokinetics of carprofen in the cat, *Res Vet Sci* 60(2):144-151, 1996. Lascelles BD, Cripps P, Mirchandarin S, et al: Carprofen as an analgesic for postoperative pain in cats: dose titration and assessment of efficacy in comparison to pethidine hydrochloride, *Small Anim Prac* 36:535-541, 1995.Balmer TV, Irvine D, Jones RS, et al: Comparison of carprofen and pethidine as postoperative analgesics in the cat, *J Small Anim Pract* 39(4):158-164, 1998. Runk A, Kyles AE, Downs MO: Duodenal perforation in a cat following the administration of nonsteroidal antiinflammatory medication, *J Am Anim Hosp Assoc* 35:52-55, 1999.

NEW ANALGESICS: WHAT IS IN THE PIPELINE FOR DOGS & CATS?

Tamara Grubb, DVM, MS, DACVA
Pullman, WA

With so much research focused on new analgesic drugs for humans, our hope would be that we could say the same about analgesic drugs for animals. Unfortunately, there doesn't seem to be much activity from the pharmaceutical companies, at least not activity that will result in new advances anytime soon. There are a few new products that have been launched in other countries (outside the US) and a few methods of drug delivery that may be clinically available in the not too distant future. Also, there are drugs currently used in human medicine (both new drugs and reformulations of old drugs) that are new to veterinary medicine and most of these are just entering clinical trials for analgesic efficacy in our patients. There are other new drugs on the market and drugs in Phase III trials in human medicine that we have not yet begun to evaluate. Also, there are many drug targets that are still in Phase I and II human trials and should be discussed, although we may not be utilizing these products any time soon. Fortunately for the immediate relief of pain in our patients, when we talk about analgesics, we don't have to limit our thoughts to drugs (although this manuscript is focused on drugs) and many new non-pharmacologic techniques are being used and /or developed in both human and veterinary medicine.

WHAT IS ACTUALLY NEW IN VETERINARY MEDICINE?

The only two new analgesic drugs that have been recently launched for veterinary patients are both coxibs and both are only available outside the US. Robenacoxib (Onsior®)[1] has been launched outside the US for the treatment of chronic pain in dogs and acute pain in cats. Also launched outside of the US is the coxib, mavacoxib (Trocoxcil), which is a sustained release NSAID approved for chronic pain in dogs. A 2010 search for literature yielded an abstract on the pharmacokinetics of the drug[2] but no information on efficacy. Information at the marketing company's website reveals that one dose of mavacoxib appears to control pain for at least 30 days. No other veterinary specific drugs have been launched but a new delivery method for opioids is currently in the final phase of approval in the US and more details should be available soon. If other drugs or drug delivery systems are in the pipeline for veterinary patients, they are likely several years from clinical use or are an extremely closely guarded secret.

WHAT ARE OUR VETERINARY COLLEAGUES WORKING ON?

The results of a January 2010 query of the International Veterinary Academy of Pain Management (IVAPM) regarding the research focus of our pain management colleagues revealed that, rather than focusing on new drugs, most of our colleagues are spending more time on reformulating drugs that we currently use and even more time on the use of non-pharmacologic therapies. Reformulation of currently used drugs is also the trend in human medicine and this can lead to 'new' products in a relatively short period of time. The primary focus of many of our colleagues appears to be methods to extend the duration of currently used drugs (through liposome encapsulation, etc..) like buprenorphine and bupivacaine. Research on acupuncture, therapeutic ultrasound, low-level laser, orthotics and a variety of physical therapy modalities is also underway. Recent publications that focus on drugs or drug delivery systems in veterinary medicine include, but are not limited to, investigations of tramadol, gabapentin, pamidronate, levobupivacaine, long-term catheters for delivery of analgesic drugs and a variety of drugs delivered in the epidural space.

DRUG TARGETS AND CURRENTLY AVAILABLE DRUGS IN HUMAN MEDICINE

Although not a complete list (not even close to being a complete list), a sampling of drugs that have recently been approved for pain management in human medicine[3,4] that have not yet reportedly been tried in veterinary medicine include, pregabalin (a select calcium channel blocker), duloxetine & milnacipran (serotonin and norepinephrine reuptake inhibitors), cyclobenzaprine (a muscle relaxant with a chemical structure similar to tricyclic antidepressants), tapentadol (a centrally acting analgesic with a dual mode of action similar to that of tramadol), topiramate (an anticonvulsant drug with efficacy against migraine headaches), tizanadine (a centrally acting alpha-2 agonist) and ziconotide (an N-type voltage-gated calcium channel blocker for intrathecal use). Drugs not yet clinically available but in the final phase of clinical research include the cyclooxygenase-inhibiting nitric oxide-donating drugs (CINODs; eg, naproxcinod) [5] and the capsaicin derived drugs. These drugs will be used as examples to discuss mechanisms of action for a variety of analgesic mechanisms but do not represent all of the drugs in any specific category. All drugs listed below that have FDA approval are approved for use in humans but not in animals (unless otherwise stated). Specific drug information can be obtained at the website for each drug. Specific scientific studies are referenced.

Pregabalin (Lyrica®) and gabapentin bind to the alpha2-delta subunit of voltage-gated calcium channels in the CNS. This reduces calcium influx into the nerve terminals with the subsequent reduction of the synaptic release of excitatory neurotransmitters, like glutamate, noradrenaline and substance P. Pregabalin isn't really 'new' in human medicine but it has continued to be approved for new analgesic indications and is used for a number of off-label indications as well. In 2005, the FDA approved pregabalin for treatment of diabetic neuropathy & post-herpetic neuralgia pain (gabapentin was approved for the treatment of post-herpetic neuralgia pain in 2002) and the drug was approved for the treatment of fibromyalgia syndrome pain in 2007. Pregabalin was the first analgesic drug approved for this type of pain. Other calcium channel blockers are being explored but none are yet in Phase III trials and no drugs from this class are currently licensed in veterinary medicine.

Duloxetine (Cymbalta®) is a serotonin and norepinephrine reuptake inhibitor (SNRI) labeled for major depressive disorder, generalized anxiety disorder, diabetic peripheral neuropathic pain (approved in 2004; first drug approved for this indication), and fibromyalgia syndrome (approved in 2008). Increases in norepinephrine and serotonin within the CNS appear to enhance the activation of descending inhibitory neurons. Duloxetine has been used off-label for a number of painful conditions, including osteoarthritis.[5] Milnacipran (Ixel, Savella), is a newer SNRI is also approved for the treatment of clinical depression and fibromyalgia syndrome pain (approved in 2009 for fibromyalgia). There are no SNRIs approved in veterinary medicine but the selective serotonin reuptake inhibitor (SSRI) fluoxetine (Reconcile®) is FDA approved for the treatment of separation anxiety in dogs. The analgesic efficacy of SSRIs in humans is controversial and no studies on SSRI analgesic properties have been conducted in dogs or cats.

Cyclobenzaprine (Apo-Cyclobenzaprine, Flexeril, Fexmid, etc…and a once-a-day extended-release formulation Amrix®) is interesting because it is a centrally-acting muscle relaxant with a chemical structure similar to tricyclic antidepressants. Indications listed on the label of the sustained release product are, 'AMRIX® is indicated as an adjunct to rest and physical therapy for relief of muscle spasm associated with acute, painful musculoskeletal conditions'. As with other tricyclic antidepressants, at least part of the mechanism of action is through noreipnephrine reuptake.

Tapentadol (Nucynta™) is a centrally acting analgesic with a dual mode of action similar to that of tramadol (mu-opioid receptor agonist and norepinephrine reuptake inhibitor).[6] Tapentadol is a Class II DEA controlled drug that was FDA-approved in November 2008 for the treatment of moderate to severe acute pain. Opioids are becoming more mainstream for treatment of chronic pain and we can expect newer opioids, new opioid delivery methods and new 'tricks' for using opioids. For instance, the use of ultra-low dosages of opioid antagonists with opioid agonists is being selectively used in human medicine. Results show that the incidence of opioid-mediated side effects are definitely reduced using this technique and analgesia may be enhanced in some patients, although the latter fact is controversial and results are inconsistent.[7]

Topiramate (Topamax®) is FDA approved for the treatment of epilepsy and the prevention of migraine headaches and has been used off-label for diabetic neuropathy. The drug has been approved since 2004 but a 2009 approval of a generic form of the drug has brought a resurgence of interest. Topiramate is intriguing because, although it's mechanism of analgesia isn't totally clear, it does purportedly have a number of mechanisms including, sodium channel blockade, potentiation of the inhibitory effects of gamma-aminobutyrate (GABA), antagonisim of the ability of kainate to activate the kainate/AMPA (non-NMDA) subtype of glutamate receptors and it may also cause calcium channel blockade at higher dosages.

Tizanidine (Zanaflex®, Sirdalud) is an orally-administered centrally acting alpha-2 adrenergic agonist that is licensed for short-term relief of muscle spasticity and cramping. The alpha-2 agonists work presynaptically to decrease the relase of presynaptic neurotransmitters.

Ziconotide (Prialt®) is an N-type voltage-gated calcium channel blocker that is used intrathecally for treatment of severe chronic pain. These calcium channels are located in both the brain and the spinal cord and blockade of these specific calcium channel decreases the release of pain-inducing neurochemicals like glutamate, calcitonin, and substance P.

The not-too-distant future

Currently in phase III trials are a variety of drugs, including Naproxcinod and Adlea.™ Naproxcinod (nitronaproxen) is a cyclooxygenase-inhibiting nitric oxide-donating drug (CINOD) that appears to have the analgesic efficacy of the parent NSAID but with a greatly reduced incidence of negative side effects because of the positive effects of the NO.[8] Naproxcinod has been shown to control the pain of osteoarthritis. Adlea is a purified form of capsaicin, which acts on the transient potential vanilloid 1 receptor (TRPV1 receptor – or 'capsaicin receptor') and has been shown to be effective for the management of post-surgical orthopedic pain and osteoarthritis of the knee,as well as pain from tendonitis of the elbow and certain types of post-trauma neuropathic pain. Resiniferatoxin (RTX) has the same mechanism of action but is more potent than capsaicin and is currently in development as a commercial product. Capsaicin is also used in a wide variety of topical creams. These creams have been on the market for a fairly long time and are not new, but have not yet been extensively used in the veterinary market and may have some application.

The more distant future

There are many other targets that are being investigated for analgesic drugs. Targets that might result in drugs that are developed relatively soon include, specific subunits of the NMDA-receptors (eg, NR2B receptor antagonists), cannabinoid receptor agonists, neurokinin-1 (NK-1) antagonists and nerve growth factor (NGF antagonists).

REFERENCES

[1.] King JN, et al. **J Vet Pharmacol Ther** 2009;32(1):1. [2.] Lees P, et al. **J Vet Pharmacol Ther** 2009;32(S)1. [3.] Kroenke K, et al. **Gen Hosp Psych** 2009;31:206. [4.] Mao J. **Journal of Pain,** epublished ahead of print 2009 [5.] Chappell AS, et al. **Pain** 2009;146(3):253. [6.] Hale M, et al. **Curr Med Res Opin** 2009;25(5):1095. [7.] Bates JJ, et al. **Anesth Analg** 2004;98:116-122. [8.] Wallace JL, et al. **Trends Pharmacol Sci** 2009;30(3):112.

CREATING AN EXCEPTIONAL SPECIALTY PATIENT & CLIENT EXPERIENCE

Robin Brogdon, MA
Huntington Beach, CA

IS MY COAT SHINY? – HOW DO I LOOK?

The health of your practice and the health of your patients should be one and the same. You get a shiny coat by being healthy. So let's look at what makes for a healthy practice. It starts with a healthy brand. We start with the big picture and dig into the details to create that exceptional patient and client experience that leads to brand equity. For good health, your brand needs to be an asset.

PERCEPTION

Your brand is more than your logo or mission statement. It should drive how you do business and thus, how others see you. This is really about perception – but perception needs substance behind it if you are to successfully deliver on your brand promise.

Brand, in its simplest definition, is a promise waiting to be fulfilled. Often a brand experience is intangible and subconscious. You can feel good or bad about a company, product or service and not necessarily know why. If your experiences are positive with a company, and then repeated often and consistently, the end result is some level of brand loyalty. But there is always confusion about 'brand'. To be certain, a brand is not a logo, tagline or a name.

WHAT'S LACKING IN MOST BRANDS?

What most marketing campaigns lack is a focus on those carrying out the brand day in and day out.

The People! Everyone needs to drink the cool-aid.

- Reception - The people who answer the phone and greet your clients
- The technicians who provide information and services to your clients
- The tech assistant who brings the pet back to the owner at the end of a visit
- The doctors who provide the medical expertise

Every single word and action your people make or take that is visible or audible to a client or potential client is part of your brand. Are they aware of that? Are you?

1. When your practice makes a promise and posts it, the mission statement and core values for instance, it is the equivalent of an advertisement, and they are standards that you are *choosing* to uphold yourself to. If everyone on your team doesn't know these by heart and what they really mean, don't post them!

2. In living up to your brand, get your client's information right. Don't inconvenience them. Be accessible – make your contact information prominent. Spell their name correctly in correspondence. Respect them. Remember, this is about *their experience.*

3. Make sure there is someone higher up the food chain to correct problems WHEN they need it. It's terribly frustrating as a customer to not get the support you need when you need it.

4. Stand behind your word. Whatever you tell a client is what you will be held to. Whatever your staff tells a customer represents not just their personal word but that of the company too. It must mean something. It's part of your company's brand. If you can't deliver, don't say it.

What we're really talking about here is building trust and credibility. Consistency and unity of message breeds confidence and trust. This is only done successfully if everyone on the team sings the same mantra – understands the practice's brand message – mission and employs the core values in the course of their daily job by the behaviors they exhibit in carrying out the brand promise.

BUILDING TRUST IN REFERRAL MEDICINE

This is not the same as buying a widget from a sales person. In a service business, people buy from people, not companies. So your people are the most important aspect of your practice. Likewise, people leave jobs because of people, not companies. Let's face it, people don't want to need your services (rDVMs or pet owners). But when people need to choose a doctor, it is done by recommendation more than any other source – from people they trust.

The best way to build trust is to humanize the practice. A surefire strategy is to involve your senior level staffers and all levels of the practice in the community. You have to be part of the community to reach those that need you. It should be part of your job. Some should be at VMA meetings, chamber of commerce, rotary club, local charities, etc. This makes a difference because the community sees that real human beings are behind the hospital doors. It's also a way to inform and educate the community that will not be taken as poaching the rDVM clients.

CUSTOMER MAPPING: FIVE STEPS TO IMPROVE SATISFACTION AND RETENTION

For many companies, their product or service no longer is what differentiates them, so they must do so in another way. Research shows that the experience is a key driver in satisfaction and retention. The client experience likely involves almost everyone on your team…from the moment they hear about you from the rDVM through receiving that final invoice.

Research by QCI International found:

- 41% of companies surveyed do not record customer contact channel preferences.
- In more than 90% of companies, staff who are responsible for talking to clients could not articulate why customers should buy from them.
- Only 13% of senior management has regular contact with customers.

It's important to establish an internal organization that will serve as the customer experience champion – the client advocate. To measure the client experience, you need to follow a disciplined process that will measure what really matters to them. The purpose is to:

- Define a set of internal metrics that define a quality customer experience.
- Identify all the contact points that are critical to that experience.
- Create a methodology for continuous improvement.

1. Define the client experience - Draw it out in diagram format.

Be sure it includes steps, contact points, interactions, relationship initiation, delivery, problem resolution, payment and client relationship management. You may even want to include the client in this exercise as you are defining their experience and perceptions, not your own internal perspective.

2. Establish expectations and priorities

You'll need to talk to your clients here – both rDVM and pet owner because you need to define both experiences in order to deliver to them well, respectively. Capture the expectation, importance and perception of your practice's performance at each step or contact point.

3. Score and even map the experience

It's also a good idea to have the internal team do this as well while you are gathering client data. Create a scoring methodology for importance and experience. Combine the scores from the clients and internal team, map them and see where the gaps are – or in essence where in the experience process you are at risk.

4. Identify opportunities for improvement

Look at the customer experience by contact point and perceptual gaps from the analysis of internal staff.

5. Create your experience scorecard

To create the scorecard, you need to establish some performance targets against which you will measure your progress – and decide which contact points and interactions are the most critical for creating a competitive advantage and building loyalty. The scorecard will help you identify the internal metrics that link directly to the client as experience. Creating a customer experience map compels an organization to take a customer-centric view.

Process adapted from Laura Peterson of VisionEdge Marketing

Let s go back to the beginning for a moment and discuss who are you as a practice? What do you stand for? How do you want to be perceived? What do you want an rDVM to experience or feel about the experience they have with your practice? Presumably you have developed a mission statement and articulated several core values.

How do you use them in your practice? Can anyone here be sure that all team members know them, understand them, and know how to employ them in their daily duties? If not, you are failing at delivering on your brand promise. If they are really what you want your practice to use as its core guideposts then you have to hire like you mean it, train continuously, hold those accountable that do not live the message. Decide what you want your business to be, then ask staff if they can live up to it – it's their choice – because this is how the practice operates.

And every business decision you make should be preceded by a question of:

Will this decision align with our mission and will we be able to act in accordance with our core values to carry it out? Remember, Brand is an asset.

WHAT DO YOU NEED TO DO TO BUILD TRUST IN YOUR PRACTICE?

Be transparent - This is not just about quality and price transparency. Do you publish the email addresses of senior staff so people can contact them directly? Take down the barriers.

Collect information - I am a preacher of how practice's can leverage word-of-mouth for building trust. Consider collecting data – what do you really know about your customers (rDVMs).

Monitor - People out there are talking about you – blogging, twitter, yelp and so on. Assign someone to monitor the media on a regular basis, address issues, participate in the online communities. Doing so creates trust – they know you are listening and want to engage with them.

Create community - It's not really about the practice, it's about bringing people together. Shouldice Hospital in Canada does this very well. Once a year they host an annual patient reunion. There's nothing like sharing an evening with people who understand what you've been through.

Adopt causes - That are strategic to your service. Trust is built by seeing you in action, serving your community. Especially by the rDVMs.

So how do you build trust? It is built and maintained over time by many small actions forming a tipping point for choice. Every step in the process, every touch point your practice has with a client or prospective client, must reinforce the message. And you can see that if there is one broken cog, it can and will damage the experience.

There is a perception that specialists are hard to reach. In many cases it would be wise to invest in systems that improve services to referral sources to be sure you can deliver on the brand promises.

Review your mission statement and core values. Adjust if needed. After you've written them, use them as a guide, a blueprint if you will, for operating your practice.

So how many of you have a shiny coat? If I've just exposed some flaws or 'hot spots' you have the opportunity to make changes. Having a health practice is all about having a healthy brand, that the inside (your staff) and your outside (rDVMs and clients) share the same view and perception of your practice because the experience they have as employees or clients must be the same as what you tell them it will be. That builds trust, loyalty, sustainability and value. A well developed brand is a true asset.

ENGAGING THE SPECIALTY VETERINARY TEAM TO BUILD LOYALTY & REVENUE

Robin Brogdon, MA
Huntington Beach, CA

IS MY BLOOD PRESSURE WNL? WHAT'S REALLY HAPPENING ON THE INSIDE?

We will explore the most under-utilized yet most cost effective form of marketing and discuss ways to engage the team, build brand champions and keep your internal processes running smoothly. The result will be more revenue and satisfied and loyal clients and referring veterinarians. It's called internal marketing - a low risk, high return proposition. In today's challenging economy, never has it been more important to attract, nurture, and enrich the employment experience of your team. We will address the keys to building and sustaining a successful practice – by focusing your efforts on your staff (the inside) – It's marketing made simple, and very cost-effective!

INSIDE OR OUTSIDE?

When most people think of marketing, they think of marketing to prospective customers or to existing ones. Essentially they think of it as an externally-focused business function. But what about marketing to your internal customers – your employees? They are, or should be, walking, talking advocates for your brand. However, most companies spend little time, attention or money on internal marketing and branding, such as making sure employees understand the brand and communicate it well. Do your employees know and live the brand? Are they brand champions?

In many ways, selling the brand on the inside is more important than selling the brand on the outside. And there's no shortage of reasons why companies should embrace the concept of internal branding. It leads to:

- Increased company loyalty and job longevity
- Employees are better able to serve customers because they understand the brand promise
- Employees who believe in the brand work harder and better
- Most importantly, when employees are brand champions, they create brand differentiation, something hard for your competition to replicate. This differentiation becomes part of your brand's edge on the competition and employees add the edge. All employees must be brand champions.

Of the traditional 4 P's - Product, Price, Promotion and Placement - are all replicable by competitors. The only thing not replicable is the 5th P – the Personality of the organization or the People.

Granted, finding and keeping employees that are brand champions is more difficult than just finding an employee. They need to be systematically found and created from those with promise. At a minimum, a company must ensure that the client and employee's perception of the brand are in sync. With increased competition, it's crucial that the company be brand focused – the employees make the difference between the brand being just an idea vs. a reality for your clients.

Internal branding is not a one-way street. To get the greatest benefits there must be a dialogue. Employees are a great source of information on the brand. You can use their perceptions, their experiences and their interactions with clients to hone the brand message. After all, the employees are the heart of the company and who better to offer valuable input into the brand.

External marketing should be no different than internal marketing. The message should be the same. Just as an organization should be listening to market forces when creating and evolving the brand, so should they listen to the internal forces. Take away: Employees must hear and experience the same messages you tell the public and must not only understand them, but be empowered to uphold them (responsibility).

Employees who like their jobs have strong opinions and knowledge of the brand and communicate this to clients. So rather than telling the employee what the brand is, use dialog and allow them to feel ownership towards the brand, to be part of defining the vision and action of the brand and interpret is as it applies to their job function. Employees should feel they "own" problems. If you have successfully branded internally, and truly conveyed the brand promise to your employees, they will naturally consider whether or not every action they take supports the brand or erodes it. When a company has an internal branding plan, the employees relate better to the brand, to the clients and each other.

So, just what is a brand champion or brand ambassador? It is a company's first and last line of defense when it comes to keeping customer loyalty.

Customers surveyed said the most important factors in staying loyal to a particular brand are 1) consistent quality, 2) customer service and 3) price. Other research shows that approx 70% of the purchase decision is made for emotional, not economic reasons, especially when buying services. Clearly, quality and service are key to a brand relationship with its customers and by its customers, I am referring to both rDVMs and pet owners. And quality and service can be controlled by your staff – your brand ambassadors. People form relationships with people and it is people who are the human, interactive face of a brand.

People buy brands for the brand experience and the employees of a company and their interactions with them are a big part of that experience – and a large part of what builds customer loyalty. Note that I said loyalty, not satisfaction. Satisfied customers change, and frequently for no other reason than to try something new. But loyal customers are much less likely to and loyal customers mean profits.

A 1996 article by Jeff Resnick stated that companies that excel at creating and maintaining loyal customers can command prices that are four to seven percent higher than weaker competitors and that they can generate three times more profit. Again, brand ambassadors are needed to make this happen. If the brand doesn't deliver on the promise, then the loyalty doesn't get created or worse yet, it gets eroded. In many ways a brand is a living breathing thing – much more than a visual – it's emotional and behavioral. Behavior reinforces the brand and employees have a great and important role to play here. When an employee and a company live the brand, they then can realize real gains and a competitive edge in the marketplace.

In order to 'live the brand' employees must truly understand the brand:

- What it is
- Why it is
- What it promises
- How it delivers
- Why you are better/different than any other brand in its space
- Why the consumers want it
- What their role is in relation to the brand

As a true ambassador, the employee needs to have a relationship with the brand that goes beyond customer service. They need to feel part of the company success or failure. You must make them understand how their actions impact the success of the company. Companies that succeed do this: Wal-Mart has its Associates; Disney has its Cast Members. They work hard at making their employees not only talk the talk but walk the walk. They make them vested in the brand and feel part of the team.

Is asking the employee to have a relationship with the brand any different than asking the clients? And knowing that people still prefer to deal with people, doesn't it make sense to have brand ambassadors who engage customers and foster relationships that lead to loyalty?

Having brand evangelists/ambassadors working for you instead of ordinary employees does 2 things:

1. Those who understand the brand are better able to deliver on brand work each day. The result is a more consistent brand message and positive experience for your customers and prospects.
2. Brand ambassadors build your brand both inside and outside your work environment. They talk to friends and associates about the brand.

Essentially, they are human billboards communicating your company message to those around them. And they do this for their own conviction and pride – not extra pay.

So how do you do this? How do employees learn to be brand ambassadors? By being shown? Not told. Brand ambassadorship must be demonstrated by the top all the way down the ranks. It isn't just a role for a few, it's for ALL employees. Ask yourself, are your employees aware of the part they play in creating a relationship and engendering loyalty with customers.

You may also see a training or human resources position in your practice as simply an administrative expense, a necessary evil. I would argue it's the key to revenue and building a successful, sustainable practice that operates to advance the mission and vision of the practice.

THE DOLLARS AND CENTS OF CUSTOMER SERVICE

Every staff member that encounters a customer can make or break the brand. When you provide consistently quality customer service you help connect the dots for your practice's brand. Meeting or exceeding your clients' expectations gives your practice credibility and helps build its reputation in your market. You have a chance of retaining your customers for the long run as well as keeping your employees engaged and committed. The lowered cost of acquiring new clients and recruiting, hiring and training new staff can be exceptional.

Another important reason to provide quality customer service answers the "What's in it for me?" question which many if not most staff will never articulate. Good customer service helps generate more revenue, which enables the practice to live its mission. The end result from this internal marketing perspective is: More revenue, which creates progress towards accomplishing the mission which begets growth and innovation which begets increased salaries, benefits, bonuses and/or promotions which begets happy employees, loyal clients.

It's really all about the internal environment or culture. Employee satisfaction, customer loyalty, and profitability are interconnected. Everyone plays a role in the never-ending chain of activity that transpires daily. When one link in the chain breaks or falters, it can and will affect the entire chain.

If you have documented customer service standards, it may be time to see if they're still meeting your needs. Are they written in manual form for everyone to have the same knowledge and same standards of accountability? If you are starting from the beginning, it's probably helpful to identify the current weaknesses before developing new customer service standards via a SWOT analysis.

Fixing the corporate culture takes transparency and communication. The more informed they are the less anxiety and greater sense of comfort they experience.

Rewards and Appreciation - Rewarding positive behavior shows employees they are valued and honored and contributes to your bottom line. Our need for approval is a common human emotion. And overwhelmingly, employees resign due to what they perceive as a lack of appreciation for their efforts or unhappiness in the workplace. A 2003 study out of the UK Chartered Institute for Personnel and Development found that it is relatively rare for people to leave jobs in which they are happy, even when offered higher pay elsewhere.

THE DANGERS OF NOT PAYING ATTENTION TO THE INSIDE

According to Gallup, there are three types of employees:

1. Engaged - They help move the organization forward.
2. Not Engaged - They put in the time but with no passion or energy – they are disconnected and it shows.
3. Actively Disengaged - Act out their unhappiness and undermine what their engaged teammates accomplish.

According to the US Employee Engagement Survey published in Gallop Management Journal, a startling 69% of workers are either not engaged or actively disengaged. Further research estimates that $370 billion is lost each year in lower productivity.

Internal Marketing matters. All that you do that touches your patients, clients and referral sources should be deliberate, planned and deliver the experience that is a reflection of your vision. The point here is that your staff does this – so make sure they are the right staff, the best trained staff and the people who you want to represent the very best of you. Marketing your practice begins with your vision. I hope through today's presentation you learned that your vision is carried out by others and that I was able to expand your vision and perhaps help you look at your staff and internal marketing activities in a new light.

The moral: Branding is like herding cats. You'll get them going in the same direction only if they're all chasing the same prey!

COMMUNICATION IN SPECIALTY CARE: CRAFTING MESSAGES FOR OUR AUDIENCES

Robin Brogdon, MA
Huntington Beach, CA

TEMPERAMENT – TAME OR FERAL?

A frequently overlooked aspect of communicating with referring vets and clients is carefully crafting what you say and how you say it. We will evaluate what messages are important for your respective audiences to hear, and determine if they are in alignment with who you really are. We'll also look at how your messages are being perceived – as tame or feral.

The goal with branding, which encompasses every method of communication, is to be seen as the solution for advanced or specialty care. We'll cover the most important aspects of what you need to do to *position* yourself in the marketplace and *communicate* effectively to the *right audience*(s). We cover all the basics, and then tackle the steps to achieve this. Once these are established, you will have the foundation in place to successfully market your practice and create a memorable brand by communicating relevant and meaningful messages.

The goal in communicating to your current and prospective clients is to create a framework for acquiring and nurturing referral relationships to fuel your practice.

We start at the beginning:

A) ASSESSMENT/DISCOVERY – Develop brand position

- Why did you create your practice?
- What did you plan to do differently than others in your area?
- What makes you special, unique or different?
- What's the mission of your practice?
- What core values have you established for your team to live by in order to carry out the mission?
- What's the best way to communicate this?
- How do you want rDVM's and pet owners to feel/think about your practice?

You need to know what you want to be in order to figure out how to be it.

B) BUILD YOUR BRAND

Leading companies do not subscribe to the common misconception that marketing is a discretionary expense. They know that it is as important as the inventory on the shelf.

Brand equity is how your clients recognize why you are different and better than the alternative. Brand equity is built on that client's direct experience with your practice. This experience, repeated over time, creates equity or value in your brand. And it serves as a kind of shorthand in the buyer's mind that separates you from everyone else. Some might call this reputation or goodwill, which CPA and valuation specialists will tell you is or can be a considerable portion of the total value of your practice. But it's common for you to think of marketing as an expense instead of as building an asset.

Brand equity is what creates loyalty. It is the quality that motivates your clients to recommend their friends or colleagues to you. Everyone wants brand equity, but building it can seem overwhelming, especially when you look at some iconic brands as role models like Coca Cola, Disney or Apple – hardly your peers and hardly comparable when it comes to your marketing budget.

Here are Five Steps to Building Brand Equity for the Small Business

1. Know who you are - Develop a brand identity that suits you today and tomorrow
2. Identify Your Target Audience(s) - Segmentation - Develop tactics and tools to reach out to each audience
3. Speak Their Language - Develop a succinct, meaningful, relevant message - Write customer-centric copy
4. Get the entire team on board – Sing the same song – Hire those that sing in harmony
5. Communicate the message creatively, consistently, professionally and frequently
   Plan, schedule, assign, execute, measure – repeat

C) COMMUNICATE

How to communicate and develop relationships with primary care veterinarians is not a secret, it's just something few businesses really invest in and most, therefore, never reach their potential. Cast as broad a net as possible by communicating with all of your audiences by using many different mediums.

STEPS IN THE PROCESS

- You've got 1 team - Make sure you've got the right team in place and that they are unified in delivering the message.
- It's got to be done 2gether - Everyone needs to believe that collaboration with the healthcare team benefits the patient and pet owner (rDVM and specialist).
- There are 3 parts to the equation – Acknowledge and respect all - rDVM, Pet Owner and Specialist.

Ultimately what you are trying to do is: Attract–Retain –Expand referral sources to grow your practice.

Turn Satisfied Clients Into Loyal Clients: ***When Times are Tough, Start Asking & Then Listen***

Perhaps you've noticed more coupons in your mailbox lately, or more door hangars appearing up on your doorstep at home. When times get tough, it seems natural to crank out the promotions. But in this world of media overload, unless there is something uniquely memorable or valuable about a particular form of outreach, you may not be spending your money in the most efficient manner.

When things get lean, it's typical for businesses to stop thinking clearly and start reacting instead of the more appropriate execution chronology of: plan, budget, schedule, execute, measure and repeat. Increasing the frequency and volume of your marketing efforts blindly is simply a desperate move, and sometimes the opposite of what you should to do improve revenues.

When your clients are under economic pressure, you risk making it worse for them by throwing everything you've got their way – which can lead to overwhelm and even confusion. So don't bombard them and risk losing them altogether. As an alternative, consider using the extra time you have to create stronger relationships. As a specialist, you've got two audiences to attend to. Here are some suggestions for turning that occasional client into one singing your praises.

Go Hunting – For information that is. Take your clients' pulse. Get their feedback. Offer incentives for them to give you answers to questions that will help you serve them better. This is powerful and if you are committed to actually doing something with the information you obtain, you will get results. The more your clients feel like you really care about them, and are sincerely working to ensure their satisfaction, the more loyal they will become, regardless of the economy.

Can We Talk? – Solicit real conversation with your clients. Don't just talk, find reasons to answer. That means you have to ask! And then listen.

When relationships are the key to prosperity, it's the quality of those relationships that equates to the success or failure of your practice. And if you are not going to make the effort during the slow periods to improve, then you probably aren't really serious about running a quality practice then either.

There has been so much hype today about 'Branding' that many have lost sight of what it really means and what it is. Branding is a long-term initiative that is predicated on building a relationship, based on trust, respect and consistency. Achieving success in building your brand is a process that takes time and patience. It takes dedication and persistence – and most importantly, a Brand Champion; someone who takes responsibility to make sure the 'brand promise' is always fulfilled.

8 Steps to a Winning Brand

1. Start with the Facts. Where have you been? Review the past history of the brand, heritage, values, beliefs, roots, etc.
2. Create Your Brand Vision/Mission Statement. Your brand vision identifies your practice's purpose for existing beyond solely that of creating profits.
3. Define Your Brand Personality. Personality helps your brand come alive. It makes your brand accessible and touchable. It helps differentiate you and gives depth and dimension to your enterprise(s).
4. Establish Your Brand Character. Brand character is all about the culture of the brand. It is the value system that directs every aspect of the enterprise(s), the principles, attitudes, and characteristics of the company.
5. Build the Relationship Between Your Brand and Your Customer. Map out the details.
6. Define Your Brand Image. There are two aspects of brand image - how you want to be seen, and how you are seen.
7. Decide How You Want Your Brand Positioned in Your Clients' Minds.
8. Deliver it in all you do. Being 100% consistent in delivering the brand experience is critical to its long-term success.

Branding is about emotional impact. Everything that represents your brand visually should resonate with your clients, sending the same primary message and triggering the same emotions. It's essential that your logo, website, marketing collateral, advertising and packaging work together to inspire your clients to act. Here are four tips for cleaning up, coordinating, refreshing and uncluttering your brand ID:

- Assess your brand promise.
- Evaluate your visual image.
- Ensure consistency.
- Update content.

Color Considerations **What's the Color of Your Brand?**

You are likely aware of the psychological impact of color. Color is one tool in our branding toolbox to help express brand attributes and create emotional connections with your constituencies. But color is not just for large corporations or products on the grocery-store shelf; it's an important aspect of your brand as well.

- Color supports differentiation

The ability to identify a company by its brand color is amazing. When you're standing at the rental-car bus stop at an airport, waiting for a shuttle van to pick you up, you know whether you're looking for a yellow (Hertz), red (Avis), or green (National) van. It's more difficult to identify Budget (orange and blue) and Alamo (yellow and blue) vans. Multiple colors seem to be harder to own. But that's not something that worried Google or eBay. Both companies have four-color logos, and it's their multicolored logos that set them apart from their competitors.

- Color can have a double effect

Some organizations and products have color in their names: JetBlue, Green Mountain Coffee Roasters, the Yellow Pages, and Blue Cross/Blue Shield. They all benefit doubly from the emotion-creating power of color.

- Color supports a renaissance

  Mars, one of the world's largest food companies, certainly knows the value of color. The company revitalized its M&M brand, which was at the end of its product life cycle, by holding a contest to find the next M&M color. It turned out to be blue, and Mars launched an extremely successful ad campaign that depicted the other M&Ms, who were jealous of the new blue M&M, holding their breath so that they, too, would turn blue.

- Color identifies your brand

  Color can be used to express your brand attributes, evoke emotion, and build that all-important connection with the people who surround your brand. Color is a key decision in the design process and one of the first priorities when working on a brand identity. The designer's job is to guide you towards a mix of colors that project your personality while being attractive to the target audience. Color is powerful. It's an important personal-branding tool, so use it wisely and consistently to support your branding activities. To make the most of your color, follow the five rules listed below.

Ensure the color you choose is...

1. Accurate
2. Relevant
3. Culturally correct
4. Applied consistently
5. Repetitive

Know who you are - Know what your audience wants and needs - Develop client-centric messaging - Deliver valuable information and engage in communication exchanges frequently and consistently. You will be seen as approachable, likeable, partnership material, trustworthy – essentially, a tame member of the healthcare team.

THE SPECIALTY PRACTICE IN THE COMMUNITY – HOW & WHERE TO BECOME INVOLVED

Robin Brogdon, MA
Huntington Beach, CA

ARE YOU A PURE BREED OR A MUTT? – WHERE DO YOU FIT INTO THE COMMUNITY?

If your vision includes more than providing specialized veterinary care, and you want to make a valuable contribution to society, then become an active member of your community. It's good for business and provides a great outlet for you and your team. We'll help you determine where to place your efforts and evaluate how you measure up through multiple sets of eyes.

What type of experiences are you delivering each day? And how do you measure your success? With your practice management software, you should be able to monitor such things as where your business is coming from, specific categories of business, etc, and it's important to do so. If you are not regularly evaluating reports about your referrals, it's analogous to getting financial statements every month from your accountant and not using the data to make business decisions. You need to use the information it gives you to improve your practice.

WHO'S OPINION MATTERS?

How do you measure it? Who is to judge? Is it the practice owners' version of success: If you are profitable? If the appointment slots are full each day? If you are paying all your personal expenses through the business and are still in the black? Whose opinion matters and what should you be measuring?

First you need to determine what constitutes success. There is no right or wrong answer – it's up to you to define – but you must, and then measure regularly exactly what you want to know at defined intervals. After measuring, it's what do you do with the information that is even more important.

Today we'll take a look at what success means from 5 different perspectives - Staff, Clients, Colleagues, Community, Financial Institutions. You'll not only get new ideas on how to evaluate your own practice, but also look at ways you can positively impact your community at the same time.

Before we move onto the 5 categories for evaluation, let's look at one of the most common ways to take a broad view of what's happening at the practice, a SWOT analysis. Not only is the analysis a good exercise, but carving out the time to think about your practice from many perspectives is valuable itself. Involving your team is essential. The purpose of a SWOT analysis is for the practice to get a sense of its *internal* strengths and weaknesses, and then determine the *external* opportunities and threats that might affect the operation of the practice.

There are two basic ways to evaluate strengths:

Internal audit – simply surveying the employees
External audit – this usually means doing some sort of rDVM/client survey

Strengths are important to identify because they can be the revenue drivers. Often they can be enhanced or showcased more prominently since you have the confidence in your ability to deliver based on these strengths.

Weaknesses are important to identify because you can't fix what is not acknowledged. You must identify the weaknesses, create an action plan to improve them, assign ownership, timelines and monitoring systems to ensure that progress is being made. If your weaknesses are exposed or grow larger, they can and will impact your bottom line, and quite possibly your reputation which will further affect your bottom line and future plans.

Opportunities and Threats - Here it becomes incumbent on the veterinarians and practice managers to monitor the external environment by keeping track of industry trends, attending professional meetings, and reading trade and business publications. The most effective managers continually ask themselves three questions:

1 - What changes are going on in the environment I need to know about?
2 - How do these changes affect me?
3 - What modifications do I need to make in my practice to account for these changes?

Opportunities are important to identify because these areas are where you can grow your practice. With the right tools and allotment of time and resources, these are revenue generators just waiting to happen.

Threats are important to identify because you want to prepare your practice for any and all scenarios to protect it.

STAFF

Measuring success from the Staff's perspective is one of the most enlightening avenues to evaluate your practice. The staff's happiness signifies many things and is the benchmark by which you could be very safe to evaluate your own practice – if you use no other method.

Job Satisfaction – Retention - Professional Development – Recruitment - Delivering the Brand Promise

- Engaged Employees = Job Satisfaction

- Job Satisfaction = Retention
- Professional Development = They care about me
- Recruitment is Easier = You become employer of choice

Your team is the unsung hero of all the marketing tools. Your people are the individuals that live out your vision most visibly. If they embrace it, believe it and live it, their satisfaction alone creates tremendous momentum and positive value for everyone you touch. Focusing on them, as your greatest asset, makes sense both ethically and financially.

CLIENTS

They are the revenue drivers – the referral sources – and the means to expand your business

- Educating clients = Increases your value as a veterinarian
- Good revenue drivers = The difference in profitability or treading water or worse
- Satisfied clients = Retention of clients and can become a referral source
- Clients who are connected to you are more compliant = Will buy recommended services
- The trusted resource for their pet = Your value is elevated making your fees palatable

If clients are happy and become a loyal source of business, they will gladly recommend your services.

COLLEAGUES

Reputation – Collegiality – Recruitment - Increased value of veterinary medicine - Alumni programs

You may not think to evaluate your practice from the perspective of your colleagues, because after all, many of them are your competitors. However, isn't it important to have a good reputation, be a positive, contributing member of your profession and know that your actions help improve the value of veterinarians which of course translates into more pets receiving the proper care. Do your colleagues find you to be collegial – open to collaboration and camaraderie. Do you attend and participate in your local and regional meetings. Do you volunteer your time to serve their boards and committees. Do you stay in touch with your alma mater and practices where you may have done an internship? Do you network at conferences and exchange information with new colleagues to build a community of resources. All of these things increase your value and the value of veterinary medicine.

COMMUNITY

Education – Sponsorships – Intern/Externships - Write a column

Evaluating your practice in the eyes of the community may also not be something you are accustomed to doing. After all, if you have a lot of clients and are making a good living, you may assume your community must like you. But it goes beyond your clients. Most of you went in to veterinary medicine with some altruistic thoughts, as well as being fascinated by the science of animal medicine and wanting to care for them. And most of you have reached a certain level of success that you are grateful for. So it's more than that. What kind of community member are you? And are you making sure that what you contribute gets noticed? And are you a good neighbor?

Partner for Productivity – Surely there are members of your community that have common interests, goals and audiences. Partner with a related business to do double duty for half the hassle. Consider co-hosting an event where you demonstrate your expertise. It can be fun, free and is a great investment in client relations.

Community + Karma can be Contagious

Everyone is much more socially aware these days. And most people really do want to give back. Talk to your team and decide where you want to contribute to your community. You'll be amazed at what it will do for your business profile as well. Your staff will respect you.

Tending to the community means you are tending to your clients too – aren't they part of the community? Ask your clients to participate with donations, time, or better yet, ideas. Not only will you feel great, but you'll be gaining great exposure and adding to the strength of your brand.

When relationships are the key to prosperity, it's the quality of those relationships that equate to the success or failure of your practice. If you are not going to make the effort during the slow periods to do a better, then you probably aren't really serious about running a quality practice then either.

FINANCIAL INSTITUTIONS

Buy/Sell - Practice valuations – Partnerships - Access to Capital

If you work with one of the financial services firms in our industry, they will tell you that it is always important to look at and manage your financials from the perspective of a lending institution. Life happens and you always want to be in a position to leverage capital, be deemed a valuable asset worthy of acquisition or have a favorable outlook to take on a new partner.

You build your businesses and reputation with the same tools you use to measure your success.

Staff – Clients – Colleagues – Community – Financial Institutions

Building a new facility and hanging out your shingle on a busy corner is not enough to accomplish success in the eyes of these 5 sectors. Without focusing on each sector, your business will not be sustainable. You may stay in business, but I'll bet you have higher turnover in clientele than you'd like, a revolving door of employees, a hit or miss reputation in the community, not be widely known and likely not look very valuable in the eyes of a lending institution.

SPECIALTY PRACTICE FINANCIAL HEALTH: WHAT METRICS SHOULD YOU EVALUATE

Karen E. Felsted, CPA, MS, DVM, CVPM
Schaumburg, IL

INTRODUCTION

Before discussing what data should be analyzed, it is first important to discuss why benchmarking is important. The term "benchmarking" is used loosely and some feel that other terms would be better. Regardless of what the activity is called, however, it is critical that it be performed. Review and analysis of critical financial and operational data should be performed on a regular basis. Analysis of data trends within the hospital as well as metric comparison to outside sources should both be done.

There is an old cliché that says "If you can't measure it, you can't manage it." While it is easy to ignore clichés because you've heard them so many times, these phrases are generally clichés for a reason—they are widely applicable truths that should not be ignored. Obviously these specific words have to be taken with a grain of salt—it's not necessary to count the paperclips on a daily basis in order to manage the purchase of office supplies but managing most areas of the hospital well means you need to measure the activities.

Improved management of the hospital is critical to better patient care and greater financial success. Quality patient care requires a sound economic foundation. A practice that is not financially successful will find it difficult to invest in the facilities, equipment, medications and staff necessary to provide the best patient care much less provide for the financial security of the practice, its owners and the staff. Results from the AVMA-Pfizer Business Practices Study released in 2006 clearly demonstrated the connection between the measurement of data and those hospitals that are financially successful. In a nutshell, the study showed that the majority of practice owners don't use financial concepts to manage their businesses and those that do make more money. For example, only 38% of the survey respondents indicated they used financial concepts to manage their business. The number of respondents who performed a monthly review of financial data ranged from 22% to 77% depending on the individual items analyzed. And yet the study showed that there was almost a $50,000 difference in the average earnings of veterinarians who used financial concepts to run their businesses versus those who did not.

KEY AREAS IN WHICH BENCHMARKS SHOULD BE REVIEWED

Listed below are the critical areas in which data review and analysis should be done on a regular basis. Within each key area are some data points that should be reviewed by all practices every month.

- Revenue
  - Revenue by service (internal medicine, surgery, etc)
  - Revenue by FTE doctor and by individual doctor
  - Revenue drivers—transactions and ATC
  - Discounts and missed charges
- Expenses—should be done by service
  - Doctor labor costs
  - Support staff labor costs
  - Costs of goods sold
  - Facility costs
- Profitability—should be done by service
- Clients
  - New clients
  - # of RDVMs—who are they
- Fees
- Facility
  - Revenue and profits per sf
  - Revenue and profits per exam room
  - Revenue and profits per surgery or special use room
- A/R
  - Total
  - Aging
- Marketing activities
- Management staff, hours and activities
- Human resources
  - # staff on a FTE basis and compared to FTE doctors
  - Staff and doctor hours
  - Overtime
  - Staff efficiency
  - Turnover
  - Compensation and benefits

It will often be necessary for the practice to gather additional information beyond that discussed here. The KPIs listed in this paper are the starting points. When a trend is discovered or an area that appears problematic, further analysis will be necessary. For example, let's say that the practice owner or manger is analyzing the revenue of the practice at the doctor level and finds that the average revenue per doctor is lower than that seen in most practices and that there is a great deal of variation in productivity amongst doctors. Some of the information you may want to gather to help determine how to address this situation includes the following:

- Number of hours worked each week by the doctors—revenue variability may be a function simply of the time spent in the practice
- Support staff help utilized by each doctor—some doctors may be able to produce more because they have access to and use more support staff
- Number of key procedures performed by each doctor in relation to the number of transactions they generate—revenue may vary because of different approaches to cases which should be more consistent

SOURCE OF DATA

Where does this data come from? Your practice most likely has two sets of software used in the operational and accounting functions of the practice. The first is the veterinary practice management system which helps manage and analyze revenue collection, inventory, receivables and other revenue related information. The second system is an accounting package such as QuickBooks or Peachtree. The accounting software is used to manage the payment of bills and payroll, prepare financial statements and aid the accountant in the preparation of the practice's tax return.

"Key performance indicator" or "KPI" is a term often used for the financial or non-financial practice operating statistics or metrics analyzed: KPIs:

- Provide information about the operations of the practice not normally available in traditional financial statements
- Help explain changes seen in the financial statements
- Measure performance of the practice
- Indicate areas in which improvement is necessary

GUIDELINES FOR EFFECTIVE FINANCIAL ANALYSIS

Standardization of revenue and expense categories

In order to compare your practice to others, it is necessary to set up your revenue and expense categories in a fashion similar to those commonly used in veterinary medicine and used in the published studies you will be comparing the practice to. In general, revenue should be classified by specialty service—internal medicine, surgery, dermatology, etc. General practice or emergency clinic revenue should be kept separately. The expense accounts in the AAHA chart of accounts are an excellent starting point for expense categories. All practices are a little different from one another and ultimately you will need to include all of the service codes and categories that best reflect how you operate; however, you will have a greater ability to compare yourself to others if your setup is similar to above and you will not lose any flexibility in internal analysis be staying consistent with the industry standards.

Consistency in the recording of information

It is important to record information in the same categories or perform calculations in the same way each time you do them in order for your numbers to be comparable over time. For example, if you started out by classifying the anesthesia purchased in a "Surgery Expense" category and then later switched it to an "Anesthesia Expense" category, these numbers will not be comparable before and after the change and your trends analysis and related conclusions may be inaccurate. In order to achieve this comparability, it is important to carefully set up your categories and define your calculations when you set up your system and, when it is necessary to add a service code to be careful that you put it in the right place. Sometimes it will be necessary for you to make a big change in order to improve your analysis. This is acceptable; just remember that in the month of the change, you will lose comparability. You will regain this comparability as time goes on and you add more data.

Common basis for comparison

Comparison of raw numbers is useful to a point in financial analysis but, in general, it is necessary to do some kind of ratio analysis or use a common basis of comparison to get the best results.

Ratios can be calculated from information available in the balance sheet, profit and loss statement or from other reports available in your practice management system. Ratios are more useful in analysis and planning compared to actual numbers for several reasons:

- Relationships between various pieces of information are often more clearly apparent
- Trends over months or years can be more easily analyzed as the practice grows
- Businesses of different sizes can be more easily compared

The most common ratio used in veterinary financial analysis is that of expressing types of revenue or expenses as a percentage of total gross revenue. For example, support staff expense may be $200,000 in the current year. If that is compared to support staff expense in the prior year during which the practice also incurred $200,000 you might think the practice was continuing to use staff as efficiently as previously. However, if the total revenue declined from last year to this year, the support staff expense as a percentage of total revenue actually increased. This means the practice isn't managing its staff costs as well as before. This is also important when comparing your practice to others. Let's say you compare your $200,000 of support staff costs to a study reporting this information for other referral practices, and that study told you that the average support staff expense in a practice was $240,000 per year. Did this tell you anything useful? Not really because you aren't necessarily comparing yourself to a practice of similar size. However, if you calculate your support staff expense as a percentage of total gross revenue (say 24%) and compare that to outside metrics that tell you that the average practice has support staff expense of 22% of gross revenue, then you know that you are doing less well than the average practice in this area. This kind of ratio analysis is critical to accurate comparison over time or between practices of different sizes.

Another frequently used common basis for comparison is the calculation of doctors on a full-time-equivalent basis. This makes it possible for you to compare your numbers to practices that have different numbers of doctors or to compare the figures in your own practice from year to year as you add doctors. A single FTE doctor is one who works 40-45 hours per week, 48+ weeks per year. If you have a doctor who works 30 hours/week 50 weeks per year, this is 0.75 of a FTE doctor. If you have a doctor who works 40 hours per week but started July 1, 2007 then for 2007, this will be a 0.50 FTE doctor.

Comparison to industry studies

It is difficult to find sources of publically available industry metrics for referral practices. Previous studies by AAHA and CareCredit yielded some information but a comprehensive set of metrics doesn't exist outside of private study groups. The National Commission on Veterinary Economic Issues (www.ncvei.org) recently created a series of benchmarking tools for referral practices (specialty and emergency) in order to help collect this data. Data being collected includes:

- Practice Productivity: Revenue, transactions and ATC
- Doctor Compensation and Benefits
- Staff Efficiency: # of Staff, Expenses, Turnover
- Practice Expenses
- Using Your Space Efficiently: revenue per sf, surgery suite, exam room
- Where Do Your Clients Come From?: sources of clients, #s of RDVMs
- Accounts Receivable and Client Payment Methods
- Marketing Your Practice
- Management Time and Tactics
- Doctor Production, Salary and Benefits
- Profitability Estimator
- Fees by Specialty

Key factors to remember when comparing your practice to published data:

- Is the data meant to represent an average practice or "best practices"? Different studies have different purposes.
- Is the data reliable? Reviewing the methodology section of the reports will give you a feel for how many practices responded and how comparative the data is. It is also important to understand how certain calculations were performed in order to know if that metric is comparable to your practice. As data is cut into ever smaller units (for example, by region or state), sample sizes get smaller. While smaller sample sizes may not be perfectly representative of the profession, you can still gain some useful information from them as long as you are careful about the reliance you place on the information.
- No study will be perfectly comparable to your practice. For example, you may be comparing 2008 data to a study with 2006 data. You may include different items in the measure of personal doctor revenue production than the study does. Or there may be a small number of respondents to a particular question in a study that you are interested in. This doesn't mean the study is useless to you. You can still get very valuable information to help you operate your business more effectively. It simply means you need to take these comparisons with a grain of salt and use them as one tool in running your business, not as the final word about how well your practice is doing.
- Just because you are different from the comparative practices doesn't mean there is something wrong with your practice. All practices are unique in one way or another and, occasionally, the published benchmarks may not be the standards you want to shoot for. However, if your numbers are significantly different, it is strongly recommended that you investigate why and determine if you have a financial or operational problem that needs to be resolved. In most cases, if your numbers are significantly different from the various studies, you have a problem.
- Comparison to industry averages can be very useful in identifying potential problems in the practice, but comparison of your own data from the current period to previous periods is better for identifying less obvious changes and monitoring corrective action.

SPECIALTY PRACTICE FINANCIAL HEALTH: WHAT TO DO IF THE NUMBERS AREN'T HEALTHY

Karen E. Felsted, CPA, MS, DVM, CVPM
Schaumburg, IL

Benchmarking, or the practice of comparing your practice's numbers to previous periods or to outside comparative metrics was discussed in the proceedings for the previous session. However, identifying trends or problem areas is not enough. If your profits aren't what you want them to be, what can you do about it? A lack of profitability either comes from revenues that are too low, expenses that are too high, or a combination of the two.

Common revenue problems in a referral practice include:

- Inconsistent fee increases
- Excessive discounting
- Missed charges
- Inefficient use of doctors and staff (low production)
  - Minimum wage employees
  - Too many part-time employees
  - Poor hiring and training
  - Little correlation between compensation and production
- Inefficient use of building
- Poor client service
- Adequate scheduling and doctor availability
- Low numbers of clients
- Poor communication of value to pet owners and RDVMs

Common expense problems include:

- Poor inventory control
- Doctor and staff compensation and benefits
- Expensive facilities
- Expensive equipment
- Poor management systems
- Upward creep in all expenses

At first you may not know how to interpret these reports or trends. Sources of information include the following:

- If you used the NCVEI benchmarking website, there is a "Recommended Treatment" section for each benchmarking tool which gives you some basic information about improvements you should consider in your practice. The Resource Library has additional studies, articles and other information.
- AAHA
- Dvm360.com
- VetPartners (formerly the Association of Veterinary Practice Management Consultants and Advisors (www.vetpartners.org)

One of the most important areas for a practice to focus on to improve profitability is staff utilization. Practices that want to become more profitable can no longer just count on fee increases or demand for services to achieve this goal. Improved profitability must come from improved business practices. In addition to contributing to profitability, employees who are allowed to learn and grow and use more skills are generally happier in their jobs and more likely to stay with the practice than those who are only allowed to do less interesting tasks. Effective leveraging of employees is also critical to the productivity of the veterinarians and the practice as a whole. Veterinarians who delegate duties to appropriate staff members are able to see more clients and generate more gross revenue and profits. This increased profitability is essential to providing good quality medicine and surgery and to continual investment in team members in the form of increased salaries and benefits and increased continuing education.

In order to achieve optimal staff utilization, the practice must:

- Have an owner or practice manager with a strong set of financial, managerial and HR skills—the title isn't enough
- Have a detailed, understanding of the skills staff members must have to provide outstanding patient care and client service
- Hire effectively—find and keep the employees with the right skills and attitudes necessary to achieve the practice's goals
- Provide high quality, effective and ongoing training programs to both new team members as well as those who have been in the practice for awhile
- Have high levels of employee retention—revolving door employees aren't around long enough to be efficient and effective
- Design and implement efficient policies, procedures and systems for getting things done

- Schedule staff, appointments, and surgeries effectively and in synch with each other
- Delegate effectively—tasks should be done by the lowest level person who can do the job properly
- Monitor staff activities frequently—in most practices staff are always busy doing something—what they are doing, however, is the key point—is it the most important activity that should be done?
- Regularly review staff utilization metrics

Financially successful practices are, almost by definition, well-managed practices. It used to be possible to have a successful practice without a great deal of business skills, but this is no longer true. Clients have higher expectations in both the medical and client service arenas, veterinarians face increased competition, and the complexity of business regulation has increased.

The first item in the above list is perhaps the most important. Most people aren't born with a full set of good management skills. Just as it took training and practice to become a board certified specialist, it takes similar dedication to become a skilled manager. One difficulty in veterinary medicine is that most veterinarians do not graduate from veterinary school with a solid grounding in business skills. Veterinary school curriculums are already strained by the burgeoning amount of veterinary knowledge that needs to be passed on to students and most students didn't take business courses in their undergraduate years. Another difficulty is that many veterinarians aren't inherently interested in business management. They went to veterinary school to learn to practice medicine, not to be a business manager. Many practices have hired practice managers to take on the business side of the practice. However, not all managers have been given the training and resources they need to do their jobs well. Fortunately, these are correctible problems.

Management duties are handled differently in different practices depending on the size of the practice, the interests of the owner veterinarians and the money available to hire managerial help. In some practices, the owner veterinarians do all of the management work either out of necessity or because they enjoy it. In other practices, the owners have hired a full-time, experienced practice manager with the result that the veterinarians primarily practice medicine. The owners make the high-level decisions, but leave day-to-day management to those paid and trained to do it. In other practices, management duties are divided between the owners and other individuals, including support staff, practice managers or associate veterinarians. As practices get bigger and as management duties become more complex, it becomes more necessary to have highly skilled, professional managers in place. Veterinarians will contribute more to the profitability of the clinic through the practice of medicine and surgery than by being managers.

While owners certainly don't have to be involved in the day-to-day management activities of a practice, it is critical that they provide vision and leadership to the business. In addition, they need to set the framework for the decisions to be made, direct and approve the overall activities, and support the management personnel in their responsibilities.

One problematic area that affects referral practices more than general practices involves the investment in equipment and facilities. What practices don't do when making the decisions to invest in equipment, staff or facilities is to make sure that the costs will lead to increased levels of revenue and profits.

For example, how much space and what kind of building is really necessary to practice veterinary medicine? Operating out of the Taj Mahal can be very psychically rewarding but may not be good financially. For example, if a practice moves into a beautiful, new facility and the rent doubles, will there be a sufficient increase in revenue (and more importantly profits) to cover this rent increase?

Lack of attention to discounts and missed charges can also lead to declining revenue. Even a small amount of products or services given away by well-meaning doctors or other team members can significantly decrease revenue and profitability. Missed charges, those not deliberately given away, can also dramatically reduce the profit margin. Capturing charges is generally about having good systems in place and is essential to efficient operations. It is a rare practice who doesn't experience these problems on a fairly regular basis.

Understanding not only the profitability of the practice but the kinds of factors that lead to this state is critical. Until the practice has an idea of the root causes of the problem, it is difficult to determine what the correct solution is. Working with a financial advisor or practice consultant may help in not only gaining a greater understanding of the issues impacting profitability but in identifying and implementing solutions.

HOW TO MEASURE & POSSIBLY REVIVE PRACTICE PROFITABILITY?

Karen E. Felsted, CPA, MS, DVM, CVPM
Schaumburg, IL

The Brakke Management and Behavior Study empirically demonstrated the importance to practice owners of understanding the finances of their practices and how few owners really do. Bottom line: The majority of practice owners don't understand financial terms and those that do, make more money. Only half of the group understood "pre-tax profits" and "cash flow." Only 10-20% of the respondents could choose the correct definition of the other terms in a multiple choice format. And it makes a big difference in earnings--male owners who answered three or more questions right had personal incomes of 7% greater than those who didn't—female owners who answered three or more questions right had incomes that were 19% higher than those who didn't. Participants in the Brakke study were not exclusively specialists but no evidence exists, either anecdotally or from formal studies, to indicate that these issues don't equally apply to specialists and general practice veterinarians.

In addition to the obvious impact on current cash flow, profitability also is a critical determinant of practice value. And while specialty practices are able to charge more for their services, their costs are also much higher and maintaining a high level of profits is arguably more difficult than in a general practice. In addition to the profitability issue, specialty practice owners also face other unique issues in succession planning including a limited buyer pool, difficulty in obtaining financing for the sale of very large practices and a lack of management skills in the potential buyers that will insure their success as owners. Profitability, at least, is an issue over which practice owners have control—if you don't like the financial performance of your practice, you can change it.

In the last few years, it has become apparent that the number of general practices with no or little value has been increasing—to the point where the Veterinary Valuation Resource Council of VetPartners™ (formerly the Association of Veterinary Practice Management Consultants and Advisors) coined the term "No-Lo Practice℠" to describe these practices. More and more practices, when appraised, did not have the value that would normally have been expected. And, in almost all cases, the owners of these practices were totally unaware of the problem. This is also true with specialty practices. On the surface, many of these practices would appear to be doing very well. They are located in attractive facilities, practice very advanced levels of medicine, have all the latest equipment and a large support staff, offer comparatively high compensation and benefits to their employees and charge well for what they do. However, practice value is largely based on profits and the very factors that make these practices look attractive on the surface are those that are reducing profitability.

Understanding the profitability of a practice is one of the most important concepts necessary to manage a veterinary hospital well. Profitability is the one single number which tells you how you are doing financially. Calculating the true operating profits of a practice is not a simple task. None of the standard financial or management reports a practice usually gets show this figure. Neither the taxable income from the tax return nor the net income from the profit and loss statement represents true profitability. This doesn't mean those reports are improperly prepared; it simply means the reports required by the IRS or accounting standards for small businesses weren't designed to determine profitability. No one report will give a practice all of the financial information it needs to make intelligent operating decisions; unfortunately, the report that seems to be prepared least often is the one that calculates true practice profitability. Because practice owners and managers aren't used to getting this kind of information, they generally don't know what the true profitability of their practice is. The first time many owners realize their true profitability is when their appraiser talks to them about it.

Operating profit is the difference between the operating revenues and expenses of a practice. Operating revenue and expenses include only items normally and necessarily seen in the day to day operations of the practice such as fees for professional services and drugs and medical supplies expense. These items should be stated at fair market value rates. For ease of comparison with other practices, the profit margin is generally stated as a percentage—this is calculated as practice profits divided by gross revenue. Some of the items that must be calculated differently to determine operating profit versus taxable income or net income include: practice owner payments, facility and equipment rent if these items are owned by the practice owner and leased to the practice, services provided by family members to the practice, depreciation, interest on debt and perks.

Owner compensation is one of the most significant adjustments and almost always has to be calculated differently in determining operating profits than would be done for the tax return or other reports. Owners often arbitrarily determine an amount they will be paid through their payroll system; this amount often has no correlation to the actual medical, surgical and management work the owner does in the practice and therefore the tax return or income statement looks as if the practice is more or less profitable than it really is. IRS regulations also dictate how some aspects of owner payments must be handled and these regulations vary by entity type. For example, owner compensation must be reported differently for a C corporation than for a partnership. A practice may appear to be more or less profitable than it really is simply because of these regulations.

So how do you calculate operating profit? Net income per the tax return is the starting point. Various adjustments are made from there.

- Add back: depreciation, amortization and interest on debt
- Deduct the estimated average amount spent on equipment per year—purchasing equipment is a true operating expense of the practice but depreciation as determined by tax law is not the best estimate
- Determine how much the owner was paid during the year and what it was comprised of (salary, rent, etc)
- Adjust owner compensation to represent a fair compensation for medical/surgical work—compensation can be estimated either via a percentage of production or as a salary comparable to what other similar specialists in your practice make
- Adjust owner compensation for management work—management expense generally averages 3-4% of gross revenues—

if you have a practice or office manager, the owner should get less than 3% of revenues as management compensation

- Adjust rent expense to fair market value if paid to the owner at a rate greater or less than fair market value
- Adjust equipment lease expense to fair market value if paid to owner at a rate greater or less than fair market value
- Determine the $ amount of personal perks paid by the practice and remove this expense—perks would be items not necessary to the operation of the practice but paid by the practice generally to gain a tax advantage (examples include excess meals and entertainment, excess auto costs, swimming pool payments, personal furniture, trips to Tahiti, etc)
- Deduct the cost associated with free services provided to the practice—family members may provide bookkeeping or other services to the practice at no charge—if the practice had to hire someone to do this work, there would be a cost involved and this should be included as an expense
- Remove any true non-recurring income or expenses such as one-time insurance proceeds or expenses related to a natural disaster
- Recalculate net income
- Divide the new net income by gross revenue

The above may sound a little daunting but there are resources available to help you. The NCVEI and VetPartners™ have joined forces to make an electronic version of a referral practice profitability estimator available on the NCVEI website (www.ncvei.org.)

If your profits aren't what you want them to be, what can you do about it? A lack of profitability either comes from revenues that are too low, expenses that are too high, or a combination of the two. Expense management is often the easiest to understand so it will be discussed first.

What practices don't do when making the decisions to invest in equipment, staff or facilities is to make sure that the costs will lead to increased levels of revenue and thus profits.

For example, how much space and what kind of building is really necessary to practice veterinary medicine? Operating out of the Taj Mahal can be very psychically rewarding but may not be good financially. For example, if a practice moves into a beautiful, new facility and the rent doubles, will there be a sufficient increase in revenue (and more importantly profits) to cover this rent increase? This is a particularly important issue for specialty practices to understand and manage due to the space requirements of some of the equipment used and the services offered.

The same goes for the addition of staff. A doctor's work life may be much easier and personally rewarding with 3 techs trailing behind him or her during the day, but does this doctor actually produce more revenue with this additional support staff? If not, the cost of the staff is eating into the profitability of the practice. Other staff problems seen in practices include the hiring of low-level, minimum wage staff that can't do the job properly, too many part-time employees, too much overtime, and a lack of training and supervision. All of these lead to inefficiencies and excess costs in getting the job done.

Declining revenue or a lack in growth of revenue is the other factor contributing to a lack of profitability. Practices often don't focus specifically on growing revenue because it's harder for them to determine exactly what to do to increase revenues. In many cases they've also been fortunate in that the revenue has just seemed to be there—this has been especially true with the high demand for specialty services.

Fees should be reviewed regularly even in these difficult economic times. It's not uncommon to see a practice that hasn't increased fees in two years or has only increased a few of them by a small percentage. Most expenses in a practice rise annually because the providers of those goods and services raise their prices—this is true of staff costs, drugs and medical supplies and the various other goods and services used by a typical practice. If the practice isn't raising its fees at least 5% per year, profitability will suffer even if nothing else changes in the practice.

Lack of attention to discounts and missed charges can also lead to declining revenue. Even a small amount of products or services given away by well-meaning doctors or other team members can significantly decrease revenue and profitability. Missed charges, those not deliberately given away, can also dramatically reduce the profit margin. Capturing charges is generally about having good systems in place and is essential to efficient operations. It is a rare practice who doesn't experience these problems on a fairly regular basis.

Understanding not only the profitability of the practice but the kinds of factors that lead to this state is critical. Until the practice has an idea of the root causes of the problem, it is difficult to determine what the correct solution is. Working with a financial advisor or practice consultant may help in not only gaining a greater understanding of the issues impacting profitability but in identifying and implementing solutions.

SUCCESSION PLANNING: ASSOCIATE BUY-INS TO A SPECIALTY PRACTICE

Karen E. Felsted, CPA, MS, DVM, CVPM
Schaumburg, IL

Bringing a new associate into practice ownership has even more challenges to it than adopting a new puppy or kitten. (They can't be locked in a room if they bark too much!) And yet, succession planning is critical to the life and health of the practice. When practice owners and associates start to talk about a partnership, they tend to focus on the specific tasks that need to be completed in order for the buy-in to occur. For example—how much is the practice worth? Who is going to draw up the legal documents? How will each person be compensated in the new ownership structure? No question these are important issues; however, it is first critical to determine if the seller and potential buyer are a good match and will work together well as a team. Partnerships generally don't fail because the practice value wasn't correctly established or the legal documents were improperly drawn up. They fail because for the same reasons marriages fail—the individuals don't have the same long-term vision for the partnership, they can't communicate well, and they don't have a mechanism for solving differences.

Understanding each other's philosophies and goals will help the potential partners decide if they are compatible enough to allow for a strong long term relationship. While it is certainly possible that opinions can change down the road, a partnership has a greater likelihood of succeeding if the potential partners are initially in accord about how critical issues will be handled by the practice in the future.

The first task in this analysis process is for the practice owner(s) to think through why they want to sell all or a part of their practice. Common reasons include:

- Ready to fully retire
- Want to decrease workload
- Desire to change careers
- Desire to practice more medicine
- Health issues
- Divorce
- Location/demographics
- Management frustrations
- Financial issues
- Want to bond associate to practice

And both the buyer and seller need to spend time understanding what each other wants from the transaction and how they do or would deal with certain critical situations. Topics for discussion include:

- Vision for the future
    - Size of practice
    - Range of specialists
    - # of locations
    - # of owners
- Decision making
    - % ownership
    - Types of decisions that require all owners to agree to, majority, 75%, etc
    - Sharing of duties
    - How will day to day decisions get made
- Ethics
    - Compliance with laws and regulations
    - Compliance with tax code
    - How clients are treated
    - How staff are treated
    - General honesty
- Risk tolerance
- Pricing
    - General level
    - Discounts
    - Pro bono services
- Personal goals
    - Money
    - Work hours/time off
    - Retirement
    - Outside participation in professional activities
- Spouse/family involvement in practice

- Communication
  - What info should be shared between partners
  - Frequency of communication
- Expectations of each other
  - Time
  - Money
  - Mentoring
  - Other
- Quality of medicine
- Operational changes
  - HR
  - Client service
  - Facilities
  - Other
- How will disagreements be handled

Selling part of the practice to an associate is one of the most common sales transactions and one of the easiest in some ways if you have the right practice owner and associate. It offers the greatest opportunity to know if the buyer is the right fit and has the right skills to be an owner. Involving associates in the "business" side of the practice and giving them the appropriate compensation for their ownership can be an excellent way of retaining them long term. Some important points about this kind of transaction include:

- Some financial investment from the purchaser is good, but don't count on it—associates often have limited financial resources
- Must be willing to provide full disclosure, honesty, and allow participation in management and financial decisions
- Associate often has perception of buying something they helped create—sometimes see discounted sales prices as something they should receive
- Usually lock in timeframe for rest of practice purchase as well as real estate purchase at time of initial transaction
- Usually best option for C corps because is almost always a stock sale
- Seller generally carries note because doesn't want to give lender lien on 100% of practice assets
- Need to do all the same documents you would do if you didn't know the buyer—don't get lax because it's a friendly deal—these can go sour

You will need advisors to help you; listed here are some of the common ones: practice appraiser, practice consultant, CPAs, attorneys. Each side needs their own CPA and attorney but a good quality appraiser should be able to prepare the valuation on a "non-advocacy" basis

A valuation of the practice is essential. A feasibility analysis done as if from the buyer's perspective will help you understand whether your practice will be attractive to prospective buyers or whether your price is too high or (occasionally) too low.

Valuing a business properly takes skill, experience, knowledge and training. In order to determine a fair value, a competent appraiser is needed—one trained in accepted valuation techniques and experienced in valuing veterinary practices. Determining a FAIR value (versus one that favors either a buyer or seller) is important for several reasons:

- A value that is too high decreases the chances of finding a buyer for the practice and, should it be sold, increases the likelihood the buyer may be unable to meet the debt obligations resulting in business failure and the seller being forced to "take back" what is left of the practice.
- A value that is too low fails to reward the seller for efforts put into building the practice.
- An inaccurate value diminishes the chance of realistic personal or business planning.

Many people currently perform appraisals of veterinary practices including licensed or certified appraisers, accountants, brokers and consultants. Appraisers use varying methodologies to determine practice value. Because of the significant variations seen amongst veterinary practice appraisals, it is critical that the user of an evaluation understand what they are getting. The areas a valuation user should know something about are:

- Method(s) used to value a practice
  - Rules of thumb
  - Income based methods
- Kind and amount of adjustments made to a practice's financial statements
- Capitalization rate determination
  - Problems with standard capitalization rates
  - Risk factors
- Feasibility analysis
- Using qualified consultants
  - Qualifications of the appraiser

Seek a practice appraiser who is familiar with the veterinary profession. An appraiser who specializes in veterinary practices will be able to help you use a valuation report in order to improve your fiscal health, and to decrease your hospital's risk as an investment for future buyers.

As a part of the valuation process, you will also need to determine exactly what will be included in the sale and how will the sale be structured. Is this a stock or an asset sale? Are you also selling the real estate where the clinic is located? There are two general types of sale:

Corporate stock sale

- Stock represents an interest in the assets + liabilities of the business, but the buyer is buying a % of the business as a whole, not in the individual assets and liabilities
- Buyer becomes responsible for known and unknown past liabilities of organization

Asset sale

- Buyer purchases individual assets—tangible or intangible
- Buyer DOES NOT purchase liabilities

In general, sellers want to sell stock because of better tax advantages and buyers want to buy assets, both because of better tax advantages and the fact that they don't inherit the liabilities of the old business. From a practical perspective, however, if the sale is a partial sale to an associate, it will be a stock sale. Almost all sales of 100% of a practice will be asset sales.

Obtaining the real estate associated with the practice is a critical issue for most buyers that must be resolved at the time of purchasing the practice. When and at what price will they get to buy it? If you're going to lease the real estate for awhile, you must deal with these issues:

- FMV rent
- Term & renewal options
- Who pays for taxes, maintenance, insurance, major repairs

Heed the advice of your experts when setting a price for your practice. Be realistic in your approach and try to look at your practice from a buyer's point of view. You will also need to consider what financing options you are willing to accept from a buyer. Are you willing to assist with the financing? In general there is good financing available from outside lenders if the practice has a strong cash flow and the buyer is a good credit risk. However, if you are selling just part of the practice, you may want to finance the sale instead of accepting a lien on your practice assets. In general you want to finance as little as possible in order to avoid the risk of non-payment.

An attorney can advise you on the preparation of legal documents, from the pre-purchase agreement through the sales contract. Do not skimp on legal services. Common documents include: prepurchase confidentiality agreement, letter of intent, letter of offer, memorandum of understanding, the purchase agreement, and a buy-sell agreement.

CELL BIOLOGY OF DEGENERATIVE (MYXOMATOUS) MITRAL VALVE DISEASE: IS THIS DISEASE INEVITABLE?

Christopher Orton, DVM, PhD, DACVS
Fort Collins, CO

Valvular heart disease in humans accounts for 20,000 deaths and over 90,000 hospitalizations at a cost of over $9 billion annually. Degenerative (age-related) disease of the mitral and aortic valve accounts for approximately 75% of valvular heart disease in humans. Degenerative (myxomatous) mitral valve disease (DMVD) is the most important heart disease in dogs accounting for approximately 75% of heart disease in this species.

DMVD in dogs appears pathologically identical to primary degenerative (myxomatous) disease of the mitral valve in humans suggesting that these conditions share a common pathogenesis. Emerging evidence suggests that DMVD is not merely an age-related degenerative process, but rather is an active disease process with identifiable initiating mechanisms, clinical and heritable risk factors, cellular and molecular mechanism that mediate a disease process. Better understanding of cellular and molecular processes that mediate DMVD could point toward pharmacologic or molecular strategies to slow or prevent disease progression.

Structural and Cellular Environment of Heart Valves

Heart valves function in an extraordinarily demanding and complex biomechanical environment. Their loading is repeated every 0.5 to 1 second over a lifetime. The gross mechanical forces on heart valves consist of 3 principal loading modes: *flexure*, *shear*, and *tension/compression*. Flexure is created by the opening and closure action of the valve. Shear is created on the surface of the valve by the flow of blood when the valve is open. Tension and compression are generated when the valve is closed and subjected to hemodynamic loading. These forces place a continuous biomechanical load on heart valves. As a result, the demand on the cellular component of heart valves to turn over and maintain a healthy extracellular matrix (ECM) exceeds that of any other connective tissue.

Heart valve leaflets have a tri-layer structure that consists of the *atrialis* (atrioventricular valves) or *ventricularis* (semilunar valves), the *spongiosa*, and the *fibrosa*. The atrialis/ventricularis layer is composed of a dense network of collagen and elastin fibers. The elastin fibers are radially aligned and thought to absorb radial strain on the leaflets created by shear stress when the valve is open. The fibrosa is the thickest layer and consists of a dense network of type I collagen. Functionally the fibrosa appears to be the main load bearing layer for tensile forces when the valve is closed. The central spongiosa layer contains a high concentration of highly hydrated glycosaminoglycans (GAG) and proteoglycans (PG). The spongiosa may have several biomechanical functions including lubrication between the fibrosa and atrialis/ventricularis as they shear and deform relative to each other, and absorption of compressive forces on the leaflets. The spongiosa may have biological functions such as allowing diffusion of oxygen and nutrients within the largely avascular leaflets.

The cellular component of heart valves consists of 3 principal cell types: *valve interstitial cells (VIC), valve endothelial cells (VEC),* and *smooth muscle cells (SMC).* VIC are fibroblast-like cells thought to be primarily responsible for maintenance and turn over (synthesis and degradation) of the ECM component of heart valves. VIC are the most abundant cells in heart valves and are present in all three layers of valve leaflets. VEC represent a single layer of cells that line the surface of heart valves analogous to vascular endothelial cells. However VEC are morphologically and functionally distinct from vascular endothelial cells. SMC are most prominent at the base of valve leaflets and extend from the base of the leaflet into the atrialis/ventricularis as a thin layer of cells just beneath the VEC. SMC express receptors to several vasoactive substances and are contractile. They are speculated to play a role in tensioning the valve leaflets however their precise role in normal heart valve function is not fully understood.

Pathology of DMVD

Gross pathologic changes associated with DMVD include progressive nodular thickening of the valve leaflets that begins at the free margin of leaflets in early-stage disease and progresses toward the leaflet base in later stages. In late-stage disease, the leaflets become scalloped and this contributes to improper leaflet coaptation. The chordae tendineae thicken, weaken, elongate, and eventually rupture allowing leaflet prolapse. Leaflet prolapse is the principal mechanism of valve insufficiency. Progressive dilation of the valve annulus contributes to improper leaflet coaptation and valve insufficiency.

A prominent microscopic feature associated with DMVD is exuberant disposition of GAG (acid mucopolysaccharides) and PG (decorin, biglycan, versican) in all valve layers, but most especially in the spongiosa layer. This aspect of the pathology accounts for the name "myxomatous" or "myxoid" pathology, and causes the gross nodularity and thickening of the valve leaflets. A distinguishing feature of myxomatous pathology is the notable absence of an inflammatory cellular component. The other principal microscopic features of DMVD are disruption and disarray of the collagenous matrix in all valve layers, and fragmentation of the elastin fibers in the atrialis. These changes reflect a shift toward predominate matrix degradation (catabolism) over matrix synthesis.

Most pathologic descriptions report an increase in cellular density associated with DMVD. Increased cellular density is generally presumed to be the result of VIC proliferation, but the precise mechanism has not been fully demonstrated in vivo. We and others have reported an increase in cellular density in both canine and human myxomatous mitral valves that is most prominent in the atrialis layer. We were unable to demonstrate a robust increase in VIC proliferative indices raising the possibility that other mechanisms such as cell migration and/or decreased apoptosis may be contributing to increases in cellular density.

Myxomatous pathology in heart valves is not restricted to DMVD. Serotoninergic valvulopathies associated with carcinoid syndrome, serotoninergic drugs (anorexigen drugs, ergot-derived drugs, antidepressant drugs, ecstasy) or exogenous administration of serotonin are also characterized by myxomatous or myxoid-like pathology in heart valves. In serotoninergic valvulopathies however, myxomatous change occurs predominately as plaque-like lesions on the surface of leaflets and chordae of affected heart valves.

Interstitial Cell Phenotype Transformation

VIC are thought to maintain normal heart valve architecture through balanced degradation and synthesis of the ECM. VIC from normal canine and human mitral valves exhibit a "quiescent fibroblastic" phenotype characterized by expression of vimentin, but not other cytoskeletal markers. Several lines of investigation demonstrate that interstitial cells are phenotypically plastic and undergo transdifferentiation to mediate developmental and pathological processes within the cardiac valves. Rabkin et al. were the first to report that VIC in human myxomatous mitral valves undergo transformation to an 'activated myofibroblast" phenotype characterized by expression of α-smooth muscle actin (α-SMA) and other cytoskeletal markers. These activated myofibroblast VIC strongly express matrix catabolic enzymes including collagenases (matrix metaloproteases [MMP]-1, MMP-13), gelatinases (MMP-2, MMP-9), and cysteine endoproteases (cathepsin S and K). Transformed "activated myofibroblasts" are now implicated in several developmental and pathological remodeling processes in cardiac valves, including myxomatous mitral valve disease.

Most attention has focused on expression of α-SMA as the signature marker of transformation of VIC to an activated or pathologic phenotype. However, we have demonstrated that that VIC expression of α-SMA in canine and human myxomatous mitral valves is largely restricted to atrialis layer, whereas myxomatous pathological change is clearly present in all valve layers. This raises questions about whether α-SMA is actually a marker of a single cellular phenotype or merely reflects a response of local VIC to an abnormal mechanical environment. Non-muscle embryonic myosin (SMemb) is a cytoskeletal element that has been associated "mesenchymal cell activation". Its expression has been demonstrated in neointimal cells after vascular injury and during atheroma formation, and in VIC of myxomatous mitral valves. We have shown that SMemb is widely expressed by VIC in all valve layers of human and canine myxomatous mitral valves, and that its expression is closely associated with MMP-1 and MMP-13 expression. We thus believe that expression of SMemb by mitral VIC may be a more meaningful marker of transformation to a "myxomatous phenotype". Little is known about the specific events that trigger VIC phenotype transformation and the pathologic events that they are mediate.

Cell Signaling in DMVD & the Autocrine Serotonin Hypothesis

Better understanding of the cell signal transduction events that mediate VIC phenotype transformation and myxomatous pathology offers the prospect of identifying therapeutic targets to slow or prevent DMVD. Several cell signaling pathways have been linked to myxomatous pathology in heart valves including serotonin (5-HT), transforming growth factor β-1 (TGFβ1), endothelin B, insulin-like growth factor, and nitric oxide.

Several lines of evidence implicate serotonin as a cause of myxomatous-like valvulopathy in humans and other species. Carcinoid syndrome, caused by high circulating serotonin produced from neoplasitic enterochromaffin cells, is associated with myxomatous-like valvulopathy primarily of right-sided heart valves. Rats injected daily with serotonin develop pathological changes in aortic valves similar to those described in human carcinoid disease in a dose-dependent manner. Several serotoninergic drugs including flenfluramine-phentermine (anorexigen), pergolide (Parkinson's disease) and ecstasy (recreational) have been associated with carcinoid-like valvulopathy. Activation of the 5-HT_{2B} receptor ($5HT_{2B}R$) has been shown to specifically mediate valve disease induced by serotonin or serotonergic drugs. Knockout of the serotonin transmembrane transporter (SERT), which together with monoamine oxidase is necessary for serotonin metabolism, causes cardiac valvulopathy in mice. Treatment of cultured VIC with serotonin increases TGFβ1 mRNA expression which in turn mediates cellular processes associated with myxomatous degeneration including expression of α-SMA and GAG production. Despite this evidence, serotoninergic valvulopathy has largely been regarded as clinically and etiologically distinct from DMVD.

We and others have shown up-regulation of the $5HT_{2B}R$ in canine myxomatous mitral valves. We have also shown down-regulation of SERT in canine myxomatous mitral valves. Both changes would be expected to enhance a serotonin signaling mechanism. Extracellular signaling-regulated kinases 42 & 44 kDa (ERK 1/2) are members of the mitogen activated protein (MAP) kinase family of serine-threonine kinases provide important signal transduction link between cell membrane receptors and nuclear gene activation. ERK 1/2 are activated by phosphorylation mediated by upstream signaling pathways. Studies in cultured VIC demonstrate that ERK are downstream phosphorylation targets of G protein-coupled serotonin receptors. We have demonstrated increased ERK 1/2 phosphorylation in canine myxomatous mitral valves compared consistent with enhanced serotonin signaling. This evidence implicates serotonin as a potential mediator of canine DMVD.

Ninety-five percent of peripheral serotonin synthesis occurs in the enterochromaffin cells of gut. Serotonin is secreted into the plasma and undergoes rapid uptake by circulating platelets in a process that is dependent on SERT. Circulating serotonin that is not absorbed by platelets is rapidly metabolized by cells in the lung and liver resulting in low circulating serotonin levels in normal plasma. Peripheral serotonin synthesis depends on the specific and rate-limiting action of tryptophan hydroxylase 1 (TPH1). We have demonstrated a 4- to 5-fold increase in protein expression of TPH1 in human and canine myxomatous mitral valves compared to normal mitral valves. We have demonstrated increased THP1 expression in all valve layers of canine and human myxomatous mitral valves and that its expression is closely associated with SMemb VIC phenotype transformation. Lastly, we have shown that increased expression of TPH1 occurs in both early- and late-stage canine DVMD. These findings implicate an autocrine serotonin signaling mechanism in the pathogenesis of canine and human DMVD, and suggest a mechanistic link between primary degenerative and serotoninergic valvulopathies. They also suggest possible therapeutic targets to slow progression of DMVD including selective serotonin receptor antagonists or inhibitors of TPH1. The specific mechanisms that trigger increased expression of TPH1 in DMVD are not known.

Role of Altered Biomechanical Environment in DMVD

The initiating causes of DMVD are poorly understood but several lines of clinical evidence implicate alterations in the biomechanical microenvironment of VIC and VEC as playing a possible causative role. In humans, heritable connective tissue disorders (Marfan syndrome, Ehlers–Danlos syndrome, osteogenesis imperfecta) are known to predispose to DMVD. The common denominator for these conditions could be weakening of the ECM component of heart valves that in turn subjects VIC to greater biomechanical strain, particularly tensile strain. In humans, hypertension is an independent risk factor for DMVD. Hypertension is also implicated as the reason why more human males with mitral valve prolapse syndrome go on to develop DMVD despite the overall higher incidence of this syndrome in females. Hypertension could certainly be reasoned to subject left-sided heart valves to greater than normal cyclic tensile stress. The role that hypertension might be playing in canine DMVD is largely unexplored because measurement of blood pressure in younger dogs prior to the onset of structural heart is not routine.

Experimental evidence further supports a hypothesis that abnormal tensile strain on VIC, either as a result of weakened ECM or hypertension, plays a causative role in DMVD. Transgenic overexpression of MMP-2 in mice causes changes consistent with mitral valve prolapse and myxomatous mitral valve disease, and has thus been proposed as an experimental animal model for these conditions. This model supports a hypothesis that any condition that weakens the ECM component increases the likelihood of DMVD. Both acute and chronic hypertension in rats cause VIC proliferation and increased collagen turnover in left-sided heart valves further implicating hypertension in the pathogenesis. Cyclic stretch has been shown to increase GAG an PG synthesis porcine VIC cultured in 3-D collagen gels. We have demonstrated that cultured canine mitral valves subjected to tensile stretch show increased expression of markers of GAG synthesis and TPH1 expression. These studies suggest that VIC have mechanoreception and mechanosignaling mechanisms capable of sensing cellular strain that could play a significant mediating role in DMVD. Experimental studies in sheep demonstrate that mitral regurgitation alone is sufficient to cause progressive myxomatous pathology in adjacent valve leaflets. These studies support the long held clinical axiom that "mitral regurgitation begets mitral regurgitation" and implicate abnormal biomechanical shear stress as a likely mediator of myxomatous pathology in mitral valves. VEC are the most likely cellular candidates for housing a mechanoreception mechanism capable of detecting abnormal shear. Specific mechanoreceptor mechanisms in heart valves have not yet been identified, but are the subject of ongoing research. Mechanoreceptor candidates include both stretch-activated ion channels and signaling pathways. Recognition of the importance of abnormal biomechanics in the pathogenesis of DMVD and identification of specific mechanoreceptor mechanisms offer the possibility of further therapeutic strategies to slow or control DMVD.

REFERENCES

[1.] Black, A., A. T. French, Dukes-McEwan, J, Corcoran, B. M. Ultrastructural morphologic evaluation of the phenotype of valvular interstitial cells in dogs with myxomatous degeneration of the mitral valve. **Am J Vet Res** 66:1408-1414, 2005. [2.] Disatian S, Ehrhart EJ, Zimmerman S, Orton EC. Interstitial cells from dogs with naturally-occurring myxomatous mitral valve disease undergo phenotype transformation. **J Heart Valve Dis** 17:402-412, 2008. [3.] Disatian S, Orton EC. Autocrine serotonin and transforming growth factor β 1 signaling mediates spontaneous myxomatous mitral valve disease. **J Heart Valve Dis** 18:44-51, 2009. [4.] Disatian S, Lacerda C, Orton EC. Tryptophan hydroxylase 1 expression is increased on phenotype-altered canine and human degenerative myxomatous mitral valves. **J Heart Valve Dis** 19: 2010 [5.] Grande-Allen, K. J., B. P. Griffin, et al. Glycosaminoglycan profiles of myxomatous mitral leaflets and chordae parallel the severity of mechanical alterations. **J Am Coll Cardiol** 42:271-277, 2003. [6.] Gupta, V., H. Tseng, et al. Effect of cyclic mechanical strain on glycosaminoglycan and proteoglycan synthesis by heart valve cells. **Acta Biomater** 5:531-540, 2009 [7.] Jian B, Xu J, Savani R, Narula N, Liang B, Levy RJ. Serotonin mechanisms in heart valve disease I: Serotonin induced upregulation of TGF-β1 via G-protein signal transduction in aortic valve interstitial cells. **Am J Pathol161**:2207-2216, 2002 [8.] Lacerda CMR, Disatian S, Orton EC. Differential protein expression between normal, early-stage, and late-stage myxomatous mitral valves from dogs. Proteomics – **Clinical Applications** 3:1422-1429, 2009. [9.] Liu AC, Joag VR, Gotlieb AI. The emerging role of valve interstitial cell phenotypes in regulating heart valve pathobiology. **Am J Pathol** 171:1407-1418, 2007. [10.] Mahimkar, R., A. Nguyen, et al. Cardiac transgenic matrix metalloproteinase-2 expression induces myxomatous valve degeneration: a potential model of mitral valve prolapse disease. **Cardiovasc Pathol** 18(5): 253-261, 2009 [11.] Olsen, L. H., K. Mortensen, et al. Increased NADPH-diaphorase activity in canine myxomatous mitral valve leaflets. **J Comp Pathol** 129:120-130 2003 [12.] Oyama, M. A. and S. V. Chittur. Genomic expression patterns of mitral valve tissues from dogs with degenerative mitral valve disease. **Am J Vet Res** 67: 1307-1318, 2006 [13.] Pedersen, L. G., J. Zhao, et al. Increased expression of endothelin B receptor in static stretch exposed porcine mitral valve leaflets. **Res Vet Sci** 82:232-238, 2007 [14.] Rabkin E, Aikawa M, Stone JR, et al. Activated interstitial myofibroblasts express catabolic enzymes and mediate matrix remodeling in myxomatous heart valves. **Circulation**104:2522-2532, 2001 [15.] Sacks, M. S., W. David Merryman, et al. On the biomechanics of heart valve function. **J Biomech** 42:1804-1824, 2009 [16.] Stephens, E. H., T. C. Nguyen, et al. The effects of mitral regurgitation alone are sufficient for leaflet remodeling. **Circulation** 118(14 Suppl):S243-249, 2008 [17.] Xu J, Jian B, Chu R, et al. Serotonin mechanisms in heart valve disease II: The 5-HT_2 receptors and its signaling pathway in aortic valve interstitial cells. **Am J Patho** l161:2209-2218, 2002.

MARKERS OF THROMBOTIC RISK & FUTURE DIRECTIONS IN THERAPEUTICS

Daniel F. Hogan, DVM, DACVIM-Cardiology
West Lafayette, IN

THROMBOTIC RISK

The risk for developing Thrombotic or thromboembolic events (TTE) is dependent upon many things but can classified under the three arms of Virchow's triad which include vascular stasis, endothelial injury, and hypercoagulability. While dramatic abnormalities in any of these arms can result in clinical thrombosis, a more practical approach would be to consider the cumulative thrombotic risk where each arm adds or reduces the risk for the development of thrombosis. However, while blood stasis and endothelial injury appear to be most related to an underlying disease condition which can occur in any individual within a population and quantified in terms of degree of severity, the hypercoagulability arm may be more of a specific characteristic of an individual. In fact, there may be a population of individuals that exhibit a particular characteristic that confers hypercoagulability although the degree of this hypercoagulability may be quite variable from one individual to the next. The variability in the hypercoagulable arm most likely results in the wide spectrum of TTE within a disease population. A pertinent example would be why some cats with extremely dilated atria with myocardial disease do not develop cardioembolic events (CE) while cats with much smaller atria experience such events. This has resulted in the mixed data from the literature where some studies conclude that left atrial size is a risk factor or CE while others do not see this relationship. Accurate prediction of hypercoagulable states would have a dramatic impact on patient prognosis and therapeutic interventions where a more accurate risk-to-benefit ratio could be determined. This would allow clinicians to decide if the possible bleeding risk from an antithrombotic agent is outweighed by the benefit of preventing a TTE. Unfortunately, these studies are currently lacking in veterinary medicine.

Inherited Hypercoagulable Disorders

The determination of the genetics for all of the proteins involved in the coagulation process in humans has allowed identification of multiple genetic mutations.

Prothrombin

There is a polymorphism of the prothrombin gene that results in mildly elevated plasma levels. This mutation has been identified as a risk factor for venous thrombosis. The exact mechanism is unknown, however it has been suggested that the mutation results in a higher transcription rate.

Factor V

By far, the most common inherited thrombotic risk factor in humans involves a mutation of factor V which has been called factor V_{Leiden}. A point mutation results in increased resistance to activated protein C (APC) inactivation and was first termed APC-resistance. Factor V_{Leiden} has been identified as the most common genetic risk factor for venous thrombosis conferring a relative risk of 2-4.

Protein C

Protein C deficiency (or loss of function) has been identified as a common inherited risk factor for venous thrombosis. Over 300 different mutations of the protein C gene have been identified to date. These reduced levels of protein C (or production of dysfunctional protein C) results in decreased inactivation of Va and VIIIa.

Protein S

Protein S acts as a cofactor for APC-mediated inactivation of Va and VIIIa. Seventy-one different point mutations have been recognized in the protein S gene leading to protein deficiency. These mutations lead to a mild increased risk for venous thrombosis. One particular mutation, protein S Heerlen, may provide a synergistic prothrombotic effect in patients with V_{Leiden}.

Antithrombin III

There are a number of inherited antithrombin III (AT III) deficiencies that carry variable increased risk for venous thrombosis. The resulting AT III function in these patients commonly range from 40%-60% of normal. The antigen levels of AT III can be normal, low or elevated depending upon the type of inherited AT III abnormality. Type I AT III deficiencies arise from reduced AT III synthesis or release so antigen levels and function are both reduced. Type II AT III deficiencies are caused by missense mutations and N-glycosylation consensus sequence mutations where the AT III molecules produced have variably reduced function but the measured AT III antigen levels are higher and may be normal.

There are no known inherited hypercoagulable states in our domestic animal species. However, there are certain breeds of cats that have an increased risk for CE which could suggest an inheritable link.

There are a number of other biomarkers that have been correlated to an increased risk for clinical thrombosis. These include elevations in **factor VIII** (DVT/PE), **homocysteine** (ACS/AMI, thrombotic stroke, DVT/PE), **lipoprotein(a)** (ACS/AMI, thrombotic and cardioembolic stroke), **plasminogen activator inhibitor-1 (PAI-1)**(ACS/AMI, DVT/PE), **thrombin-activatable fibrinolysis inhibitor (TAFI)**(DVT/PE), and **fibrinogen** (ACS/AMI, thrombotic stroke). **Hyperreactive platelets** have also been associated with an increased risk for ACS/AMI, as well as thrombotic and cardioembolic stroke.

Clinical Conditions Associated with Thrombotic Events

Immune-Mediated Hemolytic Anemia (IMHA)

Pulmonary embolism has been reported in 29%-80% of dogs at necropsy. Some dogs with IMHA have been shown to have altered hemostatic parameters including increased fibrinogen levels as well as reduced AT III levels. While these markers certainly indicate significant inflammation and an active thrombotic setting, they do not necessarily suggest the underlying cause of the prothrombotic environment. Thrombosis secondary to IMHA in humans may be related to oxidative stress and platelet hyperactivation. One study demonstrated an 8-fold increase in median P-selectin expression in dogs with IMHA suggesting the presence of circulating activated platelets. However, this may simply reflect that the prothrombotic condition results in platelet activation secondary to widespread thrombosis.

Hyperadrenocorticism

The occurrence of pulmonary thrombosis/embolism associated with hyperadrenocorticism appears to be around 18%. Studies have identified alterations in some hemostatic parameters including increases in the procoagulant factors II, V, VII, IX, X, XII and fibrinogen. Increases in thrombin-antithrombin complexes (TAT) have also been identified indicating subclinical thrombosis. There have been variable changes in AT III where both elevations and reductions have been seen. There is a 4-fold increase in the frequency of pulmonary embolism in human patients with hyperadrenocorticism than in the general population with a parallel increase in mortality. There have been elevations identified in the procoagulant factors II, V, VII, VIII, IX, X, XI and XII. Levels of factor VIII have been shown to return to normal after resolution of the hyperadrenocorticism.

Protein-Losing Nephropathy

Thromboembolic complications have been reported to occur in over 20% of dogs with protein-losing nephropathy. Reduction in circulating AT III levels and a possible correlation between platelet hypersensitivity and reduced serum albumin levels have been proposed as possible causes. In humans the incidence of thromboembolic disease in nephrotic syndrome has been reported to be approximately 1% per year with an incidence of approximately 10% and 6% for venous thromboembolism (VTE) and arterial thromboembolism, respectively within the first six months. Possible causes include loss of AT III, increased TAFI and reduced protein C activity.

Cardiogenic Embolism

The mean frequency of CE in cats with myocardial disease ranges from 6%-17%. Cats with cardiac disease have been reported to have altered platelet reactivity which could contribute to CE disease. Reductions in ATIII and protein C activity as well as increases in factor VIII and fibrinogen have been identified in cats with either a history of CE or during an acute CE event. In humans, CE occurrence rates are variable depending upon the underlying cardiac disease. Cardiogenic thromboembolic stroke, the most common and clinically significant form of CE, is seen in approximately 15% of all cardiomyopathic patients. There are few studies evaluating potential underlying hypercoagulable states with CE in humans. The focus appears to be risk stratification based upon population data from large database analyses. However, changes in platelet function similar to cats have been identified in humans with HCM.

Neoplasia

While there are no retrospective studies determining the frequency of thrombosis in animals with neoplasia, there are retrospective studies of arterial thrombosis in dogs and cats with neoplasia where there is a reported frequency between 2.5% and 30%. Clinical thrombosis is seen in up to 15% of human cancer patients at some point during their clinical course. Thrombotic mortality is the second most common cause of death in human cancer patients. A cysteine proteinase, "cancer procoagulant factor" is able to activate factor X independently of factor VIIa and has been identified in up to 85% of cancer patients. Acquired protein C resistance, and elevations in tissue factor and soluble P-selectin have also been associated with thrombosis.

Measurement of Hypercoagulability with Hemostatic assays

Platelet aggregation and thromboelastography have been used to estimate hypercoagulability in veterinary patients. Both of these methodologies provide valuable hemostatic information but appear to be better for within-animal than between-animal comparisons. While an exaggerated response to an agonist is sometimes used as evidence of hypercoagulability, a reduced inhibitory response might actually be a better measure of a hyperreactive coagulation system.

FUTURE DIRECTION IN THERAPEUTICS

It is interesting to speak of the future direction of antithrombotic therapy as there was very little progression past aspirin therapy in veterinary medicine for decades. However, there have been constant progression in antithrombotic therapy within human medicine and some of this has made its way into veterinary medicine. The focus on antithrombotic therapy has increased in veterinary medicine and this is likely to result in exciting changes over the next 5-10 years.

Antiplatelet Agents

It seems unlikely that there will be dramatic changes in the area of antiplatelet therapies for the foreseeable future. Currently, **aspirin** and the ADP_{2Y12} antagonist **clopidogrel** are the most commonly used drugs. Clopidogrel is a more potent antiplatelet agent but aspirin is dramatically cheaper. Both drugs appear to be well tolerated with minimal adverse effects. While there is no clinical evidence of superior efficacy of one drug over the other, the Feline Arterial Thromboembolism; Clopidogrel vs. Aspirin Trial (FAT CAT, www.vin.com/fatcat) is ongoing and hopes to provide this information. Another potential drug category is the glycoprotein IIb/IIIa **(GPIIb/IIIa) antagonists**. These drugs are potent antiplatelet agents but must be given IV as multiple studies have demonstrated that the oral formulations of these drugs are not effective. Additionally, one of the most common drugs in this class, **abciximab**, has been shown to not induce an *in vitro* antiplatelet effect in dogs and cats. There are newer ADP_{2Y12} agents that are less susceptible to pharmacokinetic resistance than clopidogrel and are reversible. However, clopidogrel resistance has not been clearly identified in dogs or cats and acute reversal for surgical procedures may not be important enough in veterinary medicine to overcome the dramatically higher price.

Anticoagulant Agents

While not a new drug class to veterinary medicine, the low-molecular weight heparins (**LMWH**) have generated conflicting opinions on their appropriateness in dogs and cats. Like all antithrombotics in veterinary medicine, there are currently no clinical trials evaluating the effectiveness of these agents and it is unclear as to what dose or dosing frequency is ideal. Monitoring of drug therapy with serum anti-Xa activity assays has recently been shown to not accurately reflect the antithrombotic effect of these drugs. Similar data has been seen in humans. This class of drugs seems to be a good choice for many of our patients as they can be administered chronically with minimal adverse effects. The major drawbacks appear to be the route of administration (SQ injection), and cost. Future studies to look for include clinical trials to evaluate efficacy at different dosing protocols.

A relatively new class of antithrombotic drugs, the **synthetic Xa inhibitors**, selectively inhibit Xa through the potentiation of ATIII. They have no effect on routine coagulation tests such as the activated clotting time, aPTT or PT. However, given their selective Xa inhibition, measurement of anti-Xa activity levels is an acceptable and clinically proven method for monitoring the clinical use of these drugs. Unlike the LMWH, these agents are homogeneous preparations of synthetic pentasaccharide units that bind exclusively to AT III so they have complete bioavailability when administered subcutaneously and with less inter-patient variability. **Fondaparinux (Arixtra®)** is currently approved for clinical use in humans and potentiates the innate neutralizing ability of AT III on Xa by approximately 300 fold. Multiple clinical trials have demonstrated that fondaparinux is at least as effective, or more effective, than enoxaparin in the prophylactic prevention of thrombosis associated with hip or knee replacement surgery or in the treatment of deep venous thrombosis or pulmonary embolism. While there is currently no clinical experience with fondaparinux in dogs, a recent pharmacokinetic study in cats suggested that a dosing protocol of 0.2 mg/kg SQ q 12 hrs results in anti-Xa activities similar to those found in humans. The major drawback for this class of drugs is cost as they are more expensive than the LMWH. There are newer Xa inhibitors including some that can be administered orally. These drugs may find a future place in veterinary medicine.

One of the most recent and exciting new categories of antithrombotics is the **direct thrombin inhibitors**. Recently, the Randomized Evaluation of Long term Anticoagulant therapy (RE-LY) study was completed that compared the direct thrombin inhibitor **dabigatran** to warfarin in patients with atrial fibrillation at risk for stroke or systemic embolism. Designed to detect non-inferiority, a lower dose of dabigatran was non-inferior to warfarin but had significantly less hemorrhagic events while a higher dose of dabigatran was superior to warfarin with a similar rate of bleeding. This is the first antithrombotic to be determined superior to warfarin for cardioembolic disease without an increase in bleeding. While cost will be an issue, this class of drugs could hold promise for veterinary patients.

REFERENCES

[1.] Dahlbäck B. **Sem Hematol** 1997;34:217-234. [2.] Griffin JH, et al. **J Clin Invest** 1981;68:1370-1373. [3.] vanBoven H, et al. **Semin Hematol** 1997;34:188-204. [4.] Carr AP, et al. **J Vet Intern Med** 2002;16:504-509. [5.] Feldman BF, et al. **Res Vet Sci** 1986;41:228-230. [6.] Smith SA, et al. **J Vet Intern Med** 2003;17:73-83. [7.] Hogan DF, et al. **J Am Vet Med Assoc** 2004;225:1406-1411. [8.] Van De Wiele CW, et al. **J Vet Intern Med** 2010;24:185-191. [9.] Connolly SJ, et al. **N Engl J Med** 2009;361:1139-1151.

DIAGNOSTIC & PROGNOSTIC IMPLICATIONS OF GENETIC TESTING FOR CARDIOMYOPATHIES

Kathryn M. Meurs, DVM, PhD DACVIM
Pullman, WA

INTRODUCTION

There is increasing evidence that many forms of canine and feline cardiomyopathy are inherited. 1-2,4-6 This discussion will focus on cardiomyopathies with known genetic causes.

DILATED CARDIOMYOPATHY

Portuguese Water Dog

A juvenile form of familial DCM has been reported in the Portuguese Water Dog. It is thought to be inherited as an autosomal recessive trait that is linked to a region on canine chromosome 8. [1] Affected puppies were from apparently unaffected parents and usually died between 2 - 32 weeks of age from sudden death or congestive heart failure. A test for a genetic marker highly associated with this disease is available through PennGen, http://research.vet.upenn.edu/penngen. Since this is a recessive trait, dogs that have two copies of the mutation are most likely to develop the disease. Silent carriers exist so dogs that only have one copy of the trait may pass the gene on, but will not show the trait.

HYPERTROPHIC CARDIOMYOPATHY

Maine Coon

In the Maine Coon, HCM is inherited as autosomal dominant trait. A genetic mutation has been identified in the myosin binding protein C (MYBPC3) gene. [2] The myosin binding protein C is an important cardiac sarcomeric protein and has been associated with the development of familial HCM in human beings as well. [3] In the Maine Coon cat, the mutation is a single base pair change that changes the structure of myosin binding protein C and alters its ability to interact with other contractile proteins.

The Maine Coon mutation appears to be quite breed specific. It does not appear to be associated with familial HCM in other breeds of cats. The mutation is inherited with incomplete penetrance and variable expressivity meaning that not all cats with the mutation will show the disease, or show the same severity of the disease. Cats that are homozygous for the mutation appear to be more likely to show the disease and perhaps have a more severe form. [2,4] However, it would appear that not all Maine Coon cats with HCM have this mutation. In people there are now over 400 reported HCM mutations.[3] Therefore, it is likely that there is more than one mutation in the Maine Coon cat as well.

Mutation screening can now be used to determine if a Maine Coon has the known HCM mutation. If a cat is positive for the mutation, one should carefully consider continued use as a breeding animal. It appears that approximately 30% of all Maine Coon cats tested for the mutation at Washington State University are positive for the mutation. Due to the high prevalence of the mutation in this breed, it would seem to be unwise to recommend that all cats with the mutation be removed from the breeding programs since altering the gene pool by 30% could increase the prevalence of other genetically determined disorders. Current recommendations might be to remove cats that are homozygous for the mutation from the breeding pool unless they have exceptional attributes. Heterozygous cats should be evaluated by echocardiography. If they do not show signs of hypertrophy it may suggest that they have low disease penetrance. If the cat has many other positive breed attributes and is disease negative at time of breeding they could possibly be bred to a mutation negative cat. The offspring of that mating should be screened, and if possible, a mutation negative kitten with desirable traits selected to replace the mutation positive parent in the breeding pool. Over a few generations this will decrease the prevalence of the disease mutation in the population, hopefully without greatly altering the gene pool. Disease negative but mutation positive cats should be evaluated annually for presence of disease and removed from use if they develop the disease.

Ragdoll Cardiomyopathy

A substitution mutation has also been identified in the myosin binding protein C gene in the Ragdoll cat.[5] However, the Ragdoll mutation is located in a different region of the gene that the Maine Coon.

The mode of inheritance of this mutation in the Ragdoll cat has not been identified. In the Ragdoll, homozygous cats appear to be very severely affected with development of heart failure and thromboembolic episodes often before 2 years of age. Heterozygous cats appear to have a much more mild form of the disease that may include only mild papillary muscle hypertrophy.

The Ragdoll mutation also appears to be breed specific to the Ragdoll and it has not been identified in other breeds of cats. Although there is no evidence yet that there is more than one mutation in the Ragdoll, annual clinical screening is still recommended. Echocardiographic criteria for clinical clearance should be the same as indicated above for Maine Coons.

At this time, approximately 20% of all Ragdolls tested at Washington State University have the Ragdoll mutation. Therefore, removal of all positive breeding animals from the gene pool is not without possible detriment to the breed. Breeding recommendations are the same as described above for the Maine Coon.

ARRHYTHMOGENIC RIGHT VENTRICULAR CARDIOMYOPATHY IN THE BOXER

Arrhythmogenic right ventricular cardiomyopathy (ARVC) in the boxer appears to be inherited as an autosomal dominant trait in at least some families. It is a disease of variable and age related incomplete penetrance meaning that not all affected dogs will show the same severity of disease. An eight base pair deletion associated with the development of canine ARVC has been identified in an untranslated region of a desmosomal gene.[6] Homozygous dogs appear to have a more severe form of the disease with a statistically greater number of ventricular premature complexes/24 hours than heterozygous. In human beings with familial ARVC, there are 144 known genetic mutations identified in eight genes at this time therefore, it is likely that there will be multiple mutations in the dog as well. [7] It is not yet known if this is the only causative mutation for canine ARVC therefore clinical screening is still recommended.

Genetic testing can now be performed for the ARVC mutation. It is not yet known what percentage of Boxers have the ARVC mutation but it is likely to be high enough that removal of all dogs with the mutation from the breeding pool could have a negative impact on the breed. It should be remembered that because of the incomplete penetrance of the disease not all dogs with the mutation will ever develop clinical signs and many may live a normal lifespan. It is likely that there are multiple modifying factors that may influence which dogs develop the clinical form of the disease. Our current recommendations are that dogs that are positive heterozygous for the mutation should be carefully evaluated for signs of disease (Holter monitor and possibly an echocardiogram). If they do not show signs of disease and do have other positive breed attributes they could potentially be bred to dogs that are negative for the mutation. Puppies from that litter may be screened for the mutation and over a few generations, mutation negative puppies may be selected to replace the mutation positive parent and gradually decrease the number of mutation positive dogs in the population. Boxers that are homozygous for the mutation should probably not be used for breeding unless they have very positive attributes that need to be maintained in the breed and then they should be bred to a negative dog.

REFERENCES

[1.] Werner P, et al. **Genomics** 2008;91:517. [2.] Kittleson MD, et al. **Circulation** 1999;99:3172. [3.] Keren A, et al. **Nat Clin Pract Cardiovasc Med** 2008;5:158. [4]. Sampedrano CC, et al. **JVIM** 2008;23: 91. [5.] Meurs KM,et al. **Genomics** 2007; 90:261.[6.] Meurs KM et al. **JVIM** 2009;23:674 [7.]Basso CB et al. **Lancet** 2009;373:1289.

BRACHIAL ARTERY DOPPLER WAVEFORM ANALYSIS AND MARKERS OF ENDOTHELIAL FUNCTION IN HEALTHY DOGS AND DOGS WITH CONGESTIVE HEART FAILURE. SM Cunningham, JE Rush, LM Freeman. Tufts Cummings School of Veterinary Medicine, North Grafton, MA.

Impaired endothelial function occurs in people with CHF and is a strong predictor of adverse outcome. Non-invasive measurement of endothelial function is commonly evaluated in people by ultrasound assessment of endothelium-dependent flow-mediated vasodilation (FMD). Assessment of FMD in dogs of various sizes is limited by the need for high precision to detect small changes in brachial arterial (BA) diameter. Thus other non-invasive means of assessing endothelial function, such as changes in BA flow profiles, could be of utility in dogs. Alterations in L-arginine (Arg), ADMA, SDMA, vWF, and plasma nitrate and nitrite (NO_x) may also provide information regarding endothelial function. The aims of this study were to characterize plasma markers of endothelial function and brachial artery velocity (BAV) waveforms at baseline and during ischemia-induced reactive hyperemia (RH) in healthy dogs and those with CHF.

Dogs were judged as healthy (n=7) or to have CHF (n=7) based on physical examination, echocardiogram (2D, color and Doppler), blood pressure, CBC, chemistry profile, radiographs, and plasma NTproBNP. The CHF group included dogs with ISACHC Class II or IIIa CHF due to CVD or DCM. Fasting plasma concentrations of Arg, ADMA, SDMA, NO_x, and vWF were measured. Brachial artery ultrasound was performed to record arterial diameter and BAVs at baseline. BAV changes during RH were then assessed at 10, 15, 20, 30, 45 and 60 sec following a 3 min period of vessel occlusion. Measurements were compared between groups using Mann-Whitney-U tests while potential relationships between continuous variables were assessed using Spearman rank correlations.

At baseline, CHF dogs had a lower mean velocity time integral over the entire cardiac cycle (VTI-M) compared to normal dogs, however peak forward and flow reversal BAVs were not different between groups. The peak hyperemic response was observed at the 10 sec interval in both groups. At 10 sec the peak systolic velocity (S10Vmax), systolic VTI (S10VTI), VTI-M, and the change in S10VTI (ΔS10VTI) and VTI-M from baseline were all lower in dogs with CHF compared to controls. Most of these differences were not apparent at later time points, although BAVs took longer to return to baseline values in the CHF group. No differences were noted in plasma Arg, ADMA, SDMA, or vWF. Dogs with CHF had higher plasma NO_x compared to controls (p=0.04) and NO_x was positively correlated with 2D LA:Ao (r=0.56; p=0.04). Negative correlations were found between 10 sec BAVs (S10Vmax, S10VTI, and ΔS10VTI) and ISACHC, NTproBNP, plasma NO_x, and 2D LA:Ao (r= -0.54 to -0.75; p=0.002 to 0.045).

BAV parameters during RH are different between normal dogs and dogs with CHF. Further studies are indicated to evaluate BA Doppler waveform analysis as a potential indicator of endothelial function in a larger number of dogs at varying stages of cardiac disease.

DECREASED FLOW-MEDIATED VASODILATION IN DOGS WITH MODERATE-SEVERE MYXOMATOUS MITRAL VALVE DISEASE. SG Moesgaard[1], C Klostergaard[1], M Molin[1], NE Zois[1], TB Falk[1], CE Rasmussen[1], V Luis Fuentes[2], ID Jones[2], LH Olsen[1]. 1. Department of Basic Animal and Veterinary Sciences, The Faculty of Life Sciences, University of Copenhagen, Denmark. 2. Department of Clinical Veterinary Sciences, Royal Veterinary College, University of London, UK.

Myxomatous mitral valve disease (MMVD) is a common heart valve disease in dogs. Cavalier King Charles Spaniels (CKCS) are highly predisposed to development of MMVD. Studies of plasma biomarkers of endothelial function suggest that CKCS with MMVD have endothelial dysfunction. However, biomarkers measure endothelial function indirectly and are affected by other factors. The aim of this study was to measure flow-mediated vasodilation (FMD, an ultrasound technique used for measuring human endothelial function) in dogs with MMVD.

An initial pilot study was performed in 5 healthy dogs (different breeds) where FMD was measured 5 times over a period of 5 weeks. In the clinical study, FMD was measured at 4 stages of MMVD: Group (1): 11 CKCS with no to minimal mitral regurgitation (MR); Group (2): 8 CKCS with mild MR; Group (3): 12 CKCS with moderate-severe MR: Group (4): 8 CKCS with clinical signs of heart failure due to MMVD. All dogs underwent a physical examination, blood sampling for complete blood count and serum biochemistry and echocardiography prior to FMD measurement. FMD was measured using a 13MHz linear probe to record 2D images of the brachial artery. A blood pressure cuff was inflated distal to the imaging site to a pressure of >200mmHg for 5 minutes. 2D imaging was performed 3 minutes post-occlusion. Commercially available software was used to measure vessel diameter and the peak of post-occlusion was compared with baseline to calculate percentage increase (FMD).

The pilot study showed that it is possible to measure an increase in post-occlusion vessel diameter in healthy dogs. In the clinical study a decreased FMD response was found with group (3) (5.86±3.46%) and group (4) (4.98±2.44%) compared with group (1) (8.26±5.40%) and group (2) (8.98±3.41%). When the FMD responses in groups 1 and 2 were pooled and compared to pooled responses of groups 3 and 4, the FMD was significantly decreased in groups 3+4 (8.56±4.57% vs groups 1+2, 5.51±3.05%, P=0.02). The FMD response tended to decrease with increasing MR (P=0.055). The FMD response did not correlate with body weight, age or ambient temperature.

These results support the previous findings in plasma biomarker studies indicating that CKCS with severe stages of MMVD have endothelial dysfunction. However, the FMD technique is challenging and larger number of dogs should be studied in order to evaluate the effect of increasing age and the correlation between the FMD response and relevant plasma biomarkers.

CARDIAC TROPONIN-I PREDICTS MYOCARDIAL ARTERIOSCLEROSIS AND FIBROSIS IN DOGS WITH CONGESTIVE HEART FAILURE DUE TO MYXOMATOUS MITRAL VALVE DISEASE

Torkel Falk[1], Ingrid Ljungvall[2], Nora E Zois[1], Lennart Jönsson[2], Lisbeth H Olsen[1], Henrik D Pedersen[3], Katja Höglund[2], Jens Häggström[2]

1. Faculty of Life Science, University of Copenhagen, Frederiksberg, Denmark 2. Swedish University of Agricultural Sciences, Uppsala, Sweden 3. Zeeland Pharma, Copenhagen, Denmark

An association between arteriosclerosis/fibrosis and myxomatous mitral valve disease (MMVD) in dogs has previously been demonstrated. We have also demonstrated an association between arteriosclerotic changes and fibrosis, in particular replacement type fibrosis. Cardiac troponin-I (cTnI) has been found to increase with increasing severity of MMVD.

The aim of this study was to investigate whether plasma concentrations of cTnI *in vivo* were associated with the degree of histopathological changes post mortem, particularly the extent of arteriosclerotic and fibrotic changes in the myocardium, in dogs with naturally occurring congestive heart failure (CHF) due to MMVD.

Fifty client-owned dogs with CHF due to MMVD were prospectively recruited. Clinical examination, echocardiography, and blood sampling were performed, and included dogs were re-examined every six months, if alive. Serum cTNI was analysed by a high sensitivity cTnI assay (Beckman Coulter AccuTnI). After euthanasia due to therapy failure or sudden cardiac death, a standardized pathological examination was performed, and the presence and severity of myocardial fibrosis and arteriosclerotic changes were assessed by histopathology.

Included dogs had a median cTnI of 0.08 (interquartile range 0.05-0.15) ng/mL. Multiple regression analysis confirmed a major effect of myocardial fibrosis (P>0.001), severity of arterial narrowing (P>0.001), left ventricular end diastolic dimension (P=0.001) and proximal isovolumetric surface area of the mitral insufficiency jet (P=0.004). Fibrosis score was the variable that contributed most to the model ($R^2 = 0.36$).

In conclusion, analysis of cTnI might be predicitive of myocardial fibrosis and arteriosclerosis, and this finding might contribute to increased insight in the patophysiology of MMVD.

LONGITUDINAL STUDY OF THE CARDIAC BIOMARKER NT-PROBNP IN CANINE CLINICAL HEART DISEASE. A PROGNOSTIC FACTOR FOR DOGS? Scott Forney, DVM, Stephen Ettinger, DVM, Justin Allen, DVM, Kirstie Barrett DVM, Andrew Beardow, BVS, Giosi Farace, PhD. California Animal Hospital Veterinary Specialty Group, Los Angeles, Ca. 90025; Idexx Laboratories, Westbrook, ME 04092.

Cardiac biomarker levels, traditional cardiovascular testing (physical examination, echo-doppler studies, radiography, ECG, VHS, and routine chemical analyses) were examined in dogs.

To statistically compare the relevance of cardiac biomarkers in the presence of disease, both cardiac and non-cardiac. To determine if these patients trend their biomarker levels with changes in disease.

1264 clinical CAHVSG canine patients were studied over 3 years.

Dogs were classified for their CAHVSG Heart score (from 0-9) based on the cardiac studies other than the biomarker levels (ACVIM Proc, 2008). The scores were narrowed to four grades after review. Scoring was completed prior to knowledge of the cardiac biomarker levels. Dogs seen on follow up examination had biomarker levels repeated (175 twice; 233 three times or more). A single biomarker was taken in 856 dogs.

A preliminary evaluation of mortality shows that nearly 30% of the dogs died and the mean time to death was 593 days. Grouping dogs according to their NT-proBNP level shows that dogs with higher initial NT-proBNP levels (>1500pmol/L) have lower survival than those with low NT-proBNP (<1000pmol/L); mean time was 469 days vs. 626 days. ($p<0.0001$)

There is no statistical significance between those dogs that died due to cardiac or noncardiac causes ($p=0.07$). The median time to death is twice as long for the non-cardiac cases than the cardiac cases (67 vs. 34 days). A single point NT-proBNP level might be useful for prognostic purposes.

THREE DIMENSIONAL ECHOCARDIOGRAPHY

Thomas Ryan, MD
Columbus, OH

GENERAL

1. Hung J, Lang R, Flachskampf F, et. al. 3D Echocardiography: A review of the current status and future directions. **J Am Soc Echocardiogr** 2007;20:213-233.

2. Yang HS, Bansal RC, Mookadam F, Khanderia BK, Tajik AJ, Chandrasekaran K. Practical guide for three-dimensional transthoracic echocardiography using a fully sampled matrix array transducer. **Am Soc of Echocardiog** 2008;21: 979-989.

3. Schneider AT, Hsu TL, Schwartz SL, et al. Single, biplane, multiplane, and three-dimensional transesophageal echocardiography. Echocardiographic-anatomic correlations. **Cardiol Clin** 1993;11:361–387.

RIGHT HEART

4. Fukuda S, Saracino G, Matsumura Y, et al. Three-dimensional geometry of tricuspid annulus in healthy subjects and in patients with functional tricuspid regurgitation : a real-time, 3-dimensional echocardiographic study. **Circulation** 2006;114:1492-1498.

5. Keller AM, Gopal AS, King DL. Left and right atrial volume by freehand three-dimensional echocardiography: in vivo validation using magnetic resonance imaging. **Eur J Echocardiogr** 2000;1:55–65.

6. De Castro S, Cavarretta E, Milan A, et al. Usefulness of tricuspid annular velocity in identifying global RV dysfunction in patients with primary pulmonary hypertension: A comparison with 3D echo-derived right ventricular ejection fraction. **Echocardiography:** 2008;25:289-293.

7. Fujimoto S, Mizuno R, Nakagawa Y, et al. Estimation of the right ventricular volume and ejection fraction by transthoracic three-dimensional echocardiography. A validation study using magnetic resonance imaging. **Int J Card Imaging** 1998;14:385–390.

8. Grison A, Maschietto N, Reffo E, et al. Three-dimensional echocardiographic evaluation of right ventricular volume and function in pediatric patients: validation of the technique**. J Am Soc Echocardiog** 2007;20:921-929.

9. Jiang L, Siu SC, Handschumacher MD, et al. Three-dimensional echocardiography. *In vivo* validation for right ventricular volume and function. **Circulation** 1994;89:2342–2350.

10. Niemann P, Pinho L, Balbach T, et al. Anatomically oriented right ventricular volume measurements with dynamic three-dimensional echocardiography validated by 3-Tesla magnetic resonance imaging**. J Am Coll Cardiol** 2007;50:1668-1676.

11. Papavassiliou DP, Parks WJ, Hopkins KL, et al. Three-dimensional echocardiographic measurement of right ventricular volume in children with congenital heart disease validated by magnetic resonance imaging. **J Am Soc Echocardiogr** 1998;11:770–777.

12. Shiota T, Jones M, Chikada M, et al. Real-time three-dimensional echocardiography for determining right ventricular stroke volume in an animal model of chronic right ventricular volume overload. **Circulation** 1998;97:1897–1900.

13. Maddukuri PV, Vieira ML, DeCastro S, et al. What is the best approach for the assessment of left atrial size? Comparison of various unidimensional and two-dimensional parameters with three-dimensional echocardiographically determined left atrial volume. **J Am Soc Echocardiog** 2006;19:1026-1032.

14. Sug IW, Song JM, Lee EY, et al. Left atrial volume measured by real-time 3-dimensional echocardiography predicts clinical outcomes in patients with severe left ventricular dysfunction and in sinus rhythm. **J Am Soc Echocardiog** 2008;21:439-445.

AORTIC VALVE

15. Handke M, Jahnke C, Heinrichs, G, Schlegel J, Vos C, Schmitt D, et al. New three-dimensional echocardiographic system using digital radiofrequency data: visualization and quantitative analysis of aortic valve dynamics with high resolution: methods, feasibility, and initial clinical experience. **Circulation** 2003;107:2876-9.

ENDOCARDITIS

16. Jain R, Kolias TJ. Three-dimensional transesophageal echocardiography of pacemaker endocarditis. **J Am Coll Cardiol** 2009;53:1241.

PROSTHETIC VALVES

17. Suseng L, Shernan SK, Weinert L, et al. Real-time three-dimensional transesophageal echocardiography in valve disease: comparison with surgical findings and evaluation of prosthetic valves. **J Am Soc Echocardiogr** 2008;21:1347-1354.

18. Kronzon, I, Sugeng L, Perk G, et al. Real-time 3-dimensional transesophageal echocardiography in the evaluation of post-operative mitral annuloplasty ring and prosthetic valve dehiscence. **J Am Coll Cardiol** 2009;53:1543-1547.

CONGENITAL HEART DISEASE

19. van den Bosch AE, Robbers-Visser D, Krenning BJ, et al. Real-time transthoracic three-dimensional echocardiographic assessment of left ventricular volume of ejection fraction in congenital heart disease**. J Am Soc Echocardiogr** 2006;19:1-6.

20. Chen G, Huang G, Tao Z, et al. Value of real-time three-dimensional echocardiography sectional diagnosis in complex congenital heart disease evaluated by receiver operating characteristic analysis. **J Am Soc Echocardiog** 2008;21:458-463.

21. Grison A, Maschietto N, Reffo E, et al. Three-dimensional echocardiographic evaluation of right ventricular volume and function in pediatric patients: Validation of the technique. **J Am Soc Echocardiog** 2007;20:921-929.

22. Abdel Massih, T, Dulac Y, Taktak A, et al. Assessment of atrial septal defect size with 3D-transesophageal echocardiography: comparison with balloon method. **Echocardiography** 2005;22:121-127.

23. van den Bosch AE, Ten Harkel D, McGhie JS, et al. Feasibility and accuracy of real-time three-dimensional echocardiographic assessment of ventricular septal defects. **J Am Soc Echocardiog** 2006;19:7-13.

HEMODYANMICS
Thomas Ryan, MD
Columbus, OH

GENERAL CONCEPTS

1. Maeder, MT, Kaye DM. Heart failure with normal left ventricular ejection fraction. **J Am Coll Cardiol** 2009;53:905-918.

2. Quinones MA, Otto CM, Stoddard M, et al. Recommendations for quantification of Doppler echocardiography: a report from the Doppler Quantification Task Force of the Nomenclature and Standards Committee of the American Society of Echocardiography. **J Am Soc Echocardiogr** 2002;15:167–184.

3. Thomas JD, Weyman AE. Fluid dynamics model of mitral valve flow: description with in vitro validation. **J Am Coll Cardiol** 1989;13:221–233.

4. Yoganathan AP, Cape EG, Sung HW, et al. Review of hydrodynamic principles for the cardiologist: applications to the study of blood flow and jets by imaging techniques. **J Am Coll Cardiol** 1988;12:1344–1353.

GRADIENTS & STENOSIS

5. Baumgartner H, Khan S, DeRobertis M, et al. Discrepancies between Doppler and catheter gradients in aortic prosthetic valves *in vitro*. A manifestation of localized gradients and pressure recovery. **Circulation** 1990;82:1467–1475.

6. Baumgartner H, Stefenelli T, Niederberger J, et al. "Overestimation" of catheter gradients by Doppler ultrasound in patients with aortic stenosis: A predictable manifestation of pressure recovery. **J Am Coll Cardiol** 1999;33:1655-1661.

7. Baumgartner H, Hung J, Bermejo J et al. Echocardiographic Assessment of Valve Stenosis: EAE/ASE Recommendations for Clinical Practice. **J Am Soc Echocardiog**, 2009;22:1-23.

8. Callahan MJ, Tajik AJ, Su-Fan Q, et al. Validation of instantaneous pressure gradients measured by continuous-wave Doppler in experimentally induced aortic stenosis. **Am J Cardiol** 1985;56:989–993.

9. Currie PJ, Hagler DJ, Seward JB, et al. Instantaneous pressure gradient: a simultaneous Doppler and dual catheter correlative study. **J Am Coll Cardiol** 1986;7:800–806.

10. Hatle L, Angelsen B, Tromsdal A. Noninvasive assessment of atrioventricular pressure half-time by Doppler ultrasound. **Circulation** 1979;60:1096–1104.

11. Hatle L, Brubakk A, Tromsdal A, et al. Noninvasive assessment of pressure drop in mitral stenosis by Doppler ultrasound. **Br Heart J** 1978;40:131–140.

12. Oh JK, Taliercio CP, Holmes DR Jr, et al. Prediction of the severity of aortic stenosis by Doppler aortic valve area determination: prospective Doppler-catheterization correlation in 100 patients. **J Am Coll Cardiol** 1988;11:1227–1234.

13. Richards KL, Cannon SR, Miller JF, et al. Calculation of aortic valve area by Doppler echocardiography: a direct application of the continuity equation. **Circulation** 1986;73:964–969.

14. Rifkin RD, Harper K, Tighe D. Comparison of proximal isovelocity surface area method with pressure half-time and planimetry in evaluation of mitral stenosis. **J Am Coll Cardiol** 1995;26:458–465.

15. Stamm RB, Martin RP. Quantification of pressure gradients across stenotic valves by Doppler ultrasound. **J Am Coll Cardiol** 1983;2:707–718.

16. Thomas JD, Wilkins GT, Choong CY, et al. Inaccuracy of mitral pressure half-time immediately after percutaneous mitral valvotomy. Dependence on transmitral gradient and left atrial and ventricular compliance. **Circulation** 1988;78:980–993.

QUANTITATIVE FLOW

17. Dubin J, Wallerson DC, Cody RJ, et al. Comparative accuracy of Doppler echocardiographic methods for clinical stroke volume determination. **Am Heart J** 1990;120:116–123.

18. Goldberg SJ, Sahn DJ, Allen HD, et al. Evaluation of pulmonary and systemic blood flow by 2-dimensional Doppler echocardiography using fast Fourier transform spectral analysis. **Am J Cardiol** 1982;50:1394–1400.

[19] Meijboom EJ, Rijsterborgh H, Bot H, et al. Limits of reproducibility of blood flow measurements by Doppler echocardiography. **Am J Cardiol** 1987;59:133–137.

[20] Miller WE, Richards KL, Crawford MH. Accuracy of mitral Doppler echocardiographic cardiac output determinations in adults. **Am Heart J** 1990;119:905–910.

[21] Moulinier L, Venet T, Schiller NB, et al. Measurement of aortic blood flow by Doppler echocardiography: day to day variability in normal subjects and applicability in clinical research. **J Am Coll Cardiol** 1991;17:1326–1333.

[22] Sanders SP, Yeager S, Williams RG. Measurement of systemic and pulmonary blood flow and QP/QS ratio using Doppler and two-dimensional echocardiography. **Am J Cardiol** 1983;51:952–956.

[23] Valdes-Cruz LM, Horowitz S, Mesel E, et al. A pulsed Doppler echocardiographic method for calculating pulmonary and systemic blood flow in atrial level shunts: validation studies in animals and initial human experience. **Circulation** 1984;69:80–86.

PRESSURE RECOVERY

[24] Levine RA, Jimoh A, Cape EG, et al. Pressure recovery distal to a stenosis: potential cause of gradient "overestimation" by Doppler echocardiography. **J Am Coll Cardiol** 1989;13:706–715.

[25] Niederberger J, Schima H, Maurer G, et al. Importance of pressure recovery for the assessment of aortic stenosis by Doppler ultrasound. Role of aortic size, aortic valve area, and direction of the stenotic jet *in vitro*. **Circulation** 1996;94:1934–1940.

REGURGITATION

[26] Chen C, Koschyk D, Brockhoff C, et al. Noninvasive estimation of regurgitant flow rate and volume in patients with mitral regurgitation by Doppler color mapping of accelerating flow field. **J Am Coll Cardiol** 1993;21:374–383.

[27] Enriquez-Sarano M, Miller FA Jr, Hayes SN, et al. Effective mitral regurgitant orifice area: clinical use and pitfalls of the proximal isovelocity surface area method. **J Am Coll Cardiol** 1995;25:703–709.

[28] Enriquez-Sarano M, Seward JB, Bailey KR, et al. Effective regurgitant orifice area: a noninvasive Doppler development of an old hemodynamic concept. **J Am Coll Cardiol** 1994;23:443–451.

[29] Enriquez-Sarano M, Sinak LJ, Tajik AJ, et al. Changes in effective regurgitant orifice throughout systole in patients with mitral valve prolapse. A clinical study using the proximal isovelocity surface area method. **Circulation** 1995;92:2951–2958.

[30] Flachskampf FA, Weyman AE, Gillam L, et al. Aortic regurgitation shortens Doppler pressure half-time in mitral stenosis: clinical evidence, *in vitro* simulation and theoretic analysis. **J Am Coll Cardiol** 1990;16:396–404.

[31] Otsuji Y, Toda H, Ishigami T, et al. Mitral regurgitation during B bump of the mitral valve studied by Doppler echocardiography. **Am J Cardiol** 1991;67:778–780.

[32] Samstad SO, Hegrenaes L, Skjaerpe T, et al. Half time of the diastolic aortoventricular pressure difference by continuous wave Doppler ultrasound: a measure of the severity of aortic regurgitation? **Br Heart J** 1989;61:336–343.

[33] Utsunomiya T, Doshi R, Patel D, et al. Calculation of volume flow rate by the proximal isovelocity surface area method: simplified approach using color Doppler zero baseline shift. **J Am Coll Cardiol** 1993;22:277–282.

RIGHT HEART

[34] Abbas AE, Fortuin FD, Schiller NB, et al. A simple method for noninvasive estimation of pulmonary vascular resistance. **J Am Coll Cardiol** 2003;41:1021–1027.

[35] Lee RT, Lord CP, Plappert T, et al. Prospective Doppler echocardiographic evaluation of pulmonary artery diastolic pressure in the medical intensive care unit. **Am J Cardiol** 1989;64:1366-1370.

[36] Ryan T, Petrovic O, Dillon JC, et al. An echocardiographic index for separation of right ventricular volume and pressure overload. **J Am Coll Cardiol** 1985;5:918–927.

[37] Silbert DR, Brunson SC, Schiff R, et al. Determination of right ventricular pressure in the presence of a ventricular septal defect using continuous wave Doppler ultrasound. **J Am Coll Cardiol** 1986;8:379–384.

[38] Verdejo HE, Castro PF, Concepcion R, et al. Comparison of a radiofrequency-based wireless pressure sensor to swan-ganz catheter and echocardiography for ambulatory assessment of pulmonary artery pressure in heart failure. **J Am Coll Cardiol** 2007;50:2375-2382.

[39] Yock PG, Popp RL. Noninvasive estimation of right ventricular systolic pressure by Doppler ultrasound in patients with tricuspid regurgitation. **Circulation** 1984;70:657–662.

MISCELLANEOUS

[40] Bargiggia GS, Bertucci C, Recusani F, et al. A new method for estimating left ventricular dP/dt by continuous wave Doppler-echocardiography. Validation studies at cardiac catheterization. **Circulation** 1989;80:1287–1292.

[41] Bordacher P, Lafitte S, Reuter S, et al. Echocardiographic parameters of ventricular dyssynchrony validation in patients with heart failure using sequential biventricular pacing. **J Am Coll Cardiol** 2004;44:2157-2165.

[42]. Tei C, Ling LH, Hodge DO, et al. New index of combined systolic and diastolic myocardial performance: a simple and reproducible measure of cardiac function—a study in normal and dilated cardiomyopathy. **J Cardiol** 1995;26:357–366.

[43] Waggoner AD, Faddis MN, Gleva MJ, et al. Improvements in left ventricular diastolic function after cardiac resynchronization therapy are coupled to response in systolic performance. **J Am Coll Cardiol** 2005;46:2244-2249.

ECHOCARDIOGRAPHIC ASSESSMENT OF CANINE DEGENERATIVE VALVULAR DISEASE

John D Bonagura, DVM, DACVIM
Columbus, OH

Mitral regurgitation (MR) in dogs is most often caused by chronic degenerative valvular disease (CVD). This lesion, also called valvular endocardiosis and myxomatous valvular disease, develops slowly and increases in severity over a period of years.[1,2] Progressive MR induces cardiac remodeling, characterized by left atrial (LA) and left ventricular (LV) dilatation, eccentric myocardial hypertrophy, and alteration of the intercellular matrix. Neurohormonal activation and the narrowing of small (intramural) coronary arteries may play roles in the pathogenesis of LV remodeling. Despite preservation of global LV shortening, LV dysfunction can eventually develop in CVD[3], along with increases in LA, pulmonary venous, and pulmonary arterial pressures. By this stage, many dogs are symptomatic for heart failure. Echocardiography (Echo) can identify and track many of these developments noninvasively[4] and offers useful information about the heart and hemodynamic changes that accompany chronic MR. This brief review focuses on echocardiographic assessment of dogs with CVD.

PRINCIPLES

A comprehensive clinical approach to assessing the severity of MR is optimal. This should consider the historical, physical, blood pressure, radiographic, ECG, and clinical laboratory findings along with changes evident by integrated 2D, M-mode, and Doppler echocardiographic studies. Echocardiographic abnormalities of importance include understanding the nature of the valvular lesion, the degree of cardiac remodeling, severity of MR, presence of systolic and diastolic ventricular dysfunction, and findings indicating elevations of pulmonary venous and arterial pressures.[5,6] Recognition of LV systolic or diastolic dysfunction in the setting of CVD is challenging, and there is no totally-load independent measure available. In human patients, quantitation of LV function and ventricular filling pressures is pivotal as these guide surgical intervention. Surgical treatment of canine CVD is rarely undertaken currently. However, information about LV function in CVD can be important in understanding the natural history of this disease and guide follow-up care. Potentially, the medical therapy of affected dogs could be influenced by an echocardiographic assessment of LV function or identification of elevated filling pressures. Stratification of risk can also be relevant to entry into and interpretation of clinical trials in dogs with MR.

The traditional measures of global LV systolic function are the ejection and shortening fractions. Ejection fraction (EF) is commonly measured by two-dimensional (2D) Echo using a single-plane (long axis or apical) image of the LV. Shortening fraction (SF) is measured in most situations from the M-mode echocardiogram, but also can be determined from frozen 2D images at end systole and end diastole. Additionally, experienced examiners may subjectively evaluate radial shortening of the LV (i.e., from diastole to systole) in the short-axis plane at the level of the papillary muscles. Systolic heart failure is classically characterized by reduced EF and SF and subjectively by reduced radial shortening (fractional area shortening); however, in CVD these measures can be very misleading.[7] Significant MR increases LV preload and reduces afterload. These alterations foster a normal to hyperdynamic ventricular contraction, even in the presence of impaired myocardial cell contraction.[8]

The assessment of LV diastolic function is possibly more difficult in the dog with MR. Diastolic function is complicated with critical components including active myocardial relaxation, chamber stiffness (compliance or distensibility), and the instantaneous LA to LV pressure gradient driving blood across the opened mitral valve.[9] The LA pressure directly relates to regurgitant volume, LA compliance, and pulmonary venous pressure. LV relaxation can be impaired as a function of age or associated conditions such as systemic hypertension. With severe volume overload, higher LV chamber stiffness can develop reducing distensibility. If cardiac output is limited, volume retention may occur. Additionally, moderate to severe MR produces an abrupt increase in LA pressure related to the simultaneous arrival of the regurgitant stroke volume and pulmonary venous return. This prominent v-wave transiently increases LA pressure, and contributes somewhat disproportionately to early LV filling, a period often used to assess diastolic function. Thus, the Echo assessment of diastolic ventricular function in heart failure becomes a combined assessment reflecting the interplay of muscle relaxation, chamber compliance (or distensibility), regurgitant volume, atrial function, and venous pressures.[10] Even with a full complement of Echo data, the diastolic assessment can be ambiguous.

Currently the Echo evaluation of LV function and filling pressures in dogs with CVD is based on surrogate measures that relate to ventricular remodeling, pumping function, and filling. In the future, the application of stress or exercise Echo may be useful for identifying contractile reserve (or its absence). Echo evaluations used in the author's practice are considered in the following sections. A multi-modality approach for assessment of systolic LV function in MR is suggested.

2D AND M-MODE ECHOCARDIOGRAPHY IN CHRONIC VALVULAR DISEASE

The 2D (and 3D) examination is fundamental for recognition of mitral and tricuspid valvular anatomy, including abnormal valve thickness, abnormal valve motion (prolapse; flail leaflet), and chordal rupture. It is likely that 3D (4D) imaging will provide greater insight into valvular anatomy, particularly as 3D pediatric transducers become more widely available. For practical purposes, the 2D, M-mode, color-Doppler, spectral-Doppler, and tissue-Doppler studies generally suffice. These confirm the diagnosis of chronic degenerative valvular disease and allow for a relatively comprehensive assessment of the situation.

The LV volumes and diameters needed for estimation of cardiac size and for calculation of ejection and shortening fractions can be gleaned from 2D or M-mode echocardiographic images. In MR more than half of the total stroke volume may be ejected retrograde (into the LA), so that LV volume changes over the cardiac cycle are not equivalent to cardiac output. Volumes are calculated with 2D Echo from an area tracing of the internal LV area using a single long-axis plane or two right-angled (orthogonal) apical image planes. Area is usually calculated from a length-area method (which assumes the LV is an ellipsoid chamber) or from a Simpson's rule method (of summed discs). Normal values for EF in healthy dogs using single-plane methods are approximately 45 to 55%. Volume estimates from the M-mode echo are generally made by cubing the minor dimension and adjusting for changes from normal geometry; however, the Teicholz adjustment used in most Echo systems likely results in inaccurate volumetric estimates in many dogs. Shortening fraction calculated from M-mode studies are generally normal to increased in small breed dogs with chronic MR. Normal values for shortening fraction are approximately 30 to 40% for small breed dogs but probably as low as 22 to 25% in larger breed dogs. The examiner can also evaluate the movements of multiple short-axis dimensions by tracing the internal LV area to calculate fractional area shortening at the dorsal papillary level. This value approximates the ejection fraction, and in our laboratory exceeds >45% in most healthy dogs. The actual percent of ventricular shortening identified using any of these methods relates to technical aspects of recording and measurement; physiologic state; body size; and importantly the volume of MR as described below.

These systolic function indices are unreliable in the setting of moderate to severe MR due to CVD. In these situations, both EF and SF achieve values exceeding normal and the shortening area appears hyperdynamic when observed in real time. Cardiologists often remark that a dog with severe MR and preserved LV function should have a hyperdynamic chamber, but it is difficult to quantify how high that value should achieve in a given case. Values for SF are often in the 45 to 55% range, even in dogs with overt CHF.[11] The corresponding values for EF often exceed 70%. For this reason, finding a normal EF or SF in the presence of advanced MR suggests LV myocardial failure, and perhaps a poorer prognosis. The situation is slightly different in larger breed dogs wherein normal to low shortening fraction is more commonly encountered in the setting of advanced CVD.

As MR worsens, LV diastolic volume, as estimated by the internal (minor) diameter, increases in value and it would be surprising to identify a normal LV diastolic dimension in a dog with advanced mitral valve disease. This finding of increased diastolic dimension is compatible with ventricular dilatation and volume overload. Secondly, as systolic function becomes impaired, the end-systolic dimension increases, despite the escalating volumes ejected into the lower resistance LA. Thirdly, if LV contractile function is preserved, the fully-compensated ventricle will shorten to the near- normal end-systolic volume (or diameter). Kittleson and colleagues[8] advanced this concept in small breed dogs with MR, suggesting that an increased end-systolic volume index correlated to impaired LV systolic function. In a study by Borgarelli et al.,[11] the end-systolic volume index was also abnormal in dogs with CHF caused by CVD. In large-breed dogs with MR, even EF and SF can be reduced. The indexing or modeling of LV internal measurements to body surface area or mass is very logical, and most studies demonstrate a general linear relationship between chamber size and body mass. However, multi-breed data sets suffer from relatively small sample sizes, gaps in data across body sizes, and a relative lack of points at the extreme ends of very small and very large breeds. Accordingly, the 95% confidence limits defining "normal", though statistically sound, are very wide, making it very difficult to identify mild or even mild-moderate ventricular dilatation.

A common feature of advanced CVD in dogs is apparent reduction in regional LV wall motion relative to ventricular septal motion. In normal dogs, the septal excursion is less than that of the free wall. This probably relates to both wall and global cardiac (translational) movements. However, in dogs with severe CVD the opposite is often observed. This has been interpreted as regional LV systolic dysfunction, altered translational motion from cardiomegaly, or perhaps both. However, experimentally this change has been related simply to alterations in LV geometry with deviation of the septum to the right promoting greater leftward excursion of the septum during systole.[13] Thus, the clinical relevance of this finding is unresolved (aside from supporting significant volume overload of the LV). Regional wall abnormalities also may be evident with advanced echocardiographic imaging such as myocardial strain. Additionally, one can measure the LV wall thickness along with diastolic LV dimension to estimate LV mass. It has been suggested reported that insufficient hypertrophy, as assessed by the wall thickness to chamber radius ratio, may be a risk factor for development of ventricular dysfunction and heart failure in larger breed dogs with CVD.[11]

DOPPLER ECHOCARDIOGRAPHIC STUDIES IN CHRONIC VALVULAR DISEASE

The assessment of MR severity by color Doppler Echo[6,14,15] can involve receiving chamber analysis (jet length or area relative to LA area); measurement of the vena contracta (jet width), and analysis of the proximal region of flow convergence. The former is most often used, but it should be noted that validation studies have generally suggested the need for orthogonal image planes for assessment by this method. The width of the vena contracta is often evident on color studies and corresponds to lesion severity, but different canine body sizes make analysis of this diameter problematic. The radius drawn from the regurgitant orifice to the hemisphere of the proximal isovelocity area from a shifted color baseline can be used to calculate regurgitant flow rate and effective regurgitant orifice area; however, this must also be evaluated relative to the size of the patient.

Continuous wave (CW) Doppler studies of MR can be used to identify elevated LA pressures and LV systolic and diastolic dysfunction. MR is characterized by a high velocity, turbulent, systolic jet recorded at the valve orifice and within the left atrium. The RBC velocity at the envelope of the jet relates to instantaneous pressure difference or gradient and the peak CW velocity of MR in most normotensive dogs falls into the 5 to 6 m/s range. In cases of systemic hypertension (or left ventricular outflow obstruction), the peak regurgitant velocity may exceed 6.5 m/s owing to elevated LV systolic pressure. Conversely, when there is severe systolic dysfunction, systemic hypotension, volume depletion secondary to excessive diuresis, or when pressures in the LA are very high, both the peak velocity and the rate of velocity change may be reduced. These changes reflect reductions in maximal instantaneous pressure gradient and diminished vigor of LV contraction. The subjective contour of the MR jet may also change from a rounded peak to a more abrupt peak with a rapid velocity deceleration if there is an abrupt and severe increase in LA pressure associated with MR. The first derivative of LV pressure change, dp/dt, can also be obtained from the MR jet and is considered a load dependent estimate of global LV systolic function; the value does decline in some dogs with advanced MR. Canine studies have demonstrated a close relationship between catheter-measured and Doppler estimated dp/dt.[16,17] Reference values for dp/dt using this method in small breed dogs with mild MR and normal LV function have not been published, but in our experience, these values are lower than those of instantaneous dp/dt recorded from instrumented dogs that often exceed 3000 mm Hg/s. As with other LV function indices, downward trends will be more instructive than one single measurement.

Many dogs with CVD also develop tricuspid regurgitation due to CVD and to pulmonary hypertension (PH), a frequent consequence of moderate to severe MR. Pulmonary hypertension can be explained firstly by the increase in peak and mean left atrial pressures that occur with chronic MR[18] and other pathophysiologic mechanisms are probably important. TR velocities exceeding >3 m/s (in the absence of RVOT obstruction) generally indicate PH.[19,20] Other methods for identifying PH, such as the ratio of PA acceleration time to ejection time can be useful when TR is not evident. This ratio decreases as pulmonary artery resistance increases; a ratio of <0.31 was 87% specific for the diagnosis of pulmonary hypertension[19] and in another study[20] acceleration time/ejection time ratio averaged 0.46 in the controls but was reduced (to a mean of 0.37) in dogs with moderate to severe pulmonary hypertension. Following this ratio in dogs with progressive MR may be useful in identifying onset of significant PH and in evaluating its treatment.

Doppler recordings of transmitral flow velocity and the diastolic movements of the ventricular wall and septum can be quantified to identify diastolic dysfunction and elevated LV filling pressures, but this issue is very complicated in the setting of normal to hyperdynamic LV systolic function. Mild diastolic dysfunction is characterized by impaired myocardial relaxation with E/A reversal of transmitral flow and in tissue Doppler recordings. However, increasing LA pressure caused by MR can normalize the transmitral flow pattern ("pseudonormal filling"), reproducing a normal transmitral E/A ratio. This concept that increased LA pressure will normalize early filling velocity (transmitral E-wave velocity), but that tissue velocities (E') will be minimally affected, constitutes the basis for using the ratio of E/E' as an estimate of LV filling pressures as the E' of the tissue "corrects" for the effects of relaxation on transmitral E-wave velocity. However this ratio of E/E' is only valid when E' is reduced and relatively independent of the effects of preload. Unfortunately, in advanced MR, volume overload is evident, and the high filling pressures characteristic of CHF are more difficult to identify. Thus proposed diagnostic ratios in dogs with MR are considerably higher than for patients with impaired LV systolic and diastolic function (e.g., dilated cardiomyopathy). Once the peak E-wave velocity exceeds 1.25 to 1.5 m/s, the likelihood of elevated filling pressures is much higher and the prognosis worse[21]. Additionally a high E/E' ratio also predicts elevated LA pressure. In one acute experimental study, a ratio of E/E' of >9 was associated with a pulmonary wedge pressure of >20 mm Hg 95% of the time[22]. Data from our institution (Schober and colleagues) suggest that an E/E' ratio >12, along with a high E-wave velocity, can predict findings of CHF on thoracic radiographs in dogs with chronic spontaneous CVD. This is similar to the ratio of 13 found to be about 80% sensitive and specific for the diagnosis of CHF in dogs[23]. The precise cutoffs of E/E' predicting heart failure are likely different for acute versus chronic MR.

Other Doppler measures such as pulmonary venous flow patterns are potentially useful, but often confusing due to projection of MR into the pulmonary veins. Another variable that has tracked LA pressure well in a number of studies in different species is the isovolumetric relaxation time. Normal isovolumetric time in healthy older dogs ranges from approximately 41 ms to 73 ms; impaired relaxation prolongs this interval while elevated left atrial pressure shortens it. While true isovolumetric relaxation does not occur in MR, this time is highly influenced by the combined effects of LV relaxation, LV compliance, and LA pressure. The latter effect becomes the most prominent in dogs with advancing mitral valve disease. Progressive reduction in this interval on serial examination is suggestive of elevated filling pressures and impaired LV diastolic function,[10,24] and is useful even in the setting of atrial fibrillation. In an acute study of volume loading, a ratio of peak E wave velocity (in m/s) to isovolumetric relaxation time (in ms) of >2.2 suggested mild to moderate elevations in left atrial pressure (> 15 mm Hg) in dogs.[25,2] In our practice, values of IVRT < 45 ms and values of E:IVRT >2.5 in dogs with moderate to severe MR were usually associated with radiographic signs of CHF (Schober et al. in press). Another potentially useful variable is the mitral deceleration time, which can be evaluated even in the setting of atrial fibrillation (provided short diastolic cycles are not chosen). A high velocity E wave with an abbreviated deceleration time generally indicates the combination of a high LA pressure and a noncompliant LV; however, in our studies of dogs with MR, mitral deceleration time has been a relatively weak predictor of elevated filling pressures.

Aside from providing a detailed analysis of how the LV contracts and fills, the assessment of LV function and filling pressures by Echo will become really useful when these data assist with short- and long-term prognoses, guide follow-up intervals, and help to direct cardiac therapy. It may become reasonable to perform directed Echo studies specifically to estimate the CHF risk in dogs with CVD, provided suitable cut-offs are identified. Since radiography often requires a need for heavy sedation, this makes Echo estimates attractive, especially as availability of Echo systems increases. How Doppler estimates of filling pressures measure up against radiography and biomarkers such as NT-pBNP in dogs also needs more study

OTHER METHODS FOR ASSESSING LV FUNCTION IN CVD

A number of tissue-centered and computer-based methods have become popularized for assessing LV global and regional function, and it is likely that some of these may be useful in dogs with CVD. Aside from the tissue Doppler methods discussed above, measurements of segmental contraction velocity, deformation (or strain), and rate of deformation (strain rate) have been shown to be a sensitive indicators of ventricular function in some canine studies[27]. The major promise of these examinations is found in the potential to identify LV dysfunction before simpler conventional methods are diagnostic. These variables are computer generated from proprietary software and both validation and repeatability studies are needed before these are applied widely to veterinary situations. The strain data are derived from either tissue Doppler imaging or "speckle tracking" of 2D gray scale images.[28] Myocardial function can be measured orthogonally in longitudinal, radial, or circumferential orientations. Myocardial torsion or twist can also be calculated as tests of global systolic and diastolic. In models of canine MR, early diastolic twist becomes impaired indicating diastolic dysfunction[29].

REFERENCES

[1]Haggstrom J. **Journal of Veterinary Cardiology** 2004; 6:6. [2]Pedersen HD and Haggstrom J. **Cardiovascular Research** 2000; 47:234. [3]McGinley JC et al. **European Journal of Heart Failure** 2007; 9:857. [4]Serres F et al. **J Vet Internal Med** 2008; 22:566. [5]Mazur W and Nagueh SF. Current Opinion in Cardiology 2001; 16: 246. [6]O'Gara P., et al. **J Am College Cardiology**; Cardiovascular Imaging 2008; 1:221. [7]Lee R and Marwick TH. **European Journal of Echocardiography** 2007; 8:175. [8]Kittleson MD et al. **Journal of the American Veterinary Medical Association** 1984; 184:455. [9]Nagueh SF, et al. **European Journal of Echocardiography** 2009; 10:165. [10]Ohno M et al. Circulation 2004; 89:2241. [11]Borgarelli M, et al. **J Vet Internal Med** 2007; 21:61. [12]Haggstrom J, et al. **Veterinary Clinics of North America**, Small Animal Practice 2004; 34:1209. [13]Young AA, et al. **American Journal of Physiology** 1996; 271:H2689. [14]Kittleson MD and Brown WA. J Vet Internal Med 2003; 17:84. [15]Gouni V, et al. **Journal of the American Veterinary Medical Association** 2007; 231:399. [16]Chen C, et al. Circulation 1991; 83:2101. [17]Asano K et al. **Australian Veterinary Journal** 2002; 80:25. [18]Chiavegato D, et al. **Veterinary Radiology and Ultrasound** 2009; 50:253. [19]Schober KE, et al. **J Vet Internal Med** 2006; 20:912. [20]Serres F et al. J Vet Internal Med 2007; 21:1280. [21]Borgarelli et al. **J Vet Internal Med** 2008; 22:120. [22]Oyama et al. **J Vet Internal Med** 2004; 18:667. [23]Teshima K et al. **Journal of Veterinary Science** 2005; 67:1207. [24]Diwan et al. **Circulation** 2005; 111:3281. [25]Schober KE, et al. **American Journal of Veterinary Research** 2008; 69:1034. [26]Schober KE et al. **J Vet Internal Med** 2008; 22:578. [27] Wang J et al. **Circulation** 2007; 115:1376.[28] Chetboul V, et al. **Journal of Veterinary Cardiology** 2007; 9:69. [29]Tibayan FA, et al. **The Journal of Heart Valve Disease** 2002; 11:39.

EVALUATION OF PULMONARY HYPERTENSION & DISORDERS OF THE GREAT VESSELS BY COMPUTED TOMOGRAPHY

David Sisson, DVM, DACVIM-Cardiology
Corvallis, Oregon

INTRODUCTION

Cardiac imaging can be accomplished by a variety of methods including traditional and digital radiography, echocardiography, fluoroscopy (cardiac catheterization), nuclear imaging, magnetic resonance imaging (MRI), and computed tomography (CT). These modalities differ with respect to cost of the procedure and type of information provided. Their value often resides in the structural and functional information they provide which is largely a consequence of the spatial, temporal, and contrast resolution achievable with each of these modalities. Computerized tomography has certain attributes that make this imaging modality particularly useful for the evaluation of the great vessels, coronary arteries, pulmonary vasculature and pulmonary parenchyma. Recent improvements in spatial and temporal resolution are broadening the application of CT to include the assessment of myocardial perfusion, evaluation of left and right ventricular function, quantification of the severity of mitral and aortic valvular heart disease, and the elucidation of the anatomic details of certain congenital heart defects that are difficult to image using echocardiography.[1] Unlike echocardiography and MRI, CT does not display direct estimates of regional blood flow.

DEVELOPMENT OF CARDIAC CT

Alan Cormack and Sir Godfrey Hounsfield, working independently, developed the first CT systems in the early 1970s and the significance of their achievement was recognized when they were jointly awarded the Nobel Prize in Medicine in 1979.[2] These first primitive CT imaging devices utilized a moving x-ray tube and detector and took over four minutes to acquire the data required to display a single axial tomographic slice. Somewhat later (1976), the first gated cardiac images were produced by Dr. Yo-ichirou Um`egaki and coworkers using an innovative steerable electron beam that eliminated the requirement for moving parts, resulting in a marked reduction in slice acquisition time. Unfortunately, this novel approach suffered crucial limitations of spatial resolution and low signal/noise ratios and the method has given way to the originally-conceived design using rotating tubes and detectors. In 1996, a single slice helical CT was introduced with a 1 second gantry rotation speed. Since then, system designs incorporating multiple rows of detectors (MDCT) have led to the rapid sequential introduction of 4-slice (1998), 16-slice (2001), and 64-slice (2004) CT scanning systems. Exploitation of this design element culminated with the recent introduction of an impressive 320-detector row helical CT by Toshiba in 2007. Other design innovations crucial to the development of modern CT scanners have included the development of slip-ring technology which allows for continuous high speed gantry rotation and improved temporal resolution; the use of 180^0 reconstruction algorithms that effectively double temporal resolution; the development of dual source CT imaging (DSCT) which improved temporal resolution to one-fourth of the gantry rotation time; substantial reductions of collimator slice thickness that have improved spatial resolution; novel applications of ECG-gating to aid the retrospective assembly of cardiac images and as a guide for prospective data acquisition; and the introduction of segmentation algorithms that permit phase-dependent functional analysis of cardiac images.[2-4]

SCAN ACQUISITION

When imaging the heart and lungs, temporal resolution can be optimized and motion artifacts minimized via breath holding and ECG-gating during the CT acquisition. Animal patients are usually anesthetized and intubated in order to accomplish involuntary breath holding. The CT study is accomplished in 3 stages when contrast media is administered.[5] Individual imaging centers use customized protocols that are specific to the intended application and which are frequently modified in response to individual patient circumstances. Initially, a non-contrast topogram of the chest is obtained to define the boundaries of the scan. Bolus tracking, utilizing a preset threshold of a defined area of interest, is then used to coordinate the timing of contrast injection and the onset of the image acquisition. In some circumstances where the transit time of the contrast is hard to predict, a test bolus can be administered and tracked during an abbreviated scan to guide the final contrast-enhanced, full-scan protocol. Most often, highly iodinated contrast is administered at a dose of 0.5 to 1.0 ml/kg, depending on the precise application; and the contrast is usually followed by a saline chaser of equal volume delivered at a similar rate. Typical settings of the CT scan include a tube voltage of 100 to 120 kV, depending on patient size and a tube current of 100 mA/slice; slice thicknesses varying from 0.5 to 1.0 mm depending on the application and required resolution, and a helical pitch factor approximating 0.85, which indicates some overlap of slices. Data acquisition times vary from 5 to 15 seconds. Some imaging protocols mandate the acquisition of separate arterial and venous phase data sets, doubling the acquisition time and the radiation dose to the patient. Our current system does not incorporate ECG-gating, limiting our ability to obtain high resolution images of the coronary vasculature and intra-cardiac anatomy and precluding quantitative estimates of systolic and diastolic chamber volume and ejection fraction.

IMAGE PROCESSING AND DISPLAY

After acquiring a volumetric data set in the transverse plane, they are loaded into the multiplanar reformation (MPR) program that displays the information in transverse, sagittal, and coronal image data sets at the desired image thickness, typically varying from 0.5 to 2 mm. Three-dimensional rendering is a useful means of conveying information and common processing formats include maximum intensity projections (MIP) and shaded surface display. Volume rendering is required for most of the robust imaging applications that are now used to display CT imaging data sets.[6,7] Volume rendering uses technology developed for motion picture computer animation and it requires substantial processing power. Volume rendering combines the use of opacity values and lighting effects to allow appreciation of spatial relationships between structures.[6] Arbitrary color assignments are made to create desired visual effects that are particularly dramatic when displayed using perspective volume display techniques.

CLINICAL APPLICATIONS

We are currently involved in three CT imaging projects in dogs. One is designed to evaluate the utility of contrast-enhanced CT imaging in dogs with pulmonary hypertension. Contrast-enhanced CT imaging has particular merit in the evaluation of patients with pulmonary arterial hypertension due to its utility for evaluating the pulmonary vasculature, lung parenchyma, and the heart.[8] With this methodology, it is quite easy to obtain diameter measurements of the main pulmonary artery together with measurements of its segmental and subsegmental branches. In addition, the pulmonary arterial tree can be systematically evaluated for evidence of chronic thromboembolism by noting any in situ thrombi, eccentric laminated filling defects, or occlusions. Pulmonary diseases, such as idiopathic pulmonary fibrosis, can often be identified by the distinctive CT imaging patterns that result; and a variety of airway-centered and parenchymal infiltrative disorders can be detected and quantified.

We are also evaluating the utility of CT imaging in a select group of congenital heart disorders. We believe that CT imaging will be particularly useful when echocardiography does not provide sufficient anatomic detail and magnetic resonance imaging cannot be used. Computerized tomography has been used to identify a variety of congenital heart defects including tetralogy of Fallot, patent ductus arteriosus, vascular ring anomalies, transposition of the great vessels, left and right ventricular outflow tract obstructions, and coronary artery anomalies.

Finally we are evaluating the value of CT in dogs with pericardial disease. CT is likely to have particular merit for the evaluation of patients with pericardial effusion given the common occurrence of hemagiosarcoma, heart base tumors and mesothelioma. Moreover, the sensitivity of CT for detecting pulmonary metastasis in dogs with malignant tumors is substantially greater than with traditional radiography.

SAFETY CONSIDERATIONS

There are a number of significant patient safety issues involved in contrast-enhanced CT imaging.[1] One important clinical risk is that of contrast-induced nephropathy. The requirement for general anesthesia in animal patients adds another element of low but unavoidable risk. Importantly, the estimated patient radiation dose received during contrast-enhanced CT imaging is generally in the range of 5 to 20 mSv, which is equal to the dose received from several hundred thoracic radiographs. While this level of radiation exposure is in the same range as other cardiac diagnostic studies, it is important to employ strategies that conform to ALARA principles, particularly in young animals.

REFERENCES

1. Schroeder S, Achenbach S, Bengel F, et al. Cardiac computed tomography: indications applications, limitations, and training requirements. **Eur Heart J** 2008; 29:531-556. 2. Hurlock GS, Higashino H, Mochizuki T. History of cardiac computed tography: single to 320-detector row multislice computed tomography. **Int J of Cardiovasc Imaging** 2009;25:31-42. 3. Choi SI, George RT, Schuleri KH, et al. Recent developments in wide-detector cardiac computed tomography. **Int J Cardiovasc Imaging** 2009;25:23-29. 4. Flohr TG, Ohnesorge BM. Imaging of the heart with computed tomography. **Basic Res Cardiol** 2008;103:161-173. 5. Wilson GT, Gopalakrishnan P, Tak T. Noninvasive cardiac imaging with computed tomography. **Clin Med & Research**. 2007;5:165-171. 6. Lawler, LP, Fishman, EK. Multi–Detector Row CT of Thoracic Disease with Emphasis on 3D Volume Rendering and CT Angiography. **Radiographics** 2005; 21:1257-1473. 7. Neal C. Dalrymple, NC, Prasad, SR, Freckleton, MW, et al. Informatics in Radiology (infoRAD): Introduction to the Language of Three-dimensional Imaging with Multidetector CT. **Radiographics** 2005; 25:1409-1428. 8. Hoey ETD, Gopalan D, Agrawal SKB, et al. Cardiac causes of pulmonary arterial hypertension: assessment with multidetector CT. **Eur Radiol** 2009;19:2557-2568.

OPTIMAL NUTRITION FOR FELINE CARDIAC PATIENTS

Lisa M. Freeman, DVM, PhD, DACVN
North Grafton, MA

INTRODUCTION

Cardiac disease is one of the most common medical problems in cats, and can be one of the most challenging to treat. Optimal treatment of the cat with cardiac disease requires balancing multiple medications in order to achieve beneficial effects while minimizing side effects, modifying diet, and educating the owner in order to achieve and maintain good quality of life. Many scientific advances have improved our knowledge of cardiac disease and congestive heart failure (CHF). It is becoming increasingly apparent that we not only need to treat the symptoms of CHF (eg, fluid accumulation, arrhythmias), but also need to treat the underlying neurohumoral mechanisms to achieve optimal management of CHF patients. In addition to newer cardiac medications, recent advances have improved our understanding of nutritional interventions and nutritional pharmacology. Addressing both medical and nutritional therapy is critical for the optimal management of these patients.

Optimal Body Condition

Cardiac cachexia is the loss of lean body mass that occurs in animals with CHF. Cachexia is often mistakenly viewed as an end-stage syndrome manifested by an emaciated cat but, in fact, cachexia is a slowly progressive process of lean body mass loss that can be very subtle initially and can even occur in an obese animal. Recognizing the process of cachexia at an early stage provides improved opportunities to manage it effectively. Even if the relatively insensitive criterion of weight loss (rather than muscle loss) is used to identify cachexia, this is a relatively common problem in cats with CHF. In one study, one-third of cats lost $\geq$5% of their body weight in less than 12 months (the criteria currently used for cardiac cachexia in people). This number would likely be much higher if the affected body compartment (ie, muscle or lean body mass) were more specifically assessed.

The loss of lean body mass in cardiac cachexia is a multifactorial process affected by reductions in appetite and increased energy requirements. However, it is the increased production of inflammatory cytokines, such as tumor necrosis factor (TNF) and interleukin-1 (IL-1), that are the primary mediators of cachexia. These inflammatory cytokines are known to directly cause anorexia, to increase energy requirements, and to increase the catabolism of lean body mass. These inflammatory cytokines also cause cardiac myocyte hypertrophy and fibrosis.

The deleterious effects of cachexia have been emerging and recent studies have emphasized the role of body weight and body composition in heart failure. While obesity is a risk factor for development of heart disease in people, obesity may actually be associated with a protective effect once heart failure is present - this is known as the obesity paradox. A recent large meta-analysis on body condition in people with heart failure concluded that obesity and overweight were associated with lower all-cause and cardiovascular mortality and that underweight patients consistently had a higher risk of death. Given the adverse effects associated with cachexia, the association between obesity and improved survival in heart failure appears to be due to a *lack* of cachexia, rather than to the obesity per se. This is likely due to the increased reserve of lean body mass in overweight and obese people. The obesity paradox also has been demonstrated in dogs and cats with heart failure, and recent studies in people and cats suggest a "U-shaped" curve with the worst survival for those with the lowest and highest weights. These data emphasize the importance of avoiding weight (and muscle) loss, as well as severe obesity, in the cat with heart failure. However, further research is needed to understand the effects of mild overweight or obesity in cats with CHF.

Anorexia is a common problem in cats with CHF that contributes to cardiac cachexia and is one of the important factors involved in an owner's decision of euthanasia. Although many cats with cardiac disease are not completely anorectic, many have a reduced appetite or alterations in food preferences. In one study of cats with cardiac disease (not all of which had CHF), a reduced or complete loss of appetite was present in 38% of cats. Anorexia can be the result of medical therapy, diet, or the underlying disease (the inflammatory cytokines, TNF and IL-1, which are elevated in CHF cause anorexia). Recommendations for anorectic cats with CHF include the following:

1. Make sure the CHF is controlled. Anorexia is often an early sign of worsening CHF
2. Is there a potential drug interaction or toxicity? Check renal function and electrolytes. Consider a reduction in drug dosages if the CHF is controlled.
3. Was there an abrupt diet change? Animals should be transitioned onto new diets slowly. Dietary adjustments are probably not critically important for the short-term management of CHF unless the animal is eating very high sodium foods. Drug therapy should be initiated and stable for several days, and animals should be tolerating their drug therapies well and eating well before significant dietary changes are implemented. It is more important for an animal to be eating initially than for it to be eating the "ideal" diet. Nevertheless, during this period of diet transition, high sodium diet and high sodium treats should be avoided.
4. Try switching to a more palatable food. For some cats, changing from dry to canned food (or vice versa) often helps. For others, a balanced homemade diet may be necessary. Warming the food, feeding smaller, more frequent meals, or adding flavor enhancers (eg, meats or fish cooked without sodium for cats) also may improve appetite.
5. Consider fish oil supplementation. One method of decreasing cytokine production is with omega-3 fatty acid supplementation (see below). Supplementation of fish oil, which is high in omega-3 fatty acids, can improve food intake in some animals with CHF.
6. Consider a feeding tube. Many owners appreciate the ease of feeding once a tube is placed. This allows body weight to be more easily maintained and also for the owner to feed diets with an appropriate nutrient profile, rather than to rely on what the cat will

eat voluntarily (which is not always the most appropriate).

Sodium Restriction has been recommended for heart failure for many years. Although healthy animals can easily excrete excess dietary sodium in the urine, this response is blunted in animals with cardiac disease. Therefore, some authors have recommended changing to a low sodium diet when a heart murmur was first detected, even before clinical signs are present but the authors do not recommend severe sodium restriction in animals with asymptomatic cardiac disease or with early CHF. Very low sodium diets in animals with asymptomatic disease further activates the RAA and is not associated with improvements in cardiac size or function (this has been studied in dogs but not in cats). As heart disease progresses and CHF ensues, however, sodium restriction becomes more important although the use of newer and more effective medications has diminished the need for severe sodium restriction in many patients. For an animal with cardiac disease but without CHF [i.e., an asymptomatic cat with hypertrophic cardiomyopathy (HCM)], the author counsels the owner to avoid diets high in sodium (sodium <100 mg/100 kcal) and to avoid treats or table food which are high in sodium. When CHF first arises, additional sodium restriction is recommended (sodium <80 mg/100 kcal). This does not necessarily require a commercial cardiac diet. Be careful in selecting diets designed for cats with renal disease for cats with CHF (unless severe renal dysfunction accompanies CHF) as protein restriction can contribute to muscle loss (cachexia) which is common in CHF. As CHF progresses, more severe sodium restriction may allow lower dosages of diuretics to be used to control clinical signs. Be aware, however, that the cat food is only one source of sodium and other nutrients in the diet: treats, table food, and foods used to administer medications also can be important sources for sodium in a cat's diet.

Nutritional Deficiencies/Nutritional Pharmacology

The incidence of feline dilated cardiomyopathy (DCM) secondary to taurine deficiency has decreased dramatically over the last decade, but occasional cases are still seen. Most current cases appear to be independent of taurine status. Still, taurine deficiency should be suspected whenever feline DCM is diagnosed, especially if cats have been eating a vegetarian, homemade, or otherwise unbalanced diet.

L-Carnitine is critical for fatty acid metabolism and energy production. Carnitine deficiency is associated with primary myocardial disease in a number of species, however primary myocardial carnitine deficiency is unlikely to play a role in any of the cardiomyopathies seen in cats. Nonetheless, carnitine may have some beneficial effects by improving myocardial energy metabolism.

Furosemide use has been implicated in the development of hypokalemia and hypomagnesemia due to increased urinary loss of potassium and magnesium, respectively. Hypokalemia is deleterious because it can contribute to arrhythmias. While hypokalemia is relatively common in people using furosemide, it is less common in cats. Additionally, ACE inhibitors and potassium-sparing diuretics decrease potassium excretion and can cause *hyper*kalemia. Although clinically significant hyperkalemia is uncommon, potassium supplementation or feeding diets with a high potassium content to animals receiving an ACE inhibitor increases the risk for hyperkalemia. Monitoring serum potassium concentrations for both increases and reductions is important for optimal medical and dietary intervention in cardiac patients. Like potassium, hypomagnesemia can be associated with furosemide use and can also cause arrhythmias. Serum magnesium should be monitored and supplementation should be instituted when necessary.

The omega-3 fatty acids, eicosapentaenoic acid (EPA) and docosahexaenoic acid (DHA), are long-chain fatty acids that are usually found in very low concentrations in plasma membranes compared to the n-6 fatty acids, but can be increased via diet or a supplement. Cats with CHF may have a relative plasma concentrations of EPA and DHA, as has been seen in other species with CHF. In addition to the correction of fatty acid deficiencies, a benefit of n-3 fatty acids is that breakdown products of the omega-3 fatty acids (eicosanoids) are, in general, less potent inflammatory mediators than eicosanoids derived from omega-6 fatty acids. Thus, supplementation of fish oil, which is high in omega-3 fatty acids, can decrease the production of cytokines and other inflammatory mediators.

Omega-3 fatty acids have been shown to have anti-arrhythmic effects in a number of species. Many of the beneficial effects of omega-3 fatty acids in people with CHF have been presumed to be the result of reduced arrhythmias. However, more recent studies are beginning to delineate additional positive effects, such as reduced cardiac remodeling, reduced heart rate and blood pressure, improved endothelial function, and enhanced baroreceptor function and heart rate variability. Omega-3 fatty acids can alter immune function which may contribute to the cardiovascular benefits of omega-3 fatty acids. Finally, omega-3 fatty acids reduce platelet aggregation which might be particularly useful in cats with cardiac disease due to increased platelet aggregability.

While the optimal dose of omega-3 fatty acids for cats is not yet known, the author currently recommend a dosage of 40 mg/kg EPA and 25 mg/kg DHA in cats with CHF. Fish oil supplements vary widely in the amount of EPA and DHA they contain so it is important to know the exact amount in supplements recommended. Capsules that contain approximately 180 mg EPA and 120 mg DHA can be purchased over the counter at most human pharmacies and can be administered at a dose of 1 capsule per 10 pounds body weight. Fish oil should contain vitamin E as an antioxidant, but supplements which contain other nutrients are not recommended to avoid toxicities. Cod liver oil and flax oil should not be used.

Antioxidants have received a great deal of attention for their potential role in the prevention and treatment of human cardiac disease. Reactive oxygen species are a by-product of oxygen metabolism. The body normally compensates for these damaging oxidants through the production of endogenous antioxidants. Under certain circumstances, however, oxidant levels can rise above the body's endogenous antioxidant capacity. Dogs with CHF have increased oxidative stress compared to normal dogs and also have reduced levels of lipid soluble antioxidants, creating an imbalance between oxidative stress and endogenous antioxidant protection. However, the balance of oxidants and antioxidants has not yet been studied in cats with CHF.

MONITORING THE CAT WITH CHF

The authors recommend frequent re-evaluation of cats with CHF. At recheck visits, a physical examination is done and renal function and electrolytes are evaluated. Body weight is always recorded to assess for both weight loss from altered food intake and weight gain due to fluid accumulation. Subjective assessment of muscle mass also is important because an animal can retain fat stores but lose lean body mass. Early detection of muscle loss is important to identify this syndrome (ie, cachexia) at a stage at which intervention is more likely to be successful. A complete diet history should be obtained at each visit, including cat food, treats, table food, dietary supplements, and the method the owner uses to administer medications (in one study, 34% of cat owners administered cardiac medications in cat food or human foods that were often high in sodium). In most instances, adjustment of the doses or types of medications is required, especially during the early phases of therapy. The authors believe that careful adjustment to medical and nutritional therapy can help to maintain good quality of life.

REFERENCES

[1.] Freeman LM, Rush JE. Nutritional modulation of heart disease. In: Ettinger SJ, Feldman EC (eds). **Textbook of veterinary internal medicine, 7th ed**. St Louis: Saunders Elsevier, 2010: 691-696. [2.] Oreopoulos A, Padwal R, Kalantar-Zadeh K, et al. Body mass index and mortality in heart failure: A meta-analysis. **Am Heart J** 2008; 156: 13-22. [3.] Rush JE, Freeman LM, Fenollosa N, Brown DJ. Population and survival characteristics of cats with hypertrophic cardiomyopathy: 260 cases (1990-1999) **J Am Vet Med Assoc** 2002; 220: 202-207. [4.] Torin DS, Freeman LM, Rush JE. Dietary patterns of cats with cardiac disease. **J Am Vet Med Assoc** 2007; 230: 862-867. [5.] Yang VK, Freeman LM, Rush JE. Morphometric measurements and insulin-like growth factor in normal cats and cats with hypertrophic cardiomyopathy. **Am J Vet Res** 2008; 69: 1061-1066.

THE PREVALENCE & EFFECTS OF CANINE CARDIO-RENAL-ANEMIA SYNDROME

Dan G. Ohad, DVM, PhD, DACVIM (Cardiology), Jonathan Berkowitz, DVM, Tali Bdolah-Abram, M.Sc.
Rehovot, Israel

INTRODUCTION

Human congestive heart failure (CHF) and human chronic kidney disease (CKD) often deteriorate to become terminal conditions despite advanced and even optimal medical care. Many patients with one of these conditions also develop the other: Sixty four percent of CKD patients were also diagnosed with CHF of moderate-to-severe magnitude[1], while 41% of CHF patients were found to have CKD, defined as having a serum creatinine concentration of >1.5 mg/dl[2]. Anemia is one of the more prevalent disturbances common to these two disease entities. Congestive heart failure carries the risk of either triggering or exacerbating either anemia or CKD. Similarly, CKD patients are at risk of developing either CHF or anemia.[3] A positive feedback process may, therefore, develop where all three disease entities exacerbate each other and compromise both patient well-being and the ability of care givers to stabilize his/her condition. This scenario has been termed the cardio-renal-anemia syndrome (CRAS).[3] Anemia was found to be present in 40% of CKD patients who had CHF of various severities, while the incidence of anemia among those who only had CHF with no evidence of CKD was 15-61%, depending on the definition of anemia and on blood sampling technique.[4]

One possible mechanism whereby renal function deteriorates in patients with CHF may be chronic efferent glomerular arteriolar constriction, teleologically "designed" to accommodate compromised glomerular filtration rate (GFR). This may lead to renal parenchyma hypoxia which, in turn, may trigger renal cell death followed by fibrosis.[5] It is often assumed that the major trigger of anemia in CHF patients is chronic renal injury mediated by these mechanisms, compromising erythropoietin production.[3,6] A murine study, however, demonstrated CHF-related anemia without any evidence of CKD.[7] An alternative mechanism to explain the development of CRAS may involve the liberation of cytokines from the compromised heart, which may trigger anemia through any of the following three fashions: decreasing renal erythropoietin production, interfering erythropoietin activity in bone marrow, or compromising the availability of free iron for hemoglobin production.[8] In theory, by compromising tissue oxygen delivery, anemia may lead to structural and functional cardiac and renal damage even in the absence of a background heart disease. In addition to the potentially direct ischemic injury to these highly oxygen-dependent organs, compromised oxygen delivery may also trigger peripheral vasodilation with a resultant decrease in perfusion pressure. This, in turn, may prove more harmful specifically in these two vascular beds that are, in fact, more perfusion-pressure-dependent than others. If, as a consequence, both the renin-angiotensin-aldosterone system (RAAS) and the arginine-vasopressin systems are being activated, further renal sodium chloride and water retention may be stimulated, along with a decrease in serum potassium concentration.[9] Water retention may dilute hemoglobin concentration, present a hemodynamically challenging volume overload on the heart, and thereby increase myocardial workload with elevated oxygen consumption while oxygen delivery may be compromised. This, in turn, may exacerbate ischemic cardiomyocyte injury even further.[10] Additionally, red blood cells incorporate anti-oxidants. Anemia may, therefore, result in elevated renal and cardiac tissue injury inflicted by reactive oxygen species.[11] One also has to take into account the possible renal-related side effects of traditional and common cardiac drug therapy, such as volume-contraction related pre-renal azotemia, and/or diuretic therapy-related or anorexia-related hypokalemia, which may further worsen patient condition.[12]

Canine heart disease may, too, compromise renal function, as has been demonstrated in an experimental canine tachy-cardiomyopathy model.[13] Spontaneously occurring heart disease, on the other hand, has only been scarcely associated with renal injury according to the veterinary literature. A gradual elevation in serum urea has been reported in dogs with degenerative chronic valve disease (CVD) without evidence of CHF, when treated with an angiotensin converting enzyme (ACE) inhibitor.[14] Another study demonstrated a disease-severity-related elevation in blood urea, but not creatinine, in a variety of canine heart disorders.[15] In a study conducted in 2007, 50% of dogs with degenerative CVD had azotemia and demonstrated a disease-severity related drop in GFR, paralleling an elevation in both plasma creatinine and urea.[16]

Seventy six percent of 305 cats with heart disease were found to be at least moderately azotemic (with serum creatinine levels of 1.6-2.8 gr/dl), showing no correlation between echocardiographic disease severity or between blood pressure abnormality, and azotemia severity.[17]

The use of ACE inhibition for blunting the RAAS may improve clinical signs and even delay the onset of therapy failure in both people and pets with CHF.[18] Long-lasting ACE inhibiting therapy did not significantly alter serum creatinine or urea levels, nor did it change the prevalence of azotemia among dogs with mitral regurgitation.[14] Because both ACE inhibitors and spironolactone have the potential of increasing serum potassium concentration, caution has been advised when co-administrating these agents, especially to patients with impaired renal function, so as to avoid life threatening hyperkalemia.[19,20] When co-administered to a small sample of dogs with CVD in which renal function was normal, no significant changes were documented in serum potassium levels.[18]

When human serum creatinine is progressively increasing, prognosis and final outcome are poorer especially when occurring during hospitalization for CHF.[21] There is evidence to support the fact that anemia, too, is associated with poorer prognosis and outcome in human patients treated for CHF.[22] The diagnosis of and therapy designed to address anemia in human CRAS patients (e.g., erythropoietin administration along with iron supplementation) are said to have improved cardiac function, to have improved patient general well-being and quality of life, and to have decreased the rate of hospital readmission, as well as the diuretic therapy dose.[23]

To our knowledge, there are no reports regarding the prevalence or even the existence of CRAS in veterinary patients. Neither are there available data regarding its possible association with variables such as signalment, heart disease entity, its severity, or its treatment. It should be important to investigate not only these topics but also to quantify the magnitude of CRAS, and to develop means to better identify patients at risk of acquiring it. Furthermore, if CRAS is indeed identified in veterinary patients it would justify an attempt at perfecting a comprehensive therapeutic strategy to specifically address each of its three components without compromising any of the other two.

AIM

To facilitate future prospective studies addressing the need of and the ability to diagnose, prognosticate, and treat pets with CRAS, our major goal was to examine the prevalence of dehydration, azotemia, anemia, and serum potassium changes in dogs with diagnosed heart disease of any kind. A secondary goal was to test the relationship between demographic data such as patient breed, age, and reproductive status, between heart disease entity and characteristics, and between its severity, and packed cell volume, serum creatinine, and serum potassium level deviations. Another secondary goal was to attempt a longitudinal follow up of those parameters that may potentially change throughout a chronic therapeutic regimen.

METHODS

Physical examination, echocardiography and selective blood work parameters were retrospectively collected from 223 consecutive dogs with heart disease, and compared to those from 248 dogs with no cardiac or renal dysfunction.

RESULTS

While remaining within normal range, the mean creatinine levels in the study group were lower than those in the control group. The prevalence of azotemic dogs among heart patients was 24.1%. Mean hematocrit levels were higher in the study group compared to the control group, while remaining within normal range as well. The prevalence of anemia among the cardiac study group was 22.8%, surprisingly lower than that in the control group. Six out of 156 cardiac patients (3.8%) were both anemic and azotemic.

Cardiac disease severity and chronic cardiac therapy did not affect hematocrit, creatinine or potassium levels.

CONCLUSIONS

Impaired renal function and anemia is uncommon in canine heart disease. It remains important, however, to identify these conditions so that adjustment of an appropriate treatment regimen, as well as an appropriate prognosis is facilitated. Traditional medical therapy can likely be followed as long as these parameters remain stable and within normal range.

REFERENCES

[1.] Avorn J *et al.* **Arch Intern Med**. 2002;162:2002. [2.] Silverberg DS *et al.* **J Am Coll Cardiol**. 200;35: 1737 [3.] A Laina, D *et al.* **Nature Clinic Practice Cardiovascular Medicine** 2005;2:95. [4.] Silverberg DS *et al.* **Nephrol Dial Transplant** 2003;18:141. [5.] Fine LG *et al.* **Kidney Int** 2000;57 (Suppl 75):22 [6.] Tong EM *et al.* **Semin Nephrol**.2001;21:190–203. [7.] Iverson PO *et al.* **Am J Physiol Regul Integr Comp Physiol** 2002;282:166 [8.] Means RT. **Int J Hematol**. 1999;70:7 [9.] Saito M *et al.* **J Clin Pharm and Therapeut** 2005;30:603 [10.] Anand IS *et al.* **Br Heart J**. 1999;70:357 [11.] Grune T *et al.* **Clin Nephro** 2000;53:18 [12.] Boswood A *et al.* **J Vet Cardiol** 2006;8;1 [13.] Seymour A *et al.* **Lab Anim Sci** 1994;44:443. [14.] Atkins CE *et al.* **J Am Vet Med Assoc** 2002;221:654 [15.] Pouchelon JL *et al.* **J Vet Cardiol** 2004;6:7 [16.] Nicolle AP *et al.* **J Vet Intern Med**.2007;21(5):943 [17.] Gouni V *et al.* **ECVIM Proceedings**, 2006. [18.] Thomason JD *et al.* **J Vet Cardiol** 2007; 9:103 [19.] Schepkens H *et al.* **Am J Med** 2001;110:438 [20.] Kvart C *et al.* In: Ettinger SJ, Feldman EC, editors. **Textbook Of Veterinary Internal Medicine**. 5th ed. 2000:787 [21.] Smith GL *et al.* **J Card Fail** 2003;9:13 [22.] Horwich TB *et al.* **J Am Coll Cardiol** 2002;39:1780 [23.] Silverberg DS *et al.* **Semin Nephrol**. 2006;26(4):296

CLINICAL ASSESSMENT OF CARDIOVASCULAR AUTONOMIC NERVOUS SYSTEM FUNCTION IN THE DOG

Etienne Côté DVM, DACVIM
Charlottetown, PE, Canada

INTRODUCTION

The heart beats spontaneously, but the rate of the heartbeat is controlled principally by the autonomic nervous system (ANS). This branch of the nervous system is clinically important in many deleterious ways, including induction of structural heart disease[1] and of cardiac arrhythmias.[2] The ANS also provides therapeutic target opportunities, as with beta-blocking drugs,[3] and cardiovascular autonomic function is an important marker of cardiovascular health.[4-8]

In veterinary cardiology, the effects of the ANS on the heart have been investigated at various levels.[9] In German shepherd dogs with familial arrhythmic disease, the role of adrenergic stimulation in the genesis of ventricular arrhythmias has been investigated,[10] as has the role of beta-1 receptors.[11] In a prospective observational study of privately-owned pet dogs with heart disease, Little et. al. found a 15.8-fold greater 1-year mortality in dogs with heart disease and low vagal tone (as measured with an instrument calculating beat-to-beat changes in heart rate in real time) compared to dogs with high vagal tone.[12] A high vasovagal tonus index, as calculated from heart rate variability on the electrocardiogram (ECG), conferred shorter survival times and was associated with a higher ISACHC class of heart disease in Doberman dogs with dilated cardiomyopathy as compared to a lower index,[13] and heart rate variability has been evaluated in several other studies.[14] Scant information is available, however, regarding the value to canine patients of two forms of clinical assessments of autonomic function that are commonly performed in humans: vagal maneuvers and exercise testing.

VAGAL MANEUVERS

The concept of reducing the heart rate through extracardiac physical manipulations has been recognized for centuries. Such vagal maneuvers as carotid sinus massage and ocular pressure are in routine use in human cardiology,[15] and these also are part of the standard practice of cardiology in small animals.[16] However, the response of the heart to such maneuvers is poorly quantified in dogs and cats.

We conducted a prospective, blinded evaluation of 64 volunteered healthy pets (32 dogs, 32 cats) that underwent vagal maneuvers during ECG recording. While in right lateral recumbency and under gentle restraint after acclimation, each animal underwent 4 procedures. Each of the 4 procedures lasted 15 seconds and was preceded by a washout period of at least 30 seconds. The procedures consisted of: real carotid sinus massage (moderate but comfortable digital pressure applied to the region of the carotid sinus, cranial and dorsal to the larynx), sham carotid sinus massage (hand cupped around the ventral neck but not touching the animal), real ocular pressure (moderate but comfortable digital pressure applied to both eyes through closed eyelids), and sham ocular pressure (hand cupped over the eyes, without touching the patient), and the order of procedures was varied according to a modified Latin square design to minimize bias. The investigator performing the vagal maneuvers was blinded to the ECG display. *Post-hoc* analysis included measurement of heart rate, PR interval, and QT interval prior to, during, and after the vagal maneuvers were performed. This Forum presentation will discuss the rationale behind, and execution of, the study; its results; and implications for dog and cats undergoing vagal maneuvers.

EXERCISE TESTING

Although routinely performed in human cardiovascular subjects,[17] exercise testing is virtually unknown in clinical veterinary cardiology. Kittleson et. al. evaluated normal dogs and dogs with experimentally-induced heart failure with a treadmill exercise protocol.[18] Dogs with heart failure reached anaerobic threshold sooner, and demonstrated intolerance of exercise earlier, than healthy counterparts.

Exercise testing consisting of treadmill walking or stationary bicycle riding is an integral part of cardiovascular evaluations of human patients. An emerging aspect of exercise testing in humans, where protocols have typically involved gradual increases in effort and termination of the test when an endpoint is reached, is prolonged assessment to include the rate and pattern of return to resting heart rate after the termination of exercise. Performance athletes have been known for decades to have a rapid recovery of resting heart rate after maximal exercise.[5] More recently in patients with heart disease, a delayed return to resting heart rate has been identified as a simple, independent prognostic indicator of cardiovascular mortality.[6-8]

We sought to evaluate the normal heart rate response of healthy dogs subjected to a simple exercise protocol consisting of stair climbing and descending. Sixteen dogs (4 dogs of 4 different somatotypes: cavalier King Charles spaniels, Australian shepherds, Labrador retrievers, and Newfoundland dogs) were deemed healthy via standard, complete clinical cardiovascular examinations. The dogs were instrumented with event-type ECG monitors to record 5 minutes of ECG consisting of: a 1-minute acclimation period at the base of a stairwell; climbing and descending 6 flights of stairs once while accompanied by a handler; and recovery at the base of the stairs. All dogs performed the test willingly in quintuplicate with at least 30 minutes of recovery time between sessions. Initial results have been shown,[19] and will be presented in greater detail during this Forum talk.

The protocol for exercise testing lends itself well to patients who are mobile, do not have identifiable comorbidities, and are hemodynamically stable. Accordingly, a second application of this protocol involved the recruitment of 18 dogs with naturally-occurring obesity. In these dogs, a series of cardiovascular performance tests, including 6-minute walk, tilt test, and the staircase exercise protocol, was performed at the time of recruitment and again after a 4-month period of weight loss (median weight loss: 9% body weight). Initial results of the exercise-based aspects of this experiment will be presented at the Forum.

Finally, we have been investigating a novel form of juvenile dilated cardiomyopathy in the toy Manchester terrier dog. Clinically, this disorder is characterized by a young age of onset (all known cases <1 year of age), equal distribution between genders, and a chief complaint of sudden cardiac death, either diurnally (typically preceded by a cry, and then collapse) or nocturnally. Several deaths have occurred shortly after recovery from general anesthesia for elective surgery. Evaluation of a small number of these dogs, both close relatives and non-relatives of the deceased dogs, using routine cardiovascular diagnostic tests, has not revealed any convincing abnormalities. Using the stair-based exercise protocol, we have identified cardiac arrhythmias that were not present or suspected during the standard 10-lead ECG or ECG display of the echocardiogram prior to exercise. Cumulatively, these results suggest that this form of exercise testing may provide information that supplements data obtained with other in-hospital cardiovascular tests.

REFERENCES

[1]Wittstein IS, et. al. **N Engl J Med** 2005;352(6):539. [2]Liu N, et. al. **Prog Cardiovasc Dis** 2008;51(1):23. [3]Bristow M. **Circulation** 2003;107(8):1100. [4]Gibbons RJ, et. al. **Circulation** 2002;106(14):1883. [5]Borysewicz E. **Bicycle Road Racing** 1986:92. [6]Jouven X, et. al. **N Engl J Med** 2005;352(19):1951. [7]Cole CR, et. al. **Ann Intern Med** 2000;132(7):552. [8]Morshedi-Meibodi A, et. al. **Am J Cardiol** 2002;90:848. [9]Moïse NS. **J Small An Pract** 1998;30(10):460. [10]Sosunov EA, et. al. **Cardiovasc Res** 2004;61(4):715. [11]Steinberg SF, et. al. **Am J Physiol Heart Circ Physiol** 2002; 282(4):H1181. [12]Little CJ, et. al. **Vet Rec** 2005;156(4):101. [13]Martinez Pereira Y, et al. **J Small An Pract** 2008;49(11):587. [14]Minors SL et. al. **Can J Vet Res** 1997;61(2):134. [15]Miller JM, et. al. In Libby P, et. al., eds. **Braunwald's Heart Disease**, 8th ed. 2008:763. [16]Moïse NS. In Fox PR, et. al., eds. **Textbook of Canine and Feline Cardiology**, 2nd ed. 1998:343. [17]Chaitman BR. In Libby P, et. al., eds. **Braunwald's Heart Disease**, 8th ed. 2008:195. [18]Kittleson MD, et. al. **J Vet Intern Med** 1996;10(1):21. [19]Manley RM et al. **J Vet Intern Med** 2009;23(3) :746.

ADVANCES IN INTERVENTIONAL THERAPIES FOR VETERINARY PATIENTS

Dr. Robert Prošek, Dr. Herbert Maisenbacher, Dr. Amara Estrada
Gainesville, FL & Homestead-Miami, FL

Interventional therapy is the use of diagnostic imaging to guide minimally invasive procedures. The diagnostic imaging modalities most commonly utilized in these procedures are ultrasound and fluoroscopy, but may also include endoscopy, computed tomography (CT) and magnetic resonance imaging. Specialized equipment such as introducers, catheters, and guidewires are used to access the body via minimally invasive techniques and complete the interventional procedure. Some basic examples of interventional procedures practiced routinely by veterinarians include ultrasound guided biopsies and percutaneous fluid drainage; however, interventional therapies have far reaching applications across many organ systems and disease processes and have become quite elaborate. In human medicine, interventional therapies are divided between interventional radiology and interventional cardiology. Interventional radiologists are trained in interventional vascular, respiratory, urogenital, hepatobiliary, gastrointestinal, neurologic, and oncologic therapies while interventional cardiologists are trained in interventional cardiac and vascular therapies. Although the fields of interventional radiology and interventional cardiology are relatively young, they are rapidly growing with the continual development of new devices and techniques. Many of the procedures now done commonly and considered standard of care in human medicine were only pioneered in the last few decades. Veterinary medicine clinicians have embraced many cardiac and stent procedures and are currently evaluating other interventions, and with this process building the veterinary interventional armamentarium.

ADVANTAGES AND DISADVANTAGES

Potential advantages of interventional therapies over surgical alternatives include reduced perioperative morbidity and mortality, reduced anesthesia times (once a technique is mastered; initially these procedures will take significantly longer due to a steep learning curve) and usually shorter post-op hospitalization times. In fact, most procedures are completed percutaneously (Seldinger technique) or via a minor surgical exposure of a peripheral vessel. Importantly, these therapies offer alternative treatments for patients with conditions that may not be amenable to standard therapy, the standard treatment involves risks that are unacceptable to the client/patient, or where palliation is the goal. The primary disadvantages of advanced interventional therapies is that they are often complex and require considerable technical expertise with specialized equipment. Specialized equipment such as a high resolution fluoroscope, stents, balloons, guidewires and catheters require a significant amount of initial capital investment. However, upon attainment of necessary equipment and experience, the advantages of many interventional therapies far outweigh the possible disadvantages.

CURRENT VETERINARY APPLICATIONS

Some interventional therapies utilized in human medicine have been applied to veterinary patients and several are now widely accepted in clinical practice. Procedures commonly perfomed by veterinary cardiologist include transarterial coil occlusion for patent ductus arteriosus (PDA), balloon valvuloplasty for pulmonic stenosis (PS), transvenous pacemaker therapy for bradyarrhythmias, and transvenous heartworm extraction for caval syndrome. Endoluminal tracheal stenting for tracheal collapse is commonly performed by various veterinary specialists and general veterinarians. More recently, newer interventional therapies have been described in veterinary patients. We are now almost exclusively utilizing the Amplatz Canine Duct Occluder (ACDO) for transarterial PDA occlusion. This device is based on the same technology as a device which is available for use in the human PDA, but was specifically designed for the anatomy of the canine PDA. The ACDO design allows better occlusion with considerably decreased risk of malposition than coil occlusion but at a higher expense. Other interesting cardiology interventional therapies include atrial or ventricular septal defect occlusions with the Amplatzer Septal Occluder, occlusion of an acquired aortic-to-right ventricular fistula in horses with the Amplatzer Septal Occluder, and high pressure balloon valvuloplasty for recurrent or persistent pulmonic stenosis. Intrahepatic portosystemic shunts (IHPSS) remain a surgical challenge with historically high morbidity and mortality (~ 60%) rates. Transvenous coil occlusion of IHPSS offers lower perioperative morbidity with similar long-term success. In this procedure, coils are deployed into the shunt to occlude it after a stent has been placed in the caudal vena cava where the shunt enters to prevent migration of the coils. To avoid the development of portal hypertension, complete occlusion of the IHPSS is not the initial goal, although further occlusion is possible in a subsequent procedure if necessary.

Embolization for treatment of hemorrhage or vascular malformations

Life-threatening uncontrollable hemorrhage can lead to hypovolemic shock, discomfort to patient and owner and possibly death as noted with severe epistaxis due to various etiologies. With fluoroscopic guidance, maxillary artery is isolated via the femoral artery. Polyvinyl alcohol particles or small coils are injected through the catheter to embolize the nasal supply. Other clinical problems corrected with embolization include refractory bleeding due to various causes (neoplasia, damage of vessel due to trauma or failed orthopedic surgery) and congenital arteriovenous malformations in the liver. Congenital arteriovenous hepatic malformations injected with cyanoacrylate glue can result in complete occlusion and has been performed in veterinary patients.

Interventional oncology

Interventional techniques permit fluoroscopically-guided, selective catheterization of the arteries feeding the actual tumor or body region. Intra-arterial administration of the systemic dose of chemotherapy permits significantly greater concentrations of the drug to reach the tumor without the systemic side effects that would be encountered had an equivalent IV dose been used to achieve the same local chemotherapy concentrations. Experimental studies in dogs and in rabbits have documented the improved efficacy of these techniques in achieving higher local concentrations and improved tumor remission rates, for certain types of tumors. These techniques have been used in veterinary patients for osteosarcoma, transitional cell tumors of the urinary bladder, urethra, and prostatic carcinomas.

Peri-vascular tumors can occasionally invade a vessel lumen resulting in vascular occlusion and peripheral edema (e.g. vena cava obstruction with adrenal tumors), ascites (e.g. portal hypertension due to hepatic vein obstruction due to sarcomas). Self-expanding metallic stents can be placed under fluoroscopic guidance across malignant obstructions to relieve vascular occlusions and reduce venous congestion resulting in profound clinical improvement. Animals are routinely euthanized due to the local effects of a tumor rather than the systemic effects associated with a large cancer burden. For example, malignant obstructions of the urinary tract associated with transitional cell carcinomas or prostatic tumors can result in life-threatening signs associated with complete urinary tract obstruction. Malignant obstructions in the gastrointestinal tract often cause patients to present with chronic vomiting, weight loss, tenesmus, and constipation depending upon the location of the obstruction. Palliative stenting procedures in the upper and lower urinary tract, respiratory tract, cardiovascular system, and upper and lower gastrointestinal tracts to relieve luminal obstructions due to neoplasia or various strictures can be attempted and have been described in veterinary patients.

Various tumors can be approached with interventional techniques but hepatic tumors have currently gathered the most attention as metastatic and non-resectable liver tumors pose many challenges in veterinary patients. These animals traditionally have been treated conservatively due to limited efficacy of intravenous chemotherapy for macroscopic disease and the potential deleterious side effects associated with radiation therapy along with substantial cost have led investigators to evaluate other modalities. Interventional catheter based techniques offer the potential for improved tumor response rates, prolonged survival times, and enhanced quality of life in patients with minimal systemic toxicity risks.

Other novel interventional therapies

Local thrombolytic therapy for pulmonary thromboembolism and other embolic disease, thrombectomy in cats with aortic thromboembolism, retrieval of tracheobronchial foreign bodies, antegrade urethral catheterization, embolization for treatment of idiopathic chylothorax, and catheter based delivery of mesenchymal stromal cells or myoblasts into heart and other organs to aid in repair have been attempted and performed in veterinary medicine.

CONCLUSIONS

Beyond the procedures listed above, many other interventional techniques have been described in humans and with creative thinking the potential for novel techniques is almost endless. However, many procedures have yet to be critically evaluated and deserve further research before wide acceptance. As more advances are made, there is an ever growing potential to expand the application of interventional therapy in veterinary medicine and this will translate into better care for our patients. Many cases may be found where traditional treatments have failed, were declined due to morbidity, or are contraindicated because of coexisting conditions. Take a look around, and let your imagination create a way to fix current interventional problems. "Imagination is more important than knowledge"-A. Einstein

REFERENCES

[1.] Weisse C, et al. Use of percutaneous arterial embolization for treatment of intractable epistaxis in three dogs. **J Am Vet Med Assoc** 224: 1307-1311, 2004. [2.] Reimer SB, et al. Use of rheolytic thrombectomy in the treatment of feline distal aortic thromboembolism. **J Vet Inter Med** 20(2):290-6, 2006. [3.] Weisse C, et al. Coil embolization of canine intrahepatic shunts: 42 cases. **Proceedings, 2006 ACVIM Forum**, Louisville, KY, 2006. [4.] Moritz A, et al. Management of advanced tracheal collapse in dogs using intraluminal self-expanding biliary wallstents. **J Vet Inter Med** 18:31-42, 2004. [5.] Withrow SJ, et al. Intra-arterial cisplatin with or without radiation in limb-sparing for canine osteosarcoma. **Cancer** (71(8):2484-90, 1993. [6.] Weisse C,et al. Percutaneous bland arterial embolization and chemoembolization for the treatment of benign and malignant diseases. **J Am Vet Med Assoc** 221: 1430-1436, 2002. [7.] Weisse C, et al. Evaluation of palliative stenting for management of malignant urethral obstructions in dogs. **J Am Vet Med Assoc**229:226-234, 2006. Nagata Y, Nakano Y, et al. Osseous metastases from heptocellular carcinoma: embolization for pain control. Cardiovascular Interventional **Radiol** 12(3):149-53, 1989. [8.] Hoshi S, et al. Internal iliac arterial infusion chemotherapy for rabbit invasive bladder cancer. **Int J Urol.** 4(5):493-9, 1997. [9.] Mallery KF, et al. Percutaneous ultrasound-guided radiofrequency heat ablation for treatment of hyperthyroidism in cats. **J Am Vet Med Assoc** 223(11):1602-7, 2003. Culp WT, et al. Intra-luminal tracheal stent placement in 3 cats. **Vet Surg** 2007;36:107-113. [10.] Weisse CW, et al. Potential applications if interventional radiology in veterinary medicine. **J Am Vet Med Assoc**, 223(10), 1564-1574, 2008.

ADIPONECTIN AND C-REACTIVE PROTEIN IN NORMAL AND CONGESTIVE HEART FAILURE DOGS. RM Wood, OL Nelson. Washington State University College of Veterinary Medicine, Pullman, WA.

Adiponectin (ADPN) is a cytokine produced by fat cells which has been shown to be negatively correlated with adverse cardiac conditions in humans. In the heart, adiponectin activates several pro-survival reactions, including the AMPK pathway and COX2 receptors, which protect the heart following ischemic injury. Recent studies have shown that higher levels of ADPN influence cardiac remodeling signaling, inhibiting protein synthesis and suppressing pathological cardiac growth. In humans, ADPN plasma levels rise with decreased activity of the sympathetic nervous system and β-adrenergic agonists inhibit ADPN at the level of gene expression. In contrast C-reactive protein (CRP), a marker of systemic inflammation is elevated in humans with congestive heart failure (CHF) and correlates to the severity of disease.

We hypothesized that dogs with CHF would have reduced ADPN and elevated CRP compared to normal dogs and that cytokine concentrations would predict severity of CHF. We collected serum from 23 dogs (15 healthy and 8 CHF with varying degrees of severity). Circulating adiponectin and CRP levels were quantified using a Mouse/Rat Adiponectin ELISA and a Canine CRP ELISA. mRNA expression of ADPN receptor-1 (R1) was quantified using real-time PCR in myocardium of normal and cardiac disease dogs. We found lower mean CRP concentrations in normal dogs (3.5 ± 2.3 μg/ml) than dogs with CHF (17.9 ± 17.1 μg/ml), however, the results were not statistically significant due to the large variability seen among the CHF dogs ($p=0.07$). When separated into severity of CHF, CRP concentrations in normal versus severe CHF was significant ($p < 0.05$). We found greater mean ADPN concentrations in normal dogs (13.4 ± 2.4 μg/ml) than CHF dogs (9.9 ± 3.2 μg/ml) ($p= 0.05$). In general, the greater the severity of the heart failure, the lower the level of serum adiponectin. When the 2 tests were combined, CHF dogs (all severity) had significantly lower ADPN:CRP ratios compared to normal dogs ($p < 0.001$), suggesting that the combination of lower ADPN and higher CRP concentrations has a greater correlation with CHF. mRNA expression of ADPN R1 was not significantly different in the few samples analyzed. As in humans, ADPN and CRP may play a role in the pathophysiology of heart failure in the dog.

CYTOKINE AND MATRIX METALLOPROTEINASE EXPRESSION IN MYOCARDIUM OF DOGS WITH CARDIAC DISEASE. Sonja Fonfara, Simon Tew, Udo Hetzel, Peter Clegg, Joanna Dukes McEwan. Faculty of Veterinary Science, University of Liverpool, Neston, UK.

The progression of cardiac disease and congestive heart failure (CHF) is still poorly understood. Extracellular matrix (ECM) turn-over and inflammation are important parts in progression of cardiac disease and development of CHF.

mRNA was analysed in myocardial samples (interventricular septum, right and left atrium and ventricle) of 6 dogs with no cardiac abnormalities (controls) and 5 dogs with cardiac diseases [endstage dilated cardiomyopathy (DCM): n=2, endstage degenerative valvular disease: n=1, ventricular tachycardia: n=1, severe aortic stenosis: n=1]. The house keeping gene GAPDH, interleukin (IL) 1, -2, -4, -8, -10, tumour necrosis factor α (TNF), interferon γ (IFN), transforming growth factor (TGF) β1 and β2, lysyl oxidase, matrix metalloproteinase (MMP) 1, 2, 3, 9, 13 and the tissue inhibitors of metalloproteinase (TIMP) 1, 2, 3, 4 were analysed using real time RT-PCR.

MMP- and TIMP-mRNA amounts varied in different myocardial regions with left sided increase of MMPs in DCM/CHF, whereas cytokines seem to be more evenly distributed. Pro-inflammatory cytokines [IL1 ($p=0.035$), IL2 ($p=0.01$), TNF ($p=0.057$)], MMP1 ($p=0.028$), TGFβ2 ($p=0.014$) and TIMP3 ($p=0.009$) were increased in myocardial samples of dogs with cardiac disease in comparison to control specimens.

An increase in pro-inflammatory cytokines and altered MMP/TIMP expression suggest that inflammation and ECM changes contribute to the progression of cardiac disease. TGFβ2, which is involved in embryonic cardiac development, and TIMP3, a regulator of resident fibroblasts and MMP inhibitor, might be important in myocardial fibrosis/remodelling.

MATRIX METALLOPROTEINASE-9 IS ASSOCIATED WITH SYSTOLIC FUNCTION IN DOGS WITH MYOXOMATOUS MITRAL VALVE DISEASE. I. Ljungvall[1], M. M. Rajamäki[2], S. Crosara[3], L.H. Olsen[4], C. Kvart[1], M. Borgarelli[5], K. Höglund[1], J. Häggström[1]. [1]Swedish University of Agricultural Sciences, Uppsala, Sweden, [2]University of Helsinki, Finland, [3]University of Torino, Italy, [4]University of Copenhagen, Denmark, [5]Kansas State University, USA

The breakdown and accumulation of extracellular components are regulated by a family of proteolytic enzymes, the matrix metalloproteinases (MMPs). In particular MMP-2 and MMP-9 are known to play important roles in cardiac remodeling processes. The aims of the study were to investigate whether circulating blood levels of MMP-2 and MMP-9 were associated with severity of myxomatous mitral valve disease (MMVD) in dogs, and investigate potential associations of dog characteristics, cardiac Troponin I (cTnI) and C-reactive protein (CRP) concentrations on MMP levels.

The severity of MMVD was assessed in 75 privately owned dogs. Plasma levels of the latent (pro-MMP) and active forms of MMP-2 and-9 were analyzed by zymography. Plasma concentration of cTnI was analyzed by an immunoenzymatic assay and CRP was analyzed by a canine-specific ELISA-assay.

In the included dogs, pro MMP-9, active MMP-9 and pro MMP-2 were detected, but active MMP-2 was not. No overall significant differences were found between the different MMPs and the four MMVD severity groups. Pro MMP-9 decreased with decreasing systolic blood pressure, and was higher in male dogs than in females. Active MMP-9 decreased with increasing left ventricular end systolic dimension, and with decreasing systolic blood pressure and cTnI. Left ventricular end systolic dimension was the variable most strongly associated with MMP-9. No associations were found between pro MMP-2 and the investigated variables.

In conclusion, MMP-9 decreased with worsening systolic myocardial function; hence MMP-9 has the potential to give unique information of the myocardial remodeling process in dogs with MMVD.

MYOCARDIAL PROTEIN LEVELS IN CANINE AND FELINE HEART DISEASE.

MM Sleeper[1], LT Bish[2], CA Reynolds[1], GE Singletary[1], ET Davison[1], HL Sweeney[2]. University of Pennsylvania Veterinary School[1] and Medical School[2], Philadelphia, PA.

Therapeutic gene transfer is a promising approach to address heart disease in human and veterinary patients. We have developed a procedure which results in transduction of the majority of canine and feline myocardial cells and therapeutic studies are currently underway. However, knowledge regarding alterations of various candidate transgenes in dogs and cats with various naturally occurring cardiac diseases is lacking. This information would allow more efficient prioritization of candidate transgene studies in animal populations with naturally occurring disease.

Myocardial samples were obtained from animals which were euthanized at the Veterinary Hospital of the University of Pennsylvania. Left ventricular samples were frozen in liquid nitrogen immediately after death and stored at -80 degrees C until protein extraction and immunoblot analysis. The following myocardial proteins were analyzed: phospholamban (PLB), phospho-PLB, s100a1, myostatin, ryanodine receptor, sarcoplasmic reticulum Ca^{2+} ATPase2a (SERCA2a), activin IIB receptor (ActIIBR), Bcl-2, apoptosis repressor with caspase recruiting domain (ARC). Samples from 6 normal dogs and cats were included in the analysis as well as 5 dogs with degenerative valve disease (DVD), 5 dogs with dilated cardiomyopathy (DCM), 4 cats with hypertrophic cardiomyopathy (HCM) and 5 cats with various other forms of cardiomyopathy (restrictive or unclassified; OCM).

Significant differences were noted in myostatin, phospho-PLB, s100a1, and SERCA2a expression between normal and cardiomyopathic cats and between phospho-PLB and s100a1 when considering the various groups of dogs. Cats with OCM had increased levels of full length myostatin compared to normal, whereas cats with HCM had reduced levels of full-length myostatin and myostatin C-dimer expression compared to cats with OCM. Both abnormal groups of cats had reduced s100a1 expression compared to normal cats whereas OCM cats had increased phospho-PLB expression compared to the HCM cats. SERCA2a expression was also reduced in the OCM group of cats compared to normal. There was a trend for DVD dogs ($p=0.07$) to have increased expression of s100a1 and the 20 kDa form of phospho-PLB compared to normal. Juvenile onset DCM dogs had a significant increase in the 20 kDa form of phospho-PLB compared to normal adult dogs. These results will allow scientists using therapeutic gene therapy for veterinary patients with heart disease to target more specific transgenes based on the patient's individual diagnosis.

COMPARISON OF NEW REFERENCE VALUES OF BIPLANE SIMPSON RULE AND NEW M-MODE REFERENCE VALUES TO DETECT EARLY ECHOCARDIOGRAPHIC CHANGES IN DOBERMAN PINSCHERS WITH DCM

G. Wess, J. Mäurer, K. Hartmann. Clinic of Small Animal Medicine, LMU University, Germany

Currently M-mode is considered gold standard to diagnose echocardiographic changes in Doberman Pinschers with dilated cardiomyopathy (DCM). The aim of this study was to establish reference values for Simpson method of discs (SMOD) and for M-mode values and to compare both methods with currently used M-Mode values using a prospective longitudinal study design.

Reference values for M-Mode and SMOD were calculated using 110 healthy Doberman Pinschers >5 years old with <50 VPCs/24-hours and normal echocardiographic M-mode measurements (LVIDd ≤47 mm, LVIDs ≤38 mm). A multivariate analysis including weight, age, and gender revealed that only gender had a significant influence on M-mode values. Therefore, separate M-mode cut-off values were calculated for male and female dogs. In the second part of the study, the new and currently used ("old") reference values were compared using 481 examination of healthy Doberman Pinschers and dogs with DCM (LVIDd >49 mm, LVIDs >39mm). 88 dogs were diagnosed with DCM according to all 3 methods. Additionally, 58 were normal using the established M-Mode values, but abnormal according to SMOD; 23 of these dogs were also abnormal using the new M-mode values. The prospective longitudinal study design allowed to follow-up on theses dogs and showed that 38 dogs with abnormal SMOD and all 23 with abnormal new M-Mode values became also abnormal at later examinations using the "old" M-Mode values. 20 dogs with increased SMOD values remain so far healthy (false positive).

	Cut-Off	sensitivity	specificity
EDV/BSA	>95ml/m2	94.3%	96.6%
ESV/BSA	>55ml/m2	92.9%	96.0%
EDD ♂ old	>49mm	70.6%	100.0%
new	>48mm	78.4%	100.0%
EDD ♀ old	>49mm	55.0%	100.0%
new	>46mm	86.7%	99.3%
ESD ♂ old	>39mm	62.7%	100.0%
new	>36mm	82.4%	99.0%
ESD ♀ old	>39mm	56.7%	100.0%
new	>36mm	88.3%	99.7%

SMOD is a new, valuable diagnostic test. Echocardiographic changes in Doberman Pinschers with DCM can be detected earlier using SMOD than using established reference M-mode values.

HEART RHYTHM CHARACTERIZATION DURING TRANSIENT LOSS OF CONSCIOUSNESS IN THE DOG

R.A. Santilli[1], L.V. Bontempi[2], L. Ramera[3], M. Perego[1]. [1]Clinica Veterinaria Malpensa, Samarate, Varese, Italy, [2]Ambulatorio Veterinario Associato, Bareggio, Milano, Italy, [3]Clinica Veterinaria Città di Chiari, Chiari, Brescia, Italy.

Transient loss of consciousness (T-LOC) is defined as a spontaneous, self-limited, loss of consciousness with complete recovery, irrespectively of the underlining mechanism. A common cause of T-LOC in people is syncope, a clinical entity characterised by transient global cerebral hypoperfusion. Primary and secondary disturbances of brain function such as seizures, concussion and intoxication are others possible causes of T-LOC. The aim of our study was to provide information on the electrocardiographic (ECG) findings during T-LOC in dogs.

Two hundred sixty-six dogs of different breeds were referred for rhythm monitoring because of the occurrence of T-LOC. One hundred seventy-one out of 266 dogs were males, with a median age of 8 years, and a mean weight of 24.7 $\pm$ 15.6. Two hundred-six Holter (group A) and 60 external loop recorder (R-Test) (group B) tracings were retrospectively analysed to characterise heart rhythm during T-LOC. Twelve percent of dogs of group A and 21.6 % of dogs of group B experienced T-LOC during rhythm recording, for a total 101 T-LOC (72 group A and 29 group B).

During 17 T-LOC the underlining rhythm was sinus with a mean heart rate of 151.9 $\pm$ 61.9 bpm. In 52 T-LOC paroxysmal atrioventricular (AV) complete block occurred with a mean ventricular arrest duration of 11.09 $\pm$ 5.22 sec. The ventricular arrests were preceded in 35 cases by sinus rhythms, in 6 cases by 2:1 AV block, in 10 cases by atrial fibrillation and in 1 case by a 3:1 AV block. The ventricular arrests were followed in 48 cases by escape rhythm (40 wide and 8 narrow QRS complex) with a mean rate of 33.35 $\pm$ 13.32 bpm and in 4 cases by sinus rhythm. Mean atrial rate during paroxysmal 3° AV block and ventricular arrest was 137.74 $\pm$ 37.76 bpm . In 30 T-LOC paroxysmal sinus arrest was recorded with a mean ventricular pause of 7.38 $\pm$ 4.15 sec. The sinus arrests were preceded in 21 cases by sinus rhythms, in 1 case by supraventricular tachycardia with a ventricular rate of 220 bpm and were followed in 22 cases by escape rhythm (14 narrow and 8 wide QRS complex) with a mean rate of 38.55 $\pm$ 26.31 bpm. In 8 cases of group B neither the preceding nor the subsequent heart rhythm could be analysed.

In people the reported prevalence of arrhythmic-related T-LOC is 47.9 %, in our population of dogs resulted 58.53 %. The ECG tracings presented paroxysmal bradyarrhythmias leading to a diagnosis of syncope either arrhythmic or induced by vagal cardio-inhibition. In the remaining dogs with T-LOC a sinus rhythms appeared and the cause of loss of consciousness remained unknown.

In conclusion most of T-LOC reported in our dogs are caused by paroxysmal bradyarrhytmias, further studies are needed to evaluate potential electrocardiographic findings that could differentiate arrhythmic from vagal cardio-inhibitory T-LOC.

THE USE OF PIMOBENDAN IN THE TREATMENT OF HEART FAILURE IN 28 CATS.

SG Gordon, AB Saunders, RM Roland, R Winter, LT Drourr, SE Achen, C Hariu, M Boggess and MW Miller. The Michael E. DeBakey Institute, College of Veterinary Medicine and Biomedical Sciences, Texas A&M University, College Station, TX.

There is interest in the use of Pimobendan (Vetmedin® ;Pimo) for feline heart failure (HF). This study reports the use of Pimo in 28 cats with HF. 17/28 were male and the mean age ± SD was 8.9 ± 5.2 years. 19/28 were mixed breed, there were 2 each of Burmese, Sphynx, Maine Coon and 1 each of Siamese, Birman and Persian. HF etiology was unclassified cardiomyopathy (CM) in 12, DCM in 7, right ventricular CM in 5, hypertrophic CM in 1 and congenital in 3. The baseline characteristics of all cats will be described including: clinical presentation, physical examination, systemic BP, echocardiographic parameters (2D, M-mode, Doppler & selected tissue Doppler parameters), thoracic radiographs, available blood work, concomitant diseases, concomitant medications and changes in medications at the time of initiation of Pimo. Selected parameters will be evaluated for prognostic significance relative to survival. Baseline parameters will be compared to recheck parameters in cats that underwent re-evaluation within 14-90 days of Pimo initiation. Adverse event will be reported. Overall survival will be reported. Etiology dependant survival will also be reported and compared between groups.

PREVALENCE OF HEART MURMURS AND OCCULT HEART DISEASE IN APPARENTLY HEALTHY ADULT CATS. LT Drourr, SG Gordon, RM Roland, M Boggess. The Michael E. DeBakey Institute, College of Veterinary Medicine and Biomedical Sciences, Texas A&M University, College Station, TX.

There is wide interest in the prevalence of occult heart disease (OHD) in cats with and without murmurs. The goal of this prospective study was to report the incidence of auscultatory abnormalities and OHD in a well described cohort of apparently healthy adult cats of various ages. 100 cats without history of heart disease, murmur, or other systemic diseases were recruited. Cats underwent a physical examination including auscultation by 2 investigators. Gallop rhythms, arrhythmias, and murmur timing, location, and intensity were recorded. Blood pressure, electrocardiogram, PCV/TS, BUN, creatinine, HW antibody, T4, thoracic radiographs, and a complete echocardiogram were contemporaneously performed on all cats. Cats were not sedated. Some cats had an NT-proBNP (71/100).

The median age of all cats was 3.7 (IQR; 2.7-7.0). Murmurs were detected in 44/100 cats, 26/44 were labile. No gallop rhythms or arrhythmias were detected. OHD was documented in 14/100 cats (10/14 had murmurs). Murmur etiology was not documented in 35/44 cats. The incidence of heart murmurs and OHD increased with age.

Many apparently healthy cats have murmurs and the prevalence increases with age. The prevalence of OHD is > in cats with murmurs as compared to those without murmurs, however auscultation is neither sensitive nor specific for the detection of OHD. This study and others underscore the relative importance of novel screening methods for detecting OHD in apparently healthy cats as well as the need for consensus on what constitutes a diagnosis of echocardiographic feline OHD and the incidence of progression of feline OHD.

CARDIAC EVALUATION OF THE EXERCISING HORSE

Sophy A. Jesty, DVM, DACVIM
Ithaca, NY

INTRODUCTION

Exercise necessitates that the demands of increasing oxygen consumption (VO_2) are met by the cardiac system. Changes mediated predominantly by shifts in autonomic tone ensure increased cardiac output (CO). Two large retrospective studies have been performed on horses presenting for poor performance.[1,2] Conclusions included that diagnoses were subtle and that many horses probably suffer from a summation of abnormalities. Additionally, the latter study strongly advocated the use of a high speed treadmill evaluation as a means to approximate the level of exercise at which a trained horse might be expected to work. At this time, the cardiac evaluation of poor performing horses should include at a minimum a resting and stress ECG, a resting and stress echocardiogram, and cardiac troponin evaluation. In addition, there are a number of studies showing alternative means of evaluating the cardiac system during exercise and there are a number of modalities being actively researched to improve our diagnostic acumen for this population.

WORKUP

Electrocardiogram

Heart rate (HR) is an important component of CO (HR x stroke volume or SV). HR increases with exercise such that it is proportional to the horse's speed,[3] but since horses can reach a HR of 220-240 bpm regardless of fitness level, HRmax is not an indicator of exercise capacity (HRmax does not increase with training).[4] It is the speed associated with HR that is correlated with exercise capacity and race earnings.[3,4] The two most common variables relating speed to HR are the speed at HRmax (V_{HRmax}) and the speed at a HR of 200 bpm (V_{HR200}).[5] HR recovery is mediated by autonomic function and influenced by cardiovascular fitness level.[6] In horses, the time constant of HR recovery is inversely correlated with VO_2max.[7]

21% of all horses undergoing a thorough poor performance workup had a cardiac abnormality documented; of these 75% were arrhythmias.[2] Another study found ~65% incidence of arrhythmias either during or immediately after exercise, but did not distinguish between the two.[8] Arrhythmias occurring during the post-exercise period are likely mediated by the shifting balance between sympathetic and parasympathetic tone, and are benign to a certain degree. On the other hand, arrhythmias during exercise can be a cause of decreased CO at a time that requires maximal CO, can be dangerous to the horse and rider if they degenerate into an unstable arrhythmia, and can be an indicator of cardiac damage which might have additional manifestations such as decreased systolic function. One study documented arrhythmias during and after exercise in Thoroughbred racehorses without poor performance.[9] It showed a high incidence of post-exercise arrhythmias, but only a 3% incidence of exercising arrhythmias.[9]

Equine sudden death during exercise is rare and not often proven to be cardiac sudden death,[10,11] but this does not absolve the veterinarian from the responsibility of considering a horse with exercising ventricular arrhythmias unsafe to ride until thearrhythmia resolves.

Echocardiogram

A number of studies have documented the changes that occur in mechanical function with exercise in horses. Left sided cardiac catheterization studies in horses reveal that during exercise, there is an increase in HR, an increase in left ventricular (LV) peak pressure generation, an increase in the first derivative of LV pressure generation (dP/dt), an increase in left ventricular internal diameter and volume at end diastole, an increase in left ventricular end diastolic pressure, an increase in myocardial segment velocity, an increase in SV, an increase in CO, and an increase in VO_2.[12-14] Right sided cardiac catheterization is studies in horses reveal that during exercise, there is a sharp increase in right atrial pressure (RAP) and pulmonary capillary wedge pressure (PCWP) in the first minute of hard exercise, an increase in PCWP with increasing levels of exercise, an increase in dP/dt_{max}, a decrease in dP/dt_{min}, and a decrease in the time constant of relaxation (tau) and the time of relaxation.[15-19]

There is a correlation between various indices of LV size on resting echo and speed and VO_2max.[20-22] Murmurs and regurgitation on echocardiogram at rest are generally not proven to decrease performance[23,24] Stress echocardiography is used to assess cardiac structure and function during exercise (although it is performed immediately post-exercise). Stress echocardiography should be performed with the HR as close to HRmax as possible, but horses recover after exercise incredibly quickly. Traditionally, the window of time recommended to acquire post-exercise stress echocardiographic views has been considered 30-120 seconds. Right sided pressure studies during exercise, however, have shown the rapidity with which hemodynamics change after cessation of exercise.[19] By just 60 seconds post-exercise for example, dP/dt_{max}, dP/dt_{min} and tau were already significantly different than during exercise.[19] For this and other reasons, pharmacologic stress testing using dobutamine and atropine has been investigated as a means to assess cardiac function during stress.[18, 25-28] These studies have shown that many cardiac parameters do not increase with pharmacologic stress testing to the same degree as with exercise stress testing, but that they are often similar to the parameters associated with post-exercise stress testing.[18,28]

Post stress echocardiography is insensitive for subtle changes in cardiac function that might be the cause of poor performance for two reasons: 1. echocardiography is insensitive for subtle changes in cardiac function even at rest, and 2. the immediate post-exercise period does not approximate exercise well. Despite these shortcomings, it is at the moment the best option available to assess mechanical cardiac function in poor performing horses. On the post-exercise stress echocardiogram, there should be an increase in fractional shortening (FS), although research has shown a wide coefficient of variability in FS in the 2 minutes after exercise, making it difficult to interpret the absence of an increase in FS.[12,19]

Biomarkers

Biomarkers can provide additional information about the structure and function of the heart (NT-proBNP, e.g.) and cardiomyocyte integrity (cTnI, e.g.). They can provide information about the presence or severity of disease, the extent of injury, and the prognosis.

The most common biomarker evaluated in horses at this time is cTnI, which is both sensitive and specific for cardiomyocyte damage. There are many case reports in the literature showing an increase in cTnI in horses with known cardiac damage. Additionally, there is a prospective study that shows an increase in cTnI in horses with induced myocardial damage from monensin; this study showed that the degree of the increase seemed associated with the degree of myocardial damage (as is appreciated in people).[29] Because different analyzers use different sequences as epitopes, different antibodies, and may recognize different forms of circulating cTnI, their values cannot be compared directly.[30] Reference ranges have been published for horses using a variety of analyzers.[32-35] Increases in cTnI in horses with poor performance could be a result of true underlying myocardial damage (myocarditis, e.g.) or could be a result of strenuous exercise. While some studies have shown no effect of exercise on cTnI concentration,[18,31,33] other studies have shown an increase in cTnI post-exercise in horses performing normally.[34,36] Another caveat is that an increase in cTnI represents active myocardial necrosis; a horse can have old ('quiet') cardiac damage without an increase in cTnI.

Future directions

The assessment of cardiac mechanical function in exercising horses should be improved in the future. Standard echocardiography will only ever be an indirect measure of cardiac function. Advances in equine echocardiography (such as the use of transesophageal echocardiography during exercise or the use of Tissue Doppler Imaging for myocardial motion) might improve the sensitivity of this modality. The measurement of CO greatly contributes to our assessment of cardiovascular status in other situations; this would be the same in the case of the exercising horse. CO is the bottom line in the evaluation of cardiac mechanical function. While there have been many publications on various methods of measuring CO in horses at rest, there has been only one publication on a clinically relevant means to measure CO in exercising horses.[37] Further refinements in measuring CO in exercising horses are necessary before its use is widespread in the workup of poor performing horses.

CONCLUSIONS

Stress ECG and biomarkers will be used for a long time in the workup of poor performing horses. Stress echocardiography might be surpassed by other methods to assess mechanical function in exercising horses in the future. It is likely inappropriate to conceive of a single test able to screen for the health of the entire cardiac system in poor performing horses; rather, tests used should be considered complementary to one another as various diagnostic tests evaluate different aspects of the cardiac system.

REFERENCES

[1]Morris EA, et al. **Eq Vet J** 1991; 23: 169. [2]Martin BB, et al. **J Am Vet Med Assoc** 2000; 216: 554. [3]Vermeulen AD, et al. **Eq Vet J Suppl** 2006; 36: 113. [4]Gramkow HL, et al. **Eq Vet J Suppl** 2006; 36: 118. [5]Ross RJ, **The Athletic Horse** 1994. [6]Dimka U. **J Exerc Physiol** 2009; 12: 10. [7]Hada T, et al. **Eq Vet J Suppl** 2006; 36: 141. [8]Jose-Cunnilaris E, et al. **Eq Vet J Suppl** 2006; 36: 153. [9]Ryan N, et al. **Eq Vet J** 2005; 37: 265. [10]Kiryu K, et al. **J Vet Med Sci** 1999; 61: 921. [11]Boden LA, et al. **Eq Vet J** 2005; 37: 269. [12]Rugh KS, et al. **Eq Vet J** 1989; 21: 39. [13]Hiraga A, et al. **Eq Vet J** 1999; 30; 122. [14]Pascoe JR, et al. **Eq Vet J Suppl** 1999; 30: 148. [15]Manohar M, et al. **Eq Vet J** 1994; 26: 51. [16]Manohar M, et al. **Eq Vet J Suppl** 1999; 30: 117. [17]Langsetmo I, et al. **Eq Vet J Suppl** 1999; 30: 170. [18]Durando MM, et al. **Eq Vet J Suppl** 2006; 36: 183. [19]Durando MM et al, **Eq Vet J Suppl** 2002; 34: 408. [20]Rueca F, et al. **Eq Vet J Suppl** 1999; 30: 163. [21]Young LE, et al. **Eq Vet J Suppl** 2002; 34: 467. [22]Buhl R, et al. **J Am Vet Med Assoc** 2005; 226: 1881. [23]Buhl R, et al. **Eq Vet J Suppl** 2006; 36: 178. [24]Young LE, et al. **J Vet Intern Med** 2008; 22: 418. [25]Frye MA, et al. **J Vet Intern Med** 2003; 17: 58. [26]Sandersen CF, et al. **Eq Vet J** 2005; 37: 128. [27]Sandersen CF, et al. **J Vet Intern Med** 2006; 20: 575. [28]Gehlen H, et al. **J Vet Intern Med** 2006; 20: 562. [29]Divers TD, et al. **J Vet Diagn Imag** 2009; 21: 338. [30]Adin DB, et al. **J Vet Intern Med** 2006; 20: 1136. [31]Phillips W, et al. **J Vet Intern Med** 2003; 17: 597. [32]Slack JA, et al. **J Vet Intern Med** 2005; 19: 577. [33]Begg LM, et al. **Aust Vet J** 2006; 84: 336. [34]Nostell K, et al. **J Vet Cardiol** 2008; 10: 105. [35]Kraus MS, et al. **Am J Vet Res** 2010; 71: 55. [36]Holbrooke TC, et al. **Eq Vet J Suppl** 2006; 36: 27. [37]Durando MM, et al. **Am J Vet Res** 2008; 69: 1054.

THERAPEUTIC OPTIONS IN ATRIAL FIBRILLATION

M. Kimberly J. McGurrin, BSc, DVM, DVSc, DACVIM
Guelph, ON, Canada

In the majority of cases, horses with atrial fibrillation (AF) have no apparent structural heart disease and no accompanying systemic illness. Establishment of sinus rhythm in these animals usually completely restores cardiovascular function and allows the animal to be returned to performance. Quinidine has been the primary option in the management of equine AF. Quinidine has low efficacy in chronic AF and has numerous side effects, additionally the agent is becoming more difficult to obtain. Over the last 5-7 years there has been increased interest into alternative or adjunctive options.

FLECAINIDE

Flecainide has been investigated as a potential pharmacologic alternative treatment. It has a relatively higher efficacy and lower side effect profile in recent onset AF in humans. In horses, favourable response has been noted in short duration (15 minutes) experimentally induced AF following intravenous administration. However, clinically, the agent has not been reported as effective intravenously (none of 10 horses with naturally occurring AF responded and two had potentially dangerous dysrhythmias). There has been only one report of its successful use orally. This agent has proven pro-arrhythmic in clinical use and caution should be considered if this agent is nevertheless elected.

AMIODARONE

Amiodarone has been considered to be moderately effective in the management of human AF and has been noted to control rate even if it does not restore SR. It has been associated with worsening of pulmonary disease and causing pulmonary fibrosis. Neuropathies, bradycardia and hepatotoxicity have also been reported. Most of these concerns are heightened in humans because of chronic use. Oral bioavailability is low in the horse. There have been two reports on its (IV) administration in horses. Sinus rhythm was restored in 4/6 and 3/6, respectively. However, 3/6 horses had hindlimb weakness and 3/6 horses had transient increases in serum bilirubin levels in one study and 3/6 had diarrhea in the second. A treatment period of greater than 36 hours was associated with the adverse effects in both studies.

PROPAFENONE

Propafenone has been reported to have high efficacy in recent onset AF in humans. A recent study investigating the intravenous use of propafenone in horses with AF of approximately 3 months duration (2 horses with natural AF, 4 with induced AF) reported no horses responded to treatment, despite levels within the low therapeutic range, and changes in the atrial fibrillation cycle length (indicating slowed conduction). The horses were subsequently treated successfully with quinidine.

DILTIAZEM

This agent slows conduction velocity at the AV node. Diltiazem is not proposed as an antiarrhythmic to treat atrial fibrillation but as an adjunctive therapy for quinidine. Quinidine increases conduction through the AV node. The addition of diltiazem to a treatment regimen may aid in the control of the ventricular rate. This agent has been evaluated in healthy horses but has not yet been evaluated in horses with atrial fibrillation.

ELECTRICAL CARDIOVERSION

Presently, the only non-pharmacological option for the treatment of atrial fibrillation in the horse is electrical cardioversion. A successful shock depolarizes sufficient atrium to disrupt the fibrillating waves from perpetuating. Sinus rhythm is then restored. Placing electrodes so that sufficient atrial tissue is encompassed is therefore important, as well as generating sufficient energy so that the "targeted" myocardium is completely depolarized (influenced by electrode design as well as energy selection).

Successful transthoracic cardioversion in a horse has been reported. However, the horse was relatively small and concurrent antiarrhythmic pharmaceutical therapy was required. Energy levels of 2-4J/kg or higher are commonly used in humans and several authors based on efficacy data have recommended a directive towards this "high initial energy". The maximum energy level for commercial defibrillators is 360J. Extrapolation to horses would suggest that to achieve high success rates, multiple defibrillators would be needed (720J has been used in humans).

Transvenous electrical cardioversion has far lower energy requirements in humans with reported successful energies of less than 10J (0.14J/kg). These catheters are generally placed through a femoral vein approach under fluoroscopic guidance. The procedure is more technically challenging than transthoracic cardioversion and requires specialized catheters and is therefore considered only for refractory cases. Based on energy requirements this technique was considered for the horse and has since been adapted for use.

Cardiac catheterization is achieved in the standing horse. Introducer catheters (10g) are placed in the right jugular vein approximately 15 cm apart. Cardioversion catheters are then placed through the introducer catheters. The pulmonary artery (PA) catheter is placed first through the proximal introducer. It is advanced under intracardiac pressure and ultrasound guidance into the right heart and into the pulmonary artery (PA). Ultrasound guidance alone is used to guide placement into the left branch of the PA (approximately 125 cm in depth). The second catheter is then placed through the distal introducer and is advanced until its distal end is in the right ventricle (on pressure profile and on ultrasound). This catheter is then withdrawn until the ventricular waveform on the pressure profile is lost. This ensures that the electrode is positioned such that the electrode is in the right atrium (approximately 70 cm in depth). The catheters are sutured to the horse.

Once the cardioversion catheters are secured, the horse is induced under general anesthesia. The horse is positioned in left lateral recumbency. Lidocaine infusion is avoided, as it will increase cardioversion threshold. Thoracic radiography is performed to confirm electrode positioning. The PA catheter must not overly the cardiac silhouette, as this position would increase the risk of shock to the ventricle. Successful cardioversion has been achieved when the PA catheter either appears to curve behind the cardiac silhouette, or when it traverses slightly dorsad. The latter has been associated with lower energy levels.

Once the cardioversion catheters are confirmed as correctly positioned then cardioversion can be attempted. Vasopressive agents such as dobutamine should be withdrawn at least 5 minutes before shock delivery (these agents can be arrhythmogenic and could potentiate adverse response to shock application). The shocks should be synchronized with the QRS complex. Delivery of shock on the T wave has high risk of initiating ventricular fibrillation.

The author uses an initial energy level of 50J and a maximum energy of 300J, with 2 minute intervals between shocks to allow monitoring while allowing for a possible potentiating effect from the previous shock. At the Ontario Veterinary College, the author has treated 155 episodes of atrial fibrillation with this method; 153 responded to treatment. The mean energy level at conversion in these episodes was 156.8J (range 50-300J, 0.1-0.7 J/kg).

REFERENCES

[1.] Ohmura H, et al.**J Vet Med Sci.** 2000;62(7):711. [2.] Frye MA et al **J Am Vet Med Assoc.** 2002;220(7):1039. [3.] van Loon G et al **Equine Vet J.** 2004; 36(7):609. [4.] McGurrin MK et al **J Vet Intern Med**. 2005;19(5):695. [5.] Schwarzwald CC et al; **J Vet Intern Med**. 2005;19(5):703. [6.] McGurrin MK et al **J Vet Cardiol.** 2005 Nov;7(2):109-19. [7.] Risberg AI et al **J Vet Intern Med**. 2006 ;20(1):207. [8.] Schwarzwald CC et al **J Vet Intern Med.** 2007;21(1):166. [9.] De Clercq D, et al **Equine Vet J**. 2007;39(4):344. [10.] De Clercq D et al **Vet J.** 2008;177(2):198-204. [11.] McGurrin MK, et al **J Vet Intern Med** 2008;22(3):609. [12.] De Clercq D et al **Am J Vet Res**. 2009;70(2):223.

VASCULAR RINGS & SIMILAR THINGS IN DOGS: DIFFERENTIAL DIAGNOSIS

James W. Buchanan, DVM, MMedSci, DACVIM (Cardiology)
Philadelphia, PA

INTRODUCTION

Vascular rings are congenital cardiovascular anomalies that compress the esophagus and trachea and impair swallowing in dogs and cats. In human infants, they typically cause respiratory signs. The most common form accounting for 95% of the cases in dogs is persistent right aortic arch (PRAA) with or without associated vascular abnormalities such as double aortic arches, and retroesophageal subclavian arteries (Fig. 1) (Table). (1)

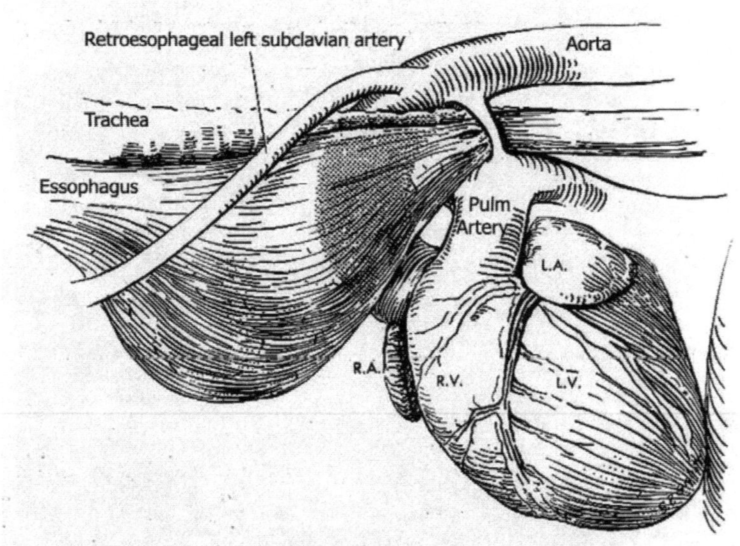

FIG. 1. Diagram of the heart, vessels, esophagus and trachea in a dog with persistent right aortic arch viewed from the left side. The esophagus and trachea are incarcerated in a vascular ring consisting of the abnormal aortic arch on the right, the ligamentum arteriosum dorsolaterally, the pulmonary artery on the left and the right pulmonary artery ventrally. From the author's article in *Current Veterinary Therapy* 1966–67.

TABLE

Vascular anomalies in 55 dogs with vascular rings

	Number	%
Persistent right aortic arch	52	95
Number with coexisting anomalies:		
Retroesophageal left subclavian artery	17	33
Double aortic arch	6	12
Patent ductus arteriosus	6	12
Persistent left cranial vena cava	6	12
Left hemiazygos vein	3	6
Retroesophageal right patent ductus arteriosus	2 *	4
Retroesophageal right subclavian artery and right ligamentum arteriosum	1 *	2

* Left aortic arch

A classic case is a weanling puppy that regurgitates undigested food shortly after eating solid food. The diagnosis is traditionally confirmed by a barium esophagram revealing dilation of the esophagus cranial to the heart. Treatment is surgical division of the ligamentum arteriosum to relieve esophageal constriction (2). The first successful management of a dog with PRAA was published in 1955 (3). I have dealt with many cases in the last 50 years and encountered several dogs that did not follow the classic course and/or proved not to have vascular rings. These will be the subjects of this presentation along with a review of normal and abnormal aortic arch embryology, old diagnostic methods, a new diagnostic tracheal sign and radiographic features of computed tomography (CT) angiography demonstrating anomalous vessels in dogs with vascular rings.

METHODS

Physical examination frequently yields evidence of non-specific megaesophagus when the throat bulges with expiration or when the chest is compressed. Endoscopy has been proposed for use in diagnosing megaesophagus but has little value in determining the cause or treatment of the megaesophagus.s. Survey radiography revealing tracheal position in DV or VD radiographs is sufficient to diagnose PRAA. The trachea is characteristically deviated leftward at the cranial border of the heart of dogs with PRAA (Fig 2). Barium esophagrams or fluoroscopy in dogs with vascular rings typically reveals dilation of the esophagus cranial to the heart but sometimes there is minimal dilation and/or barium obscures the trachea. Angiocardiography has been used for many years to visualize cardiac chambers and blood vessels but 3-dimensional analysis of vessel anatomy from 2-dimensional images is difficult and can be misleading (4,5). In non-selective angiocardiography contrast material is injected intravenously near the heart and radiographic exposures are made as the contrast material circulates through the heart and great vessels. In selective angiocardiography contrast material is injected through catheters into specific chambers or vessels to avoid superimposition and to demonstrate valvular incompetence or intracardiac shunts. It is seldom done to evaluate suspected vascular rings because it is an unnecessary expense and often is not clearly diagnostic. Non-selective CT angiography can be done following intravenous injection of contrast material and yields distinct color coded images of the esophagus, trachea and blood vessels (6). Their relationships can be seen better with this technique than with any of the preceding methods.

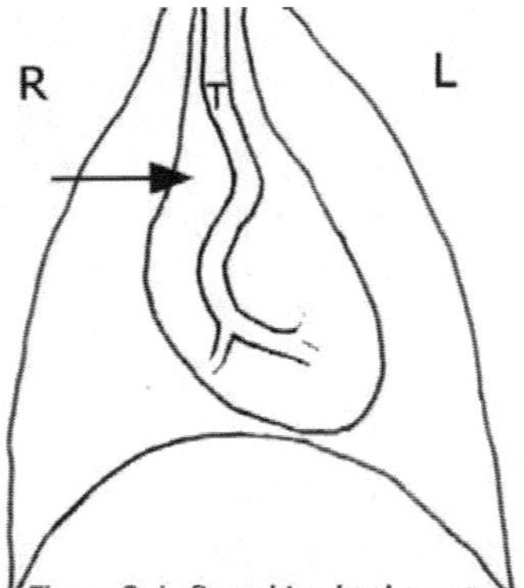

Figure 2. Leftward tracheal curve in dogs with PRAA

RESULTS

Misleading appearance of a dilated esophagus cranial to the heart has been observed in dogs with incompletely filled megaesophagus, overlong esophagus, primary esophageal stricture, and drooping esophagus between craniad vessels. Accurate diagnosis of these conditions avoids inappropriate surgery. A moderate or marked leftward curve of the trachea at the cranial border of the heart was observed in 27/27 dogs with PRAA and in none of the cases above or in 63 normal dogs or 30 dogs with generalized megaesophagus (1).

SUMMARY

Esophageal dilation cranial to the heart occasionally occurs for reasons other than vascular ring constriction. Barium esophagrams are not necessary for the diagnosis of PRAA and are less reliable than recognition of leftward tracheal deviation. CT angiography is the most accurate method for visualizing vascular rings but is not necessary if a surgeon knows the possible malformations of aortic arch pathology and how to recognize and deal with all of them intraoperatively.

REFERENCES

[1]Buchanan JW, **J Vet Intern Med** 2004;18:510-514. [2]Buchanan JW, **J Sm An Pract** 9;1968:409-428. [3]Detweiler DK and Allam M. **Cornell Vet** 45;1955:209-229 [4]Buchanan JW, **J Am Vet Radiol Soc** 1963;4:1-8. [5]Buchanan JW, and Patterson DF, **J Am Vet Radiol Soc** 1965;6:21-35. [6]Pownder S, and Scrivani P. **J Vet Card** 2008;10:125-128. [7]Buchanan JW. http://cal.vet.upenn.edu/projects/cardiosf/project/embprs/embryo.htm 2003 **[8]Buchanan JW. http://www.vin.com/library/general/JB102congAngio.htm 2006.**

FEASIBILITY OF STENTS FOR CONGENITAL VALVULAR PULMONARY STENOSIS IN DOGS

Brian A Scansen DVM, MS, DACVIM (Cardiology)
Columbus, Ohio

INTRODUCTION

Congenital valvular pulmonary stenosis (PS) is the third most commonly diagnosed congenital heart defect of dogs in North America.[1] As in children,[2] valvular PS in dogs is described in three principal forms – (1) a dome-shaped valve with commissural fusion, (2) a dysplastic and markedly thickened valve with or without hypoplasia of one or more leaflets, and (3) hypoplasia of the pulmonary annulus.[3,4] While the human form of valvular PS is predominately that of commissural fusion,[2] the canine form is more commonly characterized by thickened, dysplastic valves.[3,5,6] Without therapy, dogs with valvular PS are at risk for symptoms including exercise intolerance, syncope, sudden cardiac death, congestive heart failure, and cyanosis from a right-to-left shunt.[4,7,8]

THE EVIDENCE FOR BALLOON PULMONARY VALVULOPLASTY

Balloon pulmonary valvuloplasty (BPV) was first performed in a dog in 1980 and reported in a child in 1982.[9,10] There is evidence that BPV improves the clinical outcome of human and canine patients with valvular PS, both with a reduction in clinical symptomatology and an improvement in survival.[5, 8, 11, 12] The procedure is now routinely performed in clinical canine practice with low morbidity and mortality for those patients with a severe gradient or the presence of clinical signs referable to their disease.

Although effective for many patients, human and canine studies suggest that patients with valvular dysplasia and a hypoplastic annulus show less improvement following BPV than those with purely valvular fusion.[6,13] This stands to reason as the principal benefit of BPV is tearing of commissural fusion; it is unable to alleviate obstruction secondary to annular hypoplasia or thick, redundant valve tissue. In a study of 30 dogs with valvular PS, 94% of the 18 dogs with a normal pulmonary annular diameter and minimal valve thickening had resolution of their clinical signs and survived more than 1 year after the procedure; however, for the 12 dogs with marked leaflet thickening and a hypoplastic pulmonary annulus, only 66.6% had a favorable outcome and only 50% showed a resolution in their clinical signs.[6]

NOVEL INTERVENTIONAL THERAPIES FOR PS

Newer treatment options have been developed in children to address the variable and frequently disappointing results for BPV in the setting of a highly dysplastic pulmonary valve and/or annular hypoplasia. These include cutting balloons and intravascular stent placement.[14,15] Cutting balloons incorporate small blades, or atherotomes, which are bonded to the balloon and serve to make small incisions in a vessel resistant to standard balloon angioplasty. A small case series recently described the use cutting balloons in 4 children with variable degrees of valvar PS and infundibular narrowing/hypertrophy in the setting of Fallot's tetralogy.[14] These authors found this technique to be safe and effective in this population. More recently, a report of nine children with unsalvageable pulmonary valves and tetralogy of Fallot underwent stenting of their right ventricular outflow tract and pulmonary valve annulus.[15] The short-term results in these patients were promising and obviated immediate surgical intervention.

PRESENTATION

The purpose of this presentation is to review the literature on medical, interventional, and surgical options for canine patients with valvular PS. Additionally, novel interventional techniques coming into practice in pediatric interventional cardiology will be highlighted with a discussion of the potential to transfer these techniques to the care of canine patients. We have recently attempted pulmonary valve stenting in dogs with severe congenital valvular PS. Our initial experience, potential for widespread application, and the currently unresolved issues will be presented.

REFERENCES

1. Buchanan JW. **Current Veterinary Therapy XI: small animal practice.** 1992:647-654. 2. Waller BF, et al. **Clin Cardiol** 1995;18:45-50. 3. Patterson DF, et al. **Am J Cardiol** 1981;47:631-641. 4. Fingland RB, et al. **J Am Vet Med Assoc** 1986;189:218-226. 5. Ristic JM, et al. **J Vet Cardiol** 2001;3:13-19. 6. Bussadori C, et al. **J Vet Intern Med** 2001;15:553-558. 7. Lombard CW, et al. **J Am Vet Med Assoc** 1989;194:71-75. 8. Johnson MS, et al. **J Vet Intern Med** 2004;18:656-662. 9. Buchanan JW, et al. **J Vet Intern Med** 2002;16:116-117. 10. Kan JS, et al. **N Engl J Med** 1982;307:540-542. 11. Fawzy ME, et al. **Am Heart J** 2007;153:433-438. 12. Rao PS. **Catheter Cardiovasc Interv** 2007;69:747-763. 13. McCrindle BW. **Circulation** 1994;89:1751-1759. 14. Carlson KM, et al. **Catheter Cardiovasc Interv** 2005;64:507-512. 15. Dohlen G, et al. **Heart** 2009;95:142-147.

WHAT ARE BIOMARKERS & HOW MIGHT WE USE THEM?

Professor Adrian Boswood DipECVIM
The Royal Veterinary College, University of London

INTRODUCTION

One working definition of biomarkers was suggested as follows: a biomarker is "a characteristic that is objectively measured and evaluated as an indicator of normal biologic processes, pathogenic processes, or pharmacologic responses to a therapeutic intervention."[1] This definition could cover numerous measurable characteristics in patients with heart disease from heart rate to bioelectrical impedance! The term is now being used in a more restricted sense to indicate a substance measured, often in the blood, indicative of the presence, stage or severity of an underlying disease process.

CLASSIFICATION OF MARKERS

There are various classes of marker that have been used either clinically or in clinical research to characterize abnormalities of the cardiovascular system. They have been broadly categorized as follows,[2]

- Markers of myocyte injury,
- Markers of myocyte stress,
- Markers of remodeling,
- Markers of endothelial dysfunction,
- Markers of inflammation, and
- Neurohormonal markers.

These categories, although helpful, are not mutually exclusive; for instance NTproBNP is both a Neurohormonal marker and increased in the face of myocyte stress. The classification remains useful for considering the different circumstances in which concentrations of markers may be elevated.

POTENTIAL USES OF MARKERS

It has been suggested that in order to be useful a new biomarker should fulfill three criteria; firstly the clinician should be able to measure it, secondly it should add new information (i.e. information that is not already known about the patient's condition) and finally it should assist the clinician in managing cases.[3] Although at first glance these are relatively straightforward criteria the latter one in particular is challenging to prove. In human patients although a plethora of biomarkers has been studied in heart failure patients there are really only two or three markers that can be considered to meet all these criteria; these include natriuretic peptides and troponins, the two biomarkers that have been evaluated most extensively in veterinary patients.

Markers can used to assist in multiple aspects of the diagnosis and management of cardiovascular disease. Their potential areas of use can be summarized as follows

1) Diagnosis of acute or chronic syndromes
2) Detection of sub-clinical diseases (screening)
3) Risk stratification of patients
4) Monitoring disease progression or response to therapy
5) Selection of therapy.

When considering current and potential future uses of markers in veterinary patients I will concentrate on the evidence to support their use in these different areas and the types of studies from which evidence has been gathered. I will also describe additional studies that would need to be conducted to establish whether or not markers would be useful in other circumstances.

Cross sectional studies are usually sufficient to determine the value of markers in situations 1) and 2) above whereas demonstration of the value of biomarkers in situations 3) to 5) would require longitudinal and/or interventional studies. Cross sectional studies are often more straightforward to undertake and may require less time to complete than longitudinal or interventional studies and therefore for novel markers evidence tends to initially accrue to support their use in the earlier categories and evidence to support their value in later categories takes more time to accumulate. For this reason much of the evidence available in veterinary patients with the markers we have currently derives from cross-sectional studies.

EVIDENCE FROM VETERINARY STUDIES

Evidence for diagnosis

Numerous studies have described the potential diagnostic value of measuring cardiovascular biomarkers in different situations. A large number of studies have evaluated natriuretic peptides and troponins in both dogs and cats. There is good evidence to suggest that natriuretic peptide concentrations can discriminate dogs and cats with heart disease and heart failure from normal dogs and those that have respiratory disease.[4-6] Similar value has been demonstrated for cats.[7-8]

With the assays that are currently commercially available for measurement of natriuretic peptides in dogs and cats there has been controversy over what cut-off values to employ. It has also been shown that non-cardiac, or concurrent disease can affect the measured concentrations of natriuretic peptides.[9]

In my opinion there is little doubt that concentrations of NTproBNP can be used to differentiate, with reasonable accuracy, dogs and cats with heart failure from those not suffering from this syndrome. Whether or not they are superior to established methods of making this differentiation has not been established. These tests should therefore be regarded as complementary to conventional methods of assessment of patients and should not be regarded as an alternative.

The other biomarkers that have been studied for the differentiation of animals in heart failure from those without, and differentiating patients with respiratory distress due to heart failure from those with respiratory distress due to other causes, are the troponins I and T. Studies have tended to show that these are less effective at diagnosis and differentiation than the natriuretic peptides.[10-12]

Recently more sensitive troponin assays with lower limits of detection have been developed for use in human patients. Studies in human patients have shown that clinically valuable information can be derived from concentrations previously beneath the detection limit of the old assays.[13] One recent study has demonstrated that concentrations of troponin I measured with a high-sensitivity assay correlate with severity of disease in dogs with mitral regurgitation[14] and it will be interesting to see if additional useful clinical information can be derived with this assay.

Evidence for detection of sub-clinical disease

The use of biomarkers for the screening of sub-clinical disease in a population is much more controversial than their use in diagnosis. When screening for sub-clinical disease one is, by definition, looking for patients with an earlier stage of disease than when using a test for diagnostic or discriminatory purposes. The value of a test in this circumstance is going to be largely determined by knowledge of the prevalence of the condition of interest in the population being screened and such detailed epidemiological information is lacking for many of the diseases we frequently encounter. Some common diseases, such as valvular heart disease, manifest with an obvious heart murmur in the early stages of the disease. The value of a screening test would therefore probably be restricted to diseases that have a clinically silent occult phase such as dilated cardiomyopathy.

In dogs studies have examined the ability of natriuretic peptides and troponin to screen for occult myocardial disease and shown some ability to discriminate normal from abnormal individuals.[15-16] However the moderate sensitivity and/or specificity demonstrated in these studies would lead to very high numbers of false positive or false negative diagnoses and currently the use of tests in this way in a broad population cannot be recommended. Using the tests in a more targeted way in a high-risk population may still be of value.

A recent study in cats has also concluded that measurement of natriuretic peptides would not be useful when screening for occult myocardial disease in cats.[17] The value of screening human populations for evidence of systolic dysfunction with natriuretic peptides has not been established and currently this practice is not recommended.[18]

Evidence for risk stratification

Risk stratification involves the differentiation of a population, either ostensibly normal or diseased, into those who are at greater risk of developing a particular disease, complication of a known disease or outcome of interest. For instance patients with a disease can be stratified into those at higher risk of death compared to others. There is evidence from at least four studies in dogs to suggest that biomarkers may be able to differentiate dogs at high risk of succumbing to heart disease from those with a lower risk.[12,19-20]

Our recent study[19] shows that dogs with degenerative mitral valve disease and elevated concentrations of NTproBNP are more likely to succumb to their disease. This predictive value is strengthened by knowledge of the dog's left ventricular diameter in diastole and these two measurements are independently predictive of outcome. Their predictive value is also independent of whether or not the dog is demonstrating outward clinical signs attributable to their heart disease.

These findings show considerable promise in the differentiation of dogs suffering from a common chronic condition into those at high risk of complications in the near future from those with less significant disease.

Evidence for monitoring progression and response to therapy

Currently there are few longitudinal studies in the literature demonstrating the changes in levels of biomarkers over time or their response to therapy. Our, currently unpublished, observations suggest that NTproBNP concentrations increase more rapidly in dogs with degenerative mitral valve disease that develop signs of heart failure compared to those that do not. One recent study has shown that NTproBNP concentrations decrease in dogs following the introduction of Pimobendan.[21] Another study demonstrated that the proportion of hyperthyroid cats with detectable cTnI concentrations is higher prior to therapy than following successful therapy with radioactive iodine suggesting a reduction in ongoing myocardial damage.[22]

Selection of therapy

To my knowledge there are currently no studies in the veterinary literature which demonstrate the value of biomarkers in the selection or modification of therapy. In the human field several studies have demonstrated the value of biomarkers in guiding heart failure therapy,[23] in the detection of a sub-population of patients likely to respond favourably to a particular therapy[24] and in indicating a population of patients who will benefit from therapy.[25] All of these areas are currently unexplored in veterinary patients.

CONCLUSIONS

The ultimate proof of the value of measurement of a biomarker cannot be simply measured in terms of sensitivity, specificity or accuracy. Their genuine value would be if the measurement of a biomarker makes clinicians better at managing cases and results in improved patient outcomes. In order to demonstrate this a randomized clinical trial would need to be undertaken where patients are randomized to either have a biomarker measured and have this information contribute to patient management or to be managed optimally without knowledge of the biomarker. This would require the design and execution of a large, multicentre longitudinal study and, as veterinarians we have yet to rise to this challenge.

REFERENCES

[1.] Biomarkers and surrogate endpoints: preferred definitions and conceptual framework. **Clin Pharmacol Ther** 2001;69:89-95. [2.] Braunwald E. Biomarkers in heart failure. N Engl J Med 2008;358:2148-2159. [3.] Morrow DA, de Lemos JA. Benchmarks for the assessment of novel cardiovascular biomarkers. **Circulation** 2007;115:949-952. [4.] Boswood A, Dukes-McEwan J, Loureiro J, et al. The diagnostic accuracy of different natriuretic peptides in the investigation of canine cardiac disease. **J Small Anim Pract** 2008;49:26-32. [5.] Fine DM, Declue AE, Reinero CR. Evaluation of circulating amino terminal-pro-B-type natriuretic peptide concentration in dogs with respiratory distress attributable to congestive heart failure or primary pulmonary disease. **J Am Vet Med Assoc** 2008;232:1674-1679. [6.] Oyama MA, Fox PR, Rush JE, et al. Clinical utility of serum N-terminal pro-B-type natriuretic peptide concentration for identifying cardiac disease in dogs and assessing disease severity. **J Am Vet Med Assoc** 2008;232:1496-1503. [7.] Connolly DJ, Soares Magalhaes RJ, Fuentes VL, et al. Assessment of the diagnostic accuracy of circulating natriuretic peptide concentrations to distinguish between cats with cardiac and non-cardiac causes of respiratory distress. **J Vet Cardiol** 2009;11 Suppl 1:S41-50. [8.] Fox PR, Oyama MA, Reynolds C, et al. Utility of plasma N-terminal pro-brain natriuretic peptide (NT-proBNP) to distinguish between congestive heart failure and non-cardiac causes of acute dyspnea in cats. **J Vet Cardiol** 2009;11 Suppl 1:S51-61. [9.] Raffan E, Loureiro J, Dukes-McEwan J, et al. The cardiac biomarker NT-proBNP is increased in dogs with azotemia. **J Vet Intern Med** 2009;23:1184-1189. [10.] Connolly DJ, Brodbelt DC, Copeland H, et al. Assessment of the diagnostic accuracy of circulating cardiac troponin I concentration to distinguish between cats with cardiac and non-cardiac causes of respiratory distress. **J Vet Cardiol** 2009;11:71-78. [11.] Herndon WE, Rishniw M, Schrope D, et al. Assessment of plasma cardiac troponin I concentration as a means to differentiate cardiac and noncardiac causes of dyspnea in cats. **J Am Vet Med Assoc** 2008;233:1261-1264. [12.] Oyama MA, Sisson DD. Cardiac troponin-I concentration in dogs with cardiac disease. **J Vet Intern Med** 2004;18:831-839. [13.] Latini R, Masson S, Anand IS, et al. Prognostic value of very low plasma concentrations of troponin T in patients with stable chronic heart failure. **Circulation** 2007;116:1242-1249. [14.] Ljungvall I, Hoglund K, Tidholm A, et al. Cardiac Troponin I Is Associated with Severity of Myxomatous Mitral Valve Disease, Age, and C-Reactive Protein in Dogs. **J Vet Intern Med** 2009. [15.] Chetboul V, Tessier-Vetzel D, Escriou C, et al. Diagnostic potential of natriuretic peptides in the occult phase of golden retriever muscular dystrophy cardiomyopathy. **J Vet Intern Med** 2004;18:845-850. [16.] Oyama MA, Sisson DD, Solter PF. Prospective screening for occult cardiomyopathy in dogs by measurement of plasma atrial natriuretic peptide, B-type natriuretic peptide, and cardiac troponin-I concentrations. **Am J Vet Res** 2007;68:42-47. [17.] Hsu A, Kittleson MD, Paling A. Investigation into the use of plasma NT-proBNP concentration to screen for feline hypertrophic cardiomyopathy. **J Vet Cardiol** 2009;11 Suppl 1:S63-70. [18.] Tang WH, Francis GS, Morrow DA, et al. National Academy of Clinical Biochemistry Laboratory Medicine practice guidelines: Clinical utilization of cardiac biomarker testing in heart failure. **Circulation** 2007;116:e99-109. [19.] Moonarmart W, Boswood A, Fuentes VL, et al. N-terminal pro B-type natriuretic peptide and left ventricular diameter independently predict mortality in dogs with mitral valve disease. **J Small Anim Pract** 2010;51:84-96. [20.] Serres F, Pouchelon JL, Poujol L, et al. Plasma N-terminal pro-B-type natriuretic peptide concentration helps to predict survival in dogs with symptomatic degenerative mitral valve disease regardless of and in combination with the initial clinical status at admission. **J Vet Cardiol** 2009;11:103-121. [21.] Atkinson KJ, Fine DM, Thombs LA, et al. Evaluation of Pimobendan and N-Terminal Probrain Natriuretic Peptide in the Treatment of Pulmonary Hypertension Secondary to Degenerative Mitral Valve Disease in Dogs. **J Vet Intern Med** 2009. [22.] Connolly DJ, Guitian J, Boswood A, et al. Serum troponin I levels in hyperthyroid cats before and after treatment with radioactive iodine. **J Feline Med Surg** 2005;7:289-300. [23.] Felker GM, Hasselblad V, Hernandez AF, et al. Biomarker-guided therapy in chronic heart failure: a meta-analysis of randomized controlled trials. Am Heart J 2009;158:422-430. [24.] Zannad F, Alla F, Dousset B, et al. Limitation of excessive extracellular matrix turnover may contribute to survival benefit of spironolactone therapy in patients with congestive heart failure: insights from the randomized aldactone evaluation study (RALES). Rales Investigators. **Circulation** 2000;102:2700-2706. [25.] Ridker PM, Danielson E, Fonseca FA, et al. Rosuvastatin to prevent vascular events in men and women with elevated C-reactive protein. **N Engl J Med** 2008;359:2195-2207.

NT-proBNP: SUMMARY OF A CANINE COHORT STUDY

Andrew W Beardow, BVM&S, MRCVS, DACVIM(Cardiology)
Portland, ME

Cardiac Biomarkers represent a new and potentially powerful tool in the assessment of patients with cardiovascular disease. As with any new biomarker, questions are raised about the utility of such biomarkers in answering a number of clinical questions. Can the marker help definitively diagnose the disease? What is their utility in staging the disease? Do they respond to successful therapeutic intervention, or conversely do they predict a failure of therapeutic efficacy? The question most often posed to the clinician is that of expectations regarding both quantity and quality of life. In this presentation we will discuss several of these questions. The cornerstone of any assessment of biomarker response is the definition of a normal reference range. Several cardiac biomarkers have been made commercially available to veterinarians. Troponin I, NTproANP, and latterly NTproBNP have been or are currently available as serological tests. Many publications exist on the utility of these markers. Review of these publications may be confusing when trying to define an appropriate reference range or defining appropriate interpretive criteria.

In this presentation we will review the current literature and also present the results of a study where 1321 patients were evaluated as part of the California Animal Hospital Study. Many of these patients had been evaluated on multiple occasions. Over a third of these dogs have been seen at least twice and 24 dogs have been seen at least 10 times. A summary of these findings will be presented. We will discuss interpretation of changes in these biomarkers as disease progresses, or during therapeutic intervention, and also interpretation of the data in terms of outcome or death where such data is available.

We will also discuss the response of NTproBNP to extra-cardiac disease, where the results obtained were of significant clinical interest. In particular we will discuss the effects of changes in renal function, and the presence of pericardial effusion. Pericardial effusion (PE) is of interest because the underlying pathophysiology is contrary to that seen in other forms of heart failure which are characterized by increased wall stress. Increased wall stress is one of the primary inciting causes of an elevated NTproBNP concentration. Extracardiac compression as seen in dogs with PE would be expected to minimize increases in wall stress. We show that dogs with right heart failure secondary to chronic valve disease have significantly higher NTproBNP concentrations than those in which right heart failure was induced by PE.

Disease progression was assessed in the California Animal Hospital Study using the grading system developed and presented on several occasions by Dr Ettinger. This system allowed assessment of progression from clinical signs (murmur) without evidence of pathophysiologic change through overt and uncontrolled congestive heart failure. This system will be reviewed briefly during the presentation.

REFERENCES

[1.] Boswood A, et al. The diagnostic accuracy of different natriuretic peptides in the investigation of canine cardiac disease **JSAP** 2007 1-7 [2.] Oyama MA, et al. Assessment of serum N-terminal pro-B-type natriuretic peptide concentration for differentiation of congestive heart failure from primary respiratory tract disease as the cause of respiratory signs in dogs**. JAVMA**. December 2009;235(11):1319-25.

THE CAT IS NOT A DOG- CARDIAC BIOMARKERS

Stephen Ettinger, DVM, Kristine Yee, DVM, Andrew Beardow, B.V.Sci.,, Scott Forney, DVM, Giosi Farace, PhD, Justin Allen, DVM

Longitudinal information from over 630 biomarker studies will be presented to identify the ability to establish normal and abnormal BM levels in the cat with cardiac disease, with cardiac and other disease and without cardiac changes. The differences between dog and cat will be identified, as will the variations from within the feline species in the presence of cardiac changes. NT-pro-BNP levels separate the four grades of cardiac disease in the cat with a high level of reliability.

In many smaller studies identifying cardiac biomarkers in cats, the markers chosen and the results have varied, the conditions under which the studies were completed varied and the ability to utilize the results have not been clear. In this report, cats were studied under clinical conditions but the biomarkers themselves were taken with research level consistency. Samples were shipped to the laboratory for analysis under dry ice conditions, always arriving frozen. Measures were undertaken prior to the study to guarantee sample safety and consistency.

We will report on the normal levels of troponins (cTnI), NT-proANP and NT-proBNP from more than 630 cats from one hospital, evaluated over several years. We choose to measure biomarkers in a large number of cats in a manner that would permit us to determine the true normal levels for these animals and to do so under circumstances that allowed us to simultaneously evaluate the cats utilizing methods accepted by veterinary cardiologists. Longitudinal studies were made on some of the cats so that repeatability and or changes in biomarker levels could be evaluated to test the reliability of the data or in cases of advancing medical disease, the effects of advancing disease and or medication on the data. This report will identify these results as well as noting the significant differences between the canine and feline data.

We will show what we believe to be the repeatable and reliable normal levels of these biomarkers as well as the levels of these markers in the various states of disease as determined through clinical testing. We will show the variations of these biomarkers, where they exist, by age, breed, sex, and weight. We also looked at normal cats and those with cardiac and non cardiac disease. We will note the correlations made with other diseases, the relationship to specific cardiac and non cardiac conditions that also became apparent after the blinding period was ended. The repeatability of the studies in normal and cardiac states was determined. Other than the in house traditional cardiac information we had no access to the biomarker results until the results were unblinded.

We did not begin the project with the intent of defining pre-answered questions that we needed to validate but instead entered the project asking questions. We are still doing so and anticipate that the collection of such a large amount of data (and continuing) will provide us with many answers regarding the relevance of cardiac biomarkers as it relates to the knowledge of cardiac disease in the feline.

REFERENCES

[1.] Lalor, S. et. al.: Plasma Concentrations of Natriuretic peptides in normal cats and normotensive and hypertensive cats with chronic kidney disease. **J. Vet. Card.** (2009) 11, 571; [2.] Hsu, A. et. al.: Investigation into the use of plasma NT-proBNP concentration to screen for feline hypertrophic cardiomyopathy **J. Vet Card.** (2009) 11, 563. [3.] Fox, P. et. al.: Utility of plasma N-terminal pro-brain natriuretic peptide (NT-proBNP) to distinguish between congestive heart failure and non-cardiac causes of acute dyspnea in cats. **J. Vet. Card** (2009) 11, 551. [4.] Connolly,D. et al.: Assessment of the diagnostic accuracy of circulating natriuretic peptide concentrations to distinguish between cats with cardiac and non-cardiac causes of respiratory distress. **J. Vet. Card.** (2009) 541.

BEYOND NT-PROBNP: ALTERNATIVE BIOMARKERS OF CARDIAC DISEASE

Robert Prošek, DVM, MS, DACVIM, DECVIM-CA
Gainesville, FL

The focus of the previous lectures on NT-proBNP highlights the excitement of cardiac biomarkers along with the challenges of finding the ideal biomarker or the ideal cut-off point(s). Currently the natriuretic peptides are regarded as the most promising circulating biomarkers in our patient populations, but there are many other markers that can be used in the evaluation of these patients. There are markers of myocyte stress like adrenomedullin and ST2, markers of myocyte injury including the cardiac troponins, markers of inflammation (e.g. C-reactive protein, TNF-α, interleukin 1 and 18), markers of remodeling including the matrix metalloproteinases and tissue inhibitors of metalloproteinases, and many others including the neurohumoral markers. The various application and discovery of these markers in veterinary patients has not been fully appreciated. The latest evidence in human studies has shown that a multimarker strategy might be superior to any marker alone and this will be discussed in regards to veterinary patients based on data from the California Animal Hospital study. Due to different diseases states accounting for variances of blood levels of natriuretic peptides, a multimarker strategy might be necessary for our patients also, along with further understanding of the various markers in the pathophysiology of heart disease and failure. With better understanding of current biomarkers, along with discovery of new markers, might lead us to individualized therapy based on individual biomarker levels. Longitudinal studies are mostly lacking in veterinary medicine and will be needed to understand the full value of cardiac biomarkers, as will an objective assessment if there is a need for more single patient or low number patient studies of cardiac troponins with disease x.

MARKERS OF INFLAMMATION

Markers of inflammation were one of the first biomarkers studied in co-existence with heart failure over 50 years ago with C-reactive protein (CRP) leading the pack of inflammatory markers measured. C-reactive protein is described as an acute-phase reactant synthesized by hepatocytes in response to the proinflammatory cytokine interleukin-6. Other proinflammatory cytokines that appear to be produced by nucleated cells in the heart include TNF-α, interleukin 1 and interleukin 18. In the cytokine hypothesis of heart failure, proinflammatory cytokines are produced by a damaged myocardium and the production can be enhanced by stimulation of the sympathetic nervous system. The proinflammatory cytokines can cause myocyte apoptosis and necrosis along with activation of monocytes leading to more production of the same cytokines, which than act to further impair myocardial function. Once in heart failure, some of the detrimental effects of cytokines include further deterioration of myocardial function, endothelial dysfunction, remodeling and cachexia. In human patients, increased concentrations of cytokines have been shown to coincide with a worse outcome when in heart failure. Several studies in veterinary patients with naturally occurring heart disease and heart failure have assessed markers of inflammation, and in some circumstances revealing significant changes in these patients.

Fas, a member of the TNF-α receptor family is expressed on a variety of cells, including myocytes. The activation of Fas by the Fas ligand mediates apoptosis and appears to play an important role in the progression of heart failure. Inhibition of soluble Fas in mice reduces post-infarction ventricular remodeling and improves survival. A novel and exciting treatment option of using a nonspecific immunomodulating agent (intravenous immunoglobulins or pentoxifylline), reduces plasma levels of Fas as well as C-reactive protein and is reported to improve left ventricular function in patients with dilated or ischemic cardiomyopathy. There is growing evidence of inflammatory stimulation in our veterinary patients with heart disease.

MARKERS OF MYOCYTE STRESS

Adrenomedullin

Adrenomedullin is a peptide (52 amino-acid) produced in larger quantities by cardiac myocytes under conditions of stress and is a component of a precursor, pre-proadrenomedullin, which is synthesized in the heart, lungs, kidneys and adrenal medulla. It is a potent vasodilator with natriuretic properties and production appears to be stimulated by cardiac volume and pressure overloads. In humans, concentrations of the mid-regional proadrenomedullin peptide have been shown to independently predict outcome in patients following myocardial infarction adding information to that of derived from NTproBNP. Circulating and ventricular adrenomedullin concentrations have been shown to increase in dogs with experimentally induced heart failure.

ST2

A member of the interleukin-1 receptor family, is a circulating soluble protein for which the ligand appears to be interleukin-33 that is induced and released by stretched myocytes, much like BNP and adrenomedullin. In humans, concentrations appear to increase with heart failure class and in a separate study in human patients with heart failure, an increase in ST2 during a 14 day period was an independent predictor of death or the need for cardiac transplantation. Currently, the overall excitement in human medicine over the two myocyte stress markers ST2 and adrenomedullin, is that they appear to yield information supplementary to that provided by natriuretic peptides. It is premature to identify their benefit in clinical veterinary medicine.

NEUROHUMORAL MARKERS

Endothelin

Endothelin, the most potent pressor and vasoconstrictor substance discovered to date, was first identified in the supernatant of cultured porcine aortic endothelial cells. The family of endothelins is composed of endothelin-1 (ET-1), endothelin-2 (ET-2) and endothelin-3 (ET-3), each containing 21 amino-acid s and two disulfide bridges. Thrombin, a major factor in the pathogenesis of thrombosis, is a relatively potent inducer of ET-1 expression. Both in vitro and vivo, ET-1 causes coagulation in microcirculation of rats, and causes endothelial expression of von Willebrand factor in animals models. An increased ET-1 expression has been found in lung tissues from patients with primary pulmonary hypertension, a condition in which pulmonary vascular thrombosis is a common feature. In congestive heart failure, as myocardial performance deteriorates, systemic neurohumoral reflexes function to maintain cardiac output, organ perfusion, and circulatory homeostasis with the activation of the renin-angiotensin-aldosterone axis, vasopressin, and the sympathetic nervous system. However, these adaptive responses have the unfortunate effect of augmenting peripheral vascular tone and increasing afterload in the failing heart. These neurohumoral factors also enhance the production of the potent vasoconstrictor ET-1, which can be both beneficial and harmful, adding to the quagmire of CHF. Endothelin-1 enhances contractility in the failing heart, and up-regulation of endothelin pathways may be beneficial in providing short-term inotropic support for the failing myocardium in which beta-adrenergic responsiveness is frequently attenuated. However, in the long term, it is unclear whether up-regulation of the myocardial endothelin system in CHF is beneficial or harmful. Elevations of plasma ET-1 concentrations in experimental canine models of CHF are reported, as are elevations in naturally occurring congestive heart failure in dogs and cats.

Arginine Vasopressin

Arginine vasopressin (AVP), also referred to as antidiuretic hormone, is a nonapeptide that is synthesized in the hypothalamus and stored in the posterior pituitary gland and has vasoconstrictor and antidiuretic properties. Release of AVP from the neurohypophysis is stimulated by increased plasma osmolality or hypovolemia, and can also be induced to release by angiotensin II or sympathetic stimulation. Arginine vasopressin reacts with V_1 receptors in the heart and vasculature, and with V_2 receptors in the kidney stimulating water reabsorption via aquaporin-2 water channels in the collecting ducts. Excessive release of AVP intensifies heart failure associated with dilutional hyponatremia, fluid accumulation, and systemic vasoconstriction. A combined V_1/V_2 blocker has shown efficacy in dogs with experimentally induced heart failure and in humans with severe congestive heart failure.

MARKERS OF MYOCYTE INJURY

Various myocyte injury markers have been used in investigation of human patients with cardiac disease with cardiac specific troponins (cTn) emerging as the leading marker in this category and their measurement also widely reported in veterinary patients. It is the authors' opinion, that cardiac troponins by themselves will have little benefit as many variable diseases increase circulating troponins. Also, it might be time to move away from publications where one patient with disease x has an elevation in troponins. Likely, their place will be in a multimarker strategy or possibly mortality risk stratification (will that change anything you will do clinically?).

CONCLUSION

There is no doubt that cardiac biomarkers in veterinary medicine will find their role and become a bigger part of the clinician armamentarium. Currently, only natriuretic peptides and troponins have been incorporated into veterinary clinical practice, and even with these markers much work remains. Regardless, if it's one of the new biomarkers being studied in people such as adinopectin or chromogranin or yet undiscovered markers, the future of cardiac biomarkers is exciting. It is also very likely that not one marker will be as good as a combination marker strategy.

REFERENCES

[1.] Biondo AW et al. Genomic sequence and cardiac expression of atrial natriuretic peptide in cats. **Am J Vet Res** 63:236, 2002 [2.] Levin ER, et al. Natriuretic peptides. **N Engl J Med** 1998;339:321-328. [3.] Takemura N, et al. Plasma Atrial natriuretic peptide (ANP) in dogs with mild heartworm infection. **J Vet med Sci** 1991;53:323-325. [4.] Vollmar AM, et al. Atrial natriuretic peptide concentrations in dogs with congestive heart failure, chronic renal failure, and hyperadrenocorticism. **Am J Vet Res** 1991;52:1831-1834. [5.] Asano K, et al. Plasma atrial and brain natriuretic peptide levels in dogs with congestive heart failure. **J Vet Med Sci** 1999;61:523-529. [6.] Oyama MA, Fox PR, Rush JE, et al. Clinical utility of serum N-terminal pro-B-type natriuretic peptide concentrations for identifying cardiac disease in dogs and assessing disease severity. **J Am Vet Med Assoc** 2008;232:1496-1503. [7.] Häggström J et al: Relationship between different natriuretic peptides and severity of naturally acquired mitral regurgitation in dogs with chronic myxomatous disease. **J Vet Cardiol** 2:7, 2000. [8.] Schmidt MK, et al. Effect of azotemia on serum N-terminal proBNP concentrations in dogs with normal cardiac function: A pilot study. **J Vet Cardiol** 2009; 11, S81-S86. [9.] Fox PR, et al. Utility of plasma N-terminal pro-brain natriuretic peptide (NT-proBNP) to distinguish between congestive heart failure and non-cardiac causes of acute dyspnea in cats. **J Vet Cardiol** 2009; 11, S51-61. [10.] Khan SQ, et al. Prognostic value of midregional pro-adrenomedullin in patients with acute myocardial infarction: the LAMP (Leicester Acute Myocardial Infarction Peptide) study. **J Am Coll Cardiol** 2007;49:1525-1532. [11.] Jougasaki M, et al. Regulation of cardiac adrenomedullin in heart failure. **Peptides** 2001;22:1841-50. [12.] Rehman SU, Mueller T, Januzzi Jr JL. Characteristics of the novel interleukin family biomarker ST2 in patients with acute heart failure. **J Am Coll Cardiol** 2008;52:1458-1465. [13.] Shimpo M, et al. Serum levels of the interleukin-1 receptor family member ST2 predict mortality and clinical outcome in acute myocardial infarction. **Ciculation** 2004;109:2186-90. [14.] Yanagisawa M, et al. A novel potent vasoconstrictor peptide produced by vascular endothelial cells. **Nature** 1988; 332:411-415 [15.] Inoue A, et al. The human endothelin family:three structurally and pharmacologically distinct isopeptides predicted by three separate genes. **Proc Natl Acad Sci** USA 1989; 26:2863-2867 [16.] Prosek R, et al. Plasma endothelin-1 immunoreactivity in normal dogs and dogs with acquired heart disease. **J Vet Intern Med** 2004;18:840-844. [17.] Prosek R, et al. Measurements of plasma endothelin immunoreactivity in healthy cats and cats with cardiomyopathy. **J Vet Intern Med** 2004;18:826-830. [18.] Seta Y, Shan K, Bozkurt B, et al. Basic mechanisms in heart failure: the cytokine hypothesis. **J Card Fail** 1996; 2:243-9. [19.] Anker SD, von Haehling S. Inflammation mediators in chronic heart failure: an overview. **Heart** 2004;90:464-70.

DYSTOCIA: OUTCOME & SURVIVAL OF THE NEONATE

Katherine Cole MacGillivray, VMD, DACVIM & Peggy S. Marsh, DVM, DACVIM, DACVECC
Lexington, KY

INTRODUCTION

Dystocia in mares is defined as prolonged Stage 2 labor usually lasting longer than 30 minutes. There can be many causes of dystocia which include, although are not limited to, angular limb deformities of the fetus, fetus to mare size disproportion, mal-positioning, inadequate uterine contractions, congenital deformities or anomalies, placental abnormalities or any combination of these stated causes. Dystocia is considered a true emergency in equine practice as its duration can have a profound effect on the foal's survival as well as both the mare's reproductive and general health.

The sequelae of a dystocia for the mare include retained placenta, uterine trauma or tears, metritis, trauma and tears of the caudal portion of the reproductive tract, bruising or rupture of the intestinal tract and peritonitis. In foals, conditions which can be seen after birth by dystocia include orthopedic injury such as fractured ribs, tearing or rupture of ligaments or tendons of the limbs, fetal diarrhea with aspiration of feces into the lungs, septicemia and oxygen injury to various organ systems, such as the brain, kidneys and intestinal tract.

STUDY OBJECTIVES

For the purpose of this retrospective study, hospital records of mares that presented to Hagyard Equine Medical Institute for dystocia management during the years 2007-2009 were examined. There were several study objectives which included determination of total number of dystocias that presented to the hospital for each year and comparison of dystocia resolution procedures in regards to mare and foal survival. Other study objectives included determining the clinical conditions of foals born by dystocia, calculating the mean and median days of hospitalization and the common clinical conditions seen in foals following a dystocia birth.

Dystocia resolution procedures include assisted vaginal delivery (AVD), controlled vaginal delivery (CVD), caesarean section (CS) and fetotomy. During assisted vaginal delivery, the mare is conscious and the veterinarian provides manual assistance in vaginal delivery of the fetus. With controlled vaginal delivery, the mare is unconscious/under general anesthesia so that the veterinarian is in control of the delivery of the fetus. During CVD, the hind limbs are often hoisted. Caesarean section refers to when the fetus is removed during a celiotomy through a uterine incision. To remove a dead fetus, a fetotomy is reduced to allow vaginal removal in an awake or anesthetized mare. In review of our records, CVD is the primary dystocia resolution method utilized by the surgeons/veterinarians at the Hagyard hospital when presented with a mare with dystocia.

Additionally, the results of our study were compared to previous studies describing dystocia with regards to mare and foal outcomes[1-3]. In previous reports, mare survival following a dystocia has been high (86-91%).[1-3] Foal survival rates have been rather low ranging from 10% to 29%.[1-3] In review of Hagyard Equine Institute hospital records, mare survival rates were comparable to previous reports. Interestingly, the overall foal survival rate was higher than what has been previously reported. Not unexpectedly, greater than 50% of foals born via dystocia did require initial resuscitation. The type of resuscitation included assisted ventilation only, injectable mediation only or a combination of both. Common clinical conditions seen in foals born via dystocia included angular limb deformities, failure of passive transfer, neonatal encephalopathy (maladjustment syndrome), sepsis and rib fractures.

What is the outcome and survival of the neonate and mare subsequent to dystocia? Should an owner bring a mare to the hospital/clinic to aid in resolution? Based on the findings of our study, there is strong evidence foal survival rates have improved providing encouraging data to horse owners and farm managers even in these economic times.

REFERENCES

[1]Byron CR, et.al. **Equine Vet J.** 2003 Jan;35(1):82-85. [2]Norton JL, et.al. **Equine Vet J.** 2007 Jan;39(1):37-41. [3]Freeman DE, et.al. **Equine Vet J.** 1999 May;31(3):203-7.

INFECTIOUS ORTHOPEDIC DISEASE: ANCILLARY DIAGNOSTICS – WHAT IS NEW?

Carter E. Judy, DVM DACVS
Los Olivos, CA

INTRODUCTION

Orthopedic infections are often a much more serious clinical problem in horses compared to other species because of the potential consequences of permanent lameness and subsequent loss of use or even potentially death. Infectious orthopedic disease is a prevalent problem in all age groups of horses. While most commonly associated with neonatal and young foals, it occurs with great frequency in older animals as well. Infections involving both synovial structures or the bones of the horse can have life long implications if not identified early and accurately. Traditional techniques at identifying and localizing infectious orthopedic disease have relied upon the effective use of a combination of diagnostic tests including clinical examination, peri-neural anesthesia, blood evaluation, radiographs, arthrocentesis, wound culture, ultrasound and response to therapy. Many times these techniques fail to fully identify a problem or may under-represent its severity and diffuse nature.

While the gold standard to identify a septic structure is identification of a pathologic organism within the affected region, many times knowing where or what is affected is difficult to determine in the horse making obtaining a culture difficult. This can be compounded in young animals that are recumbent and not amenable to traditional lameness evaluation and regional localization. Additionally, the identification of a pathologic organism can take days to grow using traditional culture techniques.

The purpose of this presentation is to introduce the use of more advanced imaging diagnostic techniques such as Magnetic Resonance Imaging (MRI), Computed Axial Tomography (CAT scan) and Nuclear Scintigraphy (bone scan) in the diagnosis of infectious orthopedic disease in both young and old horses. These modalities have often been considered out of reach of many large animal veterinarians, but they have increased in use and availability in the past decade and many of these machines are available within a short distance at referral practices and universities. They are no longer only limited to use in small animals.

Computed Axial Tomography

Cross-sectional imaging methods, such as computed tomography (CT), can provide unique, diagnostically important information, which is not obtainable by traditional techniques. The history of CT in both humans and horses is relatively recent. Computed tomography is celebrating its 38th anniversary in clinical application in humans this year (2010) and has been utilized in horses for almost 25 years.

Traditional techniques such as radiography and ultrasound can have limitations when evaluating complex regions such as the skull, sinuses, nasal passages, joints and bones. Radiographs are two-dimensional representations of three-dimensional structures. Multiple objects are superimposed on each other and interpretation can be difficult. Ultrasound is excellent for soft tissue imaging, but has limitations. Deep structures, bone, and gas containing organs are poorly visualized. CT imaging can help reduce and or eliminate many of the limitations of these techniques and provides a method for better evaluating structures that are difficult to image with traditional techniques.

Computed Tomography images are a slice-by slice gray scale map of a tissue's ability to attenuate x-rays. CT images are created by the computerized reconstruction of data obtained by radiographing the animal, using a rotating, highly collimated, thin, fan shaped x-ray beam.

The patient is moved through the gantry to obtain sequential images (slices). Slice thickness is determined by the thickness of the collimated x-ray beam. The movement of the patient through the gantry determines the center between adjacent slices. The portion of the x-ray beam that escapes the patient is then received by a series of detectors in the gantry. The computer calculates the amount of x-rays attenuated by the patient for each slice. This attenuation factor is termed a CT number or Hounsfield unit. This number is unit-less and is related to the attenuating power of pure water. This number is assigned a gray scale value that determines the shade of gray for each pixel within each slice matrix. This pixel represents the density of a small volume within the patient and is termed a voxel. This gray scale image can then be manipulated digitally to enhance the appearance of different tissues such as brain, bone, lung and soft tissue and is projected onto a computer monitor or radiographic film for interpretation.

The interpretation of CT images is relatively straightforward to learn for anyone familiar with the reading of radiographs. Metallic and bony structures are shades of white to light gray, whereas fluids and soft tissues are varying shades of gray.

Small differences in tissue density can be detected using CT. Radiographs require at least a 10% difference in tissue density to discriminate between different tissues. Computed tomography can discriminate tissues with only a 0.5% difference in density. This provides for better contrast definition and the potential for seeing small changes in tissue mineral content. Gray brain matter can be differentiated from white matter and types of fluid can be distinguished such as acellular fluid, purulent exudate or blood.

The application of computed tomography has been utilized in two major ways in the horse. Both require general anesthesia. The first is through a stationary gantry with a moving table. The second is through a moving gantry and stationary table. The first technique requires a strong, durable table, and that the horse be positioned on this movable table to obtain the images. The second method allows for use within an operating theater and does not require excessive manipulation of the horse. The gantry can be moved around the horse to provide for optimal positioning of the patient and used in multiple environments.

Whereas only five fundamental opacities (air, fat, soft tissue, bone and metal) make up the basis of the conventional gray-scale radiographic image, the CT system is capable of detecting thousands of slight differences of x-ray absorption within tissues. The CT eliminates the superimposition of soft tissues and bones. Viewing and integration of this information from a series of individual CT slices lead to a three-dimensional perception and a better understanding of the anatomical relationships between structures. This is extremely helpful in examining complex regions and joints, where overlapping bones and articular surfaces are difficult to evaluate on standard radiographs. This is exceptionally useful when trying to differentiate between subtle areas of change which is often the case in early infectious processes

CT is exceptionally useful for evaluation of bone and, to a lesser degree, for evaluation of soft tissues. CT imaging can reveal important characteristics about osseous lesions. For example, the proximity of a fracture to an articular surface can be precisely defined. Osteolysis and osteogenesis can be detected before any changes are perceived on conventional radiographs. Significant changes in bone density can be documented utilizing CT. Soft tissues and fluids can be distinguished with CT better than with radiography. The tomographic slices eliminate superimposition inherent to standard radiographs. Individual muscles, tendons and ligaments can be identified. Abnormal fluid accumulations and dystrophic mineralization can be recognized within most soft tissues, as can changes in anatomical relationships within a tendon or ligament.

CT can be especially useful for the planning and application of complex fracture repair. Fractures with complex configurations requiring aggressive surgical management can be imaged, reconstructed and understood to provide the most effective and least invasive method of repair. In addition, simple fractures can be evaluated to ensure accurate placement and positioning of internal fixation devices.

More advanced techniques in CT scanning include the use of intra-arterial and intravenous contrast agents to help delineate soft tissues better as well as to determine the relative vascularity of a structure. This helps to determine the relative activity of a soft tissue lesion. This technique has helped to make CAT scan more sensitive to soft tissue and infectious injuries. In addition it can help to stage the progression of a soft tissue lesion over time.

MAGNETIC RESONANCE IMAGING

Magnetic Resonance imaging has been available in the human field since 1977 and has been used in horses since 1997. Many of the same concepts of cross-sectional imaging that CAT scan images utilize can be considered with MRI. However MRI has many unique characteristics that allow for varied examinations. Magnetic Resonance Imaging takes advantage of the unique characteristics of hydrogen protons within the body to produce a very detailed image of both the hard and soft tissues. By alternating how the tissue is magnetized, different characteristics of the area can be examined. A typical examination is composed of a variety of sequences that magnetize the tissue in different planes as well as with different magnetization protocols. This allows the veterinarian to examine for inflammation, edema, anatomical abnormalities or other irregularity in the signal for both bone and soft tissues.

Physics of MRI

Magnetic Resonance results from the effect of magnetic properties of atoms found in tissues. Protons consist of a nucleus containing both protons and neutrons. Electrons orbit the nucleus resulting in a spinning motion that is specific to each element. Since hydrogen is the most common element located within the body of mammals, it is the element with the most recognizable and useful magnetic moment. When an animal is placed in a strong magnetic field, the magnetic spins, which are normally randomly oriented, align in the direction of the magnetic field. A short pulse of high energy radio-frequency is applied and this disrupts the alignment of the hydrogen atoms. As the atoms re-align in the magnetic field, they emit a very small radio-frequency wave, that is captured and interpreted as the signal from the tissue. This is then transferred to a computer which transforms the raw data into a two dimensional representation of the tissue.

Application in the horse

There are currently two main ways that MRI is applied to the horse. This can be broken into high and low-field systems and standing and recumbent positioning. A "high-field" system is defined as a system with >1.0 T (Tesla) magnetic field. A "low field" system is one less than 0.5 T. Mid –range magnets exist but are not commonly used in equids. High field systems produce a higher detail image in a more rapid time than low field systems, but currently all require general anesthesia. Low field systems can be used from certain manufacturers in a standing configuration, but have lower quality images and can require much longer acquisition times for the same sequences as the high field magnet. In general the stronger the magnetic field, the better quality the image and the more quickly it can be acquired allowing for more sequential imaging of the patient.

Currently the regions that can be examined with the MRI in the adult horse include from the carpus and hock distally. Some scanners are able to image more proximally on the limbs in select patients. In young animals the entire horse up to about 500 lbs can be reliably imaged in many high field systems.

Imaging Protocols

Imaging protocols or sequences are the basic building blocks of the MRI exam. There are three fundamental examination sequences with some minor variations that are applied to each protocol that make up almost all of the examinations. These are:

- Proton Density – show clear anatomic structure
- T2 weighting – show high fluid contrast
- T1 weighting – good definition of bone structure

Variations to each sequence type can be made to highlight specific tissues, or to retard specific tissues types allowing better visualization of other structures. Examples of these are (not an all inclusive list, but these are what are used in horses):

- Short Tau Inversion Recovery (STIR) – High contrast, low clarity with a reduction in the signal returned from fat and an increase in the signal from water.
- Fat-suppression (FS) – eliminates signal from fat from within the structures examined allowing for differentiation of other tissues that may be obscured by the fat (i.e. bone marrow)

- ✦ Water-excitation (WE) – excites water within the structures and results in a net loss of fat signal and an accentuation of water. Provides increase clarity.

Many other variations for sequence generation exist and there are numerous sequences and types of scanning protocols available that go beyond the scope of this lecture. Further information can be found online at http://www.cis.rit.edu/htbooks/mri/ for the advanced reader or in "Equine Surgery - third edition" pg 946-963.[1]

By utilizing the unique characteristics of MRI it is possible to clearly identify regions of severe inflammation, fluid increase and anatomic disruption. The sensitivity of MRI allows for a very clear identification of septic orthopedic structures.

Nuclear Scintigraphy (Bone Scan)

The concept of nuclear scintigraphy is to inject a radioactive labeled substance into the body and then to trace the progression of that substance as it interacts with the target tissues. It has been used in horses in excess of thirty years to detect subtle regions of inflammation and bone turnover which most times correlates with the cause of lameness in the horse. Scintigraphy is often used to localize a region of lameness when other techniques such as nerve blocks and clinical examination fail to identify the region causing the problem. There are two main advantages to scintigraphy. First it is done in the standing sedated horse. The second is that the entire animal can be imaged with a single injection.

The most common radionuclide used in equine scintigraphy is technetium 99-m. This is considered a gamma isotope with a half life of 6 hours. This radionuclide is bound to a radiopharmaceutical which is dependent on the tissue of interest. In the horse this radiopharmaceutical is almost always Methylene Diphosphonate (MDP). This compound has a high affinity for bone and inflammatory cells. Other possible radiopharmaceuticals include:

- ✦ Labeled cells – White blood cells – to direct to areas of inflammation and infection.[2]
- ✦ Macroalbumin aggregates – pulmonary perfusion and distribution.
- ✦ Labeled antibodies – target specific tumors
- ✦ Simple salts – e.g. Na123I to target thyroid gland

The horse is imaged by acquiring the gamma rays through a collimator to prevent scatter onto a gamma camera much like a collimator on a radiograph machine. A computer to then processes this information to produce a diagnostic image.

There are three main phases to distribution of the radiopharmaceutical MDP that are of interest to equine veterinarians. They are:

- ✦ Vascular phase – Occurs immediately after injection and represents vascular blood flow. Has limited uses in the horse, but laminitis has been examined as such.
- ✦ Early pool phase (aka soft tissue phase) – This phase lasts about 15 minutes and represents diffusion from the vascular compartment to the extracellular compartment (soft tissues) of the animal. There is often early bone uptake in the end of this phase and may be over interpreted. Given the limited amount of time in this phase, select regions can be imaged.
- ✦ Late pool phase (aka bone phase) – This is the most useful phase and begins 2-3 hours after injection and typically lasts for up to 24 hours.

Interpretation of scintigraphic exams requires a basic understanding of the normal uptake patterns of different breeds, ages and uses of each horse. For example dorsal p1 uptake would be considered normal in a warmblood used as a jumper or dressage horse while it would be considered abnormal in a young racehorse. In young animals the high affinity for growth plates can complicate the interpretation for septic physitis. But, by comparing the contralateral growth plate, the likelihood of a growth plate to be affected can often be determined.

CONCLUSION

The use of advanced imaging techniques can greatly improve the veterinarian's ability to detect the presence and anatomical location of infectious processes within the horse. While these techniques are not the "first line of defense" in determining the presence of an orthopedic infection, they can be very helpful in identifying these problems in their early stages and allow for a rapid more effective treatment before the infection becomes advanced and possibly life threatening.

REFERENCES

[1]Sampson SN, et al. **Equine Surgery 3rd edition**, ed Auer and Stick, 2006, pg 946-963, [2]Long CD, et al **Vet Radiol Ultrasound** 2000;41[4], 354-9, [3] Kaser-Hotz, et al. **Equine Surgery 3rd edition**, ed Auer and Stick, 2006, pg 922-945.

COMPLICATIONS OF LATE GESTATION: ULTRASOUND ASSESSMENT OF THE EQUINE PREGNANCY

Stefania Bucca DVM,
Kildare Ireland

INTRODUCTION

Ultrasound technology offers safe and extensive investigation of the equine pregnancy throughout gestation, by the use of a combined transrectal and transabdominal approach. Ultrasound monitoring of feto-placental parameters provides valuable information in the identification and management of high risk pregnancies. Most pathological conditions of the equine pregnancy are commonly diagnosed during the third trimester of gestation, although the underlying pathology may have been present for a prolonged period of time prior to detection. The lack of evident clinical signs in the early stages of most conditions delays intervention and oftentimes hampers the ability to successfully manage a high risk pregnancy. Monitoring the equine pregnancy at regular intervals throughout gestation would allow early identification of problems and implementation of corrective strategies in a more timely fashion.

TRANSRECTAL APPROACH

Linear, convex or sector ultrasound technology may be employed to image the caudal aspects of the feto-placental unit in the mare, through a transrectal approach. Linear technology is the most commonly used, due to its wide popularity amongst clinicians in the reproductive field, ease of transducer orientation within the narrow equine pelvis and image interpretation. Dedicated 5 to 8MHz transducers or multi-frequency (6-10MHz) probes provide adequate transrectal imaging. High frequencies (8-10MHz) are required when detailed structural studies are carried out. Imaging of the cervix, allantochorion at the cervical pole, amnion, amniotic fluid and fetal parts may be obtained by this route. Furthermore, Doppler ultrasound technology may be applied to the uterine arteries, close to their origin in the dorsal pelvis, through a transrectal exam, providing a reliable means of monitoring uterine blood flow to the pregnant uterus. Amniotic fluid commonly appears more echogenic than allantoic, due to the closer proximity to fetal parts in motion, that impart swirling motion to free floating particles (vernix), within fetal fluids. The presence of an excessive amount of vernix could be consequent to a recent episode of strong fetal activity. But, particles remaining suspended within the fluids, unrelated to fetal activity, should be interpreted as a sign of feto-placental pathology. A similar condition has been reported as a common feature preceding abortion during MRLS (Mare Reproductive Loss Syndrome) in Kentucky. The utero-placental unit has been the subject of intense structural studies, aiming at detecting early changes and identifying abnormal conditions leading to ascending placentitis. The literature reports reference values of the combined utero-placental measurements at different stages of gestation. Measurements of the utero-placental unit should be taken close to the cervix, in an area of the ventro-caudal body where the middle uterine artery is used as landmark. A considerable difference in size may be observed during the course of the same examination, according to the volume of fetal fluids building up behind to the brim of the pelvis and the proximity of fetal parts. In the presence of ascending placentitis, a mild increase in utero-placental thickness is initially observed, commonly followed by loss of integrity and contiguity. Areas of detachment will subsequently become visible, with or without collection of anechoic fluid, blood or exsudate. Peracute insults to the cervical pole will result in sudden fetal demise and abortion, with surprisingly few detectable US abnormalities, as shown by work of LeBlanc et al., in their experimental model of placentitis. Fetal presentation, size of the fetal orbit/eye and assessment of peripheral pulses may also be carried out by trans-rectal ultrasonography of the pregnant mare, providing useful information on fetal orientation, growth and vitality, on a single field assessment of feto-placental parameters.

TRANSABDOMINAL APPROACH

Sector and convex US technologies best suit this approach. Linear technology may be employed, but linear transducers are less likely to obtain good contact to the mare's abdominal wall. The frequencies required are generally 2-6MHz and 7.5-8MHz for more detailed studies of the utero-placental unit. Proper patient preparation is essential for good quality imaging. Excessive maternal anxiety may have fetal implications in terms of heart rate accelerations. On the other hand, sedation of mares subjected to US examination for fetal well-being assessment should be avoided, as administration of sedatives to the mare will depress fetal cardiac activity. Offering the mare hay or a bucket of feed may provide sufficient diversion to allow for a full examination to take place uneventfully. Scanning of the abdomen usually initiates over midline, on a caudo to cranial direction and continues on parasagittal lines to delineate the boundaries of the pregnancy within the mare's abdomen. Fetal presentation and position are determined. Once fetal orientation has been accomplished, US investigations of anatomical structure and physiological events may commence. Parameters to be recorded include: fetal activity, fetal heart rate(FHR) and FHR reactivity, aortic diameter, fetal fluid depth and quality, thickness and contiguity of utero-placental unit.

Fetal Activity

Fetal activity is a reflection of CNS function, with decreased activity resulting from depressed CNS function. Activity is required to ensure satisfactory muscular development and skeletal joint function and promote successful postnatal adaptation. Dormant (inactive) phases are observed in fetuses of all ages but are more common in late gestation, where they can last up to 60 minutes or longer on occasion. Therefore, caution should be used when interpreting fetal activity, and reassessments are advisable. Lack of fetal movement has been associated with a negative outcome; however, sudden bouts of excessive activity followed by abrupt cessation have also been recorded in fetuses that subsequently died. Healthy equine fetuses are usually very active and major and fine tuned movements can be recognized during episodes of fetal activity. The number of major movements per hour tends to decrease as gestation advances, although a sudden increase in fetal activity is consistently observed 24-48 hrs prior to foaling.

Fetal Heart Rate

Recordings of FHR are obtained by M-mode echocardiography.The fetal cardiac area is identified within the thorax and multiple assessments of FHR and rhythm are made by placing the M-mode cursor over cardiac structures displaying the most dynamic range of excursion. FHR tracings are recorded both at rest and after activity throughout the course of the examination. Recordings during fetal activity may be quite challenging to obtain, as the fetus tends to move in and out of the angle of incidence of the M-mode cursor. FHR and FHR reactivity represent two of the most sensitive indicators of fetal well-being. FHR declines as gestation progresses. FHR variations (reactivity) depend on a functional central nervous system (CNS) and occur in response to environmental alterations and fetal activity. FHR accelerations of 25 to 40 beats per minute (bpm) for approximately 30 seconds are frequently observed in association with normal episodes of fetal activity. Movements without accelerations are commonly identified and accelerations in the abscence of stimuli are occasionally detected (5% incidence).Tachycardia or a large range of FHRs may indicate excessive fetal stress and could be brought on by painful maternal systemic problems. Bradycardia, inappropriate FHR for gestational age, or lack of heart rate variation suggests CNS depression, probably attributable to hypoxia and possible impending fetal demise. Asystole may be detected by transabdominal ultrasound in association with an immobile fetus. Equine fetal cardiac rhythm is usually regular, and cardiac arrhythmias are commonly associated with a negative outcome.

Fetal Aortic Diameter

The aortic diameter correlates to fetal size more efficiently than the sum of the fetal eye/orbit diameters or fetal eye volume. The aortic diameter should be assessed close to the base of the heart, as it emerges from the left ventricle. Measurements should be taken during systole. Although the two vessels both traverse the fetal chest in a cranio-caudal direction, the aorta should be distinguished from the anterior vena cava by its echodense walls, marked pulsatile activity and the more dorsal course within the thorax, close to the fetal spinal cord. As many as 10 measurements should be obtained during the course of one examination and the average calculated.

Fetal Fluids

Minimal and maximal allantoic and amniotic fluid depth measurements are reported in the literature. Free floating particles (vernix) are commonly observed swirling within fetal fluids and become more visible during episodes of fetal activity, particularly in the amniotic sac. The hyppomane is a feature of the allantoic sac and can often be visualised floating within the allantois, with its typical oval shape and onion like, concentrical structure. Sudden release of meconium (fetal diarrhea) may sometimes be observed in highly distressed fetuses just prior to birth, stillbirth or abortion. Accumulation of a thick layer of sediment in the most ventral areas of the allantois is sometimes observed in association with hydroallantois and when fetal activity is profoundly depressed. Pathological increases in fetal fluids have been commonly reported (hydramnios and hydroallantois). Markedly reduced volumes of amniotic fluid (olygohydroallantois) may be observed in mares with severe systemic illness. Objective assessment of fetal fluid depth requires extensive scanning of the mare's abdomen and is best carried out during phases of fetal quiescence.

Utero-Placental Unit

The diffuse, complete nature of equine placentation suggests the necessity of extensive and intimate contact between uterus and allantochorion for appropriate exchange of nutrients, oxygen and wastes. The need for adequate uteroplacental contact is highlighted by the difficulty mares have in maintaining pregnancies involving twins. However, small areas of separation of the placental membranes and uterus are commonly observed in normal pregnancies without any apparent effect on the health of the fetus. An average combined thickness of the uteroplacental unit of 1.26 ± 0.33 cm has been reported in mares with normal pregnancies in late gestation, based on vertical measurements recorded from transverse scans along sagittal or parasaggittal planes on the mare's ventral abdomen. The thickness of the uteroplacental unit is affected by numerous factors, which may reduce the efficiency of placental function. Such conditions include placentitis, placental edema and premature placental separation. Large and/or progressively enlarging areas of placental separation may lead to inefficient exchange of nutrients and adversely affect fetal growth and well-being. In addition to placental thickening, an abnormally thin uteroplacental unit has also been associated with a poor fetal outcome. Since only the parts of the uterus and placenta in contact with the ventral abdominal wall can be examined (using transabdominal ultrasound), focal areas of placental abnormality may be missed. The non-pregnant horn has a folded, thickened appearance and, although normal, could be mistakenly interpreted as utero-placental thickening.

Fetal Breathing

Rhythmical breathing movements may be observed in all fetuses in advanced gestation, when the diaphragm can be visualised. Nevertheless, fetal breathing has not been included in Reef's equine fetal biophysical profile, mostly because it cannot be consistently evaluated. Fetal breathing implies the synchronous movement of chest wall and diaphragm. Excessive convexity of the diaphragm associated with a disproportionate chest excursion is sometimes observed in distressed foetuses.

Umbilical Cord

The umbilical cord can be easily visualised in mid to advanced gestation, but rarely in its entirety. Its attachment to the dorsal wall of the gravid uterus makes an US approach rather difficult, due to the depth of the mare's abdomen and the presence of large loops of intestine with gas content. Assessing the length and blood flow within the umbilical cord is therefore rather difficult. Urachal dilatation and pouching can be observed and should be carefully evaluated when excessive cord twisting is suspected. Under these circumstances, the state of distension of the fetal bladder should be assessed, to confirm if an obstruction to urine flow exists.

Fetal Abdomen

Great anatomical detail may be obtained by serial US examination of the equine fetus. Spleen and liver can be visualized in the cranial abdomen, behind the diaphragm. The two organs can be easily distinguished by the hyperechoic appearance of the spleen and its close proximity to the gastric great curvature. The stomach is a landmark for fetal orientation and its anechoic, bean shaped silhouette facilitate fetal orientation in a cranio-caudal fashion and identifies lateral recumbency, due to its location in the left cranial abdomen. Fetal rotation along its longitudinal axis, is confirmed by the sudden change of proximity of the gastric silhouette to the US transducer. Gastric volume increases as gestation advances, to reach a plateau by term. Abnormal gastric filling has been observed in association with hydramnios and has been attributed to a disrupted swallowing pattern, often associated with abnormalities of the head. Loops of small and large intestine may be identified within the fetal abdomen and their content assessed as gestation progresses. Excessive segmental dilation, with swirling content is observed in association with congenital obstructions of the digestive tract. The fetal gonads occupy the caudo-ventral abdomen and their structural distinction between male and female fetus represent a mainstay in fetal gender determination in advanced gestation. Primary fetal sex organs are also easily assessed in advanced gestation and fetal sexing can be carried out by a transabdominal approach up to 8 months.

Biophysical Profile

An equine biophysical profile has been developed, using dynamic ultrasound techniques in pregnancies in late gestation. Six ultrasound parameters are included, chosen in relation to their ability to predict the delivery of an abnormal foal: fetal heart rate (FHR), aortic diameter, fetal activity level, uteroplacental thickness, uteroplacental contact, and maximal allantoic fluid depth. Each parameter is given a score value ranging from 0 to 2. Normalcy scores for these six variables are summed to give a biophysical profile value that ranges from 0 to 12. The results of a study by Reef and colleagues (1996) indicated that a biophysical profile score of 8 or less ensures a negative fetal outcome. A maximal score does not ensure a positive outcome, however, because unpredictable events may occur. These are events that could not realistically be predicted by a biophysical profile (ie, uterine inertia, premature placental separation at birth).

In summary, identification of a compromised pregnancy in the mare requires the collection of a database that includes a variety of ultrasound parameters indicating fetal distress and possibly suggesting neonatal compromise. Judicious interpretation of the findings and serial recording of data throughout gestation may help in the early detection of abnormal feto-maternal exchange pathways. Some sources of compromise may be readily identifiedad the impact on eto-maternal wellbeing calculated.Appropriate preventive and corrective measures may be implemented to minimize the risk of an unfavorable outcome.

REFERENCES

[1.] PipersFS et al, **J Am Vet Med Assoc**, 1984;185:766-71. [2.] Renaudin CD et al, **J Reprod Fertil Suppl** 2000;56:651. [3.] BuccaS et al, **Theriogenology** 2005;64:542. [4.] Ginther et al, **Theriogenology**,1993;40:1. [5.] Reef VB et al ***EVJ*** 1996;28:200. [6.] ReimerJM **Proceedings 43rd AAEP** Convention 1997;259. [7.] Bucca S et al, **Ippologia** 2000;11:35. [8.] McGladderyAL et al, **Proceedings 3rd Conference International Veterinary Perinatology Society**, 1993:37. [9.] Renaudin CD et al, ***EVE*** 1999;11:69. [10.] Renaudin CD et al, ***EVE*** 1999;11:75. [11.] LeBlancMM et al,**Proceedings of the 50th AAEP meeting**, 2004:127. [12.] Yamamoto K et al, ***EVJ*** 1992;24:169 [13.] Sheerin PC et al, **Proceedings of the Focus on Equine Reproduction Meeting**, 2003:22..

ADVANCES IN UNDERSTANDING NATIVE DEFENSE MECHANISMS

James A. Roth, DVM, PhD, DACVM
Ames, IA

INTRODUCTION

Native defense mechanisms are those that are always active and do not require previous exposure to protect the animal from infection. They can be activated within seconds to minutes after microbial invasion. The major native defense mechanisms include phagocytic cells (neutrophils and macrophages), the complement system, natural killer cells, antimicrobial peptides, and production of native defense cytokines by sentinel cells. Various stressors tend to reduce the efficacy of the native defense mechanisms leading to increased susceptibility to infection. The activity of all of these native defense mechanisms is enhanced if the animal has developed acquired immunity (antibody and T cell-mediated immunity) to a specific pathogen. Since this series of presentations focuses on the immunologic basis for vaccine safety and efficacy, the only aspect of native immunity to be discussed in detail will be the role of sentinel cells in early detection of microbial invasion, activation of native defense mechanisms and induction of acquired immunity.

Production of pro-inflammatory cytokines by Sentinel Cells

Macrophages, dendritic cells, and mast cells are important sentinel cells which play major roles in early detection of microbial invasion and the induction of the immune response. These cells are concentrated at or under epithelial surfaces and have receptors, called toll-like receptors (TLRs), which are capable of binding molecules that are unique to infectious agents. These molecules include bacterial lipopolysaccharide (endotoxin), flagellin, bacterial DNA (which is rich in CpG motifs), and bacterial lipoproteins. These microbial molecules are considered to be danger signals and the sentinel cells respond rapidly by secreting pro-inflammatory cytokines (Interleukin-1 (IL-1), IL-6, and tumor necrosis factor (TNF)). These molecules produce local inflammation at the site of production and activate neutrophils, macrophages, natural killer cells, and endothelial cells to have enhanced activities required for efficient elimination of a local infection.. If the proinflammatory cytokines are secreted in high enough concentrations, they will begin to circulate in the blood stream. When they reach the hypothalamus, they induce fever, loss of appetite, and lethargy. They act on the bone marrow to induce neutrophilia and cause the liver to produce large quantities of acute phase proteins. These actions enhance the ability of the native defense mechanisms to control an infection. The lethargy and loss of appetite reduce the animal's interaction with other members of the same species, thereby reducing the spread of infection. Fever slows the replication of some viruses and enhances some native defense killing mechanisms. This response requires extra energy and protein synthesis. Since the animal has a reduced appetite, the amino acids for the response are mobilized from the muscle and other tissues. Very high concentrations of the pro-inflammatory cytokines can be induced by high concentrations of microbial molecules, such as endotoxin, leading to septic shock.

Induction of the proinflammatory cytokines is essential to induce an immune response. Therefore, every time that an animal is vaccinated, proinflammatory cytokines are produced. Typically, a single vaccine will only induce low levels of cytokines and no clinically apparent reaction is observed. However, if multiple vaccines are administered at the same time, especially if some of them are Gram negative bacterins containing free endotoxin, enough cytokines may be released to cause fever, lethargy and loss of appetite lasting for a few days. The animal may already have a low grade infection causing low levels of the cytokines to be induced. The additional small amount produced by a vaccine may be enough to induce clinical signs.

Role of dendritic cells, macrophages, and B cells in Antigen Processing and presentation to T lymphocytes

T helper lymphocytes can only recognize peptide antigens presented on major histocompatability class II (MHC II) molecules. These peptides come from proteins that are endocytosed, enzymatically degraded into short peptides, and bound to MHC II molecules which are then transported to the cell surface for presentation to T helper cells. T cytotoxic lymphocytes can only recognize peptide antigens presented on MHC I molecules. These peptides come from proteins that are synthesized in the cytoplasm of the antigen presenting cell. A sampling of all proteins synthesized within a cell are processed through a proteosome in the cytoplasm and transported into the endoplasmic reticulum where short peptide fragments become bound to MHC I molecules. There are no cytotoxic T cells that recognize normal self proteins because the healthy animal is tolerant to normal self proteins. If a cell becomes infected with a virus, or mutates and starts producing foreign proteins in its cytoplasm, peptides from those proteins will be presented to cytotoxic T cells. The cytotoxic T cell will respond by killing the cell that is synthesizing foreign protein. It takes 7 to 10 days for the cytotoxic T cells to become efficient at controlling virus infection after initial exposure. They respond very rapidly on second exposure to the virus.

Antigens from both killed vaccines and live vaccines can be endocytosed, processed, and presented on MHC II molecules to T helper cells. However, only live vaccines can replicate and cause the production of pathogen proteins in the cytoplasm of infected cells. Therefore, live vaccines are typically considered to be more effective at inducing cytotoxic T cells in response to vaccination. Now that this endogenous pathway of antigen presentation is understood, researchers have developed methods to deliver vaccine antigens into the cytoplasm with out using typical modified live vaccines. One method is to use adjuvants, such as ISCOMS (immune stimulating complexes) which fuse with the cell membrane and can deliver protein antigens into the cytoplasm for processing and presentation on MHC I molecules to cytotoxic T cells. Another method is to use recombinant virus vectors which have one or more genes of the pathogen inserted into their genetic material so that when the vector infects a cell, it causes the production of pathogen proteins in the cytoplasm for processing and presentation on MHC I molecules. The advantages of using a viral vector to deliver the protein, instead of a typical modified live virus are that the vector may be easier to grow, or safer to use than the MLV vaccine. It may also break through maternal antibody more easily.

Dendritic cells, macrophages, and B lymphocytes are the cell types that play a major role in processing and presenting antigen on MHC II molecules to T helper cells. Dendritic cells are the major antigen presenting cells in all primary immune responses. Macrophages and B cells play a major role in antigen presentation to T helper cells in secondary immune responses. Dendritic cells are found in many tissues of the body, especially in the epithelium and in lymphoid tissues. Their primary function is to trap antigen, process it and present it to T helper lymphocytes. Dendritic cells in the tissues will capture antigens on their dendrites, phagocytose the antigens, break the proteins associated with those antigens down into small peptides, bind the peptides to MHC II molecules in endosomes and present those peptides attached to MHC II molecules on their cell surface. While they are doing this, they will migrate from the tissues where they found the antigen and drain with the lymphatics into lymph nodes where they will interact extensively with T helper lymphocytes capable of recognizing the foreign peptides bound to their MHC II molecules.

Initial exposure to antigen will cause the B cells and T helper cells which recognize that antigen to undergo mitosis resulting in the production of more B and T cells that recognize the antigens. This clonal expansion occurs in the lymphoid tissues. After clonal expansion, B cells become a major antigen presenting cell to T helper cells. The B cell will capture antigen using its B cell antigen receptor, which is a membrane bound form of the type of antibody molecule that it will secrete in response to antigenic stimulation. It has the same antigenic specificity. The B cell will endocytose the antigen it has captured, process the proteins associated with that antigen into small peptides, bind the peptides to MHC II molecules, and present those molecules on its surface for T helper cells to find. T helper cells which are capable of recognizing the specific peptides presented by the B cell will use the T cell receptor to bind to the MHC II molecule and peptide in an immunological synapse which also involves accessory and adhesion molecules. This interaction occurs in the lymphoid tissues. The T helper cell will then secrete cytokines which direct the B cell to undergo mitosis, switch from IgM to one of the other classes of antibody (IgA, IgE, or IgG depending on the characteristics of the pathogen), and eventually to become a plasma cell which secretes large quantities of antibody.

The initial antigen processing and presentation to cytotoxic T cells in a primary immune response occurs predominantly in the lymphoid tissue where naïve T cells and antigen presenting cells are concentrated. After the cytotoxic T cells have undergone clonal expansion and maturation in the lymphoid tissues, they circulate between the blood and the tissues monitoring for the presence of the foreign peptide from the virus which they recognize on tissue cells. If they do not find the antigen, they drain with the lymphatics to lymph nodes and eventually return to the blood through the thoracic duct. If they do find the antigen that they recognize, the will kill the cell synthesizing the foreign peptide. Thereby preventing the virus infected cell from producing more virus.

There are a complex set of interactions and signals which act on T helper cells to allow them to direct the induction of the best type of immune response for that particular pathogen. Natural infection generally induces the correct set of signals to produce the correct type of immune response. Modified live vaccines which replicate at the natural site of infection generally induce signals similar to those induced by natural infection, resulting in the correct type of immune response. Killed vaccines with adjuvants that are delivered by injection induce mostly artificial signals for the T helper cell. Those artificial signals may be the correct signals and induce good immunity, or they may be the wrong signals and cause the induction of the wrong type of immune response which is only partially protective or not protective. There are even examples of killed vaccines that enhance the disease process through induction of the wrong type of immune response.

REFERENCES

[1.] Abbas, AK, Lichtman, AH, Pillai, S. 2007. **Cellular and Molecular Immunology**. 6th edition. Saunders, Philadelphia, PA. 566 pages.; [2.] Tizard, I. R. **Veterinary Immunology: An Introduction**, 8th edition. 2009. Saunders, Philadelphia, PA 574 pages; [3.] Murphy, K, Travers, P, Walport, M. 2008. Janeway's **Immunobiology**. 7th Edition. Garland Science, New York. 887 pages.

ADVANCES IN UNDERSTANDING CELL MEDIATED IMMUNITY

James A. Roth, DVM, PhD, DACVM
Ames, IA

INTRODUCTION

Cell-mediated immunity is a rather simple label for a broad and complex set of mechanisms that require T-cell activity. Effecter T cells are induced against pathogens in response to infection or vaccination; some activated T cells eventually differentiate into memory T cells that can expand and exert antimicrobial activity very rapidly in the event of secondary exposure. The major subsets of T cells are the T helper 1 and 2 cells, cytotoxic T cells, and gamma delta T cells. B cells and the subsets of T cells recognize antigen differently and respond to antigen differently (see tables).

Lymphocyte Type	Antigen Recognition
B cells	Intact, unprocessed molecules
T helper 1 and 2 cells	Exogenous peptides processed and presented on MHC II
T cytotoxic cells	Endogenous peptides processed and presented on MHC I
Gamma delta T cells	Intact, unprocessed cell surface molecules

Lymphocyte Type	Response to Antigen
B cells	Produce Antibodies
T helper 1 cells	Secrete cytokines which activate phagocytes and natural killer cells
T helper 2 cells	Secrete cytokines which help antibody production (primarily IgA and IgE)
T cytotoxic cells	Cytotoxic for cells synthesizing foreign proteins
Gamma Delta T cells	Secrete cytokines and/or cytotoxic for cells with abnormal surface antigens

T helper 1 & 2 cells

Helper T cells coordinate the overall adaptive immune response by modulating the activities of many other immune cells, both through direct cell-cell contact and secretion of cytokines. In recent years, the balance of immune regulation between T helper 1 and T helper 2 cells has been a major area of investigation in immunology. Successful defense against the majority of bacterial and viral infections requires an early T helper (Th) 1 cell-mediated inflammatory response. Interferon-γ, interleukin-2 (IL-2), and tumor necrosis factor-α (TNF) are particularly important Th 1 cytokines. T helper 2 responses are generally characterized by IgA or IgE antibody production and often by recruitment of eosinophil and mast cell activity. Of the Th2 cytokines, IL-4, IL-10, and IL-13 are among the most dominant and best characterized. One purpose for Th2 responses appears to be mediation of immunity against helminth infections. However, when Th2 cell-driven mechanisms dominate early in infection, microbial replication may proceed unchecked. This, as well as antibody-mediated hypersensitivity, can enhance the severity of disease. Importantly, cytokines from Th1 responses tend to suppress Th2 responses, and vice versa, so that one or the other response predominates at a given time in an animal. This may have an impact on vaccine efficacy. If an animal is mounting a strong Th2 response to internal or external parasites when it is vaccinated; it may not be able to mount an effective Th1 response to the vaccine. The primary immune response seems to be more susceptible to misdirection than a secondary response. Therefore, it would be best to treat young animals for parasites, then wait two or three weeks before vaccinating them to give their immune system the best chance of mounting a strong Th1 response to the vaccines.

Measurement of cytokine production by Th cells is a useful measure of vaccine efficacy against infectious agents of animals. Much of the research that has shaped the Th1 versus Th2 theory has involved studies in mice. Human helper T cell responses also polarize toward Th1 or Th2 profiles, though the definitions seem to be less rigid [8]. There is considerable evidence to support this paradigm in numerous domestic animal species, as well [10-13]. However, it is important to be aware that the effects exerted by a cytokine in 1 species may not be identical to those exerted by its homologue in another species.

Cytokines can be categorized by the many different functional activities they mediate, and any cytokine may have multiple effects. Some cytokines activate innate immune cells such as phagocytes and natural killer cells and stimulate their participation in inflammatory processes, whereas others suppress or resolve inflammation. Several cytokines promote the activation of B cells and antibody production, as well as antibody class switching by B cells. Certain cytokines have effects further upstream in the ontogeny of cells, in that they direct the maturation and mobilization of distinct classes of cells. Chemokines control immune cell traffic in at least 2 ways: they attract the specific sets of cells that bear corresponding receptors and activate the integrin molecules that mediate leukocyte adhesion to vascular endothelium around the site of an infection.

Cytotoxic T cells

Cytotoxic T lymphocytes (CTLs), which express the CD8 protein on their surface, have an important role against a wide range of intracellular pathogens. These lymphocytes are programmed to kill infected cells when their T cell receptors (TCRs) recognize pathogen-derived peptides that have been processed and displayed on major histocompatability complex (MHC) class I molecules via the endogenous pathway. The endogenous pathway is most efficient at processing and presenting peptides which are synthesized in the cytoplasm of the infected cell. Cytotoxic T cells require seven to ten days after initial exposure to develop an effective response capable of controlling a viral infection. If an animal was previously vaccinated with a vaccine that stimulated CTL formation, those CTLs should rapidly respond and kill virus infected cells early in infection and help to stop virus replication before clinical signs appear or before they become severe.

Gamma delta t cells

Gamma delta (γδ) T cells are the other major T cell subset. The T cell receptor for this subset of T cells is a heterodimer comprised of a γ chain and a δ chain, which are structurally distinct from the α and β chains that form the T cell receptors of $CD4^+$ and $CD8^+$ T cells. Antigen recognition by γδ T cells is not usually MHC-restricted. Certain host cell markers, such as stress-induced heat-shock proteins, can activate γδ T cells. The functions of γδ T cells in response to pathogens are not well understood, but their prominent distribution in mucosal surfaces is consistent with a role in surveillance and rapid response to invaders. There is also evidence that γδ T cells modulate inflammatory responses and promote tissue repair.

Past efforts to evaluate immune responses in domestic animals typically focused on humoral immunity. One reason for this is the relative ease with which serum antibody responses can be monitored, by use of ELISAs and virus neutralization and hemagglutination inhibition assays. Immunological studies in recent years have highlighted the substantial roles of cell-mediated immunity in defense against human and animal infections. This has accentuated the importance of examining cell-mediated immunity as part of any thorough effort to characterize an immune response to infection or vaccination.

REFERENCES

[1.] Abbas, AK, Lichtman, AH, Pillai, S. 2007. **Cellular and Molecular Immunology.** 6th edition. Saunders, Philadelphia, PA. 566 pages; [2.] Tizard, I. R. **Veterinary Immunology: An Introduction**, 8th edition. 2009. Saunders, Philadelphia, PA 574 pages; nMurphy, K, Travers, P, Walport, M. 2008. [3.] Janeway's Immunobiology. 7th Edition. Garland Science, New York. 887 pages.

PRINCIPLES OF VACCINOLOGY: THE UTILIZATION OF THE IMMUNE SYSTEM TO MAKE A VACCINE EFFECTIVE

James A. Roth, DVM, PhD, DACVM
Ames, IA

MECHANISMS OF PROTECTIVE IMMUNITY

When evaluating vaccines used in veterinary medicine, serum antibody titer is often cited as an important measure of vaccine efficacy. Yet, in a number of diseases, serum antibody levels are not a good predictor of protective immunity. Specific antibodies to some infectious agents do not adequately protect the host from infection and disease. In addition, antibodies to certain infectious agents can enhance disease. One mechanism for this antibody-dependent enhancement is when immunoglobulin-bound pathogens rely on Fc receptor-mediated phagocytosis for entry to host cells. Immune complexes can also contribute to the pathogenesis of infectious diseases by triggering complement activation and inducing a type III hypersensitivity response. Another factor to consider is that antibodies may not always be detectable following an adaptive immune response. For several viral agents, adaptive cell-mediated immune responses can be mounted in the presence of passively transferred maternal antibody, even while humoral responses are suppressed. Thus, animals with no detectable antibody titers can be fully resistant to challenge.

If the various pathogenic mechanisms used by an infectious agent to invade and produce disease are understood, it is possible to make reasonable predictions as to the immune defense mechanisms which will help to protect from infection and clinical signs (see table below).

Pathogenic Mechanisms	Defense Mechanisms
• Adherence to Mucosa	Mucosal Antibody (IgA)
• Invasive parasite	IgE
• Exotoxin/Endotoxin	Neutralizing Antibody
• Viremia	Neutralizing Antibody
• Septicemia	Opsonizing Antibody
• Intracytoplasmic Growth	Cytotoxic T cells
• Growth in phagosome	T Helper 1 Cytokines
• Virus replication	Types 1 and 2 Interferon
• Infect epithelial cells	Gamma Delta T Cells

If the pathogen has to adhere to a mucosal surface to cause disease, then mucosal IgA which can block adherence should provide protection. If the pathogen is an invasive parasite, then IgE adhered to mast cells under the epithelium should cause mast cell degranulation, attraction of eosinophils, and destruction of the invasive parasite. If the clinical signs from infection or pathogenesis are due to exotoxins (like *Mannheimia haemolytica* leukotoxin) or to the effects of Gram negative bacterial endotoxin, then antibodies that neutralize the exotoxin or endotoxin should be effective. If there is a viremia early in infection, then neutralizing antibody in the serum should provide some protection. If bacteria or circulating in the blood stream (septicemia), then opsonizing antibodies which coat the bacteria and facilitate phagocytosis should help remove those bacteria. If the microorganism causes the synthesis of microbial proteins in the cytoplasm of the infected cells, then cytotoxic T cells are needed to kill the infected cells before the organism can replicate. If the pathogen grows in the phagosome of macrophages, then Th1 cytokines (gamma interferon and interleukin2) can activate the macrophages and natural killer cells to have enhanced killing activity. If the virus replicates and spreads from cell to cell rapidly, then type 1 or type 2 interferons may inhibit virus replication and provide some protection. Finally, if the microorganism infects epithelial cells of the gut or the lung, then gamma delta T cells may help to control the infection.

As illustrated in the paragraph above, there are many defense mechanisms which contribute to protection from infectious diseases. A natural infection may stimulate many of these mechanisms. A respiratory infection for example may stimulate mucosal IgA production, virus neutralizing antibody, cytotoxic T cells, Th1 cytokine production, interferon production, and gamma delta T cells in the respiratory epithelium. An ideal vaccine should also stimulate all of these protective immune mechanisms.

Evaluation of T Cell-Mediated Immunity

Thorough evaluation of vaccine efficacy requires the evaluation of T-cell responses to vaccination in populations of animals over time. This requires assays that can be performed on large numbers of out bred animals. Some assays for T cell activity are too cumbersome to use for this purpose, for example, assays for T cell cytotoxicity. Other assays for T cell activity are inherently more practical for this purpose. Evaluation of T cell activation using monoclonal antibodies and flow cytometry has proven to be a useful tool for monitoring the ability of a vaccine to induce T cell-mediated immunity.

Multi-color flow cytometry has been used to simultaneously detect the three major T cell subsets (CD4+, CD8+, and gamma delta T cells), the activation marker CD25 on the surface of activated cells, intracellular interferon gamma (a T helper 1 cytokine) and intracellular interleukin 4 (IL4, a T helper 2 cytokine). Assays are conducted by isolating peripheral blood mononuclear cells from experimental animals, putting a standardized number of cells in microtiter wells with or without the antigen of interest, incubating the cells for 4 or 5 days, then processing the cells by washing them, permeabolizing the cell membranes to allow antibody into the cytoplasm, then adding the mixture of fluorescent antibodies in the appropriate order to the cell preparations. The antibodies against

CD4, CD8, and the gamma delta T cell receptor will identify the type of T cell subset. The antibodies against CD25, interferon gamma, or IL4 will identify those T cells with each of the activation markers. The results provide the percentage of each T cell subset expressing CD25, interferon gamma, and/or IL4. My laboratory has used these techniques to monitor T cell mediated immunity in cattle, swine, and horses.

Cell-mediated immunity to Equine Herpes virus

The cell-mediated immune (CMI) response of foals to virulent equine herpesvirus-1 (EHV-1) infection was evaluated by multi-parameter flow cytometry (FCM). Ten 7-8 month old EHV-1 seronegative foals were infected intranasally with virulent EHV-1 and 10 foals served as uninfected controls. Blood samples were collected 6 and 7 weeks after infection to test for specific CMI responses to live heterologous EHV-1 recall antigen. The activation markers included major histocompatibility complex class II (MHC II), intracellular interferon gamma (IFN-gamma) and interleukin 4 (IL-4). The results from both tests were averaged before statistical analysis. Following EHV-1 stimulation, the MHC II expression index (EI) increased significantly in CD2+CD4+CD8- and CD2+CD4-CD8+ subsets of the infected group. At 4 days after incubation, the non-antigen stimulated CD2+CD4-CD8- subset of the infected group expressed a high percentage (61.1%) of MHC II. When stimulated with EHV-1, the MHC II expression declined significantly but remained at a relatively high percentage (34.4%). The IFN-gamma EI was significantly higher in infected foals in all major T cell subsets (CD2+) while only the CD2+CD4+CD8- subset showed a significant increase in intracellular IL-4 EI. The FCM results showed strong specific CMI responses to EHV-1 by all 3 tested parameters compared to the control group ($p<0.01$). The high MHC II expression in the CD2+CD4-CD8- subset suggests that this T cell subset may represent a gamma delta TCR repertoire and thereby play an important role as antigen presenting cells in the horse, as reported in other species. Being able to simultaneously quantify the frequency of specific lymphocyte subsets and the expression of cytokines that characterize activation of lymphocytes and protective CMI by multi-parameter FCM enables evaluation of subset-specific CMI responses to EHV-1 infection. This system can be applied to measure CMI responses to other equine vaccines and pathogens.

Limitations to using flow cytometry for detection of antigen specific T cell subset responsiveness

Flow cytometry has proven to be a useful tool for analysis of the induction of cell mediated immunity by vaccines through detection of T cell subset responsiveness to specific antigens. However, this technology has some important drawbacks and limitations. Since these are live cell assays, they have significant day-to-day and animal-to-animal variability. This requires that large numbers of animals be used and that they be sampled repeatedly. The procedure is labor and equipment intensive and is therefore expensive. Lymphocytes should be set up in the assay as soon as possible after blood collection. Overnight shipment of blood samples to the laboratory can work, but it may introduce additional variables. It is best to collect the blood early in the morning and have the blood in the laboratory by mid morning. Because of the day to day variability inherent in the assay, valid comparisons can only be made between lymphocytes from groups of animals assayed in parallel on the same day.

REFERENCES

[1.] Sandbulte, M.R., J.A. Roth. 2004. Methods for analysis of cell-mediated immunity in domestic animal species. **J Am Vet Med Assoc**, 225:522-530; Platt, R., B. A. Sponseller, Y.W. [2.] Chiang, and J.A. Roth. 2009. Cell-mediated immunity evaluation in foals infected with virulent equine herpesvirus-1 by multi-parameter flow cytometry. **Vet. Immunol. Immunopathol**. Accepted for Publication; [3.] Roth, J.A and L.M. Henderson. 2001. New technology for improved vaccine safety and efficacy. **Vet Clin North Am Food Anim Pract**, J.A. Roth (ed), 17:585-597

PRINCIPLES OF VACCINOLOGY: ADVERSE REACTIONS & WHY THEY SOMETIMES FAIL

James A. Roth, DVM, PhD, DACVM
Ames, IA

INTRODUCTION

Vaccines have proven to be very beneficial for controlling diseases in domestic animals. Their widespread use has dramatically reduced the incidence of severe and fatal diseases in companion animals and horses. They have also enabled the intensification of livestock production thus enabling great increases in efficiency in animal origin food and fiber production. In addition, animal vaccines have improved human health through control of zoonotic diseases such as rabies, brucellosis, and leptospirosis. Indeed, it can be argued that animal vaccines have had a profound impact on modern society. Without effective rabies vaccines many people would not opt to keep companion animals in their homes, and without effective vaccines for controlling major diseases in food producing animals the availability of animal proteins for human consumption would be greatly reduced. However, in spite of the success of animal vaccines, vaccines sometimes induce adverse reactions in animals and sometimes they fail to protect animals. When making decisions regarding vaccination programs for animals, veterinarians and animal owners must weigh the risks of vaccinating vs. the risks of not vaccinating. They must also use vaccines in a manner that induces optimal protection. This presentation will provide an overview of some of the reasons why vaccines occasionally produce adverse reactions and reasons why vaccines sometimes fail to protect animals from disease.

To produce protective immunity, a vaccine must stimulate a reaction in the animal. There usually must be a reaction both at the site of injection and systemically in order to produce an effective immune response. This reaction involves extensive activity by antigen-presenting cells, production of a variety of cytokines, and alterations in the trafficking of lymphocytes within the body. In addition, if the vaccine contains live organisms, they probably need to replicate to induce effective immunity. Live viruses must infect and replicate within cells. These essential reactions to a vaccine may induce observable clinical signs. Hopefully, the reaction to the vaccine will be mild and either unnoticeable or acceptable to the animal owner.

Regulation of veterinary vaccines

To understand vaccine safety and efficacy, it is important to understand the process by which vaccines are developed and tested by vaccine producers, and licensed by the United States Department of Agriculture (USDA), Animal and Plant Health Inspection Service (APHIS), Center for Veterinary Biologics (CVB). The federal government regulations for the United States of America regarding veterinary vaccines are found in the Virus Serum Toxin Act (VSTA) in Title 9 of the Code of Federal Regulations (9 CFR). The VSTA gives the USDA the authority to regulate veterinary vaccines in the United States. According to the 9 CFR a USDA licensed biological must be "pure, safe, potent, and efficacious, and not be worthless, contaminated, dangerous, or harmful". To understand this statement, it is important to understand what is meant by *safe* and *efficacious*. The definition found in the 9 CFR for safe or safety regarding veterinary biologics is "freedom from properties causing undue local or systemic reactions when used as recommended or suggested by the manufacturer." This definition has two important qualifiers for the term *safety*. It does not state that a vaccine should produce no reaction; rather it states that a vaccine should not cause "undue local or systemic reactions." This is recognition of the fact that stimulating a potent immune response is likely to produce at least a mild local and systemic reaction in the animal. The second important point is that according to the definition the safety of the vaccine is only ensured when it is used as recommended or suggested by the manufacturer. The recommendations and suggestions can be found on the label for the vaccine. Most vaccine label statements will indicate that a particular vaccine is only for use in healthy animals of a particular species. Healthy is defined as "apparently normal in all vital functions and free of signs of disease."

The 9 CFR definition for efficacious or efficacy is "specific ability or capacity of the biological product to effect the result for which it is offered when used under the conditions recommended by the manufacturer." The label found on the vaccine will indicate the "result for which the vaccine is offered" and will also indicate the conditions under which the vaccine is recommended for use. Therefore, it is very important to read and follow label instructions in order to achieve maximum safety and efficacy from vaccine usage.

Vaccine safety

Although vaccination has an excellent safety record, no medical procedure is completely innocuous. When animals develop adverse clinical signs within a few days to weeks of vaccination it is important to determine whether those clinical signs were vaccine induced or were not due to vaccination and only coincidentally occurred after the vaccine was administered. Animals commonly experience adverse clinical signs from a wide variety of causes and animals are commonly vaccinated. Therefore, it is to be expected that occasionally adverse clinical signs will occur after animals have been vaccinated for reasons unrelated to vaccine administration. There are also many reasons why vaccines may induce adverse reactions in the animal. It is important to differentiate true adverse vaccine reactions from false adverse vaccine reactions. Some of the causes of true adverse vaccine reactions are summarized in Table I.

TABLE I

POTENTIAL MECHANISMS RESPONSIBLE FOR ADVERSE VACCINE REACTIONS
Contamination with extraneous agents
Failure to inactivate agent in killed vaccine
Residual virulence of vaccine organisms
Vaccination of immunosuppressed animal
Immune suppression induced by the vaccine
Excessive induction of cytokine release
Multiple vaccines administered concurrently
Hypersensitivity to vaccine antigens
Type I - immediate type
Type II - cytotoxic type
Type III — immune complex type
Type IV — delayed type
Triggering or exacerbation of hypersensitivity to nonvaccine antigens
Allergies
Autoimmune disease
Induction of neoplastic changes
MLV BVD vaccine triggering mucosal disease in persistently infected cattle

Vaccine efficacy

Vaccines that are licensed by the USDA have been tested to determine that they are safe and effective. However, "effective" is a relative term. It does not mean that the vaccine must be able to induce complete immunity under all conditions which may be found in the field. This would not be realistic since the immune system is not capable of such potent protection under adverse conditions.

To be federally licensed, the vaccine must have been tested under controlled experimental conditions. The vaccinated group must have had significantly less disease than the nonvaccinated control group. This testing is typically done on healthy, nonstressed animals under good environmental conditions and with a controlled exposure to a single infectious agent. Vaccines may be much less effective when used in animals that are under stress, incubating other infectious diseases, or exposed to a high dose of infectious agents due to overcrowding or poor sanitation.

It is important to remember that for most diseases the relationship between the infectious agent and the host is sufficiently complicated that vaccination cannot be expected to provide complete protection. The vaccine can increase the animal's resistance to disease, but this resistance can be overwhelmed if good management practices are not followed. Some of the causes for vaccine failure are summarized in Table II.

TABLE II

POTENTIAL REASONS FOR VACCINE FAILURE
Insufficient time after vaccination to develop immunity
Something happened to the vaccine to make it ineffective
The physiologic status of the animal impaired the response to the vaccine
The animal was immunosuppressed at some point after vaccination
The animal was exposed to an overwhelming challenge dose of infectious agent
The duration of immunity after vaccination was not adequate
Important antigenic differences exist between the vaccine and field strains
Interference when multiple vaccines are administered concurrently

Summary

Mild local and systemic reactions to vaccines are to be expected as a natural consequence of vigorously stimulating the immune system. Dramatic adverse reactions to vaccines are occasionally due to mistakes during the production or handling of vaccines. More often, they are due to not following label instructions, particularly the restriction to only use vaccines in healthy animals. Vaccine failure to protect from disease is usually due to problems with either client education or compliance with good animal management practices. It is important for clients to understand the proper timing and method of vaccine administration, what to realistically expect for vaccine efficacy, and the importance of minimizing immunosuppressive factors and exposure to high doses of infectious agents in vaccinated animals. Veterinary vaccines have produced dramatic benefits in terms of animal health, human health, and efficiency of food production. Advances in research and the accumulating experience with vaccines are leading to safer and more effective vaccines. Proper usage of vaccines and adherence to good management practices will continue to be essential to achieve maximal vaccine safety and efficacy.

REFERENCES

[1.] Roth, J.A. 1999. Mechanistic bases for adverse vaccine reactions and vaccine failures. Advances in Veterinary Science and Comparative Medicine, 41:681-700; Code of Federal Regulations, Title 9: Animals and Animal Products, CHAPTER I--ANIMAL AND PLANT HEALTH INSPECTION SERVICE, DEPARTMENT OF AGRICULTURE **http://www.access.gpo.gov/nara/cfr/waisidx_00/9cfrv1_00.html**; [2.] Roth, J.A and L.M. Henderson. 2001. New technology for improved vaccine safety and efficacy. **Vet Clin North Am Food Anim Pract, J.A. Roth (ed),** 17:585-597.

LAWSONIA INTRACELLULARIS: MOLECULAR EPIDEMIOLOGY IN HORSES

Connie Gebhart, MS, PhD & Nicola Pusterla, DVM, ACVIM
St. Paul, MN & Davis, CA

INTRODUCTION

Lawsonia intracellularis is the causative agent of equine proliferative enteropathy (EPE), a recently recognized and/or emerged intestinal disease of horses as well as other species.[1-3] The first report of EPE in a foal occurred in 1982 and was then termed intestinal adenomatosis.[4] Since then, many reports of sporadic cases and outbreaks in foals on breeding farms have been described.[5] Reports of EPE are escalating, approaching a worldwide distribution, and the disease is increasingly being diagnosed in adult horses. Porcine PE, a pathologically identical intestinal disease also caused by *L. intracellularis*, is well-studied in swine due to its ubiquity and considerable economic importance to the swine industry.[1]

THE DISEASE

Etiology

L. intracellularis is a small, curved, Gram-negative, obligately intracellular bacterium, which resides within the apical cytoplasm of infected enterocytes. This bacterium causes proliferation of these infected enterocytes, resulting in a thickened small and sometimes large intestine.[1] The genus and species were newly identified and described in 1993 as being related genetically to the *Desulfovibrionaceae* but having numerous unique phenotypic characteristics.[6]

Clinical Disease

EPE is generally manifested in foals under one year of age and is seasonal in these foals according to weaning ages at the respective geographical locations.[7] Clinical aspects of the disease are related to a decreased absorptive capacity and are well-described.[3,5] Common clinical findings include peripheral edema, weight loss (or slow growth rate), lethargy, inappetence, fever, colic and diarrhea. An acute form, however, occasionally affects adults of several species, resulting in hemorrhagic diarrhea and often sudden death.[1] In most affected species the more common, chronic form of PE is self-limiting, and animals generally recover spontaneously after a subclinical disease course. Clinically affected and successfully treated foals, however, sell for an average of 68% less than the average price of unaffected foals by the same stallion.[7] Moreover, similar to PE in other species, EPE may be sub-clinical in foals, transiently affecting growth rate and/or causing unthriftiness.

Diagnosis

The diagnosis of EPE is challenging and relies on the presence of typical but subtle clinical signs, hypoproteinemia due to hypoalbuminemia, thickening of the small intestinal wall on abdominal ultrasonography, positive serology, or molecular detection of *L. intracellularis* in feces.[5,7] Exclusion of other enteric diseases may aid in obtaining a diagnosis of EPE; however PE very often occurs concurrently with other enteric diseases in pigs.[8]

Antemortem laboratory diagnosis of EPE relies on the use of serology for the detection of specific IgG antibodies and/or fecal PCR for shedding of the bacteria in feces. PCR of feces is very specific but its use is limited because *L. intracellularis* can only be demonstrated in the feces early in the course of the disease, subclinically affected foals may shed undetectable levels, and many factors, such as fecal inhibitors and antimicrobial treatment, may inhibit the PCR test.[5] Results of serology testing for *L. intracellularis*, using the immunoperoxidase monolayer assay (IPMA), showed that humoral IgG titers in serum are detectable longer than fecal shedding.[9] However, a positive serological result may represent exposure or subclinical infection rather than clinical disease. Serology has proven useful for routine diagnosis of EPE when combined with clinical signs and/or detection of *L. intracellularis* in feces.[5] More widespread serological testing will allow epidemiological surveys to more accurately measure the prevalence of EPE within populations.

During postmortem diagnosis, gross lesions are most commonly seen in the ileum near the ileal-cecal junction and appear as a thickening of the mucosa. Gross lesions are not evident in all cases of EPE, though, as they resolve spontaneously, often without clinical recovery. Histological lesions of EPE are uniquely proliferative and inflammation is not normally a hallmark of the disease. Definitive diagnosis can be best achieved by the use of special stains that incorporate an antibody that specifically identifies *L. intracellularis* in the tissues collected postmortem or through biopsy.[1,10]

Treatment and Prevention

Foals infected with *L. intracellularis* respond quickly to a number of different antimicrobial protocols if diagnosed early in the course of disease, before secondary problems cause further pathology. Treatment includes the use of antimicrobials such as macrolides alone or in combination with rifampin, chloramphenicol, oxytetracycline or doxycycline administered for three weeks. Supportive care includes intravenous fluids, plasma transfusion, parenteral nutrition and anti-ulcer drugs.

A commercial, modified-live *L. intracellularis* vaccine is used for the prevention of PE in the swine industry.[11] Protective immunity for PE is based on local IgA and cell mediated responses with only very low levels of systemic humoral antibody detected.[12] Humoral antibody response was detected in foals that were administered vaccine, whether vaccinated orally or intra-rectally.[13,14] The efficacy and protection conferred by vaccination is under investigation via experimental challenge studies and field efficacy trials. Hyperimmunized chicken egg antibodies incorporated into feed is another approach used for the control of PE in the swine industry that may have application to EPE control post-exposure or infection.

EPIDEMIOLOGY

Field Epidemiology

The route of infection in studied PE-affected species is fecal-oral through contact with newly infected or subclinically affected animals or due to contaminated feed, water, or environment.[1] In pigs the incubation period is two to three weeks depending on exposure dose[15] and a similar timecourse may occur in infected horses. Predisposing factors in pigs include a large herd size, recent weaning, transport to a new location, diet change and mixing.[1] For horses, proposed factors that contribute to clinical signs include stressors such as weaning, changes in diet and stabling protocol, recent transport, overcrowding, decreased immune system health, and concurrent disease such as parasitism.[3]

Typical proliferative lesions of PE with intracellular *L. intracellularis* have been described in many species of domestic and farmed animals, including pigs, hamsters, rabbits, foxes, deer, ferrets, ostriches and nonhuman primates.[1] Furthermore, *L. intracellularis* has been detected by PCR in the feces of a variety of wild animals including skunks, opossums, jackrabbits, and coyotes.[16,17] Despite an ever-increasing range of host species, *L. intracellularis* is currently not considered a zoonotic disease. Numerous potential reservoir hosts exist because of this wide and growing host range of PE. None of the epidemiological evidence supports transmission of *L. intracellularis* from domestic pigs to horses. Subclinical disease (ill thrift, slow growth rate) with typical intestinal lesions of PE has been demonstrated in cross-species infection studies in which hamsters were challenged with pig or horse isolates.[18,19] Clinical disease, with severity depending on the dose,[20] is consistently reproduced by transmission of isolates within species. This dichotomy of results suggests that *L. intracellularis* isolates may be somewhat species-specific.

The role of clinically and subclinically infected horses in the transmission of *L. intracellularis* requires further investigation. Epidemiological investigations on premises from which clinical cases have been identified have revealed that 10-65% of unaffected foals and adult horses test seropositive for *L. intracellularis*.[7,21,22] Low clinical incidence with such high seroprevalence in herdmates suggests that immune status may be an important factor in moderating clinical presentation of the disease. There is evidence that a genetic predilection may exist as immune response-related genes in horses are associated with fecal shedding of *L. intracellularis*.[23] In pigs, the timing of infection commonly corresponds with the loss of protective maternal antibodies.[24] Age predisposition in weanling foals may also be due to the concurrence of the weaning with waning of maternal antibodies.[7] Anecdotal evidence suggests that seropositive herdmates often show lethargy, inappetance, and slowed growth rates rather than overt clinical signs, signifying a milder but specific disease process. Subclinical or milder clinical disease is likely underdiagnosed in horses and, as in the swine industry, may have an economic impact.

Molecular Epidemiology

Little information is available on the source or spread of *L. intracellularis* within or between species or outbreaks because of the lack of isolate differentiation techniques. Due to the obligate intracellular nature of this organism, isolation and cultivation is extremely difficult, making the identification of subtypes by traditional methods unfeasible. Isolates from various animal species are antigenically indistinguishable, and standard phenotypic and genomic subtyping methods have not been able to provide an acceptable level of discrimination between isolates within species.[19] These preliminary investigations do suggest that isolates from pigs and horses may represent different genotypes and therefore may be species-specific.[19]

Molecular investigations of *L. intracellularis* isolates from a variety of animal species, including horses, hamsters, deer and ostriches, showed near-identical homologies of the 16S-rDNA genes to those of pig isolates.[25] Moreover, phenotypic characterization of outer membrane proteins and immunoblots of different *L. intracellularis* isolates and more sensitive molecular characterizations of the *L. intracellularis* genome demonstrated only minor differences among isolates.[19] None of these differences appears to be antigenically relevant.

Recently, the whole genome of a porcine *L. intracellularis* isolate was sequenced and, not surprisingly, found to contain numerous uncharacterized genes unique to this organism. This genome was subsequently analyzed for the presence of tandem repeats and this analysis identified very few (four) genomic regions containing variable number tandem repeat (VNTR) sequences. The monotypic nature of this organism was further confirmed. VNTR sequences in the genomes of prokaryotes are often associated with a high level of polymorphism and enable bacterial strain differentiation with substantial discriminatory power. Nevertheless, use of these *L. intracellularis* VNTR sequences did provide a sensitive method for analysis of the genetic relatedness of *L. intracellularis* bacteria or DNA obtained from various temporal and geographical locations and from various animal species. This provided insight into the phylogenetic relatedness of these isolates. Furthermore, the technique is applicable to PCR-positive feces, tissue and formalin-fixed paraffin-embedded tissue without the necessity of cultivating the bacteria. Molecular VNTR sequence profiles of *L. intracellularis* isolates from various documented PE outbreaks occurring in pigs, horses, ostriches, spider monkeys, ferrets and hamsters were analyzed.

Patterns emerged during evaluation of these diverse samples that provide some insight into the sources and phylogenetic relatedness of field isolates. VNTR sequence types obtained from pigs with PE were very different from those obtained from horses or other non-pig species. Little or no genetic variation was found between isolates from within outbreaks for any animal species or in multiple temporal samples taken from the same outbreak site. Slight variations between isolates obtained from outbreaks at different geographic locations were found, but these differences were no more pronounced among isolates from different continents than among isolates from neighboring states. Marked variation in VNTR types were found, however, between isolates from pig sources and those obtained from non-pig sources including horses (Figure 1).

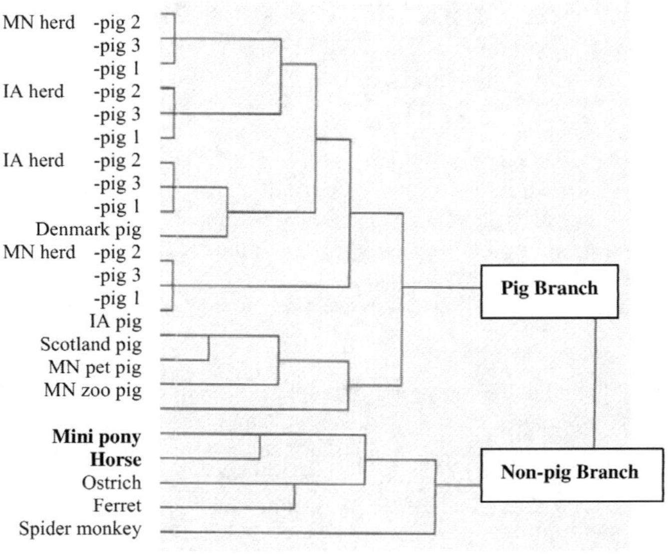

Figure 1- Dendrogram showing representative VNTR sequence relationships between *L. intracellularis* isolates from various animal species and geographical sources.

In conclusion, although *L. intracellularis* is antigenically conserved, its genome sequence contains VNTR markers useful for the demonstration of variance among isolates and may even be used to identify species-specificity of certain strains. The utility of this epidemiologic profiling method lies in the possibility of tracking different isolates of *L. intracellularis* obtained from varying animal species, geographical locations, and field outbreaks of PE.

SUMMARY

L. intracellularis is a unique organism which causes an unusual pathology in infected animals, specifically the proliferation of enterocytes as a direct result of the intracellular growth of the organism. Increased awareness of EPE in the field and the availability of diagnostics for detecting the organism in horses will result in earlier detection and prevention of disease. A preliminary epidemiological investigation into the relationships between *L. intracellularis* isolates from pigs and those from other species, including horses, suggests that they represent different strains and therefore may exhibit some species-specificity. The ability to type isolates of *L. intracellularis* will further enhance our understanding of the transmission dynamics and epidemiology of PE in pigs, horses, and other species. Advancements such as this will assist in identifying the natural ecology of this unique pathogen.

Limited knowledge of the pathogenesis of *L. intracellularis* suggests that this organism has adopted mechanisms of survival and pathogenesis that are unique among bacterial pathogens. Infection models, along with the whole genome sequence, will enable prospective studies on the pathogenesis of this organism. *L. intracellularis* is genetically unrelated to any other intracellular pathogen and may possess features that are novel in bacterial pathogenesis; thus, understanding the pathogenic properties of this organism is an intriguing goal.

REFERENCES

[1]Lawson GHK and Gebhart CJ. **J Comp Pathol** 2000;122:77. [2]Lavoie JP, et al. **Equine Vet J** 2000;32:418. [3]Lavoie JP and Drolet R. In: **Equine Infectious Diseases** 2007;313-316. [4]Duhamel GE and Wheeldon EB. **Vet Pathol** 1982;19:447. [5]Pusterla N and Gebhart CJ. **Equine Vet Educ** 2009;21:415. [6]Gebhart CJ, et al. **Int J Syst Bacteriol** 1993;43:533. [7]Frazer ML. **J Vet Intern Med** 2008;22:1243. [8]Jacobson M, et al. **Res Vet Sci** 2003;74:163. [9]Guedes RM, et al. **J Vet Diagn** 2002;14:528. [10]Guedes, RM et al. **Can J Vet Res** 2002;66:99. [11]Kroll JJ, et al. **Am J Vet Res** 2004;65:559. [12]Guedes RM and Gebhart CJ. **Can J Vet Res** 2010; in press. [13]Pusterla N, et al. **Vet J** 2009;102:458. [14]Pusterla N, et al. **Vet Rec** 2009;164:783. [15]Guedes RM and Gebhart CJ. **Vet Microbiol** 2003;91:135. [16]Tomanova K, et al. **J Wildl Dis** 2003;39:407. [17]Pusterla N, et al. **J Wildl Dis** 2008;44:992. [18]Cooper DM. PhD Thesis, 1996; University of Minnesota. [19]Al-Ghamdi G. PhD Thesis, 2003; University of Minnesota. [20]Guedes RM, et al. **Vet Rec** 2003;153:432. [21]Pusterla N, et al. **Vet Rec** 2008;163:156. [22]Pusterla N, et al. **Vet Microbiol** 2009;136:173. [23]Horín P, et al. **Mamm Genome** 2004;15:843. [24]Guedes RM, et al. **Can J Vet Res** 2002;66:258. [25]Cooper DM and Gebhart CJ. **J Am Vet Med Assoc** 1998;212:1446.

PATHOPHYSIOLOGY, CLINICAL PRESENTATION & TREATMENT OF SYSTEMIC FUNGAL DISEASE IN HORSES

Allison J. Stewart, BVSc(hons), MS, DACVIM, DACVECC
Auburn, AL

INTRODUCTION

Although the fungal kingdom comprises of >70,000 species of molds, yeasts, rusts and mushrooms, only 50 species are pathogenic to mammals. Fungi are eukaryotic organisms with a cell wall composed of chitins, mannans and glucans. The plasma membrane contains ergosterol, which is targeted by several antifungal agents. Systemic fungal infections in horses are sporadic and relatively uncommon, although geographic prevalence is quite variable. Until recently, the prognosis for systemic fungal infections has been guarded to poor. However, pharmacokinetic profiles of several antifungal agents have been reported for horses, and as affordable formulations become increasingly available, successful outcomes are likely to become more common.

CLINICAL SIGNS

Horses with discrete or diffuse pulmonary fungal granulomas, or fungal pleuropneumonia can present with signs similar to bacterial pneumonia. Fever, cough, nasal discharge, tachypnea, respiratory distress, hemoptysis and, if chronic, weight loss may be observed. Radiographic appearance of fungal pneumonia is variable, with a miliary pattern or patchy bronchopneumonia most common. Differentials include bacterial pneumonia, neoplasia, RAO, granulomatous disease complex or silicosis. Fungal infections can affect multiple organ systems and body cavities, with abdominal infection associated with signs of weight loss, colic or diarrhea.

DIAGNOSIS

Fungal pneumonia may be diagnosed from samples obtained by tracheal wash, bronchoalveolar lavage, or lung biopsy. Biopsy should be performed ideally after radiographic evaluation or with concurrent ultrasound guidance and should be obtained from the lung periphery. The lung is rich in plasminogen, so bleeding complications may be severe. Spring-loaded biopsy needles are safer than Tru-Cut biopsy instruments. Ultrasound evaluation can be used to monitor for bleeding after the procedure.

Fungal hyphae may be identified in impression smears obtained from biopsied masses or in body fluid. The presence of fungal organisms may be insignificant when cellular cytology is normal. Non-pathogenic barn fungus *Alternaria*, are often present in tracheal aspirates from healthy horses. Some fungi have characteristic morphological features that can provide a rapid cytological diagnosis (**table 1**). Histopathology slides should be stained with periodic acid Schiff (PAS), Gridley's fungus and Grocott-Gomori methenamine-silver nitrate (GMS) to help identify fungal hyphae. Immunohistochemistry, fluorescent *in situ* hybridization and DNA probes can be used to positively diagnose fungal organisms in histopathogy sections.

Some fungi have fastidious growth requirements and may be outcompeted by contaminant bacteria. Samples should be transported at room temperature in commercially available microbiological transport media. Specific culture media such as Sabouraund's dextrose agar, inhibitory mold agar containing cycloheximide and chloramphenicol are useful, but some fungi take several weeks to grow. A Pan-fungal real-time PCR assay can be used to detect a variety of fungal organisms in body fluids and fungal isolates, followed by species specific real-time PCR to positively identify the organism.[1]

Serological tests which utilize radioimmunoassays, immunodiffusion, complement fixation and ELISA are available to detect circulating antibodies against several fungal organisms. Titers often decrease with disease resolution, therefore serology can be used to monitor response to treatment and disease recrudescence.

Immune function testing

Primary pathogenic fungi such as *Blastomyces, Histoplasma, Coccidioides* and *Cryptococcus* species usually infect immunologically normal horses. Opportunistic fungi including *Aspergillus*, *Candida*, *Fusarium*, *Emmonsia* and *Pneumocystis* species have caused disease in horses that are immunocompromised, neutropenic, have neoplasia, colitis, enteritis, bacterial pneumonia or have been treated with corticosteroids. Fungal infections have been associated with host immune suppression caused by severe malnutrition, congenital immunodeficiency, or acquired immunodeficiency. Serum immunoglobulin quantification, lymphocyte subpopulation phenotyping and lymphocyte proliferation assays can be performed to test for immunodeficiency.[2, 3]

DIFFERENTIALS

Aspergillosis *Aspergillus* spp. are ubiquitous in the environment especially in moldy feed and bedding. As opportunistic pathogens they cause disease in horses that are immunosuppressed from debilitating disease, or that have been treated with immunosuppressive drugs. Infection is by inhalation of spores or by translocation of organisms across an inflamed gastrointestinal tract. *Aspergillus* spp. pneumonia in horses is almost uniformly fatal, often with no or mild respiratory signs. In humans 50-90% of patients with invasive aspergillosis die despite treatment. Traditional treatment in humans has been amphotericin B, but reversible nephrotoxicity occurs in ~50 % of cases. Voriconazole is now considered the drug of choice against human aspergillosis. A foal with localized pulmonary aspergillosis was successfully treated with surgical resection of the lesion and oral voriconazole.[4]

Treatment of Aspergillus spp. sinusitis and rhinitis in horses has been successful using oral itraconazole, topical natamycin or enilconazole (lavaged via an endoscope or indwelling catheter placed into the sinus) and insufflated nystatin powder.[5]

Candidiasis Systemic candidiasis was diagnosed and successfully treated in 4 neonatal septic foals that had been aggressively treated with numerous antibiotics and parenteral nutrition. Two of the foals were treated with amphotericin B IV, while oral fluconazole was used on the other two foals.[6] Glossitis, panophthalmitis and cystitis can be associated with systemic candidiasis.

Coccidiomycosis *Coccidioides immitis* is a soil saprophyte that grows in sandy, alkaline soils in semiarid areas. Inhaled arthroconidia enlarge to form non-budding spherules, which incite an inflammatory reaction in the lungs and lymph nodes resulting in weight loss, fever, colic and signs of respiratory disease. Compared with localized, nasal granulomas, diffuse infections with granulomas in the lungs, liver, kidney or spleen have a grave prognosis. Przewalskii horses may be more susceptible.[7]

C. immitis is difficult to culture, and spherules may not be observed histologically from antemortum lung biopsies. However, serology is very useful to diagnose infection, and decreasing titers are associated with clinical improvement.[8] Serum antibodies are detected rarely in healthy horses.[8] Infected horses have been successfully treated with fluconazole and itraconazole.[8, 9]

Pneumocystosis *Pneumocystis carinii* was reclassified from protozoan to saprophytic fungus, but due to lack of ergosterol it may actually be a plant. *P. carinii* cannot be cultured, and diagnosis is based on BAL cytology. Pneumocystosis is common in human AIDs patients, and people undergoing immunosuppressive therapy. *P. carinii* causes diffuse interstitial pneumonia, especially in Arabian foals with severe combined immunodeficiency. It has been reported in immunocompromised adult horses,[2, 10] as well as an immunocompetent foal. Trimethoprim- sulfamethozazole is the treatment of choice. Dapsone can also be used.[3]

Histoplasmosis *Histoplasma capsulatum* is most prevalent in moist soil containing bird or bat waste. Enteric, pulmonary or disseminated histoplasmosis may occur. Amphotericin B was curative in a filly with pulmonary histoplasmosis diagnosed by TTW and lung aspirate cytology. Serology is also useful for diagnosis.

Blastomycosis Blastomycosis is caused by inhalation of *Blastomyces dermatitidis* conidiae. Systemic blastomycosis has been reported in two horses, both of which were euthanized.[11, 12] Serology is useful for diagnosis.

Cryptococcosis Cryptococcosis is caused by *C. neoformans* var *grubii* (serotype A), *C. neoformans* var *neoformans (*serotype D) and *C. gattii* (serotypes B and C), which are ubiquitous, saprophytic, round, yeast-like fungi. There is an epidemiologic relationship between *C. gattii* and the Australian river redgum tree, and *C. neoformans* var *neoformans* and bird (particularly pigeon) excreta. Serologic testing is useful, with resolution of lesions correlated with declining titers.[13]

Cryptococcosis in horses primarily causes pneumonia, rhinitis, meningitis and abortion. Surgical resection of a jejunal lesion was successful in 1 horse. *C. gattii* pneumonia diagnosed by TTW and lung mass aspirates, was treated successfully with IV amphotericin B.[13] Fluconazole was successful in the treatment of cryptococcus meningitis[14] and sinonasal granulomas.[15,16]

Rare Causes of Fungal Pneumonia *Scopulariopsis* pleuropneumonia was diagnosed in a horse by culture of BAL fluid. The infection was successfully treated with a combination of oral ketaconazole and aerosolization of enilconazole.[17]

Acremonium strictum pneumonia was diagnosed by cytologic evaluation, culture and PCR testing of BAL fluid in a horse. Supportive treatment, including fluconazole was successful, although fluconazole has poor *in vitro* activity.[1] Adiaspiromycotic miliary fungal pneumonia caused by the soil mold *Emmonsia crescens* was diagnosed by percutaneous lung biopsy, resulting in euthanasia.[18]

TREATMENT

Treatment options are dependent on the site and extent of the infection, surgical accessibility, etiologic agent, known antifungal susceptibilities, evidenced based studies from human medicine and financial resources of the owner (**table 1 and 2**).

Table 1. Characteristic Morphologic Features, Diagnostic Tests and Recommended Treatment for Systemic Fungal Infections

Agent	Cytologic Appearance	Diagnostic Tests	Recommended treatment
Cryptococcus Spp.	Round, thin-walled, yeast-like fungus (5-10 μm) with a large unstained capsule (1-30 μm). Stain capsule using Mayer's mucicarmine or India ink. Narrow based budding and lack endospores.	Cytology and histopathology is very reliable. Culture Serology: capsular antigen ELISA or latex agglutination.[13]	Amphotericin B,*[13] fluconazole, meningitis[14-16] itraconazole or voriconazole
Coccidioides immitis	Spherules have a double-contoured wall and variable size (20-200 μm). Mature spherules (sporangia) contain 2-5 μm endospores (sporangiospores). In the environment, mycelium are thick-walled with barrel shaped arthroconidia.	Cytology, histopathology, culture and serology. IgM and IgG antibodies by AGID,[8] enzyme immunoassay or latex agglutination. CF (antibody) has low specificity.[8,28]	Fluconazole,[8] itraconazole [9] (ketoconazole), voriconazole or amphotericin B.
Aspergillosis Spp.	Broad (2-4 μm) septate hyphae with parallel sides and acute right angled branching.	Cytology, histopathology and culture. Serology not useful.	Amphotericin B*, voriconazole or itraconazole.[9] Minimal activity to fluconazole.
Histoplasma capsulatum	Yeasts (2-4 μm) have a thin clear halo surrounding a round or crescent- shaped basophilic cytoplasm.	Cytology, histopathology, culture and serology. AGID (antibody).	Amphotericin B,* itraconazole or voriconazole. *In vitro* susceptibility to fluconazole is poor.
Blastomyces dermatitidis	Spherical yeasts (15-17 μm) with basophilic protoplasm and unstained, uniformly shaped refractile walls. Unilateral, broad-based budding. Often located within multinucleated giant cells.	Cytology, histopathology, culture and serology. AGID (antibody).[11]	Amphotericin B*, Itraconazole or voriconazole.
Candida spp	Ovoid budding yeasts (2-4 μm) with thin walls, or chains of blastospores with septate pseudohyphae. Filamentous, regular, true hyphae also may be visible.	Cytology, histopathology and culture.	Fluconazole[6] (*C. krusei* resistant), amphotericin B[6] and voriconazole. *Candida* glossitis – mouth wash of $KMnO_4$ (0.025% q 24h) or nystatin

			(0.3 g in 10 mls of water, q 8h).
Pneumocystis carinii	Trophozoite (yeast form) is ameboid (2-5 μm) with filopodia. Sporangia (cystic form) are encapsulated spores (4-6 μm) containing 8 uninuclete spores (intracytic bodies).	BAL fluid is preferred to TTW for cytology. Can not be cultured. PCR on BAL or TTW fluid or lung. Perform immune function testing.	Trimethoprim- sulfamethoxazole (25-30 mg/kg PO q 24h) or dapsone (3 mg/kg PO q 24h for 2 months).[3]
Acremonium strictum	Mononuclear cells contain single, spherical, intracytoplasmic encapsulated spores 3-5 μm in diameter[1]		Fluconazole possibly successful in 1 horse,[1] however fluconazole has poor activity *in vitro*.

*In people intravenous amphotericin B is often administered for 2 weeks prior to prolonged oral azole therapy.

Table 2: Antifungal Drugs Used in Horses

Drug and Relative Cost	Class and Mechanism of Action	Dosage	Susceptible Organisms and Equine Reports	Side effects and Comments
Ketoconazole $$	Azole antifungals are fungistatic, by inhibiting ergosterol biosynthesis. They inhibit cytochrome P450-dependent 14α–sterol demethylase, which is essential for the formation of ergosterol.	30 mg/kg q 12h with 0.2 N HCl via NG tube.[19]	*Scopulariopsis,*[17] *Candida, Coccidioides, Cryptococcus, Histoplasma.*	Absorbed poorly in the non-acidified form.
Itraconazole $$$$		5 mg/kg q 24h PO: bioavailability of solution 60%, compared to 12% for capsules.[20]	*Aspergillus,*[21] *Candida, Coccidioides,*[9] *Cryptococcus, Scopulariopsis, Histoplasma, Bastomyces,* not *Fusarium, Mucor.*	Very unstable (requires low pH) and lipophilic. Use Sporanox® solution, not compounded formulations.
Fluconazole $$		14 mg/kg PO loading followed by 5 mg/kg q 24h PO.[22]	*Coccidioides,*[8] *Candida,*[6] *Cryptococcus,*[14-16] *Bastomyces, Histoplasma, Scopulariopsis.* Not *Aspergillus* and *Fusarium.*	High levels in plasma, CSF, synovial fluid, aqueous humor and urine.[22]
Voriconazole $$$$$		3 mg/kg PO q12h.[23] 4.0 mg/kg PO q 24h.[24,25]	*Aspergillus, Candida, Cryptococcus, Scedosporium* and *Fusarium.*	High levels in plasma, CSF, tear film, aqueous humor, urine, synovial and peritoneal fluids.[23,25]
Enilconazole $		Aerosolisation: 1.2 mg/kg in 125ml saline q12h.[17]	*Aspergillus, Scopulariopsis*[17]	*Aspergillus* plaques in the nasopharynx.
Amphotericin B $$$	Polyene macrolide. Fungicidal by irreversibly binding to ergosterol, increasing membrane permeability, leading to cell death.	0.3 mg/kg in 1 L 5% dextrose IV, increasing q3 d by 0.1 mg/kg until 0.9 mg/kg reached; for 30d. Reduce nephrotoxicity with oral NaCl.	*Aspergillus, Candida,*[6] *Coccidioides, Cryptococcus,*[13, 16] *Blastomyces, Histoplasma, Conidiobolus.*	Broadest spectrum. Poor bioavailability, limited penetration into CSF, ocular and synovial fluids. Risks- phlebitis and nephrotoxicity
20% NaI $	Exact mode of action of iodides is unknown. Very little, if any direct *in vitro* antibiotic effect. Beneficial effect on the granulomatous inflammatory process.	20-40 mg/kg IV for 7-10d or 65 mg/kg twice/week.	Although several successful cases are reported in which iodides were used as primary or adjunctive therapy, overall efficacy is considered limited at best.	Toxicity: lacrimation, cough, respiratory secretions, dermatitis. Administration during pregnancy may cause congenital hypothyroidism.
Ethylenediamine dihydro-iodide (EDDI) 80% iodine $ 76% KI $		1-2 mg/kg PO (or 20-40 mg/kg of 4.57% dextrose powder) q 12-24h for 1 week, then halve dose for weeks to months.		

REFERENCES

1. Pusterla N, et al. **Vet Clin Pathol** 2005;43(4):413. 2. MacNeill AL, et al. **Vet Clin Pathol** 2003;32(2):73. 3. Clark-Price SC, et al. **J AmVet Med Assoc** 2004;224(3):407. 4. Hilton H, et al. **J Vet Int Med** 2009;23(2):375. 5. Kendall A, et al. **J Vet Int Med** 2008;22(5):1239. 6. Reilly LK, et al. **J AmVet Med Assoc.** 1994;205:464. 7. Terio KA, et al. **J Zoo Wild Med** 2003;34(4):339. 8. Higgins JC, et al. **Vet Rec.** 2006;159(11):349. 9. Foley JP, et al. **J Vet Int Med** 1992;6:333. 10. Franklin RP, et al. **J Vet Int Med** 2002;16(5):607. 11. Toribio RE, et al. **J Vet Int Med** 1999;214(9):1357. 12. Dolente BA, et al. **Eq Vet Educ** 2003;15(3):139. 13. Begg LM, et al. **Aust Vet J** 2004;82(11):686. 14. Hart KA, et al. **J Vet Int Med** 2008;22(6):1436. 15. Cruz C, et al. **J AmVet Med Assoc** 2009;234(4):509. 16. Stewart AJ, et al. **J AmVet Med Assoc** 2009;235(6):723. 17. Nappert G, et al. **Eq Vet J.** 1996;28(5):421. 18. Pusterla N, et al. **Eq Vet J.** 2002;34(7):749. 19. Prades M, et al. **Eq Vet J.** 1989;21:211. 20. Davis JL, et al. **Am J Vet Res** 2005;66(10):1694. 21. Davis EW, et al. **J Vet Int Med** 1994;8(4):304. 22. Latimer FG, et al. **Am J Vet Res** 2001;62(10):1606. 23. Colitz CMH, et al. **Am J Vet Res** 2007;68(10):1115. 24. Davis JL, et al. **Am J Vet Res** 2006;67(6):170. 25. Passler NH, et al. **J Vet Pharmacol Therap** 2010;33(1):35.

CORRELATION OF ADRENOCORTICOTROPIN AND CORTISOL WITH PERIPHERAL LEUKOCYTE CYTOKINE EXPRESSION IN SEPTIC AND NON-SEPTIC NEONATAL FOALS. JR Gold[1], ND Cohen[1], TH Welsh Jr[2]. 1.Texas A&M University, Department of Large Animal Clinical Sciences, College Station, TX. 2.Texas A&M University, Department of Veterinary Integrative Biosciences, College Station, TX.

Studies in septic human infants have examined serum cytokine responses to understand inflammatory mechanisms and to discern their diagnostic and prognostic values. Persistent increases in interleukin (IL)-6 in septic infants are associated with increased length of hospitalization and increased risk of dying. Similar studies have been conducted in neonatal foals that suggest septic foals may be immunosuppressed relative to healthy foals. The hypothalamic-pituitary-adrenal (HPA) axis is influenced by the cytokines IL-6 IL-1β and tumor necrosis factor-α (TNF-α) in critically ill humans. No publications exist regarding the association cytokines have on the HPA axis in septic or healthy neonatal foals. The purpose of this study was to determine if a correlation exists between the HPA axis and cytokine expression in hospitalized term, neonatal septic, and non-septic neonatal foals at a tertiary referral hospital.

Blood samples were obtained from term septic foals (n=15) < 7 days of age admitted to Texas A&M University, Large Animal Teaching hospital at admission (T=0), 24(T=1), 48(T=2), 72(T=3) and 96 (T=4) hours. Total leukocyte populations were obtained using the (LeukoLOCK) total RNA isolation system and then frozen at -80 C until analysis. Plasma was separated by centrifugation, and also stored at -80C until assayed. Foals were defined as septic based upon having a positive blood culture, a sepsis score of $\geq$ 11 or both. Blood was sampled similarly in healthy term foals < 7 days of age obtained from the Texas A&M Horse Center (n=4) or from term foals (n=11) admitted to the teaching hospital for reasons other than sepsis. Endogenous basal ACTH and cortisol concentrations were measured using validated immunoassays. Cytokine mRNA expression of IL-1*β*, IL-6, IL-8, IL-10, and TNF-α was determined using real-time RT-PCR.

Data were analyzed using linear mixed-effects models to account for the repeated measures among foals. Significance was set at $P < 0.05$.

There was no significant correlation or pattern of association between either ACTH or cortisol and cytokine expression in the overall study population. ACTH had a significant correlation with cortisol ($P<0.0001$) which was not unexpected finding. Among septic foals, ACTH had a significant ($P<0.003$) positive association with IL-6 expression, and a tendency ($P = 0.070$) for a positive association with IL-8 expression. Cortisol had no significant associations with any cytokines evaluated, although IL-6 expression tended to be positively associated ($P<0.06$). Septic foals had a significant association for ACTH with cortisol ($P<0.001$). These findings indicate that a relationship exists between the HPA-axis and IL-6. Further studies are needed to determine the clinical significance of this association.

CT AND MR IMAGING OF EQUINE PITUITARY GLANDS.

H Schott II, A Pease, J Patterson, E. Howey. Michigan State University College of Veterinary Medicine, East Lansing, MI.

Pituitary pars intermedia dysfunction (PPID) affects 15-20% of the older equid population and develops slowly over years as degeneration of hypothalamic dopaminergic neurons leads to increasing size of the pars intermedia through melanotrope hyperplasia and adenoma formation. Our aims were: 1. to compare size of equine pituitary glands determined by MRI with size measured at necropsy, and 2. to assess whether treatment with pergolide causes the pituitary gland to decrease in size in PPID-affected horses. CT and MR images were acquired during the same terminal episode of general anesthesia and pituitary glands were removed immediately after euthanasia. MRI studies were performed on four untreated PPID-affected horses (28±4[SD] years, group A), four aged (25±5 years, group B) horses and four young (5±2 years, group C) horses without signs of PPID, and six PPID-affected horses (25±3 years, group D) that had been treated with pergolide for 6 months. PPID status was confirmed by dexamethasone suppression test results. CT images were also acquired on group D horses before and after 6 months of treatment. Pre-treatment CT scans were performed in winter (Jan-Feb, 2009) and all other CT and MR scans were performed from Aug-Nov, 2009.

Pituitary gland weight was 7.9±2.1, 3.1±1.3, 2.2±0.2, and 5.3±1.1 g for groups A, B, C, and D, respectively ($p<0.05$, A>B,C,D and D> B,C). Adrenal gland weights were not different between groups. Pituitary glands measurements determined at necropsy and by MRI are listed in the table:

	Group	length (mm)	width (mm)	height (mm)
	A – untreated PPID, n=4	27 ± 2^{a}	23 ± 3^{ab}	23 ± 8^{a}
Necropsy	B – aged non-PID, n=4	22 ± 4^{b}	21 ± 2^{b}	12 ± 1^{b}
	C – young non-PPID, n=4	18 ± 1^{c}	22 ± 1^{ab}	10 ± 1^{b}
	D – treated PPID, n=6	27 ± 2^{a}	24 ± 2^{a}	14 ± 3^{c}
	A – untreated PPID, n=4	26 ± 2^{a}	26 ± 2^{a}	22 ± 2^{a}
MRI	B – aged non-PID, n=4	20 ± 3^{bc}	22 ± 2^{b}	12 ± 2^{b}
	C – young non-PPID, n=4	19 ± 2^{b}	22 ± 1^{b}	9 ± 0^{b}
	D – treated PPID, n=6	23 ± 1^{c}	24 ± 2^{ab}	18 ± 2^{c}

There were no differences between necropsy and MRI measurements, although length tended ($p<0.10$) to be greater when measured at necropsy. Necropsy and MRI measurements were significantly correlated (r=0.84 length, r=0.49 width, and r=0.70 height). There was no decrease in CT measurements in group D after pergolide treatment. In fact, pituitary gland length was almost 2 mm greater ($p<0.05$) after treatment, possibly a consequence of season.

INSULIN AND GLUCOSE RESPONSES TO FEEDING GLUCOSE, FRUCTOSE AND INULIN TO NORMAL AND PREVIOUSLY LAMINITIC PONIES. KE Borer[1], SR Bailey[2], NJ Menzies-Gow[1], Y Berhane[1], PA Harris[3], J Elliott[1]. [1]Royal Veterinary College, London, UK; [2]University of Melbourne, Australia; [3]WALTHAM Centre for Pet Nutrition, Leics. UK.

Ponies prone to laminitis (PLP) may be insulin resistant. Differences in serum [insulin] between PLP and normal ponies (NP) have been shown after inclusion of inulin in the diet at 3g/kg/day[1]. The aim of this study was to assess the effect of feeding glucose, fructose and inulin on serum [glucose] and [insulin] in NP and PLP during the summer.

Six NP and 6 PLP randomly received glucose, fructose and inulin each at 1 g/kg/day once daily for 3 consecutive days, mixed with a commercial chaff-based feed whilst housed and fed ad-lib Timothy hay. Each carbohydrate was fed under 2 management conditions: after the ponies had been eating only grass for at least 7 days and again after eating only Timothy hay for at least 7 days. A wash-out period of at least 7 days was allowed between each carbohydrate feeding. Blood was collected over a 12 hour period following feeding on each day. Glucose was analysed by a commercial laboratory and insulin using a radioimmunoassay (Coat-A-Count, Siemens) previously validated in the authors' laboratory. Responses of NP and PLP to each carbohydrate were compared using a 2-way ANOVA and changes over time using a 1-way ANOVA. Results are reported as mean±SD and significance was set at $p<0.05$.

Following either of the preliminary basal diets (grass or hay), plasma [glucose] increased significantly to peak 2 hours after feeding glucose or fructose in both NP and PLP. Peak [glucose] after feeding glucose was 1.8-fold higher than baseline in PLP and 1.6-fold higher in NP. After feeding fructose, peak [glucose] was 1.2-1.3-fold higher than baseline in both NP and PLP. Minimal increases in plasma [glucose] occurred after feeding inulin at this dose.

After becoming accustomed to either grass or hay diets, serum [insulin] increased significantly above baseline after feeding glucose or fructose in both NP and PLP. Serum insulin responses after glucose feeding were significantly greater in PLP compared with NP, from 2 to 4 hours after feeding (days 2 and 3). Peak [insulin] after feeding glucose (2 hours after feeding; day 3) was 2.5 times higher in PLP compared to NP (425±194μIU/ml vs. 167±89μIU/ml). This represented an increase over baseline [insulin] of 33-fold and 15-fold in PLP and NP, respectively. The AUC for [insulin] was significantly greater in PLP vs. NP when feeding glucose after ponies had eaten only hay for 7 days previously (8616±5230 vs. 3983±3005). Serum [insulin] increased after feeding fructose, and was significantly higher in PLP vs. NP 2 hours after feeding. Peak [insulin] after feeding fructose (2 hours after feeding; day 3) was 1.7 times higher in PLP compared to NP (208±151μIU/ml vs. 124±110μIU/ml). This represented an increase over baseline [insulin] of 10-fold and 5-fold in PLP and NP, respectively. Minimal increases in serum [insulin] occurred after feeding inulin at this dose.

In conclusion, an insulin response after feeding glucose or fructose occurs in both NP and PLP, with exaggerated responses occurring in PLP. Differences in serum [insulin] after feeding 1g/kg/day glucose or fructose may distinguish between NP and PLP. Inulin fed at a low dose in this study had no significant effect on serum [insulin].

[1]Bailey et al. (2007) *J Am Vet Med Assoc* 231: 1365-1373.

REGULATION OF GLUCOSE TRANSPORT: NOVEL INSIGHTS INTO THE PATHOGENESIS OF INSULIN RESISTANCE IN HORSES. AP Waller, K Kohler, TA Burns, M Mudge, J Belknap, VA Lacombe. The Ohio State University, Columbus, OH 43210, USA.

Insulin resistance (IR), an impaired ability of insulin to stimulate glucose disposal into insulin-sensitive tissues (i.e., muscle and adipose tissue), has been widely recognized as an epidemic disease in humans, and there has recently been an increased clinical awareness of IR in horses. However, the underlying pathophysiological mechanisms of IR are still not well understood in any species. The translocation of glucose transporter 4 (GLUT4) protein (the main isoform) to the cell surface is the limiting step for glucose uptake from the bloodstream into insulin-sensitive tissues. However, the quantification of active GLUT recruited at the cell surface has proved very challenging. AS160 has recently emerged as a potential key player for GLUT translocation through an insulin-dependent pathway. However, its role is unknown during IR, as well as the role of GLUT12, a recently identified GLUT. We hypothesized that GLUT translocation and expression will be decreased in insulin-sensitive tissue, through an AS160-dependent pathway, in an equine model of IR.

Biopsies of skeletal muscle, subcutaneous (nuchal ligament and tail head) and visceral (omental, retroperitoneal and mesocolonic) adipose tissue were collected from mares, which were characterized as insulin sensitive (IS, n=5) or IR (n=5), based on the results of an insulin-modified frequently-sampled intravenous glucose tolerance test. Active GLUT transporters that translocate to the cell surface were quantified using a state-of-the art biotinylation technique with a cell-impermeant photolabeled reagent; and total GLUT-4 and -12, and AS160 protein expression were measured by Western blot technique in these biopsies. Analysis of variance or t-test was performed, as appropriate.

In skeletal muscle, GLUT4 translocation to the cell surface, but not its protein content, was significantly reduced in IR compared to IS horses ($P = 0.027$) and was not further enhanced after *in vitro* insulin stimulation. GLUT4 protein expression was significantly increased in visceral compared to subcutaneous adipose sites and skeletal muscle of both IS and IR horses ($P<0.005$). However, GLUT4 protein expression was significantly reduced in omental tissue of IR vs. IS horses ($P=0.009$). There were no significant changes in expression levels of GLUT12, total AS160 and its phosphorylation sites between groups and sample sites.

This study demonstrated that IR impaired GLUT4 recruitment to the cell surface in insulin-sensitive tissue. Our data suggested that GLUT4 in omental fat plays a key role in the regulation of glucose transport in both healthy and IR horses, while GLUT12 and AS160 are not major pathogenic factors of insulin resistance.

BLOOD RESTRICTION GUIDELINES IN THE MANAGEMENT OF CRITICALLY ILL AND INJURED HUMAN PATIENTS

R. Miller
Nashville, TN

Over 15 million units of blood are transfused per year in the United States. Many of which are used in the care of the severely injured and critically ill patient.[1]

No one will argue with the fact that red blood cell (RBC) transfusion is indicated for patients with hemorrhagic shock and hemodynamic instability. Most of the recent literature on damage control hematology during hemorrhagic shock demonstrates reduced complications, improved blood utilization and most importantly a survival benefit when an appropriate ratio of blood to blood products are utilized early in the trauma resuscitation.[2-4]

However, only a small percentage of the overall transfusions in critically ill patients are used for the treatment of hemorrhagic shock (approximately 10%). The vast majority (90%) of RBC transfusions are used to treat anemia in the ICU. Over 50% of all patients admitted to the ICU receive at least of unit of allogenic RBC with an average of about five units during their ICU admission. Eighty-five percent of patients with ICU length of stay greater than one week receive at least one unit of RBCs.[5]

Anemia in the critically ill is very common and persists throughout the duration of ICU and hospital stay with or without RBC transfusion. Its etiology is multifactorial and includes bleeding, phlebotomy, chronic disease, inflammatory response, impaired erythropoietic response, and nutritional deficiencies.

As will be described, based on a wealth of literature, RBC transfusion may be associated with an increased incidence of both infectious and non-infectious complications, longer ICU and hospital length of stay and increased mortality.

Part of the risk of RBC transfusion may be related to a "storage lesion" and an immuno-modulating effect of the RBC transfusion. The current shelf life of packed red blood cells is 42 days. Numerous abnormalities have been associated with the storage of RBCs and some studies have suggested that transfusions of older RBCs may be associated with adverse effects.[6]

Shapiro et al, reviewed anemia and blood transfusion in trauma patients admitted to the ICU. Of 586 from 111 ICU, 55% of the patients admitted to the Trauma ICU were transfused during their ICU stay. The mean pre-transfusion hemoglobin was 8.9 g/dl and mean age of the blood transfused was 20 days, ± 11.4 days.[7]

Spinella found that trauma patients transfused RBCs of increased storage age (greater than 28 days) had an independent association with mortality (odds ratio of 4) and an increased incidence of death from multi-organ failure compared to those that received fresher blood (27 days or less). If the age of transfused packed red blood cells is in fact important, it would have major ramifications on the already limited blood supply.[8]

Hill performed a metanalysis of over 13,000 surgical and trauma patients from 20 peer-reviewed articles and found overwhelming evidence that allogenic blood transfusions are associated with a significant increased risk of post-operative bacterial infections in surgical patients with the greatest risk being in the subgroup of trauma patients.[9]

Croce et al, studied trauma patients with moderate degrees of injury (ISS less than 25) where no blood was transfused in the initial 48 hours after injury. They found that delayed transfusion was independently associated with ventilator-associated pneumonia, ARDS, and deaths regardless of ISS. They concluded from their data that a judicious transfusion policy should be mandated after resuscitation.[10]

The first and only prospective multicenter randomized controlled clinic trial of transfusion requirements in critically care (TRICC) was published in 1999 by Hebert, et al, in the New England Journal of Medicine. 838 critically patients with euvolemia and a hemoglobin of less than 9 g/dl within the first 72 hours after admission were enrolled from 22 ICUs in Canada. Patients were randomly assigned to either a restrictive strategy of transfusion where RBC were transfused only if the hemoglobin was less than 7 g/dl or to a liberal strategy were RBC transfusions were given if the hemoglobin was less than 10 g/dl. Overall and 30 day mortality was similar between the two groups. However, both hospital and 30 day mortality was significantly lower among patients who were less acutely ill (APACHE score less than 20) and in those younger than 55 years of age. This study clearly demonstrated that a restrictive strategy of RBC transfusion is at least as effective and possibly superior to a liberal transfusion strategy in critically ill patients.[11]

The ABC trial (Anemia and Blood Transfusion in Critically Ill Patients) was a European prospective observational study following 3543 patients in 146 ICUs through 28 days, discharge and death. The mean admission hemoglobin was 11.3 g/dl. The mean transfusion rate in this study was 37%. ICU and overall mortality was significantly higher in patients receiving blood. (ICU 18.5% versus 10.1%, P>0.01; overall 29% versus 14.9%, P>0.001).[12]

The CRIT study was the first major US prospective multicenter observational study evaluating anemia in blood transfusions in critically ill patients. The study included almost 5,000 patients in 284 medical and surgical ICUs. The mean hemoglobin at baseline was 11g/dl. Overall 44% of the patients received one or greater units of RBCs while in the ICU (mean of 4.6 units). Mean transfusion hemoglobin was 8.6 g/liter and the mean time to the first ICU transfusion was between two and three days. The number of RBC transfusions a patient received was independently associated with a longer ICU and hospital length of stay, more total complications and a higher mortality.[13]

More recently, Marik et al, performed a systematic review of the literature on the efficacy of RBC transfusion in the critically ill. Forty-four studies including 272,596 patients were identified. Outcome measures included mortality, infection, multi-organ dysfunction syndrome and ARDS. In 42 of the 45 articles the risk of RBC transfusion outweighed the benefits. Only one study showed that the benefits of RBC transfusion outweigh the risks in a subgroup of elderly patients with acute myocardial infarction and a hematocrit of less than 30%. Twenty-two studies examine the association between RBC transfusion and nosocomial infections; in all these studies blood transfusion was independent risk factor for infection. Additionally, there were 10 trauma articles that reviewed almost 44,000 injured patients. In all 10 articles, the risk of RBC transfusion outweighed its benefit in all categories including infection, MODS, SIRS, and mortality.[14]

Due to this striking evidence of the detrimental effects of RBC transfusion, Earley et al, implemented an anemia clinical practice guideline aimed to reduce transfusions after the acute resuscitation phase in the management of trauma. The guideline was based on transfusion trigger of 7 g/dl once hemodynamic stability was achieved. After implementation of the guideline the mean RBC per patients transfused decreased from 23 units to 17 units with no difference in length of stay, risk of death or myocardial infarction and had a significant decrease in the incidence of ventilator-associated pneumonias.[15]

Zilberberg and Shorr created an analytical model to determine the number of transfusion-attributable severe acute complications (TSACs) avoided and a direct cost savings incurred annual in the United States based on a universal adoption of a restrictive transfusion strategy with a trigger less than 7 g/dl. The model only included patients at risk for a late transfusion ($\geq$ ICU day 3) in the absence of active bleeding. They estimated that about a million adult patients in ICUs in the Unite States annually are at risk for a late transfusion. With the current transfusion practices, about 3.1 million packed red blood cell units are infused to these patients. With universal institution of the restrictive practice guideline, the transfusion burden would be reduced by 42% to 1.8 million units. This saves about 1.3 million units of packed red blood cells annually and prevents about 40,000 TSACs annually at a cost savings of about $820 million. This does not include the cost savings in avoiding TSACs.[16]

Most recently a superb clinical practice guideline for red blood cell transfusion in adult trauma and critical care was developed through a collaborative effort between the Society of Critical Care Medicine and Eastern Association for the Surgery of Trauma Practice Management Workgroup.[17] This collaborative effort was led by Dr. Lena Napolitano and a summary of the key recommendations include the following:

1. A restrictive strategy of RBC transfusion (transfuse when hemoglobin is less than 7 g/dl) is as effective as a liberal transfusion strategy. (Transfusion when hemoglobin is less than 10 g/dl).
2. Using only hemoglobin level as a trigger for transfusion should be avoided. The decision for RBC transfusion should be based on an individual patient's intravascular volume status, evidence of shock, duration extent of anemia, and cardiopulmonary physiologic parameters.
3. There is no evidence that blood transfusions increase tissue oxygenation and optimal transfusion triggers in sepsis are not known.
4. Red blood cell transfusion should be avoided in patients at risk for acute lung injury and ARDS after completion of the resuscitation.
5. There is no evidence that red blood cell transfusion helps weaning from mechanical ventilation or benefits in moderate to severe closed head injury.
6. Red blood cell transfusions are associated with an increase in nosocomial infection rates independent of other factors.
7. Red blood cell transfusions are associated with a longer ICU and hospital length of stay and are an independent risk factor for multi-organ failure, SIRS, and increased mortality

REFERENCES

[1.] National Blood Data Resource Center: Comprehensive report on blood collection and transfusion in the Unites States available at http://www.aabb.org. [2.] Gunter OL et al. Optimizing outcomes in damage control resuscitation: identifying blood product ratios associated with improved survival. **J Trauma** 2008 Sep;65(3):527-534. [3.] Nunez TC et al. Early Prediction of massive transfusion in trauma: simple as ABC (Assessment of Blood Consumption)? **J Trauma** 2009 Feb; 66:346-352. [4.] Cotton BA et al. Predefined massive transfusion protocols are associated with a reduction in organ failure and postinjury complications. **J Trauma**. 2009 Jan; 66(1):41-8; discussion 48-9. [5.] Nichol AD. Restrictive red blood cell transfusion strategies in critical care: does once size really fit all: **Crit Care Resus** 2008; 10:323-327. [6.] Ho J etal. Effects of storage on efficacy of red cell transfusion: when is it not safe? **Crit Care Med** 2003, 31(12 Suppl): 5687-5697. [7.] Shapiro MJ et al. Anemia and blood transfusion in trauma patients admitted to the intensive care unit. **J Trauma** 2003; 55:269-274. [8.] Spinella PC et al; Duration of red blood cell storage is associate with increased incidence of deep vein thrombosis and in hospital mortality in patients with traumatic injuries. **Crit Care** 2009; 13:R151. [9.] Hill GF et al. Allogeneic blood transfusion increases the risk of postoperative bacterial infection: A meta-analysis. **J Trauma** 2003; 54:908-914. [10.] Crorce MA et al. Transfusions result in pulmonary morbidity and death after a moderate degree of injury. **J Trauma** 2005; 59:19-24. [11.] Hebert PC et al. A multicenter, randomized, controlled clinical trial of transfusion requirements in critical care. **N Eng J Med** 1999; 340:409-417. [12.] Vincent JL et al. Anemia and blood transfusion in critically ill patients. **JAMA** 2002; 288:1499-1507. [13.] Corwin HL et al. The CRIT study: Anemia and blood transfusion in the critically ill –Current clinical practice in the United States. **Crit Care Med** 2004, 32:39-52. [14.] Marik PE. Efficacy of red blood cell transfusion in the critically ill: A systematic review of the literature. **Crit Care Med** 2008; 36:2667-2674. [15.] Earley AS et al. Anemia management program reduces transfusion volumes, incidence of ventilator-associated pneumonia, and cost in trauma patients. **J Trauma** 2005; 61:1-7. [16.] Zilberg MD, Shorr AF. Effect of a restrictive transfusion strategy on transfusion-attributable severe acute complications and costs in the US ICUs: A model simulation. **BMC Health Ser Rsrch** 2007; 7:138. [17.] Napolitano LM et al. Clinical practice guideline: Red blood cell transfusion in adult trauma and critical care. **Crit Care Med** 2009; 37:3124-3157.

DAMAGE CONTROL HEMATOLOGY: TRAUMA EXSANGUINATION PROTOCOL IN THE HUMAN PATIENT

R. Miller
Nashville, TN

Trauma is the leading cause of death and disability in the human population and exceeds all other causes of death combined in persons less than 36 years old. Hemorrhage accounts for about 40% of all trauma deaths and control of bleeding is extremely challenging in the presence of an established coagulopathy. Uncontrolled non-surgical hemorrhage has caused early termination of operative intervention, damage control techniques and increases the incidence of multi-organ failure and death.

The majority of deaths from trauma occur within the first few hours after injury. Hemorrhage causes the largest percent of deaths within the first hour and hemorrhagic shock and exsanguination account for 80% of the deaths in the operating room and 50% of the deaths within the first 24 hours after injury. Very few hemorrhagic deaths occur after the first 24 hours and only central nervous system injury is more lethal. Most preventable trauma deaths occur early and involve airway management, thoracic injuries, traumatic brain injuries, and most importantly, control of hemorrhagic shock.

Damage control was a term initially used by the US Navy to describe the capacity of the ship to absorb damage and maintain mission integrity. It is now the standard descriptor of the three-phase surgical approach to a severely injured trauma patient. In stage one, an initial trauma celiotomy is performed controlling hemorrhage and contamination and then curtailing the abdominal operation with intraabdominal packing and application of a temporary abdominal closure. Stage two includes transport to the intensive care unit for secondary resuscitation with rewarming, correction of coagulopathy, and ventilatory support, continually reexamining and planning definitive operation. The third stage entails reoperative trauma celiotomy, removal of packing, definitive repairs, and decisions on abdominal wall closure. The decision to abort operative intervention and initiate damage control must be made early in order to prevent depletion of physiologic reserve which can culminate in physiologic exhaustion and death. Physiologic reserve is defined as each individual's ability to tolerate injury and a function of unique host factors including age, gender, preexisting disease, and immunocompetence.

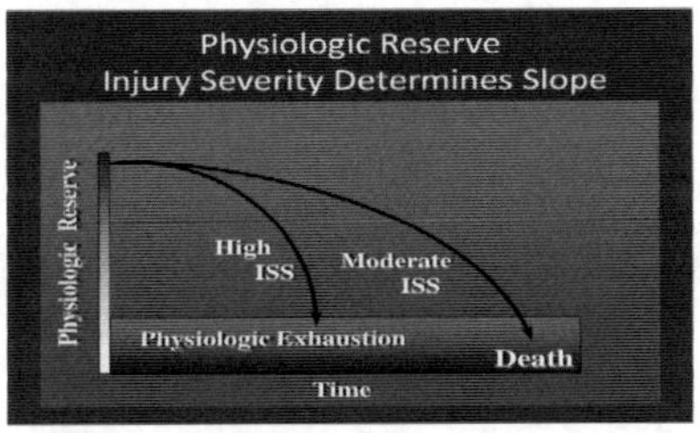

The mechanism contributing to coagulopathy following injury is multifactorial and includes initiation of both the inflammatory and coagulation cascades leading to what is now termed the acute coagulopathy of trauma-shock (ACoTS). ACoTS is present upon admission in about ¼ of trauma patients. It has been shown to have a four-fold increase in mortality and worsens the outcome from traumatic brain injury. There is increasing evidence that an aggressive correction of this coagulopathy dramatically reduces mortality.

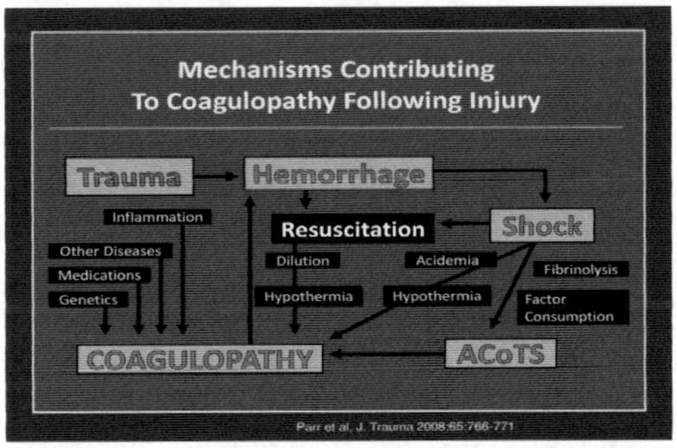

Damage control resuscitation and the direct treatment of coagulopathy have been relatively neglected until most recently. It was viewed as a product of resuscitation, hemodilution, hypothermia, and delayed blood bank logistics. It is essential to address the deadly triad of shock which includes hypothermia, acidosis and coagulopathy immediately upon admission. Damage control resuscitation also involves a limited resuscitation, maintaining the systolic blood pressure of about 90mm of mercury and utilizing blood and blood products as the primary source of volume resuscitation. Although the exact ratio of blood to blood products infused is controversial; both military and civilian resuscitation data reveal that a FFP to packed red blood cells ratio of about 2:3 achieves the best survival rates.

Vanderbilt University Medical Center initiated a trauma exsanguination protocol(TEP) in February 2006. Through multiple peer-reviewed publications we have shown a significant reduction in the odds of mortality after the initiation of the TEP with a significant decrease in overall blood product consumption. Since the inception of this protocol, we have seen a significant reduction in severe sepsis, septic shock, and multiorgan failure. Additionally, ventilator-associated pneumonia, pulmonary failure, open abdomen, and the development of abdominal compartment syndrome all have been lower with this protocol in place. Currently a scoring system is being tested across the country to aid the trauma resuscitation team in activating the TEP protocol promptly.

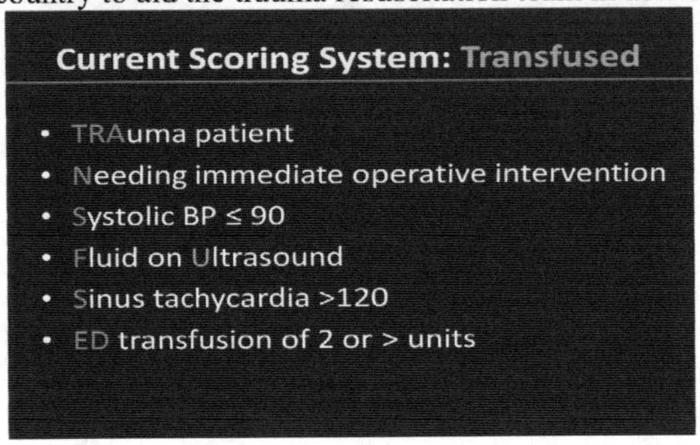

REFERENCES

[1.] Kauvar et al. Impact of Hemorrhage on Trauma Outcome: An overview of Epidemiology, Clinical presentations and therapeutic considerations. **J. Trauma** 2006;60:S3-S11 [2.] Borgman et al. The ratio of blood products transfused affects mortality in patients receiving massive transfusions at a combat support hospital. **J. Trauma** 2007; 63:805-813 [3.] Duchesne et al. Review of Current Blood Transfusion Strategies in a Mature Level 1 Trauma Center: Were we wrong for the last 60 years? **J. Trauma** 2008; 65:272-278 [4.] Cotton et al. Damage Control Hematology: The impact of a Trauma Exsanguination Protocol on survival and blood product utilization. **J. Trauma** 2008;64:1177-1183 [5.] Gunter et al. Optimizing outcomes in Damage Control Resuscitation: Identifying blood product ratios associated with improved survival. **J. Trauma** 2008; 65:527-534 [6.] Nunez et al. Earl prediction of Massive Transfusion in Trauma: Simple as ABC (Assessment of Blood Consumption)? **J. Trauma** 2009; 66:346-352 [7.] Cotton et al. Predefined Massive Transfusion Protocols are associated with a reduction in organ failure and postinjury complications. **J. Trauma** 2009;66:41-49 [8.] Cotton et al. room for (Performance) Improvement: Provider-related factors associated with poor outcomes in massive transfusion. **J. Trauma** 2009;67:1004-1012

MAINTENANCE FLUID THERAPY: ARE WE OVERESTIMATING OUR EQUINE PATIENTS' NEEDS?

Harold C. Schott II, DVM, PhD, DACVIM
East Lansing, MI

INTRODUCTION

Unlike people and small animals, massive trauma leading to acute hemorrhagic, hypovolemic shock is a rare event in equine medicine. Rather, the majority of horses in need of high volume intravenous (IV) fluid support are afflicted with gastrointestinal disease resulting in large volume nasogastric reflux or profuse diarrhea. Although affected horses can develop profound fluid deficits, they develop over a period of one or more days allowing greater time for homeostatic responses to partly re-establish effective circulating volume. Thus, emergent resuscitative fluid therapy is not commonly indicated in horses and when aggressive replacement fluid therapy is pursued, unintended adverse consequences may develop. Examples can include foals with hypoxic-ischemic encephalopathy and horses with enterocolitis of several days duration. In both examples, acute administration of replacement fluids often results in a sodium (Na^+) load that greatly exceeds the estimated Na^+ deficit. In addition to promoting edema formation, increased urine output to excrete excess Na^+ is accompanied by an obligate urinary K^+ loss that can exacerbate a total body K^+ deficit.

Next, one of the major limitations in our current equine fluid therapy arsenal is lack of a commercial maintenance fluid packaged in 3 or 5 liter bags. Consequently, horses are nearly exclusively treated with replacement fluid products that have a composition similar to extracellular fluid (ECF). These fluids are high in Na^+, low in K^+, and provide no nutritional support. Realistically, this is not a major problem for horses that are receiving IV fluids for 48 hours or less but can become a limiting factor in providing the highest quality care to patients with substantial, ongoing fluid losses of several days duration (again, horses with nasogastric reflux due to ileus or profuse watery diarrhea). Before specific recommendations for maintenance fluid therapy will be made, a brief review of the indications for and consequences of administration of large volumes of replacement fluids will be addressed.

ASSESSMENT OF NEED FOR FLUID SUPPORT

Assessment of a patient's need for fluid support involves a combination of physical exam findings and rapidly available laboratory data. Unfortunately, some clinical measures of hydration status including mucous membrane dryness and skin tent duration may be less reliable than traditionally taught due to variation with age, skin site used, and rapidity of fluid depletion and replacement.[1] Further, parameters such as heart rate and indirect blood pressure measurement are also now recognized to be less reliable indicators of impending cardiovascular collapse in acute trauma patients as compared to measures of tissue oxygenation.[2] Nevertheless, in horses with gastrointestinal fluid losses, heart rate (in the absence of arrhythmias or intermittent moderate to severe pain) continues to provide a practical index of effective circulating volume and should remain an important parameter when assessing both the need for and rate of replacement of a fluid deficit. For example, horses with enterocolitis and a total protein concentration of 3.0 g/dl may have heart rates ranging from normal to >100 beats/min, depending on duration of disease and accompanying fluid shifts. Administration of replacement fluids and possibly oncotic agents (plasma and/or hetastarch) is clearly indicated in horses with persistent tachycardia and ongoing fluid losses but may be of little benefit to horses with normal heart rates and in which excessive fluid loss has ceased.

Rapidly available laboratory data generally includes packed cell volume (PCV), total solids (TS) as an estimate of total protein concentration, venous pH, and lactate and electrolyte concentrations. In horses without acute hemorrhage, PCV can be used to assess the degree of hemoconcentration but accurate interpretation can be complicated by splenic contraction and a likely increasing measurement error as values exceed 50-55%. An increase in TS can be a useful indicator of loss of plasma water (and volume) but is complicated by protein loss into the gut lumen in many patients with gastrointestinal disease. Next, lactate concentrations at hospital admission and rate of change during treatment have received considerable attention as indirect indicators of tissue perfusion and oxygenation and have been associated with outcome in foals[3] and adult horses.[4] In general, the higher the lactate concentration the greater the indication for IV fluid support; however, it is important to remember that lactate concentration can also increase in patients with increased muscle activity (e.g., after a seizure) and is not directly correlated with the need for aggressive fluid therapy. Realistically in the end, clinical experience usually becomes the primary factor influencing a clinician's decision about need for initiation of and rate of administration of intravenous fluid therapy.

CONSEQUENCES OF REPLACEMENT FLUID SUPPORT

Use of large volumes (100 liters or more over 48 hours in a 500 kg horse) of replacement fluids has become a standard practice in many hospitals for support of horses with moderate to severe gastrointestinal diseases. Again, these fluids are high in Na^+, low in K^+, and provide no nutritional support. Since total ECF volume of a 500 kg horse is 100 liters, replacement of the entire ECF volume also replaces 100% of exchangeable Na^+ (14,000 mEq). Despite loss of substantial amounts Na^+ in reflux or diarrhea fluid, this type of therapy results in excessive Na^+ replacement with consequences of promoting edema formation and increasing urinary K^+ loss (**Figure 1**). Assuming that one measure of

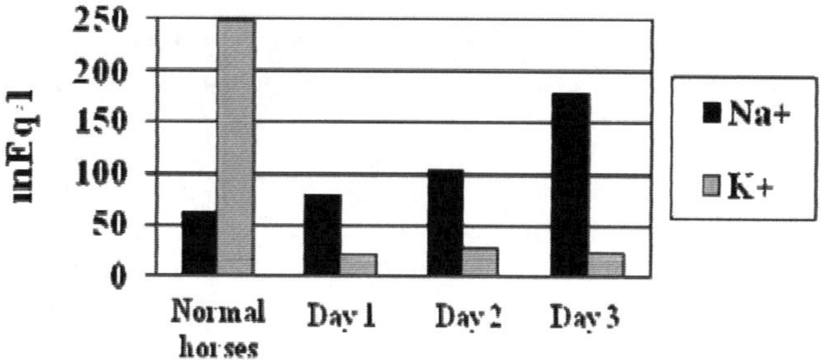

Figure 1. Urinary sodium and potassium concentrations in hospitalized horses that received replacement intravenous fluid therapy over a period of several days during which they had no or minimal intake of feed and water (Schott, unpublished data).

effective fluid replacement is an increase in urine production and that normal urine output is 6-8 liters/day, a doubling of urine production would equate to 15 liters/day and an obligate K^+ loss of 375 mEq/day (assuming a lowest achievable urinary K^+ concentration of 25 mEq/l). Most replacement fluids have a K^+ concentration of 5 mEq/l; thus administration of 50 liters/day provides replacement of only 250 mEq of K^+. Even without considering the K^+ loss in gastrointestinal fluid losses, use of replacement fluids alone generally exacerbates total body K^+ depletion in a completely anorectic horse. Clearly, this approach to fluid therapy overestimates a patient's Na^+ needs and fails to maintain adequate body K^+ stores.

NEED FOR MAINTENANCE FLUID SUPPORT

From the above discussion, the need for maintenance fluid products in equine practice, especially for completely anorectic horses, should be clear. A maintenance fluid should contain less Na^+ and more K^+ and, ideally, a source of calories for nutritional support. Several such commercial fluid products designed for use in people are available and have been used in small animal patients. Unfortunately, the largest volume available is a 1 liter bag, impractical for use in horses requiring high volume fluid support. As a consequence, most equine clinicians currently add a variety of supplemental products (e.g., KCl, calcium gluconate, dextrose, etc.) to larger volume replacement fluid bags to address the limitations of replacement fluids. This practice fails to address the problem of excessive Na^+ replacement and often leads to increased tonicity of the final product. Another common practice is to supplement KCl as a concentrated oral slurry in horses with diarrhea but this route is not available for horses with persistent gastrointestinal reflux. Further, administration of hypertonic oral electrolyte slurries may negatively affect appetite and repeated doses have been reported to exacerbate gastric ulcers.[5]

MAINTENANCE FLUID SUPPORT STUDIES

The author and colleagues have compared the effects of fluid support of healthy horses deprived of food and water with a commercially available replacement fluid (Normosol-R™ = R), a maintenance fluid (Normosol-M™ = M) as well as a modification of Normosol-M™ (M+) with additional dextrose, potassium chloride, potassium acetate, calcium gluconate, and magnesium sulfate (**Table 1**).[6,7] Water and electrolyte balance; serum concentrations of glucose, triglycerides, and insulin; and glycogen content of hepatic and skeletal muscle were compared in six horses that were deprived of feed and water while supported with these three fluids (1.5 ml/kg/h). Clinical hydration status, body mass (BM), and water and electrolyte balance were measured daily for a 3 day baseline period (with hay and water intake), during a 5 day period of feed and water deprivation (fluid infusion period), and during a 3 day recovery period (with hay and water intake).

Solutions	dextrose (g/l)	Na^+	K^+	Cl^-	Ca^{++}	Mg^{++}	sulfate	acetate	gluconate	Osmolarity (mOsm/L)
Normosol-R	0	140	5	98		3		27	23	294
Normosol-M + 5% dextrose	50	40	13	40		3		16		363
"modified" Normosol M + 5% Dextrose	68	36	39	36	24	21	18	24	24	574

* Na^+, K^+, Cl^-, Ca^{++}, Mg^{++}, acetate, and gluconate all provided as mEq/L

Table 1. Composition of the commercially available replacement and maintenance solutions and the modified maintenance solution.

Horses experienced a 3-9% BM loss during the fluid infusion period and several developed mild diarrhea from day 3-5 of the infusion period; however, clinical examination revealed normal hydration parameters in all horses during all infusion periods. Of interest, no horse drank water when it was offered once daily during each infusion period. Although there were mild variations in plasma electrolyte concentrations and acid-base balance with the different treatments, none were significant. However, both urine and fecal Na^+ and chloride (Cl^-) excretion were significantly greater with R than with M or M+. All horses experienced a negative K^+ balance during infusion with all fluid products but the magnitude of the estimated K^+ deficit was less with M or M+ than with R. Increased urine output with hyposthenuria was seen in all horses with all fluids but urinary losses of calcium and magnesium were attenuated with M+ as compared to R or M.

Glucose and insulin concentrations increased significantly during the infusion period with M and M+ but remained unchanged from the baseline period with R. There was no significant difference in the magnitude of the increases between M and M+ but glucosuria was detected on five infusion days with M+ as compared to only once with M. In contrast, triglyceride concentrations increased significantly with R but remained unchanged with M or M+. By the end of the infusion period, liver glycogen content increased by 30% and 47% with M and M+, respectively, but decreased by 86% with R. Changes in muscle glycogen content were not detected.

Although no differences in clinical examination findings were observed, changes in Na^+, K^+, and Cl^- balance during the infusion period were attenuated with M and M+, as compared to R. Further, M+ was superior to M with regard to calcium and magnesium balance. Although insulin secretion and tissue glucose uptake were increased with both the 5% (M, 75 mg/kg/h) and 6.8%

(M+, 102 mg/kg/h) dextrose solutions, they were inadequate to prevent mild to moderate hyperglycemia and glucosuria. However, both dextrose-containing solutions were effective in preventing an increase in triglyceride concentration during a period of food deprivation.

All in all, both maintenance solutions were superior to R for water and electrolyte balance. However, the horses were unable to utilize all of the dextrose (glucose) in the maintenance solutions at the rate at which they was administered. Thus, an ideal maintenance solution for intravenous use in horses should have a dextrose concentration less than 5%. Further, due to the variable rates at which IV fluids may be administered, addition of 50% dextrose to fluids to produce a final concentration of 1-3% may be the most logical approach in development of a maintenance fluid for horses.

CONCLUSION

Our studies have demonstrated clear improvements in electrolyte balance with maintenance solutions as compared to a replacement solution. The next steps are to refine the composition of the maintenance solution for horses and to compare the effects of a replacement solution with a maintenance solution designed for horses in a large group of horses requiring fluid support for 4 or more days. Ideally, such a study would be a prospective, multi-center, randomized clinical trial with an outcome of a commercial maintenance fluid for horses in a large volume bag.

REFERENCES

[1.] Pritchard JC, et al. **Equine Vet J** 2008;40:558. [2.] Convertino VA, et al. **Trauma** 2008;64(Suppl 4):S342. [3.] Wotman K, et al. **J Vet Intern Med** 2009;23:598. [4.] Tennent-Brown BS, et al. **J Vet Intern Med** 2010;24:198. [5.] Holbrook TC, et al. **Equine Vet J** 2005;37:501. [6.] Schott HC, et al. **J Vet Intern Med** 2007;21:602. [7.] Schott HC, et al. **J Vet Intern Med** 2007;21:619.

REPLACEMENT FLUID THERAPY FOR THE CRITICAL EQUINE PATIENT

K. Gary Magdesian, DVM, DACVIM, DACVECC, DACVCP
Davis, CA

INTRODUCTION

Replacement of blood volume deficits is the cornerstone of treating the critically ill horse. Hypoperfusion, associated with hypovolemia, hypotension and reduced cardiac output, is associated with multiple organ dysfunction and mortality if unchecked. Interstitial, and then intracellular, fluid losses must also be addressed, although their priorities fall below those of circulating volume. Acute abdominal disease (colic and colitis), acute hemorrhage and sepsis are among the most common disorders requiring rapid fluid replacement in horses and foals.

The decision for optimal fluid type and rate of replacement in horses is subjective and without scientific consensus, making fluid therapy difficult at best. Detection of dehydration is highly insensitive, with earliest clinical detection occurring at 3-5 % of body weight loss in fluid. Even more challenging is to estimate the percent loss from each of the fluid subcompartments, including plasma volume, interstitium, and intracellular fluid (ICF). Purely extravascular losses are particularly difficult to detect. Little data are available to guide choice of fluid in equine patients, including isotonic crystalloids, hypertonic crystalloids, or colloids, as well as rates and volumes needed.

There are 4 major decisions in formulating replacement fluid therapy:[1]

1. Goals of fluid therapy
2. Choice of fluid type
3. Rate of administration
4. Limits to fluid administration

1. GOAL OF REPLACEMENT FLUID THERAPY

In the acutely ill horse, the primary goal of fluid administration is to reestablish tissue perfusion. The immediate priority is therefore restoration of circulating volume and fluids are primarily administered for reversal of hypovolemia. Hypovolemia may be due to external losses, as with diarrhea or hemorrhage, or internal losses due to transudation or exudation of fluids. Relative hypovolemia occurs as venous capacitance increases due to release of inflammatory mediators in sepsis or systemic inflammatory response syndrome (SIRS), anesthetic agents, or vasodilatory pharmaceuticals. In all of these cases the goal of volume repletion is to restore venous return, cardiac output and arterial pressure, in order to normalize organ perfusion.

Replenishment of interstitial losses is next in priority, because of the communication of the interstitium with plasma volume across the vascular membrane. Intracellular losses can be replaced more slowly, often over several hours. Additional immediate goals include correction of acid-base and electrolyte abnormalities. In adult horses, the primary acid-base derangement is lactic acidosis, although sodium and chloride abnormalities can also contribute. Colloid replacement is another potential goal in the concurrently hypovolemic and hypoproteinemic animal.

Specific goals include normalization of clinical perfusion parameters: mentation, heart rate, extremity temperature, pulse quality, jugular fill, and urine production. Laboratory perfusion parameters include lactate concentration, venous oxygen saturation, urine specific gravity, and serial packed cell volume measurement and total protein concentrations. Optimizing filling pressures, most practically in the form of central venous pressure (CVP), is a controversial goal of fluid replacement. Ultimately, improvement in arterial blood pressure is sought as a marker of the potential for adequate tissue perfusion.

II. FLUID TYPE

Fluids choices available for volume resuscitation include isotonic and hypertonic crystalloids, as well as colloids. Hypotonic fluids (0.45 % NaCl/2.5 % dextrose [hypotonic in vivo]; PlasmaLyte 56 [Baxter Healthcare Corporation, Deerfield, IL]; Normosol M ; [Abbott Laboratories, North Chicago, IL], 5% dextrose) are not replacement fluids and should never be bolused.

There is currently no consensus in the veterinary or human medical communities regarding the optimal fluid choice for resuscitation of hypovolemia. Several meta-analyses and systematic reviews have not shown a benefit of either isotonic crystalloid or colloid for use in human patients.[2] Based on cost differences, the conclusion in most of these studies is that crystalloids should be administered over colloids under most circumstances. In addition, there is some evidence for an increased risk of side effects (anaphylaxis, coagulopathies, renal dysfunction) with the use of colloids. Isotonic crystalloids distribute among the entirety of the ECF, with approximately 20-25 % of the administered volume remaining within the plasma volume, and 75-80 % distributing extravascularly to the interstitium. Due to this distribution pattern, the maximum hemodynamic effects of crystalloids occur immediately after infusion. Traditionally, three to four times the amount of isotonic crystalloids has been recommended to replace a deficit within the intravascular compartment because of this extravascular distribution. However, due to concerns over interstitial and tissue edema, some authors have questioned the need for this practice.[3] Colloids distribute virtually entirely to the vascular compartment in normal animals, however in animals with SIRS or severe trauma they may extravasate to the interstitium due to reductions in the capillary reflection coefficient associated with altered vascular permeability. In doing so, they would potentiate edema by raising interstitial oncotic pressure. The use of both crystalloids and colloids together has advantages in some circumstances. The author commonly administers isotonic and hypertonic crystalloids in conjunction with hetastarch as a rapid volume expander in adult horses.

Isotonic crystalloids are conveniently available in 5 L bags. These include 0.9% saline, lactated Ringer's solution (LRS), PlasmaLyte A (Baxter Healthcare Corporation, Deerfield, IL) and Normosol R (Abbott Laboratories, North Chicago, IL). Saline has little application for use in horses except for those with marked hyperkalemia, metabolic alkalosis or hypochloremia; the reason for its limited applicability is its tendency to produce a hyperchloremic metabolic acidosis in both humans and horses.[4-5] Hyperchloremic acidosis causes afferent renal artery vasoconstriction in animal models, and is associated with renal dysfunction, inflammatory cytokine release and possibly additional organ dysfunction.[6] LRS is a commonly administered balanced electrolyte solution that has long been used in human and veterinary medicine. It contains less chloride and sodium than saline; however it is still mildly hyperchloremic relative to equine plasma, and it is therefore theoretically possible to increase plasma chloride with large volume administration of LRS. Normosol and PlasmaLyte fluids are the closest to equine plasma in terms of electrolyte composition and may be the most physiologic fluid available for horses. They contain acetate and gluconate rather than lactate as the alkalinizing salts, which is an advantage for horses with liver dysfunction.

Colloids are the other major fluid choice for volume resuscitation in the horse. They have the theoretical advantage of primarily intravascular distribution, which could lead to more rapid volume expansion. However, in animals with SIRS or sepsis this may not hold true due to altered vascular integrity. This makes the use of colloids challenging in critically ill horses. For example, large volumes of crystalloids could potentiate pulmonary edema in foals or horses with acute respiratory distress syndrome, making colloids an attractive choice for these cases. However, without knowledge of the specific alterations in the Starling equation for the individual horse, it is impossible to know whether colloids would also extravasate into the lungs and worsen pulmonary edema. Deterioration of clinical or blood gas parameters post-colloid administration warrants reconsideration of colloid use. Horses with concurrent hypovolemia and hypoproteinemia (relative or absolute) are potential candidates for colloid administration as resuscitative fluids; however, they should be serially monitored for development or worsening of edema.

Plasma and hetastarch are the two primary colloids available for use in horses and foals. Dextrans are associated with an increased risk of side effects (coagulopathies, allergic reactions and renal failure) as compared to hetastarch and are similarly expensive, therefore offering no advantage over hetastarch. Plasma is not suitable as a replacement fluid; it must be thawed slowly and rapid bolus administration should be avoided due to anaphylaxis or allergy risks. Currently, the use of plasma in critical care is controversial; it offers the potential benefit of providing albumin, antithrombin and antibodies. Potential disadvantages include anaphylaxis or other inflammatory reactions and provision of additional clotting factors to already prothrombotic patients. Hetastarch has dose dependent effects on coagulation, interfering with platelet function and producing a von Willenbrand-like state. In humans, hetastarch rarely has been reported with hyperoncotic renal dysfunction in sepsis. Although there are no reports of this in horses, it should be used with caution in patients with renal dysfunction. In a study evaluating hetastarch in hypoproteinemic horses with GI diseases, COP exhibited an increase for up to 24 h after administration. Pentastarch is a more homogeneous, in terms of molecular weight variability, derivative of hetastarch. The COP of pentastarch is 40 mmHg, as compared to 30 mmHg for hentastarch, providing more efficient volume expansion. Pentastarch is associated with fewer effects on coagulation in humans, and has less small molecular weight particles that could escape from a compromised vasculature. Pentastarch is not currently available in the US.

The primary hypertonic solution available for fluid replacement is hypertonic saline (7-7.5 %). Hypertonic saline results in a shift of first interstitial and later cellular fluids into the intravascular space. In addition to causing plasma volume expansion, hypertonic saline has mild inotropic, anti-inflammatory and anti-edema effects, causing renewed interest in its use in critical care. It is protective against edema and compartment syndromes. Administered doses should not exceed 4 ml/kg. It should be used concurrent to isotonic crystalloids, as the plasma expansion it produces is short-lived. Contraindications to administration of hypertonic saline include preexisting hypernatremia or hyperosmolarity, and uncontrolled hemorrhage. Hypertonic saline should be used with caution in neonatal foals, as they appear to be less tolerant of rapid swings in plasma sodium concentrations.

III. RATE OF ADMINISTRATION

Predicting the volume of fluid required to treat hypovolemia and dehydration is difficult. Signs of dehydration (reduced skin turgor, tacky membranes) can be misleading, and deciphering the distribution of the deficits is even less accurate. Clinical signs of hypovolemia, such as tachycardia and decreased extremity temperature, are also insensitive and nonspecific. Fluid overload and persistent hypovolemia can both result in multiple organ dysfunction syndrome. A simpler approach than trying to calculate absolute fluid deficits and one that is likely subject to fewer complications is that of the "fluid challenge", a method for guiding fluid repletion in human critical care.[1] This technique allows for prompt correction of fluid deficits in critically ill patients, while minimizing the risks of fluid overload and edema. With this method, incremental volumes of fluid are administered rapidly over defined intervals until clinical signs and laboratory indicators of hypoperfusion abate or plateau, or central venous pressures increase to maximum. It should be noted that central venous pressure (CVP) does not always accurately predict intravascular preload, because it is also dependent on ventricular compliance and afterload. However, it can serve as a "cap" to fluid loading in horses that do not have heart disease. Clinical parameters that should be serially monitored as markers of improvement in volume status include heart rate (less reliable in neonatal foals), extremity temperature, capillary refill time, pulse quality, jugular refill, mentation, blood lactate, packed cell volume, total protein concentration, and arterial blood pressure. Urine output is an important marker of adequacy of end organ perfusion, and the production of dilute urine is one end point to fluid bolus administration.

Guidelines for fluid administration rates includes 10-20 mL/kg of isotonic crystalloid over 20-30 minutes followed by reassessment of the clinical and laboratory perfusion parameters.[1,2] These are repeated over 30-60 minutes until perfusion parameters improve or plateau (no further improvement, indicating a possible need for inotrope or vasopressor therapy), urine is produced, or limits to fluid volume administration are reached.[7] Most hypovolemic foals and horses need 2-4 and 2-3 of such boluses, respectively, and the number is based on serial reassessment. Excessive administration of isotonic crystalloids should be avoided, because that would facilitate tissue edema, worsen oxygenation, and possibly allow for compartment syndrome.[2] If colloids are selected for replacement, a smaller volume should be administered due to relative greater retention within the vascular space. Doses of 3-10 mL/kg can be used for hetastarch, although a total daily dose of 10 mL/kg should not be exceeded due to risk of coagulopathies at higher doses. Colloids should be administered along with isotonic fluids, as they do not rehydrate the interstitial space. The dose of hypertonic saline (7-7.5 %) for fluid resuscitation is 4 mL/kg, once, which should be administered concurrent to or followed by isotonic crystalloids.

IV. LIMITS

Fluid overload is an area receiving increasing attention in human critical care. The primary limit to fluid administration is edema. Edema can be immediately life threatening in the form of pulmonary edema, particularly in septic neonatal foals and adult horses with SIRS that have altered capillary permeability. Longer term edema can have consequences on tissue oxygenation and organ dysfunction through secondary compartment syndromes.[4] The risks of edema and compartment syndrome are greater than previously recognized; they are associated with worsening outcomes in human patients.[4,8,9] Secondary compartment syndrome is abdominal compartment syndrome in the absence of abdominal injury, and is thought to be associated with early and overzealous crystalloid administration. Another potential drawback of excessive fluid administration is induction of inflammatory cytokine release, as this has been shown to occur with administration of some types of fluids. Under-resuscitation of hypovolemia also has its own risks of organ injury through inadequate oxygen delivery, again warranting close and serial monitoring of patients undergoing replacement fluid therapy.

Colloids have erroneously been associated with a reduced predilection for interstitial edema. During volume loading, if more colloid is administered than the actual blood volume deficit, colloids do not completely remain within the circulatory compartment. Up to 60 % of the infused volume can load the interstitial space.[10,11] Therefore, colloids should not be administered without close monitoring of end points and limits.

Diagnosis of under- or over-loading of fluid therapy in the clinical setting is difficult. Filling pressures, usually measured as CVP, can be misleading as indicators of volume repletion. However, they are likely more accurate as ceilings to fluid loading. That is, once maximal CVPs are reached (10 cm H_20 [7.4 mmHg] in foals, 10-15 cm H_2O [7.4-11.0 mmHg] in adult horses), fluid boluses should cease. At that point, fluid therapy should switch to maintenance + on-going losses or slightly more for fluid diuresis. In addition, a decrease in arterial oxygen saturation during fluid loading warrants reassessment of fluid volume, and this is particularly useful in foals. An increase in respiratory rate or character is another possible indicator of fluid overload. Urine production is an excellent end point to fluid bolus administration, as dilute urine indirectly indicates renal perfusion.

CONCLUSION

Fluid resuscitation of hypovolemia in horses is challenging and dynamic. Optimal fluid delivery will vary from patient to patient. In general, the "fluid challenge" method with serial reassessment used in human critical care is a safe means of providing volume replacement rapidly. The long time debate over whether crystalloids or colloids are more effective for fluid resuscitation has not yielded definitive support for either. Because of increased cost and potential for side effects, isotonic crystalloids should remain the main stay of replacement fluid therapy. There has been a movement away from 0.9 % saline as the crystalloid of choice for treatment of hypovolemia, due its propensity to cause hyperchloremic metabolic acidosis and the associated deleterious effects. Balanced electrolyte solutions appear to be more physiologic for horses, especially the commercial acetated fluids, as the electrolyte composition is closer to that of equine plasma. Colloids may be used concurrently but at lower doses. Hypertonic saline has several hemodynamic and physiologic advantages and is a useful adjunct to isotonic fluids in volume resuscitation of the adult horse. Colloids may not be retained in the intravascular space in horses with an impaired endothelial barrier, such as those with severe SIRS. This may generate an increased oncotic pressure in the interstitium with fluid accumulation and subsequent edema.

Table 1. Composition of Replacement Crystalloids

Crystalloid	Sodium	Potassium	Chloride	Calcium	Magnesium	Osmolarity	Organic
LRS	130	4	109	3	0	272-273	28 lactate
Saline (0.9%)	154	0	154	0	0	308	0
Ringers soln	147-148	4	156	4.5	0	310	0
Plasma-Lyte148	140	5	98	0	3	294	27 acetate 23 glucon.
Normosol R	140	5	98	0	3	294	27 acetate 23 glucon.

REFERENCES

[1]Vincent JL, et al. **Crit Care Med** 2006:34(5):1333; [2]Ribeiro Jr MAF, et al. **Turk J Trauma Emerg Surg** 2009;15(4):311; [3]Bauer M, et al. **Best Pract Res Clin Anaesth** 2009; 23:173.[4]Roche AM, et al. **Curr Opin Crit Care** 2009;15:520-524; [5]Stewart PA. **Can J Physiol Pharmacol** 1983;61:1444; [6]Kellum JA, et al. **Chest** 2006; 130:962; [7]Corley KTT, **Eq Vet Educ** 2002;14(6):328; [8]Sakr ,Y et al. **Chest** 2005;128:3098; [9]Van Biesen W, et al. **J Nephrol** 2005;18:54; [10]Jacob M et al. **Best Pract Res Clin Anaesth** 2009;23:145; [11]Rehm M, et al. **Anesth** 2001;95:849.

CHEMOPROPHYLACTIC EFFECTS OF GALLIUM MALTOLATE AGAINST *RHODOCOCCUS EQUI* PNEUMONIA AMONG FOALS AT ENDEMIC EQUINE BREEDING FARMS

M. Keith Chaffin,[1] Noah D. Cohen,[1] Ronald J. Martens,[1] Michael S. O'Conor,[1] Lawrence R. Bernstein,[2]
1. Department of Large Animal Clinical Sciences, College of Veterinary Medicine & Biomedical Sciences, Texas A&M University, College Station, TX. 2. Terrametrix, Menlo Park, CA.

Rhodococcus equi is an important cause of pneumonia in foals worldwide. We previously reported the results of a randomized, non-blinded, clinical trial demonstrating that chemoprophylaxis with azithromycin reduced the incidence of *R. equi* pneumonia. This prophylactic strategy has limited practical application due to the potential for development of antimicrobial resistance. Gallium maltolate (GaM) has potential for control of *R. equi* pneumonia. The purpose of this study was to determine the chemoprophylactic effects of GaM on the incidence of *R. equi* pneumonia among foals at *R. equi* endemic equine breeding farms.

A controlled, randomized, double-blinded, clinical trial was performed at 12 equine breeding farms with a history of endemic *R. equi* infections among their foal populations. Foals assigned to group 1 were treated with a placebo (indistinguishable from GaM formulation) and foals assigned to group 2 were treated with a GaM formulation (approximately 30 mg/kg, PO, q 24 hrs) during the first 2 weeks of life. Enrolled foals were monitored for development of *R. equi* pneumonia. Data were compared between treatment groups.

There were 483 foals studied. There was no significant difference between group 1 (32%) and group 2 (35%) foals in the proportion of foals that developed *R. equi* pneumonia.

The results of this study demonstrate that chemoprophylaxis using GaM, administered at this dosage regimen, does not effectively reduce the incidence of *R. equi* pneumonia among foals at endemic breeding farms. Future investigations are needed to assess the prophylactic effectiveness of expanded GaM doseage regimens and the therapeutic effectiveness of GaM in foals with *R. equi* pneumonia.

IMMUNE MODULATION BY A VAPA/CPG ODN VACCINE IN FOALS UNDERGOING EXPERIMENTAL *R. EQUI* CHALLENGE. K.L. Lohmann[1], A.M. Lopez[2], S. T. Manning[1], F. J. Marques[1], R. Brownlie[2], A. L. Allen[1], A. E. Sangster[1], G. Mutwiri[2], V. Gerdts[2], A. Potter[2], and H.G.G. Townsend[1,2]. 1. University of Saskatchewan, Western College of Veterinary Medicine, Saskatoon, SK. 2. University of Saskatchewan, Vaccine and Infectious Disease Organization, Saskatoon, SK.

Foal pneumonia caused by *Rhodococcus equi* is an important disease for which an effective, safe vaccine is not currently available. We evaluated the immunogenic and protective potential of a recombinant VapA/CpG ODN vaccine in neonatal foals. Foals (n=8) were vaccinated between 1 and 6 days of age (Day 1) and again 14 days later; control foals (n=7) received an injection of PBS. All foals were challenged by intrabronchial administration of 5×10^6 *R. equi* 103+ on Day 29. Bronchoalveolar lavages were performed on Days 15, 29, and 36 and total cell count, differential cell count, rVapA-stimulated cell proliferation and IFNγ mRNA expression determined. Clinical examination, complete blood counts, serology for VapA-specific total IgG, IgG_a, IgG_b and IgG_t, and collection of nasal and fecal swabs for culture were performed on Days 1, 15, 29, 36, 43 and 50. Foals were humanely euthanized on Day 50 and severity of pneumonia scored on a 4-point scale. Clinical, hematologic, and immune response data were analyzed using repeated measures analysis over time; differences between groups at specific time points of biological interest and post-mortem lung scores were compared using the Wilcoxon Rank Sum test. None of the foals developed clinically apparent pneumonia. Two foals shed virulent *R. equi* in feces, however, real-time PCR revealed the isolates to be different from the challenge strain. Vaccination resulted in a significant increase in VapA-specific immunoglobulin production and proliferation of BAL cells prior to intrabronchial *R. equi* challenge, and in an increased expression of VapA-specific IFN-γ mRNA by BAL cells following challenge. Post-mortem lung severity scores did not differ significantly between groups. While protection against experimental challenge was not achieved, these findings support further work to optimize the vaccine formulation and/or vaccination protocol.

PREVALENCE OF GASTRODUODENAL ULCERATION IN NON-SURVIVING NEONATAL FOALS: 354 CASES (1995-2006). LC Sanchez, JR. Elfenbein. Hoffman Neonatal Intensive Care Unit, University of Florida College of Veterinary Medicine, Gainesville, FL.

Gastric ulcer disease is reportedly common in sick neonatal foals and has been associated with rupture without prior clinical signs. The purpose of the current study was to evaluate factors associated with gastroduodenal ulceration in non-surviving neonatal foals.

All equine necropsy reports from 1995-2006 from the University of Florida Veterinary Medical Center were reviewed. Foals less than 30 days of age were included. Year, age, breed, sex, diagnosis, and presence of glandular, non-glandular, and/or duodenal ulceration were recorded. Diagnoses were grouped as follows: sepsis, prematurity, respiratory, gastrointestinal, musculoskeletal, neurological, urogenital, or other. Multiple diagnoses were possible. Unless gastric and/or duodenal ulceration was the only lesion identified, it was not included as a gastrointestinal lesion. Factors associated with ulceration were identified via backward stepwise logistic regression. In order to characterize the hospital usage of anti-acid medications (ranitidine, cimetidine, omeprazole), a computerized database of all neonatal foal admissions was searched for use of anti-acid therapy in foals <30 days during the study years. Relative frequency of anti-acid therapy and prevalence of proximal gastrointestinal ulceration were group by decade (1995-2000 vs. 2001-2006) and compared by Chi square analysis.

Three hundred fifty-four foals were included in the study. Fifty-four (15%) foals had ulceration (43 non-glandular, 24 glandular, 3 duodenal). Only one foal in the study had a perforated gastric ulcer. Gastrointestinal disease had a significant positive association with ulceration, regardless of site. Increasing age was significantly associated with squamous, duodenal or any ulceration. Year was not retained in the model regardless of ulcer location, nor was ulcer prevalence significantly different by decade. Anti-acid therapy was less common in the overall hospital population from 2001-2006 relative to 1995-2000.

In conclusion, the prevalence of gastroduodenal ulcers at necropsy was fairly low in the neonatal foal population in the current study. Ulcer prevalence did not change significantly over time, although the usage of anti-acid therapy decreased over time in this population.

DEXAMETHASONE AND FLUTICASONE ARE EFFECTIVE TO TREAT AIRWAY HYPERSENSITIVITY IN HORSES WITH INFLAMMATORY AIRWAY DISEASE. T. Tohver, R. Léguillette. University of Calgary Faculty of Veterinary Medicine, Calgary, AB, Canada

The purpose of the study is to determine if dexamethasone (0.05 mg/kg IM SID) and inhaled fluticasone administered with the AeroHippus® (3000 mcg BID) are effective in decreasing airway hypersensitivity and reactivity (expressed as the provocative concentration of histamine (PCR_L75) inducing a 75% increase in lung resistance), airway inflammation (measured in the bronchoalveolar lavage fluid (BALF)) and clinical signs in horses with inflammatory airway disease (IAD).

Eight horses with IAD were randomly assigned to a treatment group with either dexamethasone (DEX) or fluticasone (FLUT) in a cross-over design study. Each group was treated with one medication for 2 weeks, followed by a washout period before being treated for 2 weeks with the other medication. All horses included in the study showed: 1- respiratory clinical signs during exercise 2- no increase in lung resistance at rest after a challenge with moldy hay 3- airway hypersensitivity (i.e. low PCR_L75) 4- increased percentage of mast cells or/and eosinophils or/and neutrophils in the BAL. Lung mechanics and histamine challenges were performed at baselines, then weekly during treatment, and BALs were done at baselines and at the end of every treatment phase. Clinical signs at exercise were scored every other day. Statistical analysis to compare the effect of treatments on PCR_L75, BALF cells and clinical signs scores were performed using a Wilcoxon signed rank test.

There was no significant difference in PCR_L75 between groups at baselines. The PCR_L75 values decreased significantly after two weeks of DEX and FLUT treatment and after one week of DEX treatment. The airway reactivity decreased significantly after one and two weeks of DEX and FLUT treatment. The doses of histamine used up to a maximum of 32mg/ml did not induce a 75% increase in R_L after two weeks of treatment in all but one horse in each treatment group. There was no statistical difference in the clinical signs' scores or in the BALF differential cell counts before and after each treatment.

In conclusion, both DEX and FLUT treatments significantly inhibited airway hypersensitivity and reactivity in horses with IAD. There are no significant effects on the clinical signs or the inflammatory cells in the BALF.

DOSE TITRATION OF THE CLINICAL EFFICACY OF INTRAVENOUS PHENYLBUTAZONE IN A REVERSIBLE MODEL OF FOOT LAMENESS IN HORSES.

JH Foreman. University of Illinois, Urbana, IL.

Recently-approved Fédération Equestre Internationale rule changes allow the use of "half-doses" of phenylbutazone (PBZ) up to 12 hrs prior to competition, but the clinical analgesic efficacy of half-doses has not been reported previously.

The objective of this experiment was to compare the clinical efficacy of various doses of PBZ with a negative control. The hypothesis was that higher PBZ doses would result in improved efficacy in a reversible model of foot lameness.

Eight horses were shod with adjustable heart bar shoes on one front foot. Weekly AAEP Grade +4.0 lameness was induced by tightening a set screw against the heart bar. HR and lameness score were monitored at rest; every 20 min after lameness induction for 5 hrs; and hourly for another 8 hrs. One hr after lameness induction, treatment was administered IV in a randomized blinded manner and included negative control (isotonic saline: SAL) or PBZ at 2.2 (half-dose), 4.4, or 8.8 mg/kg. Results were compared using RM ANOVA and Student-Newman-Keul's test with the level of significance set at $P<0.05$.

HR and lameness score changes mirrored one another. Half-dose PBZ reduced HR compared to SAL only from 3.67-5.0 hrs after administration ($P<0.05$). Full and double-dose PBZ reduced HR from 3.0-12.0 hrs post-administration ($P<0.05$); there was no difference in response between full and double-dose PBZ.

The conclusion was that half-dose PBZ was clinically effective for only a brief period and was not effective for 6 hrs prior to the 12-hr point at which competition is legal under new FEI rules.

IN VIVO EXPANSION OF CIRCULATING RED CELL MASS IN HORSES CAUSES HYPOCOAGULABILITY AS MEASURED *IN VITRO* BY WHOLE BLOOD THROMBOELASTOMETRY. SA Smith,[1] PA Wilkins,[2] M McMichael,[2] SC Clark-Price,[2] KM Lascola[2], P DePedro[2] Departments of Biochemistry[1], Veterinary Clinical Medicine[2], University of Illinois, Urbana, IL

Studies report a correlation between indicators of red cell mass and results of thromboelastography/thromboelastometry (TE) for humans and experimental animals. *In vitro* manipulation of packed cell volume (PCV) impacts TE, with higher PCV causing relative hypocoagulability. Our objective was to evaluate the effect of increased circulating red cell mass on TE.

Baseline (B) whole blood samples from 6 normal horses were collected into citrate or EDTA. Citrated blood was immediately subjected to TE with either a tissue factor or ellagic acid trigger and also used to measure plasma PT, aPTT, and fibrinogen. CBC's were performed on EDTA blood. In order to increase PCV via splenic contraction, horses were infused with 6 $\mu g \cdot kg^{-1} \cdot min^{-1}$ phenylephrine for 15 min. Blood was collected and assays repeated 5 min (P) and 2 hr after completion of infusion. TE parameters included clot time (CT), clot formation time (CFT), alpha angle (α), and maximum clot firmness (MCF). Statistical comparisons were by repeated measures ANOVA.

Red cell count, hemoglobin, and PCV (B:40+/-1, P:55+/-1) significantly increased 5 min post-infusion compared to baseline. 5 min post infusion samples were hypocoagulable on TE as indicated by significantly increased CT (B:332+/-16, P:389+/-21), and CFT (B:115+/-8, P:183+/-14), and decreased α (B:68+/-1, B:57+/-2), and MCF (B:49+/-2, B:40+/-3). Routine coagulation assays (PT, aPTT, Fib) and platelet counts were unchanged. Red cell mass and TE parameters returned to baseline at 2 hr post-infusion.

Increased red cell mass *in vivo* causes apparent hypocoagulability. Coagulation in horses with polycythemia or hemoconcentration will appear abnormal when measured by TE.

ULTRASOUND CERVICAL PARAMETERS IN THE MARE THROUGHOUT GESTATION

Stefania Bucca DVM, Ursula Fogarty MVB, PhD, MRCVS

Placentitis is one the leading causes of fetal and neonatal death in the horse, worldwide. Placentitis in mares is reported to be most commonly caused by bacteria ascending through the vagina. A key role of the cervix in the pathogenesis of ascending placentitis could therefore be speculated. In order to maintain its vital role as a physical barrier, a high level of tightness has been attributed to the cervical canal throughout gestation. Transient cervical relaxation has been anecdotally_reported to occur during the last month of pregnancy, based on rectal palpation and vaginoscopy findings. Objective evidence of cervical physiologic changes during pregnancy is currently lacking. The present study aims at identifying US parameters to characterize size and echo-texture of the equine cervix throughout gestation in normal pregnancies. For the purpose of this study the cervix of 42 light breed horse mares were examined by ultrasound at by-weekly intervals throughout gestation from day 60 to parturition. Measurements at three designated areas and a quality score were recorded on each examination.

NEUROAXONAL DYSTROPHY IN QUARTER HORSES. CJ Finno, M Aleman, JE Madigan, R Higgins, DL Bannasch. University of California, Davis, CA.

In many species, an underlying genetic basis for Neuroaxonal Dystrophy (NAD) has either been identified (humans) or suspected (horses, sheep, cats, dogs). Equine NAD is considered the basic disease underlying equine degenerative myeloencephalopathy (EDM), with a high probability that the pathophysiology of the two diseases is similar. The purpose of this study was to describe clinical (n=148) and histopathologic (n=5) findings of NAD in Quarter horses, to determine mode of inheritance and perform a genome wide association study (GWAS) to determine a candidate chromosomal region associated with NAD in a family of Quarter horses. Phenotyping was based upon three investigators performing an independent neurologic assessment and a mean ataxia score (MAS from 0-5) was utilized. Of the 148 horses examined, 60 (41%) were assigned a MAS of <1, 52 (35%) were assigned a MAS between 1 and 1.99, 26 (18 %) were assigned a MAS between 2 and 2.99, 8 (5%) were assigned a MAS between 3 and 3.99 and 2 (1%) a score of 4. Five horses (3 had also undergone neurologic evaluation) were diagnosed on histopathologic evidence of dystrophic neurons and axonal spheroids occurring bilaterally and symmetrically in the lateral accessory cuneate, medial cuneate and gracilis nuclei in the brainstem and in Clarke's nucleus in the spinal cord. A dominant mode of inheritance is supported through pedigree analysis. Genome wide association was performed on 85 horses (30 affected, MAS≥2; 24 intermediate, MAS>0.17 but <2; and 31 unaffected, MAS≤0.17) and three candidate chromosomal regions were identified (*p* genome 0.015, 0.019, and 0.034 after 50,000 permutations). These candidate regions are being investigated further to attempt to identify a causative mutation for NAD and provide a future genetic screening test for diagnosis.

PATHOPHYSIOLOGY OF & TREATMENT OPTIONS FOR EQUINE FUNGAL KERATITIS

Alison B. Clode, DVM, DACVO
Raleigh, NC

GENERAL CONSIDERATIONS

Although significant geographic variability exists, fungal infections of the cornea are commonly encountered in equine ophthalmology. The eastern, southern, and Midwestern portions of the United States have high disease prevalence, ranging from 6 to 40% in horses with ulcerative keratitis,[1-3] while occurrence in the western and central northern regions is rare.

CLASSIFICATION OF KERATITIS

Keratitis, or corneal inflammation, may be characterized as ulcerative or nonulcerative, infiltrated or uninfiltrated, and superficial or deep. Ulcerative keratitis is diagnosed by fluorescein dye uptake, while nonulcerative keratitis manifests as vascularization and edema. Infiltrated keratitis may be ulcerative or nonulcerative, and is characterized by infiltration of inflammatory cells and/or infectious organisms. Clinically, infiltration appears as yellowish corneal opacification, with more marked vascularization, anterior uveitis, and pain. With or without ulceration or infiltration, keratitis may be focused in the superficial or deep portions of the cornea. Examination, fluorescein dye test, and corneal cytology and cultures are critical for correctly and completely characterizing keratitis, which in turn determines appropriate medical and/or surgical therapy and prognosis.

CLINICAL SIGNS OF FUNGAL KERATITIS

Clinical presentations of fungal keratitis include superficial ulcerative plaque formation, stromal ulceration, or stromal abscess. In contrast to bacterial disease, which tends to produce circular, uniform lesions, with or without true keratomalacia ('melting'), ulcerative fungal disease often results in more irregular infiltration with stromal loss of variable depths within a lesion. Nonulcerative keratomycosis (stromal abscess) may manifest as discrete yellowish foci at any depth within the stroma, or as a more diffuse infiltration with often marked surrounding edema. Regardless of presentation, corneal vascularization is often prominent and anterior uveitis (miosis, flare, ocular hypotony) may be severe. The degree of anterior uveitis and the degree to which it is controlled by therapy serve as good markers for the severity of corneal disease, with more intense uveitis indicating more severe corneal disease.

DIAGNOSIS OF FUNGAL KERATITIS

Ulcerative keratomycosis is diagnosed by a combination of clinical signs, corneal cytology, and fungal culture. Cytology and cultures may be collected using a sterile Kimura spatula or sterile scalpel blades, and cytology samples are placed directly onto a slide and stained with Diff-Quik or Gram stain – hyphae and variable numbers of neutrophils are diagnostic. Cytology is both sensitive and specific for fungal infection, and provides results within minutes (versus days to weeks for cultures). Typical agents include *Fusarium* spp, *Aspergillus* spp, and *Penicillium* spp., all of which may be part of the normal ocular flora (although *Fusarium* spp are more likely to be acquired from the environment).

MEDICAL TREATMENT

General principles

Medical therapy consists of topical and oral medications to control the infection, inflammation, and pain. In addition to choosing antimicrobial medications which have an appropriate spectrum for the agents involved, other principles of topical ocular therapy to consider include the transcorneal penetration of a drug, the effect of a drug on wound healing, and interactions among individual drugs administered. When ulcerative disease is present, most medications will more effectively penetrate into corneal tissue, potentially affording greater efficacy, while fewer medications can penetrate the cornea through intact epithelium. Individual drugs also may negatively impact epithelial wound healing. Additionally, if drugs are administered in either combination or rapid succession, they may either wash each other out or may inactivate one another based on formulation properties (pH, protein binding, etc). For these reasons, it is recommended that all topical drugs be individually administered through a subpalpebral lavage catheter, pushed through gently with air, and not be allowed to sit within the tubing where they may interact with and bind to one another, or be exposed to inappropriate temperature or lighting conditions.

Topical medications

Topical medications include antifungals, therapeutic antibiotics (to treat or avoid potential concurrent bacterial infections), atropine, and antiprotease medications (see Tables 1 and 2). Antiproteases are used to control corneal digestion which occurs with excessive protease production by inflammatory cells, infectious organisms, and reactive keratocytes. Initially, antimicrobials and antiproteases are administered every 2-4 hours, and the frequency is decreased as a positive response to therapy occurs. Atropine (1%) may be administered every 6 to 12 hours initially, depending upon the degree of flare/hypopyon and miosis, and should be decreased as soon as adequate mydriasis is achieved and maintained. Treatment frequently continues for a minimum of 6 weeks, and owners should be made aware of this duration at the onset of therapy.

Table 1. Topical Ophthalmic Antifungal and Antibiotic Medications (* = preferred therapy at NCSU)

Drug	Spectrum	Transcorneal Penetration
* Natamycin 5% suspension	Excellent	Poor
* Voriconazole 1% solution (reconstitute according to package instructions)	Excellent	Excellent[4]
Itraconazole 1% (compounded in DMSO)	Good	Good[5]
Miconazole 1% (compounded or vaginal cream)	Good	Good
* Moxifloxacin 0.5%	Excellent (G+ and G-) Less effective versus *Pseudomonas* Therapeutic only	Excellent[6]
* Ciprofloxacin 0.3%	Good (G-) Good versus *Pseudomonas* Therapeutic only	Poor[6]
Tobramycin 0.3%	Good (*Staph, Strep, Pseudomonas*) Increasing resistance possible	Good
Triple antibiotic preparations (neomycin, polymyxin B, bacitracin or gramicidin)	Good (G+ and G-) Prophylactic or therapeutic	Poor

Table 2. Topical Antiprotease Medications[7] (* = preferred therapy at NCSU)

Agent	Mechanism of Action	Active Versus
* Autologous Serum -- α_2-macroglobulin -- α_1-antitrypsin	Entrapment of protease	-- serine proteases and MMPs -- serine proteases
* Tetracyclines	Zinc and calcium chelation	MMPs
EDTA 0.2%	Zinc and calcium chelation	MMPs
N-Acetylcysteine 5-10%	Zinc and calcium chelation	MMPs
Ilomostat 0.1%	Zinc and calcium chelation	MMPs

Systemic Medications

Oral (or injectable) medications include antifungal and anti-inflammatory agents, and potentially antibiotics (see Table 3). Antifungal agents and antibiotics are generally administered for a minimum of 3 weeks, depending upon the response to therapy, and antiinflammatories are administered as needed, generally for 2-3 weeks. Gastroprotectants may be administered as necessary as well.

Table 3. Systemic Medications (* = preferred therapy at NCSU)

Drug	Dose (Route)	Key Points
* Fluconazole	14 mg/kg once, then 5 mg/kg q24h (PO)	Good spectrum Good intraocular penetration[8] $
Itraconazole	1.5 mg/kg IV q24h 5 mg/kg PO q24h	Good spectrum Poor intraocular penetration[9] $$
Voriconazole	3 mg/kg PO q12h[10]	Excellent spectrum Excellent intraocular penetration $$$
* Flunixin meglumine	1.1 mg/kg q12h (PO)	Better for ocular disease than phenylbutazone
* TMS	20-30 mg/kg q12h (PO)	
Potassium penicillin	20,000 U/kg q6h (IV)	Administered prior to surgery
Gentamicin	4-6.6 mg/kg q24h (IV)	Administered prior to surgery
Doxycycline	10-20 mg/kg q12h (PO)[11]	Antiprotease activity

Subconjunctival Therapy

While not evaluated for overall efficacy or for potential secondary complications or local toxicities, subconjunctival injections of amphotericin B have been utilized in refractory cases of keratomycosis for which surgical therapy is not an option. Injection of 0.2 ml of a 5 mg/ml solution into the subconjunctival space, once every 48 hours for a total of 3 doses, may be performed in an attempt to control stromal abscesses. Topical administration is not useful due to minimal epithelial penetration.

SURGICAL TREATMENT AND PROGNOSIS

General Principles

Surgical therapy is indicated if the infection either does not improve or worsens; if stromal loss is >50%; or if anterior uveitis is not controlled with medical therapy. Surgical options include keratectomy with adjunctive grafting procedures (conjunctiva, amnion), or keratoplasty procedures (penetrating, posterior lamellar, deep endothelial lamellar). These procedures are best performed under general anesthesia with an operating microscope and microsurgical instruments. Although general anesthesia carries its own set of risks, surgical therapy for infected ulcers generally improves the prognosis and may ultimately shorten the total duration of treatment. A survey of 86 eyes receiving PK for a variety of reasons indicated a positive visual outcome in nearly 80%, with an average healing time of just under 7 weeks.[12] A favorable visual outcome and globe retention were reported in over 90% and 95%, respectively, of 39 cases of ulcerative keratomycosis receiving either medical or medical and surgical therapy, treated for a mean duration of 48 days.[3] Regardless of the projected outcome, owners must be prepared for a long duration of therapy with an uncertain prognosis.

REFERENCES

[1]Moore C, et al. **J Am Vet Med Assoc** 1983;182:600. [2]Utter M, et al. **Eq Vet J** 2009;21:321. [3]Andrew S, et al. **Eq Vet J** 1998;30:109. [4]Clode A, et al. **Am J Vet Res** 2006;67:296. [5]Ball M, et al. **J Am Vet Med Assoc** 1997;211:199. [6]Clode A, et al. **Am J Vet Res** in press. [7]Ollivier F, et al. **Am J Vet Res** 2003;64:1081. [8]Latimer F, et al. **Am J Vet Res** 2001;62:1606. [9]Davis J, et al. **Am J Vet Res** 2005;66:1694. [10]Colitz C, et al. **Am J Vet Res** 2007;68:1115. [11]Davis J, et al. **Am J Vet Res** 2006;67:310. [12]Brooks D, et al. **Vet Ophthalmol** 2008;11:123.

PATHOPHYSIOLOGY & TREATMENT OPTIONS FOR EQUINE GLAUCOMA

Alison B. Clode, DVM, DACVO
Raleigh, NC

BACKGROUND

Glaucoma is a group of diseases, for which an elevated intraocular pressure (IOP) is a risk factor, that produce retinal and optic nerve damage, and subsequent blindness. Although primary glaucoma due to inherent iridocorneal angle (ICA) defects may occur in horses, they are not officially recognized or characterized. Most commonly, glaucoma occurs secondary to anterior uveitis due to a variety of causes, and occasionally in associated with intraocular tumors.

PATHOGENESIS

Aqueous humor (AH) is produced by the ciliary body through three mechanisms: passive diffusion, ultrafiltration, and active secretion, with active secretion by the carbonic anhydrase (CA) enzyme system the most significant contributor. Following formation in the ciliary body, AH flows from the posterior chamber, through the pupil, to the anterior chamber, with integrity of this flow pattern critical for maintaining normal intraocular pressure. AH then exits the eye through the ICA (conventional outflow) or the suprachoroidal and supraciliary spaces (uveoscleral, unconventional outflow), to be resorbed by local vasculature. Normal IOP in the horse ranges from 15 to 25 mmHg, although higher values may be observed in individual horses with otherwise normal ophthalmic exams. Increases in IOP occur when either AH flow within or exit from the eye are obstructed. Blockage of AH flow through the pupil may result when the lens or vitreous protrude through the pupil, or when pre-iridal fibrovascular membranes (PIFVM) or posterior synechiae develop in association with intraocular inflammation. Blockage of AH drainage from the eye may occur when the ICA is obstructed by protein, inflammatory or neoplastic cells, PIFVM, peripheral anterior synchiae, or trabecular meshwork swelling or compression, all of which also may occur in association with intraocular inflammation. Regardless of the underlying cause, these conditions increase IOP, which decreases ocular blood flow and compresses the optic nerve, ultimately perpetuating inner retinal (i.e., ganglion cell) death and associated vision loss.

CLINICAL SIGNS & DIAGNOSIS

Diagnosing glaucoma in the horse is difficult in comparison to dogs, as horses may present with minimal signs of pain or blindness even with significant IOP elevations. It is therefore imperative that clinicians combine information regarding the signalment, history, clinical signs, and ophthalmic exam with a high index of suspicion for the disease. The most common clinical signs include buphthalmos, corneal edema, Haab's striae, and mydriasis, although multiple other signs may be present which support the diagnosis of glaucoma. Tonometry may be readily performed in the horse utilizing either applanation or rebound tonometry, both of which have been validated for use in the horse.[1] Factors that may influence IOP measurements include diurnal variation, sedation, or excessive manipulation of the globe while performing tonometry. IOP is lower during the light phase and higher during the dark phase,[2] although the clinical significance of such variation is unclear. Sedation with xylazine has been documented to decrease IOP by 23 to 27%,[3] while detomidine increased the IOP by an average of 8 mmHg due to the head being below the level of the heart.[4] Excessive pressure on the globe associated with opening the eyelids may also result in a false elevation, indicating that it is often appropriate to perform an auriculopalpebral nerve block prior to measuring IOP.[3] While potential alterations in IOP associated with each of these factors may not be clinically significant when evaluating an individual patient, they emphasize the need for consistency in evaluation technique.

MANAGEMENT

Management of glaucoma depends on the ability to decrease AH formation and increase AH outflow, both of which can be achieved either medically or surgically, in conjunction with identification and correction of the underlying cause. Because of the invasiveness and potential complications associated with available procedures, surgical therapy is generally not performed prophylactically, and is instead reserved for cases refractory to medical therapy.

Medical Therapy (NEW PUBMED SEARCH)

Evaluation of antiglaucoma medications in horses has primarily been performed utilizing normal, nonglaucomatous horses. While such studies serve an important purpose in determining potential utility of individual medications, it is important to note that in most species, the degree of IOP lowering is greater in glaucomatous individuals, indicating that investigation of individual medications in horses with glaucoma is indicated to determine the most effective therapeutic regimens.

Currently, beta blockers and CA inhibitors (CAI) are the primary options for topical medical management of glaucoma in horses. Beta-blockers (timolol maleate 0.5%) decrease AH formation through inhibition of intraocular cAMP production, lowering IOP by 27% (~7 mmHg) in normal horses on a twice-daily dosing schedule.[5] CAIs (dorzolamide 2%, brinzolamide 1%) also decrease formation by inhibiting CA (isoenzyme II) in the ciliary body epithelium. Twice-daily topical administration of dorzolamide, alone or in combination with timolol, reduced IOP by 10% (2 mmHg),[6] while topical brinzolamide decreased IOP by 5 mmHg (21%).[7]

Topical prostaglandin analogues (i.e., latanoprost 0.005%) increase uveoscleral outflow, theoretically having a potentially large effect on IOP in heavily uveoscleral outflow-dependent equine eyes. Once-daily latanoprost, administered to normal horses, decreased IOP by 3 mmHg (17%) in mares and 1 mmHg (5%) in geldings, however side effects (blepharospasm, epiphora, conjunctival hyperemia) were significant.[8] In consideration of the side effects, as well as the likelihood that glaucoma in horses is secondary to uveitis, use of latanoprost in horses is therefore generally not advised. To date, other prostaglandin analogues have not been evaluated in the equine eye.

While administration of mydriatics is generally contraindicated in cases of primary glaucoma, atropine (a therapeutic mydriatic-cycloplegic) has variable effect on IOP in horses. Mydriasis may lead to occlusion of the ICA by the peripheral iris, thereby obstructing drainage, while cycloplegia may relax the uveal tissue and increase uveoscleral outflow. One study of atropine in normal horses documented an 11% decrease in IOP following treatment with topical atropine,[9] while another documented no significant change.[10] Due to these variations, atropine is not recommended as a primary therapy for horses diagnosed with glaucoma; however, it may be administered in cases of glaucoma secondary to uveitis.

Additional medical therapy may include administration of topical corticosteroids or non-steroidal anti-inflammatories, particularly in cases of glaucoma secondary to uveitis. Oral corticosteroids or non-steroidal anti-inflammatories may also be indicated.

Drug(s)	**Indication**	**Recommended Dosing**
Timolol 0.5%	Glaucoma	q8-12h
Dorzolamide 2%	Glaucoma	q8h-12h
Brinzolamide 1%	Glaucoma	q8-12h
Systemic NSAID	Glaucoma (analgesia) Uveitis (anti-inflammatory)	q12h
Topical corticosteroid	Uveitis with Secondary glaucoma	q6-8h
Topical NSAID	Uveitis with Secondary glaucoma	q6-8h
Topical atropine	Uveitis with Secondary glaucoma	q12-24h

Surgical Therapy

Surgical management may be utilized to maintain a positive response to medical therapy and preserve vision, or to restore comfort to a blind eye.

Visual patients may be candidates for transscleral laser cycloablation (TSCP) to decrease AH production by selective destruction of CB epithelium and stroma. Localization of the pars plicata is necessary for effective laser placement, identified at 4-6 mm posterior to the limbus in the dorsotemporal and ventrotemporal quadrants.[11,12] The diode laser wavelength of 780-850nm is better absorbed by melanin and can be utilized with lower total energy doses (~**150J/eye**) than the Nd:YAG laser wavelength of 1064 nm (~**250J/eye**) to achieve similar clinical effect. Based on desired histologic damage produced by diode TSCP in normal eyes, a starting energy of 2.25J/site (1500 mw x 1500 msec) for 60 sites (avoiding 3 and 9 o'clock on the globe) is considered appropriate.[12] Clinically, 42 eyes of 36 horses treated with diode TSCP (mean 183J/eye, range 99-375J) followed as long as 68 weeks, had good IOP control and 59% were sighted at final follow-up, however 100% (8/8) of eyes surviving to 20-68 weeks required medical therapy.[13] Evaluation of Nd:YAG TSCP in 23 eyes of 16 horses identified a 70% success rate, defined as IOP <30 mmHg, for eyes surviving longer than 20 weeks post-procedure.[14] Of those eyes, 60% were visual, and 45% were not receiving topical ocular anti-glaucoma medications. Common complications of both lasers include hyphema (11%), conjunctival hyperemia (22%), and corneal ulceration (13%). As determined by this study, an appropriate protocol was a power setting of 11W, duration 0.4 sec, for 60 sites (264 J/eye).

Additional procedures which have not yet been clinically evaluated in horses but have shown promise in canine patients include placement of gonioimplants and use of endoscopic cyclophotocoagulation (ECP). Gonioimplants are shunts which divert AH from the anterior chamber to the subconjunctival space, from where it is absorbed into peripheral circulation. These shunts are commonly placed in dogs during the same surgical procedure as TSCP is performed, ideally increasing the efficacy of both procedures. Complications associated with gonioimplants as reported in dogs include shunt occlusion, cataract development, and endothelial touch.[15] No published reports exist on use of this procedure in the horse, however. ECP combines an intraocular camera with a laser, allowing precise targeting and visualization of the ciliary processes as they are ablated with the laser. Complications may include intraocular hemorrhage, cataract development, and phthisis, however no published reports exist on use of this procedure in veterinary patients.

Intravitreal gentocin injection, enucleation, or evisceration may be performed in permanently blind, painful eyes.

PROGNOSIS

The prognosis for vision and globe retention is largely dependent upon the ability to identify and control the underlying cause, which is commonly uveitis. In such cases, accompanying ocular changes such as cataracts, lens luxation or subluxation, vitreal degeneration, and retinal inflammation, may have an equally deleterious impact on vision, as well as on the ability to control underlying uveitis.

REFERENCES

[1]Knollinger AM, et al. **Am J Vet Res** 2005;227(2):244. [2]Bertolucci C, et al. **Chrono Int** 2009;26(2):348. [3]van der Woerdt A, et al. **Am J Vet Res** 1995;56:155. [4]Komaromy AM, et al. **Am J Vet Res** 2006;67:1232. [5]van der Woerdt A, et al. **Vet Ophthalmol** 2000;3:165. [6]Willis AM, et al. **Am J Vet Res** 2001;62:709. [7]Germann SE et al. **Eq Vet J** 2008;40:662. [8]Willis AM, et al. **Am J Vet Res** 2001;62:1945. [9]Herring IP et al., **Vet Ophthalmol** 2000;3:139. [10]Mughannam AJ, et al. **Vet Ophthalmol** 1999;2:213. [11]Miller TL, et al. **Vet Ophthalmol** 2001;4:183. [12]Morreale RJ, et al. **Vet Ophthalmol** 2007;10:84. [13]Annear MJ, et al. **Vet Ophthalmol** 2008;11:424. [14]Whigham HM, et al. **Vet Ophthalmol** 1999;2:243. [15]Sapienza JS, et al. **Vet Ophthalmol** 2005;8:121.

PATHOPHYSIOLOGY & TREATMENT OPTIONS FOR EQUINE INFLAMMATORY CORNEAL DISEASE

Caryn E. Plummer, DVM, DipACVO
Gainesville, FL

BACTERIAL ULCERS

Corneal ulcers infected with a bacterial agent can be very serious and may manifest with extensive cellular infiltrate, melting or sloughing of corneal tissue and occasional abscessation. The most important route of therapy is antimicrobial medication directly aimed at the causative or complicating microbe. Culture and sensitivity testing is imperative, since so many "bugs" are exhibiting resistance to even our "big gun" back-up drugs. Initial empiric therapy should be chosen after samples from a corneal scraping have been examined for cytologic evidence of infection and the morphologic appearance of the infectious agents. Anti-collagenolytic agents are also critical to successful treatment of these infected wounds due to the upregulation of corneal destructive/resorptive house-keeping enzymes. Autologous serum or plasma, EDTA, acetylcysteine, doxycycline and ilomostat are the current favorites.

FUNGAL KERATITIS

Please see discussion by Dr. Clode.

HERPETIC KERATITIS

Multiple, superficial, white, punctate or linear opacities of the cornea, with or without fluorescein dye retention, are found associated with equine herpes virus 1 and 2. The focal punctate corneal opacities may be found at the end of superficial corneal vessels, and may retain rose Bengal stain. Varying amounts of ocular pain, conjunctivitis, and iridocyclitis are present. Multiple animals in a herd or individuals that regularly congegrate with other horses such as on the show circuit may be affected. Topically administered idoxuridine and trifluorothymidine (TID) have been used with topical NSAIDs for treatment of equine herpes ulcers, but recurrence is common. Lysine administered 20-30gm/day PO may be helpful.

STROMAL ABSCESSES

Focal trauma to the cornea can inject microbes and debris into the corneal stroma through small epithelial ulcerative micropunctures. A corneal abscess may develop after epithelial cells adjacent to the epithelial micropuncture divide and migrate over the small traumatic ulcer to encapsulate infectious agents or foreign bodies in the stroma. Epithelial cells are more likely to cover a fungal than a bacterial infection. Reepithelialization forms a barrier that protects the bacteria or fungi from topically administered antimicrobial medications. Reepithelialization of stromal abscesses interferes with both routine diagnostics and treatment.

Corneal stromal abscesses can be a vision threatening sequelae to apparently minor corneal ulceration in the horse. A painful, blinding chronic iridocyclitis may result. Most stromal abscesses involving Descemet's membrane are fungal infections. The fungi seem "attracted" to the type IV collagen of Descemet's membrane. Both superficial and deep stromal abscesses do not heal until they become vascularized. The patterns of corneal vascularization are often unique suggesting that vasoactive factors are being released from the abscess that influences the vascular response.

Medical therapy consists of aggressive use of topical antifungals and antibiotics, topical atropine, and systemic NSAIDs. Superficial stromal abscesses are more likely to respond positively to medical therapy than deeper lesions. If reduced inflammation of the cornea and uvea are not found after several days of medical treatment or if the uveitis worsens despite appropriate therapy, surgical removal of the abscess should be considered. Deep lamellar (DLEK), posterior lamellar (PLK) and penetrating keratoplasties (PK) are utilized in abscesses near Descemet's membrane, and eyes with rupture of the abscess into the anterior chamber. Surgical removal and corneal transplantation eliminates sequestered microbial antigens, and removes necrotic debris, cytokines and toxins from degenerating leukocytes in the abscess.

Eosinophilic keratoconjunctivitis

Has an unknown etiology, but may be an immune-mediated disease. All ages and breeds of horses can be affected with many cases reported in the spring. Clinical signs include corneal granulation tissue, blepharospasm, chemosis, conjunctival hyperemia, mucoid discharge, and corneal ulcers covered by raised, white, necrotic plaques. Eosinophilic keratoconjunctivitis can resemble a corneal tumor (e.g. SCC) in appearance. KCS may develop in affected horses due to lacrimal gland inflammation. The lacrimal gland should be palpated to detect swelling.

Corneal cytology typically contains numerous eosinophils and a few mast cells to rule out similar appearing infectious and neoplastic causes. Superficial lamellar keratectomy to remove plaques speeds corneal healing. Topical corticosteroids (1% prednisolone acetate or 0.1% dexamethasone) 4 to 6 times a day in early stages (in spite of corneal ulcerations – but with cautious monitoring), antibiotics (e.g., bacitracin-neomycin-polymyxin), 1% atropine, and 0.03% phospholine iodide (BID) in combination with systemic nonsteroidal antiinflammatory drugs are indicated. Topical cromolyn sodium (4.0% TID) or lodoxamide (0.1% TID), mast cell stabilizers, can also aid healing. Systemic corticosteroids may be necessary. Horses with EK should be dewormed twice with ivermectin 10 days apart. These lesions are typically slow to heal. Scarring of the cornea occurs.

IMMUNE-MEDIATED KERATITIS

Nonulcerative keratitis is a recognized primary corneal disease syndrome with variable subsets of clinicopathogenesis and clinical appearance. Recent convention describes a group of inflammatory conditions that are seemingly immune-mediated in origin or progression referred to as immune-mediated keratitis or IMMK. It seems that even within the subsets that will be introduced below, there is considerable variation in the clinical appearance and response to therapy in different geographic regions. This may indicate that different populations have different genetic predispositions or that there are different environmental factors influencing or precipitating the disease process. It is presumed that IMMK is a non-infectious process; however, evidence to support or refute this claim is lacking. Thus far histologic examination and microbial cultures have failed to identify an infectious agent. In all types of IMMK, signs of uveitis or severe discomfort are not usually observed.

Superficial IMMK is characterized by a non-painful waxing and waning course of chronic opacity that is subepithelial, superficial, white to yellow infiltrate surrounded by superficial branching corneal vascularization. This form can usually be controlled with topical steroids or cyclosporine. Superficial keratectomy appears to be useful to decrease autoantigen load and decrease recurrence.

The mid-stromal form appears as midstromal cellular infiltrate with mild, surrounding corneal edema and vascularization. The cellular infiltrate is deeper and denser than the superficial form and the cornea is generally more opaque. Vascularization tends to be less branching and straighter. Treatment is the same as for superficial IMMK, but tends to be less responsive.

Deep or endothelial disease appears as endothelial cellular infiltrate with diffuse corneal edema. Also called endotheliitis, this form is characterized by a chronic, slowly progressive, nonpainful patch or full-thickness corneal edema. Pale green lacunae may develop and bullous keratopathy may also occur and result in superficial focal or multifocal ulcerations. Endothelial disease is the least amenable to therapy. There have been recent anecdotal reports of response to the topical NSAID bromfenac. Topical hypertonic saline may help decrease the accumulation of edema.

REFERENCES

[1.] Matthews A, Gilger BC. **Vet Ophthalmol** 2009; 12: 10. [2.] Andrew SE, Willis AM. **Equine Ophthalmology** 2005: 157. [3.] Gilger BC, Michau TM. **Vet Ophthalmol** 2005; 8: 233. [4.] Baker A, Plummer CE. **Vet Ophthalmol** 2008; 11: 381. [5.] Ollivier FJ. **Vet Ophthalmo** 2007; 10: 199. [6.] Brooks DE. **Vet Clin North Am** 2004; 20: 345. [7.] Yagamata M. **J Am Vet Med Assoc** 1996; 209: 1283. [8.] Matthews A. **Equine Vet Ed** 2000; 12: 271.

PATHOPHYSIOLOGY & TREATMENT OPTIONS FOR EQUINE RECURRENT UVEITIS

Caryn E. Plummer, DVM, DACVO
Gainesville, FL

By definition, uveitis is inflammation of the uvea of the eye (iris, ciliary body and choroid). The complex of diseases known as "Equine Recurrent Uveitis" (ERU) refers to intraocular inflammation that recurs or persists causing various degrees of inflammation, scarring, degeneration and dysfunction of multiple components of the eye. Several classification schemes are used to differentiate subsets of clinically observed disease. Some experts prefer to refer to the syndrome as "Equine Persistent Uveitis" as inflammation is detectable on a cellular basis in all of these conditions, including quiescent and insidious cases. Miosis is a hallmark of most acute forms of recurrent or persistent uveitis. Aqueous flare may be the first sign to appear and the last clinical sign to disappear.

CLASSIFICATION BY OBSERVED INFLAMMATION OVER TIME

Classic uveitis

Horses demonstrate repeated bouts of severe inflammation and pain in one or both eyes. In between bouts, the eye(s) appear comfortable. Horses that are experiencing their first bout of classic uveitis are called *primary* cases. As not all uveitis is recurrent, a case is not termed "recurrent" until two or more classic episodes have occurred.

Insidious uveitis

Horse appears normal to the owner and the animals do not exhibit overt ocular pain. Most often seen in Appaloosas, draft horses or European warmbloods. Examination of the globe shows progressive deterioration of numerous ocular structures that is progressive over time. Pupils may be slightly miotic. Subtle signs include corneal haze, slight aqueous flare, muddy iris color, slight miosis, iris rim atrophy, corpora nigrans atrophy, cataract, low grade vitreous haze and retinal scarring.

CLASSIFICATION BY STAGE OF DISEASE AT TIME OF EXAMINATION

Acute cases are horses who are suffering from a flare up of classic uveitis. Signs of an acute bout include pain, lacrimation, photophobia, chemosis, conjunctival hyperemia, corneal edema, corneal vascularization, dot like keratic precipitates on the corneal endothelium, aqueous flare, hyphema or hypopyon, miosis, vitritis and ocular hypotension (pressure of 10-12 mm Hg).

Quiescent cases are horses in a "calm" time of their cycle of repeat inflammation. Slit lamp examination will show subtle flare in the anterior chamber, indicating low grade persistent inflammation. Ocular examination may reveal scarring or signs of previous disease including chronic corneal edema, iris atrophy, iris color change, synechiae, pigment rests on the lens, cataract, densities or haze in the vitreous, or scarring around the optic disc. Optic disc scarring is seen as either a focal "bullet hole" pattern of numerous tiny scars, or a "butterfly" pattern where wing shaped islands of depigmentation flank either side of the optic disc.

End stage uveitis cases have eyes that have undergone severe degeneration and chronic scarring. They may show extensive corneal scarring, circumfrential synechiae, dense cataract, lens luxation or subluxation, secondary glaucoma, retinal detachment or phthisis bulbi. End stage uveitis is associated with vision loss.

PATHOPHYSIOLOGY

Uveitis begins with compromise of the blood ocular barrier. This barrier normally functions to keep the aqueous and vitreous clear as tight fenestrations between the cells of ocular capillary walls prevent cells and large molecules from passing through the blood vessels of the iris and the choroid. The blood ocular barrier also serves to isolate intraocular structures from the normal traffic of immune surveillance, making the tissue of the inside of the eye an immune privileged site.

In acute uveitis, the uveal blood vessels in the iris, ciliary body and choroid thicken and become congested. Soon they become "leaky", and cells and inflammatory mediators cross the compromised blood ocular barrier and enter the inside of the eye. Most of the cells that cross the barrier initially are neutrophils. This is seen grossly as hypopyon, aqueous flare, and vitreous haze. Neutrophils that enter the eye are soon replaced by large numbers of lymphocytes, some of which infiltrate the connective tissue of the ciliary body and iris, forming large follicle like clusters. Antibodies and inflammatory cytokines are detectable inside the eye and within ocular tissues. These substances probably react with host and (in some cases) infectious factors to contribute to ongoing pathologic change. Numerous heavy exudates appear on intraocular tissues, most notably on the posterior surface of the iris and ciliary body, on the capsule of the lens, and in the layer between the retinal pigmented epithelium (RPE) and the photoreceptors of the retina. The exudates interfere with the function of the ocular tissue they adhere to. Cytokine activity mediates tissue destruction. With repeated or persistent inflammation the various chronic changes described in blue above occur. Vision loss results when dense cataract and synechiae obscure acuity, when the retina detaches or degenerates and no longer can transmit processed light signals to the brain or when glaucoma causes ischemic damage and degeneration of the optic nerve.

ETIOLOGY

Decades of research have substantiated that recurrent uveitis is an immune mediated disease. However some bacterial, viral and parasitic infections, as well certain host conditions have been associated as triggering events for the syndrome. Factors include bacterial infections (leptospirosis, *Borrelia burgdorferi*, brucellosis, *Streptococcus*, *Rhodococcus equi*, generalized septicemia), viruses (influenza, equine viral arteritis, parainfluenza), parasites (Onchocerciasis, *Strongylus*, Toxoplasmosis) and certain host conditions (tooth root abscess, severe hoof abscess, septicemia, and severe trauma). Of all the possible infectious triggers, leptospirosis is the most significant worldwide.

Leptospirosis and Uveitis: The most significant serovars associated with disease are *L. interrogans* serovar *Pomona* (seen often in the USA) and *L. interrogans* serovar *Grippotyphosa* (seen often in Germany and central Europe) Factors that increase the risk of leptospirosis in horses include pasture access to cows, pigs or deer, close proximity to streams or ponds frequented by the same, use of piped pond water for drinking water, heavy infestation of stable with rats and a rainy season with persistence of ground water. Horses become infected when they drink water contaminated by the urine of a carrier animal (often a cow, deer, pig or rat). The spirochete gains access to the horse's bloodstream by mechanical penetration of mucous membranes. Bacteremia results in clinical illness, manifested by anemia, fever and flu like symptoms. Clinical disease is mild and self limiting, and rarely diagnosed as a leptospiral infection when acute. Resolution of signs does not mean elimination of the bacteria from the body—the spiral organisms often colonize the kidneys of the horse and may persist for a few months, being shed in the urine. Ocular signs of classic uveitis associated with leptospirosis rarely occur during acute leptospiral infection. Rather they begin months later. Initial signs subside with or without therapy but recur at unpredictable intervals. Subsequent episodes of inflammation may be more or less severe than the initial one. A unified theory is yet to appear to explain all the ocular events that accompany leptospiral associated uveitis.

- Is it a direct toxicity of an intraocular infection with the spirochete?
- Is it an autoimmune disease against "self" that is triggered by molecular mimicry between leptospira and host tissue?
- Are leptospira somehow modulating the immune response of the eye?

These questions continue to challenge researchers. Although systemic infection with pathogenic strains of leptospirosis is clearly a common trigger for vision threatening recurrent uveitis, It is likely that the genetic makeup, specifically that of the MHC complex, of the individual horse that contracts the disease plays a role in determining both susceptibility to leptospirosis as an inciting trigger, and severity of subsequent inflammatory episodes.

BREED AND UVEITIS

Recent work has also shown that certain breeds are at risk for uveitis, most notably Appaloosas, European warmbloods and draft horses. The Appaloosa breed is 8.3 x more at risk than other breeds for uveitis. Appaloosas that have insidious disease often have overall roan or light coat colors rather than dark coats with a rump blanket. The skin around the lids of affected Appaloosas is often mottled or pink in pigmentation. Mane and tail hair may be sparse. It is theorized that these horses have genetic proclivity to uveitis due to aberrations in the Major Histocompatibility Complex, specifically in their Equine Lymphocyte Antigen subtype. Recent research from Germany has supported this concept in German warmbloods susceptible to disease.

THERAPY OF UVEITIS

- Mydriasis is essential therapy for all cases of acute uveitis. Initial application of atropine should be frequent until pupil is fully dilated. Severe cases may show poor response to the action of mydriatics.
- Topical corticosteroids and systemic NSAIDS are the core elements of anti-inflammatory field therapy for acute attacks. Therapy should be intense for about two weeks and may be tapered over another two weeks. Subconjunctival and/or systemic corticosteroids are indicated in severe cases.
- Corticosteroid topical therapy may induce calcific band keratopathy, especially in horses with leptospiral associated uveitis. Treat wi†h EDTA chelation/keratectomy.
- Horses with acute uveitis are very painful and may suffer secondary corneal ulcers from self trauma. Corticosteroids are contraindicated in these cases—use topical NSAIDs instead and treat for any infection/collagenolysis of the ulcer.
- Anecdotal reports indicate that a one month course of oral doxycycline or enrofloxacin may help cases that are associated with leptospiral infection.
- Acupuncture therapy may help moderate the frequency or severity of episodes
- Therapy does little to alter the progression of disease in horses with insidious uveitis.
- Some horses (especially Appaloosas) develop secondary glaucoma late in their disease. Glaucoma therapy is difficult, but Timolol, Dorzolanide or a combination of these drugs (Cosopt) may be tried. Judicious topical steroids may help as well.
- Consider referring ERU cases to a specialty equine ophthalmology referral center for suprachoroidal cyclosporine implant surgery. Best candidates are classic cases: horses early in the course of the disease that have experienced only a few "attacks" and show little or no permanent ocular scarring. Cost is around $2500.
- Although horses in central Europe routinely undergo pars plana vitrectomy for ERU, results have been poor with this procedure in the US and the procedure is not often advised here.
- Intravitreal injection with an antimicrobial agent like gentamicin or an immune-suppressive agent (dexamethasone, sirolimus)

PROGNOSIS OF UVEITIS

Visual prognosis for horses suffering from multiple acute attacks of uveitis or insidious chronic disease is ALWAYS guarded. Data on the incidence of blindness in uveitic horses is lacking, but it is clear that uveitis is the leading cause of blindness in horses worldwide.

Future understanding of recurrent uveitis in all species will revolve around the investigation of several key questions:

1. What genetic makeup predisposes individuals to develop disease?
2. If the disease is autoimmune, which autoantigens participate in the initiation and perpetuation of inflammation?

3. What immune mechanisms initiate the immune response and mediate tissue destruction?

Study continues to determine the genetic predisposition for the syndrome in certain horse breeds (i.e., Appaloosas and German warmbloods). If a genetic marker is associated with susceptibility, then horses with this genotype could be identified and excluded from use for reproduction thus decreasing the prevalence of the disease. New immunosuppressive therapies, such as tacrolimus or rapamycin (sirolimus), may offer hope in the medical management. Perfecting a device to deliver such a medication may also be feasible. Studies are also being done to determine the role of leptospirosis or other microorganisms in the initiation and pathogenesis of ERU. Effort continues to further quantify the immune events that characterize inflammation and mediate recurrence.

REFERENCES

[1.] Dwyer A, Gilger BC. **Equine Ophthalmology** 2005: 285. [2.] Douglas LC. **J Vet Pharmacol Therap** 2008; 31: 511. [3.] Deeg CA. **Vet Ophthalmol** 11: 61. [4.] Deeg CA. **Ophthalmic Res** 2008; 40: 151 [5.] Divers TJ. **Equine Vet J** 2008; 40: 167. [6.] Gilger BC. **Am J Vet Res** 2008; 69: 1329. [7.] Gilger BC. **Invest Ophthalmol Vis Sci** 2006; 47: 2596.

EVALUATION OF HEART RATE & RHYTHM IN THE EXERCISING HORSE

M. Kimberly J. McGurrin, BSc, DVM, DVSc, DACVIM
Guelph, ON, Canada

The cause of poor performance in the equine athlete is often difficult to determine. This is especially true for racehorses where the poor performance may be detailed by slowing of a few seconds or is intermittent in nature. Multiple findings commonly present themselves on the performance exam and the relative contribution of each is difficult to elucidate. Often heart rhythm abnormalities are investigated based on elimination of other causes or suspicion based on performance analysis. Detailed investigation in conditions that are as near an approximation of precipitating conditions may be required.

Heart monitors, telemetry units and Holter monitors may all be used in the monitoring and investigation of fitness and poor performance. Each has its advantages and detriments, and each may be more suitable for certain applications than others.

HEART RATE MONITORS

Heart rate monitors use proprietary algorithms to detect the heart rate signal. Most of their data is based on human ECG morphology. These algorithms include such information as maximum or minimum signal in a section of ECG (presumed to be a "QRS"), slope calculations and filters to eliminate noise and detection of T waves. Particularly in the past, there were serious concerns about the accuracy of heart rate monitors for equine patients. In part, this was associated with the algorithm interpreting the equine T wave as a QRS complex. Recent evaluations comparing heart monitors that are adapted for equine use to simultaneous ECG recordings show them to be relatively accurate, especially at mid-range heart rates. A number of trainers now have access to heart monitors and use them for fitness monitoring of their horses. Some use monitors that record heart rate response to wrist recorder and then analyze post-work, others will monitor the horse as it works (via telemetry). Many don't use a V200 model to follow fitness, but rather compare response to previous training demands and compare horses to each other. Some heart rate monitors have GPS capabilities, which will allow velocity calculations, eliminating stopwatch velocity calculations.

In the clinical setting heart rate monitors are best used as an adjunct to other means of evaluation for performance. They cannot be used to evaluate cardiac abnormalities without further investigation. During an exercise test, heart rate monitors prove invaluable in ensuring that the horse is exercising to a specified level and allows for standardization of testing, more so between tests on the same animal. They are valuable in fitness monitoring and in evaluating heart rate response to exercise. They are useful in treadmill airway evaluations to ensure that the horse is exercised sufficiently. Obvious variations from anticipated ranges should prompt investigation with other devices (e.g. heart rate of 200 at a jog). They are not of value in detecting arrhythmia, where abnormal beats will be interpreted as noise and the heart rate will not register within the anticipated range.

TELEMETRY (ECG)

Telemetry ECG recordings have the added benefit of recording the actual ECG signal. This allows rate and rhythm evaluation. Computer recording permits later detailed analysis. Multiple monitors may be able to simultaneously record to the same computer (4-6). Some of these systems have software that can be used on hand-held computers, which would allow for more portability and field applications. The software often has difficulty in interpreting abnormal beats and visual evaluation of the entire recording is best. Telemetry units are excellent devices for treadmill application and for field application where the horse can be followed or is in relatively close proximity. Otherwise the range of these devices needs to be considered. Most have a range of 100 meters.

In the clinical setting they have many of the same capabilities of heart rate monitors, and since they do not have to print and yet can be continuously monitored they can be used as a heart rate monitor. They therefore can be used in exercise tests and in airway evaluations to ensure adequate exercise levels. At the same time heart rhythm is recorded and the accuracy of heart rate detection can be checked. These units are especially valuable when dysrhythmia is considered as a potential cause for poor performance, or for performance evaluations when the cause is unknown, where multiple tests can be performed at once.

HOLTER MONITORS

Holter monitors similarly will record the ECG signal. They are not constrained by range. They are limited by the fact that for most devices, once they are recording one is unable to check the quality of the ECG and therefore are uncertain whether the ECG recording has remained diagnostic. The units are self-contained and therefore there is no constraint on how many devices are recording at one time, which can be a concern with telemetry devices (more of a concern for research applications). The issues with software identification of abnormal beats remain; particularly with the software that is designed for human application, which will misinterpret forms. Therefore one has to evaluate the entire recording "manually".

TREADMILL OR "TRACKSIDE"

Treadmill testing is valuable because the testing can be standardized. The speed can be controlled, there are no variations in track conditions and all animals are subjected to the same environmental conditions. In this situation a fixed ECG could be used if one did not have access to telemetry or to Holter equipment. Questions remain whether the test is sufficiently similar to track conditions for poor performance analysis, and how much of a role the many factors associated with race environment play. For example, in evaluating heart rate response to exercise a rider may influence how the horse responds. So might the track condition, and the conditioning of the horse to the track or treadmill, and the excitability of the individual horse. All of these have been shown to lower V200. This also demonstrates the importance of not using a single test to evaluate fitness.

SIGNAL QUALITY

With all modalities artifact due to movement is a major factor in determining signal quality. With heart rate monitors the heart rate will be inaccurate. With telemetry and Holter recordings poor electrical contact resulting in artifacts makes interpretation of the tracing difficult. Electrodes should be positioned to minimize motion artifact. Options for electrode positions include placement under a girth in pouches or the use of glue or suture.

A routine base apex lead may be used if the electrodes are glued. With this positioning the wires are more exposed than if the electrodes are under the girth. This may be a factor when considering placement of rider and harness. If electrodes are placed under the girth there is the possibility of the girth moving and causing artifact. The girth needs to be sufficiently tight to hold the electrodes in place. Individual variation due to conformation of the animal will play a role and ensuring electrode contact may necessitate different positioning especially for electrodes near the withers. Larger surface area electrodes may help because they will maintain some contact even with motion.

Selection of electrode material may aid in eliminating artifact. While some of the recording devices come with their own electrodes, some are disposable, and this may be costly or these electrodes may not be the most efficient at obtaining high quality signals. Adhesive electrodes are small and the adhesive material often has to be modified by the addition of stronger materials.

HEART RATES WHILE RACING

In 2007 at a Standardbred track in Ontario, the author co-conducted a study investigating heart rate and rhythm responses during live race conditions. Purpose designed surcingles with repeated-use electrodes were placed under the race harness. Holter monitors were placed on all horses in two races a night two nights a week during the summer race meet and 345 Holter recordings were obtained. Several horses were repeatedly monitored resulting in 288 horses being sampled in total. The monitors were placed from the time the horses were initially harnessed before first warm-up until the harness was removed post-race. All horses completed the scheduled races and all horses raced subsequently.

Analysis is ongoing at the time of proceedings submission. All recordings are being evaluated for heart rate response and heart rhythm both during and after exercise periods.

REFERENCES

[1.] Scheffer CW et al. **Vet Rec** 1995;137(15):371. [2.] Martin, B.B. et al. **J Am Vet Med Ass** 2000 ;216: 554. [3.] Kingston JK et al. **Eq Vet J Suppl**. 2006 ; 36:106-9. [4.] Jose-Cunilleras E, et al. **Eq Vet J Suppl.** 2006; 36:163. [5.] Linder AE. **J Anim Sci** 2009 Nov E-pub. [6.] Buhl R. **Vet J.** 2009 Dec 24. E-pub.

SPINAL CAUSES OF POOR PERFORMANCE IN THE SPORT HORSE

Carter E. Judy, DVM DACVS
Los Olivos, CA

INTRODUCTION

Back and spinal problems in the horse are a significant cause of poor performance and lost use in horses. The incidence of back pain in the horse is really unknown and the degree it affects horses has not been well studied. Since many horses with primary back pain present without an overt lameness, the condition can be confusing and difficult to diagnose for the veterinarian. Definitive diagnosis of back problems can be difficult given the vague clinical signs and lack of good pathologic reports.[1-4] This has led to a lot of misinformation, misunderstanding and confusion about the prevalence, diagnosis, and management of spinal disorders and back pain in horses. Given the large number of people with back pain, many owners correlate their own experiences to their animal's clinical signs causing for even further confusion and misinterpretation. Many horses perform well with underlying, undiagnosed back problems, while others perform poorly with similar issues. The purpose of this presentation is to discuss the diagnosis, and treatment options for spinal causes of poor performance in the sport horse.

ANATOMY

The axial skeleton of a typical horse follows the formula (C7, T18, L6, S5, Cd15-21). Variations in anatomy do occur with different breeds (e.g. some Arabians will have 5 lumbar vertebrae). The purpose of the axial skeleton is threefold; 1) protect the spinal cord, 2) provide support for soft tissue attachments, 3) flexibility for movement and locomotion. Failure of any one of these purposes can result in a reduction in performance and function resulting in clinical signs.

DIAGNOSIS

Challenges

The diagnosis of back pain in the horse can be challenging. This is mostly due to the fact that back pain is most commonly manifested as decreased performance rather than overt lameness. Since many horses perform adequately with back pain, it is often underreported or not noticed as a clinical entity easily. On the other hand, given the nature of some horses to be flighty and "thin-skinned" and hyper-responsive to palpation to their backs, an evasive response may be falsely interpreted as being back pain instead of a behavioral abnormality. This duality in clinical signs makes interpretation difficult to determine if the problem is real or simply temperamental. Another example is the "cold backed" horse who has transient stiffness when saddled or cinched. This apparent hypersensitivity over the back results in hyper-flexion and stiffness that resolves after several minutes of riding, in some cases the horses may even buck or throw their riders. Most times there are no other clinical signs associated with the back pain to explain it. Whether this is simply temperament or a clinical entity is a challenge to determine and can require of effort to differentiate pathology from personality.

Types of Back Pain

Back problems can be broken into three major categories; Primary back problem, secondary back problem and presumed back problem.[4] A primary back problem is one directly related to a pathologic problem within the axial skeleton or associated soft tissues. Examples of primary back problems include:

1) Soft tissue injuries: Supraspinous ligament desmitis, sacroiliac desmitis, etc…
2) Osseous injuries: impinging dorsal spinous processes, vertebral fracture, spondylosis, osteoarthritis, etc…
3) Neurologic disorders: EPM, EDM, EHV-1, EMND
4) Tack associated
5) Idiopathic

Most primary back problems do not result in overt lameness in the traditional sense of being consistently lame so as to be graded using conventional grading systems. Most primary back lesions result in poor performance or decreased performance compared to previous sporting events.

A secondary back problem is one that is the result of another issue within the horse. A common cause of secondary back pain is osteoarthritis of the lower hocks joints (distal inter-tarsal and tarso-metatarsal joints). Since lameness is not a typical feature of primary back problems, the presence of lameness with a suspected back injury should be carefully evaluated for concurrent pathologies in other regions besides the back.

Presumed back problems may be the most difficult to deal with since many times they are a perception of a problem by the owner, rider or trainer that may not be a problem at all. Lack of ability, poor training and lack of fitness are often perceived as a spinal problem or back pain when in reality no pathology exists. All equine athletes are horses, but not all horses are necessarily athletic. Since back problems are difficult to decipher to begin with, this conundrum of the belief of a back problem being present causes a real challenge to the veterinarian.

Clinical Evaluation

The clinical examination is often one of the most telling aspects for identifying back issues in performance horses. Many horses will exhibit temperament changes that cause them to respond erratically to previously innocuous stimuli. Pinning of the ears, swishing of the tail, erratic behavior or other strange behavior are often telltale signs of back pain. Conformation should be thoroughly evaluated. Horses with alterations in posture, muscle asymmetry, muscle atrophy, osseous asymmetry or generalized back atrophy often are indicative of an underlying primary back or spinal problem. The tack should be carefully evaluated for proper fit. A properly fitted saddle needs little extra padding. If multiple pads and wedges are used to alter the fit of a saddle, it may likely be the cause of the back discomfort. Gait evaluation should be performed under all gaits both with and without a rider. A complete orthopedic and neurologic evaluation should be performed to rule out secondary causes of back pain since many hind limb lameness conditions result in secondary back pain. A rectal examination should be included in all horses with back pain to evaluate for osseous changes, hypertrophy and atrophy of the sub-lumbar musculature as well as for regions of pain and discomfort.

The use of diagnostic anesthesia can be utilized, but care should be taken to avoid anesthetizing vital nerves. Inter-spinous anesthesia, sacroiliac anesthesia and deep epaxial muscular anesthesia are the most commonly performed and can be enlightening as to the source of discomfort. It can be difficult in many cases of overriding dorsal spinous processes to place the needle between the dorsal spinous processes. A para-median approach can be utilized to access these regions of question. Dorsal articular facet anesthesia (using ultrasound guidance) and sacroiliac anesthesia can also be useful in localizing the specific region of pain in the cervical and sacro-iliac regions.

Diagnostic Imaging

Radiographic evaluation of the cervical spine is relatively straightforward and can be performed standing. Oblique projections of the cervical spine can be useful in identifying dorsal articular facets that may have significant osteoarthritis. Standing radiographs of the thoracic vertebrae are possible and best achieved utilizing high kvP and mas protocols to image the thoracic spine. However portable units with high quality digital or high-speed plain film systems can routinely image the dorsal spinous processes as well. Complete radiographic evaluation of the caudal thoracic, lumbar spine, sacroiliac region and pelvis is extremely challenging standing. It can be performed under general anesthesia (as well as the other locations), but the value of these radiographs is questionable in cases with subtle clinical signs and should likely be reserved for severe cases or those where fracture is suspected.

Nuclear scintigraphy is a very useful diagnostic modality to identify issues with the spine of the horse. This involves the injection of a radiopharmaceutical with a high affinity for bone remodeling and inflammation. The benefit of scintigraphy is that is performed standing and can image the entire horse. The negatives of scintigraphy are the cost ($1000-$2000), relative availability of the machines, and the lack of sensitivity for discrete soft tissue injuries. In this author's opinion, scintigraphy is an excellent way to image a horse with back, spinal or sacroiliac pain and can be a very sensitive technique.

Ultrasound of the back can be useful in imaging those structures that are visible. Many ligaments and soft tissues lie behind bony structures that make interpretation of some structures quite difficult. While the dorsal portions of the sacroiliac ligament and supraspinous ligaments are visible with ultrasound, the deeper structures are much more difficult to image and interpretation should be made with caution. Many pelvic and vertebral articulations can be visualized as well.[5,6]

Thermography is often used in the evaluation of back pain and spinal problems. Its usefulness is limited to those structures superficial to the axial skeleton. Interpretation of thermographic images takes serial evaluations utilizing the proper equipment and with a person familiar with the interpretation of thermographic images. Thermography is quick, safe and non-invasive. It can provide useful information and some research indicates its usefulness in the diagnosis and management of back pain in the horse.[7]

CLINICAL ENTITIES

The list of possible injuries and diagnoses is long, but several conditions seem to occur with the most prevalence. Below is a partial summary of common injuries based on broad categories. This list is not all-inclusive:

1) Muscle strains: damage to the epaxial musculature is reported to be the most common cause of back pain in the horse.[4]
2) Ligamentous damage: these injuries can be mild, moderate or severe and can affect any ligament involving the spine. The supraspinous ligament is commonly injured in jumpers most commonly in the cranial lumbar region and has a high recurrence rate. Sacroiliac desmitis is also a common condition resulting in poor performance and often lameness in the horse and can be unilateral or bilateral.
3) Superficial soft tissue injuries: Disruptions in the skin and irritation of the skin cannot be excluded as causes of back pain in horses. The clinical signs of back pain can be very similar to deeper injuries when the skin has been injured and is subject to direct pressure by both well fitting tack along with poor fitting tack.
4) Osseous pathology: Vertebral anomalies and deformities can occur and lead to increased strain on other structures resulting in secondary injuries. Inter-vertebral disk disease is very uncommon in the horse and is infrequently diagnosed. Fractures and other traumatic events result in significant lameness in the acute phase and can result in poor performance once they become chronic.
5) Articular pathology: Dorsal articular facet osteoarthritis, zygapophyseal arthritis and sacroiliac synostitis can all be causes of significant performance problems in the horse. Impinging or over-riding spinous processes are a common occurrence as well and can be a cause of subclinical issues as well as overt lameness problems. This occurs most commonly between T13 and T18. Jumpers and dressage horses tend to be over-represented.[3]
6) Neurologic disease: Cervical dynamic stenosis can manifest as poor performance in mild cases without overt signs of ataxia. Lower motor neuron disease can cause muscle fasiculations that are often interpreted as back muscle issues. Lyme disease has been reported to cause polyarthritis that can affect the spinal articulations and produce odd clinical signs.[4]

TREATMENT

The treatment of spinous problems in horses involves more disciplines than most other injuries in the horse. The treatment of back problems can be broken down into three broad categories; Conservative medical management, surgical management, and "alternative" management therapies.

Conservative medical management includes the use of medications, therapeutic exercise programs, and corrective management strategies including tack fit and equitation alterations. The medications most commonly used include non-steroidal anti-inflammatories such as flunixin meglumine, phenylbutazone and firocoxib with various dosage regimens. In chronic low-grade cases of back pain such as over-riding dorsal spinous processes, the use of a course of non-steroidal medication along with a therapeutic exercise program can be quite effective at managing the problem. The use of muscle relaxants has been advocated by some veterinarians for muscular spasms. Medications such as dantrolene, methocarbomal and thiocolchicoside have all been utilized. In the author's experience the response to muscle relaxants is highly variable and does not seem to have a consistent, uniform effect across horses with similar problems when used systemically. The use of corticosteroid either systemically or more commonly via direct injection is commonly used in a wide variety of cases including osteoarthritis of the dorsal articular facets to direct injection into regions of inflammation and chronic pain. They are often combined with products such as Serapin (derivative of the pitcher plant (*Sarraseniaceae*) for injection into regions with low-grade pain and chronic inflammation or are combined with hyaluronic acid in cases where articular joints are being injected. Sclerosing agents and iodine have been utilized to treat many conditions of the back and sacroiliac ligaments in an attempt to cause scarring and "stabilization" of the affected regions. Scientific reports on the benefit of these types of procedures are limited or non-existent and it is based mostly on anecdotal use. In some mares, altrenogest is utilized to suppress estrus due to the belief that ovarian activity may lead to a sore back or "cold-back." Again the scientific evidence is limited and its use is based on anecdotal benefits. The use of "regenerative" techniques such as bone marrow derived cultured mesenchymal stem cells, fat derived processed mesenchymal stem cell, platelet rich plasma, etc… are being explored as potential alternatives to other medical treatments. Controlled studies are not available on their use in this location, but anecdotally it does appear to have some benefit.

Mesotherapy is a technique that can be used to reduce chronic back pain as well as more acute signs secondary to over-riding dorsal spinous processes in horses. It involves injecting a mixture of local anesthetic, corticosteroid and muscle relaxant into the dermis utilizing very short needles. The mechanism of action involves type I and type II nerve fibers from the skin that have collateral fibers that inhibit pain conduction in the spinothalamic fasciculus from deeper structures of the same spinal segment to the thalamus and cerbrum.[8] These injections are made into the affected dermatomes and those caudal to it to ensure the caudal orientation of the segmental nerves are treated. Significant improvement in clinical signs will occur within 10-14 days and can last from 3-12 months depending on severity, location and underlying pathology.

Surgical treatment of spinal and back pain in the horse is limited in scope. While cervical fusion is often considered for horses with neurologic signs only, it can be considered in cases with severe osteoarthritis of the cervical spinal segments. Most surgical therapy of the spine and for back pain is focused on horses with over-riding dorsal spinous processes. Prior to consideration for surgical resection, it is imperative that the diagnosis be confirmed as the cause of the problem since the surgery is fairly aggressive and invasive. Radiographic signs alone of impingement do not necessarily equate to a clinical problem. Other supporting information such as scintigraphy, result to regional anesthesia and response to conservative medical therapy are necessary to ensure the diagnosis is correct. Horses who have failed conservative therapy do appear to have a good response to surgical resection with 72% returning to full work.[9] Horses with fractures of the dorsal spinous processes may also be considered for surgical removal, but many horses with this problem will respond favorably to conservative therapy.

"Alternative" therapies can include chiropractic manipulation, acupuncture, physical therapy and massage. Chiropractic evaluation focuses on evaluating vertebral dysfunction and may be considered when conventional veterinary care has not aided in the problem. Acupuncture therapy is believed to help in chronic pain by affecting opiod and non-opiod neurological pathways.[4] Pain relief is often rapid, but not long lasting. Physical therapies such as electro stimulation for direct pain control and alterations of nerve conduction have been advocated but little research is available on its benefit in the horse.

CONCLUSION

The identification, diagnosis and treatment of back and spinal problems can be quite challenging in the horse. Given the vague clinical signs, limited ability to image the region and wide variety of traditional and non-traditional therapies utilized currently, the success of such cases is highly variable. Further research into this area is needed to help clarify the extent of this problem and to help focus more specific, problem oriented therapies.

REFERENCES

[1]Cauvin E., **In Practice**, 1997; 19 522-533, [2]Jeffcott LB., **Equine Vet J**, 1979; 11(3): 129-136, [3]Jeffcott LB., **Equine Vet J**, 1980 12(4): 197-210, [4]Jeffcott LB. et al, **Equine Sports Medicine and Surgery** 2004, pg 433-474, [5]Denoix J-M, **Vet Clin North Am Eq Pract** 1999; 15(1):131-160, [6]Denoix J-M, **Dubai International Equine Symposium**, 1996; 273-286, [7]von Schewinitz DG, **Vet Clin North Am Eq Pract**, 1999; 15(1):161-177, [8] Denoix J-M, **Diagnosis and Management of Lameness in the Horse** 2003, 509-521, [9] Walmsley JP, et al, **Equine Vet J** 2002; 34(1): 23-28

MUSCLE DISORDERS RELATED TO POOR PERFORMANCE

Stephanie J Valberg, DVM PhD, DACVIM, Anna Firshman DVM PhD, DACVIM
St Paul, MN

Optimal function of skeletal muscle is essential for successful athletic performance. Even minor derangements in locomotor muscle function will impact power output, coordination, stamina and desire to work during exercise. While many myopathies are easy to recognize in the acute stages, low grade muscle strains or weakness and chronic myopathies may be difficult to diagnose. The challenges of identifying their contribution to poor performance in horses include;

1) differentiating between true pain with exercise and an uncooperative attitude
2) locating and assigning significance to focal muscle strain
3) determining the degree to which orthopedic pain contributes to muscle pain (especially for internal medicine specialists)
4) determining the degree to which myopathies contribute to orthopedic pain (especially for orthopedic specialists)
5) identifying muscle weakness as a component of gait alterations
6) differentiating myogenic from neurogenic weakness

DIAGNOSTIC APPROACH

The diagnostic approach to poor performance involves by necessity a detailed evaluation of all body systems. Frequently poor performance in athletes is not due to one specific cause but often a combination of disorders involving pain in multiple joints as well as in skeletal muscle. Trying to determine the source of orthopedic and muscle pain requires a thorough history, physical, neurologic and lameness examination as well as ancillary diagnostic tests.

History

The history should include the diet, supplements, performance level, fitness level, exercise schedule, duration, severity and frequency of poor performance, previous lameness, medications and any factors that initiate the performance problem.

Physical exam

A detailed physical evaluation of the muscular system includes careful inspection of the horse standing perfectly square for symmetry of muscle mass and evidence of fasciculations. The entire muscle mass should be palpated for heat, pain, swelling or atrophy by comparing with contralateral muscle groups. Running a blunt instrument over the lumbar and gluteal muscles should illicit extension (swayback) followed by flexion (hogback) in healthy animals. Guarding against movement may reflect abnormalities in the pelvic or thoracolumbar muscles, or pain associated with the thoracolumbar spine or sacroiliac joints. The horse should then be observed at a walk and trot for any neurologic or gait abnormalities, followed by flexion tests for lameness. Blood samples should be drawn prior to any exercise for a CBC and serum biochemistry profile.

Exercise Challenge

An exercise challenge can be helpful in detecting subclinical exertional rhabdomyolysis and in observing the progression of a gait abnormality with exercise. Blood samples to evaluate peak changes in CK activity should be taken before exercise and about 4-6 hrs after, but NOT immediately, after exercise. The exercise test in unfit horses may involve 2 minute intervals of walk and trot for up to 15 min. Clinical judgment should be used and horses should not be pushed if they seem reluctant to continue to exercise after a brief rest. For fit horses, 4 min of walk and 11 min of continuous trot can be used. During the test, horses should be observed carefully for exacerbation of lameness, changes in impulsion, stiffness, shortened stride and development of a sour attitude. In addition to quantifying the extent of rhabdomyolysis during mild exercise, the test is helpful in deciding how rapidly to put a horse back into training. Horses with marked elevations of CK 4 hrs following light exercise should be very gradually reintroduced to exercise.

Ancillary diagnostic tests

Depending on the breed of horse, EDTA blood for genetic tests for MH and PSSM1 (see applicable breeds below), serum vitamin E concentration (particularly if there is reduced muscle mass or weakness), whole blood selenium concentration (if horses are not receiving supplementation and are in a deficient area) and muscle biopsy may be indicated. For horses with atrophy, biopsies of the atrophied muscle +/- sacrocaudalis dorsalis muscle should be considered. For horses with exercise intolerance or exertional rhabdomyolysis, gluteal or semimembranosus biopsies are preferable in order to look for generalized myopathies. Fresh muscle that is shipped on ice packs and then frozen in specialized laboratories, rather than formalin-fixed muscle, is preferred because muscle fiber type composition (% of type 1, 2A and 2B), fiber oxidative capacity (degree of oxidative staining), lipid storage, and many more metabolic parameters can only be assessed in frozen sections.

In complex cases, repeated evaluations may be required, including examination of the horse under saddle, use of bone scans to localize inflammation, nerve blocks to confirm the source of pain and radiography and ultrasound examination to characterize lesions. In addition, if horses have exercise intolerance, a standardized incremental exercise test that incorporates monitoring heart rate, blood lactate and serum free fatty acids may be of value to assess horses for metabolic derangements.

DIFFERENTIAL DIAGNOSES

Following a complete evaluation, the origin of muscle dysfunction should fall into one or more of the following categories;

1) Pain
   - Muscle contusions or tearing
   - Primary muscle strain either focal or generalized from overuse
   - Muscle strain secondary to skeletal pain and altered gait
   - Neuritis
2) Weakness
   - Myogenic atrophy
     - Disuse
     - Deficiency of Vitamin E
     - Immune mediated myopathy
     - Malnutrition/malabsorption
     - Cushing's disease
   - Neurogenic Atrophy
   - Lower Motor Neuron lesions - trauma, chronic Vitamin E deficiency, EPM, polyneuropathy
3) Inadequate oxidative metabolism
   - Insufficient training of those myofibers that contribute to the expected performance leading to fatigue
   - Skeletal muscle oxidative enzyme defects[1]
4) Disruption of glycogen/glucose metabolism and energy regulation
   - Polysaccharide storage myopathy (PSSM) type 1 due to *GYS1* mutation[2,3]
   - PSSM type 2[4,5]
5) Inadequate fat metabolism
   - Inadequate supply of lipid
   - Lipid storage disorders
6) Disruption of excitation contraction coupling
   - Malignant hyperthermia (MH)[6,7]
   - Recurrent exertional rhabdomyolysis (RER)[8-10]

Heritable myopathies and association with specific breeds

PSSM1 is associated with a mutation in the glycogen synthase 1 gene (*GYS1*) and a DNA test is available.[3] http://www.vdl.umn.edu/guidelines/equineneuro/home.html PSSM1 predominates over PSSM2 as a cause of muscle disease in Quarter Horses and also occurs in 20 other equine breeds. PSSM1 is very common in Belgian and Percheron horses but infrequent in Clydesdale and Shires. PSSM2 is more common than PSSM1 in Warmblood breeds, and also occurs in Arabians, Morgans, a subset of Quarter Horses and other light breeds. MH due to a *RYR1* genetic mutation has only been found in Quarter Horse-related breeds.[7] MH testing is also performed at the University of Minnesota. RER occurs more commonly in horses with Thoroughbred breeding, as well as in Standardbreds, Arabians and Warmbloods that have a more excitable temperament.[11] A diagnosis is made based on ruling out other myopathies via genetic testing and muscle biopsy. A DNA based test is not yet available for RER. Lipid storage disorders are diagnosed through Oil red O staining of frozen sections of muscle biopsies and by plasma acylcarnitine and urine organic acid profiles (http://www.baylorhealth.edu/imd/clinicaltests.htm).

PRINCIPLES OF TREATMENT

Managing muscle disorders in sport horses involves a combination of; (1) rest (2) selection of medications to reduce pain, inflammation, address deficiencies or counteract pathophysiologic mechanisms, (3) diet regimes designed to counteract any metabolic deficiencies, (4) rehabilitation therapy programs, (5) gradual reintroduction of the intended performance. Although new to North American veterinary practice, rehabilitation therapy in horses has been used extensively in UK and Europe with a dramatic impact on restoring muscle function and preventing recurrence of muscle disorders. Rehabilitation therapy focuses on alleviating sensory, neuromotor, and mechanical abnormalities through pain modulation, manual therapy, proprioceptive and motor retraining, therapeutic exercise, and functional and sport-specific activities. Modalities currently used in equine medicine include massage, stretching, hot and cold therapy, therapeutic ultrasound, neuromuscular stimulation and therapeutic excercises including aquatreadmill exercise, and graduated exercise programs.[12]

SPECIFIC MANAGEMENT

1) **Underlying lameness.** Many horses with either myopathies secondary to orthopedic pain or primary myopathies may not fully respond until underlying issues with hoof balance, ligament, tendon and joint pain are fully addressed. A holistic approach combining treatment of orthopedic disorders and rehabilitation therapy may take several months to restore complete soundness; however, it is the best long term strategy for horse and rider.
2) **Muscle strains, tears.** Adequate rest, cold therapy and nonsteroidal anti-inflammatory medication form the basis for initial treatment. To prevent permanent changes in gait, therapeutic ultrasound and stretching of the muscle should begin once the acute pain has resolved. Strength building exercises prior to resuming full exercise such as hand walking once the initial

stiffness and swelling has dissipated and progression to ground poles, caveletti and work under water can prevent permanent changes in stride and future re-injury.

3) **Weakness.** Once the primary cause of muscle weakness has been addressed, physical therapy can be used to build muscle mass and strength. One consideration for these horses should be assessment of serum vitamin E. We have recognized hypovitaminosis E in conjunction with moderate generalized muscle weakness and decreased muscle mass prior to development of recognizable signs of equine motor neuron disease. Dietary supplementation with natural vitamin E combined with resistance training such as aquatreadmill training, and work on a slope as well as use of a theraband and ground poles can help build strength.
4) **Insufficient training.** The degree to which the muscles necessary for a given type of performance are recruited during a fitness program should be evaluated and a new program designed to provide gradually increasing intervals of recruitment of these muscle groups. For example, horses competing over cross country courses should be exposed to intervals of speeds that replicate speeds attained during competition. Horses competing in dressage need to gradually develop the strength and coordination necessary to perform specific movements such as passage or piaffe.
5) **PSSM.** PSSM1 is due to disruption of energy metabolism originating from a persistently active glycogen synthase enzyme which appears to inhibit oxidative glycogen metabolism during aerobic exercise.[13] The specific cause of PSSM2 is not known however these horses appear to respond to similar diet and exercise regimes as PSSM1 horses.[4;14] The approach to managing both forms of PSSM is directed at providing as much movement as possible through daily turn out and daily exercise. Exercise on a longe line or under saddle must be introduced gradually after an acute episode of exertional rhabdomyolysis. The duration of exercise, not just the intensity, should be considered in beginning exercise. Many times this means beginning with 5 minutes or less of walk and trot. Prolonged hand walking should initially be avoided as it can precipitate another episode of rhabdomyolysis.

   The amount of digestible energy (DE) provided as starch and sugar should be minimized in PSSM horses by avoiding lush pastures, providing hay with <12% nonstructural carbohydrate (NSC) content and avoiding grain.[15] Many horses will also require an alternate energy source in the form of fat. For overweight horses, this can initially be accomplished by exercising after a 6-12 h fast. For horses with normal weight, a low starch (<15% NSC) feed which provides 10-15% of daily energy in the form of fat should also be fed. Owners usually report a change in attitude and energy within 3-4 weeks of adding fat to the diet.
6) **Lipid storage myopathy.** Recently we have identified horses that have muscle pain and exercise intolerance when fed high fat feeds and these horses have excessive lipid storage in muscle biopsies. When the diets of these horses are gradually switched from fat supplemented feeds to oats and a vitamin/mineral ration balancer they have shown marked improvement in energy levels and performance.
7) **MH.** To date there are no tested management strategies for horses with MH, although dantrolene seems like a reasonable but expensive approach. Horses with both MH and PSSM have been shown to respond to high fat low starch diets although their response is not as favorable as horses with PSSM alone.[16]
8) **RER.** While exertional rhabdomyolysis is the most common presentation in young horses, RER often presents in older horses as muscle soreness and reluctance to collect. Thoroughbreds susceptible to RER that are in moderate to high intensity training should be fed concentrate feeds that are relatively low in starch (<20% DE) and high in fat (20% DE) to minimize post –exercise muscle damage.[17] Management changes should be instituted to minimize excitement and nervousness. Dantrolene (2-4mg/kg PO) given 1 hour before exercise has proven to be effective in preventing RER in some horses. Rehabilitation therapy practices that are directed at muscle soreness and stiffness such as massage, stretching, long and low head carriage warm up periods are also beneficial.

Conflict of Interest statement: Drs. Valberg, Mickelson and McCue own the license for PSSM testing and receive sales income from its use. Their financial and business interests have been reviewed and managed by the University in accordance with its conflict of interest policies.

REFERENCES

[1] S.J.Valberg, G.P.Carlson, G.H.Cardinet, III, E.K.Birks, J.H.Jones, A.Chomyn and S.DiMauro, **Muscle Nerve** 17 (1994) 305-312. [2] S.J.Valberg, G.H.Cardinet, III, G.P.Carlson and S.DiMauro, **Neuromuscul.Disord.** 2 (1992) 351-359. [3] M.E.McCue, S.J.Valberg, M.B.Miller, C.Wade, S.DiMauro, H.O.Akman and J.R.Mickelson, **Genomics** 91 (2008) 458-466. [4] M.McCue, A.Armien, M.Lucio, J.Mickelson and S.Valberg, **Vet Pathol.** (2009). [5] M.E.McCue, S.J.Valberg, M.Lucio and J.R.Mickelson, **J Vet Intern.Med** (2008). [6] M.Aleman, J.Riehl, B.M.Aldridge, R.A.Lecouteur, J.L.Stott and I.N.Pessah, **Muscle Nerve** 30 (2004) 356-365. [7] M.Aleman, J.E.Nieto and K.G.Magdesian, **J Vet Intern.Med** 23 (2009) 329-334. [8] L.R.Lentz, S.J.Valberg, L.V.Herold, G.W.Onan, J.R.Mickelson and E.M.Gallant, **Am.J.Vet.Res.** 63 (2002) 1724-1731. [9] L.R.Lentz, S.J.Valberg, E.M.Balog, J.R.Mickelson and E.M.Gallant, **Am.J.Vet.Res.** 60 (1999) 992-999. [10] S.J.Valberg, J.R.Mickelson, E.M.Gallant, J.M.MacLeay, L.Lentz and C.F.de la, **Equine Vet.J.Suppl** 30 (1999) 533-538. [11] J.M.MacLeay, S.A.Sorum, S.J.Valberg, W.E.Marsh and M.D.Sorum, **Am.J.Vet.Res.** 60 (1999) 1562-1566. [12] R.Paulekas and K.K.Haussler, **J Equine Vet Sci** 29 (2009). [13] L.Borgia, Valberg S.J., McCue M.E., J.Pagan and C.R.Roe, in: The effect of varying dietary starch and fat on metabolic response to exercise and plasma creatine kinase activity in Quarter Horses with Polysaccharide Storage Myopathy , in press. [14] M.E.McCue, W.P.Ribeiro and S.J.Valberg, **Equine Vet J Suppl** (2006) 340-344. [15] W.P.Ribeiro, S.J.Valberg, J.D.Pagan and B.E.Gustavsson, **J.Vet.Intern.Med**. 18 (2004) 887-894. [16] M.E.McCue, S.J.Valberg, M.Jackson, L.Borgia, M.Lucio and J.R.Mickelson, **Neuromuscul.Disord.** 19 (2009) 37-43. [17] E.C.McKenzie, S.J.Valberg, S.M.Godden, J.D.Pagan, J.M.MacLeay, R.J.Geor and G.P.Carlson, **J.Vet.Intern.Med**. 17 (2003) 693-701.

DOPPLER ECHOCARDIOGRAPHY IN EQUINE PATIENTS: PULSED WAVE, CONTINUOUS WAVE, & COLOR DOPPLER

Sophy A. Jesty, DVM, DACVIM
Ithaca, NY

INTRODUCTION

Doppler echocardiography has greatly increased the information gained during cardiac ultrasound. Doppler echocardiography, unlike 2D and M-mode echocardiography, can provide information regarding the direction, velocity, character, and timing of blood flow within the heart and great vessels. It thus provides hemodynamic information which would otherwise be absent form a noninvasive cardiac workup. A full echocardiogram includes the use of some if not all of the commonly used Doppler modalities (pulsed wave, continuous wave, color flow) as they are complementary to each other. Articles and chapters for review are included in the reference section.[1-4]

Once velocity of blood flow is known, a pressure gradient between the two areas over which flow is occurring can be calculated using the modified Bernoulli equation: $\Delta P = 4V^2$, where P is pressure and V is velocity. For example, if the velocity of a mitral regurgitation jet is known, the pressure gradient between the left ventricle and left atrium during systole can be calculated. Knowing pressure gradients can allow for an estimation of pressures within specific cardiac chambers, which greatly aides in the assessment of the normalcy of cardiac physiology; in this way knowing pressure gradients allows for a better understanding of the hemodynamic significance of a specific cardiac lesion.

DOPPLER MODES

Pulsed wave Doppler

Pulsed wave (PW) Doppler works by transmitting an ultrasound pulse into the tissue and then waiting until that pulse returns from a specified tissue depth (set by the operator using a sample gate). The velocity of blood flow within that specific region can then be calculated using the ultrasound frequency shift (this is termed range resolution). PW Doppler is therefore very specific for location, but it is incapable of resolving higher frequency shifts (i.e. higher velocities). When the PW Doppler is used to interrogate low velocity, laminar flow the profile is hollow. When PW Doppler is used to determine high velocity, turbulent flow the profile fills (spectral broadening).

Continuous wave Doppler

Continuous wave (CW) Doppler continuously transmits and receives ultrasound signals. It therefore cannot have range resolution because it cannot detect the depth of the reflected signal. Velocities all along a cursor line are displayed, so location is less specific than using PW Doppler. The continuous transmission and reception of ultrasound signals, however, allow for resolution of very high velocities. When CW Doppler is used to interrogate flow of any kind (laminar or turbulent), spectral broadening of the profile is seen because CW Doppler samples many velocities along the cursor line.

Color Doppler

Color flow Doppler is a form of PW Doppler, so it can differentiate low from high velocities, but cannot resolve higher velocities from one another. Blood flow is coded as different colors and intensities, depending on the direction of blood flow, velocity, and character of flow (laminar vs. turbulent). Because color Doppler is graphic, it is easy to interpret and is a good screening tool, but it cannot provide specific velocity information. When a color Doppler map is placed over a 2D echocardiographic image, flow is coded blue if it is moving away from the transducer and flow is coded red if it is moving toward the transducer (BART). Enhanced color maps will further display velocity information, with colors of flow moving away from the transducer ranging from a deep blue to a light blue and colors of flow moving toward the transducer ranging from deep red to orange/ yellow depending on velocity. In color Doppler, aliasing involves a color reversal between the Nyquist limit and 2x the Nyquist limit. If velocity exceeds 2x the Nyquist limit, blood flow will be coded using a variance map (colors other then blue and red).

PHYSICS

Angle of interrogation

The calculated velocity is directly proportional to the cosine of the angle of interrogation. Therefore, Doppler echocardiography provides accurate velocity of blood flow, *as long as the angle of interrogation is close to parallel to the direction of flow*. The importance of the criterion cannot be overstated and in many instances it is impossible in equine echocardiography to have a near parallel angle of interrogation; in fact angles of interrogation > 15° are considered unacceptable.[1] Blood flow through the heart and great vessels is normally low velocity, laminar flow (in the range of 1 m/sec). Higher velocity, turbulent flow is abnormal blood flow.

Transducer frequency

The measured velocity is inversely proportional to the frequency of the transducer. Therefore, Doppler echocardiography allows for the highest measured velocity of blood flow *when the transducer frequency is low*. The time interval between pulses (reciprocal of pulse repetition frequency or PRF) must be two times the sample depth in order for the ultrasound to be transmitted and received. So PRF must decrease as sample depth increases, but PRF is proportional to the velocity that can be accurately measured. This means that as sample depth increases, the velocity that can be accurately measured decreases. The PRF must be 2x the velocity to accurately measure it. Therefore, the maximum velocity that can be measured accurately is ½ PRF (this is referred to as the Nyquist limit). In other words, if the Nyquist limit is exceeded, the velocity will become ambiguous; this is referred to as aliasing.

Depth of sampling

The measured velocity is inversely proportional to the sampling depth. Therefore, Doppler echocardiography allows for the highest accurately measured velocity of blood flow *when the sampling depth is small.* In other words, aliasing is inversely proportional to the depth of sampling if the transducer frequency remains constant. The Nyquist limit is surpassed (and therefore aliasing occurs) at much lower velocities at deeper depths of sampling than shallower depths of sampling.

Optimization of Doppler

A general understanding of the physics behind Doppler echocardiography can aid in the optimization of this modality. The following is a list of means by which to improve the information provided by Doppler:

1. Decrease the angle of interrogation
2. Decrease transducer frequency
3. Decrease sampling depth
4. Move the PW baseline up or down to accommodate a higher velocity of blood flow in one direction (at the expense of blood flow in other direction)
5. Increase the frame rate during color Doppler imaging by narrowing the sector width and eliminating the 2D image outside of the color Doppler map
6. Decrease the packet size during color Doppler imaging to increase the frame rate

REFERENCES

[1]Boon JA. **Manual of Veterinary Echocardiography,** 1998; p.1. [2]Reef VB. **Equine Diagnostic Ultrasound** 1998; p.215. [3]Bonagura JD, et al. **Eq Vet J Suppl** 1995; 19: 5. [4]Marr CM, et al. **Br Vet J** 1994; 150: 527.

THE USE OF DOPPLER ECHOCARDIOGRAPHY IN THE WORKUP OF EQUINE CARDIAC CASES: CASE DISCUSSIONS

Sophy A. Jesty, DVM, DACVIM
Ithaca, NY

INTRODUCTION

Doppler echocardiography provides useful information in the workup of equine cardiac cases and can help in further characterizing the significance of disease, thereby providing the clinician a more specific prognosis for performance and life expectancy. As introduced in the preceding hour, Doppler echocardiography has three common modes: pulsed wave, continuous wave, and color. These modes are complementary and should be used in conjunction with one another to ascertain as completely as possible the hemodynamic relevance using noninvasive means.

This hour will focus on a variety of cardiac cases for which the use of Doppler echocardiography adds valuable information beyond what 2D and M-mode echocardiography can give.

DOPPLER ECHOCARDIOGRAPHY

Doppler echocardiography provides velocity of blood flow, *as long as the angle of interrogation is close to parallel to the direction of flow*. The importance of the criterion cannot be overstated and in many instances it is impossible in equine echocardiography to have a near parallel angle of interrogation, nonetheless the following discussion assumes valid signal acquisition.

Once velocity of blood flow is known, a pressure gradient between the two areas over which flow is occurring can be calculated using the modified Bernoulli equation: $\Delta P = 4V^2$, where P is pressure and V is velocity. For example, if the velocity of a mitral regurgitation jet is known, the pressure gradient between the left ventricle and left atrium during systole can be calculated. Knowing pressure gradients can allow for an estimation of pressures within specific cardiac chambers, which greatly aides in the assessment of the normalcy of cardiac physiology; in this way knowing pressure gradients allows for a better understanding of the hemodynamic significance of a specific cardiac lesion.

CASES

Equine cardiac cases which demonstrate the utility of Doppler echocardiography will be reviewed. The principles covered by the case presentations are the following:

Congenital Intracardiac shunts

Ventricular septal defects (VSDs) are the most common congenital cardiac lesion in horses. Normally, the echo dropout of the VSD is easily identified and the size of the defect is correlated with prognosis for performance and life expectancy.[1-3] The measurement of VSDs might not be accurate in each case however, given that they are sometimes irregular in shape. Additionally, the significance of the shunt between the two ventricles might not be proportional to the size of the VSD if other structures (such as the aortic valve) mechanically obstruct shunt flow. The true hemodynamic significance of the VSD shunt can be additionally assessed using Doppler echocardiography to evaluate the velocity of the shunt flow.

Continuous wave Doppler can be used to calculate the pressure gradient between the left ventricle (LV) and right ventricle (RV) during systole by measuring the peak velocity of the shunt flow. LV pressure (LVP) will not be measured directly, but normally the pressure generated by the LV during systole is transmitted to the systemic arteries and is very close to systolic arterial blood pressure (SAP), which can be measured. Therefore, $\Delta P = 4V^2$ or (LVP-RV pressure or RVP) $= 4V^2$ can be rearranged such that $RVP = SAP - 4V^2$. Normally, VSD shunt velocity should be 4.0-6.0 m/sec; this indicates a physiologic pressure gradient between the ventricles and minimal hemodynamic significance. If the shunt velocity is low (< 4.0 m/sec), either the RV pressure is abnormally high (as can occur due to a large shunt volume, pulmonary hypertension, or when the RV faces systemic blood pressures) or the LVP is abnormally low (as can occur with systolic dysfunction during heart failure).

If the VSD size is < 2.5cm and the shunt velocity is > 4.0 m/sec, horses have a good prognosis for life expectancy and some level of performance.[1-3] If the VSD size is > 2.5cm but < 3.5cm and the shunt velocity is > 3.0 m/sec but < 4.0 m/sec, exercise intolerance is likely although they may be used as pleasure horses.[2,3] If the VSD size is > 3.5cm and the shunt velocity is < 3.0 m/sec, horses have a decreased life expectancy due to congestive heart failure.[1-3] If VSD shunt velocity is low, this is a clue that something in addition to a simple inconsequential VSD is present, whether it be secondary effects of a large VSD or evidence of a more complex congenital malformation.

Mitral and tricuspid regurgitation

Acquired degenerative valve disease occurs commonly in middle aged and older horses, resulting in insufficient valves. Mitral regurgitation (MR) is the most likely regurgitation to limit performance and life expectancy.[3-5] The clinical significance of MR depends on the severity. Mild MR can exist for many years without causing clinical signs, whereas more severe MR can cause congestive heart failure and premature death.[6] Determining the prognosis for horses with MR requires Doppler assessment in addition to left ventricular and left atrial chamber measurements.

Color flow Doppler echocardiography can be used to semi-quantitatively grade the amount of regurgitation in the left or right atria. The regurgitant jet will be coded a mosaic of colors *other* than red and blue due to its high velocity and turbulence. The width of the jet's base at the level of the mitral or tricuspid valve and the extent to which it fills the atrium during systole are indicative of the severity of the regurgitation. Additionally, proximal isovolumic surface area quantification and vena contracta width may prove useful in quantifying equine mitral and tricuspid regurgitation, as they have been in small animals and people.[7,8]

Continuous wave Doppler echocardiography can be used to calculate the pressure gradient between the LV and the LA during systole by measuring the velocity of the MR jet. Again, SAP can be used as a measure of LVP, so $\Delta P = 4V^2$ or $(LVP-LAP) = 4V^2$ can be rearranged such that $LAP = SAP - 4V^2$. Normally, the velocity of MR should be approximately 5.0-6.0 m/sec; this indicates a physiologic pressure gradient between the LV and LA. If velocity of MR is low (< 5.0 m/sec), either the LVP is abnormally low (systolic dysfunction associated with heart failure) or the LAP is abnormally high (which occurs due to MR) or both. If the velocity of MR is high (> 6.0 m/sec), the LVP is abnormally high (which occurs with systemic hypertension). Also, the intensity of the continuous wave signal of mitral and tricuspid regurgitations increases as the severity increases.

Continuous wave Doppler echocardiography can be used to calculate the pressure gradient between the right ventricle (RV) and the right atrium (RA) during systole by measuring the velocity of the tricuspid regurgitation (TR) jet. The right atrial pressure (RAP) can be measured using central venous pressure (CVP). The equation $\Delta P = 4V^2$ or $(RVP - RAP) = 4V^2$ can be rearranged such that $RVP = CVP + 4V^2$. Normally, the velocity of TR is < 2.8 m/sec. If the TR velocity is high (> 2.8 m/sec), the RVP is abnormally high (which occurs with pulmonary hypertension). Note that this is the only noninvasive means to diagnose pulmonary hypertension.

Horses with severe regurgitation on color flow Doppler echocardiography but without significant atrial dilation are more likely to progress faster. Other indicators of mitral regurgitation severity, which can affect performance and life expectancy, include a regurgitant jet area > 50% of the LA, a decrease in systolic function, and pulmonary hypertension.[3-5,9,10] Tricuspid regurgitation is far less likely to affect performance or life expectancy.[3,4,11]

Aortic insufficiency

The aortic valve is the most likely to be affected by degenerative valve disease seen in middle aged and older horses, so aortic insufficiency (AI) has increasing prevalence in these cohorts.[3,12] AI is usually very slowly progressive and does not cause clinical signs or decreased life expectancy, but it carries a risk of exercising ventricular arrhythmias when severe.[13] Since this poses a risk to horse and rider, it becomes important to qualify the significance of AI in these patients. In conjunction with systemic arterial pulse quality and aortic and left ventricular diameter on 2D echo, Doppler echocardiography can aid in assessing the hemodynamic significance of the insufficiency.

Color flow Doppler echocardiography can semi-quantitatively assess the severity of AI. When the color jet extends beyond the level of the mitral valve in the LV and occupies > 2/3 the space of the left ventricular outflow tract, the AI is severe.[3] The area of the color jet at its origin standardized to the area of the aortic valve, can also provide a measure of severity. Finally, if late diastolic mitral regurgitation is recognized, it indicates a severe end diastolic increase in LVP secondary to the AI.

Continuous wave Doppler echocardiography can be used to calculate the pressure gradient between the aorta and the LV during diastole by measuring the velocity of the AI jet. Diastolic arterial blood pressure (DAP) can be measured, so the equation $\Delta P = 4V^2$ or $(DAP - LVP) = 4V^2$ can be rearranged such that $LVP = DAP - 4V^2$. Normally, the peak velocity of AI is > 4.0 m/sec. If The AI velocity is low (< 4.0 m/sec), either the diastolic blood pressure is abnormally low (as can occur with diastolic runoff due to AI) or the LVP during diastole is abnormally high (as can occur due to increasing LVP during diastole because of the AI regurgitant volume) or both. Additionally, the calculation of pressure half time (PHT) can be useful in the assessment of severity of AI.[3] PHT is the time is takes for the velocity to decrease by half. A long, gradual PHT (> 500 msec) indicates that throughout diastole, the pressure gradient between the aorta and LV is maintained, which is physiologic and an indicator of minimal hemodynamic significance. A short, steep PHT (< 500 msec) indicates that the pressure gradient between the aorta and LV equilibrates quickly in early diastole, which is not physiologic and an indicator of significant hemodynamic significance.

REFERENCES

[1]Pipers FS, et al. **J Am Vet Med Assoc** 1985; 187: 810. [2]Reef VB. **Eq Vet J Suppl** 1995; 19: 86. [3]Reef VB. **Eq Vet J Suppl** 1995; 19: 71. [4]Reef VB, et al. **Eq Vet J** 1998; 30:18. [5]Bonagura JD. **Eq Vet Educ** 1990; 2: 31. [6]Imhasly A, et al. **Vet J** 2009; 183: 166. [7]Muzzi RAL, et al. **J Vet Cardiol** 2003; 5: 33. [8]Thomas JD. **Heart** 2002; 88: 651. [9]Long K. **Eq Vet Educ** 1990; 2: 15. [10]Long K. **Eq Vet Educ** 1993; 5: 161. [11]Patteson MW, et al. **Eq Vet J** 1993; 25: 409. [12]Reef VB, et al. **Am J Vet Res** 1987; 48: 904. [13]Reef VB. Cardiovascular ultrasonography. In: **Equine Diagnostic Ultrasound**, Ed: Reef VB. 1998; 215.

WHICH ENDOCRINE DISORDER ARE WE DEALING WITH? PITUITARY PARS INTERMEDIA DYSFUNCTION VERSUS EQUINE METABOLIC SYNDROME: DIAGNOSTICS

Nicholas Frank, DVM, PhD, DACVIM
Knoxville, TN

INTRODUCTION

It is sometimes challenging to determine whether a horse suffers from pituitary pars intermedia dysfunction (PPID), Equine Metabolic Syndrome (EMS), or both disorders concurrently. This clinical workshop focuses upon diagnostic testing for PPID and EMS, with the aim of improving recognition of these important endocrine disorders. Advanced PPID is relatively straightforward to diagnose, so this discussion will primarily focus upon recognition of early PPID.

EARLY PPID

Pituitary pars intermedia dysfunction (also called equine Cushing's disease) is associated with laminitis, muscle atrophy, immunosuppression, and regional fat redistribution.[1] It is particularly important to diagnose PPID before laminitis develops in order to prevent this devastating disease. Unfortunately, PPID is more difficult to recognize in its earlier stages and diagnostic tests have lower accuracy at this point. The following points may therefore be helpful for recognizing early PPID:

Age – Since PPID is more common in older horses, this diagnosis should be considered in any horse > 15 years of age.

Hair coat abnormalities – Hirsutism is an easily recognized clinical sign of advanced PPID, but hair coat abnormalities associated with early PPID are less specific and harder to detect. Affected horses exhibit delayed shedding of the winter hair coat, increased hair length, and dullness of the hair coat. If this problem is suspected, owners should record the time that their horse sheds its winter hair coat and compare it with other horses in the same barn. A horse with early PPID may retain its winter hair coat for a several weeks longer than other horses. Examination of control horses from the same barn is recommended because the timing of hair coat shedding depends upon day length and therefore geographical location.

Retention of winter hairs in certain regions of the body – Some horses with early PPID shed most of their winter hair coat, but retain hairs along the palmar or plantar aspects of the legs, behind the elbow, or beneath the mandibles. A thorough physical examination is required to detect this problem and the client must be questioned about the horse's normal hair coat. For example, clients must be asked whether longer hairs along the palmar/plantar aspects of the legs are normal for their animal.

Shift in metabolism – Owners sometimes report that their horse has undergone a "shift in metabolism" (from easy keeper to a hard keeper) and is now losing weight and body condition. This problem is difficult to confirm unless a complete dietary history has been recorded, but clinicians should suspect PPID when a middle-aged horse begins to lose muscle mass and requires more calories than it did in the past. Owners may attribute this to aging, but the clinician should judge whether the decrease in muscle mass is appropriate or accelerated in an individual horse. In many cases, loss of muscle mass has occurred over a few months, instead of the gradual process associated with aging.

Regional adiposity – This clinical sign is associated with both EMS and PPID. In the case of EMS, regional adiposity first develops at an earlier age (< 15 years) and then persists. In contrast, detection of regional adiposity for the first time in an older (> 15 years) horse indicates the development of PPID. As discussed below, horses can develop EMS at an earlier age and then concurrent PPID as they get older. These patients have fat deposits that were previously associated with EMS and now become a component of PPID. Regional adiposity refers to enlarged adipose tissues within the neck region ("cresty neck") and fat pads cranial to the tail head.

Poor performance, change in attitude, and declining general appearance – These vague presenting complaints sometimes lead to the diagnosis of early PPID. Poor performance has not been documented in horses with PPID, but this problem prompts assessment of the horse's appearance and body condition, which leads to consultation with a veterinarian.

Lameness – This problem is a presenting complaint for PPID in early and advanced cases. Pituitary pars intermedia dysfunction has been associated with laminitis[1] and is a predisposing factor for this condition. There are several potential mechanisms linking PPID with increased laminitis risk, including altered vascular dynamics, insulin resistance (IR), and weakening of dermal-epidermal attachments; all of which have been comprehensively reviewed by Johnson et al.[2] Horses with PPID are also at greater risk for developing sole abscesses and white line disease as a result of immunosuppression. These problems may be identified during routine hoof care and the farrier may be the first person to suspect an underlying endocrine disorder. Examination of the hoof may also reveal divergent hoof rings (founder lines) that indicate previous laminitis events. Mild laminitis may go undetected in pastured horses, particularly in horses that are rarely ridden. Divergent hoof rings are not definitive evidence of laminitis, but should raise concern about the problem. Third phalanx rotation may be identified on lateral radiographs of the foot.

Infertility – Pituitary pars intermedia dysfunction should be considered when aged mares develop fertility problems. More research is required to determine the effects of PPID on the reproductive cycle and uterine environment. At present, only anecdotal evidence is available to suggest that reproductive performance improves in mares with PPID that receive treatment.

DIAGNOSTIC TESTING FOR EARLY PPID

Only two studies to date have attempted to correlate test results with different stages of PPID. Beech et al.[3] examined horses with clinical signs of PPID, clinically normal animals, and 4 horses with an equivocal clinical signs. In that study, the thyrotropin-releasing hormone (TRH) stimulation test confirmed the diagnosis of PPID in horses with clinical signs of the disorder and positive test results were detected in clinically normal older horses with histological evidence of pituitary hyperplasia. The latter finding may be important if future studies show that pituitary hyperplasia is associated with early PPID. However, it is also conceivable that pituitary hyperplasia is a normal finding in older horses and becomes more pronounced at different times of the year. If this is the case, ante mortem detection of pituitary hyperplasia will have little clinical significance. Miller et al.[6] assessed pituitary lesions in relation to oral domperidone challenge test results. Results of this study showed that horses with early (grade 3; 1 to 5 scale) and advanced (grades 4 and 5) pituitary lesions had abnormal oral domperidone test results, and that adrenocorticotropin hormone (ACTH) responses to domperidone increased in accordance with histological grade. Both the TRH stimulation test and oral domperidone challenge test are appropriate diagnostic tests for PPID, but are not widely used at this time.

Currently recommended tests – Resting ACTH concentrations and the overnight dexamethasone suppression test (DST) are easily performed, so these tests are currently recommended. However, the sensitivity of these tests must be questioned for early PPID. In the study performed by Miller et al.,[6] resting ACTH concentrations were within reference range for 5 of 5 horses with grade 3 lesions, 6 of 12 horses with grade 4 lesions, and 1 of 4 horses with grade 5 lesions. The DST had high specificity when examined by Beech et al.,[3] but sensitivity ranged from 23% to 66% depending upon the groups compared. *If early PPID is suspected, the clinician must often decide whether to institute treatment on the basis of clinical judgment alone or leave the patient untreated until positive test results are obtained.* Pergolide can be administered for 6 months (1 mg total dose orally per day) as a therapeutic trial.

There are two other approaches to diagnosing PPID using resting ACTH concentrations: 1) targeting the seasonal increase in hypothalamic-pituitary-adrenal activity or 2) testing animals on multiple occasions throughout the year. Plasma ACTH concentrations increase in the late summer and autumn in healthy horses and those with PPID,[3, 7] and are higher in affected animals. The magnitude of the seasonal increase in ACTH concentration might therefore serve as a diagnostic test. Preliminary data from a small number of horses studied by our research group indicate that a seasonal cut-off value of 100 pg/mL is appropriate for diagnosing PPID in September and October, instead of the 35 pg/mL value used at other times of the year. If this cut-off value is supported by additional results, horses with early PPID may be identified in the future by measuring ACTH concentrations in September or October. Our study also showed that ACTH concentrations fall above and below the 35 pg/mL cut-off value during the year in some PPID horses (outside of the seasonal increase), indicating that test accuracy can be improved by testing horses at multiple times throughout the year.

EQUINE METABOLIC SYNDROME

Insulin resistance is the endocrine component of EMS and some clinicians prefer to focus on this problem because it can be measured and defined. However, the situation in horses extends beyond IR and involves the metabolic status of the animal and occurrence of laminitis. It is therefore preferable to group several problems together as EMS, so that owners and veterinarians recognize that the metabolic efficiency of the animal contributes to the development of obesity, and therefore IR and increased laminitis risk. Limited data are available for horses, but obesity and insulin resistance have both been associated with increased risk of laminitis in ponies.[8, 9] Equine Metabolic Syndrome is therefore a group of conditions that occur concurrently in association with increased laminitis risk. Components include increased metabolic efficiency, obesity, regional adiposity, hyperinsulinemia, and insulin resistance (IR). Obesity, hyperinsulinemia, and IR are all alterations in normal physiology that potentially contribute to the development of laminitis.

CLINICAL PRESENTATION

Age – Obesity is a component of EMS that first develops in younger horses (< 15 years of age). Affected horses may be genetically predisposed to obesity because of enhanced metabolic efficiency and researchers at the University of Minnesota and Cornell University are searching for this genetic link. Some horses become obese when they reach maturity whereas others develop obesity later in life when they are overfed and under-exercised. In general, EMS should be suspected in any obese horse with regional adiposity that the owner describes as an "easy keeper".

Obesity and regional adiposity – Most horses with EMS are obese and show enlargement of adipose tissues within the neck crest ("cresty" neck). Fat pads also develop close to the tail head and within the prepuce or mammary gland region, so some horses present with the complaint of preputial swelling or mammary gland enlargement. Affected horses occasionally have randomly distributed subcutaneous adipose tissue deposits. If regional adiposity first develops in an older horse (> 15 years), it is more likely to be associated with PPID .

Laminitis – Acute laminitis is the presenting complaint for many horses with EMS. Laminitis appears to occur spontaneously without a history of grain overload, bacterial disease, or retained placenta. However, further questioning of the client or examination of the surroundings reveals that laminitis occurred in association with changes in the pasture grass, including rapid growth in the spring, drought in the summer, frost in the winter, or transfer to a new pasture. Laminitis can also occur at a subclinical level and lead to structural changes in the hoof that may be recognized as divergent hoof rings or radiographic evidence of third phalanx rotation.

Infertility – Impaired reproductive performance may be a presenting complaint for EMS. Alterations in reproductive cycling have been detected in obese insulin-resistant mares.[10]

DIAGNOSTIC TESTING FOR EMS

Resting glucose and insulin concentrations – This testing is easy to perform because it only requires collection of a single blood sample, but should only serve as a screening test for IR. Horses must undergo a short fast before samples are collected and this can be accomplished by leaving only one flake of hay with the animal after 10 PM the night before and then collecting a blood sample the following morning. Reference ranges for insulin vary among laboratories according to the assay used and control population. With the radioimmunoassay used in our laboratory, a cut-off value of 20 μU/mL defines hyperinsulinemia. Most insulin-resistant horses maintain blood glucose concentrations below 100 mg/dL, so hyperglycemia raises additional concern. Persistent hyperglycemia signals loss of glycemic control, which is seen in some patients with pancreatic insufficiency. Resting insulin concentrations may be low, high, or within reference range in patients with pancreatic insufficiency, so the key finding is persistent hyperglycemia. Detection of persistent hyperglycemia indicates the development of diabetes mellitus.

Combined glucose-insulin test – There are two indications for performing this test: 1) resting insulin concentrations are within reference range, yet EMS is suspected, and 2) a test is needed to document and monitor IR. This test involves insertion of an intravenous catheter and administration of dextrose followed immediately by insulin. This procedure has been described in previous publications.[11, 12]

Oral sugar test – An easy test for detecting IR involves administration of Karo® Light Syrup (50 mg/kg bwt; 15 mL per 100 kg) orally via dose syringe and then collection of a blood sample 60 to 90 minutes later. The test is performed under short-term fasting conditions and both glucose and insulin concentrations are measured. Results of a pilot study indicate that IR can be diagnosed by detecting a glucose concentration > 110 mg/dL and insulin concentration > 60 μU/mL between 60 to 90 minutes post-treatment.

New tests in development – In the future it is likely that practitioners will submit blood for resting leptin, adiponectin, and C-peptide concentrations, but these tests are not commercially available yet.

EMS/PPID CASES

Horses with EMS can develop PPID as they get older. Obesity has been associated with increased inflammatory cytokine expression in the blood of horses[13] and chronic inflammation within the adipose tissues of obese mice.[14] Oxidative damage has also been associated with obesity in humans,[15] so EMS may predispose horses to PPID by increasing oxidative damage to dopaminergic neurons.[16] If this is the case, PPID may develop at a younger age in horses with EMS.

CLINICAL PRESENTATION

Age – Although it was previously stated that PPID affects horses > 15 years of age, it is advisable to begin monitoring horses with EMS for early signs of PPID once they reach 12 years of age.

Body condition – The obese body condition associated with EMS is maintained in most cases, with epaxial muscle atrophy developing over time. Rounding of the abdomen may become more pronounced when the abdominal musculature weakens in obese horses.

Hair coat abnormalities – Detection of hair coat abnormalities in a horse with EMS signals the onset of PPID. Shedding patterns should be monitored to detect retention of the winter hair coat.

Alteration in metabolism – Horses with EMS may require more calories when PPID develops.

Exacerbation of hyperinsulinemia – A spontaneous increase in glucose or insulin concentrations can signal the development of PPID in a horse with EMS that was successfully managed in the past. The author hypothesizes that IR is exacerbated by the development of PPID in horses with EMS, which is associated with increased risk of laminitis. This hypothesis is based upon clinical experience, rather than scientific evidence, so studies are required to examine this question further. As more horses with PPID have undergone testing for IR, it has become apparent that relationships between these conditions are more complicated than previously reported.[17] When hyperinsulinemia and IR are detected in a horse with PPID, it should first be determined whether the animal has a history of EMS. Insulin resistance is likely to be a pre-existing problem that has been exacerbated by PPID, rather than a component of the disorder.

Increased incidence or severity of laminitis – In accordance with the previously stated hypothesis, concurrent PPID should be suspected when a well-managed EMS horse develops laminitis again after being free of this problem for several years. This emphasizes the need for early detection of PPID in horses with EMS. Medical management of PPID can facilitate control of IR in affected horses and may lower the risk of laminitis events.

REFERENCES

[1]Donaldson MT et al. **J Am Vet Med Assoc** 2004;224:1123. [2]Johnson PJ et al. **Vet Clin North Am Equine Pract** 2002;18:219. [3]Beech J et al. **J Am Vet Med Assoc** 2007;231:417. [4]Beech J and Garcia M. **Am J Vet Res** 1985;46:1941. [5]McFarlane D, et al. **Domest Anim Endocrinol** 2006;30:276. [6]Miller MA et al. **Vet Pathol** 2008;45:26. [7]Donaldson MT et al. **J Vet Intern Med** 2005;19:217. [8]Treiber KH, et al. **J Am Vet Med Assoc** 2006;228:1538. [9]Carter RA et al. **Equine Vet J** 2009;41:171. [10]Vick MM, et al. **Reprod Fertil Dev** 2006;18:609. [11]Eiler H, et al. **Am J Vet Res** 2005;66:1598. [12]N. Frank. **J Equine Vet Sci** 2009;29:259. [13]Vick MM et al. **J Anim Sci** 2007;85:1144. [14]Xu H, et al. **J Clin Invest** 2003;112:1821. [15] Vincent, HK, et al. **Diabetes Obes Metab** 2007;9:813. [16]McFarlane D and Cribb AE. **Am J Vet Res** 2005;66:2065. [17]Garcia MC and Beech J. **Am J Vet Res** 1986;47:570.

WHICH ENDOCRINE DISORDER ARE WE DEALING WITH? PITUITARY PARS INTERMEDIA DYSFUNCTION VERSUS EQUINE METABOLIC SYNDROME: TREATMENT OPTIONS

Nicholas Frank, DVM, PhD, DACVIM
Knoxville, TN

INTRODUCTION

In this second clinical workshop we will discuss medical treatment of pituitary pars intermedia dysfunction (PPID) and Equine Metabolic Syndrome (EMS). These endocrine disorders require management as well as medical treatment, so it is important to institute appropriate housing, diet, and exercise changes. These management changes have been reviewed elsewhere, so this discussion will focus upon treatment options.

TREATMENT OPTIONS FOR PPID

Pituitary pars intermedia dysfunction, which is also called equine Cushing's disease, develops when melanotropes within the pars intermedia of the pituitary gland secrete larger amounts of hormones derived from the prohormone proopiomelanocortin (POMC). In the pars intermedia of a healthy horse, the end products of the POMC pathway are alpha melanocyte-stimulating hormone (αMSH), beta-endorphin, and corticotrophin-like intermediate peptide (CLIP). When PPID develops, secretion of αMSH and CLIP increases and other hormones derived from POMC are also released. These hormones include adrenocorticotropin hormone (ACTH), which is the primary product of the POMC conversion pathway in the pars distalis. Secretion of ACTH from the pars intermedia induces hyperadrenocorticism because production of this hormone is unregulated in this region of the pituitary gland. The pars intermedia does not secrete ACTH under normal circumstances, so it is not under negative feedback control through the hypothalamic-pituitary-adrenal axis.

Loss of dopaminergic inhibition creates a permissive environment for hyperplasia or neoplasia within the pars intermedia. Dopamine is an inhibitory neurotransmitter that acts on D2 dopamine receptors and reduces the activity of melanotropes. This neurotransmitter is secreted by dopaminergic neurons that have their cell bodies within the paraventricular nuclei of the hypothalamus and extend down to the pars intermedia. These dopaminergic neurons degenerate over time as a result of oxidative damage.[1, 2] It is assumed that dopaminergic neuron degeneration is enhanced in horses with PPID, which leads to pituitary hyperplasia, hypertrophy, and the formation of functional pituitary adenomas.

Pergolide mesylate – This ergot alkaloid dopamine receptor agonist is administered to horses with PPID to restore dopaminergic inhibition of melanotropes. Interaction with D2 receptors inhibits hormone secretion and is therefore associated with improvement in clinical signs. It has not been determined whether pergolide treatment also inhibits the development of pituitary hyperplasia or reduces the size of pituitary adenomas, but these beneficial effects are plausible considering the mechanism of action of this drug. Pergolide was previously marketed as Permax® (previously manufactured by Eli Lilly Co) and used for the treatment of Parkinson's disease. However, this product was voluntarily withdrawn from the market in March 2007 after the Federal Drug Administration issued a warning that pergolide was associated with increased incidence of valvular regurgitation in people.[3, 4] Since that time, pergolide has only been available through compounding pharmacies. This has lowered the price of the drug, but also raised concerns about the stability and efficacy of compounded pergolide. At present, the recommended approach is to purchase pergolide in granule form and assume a 60-day shelf life or as an aqueous suspension and assume a 30-day shelf life. In a recent study, Davis et al.[5] demonstrated that pergolide lacked stability over a 35-day period when prepared as an aqueous suspension. This group also showed that higher temperatures and exposure to light enhanced degradation, so aqueous preparations should be stored in dark containers and refrigerated. Even under these conditions, a maximum 30-day shelf life should be assumed for aqueous preparations. Pergolide is prescribed on a total dose basis, using a starting dose of 1.0 mg/day for horses and ponies with PPID. If the horse owner is willing to administer half the dose twice daily, this may be preferable on the basis of recent pharmacokinetic data.[6] Some owners report loss of appetite when pergolide is first started. If anorexia develops, treatment should be halted for 2 days or until appetite improves, and then restarted at 0.25 mg/day for 2 days, 0.5 mg/day for 2 days, and 0.75 mg/kg for 2 days. Another side effect of pergolide treatment is temporary dullness after initiating therapy and this can be addressed using the same strategy. Pergolide is currently purchased from compounding pharmacies as a 1.0 mg per 2 mL suspension and costs approximately $1.60/day for a horse receiving 1 mg/day.

Cyproheptadine – Both pergolide and cyproheptadine lower plasma ACTH concentrations in PPID horses,[7] but pergolide is more effective.[8] Donaldson et al.[8] reported that 17 of 20 horses (85%) with PPID improved with pergolide treatment, compared with only 2/7 horses treated with cyproheptadine. Cyproheptadine antagonizes serotonin, which is thought to be a stimulatory neurotransmitter for pars intermedia melanotropes. Treated horses occasionally exhibit sedation when treatment is initiated. A dosage of 0.25 mg/kg PO q12h is recommended and the drug is available as 4 mg tablets in labeled (Periactin®, Merck & Company, Inc.) and generic forms. At present, tablets cost approximately $0.50 each ($14/day for a 450-kg horse), so most practitioners order the drug from compounding pharmacies. Cyproheptadine is compounded into an aqueous suspension at 50, 100, or 150 mg/mL concentrations and costs approximately $1.30/day for a horse weighing 450 to 525 kg.

Trilostane – This drug is a 3-beta hydroxysteroid dehydrogenase inhibitor, so it acts by inhibiting cortisol synthesis within the adrenal cortex. Trilostane is commonly used in dogs for the management of pituitary-dependent hyperadrenocorticism when increased ACTH secretion from a pars distalis adenoma induces adrenal hyperplasia. However, the reported incidence of adrenal hyperplasia is low (20%) in horses with PPID,[9] so this may limit the usefulness of trilostane in this species. Despite this theoretical assumption, McGowan and Neiger[10] reported improvement in clinical signs in 20 horses with PPID that were treated with trilostane at a mean dosage of 0.5 mg/kg. Lethargy resolved, polyuria/polydipsia decreased, and recurrent or chronic laminitis improved in 13/16 affected horses. Combined dexamethasone/thyrotropin-releasing hormone test results improved after treatment, but did not return to normal. The recommended dosage of trilostane has subsequently been increased to 1.0 mg/kg (q24h; given in the evening). Trilostane is now approved by the Federal Drug Administration for the treatment of hyperadrenocorticism in dogs and is marketed in the United States as Vetoryl® (Dechra Veterinary Products). This product is only available as 60 mg capsules and is very expensive for horses (approximately $100 for 30 capsules; $26/day for a 500-kg horse; $800/month). Trilostane treatment is therefore reserved for cases where ACTH stimulation test results indicate significant adrenal involvement.

Combination therapy – Some horses with PPID respond to combination treatment with pergolide and cyproheptadine. One approach is to reach the maximum pergolide dosage of 5 mg/day and then add cyproheptadine treatment (0.25 mg/kg PO q12h) to the regimen. Other clinicians with experience in this field add cyproheptadine when the pergolide dosage reaches 3 mg/day.[11]

RESPONSE TO PPID TREATMENT

Clinical signs and behavior – Although pergolide sometimes induces transient mental dullness when treatment is started, most horses respond positively by exhibiting increased physical activity and energy. Polyuria and polydipsia often resolve with treatment and skeletal muscle mass can improve over time, particularly in animals that are still being ridden. Effects of treatment on hair shedding take longer to recognize, but many horses show improvement in shedding the following season. Improvement in laminitis has also been reported.[7] Horse owners should always leave precautionary measures in place to prevent laminitis, but laminitis risk appears to decrease with pergolide treatment. Horses with pre-existing insulin resistance that has been exacerbated by PPID should be monitored by rechecking insulin concentrations. One commercial laboratory (Cornell University Animal Health Diagnostic Laboratory) offers an ACTH and insulin panel ($38/sample) that can be run on a single EDTA plasma sample. In the author's experience, plasma insulin concentrations decrease when pergolide treatment is initiated in horses with PPID and hyperinsulinemia.

Plasma ACTH concentrations and dexamethasone suppression test (DST) results – Whichever test was performed to diagnose PPID should be repeated 4 weeks after treatment is initiated. If PPID was diagnosed on the basis of an elevated plasma ACTH concentration, a goal of treatment should be to return concentrations to within reference range (< 35 pg/mL by chemiluminescent assay).[7] However, ACTH concentrations vary over time in horses, so small alterations should not be over interpreted. If the plasma ACTH concentration is still significantly (> 10 pg/mL) above reference range at the 4-week recheck, the pergolide dosage should be increased (0.5 or 1.0 mg/day). This procedure is repeated until a maximum pergolide dosage of 5 mg/day is reached. If plasma ACTH concentrations remain elevated at the maximum dosage, combination therapy should be considered. Alternatively, the clinician should take a pragmatic approach and decide upon a dosage that is economically feasible, yet still beneficial to the patient. If the DST is used to diagnose PPID, the pergolide dosage should be increased until cortisol concentrations < 10 ng/mL (1.0 μg/dL) are detected 19 or 24 hours after dexamethasone administration. Failure to suppress after 4 weeks indicates that the dosage should be increased. Both plasma ACTH concentrations and the DST are affected by season, so this should be considered when scheduling follow-up testing. Higher ACTH concentrations and false positive DST results are detected in the late summer/autumn period (September and October in Tennessee).[12]

TREATMENT OPTIONS FOR EMS

Equine metabolic syndrome is a disorder that should be managed with diet, housing, and exercise changes, rather than drugs in the majority of cases. Veterinarians have a responsibility to recommend management changes and discourage horse owners from administering drugs as a substitute. However, there are two indications for pharmacological intervention: 1) short-term (3 to 6 months) treatment while management changes are taking effect and 2) refractory cases. Two drugs have been examined to date:

Levothyroxine sodium – The Thyro L® product manufactured by Lloyd, Inc. has been evaluated in several studies conducted by the author's research group.[13-16] Administration of levothyroxine at higher dosages has been associated with weight loss and increased insulin sensitivity,[13-15] and pretreatment with levothyroxine for 14 days prevented horses from developing insulin resistance after endotoxin infusion.[16] Levothyroxine has been administered at an approximate dosage of 0.1 mg/kg, which has been rounded to 48 mg/day for horses weighing 450 to 525 kg. It is assumed that levothyroxine induces weight loss by raising circulating thyroxine concentrations and stimulating basal metabolic rate. Serum thyroxine concentrations rise to between 40 and 80 ng/mL (reference range: 10 to 40 ng/mL) in treated horses, but clinical signs of hyperthyroidism are not observed. Weight loss is enhanced by restricting caloric intake and increasing exercise at the same time that levothyroxine is administered. Horses should not be allowed free access to pasture during treatment because levothyroxine is likely to induce hyperphagia, which offsets the effects of treatment. Levothyroxine is primarily administered for the purpose of accelerating weight loss in obese horses and can be prescribed for 3 to 6 months while other management practices are instituted. In a recent study obese horses with chronic EMS were treated with levothyroxine without calorie restriction to determine whether the drug was still effective in animals with normal caloric intake. Interestingly, a higher dosage of 72 mg/day (approximately 0.15 mg/kg) was required to induce significant weight loss and improvement in insulin responses in these severely affected horses. Current recommendations are therefore to initiate treatment at 48 mg/day (with restricted caloric intake) and then increase to 72 mg/day after 3 months if the horse remains obese. Most horses are treated for 6 months and then taken off the drug by administering 24 mg/day for 2 weeks and then 12 mg/day for 2 weeks. A level scoop of Thyro-L® powder contains 12 mg levothyroxine. Some horse owners and veterinarians leave horses on a "maintenance" dose of levothyroxine (12 to 24 mg/day), but there is no scientific evidence to support this approach. At the time of writing, Thyro-L® can be purchased for $25 (1 lb; 1,000 mg; 83 teaspoons) or $175 (10 lb). Each pound of Thyro-L® contains 1,000 mg levothyroxine (approximately 80 teaspoons), so this treatment costs < $1 per day for 48 mg/day and < $1.50 per day for 72 mg/day if a 10-lb container is purchased.

Metformin hydrochloride – Glucophage® (Bristol-Myers Squibb) and generic forms of metformin have been used for decades in human medicine. Metformin is a biguanide drug that is administered to control hyperglycemia and increase tissue insulin sensitivity in humans with diabetes mellitus. This drug suppresses hepatic glucose production by activating AMP-activated protein kinase, which inhibits gluconeogenesis and lipogenesis while increasing fatty acid oxidation and lipolysis.[17] Two key gluconeogenesis enzymes, phosphoenolpyruvate carboxykinase (PEPCK) and glucose-6-phosphatase are inhibited by metformin through this mechanism. The insulin-sensitizing effects of metformin may also be mediated by skeletal muscle AMPK, causing increased glucose transporter 4 abundance within cell membranes and enhanced glucose uptake.[18] One study also describes AMPK-independent effects of metformin on cardiac muscle, with results indicating that p38 mitogen-activated protein kinase (p38 MAPK) and protein kinase C pathways are activated.[19] Only a small number of studies have been performed to date in horses. Durham et al.[20] reported that resting insulin concentrations and proxy measures of insulin sensitivity improved in horses and ponies with presumed insulin resistance when treated with metformin at a dosage of 15 mg/kg PO q12h. Administration of metformin at this dosage was associated with positive clinical outcomes, but Hustace et al.[21] subsequently reported that the oral bioavailability of 3 grams metformin was 7.1 ± 1.5% in fasted horses and 3.9 ± 1.0% in fed animals. Our current recommendation for metformin is therefore 30 mg/kg PO q12h. This drug is available as generic tablets and at the time of writing is available from a veterinary pharmacy at a cost of $140 for 1,000 tablets (1 gram/tablet). A 500-kg horse receives 15 tablets twice daily, so 900 tablets are required per month (approximately $125/month).

Response to treatment – Weight loss can be monitored using a weight tape or walk-on scale and neck circumference measurements may be useful for assessing improvement in regional adiposity.[22] Fasting blood glucose and insulin concentrations and combined glucose-insulin test results should return to normal with treatment. However, there are some patients that remain insulin-resistant in the face of management changes and medical treatments. These animals remain at higher risk for laminitis, so every effort should be made to avoid triggers for this disease. Severely affected horses and ponies should be held off pasture and housed in dry lots or small grass paddocks.

DIABETES MELLITUS

Affected horses suffer from persistent hyperglycemia and sometimes glucosuria. This condition becomes a greater concern when glucosuria causes electrolyte depletion and anorexia. Diabetic horses are more susceptible to hypertriglyceridemia when stress and negative energy balance stimulate lipid mobilization, and this can occur when other medical problems develop such as gastrointestinal disease. Horses with diabetes mellitus that are in good general heath should be managed with dietary changes and administration of metformin, whereas patients in crisis require exogenous insulin. Short-acting regular (soluble) insulin can be administered intravenously as a constant rate infusion or long-acting insulin can be given via subcutaneous injection. The decision to administer insulin is based upon the degree of glucosuria and hypertriglyceridemia. Diabetes mellitus is sometimes associated with PPID, so pergolide should be administered to these patients.

REFERENCES

[1]McFarlane D, et al. **J Neuroendocrinol** 2005;17:73. [2]McFarlane. **Ageing Res Rev** 2007;6:54. [3]Zanettini R, et al. **N Engl J Med** 2007;356:39. [4]Schade R, et al. **N Engl J Med** 2007;356:29. [5]Davis JL, et al. **J Am Vet Med Assoc** 2009;234:385. [6]Wright A, et al. **J Vet Intern Med** 2008;22:710 (abstract). [7]Perkins GA, et al. **Equine Vet J** 2002;34:679. [8]Donaldson MT, et al. **J Vet Intern Med** 2002;16:742. [9]Schott HC. **Vet Clin North Am Equine Pract** 2002;18:237. McGowan CM and Neiger R. **Equine Vet J** 2003;35:414. [11]Schott HC. **Proc AAEP** 2006;52:60. [12]Donaldson MT, et al. **J Vet Intern Med** 2005;19:217. [13]Frank N, et al. **Am J Vet Res** 2008;69:68. [14]Frank N, et al. **Am J Vet Res** 2008;69:76. [15]Sommardahl CS, et al. **Am J Vet Res** 2005;66:1025. [16]Tóth F, et al. **Am J Vet Res** 2010;71:60. [17]Kim YD, et al. **Diabetes** 2008;57:306. [18]Musi N, et al. **Diabetes** 2002;51:2074. [19]Saeedi R, et al. **Am J Physiol Heart Circ Physiol** 2008;294:H2497. [20]Durham AE, et al. **Equine Vet J** 2008;40:493. [21]Hustace JL, et al. **Am J Vet Res** 2009;70:665. Frank N, et al. **J Am Vet Med Assoc** 2006;228:1383.

VERTICALLY TRANSMITTED INFECTIOUS DISEASES OF CATTLE

Simon Peek BVSc, MRCVS, PhD, DACVIM, DECEIM,
Madison, WI

INTRODUCTION

The possibility for vertical transmission of specific infectious diseases in both beef and dairy cattle is rightly a concern for veterinarians and producers alike. Included within the list of infectious agents for which there has been avid discussion of the relative risk of vertical transmission are bovine viral diarrhea virus (BVDV), bovine leukosis virus (BLV), *Mycobacterium paratuberculosis*, blue tongue virus (BTV), *Neospora caninum* and bovine spongiform encephalopathy. Veterinarians and producers that deal with predominantly grade cattle of lesser individual value are necessarily concerned about the production and economic implications of vertically acquired infectious agents, whereas those who work with individually valuable animals, particularly those involved in assisted reproductive technologies such as conventional *in vivo* embryo transfer, oocyte recovery and *in vitro* fertilization have greater concerns about product quality and the pure infectious disease status of the dam and progeny produced. For several of the diseases named above international export and import concerns are prominent.This presentation will focus on the current knowledge regarding vertical transmission of several of these key diseases and the implications for veterinarians working with all types of cattle.

BOVINE LEUKOSIS VIRUS

The ability of BLV to effect both horizontal and vertical transmission has long been recognized. The proportionate importance of vertical transmission is far less than horizontal, with the majority of infections spread being via transfer of BLV infected lymphocytes in blood, colostrum and milk. Iatrogenic transmission during cattle handling, processing and routine husbandry is a frequent means of spread that is commonly targeted by control programs. Serologic surveys within the US show wide variation in state by state infection levels but a consistent finding amongst studies is that dairy breeds have a higher seroprevalence than do beef breeds[1]. The possibility of transplacental infection from a BLV positive cow to her calf *in utero* has long since been verified but there is little recent data that adds to the literature from the 1980s and 1990s. Epidemiologic studies examining *in utero* spread in BLV positive dams via pre-colostral blood testing of neonates have demonstrated vertical transmission rates that vary from 0.5% to 18%[1]. Maternal lymphocytosis (>12,000/μL) and a high herd seroprevalence appear to be factors that may increase the vertical transmission rates in these epidemiologic studies[1,2]. The dam's age, parity and whether or not maternal infection takes place during pregnancy are factors that do not appear to increase the likelihood of *in utero* transmission[1]. Studies on commercial dairies in the US would suggest that 4% is a reasonable estimate for the proportion of progeny that will be born infected from seropositive dams. Semen from BLV infected bulls is not considered infective, unless there is significant leucopsermia and infected lymphocytes are transferred. Similarly, ova and embryos from BLV positive donors are not considered sources of infection for recipient animals.

BOVINE VIRAL DIARHHEA VIRUS

Of all bovine infectious diseases, the pathogenetic significance of vertical transmission is perhaps best elucidated for BVDV. The relevance of transplacental infection in early gestation (prior to 120 days) as a means of viral persistence within cattle populations is manifest. Infection with non-cytopathic strains of BVDV (ncp BVDV) before fetal immunocompetence can result in a persistently infected calf if the pregnancy goes to term. These PI calves represent a major reservoir of infective virus on farms and propagate the disease within cattle populations. The mechanisms by which ncp BVDV establishes the PI status are complex and appear to be associated with avoidance of an appropriate host innate immune response to viral infection within the fetus, despite normal concurrent anti-viral responses in the dam[3]. Interference with normal type 1 interferon responses appears to be important in viral persistence and establishing the PI status[4]. BVDV infected, cryopreserved or fresh semen have long been documented as potential means of horizontal transmission of infection with subsequent consequences for *in utero* transmission before fetal immunocompetence occurs; however current USA and international requirements for infectious disease screening of sires should reduce this risk to near zero if appropriately followed. Natural service or herd bulls should therefore be screened in a comparable way for biosecurity reasons. The risks for vertical transmission of BVDV via traditional *in vivo* or *in vitro* embryo transfer technologies appear to be very low, but not zero[5,6]. Typical embryo washing and trypsin treatment appear to remove all but a few very high affinity strains of BVDV but under certain experimental conditions virus may still be retained on *in vivo* produced embryos[7]. Whether or not this is ever sufficient for infectivity is uncertain. Consequently, concurrent vaccinal protection of donors and recipients is suggested. The resistance of the *zona pellucida* to virus infection may well serve to protect IVF embryos from internal infection but there is evidence that this technique can result in greater embryo associated virus being introduced via the surface of the *zona pellucida* than is the case with conventional *in vivo* transfer[7]. Subsequent uterine tubule infection could result but the pathogenetic significance of this in natural disease transmission is uncertain at this time for IVF embryos. Embryo donor, recipient and semen donor screening are of obvious critical importance in preventing spread of BVDV by this means, as is the quality control of all sera and bovine fluids used in the process. It is probable that similar concerns exist for embryos derived via somatic cell nuclear transfer as do for IVF embryos[7,8].

BOVINE SPONGIFORM ENCEPHALOPATHY

The etiologic and pathologic similarities between BSE and scrapie alongside the documented cases of BSE in progeny of affected dams and beyond the temporal institution of regulations banning animal by-product feeding of cattle has led to considerable interest in the possibility of vertical transmission of BSE in cattle. However, at this time there is no definitive proof that this occurs. Similarly, embryo transfer has not been proven to be a means of BSE transmission when infected donors or bulls were used[9,10,11].

BLUE TONGUE VIRUS

Interest in vertical transmission of BTV is not so much driven by the clinical severity or relevance of the infection in cattle as it is by the reportable nature of the disease and its major importance in the international trade of animals and animal products. This is particularly relevant for producers who deal with the European and Asian export markets. Up until very recently experimental attempts to demonstrate transplacental infections of calves with wild type virus had been unsuccessful in pregnant cattle, although there are numerous studies that document successful transplacental transmission with cell attenuated BTV strains[12]. During early gestation the most likely reproductive consequences of fetal infection are fetal death and resorption with impaired fertility in the dam. Some aborted bovine fetuses following BTV exposure will demonstrate similar teratogenic defects involving the CNS and ocular system that can also be seen with BVDV infection. However later term experimental infection has been associated with the birth of weak, low birth weight calves that may also be viremic. Clinical experience in Europe during recent outbreaks of the disease have heavily suggested that transplacental infection might be a significant means by which the virus persists in cattle populations but it was not until very recently that the possibility that wild type BTV could be transmitted transplacentally following oral inoculation of a pregnant dam was verified[12,13,14]. Interestingly, that particular study utilizing BTV type 8 (a strain not yet identified in North America) was performed using cattle that were late pregnant, well beyond the time in gestation when fetal immunocompetence develops[12]. Although BTV can be transmitted in semen and this can represent a potential means of horizontal infection, it does not appear that embryo transfer is a significant mode of transmission.

Neospora caninum

Transplacental infection with *Neospora caninum* is a major means of spread for the organism. Recently there has been an effort to adopt the terms endogenous and exogenous transplacental infection so as to distinguish between vertical transmission that arises due to a novel, acute infection of a pregnant dam (exogenous) versus that occurring in a pre-existent, chronically infected dam. (endogenous)[15,16]. Recent studies have demonstrated that oocytes and early embryos are not exposed to the organism in the uterus of seropositive dams such that embryo transfer can be an authentic technique for eliminating vertical transmission[16]. Current embryo transfer techniques are therefore an effective way of deriving replacement stock that are seronegative regardless of the status of the dam.

Mycobacterium paratuberculosis

The fact that the majority of cattle become infected with this agent in the immediate neonatal period can often make the issue of whether or not true, *in utero*, vertical transmission occurs moot, particularly in grade cattle. The feco-oral route is the predominant means of spread and the mode of transmission that producers should be focused on for control purposes. However it is not usual for clients and veterinarians involved in the field of embryo transfer to seek information as to whether or not a donor or recipient could potentially transmit the disease *in utero* to a valuable calf. This question is most commonly asked from the perspective of the older, seropositive embryo donor, or in the context of screening recipients prior to their use. The individually valuable mature cow with clinical Johne's disease from whom a client is trying to acquire "one last calf" is another scenario in which we may be asked to give an opinion as to the likelihood of vertical transmission. The literature provides some guidance but is by no means declarative at this point in time. Many of the published studies on *in utero* transmission of *M.paratuberculosis* are epidemiologic in nature and can be best summarized by the sparingly useful comment that the incidence of calves infected *in utero* on any farm increases as the ratio of clinical cases to sub-clinical ones goes up. However, epidemiologic studies suggest that the estimated incidence of true vertical transmission in herds with a within herd prevalence of 5% is about 0.44-1.2 infected calves per 100 cows compared to 3.5-9.3 calves per 100 cows when the herd prevalence rises to 40%[17]. The risk of *in utero* transmission markedly increases in such studies when one considers dams showing clinical signs compared to those as yet only subclinically infected. Estimates place approximately 9% (6-14%, 95% CI) of calves from subclinically infected dams as being likely to be infected *in utero* versus approximately 39% (20-60%, 95% CI) of those born to dams with clinical signs[17]. Studies examining the safety of embryo transfer or IVF techniques suggest that the zona pellucida is an effective barrier to the organism and that neither technique represents a risk for transmission of the infection from donor to fetus or donor to recipient, provided that the sequential washing procedures suggested by the IETS are observed[18].

REFERENCES

[1.] Hopkins S, and DiGiacomo R. **Vet Clin Nth Amer Food Animal** 1997,;13,1:107. [2.] Lassauzet M, et al **Can J Vet** Res 1991;55:264. [3] Shoemaker M, et al **J Interfer Cyto Res** 2009;29:23. [4.] Smirnova N, et al. **Virus Res** 2008;132:49. [5.] Gard J, et al. **Theriogenology** 2007;67:1415. [6.] Bielanski A, et al. **Theriogenology** 2009;71:499. [7.] Gard J, et al **Theriogenology** 2007;68:434. [8.] Van Soom A, et al. **Reproduction Fert Dev** 2010;22:21. [9.] Harman J and Silva C. **JAVMA** 2009;1:234:59. [10.] Simmons M, et al.**Vet Res** 2008;39:34. [11.] Wrathall A, et al. **Theriogenology** 2008;70:725. [12.] Backx A, et al. **Vet Microbiol** 2009;138:235. [13.]Menzies F, et al **Vet Rec** 2008;163:203. [14.] De Clercq K, et al **Transb Emerging Dis** 2008;55:352. [15.] More G, et al. **Vet Parasitol** 2009;160:51. [16.] Moskwa B, et al. **Vet Parasitol** 2008;158:370. [17.] Whittington R and Windsor P. **Vet J** 2009;179:60. [18.] Bielanski A, et al. **Theriogenology**2006;66:260.

GOAT SCRAPIE: UPDATE & COMPARISONS WITH SHEEP SCRAPIE

Stacey R. Byers, DVM, MS, DACVIM
Fort Collins, CO

INTRODUCTION

Scrapie is a form of a transmissible spongiform encephalopathy (TSE). Other examples of TSE's include bovine spongiform encephalopathy (BSE), chronic wasting disease in cervids, and Creutzfeldt–Jakob disease in humans. Scrapie is primarily a disease in sheep and does infect goats; it is not infectious to people. Research indicates scrapie was not responsible for the BSE outbreak in Great Britain that occurred in the late 1980's.

The purpose of this presentation is to provide current information regarding scrapie in sheep and goats, particularly within the United States.

History

Sheep were first recognized to be infected with scrapie in the mid 1700's. From England, the disease slowly spread to other parts of Great Britain and Europe until after World War II with a large exportation of preclinical sheep around the world. The first case in North America occurred in Canada following importation from Great Britain in 1938. Scrapie was introduced in to the United States in 1947 from sheep imported from Canada. Australia and New Zealand were also recipients of scrapie-infected sheep but underwent a massive eradication program in the 1950's that eliminated the disease. Currently these two countries are designated "scrapie free". Scrapie in goats was first identified in the 1940's and is still relatively rare but has been identified in many countries.

Agent

There have been multiple proposals regarding the agent responsible for TSE's. The most accepted hypothesis is that a prion protein is responsible for the TSE's in the various host species. The prion protein is a normal constituent of cell membranes with a myriad of functions still being discovered. For example, the protein is associated with synaptic vesicles in neuronal cells, binds copper with high affinity, and is involved in cell survival. Similar proteins have been identified in birds, fish, amphibians, and reptiles. A single gene codes for the prion protein and all mammals carry a pair of these genes. The gene is comprised of a variable number of codons or amino acid sequences depending on species. Gene arrangement is highly conserved, with variations in length between species due to size of the untranslated regions. Polymorphisms are possible for some of the codons which alter susceptibility and incubation period to scrapie and other TSE's.

The normal prion protein (PrP^{C}) is comprised of multiple alpha helices, beta sheets, coils, and beta turns, and is protease sensitive. The abnormal prion protein (PrP^{SC}) is an isoform of the PrP^{C} and has an increased number of beta sheets and other structural changes which render it partially proteinase- K resistant. This protease resistance contributes to the accumulation of the PrP^{SC} form. The PrP^{SC} form is capable of inducing conformational changes in normal prion proteins resulting in more PrP^{SC}. The mechanism that initiates the conversion of PrP^{C} to PrP^{SC} is unknown.

Classical and Atypical Scrapie

Previously it was thought there was only one form or strain of scrapie and only sheep were infected. Research and testing of infected herds and suspicious animals have altered these ideas. Scrapie is now divided into 2 broad categories, classical and atypical scrapie. Both sheep and goats can be infected by either type. Classical scrapie is the traditional form of scrapie and has been well studied over the years. It is now recognized that there are multiple "strains" that vary in virulence, pathogenesis, incubation periods, and neuropathology. There is variability in susceptibility within and between breeds due to genetic variations. The combination strain and genetic sequencing of the prion gene are likely the primary factors affecting transmission.

Atypical scrapie was identified in Norway in 1998 (aka Nor98 scrapie), and since then has been identified in multiple countries. There is only one strain to date, and this form of scrapie is suspected to be a spontaneous event. Atypical scrapie has been reported in both sheep and goats. Clinical signs and brain lesions vary between the two types of scrapie and are discussed in more detail below.

Research on classical scrapie in sheep, has identified 3 codons or amino acid (AA) sequences from the prion protein gene which are strongly linked to classical scrapie. The AA sequences alters susceptibility or resistance and incubation period following infection. Codon 136 affects susceptibility to some scrapie strains. Valine increases susceptibility but alanine confers higher resistance to scrapie. Codon 154 plays a minor role in classical scrapie, and is not often used in US for genetic testing. Codon 171 is the major determinant of susceptibility. Possible AA's are glutamine, arginine, histadine, and lysine. Glutamine and histadine have the same susceptibility and lysine is unknown. Arginine confers a degree of resistance so that homozygous sheep are resistant and heterozygous are only rarely susceptible to scrapie.

The genetics for classical scrapie in goats are not clearly identified due to the more recent emergence in this species and the increased number of polymorphisms in the gene. There appears to be a weaker link between resistant and susceptible genotypes with none identified as totally protective. The low prevalence of some genotypes in different breeds and regional variations complicates the genetic analysis and susceptibility to scrapie. The currently relevant codons appear to be 37, 142, 143, 146, 154, 211, and 222 and a 3-octapeptide repeat.[1-3] Codon 142 (methionine) appears to increase incubation period before clinical signs manifest. The effects are also seen in goats heterozygous for methionine. Codons 146 (serine or aspartate) and 222 (lysine) show promise as resistant genotypes; codons 143 (arginine) and 154 (histadine) appear to decrease susceptibility. Homozygotes at codon 146 (asparagine) appears to be linked to increased susceptibility. However not all polymorphisms appear in all breeds which complicates identification of resistant and susceptible genotypes.

The genetic role in atypical scrapie is less well understood due to the random nature and low prevalence. Cases have been found in sheep with known classical scrapie resistant genotypes. Most cases have been found incidentally during screening processes. In sheep, variations at codons 141 and 154 are observed and at codon 154 in goats with this form of scrapie.[4]

TRANSMISSION

Classical

Classical scrapie is not heritable but susceptibility to the disease is. The protein is infectious and can be transmitted to both susceptible and resistant genotypes; however disease only manifests in the susceptible genotype.[5] Infection with classical scrapie requires exposure to the abnormal prion protein therefore not all genetically susceptible sheep and goats will develop scrapie. As with any infectious disease, other factors such as breed, unknown genetic factors, infectious dose, scrapie strain, and exposure levels, play a role in whether an individual develops scrapie. All ages can become infected, however genetically susceptible neonatal lambs and kids having an increased risk and requiring a smaller infectious dose than older animals.[5]

Transmission primarily occurs through oral exposure to placental tissues or fetal fluids. The PrP^{SC} only accumulates in the placentome of susceptible genotype fetuses but does not transfer to the fetus.[4,6] Multiple fetus pregnancies can result in susceptible and resistant genotypes and variable deposition in placentomes regardless of fetal genotype.[6] Transmission by milk or colostrum has been documented which increasing suspicion of transmission in other body fluids.[7,8] Lateral and horizontal infections occur through both direct and indirect contact with the PrP^{SC} protein. Scrapie is not transmitted in utero, through semen, or properly washed embryos.[6,9] As research increases in goat scrapie, difference are being discovered . Goat placentas contain less PrP^{SC} than sheep placentas, and accumulation increases over successive pregnancies (personal communication).

Two probable pathways for infection exist following oral exposure to the abnormal protein. One pathway involves early replication in lymphoreticular tissue of the gastrointestinal tract with spread to the local draining LNs and non-neural tissues. Immunohistochemical (IHC) testing has found accumulation particularly in the palantine tonsils and Peyer's patches within the follicular dendritic cells and macrophages in the follicles. At some point, the PrP^{SC} is transported to the central nervous system via the dorsal vagus nerve and/or blood, and accumulates in specific CNS tissues. Clinical signs develop after sufficient neuronal damage occurs. The alternative pathway involves direct transport of the PrP^{SC} from the enteric nervous system to the CNS.

Shedding of the abnormal protein occurs before clinical signs develop. Since the incubation period is variable the shedding time frame is not well defined. Current research indicates goats typically do not start shedding until the year before clinical signs (personal communication).

Atypical

Less is known and understood regarding the atypical form of scrapie. Currently, atypical scrapie is presumed to be a spontaneous form of scrapie. Similar spontaneous forms of TSE's have been identified in cattle and humans. Due to a few reported cases of multiple affected animals in the same group, it is speculated that it has low transmissibility or is not transmitted under natural conditions.[10] Research has found that if lymphoid tissue is involved, horizontal transmission of scrapie is efficient, however when the peripheral lymphoid tissues are not substantially infected, such transmission is relatively inefficient.[10] Since no detectable PrP^{SC} is found in peripheral lymphoid tissues of atypical scrapie animals, transmission to other animals would be less common.[10,11] Also transport to the CNS takes place directly following ingestion of the PrP^{SC} or the conversion to PrP^{SC} occurs spontaneously in the brain.

CLINICAL SIGNS

Classical

Clinical animals are afebrile and the course is variable but progressive over 1-6 months. Typically sheep and goats start showing signs between 3-5 years old but cases have been seen in sheep less than 18 month old. Sheep exhibit behavior changes including apprehension, separation from flock, reluctance to approach the feed bunk, and aggression. Weight loss occurs despite a good appetite. Pruritis occurs with the commonly associated wool loss from rubbing or chewing. Characteristic lip smacking can be elicited with back scratching. Neurologic signs include star gazing, head pressing, tremors especially in the head and neck, ataxia, and gait abnormalities. Sheep with gait abnormalities will high step on the front limbs and /or bunny hop on the rear limbs. Eventually the animal is unable to rise, and if the disease is allowed to progress, found dead. Some affect sheep are simply found dead, typically after a stressful event.

Goats show signs similar to that observed in sheep. However, goats are less likely to rub and instead use horns and feet to scratch. Dairy goats can become difficult to milk. Goats also can become aggressive and cannibalistic. Pica has been observed.

Atypical

Brain lesions are different than that identified in classical scrapie, likely explaining the variability in clinical signs and onset. Atypical scrapie is seen in older animals, on average, 6.5 years old. More than 60% of affected sheep and goats are greater than 5 years old and animals up to 20 years old have been detected. Clinical signs are uncommon but a few cases have reported ataxia, weight loss, and anxiety; pruritis is not typically observed.[10] Most cases have been identified incidentally during post mortem screening.

DIAGNOSIS

Classical Scrapie

Ante mortem testing is based on IHC staining of lymphoid follicles. The more popular sites for collecting are the third eyelid and rectal mucosa.[12] Test sensitivity is depended on the number of lymphoid follicles present and species differences. Rectal biopsy results in goats have been unreliable because of the low number of positive follicles (personal communication). Blood tests are under development, however there is an inherent difficulty due to the lack of a host immune response, no foreign antigen, and the intracellular presence of the PrP^{SC}.

Post mortem testing is performed on various lymphoid tissues and brain sections. Abnormal prion protein accumulation varies with breed, age, length of infection, genotype, and type of scrapie so multiple sections must be examined. All prion diseases share several neuropathological features including neuronal loss, gliosis, deposition of amyloid plaques, and presence of large vacuoles that generates a spongiform appearance in brain tissue. Generally characteristic brain lesions are similar between sheep and goats. Immunohistochemical staining for PrP^{SC} accumulation in tissue sections has confirmed a consistent involvement of the caudal medulla in both species. The presence PrP^{SC} in the dorsal motor nucleus of the vagus nerve (DMNV) is used categorize classical scrapie in both species. Both species typically demonstrate accumulation of PrP^{SC} in a wide variety of tissues outside the CNS. Accumulation appears to be linked to the incubation period.

Western blot and ELISAs are also available for detection of scrapie. Rapid screening tests using ELISA and Western blot have been evaluated in Europe and have high specificity compared to IHC, however the sensitivity is difficult to assess due to the low prevalence of scrapie.[13]

Atypical Scrapie

Atypical scrapie shows different locations for PrP^{SC} accumulation in the CNS as compared to classical scrapie and variations are even seen between cases of atypical scrapie. Research has found spongiform changes and gliosis in more rostral brain structures, such as thalamus, midbrain and areas of the cerebral cortex and an absence in the medulla oblongata. Immunohistochemistry does not identify PrP^{SC} in lymphoid tissues typically seen with classical scrapie. No intraneuronal staining has been reported. Some cases display minimal staining restricted only to the spinal tract nucleus of the trigeminal nerve and the DMNV is not affected. Cerebellar and cerebral cortices are often the most immunostained, but staining of the cerebellum is variable.[10]

Western blot can differentiate between classical and atypical scrapie due to variations in band profiles. The PrP^{SC} found in atypical scrapie is slightly more susceptible to proteinase-K digestion creating the differences.[4,10] Western blot for atypical scrapie include a multiple protein band pattern with an unglycosylated band that measures 15 kDa, whereas classic scrapie produces a 3 protein band pattern that ranges from 20 to 30 kDa.[10]

PREVENTION AND CONTROL

No treatments are currently available. Scrapie is a notifiable disease and the national scrapie eradication program is still underway in the United States. The outbreaks in goats will likely delay the eradication deadline set for 2010. Management is aimed at decreasing the flock/herd risk through management, genetics, and preclinical testing. Management practices are common sense strategies involving biosecurity, separation of susceptible or new animals during parturition, and sanitation. Genetic tests can be used in sheep to select for genotypes resistant to classic scrapie. However this does not provide protection against atypical scrapie, and no tests are available for either type of scrapie in goats. Preclinical testing using third eyelid or rectal mucosal biopsies can be used to identify positives animals however rectal mucosa biopsies are variable in goats.[12]

Post mortem testing is recommended for all sheep and goats dying of unknown causes, in particular if animals are from groups of unknown scrapie status, animals displaying abnormal behavior prior to death, or goats housed with sheep.

RESOURCES

www.eradicatescrapie.org
www.aphis.usda.gov/vs/scrapie and www.aphis.usda.gov/animal_health
www.animalagriculture.org/scrapie

REFERENCES

[1]Arrizubieta MJ, et al. **Anal Biochem** 2009;393:229. [2]Sofianidis G, et al. **J Comp Pathol** 2008;138:90. [3]White S, et al. **Genet Sel Evol** 2008;40:553. [4]Vaccari G, et al. **Vet Res** 2009;40. [5]Hamir AN, et al. **J Vet Diagn Invest** 2009;21:64. [6]Wrathall AE, et al. **Theriogenology** 2008;70:725. [7]Maddison BC, et al. **J Virol** 2009;83:8293. [8]Konold T, et al. **BMC Vet Res** 2008;4:14. [9]Low JC, et al. **Theriogenology** 2009;72:809. [10]Benestad SL, et al. **Vet Res** 2008;39:19. [11]Loiacono CM, et al. **J Vet Diagn Invest** 2009;21:454. [12]Dennis MM, et al. **Am J Vet Res** 2009;70:63. [13]Seuberlich T, et al. **J Vet Diagn Invest** 2009;21:97.

LESIONS OF THE HOOF HORN CAPSULE & THE RELATIONSHIP TO ANATOMY & WEIGHT BEARING

Sarel R van Amstel BVSc, MMedVet, DABVP, DACVIM. & Jan K Shearer DVM, MS.
Knoxville, TN

INTRODUCTION

The anatomy of the internal structures within the claw capsule as well as that of the claw horn plays a unique and vital role in maintaining claw health during weight bearing. Important functions such as maintaining normal horn growth and claw conformation can become compromised by a variety of external and internal factors including the normal biomechanics of weight bearing. Common claw lesions and their relationship to pertinent anatomical structures and weight bearing where applicable will be discussed.

VERTICAL WALL FISSURE

Vertical wall fissure occurs most commonly on dorsal wall (toe) and consists of a partial or full thickness fracture of the horn of the wall. Front claws seem to be more commonly affected.

Horn cells are arranged in 2 main structural forms in equal amounts: tubular and inter-tubular horn. Tubular horn cells are arranged in a steep spiral around axes of the tubules. Horn tubules differ in size, number and shape in various parts of the claw. Tubules are round near the inside of wall and oval near the surface. Horn tubules provide structural strength to horn thus the number of horn tubules are directly related to horn hardness. There are approximately 80 tubules/mm^2 in the wall, while the sole and heel has only20/ mm^2.

Inter tubular horn consists of sheets of elongated polygonal cells arranged parallel with bearing surface. Horn cells are connected by intercellular cementing substance (ICS) consisting of ceramids, fatty acids and cholesterol.

Vertical wall cracks relates to the physical properties of horn particularly fracture toughness. The horn of the wall shows greatest fracture resistance for cracks running proximally because of the parallel orientation of intertubular horn cells to the bearing surface. Vertical cracks originate at the ICS junction between tubular and inter-tubular horn. Progression of a vertical crack is diverted along the horizontal plane of the inter-tubular cells which is the least fracture resistant direction. This shift in direction of the horn fracture limits its progression thus the size of the fissure.

The hydration level of horn is reported not to have a major effect on fracture properties. Cornified horn cell membrane (plasmalemma) is permeable to water and solutes but not large molecules (protein). The outward hydrostatic force and osmotic gradient from the dermis to the cornified keratinocytes influences the water content of horn cells. Horn that is immersed in water will loose electrolytes to the outside thus affecting the osmotic gradient. Fracture toughness is reported to be best at ± 18% mid-wall moisture. Desiccated horn is reported to predispose to vertical wall fissure. Horn quality is another factor which will predispose to horn fracture. Biotin is one important substance which may influence horn quality since it plays an important role in the formation and composition of the inter-cellular cementing substance. Biotin improves cell to cell adhesion as well as keratinocyte protein metabolism thus keratinization. Laminitis also predispose to the formation of inferior quality horn because of interference with blood and nutrient flow to the basement membrane and basal layer of the epidermis.

Body weight and a steep dorsal wall angle (normal angle is $50 - 55^0$) may also predispose to vertical wall crack in the front claws due to weight bearing.

HORIZONTAL WALL FISSURE

Horizontal wall fissure represents a partial or full thickness fissure of the wall of the claw commonly seen on the abaxial wall and denotes partial or complete interruption in horn growth.

Horn of the wall is produced in the area of the coronary corium which is situated between the perioplic corium (coronary band) and the sensitive laminae. The corium consists of a rich vascular network, which terminates in dermal papillae or vascular pegs. A vascular peg consists of a main arteriole connecting directly to a venule at the tip with an extensive capillary network in between.

The basement membrane (BM) is situated between the epidermis and corium and is a key structure in bridging the cytoskeleton of the keratinocytes in the basal layer of the epidermis via the corium to the connective tissue of the distal phalanx. The BM consists of a complex lattice composed of both collagen and glycoprotein (laminin, fibronectin, amyloid P, entactin and heparin sulfate proteoglycan). It provides both anchorage and orientation to keratinocytes in the basal layer and damage of the BM leads to loss of organized structure. Keratinocyte proliferation and differentiation are also controlled by signals from the BM. Pathological changes within the BM result in inhibition of growth or activation and hyperproliferation of keratinocytes and an abnormal keratinocyte growth pattern and expression of keratin protein subtypes thereby producing inferior quality horn.

The basal layer of the epidermis consists of keratinocytes undergoing active proliferation and differentiation. Neighboring keratinocytes are tightly bound by desmosomal intracellular junctions and imbedded in the intercellular cementing substance (ICS). Keratinocytes become large, polygonal and flatter upon transition into the Stratum spinosum, and cell contents replaced with keratin proteins (keratinization). These cells then undergo programmed death in the final stages of keratinization and differentiation (cornification).

Keratin protein (Cyto-keratins) is a fiber-reinforced composite material consisting of various subtypes. Keratin provides structural support to individual horn cells in order to maintain shape and proportionate dimensions of the claw horn components. Keratin proteins consist of long slender fibers (tonofilaments or microfibrils), which are aligned parallel to the long axis of the cell. Fusion of two filaments takes place to form a rope-like coiled coil. Individual fibers consist of stable and unstable domains. Cysteine–cysteine disulphide in the unstable domain cross-links by means of a binding protein to the intracellular matrix. The hydration level inside the cell stabilizes the remainder of the intracellular protein matrix. During conditions of low hydration of the epidermis, extensive secondary hydrogen cross-linking through glycine and tyrosine makes the protein matrix less flexible thus increasing horn hardness. High levels of epidermal moisture results in greater distances between secondary cross-linking, which makes the protein matrix more flexible thus decreasing horn hardness.

` Changes in the blood supply and nutrient flow caused by laminitis may adversely affect the basement membrane. In addition proliferation, differentiation and keratinization of the horn cells in the epidermis is interrupted resulting in horizontal fissures which if severe may involve the full thickness of the wall. In addition, because of poor quality horn changes in claw horn conformation is commonly present. Thus any compromise in blood flow will have a negative effect on horn production. Horn production is also dependent on the supply of nutrients which include adequate levels of protein, energy, lipids, vitamins A, D and E, calcium and phosphorous. Micro-nutrients such as the sulfur containing amino acids like cysteine and methionine are essential for cross-linking of keratin filaments. Trace minerals particularly zinc and copper, as well as biotin have very important roles in the keratinization of horn cells and integrity of the ICS of claw horn.

SOLE ULCER (ZONE 4: TYPICAL ULCER; ZONE 6: HEEL ULCER; ZONE 5: TOE ULCER)

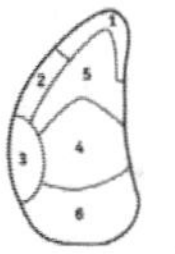

Sole ulcer represents a full thickness break in the sole horn resulting from traumatic damage to the sole corium following failure of the suspensory system for the third phalanx.

The third phalanx is suspended within the claw horn capsule by the laminar corium and a series of collagen fiber bundles that stretch from the insertion zone on the surface of P_3 to the basal layer of the epidermis via the basement membrane. The interface between dermal and epidermal components is the interdigitating dermal (sensitive laminae) and epidermal laminae (horn leaflets). The result is that P_3 hangs within the claw capsule and weight is transferred as tension onto the wall of the claw capsule. The interface between dermal and epidermal components (interdigitation of sensitive laminae and horn leaflets) in cattle differs significantly from that of horses. First, the laminar corium (sensitive laminae) is much less extensive in cattle as compared to horses. Secondly, there are no secondary laminae in the laminar corium of cattle. Therefore, capabilities with respect to mechanical load carried on the claws of cattle, varies significantly. In the horse load bearing is primarily on the wall. Cattle, on the other hand, simply cannot handle the same amount of mechanical load on the walls of their claws. Instead, weight bearing in cattle requires displacement of load to alternate support structures within the sole and particularly the heel. The caudal aspect of P3 has a distinct bony protrusion (flexor tuberosity) where the deep flexor tendon attaches. This part of P3 has limited support from the interdigitating segment of the laminar corium and horn leaflets and therefore has its own support apparatus.

The primary structures within the support apparatus of the caudal aspects of P3 in cattle are:
Firstly the **digital cushion**, which consists of loose connective tissue and adipose tissue. The digital cushions are arranged in a series of 3 parallel cylinders. The volume of all three pads in a single claw adds up to ~ 5.7ml. All three cushions extend from the skin/ horn junction at the heel towards the tip of the third phalanx. The abaxial and middle fat pads are shorter than the axial pad. The shorter pads lie superficial to the deep digital flexor tendon and do not reach further distal than the insertion of the tendon. The axial fat pad courses from the axial border of the heel bulb towards the middle of the sole surface where it ends in the middle third. Recent studies by Swiss researchers have shown that the amount of fat (and thus cushioning capacity) increases with increasing age. This is believed to have significant implications for animals relative to their susceptibility to claw disorders.

The second component of the suspensory system for the caudal aspect of P3 is the **corium and associated connective tissue** attachment to the digital cushion and the inside of the abaxial claw wall. The axial aspect of P3 is supported by attachments to the inside of the axial claw wall and the distal interdigital cruciate ligament

The pathogenesis of sole ulceration (typical ulcer and heel ulcer) is primarily based on mechanical injury to the solar corium, in particular the basement membrane and basal layers of the sole epidermis (sole horn) (zones 4and 6). Such mechanical injury has been associated with laxity or break down of the collagen fiber attachment (suspensory apparatus) between the caudal part of the third phalanx and the basal layers of the wall epidermis, resulting in sinking of the bone. Collagen is a structural protein and the main component of the connective tissue attachment (suspensory apparatus). Laxity or break down of this collagen support may be hormonal or enzymatic and is associated with calving and /or laminitis.

Toe ulcer (zone 5) represents a full thickness break in the sole horn at the apex of the toe resulting from traumatic damage to the sole corium following failure of the suspensory system (interdigitation of sensitive laminae and horn leaflets) for the cranial and middle aspects of the third phalanx at the toe. The apex of the third phalanx causes pressure on the corium and epidermis.

Blood supply to dermal folds (sensitive laminae) in the laminar corium consists of numerous centrally situated arterioles and venules, which branch to form a dense arcade of capillaries and venules at the base and ridge of the dermal folds. Short hook-shaped direct or indirect arterio-venous anastomoses occur irregularly at the base of the dermal folds.

Interference of the blood supply to the dermal folds can result in separation of the laminar corium and horn leaflets with rotation of the apex of P3 with traumatic damage to the cranial aspect of the solar corium. This may result in the development of a toe ulcer.

Toe ulcer may also occur in the sole adjacent to the white line in zone 2 as result of overly thin soles. Abrasive walking surfaces may cause thinning of the sole to the extent that the sole becomes separated from the white line with the formation of an ulcer.

WHITE LINE DISEASE

White line disease goes through various stages and includes the following progression: White line hemorrhage; White line separation; White line abscess (subsolar and or wall/coronary band abscess); Complicated white line disease involving structures in the heel and eventually the distal interphalangeal joint.

Eighty percent of the horn of the white line (WL) consists of laminar horn, which is produced from a) germinal epithelium overlying the laminar corium just below the coronary corium. This forms the outer zone of the WL and consists of soft non-tubular horn b) germinal epithelium overlying the dermal papillae (also referred to as dermal caps or cap-papillae), which protrude from the dermal folds on the laminar corium and produce non-tubular cap-horn. This horn represents the middle zone of the WL and c) terminal papillae which form large loosely arranged tubules (inner zone). The remainder of the white line is made up of coronary horn (horn leaflets). Laminar horn is non-tubular horn (outer and middle zones), but has large, short and hollow tubules toward the bearing surface (inner zone). This makes it structurally soft, flexible, with a high turnover rate. Because of the high turnover rate the cells are generally less mature and therefore softer and less resistant to wear and other environmental factors.

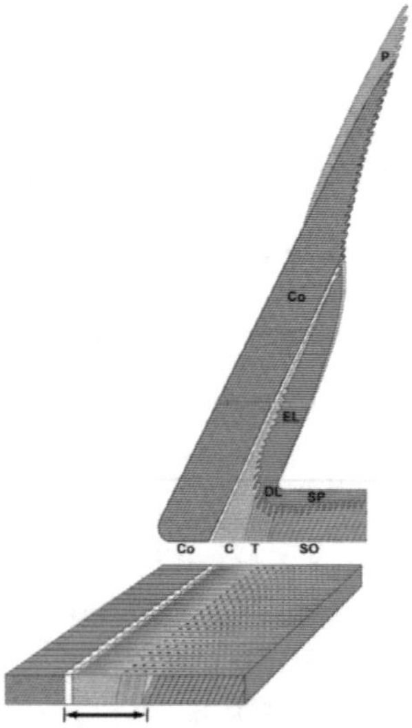

Arrow indicates 3 zones of the WL
Co = wall
C = middle zone of WL
T= Inner zone of WL
SO = sole
Thin line between Co (wall) and C (WL middle zone) = outer zone of WL

The causes of WLD includes overgrowth in which case the condition starts from the outside or subclinical laminitis and breakdown of the suspensory system in which case the condition starts on the inside and works its way to the outside. Horn overgrowth is particularly evident in the outside claw of the rear leg. Force plate studies have shown that the ground reaction force associated with weight bearing is highest at the abaxial heel/wall/sole junction (zone 3) of the rear outer claw during the initial phase of the stride (heel strike). Mechanoreceptors stimulation in the basal layers of the epidermis results in accelerated epidermal cell proliferation and differentiation thus overgrowth. Increased weight bearing will predispose to mechanical breakdown of the soft horn of the white line particularly evident in zone 3.

WLD may occur in any of the following zones: axial 1; abaxial 1; abaxial 2 and abaxial 3. Causes of WLD can be summarized as follows: **Axial zone 1**: 1. Screw claw. The WL is non-weight bearing because of rotation of the toe leading to abnormal cell turn over and separation. 2. Laminitis resulting in separation of dermal folds and horn leaflets thus interfering with laminar horn formation. **Abaxial zone 1and 2**: 1. Screw claw and 2. Laminitis (see previous). 3. Extension of toe ulcer due to abrasive surface (concrete) or P3 rotation. **Abaxial zone 3**: 1. Laminitis 2. Weight bearing (see above)

CONCLUSION

An intimate knowledge of the structural anatomy of the claw and associated structures as well as the normal biomechanics of weight bearing greatly facilitates understanding the pathogenesis and progression of claw horn lesions. This in turn will facilitate identification of possible predisposing causes leading to appropriate lameness control measures.

REFERENCES

Baillie, C., C. Southam., A. Buxton, and P. Pavan. 2000. Advanced Composites Letters. 9(2):101-113. Greenough P, Laminitis and Lameness: A hands on approach 2007. Hendry KAK, Knight CH, Galbraith H, Wilde CJ. 2000 In 11th International Symposium on Disorders of the Ruminant Digit and 3rd International Conference on Bovine Lameness. Parma, Italy, pp. 128–129. Hirschberg RM, Mulling ChKW. 2002 In Proceedings of the 12th International Symposium on Lameness in Ruminants, Orlando, FL, pp. 223– 226. Lischer ChJ, Ossent P, Raber M, Geyer H 2002, Vet Rec151, (23) 694 – 8. Tarlton JF, Holah DE, Evans KM, Jones S, Pearson GR, Webster AJ. 2002 Vet J 163 (2) 196-204. Tarlton JF, Webster AJF. 2002, In Proceedings of the 12th International Symposium on Lameness in Ruminants, Orlando, FL, pp. 395–398. Bertram, J.E.A, and J.M. Gosline. 1987. J. exp. Biol. 130:121-136. van Amstel SR, Shearer JK. Review of Pododermatitis Circumscripta (Ulceration of the Sole) in Dairy Cows. J Vet Intern Med 2006: 805-811 Van Amstel SR, Shearer JK. Manual for the Treatment and Control of Lameness 2006 Blackwell publichers. V an der Tol PP, Metz JH, Noord huizen – Stassen EN, Back W, Braam CR, Weijs Wa. 2003, J Dairy Science.Sep 86, (9) 2875-83. Wagner, I.P., and D.M. Hood. 2002. AJVR, 63(8):1140–1144.

PATHOPHYSIOLOGY & DIFFERENTIATION OF TOE LESIONS IN DAIRY CATTLE

Jan K. Shearer, DVM, MS and Sarel R. van Amstel BVSc, MMedVet(Med), DABVP, DACVIM
Ames, IA

INTRODUCTION

For the past several years these authors have been studying toe lesions in cattle with a specific interest in their pathogenesis and underlying causes. Our work to date indicates that there are at least 7 conditions that might predispose to toe lesions. These include: toe ulcer (zone 5), white line disease (zones 1 & 2), thin soles and thin sole toe ulcer (TSTU) in zone 5 adjacent to the junction of zones 1 & 2, corkscrew claw (zones 1 & 2), wall cracks (especially horizontal wall cracks), trauma-related lesions affecting the tip or apex of the claw and iatrogenic causes associated with over-trimming or improper trimming technique. In our experience, toe ulcers associated with laminitis and rotation of the apex of the third phalanx (P3) are relatively rare compared with ulcers of the heel or sole. White line disease lesions in the toe are slightly more common and sometimes associated with inappropriate trimming techniques (i.e. excessive removal of the axial wall at the toe in zone 1). Laminitis is believed to be a potential contributor to white line lesions in the toe as a result of its effects on horn quality and the production of dyskeratotic horn. However, the pathogenesis of this lesion appears to be different than that observed for white line lesions occurring in the abaxial heel (zone 3). Corkscrew claw is relatively common and predisposes to toe lesions by a couple of different routes. Trauma-related lesions that result in toe abscess formation are more common in feedlot cattle where hyper-excitable animals are prone to traumatic injuries of the toe during processing or hauling. The experience of these authors suggest that one of the most common causes of toe lesions in dairy cattle, particularly on large dairies, is thin soles and TSTU. These disorders are a consequence of abrasive flooring conditions and complicated by the requirement of cows to walk long distances on concrete surfaces in large operations. They may also be exacerbated by over-trimming and/or by the inappropriate scheduling of preventive maintenance trimming activities.

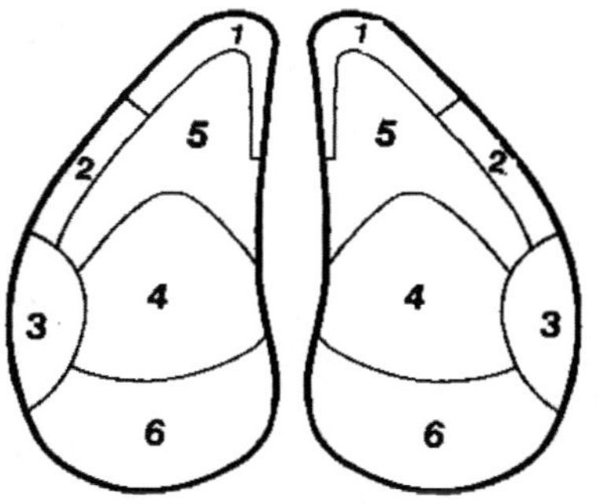

Claw Zones

TOE ULCER

Breakdown of the suspensory apparatus associated with the release of matrix metalloproteinases leads to sinking and rotation of the third phalanx. For reasons that are not totally clear, in some animals rotation of the apex of P3 surpasses sinking of the caudal portion of the bone. The result is pressure on the solar corium in zone 5 that leads to hemorrhage that may be most noticeable in the white line region of zone 1. Dysfunction of the solar corium at the point of injury can lead to development of a full thickness defect in the sole. In the worst case scenario, the solar corium and P3 may actually prolapse through the sole. We speculate that one reason why the more severe form of this lesion is less commonly observed in cattle may be due to differences in the suspensory apparatus of the bovine compared with the equine. In cattle breakdown of the suspensory apparatus by metalloproteinase activity leads to weakness and elongation of collagen fiber bundles, whereas in equines a true separation of the dermal-epidermal junction is more likely to occur. Regardless, the differentiation of a toe ulcer from other toe lesions may be difficult since lesions are often well-advanced by the time they are first observed.

WHITE LINE DISEASE (ZONES 1 & 2)

The white line is produced by the laminar corium. It is a 3-part structure consisting of an outer, intermediate and inner zone. The outer and intermediate zones consist of laminar horn, whereas the inner zone (adjacent to the sole) is a heterogeneous combination of laminar horn and loosely arranged tubules. These structural features make it the softest and least resistant part of the claw capsule and subject to damage by mechanical shearing forces and the penetration of bacteria and foreign bodies such as coarse dirt and gravel. The area of the white line most commonly affected is the abaxial heel-sole-wall junction of the lateral claw (zone 3 on the claw zone diagram). The white line in this region is naturally predisposed to greater mechanical impact and wear during locomotion since this area bears the impact of heel strike during the foot placement phase of the stride. Overgrowth and overloading of the outer claw tends to exacerbate load bearing and may increase white line disease problems in zone 3.

White line lesions in zones 1 & 2 tend occur less commonly and may have a slightly different pathogenesis. These areas of the white line do not withstand mechanical impact to the same degree as zone 3 and thus, claw horn wear and turnover is slower in these zones. Nonetheless, lesions normally begin as small cracks or spaces within the white line that become infiltrated with organic matter. The entrapment of extraneous material within these spaces may be visualized as one or more dark lines, cracks or often as a small dark spot within the white line. In other cases where separation of the white line is advanced and complicated by infection, the lesion may appear as a large area of loose necrotic horn. Abscess formation associated with white line disease creates severe lameness. These abscesses and associated purulent material may accumulate in the subsolar region of the toe or in many cases migrate upward beneath the wall. Upon reaching the skin-horn junction, these may rupture forming a sinus tract at the coronet. In the worst case scenario, infection may extend into P3 where the resulting osteitis creates a very painful and often times chronic condition. Proper treatment requires the debridement of all necrotic tissue including affected portions of P3.

CORKSCREW CLAW

Corkscrew claw is most commonly observed in the lateral claw of rear feet in cattle. It is reported to be a heritable condition, however factors such as claw disease and housing conditions are likely to influence occurrence. It is characterized by an abaxial to axial rotation of the toe that displaces the sole, axial wall and white line into an axio-dorsal position. Internally, P3 rotates in similar fashion such that the dorsoplantar plane of the distal interphalangeal joint may be rotated by as much as 11 degrees. In the heritable form of this disorder, there are a couple of important anatomical abnormalities that contribute to development of this problem: 1) a misalignment of the second and third phalanges that positions the abaxial surface of P3 in close proximity to the weight bearing surface, and 2) an elongated and narrow third phalanx. Curving and upward rotation of the apex of the toe results in weight bearing on the mid to caudal portion of the abaxial wall. Corkscrew claws are normally larger and typically bear the majority of weight in the foot. The corresponding lack of weight borne on the unaffected digit usually results in significant atrophy of that claw.

Toe lesions in corkscrew claws develop in one of two ways: 1) white line separation in zones 1 and 2, and 2) as the axial wall rotates and repositions itself dorsally, it has a tendency to become folded over on itself trapping organic matter within the fold. This is a near perfect environment for anerobic bacteria that eventually cause further necrosis ultimately exhibiting itself as a toe abscess. Horn of the white line in zones 1 & 2 of corkscrew claws often shows evidence of hemorrhage and degeneration. This increases its vulnerability to separation and increases the possibilities of white line disease in these zones. Routine trimming at 3-4 month intervals helps to maintain normal shape and function of the claw capsule, but must be done carefully to avoid excessive thinning of the sole in the toe (zones 1 & 2).

TRAUMA-RELATED LESIONS OF THE TOE

Cattle working facilities, certain types of flooring conditions and rough or careless animal handling contribute to an increased risk of trauma-related lesions of the toe. These are normally more common in feedlot cattle or younger animals that may be less accustomed to human contact and handling. Fractures of the claw capsule, and on occasion P3, lead to severe lameness and complications such as the formation of a sequestrum. A sequestrum often occurs when injury results in fracture of P3 with subsequent loss of blood supply to that portion of the bone. It is not uncommon to find a loose necrotic piece of bone in chronic toe lesions. In many cases, removal of this necrotic bone will permit complete healing of an otherwise chronic lesion.

IATROGENIC - OVERTRIMMING AND IMPROPER TRIMMING TECHNIQUE

Until recent time foot trimming in North America was learned largely through experience or by working as an apprentice with an experienced person. In most cases, the trimming techniques applied to cattle were similar to those used on horses. However, as we have learned more about the anatomy and physiology of the bovine foot, our approach to trimming has changed. Toussaint Raven with his book "Cattle Footcare and Claw Trimming" is credited with introducing the greatest advancements in foot trimming technique in modern time. Properly applied, his functional claw trimming procedures prevents over-trimming of the foot, balances weight bearing within and between the claws of each foot, and provides the foot with stable weight bearing surfaces.

One of the most common trimming errors is over-trimming. Different trimming methods can result in significant differences in sole thickness. For example, one study found that the trimming method described by Toussaint Raven (also referred to as the Dutch method) resulted in fewer over-trimmed feet as compared to another popular method. The Dutch method uses length of the dorsal wall as a guide to estimate sole thickness. For the average adult Holstein cow a toe length of 3 inches (7.5cm) corresponds to a sole thickness of ~ 0.25 inches (~6.25 mm), which under normal circumstances of growth and wear provides enough sole horn to protect the solar corium. However, trimming the sole to a thickness of 3-4 mm or less creates conditions whereby the sole may not be to protect the underlying solar corium in an animal housed on a hard flooring surface. Over-trimming typically leads to lesions in zone 5 characterized by separation of the sole adjacent to the white line in zones 1 and 2. These are described in greater detail in the section under thin soles.

There are several improper trimming techniques that may result in toe lesions: trimming the toe too short, excessive removal of axial wall and white line, and excessive removal of wall horn on the apex of the claw and others. Many of these techniques create lesions by direct exposure of the corium or by causing weakness of the corresponding portion of the claw horn capsule affected.

Finally, as keratinocytes migrate outward from the basal cell layer of the epithelium, they incorporate keratin within their cellular structure. The most mature horn cells and those with the greatest degree of keratinization are those within the stratum corneum layer. This corresponds to the most external layer or horn that has reached the surface of the claw capsule. Trimming removes the most mature and therefore the hardest horn layers. Horn that remains is less mature and thus softer (i.e. wears more rapidly). While regular trimming that corrects claw horn overgrowth and altered weight bearing is a useful adjunct to foot and leg health, trimming schedules that are too aggressive or frequent, coupled with flooring conditions that contribute greater than average wear rates, increases the potential for thin sole problems. Trimming schedules must be developed on a farm-by-farm basis in consideration of claw horn wear and growth characteristics unique to each operation.

WALL CRACKS

Cracks or fissures in the hoof wall are common in cattle. Those which run in a vertical direction (from the coronet to the weight bearing surface) are referred to as vertical wall cracks or sandcracks. Incidence rates as high as 64% have been reported in beef cattle, compared with less than 1% in dairy cattle. Therefore, for the purposes of this discussion, we will focus our attention on horizontal wall cracks.

Cracks or fissures that run in a horizontal direction (parallel to the coronet) are referred to as horizontal wall cracks. These are common in both beef and dairy cattle and when severe may result in profound lameness. The causes of horizontal wall cracks are better understood. In some cases they simply signal a "physiological change" that has resulted in a mild to moderate interruption of horn growth and formation in the basal cell layer of the coronary corium. In others, they are representative of conditions (often related to disease disorders) that have led to significant "physiological stress" and severe interruption of horn formation in the hoof wall. These severe disruptions in horn formation are exhibited as very distinct ridges and grooves that run in a horizontal direction on the hoof wall. They are often referred to as "hardship grooves" or "stress lines". In the most extreme cases where the fissure is sufficiently deep to result in a full thickness defect of the wall, the lesion is often referred to as a "thimble". As these fissures migrate downward toward the weight bearing surface, the segment of the wall closest to the weight bearing surface becomes moveable as it reaches or nears the weight bearing surface. When this occurs, it becomes extremely painful as it pinches and damages the underlying sensitive tissues. The objective of treatment is to stabilize the lose sections of the wall by removal of loose portions or by other means including the use of a foot block on the opposite claw assuming it is more stabile and will support a block.

THIN SOLES AND THIN SOLE TOE ULCER

In housing systems where the rate of sole horn wear exceeds the rate of growth excessive thinning of the sole is likely to occur. Previous work has demonstrated that claw horn hardness is influenced by nutrition, contact with manure slurry and moisture content of claw horn. Claw horn is continually exposed to high moisture conditions particularly during the hot and humid summer months. Heat stress abatement procedures require that cows have access to sprinklers and fans, misters or high pressure fogging systems. Claw horn moisture content is also affected by manure management systems based on flushing of fresh or recycled water to clean floors in barns, holding areas and travel lanes. Wear rates are also affected by the spatial layout of facilities which require cows to walk long distances to and from barns and milking areas. This is exacerbated by abrasive flooring conditions that include sharp turns and sloped walkways. Excessive sole horn wear is especially common in new installations where freshly hardened concrete creates a particularly abrasive surface as a consequence of the presence of surface aggregate which naturally forms on the flooring surface as the concrete cures. This observation has become as commonplace as to have its own name "New Concrete Disease".

The occurrence of thin soles is also influenced by conditions contributing to poor cow comfort such as overcrowding and reduced stall use due to improper stall design and insufficient bedding. Issues of dominance also affect stall use. When stall numbers are equivalent or less than the total number of cows in the barn, timid animals such as heifers may have less opportunity to rest. Some recommend that there be at least 10% more free stalls than cows to permit greater choice and encourage resting. Generally speaking, stall design should accommodate the following resting behaviors: 1) Ability for the cow to stretch front legs forward. 2) Ability to lie on its side with sufficient space for the head and neck. 3) Ability to rest its head on its side. 4) Sufficient room for the cow to rest its legs, udder, and tail on the free stall platform and have a clean, dry and soft bed. Some US recommendations for Holstein cattle include construction of a free stall 8 ft (2.5m) long [7ft 6in. (2.28m)] for two facing rows) and 4 ft (1.25m) wide with a brisket locator 15in. (38cm) high and located 5ft. 8in. (1.72m) from the stall curb. For free stalls facing a wall a length of 9 ft. 8 in. (2.94 m) are recommended by some.

Sole thickness should be a minimum of ¼ inch (~ 7 mm) to properly protect the corium and support a cow's weight on a hard surface such as concrete. When sole thickness is 1/8 inch (~ 3-4 mm) or less, the sole will flex with finger pressure. Continued wear and thinning of the sole leads to a separation of the sole from the white line at the junction of zones 1 & 2. This exposes the solar corium and therefore by definition is properly termed an ulcer. These authors have chosen to call it a thin-sole-toe-ulcer (TSTU) to distinguish it from toe ulcer due to laminitis. It is a very prevalent lesion in the southeastern US and observation in recent years suggests that the incidence of thin soles and TSTU is increasing as herds continue to enlarge.

SUMMARY

The differentiation of toe lesions is important since the predisposing causes of each are different. Toe lesions represent some of the most painful of lameness conditions, treatment can be complicated and complete recovery is often prolonged. Misdiagnosis of toe lesions is common since most are observed in an advanced state where the underlying cause may be difficult to determine. Nonetheless, an accurate diagnosis and determination of underlying cause are important to the development of effective management recommendations.

REFERENCES

AABP Lameness Committee, AABP Fact Sheet. 2006. Barnes MM: **Br Vet J.** 145 (5): 436-445, 1989. Budras KD, Habil Dr, Mulling Ch, Horowitz: A. **AJVR** 57(4): 444-4551, 1996. Cermak J: **Bovine Pract** 23: 76-78, 1998. De Belie N, Rombaut E: **Biosystems Engineering** 85 (3): 339-346, 2003. Kempson SA, Langridge A, Jones JA: **Proceedings of the 10th International Symposium on Lameness in Ruminants**, Lucerne, Switzerland: 216-217, 1998. Mulling Ch, Budras KD: **Proceedings of the 10th International Symposium on Lameness in Ruminants**, Lucerne, Switzerland: 214-215, 1998. Mulling Ch KW: **Proceedings of the 12th International Symposium on Lameness in Ruminants,** January 9-13, Orlando, Florida: 90-98, 2002. Ossent P, Lischer C: In Practice. September 1998: 415-427. Oxford concise Medical Dictionary. Fourth Edition. Oxford University Press, Walton street, Oxford ox2 6dp: 684, 1994. Paulus N, Nuss K: **Proceedings of the 12th International Symposium on Lameness in Ruminants**, January 9-13, Orlando, Florida: 428-430, 2002. Shearer JK, van Amstel SR: **Proceedings of the Western Dairy Management Conference**. Reno, Nevada, March 7-9: 149-159, 2007. Shearer JK van Amstel SR, Benzaquen M, Shearer LC: **Proceedings of the14th International Symposium on Lameness in Ruminants**, Colonia, Uruguay, November 7-11: 110-111, 2006.Thompson PN: Jnl of the S Afr Vet Assoc. 69 (1): 23-26, 1998. Toussaint Raven E: Chapter 3: 75-106, 1989. Valentino LW, St.Jean G, Anderson DE, Desrochers A, Kersting K, Lopez M, Adams SB, Huhn J, Mueller EPO, Cohen ND: 110 cases (1987-1997). JAVMA 217 (3): 376-383, 2000. van Amstel SR, Palin FL, Shearer JK, Robinson BF:**Bovine Pract** 36 (2): 136-140, 2002. van Amstel SR, Shearer JK, Palin FL: J. **Dairy Sci**., 87:757-763, 2004. van Amstel SR, Palin FL, Shearer JK: **Veterinary Record** 155: 630-633, 2004. van Amstel SR, Shearer JK, Palin FL, Cooper J, Rogers GW: **Proceedings of the14th International Symposium on Lameness in Ruminants**, Colonia, Uruguay, November 7-11: 142-143, 2006. van Amstel SR, Shearer JK: **Manual for Treatment and Control of Lameness in Cattle**.. Blackwell Publishing Professional, Ames, Iowa: 59-62, 2006. Van der Tol PP, Metz JH, Noordhuizen-Stassen EN, Back W, Braam CR, Weijs WA: **J Dairy Sci**., 86 (9): 2875-83, 2003. Van der Tol , Metz JHM, Noordhuizen-Stassen EN, Back W, Braam CR, Weijs WA: **Proceedings of the 12th International Symposium on Lameness in Ruminants**, Orlando Florida: 202-205, 2002.

UPPER LEG LAMENESS: DIAGNOSTIC & THERAPEUTIC APPROACHES

André Desrochers DMV, MS, DACVS, DECBHM
Saint-Hyacinthe, Québec, Canada

INTRODUCTION

Upper leg lameness conditions are difficult to diagnose in adult cattle. Depending on the clinician, definition of upper leg lameness conditions will be either above the fetlock or above the carpus or tarsus. Lameness above the carpus and tarsus are difficult to diagnose and treat because of the heavy surrounding muscles. Even if you suspect upper leg lameness the foot of the affected leg should be carefully examined except for obvious reason of lameness (e.g. fracture, tendon laceration, and septic arthritis). A thorough lameness exam is often needed to specify the area affected followed by ancillary tests like arthrocentesis and medical imaging. The most frequent conditions causing upper leg lameness are listed here:

Fractures: Tibia, radius, femur (calves), femoral head (adult), humerus
Septic arthritis: carpus, tarsus, stifle
Ligament injuries: Stifle, coxofemoral luxation, patellar luxation, upward fixation of the patella
Tendon injuries: gastrocnemius, peroneus tertius
Osteochondrosis: stifle, tarsus, shoulder
Neurologic: sciatic nerve (peroneal and/or tibial branches), femoral nerve, obturator nerve

LAMENESS EXAMINATION

Observation at a distance

A great deal of information is obtained by taking a few minutes to look at the animal quietly in its stall or in its normal environment (e.g. pasture, dry lot, group housing). Lameness is most easily assessed when the cow is observed in motion. In tie stall barns, lameness evaluation is more challenging because of limited space and ease of movement. Displacement of cows in the hallway could be difficult and hazardous.

Cattle lameness is generally obvious by observing the cow's stance. Attention should be paid to the posture of the cow, including the back, shoulders, pelvis, and major limb joints. With the animal standing, the general stance is observed first and more specifically at each limb and digit. Compare one region to the opposite side and determine if obvious swelling, wounds, shifting of weight, and foot posture such as toe touching or displacement of weight bearing on the medial or lateral claw are present.

Examination of the claws reveals excess wear of the wall and sole of the healthy digit. In long standing diseases with severe lameness, the heels are taller and the wall longer on the affected claw compared with that of the healthy claw. A dropped fetlock (e.g. hyperextension of the fetlock joint) may be noticed on the sound limb because of excessive load on the flexor tendons and suspensory ligament. In young animals, angular limb deformities secondary to uneven weight bearing occurs rapidly with chronic lameness. When chronic lameness occurs in a hind limb, the contralateral tarsus typically will develop varus deformity. When chronic lameness occurs in a forelimb, the carpus typically develops a valgus deformity.

Differential diagnoses for non-weight bearing lameness should always include: sole abscess, fracture, major joint luxation (e.g tarsus), critical weight bearing ligament or tendon injury (e.g. gastrocnemius muscle), critical nerve injury (e.g. radial nerve, femoral nerve, sciatic nerve), septic arthritis, and septic tenosynovitis. An abnormal deviation of the limb is usually related to a fracture or joint luxation. The stance and position of the limb is abnormal with nerve damage, tendon rupture, or a severe ligament injury. Cattle affected with a radial nerve paralysis will have a dropped elbow, but this must be differentiated from a humerus fracture, radius / ulna fracture, or septic arthritis of the elbow joint. A rupture of the tendinous portion of the gastrocnemius muscles is shown by a hyperflexion of the hock and a drop calcaneus, but this must be differentiated from a fractured calcaneus or sciatic nerve paralysis. Careful attention should be paid to muscle atrophy because this could be caused by nerve injury or disuse. Neurogenic muscle atrophy occurs very rapidly (7 to 10 days) and is severe. Muscle atrophy caused by disuse occurs over a longer period of time (weeks). Chronic lameness of the front limb will usually bring atrophy of the triceps, biceps and scapular muscles. The consequence of this atrophy is a more apparent shoulder with joint instability and the animal may be falsely diagnosed with shoulder joint diseases. Similarly, atrophy of the muscles of the rear limb causes pronounced greater trochanter of the femur which may be misdiagnosed as a coxofemoral joint luxation.

Examination during movement

In certain cases, lameness is subtle and other procedures are necessary to localize the lesion. Lameness' characteristics are easier to assess with the animal walking. The observer looks at the severity of the lameness and assesses the individual components of the gait including the arc of flight, position of the digit when it touches or leaves the ground, and the relative time spent in each phase of the stride.

When diseases of the proximal limb, such as the hip, stifle, or shoulder are suspected, the individual structures should be palpated as the animal walks. Bone-on-bone crepitation could be felt if a luxation or fracture is present. Soft tissue crepitation may be felt if tendon or ligament injury is present. Gas crepitation may be felt if emphysema is present. It is sometimes difficult to pinpoint the location of the crepitation because it can be felt some distance from the lesion. Marked bone-on-bone crepitation originating from the stifle and coxofemoral joint often feel similarly and can be difficult to distinguish. Rectal palpation of the pelvis and region of the coxofemoral joint during ambulation may help localize the lesion to the coxofemoral joint. Alternatively, a stethoscope may be used to auscultate the suspected regions during walking or during manipulation of the limb in an attempt to localize the point of maximum intensity of the crepitus.

Examination of the affected limb

At this point, the clinician should have an idea of which leg is affected and an estimation of the affected region of the limb. Now, the affected limb must be examined carefully. Unless an obvious lesion is apparent, we start by a palpation of the limb from the digit working up the leg. The clinician should watch for pain reaction and determine if swelling, deformation, crepitation, warmth, and wounds are present. A hoof tester is used to evaluate pain of the claw. A hoof tester is often used to rule out any pain originating from the foot. Sedatives and tranquilizers should be avoided whenever pain responses are desired. After localization of the lesion, local anesthesia or sedation may be required to complete the examination.

Examination of long bones is performed by applying firm pressure in regions of minimal soft tissue covering (e.g. medial aspect of tibia and radius, greater trochanter of femur, greater tubercle of humerus, etc). If the animal has an adverse response, evidenced by withdrawal, avoidance, attempts to kick the evaluator, or muscular flinching, then the opposite leg should be palpated for comparison. Most fractures are obvious, but incomplete non-displaced fractures can be suspected if there is a painful reaction after deep palpation of the limb. Each joint should be palpated separately and complete flexion, extension, abduction and adduction of the limb done. Isolation of the shoulder and elbow or of the stifle and tarsus are difficult when flexion or extension movements are performed because these joints as united by muscle tendon units.

Specific physical exams are used when injury to the coxofemoral joint or cruciate ligaments are suspected. Examination of the coxofemoral joint requires manipulation of the rear limb which can be difficult in heavy or stubborn animals. The cranial drawer test is performed with the animal standing but coxofemoral joint assessment is easier on the animal in lateral recumbency with the affected limb uppermost. The relative position of the greater trochanter to that of the tuber coxae and the tuber ischii should be determined before than animal is laid down. The normal position of the greater trochanter is ventral to both of these bony prominences and imaginary lines drawn between them will create a "triangle". Failure to palpate the greater trochanter may suggest a cranioventral luxation of the coxofemoral joint. Positioning of the greater trochanter in-line with the tuber coxae and tuber ischii suggests caudodorsal luxation of the coxofemoral joint. After the animal is laid down with the affected leg uppermost, the foot or the metatarsus III/IV is grasped and the entire limb rotated while performing repeated abduction and adduction motions. Fracture of the physis of the head of the femur (capital physeal fracture) should elicit crepitation of the hip that can be felt and occasionally heard. Coxofemoral joint luxation might elicit more crepitation (ventral luxation), excessive movement of the greater trochanter, and ease of abduction if the luxation is cranioventral.

Cranial cruciate ligament (CCL) rupture is more difficult to diagnose. Typically, the stifle is swollen and painful to palpation. A "drawer" test can be performed with the animal standing and is easier if weight bearing on the injured limb. The examiner stands immediately behind the affected leg and places both hands on the tibial crest by encircling the limb. Then, the examiner's knee is placed on the back of the calcaneus. A drawer test is positive if displacement or a crepitation can be felt after firm caudal traction on the tibial crest followed by a sudden release. The anatomy and function of the rear limb of cattle is such that the tibia is already displaced cranial to the femur when the CCL is ruptured. Thus, movement of the tibia caudal is a sign of "positive drawer". An alternative technique is for the examiner to stand cranial and lateral to the affected limb and place both hands on the tibial crest. Then, a firm, rapid thrust is applied to the proximal tibia and the limb observed for displacement and felt for crepitus.

SOME SPECIFIC UPPER LAMENESS

Long bone fractures

Humeral, femoral, tibial and radial fractures are usually easy to diagnose but difficult to treat. Humeral fracture is the only one that can be treated successfully with stall rest only. Internal fixation of external immobilisation must be used for the other long bone fractures. Tibial and radial fractures are successfully treated with plates, pin casting or Thomas splint. Diaphyseal fermoral fracture is more frequent in calves and is usually treated with IM stack pinning. Femoral head fracture can be fixed with canulated screws.

Cranial cruciate ligament rupture

Cranial cruciate ligament (CCL) rupture is not a common disease in cattle. However it can be misdiagnosed or not considered on a cow with obscured hind limb lameness. Meniscus injury is reported to be common with CCL injury. Its rupture is caused by an acute injury or secondary to a degenerative arthritis with progressive fraying of the ligament. Clinical presentation is pretty much the same whatever is the origin of the rupture. The cow is lame and the stifle is swollen. If the instability is severe, a popping noise can be heard at walk. Palpation of the stifle revealed swelling and pain. A drawer movement can be performed on the animal standing. A drawer is considered positive if crepitation and movement are felt during the manipulation. Radiographic views of the stifle will confirm your diagnosis.

The decision to treat or not will depends of the animal value, its weight, and the chronic state of the disease (degenerative changes on radiographs). Cattle with severe osteoarthritis are poor candidate for a surgery. Conservative treatment is proposed if the animal weight is lower than 500 kg or cattle with severe osteoarthritis. Two surgical treatments are commonly used: articular capsule imbrications or intraarticular techniques. Both surgical approaches have to be performed under general anesthesia. Imbrication only is used on cattle between 500 and 600 kg. Intraarticular approaches are applied on cattle weighing 600 kg and more. Short-term prognosis is good. Seventy to 80 % will return to previous production but depending of the facilities and the type of production, osteoarthritis may shorten life expectancy. The goal of this surgery is to prolong production life of an animal by retarding the degenerative process, allowing collecting embryos or semen.

Coxo-femoral luxation

Most coxo-femoral luxation occurred during the peripartum period following milk fever or paresis of the hind limb (obturator nerve paralysis). Other causes of luxation are accidental, like mounting during oestrus of slipping on a wet floor. Typically there are two kind of clinical presentation: standing or downer cows. Downer cows have a poor prognosis because of high probability of recurrence after reduction and concomitant nerve and muscle damages. Standing cows with coxofemoral luxation are quite rare in our clinics. Nevertheless, those animals have a good prognosis if the reduction is performed within 48 hours after the beginning of clinical signs. Clinical diagnosis is difficult and should be based on 3 criteria. The first one is external rotation of the digits and internal rotation of the hock on a standing animal. The second is the asymmetric position of the greater trochanter compared to the opposite side. Finally, the third criterion is a crepitating joint when rotation of the hip is performed. If the animal is down, palpation and manipulation is easily performed on lateral recumbency. Femoral neck fracture or slipped capital physis should be included in the differential diagnosis. Radiographic views of the hips are helpful and diagnostic but not available to most veterinarians because powerful radiographic equipment is necessary.

Most cattle with luxation seen in our clinics are ventral and fatal because of the impossibility of reducing the luxation, the extensive muscle damage and the cow being almost always down. In the literature, cranio-dorsal luxations are reported more frequently. If the luxation occurred less than 48-72 hours, a manual reduction is indicated. However, if the luxation is chronic, internal reduction is necessary. Contracted muscles of the hip and the presence of a hematoma in the acetabular cavity make manual reduction, in chronic cases, quite difficult. Reduction manoeuvres are performed on a sedated animal lying down on its sound hip.

Osteochondrosis

Osteochondrosis is a common disorder in horses and pigs but infrequently reported in cattle. An abnormal endochondral ossification of the articular cartilage results in fragmentation of the cartilage or a sub-chondral cyst. It's a disease related to fast growing animals, fed with high-energy ration with mineral imbalance. The joints most frequently affected in cattle are reported to be the proximal humerus, distal radius and the lateral femoral trochlear ridge. This was based on slaughterhouse finding. These lesions could be clinical or not. However, in our clinics we see OCD of the stifle, tarsus and distal interphalangeal joint. Animal with clinical signs will show a moderate lameness with distended joint (s). Clinical diagnosis is based on the history (rapid growth, mild chronic lameness unresponsive to symptomatic treatment) and physical exam. Radiographic views of the affected joints are necessary to confirm the diagnosis and to differentiate from other joint diseases. Certain radiographic lesions are difficult to differentiate from septic arthritis. A sclerotic area usually surrounds the lysis in the sub-chondral bone. Arthrocentesis is indicated to rule out septic arthritis.

Conservative treatment should be considered if radiographic lesions and clinical signs are moderate. Conservative treatment for osteochondrosis is stall rest, ration adjustment, and NSAIDs if needed. If a cartilage fragment is loose and large, arthroscopic debridement is indicated. Prognosis is generally good depending of the lesion's size and the severity of the clinical signs.

Patellar luxation and femoral nerve paralysis

Following dystocia, the femoral nerve or the medial femoropatellar ligament may be damaged secondary to excessive traction. Clinical signs are immediate and the calf has difficulty to raise and stand. Their stance and gait are typical. The patella is displaced laterally to the lateral ridge of the femoral trochlea. It is easy to replace the patella at its normal position but usually it does not stay in place very long. Biomechanically, patellar luxation or femoral nerve paralysis will have the same consequences: voluntary extension of the stifle is impossible. If the nerve is responsible of the patellar luxation, muscles atrophy should be more rapid (3 days). It is not always easy to evaluate this on a thin Holstein calf. Femoral nerve paralysis could be bilateral. Surgical treatment concentrates on the patellar luxation. Medial imbrication of the patella is the treatment of choice. If the imbrication is not sufficient to keep the patella in place, trochlear sulcoplasty should be performed. The goal of the surgery is to keep the patella in its anatomical position and ready to go for when the quadriceps is functional again. It may take a few months depending of the damage. Prognosis is always guarded. Calf with a bilateral femoral nerve paralysis has a poor prognosis.

REFERENCES

[1.] Greenough PR. **Lameness in cattle** W.B Saunders Philadelphia 1997 3rd edition Basic concepts of bovine lameness. 3-13 [2.] Hull BL. Fractures and luxations of the pelvis and proximal femur. **Vet Clin North Am Food Anim Pract** 12:47, 1996. [3.] Jubb TF et al. Prognostic factors for recovery from coxo-femoral dislocation in cattle. **Aust Vet J** 66:354, 1989. [4.] Ducharme NG. Stifle injuries in cattle. **Vet Clin North Am Food Anim Pract** 12:59, 1996. [5.] Nelson DR et al. Peripheral detachment of the medial meniscus with injury to the medial collateral ligament in 50 cattle. **Vet Rec** 127:59, 1990. [6.] Healy A. Dystocia and femoral nerve paralysis in calves. **Compendium continuing education** 19:1299, 1997. [7.] Trostle SS et al. Clinical and radiographic findings, treatment, and outcome in cattle with osteochondrosis: 29 cases (1986-1996). **J Am Vet Med Assoc** 211:1566, 1997.

CARDIAC DISEASES OF CATTLE

Simon Peek BVSc MRCVS, PhD, DACVIM, DECEIM & Sheila McGuirk DVM, PhD, DACVIM
Madison, WI

INTRODUCTION

Congenital, infectious, neoplastic, nutritional and toxic causes of cardiac disease are occasionally encountered in primary and referral bovine practice. This presentation will review the most commonly encountered conditions with an emphasis on dairy cattle in the Northern United States. In addition the results of a small retrospective study on cases of idiopathic hemorrhagic pericarditis will be presented describing the diagnosis, management and outcome for adult cattle with this condition, based upon clinical cases seen at the University of Wisconsin's Veterinary Medical Teaching Hospital in recent years.

PERICARDIAL DISEASE

Septic pericarditis remains the best documented and most commonly reported pericardial disease in cattle. Within the literature it is most commonly attributed to foreign body penetration through the diaphragm, originating from the reticulum, although the potential does exist for extension of septic intra-thoracic disease independent of reticuloperitonitis. Clinical experience would suggest the latter is far less common, although occasional cases of septic pleural or pulmonary disease will extend to involve the pericardial sac. Aseptic pericardial effusions are most commonly associated with cardiac neoplasia (typically lymphosarcoma) or the recently documented condition of idiopathic hemorrhagic pericardial effusion[1,2]. Rare cases of cardiac neoplasia other than lymphosarcoma such as mesothelioma and chemidectoma are documented and these can lead to clinically significant accumulations of pericardial fluid and resultant tamponade[3].

IDIOPATHIC HEMORRHAGIC PERICARDIAL EFFUSION

There have been 2 reports in the literature[1,2] of this condition and we have seen a number of cases admitted to our hospital during the last 5 years at the University of Wisconsin. It is an unusual condition but over the last 5 years admissions with this disease have approximately equalled the number of septic pericarditis cases in our hospital. It is most noteworthy from the perspective of outcome and longevity for affected cattle following treatment. As has been previously noted by Firshman et al (2006) pericardial drainage and anti-inflammatory medication can bring about prolonged (>1 year) remission. Our experiences have been similar suggesting that the common presenting signs and examination findings of cardiac tamponade and congestive failure can be reversed for sufficient time frames to allow for reproductive salvage and even a return to productivity in some mature dairy cattle. Part of the conference presentation will detail the case presentation, diagnosis, and outcome for cattle affected with this condition.

ENDOCARDIAL DISEASE

The most common endocardial disease encountered in cattle is septic endocarditis. Most lesions are vegetative and valvular in location although concurrent mural endocardial lesions may also be seen. Mural endocarditis in the absence of valvular lesions is extremely rare. The retrospectives that have been published list the right atrioventricular (AV) valve as the most commonly affected , and the left AV valve being the next most likely to be involved[4]. Microbiological investigation of valvular lesions at *post mortem* and the blood of affected cattle *ante-mortem* have demonstrated that *Arcanobacterium pyogenes* is the most commonly cultured organism[3].

CONGENITAL HEART DISEASE

The largest retrospective study of congenital cardiac disease in calves published to date has documented that septal defects are most frequently encountered with ventricular septal defects (VSD) (198 out of 704) outnumbering atrial septal defects (ASD) (148 out of 704)[5]. As ultrasound has become a more commonly employed diagnostic tool for the evaluation of neonates with murmurs, or other clinical signs consistent with a potential cardiac defect (cyanosis, persistent tachycardia, stunted growth), it has become evident that a small number of individuals with small, solitary defects may survive and live functional, productive lives. Small VSDs without concurrent evidence of shunting, cardiac chamber dilation or notable growth retardation would be the best examples of this in the author's experience. Inevitably, ethical questions regarding the reproductive use of animals that have been diagnosed with a congenital defect swiftly follow the diagnosis if owners are debating whether or not the animal is to be culled. Categoric evidence linking common congenital heart defects in cattle with specific patterns of genetic inheritance is lacking although it is appropriate to broach the possibility of a hereditary component with owners.

MYOCARDIAL DISEASE

Within the US, primary myocardial conditions are rare. Geographic areas that are selenium deficient have been historically notorious for sporadic cases of nutritional myodegeneration (white muscle disease), in which myocardial lesions can contribute to sudden death. The majority of cattle that are affected with white muscle disease have clinical and biochemical evidence of skeletal muscle injury that lead to a suspicion of the disease ahead of cardiac signs. However, in endemic areas one should be aware of the potential for the condition to cause sudden death, or a brief period of respiratory distress alongside clinical evidence of cardiac disease prior to death, in young cattle. The potential for involvement of the myocardium and epicardium following *Histpohilus somni* infection will be familiar to veterinarians working with feedlot cattle, but seems to be very unusual in dairy cattle. Most frequently the myocardial lesions will be noted during *post mortem* examination of a feedlot animal in which the predominating clinical and pathologic findings related to the respiratory tract.

Lymphosarcoma represents an important differential for cattle over 2 years of age that present with signs of cardiac disease. The right atrium is a documented predilection site for tumors in bovine leukosis virus (BLV) positive cattle[4]. Tachyarrythmias, tachycardia with sinus rhythm and outward signs of predominantly right sided congestive heart failure (brisket edema, distended jugular veins and/or true jugular pulsation, prominent, turgid milk veins) would be the common physical examination findings. Occasional cases may auscult with muffled heart sounds similar to a pericarditis case due to effusion within the pericardial sac. During echocardiography it is frequently much easier to determine the presence and extent of the cardiac failure than it is to visualize the neoplastic infiltration of the myocardium.

Dilated cardiomyopathy is a rarely encountered primary myocardial disease in Holstein-Freisian cattle[6]. It is hereditary and the majority of reports of the condition are from Europe[6,7] and Japan[8]. Affected cattle develop clinical signs of congestive heart failure, typically in young adulthood. Recently the gene associated with this condition has been identified and mapped in the Holstein-Freisian breed[9]. There are reports of the condition in other European breeds that are not commonly encountered in North America.

Cor pulmonale secondary to high altitude and chronic pulmonary hypertension is a geographically distinct cause of congestive heart failure in North America with which the authors are clinically unfamiliar.

Toxic causes of cardiomyopathy are extremely rare in cattle, although documented cases of ionophore toxicity and plant ingestion do exist in the literature. The greater safety margin for ionophores in cattle compared to horses means that it is extremely unlikely that inadvertent mixing errors will lead to toxicity in adult cattle although it would appear that the safety margin may be lower in calves. Plants such as white snakeroot, coffee senna and hairy vetch are theoretically capable of inducing myocardial injury and resultant cardiac dysfunction. However, the increased confinement of dairy cattle means that these causes are highly unlikely to ever be encountered in dairy practice and still rare in beef practice.

ANCILLARY DIAGNOSTICS FOR CARDIAC DISEASE IN CATTLE

Electro- and echocardiographic evaluation of the bovine heart are typically confined to the referral or university teaching hospital setting but undoubtedly represent the most useful and eloquent diagnostic information available to us. Referral to a facility in which they can be performed should certainly be considered in individually valuable cattle, whether it be for diagnostic, therapeutic or prognostic purposes. In recent years, possibly fuelled by an increased awareness of their diagnostic utility in other species, there have been a small number of publications describing biochemical markers of cardiac disease in cattle, specifically cardiac troponins I (cTnI) and T (cTnT). Publications so far have been retrospective case studies involving cattle with a variety of cardiac disorders[10], traumatic pericarditis[11], as well as individual case reports on a calf with myocarditis due to foot and mouth infection[11], and an idiopathic pericarditis case[2]. Human assays for cTnI and cTnT have also been used to measure myocardial injury during experimental endoxtemia in calves[12]. The diagnostic utility of the troponins would be best served at this point by the establishment of normal reference ranges for adult and neonatal cattle using modern immunoassays that were simultaneously validated for the bovine species.

REFERENCES

1. Firshman A, *et al.* **J Vet Intern Med** 2006;20:1499. 2. Jetsy S, *et al.* **JAVMA** 2005;226(9):1555. 3. Buczinski S, *et al.* **Vet J** 2009;ePub ahead of press. 4. Peek S and McGuirk S. Rebhun's Diseases of Dairy Cattle, 2nd ed,; 2008;43. 5 Ohwada K and Murakami T. **J Jpn Vet Med Assoc** 2000;53:205. 6. Nart P *et al* **Vet Rec** 2004;155:355. 7. Leifsson P and Angerholm J. **J Vet Med Series** 2004;51:332. 8. Furoka H *et al* **J Comp Path** 2001;125:159. 9. Owczarek-Lipska M *et al* **Mamm Genome** 2009;20:187 10. Mellanby R *et al* **Vet Rec 2007**;161:454. 11. Gunes V *et al* **Vet Rec** 2008;162:514 12. Gunes V *et al.* **Vet Rec** 2005;156:714. 13. Peek S *et al.* **Can J Vet Res** 2008;72:356.

CHALLENGING CASE EXAMPLES OF BOVINE INTESTINAL DISEASE

André Desrochers DMV, MS, DACVS, DECBHM,
St-Hyacinthe, Québec, Canada

DECISION MAKING

Intestinal surgery is very difficult to perform in field condition. An experienced assistant is necessary for most intestinal procedures in cattle. Fortunately, most intestinal pathologies are treated medically rather than surgically. Before taking the decision to favour a surgical or medical approach, you need to gather information and approach the case in a logical manner (1) Review the possible causes of abdominal pain relevant to the particular case, (2) Recognize the indications for immediate surgery, (3) Is surgery an option (economics, facilities, surgical abilities) (4) Establish medical treatment prior to or during surgery, (5) If surgery is postponed, determine in advance a precise time and list criteria to be monitored help decision making, (6) Establish and present to the client a most realistic prognosis and cost estimate. A systematic approach based on adequate signalment and history, complete physical examination and judicious choice of ancillary tests represents the tools available to the clinicians.

There is no definitive criteria that will help you decide if you must do a surgery or not. Frequent clinical signs and physical findings are: colic, distended abdomen, tachycardia, and shock. Very distended abdomen needs assistance rapidly. A stomach tube should be passed to decompress the rumen. If no free gas is coming out of the tube, it's maybe a frothy bloat or intestinal volvulus. Abnormal findings at rectal palpation would be presence of fresh blood, absence of feces, tension band, distended small bowel, distended cecum, a mass in the middle of small bowels. Ileus and peracute enteritis can be easily confused with jejunal obstruction. Usually the small bowel will be distended primarily with fluids and easily palpable per rectum. Tension bands maybe palpated as well. In volvulus, intestines are severely distended with gas occupying the pelvic cavity. Intussusception is difficult to identify precisely but usually we suspect it if we feel a hard mass in the middle of distended small bowels. Hemorrhagic jejunal syndrome might have similar rectal examination finding except that the rectum is filled with dark raspberry like feces in variable volume. Animals with enteritis may run a fever and show other clinical signs or lab results compatible with it. Finally, the heart rate is a good indicator of the severity of the condition but should be correlated with other findings. A cow with mildly distended small bowel and a heart rate of 80 does not need immediate assistance. However, a cow with a distended organ and heart rate of 100 and dehydrated needs immediate assistance.

SURGICAL TECHNIQUE

The most common surgical conditions reported are: intussusception, volvulus of the jejunum, incarceration and hemorrhagic jejunal syndrome. Other less frequent conditions are: atresia coli, gut tie in steers, duodenal obstruction, trichobezoar, and incarceration. Recently, ileal impaction has been described.Intestinal surgeries are performed through a right flank incision while the animal is standing. Preoperative analgesia is given to make the animal comfortable for the surgery. Sedation should be administered with care because the animal might go down during the procedure. However, if the animal is weak, it is better to plan to do the surgery with the cow in left lateral recumbency. Intestinal resection and anastomosis principles are the same as described in other species. However, in cattle, the mesentery is short and fatty rendering vessels ligation difficult. The surgeon should pull gently on the mesentery otherwise the animal might go down because of the pain. It is recommended to infiltrate the mesentery with lidocaine before resection to alleviate the pain from pulling on the mesentery. End-to-end anastomosis is more commonly done than side-to-side. The anastomosis is sutured with one layer of full thickness simple interrupted sutures. Simple continuous suture can be used but should be interrupted at one point to avoid a purse string effect (2X 180°). The mesentery should always be sutured to avoid incarceration. Postoperative treatment is similar to other clean contaminated gastrointestinal surgery: broad spectrum antibiotics, analgesic, fluids. Intussusception, trichobezoar and ileal impaction has a better prognosis than the other aforementioned surgical conditions.

SPECIFIC INTESTINAL SURGICAL DISEASES

Duodenal obstruction

Only the oral part of the cranial duodenum and a portion of the descending duodenum can be exteriorized, therefore any intervention on the duodenum is difficult. Duodenal obstruction is rarely reported in the literature. Two types of obstruction are reported: functional, intraluminal or extraluminal obstruction and strangulation. Phytobezoar/trichobezoar and gravel were reported to cause intraluminal obstruction while adhesions and liver abscesses will cause extraluminal obstruction. Strangulation of the caudal flexure of the duodenum was reported in 2 cows during late pregnancy because the uterus has passed through a rent in the mesoduodenum.

Atresia coli

Origin of atresia coli is not well understood and it is most likely from multifactorial origin. Clinical presentation is typical with absence of feces since birth, bloating, colic and depression. Affected calves are rarely diagnosed before 48 hours of age but no later than 5 days because the condition is fatal. However, we have seen calves diagnosed with atresia coli at 10 days of age that were successfully treated. The age at presentation has never been associated with survival rate even if they have peritonitis. The surgery should be performed as soon as possible but the dehydration or hypopoproteinemia is addressed first. Therefore, if needed, fluids, plasma, antibiotics should be administered before going in surgery. If surgery is elected, two options are possible depending of the animal future. The two options are: bypass of the atresic segment and anastomosis to the descending colon or typlostomy/cecostomy into the right flank. The typhlostomy is a salvage procedure that allows the animal to grow until market weight is achieved. Bypass of the atresic segment is a very delicate surgery and in the authors opinion it must be performed under general anesthesia. The right paralumbar fossa laparotomy is the approach of choice to expose intestines. The proximal portion of the incision starts just ventral to the transverse apophysis. A typlotomy is strictly performed prior to the anatomosis to decompress the large colon. The spiral colon might be decompressed as well if needed. A 10-15 cm side to side anastomosis is performed between the descending colon and the blind loop of the spiral colon keeping it in an anatomic position. Postoperative complications are common from functional stenosis to peritonitis. Prognosis of colocolic anastomosis is approximately 40%, with no difference between end-to-side and side-side anastomosis technique.

REFERENCES

[1.] Anderson, D.E. and J.M. Ewoldt, Intestinal surgery of adult cattle. **Vet Clin North Am Food Anim Pract,** 2005. 21(1): p. 133-54. [2.] Lores, M., J.A. Haruna, and A. Ortenburger, Bilateral 'gut-tie' in a recently castrated steer. **Can Vet J**, 2006. 47(2): p. 155-7. [3.] Braun, U., A. Steiner, and M. Gotz, **Clinical signs, diagnosis and treatment of duodenal ileus in cattle.** Schweiz Arch Tierheilkd, 1993. 135(11-12): p. 345-55. [4.] Abutarbush, S.M. and J.M. Naylor, Obstruction of the small intestine by a trichobezoar in cattle: 15 cases (1992-2002). **J Am Vet Med Assoc**, 2006. 229(10): p. 1627-30. [5.] Deprez, P., et al., Incarceration of the small intestine in the epiploic foramen of three calves. **Vet Rec,** 2006. 158(25): p. 869-70. Nuss, K., et al., Ileal impaction in 22 cows. **Vet J**, 2006. 171(3): p. 456-61. [6.] Syed M, Shanks RD. Incidence of Atresia coli and relationships among the affected calves in one herd of Holstein cattle. **J.Dairy.Sci.** 1992;75:1357-1364. [7.] Ducharme NG, Arighi M, Horney FD, Barker IK, Livesey MA, Hurtig MH, Johnson RP. Colonic atresia in cattle: a prospective study of 43 cases. **Can Vet J** 1988;29(10):918-924. [8.] Dreyfuss DJ, Tulleners EP. Intestinal atresia in calves: 22 cases (1978-1988). **J Am Vet Med Assoc** 1989;195(4):508-13. [9.] Smith DF, Ducharme NG, Fubini SL, Donawick WJ, Erb HN. Clinical management and surgical repair of atresia coli in calves: 66 cases (1977-1988). **JAVMA** 1991;199(9):1185-90. [10.] Constable PD, Huhn JC, Morin DE, Nelson DR. Atresia coli in calves: Etiopathogenesis and surgical management. **Bov Pract** 1999;33(1):70-3. [11.] Koller U, Lischer C, Geyer H, Dressel C, Braun U. Strangulation of the duodenum by the uterus during late pregnancy in two cows. **Vet J** 2001;162:33-37

HORIZONTAL TRANSFER OF ANTIBIOTIC RESISTANCE GENES IN NATURE: IS AGRICULTURAL USE OF ANTIBIOTICS HAZARDOUS TO HUMAN MEDICINE?

Abigail A. Salyers & Nadja B. Shoemaker
Urbana, IL

The Issues

In the debate over possible consequences for human health of widespread use of antibiotics, two different concerns have emerged. The first was whether there were sufficient antibiotic residues in meat and animal products such as eggs and milk to pose a problem, especially in children. This concern has been allayed successfully by the imposition of stringent limitations on the amount of antibiotic residues that are permitted in meat, milk and eggs.

One aspect of the antibiotic residues issue, however, has not been adequately addressed, because it was discovered only recently. This is the entry of antibiotics into various water supplies. Easily measurable concentrations of some antibiotics such as tetracyclines have been found in streams and ground water (Barret, 2004). This will be a more difficult problem to solve than the problem of residues in milk or other animal products because removal of antibiotics from urine and manure is probably not feasible economically. Also, some antibiotics that end up in water sources arise from human waste. It may be possible to sort this out to some extent by using forms of antibiotics that are used primarily in agriculture rather than human medicine serve as markers of the origin of the antibiotic residues. Since this is an issue that has only been identified recently and since the impact of water-borne antibiotics on human health has not been evaluated, it will not be considered further in this article. Another concern about water has been that antibiotic-resistant bacteria found on plants irrigated with "grey water" from animal production farms may affect human health because, like meat, they move through the food supply and enter the human body.

At first in the debate over agricultural use of antibiotics and human health, attention focused on gastrointestinal pathogens such as *Salmonella* and *Campylobacter*. One can argue that this is not as big a concern as it might seem at first, because antibiotics are generally not used to treat gastroenteritis caused by these pathogens. The flaw in this argument is that a small fraction of cases of gastroenteritis caused by these bacteria progress to invasive, systemic disease that must be treated by antibiotics. The argument that cooking solves this problem is belied by the continued widespread incidence of gastroenteritis caused by these pathogens despite intense efforts to educate the public about proper handling of food. Also, plants have been responsible for some outbreaks of this type of disease.

This article will not attempt to review the vast literature on antibiotic resistant gastrointestinal pathogens in the food supply because this has already been much reviewed and analyzed. Instead, this article will focus on a more recently recognized, and potential more dangerous, consequence of antibiotic-resistant bacteria moving from the farm to the human food supply: the possibility that animal bacteria, even those that are not usually capable of causing disease, could transfer their antibiotic resistance genes to bacteria such as *Bacteroides* spp. that are normally resident in the intestinal tract but can on occasion cause serious post-surgical infections. Moreover, these resident colonic bacteria might in turn transfer resistance genes to bacteria from the human mouth and upper respiratory tract that are constantly swallowed and pass through the human colon. These include such potential human pathogens as *Streptococcus pneumoniae* and *Staphylococcus aureus*. The residence time in the colon of bacteria that are only passing through, at least 24 hr, is more than enough time to allow gene transfers to occur.

Critical Analysis of the Gene Transfer Argument

The scenario for a connection between antibiotic use on farms and eventual appearance of resistance genes from animal bacteria in human pathogens is summarized in Table 1. A detailed discussion of each step in this scenario follows.

Table 1. Theoretical scenario for a possible connection between antibiotic use on farms and the eventual appearance of newly resistant human pathogens.

Step 1. Antibiotic bacteria that are not directly acting pathogens ("commensals") become resistant in the animal due to selection for them by antibiotic use.

Step 2. Resistant bacterial strains from animals or plants (due to irrigation with grey water) enter the human body via the food supply and pass through the human colon, where the highest concentrations of bacteria are found and the residence time is at least 24 hr. This is enough time for gene transfer to human strains to occur.

Step 3. Human strains that are capable of causing opportunistic infections may later cause such infections. Also resistant bacteria residing in the colon could transfer their newly acquired resistance genes could transfer their genes to swallowed opportunistic pathogens such as *Staphylococcus* and *Streptococcus* species, which are then released into the environment and could reenter the body to cause opporotunistic infections in other areas of the body.

Step 1. Selection of antibiotic resistant strains. The first step in the scenario is the rise of resistant bacteria on farms that use antibiotics. Many people accept without question that using antibiotics on farms is the main, if not the sole, cause of the presence of antibiotic bacteria in the intestines of animals. There is no question that the use of antibiotics selects for antibiotic resistant bacteria. Even low concentrations of antibiotics (subtherapeutic levels) can act as a selection pressure if it allows resistant bacteria to have even a small fitness advantage.

The question remains, however, whether antibiotics are the only selection pressure. This question was first raised by studies that found antibiotic resistant bacteria on farms that do not use antibiotics. Of course, this could be explained by antibiotics in ground water or streams moving from one farm to another, but there may be other explanations. One is that antibiotics are not the only selection for resistant bacteria. Two studies we conducted provide support for considering more seriously the question of whether antibiotics are the only selection pressure for antibiotic resistance genes.

In a study of antibiotic resistance genes in human colonic *Bacteroides* strains isolated before 1970 and after 1995 showed that although the incidence of a tetracycline resistance gene, *tetQ,* rose significantly in the intervening period, when tetracycline use in human medicine was most intense, 20% of the strains isolated before 1970, some well before that date, carried this resistance gene. It is hard to justify this level of carriage on contact with antibiotic producer in soil because these bacteria and fungi produce extremely low concentrations of antibiotics in nature (Shoemaker et al., 2001).

Even more surprising were the results of a study published in 2009, in which we used PCR amplification and sequencing to detect tetracycline and macrolide resistance genes in vaginal swab samples taken from captive and wild baboons. This study was done as an adjunct to a project designed to define the resident vaginal microbiota of baboons, animals used as models for human female reproductive problems (Jeters et al., 2009). Two of the tetracycline resistance gene included in the screen, *tetM* and *tetW,* were found in virtually all of the samples, even those from wild baboons in Africa. Some erythromycin resistance genes were found less commonly but were present in a number of the animals. We were screening mainly for *tet* and *erm* genes because these are widespread in human *Bacteroides* and the Gram-positive bacteria, the two groups that are numerically dominant in the human colon and in the baboon vaginal tract. Given that the captive baboons were housed in an isolated location in Texas, where only tap quality water was used for watering and cage cleaning, and given that the African baboons were living far from human populations and farms, the ground water argument does not hold. Finally, it is worth noting that phylogenetic studies of antibiotic resistance genes have shown that these genes arose long before antibiotics were used and in some cases long before humans appeared on the evolutionary scene (Nikolich and Salyers., unpublished data).

Learning whether there are other pressures selecting for resistance genes besides antibiotics is important, because if so, this information needs to be incorporated in future strategies for stopping or slowing the rise in carriage of antibiotic resistance genes. The possibility that this is the case should not, however, be used to argue that unrestricted use of antibiotics either in agriculture or human medicine is justified. There is no question that once a resistant bacterial strain is present in an environment, antibiotic use can rapidly and drastically increase its numbers. The question is where resistance genes originally arose and why they seem to be so stably maintained in the absence of selection (Salyers and Amabile-Cuevas, 1997)

Step 2. Resistant bacteria enter the human body via the food supply. Whatever the origin of the resistant strains, there is little question that such strains from agricultural sources enter the food supply and, through it, the human body. We know this from studies in which bacteria have been cultivated from various meats, especially chicken, that are on sale in grocery stores. Since these bacteria can be cultivated, they are clearly still viable.

There are two limitations of the studies that have been done. The first is that virtually all studies focus on cultivation of a very limited number of bacterial species, principally *E. coli, Salmonella, Campylobacter* and *Enterococcus* spp. These were undoubtedly chosen because they are easy to cultivate and are known to cause human disease. It would be interesting to take a cultivation-independent approach to the question of what bacteria are on food. That is, 16S rRNA gene amplification and DNA sequencing of the amplicons (cultivation-independent approach) could be used to get a complete picture of what is present on various foods found in supermarkets. The answer could change the way we think of the "microbiota" of food.

Two objections have been made to taking a cultivation-independent approach. First, such a study could be expensive and technically difficult, and second, detecting bacterial DNA does not mean the bacterium is alive. The first of these objections was valid 10 years ago, but the technology has now developed to the point that it is now as cheap or cheaper to take the cultivation-independent approach. The same approach could be used to sample antibiotic resistance genes carried by bacteria on the food. Such studies can now be done quantitatively using quantitative PCR. The second objection has been solved in the form of reverse transcriptase PCR (RT-PCR), which can be used to detect the transcript of the gene being amplified. If a bacterium is producing 16S rRNA, it is certainly not dead.

A more difficult question to answer is whether all these bacteria reach the intestine alive after passing through the stomach and small intestine. We know from hard experience that the gastrointestinal pathogens do this. And since it appears that infants obtain their intestinal microbiota from their parents, other bacteria can clearly transit the stomach in high enough numbers to colonize the colon.

Step 3. Does resistance gene transfer occur among bacteria, both normal colonizers and swallowed bacteria, in the human colon. There are three known mechanisms of bacterial gene transfer: natural transformation, bacteriophage transduction and conjugation. Since bacteria passing through the colon would be members of different species and genera than those that normally colonize the colon, the gene transfers of greatest interest would be those that cross species and genus lines. Natural transformation involves the binding of a double stranded DNA segment, degradation of one of the strains and uptake of a single strand of the foreign DNA. Since this DNA would have to have regions of identity with the chromosome in order to integrate, natural transformation strongly favors transfers between very similar bacteria. This attribute, together with the fact that so far none of the numerically predominant colon bacteria are known to be capable of natural transformation, makes it unlikely that natural transformation would make much of a contribution in the scenario being considered here. Bacteriophage transduction is similarly restricted because there must be a bacteriophage receptor on the recipient bacterium's surface, and such receptors are generally limited to strains of the same species. Conjugation, the direct cell-to-cell transfer of mobile elements, is the type of transfer most likely to be responsible for trans-species and trans-genus DNA transfers. In fact all of the genes mentioned in **Figure 1** are associated with conjugative elements.

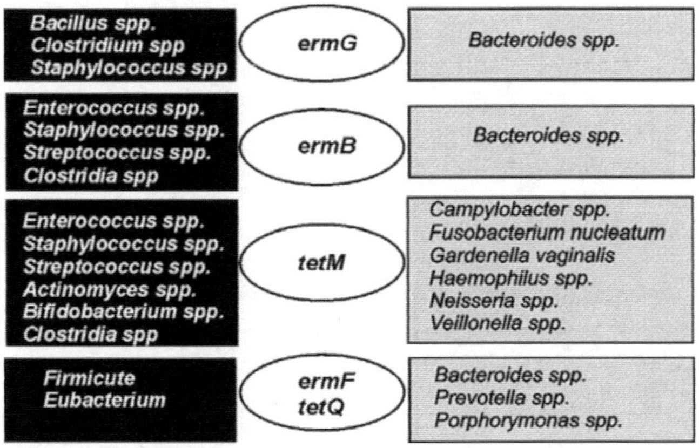

Figure 1. Examples of possible cases of horizontal gene transfer in nature. Bacteria in the boxes on the left are Gram-positive. Bacteria in the boxes on the right are Gram-negative. The resistance gene in the ovals between the rectangular boxes are antibiotic resistance genes that have been found by DNA-DNA hybridization or DNA sequencing in the bacteria in boxes to which the oval is connected.

There is a growing amount of evidence that cross-species and cross-genus transfers occur in the human colon. This evidence takes the form of finding nearly identical copies of resistance genes (more than 95% DNA sequence identity in members of different species (Salyers et al., 2007a and 2007b).

Two particularly interesting examples of transfers that may have occurred between bacteria normally found in animals and bacteria isolated from the human intestinal tract involve the tetracycline resistance gene *tetQ* and the erythromycin resistance gene *ermB*. The *tetQ* gene was found both on a plasmid in a rumen bacterium, *Prevotella ruminicola,* and on a conjugative transposon in human colonic *Bacteroides* spp. We have shown that under laboratory conditions, a mobile element, CTn12256, can transfer *tetQ* from *Bacteroides thetaiotaomicron* to *P. ruminicola B14*. In the case of the human colonic species, *tetQ* is now widely disseminated, a process that seems to have accelerated over the past several decades. In the human *Bacteroides* spp., *tetQ* is normally found not on plasmids but on integrated mobile elements called conjugative transposons. The excision and transfer of a conjugative transposon is illustrated in **Figure 2**. It is interesting to note that in the case of the most widespread *tetQ*-carrying *Bacteroides* conjugative transposon, CTnDOT, excision and transfer are stimulated considerably by exposure of the donor to tetracycline. However, the only CTn we have found that transfer *tetQ* to *P. ruminocola* does not exhibit tetracycline-stimulated transfer. This raises the possibility that if *tetQ* from an animal strain of bacteria made its way into *Bacteroides* as part of a tetracycline-stimulated conjugative transposon, that human use of tetracycline could contribute to its spread to other *Bacteroides* strains. *Bacteroides* is a significant opportunistic pathogen of humans, so the spread of antibiotic resistance in this genus is a cause for concern. Tetracycline is no longer used to treat *Bacteroides* infections because so many strains are resistant to it, but the same conjugative transposons may also carry other resistance genes that go along for the ride when the conjugative transposon is transferred. Also of interest that in our study of resistance genes in the baboon microbiota, we were able to link some of the widespread genes with known conjugative transposons by using PCR to amplify the region between the conjugative transposon and the resistance gene.

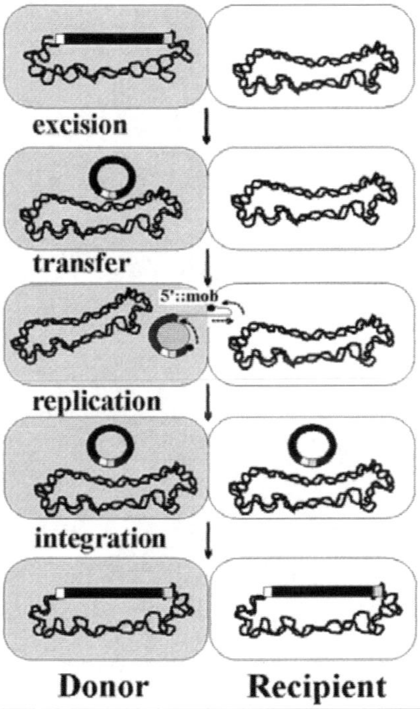

Figure 2. Steps in the transfer of a conjugative transposon. The element is normally integrated into the chromosome. To transfer, it first excises to form a circular intermediate. Then a single-stranded copy of the circular form is transferred to a recipient, where it reintegrates into the chromosome.

More recently, we found a second example of a possible human-animal resistance gene transfer (Schlessinger et al.). In this case, a conjugative transposon was also involved. The conjugative transposon, CTnBST, was found in a human colonic *Bacteroides* strain. It carried a 13 kbp insert, which contained an erythromycin resistance gene (*ermB*). This segment was identical at the DNA sequence level to a segment of DNA found in a rumen bacterium, *Arcanobacterium pyogenes*. It is not known whether the *ermB* DNA segment found in the rumen bacterium is part of a mobile element, but somehow this segment, which has a %G+C content that is characteristic of low %G+C gram-positive bacteria, not of *Bacteroides* spp., made its way into *Bacteroides* and ended up as part of an integrated mobile element. How this merger of the 13 kbp segment with what appears to be a *Bacteroides* conjugative transposon is not clear.

Although it might seem at first that this example shows that the DNA segment in question came from a rumen strain directly into a human colonic microbe during passage of the rumen microbe through the human intestinal tract, this is not necessarily the case. The DNA segment might have moved through one or more intermediate microbe. Defining direction of transfer and participants in the transfer process is currently virtually impossible to do unless the transfer occurs under laboratory or controlled field conditions. Yet another complication in establishing the contribution, if any, of agricultural use of antibiotics to increasing resistance of human pathogens is that bacteria in food are not the only external-source bacteria that move through the human intestinal tract is that bacteria from the mouth and bacteria inadvertently ingested from aerosolized soil and water also move constantly through the intestine.

Probiotic use, a practice that is very common in Asia and increasingly common in Western countries, also complicates establishing cause and effect of gene transfers originating from animal strains of bacteria because some of these strains are resistant to antibiotics Little is known about resistance genes in the probiotic strains of *Lactobacillus* and *Bifidobacterium*. This possible source of resistance genes that enter human colonic bacteria would have to be considered before concluding that bacteria from the food supply are solely responsible for providing such resistance genes.

The available evidence shows that trans-genus transfers of resistance genes by conjugation is the main source of resistance gene spread in the human colon. The fact that the same resistance gene is found in bacteria of the sort that would be swallowed as well as in bacteria that normally colonize the colon suggests that transfers may occur not only among members of the major populations of bacteria in the colon but also between the normal microbiota and bacteria that are just passing through. It should be noted that the evidence discussed only shows that transfers seem to be occurring between members of the major colonic bacteria and bacteria in the colon and transient bacteria. As already mentioned, it is not possible at this time to be sure of the direction of transfer. It is also not possible to state with confidence that bacteria entering the human body through the food supply are participating in these transfers. More work is needed to determine if this is the case. It is now feasible, however, to do such a survey. The problem continues to be that there are relatively few laboratories working in this area and there is little funding, so progress will continue to be slow. Nonetheless, the potential for a contribution to the spread of resistance genes by agricultural use of antibiotic exists and should be considered in any assessment of risk and consequence of this type of antibiotic use.

REFERENCES

Barrett, J, **Environ Health Perspect.** 2004; 112: A736. Jeters R et al, **Microb Drug Resist.** 2009; 15, in press. Salyers, A and Amabile-Cuevas, C., **Appl.. Environ. Microb**., 1997: 41: 2321; Salyers, A. et al., **Clin. Microb. Newsletter** 2007a; 29: 17. Salyers, A. et al.. **Clin. Microb. Newsletter** 2007b; 29: 25. Schlesinger, D. et al., **Appl. Environ. Micro**. 2007; 73: 4226; Shoemaker, N. et al. **Appl. Environ. Microb.** 2001; 67: 3488.

DIAPHRAGMATIC PARALYSIS IN JUVENILE ALPACAS. SR Byers[1], GM Barrington[2], G Haldorson[3], T Holt[1], RC Callan[1]. 1. Colorado State University College of Veterinary Medicine, Fort Collins, CO. 2. Washington State University College of Veterinary Medicine, Pullman, WA. 3. Washington Animal Disease Diagnostic Laboratory, Pullman, WA.

Diaphragmatic paralysis is a rarely reported condition in alpacas; however over a 6 month period a single farm referred 3 juvenile alpacas to the Washington State University Veterinary Teaching Hospital for respiratory issues and ataxia. Through testing and post mortem examinations, diaphragmatic paralysis was diagnosed. The purpose of this retrospective study was to determine minimally invasive diagnostic procedures to confirm the condition as compared to postmortem examinations, and identify potential etiologies.

Review of medical records from WSU and Colorado State University Veterinary Teaching Hospitals identified 12 alpacas (11 juvenile alpacas and 1 adult) with presumptive diaphragmatic paralysis (2002-2009). Only the juveniles were evaluated in this study. The alpacas were less than 6.5 months of age and almost evenly divided between male and female. Test methods compared included thoracic radiography; fluoroscopy; echocardiography; and arterial blood gas, complete blood count, serum chemistry, cerebrospinal fluid, micro and macronutrients, toxin, and heavy metal measurements. Alpacas that died or were euthanized received full post mortem examinations.

Diaphragmatic paralysis was confirmed by fluoroscopic examination in 7of 7 alpacas. Histopathologic evaluation revealed phrenic nerve degeneration in all 6 alpacas that were submitted for post mortem examination. Owners declined fluoroscopic imaging in 4 cases due to a presumptive diagnosis based on clinical signs, prognosis, and response to treatment.

Fluoroscopy proved to be the superior antemortem diagnostic procedure. The alpacas tolerated the procedure well with little or no sedation required to maintain the animal in sternal recumbency. Histopathologic examination showed degeneration of the phrenic nerve as well as other long nerves in several of the juvenile alpacas. No single etiologic agent was identified, however genetics or micronutrient levels were suspected as contributing factors in 8 of the 11 cases.

EFFECTS OF PLASMA AND HETASTARCH ON BLOOD COLLOID OSMOTIC PRESSURE AND COAGULATION TIMES IN HEALTHY ALPACAS

E McKenzie,[1] M Payton[2], J Riehl,[1] R Garrett.[1] 1.Oregon State University, Corvallis, OR; 2.Oklahoma State University, Stillwater, OK.

The objective of this study was to compare the effects of equivalent volumes of hetastarch and llama plasma on blood colloid osmotic pressure (COP), and specific hematologic variables in healthy alpacas.

In a randomized cross over trial, five alpacas were administered 15ml/kg of llama plasma or 6% hetastarch intravenously six weeks apart. COP, hemoglobin, total protein, albumin, hematocrit and platelet count were measured before, immediately after, and 1, 2, 3, 4, 5, 6, 12, 24, 36, 48, 72 and 96h after treatment. Prothrombin time (PT) and partial thromboplastin time (PTT) were measured before and 2 and 7h after treatment. Data were analyzed by RMANOVA ($p<0.05$).

There was no difference between any variables before treatments. Mean COP was 21.1±0.5 and 21.1±1.5mmHg before plasma and hetastarch respectively. Plasma significantly increased COP at 12h (23.2±0.8) through 96h (24.2±0.5). Hetastarch significantly increased COP immediately after treatment (23.7±1.3) through 72h (23.8±0.9). Mean change in COP after treatment was greater for hetastarch versus plasma at 1 through 12h; at all subsequent timepoints the treatments induced equivalent increases in COP.

Hetastarch mildly increased PT and PTT 2h after treatment, and plasma decreased PT and PTT 7h after treatment. Both treatments decreased hemoglobin and hematocrit for 12h. Hetastarch decreased total protein and albumin for 24h, and plasma significantly increased these variables from 24 through 96h after treatment.

In conclusion, hetastarch causes an immediate increase in COP in alpacas with minimal effects on coagulation and other hematologic variables. Plasma induced an equivalent increase in COP 12h after treatment.

ORAL PHARMACOKINETICS OF PONAZURIL IN HEALTHY LLAMAS (*LAMA GLAMA*).

M.E. Prado[1], J.T. Ryman[2], T. Martin-Jimenez[1], B. Meibohm[2]. [1]College of Veterinary Medicine, University of Tennessee, Knoxville, TN and [2]College of Pharmacy, University of Tennessee, Memphis, TN.

Eimeria macusaniensis is a major cause of diarrhea in New World Camelids (NWCs) in the United States and worldwide. Currently, infected NWCs are treated empirically with different anti-protozoal chemotherapeutics such as Ponazuril and amprolium; however, information regarding the bioavailability of these drugs is lacking. Ponazuril is a triazine coccidial drug related to toltrazuril that has been shown to be effective against different members of the phylum Apicomplexa. The purpose of this study was to determine the pharmacokinetics of oral Ponazuril in healthy llamas. Ponazuril was administered once orally to 6 llamas (20 mg/kg of body weight) and serum was collected on days 0, 0.5, 1, 2, 3, 4, 5, 6, 7, 9, 11, 14, 21, 28, 35, 42 and 49. Serum samples were analyzed by a validated reversed phase High Performance Liquid Chromatography (HPLC) assay with UV absorbance detection. Ponazuril was readily absorbed by healthy llamas following a single oral dose. The mean elimination half-life, Tmax, Cmax and Area Under the Curve (AUC) for Ponazuril were 135±16.7 h, 84±25.2 h, 23.6±6.03 mg/L and 7515±2750 hr*mg/L, respectively. Plasma concentrations of Ponazuril peaked at 23.60±6.0 µg/mL at 84 (range 48-120) hours post-administration. Thereafter, plasma concentrations gradually decreased but remained detectable up to 35 days post-administration. No adverse effects were observed during the duration of the study. In summary, results from this study showed that oral administration of Ponazuril to healthy llamas at 20 mg/kg resulted in plasma concentrations that were high enough to be effective against different Apicomplexans based on data for other species The effective minimal inhibitory concentration of Ponazuril for *E. macusaniensis* in NWCs is currently unknown. Studies are underway to determine the efficacy of this drug and other anti-coccidials against *E. macusaniensis* in NWCs.

INTEGRATION OF PHARMACOKINETIC AND PHARMACODYNAMIC DATA FOR OXYTETRACYCLINE IN HEALTHY AND PNEUMONIC CALVES.

T Potter, J Illambas, A Rycroft and P Lees. Royal Veterinary College, London, UK.

The objective of this project was to validate pharmacokinetic-pharmacodynamic (PK-PD) integration and modelling approaches to dose schedule design, by generating and correlating data on the pharmacokinetics (PK) and pharmacodynamics (PD) of oxytetracycline.

Minimum inhibitory concentrations (MIC) for oxytetracyline were determined in Mueller Hinton Broth (MHB) and calf serum, using 5 overlapping sets of doubling dilutions for 6 clinical isolates of each of two organisms, *M.haemolytica* and *P.multocid*a. Pharmacokinetic data and *ex vivo* bacterial kill curves were established for three biological fluids; serum, exudate and transudate using a tissue cage model. The PK study used 10 healthy female Aberdeen Angus Holstein Friesian Cross calves aged 3-5months. Oxytetracycline (Alamycin LA) was administered intramuscularly at 20mg/kg. In a further trial serum pharmacokinetics, clinical response and lung pathology were examined in a *M.haemolytica* model of calf pneumonia. Eight Holstein Friesian bull calves, aged 12 weeks with induced pneumonia were treated with oxytetracycline (Alamycin LA) administered intramuscularly at 20mg/kg. Serial blood samples were collected to establish serum pharmacokinetics and clinical parameters were monitored over a 48h period. Animals were then euthanased and *post mortem* examinations were performed. Clinical and pathological findings were compared to a group of 8 untreated control calves.

Mean MICs in MHB for *M.haemolytica* and *P.multocida* were 0.24µg/ml and 0.25µg/ml respectively, and were significantly different to those measured in serum; 7.13µg/ml and 7.20µg/ml. Pharmacokinetic data are presented in table 1. In the pneumonia model study, no significant difference was observed in clinical parameters and percentage lung consolidation (23.1% vs 21.7%) between the oxytetracycline group and the untreated control animals.

The C_{max} values from both PK studies were lower than the average MIC for the two pathogens in serum. Targeted partial areas were used to express the actual ratio of average blood concentration relative to a targeted MIC. Based on the MIC in serum for the first 24h, targeted partial areas were 0.51 and 0.36 for the healthy and pneumonic animals respectively. The alteration in drug PK in diseased animals together with the matrix effect on the MIC need to be taken into account in designing appropriate dose schedules. These discrepancies may account for the lack of clinical efficacy observed in parts of this project.

Animals	Cmax (µg/ml)	AUC_{24} (h.µg/ml)	Using serum MIC 7.17µg/ml		Using broth MIC 0.25µg/ml	
			C_{max}/MIC	AUC_{24}/MIC	C_{max}/MIC	AUC_{24}/MIC
Healthy	5.26	87.52	0.73	12.21h	21.04	350.08h
Pneumonic	3.41	61.21	0.48	8.54h	13.64	244.84h

Table 1: Pharmacokinetic variables and surrogate markers of antimicrobial activity for oxytetracycline in healthy and pneumonic calves.

BIOAVAILABILITY OF ORAL SELENIUM SUPPLEMENTS IN SHEEP. JA Hall[1], RJ Van Saun[2], WC Stewart[1], WR Vorachek[1], W Mosher[1], G Pirelli[1]. [1]Oregon State Univ. College of Veterinary Medicine, Corvallis, OR. [2]Pennsylvania State Univ. College of Agricultural Sciences, University Park, PA.

The objective of the study was to compare bioavailability between oral selenium (Se) sources at three levels of supplementation in sheep. Bioavailability was assessed by measuring Se concentration in whole blood (WB) and serum at various time intervals over a 12-month period.

This study was part of a larger study evaluating the role of Se supplementation (source, supplementation rate) on Se status, immune responses, and susceptibility and recovery from foot rot (FR) in sheep. The study population included 240 mature ewes from three genotypes (Polypay, Suffolk, Crossbred). Healthy and FR-affected sheep were randomly assigned to 1 of 8 treatment groups based on Se supplementation rate (0, 0.3, 0.9 and 1.5 ppm) and source (selenite, selenate (.3 ppm only) and selenomethionine). All treatment groups were balanced for number of healthy and FR-affected sheep and total FR scores. All sheep were dosed weekly with the calculated amount of Se supplement or water (controls) that would be equivalent to the weekly consumption. Intake rates were based on NRC estimations using measured body weights. Blood samples were collected prior to initial dosing and then monthly throughout the study. WB Se concentrations were measured monthly in controls and all 0.3 ppm groups. WB and serum Se concentrations were measured quarterly for all groups. Se concentration data were analyzed by ANOVA for repeated measures with main effects of group and month, and with initial Se concentration as a covariate.

Treatment group, month, and their interaction highly influenced ($P<.0001$) WB and serum Se concentrations throughout the study. Control sheep showed a 78.8% and 58.8% decline ($P<.0001$) in WB (250 to 53 ng/ml) and serum (97.4 to 40.1 ng/ml) Se concentrations, respectively, over the duration of the study, consistent with consuming a low Se diet without supplementation. Comparing WB Se concentrations across Se sources at the 0.3 ppm supplementation rate, selenite and selenate supplemented sheep were not different; both had higher ($P=.0001$) WB Se concentrations compared to controls starting at 2 months and remained higher ($P<.0001$) for the study duration. At the 0.3 ppm supplementation rate, selenomethionine supplemented sheep had higher WB Se concentrations compared to controls ($P<.0001$) at all time points, and after 2 months compared to selenite and selenate supplemented sheep. Both WB and serum Se concentrations showed increasing ($P<.0001$) dose-response curves to higher selenomethionine supplementation rates. In contrast, mean WB Se concentrations were not different between 0.9 and 1.5 ppm selenite groups and did not differ from 0.3 ppm selenomethionine supplementation. Inorganic Se sources seemingly have limited ability to increase Se status compared to selenomethionine, most likely a result of increased excretion. The lowest rate of Se supplementation from selenomethionine was equal to higher rates of selenite supplementation suggesting a greater bioavailability of selenomethionine.

POTASSIUM HOMEOSTASIS AND HYPOKALEMIA IN LACTATING DAIRY COWS WITH LEFT DISPLACED ABOMASUM. P Constable[1], W Grunberg[2], U Schroder[3], R Staufenbiel[3], M Rohn[3]. [1]School of Veterinary Medicine, Purdue University, West Lafayette, IN; [2]Utrecht Universitat, The Netherlands; [3]Klinik fur Klauentiere, Diagnostischer Dienst am Rind, Freie Universitat, Berlin, Germany.

Abnormal potassium homeostasis and hypokalemia occurs commonly in dairy cows with left displaced abomasum (LDA). The major objective of this study was to identify potential mechanisms for hypokalemia in cows with LDA. Accordingly, the results of preoperative clinicopathologic analyses for 1,160 dairy cows >22 months of age with an LDA presented at the Clinic for Ruminants and Swine of the Freie Universitat, Berlin Germany were retrieved. The following serum biochemical parameters were assayed using automated methods: concentrations of potassium ([K]), sodium ([Na]), calcium ([Ca]), magnesium ([Mg]), chloride ([Cl]), phosphorus ([Pi]), urea ([urea]), and total bilirubin ([TBil]), as well as aspartate amino transferase (AST) activity. Analysis of the erythron and leukon was performed on blood collected into EDTA-containing tubes. Pearson correlation coefficients were calculated on raw or log-transformed values and only variables with $r>0.2$ or <-0.2 are reported. Effects of measured parameters on serum [K] were tested by multivariable linear regression analysis. $P < 0.05$ was significant.

The mean [K] was 3.7±0.7 mEq/L (range 1.7 to 5.9 mEq/L) and 55% percent of cattle with LDA (636/1,160) were hypokalemic (serum [K] <3.9 mEq/L). Severe hypokalemia (serum [K] <2.5 mEq/L) was present in 6% of cattle with LDA. Serum [K] was positively correlated with serum [Cl] ($r=+0.54$), [Na] ($r=+0.33$), and [Ca] ($r=+0.22$), and negatively correlated with serum log[TBil] ($r=-0.47$), packed cell volume ($r=-0.37$), blood [hemoglobin] ($r=-0.34$), and log[urea] ($r=-0.33$) and logAST activity ($r=-0.31$) in serum. Multivariable linear regression analyses indicated that serum [Cl], log[TBil], [Ca], log[Pi], log[Mg], log[urea], and packed cell volume were independent predictors of serum [K] ($R^2=0.45$).

We conclude that hypokalemia occurs commonly in dairy cows with LDA, and that hypokalemia most likely results from sequestration of abomasal fluid within the gastrointestinal tract (low serum [Cl]), decreased feed intake relative to the cows level of milk production (high serum log[TBil], low serum [Ca], log[Pi], and log[Mg], and hypovolemia (high packed cell volume, high serum [urea]). The negative association between serum [K] and packed cell volume and serum [urea] suggests that aldosterone secretion plays a role in the development of hypokalemia. Because the majority of cattle with LDA are hypokalemic and serum [K] is most closely associated with serum [Cl], oral KCl should be routinely administered to cattle with LDA. Treatment of hypokalemia in dairy cattle with LDA should also include surgical correction of the LDA, increasing dry matter intake, and rehydration. We believe that this is the first large scale study to identify potential mechanisms for hypokalemia in lactating dairy cattle.

LACTOFERRIN: INTRACELLULAR EFFECTS THROUGH MAPK SIGNALING A FURTHER EXPLAINATION OF ITS ANTI-INFLAMMATORY PROPERTIES. Maisie E Dawes, Western University of Health Sciences, Pomona, CA.

The heme-associated iron-binding apoprotein lactoferrin (LF) is known for its, anti-inflammatory, anti-parasitic, antimicrobial and bactericidal effects. Lactoferrin demonstrates ubiquity throughout mammalian host biological fluids: Saliva; tears; mammary secretions, as well as at the level of mucosal surfaces. It is also released from immune cells under pathogenic stimulation. Numerous studies speak to its ability to alter leukocyte function, inhibit cytokine production, and bind lipopolysaccharide (LPS); significant mechanisms through which it is believed to achieve its anti-inflammatory effects. More recently, investigators have demonstrated lactoferin's ability to interact with host DNA; and have elucidated its role in the apoptotic signaling pathway. The purpose of this study was to further characterize the mechanisms through which it modulates inflammation in the face of bacterial endotoxin. In earlier studies, we demonstrated lactoferrin's ability to limit the expression of inducible cyclooxygenase-2 and the gelatinase, matrix metalloproteinase -9 by LPS-induced macrophages. The generation of these inflammatory mediators is modulated by pro- inflammatory cytokines such as interleukin-1 β (IL-1 β) and tumor necrosis factor-alpha (TNF-α), the production of both being dependent on signaling through the p38 mitogen activated protein kinase (MAPK) pathway. We hypothesized that LF would inhibit p38 phosphorylation. This study was approved by Western University's Institutional Animal Care and Use Committee.

Peripheral mononuclear cells (5 x 10^6) isolated from buffy coat cells of healthy neonatal to 4-month old Holstein calves were cultured in the presence and absence of LF (200 ng/ml), LPS (1μg/ml), anisomycin (25 μg/ml), a known p38 activator - the positive control, and 50 mM of SB203580, a known p38 inhibitor - the negative control. Sample lysates obtained post culture was subjected to immunoprecipitation and kinase reactions. Reactions were terminated under reducing conditions and evaluated using western immunoblotting. Phosphorylation of activated transcription factor-2 (ATF-2) by phosphorylated p38 served as the marker of investigation. Immunologically reactive ATF-2 expression by LPS and anisomycin-treated cells was compatible with a prominent band at 40 kD. Evidence of LF-induced inhibition of LPS-induced p-38 activation was observed in lanes representative of co-cultures of LF + LPS; LF + anisomycin; and anisomycin + SB203580, which was demonstrated by decreased immunological reactivity at 40kD. Our findings suggest that LF interferes with LPS-induced p-38 activation of transcription factor ATF-2, *in vitro*. This serves as additional proof of its potential use in attenuating the systemic effects of LPS.

CLINICAL, MORPHOLOGICAL AND MOLECULAR STUDY OF A NEW FORM OF HEREDITARY MYOTONIA IN MURRAH WATER BUFFALO

AS Borges[1], JD Barbosa[2], LAL Resende[1], LSLS Mota[1], RM Amorim[1], JF Garcia[1], TL Carvalho[1], NJ Winand[3]. 1. São Paulo State University, Botucatu, Brazil. 2. Pará Federal University, Pará, Brazil. 3. Cornell University, College of Veterinary Medicine, Ithaca, NY.

Hereditary myotonia caused by mutations in the skeletal muscle chloride channel (CLCN1) has been described in man, goats, dogs and mice. Recently we identified a hereditary form of myotonia in Murrah water buffalo. The purpose of this study was to characterize the clinical, morphological, and molecular genetic features of this disorder.

Clinical examination and EMGs were performed on 8 affected animals. Muscle biopsies were obtained for RNA extraction and histological evaluation. The CLCN1 cDNAs from normal and affected animals were compared. The relevant genomic fragment consisting of partial exon 3, intron 3 and the 5' portion of exon 4, was also amplified and sequenced.

Affected animals exhibited muscle hypertrophy and stiffness. In some animals stiffness was severe, leading to prolonged recumbency after startle. Myotonic discharges were observed on EMG and dystrophic changes were not evident. The cDNA from myotonic buffalo uniformly lacked 43 nucleotides at the 3' end of exon 3 relative to wild-type. Analysis of the corresponding genomic fragment revealed that aberrant splicing caused the absence of this exonic sequence from the transcript. The abnormal splicing pattern was uniformly associated with homozygosity for a synonymous SNP in exon 3. The predicted impact of this aberrant splicing event is alteration of the translational reading frame and premature termination of translation. Genotype/phenotype information from a limited number of animals suggests a recessive mode of inheritance.

Cryptic splicing associated with synonymous substitutions in coding sequences have been described in several human diseases, but this is the first report associating this mechanism with congenital myotonia. This is also the first description of congenital myotonia in the water buffalo.

NOCIEPTION & PHYSIOLOGY OF PAIN IN FOOD ANIMALS

Hans Coetzee, BVSc, Cert CHP, PhD, DACVCP
Manhattan, KS

Pain is defined as an aversive feeling or sensation associated with actual or potential tissue damage resulting in physiological, neuroendocrine, and behavioral changes that indicate a "stress" response". Nociception is the process of pain perception and involves 5 distinct stages:

1. Tissue Damage: Releases bradykinin that in turn releases substance P and prostaglandin at the site of injury.
2. Transduction: Conversion of the chemical signal to electrical energy
3. Transmission: Transmission of the electrical impulse occurs via 2 distinct nerve fiber tracts. A-fibers transmit sharp, distinct pain sensations. On the other hand, C-fibers transmit poorly localized chemical signals. In both cases, signal ascend via spinothalamic tracts to the brain.
4. Pain Perception: This process localizes the pain and intensity
5. Pain Modulation: This is brought about via endogenous analgesic pathways through the release of endorphins and encephalins.

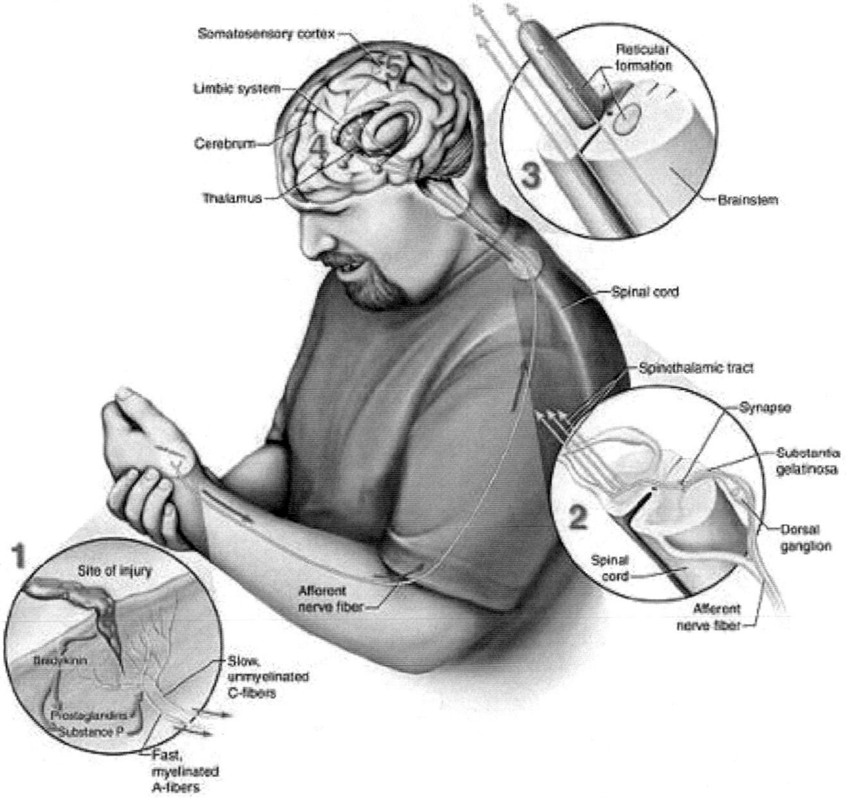

Available at http://wildiris3.securesites.net/cms_prod/files/course/91/pain-pathways.jpg

Pain and distress inflicted by routine animal husbandry procedures such as castration are major animal welfare concerns in beef cattle production. In clinical veterinary practice, behavior is used to identify animals in pain, while in research both physiological responses (typically plasma cortisol responses) and behavior are used.[1] Assessment of individual animal behavioral changes in response to pain is highly subjective and may be influenced by differences in individual perception and interpretation. Acute cortisol response is not always an accurate indicator of distress since the adrenal system is also responsible for maintaining homeostasis. Therefore, the need for a robust, repeatable physiological indicator of distress for use in the assessment of production procedures and environments has been recognized.

Surgery-induced pain consists of two phases: an immediate incisional phase and a prolonged inflammatory phase that arises primarily due to tissue damage. [2] Several methods have been developed to directly or indirectly assess pain associated with dehorning and castration in calves. Measurements of physiological changes include assessment of heart rate[3] and heart rate variability[4] and changes in peripheral vascular perfusion determined by thermography[4]. Physiological effects of dehorning have also been evaluated by assessing changes in bodyweight and performance[5]. Neuroendocrine effects have been studied using circulating biomarkers such as cortisol[6] and substance P[7]. Behavioral responses to painful events have been assessed using constant video surveillance, chute exit speed determination[8], step counts and stride length[9]. Remote accelerometer sensors can objectively monitor cattle behavior, and this system has been validated with high accuracy, compared to video observation, in predicting cattle activity[10]. Accelerometers have been used to illustrate behavioral changes in cattle after castration[11].

ASSESSMENT OF PAIN IN CATTLE ASSOCIATED WITH CASTRATION

Assessment of individual animal behavioral changes in response to pain is highly subjective.[1] Escape behaviors that are present during castration but not seen afterwards may reflect a specific acute pain[4] or may be a behavioral response indicative of a desire to escape restraint. Studies have reported that surgically castrated calves struggle and kick during the operation but calves castrated with rubber rings do not. Macauley and others[12] reported that calves castrated surgically moved around much less than control calves or calves castrated using a Burdizzo. Fisher and others[13] found that 14 month old bulls castrated surgically stamped their hind feet, swished their tails and grazed less in the afternoon following castration than control bulls and bulls castrated using bands. Behaviors indicative of a painful sensation such as turning the head towards the hindquarters, alternate lifting of the hindlegs, abnormal postures and slow movement of the tail has been reported weeks after rubber-ring castration. The relative level of pain signified by the different behaviors seen after castration is difficult if not impossible to quantify.

Production parameters are imprecise measurements of pain experienced by animals following castration.[13] Weight gain following castration may also be negatively influenced by decreased testosterone following castration.[14] Assessment of production parameters is critical if animal well-being research is to have relevance to livestock producers. The ADG of 7 week-old calves over a 5 week period following castration using rubber rings, clamp or surgery has been found to be lower than non-castrated calves but similar between the different castration methods.[15] Rubber ring and surgical castration were reported to cause a decrease in ADG of 50% and 70% respectively in cattle aged 8 to 9 months.[16] When 8, 9 and 14 month old cattle were castrated surgically or using latex bands, cattle castrated later had poorer growth rates than those castrated at weaning. Cattle castrated with latex bands also had lower growth rates than those castrated surgically during the following 4 – 8 weeks.

Several studies have evaluated acute cortisol response as a method to determine the extent and duration of distress associated with castration in cattle have recently been reviewed.[1] Surgical and latex band castrations, especially when performed in older cattle, appear to elicit higher plasma cortisol responses that remain above pre-treatment levels for longer. Peak cortisol concentration following surgical castration occurs within the first 30 minutes after castration and range from 45 nmol/L following rubber ring castration to 129 nmol/L following surgical castration. The duration of plasma cortisol response above pre-treatment levels typically ranges from 60 minutes following Burdizzo castration to 180 minutes following surgical castration. Fisher and others[16] reported that plasma cortisol response was significantly reduced during the first 90 minutes following surgical or Burdizzo castration in calves when the lidocaine was given suggesting an association between cortisol response and pain.

Table 1 shows the Cmax, which is the maximum cortisol concentration in serum, and the Tmax which is the time after castration when maximum cortisol concentration occurred. When we consider rubber ring castration in 6-day old calves we see the cortisol concentration was much lower and occurred much earlier than in 2-4-month-old animals. However, that association doesn't really hold here for surgical castration. We see a much higher cortisol concentration in 6-day-old calves compared to 2-4-month-old calves. This does not really fit with our hypothesis that surgical castration should be less stressful in younger calves. The literature suggests those 6-day-old calves have a higher cortisol concentration than those older calves. This does not imply that we should wait and castrate them older. What I'm trying to point out is cortisol responses are an imperfect measure of pain associated with castration.

Table 1

	Age				
Method	**6 days**	**21 days**	**42 days**	**2 – 4 months**	**5.5 months**
Rubber ring	**60** (36 min)	45 (48 min)	45 (60 min)	**76** (90 min)	
Latex Band				101 (30 – 60 min)	
Burdizzo	**80** (24 min)	50 (24 min)	60 (24 min)	**64** (30 min)	87 (30 min)
Surgery (Pull)	**105** (24 min)	65 (24 min)	110 (24 min)	**68** (30 min)	
Surgery (Cut)					129 (30 min)

Adapted from: Stafford K and Mellor D. The Welfare significance of the castration of cattle: a review. New Zealand Veterinary Journal 53(5) 271-8, 2005

Table 2

	Age				
Method	**6 days**	**21 days**	**42 days**	**2 – 4 months**	**5.5 months**
Rubber ring	**132 min**		96 min	**132 min**	180 min
Latex Band				**180 min**	
Burdizzo	**60 min**	60 min	72 min	**90 min**	90 min
Surgery (Pull)	**132 min**	84 min	132 min	**180 min**	
Surgery (Cut)					360 – 600 min

Adapted from: Stafford K and Mellor D. The Welfare significance of the castration of cattle: a review. New Zealand Veterinary Journal 53(5) 271-8, 2005

Table 2 contains an interesting data set examining the duration of plasma cortisol response elevation above pre-treatment levels. With a rubber ring castration in 6-day-old calves and 2-4-month-old calves the time above baseline cortisol response is identical. The duration of cortisol response was the same in those two age groups of calves even though we currently recommend doing them as early as possible. The take home message is that measurement of plasma cortisol is not a perfect measure of pain in animals. Cortisol measurement will not answer the questions we need answered in order to address animal welfare concerns. In Table 2 we see that surgical castration in 6-day-olds produces a much shorter duration of cortisol response than surgical castration in 2-4-month-old animals. This is typically what we would expect. However, there are also aspects of these data that do not fit with our hypothesis. For example, the AVMA backgrounder states that elastrator rubber ring techniques have been associated with chronic pain and should be discouraged, but 6-day-old calves had the same duration of cortisol as 2-4-month-old calves.

Substance P is an 11-amino acid prototypic neuropeptide that regulates the excitability of dorsal horn nocireceptive neurons and is present in areas of the neuroaxis involved in the integration of pain, stress, and anxiety. Studies have shown that plasma SP levels are up to 27-fold greater in human patients with soft tissue injury than healthy controls.[17] Our research group recently conducted a study to evaluate plasma substance P (SP) and cortisol response following castration.[7] Calves were acclimated for 5 days prior to random assignment based on scrotal circumference to a castration or uncastrated control group. Blood samples were collected at -24, -12, and 0 hours pre-castration and 3, 10, 20, 30, 45, 60, 90, 120, 150, 180, and 240 minutes post-castration. Vocalization and attitude scores were determined at the time of castration or simulated castration. Plasma SP and cortisol were determined using competitive and chemiluminescent enzyme immunoassay, respectively. Data were analyzed by repeated measures analysis using a Mixed Effects model allowing for unequal variances. No significant difference in plasma cortisol response between castrated and uncastrated calves was observed over time ($p=0.644$). In contrast, mean plasma SP concentrations were significantly higher in castrated calves compared to uncastrated controls over the course of the study ($p=0.042$). Cortisol responses over time in calves with vocalization scores of 0 were not significantly different from calves with vocalization scores of 3 ($p=0.17$). However, calves with vocalization scores of 3 had significantly higher SP levels when compared to calves with scores of 0 ($p=0.033$). These findings contradict previous reports that show an increase in plasma cortisol relative to pain post-castration. Significant increases in plasma SP concentration post-castration suggest that this measurement may be associated with nociception; however, further investigation is necessary.

Accelerometers have been used in other species to detect lameness and remotely monitor level of animal activity. Our research group has utilized video observations to determine the accuracy of accelerometers to measure behavior changes in cattle and to determine differences in beef bull behavior post-castration.[11] Three Holstein calves and 12 healthy beef bulls had two-dimensional accelerometers placed on three animals and data was logged simultaneous to video recording of animal behavior. The subsequent data set was used to generate and validate a predictive model to classify animal posture (standing or lying) and type of activity (standing in place, walking, eating, getting up, lying awake, or lying sleeping). The algorithms developed were used to conduct a prospective trial to determine differences in bull behavior in the first 24 hours post-castration ($N=6$) compared to both control animals (non-castrated) ($N=6$) and pre-castration readings from the same bulls. Based on the analysis of the 2-D accelerometer signal, posture can be classified with a high degree of accuracy (98.3%) and the specific activity can be estimated with a reasonably low misclassification rate (23.5%). Employing the system to compare behavior post-castration revealed that castrated calves spent a larger ($P < 0.05$) amount of time standing (79.3%) compared to either pre-castration readings (51.2%) or control calves after castration (64.3%). Animals also spent a lower percentage of the time eating in the post-castration phase. The 2-D accelerometers provided accurate classification of animal posture and reasonable classification of animal activity. Collected data allowed quantification of behavioral differences between animals after a surgical procedure and provides a valuable tool to compare research with behavioral endpoints.

Australian researchers used radar speed cameras to measure the speed of cattle exiting a squeeze chute.[8] Cattle with faster exit speeds had lower weight gains, more sickness, and more dark cutting meat. The major problem with chute exit speed as a means to determine pain is that it doesn't work well for highly acclimated, very tame animals such a dairy cattle. There are also certain breeds such as *Bos indicus* breeds that are more prone to demonstrate rapid chute exit speeds when compared with *Bos taurus* breeds.

REFERENCES

[1]Stafford, K.J. and Mellor, D.J. **New Zealand Veterinary Journal**. 2005. 53(5):271–278. [2] Kissin, I. **Anesthesiology**. 2000. 93:1138 – 43. [3] Heinrich, A. et al. **J Dairy Sci** 2009. 92(2): 540-547.[4] Stewart, M. et al. **J Dairy Sci**. 2009. 92(4): 1512-1519. [5] Faulkner, P.M., Weary, D.M. **J Dairy Sci.** 2000. 83:2037-2041. [6]Stafford, K.J., D.J. Mellor. **The Veterinary Journal**. 2005. 169: 337 – 349. [7] Coetzee et al. **Amer J Vet Res.** 2008. 69(6): 751 – 762. [8]Burrows, H.M., R.D. Dillon. Aust. **J. Exper. Agric.** 1997. 37:407-411. [9] Currah, J.M et al. **Can Vet J.** 2009. 50(4): 375–382. [10] Robert, B. et al. Comput Electron Agric. 2009. 67: 80-84. [11]White, B. J et al.. Amer **J Vet Res**. 2008. 69 (8): 1005-1012. [12] Macauley, A.S.et al. **Journal of Animal Science**. 1986. 63:166. [13] Robertson, I.S et al. **Research in Veterinary Science**. 1994. 56: 8–17. [14] Knight, T.W, et al. **New Zealand Journal of Agricultural Research**. 1999. 42:255–268. [15] Fenton, B.K. et al. **Veterinary Record.** 1958. 70:101-102.[16.] ZoBell, D.R et al. **Canadian Journal of Animal Science.** 1993. 73:967-970. [16] Fisher, A.D. et al. J. **Anim. Sci.** 1996. 74:2336-2343. [17] Onuoha GN, Alpar EK**. Life Sci** 1999;65(13):1351-1358.

PHARMACOLOGY OF PAIN MANAGEMENT IN FOOD ANIMALS

Hans Coetzee, BVSc, Cert CHP, PhD, MRCVS, DACVCP
Manhattan, KS

The 21st century consumer is wealthier but also more detached from production agriculture than any in history. Therefore, animal welfare concerns are becoming an important issue to our clients and customers. We recognize that pain is an inevitable consequence of many routine animal husbandry procedures in farm animals. However, how can we assess if cattle are in pain and if our analgesic interventions are working? In this session we will examine novel approaches to measuring pain and analgesic drug efficacy in cattle. We will discuss a practical sub-anesthetic/ analgesic drug combination you can use to take the edge off the fractious cases you encounter.

Nociception is an inevitable consequence of many routine management procedures in farm animals. Castration is considered one of the most stressful experiences for livestock by the American Veterinary Medical Association (AVMA)[1] and is performed on approximately 15 million calves in the U.S annually. Dehorning and castration are especially significant in terms of animal welfare because preemptive analgesia can be applied in advance of the painful stimulus, thereby preventing sensitization of the nervous system to subsequent stimuli that could amplify pain. The AVMA "supports the use of procedures that reduce or eliminate the pain of dehorning and castrating of cattle" and proposes that "available methods of minimizing pain and stress include application of local anesthesia and the administration of analgesics."[1] In spite of this, a recent survey of 184 bovine veterinarians conducted by our research group found that only 1 in 5 U.S veterinarians use anesthesia or analgesics at the time of castration. One reason for this discrepancy is the lack of objective methodology to quantify the most effective pain mitigation strategies.

It is remarkable that although administration of local anesthesia prior to castration and dehorning is legislated in several European countries, there are currently no analgesic drugs specifically approved for pain relief in livestock by the U.S Food and Drug Administration (FDA). FDA Guidance Document 123 for the development of effectiveness data for non-steroidal anti-inflammatory drugs (NSAIDs) states that "validated methods of pain assessment must be used in order for a drug to be indicated for pain relief in the target species."[2] The identification and validation of robust, repeatable pain measurements is therefore fundamental for the development and approval of effective analgesic drug regimens for use in livestock. Research to address our limited knowledge in this area is therefore essential to formulating science-based recommendations.

In practical terms, resistance to requiring injectable analgesia for routine castration and dehorning is largely based on time and logistical issues. However, a recent study evaluating novel methods of analgesia for tail docking in lambs demonstrated that castrating by banding in 1-2 day old lambs required an average of 28 seconds without analgesia and 68 seconds when an injectable local anesthetic was administered by jet-injector.[3] While the United States has not followed other countries in legislating the use of local anesthesia during castration and dehorning, it is appropriate for the veterinary profession to pursue practical, rapid, and effective methods for the relief of pain related to these procedures. The authors are aware of practicing veterinarians that have adopted local anesthesia regimens for these procedures.

Challenges associated with providing analgesia in food animals

(1) Absence of FDA approved analgesic compounds in the U.S requiring ELDU under AMDUCA
(2) Time delay between administration of a local anesthesia and the onset of activity that may add significant time to the procedure
(3) Technical demands associated with analgesic drug administration (for example, flunixin is only labeled for IV administration)
(4) Most available analgesic compounds have a short half-life necessitating frequent administration
(5) DEA regulation of narcotic analgesics
(6) The cost associated with providing analgesia
(7) The willingness (or unwillingness may be more accurate) of producers to adopt these measures.

Compounds available to provide analgesia

Nociception involves the transduction of chemical signals at the site of injury into electrical energy and the transmission of these impulses via the spinothalamic tracts to the brain where pain perception occurs.[4] Several compounds including opioids (eg. butorphanol), local anesthetics (eg. lidocaine), α2 - adrenergic receptor agonists (eg. xylazine) and NMDA receptor antagonists (eg. ketamine) exert direct analgesic effects by targeting specific receptors in the central and peripheral nervous system. In contrast, NSAIDs produce analgesia and reduce inflammation by inhibiting the enzyme cyclooxygenase (COX) and subsequent prostaglandin production in the peripheral tissues and CNS. The observed effect of NSAID administration in modulating the effect of pain associated with dehorning is therefore likely due to a combination of direct analgesic effects in the CNS and anti-inflammatory effects at the site of injury.

Multimodal analgesia

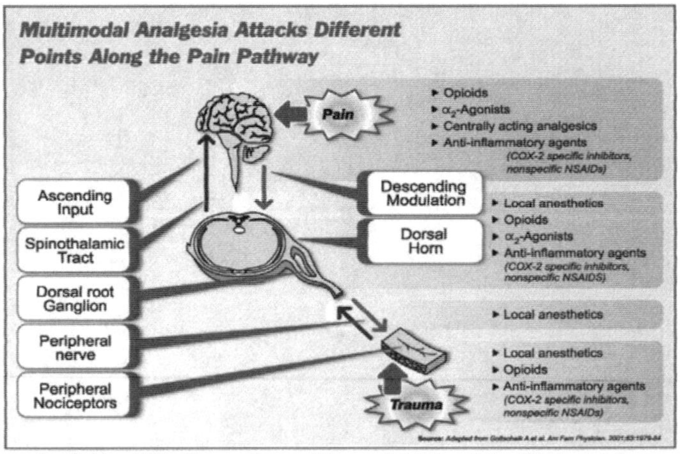

Available at http://www.physweekly.com/picts/specialxxi03/chart-b.jpg

Pharmacokinetic/Pharmacodynamic Modeling of Analgesic Drugs

Our research group recently conducted a study to examine the effect of oral aspirin and intravenous sodium salicylate on acute plasma cortisol response following surgical castration.[4] Twenty bulls, randomly assigned to the following groups; 1) uncastrated, untreated controls, 2) castrated, untreated controls, 3) 50mg/kg sodium salicylate IV pre-castration and 4) 50mg/kg aspirin (acetylsalicylic acid) per os pre-castration, were blood sampled at 3, 10, 20, 30, 40, 50 minutes and 1, 1.5, 2, 4, 6, 8, 10 and 12 hours post-castration. Samples were analyzed by competitive chemiluminescent immunoassay and fluorescence polarization immunoassay for cortisol and salicylate respectively. Data were analyzed using noncompartmental analysis, a simple cosine model, ANOVA and t-tests. Intravenous salicylate Vd_{ss} was 0.18 L/kg, Cl_B was 3.36 mL/min/kg and $T_{½λ}$was 0.63 hours. Plasma salicylate concentrations above 25 µg/mL coincided with significant attenuation in peak cortisol concentrations (p=0.029). Peak salicylate concentrations following oral aspirin administration was less than 10 µg/mL and failed to attenuate cortisol response. Once salicylate concentrations decreased below 5 µg/mL, cortisol response in the castrated groups were significantly higher than uncastrated controls (p=0.018). To our knowledge this is the first study relating plasma analgesic drug concentrations directly to mitigation of plasma cortisol response post-castration. These findings have important implications for designing effective analgesic regimens to alleviate the stress response associated with painful routine animal husbandry procedures.

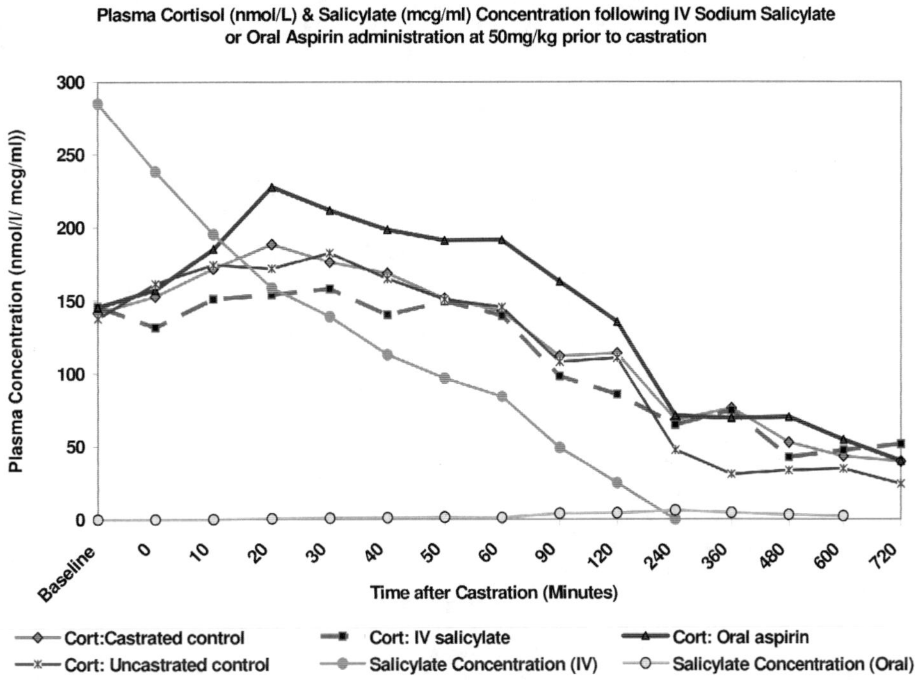

A protocol for use of IM butorphanol/xylazine/ketamine (BXK) has been evaluated. The regimen consists of butorphanol (0.01 – 0.025 mg/kg) + xylazine (0.02 – 0.05 mg/kg) + ketamine (0.04 – 0.1 mg/kg). Thus a 450 kg animal, 5 mg butorphanol, 10 mg xylazine, and 20 mg ketamine would constitute the low end of the dosing range. **Note that the calculation should involve 2x xylazine as compared to butorphanol and 2x ketamine as compared with xylazine.** We have noted up to an hour of cooperation using this protocol but more fractious patients may require increased doses.

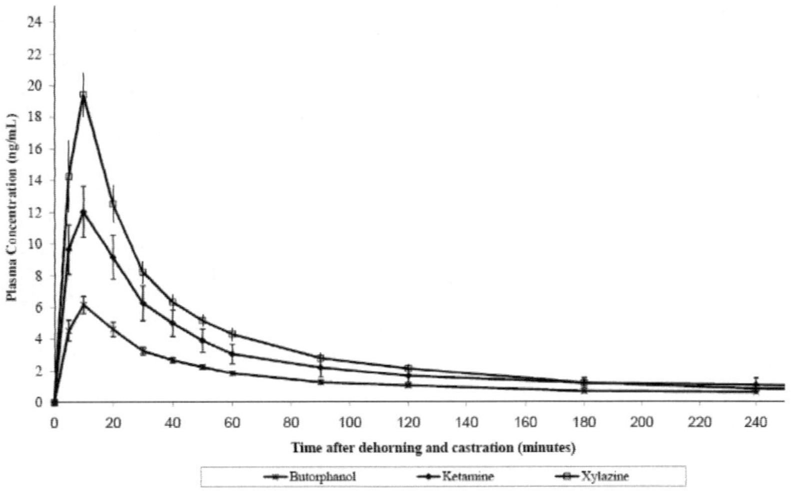

Meloxicam

Meloxicam is a NSAID of the oxicam class that is approved in the European Union for adjunctive therapy of acute respiratory disease; diarrhea and acute mastitis when administered at 0.5 mg/kg IV or SC. Meloxicam is considered to bind preferentially to cyclooxygenase-2 (COX-2) inhibiting prostaglandin synthesis although definitive evidence of COX-selectivity in calves is deficient in the published literature. Heinrich et al. (2009)[5] demonstrated that 0.5 mg/ kg meloxicam IM combined with a cornual nerve block reduced serum cortisol response for 6 hours in 6-12 wk old calves compared with calves receiving only local anesthesia prior to cautery dehorning. Furthermore, calves receiving meloxicam had lower heart rates and respiratory rates than placebo treated control calves over 24 hours post-dehorning. Stewart et al. (2009)[6] found that meloxicam administered IV at 0.5 mg/kg mitigated the onset of pain responses associated with hot-iron dehorning in 33 ± 3 day old calves compared with administration of a cornual nerve block alone as measured by heart rate variability and eye temperature. These findings indicate that administration of meloxicam at 0.5 mg/kg IV or IM decreases physiological responses that may be linked to pain and distress associated with cautery dehorning in preweaning calves.

The purpose of this study was to investigate the pharmacokinetics and oral bioavailability of meloxicam in ruminant calves[7]. Six Holstein calves (145 – 170 kg) received either meloxicam IV at 0.5 mg/kg or oral meloxicam at 1 mg/kg in a randomized cross-over design with a 10-day washout period. Plasma samples collected up to 96 hours post-administration were analyzed by LC-MS followed by noncompartmental pharmacokinetic analysis. A mean peak plasma concentration (Cmax) of 3.10 ug/mL (Range: 2.64 – 3.79 ug/mL) was recorded at 11.64 hours (Range: 10 – 12 hours) with a half-life (T ½ λz) of 27.54 hours (Range: 19.97 – 43.29 hours) after oral meloxicam administration. The bioavailability (F) of oral meloxicam corrected for dose was 1.00 (Range: 0.64 – 1.66). These findings indicate that oral meloxicam administration could be an effective and convenient means of providing long-lasting analgesia to ruminant calves.

Mean +/- SD Meloxicam Plasma Profile

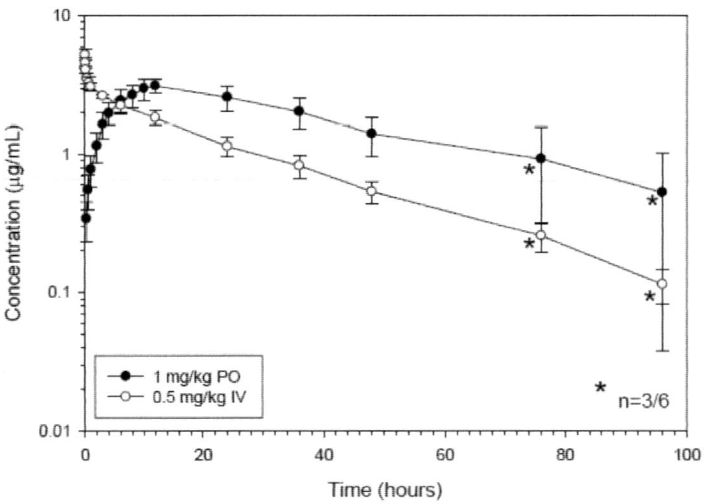

In the United States, meloxicam administered to cattle by any route constitutes extra-label drug use (ELDU). Under the Animal Medicinal Drug Use Clarification Act (AMDUCA), ELDU is permitted for relief of suffering in cattle provided specific conditions are met. These conditions include that (1) ELDU is permitted only by or under the supervision of a veterinarian, (2) ELDU is allowed only for FDA approved animal and human drugs; (3) ELDU is only permitted when the health of the animal is threatened and not production purposes; (4) ELDU in feed is prohibited and (5) ELDU is not permitted if it results in a violative food residue. Therefore, use of oral meloxicam to alleviate suffering associated with dehorning and castration in calves in the United States would be required by law to comply with these regulations. Currently the only NSAID approved for use in cattle in the United States is flunixin meglumine. The plasma elimination half-life of flunixin is reported to be 3 – 8 hours therefore requiring once daily administration. Although this drug class is recognized as having analgesic properties, flunixin is only indicated for control of fever associated with respiratory disease or mastitis, and fever and inflammation associated with endotoxemia, rather than for control of pain. Studies demonstrating the analgesic effects of flunixin at the approved dose of 2.2 mg/kg are deficient in the published literature. Use of flunixin meglumine is further complicated by the requirement for intravenous administration which is more stressful on the animal and involves more skill and training on the part of the operator. Several reports have suggested that the IM administration of flunixin may result in significant myonecrosis and tissue residues. In the absence of data demonstrating that flunixin reduces signs of pain and distress associated with dehorning and castration in calves, it could be argued that use of oral meloxicam for this purpose can be justified under AMDUCA. Meloxicam (20 mg/ml) is approved for use in cattle in several European countries with a 15 day meat withdrawal time and a 5 day milk withdrawal time following administration of 0.5 mg/kg IM or SC. An oral meloxicam suspension (1.5 mg/mL) and injectable formulation (5 mg/mL) are approved in the United States for the control of pain and inflammation associated with osteoarthritis in dogs. Furthermore, an injectable formulation (5 mg/ml) is approved for the control of post-operative pain and inflammation in cats. Several generic tablet formulations containing meloxicam (7.5 and 15 mg) have recently been approved for relief of signs and symptoms of osteoarthritis in human medicine. The cost of administering IV meloxicam to calves in the present study was approximately US $58.00/100 kg bodyweight and the cost of administering oral meloxicam was US $0.30/100 kg bodyweight.

REFERENCES

[1] American Veterinary Medical Association. Available at http://www.avma.org/reference/backgrounders/castration_cattle_bgnd.asp. Accessed 5 September 2007.2. [2] U.S. Food and Drug Administration, Center for Veterinary Medicine. Available at http://www.fda.gov/cvm/Guidance/guide123.htm. Accessed 5 September 2007. [3] French NP et al. **Vet Rec.** 1994; 134(18):463-7. [4]Coetzee JF et al: **J Vet Pharmacol Ther** 2007. 30:305–31. [5]Heinrich, A. et al. **J Dairy Sci** 2009. 92(2): 540-547. [6] Stewart, M. et al. **J Dairy Sci**. 2009. 92(4): 1512-1519. Coetzee, J.F.et al. **Vet Ther**. 2009. 10(4): E1 – E8.

USING BEHAVIOR FOR CLINICAL ASSESSMENT OF PAIN IN BOVINES

Suzanne T. Millman, PhD
Ames, IA

INTRODUCTION

Animal welfare is an important facet of bovine veterinary practice, and practices that cause pain are of particular concern. In a national survey commissioned by the American Farm Bureau[1,2], 81% agreed that "(f)arm animals have roughly the same ability to feel pain and discomfort as humans". Further, 68% of respondents believed that the government should take a more active role in promoting farm animal welfare, and 76% did not consider food prices to be more important than farm animal well-being.

Amongst bovine practitioners, there are regional and generational differences in attitudes about pain management. In a national Canadian survey[3], analgesia was less likely to be provided for dehorning beef calves versus dairy calves of the same age (beef: 60%, dairy: 85%, P<0.001). Veterinarians were more likely to provide analgesia if they believed dehorning was painful, and they were three times more likely to provide analgesia if concerns for their personal safety[4]. In Ontario, Canada 92% of bovine veterinarians and 22% of dairy producers use local anesthetic for dehorning some or all calves[5]. Both veterinarians and producers cited pain relief as primary reason, whereas time and cost were cited as reasons for not using analgesia. Producers that used analgesia were 6.5 times more likely to have a veterinarian involved with their decision making about dehorning, whereas 13% of producers were unaware of opportunities for pain management. A recent survey of dairy producers in Northeast and North Central USA similarly found 12.4% of dairy owners used anesthetic when dehorning calves[6], suggesting that bovine practitioners should offer information about pain management options.

American Veterinary Medical Association and Canadian Veterinary Medical Association animal welfare policies recommend pain management, particularly for surgical procedures. Pain is defined as "an unpleasant sensory and emotional experience associated with actual or potential tissue damage"[7]. Pain experience involves sensation and transduction of noxious stimuli in afferent A and C nerve fibers, modulation of the signal in the dorsal horn of the spinal cord and finally perception of the signal by the brain[8], specifically the pre-frontal cortex[9]. Affective states, such as pain, can only be measured indirectly, and hence, present challenges for both human and veterinary medicine. How can we know if an animal is experiencing pain and whether our interventions are effective? Multidisciplinary approaches are warranted, and behavior is a key component to help us answer these questions.

PAIN-RELATED CHANGES IN ACTIVITY AND POSTURE

Animal behavior is a tool that is central to detection of ill, injured and unthrifty livestock. Since there is no "Gold Standard" for bovine pain assessment, such as self-reporting, animal behavior provides the most integrated response of an animal to its situation. However, interpretation of behavior responses rests on knowledge about evolved species-specific responses, animal cognition and how behavior is integrated with immune and endocrine systems.

Subjective pain scales have been used for pain assessment in humans, in which patients are asked to numerically rate their pain on a scale from 1 to 10 or on a visual analogue scale with lowest scores associated with no pain at all and highest scores associated with worst pain imaginable. These subjective scoring systems have been used in veterinary practice for assessing bovine pain[10], but since the animal cannot self-report, validity of pain assessment by proxy has been criticized due to its dependent on the knowledge and interpretation of the owner or veterinarian. However, subjective scoring is often the most practical option for assessing animal pain in field conditions, and can be strengthened with clearly articulated and/or graphically illustrated categories for assessment, observer training and integration of knowledge from primary research.

Several postural changes provide useful diagnostics and specific types of pain. Visceral pain associated with castration results in characteristic immobile postures ("statue-standing"), abnormal standing postures such as standing still with hind-legs stretched and apart, easing of hindquarters and lying in sternal or lateral recumbency with full or partial extension of the hind-legs[11]. Similarly, abnormal tail postures occur with diarrhea in calves, likely due to visceral discomfort[12]. In contrast, a number of characteristic postures are used for detection of lameness in dairy cattle, such as perching (in which dairy cattle stand with front hooves in the stall and in the alley with the rear hooves), crossed legs, claw rotation and back arch[13]. Hock-hock distance has been used to identify mastitis pain[14]. However, caution is warranted since changes in posture do not necessarily reflect pain. Current research by PhD candidate Amy Stanton (unpublished data) suggests that ear droop and head tilt that occur during otitis are not associated with increased pain sensitivity when measured using a mechanical nociception test.

Changes in behavioral activities can also provide insights into animal pain, and can be measured using direct observation, video analysis and increasingly with biotelemetry equipment, such as accelerometers. To quantify behavioral responses, applied ethologists develop ethograms to provide mutually exclusive definitions for each behavior that can be analysed in terms of latency, duration and frequency of occurrence. Video technology has advantages for blinding of observers to treatment groups to avoid bias, to train observers on the ethogram and to test intra- and inter-observer reliability. Pain-related changes in behavior can be general, such as increased restlessness or decreased feeding, and are associated with convalescence. However, some activities appear to be characteristic and relatively sensitive indicators of pain. Ear-flicking, head shaking and head scratching are associated with postsurgical dehorning pain[15], and these behaviors are reduced when pain is mitigated by local anesthetics and nonsteroidal anti-inflammatory drugs[16-19]. Foot stamping, easing of quarters and scrotal licking are associated with postsurgical castration pain, and can be used assess differences in pain associated with castration techniques[11]. During first 3h following castration, calves castrated with rubber rings displayed more active behaviors than calves that were only handled, castrated with a Burdizzo, surgical castrated or castrated with a Burdizzo and rubber ring. This difference was particularly evident for foot stamping/kicking and easing quarters. Rubber ring castration was also associated with significantly more lesion licking during the first 48 hours, with 29.3 counts rubber ring versus 8.6 counts with surgical castration.

EMERGING TECHNIQUES FOR "ASKING" ANIMALS ABOUT PAIN

In human medicine and biomedical research, mechanical tests have been used to quantify pain sensitivity and allodynia. Pressure algometry measures mechanical nociception, and a more sensitive version of this technology are von Frey's Filaments that can assess nociception thresholds on acutely sensitive tissues, such as wounds. In our research group, we are using pressure algometry to quantify changes in nociception thresholds associated with dehorning. Following sham and actual dehorning, calves show greater sensitivity following actual dehorning and did not become sensitized to the procedure in the sham treatment. Significant differences in nociception threshold by dehorned calves were detected for 59 hours (unpublished data). In another study, pain sensitivity after dehorning was decreased in calves that received a single dose of meloxicam (unpublished data). Algometry is also being used by our research group and others, to assess visceral pain, and for pain associated with bovine lameness[20-22].

Thermal sensitivity has been applied to assess parturient pain in cattle, using a radiating heat stimulus directed at the coronary band of the unshaven forefoot and a foot-withdrawl response[23]. Cows that consumed placenta or amniotic fluid tolerated more heat at 1h post partum than they did at 6h and 36h (P<0.004), whereas control cows did not. A laser foot lift test was developed and validated, aiming at the caudal aspect of the metatarsis[24], and infrared thermography indicated that cattle lifted the affected leg when skin temperature reached 45-55C. Results were more reliable at higher power settings (4.5W vs. 2.25W), during which animals were more likely to kick than move legs. These researchers found that at least three tests were needed to control for individual variation. Laser foot lift tests have also been validated with dairy cows, and was sufficiently robust to detect hypoalgesia associated with environmental stressors[25].

In many species, distress calls are emitted at particular pitch and volume[26]. Hot-iron branding caused 58 of 95 calves to vocalize versus 7 of 94 calves during sham branding (P<0.0001)[27]. Furthermore, hot-iron branding resulted in calls that were higher in frequency (branded: 186.7 Hz +/- 5.2; sham branded: 141.6 Hz =/- 6.6, P<0.01), and of higher relative intensity (P<0.05). However, although vocalizations may be a very sensitive measure of painfulness, it may not be practical for cattle since vocalizations are relatively rare, with only 50% of cattle vocalizing during branding[28], 14% during castration[29]and 13% during disbudding[30] procedures, versus goats for which 100% vocalized during disbudding[31]. There are also phenotypic differences between beef steers in terms of likelihood to vocalize[32]. Approximately, 36% of steers vocalized during processing, but 45% of steers with completely black hides called versus 30% of cattle with completely white hides (P<0.014). Calves that vocalized in restraint tests were more likely to also call during isolation (P<0.037), suggesting differences in temperament. Acoustic properties differed with short calls at lower fundamental frequencies given during isolation versus during restraint (P<0.001).

Opportunities to quantify escape behaviors are possible using exit speed, during which an infrared beam is placed at a set distance from the restraint chute. As each animal exits the chute and crosses the beam, the timer automatically stops and latency recorded. Similarly, strain gauges and load cells have been used to measure the pressure that cattle exert to holding chute during branding, and digital video image analysis to quantify escape behavior[33]. Hot-iron branding was associated with greater maximum exertion force, greater head movement distances and greater head movement velocities than freeze branding or sham branding. In this experiment, image analysis was a more sensitive technique for detecting treatment differences than frequency counts or exertion forces, it was quick and inexpensive, but its application depends on animal restraint methods.

Last, cognitive tests are the most promising techniques for "asking" animals about their subjective experiences, such as pain. In a pivotal study, lame broiler chickens were found to preferentially consume feed containing the NSAID carprofen, whereas non-lame birds did not[34]. Rushen[35] used this technique on sheep to assess pain associated with electro-immobilization. After receiving four treatments of electro-immobilization, sheep took twice as along to move through the race when compared with sheep that received physical restraint, and this aversion was still evident 12 weeks after the treatments, suggesting strong affective response to this handling procedure. However, these cognitive tests depend on the capacity of the cow or calf to learn associations between events. Cattle must have sufficient cognitive development and function to learn to predict events based on signs and signals. Hence this technique is time consuming and may not be amendable to particular research questions.

CONCLUSIONS

Animal behavior is a tool that is central to detection of ill, injured and unthrifty livestock. Affective states such as pain and malaise can only be measured indirectly, and when assessing these states in animals, there is no "Gold Standard", such as self-reporting. Animal behavior provides the most integrated response of an animal to its situation. Interpretation of behavior responses rests on knowledge about evolved species-specific responses, animal cognition and how behavior is integrated with immune and endocrine systems. Greater understanding of sickness motivation may provide opportunities to facilitate convalescence in compromised cattle

REFERENCES

[1]Norwood B, et.al. **Feedstuffs** 2007;42:14. [2]Lusk JL et.al. **JAVMA** 2008;233:1121. [3]Hewson CJ, et.al. **Can Vet J** 2007;48:155. [4]Hewson CJ, et.al. **Can Vet J** 2007;48:1129. [5]Misch LJ, et.al **Can Vet J** 2007;48:1249. [6]Fulwinder WK, et.al. **J Dairy Sci** 2008;91:1686. [7]International Association for the Study of Pain (ISAP). **Pain** 1979;6:249. [8]Anderson DE, et.al **Vet Clin N Am - Food A** 2005;21:623. [9]Anil SS, et.al. **JAVMA** 2002;220:313. [10]Rutherford KMD, **Anim Welfare** 2002;11:31. [11]Molony V, et.al. **Appl Anim Behav Sci** 1995; 46:33. [12]Todd CG, et.al. **J. Dairy Sci** 2007;85:369. [13]Zurbrigg K, et.al. **J Dairy Sci** 2005;88:3201. [14]Milne MH, et,al, **Cattle Practice** 2003;11:289. [15]Weary DM, et.al. **Appl Anim Behav Sci** 2006;100:64. [16]McMeekan CM, et.al. **NZ Vet J** 1999;47:92. [17]Faulkner PM, et.al. **J Dairy Sci** 2000;80:2037. [18]Milligan BN, et.al. **Can Vet J** 2004:45:140. [19]Stewart M, et.al. **J Dairy Sci** 2009;92:1512. [20]Whay HR, et.al. **Vet J** 1998;156:23. [21]Dyer RM, et.al. **J Dairy Sci** 2007;90:4592. [22] Laven RA, et.al. **NZ Vet J** 2008;56:210. [23]Pinheiro Machado LC, et.al. **Physiol Behav** 1997;62:1339. [24]Veissier I, et.al. **Appl Anim Behav Sci** 2000;66:289. [25]Herskin MS, et.al. **J Anim Sci** 2007;85:240. [26]Weary DM, et.al. **Anim Behav** 1995;50:1047. [27]Watts JM, et.al. **Appl Anim Behav Sci** 1999;62:125. [28]Lay DC, et.al. **Appl Anim Behav Sci** 1992;33:137. [29]Currah JM, et.al. **Can Vet J** 2009;50:375. [30]Stewart M, et.al. **Physiol Behav** 2008;93:789. [31]Alvarez L, et.al. **Anim Welfare** 2010;19:55. [32]Watts JM, et.al. **Appl Anim Behav Sci** 2001;66:289. [33]Schwartzkoff-Genswein KS, et.al. **J Anim Sci** 1998;76:972. [33]Danbury TC,et.al. **Vet Record** 2000;146:307. [34]Rushen J, **Appl Anim Behav Sci** 1986;15:315.

SICKNESS BEHAVIOR & THE CONVALESCENT FOOD ANIMAL

Suzanne T. Millman, PhD
Ames, IA

INTRODUCTION

In veterinary medicine, changes in an animal's behavior are often used as the first clinical signs of illness or injury. Good stockpeople develop an "eye" for the animals in their care, and can become highly skilled at picking up subtle changes in behavior patterns at the individual animal or pen level. Furthermore, emergence of technologies that automatically record feed intake, water intake and activity, particularly in the dairy industry, provide opportunities to improve detection of compromised individuals. The behavior of unthrifty animals is also of importance when designing a management plan for convalescence. Understanding of the mechanisms associated with sickness behavior and recognition of the species-specific behavioral needs of compromised livestock are important for designing effective management protocols to aid recovery and to care for individuals that fail to respond to conventional treatments.

WHAT IS "SICKNESS BEHAVIOR"?

Vertebrate species display characteristic changes in behavior during infection, such as reduced feeding, increased huddling, shivering, lethargy, sleep and social isolation. Hart[1] postulated that these behavioral changes represent an evolutionary strategy to facilitate recovery from disease. Fever is an important component of the immune response to reduce pathogen proliferation[2,3], and to increase core body temperature by 1C, a mammal must increase its metabolic rate by 13%. Hence, sparing energy through reduced activity, huddling and increased sleep facilitates the febrile response, resulting in an internal environmental that is less favorable to pathogen proliferation. Anorexia results in animals that expend less effort foraging, are less conspicuous to predators and devote more resources to mounting an immune response. Subsequent research in the interdisciplinary field of psychoneuroimmunology indicates that sickness behavior is mediated by pro-inflammatory cytokines, IL-1, IL-6 and TNFα, within specific locations in the central nervous system. As part of the innate immune response, these cytokines are released by macrophages and can cross the blood-brain barrier, but they are also produced directly in the brain by glial cells in response to vagal nerve stimulation[4].

However, expression of sickness behavior is not straightforward, and varies between and within individuals according to the environmental context. Sickness behavior is organized as a motivational state, and hence it competes for expression with other motivational states, such as hunger, defense or sex[5]. Laboratory experiments using lipopolysaccharide (LPS) evoke a proinflammatory cytoine cascade and characteristic changes in behavior and physiology. However, when LPS is administered to lactating female mice, some aspects of maternal behavior are suppressed and others are not. If nests and rat pups are dispersed, female rats will forfeit convalescence in favor of pup retrieval. However, effort to rebuild the nest is only performed in rooms at 6C, whereas convalescence is favored at 22C[6]. Similarly, LPS does not affect nest defense because dams display aggression against a male intruder, but social exploration is suppressed[7]. Although sickness normally suppresses activity and exploration, these behaviors are expressed when animals are placed in novel environments[8]. Conversely, sickness impairs learning of new tasks, but performance of tasks previously learned[9]. Interestingly, gender differences exist since LPS reduces sexual behavior in female rats, but not in males[10].

BOVINE SICKNESS BEHAVIOR?

Our research group, and others, are exploring the sickness and pain-related behavior of livestock, using biotelemetry and video techniques, together with physiologic and performance parameters. Detailed data collection of behavior and physiologic parameters is time consuming and labor intensive, but technologic advances provide increasing opportunities to collect some of this information automatically in the field and in the laboratory. Accelerometers can be attached to a hindleg to measure bouts of activity and rest, and can detect increased activity associated with estrus. This is a promising technique for detecting injury or illness. Cattle normally display crepuscular circadian patterns of activity, with active periods corresponding with dawn and dusk. Todd[11] found increased activity by veal calves experiencing diarrhea due to viral agents relative to calves affected by diarrhea due to other (bacterial, parasitic) causal agents (p=004). Increased activity has also been observed in cows 8-9 days prior to clinical signs of ketosis, displaced abomasum and digestive disorders[12]. Dairy heifers that fail to gain weight during the immediate post-weaning period also display high levels of activity relative to calves with medium or high growth rates, perhaps due to social stress, failure to transition to bunk versus pail feeding or subclinical illness (unpublished data). Restlessness during onset of illness may arise from discomfort or changes in neurotransmitters[13].

Sleep is an important component of sickness behavior, positively associated with recovery from bacterial infections[14]. An experiment involving free-ranging kudu in South Africa provide insights into demands for rest during illness[15]. The kudu were surgically implanted with dataloggers to collect activity and body temperature patterns. After release, the kudu developed pneumonia for 6-10 days and died. Kudu became febrile, and during this period daily activity during fever decreased to 40% of activity during afebrile days. Furthermore, during fever they actively sought microclimates that were warmer than afebrile periods. For cattle, demands for rest during illness and injury can be expected to surpass 12-13 hours of rest per day required by healthy heifers[16]. In one study, lame cows spent more time lying out of the cubicles when housed in a freestall barn with a 1:1 cubicle: cow ratio, which could arise due to competition for resting sites or due to poor cubicle design that fails to accommodate the space envelope required for standing-lying transitions[17]. Opportunity to rest may be thwarted by discomfort and ability to transition between standing and lying. During ruminal acidosis, heifers display greater difficulty lying down and spend more time performing lying intention movements[18]. Rest is most likely to be facilitated in thermal-neutral environments that provide comfortable surfaces[19], good footing and protection from bullying.

Changes in feeding behavior are also useful indicators of compromised cattle, often appearing as a first clinical sign of disease. Direct observations are best made immediately after fresh feed is provided, when motivation to feed should be highest. In some operations, including dairy units, electronic identification allows monitoring of individual animal feed intakes Ketotic dairy cattle display a rapid decrease in feed intake (10.4 kg of fresh matter per day), feeding time (45 minutes less per day) and feeding rate (25 g of fresh matter per minute less) days prior to diagnosis by farm staff[20]. Conversely, locomotion disorders produced smaller changes in feed intake (1.6 kg less fresh matter per day), feeding time (19 minutes less) and increased feeding rate (22 g FM more per minute). This group also examined feeding behavior of cattle 30d prior to trimming and 30d after trimming. Cattle classified as lame at trim showed decreased feeding time and visits during the 30 days prior to trim, and then feeding time and visit increased during the 30 days post-trim, suggesting recovery from painful condition. These changes in feeding behavior were consistent for cows fed high or low forage diets. Similarly, changes in feeding behavior occur two weeks prior to onset of clinical signs of metritis, and these cows are also less likely to compete at the feed bunk[21,22]. Changes in feeding behavior of beef cattle have also been detected days before detection of bovine respiratory disease[23,24].

Although feeding motivation is generally depressed during illness, decreased appetite may result from specific food aversions or preferences that may be innate or learned. When offered a variety of foods, beef cattle differ markedly in preferences for food high in protein or energy[25], with transient taste aversions among the reasons. Dairy calves injected with LPS decrease consumption of free choice hay, but milk consumption is not affected if calves are limit fed[26,27]. Appetite in compromised animals may be stimulated by frequent provision of a variety of fresh feedstuffs that are easy for the animal to access, in terms of competition from other cattle and bunk access. There is evidence that some animals can self-medicate, preferentially selecting food items that previously have been associated with alleviation of illness. Sheep treated with grain, tannins and oxalic acid preferentially consume sodium bentonite, polyethylene glycol and dicalcium phosphate respectively[28]. Lame broiler chickens preferentially consume feed containing the NSAID carprofen[29], and placentophagia by cows is associated with decreased pain in a thermal sensitivity test[30].

The compromised animal has a different set of behavioral priorities than its healthy herd-mates, and hence hospital pens can provide specialized care for recuperation. Unfortunately there has been scant scientific scrutiny about hospital pen design and management. Identifying and tracking individual animals is a challenge within feedlot and large dairy operations, and hence placing an animal in the hospital pen ensures treatment protocols are followed. However, due to the effects of novelty and stress associated with changes to social groups, some compromised animals may fare better in their home pen if they can maintain access to resting, feeding and drinking site while avoiding bulling. The association between sickness and buller-rider syndrome[31] suggests that in feedlots, pulling infirm individuals to the hospital pen is warranted, whereas cow-calf and dairy operations may choose to attend to compromised cows and calves within the home pen or pasture. Dominant individuals are generally able to prioritize sickness behavior, whereas subordinate individuals prioritize defensive behaviors. Convalescent individuals attempt to avoid agonistic encounters by initiating less aggression, and this strategy is most successful for dominant individuals within a group[32]. However, amongst small, stable social groups, there is some evidence of care-giving behavior amongst familiar herd-mates that could mitigate stress and facilitate recovery. Lame cows receiving more social licking than their non-lame herd-mates[17].

CONCLUSIONS

The behaviors that an animal displays during acute illness, such as anorexia, dullness and lethargy, may be more than a simple consequence of disease and may in fact aid in recovery. Understanding sickness behavior of livestock can provide opportunities for improved detection and care of compromised individuals.

REFERENCES

[1]Hart BL, **Neurosci Biobehav Rev** 1988;12:123. [2]Kluger MJ, et.al. **Science** 1975;188:166; [3]Vaughn LK, et.al **Brain Res Bull** 1980;5:69; [4]Dantzer R, **Brain Behav Immun** 2003;178:1; 5Aubert A, **Neurosci Biobehav Rev** 1999;23:1029; [6]Aubert A, et.al. **Brain Behav Immun** 1997;11:107; [7]Weikl ZM, et.al. **Physiol Behav** 2006;87:694; [8]Engeland CG, et.al. **Physiol Behav** 2001;72:481; [9]Aubert A, et.al. **Brain Behav Immun** 1995;9:129; [10]Yirmina R, et.al. **Brain Behav Immun** 1995;9:220; [11]Todd CG, et.al. **J Anim Sci** 2007;85:369; [12]Edwards JL, et.al. **J Dairy Sci** 2004;87:524; [13]Zager A, et.al. **Euro Neuropharm** 2009;19:589; [14]Toth LA, et.al. **Proc Soc Exp Biol Med** 1993;203:179; [15]Hetem RS, et.al. **Am J Physiol Regul Integr Comp Physiol** 2007;294:R246; [16]Jensen MB, et.al. **Appl Anim Behav Sci** 2005;90:207; [17]Galindo F, et.al. **J Appl Anim Welfare Sci** 2002;5:193; [18]Niss DB, et.al. **J Dairy Sci** 2009;92:617; [19]Rushen J, et.al. **J Dairy Sci** 2007;90:3647; [20]Gonzales LA, et.al **J Dairy Sci** 2008;91:1017; [21]Urton G, et.al. **J Dairy Sci** 2005;88:2843; [22]Huzzey JM, et.al. **J Dairy Sci** 2007;90:3220; [23]Buhman MJ, et.al. **Am J Vet Res** 2000;61:1163; [24]Quimby WF et.al. **Can J Anim Sci** 2001;81:315; [25]Atwood SB, et.al. **J Anim Sci** 2001;79:3034; [26]Borderas TF, et.al. **J Anim Sci** 2008;86:2920; [27]Borderas TF, et.al. **J Dairy Sci** 2009;92:4549. [28]Villalba JJ, et.al. **Anim Behav** 2006;71:1131-1139; [29]Danbury TC, et.al. **Vet Record** 2000;146:307; [30]Pinheiro Machado, LC, et.al **Physiol Behav** 1997;62:1339; [31]Taylor LF, et.al. **Aust Vet J** 1997;75:732; [32]Cohn DWH, et.al. **Physiol Behav** 2006;87:932;

RUMEN ACIDOSIS: PATHOPHYSIOLOGY & ETIOLOGY

Paul J. Plummer, DVM, PhD, DACVIM
Ames, IA

INTRODUCTION

Rumen acidosis is a complex mutifactorial disease process of domesticated ruminants. In recent years it has been increasingly recognized as a significant economic burden to both the dairy cattle and beef cattle industries. By definition the process is characterized by a decrease in rumen pH below physiological levels normally seen in a fully functional rumen. Normal rumen pH is normally considered to range between 5.8 and 6.2 with some variance based on diet being consumed by the animal. If novel highly fermentable food sources are provided this can result in a significant drop in pH associated with rapid rumen fermentation of the feedstuffs.

Classifications

Rumen acidosis is generally subdivided into severe classifications based on severity and prognosis. Sub-acute rumen acidosis (SARA) is also sometimes referred to as sub-clinical or chronic rumen acidosis and is associated with intermittent bouts of mild to moderate acidosis that generally does not result in severe disease manifestations. In contrast, acute rumen acidosis is associated with a rapid onset of fulminate disease and death in many cases. The use of these differentiating criteria is useful for helping to communicate the severity and expected sequella when talking to peers. Having described these classifications it is often helpful to understand that the physiological processes of acidosis occur over a period of time and thus represent a changing pH profile that is not static. This issue is particularly important in terms of making a diagnosis since normal variation occurs in pH over the course of a 24-hour period. Hence, the time of day in relation to feeding can have a significant impact on the pH of rumen fluid obtained. Subacute rumen acidosis is typically diagnosed when the pH of a rumen fluid sample measures between 5.0 and 5.5. When the rumen pH drops below 5.0 these cases are typically defined as acute rumen acidosis.

Clinical Observations

In many cases cattle with sub-acute rumen acidosis may have few or no clinical signs during an episode. As mentioned above diagnosis is often complicated by the variability of pH and requires sample collection to be consistent. In most cases the diagnosis of SARA is based on multiple samples collected from a group of animals in the same management group. In cases where diagnosis of an individual animal is needed it is important to rely on more than rumen pH to make this diagnosis. In contrast acute rumen acidosis is often severe and has rapid onset of clinical signs consistent with severe lactic acidosis and cardiovascular collapse. Acute death may be the first indication of acute rumen acidosis in a group of cattle.

Pathophysiology

The normal rumen buffering system utilizes ammonia, bicarbonate and phosphate to maintain physiologic pH. When the rumen pH drops below 5.0 the volatile fatty acids (VFAs) also provide some buffering activity since their pKa values are roughly 4.8. In the case of rumen acidosis an excess of lactate is often formed. Lactate is a 10- fold stronger acid with a pKa of roughly 3.8 which results in a precipitous drop in rumen pH.[1] Following the ingestion of readily fermentable starch the rumen microbes produce significant amounts of amylase which consequently results in the release of large quantities of glucose. The resulting elevations in rumen glucose result in three primary consequences. First, increased rumen glucose allows for an overgrowth of *Streptococcus bovis* and increased lactate concentrations as a byproduct of pyruvate fermentation by lactate dehydrogensase. Secondly the rumen glucose concentrations allow for an overgrowth of coliforms and other amino acid decarboxylaters. These organisms can release significant concentrations of endotoxin and histamine that can be absorbed and result in endotoxic shock and cardiovascular compromise. Finally, the increased osmolality associated with the high glucose concentrations inhibits VFA production that further exacerbates the high osmolality of the rumen ingesta. As the rumen pH continues to drop and the osmolality increases the rumen protozoa are rapidly killed and there is an increase in the rumen populations of lactobacilli. Interestingly, in SARA as rumen pH continues to drop lactate fermentation is shut down below pH 5.0. This results in a situation where the lactic acid concentrations in the rumen fluid peak at a pH between 5.0 and 5.5 and drop back down to baseline levels below pH 5.0.[2] Despite this drop in lactic acid the total organic acid concentration of the rumen continues to increase as pH declines. In contrast, cases of acute acidosis have continuing elevations of lactic acid even as pH drops to 4.0. Data suggest that L-lactate and D-lactate are the primary sources of the lowered rumen pH concentrations.[3] Increasing rumen osmolality results in decreased VFA absorption. This results in a net movement of fluid into the rumen and an associated drop in vascular volume. Vascular volume contraction in concert with increased blood concentrations of VFAs and lactate results in increased tonicity of the blood. This hypertonicity is sensed by osmo-receptors and results in decreased feed intake and decreased salivation. In severe cases there is a precipitous drop in blood pH associated with accumulating concentrations of blood acetate. The contribution of the other VFAs to blood pH is minimized since proprionate is rapidly metabolized on first pass through the liver and butyrate is converted to BHBA to rumen epithelium. The pH and osmolality of the rumen contents result in edema formation of the rumen epithelium and subsequent sloughing in severe cases.

Etiology

The etiology of both acute and chronic rumen acidosis is associated with the consumption of highly fermentable food sources, especially in cases where these are newly introduced or altered in their presentation. When large amounts of fermentable feeds enter a rumen the resulting microbial fermentation leads to lowered rumen pH. It is important to remember that the ration on paper is not always indicative of the ration consumed. Especially where total mixed rations are being utilized it is important to evaluate the feed bunks for sorting that may be occurring. Such a situation often results in cattle that are being offered an appropriately balanced ration but sort out the long-stem fiber and preferentially consumer the fermentable grains. In acute rumen acidosis the history often includes rapid introduction of novel food sources either accidently (ie. animals break into feed storage area or hired help make drastic changes) or intentionally. In some cases animals that have been unable to consume food for some period of time due to inclement weather preventing feed consumption or illness resulting in lower feed intakes will ingest large volumes of feed upon reintroduction to feed access and develop rumen acidosis.

Conclusions

Acute rumen acidosis is a life threatening medical condition that occurs in beef and dairy cattle as well as small ruminants. The pathophysiology of this process results in rapid development of clinical signs and in severe cases acute death. Most cases of acute rumen acidosis are associated with rapid introduction or reintroduction of new food sources that are highly fermentable. Subacute rumen acidosis is most commonly seen in dairy cattle and is much less specific in its clinical signs. In these cases the rumen pH is maintained in low-grade acidotic states (pH 5.0-5.5) for significant periods of time over weeks to months. Most often SARA is diagnosed as a result of diagnostics performed to evaluate decreased productivity or increased clinical signs of sequella.

REFERENCES

[1]Merck Veterinary Manual [2]Nagaraja et.al. **Vet Clinics of NA:Food Animal Practice** 2007; 23(2):333-50 [3]Owens et. al. **J Anim Sci** 1998: 76(1):275-86

RUMEN ACIDOSIS: SEQUELLA & MANAGEMENT

Paul J. Plummer, DVM, PhD, DACVIM
Ames, IA

INTRODUCTION

Rumen acidosis is a common physiological condition of both dairy and beef cattle. As the name would suggest this condition occurs as a result of a drop in pH of the rumen contents following the ingestion of highly fermentable feeds. The pathophysiology and etiology were described previously. This article will focus on the clinical signs, sequella and management of the disease process.

SEQUELLA

Feed Intake

Similar to the situation with rumen pH the sequella of rumen acidosis are likely best viewed as a continuum of disease. Ultimately the severity of the sequella is in a large part determined by the initial severity of the acidic episode. It is also important to remember that the clinical signs may not be observed for weeks to months following the episode of acidosis. One of the most consistent changes observed is a variable or decreased food intake. In acute rumen acidosis complete anorexia is commonly observed while less severe depressions in feed intake may be observed in subacute rumen acidosis (SARA) cases. The changes are typically the result of elevated rumen VFA concentrations resulting in decreased rumen motility and concurrent blood hypertonicity resulting in osmoreceptor triggering with appetite depression.

Milk Production and Components

Changes in milk production or milk fat may also be observed in dairy cattle where this parameter is more commonly monitored then in beef cattle. The decreased milk production is often a manifestation of the decreased feed intake already discussed. Traditionally the low milk fat production has been associated with low acetate concentrations. Recent research suggest however that incompletely saturated fatty acids can actually inhibit milk fat production in the udder.[1,2,3] When evaluating samples for milk fat depression in dairy cattle the use of individual animal samples is typically better then bulk tank milk. It is also important to remember that other factors such as excessive dietary levels of unsaturated fatty acids and over-dosing of ionophores can result in similar changes to milk fat concentrations. One other parameter of milk production that may be of use in evaluating herd level issues is to look for relatively lower milk production in older cattle while adjusting for age. This is perhaps best done by comparing the average ME (mature equivalent) of 1st lactation cows to that of older cattle. Herds with chronic SARA issues will often have significant drops in milk production in the older cattle associated with the chronic negative effects of pH.

Immunology

Rumen acidosis and SARA have significant impacts on immune function of cattle. See the associated proceedings article from this session.

Endotoxin

The mass killing of gram negative bacteria in the rumen as a result of the dropping pH results in the release of large amounts of endotoxin. Additional endotoxin is released during bacterial replication of the organisms that overgrow in these conditions further exacerbating the problem. Research has demonstrated that rumen endotoxin levels significantly increase even in cases of SARA.[4] The rising endotoxin levels weaken the rumen epithelial barrier and results in LPS leakage into the vasculature. Once in the vasculature the endotoxin results in a cascade of physiological events including the release of TNF, vasculature leakage and the expression of tissue factor with subsequent potential for coagulopathies. This inflammatory response results in the release of acute phase proteins including haptoglobin and serum amyloid-A by the liver.

Laminitis

Decades of experience have indicated a role of rumen acidosis in the subsequent development of laminitis in cattle. Although the pathophysiology of this process is still somewhat unclear we are starting to get some clues as to how this may occur. Research suggests that rumen acidosis is associated with elevated concentrations of histamine in the rumen fluid as a result of the overgrowth of decarboxylating bacteria. There also appears to be an increased ability of histamine to translocate into the vasculature at pH level below 5.1. This increased translocation is associated with a drop in epithelial catabolism of histamine by diamine oxidase. Increased vascular concentrations of histamine work in concert with the increased expression of tissue factor to result in coagulopathies and vascular disturbances that ultimately precipitate the development of laminitis.

Parakeratosis – Rumenitis and Liver Abscesses

Initial elevations in ruminal proprionic and butyric acid result in ruminal papilla elongation to increase surface area and increase absorption of VFAs. With prolonged exposure to these elevated VFA levels rumen parakeratosis develops. The collective activity of increased rumen osmolality, parakeratosis, epithelial edema and endotoxin results in significantly impaired rumen epithelium. The impaired mucosa allows for translocation of *F. necrophorum* and *Arcanobacterium pyogenes*. Carriage of these bacterial organisms via the portal circulation to the liver results in bacterial seeding of the liver parenchyma, the caudal vena cava, and bacterial embolism throughout the body. Over a period of months disseminated abscesses form throughout the body. DNA finger printing of isolates obtained from these abscesses confirms that they match those of isolates obtained from the rumen wall, confirming that ruminal translocation is the likely source of bacteria.[5]

Polioencephalomalacia

Rumen acidosis has long been associated with an increased incidence of polioencephalomalacia. Traditionally this has been associated with an overgrowth of rumen bacteria that produce thiaminases. These thiaminases are then presumed to act on ruminal thiamine and result in decreased absorption of thiamine ultimately resulting in the development of thiamine deficiency polioencephalomalacia. In more recent years there is a growing interest in the role of rumen acidosis associated with sulfur toxicity. As the pH of the rumen contents decreases there is a significant increase in the gas cap concentrations of H_2S. This increases the exhalation of hydrogen sulfide and may result in increased cerebral concentrations where sulfur is thought to interfere with cytochrome oxidase and cellular respiration.

MANAGEMENT

The key management factors needed to minimize the risk of rumen acidosis can basically be summed up into two principles. The first principle is consistency. Maintaining a consistent diet, fed in a consistent manner on a consistent schedule with consistent management will go a long way towards minimizing both acute and subacute rumen acidosis. In many cases of acute rumen acidosis the inciting cause is generally a change from the normal. The second primary principle is to make feed changes gradually. Introduction of novel food sources, especially those that are highly fermentable, will routinely result in some degree of rumen acidosis. Hence, when feed changes are made they should occur gradually over a period of weeks to months depending on the degree of change. Specifics examples of these management principles will be discussed in the presentation.

CONCLUSIONS

Rumen acidosis and subacute rumen acidosis are associated with a gradient of low pH values between 4.5 and 5.5. The severity and time of onset of sequella are variable and primarily associated with the severity of the initial rumen insult. In many cases the sequella are not observed for weeks to months after the episode making prognostication of acidosis bouts difficult. Effective management of acidosis relies on consistency of feed management and minimizing drastic changes in feed sources.

REFERENCES

[1]Gervais et. al. **J Dairy Sci** 2009; 92(10)4876-88. [2]Alzahal et al. **J Dairy Sci** 2009; 92(3):1108-16 [3]Mutsvangwa et al. **J Dairy Sci** 2003; 86(12):4043-46 [4] Emmanuel et. al. **J Dairy Sci.** 2007 Dec;90(12):5552-7. [5]Nagaraja and Lachtenberg, **Vet Clin North Am Food Anim Pract.** 2007 Jul;23(2):351-69.

ALTERATIONS IN BOVINE IMMUNOLOGY ASSOCIATED WITH SUBACUTE RUMEN ACIDOSIS & KETOSIS

Christopher CL Chase DVM, MS, PHD, DACVM & Nicolas Villegas DVM, MS
Brookings, SD

INTRODUCTION

The common practice of vaccinating during the transition period (15-45 days in milk) is an immunological challenge for the cows due to the negative energy balance (NEB) associated with the high energy demands and the low dry matter intake typically observed post partum. This NEB is generally observed in all dairy herds and explains the loss of body condition after calving. The effects of two specific metabolic diseases along with generalized immune suppression in the peripartum period will be discussed.

KETOSIS

Approximately 30% of dairy cows suffer sub-clinical ketosis during the fresh period as a result of the negative energy balance.[1] The pathogenesis of this phenomenon is explained by the metabolic changes that occur when nutrient intake, particularly energy, does not meet production demands. In high-producing cows this metabolic disorder usually occurs from a few days up to six weeks post-calving, with the highest incidence occurring at about three weeks post partum. Most high-producing cows experience sub-clinical ketosis in early lactation when they are unable to consume enough energy to meet demands.[2] Cows with this disorder are known to have primary ketosis. The body temperature in cows with primary ketosis is normal. An elevated body temperature accompanying primary ketosis suggests additional complications. Ketosis can be the also result of other concurrent problems. Complications such as retained placenta, displaced abomasums, or any disease, which reduces appetite, may result in secondary ketosis.

Cows that develop ketosis are usually in negative energy balance and, therefore, are utilizing body fat and protein stores as a result of a drop in blood glucose concentration. When fat molecules reach the liver, they are converted to ketones, and elevated ketone levels in the blood result. High levels of blood ketone bodies in blood interfere with the production of T cells,[3] T cell proliferative response[4] and impair the chemotactic response of leukocytes.[5] This impaired chemotactic response may explain the link between elevated ketone levels and the risk of mastitis. Since sub-clinical ketosis is present in nearly 30% of fresh dairy cows, vaccination during this the freshening period may be not the best approach and vaccinating during the dry period might be a better alternative.

SUBACUTE RUMEN ACIDOSIS

Subacute ruminal acidosis (SARA) affects 19% of early lactation and 26% of mid-lactation cows in the US.[6] Early lactation cows and cows at peak dry matter (DM) intake are at most risk from SARA. The early lactation cows are at more at risk to SARA due to reduced rumen capacity and microbial function along with the abrupt dietary change to higher energy density diets. Cows at peak dry mater intake are at risk of higher ruminal acid production.[6] SARA results in increased acute inflammation as measured by serum acute phase proteins.[7] SARA also resulted in increased immunosuppression as measured by increased cortisol secretion, reduced phagocytosis and migration of neutrophils.[8] Our work with SARA indicated lymphocyte immunosuppression following the induction of SARA using either 1) the basal diet supplemented with 700 mL/d of butylene glycol (BG) to induce ketoacidosis by increasing blood ß-hydroxybutyate (BHBA); 2) the basal diet supplemented with 1.2 ± 0.1 kg/d of anionic salts (AS; Soychor 16.7, West Central Soy, Ralston, IA) to induce a metabolic acidosis; and 3) the basal diet with all forage replaced by finely ground corn and soybean meal blended to provide 15% CP (HG), to induce lactic acidosis.[9] Following stimulation with the mitogen, *Staphylococcus aureus* exotoxin B, lymphocyte responsiveness was decreased. There was also decreased proliferation and memory response to bovine herpesvirus 1 (BHV-1; IBR) and bovine viral diarrhea virus (BVDV). We found no effect on antibody titers following induction of SARA.

IMMUNITY DURING THE PERIPARTUM PERIOD

Dairy cows are continuously challenged with the need to increase milk production. The alteration of the immune system and the innate host resistance mechanism in dairy cows usually begins 3 weeks before parturition, and it is maximized after calving, when milk yield peaks and the energy balance begins to improve. Some alterations in the host defense mechanisms that occur during the periparturient period are associated with changes in hormone profiles and the metabolic and physiological stress of parturition. These changes can contribute to the high incidence of disease and the low immune response to vaccination experienced by the periparturient cow. Evidences of the changes in the immune system and the non-specific host defense mechanism occur in the periparturient dairy cow.[10, 11, 12]

Hormonal changes in the endocrine system of cows later in pregnancy are well known and are characterized by higher circulating levels of estrogen and progesterone. These hormones can depress the immune system function. High estrogen concentrations, for example, can reduce neutrophil function, and high levels of progesterone can suppress lymphocyte function. Cortisol produced in the fetal adrenal glands can affect cell- mediated response in the cow, such as the production of interleukin-2 and interferon gamma.[13] These changes may explain why the dairy cow is more susceptible to mastitis during the first 2-3 weeks of lactation.[14, 15] Among thes changes at calving time are high cortisol levels, which produce immunosuppression and an increase in estrogen that can also suppress lymphocyte function. Depression in postpartum leukocyte function has been correlated to shifts in leukocyte trafficking patterns at this time.[16,17] The alteration in the proportions of the peripheral blood lymphocyte subset has been monitored in dairy cows during the pre-partum and post-partum periods.[18] The variation of T cells was significant during the peripartum period, particularly around parturition. A decrease in the total number of T lymphocytes and changes in the T subpopulation have been reported in peripheral blood.[19] B cell and MHC II + populations remained constant until after calving and then decreased, returning to the initial subset proportion by Week 16.

SUMMARY

The fresh cow experiences several metabolic disorders that often occur at a subclinical level, the most common of these being ketosis and acidosis. Both of these metabolic disorders cause immunosuppression. This, along with the negative energy balance due to increase in milk production and lower feed intake, makes the postpartum cow a poor target for a strong immune response. Timing of booster vaccinations should take these metabolic factors into consideration resulting in vaccination later in the postpartum period or earlier in the dry cow period.

REFERENCES

[1.] Duffield TF. **Vet Clin N Am Food Animal Practice** 1999;16:231-253. [2.] Schultz LH, et al. **The Ruminant Animal: Digestive Physiology and Nutrition** 1988; p. 493-528. [3.] Suriyasathaporn W, et al. **Vet Immunol Immunopath** 1999;68:177-186. [4.] Donovan D. PhD Dissertation. .South Dakota State University, 2003. [5.] Van Werven T., The role of leukocytes in bovine *Escherichia coli* mastitis, PhD thesis,Utrecht University, Utrecht, The Netherlands, 1999. [6.] Mulligan FJ, et al. **Vet J** 2008;176:3-9. [7.] Plaizier JC, et al. **Vet J** 2008;176:21-31. [8.] Enemark JMD. **Vet J** 2008;176:32-43. [9.] Donovan DC, et al. **J An Sci** 2003;81:3088-3094. [10.] Ishikawa H. **Jpn J Vet Sci** 1987;49:469-475. [11.] Kashiazake Y, et al. **Jpn J Vet Sci** 1985;47:337-339. [12.] Kehrli M, et al. **AJVR** 1989;50:215-220. [13.] Corbeil LB. **Vet Clin N Am Food Animal Practice** 2001;17:567-583. [14.] Oliver SP, et al. **J Dairy Sci** 1988;71:2584-2606. [15.] Sordillo LM, et al. **J Dairy Sci** 1997;80:1851-1865. [16.] Shafer-Weaver KA, et al. **J Dairy Sci** 1999;82:1696-1706. [17.] Kayoko K, et al. **J Dairy Sci** 2002;85:1437-1444. [18.] Van Kampen C, et al. **Vet Immunol Immunopath** 1997;59:79-92. [19.] Park YH, et al. **Vet Immunol Immunopath** 1993;36:137-151.

PRACTICAL VACCINOLOGY OF DAIRY CALVES

Christopher CL Chase DVM, MS, PHD, DACVM
Brookings, SD

FREQUENCY OF VACCINATION AND INTERVAL BETWEEN VACCINATIONS- CAN WE OVERVACCINATE?

Many vaccine protocols have been developed to vaccinate the young calf at frequencies as often as weekly during the first and second month of age. Finding any experimental studies that support this frequency is difficult. From other immune systems, it is clear that too frequent vaccination in young animals can lead to antigen specific tolerance (i.e. the lack of any immune response to the antigen).[1] This is the result of suppressive T cells and the deletion of T and B clones specific for those antigens. Another possible adverse outcome of over-vaccination is autoimmunity. Development of autoimmunity is based on priming against the animal's own antigens (self) or by the close mimicry of vaccine antigens to the animal's own antigens. An example of this mimicry is that antibodies against one of the surface proteins of IBR cross-react with a surface protein of the immune cells.[2] Stimulation of inflammatory mechanisms associated with vaccination, such as adjuvants like alum or microbial ligands or repeated re-exposure to vaccine components that drive expansion of autoimmune B and/or T cells clones may occur with frequently repeated vaccine stimulation.[3,4] In all animals following vaccination there is expansion in the populations of responding T and B cell clones. There is a requirement for a good immune response that this clonal expansion not only stop but that an active process of cell death (apoptosis) must also occur. This "waning process" allows "culling" T or B cells that may be poor responders or even cause autoimmunity to be removed by apoptosis.[5] This whole process from vaccination to achieving homeostasis takes at least 3 weeks for the development of a primary response that can then be boosted to get a true anamnestic secondary response. This short review is extracted from a full review published in 2008.[6]

DEVELOPING A VACCINATION PROGRAM

The first necessity of planning a calf vaccine program is to assess the disease risks at the production site. Often "blanket vaccination" programs are suggested for many pathogens, which may or may not be a threat to young calf health. Carefully review the antigens that are being used to make sure they make sense for the operation. Secondly the effect of maternal immunity and the age of the calf must be considered. The relationship is linear: the younger the calf, the poorer the response, the older the calf, and the better the response. However, the inverse relationship is true from the standpoint of protection afforded by maternal immunity: the younger the calf, the better protection due to high levels of maternal antibody, the older the calf, the more susceptible to disease due to waning maternal antibody.

Management factors also come into play. In dairy operations, isolation of calves from exposure to pathogens and good bio-security can provide a window of enhanced protection by maternal immunity giving an extended window before vaccination is necessary. Some practices, such as feeding waste milk on dairies, may "break" the isolation by introducing both pathogens and antibiotics that alter the natural flora developing in the calf to make them more susceptible. This may warrant a more aggressive vaccination program.

BRSV Vaccines

Active immunization for BRSV is probably the most difficult because of maternal interference. This requires careful monitoring and frequent revaccination so that the calves can be protected before the animals reach the "window of susceptibility", the time frame where animals are no longer protected by passive immunity and active immunity has not been stimulated. A promising development has been an intranasal BRSV vaccine that has been licensed in Europe.[7] IN vaccination takes advantage of the poor penetration of antibody from colostrum onto mucosal surfaces providing less interference with the function of vaccines. This vaccine reduced clinical disease in 3-week-old colostrum-fed calves challenged 66 days following vaccination. Another study administered a commercial 4-way viral MLV vaccine licensed for parenteral use, intranasally, in 2-week-old colostrum-fed calves and found it protective 8 days post vaccination.[8] A single dose of inactivated 4 way viral vaccine in 4-5-week-old calves protected calves against a BRSV challenge 3 weeks post vaccination.[9]

BVDV Vaccines

The risk for BVDV infection and disease in the young calf appears to be much lower than BRSV. The additional immunosuppression in the calf as the result of the BVDV component of many parenterally administered BVDV MLV[10] deserves consideration. Studies in calf ranches have indicated that there is little advantage gained from vaccinating calves under 60 days of age[11] and that maternal antibody protection lasts from 70-110 days.[12] The greatest risk for BVDV infection was from 4-9 months of age as animals were group housed in larger groups.[13] Vaccination programs that vaccinated calves in the first 60 days, and then much later at 4 to 9 months, were not effective in preventing infection.[13] Based on this information, BVDV control programs that use MLV probably should begin around 2-3 months of age followed by revaccination after 3-4 weeks.

BHV-1 Vaccines

BHV-1 infections, like BVDV, are unlikely in young calves. There are several positive aspects of IN BHV-1 vaccines: maternal interference is less likely at mucosal sites in young animals so the vast majority develop active immunity.[14,15] After IN vaccination high levels of mucosal and serum interferon are also produced[16] that both direct an antiviral effect, and may also aid in the development of the neonatal immune response. BHV-1 is also immunosuppressive.[17] However, intranasal BHV-1 vaccines, because of their localized and limited replication in the nasal epithelium, are less of a danger for development of immunosuppression than parenteral vaccines. All BHV-1 vaccines including IN vaccines[18] can result in latency- the ability of the virus to reactivate (recrudesce) and be shed. The biological relevance of this reactivation and shedding of BHV-1 is unclear, as there is no lesions or disease syndromes associated with recrudescence, and the virus can only be reactivated experimentally following several days of treatment with high doses of dexamethasone.[18.]

Clostridial Vaccines

Multistrain *Clostridium* spp. bacterin-toxoids are a frequent part of the vaccination program of heifers. These pathogens represent the cause of sporadic enteric and musculoskeletal disease. Localized vaccine reactions are the most frequent and serious side effect.[19] Maternal interference significantly inhibited *Clostridial* spp. antibody responses in calves vaccinated at 3[20] or 50 days of age.[21]

***Mannheimia* and *Pasteurella* Vaccines**

These bacterial pathogens are frequently isolated from many cases of calf pneumonia, and pose a significant threat to heifer development. Bacterin-toxoids and avirulent MLVs are commonly used. Bacterin-toxoids have been shown to be inhibited by maternally transferred immunity prior to 6 weeks of age.

***Mycoplasma bovis* Vaccines**

This emerging pathogen is also frequently isolated from pneumonia in the heifer. The current vaccine in common use is a bacterin. A single experimental vaccine administered at 3 weeks to calves with low *M. bovis* antibodies was shown to be protective when calves were challenged 21 days later.[22]

***Salmonella* spp Vaccines**

Although *Salmonella* spp are important pathogen of calves, there are few well-designed control studies in the literature other than field observational studies of *Salmonella* vaccine efficacy. There are three major groups of *Salmonella* vaccines: inactivated with gram-negative core antigens (J-5, J-Vac and Endovac-Bovi), attenuated live vaccine (Entervene-D) and the subunit siderophore receptor protein (SRP) vaccines. All three vaccine groups have been shown to have efficacy in the field. Only the attenuated live vaccines are labeled for use in calves (2 weeks of age or older).

Rotavirus & Coronavirus Vaccines

These two viruses are common causes of neonatal diarrhea. Maternal antibody dramatically decreases vaccine efficacy.[23,24] Onset of protection against disease and shedding in calves occurs before secretory IgA is produced 10 days after vaccination,[23] implying that the oral administered vaccine may activate innate immune system in the gastrointestinal tract and decrease disease. These vaccines would be of greatest value for herds with colostrum low in levels of rotavirus and/or coronavirus antibody.

SUMMARY

Vaccination of the heifer calf is complicated by the presence of significant levels of maternal antibody that persist in the calf, colostral and neonatal hormonal factors, the lack of full immune competence, and interference in the function of vaccines by the presence of maternal immunity. The first necessity of planning a calf vaccine program is to assess the disease risks at the production site. Carefully review the antigens that are being used to make sure they make sense for the operation. The affect of maternal immunity and the age of the animal must be carefully considered in determining the vaccination schedule. The use of MLV containing immunosuppressive agents should be planned to avoid giving them at times when the animals have low immune competence and/or are immune suppressed, or by using mucosal vaccination routes that minimize induced immunosuppression and interference by maternal antibody.

REFERENCES

[1.] Raimondi G, et al. **Methods Mol Biol** 2007;380:1-24. [2] Fitzpatrick DR, et al. **Immunology** 1990;70:504-512 [3] Siegrist CA. **J Comp Pathol** 2007;137:S46-50. [4] Molitor ML et al. **Front Biosci** 2007;12:3302-11. [5] Wagner DH. **Clin Immunol** 2007;123:1-6. [6] Chase et al. **Vet Clin N Am Food Animal Practice** 2008;24:87-104. [7] Vangeel I. **Vet J** 2007;174:627-35. [8] Ellis J, et al. **J Am Vet Med Assoc** 2007;230:233-43. [9] Mawhinney IC, et al. **Vet Rec** 2005;156:139-43. [10] Kelling CL. **Vet Clin N Am Food Animal Practice** 2004;20:115-29. [11] Thurmond MC, et al. **J Am Vet Med Assoc** 2001;219:968-75. [12] Munoz-Zanzi CA, et al. **J Am Vet Med Assoc** 2002;221:678-85. [13] Rush DM, et al. **J Am Vet Med Assoc** 2001;219:1426-31. [14] Todd JD. **Dev Biol Stand** 1976;33:391-5. [15] Lemaire M, et al. **J Clin Microbiol** 2000;38:1885-94. [16] Todd JD, et al. **Infect Immun** 1972;5:699-706. [17] Ambagala APN, et al. **Vet Immunol Immunopath** 2005;107:1-15. [18] Lemaire M, et al. **Vaccine** 2001;19:4795-804. [19] Stokka GL, et al. **J Am Vet Med Assoc** 1994;204:415-9. [20] Troxel TR, et al. **J Anim Sci** 2001;79:2558-64. [21] Troxel TR, et al. **J Anim Sci** 1997;75:19-25. [22] Nicholas RAJ, et al. **Vaccine** 2002;20:3569-75. [23] van Zaane D, et al. **Vet Immunol Immunopath** 1986;11:45-63. [24] de Leeuw PW, et al. **Res Vet Sci** 1980;29:142-7.

CALF PNEUMONIA: CLINICAL ASSESSMENT & MONITORING IN CALF FACILITIES

Sheila M. McGuirk, DVM, PhD, DACVIM & Simon F. Peek, BVSc, MRCVS, DACVIM
Madison, WI

INTRODUCTION

Pneumonia is an important disease of preweaned dairy calves, with reported incidence rates that vary between 7.5 and 25.6%.[1-5] Blamed for 1.5 to 4.2% of calf deaths[6,7], the percent of preweaned dairy calf deaths attributed to pneumonia (21.3%)[8] is second only to deaths due to diarrhea. In addition, survivors of dairy calf pneumonia have long term consequences of reduced growth rate, a higher age at calving and shortened lifespan in the herd.[9,10,11] Considering death loss, impaired productivity, and treatment costs, calf pneumonia is a disease of significant economic importance. Typically viewed as a problem of calves after weaning, pneumonia is relatively common in preweaned calves, affecting individuals as young as 2 weeks and peaking in incidence at 5 to 6 weeks of age.[2,12,13] The diagnosis of pneumonia in young calves is challenging and, particularly with appetite-based screening, producer diagnosis fails to identify as many as 50% of affected calves.[2,14] Without early detection, effective treatment intervention is less likely and endemic calf pneumonia becomes a major herd problem, with otitis media, chronic pneumonia, and recurrent bloat being commonplace amongst calves of weaning age.

The relationship between dairy calf pneumonia and calf housing risk factors has been identified.[5,6,13] Frequently cited as the ideal housing for young dairy calves, calf hutches can present seasonal challenges under conditions of extreme heat or cold and are being supplanted by a growing number of indoor rearing facilities. Seasonal respiratory disease challenges common to calf barns with calves in individual pens are addressed by managing pen design, size, number, type of bedding, barn filling and emptying pattern, and air quality within the pen.[13, 15] With many recognized advantages, indoor group housing puts young calves at risk for increased respiratory disease.[5,16,17] With the formation of small, stable calf groups that provide at least 28 square feet of space per calf, offer increased milk allowance or ad-libitum feeding and all-in, all-out management of group pens, group pen respiratory disease risk factors can be minimized.[12,18-22]

CLINICAL ASSESSMENT OF DAIRY CALF PNEUMONIA PROBLEMS

With each herd investigation, we aim to calculate calf pneumonia morbidity and mortality rates for the previous year. With access to good calf health records, a monthly analysis of the number of cases can be performed and related to number of calving cows, serum total protein concentration, ambient temperature, or other management practices of interest to the investigation. The data can be used to find seasonal trends, months of concern, age at first diagnosis, or to construct survival curves for age at first treatment, treatment type or other management relationships of importance. With limited, inconsistent or inaccurate calf health records, greater emphasis is placed on recording accurate information regarding routine calf care (herd vaccinations, navel care, colostrum, feed types, amount, times and additives), pneumonia definition, pneumonia treatment protocols and calf caregivers. Information gathered is assessed for accuracy by observations made during a tour of facilities. Close observation of the maternity pen, feed preparation area, calf holding pens, stored colostrum, feeding equipment, medicine inventory, supplies, calf housing, calves and people are made at this time. Make note of the cleanliness of equipment, facilities, animals, feed, water and people. Comfort is evident when preweaned calves spend 70 to 80% of non-feeding time lying down.[23] Calves should be calm but responsive and curious during the tour. Consistency and follow through in feeding and treatment protocols may be evident by identifying markers on the pen or the calf.

Record retrieval, questions and a tour of facilities is followed by individual calf assessment that determines the calf pneumonia prevalence that day. Using a respiratory scoring system previously described[13] and available our website: http://svmweb.vetmed.wisc.edu/dms/fapm/fapmtools/8calf/calf_respiratory_scoring_chart.pdf, examined calves are classified as having respiratory disease (total respiratory score >4) or not (total respiratory score <5). For small calf groups, all calves are scored but when the number of calves is large, a non-randomized representative sample that includes all calf ages is scored. For farms with more than one type of calf housing, a non-randomized representative sample of calves from each housing type is assessed. From these respiratory scoring data, the prevalence and distribution of respiratory disease by age is determined. Even with a post weaning calf pneumonia complaint, the preweaned calves are assessed to determine whether the post weaning respiratory disease is the result of uncured calves entering group pens.

The respiratory scoring system, which is based on the assessment of nasal discharge, ocular discharge, ear position and spontaneous coughing from outside the calf pen and rectal temperature and tracheal compression to induce coughing during a hands-on exam inside the calf pen, classifies a calf as having respiratory disease when 2 or more clinical signs or respiratory disease are present. For calves that attain a total respiratory score of 5 or higher from outside the pen, the pen is not entered. Treatment records determine which of the calves with a respiratory clinical score >4 have been treated previously and the farm's pneumonia detection rate is calculated as follows:

$$\frac{\text{Number of calves with respiratory score >4 that were treated within the last 7 days}}{\text{Total number of calves with a respiratory score >4}}$$

Multiply the result by 100 to determine the farm's pneumonia detection rate. The goal of 85% is readily achieved when trained individuals (farm personnel or veterinary technicians) are assigned to score calves twice weekly. In addition to determining prevalence, apparent age of onset of pneumonia and farm detection rate, the respiratory scoring results can also be used to determine problem barns, treatment efficacy, recurrence rates and to select calves for further diagnostic testing.

HERD BASED DIAGNOSTIC TESTING FOR CALF PNEUMONIA

Depending on the specific objectives of a calf pneumonia investigation, diagnostic testing is useful to determine etiology, to refine vaccination schedules or to develop more effective treatment protocols. For herd investigations, we have suggested that a minimum of 6 affected and untreated calves from the at-risk age group be sampled.[24] To find bacterial or viral pathogens, transtracheal wash (TTW) fluid, deep nasopharyngeal swabs or bronchoalveolar lavage (BAL) fluid are candidate tests. The selected test must be performed efficiently, economically and effectively on the farm. With a desired minimum sample size of 6 calves, we have not found TTW fluid collection to be applicable for farm use. Deep nasopharyngeal swabs have been used effectively in herd-based investigations (DC Sockett, WI Veterinary Diagnostic Laboratory, personal communication), using 33 inch length double-guarded culture swabs. Polyester viral culture swabs and bacterial and viral transport media are required. Both nostrils are swabbed to obtain 2 viral and 1bacterial sample per calf. Another approach is to collect BAL fluid from calves as previously described[24] and used to obtain bacterial, viral and cytologic results useful to the investigation of a dairy calf pneumonia problem. A comparison of nasopharyngeal swab and BAL fluid analysis results from the same calves are shown below.

Table 1. Nasopharyngeal swab and bronchoalveolar lavage fluid results from a dairy calf pneumonia herd investigation

Calf ID	Birth Date	Total Respiratory Score	BAL - % macrophages	BAL % Neutrophils	Bacterial results	*Mycoplasma* results	Virus PCR Results	Interpretation and commemts
5175	6/22/09 (weaned)	Not done; farm requested sample	98%	2%	NSG - BAL	Negative	BAL: corona, BRSV, BVD, IBR neg	Low cellularity; no inflammatory changes
5188	6/30/2009	5	Not done	Not done	NSG - BAL and pharyngeal swab	Negative	BAL: corona, BRSV, BVD, IBR neg; pharyngeal swab: corona pos, others neg	Small sample - no cytology done
5189	6/30/2009	6	99%	1%	NSG - BAL and pharyngeal swab	Negative	BAL: corona, BRSV, BVD, IBR neg; pharyngeal swab: corona pos, others neg	Moderately cellular but no inflammatory changes
5203	7/5/2009	5	86%	13%	NSG - BAL and pharyngeal swab	Negative	BAL and pharyngeal swab corona virus pos; BRSV, BVD, IBR neg	Highly cellular but no significant inflammation
5206	7/8/2009	5	95%	5%	NSG - BAL and pharyngeal swab	Negative	BAL and pharyngeal swab corona virus pos; BRSV, BVD, IBR neg	Moderately cellular but no inflammatory changes
5207	7/8/2009	6	84%	15%	NSG - BAL; pharyngeal swab - *P. multocida*	Negative	BAL and pharyngeal swab corona virus pos; BRSV, BVD, IBR neg	Moderately cellular but no significant inflammatory changes
5220	7/13/2009	6	24%	74%	NSG - BAL; pharyngeal swab - *P. multocida*	Negative	BAL and pharyngeal swab corona virus pos; BRSV, BVD, IBR neg	Low volume sample; highly cellular with purulent inflammation

To develop an effective antibiotic treatment protocol for a dairy with calf pneumonia problems, nasal swab results can be used as described.[24] From each of 6 affected, untreated calves, 2 culturettes are submitted, one for bacterial and the other for *Mycoplasma bovis* culture. The antibiotic susceptibility patterns exhibited by nasal *Pasteurella multocida, Mannheimia haemolytica* and *Histophilus somnus* isolates from each calf are used to predict the susceptibility of bacterial isolates from the lung.[25] If more than 1 of 6 or more than 15% of the tested calves culture *Mycoplasma bovis* from the nose[26], the antibiotic protocol is adjusted to reflect a potential role of this organism. From the nasal swab antibiotic susceptibility results shown in Table 2, ceftiofur, enrofloxacin, florfenicol, tilmicosin or tulathromycin might be chosen for the herd-based calf pneumonia treatment protocol. If *M.bovis* is found in more than 15% if the nasal swab cultures, tulathromycin, tilmicosin, florfenicol or enrofloxacin might be considered.

Table 2. Nasal swab antibiotic susceptibility results from 6 calves

Antibiotic	*Pasteurella multocida*	*Mannheimia haemolytica*	*Histophilus somnus*
Ampicillin/Amoxicillin	Sensitive	Resistant	Resistant
Ceftiofur	Sensitive	Sensitive	Sensitive
Enrofloxacin	Sensitive	Sensitive	Sensitive
Florfenicol	Sensitive	Sensitive	Sensitive
Spectinomycin	Sensitive	Sensitive	Resistant
Sulphadimethoxine	Sensitive	Resistant	Resistant
Tetracycline	Resistant	Resistant	Resistant
Tilmicosin	Sensitive	Sensitive	Sensitive
Tulathromycin	Sensitive	Sensitive	Sensitive

CALF HOUSING EVALUATION

Calf pneumonia investigations should identify the calf housing factors that might increase the risk of respiratory disease. From a study in naturally ventilated calf barns in winter, the factors associated with the lowest respiratory disease prevalence were solid panels between calves that prevented calf to calf contact, deep bedding sufficient to completely cover the legs of calves lying down, and low airborne bacteria counts.[13] To lower aerosolized bacterial counts in individual calf pens, increase pen size to the recommended 32 square feet per calf[20], limit the number of solid barriers in individual calf pens to the 2 that prevent calf to calf contact, lower barn temperature and use outside air to dilute stale air in the calf pen microenvironment.[27] There should be 15% more calf pens than calves at maximum occupancy to provide adequate time to clean, disinfect and rest pens between successive calf occupants.[28] Being able to completely empty, clean, disinfect and rest the calf barn or large areas of individual calf pens at least once each year presents a significant opportunity to maximize respiratory health of calves by lowering the population and survival time of resident microorganisms within the calf environment.

CONCLUSIONS AND SUMMARY

Respiratory disease is under diagnosed but an effective system to screen preweaned calves twice weekly will result in earlier detection and implementation of a treatment protocol when it is most likely to be successful. Ear infections, recurring bloat and chronic pneumonia calves in post weaning calf groups provide evidence that untreated pneumonia is a problem. Nasal swab, deep nasopharyngeal swab or bronchoalveolar lavage fluid culture results can guide treatment decisions, clarify the role of potential etiologic agents and refine vaccination strategies. Calf housing concerns must be identified and limited to reduce the calf pneumonia. Scoring weaned calves to identify uncured calves that should be held and treated before moving to group housing will limit post weaning respiratory disease.

REFERENCES

[1]Waltner-Toews D, et al. **Prev Vet Med** 1986; 4:103. [2]Virtala A-M, et al. **J Am Vet Med Assoc** 1996;208:2043. [3]Wells SJ, et al. **Prev Vet Med** 1996; 69:185. [4]Donovan GA, et al. **Prev Vet Med** 1998;34:41. [5]Svensson C, et al. **Prev Vet Med** 2003;58:179. [6]Ames TR. **Vet Clin NA Food Anima Pract** 1997;13:379. [7]Andrews AH. **Cattle Pract** 2000;8:109. [8]USDA, 2009. USDA:APHIS:VS, CEAH, Fort Collins, CO #550,1209. [9]Waltner-Toews D, et al. **Can J Vet Res** 1986;50:314. [10]Virtala AK, et al. **J Dairy Sci** 1996;208:2035. [11]Correa MT, et al. **Prev Vet Med** 1988;6:253. [12]Sivula NJ, et al. **Prev Vet Med** 1996;27:155. [13]Lago A, et al. **J Dairy Sci** 2006;89:4014. [14]Quimby WF, et al. **Can J Anim Sci** 2001;81:315. [15]Lundborg GK, et al. **Prev Vet Med** 2005;68:123. [16]Waltner-Toews D, et al. **Prev Vet Med** 1986;4:137. [17]Curtis CR, et al. **Prev Vet Med** 1988;5:293. [18]Chua B, et al. **J Dairy Sci** 2002;85:360. [19]Svensson C, et al. **Prev Vet Med** 2006;73:43. [20]Stull C, et al. **Vet Clin NA Food Anim Pract** 2008;24:191.[21]Pedersen RE, et al. **Livestock Sci** 2009; 121:215. [22]Assié S, et al. **Prev Vet Med** 2009;91:218. [23] Panivivat R, et al. **J Dairy Sci** 2004;87:3736. [24]McGuirk SM. **Vet Clin NA Food Anim Pract** 2008;24:139. [25]DeRosa DC, et al. **J Clin Microbiol** 2000;38:327. [26]ter Laak EA, et al. **Zentralbl Veterinarmed B** 1992;39:610. [27] Nordlund KV. **Vet Clin NA Food Anim Pract** 200824:41. [28]Heath SE. **Comp Cont Educ Pract Vet** 1992;14:385.

PHARMACOLOGY OF MEDICATIONS USED TO TREAT CALF PNEUMONIA

Hans Coetzee, BVSc, Cert CHP, PhD, DACVCP & Mike Apley DVM, PhD, DACVCP
Manhattan, KS

In global terms, BRD treatment and metaphylaxis represents the most lucrative market for the veterinary pharmaceutical industry. The Holy Grail is the development of antimicrobials that require a single administration that provides an extended duration of activity. But how does plasma drug concentration relate to clinical outcome? How long should we wait to re-treat animals that appear to be non-responders? Also, is there a risk that these long acting drugs will encourage development of antimicrobial resistance? In this session we will address these questions and dispel 5 antimicrobial treatment myths relating to combination therapy, route of drug administration, class switching and using body temperature as a basis for treatment.

FIVE ANTIMICROBIAL TREATMENT MYTHS

1. Two Antimicrobials work twice as well as one.
2. You must give something IV at the same time as a long-acting antibiotic to get a quick response.
3. If calves with BRD haven't responded to the first drug, switch to another drug class.
4. Aggressive is good. If they don't look better in 24 hours, add another drug to the treatment rotation.
5. The hotter they are, the sicker they are. Therefore, base your drug choices for respiratory disease on rectal temperatures.

MYTH 1:- TWO ANTIMICROBIALS WORK TWICE AS WELL AS ONE

The Fractional Inhibitory Concentration (FIC) is one way to look at antimicrobial "synergy"[1]. The FIC is defined in the following equation:-

$$\mathbf{FIC} = \frac{\textbf{MIC of drug A + B}}{\textbf{MIC of drug A alone}} + \frac{\textbf{MIC of drug B+A}}{\textbf{MIC of drug B alone}}$$

The FIC is interpreted using the criteria detailed in Table 1.

FIC	Interpretation
< 0.5	Synergism
0.5 to 1.5	Additive
2.0	Indifferent
> 2.0	Antagonistic

Synergistic Combinations

If we consider synergism using the FIC criteria there are very few antimicrobial combinations in veterinary medicine that are truly synergistic. In some cases, synergism may be pathogen specific and cannot be universally applied across all species or all isolates. Some examples of drug combinations that demonstrate synergism are:-

- Combination of a beta-lactam with sulbactam or potassium clavulanate
- Combination of a sulfa with a diaminopyrimidine (potentiated sulfa)
- Aminoglycosides and beta-lactams (some combinations)

Antagonism Examples

In some cases, the veterinary tendency towards polypharmacy and the belief that "more is better" may in actual fact be harmful to our patients. One such example is the combination of chlortetracycline (bacteriostatic protein synthesis inhibitor) and penicillin (bacteriocidal cell wall synthesis inhibitor). Penicillin requires actively dividing bacteria that are producing cell walls to exert a bacteriocidal (killing) effect. Chlortetracycline reversibly binds to the 30S ribosomal subunit thereby inhibiting bacterial multiplication. Chlortetracycline therefore inhibits optimal penicillin activity. This antagonism has been demonstrated in controlled clinical trials in humans with pneumococcal meningitis. In one study, children with meningitis responded better to ampicillin alone than a combination of ampicillin, chloramphenicol and streptomycin. Burrows and Ewing (1989)[2] demonstrated mixed synergism, additive effects and indifference for combinations of erythromycin and oxytetracycline or spectinomycin against *Pasteurella* (*Mannheimia*) *haemolytica* .For example, the combination of erythromycin and spectinomycin was found to be synergistic, additive and indifferent against 15, 13 and 6 isolates respectively. Ose and others[3] demonstrated in-vitro synergism between tylosin and oxytetracycline against *Pasteurella* (*Mannheimia*) *haemolytica* but the clinical relevance of this is not known.

TAKE HOME MESSAGE

Synergism between antimicrobials is extremely variable. There is also a high possibility of antagonism between some combinations, such as β- lactams with bacteriostatic agents like oxytetracycline and the macrolides.

MYTH 2:- YOU MUST GIVE SOMETHING IV AT THE SAME TIME AS A LONG-ACTING ANTIBIOTIC TO GET A QUICK RESPONSE

Table 2 illustrated that many of the BRD therapies considered "long acting" achieve peak plasma drug concentrations in < 1 hour. The exceptions would be oxytetracycline, florfenicol and procaine penicillin. Therefore in many cases, our desire to give something IV is more to placate the producer who expects us to "do something" rather than actually having a scientific reason for doing something.

Myth 3 and 4:- When should we switch drugs; The concept of Post-Treatment Intervals

The following list contains examples of single injection (Long Acting) BRD antimicrobials:-

- Enrofloxacin (Baytril 100®, Bayer)
- Florfenicol (Nuflor®, Schering-Plough)
- Tilmicosin (Micotil®, Elanco)
- Tulathromycin (Draxxin®, Pfizer)

- Oxytetracycline 200 mg/ml (Terramycin LA-200®, Pfizer,and generics)
- Oxytetracycline 300 mg/ml (Alamycin 300, NBK)

These antimicrobials are typically administered q24 or q48H:-

- Ceftiofur hydrochloride (Excenel® RTU, Pfizer)
- Danofloxacin (Advocin®, Pfizer)
- Marbofloxacin (Marbocyl)
- Cefquinome (Cephaguard)

Table 2

Antimicrobial	**Serum or plasma Tmax - reported range or mean ± SD (hours)**	
Ampicillin trihydrate - 7.7 mg/kg IM	1.1 ± 0.63	
Ceftiofur sodium - 2.2 mg/kg SC	0.67 - 3.0	
Ceftiofur hydrochloride - 2.2 mg/kg SC	1 - 5	
Danofloxacin - 5 mg/kg SC	1 ± 0	
Enrofloxacin - 5 mg/kg SC	Calves 1.9 ± 0.21	Adults 3.5 ± 0.46
Florfenicol - 40 mg/kg SC	4 - 6	
Florfenicol - 20 mg/kg IM	3.17 ± 1.33	
Oxytetracycline (100 mg/ml) - 20 mg/kg IM	4	
Oxytetracycline (200 mg/ml) - 20 mg/kg IM	4 - 8	
Procaine penicillin G - 30,000 IU/kg IM	2 - 4	
Tilmicosin - 30 mg/kg SC	1.0 serum (8 hrs lung)	
Tulathromycin - 2.5 mg/kg SC	**0.5 plasma (24 hrs lung)**	

The Million Dollar Question

Do we know enough about pharmacodynamics to predict when therapy has "expired"?

There are several key questions that must be considered when we are examining the interaction between the antimicrobial and the target organism.

- How should we describe drug concentrations?
- How antimicrobials are best presented to the target pathogen?
- What is the target MIC for the pathogen?
- Then...can we put these together to predict when we should retreat a non-responding animal?

1. How should we describe drug concentrations?

The pharmacokinetic parameters needed to describe drug concentration in the animal are straightforward.
1. How **high** does the plasma concentration get with the route of administration we are using?
2. How **long** does the drug remain in the plasma?
3. Any idea what's going on in the tissues as far as concentration?

Terms describing or making inferences about concentrations

Cmax is the highest concentration reached in the plasma or a specific tissue.

The **volume of distribution (Vd)** may be used as an indication of the relationship between vascular and extravascular concentrations of a drug. The apparent volume of distribution is the volume of fluid (expressed as l/kg of body weight) necessary to contain the total amount of drug in the body if it were uniformly distributed and the concentration in this hypothetical fluid were equal to the plasma concentration. Drugs that tend to stay in the plasma have a Vd much less than 1, drugs that have wide distribution have a Vd near 1, and drugs with very wide distribution have a Vd much greater than 1.

Area under the curve (AUC) refers to the total area under the plasma concentration curve. The AUC following an IV injection of a drug essentially represents "all of the drug". Comparing this AUC to the AUC following IM, SC, or oral administration allows the calculation of bioavailability. Bioavailability is the percent of a drug available after administration by a specified route (other than IV) compared to IV administration of the same amount. Adjustments are necessary for flip-flop pharmacokinetics.

Terms describing rates

Tmax describes time to peak concentration in the plasma, this may also be described for a tissue. In the plasma, it is at the time of injection for an IV bolus, and reflects the rate of absorption from an IM injection.

$T\frac{1}{2}\alpha$ is the time required for the plasma concentration to decrease by ½ during the distribution phase of the plasma concentration curve (**distribution half-time**), and estimates the rate of distribution to the tissues. This phase consists primarily of distribution to the tissues, but also includes some elimination processes. It is greatly confounded by absorption following an IM injection.

$T\frac{1}{2}\beta$ is the time required for the plasma concentration to decrease by ½ during the elimination phase of the plasma concentration curve (**elimination half-time**). Elimination from the plasma and tissues predominates in this phase. Although the plasma **$T\frac{1}{2}\beta$** may give an indication of the tissue **$T\frac{1}{2}\beta$,** they are not necessarily equal.

2. How antimicrobials are best presented to the target pathogen?

What is the most basic thing to know about an antimicrobial? How it kills microbes! Application of pharmacokinetics is futile without knowledge of pharmacodynamics. A basic knowledge of bacteriostatic and bactericidal classifications along with site of action is a start. First, just what is this static/cidal thing anyway?? Take home message on 'cidal vs. 'static: It is not a hard and fast classification, but based on the relationship of the MIC and MBC. Bacteriostatic compounds can kill microbes at the proper concentration in-vitro, but the question is if this concentration can be practically obtained in-vivo.

Bacteriostatic	Bactericidal
Tetracyclines	β-lactams
Aminocyclitols	Aminoglycosides
Macrolides	Fluoroquinolones
Sulfonamides	
CHPC	
Lincosamides	

Site of action of antimicrobials:

Ribosome* - 30S subunit	Tetracyclines. Aminoglycosides, Aminocyclitols
Ribosome* - 50S subunit	Macrolides, Lincosamides, Chloramphenicol, Florfenicol
Cell wall	β-lactams (Penicillins and Cephalosporins)
Purine synthesis	Sulfonamides, Trimethoprim
DNA gyrase inhibition	Fluoroquinolones

*Protein synthesis inhibitors

The next step is to classify the antimicrobials as to the type of dosing regimen required for maximum efficacy.

Bacteriostatic antimicrobials and β-lactams require a substantial portion of the dosing interval above the MIC of the pathogen (usually relating to the plasma concentration). The bactericidal β-lactams are in this group because the cell wall of the bacteria become saturated with the β-lactam and increased concentrations do not help. In-vitro pharmacokinetic modeling has supported this theory. Increasing the in-vitro concentration of a β-lactam more than about 4x the MIC (in-vitro) does little to increase efficacy. The tetracyclines are actually considered to be more dependent on the AUC:MIC ratio than on time above MIC for efficacy, although experimental support of this view is thin.

Fluoroquinolones and aminoglycosides are considered to be concentration-dependent, bactericidal antimicrobials. They respond favorably to increasing concentrations with little positive response to extended duration of drug concentrations. Some studies suggest that AUC may be the best predictor of fluoroquinolone efficacy, but remember that cranking the Cmax also has a pronounced effect on AUC. Fluoroquinolones are sometimes thought of as responding to "total exposure" of the pathogen to the antimicrobial. Published work in humans has suggested that a fluoroquinolone peak plasma concentration of at least 8-10x the pathogen MIC is required to prevent resistance in refractive pathogens such as *Pseudomonas*, this is referred to as the Mutant Prevention Concentration (MPC) and will vary for different fluoroquinolones.[4]

So, when we put this all together, what kind of predictions can we make about how long we can wait before we retreat an animal?

Figure 5 depicts the current suggested minimum and maximum times before moving to additional therapy in non-responding BRD cases. Suggested time periods are days (24 hour periods) after the only or last administration.

Figure 5

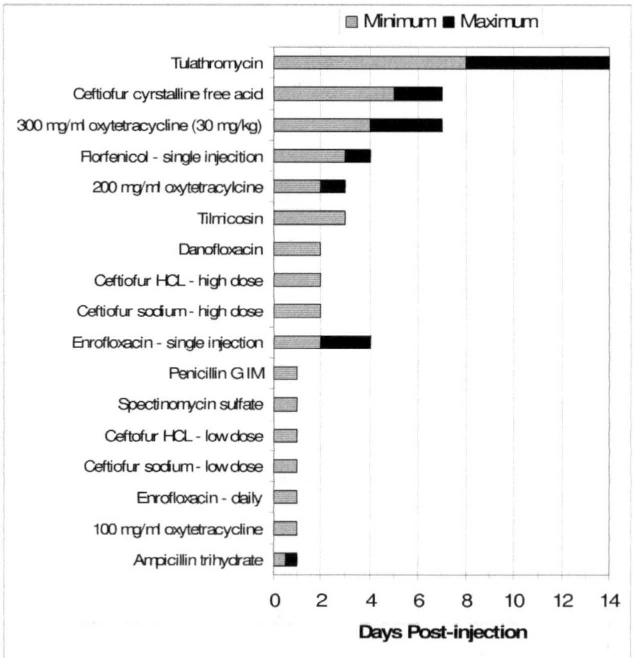

REFERENCES

[1] Hall MJ et al. Journal of Antimicrobial Chemotherapy 1983 **11**, 427-433. [2] Burrows, GE et al.. Journal of Diagnostic Investigation 1989. (4)::299-304. [3] Ose EE et al. VM SAC Veterinary Medicine and Small Animal Clinician, 1976 .71:92-95. [4] Blondeau JM et al. J Chemother. 2004;16 Suppl 3:1-19.

PRACTICAL VACCINOLOGY OF DAIRY CALVES

Christopher CL Chase DVM, MS, PHD, DACVM
Brookings, SD

FREQUENCY OF VACCINATION AND INTERVAL BETWEEN VACCINATIONS- CAN WE OVERVACCINATE?

Many vaccine protocols have been developed to vaccinate the young calf at frequencies as often as weekly during the first and second month of age. Finding any experimental studies that support this frequency is difficult. From other immune systems, it is clear that too frequent vaccination in young animals can lead to antigen specific tolerance (i.e. the lack of any immune response to the antigen).[1] This is the result of suppressive T cells and the deletion of T and B clones specific for those antigens. Another possible adverse outcome of over-vaccination is autoimmunity. Development of autoimmunity is based on priming against the animal's own antigens (self) or by the close mimicry of vaccine antigens to the animal's own antigens. An example of this mimicry is that antibodies against one of the surface proteins of IBR cross-react with a surface protein of the immune cells.[2] Stimulation of inflammatory mechanisms associated with vaccination, such as adjuvants like alum or microbial ligands or repeated re-exposure to vaccine components that drive expansion of autoimmune B and/or T cells clones may occur with frequently repeated vaccine stimulation.[3,4] In all animals following vaccination there is expansion in the populations of responding T and B cell clones. There is a requirement for a good immune response that this clonal expansion not only stop but that an active process of cell death (apoptosis) must also occur. This "waning process" allows "culling" T or B cells that may be poor responders or even cause autoimmunity to be removed by apoptosis.[5] This whole process from vaccination to achieving homeostasis takes at least 3 weeks for the development of a primary response that can then be boosted to get a true anamnestic secondary response. This short review is extracted from a full review published in 2008.[6]

DEVELOPING A VACCINATION PROGRAM

The first necessity of planning a calf vaccine program is to assess the disease risks at the production site. Often "blanket vaccination" programs are suggested for many pathogens, which may or may not be a threat to young calf health. Carefully review the antigens that are being used to make sure they make sense for the operation. Secondly the effect of maternal immunity and the age of the calf must be considered. The relationship is linear: the younger the calf, the poorer the response, the older the calf, and the better the response. However, the inverse relationship is true from the standpoint of protection afforded by maternal immunity: the younger the calf, the better protection due to high levels of maternal antibody, the older the calf, the more susceptible to disease due to waning maternal antibody.

Management factors also come into play. In dairy operations, isolation of calves from exposure to pathogens and good bio-security can provide a window of enhanced protection by maternal immunity giving an extended window before vaccination is necessary. Some practices, such as feeding waste milk on dairies, may "break" the isolation by introducing both pathogens and antibiotics that alter the natural flora developing in the calf to make them more susceptible. This may warrant a more aggressive vaccination program.

BRSV Vaccines

Active immunization for BRSV is probably the most difficult because of maternal interference. This requires careful monitoring and frequent revaccination so that the calves can be protected before the animals reach the "window of susceptibility", the time frame where animals are no longer protected by passive immunity and active immunity has not been stimulated. A promising development has been an intranasal BRSV vaccine that has been licensed in Europe.[7.] IN vaccination takes advantage of the poor penetration of antibody from colostrum onto mucosal surfaces providing less interference with the function of vaccines. This vaccine reduced clinical disease in 3-week-old colostrum-fed calves challenged 66 days following vaccination. Another study administered a commercial 4-way viral MLV vaccine licensed for parenteral use, intranasally, in 2-week-old colostrum-fed calves and found it protective 8 days post vaccination.[8] A single dose of inactivated 4 way viral vaccine in 4-5-week-old calves protected calves against a BRSV challenge 3 weeks post vaccination.[9]

BVDV Vaccines

The risk for BVDV infection and disease in the young calf appears to be much lower than BRSV. The additional immunosuppression in the calf as the result of the BVDV component of many parenterally administered BVDV MLV[10] deserves consideration. Studies in calf ranches have indicated that there is little advantage gained from vaccinating calves under 60 days of age[11] and that maternal antibody protection lasts from 70-110 days.[12] The greatest risk for BVDV infection was from 4-9 months of age as animals were group housed in larger groups.[13] Vaccination programs that vaccinated calves in the first 60 days, and then much later at 4 to 9 months, were not effective in preventing infection.[13] Based on this information, BVDV control programs that use MLV probably should begin around 2-3 months of age followed by revaccination after 3-4 weeks.

BHV-1 Vaccines

BHV-1 infections, like BVDV, are unlikely in young calves. There are several positive aspects of IN BHV-1 vaccines: maternal interference is less likely at mucosal sites in young animals so the vast majority develop active immunity.[14,15] After IN vaccination high levels of mucosal and serum interferon are also produced[16] that both direct an antiviral effect, and may also aid in the development of the neonatal immune response. BHV-1 is also immunosuppressive.[17] However, intranasal BHV-1 vaccines, because of their localized and limited replication in the nasal epithelium, are less of a danger for development of immunosuppression than parenteral vaccines. All BHV-1 vaccines including IN vaccines[18] can result in latency- the ability of the virus to reactivate (recrudesce) and be shed. The biological relevance of this reactivation and shedding of BHV-1 is unclear, as there is no lesions or disease syndromes associated with recrudescence, and the virus can only be reactivated experimentally following several days of treatment with high doses of dexamethasone.[18.]

Clostridial Vaccines

Multistrain *Clostridium* spp. bacterin-toxoids are a frequent part of the vaccination program of heifers. These pathogens represent the cause of sporadic enteric and musculoskeletal disease. Localized vaccine reactions are the most frequent and serious side effect.[19] Maternal interference significantly inhibited *Clostridial* spp. antibody responses in calves vaccinated at 3[20] or 50 days of age.[21]

***Mannheimia* and *Pasteurella* Vaccines**

These bacterial pathogens are frequently isolated from many cases of calf pneumonia, and pose a significant threat to heifer development. Bacterin-toxoids and avirulent MLVs are commonly used. Bacterin-toxoids have been shown to be inhibited by maternally transferred immunity prior to 6 weeks of age·

***Mycoplasma bovis* Vaccines**

This emerging pathogen is also frequently isolated from pneumonia in the heifer. The current vaccine in common use is a bacterin. A single experimental vaccine administered at 3 weeks to calves with low *M. bovis* antibodies was shown to be protective when calves were challenged 21 days later.[22]

***Salmonella* spp Vaccines**

Although *Salmonella* spp are important pathogen of calves, there are few well-designed control studies in the literature other than field observational studies of *Salmonella* vaccine efficacy. There are three major groups of *Salmonella* vaccines: inactivated with gram-negative core antigens (J-5, J-Vac and Endovac-Bovi), attenuated live vaccine (Entervene-D) and the subunit siderophore receptor protein (SRP) vaccines . All three vaccine groups have been shown to have efficacy in the field. Only the attenuated live vaccines are labeled for use in calves (2 weeks of age or older).

Rotavirus & Coronavirus Vaccines

These two viruses are common causes of neonatal diarrhea. Maternal antibody dramatically decreases vaccine efficacy.[23,24] Onset of protection against disease and shedding in calves occurs before secretory IgA is produced 10 days after vaccination,[23] implying that the oral administered vaccine may activate innate immune system in the gastrointestinal tract and decrease disease. These vaccines would be of greatest value for herds with colostrum low in levels of rotavirus and/or coronavirus antibody.

SUMMARY

Vaccination of the heifer calf is complicated by the presence of significant levels of maternal antibody that persist in the calf, colostral and neonatal hormonal factors, the lack of full immune competence, and interference in the function of vaccines by the presence of maternal immunity. The first necessity of planning a calf vaccine program is to assess the disease risks at the production site. Carefully review the antigens that are being used to make sure they make sense for the operation. The affect of maternal immunity and the age of the animal must be carefully considered in determining the vaccination schedule. The use of MLV containing immunosuppressive agents should be planned to avoid giving them at times when the animals have low immune competence and/or are immune suppressed, or by using mucosal vaccination routes that minimize induced immunosuppression and interference by maternal antibody.

REFERENCES

[1.] Raimondi G, et al. **Methods Mol Biol** 2007;380:1-24. [2] Fitzpatrick DR, et al. **Immunology** 1990;70:504-512 [3.] Siegrist CA. **J Comp Pathol** 2007;137:S46-50. [4] Molitor ML et al. **Front Biosci** 2007;12:3302-11. [5] Wagner DH. **Clin Immunol** 2007;123:1-6. [6] Chase et al. **Vet Clin N Am Food Animal Practice** 2008;24:87-104. [7] Vangeel I. **Vet J** 2007;174:627-35. [8] Ellis J, et al. **J Am Vet Med Assoc** 2007;230:2 33-43. [9] Mawhinney IC, et al. **Vet Rec** 2005;156:139-43. [10.] Kelling CL. **Vet Clin N Am Food Animal Practice** 2004;20:115-29. [11.] Thurmond MC, et al. **J Am Vet Med Assoc** 2001;219:968-75. [12.] Munoz-Zanzi CA, et al. **J Am Vet Med Assoc** 2002;221:678-85. [13.] Rush DM, et al. **J Am Vet Med Assoc** 2001;219:1426-31. [14.] Todd JD. **Dev Biol Stand** 1976;33:391-5. [15.] Lemaire M, et al. **J Clin Microbiol** 2000;38:1885-94. [16.] Todd JD, et al. **Infect Immun** 1972;5:699-706. [17.] Ambagala APN, et al. **Vet Immunol Immunopath** 2005;107:1-15. [18.] Lemaire M, et al. **Vaccine** 2001;19:4795-804. [19.] Stokka GL, et al. **J Am Vet Med Assoc** 1994;204:415-9. [20.] Troxel TR, et al. **J Anim Sci** 2001;79:2558-64. [21.]. Troxel TR, et al. **J Anim Sci** 1997;75:19-25. [22.] Nicholas RAJ, et al. **Vaccine** 2002;20:3569-75. [23.] van Zaane D, et al. **Vet Immunol Immunopath** 1986;11:45-63. [24.] de Leeuw PW, et al. **Res Vet Sci** 1980;29:142-7.

FIVE OF THE BEST FOOD ANIMAL INTERNAL MEDICINE JOURNAL ARTICLES PUBLISHED IN 2009

Peter Constable, BVSc(Hons), MS, PhD, DACVIM
West Lafayette, IN

THE OTHER FIVE BEST FOOD ANIMAL INTERNAL MEDICINE JOURNAL ARTICLES PUBLISHED IN 2009

Allen Roussel, DVM, MS, DACVIM, DECBHM
College Station, TX

Peer-reviewed food animal medicine and surgery articles published in 2009 were examined. English-language articles published in the top 50 veterinary journals and top 10 animal agricultural journals based on 5-year impact factor (ISI Web of Knowledge) and articles published in other journals based on recommendations from colleagues were considered. The title and abstract of eligible articles were reviewed and articles selected for further consideration. Ten evidenced-based articles that dealt with an important topic were subsequently selected for presentation and discussion. Criteria used for selection included diversity, excellence in study design and statistical analysis, and a high likelihood that the results of the study will impact the way we practice food animal medicine and surgery. The papers selected contained a clearly stated rationale and hypothesis, and a well described materials and methods section that will permit anyone, anywhere, to reproduce the study.

Study designs used in food animal medicine and surgery include descriptive studies (case reports, case series, surveys) and analytic studies. The latter category includes observational studies (case-control, cross-sectional, and cohort studies) and experimental studies (laboratory studies and randomized clinical trials) (Dohoo et al 2003). Meta-analysis is a subset of systematic review and provides a very helpful evidenced-base approach as meta-analyses should have a higher external validity than a single study (Lean et al 2009).The 10 papers selected used an appropriate method of randomization (where appropriate) and had appropriate statistical power as well as internal validity and external validity (generalizability). In general terms, large multi-site studies have greater external validity and small controlled laboratory or experimental station studies have greater internal validity (Lean et al 2009). Although the ideal study has excellent internal and external validity, completion of the ideal study usually requires extensive resources which are usually unavailable.

The intent of limiting the list to 10 studies was to focus attention and discussion on a small number of evidenced-based papers that are likely to become "game changers". A similar approach was used to select 10 papers from 2008 for discussion at the 2009 ACVIM Forum (Constable 2009). The selected papers for discussion at the 2010 ACVIM Forum increase our understanding of the pathophysiology of disease or change the way we diagnose, treat, and control disease or perform surgery. Many excellent papers were not able to be listed because of the desire for diversity in species and organ system(s) affected, age of animal, causative agent, and geographic location. The strong emphasis on diversity meant that the 10 selected papers were based on 3 food animal species (cattle, sheep, goats) and were published in 6 different journals. The 5-year impact factor rank for the Veterinary Sciences subject category was 18, *Vet J*; 21, *Prev Vet Med*; 22, *J Am Vet Med Assoc*; 40, *Am J Vet Res*; 41, *Vet Rec*, and for the Agriculture, Dairy and Animal subject category was 1, *J Dairy Sci*. The 10 papers originated from 4 countries (Germany, The Netherlands, Switzerland, USA) and are listed below in alphabetical order within each category, which are arranged from the highest level of evidence (meta-analyses) to the lowest level of evidence (descriptive studies):

Meta-analyses

1. Halasa T, Nielen M, Whist AC, Østerås O. Meta-analysis of dry cow management for dairy cattle. Part 2. Cure of existing intramammary infections. *J Dairy Sci* 2009; 92:3150-3157. An interesting companion paper is: Halasa et al. Meta-analysis of dry cow management for dairy cattle. Part 1. Protection against new intramammary infections. *J Dairy Sci* 2009; 92:3134-3149.

Experimental studies

1. Alinovi CA, Ward MP, Lin TL, Wu CC. Sample handling substantially affects Johne's ELISA. *Prev Vet Med* 2009; 90:278-283 – laboratory study.

2. Jones ML, Streeter RN, Goad CL. Use of dietary cation anion difference for control of urolithiasis risk factors in goats. *Am J Vet Res* 2009; 70:149-155 – experimental study.

3. Pithua P, Godden SM, Wells SJ, Oakes MJ. Efficacy of feeding plasma-derived commercial colostrum replacer for the prevention of transmission of *Mycobacterium avium* subsp *paratuberculosis* in Holstein calves. *J Am Vet Med Assoc* 2009; 234:1167-1176 - randomized clinical trial.

4. Washburn KE, Bissett WT, Fajt VR, Libal MC, Fosgate GT, Miga JA. Comparison of three treatment regimens for sheep and goats with caseous lymphadenitis. *J Am Vet Med Assoc* 2009; 234:1162-1166 - randomized clinical trial.

Descriptive studies

1. Bicalho RC, Machado VS, Caixeta LS. Lameness in dairy cattle: A debilitating disease or a disease of debilitated cattle? A cross-sectional study of lameness prevalence and thickness of the digital cushion. *J Dairy Sci* 2009; 92:3175-3184 – cross sectional study.

2. Braun U, Rauch S, Hässig M. Ultrasonographic evaluation of reticular motility in 144 cattle with vagal indigestion. *Vet Rec* 2009; 164:11-13 – case series.

3. Heppelmann M, Rehage J, Kofler J, Starke A. Ultrasonographic diagnosis of septic arthritis of the distal interphalangeal joint in cattle. *Vet J* 2009; 179:407-416 - case series.

4. Hessman BE, Fulton RW, Sjeklocha DB, Murphy TA, Ridpath JF, Payton ME. Evaluation of economic effects and the health and performance of the general cattle population after exposure to cattle persistently infected with bovine viral diarrhea virus in a starter feedlot. *Am J Vet Res* 2009; 70:73-85 - observational study.

5. Janke JJ, Osterstock JB, Washburn KE, Bissett WT, Roussel AJ, Hooper RN. Use of Walpole's solution for treatment of goats with urolithiasis: 25 cases (2001-2006). *J Am Vet Med Assoc* 2009; 234:249-252 - case series.

REFERENCES

[1.] Constable PD. Top ten evidenced-based papers in food animal medicine and surgery for 2008. **Proceedings of the 27th ACVIM Forum**, June 2009. [2.] Dohoo IR, Martin W, Stryhm H. Introduction to observational studies. **In Veterinary Epidemiologic research**. AVC Inc, Charlottetown, Prince Edward Island, Canada, 2003, pp 139-150. [3.] Lean IJ, Rabiee AR, Duffield TF, Dohoo IR. Invited review: Use of meta-analysis in animal health and reproduction: Methods and applications. **J Dairy Sci** 2009; 92:3545-3565.

ANTIEPILECTIC DRUG THERAPY: THE NEXT GENERATION

Michael Podell MSc, DVM, DACVIM (Neurology)
Northbrook, IL

Over the past two decades, the number of antiepileptic drugs (AEDs) available to veterinarians has quadrupled. Coupled with this increase is the ability to rapidly and accurately diagnose underlying brain disease with readily accessible MRI scanning. As a result, the veterinary community is attuned to the need for early treatment intervention. As more treatment choices become available, the unrelenting questions still arise are when should treatment begin, which initial drug therapy is best for our patients, when should treatment changes be considered, and finally, what advantages do newer drugs provide for our patients. The *purpose* of this presentation is three-fold: 1. Review decision making strategies for AED therapy; 2. Provide an overview of the potential applicability of several newer AEDs currently available; and 3. Present information on the next generation of therapeutic advances in epilepsy.

DECISION MAKING STRATEGIES FOR AED THERAPY

When to start treatment

The decision to start AED treatment is based on a number of factors, including etiology, risk of recurrence, seizure type and effect on patient, and risk of treatment. Risk factors for seizure recurrence are not well-established for cats and dogs. A number of relative risk factors have been identified in epileptic people, to include a diagnosis of current or previous defined cerebral lesions or trauma, presence of interictal EEG epileptic discharges (up to 90 % recurrence rate) and a history of marked post-ictal adverse effects (Todd's paralysis).[1] Evidence-based guidelines through several international groups are well established for people based on risk: benefit ratio and predictability factors of drug effect.[1,2] From these guidelines, several commonalities exist in guiding clinical practice to include confirmation of an epileptic seizure event and seizure type, obtaining a definitive diagnosis, knowledge that recurrent seizure activity is correlated with poorer long term treatment success, and the influence of treatment on quality of life factors.[3] Thus, the decision to treat is a reflection of the treatment goals to reduce or eliminate epileptic events, reduce seizure severity, avoid adverse effects, and reduce seizure-related mortality and morbidity.[4]

While similar information is not as readily available in our patient population, extrapolation is possible to provide rationale treatment guidelines. *Overwhelming evidence exists in people that there is no benefit to start treatment after a single unprovoked event.*[5] The earlier AED therapy is started, however, the better the potential outcome may be for seizure control. Reasons to start AED include: 1) Identifiable structural lesion present or prior history of brain disease or injury; 2) Status epilepticus has occurred (ictal event $\geq$ 10 minutes; 3) Two or more generalized seizures occur within a 24 hour period; 4) Two or more isolated seizure events occur within a 6 month period; and 5) Prolonged, severe, or unusual post-ictal periods occur.

Drug selection

AED selection is based on a number of factors, including seizure type, efficacy, and tolerability. Again, evidence-based guidelines established by the International League Against Epilepsy, American Academy of Neurology, and Standard and New AED Trials (SANAD) provide probability based recommendations.[1,2,5] Despite these guidelines, *no evidence exists that any single AED provides a better outcome for adults with unprovoked epilepsy when early treatment is started.*[3] Drug selection, therefore, is often based on tolerability. Adverse effects can be divided into transient, persistent and life-threatening categories (idiosyncratic or predictable). Most transient adverse effects are avoidable with titration dosing and dissipate within several weeks. Persistent effects are either CNS dose dependent effects associated with sedation, ataxia, vertigo or cognitive impairment or metabolic related with hormonal imbalances, metabolic syndromes and degenerative effects (e.g., osteoporosis). Severe life-threatening effects are mainly associated with either idiosyncratic bone marrow disease (e.g., aplastic anemia) or predictable organ damage over time (e.g., hepatotoxicity).

Success parameters

Epilepsy treatment should be goal-oriented and approached in an objective fashion. Seizure elimination or significant seizure reduction, reducing seizure severity and maintaining normal lifestyle for patient and owner are all important considerations. While many drugs may provide initial improvement in seizure control, long-term efficacy is dependent on many factors. With only approximately 60 – 80% of both human and canine epileptic patients responsive to treatment, [6, 7] evaluation of reasons for recurrent seizure activity (refractory epilepsy) is important. These factors can be broken down into three main dependent variables: disease related, drug-related, and patient-related problems. Disease related factors include presence of an undiagnosed underlying brain disease, such as cortical malformation, prior trauma, or active disease process. Occult conditions can lead to localization related epilepsy where epileptic foci develop drug resistance due to architectural brain changes. Drug-related mechanisms include ineffective mechanism of action, development of tolerance, and alteration of drug target or uptake over time.[8] Seizure specific therapy targets a drug for a specific seizure type, and inappropriate drug selection may result in poor control. Patient-related issues are now being discovered in relation to gene polymorphisms that affect AED pharmacokinetic or pharmacodynamic properties. As a result, altered drug metabolism or action is no longer predictable as compared to the general patient population.

THE GENERATION GAP

History of AED Development

From the introduction of bromide in 1857 by Sir Charles Locock to treat "hysterical" epileptic fits in women, to the development of phenobarbital in early 1900's and the introduction of phenytoin, valproate and carbamazepine in the early 1970s, the use of these first generation AEDs was limited by mechanism, benefit, and adverse effects. Not until the early 1990's did a revolution in drug development occur with the introduction of felbamate. The enhancement of rodent models for drug discovery screening was a driving factor in this expansion. The succession of the second generation drugs over the next two decades resulted in greater treatment success with significant improvement in tolerability. As the second decade of the 21st century starts, a third generation of AED therapy enters with more advances in improving seizure control and patient quality of life. **(Table 1)**

Mechanisms of action

Antiepileptic drugs can be classified into three broad mechanistic categories **(Table 2)**: 1. Enhancement of inhibitory processes via facilitated action of gamma amino-butyric acid (GABA); 2. Reduction of excitatory transmission; and 3. Modulation of membrane cation (Na or Ca^{+2}) conductance. Unfortunately, several limitations exist in the selection of AEDs for use in veterinary medicine.[9] In the past, many of the AEDs useful in people could not be prescribed to small animals either due to inappropriate pharmacokinetics (too rapid of an elimination), ineffectiveness, or the potential hepatotoxicity. The result was that the most commonly used AEDs in veterinary medicine were from the same mechanistic category, that of enhancing brain inhibition. The advent of new AEDs with alternative mechanisms of action is now available to allow a broader selection of treatment options.

Table 1. Categorization of Antiepileptic Drugs by Generation of Drug Development

FIRST GENERATION	SECOND GENERATION	THIRD GENERATION	NEXT GENERATION
Bromide	Felbamate	Lacosamide	Flurofelbamate
Phenobarbital	Gabapentin	Rufinamide	Brivacetam
Benzodiazepines	Oxcarbazine		Carisbamate
Phenytoin	Zonisamide		Retigabine
Carbamazepine	Lamotrigine		Tiagabine
Valproate	Levetiracetam		Losigamone
	Pregabalin		Remacemide
	Topiramate		Seletracetam

Clinical AED pharmacology

The efficacy and safety profiles of AEDs are determined in large part by their pharmacokinetic properties. Drugs that are the easiest to use by the general population are ones that have the most favorable pharmacokinetic properties. For this reason, phenobarbital (PB) has been widely used in veterinary medicine for generations. Ultimately, the most desirable pharmacokinetic profile of an AED is one that has complete bioavailability, availability as a parenteral formulation, an elimination half-life suitable for daily or twice daily dosing, linear elimination kinetics, no autoinduction of enzymatic biotransformation, no pharmacokinetic interactions with other drugs, rapid brain penetration, a volume of distribution with a single compartment, low and nonsaturable protein binding, and no active metabolites. Unfortunately, the ideal AED has not been formulated for any species.

Table 2. Summary of the Mechanism of Actions of Currently Available Antiepileptic Drugs.

	MECHANISM OF ACTION				
	Decrease seizure onset		Decrease seizure spread		
Drug	**Enhanced Na+ channel inactivation**	**Enhanced GABA activated Cl- conductance**	**Reduced Ca^{+2} channel current**	**Reduced glutamate excitation**	**Novel mechanism**
Phenobarbital		++	+	?	
Bromide		++*			
Felbamate	+	+		+	
Benzodiazepines		++			
Gabapentin	?	+	++		
Topiramate	+	+		+	
Zonisamide	+		++		
Levetiracetam		?	?	?	SV2a receptor
Pregabalin			++		
Lacosamide	+				CRMP
Rufinamide	++				

*Competitive displacement of Cl- through activated GABA receptors; ? = Possible mechanism; SV2a=synaptic vesicle 2A binding; CRMP=collapsin response mediator protein 2 binding.

First Generation AEDs

Phenobarbital is well-tolerated and is still the standard of care for first generation AED use in the cat and dog, as it is highly efficacious and well tolerated. The drug has high bioavailability, is over 50% protein bound, has a long elimination half-life, with predominant hepatic metabolism and autoinduction of hepatic microsomal enzymes (p-450 system), which can progressively reduce the elimination half-life with chronic dosing. *Bromide* is an inorganic salt with a prolonged elimination half-life (15-20 days), is solely renal excreted unchanged, has proven efficacy, but does have greater dose-dependent adverse effects.[10]

Second Generation AEDs

Felbamate is believed to increase seizure threshold and prevent seizure spreading by reducing excitatory neurotransmission in the brain and increasing GABA activity. Felbamate is most useful as monotherapy in the treatment of uncontrolled partial epilepsy and Lennox-Gastaut syndrome with atonic seizures. In dogs, the drug has a high bioavailability and protein binding capability and is metabolized by the hepatic microsomal p450 enzymes.[11] Effective control of focal seizure activity (automatisms) with documented therapeutic serum concentrations has been shown with felbamate therapy in dogs.[12] Felbamate is a non-sedating drug, but has been correlated with a higher incidence of idiosyncratic aplastic anemia and liver toxicity in people. Similar, but rare and reversible blood cytopenias and hepatotoxcity have been observed in the dog. Serial monitoring of the complete blood count and chemistry panel is recommended at 1 month and every 3 months during treatment. Initial dose is 5-10 mg/kg q 12 hours with titration to q 8 hr dosing. Trough serum drug concentration is typically done 1 –2 weeks after initiation of treatment, with a therapeutic range between 25-100 mg/L.

Topiramate is a sulphamate-substituted monosaccaride with multiple mechanisms of action of rapidly potentiated GABA activity, sodium channel blockade and post-synaptic glutamate receptor blockage that is efficacious against a wide variety of seizure types.[12] In people, topiramate is well-absorbed, is primarily renal excreted as an unchanged drug and has a relatively long-half life of 20-30 hours. In the dog, however, topiramate has a relatively short elimination half-life in the dog of 2-4 hours,[13] but may exhibit prolonged pharmacodynamic activity in the brain due to high affinity receptor binding. Initial dosing is 2- 5 mg/kg q 12 hours with gradual adaption to 10 mg/kg q 12 hr. Loss of appetite, irritability, GI upset and sedation are potential adverse effects.

Zonisamide is a substituted 1,2-benzisoxazole derivative that works via multiple mechanisms to block the propagation of epileptic discharges and suppress focal epileptogenic activity through sodium and calcium channel blockade and GABA activity potentiaion .[14] In the dog, zonisamide is well-absorbed, has a relatively long-half life (15-20 hours), has high protein-binding affinity and has both hepatic and renal metabolism.[15] The drug is highly concentrated in red blood cells due to high binding to carbonic anhydrase and other red cell protein components. Broad-spectrum antiepileptic activity has been reported against a variety of seizure types, with particular improvement in the treatment of adult myoclonus epilepsy.[16]. Since serum drug levels are lowered by phenobarbital, phenobarbital doses should be decreased by 25%. . Initial dose is 5 mg/kg q 12 hr with a therapeutic range of 10-40 mcg/ml, with greater efficacy at 20 mcg/ml.[17] Major adverse effects include a higher incidence of sedation, GI upset, metabolic acidosis and renal calculi formation.

Levetiracetam is the S-enantiomer of the ethyl analogue of piracetam with a unique mechanism of action mediated by binding to the synaptic vesicular protein, SV2A, that decreases neurotransmitter release. In dogs, the drug is well-absorbed, is rapidly metabolized with an elimination half-life estimate of 4-8 hours and is predominantly renal excreted (> 80%).[18] Wide fluctuations of drug metabolism, however, occur in the dog. Initial dose is 10- 20 mg/kg q 12 hours with gradual increment to 20 or higher mg/kg q 8 hour dosing. The therapeutic range is not well-defined and drug monitoring is recommended only to establish individual patient pharmacokinetic pattern. The drug is well-tolerated, with sedation noted as the most common adverse effect. Parenteral formulation is available for IV loading at 40-60 mg/kg over 10 minutes.

Lamotrigine is a novel drug that is chemically unrelated to current AEDs. Although efficacious in human epileptic patients, the drug is converted to a cardiotoxic 2-N methyl metabolite in dogs, [19] not found in people, and is not recommended for use in dogs.

Pregabalin is a structural analogue of GABA with a mechanism of action of voltage-gated calcium channel modulation that decreases depolarization induced calcium influx at nerve terminals, ultimately, to reduce excitatory neurotransmitter release. The mean elimination half-life is estimated at 7 hours in dogs.[20] Metabolism appears to be predominantly renal excretion with minimal protein binding and drug interaction. Efficacy as an add-on for refractory partial seizures was found in several studies in people and in a recent study in dogs.[21] A definitive therapeutic range has yet to be determined for people or dogs. Initial dose recommendation is 4 mg/kg q 8 to 12 hours. Adverse effects appear limited to sedation and ataxia.

Third Generation AEDs

Lacosamide (Vimpat®) is a functionalized amino acid proven to decrease neuronal discharge frequency and synaptic excitability.[22] The postulated mechanisms of action includes selective slow inactivation of sodium channels and novel binding to collapsin response mediator protein-2. In people, the drug is well-absorbed, has minimal first pass effect with predominant renal excretion, low protein binding, favorable drug-drug interactions with other AEDs, and is well-tolerated. Clinical trials demonstrate comparable decrease in seizure frequency as that of levetiracetam and zonisamide at a dose of 100-200 mg q 12 hours. A parenteral formulation is available for IV loading. No data was found regarding clinical use in dogs or cats.

Rufinamide (Banzel®) is a novel AED that is structurally unrelated to any other AED. The main mechanism of action is related to prolongation of the inactive state of the sodium channel to prevent neuronal depolarization. In people, the drug is slowly absorbed with low bioavailability. Renal excretion is high and no induction of the hepatic p450 system was found, although other hepatic metabolized drugs decrease serum concentration. Of a total of 9 double-blinded studies in people, 5 revealed a positive effect of rufinamide to treat refractory partial seizures but not generalized seizures.[23] Initial dosing ranged from 10 to 40 mg/kg/day, with dose dependent adverse effects of sedation, fatigue and dizziness noted. A parenteral formulation is not available. No data was found regarding clinical use in dogs or cats.

THE NEXT GENERATION

The future of epilepsy treatment is undergoing a multifaceted, exponential growth. Over 15 new AED are currently in clinical trials throughout the world. A common denominator for drug development is the ability to define pharmacoresistant therapy by both drug action and patient pharmacogenomics. With the advent of more advanced diagnostic capabilities to map localization epilepsy, conventional surgical and radiosurgical intervention can provide curative outcomes previously never imagined. Vagal and brain stimulators introduce physiologic methods of altering the brain's baseline seizure threshold. While veterinarians may not incorporate all of these modalities into our clinical practice yet, the future potential to help our patients grows every day as well.

REFERENCES

[1.]SIGN Guideline 70. (2003) Scottish Intercollegiate Guidelines Network. Edinburgh . www.sign.ac.uk; [2.] American Academy of Neurology, (2004) www.aesnet.org/go/practice/guidelines; [3.] Stephen LJ and Brodie MJ (2009), **Neurol Clinics** 27:967-992; [4.] Perucca E. (2009) In: **The Treatment of Epilepsy** (ed, Shorvon, Perucca, Engel), p 121; [5.] Glauser T, et al. ILAE Guidelines (2006), **Epilepsia** 47:1094-1020; [6.] Brodie MJ, et al (2007) **Neurology** 68:402-408; [7.] Podell M (2005) **Manual of Canine and Feline Neurology**, Platt SR and Olby N. BSAVA, 97-112; [8.] Loscher and Schmidt D. (2006) **Epilepsia** 47:1253-1284; [9.] Podell, M. (1998): **Clin Tech Small Animal Practice** 13, 185-192; [10.] March, P. A., et al. (2002):. **J Vet Pharmacol Ther** 25, 425-432; [11.] Adusumalli V. A., et al. (1992). **Epilepsia** 33, 955-960; [12.] Ruehlmann, D. S et al (2001):. **J Small Animal Practice** 42, 403-408.; [13.] Petroff, Oa, et al (2001). **Epilepsia** 42, 543-548; [14.] Streeter AJ et al (1995). **Drug Metab Dispos** 23:90-93; [15.] Ito, T., et al.. (1980):. **Arzneimittel-Forschung** 30, 603-609; [16.] Walker, R. M., et al.

(1988): **Fundam Appl Toxicol** 11, 333-342; [17.] Dewey C (2004) **JAAHA** 40:2850291; [18.] Volk HA et al. (2008) **Vet J** 176:310-319; [19.] Wong I., and Lhatoo S. (2000):. **Drug Safety** 23, 35-56; [20.] Salazar V, et al. (2009) **Vet Anesth Analg**, 36:574-580; [21.] Dewey C et al. (2009) **JAVMA** 235:1442-1449); [22.] Halford JJ and Lapointe M (2009) **Epilepsy Currents** 9: 1-9; [23.] Biton V. (2009) In: **The Treatment of Epilepsy** (ed, Shorvon, Perucca, Engel), p. 651.

CLINICAL APPLICATIONS OF ELECTROENCEPHALOGRAPHY IN VETERINARY MEDICINE

D. Colette Williams, PhD candidate
Davis, CA

INTRODUCTION

Electroencephalography is defined as "1) the science related to the electrical activity of the brain and 2) the practice of recording and interpreting EEGs".[1] Thus, an electroencephalogram is a "record of electrical activity of the brain taken by means of electrodes placed on the surface of the head, unless otherwise specified".[1] A similar, but far more invasive technique, that of electrocorticography (ECoG), involves the placement of electrodes directed on the cerebral cortex or the dura mater. Ion fluctuations in cortical pyramidal neurons, resulting from numerous excitatory post-synaptic potentials (EPSPs) and inhibitory post-synaptic potentials (IPSPs), are the primary basis for the waveforms depicted in the normal EEG. The background slowing and spindles associated with slow wave sleep are the result of synchronized activity from thalamocortical projection neurons.

The field has a long history, dating back to 1875 when Richard Caton published recordings from monkeys, rabbits and cats obtained via galvanometer. Fifty years later, Hans Berger performed the first EEGs on humans. Dietsch, in 1932, described the application of Fourier analysis to EEG data for the purpose of obtaining spectral information. Findings from children with petit mal (absence) epilepsy were published in the mid 1930s, as were Bremer's studies on the affects of neural generators on sleep in cats. Human sleep research followed in the early 1950s, and at the end of the decade, Brass described his findings on canine EEG. Additional studies pertaining to veterinary EEG or ECoG have made their way into print, by authors such as Pampiglione, Ruckebusch, Dallaire, Knecht, Redding, Klemm, Jaggy, Strain, Steiss, Lewin, Bergamasco and Holliday, to name a few.

INDICATIONS

Electroencephalography has been used extensively in human medicine for the diagnosis of seizure disorders. Some epileptic syndromes are classified, in part, by specific epileptiform discharges (ED, Fig.1) found on EEG. This test is best used to answer specific, clearly defined questions regarding management of seizure disorders.[2] Recordings can be beneficial in the assessment of anticonvulsant therapy efficacy both acutely (status epilepticus) and in chronic cases. In human and veterinary medicine, EEG can be useful in differentiating epilepsy from psychiatric (behavioral) events, syncopal episodes or paroxysmal dyskinesias.

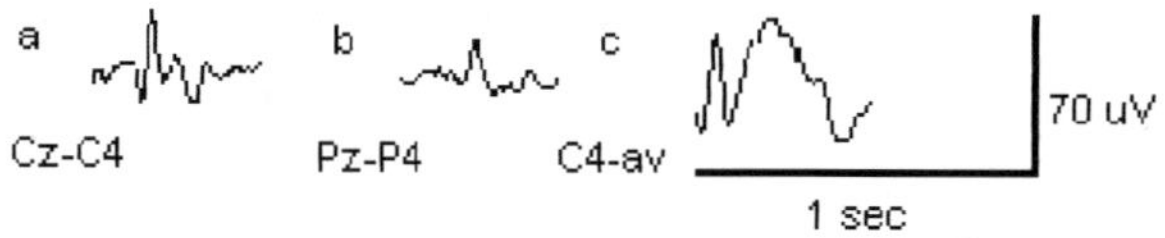

Figure 1. Examples of epileptiform discharges: a) spike (<70 msec), b) sharp wave (70-200 msec) and c) spike-and-wave (slow wave >200 msec)

The technique can also be a tool for diagnosing various sleep disorders. Cerebral function can be readily determined in comatose or obtunded patients and critical cases can be monitored long term. Quantitative EEG (qEEG) is based on changes in EEG spectral data (Fig 2) and has been utilized as a means of assessing depth of anesthesia in both large and small animals.

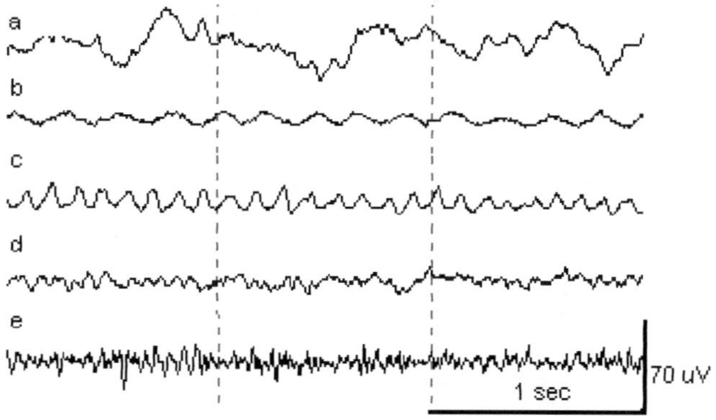

Figure 2. Waveform spectrum. Examples of epochs containing large amounts of a) Delta 0-4 Hz , b) Theta 4-8Hz, c) Alpha 8-13 Hz, d) Beta 13-30 Hz and e) Gamma >30Hz.

LIMITATIONS

Abnormalities in EEG background activity are often nonspecific and similar findings can result from toxic, metabolic, degenerative, developmental, neoplastic and infectious/inflammatory insults. Though subtle differences have been described in EEGs obtained from humans, documentation in animals is limited. Periodic lateralized epileptiform discharges (PLEDs) have been associated with lateralized encephalopathies in people and are thought to be related to focal seizures (Fig 3).[3]

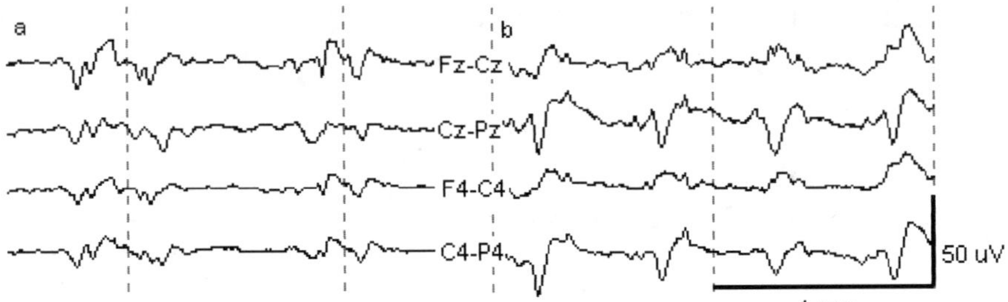

Figure 3. Probable PLEDs in a dog with *Aspergillus fumigatus* meningitis/ventriculitis recorded: a) before and b) after a focal seizure.

Despite its value in identifying ED in seizure disorders, EEG does not differentiate between symptomatic, cryptogenic and idiopathic epilepsy. Other diagnostic tests employed in these cases, namely imaging (MRI/CT) and CSF analysis, can be used to distinguish the former from the latter two. Even in epileptic patients, interictal recordings may be normal.

Activation Techniques

To increase the diagnostic yield of an EEG, a number of protocols have been employed. In several human epileptic syndromes, specific stimuli can trigger the appearance of ED. Most notably are 3/second spike-and-wave discharges resulting from a period of hyperventilation in absence epilepsy patients and variable types of ED associated with photic stimulation (some frequency specific) in several other seizure disorders. The author recently had a feline patient in which ED (along with partial or generalized seizures) could be consistently induced by a click sound.

Two additional activation techniques, applicable to veterinary patients as well as humans, are the use of sleep during an EEG and a period of sleep deprivation prior to recording. Slow wave sleep (SWS) increases the likelihood of detecting ED regardless of whether it is natural or drug induced (Fig 4).[4,5,6] The use of an appropriate dose of an α_2 adrenergic agonist (xylazine, detomidine, metedetomidine or dexmetedetomidine) for sedation virtually guarantees a period of SWS, regardless of the species. Though these drugs induce SWS, it is thought that they suppress rapid eye movement (REM) sleep, thus should not be used if the EEG is indicated for a sleep disorder.

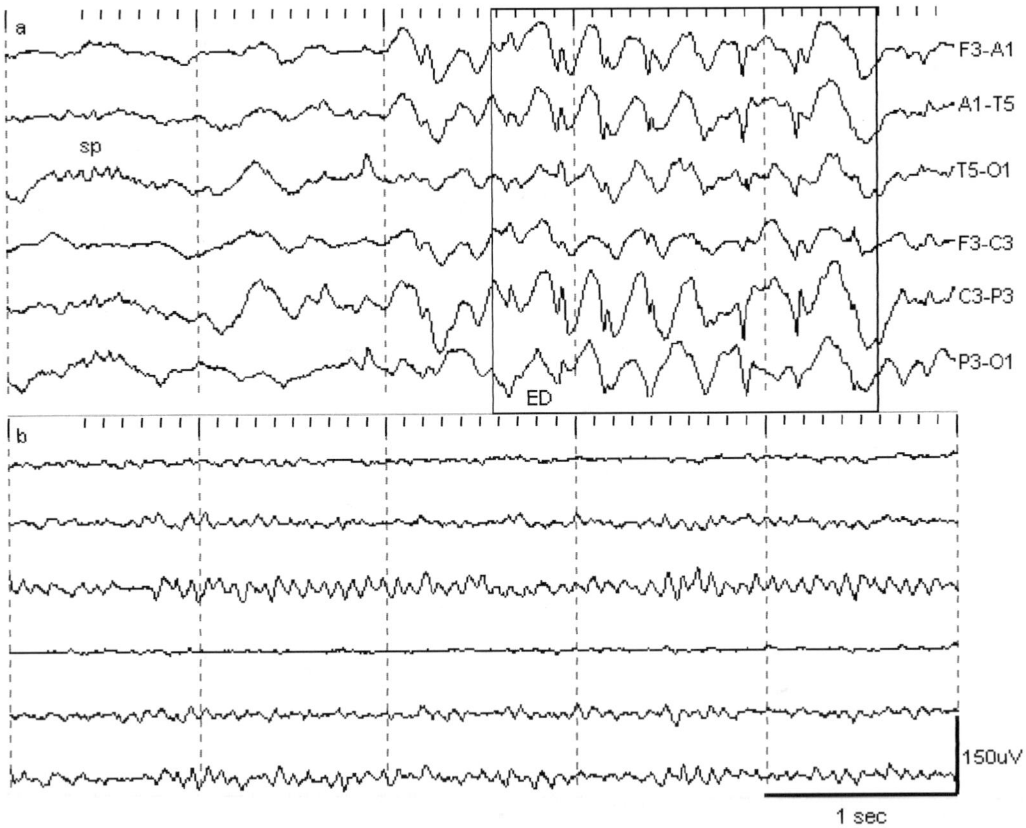

Figure 4. a) spike-and-wave discharges (boxed) during metedomine induced SWS, b) normal wakefulness after reversal with atipamazole in a young CA sea lion with domoic acid toxicosis. sp=sleep spindle, ED=epileptiform discharges.

RECORDING TECHNIQUE

Equipment used in EEG recording can be either analog or digital. The advantages of the latter have been well described.[7] Some manufacturers have telemetric units on the market and many offer EEG as an add-on feature with certain models of EMG/evoked potential systems.

For EEG, needle electrodes are commonly used in veterinary medicine, though the use of surface electrodes or alligator clamps has also been described (a variety of other types are used for ECoG recordings). Electrode placement varies with species and size but designation is generally based on human descriptions (F=frontal, C=central, P=parietal, O=occipital, T=temporal, left=odd numbers, right=even numbers and z=midline). For localizing events, via identification of instrumental phase reversals in bipolar montages (Fig 5), a minimum of 3 electrodes arranged in a chain (ideally, in both sagittal and transverse planes) is required. The information presented here is based on techniques as described by Holliday.[8]

INTERPRETATION CONSIDERATIONS

A variety of factors will influence the appearance of the EEG, even in normal animals. Breed and size will have an affect on the amplitude of the EEG signal, as increased skull thickness and scalp musculature may necessitate the use of higher amplifier sensitivities. Post-natal brain maturation takes time, so age needs to be taken into account as well. Recent data recorded from normal 2 week old kittens, have shown similarities to the quiet sleep of infants in that both demonstrate discontinuity, periods of quiescence which alternate with bursts of activity.[9] Amplitude (and power) also vary with age.[9]

State of vigilance must be considered as background activity varies between SWS (also referred to as delta sleep) and quiet wakefulness/drowsiness or REM sleep (the latter can be relatively similar in appearance).[10] Normal transient events, consisting of sleep spindles, vertex sharp waves and K-complexes (Fig 5), frequently occur during SWS (and occasionally drowsiness) and should not be interpreted as ED (especially in young animals, when sleep spindles often look "spiky").

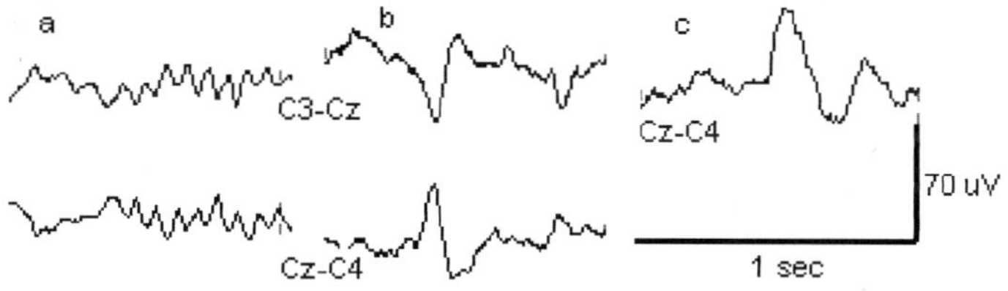

Figure 5. Normal transients of SWS: a) sleep spindle, b) vertex sharp wave and c) K-complex.

Various drugs, including those administered for sedation/anesthesia, as well as, medications currently used in the patient's treatment regimen, can have an affect on the EEG, most in a dose dependant manner. Barbiturates and benzodiazepines are associated with fast beta in some species.[4,11,12] The former will induce burst-suppression patterns at high doses.[12] The anticonvulsant properties of both drugs may decrease the likelihood of finding ED in an EEG. In humans, propofol administration has been associated with ED in both epileptics and non-epileptics.[12] Isoflurane reviews are mixed, some report ED frequency is decreased while others describe no change in ED when used along with nitrous oxide.[12] It is well known for its ability to produce burst-suppression EEG patterns at surgical planes of anesthesia in both humans and animals (Fig 6).[12,13]

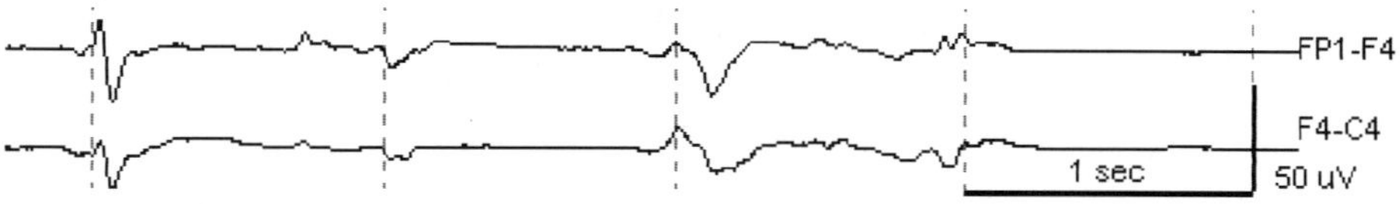

Figure 6. Burst-suppression pattern in a horse under isoflurane anesthesia.

SUMMARY

Electroencephalography can provide the veterinary clinician with valuable information regarding a patient's cerebrocortical function. Not intended to be used as a "stand alone" test, it is best utilized in combination with other diagnostic modalities.

REFERENCES

[1]Noachtar S, et al.**Electroencephogr Clin Neurophysiol Suppl** 1999;52:21. [2]Binnie CD, et al. **Clin Neurophysiol** 1999;110:1671. [3]Zifkin BG, et al. **Practice of Clinical EEG, 3rd ed.** 2003;288. [4]Aleman M, et al. **J Vet Intern Med** 2006;20:1443. [5]Goldstein T, et al. **Proc Biol Sci** 2008;275(1632):267. [6]Talke P, et al. **J Neurosug Anesthesiol** 2007;19(3);195. [7]Holliday TA, et al. **Vet Neuro Neurosurg J** 2001;3(1) and 2003;5(1). [8]Holliday TA, et al. **Clin Tech in Sm An Prac** 1998;13(3):132. [9] Lewis MJ, et al. **Am J Vet Res** 2010 (in press); [10]Williams DC, et al. **J Vet Intern Med** 2008;22:630. [11]Jeserevics J, et al. **J Vet Intern Med** 2007;21:1299. [12]Van Cott, et al.**Current Practice of Clinical EEG, 3rd ed.** 2003;463. [13]Scheller MS, et al. **Br J Anaesth** 1990;65:388.

ENDOSCOPIC TRANSSPHENOIDAL PITUITARY SURGERY

Adam N. Mamelak[1], MD, David Bruyette[2], DVM, DACVIM, Tina Jo Owen[3], DVM, DACVS
[1]Department of Neurosurgery, Cedars-Sinai Medical Center, Los Angeles, CA
Departments of [2]Medicine and [3]Surgery, VCA West Los Angeles Animal Hospital, Los Angeles, CA

The human pituitary gland is uniquely situated at the base of the skull, posterior to the nasal passages and sphenoid sinus, and inferior to the brain. Due to this location the pituitary gland and associated tumors can be approached from a nasal route rather than a cranial route utilized for most brain lesions. Reaching the pituitary region via the sphenoid sinus located directly behind the nose, the so called transsphenoidal (TS) approach offers many advantages, including direct exposure of the pituitary fossa, minimal tissue destruction, and protection of the intracranial contents.

The transsphenoidal approach to the brain has been recognized since the times of the Egyptians, who utilized this technique to remove the brain prior to mummification. The modern version of the TS was initially performed in the early 1900s by Shloffer, and popularized by Cushing[1, 2]. In these early days, without the advantages of magnification or illumination beyond that of a headlight, the procedure was fraught with difficulties. It was eventually abandoned in favor of a transcranial approach. In the mid 1960's the operating microscope was introduced to neurosurgery and related disciplines. The microscope provided the light source, magnification, and depth of field needed to re-address the previous limitations of transsphenoidal route. With the microscope providing the needed magnification and illumination, surgeons such as Giout in France perfected the sub-labial trassnsphenoidal approach, with a resultant rebirth of the field[2]. Utilizing X-ray fluoroscopy to identify the sella tursica and assure an accurate trajectory, the microscope allowed surgeons excellent visualization, which in turn resulted in safe and more complete operations. The TS rapidly gained in popularity, and has subsequently become the dominate technique for the treatment of sellar and parasellar pathology for the past 50 years. At present well over 95% of pituitary tumors are treated utilizing the TS approach. Microscope based TS is very safe, with a mortality rate of well below 1%, and low morbidity.

Despite these obvious advantages, the sub-labial TS has several shortcomings. Firstly, it is a destructive approach, resulting in the removal of the inferior third of the cartilaginous septum. The incision in the upper gum invariably results in numbness to the upper gum line, and 5-10% of patients develop longer term rhinological complication such as saddle-nose deformity[3, 4]. Efforts to minimize these side effects include the collumellar approach, in which incisions are made in the anterior nasal mucosa, a tissue plane is developed to separate the nasal mucosa from the cartilaginous septum, and the septum is fractured out to allow insertion of the speculum. Although slightly less destructive, this approach provides lesser exposure and freedom of movement than the sub-labial method[3].

With the advent of modern day endoscopes, otorhinological surgeons began to perform functional sinus surgery, including endoscopic sphenoidotomy for chronic sinus infections. Their work demonstrated the safety of opening the sphenoid sinus without a sub-mucosal plane. Subsequent to this, endonasal speculums were developed that facilitated direct endonasal microscopic transsphenoidal surgery. This approach proved to be extremely efficient and has risen in popularity amongst neurosurgeons to the point that it has largely replaced the sub-labial approach for many high volume pituitary surgeons[5]. The endonasal microscopic technique has many advantages including lack of upper gum numbness or saddle nose deformity, rapid tissue healing, less post-operative pain and fewer nasal complications. However, the microscope based endonasal TS suffers one major disadvantage over more traditional methods, namely a markedly reduced field of view and range of motion for visualization. Since the microscope objective has a focal point of approximately 350-400 mm, the field of view through the nasal speculum is typically centered directly on the sella floor, but much of the surrounding structures cannot be seen. Further, all surgical instruments must be introduced through this same narrow corridor, resulting in a limited range of motion for the surgeon, and often a reliance on "one-handed" techniques.

More recently, surgeons have begun to employ the endoscope as an alternative to the microscope, with quite impressive results. Unlike the large microscope, the endoscope is passed directly into the sphenoid sinus. Visualization of the operative field is accomplished via a video camera attached to the end of the endoscope, with images displayed on a video monitor directly behind the patient's head. With a focal length of 1 -2 cm, the endoscope provides a panoramic view of the entire base of the skull at the sphenoid sinus (Fig 1A). Cadaveric research studies have documented the marked increase in the field of view obtained by the endoscope when compared to the microscope, regardless of the approach utilized [6](Fig 1B). Technical improvements have exploded in this field over the past 5- 10 years, and include the addition of high definition video cameras, improved surgical tools designed for direct endonasal approaches, and the use of a two-surgeon technique. Both single nostril and two nostril approaches can be employed, although we and most other experienced groups favor a bi-nasal approach which permits much greater flexibility and broader access with no significant changes in nasal morbidity[7].

A

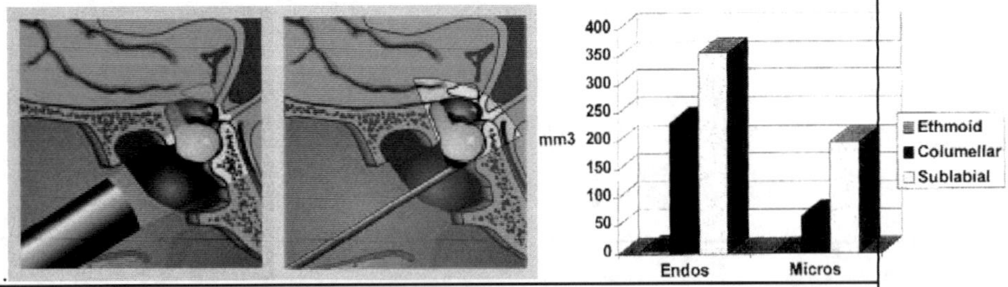

Figure 1: A) comparing the field of view of the microscope and endoscope. The microscopic view is limited by the speculum while the endoscope view is more panoramic. B) The endoscopic field of view is almost twice as large as that of the microscope, regardless approach (reproduced with permission from Spencer et al, 1999).

At our institution, we routinely perform endoscopic endonasal transsphenoidal surgery utilizing a single surgeon and a pneumatic endoscope holder rather than a second surgeon. In our hands this is a more rapid and equally safe technique. The basic technique includes the following steps: 1) Rigid fixation of the head in a 3-point head clamp in a semi-sitting “lounge chair” position. 2) Registration of a previously acquired MRI to the patient’s facial landmarks, and use of a frameless stereotactic navigation tool to identify skull base landmarks, 3) insertion of a rigid endoscope (4mm, 0 degree) into the nostril for visualization via an HD camera and monitor, 4) removal of the anterior wall of the sphenoid sinus and posterior thirds of the midline septum 5) similar dissection in the left nostril 6) securing of the endoscope in place with a pneumatic holder and 7) tumor removal using bimanual microsurgical dissection techniques. At the completion of the surgery the nasal turbinates are simply re-aligned with no other closure needed. Nasal packing is rarely needed. Typical hospital stays are 1-2 days, and most people resume work and other normal activities within 2 weeks of surgery.

The endoscopic endonasal approach satisfies all of the requirements and demands of micro-neurosurgery including wide access to all anatomy, visualization of lateral structures such as the cavernous sinuses not easily seen through a microscope, and the ability to reach all structures for dissection. However, present day endoscopes provide a 2-dimensional non-stereoscopic view, as compared to the three-dimensional view of the microscope. Adjusting to this lack of stereoscopy is initially difficult for many neurosurgeons, although this limitation is rapidly overcome; an experienced echoed by general surgeons using laparoscopy and thoracoscopy. Second, utilization of the endoscope requires the development of a unique skill set, including frequent cleaning of the lens, and orientation to the endoscopic anatomy. Finally, clouding of the endoscope with debris can result in sub-optimal image quality. The addition of HD technology, brighter light sources, and endoscope cleaning devices has greatly minimized these issues. Thus, with a bit of patience and a commitment to the development of endoscopic skills, most neurosurgeons that convert to the endoscopic approach stick to it with fervor. At our institution we have performed endoscopic transsphenoidal surgery on over 300 patients. The endoscopic approach was abandoned on only 2 cases early in our experience, and none in the past 4 years.

Reported outcomes with fully endoscopic pituitary surgeon are encouraging. These reports generally indicate shorter hospital stays, greater patient post-operative comfort (especially in the absence of nasal packing), a complication rate similar to that observed with microscope based surgery, and surgical outcome that equal or surpass microscope results in terms of extent of resection or hormonal cure[5, 8]. Further, the wide field of view provided by the endoscope has resulted in dramatic expansion in the use of the TS as an approach to more deeply seated midline intracranial lesions such as craniopharyngioma, meningioma, or chordoma[9]. Until recently these lesions were almost exclusively treated via a trasncranial or complex skull base approach. This concept is rapidly changing, with the advent of techniques for transnasal repair of the skull base. The development of surgical technologies such as endoscopic lasers and tissue emulsifiers is rapidly evolving, leading to better results. Finally, the addition of true stereoscopic endoscopies may overcome one of the final barriers to endoscopy.

The application of endoscopy to human pituitary surgery is spilling over to veterinary medicine. Pituitary dependent hyperadrenocorticism (PDH) caused by pituitary tumors is extremely common in canines. Surgical management of this disease via the TS is the most effective treatment, resulting in cure or remission in 75-85% of dogs. Meij and colleagues in the Netherlands have developed and extensively

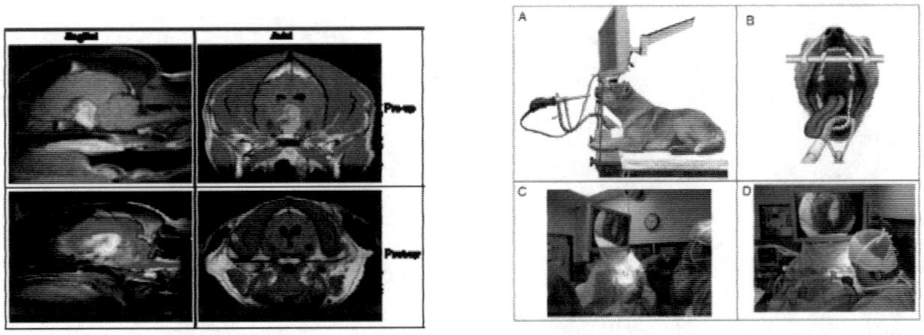

Figure 2: Left panel- Pre- and post-operative MRI scans of a dog with PDH that underwent removal using the technique described. Right panel- Surgical set up. In A&B) dog is placed in open-mouthed sternal recumbancy with the telescope placed in front of the mouth and images projected on an overhead HD screen. In C &D) surgeon operates from an image

reported on a trans-oral, transsphenoidal approach to these tumors with a mortality rate of approximately 8%[10, 11]. However, their results have not been adopted on a widespread basis by veterinary surgeons, likely due to a lack of a means to adequately magnify and illuminate the operative field, resulting in a technically demanding procedure. Our group, a collaboration between human neurosurgeons, veterinary endocrinologists and veterinary surgeons, has modified the surgical technique and introduced telescope based illumination and magnification to improve the safety and efficacy of these procedures, while potentially making the procedure accessible to a larger number of veterinary surgeons. The operative field is visualized using a novel rigid telescope and high definition camera system (VITOM™, Karl Storz Endoscopy, Germany) that provides up to 40 x magnification, excellent illumination, and HD quality (1080p) resolution of the surgical anatomy[12]. The scope looks identical to an endoscope, but has a focal length of 250mm, and therefore sits away from the operative field. This working distance provides magnification and illumination of the operative field without restricting the introduction of surgical instruments. The HD monitor is placed directly in front of the surgeon and just above the dog's head, with the surgical dissection carried out from the monitor (fig 2). This allows all surgeons and operating room personnel to visualize the procedure with equal clarity. To date we have performed this procedure in over 40 cadaver canines and 9 live animals, with very promising results. A trial to determine the safety and efficacy of this method is underway. It is our hope that with the addition of the endoscope based technology, the surgery will become much more accessible to veterinary patients.

Endoscopic pituitary surgery is rapidly becoming the gold standard method in humans, in large part due to the many advantages it brings over traditional microscope based methods. Translation of these tools into the veterinary setting is underway. Long focal distance endoscopes (or "exoscopes") are currently being utilized to refine the technique of transsphenoidal surgery for pituitary masses in dogs, with promising early results. These methods are likely to further develop and play a significant role in veterinary surgery in the next decade.

REFERENCES

[1.] Rosegay H. Cushing's legacy to transsphenoidal surgery. **J Neurosurg.** Apr 1981;54(4):448-454. [2.] Liu JK, Das K, Weiss MH, Laws ER, Jr., Couldwell WT. The history and evolution of transsphenoidal surgery. **J Neurosurg.** Dec 2001;95(6):1083-1096. [3] Dew LA, Haller JR, Major S. Transnasal transsphenoidal hypophysectomy: choice of approach for the otolaryngologist. **Otolaryngol Head Neck Surg.** Jun 1999;120(6):824-827. [4.] Nabe-Nielsen J. Nasal complication after transsphenoidal surgery for pituitary pathologies. **Acta Neurochir (Wien)**. 1989;96(3-4):122-125. [5.] Sanai N, Quinones-Hinojosa A, Narvid J, Kunwar S. Safety and efficacy of the direct endonasal transsphenoidal approach for challenging sellar tumors. **J Neurooncol**. May 2008;87(3):317-325. [6.] Spencer WR, Das K, Nwagu C, et al. Approaches to the sellar and parasellar region: anatomic comparison of the microscope versus endoscope. **Laryngoscope**. May 1999;109(5):791-794. [7.] Snyderman CH, Pant H, Carrau RL, Prevedello D, Gardner P, Kassam AB. What are the limits of endoscopic sinus surgery?: the expanded endonasal approach to the skull base. **Keio J Med**. Sep 2009;58(3):152-160. [8.] Jho HD, Carrau RL. Endoscopic endonasal transsphenoidal surgery: experience with 50 patients. **J Neurosurg**. Jul 1997;87(1):44-51. [9.] Gardner PA, Kassam AB, Snyderman CH, et al. Outcomes following endoscopic, expanded endonasal resection of suprasellar craniopharyngiomas: a case series. **J Neurosurg**. Jul 2008;109(1):6-16. [10.] Meij BP, Voorhout G, van den Ingh TS, Hazewinkel HA, Teske E, Rijnberk A. Results of transsphenoidal hypophysectomy in 52 dogs with pituitary-dependent hyperadrenocorticism. **Vet Surg.** May-Jun 1998;27(3):246-261. [11.] Meij BP, Voorhout G, Van den Ingh TS, Hazewinkel HA, Van't Verlaat JW. Transsphenoidal hypophysectomy in beagle dogs: evaluation of a microsurgical technique. **Vet Surg**. Jul-Aug 1997;26(4):295-309. [12.] Mamelak AN, Danielpour M, Black KL, Hagike M, Berci G. A high-definition exoscope system for neurosurgery and other microsurgical disciplines: preliminary report. **Surg Innov**. Mar 2008;15(1):38-46.

EQUINE MYOPATHIES: IT'S NOT JUST TYING UP ANYMORE

Stephanie Valberg DVM PhD, Diplomate ACVIM
St Paul MN

Research has provided a great deal of new information regarding specific causes of myopathies in horses. This presentation will provide a review of specific equine myopathies organized according to their etiopathology; metabolic myopathies; disorders of sarcolemma and excitation contraction coupling; and immune-mediated myositis.

METABOLIC MYOPATHIES

Two specific genetic mutations associated with glycogen storage disorders have recently been identified, glycogen branching enzyme deficiency[1] which affects developing feti and foals and polysaccharide storage myopathy type 1 [2] which affects adult horses primarily. A second form of PSSM [3] exists, however, the genetic basis for this form of PSSM is not known.

GLYCOGEN BRANCHING ENZYME DEFICIENCY (GBED)

Breeds: Quarter horse-related bloodlines, particularly descendants from stallions Zantanon and King.
Inheritance: Autosomal recessive
Prevalence: In general 8% of the Quarter Horse breed are carriers, however, this increases to 28% within Western pleasure horses.[4] About 3% of abortions may be due to GBED.[5]
Age affected: Signs usually present in utero or at birth.
Clinical signs: Abortion or stillbirth are common, however, foals may be born alive and are weak at birth. With supportive care they may live to up to 18 weeks of age. Death may be sudden when exercised on pasture, associated with weak respiratory muscles or the result of euthanasia due to persistent recumbency. Treatable flexural deformities of all limbs and recurrent hypoglycemia and seizures occur in some affected foals.
Treatment: none
Mutation: A point mutation in exon 1 changes a tyrosine to a premature stop codon in the glycogen branching enzyme gene (*GBE1*), which is expressed in numerous tissues.[6]
Diagnosis: Samples of muscle and heart stained for Periodic acid Schiff's (PAS) show a variable amount of abnormal PAS positive globular and crystalline intracellular inclusions. Genetic testing is done by Veterinary Genetics Laboratory at the University of California, Davis or Vetgen in Michigan on mane or tail hair roots.

POLYSACCHARIDE STORAGE MYOPATHY TYPE 1 (PSSM1)

Breeds: > 20 breeds including Quarter horse-related, Belgians, Percherons, Morgans, Mustangs and some Warmbloods. Likely dates back 1200 years to the "Great Horse".[2,7]
Inheritance: Autosomal dominant
Prevalence: 8% of the Quarter Horse related breeds, 30% of halter Quarter Horses and 36-50% of Belgians and Percherons are affected.[4]
Age affected: Signs usually begin by 2 to 3 years of age but may occur in weanlings. A proportion of horses are subclinical.
Clinical signs: Firm painful muscles, stiffness, fasciculations, sweating, weakness and reluctance to move with light exercise. Occasionally gait abnormalities, mild colic and muscle wasting are seen. Serum CK and AST activity are usually elevated, however this may not be the case in Draft breeds.
Treatment: Diet is modified to minimize soluble starch and sugar and provide 15% of daily digestible energy in the form of fat. This is combined with regular daily exercise and turn out.[8]
Mutation: A gain of function point mutation in the *GYS1* gene results in an arginine to histidine substitution in the skeletal muscle glycogen synthase enzyme.[2]
Diagnosis: Muscle biopsy samples can be evaluated for the presence of amylase-resistant PAS positive polysaccharide. A more specific diagnosis can be made through genetic testing on mane or tail hair roots, or unclotted blood samples. Testing is performed at the University of Minnesota. http://www.vdl.umn.edu/vdl/ourservices/neuromuscular.html

POLYSACCHARIDE STORAGE MYOPATHY TYPE 2 (PSSM2)

Breeds: Quarter Horse-related breeds, a few Arabians and possibly other light breeds.[7]
Inheritance: Unknown.
Age affected: Signs usually begin by 2 to 3 years of age but may occur in weanlings.
Clinical signs: Rhabdomyolysis with or without exercise.
Treatment: Diet is modified to minimize soluble starch and sugar and provide 15% of daily digestible energy in the form of fat. This is combined with regular daily exercise and turn out.
Mutation: Unknown.
Diagnosis: Muscle biopsy samples are evaluated for presence of amylase sensitive and/or resistant polysaccharide at the Neuromuscular Diagnostic Laboratory at the University of Minnesota.[3]

One lipid storage disorder has been described in horses due to multiple acyl CoA dehydrogenase deficiency.[9] The term atypical myopathy, atypical myoglobinuria and pasture myopathy[10] have been used previously for this disorder.

MULTIPLE ACYL CoA DEHYDROGENASE DEFICIENCY (MADD)

Breeds: A variety of breeds are affected.

Inheritance: Appears to be an acquired disorder

Prevalence: Occurs seasonally in pastured horses in the spring and fall particularly when wet windy weather is preceded by drought. The disorder has been reported in northern Europe, the United Kingdom and the Midwestern United States.[11]

Age affected: A wide range of ages are affected, young horses may be more susceptible.

Clinical signs: Sudden onset of weakness, muscle fasciculations, recumbency, elevated respiratory rate, heart rate, pulmonary edema and/or esophageal choke. Marked elevations in serum CK and AST activity are present.

Treatment: Fluid therapy to prevent renal failure, antioxidants such as vitamin E, vitamin C and dimethylsulfoxide as well as ribloflavin are recommended. Prognosis is poor.

Etiology: An unknown toxin potentially in the soil or in plants appears to inhibit beta oxidation of lipids

Diagnosis: Urine organic acid profiles show marked elevations in lactic acid, ethylmalonic acid, 2-methylsuccinic acid, butyrylglycine (iso)valerylglycine, and hexanoylglycine. Plasma acylcarnitine profiles are also abnormal. Frozen sections of muscle biopsies of postural and respiratory muscles show acute myodegeneration and excessive accumulation of lipid droplets in the majority of myofibers.

There is one report in the literature of a mitochondrial myopathy in an Arabian mare that caused marked exercise intolerance and lactic acidosis with light exercise. The defect was due to a deficiency in Complex I of the respiratory chain.[12] No pattern of inheritance or genetic mutation was identified.

DISORDERS OF SARCOLEMMAL DEPOLARIZATION AND EXCITATION CONTRACTION COUPLING

Several disorders of muscle cell depolarization and muscle cell contraction exist in horses. The genetic mutations for two of these (HyPP and MH) have been identified and the chromosomal locations of causative genes for recurrent exertional rhabdomyolysis may have been identified. The specific defect causing myotonia is unknown in the horse.

HYPERKALEMIC PERIODIC PARALYSIS (HyPP)

Breeds: Quarter Horse-related bloodlines that are descendant from the stallion Impressive.

Inheritance: Autosomal dominant[13]

Prevalence: In general 3% of the Quarter Horse breed is affected, however, in halter horses 60% are affected.

Age affected: Signs usually begin by 2 to 3 years of age.

Clinical signs: Vary from asymptomatic to intermittent myotonia, third eyelid prolapse, muscle tremors, weakness to temporary paralysis. Horses homozygous for HyPP may present with difficulty swallowing or respiratory distress.

Treatment: Intravenous administration of calcium gluconate slowly usually stops episodes. Prevention involves a diet with < 1% potassium and if necessary acetazolamide.

Mutation: A point mutation in the sodium channel gene *SCNA4* causes a phenylalanine/leucine substitution in a key part of the voltage-dependent skeletal muscle sodium channel alpha subunit which controls channel activity.[13]

Diagnosis: Veterinary Genetics Laboratory at the University of California, Davis on mane or tail hair roots.

MALIGNANT HYPERTHERMIA (MH)

Breeds: Quarter horse-related bloodlines[14]

Bloodlines: High frequency in some halter and western pleasure horse families many of which also segregate for PSSM[15]

Inheritance: Autosomal dominant.

Prevalence: < 1% of the Quarter Horse breed is affected.

Age affected: Adults

Clinical signs: High temperature, metabolic failure and death under anesthesia. Tying up and fever may occur alone or if horses concurrently have PSSM1 episodes of rhabdomyolysis can be very severe. Sudden death is reported.

Treatment: Oral dantrolene sodium may be preventative, however, an affordable intravenous form is not available to provide to horses during a fulminant episode.

Mutation: Point mutation in the *RYR1* gene that results in an arginine to glycine substitution in the calcium release channel.[14]

Diagnosis: Genetic testing at Neuromuscular Diagnostic Laboratory at the University of Minnesota.

RECURRENT EXERTIONAL RHABDOMYOLYSIS

Breeds: Thoroughbreds and potentially Standardbreds, Arabians, Warmbloods, racing Quarter Horses worldwide

Prevalence: 5-10 % of racing Thoroughbreds develop clinical signs. Much more prevalent amongst 2 and 3 year old fillies especially if they have a nervous temperament.[16]

Age affected: Onset when horses are physically fit and in an environment where they are nervous/excitable. Adult riding horses usually display persistent muscle soreness and infrequent clinical episodes of exertional rhabdomyolysis

Clinical signs: Muscle pain, cramping, sweating, stiffness with exercise often in combination with excitement or a few days off prior to exercise.

Treatment: The disorder can be managed by decreasing excitement, feeding a low starch high fat diet, regular exercise and potentially dantrolene provided 60-90 minutes before exercise.[17]
Inheritance: Potentially autosomal dominant but there may be more than one gene influencing expression.
Mutation: Unknown. Two potential chromosomal loci associated with RER in Thoroughbreds have been located by genome mapping.
Diagnosis: Currently based on clinical signs and ruling out other myopathies through muscle biopsy.

MYOTONIA
Breeds: Relatively rare but found in a wide variety of breeds and no known heritable basis has been established.
Clinical signs: Well developed musculature is apparent within weeks of birth. In some animals, nonprogressive muscle hypertrophy and myotonic dimpling and stiffness that improves with exercise is observed. In dystrophic cases, muscle hypertrophy and mytotonic dimpling progresses to focal muscle atrophy and stiffness which worsens with exercise.
Treatment: Signs may improve with phenytoin but it is not a practical long term solution
Diagnosis: Currently based on identification of myotonic discharges on electromyography. In dystrophic cases, biopsies of affected muscles contain myofibers with numerous centrally located nuclei, muscle atrophy, fiber type grouping, ring fibers, sarcoplasmic masses and fibrosis.

INFLAMMATORY MYOPATHIES

Horses develop inflammatory myopathies in association with infarctive purpura hemorrhagica and immune mediated myositis.

IMMUNE MEDIATED MYOSITIS (IMM)
Breeds: Primarily Quarter Horse related breeds especially western pleasure and cutting horses.
Age of onset: Bimodal distribution of either less than 8 or over 17 years of age. In about 40% of cases the onset is associated with a respiratory infection. Vaccinations have been suggested, but not proven, to trigger muscle wasting.[18]
Clinical signs: Rapid onset of atrophy of gluteal and lumbar muscles with progression to involve neck and proximal forelimb muscles. Weakness and malaise occur. Without treatment muscle mass gradually returns over months.
Treatment: Corticosteroids
Diagnosis: Currently based on identification of macrophages and CD4+ T lymphocytic infiltrates in myofibers and blood vessels within biopsies of epaxial or gluteal muscles. Lesser numbers of CD8+ cells and small clusters of B lymphocytes are seen and myofibers do not stain for equine IgG.

REFERENCES

[1.] S.J.Valberg, T.L.Ward, B.Rush, H.Kinde, H.Hiraragi, D.Nahey, J.Fyfe and J.R.Mickelson, **J.Vet.Intern.Med.** 15 (2001) 572-580. [2.] M.E.McCue, S.J.Valberg, M.B.Miller, C.Wade, S.DiMauro, H.O.Akman and J.R.Mickelson, **Genomics** 91 (2008) 458-466. [3.] M.McCue, A.Armien, M.Lucio, J.Mickelson and S.Valberg, **Vet Pathol.** (2009). [4.] R.C.Tryon, M.C.Penedo, M.E.McCue, S.J.Valberg, J.R.Mickelson, T.R.Famula, M.L.Wagner, M.Jackson, M.J.Hamilton, S.Nooteboom and D.L.Bannasch, **J Am Vet Med Assoc.** 234 (2009) 120-125. [5.] M.L.Wagner, S.J.Valberg, E.G.Ames, M.M.Bauer, J.A.Wiseman, M.C.Penedo, H.Kinde, B.Abbitt and J.R.Mickelson, **J.Vet.Intern.Med**. 20 (2006) 1207-1211. [6.] T.L.Ward, S.J.Valberg, D.L.Adelson, C.A.Abbey, M.M.Binns and J.R.Mickelson, **Mamm.Genome** 15 (2004) 570-577. [7.] M.E.McCue, S.J.Valberg, M.Lucio and J.R.Mickelson, **J Vet Intern.Med** (2008). [8.] W.P.Ribeiro, S.J.Valberg, J.D.Pagan and B.E.Gustavsson, **J.Vet.Intern.Med.** 18 (2004) 887-894. [9.] C.M.Westermann, L.Dorland, D.M.Votion, de Sain-van der Velden MG, I.D.Wijnberg, R.J.Wanders, W.G.Spliet, N.Testerink, R.Berger, J.P.Ruiter and J.H.van der Kolk, **Neuromuscul.Disord.** 18 (2008) 355-364. [10.] C.J.Finno, S.J.Valberg, A.Wunschmann and M.J.Murphy, **J.Am.Vet.Med.Assoc.** 229 (2006) 1134-1141. [11.] D.M.Votion and D.Serteyn, **Vet J** 178 (2008) 185-190. [12.] S.J.Valberg, G.P.Carlson, G.H.Cardinet, III, E.K.Birks, J.H.Jones, A.Chomyn and S.DiMauro, **Muscle Nerve** 17 (1994) 305-312. [13.] J.A.Rudolph, S.J.Spier, G.Byrns, C.V.Rojas, D.Bernoco and E.P.Hoffman, **Nat.Genet.** 2 (1992) 144-147. [14.] M.Aleman, J.E.Nieto and K.G.Magdesian, **J Vet Intern.Med** 23 (2009) 329-334. [15.] M.E.McCue, S.J.Valberg, M.Jackson, L.Borgia, M.Lucio and J.R.Mickelson, **Neuromuscul.Disord.** 19 (2009) 37-43. [16.] J.M.MacLeay, S.A.Sorum, S.J.Valberg, W.E.Marsh and M.D.Sorum, **Am.J.Vet.Res.** 60 (1999) 1562-1566. [17.] E.C.McKenzie, S.J.Valberg, S.M.Godden, C.J.Finno and M.J.Murphy, **Am.J.Vet.Res.** 65 (2004) 74-79. [18.] S.S.Lewis, S.J.Valberg and I.L.Nielsen, **J Vet Intern.Med.** 21 (2007) 495-503.

CLINICAL & MOLECULAR ASPECTS OF MALIGNANT HYPERTHERMIA IN THE HORSE

Monica Aleman, MVZ, PhD, DACVIM
Davis, CA

INTRODUCTION

Malignant hyperthermia (MH) is a potentially fatal pharmacogenetic disorder described in humans, pigs, dogs and horses.[1-5] Naturally occurring MH-like episodes have also been documented in wildlife, cats and cattle. Descriptions of clinical manifestations of MH in people date to the early 1900's. An autosomal dominant predisposition to the disorder triggered by anesthetics was first reported by Denborough in 1960.[1] Since the introduction of inhalation anesthesia in horses in the 1960's, cases have been suspected but seldom reported. The disorder was first documented in the horse in 1975 and since then, MH or MH-like episodes have been described in Quarter Horses, Paints, Appaloosas, Thoroughbreds, Arabians, and Pony breeds.[6,7]

Malignant hyperthermia can be triggered in susceptible individuals by halogenated anesthetics, depolarizing muscle relaxants, and stress.[1] Mutations in the ryanodine receptor 1, also known as the calcium release channel of the sarcoplasmic reticulum of the skeletal muscle, have been implicated in over 50% of cases of MH in people and in all cases in pigs, dogs and horses.[1-5] This receptor plays a crucial role in skeletal muscle excitation-contraction coupling. The sarcoplasmic reticulum is the major source of intracellular calcium in skeletal muscle.[8,9] Alterations in the structure and function of the receptor will result in the excessive release of calcium from the sarcoplasmic reticulum with subsequent triggering of hypermetabolic events that can result in death.[1]

RYANODINE RECEPTOR

The ryanodine receptors (RyRs) are transmembrane calcium-release channels that bind specifically the plant alkaloid ryanodine, which is the reason for its name.[8] Mammalian tissues express 3 isoforms of ryanodine receptors that are structurally related but possess distinct functional and regulation properties and include the following: skeletal muscle ryanodine receptor (sRyR or RyR1), cardiac muscle (cRyR or RyR2), and brain isoform (bRyR or RyR3).[8] However, these isoforms are not exclusive of these tissues.[8,9] Mammalian skeletal muscle also contains a small amount of RyR3 which concentration varies among different muscles.

The ryanodine receptor 1 is a 2.3 MDa homotetramer with a central channel for the release of calcium. The N-terminus of the tetramer forms a large foot region that projects to the sarcoplasm and contains most of the ligand binding sites of the receptor; while the C-terminus forms the transmembrane channel. Additional proteins encoded by different genes are bound to this large and complex receptor. The RyR1 is highly concentrated at the triad junctions between the sarcoplasmic reticulum terminal cisternae and sarcolemmal T-tubules.[8,9] These specialized domains contain a key receptor in the T-tubule membrane called the dihydropyridine receptor (DHPR) which is an L-type voltage-sensitive calcium channel.[8,9] Upon membrane depolarization and the presence of calcium, the DHPRs undergo conformational changes and through protein to protein interactions with the RyRs, the RyRs open its channel and release calcium during excitation-contraction coupling.

Endogenous and exogenous modulators can alter the function of the ryanodine receptor.[8] Examples of endogenous modulators include ATP, Ca^{2+}, Mg^{2+}, cADP ribose, nitric oxide, and calmodulin among others. The activating modulators include μM concentrations of Ca^{2+}, calcium-depleted calmodulin, and ATP. Among the endogenous modulators, Ca^{2+} is considered the physiological activator because other ligands cannot activate the channel in its absence or may require Ca^{2+} for maximal effect.[8] The inhibitory modulators include mM concentrations of Ca^{2+}, calcium-saturated calmodulin, and Mg^{2+}.[8] Exogenous modulators include drugs such as halogenated anesthetics (mainly halothane), depolarizing muscle relaxants (succinylcholine), and others (i.e. 4-chloro-m-cresol; 3,4-methylenedioxymethamphetamine ["ecstasy"]).[8]

CLINICAL SIGNS AND TRIGGERING EVENTS

Clinical manifestations of MH are the result of the increased intracellular aerobic and anaerobic metabolism.[1] A marked increase in muscle metabolism results in rapid increase in body temperature that may exceed 43°C (109.4°F). Affected species develop hyperthermia, hyperhidrosis, tachycardia, dysrhythmias, tachypnea, hypertension, muscle rigidity, rhabdomyolysis, myoglobinuria, and death.[1,4,10] However, there may be differences in the severity of signs among species with humans, pigs and horses showing more profound signs than those reported in dogs.[1,3,4,10] Affected breeds of pigs include Duroc, Landrace, Petrain, Poland China, and Yorkshire.[3] In dogs, MH and MH-like events have been described in Pointers, Greyhounds, mixed-breed Doberman Pinscher-German Shepherds, and Labrador retrievers.[4] Horses with the *RyR1* gene mutation include Quarter Horses and Paints.[10] However, the disorder has been suspected in other breeds as previously mentioned. There is no apparent gender or age predisposition in horses. Horses may appear normal in the absence of triggering events or may have intermittent signs of hyperthermia, rhabdomyolysis, stiffness, and sweating.[10] Exposure of MH-susceptible individuals to triggering agents may not always lead to an MH crisis. The prevalence of MH in the horse appears to be low (< 1% in the Quarter Horse population).

Both, anesthetic- and nonanesthetic-triggered MH have been reported in susceptible individuals (humans, pigs, dogs and horses).[10] All halogenated anesthetics (halothane, isoflurane, desflurane, and sevoflurane) can induce MH in susceptible humans and pigs, but most cases have been associated with halothane. Studies in mutated pigs have shown that halothane exposure results in the fastest onset of an MH crisis followed by isoflurane and desflurane.[11] Fatal MH associated with halothane and isoflurane has been documented in horses.[10]

Nonanesthetic-triggered MH has been associated with physical and environmental stress and exertion in people, though less common than the anesthetic form. Nonanesthetic form of MH such as porcine stress syndrome (PSS) and pale soft exudative (PSE) muscle has been documented in pigs. This form can be induced by stress from transportation, unfamiliar environment, fights, exertion, weaning, coitus, heat or extreme cold, branding, and preslaughter practices. In horses, clinical manifestations of nonanesthetic-associated MH may include rhabdomyolysis, "colic-like" episodes, muscle rigidity, stiffness, intermittent hyperthermia and sweating.[10] These signs may be induced or associated with exercise, stress, breeding, and concurrent illness or myopathy.[10] More recently, it was shown that Quarter Horses and Paints with polysaccharide storage myopathy (PSSM) type 1 caused by a glycogen synthase 1 (*GYS1*) gene mutation had a more severe clinical phenotype occasionally resulting in death, if a concurrent *RyR1* gene mutation was present.[12]

Exogenous triggers such as the administration of succynilcholine and caffeine in the research setting, consumption of recreational drugs (ecstasy), and nerve stimulation studies, have induced clinical MH.[8,9]

CLINICOPATHOLOGICAL FINDINGS

The most common clinicopathological findings in susceptible species consist of respiratory and metabolic acidemia, hypercapnia, hyperlactatemia, electrolyte derangements (hyperkalemia, hyponatremia, hypocalcemia or hypercalcemia, hyperphosphatemia), increased serum creatine kinase (CK) activity, and myoglobinuria.[10] Other common findings include azotemia and hyperproteinemia. Horses that die peracutely may have serum CK activity within reference values due to insufficient time for levels to peak. Horses that survive an episode may have serum CK activity within several hundred thousands IU/L. Diseased horses may have intermittent or persistent elevations of muscle enzymes (CK 600-36,000 IU/L, reference range 119-287 IU/L; aminotransferase [AST] within reference range to a few thousands, reference range 168-494 IU/L).[10]

Histochemical evaluation of skeletal muscle specimens from MH susceptible species does not show specific alterations. Common but nonspecific histochemical features of muscle samples collected during fatal episodes in horses include multifocal interfascicular and myofiber edema, myonecrosis, hypercontracted fibers, glycogen depletion, and decreased myophosphorylase and oxidative activities.[10]

DIAGNOSIS

A presumptive diagnosis of MH has been based on the development of hypercapnia, acidosis, hyperthermia, profuse sweating, tachycardia, tachypnea, and muscle rigidity triggered by anesthetic and nonanesthetic events. Because clinical manifestations of MH are triggered such as in the case of anesthesia, diagnosis before anesthesia is difficult. Six different genetic loci have been identified to confer MH susceptibility in people. One of the loci includes the *RyR1* gene which accounts for more than 50% of the reported cases Worldwide. Specific mutations in the DHPR (*CACNA1SA*) gene also confer MH susceptibility in people. In addition, the number of different mutations within the *RyR1* gene makes genetic testing difficult. Therefore, the *in vitro* contracture test (IVCT) is still the gold standard for the diagnosis of MH in humans.[13]

The presence of a single missense point mutation in the *RyR1* gene that results in an amino acid substitution has been associated with the disease in pigs (autosomal-recessive, C1843T→R615C), dogs (autosomal-dominant, T1640C→V547A), and horses (autosomal-dominant, C7360G→R2454G).[3-5] The IVCT has also been used in these species for the diagnosis of MH.[14] The IVCT is based on the differential contractile response of unaffected and MH susceptible skeletal muscle to caffeine and halothane. Skeletal muscle from MH susceptible animals exhibits hypersensitivity to caffeine- and halothane-induced contracture, whereas muscle from unaffected animals is not hypersensitive to either agent. However, this test is not readily available in veterinary medicine, and requires trained personnel, appropriate facilities, standard protocols, and fresh muscle samples.[14] Because of the limited availability of IVCT and presumed presence of a single point mutation in Quarter Horses and related breeds, genetic testing is more suitable for the diagnosis of MH.[15] It is uncertain whether the *RyR1* (C7360G) gene mutation is responsible for all cases of MH in horses or if it is specific to the Quarter Horse and related breeds. Genetic testing is available at the Neuromuscular Disease Laboratory at the University of California at Davis (USDA International permit for importation of samples is available).

TREATMENT AND PREVENTION

Treatment of an MH crisis consists of immediate discontinuation of the triggering agent (i.e. anesthesia), oxygen administration, cold therapy (cooled intravenous fluids, body ice packs, hydrotherapy), and dantrolene sodium administration (Dantrium sodium, Proctor and Gamble Pharmaceuticals, Cincinnati, OH). The administration of intravenous dantrolene sodium is considered the gold standard for the treatment of malignant hyperthermia in human medicine and across species. However, due to expense, dose (unknown for MH crisis), shelf life, and availability; most equine veterinary hospitals do not have injectable dantrolene. If the patient survives, supportive therapy and care must be provided. The resulting rhabdomyolysis and associated complications must be addressed or prevented.

Caution must be practiced in horses with preexisting conditions such as myopathies or with a history of anesthesia-associated myopathy that may increase their risk of anesthesia. Family history may provide relevant information. Recommendations are to avoid anesthesia or use alternative anesthetics in susceptible horses but still exercise caution. Oral administration of dantrolene sodium on an empty stomach, 1.5 to 2 hours prior to anesthesia may be helpful. Bioavailability of dantrolene is approximately 39% after oral administration with highest absorption when administered after at least 3 hours of fasting.

REFERENCES

[1]Denborough MA. **Lancet** 1998;352(9134):1131. [2]McCarthy TV et al. **Nature** 1990;343(6258):562. [3]Fujii J, et al. **Science** 1991;253(5018):448. [4]Roberts MC, et al. **Anesthesiology** 2001;95(3):716. [5]Aleman M, et al. **Muscle Nerve** 2004;30(3):356. [6]Klein L. **Vet Anesth** 1975;2:41. [7]Aleman M et al. **J Vet Intern Med** 2005;19(3):363. [8]Zucchi R et al. **Pharmacol Rev** 1997;49(1):1. [9]Endo M. **Physiol Rev** 2009;89(4):1153. [10]Aleman M et al. **J Vet Intern Med** 2009;23(2):329. [11]Wedel DJ, et al. **Anesthesiology** 1993;78(6):1138. [12]McCue ME, et al. **Neuromuscul Disord** 2009;19(1):37. [13]The European Malignant Hyperpyrexia Group. **Br J Anaesth** 1984;56(11):1267. [14]Hildebrand SV, et al. **J Am Vet Med Assoc** 1990;196(7):1077. [15]Nieto JE, et al. **J Vet Intern Med** 2009; 23(3):619.

APPROACH TO LYSOSOMAL STORAGE DISEASES IN THE DOG: THE NEURONAL CEROID LIPOFUSCINOSES

Dennis O'Brien DVM PhD DACVIM-Neurology
Columbia, MO

Introduction

The lysosomal storage diseases are a heterogeneous group of progressive neurodegenerative diseases. Most are inherited as autosomal recessive traits. While the individual diseases may be uncommon and limited to single breeds, as an aggregate they are not an uncommon cause of neurologic disease in dogs. In this paper, we will review the pathogenesis of these conditions with a focus on a subgroup, the neuronal ceroid lipofuscinoses, as an example of the conditions.

Lysosomes: the cell's recycling centers

Proteins within cells can become non-functional from a variety of processes. Sometimes they will be synthesized abnormally, either as a result of a mutation in the gene for that protein or simply due to an error in synthesis or folding. Once formed, proteins may undergo oxidative damage or become misfolded rendering them non-functional. Such proteins need to be catabolized to remove them from the cytoplasm and recycle the materials involved for synthesis of new proteins. In addition, normal proteins may need to be removed in order to regulate homeostasis within the cell. Both of these processes can be served by the lysosome.

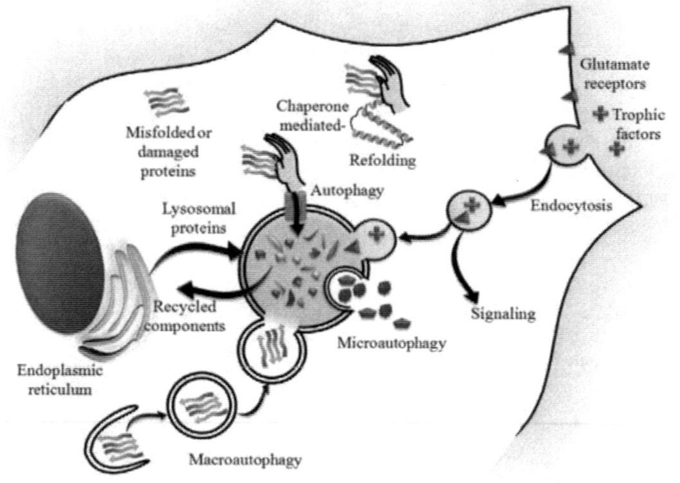

Figure 1: Lysosomal enzymes, ion pumps, and other proteins are manufactured in the endoplasmic reticulum and must be targeted to the lysosome. Large macromolecules are incorporated into autophagasomes which fuse with the lysosome (macroautophagy). Smaller molecules may be taken directly into the lysosome (microautophagy), a process often used to control regulatory signals. Chaperones may refold damaged proteins into correct configuration. If damaged proteins cannot be salvaged, chaperones may unfold them so that they can be degraded through chaperone mediated autophagy via lysosome associated membrane proteins (LAMPs). Endocytosis brings surface receptors such as glutamate receptors or trophic factors into the cell where they either trigger second messenger signals or are sent to the lysosome for recycling.

The lysosome is a membrane bound organelle within the cytoplasm (Figure 1). Enzymes manufactured in the endoplasmic reticulum are transported to the lysosome where the lower pH within the lysosome or specific activators will enable them to degrade complex macromolecules. Keeping these enzymes enclosed prevents damage to functional proteins within the cytoplasm. Other proteins such as ion pumps and membrane channels also must be incorporated into the lysosome to function properly.

The process of degrading cellular components is called autophagy to contrast it with phagocytosis of extracellular components (e.g. virus phagocytized by leukocytes). Autophagy occurs through three main mechanisms: macroautophagy, microautophagy and chaperone mediated autophagy. In macroautophagy, a region of the cytosol will be sequestered within a limiting membrane which fuses to form the autophagosome. Often this region will contain aggregates of malfolded proteins. The autophagosome lacks enzymes and thus must fuse with the lysosome to degrade the contents. Alternatively, smaller portions of the cytosol can be directly incorporated into the lysosome through microautophagy. Microautophagy is more commonly involved in basal cell component turnover and regulating homeostasis by removing components no longer needed. Cell surface proteins such as neurotransmitter receptors and extracellular signaling molecules may be brought into the cell through endocytosis. The endosomes produced may also fuse with the lysosome to degrade the contents thus regulating their availability.

Chaperones are proteins that guide the folding of other proteins during synthesis and help maintain them in the functional conformation, but they can also refold misfolded proteins back into their normal conformation. For proteins that cannot be salvaged, chaperones can assist in their degradation by unfolding the proteins to allow easier access to catalytic sites. These chaperone altered proteins can be fed into the lysosome through lysosomal associated membrane proteins such as LAMP2A.

Lysosomes storage diseases:

The complex macromolecules that make up neurons and myelin are broken down into reusable or excretable fragments through a series of steps. Each step is mediated by a separate enzyme and depends on the preceding step to efficiently catabolize the substrate. A mutation in the gene for one of these enzymes can lead to decreased or absent enzyme activity. The catabolism cannot proceed beyond this roadblock and the substrate immediately upstream accumulates within the lysosome producing a membrane bound storage body (Figure 2). Alternatively other proteins necessary for maintenance of lysosomal pH or membrane proteins may be altered affecting overall lysosomal function. Finally some proteins may be manufactured normally but not targeted to the lysosome appropriately (e.g. the mucolipidioses) or not activated (e.g. GM2 gangliosidosis form AB).

Regardless of the underlying reason, accumulation of materials within a membrane enclosed vesicle within the cytoplasm is a hallmark of the lysosomal storage diseases. The traditional view of their pathogenesis is that the accumulation of the abnormal materials is somehow toxic to the cell. This view is probably an oversimplification of a very complex process. In addition to the physical accumulation of byproduct, the associated dysfunction of the lysosome can have wide-ranging ramifications. The breakdown products of autophagy may be used to synthesize new proteins. This process can be especially important when the animal is in a catabolic state as the need for recycled materials becomes more acute. The role of the lysosome in regulating homeostasis may also be impaired by the dysfunction, leading to altered cell signaling and apoptosis. Though other organs such as viscera, bone and connective tissue may be involved, over 2/3 of the lysosomal storage diseases affect the brain, and some such as the neuronal ceroid lipofuscinoses affect primarily brain. While in some diseases, all neurons or oligodendroglia are affected, other can be selective. For example research by Dr. Phil March has shown that GABA cells are selectively lost in one canine lysosomal storage disease.

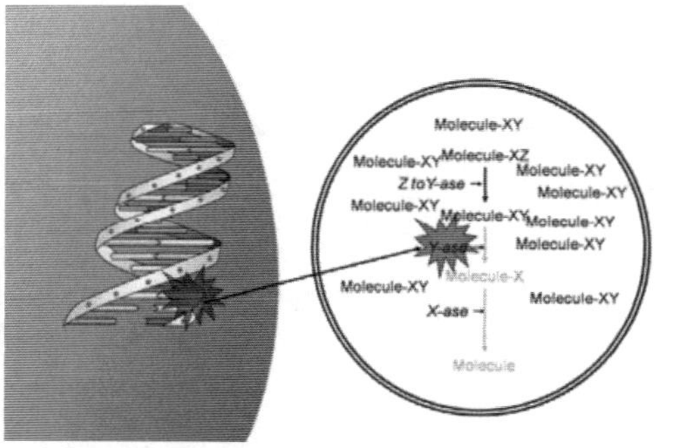

Figure 2: Mutations that lead to a deficiency of a lysosomal enzyme can block a step in the degradation of complex macromolecules. The molecule upstream from the deficit then builds up within the lysosome.

The Neuronal Ceroid Lipofuscinoses (NCLs)

The terms ceroid and lipofuscin refer to lipopigments which autofluoresce with certain spectra of light. Lipofuscin accumulation is common in aging cells. The term ceroid was originally applied to similar pigments in cirrhotic livers, but has subsequently been applied to all autofluorescent inclusions that accumulate for reasons other than aging. The NCLs have the common denominator of autoflourescence makes them readily recognizable as a class of diseases on histopathology. The nature of the accumulated storage product varies with the specific type of NCL. The clinical signs associated with some recently studied NCLs are summarized in table 1; see Jolly 1994 for additional information.

The clinical signs of NCL are variable, but there are some common themes. All the NCLs of dogs have a delayed onset of signs, though neonatal disease has been seen in other species. This can vary from less than a year of age (e.g. CLN1 in dachshunds) to over 5 years of age (e.g. Tibetan terriers). Once signs begin, they are progressive and the disease is typically fatal. Vision loss is a common, but not universal, characteristics of the NCLs. In some cases, such as CLN2 in dachshunds, retinal cell degeneration occurs early in the course of the disease. Sluggish pupillary light reflexes and abnormal ERGs will progress to complete blindness. In the Tibetan terriers, retinal function is spared but cortical vision abnormalities with variable menace response and difficulty tracking objects or negotiating obstacles develop late in the disease. Unrelated progressive retinal atrophy also occurs in Tibetan terriers which could explain the early onset of blindness reported in some of the early reports of NCL in the breed. Other breeds, such as American Staffordshire Terriers and American Bulldogs do not show visual deficits. Progressive ataxia is probably the most consistently reported finding. In the Am Staff the ataxia is primarily cerebellar with dysmetria, balance problems, and downbeat nystagmus, while in the American Bulldog it is primarily a sensory ataxia with proprioception deficits. This can be accompanied by spastic paresis or more rarely diminished reflexes suggesting peripheral nerve involvement. The most disturbing sign of NCL is the progressive dementia that occurs in some breeds such as the Border Collie and Tibetan Terrier. Affected dogs lose their prior training and often appear confused. They can become hyperesthetic to sounds, touch, and movement and will often become withdrawn and fearful. Attempts to pet or handle them can result in aggression and it is not unusual for owners to be bitten by their once friendly pet. When seizures occur, they can either be generalized tonic-clonic seizures or myoclonus, and are often refractory to therapy. Generally speaking the signs will be symmetrical though occasional exceptions occur such as a horizontal nystagmus.

While the constellation of signs within a breed can be fairly characteristic for the NCLs, other diseases could produce similar clinical signs. Thus a complete neurologic workup would be needed to rule out other causes. Brain imaging typically shows non-specific brain atrophy (Figure 3). Since storage material is confined to the nervous system, brain biopsy would be necessary for antemortem confirmation of NCL, but with the increasing availability of DNA testing, many breeds can be genotyped to assess the relative risk of NCL for that animal.

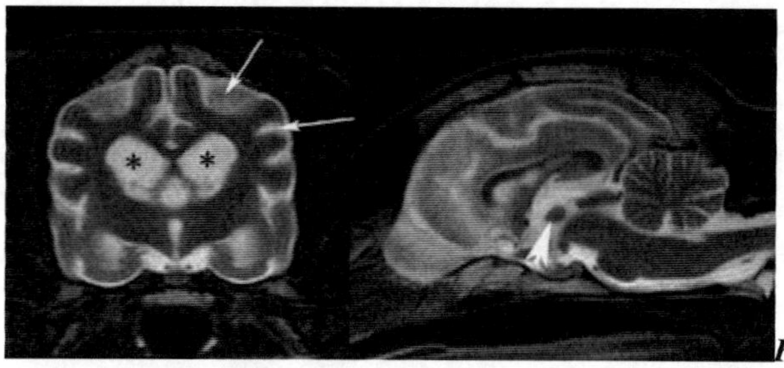

***Figure 3:** Brain atrophy can be seen as enlargement of the ventricles (asterisks), prominent sulci (arrows), and a small intrathalamic adhesion (arrowhead) in this T2 weighted MRI of a Tibetan terrier with NCL*

Genetic tests for NCLs

The mutations responsible for the NCLs in several breeds have been identified (See Table 1) through either candidate gene or gene mapping approaches (see O'Brien 2006 for further discussion of these approaches). Where the mutation has been well characterized and correlated with biochemical abnormalities, the sensitivity of the tests for detecting the mutation is virtually 100%. In diseases with a delayed onset, however, a dog could be homozygous for the mutant allele, but still be afflicted with a different disease. Thus other potential differential diagnoses need to be entertained, but the risk of a dog who carriers the mutant alleles of developing the disease in the future also needs to be weighed into the clinical decision making process. DNA testing is most valuable for breeders. By testing breeding stock, breeders can use wise breeding strategies to avoid producing affected dogs and decrease the frequency of the mutant allele in the breed. Often such strategies continue to utilize carriers as breeding animals to avoid further genetic bottlenecks. When NCL is diagnosed in breeds where the gene has not yet been identified, DNA samples should be collected to permit identification of the gene in the future.

Therapy: hope for the future

Currently there is no effective therapy for NCLs. Symptomatic therapy such as anticonvulsants, antianxiety drugs, and behavioral modification can help alleviate signs, but these diseases are inevitably progressive. Often helping the owner decide when to elect euthanasia in a dog who is undergoing a slowly progressive dementing disease is the most challenging aspect of these diseases. Some of the enzymes involved are soluble and can be taken up from interstitial fluids by neurons. Thus gene therapy would not necessarily need to transfect every neuron to produce improvement and enzyme replacement into CSF could potentially deliver enzyme to neurons. Trials are under way at several institutions to develop such therapies in dogs. Such approaches will inevitably be applied to human medicine first, but when the therapies have been developed in the dog, they should be more readily available for veterinary applications than if they had only been studied in laboratory rodents.

Table 1: Characteristics of some recently investigated Neuronal Ceroid Lipofuscinoses

Breed	Gene	Onset & progression	Vision	Cognitive	Aggression	Ataxia	Cerebellar signs	Seizures	Other
American Staffordshire Terrier	In press	1.5-9 years Protracted	Normal			Yes	**Hypermetria, nystagmus, imbalance**	No	
American Bulldogs	Cathepsin D	0.9-3 years Protracted	Normal			**Yes** CP loss	Hypermetria,	No	**Paraparesis** → non-ambulatory tetraparesis
Australian Shepherd	Unknown	18 months	Blind	Behavior changes, hyperesthesia, anxiety		Yes	Hypermetria	Fly-biting	Paresis
Border Collies	CLN5	15-23 months	Blind	**Behavior changes**, hyperesthesia, anxiety	Yes	Yes		Possible	
Dachshund	TPP1 (CLN2)	7-9 months Acute	Blind	Loss of training	Yes	Yes	Hypermetria	Myoclonus	
Dachshund	In press	9 months Acute	Blind	Behavior changes, hyperesthesia, anxiety, confusion		Yes			Persistent head movements
English Setters	CLN8	1-1.5 years	Blind	Mentally dull		Yes		Yes	
Labrador Retriever	Unknown	7 years	Normal			Yes	Hypermetria	Yes	Tetraparesis
Polish Owczarek Nizinny (PON)	Unknown	0.5-4 years Protracted	Blind	Nervousness, confusion	Yes	Yes		No	
Tibetan Terrier	In Press	4-8 years Protracted	Central deficits	**Behavior changes, hyperesthesia,** anxiety	**Yes**	Yes		Yes	Rare LMN signs, nystagmus

REFERENCES

1. S. U. Walkley. **J Inher Metab Dis** 32 (2):181-189, 2009; 2. Olby NJ et al **Proc ACVIM Forum** 27:304, 2009; 3. Abitbol M et al **Proc ACVIM Forum** 27:305, 2009; 4. Mole SE et al ed. The neuronal ceroid lipofuscinoses (Batten disease). 2nd ed. **Biomedical and Health Research**. Amsterdam: IOS Press, 2010: 5. S. A. Melville et al. **Genomics** 86 (3):287-94, 2005; 6. T. Awano et al. **Molecular Genetics and Metabolism** 87 (4):341-8, 2006: 7. T. Awano et al. **Molecular Genetics & Metabolism** 89 (3):254-260, 2006; 8. M. L. Katz et al. **Biochemical & Biophysical Research Communications** 327 (2):541-7, 2005; 9. R. D. Jolly et al. **Journal of Small Animal Practice**. 35 (6):299-306, 1994; 10. Mitchison HM et al. **Brain Pathology** 2004; 14: 86-96; 11. D. P. O'Brien and M. L. Katz **J Vet Int Med** 22 (2):472-475, 2008; 12. D. P. O'Brien, **ACVIM Proceedings**, Louiseville KY:308-310, 2006; 13. J. H. Rossmeisl et al. **Journal of Veterinary Diagnostic Investigation** 15 (5):457-460, 2003

ACUTE LOWER MOTOR NEURON PARALYSIS

Andrea Fischer, DVM, DrMedVet, DrHabil, DECVN, DACVIM
Munich, Germany

INTRODUCTION

The lower motor neuron comprises the motor neuron in the spinal cord, the ventral nerve root and the peripheral motor nerve. Neuromuscular endplates and muscle fibres form a functional unit (motor unit) with the lower motor neuron. Thus patients with acute Motoneuron, ventral nerve root, peripheral nerve, neuromuscular endplate or muscle disorders show identical clinical signs characterized by loss of motor function, muscle strength and spinal reflexes (e. g. LMN paralysis) but normal sensation and normal mental status. In people, acute lower motor neuron paralysis is covered by the term acute motor neuropathy. In clinical veterinary medicine, acute lower motor neuron paralysis is typically seen with acute polyradiculoneuritis, which may occur with or without exposure to racoon saliva, tick paralysis, botulism and myasthenic crisis. Time from onset to peak is typically less than two weeks in polyradiculitis and may be as short as one day in the others. From a VITAMIN D / DAMNIT perspective, most diseases fall into the inflammatory/immune mediated and metabolic-toxic disease category. Fortunately, acute LMN paralysis is often associated with a good prognosis despite its severe appearance with the notable exception of motoneuron disorders and severe axonal degeneration.

POLYRADICULITIS

Acute canine polyradiculoneuritis is characterized by ventral nerve root inflammation and rapid development of non-ambulatory LMN tetraparesis or tetraplegia. Hyperalgesia of distal limbs is a frequent feature but may not be present in each case. Rarely, spinal pain may be observed. The term Coonhound paralysis infers that polyradiculitis is caused by an immune-mediated reaction to antigen present in the saliva of raccoons. A raccoon bite, however, is not a requirement for this diagnosis, because acute polyradiculitis occurs worldwide in dogs and is considered the counterpart of Guillain-Barré syndrome (GBS) in people. GBS syndrome in people is now considered a post-infectious immune-mediated disease. Various combinations of anti-ganglioside antibodies are strongly associated with clinically and electrophysiologically defined GBS variants in people. In the acute axonal motor neuropathy (AMAN) subtype, antibodies to gangliosides, initiated through molecular mimicry by antecedent intestinal *Campylobacter jejuni* infection, attack gangliosides expressed in human peripheral nerve axolemma, inducing a primary axonal damage. Molecular mimicry (the sharing of structurally homologous antigenic determinants) between gangliosides and *C. jejuni* lipooligosaccharide is central to the induction of this response. Other bacteria (*Mycoplasma pneumonia, Haemophilus influenza)* and viruses (cytomegalovirus, Epstein-Barr virus, HIV) have also been incriminated. Correlations have been established between anti-ganglioside antibodies and specific clinical phenotypes in people, notably between anti-GM1/GD1a antibodies and the acute motor axonal variant of GBS and anti-GQ1b/GT1a antibodies and Miller-Fisher syndrome characterized by opthalmoplegia, ataxia and areflexia. The presynaptic motor nerve terminal lying outside the blood-nerve barrier has been recognized as a potential target site for a subset of antiganglioside antibodies. Canine polyradiculitis has been compared to the motor axonal form of GBS which is associated with a worse prognosis in people. Yet, dogs will often recover within 3–5 weeks with just physical rehabilitation and supportive care. Similar to GBS syndrome in people, corticosteroid treatment appears not to influence recovery rate. On the other hand, prolonged courses up to 3 months, incomplete recovery and lack of improvement were recognized. In humans with GBS, treatment with either plasmapheresis or high-dose human immunoglobulins (IVIg) has proven effective in controlled clinical trials. Treated patients had a significantly more rapid onset of motor recovery, a shortened time to recover walking without aid and less frequently required artificial ventilation. Because IVIg therapy is much simpler to administer and less likely to cause complications, it has been adopted as the favored treatment in people. While removal of antibodies and complement components is likely to be underlying the effect of plasmapheresis, the mechanism of action of IVIg is less clear and presumably pleiotropic, possibly including antiidiotypic neutralization of pathogenic antibodies. In a murine model, IVIg inhibited binding of anti-GQ1b antibodies to the neuromuscular junction and also displaced GQ1b-bound antibody. Currently, there is only preliminary evidence from non-controlled clinical trials supporting the use of human immunoglobulins in dogs with acute polyradiculitis.

BOTULISM

Botulism is an intoxication caused by neurotoxins produced by *Clostridium botulinum* and rarely by other clostridia like *Cl. butyricum* and *Cl. baratii* in people. These anaerobic, gram-positive, spore-forming bacteria are distributed in the soil worldwide. There are several types of botulinum neurotoxins (A, B, C, D, E, F and G). Almost all human cases are caused by the A, B or E serotype, whereas all canine cases to date have been caused by toxin type C and rarely (two cases in Senegal) type D. The most common mode of intoxication in people and animals is ingestion of the preformed toxin (food-borne-botulism). Most reported cases in dogs are associated with the ingestion of carrion. Cadavers of birds are frequent sources of intoxication, because their intestines often contain C. botulinum type C. After the birds' death, bacteria can multiply and produce large amounts of toxin and spores. Thus, outbreaks of canine botulism are often related to areas of avian epizootics of botulism. The only reported natural outbreak of botulism in a group of cats was also traced back to the ingestion of a pelican carcass and botulinum toxin type C. Wound botulism and infant botulism represent forms of botulism in which the toxin is produced by clostridia organisms multiplying within a wound or the GI tract of infants (toxico-infectious botulism), but these have not yet been described in dogs or cats. Intestinal colonization with toxigenic clostridia can exist in adult people and dogs without any clinical signs unless the normal intestinal micro flora is disrupted (antibiotic treatment, immunosuppression, GI surgery). With food borne botulism, botulinum toxin is absorbed from the upper small bowel into the lymphatic system and thereafter distributed haematogenously to axon terminals of cholinergic nerves e. g. the motor nerve terminal and the peripheral autonomic system. Botulinum toxin does not cross the blood brain barrier. The H chain is responsible for the specific binding to a high affinity ganglioside surface receptor complex on the presynaptic peripheral nerve terminal, internalization of the toxin (receptor-mediated endocytosis) and translocation into the cytosol of the peripheral nerve. The L chain represents the active neurotoxin. It acts as a zinc-dependent endopeptidase which cleaves and thus inactivates SNARE- (soluble N-ethylmaleimide-sensitive-factor attachment protein receptor) proteins within the nerve terminal. SNARE proteins mediate docking and fusion of neurotransmitter vesicles to the presynaptic plasma membrane and thus promote calcium-mediated neurotransmitter release into the synaptic cleft. Unlike tetanus neurotoxin there is no further intracellular trafficking of the L chain. Thus, all botulinum toxin types interfere with acetylcholine release at the neuromuscular endplate resulting in acute lower motor neuron paralysis. Autonomic dysfunction may also occur. Toxins differ by their specific cleavage sites. Botulinum toxin C1 has a dual action directed against SNAP-25 and syntaxin while botulinum toxin type A cleaves SNAP-25 and botulinum toxin types B and D cleave VAMP at specific amino-bonds. The clinical diagnosis of botulism is based on the combination of acute lower motor neuron paralysis and a history which suggests toxin ingestion with the food. Botulism may be suspected if multiple animals in a group or from the same household are affected. Proof of botulism requires the demonstration of botulinum toxin by mouse bioassay followed by determination of the specific toxin type by the mouse protection assay with specific antitoxins. Other diagnostic tests have been developed but are still of less sensitivity than the mouse bioassay. PCR is based on amplification of genes encoding the various botulinum neurotoxins and thus will only detect toxigenic C. botulinum organisms but not the neurotoxin itself. Therefore, it is only helpful in environmental investigations and detection of toxigenic bacteria. Enzyme-linked immunosorbent assay (ELISA) can assist with detection of the toxin, but each type of botulinum toxin needs to be addressed separately. ELISA may also respond to inactive toxin and its sensitivity is still not as good as the mouse bioassay. Yet, it may be a useful adjunct for investigation of clinically suspect botulism. Avian eyelid closure test was suggested as another sensitive bioassay. Antibodies directed against the various botulinum toxins can be measured in serum. Positive antibody titres support contact to toxigenic C. botulinum, but toxin-specific antibodies are also present in healthy dogs Antibodies directed against the various botulinum toxins can be measured in serum. Positive antibody titres support contact to toxigenic C. botulinum, but toxin-specific antibodies are also present in healthy dogs. Thus, botulism can only be proven by repeated measurements for increasing antibody concentrations after three weeks. Electrodiagnostic examination has the advantage that the results are immediately available to the clinician and has therefore become a standard test in the initial assessment in human botulism. Electrodiagnostic testing of dogs with botulism has rarely been reported. Botulism has a self-limiting course and many animals will recover within two to three weeks. Treatment is mainly supportive, because spontaneous recovery will occur in most dogs within two to three weeks if the amount of ingested toxin is not too large and respiratory failure and complications like pneumonia or urinary tract infections are avoided. Depending on the degree of symptoms, patients should be hospitalized and monitored closely for respiratory failure. Aggressive treatment with intubation and mechanical ventilation should be considered for severely affected animals. Clindamycin, lincomycin and aminoglycosides are neuromuscular blocking antibiotic agents and should be avoided. Administration of botulinum antitoxin is discussed controversially due to risk of anaphylaxis, its action which is restricted to free circulating toxin but not to internalized toxin and its toxin-specific mode of action, thus only antitoxin C would be indicated in dogs. It's our experience that even dogs severely affected with botulism will recover if appropriate care and mechanical ventilation are instituted.

ACUTE MYASTHENIC CRISIS

Myasthenia gravis typically causes exercise-induced weakness and a waxing and waning course. By definition, all patients with myasthenic crisis are in respiratory failure due to muscle weakness and require ventilatory assistance. Myasthenic crisis differs from botulism and polyradiculitis, because spinal reflexes tend to be preserved. Myasthenia gravis is more commonly associated with facial muscle weakness and depressed palpebral reflexes and also cervical muscle weakness resulting in ventroflexion of the head and neck. It is typically precipitated by poor control of generalized disease, drugs, anesthesia, aspiration or surgery, stress, hot environment or hyperthyroidism in humans.

OTHERS

Tick paralysis and coral snake poisoning need to be considered depending on the geographic. Motoneuron disease (poliomyelitis e. g. rabies, motoneuron degeneration), polymyositis, acute necrotizing polymyopathy, exposure to toxins (ionophores e. g. salinomycin, monensin; organophosphates), or certain metabolic disorders (hypokalemia, hyperkalemia, hypercalcemia, hypoglycemia) are incriminated less commonly. Metabolic disorders and ion channelopathies (e. g. episodic hyperkalemic paralysis from sodium channelopathy) are more typically associated with a waxing and waning course or short repetitive episodes of acute LMN paralysis than with sustained LMN motor paralysis.

CLINICAL APPROACH

Acute generalized lower motor neuron paralysis is an easy to recognize clinical entity. At the peak of the disease, neurologic examination is characterized by non-ambulatory tetraparesis or tetraplegia with areflexia or hyporeflexia of spinal reflexes. The most characteristic finding and the baseline for establishing a diagnosis are reduced or absent flexor reflexes of all limbs and absent patellar reflexes. Muscle tone is clearly reduced and flaccidity may be impressive in some dogs. Most patients show also weakness of the cervical muscles resulting in inability to lift the head. Mentation tends to be unimpaired. Cranial nerve signs with the exception of a loss of bark are uncommon in polyradiculitis, but in people facial paralysis may occur in some cases of GBS. Literature states that oculomotor paralysis including mydriasis is a frequent feature of botulism. In six dogs with botulism a variety of clinical signs were observed ranging from non-ambulatory LMN tetraparesis without cranial nerve signs to flaccid tetraplegia with multiple cranial nerve signs (mydriasis, facial, tongue and jaw paralysis) in one and mydriatic pupils as the only cranial nerve sign in another tetraplegic dog. In any dog with acute lower motor neuron paralysis anamnesis should aim, firstly, at confirmation of the acute onset and exclusion of acute exacerbation of chronic progressive LMN paresis, and, secondly, to detect exposure to snakes, ticks, botulinus toxins and racoons. Diagnostic injections of edrophonium chloride (0.1 - 0.2 mg/kg i. v. or 1 – 10 mg/dog) may be performed as a readily available bedside tests for myasthenia gravis and if not available neostigmine 0.04 mg/kg i. m-. has been recommended. False positive and false negative results and even worsening weakness can occur, thus acetylcholine receptor antibodies are preferred. Thoracic radiographs and contrast swallows are recommended to recognize megaesophagus which may severely impact management of these patients. CSF analysis is recommended in any case of acute lower motor neuron paralysis to recognize CNS inflammation as an underlying cause and also albuminocytologic dissociation characteristic for polyradiculitis.

Electrodiagnostics

EMG and nerve conduction studies are used as the neurologists` readily available bedside tests to substantiate the suspicion of generalized LMN disease and to recognize disease categories. Spontaneous electric activity in multiple muscles of several limbs is indicative of widespread denervation and most characteristic for acute polyradiculoneuritis. Diagnostic problems arise from the delay in the appearance of spontaneous electric activity. Usually it will take at least five days in dogs and uncommonly up to seven days from disease onset until EMG changes can be fully appreciated. In some cases of acute polyradiculoneuritis, we repeated the EMG on day 7 after onset to demonstrate severe generalized denervation which had not been present on day 5. Yet, polymyopathy/polymyositis can also lead to spontaneous activity on EMG, and definite exclusion of muscle disease would require biopsies. Spontaneous electric activity has also been observed in the course of canine botulism. Dogs with botulism tended to recover more rapidly within two weeks and paralysis peaked more early in the course of the disease compared to polyradiculits. Consequently, we failed to record SPA in any dogs with suspected botulism because electrodiagnostic examination was performed early when paralysis was most evident and not during recovery. Small amplitude, polyphasic CMAP were frequently found in polyradiculitis after distal nerve stimulation and recording of CMAP with alligator clamps serving as surface electrode. Similarily, conduction failure of the most distal aspect of the peripheral nerves has been described in people with GBS. Further evidence for polyradiculoneuritis is provided by delayed appearance of F-waves or non recordable F-waves. Yet, evaluation of F-waves may not be feasible in some dogs with small amplitude CMAPs. Neuromuscular transmission is usually investigated by repetitive nerve stimulation at 3 Hz or single fibre electromyography. A decremental response indicates postsynaptic neuromuscular transmission failure as in myasthenia gravis. An incremental response is indicative of presynaptic defects of neuromuscular transmission (e. g. acetylcholine release) and has been described in people with botulism and the paraneoplastic Lambert-Eaton myasthenic syndrome. Repetitive stimulation in a posttetanic setting may be useful for differentiation of postsynaptic and presynaptic disorders of neuromuscular transmission.

Muscle and nerve biopsies

Biopsies are needed to exclude an acute exacerbation of a more chronic polyneuropathy and to eliminate polymyositis / polymyopathy as the cause of LMN paralysis. Muscle and nerve biopsies will be normal or only indicate neurogenic muscle atrophy in most cases of acute polyradiculoneuritis.

REFERENCES

Cencioni et al. **Int J Immunopathol Pharmacol**. 2009;22(4):1043. Hughes RA et al. **Lancet** 2005;366:1653. Douglas MR, Winer JB. **Exp Rev Neurother** 2006;6:1569. Hughes RA et al. **Cochrane Database Syst Rev** 2006:CD002063. Dalakas MC. **JAMA** 2004;291:2367. Hughes R. **J Neurol** 2008;255 Suppl 3:7. Gold R et al.. Nat Clin Pract Neurol 2007;3:36. Nobile-Orazio E, Terenghi F. **J Neurol** 2005;252 Suppl 1:I7. Hughes RA. **J Anat** 2002;200:331. Schröder A et al. **Rev Neurother.** 2009;9(9):1331. GOLD, RM et al. **Nat. Clin. Pract. Neurol**. 2007;3: 36. Cuddon PA. **J Vet Intern Med** 1998;12:294-303. Cai S, Singh BR. **Crit Rev Microbiol 2007**;33:109-125. Bruchim Y et al. **Vet Rec** 2006;158:768. Chaudhuri A, Behan PO. QJM. 2009;102(2):97.Elad D et al.**J Clin Microbiol** 2004;42:5406. Gophen M et al.. **Environ ToxicolWater Qual** 2006;6:77. Wang J et al. **Toxicon** 2007;49:1019.

PITUITARY SURGERY IN DOGS: MOLECULAR PATHOGENESIS OF CANINE PITUITARY DEPENDENT HYPERADRENOCORTICISM

David Bruyette[1], DVM, DACVIM, Adam. Mamelak, MD[2], Tina Jo Owen[3], DVM, DACVS.
[2]Department of Neurosurgery, Cedars-Sinai Medical Center, Los Angeles, CA; Departments of [1]Medicine and [3]Surgery, VCA West Los Angeles Animal Hospital, Los Angeles, CA.

While CD is rare in humans, it is an extremely common endocrinopathy in dogs. CD, more commonly referred to as pituitary dependent hyperadrenocorticism (PDH), has an incidence of 0.2% of all dogs, or approximately 100,000 dogs yearly in the US. Over 90 % of these dogs are subsequently diagnosed with CD while the remaining 10% have adrenal dependent hyperadrenocorticism (ADH). Most canines that are diagnosed with CD present with central obesity, a pendulous and distended abdomen, ravenous appetite, bilaterally symmetrical endocrine alopecia, hair loss, skin changes, and severe polyuria/polydipsia. Untreated dogs are at risk for the development of infections, hypertension, and diabetes mellitus. MRI and CT protocols to image pituitary tumors are widely used in veterinary medicine. The use of these imaging modalities along with biochemical assays and provocative tests of ACTH, cortisol, and other metabolites levels allow accurate determination of the success of therapy.

Several medical treatments are available for treatment of PDH in dogs. Medical options include drugs that chemically destroy the adrenals (lysodren; op-DDD), inhibit enzymes in the adrenal leading to the synthesis of cortisol (ketoconazole, trilostane) or inhibit the release of ACTH from the pituitary gland (selegiline, cabergoline). While these treatments can improve the clinical signs in 40-80% of dogs, they need to be chronically administered, necessitate frequent monitoring and do not cure or address the primary cause of the disease (the pituitary tumor). Unlike human disease, tumors in canines are often very large by the time they are detected. Further, these tumors tend to grow intracranially, probably due to the thick skull base of dogs.

Analysis of the receptor subtypes in PDH shows many similarities to human CD, but also key differences. For example, while both species express the dopamine D2 receptor subtype and leukocyte inhibitory factor receptor (LIFR), canine tumors preferentially express the somatostatin sst2 receptor subtype rather than the sst5 subtype observed in humans. Thus ACTH can be potentially suppressed release with a combination of cabergoline and octreotide or BIM23A760 in dogs, while human tumors are less responsive to octreotide relative to pasireotide. In addition, recent evidence has shown efficacy in treating canine CD with 9-cis-retinoic acid as well as the D2 agonist cabergoline, when neither of these therapies has proven particularly effective in humans.

POMC gene expression can be suppressed in vitro with retinoic acid while in vitro antagonism of the PPAR-γ pathway has implicated a potential role for rosiglitazone, but again limited clinical trials have not demonstrated long-term efficacy in humans. While laboratory studies have provided important guidance, clinical results with agents specifically blocking these pathways has been generally disappointing to date, suggesting that better targeting or other pathways need to be identified.

We have recently embarked on a study utilizing endoscopic transsphenoidal hypophysectomy for the treatment of PDH in dogs. We will be utilizing the tissue obtain from affected and normal dogs to continue work on the molecular pathogenesis of PDH in dogs and to evaluate novel therapeutics. Tissue will be used for histopathology, immunohistochemistry, protein analysis, and cell culture.

Tissue Preservation

Tissue for RNA and DNA analysis was placed in RNAlater™ and kept at 4°C for 1-4 days after which tissue was placed in a new microcentrifuge tube and kept in -80°C until analysis. Tissue for Western blot analysis is snapped frozen in liquid nitrogen and kept in-80°C. Tissue for immunohistochemistry is stored in formalin 10% at 4°C. Tissue for primary cultures is placed in 10ml of Dulbecco's Modified Eagle Medium with 1% penicillin-streptomycin and immediately processed.

Primary Tissue Culture

Tumor or whole pituitaries are placed into dish with 10ml room temperature DMEM 0.3% BSA. They are washed 3 times, placed in a 50 ml tube, and centrifuged at room temperature for 5 minutes at 1000rpm. The tissue is transferred to a Petri dish with very little media, chopped very finely with a sterile scalpel, transferred to a tube with 10ml media and centrifuged at 2000rpm, for 5 min at room temperature. The supernatant is discarded. In a separate tube 35mg collagenase, 15mg hyaluronidase and 9.5ml DMEM 0.3% BSA, are shaken to mix and filtered into the tube with tissue . The tissue is incubated at 37°C for 15-30 min, and shaken well every 5 min. 15-20 ml DMEM 10% FBS is added to stop enzyme activity, and then centrifuged 1500-2000 rpm for 5-10 min. Cells are then re-suspended in a small volume and plated in tissue culture dishes.

We are utilizing Taqman Realtime PCR to measure the mRNA expression levels of EGFR and POMC in each canine CD tumor. Immunostaining of all tumors for EGFR, ACTH, Tpit, and Her2 antibodies will be performed. These data will characterize some of the constitutive mechanisms underlying ACTH production and regulation in CD. mRNA expression levels will be correlated to response to gefitinib in vitro, via measures of ACTH levels and POMC expression. Following chronic exposure to gefitinb we will also measure cell death via an apoptosis assay, cellular turnover with BUDR labeling, and histological response via H&E staining. These experiments are aimed to better elucidate the role of the EGFR pathway and its involvement in the growth and regulation of CD tumors. These experiments will clarify a potential signaling pathway important in CD, and provide a rational for clinical testing. Utilizing the relatively abundant canine tissue provides a means to more rapidly and efficiently define the role of the EGFR pathway in CD than with smaller and less abundant human tumors, while avoiding the many pitfalls of established cell lines such as Att20.

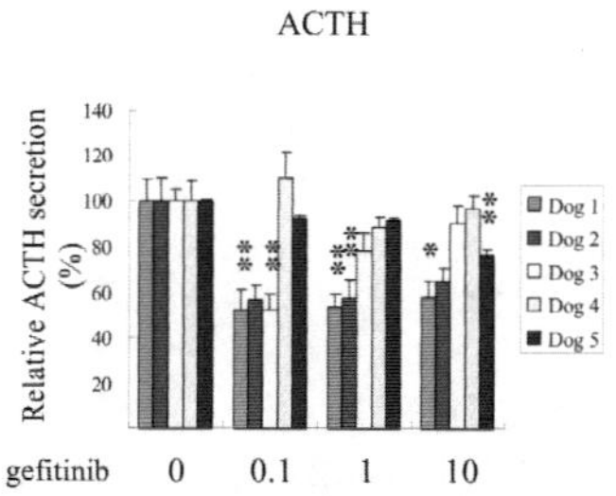

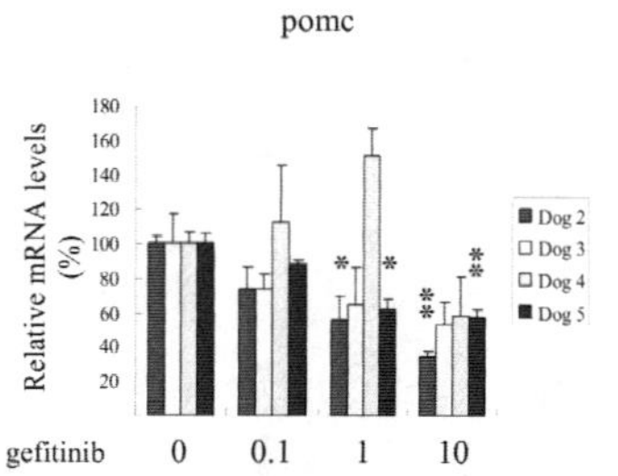

If in vitro experiments continue to support the view that EFGR blockade results in prolonged reduction n ACTH production and/or cell death, a "pre-clinical" trial of this agent will be initiated in canines that have failed or refused medical therapy. Prior to therapy initiation canines will undergo full hormonal screening and MRI imaging to document PDH and tumor size. While receiving the drugs animals will be monitored for side effects, and also undergo monthly measurement of ACTH, UCCR, and response to LDDS. MRI imaging will be performed at 1 month and 3 months after treatment. After 3 months animals with residual tumor will undergo transsphenoidal tumor removal with curative intent. The acquired tumor tissue will undergo analyses similar to non-treated tumor as described above, including EGFR and POMC mRNA expression levels, immunoassaying, formal histology, and response to in vitro gefitinib. These data will provide critical additional information as to the effect of EGFR blockade on tumor growth. Depending on the efficacy of this agent, further trials and a possible human clinical trial can be considered.

REFERENCES

[1.] Hanson JM, et al: **Domest Anim Endocrinol** 2008,4:217. [2.] Van Rijn SJ, et al: **Domest Anim Endocrinol** 2009, Dec 6 (Epub). [3.] Hanson JM, et al: **Domest Anim Endocrinol** 2009, Dec 6, 2009 (Epub). [4.] deBruin C, et al: **Endocrinology** 2008, 9: 4357. [5.] Castillo V, et al: **Endocrinology** 2006, 9: 4438. [6.] Castillo V, et al: **Res Vet Sci** 2010, 88: 26. [7.] Castillo V, et al: **Res Vet Sci** 2008, 85: 26.

CYBERKNIFE RADIOSURGERY FOR IRRADIATION OF BRAIN TUMORS IN DOGS AND CATS. SC Charney, MR Witten, JM Berg, RJ Joseph; Animal Specialty Center, Yonkers, NY.

Radiation therapy is often used for the treatment of brain tumors in dogs and cats. Conventional radiation therapy requires the use of multiple fractions over many weeks to minimize damage to normal tissue. The CyberKnife® Radiosurgery System[a] (CyberKnife) is a frameless stereotactic radiosurgery system that uses multiple non-isocentric, non-coplanar beams to deliver single fraction or hypofractionated radiation with sub-millimeter accuracy. The large number of beams combined with continual image guidance throughout treatment results in a high degree of conformality and steep dose gradients.

CyberKnife couples an orthogonal pair of x-ray cameras to a dynamically manipulated robot-mounted linear accelerator that guides the therapy beam to the treatment target. CyberKnife uses skeletal landmarks or implanted fiducial markers to track the treatment target during planning and treatment delivery. Proprietary inverse planning software is used to create a treatment plan.

The purpose of the study was to evaluate the clinical outcome of dogs and cats with brain tumors treated with CyberKnife radiation therapy.

Forty dogs and cats with brain tumors (presumptive diagnosis based on Magnetic Resonance Imaging (MRI) or Computed Tomography (CT); meningiomas, gliomas, cranial nerve tumors, pituitary tumors, a choroid plexus tumor, and two tumors of unidentifiable origin) and treated with CyberKnife were included. Tumor volumes ranged from $284 mm^3$ to $39,777 mm^3$ (mean $6,412.5 mm^3$). Total treatment doses ranged from 1200 to 3000 cGy given in one to three fractions. Overall median survival time was 347 days. Fifteen patients are still alive with a median time to follow-up of 504 days. All animals with presumed gliomas are dead.

These results indicate that the CyberKnife radiation provides comparable outcomes to conventional radiation with fewer required fractions and that glioma tumor type may have a poorer prognosis than other tumor types.

[a] Accuray, Inc., Sunnyvale, CA

EXTRACELLULAR HSP70 RELEASE IN CANINE STEROID RESPONSIVE MENINGITIS-ARTERITIS.

S. Moore[1], M. Kim[1], A. Tipold[2], M. Oglesbee[1]. 1. The Ohio State University College of Veterinary Medicine, Columbus, OH. 2. School of Veterinary Medicine, Hanover, Germany.

The major inducible 70 kDa heat shock protein (hsp70) is known for both its intracellular role in protecting against protein denaturing insult and for its ability to stimulate innate and adaptive immune responses when released from cells. Despite the potential biological activity of extracellular hsp70 (ehsp70), release into the central nervous system (CNS) has been largely ignored, having being examined only in cases of trauma or ischemia reperfusion injury. The purpose of this study was to evaluate the ehsp70 response in a naturally occurring canine model of CNS immune-mediated inflammatory disease. A commercially available hsp70 ELISA was used to quantify ehsp70 in serum and CSF of healthy dogs (n=8), dogs with acute, untreated steroid responsive meningitis-arteritis (SRMA) (n=30) and dogs previously diagnosed with SRMA who were receiving corticosteroid treatment (n=30). Results were analyzed using a Kruskal-Wallis ANOVA followed by Dunn's multiple comparison test. Mean ehsp70 concentrations in CSF were significantly elevated in dogs with acute SRMA when compared to controls (1.27 and 0, respectively; p=0.029), while ehsp70 concentrations in CSF of treated dogs and controls were indistinguishable (p=0.12). Additionally, while serum ehsp70 concentrations were also significantly elevated compared to controls, no correlation existed between serum and CSF concentrations in the acute SRMA dogs, and in general, ehsp70 concentrations were lower in serum relative to CSF. These findings are consistent with local release of hsp70 in the CNS and not leakage in to the CSF from blood brain barrier disturbance. Results suggest that dogs with SRMA may provide an important model in which the biological significance of ehsp70 may be established, addressing the degree to which CNS inflammation may be regulated through modulation of ehsp70 levels.

DEGENERATIVE MYELOPATHY OF THE GERMAN SHEPHERD DOG - I

Dr. Roger Clemmons
Gainesville, FL

Degenerative Myelopathy (DM) was first described as a specific degenerative neurologic disease in 1973. Since then, much has been done to understand the processes involved in the disease and into the treatment of DM. Although a few cases have been reported in other large breeds of dogs, the disease appears with relative frequency only in the German Shepherd breed, suggesting that there is a genetic predisposition for German Shepherd dogs (GSD) in developing DM. The purpose of this presentation is to review the past findings in GSDM (German Shepherd Dog Myelopathy) to provide a historical perspective about the disease.

The age at onset is 5 to 14 years, which corresponds to the third to sixth decades of human life. Diagnosis of DM is made by a history of progressive spinal ataxia and weakness that may have a waxing and waning course or be steadily progressive. This is supported by the neurologic findings of a diffuse thoracolumbar (TL) spinal cord dysfunction. Clinical pathologic examinations are generally normal except for an elevated cerebral spinal fluid (CSF) protein in the lumbar cistern. Electromyographic (EMG) examination reveals no lower motor unit disease, supporting the localization of the disease process in the white matter pathways of the spinal cord. Spinal cord evoked potentials recorded during the EMG do show changes which help determine the presence of spinal cord disease. Imaging of the spinal column including MRI is normal (other than old age changes) in uncomplicated DM. However, analysis of the size of the TL spinal cord has suggested that there may be diminished spinal cord volume.

The gross pathologic examination of dogs with GSDM generally is not contributory toward the diagnosis. The striking features being the reduction of rear limb and caudal axial musculature. The microscopic neural tissue lesions consist of widespread demyelination and axonal loss of the spinal cord, with the greatest concentration of lesions in the TL spinal cord region. In severely involved areas, there is also a reduced number of axons, an increased number of astroglial cells and an increased density of small vascular elements. In the thoracic spinal cord, nearly all funiculi are vacuolated. Similar lesions are occasionally seen scattered throughout the white matter of the brains from some dogs, as well. Many patients have evidence of plasma cell infiltrates in the kidneys on throughout the gastrointestinal tract, providing a hint to the underlying immune disorder causing DM.

Dogs afflicted with GSDM have depressed lymphocyte blastogenesis to plant mitogens. The depression of their cell mediated immune responses correlates with the clinical stage and severity of the disease. Furthermore, this suppression has been shown to be due to the genesis of a circulating suppressor cell. Some dogs with DM exhibit antigen-binding cells specific to canine myelin basic protein. Lumbar CSF of GSDM dogs also contain increased concentrations of IgG with oligocloncal bands and myelin basic protein. Complement and IgG have been shown to be bound within lesions within the spinal cords of dogs with GSDM. These patients also show increased circulating immune-complexes in their sera.

Treatment of GSDM includes exercise, diet, supplements and medications. Each of these has an impact upon health and upon the disease. The components of the treatment work together to reduce the progression of GSDM. They target the processes which we have uncovered as the causes of the pathologic changes we see. We have seen few side-effects (mostly GI upset) in the patients diagnosed with GSDM. Current medication recommendations are aminocaproic acid (10-15 mg/kg q8h) and acetylcysteine (25 mg/kg q8h, qod). Other approaches will be discussed.

DEGENERATIVE MYELOPATHY OF THE GERMAN SHEPHERD DOG - II

Dr. Roger Clemmons
Gainesville, FL

Degenerative Myelopathy of the German Shepherd dog (GSDM) may be a unique neurodegenerative disease seen in the German Shepherd breed and similar dogs. Genomic analysis with random amplified polymorphic DNA (RAPD) methods has suggested that GSDM may have a genetic basis. The test is used to help determine which GSDM possible dogs can be separated into those who are probable GSDM patients from those who probably do not have the condition, since up to 75% of those with clinical signs actually do not have GSDM.

On the other hand, the DM Flash test is not a good screening test for this reason…there are too many false positive results due to a high incidence of the change in the breed while only a limited (10-20%) of the positives get the disease. However, that said, it does say that the risk of developing GSDM is 12 times higher in dogs who are positive on the DM Flash test than in dogs who are negative. A negative test suggests there is a 99% chance the dog will not get GSDM. However, a positive test in normal dogs has only an 11% predictive value.

In dogs with clinical signs of possible DM, the DM Flash test is a good diagnostic test. It is 96% sensitive and 99% specific and is generally going to separate those who have GSDM from those who do not have it (only some other neurologic disease). It still does not replace other tests since there may be more than one disease, but it probably is sufficient to confirm the diagnosis of GSDM and move the patient to the "probably GSDM" category. What this means is that a positive DM Flash test means there is less than a 1% chance the dog does not have DM as part of the clinical picture, while a negative DM Flash test means that there is a 4% chance that the dog may still have GSDM even though the test was negative. With a negative result, only additional tests can be sure that GSDM is part of the disease process or whether another disease is more likely.

So, while 91% of the proportional risk of developing GSDM is accounted by having a positive DM Flash test (genetics accounts for most of the reason the patient gets GSDM). A positive test in a normal dog does not mean they will inevitably get GSDM; just that they are at a greater risk. A negative test means it is highly unlikely that the dog will get GSDM, but not impossible. That is about as good as any test. Those who test positive are the ones who should be watched and extra care taken with. We may ultimately find out what other factors influence the genetics to lead to GSDM and then we might be able to prevent the disease from happening. For now, all you can say is there is a risk of developing GSDM and that might be a reason to alter breeding, but only if other things are factored in as well.

Several regions of gene variation have been found to be altered in GSDM; but none appears to be the high risk target explaining which dogs, who are positive in the DM Flash test, will develop the disease. Like families of MS patients, there is a risk that others might get MS because of increased genetic risk. However, genomic wide investigations have only led to finding common mutations which, while associated with the disease, are not the genetic cause. Further work is needed to elucidate the genetic and environmental issues which lead to the development of GSDM.

SOD1-ASSOCIATED CANINE DEGENERATIVE MYELOPATHY: A PROGRESS REPORT.

JR Coates[1], R Zeng,[1] GC Johnson[1], ML Katz[2], JF Taylor[3], T Awano[1], S Kanazono[1], S Khan[1], L Hansen[1], JA Villamil[1], AW Hahn[1], GJ Haldorson[4], D Righter[4], K Lindblad-Toh[5], GS Johnson[1]. 1. College of Veterinary Medicine, 2. School of Medicine, 3. College of Agriculture and Natural Resources, University of Missouri, Columbia, MO. 4. College of Veterinary Medicine and Washington Animal Disease Diagnostic Laboratory, Washington State University, Pullman, WA. 5. Broad Institute of Harvard and MIT, Cambridge, MA.

Canine degenerative myelopathy (DM) is an adult onset fatal neurodegenerative disease that occurs in many breeds. The histopathology is classified as a multisystem central-peripheral axonopathy. Recently, a missense mutation in the superoxide dismutase 1 (*SOD1*) gene was found to be a cause for DM suggesting that DM is similar to some forms of human amyotrophic lateral sclerosis (ALS).

This report includes genotypes at SOD1:c.118A>G for DNA samples from 11,736 purebred dogs genotyped at the University of Missouri on or before December 22, 2009. The mutant *SOD1:c.118A* allele was found in 91 of the 192 examined dog breeds or varieties. One hundred of 101 dogs from 14 different breeds or varieties with histopathologically confirmed DM were homozygous for the *SOD1:c.118A* allele; one Chesapeake Bay Retriever with histopathologically confirmed DM was an A/G heterozygote.

The relationship between breed-specific *SOD1:c.118A* allele frequencies and breed-specific DM prevalences were ascertained from the records in the Veterinary Medical Database (VMDB). We found a highly-significant correlation between breed-specific *SOD1:c.118A* allele frequency and breed prevalence of DM.

The occurrence of the *SOD1:c.118A* allele in a wide variety of dog breeds indicates that DM should be considered among the differential diagnoses for middle aged or older dogs of any breed presenting with a chronic progressive onset of upper motor neuron signs in their pelvic limbs. Although considered rare, it is possible that carrier dogs may develop DM.

Potential biomarkers are being evaluated in neural tissue morphometry, cerebrospinal fluid, gait, imaging and electrophysiologic studies, and genome-wide association mapping.

IDENTIFICATION OF DISEASE ASSOCIATED LOCI BY GENOME-WIDE ASSOCIATION OF SINGLE NUCLEOTIDE POLYMORPHISMS IN PUG DOGS WITH NECROTIZING MENINGOENCEPHALITIS. RM Barber[1], SJ Schatzberg[1], BF Porter[2], AN Allen[3], JJ Corneveaux3, SR Platt[1], M Kent[1], MJ Huentelman[3]. 1. University of Georgia College of Veterinary Medicine, Athens, GA. 2. Texas A&M College of Veterinary Medicine and Biomedical Sciences, College Station, TX. 3. The Translational Genomics Research Institute (TGen), Phoenix, AZ.

Necrotizing meningoencephalitis (NME) is a non-suppurative inflammatory disorder of the canine central nervous system. Overrepresented in Pug dogs, NME also occurs in other small breeds including the Maltese and Chihuahua. The complete etiopathogenesis of NME is unknown but non-Mendelian inheritance has been demonstrated in Pug dogs, suggesting a role for genetic risk factors in the development of disease. The purpose of this investigation was to identify disease susceptibility loci in Pug dogs with NME through genome-wide association (GWA) of single nucleotide polymorphisms (SNPs) in affected and non-affected Pug dogs.

Genomic DNA was isolated from 30 Pug dogs with histopathologically confirmed NME and 68 healthy, control Pug dogs without evidence of neurological or autoimmune disease. Genomic DNA quality was assessed with 2% agarose gel electrophoresis and quantity with fluorometric dsDNA quantification. Genome-wide association of > 100,000 SNPs was performed using the Illumina Canine Infinium® HD BeadChip, and SNPs were analyzed with PLINK (http://pngu.mgh.harvard.edu/purcell/plink/) with a minor allele frequency of > 5% and call rate of > 98%.

Three loci strongly associated with the development of NME were identified. Genome-wide significance with Bonferroni correction was reached for an approximately 7 MB region of dog leukocyte antigen (DLA) class II on chromosome 12 ($p_{raw} = 1.9 \times 10^{-8}$, $p_{genome} = 0.03$), including genes *DSCR4*, *HLA-DPB1*, *PACSIN1*, *PHF1* and *SYNGAP1*. Disease associated genes on other chromosomes included the *STYX* gene on chromosome 8 and the *AFAP1L2* gene on chromosome28

The results indicate that a mutation in the DLA class II region of Pugs dogs contributes to disease development. Additionally, the finding of several disease associated loci is in Pug dogs with NME is consistent with non-Mendelian inheritance and suggests that NME may be a polygenic disorder. Fine mapping studies of additional NME-affected breeds are currently underway to further characterize the identified loci and will be followed by sequencing to identify exonic mutations.

ARE CERVICAL VERTEBRAL RATIOS USEFUL IN THE DIAGNOSIS OF CERVICAL SPONDYLOMYELOPATHY IN DOBERMANS? Ronaldo C. da Costa[1], Jacqueline Johnson[1], Joane Parent[2]. [1]Department of Veterinary Clinical Sciences, The Ohio State University, Columbus, OH, [2]Université de Montréal, St-Hyacinthe, QC, Canada.

Cervical vertebral ratios are routinely used in horses and humans. Use of cervical vertebral ratios has rarely been reported in dogs. Our goal was to compare the cervical inter- and intravertebral ratios of clinically normal Dobermans with those of Dobermans with cervical spondylomyelopathy (CSM) to establish if consistent differences exist between normal and CSM affected dogs.

Twenty four Doberman pinschers were studied, 6 clinically normal dogs and 18 with CSM. Cervical radiographs and magnetic resonance imaging were performed in all dogs. Lateral cervical radiographs were used to perform 8 measurements of the vertebral bodies and vertebral canal from C3 to C7. Measurements were performed blindly by the same investigator. Six inter- and intravertebral ratios were calculated: 1) minimum sagittal intravertebral ratio (minimum intravertebral canal height divided by vertebral body height), 2) modified sagittal intravertebral ratio (minimum intravertebral canal height divided by modified vertebral body height), 3) sagittal intervertebral ratio (intervertebral distance divided by vertebral body height), 4) modified sagittal intervertebral ratio (intervertebral distance divided by modified vertebral body height), 5) sagittal diagonal intervertebral ratio (diagonal intervertebral distance divided by vertebral body height), 6) modified sagittal diagonal intervertebral ratio (diagonal intervertebral distance divided by modified vertebral body height). Paired t-test was performed to compare all ratios between clinically normal and CSM-affected Dobermans.

Four cervical vertebral measurements were found to be significantly different between normal and CSM affected dogs: modified sagittal intravertebral ratio at C4 ($P = 0.043$) and at C5 ($P = 0.033$), modified sagittal intervertebral ratio at C5-6 ($P = 0.016$), and modified sagittal diagonal intervertebral ratio at C5-6 ($P = 0.036$). In all cases the ratios of CSM-affected Dobermans were smaller. In conclusion, significant differences in inter- and intravertebral cervical ratios were identified between normal and CSM-affected Dobermans and they may be used as a screening tool for the diagnosis of CSM in Dobermans.

EVALUATION OF LEVETIRACETAM AS ADJUNCTIVE TREATMENT FOR CANINE EPILEPSY. KR Muñana[1], WB Thomas[2], KD Inzana[3], JA Nettifee-Osborne[1], KJ McLucas[2], NJ Olby[1], CL Mariani[1], PJ Early[1]. 1. North Carolina State University College of Veterinary Medicine, Raleigh, NC. 2. University of Tennessee College of Veterinary Medicine, Knoxville, TN. 3. Virginia-Maryland Regional College of Veterinary Medicine, Blacksburg, VA.

The purpose of this blinded clinical trial was to evaluate the safety and efficacy of levetiracetam (LEV) as adjunctive therapy for refractory canine epilepsy. Idiopathic epileptic dogs with $\geq$ 4 seizures per month or cluster seizures despite appropriate antiepileptic drug (AED) treatment were enrolled. Following an 8-week baseline assessment, dogs were randomized to receive LEV (20 mg/kg PO q8h) or placebo for 16 weeks, and then were crossed over to the alternate treatment for 16 weeks after a 4-week washout. Owners recorded information on seizure frequency and adverse events on standard forms. Hemogram, chemistry profile, urinalysis and serum AED concentrations were evaluated at established intervals. Mixed models were fit to compare data over the course of the study. Twenty-two of 34 (64.7%) randomized dogs completed the study. Mean reduction in seizures from baseline during LEV treatment was 50.86%. Thirteen dogs (59%) experienced a $\geq$50% reduction in seizures. A decrease in average weekly seizure frequency during LEV administration (1.092) compared to baseline (1.922) was identified ($p = 0.015$), but this difference was not significant when the reduction in seizures with placebo was considered ($p=0.310$). The most common adverse event was ataxia, but the increased incidence during LEV administration did not differ significantly from placebo. No changes in laboratory parameters were identified. Findings suggest that treatment with LEV is safe in epileptic dogs. The trial failed to demonstrate efficacy of LEV over placebo, although the power of the study was limited. Further evaluation of LEV as treatment for canine epilepsy is warranted.

MAGNETIC RESONANCE IMAGING OF DISKOSPONDILITIS IN 23 DOGS

I Carrera[1], R Gonçalves[2], F McConnell[2], M Sullivan[1]. 1. Faculty of Veterinary Medicine, University of Glasgow, Glasgow, UK. 2. Small Animal Teaching Hospital, University of Liverpool, Neston, UK.

The diagnosis of diskospondylitis is based on clinical signs, diagnostic imaging and laboratory results. To date radiography has been the imaging technique most widely used. This retrospective study aimed to describe the Magnetic Resonance Imaging (MRI) findings observed in dogs with confirmed diskospondylitis and evaluate the advantages and limitations of MRI in the management of this condition.

Inclusion criteria comprised: complete clinical history and neurological examination, MRI study of the region of the vertebral column affected, including in all cases sequences pre- and post-contrast sequences; urine, blood and/or intervertebral disk cultures, and favourable progression of the clinical signs in response to antibiotic treatment.

Twenty-three dogs and 28 sites of infection were evaluated. In all sites, there was involvement of two adjacent vertebral end-plates and the associated disk. T1W images revealed loss of the normal bright marrow signal on the vertebral bodies, manifested as hypointense areas involving the end-plates and vertebral body marrow adjacent to them. These areas corresponded in T2-weighted images to hyperintense areas in 2 sites (7.1%), being hypointense in the remainder (92.9%); while fat suppression sequences always showed high signal intensity to normal bone on the affected areas. Erosion and irregularity of the end-plate was seen in 100% of the sites. All intervertebral disks were misshapen and distorted, 26 (89.2%) showed very high signal intensity in T2. Post-contrast sequences revealed enhancement of the end-plates (100%), intervertebral disks (89.2%), and paravertebral soft tissues (92.8%). Epidural extension of the infection was very well-delineated in post-contrast images at 26 sites (92.8%). Cord compression was present at 21 sites (75%). Subluxation was observed in 2 cases (8.3%).

MRI depicts clearly the exact location and extension of diskospondylitis. The capability to assess the paravertebral soft tissues, epidural space and spinal cord, greatly helps to manage dogs with neurological clinical signs. Moreover, early foci of infection are promptly delineated allowing aggressive early treatment.

OBSERVATIONAL INVESTIGATION OF THE CLINICAL COURSE OF POLYRADICULITIS AFTER TREATMENT WITH HUMAN INTRAVENOUS IMMUNOGLOBULINS

A Fischer[1], TA Steinberg[1], K Jurina[2], K Matiasek[3], K Hirschvogel[1]. 1. Clinic of Small Animal Medicine, Ludwig-Maximilians-University of Munich, 2. Tierklinik Haar, Germany. 3. Animal Health Trust, Newmarket.

The purpose of this study was to describe the clinical course of acute canine polyradiculoneuritis (ACP) following treatment with intravenous immunoglobulin (IVIg).

Study design was a combined retrospective and prospective clinical case series

Twelve client-owned dogs diagnosed with polyradiculoneuritis were treated with human IVIg (2000–2008).

Dogs with ACP were treated with IVIg (0.3–2 g/kg [0.14-0.9 g/Ib], IV). Diagnosis of ACP was confirmed using clinical features, electrodiagnosis, CSF analysis, and muscle/nerve biopsies. A simple 6 point gait score was applied. Duration of initial progressive phase, the time from IVIg administration until dogs walked without support and the duration of the complete episode (onset to walk without support) was evaluated.

Neurologic signs progressed for 2-11 days (median 6 days). All dogs but one (dog 12) were treated with IVIg within 3-15 days (median 11 days) of the first evidence of weakness. Eight dogs were considered as responders and walked without support within thirteen days (range 3-13 days; median 7 days) after IVIg treatment. Less improvement of motor function was noted in another four dogs, considered as non-responders. These walked without support between 21-120 days (median 61 days). The complete episode in the responders lasted between 15-58 days (median 18.5 days) and in the non-responders between 29-123 days (median 68.5 days). An adverse reaction was observed in two dogs.

It was concluded that human IVIg infusion may promote recovery in polyradiculitis. A prospective blinded treatment study is thus recommended.

LOW FIELD MAGNETIC RESONANCE IMAGING (MRI) AND TRANSCRANIAL MAGNETIC STIMULATION (TMS) IN DOGS WITH AND WITHOUT CLINICAL SIGNS OF DISK ASSOCIATED WOBBLER SYNDROME (DAWS). S De Decker[1], I Gielen[1], L Duchateau[1], J Lang[2], R Dennis[3], N Corzo[4], H van Bree[1], I Van Soens[1], L Van Ham[1]. [1]Faculty of Veterinary Medicine, Ghent University, Belgium, [2] Vetsuisse Faculty Bern, Bern University, Suisse, [3]Animal Health Trust, Newmarket, United Kingdom, [4]Davies Veterinary Specialists, Hertfordshire, United Kingdom.

The purpose of this study was to investigate the spectrum, incidence, repeatability, and consequences of cervical MRI abnormalities in dogs with and without DAWS. Additionally, it was evaluated if TMS could differentiate between relevant and irrelevant cord compressions.

Twenty clinically normal Doberman Pinschers, 17 clinically normal Foxhounds, and 21 dogs with DAWS underwent cervical MRI (0.2T). Disk degeneration, ventral and dorsal spinal cord compression, vertebral body abnormalities, and intraspinal signal intensity (ISI) changes were evaluated. In part 1 of this study, the MRI studies of all clinically normal dogs were reviewed. In part 2, MRI studies of dogs with DAWS were randomly mixed with those of age-matched clinically normal Doberman Pinschers (n=12) and Foxhounds (n=11) and were presented to 4 independent board-certified radiologists. In part 3, Doberman Pinschers with clinical signs of DAWS (n=16), and clinically normal Doberman Pinschers with (n=6) and without (n=11) spinal cord compression underwent TMS with recording of latencies and amplitudes from the extensor carpi radialis (ECRM) and cranial tibial (CTM) muscles.

Part 1 of this study revealed a high incidence of MRI abnormalities in clinically normal Doberman Pinschers and Foxhounds. Only the occurrence of vertebral body abnormalities was significantly higher in Doberman Pinschers. Disk degeneration and ventral compression were significantly associated with higher age. Disk degeneration, ventral and dorsal compression were significantly associated with the more caudal disk spaces. Part 2 demonstrated a good intraobserver agreement for dorsal compression and new bone formation. The other parameters were associated with a very good intraobserver agreement. Interobserver agreement was considered good for disk degeneration and vertebral body abnormalities and moderate for the other assessed parameters. Two of the 21 patients and 4 of the 23 clinically normal dogs were erroneously categorized as clinically normal and clinically affected, respectively. Part 3 revealed significant differences in ECRM latencies, CTM latencies, and CTM amplitudes between the 3 different groups globally and between clinically affected Doberman Pinschers and the two groups of clinically normal dogs seperately. There were no significant differences between both groups of clinically normal Doberman Pinschers.

Degenerative MRI abnormalities are commonly seen in the caudal cervical region of older clinically normal Doberman Pinschers and Foxhounds. These findings are generally associated with a very good intraobserver agreement, moderate interobserver agreement and the occurrence of false positive and false negative clinical assessments. TMS can be use to differentiate between clinically relevant and irrelevant spinal cord compressions seen on MRI.

ASSOCIATION BETWEEN THE ABCB1 (MDR1) GENE AND SEIZURE CONTROL IN CANINE EPILEPSY.

KR Muñana[1], JA Nettifee-Osborne[1], RL Bergman Jr[1], KL Mealey[2]. 1. North Carolina State University College of Veterinary Medicine, Raleigh, NC. 2. Washington State University College of Veterinary Medicine, Pullman, WA.

Medically refractory seizures are a significant problem in both human and canine epilepsy. Altered expression of ABCB1 (MDR1), the gene encoding for p-glycoprotein, has been proposed to play a role in drug-resistant epilepsy. The aim of this study was to determine whether heterogeneity of the ABCB-1 gene is associated with seizure control in canine epilepsy. Twenty-nine epileptic Collies being treated with antiepileptic drugs (AEDs) were evaluated. Based on owner records, dogs were classified as having good control ($\leq$1 seizure/month, no cluster seizures) or poor control (>1 seizure/month, and/or clusters). Serum AED concentrations were obtained, and ABCB1 genotyping was performed on buccal samples using a previously described technique. Data was analyzed for associations between genotype and seizure control, number of AEDs administered, and serum AED concentrations, using a Fisher's exact test for categorical variables and the Kruskal-Wallis test for continuous variables. Fourteen dogs (48%) were homozygous for the mutation (M/M), 11 dogs (38%) were heterozygous (M/N) and 4 dogs (14%) had the wild-type genotype (N/N). Dogs with the M/M genotype were significantly more likely to have good seizure control than M/N or N/N dogs (p=0.003). Serum phenobarbital and bromide concentrations did not differ between groups, but dogs in the M/N and N/N groups received a larger number of AEDs than dogs in the M/M group (p=0.014). The ABCB1 mutation, which results in loss of functional p-glycoprotein, is associated with good seizure control in epileptic Collies. These findings suggest that p-glycoprotein may play a role in treatment response with AEDs.

KINEMATICS OF THE CRANIAL AND CAUDAL CERVICAL SPINE IN LARGE BREED DOGS.

Jacqueline Johnson[1], Ronaldo C. da Costa[1], Matthew J. Allen[1], S. Bhattacharya[2], Vijay K. Goel[2] 1. Department of Veterinary Clinical Sciences, The Ohio State University, Columbus, OH. 2. Department of Bioengineering, The University of Toledo, Toledo, OH.

The diagnosis and treatment of disorders of the canine cervical spine is hindered by a lack of understanding of the biomechanical features of the cervical spine. The goal of this study was to define kinematic motion patterns of the canine cervical spine, identify differences between the cranial (C_2-C_4) and caudal (C_5-C_7) segments, and to distinguish primary (PM) and coupled motions (CM).

Cadaveric cervical spines were collected from 8 Foxhounds (29-33kg). Spines were deemed free of pathology using radiography, CT and MRI. Musculature was removed without damaging underlying ligaments and joints capsules. Spines were mounted in a custom-made pure-moment testing jig, data was collected using an optoelectronic camera, and analyzed using a custom software program. Range of motion (ROM), neutral zone (NZ), CM and CM/PM was calculated for flexion/extension, left/right lateral bending and left/right axial rotation.

There was no difference between flexion and extension, but total flexion/extension ROM was reduced at C_5-C_6. Axial rotation was approximately 3 times greater in the caudal segments (15.8° vs 4.2°). Lateral bending and axial rotation were strongly coupled with both movements occurring simultaneously in the same direction. In lateral bending, CM/PM was 0.46 and 1.10 in the cranial and caudal segments respectively; in axial rotation, CM/PM was 1.0 and 0.8 respectively.

The cranial and caudal cervical segments have significantly different motion patterns. Bending/rotation are coupled throughout the cervical spine. Axial rotation is greatest in the caudal cervical spine. This may have implications in the development of disc-related spinal compressions and cervical spondylomyelopathy in large-breed dogs.

NEW QUANTITATIVE ASSAYS FOR THE DIFFERENTIAL DIAGNOSIS OF EQUINE PROTOZOAL MYELOENCEPHALITIS (EPM). SM Reed[1], DK Howe[2], MR Yeargan[2], JK Morrow[3], AJ Graves[3] and WJA Saville[4]. 1. Rood and Riddle Equine Hospital, Lexington, KY. 2. University of Kentucky, Lexington, KY. 3. Equine Diagnostic Solutions, Lexington, KY. 4. The Ohio State University, Columbus, OH.

A field study was conducted on neurologic horses to evaluate new quantitative assays and to re-establish the importance of performing cerebrospinal fluid (CSF) taps for accurate diagnosis of EPM.

Two new ELISAs, designed to be run concurrently, have been developed at the Gluck Equine Research Center. These assays incorporate three unique antigens of *Sarcocystis neurona* to measure the IgG response to parasite infection. Paired serum and CSF samples collected from 131clinical cases examined at Rood and Riddle Equine Hospital (RREH) over the past three years were tested with the two complementary assays. The ELISA results were compared to standard WB results performed at Equine Diagnostic Solutions (EDS) laboratory. Neurologic status was monitored at presentation and on follow-up examination for most patients. Histopathological results were available for 14 horses with paired samples and another 5 horses with only CSF available.

Qualitative WB and quantitative ELISA results, clinical histories and outcomes, and necropsy results (where available) were tabulated. Cases were grouped by diagnosis into EPM, CVM and other, and the compiled data were then analyzed. Strong correlation was observed between ELISA and WB results. Endpoint serum titers alone were not predictive of an EPM diagnosis. However, endpoint CSF titers were correlated strongly with an EPM diagnosis. Furthermore, a serum-to-CSF titer ratio was the most informative criterion for the differential diagnosis, consistent with antigen-specific antibody indices that are used for determining intrathecal antibody production. Serum:CSF ratios <100 were observed for 83% (24/29) of EPM cases, while 97% (36/37) of CVM diagnoses and 88% (57/65) of other neurologic diagnoses had serum:CSF ratios of >100. In total, the serum:CSF titer ratio obtained with the new ELISAs provided an accurate diagnosis for 89% (117/131) of these neurologic cases. Importantly, a 1:1000-titer blood sample spiked into pristine CSF had no appreciable effect on ELISA results until there was greater than 10,000 red cells per microliter, thereby demonstrating the utility of these assays even when there is significant blood contamination.

This study verified that analysis of both serum and CSF from horses suspected to have EPM is more beneficial than examination of serum alone. The results support the use of these quantitative assays as valuable new tools for the diagnosis of EPM in horses showing clinical signs.

DIAGNOSIS & MANAGEMENT OF CRANIOCERVICAL JUNCTION ANOMALIES: "IT'S COMPLICATED"

Sofia Cerda-Gonzalez, DVM, DACVIM (Neurology)
Ithaca, NY

INTRODUCTION

The craniocervical junction (a.k.a. craniovertebral junction) includes the occipital bone, the atlas, and axis, the single joint cavity that unites them, and the ligamentous attachments of the atlantooccipital and atlantoaxial junctions. This area allows more motion than any other portion of the axial spine. Its mobility and complex anatomy present many opportunities for errors in development leading to instability and damage to the underlying neural structures. Disorders of the CCJ are most frequently seen in toy and small-breed dogs. In particular, Cavalier King Charles Spaniels and Griffon Bruxellois are predisposed to CLM, while atlanto-axial instability is most frequently seen in the Chihuahua, Yorkshire terrier, miniature and toy Poodle, Pomeranian, and Pekingese breeds.

Commonly recognized disorders include Chiari-like malformations, atlanto-axial instability/subluxation, abnormalities of the dens (hypoplasia/aplasia/non-union), and occipito-atlanto-axial malformations. Traumatic subluxation of the atlantooccipital or atlantoaxial junctions and atlas/axis fractures may also be seen. Recently, non-traumatic atlantooccipital overlapping, atlantoaxial dural/fibrous band formation, and incomplete ossification of the atlas have been described in the dog. Multiple craniocervical junction anomalies are frequently present concurrently and overlap in their clinical manifestation, complicating diagnosis and treatment decisions by blurring the contribution of each disorder to the development of syringomyelia and/or clinical signs. Consequently, it may be preferable to consider these as part of a continuum of "craniocervical junction disorders" when considering diagnostic and therapeutic options rather than considering individual disorders alone.

This presentation aims to provide an overview of the clinical presentation, diagnostic and treatment options available for dogs with congenital disorders of the craniocervical junction. Diagnostic challenges and strategies to overcome these will be discussed, along with therapeutic options for these diseases.

CLINICAL SIGNS

Disorders of the craniocervical junction overlap in their clinical manifestation, reflecting compression and/or concussion of the caudal brainstem, and cerebellum, and cervical spinal cord. Compromise of the vertebro-basilar vascular system and compression of cervical nerve roots can also play a role in the development of clinical signs and/or syringomyelia. Clinical signs are typically seen in young to middle-aged dogs. Their onset may be gradual or may develop acutely following minor trauma such as falling from furniture, hitting objects head-on, rough play, or dogfights, particularly in dogs with instability of the atlantooccipital and atlantoaxial junctions. In these dogs respiratory paralysis and sudden death can result if instability is severe.

Affected dogs may show signs of a cervical myelopathy, including tetraparesis, upper motor neuron/sensory ataxia, and proprioceptive deficits in all limbs. Signs are typically most pronounced in the pelvic limbs except in dogs with central cord syndrome. Dogs with cervical spinal cord involvement may also develop scoliosis. Cerebellar involvement can manifest as a hypermetric gait, a wide-based posture, and menace deficits despite intact vision and intact palpebral reflexes. Vestibular signs, facial paresis, and seizures are also reported.

Hyperesthesia (abnormal sensitivity of normal skin to light pressure, touch, and temperature changes), allodynia (pain caused by a non-painful stimulus), and paresthesias (spontaneous sensations of tingling, creeping or pricking of the skin) can develop. These dogs demonstrate pain on head and cervical palpation, scratching of the neck and shoulder region without making skin contact ("phantom scratching"), rubbing of the face and neck on furniture, and shoulder/axillary pain when being picked up. Severely affected dogs cry out spontaneously or during play and may avoid interacting with their owners and/or littermates.

DIAGNOSIS

Chiari-like malformation

A diagnosis of a Chiari-like malformation is obtained using magnetic resonance (MR) imaging of the craniocervical junction and cervical spinal cord. Findings consistent with this diagnosis include: cerebellar indentation and herniation through the foramen magnum, attenuation or obliteration of the cerebellomedullary cistern, and medullary kinking. Ventriculomegaly is frequently present as well. Ultrasonography of the atlantooccipital junction can be used to identify cerebellar herniation, but does not allow visualization of the spinal cord to determine if syringomyelia is present.

It is important to remember that a Chiari-like malformation may be present without clinical signs, particularly in the Cavalier King Charles Spaniel. Similarly, in people with Chiari type 1 malformations up to 30% of cases remain asymptomatic. Syringomyelia, an accumulation of fluid within the spinal cord parenchyma, is frequently seen in dogs with Chiari-like malformations as an area of T2-hyperintensity and T1-hypointensity within the spinal cord parenchyma. Hydromyelia, defined as dilation of the central canal within the spinal cord, may otherwise be seen. Although the presence of syringomyelia is correlated to the presence of clinical signs this finding may also be seen in asymptomatic dogs. Cerebrospinal fluid flow studies and volumetric measurements have been proposed as methods of predicting which dogs with Chiari-like malformations will develop clinical signs. In patients diagnosed with a Chiari-like malformation the brain, cervical spinal cord, and craniocervical junction should be assesses for additional potential causes of clinical signs.

Atlantooccipital overlapping

Non-traumatic atlantooccipital overlapping is seen as a decreased distance between the dorsal arch of the atlas and the supraoccipital bone, with the former located either immediately ventral to the foramen magnum or within the caudal fossa. The atlas may obliterate the cerebellomedullary cistern and indent the caudal aspect of the cerebellum. Medullary kinking can be seen concurrently. These changes are inconspicuous on cervical radiographs; advanced imaging is necessary to obtain a diagnosis.

Atlantooccipital overlapping can be dynamic, worsening with extension of the craniocervical junction. This is best evaluated by imaging the craniocervical junction in neutral and mildly extended positions. Magnetic resonance imaging best demonstrates parenchymal changes secondary to atlantooccipital overlapping, such as cerebellar and medullary compression. Computed tomography is necessary to confirm the position and alignment of the atlas in relation to the foramen magnum. Atlantooccipital overlapping resembles the human condition basilar impression/invagination (terms used interchangeably) (BI), seen as displacement of the atlas and axis towards or into the foramen magnum. In both the human and canine conditions other disorders of the craniocervical junction are frequently present, including atlantoaxial instability, dens dysplasia/hypoplasia, and Chiari malformations.

Dural/fibrous band overlying C1-C2 junction

Using MR imaging a dorsal compressive lesion may be seen obliterating the dorsal subarachnoid space and causing varying degrees of spinal cord compression at the level of the atlantoaxial junction. This dural/fibrous band is also visible at surgery as a thick white membrane, and is frequently present in dogs with a Chiari-like malformation. The dorsal subarachnoid space may appear dilated immediately cranial to the lesion, while syringomyelia may begin immediately caudal to the dural/fibrous band. Although a relationship between this lesion and syringomyelia formation has not yet been documented, obstruction to cerebrospinal fluid flow secondary to this lesion is suspected in people and dogs.

Atlantoaxial instability/subluxation, dens hypoplasia/aplasia

Atlantoaxial instability/subluxation may be diagnosed on cervical radiographs by demonstrating dorsal displacement of the axis relative to the atlas and increased distance between the dorsal arch of the atlas and the spinous process of the axis. Abnormalities of the dens, including dens hypoplasia or aplasia, may also be apparent. If instability is not seen with the neck in a neutral position, it may be demonstrated by positioning the neck in mild flexion, preferably using fluoroscopy. In a subset of cases instability is not demonstrable using radiography alone. Dens abnormalities, such as dorsal angulation of the dens or congenital non-union of the dens may also not be visible on plain radiographs. In these cases MR imaging can elucidate the diagnosis.

In dogs with atlantoaxial instability that is apparent on cervical radiographs advance imaging is also necessary to assess degree of neural compression and to determine whether other disorders of the craniocervical junction are present, such as syringomyelia. Magnetic resonance imaging is preferred over myelography, as it allows evaluation of neural parenchyma in addition to revealing degree of spinal cord compression.

TREATMENT

Medical and surgical treatment options are described for disorders of the craniocervical junction. Medical options overlap for many of these conditions. Corticosteroids may be used for their anti-inflammatory properties, their effect on CSF production, and their ability to limit sympathetically mediated pain. Other commonly prescribed pain medications include gabapentin, pregabalin, cyclooxygenase 2 inhibitors such as carprofen, and opioid analgesics such as fentanyl and butorphanol. Anti-epileptic medications are necessary if seizures are present. Although treatment options can improve clinical signs, in many cases treatment is merely symptomatic and allows disease progression to continue.

In dogs with atlantoaxial instability immobilization of the head and neck using a ventrally reinforced cervical splint for a minimum of 6 weeks is essential, along with strict exercise restriction. This is used to encourage fusion of the atlantoaxial junction through fibrosis. Splinting and exercise restriction have also improved the clinical signs of dogs with atlantooccipital overlapping.

Surgical options depend on the condition(s) present. Specific options include surgical stabilization of the atlantoaxial and/or atlantooccipital junctions, suboccipital craniectomy (a.k.a. foramen magnum decompression), dural/fibrous band resection, and odontoidectomy. The most challenging aspect of making treatment decisions for dogs with disorders of the craniocervical junction is deciphering the role of each abnormality in the patient's clinical presentation. As we begin to recognize a larger number of anomalies in the craniocervical junction we increasingly need to address more than one disorder per patient. In formulating a treatment plan, a number of factors must be considered:

- Presence and etiology of neural compression/concussion
- Presence of unstable or dynamic compressive lesions
- Presence of secondary pathological conditions, such as syringomyelia
- Patient factors, such as skeletal maturity and size
- Owner preference, financial concerns

REFERENCES

1. Barnes PD, et al. **Magn Reson Imaging Clin N Am** 2000;8(3):651-674. 2. Cerda-Gonzalez S, et al. **Vet Radiol Ultrasound** 2009;50(3):264-268. 3. Cerda-Gonzalez S, et al. **Vet Radiol Ultrasound** 2009;50(5):467-476. 4. Cerda-Gonzalez S, et al. **Vet Clin North Am Small Anim Pract** 2010;40(1):121-124. 5. Chamberlain WE, et al. **Yale J Biol Med** 1939;11:487-96. 6. Couturier J, et al. **J Small Anim Pract** 2008;49(9):438-443. 7. Cross HR, et al. **J Small Anim Pract** 2009;50(8):399-405. 8. Dewey CW, et al. **Comp Cont Ed Pract Vet** 2009 (web only). 9. Goel A, et al. **Neurol India** 2009;57(3):253-246. 10. Levine DN, et al. **J Neurol Sci** 2004;220:3-21. 11. March PA, et al. **J Vet Int Med** 2005;19:418-419. 12. Nakamura, et al. **Neuropathology** 2000;20(1):38-43. 13. Owen MC, et al. **Vet Radiol Ultrasound** 2009;49(6):570-572. 14. Rusbridge C, et al. **J Small Anim Pract** 2009;50(8):386-393. 15. Schmidt MJ, et al. **Am J Vet Res** 2009;70(4):508-512. 16. Schmidt MJ, et al. **Vet Radiol Ultrasound** 2008;49(5):472-476. 17. Stalin CE, et al. **Am J Vet Res** 2008;69(1):89-93. 18. Vermeersch K, et al. **Vet Surg** 2004;33(4):355-360. 19. Warren-Smith CM, et al. **Vet Radiol Ultrasound** 2009;50(6):635-638.

SUCCESSES & FAILURES IN THE SURGICAL MANAGEMENT OF CAUDAL OCCIPITAL MALFORMATION SYNDROME IN DOGS

Andy Shores, DVM, MS, PhD, DACVIM (Neurology)
Auburn, AL

Introduction

Surgical and medical management of COMS / Chiari malformation in dogs is, in the grand scheme of things, relatively new in veterinary neurology and neurosurgery. Many of us have set out to answer the basic questions regarding this syndrome and some of the basic answers are known; however, the more complex questions await the benefits of experience and controlled and long-term studies. This presentation will capsualize my approach to COMS / Chiari syndrome management, the lessons I have learned in management through successes and failures, and my intended directions for improving long term outcomes.

Defining COMS / Chiari Syndrome (Chiari-Like Malformation/ Syringomyelia [CM/SM] Syndrome)

Chiari-Like Malformation (CM) in the Cavalier King Charles Spaniels (CKCS) has recently been described as a *mismatch* of caudal fossa skull volume and caudal fossa parenchyma volume. This mismatch, in some, changes the dynamics of CSF flow and results in the development of syringomyelia (SM). The pathogenesis of CM and SM remain under investigation; however, the consequence of these conditions can result in patients with a variety of clinical signs including phantom scratching, cervical pain or hypersensitivity, paresis, and scoliosis. Characteristic MRI changes have been described.

Some estimate the incidence of CM in Cavalier King Charles Spaniels is 95% (CKCS), while as many as 50% of these have an accompanying syringomyelia (SM). Other breeds reported with CM (but with a lower incidence) include the Boston Terrier, Brussels Griffon (Griffon Bruxellois), Bull Terrier, Chihuahua, King Charles Spaniel (the English toy spaniel), Maltese Terrier, Pomeranian, Toy Poodle, and the Yorkshire Terrier. CM and SM have been diagnosed at Auburn University in a Welch Corgi and a Pug.

Indications for Surgery

Surgery is often recommended in dogs with CM and SM at Auburn University using the following criteria: 1) MRI evidence of CM and SM; 2) a syrinx measuring $\geq$ 3 mm diameter; 3) clinical signs of phantom scratching, cervical pain or hypersensitivity, or thoracic limb paresis (and without MRI / CSF evidence of other changes that could produce the same clinical signs). Owners are given the option of medical therapy, although surgical therapy is the recommended option in patients matching the aforementioned criteria. Younger patients are not excluded from this recommendation when they meet the criteria.

Surgical Procedure

The surgical technique includes foramen magnum decompression (FMD) using Lempert and Micro-Kerrison ronguers, a durotomy using a #12 scalpel blade and tenotomy scissors, duroplasty using swine intestinal submucosa (SIS) and a free autogenous adipose tissue graft (FAATG). A dorsal laminectomy of the cranial portion of C1 is sometimes necessary. After the decompression and the durotomy, adhesions of the dura to the nervous tissue parenchyma are broken down using triangular shaped cellulose sponge. The dura is marsupialized by suturing the cut edges to surrounding fascia using 5-0 interrupted synthetic absorbable monofilament sutures. The SIS is sutured similarly over the durotomy defect in a tented fashion. The FAATG is typically obtained from the gluteal region and placed over the site prior to routine closure.

Immediate postoperative care typically consists 1 to 2 nights' stay in the ICU with maintenance IV fluids. Medications administered include gabapentin, prednisolone, and omeprazole *per os*. Most patients are discharged to the owner's care 2 to 3 days after surgery. At home care includes rest and restricted activity for 3 weeks with a gradual return to normal activity after this time. Gabapentin and omeprazole therapy are continued for at least 6 months and often for life while prednisolone therapy is continued for either 7 days (without tapering) or over a course of 3 weeks (tapering dose regime).

Lessons Learned

In the genesis of our technique, I think several observations have improved the surgical procedure. The general principles for all surgeries (exposure and hemostasis) apply. In addition, judicious use of the electro-scalpel during the muscle dissection reduces the time of the procedure. The use of a surgical drill is not necessary for the FMD. The occipital bone is usually *eggshell* thin in the CKCS. In the Chihuahua, the occipital bone often does not extend over the area requiring decompression and boney decompression is not required. The area of the FMD only needs to extend to the medial edges of the cerebellar hemispheres (this can be judged by the thickness of the dura in this area). At least some portion of the dorsal lamina of C1 should be removed as evidenced by the thickness of the dura in this region and the presence of adhesions to the underlying parenchyma. It is not necessary to completely resect the fibrous tissue overlying the cisterna magna. Once a portion of the dura can be defined by surgical dissection, the durotomy is performed and any overlying, thickened tissue is sutured along with the dura to the surrounding fascia in the marsupialization. Excessive dissection of this thickened fibrous tissue usually results in additional hemorrhage and tissue trauma.

The FAATG should be about 150% of the area covered to allow for atrophy and loss of tissue before adequate blood supply develops. Grafts more than 5 mm thick do not revascularize well, and less than 2 mm have not been evident over the marsupialization site on 6-month follow-up MRIs.

All patients do not appear to respond equally to gabapentin therapy. We have begun assaying gabapentin levels in an attempt to optimize this therapy.

Older patients and patients with concomitant primary secretory otitis media (PSOM) or generalized ATOPY or food allergies do not respond to surgery as well as patients not exhibiting these additional problems. The author believes the older patients may have permanent changes in the dorsal root ganglia of the upper cervical spinal nerves. Uncontrolled allergies and middle ear disease can continue to drive the hypersensitivity issue and promote the phantom scratching and sometimes continued episodes of pain and discomfort. Younger patients (3 to 6 months of age) matching the earlier criteria for surgical candidacy have had a reduction in the syrinx diameter and appear to improve more rapidly and more completely than older patients.

Summary and Conclusions

Surgical management of CM and SM is an evolving process. Obviously, a controlled randomized study has the potential of answering some of our questions; however, this is not very practical in the clinical setting. Some tailoring of the surgical management to the individual has seemed necessary in the development of our technique, although being as consistent as possible has given us some level of data to support what we do. To date, none of our patients have required re-operation, although follow-up times have not exceeded 2 ½ years. The morbidity of this procedure has been minimal and there have been no deaths. Initial recoveries previously reported in 14 patients with follow-up times of 1 week to 1 year have been judged good to excellent by the owners. Follow-up MRI studies have demonstrated survival of the FAATG over the durotomy site. Consultations with the dermatology service has assisted in management of many of our patients have improved outcome and client satisfaction. Younger surgical candidates need to be recognized early by the referring veterinary population to afford a better chance for a more complete recovery. Investigations of efficacious gabapentin levels and of newer drugs to manage neuropathic pain and hypersensitivity may greatly augment surgical management. At this time, diligence in minimizing tissue trauma and the use of the FAATG seemingly has resulted in initial improvement in all patients and to date no repeat surgeries have been necessary.

REFERENCES

[1.] CAPPELLO R, et al: **VetSurg** 36:509, 2007 [2.] http://www.cavalierhealth.org/syringomyelia.htm [3.] CERDA-GONZALEZ, S, et al.: **VetRadUltra** 50 (1): 37, 2008. [4.] CROSS HR, et al: **J Small Anim. Pract**. 2009 Aug; 50(8):399. [5.] DRIVER C, et al:. **22nd ESVN** Annual Symposium, Sept. 2009. [6.] AKIN EY, et al: **Veterinary Surgery Forum** (abstr), 2009. [7.] CARRERA I, et al: **Am J Vet Res**; 2009 Mar;70(3):340.

HISTOPATHOLOGIC CHANGES & MRI FINDINGS IN DEGENERATIVE FORAMINAL STENOSIS

Thomas Gödde, Tierärztliche Gemeinschaftspraxis Piding, Germany & Kaspar Matiasek, Neuropathology Laboratory, The Animal Health Trust, Newmarket, UK

L7-S1 foraminal stenosis as part of a degenerative lumbosacral stenosis is a quite common disease entity in our patients presented with clinical signs of a cauda equina – syndrome as well as lameness of unknown origin of the hind limbs In our study about the surgical treatment of lumbosacral foraminal stenosis using a lateral approach in 20 dogs with degenerative lumbosacral stenosis, recognition of L7-S1 foraminal stenosis was defined as absence of the physiological fat signal in one or more of the three foraminal zones on sagittal and transversal imaging planes respectively. Images were obtained by means of a 1.5 Tesla magnet and a human spinal cord coil or a 0.23 Tesla low field system in combination with a solenoid coil and a SpinEchoT13D technique with thin slices <2mm. During both exams, patients were positioned with extended limbs in ventral or lateral recumbency. Decompression of the affected nerve roots at the level of the L7-S1 neuroforamen was achieved by means of a uni- or bilateral dorsolateral foraminotomy and nerve biopsies were harvested in selected cases. The current investigation was undertaken

-to define histopathologic changes of the affected L7-S1 nerve roots to prove our MRI based definition of a foraminal stenosis
-to investigate pathophysiologic mechanisms of radicular pain
-to adopt our imaging protocol to improve recognition of foraminal stenosis and concurrent L7 nerve root pathology

Histopathology

L7 nerve roots were evaluated in four dogs. All animals showed an extensive concentric perineurial fibrosis. The density of large myelinated Ia/b afferents was markedly reduced in all samples while small myelinated fibres were less affected. About 60% of remaining large fibers showed multisegmental hypomyelination and occasional active myelin sheath breakdown with axonal preservation. The endoneurial sheath of affected fibres was remarkably enlarged and accompanied by a general increase in endoneurial collagen. Moreover all nerve roots showed numerous subperineurial and endoneurial Renaut bodies of 40 to 400 μm diameter.

Inflammatory changes were not observed. Two samples exhibited a moderate media hyperplasia in most endoneurial arteries. Further vascular changes included meander like course of epineurial veins. Electron microscopy confirmed the demyelinating character of nerve fiber damage. Further, multiaxonal collagen pockets were occasionally seen.

Pathophysiology

Both the abundancy of Renaut bodies and nerve fiber pathology are consistent with a chronic compression neuropathy. Due to direct mechanical influence and/or vascular impairment, the replacement fibrosis is complicated by fibroplasia of the perivascular endoneurial interstitium and concentric perineurial hyperplasia with further impact on the endoneurial milieu. Due to its larger diameter intraforaminal aspects of the dorsal root ganglia are particularly prone to pressure. This influence as well as the incomplete differentiation result in dynamic changes and phenotypic switching with suspected hyperexcitability (hyperalgesia) and conveyance of non-noxious signals on the pain pathways (allodynia). The damage of proprioceptive fibers may furthermore lead to misinformation on the actual muscle tone and tension.

Histopathologic implications to improve the imaging protocol

The preforaminal zone of the spinal canal representing the lateral recesses has to be added to the observation protocol on transversal SE T1 images, as visible changes of nerve root thickening, sometimes in form of a number 8 figure start already there and may be followed through the neuroforamen. Application of dorsal GESTIR protocols with 4mm slice thickness, tilted about 7-10 degree from craniodorsal to caudoventral allow to follow the long axis of the spinal nerve from L7 at least to the level of the entrance zone of the neuroforamen. The normal low field MRI image is a slightly whitish dotted line representing the exiting spinal nerve in combination with an oval shaped, hyperintense structure representing the spinal ganglion. Recognition of the nerve is facilitated as the epidural fat appears black and consequently contrast to the exiting nerve is high. Pathologic changes include preforaminal and foraminal, hyperintense thickening of the spinal nerve root, probably due to embedding of the Renaut bodies. The spinal ganglion is usually no longer detectable in this surrounding. At the middle to exit zone of the neuroforamen, the exiting nerve is usually represented by a hyperintense, round to oval structure, significantly smaller in diameter due to the stenotic changes.

Conclusions

-according to the histopathologic changes of the L7 nerve root in our biopsies, absence of the foraminal fat signal in two or more foraminal zones on sagittal as well as transversal images is considered pathognomonic for a L7-S1 foraminal stenosis

-the dorsolateral approach to the neuroforamen allows decompression of the nerve roots as well as establishing a diagnosis by means of a nerve root biopsy

-histopathologic changes include severe fibrosis with increase in nerve root diameter, embedding of Renaut bodies, abnormal vessels, loss of axons and demyelination at the level as well as proximal and distal to the foraminal border lines

-occurrence of Renaut bodies is a diagnostic prerequisite of foraminal stenosis and allows differentiation from other inflammatory, infectious and neoplastic changes

-MRI recognition of a foraminal stenosis includes loss of fat signal due to nerve root thickening and/or obstruction by bony or connective tissue proliferations and less frequently disc material

-preforaminal nerve root thickening – "number 8 figure" - at the level of the spinal recess cranial to the foraminal entrance zone may be observed on transversal SE T1 weighted images, especially in chronic cases

-a slightly tilted craniodorsal to caudoventral GESTIR protocol allows definition of a hyperintense nerve root ; recognition of the dorsal root ganglion of the contralateral side may serve as a control

-both fibre damage and neuroplastic changes explain the neurological deficits and the neuropathic, NSAID non-responsive pain in animals, affected with this condition

THE ULTRASONIC ASPIRATOR: CURRENT & POTENTIAL NEUROSURGICAL APPLICATIONS

Patrick J. Kenny BVSc DACVIM (Neurology) DECVN
London, UK

The ultrasonic aspirator was first developed in the late 1960's as an ophthalmic surgical tool to aid in the removal of cataracts (phaecoemulsification)[1]. It's potential as a neurosurgical tool was quickly recognized and within a decade was being used in the excision of various tumors of the nervous system[2].
Several models of surgical ultrasonic aspirators are currently commercially available (Cavitron Ultrasonic Surgical Aspirator (CUSA™), Integra Selector®, Sonopet Omni®, etc). Most early experimental work and clinical reports were performed using the CUSA™.

Principles of Ultrasonic Aspiration & Dissection

At a basic level the hand piece of an ultrasonic aspirator consists of a hollow tube that vibrates ultrasonically in a longitudinal plane. Irrigation is directed to the tip with an outer sleeve. Suction is applied via the center of the tube. Suction pressure, irrigation rate and power of ultrasonic vibration are controlled at the main console. Operation is controlled by the surgeon with a foot pedal. When in operation all three functions (vibration, irrigation and suction) occur simultaneously.

The ultrasonic vibration causes fragmentation of tissues in two ways. Tissues with a high water content (such as neoplastic cells) are fragmented by cavitation. Fibrous tissues are fragmented by direct mechanical action of the tip. Tissues with a high collagen or elastin content (e.g. blood vessels, dura, tumor capsules) are relatively spared. Ultrasonic dissection therefore allows selective tissue fragmentation[3]. Tissue/tumor within 1-2mm of the tip is fragmented[2] and immediately removed by suction. Aerosolation of tumor cells has been reported[4] but the clinical significance of this finding has not been investigated. An important advantage of ultrasonic dissection is decreased traction and transmitted movement to surrounding tissues compared to conventional surgical techniques[5]. Encapsulated masses can be evacuated from the inside before removing the capsule using conventional techniques. Sparing of vascular structures facilitates a cleaner surgical field, but the potential for vessel damage and hemorrhage has been noted at higher powers[3]. Aspirated tumor tissue has been reported to be successfully examined histologically[6], but it has been recommended to remove a small sample by conventional methods for evaluation[7].

Effect of Ultrasonic Dissection on Neural Tissue

Experimental studies on cats show ultrasonic aspiration at low powers produced localized lesions, sparing function and blood flow of adjacent cord, with histologically sharp demarcations between the aspiration cavity and normal tissue[8]. Prolonged exposure (30 minutes) of cat spinal cord to ultrasonic vibrations without physical contact resulted in functional (loss of somatosensory evoked potential, paraplegia) and histological changes (white matter edema)[2]. These changes were more severe in cats where the dura was opened. Experiments on rat sciatic nerve have shown that contact with the probe tip at low intensities will abolish the conduction of action potentials, but contact with the vibrating probe shaft has little effect[8]. Rat facial nerves have also been experimentally subjected to ultrasonic vibration[9]. An acute decrease of the conducted nerve action potential and increase in latency of muscle action potential was detected. In the majority of rats facial movements returned within two weeks. At two weeks there was histological evidence of perineurial disruption, fibre degeneration and of regeneration. Multiple unilateral cranial nerve palsies have been reported after the removal of an acoustic neuroma[10]. The suspected pathogenesis was transmission of high power ultrasonic vibration through the petrous bone. Aspiration power of <20% created minimal acute morphologic damage to the cerebral cortex in a dog model[11].

Current Neurosurgical Applications in Human and Veterinary Neurosurgery

Ultrasonic aspirators are widely used in human neurosurgery for the resection of intracranial (meningioma[5], acoustic neuroma[5], glioma[5], craniopharyngioma[5]) and spinal neoplasms (meningioma[5], neurofibroma[5], ependymoma[5,12], astrocytoma[5,12]), and in epilepsy surgeries such as temporal lobe resections[13].

There are several literature reports documenting the use of ultrasonic aspirators in veterinary neurosurgery, including the excision of intracranial meningiomas[3,14,15], a choroid plexus papilloma[3], a cerebrovascular malformation[3], and for hypophysectomy[7].

Recent Developments

Manufacturers of ultrasonic aspirators have trended towards developing smaller, lighter hand-pieces to decrease surgeon fatigue and improve operation in deep or confined areas[16]. Modifications have also been made to aspirator tips to allow easier resection of fibrotic tumors[17].

In the last decade a variation of ultrasonic aspirator technology has led to the development of an ultrasonic bone curette (Sonopet Omni®). The design of the hand-piece is similar in that irrigation and suction is supplied, but the tip has various configurations of an abrasive surface and is modified to vibrate ultrasonically in a torsional as well as longitudinal plane. This allows the cutting and scraping of bony structures with an oscillating tip as an alternative to the conventional rotating burr, and therefore a decreased possibility of catching or being dragged into surrounding tissues. The ultrasonic bone curette is reported to be safe and effective when used adjacent to dura, nerves and vessels[18]. It has been used successfully used in several spinal and skull-base surgeries including foramen magnum decompression, cervical laminoplasty, laminotomy[19], anterior clinoidectomy, and opening of the internal auditory canal[20].

Potential Veterinary Neurosurgical Applications.

Ultrasonic dissection and aspiration has the potential to improve resection of many extra- and intra-axial tumors of the nervous system in veterinary patients. The reported safety factor inherent to the ultrasonic bone curette lends itself to procedures that require the removal of bone in close proximity to vital neural or vascular structures, such as cranectomies, caudal fossa decompressions, laminectomy for cervical spondylomyelopathy, decompression of lumbosacral and foraminal stenosis, and intervertebral disc disease. Post-operative outcomes should be studied and compared to conventional techniques.

REFERENCES

[1]Kelman CD. **Am J Ophthalmol** 1967:64;23-35. [2]Flamm ES. et al **Neurosurgery** 1978:2;240-5. [3]Shores A. et al **Prog Vet Neurol** 1991:2;89-94. [4]Preston JK. et al **Br J Neurosurg** 1999:13;486-9. [5]Epstein F. **Clin Neurosurg** 1983:31;497-505. [6]Malholtra V. et al **Acta Neurochir (Wien)** 1986:81;132-134. [7]Axlund TW. et al **Vet Surg** 2005:34;179-89. [8]Young W. et al **Neurosurgery** 1981:8;689-94. [9]Gleeson MJ. et al **Clin Otolaryngol** 1986:11;177-87. [10]Ridderheim PA. et al **Acta Neurochir (Wien)** 1987:89;84-86. [11]Bagley RS. et al **Am J Vet Res** 2001:62;248-51. [12]Garces-Ambrossi GL. et al **J Neurosurg Spine** 2009:11;591-9. [13]Harkness W. et al **Childs Nerv Syst** 2006:22;936-944. [14]Greco JJ. et al **J Am Vet Med Assoc** 2006:229;394-400. [15]Cavanaugh RP. et al **J Small Anim Pract** 2008:49;244-8. [16]Sawamura Y. et al **Neurosurgery** 1999:45;1192-6. [17]Klopfenstein JD. et al **Barrow Quarterly** 2004:20. [18]Nakagawa H. et al **International Congress Series** 2004;445-9. [19]Nakagawa H. et al **J Neurosurg: Spine** 2005;2:431-435. [20]Hadeishi H. et al **Neurosurgery** 2003:52;867-70.

A MITOCHONDRIALLY INHERITED NEUROLOGICAL SYNDROME IN THE GOLDEN RETRIEVERS

Karin Hultin Jäderlund, DVM, PhD, DECVN
Oslo, Norway

INTRODUCTION

In 2002, three related Golden Retrievers with gait abnormalities were presented to the author. The clinical picture of those dogs was not in agreement with any other previously described canine disease. Since then, more than 20 Golden Retriever dogs have been diagnosed with this novel neurological syndrome by the author. Both the phenotype and the genotype have been published (1, 2). The disease was characterized as a mitochondrially inherited sensory ataxic neuropathy owing to the neurological signs, in agreement with a similar clinical syndrome in humans. Sensory ataxic neuropathies present with loss of proprioceptive sensations and tendon reflexes, with preservation of muscle strength (3). This syndrome is associated with a number of aetiologies in humans and mitochondrial disorders are recognized among them. The canine disease is suitable to serve as a naturally occurring animal model for mitochondrial diseases.

CLINICAL CHARACTERISTICS

All owners of affected dogs had noticed an insidious onset of gait abnormalities in puppyhood. The signs were slowly progressive over months to years. Routine physical examination did not identify any abnormalities except an ataxic gait and, in most cases, hyperextension of the carpi. Affected dogs had no overt muscle atrophy. At neurological testing, postural reactions and spinal reflexes were generally decreased. The patellar reflexes were most prominently affected and were completely absent in a majority of the dogs. The pelvic limbs were more affected by neurological deficits than the thoracic limbs. Pain perception was within normal limits. The constellation of signs was similar from dog to dog, but the severity of signs differed between dogs, apparently irrespective of age or duration from onset of signs. The neuroanatomical diagnosis in these dogs referred to the sensory compartment of the spinal reflex arcs in both pelvic and thoracic limbs, and to afferent proprioceptive pathways in sensory nerves and spinal cord. Results from routine blood work, cerebrospinal fluid analyses and radiographic examinations were not diagnostic. At electrophysiology, affected dogs had significantly lower sensory nerve conduction velocities than control dogs in all nerves tested (superficial radial, ulnar, peroneal and tibial nerve). Concerning motor nerve conduction studies, amplitudes in sensory nerve testing and electromyography, no conclusive differences from control dogs were found. About half of the affected dogs were euthanized before attaining four years of age, owing to neurological impairment. So far, no dog has become non-ambulatory or has died spontaneously as a result of the disease. The oldest affected dog hitherto was euthanized at ten years of age because of an unrelated disorder.

NEUROPATHOLOGY

In both the central and the peripheral nervous system, a typical symmetrical spatial distribution of different types of neurodegenerative changes was seen at post mortem examination. In the spinal cord and peripheral nerves, the changes were consistent with a chronic progressive central and peripheral sensorimotor axonopathy affecting in particular the proprioceptive pathways with very mild involvement of peripheral motor axons (1, 4).

GENETICS

A common ancestor, a female born in 1976, was in the maternal lineages of all affected dogs, for at most eight generations back in time. Analysis of the pedigrees excluded a nuclear inheritance, and hence transmission was proven to be mitochondrial (2). The mitochondrial (mt) genome is a circular double-stranded molecule containing 37 genes. The 37 genes encode 13 protein subunits of the respiratory chain enzymes together with 22 transfer RNAs (tRNAs) and two ribosomal RNAs. All the other subunits of the respiratory chain enzymes are encoded from the nuclear genome. In every somatic cell, there are thousands of copies of the mtDNA. Because of the polyploid nature of the mtDNA content in any one cell, both mutant and wild type mtDNA can coexist in the cell, a condition called heteroplasmy. The proportion of mutant mtDNA to total mtDNA is called the mutant load. With just one type of mtDNA present – wild type or mutant – the condition is called homoplasmy.

The complete mtDNA was re-sequenced in affected and control dogs. A single base-pair deletion was identified in the mt $tRNA^{Tyr}$-gene of the mtDNA from affected dogs and maternal relatives. Mutant load was estimated by a quantitative oligonucleotide ligation assay (qOLA) method. All the affected dogs and all their tested clinically unaffected 1st, 2nd, 3rd and 4th degree maternal relatives had a mutant load of >88% in blood. Concerning the degree of heteroplasmy, the group of affected dogs was not possible to distinguish from these close maternal relatives. In more distant maternal relatives, the degree of heteroplasmy varied from 5 to 100% wild type among dogs (2, 4).

A northern blot analysis was applied to measure the steady-state levels of the mutated $tRNA^{Tyr}$-molecule compared to control mt tRNA-molecules, in muscle tissue from three affected dogs and two control dogs. All three affected dogs showed significantly reduced steady-state levels of $tRNA^{Tyr}$ compared to the control dogs and the control tRNAs. These data indicated an increased degradation and hence an impaired stability of mutated $tRNA^{Tyr}$ (2).

STUDIES OF MITOCHONDRIA

Fresh muscle biopsies were harvested from five affected dogs and five age-matched control dogs of the same breed. Muscle mitochondria were prepared, and respiratory chain enzyme activities and rates of mitochondrial ATP production were determined pair-wise in a blinded fashion according to Wibom et. al. (5). In affected dogs, the ATP production rate and the enzyme activities in the respiratory chain complexes were reduced in a mode that indicated dysfunction of the respiratory chain units encoded by mtDNA

(2). Moreover, the reduced activities of multiple enzymes with subunits encoded by mtDNA were in accordance with mutations in tRNA-genes. Complex II of the respiratory chain showed no significantly reduced activity, and is the only complex entirely encoded by the nuclear DNA.

Freshly frozen muscle biopsies from the same animals were used for histochemistry and electron microscopy. On combined staining for succinate dehydrogenase (SDH) and cytochrome oxidase (COX) activity, affected dogs had a bluish staining of the muscle fibres compared with their matched controls (2). This indicated a shift towards a lower COX activity compared with the SDH activity in the muscles of affected dogs. The SDH and COX stain, respectively, reflect the activity of complex II and complex IV of the respiratory chain. On electron microscopy, one affected dog had paracrystalline inclusions in the mitochondria, a finding characteristic for human adult mitochondrial muscle pathology (6).

CONCLUSIONS

Mitochondrially inherited sensory ataxic neuropathy in Golden Retriever dogs is a novel neurodegenerative syndrome. The disease has a chronic, slowly progressive clinical course and the prognosis is guarded with respect to a full-length dog life. The disease is caused by a deletion in the mt $tRNA^{Tyr}$-gene. The phenotypic expression of this disease is associated with a high mutant load, but the disease seems to penetrate only at a low frequency. Therefore, the nuclear genetic background and/or some environmental factor may also be involved in the disease presentation. By not using carrier bitches for breeding, the disease incidence can be reduced.

REFERENCES

[1]Hultin Jäderlund K, et. al. **JVIM** 2007;(21)1307. [2]Baranowska I, et.al. **PLoS Gen** 2009 5(5):e1000499. doi:10.1371/journal.pgen.1000499. [3]Illa I, et. al. **Rev Neurol (Paris)** 2001;(157)517. [4]Hultin Jäderlund K. Acta Universitatis Agriculturae Sueciae, **Doctoral Thesis No.** 2009:73. [5]Wibom R, et. al. **Anal Biochem** 2002;(311)139. [6]Bourgeois et **Tarnopolsky. Mitochondrion** 2004;(4)441.

ALASKAN HUSKY ENCEPHALOPATHY: AN UPDATE

Karen M. Vernau DVM, MAS, DACVIM (Neurology)
JA Runstadler, HJ Husan, RJ Higgins, C Ackerley, BK Sturges, PJ Dickinson, C Giulivi, GD Shelton, JM Cameron, BH Robinson.
Davis, CA

INTRODUCTION

Alaskan Husky Encephalopathy (AHE) is a devastating, fatal disease, reported in 13 Alaskan huskies from different regions in the northern United States, including Alaska, Massachusetts, New York, Wyoming, Maine and Minnesota. The disease affects both male and female dogs; multiple dogs from the same litter are often, but not always, affected.[1,2] Most affected Alaskan husky dogs were juvenile (range 6 months to 6 years of age), racing sled dogs. Because the Alaskan husky mainly serves as a working dog in remote regions, many affected dogs are euthanized or die without a clinical diagnosis and/or a necropsy; anecdotal reports suggest that AHE is more common than reported. A similar disease is reported in 11 European Yorkshire terriers.[3]

Affected dogs may have an acute or chronic progressive history, usually with multifocal central nervous system deficits. Common neurological problems include seizures, changes in mentation, difficulty prehending food, absent menace response, hypermetria, proprioceptive positioning deficits, facial hypalgesia, ataxia and tetraparesis. Some dogs have had a waxing and waning clinical history, with dramatic changes in clinical signs. Diagnostic work-up was limited in the 13 dogs already described. One dog had normal serum pyruvate and lactate concentrations, and two dogs had normal cerebrospinal fluid pyruvate and lactate concentrations compared to 4 normal dogs. Two affected dogs (26 and 29 months of age) from the same litter in Alaska had intracranial imaging (computed tomography (CT) or magnetic resonance (MR)). On the CT images of one dog, there were bilateral hypoattenuating lesions in the thalamus. On the T2 weighted MR images of the second dog, there were bilateral hyperintense areas in the brainstem, from the thalamus to the medulla, as well as bilateral lesions in the putamen, caudate nucleus and claustrum. All affected dogs died or were euthanized, most within 6 months of onset of clinical signs; one dog lived for 1 year after onset of clinical signs when it died of "natural causes".[1,2]

On necropsy, neuropathological findings consisted of grossly visible bilateral, symmetrical, cavitated lesions in the brainstem. These lesions were consistently present in the thalamus, and variably present in the more caudal aspect of the brainstem. On microscopic examination, lesions consisted of neuronal necrosis, vascular hypertrophy, vascular hyperplasia and spongiosis, with neuronal sparing in other areas. Affected areas included multifocal lesions in the base of sulci of the cerebral cortex, gray matter of the cerebellar folia, the ventral cerebellar vermis, basal nuclei, and brainstem and cervical spinal cord. Pathologically, these lesions were described as subacute, necrotizing encephalomyelopathy, and the disease was called AHE.

Recognized causes of symmetrical brain lesions include metabolic, nutritional and toxin-induced etiologies. Since AHE in Alaskan huskies, and a similar disease Leigh-like subacute necrotising encephalopathy in Yorkshire terriers[3] occur in these specific breeds, it is likely that AHE is an inherited metabolic disorder, while a nutritional or toxic cause is less likely. Additionally, the clinical signs, MRI imaging features, and neuropathology, differ from descriptions of previously reported nutritional or toxin-induced central nervous system diseases.

LEIGH SYNDROME IN PEOPLE

Alaskan Husky Encepathalopathy resembles Leigh syndrome (LS) in people.[2] Leigh syndrome was first reported in 1951, in a human infant where the gross and histopathogical findings of subacute necrotizing encephalomyelopathy were described in detail.[4] Today, the diagnosis is based on characteristic topography of lesions in neuroradiological or neuropathological studies, such as bilaterally symmetrical lesions with increased signal intensity in the brainstem, spinal cord, cerebellum and cerebral cortex, with bilateral calcifications in the thalamus. Leigh syndrome most commonly affects infants and children; rarely adults; patients often have heterogeneous clinical signs. Infants with LS commonly have feeding problems, hypotonia, seizures, ataxia, and usually die before 2 years of age.[5]

Leigh syndrome is the clinical manifestation of a defect in mitochondrial energy metabolism. A number of biochemical deficiencies cause LS, such as pyruvate dehydrogenase complex deficiency and mitochondrial respiratory chain complex deficiency.[6] Genetic mutations causing LS are heterogenous, and may be caused by both mitochondrial DNA mutations (e.g.T8993G) and nuclear DNA mutations (affecting enzymes such as cytochrome c oxidase).[5,7] However, it up to 50% of people with LS, a molecular diagnosis is not made, likely because many of the nuclear DNA proteins for mitochondrial function have not yet been identified.[8]

MULTICENTER COLLABORATIVE STUDY

The hypothesis of this study is that AHE is caused by an inherited disorder affecting mitochondrial energy generation. Specific aims were to define the signalment, clinical signs, diagnostic features and long-term disease course in Alaskan huskies with AHE, and to evaluate the neuropathology, ultrastructure and function of the mitochondria in Alaskan huskies with AHE compared to age-matched clinically normal Alaskan husky dogs.

Since February 2008, we have examined 6 juvenile dogs from an Alaskan kennel, at the UC Davis William R. Pritchard Veterinary Medical Teaching Hospital (VMTH). All dogs were between 6-12 months of age; 4 were female and were 2 were male. All dogs had multifocal brain disease; clinical signs were similar to the previously reports in Alaskan huskies. Results of extensive extracranial diagnostic work-ups were normal. MRI of the brain was abnormal and remarkably similar in all 6 dogs, with symmetrical, multifocal lesions in the subcortical gray matter, thalamus, caudate nucleus, claustrum, putamen, midbrain, cerebellar vermis and brainstem.

Three dogs were euthanized immediately after diagnostic work-up, and 3 dogs were neutered and placed into foster homes for long-term follow up. One male dog had static neurological deficits for 5 months after presentation, when he developed progressive multifocal brain disease, and was euthanized. Repeat MRI four months after the initial MRI showed progression of the previously identified lesions. The two remaining dogs (one male and one female) are alive 1.5 years and 2 years after initial presentation. The MRI was repeated in one dog 1 year after the initial MRI and no changes were noted, compared to the original MRI. The other dog continues to have infrequent generalized seizures, and receives oral phenobarbital, but otherwise both dogs have static neurological deficits without a waxing and waning history.

To date, we have euthanized 4 dogs. On gross pathology, bilaterally symmetrical areas of cavitation were noted within the thalamus. Microscopically, lesions were widespread, and included foci of bilaterally symmetrical degeneration and encephalomalacia in the basal nuclei, thalamus, and brainstem (pons and medulla). In addition, there were extensive multifocal lesions noted within the subcortical gray matter of the cerebral cortex confined to the base of sulci and symmetrical lesions within the gray matter of cerebellar folia in the ventral vermis. The clinical, neuroradiological and neuropathological findings were almost identical to those of previous reports of AHE in dogs.[1,2]

Biochemically, no significant abnormalities were found in serum/CSF lactate and pyruvate levels, in pyruvate dehydrogenase enzyme activity or 2-^{14}C pyruvate whole cell oxidation in fibroblasts, when compared with controls. Electron transport chain enzyme activities were within normal limits in muscle and liver (complexes I, II, I+III, II+III, IV and citrate synthase) but a reduction in complex IV and V activity was noted in the cerebral cortex and thalamus compared to controls. No genetic mutations were present in eleven candidate nuclear and mitochondrial genes known to cause LS in humans.

We conclude that there is no evidence to support a systemic defect within the TCA cycle, pyruvate dehydrogenase complex, or respiratory chain enzymes as the cause of AHE, but mitochondrial involvement is still suspected, suggesting the disease is unlike any known cause of Leigh Syndrome. In addition, the similarity of this condition to LS in humans, suggest that further investigation of AHE in Alaskan huskies may lead to the development of a canine model for elucidating other mechanisms of LS.

REFERENCES

1. Wakshlag JJ, de Lahunta A, Robinson T, et al. Subacute necrotising encephalopathy in an Alaskan husky. **J Small Anim Pract** 1999;40:585-589. 2. Brenner O, Wakshlag JJ, Summers BA, et al. Alaskan Husky encephalopathy--a canine neurodegenerative disorder resembling subacute necrotizing encephalomyelopathy (Leigh syndrome). **Acta Neuropathol (Berl)** 2000;100:50-62. 3. Baiker K, Hofmann S, Fischer A, et al. Leigh-like subacute necrotising encephalopathy in Yorkshire Terriers: neuropathological characterisation, respiratory chain activities and mitochondrial DNA. **Acta Neuropathol** 2009;118:697-709. 4. Leigh D. Subacute necrotizing encephalomyelopathy in an infant. **Journal of Neurology, Neurosurgery and Psychiatry** 1951;14:216-221. 5. Hirano M, Kaufman P, De Vivo D, et al. Mitochondrial neurology I: encephalopathies. In: DiMauro S, Hirano M, Schon EA, eds. Mitochondrial Medicine. Abingdon, Oxon: **Informa Healthcare**; 2006:27-44. 6. Desnuelle C, Richelme C, V. P-F. Neurological Features of Genetic and Aquired Metabolic Mitochondrial Defects. In: Desnuelle C, DiMauro S, eds. **Mitochondrial Disorders** from pathophysiology to acquired defects. Paris: Springer; 2002:193-210. 7. Rahman S, Blok RB, Dahl HH, et al. Leigh syndrome: clinical features and biochemical and DNA abnormalities. **Ann Neurol** 1996;39:343-351. 8. Staley KJ, Sims KB, Grant PE, et al. Case records of the Massachusetts General Hospital. Case 28-2008. An 8-day-old infant with congenital deafness, lethargy, and hypothermia. **N Engl J Med** 2008;359:1156-1167.

SUBACUTE NECROTIZING ENCEPHALOPATHY (SNE) IN YORKSHIRE TERRIERS: AN UPDATE

Kaspar Matiasek, DVM, DrMedVet, FTA-Neuropathol
Newmarket, Suffolk, UK

INTRODUCTION

Human Leigh disease is a syndromic encephalopathy of mitochondrial origin that results from various respiratory chain abnormalities, all of which are associated with multiple clinical and pathological phenotypes. A canine resemblance of this fatal neurodegenerative disease has been described in Alaskan Huskies[1] and Yorkshire Terriers[2].

CASE SERIES

The pathological picture of SNE in both Huskies and Yorkshire Terriers is characterised by (1) a V-shaped, non-contiguous malacia extending from the striatum towards the medullospinal transition, (2) infarct-like cerebrocortical lesions surrounding the deep aspects of parietotemporal neopallial sulci and (3) Purkinje cell loss within the central parts of the cerebellar vermis. Corresponding to the individual disease course, the brain pathology in the different areas is composed of active and chronic lesions but the distribution pattern and site-specific mode of damage are identical amongst affected individuals. Most of the Yorkshire Terrier cases have been presented to the clinician with a history of either acute seizures or gait abnormalities and compulsive walking within the first year of life. Neurological examination always shows mental deficits, tetraataxia and dysmetria, delayed or absent conscious proprioception, hyperreflexia and hyper/dystonia of appendicular muscles. Cranial nerve testing indicates central visual deficits and the menace response is absent. Occasionally, the dogs are deaf and show other cranial nerve signs. In contrast to the Huskies, a monophasic course lead to rapid deterioration within 6 months after diagnosis in far most of the Terriers.

The mitochondrial origin was suspected due to the similarity to human Leigh syndrome and the presence of mild electron microscopical changes to the mitochondria. Ragged red fibers were not present. Preliminary enzymological data uncovered a respiratory chain 1 and 4 problem in two of the dogs. Sequence analysis of the mitochondrial genes encoding for mitochondrial t-RNAs just showed non pathogenic polymorphisms. Leigh syndrome relevant ATP-6 and major rearrangements were excluded.

CURRENT DEVELOPMENTS

Apart from the juvenile manifestation, late onset cases and remittent-relapsing courses have now also been recognised in mainland Europe, similar to Alaskan husky form,. In one of the cases a high concurrent urinary excretion of L-2 hydroxyglutaric acid was measured. Furthermore, absolute lactate levels and lactate-pyruvate relationship in blood and csf showed individual changes.

Since combined RCC defects may indicate an impairment of mitochondrial gene transcription further nuclear players, such as the elongation factor are in focus of our genetic screen. In the cases without muscle enzymology we investigate the expression of RCC in the brain in order to narrow the panel of candidate genes. Sequencing of the mitochondrial genome of 12 affected dogs is in progress.

REFERENCES

[1.] Brenner, O. et al. **Acta Neuropathol.** 2000; 100:50. [2.] Baiker, K. et al. **Acta Neuropathol.** 2009; 118(5):697

DIAGNOSIS & TREATMENT OF MITOCHONDRIAL MYOPATHIES

G. Diane Shelton, DVM, PhD, DACVIM (Internal Medicine)
La Jolla, CA

INTRODUCTION

The clinical recognition of mitochondrial disease is often challenging with an ever increasing spectrum of documented phenotypes. In addition to a predominantly "myopathic" or "encephalomyopathic" clinical presentations, patients may present with multi-organ system manifestations in which muscle involvement is only a part. Mitochondrial disease can occur at any age. Organs with a high energy demand are typically affected, including skeletal and cardiac muscle, endocrine organs, kidney, non-mucosal components of the intestinal tract, retina, and the central nervous system. Although effective treatments remain elusive, definitive diagnosis is crucial for determining appropriate symptomatic management as well as an accurate prognosis. Diagnostic difficulty results not only from the broad spectrum of clinical signs and heterogeneity of the diseases, but also from the absence of reliable screening or diagnostic biomarkers that are both sensitive and specific in all cases of mitochondrial disease. Currently extensive laboratory testing needs to be performed. Primary mitochondrial respiratory chain disease is a heterogeneous group of disorders characterized by impaired energy metabolism due to presumed genetically-based oxidative phosphorylation (OXPHOS) dysfunction. To add an additional level of complication, there is a dual genome origin with genetic abnormalities in either mitochondrial DNA (mtDNA, having a maternal pattern of inheritance) or nuclear DNA (nDNA, having classical Mendelian inheritance). More than 150 mtDNA pathogenic point mutations and 100 mtDNA deletions have been identified in symptomatic human patients.[1, 2] We are just at the tip of the iceberg in the recognition and diagnosis of mitochondrial disease in our veterinary patients. The specific aims of this discussion are to describe the clinical features most indicative of mitochondrial disease, identify baseline diagnostic testing for this group of diseases, describe the differences between primary versus secondary mitochondrial disease, and discuss available treatment options.

WHEN TO SUSPECT MITOCHONDRIAL DISEASE (RED FLAGS)

- Neurologic - Basal ganglia disease, neurodegeneration, myoclonus, ataxia, MRI findings consistent with Leigh disease
- Cardiovascular - Hypertrophic cardiomyopathy with rhythm disturbance, cardiomyopathy with lactic acidosis, dilated cardiomyopathy with muscle weakness, Wolff-Parkinson-White arrhythmia
- Ophthalmologic - Retinal degeneration, ophthalmoplegia/paresis, ptosis, sudden onset optic neuropathy/atrophy
- Gastroenterologic - Unexplained or valproate-induced liver failure, severe dysmotility, pseudo-obstructive episodes
- Other including myopathic presentation - Young patient with unexplained hypotonia, weakness, failure to thrive and metabolic acidosis, exercise intolerance not in proportion to weakness, hypersensitivity to general anesthesia, episodes of acute rhabdomyolysis

Other non-specific(vague) clinical signs such as unexplained chronic or cyclic vomiting, endocrinopathies, or failure to thrive, particularly if they occur in isolation, could be consistent with an underlying mitochondrial disease, but do not necessarily indicate a mitochondrial problem per se. When these diverse clinical signs are present in combination and a specific underlying cause not identified, the likelihood of a mitochondrial disorder increases. This is particularly true if the nonspecific features involve different organ systems. Such findings should prompt initiation of appropriate baseline diagnostic investigations for mitochondrial dysfunction.

BASELINE SCREENING TESTS FOR MITOCHONDRIAL DISEASE: INITIAL EVALUATION (adapted from[3])

Metabolic screening of blood and urine for all patients

- Basic chemistries, CBC
- Liver enzymes and ammonia
- Creatine kinase
- Blood lactate, pyruvate, and lactate to pyruvate ratio at rest and post-exercise
- Quantitative plasma amino acids
- Quantitative urine organic acids
- Plasma acylcarnitine analysis

Metabolic screening of spinal fluid for the patient with neurologic signs

- Lactate and pyruvate
- Quantitative amino acids
- Routine studies including cell count, glucose, and protein measurement

Characterize systemic involvement in all patients

- Echocardiogram
- Electrocardiogram
- Ophthalmologic examination
- Audiology testing
- Brain MRI

LACTIC ACIDOSIS

Lactate, the product of anaerobic glucose metabolism, accumulates when aerobic metabolism is impaired, causing a shift in the oxidized-to-reduced NAD+/NADH ratio within mitochondria. Elevated plasma lactate and/or pyruvate levels may occur in a wide range of conditions including anaerobic exercise, systemic diseases that increase blood lactate levels (including but not limited to hypoxia, hypotension, shock and sepsis), and in many metabolic diseases (amino acid disorders, organic acidemias, urea cycle defects, pyruvate metabolism defects, Krebs cycle defects, mitochondrial OXPHOS disorders, fatty acid oxidation disorders and others). The accurate interpretation of abnormal lactate and pyruvate concentrations is not always straightforward. Spurious elevations of plasma lactate commonly occur as a result of physical exercise before collection, poor blood collection techniques, or the animal struggling during blood collection. However, lactic acid accumulation in blood or CSF can be an important, albeit non-specific, marker of mitochondrial disease.[3, 4]

Blood for lactate analysis is usually collected into sodium fluoride tubes stopping the erroneous production of lactate by red blood cells. As an alternative, a blood drop can be measured with a hand held lactate analyzer. Determination of pyruvate levels can also be a challenge but is necessary to quantify blood lactate to pyruvate ratios, which indirectly reflect the NADH/NAD+ cytoplasmic redox state.[5] Pyruvate is quite unstable and requires proper collection techniques and handling. Blood for pyruvate analysis requires collection into 8% perchloric or trichloroacetic acid to precipitate proteins, immediate placement on ice, and either rapid analysis or immediate freezing of the supernatant. Mitochondrial disease patients may have normal lactate and pyruvate levels unless undergoing a metabolic crisis or following exercise so normal results cannot rule out an abnormality. Pyruvate may be elevated in pyruvate metabolism defects such as pyruvate dehydrogenase deficiency, pyruvate carboxylase deficiency, or biotinidase deficiency, but can also be elevated in the immediate postprandial period. Elevated plasma alanine levels (from amino acid quantification), may be a useful indicator of long-standing pyruvate accumulation. [3, 4]

Amino Acid Analysis

Amino acid analysis can be a useful adjunct when alanine elevations are present in both plasma and CSF, especially when hyperalaninemia occurs in a fasting specimen (mild elevations can occur from recent ingestion of a carbohydrate-rich meal). Similar to other biochemical markers of mitochondrial dysfunction, the sensitivity of an elevated alanine for mitochondrial disease is low, as it may only be elevated at certain times during a physiologic stress or regression. Thus, normal alanine levels do not exclude the diagnosis of mitochondrial disease. Elevations of proline, glycine and sarcosine have also been associated with mitochondrial dysfunction.[4]

Organic acid analysis

Organic acids (OAs) are byproducts of protein, carbohydrate and fat catabolism. Urine is the preferred specimen for OA screening owing to the ready excretion and greater extraction efficiency in urine compared to plasma, but can also be performed on CSF and plasma in select clinical scenarios. Urine OA analysis is routinely obtained in infants with sudden-onset encephalopathy in an effort to identify amino and organic acidopathies, as well as certain fatty acid oxidation disorders. During a period of clinical stability, urine OAs may have a low sensitivity for detecting mitochondrial disease, as abnormalities are often only present when a patient is acutely symptomatic. Increased excretion of tricarboxylic acid (TCA) cycle intermediates, ethylmalonic acid, and 3 methyl glutaconic acids commonly occur in mitochondrial disease, but are rarely diagnostic of a specific mitochondrial disorder.[4]

Carnitine analysis

Carnitine serves as a mitochondrial shuttle for free fatty acids and acts as a buffer for potentially toxic coenzyme A (CoA) esters. It permits restoration of intramitochondrial CoA and removal from the mitochondrion of esterifed intermediates by enabling their urinary excretion. In addition, this analysis enables identification of secondary fatty acid oxidation defects and carnitine deficiency which may occur in primary OXPHOS disorders. Quantification of blood total and free carnitine levels, along with acyl-carnitine profiling permits identification of fatty acid oxidation defects, as well as some primary amino acidemias and organic acidemias. [4]

TISSUE INVESTIGATIONS

For the diagnosis of both mitochondrial myopathies and encephalomyopathies, the muscle biopsy has a pivotal role. [6] The value of accurate histochemical analysis (in frozen tissue free from artifacts) should not be underestimated. Muscle biopsy allows for pathological (light and ultrastructural), biochemical and molecular studies. The biopsy is evaluated morphologically for evidence of excessive mitochondrial proliferation (modified Gomori trichrome stain for visualization of "ragged-red" fibers and succinic dehydrogenase [SDH] reaction for "ragged-blue" fibers) suggestive of primary OXPHOS disease. The SDH reaction reveals an entirely nuclear-encoded component of the respiratory chain and can be compared with the histochemical stain for cytochrome C oxidase, which is usually affected by mtDNA mutations. Muscle pathology often includes excessive storage of lipid droplets and glycogen, reflecting a general impairment of intermediate metabolism secondary to a defect of terminal oxidation. Muscle biopsies also demonstrate other pathological changes in muscle that could produce secondary abnormalities in muscle biochemistry. Muscle tissue can further be used for assay of individual respiratory chain enzyme complexes, for high resolution respirometry studies of OXPHOS activity, and for extraction of DNA for genetic testing. Mitochondrial proliferation and ultrastructural abnormalities can also be studied in the liver. It is important to perform a skin biopsy at the time of tissue biopsy to obtain cultured fibroblasts for possible later study.

DISTINGISHING PRIMARY FROM SECONDARY MITOCHONDRIAL DYSFUNCTION

Even when mitochondrial dysfunction is confirmed by sophisticated biochemical testing, it can be challenging to distinguish whether the cause for the dysfunction is a gene that directly impacts the electron transport chain (ETC) or is secondary to an unrelated genetic or environmental cause. The term "primary mitochondrial disease" refers specifically to mitochondrial dysfunction caused by genetic mutations (both mtDNA and nDNA genomes), directly impacting the composition and function of the ETC. However, mitochondrial dysfunction may occur secondarily in which respiratory chain dysfunction and oxidative stress (excessive accumulation of reactive oxygen radicals) are presumably the final common pathways of diverse external (i.e.: drugs) or intrinsic causes with defects in another energy related metabolic pathway (i.e.: fatty acid oxidation or amino acid metabolism). Currently, there is no accepted gene-based diagnostic algorithm for the diagnosis of primary mitochondrial disease that is useful for all patients or pursued by all metabolic specialists.[4] Testing for nDNA mutations can be performed on any tissue, including blood. However, most diagnostic nDNA gene testing should not be performed a priori but, rather, guided by the clinical picture, tissue-specific signs, and morphological and biochemical findings in a given patient.[4] In contrast, testing for mtDNA mutations is frequently most informative when performed on a muscle biopsy specimen.[4]

TREATMENT OF MITOCHONDRIAL DISEASES

The treatment of mitochondrial disease is still in its infancy. Aside from symptomatic management, treatment of mitochondrial disease focuses on maintaining optimal health, using preventative measures to mitigate worsening during times of physiologic stress (such as infection, dehydration, or surgery), and avoiding mitochondrial toxins.[7] Scientific support for the use of vitamin-based and cofactor-based mitochondrial therapies is accumulating. Such pharmacologic supplements are intended to promote critical enzymatic reactions, reduce putative damage from of excess free radicals, and scavenge toxic acylcoenzme A (acyl CoA) molecules, which accumulate in mitochondrial disease. Exercise also has an important role in mitochondrial disease therapy. Exercise has been shown to reduce the burden of abnormal mitochondria and increase the percentage of healthy, nonmutated mtDNA and improve endurance and muscle function. [8]

- Healthful nutrition – secondary mitochondrial dysfunction occurs with extreme malnutrition, anorexia, starvation and illness-related cachexia. Some patients may need restriction of carbohydrate, protein or fat.
- Avoiding mitochondrial toxins – drugs with reported mitochondrial toxicity include valproic acid, antiretrovirals, statins, aspirin, aminoglycoside antibiotics, acetaminophen, beta-blockers and possibly corticosteroids.[7]
- Ketogenic diet – Recent animal studies have shown that the high-fat ketogenic diet may be beneficial in optimizing mitochondrial function.[9] However, many mitochondrial disease patients have secondary fatty acid oxidation disorders and such a diet may be contraindicated. The ketogenic diet is the standard of care for pyruvate dehydrogenase deficiency, but is contraindicated in patients with known fatty acid oxidation disorders or with pyruvate carboxylase deficiency.
- Pharmacological treatment and supplements[7]
  - Coenzyme Q10 – Integral component of ETC, important intracellular signaling role, antioxidant and pro-oxidant roles
  - Riboflavin (B_2) – Flavoprotein precursor, key building block in complex I and II, and cofactor in key enzymatic reactions involving fatty acid oxidation and the Krebs cycle.
  - L-creatine – Source of high-energy phosphate, intracellular buffer for ATP, energy shuttle
  - L-arginine – Produces nitric oxide which has neurotransmitter and vasodilatory properties
  - L-carnitine – Critical role in mitochondrial fatty acid oxidation and esterification of free fatty acids
  - B-vitamin complexes - Antioxidant
  - Vitamins E and C - Antioxidant
  - Alpha-lipoic acid - Antioxidant

REFERENCES

[1]Dimauro S, et al. **Ann Med** 2005;37:222-232, [2]Greaves LC, et al. **IUBMB Life** 2006;58:143-151, [3]Hass RH, et al. **Pediatrics** 2007;120:1326-1333, [4]Haas RH, et al. **Mol Genet Metab** 2008;94:16-37, [5]Debray FG, et al. **Clin Chem** 2007;53:916-921,
[6]DiMauro S, et al. **Myology 3rd Ed**, 2004, [7]Parikh S, et al. **Current Treatment Options in Neurology** 2009;11:414-430, [8]Jeppesen TD, et al. **Brain** 2006;129:3402-3412, [9]Bough, KJ et al. **Ann Neurol** 2006;60:223-235.

THE APPLICATION OF GENE ONTOLOGY ANALYSIS TO FELINE INJECTION SITE SARCOMA GENE EXPRESSION DATA. ML Hauck[1], JC Fisher[1], DM Nielsen[1], C Cassidy[1], SL Gillings[2] F McCarthy[3]. 1. North Carolina State University, Raleigh, NC. 2. Seattle, WA. 3. Mississippi State University, Starkville, MS.

The genetic changes responsible for the malignant transformation in injection site sarcomas (ISSa) are unknown. The purpose of this study was to evaluate differentially expressed genes utilizing Gene Ontology (GO) to identify biological processes important in this transformation.

Gene expression data was collected from sixteen ISSa utilizing a human cDNA array. Twenty-seven thousand genes were interrogated. Differentially expressed genes were determined at the 5% false discovery rate using JMPgenomics. Fibrosarcomas (FSA), normal muscle and normal lymph node were used for comparison.

GO analysis of differentially expressed transcripts yielded the biological processes associated with the upregulated genes in all pairwise comparisons. In comparing differentially expressed genes between ISSa and FSA, ISSa had biological processes upregulated from DNA repair and telomerase maintenance, multiple metabolic pathways, cell communication and cell adhesion. FSA had multiple metabolic pathways, cell motion, cell cycle and cell communication. Although functional groups may overlap between samples, the genes represented are differentially expressed, suggesting these two tumor types utilize different pathways to be successful as neoplastic cells. For example, FSA express the adenosine A2b receptor, which plays a role in cell growth and gene expression, and ISSa express SMAD2, important in TGF-beta signaling. Both of these proteins fall in the category of cell communication.

GO analysis offers another means of understanding complex data sets such as transcriptome or proteomic data. In conjunction with pathway analysis, GO analysis offers an overview of cellular processes and can lend direction to further work in evaluation of tumor specific abnormalities.

PHASE I STUDY OF *CLOSTRIDIUM NOVYI-NT* SPORES IN TUMOR-BEARING DOGS. E. Krick[1], K. Sorenmo[1], S. Zhou[2], K. Kinzler[2], B. Vogelstein[2], L. Diaz[2]. 1. University of Pennsylvania School of Veterinary Medicine, Philadelphia PA. 2. Sidney Kimmel Comprehensive Cancer Center at Johns Hopkins University, Baltimore, MD.

The study purpose was to establish the maximum tolerated dose (MTD) of *Clostridium novyi-NT* (*C. novyi-NT)* spores in tumor-bearing dogs. Secondary purposes included pharmacokinetic analysis/documenting spore germination within tumors and tumor response. A standard dose-escalation scheme was planned, with MTD defined as the highest dose at which 0 or 1/6 dogs experienced dose-limiting toxicity (DLT). Dogs received one dose of *C. novyi-NT* spores intravenously and were hospitalized for monitoring. Toxicities were graded according to VCOG criteria and interventions performed according to specific endpoints/guidelines. Toxicity > Grade 2 or any toxicity combination that significantly affected patient status were considered DLT. Response, using RECIST criteria, was measured at 28 days.

Six dogs were enrolled. The first two experienced DLT, and the dose was de-escalated by 90%. Two of the next four dogs experienced DLT; therefore, further dose escalation was stopped, as the study endpoint was reached. Fever (medically manageable) was the most common toxicity (n=6). Two dogs developed *C. novyi-NT* abscesses: one within a nasal carcinoma and one splenic abscess. Both dogs required surgical intervention. *C. novyi-NT* was cultured from 1/6 tumors. Five dogs were available for response assessment: four had stable disease, and one had progressive disease.

The results of this study indicate that *C. novyi-NT* can germinate within canine tumors, and, although toxicity is common, it is medically and/or surgically manageable. Further studies in dogs with superficial tumors may allow for continued dose escalation, and results from these studies will inform future clinical trials in veterinary and human oncology.

PARR & FLOW CYTOMETRY FOR HEMATOLOGIC MALIGNANCY: ARE THEY CLINICALLY USEFUL?

Susan E Lana, DVM, MS, DACVIM (Oncology)
Fort Collins, CO

INTRODUCTION

With the advancement of molecular diagnostics in cancer medicine in recent years clinicians find themselves with a variety of options when it comes to selecting tests for their patients. In many cases the assays had previously been available only in human medicine or in a veterinary research setting making it difficult to know which clinical scenarios warrant such diagnostics. In many cases advanced diagnostics have proven to be more sensitive, more specific, and better able to characterize or sub classify disease than traditional methods such as cytopathology or histopathology. These test have provided a wealth of information and insight into the hematologic diseases we commonly see in veterinary medicine, however it can be difficult to know when the results of a test can guide a clinical decision such as when to continue or not continue treatment or when a test might only provide academic information that for some clients might be a waste of resources. The purpose of this review is to provide information regarding two widely available diagnostic test, PARR and flow cytometry. The intent is to provide insight into clinically relevant information that can be obtained from these assays when applied to common hematologic malignancies.

PCR FOR ANTIGEN RECPTOR REARRANGEMENT

Molecular characterization of hematologic malignancies has lead to the development of useful diagnostic and prognostic tests. In many human hematologic malignancies disease specific genetic alterations have been identified and provide an appropriate target for sensitive PCR based assays. For example in the case of in chronic myelogenous leukemia, the translocations of the bcr-abl gene produces a specific genetic product which when amplified by PCR allows for detection of as few as 1 neoplastic cell in 10^6 cells[1]. Sensitive detection of circulating tumor cells before treatment to determine the extent of disease, and after treatment to determine the efficacy of therapy, has resulted in improved outcome and monitoring of many types of leukemias[1]. Because lymphoma is the result of clonal lymphocyte expansion, neoplastic populations of lymphocytes can be identified by establishing the presence of clonal populations of lymphocytes and the PARR test is based on the premise that hematologic malignancies are clonal.

Detection of clonality relies on the fact that lymphocytes contain DNA regions that are unique in length and sequence. This is a result of the recombination of the V, D, and J gene segments (B cell) or the V and J segments (T cell) to produce the CDR3 region that encodes the antigen-binding portion of the immunoglobulin (Ig) heavy chain or T cell receptor gamma (TCRγ).[2,3] Current methodology uses polymerase chain reaction (PCR) to detect antigen receptor rearrangements (PARR), with primers for conserved regions of the V and J genes used to amplify the desired CDR3 region. Detection of clonal lymphocyte populations by this method is well-established in human medicine[4], and is routinely used in several veterinary laboratories to aid in the diagnosis of lymphoma and leukemia.[5,6] It is important to note that the conditions for the assay have a profound effect on the sensitivity and specificity of the results; therefore each laboratory conducting the assay should report these numbers when they report results to allow proper interpretation of the findings. It has been established that the PARR test is much more sensitive than routine light microscopy at detecting neoplastic populations in the blood. We recently reported that 73% of patients with malignant clones in their lymph nodes also had malignant cells in their peripheral blood.[7] Only 21% of the patients had malignant cells detectible by light microscopy indicating that the PARR test is sensitive and that the majority of patients have previously undetected disease in the peripheral blood.[6,7]

Most canine patients treated for lymphoma (>80%) will achieve complete clinical remission. Cures however, are rare and the majority will relapse and ultimately die of their disease.[8] The cause of this relapse is the presence of residual malignant cells that are not detected by relatively insensitive means (physical exam, light microscopy) which proliferate over time and lead to relapse. With the advent of more sensitive molecular techniques, methods to measure minimal residual disease (MRD) are being incorporated into clinical case management.[1,9] The rationale for studying MRD is to improve the measurement of treatment response, provide prognostic information, and to ultimately optimize treatment strategies.[4] In childhood acute lymphoblastic leukemia (ALL), the clinical usefulness of measuring MRD was conclusively demonstrated in the late 1990's.[10] In several studies initial response to treatment evaluated via MRD methods has been found to be a strong prognostic factor, more so than other presenting factors such as age or total white blood cell count. .[4, 10-14] In fact depending on the study examined, 42-75% of children who are MRD negative at the end of chemotherapy induction, have an excellent outcome with relapse rates of only 2-8%. Conversely, high levels of MRD at this same time point have been associated with a 50- 100% relapse rate.[4, 10-14] Measurement of MRD and altering treatment course based on the results is now the standard of care in childhood ALL. Studies to determine the utility of monitoring MRD in other childhood hematologic malignancies such as lymphoblastic lymphoma are ongoing.[10,15]

The presence of MRD has also been monitored in human follicular lymphoma and mantle cell lymphoma, both of which are B cell tumors.[16] Treatment of follicular lymphomas can result in long term remission and favorable prognosis, whereas mantle cell lymphomas are less responsive to treatment and are often fatal. The level of detectable MRD in the blood of patients with these two diseases correlates with clinical course, in that most mantle cell lymphoma patients continue to have MRD in their blood after chemotherapy, whereas follicular lymphoma patients treated with a combination of chemotherapy and the monoclonal antibody rituximab do not.[17,18] These findings suggest that the biologic behavior and subsequent ability to eliminate MRD by chemotherapy varies across the same histologic subtype, adding valuable prognostic information to the clinical management of these diseases. In other human lymphoid malignancies such as non Hodgkin's lymphoma it has yet to be determined if the ability to clear MRD will translate into improved patient outcome.[4,10] In canine hematologic malignancies, one report monitoring MRD in 8 canine lymphoma patients has been published, and 50% of the dogs achieved a molecular remission as determined by qualitative PCR in pre treatment and post treatment lymph node biopsy samples.[20] This study was not quantitative, however and did not use tumor specific primers, so no conclusions about the correlation between minimal residual disease and survival could be drawn.

A more sensitive quantitative assessment of MRD is now routinely performed on human leukemia patients using SYBR green real time PCR.[16] Tumor and patient specific primers that bind either to antigen receptor genes or to novel genes created by tumor specific translocation can be used to determine the number of neoplastic cells/ul of blood, or the number of neoplastic cells/white blood cells.[16] Quantitative real time PCR (RQ-PCR) has become an accepted and highly sensitive tool for monitoring MRD in many hematologic malignancies.[20] Current and future uses include stratifying patients at the time of diagnosis based on the degree of tumor burden or expression levels of gene transcripts. The response to traditional and targeted therapies can also be monitored and the kinetics of this response can provide important prognostic information, even alter the intensity of therapy. Additionally monitoring MRD by RQ-PCR may also allow detection of early disease progression often times before clinical indicators are present.[16] The overreaching goal is to tailor therapy to the individual and improve patient outcome and studies are ongoing in veterinary medicine to validate the best clinical use of this technique.

FLOW CYTOMETRY

Simplistically, flow cytometry is an analytical test that can be used to evaluate cell suspensions based on staining the cells with specific flurochromes and passing them through a cytometry chamber. Cells are sorted or gaited according to the staining characteristics of the different populations. Cells can also be sorted by size. Almost any single cell suspension can be analyzed by flow cytometry from whole blood to LN aspirates. With respect to lymphocytes this technique can also establish if the lymphocyte population is dominated by a single lymphocyte phenotype (neoplasia) or heterogenous expansion (reactive hyperplasia) – it has the advantage that a large panel of antibodies can be used on freshly aspirated samples, but the disadvantage of not being able to examine cells within the context of the lymph node architecture. A recent study (Williams et al 2008) using flow cytometry in patients with lymphocytosis demonstrates the use of this diagnostic test to provide prognostic information. Information from 96 patients was available for analysis. Patients were grouped into 4 classifications including CD 21 positive (B cell); CD 3 positive (T cell); aberrant T cell; and CD 34 positive (undifferentiated progenitor cells). Outcome could be predicted regardless of the treatment given in 3 of the categories. CD 34 positive lymphocytosis patients had a very poor prognosis with median survivals of 16 days. T cell cases were separated by cell count with patients having counts > 30,000 cells having a worse survival (131 days) compared to those with counts < 30,000 (>1000 days). For those patients with B cell lymphocytosis cell size was predictive of survival with those patients having large cells doing far worse with median survivals of 129 days compared to those with small cells (median not reached at 1000 days).[21] Flow cytometry classification is also being used to provide prognostic information in B cell lymphoma patients. Preliminary results indicate that differences in size of cell as well as amount of expression of surface markers such as MHC II can help group patients into those that do well and those that do poorly. In a series of B cell lymphoma patients who had LN aspirates evaluated by flow cytometry those with large sized cells had a median survival of 185 days. Those with small to medium sized cells were classified according to amount of class II MHC expression on the surface. Patients with low expression had a decreased survival at 222 days and those with high expression had a median survival of 325 days.

Another clinical setting in which flow cytometry has been helpful is diagnosing mediastinal masses. In most cases the major differentials include thymoma or lymphoma. The majority of thymocytes should express CD4 and CD8 surface markers. This unique phenotype allows differentiation of the cells from almost all other lymphoid cell populations in the body, including malignant ones. In a previous report 16 canine patients with a definitive diagnosis that presented with a mediastinal mass had samples evaluated by flow cytometry. Of those, 6/6 case with thymoma had greater than 10% CD4+/ CD8+ cells. Of the 7 with confirmed lymphoma, 6 had < 2% CD4+/ CD8+ cell populations making this phenotype highly supportive of a diagnosis of thymoma when present.[22] As more information is obtained from flow cytometry samples and more markers validated in veterinary species the utility of this test will grow.

REFERENCES

1. Morley A. **New England Journal of Medicine** 1998;339:627. 2. Raulet DH. **Annu Rev Immunol** 1989;7:175. 3. Schatz DG, et al **Annu Rev Immunol** 1992; 10:359. 4. Bruggemann, M, et al. **Acta Haematol** 2004;112:111. 5. Burnett RC, et al. **Vet Pathol** 2003;40:32. 6 Keller RL, et al. **Vet Clin Path** 2004; 33:145. 7. Lana SE, et al. **J Vet Intern Med** 2005 20(2):329. 8. Vail DM, et al. In: Withrow SJ, MacEwen EC, eds. **Small Animal Clinical Oncology**, 3rd ed. 2001: 558. 9. van Dongen JJ, et al. **Leukemia** 2003;17:2257. 10. Cairo MS. **Clinical Advances in Hematology & Oncology** 2007;5:876. 11. Cave H, et al. **New England Journal of Medicine** 1998;339:591. 12. Coustan-Smith E, et al. **Blood** 2000;96:2691. 13. Nyvold C, et al. **Blood** 2002;99:1253. 14. van Belzen N, et al. **Leukemia** 1997:11:1742. 15. Zhou J,et al. **Blood** 2007;110:1607. 16. Schuler F, Dolken G. **Clinica Chimica Acta** 2006; 363:147. 17. Corradini P, et al. **J Clin Oncol** 2004; 22: 1460. 18. Czuczman MS, et al. **Ann Oncol** 2001;12: 109. 19. Calzolari C, et al. **Veterinary Research Communications** 2006;30(Suppl 1):285. 20. van der Velden VHJ et al. **Leukemia** 2007;21:604. 21. Williams M, et al. **J Vet Intern Med.** 2008;22:596. 22. Lana SE, et al. **J Vet Intern Med** 2006;20:1161.

PROGENITOR CELL TRANSPLANTATION FOR THE TREATMENT OF CANINE B-CELL LYMPHOMA

Steven E. Suter VMD, MS, PhD, DACVIM (Oncology)
Raleigh, NC

INTRODUCTION

For close to 5 decades, the domestic dog (*Canis familiaris*) has served as a preclinical model for hematopoietic cell transplantation (HCT).[1] Based on limited knowledge gained in early, uniformly unsuccessful, human clinical trials and later in murine studies, dogs were identified as a suitable species for study based on their large body size, longer life span, genetic diversity, random-bred nature, ease of handling, and relatively economical colony maintenance costs.[2] For these reasons, there is a tremendous body of literature documenting the utility of dogs for studies of HCT, total body irradiation (TBI), myeloablation, cell dose requirements, graft manipulation, engraftment, and GVHD. Therefore, the vast majority of clinical HCT problems encountered in people were solved using dogs. It is estimated that >95% of human transplant procedures were first perfected using canines.[1]

Bone marrow transplantation was revolutionized by a number of important discoveries in both humans and dogs dating back to the 1990s. First, the cloning and large-scale production of recombinant hematopoietic cytokines, including granulocyte colony-stimulating factor (rhG-CSF, recombinant human G-CSF and rcG-CSF, recombinant canine G-CSF), stem cell factor (rhSCF and rcSCF), and granulocyte macrophage colony-stimulating factor (rhGM-CSF and rcGM-CSF), that could increase circulating hematopoietic progenitors and support their *in vivo* growth.[3,4] Second the finding that peripheral blood contains small numbers of hematopoietic progenitor cells and that recombinant cytokines dramatically increase these numbers via recruitment from the bone marrow.[5,6] Third, the production and characterization of anti-CD34 monoclonal antibodies directed against human[7], mouse[8], primate[9], and canine mononuclear cells[10] that allowed the isolation of a population of cells that can reconstitute the peripheral blood cell lineages after lethal marrow ablation. And finally, the development of sophisticated continuous-flow blood separators that are able to remove and separate blood components via centrifugation on the basis of specific cell densities.[11,12]

We recently obtained 3 Baxter-Fenwal CS-3000 Plus with Access Management System continuous blood separators (Fenwal, Baxter Healthcare Corporation, Deerfield, IL) which were kindly donated by the Mayo Clinic, Rochester, MN. These machines were initially tested using normal research dogs to determine if an ample number of CD34+ peripheral blood progenitor cells could be harvested to ensure complete hematopoietic reconstitution following lethal total body irradiation. Based on these initial successful studies, North Carolina State University Veterinary Teaching Hospital began treating client-owned dogs diagnosed with B and T cell lymphoma using peripheral blood stem cell transplantation (PBSCT).

INITIAL STUDIES

Apheresis Procedure

Eight normal dogs (2 unprimed and 6 primed with rhG-CSF [(Neupogen, Amgen, USA)] initially underwent the apheresis procedure to collect peripheral blood CD34+ progenitor cells via collection of peripheral blood mononuclear cells (PBMN).[13] For jugular catheter placement, sedation was instituted using midazolam (0.2-0.4 mg/kg IV), butorphanol (0.2-0.4 mg/kg IV), and propofol (1-2 mg/kg IV as needed). The dogs were intubated and placed on servoflurane. A 12-gauge, 20-cm central venous dual-lumen catheter (Arrow International, Reading, PA) was aseptically placed in a jugular vein for venous access and return. Three additional 18-gauge 2.9 cm catheters were placed in a cephalic vein and both saphenous veins for 1) medetomidine HCl (Dormitor, Pfizer, NY) constant rate infusion (CRI), 2) 10% calcium gluconate CRI, and 3) serum ionized Ca (iCA) monitoring. Once the apheresis procedure was under way for ~1/2 hr, the servoflurane was discontinued and the dogs were extubated when appropriate mentation was noted. For the rest of the procedure, a light plane of anesthesia was maintained using a medetomidine HCl CRI (0.2-0.5 ug/kg/min) with additional boluses of butorphanol or Propofol if needed. PBMC apheresis was performed using the Baxter-Fenwal CS-3000 Plus cell separator with the small volume separation container holder (SVSCH) and small volume collection chamber (SVCC) and a disposable double needle open system apheresis tubing kit (Baxter-Fenwal, Access Open System Apheresis Kit, #4R2224) as suggested by the manufacturer for patients with body weights of 50 kg or less (Baxter Fenwal CS-3000 Mononuclear Cell Collection, CLE005 rev.2-1/1/01). Utilization of the SVSCH and SVCC can negate the need for blood priming in small patients since the extracorporeal volume of these containers and the tubing kit is ~ 240 mls. The centrifuge speed was set at 1600 rpm, while the machine inlet flow rate was 1500 mls/hr. Acid-citrate-dextrose formula A (ACD-A, Baxter-Fenwal, Deerfield, IL) anticoagulant infusion was overseen by the Access Management System which infuses anticoagulant at a WBC:ACD-A ratio of 10:1. The disposable tubing kit and collection chambers were primed with ACD-A and 0.9% saline. During the apheresis procedure, the dogs were monitored using continuous electrocardiogram, indirect blood pressure measurement, and body temperature measurement. Pulse and respiratory rate were also continuously monitored. Serial blood iCa, sodium (Na), potassium (K), pH, oxygen partial pressure (PO2), oxygen saturation (SO2) and glucose concentrations were evaluated every ½ hr using a handheld blood analyzer (iSTAT, Abbott Point of Care, Inc., Princeton, NJ). Hot water and air heating pads were used to maintain body temperature in the physiological range, while a CRI of 10% Ca gluconate solution at 20 mls/hr was used to control citrate-induced hypocalcemia. If serum iCa levels dropped by >0.2 mmol/L the infusion rate was increased to 30 mls/hr, with incremental increases of 10 mls/hr if needed. Upon completion, a final iCa was recorded, blood was drawn for a CBC, and the dogs were given Antipamezole HCl (Antisedan, Pfizer) at five times the total medetomidine dose per the manufacturer's instructions as a reversing agent. Once these dogs were awake and able to walk, all their catheters were removed and the dogs were returned to the general research colony. For the first two dogs, the aphereresis procedure was continued for only 2 hours. When harvesting the remaining 6 dogs who received Neupogen, the procedure was continued until ~2.5-3 total blood volumes (TBV) were cycled through the machine. Therefore, the total time of the procedure for these dogs varied based on the dog's body weight and the calculation of TBV = BW (kg) x 90mls blood/kg.

Flow Cytometry

Immediately upon completion of the apheresis, 1 ml of the harvested product (of 57 ml total) was used for complete blood count analysis (CBC) using an automated hematology analyzer (Advia 2120, Siemens, Deerfield, IL). An additional 1 ml was assayed for CD34+ progenitor cell content using a commercially available mouse anti-canine CD34 monoclonal antibody.[10] Flow cytometry was performed using a FASCan (Becton-Dickinson, Franklin Lakes, NJ) equipped with a 15-mW argon laser tuned to 488 nm (Spectra-Physics, Montain View,CA). The software for data acquisition was CELLQuest (Becton-Dickinson). An isotype control (Jackson Immunoresearch, West Grove, PA) was used to control for nonspecific antibody staining. The percentage of CD34+ mononuclear cells x the total white blood cell count was used to calculate the total number of harvested CD34+ progenitor cells. This number was then divided by the weight of the dog (kg) to determine the total number of CD34+ cells/kg harvested.

Results

For the first normal dog, who did not receive Neupogen, the apheresis procedure was continued for a total of only 2 hours and CD34+ enumeration was not attempted. Our goal with this dog was to optimize the apheresis procedure in our hands and evaluate the systemic effects of ADC-A infusion and the harvesting procedure itself. This dog tolerated the procedure well with no evidence of hypocalcemia or hypothermia. The second dog who did not receive Neupogen was harvested for a length of time based on its TBV. Although we were able to document mononuclear cell collection, the PBMN CD34+ % of this collection was extremely low (0.35%).

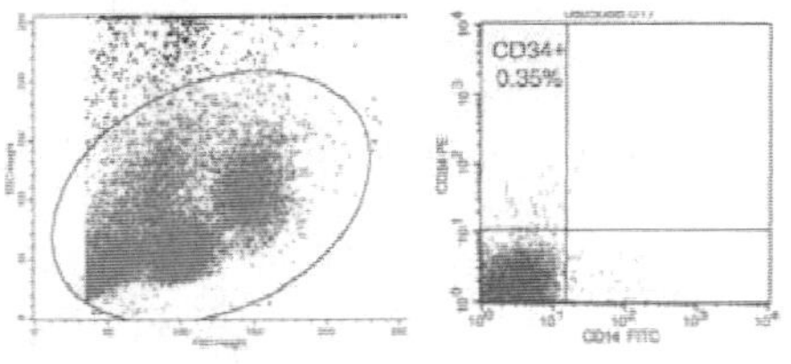

Fig 1. Flow cytometric analysis of apheresis product. The left dot plot represents forward scatter (x-axis) vs side scatter (y-axis) of the harvested product with gating (circle) on the cells that are analyzed for CD34/CD14 content in the right dot plot. The CD34+ percentage was 0.35% (upper left quadrant).

Approximately 3 x 10^5/kg total CD34+ progenitor cells were harvested and based on previous reports documenting complete hematologic recovery of 100% of patients who receive $\geq$2 x 10^6 CD34+ cells/kg[14,15], we concluded this target dose was not achieved. The following 6 normal dogs received Neupogen for 6 days as previously described at 5 ug/kg SQ BID, with a goal of reaching >30,000/ul total white blood cells.

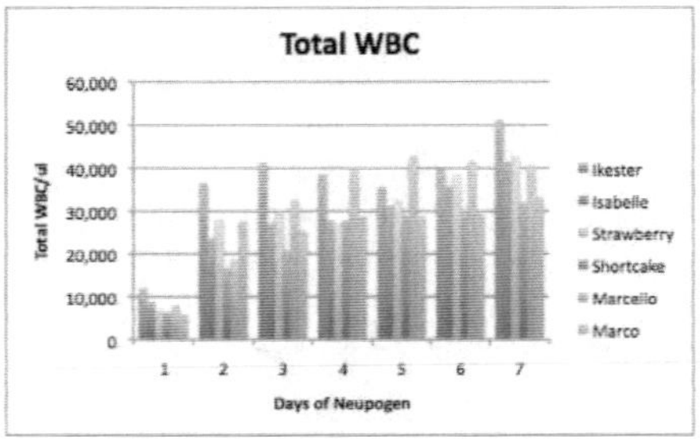

Fig 2. Total white blood cell count of 6 normal dogs receiving Neupogen for 6 days. In all dogs, the target dose of 30,000/ul was reached. All dogs received a double dose of Neupogen (10 ug/kg) 2 hours before the apheresis procedure.

Apheresis of these 6 dogs was performed with the intent of harvesting enough CD34+ progenitor cells to perform a PBSCT, therefore, the length of the procedure was based on the patients body weight, the calculated TBV, and an inlet flow rate of 1500 ml/hr. The average procedure length of these dogs was 4.5 hours, with a range of 3.5 hrs.-5.25 hrs. Fig 3 shows the flow cytometric CD34+ analysis of the harvested product from one of these dogs.

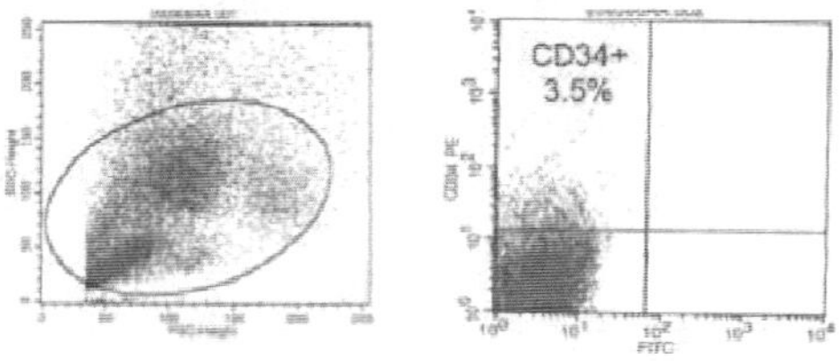

Fig 3. CD34+ enumeration of a Neupogen primed normal dog apheresis product. The CD34+ % of this product was 3.5% (upper left hand quadrant). Based on the total number of cells harvested in 50 mls of product, ~1.94 x 10^7 CD34+ cells/kg were harvested.

In all 6 dogs receiving Neupogen, the target dose of 2 x 10^6/kg CD34+ cells/kg was easily achieved, with a range of 7.8 x 10^6/kg to 3.3 x 10^8/kg. In addition, all dogs tolerated the procedure extremely well, with no evidence of any clinical abnormalities with appropriate monitoring and support.

CLIENT-OWNED B-CELL LYMPHOMA PATIENTS

Based on our results and previous reports documenting the utility of this procedure when treating dogs with hematologic malignancies[16,17], we began treating client-owned patients diagnosed with B and T-cell lymphoma with PBSCT beginning in November, 2008. Only the results of the B-cell patients are presented here. All patients were diagnosed by their referring veterinarians and received standard multi-agent systemic CHOP-based chemotherapy (cyclophosphamide, adriamycin, vincristine, prednisone) and were in complete clinical remission before PBSCT. All patients received high-dose cyclophosphamide (500mg/m^2 IV) in an effort to clear their blood and bone marrow of all cancer cells 2 weeks before starting Neupogen for 6 days.[17] Blood and bone marrow PARR (PCR for antigen receptor rearrangements)[18] was performed upon presentation to NCSU and all samples were negative using B cell primers. In all dogs, the target dose of >30,000 WBC/ul after 6 days of Neupogen was achieved (6 of these dogs are seen in Fig. 4).

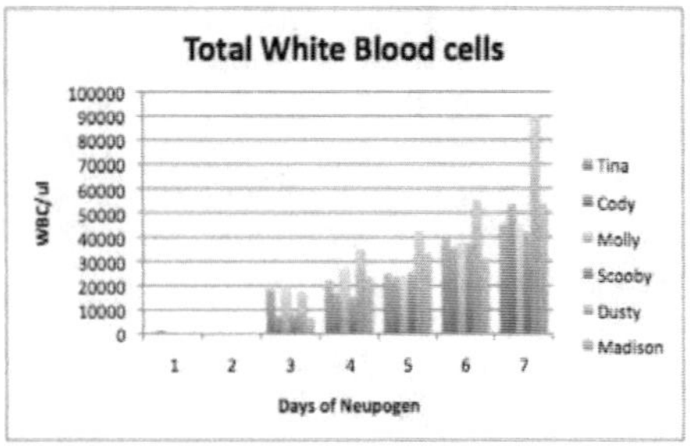

Fig 4. Total white blood cell count of 6 dogs in complete remission for B cell lymphoma receiving Neupogen for 6 days. Day 7 is the total white blood cell count after the apheresis procedure (~ 7 hrs after the double dose of Neupogen).

All dogs underwent peripheral blood mononuclear cell apheresis as described above with no complications noted. Below is the flow cytometric analysis for CD34+ content (3.0%) from one of these dogs.

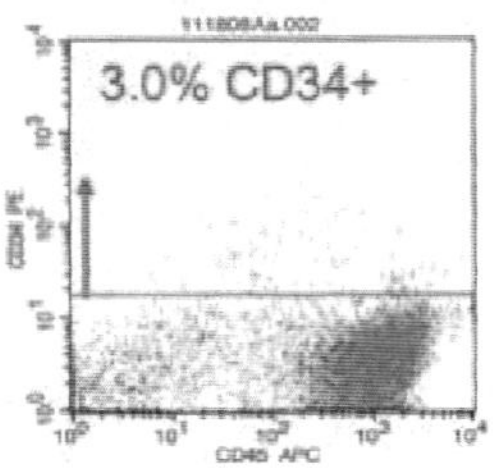

Fig 5. CD34+ enumeration of apheresis product using flow cytometry. A total of 4.726×10^9 total cells were harvested. Based on 3% CD34+ cells, a total of 3.9×10^6 CD34+ cells/kg were harvested, well above the target of 2×10^6 CD34+ cells/kg.

After CD34+ cell enumeration, the apheresis products were refrigerated overnight. The following day, all patients underwent 10 Gy total body irradiation (TBI) using a previously described radiation protocol.[17] The harvested cells were then returned to the patient intravenously immediately after TBI. The patients remained in the hospital for 2 weeks post-TBI for supportive care relating to TBI toxicities, which included variable gastrointestinal signs and severe cytopenias. All patients showed significant hematologic recovery during this time (Fig 6).

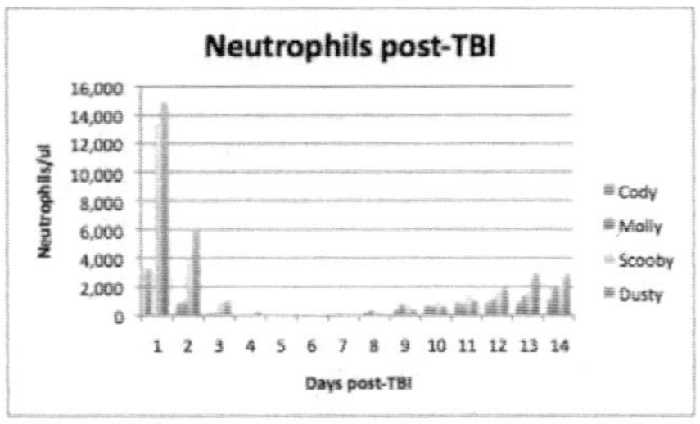

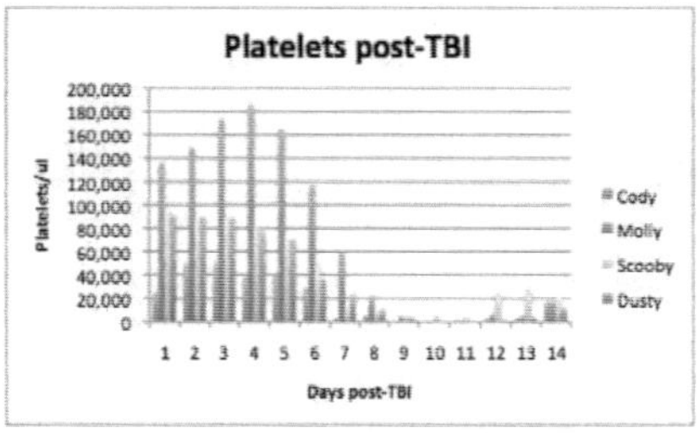

Fig 6. Post-TBI/PBSCT hematologic recovery in 4 dogs. The neutrophil nadir in most dogs occurred ~day +4-7, while the platelet nadir occurred ~day +9-11. All dogs cytopenias either normalized (neutrophils) or increased significantly (platelets) within 14 days post-TBI.

A total of 8 dogs underwent PBSCT using mononuclear cells harvested using the Baxter-Fenwal CS-3000 Plus cell separator. One dog, with only one kidney that was bordering on insufficiency prior to the procedure, went into acute renal failure and died in the hospital 9 days after TBI. Another dog, which engrafted appropriately while in the hospital, died of unknown causes approximately 3 weeks after the procedure. On necropsy, this dog had no evidence of lymphoma, but had systemic pathology suggestive of DIC. 2 dogs relapsed with disease ~3.5 months after the procedure and were eventually euthanized. And finally, one dog who was ~9.5 months post PBSCT, developed inoperable liver cancer and was euthanized with no evidence of lymphoma on necropsy. The 3 remaining dogs, none of which are receiving systemic chemotherapy, are 13 months, 11 months, and 9 months post-PBSCT with no evidence of disease and clinically normal at home.

These initial results support the notion that canine patients with lymphoma can undergo PBSCT in a clinical setting and become long-term survivors. Although an ~30% cure rate of canine patients has been reported previously[16], these transplants were performed in a research setting approximately 25 years ago. Therefore, more patients need to be transplanted and more time needs to elapse before we can begin to determine the true cure rate (surviving >2 yrs) of lymphoma using this technique.

REFERENCES

[1]Lupu M, et al. **Vet Comp Oncol** 2007;5(1):14. [2]Ostrander EA, et al. **Genome Res** 2005;15:1706. [3]de Revel T, et al. **Blood** 1994;83:3795. [4]Sandmaier BM, et al. **Blood** 1996;87:3508. [5]Appelbaum FR. **Exp Hematol** 1979;7:7. [6]Storb R et al. **Blood** 1967;30:805. [7]Ratajczak MZ, et al. **Leukemia** 1998;12:942. [8]Morel F, et al. **Blood** 1996;88:3774. [9]Andrews RG, et al. **Blood** 1992;80:1693. [10]McSweeney PA, et al. **Blood** 1998;91:1977. [11]Buckner D, et al. **Blood** 1969;31:653. [12]Herbst EW, et al. **Blut** [German] 1975;30:265. [13]Storb R, et al. **J Vet Intern Med** 2008;22:74. [14]Singhal S, et al. **Bone Marrow Transplant** 2000;26:489. [15]Ketterer N, et al. **Blood** 1998;91:3148. [16]Appelbaum FR, et al. **Transplantation** 1986;41(1):19. [17]Lupu M, et al. **J Amer Vet Med Assoc** 2006;228:728. [18]Burnett RC, et al. **Vet Pathol** 2003;40:32.

GENOMIC CHARACTERIZATION OF CANINE LYMPHOID CELL LINES

Steven E. Suter VMD, PhD, DACVIM, Eric L. Seiser, Rachael Thomas PhD, Kristy L. Richards MD, PhD, Kathryn K. Byler, Tessa E. Breen, Peter Moore BVSc, Ph.D, Matthew Breen PhD
Raleigh, NC

INTRODUCTION

Human tumor cell lines have played a fundamental role in our understanding of cancer genetics and physiology and have been used extensively in the discovery and characterization of new chemotherapeutic agents. Using a resource panel of 60 cell lines to represent the most common adult tumors, the Developmental Therapeutics Program (DTP) of the United States National Cancer Institute (NCI) has screened over 500,000 compounds for their ability to inhibit growth *in vitro*.[1] Molecular characterization of these cell lines has resulted in the generation of a large volume of genetic and molecular screening data that add substantially to their utility. It is accepted that the clinical predictive value of a cell line, as defined by *in vitro* drug activity translating into Phase II clinical trial efficacy in patients, may be limited due to incomplete representation of the primary tumor by the cell line.[2] However, when used under the right framework, cell line-derived data have been shown to be of predictive value for tumors such as non-small cell lung cancer, breast, ovary, and colon cancers.[3] In addition, data suggest that preclinical compound-oriented screens may be useful predictors in hematological malignancies.[2]

Although over 1,000 human leukemia and lymphoma (LL) cell lines have been described, only a very limited number of canine lymphoid tumor cell lines have been reported previously of which only one, OSW, has been characterized in detail at the genomic level.[4] Five other canine LPD cell lines, GL-1 (B-cell leukemia), CLGL-90 (chronic large, granular lymphocytic T-cell leukemia), 17-71 (B-cell lymphoma), CL-1 (T-cell lymphoma) and CLL-1390 (primitive leukocytic round cell neoplasia)[5-8] have been used extensively for a variety of *in vitro* studies examining canine tumor biology and drug response[9-14] Despite having been in circulation for up to 20 years these five cell lines have received little or no genomic characterization; consequently their precise relevance to the *in vivo* diseases, along with their clinical predictive values, has yet to be established.

The completion of the 7.6x genome sequence assembly and haplotype structure of the domestic dog[15] has facilitated the development of a panel of high-resolution tools for canine genomic investigation. We used high-resolution array-based comparative genomic hybridization (aCGH) analysis to identify DNA copy number aberrations in the five canine LL cell lines, in combination with high-density expression array analysis for evaluation of transcriptional status. These data were compared directly with cytogenetic characteristics and gene expression profiles of spontaneous canine LL tumors to elucidate shared cytogenetic and gene expression abnormalities and to assess their relevance to primary canine lymphoid malignancies. Ultimately, the characterization of abnormal DNA copy number and differential regulation of key genes in canine LL cell lines will help determine their clinical predictive value. These data also will extend opportunities for comparative investigations of lymphoid neoplasia between both species.[16,17]

RESULTS

DNA Ploidy Analysis

Peripheral blood mononuclear cells (PBMN) isolated from a normal dog and the cell lines were stained with propidium iodide (PI) and analyzed using flow cytometry (FACScan, Becton Dickinson, USA).[18] The PMMN were used as a normal diploid (n = 2) control. Fig. 1 shows a histogram of PI stained PBMN, 17-71, and GL-1 cells. These results show that both of the cell lines exhibit aneuploidy-both being hyperdiploid. Interestingly, all 5 canine cell lines exhibit hyperdiploidy.

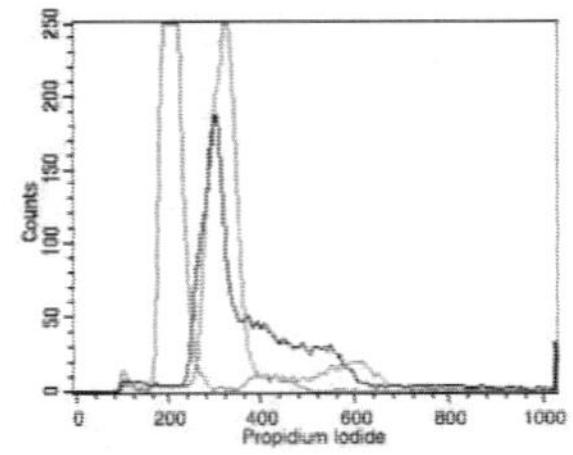

Fig 1. PI staining of normal PMNC and the canine lymphoid cell lines, 17-71 and GL-1. A DNA index (DI) was calculated.[18] The DI of the 17-71 cells was 1.5, while the DI of the GL-1 cells was 1.4. The DI of the remaining 3 cell lines (CL-1, CLGL-1390, and CLL-1390) was also in the 1.4-1.5 range.

Array CGH Analysis

aCGH profiles of each of the five dog LL cell lines demonstrated distinct distributions of DNA copy number changes. All five cell lines presented with numerous DNA copy number aberrations, comprising both gains and losses. The size of these aberrations ranged from 20 kb to entire chromosomes. In both of the leukemia derived cell lines (GL-1 and CLGL-90), the total number of genomic losses was more frequent than gains. While the T-cell lymphoma CL-1 also presented with more DNA copy number losses as compared to gains, the B-cell lymphoma 17-71 displayed an opposing bias. When comparing the aCGH profiles between the cell lines it was apparent that multiple regions of DNA copy number imbalance were shared between cell lines of common origin and also common immunophenotype. The two lymphoma cell lines (17-71 and CL-1) exhibited 17 regions of shared genomic imbalance across 14 chromosomes: gains within CFA 13, 31, 32, and 34, as well as shared losses in CFA 3, 6, 11, 12, 13, 14, 16, 19, 26 and 32. Three of these imbalances, gain in CFA 13, gain in CFA 31 and loss in CFA 14, have been defined previously as recurrent aberrations in spontaneous canine lymphoma {Thomas, 2003 #4}. The leukemia derived cell lines (GL-1 and CLGL-90) showed 10 common imbalances across eight chromosomes, with gains in CFA 2 and 8 as well as losses in CFA 6, 8, 16, 18, 26 and 28. Immunophenotype comparison also revealed shared regions of copy number imbalance. Specifically the two B cell lines (GL-1 and 17-71) shared four regions of DNA copy number imbalance, including losses in CFA 19, 26, 27 and 32. The two T-cell lines (CL-1 and CLGL-90) displayed 25 common genomic imbalances across nine chromosomes: gain in CFA 37 and losses in CFA 4, 6, 8, 16, 18, 23, 28 and 32. The primitive leukocytic round cell neoplasm cell line (CLL-1390) shared genomic imbalances with the each of the other four cell lines, including gain of the proximal half of CFA 13 (shared with the two lymphoma cell lines) and loss of a region of CFA 28 (shared with the leukemia cell lines).

Gene Expression Analysis

Gene expression microarray data were filtered to remove probe sets that displayed limited variation among the five canine LL cell lines, six primary canine lymphomas and six non-neoplastic lymph nodes, yielding 2661 probe sets that were used for all subsequent analyses. Unsupervised hierarchical clustering analysis produced four discrete groups: i) the six normal lymph nodes, ii) the three primary B-cell canine lymphomas, iii) the three primary T-cell canine lymphomas, and iv) the five dog LL cell lines (Figure 2). Within the cluster of cell lines the samples grouped according to disease type (lymphoma vs leukemia) rather than B-cell vs. T-cell phenotype. The leukemia lines GL-1 and CLGL-90 and the leukocytic round cell line CLL-1390 formed a separate cluster, while the lymphoma lines 17-71 and CL-1 segregated from all other lines. Clustering analysis segregated the cell lines from both the normal lymph node samples and the primary tumor samples, but differential expression analysis revealed that there are numerous cancer-associated genes differentially expressed between each of the cell lines. Assessment of genes differentially expressed between cell lines and immunophenotype-matched primary lymphomas identifies multiple cancer-related genes with dissimilar expression in each cell lines. While expression differences were observed between cell lines and immunophenotype-matched primary lymphomas, a small subset of cancer associated genes that were differentially expressed between cell lines and normal lymph nodes exhibited similar expression with the immunphenotype-matched primary tumor.

Cytogenetic Analysis

The DNA copy number of ten BAC clones was evaluated in each of the five cell lines using multicolor FISH analysis. CLGL-90 exhibits a homozygous deletion of the *CDKN2* locus in CLGL-90. Both CLL-1390 and CL-90 show a homozygous deletion of the *PTEN* locus. The GL-1 cells show an amplification of the *FLT3* locus (n=4), while the 17-71 cells have an amplification of the *KIT* locus (n>4). Application of the FISH panel identified structural chromosome aberrations in the CLGL-90, 17-71 and CLL-1390 cell lines. The T-cell leukemia line CLGL-90 presented with a balanced copy number (n = 2) of the *MYC* and *IGH* loci, with one copy of each locus located on separate autosomes, in agreement with the normal dog karyotype. The other copy of each locus was, however, located at either end of the same derivative chromosome in CLGL-90, indicating that an abnormal fusion event had occurred between regions of CFA 13 (*MYC*) and CFA 8 (*IGH*) in this cell line. FISH analysis of *MYC* and *KIT* revealed the existence of a putative isochromosome of CFA 13 in both 17-71 and CLL-1390. The 17-71 cell line displayed two copies of the putative isochromosome in ~90% of the metaphase spreads analyzed, while CLL-1390 typically displayed only one isochromosome per cell. aCGH analysis of GL-1 identified amplification of a ~20Mb region of CFA 25 extending from the centromere and encompassing the region encoding the gene *FLT3*. Initial FISH data for the BAC containing *FLT3* confirmed the presence of 4 copies of this locus in GL-1. Subsequent FISH analysis with a panel of five clones from CFA 25 revealed that the amplified region was extensive, representing a duplication of the region spanning 3.1 – 14.7 Mb of CFA 25.

Quantitative Real-Time PCR

Quantitative RT-PCR was performed using primers specific for canine *PTEN*, *FLT3*, *MYC* and *KIT*, which were selected due to observed homozygous deletions or extensive copy number increases in aCGH and FISH analysis within one or more of the five cell lines. The fold changes of each of these four loci were normalized to the expression level of the reference gene RLP32. There were no detectable copies of the *PTEN* locus in CLL-1390 and CL-1. Expression analysis using qRT-PCR indicated that while the *PTEN* expression levels of 17-71and CLGL-90 were similar to each other and GL-1 slightly elevated, there was massive down regulation of *PTEN* both in CL-1 and CLL-1390. Analysis of *FLT3* expression levels demonstrated a large increase in expression in GL-1, at least 70-fold higher, when compared to the other cell lines and the normal lymph node. Interestingly, the T-cell lines CLGL-90 and CL-1 had lower levels of expression than the normal lymph node whereas the round cell neoplasm had *FLT3* levels comparable to the normal lymph node. *MYC* qRT-PCR showed only very small differences in expression levels across the cell lines, with the gene being expressed approximately two-fold higher in 17-71, CLGL-90 and GL-1 than in CL-1 and CLL-1390. Lastly, when *KIT* expression was examined by qRT-PCR it was evident that while four of the five lines had 10-75 fold increase in *KIT* expression compared to the normal lymph node, the line CL-1 had over a 200-fold decrease in expression.

Integrated CGH and GEP analysis

Investigation of genome-wide distribution of DNA copy number imbalance and aberrant gene expression revealed associations between the genome and the transcriptome in the cell lines. For all five cell lines, the distribution of expression fold changes (increase, decrease, no change) between a cell line and the normal lymph nodes differed between probe sets that fell within regions of DNA copy number loss and regions of normal copy number (p-value <0.01). Specifically, in regions of DNA copy number loss there was a greater proportion of probe sets with decreased gene expression, as well as a smaller proportion of probe sets with increased gene expression. Four of the five cell lines (GL-1, CL-1, CLGL-90, and 17-71) displayed discrete distributions of expression fold changes in regions of genomic gain when compared with regions of normal DNA copy number (p-value <0.01). These regions of genomic gain displayed a greater proportion of probe sets with increased expression fold change and a smaller proportion of probe sets with a decreased expression fold change. To further assess the association between genomic copy number and differential gene expression, two chromosomes were examined in greater detail: CFA 13 in 17-71 and CFA 25 in GL-1. aCGH and FISH analyses both confirmed gain of CFA 13 in the B-cell lymphoma cell line 17-71. Since CFA 13 contains two oncogenes (*MYC* and *KIT*) that are relevant to lymphoid tumor biology in both human and canine populations, we selected this chromosome to examine the association between chromosomal gain and dysregulation of gene expression. There is a significant difference (p-value <0.01) in the distribution of expression fold changes between CFA 13 and the distribution of expression fold changes in balanced regions of the genome. Specifically, increased fold changes are observed in 25.07% of the probe sets in CFA 13 compared to 16.57% within the copy number neutral regions of the genome. The B-cell leukemia line GL-1 displays a gain in the proximal half of CFA 25, which contains the oncogene *FLT3*, and therefore we examined the association of DNA copy number and gene expression across a chromosome with a regional DNA copy number imbalance. Probe sets within the proximal gain in CFA 25 demonstrated a statistically significant difference (p-value <0.01) in the distribution of expression fold changes compared to the copy number neutral region of CFA 25. Increased expression was observed in 33.78% of the probe sets in the gained region of CFA 25, compared to 17.4% within the remainder of the chromosome.

Discussion

The depth and breadth of utility of cancer cell lines as *in vitro* models is dependent on the extent to which they represent the primary malignancy from which they were derived. The five canine lymphoid cells used in this study have been in circulation for up to 30 years during which time they have been considered to be representative of the *in vivo* disease from which they were initiated. Using a series of complementary techniques and resources we have assessed the genomic characteristics of these five cell lines towards determining the degree to which they represent the *in vivo* disease, and thus beginning to assess their clinical predictive value. We defined their genomic DNA copy number profiles using a combination of high-resolution aCGH analysis and multicolor SLP FISH analysis, and assessed their gene expression profiles using genome-wide expression array analysis in combination with qRT-PCR. Our findings revealed instances of DNA copy number imbalances that are shared between the lymphoma cell lines and also with primary canine lymphomas, as well as genomic aberrations that are not. In addition, gene expression data suggests that there are global transcriptional differences from primary tumors that may limit the utility of these cell lines as models for *in vivo* disease.

REFERENCES

[1]Shoemaker RH. **Nat Rev Cancer** 2006;6(10):813. [2]Baguley BC, et al. **Eur J Cancer** 2004;40(6):794. [3]Voskoglou-Nomikos T, et al. **Clin Cancer Res** 2003:9(11):4227. [4]Kisseberth WC et al. **Leuk Res** 2007;31(12):1709. [5]Nakaichi M, et al. **J Vet Med Sci** 1996;58(5):469. [6]Momoi Y, et al. **Vet Immunol Immunopathol** 1997;59(1-2):11. [7]Steplewski Z, et al. **Cancer Immunol Immunother** 1987;24(3):197. [8]Suter SE, et al. **Clin Cancer Res** 2005;11(4):1579. [9]Sano J, et al. **J Vet Med Sci** 2004;66(6):709. [10]Steplewski Z, et al. **In Vivo** 1990;4(4):231. [11]Tamura K, et al. **J Vet Med Sci** 2007;69(9):925. [12]Tamura K, et al. **Vet Immunol Immunopathol** 2006;110(1-2):163. [13]Uozurmi K, et al. **Res Vet Sci** 2005;78(3):217. [13]McCall JL, et al. **Leuk Res** 2009;34(4):69. [14]Jamadar-Shroff, et al. **Clin Cancer Res** 2009;15(4):1269. [15]Lindblad-Toh K, et al. **Nature** 2005;438(7069):803. [16]Breen M, et al. **Chromosome Res** 2008;16(1):145. [17]Paoloni M, et al. **Nat Rev Cancer** 2008;8(2):147. [18]Abu Talab FM, et al. **J Egyptian Nat Cancer Inst** 2002;14(4):303.

MULTIDRUG RESISTANCE TRANSPORTERS: BEYOND MDR1

Katrina L. Mealey DVM, PhD, DACVIM, DACVCP
Pullman, WA

INTRODUCTION

ABC (ATP Binding Cassette) transporters, also called multidrug resistance transporters, are one of the most important causes of multidrug resistance in tumor cells. The ABC transporters belong to the ABC Superfamily of proteins, one of the largest known protein families. Members of this family are primarily active transport proteins that are expressed on cell membranes. In prokaryotes these proteins can serve import and export functions but in eukaryotic cells, only export functions have been described. The ABC transporter family contains over 40 members which have been subdivided into 7 subfamilies (designated ABC 'A' through ABC 'G'). Members of these subfamilies participate in antigen presentation, cholesterol and lipid transport, mitochondrial homeostasis, regulation of several ion channels and multidrug resistance. Only 15 ABC transporters have been observed to export chemotherapy drugs, but most of these observations are limited to in vitro conditions. Only 3 members of the ABC transporter family have been shown to be relevant to clinical chemotherapeutic drug resistance in cancer patients: P-glycoprotein (P-gp), Multidrug resistance-related protein, and Breast Cancer Resistance Protein. Information gleaned from research in human tumor cell lines and cancer patients has, to date, been applicable to veterinary patients. Specifically, substrates for a particular ABC transporter in humans are also substrates for the canine and feline orthologues. Similarly, compounds that inhibit a particular ABC transporter in humans also appear to inhibit the orthologous canine and feline transporters. This article will briefly review the role of P-glycoprotein in canine and feline cancer patients (since this topic has been reviewed elsewhere) and will provide a more in-depth review of the literature regarding Multidrug resistance-related protein and Breast Cancer Resistance Protein in human as well as veterinary patients.

ABCB1 (P-glycoprotein; P-gp)

P-gp has been fairly well characterized in dogs. Tissue distribution of P-gp in dogs is similar to that in humans, suggesting a similar physiologic role. Additionally, P-gp has been shown to contribute to chemotherapeutic drug resistance both in vitro (canine cell culture studies) and in vivo (canine cancer patients). For example, in dogs with lymphoma P-gp expression was higher in relapsed samples than in pre-treatment samples[1]. The expression of P-gp in feline lymphoma was also examined[2]. In contrast to dogs, P-gp expression was not predictive of remission or survival time.

P-gp substrates include many anticancer drugs (anthracyclines, vinca alkaloids, epipodophyllotoxins), macrocyclic lactones (ivermectin, selamectin, milbemycin, etc), loperamide, and other drugs. A number of studies have addressed P-gp's role in drug absorption, distribution and excretion and how these pharmacokinetic parameters affect drug safety and efficacy. Much of this information has been generated from studies and clinical observations of dogs affected by a mutation in the ABCB1 gene. The ABCB1 polymorphism in dogs consists of a 4 base-pair deletion mutation. This deletion results in a shift of the reading frame that generates several premature stop codons. Because protein synthesis is terminated before even 10% of the protein product is synthesized, dogs with two mutant alleles exhibit a P-gp null phenotype, similar to mdr1 (-/-) knockout mice[3]. Affected dogs include many herding breeds. For example, roughly 75% of collies in the United States, France, and Australia have at least one mutant allele. Other affected herding breeds, albeit at a lower frequency, include Old English Sheepdogs, Australian Shepherds, Shelties, English Shepherds, Border Collies, German Shepherds, Silken Windhounds, McNabs, and Long-haired Whippets.

Drug distribution, the delivery of drugs from the systemic circulation to tissues, can be dramatically affected by alterations in P-gp function. P-gp is a component of the blood-brain barrier, the blood-testes barrier, and the placenta. Therefore, distribution of P-gp substrate drugs to these tissues is greatly enhanced in ABCB1(mut/mut) dogs. ABCB1(mut/mut) dogs experience adverse neurologic effects after a single dose of ivermectin (>120 μg/kg). Affected dogs also appear to have increased susceptibility to neurologic adverse effects of other avermectins including milbemycin, selamectin, and moxidectin. Avermectins are not the only drugs that P-gp normally excludes from the brain. Loperamide is an opioid that is generally devoid of CNS activity because it is excluded from the brain by P-gp. At routinely used therapeutic doses, loperamide causes profound CNS depression in ABCB1 mutant/mutant dogs[4]. Less information is available regarding P-gp and the blood brain barrier in cats. The author has received anecdotal reports of ivermectin toxicity in cats after standard doses, but whether or not the underlying cause is a result of altered P-gp expression or function is not currently known.

P-gp can also affect the elimination of drugs, including anticancer agents, from the body. Renal and biliary excretion are the most important pathways for drug elimination. P-gp is expressed on renal tubular cells and biliary canalicular cells, suggesting that it plays an important role in drug excretion. Concurrent administration of a P-gp-inhibitor decreases the biliary and renal clearance of doxorubicin in rats. Studies comparing biliary excretion of the P-gp substrate ^{99m}Tc-sestamibi in ABCB1 WT/WT and ABCB1 mut/mut dogs demonstrate that ABCB1 mut/mut dogs do not excrete this drug into bile. ^{99m}Tc-sestamibi is essentially undetectable in gallbladders of ABCB1 mutant/mutant dogs but is highly concentrated in gallbladders of ABCB1 WT/WT dogs. Diminished biliary drug excretion may play a role in the apparent increased sensitivity of herding breeds to chemotherapeutic drugs that are P-gp substrates. For example, ABCB1(mut/mut) and ABCB1(WT/mut) dogs are significantly more likely to develop hematologic toxicity (neutropenia and thrombocytopenia) after treatment with the P-gp substrate vincristine than ABCB1 (WT/WT) dogs[5]. However, these dogs tolerate cyclophosphamide (a drug that is not a P-gp substrate) at full doses. In dogs, vincristine is eliminated primarily via biliary excretion of parent drug with some urinary excretion of parent drug and metabolites. Therefore, reduced doses of P-gp substrate chemotherapeutic agents are recommended in ABCB1 mut/mut and ABCB1 WT/mut dogs.

ABCC1 (Multidrug resistance related protein)

MRP1 was first identified as being responsible for multidrug resistance in a human lung carcinoma cell line. Subsequent studies have shown that MRP1 contributes to multidrug resistance in a variety of tumor types in human patients. Substrates for MRP1 include methotrexate, doxorubicin, etoposide, vinca alkaloids, paclitaxel, and irinotecan, as well as glutathione (GSH)-, glucuronide-, and sulphate-conjugates of these compounds[6]. Like all ABC transporters, MRP1 evolved not to provide multidrug resistance to cancer cells, but to serve as part of a complex cellular defense system that recognizes and removes potentially toxic compounds from within cells. MRP1 is expressed at basal levels in many tissues throughout the body but is expressed at high levels in adrenal gland, choroid plexus, kidney, placenta, helper T-cells, testis and muscle. Expression of MRP1 in Sertoli cells of the testis protects germline cells from potentially toxic xenobiotics in blood. Similarly, MRP1 expressed at the choroid plexus transports drugs from the cerebrospinal fluid to the blood to protect sensitive neural tissue. MRP1 also serves to regulate inflammation by virtue of its ability to transport a variety of inflammatory mediators. Endogenous substrates for MRP1 include glutathione, glutathione disulfide, glucuronide conjugates of hormones, leukotrienes, and prostaglandins. Thus, it appears that MRP1 normally functions to export endogenours intermediates of cellular metabolism and to protect the cell from toxic insult. The influence MRP1 has on intracellular GSH-GSSG balance implies that MRP1 also plays an important role in regulating oxidative stress[7].

Compared to data in humans, information regarding MRP in veterinary patients is lacking. The author was unable to locate any literature on MRP expression in feline tumors. Several groups of investigators have documented MRP expression by a variety of canine tumor types including transitional cell carcinoma[8], mast cell tumors[9], osteosarcoma[10], and mammary cancer[11]. Unfortunately, the effect of MRP expression on prognosis or chemotherapeutic drug resistance was not investigated in these studies. Thus, our understanding of the role of MRP in canine and feline multidrug resistance is incomplete. Future investigations should address substrate similarities and differences between canine/feline MRP and human MRP, whether or not compounds that inhibit human MRP also inhibit canine/feline MRP, and whether or not expression of MRP (with and without other ABC transporters) correlates with prognosis or clinical multidrug resistance.

ABCG2 (Breast Cancer Resistance Protein or BCRP; Mitoxantrone Resitance Protein MXR)

Like P-gp and MRP, BCRP was initially discovered as a protein that was overexpressed in a multidrug resistant cell line. This cell line was selected for resistance by exposure to mitoxantrone and was found not to express P-glycoprotein or MRP1, but a new resistance pump, BCRP. The ABCG2 gene is highly conserved among species including mammals, birds, reptiles and fish, suggesting that the protein serves an important physiologic function. Like other ABC transporters, BCRP is expressed in many normal tissues in including placenta, brain, mammary gland, testis, gastrointestinal tract, liver, hematopoietic stem cells and others. A long list of substrates and inhibitors of BCRP already exists, and the list continues to expand. With regard to cancer therapy, important substrates of BCRP include mitoxantrone, camptothecin derivatives, etoposide and methotrexate. Of particular interest, due to the recent FDA approval of toceranib for treatment of canine mast cell tumors, many tyrosine kinase inhibitors have been shown to be substrates for BCRP[12]. In fact tyrosine kinase inhibitors are reported to be both substrates for and inhibitors of BCRP. Thus, there appear to be complex interactions between tyrosine kinase inhibitors and BCRP which could affect the disposition of other drugs concurrently administered to the same patient.

In a recent report involving canine mammary adenocarcinomas and adenomas, BCRP was expressed in a high percentage of adenocarcinomas (> 85%) and a small percentage of adenomas (~ 28 %)[13]. Tumor samples with a higher malignancy grade were reported to have a moderate correlation with the degree of BCRP expression. Research in the author's laboratory has also demonstrated BCRP expression in a variety of canine solid tumors (both sarcomas and carcinomas). Most of these specimens consisted of pre-treatment biopsies suggesting that BCRP may contribute to intrinsic chemotherapy resistance in these particular tumors. Ongoing work in the author's laboratory involves determining which chemotherapeutic drugs are substrates for canine and feline BCRP. Additionally, we are investigating differences in function and substrate specificity between canine and feline BCRP.

Potential Role of Inhibitors of ABC Transporters

There is abundant literature supporting a role for P-gp, MRP1 and BCRP as contributors to multidrug resistance in human cancer patients, although the role of MRP1 in multidrug resistant tumors may be more controversial. It is important to note, however that even though each of these transporters contributes to multidrug resistance, they are not likely to be sole cause of chemotherapeutic failure in most cancer patients. Tumor cells often develop many resistance mechanisms such that they are resistant to drugs that are not substrates for any of the ABC transporters. Clinical trials of P-gp inhibitors and ABCG2 inhibitors (ABC reversal agents) have often had disappointing outcomes[14]. A number of reasons have been suggested for the less-than-desirable outcomes observed in clinical trials of ABC transporter reversal agents. Nonspecific toxicity of ABC reversal agents to cancer patients is a common problem. Another reason is the unexpected and undesired pharmacokinetic interactions between the reversal agents and the chemotherapeutic agents used for treating cancer patients. For example, the reversal agents often block excretion of the anti-cancer drugs resulting in their accumulation in plasma which then increases their toxicity. An additional reason for poor response to ABC transporter reversal agents in some, but not all, cancer patients is the presence of single nucleotide polymorphisms (SNPs) amongst the ABC drug transporters. These SNPs might affect binding of the reversal agent to the transporter, limiting the efficacy of the ABC reversal agent. The expression of multiple ABC drug transporters by a single tumor cell is well documented; therefore use of a reversal agent effective for only one ABC transporter would not prevent the multidrug resistant phenotype. Thus, a great deal more research is needed before ABC reversal agents are likely to have a significant impact on treatment outcomes in cancer patients.

REFERENCES

[1.] Bergman et al, **J Vet Int Med**, 1996;10:354-359. [2.] Brenn et al, **Vet Comp Oncol**, 2008;6;201-211.[3.] Mealey et, **Pharmacogentics,** 2001;11:727-733. [4.] Mealey et al, **Drug Metab Disp**, 2008;36:1073-1079. [5.] Mealey et al, **J Vet Int Med**, 2008;22:996-1000. [6.] Munoz et al, **IUMB Life,** 2007;59:752-757. [7.] Eckford and Sharom, **Chem Rev**, 2009;2989-3011.[8.] Lee et al, **Res Vet Sci** 2007;83:210-216. [9.] Tojo et al **J Vet Med Sci** 2002; 64:531-533. [10.] Mealey et al, **Vet Comp Onc**, 2003;1:67-75. [11.] Honscha et al, **Reprod Domest Anim** 2009;44:218-223. [12.] Lemos et al, **Br J Cancer** 2008;98:857-862. [13.] Nowak et al, **In Vivo** 2009;23:705-709. [14.] Lee CH. **Methods Mol Biol** 2010;596:325-340.

IMMUNE RESPONSE IN DOGS WITH MELANOMA RECEIVING A GD3 VACCINE CONTAINING THE ADJUVANT MONTANIDE ISA 70®. Rowan J. Milner[1], Jim P. Farese[1], Marc Salute[1], Nick Bacon[1], Karri Barabas[1], Shannon Parfitt[1], Sarah Lyles, Monica Clemente, Marc Salute, Kelvin Kow. 1. Department of Small Animal Clinical Sciences, University of Florida. Gainesville, Florida, USA.

We report on the results from a prospective IACUC approved clinical trial in dogs with melanoma that received a GD3 based melanoma vaccine containing the adjuvant Montanide ISA 70®. Montanide ISA 70® is an oil/surfactant-based adjuvant where surfactants are combined with non-metabolizable and/or metabolizable oil. Side effects, survival and immune response was compared to a previously reported Phase-1 GD3 trial (n=65).

Twenty-two dogs with melanoma were recruited. Blood was collected pre- and post vaccination for GD3 specific IgG and IgM titers and NK-cell assay. Vaccination occurred monthly for 3 months followed by restaging at the 4th visit. 50% were spayed females.

Mixed-breeds were most common at 27%, Golden Retrievers 13% and others breeds at 60% (≤ 2 individuals). Melanoma sites included 73% oral, 27% pedal. Sixteen (72%) were graded, 25% were low grade (< 3MF/10hpf), 25% intermediate (3-14 MF/10hpf) and 50% high grade (>14 MF/10hpf). Antibody titers results were available for 63% of dogs and NK Assay for 68%. For IgG titers, 36% dogs showed increasing titers, 57% no change and 7% decreasing titers. Only 13% of dogs showed a positive NK Assay. Fifty percent of dogs showed significant local reactions to the vaccine, ranging from indurate lesions to abscesses. The median survival for this group was 263 days (Inter-quartile range of 213 to 640 days), this was not statically different from the dogs in a Phase-1 GD3 trial. Because of the small number, further analysis was not possible. However, the immune response in the Montanide ISA 70® group appears to differ significantly from the Phase-1 GD3 group with only 13% showing a positive NK response compared to 67% for the Phase-1 group. More Montanide ISA 70® dogs also had a positive IgG response.

These results appear to reflect a Th-2 response. To verify this, changes in serum cytokines and chemokines are currently under investigation by our laboratory using MILLIPLEX MAP technology. The significant local vaccine reaction (50%) to the Montanide ISA 70® in the GD3 vaccine compared to Phase-1 dogs was unacceptable and cannot be recommended.

THERMOCHEMOTHERAPY FOR THE TREATMENT OF SOLID TUMORS. J Phillips[1], D Clark[2], G Aknine[2], O Smrkovski[1], M Wilson[1]. 1. University of Tennessee College of Veterinary Medicine, Knoxville, TN. 2. Biothermatics, Inc., Franklin, TN.

The role of chemotherapy and hyperthermia has been studied in veterinary medicine for many years. Regional hyperthermia in combination with such drugs as bleomycin and specifically formulated doxorubicin has proven successful for the treatment of small cutaneous tumors. The *in vivo* activity of regional hyperthermia in combination with carboplatin chemotherapy, however, has not been investigated in veterinary patients. Here we describe the use of systemic carboplatin chemotherapy in combination with regional microwave hyperthermia for the treatment of solid tumors in veterinary patients. Patients eligible for therapy included cats, dogs, and horses diagnosed with a measureable malignant solid tumor that was amenable to regional hyperthermia. Hyperthermia was applied using the iThermTWS (Biothermatics) compact microwave hyperthermia unit. Treatments were monitored using the thermosensors designed within the applicator and verified using invasive thermometry. Target skin temperatures ranged from 41-43°C and were maintained for 45 minutes per treatment. Patients were treated every other day pending response to therapy. A total of 25 patients have been accrued of which 13 are evaluable for response and toxicity. These patients include 3 horses, 4 cats, and 18 dogs. Cutaneous toxicities have varied from none (Grade 0) to moderate/severe (Grade III/IV) and appear to be limited to the tumor tissue within the treatment field. Responses have included 7 complete clinical responses, 4 partial responses, and 2 patients with stable disease. Subjective palliative improvement in pain scores was also noted in 3 patients with extensive bone involvement. Overall, the iThermTWS proved simple to use, safe, and reliable. When combined with carboplatin chemotherapy it appears to be highly effective for the treatment of solid tumors.

INVESTIGATION OF A PROCASPASE-3 ACTIVATING COMPOUND (PAC-1) FOR THE TREATMENT OF CANINE LYMPHOMA. LD Garrett[1], PW Lucas[1], L Dirikolu[1], J Schmit[1], PJ Hergenrother[2], TM Fan[1]. University of Illinois Department of Veterinary Clinical Medicine[1] and Department of Chemistry[2], Urbana, IL.

PAC-1, a synthesized small molecule activator of procaspase-3, preferentially induces apoptosis of cell types containing elevated concentrations of procaspase-3. Higher concentrations of procaspase-3 have been found within cancer cells in comparison to their normal cellular counterparts. Intracellular concentrations of procaspase-3 have been identified within canine malignant lymphocytes, both in cell lines and in spontaneous-tumor samples, lending support to the therapeutic potential of procaspase activators for canine lymphoma. The purpose of this study was to determine the pharmacokinetics of oral and intravenous PAC-1 administration in healthy dogs and to ascertain if a pro-apoptotic effect could be induced in canine lymphoma cell lines at PAC-1 concentrations achievable *in vivo*.

Single-dose PAC-1 (dissolved in B-cyclodextran) pharmacokinetics (PK) was investigated using six healthy dogs in an oral/intravenous cross-over study. Blood was drawn for PK analysis at time points 0, 15, and 30 minutes and at 1, 2, 4, 6, 8, 12, and 24 hours post administration. Subsequently, a 48 hour constant rate infusion (CRI) of PAC-1 after a 10 minute bolus loading dose was evaluated in four healthy dogs. Blood was drawn for PK analysis at 1, 6, 12, 24, 36, and 48 hours after the initiation of the infusion. PAC-1 activation of procaspase-3 leading to apoptosis was evaluated in two canine B-cell lymphoma cell lines *in vitro*. The canine lymphoma 17-71 and GL-1 cell lines were incubated with increasing concentrations of PAC-1 for 24 and 48 hours in an attempt to mimic a PAC-1 concentration and duration achievable in vivo. Flow cytometry was used to assess apoptosis at the 24 and 48 hour time points.

Oral PAC-1 bioavailability was extremely variable with an average of 41%. The PK data for the single-dose intravenous PAC-1 showed very consistent blood concentrations between the dogs with a rapid decrease in PAC-1 concentrations to less than 10 uM by the 2 hour time point. The 48 hour CRI administration of PAC-1 achieved a concentration of PAC-1 at or greater than 10 uM from the 6 hour time point on. Time points 24, 36, and 48 hours had PAC-1 concentrations equal to 18uM. *In vitro*, the lymphoma cell lines exhibited apoptosis at PAC-1 concentrations of 17.5 and 10 uM for incubation durations of 24 and 48 hours, respectively. The percent apoptosis ranged from 20-70%, depending on the cell line and PAC-1 concentration and exposure duration.

A CRI of PAC-1 can be maintained for 48 hours in healthy dogs and achieves a blood concentration of an amount and duration that induces apoptosis in malignant canine lymphocytes in cell culture. Thus, PAC-1 is a viable therapeutic for an anti-cancer trial in dogs with lymphoma.

EXPRESSION AND FUNCTION OF SURVIVIN IN CANINE OSTEOSARCOMA. J Shoeneman, BJ Rose, EJ Ehrhart III, B Charles, DH Thamm. College of Veterinary Medicine and Biomedical Sciences, Colorado State University, Fort Collins, CO.

Osteosarcoma (OSA) is a common and highly malignant canine tumor, for which new therapies are desperately needed. Apoptosis resistance is one of the hallmarks of cancer. Survivin, a member of the Inhibitor of Apoptosis Protein (IAP) family, is critical in apoptosis resistance and cell proliferation, and is over-expressed in a variety of human cancers. Survivin expression is an independent negative prognostic factor in many human malignancies, and evidence indicates a potential pathogenic and prognostic role in human OSA. We previously demonstrated that survivin protein is commonly expressed in canine lymphoma, and that strong expression is associated with an inferior outcome following chemotherapy; however, we are unaware of any studies evaluating the potential pathologic or prognostic role of survivin in canine OSA. We sought to examine the effects of survivin inhibition *in vitro* on proliferation, apoptosis, cell cycle distribution and chemotherapy sensitivity.

The canine OSA cell lines Abrams and D17 were utilized for these investigations. Survivin inhibition was accomplished via a custom siRNA designed against the canine survivin mRNA sequence. A scrambled siRNA was utilized as a specificity control. siRNA transfection was accomplished via lipofection. Western analysis was used to confirm gene knockdown. Alterations in viable cell number and apoptosis were evaluated via manual cell counting and an activated caspase-3/7 ELISA respectively. Changes in cell-cycle distribution were assessed using ethidium bromide staining and flow cytometry. Apoptotic responses to the antineoplastic drugs doxorubicin and carboplatin were assessed using the caspase-3/7 ELISA above.

siRNA transfection resulted in approximately 85% knockdown of survivin expression in both cell lines. This was associated with a dramatic reduction in total and viable cell number, and a significant enhancement of basal apoptosis. Survivin knockdown sensitized canine OSA cells to carboplatin and doxorubicin associated cell death, and knockdown of survivin was associated with an accumulation of 4N (hyperdiploid) cells and an increased sub-G1 (apoptotic) population as assessed cytometrically.

In conclusion, siRNA-mediated inhibition of survivin resulted in significant growth arrest, apoptosis induction and chemosensitization in 2 canine OSA cell lines. Future and ongoing studies will evaluate the efficacy of a small-molecule survivin inhibitor *in vitro* and in a murine xenograft, and will assess the prognostic significance of survivin expression in a series of primary canine OSA tissues.

CLONING AND EXPRESSION OF CANINE CD25 FOR VALIDATION OF AN ANTI-HUMAN CD25 ANTIBODY TO IDENTIFY T REGULATORY LYMPHOCYTES IN DOGS WITH OSTEOSARCOMA. Rissetto KC[1], Rindt H[2], Selting KA[1], Henry CJ[1, 3], Villamil JA[1], Reinero CR[1] [1]Department of Veterinary Medicine and Surgery, University of Missouri, Columbia, MO. [2] Department of Veterinary Pathobiology. University of Missouri, Columbia, MO. [3]Department of Internal Medicine, Hematology/Oncology Division, University of Missouri School of Medicine, Columbia, MO

T regulatory cells (Tregs) suppress anti-tumor immunity and can increase with some cancers. Phenotypically, CD4+CD25+FoxP3+ cells better represent Tregs, although only CD4+FoxP3+ Tregs have been evaluated in dogs. We hypothesized that the anti-human CD25 antibody cross-reacts with canine CD25 protein and that dogs with osteosarcoma (OSA) would have an increased %Tregs compared to healthy controls.

The canine CD25 gene was cloned and transfected into human HeLa cells. Mouse-anti human CD25 monoclonal antibody was evaluated via flow cytometry to validate its use in dogs. Blood and lymph node (LN) aspirates from 35 healthy dogs and 10 dogs with untreated appendicular OSA were labeled for flow cytometry using anti-canine CD4, anti-human CD25, and anti-rat/mouse FoxP3 antibodies. Populations analyzed included CD4+CD25+ lymphocytes, CD4+CD25+FoxP3+ lymphocytes, and CD4+CD25+ cells expressing FoxP3; these were compared using non-parametric statistical analysis with $p < 0.05$ considered significant.

No significant differences were detected in circulating or nodal %Tregs between healthy dogs and dogs with OSA, or between healthy dogs of different ages. The %Tregs in control blood and LNs, respectively were (median): CD4+CD25+ cells, 1.98% and 7.15%; CD4+CD25+FoxP3+ cells, 0.98% and 5.36%; and CD4+CD25+ cells expressing FoxP3, 47.01% and 71.68%. For dogs with OSA, Treg populations in blood and tumor-draining LNs, respectively, were: CD4+CD25+ cells, 2.2% and 7.67 %; CD4+CD25+FoxP3+ cells, 0.84% and 3.95%; and CD4+CD25+ cells expressing FoxP3, 39.68% and 60.38%.

The anti-human CD25 antibody definitively recognizes canine CD25 protein. Percentages of Tregs in dogs with OSA are not significantly different from control dogs.

VCS Award Winning Abstract

COMBINED IN VITRO EFFECTS OF CALCITRIOL WITH CCNU, VINBLASTINE, IMATINIB OR TOCERANIB ON C2 CANINE MASTOCYTOMA CELLS. Erin Malone, Kenneth Rassnick, Joseph Wakshlag. Cornell University, Ithaca, NY.

CCNU, vinblastine, and tyrosine kinase inhibitors are used to treat non-resectable, metastatic, and high-grade canine mast cell tumors. Calcitriol, the biologically active form of vitamin D, exerts potent antineoplastic activity in a wide range of human and canine tumors. The combination of calcitriol with a diverse array of antineoplastic agents demonstrates even greater activity. In this study, we examined whether calcitriol increased chemotherapy or tyrosine kinase inhibitor cytotoxicity in C2 canine mast cells.

C2 cells were incubated with increasing concentrations of calcitriol (0.1-25nM). Twenty-four hours later, cells were treated with either vehicle control alone or increasing concentrations of CCNU (2.5-20uM), vinblastine (1.25-10nM), imatinib (1.6-12.5nM) or toceranib (3.13-25nM). After 48h, cell survival was assessed with an MTT assay and data was used to derive a combination index (CI; values <1, 1, >1 indicate synergism, additivity, antagonism, respectively).

CI values <1 were obtained for calcitriol (0.1-6.25nM) combined with CCNU (5-10uM), vinblastine (2.5-5nM), imatinib (1.67-12.5nM) and toceranib (3.13-25nM), indicating the interactions are synergistic. The concentrations of calcitriol and either CCNU, vinblastine, imatinib or toceranib required for 50% growth inhibition were generally 2-4 fold lower when the drugs were used in combination than when used individually. There was potential antagonism when calcitriol concentrations >6.25nM were combined with any of the drugs.

Calcitriol and various chemotherapeutics had synergistic, antiproliferative effects on C2 mast cells. Findings suggest that these combinations may have significant clinical utility in dogs with mast cell tumors.

VCS Award Winning Abstract

IMMUNOPRECIPITATION TO SCREEN LYMPHOMA CELLS FOR HYPERMETHYLATION. JN Bryan, M Jabbes, Tripp CD. Department of Veterinary Clinical Sciences, Washington State University, Pullman, WA.

Hypermethylation silences expression of genes in human non-Hodgkin lymphoma (NHL).The gene *DLC1* has been shown to be hypermethylated in canine NHL. This purpose of this study was to screen candidate genes, hypermethylated in human NHL, for hypermethylation in canine NHL cells.

DNA and RNA were isolated from two normal lymph nodes, lymphocytes from a dog with chronic lymphocytic leukemia (CLL), and OSW cells cultured *in vitro*. DNA was sheared by sonication to fragments averaging 300bp in length. An anti-methylcytosine antibody was used to precipitate (MeDIP) fragments containing methylcytosine. Gene-specific primers were used to amplify desired genes and a housekeeping gene (*GAPDH*) using real-time PCR. Genomic DNA was compared with MeDIP to identify genes containing methylcytosine. cDNA was prepared for each sample and hybridized with an Affymetrix GeneChip Canine Genome 2.0 Array to quantify genome-wide gene expression. Genes examined were *DLC1, CCND1, CDKN1A, MGMT, CDKN2B, LHX2, POU3F3, PRKCE,* and *RARB2*.

Hypermethylation was most consistently identified in OSW. The genes *DLC1, CCND1, CDKN1A, MGMT, LHX2, POU3F3,* and *RARB2* contained methylcytosine by real-time analysis in OSW. The genes *CDKN1A, MGMT, LHX2, POU3F3,* and *RARB2* contained methylcytosine in CLL. Expression was reduced in *DLC1, LHX2*, and *MGMT* in both cell types and in *CDKN1A* in CLL. MeDIP generates a pool of methylcytosine-rich DNA which can be queried for the presence of gene promoters. This study demonstrates that hypermethylation is present in NHL of dogs in a similar pattern to humans. Further evaluation of this as a therapeutic target is warranted.

INHIBITION OF GLYCOLYSIS WITH 2-DEOXYGLUCOSE INHIBITS SPONTANEOUS METASTASIS IN MURINE OSTEOSARCOMA. Douglas H. Thamm, Janet Lori, Joseph L. Sottnik; Colorado State University, Fort Collins, CO.

We recently reported on a re-analysis of previously published gene expression data from a large set of primary solid and metastatic tumors originating from different tissues using the latest available tools for normalization, identification of differentially expressed genes and pathway analysis (Ptitsyn A., et al, BMC Bioinformatics 2008;9 Suppl:S8). Analysis of multiple independent datasets indicated significantly reduced oxidative phosphorylation in metastases compared to primary solid tumors, implying a potential enhancement of the Warburg phenotype in metastatic progression. We thus sought to determine if inhibition of anaerobic glycolysis with the competitive inhibitor 2-deoxy-D-glucose (2-DG) would reverse this metastatic bioenergetic phenotype in highly metastatic cancer cells. We evaluated the effects of 2-DG on DLM8 osteosarcoma cell migration and Matrigel invasion via Boyden chamber assays, and modulation of *in vitro* production of matrix metalloproteases, VEGF, and cathepsin L were evaluated by gelatin zymography, ELISA and western analysis respectively. The effects of 2-DG administration *in vivo* was evaluated in an orthotopic, spontaneously metastatic, luciferase-transfected postsurgical model of murine osteosarcoma utilizing DLM8, which is syngeneic to C3H mice. DLM8 cells were implanted into the distal tibias of C3H mice, followed by amputation of the affected limb 16 days later. 2-DG or saline were administered postoperatively (500 mg/kg IP 3x/week, n=5 per group) until sacrifice. Presence of metastasis was evaluated using biweekly bioluminescence imaging and metastasis-free interval calculated using the Kaplan-Meier method. Concentrations of 2-DG associated with modest antiproliferative activity dramatically inhibited DLM8 cell migration and invasion. VEGF production was inhibited at high 2-DG concentrations, *but there were no repeatable changes in cathepsin L or gelatinase activity within the concentration range associated with inhibition of migration and invasion.* Postoperative administration of 2-DG significantly delayed DLM8 metastasis (median metastasis-free interval 8 vs 34 days, $p = 0.049$). In conclusion, glycolysis inhibition with 2-DG is associated with an inhibition of osteosarcoma migration, invasion and *in vivo* metastasis, at concentrations associated with modest antiproliferative activity.

EVOLUTION OF THE VETERINARY ONCOLOGY PARADIGM: SHOULD A CURE BE OUR GOAL?

Chand Khanna, DVM, PhD, DACVIM
Bethesda, MD

What "C" Word Are We Most Afraid Of?

From a historical perspective the practice of veterinary oncology emerged from a realization that cancers in dogs and cats are similar to cancers in humans and indeed were responsive to similar types of treatment. These treatment options were largely cytotoxic in nature. Selection of specific agents were not matched to a detailed understanding of cancer biology and was often based on their use in the human clinic. The cytotoxic nature of standard chemotherapy regimes generally provides a clear and consistent relationship between the dose-intensity of a therapy, the chance of therapeutic benefit and the risk for toxicity. This relationship continues to define the practice of veterinary oncology. For the most part dose-limiting toxicities in dogs and cats are gastro-intestinal and or hematologic. The risk of these toxicities frames and limits the success of most chemotherapy protocols used in veterinary practice and reduces effectiveness compared to their use in human patients. Without necessary intensive supportive care options or alternative affordable treatments the goal of short-term disease control with parallel maintenance of quality of life has been prioritized. This model of treatment in veterinary oncology has become the convention and is embraced by most pet owners who have anxiety and concerns about the impacts of chemotherapy on their pet's quality of life. We have been correctly forced to emphasize control of disease and relatively short-term remissions rather than cure. Our reporting and discussion of outcomes for the treatment of cats and dogs is largely limited to response rates and response durations. Nonetheless, it is important to note that current treatment protocols do cure a proportion of our veterinary patients. These rates of cure are not rigorously reported in the veterinary literature and infrequently discussed with owners. It is reasonable to ask if the "C" word that veterinary oncologists are most concerned about speaking is "Cure" and not "Cancer".

Should We Begin To Imagine Discussions That Include The Term "Cure"

As our growing understanding of cancer biology expands and is translated to novel treatment approaches, significant improvement in long-term remissions and cures have been seen in human cancer patients. This has come from earlier disease detection strategies, successful management of many previously dose-limiting toxicities related to cytotoxic drugs, identification and targeting of specific drivers of the cancer phenotype. The completion and release of a publicly accessible draft of the canine genome and advances in our understanding of feline genomics have recently provided the opportunity to understand cancer in companion animals in similar ways. This includes detailed studies of chromosomal gains and losses, examination of cancer-associated genes, study of gene amplification, mutation or loss. The translation of this basic understanding of cancer biology has fueled drug and disease management strategies in humans, dogs and cats. It is reasonable that new cancer drugs made for cats and dogs will be developed, that existing molecular-based or other targeted drugs developed for human use will have a clear rationale for use in dogs and cats, and that existing conventional cytotoxic agents and radiation therapy will be more effectively used or combined in the treatment of dogs and cats with cancer. Will the emergence of more effective treatments that lead to higher cure rates for cancer lead to a change in our perspective or should we begin to imagine discussions that include the term "cure"?

We Cure Patients Today But Should We Talk About It?

The term *cure* or *cured* is complicated by a reasonable concern that we inflate expectations of pet owners. Distinct from cure or cured is the notion of cure or survival rate, which describes the chance that an individual is free of disease at a certain point in time. The point in time used to define cure or survival rate must be sufficiently long such that patients who exceed such a time-point have a reasonable expectation of cure and that patient attrition beyond that point as a results of late recurrence will be rare. In human patients no specific convention for a time at which cure rate is reported is available. However, for many adult cancers, 5-year survival rates are often cited with an expectation that recurrence after 5 years is rare. The definition of an equivalent of a 5-year survival rate in dogs and cats is complicated by heterogeneity in the biologies of recurrence for many veterinary cancers and the impact of differences in life-span between large dogs, small dogs and cats. Despite, this variability, similar to the human convention the use of a 2-year survival rate is often considered, but not discussed, in the same way. Based on the published data, the use of a 2-year cure or survival rate is reasonable and yields a definable population of cured animals for a number of cancers. The consistent reporting of 2-year survival or cure rates in the veterinary literature may be an important part of our evolution out of the era of palliation. More sophisticated determinations of the times to report cure or survival rates in animals will require consideration of the size of the animal, age of cancer diagnosis, and biology of specific cancer.

How Are Cures Achieved?

As discussed above, we currently do cure veterinary patients afflicted with a number of distinct cancers, including those who require systemic therapy. It is reasonable that these cures are based on a number of influences related to the specific tumor, the host and treatment modality. As suggested earlier, features of each of these determinants are increasingly becoming understood.

Tumor: From the standpoint of the tumor, the ability to conduct next generation sequencing and parallel genomic and proteomic studies on specific cancers provides an unparalleled opportunity to define all genomic variations, amplifications, mutations, and losses in the tumor of an individual or a population of individuals. Such data will identify recurrent and so-called driver alterations that are responsible for the cancer phenotype in an individual. Beyond mere expression, driver alterations are essential for cancer development and progression. When identified, appropriate targeting of these driver alterations have been consistently shown to lead to the most successful treatment outcomes. Detailed and multi-level platform studies in canine cancers are currently underway at a number of institutions around the world and will provide a new view from which to consider cancer in dogs.

Host: Studies of the host are not restricted to the identification of genetic risk factors for cancer development. Indeed, host genomic studies in women with breast cancer have identified those patients with the greatest risk for cancer spread (unrelated to tumor development). Studies of the host have also included an individual's tolerance or response to a specific therapy, so called pharmacogenomics. Collectively the ability to define the features of the host risk for cancer development, progression and effective treatment will allow earlier diagnosis of cancer in at risk groups, more aggressive treatment of patients at risk for rapid cancer progression and provide improved opportunities to select optional treatments for individual patients. The ability to describe the host by breed determinants will hasten the identification and application of such tailored approaches to veterinary cancer patients.

Treatment: The understanding of the cancer and host biologies discussed above have allowed a departure from the sole use of non-targeted cytotoxic agents in the treatment of cancer in both human and veterinary patients. The use of agents that are specifically directed against a cancer target of interest is now a reality. As discussed above, an important opportunity in providing cancer cures will be the identification of "driving" events in a cancer and the successful targeting of these events with novel therapies.

The future veterinary oncology patient will be managed based on considerations of the tumor, host and available drugs. It is reasonable that discrete subsets of patients will be managed with combinations of rationally selected therapies. This view of veterinary cancer treatment will seek cures as an outcome. An approach to cancer treatment in dogs and cats that includes early diagnosis by screening at risk individuals, providing specific treatments based on a biological understanding of the cancer, combining these treatments with existing and effective drugs, and monitoring responses using a biological understanding of disease will be the art of a new veterinary oncology paradigm. This evolution of the field will lead to unique challenges and opportunities for the veterinary oncologist and growing interest from the pharmaceutical industry in veterinary oncology. The outcome will be an increasing number of cured patients. It is reasonable that we begin to consider this discussion as our paradigm begins to change.

STATISTICS & CLINICAL TRIAL DESIGN: AN ONCOLOGY PERSPECTIVE

Melissa C. Paoloni, DVM, DACVIM (Oncology)
Bethesda, MD

INTRODUCTION

Clinical trials in veterinary oncology are growing in scope and importance. Trials that evaluate novel therapies for pet animals with cancer have the opportunity to improve the outcome and quality of life (QOL) for veterinary patients, as well as human patients with cancer. The demand for clinical trials has been fueled by a sustained increase in the number of pet animals diagnosed with cancer and the pet-owning public's interest in finding effective and well-tolerated treatments for their pets. Interestingly, the outcome for patients who are managed within a clinical trial has been consistently shown to be superior to that of patients who receive the same agents outside the setting of a clinical study. This suggests that clinical trials are of value not only for improving care of future patients, but also for those patients enrolled in a study. Through the conduct of clinical trials, our opportunities to prevent, diagnose, and treat cancer will continue to improve. It is likely that the availability of clinical trials in veterinary oncology will also continue to expand. This expansion will require and welcome greater direct and indirect participation from both primary care veterinarians and veterinary specialists. The goals of this lecture are to introduce concepts of clinical trial design and conduct and to review the statistical tests readily used in their analysis.

CHOICE OF STUDY QUESTION

Each clinical trial must have a primary question or objective. The primary question, as well as any secondary or subsidiary questions, should be carefully selected, clearly defined and stated in advance.

ENDPOINTS

Endpoints are selected to answer the primary and secondary objectives of a trial and reflect the specific measures that will be followed in a study. Choice of a primary endpoint for a drug or a device to be considered efficacious must demonstrate tangible clinical benefit, generally defined as an improvement in survival or improvement in symptoms (Randomized Control Trials). The primary endpoint is the key decision in designing a trial and the major determinant of sample size.

Primary endpoints should be:

- Consistent with the primary study question and the hypothesized effect of the intervention
- Clearly defined and specified in advance
- Capable of being ascertained as completely as possible (ideally in every subject)
- Measured in the same way for all subjects
- Capable of unbiased assessment

Types of Primary Endpoints

- Binary (e.g., objective response rate,30-day mortality, 30 day overall mortality)
- Ordinal (e.g., toxicity- graded from 0 [none] to grade 5 [death]; two or more seizures)
- Continuous (e.g., quality of life, visual analog scale, CD-4, lymphocyte count)

Common endpoints in clinical trials include defining a maximally tolerated dose (MTD) to assess safety and evaluating tumor response, disease free interval (DFI) and survival time (ST) to measure activity. Criteria for safety and toxicity have been recently established for veterinary patients through the Veterinary Cooperative Oncology Group (VCOG-CTCAE). These published guidelines list a number of maladies that may be noted in the evaluation of an animal in a clinical trial and allow the categorization of the severity of these maladies using a grade from I (least severe) to IV (most severe). These guidelines allow consistency in the reporting of safety/toxicity endpoints in clinical trials. Reporting of the safety/toxicity of a new agent assessed in a clinical trial is of significant importance. The effectiveness of a new cancer agent can be measured in several ways. These include response rates in measurable/macroscopic disease, DFI and overall survival as mentioned above and quality of life (QOL) assessments. Endpoints differ depending on the type and objectives of a study and are defined prospectively in the study protocol. Biological or correlative endpoints, such as pharmacokinetic and pharmacodynamic assessments, are becoming increasingly important in veterinary oncology clinical trials. These biologic endpoints have been incorporated into study designs to better understand the mechanisms and biological effects of novel therapies/interventions.

RANDOMIZED CONTROLLED TRIALS

A randomized controlled trial (RCT) is a type of scientific experiment, in this case clinical trial, most commonly used in testing the efficacy or effectiveness of a novel drug or health technologies (such as pharmaceuticals, medical devices or surgery). It is the ideal way to determine the validity of the study's null hypothesis. Patients are randomized to receive a particular intervention, often with both the patients and investigators "blinded" (in this case double blinded) to the intervention administered.

RANDOMIZATION

There are two processes involved in randomizing patients to different interventions. First is choosing a randomization procedure to generate a random and unpredictable sequence of allocations-this may be a simple random assignment of patients to any of the groups at equal probabilities, or may be complex and adaptive. A second and more practical issue is allocation concealment, which refers to the stringent precautions taken to ensure that the group assignment of patients are not revealed to the study investigators prior to definitively allocating them to their respective groups.

There are a couple of statistical issues to consider in generating randomization sequences:

- Balance: since statistical tests are most powerful when the groups being compared have equal sizes, it is desirable for the randomization procedure to generate similarly-sized groups.
- Selection bias: depending on the amount of structure in the randomization procedure, investigators may be able to infer the next group assignment by guessing which of the groups has been assigned the least up to that point. This breaks allocation concealment and can lead to bias in the selection of patients for enrollment in the study.
- Accidental bias: if important covariates that are related to the outcome are ignored in the statistical analysis, estimates arising from that analysis may be biased. The potential magnitude of that bias, if any, will depend on the randomization procedure.

Simple Randomization

- Unless otherwise stated, focus is on randomization to two treatments A and B
- Each participant enters a trial and is assigned treatment A or B with probability (p). Most often p is chosen to be 1/2. The number of patients assigned treatment A follows a binomial distribution with sample size n.
- Can lead to imbalance between groups.

Blocked Randomization

- Blocked randomization is a common solution for the problem of imbalance
- As patients enter a trial we define a block size and a proportion to each treatment
  - Example: 2 treatments, block size b=4 patients, 2 patients to receive treatment A, 2 receive treatment B, that is proportional to each treatment ½
  - Example: Within a block of size b=4 there are 6 orderings of 2 A's and 2 B's: AABB, ABAB, ABBA, BAAB, BABA, BBAA

Stratified Randomization

- Patient response may depend on prognostic factors (e.g., previous MI, WBC count level)
- Randomization helps balance prognostic factors across treatments. But some imbalances may occur by chance.
- A suggested strategy is blocked randomization within pre-defined combinations of prognostic factors (stratification)

Adaptive Randomization

- Biased Coin
  - This is a modification to simple randomization where the chance of allocation is biased in favor of the under-represented treatment when the imbalance passes some threshold. For example, if after 10 randomizations 7 patients had been randomized to A and 3 to B, then the coin might be biased to give a 2/3 chance that the next patient is allocated to B and only 1/3 chance to A.
- Urn Model
  - Start with m A balls and m B balls in an urn and choose a ball at random. If an A ball is chosen assign that treatment, put the A ball back in the urn and add an additional B ball to the urn. Continue the process.

Outcome-adaptive or response-adaptive randomization is a variation on traditional randomization to address the issue of a more "desirable" or "efficacious" treatment arm developing over time. Randomization probabilities are adjusted continuously throughout the trial in response to the data. The probability of a treatment being assigned increases as the probability of that treatment being superior increases. The statistical advantages of randomization are retained, while on average more patients are assigned to superior treatments.

POWER:

Power is a measure of the ability of a study to identify statistically significant differences between treatment groups. If the difference between treatment groups is small, the number of patients included in a trial needs to be large. The size of the patient population is a primary determinant of its power. By increasing the statistical power in a study, the chance of demonstrating statistically significant differences between treatment groups is increased. It is important to note that the size of a study (i.e. number of participants) should be defined before a clinical trial is initiated.

SURVIVAL ANALYSIS:

Survival analysis is the basis of oncology clinical trial outcome assessment. It is defined as a collection of statistical procedures for the analysis of data that corresponds to the time from a well-defined origin until the occurrence of some particular event. Survival data is generally not symmetrically distributed, so it is not reasonable to assume that data have a normal distribution, and censoring of data is required for analysis. Survival time of an individual is censored when the endpoint has not been observed for that person. Censoring occurs when we have some information about individual survival time, but we do not know the survival time exactly.

Goals of Survival Analysis

- To estimate and interpret survivor and/or hazard functions
- To compare survivor and/or hazard functions
- To assess the relationship of explanatory variables to time

Each of the below statistical concepts will be discussed with examples given as to how to analyze and better interpret clinical trial data.

Survivor Function

- The survivor function, S(t), gives the probability that a person survives longer than some specified time t.
- Estimating S(t): The Kaplan-Meier Method S(t) can be expressed as a product of conditional probabilities and that each observation, whether censored or uncensored, can help to estimate some of these conditional probabilities.

Hazard Rate

- 1) Hazard rate at time t (or hazard function at time t) is the instantaneous failure rate for an individual who has not yet failed at time t.
- 2) Hazard rate at time t is the negative slope of the survival function at time t on a log scale.

REFERENCES

[1]Paoloni, M and Khanna C. Chapter 5: Clinical Trial Design. *Cancer Management in Small Animal Practice*. Eds. Henry and Higginbotham.

MOLECULAR & BIOLOGIC ACTIVTIY OF RECEPTOR TYROSINE KINASES IN CANINE CANCER

Douglas H. Thamm, VMD, DACVIM (Oncology)
Fort Collins, CO

INTRODUCTION

Cancer is a disease characterized by dysregulated growth, abridged cell death, and enhanced cell migration, invasion and angiogenesis. While the molecular mechanisms responsible for conferring this phenotype are very diverse, one of the classes of molecule that has been receiving a great deal of recent attention as a potential target for therapy are the **receptor tyrosine kinases (RTKs)**. These are cellular receptors for extracellular growth factors that facilitate communication from the extracellular milieu to the cell interior, mediating functions such as growth, survival, invasion and angiogenesis. A panoply of different RTKs are expressed by almost all tumor cells, but they can play variable roles in the pathogenesis of the disease.

Most RTKs exist as monomers, which homodimerize or heterodimerize upon contact with the appropriate ligand, inducing a conformational shift that allows phosphorylation of tyrosine residues in the intracellular domain. This then triggers several different intracellular second messenger cascades, culminating in altered gene expression and, often, a more malignant phenotype **(Fig. 1)**.

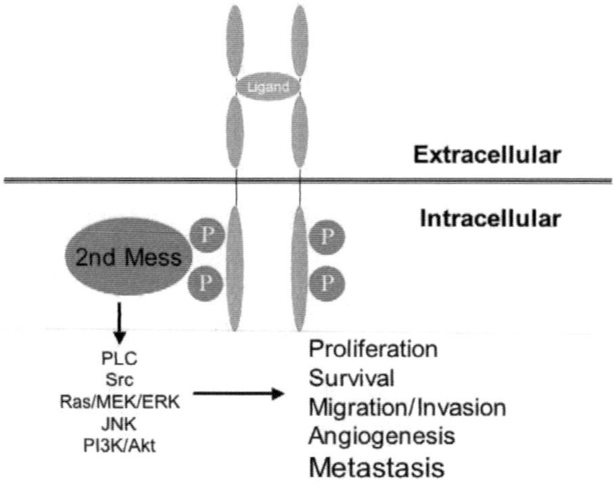

Figure 1. Diagram of a generic receptor tyrosine kinase

There are several new cancer drugs currently available on the human market that act by inhibiting signaling through one or more RTKs, and many more are in late-stage clinical development (see below). In addition to their potential for off-label use in veterinary cancer patients, **RTK inhibitors specifically labeled for use in veterinary patients are now available.** Thus, an understanding of this class of protein and the drugs designed to inhibit RTK signaling will be critical as veterinary oncology advances in the next decade.

RTK ACTIVATION

Under most normal circumstances, activation of RTKs occurs only when the receptor encounters the appropriate ligand in the extracellular milieu. However, there are a variety of mechanisms by which RTKs can be inappropriately activated. These include: (1) Production of the ligand by the tumor cells themselves (autocrine stimulation); (2) Overexpression of the RTK, leading to spontaneous dimerization; (3) Mutations in the RTK, leading to constitutive activation in the absence of bound ligand. Any of these alterations can result in inappropriate signaling through the receptor, leading to an enhanced oncogenic phenotype.

RTK INHIBITION AND HUMAN CANCER

One of the first tyrosine kinases to be definitively implicated in human tumorigenesis and successfully inhibited was the ***bcr-abl*** fusion protein (the product of the so-called Philadelphia chromosome translocation), demonstrated to be present and functional in many cases of chronic myelogenous leukemia (CML).[1] This chimeric protein fuses the activation domain of the Bcr protein to the kinase domain of the Abl protein, rendering it constitutively active and capable of signaling at all times, providing an unending stimulus for cell proliferation and protection against cell death (apoptosis). Given the dependence on this fusion kinase for proliferation and survival in CML cells, an effort was made to find a drug capable of specifically inhibiting this signaling. This culminated in the discovery of a small molecule called STI571 (also called imatinib mesylate or Gleevec®). This molecule was capable of potently inhibiting signaling through the Abl kinase, resulting in diminished CML cell proliferation and enhanced cell death.[2] This molecule showed excellent antitumor activity and a good safety profile in humans with CML, and is currently approved for that indication.

Since the successful approval of imatinib several additional RTK inhibitors have been approved for use in human cancer **(Table 1)**.

Table 1. RTK Inhibitors Approved for Human Use

Name	Class	Targets	Indication
Imatinib (Gleevec®)	Small molecule	Bcr-Abl, Kit, PDGFR	Chronic myelogenous leukemia (CML), gastrointestinal stromal tumor (GIST)
Trastuzumab (Herceptin®)	Monoclonal antibody	HER2 (erbb2)	Breast cancer
Gefitinib (Iressa®) *Erlotinib (Tarceva®)*	Small molecules	EGFR	Lung cancer
Cetuximab (Erbitux®)	Monoclonal antibody	EGFR	Colorectal cancer
Bevacizumab (Avastin®)	Monoclonal antibody	VEGFR2	Colorectal cancer
Sorafenib (Nexavar®)	Small molecule	Kit, FLT3, VEGFR2/3, PDGFR	Renal cell carcinoma (RCC)
Sunitinib (Sutent®)	Small molecule	Kit, FLT3, VEGFR1-3, PDGFR, CSF-1R, RET	RCC, GIST
Dasatinib (Sprycel®)	Small molecule	Bcr-Abl, Src, KIT, EPHA2, PDGFR	CML, ALL
Nilotinib (Tasigna®)	Small molecule	Bcr-Abl, Kit, PDGFR	CML
Lapatinib (Tykerb®)	Small molecule	EGFR, HER2	Breast cancer
Panitumumab (Vectibix®)	Monoclonal antibody	EGFR	Colorectal cancer

RTKs AND ANGIOGENESIS

There are a group of RTKs that mediate the growth of new blood vessels (angiogenesis) in tumors. These include receptors for vascular endothelial growth factor, basic fibroblast growth factor, the angiopoietins, and others. As in tumor cells themselves, signaling through these RTKs are of prime importance in the growth and maintenance of tumor vasculature, and therapies designed to inhibit signaling through these angiogenic growth factor receptors (e.g. bevacizumab, sorafenib, sunitinib) are showing great promise in human clinical trials, and veterinary clinical trials using toceranib (Palladia®), which targets VEGFR2 as well as KIT (see below), are eagerly anticipated.

RTKs AND VETERINARY CANCER

Canine Mast Cell Tumor and KIT

While in general the patterns of expression and function of RTKs in veterinary cancer is less well understood that in human oncology, there are some examples where presence, activation and function have been well elucidated and treatment directed at the RTK successfully employed. The best example involves the RTK **KIT**, the receptor for stem cell factor, in canine mast cell tumor (MCT). London et al and other investigators have demonstrated in multiple studies that KIT is expressed on most canine MCT, and more interestingly, that many canine MCT harbor a mutation in the gene coding for KIT that renders it constitutively active.[3] This mutation may be more common in higher grade tumors.[4] Furthermore, the small molecules toceranib (Palladia®) and masitinib (Kinavet®, Masivet®) that inhibit signaling through KIT can potently inhibit the growth of canine MCT harboring an activating KIT mutation *in vitro*.[5] This elegant work culminated in clinical trials of toceranib and masitinib in canine MCT, in which significant antitumor activity were demonstrated.[6-9] These drugs, and their use in canine MCT will be discussed in greater detail in additional sessions at this conference.

Feline Vaccine-Associated Sarcoma and PDGFR

Recent work has demonstrated that a majority of feline vaccine-associated sarcomas (VAS) express the RTK **PDGFR**, the receptor for platelet-derived growth factor (PDGF). Additionally, PDGF stimulates the proliferation of feline VAS cells in culture and protects VAS cells from the antiproliferative and pro-apoptotic effects of doxorubicin. Furthermore, inhibition of PDGFR signaling with imatinib eliminates the oncogenic effects of PDGF and inhibits tumor growth in a nude mouse model of VAS.[10] Therapy with imatinib has been pursued in a small number of tumor-bearing cats to date, and it appears to be tolerated at doses approaching those associated with antitumor activity in humans.[11]

RTKs and Canine Osteosarcoma

A variety of RTKs have been identified in canine osteosarcoma (OSA) cells, including PDGFR, HER2 (one of several receptors for epidermal growth factor), IGF-1R (the receptor for insulin-like growth factor 1), and MET (the receptor for hepatocyte growth factor). IGF-1R and MET have been shown to mediate important oncogenic functions such as proliferation, anti-apoptosis, invasion, motility and chemoresistance in canine OSA cells,[12,13] and interestingly, constitutive activation of the MET receptor has been demonstrated in canine OSA as well.[13,14] Furthermore, MET activity can be inhibited in canine OSA and other canine tumor cells by a small molecule *in vitro*.[14] A randomized, placebo controlled trial in dogs with OSA using a somatostatin analog designed to reduce IGF-1 levels demonstrated no difference between dogs receiving chemotherapy and somatostatin analog and dogs receiving chemotherapy and placebo,[15] however serum IGF-1 concentrations were reduced by less than half in the treated dogs. It is quite possible that more potent inhibition might result in different results.

A more complete list of RTKs identified in veterinary cancer and levels of evidence for their utility as targets for therapy is presented in **Table 2**.

Table 2. Receptor Tyrosine Kinase Expression, Function and Inhibition in Veterinary Cancer.

RTK	Ligand	Tumor Type	Level of Evidence
KIT	Stem Cell Factor	Canine mast cell tumor, canine hemangiosarcoma	Expressed on most MCT and HSA: ~30% of MCT carry a functional activating mutation. Small molecules capable of inhibiting signaling and antitumor effects *in vitro* and in patients have been identified.
		Feline mast cell tumor	Expressed on most MCT. No functional mutations identified.
MET	Hepatocyte Growth Factor / Scatter Factor	Multiple canine tumors, including osteosarcoma, melanoma, histiocytic sarcoma, mast cell tumor	Expressed in most tumors of the histotypes examined. Constitutive activation detected in osteosarcoma. Receptor is functional and signaling inhibitable *in vitro* with small molecules.
IGF-1R	Insulin-like Growth Factor-1	Canine osteosarcoma, melanoma	Expressed in most tumors of the histotypes examined. Receptor is functional and signaling inhibitable *in vitro* with small molecules.
EGFR	Epidermal Growth Factor, Transforming Growth Factor-alpha, others	Canine mammary, lung, transitional cell, and nasal carcinoma, feline oral SCC	Expression documented most tumors of the histotypes examined. Receptor is functional in canine mammary carcinoma and signaling inhibitable *in vitro* with small molecules.
HER2	Epidermal Growth Factor, Transforming Growth Factor-alpha, others	Canine and feline mammary carcinoma, canine osteosarcoma	Expressed in a large minority of canine and feline mammary tumors, appears overexpressed in some canine osteosarcomas. Coreceptor for EGFR so may contribute to biologic effects of EGF in mammary carcinoma.
PDGFR	Platelet-derived growth factor	Canine osteosarcoma and hemangiosarcoma, feline vaccine-associated sarcoma	Expressed in canine osteosarcoma. Expressed, functional and inhibitable with small molecules *in vitro* and in mouse models of vaccine-associated sarcoma.
VEGFR2	Vascular endothelial growth factor	Canine hemangiosarcoma, melanoma, nasal carcinoma Tumor vasculature	Expressed in some melanoma and nasal carcinomas and most HSA evaluated. Some evidence of functionality in canine HSA cells. Increased concentrations of ligand detected in serum of HSA patients.
FGFR1 and 2	Basic fibroblast growth factor	Canine hemangiosarcoma Tumor vasculature	Expressed in most HSA evaluated. Evidence of functionality in canine HSA cells. Increased concentrations of ligand in urine of HSA patients.
TrkA	Nerve growth factor	Canine osteosarcoma	Expressed in most OSA. Blockade of TrkA induces apoptosis.

In conclusion, receptor tyrosine kinases represent an intriguing and potentially very powerful target for therapy in canine and feline malignancy. New drugs targeting these RTKs are being introduced at an ever-increasing pace, and it is virtually assured that this exciting new form of treatment will find an important place alongside surgery, radiation therapy and chemotherapy in the veterinary cancer clinician's armamentarium.

REFERENCES

[1]Giles, F.J., et. al. **Curr. Mol. Med.** 2005;5(7):615. [2]Buchdunger, E., et. al. **Cancer Res.** 1996;56(1):100. [3]Downing, S., et. al. **Am. J. Vet. Res.** 2002;63(12):1718. [4]Zemke, D., et. al. **Vet. Pathol.** 2002;39(5):529. [5]Liao, A.T., et. al. **Blood** 2002;100(2):585. [6]London, C.A., et. al. **Clin. Cancer Res.** 2003;9(7):2755. [7]Pryer, N.K., et. al. **Clin. Cancer Res.** 2003;9(15):5729. [8]London, C.A., et. al. **Clin Cancer Res.** 2009;15(11):3856. [9]Hahn, K.A., et. al. **J Vet Int Med** 22:1301. [10]Katayama, R., et. al. **Cancer Chemother. Pharmacol.** 2004;54:25. [11]Lachowicz, J.L., et. al. **J. Vet. Intern. Med.** 2005;19(6):860. [12]MacEwen, E.G., et. al. **J. Cell. Biochem.** 2004;92(1):77. [13]MacEwen, E.G., et. al. **Clin. Exp. Metastasis** 2003;20(5):421. [14]Liao, A.T., et. al. **Vet. Compar. Oncol.** 2005;3(2):61. [15]Khanna, C., et. al. **Clin. Cancer Res.** 2002;8(7):2406.

TYROSINE KINASE INHIBITOR (TKI) THERAPY IN COMPANION ANIMALS: YEAR ONE

Cheryl London , DVM, PhD, ACVIM (Oncology), Associate Professor, Ohio State University
Columbus, OH

INTRODUCTION

Inhibitors of tyrosine kinases (TKIs) have become an integral part of human cancer therapy over the past 5-10 years. Drugs such as Gleevec, Sutent, Sorafenib, Tarceva, among others, are now routinely used to treat a variety of cancers including chronic myelogenous leukemia, gastrointestinal stromal tumor, renal cell carcinoma, lung cancer, among others. A number of inhibitors are now in clinical trials as well. Despite the widespread use of such inhibitors in the human arena, limited data exists on the clinical efficacy of TKIs in veterinary medicine. In part, this is due to the fact that tumor types in which such inhibitors may work are only beginning to be defined. Additionally, many of the human TKIs are currently cost-prohibitive, preventing their widespread use. In June 2009, the first TKI for use in canine cancer patients, Palladia (toceranib phosphate) was approved by the FDA. The following review will summarize the recent clinical experience with Palladia in treating dogs with cancer.

Palladia: Mechanism of action

Palladia (toceranib phosphate) is a small molecule inhibitor that blocks a variety of tyrosine kinases expressed on the cell surface; these are known as receptor tyrosine kinases (RTK). It works as a competitive inhibitor of ATP, thereby preventing receptor phosphorylation and subsequent downstream signal transduction. It is known to block the activity of several members of the split kinase family including VEGFR, PDGFR, Kit, CSF-1, and Flt-3 and likely also inhibits the activity of another RTK called Ret. Palladia was originally developed as an anti-angiogenic agent as an inhibitor of VEGFR and PDGFR family members, and is known to inhibit angiogenesis in a variety of mouse tumor models. However, inhibition of other members of the split kinase family such as Kit result in direct anti-tumor activity as well. This combination of anti-angiogenic and anti-tumor activity likely provides broader clinical activity than that observed with more specific TKIs. The structure of Palladia and its mechanism of action are shown below:

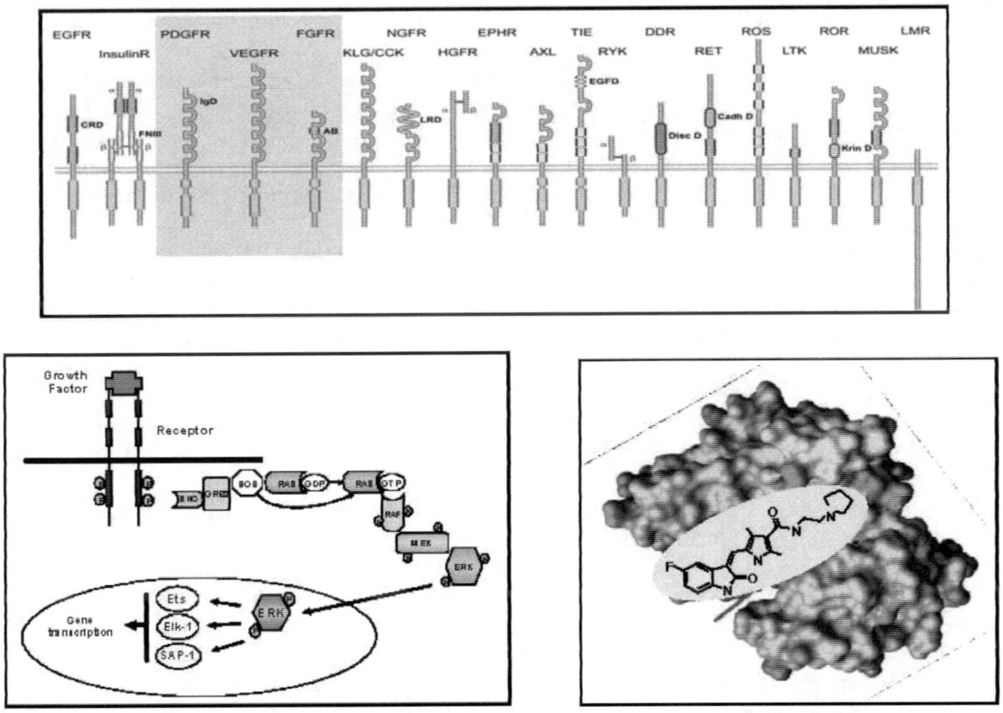

Figure 1: A) Receptor tyrosine kinases are grouped into families based on their structure. All RTKs possess an extracellular ligand binding domain, a transmembrane domain, and a cytoplasmic domain that is responsible for binding ATP and mediating signal transduction. B) Binding of RTKs by their growth factors induces receptor dimerization, phosphorylation, and the subsequent activation of cytoplasmic signaling elements, eventually influencing gene expression and cell cycling to regulate cell growth and survival. C) Small molecule inhibitors of RTKs typically act as competitive inhibitors of ATP, displacing this molecule and thereby preventing receptor phosphorylation and activation.

Review of Palladia clinical trials

The first evaluation of Palladia in veterinary patients was a phase I clinical trial exploring the safety and activity of this multi-targeted TKI (*36*). In 57 dogs with a variety of cancers (carcinomas, sarcomas, MCTs, melanomas, and lymphomas) objective responses occurred in 16 dogs (RR=28%), with stable disease in an additional 15 dogs, for an overall biological activity of 54%. Responding tumors included sarcomas, carcinomas, melanomas, myeloma, and MCTs. The highest RR was in MCTs, with 10/11 dogs with Kit mutations exhibiting responses (n=9) or stable disease (n=1). A placebo controlled randomized registrational study of Palladia was then performed in dogs with non-resectable Grade II and III MCTs (*37*). During the blinded phase, the RR in Palladia treated (n=86) dogs was 37.2% (7 CR, 25 PR) versus 7.9% (5 PR) in placebo-treated (n=63) dogs. Of 58 dogs that received Palladia following placebo-escape, 41.4% (8 CR, 16 PR) experienced an objective response. The overall RR for all 145 dogs receiving Palladia was 42.8% (21 CR, 41 PR). Dogs with Kit mutations were much more likely to respond (SD≥10 weeks, CR, or PR) to Palladia than those without (82% vs 55%) and dogs without lymph node metastasis had a higher response rate than those with involvement (67% vs. 46%). These two clinical trials demonstrated that Palladia has biologic activity in canine MCTs and likely has activity in a variety of other tumor types either through inhibition of angiogenesis, direct anti-tumor effects, or a combination of both.

The Palladia experience

Palladia was approved for use in June 2009 and since that time over 1000 dogs have been treated with drug. Data will be presented that summarizes the types of tumors treated, spectrum of toxicities, and dosing regimens. While the label indicates use of Palladia for recurrent Grade II and III MCTs, it has also been used quite extensively off-label to treat a variety of tumors as the Phase I data indicated that additional tumor types may respond to therapy. The discussion below addresses several important clinical considerations including dosing, optimal treatment of MCTs, and tumor types in which additional activity has been observed.

Palladia dosing

The Palladia label indicates dosing should be initiated at 3.25 mg/kg every other day (EOD) and dose reduced as needed based on clinical side effects; the lowest recommended dose is 2.2 mg/kg. While this has been established as the maximum tolerated dose, evidence exists that good biologic activity occurs when doses are initiated below the 3.25 mg/kg rate and flexible dosing schedules may also be effective. For example, in the Phase I study, of 16 dogs treated with Palladia at 2.5 mg/kg EOD, 6/16 (37.5%) exhibited response to therapy (4 complete and 2 partial) while an additional 5 dogs had stabled disease for greater than 10 weeks. This compares favorably with 20 dogs treated with 3.25 mg/kg EOD in which 8 (40%) had objective responses (2 complete and 6 partial) and an additional 4 dogs had stable disease for 10 weeks. Therefore, the overall biologic activity in the 2.5 mg/kg group was 68% compared to 60% in the 3.25 mg/kg group. Based on these data, it is reasonable to assume that for many dogs lower doses of Palladia result in significant clinical activity. This is important as the recent clinical experience with Palladia in dogs suggests that dosing at 2.5-2.75 mg/kg EOD is better tolerated than the higher dose, resulting in less toxicity, better owner compliance, and, importantly, fewer drug holidays. Indeed, over the past 8 months, dogs with a variety of tumors may respond to Palladia at this dosing regimen. Responding tumor histologies are discussed below.

With respect to the actual dosing regimen, there again is some flexibility in how this drug is administered. The label indicates that the drug should be administered EOD. However, some dogs do not tolerate this dosing regimen even at the 2.5 mg/kg dose rate. The author of this review and other medical oncologists who have worked with Palladia over the past year have found that a Monday/Wednesday/Friday schedule of dosing may be better tolerated by some dogs. This schedule may be particularly useful when Palladia is combined with other therapeutics, such as cyclophosphamide or NSAIDs (discussed below) as part of a metronomic treatment protocol. In the uncommon circumstance when a dog cannot tolerate the M/W/F schedule, it may be possible to administer drug every third day and maintain efficacy in certain cases. However, it is important to note that this is not ideal and may result in sub-therapeutic drug exposure.

Concomitant medications for management of Palladia side effects

Patients with a variety of cancers may have underlying conditions that predispose them to some of the toxicities induced by Palladia. This is particularly true for MCT patients who may have subclinical (or clinical) GI ulceration making them more susceptible to vomiting, diarrhea and GI bleeding. For such patients, it is highly recommended that pretreatment with appropriate supportive care be initiated. For MCT patients this may include the use of antacids (famotidine, omeprazole), anti-histamines (diphenhydramine), prednisone to help with tumor inflammation and potentially decrease MCT mediator release, and sucralfate in dogs with a positive hemoccult. In the author's experience, the use of these drugs 4-7 days prior to initiation of Palladia therapy does decrease the likelihood of toxicity. For dogs that experience inappetence or vomiting, the author has found that either metoclopramide or ondansetron work quite well and can be administered daily. Maropitant (Cerenia) may also be of benefit. With respect to diarrhea, the author has found the loperamide (Imodium) can be given on Palladia days to help prevent or lessen this side effect. Others have also found metronidazole useful. It is important to note that NSAIDs should be used with extreme caution in dogs receiving Palladia; they should never be given on the same day, and may significantly enhance GI toxicities. In general, dogs with MCTs tend to be more sensitive to Palladia side effects than those with non-MCTs. Lastly, Palladia is likely to enhance the magnitude of clinical illness in patients that are already sick and it is therefore highly recommended that every effort be made to increase the health status of dogs prior to treatment initiation.

Palladia as part of metronomic therapy

Over the past few years there has been significant interest in using low doses of chemotherapy and/or other anti-cancer agents to inhibit tumor angiogenesis. Metronomic therapy is generally designed to prevent tumors from continuing to grow (induce stable disease) although in some cases tumor regression can be noted. In the human arena, metronomic therapy has shown surprising success in some studies where patients had failed all conventional studies. In these trials, low dose chemotherapy (generally cyclophosphamide and capcitabine or 5-FU) were combined with anti-angiogenic agents such as Avastin (monoclonal antibody directed against VEGF). The importance of combining a cytotoxic with an anti-angiogenic agent was highlighted by work demonstrating increases in VEGF levels and circulating endothelial cell precursors (CEPs) in patients receiving anti-angiogenic agents as sole therapy, resulting in early drug failure. The addition of the cytotoxic agents helps to reduce the CEPs and in some cases, circumvent resistance. Metronomic therapy has been used in veterinary medicine, usually through combination of low dose cyclophosphamide administered daily with an NSAID, and evidence suggests it may have activity in certain settings such as hemangiosarcoma and soft tissue sarcomas. Recently, Palladia has been combined with low dose cyclophosphamide (10 mg/m2) in a dogs with a variety of metastatic tumor types. In general, the cyclophosphamide is administered on the Palladia "off" days (i.e., alternating with Palladia). The author's experience and the experience of other medical oncologists suggest that this treatment regimen is very well tolerated, does not exacerbate clinical toxicities from Palladia, and may result in greater biologic activity than with either drug alone.

Biologic activity of Palladia in non-MCT patients

Over the past year dogs with a variety of non-MCT histologies have been treated with Palladia, either alone or as part of a metronomic protocol. Tumor types in which biologic activity (stable disease, partial response) have been observed include anal sac adenocarcinoma, thyroid carcinoma, melanoma, soft tissue sarcoma, gastric carcinoma, gastrointestinal stromal tumor, nasal carcinoma, osteosarcoma, and transitional cell carcinoma. Anal sac tumors appear to be particularly sensitive to Palladia and several dogs have experienced significant disease regression after treatment. It is important to note that most of these patients had failed standard therapies, often had metastatic disease, and were therefore being treated with Palladia as a salvage therapy. Additionally, several patients also received low dose cyclophosphamide in addition to the Palladia, in an effort to enhance the anti-angiogenic effect. Importantly, most patients tolerated these therapies well and were able to maintain therapy for several months.

Summary and ongoing/future work

The use of Palladia should be considered when standard therapies have not been of benefit, or have failed, or if no good therapy exists for a particular cancer. The benefits of Palladia may be enhanced through combination with low dose cytotoxic therapy, particularly cyclophosphamide. Several challenges remain including continuing to define tumor types (other than MCT) which may respond to Palladia, determining appropriate dosing regimens that maintain drug efficacy, and evaluating activity in the setting of microscopic disease. Additionally, clinical trials are underway to explore the combination of Palladia with standard chemotherapeutics and local radiation therapy that hopefully will enhance objective response rates in a variety of cancers.

REFERENCES

[1.] C. A. London et al., Phase I dose-escalating study of SU11654, a small molecule receptor tyrosine kinase inhibitor, in dogs with spontaneous malignancies. **Clin Cancer Res** 9, 2755 (2003). [2.] C. A. London et al., Multi-center, placebo-controlled, double-blind, randomized study of oral toceranib phosphate (SU11654), a receptor tyrosine kinase inhibitor, for the treatment of dogs with recurrent (either local or distant) mast cell tumor following surgical excision. **Clin Cancer Res** 15, 3856 (2009). [3.] Ebos JM, Lee CR, Kerbel RS. Tumor and host-mediated pathways of resistance and disease progression in response to antiangiogenic therapy. **Clin Cancer Res**, 15, 5020 (2009)[4.] Kerbel RS. Issues regarding improving the impact of antiangiogenic drugs for the treatment of breast cancer. **Breast** 18, S41, 2009.

MEDICAL & SURGICAL ONCOLOGY: DIAGNOSIS & MANAGEMENT OF LYMPHOMA IN DOGS & CATS

J. Paul Woods, DVM, MS, DACVIM (Internal Medicine, Oncology); Sarah Boston, DVM, DVSc, DACVS
Guelph, ON, Canada

INTRODUCTION

Lymphoma presents in a variety of anatomical distributions with a variety of morphologically and functionally different cell types. This cancer is actually a heterogeneous group of malignancies with different tissue tropisms and resulting clinical behaviours. This heterogeneity results from the diversity of normal cellular differentiation exhibited by lymphocytes in their normal reaction to disease, i.e. development of helper T cells, suppressor T cells, memory T cells, early B cells, plasma cells, memory B cells, etc. A lymphoma consists of a proliferation of a clone (or rarely clones) of cells arrested at a particular stage of normal lymphocyte proliferation.[1-3]

Morphologically this results in tumors which have many different appearances. Clinically, each case has a very homogeneous population of cells (a clone) but there can be a wide variation between cases. Thus, there is a wide range in response to a given standard treatment regime. Malignant cells may express exaggerated or altered behaviour based on their benign functional counterparts, resulting in such paraneoplastic syndromes as hypercalcemia and monoclonal gammopathies. Canine and feline lymphoma is therefore a malignancy of lymphocytes which is extremely variable in the type of cell differentiation, resulting in variable clinical presentations and response to treatment.

Malignant lymphoma is the most common hematopoietic neoplasm in both the dog and cat. In dogs it has been reported as the fourth most common canine neoplasm with an occurrence of 24-30 cases/100,000 dogs per year (approximately twice the human incidence and 2/3 the incidence in cats). The cat has the highest incidence of lymphoreticular malignancy of any species with 200 cases/100,000 cats at risk. One third of all cat tumors are hematopoietic tumours of which 90% are lymphoma. Some cases of lymphoma in the cat are caused by the feline leukemia virus (FeLV). There is speculation for a contagious retrovirus for lymphoma in dogs, howevere direct evidence is lacking. Genetic predisposition might be an important contributing factor to the development of lymphoma in the dog because some breeds are at high risk for development of lymphoma while other breeds have a low incidence. Breeds of dogs reported to have a higher incidence include boxers, basset hounds, St. Bernards, Scottish terriers, Airedales, and bulldogs. (Breeds at low risk include dachshunds and Pomeranians.)[1-3]

DIAGNOSIS

The minimum requirement for making the diagnosis is a confirmatory fine-needle or tissue biopsy. Select an enlarged lymph node other than submandibular lymph nodes which are usually scarred and hyperplastic due to oral antigenic stimulation. The absolute minimum sample is a well-prepared fine needle aspirate (FNA) smear (not a crush preparation: one option is to prepare the smear just as if it were a blood film, drawing the cells along behind the edge of the slide). Cytology preparations must be gently done and well stained: the most common problems with samples are ruptured cells and inadequate staining density. If lymphadenopathy is not evident, then bone marrow, liver, spleen, kidney, or mediastinal involvement must be considered and biopsy of identified lesions in such places may be required to demonstrate the disease.[1-3]

Classification of lymphoma is based on anatomic, histologic, and phenotypic characteristics. Clinical signs are related to anatomic forms. Major anatomic types are: multicentric: most common (80%); gastrointestinal with little nodal involvement (8%); skin: uncommon; thymic: uncommon; and miscellaneous extranodal (e.g. splenic, CNS): rare. Histologically, regardless of classification systems, most lymphoma in both species are considered to be of high grade malignancy.[1-3] Immunophenotyping (B- vs T-cell) can be determined with flow cytometry, immunocytochemistry, or immunohistochemistry.[4,5] In cases where the diagnosis is ambiguous, reactive (polyclonal) lymphocytes can be distinguished from neoplastic (monoclonal) lymphocytes using a PCR for antigen receptor rearrangements (PARR) assay.[6] Sensitivity varies by lab but is ~75% for detecting lymphoma (although Ehrlichia may also produce a lymphocyte clonal expansion).[7]

STAGING

Staging indicates the extent of involvement of lymphoma. Stage in dogs is as follows: Stage I: single lymph node involved (rare); Stage II: regional nodal involvement; Stage III: generalized lymphadenopathy (common); Stage IV: liver and/or splenic involvement; Stage V: bone marrow involvement or extranodal lymphoma. Each stage is further subclassified by physical performance status, qualitatively, as: (a) without systemic signs, or (b) with systemic signs. Important related abnormalities include anemia, hyperglobulinemia, and hypercalcemia. In general, the prognosis becomes worse for more advanced stages, howeverperformance subclassification (a or b) may be more important. Although advanced stage disease and the presence of hypercalcemia have been shown to adversely correlate to prognosis in the dog, response to therapy may alter these abnormalities and allow for long-term survival in any individual instance. Feline leukemia infection may be detrimental in general to long-term survival in feline lymphoma but it is not predictive of initial response to therapy.[1-3]

taging should be done to determine clinical stage of disease and ability to handle chemotherapeutic drugs (e.g. liver function).[1-3] Staging requires: Complete blood count (CBC) to look for leukemia, marrow compromise (phthisis), or splenic malfunction; biochemical profile to evaluate calcium, protein, liver enzymes, bilirubin, creatinine, and urea for evidence of organ dysfunction; bone marrow aspirate to evaluate for involvement of the bone marrow; and radiographs of thorax and abdomen (or abdominal ultrasound) to evaluate for lymphadenomegaly or organomegaly; .

TREATMENT

Without treatment lymphoma is a uniformly fatal disease with usually rapid rate of progression, as most dogs and cats present in a late stage of disease and have rapidly dividing aggressive cell types. Average survival without treatment is about 30 days. With treatment, lymphoma survival times are usually increased, and well-managed patients that establish complete remissions typically enjoy excellent quality of life until their disease is no longer responsive to treatment. The costs of treatment are significant and must be considered on an individual basis. The aim of treatment is to completely eradicate the tumour. In practice, the clinician attempts to reduce the tumor burden to non-detectable levels for as long as possible, thereby increasing the pet's quality of life and prolonging survival. There is great variation in the response to standard treatment protocols.

Individual response and ultimate survival can be highly variable; however, reports of treated dogs indicate complete remissions rates of 60-80% with median survivals of 6-12 months,. In cats, the complete remission rate is approximately 60% with a median survival time of 6-7 months. These figures are representative of relatively aggressive protocols reported by teaching clinics. Known variables include treatment protocol, clinical stage, size of pet, sex of pet, owner compliance, and the clinician's commitment and skills.[1-3]

The major problem is the development of tumor resistance to the drugs used. The mechanisms of this are: (1) genetic diversity and selection for altered metabolic pathways; (2) active transport of the drugs out of the cell after entry by membrane carrier proteins, shown for vincristine and adriamycin, and (3) cells not in appropriate phase of cell cycle for drug to act. The toxicity of the drugs is a related problem, in that escalating the dosage to overcome resistance is limited by toxicity to normal tissues. Another related problem is lack of efficacy. Many drugs are effective only when combined with others or when used at unacceptably high dosage levels.[1-3]

Radiation has been used either as palliative for bulky localized disease (e.g. mediastinum) or signs due to CNS lymphoma, or as half-body radiation in addition to chemotherapy.[8-10] There are also clinical trials investigating bone marrow transplantation to treat lymphoma.[11,12]

CANINE LYMPHOMA PROTOCOLS[13]

Single Agent

Prednisone is the simplest and cheapest therapy which may provide some degree of relief for a few weeks; however, with prednisone therapy alone, the average pet only lives 2 months (compared to 1 month with no therapy). Prednisone will induce a complete remission in one third of the pets, a partial remission in one third, and no response in one third. Treating with corticosteroids alone, however, may jeopardize the ability to attain a durable remission with subsequent combination therapy, as pre-treatment with corticosteroids can induce the phenomenon of pleiotropic drug resistance by enhancing the expression of P-glycoprotein cell membrane pumps for removal of drugs from cells.

Doxorubicin (Adriamycin) is a very effective drug all by itself. It has been used as an induction agent and is at least as effective, if not more so, than multi-agent protocols that do not contain doxorubicin, such as COP(cyclophosphamide, vincristine, prednisone). Alternatively, it can be used as a rescue agent after COP-type protocols have failed. As a single-agent, doxorubicin is given at a dose of 30 mg/m^2 or 1 mg/kg intravenously every 21 days for a total of 5 treatments. Administration of this drug is limited by cumulative cardiotoxic effects, requiring a maximum of about 8 lifetime treatments per patient.
There is a 5% risk of irreversible cardiotoxicity (dilated cardiomyopathy) at a total cumulative dose of 180 mg/m^2, and a 10% risk at 240 mg/m^2 (8 treatments). Complete remission (all lymph nodes back to normal size) usually occurs by the end of the first three weeks, and should definitely occur by the second treatment. If there is a questionable response, a different protocol should be attempted.[14]

Multidrug Protocols

Multidrug protocols are the standard-of-care for treatment of lymphoma, and have the advantage of using multiple mechanisms to eradicate tumor cells. Common protocols are outlined in many references. Many use an initial combination of cyclophosphamide, vincristine and prednisone. Some reserve l-asparaginase (Elspar) and doxorubicin for consolidation or rescue roles. Others add cytosine arabinoside or methotrexate. However, the most common multidrug protocol combination typically includes vincristine, cyclophosphamide, doxorubicin, and prednisone, such as the Madison-Wisconsin Short Protocol, listed below.

COP

(**C**ytoxan® [cyclophosphamide], **O**ncovin® [vincristine], **P**rednisone)
These protocols are generally of two types: frequent low doses, or infrequent high dose blasts, both using the same drugs.[15,16]

COAP

(**C**ytoxan® [cyclophosphamide], **O**ncovin® [vincristine], cytosine **A**rabinoside, **P**rednisone)[17]

L-VCMP

(**L**-asparginase, **V**incristine, **C**ytoxan® [cyclophosphamide], **M**ethotrexate, **P**rednisone)[18]

L-VCAMP

(**L**-asparginase, **V**incristine, **C**ytoxan® [cyclophosphamide], **A**driamycin® [doxorubicin], **M**ethotrexate, **P**rednisone)[19]

L-VCAMP (Madison-Wisconsin Protocol)

(**L**-asparginase, **V**incristine, **C**ytoxan® [cyclophosphamide], **A**driamycin® [doxorubicin], **M**ethotrexate, **P**rednisone)[20]

L-VCAP-Short (Madison-Wisconsin Short Protocol)

(**L**-asparginase, **V**incristine, **C**ytoxan® [cyclophosphamide], **A**driamycin® [doxorubicin], **P**rednisone)[21]

Ontario Veterinary College Protocol[9]

Similar to intensified Madison-Wisconsin dose protocol with addition of two half body radiation treatments and substitution of epirubicin (30 mg/m^2) for doxorubicin. No maintenance phase.

FELINE LYMPHOMA PROTOCOLS[13]

COP

(**C**ytoxan® [cyclophosphamide], **O**ncovin® [vincristine], **P**rednisone)[22,23]

COPA

(**C**ytoxan® [cyclophosphamide], **O**ncovin® [vincristine], **P**rednisone, **A**driamycin® [doxorubicin])[24]

VCM

(**V**incristine, **C**ytoxan® [cyclophosphamide], **M**ethotrexate)[25]

L-VCMP

(**L**-asparginase, **V**incristine, **C**ytoxan® [cyclophosphamide], **M**ethotrexate, **P**rednisone)[26]

L-VCAMP

(**L**-asparginase, **V**incristine, **C**ytoxan® [cyclophosphamide], **A**driamycin® [doxorubicin], **M**ethotrexate, **P**rednisone)[19,27]

L-VCAM (Madison-Wisconsin Protocol)

(**L**-asparginase, **V**incristine, **C**ytoxan® [cyclophosphamide], **A**driamycin® [doxorubicin]), **M**ethotrexate)[28]

REFERENCES

1. Vail DM et al. In: Withrow SJ, Vail DM eds. **Withrow & MacEwen's Small Animal Clinical Oncology**, 4th edition, Saunders Elsevier, St. Louis, 2007; 699. 2. Morrison WB. **Lymphoma in Dogs and Cats**, Teton New Media, Jackson, WY, 2005. 3. Bryan JN. In: Henry CJ, Higginbotham ML eds. **Cancer Management in Small Animal Practice**, Saunders Elsevier, St. Louis, 2010;343. 4. Gauthier MJ, et al. **J Vet Intern Med** 2005;19:193. 5. Gibson D, et al. **J Vet Intern Med** 2004;18:710. 6. Burnett RC, et al .**Vet Pathol** 2003;40:32. 7. Avery A. Top **Comp Anim Med** 2009;24(3):144. 8. Straw RC, et al. **J Am Vet Med Assoc** 1986;189:927. 9. Abrams-Ogg ACG, et al. **Proceedings of Veterinary Cancer Society** 1999;19:8. 10. Gustafson NR, et al. **Vet Comp Oncol** 2004;2:125. 11. Frimberger AE et al. **J Vet Intern Med** 2006;20:355. 12. Lupu M, et al. **J Am Vet Med Assoc** 2006;228:728. 13. Woods JP. In: Allen DG et al. eds. **Handbook of Veterinary Drugs**, 3rd edition, Lippincott Williams & Wilkins, Philadelphia, 2005;72. 14. Postorino NC, et al: **J Am Anim Hosp Assoc** 1989;25:221. 15. Cotter SM. **J Am Anim Hosp Assoc** 1983;19:159. 16. Madewell BR. **Am J Vet Res** 1975;36:1525. 17. Theilen GH, et al. **J Am Vet Med Assoc** 1977;170:607. 18. MacEwen EG, et al. **J Am Vet Med Assoc** 1987;190:564. 19. Matus RE. In Kirk RW ed. **Current Veterinary Therapy X.** Philadelphia: WB Saunders, 1989;482. 20. eller ET, et al. **J Vet Intern Med** 1993;7:289. 21. Chun R, et al. **J Vet Intern Med** 2000;14:120. 22. Cotter SM. **J Am Anim Hosp Assoc** 1983;19:166. 23. Teske E, et al. **J Vet Intern Med** 2002;16:179. 24. Moore SA, et al. **J Vet Intern Med** 1996;10:372. 25. Jeglum KA, et al. **J Am Vet Med Assoc** 1987;190:174. 26. Mooney SC, et al. **J Am Vet Med Assoc** 1989;194:696. 27. Zwahlen CH, et al. **J Am Vet Med Assoc** 1998;213:1144. 28. Vail DM, et al. **J Vet Intern Med** 1998;12:349.

MEDICAL & SURGICAL ONCOLOGY: DIAGNOSIS & MANAGEMENT OF MAST CELL TUMORS IN DOGS & CATS

J. Paul Woods, DVM, MS, DACVIM (Internal Medicine, Oncology); Sarah Boston, DVM, DVSc, DACVS
Guelph, ON, Canada

INTRODUCTION

Mast cell tumors (MCT) are the most common cutaneous tumors in the dog representing 20-25% of all canine skin tumors. In the cat, MCT represent 15% of all tumors (second most common cutaneous tumor). No sex predilection has been found in the dog, but in the cat the male:female ratio is 2:1. MCT occur most commonly in middle-aged dogs and cats with a mean age of about 8 years. Siamese cats, boxers, Boston terriers and other breeds of bulldog descendants (such as the Pug) are at higher risk to develop MCT.[1,2]

Mast cell tumors may exist in cutaneous or extracutaneous sites. In dogs the most common sites are the skin of the trunk and perineal region (50%). the skin of the extremities (40%), and the cutaneous head and neck (10%). MCT may arise as multifocal tumors in about 11% of dogs . In cats, most MCT are found in the head and neck region. Occasionally MCT are located only in a visceral location in cats, such as the spleen.[1,2]

The biologic behavior of MCTs is extremely variable. In dogs, histologic grading can help determine the likelihood of recurrence or metastasis. High-grade mast cell tumors have a high metastatic potential, and dogs may present with metastasis to regional lymph nodes or, rarely, distant metastases to blood, bone marrow, or abdominal organs such as the liver or spleen. . Dogs with advanced disease may have gastric or duodenal ulcers (related to elevated systemic histamine levels with secondary excess of hydrochloric acid secretion in the stomach), and present with vomiting, melena, and anemia. Occasionally mechanical manipulation during examination will result in degranulation of mast cells which results in erythema and wheal formation (Darier sign).[1,2]

Diagnosis of MCT can often be made by a fine needle aspirate (FNA) cytology, however excisional biopsy is required for accurate histologic grading. Establishing a diagnosis prior to surgical treatment can be very helpful because the treatment should be aggressive and precaution may need to be taken to prevent or control heparin and histamine release. Diagnostic work-up may consist of staging with a complete blood count (CBC) (eosinophilia, basophilia, mast cells), buffy coat smear, bone marrow aspirate, FNA of the lesion and regional lymph node, abdominal radiographs or ultrasound (lymphadenomegaly or organomegaly). In advanced canine MCT, a coagulogram (PT, PTT, FDP) may be performed to evaluate systemic coagulation status.[1,2] The presence of the occasional mast cell in the buffy coat or in the draining lymph node must be interpreted with caution, because non-neoplastic diseases can also result in mastocytosis.[3] The tumor biopsy may be submitted for additional tests, such as mitotic index or immunohistochemistry (IHC) for markers of Ki67, c-kit mutations (CD117), and proliferation markers such as AgNOR staining , which are prognostic for tumor recurrence and metastasis in some canine MCT.[4,5]

THERAPY

The therapeutic approach for MCT is based on the clinical stage and histologic grade of the tumor.

Surgery is the primary treatment for MCT, particularly for control of local disease.[6,7,8] Wide surgical excision (3 cm margins and one fascial plane deep) is the best approach local therapy for tumors that have not spread, and this therapy may be curative. However, recent evidence suggests that 2 cm margins laterally may be sufficient for complete excision of some tumors.[7] All surgical margins should be histologically evaluated.

Radiation therapy can be used as adjuvant therapy after incomplete surgical excision or as primary treatment if surgery is not an option .[9,10,11,12] The majority (96%) of local intermediate grade (grade 2) tumors that are incompletely excised and followed with definitive radiation therapy can be controlled for 5 years or more. Radiation therapy can also be used as a treatment for bulky disease when surgery is not an option, and has a reported 5-075% control rate at one year. Chemotherapy is less effective than surgery or radiation. However, patients that have multiple tumors or advanced local disease where local treatment is not an option may be managed by chemotherapy. Although remissions may be achieved, cure is unlikely. Remission times with chemotherapy are usually 10-20 weeks. Multiagent protocols (e.g. prednisone and vinblastine) may achieve a higher response rate than single agent therapy.[1,2,13] In dogs, glucocorticoid therapy (oral prednisone [1-2 mg/kg PO or 20-40 mg/m^2 PO q24 hr] or triamcinolone [1 mg/cm diameter of tumor intralesionally q 2 weeks]) may result in partial remission (PR) or occasionally complete remission (CR); however feline MCT appears to be less responsive to glucocorticoids. Recently, tyrosine kinase inhibitors (TKI) (toceranib [Palladia], masitinib [Kinovet], imatinib [Gleevec]) have become more available in veterinary medicine.[14-21] Kit, a tyrosine kinase, has been found to be mutated in 9-30% of high grade MCT in dogs.[15] In the future, TKIs in combination with other chemotherapy may result in higher response rates.

Ancillary drug therapy is important with canine MCT. Dogs with mastocytosis, measurable MCT, or evidence of gastrointestinal bleeding should receive H_2 antagonists (i.e. famotidine, ranitidine) to reduce gastric acid secretion and prevent gastrointestinal ulceration. Sucralfate (0.5-1 g PO q 8h) may be given in cases with evidence of gastrointestinal ulceration and bleeding. H_1 antagonists (e.g. diphenhydramine 2-4 mg/kg PO q 12 h) are used along with famotidine prior to and following surgical removal of canine MCT to help prevent the negative effects of local histamine release on fibroplasia wound healing.[1,2]

PROGNOSIS

In general, cutaneous MCT carry a more guarded prognosis in the dog than in the cat. MCT in the Boxer are usually of a lower histologic grade than those found in other breeds. MCT in Siamese are of the less malignant histiocytic type. Prognosis is based on species, breed, tumor location, growth rate, extent of disease, and the histologic grade. The grade is based upon cellular differentiation, cellular pleomorphism, cytoplasmic granules, mitotic figures, and depth of invasion. The more undifferentiated tumor then the higher the grade and the poorer the prognosis.[1,2,22] Tumors located in the perineal or preputial area are likely to metastasize both locally and to deep lymph nodes, and prognosis for tumors in these areas may be worse. Rapid growth rate is also an important prognostic indicator. Dogs with MCT growth rates of greater than 1 cm/week have only a 25% chance of surviving an additional 30 weeks.

Regardless of local therapy chosen, even dogs with low to intermediate grade tumors should be evaluated regularly for local recurrence and possible systemic spread.

CHEMOTHERAPEUTIC PROTOCOLS FOR CANINE MAST CELL TUMOR[23]

Prednisone[24,25]
Prednisone 1 mg/kg PO q 24 h
or
Prednisone 40 mg/m^2 PO q 24 hr for 1 week, then 20 mg/m^2 PO q 24 hr for 3 weeks, then 20 mg/m^2 PO q 48 hr for 3 weeks, then reduce dose by 50% every 3 weeks to lowest dose that maintains remission or stable disease.

Triamcinolone[26,27]
Triamcinolone 3-6 mg/dog intralesionally once/week
or
Triamcinolone 1mg / cm tumor diameter, maximal dose 10-40 mg, q 14 days

CCNU (lomustine)[27]
CCNU (lomustine) 70-90 mg/m^2 PO every 3 weeks

Prednisone/Vinblastine[28]
Prednisone 2 mg/kg PO q 24 h; Vinblastine 2 mg/m^2 IV q weekly for 4 weeks followed by 4 treatments every 2 weeks
Continue Vinblastine 2 mg/m^2 IV every 2 weeks for as long as appears to have an objective response. Taper and discontinue prednisone over 12-26 weeks.

CVP [29,30,31]
(**C**yclophosphamide/**V**inblastine/**P**rednisone)
Cyclophosphamide 50 mg/m^2 PO q 48 h or for 4 days per week; Vinblastine 2 mg/m^2 IV once a week; Prednisone 20-40 mg/m^2 PO q 48h
or
Day 1 Vinblastine 2-2.2 mg/m^2 IV ;
Days 8,9,10,11 Cyclophosphamide 50-75 mg/m^2 PO OR Day 8 Cyclophosphamide 200-250 mg/m^2 IV (Give furosemide 2 mg/kg IV if giving IV Cyclophosphamide);
Daily Prednisone 1 mg/kg PO q 24 h
Repeat cycle every 3 weeks for 6 months. Prednisone tapering began at 4 months and discontinued by 7 months.

Tyrosine Kinase Inhibitors (TKI)
Toceranib (Palladia) [14-16]
3.25 mg/kg PO q 48 h

Masitinib (Kinavet)[17]
12.5 mg/kg PO q 24 h

Imatinib (Gleevec)[18,21]
5 to 10 mg/kg PO q 24 h

REFERENCES

1. Thamm DH, et al. In: Withrow SJ, Vail DM eds. **Withrow & MacEwen's Small Animal Clinical Oncology**, 4th edition, Saunders Elsevier, St. Louis, 2007; 402. 2. Fox LE. In: Morrison WB ed. **Cancer in Dogs and Cats Medical and Surgical Management**, 2nd edition, Teton New Media, Jackson, WY, 2002;451. 3. Bookbinder PF, et al, **J Am Vet Med Assoc** 1992;200:1648. 4. Scase TJ, et al. **J Vet Intern Med** 2006;20:151 5. Romansik EM, et al. **Vet Pathol** 2007;44:355. 6. Murphy S, et al. **Vet Rec** 2004;154:743. 7. Simpson AM, et al. **J Am Vet Med Assoc** 2004;224:236. 8. Seguin B, et al. **J Am Vet Med Assoc** 2001;218:1120. 9. Al-Sarraf R, et al. **J Vet Intern Med** 1996;10:376. 10. Turrel JM, et al. **J Am Vet Med Assoc** 1988;193:936. 11. Dobson J, et al. **Vet Comp Oncol** 2004;2:132. 12. Frimberger AE, et al. **J Am Anim Hosp Assoc** 1997;33:320. 13. Davies DR, et al. **J Am Anim Hosp Assoc** 2004;40:124. 14. London CA, et al. **Clin Cancer Res** 2003;9:2755. 15. Downing S, et al, **Am J Vet Res** 2002;63:1718. 16. London CA. **Top in Comp An** 2009;24(3):106. 17. Pryer NK, et al. **Clin Cancer Res** 2003;9:5729. 18. London CA, et al. **Clin Cancer Res** 2009;15(11):3856. 19. Hahn KA, et al. **J Vet Intern Med** 2008;22:1301. 20. Isotani M, et al. **J Vet Intern Med** 2008;22(4):985. 21. Lachowicz JL, et al. **J Vet Intern Med** 2005;19:860.. 22. Bostock DE. **J Small Anim Pract** 1973;14:27. 23. Woods JP. In: Allen DG et al. eds. **Handbook of Veterinary Drugs**, 3rd edition, Lippincott Williams & Wilkins, Philadelphia, 2005;72. 23. McCaw DL, et al. **J Vet Intern Med** 1994;8:406. 24. O'Keefe DA. **Vet Clin North Am** [Small Anim Pract] 1990;20:1105. 25. Hess PW, et al. **J Am Anim Hosp Assoc** 1976;12:350.26. Tams TR, et al. **Comp Cont Educ Pract Vet** 1981;3:869. 27. Rassnick KM, et al. **J Vet Intern Med** 1999;13:601. 28. Thamm DH, et al. **J Vet Intern Med** 1999;13:491. 29 O'Keefe DA, **Vet Clin North Am** [Small Anim Pract] 1990;20:1105. 30. Elmslie R. **Vet Cancer Soc Newsl** 1997;20:6. 31. Camps-Palau M. **Vet Comp Oncol 2007;5:156.**

MEDICAL & SURGICAL ONCOLOGY: DIAGNOSIS & MANAGEMENT OF OSTEOSARCOMA IN DOGS & CATS

Sarah Boston, DVM, DVSc, DACVS; J. Paul Woods, DVM, MS, DACVIM (Internal Medicine, Oncology)
Guelph, ON, Canada

INTRODUCTION

Osteosarcoma is the most common primary bone tumor in dogs and makes up approximately 85% of the tumors of the skeleton. This is primarily a disease of large and giant breed dogs. Older dogs are more commonly affected, with a median age of 7 years. Osteosarcoma has a predilection for the metaphyseal regions of long bones. The most common sites of osteosarcoma include the metaphyseal region of the distal radius, proximal humerus, distal femur and distal and proximal tibia.[1,2]

DIAGNOSIS AND STAGING

A presumptive diagnosis of osteosarcoma is based on the signalment, tumor location and radiographic signs. A decision to biopsy the bone prior to treatment will depend on several factors. If the presentation of the case is very typical of a primary bone tumor and the owner's willingness to treat will not be altered by the tumor type, the clinician may opt to skip this step and move forward with definitive therapy. Cases with an atypical presentation for osteosarcoma should be biopsied prior to definitive therapy.[1,2]

Bone biopsy is performed using a Jamshidi needle or Michele trephine and a closed technique in most cases.[3] When planning the site of the bone biopsy, it is also important to take the options for definitive therapy into consideration. The biopsy tract is considered contaminated with tumor cells and will need to be removed during definitive therapy. This becomes important in cases where limb spare surgery is an option or in cases of flat bone osteosarcoma. If stereotactic radiosurgery is a consideration for limb spare, a bone biopsy should not be performed in cases that are typical of osteosarcoma. The reason for this is that one of the major complications associated with stereotactic radiosurgery is pathologic fracture and this has been seen at the bone biopsy tract sites. Another option for determining a pretreatment diagnosis is cytology via a fine needle aspirate. This technique has been used in concert with ultrasound in an attempt to improve the yield. The ultrasound can be used to look for a break in the cortex through which to insert a large gauge needle. A diagnostic sample was obtained in that study in 32/36 cases.[4]

Staging is performed to determine if there is gross spread of disease. Almost all (90-98%) cases of osteosarcoma have metastasized at the time of diagnosis. However, this is micrometastatic disease in most cases, as the rate of diagnosing gross metastasis at the time of presentation is only ~15%. The two most common sites of metastasis are the lungs and another bone site.[1,2] Three-view thoracic radiographs are an essential part of tumor staging. CT scan is the modality of choice for evaluating for lung metastasis in human osteosarcoma patients. This test has been shown to be highly sensitive but not specific for diagnosing gross metastasis to the lung in humans. As CT scans become more common-place in staging for osteosarcoma in canine patients, more information will become available as to how to interpret positive lesions on the CT scan.

The probability of diagnosing gross bone metastases at the time of diagnosis is low (~ 8%). Evaluation for metastases to other bones involves a complete orthopedic examination and long bone survey radiography or nuclear scintigraphy. A complete orthopedic examination including palpation of the spine is important to determine if the dog is a good candidate for an amputation and to evaluate for possible metastatic disease. Areas that are painful should be radiographed.

The rate of metastasis to regional lymph nodes has been reported to be 4.4%.[5] Dogs with lymph node metastases have a much shorter survival time (approximately 2 months). The regional lymph nodes should be palpated and aspirated if enlarged. If the diagnosis of lymph node metastasis is not made prior to amputation, it is important that the lymph node is specifically examined histologically post-amputation.

SURGICAL TREATMENT

Amputation

The goal of surgical treatment is complete resection of the primary tumor to prevent further tumor progression and remove the source of pain. Most commonly this is achieved by amputation of the affected limb. In most cases there is a significant reduction in weight bearing in the affected limb prior to surgery and the patient will adjust easily to walking on three legs post-operatively. A thorough orthopedic examination should be performed prior to amputation. In most cases, dogs will cope well after amputation, even if there is evidence of osteoarthritis. Dogs treated with amputation alone will generally live for 3-4 months until they succumb to metastatic disease.

Limb Spare

The distal radius is the site that has been most successful for limb spare surgery. However, there are options available for other sites. Owner selection for this procedure to be successful is as important as case selection. The owners who choose to move forward with this option must be fully educated that this option carries with it a significantly greater financial and time commitment than amputation and that there is a much higher risk of post-operative complications. Common complications of limb spare surgery include infection and implant failure. Local recurrence is also a potential complication of limb spare surgery. In a small percent of limb spare cases, the complications are catastrophic and revision surgery or amputation is required. Guidelines for selecting candidates for distal radial limb spare are that the patient is negative for gross metastasis, has minimal soft tissue involvement and that less than 50% of the length of the bone is affected radiographically.

There are many limb spare techniques that are currently employed by surgical oncologists to treat canine distal radial osteosarcoma. These include: cortical allograft,[2] pasteurized autograft,[6] intraoperative radiation (autograft),[7] endoprosthesis, vascularized ulnar transposition,[8] stereotactic radiosurgery,[9] bone transport osteogenesis,[10] and intraosseous transcutaneous amputation prosthesis (ITAP). With the exception of intraoperative radiation, stereotactic radiosurgery and ITAP, these techniques involve the resection of the distal radius with appropriate margins and the replacement of the defect that is created with an autograft, allograft, prosthesis or regenerate bone.[11] As with many surgical procedures with multiple techniques available, there is not one perfect technique that is without its own set of complications. Currently the endoprosthesis technique appears to be used most commonly by surgical oncologists.

The endoprosthesis limb spare technique uses a 316L stainless steel endoprosthesis that is commercially available. The original endoprosthesis was developed by Charles Kuntz and has been modified over time. In a study by Liptak et al that prospectively compared clinical cases of allograft limb spare and endoprostheses, significantly more of the radius was resected using the 122mm endoprosthesis.[11] The authors postulated that this might have increased the incidence of proximal radial screw pullout. The newer generation VOI endoprosthesis is available in two lengths and is coated with hydroxyapatite to encourage bone ingrowth. The screws are locking screws to help reduce screw pullout. In Liptak's study comparing allograft and endoprosthesis in clinical cases, there was no significant difference in infection rate, infection severity, median time to infection, surgical time, limb use, implant failure rate or oncologic outcome.[12] A biomechanical study of allograft vs. endoprosthesis in cadaveric limbs showed the endoprosthesis to be biomechanically superior to the allograft. This study also showed that there was no biomechanical advantage to preserving the ulna in either the allograft or the endoprosthesis group.[12,13]

SRS was reported by Farese et al as a limb spare technique that involves a single dose of radiation that is accurately delivered to the bone tumor with minimal radiation delivered to the surrounding tissues.[9] This delivery technique involves creating a frame around the limb that is used to ensure accurate positioning during the planning CT and the delivery of radiation. A contrast-enhanced CT is then obtained, which is used to plan a radiation dose that will conform to the shape of the tumor target by the use of a linear accelerator. After treatment planning, the patient was transferred to the radiation suite and a single dose of 3000cGy was delivered to the tumor volume. There is a steep dose gradient between the tumor and surrounding tissues to minimize damage to healthy tissues. This technique was performed in 11 dogs, with 9/11 having an osteosarcoma of the distal radius. Four of the dogs developed pathological fractures. The advantage of this technique is that it is non surgical and therefore infection and implant failure are not a concern. The main disadvantages are that SRS is not readily available and there is a risk of pathologic fractures.

ITAP is a new method of limb spare that employs a custom metal prosthesis that is fitted into the medullary cavity of the amputated bone. It is a new treatment technique that has been reported in the USA and the UK. It is rarely performed due to the expense and the expertise necessary, but it is likely to become more common in the future. The fundamental part of the development of this technique is creating a material that is of a type and porosity that allows for ingrowth of the subcutaneous and cutaneous tissues into the prosthesis-skin interface. Early results with this technique are promising and will provide another technique for limb spare in the future for osteosarcoma and other malignant tumors of the distal extremities.

The complications rate for limb spare surgery is much higher than for other orthopedic procedures. The three major complications that occur are infection, local recurrence and implant failure.

The reported infection rate for limb spare varies widely but is 40-75%. Limb spare infections are difficult to manage because of the relative lack of blood supply to the distal antebrachium and because of the presence of a large implant and/or graft. With severe infections that fail to respond to antibiotics and result in poor limb use, amputation may have to be performed. The high infection rate has become a major pitfall in limb spare surgery. However, the silver lining is that dogs that develop infected limb spares with both cortical allografts and endoprostheses appear to have a survival benefit over dogs that do not develop infection.[12,14]

Local recurrence rates for limb spare are reported at 15-25%. The risk of local recurrence is increased when the tumor capsule is encountered intraoperatively and care must be taken to avoid this, especially during the dissection between the radius and ulna. Removal of the ulna en bloc with the radius, either proximal to the styloid process as in ulnar transposition or at the level of the radial osteotomy, as reported by Liptak et al, may help to decrease this complication. Dogs with evidence of a pathological fracture or a large amount of soft tissue involvement preoperatively also have an increased risk of recurrence.

The third major complication associated with limb spare is implant failure. Liptak reported a construct failure rate of 40% when 20 dogs that had either allograft or endoprosthesis limb spare were evaluated. The mode of construct failures will vary depending on the limb spare technique used. Liptak found that the failures in the allograft limb spare dogs tended to occur distally, whereas the endoprosthesis construct failures occurred proximally. After SRS or intraoperative radiation, the construct fails at the level of the irradiated bone. LBTO can result in failure of one of the wires, and this is generally managed by replacement of the wire and is a relatively minor complication. Recently, dogs that developed a construct failure after limb spare were reported to have a survival advantage.[13]

ADJUNCTIVE TREATMENT

Chemotherapy[15]

Chemotherapy has been shown to significantly increase the survival times of dogs undergoing surgical treatment of osteosarcoma. The median survival time varies in the literature from 202-540 days. Most reported median survival times are around 1 year. The options for chemotherapeutic agents are doxorubicin, carboplatin and cisplatin.[16-20] These drugs can be given as single agent therapy or in combination (doxorubicin and carboplatin or cisplatin). There has not been definitive evidence to suggest that one protocol produces a significantly longer survival time.

Radiation

Palliative radiation is another treatment option for dogs with osteosarcoma in which surgery is not an option.[21] This treatment involves treating the affected bone with 2-4 doses of radiation. This type of radiation is low typically does not cause side effects to the skin or surrounding tissues. This option can be combined with other medications that may help with pain control and possibly slow the progression of disease. The goal of palliative radiation is to reduce the pain and inflammation associated with the tumor and thereby to improve quality of life. Advantages of this treatment are that it is significantly less expensive ($1,000-1,800) and it is a good option for owners who do not want their dog to undergo surgical treatment. It is also a viable option for patients with evidence of metastatic disease where amputation is not practical. Dogs with more than one bone lesion can receive radiation to both sites at the same time. The disadvantage of this treatment is a relatively short survival time (the average is 3 months) and that not all dogs will respond to therapy (approximately 70% have a good response) as well there is a risk of pathological fracture. Full course, curative intent radiation is generally not performed due to the side effects that can occur to the skin, the duration of radiation and the expense involved.

Palliative Therapy

If analgesics are the sole therapy for osteosarcoma, this treatment should only be employed for a short period. Most patients will have some improvement with an NSAID, gabapentin, and/or opioid analgesics. Bisphosphonates are a relatively new palliative therapy for osteosarcoma. This class of drugs are osteoclast inhibitors and therefore will decrease the rate of osteolysis. They may also have other antineoplastic effects. These medications have a poor oral absorption in dogs and the preferred route of administration is intravenously. The most commonly used bisphosphonate at Colorado State University (CSU) and OVC is pamidronate at a dose of 1mg/kg as an infusion with 250cc of saline. This medication has the theoretical risk of causing renal toxicity and monitoring of renal function is recommended. The infusion is repeated every 30 days. In most cases, this therapy is combined with palliative radiation.

Metronomic chemotherapy is another relatively new form of chemotherapy.[22] This is an oral, regularly administered (every day or every other day) low dose of chemotherapy that may slow the progression of disease. One combination of drugs includes an NSAID, low dose doxycycline, and low dose cyclophosphamide. Potential side effects include bone marrow suppression and sterile hemorrhagic cystitis from cyclophosphamide. The efficacy of this therapy is currently unknown. This therapy should not replace standard chemotherapy but can be considered in cases of metastatic disease or when palliative radiation is used to treat osteosarcoma.

REFERENCES

[1.] Dernell WS, et al. In: Withrow SJ, Vail DM, eds. **Small Animal Clinical Oncology** 4th ed, St. Louis, Missouri, 2007:540. [2.] Dernell WS. In: Slatter D ed. **Textbook of Small Animal Surgery** 3rd ed. Philadelphia, PA: Saunders, 2003:2272. [3.] Powers BE, et al. **J Am Vet Med Assoc** 1988;193:205 [4.] Britt T, et al. **J Small Anim Pract** 2007;48:145. [5.] Hillers KR, et al. **J Am Vet Med Assoc** 2005;226:1364 [6.] Morello E, et al. **Vet Surg** 2003;32:539. [7.] Boston SE, et al. **Vet Surg** 2007; 36:314. [8.] Seguin B, et al. **Vet Surg** 2003;32:69. [9.] Farese JP, et al. **JAVMA** 2004;225:1567. [10.] Ehrhart N. **Vet Surg** 2005;34:24. [11.] Jehn CT, et al. **Vet Surg** 2007;36:324. [12.] Liptak JM, et al. **Vet Surg** 2006;35:518. [13.] Liptak JM, et al. **Vet Surg** 2006;35:150. [14.] Lascelles BDX, et al. **Ann Surg Onc** 2005;12: 1073. [15.] Woods JP. In: Allen DG et al. eds. **Handbook of Veterinary Drugs**, 3rd edition, Lippincott Williams & Wilkins, Philadelphia, 2005;72. [16.] Straw RC, et al. **J Vet Intern Med** 1991;5:205. [17.] Ogilvie GK, et al. **Am J Vet Res** 1988;49:1076. [18.] Berg J, et al. **J Am Vet Med Assoc** 1995;206:1555. [19.] Bergman PJ, et al. **J Vet Intern Med** 1996;10:76. [20.] Kent MS, et al. **J Vet Intern Med** 2004;18:540. [21.] LaRue SM, et al. **J Am Vet Med Assoc** 1989;195:1734. [22.] Mutsaers AJ. **Top Comp Anim Med** 2009;24(3):137.

SOFT TISSUE SARCOMAS IN DOGS & CATS

Sarah Boston, DVM, DVSc, Diplomate ACVS
Guelph, ON, Canada

SOFT TISSUE SARCOMAS IN DOGS

Introduction

Soft tissue sarcomas (STS) are a heterogeneous group of tumors that are named after their presumptive tissue of origin. The focus of this discussion will be tumors that are histologically diagnosed as soft tissue sarcomas, spindle cell tumors, nerve sheath tumors, hemangiopericytomas and mesenchymal cell tumors. More broadly, soft tissue sarcomas can include other tumors of connective tissue origin, including fibrosarcoma, liposarcomas, leiomyosarcoma, malignant fibrous histiocytoma, myxosarcoma, myxofibrosarcoma, lymphangiosarcoma and hemangiosarcoma. Despite the varying tissues of origin, soft tissue sarcomas have many features in common. They are locally aggressive and recurrence is common after a marginal excision. They tend to metastasize hematogenously, rather than via the lymphatics. Overall the metastatic rate for these tumors is low and obtaining local control is the most important component of therapy.

Diagnostic Workup

Soft tissue masses should be approached with a FNA as a first diagnostic step. Even with a convincing cytological diagnosis of spindle cell tumor, mesenchymal cell tumor, or sarcoma, an incisional biopsy of the mass for histologic confirmation and tumor grading is recommended prior to definitive surgery. This is especially true if a wide or radical excision is the planned course of action, as this will result in considerable expense and morbidity to the patient. It is important to keep in mind the eventual surgical planning that will take place when performing an incisional biopsy. For subcutaneous masses, the skin incision should be made in the central portion of the skin involved. The skin incision should be small relative to the mass and should never approach the margin of tumor and normal tissue, as this may compromise margins of excision. A wedge of tissue can be removed or skin punch tool can be used to take the biopsy. A Tru-cut biopsy is another potential method for incisional biopsy and may be useful for sarcomas that are not in the skin or subcutaneous tissue. The disadvantage of this technique is that it provides a much smaller sample, making definitive diagnosis and grade assessment more difficult. It is important to note that although an incisional biopsy will help to give more definitive information regarding diagnosis and grade, the pathologist is only provided with a small amount of tissue with the biopsy and it is possible that the tumor will be upgraded or downgraded when the entire mass is evaluated. Prior to definitive surgical removal, staging should be performed with three-view thoracic radiographs, evaluation of the draining lymph node(s), and in some cases, evaluation of internal organs such as the liver and spleen. The metastatic rate is dependent on the grade. Reported metastatic rates are 0-15% for a Grade I STS, 10-20% for a Grade II STS and 40-50% for a Grade III STS[1]. Local staging (identifying the exact structures involved and extension of the mass) should be performed in many cases prior to definitive surgical resection. This can be done with either CT or MRI, depending on availability.

Surgical Treatment

Surgery remains the mainstay of treatment for soft tissue sarcoma in dogs. Although chemotherapy and radiation may play a role, this class of tumors is not very responsive to these modalities in the gross disease setting. Because of the high recurrence rate with STS that are not removed with adequate margins, a wide or radical excision is recommended. Wide excision refers to removal of the tumor with 3cm margins laterally and one fascial plane deep to the tumor. A radical excision refers to the removal of an anatomic segment to achieve clear margins of excision. The margins of excision should be inked with tissue ink and evaluated by a pathologist for completeness of excision. It is very important to communicate with the pathologist regarding margins of excision. It is also important that the deep margin is evaluated by the pathologist, not just the lateral margins of excision.

Soft tissue sarcomas tend to form a pseudocapsule of compressed tumor cells at the periphery. This can give the impression that they are mobile and can be easily shelled out. A marginal excision with minimal to no normal tissue surrounding the mass will leave tumor cells behind and a recurrence is very likely with this treatment. An unplanned excision is an excision that was performed without knowledge of the tumor type prior to removal and is generally a marginal or incomplete excision if the mass is then diagnosed as a STS. This scenario might then prompt the clinician who performed the unplanned excision to refer the case to a specialist in oncology. A recent study by Bacon et al evaluated the success of primary re-excision after recent inadequate resection of soft tissue sarcomas in dogs.[2] In 41 dogs, re-excision of the surgical scar resulted in a long-term favorable prognosis without the need for radiation or amputation. The local recurrence rate in that study was 6/39 (15%) and distant metastasis occurred in 4/39 dogs (10%). Residual tumor was identified histologically in only 22% of cases and this was not found to be predictive of local recurrence. If possible, re-excision of the surgical scar should be performed after unplanned excision of STS. Re-excision should be performed with wide margins, removing the fascial plane below the scar.

Kuntz et al evaluated prognostic factors of soft tissue sarcoma and found that the number of mitotic figures and necrosis on histopathology was predictive of survival time and the number of mitotic figures was predictive of distant metastasis.[1] This suggests that tumor grade may play a role in how we treat these tumors both systemically and locally. A recent study in Veterinary Pathology evaluated whether or not tumor grade was a predictor of local recurrence in marginally excised subcutaneous soft tissue sarcomas.[3] This study evaluated 138 tumors, of which 34 were completely excised and 104 were marginally excised. None of the completely excised tumors with follow up information (0/30) recurred. Of the incompletely excised tumors, 3/41 (7%) of the grade I, 14/41 (34%) of the grade II and ¾ (75%) of the grade III tumors recurred after marginal excision. These comparisons were statistically significant. Overall this study concluded that histological grade is a strong predictor of recurrence of marginally excised subcutaneous STS. Clean histologic margins were also a strong predictor of nonrecurrence.

Another recent retrospective study by Stefanello et al evaluated the recurrence rate for low-grade soft tissue spindle cell sarcoma of the extremities and found that with 37 tumors in 35 dogs.[4] The margins of excision in this study were dirty (12), clean but close (12) and clean (11). Follow up ranged from 210-2202 days. Median survival time (MST) and disease-free interval (DFI) were not reached because <50% of the dogs died of disease-related events by the end of the study. There was not a significant difference between survival in the three groups of dogs. This could be due to the small number of dogs in each group. The DFI and MST were 697.8 and 703.5 days. This study does suggest that distal extremity, low grade soft tissue sarcoma may be amenable to a less aggressive treatment protocol for both local and systemic therapy.

Another recent study by Chase et al evaluated the outcome after removal of canine spindle cell tumors in first opinion practice in the UK.[5] In 104 canine spindle cell tumors, the method of resection was described as marginal in 45 (43.3%), narrow in 18 (17.3%), wide in 5 (4.8%) and radical in 6 (5.8%) and not recorded in 30 (28.8%). Tumor recurrence occurred in 29 dogs (27.9%). The distribution of local recurrence was 13/45 (28.9%) of the marginal excision group, 3/18 of the narrow excision group (16.7%) and 1/5 (20%) in the wide excision group. The median survival time was 1,013 days. Of the 83 dogs that were dead at the time of the study, 18 (21.7%) of the deaths were tumor related. Of the 18 dogs that died, the MST was 878 days. An assessment of tumor invasiveness by palpation was predictive of DFI and ST.

The findings that some cases can do very well with a low-grade spindle cell tumor, even with marginal excision, does not suggest that local therapy is not the most important part of treatment. It does, however, highlight the need to assess every patient as an individual and to take multiple factors into account when determining the best treatment approach. The tumor grade, size, location and how amenable the tumor is to wide excision are very important. In cases where a wide excision can be carried out with minimal morbidity to the patient, this should be the treatment pursued, regardless of grade. When the tumor is not amenable to wide excision without the need for amputation, other factors should come into play. The two important factors to consider are the tumor grade and the age of the patient. A low grade tumor in a geriatric patient removed with a marginal excision may not recur in that patient's life time, as was seen in Chase and Stefanello's studies, where a large proportion of the patients died of non-tumor related causes. However, a higher grade in a younger patient will increase the risk of a local recurrence with a marginal surgical resection. When considering a surgical approach that is marginal, it is important to have a plan for long-term local control and/or for managing a recurrence. This may include amputation or re-excision and radiation therapy.

Adjunctive Treatments

Because local control is crucial in soft tissue sarcoma, radiation is often used as adjunctive treatment after a marginal excision or when a wide excision is performed with inadequate histological margins. It is very important that the surgeon performing the excision has a plan for either a curative intent surgery (wide excision) or a cytoreductive surgery (marginal excision) that will be followed with radiation therapy. The worst-case scenario is an attempted wide excision with dirty histological margins. This creates the most surgical morbidity and cost, increases the chance of healing complications and creates the largest possible radiation field. Although this unfortunate circumstance will occur occasionally, it can be avoided in most cases with appropriate preoperative planning. If clean margins are unlikely to be achieved based on preoperative imaging, a better approach would be to plan for a marginal excision and follow with radiation. Radiation should not be used as a mop to clean up after a messy surgeon. Although it can be a safety net to fall back on when the margins of excision are not clear, it should not be relied upon in all cases. Full course radiation therapy has been shown to be effective adjunctive therapy to achieve local control after a marginal excision. However, local control is not achieved in all cases.

Chemotherapy is generally not recommended for grade I or II STS because of the relatively low metastatic rate. Systemic chemotherapy for the treatment of high grade STS is controversial. Selting et al retrospectively evaluated 39 dogs with high grade STS that were treated with adjuvant doxorubicin.[6] The overall DFI was 724 and the MST was 856. A significant difference was not found when dogs that received doxorubicin (21 dogs) were compared to dogs that did not (18 dogs). Although this data would suggest that a survival benefit is not provided by doxorubicin in cases of high grade STS, a prospective, randomized clinical trial is necessary to truly determine if there is a role for doxorubicin in cases of high grade STS. Such a study would be difficult because of the paucity of cases of high grade STS and the relatively long survival time.

Elmslie et al recently reported the beneficial effects of metronomic chemotherapy (continuous, low dose chemotherapy) in dogs with incompletely resected soft tissue sarcomas.[7] Thirty dogs with incompletely resected STS were treated with cyclophosphamide and piroxicam. These dogs were compared with 55 contemporary controls. The DFI in the control dogs was 211 days, versus the treatment group, where the DFI was not reached. Metronomic chemotherapy appears to play an important role in local control of incompletely resected STS.

SOFT TISSUE SARCOMAS IN CATS: FELINE VACCINE ASSOCIATED SARCOMA

Although there may be a trend towards a less aggressive approach to the management of some low-grade soft tissue sarcomas in dogs, the opposite is true for the approach to vaccine-associated sarcomas (VAS) in cats. VAS, also known as injection site sarcomas (ISS), were first reported in the early 1990s. A strong association has been noted with administration ofinactivated (killed) vaccines such as FeLV and rabies vaccinations and VAS development. Tumor development has also been noted with other subcutaneous injections. These tumors are thought to arise from sites of chronic, intense inflammation that leads to the proliferation and transformation of fibroblasts. Most commonly these tumors are high-grade fibrosarcomas that display an aggressive biologic behavior. Recommendations from the Vaccine Associated Feline Sarcoma Task Force (VAFSTF) have been put in place in an attempt to provide epidemiologic information regarding injections and tumor development, and to improve the chance of tumor detection and resection by providing guidelines on injection location. It is now recommended that cats be vaccinated over the lower limb to allow early detection and radical resection of VAS, should they occur. Specifically, cats should receive rabies vaccine in the distal right pelvic limb, the feline leukemia virus vaccine in the distal left pelvic limb, and all other injections in the distal right thoracic limb. Also, it is recommended that injections be administered subcutaneously rather than intramuscularly. A recent study by Shaw et al evaluated the changes in VAS sites from 1990-2006 to determine if the recommendations of the VAFSTF resulted in different locations of VAS.[8] When sites were compared before and after the vaccination recommendations in 1996, there was a decrease in the percentage of inter-scapular and right and left thoracic VAS and a corresponding increase in the number of VAS over the right thoracic limb and the combined regions of the pelvic limb and abdominal wall bilaterally. The authors hypothesized that the VAS over the lateral abdominal wall were likely a result of an aberrant injection that was intended for the proximal pelvic limb but resulted in injection over the abdominal wall due to the position of the cat at the time of injection. The proximal limb and lateral abdominal wall remain problematic areas to treat with wide surgical resection and this study highlights the need for continued education of general practitioners to vaccinate over the lower extremity and for continued development of vaccines with a limited inflammatory response.

Like other sarcomas, aggressive local control remains the mainstay of treatment for VAS. Surgery is the primary method of local control, and is often combined with radiation therapy pre- or post-operatively. Cohen et al evaluated 78 cats with VAS treated with conservative or wide excision followed by full course radiation therapy.[9] The MST for all cases was 730 days. 41% of the cats had a local recurrence and 12% developed metastasis. Whether the cats had a wide or conservative excision did not affect the recurrence rate. 26 cats received chemotherapy. There was no difference in the recurrence rate, metastastic rate or survival times in cats that received chemotherapy compared with cats that did not. This study suggests that conservative excision and radiation therapy to 3cm margins may be equivalent to wide excision with 3cm margins and radiation therapy. However, the overall recurrence rate in this study was high, and local control was not achieved in 41% of the cases. More aggressive local control appears to be necessary. Cohen's study found that the number of surgeries was inversely related to a successful outcome. This was also found in a retrospective study by Eckstein et al.[10] Kobayashi et al reported 92 cats with VAS treated with preoperative radiation therapy and found that clean margins of excision had a significant survival benefit, with a mean time to first event (local recurrence or metastasis) being 986 days in cases with clean margins, compared with 292 days in cases that did not.[11] This highlights the importance of an incisional, rather than an excisional biopsy, followed by expedient referral to an oncologic surgeon for aggressive local resection.

A recent abstract presentation at ACVS by Phelps et al described wide or radical excision of VAS using 5cm lateral margins and two muscle planes or bone as the deep margin of excision in 99 cats.[12] The recurrence rate was low compared with other studies at 15.4%, with a median survival time of 901days. Incisional dehiscence was the most common complication. However, the overall complication rate was low. This study suggests that surgery alone can be an effective method of treatment of VAS. However, the margins of excision need to be extensive and this will require a surgeon with additional training and experience in surgical oncology. Even a 1cm mass will require an excisional diameter of 11cm, which is considerable in a small patient. Most VAS are much larger than this at the time of diagnosis and excision. Post operative care in an ICU is essential to provide adequate support and analgesia. Many of these patients will require a blood transfusion intraoperatively.

Radiation therapy is often used in combination with surgery. It is difficult to determine if this should be performed before or after surgery, if at all. The benefit of performing radiation therapy first is that the margins of excision will be sterilized, and the radiation field will be as small as possible. The disadvantage of this approach is that there may be an increase in incisional complications due to the fact that the surgical site will be within the radiation field. Kobayashi et al reported 92 cats with VAS treated with preoperative radiation therapy and concluded that this is an effective treatment plan in cases of VAS, especially if clean margins of excision can be achieved post-irradiation.[11] Each case must be evaluated on an individual basis.

A retrospective by Bregazzi et al evaluated cats treated with surgery and radiation, with and without doxorubicin also failed to show a survival benefit with the addition of chemotherapy to the treatment protocol.[13] It is difficult to interpret retrospective studies because there may be a selection bias in the cases that are selected for chemotherapy. Eckstein et al did show a significant benefit in the PFI when chemotherapy was used in the gross disease setting in cases treated with coarse-fractionated radiation therapy compared with cases that did not receive chemotherapy.[10]

REFERENCES

1. Kuntz CA et al. **J Am Vet Med Assoc** 1997;211:1147. 2. Bacon NJ et al. **J Am Vet Med Assoc** 2007;230:548. 3. McSporran KD. **Vet Pathol** 2009;46:928 4. Stefanello D et al. **Vet Surg** 2008;37:416-465. 5. Chase D et al. **J of Sm Anim Pract** 2009;50:568. 6. Selting KA et al. **J Am Vet Med** Assoc 2005;227:1442. 7. Elmslie RE et al. **J Vet Intern Med** 2008; 22:1373. 8. Shaw SC et al. **J Am Vet Med Assoc** 2009;234:376. 9. Cohen M et al. **J Am Vet Med Assoc** 2001;219:1582. 10. Eckstein et al. **J of Vet and Comp Oncology** 2009;7:54. 11. Kobayashi T et al. **Vet Rad & Ultrasound** 2002;43:473. 12. Phelps HA et al. **ACVS Symposium** online abstract Oct 2009. 13 Bregazzi VS et al. **J Am Vet Med Assoc** 2001;218:547

CLINICAL APPLICATION OF RETINOIDS FOR THE TREATMENT OF CANINE & FELINE CANCER

Carlos H. de M. Souza, DVM, MS. DACVIM(Oncology)
Columbia, MO

INTRODUCTION

Retinoids are natural or synthetic derivatives of vitamin A that exert profound effects on the growth and differentiation of many cell types both *in vivo* and *in vitro*. Retinoids are essential for epithelial differentiation; they have been demonstrated to induce growth inhibition and cell cycle arrest in cancer cells by induction of terminal differentiation and apoptosis.[1] This effect occurs through interaction with specific retinoid nuclear receptors that are part of the steroid superfamily of nuclear receptors.[2,3] Retinoid receptors function as ligand-dependent transcription factors, and with retinoid binding and activation of the receptor leading to transcription of responsive genes. Retinoid receptors are divided in two families: the retinoic acid receptors (RARs); and the retinoid X receptors (RXRs, also called rexinoids). Each family has three receptor subtypes (α, β and γ), with multiple isoforms and specific genes that encode each of them. In the physiologic setting, the natural diet-derived retinoids (all-trans retinoic acid-ATRA and 9-cis-RA), have different binding affinities to the receptors. Specifically, ATRA and 9-cis-RA bind RAR, but only 9-cis-RA binds RXR).[2-5] Since both therapeutic and adverse effects vary depending on the class and isoform of receptor bound, a variety of new synthetic retinoids with exquisite isoform specificity are under investigation.[6] Additional complexity in this signaling pathway comes from the fact that RAR can form heterodimers with RXR. Furthermore, RXR can form homodimers and can also form heterodimers with different steroid receptors, such as the thyroid receptor, vitamin D receptor, peroxisome proliferator-activated receptor, and a number of orphan receptors. The activated dimer binds to specific segments of DNA called retinoic acid response elements (RARE) and retinoid X response elements (RXRE), which are embedded within the promoters of retinoid-regulated genes.[3] Transcriptional regulation by retinoid receptors requires activation of the receptor. The inactive receptor exists as a complex formed by retinoid receptors and corepressor proteins. Corepressors link the dimerized RAR or RXR to histone deacetylases (HDAC). These HDACs remove acetyl groups from histones that are attached to nucleotides. Deacetylation causes chromatin condensation and gene silencing. Binding of the retinoid affects the structural conformation of the receptor ligand-binding domain, leading to dissociation of the corepressor. Histone acetyl transferases (HAT) are then recruited by another group of proteins called coactivators. Acetylation of the histones by HAT causes nucleosomal repulsion, and chromatin decondensation. Transcription of responsive genes can then be initiated.[3-6] In addition to directly influencing gene transcription, RAR and RXR can inhibit cell proliferation signals by inhibition of the transcription factor AP-1 (via c-Jun and c-Fos pathway). The interaction between retinoid receptors and AP-1 has been demonstrated to inhibit cancer formation in studies of chemical carcinogenesis.[3-6]

Retinoids have been shown to suppress pre-malignant lesions such as oral leukoplakia, cervical dysplasia and xeroderma pigmentosum in people, and also actinic keratosis in dogs.[1,2,7] An epidemiological link between low vitamin A intake and development of squamous cell carcinoma (SCC) has been identified in people.[1] In human medicine, retinoids have been used successfully as single agent therapy in the treatment of promyelocytic leukemia (PML), juvenile chronic myelogenous leukemia, and in cutaneous lymphoma.[2,3,8-11] When administered in conjunction with interferon-α2A, retinoids aid in the treatment of human SCC of the skin, uterine cervix, and also renal carcinoma. Clinical trials have demonstrated that administration of retinoids can inhibit the development of second primary cancers after treatment of squamous cell carcinoma of the head and neck, lung tumors, and hepatic carcinomas.[2,8-11] Retinoic acid has also been shown to induce differentiation of thyroid carcinomas, leading to increased radiation sensitivity to iodine-131.[12,13]

In dogs, retinoids have been used mainly in dermatology as a differentiation agent for cases affected by actinic keratosis, sebaceous adenitis, and benign pilomatrixomas.[7,14,15] Few studies have demonstrated the effects of retinoids against canine and feline cancer. Retinoids such as ATRA and etretinate have been used successfully in dogs with cutaneous lymphoma and cats with cutaneous SCC and lymphoma.[16-18] Natural retinoids (ATRA and 9-cis-RA) have also been shown to induce differentiation, growth inhibition, and apoptosis in canine osteosarcoma cells. Histologic differentiation towards a less malignant phenotype was also observed after *in vitro* treatment with retinoids in canine osteosarcoma and feline mammary tumor cells.[19-24] The effects of retinoids on canine melanoma cells have also been studied, where changes in growth rate or differentiation were not observed when cells were treated with the natural retinoids (ATRA and 9-cis-RA). However, additional studies showed that when a substance acting as a retinoid synergist was added to this cell line, significant effects in both growth inhibition and differentiation were induced.[25,26] In addition, two recent studies have shown that MCT cell lines express retinoid receptors and that the level of expression of RARα mRNA correlates well with growth inhibition caused by ATRA.[27,28] With development of synthetic retinoids, specific receptor targeting has led to superior results in both cancer prevention and treatment in human medicine. Additional understanding of the patterns of retinoid receptor expression suggests that specific retinoid receptor interacting agents may enhance responses in canine cancer patients.

REFERENCES

1. Klaassen I, et al. **Oral Oncol** 2002;38:532. 2. Hansen LA, et al. **Carcinogenesis** 2000;21(7):1271. 3. Altucci L, et al. **Nature Rev Can** 2001;1:181. 4. **Lippman SM, et al.** J Nutr **2000;130: 479S.** 5. **Chambon P.** J FASEB **1996;10:940.** 6. **Fontana JA, et al.** Leukemia **2002;16:463.** 7. **Marks SL, et al.** J Am Acad Dermatol **1992;27(1):11.** 8. **Duvic M, et al.** J Clin Oncol **2001;19:2456.** 9. **Dragnev KH, et al.** The Oncologist **2000;5:361.** 10. **Sun SY, et al.** Clin Can Res **2000;6:1563.** 11. **De Thé H, et al.** Oncogene **2001;20:7136.** 12. **Schmutzler C, et al.** Thyroid **2000;10(5):393.** 13. **Grumwald F, et**

al. Nuc Med **1998;38(11):1903.** [14.] **White SD, et al.** J Am Vet Med Assoc **1995;207(2):197.** [15.] **Toma S, et al.** Vet Dermatol **2005;16:346.** [16.] **White SD, et al.** J Am Vet Med Assoc **1993;202(3):387.** [17.] Messinger LM In: Kunkle G: Feline dermatology. **Vet Clin N Am Sm Anim Pract** 1995; 25(4):996. [18.] Evans AG, et al. **Am J Vet Res** 1985; 46(12):2553. [19.] **Barroga EF, et al.** J Vet Med Sci **1998;60(11):1269.** [20.] **Barroga EF, et al.** Res Vet Sci **1999;66:231.** [21.] **Hong SH, et al.** Am J Vet Res **2000;60(12):69.** [22.] **Hong SH, et al.** J Vet Med Sci **2000;62(4):469.** [23.] **Hong SH, et al.** Res Vet Sci **2000;68:57.** [24.] Ivanyi D, et al. **Cancer Lett** 1993; 73(2-3):191. [25.] **Ohashi E, et al.** J Vet Med Sci **2001;63(1):83.** [26.] **Ohashi E, et al.** J Vet Med Sci **2002;64(2):169.** [27.] **Ohashi M, et al.** J Vet Med Sci **2006; 68(8):797.** [28.] **Miyajima N, et al.** J Vet Int Med **2006;20:348.**

SRS, IMRT, IGRT & OTHER NOVEL RADIATION TECHNOLOGIES NOW BEING USED IN VETERINARY MEDICINE

Dr. Susan M. LaRue
Fort Collins, CO

In the 1990s the field of Radiation Oncology underwent a renaissance due to the development of new technologies. These technologies were not just elegant physics-based devices looking for medical applications. These developments, particulary IMRT and PET/CT, enhanced and enabled better application of known radiation biology principles. In the early part of this century advancement in stereotactic imaging technology (image guided radiation therapy, IGRT) made it possible to oligofractionate radiosurgery. Called stereotactic radiation therapy (SRT) or stereotactic body radiation therapy (SBRT), this technique uses large doses per fraction targeted at the tumor, sparing normal tissues by avoidance. Improved imaging has also made dynamic adaptive radiation therapy (DART) possible. DART is where the planning target volume is modified as the patient responds to treatment.

These technologies are becoming increasingly available to the veterinary profession. However, it is important to distinguish when these advanced technologies are indicated. For many types of tumors currently used methodologies, like 3-D conformal radiation therapy (3D-CRT) remain the optimal method of treatment. The purpose of this lecture is to describe the new technologies and compare them to current standard-of-care radiation techniques. This proceeding will provide a glossary annotated by the author of terminology relating to current and emerging radiation therapy technologies.

Conventional external beam radiotherapy (2DXRT)

Conventional external beam radiotherapy (2DXRT) is generally delivered with single beams or using parallel opposed portals. Dose is manually calculated to a reference point in the treatment field, usually the tumor isocenter. This type of radiation therapy is most useful when treating fairly cubic or rectangular prism like structures, where the dose the treatment field can be accurately predicted by the dose to the reference point. In veterinary medicine this is the standard of care for soft-tissue sarcomas and mast cell tumors of the extremities. It is also useful for tumors of the spine, mediastinum, and whole/partial brain treatment. This is also commonly used for palliative therapy in veterinary medicine because it is not reliant on CT or MRI based imaging, making it less expensive. Electron beam radiation therapy can be administered using a 3 dimensional treatment plan, but is more commonly manually calculated.

Electron Beam Radiation Therapy

Electrons therapy is a form of external beam radiation therapy where electrons are directed at the tumor. Electrons don't penetrate as well as x-rays, which are normally used for external beam radiation therapy. That makes them well suited for treating tumors close to the skin surface. They are particularly useful for treatment skin tumors over the abdomen, so that excess radiation dose not reach the intestinal track, and for tumors over the thorax, so that the dose of radiation to the lungs can be limited. Electrons are also useful for tumors near the eye. A special ceramic covered tungsten contact lens block can be placed over the eye to greatly diminish dose, while the prescribe dose can be delivered to the area surrounding the eye.

Three Dimensional Conformal Radiation Therapy (3DCRT)

3DCRT uses CT-based imaging to generate 3D images of specific internal structures. Both tumor and normal tissue structures are identified by the radiation oncologist and "contoured". Information relating the radiation dose to the volumes of specific structures is available as a dose volume histogram (DVH). This allows more precise description of the prescribed dose to the tumor and to normal tissue structures. Customized blocks or static multileaf collimators (MLC) are used to shape the beam. An important but sometimes overlooked aspect of 3D-CRT is the use of positioning devices such as face masks and vacu-lock positioning bags to allow precise re-positioning from fraction to fraction. With confirmation by a radiation medical physicist, these types of positioning devices can decrease the CTV (clinical target volume) to PTV (planning target volume) expansion. This is the standard of care in radiation therapy for most cancers treated in veterinary medicine. It also remains standard-of-care for a number of human cancers including breast cancer and lung cancers.

Intensity Modulated Radiation Therapy (IMRT)

The intensity of each beam is controlled through collimator leaves (MLC) that move **during** treatment, allowing the dose to be sculpted around important normal structures. This technique requires a specific treatment planning system that uses ***inverse planning.*** Inverse planning requires that the various tumor structures (Gross tumor volume [GTV], CTV and PTV) as well as critical normal tissue structures be indentified and contoured into the planning system. Optimization objectives for each structure are entered, and a sophisticated algorithm attempts to meet all objectives. This is standard of care for the treatment of prostate tumors, head and neck cancers, vertebral cancers, some brain cancers, and pelvic cancers in humans. This technology is valuable for treating geometrically complex tumors in animals such as nasal, vertebral body, bladder, prostate and perineal tumors. There are 2 major benefits associated with IMRT. First, dose to adjacent normal tissue structures can be minimized, dramatically reducing acute effects. Patients are more comfortable and require less pain medication (personal communication, Ronald Carsten). In addition, the tumor dose can be increased. Dose was increased dramatically in the treatment of human prostatic cancers while still minimizing acute effects.

Stereotactic RadioSurgery (SRS) and Stereotactic Radiation Therapy (SRT)

A non-surgical technique using highly focused radiation beams that requires some sort of stereotactic imaging system to confirm fiducial location. A frame based system or imaging system can be used to identify implanted fiducial markers, or anatomic fiducials, such as bony landmarks. Even soft tissue structures such as prostates and bladders can serve as fiducials if stereotactic imaging system includes cone beam CT (CBCT). SRS generally (but not always) refers a single treatment and is most commonly used for brain lesions. The Gamma Knife, developed by Lars Leksell in 1967, was the first device developed. The gamma knife has Still in use, it was designed for brain cancers and required a frame based system screwed into skull. The gamma knife has 201 cobalt sources that can be directed towards a single target. By the 1990s stereotactic imaging had made technological advances and other systems became available that were more conducive administering more than one fraction. For tumors not located near radiation sensitive structures, SRS may have radiobiological advantages. However for tumors near radiation sensitive structures such as spinal cord, rectum, and optic chiasm, coarsely fractionating the radiation may be advantageous to help spare these structures. SRT (also referred to as Stereotactic Body Radiation Therapy [SBRT]) is undergoing widespread evaluation in human clinical trials. In veterinary medicine it is being evaluated for the treatment of nasal tumors, brain tumors, pituitary tumors, pelvic and perineal tumors, osteosarcoma, soft tissue sarcomas, and vertebral body tumors.

A number of different systems have been developed for SRS and SRT. The **Body Gamma Knife** is an adaptation of the original gamma knife, but can be positioned around many more parts of the body. The **Cyberknife** is a robotic radiosurgery system. A robotic arm directs a small linear accelerator beam. An imaging system identifies implanted fiducial markers to ensure accuracy. Accelerator based systems, such as the **Varian Trilogy**, have on board imaging systems (OBI) including both kV and CBCT. All of these systems are accurate, but still rely on good treatment planning and input from the radiation oncologist.

Image Guided Radiation Therapy (IGRT)

Accounts for the variability in tumor location within the body between treatments. Initial treatment planning images are compared to current images taken in the treatment room to see if the patients positioning must be adjusted. A number of different technologies can be used to achieve this. While a required feature for SRS or SRT, IGRT is an extremely valuable addition to IMRT and 3DCRT as well. It can reduce the PTV expansion, reducing effects to adjacent normal tissues.

Dynamic Adaptive Radiation Therapy (DART)

DART is a technique where the radiation plan is adapted as the tumor shrinks during treatment. CBCT from the treatment can be imported back into the treatment planning system, and subtle adjustments can be made.

CASE PRESENTATIONS OF VETERINARY CANCER PATIENTS TREATED WITH NOVEL RADIATION TECHNOLOGIES

Dr. Susan M. LaRue
Fort Collins, CO

Intensity Modulated Radiation Therapy (IMRT) for treatment of nasal tumors

Sunni, an 8 year old spayed female diagnosed with a nasal adenocarcinoma

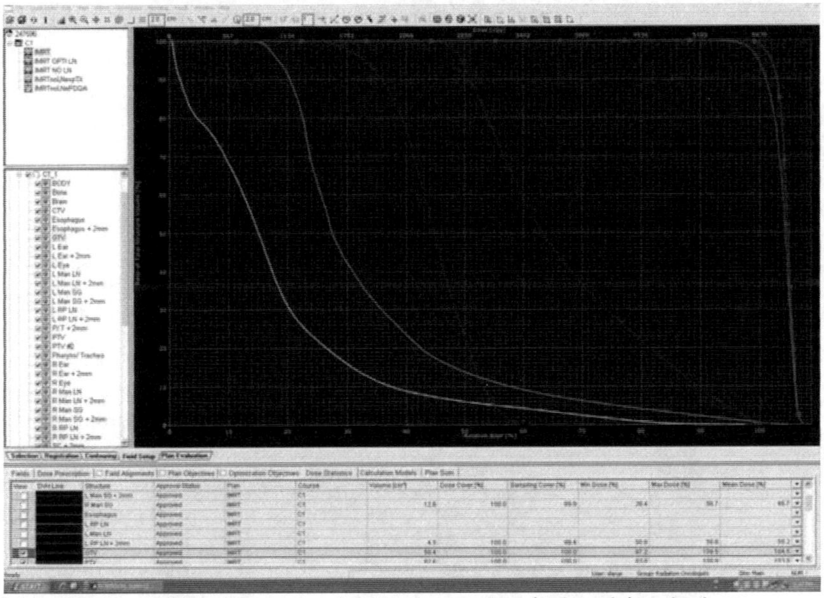

Figure 1. Dose Volume Histogram showing tumor dose (right side) and other structures (left in figure)

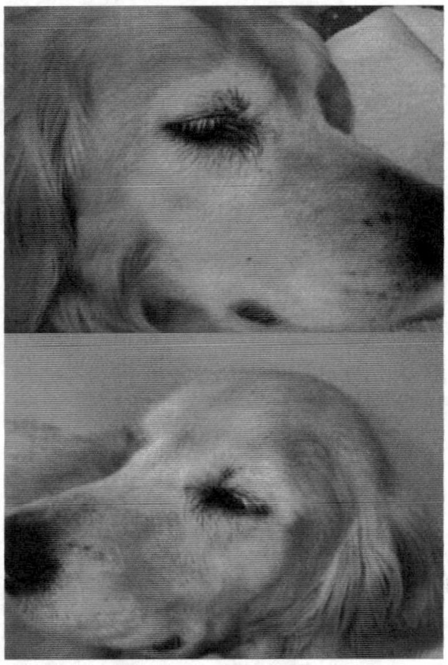

Figure 2. Minimal acute effects post treatment

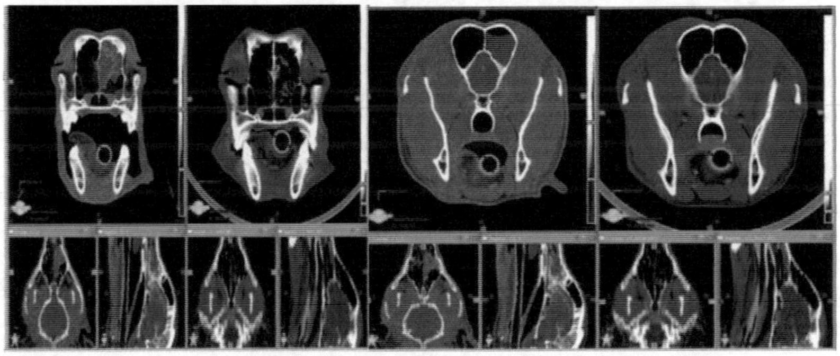

Figure 3. Pre and 6 month post treatment CT of nasal and sinus region

IMRT and Image Guided Radiation therapy (IGRT) to produce Dynamic Adaptive Radiation Therapy (DART)

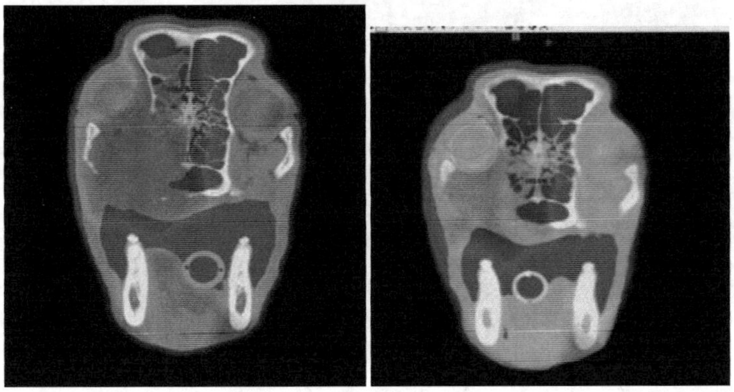

IGRT for Bladder and Prostate Tumors

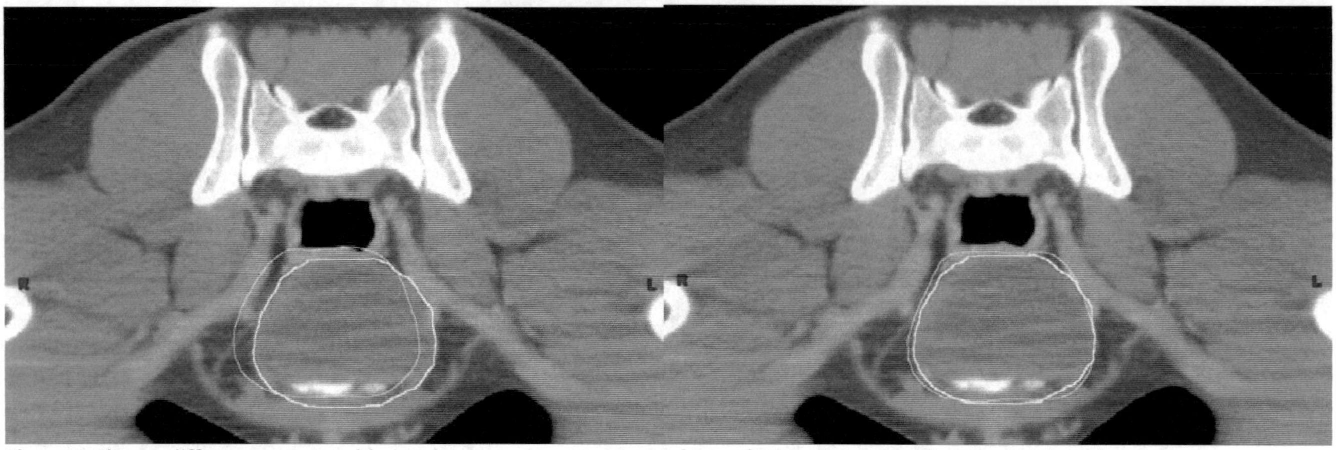

Figure 1 Shows difference to matching to bony anatomy compared to soft tissue anatomy in a dog with a bladder tumor

Stereotactic Radiation Therapy (SRT) for Treatment of Osteosarcoma

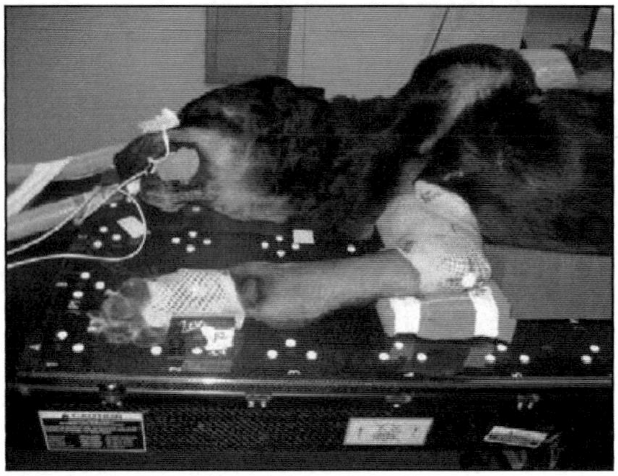

Figure 1 Patient in positioning device

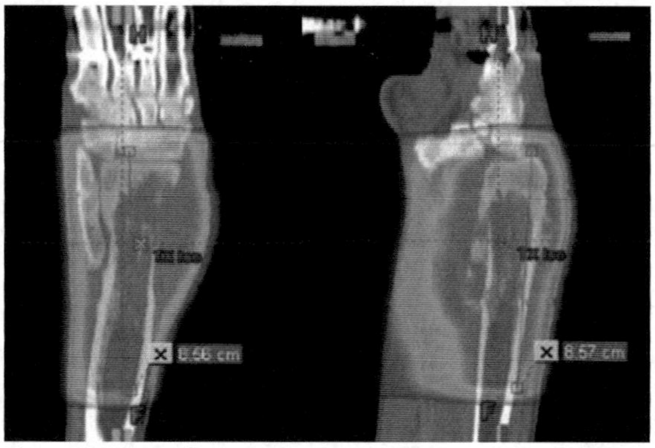

Figure 2 Dose shown as colorwash on treatment plan

SRT for Treatment of Pituitary Acromegally

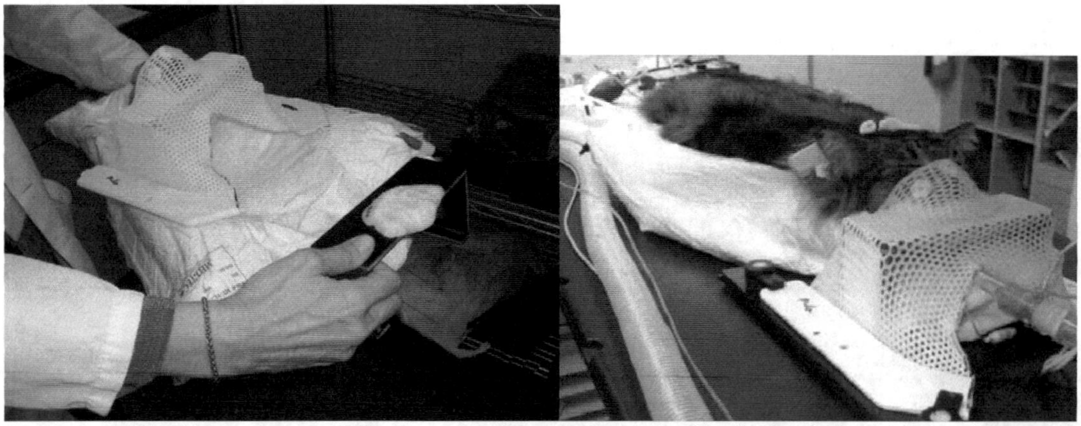

Figure 1. Positioning equipment includes carbonfiber platform, bite block, vaculock cushion and face mask. On right the cat is positioned

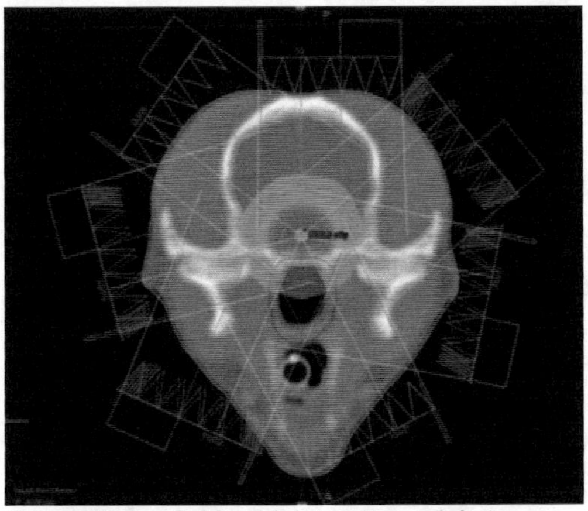

Figure 2 Seven fields with intensity modulation provides rapid dropoff between tumor and normal tissues.

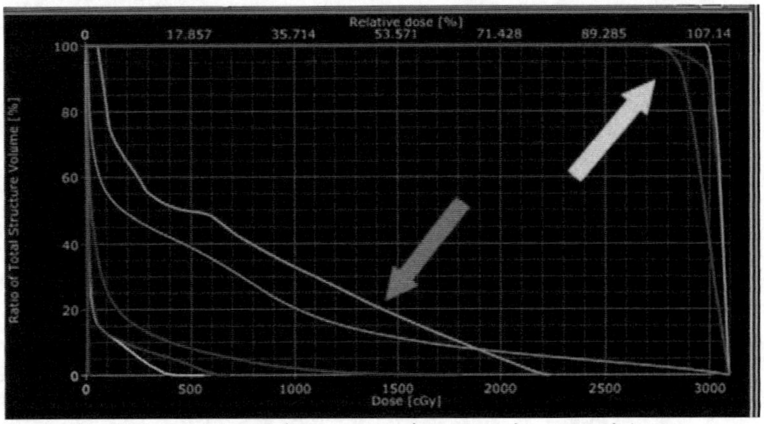

Figure 3 DVH showing tumor (green arrows) compared to normal tissue structures (red arrows)

SRT for treatment of Nasosinal tumors

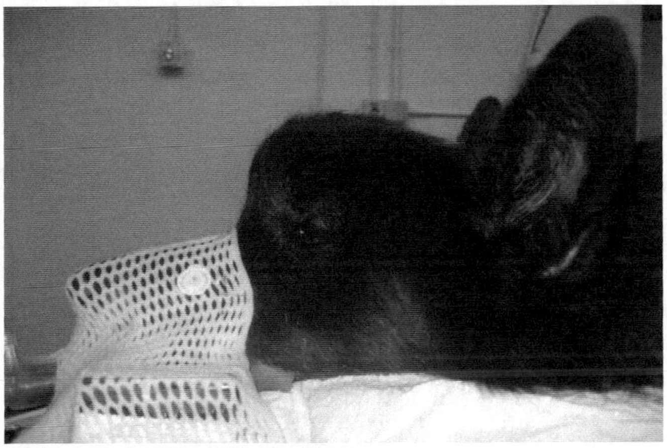

Figure 1. Patient in positioning device.

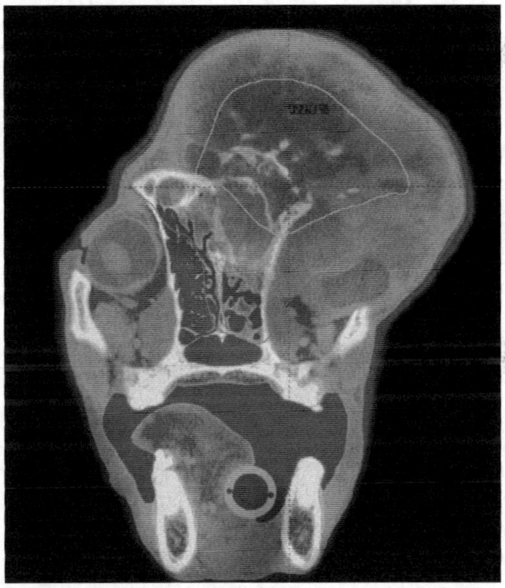

Figure 2 Intensity Modulated Plan spared eye, brain and skin.

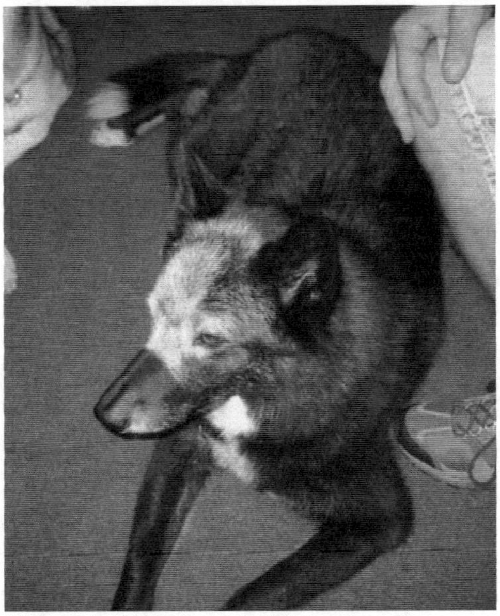

Figure 3 Patient one year after treatment

ROLE OF INFECTION IN TUMOR CONTROL

Douglas H. Thamm, VMD, DACVIM (Oncology)
Fort Collins, CO

INTRODUCTION

Infectious organisms have been associated with the pathogenesis of various tumor types, including cervical carcinoma (human papillomavirus), equine sarcoid (bovine papillomavirus), liver tumors (hepatitis virus), nasopharyngeal carcinoma and Burkitt's lymphoma (Epstein-Barr virus), Kaposi's sarcoma (human herpesvirus 8), and various retrovirus-related hematological neoplasms. A much more interesting role of infection, especially bacteria, may involve a role in the control of neoplasia.

Bacteria were one of the first nonsurgical cancer therapies. Around the turn of the last century, a New York surgeon named William Coley made that observation that the severe postoperative infections commonly encountered in cancer patients were occasionally associated with dramatic tumor regressions. Although the mechanisms underlying this antitumor effect were unknown, Coley was impressed enough with the observed activity that he began purposefully infecting patients with inoperable tumors with live bacteria. Later, a killed bacterial product was utilized (Coley's Mixed Bacterial Vaccines).[1] This product was associated with reduced, but still significant, activity and less worrisome adverse effects. With the advent of radiation therapy in the early 20th century and chemotherapy shortly thereafter, Coley's treatment largely fell out of favor.

In the last 3+ decades, there has been a resurgence of interest in the role of bacterial infection in cancer control. A case series of meaningful antitumor regression following infection has been documented in human brain tumor patients. Significantly reduced risk of recurrence and prolongation of overall associated with postoperative infection have been documented in humans with melanoma and lung cancer, respectively. A meta-analysis of cases of spontaneous regression of cancer in the human literature identified either documented infection or sustained fever of unknown origin in 62 of 224 cases.

BACTERIAL INFECTION AND OSTEOSARCOMA

It has been well known that limb-sparing surgery for canine osteosarcoma (cortical allograft and metal implant procedures) is associated with a very high incidence of chronic postoperative bacterial infection. While this unquestionably presents a clinical management challenge, a study recently demonstrated that dogs undergoing limb-sparing surgery for osteosarcoma that subsequently developed a chronic bacterial infection at the surgery site had a superior outcome to those treated identically that do not develop an infection (time to metastasis 539 vs. 239 days).[2] Subsequent to this report, a retrospective study demonstrated that this same phenomenon is true in humans with osteosarcoma undergoing limb salvage procedures.[3] In support of these clinical findings, investigators have demonstrated reduction of growth of experimental murine melanomas in mice infected with *Toxoplasma* and other agents.[4]

We recently developed a mouse model of chronic bacterial osteomyelitis to investigate the effects of osteomyelitis on tumor growth.[5] Chronic *Staphylococcus* or *Pseudomonas* osteomyelitis was established in C3H mice and the effects of infection on tumor growth of syngeneic DLM8 osteosarcoma were assessed. Experiments were repeated in other mouse strains with other tumor types. The effects of infection on tumor angiogenesis and innate immunity, including NK cell and monocyte responses, were assessed. Osteomyelitis signifcantly inhibited the growth of tumors in mice, and that the effect was independent of the infecting bacterial type, tumor type, or mouse strain (**Figure 1**). Depletion of NK cells or monocytes reversed the antitumor activity elicited by infection. Moreover, infected mice had a significant increase in circulating inflammatory monocytes and numbers of tumor associated macrophages. Infection suppressed tumor angiogenesis but did not affect the numbers of circulating endothelial cells. Thus, it appears that the antitumor effect of bacterial infection is associated with stimulation of innate immunity, potentially mediated via interferon-gamma or other cytokines produced by macrophages and NK cells following stimulation.

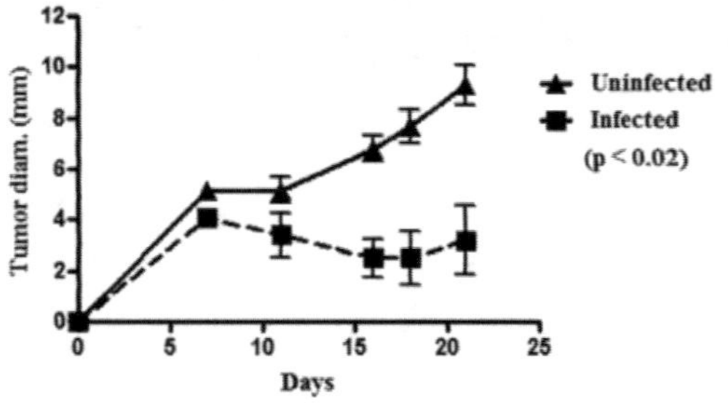

Fig. 1. *Staphylococcus osteomyelitis inhibits murine osteosarcoma growth.* Mice were infected with *S. aureus* containing biofilms three days before tumor challenge. Tumor growth kinetics were assessed by two-dimensional measurements and the longest diameter is reported as the mean (± SEM) calculated for each group. Infected C3H mice (n=5 per group) challenged with syngeneic DLM8 osteosarcoma cells have significantly inhibited tumor growth ($p < 0.02$) compared to uninfected mice.

SALMONELLA AND CANCER

One obvious extension to the observations described above would be the purposeful infection of cancer patients with bacteria to recapitulate the observations of Coley and osteosarcoma researchers.

Salmonella have been shown to infect and preferentially accumulate within implanted tumors in mice, achieving tumor/normal tissue ratios of approximately 1,000:1. A *S. typhimurium* strain has been attenuated by partial deletion of the *msbB* gene, which is responsible for addition of a terminal myristyl group to lipid A. Lipopolysaccharide derived from these lipid A mutants has diminished ability to induce tumor necrosis factor-α in human monocytes. The bacteria have been further attenuated by partial deletion of the *purI* gene, creating a growth requirement for external sources of purines, which are thought to be present in high concentrations in the interstitial fluids in the tumor microenvironment. The mutations of this attenuated form of *S. typhimurium* (VNP20009, Vion Pharmaceuticals) were accomplished by deletions of large portions of the genes, making reversion to wild-type highly unlikely. These modifications allow for significant dose-escalation in mice when compared with the native organism, and do not affect its tumor specificity.

The VNP20009 mutant *Salmonella* has shown antitumor activity in a variety of murine syngeneic and xenogeneic tumor models after intralesional and systemic administration. It has supra-additive antitumor effects when combined with external beam radiation and chemotherapy. Additionally, bacteria have the potential to serve as excellent vectors for gene delivery due to their large genome size and ease of genetic manipulation; transgenes coding for cytokines, antiangiogenic factors, enzymes and immunogens can all be expressed following bacterial infection. Tumor-directed transgene expression has been demonstrated after administration of VNP20009 carrying the genes for green fluorescent protein or *E. coli* cytosine deaminase. These findings led to a human pilot trial of administration of genetically modified *Salmonella* carrying the *E. coli* cytosine deaminase gene, in which enhanced conversion of 5-fluorocytosine to 5-fluorouracil was detected within the tumor microenvironment in 2 of 3 patients treated.[6]

A Phase-I study of VNP20009 in humans with melanoma was reported. The maximum tolerated dose was 3×10^8 cfu/m^2, with thrombocytopenia, anemia, persistent bacteremia, hyperbilirubinemia, diarrhea, vomiting, nausea, elevated alkaline phosphatase, and hypophosphatemia as the dose-limiting toxicities. No major antitumor responses were observed, and bacteria were cultured from tumor tissue in only 12.5% of patients.[7]

Several other groups have demonstrated the potential of orally delivered genetically modified *Salmonella* in tumor vaccination strategies: excellent tolerability, rapid clearance from the GI tract, and generation of potent humoral and T cell-mediated immune responses have been seen in a variety of mouse models.[8]

VNP20009 in canine tumors

We completed a Phase-I dose-escalation trial of VNP20009 in dogs with spontaneous neoplasia.[9] Forty-one pet dogs with a variety of malignant tumors received weekly or biweekly intravenous infusions of VNP20009, at doses ranging from 1.5×10^5 to 1×10^8 cfu/kg. The maximum tolerated dose was 3×10^7 cfu/kg, with fever being the dose-limiting toxicity. One treatment-related acute death occurred. Bacteria were cultured from tumor tissue in 42% of cases. 35 patients were evaluable for antitumor response. Clinically significant antitumor responses were seen in 15% (4 CR, 2 PR), and disease stabilization for at least 6 weeks occurred in another 10%. Three additional dogs experienced minor responses (<50% regression) or transient PR. Responses were seen in dogs with melanoma, soft-tissue sarcoma (STS), osteosarcoma and cutaneous lymphoma. Interestingly, 6 additional dogs had evidence of regional disease regression (5 of which met the criteria for PR based on duration) in the face of no response or progression of metastatic disease.

Seven dogs developed tumor abscessation or hemorrhage after treatment. This occurred in 2 dogs with melanoma, 2 dogs with STS, and 3 dogs with probable hemangiosarcoma (HSA). These reactions occurred at a variety of doses. In the dogs with melanoma and STS, tumor abscessation occurred within 1 week after the first treatment. VNP20009 organisms were cultured from the abscess fluid in 3 of these 4 cases.

Salmonella and angiogenesis

Given the interesting clinical activity seen in patients with vascular neoplasia, we sought to determine if VNP20009 could be exerting some of its antitumor activity through endothelial damage (**Figure 2**). Cultured bovine pulmonary artery endothelial cells (BPAEC) were exposed to varying multiplicities of infection (MOI) of VNP20009, with or without the presence of a submicron filter transwell insert, preventing the direct contact of the bacterium and the BPAEC. VNP20009 induced BPAEC apoptosis in a dose- and time-dependent fashion. In order to determine the bacterial moiety responsible for BPAEC apoptosis, VNP20009 conditioned medium (CM) was pre-incubated with 10 ug/mL polymyxin B (PMB) to inactivate LPS, or boiled for 60 minutes to inactivate protein

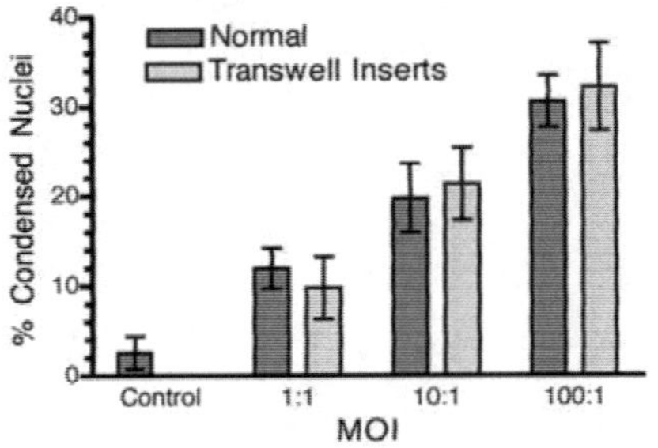

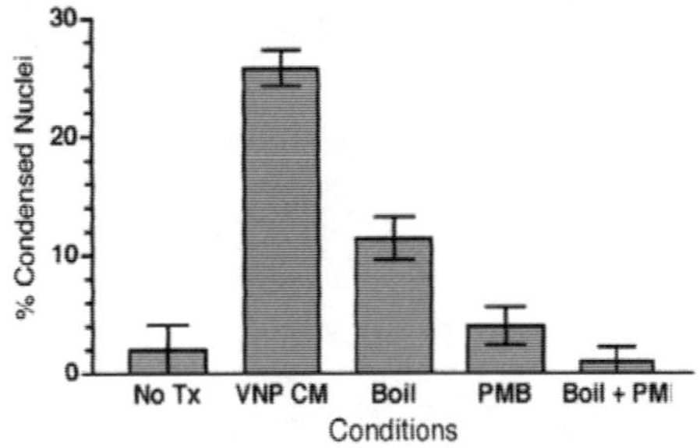

Figure 2: A. *VNP20009 induces endothelial apoptosis in a dose dependent fashion.* **A.** BPAEC were incubated for 6 hours with varying MOI of VNP20009, with or without a 0.4-um transwell filter separating BPAEC and bacteria, then chromatin condensation was assessed using Hoechst 33342 staining and fluorescence microscopy. Culture of the chamber containing EC was uniformly negative for bacteria. Physical separation did not alter the induction of apoptosis in BPAEC. **B.** *Effect of LPS and protein inactivation on VNP20009 activity.* VNP20009 conditioned medium was pre-incubated with polymyxin B (PMB) or boiled prior to being diluted 1:1 with fresh media and added to cultured BPAEC. 6 hours later, chromatin condensation was assessed as described above. Both protein inactivation by boiling and LPS inactivation with PMB diminished apoptosis, and combined LPS and protein inactivation exhibited the most profound effect.

constituents. LPS inactivation resulted in profound attenuation of apoptosis, although heat inactivation also diminished apoptosis to some degree. Thus, despite the fact that the msbB gene deletion in VNP20009 results in an attenuated LPS moiety, LPS appears to be the major contributor to apoptosis induction in BPAEC, potentially via autocrine induction of tumor necrosis factor-alpha (TNF).[10]

PEGYLATED TUMOR NECROSIS FACTOR ALPHA

A non-infectious strategy that could recapitulate some of the effects observed with active bacterial infection involves the exogenous administration of TNF, one of the primary mediators of inflammation secondary to LPS.

TNF was first identified as a cytokine capable of inducing hemorrhagic necrosis in tumors. It exhibits a range of biologic effects, including direct tumor cell cytotoxicity, immunomodulation and endothelial toxicity. Early enthusiasm for TNF as a systemic anti-cancer therapy was tempered when multiple studies indicated that TNF had a very short circulating half-life (14 minutes to 2.5 hours) and severe side effects, primarily hypotension. Data obtained from preclinical murine models demonstrated that PEGylation of human TNF (PEG-TNF) can decrease toxicity and increase antitumor efficacy in mouse models, when compared to native TNF. We hypothesized that PEGylation of TNF would result in decreased toxicity and increased circulating half-life compared with native TNF, and thus exhibit more biologic and antitumor activity than native TNF in tumor-bearing dogs. To test this hypothesis, we performed pharmacokinetically and pharmacodynamically intensive Phase 1 dose escalation study in dogs with spontaneous cancer.[11] The endpoints of this study were 1) evaluation of safety; 2) determination of plasma and intratumor pharmacokinetics; 3) determination of the maximum tolerated dose (MTD); and 4) evaluation of clinical and biological efficacy of PEG-TNF.

Fifteen dogs were enrolled at doses from 20.0 to 30.0 μg/kg. Dose-limiting toxicity at 30.0 μg/kg consisted of vascular leak in 1 and hypotension/coagulopathy in 1, establishing 26.7 μg/kg as the MTD. Mean elimination half-life was 15.3 +/- 4.9 h, markedly higher than that reported for native TNF. Biologic activity (transient fever and leukopenia, increased tumor inflammation and necrosis) was observed at all dosages. A significant increase in tumor blood flow was observed with DCE-MRI. Minor/transient antitumor responses were observed in dogs with melanoma, squamous cell carcinoma and mammary carcinoma, and a partial response was observed in a dog with subcutaneous hemangiosarcoma.

FUTURE DIRECTIONS

How else might the apparent antitumor activity associated with bacterial infection be harnessed in the clinic? It is likely that a significant component of the immunomodulatory activity observed with bacterial infection may be mediated through ligation of various pattern recognition receptors (PRR) on immune cells, innate immune response-expressed proteins that respond to pathogen-associated molecular patterns (PAMP). These include the Toll-like receptors (TLR) and peptidoglycan recognition proteins (PGRP). Interestingly, agonists of both TLR and PGRP have been evaluated in canine clinical trials, and have been associated with encouraging antitumor activity.

A recent article demonstrated that intravenous infusion of cationic liposome-DNA complexes, which are potent stimulators of TLR9, is associated with a reduction in tumor microvessel density and objective evidence of antitumor activity in dogs with spontaneous soft-tissue sarcomas.[12]

A large body of work has demonstrated that intravenous infusions of liposome muramyl tripeptide phosphatidylethanolamine (L-MTP-PE), a synthetic mycobacterial cell wall component and a ligand for PGRP-IαC, has remarkable antimetastatic activity, when used alone or in combination with chemotherapy, in canine osteosarcoma and hemangiosarcoma.[13,14]

Although neither of these 2 therapies is commercially available, it is hoped that these stimulators of innate immunity may eventually provide a safe and efficacious means to recapitulate the encouraging antitumor activity associated with bacterial infection.

REFERENCES

[1] Coley WB. **Am J Med Sci** 1906;131(3):675. [2] Lascelles BDX, et al. **Ann Surg Oncol** 2005;12(12):1073. [3] Jeys LM, et al. **Ann Surg Oncol** 2007;14(10):2887. [4] Rankin EB, et al. **Cancer Biol Ther** 2003;2(6):687. [5] Sottnik JL, et al. **Cancer Immunol Immunother** 2010;59(3):367. [6] Nemunaitis J, et al. **Cancer Gene Ther** 2003;10:737. [7] Toso, JF et al. **J Clin Oncol** 2002;20(1):142. [8] Xiang R, et al. **Cancer Res** 2005;65(2):553. [9] Thamm DH, et al. **Clin Cancer Res** 2005;11(13):4827. [10] Thamm DH, et al. **Proc Am Assoc Cancer Res** 2003, Abstr #45. [11] Thamm DH, et al. **Clin Cancer Res**, *in press*. [12] Kamstock D, et al. **Cancer Gene Ther** 2006;13(3):306. [13] Kurzman ID, et al. **Clin Cancer Res** 1995;1(12):1595. [14] Vail DM, et al. **Clin Cancer Res** 1995;1(10):1165.

AIRWAY OBSTRUCTIVE DISORDERS: ARE HUMANS GOOD MODELS FOR HORSES?

N. Edward Robinson BVetMed, PhD, MRCVS
East Lansing, MI

INTRODUCTION

The term "one medicine", attributed to William Osler, is generally now interpreted to mean that veterinarians, physicians, and other medical and non-medical personnel should work together to combat infectious diseases. As a physiologist, my definition of "one medicine" is somewhat different and refers to the commonality between species in structure, function, and disease mechanisms so that the world is enriched by investigating many species and understanding the unique ways they adapt to their environment and respond to agents that cause dysfunction. In 1667, when Robert Hooke used a dog to demonstrate that rhythmic movement of the lungs was not essential to maintain life if the lungs he and his readers clearly understood the concept of "one medicine" because it was not necessary to repeat these experiments on people to convince readers that the same principles applied to human lungs.

In the great era of investigation of lung structure and function that began during World War II (with the need to fly at high altitude) and continued into the 1970s, most of the investigations were conducted on animals and differences among species contributed to some important discoveries. Realization that brisket disease of cattle was a consequence of pulmonary hypertension induced by the hypoxia of altitude led to the understanding that species differences in the pulmonary vascular response to hypoxia are due to differences in the amount of vascular smooth muscle; species with differing degrees of lung lobulation were used to investigate factors controlling ventilation and perfusion; knowledge of the mechanical behavior of the lungs and rib cage of different species contributed to our ability to ventilate neonates, adult humans, and various mammals; and the special character of sheep lung lymphatics revolutionized the understanding of vascular permeability in the lung. Weibel and his colleagues capitalized on species differences in a large series of investigations that demonstrated the mathematical relationships between structure, gas exchange, and gas transport. Dogs were even "made into humans" by setting them upright in experiments to investigate gravitational effects on the lung.

The advent of molecular biology saw a decrease in the number of species used in investigations because it was well accepted that the majority of cellular and molecular mechanisms are common to all species. Thus we could say that the whole world now recognizes "one medicine." However, restriction of investigation of mechanisms of disease to the mouse has some disadvantages because humans and larger animals are not simply large mice! This is particularly apparent in the respiratory system. While the basic unit of the lung, i.e. the acinus (terminal bronchiole, alveolar duct and alveoli), is similar in mice and larger mammals, these gas exchange units must be joined together by generations of conducting airways to form the large lungs of people (2.3 kilomice), horses (15 kilomice), elephants (250 kilomice) and sperm whales (1.5 megamice). Thus, mice lack the multiple generations of conducting airways with their specialized epithelial cells, cartilage, smooth muscle, and mucus secreting apparatus. Only the trachea of the mouse is cartilaginous, all its intrapulmonary airways lack cartilage. Also, the trachea of a mouse is lined by cuboidal epithelium like the small bronchioles rather than with the pseudostratified columnar epithelium typical of the trachea of larger mammals. Chronic airway diseases are an affliction of the tracheobronchial tree and are associated with structural and functional changes in the mucus producing apparatus and airway smooth muscle. For this reason it would seem more reasonable to investigate these types of diseases in species with multiple generations of conducting airways rather than in the mouse with its "bronchiole-like" trachea and bronchi.

WHAT ARE AIRWAY OBSTRUCTIVE DISORDERS?

Airway obstructive disorders lead to an increase in the resistance to airflow in the conducting airways. Because the upper airway comprises most of the resistance to airflow, nasal stenosis and laryngeal paresis can dramatically increase airway resistance so that affected animals have difficulty exercising and may even have difficult breathing at rest. In the tracheobronchial tree, most of the airway resistance is caused by the trachea and larger bronchi with bronchioles contributing little. Consequently, tracheal and bronchial obstructions cause more respiratory distress than do those of the bronchioles. In the latter, there needs to be extensive obstruction before animals have respiratory distress at rest. Less severe bronchiolar obstruction impairs gas exchange during exercise and results in loss of exercise tolerance.

Tracheobronchial obstructions can be a consequence of luminal obstructions, changes in the airway wall, loss of peribronchial airway support, or masses (including enlarged atria) impinging on the airway. Many of these factors are important as causes of airway obstruction in the chronic inflammatory obstructive airway diseases such as RAO in horses, asthma in cats, bronchitis in dogs, and asthma and chronic obstructive pulmonary disease (COPD) in people. In all these diseases, exacerbations of airway obstruction are associated with worsening of airway inflammation, often as a consequence of inhaled environmental irritants. Because asthma prevalence has been increasing especially in the young and because COPD is a leading cause of lost workdays and death in older individuals, considerable research dollars have been directed to these two conditions. As veterinarians, it is therefore reasonable to ask if humans provide good models for chronic airway disease in mammals. The question is best answered in the case of horses because more research has been devoted to horse airway disease than to feline asthma and canine bronchitis.

AIRWAY DISEASE PHENOTYPES

People

When thinking of human obstructive lower airway diseases, one generally begins with asthma and COPD. Asthma is characterized by recurrent bouts of airway obstruction initiated by airway smooth muscle contraction and interspersed with periods of relatively normal lung function. Allergic asthma (a.k.a extrinsic asthma) frequently begins in childhood as part of a constellation of atopic conditions that can also include allergic dermatitis and food allergies. In many individuals, their childhood asthma wanes or disappears in adulthood. In others, the asthma may become more severe and can be a lifetime problem. Some people develop non-allergic (extrinsic) asthma as adults, usually as a consequence of working in environments that contain high concentrations of particulates, endotoxin, or industrial toxins. Asthmatic individuals have airways that are hyperresponsive to inhaled stimuli so that excessive bronchospasm occurs in response to certain triggers, which may be a specific antigen, high particulate concentrations, cold dry air, exercise, or viral infection. Trying to fit these various triggers to the inflammatory response and lung function changes they induce, quickly makes one realize that asthma is a heterogeneous disease.[1]

COPD is a degenerative lung disease of smokers in which chronic bronchitis is usually accompanied by alveolar emphysema. The airways are obstructed as a consequence of excessive mucoid accumulations and by loss of the normal tethering support provided by surrounding alveolar septa. In comparison to asthma, airway smooth muscle contraction is less important as a cause of airway obstruction in COPD. COPD generally develops in middle to old age and is usually associated with a long history of cigarette smoking.

Although these definitions of asthma and COPD appear quite distinct, features of both may occur in some people, especially as they age. In addition, there are other airway obstructive diseases that also have some features of either asthma, COPD or both. These conditions include viral wheezing, bronchiectasis, and eosinophilic bronchitis.[2]

Horses

In horses, we currently recognize two phenotypes of chronic airway disease; recurrent airway obstruction (RAO, a.k.a. heaves) and inflammatory airway disease (IAD). Horses with RAO have measurable airway obstruction or clinically obvious respiratory distress when at rest.[3] The reversible obstruction is induced by organic dust exposure and wanes when horses are removed from the triggering environment. Most of the obstruction is due to airway smooth muscle contraction. Animals that cough, lack respiratory distress at rest, have excess tracheobronchial mucoid secretions, and no evidence of acute respiratory problems are said to have IAD, but precise definition of this phenotype is proving very difficult. Initial attempts were based on the presence of visible accumulations of mucoid secretions within the trachea, and increased number of inflammatory cells in tracheal wash or bronchoalveolar lavage fluid (BALF).[4] However, because many apparently normal horses have a large percentage of neutrophils in the tracheal wash and healthy racehorses in training have more neutrophils in their tracheal wash than those not training, the newer IAD definition avoids use of tracheal cytology and requires documentation of lower airway inflammation in BALF or demonstration of low-grade changes in lung function or poor performance.[5] Other phenotypes of equine airway disease may exist in horses but these are not usually uncovered in the typical investigations of a few horses at referral centers. An extensive multi-center, longitudinal field study to characterize the natural history, BALF cytology and lung function might uncover horses with, for example, persistent eosinophilic inflammation or others with airway obstruction yet little inflammation.

LUNG FUNCTION

In both people and horses with chronic airway diseases, the lung function changes are typical of diffuse airway obstruction. Airway resistance is increased, indices such as forced expiratory volume in one second (FEV1.0) derived from flow-volume loops are decreased, and dynamic lung compliance is decreased.[3] In asthma and RAO, these changes can be largely reversed by administration of beta2-adrenergic and anticholinergic bronchodilators. Because of the structural changes in the lungs and airways, reversibility of airway obstruction is much less in human COPD and is best induced by anticholinergic drugs. When their airways are challenged with bronchoconstrictor agonists such as methacholine and histamine subjects with asthma, RAO and COPD all demonstrate airway hyperresponsiveness. That is, there is a greater degree of airway obstruction (smooth muscle contraction) at a given dose of agonist. The reasons for hyperresponsiveness include thickening of the airway mucosa luminal to the smooth muscle, facilitation of contraction by inflammatory mediators, and loss of inhibition of contraction by beta-adrenergic and nitroxidergic mechanisms.

INFLAMMATION

Structural and functional changes that occur in the airways of people and animals with chronic airway disease are a consequence of airway inflammation. In allergic asthma, the eosinophil was always described as the characteristic cell of this disease. However, acute exacerbations of asthma and especially severe asthma are frequently characterized by an influx of neutrophils. In the horse, an increased percentage of neutrophils in BALF is the most frequent finding in cases of chronic airway disease[3,6] although occasional animals have an increased percentage of mast cells or eosinophils. Investigations of BALF cytology in horses largely present snapshots of a single time interval so we don't know, for example, if eosinophilic inflammation can occur in horses that usually have many neutrophils or vice versa. This will only be uncovered in the type of longitudinal study described above.

Etiology of inflammation

The inflammatory cell influx into the airways is controlled by chemokines and cytokines released from lymphocytes, macrophages, epithelium and other cells. In mice the pattern of cytokine release has been characterized as Th1 (TNF and IFN) associated with a neutrophilic response, Th2 (IL-4, IL-5, IL-9, eotaxin, and IL-13) associated with eosinophil influx, and Th17 (IL-17 and IL-17F from Th17 lymphocytes and IL-25, IL-33 and stromal lymphoprotein (TSLP) from airway epithelium) associated with a neutrophilic or mixed response.[1,7] Neither human asthma nor RAO of horses tidily fit into any these immune categories. Although the increased concentrations of IgE in serum, increased concentrations of Th2 cytokines, and esosinophil and mast cell infiltration of the airways are typical Th2, the response of asthmatics to specific antagonists of the Th2 pathway has been disappointing[1] except in the case of mepolizumab (an IL-5 blocking monoclonal antibody) that has therapeutic benefits in exacerbations of corticosteroid-resistant asthma characterized by persistent eosinophil influx into the airway wall.[8] This drug is not beneficial to the asthma population in general, which highlights that asthma is not a single disease but a mixture of phenotypes and that new therapies may need to be targeted at specific phenotypes that are characterized not simply by symptomatology but rather by their pattern of cytokine release.

In both asthma and RAO, the airway epithelium is the target hit first by inhaled "triggers" and so it is highly likely that signaling by the epithelium is important in initiating inflammation.[1] The epithelium is intimately linked to the immune system through dendritic cells, the processes of which extend between epithelial cells to the tight junctions. Dendritic cells are involved in antigen recognition and can be programmed through innate immune pathways to shape the adaptive immune response. For example, activation of toll-like receptors (TLRs) 2, 4, 8, and 9 can drive the dendritic cells to a Th2 response.[9] Hence although asthma is a Th2-like disease, this may be initiated by activation of TLRs of the innate immune system, which may explain why respiratory virus infections in early life increase the risk of later asthma.[10] On a different note, one cannot discount involvement of the gastrointestinal immune system because changes in the gastrointestinal microflora can also reprogram immune responses from Th1 to Th2.[11]

The number of investigations of the immune reponses of RAO-affected horses is small and differences in what is measured, time of measurement, history of the horses preclude any clear conclusions (not surprising given the confusion in asthma after vast investments of money). Suffice it to say that evidence has been presented for and against Th2-mediated responses. It is clear however, there is no clinical evidence in horses for an IgE-mediated immediate airway obstruction when animals are exposed to "triggers".[12] Also, there is no evidence for a childhood form of asthma and only rarely eosinophilic inflammation. So, in these respects, allergic asthma is not a good model for RAO. The neutrophilic inflammation of RAO is more characteristic of human COPD but the reversibility of airway obstruction is greater in RAO. Neutrophilic influx is also characteristic of industrial types of asthma that are associated with exposure to endotoxin, which typically triggers a Th1 response, which should be associated with a reduced Th2 response. Studies of asthmatics indicate the situation is more complex. While exposure to high endotoxin concentrations is protective, a variant of CD14 is associated with more severe atopy.[13] However, a recent multi-center study of candidate genes for wheeze and allergy found that TLR4 (the endotoxin receptor) was one of only four genes associated with wheeze in the past year.[14] Variants in TLR4 were also related to allergen-specific IgE pointing out once again the likelihood of interactions between innate and adaptive immune responses in airway disease.[14]

TREATMENT AND PREVENTION

Despite vast amounts of research into immune responses and signaling mechanism involved in the development of asthma, bronchodilators and corticosteroids remain the mainstays of therapy. The management of asthma has shifted from use of inhaled corticosteroids for only severe exacerbations of disease to emphasis on long-term control of inflammation in order to prevent airway remodeling and the long-term deterioration of lung function. Combined therapy with a bronchodilator and a corticosteroid is quite routine. In horses, practitioners are reticent to use effective doses of corticosteroids for short-term treatment and lower doses for long-term prevention because of the fear of laminitis. The latter might be lessened by use of inhaled corticosteroids but cost and difficulty of administration have limited the development and implementation of such treatment.

With regard to other pathways to prevention and treatment, monteleukast (Singulair®), a sulfidopeptide leukotriene receptor antagonist, is effective in preventing exacerbations in some people with asthma but in the small trial done on horses, this type of drug was not beneficial. Surprisingly, none of the experimental compounds that are effective at blocking the Th2 immune pathways in rodents have been effective in trials on asthmatics, except for mepolizumab, which is only effective in a small subgroup of patients (see above). This lack of efficacy of blockers of Th2 mechanisms and the evidence for involvement of TLRs suggests that a whole new approach may be necessary to find effective therapies. The cost of such an approach rules out the possibility of new medications for horses until they have been developed for people.

Ultimately, the best treatment is prevention. In both asthmatics and horses, prevention of exposure to triggers is very effective in preventing exacerbations of airway obstruction but in both species, the practicalities of doing this prevent full compliance. In the long run, we may learn that both RAO and asthma are initiated by early life exposures to viruses or other agents so that vaccination or other preventive therapy might reduce the future prevalence of both diseases.

CAN REAL DISEASE BE MIMICKED IN THE LABORATORY?

While rodents are convenient laboratory animals, can be genetically manipulated to investigate immune mechanisms, and can be experimentally challenged to produce some of the features of asthma, in no way do they mimic real disease. It is impossible to recreate a lifetime of environmental and microbiological exposures in an animal that lives a few years. In addition, its airway anatomy is unsuited as a model for chronic airway disease. The time has come to take the information gleaned from rodents and to apply it to large-scale genetically-based epidemiological investigations such as are being undertaken in humans.[14] We will never afford to conduct enough laboratory-based investigations of immune mechanisms to understand the complexity of RAO so we should by-pass that step and begin the cooperation among laboratories and practitioners necessary to establish large data bases that can be used to find the genetic and lifestyle risk factors for equine airway disease.

REFERENCES

[1.] Holgate ST, et al. **Immunity** 2009;31:362. [2.] Wardlaw AJ, et al. **Clin exp Allergy** 2005;35:1254. [3.] Anon.. **Equine Vet J** 2001;33:5. [4.] Anon. **Equine Vet Educ** 2003;5:81. [5.] Couetil LL, et al. **J Vet Intern Med** 2006;21:356. [6.] Bedenice D, et al. **J Vet Intern Med** 2008;22:1022. [7.] Kolls JK, et al. **Immunity** 2004;21:467. [8.] Haldar P, et al. **New Engl J Med** 2009;360:973. [9.] Holgate ST. **Trends Immunol** 2007;28:248. [10.] Jackson DJ, et al. **Am J Respir Crit Care Med** 2008;178:667. [11.] Shreiner A, et al. **Adv Exp Med Biol** 2008;635:113. [12.] Tahon L, et al. **Vet Immunol Immunopathol** 2009;127:85. [13.] Simpson A, et al. **Clin Exp Allergy** 2009; e-Pub Nov 25. [14.] Genuneit J, et al. **Clin Exp Allergy** 2009;39:1875.

DON'T BLAME THE DOG: THE INFLUENCE OF HUMAN MEDICINE ON COMPANION ANIMAL INFECTIOUS DISEASES

J Scott Weese DVM DVSc DipACVIM
Guelph, Ontario

INTRODUCTION

The concept zoonotic diseases, diseases that are naturally transmitted from animals to humans, is well established. The opposite route of transmission, from humans to animals, receives much less attention, despite increasing evidence that it may be of concern from both animal and public health standpoints. As more attention is paid to the microbiological interaction between humans and their close animal contacts (i.e. pets), as tools to better characterize microorganisms become more widely available and applied to companion animals and as more scrutiny is placed on the household as a site of interspecies transmission of pathogens, evidence is emerging that microorganisms may frequently be transmitted between humans and pets, in both directions.

PATHOGENS OF CONCERN

Examples of the some of the more common, concerning or interesting pathogens are provided below.

Methicillin-resistant *Staphylococcus aureus*

Methicillin-resistant *S. aureus* (MRSA) has been one of the most prominent examples of the influence of a human infectious disease on companion animals. MRSA is a critically important pathogen in humans, being a leading cause of hospital-associated infections and the cause of an epidemic of community-associated infections in many regions internationally. MRSA is clearly of human origin, being identified in humans in the late 1950s, shortly after the introduction of methicillin. It wasn't until the 1990s and 2000s, and the phase of expansion of MRSA into a community-associated pathogen in humans that the first reports of MRSA in companion animals emerged.[1-6] Initial reports in dogs and cats, both published and anecdotal, tended to involve pets that visited hospitals or that were owned by people with close contact with the human healthcare system, including both patients and healthcare workers, however as MRSA expanded in both humans and animals, this close apparent link with the human healthcare system is no longer apparent. Residing with a human healthcare worker has been reported as a risk factor for MRSA colonization in dogs in some, but not all[7,8] studies, and the majority of MRSA cases reported do not have a clear link to the human healthcare system.

The MRSA strains found in pets provides insight into the human origin. MRSA strains found in pets are almost invariably those found most commonly in humans.[3,8-13] The predominant strains in humans vary geographically, and those differences are also reflected in pets. As changes in MRSA strain prevalence occur in a region, such as the increase in the USA300 clone in North America, those changes are also noted in pets. Therefore, the molecular epidemiology of human and pet MRSA are closely linked, and given the commonness of MRSA in humans and humans as the main reservoir of *S. aureus*, it is clear that these changes are being driven by humans.

Dogs that visit human hospitals offer a unique opportunity to study the human hospital/pet interface. An earlier cross-sectional study of visitation dogs did not find MRSA in any of 102 dogs,[14] however a subsequent longitudinal cohort study identified acquisition of MRSA by 8% of hospital visitation dogs over the course of a year compared to just 1% by control dogs that visited non-healthcare settings such as schools.[15] Contact with human hospitals and contact with children were risk factors for MRSA acquisition. Additionally, a nested case-control study involving the hospital visitation group identified patients feeding dogs treats by hand and allowing dogs to lick patients' faces as risk factors, likely as a result of close contact with colonized or contaminated human body sites. MRSA acquisition by cats in a longterm care facility has also been reported.[16]

As MRSA colonization rates have increased in the general human population, it is perhaps not surprising (at least in retrospect) that MRSA colonization and infection appear to have increased in pets in the community. Various studies have evaluated the prevalence of colonization in different populations of dogs and cats, with rates of 0-4% typically being reported.[8,17-20] While low, this prevalence is rather remarkable, considering it is similar to the prevalence in healthy humans in most regions, despite the fact that *S. aureus* is not typically considered a true commensal in dogs (and perhaps cats). Regardless, given the prevalence of MRSA in the healthy human population and little evidence that it will decrease in the near future, MRSA exposure of pets will be an ongoing risk.

Vancomycin resistant enterococci

Vancomycin resistant enterococci are important opportunistic pathogens in human hospitals, causing both sporadic disease and outbreaks. While of much less impact in the general population, VRE can be found in a small percentage of people in the community. VRE has been identified in companion animals, and there may be two main and different routes of acquisition and dissemination. Relatively high VRE colonization rates have been reported in dogs Europe in the past and this was associated with the use of avoparcin (a glycopeptides like vancomycin) in food animals.[21] Banning of avoparcin resulted in an apparent pronounced decline in VRE colonization in dogs,[22] and rates in dogs in Europe are now low.[23] In contrast, VRE was very rare in pets in North America (where avoparcin was not used) during that time, and apparently remains so.[14,15] Sporadic clinical infections with VRE are encountered in dogs and cats, both in referral hospitals and primary care practices, but these currently appear to be quite rare. The potential impact of human VRE in companion animals is unclear. VRE acquisition was reported in a hospital visitation dog,[15] likely a result of acquiring it from a patient, but clear evidence of human-animal transmission is currently lacking. Limited information is available regarding the VRE strains found in pets compared to humans, although dogs are known to carry vancomycin-susceptible enterococci lineages associated with human infections.[23] Risk factors for VRE infection in pets are poorly understood. VRE does not appear to have established as strong of a community 'foothold' in humans as has MRSA, so the potential for exposure of pets from colonized humans should be somewhat limited. Risks of human-pet transmission may therefore be highest amongst a limited number of households containing people at higher risk of VRE carriage (i.e. people returning from hospital stays) or animals that visit healthcare facilities. More information is required about the VRE in humans and animals, but the potential impact of this important opportunistic pathogens in some populations should be considered.

Multidrug resistant (MDR) Gram-negative bacteria

Multidrug resistance is an emerging problem in *E. coli* and other Gram negative organisms. While not to the extent (or profile) of MRSA, MDR Gram negatives particularly those producing extended spectrum beta-lactamases (ESBLs) or extended spectrum cephalosporinases (ESCs) are an increasing problem in both humans and companion animals.[24-30] More recently, carbapenemase-producing organisms, predominantly carbapenemase-producing *Klebsiella* species (KPC) have emerged as a serious concern in human medicine[31,32] and there are anecdotal reports of their presence in companion animals. The latter most likely represents human to pet transmission, because simultaneous emergence of this phenotype in human and animals would be unexpected, particularly given the relatively low use of carbapenems in pets compared to humans.

The pronounced diversity of many Gram negatives such as *E. coli*, the potential for common source infection (i.e. food and environmental sources), the multiple potential sources of infection and more complicated epidemiology compared to some other organisms complicates assessment of interspecies transmission. More information is available regarding transmission of *E. coli*, including commensal, verotoxigenic and uropathogenic strains. These have focused on similarity between human and animal strains, typically with an emphasis on pets as a potential source of human infection.[33-36] Characterization of actual interspecies transmission has been limited, but given the similarity in strains that can be found in humans and pets, there is reasonable assumption that transmission in one or both directions could occur, with both susceptible and resistant strains.

It is unlikely that humans are the sole or even predominant source of MDR Gram negative bacteria in companion animals, yet there is certainly potential for an influence. As these organisms become more common and problematic in people in hospitals and the community, the potential for exposure of pets increases, and given the evidence of the potential for interspecies transmission of commensal susceptible *E. coli*, it is logical to assume that this influence could extend to MDR *E. coli*, along with other MDR Gram negative organisms.

Clostridium difficile

Clostridium difficile is the most important cause of hospital-associated diarrhea in humans and *C. difficile* infection (CDI) is increasing in incidence and severity worldwide. Recent evidence suggests that CDI may be an increasing or overlooked problem in people in the community. The role of *C. difficile* in disease of companion animals is somewhat unclear. While most animals that are shedding *C. difficile* are clinically normal, an association has been made between the presence of *C. difficile* toxins in feces and the presence of diarrhea in dogs.[37,38]

The epidemiology of *C. difficile* in pets is poorly understood. Nosocomial acquisition can occur[39,40] but most cases of infection and colonization appear to be unassociated with veterinary healthcare contact. As with MRSA, hospital-visitation animals have provided some insight into the potential for human to animal transmission. A cross-sectional study of visitation dogs identified *C. difficile* colonization in 58% of dogs,[14] an astounding number considering other studies have typically reported low (0-10%) prevalences in other healthy adult dog groups.[38,39,41] Included in that group was a dog colonized with ribotype 027/NAP1, a hypervirulent outbreak clone that is a leading cause of sporadic disease and outbreaks in humans internationally. A subsequent longitudinal study of visitation dogs identified *C. difficile* acquisition by 28% of visitation dogs over the course of a year, a significantly higher rate than dogs that were involved in non-hospital visitation programs.[15] In that study, contact with human hospitals, contact with children and prior antimicrobial treatment were risk factors for *C. difficile* acquisition. Additionally, and fascinating from an ecological standpoint, antimicrobial treatment of a person in the household was a risk factor for the dog shedding *C. difficile*. Presumably, antimicrobial treatment resulted in a greater likelihood of *C. difficile* shedding by the person (something that is well known) with subsequent transmission to the pet. Among the hospital visitation group, licking patients and being allowed up on beds were associated with *C. difficile* acquisition. Another study, involving dogs in the general population in Canada, reported that dogs living with an immunocompromised person were 7.9 times as likely to shed *C. difficile* compared to other dogs.[42] Studies of *C. difficile* isolates from dogs indicate that isolates from dogs are typically, if not exclusively, strains that are also found in humans.[42,43] The animal health, as well as subsequent human health, implications of human-dog transmission of *C. difficile* are unclear but this is an area of concern and the research to date provides a strong indication that this pathogen can be transmitted from humans to animals.

Novel H1N1 influenza

The emergence of the novel H1N1 influenza pandemic in 2009 highlighted the issues of human-animal transmission of infectious diseases and the potential impact of this, both in terms of animal health (which was limited) and psychological impact (potentially greater). While of human, avian and swine origin, the pandemic H1N1 pandemic was clearly a human pandemic but one that was also associated with 'spillover' infections in animals such as dogs, cats, ferrets, swine and poultry. Reports of infected pets invariably involved pets whose owners had influenza. When one considers that fact, plus the relatively limited contact that most pets have with other animals or humans, it is clear that pet infections were associated with their household contacts, and a limited (but probably underreported) number of pet deaths occurred from infections acquired from their owners.

Herpes simplex virus (human herpesvirus-1)

While a seemingly uncommon event, human herpesvirus is an example of a presumably overlooked and potential fatal consequence of human-animal transmission. Fatal human herpesvirus infection has been transmitted from humans to pet rabbits.[44] These have involved women with active facial and labial lesions, with indirect transmission via the hands as the presumed source of infection. While the risk of transmission is probably low, people with active herpesvirus lesions should avoid direct contact of pets with infected sites and focus on good hygiene (especially hand hygiene) habits.

Tuberculosis

Mycobacterium tuberculosis is a pathogen of considerable public health concern. Humans are the host reservoir but many other species, including dogs, cats and birds are at least somewhat susceptible to infection. It is possible that subclinical transmission of tuberculosis from people to pets is underestimated, as suggested by a study that reported isolation of *M. tuberculosis* from 15% of healthy dogs and cats that were in contact with infected people.[45] However, another study did not identify the organism in any of 29 pets exposed to infected individuals. The same *M. tuberculosis* strain has been found in infected people and their pets, providing much stronger support for human-pet transmission.[46,47] Transmission of tuberculosis from people to pet birds has been more clearly established, particularly associated with close contacts such as feeding birds food that was pre-chewed by an infected person or mouth-to-beak feeding.[48,49]

A major concern in human medicine is the expansion of multidrug resistant and extensively drug resistant (XDR) tuberculosis. While these have not been identified in companion animals, the consequences would be potentially grave from the pet's standpoint. Tuberculosis can be difficult to diagnose early in pets and difficult to treat, resulting in morbidity and mortality. Further, various individuals have recommended immediate euthanasia of pets with tuberculosis because of the unknown but theoretical potential for subsequent zoonotic transmission. Certainly, pressures to euthanize animals infected with MDR or XDR tuberculosis would be significant and infection with either of those would likely be fatal, either from disease or public health intervention.

INFECTION PREVENTION AND CONTROL

Objective information on the control of interspecies transmission of these, and other, pathogens is limited. The degree of transmissibility for various pathogens is poorly understood and is presumably variable. Risk factors for human-animal transmission are also presumably variable between different animal species and pathogens, and only limited information is available for a small group of pathogens. Consideration of known or presumed routes of transmission (fecal-oral, contact, aerosol), infection/colonization sites in humans and animals, and general infection control practices can be used to establish reasonable but non-evidence-based guidelines. The main emphasis is on general household and veterinary clinic infection control principles, particularly hand hygiene. Prudent antimicrobial use is also an important factor for many of the above diseases, based on evidence or logical concern about the role of routine antimicrobial therapy in pets and people on the potential for interspecies transmission of selected pathogens. Research on the efficacy of different control measures is needed to help reduce the risk of human-pet transmission and the potential animal and public health consequences.

REFERENCES

[1]Manian FA. ***Clin Infect Dis*** 2003;36:e26-28. [2]van Duijkeren E, et al. ***Emerging Infect Dis*** 2004;10:2235-2237. [3]O'Mahony R, et al. ***Vet Microbiol*** 2005;109:285-296. [4]Rich M, et al. ***Vet Microbiol*** 2005;111:237-240. [5]Tomlin J, et al. ***Vet Rec*** 1999;144:60-64. [6]Cefai C, et al. ***Lancet*** 1994;344:539-540. [7]Boost MV, et al. ***Epidemiol Infect*** 2007:1-12. [8]Kottler S, et al. ***J Vet Intern Med*** 2009. [9]Boost M, et al. ***Clin Microbiol Infect*** 2007;13:731-733. [10]Leonard F, et al. ***Vet Rec*** 2006;158:155-159.[11]Loeffler A, et al. ***J Antimicrob Chemother*** 2005;56:692-697. [12]Moodley A, et al. ***J Antimicrob Chemother*** 2006;58:1118-1123. [13]Weese J, et al. ***Vet Microbiol*** 2006;115:148-155. [14]Lefebvre S, et al. ***J Hosp Infect*** 2006;62:458-466. [15]Lefebvre SL, et al. ***J Am Vet Med Assoc*** 2009;234:1404-1417. [16]Coughlan K, et al. ***Zoonoses and Public Health*** 2009. [17]Baptiste K, et al. ***Emerg Infect Dis*** 2005;11:1942-1944. [18]Vengust M, et al. ***Lett Appl Microbiol*** 2006;43:602-606. [19]Hanselman BA, et al. ***Vet Microbiol*** 2007;126:277-281. [20]Hanselman BA, et al. ***Can Vet J*** 2009. [21]van Belkun A, et al. ***Lancet*** 1996;348:1038-1039. [22]Wagenvoort JH, et al. ***J Antimicrob Chemother*** 2003;52:532. [23]Damborg P, et al. ***Vet Microbiol*** 2008. [24]Carattoli A. ***Clin Microbiol Infect*** 2008;14 Suppl 1:117-123. [25]Moreno A, et al. ***Vet Microbiol*** 2008;129:203-208. [26]Sidjabat HE, et al. ***J Med Microbiol*** 2007;56:426-434. [27]Kang C, et al. ***Infect Control Hosp Epidemiol*** 2004;25:860-867. [28]Manzur A, et al. ***J Clin Microbiol*** 2007;45:2365-2369. [29]Paterson D. ***Am J Infect Control*** 2006;34:S20-28. [30]Zhanel GG, et al. ***Antimicrob Agents Chemother*** 2008;52:1430-1437. [31]Deshpande LM, et al. ***Microb Drug Resist*** 2007;12:223-230. [32]Livermore DM. ***J Antimicrob Chemother*** 2009;64 Suppl 1:i29-36. [33]Hogg RA, et al. ***Vet Rec*** 2009;164:304-305. [34]Féria C, et al. ***FEMS Microbiol Lett*** 2001;202:205-208. [35]Johnson JR, et al. ***J Clin Microbiol*** 2009;47:3721-3725. [36]Stenske KA, et al. ***Am J Vet Res*** 2009;70:1401-1406. [37]Cave NJ, et al. ***J Am Vet Med Assoc*** 2002;221:52-59. [38]Weese JS, et al. ***J Vet Intern Med*** 2001;15:374-378. [39]Clooten J, et al. ***Vet Microbiol*** 2008;129:209-214. [40]Weese J, et al. ***J Vet Intern Med*** 2003;17:813-816. [41]Borriello S, et al. ***J Clin Pathol*** 1983;36:84-87. [42]Weese J. ***Epidemiol Infect*** 2010;In press.[43]Arroyo L, et al. ***J Med Microbiol*** 2005;54:163-166. [44]Müller K, et al. ***J Am Vet Med Assoc*** 2009;235:66-69. [45]Hawthorne VM, et al. ***Br Med J*** 1957;2:675-678. [46]Erwin PC, et al. ***Emerging Infect Dis*** 2004;10:2258-2210. [47]Une Y, et al. ***Comp Immunol Microbiol Infect Dis*** 2007;30:415-425. [48]Schmidt V, et al. ***Avian Pathol*** 2008;37:589-592. [49]Steinmetz HW, et al. ***Avian Dis*** 2006;50:641-645.

HEREDITARY NEPHROPATHY (ALPORT SYNDROME) IN DOGS, MICE, & PEOPLE

Dr. George E. Lees, DVM, MS, DACVIM
College Station, TX

INTRODUCTION

The concept of 'One Medicine' means somewhat different things to different people, but I have experienced One Medicine mainly in the context of cross-species application of relevant biomedical information in various settings. This comparative medicine approach to what we do is, in my opinion, deeply engrained in all veterinarians by our training and experience. We habitually seek to identify and apply the basic principles of biology and medicine that are shared across different species and systems (ie, we recognize and exploit the similarities that exist), but we also understand and accept that there often are large and important differences between the various species with which we work (eg, "Cats are not small dogs", etc).

When taken to cellular, molecular, and genetic levels, comparative medicine has great power to elucidate the underlying mechanisms of disease (and thus possibly identify new therapeutic options). One excellent example of this is the progress that has been made in understanding Alport syndrome over the last 2 decades. The molecular basis of Alport syndrome, which is a disease caused by defects in basement membrane (ie, type IV) collagen, was first reported in 1990. At Texas A&M University, we became involved with this field beginning in 1992-93 when dogs that proved to have canine Alport syndrome (aka, hereditary nephropathy) were brought to the Small Animal Hospital for evaluation of juvenile-onset renal disease. This is an overview of the investigations of Alport syndrome that I have been involved with since then. Much of the work was done at Texas A&M University; however, much work also has been done by collaborators at several other institutions in the United States, as well as in Sweden and Denmark.

Today, glomerular basement membranes are known to contain a crucial type IV collagen network composed of 3 individual peptides that must be synthesized and assembled correctly in order to maintain the structural and functional integrity of the glomerular filtration barrier. The 3 peptides are the alpha-3, alpha-4, and alpha-5 chains of type IV collagen, each of which is encoded by a single gene: COL4A3 encodes the alpha-3 chain, COL4A4 the alpha-4 chain and COL4A5 the alpha-5 chain. Gene defects that disrupt the proper synthesis of any one of the 3 component chains can disrupt the entire alpha-3-4-5 collagen IV network. Such defects are now known to be the causes of Alport syndrome. Because the COL4A5 gene is located on the X-chromosome, Alport syndrome often is inherited in an X-linked fashion. However, because the COL4A3 and COL4A4 genes are on another chromosome, Alport syndrome also can be inherited as an autosomal (usually autosomal recessive) trait.

DISCOVERY AND CHARACTERIZATION OF HEREDITARY NEPHROPATHY (HN) IN DOGS

English Cocker Spaniels

Although terms used to identify the condition have evolved over the years, a hereditary renal disease has been observed in Cocker Spaniels since the late 1940s. Krook described a juvenile nephropathy that he termed 'renal cortical hypoplasia' in 40 Cocker Spaniels that were examined at the Royal Veterinary College in Stockholm from 1946 to 1956. Additional early reports about renal cortical hypoplasia in Cocker Spaniels from Sweden, Switzerland and Australia described juvenile onset of a fatal proteinuric nephropathy. Subsequent studies showed that the disease was inherited in an autosomal recessive fashion and that the primary renal lesions involved glomeruli. With these developments, familial nephropathy or "FN" became the diagnostic term most widely used for the disease. Beginning in the 1980s, investigators using transmission electron microscopy noted that the ultrastructural appearance of the glomerular basement membranes (GBM) of affected dogs was like that of human Alport syndrome, but that clue was of little assistance while the molecular and genetic basis of Alport syndrome remained obscure. Starting in the late 1980s, however, a series of discoveries established that Alport syndrome was caused by genetic type IV collagen defects. Thus, when distinctive ultrastructural GBM changes were identified in an affected dog examined in 1992 at Texas A&M University, we began systematic investigations of the disease based on the hypothesis that it was a genetic type IV collagen disorder. These studies defined the clinical, pathologic, and ultrastructural features of the disease, including the evolving pattern of changes seen as the disease progressed.[1-3] Importantly, immunostaining kidney from affected dogs showed complete absence of the $\alpha3$(IV) and $\alpha4$(IV) collagen peptide chains that normally are present in the GBM. This finding further suggested that the mutation responsible for the disease would be located in 1 of the 2 genes encoding these proteins. We subsequently described the use of quantitative real time-PCR to identify COL4A4 as the candidate gene causative for familial nephropathy in Cocker Spaniels and then sequencing of COL4A4 to identify a single base change in exon 3 that produces a premature stop codon.[4] This is the first autosomal form of canine hereditary nephropathy in which the underlying genetic cause was identified. A genetic test for this mutation is now available worldwide, and appropriate use the test will permit breeders of Cocker Spaniels to eradicate the disease.

Mixed breed ('Navasota') dogs

The first member of a mixed breed family of dogs that proved to have an X-linked form of HN was brought to TAMU in 1993. The investigative approach that was being used in the English Cocker Spaniels was also applied to this family to characterize the clinical, pathologic, and ultrastructural features of the disease, including the pattern of changes seen as the disease progressed.[5] Additionally, we obtained and cryopreserved semen from affected male dogs, which later permitted us to establish a colony of dogs with X-linked HN that could be bred to produce more affected dogs as needed for prospective studies. The colony started in the Fall of 1997, and the first affected male progeny were produced in the Fall of 1998. Initially, immunostaining for the presence or absence of the collagen IV $\alpha5$ chain in the epidermal basement membrane within a skin biopsy taken at one week of age was used to identify the genetic status of each puppy. However, we subsequently identified the causative COL4A5 mutation in this family, and we now use a direct genetic (DNA-based) test to genotype dogs in the colony.[6]

STUDIES OF HN IN DOGS AND MICE AS MODELS OF HUMAN ALPORT SYNDROME

In humans, X-linked Alport syndrome is much more common (in about 85% of affected families) than the autosomal (usually autosomal recessive, and rarely autosomal dominant) forms of the disease. However, the first genetically engineered mouse models of Alport syndrome were autosomal recessive. For these reasons and because establishing a breeding colony of the dogs with X-linked HN was do-able (all female progeny from normal female dogs inseminated with frozen-thawed semen from affected males would be obligate carriers; ie, potential breeding stock), we created the X-linked canine HN colony at TAMU and began prospective studies to investigate the pathogenesis and treatment of Alport nephropathy. These studies included: a) an evaluation of the effects of enalapril administration to affected males,[7] b) studies of the effects of cyclosporine A administration to affected males and to carrier females, [8] and an examination of the mechanisms underlying the alterations of glomerular permselectivity that occur as the disease progresses in affected males.[9,10] We also explored various aspects of the pathogenesis of Alport nephropathy at the molecular level including: a) studies of the composition of normal and Alport GBM in several species,[11,12] b) studies of the changes in matrix metalloprotease activity in Alport kidney,[13,14] and c) a global analysis of gene expression in Alport kidney.[15] In addition, we have studied patterns of X-chromosome inactivation in dogs,[16] and the influence of X-inactivation on disease severity in genetically engineered mice with X-linked Alport syndrome.[17]

STUDIES OF HN IN DOGS AS A CANINE MODEL OF CANINE RENAL DISEASE

Although HN is a rare disease in dogs, it also is an example of a common problem – that is, existence of glomerular lesions that cause proteinuria and a chronic, progressive deterioration of renal function that leads to end-stage renal failure. For this reason, the HN dog colony at TAMU has been a valuable resource for investigation of many issues of direct relevance to the diagnosis and management of chronic kidney disease in dogs. One study examined the effects of serial ultrasound-guided renal biopsies on kidney structure and function.[18] We also examined the onset of proteinuria in dogs with HN and showed that the onset of microalbuminuria preceded the onset of overt proteinuria in both affected males and carrier females.[19,20] Because young-adult carrier females generally have mild renal disease characterized by mild to moderate but stable proteinuria and negligible or no change in other renal functions over extended periods, we used them to examine the effects of various interventions on their magnitudes of proteinuria. Both dietary protein intake and short-term prednisone administration were shown to have substantial but reversible effects on the magnitude of proteinuria in these dogs.[21,22] Further studies of proteinuria in these carrier females identified the amount of day-today variation in UPC ratios observed in dogs with stable proteinuria and provided a basis for interpreting UPC values when evaluating them serially to detect significant increases or decreases in magnitude of proteinuria over time.[23] In addition to our studies of the total quantity of protein in the urine of dogs with HN, we have used various proteomic techniques to separate and characterize many of the various proteins that appear in the urine at different stages of chronic progressive kidney disease. We started with sodium dodecyl sulfate-polyacrylamide gel electrophoresis (SDS-PAGE) analysis of proteins in serial urine samples collected over the full course of chronic progressive renal disease in affected male dogs with X-linked HN. These studies showed the development of tubular proteinuria, in addition to glomerular proteinuria, as the disease progressed.[24] We subsequently used morphometric analysis of serial renal biopsies to show that the development of tubular proteinuria correlated with the presence of structural changes in the tubulointerstitium and that the earliest detectable changes preceded the onset of azotemia.[25] This work has recently been extended by showing that disruption of megalin-mediated endocytosis of filtered proteins by tubular epithelial cells is an important cause of the tubular proteinuria that occurs as chronic glomerular disease progresses to end-stage renal failure.[26] Ongoing studies are using other proteomic techniques to separate and measure specific proteins in the urine of dogs with X-linked HN in hopes finding new ways to improve the detection and staging of chronic kidney disease in dogs.[27]

REFERENCES

[1]Lees GE, et al: **J Vet Intern Med** 1997;11:80. [2]Lees GE, et al: **J Am Anim Hosp Assoc** 1998;34:189. [3]Lees GE, et al: **Kidney Int** 1998;54:706. [4]Davidson AG, et al: **J Vet Intern Med** 2007;21:394. [5]Lees GE, et al: Am J Vet Res 1999;60:373. [6]Cox ML, et al: **Mamm Genome** 2003;14:396. [7]Kashtan CE, et al: **J Am Soc Nephrol** 2000;11:408A. [8]Lees GE, et al: **J Am Soc Nephrol** 2001;12:555A. [9]Kashtan CE, et al: **J Am Soc Nephrol** 2002; 13:539A. [10]Kashtan CE, et al: **J Am Soc Nephrol** 2004;15:430A. [11]Lees GE, et al: **Am J Vet Res** 2000;61:38. [12]Kashtan CE, et al: **J Am Soc Nephrol** 2001; 12:252. [13]Rao VH, et al: **Kidney Int** 2003;63:1736. [14] Rao VH, et al: **Ped Nephrol** 2005;20:732. [15]Greer KA, et al: **Mamm Genome** 2006;17:976. [16]Bell RJ, et al: **Cytogenet Genome Res** 2008;122:37.. [17]Rheault MN, et al: **Nephrol Dialysis Transplant** 2010;25:764. [18]Groman RP, et al: **Vet Radiol Ultrasound** 2004;45:62. [19]Lees GE, et al. **J Vet Intern Med** 2002;16:353. [20]Hsieh OF, et al: **J Vet Intern Med** 2005;19:432. [21]Burkholder WJ, et al. **J Vet Intern Med** 2004;18:165. [22]Lees GE, et al. **J Vet Intern Med** 2002;16:352. [23]Nabity MB, et al: **J Vet Intern Med** 2007;21:425-30. [24]Kashtan C, et al: **J Am Soc Nephrol** 2004;15:430A. [25] Lazaretti P, et al: **J Vet Intern Med** 2006;20:737. [26]Vinge L, et al: **Nephrol Dialysis Transplant** 2010; in press. [27]Nabity MB, et al: **J Vet Intern Med** 2009; 23:764.

PATHOPHYSIOLOGY OF STRESS-RELATED GASTROINTESTINAL DISORDERS

Adam J. Moeser, DVM, PhD
Raleigh, NC

INTRODUCTION

Disease is a multi-factorial process involving interactions between the host, pathogen, and environment. Much of the study of disease pathogenesis has been focused on pathogen interactions with host tissues. Recently, the influence of environmental factors on host defense and pathogen virulence has received increased attention. In particular the influence of life stress on disease prevalence and severity has become well-recognized in human and veterinary medicine. It is well-known that stressors (psychological and/or physical stress) predispose individuals to a variety of acute and chronic diseases[1, 2] but the pathophysiology remains largely unknown and thus treatment options are limited. The gastrointestinal (GI) system is particularly sensitive to stress and this is reflected by the increasing number of stress-related GI diseases in humans including irritable bowel syndrome (IBS), inflammatory bowel diseases (IBD), gastroesophageal reflux disease (GERD) and more. Likewise, veterinary species often undergo significant stress and develop similar GI disorders and thus a greater understanding of these disorders will benefit human and veterinary patients. The objective of this paper is to review the basic physiology of the stress response and how stress directly influences GI health and defense. In addition this paper will review animal models commonly used in stress studies and discuss the relevance of these models to veterinary and human medicine.

BIOLOGY OF THE STRESS RESPONSE

Stress can be defined as any threat to homeostasis of an organism.[3] The stress response is a highly conserved mechanism involving coordinated signaling between central and peripheral nervous systems in effort to maintain or restore homeostasis. Therefore the stress response is absolutely critical for survival. However, in conditions where the stress is overwhelming, e.g., multiple stressors imposed on one another, chronic stress, traumatic stress, etc, or when the ability of an individual to mount an adequate stress response impaired, profound disturbances in physiology and health are observed. Stressors can be physical, psychological, chemical, etc but the body's initial stress response is similar. The most well-known and studied stress response system is the hypothalamic-pituitary-adrenal (HPA) HPA axis. In this pathway, a perceived stress induces the synthesis and release of corticotropin releasing factor (CRF) by the hypothalamus. CRF is subsequently delivered to the anterior pituitary via the hypophyseal portal system where it activates CRF receptors causing release of adrenocorticotropic hormone (ACTH). ACTH binds to ACTH receptors in the adrenal cortex inducing release of cortisol into the circulation. Therefore cortisol is often used as a physiologic marker of stress. The biologic actions of cortisol have been well-studied and include potent gluconeogenic, anti-inflammatory, and immunosuppressive properties. Circulating cortisol also provides a level of control over the stress response through the negative feedback on the hypothalamus. During the stress response, the adrenal gland also releases catecholamines (epinephrine and norepinephrine) which function in the fight or flight response to divert blood flow to critical organ systems such as the cardiovascular system and muscles. The overall function of this important response is to mobilize body resources to allow the individual or animal to cope with a perceived physical threat.

THE EFFECTS OF STRESS ON GI PHYSIOLOGY AND FUNCTION

Stress has a profound influence on GI physiology resulting in a variety of clinical symptoms including diarrhea, abdominal pain, constipation, weight loss, etc. The underlying GI pathophysiology contributing to stress-related clinical disease is described below.

Increased Intestinal epithelial permeability

The intestinal epithelium forms a selective barrier to antigens, toxins, and microorganisms that reside in the intestinal lumen. Impairment of the intestinal barrier, characterized increased intestinal permeability, triggers the rapid onset of inflammation as a result of increased transmigration of luminal microorganisms, toxins, and antigenic substances into the sub-epithelial lamina propria. Several research studies have demonstrated that rodents subjected to various acute stressors display marked disturbances in intestinal barrier function.[4, 5] Intestinal barrier dysfunction is a hallmark for stress-related GI diseases most notably IBD and IBS.[6, 7]

Disorders in intestinal secretion and absorption

Stress can influence both secretory and absorptive properties in the gut. Several studies report intestinal hypersecretion, a response determined to be a result of stimulated Cl^- secretion.[4] Na glucose co-transport activity was also shown to be impaired in a stress rat model. [8]

Increased bacterial adherence

Earlier studies with rodents showed that chronic stress results in enhanced bacterial adhesion and penetration into enterocytes.[9] The mechanisms of stress-induced impairment of mucosal defenses are unclear but interactions between stress hormones, mast cells, and enteric nerves with the intestinal mucosa appear to play a major role. The catecholamine norepinephrine appears to play a role as exposure to norepinephrine to the cecal and colonic mucosa of pigs significantly enhanced the adherence of enterohemmorhagic E coli O157:H7.[10] In rat pups, maternal separation stress increased bacterial adherence and invasion that was mediated by CRF.[11]

Disorders in motility and hypersensitivity.

The enteric nervous system controls many aspects of intestinal function including motility, secretion, absorption, antigen uptake and inflammation. Specific effects of stress on the enteric nervous system include alterations in intestinal motility patterns, increased secretion, and inflammation. Stress has been shown to increased colonic motility and visceral hypersensitivty[12, 13], both thought to play a key role in the pathogenesis of IBS.

MECHANISMS OF STRESS-INDUCED INTESTINAL DYSFUNCTION

As discussed above, the role of stress in clinical GI disease is well-documented and research investigating these disorders has led to the increased understanding of the underlying pathophysiology. However, the signaling mechanism(s) by which a stress alters intestinal physiology is poorly understood. Until recently, the popular concept was that the influence of stress on peripheral organ systems was largely attributed to activation of the HPA axis and adrenal cortisol. However, the exact role of this system in GI dysfunction is unclear. For example, surgical or chemical ablation of the adrenal gland in rats did not influence GI function (intestinal permeability or ion secretion) in response to acute restraint stress[5] suggesting that the HPA axis and cortisol release does not appear to play a major role. This has led to investigations into alternative stress pathways that mediate GI physiology. One important discovery that supported a HPA axis-independent pathway was that stress peptides, including CRF and a family of related neuropeptides, the urocortins, could be synthesized locally within intestinal tissues. Furthermore, local expression of CRF receptors in the gut was determined providing evidence for a localized stress response system outside of the central nervous system and HPA axis. Since this finding, several studies have been conducted to elucidate the role of peripheral CRF/CRF receptor systems in stress-induced intestinal disorders. These studies demonstrate increased local production of intestinal CRF in stress-related intestinal diseases including IBD and infectious diarrhea.[14, 15] With the use of select CRF receptor antagonist drugs, it has been shown that activation of CRF receptor pathways in the intestine are responsible for several pathophysiologic processes including impairment in barrier function, hypersecretion, mucosal inflammation, disturbed motility, and visceral hypersensitivity.[16] The pathophysiology resulting from CRF receptor activation largely reflects the specific intestinal cell types that express CRF receptors. Specific cell types that express CRF receptors include immunocytes and enteric neurons. In particular intestinal mast cells have been shown to be highly activated by stress likely due to the expression CRF receptors on their cell surface. Mast cells play a central role in stress-induced intestinal disease as demonstrated by the increased numbers of activated mast cells in stressed individuals and studies in mast cell-deficient animals which appear to be resistant to stress-induced intestinal dysfunction.[17, 18] The exact mechanism by which mast cells contribute to intestinal dysfunction is unclear. Upon activation, mast cells are capable of releasing a variety of preformed mediators (Proteases, histamine, TNF-α, etc) and synthesized mediators (IL-6, VEGF, prostaglandins, etc) that have profound influence on intestinal physiology. The specific release of mast cell mediators and their contribution to specific GI functions is not well understood and requires further study.

ANIMALS MODELS OF STRESS-INDUCED INTESTINAL DISEASE

Our understanding of stress-related disorders has come almost exclusively from rodent models. Although these models have led to a more fundamental understanding of how stress influences intestinal physiology there are important limitations to these models to consider when making a direct comparison to larger animal species and humans. Firstly, the nature of the stress protocols (water avoidance stress, cold restraint stress) are short-term traumatic stressors that may not reflect the low grade, chronic stress that is more commonly observed in veterinary species and humans. Second, the CNS and enteric nervous system is more simplified in the rodent and may not represent the physiologic response (stress perception and clinical outcome) observed in higher species. For example, in response to stress, rodents do not develop diarrhea nor do they exhibit extensive mucosal inflammation which are characteristic of human disease. Given these limitations, the biomedical field has placed considerable emphasis on the need for large animal models of human disease.

We have worked extensively with the pig as a model for stress-induced intestinal diseases. The pig model has several features that make it an ideal model for studying stress-related disease. Pigs are omnivores and their digestive anatomy and physiology is more similar to humans than other laboratory animals. Pigs are intelligent and possess a highly-developed central nervous system which is critical for perceiving and responding to psychological stressors. In addition, in swine production systems, pigs often experience low-grade chronic stressors (increased stocking density, littermate separation, competition for food, etc) that are more comparable to human stressors. In our studies we have found that the stress response in pigs and the resulting GI pathophysiology (mucosal inflammation) and clinical presentation (diarrhea, weight loss) are more comparable to human disease compared with rodents.[19-21] Altogether, the similarity between pigs and other veterinary species and humans will be important when evaluating the efficacy of therapeutic strategies.

SUMMARY

Veterinary species and humans undergo chronic stress resulting in a variety of GI diseases, however our understanding the mechanisms involved has only begun to be elucidated. As we consider the one medicine theme, it appears necessary that we explore stress mechanisms in large animal species in order to fully understand the complexities of these disorders. This translational approach will hopefully lead to novel therapeutic approaches to lessen the impact of stress on intestinal disease.

REFERENCES

[1.] Bennett EJ, Tennant CC, Piesse C, Badcock CA, Kellow JE. Level of chronic life stress predicts clinical outcome in irritable bowel syndrome. **Gut** 1998;43:256-61. [2.] Blanchard EB, Lackner JM, Jaccard J, Rowell D, Carosella AM, Powell C, Sanders K, Krasner S, Kuhn E. The role of stress in symptom exacerbation among IBS patients. **J Psychosom Res** 2008;64:119-28. [3.] Selye H. A syndrome produced by diverse nocuous agents. 1936. **J Neuropsychiatry Clin**

Neurosci 1998;10:230-1. [4] Saunders PR, Kosecka U, McKay DM, Perdue MH. Acute stressors stimulate ion secretion and increase epithelial permeability in rat intestine. **Am J Physiol** 1994;267:G794-9. [5] Santos J, Saunders PR, Hanssen NP, Yang PC, Yates D, Groot JA, Perdue MH. Corticotropin-releasing hormone mimics stress-induced colonic epithelial pathophysiology in the rat. **Am J Physiol** 1999;277:G391-9. [6] Barbara G. Mucosal barrier defects in irritable bowel syndrome. Who left the door open? **Am J Gastroenterol** 2006;101:1295-8. [7] Porras M, Martin MT, Yang PC, Jury J, Perdue MH, Vergara P. Correlation between cyclical epithelial barrier dysfunction and bacterial translocation in the relapses of intestinal inflammation. **Inflamm Bowel Dis** 2006;12:843-52. [8] Boudry G, Cheeseman CI, Perdue MH. Psychological stress impairs Na+-dependent glucose absorption and increases GLUT2 expression in the rat jejunal brush-border membrane. **Am J Physiol Regul Integr Comp Physiol** 2007;292:R862-7. [9] Soderholm JD, Yang PC, Ceponis P, Vohra A, Riddell R, Sherman PM, Perdue MH. Chronic stress induces mast cell-dependent bacterial adherence and initiates mucosal inflammation in rat intestine. **Gastroenterology** 2002;123:1099-108. [10] Chen C, Lyte M, Stevens MP, Vulchanova L, Brown DR. Mucosally-directed adrenergic nerves and sympathomimetic drugs enhance non-intimate adherence of Escherichia coli O157:H7 to porcine cecum and colon. **Eur J Pharmacol** 2006;539:116-24. [11] Soderholm JD, Yates DA, Gareau MG, Yang PC, MacQueen G, Perdue MH. Neonatal maternal separation predisposes adult rats to colonic barrier dysfunction in response to mild stress. **Am J Physiol Gastrointest Liver Physiol** 2002;283:G1257-63. [12] Tache Y, Martinez V, Wang L, Million M. CRF1 receptor signaling pathways are involved in stress-related alterations of colonic function and viscerosensitivity: implications for irritable bowel syndrome. **Br J Pharmacol** 2004;141:1321-30. [13] Fukudo S, Nomura T, Hongo M. Impact of corticotropin-releasing hormone on gastrointestinal motility and adrenocorticotropic hormone in normal controls and patients with irritable bowel syndrome. **Gut** 1998;42:845-9. [14] Kresse AE, Million M, Saperas E, Tache Y. Colitis induces CRF expression in hypothalamic magnocellular neurons and blunts CRF gene response to stress in rats. **Am J Physiol Gastrointest Liver Physiol** 2001;281:G1203-13. [15] Kokkotou E, Torres D, Moss AC, O'Brien M, Grigoriadis DE, Karalis K, Pothoulakis C. Corticotropin-releasing hormone receptor 2-deficient mice have reduced intestinal inflammatory responses. **J Immunol** 2006;177:3355-61. [16] Martinez V, Tache Y. CRF1 receptors as a therapeutic target for irritable bowel syndrome. **Curr Pharm Des** 2006;12:4071-88. [17] Barbara G, Stanghellini V, De Giorgio R, Cremon C, Cottrell GS, Santini D, Pasquinelli G, Morselli-Labate AM, Grady EF, Bunnett NW, Collins SM, Corinaldesi R. Activated mast cells in proximity to colonic nerves correlate with abdominal pain in irritable bowel syndrome. **Gastroenterology** 2004;126:693-702. [18] Santos J, Yates D, Guilarte M, Vicario M, Alonso C, Perdue MH. Stress neuropeptides evoke epithelial responses via mast cell activation in the rat colon. **Psychoneuroendocrinology** 2008;33:1248-56. [19] Moeser AJ, Klok CV, Ryan KA, Wooten JG, Little D, Cook VL, Blikslager AT. Stress signaling pathways activated by weaning mediate intestinal dysfunction in the pig. **Am J Physiol Gastrointest Liver Physiol** 2007;292:G173-81. [20] Moeser AJ, Ryan KA, Nighot PK, Blikslager AT. Gastrointestinal dysfunction induced by early weaning is attenuated by delayed weaning and mast cell blockade in pigs. **Am J Physiol Gastrointest Liver Physiol** 2007;293:G413-21. [21] Smith F, Clark JE, Overman BL, Tozel CC, Huang JH, Rivier JE, Blikslager AT, Moeser AJ. Early Weaning Stress Impairs Development of Mucosal Barrier Function in the Porcine Intestine. **Am J Physiol Gastrointest Liver Physiol** 2009.

BIOMARKERS FOR SEPSIS IN DOGS

Amy E. DeClue, DVM, MS, DACVIM (SAIM)
Columbia, MO

INTRODUCTION

A biomarker, or biological marker, is defined as "a characteristic that is objectively measured and evaluated as an indicator of normal biological processes, pathogenic processes, or pharmacologic responses to a therapeutic intervention".[1] Another, perhaps more simple, definition is a "quantifiable measurement(s) of biological homeostasis that define(s) what is 'normal', therefore providing a frame of reference for predicting or detecting what is 'abnormal'".[2] Biomarkers include physiologic indicators (e.g., body temperature, respiratory rate, heart rate) or endogenously produced biochemical compounds found in various body fluids or tissues (e.g., TNF, IL-6, CNP). Molecular biomarkers also include nucleic acid-based biomarkers such as gene mutations or polymorphisms; and quantitative analysis of the gene expression of peptides, proteins, lipids metabolites, and other small molecules.

Sepsis, defined as the systemic inflammatory response to infection, is associated with substantial morbidity and mortality in dogs. Identifying biomarkers that would allow the rapid diagnosis of sepsis facilitating early intervention, particularly antimicrobial therapy[3], would undoubtedly benefit patients with sepsis. In addition to their role as a diagnostic tool, biomarkers may allow for temporal categorization of systemic inflammation such that tailored therapeutic interventions can be implemented. Biomarkers are also being investigated to facilitate the identification of potential therapeutic targets for immunomodulation and predict patient response to novel treatment strategies. Additionally, biomarkers that correlate with disease severity and predict morbidity and mortality are under investigation to assist with client education and patient stratification in clinical trials.

Determining the most ideal biomarkers for the diagnosis, staging, and prognostication of sepsis is complex. There are many potential biomarkers for sepsis; more than 100 distinct molecules have been proposed.[4] This discussion will review biomarkers that have been evaluated in clinical canine sepsis. It is important to be cognizant of the inconsistencies in the literature, including discordant conclusions regarding biomarker relevance in certain settings. Some of the sources of inconsistency include variable patient populations, assay methodology, reference standards, site of sampling, requirements for activation; and the presence of soluble receptors, circulating inhibitors, and carrier proteins which may interfere with measuring the biomarker of interest.

BIOMARKERS

Clinical evaluation – Rectal temperature, heart rate, respiratory rate and leukocyte count have been used as markers for the systemic inflammatory response syndrome (i.e., SIRS criteria; table 1) in dogs. In one study of 500 dogs with a variety of illnesses, not just sepsis, presenting to a veterinary teaching hospital, mortality was significantly associated with the number SIRS criteria fulfilled.[7] While these parameters are often used as part of the clinical definition of sepsis, it is important to note that these criteria are not necessarily specific for sepsis (i.e., infection) as many non-infectious inflammatory conditions may case changes in temperature, heart rate, respiratory rate and white blood cell count. The sensitivity and specificity of SIRS criteria for the diagnosis of sepsis is variable based on the study and the cut off criteria used but ranges from a sensitivity of 77-97% and specificity of 64-77% when 2 of the 4 criteria were met.[8] Multiple organ dysfunction syndrome (MODS) has recently been evaluated in a retrospective study of 114 dogs with sepsis secondary to GI leakage (defined as infection based on cytologic examination, positive bacterial culture, or gross evidence of gastrointestinal tract leakage).[6] Organ dysfunction was defined as noted in table 2. Data from this study is summarized in table 3. Approximately 60% of dogs had coagulation dysfunction, 49% had hepatic dysfunction, 31% had respiratory dysfunction, 17% had cardiovascular dysfunction and 12% had renal dysfunction.[6] Multiple organ dysfunction syndrome (dysfunction of ≥ 2 organ systems) was diagnosed in 50% of the population.[6] Dogs with MODS had a mortality rate of 70% compared to 25% for dogs without MODS.[6] There was a significant association between the presence of organ dysfunction and death.

Table 1: SIRS Criteria for dogs[5]

Temperature	> 102.6°F or < 99°F
Heart rate	>140 bpm
Respiratory rate	>40 bpm
White blood cell count	>19,500/ul or < 5,000/ul or >5% bands

Table 2: Definitions of organ dysfunction in dogs.[6]

Organ	Criteria
Renal	An increase in creatinine concentration ≥ 0.5 mg/dL from presurgical values without evidence of pre-or post renal azotemia
Cardiovascular	Hypotension severe enough to require vasopressor therapy
Respiratory	Need for supplemental oxygen administration or mechanical ventilation; determined based on clinical assessment, blood gas analysis (alveolar-arterial gradient in partial pressure of oxygen >10 mm Hg) and/or results of pulse oximetry (SpO2 < 95)
Hepatic	Plasma or serum total bilirubin of >0.5 mg/dL
Coagulation	PT or PTT > 25% above the upper reference limit and/or platelet count of ≤ 100,000/μL

Table 3: Data summary[6]

Dysfunctional organ systems	Percent of tpopulation	Survival (%)	Odds ratio for risk of death
0	21.9	84	
1	28.1	69	6.73
2	22.8	46	6.67
3	14.9	24	10.21
4	9.6	9	18.71
5	2.6	0	

Endotoxin Activity (EA)- Measurement of EA in people at the time of ICU admission has a high negative predictive value (NPV) for the diagnosis of Gram-negative infection and sepsis; such that it may have a role in the exclusion of sepsis (Se = 85.3%, Sp = 44%, NPV 98.6%).[9] Unfortunately however EA does not seem to be useful in human medicine in predicting disease severity and assessing the risk of development of severe sepsis / organ dysfunction. A small number of veterinary studies have evaluated EA; two in dogs and one in hoses. The first of the canine studies evaluated 22 healthy dogs, 11 sick dogs without evidence of SIRS, and 8 dogs with suspected sepsis (defined by the authors as SIRS with toxic neutrophils on blood smear evaluation).[10] They found that EA was significantly higher in dogs with sepsis compared to those without SIRS and healthy dogs. The authors thus concluded that the EAA is useful for endotoxin measurement in dogs and proposed a clinical cut off value of 0.346, since only one of the healthy dogs had an EA above that value. The second study evaluated the ability of the EAA to identify infection and predict outcome in 100 dogs consecutively admitted to a university teaching hospital ICU.[11] In contrast to the first study however EAA did not correlate with the presence of infection ($p>0.1$) but was useful as a part of a multivariate prognostic model to predict survival. Clearly further studies are warranted regarding the use of EA as a biomarker for sepsis in dogs.

Tumor Necrosis Factor (TNF) -TNF has been found to have prognostic significance in a variety of human studies; with higher TNF concentrations correlating with mortality and persistent TNF production being associated with a poorer prognosis.[12-14] A variety of investigators have studied the potential role of TNF as a biomarker in veterinary patients with sepsis. TNF activity has been demonstrated in hospitalized dogs with canine parvoviral (CPV) enteritis and there was significant association between increasing TNF activity over time and morality in this population.[15] TNF concentrations have also been used as an indicator of response to treatment in clinical CPV,[16] and experimental endotoxemia in dogs.[17] In addition TNF has been evaluated as a potential biomarker for dogs with septic peritonitis. In 23 dogs with septic peritonitis, there was no detectable serum TNF (measured by ELISA) in 17 dogs on day 1; whereas 16 dogs had detectable TNF in their abdominal fluid. [18] The median abdominal fluid TNF concentration was 0.69 ng/mL for surviving dogs, compared to 1.16 ng/mL for non-survivors, although this difference was not statistically significant.[18] Plasma TNF bioactivity at hospital presentation in dogs with sepsis, non-infectious SIRS and healthy controls has been evaluated by investigators at the University of Missouri, Comparative Internal Medicine Laboratory (MU CIML). We found that significantly more dogs in the sepsis or SIRS groups had measurable TNF bioactivity in their plasma compared to healthy controls, but there was no difference between the sepsis and SIRS groups.[19] Additionally, plasma TNF bioactivity was not significantly correlated with mortality.

Interleukin (IL)-6- Several studies in human medicine have demonstrated a positive correlation between plasma IL-6 concentrations both at admission and over time with subsequent morbidity and mortality. One such study revealed a direct correlation between plasma IL-6 and the risk of death in patients with intra-abdominal sepsis; with IL-6 measurement alone predicting outcome with >80% accuracy.[20] Persistent production of IL-6 seems to be more important than initial or peak concentrations in regards to association with patient outcome, although IL-6 may simply serve as a marker for disease severity in the absence of a causal relationship. [21-23] Unfortunately, however, IL-6 production is not specific for the diagnosis of sepsis as IL-6 can also be strongly induced in SIRS of non-infectious origin, such that increased plasma concentrations have been documented in human patients following surgery, those with immune mediated disease, and transplant rejection. IL-6 has documented superiority to procalcitonin (PCT) in neonates with sepsis but not adult patients, in which PCT is the diagnostic and prognostic biomarker of choice. There are 2 studies in veterinary medicine that have evaluated plasma IL-6 concentrations in dogs with sepsis and SIRS. In the first study, plasma IL-6 concentrations at presentation predicted outcome in canine SIRS and sepsis.[24] In a study conducted at the MU CIML, plasma IL-6 concentrations were evaluated at hospital presentation in dogs with sepsis, non-infectious SIRS and healthy controls. Significantly more dogs in the sepsis or SIRS groups had IL-6 in their plasma compared to healthy controls, but there was no difference between sepsis and SIRS groups. Additionally, plasma IL-6 concentration was not significantly correlated with mortality.[19] It should be noted that these 2 studies used different assay methodologies to evaluate IL-6, the first used a bioassay and the second an ELISA technique.

Nitric Oxide (NO) - Nitric oxide plays an important role in the pathophysiology of sepsis. Pathogen associated molecular patterns and pro-inflammatory mediators induce iNOS in the endothelium, vascular smooth muscle, macrophages and various parenchymal calls.[25] NO is thought to be a key mediator of the myocardial dysfunction, hypotension and vascular hyporeactivity associated with sepsis, severe sepsis and septic shock.[25, 26] Nitric oxide also has profound immunomodulatory functions during sepsis.[25, 27, 28] Inhibiting NO production in experimental animal models of sepsis and human clinical trials has been shown to have beneficial effects on the hemodynamic stability.[23, 29-31] Nitric oxide has been evaluated in dogs with critical illness previously. In this study, critically ill dogs had significantly higher plasma nitrate and nitrite concentrations (stable markers of NO production) than non-critically ill or healthy dogs. [32] Additionally, plasma nitrate/nitrite correlated with disease severity and survival.[32] Investigators at the MU CIML have evaluated NO in dogs with naturally developing sepsis.[33] We found dogs with sepsis had significantly higher plasma NO concentrations than healthy control dogs; however there was no difference between septic dogs and dogs with non-infectious causes of SIRS. Nor was there a difference in plasma NO concentrations between the survivors and non-survivors.[19, 33] Jacobson et al had similar findings in a cohort of dogs with babesiosis.[34]

IL-10 - Interleukin 10 is a potent anti-inflammatory cytokine which down regulates the inflammatory response through various mechanisms.[35] IL-10 acts to inhibit synthesis of many pro-inflammatory mediators (IFN-γ, IL-3, IL-2 and TNF) as well as enhancing production of IL-1RA by monocytes. IL-10 also blocks IFN-γ, binding at a key step in its signal transduction pathway.[35, 36] Diseases associated with increased IL-10 mRNA expression in dogs include Leishmania infantum[37] and canine distemper virus[38]; however little investigation has been carried out in clinical cases of bacterial sepsis. In a study conducted at the MU CIML, plasma IL-10 concentrations were evaluated in dogs with sepsis, non-infectious SIRS and healthy controls.[19] There was no difference in the number of dogs with measurable plasma IL-10 concentrations among groups nor were plasma IL-10 concentrations significantly correlated with survival.

Procalcitonin (PCT) - Procalcitonin is perhaps the single most useful biomarker for sepsis in adult human patients. PCT is the precursor peptide of the hormone calcitonin; it is typically expressed in the C cells of the thyroid gland, but also secreted by non-thyroidal tissues. In human patients and experimental models it appears that circulating concentrations of PCT begin to increase within 2-6hrs of induction of bacteremia and/or endotoxemia, with a circulating half-life of approximately 24 hours.[39-41] PCT concentration will also be transiently increased with trauma, such as surgery; but return to baseline within 24 hours in the absence of infection or sepsis. While the exact role of PCT is still being determined in human medicine it appears to be useful in the diagnosis of sepsis,[42, 43] to guide antibiotic therapy (both commencement, cessation and de-escalation of antibiotics)[44, 45] and for prognostication[46]. Another interesting feature of PCT as a biomarker is that it is useful in the setting of chemotherapy induced sepsis, since it is a tissue-based rather than leukocyte-based biomarker. Unfortunately the ELISA for PCT in humans cannot be used for dogs due to differences in the carboxy terminus of the PCT molecule, such that the only study evaluating PCT in dogs investigated PCT mRNA expression by qRT-PCR. [47] This study demonstrated greater PCT mRNA in sick dogs (n=56; categorized into groups with infectious, inflammatory/non-infectious, neoplastic and other 'diseases) compared to healthy controls (n=11) ($p<0.001$), but no there was no difference in PCT mRNA expression based on the primary disease process.[47] Interestingly in five dogs from each group, PCT mRNA expression was determined at two time points; and in each dog the changes in its expression were in agreement with the clinical evaluation of patient improvement or deterioration, suggestive of a possible prognostic role.[47]

Markers of Coagulation - Altered coagulation is directly correlated to disease severity and mortality in human patients with sepsis. Hypercoagulability is also a common finding in dogs with naturally occurring sepsis as evidenced by prolongation of prothrombin time, partial thromboplastin time, increased D-dimer and fibrinogen degradation products concentrations, and depletion of antithrombin (AT)and activated protein C (aPC).[6, 48, 49] It appears that aPC may act as a useful prognostic indicator in dogs with sepsis.[48] Evaluation of hemostasis in dogs with CPV enteritis also documented decreased AT activity and hypercoagulability in affected dogs.[49] Von Willebrand factor antigen concentration has been evaluated as a marker of endothelial activation and is an independent predictor of survival in humans with sepsis. In a study of 14 dogs with sepsis and 11 control dogs, Von Willebrand factor antigen concentration was significantly higher in the sepsis group compared to control ($p<0.005$) but there was no difference between the survivors and non-survivors in the sepsis group.[50]

Natriuretic Peptides (NP) - The NP system is an endocrine mechanism that regulates salt and body fluid balance. In mammals it consists of three structurally related members; atrial natriuretic peptide (ANP), brain or B-type natriuretic peptide (BNP) and C-type natriuretic peptide (CNP). ANP and BNP are made primarily in the heart, whereas CNP is made in many organs, predominately vascular endothelial cells. NPs are synthesized as prepropeptides and then cleaved inactive N-terminal (NT) peptide and a biologically active C-terminal mature peptide. In many cases, the NT-peptide has greater plasma half-life and stability compared to the biologically active c-terminal peptide and thus is a preferred assay target. ANP, BNP and CNP are promising biomarkers for sepsis in humans[51-53] To the author's knowledge, ANP has not been evaluated in dogs with sepsis. In pilot study performed at MU CIML, NT-pBNP was evaluated in a small number of dogs with sepsis and healthy controls. There was no significant difference in serum NT-pBNP concentrations between the populations. We have also investigated NT-pro-CNP as a biomarker for sepsis in dogs. We found that dogs in the sepsis group (n=11) had significantly greater serum NT-pCNP concentrations than dogs in the SIRS group (n=11) and healthy controls (n=49) ($p<0.001$).[54] Using a cut-off value of 10.2 pmol/L, the sensitivity and specificity of serum NT-pCNP for differentiating sepsis from SIRS in dogs was 82% and 94%, respectively.[54]

C-reactive protein (CRP) - CRP is an acute phase protein that has been evaluated extensively as a biomarker both in human and veterinary medicine. In human medicine its use has been surpassed given the development of more sensitive and specific sepsis biomarkers such as PCT. A CRP assay has been validated for use in dogs[55] and increased CRP has been found in dogs associated with surgery, immune mediated disease,[56, 57] pancreatitis,[58] gastrointestinal disease and a variety of infectious insults including sepsis.[59, 60] CRP concentrations appear to be higher in critically ill dogs compared with controls.[61] The change in CRP over time has also been suggested to predict survival in dogs with sepsis.[62]

CONCLUSION

Strong criteria for the rapid diagnosis and prognostication of sepsis in dogs are needed. Large scale, multi-institutional studies to critically evaluate scoring systems for sepsis, including both clinical evaluation and endogenously produced biochemical compounds, will help establish sepsis biomarkers with the potential to play an important role in rational patient management and patient stratification for clinical trials.

ACKNOWLEDGEMENTS

The author would like to thank Dr. Claire Sharp at Tufts University for her substantial contribution to these proceedings.

REFERENCES

[1]Group BDW. Clin Pharmacol Ther 2001;69:89. [2]Dalton WS, et al. Science 2006;312(5777):1165. [3]Garnacho-Montero J, et al. Crit Care 2006;10(4):R111. [4]Marshall JC, et al. Crit Care Med 2003;31(5):1560. [5]DeClue A, Sepsis and the Systemic Inflammatory Response to Infection: in *Text Book of Veterinary Internal Medicine.* (2010) pp. 523. [6]Kenney EM, et al. J Am Vet Med Assoc;236(1):83. [7]Okano S, et al. Vet Rec 2002;150:245. [8]Hauptmann J, et al. Vet Surg 1997;26:393. [9]Marshall JC, et al. J Infect Dis 2004;190(3):527. [10]Kjelgaard-Hansen M, et al. Comp Immunol Microbiol Infect Dis 2008;31(6):477. [11]Dickinson AE, et al., paper presented at the 15th IVECCS, Chicago, IL, 2009. [12]Martinez M, et al. Crit Care Med 1999;27:1303. [13]Oberhoffer M, et al. Clin Chem Lab Med 1999;37(3):363. [14]Gogos CA, et al. J Infect Dis 2000;181(1):176. [15]Otto C, et al. JVIM 1997;11:65. [16]Senturk S. J Vet Pharmacol Ther 2005;28(1):57. [17]DeClue AE, et al. Am J Vet Res 2008;69(2):228. [18]Humm KR, et al., in *14th IVECCS.* (Phoenix, AZ, 2008), pp. 819. [19]DeClue A, et al. Unpublished work 2010. [20]Patel RT, et al. Br J Surg 1994;81(9):1306. [21]Casey LC, et al. Ann Intern Med 1993;119(8):771. [22]Barriere SL, et al. Crit Care Med 1995;23(2):376. [23]Tanigawa K, et al. J Crit Care 1999;14(4):186. [24]Rau S, et al. Vet Clin Pathol 2007;36(3):253. [25]Kirkeboen KA, et al. Acta Anaesthesiol Scand 1999;43(3):275. [26]Abraham E, et al. Crit Care Med 2007;Publish Ahead of Print. [27]Clancy RM, et al. J Clin Invest 1992;90(3):1116. [28]Kubes P, et al. Proc Natl Acad Sci U S A 1991;88(11):4651. [29]Watson D, et al. Crit Care Med 2004;32(1):13. [30]el-Dwairi Q, et al. Am J Physiol 1998;274(3 Pt 1):C770. [31]Chan D, IVECCS, Atlanta, GA, 2005. [32]Jones J, et al. J Vet Emergency Crit Care 1999;9(4):195. [33]Whitehead ZR, et al.

IVECCS. Phoenix, AR, 2008, pp. 827. [34]Jacobson L, et al. Vet Parasitol 2002;104:27. [35]Gosain A, et al. J Burn Care Rehabil 2005;26(1):7. [36]Rozanski E, IVECCS, San Antonio, Texas, 2006. [37]Boggiatto PM, et al. Clin Vaccine Immunol 2009. [38]Beineke A, et al. Viral Immunol 2008;21(4):401. [39]Meisner M, et al. Shock 2003;19(2):187. [40]Dandona P, et al. J Clin Endocrinol Metab 1994;79(6):1605. [41]Nijsten MW, et al. Crit Care Med 2000;28(2):458. [42]Assicot M, et al. Lancet 1993;341(8844):515. [43]Simon L, et al. Clin Infect Dis 2004;39(2):206. [44]Christ-Crain M, et al. Lancet 2004;363(9409):600. [45]Stocker M, et al. Neonatology 2009;97(2):165. [46]Jensen JU, et al. Crit Care Med 2006;34(10):2596. [47]Kuzi S, et al. J Vet Diagn Invest 2008;20:629. [48]de Laforcade A, et al. J Vet Intern Med 2003;17:674. [49]Otto CM, et al. J Am Vet Med Assoc 2000;217(10):1500. [50]Rogers CL, et al. J Vet Intern Med 2009. [51]Morgenthaler NG, et al. Crit Care 2005;9(1):R37. [52]Chen Y, et al. Am J Emerg Med 2009;27(6):701. [53]Hama N, et al. Biochem Biophys Res Commun 1994;198(3):1177. [54]DeClue A, et al. J Vet Emergency Crit Care 2009;19 (S1):A1. [55]Kjelgaard-Hansen M, et al. J Small Anim Pract 2008;49(6):282. [56]Griebsch C, et al. Vet Clin Pathol 2009. [57]Lowrie M, et al. J Vet Intern Med 2009;23(4):862. [58]Mansfield CS, et al. J Am Vet Med Assoc 2008;233(6):936. [59]Dabrowski R, et al. Theriogenology 2009;72(4):471. [60]Nakamura M, et al. J Vet Med Sci 2008;70(2):127. [61]Chan DL, et al. J Vet Intern Med 2009;23(3):559. [62]Gebhardt C, et al. J Vet Emerg Crit Care (San Antonio) 2009;19(5):450.

USE OF MOLECULAR ASSAYS FOR THE DIAGNOSIS OF FELINE INFECTIOUS DISEASES

Michael R. Lappin, DVM, PhD, DACVIM
Fort Collins, Colorado

Clinical syndromes induced by infectious agents are common in cats. Documenting current infection is still present is the best way to make a definitive diagnosis. Commonly used techniques vary by the body system but include fecal flotation, cytology, histopathology, immunohistochemistry, culture, antigen tests, and molecular diagnostic assays. For some agents, antibody test results are also used to help make a clinical diagnosis. However, presence of antibodies may only document prior exposure, not current infection.

Sensitivity is the ability of an assay to detect a positive sample; specificity is the ability of an assay to detect a negative sample. Sensitivity and specificity vary with each assay. Positive predictive value (PPV) is the ability of a test result to predict presence of disease; negative predictive value (NPV) is the ability of a test result to predict absence of disease. Many of the infectious agents encountered in feline practice infect a large percentage of the population, resulting in positive organism detection techniques or serum antibody production. However, they only induce disease in a small number of cats in the infected group. Classic examples include coronaviruses, *Toxoplasma gondii*, and *Bartonella* spp.. For these agents, even though assays with good sensitivity and specificity are available, the predictive value of a positive test is actually very low.

One of the newest organism demonstration techniques used in practice is the polymerase chain reaction (PCR) assay. This reaction amplifies small quantities of DNA to detectable levels. By use of a reverse transcriptase step, RNA is converted to DNA; therefore the technique can also be used to detect RNA (RT-PCR). PCR assays are of great value for documentation of infections, particularly if the organism in question is difficult to culture (e.g., *Ehrlichia* spp., *Mycoplasma* spp.) or cannot be cultured (e.g., hemoplasmas). Specificity can be very high, depending on the primers used in the reaction. For example, primers can be designed to detect one genus but not others. Primers can also be designed to identify only one species. For example a PCR assay can be developed to detect all haemoplasmas or just one species such as *Mycoplasma hemofelis*.

Because of the inherent sensitivity of the reaction, PCR can give false-positive results if sample contamination occurs during collection or at the laboratory performing the procedure. False-negative results can occur if the sample is handled inappropriately. Results may also be affected by treatment. For example, many cats with hemoplasmosis become negative by PCR assay of blood while on doxycycline or fluoroquinolone therapy and so should be sampled prior to treatment. Other potential problems are that minimal standardization exists among commercial laboratories offering PCR assays and minimal external quality control exists.

Although PCR assays can be one of the most sensitive for documentation of infections, positive test results do not always prove that the infection is resulting in clinical illness. For example, because the technique detects DNA of both live and dead organisms, positive test results may be achieved even if the infection has been controlled. When the organism being tested for commonly infects the background population of healthy cats, interpretation of results for a single animal can be difficult. For example, *Bartonella henselae* frequently infects up to 20% of healthy cats in some environments and so detection of a positive test result in a clinically ill cat does not prove a disease association. In addition, for some agents the currently available PCR assays cannot discriminate between vaccine strains and field strains. For example, currently available PCR assays for feline herpesvirus 1 (FHV-1) and feline calicivirus (FCV) also amplify modified live vaccine strains, so a positive result does not indicate presence of a pathogenic strain. Real-time PCR can be used to determine the amount of microbial DNA in a sample. It is possible that the DNA load will correlate to the presence of disease for some agents. However, some agents are very host adapted and can have large amounts of DNA present in samples from healthy carrier cats. For example, the number of '*Candidatus* M. haemominutum' copy numbers/µl of blood does not correlate to the PCV. Based on these findings, it is very important that small animal practitioners carefully assess the predictive values of currently available PCR and the expertise and reliability of the laboratory that will be performing the assays. New PCR assays are being developed almost daily. The purpose of this lecture is to use several common infectious disease agents to emphasize important points concerning molecular assays in the diagnosis of feline infectious diseases.

Respiratory agents

FCV is a common differential diagnosis for cats with clinical evidence of rhinitis and stomatitis. Less commonly, FCV is associated with conjunctivitis, polyarthritis, and lower airway disease in kittens. Virus isolation can be used to document current infection but takes at least several days for results to return. Because of wide-spread exposure and vaccination, the positive predictive value of serological tests is poor. Reverse transcriptase (RT) PCR assays can be used to amplify the RNA of FCV and results can be returned quickly. However, these assays also amplify vaccine strains of FCV. FCV RNA can be amplified from samples collected from normal carrier cats as well as clinically ill cats and so have poor positive predictive value. For example, in one study in our laboratory, presence of FCV RNA failed to correlate to the presence or absence of stomatitis in cats (1). In addition, amplification of FCV RNA cannot be used to prove virulent systemic calicivirus infection. Results of FCV RT PCR can also be falsely negative and so can have poor negative predictive value.

FHV-1 is a common differential diagnosis for cats with clinical evidence of rhinitis, stomatitis, conjunctivitis, keratitis, and facial dermatitis. Because of wide-spread exposure and vaccination, the positive predictive value of serological tests is poor. FHV-1 can be documented by direct fluorescent staining of conjunctival scrapings, virus isolation, or PCR. FHV-1 DNA can be amplified from conjunctiva, nasal discharges, and pharynx of healthy cats and so the positive predictive value of conventional PCR assays is low (2). Currently used PCR assays also detect vaccine strains of FHV-1, further lessening the positive predictive value of the assays (3). In one study in our laboratory, presence of FHV-1 DNA failed to correlate to the presence or absence of stomatitis in cats. Quantitative PCR may ultimately prove to correlate to the presence or absence of disease but failed to correlate to presence of conjunctivitis in one study (4). The negative predictive value of FHV-1 PCR assays is also in question because many cats that are likely to have FHV-1 associated disease are negative. This may relate to clearance of FHV-1 DNA from tissues by a hypersensitivity reaction. Tissue biopsies have greater sensitivity than conjunctival swabs but do not necessarily have greater predictive value. FHV-1 DNA can be amplified from aqueous humor of some cats but whether this indicates FHV-1 associated uveitis is unknown.

Mycoplasma spp. *Chlamydophila felis*, and *Bordetella bronchiseptica* are other common respiratory pathogens in cats. As for FHV-1 and FCV, PCR positive test results for these organisms cannot be used to distinguish a carrier from a clinically ill cat. In addition, PCR assays do not provide antimicrobial drug susceptibility testing and so for cats with potential bordetellosis, culture and sensitivity is the optimal diagnostic technique, especially if an outbreak is occurring. *Toxoplasma gondii* DNA has been amplified from airway washings of some cats with lower respiratory tract disease and so PCR is an option for evaluation of samples from diseased animals from which the organism is not identified cytologically.

Gastrointestinal agents

The diagnosis of *Giardia* spp. infection is generally made with the combination of fecal flotation techniques and wet mount examination. Fecal antigen tests are also accurate and there are several assays available for point of care use, included one labeled for veterinary use (5). Fecal PCR assays are often falsely negative because of PCR inhibitors in stool and so PCR should not be used as a screening procedure for this agent. However, *Giardia* spp. PCR can be used to determine whether the infective species is a zoonotic assemblage which is the primary indication for this technique. However, it now appears that assemblage determination should be performed on more than one gene for most accurate results.

While *Cryptosporidium* spp. infection is common, it is unusual to find *C. felis* oocysts after fecal flotation in cats. Acid-fast staining of a thin fecal smear is cumbersome and insensitive. Antigen assays titrated for use with human feces are inaccurate when used with cat feces. Thus, PCR may be aid in the diagnosis of cryptosporidiosis in dogs and cats and has been shown to be more sensitive than IFA in cats (6). *Cryptosporidium* spp. PCR assays are indicated in IFA negative cats with unexplained small bowel diarrhea and when the genotype of *Cryptosporidium* is to be determined. However, *C. felis* infection in cats is common and so positive tests results do not always prove that the agent is the cause of the clinical disease. No drug is known to eliminate *Cryptosporidium* spp. infections and small animal strains are not considered significant zoonotic agents so PCR is never indicated in healthy animals.

PCR assays are also available for detection of DNA of *Tritrichomonas foetus, Salmonella* spp., *Campylobacter* spp., *Clostridium* spp., parvoviruses, and *T. gondii* and a RT-PCR assay is available for coronaviruses. Trophozoites of *T. foetus* can often be detected on wet mount examination of fresh feces which can be completed as an in clinic test. DNA of *T. foetus* can be detected in healthy carrier cats and so positive results do not always prove illness from the organism (7). Cases with suspected salmonellosis or campylobacteriosis should be cultured rather than assessed by PCR to determine the anti-microbial susceptibility patterns. In dogs, the PPV of *Clostridium* spp. PCR assays on feces is low and if used, should be combined with enterotoxin assays. Information in cats is currently lacking. There is no current evidence that parvovirus PCR on feces is superior to currently available antigen assays. *Toxoplasma gondii* is only shed for about 7-10 days and millions of oocysts are generally shed during this time making the organism very easy to identify. Thus, PCR assays are usually not needed to diagnosis this infection. Because virus isolation is not practical clinically, RT-PCR is used most frequently to detect coronaviruses RNA in feces. However, positive test results do not differentiate FIP inducing strains from enteric coronaviruses.

Blood borne agents

Mycoplasma haemofelis (Mhf), '*Candidatus* Mycoplasma haemominutum' (Mhm), and '*Candidatus* M. turicensis' (Mtc) all can be found in cats. In at least two studies of experimentally infected cats, Mhf is apparently more pathogenic than Mhm. It appears that Mtc has intermediate pathogenicity. Diagnosis is based on demonstration of the organism on the surface of erythrocytes on examination of a thin blood film or PCR assay. Organism numbers fluctuate and so blood film examination can be falsely negative up to 50% of the time. The organism may be difficult to find cytologically, particularly in the chronic phase. Thus, PCR assays are the tests of choice due to sensitivity (8). Primers are available that can amplify all three hemoplasmas. Real time PCR assays can be used to monitor copy numbers during and after treatment but do not have greater sensitivity, specificity, or predictive value than conventional PCR assays. PCR assays should be considered in the evaluation of cats with unexplained fever or anemia that are cytologically negative. In addition, the ACVIM recommends screening cats for use as blood donors by PCR assays for haemoplasmas (9). Many cats are carriers of the relatively non-pathogenic '*Candidatus* M. haemominutum' and so positive test results may not always correlate to the presence of disease (poor PPV).

Cats can be infected by *E. canis* like organism (10) and *Anaplasma phagocytophilum* (11). Little is known about the other agents in these genera in regards to cats. As the organisms are in different genera, serological cross reactivity is variable. Thus, while the clinical syndromes can be similar, there is no one serological test to document infection and there is currently no standardized serology for cats. In addition, some cats with *E. canis* infection do not seroconvert and so PCR assay is superior to serology in cats. PCR assays can be designed to amplify each organism. Alternately, primers are available to amplify all of the organisms in a single reaction and then sequencing can be used to determine the infective species.

Cats can be infected by *Rickettsia felis* and have been shown to have antibodies against *R. rickettsii*. Fever, headache, myalgia, and macular rash in humans have been attributed to *R. felis* infection in several countries around the world. In recent study in our laboratory, we assayed 92 pairs of cat blood and flea extracts from Alabama, Maryland and Texas, using PCR assays that amplify a region of the citrate synthase gene (gltA) and the outer membrane protein B gene (ompB). Of the 92 pairs, 62 of 92 (67.4%) flea extracts and none of the cat blood samples were positive for *R. felis* DNA (12). In another study, we showed *R. felis* and *R. rickettsii* antibody prevalence rates in cats with fever to be 5.6% and 6.6%, respectively but neither organism was amplified from blood (13).

These results prove that cats are sometimes exposed but further data are needed to determine significance of diseases associations. Whether *Rickettsia* spp. PCR assays are indicated for use in cats at this time is unknown.

Blood culture, PCR assay on blood, and serologic testing can be used to assess individual cats for *Bartonella* spp. infection. Cats that are culture-negative or PCR-negative and antibody-negative and cats that are culture-negative or PCR-negative and antibody-positive are probably not a source of flea, cat, or human infection. However, bacteremia can be intermittent and false-negative culture or PCR results can occur, limiting the predictive value of a single battery of tests. While serologic testing can be used to determine whether an individual cat has been exposed, both seropositive and seronegative cats can be bacteremic, limiting the diagnostic utility of serologic testing. Thus, testing healthy cats for *Bartonella* species infection is not currently recommended (14). Testing should be reserved for cats with suspected clinical bartonellosis. Because *Bartonella* spp. infection in cats is so common in healthy cats, even culture or PCR positive results does not prove clinical bartonellosis. For example, while we detected *Bartonella* spp. DNA in more cats with fever than pair matched cats without fever, the healthy cats were still commonly positive (15). I currently recommend combined serology with PCR in evaluation of cats with suspected bartonellosis.

Cytauxzoon felis in clinically affected cats is usually easily identified on cytological examination of blood smears or splenic aspirates. Serologic testing is not commercially available. PCR can be used to amplify organism DNA from blood from cats that are cytologically negative (16).

Antibodies against feline immunodeficiency virus (FIV) are detected in serum in clinical practice most frequently by enzyme-linked immunosorbent assay (ELISA). Comparisons between different tests have shown the results of most assays are comparable (17). Results of virus isolation or RT-PCR on blood are positive in some antibody-negative cats. False-positive reactions can occur using ELISA; hence, positive ELISA results in healthy or low-risk cats should be confirmed using Western blot immunoassay. Kittens can have detectable, colostrum-derived antibodies for several months. Kittens less than 6 months of age that are FIV seropositive should be tested every 60 days until the result is negative. If antibodies persist at 6 months of age, the kitten is likely infected. Virus isolation or RT-PCR on blood can also be performed to confirm infection. However, FIV is not present in the blood in high levels and so false negative results are common. Thus, the assay is not very accurate for distinguishing a vaccinated cat from a naturally exposed cat (18).

Most cats with feline leukemia virus infection are viremic and so molecular diagnostic assays are not usually needed in clinical practice. However, use of newer sensitive real time PCR assays have been used to accurately characterize the stages of infection (19). However, these assays are not commonly available commercially.

RNA of both FIPV and FECV can be amplified from the blood of cats and so positive test results do not always correlate with the development of FIP. Amplification of the mRNA of the M gene by RT-PCR had mixed results in two studies performed to date. In the one study, 13 of 26 apparently normal cats were positive for FECV mRNA in blood suggesting that the positive predictive value of this assay for the diagnosis of FIP was low (20).

Ocular agents

Toxoplasma gondii, *Bartonella* spp., FHV-1 and coronavirus are the organisms for which DNA or RNA has been amplified most frequently from the aqueous humor of cats with endogenous uveitis. While little is know about the predictive value of these assays when used with aqueous humor, the combination of molecular assays with local antibody production indices may aid in the diagnosis of some cases.

REFERENCES

[1]Quimby JM, et. al. **J Feline Med Surg**. 2008;10:66. [2]Veir JK, et. al. **J Feline Med Surg**. 2008;10:551. [3]Maggs DJ et. al. **Am J Vet Res.** 2005;66:1550. [4]Low HC, et. al. **Am J Vet Res.** 2007;68:643. [5]Mekaru SR, et. al. **J Vet Int Med** 2007;21:959. [6]Scorza AV, et. al. **J Parasitol** 2003;89:423-426. [7]Gookin JL, et. al. **J Clin Microbiol** 2004;42:2707. [8]Jensen WA, et. al. **Am J Vet Res** 2001;62:604. [9]Wardrop KJ, et. al. **J Vet Intern Med**. 2005;19:135. [10]Breitschwert EB, et. al. **J Vet Intern Med** 2002;16:642. [11]Lappin MR, et. al. **J Am Vet Med Assoc** 2004; 225:893. [12]Hawley JR, et. al. **J Feline Med Surg**. 2007 9:258-562. [13]Bayliss DB et. al. **J Feline Med Surg**. 2009;11:266-70. [14]Brunt J, et. al. **J Feline Med Surg**. 2006;8:213. [15]**Lappin MR** et. al. **J Fel Med Surg** 2009;11:141. [16]Haber MD, et. al. **Vet Parasitol.** 2007;146:316. [17]Hartmann K et. al: **J Feline Med Surg** 2007;9:439. [18]Crawford PC et. al. **J Am Vet Med Assoc** 2005;226:1503. [19]Levy J, et. al. **J Feline Med Surg**. 2008;10:300. [20]Can-S Ahna K et. Al. **J Feline Med Surg** 2007;9:369..

METABOLIC & CARDIAC DERANGEMENTS IN CANINE BABESIOSIS

Remo Lobetti BVSc (Hons) MMedVet (Med) PhD Dipl ECVIM
South Africa

Canine babesiosis is a tick-borne disease caused by the hemoprotozoan parasites of the genus *Babesia*. *Babesia canis* and *B. gibsoni* are responsible for canine babesiosis throughout the world. *Babesia canis* consists of a group of three different subspecies, namely *B. c. canis*, *B. c. vogeli*, and *B. c. rossi*. *Babesia gibsoni* is subdivided into 2 subspecies: the North American and Asian subspecies. Three other babesia species that can affect the dog are; (1) a small *Babesia* that is distantly related to *B. gibsoni*, but more closely related to *B. microti*, *B. rodhaini*, and *Theileria equi*; (2) a *B. microti*-like organism; and (3) *B. conradae*, which is a closely related piroplasm to isolates from wildlife and humans. Babesiosis primarily affects erythrocytes but can involve multiple organs and can range from a relatively mild to a fatal peracute disease. Canine babesiosis can be clinically classified as uncomplicated or complicated with uncomplicated babesiosis being further divided into mild, moderate, or severe disease, depending on the severity of anemia. Complicated babesiosis involves clinical manifestations that are unrelated to the hemolytic disease. Recently described complications of babesiosis are metabolic derangements and cardiac changes.

METABOLIC CHANGES

Carbohydrate

Abnormal carbohydrate metabolism appears to occur frequently in severe and complicated babesiosis as in one study, dogs with naturally occurring, severe and/or complicated babesiosis showed varying degrees of hypoglycemia, hyperglycemia, hyperlactatemia, and hyperpyruvatemia. In that study 3 of the 5 fatal cases showed concurrent hypoglycemia, hyperlactatemia and hyperpyruvatemia. In babesiosis a significant negative correlation between glucose and lactate but not between lactate and pyruvate has been reported, which is consistent with findings in malaria.

A wide range of lactate: pyruvate (L/P) ratios have been documented in babesiosis, due to variability in both lactate and pyruvate and the fact that they frequently do not change in consistent ways relative to one another. Variable pyruvate and L/P ratios have also been reported in other studies. Factors contributing significantly to variance in L/P ratio in severe adult malaria were creatinine, shock and hyperparasitemia. Normal or close-to-normal L/P ratios have been reported in malaria and sepsis.

Glucose

Hypoglycemia is a common complication of canine babesiosis with a general prevalence of 9% and increasing to 20% in hospitalized cases. In admitted cases, severe hypoglycemia (< 40mg/dl) was documented in approximately 6% of dogs. Risk factors for hypoglycemia are severe anemia, icterus, age (< 6 months of age), and vomiting. Toy breeds and pregnant bitches do not appear to be at a higher risk for hypoglycemia than other dogs. The importance of hypoglycemia is that it can result in neurological signs and thus be misdiagnosed as cerebral babesiosis, which carries a much poorer prognosis.

Hypoglycemia in babesiosis can be caused by increased consumption, decreased intake, decreased production and/or increased losses. Increased glucose consumption can result from hypermetabolism, anaerobic glycolysis, and consumption by the parasites. An increase in non-insulin-mediated glucose consumption can occur in sepsis and trauma. Inflammatory mediators such as TNF trigger increased glucose consumption, particularly in macrophage-rich tissues such as spleen, liver and lung; which then continues despite the hypoglycemic state. As canine babesiosis is a sepsis-like state, hypermetabolism could explain the hypoglycemia.

Hyperinsulinemia as a result of inappropriate insulin secretion may precipitate hypoglycemia, and has been suggested as a possible cause of hypoglycemia in human and murine malaria. In dogs with babesiosis induced hypoglycemia, it has been shown that insulin secretion was appropriately low in the majority of cases with only 2 dogs having inappropriately elevated insulin concentrations. Thus hyperinsulinemia appears to be a rare cause of hypoglycemia in canine babesiosis.

Anorexia has not typically been associated with hypoglycemia in dogs subjected to starvation or children with malaria. Suppression of gluconeogenesis has been suggested in malaria and experimental sepsis, but increased gluconeogenesis can also occur. Increased glycogenolysis and impaired glycogenesis causes depletion of hepatic glycogen in endotoxemic dogs. Hepatic dysfunction has been reported in canine babesiosis and could be associated with reduced gluconeogenesis. Hypoglycemia, however, is rarely associated with liver disease alone.

Lactate

Humans with severe disease often have persistent hyperlactatemia with blood lactate concentration correlated with the survival rate. Hyperlactatemia is an important finding in patients in septic shock and can be indicative of hypermetabolism (stress lactate), tissue hypoxia (shock lactate), or both. In dogs, blood lactate concentration has been shown to be of prognostic value in patients with gastric dilatation-volvulus and in dogs admitted to intensive care units. Serial blood lactate in dogs with naturally occurring, severe or complicated canine babesiosis has shown hyperlactatemia (> 2.5mmol/l) in 50% of dogs. Measurements significantly associated with mortality were blood lactate concentration >5mmol/l on admission; blood lactate concentration >2.5mmol/l at 8, 16, and 24 hours after admission; and increase or <50% decrease in blood lactate concentration within 8 and 16 hours after admission. Blood lactate concentration persistently >4.4mmol/l indicated a very poor prognosis. Serial blood lactate measurements are useful in predicting survival in dogs with severe and complicated canine babesiosis.

Pyruvate

Sepsis, bacteremia, febrile infections, and malaria can all result in elevated blood pyruvate levels and hyperpyruvatemia has been reported to occur in dogs with severe and complicated babesiosis. Although lactate production by malaria parasites has been suggested it is not considered quantitatively important. Hepatic dysfunction, potentially causing reduced lactate clearance, occurs in babesiosis, but organ dysfunction alone seems insufficient to explain elevated lactate in critical illness and sepsis. Impairment of pyruvate dehydrogenase activity could account for accumulation of pyruvate and hence lactate, but studies have reported both increased and decreased pyruvate oxidation.

CARDIAC CHANGES

Cardiac dysfunction in canine babesiosis has traditionally been regarded as a rare complication, with the majority of lesions being reported as incidental findings at autopsy. Recent studies have, however, demonstrated cardiac lesions in canine babesiosis that are of clinical importance.

Pro-BNP

Biochemical markers are being used more frequently in the diagnosis of myocardial diseases in animals with both atrial natriuretic peptide (ANP) and brain natriuretic peptide (BNP) being potentially useful markers. ANP is released from the atria in response to increased atrial stretch, whereas BNP is secreted by the ventricles in response to volume and pressure overload. In humans ANP and BNP concentrations are good indicators for the severity of left ventricular dysfunction that occurs with acute myocardial infarction and congestive heart failure. In veterinary medicine elevated ANP concentrations have been reported in dogs with mitral valve regurgitation, heartworm disease, and congestive heart failure. BNP is released as a precursor molecule – pro-BNP, which has a longer half-life in circulation than BNP. A pro-BNP value less than 210 pmol/l is considered negative for heart disease, between 210 and 300, considered probable for heart disease, and over 300, indicative of heart disease.

In dogs with babesiosis pro-BNP median values with mild uncomplicated, severe uncomplicated, and complicated babesiosis were 245.5 pmol/l (range 8.5-683.3), 650 (range 260-2028), and 638 (range 310-2214), respectively. These findings imply that dogs with babesiosis have reduced cardiac function and that the dysfunction becomes more severe as the disease severity increases. This myocardial dysfunction occurs either earlier and/or is independent of myocardial damage which can be more subtle.

Troponin

Cardiac troponins are thin, filament-associated proteins of cardiac muscle that act as the regulatory subunit of the troponin complex associated with the actin thin filament within the cardiac myocyte. Troponins are important cardiac biomarkers as they are highly sensitive markers for myocardial cell death. They are more sensitive and persistent indicators of cardiac injury, with high tissue specificity, than other markers even in the presence of marked skeletal muscle injury, liver disease, and renal disease. This is an important consideration in babesiosis, in which renal failure, liver impairment, and rhabdomyolysis can occur.

In canine babesiosis it has been shown that troponin I is a sensitive marker of myocardial injury with the magnitude of elevation of troponin I appearing to be directly proportional to the severity of the disease. Dogs dying of canine babesiosis with histological evidence of myocardial lesions had higher concentrations of troponin I than did dogs that survived.

Blood pressure

Dogs with severe and complicated babesiosis are frequently presented in a state of collapse and clinical shock, with the shock resembling the hyperdynamic phase of septic shock. Collapsed dogs with babesiosis may not display classic signs of shock syndrome, partially because of the hemodynamics of hemolytic anemia. The pulse may be bounding or weak; temperature elevated or subnormal; and mucous membranes pale, icteric or congested (hemoconcentration). Babesial shock, like endotoxic shock, may pass through a hyperdynamic stage followed by a hypotensive stage. It has been shown that hypotension occurs frequently in babesiosis and that the presence and severity of hypotension increases with increased disease severity. The presence of hypotension in a large proportion of dogs with complicated babesiosis is consistent with the hypothesis that inflammatory mechanisms play a major role in this disease, and can result in a sepsis-like state. It is likely that hypotension in babesiosis is due to a combination of vasodilation, reduced vascular volume due to increased vascular permeability and/or dehydration, and myocardial depression. Myocardial infarction can result in vascular pooling and myocardial depression. Thus, a feasible explanation for the hypotension in canine babesiosis would be myocardial pathology. In addition, hypotension can play a role in the pathophysiology of the disease as it has been hypothesized to facilitate parasite sequestration within the capillary beds.

ECG

Dogs with babesiosis can show a variety of arrhythmias and ECG abnormalities. Arrhythmias include SA block, sinus arrest, first and second degree AV block, ventricular tachycardia, and VPCs. Reported ECG abnormalities are low R-amplitude, prolonged QRS, ST deviation, large T waves, and notching of the R wave. These ECG changes have been described in dogs with myocardial infarction, ischemia and inflammation and in Chagas disease. Although bradycardia and irregular rhythms in dogs may be poor prognostic indicators, in general the ECG changes do not appear to be related to disease severity or to pathological changes and thus the clinical value of ECG monitoring in canine babesiosis is limited. The only exception to this is the presence of VPCs as they appear to be associated with an elevated troponin I concentration.

Cardiac pathology

Macroscopic cardiac lesions that have been described in dogs with babesiosis include pericardial effusion and pericardial, epicardial and endocardial hemorrhage, involving one or more of the chambers. On cardiac histopathology from dogs that succumbed to babesiosis, hemorrhage, necrosis, inflammation and fibrin micro-thrombi in the myocardium have been documented.

This myocardial pathology would result in the ECG changes described above as well elevations in troponin I. Myocardial damage results in left ventricular failure, which can result in hypotension and an expansion of the plasma volume due to homeostatic mechanisms.

REFERENCES

[1.] Dvir E, Lobetti RG, et al. Electrocardiographic changes and cardiac pathology in canine babesiosis. **J Vet Cardiol**. 6:15, 2004. [2.] Jacobson LS, Lobetti RG, et al. Blood pressure changes in dogs with babesiosis. **J South African Vet Assoc**. 71:14, 2000. [3.] Jacobson LS, Lobetti RG. Changes in glucose, lactate and pyruvate in canine babesiosis. **AJVR.** 66: 244, 2005. [4.] Keller N, Jacobson LS, et al. Prevalence and risk factors of hypoglycemia in virulent canine babesiosis. **JVIM**. 18:265, 2004. [5.] Lobetti RG, Dvir E, et al. Cardiac troponins in canine babesiosis. **JVIM**. 16: 63, 2002. [6.] Nel M, Lobetti RG, et al. Prognostic value of blood lactate, blood glucose and hematocrit in canine babesiosis. **JVIM**. 18: 471, 2004. [7.] Rees P, Schoeman JP. Plasma insulin concentrations in hypoglycaemic dogs with *Babesia canis rossi* infection. **Vet Parasitol.** 152:60, 2008.

IMMUNOLOGIC & CLINICAL ASPECTS OF FELINE SEPSIS

Amy E. DeClue, DVM, MS, DACVIM (SAIM)
Columbia, MO

INTRODUCTION

Sepsis is a Greek work meaning "decomposition of animal or vegetable organic matter in the presence of bacteria".[1] In the clinic, sepsis has come to mean infection with concurrent clinical evidence of the systemic inflammatory response (i.e., SIRS because of infection). At one time, sepsis and bacteremia were considered synonymous terms. However, as our knowledge of sepsis has grown we have realized that the clinical manifestations of sepsis relate to wide-spread inflammation and not necessarily wide-spread infection. There are several definitions pertaining to sepsis that you should be aware of:

- **Bacteremia** – the presence of viable bacteria in the blood stream
- **Systemic inflammatory response syndrome (SIRS)** – a clinical syndrome caused by systemic inflammation of infectious (i.e., sepsis) or non-infectious origin. In cats, the diagnosis of SIRS is based on fulfillment of at least 2 of 4 criteria: tachycardia, tachypnea, hypo-or hyperthermia, and either leukocytosis, leucopenia or bands neutrophils.
- **Sepsis** - the systemic inflammatory response to infection
- **Severe sepsis** – the systemic inflammatory response to infection associated with organ dysfunction and manifestations of hypoperfusion or hypotension.
- **Septic shock** – the systemic inflammatory response to infection with hypotension despite adequate fluid resuscitation along with the manifestations of hypoperfusion.
- **Multiple organ dysfunction syndrome (MODS)** – altered function of 2 or more organs secondary to SIRS such that homeostasis cannot be maintained without intervention.
- **Acute respiratory distress syndrome (ARDS)** – a pulmonary inflammatory disorder characterized by non-cardiogenic pulmonary edema, neutrophilic inflammation and hypoxemia.

IMMUNOLOGY

The sequence of events leading to sepsis is complex and incompletely understood. In the initial phases of in infection, microbial products induce systemic inflammation through activation of immune cells. The induction of systemic inflammation may start at a local site like an abscess in the subcutaneous space. For sepsis to develop from a local infection, inflammatory mediators and/or microbial products must enter into the systemic circulation and activate inflammatory cells throughout the body. Sepsis is not simply the induction of inflammation, but rather it is induction of proinflammatory components of the immune system to the extent that they overwhelm normal anti-inflammatory counter regulatory mechanisms allowing inflammation to go unchecked. Ultimately, the unchecked proinflammatory cascade leads to inflammatory cell infiltration, altered thermoregulation, vasodilation, vascular leakage, coagulation, hemodynamic instability and, in some cases, organ failure.

Since bacterial infection is the most common cause of sepsis in cats, we will focus on bacterial initiation of inflammation. Similar inflammatory pathways may be involved with other types of organisms as well. During sepsis, pathogen associated molecular patterns (PAMPs) like lipopolysaccharide (LPS, endotoxin) from gram negative bacteria or peptidoglycans from gram positive bacteria are released. PAMPs activate immune cells resulting in production of inflammatory mediators. For example, in the case of a gram negative bacterial infection, the lipid A portion of LPS first binds to LPS binding protein and then it is recognized via macrophage cell surface receptors like CD14. The main function of CD14, which lacks a transmembrane domain, is to transfer LPS to toll like receptor (TLR)-4 and MD-2 for subsequent cellular activation. Once LPS binds to these cell surface receptors, the macrophage becomes activated and intracellular signaling through activation of NF-kappa B is initiated. Activation and nuclear translocation of NF-kappa B results in the transcription of multiple inflammatory mediators that have been implicated in the induction and maintenance of sepsis. While there are differences in the mechanism of activation based on the microorganism involved, the resultant immune response is fairly consistent for most organisms. This may be one explanation for why, although there is a plethora of organisms that may induce sepsis, the inflammatory and clinical outcome is similar.

While there are many inflammatory mediators involved in sepsis, only a few have been evaluated in cats like tumor necrosis factor (TNF), interleukin (IL)-1β, IL-6, CXC chemokine ligand (CXCL)-8 (previously referred to as IL-8), IL-10 and leukotrienes. TNF and interleukin IL-1β are prominent in the early pro-inflammatory phase of sepsis while IL-6 and CXCL-8 are involved in maintenance of inflammation and IL-10 is important for the resolution of inflammation.

- Tumor necrosis factor alpha (TNF) is a potent mediator of systemic inflammation produced primarily by monocytes and macrophages. TNF activates multiple secondary cascades results in a prolonged inflammatory response. When given systemically, TNF administration results in several pathophysiologic changes seen in clinical sepsis including increased endothelial cell permeability, fever, hepatic production of acute phase proteins, neutrophil release from bone marrow and expression of endothelial leukocyte adhesion molecules. TNF is directly cytotoxic to some cells by induction of apoptosis and promotes the activity of Th cells, resulting in both humoral and cellular immune responses.
- IL-1β, predominantly produced by activated mononuclear phagocytes, is stimulated by bacterial products such as LPS and other pro-inflammatory cytokines such as TNF. IL-1β increases the expression of a wide variety of other pro-inflammatory mediators, cytokines, chemokines and adhesion molecules; stimulates neutrophil release from the bone marrow; induces

fever; and stimulates hepatic acute phase protein synthesis. IL-1β also has a prominent role in inducing pulmonary inflammation and hemodynamic derangement during sepsis.

- IL-6 production by monocytes/macrophages, lymphocytes and endothelial cells is stimulated by TNF, IL-1β and LPS. As with the aforementioned pro-inflammatory mediators IL-6 has multiple functions, including the induction of fever, hepatic synthesis of acute phase proteins and differentiation of B cells to produce immunoglobulins. IL-6 may also mediate gastrointestinal barrier dysfunction in shock and activate the coagulation system. However, IL-6 has some anti-inflammatory properties including like decreasing the activity of TNF and IL-1.
- CXCL-8 is produced by several cell types including macrophages, epithelial cells and endothelial cells. The main role of CXCL-8 is the recruitment and activation of neutrophils.
- IL-10 is expressed and secreted by a variety of cell types; predominantly T helper 2 (Th2) cells but also monocytes/macrophages and dendritic cells. IL-10 is a potent anti-inflammatory cytokine which down regulates the inflammatory response through various mechanisms. IL-10 acts to inhibit synthesis of many pro-inflammatory mediators (IFN-γ, IL-3, IL-2 and TNF); enhances production of IL-1ra by monocytes and blocks the effects of IFN-γ.

Pathogen associated molecular patterns from both gram negative and gram positive bacteria stimulate significant production of TNF, IL-1β and CXCL-8 from feline whole blood, *ex vivo.*[2] However, the type of PAMP stimulus alters the intensity of the inflammatory mediator response. Lipopolysaccharide and peptidoglycan are more potent inducers of TNF than lipoteichoic acid and LPS is a more potent inducer of IL-1β and CXCL-8 than either lipoteichoic acid or peptidoglycan in cats.[2] *In vivo,* Low dose endotoxin infusion results in the development of systemic inflammation demonstrated by fever, hypotension, and TNF, IL-6, CXCL-8 and IL-10 production in cats.[3] Fever develops within 4 hours of initiating endotoxin administration and cats have a biphasic hypotensive response at 1.5 and 5+ hours.[3] Peak TNF response is observed between 1-2 hours while peak IL-6 and CXCL-8 response is observed at 4 hours after initiation of endotoxin infusion.[3] In a clinical evaluation of feline sepsis performed at the University of Missouri, Comparative Internal Medicine Laboratory, cats with sepsis had significantly greater plasma TNF activity compared to control cats, but not cats with non-infectious forms of SIRS.[4] There were significantly more cats with detectable concentrations of IL-6 in the plasma than the SIRS or control group. There was no difference in IL-1β or CXCL-8 among groups. Both IL-1β and IL-6 were significantly and positively correlated with mortality in the sepsis group.

Cysteinyl-leukotrienes (cys-LT) are important mediators of inflammation and cardiovascular dysfunction during many disease processes including sepsis. Cysteinyl-leukotrienes have been implicated in the development of myocardial insufficiency, hypotension, acidemia in experimental models of feline gram negative sepsis.[5-8] Experimental cys-LT inhibitors preserve left ventricular function and alleviated hypotension and decreased cardiac output, in a model of feline sepsis.[5, 8] Zafirlukast, a cys-LT receptor antagonist, ameliorated the acute hypotensive response to low dose endotoxin infusion endotoxin in cats, but failed to significantly alter rectal temperature, heart rate or plasma TNF and IL-6 concentrations.[9]

Other immunologic changes associated with endotoxemia or sepsis in cats include altered circulating leukocyte number and histopathologic evidence of inflammation. Cats develop leucopenia and neutropenia with a nadir at 2 and 4 hours, respectively, after initiation of endotoxin infusion.[3] Lymphopenia has a slower in onset with the nadir at 6+ hours.[3] There is a rapid rebound of the leukocyte and neutrophil count with return to baseline by 6 hours. Leukocytosis, leucopenia, band neutrophils and neutrophils with toxic changes have all been reported in cats with naturally developing sepsis.[10, 11] The lungs are the most consistently affected tissue following endotoxin infusion in cats.[3] Patchy alveolar congestion, multifocal acute alveolar epithelial necrosis, small foci of neutrophils within the alveoli, as well as mild edema and moderate numbers of increased intra-capillary megakaryocytes are consistent histologic findings in the lung of cats post endotoxin administration.[3] Additionally, cats may develop mild lymphocyte apoptosis in the spleen and/or intestinal Peyer's patches.[3] A necropsy study of severe sepsis in cats demonstrated that multiorgan necrosis or inflammation was observed in all cats with the lung being the most commonly affected organ.[11]

CLINICAL FINDINGS

Sepsis has been associated with many conditions in the cat including septic peritonitis, hepatic abscessation, pyothorax, bacteremia, pneumonia, endocarditis, pyelonephritis, pyometra, viral, fungal protozoal, and parasitic infections.[11-15] Accurately describing the clinical findings associated with sepsis in cats is challenging since diagnostic criteria for sepsis in cats have not been prospectively evaluated. There are several diagnostic criteria that have been proposed ranging from simply the recognition of systemic consequences of infection to more elaborate diagnostic schemes requiring fulfillment of SIRS criteria and identification of infection. The most commonly used SIRS criteria (table 1) are based on a retrospective study of clinical and necropsy findings in 29 cats with sepsis.[11] While this was an important study and certainly added to our understanding of the clinical aspects of sepsis, caution should be used when applying criteria from a study with a 100% mortality rate. It is possible that these criteria exclude cats with less severe forms of sepsis. Nevertheless, many clinicians use these criteria to define systemic inflammation in the context of sepsis.

Table 1: SIRS Criteria[11]	
Temperature	>103.5°F or <100°F
Heart rate	>225 or <140
Respiratory rate	>40
White blood cell count	>19,500/ul or < 5,000/ul or >5% bands

The following data, which focuses on bacterial sepsis, has been compiled from a literature search using combinations of the following terms: feline, sepsis, septic peritonitis, bacteremia, abscess, pyometra, and pyothorax. The terms bacteremia, abscess, pyometra and pyothorax were selected because cats with those conditions are at risk for developing a systemic inflammatory response to infection. These data should be interpreted with caution as there are several important limitations to this analysis. First, the inclusion criteria for these studies were often focused on the source or type of infection (mainly bacterial) and criteria to demonstrate systemic inflammation (like SIRS criteria) were not always used. Thus, this information should be viewed as clinical findings in cats sepsis or cats with infections that are likely to result in sepsis. Additionally, these studies were retrospective in nature thus the information was often incomplete. Finally, each parameter was not evaluated by every study and some studies only reported group means, not the number of cats with each abnormality. In cases where the group mean was presented, data were not reported. Data are presented as the number of cats with the abnormality, the total number of cats with that particular parameter recorded and finally the percent of the population that had the finding of interest.

Table 2: Organism	Number of positive cultures	Percent
E. coli	55	23
Pasteurella	39	16
Streptococcus	17	7
Enterococcus	17	7
Staphylococcus	11	5
Enterobacter	7	3
Pseudomonas	4	2
Corynebacterium	4	2
Salmonella	3	1
Micrococcaceae	2	1
Acinetobacter	2	1

Table 3 Parameter	Number of cats affected	Total number of cats	Percent
Anorexia	141	198	71
Lethargy	142	200	71
GI signs	38	120	31
Weight loss	32	133	24

Table 4 Parameter	Number of cats affected	Total number of cats	Percent
Tachypnea	72	94	77
Pale MM	87	115	76
Abdominal pain	86	153	56
Fever	47	136	35
Hypothermia	52	189	28
Bradycardia	32	135	24

There were ten studies identified that contained clinical and/or clinicopathologic data.[4, 11-19] A total of 336 cats were included in this review. Gender was reported in 294 cats; approximately 60% of cases were male and 40% were female. Age was reported for 257 cats with an overall mean age of 4.58 years. Duration of hospitalization was reported for 135 cats with an overall mean of 3.14 days. Survival statistics were reported for 214 cats with an overall survival rate of 57.9%. The location of infection was identified as pleural space (n=131), peritoneum (95), blood (71), liver (14), systemic (7), endocardium (5), lung (5), kidney (3), subcutaneous space (2), bone (1), uterus (1) and CNS (1). The majority of cats had bacterial infections (328); 8 cats had other types of infection (FeLV (2), FIP (2), *Cytauxzoon felis* (2), FIV, *Toxoplasma*). Aerobic cultures were performed in 237/324 cats (73.1%). Some studies did not indicate if anaerobic cultures were also performed. For studies that reported growth of anerobic organisms, *Clostridium* (30), mixed anaerobes (25), *Bacteroides* (17), *Fusobacterium* (15) and *Actinomyces* (13) were the most commonly reported results. The most common organisms identified with aerobic bacterial culture are listed in table 2. The frequently reported clinical signs (table 3), physical examination findings (table 4) and clinicopatholgic abnormalities (table 5) are listed in their respective tables.

Table 5: Clinicopathologic parameters.

Parameter	Number of cats affected	Number of cats evaluated	Percent
Hypoalbuminemia	74	116	64
Acidemia	21	34	62
Band neutrophils	93	158	59
Hyperlactatemia	30	52	58
Hyponatremia	27	50	54
Hypocalcemia	36	73	49
Hyperbilirubinemia	56	121	46
Anemia	39	91	43
Leukocytosis	48	114	42
Increased BUN	31	77	40
Hyperglycemia	55	156	35
Increased ALT	28	116	24
Hypokalemia	12	50	24
Creatinine	17	77	22
Leukopenia	13	70	19
Hypoglycemia	25	137	18
Increased ALP	13	79	16
Hyperkalemia	6	50	12
Hypernatremia	1	50	2

REFERENCES

[1]Geroulanos S, et al. **Intensive Care Med** 2006;32:2077. [2]Stich A, et al. **J Vet Emergency and Crit Care** 2009. [3]DeClue A, et al. **Vet Immunol Immunopathology** 2009;132:167. [4]DeClue A. Unpublished data 2010. [5]Schutzer K, et al. **Circ Shock** 1989;29:13. [6]Schutzer K, et al. **Am Rev Resp Disease** 1993;147:1380. [7]Schutzer K, et al. **Europ Resp J** 1994;7:1131. [8]Pacitti N, et al. **Circ Shock** 1987;21(2):155. [9]Sharp C, et al. **J Vet Emergency and Crit Care** 2008. [10]Segev G, et al. **J Vet Intern Med** 2006;20(1):20. [11]Brady C, et al. J Am Vet Med Assoc 2000;217(531-5). [12]Sergeeff J, et al. JVIM 2004;18:295. [13]Costello M, et al. J Am Vet Med Assoc 2004;225:897. [14]Waddell L, et al. **JAVMA** 2002;221:819. [15]Barrs V, et al. **J Fel Med Surg** 2005;7:211. [16]Ruthrauff CM, et al. **J Am Anim Hosp Assoc** 2009;45(6):268. [17]Greiner M, et al. **J Feline Med Surg** 2007;9(5):404. [18]Demetriou JL, et al. **J Small Anim Pract** 2002;43(9):388. [19]Parsons KJ, et al. **J Small Anim Pract** 2009;50(10):518.

ADVANCES IN THE PREVENTION & MANAGEMENT OF *BORRELIA BURGDORFERI* INFECTIONS IN DOGS

Richard E Goldstein, DVM, DACVIM, DECVIM-CA
Ithaca, NY

INTRODUCTION

Lyme disease and its causative tick-borne agent *Borrelia burgdorferi* have become extremely common in many areas of the United States and Europe over the last 20 years in both humans and dogs (1). Anecdotally, it appears that 2009 stood out in terms of the emergence of high concentrations of Lyme-positive dogs in areas that previously were thought to have only occasional cases. Despite the large number of clinical and nonclinical Lyme cases seen in small animal practice in Lyme-endemic areas, the veterinary practitioner faces many unanswered questions regarding this disease. These include even the most basic question— how to treat a clinical case of Lyme disease or even how to recognize infection with *Borrelia burgdorferi* in a sick dog? Additional questions most practitioners struggle with, perhaps even more often, are if, when, and then how to monitor nonclinical dogs for evidence of *Borrelia burgdorferi* infection? And perhaps most important of all is how to prevent Lyme disease? How good is tick control? Should we or should we not vaccinate? Who should we vaccinate and do vaccines prevent or promote the most severe manifestation of Lyme disease— Lyme nephritis? After a brief overview of *Borrelia burgdorferi* pathogenesis necessary for the understanding of these issues, this discussion will focus on ways that veterinarians can answer these questions.

CLINICAL SIGNS OF LYME DISEASE

The lack or apparent lack of clinical signs in most dogs with active Lyme infection makes both the diagnosis and the study of this disease very difficult. In dogs, overt clinical signs are observed only in approximately 5% to 10% of infected cases (2). These signs tend to occur 2 to 5 months after the infection and include lameness (mono- or polyarthritis, **Figure 1**), lymphadenopathy, lethargy, and fever (3). Unlike in humans with acute infections, skin lesions are uncommon in dogs. The classic Lyme signs of arthritis, pain, fever, and lethargy typically resolve within approximately 3 days, in some cases only with antibiotic therapy (1). Some questions remain regarding more serious, less common syndromes that have been associated with Lyme infection in dogs including renal disease (Lyme nephritis), cardiac disease (myocarditis), and neurologic disease. Another question yet to be answered is whether some dogs get the devastating chronic recurrent disease seen in some infected humans.

DIAGNOSIS

Bacterial Culture or PCR Identification

This is very difficult in the case of *Borrelia* due to the small number of infecting organisms and the complicated techniques involved in their successful culturing.

Serology

Clinically we must rely on serology in conjunction with clinical signs to diagnose this disease. There are currently three types of serologic tests commercially available: nonspecific or kinetic enzyme-linked immunosorbent assay (ELISA), or KELA; C6 antibody testing; and Western blot.

Nonspecific or Kinetic ELISA (KELA)

This is a very sensitive test aimed at identifying any antibodies produced against *Borrelia* whole cell antigen. It does NOT differentiate between antibodies produced in reaction to Lyme infection versus Lyme vaccination and will be positive in both instances. Since we can usually not be sure of infection status and many times of vaccination status, a positive nonspecific ELISA should ideally be followed up with an additional test that would confirm infection, such as a Western blot (4) or a C6 antibody test (5).

C6 Antibody Testing

There are currently two commercially available tests for canine antibodies against the Lyme C6 peptide. This peptide is expressed only during infection; therefore, these tests are meant to be positive only in the event of natural exposure and negative in naive dogs or dogs vaccinated for Lyme (6). These tests include the in-house SNAP® 3Dx® or SNAP® 4Dx® Tests and the quantitative C6 antibody test available through IDEXX.

The SNAP 4Dx is an excellent test for screening dogs for Lyme infection. This can be done as part of a screening program for nonclinical dogs or when Lyme disease is suspected. A positive result is indicative of active Lyme infection (7).

The quantitative C6 antibody test checks for similar antibodies as the 4Dx but in a quantitative fashion. We still have a lot to learn about the value of knowing the quantitative C6 titer in all cases of Lyme infection, but we do know that the quantitative C6 titer does drop with therapy (8) and we have shown that the titer also correlated very well with circulating anti-Lyme immune complexes (9) and in the future will likely be a useful tool in treatment decisions of nonclinical dogs.

Western Blot

This technique involves a blot smeared with Lyme antigens located in known locations. This is a relatively expensive and labor-intensive type of test and its interpretation requires expertise, therefore this will never be available as in-house technology. As we have learned more about the 4Dx and quantitative C6 assays, they have taken over some of the role of the Western blot as a confirmatory assay. At this time I would recommend using the Western blot only in dogs where the vaccinal status is important to the veterinarian.

LYME PREVENTION

Tick Removal

It takes time for an infected tick to transmit *Borrelia* to a dog; this typically can happen no sooner than 48 hours from the beginning of the blood meal (1). Therefore, daily tick removal especially immediately after possible tick exposure is beneficial. People should handle possibly infected ticks with care though, since there is the potential for transmission to the person from the tick. Tick control is an important and indispensable part of any Lyme prevention strategy. A recent study demonstrated that tick control with fipronil was successful in preventing Lyme infection in dogs when exposed to infected ticks compared with non-treated control dogs (10).

Should We Vaccinate??

Whenever contemplating the use of a vaccine one needs to weigh the pros and cons. These are the considerations that, in my opinion, should be evaluated in such cases:

• How common is the pathogen? How common is the disease? How severe is the disease? How treatable is the disease? How good is the natural immunity? How expensive, reliable, and safe is the vaccine? Is there a zoonotic potential?

For canine Lyme disease vaccines:

• Common? The disease is very common. Even though most dogs are asymptomatic, the infection rate is so high in the Northeastern United States that it is a very common disease. Ten percent of infected dogs showing clinical signs out of 50% to 75% of the dogs in some areas is a lot!

• Severity? Lyme disease is not that bad in most cases. What we do not know is the prevalence and significance of more severe syndromes such as Lyme nephritis and chronic recurrent disease.

• Treatable? Lyme disease is treatable in most cases in terms of eliminating clinical signs. But in many dogs, based on experimental studies, we may never get rid of the bacteria even with antibiotic therapy (11). So perhaps Lyme disease is not that treatable. And then of course there is Lyme nephritis, which at best seems very hard to treat.

• Natural immunity? Not very good, because the immune response appears to dampen over time. This is likely a result of the bacteria "hiding" themselves from the immune system, mainly in tissue such as synovial membranes, and down-regulating their immunogenic surface proteins (12).

• How expensive, reliable, and safe is the vaccine? Three kinds of vaccines currently are commercially available. All appear to be relatively inexpensive. Two are whole cell bacterin based and the other a recombinant outer surface protein A (OspA) single antigen vaccine. There is very little published safety data. All three vaccines are approved for dogs and so had to go through safety testing. All three types of vaccine are likely to be effective. How do these vaccines work? One mechanism of action common to all three vaccines is via production of anti-OspA antibodies. These antibodies produced in the dog in response to vaccinal OspA antigen work in the gut of the tick to bind the bacteria during the blood meal (13), thus sterilizing the gut of the tick and preventing the transmission of the bacteria into the dog.

• Safety of Lyme vaccines? There are different aspects to this question: Regarding acute vaccine reactions, one could assume that all the dogs in the vaccine studies seemed fine so the vaccine must be safe. Obviously that is not a definitive answer. The vaccine studies are too small to pick up uncommon reactions, are too short in duration, are invariably conducted by the vaccine manufacturers, and many of these studies are never published. While safety studies that are much larger are conducted on every vaccine, these are typically looking for acute reactions and most often not published. The more pressing question is: Do vaccines contribute to clinical signs associated with Lyme disease or to the syndrome known as Lyme nephritis? There is a large amount of new and exciting data that attempts to answer this question, and it is a good question because we now know that Lyme nephritis is not caused by an inflammatory response to renal invasion of *Borrelia burgdorferi* organisms (13). Lyme nephritis is a condition associated with an accumulation of immune complexes in the kidney. We have shown that all Lyme-positive dogs have a measurable level of Lyme-specific circulating immune complexes in their blood (9). Recently we have also been able to purify renal immune complexes from kidneys of dogs that died of Lyme nephritis and to identify Lyme-specific antibodies present within those immune complexes. Studies are still ongoing regarding possible co-infections or other antigen–antibody complexes in those kidneys. Other evidence for the lack of a key role for vaccine-induced antibodies in Lyme nephritis is the fact that, despite looking very closely, we have yet to identify a dog with Lyme nephritis and solely vaccinal antibodies within the renal immune complexes. Vaccinal antibodies in the immune complexes are also fairly rare in dogs that have been infected as well as vaccinated against Lyme disease.

To answer the question directly, though, we have just completed a large study at Cornell assessing the possible effects of vaccination with either a whole cell bacterin or recombinant OspA vaccines on immune complex formation in dogs. Lyme-negative as well as Lyme-positive dogs were included in different groups in the study. Briefly, the study suggests that there is very little cause for concern when vaccinating Lyme-negative dogs. When vaccinating Lyme-positive dogs, there may be cause for some concern, especially if the C6 antibody titer is very high and the whole cell bacterin vaccine is being used. Alternative protocols involving treatment prior to vaccination and the use of recombinant vaccines in Lyme-positive dogs are currently being studied and are summarized below.

TREATMENT

I believe there is little question about treating Lyme-positive dogs with clinical signs. What should we use? There is no proven benefit in vaccinating a dog with clinical Lyme disease from a therapeutic standpoint and vaccination is **not** recommended. That leaves antibiotics: amoxicillin and doxycycline both work well and should be used for 4 to 6 weeks. Doxycycline will also likely be better at treating additional organisms that may have co-infected our patient along with Lyme disease and is therefore the therapy of choice (14).

The more difficult question is what about the nonclinical majority, that is, Lyme-positive dogs with no clinical signs? Why not treat them all? Although it is hard to look the concerned owner of a young Lyme-positive golden retriever in the eye and refuse to treat a possibly fatal disease, theoretically we should approach the treatment question from a risk/cost-benefit standpoint. At this time I am not sure we have enough information to make an informed choice in this matter. What do we know?

• Every Lyme-positive dog should be tested for proteinuria and if proteinuric (urine protein-creatinine ratio [UPC] > 0.5–1), these dogs should be treated as clinical cases. They should be given antibiotics and should not be vaccinated, at least not until after being treated.
• Lyme-positive dogs may remain infected with Lyme at low numbers with or without treatment (11).
• There is good evidence to show that treatment tends to lower Lyme titers faster than they would decline without treatment (especially C6 antibody titers)(8). This is also true for circulating immune complexes (15) and therefore may be true for bacterial load as well. Based on experimental and a growing number of field studies, IDEXX Laboratories is currently recommending treatment above a certain C6 quantitative titer and then rechecking titers 6 months after the initial positive test (or 5 months after the treatment has ended) to evaluate the success of the treatment.. The more we gain experience with this approach, and as more controlled field studies are published, the better this looks. We hope to have a more definitive answer to this question in the near future.

LYME NEPHRITIS

Current Recommendations for Screening and Treatment of Dogs with Suspected Lyme Nephritis

These recommendations are based on limited experience and theory, and not yet on strong clinical data (14):

• Monitor dogs in endemic areas for Lyme infection.
• Screen all positive dogs for signs of proteinuria or microalbuminuria.
• Screen all dogs that present with proteinuria or microalbuminuria for Lyme.
• Consider treating any dogs positive for both Lyme and proteinuria or microalbuminuria with 4 to 6 weeks of doxycycline.
• If proteinuria persists or worsens (based on urine protein/creatinine ratio):
  ○ Continue doxycycline
  ○ Consider a low-protein diet and an angiotensin-converting enzyme (ACE) inhibitor
  ○ Consider renal biopsy
  ○ If the renal biopsy is consistent with immune-mediated glomerulonephritis, consider immunosuppression with drugs such as mycophenolate, azathioprine, chlorambucil, cyclophosphamide or cyclosporine in consultation with a veterinary internist.

REFERENCES

1. Greene CE, Straubinger RK. Borreliosis. In: Greene CE (ed): **Infectious Diseases of the Dog and Cat**, 3rd ed. Philadelphia: WB Saunders Co, 2006, pp 417-435. 2. Levy SA, Magnarelli LA. Relationship between development of antibodies to *Borrelia burgdorferi* in dogs and the subsequent development of limb/joint borreliosis. **J Am Vet Med Assoc**. 1992;200:344–347. 3. Appel MJG, Allen S, Jacobson RH, et al. Experimental Lyme disease in dogs produces arthritis and persistent infection. **J Infect Dis**. 1993;167:651-664. 4. Lindenmayer J, Weber M, Bryant J, et al. Comparison of indirect immunofluorescent-antibody assay, enzyme-linked immunosorbent assay, and western immunoblot for the diagnosis of Lyme disease in dogs. **J Clin Microbiol**. 1990;28:92-96. 5. Levy S, O'Connor TP, Hanscom JL, et al. Utility of an in-office C_6 ELISA test kit for determination of infection status of dogs naturally exposed to *Borrelia burgdorferi*. **Vet Therap.** *2002;*3:308-315. 6. Liang FT, Jacobson RH, Straubinger RK, et al. Characterization of a *Borrelia burgdorferi* VlsE invariable region useful in canine Lyme disease serodiagnosis by enzyme-linked immunosorbent assay. **J Clin Microbiol**. 2000;38:4160-4166. 7. Bacon RF, Biggerstaff BJ, Schriefer ME, et al. Serodiagnosis of Lyme disease by kinetic enzyme-linked immunosorbent assay using recombinant VlsE1 or peptide antigens of *Borrelia burgdorferi* compared with 2-tiered testing using whole-cell lysates. **J Infect Dis.** 2003;187:1187-1199. 8. Levy SA, O'Connor TP, Hanscom JL, et al. Quantitative measurement of C6 antibody following treatment of *Borrelia burgdorferi* antibody-positive nonclinical dogs. **Clin Vaccine Immunol**. 2007;Nov 14 [Epub ahead of print]. 9. Goldstein RE, Atwater DZ. Serologic and circulating immune complex analysis in dogs naturally infected with *Borrelia burgdorferi* (abstract). **ACVIM Forum.** 2006. 10. Jacobson R, McCall, J, Hunter J, et al. The ability of fipronil to prevent transmission of *Borrelia burgdorferi*, the causative agent of Lyme disease to dogs. **J Appl Res Vet Med.** 2004;2:1. 11. Straubinger RK, Straubinger AF, Summers BA, et al. Status of *Borrelia burgdorferi* infection after antibiotic treatment and the effects of corticosteroids:An experimental study. **J Infect Dis**. 2000;181:1069-1081. 12. Straubinger RK. PCR-based quantification of *Borrelia burgdorferi* organisms in canine tissues over a 500-day postinfection period. **J Clin Microbiol**. 2000;38(6):2191-2199. 13. Hutton TA, Goldstein RE, Njaa BL, et al. The search for *Borrelia burgdorferi* in kidneys of dogs with suspected "Lyme nephritis." **J Vet Intern Med.** 2008;22(4):860-865. 14. Littman MP, Goldstein RE, Labato MA, et al. ACVIM small animal consensus statement on Lyme disease in dogs: diagnosis, treatment, and prevention. **J Vet Intern Med**. 2006;20(2):422-434. 15. Goldstein RE, Atwater DZ. Serology and circulating immune complexes in dogs naturally infected with *Borrelia burgdorferi* before and after doxycycline therapy (abstract). **ACVIM Forum.** 2006.

FELINE PANLEUKOPENIA – RECOGNITION AND MANAGEMENT OF ATYPICAL CASES

Katrin Hartmann, Prof., Dr. med. vet., Dr. habil., DECVIM
Munich, Germany

INTRODUCTION

Feline panleukopenia, caused by feline parvovirus (FPV), is a highly contagious disease affecting all members of *Felidae.*[1] Severity of clinical signs depends on age, immune status, and concurrent infections. Clinical disease ranges from subclinical infection to a peracute syndrome with sudden death. Typical initial signs include fever, depression, and anorexia.[2] Affected cats may initially present for vomiting and, with lower frequency, develop watery to hemorrhagic diarrhea. Patients die from complications associated with secondary bacterial infection, sepsis, dehydration, and disseminated intravasal coagulopathy (DIC). Mortality of 25 % - 90 % in acute panleukopenia and up to 100 % in peracute infections has been reported.

ETIOLOGY

FPV is a non-enveloped single-stranded desoxyribonucleic acid (DNA) virus which is highly resistant to physical influences and chemical substances. It is the prototype virus of a number of very closely related parvoviruses that can be isolated from various carnivores such as dogs, mink, raccoons, raccoon dogs, foxes, and other canids.[3] FPV is known to replicate efficiently in cats and other members of *Felidae*, as well as in raccoons, mink, and foxes.[4] In dogs, FPV replication is possible only in lymphoid tissues, such as thymus, spleen and bone marrow, but not in the gut. Infected dogs do not develop disease, and FPV is not shed from infected dogs.[5] In 1978, a very closely related new virus in dogs was detected. It was named canine parvovirus type 2 (CPV-2) to distinguish it from another parvovirus in dogs isolated in 1970, that is now called canine minute virus (CaMV). CPV-2 had derived from FPV by acquiring 5 or 6 amino acid mutations in the capsid protein gene[6] and crossing several species barriers. CPV-2 was not able to replicate in cats. However, during subsequent adaptation to the canine host, the amino acid changes that enabled the new virus to better bind to the canine receptor, also gave the virus the ability to infect and replicate in cats.[7] Thus, the canine parvoviruses, that are now circulating in the dog populations worldwide and that are called antigenic types CPV-2a, CPV-2b, and CPV-2c are able to infect cats and may also cause a disease in cats[8,9] that is not distinguishable from panleukopenia caused by FPV. CPV infections of cats still appear to be rare in Europe and the USA,[8] but seem to by common in Asia, e.g., in Taiwan.[10]

EPIDEMIOLOGY

Parvoviruses are shed in high amounts with feces. In shelters and catteries with infected cats, a high amount of virus can accumulate very quickly. In contaminated environments, parvoviruses may stay infectious for weeks, months, and up to more than a year. The virus is only inactivated by disinfectants that contain peracetic acid, formaldehyde, sodium hypochlorite, or sodium hydroxide. In addition, the virus is highly contagious. Therefore, only non-susceptible (i.e., successfully vaccinated) cats should be allowed to enter a contaminated environment.[1]

PATHOGENESIS

FPV causes a systemic infection. The virus is transmitted *via* fecal-oral route. It initially replicates in oropharynx and is then distributed *via* cell-free viremia to virtually all tissues. FPV efficiently replicates in all fastly deviding cells. It infects lymphoid tissues and causes cellular depletion. Lymphopenia may arise as a result of lymphocytolysis but also can be attributed to indirect effects, e.g., lymphocyte migration into tissues. It also replicates in bone marrow cells, including early progenitor cells and causes a significant decrease of virtually all myeloid cell populations. Both mechanisms lead to the marked panleukopenia observed in FPV-infected cats. Besides leucopenia, a typical feature of FPV infection is the shortening of the intestinal villi due to a sometimes complete loss of epithelial cells in the gut through FPV replication in the dividing cells of the epithelium, the "Lieberkühn crypts". Intrauterine transmission or perinatal infection may cause disorders of the central nervous system. A syndrome known as "feline ataxia" results from an impaired development of the cerebellum due to lytic virus replication in the Purkinje cells in infected kittens. Fetal infection may induce a form of tolerance to the virus so that kittens continue to shed virus for extended periods of time after birth. Fetuses infected between the day 35 and 45 of pregnancy have depressed T-lymphocyte-mediated immunity.

Antibodies play an important role in the protective immune response to FPV. Maternally derived antibodies (MDA) efficiently protect kittens from fatal infection. Passive immunity is replaced by an active response obtained by vaccination or mild natural infections. Natural or acquired immunity is solid and long lasting. MDA have a biological half-life of about ten days.[11] Most cats are believed to have maternal antibodies at protective titers until week 6 to 8. Kitten serum antibody titres dependent on the real intake of colostrums, and my be highly variable between kittens. Decrease of MDA leads to an "immunity gap" (in most kittens around 6 to 8 weeks of age), when the antibody level is too low to protect against natural infection, but still high enough to interfere with vaccination.[11,12]

CLINICAL SIGNS

The clinical signs in cats with panleukopenia are different from those that are seen in dogs with CPV infection. Hemorrhagic diarrhea is much less frequent in cats. In a recent study (that will be presented at the conference), hemorrhagic diarrhea was only present in 14.1% of cats with panleukopenia. Only 69.3% of cats with panleukopenia had diarrhea and 62.7% had vomiting. Some of these cats likely had a peracute form of the disease and died before gastrointestinal signs occured. However, 10.2% never developed diarrhea or vomiting despite a prolonged course of illness and had a favorable outcome in 54.2%. In addition, 34.2% of the cats never showed leukopenia during their disease, and 15 cats in the study neither had gastrointestinal signs nor leukopenia. In these cats, FPV infection was suspected either because of known contact with an infected cat, or the disease was diagnosed at necropsy. No clinical sign was significantly related to the outcome in the cats of this study. FPV infection leads to severy panleukopenia. It was shown in the study, that severity of neutropenia parallels that of clinical illness and risk to die, similar to what happens in dogs. Of the 137 cats in the study, 38.7% had lymphopenia. In contrast to neutropenia, lymphopenia was not associated with disease outcome. Lymphopenia with concomitant neutropeniapenia was observed more often than solitary lymphopenia (27.7 % *versus* 11.0%). The second most common observed blood cell abnormality was thrombocytopenia attributed to megakaryocyte destruction or increased consumption with DIC. In most cats, thrombocytopenia was not prominent. However, a lower platelet count was significantly correlated with a negative outcome. A low platelet count may also reflect severity and more advanced stage of disease due to severe viral megakaryocyte destruction, and thus, explain its role as a negative prognostic factor as detected. Anemia was not associated with negative outcome. Due to the relatively long life span of red blood cells, marked anemia is not common in FPV infection, unless intestinal blood loss is severe; however, a nonregenerative anemia can be seen with FPV. Mild anemia can be masked by severe dehydration. Hypoalbuminemia was the most frequent biochemical abnormality in the study, resulting from decreased intake and protein leakage into the gut due to severe mucosal lesions. An albumin level < 30 g/l at presentation was associated with a negative outcome. This can be explained by the association between hypoalbuminemia and decreased plasma colloid osmotic pressure, which diminishes effective perfusion at the capillary level, possibly leading to DIC, organ failure, and death. A potassium serum level < 4 mmol/l at presentation was also associated with a negative outcome. Hypokalemia can be explained by anorexia, vomiting, increased gastrointestinal potassium losses, fluid therapy, or possible refeeding syndrome and most likely reflects the severity of the enteritis.

DIAGNOSIS

FPV can be detected directly by isolation of the virus from blood or feces in cell culture. However, these methods are rarely used for routine diagnosis today. In practice, FPV antigen usually is detection in feces using commercially available enzyme-linked immunosorbent assays (ELISA) or latex agglutination or immunochromatographic tests. These tests have a relatively high sensitivity and specificity when compared to reference methods.[13] Tests marketed for the detection of FPV antigen as well as those detecting CPV may be used to look for FPV in feces.[13,14] The analytical sensitivity of the antigen tests can be compromised by the presence of antibodies in feces which may bind the viral epitopes and therefore render them inaccessible to the monoclonal antibodies of the test.[15] Specialised laboratories may offer virus detection using electronmicroscopy of fecal samples or PCR using whole blood or feces. PCR in blood is recommended in cats without diarrhea or when no fecal samples are available.

FPV antibodies can be detected by ELISA or indirect immunofluorescence. However, antibody tests have no value in the diagnosis of FPV. However, the presence of antibodies is proof of protection against FPV infection under field conditions.[16]

PROGNOSIS

A recent study (that will be presented at the conference) demonstrated that leukocyte and platelet counts, as well as albumin and potassium serum concentrations at time of presentation represent prognostic markers in cats with panleukopenia, whereas vaccination status, age, clinical signs, and housing conditions are not. The study showed that most affected cats were younger than 1 year, 56.7% were younger than 6 months. However, 25.3% were older than 1 year, and 10.7% even older than 5 years. This is important to realize in light of the thought that older cats are traditionally believed to be protected either by widespread vaccination or earlier subclinical infections. Reasons for the relatively high infection rate in older cats may include inadequate immunization and lack of virus exposure in the environment (e.g., high percentage of indoor cats resulting in a lower degree of immunity). The study was performed in Munich, a German city, in which many cats are kept strictly indoors. In the study, there was no significant correlation between age and severity of clinical signs and outcome. Based on the results, young cats are more susceptible to feline panleukopenia, but do not have an increased risk of death, which contradicts the idea that the highest morbidity and mortality occur in kittens up to 12 months of age. In the study, indoor cats were overrepresented compared to outdoor cats, and 14.5% had no contact with other cats indicating that indirect transmission is an important way of infection. Transmission can occur *via* contaminated clothing, cages, and insect vectors. There was no significant difference in vaccination status between in- and outdoor cats. Therefore, it can be assumed that natural contact with parvovirus is less likely in indoor cats, thus, their chance to build a protective immunity because of inapparent infection is decreased. Additionally, some cats were only vaccinated according to the "old" vaccination protocol (at weeks 8/9 and 12, but not 16). The lack of protective antibody titers in face of vaccination may be due to interference with maternal antibodies, which can last up to 20 weeks of age.[17] These findings support the need to follow the current guidelines of vaccinating at 6 weeks of age and then every 3 to 4 weeks until 16 weeks of age.[1,18] No significant correlation between vaccination status and outcome or duration of hospitalization was found in the study. However, none of the patients was completely vaccinated according to the new vaccination guidelines,[1,18] demonstrating the inefficiency of the "old" vaccination protocol.

TREATMENT

Restoration of fluid, electrolyte, and acid-base balance, preferably by intravenous fluids, is the most important aspects of symptomatic treatment. As the gut barrier often is destroyed in FPV-infected cats, intestinal bacteria may easily translocate into the blood stream. Bacteraemia may occur and, facilitated by the existing neutropenia, lead to sepsis in these immunocompromised patients. Thus, prevention of sepsis is essential. As bacteria usually derive from the gut, an antibiotic with a proven efficacy against gram-negative bacteria is recommended. Antibiotics should be administered parenterally. Oral intake of water and food should only be restricted if vomiting persists and should be continued as long as possible and restarted as early as possible. A highly digestible diet is preferred but if the cat does not tolerate this, any diet is better than no food intake at all. If persistent vomiting occurs, anti-emetics should be considered. Vitamin supplements, particularly of the B vitamin complex, can be given to prevent development of thiamine deficiency (which rarely occurs). Cats that develop hypoproteinemia may require plasma or whole blood transfusions to restore oncotic pressure. Plasma transfusions are also helpful to treat DIC (in combination with heparin), as they replace used anti-thrombin III and other important cytokines. In cats that are anorexic or show severe vomiting and/or diarrhoea, or in patients with persisting hypoproteinaemia, total or partial parenteral nutrition is required, preferably *via* a central venous catheter (jugular vein).

Anti-FPV serum (or antibody-containing commercial products) can be used to prevent infection in susceptible animals following exposure to FPV. No studies are available on the efficacy of anti-FPV serum in cats, although the therapeutic efficacy of serum has been demonstrated in a field study in dogs. A similar beneficial effect may be expected in cats. Interferon-ω has been shown to be effective in the treatment of CPV infections in dogs[19,20] and also inhibits the replication of FPV *in vitro*. However, only one study has been published on its use in cats with panleukopenia in which it was given in a prophylactic approach, but did not exhibit efficacy.[21]

VACCINATION

FPV vaccination is regarded as a core vaccine, and thus, every cat (independent of lifestyle and environment) needs to be protected against FPV.[1] Thus, vaccination against FPV is recommended for every cat, with exception of those cats that have recovered from panleukopenia or have had subclinical FPV infection. Earlier infections with FPV, and thus protection, may be determined by antibody detection. Even strictly indoor cats cannot necessarily avoid encountering FPV, since the virus is so stable in the environment and can easily be transmitted through fomites. The immunity induced by FPV vaccines usually leads to solid protection, and most vaccinated cats are completely protected from developing clinical disease. FPV vaccination also leads to good protection against CPV. Both modified-live virus (MLV) and adjuvanted inactivated virus vaccines are available for administration by injection, and both provide solid immunity against FPV.[1] Generally, a more rapid and more substantial protection may be afforded by using MLV vaccines. However, MLV vaccines should not be used in pregnant queens because of the risk of transfer of virus to the fetus and fetal damage (especially to the developing cerebellum). MLV FPV vaccines should also not be administered to kittens under 4 weeks of age (to avoid potential damage to the cerebellum which is still developing in young neonates).

Many kittens will be protected by MDA in the first weeks of life. However, without testing, the level of protection and the point at which a kitten will become susceptible to disease and/or become able to respond correctly to vaccination cannot be determined, as there is considerable inter-individual variation. In general, MDA will often wane to a point that will allow response to vaccination by 8 to 12 weeks of age, and an initial vaccination at 8 to 9 weeks of age followed by a second vaccination 3 to 4 weeks later has been recommended in earlier times. However, kittens with poor MDA may be vulnerable at an earlier age (and thus, in high risk environments, earlier first vaccination, e.g., at 6 weeks may be recommended), while other cats may possess so many MDA that they are incapable of responding to vaccination until sometime after 12 weeks of age (and therefore, vaccination at 16 weeks is recommended according to the new protocols). Following a primary course of vaccination, it is recommended that all cats receive a first booster no later than 12 months after the initial vaccine course, and subsequent booster vaccinations are given at intervals of 3 years or longer. Substantial data exists to suggest that cats, which respond to FPV vaccination, have solid immunity that lasts for several (7 or more) years even in the absence of any repeat vaccination or natural challenge.[1,22]

REFERENCES

[1]Truyen U, et al. **J Feline Med Surg** 2009;11:538. [2]Addie DD, et al. **Vet Rec** 1996;138:119. [3]Parrish CR. **Adv Virus Res** 1990;38:403. [4]Steinel A, et al. **J Wildlife Dis** 2001;37:594. [5]Truyen U, Parrish CR. **J Virol** 1992;66:5399. [6]Truyen U. **Vet Microbiol** 1999;69:47. [7]Hueffer K, Parrish CR. **Curr Opin Microbiol** 2003;6:392. [8]Truyen U, et al. **Virology** 1996;215:186. [9]Mochizuki M, et al. **J Clin Microbiol** 1996;34:2101. [10]Ikeda Y, et al. **J Virol** 1998;72:6932. [11]Scott F, et al. **J Am Vet Med Assoc** 1970;56:439. [12]Dawson S, et al. **J Fel Med Surg** 2001;3:17. [13]Neuerer FF, et al. **J Feline Med Surg** 2008;10:247. [14]Schunck B, et al. **J Virol Methods** 1995;55:427. [15]Lutz H, et al. **Vet Immunol Immunopathol** 1995;46:21. [16]Lappin MR, et al. **J Am Vet Med Assoc** 2002;220:38. [17]Addie DD, et al. **Vet Rec** 1998;142:353. [18]Richards JR, et al. **J Am Vet Med Assoc** 2006;229:1405. [19]De Mari K, et al. **Vet Rec** 2003;152:105. [20]Minagawa T, et al. **Vet Microbiol** 1999;69:51. [21]Paltrinieri S, et al. **Vet Immunol Immunopathol** 2007;118:68. [22]Gore TC, et al. **Vet Ther** 2006;7:213.

KEY WORDS

feline parvovirus, FPV, cat, clinical signs

IMPROVED SURVIVAL OF CATS WITH ACUTE CYTAUXZOONOSIS TREATED WITH ATOVAQUONE AND AZITHROMYCIN AS COMPARED TO IMIDOCARB DIPROPIONATE. LA Cohn[1], AJ Birkenheuer[2], JD Brunker[3], ER Ratcliff[4]. 1.University of Missouri, Columbia, MO. 2. North Carolina State University, Raleigh, NC 3.Oklahoma State University, Stillwater, OK, 4. Fair Grove Veterinary Services, Fair Grove MO.

Cytauxzoonosis is a frequently fatal tick transmitted protozoal disease of cats. Although no prospective studies support its efficacy, imidocarb dipropionate (ID) has been used as standard therapy for many years. Recently, a prospective study suggested that a combination of atovaquone and azithromycin (A&A) might have efficacy in the treatment of cytauxzoonosis. The purpose of this prospective, randomized study was to compare the efficacy of ID with that of A&A for the treatment of acute cytauxzoonosis in naturally infected cats.

Veterinary clinics in regions with endemic cytauxzoonosis were recruited for this study. A pre-determined order of randomized entry for cats into one of two (ID and A&A) treatment groups, drugs adequate to treat 3 cats, and a basic algorhythm for care including the dose regimen (ID – 3mg/kg IM, 15 minutes after administration of 0.05mg/kg atropine SC to be repeated in two weeks; A&A, atovaquone 15mg/kg PO q8h and azithromycin 10mg/kg PO q24h) were supplied to each clinic. Inclusion criteria were a clinical presentation suggestive of cytauxzoonosis and fully informed owner consent, while a temperature below 100°F or a moribund condition were exclusion criteria. Diagnosis was confirmed via PCR. All cats received heparin (200 U/kg SQ q 8h) and crystalloid fluids but additional treatments were left to the discretion of the attending veterinarian. The primary endpoint, survival, was compared via chi-square testing.

Sixteen private practices (n=81 cats) and two referral centers (n=15 cats) participated in the trial each enrolling between 1 and 20 cats. Of 96 cats enrolled, diagnosis was confirmed in 84. Four cats were excluded due to either failure to adhere to order of randomization or incomplete medical records. Twenty-six cats were treated with ID, while 54 received A&A. Gender, breed, and age did not differ between treatment groups. The most common presenting complaints for all cats were lethargy, anorexia, and prolapse of the third eyelid with duration of signs from 1 to 6 days. Fever was present in all but 5 cats, with a mean temperature of 104.5°F in both treatment groups. Initial PCV in the ID and A&A groups were 25.6±7.9%and 25.9±6.5%, respectively, while initial WBC counts were 4.5±2.7 and 5.4±2.9 K/µl. Seven of 26 cats (27%) treated with ID survived to hospital discharge, while 32 of 54 cats (59%) treated with A&A survived ($p= 0.01$). Of all the cats that died, 9 of 19 cats treated with ID and 16 of 22 cats treated with A&A died the day of or the day after hospital admission.

Cats which received supportive care plus A&A demonstrated improved survival to hospital discharge as compared with cats receiving supportive care and ID. A uniformly efficacious treatment may be impossible to find considering the rapid demise of many cats after presentation.

PATHOGENS ASSOCIATED WITH CANINE INFECTIOUS RESPIRATORY DISEASE COMPLEX (CIRDC) IN THE USA (FALL, 2009). Sanjay Kapil, OADDL, Stillwater, OK.

Canine infectious respiratory disease complex (CIRDC) in group housed dogs is clinically characterized by cough, nasal discharge and fever. The purpose of the study was to perform complete microbiological examination of CIRDC cases in the USA.

A total of 355 nasal swabs and 17 lung tissues were submitted for canine influenza A (CIV) RT-PCR; virus isolation on susceptible cell lines; and bacterial identification using biochemical tests and sensitivities using TREK diagnostic system.

Twenty-nine samples were positive for canine influenza A by CIV RT-PCR. Canine adenovirus-2 was detected in 2; canine herpesvirus in 3; canine respiratory coronavirus in none; and canine para influenza-2 in one specimen. Of the Culturette swabs, 28 were *Bordetella bronchiseptica*; 34 were *Streptococcus sp.*; 6 were *Pasteurella sp.*; and 12 were *Pseudomonas* positive. No bacteria were isolated from 174 respiratory specimens. Most of the bacterial pathogens (*Bordetella bronchiseptica*) from CIRDC were susceptible to Orbifloxicin *in vitro*.

There is no recently published complete microbiological study on the infectious agents associated with CIRDC in the USA. Canine influenza A, *Streptococcus sp.* and *Bordetella* were 3 major pathogens in CIRDC cases in the USA during fall, 2009. CIV was detected in upper- and lower-respiratory tract infections in susceptible dogs. The protocols for control, prevention and treatment of CIRDC cases need to be updated to cover the recent pathogens currently circulating the USA.

BIOLOGICAL VARIABILITY OF CANINE PANCREATIC LIPASE IMMUNOREACTIVITY AND C-REACTIVE PROTEIN CONCENTRATIONS IN CLINICALLY HEALTHY DOGS. PC Carney[1], CG Ruaux[1], JS Suchodolski[2], JM Steiner[2]. 1. Oregon State University College of Veterinary Medicine, Corvallis, OR. 2. GI Laboratory, Texas A&M University College of Veterinary Medicine, College Station, TX.

C-reactive protein (CRP) and canine pancreatic lipase immunoreactivity (cPLI) may serve as clinically useful indicators of generalized inflammation and pancreatic inflammation, respectively, but the magnitude of change over serial measurements in a given individual that is clinically meaningful (e.g., for the purposes of monitoring) is unknown.

This study investigated indices of analytical, inter- and intra-individual biological variability of CRP and cPLI in 12 clinically healthy dogs. Blood was collected daily for 7 days, weekly for 6 weeks, and monthly for 3 months. Owners were questioned about their dog's health status during the period between the current and previous blood collection, and temperature, respiratory rate, and heart rate (TPR) were recorded at each time point. Serum cPLI (Spec cPL®) and CRP concentrations were determined via commercially available, previously validated assays.

Three levels of analysis for outliers were carried out. Results identified as potential outliers were excluded if TPR or history indicated the possible presence of historical or present disease. One sample from a single dog and all samples from one dog were excluded on this basis; the latter dog had chronic inflammatory skin disease. Indices of biological variation were then calculated. All subsequent measurements were checked for correlation with the first sample concentration using Spearman's test. Components of biological variation were calculated using nested ANOVA.

For CRP, the inter-individual variability (CV_G) was 90.8%, the intra-individual variability (CV_I) was 115.5%, and the analytical variability (CV_A) was 6.3%; the index of individuality was 0.74, and the one-sided critical difference for sequential values ($p<0.05$) was 269.9%. For cPLI, CV_G=49.48%, CV_I=193.8%, and CV_A=8.4%, yielding an index of individuality of 0.24 and a one-sided critical difference for sequential values of 452.64%. Significant correlation ($p<0.05$) was likely between the first sample measurement and subsequent measurements for CRP over the short term (i.e., days), but uncommon with longer temporal separation. Subsequent cPLI values rarely correlated with the first sample. Applying one-sided critical difference values to means from each group, CRP concentrations > 9.36 mg/L and cPLI concentrations > 281 μg/L are likely to be physiologically significant.

Based upon these results, the use of a reference range for cPLI is appropriate, but subsequent measurements are unlikely to show good correlation with previous measurements; conversely, use of a reference range for CRP is of questionable utility, but agreement between repeated measures over the short term is expected. Large changes in serial measurements of cPLI, such as those observed clinically in dogs with pancreatitis, are required to be clinically meaningful in an individual, whereas more modest changes are likely to be meaningful for CRP.

EVALUATION OF DETEMIR INSULIN IN DIABETIC DOGS MANAGED WITH HOME BLOOD GLUCOSE MONITORING. SL Ford[1], JS Rand[2] TM Ghormley[1] HM Lynch[1] VCA Emergency Animal Hospital and Referral Center, San Diego, CA. 2. Centre for Companion Animal Health, School of Veterinary Science, University of Queensland, Australia.

There are no reported studies of short or long-term use of detemir in diabetic dogs. The aim of this study was to report outcomes using detemir and home blood glucose monitoring in predominantly long-term, poorly-controlled diabetic dogs.

Thirteen dogs diagnosed with diabetes mellitus were included in the study. Detemir was administered twice daily and a consistent diet fed. Initial insulin dose was 0.1-0.2U/kg/injection, which was then adjusted aiming to minimize clinical signs and achieve maximum blood glucose concentrations below 300 mg/dL measured using a portable blood glucose meter calibrated for canine blood. Owners performed an average of 4 +/- 2 blood glucose measurements per day in the stabilization period, and twice daily thereafter. Owners supplied documentation of daily insulin dosages, glucose concentrations and clinical data.

Prior to beginning detemir, most dogs (10/13) in the study were initially treated with another insulin (5/10 with biosynthetic human NPH [Humalin-NPH] and 5/10 with porcine lente [Vetsulin]) for a mean of 54 weeks (median 44 weeks), but failed to achieve glycemic control because of short duration of insulin action (6/10), or lack of blood glucose lowering effect (4/10). Three dogs (3/13) were newly diagnosed with highly motivated owners.

All dogs responded to detemir with improved glycemic control. In previously-treated, poorly-controlled dogs, mean blood glucose decreased from 383 to 279 mg/dL ($p<0.01$), fructosamine decreased from 598 to 373 umol/L ($p<0.05$), and clinical signs improved. In newly diagnosed dogs, mean blood glucose decreased from 391 mg/dl to 236 mg/dl; $p<0.01$). Dogs (10/13) continued on detemir to the end of the reporting period (median 10 months, range 4-24 months), or until death (3/10) from causes not directly related to diabetes (median 10 months; orthopedic disease, anal sac gland carcinoma, and IBD).

Half the dogs (54%, 7/13) were well regulated (mean of 2 daily pre-insulin blood glucose measurements ≤300 mg/dL), and the remaining dogs (46%, (7/13) were moderately-well regulated (mean ≤360 mg/dL). Biochemical hypoglycemia (< 60mg/dl) was detected in 1.9% of all blood glucose measurements and occurred on average 7.5 times/dog during the treatment period (median 10mths). Biochemical hypoglycemia was more common in small dogs and those with partial anorexia. Mean and median insulin dose in the last week in which data were available were 0.45 and 0.22 U/kg/injection.

In conclusion, detemir improved glycemic control in this small study of diabetic dogs and warrants further investigation for dogs which are inadequately controlled with other insulin. Home blood glucose monitoring is a valuable aid in glycemic regulation and assists in preventing severe clinical hypoglycemia. The potency of detemir limits its use in small dogs.

TITRATION OF NFE IN CANNED FORMULATIONS TO INDUCE KETOSIS AND WEIGHT LOSS IN OBESE CATS. Melendez L, Schoenherr WD, Hahn KA. Hill's Pet Nutrition Inc., Topeka, KS, USA.

Dietary-induced ketosis may alter a cat's metabolism to a state of simulated fasting even though adequate calories are being ingested. This study evaluated the ability of canned foods with varying levels of NFE (Nitrogen Free Extract; an estimate of soluble carbohydrate) to induce ketosis in obese cats and to determine the effects on weight loss.

32 obese cats (body condition score > 4; 5 point scale) were randomized to be fed one of 4 different canned formulations varying in levels of NFE (15.3, 17.1, 18.0, and 21.3%; DM) for 28 days. Food intake was restricted for the duration of the experiment to cause weight loss. Each cat received its daily food amount based on energy requirements of its initial body weight (kcal offered per day = 0.7 x (70 x initial body weight $(kg)^{.75}$)). The amount of food offered daily to each animal was calculated by dividing the amount of calories to be offered by the caloric density of the food (kcal/kg). Use of this equation was estimated to allow animals to lose body weight at a rate of 0.5 to 1.5% of their initial body weight per week.

The foods were effective for weight loss in the obese cats. Cats lost an average of 0.465 kg of weight at a rate of 1.75% of initial body weight per week. At two weeks, serum β-hydroxybutyrate (BHBA) was significantly elevated ($P<.05$) for all treatments and non-esterified fatty acids (NEFA) significantly increased ($P<.05$) in three of the four treatments when compared with the initial values. At the completion of the 28-day study, cats fed the 15.3 and 18.0% NFE foods continued to have elevated BHBA ($P<.05$). NEFA returned to the normal levels for all treatments on day 28 when compared with the initial values. Serum biochemical values (e.g., glucose, urea, creatinine, albumin, triglyceride, cholesterol) were within the normal reference ranges throughout this study for all foods.

These data indicate that at 14 days, cats shifted their metabolism from using carbohydrates to fat as an energy source and cats fed a food having >18% NFE cannot maintain the mild ketosis. This study demonstrates that obese cats fed a high protein, low carbohydrate, moderate fiber dry food lose weight at effective levels and dietary ketosis was induced and maintained when dietary NFE levels were at or below 18%.

LIFESPAN ANALYSIS OF METABALOMIC PROFILES IN CATS. S Yu and SC Zicker. Pet Nutrition Center, Hill's Pet Nutrition, Topeka, KS.

The metabolome is a full complement of molecules of low molecular weight present in cells, or biological fluids, in a particular physiological or developmental state. It is a fingerprint small molecules of an animal's metabolism at a specific point in time. To better understand temporal relationships of metabolic changes it is necessary to look at multiple timepoints across different ages. In this study 96 apparently healthy American Shorthair domestic cats (male = 29, female = 67) with ages ranging from 1.7 to 15.9 years were fed a nutritionally complete and balanced dry cat food for 3 weeks prior to sample acquisition. Healthy cats were defined by a normal physical examination by a licensed veterinarian, normal serum chemistry profile, normal complete blood count, and normal concentrations of serum thyroxine within three months prior to initiation of the study. All cats were immunized against feline panleukopenia, calici and herpes viruses, and rabies. Blood samples were drawn from all animals after overnight with-holding of food and plasma was isolated for metabolite analysis. Plasma was analyzed by a proprietary extraction and analysis system that utilized both GC-MS and LC-MS techniques. Metabolites were plotted against age and analyzed statistically for effects of age, sex, and their interactions. There was a significant effect of age ($p < 0.05$) for 3-methylhistidine, arginine, carnosine, gamma-gluatmyltyrosine, glutathione, isovalerylcarnitine, oleoylcarnitine, proline-hydroxy-proline, proline, serotonin, beta-sitosterol, glycine, myo-inositol, and pyruvate. Significant effects of gender ($p < 0.05$) were found for isovalerylcarnitine and proline-hydroxy-proline. Significant interactions of age and sex ($p < 0.05$) were present for gamma-glutamylphenylalanine, gamma-glutamyltyrosine, glutathione, isovalerylcarnitine, oleoylcarnitine, serotonin, and tyrosine. General interpretations of these changes suggest a decrease with age in muscle and bone turnover (3-methylhistidine, proline-hydroxyproline), antioxidant capacity (glutathione, carnosine), and possibly immune function (arginine). A shift in metabolism with age to possible increased fatty acid metabolism is suggested by increased acylcarnitines (isovalerylcarnitine,oleoyl carnitine). Increased concentrations of plasma pyruvate have also been associated with increased reliance on fatty acid metabolism as well as inflammatory conditions of the intestine. Serotonin displayed an interesting age effect with almost no variation from 1-5 years of age and a sudden increased variation from 6-16 years of age. The physiologic implications of this remain unknown but may indicate an age related disturbance in serotonin homeostasis. Finally, myo-inositol has been shown to increase with uremia and also may impart neuropathies when exogenously administered to animals. Metabolites that changed with age may impact biological function and also be used to develop biomarkers for aging processes. Interpretation of results will need to be evaluated against nutritional and disease factors to differentiate normal aging from sub-clinical disease. However, metabolomic profiling did prove useful in developing targets for hypothesis driven studies.

URINARY TRACT CATHETERIZATION & NOSOCOMIAL INFECTIONS IN DOGS & CATS

Jeanne A. Barsanti, DVM, DACVIM (Specialty of Small Animal Internal Medicine)
Athens, GA

INTRODUCTION

One important iatrogenic cause for UTI is urinary tract catheterization, both catheters placed through the urethra and cystostomy tubes. Dogs and cats that develop UTI during hospitalization are over 100 times more likely to have had a urinary catheter and these animals are hospitalized longer than those without UTI.[1] Twenty one per cent of dogs with intervertebral disc disease requiring bladder management during hospitalization developed UTI, as opposed to 0% of dogs requiring surgery that did not require bladder management.[2]

PATHOPHYSIOLOGY

Most catheter-induced infections involve one bacterial species.[2,3,4] Bacteria, especially gram negative organisms such *E. coli, Enterobacter, Proteus*, and *Klebsiella*, establish periurethral colonization, and these bacteria can be inoculated into the bladder during catheterization.[2,5,6] The hospital environment can be a source of the infecting organism [7,8] and the organism causing UTI can be found in rectal flora prior to the urine.[8] In humans, lack of proper hand washing by health care personnel is largely responsible for transmission of nosocomial UTI in catheterized patients.[9,10] In people, the risk of UTI with intermittent catheterization is 1% to 3% per insertion, so if a catheter must be inserted several times a day, most people become infected after a few weeks.[9] In dogs with thoracolumbar disc herniation, each day of catheterization or manual expression increased risk of UTI by 1.3-1.5 times.[2,4] The risk of UTI with an indwelling catheter in dogs with a closed system increased by 20% for each year increase in age, 27% for each day increase in catheterization, and 454% with administration of antimicrobial drugs.[4] Three days or more of indwelling urinary catheterization increases risk of UTI in both dogs and cats.[8,11] Most dogs and cats with cystostomy tubes also develop UTI with one or multiple organisms.[3]

In men and women, intermittent catheterization is less conducive to infection than indwelling catheters.[9,10,12] This may also be true for male dogs but is less likely to be true for female dogs because the urethral orifice is intravestibular. With indwelling catheters, gender does not affect risk for UTI in dogs and cats.[2,4,11,13,14] Bacteria can ascend either around or through the catheter. Closed, sterile systems can prevent bacterial access within the catheter lumen, which is the most rapid route of entry, but it is difficult to prevent access along the extraluminal side of the catheter.[15-17] Once bacteria gain access to the urinary tract, the level of bacteriuria increases rapidly within 24-48 hours.[10] This is in contrast to introducing bacteria into healthy, noncatheterized people or animals in which introduction of bacteria into the bladder rarely leads to a UTI.[10,18] Factors involved in the increased risk of UTI in catheterized individuals include presence of residual urine, damage to the bladder wall through overdistention, the presence of the catheter, and the presence of biofilm on the catheter.[10]

Biofilm formation on the catheter is important in the development of infection when catheters are indwelling. A biofilm can be deposited within hours of catheter insertion.[19] A biofilm is not static, but composed of multiple species of bacteria, their secreted matrix, and components from urine.[10] The first step in formation of a biofilm on a urinary catheter is deposition of urine proteins, electrolytes, and other organic materials.[10] Tamm-Horsfall glycoprotein facilitates attachment of uropathogens to the catheter via hairlike projections on *E. coli* and related gram negative organisms.[10] Organisms form colonies and secrete a polysaccharide matrix.[10] Slow urine flow predisposes to bacterial colonization along the catheter. Some organisms detach and become free floating. Organisms in biofilms are poorly susceptible to antimicrobial agents due to poor diffusion into the biofilm, local microenvironment, lack of host defense mechanisms, and enhanced microbial resistance due to bacterial proximity in colonies.[10] Once an infection become established on a urinary catheter, it cannot be treated successfully and the catheter must be removed.

TRANSURETHRAL CATHETERIZATION

Catheterization should always be performed as aseptically and gently as possible. A sterile, disposable catheter should be used as infection has been documented in normal dogs when clean, but not sterile, catheters are used.[6] The catheter should be passed using only sterile instruments or gloves. This method of catheterization rarely introduces bacteria into urine in male dogs, and only in low numbers(less than 10^3 per milliliter).[18,20] In contrast, catheterization of female dogs, even when being vigilant about using aseptic technique, introduces bacteria approximately 50% of the time; occasionally, large numbers (10^5 per milliliter) of organisms are introduced.[20] Because of the difficulty in distinguishing infection from contamination and because of the small risk of inducing UTI, catheterization is not an adequate substitute for cystocentesis for urine collection in female dogs. In normal cats, catheterized samples have fewer than 10^3 bacteria per milliliter in males and females.[21] However, unsedated cats may be difficult to catheterize without causing trauma.

Previously used IV fluid bags that never contained dextrose and that were stored cleanly for 0-17 days (median of 3 days) remained sterile.[22] Such bags can thus be safely used as urine collection reservoirs with indwelling urinary catheters.

CONSEQUENCES OF CATHETER INDUCED UTI

Potential complications are pyelonephritis, renal abscesses, prostatitis and epididymits.[6] UTI with organisms that alkalinize the urine can result in struvite urolithiasis. UTI are associated with endocarditis[23] and diskospondylitis, through bacteremia.[24,25] Urosepsis refers to bacteria, most commonly gram negative, infiltrating the systemic circulation from a urinary tract source. In people this is most commonly associated with urinary tract obstruction or iatrogenic manipulation, typically catheterization, in a critical care setting.[26] Of people with catheter-associated UTIs, 1% to 4% developed bacteremia, and fatality rates as a result varied from 13% to

30%, depending on the study.[19,27]

CLINICAL SIGNS

Many catheter-induced infections are asymptomatic.[2,3] Potential signs are hematuria, dysuria, cloudy urine, and foul smelling urine. Signs of septicemia are possible.

DIAGNOSIS

Cystocentesis samples are preferred for culture to document UTI. However, when an indwelling catheter or cystostomy tube is in place, samples may be obtained through the catheter or tube for quantitative culture. Low numbers of organisms in samples obtained from indwelling urinary catheters usually progress within days to high concentrations of bacteria unless the catheter is removed.[9] Culture of the tip of an indwelling urinary catheter on its removal should not be used because of poor positive predictive value.[11] It is important to culture urine either on the day of indwelling catheter removal or 24 hours after.[28]

TREATMENT

Treatment of UTI developing during indwelling urinary catheterization should be delayed until the catheter is removed unless systemic signs of infection develop. In evaluating culture results when the urinary catheter is removed, any bacteria isolated should be considered significant as long as the sample was collected from the catheter with aseptic technique or by cystocentesis. Treatment should continue for 10 days, with reculture of urine approximately 1 week after therapy is discontinued.

Antimicrobial resistance is a potential problem in treating catheter induced UTI. Use of antibiotics prior to or during urinary catheterization increases the risk of antimicrobial resistance.[8] The urinary tract was the most common site of infection with multidrug resistant *E. coli* and *Enterobacter*, and most cases were associated with urinary catheterization and/or bladder dysfunction.[29] Organisms causing UTI in dogs with thoracolumbar disc disease were resistant to approximately 50% of drugs tested.[2] Antimicrobial resistance among *E. coli* causing UTI was higher in samples from academic institutions and animal hospitals in the US [30] as compared to animal clinics in Sweden.[31] This is probably related to the higher likelihood of complicated UTI in academic institutions and hospitals. Resistance to multiple drugs by *E. coli* varied with academic institution from 9-31% of strains.[32] Factors associated with percent resistant were gender (males), previous antimicrobial use (within 30 days), use of immunosuppressive drugs (within 30 days), days of hospitalization, presence of underlying diseases (especially those requiring urinary catheterization or causing inability to urinate), and surgery.[29,33]

Athough organisms isolated from catheter induced infections may be multidrug resistant, antimicrobials to which the organism is reported resistant may still be effective once the catheter is removed due to high urine concentrations of antimicrobials when renal function is normal. Most cases of MDR *E. coli* UTI resolved with treatment with amoxicillin/clavulanate (12.5-25 mg/kg PO q12hr for 20-30 days) or chloramphenicol (50 mg/kg PO q 8 h for 11-33 days).[29] Because of the problem of antibiotic resistant UTI, the effect of bacteriophages on uropathogenic *E. coli* isolated from dogs and cats has been investigated *in vitro*.[34] Bacteriophages are viruses that infect and kill bacteria. This study found that a group of 10 bacteriophages could lyse over 90% of UPEC. Whether these bacteriophages could be used safely and with efficacy *in vivo* remains to be determined.

PREVENTION OF UTI

Urinary Tract Manipulation

A single episode of catheterization rarely results in UTI and usually does not warrant antimicrobial therapy. To prevent inducing UTI in animals without infection undergoing more invasive urinary tract procedures such as cystoscopy, urethroscopy, contrast urethrocystography, electrodiagnostic procedures, urohydropulsion, and prostatic massage, short-term administration of a broad-spectrum antimicrobial may be helpful.[35] Antimicrobial use for 24 to 72 hours, beginning a few hours before the procedure, should be sufficient. Drugs to consider include amoxicillin-clavulanate, trimethoprim-sulfonamide, macrocrystalline nitrofurantoin, a first generation cephalosporin, or enrofloxacin.

With repeated, intermittent urinary catheterization, antibiotic prophylaxis is also not indicated. In children with intermittent catheterizations, antimicrobial prophylaxis was not associated with a lower rate of bacteriuria, reduction of periurethral colonization by bacteria, or a lower rate of symptomatic UTI.[36] In adults with intermittent catheterizations, the incidence of UTIs was reduced by bladder irrigation with a solution of neomycin and polymyxin (Neosporin G.U. Irrigant, Burroughs Wellcome, Research Triangle Park, N.C.) or by oral methenamine hippurate or nitrofurantoin.[12,37]

The incidence of infection in dogs that had intervertebral disc surgery was similar whether manual expression or intermittent catheterization was used.[2]

Indwelling Urethral or Cystic (Cystostomy Tube) Catheterization

The guiding rule should be to avoid indwelling catheterization if possible and when required, to terminate use as soon as possible. Appropriate indications for indwelling catheters are: 1) temporary relief of anatomic or functional obstruction; 2) urinary incontinence in patients with periurethral wounds; 3) urine output monitoring in critically ill patients; 4) in the preoperative and postoperative management of lower urinary tract trauma. After relief of urethral obstruction, an indwelling urinary catheter should be left in place only if one of the following 4 conditions is present: 1) the obstruction was relieved with difficulty; 2) the urine stream is weak and small after relief of obstruction or a large amount of crystalline debris is evident with bladder flushing; 3) the dog or cat is uremic; 4) detrusor dysfunction is present secondary to over distention.

The following factors are most important in preventing UTI with an indwelling urinary catheter: 1) aseptic catheter placement; 2) maintenance of a closed catheter system; 3) dependent drainage; 4) most importantly, minimization of the duration of catheterization.[10,14,38] It is recommended that individuals wear gloves when manipulating or emptying urine drainage bags and wash hands after contact with every patient.

Antimicrobial therapy should not be used prophylactically to avoid infection with resistant organisms. Prophylactic use of an antimicrobial may reduce infection rates with short-term use (less than 3 days) of a closed, indwelling catheter system. Although antibiotics may delay the onset of bacteriuria, they cannot prevent infection if the catheter remains in place, especially for 3 days or longer and one usually does not know upon placement of a catheter how long it will be required. Even with short-term catheterization, urinalysis, urine culture, and treatment after catheter removal may be preferable to prophylactic antibiotics. The risk of UTI with indwelling catheters was increased by 454% with administration of antimicrobial drugs in dogs with intervertebral disc surgery.[4]

Antimicrobial impregnated urinary catheters may prevent catheter associated UTI in some situations, but are expensive.[10] Instillation of antimicrobial agents into the bladder in people with indwelling catheters is of little benefit.[9,10] In general, the short contact time plus the necessity of disconnecting the closed catheter system to infuse the antimicrobial substance abrogates any beneficial effect. Placing antimicrobial agents into the drainage bag has not been found to be beneficial.[9,10] Placing lubricants or antibiotic creams at the urethral meatus was also of no benefit in humans.[9,10]

Glucocorticoids, even at anti-inflammatory doses, should be avoided in animals with indwelling urinary catheters. In cats with indwelling urinary catheters, 5 mg/day of oral prednisolone predisposed the cats to developing bacterial pyelonephritis and did not reduce catheter-associated inflammation.[39]

A newer way to prevent symptomatic UTI in people with indwelling urinary catheters is to inoculate such individuals with an *E. coli* strain associated with asymptomatic infection.[40] In 30 colonized people, the incidence of symptomatic UTI was markedly reduced compared with the incidence in people not inoculated with this organism. This suggests that the presence of asymptomatic UTI should be tolerated in dogs and cats with indwelling catheters or tubes as long as no systemic signs of infection are present.

PUBLIC HEALTH CONSIDERATIONS

Although there are no reports of human infections associated with catheter induced UTI in dogs and cats, there is evidence for sharing of uropathogenic E. coli between people and their pets.[41-43] Some virulence factors in *E.coli* strains isolated from dogs affected with UTI are identical to virulence factors identified in human urogenital infections,[44] even though the vast majority of canine, feline, and human *E. coli* isolates associated with UTI are different.[45] Within household sharing of *E. coli* clones occurred most commonly human to human and pet to pet, but can occur between pet and human.[46]

REFERENCES

[1]McDonell J, et al. **J Vet Int Med** 2008; 22:788.[2]Bubenik L, et al. **Vet Surg** 2008;37:791.[3] Beck AL, et al. **J Am Vet Med Assoc** 2007;230:1184.[4] Bubenik LJ, et al. **J Am Vet Med Assoc** 2007; 231:893.[5]Schlager TA, et al,. **Clin Infect Dis** 1999;28:346.[6] Thomas JE. **J Am Vet Med Assoc** 1979;174:705.[7] Sanchez S, et al. **J Clin Microbiol** 2002;40:3586.[8] Ogeer-Gyles J, et al. **J Am Vet Med Assoc** 2006;229:1584.[9]Saint S, et al. **Arch Intern Med** 1999;159:800.[10]Trautner BW, et al. **Arch Intern Med** 2004;164:842.[11] Smarick SD, et al. **J Am Vet Med Assoc** 2004;224:1936.[12] Stamm WE, et al. **N Engl J Med** 1993;329:1328.[13] Barsanti JA, et al. **J Am Vet Med Assoc** 1985;187:384.[14] Lippert AC, et al. **J Am Anim Hosp Assoc** 1988;24:627.[15] Nickel JC, et al. **Urology** 1992;39:93.[16] Riley DK, et al. **Am J Med** 1995;98:349.[17] Roberts JA, et al. **Urology** 1993;41:338.[18] Biertuempfel PH, et al. **J Am Vet Med Assoc** 1981;178:989.[19] Denstedt JD, et al. **J Endourol** 1998;12:493. [20]Comer KM, et al. **J Am Vet Med Assoc** 1981;179:891.[21] Lees GE, et al **J Am Vet Med Assoc** 1984;184:449.[22] Barrett M, et al. **J Am Anim Hosp Assoc** 2008;44:2.[23] Sykes JE, et al. **J Am Vet Med Assoc** 2006;228:1723.[24] Smith JD, et al.**J Am Anim Hosp Assoc** 1996;32:29.[25] Watson E, et al.Vet Radiol Ultrasound 1993;34:397.[26] Nickel JC. J Urol 2005;173:27.[27] Liu HL, et al. Am J Med 2005;118:14.[28] Hugonnard M, et al. **J Vet Int Med** 2008;22:1464.[29] Gibson JS, et al. **J Vet Int Med** 2008;22:844.[30]Boothe D, et al. J Vet Int Med 2006;20:760.[31]Hagman R, et. al. **Vet Rec** 2005;157:193.[32]Bothe D, et.al. **J Vet Int Med**2005;19:468.[33]Boothe D, et. al. J Vet Int Med 2005:19:468. [34]Freitag T, et al. Res Vet Sci 2008;85:1.[35] Bhatia NN, et al. **Urology** 1992;39:583.[36] Schlager TA, et al. **Clin Infect Dis** 1999;28:346.[37] Cunha BA. **Urology** 1988;32:67.[38] Warren JW. **Med Clin North Am** 1991;75:481.[39] Barsanti JA, et al. **J Vet Intern Med** 1992;6:64.[40] Darouiche RO, et al. Urology 2001;58:339.[41] Murray AC, et al. **Ann Intern Med** 2004;140:848.[42] Stenske KA, et al. **Am J Vet Res** 2009;70:1108.[43] Johnson JR, et. al. **J Clin Microbiol** 2008;46:4078.[44] Siqueira AK, et al **Res Vet Sci** 2009;86:206.[45] Freitag T, et al. **J Vet Int Med** 2004;18:437.[46] Johnson JR, et al. **J Infect Dis** 2008;197:218.

FOSFOMYCIN: AN ALTERNATIVE DRUG FOR TREATMENT OF E. COLI URINARY TRACT INFECTIONS?

Dawn Merton Boothe, DVM, PhD, DACVIM, DACVCP, Patrick Hubka BS
Auburn, AL

Multidrug-resistant pathogenic E. coli is an emerging national disease in veterinary companion animal medicine.

E. coli is an organism which responds to antimicrobial selection pressure by developing resistance (Miller 2002). Multidrug resistant *E. coli* has been documented as a cause of canine urinary tract infections (Cooke 2002), nosocomial infections in a veterinary teaching hospital (Sidjabat 2006), and in fecal isolates in ICU patients (Ogeer-Gyles 2006).We have previously reported in abstract form (Boothe, 2005a) a high incidence of multidrug-resistance (MDR) in *E. coli* uropathogens isolated from canine patients admitted to Auburn University's Veterinary Teaching Hospital (n=175). Overall *E.coli* resistance to 17 antimicrobials included in our standard antibiogram was 31%. More problematic, resistance approximated 40% to 50% toward first (amoxicillin, amoxicillin-clavulanic acid, cephalexin and sulfonamide-trimethoprim) and second choice (enrofloxacin) empirical antimicrobals. A similar pattern was found in other southern Veterinary Teaching Hospitals as well (Boothe 2005b). As a follow-up to these findings, our report in progress (initial presentation in abstract form, Smaha 2006), *Escherichia coli antimicrobial Resistance in Small Animals: The scope of the Problem,* we prospectively determined the susceptibility of canine E. *coli* pathogens (n=376) causing spontaneous disease from different geographical areas of the United States. Of these isolates, ,from the urinary tract, approximately 50% were characterized by MDR (Shaheen 2010, Boothe 2010). Based on this data, we have implemented a prevalence study, funded by the Morris Animal Foundation, that focuses on the phenotyping of *E. coli* associated with spontaneous disease in dogs and cats. Isolates are subjected to C&S using custom made plates (Trek Diagonistic®). To date, we have studied close to 1000 isolates. Approximately 75% of the isolates are from the urinary tract. **Figure 1** demonstrates the distribution of those isolates collected thus far expressing NDR (no drug resistance), SDR (single drug or drug class) , MDR (multiple; 2 to 4 drugs or drug classes), XDR, (extreme; 5 to 6) and ZDR (all 7 drug classes tested).

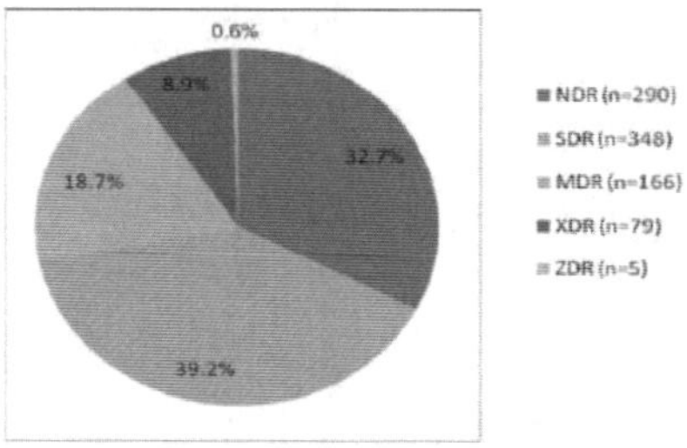

Figure 1. Number of feline or canine pathogenic *E. coli* expressing no (NDR), single (SDR), multiple (2-4 drug classes; MDR), extreme (5-6 drug classes, XDR) or total (all drug classes tested; ZDR) resistance.

Although MDR and XDR isolates largely retain susceptibility to aminoglycosides, treatment of these with aminoglycosides is costly, often requiring hospitalization. Further, aminoglycoside therapy is associated with the risk of nephrotoxicity. Although MDR E. coli continues to express susceptibility to nitrofurantoin, this drug also can be associated with undesirable side effects as well as potential adverse reactions (peripheral neuropathy).

The importance of integration of pharmacokinetic-pharmacodynamic parameters (from the species of interest) as the basis of antimicrobial therapy in the clinical patient is a widely accepted necessity among medical communities and is increasingly being applied to clinical medicine (Vinks 2009). Relevant pharmacokinetic parameters for concentration dependent drugs include maximum drug concentration (C_{max}) and for time dependent drugs, C_{max} and plasma and elimination half-life. Relevant population pharmacodynamic statistics include the MIC_{90}, which represents the 90th percentile within the range of MIC which characterize the isolate's susceptibility to the drug of interest.

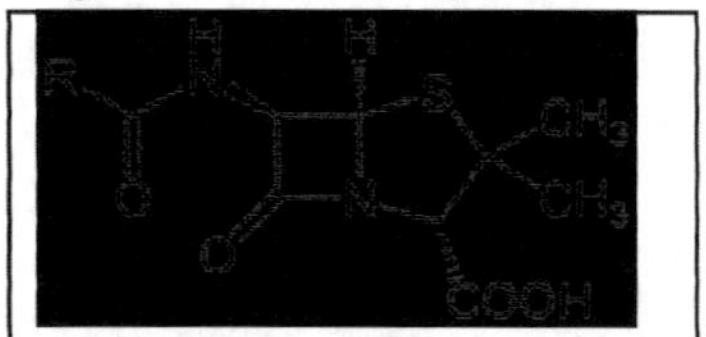

Beta lactam ring (red) of penicillin

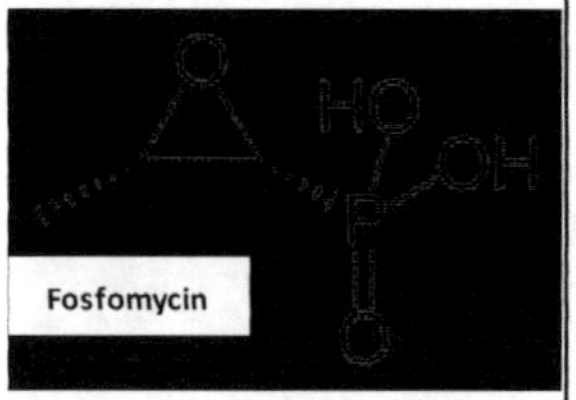

Figure 2. Structures of the cell wall inhibitors penicillin (top) and fosfomycin (bottom)

There is a need for a safe, convenient, effective alternative drug for treatment of canine multidrug resistant E. coli uropathogens.

The need is exemplified by three clinical cases presented to the author. 1. Shadow K, is a 6 yr old Scottish Terrier with a chief complaint of chronic pyoderma, who has been treated with antimicrobials (cephalexin, amoxicillin-clavulanic acid, or enrofloxacin at low to mid doses) and immunosuppressive therapy for 2 years. Her serum creatinine and blood urea nitrogen are 3.5 and 75 mg/dl, respectively. Three weeks ago, she developed clinical signs indicative of UTI and urinalysis reveals clumps of white blood cells; C&S reveals > 10^5 CFU/µl of MDR-*E. coli* susceptible only to amikacin. 2. Nanouk H. is a 12 year old female Malamute with a chief complaint of recurrent UTI, associated with poor vulvular confirmation. Her *E. coli* infection progressively developed resistance to amoxicillin, amoxicillin clavulanic acid, and most recently, enrofloxacin. Currently, the organism is susceptible only to nitrofurantoin and amikacin. Nanouk has undergone vulvoplasty and is now being re-presented for treatment. 3. Wyn Wyn L. is a 10 year old female mixed breed dog presented in her third year of treatment for pemphigus foliaceus. Immunosuppressive therapy includes glucocorticoids and cyclophosphamide. Urine C&S reveals 10,000 CFU/µl of MDR-*E. coli* susceptible only to amikacin. Although clinical signs are absent, a previous episode of septicemia has led both client and clinician to pursue treatment of the infection.

Fosfomycin tromethamine (Monurol®) offers a potentially safe, effective and convenient alternate treatment for these patients with MDR-E. coli UTI.

Fosfomycin, a phosphonic acid which contains a carbon-phosphorous bond, is a natural antibiotic produced by *Streptomyces fradiae* (**Fig 2**). Its *in vitro* spectrum is broad, with potential efficacy toward isolates expressing MDR, including *E. coli* and Gram positive isolates. Approved for human use in the USA, its indication is as a one time (or up to 3 days) treatment of *E. coli* UTI in humans (Schito 2003). Fosfomycin is a phosophonol pyruvate analogue that irreversibly inhibits phosphoenol pyruvate transferase, an enzyme that catalyzes the first step of cell wall peptidoglycan synthesis of microbial cell walls (Fillgrove 2007). Although its mechanism of action is similar to the beta-lactams, unlike the beta-lactams, fosfomycin is not susceptible to destruction by any class of beta-lactamases. Rather, resistance to fosfomycin, which is unusual, reflects the FosX or FosA enzyme which hydrolyzes the drug. The gene for this protein is chromosomally mediated. Thus, when resistance does occur, it is usually SDR, and generally not associated with MDR (Fillgrove 2007). Further, compared to susceptible strains, fosfomycin-resistant mutants are impaired, exhibiting poorer growth rates, and reduced adherence to uroepithelial cells (particularly in the presence of -acetylcystein (Marchese 2003a) and urinary catheters (Marchese 2003b)). As a cell wall inhibitor, fosfomycin is bactericidal when present at the site of infection at therapeutic concentrations. Cell wall inhibition indicates efficacy to be time dependent and facilitated if drug concentrations exceed the MIC of the infecting microbe for at least 50% of the dosing interval (T>MIC=50%). However, fosfomycin also exhibits a concentration-dependent effect (Kawaguchi 2002), which facilitates efficacy in the face of a short elimination half-life as long as concentrations are substantially above (eg, 10 X) the MIC of the infecting microbe. Studies in humans have demonstrated that fosfomycin distributes well to soft tissues, reaching therapeutic breakpoints (Sauermann 2007). Other attributes of fosfomycin that support its use for treatment of *E. coli* UTI include renal excretion, synergistic interaction with several other classes of antimicrobials (Olay 1974, 1989), and preparation as a 3 g sachet (granules) which is mixed with water to orally deliver approximately 40 mg/kg (in humans). Kill studies in our laboratory suggest that the efficacy of fosfomycin is not concentration dependent, suggesting, as might be expected of a cell-wall inhibitor, time-dependent effects. We anticipate that a week of treatment with Monurol® at 40 mg/kg bid will cost approximately $12 /kg for 1 week of therapy. Alternative therapies are based generally on injectable drugs (aminoglycosides, carbepenems) and thus hospitalization, and for aminoglycosides, intensive monitoring. Side effects of fosfomycin appear to be limited to diarrhea.

Role of fosfomycin in Canine UTI

This seminar will discuss the pharmacodynamics of fosfomycin and the kinetics after administration of the commercially available tromethamine salt . Based on these studies, fosfomycin tromethamine appears to be a drug which may reasonably be effective for treatment of canine UTI, including that caused by MDR *E. coli.* However, clinical trials are now indicated to determine the most appropriate dosing regimen (bid versus once daily) and the duration of therapy. Accordingly, until scientific evidence defines the most appropriate dose, interval and duration, the use of fosfomycin might be prudently reserved for treatment of MDRE. coli for which no other alternative exists and the risk of not treating the infection presents harm to the patient. Use of fosfomycin should be based on susceptibility testing only. Currently, susceptibility testing for fosfomycin is only available through e-testing at selected laboratories (including the Clinical Microbiology Laboratory at Auburn University). If the decision is made to use fosfomycin, reculture and supportive care are indicated (see Figure for a suggested algorithm of treatment of recurrent UTI). .

REFERENCES (Others Available Upon Request)

[1.] Boothe DM, Smaha TJ, Davis H, et al. Comparison of Gram Negative Antimicrobial Resistance Patterns at Four Universities. **Abstract**, Annual Forum, Amer Coll Vet Int Med, Baltimore, MD, June, 2005a. [2] Boothe DM, Smaha TJ, Hathcock TL, et al. Escherichia coli Resistance in Canine Urinary Tract. **Abstract,** Annual Forum, Amer Coll Vet Int Med,, Baltimore, MD, June, 2005b [3.] Boothe DM, Smaha T. Escherichia Coli Antimicrobial Resistance in Small Animals: The Scope of the Problem. **Abstract,** Annual Forum, Amer Coll Vet Int Med,, Louisville, KY.June, 2006b. [4.] Boothe DM. Principles of antimicrobial therapy.**Vet Clin North Am Small Anim Pract.** 2006 Sep;36(5):1003-47. [5.] Boothe DM, Smaha T, Shaheen BW. Antimicrobial susceptibility patterns of clinical isolates of *E. coli* isolated from dogs and cats in United States. **Submitted Veterinary Microbiology**, 2010. [6.] Cooke CL, Singer RS, Jang SS et al. Enroflxoacin resistance in *Eschericia coli* isolated from dogs with urinary tract infections. **JAVMA** 2002 220: 190-192. [7.] Fillgrove KL, Pakhomova S, Schaab MR, Newcomer ME, Armstrong RN. Structure and mechanism of the genomically encoded fosfomycin resistance protein,FosX, from Listeria monocytogenes.**Biochemistry**. 2007 Jul 10;46(27):8110-20. [8.] Kawaguchi K, Hasunuma R, Kikuchi S, Ryll R, Morikawa K, Kumazawa Y. Time- and dose-dependent effect of fosfomycin on suppression of infection-induced endotoxin shock in mice. **Biol Pharm Bull**. 2002 Dec;25(12):1658-61. [9.] Grif K, Dierich MP, Pfaller K, Miglioli PA, Allerberger F. In vitro activity of fosfomycin in combination with various antistaphylococcal substances.**J Antimicrob Chemother**. 2001 Aug;48(2):209-17. [10.] Gutierrez OL, Ocampo CL, Aguilera JR, Luna J, Sumano LH. Pharmacokinetics of disodium-fosfomycin in mongrel dogs.Res **Vet Sci**. 2008 Aug;85(1):156-61 [11.] Johnson, J. Evolultion of Pathogenic Escherichia coli. In: Donnenberg MS, (ed) Exherchia coli: Virulence Mechanisms of a Versatile Pathogen. **Elsevier Science, Academic Press**, San Diego. 2002, p 55-70. [12.] Kawaguchi K, Hasunuma R, Kikuchi S, Ryll R, Morikawa K, Kumazawa Y. Time- and dose-dependent effect of fosfomycin on suppression of infection-induced endotoxin shock in mice. **Biol Pharm Bull**. 2002 Dec;25(12):1658-61. [13.] Marchese A, Bozzolasco M, Gualco L, Debbia EA, Schito GC, Schito AM. Effect of fosfomycin alone and in combination with N-acetylcysteine on E. colibiofilms.**Int J Antimicrob Agents**. 2003a Oct;22 Suppl 2:95-100. [14.] Marchese A, Gualco L, Debbia EA, Schito GC, Schito AM. In vitro activity of fosfomycin against gram-negative urinary pathogens and the biological cost of fosfomycin resistance.**Int J Antimicrob Agents**. 2003b Oct;22 Suppl 2:53-9.

NUTRITIONAL MANAGEMENT OF ACUTE PANCREATITIS IN THE DOG & CAT

Dr Caroline Mansfield BSc BVMS MVM DECVIM-CA
Perth, Western Australia

BACKGROUND

Acute pancreatitis in dogs can be a challenging disease to manage, with a wide spectrum of severity. Many dogs will recover with minimal supportive care, whilst others will develop life-threatening complications and require intensive management. The occurrence of acute pancreatitis in cats is much less common than the chronic form, but it too is often associated with the development of life-threatening complications and the concurrent development of liver disease.

The systemic complications in acute pancreatitis are triggered by a complex chain of events starting with the actions of circulating pancreatic enzymes such as phospholipase, which are normally inhibited by circulating anti-proteases such as α-macroglobulin or α_1-protease inhibitor. When these circulating anti-proteases are exhausted the complement, kinin and fibrinogen cascades are activated with subsequent systemic abnormalities. The many cytokines activated with this systemic inflammatory disease have also been implicated in the deleterious effects. Increased concentrations of interleukin-6 (IL-6), IL-8, tumour necrosis factor alpha (TNF-α), C-reactive protein (C-RP) and poly-morphonuclear elastase (PMNe) have been shown to correlate with severity of disease in people[1].

Intestinal status is considered to be a major contributor to the inflammatory status in many conditions, including acute pancreatitis and to maintain intestinal health, intraluminal nutrition is considered ideal[2,3]. It has been shown that fasting leads to intestinal mucosal atrophy[4], increased rate of enterocyte apoptosis[5], decreased glutamine and arginine transport[6], a change in the mucin composition of goblet and deep crypt cells[7] and a breakdown in the intestinal barrier leading to increased intestinal permeability[8]. Canine experimental models have also shown decreased bacterial translocation with intra-jejunal feeding and amelioration of inflammatory response[9,10].

'Pancreatic rest' has been the cornerstone of nutritional management of pancreatitis in the dog, as well as in people until the past decade. The rationale behind this concept is that decreasing the pancreatic digestive activity aids in the resolution of inflammatory activity. Recent studies in people have shown that a decrease in protein digestive enzymes is the most important component of this rest, but the volume of pancreatic secretion is not as important. It has also been shown that reduction of protein digestive enzymes does not need to decrease below the basal level[11].

THE DEBATE ABOUT ENTERAL NUTRITION

In People

There exists a significant lack of consensus regarding nutritional recommendations in human acute pancreatitis despite a plethora of studies assessing the role of enteral nutrition[12,13]. Part of this discordance is due to the fact that the treatment groups are composed of different degrees of diseases severity. Another contributing factor is that most studies only compare interventional methods of feeding (jejunal, nasogastric or total parenteral nutrition) with each other, and not with fasting. Despite this lack of consensus there is agreement in meta-anlayses[14] that early naso-gastric feeding is well tolerated and not associated with significant complications[15]. Various studies have comparing early naso-gastric feeding to total parenteral nutrition have shown decreased hospitalization times, decreased severity, inflammatory responses and rate of septic complications in the feeding group[16,17].

In Dogs

There is a distinct lack of controlled studies comparing different nutritional modalities in dogs with acute pancreatitis. Contributors to this lack of research may include the challenges of definitive diagnosis, the unpredictable nature of the disease and a lack of ensuring dogs with similar severity are recruited to remove compounding bias. Recently, a clinical severity scoring index has been developed and published assessing what factors were most likely to contribute to mortality[18]. The single most significant negative prognostic factor was lack of enteral nutrition for 3 days or greater. It is quite possible that this factor is even more significant, as the numbers of days without enteral nutrition were only those that could be confirmed, and due to the retrospective nature of that study the days often only were counted upon admission to a veterinary clinic. The amount of 'self-fasting' that occurred prior to admission was not assessed.

Not only is there no consensus of whether it is necessary to fast or feed dogs with pancreatitis, there is also no consensus as to which diet to feed. Another study done at our institution evaluated the healthy pancreatic response to a variety of food types and determined that there was no statistical difference, but a trend towards less stimulation with fat restriction, addition of oral pancreatic enzyme extracts and medium chain triglycerides[19]. This study only assessed healthy dogs, and so may well not represent what is required in naturally occurring pancreatitis. Enteral administration of medium-chain triglycerides is thought to result in minimal CCK secretion in people, thereby reducing pancreatic stimulation, especially compared to liquid enteral diets that contain long-chain triglycerides.[20] Administration of supplemental pancreatic enzymes has been applied in the management of people and dogs with exocrine pancreatic insufficiency that is secondary to chronic pancreatitis, but did not appear to alter pancreatic adaptation in any of the healthy dogs assessed.[21,22]

Studies of inflammatory response in people show a therapeutic window of opportunity about 36 hours after the initial onset of abdominal pain[23]. It is considered that during this time, interventional measures aimed at reducing the cytokine cascade are most likely to be successful.

We have most recently finished assessing early interventional enteral feeding compared to total parenteral nutrition in dogs at our institution (unpublished study). 10 dogs of similar (severe) clinical severity, without concurrent diabetes ketoacidosis and weighing more than 10 kg bodyweight were included in the study. They were sequentially allocated to an enteral feeding group (esophasgotomy feeding of previously identified diet within 24 hours of admission) or a total parenteral nutrition group. All dogs received additional supportive care including plasma transfusion and micro-enteral nutrition. There was no difference in mortality or days of hospitalization between the two groups, but there was a quicker and greater reduction in clinical severity scores in the feeding group.

There were a greater number of more severe complications in the TPN group compared to the feeding group. There was also more regurgitation/vomiting episodes in the TPN group compared to the feeding group. This only confirms that TPN is not advantageous in acute pancreatitis and that early enteral (pre-pyloric) feeding is safe and well tolerated in dogs with severe pancreatitis. Further studies with large cohorts of dogs are necessary before making definitive recommendations regarding which dogs to feed, what to feed them and when to feed them.

In Cats

The amount of investigation into feline pancreatitis is even scarcer than the scarce canine studies. This may be attributable to historical difficulties in diagnosing this disease. However, there are some assumptions that may be made in cats and are probably sound until proven otherwise. The role of dietary fat appears to be less important in cats than in dogs, although a recent study suggests Australian Burmese cats are slower to clear triglycerides[24] following a high-lipid meal. The seriousness of concurrent disease often overrides any consideration of 'resting the pancreas' and early interventional feeding is often undertaken in cats to prevent or treat concurrent hepatic lipidosis. Interestingly, this does not seem deleterious to the resolution of the pancreatitis, and may serve as reminder that strict pancreatic rest is not always necessary in all species.

CONCLUSION

It would appear that early enteral nutrition is safe and well tolerated in dogs with severe pancreatitis, and total parenteral nutrition is not recommended. However, not all dogs with acute pancreatitis will require early enteral nutrition, and in the author's opinion this treatment modality should be reserved for the more severe cases or when there has been no enteral nutrition for greater than 3 days. Early consideration of nutritional requirements in cats with acute pancreatitis is vital to prevent development of life-threatening complications such as hepatic lipidosis.

REFERENCES

1. Al-Mofleh I. **World J Gastro** 2008;14:675-684 2. Flint RS et al.. **World J Gastro** 2003, 5(3)::69-85. 3. Niv E et al. **World J Gastro** 2009; 21:1281-1288. 4. Hernandez-Aranda JG-C et al. **Nutr Hosp** 1996;11:160-166. 5. Fukuyama K et al.. **Dig Dis Sci** 2001;46:545-549. 6. Sarac TP et al.. **Surgery** 1994;116:67-85. 7. Sharma R et al. **Dig Dis Sci** 1995;40:2532-2539. 8. Deitch E. et al. **Ann Surg** 1987;205:681-692. 9. Qin HL et al. Hepatobiliary Pancreat Dis Int 2002;1:150-154. 10. Qin HL, , et al. **World J Gastro** 2003;9:2270-2273. 11. Ioannidis O et al. **J Pancreas** 2008;9:375-390. 12. Koretz RL.. **Nutr Clin Pract** 2009;24:316-324. 13. Marik PE et al. .**British Med J** 2004;328:1407. 14. Petrov MS et al. **J Pancreas** 2008;9:440-448. 15. Eckerwell GE et al. **Ann Surg** 2006;244:959-967. 16. Windsor AC, et al. **Gut** 1998;42:431-435. 17. Kalfarentzos F, et al. **Br J Surg** 1997;84:1665-1669. 18. Mansfield CS et al. ***JAVMA*** 2008;233:936-944. 19. James FE et al. **Am J Vet Res** 2009;70:614-618. 20. Shea JC et al.. **Pancreatology** 2003;3:36-40. 21. Meier RF et al. **Best Pract Res Clin Gastroenterol** 2006;20:507-529. 22. Watson P. **In Practice** 2004;26:64-77. 23. McKay CJ et al. **Br J Surg** 2004;91:1243-1244. 24. Kluger E et al. **J Feline Med Surg** 2009;11:82-90.

A GENETIC APPROACH TO GALLBLADDER MUCOCELES IN DOGS

Katrina L. Mealey DVM, PhD, DACVIM, DACVCP
Pullman, WA

INTRODUCTION

A gallbladder mucocele is simply an enlarged gallbladder that contains an excessive amount of mucus. The first report of gallbladder mucoceles in dogs was in 1965[1]. This report described finding mucinous hypertrophy of the gallbladder incidentally in older dogs undergoing necropsy. More recently, gallbladder mucoceles (mucinous hyperplasia or mucinous cholecystitis) have been documented to be an increasingly important cause of hepatobiliary disease in dogs[2-4]. Whether or not this apparent increase in gallbladder mucoceles in dogs represents a true increase in disease frequency or is simply a reflection of increased ultrasonographic examination of the abdomen is not currently known. In contrast to the 1965 report, more recent reports involving gallbladder mucoceles describe a disease that can be associated with a high degree of morbidity and mortality, and not simply an incidental finding. Gallbladder mucoceles tend to occur in older dogs (9 years of age is the median reported in several studies)[2-4]. Clinical signs may be nonspecific (vomiting, anorexia, fever) or affected dogs may show signs of gallbladder involvement such as cranial abdominal pain and icterus. If the gallbladder has ruptured, the dog may show signs of peritonitis and/or shock. Gallbladder mucoceles have also been identified (ultrasonographically) in dogs with no apparent clinical signs as well. Clinicopathologic abnormalities may include hyperbilirubinemia, increased liver enzymes, hypercholesterolemia and hypertriglyceridemia. The key histopathologic finding in dogs with gallbladder mucoceles is mucinous hyperplasia with or without inflammatory infiltrates and necrosis. Histopathologic findings of the liver that may be associated with gallbladder mucoceles include intrahepatic cholestasis, cholecystitis, and periportal inflammation.

Proposed Etiologies and Potential Candidate Genes

The etiology of gallbladder mucoceles in dogs has not yet been identified. Suggested etiologies include:

1-A primary or secondary motility disorder of the gallbladder
2-A secondary complication of dyslipidemias (Shetland Sheepdogs in particular)
3-Extra-hepatic biliary duct obstruction
4-A primary or secondary disorder of mucus-secreting cells

Smooth muscle of the gallbladder contracts in response to cholecystokinin (CCK). CCK is released from enterocytes in response to food containing fat and protein. While CCK is the most potent secretagogue, motilin and cholinergic stimulation also stimulate gallbladder contraction. Disorders of gallbladder contraction have not been studied as extensively in dogs as they have been in people. Several disorders of gallbladder emptying have been described in people, including myotonic dystrophy type 1 (DM1)[5] and altered CCK-A receptor expression[6]. To the author's knowledge, neither of these has been investigated in dogs as a potential cause of gallbladder mucoceles. Because humans with these disorders do not develop gallbladder mucoceles, neither appears to be a strong candidate gene for gallbladder mucocele formation in the dog.

Dyslipidemias have been described as a breed-related phenomenon in Shetland Sheepdogs and Miniature Schnauzers. Gallbladder mucoceles have also been identified as a breed-associated phenomenon in Shetland Sheepdogs with coexistence of dyslipidemia and gallbladder mucocele occurring in 70 % of dogs in one study[2]. In contrast, Miniature Schnauzers have not been identified as a breed that is predisposed to gallbladder mucocele formation. Furthermore, dyslipidemia is not among the most common clinicopathologic abnormalities in dogs with gallbladder mucoceles. Thus, dyslipidemia (and its potential causative genetic disorders) does not appear to be a primary cause of gallbladder mucocele formation in most affected dogs.

Gallbladder mucoceles are uncommon in people, but when they do occur, they tend to be associated with extrahepatic bile duct obstruction. Prospective studies in dogs, in which the common bile duct or cystic duct was ligated, failed to consistently cause gallbladder mucocele formation[7]. Similarly, retrospective studies of dogs with gallbladder mucoceles do not describe extrahepatic bile duct obstruction as a consistent finding[2,3].

Primary disorders of mucin secretion by the gallbladder have been described in humans and mice. For example, two Lith genes (Lith3 and Glycam1) have been shown to affect gallbladder mucin accumulation, cholesterol crystallization and gallstone formation in mice[8]. In primary cultures of human gallbladder epithelial cells, epidermal growth factor pathway stimulation resulted in upregulation of MUC5AC, a gel-forming mucin known to accumulate in gallstone disease[9]. However, since gallstone formation does not appear to be associated with gallbladder mucoceles in dogs, these genes were not considered strong candidate genes.

While there are a number of mechanisms that could cause secondary disorders of mucus secreting cells, our laboratory has chosen to investigate bile salt-stimulated mucin secretion as an etiologic factor in gallbladder mucocele formation.

Toxicity of Bile Salts

Bile is produced by the cooperative actions of a number of transporters located on the canalicular membrane of hepatocytes. Active transport of biliary solutes creates an osmotic force that attracts water through tight junctions and aquaporins in the hepatocyte membrane, with bile acids (bile salts) being the most important biliary solute. Other important solutes of bile include cholesterol and phospholipids. Tight regulation of bile salt homeostasis is crucial because of the inherent toxicity of bile salts to hepatocytes and to epithelial cells lining the biliary tract. One of the most important factors functioning to protect the biliary tract from bile salt-induced cytotoxicity is the presence of phospholipid, phosphatidylcholine (PC) in particular, in bile. Bile salt cytotoxicity is substantially reduced in the presence of PC owing to the formation of mixed micelles (PC + bile salts) rather than simple micelles (bile salts only).

A decrease in the PC content of bile increases bile salt-induced damage to epithelial cells lining the biliary system. Since hypersecretion of mucin is the typical physiologic response of any epithelial lining to injury, gallbladder mucoceles may result from chronic bile salt-induced injury to the epithelial lining of the biliary system. Indeed, bile salts have been shown to directly stimulate mucin secretion by canine gallbladder epithelial cells[10].

ABCB4 as a Candidate Gene for Gallbladder Mucoceles in Dogs

ABCB4 functions exclusively as a phospholipid translocator[11]. ABCB4 is expressed on the canalicular membranes of hepatocytes where it translocates PC from the hepatocyte to the biliary canalicular lumen. Proper function of ABCB4 is critical for maintaining hepatobiliary health as evidenced by the myriad of diseases that occur in humans when polymorphisms of ABCB4 cause complete or partial protein dysfunction. Over thirty disease-causing mutations in human ABCB4 have been described. The disease spectrum ranges from severe (debilitating diseases of young children that require liver transplantation) to mild with disease severity dependent upon the nature of the mutation. Milder disease occurs when the ABCB4 gene mutation reduces but does not eliminate transport activity of the protein. Similarly, milder forms of disease exist in patients that are heterozygous for mutations that eliminate transporter activity (i.e., truncations). *Abcb*4-/- mice also develop severe hepatobiliarybiliary disease, providing additional evidence that ABCB4 function is essential for hepatobiliary health.

Recently Shetland Sheepdogs were identified as a breed that is predisposed to gallbladder mucocele formation, suggesting a genetic predisposition[2]. Because ABCB4 dysfunction is associated with hepatobiliary disease in other species, we postulated that a defect in canine *ABCB*4 might be associated with gallbladder mucoceles in Shetland Sheepdogs (and potentially other breeds as well). Therefore, we sequenced canine *ABCB*4 in affected and unaffected Shetland Sheepdogs as well as affected and unaffected dogs of other breeds.

A mutation in canine *ABCB*4 was identified in Shetland Sheepdogs as well as other dog breeds diagnosed with gallbladder mucocele. A diagnosis of gallbladder mucocele required one of the following: an ultrasound report by a boarded veterinary radiologist, a surgical report by a boarded veterinary surgeon, or a histopathology report. This mutation was significantly ($P < 0.05$) associated with the diagnosis of gallbladder mucocele (and its associated hepatobiliary pathologic changes) in Shetland Sheepdogs as well as other dog breeds. The disease appears to be inherited in a dominant fashion with incomplete penetrance (1 dog affected by the mutation had no evidence of gallbladder disease at 9 years of age).

Potential Benefits of *ABCB*4 Genotyping

Many of the dogs in our study were severely affected at the time of diagnosis and died of their disease despite surgical intervention. Our discovery of the *ABCB*4 mutation allows early identification of dogs predisposed to gallbladder mucocele formation. Genotyping of young dogs for the *ABCB*4 mutation would allow veterinarians to closely monitor gallbladder mucocele development and progression in affected dogs. Surgical intervention could be performed earlier in the disease process before disease-induced morbidity places the patient at higher risk for intra- and post-operative complications.

Another benefit of genotyping dogs for the *ABCB*4 mutation is the possibility of early medical or dietary management to prevent or at least delay the onset of mucocele formation. Currently, treatment options for managing dogs with gallbladder mucoceles are limited because information regarding the etiology of the disease has been lacking. There are some reports of patients with "early" gallbladder mucoceles benefiting from treatment with ursodeoxycholate, a relatively hydrophilic bile acid. Perhaps a greater percentage of dogs could benefit from earlier treatment and/or early screening (i.e., ultrasound) for the disease once they have been determined to be affected by the *ABCB*4 mutation.

REFERENCES

[1.] Kovatch RM et al. **Pathol Vet** 1965;2:574-584. [2.] Aguirre AL et al. **J Am Vet Med Assoc** 2007;231:79-88 [3.] Pike et al. **J Am Vet Med Assoc** 2004;224:1615-1622. [4.] Worley et al. **J Am Vet Med Assoc** 2004;225:1418-1422 [5.] Cardani R et al. **Neuromuscul Disord** 2008;18:641-645. [6.] Ding et al. **Hepatobiliary Pancreat Dis Int** 2005;4:295-298. [7.] Bernhoft RA et al. **Gastroenterology** 1983;85:1166-1171. [8.] Lammert F et al. **Hepatology** 2002;36:1145-1154. [9.] Finzi L et al. **Amer J Pathol** 2006;169:2031-2041. [10.] Klinkspoor et al. **Biochem J** 1998;332:257-262. [11.] Oude Elferink RP and Paulusma CC. **Pflugers Arch** 2007;453:601-610.

MINIMALLY INVASIVE APPROACHES TO THROMBOSIS IN SMALL ANIMALS

Marilyn Dunn DMV, MVSc, DACVIM
St-Hyacinthe, Quebec, Canada

INTRODUCTION

Thrombosis is a major cause of mortality in people with heart disease, cancer, and stroke which are the 3 most common causes of death in developed countries. Aspirin, heparin, and warfarin have long been the mainstays of antithrombotic therapy in human medicine. Although hereditary thrombophilias have not been identified in dogs or cats, thrombosis is recognized as a common complication of many acquired diseases, including cardiac, endocrine, inflammatory, and neoplastic disorders.

Normal hemostasis is maintained through an intricate balance between endogenous anticoagulants and procoagulants. The net effect is preservation of blood flow in the systemic vasculature with localized coagulation at sites of vessel injury[1].

Clinical signs of a thrombus can be variable and depend on the affected organ. Clinical signs result from compromised blood flow to the organ. Thrombi may obstruct flow locally or break off and lodge downstream where the vessel caliber is smaller and there is slower flow.[2,3]

Diagnosis of hypercoagulability is a great challenge in veterinary medicine. Routine coagulation tests (PT, PTT) are designed to detect hypocoagulability. Low plasma antithrombin activity and high fibrin degradation product, d-dimer, and fibrinogen levels provide laboratory evidence of hypercoagulability. Thromboelastography (TEG) is a technique that depicts global hemostasis, beginning with clot formation and ending with clot lysis. Evaluation of platelet function and TEG may prove useful in future veterinary studies to detect hypercoagulability.[1-3]

Actual detection of thromboemboli can be very difficult. A combination of Doppler ultrasonography, angiography, venography, contrast CT, MRI and ventilation perfusion scans can aid in their identification however, many cases of thromboembolism likely go undiagnosed in small animals.[1-3]

Treatment of thrombosis can be aimed at reducing thrombogenesis or dissolution of existing clots through the use of thrombolytic agents. Reducing thrombogenesis can be achieved with antiplatelet drugs or anticoagulants. Antiplatelet agents act through inhibition of platelet activation pathways or interference with membrane receptors. The 3 classes of antiplatelet drugs in current use include: nonsteroidal anti-inflammatory drugs (cyclooxygenase inhibitors), thienopyridines (ADP receptor antagonists) and GPIIb/IIIa blockers (fibrinogen receptor antagonists).[2]

Anticoagulants inhibit the generation of fibrin, but do not dissolve pre-existing fibrin clots. Anticoagulants are used for thromboprophylaxis in patients considered at risk for a first thrombotic event and to limit fibrin formation in patients with documented thromboembolic disease. There are currently 3 anticoagulants in clinical use: warfarin, heparin and low molecular weight heparins.[2]

THROMBOLYTIC DRUGS

Thrombolytic drugs are less commonly used and target the clot directly by accelerating fibrinolysis. Fibrinolytics augment fibrinolysis by converting plasminogen to plasmin and overwhelming plasmin inhibitors. Side effects reported with these agents are bleeding tendencies, pulmonary embolism (when treating venous thromboembolism) and reperfusion injury. These agents can be infused systemically or locally (agent delivered through a catheter directly into the affected vessel). In people, catheter directed thrombolytic therapy appears to be associated with better outcomes and less bleeding complications for both arterial and venous thrombosis.[4] Tissue plasminogen activator, streptokinase and urokinase have also been used with variable outcomes in veterinary patients.[3]

CASE PRESENTATION

An 8 year old female spayed Greyhound was presented to the Veterinary Teaching Hospital of the University of Pennsylvania with acute paresis and pain in the hind limbs. Initial physical examination revealed cold distal hind limbs (worse on the right side), markedly decreased femoral pulses bilaterally, lack of metatarsal pulses bilaterally and cyanotic nail beds. Following the administration of analgesics, diagnostic tests confirmed protein losing nephropathy (PLN) and hypertension. Ultrasonographic examination of the distal aorta showed a hyperechoic structure at the aortic trifurcation extending into the external iliac arteries. A distal aortic thrombosis was highly suspected. The dog received conservative management for the first 18 hours following admission. As her status failed to improve, it was decided that an intervention to re-establish blood flow to the area would be beneficial. The dog was anesthetized and her right carotid artery was accessed. An aortogram revealed marked filling defects in the caudal aorta at the trifurcation. Markedly reduced flow was seen in her external and internal iliac arteries. The left carotid artery was then accessed to allow placement of 2 stents simultaneously. Two caval stents measuring 8mm in width and 60 mm in length were deployed simultaneously in the proximal external iliac arteries and ending together in the distal aorta. An aortogram following stent placements showed much improved flow in both the internal and external iliac arteries. A catheter was inserted over a guidewire into the distal aorta and 6.5mg of tissue plasminogen activator was trickled in over a few minutes. During the post-operative period, the dog was on analgesics and anticoagulants. Over the following 3 days, she progressively improved and regained some motor function of the hind limbs. Following discharge and management of her PLN, hypertension and outpatient anticoagulant therapy, she continued to improve. Important sequelae to her thrombosis has been an irreversible quadriceps contracture preventing flexion of her knee.

MINIMALLY INVASIVE APPROACHES

A combination of vessel stenting and catheter directed thrombolysis was used to re-establish blood flow rapidly in this patient. Stenting plastered the clot up against the aortic wall while a thrombolytic agent was infused to address clots downstream within the arterial tree.

Vascular stenting

Various nitinol-based vascular stents have been developed to treat both arterial and venous thrombosis in people. Stents are non-thrombogenic and do not appear to be associated with an increased risk of thrombosis at implantation site however, as they are often used in hypercoagulable patients, anticoagulants and/or thrombolytics are used concurrently. These stents can be covered or not and are used to treat vessel occlusion secondary to atherosclerotic plaques or in the treatment of aortic dissection. Covered stents or stent grafts consist of a metal exoskeleton covered with a polytetrafluroethylene (PTFE) material which is impermeable to blood.[5] Stents can be mounted on a balloon or be self-expanding. Some complications reported in people are stent fracture in mobile sites such as upper extremity deep vein thrombosis.[5,6] Once in place, vascular stents cannot be removed. Vascular stenting is applicable to veterinary patients and in the future may improve outcome in certain thrombotic conditions as it has in human medicine.

Embolic trapping devices

Various devices ranging from baskets to 'spider-like' endovascular implants have been developed in people to trap emboli and prevent them from reaching the lungs (caval implants in the presence of deep vein thrombosis). Once trapped, emboli can undergo fibrinolysis. The most commonly used embolic trapping devices are vena cava filters which are placed in 3-11% of human patients with DVT. The use of these filters is indicated when primary therapy can not be started, must be stopped or is insufficient in protecting patients from pulmonary embolism. Filters are inserted fluoroscopically into the inferior vena cava just under the lowest renal vein. These devices can be permanent or retrieved once the thrombotic disease is under control. Filters can be made of nitinol or stainless steel and are biocompatible and nonthrombogenic. To my knowledge, these devices have yet to be used in clinical veterinary patients.[7,8]

Percutaneous mechanical thrombectomy (PMT)

PMT can be divided into wall contact devices and nonwall contact devices. A nonwall contact device, the Angiojet system has undergone evaluation in a small number of cats.[9] The Angiojet system emits high velocity saline jets that are directed backwards from the tip of the device to outflow channels. These flow jets generate a vacuum force that draws the thrombus back into the catheter. This system has been used successfully in cats with distal aortic thromboembolism . In human medicine, instead of using saline flush it has been recommended to use tissue plasminogen activator in the flush to improve thrombolysis at the clot and distal to it. This combined approach may improve outcome.[10,11]

Multi-modality approach

A multi-modality approach using systemic anticoagulant therapy, catheter-directed thrombolysis and endovascular stents may be the most appropriate approach in clinical veterinary patients. A multi-modality approach allows direct intervention to rapidly remove the clot or open the vessel lumen while lysing distal clots and preventing further clot formation. This combined approach may result in much improved morbidity and mortality. Given the variable presentation and wide range of underlying disease processes in our patients, an individual patient approach will most likely yield the best results.

REFERENCES

1. **Textbook of Veterinary Internal Medicine** 6th ed. St. Louis Missouri: Saunders; 2005. 2. Antiplatelet and Anticoagulant Therapy. In **In Kirk's Current Veterinary Therapy XIV** Saunders, Philadelphia 2009. 3. Acquired Coagulopahty V: Thrombosis. In: Schalm's **Veterinary Hematology** 5th ed.: Philadelphia; Lippincott, Williams & Wilkins 2000. 4. Endovascular Intervention of Iliofemoral Venous Thrombosis. **Endovasc Today** 2006: 60-66. 5. Endoluminal treatment of aortic dissection. **Eur Radiol** 2003:13:2521-2534. 6. Upper Extremity Deep Vein Thrombosis: Interventional Management. **Techniques in Vascular and Interventional Radiology** vol7: 2004; 86-90. 7. Embolic Protection Questions Answered. **Endovasc Today** 2006: 67-79. 8. Guidelines for the Use of Retrievable and Convertable Vena Cava Filters: Report from the Society of Interventional Radiology Multidisciplinary Concensus Conference. **J Vasc Interv Radiol** 2006:17: 449-459. 9. Use of rheolytic thrombectomy in the Treatment of Distal Aortic Thromboembolism. **J Vet Intern Med** 2006:290-296. 10. Mechanical Thrombectomy for DVT. Techniques in Vascular and Interventional Radiology, 2004: 79-85. 11. Endovascular Management of acute deep vein thrombosis. **Americ J of Med.** 2005:118: 31S-36S.

CYCLOSPORINE & THE HEMOSTATIC SYSTEM: IS THE DRUG SAFE IN IMHA PATIENTS?

Andrew Mackin, BSc, BVMS, MVS, DVSc, FACVSc, DACVIM
Starkville, MS

IMMUNE-MEDIATED HEMOLYTIC ANEMIA

Immune-mediated hemolytic anemia (IMHA), a common cause of anemia in dogs, results from the antibody-mediated destruction of circulating red blood cells.[1] Unfortunately, despite our best efforts, mortality rates for canine IMHA are still unacceptably high. Most large retrospective and prospective studies of dogs with IMHA cite patient mortality rates of around 50% (range 25% to 70%), and mortality rates have not really improved much at all over the past few decades.[2-8] The majority of canine patients that either die naturally or are euthanatized because of IMHA die within the first few weeks of diagnosis, and the largest cause of death is thrombotic disease, particularly pulmonary thromboembolism (PTE). In fact, the incidence of pulmonary thromboembolism in dogs with IMHA is estimated to be as high as 80%.[3,4] The mechanisms of PTE formation are not completely understood, but it is hypothesized that a hypercoagulable state involving both platelets and the clotting system is involved in the development of thromboembolic diseases.[9,10] Platelets play an integral role in thrombus development in IMHA patients. Dogs with IMHA have been shown to have hyperactive platelets with associated membrane alterations and increased expression of P-selectin.[9] P-selectin is a cell adhesion molecule expressed on platelets, vascular endothelium, and leukocytes. Platelet P-selectin expression has been shown to be directly related to thrombus formation, as illustrated by recent studies in mice that demonstrated that inhibition of P-selectin results in a significant reduction in deep vein thrombosis.[11] As well as changing shape and expressing P-selectin, activated platelets release vasoactive agents such as serotonin, adenosine diphosphate and prostaglandins (particularly thromboxane A_2) which mediate vasoconstriction and further predispose to thrombus formation by decreasing blood flow and increasing blood stasis.[11]

Although clinicians are acutely aware of the potential for PTE in canine IMHA patients, most therapeutic strategies that have been utilized in attempt to prevent thromboembolism have not been proven to be effective. Clinicians have tried utilizing warfarin, various subcutaneous and intravenous unfractionated heparin protocols, and low molecular weight heparins such as enoxaparin and dalteparin, all with as yet minimal documentable success. To date, only low-dose oral aspirin (0.5 mg/kg once daily) has been associated with significant reductions in IMHA mortality rates in dogs and, even with low-dose aspirin, mortality rates are still unacceptably high.[8] Given that there is no established drug therapy that is highly effective at preventing PTE in canine IMHA patients, clinicians must strive to reduce the incidence of PTE by any means possible, including rigorous avoidance of potential risk factors. Since the list of proposed risk factors for increased hypercoagulability and predisposition to PTE in dogs with IMHA is long and includes many strategies commonly associated with treatment of the disease, including glucocorticoids, intravenous catheterizations, multiple transfusions, and high dose intravenous immunoglobulins, complete 'avoidance of risk factors' is typically impractical.[12] Still, clinicians should at least be aware of those therapeutic strategies that have been associated with an increased risk of thrombosis, and should always consider the potential increased risk of PTE when making therapeutic decisions, including decisions regarding which immunosuppressive agents to use.

Immunosuppressive Therapy in IMHA Patients

Treatment of IMHA in dogs is centered on immunosuppressive therapy and concurrent supportive care. Although glucocorticoids are the cornerstone of immunosuppressive therapy, additional agents such as cyclophosphamide, azathioprine or cyclosporine are often needed, both to treat IMHA patients that are refractory to steroids alone and to reduce dependence on high doses of glucocorticoid. Traditionally, cyclophosphamide and azathioprine have been the immunosuppressive agents most commonly used in conjunction with glucocorticoids in canine IMHA patients. Cyclophosphamide, however, has to a large extent fallen out of favor because of unacceptable side-effects such as gastrointestinal disorders, myelosuppression, and sterile cystitis, and because canine IMHA patients treated with the drug appear to have higher mortality rates and shorter survival times than dogs treated with alternative immunosuppressive regimens.[6,13,14] Azathioprine has therefore become the immunosuppressive drug of choice to be utilized in combination with glucocorticoids in most IMHA patients, because the drug is relatively inexpensive, typically well-tolerated, and often effective. Azathioprine does, however, have some significant limitations, including consistent mild to moderate suppression of erythropoiesis, occasional mild gastrointestinal signs, infrequent but marked myelosuppression or hepatotoxicity, and human tablet sizes that are unsuitable for use in smaller dogs.[5,6,8] Additionally, in some dogs, even the combined immunosuppressive effects of glucocorticoids and azathioprine are insufficient to control IMHA. In such patients, in more recent years, it is becoming increasingly recommended that cyclosporine be used as either an 'add on' or alternative to the more traditionally used immunosuppressive agents. However, because cyclosporine has only relatively recently attained acceptance as an immunosuppressive agent for use in dogs with immune-mediated diseases, there is very little published in either retrospective or prospective studies in canine IMHA patients regarding the drug's efficacy or safety.[14]

CYCLOSPORINE

Cyclosporine, a powerful immunosuppressive agent originally introduced into human medicine for organ transplantation, is regarded as a promising drug for the treatment of many different immune-mediated diseases in dogs.[15-18] Cyclosporine targets T-cells by binding to cyclophilin, with the resultant cyclosporine/cyclophilin complex inhibiting calcineurin, a calcium-dependent serine/threonine phosphatase. Calcineurin is necessary for T-cell production of cytokines such as IL-2, IL-4 and IFN-γ. Cyclosporine-induced inhibition of cytokine expression inhibits proliferation of activated T- lymphocytes by arresting generation of cytotoxic T-cell clones.[17] Cyclosporine typically does not produce many of the side effects associated with other immunosuppressive agents in dogs, although gingival hyperplasia, mild gastrointestinal signs, hepatotoxicity and nephrotoxicity are occasionally reported.

Cyclosporine has the additional beneficial property of being minimally myelosuppressive. The relatively recent availability of a veterinary-approved oral cyclosporine product has rapidly increased the popularity of the drug for treatment of a wide range of immune-mediated diseases in the dog, including IMHA. Recent human studies and results of our preliminary research, however, have suggested that cyclosporine could have the potential to contribute to the development of PTE in dogs with IMHA.

In humans, cyclosporine has been shown to increase platelet procoagulant activity and accelerate thrombin generation.[19] Researchers have used flow cytometry to determine that cyclosporine causes changes in the human platelet plasma membrane, such as increased expression of platelet phosphatidylserine, which result in promotion of platelet procoagulant activity.[19] Cyclosporine also increases platelet production of thromboxane A_2, a prostaglandin generated via the enzyme cyclooxygenase (COX) that activates platelets and causes vasoconstriction.[20-22] Preliminary work several years ago in our Flow Cytometry Laboratory at the Mississippi State University College of Veterinary Medicine (MSU-CVM) also suggested that, as in humans, cyclosporine may increase platelet procoagulant activity in some dogs.[23] In our small preliminary study, flow cytometry was used to evaluate P-selectin and phosphatidylserine as surface markers of platelet reactivity in three dogs before and after the administration of cyclosporine. Our preliminary study revealed markedly increased platelet P-selectin expression and mildly decreased phosphatidylserine expression, suggesting that, at the very least, exposure to cyclosporine was causing detectable changes in platelet surface membranes.[23] Preliminary results of a small *in-vitro* companion study at MSU-CVM utilizing several in-house analyzers which enable a more 'global' assessment of the effects of cyclosporine on both platelets and coagulation, the Siemens Platelet Function Analyzer-100® (which measures platelet function using a membrane aperture) and the Sienco SonoClot® Coagulation & Platelet Function Analyzer (which utilizes viscoelastometry to evaluate clot formation), also demonstrated that exposure to cyclosporine appeared to decrease the time of canine platelet plug and fibrin clot formation.[23]

Based on both human studies and the results of our preliminary research, we became concerned that the administration of cyclosporine to dogs with IMHA might have the highly undesirable potential to further increase the risk of thromboembolic disease and lead to an even higher incidence of PTE in IMHA patients. Since veterinarians are turning to cyclosporine as a potential life-saver in dogs with IMHA, it could have disastrous consequences if, in fact, the drug predisposed such patients to PTE and actually led to an increased mortality rate. We therefore devised and recently completed a much more comprehensive study at MSU-CVM designed to determine whether cyclosporine actually predisposes dogs to thrombosis.

Cyclosporine and Canine Platelet Procoagulant Activity

In our comprehensive study, we thoroughly evaluated platelet reactivity and procoagulant activity in eight healthy dogs before and after exposure to cyclosporine.[24] Pre-treatment assessment included a complete blood count, serum biochemistry, urinalysis, prothrombin time (PT), partial thromboplastin time (PTT), plasma fibrinogen levels, antithrombin, von Willebrand factor, D-Dimer, buccal mucosal bleeding time (BMBT), flow cytometric evaluation of a battery of markers of platelet function, activation and procoagulant activity (including P-selectin, phosphatidylserine, COX-1 and COX-2), viscoelastometric analysis of clot formation and retraction using the Sienco SonoClot®, evaluation of platelet function using the Siemens PFA-100® with collagen/ADP and collagen/EPI cartridges, and measurement of urine thromboxane B_2 (a metabolite of thromboxane A_2 that is more stable). The dogs were then administered oral cyclosporine at immunosuppressive doses (starting dose 10 mg/kg bid), and doses were adjusted accordingly to reach target immunosuppressive cyclosporine levels as determined by high-performance liquid chromatography (HPLC) in our Clinical Pharmacology Laboratory. Samples were collected two hours after administration at peak plasma levels, one day after initiating treatment, and after cyclosporine blood levels had achieved steady state and been determined to be at immunosuppressive levels for each individual dog, and evaluated using the battery of tests of hemostasis used for pre-treatment assessment. After a wash-out period, the study was repeated using a lower, anti-inflammatory dose of cyclosporine (5 mg/kg sid) that is typically used for canine atopy and the long-term maintenance of IMHA patients.[24]

Unlike human studies, and in contrast to our preliminary work, results of our comprehensive study strongly suggested that cyclosporine does not cause a detectable increase in platelet reactivity or procoagulant activity during the first week of drug administration in dogs.[24] Mean platelet expression of P-selectin, phosphatidylserine, COX-1 and COX-2 all tended to decrease with cyclosporine therapy at both high and low doses, PFA-100® closure times tended to be slightly prolonged, and SonoClot® and thromboxane B_2 results were essentially unaffected.[24] Results derived from work in normal healthy dogs cannot be directly translated to predict similar effects at IMHA dogs, but at the very least we have found no experimental evidence that the effects of cyclosporine on canine platelets are potentially dangerous in such patients. Although these results are reassuring, they do not provide an absolutely complete picture of the effects of cyclosporine on the hemostatic system, particularly considering that the drug has, in other species, been shown to have increased procoagulant activity in cells other than platelets (such as blood mononuclear cells, red blood cells, and vascular endothelial cells).[22,25,26] Clinical trials evaluating the safety of cyclosporine in dogs with IMHA, including assessment of potential prothrombotic effects, are still indicated and warranted.

CLINICAL EFFICACY OF CYCLOSPORINE IN CANINE IMHA PATIENTS

Canine IMHA is associated with a high mortality rate during the first few weeks of hospitalization and treatment.[5,6] Given this high early mortality rate, it is imperative that immunosuppressive therapy be both as rapid-acting and as effective as possible in IMHA patients. Since cyclosporine is a relatively new drug used in the treatment of IMHA, clinical efficacy data regarding the use of cyclosporine in canine IMHA patients is, to a large extent, lacking. Given how challenging it can be to generate sufficient data regarding the use of immunosuppressive drugs in IMHA patients, it may be many years until solid retrospective or prospective studies evaluating cyclosporine in dogs with IMHA are generated. In the meantime, we must use whatever data is already available to make educated guesses regarding the clinical efficacy of cyclosporine in IMHA patients. Even if cyclosporine is shown to be 'safe' in that it does not have procoagulant effects on canine platelets, the drug actually has the potential to be a dangerous or unsafe treatment choice in IMHA patients if it is ineffective or too slow to effect.

Oral cyclosporine has highly unpredictable bioavailability in both people and dogs, and it is therefore widely recommended that, in order to achieve acceptable immunosuppressive effects, cyclosporine blood levels be monitored and doses altered to achieve uniform 'target' drug blood levels.[27,28] In veterinary medicine, however, ideal cyclosporine blood levels are not well-defined, and target recommendations vary widely between clinical pharmacology laboratories. Therefore, in order to better define ideal target cyclosporine blood levels, we recently evaluated the effects of cyclosporine on a comprehensive panel of biomarkers of immunosuppression in normal dogs.[29,30] In our laboratory at MSU-CVM , we measured T-cell surface antigen (CD25 and CD95) and cytokine expression (IL-2, IL-4, and INF-γ) via flow cytometry and T-lymphocyte IL-2 and INF-γ mRNA expression via quantitative polymerase chain reaction (QPCR) in dogs receiving varying oral doses of cyclosporine. As expected, we confirmed a wide variability on cyclosporine bioavailability in individual dogs and revealed that, at high drug blood levels, T-cell expression of IL-2 and INF-γ is reliably suppressed. At lower drug blood levels, however, our results raised some intriguing questions regarding cyclosporine pharmacodynamics. Specifically, we discovered that, despite comparable cyclosporine blood levels, T-lymphocyte cytokine expression was markedly suppressed in some dogs, and minimally affected in others.[30] Our results strongly suggest that variability in the immunosuppressive effects of cyclosporine in dogs is influenced by more than simple variations in bioavailability and blood drug levels and that, even at comparable cyclosporine blood levels, there appears to be marked individual dog-to-dog variation in T-cell drug responsiveness. We are currently exploring explanations for this individual animal-to-animal variability in cyclosporine responsiveness at comparable blood levels, including inherited and drug-induced alterations in the function of cell membrane efflux pumps such as P-glycoprotein.[31,32] In the meantime, our findings suggest that, except at high drug blood levels (trough levels of at least 600 ng/mL measured by HPLC), blood cyclosporine levels do not always reliably predict drug immunosuppressive effects.

What does this mean for the IMHA patient? The potential problem with relying on cyclosporine as a first-line immunosuppressive agent in critical IMHA patients is that, in some patients, immunosuppressive effects will be unreliable, particularly during the important first few weeks of therapy. Because of highly variable oral bioavailability, in many dogs the initial starting cyclosporine dose will not be sufficient to achieve high trough blood levels, necessitating a delay while blood levels are measured and original doses are adjusted upwards. For example, in our study, even though we started dogs on an aggressive oral starting dose of cyclosporine (10 mg/kg bid), we still had to increase the oral dose in many dogs in order to achieve the high blood drug levels that could reliably suppress T-cell function. Furthermore, we now know that, until this high blood drug level is achieved, T-cell function in some dogs will not be suppressed. On the positive side, the short half-life of cyclosporine means that, within a few days of starting drug therapy, blood levels have reached a relatively steady state, and blood drug levels can therefore be measured as little as two days (and, arguably, one day) after commencing therapy. Additionally, once high blood drug levels are achieved, immunosuppressive effects appear to occur rapidly: based on our work, we know that marked T-cell suppression can be reliably produced within one week of achieving high blood cyclosporine levels.

In the short term, if cyclosporine is used as a first-line immunosuppressive drug in critical canine IMHA patients, we recommend starting with an aggressive oral starting dose of 10 mg/kg twice daily, measuring trough (twelve hours post-dosing) blood cyclosporine levels two days after commencing therapy, and adjusting doses accordingly to achieve high blood drug levels. In the future, at least based on the human precedent, we may be able to establish peak (two hour post-dosing) blood drug level recommendations that more accurately predict immunosuppressive effects.[15,33-35] More promisingly, given the individual dog-to-dog variations in responsiveness to cyclosporine, establishment of truly individualized drug therapy by measuring the effects of the drug on T-cell function (via measurement of cytokine expression via flow cytometry or QPCR) may be the best method of determining rapid and effective immunosuppression in critical canine IMHA patients.

REFERENCES

[1]Scott-Moncrieff JC, et.al. **JAAHA** 2001;37:220. [2]Reimer ME, et.al. **JAAHA** 1999;35:384. [3]Balch A, et.al. **Compend Contin Educ Prac Vet** 2007;29:217. [4]Carr AP, et.al. **JVIM** 2002;16:504. [5]Piek CJ, et.al. **JVIM** 2008;22:366. [6]Reimer ME, et.al. **JAAHA** 1999;35:384. [7]Burgess K, et.al. **JVIM** 2000;14:456. [8]Weinkle TK, et.al. **JAVMA** 2005;226:1869. [9]Weiss DJ, et.al. **JVIM** 2006;20:682. [10]Sinnott VB, et.al. JVECC 2009;19:484. [11]Sobieszczyk P, et.al. **Circulation** 2002; 106:1748. [12]Tsuchiya R, et.al. **JVIM** 2009; 23:1164. [13]Mason N, et.al. **JVIM** 2003;17:206. [14]Grundy SA, et.al. **JAVMA** 2001;218:543. [15]Barden MJ, et.al. **Cell Proliferat** 2007;40:50. [16]Jensen SA, et.al. **Transpl** P 2001;33:3003. [17]Daigle JC. **JAAHA** 2002;38:205. [18]Allenspach K, et.al. **JVIM** 2006;20:239. [19]Tomasiak M, et.al. **Neph Dialysis Transpl** 2007;22:1750. [20]Oskarsson HJ, et.al. Hypertension 1997;29:1314. [21]Perico N, et.al. **Am J Physiol** 1986;251:581. [22]Zheng Y, et.al. **Chin Med J** 2009;122:1674. [23]Thomason J, et.al. 27th **ACVIM Forum** (abstract) 2009. [24]Thomason J, et.al. 28th **ACVIM Forum** (abstract) 2010. [25]Carlsen E, et.al. **Clin Exp Immunol** 1985;60:407. [26]Bombeli T, et.al. **J Lab Clin Med** 1996;127:621. [27]House AK, et.al. **Vet Surg** 2006;35:543. [28]Campbell KL, et.al. **Compend Contin Educ Prac Vet** 2005;27:339. [29]Archer T, et.al. 27th **ACVIM Forum** (abstract) 2009. [30]Archer T, et.al. 28th **ACVIM Forum** (abstract) 2010. [31]Parasrampuria DA, et.al. **Pharm Res** 2001;18:39. [32]Bonhomme-Faivre L, et.al. **Transpl** 2004;78:21. [33]Cantarovich M, et.al. **Transpl P** 2004;36:4425. [34]Cantarovich M, et.al. **Transpl P** 1998;30:1462. [35]Davies RA, et.al. **Transpl P** 2007;39:3334.

THE PATHOPHYSIOLOGY OF CANINE DIABETES MELLITUS

Rebecka S. Hess, DVM, DACVIM
Philadelphia, PA

INTRODUCTION

The pathophysiology of canine diabetes mellitus (DM) is incompletely understood. However, the pathophysiology is probably multifactorial and includes genetic and immune mediated factors. It is possible that different breeds of dogs have different genetic or immune mediated etiologies, and that the pathophysiology of DM is not identical in all breeds of dogs.

CLASSIFICATION OF DIABETES MELLITUS IN HUMAN BEINGS

In human beings, clinical and clinicopathologic features help distinguish between classic type I and type II DM. Young age of onset, lean body condition, acute hyperglycemia, a propensity for diabetic ketoacidosis (DKA), presence of pancreatic beta-cell autoimmunity and other autoimmune diseases, and the need for exogenous insulin therapy have all been considered characteristic of type I DM. Obesity and insulin resistance have been considered classic characteristics of type II DM.[1]

In between type I DM and type II DM

In recent years many intermediate forms of DM have been recognized. These forms of DM do not conform to the classic definitions of type I or type II and have been termed, depending on their specific characteristics, autoantibody-positive phenotypic type II diabetes, maturity-onset diabetes of the young (MODY), latent autoimmune diabetes in adults (LADA), slowly progressive type I diabetes, latent type I diabetes, double diabetes, or type 1.5 diabetes.[1, 2]

Age

Older age was historically considered a phenotypic characteristic of type II DM, classically recognized in older, obese individuals. However, with the obesity epidemic affecting younger patients, type II diabetes is now also recognized in obese adolescents.[2]

Ketosis

One example of a syndrome that can not be strictly defined as type I or type II DM is ketosis-prone DM.[3] Historically, human beings with DKA were classified as having type I DM which has been characterized by presence of beta-cell autoantibodies (AA+) and absent beta-cell function (B^-). However, several studies have demonstrated that patients with DKA can have beta-cell autoantibodies and beta-cell function (AA+ B^+) or have no beta-cell autoantibodies, with or without beta-cell function ($AA^- B^+$, or $AA^- B^-$).[3]

MODY

MODY is another form of DM which defies conventional definition in that it affects children, adolescents, and young adults, is not associated with obesity, but unlike type I DM, is characterized by chronic mild, asymptomatic hyperglycemia that may last for many years. Additionally, unlike type I DM, MODY is not an autoimmune disease. Several forms of MODY have been characterized, and all involve one of several possible mutations that affect beta-cell function. The mutations are autosomal dominant and involve various genes that are expressed in beta-cells, including the glycolytic enzyme glucokinase and several regulators of insulin gene transcription. The mutation leads to beta-cell dysfunction and DM which often can be treated with oral hypoglycemic agents, diet, and exercise (similar to type II DM), but may require insulin therapy (similar to type I DM).[2]

Beta-cell autoantibodies

Autoimmunity directed at beta-cells has been traditionally associated with type I DM. Beta-cell autoantibodies include antibodies directed at islet-cell cytoplasm (ICA), glutamic acid decarboxylase 65 (GAD65), the intracytoplasmatic domain of the tyrosine phosphatase-like protein IA-2 (IA-2A), insulin (IAA), and the zinc transporter ZnT8. While autoimmune diabetes should classically present in young patients, there are at least three separate populations of adults that develop autoimmune diabetes. Adult onset type I diabetes is a phenotypically classic, autoimmune type I DM that develops in adults instead of in children. Type 1.5 diabetes is phenotypically similar to type II DM in that it is observed in adult, obese patients, but these patients are antibody positive. LADA also develops in adults, but unlike adult onset type I DM it is not treated with insulin within the first six months of diagnosis.[1]

LADA

The ICA, GAD65, IA-2, insulin, and ZnT8 autoantibodies have been identified in type I as well as LADA patients, and in both of these diabetic populations there is also presence of islet-reactive T-cells. Therefore, the underlying disease process in both type I DM and LADA is considered to be autoimmune. However, the classic type I DM patient is young, whereas the typical LADA patient is older than 35 years of age. Both type I DM and LADA patients tend to be non-obese, but type I DM patients require exogenous insulin therapy at the time of diagnosis whereas LADA patients can be controlled with diet and oral hypoglycemic agents for several months, before requiring insulin therapy. LADA patients can eventually develop DKA, similar to type I DM patients.[1]

Both LADA and type I DM are autoimmune diseases characterized by presence anti-beta-cell autoantibodies and islet reactive T-Cells. However, some differences between these two forms of the disease have been documented in autoantibody clustering (the titer and number of various specific antibodies formed), T-cell reactivity (to different antigens), and genetic susceptibility and protection. Human Leukocyte Antigen (HLA) susceptibility and protection have been documented in both type I DM and LADA. HLA DR3, DR4, and DQB1*0201 and 0302 increase the risk of type I DM while DR2 and DQB1*0301 and 0602 confer protection against type I DM. Whether there are differences in HLA susceptibility alleles between type I DM and LADA, is controversial. However, it seems as if LADA patients have an increased frequency of the protective alleles DR2 and DQB1*0602 and this may partially explain the later age of onset of LADA compared to classic type I DM.[1]

CLASSIFICATION OF DIABETES MELLITUS IN DOGS

Type I DM or LADA

Canine DM resembles human type I DM in that non-obese dogs develop severe acute hyperglycemia requiring exogenous insulin therapy at the time of diagnosis, and in that they can develop DKA.[4, 5] However, unlike classic type I DM in humans, most dogs develop DM as middle aged adults.[5] Because canine DM resembled type I DM but develops at a later age, some researchers have suggested that canine DM resembles LADA. However, it has not yet been conclusively established that canine DM is an autoimmune disease. If, eventually, it is established that canine DM is a T-cell mediated autoimmune disease which resembled type I DM in human being, the title of adult onset type I diabetes (rather than LADA) may be more appropriate, because dogs require insulin therapy at the time of diagnosis.

Beta-cell autoantibodies

Reports of the presence of anti-beta-cell antibodies in dogs are small. An early study identified anti-beta-cell antibodies in nine of the 23 previously untreated, newly diagnosed diabetic dogs.[6] In a more recent study, four of 30 (13%) previously untreated, newly diagnosed diabetic dogs had significant GAD65 auto reactivity compared to controls and 3 dogs were positive for autoantibodies to insulinoma antigen-2 (IA-2). Two diabetic dogs showed autoreactivity to both antigens.[7]

While this study did document that GAD65 antibodies were significantly more common in diabetic dogs than in control dogs, the percent of diabetic dogs that have GAD65 antibodies (13%) is still small compared to diabetic human beings.[7, 8] In human beings, about 85% of newly diagnosed type I DM patients have GAD65 antibodies, and about 90% of first-degree relatives of patients with type I DM who have antibodies to GAD65, IA-2, or insulin develop type I DM within several years after the detection of the antibodies.[8] In humans, detection of anti-beta-cell antibodies in normoglycemic relatives of patients with type I DM helps predict that an individual will develop type I DM years before the onset of the disease.[9] The clinical significance of detection of such antibodies in dogs, at the time of diagnosis of diabetes is unknown. It remains to be seen whether anti-beta-cell autoantibodies are part of the etiology of canine diabetes, or whether they can predict the onset of clinical signs, or whether presence of these antibodies is a result of destruction of beta-cells.

Genetic predisposition

Although genetic predisposition to DM is apparent from the fact that certain breeds of dogs are at increased risk for the disease, while others are at decreased risk for the disease, the specific genetic determinants of canine DM remain unknown. Samoyeds, Australian Terriers, Pugs, Toy and Miniature Poodles, and Miniature Schnauzers are at increased risk for DM in the United States, whereas German Shepherds, Golden Retrievers, and American Pit Bull Terriers are at decreased risk for the disease in the United States.[10] Because the genetic pool of pure breed dogs is very small, it is possible that different breeds of dogs are at increased or decreased risk for DM in different parts of the world.

The two most important genetic determinants of type I DM in humans are the major histocompatibility complex class II (MHC-II) genes and the insulin gene region. MHC-II genes coding for HLA class II proteins are estimated to determine about 40% of type I DM heritability in humans, and the insulin gene region contributes about 10% to this heritability.[11] HLA class II proteins promote recognition of protein fragments by T-cell receptors, and facilitate subsequent activation of T lymphocytes. One of the theories regarding the action of MHC genes is that in people at increased risk for type I DM, MHC genes favor positive selection for anti-pancreatic T-cells, whereas in individuals at decreased risk for type I DM there is selection for T-cell nonresponsiveness to pancreatic beta-cells.[11]

A study of DNA from 460 diabetic dogs and 1047 control dogs identified three DLA haplotypes, DRB1*009/DQA1*001/DQB1*008, DRB1*015/DQA1*0061/DQB1*023, and DRB1*002/DQA1*009/DQB1*001, that were observed significantly more in the diabetic dogs compared to the controls. One DLA-DQ haplotype, DQA1*004/DQB1*013, was significantly reduced in cases with diabetes and may therefore be protective.[12]

T-cell dysfunction

T-cell dysfunction is an important component of the pathophysiology of type I DM in humans. In type I DM in humans, there is a misdirected T-cell-mediated attack against pancreatic beta-cells that leads to destruction of these cells, and inability to synthesize insulin. T-cell receptors are associated with several clusters of differentiation (CD) antigens on the cell surface. Among these clusters of proteins is the CD3 complex that transmits activation signals into the cell when the T-cell receptor binds an antigen. Recently, administration of Anti CD3 antibodies, which are thought to modulate the immune response by induction of tolerance to pancreatic beta-cells has shown some promise for treatment of type I DM.[13]

Single nucleotide polymorphisms (SNPs) of T-cell cytokines were therefore analyzed in diabetic and non-diabetic dogs. When SNPs were examined for an association with diabetes, significant associations were observed for IL-4 in the Collie, Cairn Terrier, and Schnauzer and for IL-10 in the Cavalier King Charles Spaniel. These findings suggest that T-cell dysfunction may also be important in canine DM.[14]

Insulitis

Insulitis is a process in which inflammatory cells infiltrate the islets of Langerhans and cause destruction of insulin secreting beta-cells. Severe destructive insulitis is observed in about 85% of female non-obese diabetic (NOD) mice and about 25% of male NOD mice. The NOD mouse develops a spontaneous form of autoimmune diabetes which resembles human type I DM in many aspects, and has therefore been extensively studies. In the NOD mouse, there is an infiltration of dendritic cells and macrophages and then T-cells (CD4 and CD8) and B cells into the area surrounding the pancreatic islets (peri-insulitis) beginning at about 3 weeks of age and culminating with selective T-cell–mediated destruction of insulin-producing beta-cells by 4–6 months of age.[15] The mechanisms by which this inflammation leads to destruction of beta-cells include lysis by CD8 T-cells, and damaging by cytokines produced by CD4 T-cells or inflammatory mediators released by other cells. [16] Insulitis that develops in humans with type I DM is not as frequent or severe and is predominated by CD8 T-cells. [16] A study of the histopathology of pancreatic biopsies from humans that were positive for antibodies that predict the onset of type I DM (ICA, GAD, IA-2) found that only 2/62 patients and none of the 62 controls had insulitis. This study suggests, that unlike the NOD mouse, insulitis may not precede the onset of type I DM, in most human cases. [17] Detailed histologic studies of the pancreas in diabetic dogs are lacking, and insulitis is reported infrequently in dogs.[18, 19] The pathophysiologic significance of this is not known.

Obesity

In human beings, insulin resistance and obesity are important features of type II DM. A number of studies have shown that in dogs with experimentally induced obesity, plasma leptin concentration is increased, whereas plasma adiponectin is decreased. [20, 21] Additionally, expression of the peroxisome proliferator-activated receptor-g (PPARg), the glucose transporter GLUT4, and lipoprotein lipase in adipose tissue and skeletal muscle are decreased in obese, insulin resistant dogs.[20] Plasma ghrelin concentration is also significantly decreased in obese dogs.[21] Some of these hormonal changes have been shown to be reversible with weight loss. Although most of these studies have investigated dogs with experimentally induced obesity, one study of naturally-occurring obesity in otherwise healthy dogs also found that fasting plasma leptin concentration was significantly higher in obese dogs, as compared to dogs with normal body conditions.[22] The clinical significance of the hormonal changes associated with obesity in dogs has yet to be determined. Similarly, the clinical significance of obesity induced insulin resistance in dogs, is not known, but it is probably not an important clinical entity in canine diabetes.

CONCLUSIONS

The clinical characteristics of canine DM suggest that this disease most resembles type I DM in human beings. While some preliminary data regarding predisposing genetics, presence of anti beta-cell antibodies, and genetics of T-cell mediators have been reported in dogs with DM, the understanding of the pathophysiology of canine DM is still limited.

REFERENCES

[1]Naik RG, et al. **J Clin Endocrinol Metab** 2009;94: 4635, [2]Fajans SS, et al. **N Engl J Med** 2001;345(13):971, [3]Balasubramanyam A, et al **Endocrine Reviews** 2008; 29(3):292, [4]Hume DZ, et al. **J Vet Intern Med** 2006;20:547, [5]Hess RS, et al. **JAVMA**;2000:217(8): 1166, [6]Hoenig M, et al. **Vet Immun and Immunopath** 1992; 32 :195, [7]Davison LJ, et al. **Vet Immun and Immunopath** 2008;126:83, [8]Jun HS, et al. **Cell. Mol. Life Sci.** 2002;59:1892, [9]Sonesko JM, et al. **Diabetes Care** 2008;31:528, [10]Hess RS, et al. **JAVMA** 2000;216(9):1414, [11]Concannon P, et al. **N Engl J Med** 2009;360:1646, [12]Kennedy LJ, et al. **Tissue Antigens** 2006; 68 :467, [13]Begum S, et al Endocrinology 2009;150: 4512, [14]Short AD, et al. **Vet Immun and Immunopath** 2009;128 :137, [15]Delovitch TL et al. **Immunity**, 1997:7;727, [16]Akirav E et al **Diabetes** 2008:57:2883, [17]In't Veld P, et al. **Diabetes** 2007;56:2400, [18]Jouvion G, et al. **Endocrine Pathology** 2006;17(3):283, [19]Gepts W, et al. **Diabetologia** 1967;3:249, [20]Gayet C, et al. **Br J of Nut** 2007;98:497, [21]Jeusette IC, et al. **Am J Vet Res** 2005:66:81, [22]Ishioka K, et al. **Res in Vet Sci** 2007;82:11.

THE RECENT DEVELOPMENTS IN THE MANAGEMENT OF DIABETES MELLITUS IN DOGS

Linda M. Fleeman BVSc PhD MACVSc
Melbourne, Australia

INSULIN THERAPY – WHAT IS NEW, WHAT IS STILL AVAILABLE, AND HOW TO BEST USE THEM

A large variety of human insulin preparations are available, as well as veterinary insulin preparations registered specifically for use in dogs and cats (Table 1). All products are potentially efficacious for the treatment of canine diabetes mellitus although there is limited veterinary information on most of the recently released human products. As improved therapeutic protocols are developed for the management of human diabetes, new products are continually being developed while others are withdrawn. The current trend in human medicine is towards long-acting 'peakless' insulin preparations that aim to provide continuous basal insulin supplementation, and which are typically combined with administration at meal times of a rapid-onset, short-acting insulin preparation. This has resulted in many changes in the market for short- and intermediate-acting insulin products and their pre-mixed combinations. It is these products that are most commonly used for the management of diabetes in dogs and so veterinarians must now adapt to the withdrawal of several familiar insulin preparations.

Which insulin preparations have been withdrawn?

IDEXX Pharmaceuticals have recently withdrawn PZI-VET, which was a 40 U/mL bovine/porcine preparation of protomine zinc insulin registered in several countries for veterinary use. The loss of this product highlights the practical problems often associated with production of animal sourced insulin. In the human market, all of the lente-type insulin preparations have now been discontinued. Monotard (produced by Novo Nordisk) and Humulin L (produced by Eli Lilly) were both 100 U/mL preparations of lente insulin marketed for human diabetics that have been discontinued. Ultratard (Novo Nordisk) and Humulin U (Eli Lilly) were longer-acting ultralente preparations that were discontinued a number of years ago. There is a similar trend with all insulin preparations containing regular crystalline insulin. For example, the convenient 10 mL vials of Mixtard 30/70 (Novo Nordisk), which is a 100 U/mL pre-mixed combination of 30% short-acting regular insulin and 70% intermediate-acting isophane (NPH) insulin, are no longer available in many countries.

Which insulin preparations are still available?

Porcine insulin zinc suspension

The only lente insulin preparation now is Vetsulin (in the USA) / Caninsulin (elsewhere in the world) made by Intervet Schering-Plough, which is a 40 U/mL porcine insulin zinc suspension specifically registered for veterinary use. Availability of Vetsulin in the USA has recently been affected by the finding that the product may have varying amounts of crystalline zinc insulin[1]; however supply of Caninsulin elsewhere is unaffected and has been reliable for many years. There is good evidence that this insulin product is effective for the treatment of diabetes mellitus in dogs[2]. Lente insulin has an intermediate duration of action and is a combination of 30% short-acting semilente insulin and 70% longer-acting ultralente insulin. This combination results in a relatively predictable and rapidly obtained peak effect. The average duration of action of porcine insulin zinc suspension in dogs is 16 hours resulting in significant lowering of blood glucose concentration in the majority for 12 hours or less[3,4]. It is particularly suited to a twice-daily dosing regimen in diabetic dogs because the meal can be fed at the time of the insulin injection. This more dilute 40 U/mL veterinary preparation is advantageous for small dogs, which might require a total insulin dose of only 1 or 2 U. Dosing is more simple and accurate if specific U-40 syringes are prescribed with this product.

Regular insulin

Products containing 100% regular crystalline insulin are still available and have long been recommended for the management of diabetic ketoacidosis and ketosis in dogs[5]. If these products are withdrawn from the human market, it will be necessary for veterinarians to use the newer short-acting insulin preparations instead (see below).

Isophane / Neutral Protamine Hagedorn (NPH) insulin

Humulin N (Eli Lilly) and Protophane (Novo Nordisk) have an intermediate duration of action and are effective for the treatment of diabetes in dogs when administered twice-daily[6]. Postprandial hyperglycemia can occur in some well-regulated diabetic dogs when meals with high insoluble fiber content are fed at the same time as the Humulin N insulin injection[6], indicating that the ideal time for feeding such meals might be 1-2 hours after the insulin injection, which is not convenient for many owners.

Pre-mixed combinations of 30% regular and 70% isophane (NPH) insulin

Humulin 70/30 (Eli Lilly) and Mixtard 30/70 (Novo Nordisk) are still available in most countries. These 100 U/mL pre-mixed combinations of 30% short-acting and 70% intermediate-acting insulin have a similar action to lente insulin and so are also suited to a twice-daily dosing regimen in diabetic dogs where the meals are fed at the same time as the insulin injections. Insulin dosing pens are available for both these pre-mixed regular and NPH insulin products and the 100% NPH insulin preparations. These pens can administer insulin doses in either 0.5 or 1 U increments, and so provide a practical option for dogs that require small doses of insulin or for owners who have difficulty measuring and administering insulin doses using a syringe and needle.

What about the new insulin preparations?

Recombinant human protamine zinc insulin

The release of ProZinc (Boehringer Ingleheim) provides a veterinary product to fill the gap in the market created by the discontinuation of PZI-VET, and efficacy of the new product[7] for treatment of diabetic cats is comparable to that of the discontinued product[8]. Protamine zinc insulin is not generally recommended as a first choice therapy for diabetic dogs as it has a less predictable action with a slower onset than lente insulin[9] or the 70% or 100% NPH preparations.

The short-acting lispro, aspart, and glulisine insulin preparations

Lispro insulin (Humalog, Eli Lilly) has been shown to be a safe and effective alternative to regular insulin for the treatment of diabetic ketoacidosis in dogs[10].

Table 1. Currently available insulin products*

Type of insulin	Concentration	Brand name(s)	Manufacturer	Multi-use vials	Injection pens
		VETERINARY INSULIN PREPARATIONS			
insulin zinc suspension (lente)	40 U/mL	(Vetsulin); Caninsulin	Intervet Schering-Plough	2.5 & 10 mL	None
protamine zinc insulin (PZI)	40 U/mL	ProZinc	Boehringer Ingleheim	10 mL	None
		TRADITIONALLY AVAILABLE HUMAN INSULIN PREPARATIONS			
regular insulin	100 (& 500) U/mL	Humulin R	Eli Lilly	3 & 10 mL	None
regular insulin	100 U/mL	ActRapid	Novo Nordisk	10 mL	NovoPens including Junior/Demi[a]
isophane (NPH) insulin	100 U/mL	Humulin N	Eli Lilly	10 mL	HumaPen Luxura incl. HD[a]; Humulin N Pen[b]
isophane (NPH) insulin	100 U/mL	Protophane	Novo Nordisk	10 mL	NovoPens incl. Junior/ Demi[a]; Innolet[a]; NovoLet[b]
50% regular & 50% isophane (NPH)	100 U/mL	Mixtard 50/50	Novo Nordisk	None	NovoPens including Junior/Demi[a]; Innolet[a]
30% regular & 70% isophane (NPH)	100 U/mL	Mixtard 30/70	Novo Nordisk	None	NovoPens including Junior/Demi[a]; Innolet[a]
30% regular & 70% isophane (NPH)	100 U/mL	Humulin 70/30	Eli Lilly	10 mL	HumaPen Luxura incl. HD[a]; Humulin 70/30 Pen[b]
		NEW HUMAN INSULIN PREPARATIONS			
insulin lispro	100 U/mL	Humalog	Eli Lilly	3 & 10 mL	Humapen Luxura incl. HD[a]; Autopen[a]; KwikPen[b]
insulin aspart	100 U/mL	NovoLog; NovoRapid	Novo Nordisk	10 mL	NovoPens including Junior/Demi[a]; FlexPen[b]
insulin glulisine	100 U/mL	Apidra	Sanofi Aventis	10 mL	OptiClik[a]; SolarStar[b]
50% lispro & 50% lispro protamine	100 U/mL	Humalog Mix 50/50	Eli Lilly	10 mL	HumaPen Luxura including HD[a]; KwikPen[b]
30% aspart & 70% aspart protamine	100 U/mL	NovoLog Mix 70/30; NovoMix 30	Novo Nordisk	10 mL	NovoPens including Junior/Demi[a]; FlexPen[b]
25% lispro & 75% lispro protamine	100 U/mL	Humalog Mix 75/25	Eli Lilly	10 mL	HumaPen Luxura including HD[a]; KwikPen[b]
insulin glargine	100 U/mL	Lantus	Sanofi Aventis	10 mL	Autopen[a]; SolarStar[b]
insulin detemir	100 U/mL	Levemir	Novo Nordisk	10 mL	NovoPens including Junior/Demi[a]; FlexPen[b]

* Availability and brand names of commercial insulin products and injection devices change over time and vary between countries.

[a] Insulin injection pen that uses 3 mL cartridge refills. All deliver 1 U dose increments, except for the NovoPen Junior/Demi and the HumaPen Luxura HD which deliver 0.5 U increments.

[b] Pre-filled disposable 3 mL insulin injection pen. All deliver 1 U dose increments, except for the NovoLet which delivers 2 U dose increments.

Pre-mixed combinations of short-acting lispro or aspart insulin with longer-acting protamine insulin

Humalog Mix 75/25 (Eli Lilly) is marketed for human diabetics as an improved product that can be used twice-daily in replacement of Humulin 70/30[11]. Compared with Humulin 70/30, Humalog Mix75/25 in human subjects has a more rapid and predictable onset of glucose-lowering activity[12], with greater reduction in postprandial glycemia when administered with a meal[13], and a similar duration of action. There is currently no published information on the use of these new pre-mixed combinations for the management of diabetes in dogs. Research in diabetic dogs is warranted as the product(s) might prove reliable and efficacious when administered twice-daily at the same times as meals. Based on the human data, it can be predicted that the risk of hypoglycemia might be greatest in the first 2 hours after injection in dogs and so it will likely be crucial that meals are always consumed at the time of injection.

Insulin glargine and insulin detemir

Lantus (Sanofi Aventis) and Levemir (Novo Nordisk) are long-acting synthetic insulin analogues developed for the human diabetic market to provide continuous basal insulin concentrations that inhibit hepatic glucose production. The therapeutic aim is to mimic the physiological pattern of insulin secretion of healthy subjects, and the basal insulin levels provided by these products are supplemented at meal times by administration of short-acting insulin preparations that act during the postprandial period[14]. Insulin glargine is efficacious for the treatment of diabetes in cats[15]. In contrast, information regarding the use of insulin glargine in diabetic dogs is limited, and initial pharmacological assessment in healthy dogs found that glargine administration resulted in an unpredictable serum insulin concentration response and failed to produce a significant glucose-lowering effect in 2 of 9 dogs after a single injection[9]. For diabetic dogs, it is important to use an insulin preparation that has a predictable and reliable glucose-lowering effect. This is because it is crucial that the period of maximal insulin action coincides with the postprandial period if clinical signs are to be controlled with minimal risk of hypoglycemia. Generally, the longer the duration of action of an insulin preparation, the greater the variability in response between dogs receiving the same dose and the less predictable the period of maximal insulin action. Until there is further evaluation of the efficacy of insulin glargine or levemir for the treatment of diabetes in dogs, it is not recommended that they are used for initial therapy.

Should we use insulin dosing pens for diabetic dogs?

Insulin dosing pens are designed to be used by people with no formal medical training with the goal of making the task of measuring and administering insulin doses easier, less painful, and more accurate and precise. They can provide the same benefit for diabetic animals and their owners. Most insulin dosing pens allow administration of doses in 1 U increments. Many owners appear to be able to measure doses in 0.5 U increments if they use a 0.3 mL U-100 insulin syringe. However it is worth noting that insulin dosing pens are much more accurate and precise than insulin syringes for small doses. For example, when administering a dose of 1 U of a 100 U/mL insulin preparation, the mean dose delivered using the AutoPen is 0.93 U (range: 0.63-1.20 U) and for the SolarStar pen is 1.02 U (range: 0.60-1.40 U). In comparison, a study evaluating the accuracy and precision of insulin dosing by pediatric nurses using 0.3 mL or 0.5 mL insulin syringes found that the mean $\pm$SD dose delivered when intending to administer 1 unit was 1.64 $\pm$0.38 (range: 0.65-2.80 U)[16]. Importantly, it can be seen that most syringe-measured doses exceed the intended dose.

The HumaPen Luxura HD pen (by Eli Lilly) and the NovoPen Junior (USA) / Demi (elsewhere) pens (by Novo Nordisk) are specifically designed for use in babies and children and deliver accurate and precise insulin doses in 0.5 U increments. As these pens can be used with the 3 mL cartridges of all the intermediate-acting 70% and 100% NPH insulin preparations, they are particularly suitable for use in diabetic dogs.

When insulin dosing pens are not in use, for example before they are opened, they should be stored in the refrigerator, however once opened and in use, the pens should *not* be refrigerated and instead should be kept at room temperature <30°C (86°F) away from direct heat and light. Before each use, insulin dosing pens typically need to be primed until a stream of insulin appears, which can seem wasteful to those not familiar with the protocol. During injection, the needle must be left in the skin for several seconds while the injection button is held down and, when the needle is removed after injection, it is normal for there to be a drop of insulin visible on the tip of the needle.

GLYCEMIC MONITORING – WHAT IS NEW?

The AlphaTRAK blood glucose meter (Abbott Animal Health) is designed specifically for veterinary patients and gives more accurate and precise blood glucose measurements for dogs than blood glucose meters designed for human samples[17]. It requires only 0.3 μL of blood and so is suitable for home monitoring of blood glucose concentration as well as for hospital use. Unlike human blood glucose meters, the AlphaTRAK meter does not consistently give results that are lower than the laboratory reference method, and so can potentially miss hypoglycemia in insulin-treated patients. If there is ever doubt about the accuracy of any blood glucose concentration result, the best approach is to repeat the measurement using the same meter or a laboratory reference method. Although rarely relevant in veterinary medicine, it is noteworthy that the AlphaTRAK test strips use the glucose dehydrogenase (GDH)-pyrroloquinoline quinone (PQQ) enzyme and thus will cross-react with non-glucose sugars and so yield an erroneously high result[18,19].

Continuous monitoring of subcutaneous or interstitial glucose concentration provides several advantages for diabetic dogs compared with traditional methods of serial blood glucose monitoring[20]. The major advantage is that glucose results are obtained every 5 minutes allowing clinicians to base treatment decisions on much more detailed glycemic data. The Guardian REAL-Time Continuous Glucose Monitoring System (CGMS) (Medtronic) provides glucose results in real time and uses wireless technology so the monitor does not need to be attached to the animal. There is strong correlation between interstitial and blood glucose concentrations in diabetic dogs[21,22] with a delay of 5-12 minutes compared with blood glucose[23]. Importantly, the CGMS system was shown to be valid for use in the home environment of diabetic dogs[22] and facilitated detection of hypoglycemia[21]. Limitations of the system include a 2 hour set up period before glucose readings are obtained and the requirement for blood sampling during monitoring 2-3 times every 24 hours to calibrate the device. Importantly, all information on this new technology is based on standard measuring conditions and so there is no information on how the system is affected in dogs by physiological and pathological states such as dehydration, acidosis, anesthesia, etc. As with all methods of monitoring glycemia, it is not advisable to rely on the CGMS as the sole monitoring tool. If there is doubt about the accuracy of interstitial glucose results, the best approach is to immediately obtain a blood glucose measurement.

REFERENCES

[1]FDA CVM alert at http://www.fda.gov/AnimalVeterinary/NewsEvents/CVMUpdates/ucm188752.htm. [2]Munroe WE, et. al. **J Vet Intern Med** 2005;19:675. [3]Graham PA, et. al. **J Small Anim Pract** 1997;38:434. [4]Fleeman LM et. al. **Vet Record** 2009;164:232. [5]Feldman EC & Nelson RW. Canine and Feline Endocrinology and Reproduction. 2004;580. [6]Palm CA, et. al. **J Vet Intern Med** 2009;23(1):50. [7]Nelson RW, et. al. **J Vet Intern Med** 2009;23:787. [8]Nelson RW, et. al. **J Am Vet Med Assoc** 2001;218:38. [9]Stenner VJ, et. al. **J Vet Intern Med** 2004;18:444. [10]Sears, KW, et. al. **J Vet Intern Med** 2009;23:696. [11]Eli Lilly prescribing information at http://www.pi.lilly.com/us/humalog7525-pi.pdf. [12]Malone, JK, et. al. **Diabetes Metab** 2000;26:481. [13]Malone, JK, et. al. **Clin Ther** 2000;22:222. [14]Quattrin T. **Expert Opinion on Pharmacotherapy** 2004;5:2597. [15]Marshall, RD, et. al. **J Fel Med Surg** 2009;11:683. [16]Casella SJ, et. al. **Pediatrics** 1993;91:1155. [17]Cohen TA, et. al. **J Am Vet Med Assoc** 2009;235:276. [18]Ginsberg B. **J Diabetes Sci Technol** 2009;3:903. [19]FDA Public Health Notification: http://www.fda.gov/MedicalDevices/Safety/AlertsandNotices/PublicHealthNotifications/ucm176992.htm. [20]Wiedmeyer CE, et. al. **Diabetes Technol Ther** 2005;7:885. [21]Davison LJ, et. al. **J Small Anim Pract** 2003;44:435. [22]Wiedmeyer CE, et. al. **J Am Vet Med Assoc** 2003;223:987. [23]Rebrin K & Steil GM. **Diabetes Technol Ther** 2000;2:461.

A MODEL OF ACID-INDUCED INJURY IN THE CANINE STOMACH. T Hill, A Blikslager. North Carolina State University College of Veterinary Medicine, Raleigh, NC.

There are many recognized contributing factors of gastric ulcers, including non-steroidal anti-inflammatory therapy, liver disease and others. Many of these predisposing factors mediate injury via increased acid secretion or increased susceptibility of the gastric mucosa to acid injury. We developed a model to study acid injury in canine pyloric mucosa and examine mucosal recovery.

Gastric antral mucosa was harvested from 9 shelter dogs immediately after euthanasia. Mucosa from each dog was mounted on Ussing chambers and treated both with acid Ringer's solution at pH 1.0 for 15 minutes on the mucosal side and normal Ringer's solution as control. Tissues were maintained for a total of 180 minutes; permeability was assessed electrophysiologically.

Permeability was significantly increased 15 minutes after application of acid Ringer's ($p=0.001$). Steady state transepithelial resistance (TER) of control tissues was $127.5 \pm 12.1\ \Omega \cdot cm^2$. Mean TER of acid injured tissues 15 minutes after acid injury was $53.2 \pm 13\ \Omega \cdot cm^2$. 150 minutes after restoration of normal Ringer's solution to the injured mucosa, mean TER of acid injured tissue had recovered to near baseline ($93.1 \pm 12\ \Omega \cdot cm^2$) which was not significantly different from control tissues.

Ex vivo models of gastric injury in dogs have not been successfully developed prior to this study. Addition of HCl caused a significant increase in mucosal permeability shortly after injury that rapidly resolved after removal of acid Ringer's. This model of acid injury and recovery may be useful for understanding the pathogenesis of peptic ulcers and mucosal repair after injury.

A ZINC L-CARNOSINE CONTAINING PRODUCT DOES NOT AMELIORATE ACID-INDUCED INJURY IN THE CANINE STOMACH. T Hill, A Blikslager. North Carolina State University College of Veterinary Medicine, Raleigh, NC.

Gastric ulcer disease is common in dogs. Multiple medical therapies have been developed for both the prevention and treatment of canine gastric ulcers. Zinc l-carnosine is a compound used in human medicine for the treatment of gastric ulcers with reported success. However the use of zinc l-carnosine in dogs has not been described. We hypothesized that Gastricalm®, a zinc l-carnosine-containing compound formulated for dogs, would prevent increased gastric permeability in response to application of HCl.

Mucosa from the gastric antrum was harvested from 6 shelter dogs immediately after euthanasia, and mounted on Ussing chambers. The tissues were equilibrated for 30-minutes prior to addition of HCl (final pH of 1.5-2.5), Gastricalm®, and acid Ringers solution plus Gastricalm®. Tissues were maintained for 180-minutes, during which permeability was assessed electrically.

Permeability of the gastric mucosa was significantly increased in a dose-dependent manner by addition of HCl, whereas control tissues maintained viability for the study period. Change in permeability was detected within the first 15-minutes of acid application and progressed over the subsequent 150-minutes. Gastricalm® had no significant effect on this increase in permeability.

Addition of HCl caused a dose-dependent increase in permeability over time. However, there was no significant effect of Gastricalm® in this study, suggesting the widespread use of this agent is unfounded.

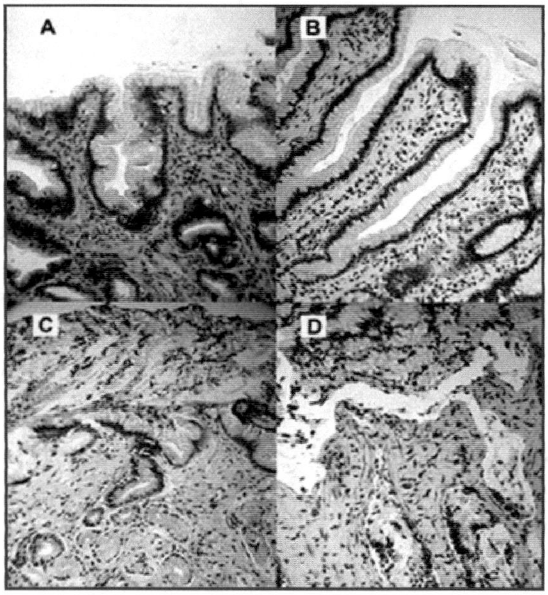

Gastric mucosa at 180 minutes, showing loss of gastric glands and mucosal epithelium with acid Ringer's solution and absence of amelioration with Gastricalm® . A: Normal Ringer's solution, B: Normal Ringers solution with Gastricalm®, C: Acid Ringer's solution, D: Acid Ringer's solution with Gastricalm®

DIAGNOSIS OF PRIMARY COPPER HEPATOPATHY IN DOGS VIA MEASUREMENT OF BILIARY COPPER CONCENTRATION. BM Pressler[1], GE Moore[1], RP Favier[2]. [1]Purdue University, West Lafayette, IN. [2]Utrecht University, Utrecht, The Netherlands.

Primary copper (Cu) hepatopathy due to defective biliary excretion of Cu has been described in a number of pure-breed dogs. Alternatively, dogs with liver disease of other etiologies oftentimes also are noted to also have hepatic Cu concentrations above reference range. Diagnosis and differentiation of primary Cu hepatopathies versus other hepatopathies with secondary Cu accumulation requires liver biopsy (a relatively invasive procedure) followed by assay of hepatic Cu concentration and special stains to determine the distribution of Cu within the hepatic lobule. Because excess Cu is normally excreted into bile, we hypothesized that decreased biliary Cu concentration would allow diagnosis of Cu-associated hepatopathy in a colony of affected dogs, and differentiation from dogs with possible secondary Cu accumulation due to other liver diseases.

Paired liver and bile samples were collected from 37 overtly healthy hound-type adult dogs (group 1A), 5 healthy, adult COMMD$^{+/-}$ Beagle-Bedlington terrier crossbreed dogs (group 1B), 18 client-owned dogs with a variety of non-primary Cu hepatopathies (group 2), and 4 adult COMMD$^{-/-}$ Beagle-Bedlington terrier dogs previously confirmed to have primary Cu hepatopathy by assay of liver Cu concentration (group 3). All dogs in group 1A, 1B, and 3 were being fed commercial dry adult maintenance dog foods, whereas group 2 dogs were fed a variety of diets based on owner and referring veterinarian preference. Group 3 dogs had been receiving penicillamine (10 mg/kg PO q12h) which was discontinued 6 days prior to obtaining liver and bile samples. Liver and bile Cu concentrations were assayed by a commercial laboratory, and differences analyzed by two-sample t-test, with significance designated at $p<0.05$.

Liver and bile Cu concentrations (in ppm) for the four groups of dogs were:

Group	Mean ± SD Liver Cu	Mean ± SD Bile Cu
1A (healthy hound dogs)	57.5 ± 28.3	31.6 ± 13.4
1B (COMMD$^{+/-}$ dogs)	127.6 ± 70.9	22.5 ± 13.7
2 (non-Cu hepatopathy dogs)	167.9 ± 114.2	19.4 ± 18.6
3 (COMMD$^{-/-}$ dogs)	4070.0 ± 1547.4	4.3 ± 1.6

Healthy dogs (Groups 1A + 1B) had significantly higher biliary Cu concentrations than dogs with hepatopathies (Groups 2 + 3; $p<0.01$) or dogs in Groups 2 or 3 when considered separately ($p<0.01$). In this population of dogs, a biliary Cu concentration of 7.5 ppm or higher was 100% specific and 50% sensitive for excluding a diagnosis of primary Cu hepatopathy.

These results suggest that biliary Cu measurement may be a less-invasive method than liver biopsy for excluding a diagnosis of primary Cu hepatopathy in dogs. However, a larger number of dogs with primary Cu hepatopathy are required to determine if these results can be extrapolated beyond this colony, and in particular, testing of young affected dogs is required to validate this as an early screening test.

SELECTED GROWTH FACTOR EXPRESSION IN *SPIROCERCA LUPI*-INDUCED ESOPHAGEAL NODULES. E. Dvir, SJ Clift. Faculty of Veterinary Science, University of Pretoria, Pretoria, South Africa.

The study aims to assess the expression of vascular endothelial growth factor (VEGF)-A, fibroblast growth factor (FGF) and platelet-derived growth factor (PDGF) in the progression of the spirocercosis-induced esophageal nodule in the dog from an early, non-neoplastic, inflammatory nodule to a neoplastic sarcoma.

Triplicate 4-μm-thick sections from 62 paraffin blocks containing *Spirocerca*-induced nodules were stained by immunohistochemistry using R&D System, polyclonal goat anti-canine VEGF and anti-human FGF and PDGF antibodies. The nodules were classified as early inflammatory (n=15), pre-neoplastic (n=27) and neoplastic/sarcoma (n=20). Additionally, 10 sections of normal distal third of the esophagus and 21 non-spirocercosis-related sarcomas were evaluated and compared with the *Spirocerca*-induced nodules. Five non-overlapping high power fields per case were evaluated under light microscopy and the fibroblasts were evaluated for percentage positive staining. The intensity of the positive staining was further divided as either strong (score 2) or weak (score 1). The intensity score was multiplied by the percentage to yield a field score and a final score was obtained by calculating an average of the 5 fields. Antibody staining was compared between the different histological grades and the controls using the Kruskal-Wallis Test followed by the Mann-Whitney Test as a post hoc testing. The level of significance was set at 0.05.

There were significant differences between the groups' score in all the growth factors that were examined. The normal esophagus showed no staining for any of the growth factors. FGF had the highest score in the non-spirocercosis-related sarcoma (median 118, 3-194) followed by spirocercosis-induced sarcoma (34.5, 0-138), pre-neoplastic nodule (8, 0-99) and early nodule (0, 0-30). All the differences among the groups were significant. VEGF had the highest score in the non-spirocercosis-related sarcoma (median 47, 1-110) followed by spirocercosis-induced sarcoma (26, 0-136), pre-neoplastic nodule (0, 0-62) and early nodule (0, 0-35). The differences among the groups were significant, except between the normal vs. early nodule, early vs. pre-neoplastic and non vs. spirocercosis-induced sarcoma. PDGF had the highest score in the non-spirocercosis-related sarcoma (median 29.2, 0-70) followed by pre-neoplastic nodule (23.4, 0-95), early nodule (13.6, 0-132) and spirocercosis-induced sarcoma (0, 0-47). The differences among the groups were significant except the normal vs. spirocercosis-induced sarcoma, early vs. pre-neoplastic nodule and non-spirocercosis-related sarcoma and pre-neoplastic nodule vs. non-spirocercosis-related sarcoma.

The expression of VEGF and FGF increased as the nodule progressed from early inflammation to sarcoma, but it is not limited to spirocercosis-induced sarcoma. The expression of PDGF in spirocercosis is restricted to the early stages of the nodule progression. Further investigation is warranted to establish if FGF, VEGF or PDGF play a role in the pathogenesis of the malignant transformation in canine spirocercosis.

THE EFFECTS OF AN INCRETIN MIMETIC, EXENATIDE, IN HEALTHY CATS

JD Brunker, TM Seyfert, LK Maxwell, D McFarlane
Center for Veterinary Health Sciences, Oklahoma State University, Stillwater, OK

The incretin effect refers to the phenomenon in which orally ingested glucose leads to a greater insulin response compared to intravenously administered glucose. Incretin mimetics, such as exenatide, represent a potentially attractive and safe therapeutic option for cats with type 2 diabetes mellitus. The purpose of the study was to determine the insulinotropic effects of exenatide during a hyperglycemic clamp in 6 healthy cats. The second phase of the study was to determine drug safety following 4 weeks of twice daily exenatide administration in 9 healthy cats using physical examinations, body weight, hematology, and serum biochemistry.

A randomized cross over design was used, with each cat receiving exenatide at 0.1 µg/kg, 0.24 µg/kg, or 1 µg/kg or saline. The hyperglycemic clamp was initiated by administering 20% dextrose IV to maintain the blood glucose between 180 mg/dl and 200 mg/dL (~10 mmol/l to 11 mmol/l). Once the blood glucose concentration and the dextrose infusion rate had remained steady for at least 30 minutes, the dose of exenatide or saline was administered subcutaneously and that was marked as time 0. Blood samples were collected for insulin measurements at times -10, -5, 0, 15, 30, 45, 60, 90, 120, 180, and 240 minutes.

For the chronic administration phase, exenatide at a dose of 1 µg/kg was administered subcutaneously every 12 hours for 28 days. The cats were monitored twice daily for adverse effects. Body weights were measured weekly throughout the chronic administration period. Insulin was expressed as the ratio of plasma insulin concentration: baseline insulin concentration.

Exenatide effectively increased insulin release in a dose-dependent manner. At a dose of 1 µg/kg of exenatide, the onset of action was approximately 15 minutes and the duration of action was approximately 90 minutes. The area under the curve was calculated for the insulin ratio from 15 to 90 minutes after the subcutaneous administration of exenatide over a range of dose rates. Exenatide appeared to conform to a sigmoidal E_{max} dose-response model. No adverse effects were noted following 28 days of exenatide at 1 µg/kg q 12 hr. All cats lost weight after 28 days on drug.

Exenatide increased insulin release in healthy cats, with maximum observed effect at a dose of at1 µg/kg. This dose when administered chronically induced significant weight loss without any apparent adverse effects.

FELINE HERPESVIRUS-1 AND *MYCOPLASMA* SPP. INFECTIONS IN CATS WITH ACUTE CONJUNCTIVITIS IN AN ANIMAL SHELTER. D. Zirofsky[1], C.C. Powell[1], W. Rekers[2], M. Brewer[1], M.R. Lappin[1]. 1. Colorado State University, Fort Collins, CO. 2. Cat Adoption Team, Sherwood OR.

In previous studies, DNA of feline herpesvirus 1 (FHV-1) and *Mycoplasma* spp. were the most likely to be amplified by PCR assays from nasal discharges of cats with rhinitis and ocular discharges of cats with conjunctivitis. However, both organisms can colonize healthy cats and in the previous studies, only small numbers of samples from normal cats were included as controls. Thus, the significance of positive PCR test results for FHV-1 and *Mycoplasma* spp. in samples from clinically ill cats is unclear. The purpose of this study is to compare the prevalence rates for FHV-1 DNA and *Mycoplasma* spp. DNA in samples from cats with and without conjunctivitis.

Cats housed in two animal shelters in Oregon were used for the study. Cats with conjunctivitis with no obvious exogenous cause, with or without rhinitis were included in the study. For every two cats with conjunctivitis, samples were to be collected from a cat housed in the shelter without conjunctivitis. Conjunctival cells were collected by rubbing a sterile cotton swab gently in the conjunctival fornix of both eyes after application of topical 1% proparacaine to induce local analgesia. Samples were stored at -20C until shipped to Colorado State University where they were processed and assayed for total DNA as well as DNA of FHV-1 and *Mycoplasma* spp. using adaptations of previously published PCR assays. Differences in prevalence rates were compared by Fisher's exact test with significance defined as $p < 0.05$.

Overall, samples were collected from 69 cats with conjunctivitis and 30 cats without conjunctivitis. The results of the PCR assays are shown in the following table.

Assay result	Conjunctivitis (n = 69)	Control (n = 30)	p value
FHV-1 alone	16 (23.2%)	8 (26.7%)	0.7917
Mycoplasma alone	21 (30.4%)	2 (6.7%)	0.0097
Both organisms	10 (14.5%)	1 (3.3%)	0.165
Neither organism	22 (31.9%)	19 (63.3%)	0.0043
Any FHV-1	26 (37.7%)	9 (30%)	0.5017
Any *Mycoplasma*	31 (44.9%)	3 (10%)	0.0006

Results of *Mycoplasma* spp. and FHV-1 PCR assay results do not always document a disease association. However, while both organisms can be amplified from cats with and without conjunctivitis, *Mycoplasma* spp. were amplified from a greater percentage of cats with conjunctivitis suggesting that this microbial group may be common primary pathogens in cats with this syndrome.

URINARY THROMBOXANE METABOLITES ARE AFFECTED BY ASPIRIN ADMINISTRATION TO HEALTHY DOGS. CM Hoh[1], SA Smith[2], M McMichael[1], JK Byron[1]. 1. University of Illinois College of Veterinary Medicine, Urbana, IL. 2. University of Illinois College of Medicine, Urbana, IL.

Anti-inflammatory dose aspirin is proven to decrease canine platelet function, but *in vivo* effects of "mini-dose" aspirin have not been determined. The primary urinary metabolite of platelet thromboxane A_2 is 11-dehydro-thromboxane B_2 (11-$dTXB_2$) for humans or 2,3-dinor-thromboxane B_2 (2,3-$dTXB_2$) for dogs. For humans, a $<60\%$ decrease in urinary 11-$dTXB_2$ indicates aspirin resistance. The study objective was to evaluate the *in vivo* effects of mini-dose aspirin in normal dogs by measuring urinary thromboxane metabolites.

Healthy dogs (n=20) received aspirin 1 mg/kg/dX7d and, after a washout period, 10 mg/kg once (n=10). Urine was collected prior to aspirin administration, then 12 hours following the last dose. Thromboxane metabolites were measured by ELISA, and normalized to urine creatinine.

Baseline 11-$dTXB_2$ levels were low (0.5±0.3 ng/mg creat) and not impacted significantly by either aspirin protocol. Baseline 2,3-$dTXB_2$ levels were higher (6.7±2.7 ng/mg creat) and were significantly ($p<0.001$) decreased by a single 10 mg/kg dose (range: 74-33% decreased) but not consistently changed by 1 mg/kg/dX7d (range 71% decreased to 13% increased). 2,3-$dTXB_2$ levels decreased by $>60\%$ in only 2 of 20 dogs receiving 1 mg/kg/dX7d.

Urinary 11-$dTXB_2$ levels are low and not impacted by either mini- or anti-inflammatory doses of aspirin, indicating that this metabolite does not accurately reflect platelet thromboxane metabolism in dogs. Urinary 2,3-$dTXB_2$ levels significantly decreased after a single anti-inflammatory dose, indicating that this metabolite is impacted by platelet cyclo-oxygenase function. However, the mini-dose did not consistently impact urinary 2,3-$dTXB_2$ levels, suggesting that this dose is insufficient to reliably inhibit canine platelets.

INFLUENCE OF BODY CONDITION ON PHARMACOKINETICS OF ORAL PREDNISOLONE AND PREDNISONE IN CLINICALLY HEALTHY CATS. SA Center, JF Randolph, KL Warner, KW Simpson, NJ Place. College of Veterinary Medicine, Cornell University, Ithaca, NY.

We sought to validate an HPLC method to measure prednisolone and prednisone in feline plasma, and examined the effect of body condition on the pharmacokinetics of a 2 mg/kg oral glucocorticoid dose in 5 normal-conditioned and 6 over-conditioned adult cats.

HPLC validation used calibration standards (prednisone, prednisolone, hydrocortisone and betamethasone [internal standard]) in pooled feline plasma, and demonstrated linearity, lack of interfering analytes, repeatability and recovery (inter-assay [n=10], intra-assay [n=8]). Lean body mass (LBM) was quantified using a validated multifrequency bioelectrical impedance analysis system. Paired T tests and ANOVA (alpha $\leq$ 0.05) determined significance.

The assay was linear ($R^2 \geq 0.99$), recovered 84%-105% of either drug, and had intra- and inter-assay CV% ranging from 1.1-15.1%. Significantly higher maximum concentration (Cmax: 3,074+/-1,364 vs 275+/-155ng/mL), greater area-under-the-curve (AUC: 8,605+/-4,644 vs 2,177+/-1,498hr-ng/mL), faster elimination rate constant (0.23 +/-0.06 vs 0.11+/-0.030), shorter time to Cmax (Tmax: 1.5+/-0.9 vs 3.0+/-1.7hrs), and shorter half-life (3.17+/-0.72 vs 7.79+/-3.53hrs) were shown for prednisolone compared to prednisone. Prednisone was readily converted (>99%) to prednisolone, but estimated prednisone bioavailability was ~26% relative to prednisolone. Over-conditioned cats received significantly more glucocorticoid / LBM (5.8+/-1.2 vs 3.1+/-0.7mg/kg), and had significantly higher Cmax (3,704+/-1,106 vs 1,907+/-650ng/mL) and AUC (11,393+/-4,525 vs 5,259+/-1,634hr-ng/mL) for prednisolone compared to normal-conditioned cats.

This HPLC method can quantify prednisolone and prednisone in feline plasma. Apparent limited oral bioavailability of prednisone cautions against its therapeutic use in cats. Conventional prednisolone dosing based on total body weight or 'per-cat-dose' in over-conditioned cats might contribute to adverse effects (e.g., diabetes mellitus, hepatic lipidosis).

OVERVIEW OF ADRENAL GLAND PHYSIOLOGY & DIAGNOSTIC TESTING

J. Catharine Scott-Moncrieff MA, MS, Vet MB, DACVIM DECVIM
West Lafayette, IN

PHYSIOLOGY OF THE ADRENAL CORTEX

The adrenal gland is made up of the adrenal medulla which secretes catecholamines, and the adrenal cortex which secretes glucocorticoids, mineralocorticoids, and androgens. Adrenal steroids are derived from cholesterol and all contain the cyclopentanoperhydrophenanthrene (CPPP) nucleus. Cytochrome P450 enzymes present in the adrenal cortex (P450scc, P450 c17, 3β-hydroxysteroid dehydrogenase, P450 c21, P450c11, and P450c11AS), catalyze formation of the different adrenal steroids (figure 294-1). The major secretory products of the adrenal glands in dogs and cats are cortisol, aldosterone, and the androgens dehydroepiandrosterone, and androstenedione. Other hormones such as progesterone and 17α-hydroxyprogesterone are synthesized in the adrenal gland, but in normal animals only small amounts are secreted into the systemic circulation. Adrenal derived androgens serve as substrates for synthesis of estrogen and testosterone in peripheral tissues.

The adrenal cortex is composed of three histopathologic layers, the zona glomerulosa (ZG), the zona fasciculata (ZF), and the zona reticularis (ZR). The ZG is the outermost layer of the adrenal gland and is the only layer capable of synthesizing and secreting aldosterone because it contains the enzyme P450c11AS (aldosterone synthase). The ZG does not contain the enzyme P450 c17, so it is incapable of synthesizing cortisol or androgens. The ZF and ZR both synthesize androgens and cortisol.

Synthesis and secretion of cortisol by the adrenal glands is regulated by the hypothalamic pituitary adrenal axis. Corticotrophin releasing hormone (CRH) secreting neurons in the hypothalamus have axons that terminate in the anterior pituitary. Corticotrophin releasing hormone stimulates secretion of ACTH from the pituitary gland. ACTH is released into the blood, attaches to receptors in the adrenal cortex and stimulates synthesis and secretion of cortisol. As the plasma cortisol concentration rises, cortisol inhibits CRH and ACTH release from the hypothalamus and pituitary by negative feedback. Increased ACTH concentration also inhibits CRH release from the hypothalamus. ACTH release in normal animals is episodic. A diurnal rhythm has not been documented in dogs and cats. Factors that stimulate CRH release include stress, hypoglycemia, and physical exercise.

Synthesis and secretion of aldosterone is regulated by the renin-angiotensin axis, the plasma potassium concentration and to a minor extent, the plasma sodium and ACTH concentration. Increased plasma potassium concentrations and angiotensin II both markedly increase aldosterone release from the adrenal cortex. Increased angiotensin synthesis is stimulated by decreased extracellular fluid volume which causes a reflex increase in renal nerve discharge and increased renal intra-arterial mean pressure. This results in increased renin secretion from the juxtaglomerular cells that surround the afferent arterioles as they enter the glomeruli. Renin acts on angiotensinogen to form angiotensin I which is then converted to angiotensin II primarily in the lungs by angiotensin converting enzyme. Angiotensin II causes increased aldosterone synthesis and secretion by increasing conversion of cholesterol to pregnenolone, and corticosterone to aldosterone. Increased plasma potassium also directly stimulates secretion of aldosterone. Total absence of ACTH can decrease aldosterone secretion, but ACTH has little effect in controlling the rate of secretion and hypophysectomy does not result in mineralocorticoid deficiency.

Glucocorticoids have a variety of effects that make them crucial to normal homeostasis. Glucocorticoids stimulate hepatic gluconeogenesis and glycogenesis, and enhance protein and fat catabolism. They have a permissive action on many metabolic reactions such as lipolysis and calorigenesis and they are important in maintaining vascular reactivity to catecholamines. Glucocorticoids are also important in maintaining normal blood pressure and counteracting the effects of stress.

Mineralocorticoids increase absorption of sodium and secretion of potassium in the kidney, sweat glands, salivary glands and intestinal epithelial cells. Thus aldosterone is critical in conserving body salt. The main site of action in the kidneys is the principal cells of the collecting tubules and to a lesser extent the distal tubules and collecting ducts. Aldosterone also causes secretion of hydrogen ions in exchange for sodium in the intercalated cells of the cortical collecting tubules. Thus a deficiency of aldosterone causes a mild metabolic acidosis.

DIAGNOSTIC TESTING OF THE ADRENAL GLAND

There are a number of different tests that may be utilized to evaluate the function of the adrenal gland in dogs and cats. The choice of test is dependent on the species of interest and the suspected disease process. Test choice is also influenced by presence or absence of concurrent illness and whether there has been prior treatment with corticosteroids.

Basal cortisol

Measurement of a basal cortisol concentration has little utility in diagnosis of hyperadrenocorticism and a low basal cortisol is also not adequate for confirmation of a diagnosis of hypoadrenocorticism, because some normal dogs have a low basal cortisol concentration but an appropriate response to ACTH administration. Measurement of a basal cortisol of >2μg/dl had a negative predictive value of 100% in a study of 123 dogs evaluated for hypoadrenocorticism. Thus a basal cortisol of >2μg/dl is a useful test to exclude a diagnosis of hypoadrenocorticism.

ACTH stimulation test

The ACTH stimulation test is used in both diagnosis of hypoadrenocorticism and hyperadrenocorticism. The ACTH stimulation test relies on the assumption that hyperplastic or neoplastic adrenals have abnormally large reserves of cortisol and therefore hyper-respond to maximal stimulation by ACTH while dogs with hypoadrenocorticism have no adrenal reserve. The ACTH stimulation test is the gold standard test for confirmation of a diagnosis of hypoadrenocorticism. It is also useful in diagnosis of hyperadrenocorticism especially in dogs with suspected exposure to exogenous glucocorticoids and with concurrent illness. It is also extensively used in the management of dogs with pituitary hyperadrenocorticism (PDH) receiving medical therapy. Unfortunately the ACTH stimulation test is less sensitive for diagnosis of hyperadrenocorticism (sensitivity approximately 85%) than other tests such as the low dose dexamethasone suppression test and it has particularly poor sensitivity in dogs with adrenal tumors (AT) (sensitivity 50 - 60%). Because of the low sensitivity of this test, a diagnosis of HAC should not be excluded based on a normal ACTH stimulation. The ACTH stimulation test does not distinguish between ADH and PDH. For an ACTH stimulation test, a baseline sample for measurement of cortisol is collected followed by administration of synthetic ACTH (cosyntropin) at a dose of either 250μg/dog or 5μg/kg IV. A second sample for measurement of cortisol is collected one hour after administration of ACTH. The lower dose of 5μg/kg has been evaluated in normal dogs dogs with hyperadrenocorticism, and dogs with suspected hypoadrenocorticism. These studies have confirmed that the two doses result in an equivalent increase in cortisol concentration. Equivalent responses to cosyntropin have also been reported with the IV and IM routes of administration. One dose of dexamethasone can be administered prior to performing an ACTH stimulation test if clinically indicated, because ACTH will override the suppressive effects of one dose of dexamethasone and dexamethasone is not detected by the assay for cortisol. Cosyntropin can be reconstituted and frozen at -20 C in plastic syringes for up to six months with no effects on bioavailability. Use of compounded ACTH is an alternate choice for performing the ACTH stimulation if cosyntropin is not available, however potential variability between compounding pharmacies and variability of the peak cortisol response makes this a less than ideal choice. If compounded ACTH gel is used two post ACTH samples should be collected at 1 and 2 hours after ACTH administration. In dogs with hypoadrenocorticism, both the pre- and post- ACTH cortisol concentrations are usually less than 1μg/dl, and both values should be less than the reference range for basal cortisol (usually 2 μg/dl) to confirm the diagnosis. In dogs with hyperadrenocorticism an excessive response to ACTH stimulation (> 25μg/dl) is expected. In cats synthetic ACTH is administered either IV or IM at a dose of 125 μg /cat. Samples are collected at baseline, and then at 30 and 60 minutes after administration of ACTH. A post ACTH cortisol of >15 μg/dl at either 30 or 60 minutes after administration is consistent with a diagnosis of hyperadrenocorticism. It is important to use the protocol and reference range established by the laboratory being used for cortisol testing.

Low-dose dexamethasone suppression test

The low dose dexamethasone suppression (LDDS) test is used for diagnosis of hyperadrenocorticism and is a sensitive screening test (sensitivity 95%). It does not however, allow differentiation of iatrogenic hyperadrenocorticism and may be falsely positive or negative in this situation. The LDDS test relies on the fact that the administration of exogenous glucocorticoids should suppress the production of ACTH from the normal pituitary and therefore the production of cortisol from the normal adrenal. This suppression persists in the normal dog for 24 - 48 hours. Since dexamethasone is not detected by the assay for cortisol, this suppression can be measured after administration of exogenous dexamethasone. Adrenal tumors function independently of ACTH and a hyperplastic or neoplastic pituitary gland is relatively resistant to negative feedback from circulating steroids, so in either cause of hyperadrenocorticism no suppression occurs. The test requires a blood sample for the measurement of a baseline cortisol, followed by the administration of dexamethasone IV at a dose of 0.01 mg/kg. The patient is then left undisturbed in a cage and a second blood sample is collected 8 hours later. In normal dogs the cortisol concentration of the 8 hour sample should be less than 1.0 -1.5 μg/ml. Dogs with HAC fail to have suppression of the cortisol concentration. Additional information may be obtained by measuring a 4-hour cortisol concentration. If suppression occurs at 4 hours but "escape" occurs at 8 hours this is diagnostic for PDH and further differentiation testing is unnecessary.

The low dose dexamethasone suppression test is the test of choice for diagnosis of feline hyperadrenocorticism and is performed using a higher dose of dexamethasone (0.1 mg/kg IV) than in the dog. A base-line blood sample is collected, and additional samples are collected at 4 and 8 hours after dexamethasone administration. Cortisol concentration will be suppressed (<1.5 μg/dl) at 8 hours in normal cats but not in cats with hyperadrenocorticism. A few cats with HAC will suppress normally on the LDDS. If the index of suspicion for hyperadrenocorticism is high, a second test using the lower dose of dexamethasone (0.01 mg/kg) can be performed but some normal cats will fail to suppress at this dose.

Urinary cortisol:creatinine ratio

The urinary cortisol:creatinine ratio is a very useful screening test for spontaneous hyperadrenocorticism. This test is performed on a voided morning urine sample collected at home, so is a very convenient initial screening test. The normal range varies between laboratories. This test is extremely sensitive for HAC but is not very specific. Diseases other than hyperadrenocorticism may increase the cortisol:creatinine ratio. The cortisol;creatinine ratio is best used as a screening test to rule out HAC as a differential diagnosis. A dog with a normal cortisol:creatinine ratio is unlikely to have hyperadrenocorticism (with the exception of atypical hyperadrenocorticism). The cortisol;creatinine ratio has some potential for identification of dogs at risk of hypoadrenocorticism due to mitotane administration but has not been evaluated for routine diagnosis of hypoadrenocorticism.

Endogenous ACTH concentration

Measurement of endogenous ACTH is used in dogs and cats to distinguish primary from secondary hypoadrenocorticism and to distinguish pituitary dependent and adrenal dependent hyperadrenocorticism. In dogs with confirmed hypoadrenocorticism that have normal electrolyte concentrations, measurement of an ACTH concentration above the reference range confirms a diagnosis of primary hypoadrenocorticism, while an ACTH concentration within or below the reference range is consistent with a diagnosis of

secondary hypoadrenocorticism. In dogs without the classic electrolyte changes of hypoadrenocorticism, differentiation of primary from secondary hypodrenocorticism allows identification of dogs that are at risk for progression to complete adrenal failure.
In dogs with hyperadrenocorticism, plasma ACTH concentrations should be high in pituitary dependent disease, while in functional adrenal tumors the plasma ACTH concentration should be low. The reference range for ACTH depends upon the laboratory and assay used. In our laboratory, ACTH concentrations >15 pg/ml are consistent with a diagnosis of PDH. ACTH concentrations <10 pg/ml are suggestive of ADH however some dogs with PDH have low values probably due to episodic ACTH secretion. The result of the endogenous ACTH should therefore never be relied on alone to confirm a diagnosis of adrenal tumor. Endogenous ACTH is susceptible to rapid enzymatic degradation in vitro, so handling recommendations from the laboratory should be carefully followed. In dogs with PDH concentrations of ACTH precursors are correlated with the size of the tumor.

Low and High dose dexamethasone suppression tests

These tests are the most commonly used routine endocrine tests to differentiate pituitary dependent hyperadrenocorticism from adrenal dependent hyperadrenocorticism. These tests rely on the assumption that adrenal tumors are more resistant to the feedback effects of dexamethasone than are hyperplastic or neoplastic pituitary glands. The protocol for the high dose dexamethasone suppresion test is the same as the low-dose test except that the dose of dexamethasone is 10 x higher (0.1 mg/kg) and that a 4 hour sample is usually not collected. It is important that dexamethasone SP is used in this test since dexamethasone in polyethylene glycol can cause CNS signs at this high dose (This should be referenced or noted as the author's experience if possible). Suppression is defined as a 4 or 8 hour plasma cortisol concentration that is < 50% of the baseline sample. Approximately 75 - 80% of dogs with pituitary dependent hyperadrenocorticism will show suppression 8 hours after dexamethasone administration, while functional adrenal tumors will fail to have suppression of the cortisol concentration. Therefore suppression after the low or high dose dexamethasone test is diagnostic for PDH. Lack of suppression however, is not diagnostic for ADH and additional testing is required. If greater than 50% suppression is seen at 4 or 8 hours on a LDDS, a HDDS test is unnecessary.

Plasma aldosterone

Diagnosis of hypoadrenocorticism has historically relied on the cortisol concentration as an indicator of adrenal reserve while plasma aldosterone concentrations have largely been ignored, primarily because plasma aldosterone assays have not been widely available. However cortisol concentrations may not necessarily parallel aldosterone concentrations particularly in dogs with glucocorticoid deficient hypoadrenocorticism, and aldosterone might theoretically be useful in distinguishing primary from secondary hypoadrenocorticism and pure glucocorticoid deficiency from classic primary hypoadrenocorticism. In one study, plasma aldosterone concentrations were not useful in distinguishing these three groups of dogs. Further work is necessary to evaluate the utility of the aldosterone concentration in this scenario.

Aldosterone-to-renin and Cortisol-to-adrenocorticotrophic hormone ratios

Measurement of cortisol:ACTH ratio (CAR) and aldosterone:renin ratio (ARR) has been proposed as an alternative diagnostic test for hypoadrenocorticism in dogs. In a study of 22 dogs with primary hypoadrenocorticism and 60 normal dogs, there was overlap between the groups for aldosterone, renin, cortisol, and ACTH concentrations as a single measurement but the CAR and ARR was consistently lower in dogs with hypoadrenocorticism than in normal dogs. This suggests that these ratios could potentially be used in place of the ACTH stimulation test for diagnosis of hypoadrenocorticism. Advantages of measuring these ratios include the need for collection of only one blood sample, and avoiding the cost of performing the ACTH stimulation test. Disadvantages include the lack of availability of the renin and aldosterone assay, and the issues of sample handling with regard to measurement of endogenous ACTH. In addition more studies are necessary to evaluate the CAR and ARR in dogs with other illnesses that might undergo testing for hypoadrenocorticism. The CAR ratio in secondary hypoadrenocorticism might also overlap with normal dogs.
Sex hormone profiles: Recent studies have demonstrated that dogs with hyperadrenocorticism also secrete excessive quantities of other adrenal steroids, in particular progesterone, 17-hydroxy-progesterone, androstenedione and DHEA. Adrenocortical carcinomas secrete larger amounts of sex hormones relative to cortisol in response to ACTH than do benign malignant tumors and dogs with PDH, although there is significant variability between dogs. Measurement of a sex hormone profile (University Tennessee Endocrinology Laboratory) before and after administration of ACTH may be useful in the diagnosis of early or atypical hyperadrenocorticism. Sex hormone testing may also be useful in diagnosis of atypical hyperadrenocorticism caused by sex hormone secreting adrenal tumors. In these cases cortisol secretion may be normal or sub-normal. The same protocol is used for measurement of sex hormones before and after ACTH stimulation as is used for the standard ACTH stimulation test.

REFERENCES

[1.] Lennon EM, et al. Use of basal serum or plasma cortisol concentrations to rule out a diagnosis of hypoadrenocrticism in dogs: 123 cases (2000-2005). **Journal of the Veterinary Medical Association** *2007;231:413-416.* [2.] Javardi S, et al. Aldosterone-to-renin and cortisol-to-adrenocorticotrophic hormone ratios in healthy dogs and dogs with primary hypoadrenocorticism. **J Vet Intern Med** *2006;20:556-561.* [3.] Frank LA, Schmeitzel LP, Oliver JW. Steroidogenic response of adrenal tissues after administration of ACTH to dogs with hypercortisolemia. **JAVMA** 2001;218:214-216. [4.] Hill KE, et.al. Secretion of sex hormones in dogs with adrenal dysfunction. **J Am Vet Med** 2005;226:556-561. [5.] Ristic JME, Ramsey IK, et.al. The use of 17-hydroxyprogesterone in the diagnosis of canine hyperadrenocorticism . **Vet Rec** 1999;144:551-554. [6.] Granger N, de Fornel P, Devauchelle P, et al. Plasma Pro-opiomelanocortin, Pro-adrenocorticotropin hormone, and pituitary adenoma size in dogs with Cushing's disease. **J Vet Int Med** *2005;19:23-28.*

HYPOADRENOCORTICISM IN DOGS & CATS: UPDATE ON DIAGNOSIS & TREATMENT

J. Catharine Scott-Moncrieff MA, MS, Vet MB, DACVIM DECVIM
West Lafayette, IN

INTRODUCTION

Hypoadrenocorticism (Addison's disease) results from failure of the adrenal glands to secrete glucocorticoids (primarily cortisol) and mineralocorticoids (primarily aldosterone). Most cases of hypoadrenocorticism are due to primary adrenal failure, resulting in deficiency of usually both cortisol and aldosterone from the adrenal cortex. More rarely, Addison's disease may be due to pituitary dysfunction resulting in a failure of ACTH secretion and pure glucocorticoid deficiency (secondary adrenal failure). In secondary hypoadrenocorticism, mineralocorticoid secretion is expected to be normal.

ETIOLOGY OF HYPOADRENOCORTICISM

Since most cases of hypoadrenocorticism either die with no necropsy or are successfully treated, the underlying cause is usually not known in individual patients. Primary adrenal failure is suspected to be immune-mediated in most cases. Other causes of primary hypoadrenocorticism include granulomatous destruction or hemorrhagic infarction of the adrenal gland, adrenalitis, neoplasia, amyloidosis and necrosis. Immune mediated destruction of the adrenal gland may also occur together with other immune mediated endocrine disorders such as hypothyroidism, diabetes mellitus, and hypoparathyroidism. Destructive lesions in the hypothalamus or pituitary due to neoplasia, inflammation, or trauma may cause secondary hypoadrenocorticism. Idiopathic ACTH deficiency may also occur.

Primary adrenal failure may also be caused by a number of drugs including mitotane, trilostane, and ketoconazole. Adrenal suppression caused by ketoconazole is reversible but adrenal failure caused by mitotane may be permanent. Although adrenal suppression caused by trilostane is reversible in most cases, adrenal gland necrosis associated with trilostane treatment has been reported in some dogs. Administration of glucocorticoid drugs may cause secondary adrenal failure. After corticosteroid administration (topical, oral, or injectable) suppression of ACTH production from the pituitary gland occurs within a few days. This results in secondary adrenal gland atrophy. How long the adrenal axis is suppressed depends on the potency and half-life of the administered glucocorticoid. Long acting depot drugs are the most potent adrenal suppressants and can cause suppression for 5 - 6 weeks or longer.

CLINICAL SIGNS

Signalment

Seventy percent of dogs diagnosed with hypoadrenocorticism are female, and most are young to middle-aged dogs (mean 4 - 5 years). The disease is heritable in the Standard Poodle, Bearded Collie, Portuguese water dog and the Nova Scotia Duck Tolling Retriever (NSDTR) and in these breeds no obvious sex predisposition is evident. In the Standard Poodle, Portuguese Water dog, and NSDTR, the disease appears to be inherited as an autosomal recessive trait. Incidence of hypoadrenocorticism in the NSDTR is estimated to affect 1.4% of the population while in the Standard Poodle 8.6% of poodles in one study were affected.

History and Physical examination

Clinical signs may be either acute or gradual in onset and often wax and wane. Owners may not realize how long their dog has been ill until treatment results in a dramatic improvement in activity level. Since 85-90 % of adrenal reserve must be depleted before clinical signs are observed, it may require a stressful event to trigger clinical illness. Clinical signs may be very vague and are rarely pathognomonic for the disease. Anorexia, vomiting, lethargy/depression, weakness, weight loss, diarrhea, shaking/shivering, polyuria, polydipsia, and abdominal pain may be observed. Most of these clinical signs can occur due to glucocorticoid deficiency alone. If mineralocorticoids are also deficient, the clinical signs tend to be more severe and polyuria, polydipsia, hypovolemic shock, collapse and dehydration are often present. Less common clinical signs include acute gastrointestinal hemorrhage, and seizures due to hypoglycemia or electrolyte derangement. The physical examination may be normal or may reveal lethargy, weakness, dehydration, bradycardia, weak pulses, decreased capillary refill time, and other evidence of hypovolemic shock

DIAGNOSTIC TESTING

Clinical Pathology

A complete blood count may reveal a nonregenerative normocytic normochromic anemia, or the hematocrit may be increased due to dehydration. Eosinophilia, neutrophilia or lymphocytosis occur in only 20-30% of dogs with hypoadrenocorticism, but lack of a stress leukogram in a dog with systemic illness is common. A chemistry profile may reveal hyponatremia, hypochloremia, hyperkalemia, hypercalcemia, and hyperphosphatemia. These changes occur due to aldosterone deficiency with a resultant failure of the kidneys to conserve sodium. This is accompanied by profound fluid loss, shift of K^+ ions to the extracellular compartment, pre-renal azotemia due to decreased renal perfusion and hypovolemia. A mild to moderate metabolic acidosis may also occur because lack of aldosterone impairs renal tubular hydrogen ion secretion. Other serum biochemical abnormalities that may occur in dogs with hypoadrenocorticism include hypoalbuminemia, hypocholesterolemia, hypoglycemia, and increased liver enzymes. Specific gravity of the urine is commonly less than 1.030 due to loss of the normal medullary concentration gradient and impaired water reabsorption by the renal collecting tubules. The changes on the minimum data base in dogs with hypoadrenocorticism may initially mimic other disorders such as renal failure, hepatic disease, gastrointestinal disease, or insulinoma.

Serum electrolyte abnormalities

The majority of dogs with hypoadrenocorticism have the classic electrolyte changes of hyponatremia and hyperkaelina due to aldosterone deficiency. It is now recognized however, that a subset of dogs with hypoadrenocorticism do not have the classic electrolyte changes typically observed in dogs with hypoadrenocorticism. In a retrospective study of dogs with hypoadrenocorticism,

24% of dogs lacked hyponatremia and hyperkalemia. In a study of 25 NSDTR, 32% lacked electrolyte abnormalities at the time of diagnosis. Reasons for normal electrolytes include secondary hypoadrenocorticism due to decreased ACTH secretion, selective destruction of the zona fasciculata and reticularis, or early stage disease in which there has not yet been complete destruction of the zona glomerulosa. Dogs with glucocorticoid deficient hypoadrenocorticism tend to be older, have a longer duration of clinical signs, and are more likely to be anemic, hypoalbuminemic and hypocholesterolemic. It is important for clinicians to recognize that an absence of the characteristic electrolyte changes does not exclude a diagnosis of hypoadrenocorticism. Conversely reliance on measurement of electrolytes alone for diagnosis of hypoadrenocorticism can be misleading, because there are many other causes of these electrolyte changes.

Na:K ratio

The Na:K ratio is usually low in dogs with hypoadrenocorticism, and this ratio may be useful to guide emergency diagnosis and treatment while waiting for definitive test results. Use of a Na:K ratio cut off of <27 or 28 for predicting a diagnosis of hypoadrenocorticism resulted in correct classification of disease state 95% of the time in a retrospective study of 76 dogs with hypoadrenocorticism and 200 dogs with other disorders. It is important to remember however, that this ratio is only useful in dogs which have electrolyte changes. As discussed above, electrolyte concentrations and therefore the Na:K ratio may be completely normal in dogs with both primary and secondary hypoadrenocorticism.

Imaging studies

Most untreated dogs with hypoadrenocorticism have one or more radiographic abnormalities on thoracic and abdominal radiographs including microcardia, small cranial lobar pulmonary artery, narrow posterior vena cava, or microhepatica. These changes reflect the degree of hypovolemia, therefore they are more likely to be present in dogs with electrolyte abnormalities. Occasional dogs may have evidence of megaesophagus. Most dogs with hypoadrenocorticism have a measurable reduction in size of the adrenal glands, and sometimes the adrenal glands cannot be identified on ultrasound. There is overlap however with the adrenal size of normal dogs. The presence of small adrenal glands on ultrasound, although supportive of a diagnosis of hypoadrenocorticism, is not adequate for confirmation.

Electrocardiogram

In dogs with hyperkalemia, abnormalities may be present on the electrocardiogram. These include a peaked T wave and shortening of the QT interval in mild hyperkalemia, widening of the QRS complex, decreased QRS amplitude, increased duration of the P wave, and increased P-R interval in moderate hyperkalemia, and loss of P waves and ventricular fibrillation or asystole in severe hyperkalemia.

Basal cortisol

Measurement of a basal cortisol concentration is not adequate for confirmation of a diagnosis of hypoadrenocorticism, because some dogs have a low basal cortisol concentration but have an appropriate response to ACTH administration. Measurement of a basal cortisol of >2μg/dl had a negative predictive value of 100% in a study of 123 dogs evaluated for hypoadrenocorticism. Thus a basal cortisol of >2μg/dl is a useful test to exclude a diagnosis of hypoadrenocorticism.

ACTH stimulation test

An ACTH stimulation test is necessary to confirm a diagnosis of hypoadrenocorticism because not all dogs with hypoadrenocorticism have the expected electrolyte changes, and because many other disorders may mimic the characteristic findings of Addison's disease. It is acceptable to base emergency treatment upon a tentative diagnosis (based on electrolyte abnormalities), however an ACTH stimulation test should always be performed prior to initiating long term treatment. In dogs with hypoadrenocorticism, both the pre- and post- ACTH cortisol concentrations are usually less than 1μg/dl, and both values should be less than the reference range for basal cortisol (usually 2 μg/dl) to confirm the diagnosis. Although there is usually a clear distinction between the response to ACTH in a dog with hypoadrenocorticism and that of a dog with adequate adrenal reserve, sometimes borderline results can occur (post ACTH cortisol concentrations between 2 and 6 μg/dl). Other causes of inadequate response to ACTH stimulation include prior glucocorticoid administration (other than recent use of IV dexamethasone), lysodren, trilostane or ketoconazole administration, loss of activity of the ACTH product administered, and errors in administration of ACTH. Rarely dogs with sex hormone secreting adrenal tumors will have a flat line response to ACTH, however these dogs usually have overt signs of hyperadrenocorticism. Interpretation of such test results is difficult if no other underlying cause of adrenal suppression can be identified. It is possible that dogs with early adrenal gland dysfunction may initially have borderline results, but this is currently poorly documented. Whether inadequate adrenal reserve in a dog with another underlying disease (relative adrenal insufficiency) can contribute to clinical signs is also not established. Recommendations for further evaluation in dogs in which the ACTH stimulation test is borderline include repetition of the ACTH stimulation test, and measurement of concurrent ACTH concentration. Measurement of aldosterone and renin activity might also be useful in dogs with electrolyte abnormalities. Whether dogs with apparent relative adrenal insufficiency benefit from administration of glucocorticoids is unknown.

Aldosterone-to-renin and Cortisol-to-adrenocorticotrophic hormone ratios

Measurement of cortisol:ACTH ratio (CAR) and aldosterone:renin ratio (ARR) has been proposed as an alternative diagnostic test for hypoadrenocorticism in dogs. Advantages of measuring these ratios include the need for collection of only one blood sample, and avoiding the cost of performing the ACTH stimulation test. Disadvantages include the lack of availability of the renin and aldosterone assay, and the issues of sample handling with regard to measurement of endogenous ACTH.

Feline hypoadrenocorticism

Hypoadrenocorticism is a rare diagnosis in the cat. The majority of reported cases have been caused by primary adrenal failure. In most cases the underlying cause is suspected to be immune-mediated, although bilateral adrenal infiltration due to lymphoma, and permanent or transient adrenal failure secondary to trauma have also been reported. There are no published reports of spontaneous secondary hypoadrenocorticism in cats although iatrogenic hypoadrenocorticism due to rapid withdrawal of corticosteroids as been reported. There is no breed or sex predisposition and affected cats have ranged in age from 1.5 to 14 years of age. Clinical signs and laboratory abnormalities associated with hypoadrenocorticism in the cat are similar to those reported in the dog. Clinopathologic abnormalities include hyperkalemia, hyponatremia, azotemia, hyperphosphatemia, anemia, lymphocytosis, and eosinophilia. Atypical hypoadrenococrticism in which electrolyte abnormalities are absent has yet to be reported in the cat. The diagnosis of feline hypoadrenocorticism is confirmed by lack of cortisol response to ACTH administration. The most common protocol utilizes a dose of 125μg of cortrosyn administered IM or SQ with samples collected at 0, 30 minutes and 60 minutes. It is important to recognize that the reference range for the post ACTH cortisol is lower in cats than in dogs. To confirm a diagnosis of hypoadrenocorticism both the pre and post cortisol concentrations should be < than 2 μg/dl. Treatment of feline hypoadrenocorticism is with either fludrocortisone at a dose of 0.02 mg/kg per day or with DOCP at a dose of 10 -12.5 mg per month.[55] Either oral prednisone or prednisolone (1.25 mg/cat PO q 24 hours) may be used for glucocorticoid replacement. In cats that are difficult to handle, IM injectable methylprednisolone acetate can be administered monthly at the same time as DOCP administration. Response to treatment is similar although clinical signs such as anorexia, lethargy, and weakness may sometimes take longer to resolve than in dogs. Prognosis for long-term survival is good with the exception of cats in which the underlying cause is neoplasia.

ATYPICAL PRESENTATIONS OF ADDISONS DISEASE

Although diagnosis of a typical case of hypoadrenocorticism is usually straight forward, many cases have a less typical presentation with normal electrolyte concentrations.

Pure glucocorticoid deficiency

Although most cases of Addison's disease present with typical electrolyte abnormalities, in some cases the biochemical profile reveals normal electrolytes. This is always the case in secondary hypoadrenocorticism, but may also occur with early primary adrenal failure. Some dogs with primary adrenal failure will initially have glucocorticoid deficiency alone and later progress to mineralocorticoid deficiency. Interestingly, in one retrospective study of 11 dogs with glucocorticoid deficient hypoadrenocorticism, only one dog ultimately developed mineralocorticoid deficiency, despite the fact that the majority of the dogs (9/11) were diagnosed as having primary hypoadrenocorticism.

Measurement of an endogenous ACTH concentration will allow differentiation of primary versus secondary hypoadrenocorticism in these dogs and thus allow the clinician to decide whether long-term monitoring of electrolyte concentrations is necessary. Measurement of an increased ACTH concentration confirms a diagnosis of primary hypoadrenocorticism, while an ACTH concentration within or below the reference range this is consistent with a diagnosis of secondary hypoadrenocorticism. Assays for measurement of aldosterone are also available, however they are unreliable in identifying those dogs that will go on to require mineralocortiocoid as well as glucocorticoid supplementation. In one study of dogs with pure glucocorticoid deficiency dogs without electrolyte abnormalities were more likely to be anemic, hypoalbuminemic, and hypocholesterolemic.

Hypoglycemia

Hypoglycemia severe enough to cause a seizure disorder may occur in hypoadrenocorticism. Other more classic electrolyte changes may or may not be present. Hypoadrenocorticism should always be considered in the differential diagnosis of dogs presenting with hypoglycemia. It is possible that some cases of hunting dog hypoglycemia reported in the past were actually cases of atypical Addison's disease.

Severe gastrointestinal hemorrhage

Severe gastrointestinal hemorrhage may be a feature of hypoadrenocorticism. Causes of gastrointestinal hemorrhage in hypoadrenocorticism may include ischemia due to severe hypovolemia, and the effect of cortisol deficiency on the mucosa of the gastrointestinal tract. Interestingly in one report, dogs with gastrointestinal hemorrhage initially had typical Addisonian electrolyte changes, but by the time of referral had either normokalemia, or hypokalemia. Electrolyte changes typical of hypoadrenocorticism may also occur in dogs with other causes of severe gastrointestinal disease due to hypovolemia and acidosis. Differentiation of Addison's from other causes of gastrointestinal disease can only be made by the ACTH stimulation test. Dogs with severe gastrointestinal hemorrhage may require blood transfusion(s), and may have prolonged recovery.

Hepatopathy

Thirty percent of dogs with Addison's disease have increased hepatic enzymes, and in addition, hypoalbuminemia, hypocholesterolemia, and hypoglycemia are common in this disease. Hypoadrenocorticism may therefore mimic hepatic failure. Abnormalities in liver function tests have also been reported in some Addisonian dogs. It has been speculated that immune mediated hepatitis may occur concurrently with hypoadrenocorticism in some dogs. Alternatively the liver could be secondarily affected due to hypotension and impaired tissue perfusion. Regardless, hepatic abnormalities detected in Addisonian dogs resolve with no specific treatment other than that for hypoadrenocorticism.

Renal failure

Azotemia is common in hypoadrenocorticism due to hypovolemia, hypotension, and decreased renal perfusion. Usually renal abnormalities are rapidly corrected with treatment, however a delay in treatment may cause secondary renal damage. Permanent renal failure is uncommon. Hypoadrenocorticism should always be considered in the differential diagnosis of animals presenting with evidence of acute renal failure.

Megaesophagus

Reversible megaesophagus associated with hypoadrenocorticism has been reported in dogs. The cause of megaesophagus has been suggested to be the effect of abnormal electrolyte concentrations on neuromuscular function, however in some affected dogs electrolyte concentrations are normal.

Multiple endocrinopathies

Hypoadrenocorticism may occur in conjunction with other endocrine deficiencies such as hypothyroidism, diabetes mellitus, and hypoparathyroidism. A series of 10 dogs with concurrent hypoadrenocorticism and hypothyroidism have been reported. Hypothyroidism should be considered in any dog with hypoadrenocorticism that has a poor clinical response to initial treatment. In some cases profound hypothyroidism may mask the typical electrolyte changes of Addison's disease. Conversely a diagnosis of concurrent hypothyroidism may be the cause of a poor response to treatment in an Addisonian patient.

ACUTE TREATMENT

Rapid treatment of dogs with suspected Addison's disease is vital especially if profound electrolyte abnormalities are present. Aims of treatment include correction of hypotension/hypovolemia, correction of electrolyte imbalances, provision of an immediate source of glucocorticoids, and correction of acidosis, hypoglycemia, and hypercalcemia.

The suggested procedure for dogs presenting with signs of hypovolemia in which Addison's disease is suspected is to immediately: place IV catheter in cephalic or jugular vein, and collect a blood sample for measurement electrolytes and cortisol. Synthetic ACTH is then administered IV, and a second blood sample for measurement of cortisol collected 1 hour later. Fluid therapy (0.9% saline IV, 30 - 80 ml/kg/24 hours plus correction for dehydration) should be started immediately. Once the second blood sample has been collected, administer prednisolone sodium succinate at a dose of 4-20 mg/kg IV, or hydrocortisone hemisuccinate or hydrocortisone phosphate at dose of 2-4 mg/kg IV or dexamethasone 0.5 to 2.0 mg/kg as an initial dose. Then add dexamethasone 0.05 - 0.1 mg/kg q 12 hours into fluids until can switch to oral glucocortiocoids. If animal is in shock, administration of steroids should be at shock doses and this should take precedence over establishing an immediate diagnosis. For dogs with hyperkalemia consider IV glucose and insulin to rapidly lower serum potassium, and calcium gluconate to protect the heart from the cardiosuppressive effects of hyperkalemia.

Options for mineralocorticoid supplementation include fludrocortisone, desoxycorticosterone pivalate, or hydrocortisone. The choice of mineralocorticoid depends upon the clinical status of patient (oral versus injectable), product availability, and confidence in diagnosis. In most cases electrolytes normalize with fluid therapy and glucocorticoids alone and immediate mineralocorticoid supplementation is not necessary. Long-term mineralocorticoid therapy can be initiated once the animal is stable and the diagnosis is confirmed. Other treatments that may be indicated in individual patients include synthetic colloids, blood transfusion, and IV dextrose. Parameters that should be monitored during treatment include serum electrolytes and acid-base status, urine output, ECG, blood pressure, and if possible central venous pressure. IV fluid therapy should be continued until the animal is fully rehydrated and oral intake is possible. Injectable medication should be continued until oral medications can be substituted.

MAINTENANCE THERAPY

Options for long term mineralocorticoid treatment include fludrocortisone (0.1 mg/5 kg body weight divided bid) or desoxycorticosterone pivalate (2.2 mg/kg IM q 25 days initially). For both of these mineralocorticoids the dose should be titrated to effect. The dose of fludrocortisone typically needs to be increased over time whereas in many cases the dose of DOCP can be decreased over time. In our clinic we reduce the dose of DOCP by 10% a month provided the electrolytes remain in the normal range at 30 days. In one study the range of doses needed for good control of hypoadrenocorticism ranged from 1.65 to 2.2 mg/kg at intervals ranging from 21-30 days. Dogs without electrolyte derangements do not require mineralocorticoid treatment, although electrolytes should be monitored frequently.

Prednisone is typically recommended for glucocorticoid replacement. The starting dose is 0.1 to 0.22 mg/kg initially, but this dose should then be tapered to the lowest dose that will control the clinical signs. It is important to avoid excess prednisone supplementation because this may result in manifestations of hyperadrenocorticism. Only 50% of dogs on fludrocortisone require supplemental prednisone, whereas most dogs on DOCP require prednisone at least every other day.

REFERENCES

[1.] Famula TR, et al. Heritability and complex segregation analysis of hypoadrenocorticism in the standard poodle. **JSAP** 2003;44:8-12. [2.] Thompson AL, et al. Comparison of classic hypoadrenocorticism with glucocorticoid-deficient hypoadrenocorticism in dogs: 46 cases (1985-2005). **Journal of the Veterinary Medical Association** 2007;230:1190-1194. [3.] Adler JA, et al. Abnormalities of serum electrolyte concentrations in dogs with hypoadrenocorticism. **J Vet Intern Med** 2007;21:1168-1173. [4.] Hughes AM, et al. Clinical features and heritability of hypoadrenocorticism in **Nova Scotia Duck Tolling Retrievers:25 cases** (1994-2006).

HYPERADRENOCORTICISM IN THE DOG & CAT: CLINICAL PRESENTATION & DIAGNOSIS

J. Catharine Scott-Moncrieff MA, MS, Vet MB, DACVIM DECVIM
West Lafayette, IN

INTRODUCTION

Approximately 80 to 85% of cases of hyperadrenocorticism (HAC) in dogs are due to pituitary dependent hyperadrenocorticism (PDH), with the remainder due to an adrenocortical tumor (AT). Cortisol is the most common secretory product of the adrenal gland in hyperadrenocorticism, but excessive production of other steroid hormones (sex hormones, mineralocorticoids) has also been reported in dogs and cats.

SIGNALMENT

Hyperadrenocorticism affects middle-aged and older dogs. The median age is 11 years. Hyperadrenocorticism has been diagnosed in numerous breeds with Poodles, dachshunds, beagles, German shepherd dogs, Boston terriers and boxers being over-represented. There is a slight increased predisposition for females (55-65%). Larger breeds may be affected more frequently with AT than smaller breeds.

CLINICAL SIGNS OF HYPERADRENOCORTICISM

In most cases of hyperadrenocorticism, secretion of abnormal quantities of glucocorticoids is believed to be the cause of Cushing's syndrome. Clinical signs include polydipsia, polyuria, polyphagia, abdominal enlargement, hepatomegaly, cutaneous changes (alopecia, cutaneous atrophy, calcinosis cutis, hyperpigmentation), muscle weakness, decreased exercise tolerance, excessive panting, hypertension, truncal obesity, lethargy, weight gain, immunosuppression, insulin resistance, and decreased sexual function. Hypertension has been documented in up to 80% of dogs with HAC. Increased circulating concentrations of other adrenal hormones such as progesterone or 17- hydroxyprogesterone may be associated with clinical signs that are indistinguishable from those due to glucocorticoids. This is hypothesized to be due to the marked intrinsic glucocorticoid activity of progestagens. In rare cases of hyperadrenocorticism, increased concentrations of adrenal androgens, estrogens, and mineralocorticoids, may cause virilization, feminization, or hypertension respectively.

LABORATORY CHANGES IN HYPERADRENOCORTICISM

Common abnormalities on the minimum data base include leukocytosis, monocytosis, eosinopenia, lymphopenia and mild polycythemia. Increased levels of alkaline phosphatase (ALP) are found in 95% of dogs with hyperadrenocorticism. Serum alanine aminotransferase (ALT) concentrations are more mildly increased. Other biochemical changes include hypercholesterolemia, hyperglycemia, hypophosphatemia, hypernatremia, and hypokalemia. Urinalysis may reveal low urine specific gravity, glucosuria, proteinuria or evidence of urinary tract infection. Urinary tract infection is identified in up to 50% of dogs with hyperadrenocorticism , although as few as 5% have clinical signs. Urine culture and sensitivity should be routinely performed in dogs with suspected HAC. Dogs with HAC also have increased risk for calcium containing urolithiasis.

DIAGNOSIS OF HYPERADRENOCORTICISM

Diagnosis of Cushing's syndrome is made by consideration of historical findings, physical examination, review of a laboratory minimum data base (CBC, serum biochemical profile, urinalysis) and performance of specific endocrine function tests (ACTH stimulation test, low dose dexamethasone suppression test, urine cortisol:creatinine ratio). Measurement of a baseline cortisol is of little value for diagnosis of Cushing's syndrome. More recently evaluation of sex hormones before and following administration of ACTH has also been used to support the diagnosis.

WHICH TEST SHOULD BE PERFORMED FIRST?

If clinical signs of hyperadrenocorticism are present and there is no history of exogenous corticosteroid administration, the LDDS is the most appropriate initial test. If this test is abnormal or borderline the ACTH stimulation test may be used to confirm or support the diagnosis and obtain a baseline for monitoring response to treatment. If hyperadrenocorticism is not confirmed by the LDDS test, an ACTH stimulation test should be performed if clinical suspicion for the disease is high. Testing should be repeated in 1-3 months if testing is negative but no other cause of the clinical signs is identified. In dogs with obvious clinical signs of hyperadrenocorticism and persistent normal cortisol testing, measurement of a sex hormone profile should be considered (see below). In animals with concurrent disease or any history of exogenous steroids, the ACTH stimulation test is the test of choice for initial testing. In dogs in which the clinical suspicion for hyperadrenocorticism is low but hyperadrencorticism needs to be ruled out (e.g. dogs with persistent increase in alkaline phosphatase), the cortisol:creatinine ratio is an excellent screening test.

SEX HORMONE PROFILES IN DOGS WITH HYPERADRENOCORTICISM

Dogs with both PDH and AT (confirmed by hypercortisolemia during ACTH stimulation or low dose dexamethasone suppression tests) have increases of other steroid adrenal hormones both before and after stimulation with ACTH. Hormones that are commonly increased include dehydroepiandrosterone, androstenedione, progesterone, and 17-hydroxyprogesterone. The specific profile varies from dog to dog but in most dogs with hyperadrenocorticism, at least one steroid hormone other than cortisol is increased, and often 2-3 are increased. Concentration of 17-hydroxyprogesterone in dogs with hyperadrenocorticism has been evaluated more extensively than other hormones and studies suggest that approximately 70% of dogs with hyperadrenocorticism have an increase in 17-hydroxyprogesterone after stimulation with ACTH. Dogs with non-adrenal illness also have increases in 17-hydroxyprogesterone. Of 35 dogs with neoplasia that had no clinical signs of hyperadrenocorticism, 11 had high post ACTH 17-hydroxyprogesterone concentrations, whereas only 3 dogs had high cortisol concentrations. This suggests that the specificity of 17-hydroxyprogesterone for diagnosis of hyperadrenocorticism is less than that of serum cortisol. There is extensive overlap between concentrations of sex hormones in normal dogs, dogs with HAC and dogs with non-adrenal illness.

ATYPICAL HYPERADRENOCORTICISM

This term is used to describe cases of hyperadrenocorticism (both PDH and ADH) in which clinical signs and response to treatment are consistent with a diagnosis of hyperadrenocorticism, but the standard screening tests (ACTH stimulation test, low dose dexamethasone suppression test) are normal. In dogs with PDH, cortisol secretion is usually normal while in sex hormone secreting adrenal tumors, cortisol secretion is typically suppressed below the reference range. A sex hormone profile panel before and after ACTH stimulation should be considered in dogs that have clinical signs and clinical laboratory evidence of hyperadrenocorticism, no evidence of other cause for their clinical signs, but normal or borderline cortisol testing. Sex hormone measurement should also be considered in dogs with clinical signs of hyperadrenocorticism and suppressed cortisol concentrations after ACTH stimulation if treatment with exogenous steroids or mitotane can be ruled out. The current profile run by the Clinical Endocrinology Service at the University of Tennessee includes androstenedione, estradiol, progesterone, 17 hydroxyprogesterone, and aldosterone. The protocol is identical to that for a standard ACTH stimulation test, but the profile requires a larger volume of serum. Marked increases in 2 or 3 adrenal steroid hormone concentrations (1.5 to 2.0 greater than the high end of the reference range) are very supportive of a diagnosis of hyperadrenocorticism in patients with appropriate clinical signs of hyperadrenocorticism but normal or borderline cortisol testing. This clinical opinion needs to be confirmed in by further retrospective and prospective studies. Sex hormone profiles are not recommended for routine diagnosis of hyperadrenocorticism. Concentrations of sex hormones vary depending on sexual status, therefore reference ranges must be established based on sex and neuter status.

DIFFERENTIATION OF PDH FROM ADRENAL TUMORS

Clinical and routine laboratory findings are not useful in distinguishing PDH from functional adrenal tumors. The ACTH stimulation test and cortisol:creatinine ratio are also unable to differentiate adrenal tumors from PDH. Endogenous ACTH concentrations, and endocrine function tests such as the high and low dose dexamethasone suppression tests, are the most useful endocrine tests for distinguishing AT from PDH. Results of the various differentiation tests can be difficult to interpret, particularly in the presence of bilateral adrenal neoplasia, concurrent AT and PDH, or concurrent pheochromocytoma. Suppression of basal cortisol by more than 50% after the low or high dose dexamethasone test is diagnostic for PDH. Lack of suppression however, is not diagnostic for ADH and additional testing is required. If greater than 50% suppression is seen at 4 or 8 hours on a LDDS, a HDDS test is unnecessary. In our laboratory ACTH concentrations >15 pg/ml are consistent with a diagnosis of PDH. ACTH concentrations <10 pg/ml are suggestive of ADH, however some dogs with PDH have low values probably due to episodic ACTH secretion. The result of the endogenous ACTH should therefore never be relied on alone to confirm a diagnosis of adrenal tumor.

Diagnostic imaging

Approximately 57% of adrenal tumors are identified on abdominal radiographs as a soft tissue mass or a mineralized opacity, compared to 72% with ultrasonography. Other radiographic findings may include hepatomegaly, osteopenia, dystrophic mineralization and distension of the urinary bladder. Thoracic radiographs may reveal evidence of metastasis, pulmonary thromboembolism or mineralization. In dogs with PDH, ultrasound of the adrenal glands typically reveals bilaterally symmetrical enlargement of the adrenal glands with preservation of normal adrenal architecture. The glands may not be identical in size but the difference in thickness is usually less than 5mm. The size of the adrenal glands may be within the normal range in some dogs with PDH. In dogs with a functional adrenal tumor there is unilateral adrenal gland enlargement with abnormal adrenal gland architecture. Evidence of tumor thrombus within the vena cava is detected on ultrasound in 11% of dogs with adrenocortical tumors and evidence of distant metastasis to the liver, kidneys, and other abdominal organs may be present. Bilateral adrenal tumors and macronodular hyperplasia of the adrenal gland in dogs with PDH may complicate the interpretation of ultrasound findings. Computed tomography is also a useful tool for evaluating the adrenal glands, especially in dogs with adrenal tumors. MRI or CT of the brain is helpful in determining pituitary tumor size in dogs with PDH. Approximately 70% of dogs with PDH have a detectable pituitary tumor on CT or MRI.

HYPERADRENOCORTICISM IN CATS

Cats with hyperadrenocorticism are middle-aged or older (median 10 years of age, range 5-16 years) and females are slightly over-represented. Most have diabetes mellitus at the time of diagnosis. Clinical signs include insulin resistant diabetes mellitus (polyuria, polydipsia, polyphagia, weight loss, and peripheral neuropathy), lethargy, abdominal enlargement or pot-bellied appearance, muscle atrophy, unkempt hair coat, bilaterally symmetrical alopecia, cutaneous fragility, and recurrent abscess formation. The physical examination may reveal hepatomegaly, seborrhea, and cutaneous lacerations. Skin fragility may be so severe that tearing of the skin occurs during routine grooming. Virilization, estrus behavior, and hyperaldosteronism may occur in cats with adrenal tumors that secrete adrenal steroids other than cortisol.

The results of a CBC, biochemical panel, and urinalysis are usually typical of cats with diabetes mellitus. Increased ALP, ALT, hypercholesterolemia, hyperglycemia, and low BUN are common. Endocrine tests used for diagnosis of feline hyperadrenocorticism include the ACTH stimulation test, the low dose dexamethasone suppression test, and the urine cortisol:creatinine ratio (C:Cr). Pituitary dependent hyperadrenocorticism occurs in 85% of cats with spontaneous hyperadrenocorticism, while 15% have functional adrenocortical tumors. Abdominal ultrasound is the most useful diagnostic test for differentiation of PDH from AT. Either the high dose dexamethasone suppression test (1 mg/kg) or measurement of endogenous ACTH concentration may be helpful in equivocal cases.

REFERENCES

[1.] Syme HM, Scott-Moncrieff JC, et al. Hyperadrenocorticism associated with excessive sex hormone production by an adrenocortical tumor in two dogs. **JAVMA** 2001;219:1725-1728. [2.] Rossmeisl JH, Scott-Moncrieff JC, et al. Hyperadrenocorticism and hyperprogesteronemia in a cat with an adrenocortical carcinoma. **JAAHA** 2000;36:512-7. [3.] Boord M, Griffin, G. Progesterone secreting adrenal mass in a cat with clinical signs of hyperadrenocorticism. **J Am Vet Med Assoc** 1999; 214:666-672. [4.] DeClue AE, Breshears LA, Pardo ID, et al. Hyperaldosteronism and hyperprogesteronemia in a cat with an adrenal cortical carcinoma. **J Vet Int Med** 2005;19:355-358. van der Vlught-Meijer RH, Meij BP, van den Ingh TSGAM, et al. Dynamic computed tomography of the pituitary gland in dogs with pituitary-dependent hypeeradrenocorticism. **J Vet Int Med** 2003;17:773-780. [5.] Sieber-Ruckstahl NS, et al. Evaluation of cortisol precursors for the diagnosis of pituitary dependent hypercortisolism in dogs. **Vet Rec** 2008;162:673-678. [6.] Forrester SD, et al. Retrospective evaluation of urinary tract infection in 42 dogs with hyperadrencorticism or both. **J Vet Intern Med** 1999;13:557-560. [7.] Hess RS, et al. Association between hyperadrenocorticism and development of calcium containing uroliths in dogs with urolithiasis. **JAVMA** 1998;212:1889-1891. [8.] Ortega TM et al. Systemic arterial blood pressure and urine protein/creatinine ratio in dogs with hyperadrenocorticism. **JAVMA** 1996;209:1724-1729. [9.] Benitah N, et al. Evaluation of serum 17-hydroxyprogesterone concentration after administration of ACTH in dogs with hyperadrenocorticism. **JAVMA** 2005;227:1095-1101. [10.] Behrend EN et al. Serum 17-α-hydroxyprogesterone and corticosterone concentrations in dogs with nonadrenal neoplasia and dogs with suspected hyperadrenocorticism. **JAVMA** 2005;227:1762-1767.

UPDATE ON TREATMENT OF HYPERADRENOCORTICISM: WHAT IS THE CURRENT RECOMMENDATION?

J. Catharine Scott-Moncrieff MA, MS, Vet MB, DACVIM DECVIM
West Lafayette, IN

In the United States, most dogs with canine pituitary dependant hyperadrenocorticism (PDH) are treated medically with either opDDD (mitotane) or trilostane. Hypophysectomy, the treatment of choice for PDH in people is not commonly performed in dogs in the US, although it is commonly performed in Europe and outcomes are reported to be superior to that of medical management. Both mitotane and trilostane may be effective in management of both pituitary dependent and adrenal dependent forms of hyperadrenocorticism, and both drugs have advantages and disadvantages.

It is ideal to identify which form of hyperadrenocorticism is present when formulating an approach to treatment. Although surgery is often considered the treatment of choice for adrenal tumors, medical therapy can also be effective. Tests that may be used to differentiate PDH from AT include the low and/or high dose dexamethasone suppression tests, measurement of endogenous ACTH concentration, and imaging studies such as ultrasound and computed tomography. If identification of the cause of hyperadrenocorticism is not possible, mitotane is the best choice for treatment because this drug has cytotoxic effects that can cause tumor shrinkage.

TREATMENT OF PITUITARY DEPENDENT HYPERADRENOCORTICISM

Prior to initiation of treatment for PDH, the diagnosis should be confirmed by appropriate endocrine testing, and patients should be systemically well with a good to excellent appetite. If the appetite is poor or other signs of systemic illness are present, a diagnostic work up for the cause of these signs is necessary before considering treatment. Because medical therapy does not address the pituitary tumor that is the underlying cause of the disease dogs that do not have clinical signs of hyperadrenocorticism should not be treated.

Mitotane

Mitotane is derived from the insecticide DDT and is a potent adrenocorticolytic agent. The drug causes progressive necrosis of the zona fasciculata and zona reticularis and at higher doses may also cause necrosis of the zona glomerulosa. Other effects of mitotane include fatty degeneration, centrilobular atrophy and congestion of the liver. Normal dogs are clinically quite resistant to the effects of the drug. Mitotane should never be administered in animals that are not eating well. Therapy with mitotane is begun at a dose of 50 mg/kg divided q 12 hours. Absorption of mitotane is increased if administered with fat. Glucocorticoids are not usually administered concurrently, but a small supply of prednisone should be made available to the owner for emergencies. Mitotane should be administered until water consumption decreases to < 100 ml/kg/day, or until a decreased appetite, depression, diarrhea, or vomiting are observed. The time for clinical response is quite variable but most dogs respond within 3-7 days. At this point the dog should be reevaluated and an ACTH stimulation test performed. Prednisone treatment (0.2 mg/kg) should be initiated in patients that are showing clinical signs of hypocortisolemia, until the results of the ACTH stimulation test are known. In patients which are not polydipsic prior to therapy, patients in which water consumption cannot be monitored, and patients in which polydipsia is due to another underlying disease (e.g. diabetes mellitus), mitotane should be administered for a maximum of 5-7 days prior to ACTH stimulation testing. The goal of treatment is to have both the pre- and post-cortisol measurement in the normal resting range (2-6 μg/dl). Maintenance therapy (50 mg/kg q 7-10 days) is started once the ACTH stimulation test shows adequate suppression and prednisone therapy (if necessary) has been discontinued. Failure to use maintenance therapy will result in re-growth of the adrenal cortex and recurrence of clinical signs. Efficacy of maintenance therapy is monitored by an ACTH stimulation test after one month of maintenance treatment and then every 3 months. The dose of mitotane required for long term maintenance is very variable (26-330 mg/kg/week). Reasons for treatment failure include incorrect diagnosis (for example iatrogenic hyperadrenocorticism), the presence of an adrenal tumor (although some adrenal tumors will respond well), loss of drug potency due to poor storage or compounding of mitotane, or the requirement for a higher dose or duration of treatment in some dogs. Side effects of mitotane include gastric irritation, Addisonian crisis, and very occasionally neurologic signs. Mean survival time of 200 dogs treated with mitotane was 2.2 years (range 10 days to 8.2 years).

Trilostane

Trilostane is a synthetic hormonally inactive steroid analogue which is a competitive inhibitor of the 3 ß-hydroxysteroid dehydrogenase system. The drug blocks production of a number of adrenal steroids including cortisol and aldosterone. Trilostane is rapidly absorbed orally (peak concentrations within 1.5 hours) although suppression of plasma cortisol concentrations is short lived (< 20 hours). The drug is well tolerated by dogs and is reportedly effective in controlling polyuria and polydipsia in 70% of cases, and dermatologic changes in approximately 60% of patients. The mean starting dose in one study was 6 mg/kg with a mean final dose of 4-16 mg/kg. Current recommendations for use of trilostane are to start at a dose of 2-6 mg/kg q 12 - 24 hrs and then increase or decrease the dose based on evaluation of ACTH stimulation tests performed 4-6 hours after drug administration. Twice a day therapy at a starting dose of 1-3 mg/kg may result in good control of clinical signs at a lower total daily dose with less risk of adverse effects than higher doses q 24 hours. ACTH stimulation testing should be performed 10, 30 and 90 days after start of treatment and 30 days after each dose adjustment. Trilostane is well tolerated in most dogs. Adverse effects that are usually mild and self limiting include diarrhea, vomiting, and lethargy in up to 63% of treated dogs. Occasional dogs have developed hypoadrenocorticism which is generally glucococorticoid deficient only, although dogs with evidence of mineralocorticoid deficiency have been reported.

Hypoadrenocorticism induced by trilostane is generally reversible, although in rare cases this may take several months likely because of adrenocortical necrosis. There have been anecdotal reports of acute death shortly after starting trilostane treatment. In a recent retrospective study of dogs treated with either mitotane (n= 25) or trilostane (n=123), long term survival was not statistically different between the groups. Median survival in the mitotane group was 708 days (range 33-1399), and in the trilostane group was 662 days (range 8-1971). Of the dogs that died in this study, 11% died due to causes that were attributed to the underlying PDH, and a further 17% died of causes that could have been related to their underlying PDH. In 38% of cases the cause of death could not be directly attributed to PDH, and in 34% the cause of death was unknown.

Other drugs that have been reported to be effective in dogs with PDH include ketoconazole, and l-deprenyl. Ketoconazole (Nizoral®) is an imidazole derivative which is an orally active antifungal agent. Ketoconazole inhibits ergosterol synthesis in fungi, and interferes with gonadal and adrenal steroid synthesis in mammals by inhibiting cytochrome p - 450 dependent enzymes. Ketoconazole has been used successfully in the management of PDH in dogs. A dose of 5-25 mg/kg q 12 hours, results in a rapid reduction of serum cortisol concentrations in most animals. Signs of toxicity are uncommon but include anorexia and vomiting. Toxicity can be minimized by gradual increase of the dose. In a study of 48 dogs with PDH treated with ketoconazole, median survival was 25 months.

L-deprenyl is a selective monoamine oxidase type B inhibitor which increases CNS dopamine concentrations. Treatment with L-deprenyl has recently been reported to be effective in less than 20% of dogs with PDH. Retinoic acid and cabergoline have also been reported to have efficacy in treatment of canine PDH. Treatment with melatonin, lignins,and flaxseed oil has been recommended for treatment of atypical hyperadrenocorticism but there are as yet no studies to support their efficacy. Bilateral adrenalectomy is an option in dogs with PDH that do not tolerate or respond to medical therapy. Disadvantages include the cost, induction of permanent hypoadrenocorticism, and risk of surgery in a dog with poorly controlled PDH. This procedure is therefore rarely used for treatment of PDH in dogs.

Radiation therapy improves outcome in dogs with PDH that have associated neurological signs due to a pituitary tumor, and in dogs with pituitary tumors greater than 0.8 to 1.0 cm diameter. Neurologic abnormalities attributed to the presence of a space occupying pituitary mass include change in behavior, inappetance, obtundation, aggressiveness, anisocoria, apparent blindness and seizures. Complications associated with radiation therapy include alopecia and leukotrichia. Resolution of associated endocrine signs after radiation therapy is not consistent and may be delayed for several months; therefore dogs with PDH may need continued medical treatment. Dogs that present with more severe neurological impairment (e.g., stupor, seizures) have a worse outcome than dogs that present with signs of mild neurological impairment. Likewise, dogs with large tumors have a poorer prognosis than dogs with smaller tumors.

TREATMENT OF ADRENOCORTICAL TUMORS

Treatment of adrenal neoplasia involves either surgical resection of the tumor or medical therapy with mitotane, trilostane or ketoconazole. Adrenalectomy should be reserved for patients that do not have evidence of extensive tumor invasion or metastasis. After unilateral adrenalectomy, normal adrenal gland tissue will be atrophic and glucocorticoid and sometimes mineralocorticoid supplementation are necessary in the perioperative and postoperative period. Dexamethasone provides primarily glucocorticoid replacement. This should be administered in IV fluids over 6 hours as soon as the tumor is identified at surgery??. Alternatively hydrocortisone can be used as a continuous infusion to provide both glucocorticoid and mineralocorticoid supplementation. Other options for mineralocorticoid support include fludrocortisone or desoxycorticosterone pivalate administered 12-24 hours before surgery. Blood pressure, central venous pressure, electrolytes and blood glucose should be carefully monitored before, during and after surgery. The merits of pre-operative anticoagulant administration to prevent thromboembolic complications have yet to be proven. After surgery, parenteral glucocorticoids should be continued for 48 - 72 hours until the patient is bright, alert and eating. Oral supplementation with prednisone can then replace parenteral therapy. Supplementation with mineralocorticoids if utilized should be discontinued after 2 doses and the patient monitored for hyperkalemia and hyponatremia. The prognosis for dogs with adrenocortical tumors without metastatic disease that survive the initial perioperative period is good. In one study 80% of dogs survived the perioperative period.

Some dogs with inoperable tumors or metastatic disease may respond to medical treatment with high doses of op-DDD (mitotane). Mitotane is often successful in controlling clinical signs and in some cases may also result in tumor shrinkage. In one study the mean dose of mitotane required to control clinical signs in dogs with adrenal tumors ranged from 35 to 1275 mg/kg/week. Although the survival times for dogs with adrenal tumors treated with mitotane (median 12 months) are shorter than those for dogs treated for PDH, occasional dogs may have prolonged survival. Dogs without metastatic disease tend to have better long-term outcome. Trilostane has also been used successfully to control clinical signs of Cushing's in dogs with adrenal tumors. The approach to treatment is similar to that outlined for management of PDH. It should be recognized that trilostane will not cause tumor shrinkage, in fact the adrenals tend to actually become larger during trilostane treatment. Ketoconazole may also be used for palliation of clinical signs prior to adrenalectomy but it has little advantage over trilostane.

TREATMENT OF FELINE HYPERADRENOCORTICISM

Medical management of hyperadrenocorticism is less successful in the cat than the dog. Moreover, cats with hyperadrenocorticism are often poor surgical candidates. Thus the long term prognosis in cats with hyperadrenocorticism is guarded. Trilostane is currently the most effective drug for medical management of cats with PDH. Treatment with trilostane may improve clinical signs but cats typically remain diabetic. The effective dose of trilostane ranges from 15 mg PO q 24 hrs to 60 mg PO q 12 hours. ACTH stimulation tests should be used to titrate the dose, similar to the protocol for dogs. Metyrapone, an 11β-hydroxylase inhibitor that blocks the conversion of 11-deoxycortisol to cortisol, has been used to successfully treat hyperadrenocorticism in the cat. The recommended dose is 65 mg/kg PO every 8 to 12 hours. Results of treatment with mitotane and ketoconazole have been disappointing with partial responses observed in only some cats. Radiation therapy is another option for cats with PDH. Resolution of neurologic signs, improved clinical signs, and improved glycemic regulation has been reported after radiation treatment. Improvement in endocrine function occurs within 1-5 months of completion of radiation therapy.

In cats with functional adrenal tumors, adrenalectomy is the treatment of choice although initial medical treatment may be necessary in severely debilitated cats. Bilateral adrenalectomy is also effective in cats with PDH that do not respond to medical therapy. As in dogs, intravenous fluid supplementation and glucocorticoid administration are vital for successful post-operative management. Cats with diabetes mellitus require careful management of insulin requirements perioperatively. Cats undergoing bilateral adrenalectomy or with evidence of hypoadrenocorticism (hyperkalemia, hyponatremia) need supplemental mineralocorticoids (deoxycorticosterone pivalate or fludrocortisone acetate). Post-operative complications include electrolyte derangements, skin lacerations, pancreatitis, hypoglycemia, pneumonia, sepsis, and venous thrombosis. The long term prognosis for cats with hyperadrenocorticism is fair to good following adrenalectomy. Of 10 cats treated by adrenalectomy, 3 died within 5 weeks of surgery of complications associated with sepsis or thromboembolism. The remaining 7 cats survived a median of 12 months.

REFERENCES

[1.] Kintzer PP, Peterson ME, et al. Mitotane treatment of 32 dogs with cortisol-secreting adrenocortical neoplasms **JAVMA** 1994; 205:54-61. [2.] Kyles AE, Feldman EC, et al. Surgical management of adrenal gland tumors with and without associated tumor thrombi in dogs: 40 cases (1994-2001). **JAVMA** 2003;223:654-662. [3.] Anderson CR, Birchard SJ, et al. Surgical treatment of adrenocortical tumors: 21 cases (1990-1996). **JAAHA** 2001;37:93-97. [4.] Kintzer PP, Peterson ME, et al. Mitotane treatment of 200 dogs with pituitary-dependent hyperadrenocorticism. **JVIM** 1991;5:182-190. [5.] Meij BP, Voorhout G, et al. Results of transphenoidal hypophysectomy in 52 dogs with pituitary-dependent hyperadrenocorticism. **Vet Surg** 1998;27:246-261. [6.] Neiger R, Ramsey I, et al. Trilostane treatment of 78 dogs with pituitary-dependent hyperadrenocorticism. **Vet Record** 2002;150: 799-804. [7.] Barker EN, Campbell S, et al. A comparison of the survival times of dogs treated with mitotane or trilostane for pituitary-dependent hyperadrenocorticism. **JVIM** 2005;19:810-815. [8.] Alenza DP, Arenas C, et al. Long-term efficacy of trilostane administered twice daily in dogs with pituitary dependent hyperadrenocorticism. **JAAHA** 2006;42:269-276. [9.] Vaughan MA, Feldman EC, et al. Evaluation of twice daily, low dose trilostane treatment administered orally in dogs with naturally occurring hyperadrenocorticism. **JAVMA** 2008;232:1321-1328. [10.] Castillo V et al. Retinoic acid as a novel medical therapy for Cushing's disease in dogs. **Endocrinology** 2006;147:4438-4444. [11.] Reusch CE et al. The efficacy of L-Deprenyl in dogs with pituitary dependent hyperadrenocorticism. **J Vet Int Med** 1999;13:291-301. [12.] Lien Y-H et al. Use of ketoconazole to treat dogs with pituitary dependent hyperadrenocorticism: 48 cases (1994-2007). **JAVMA** 2008;233:1896-1901. [13.] Galac S et al. Urinary corticoid:creatinine ratios in dogs with pituitary dependent hypercortisolism during trilostane treatment. **J Vet Intern Med** 2009;23:1214-1219. [14.] Ashley PF et al. Effect of oral melatonin administration on sex hormone, prolactin, and thyroid hormone concentration in adult dogs. **JAVMA** 1999;215:111-1115. [15.] Castillo VA, et al. Cushing's disease in dogs: Cabergoline treatment. **Res vet Sci** 2008;85:26-34.

EVALUATING RENAL FUNCTION: MORE THAN BUN & CREATININE

Linda Ross, DVM, MS, DACVIM (Small Animal Medicine)
North Grafton, MA

INTRODUCTION

A diagnostic test for any disease should ideally be both sensitive and specific. Sensitivity refers to the probability that the test yields a positive result in patients with the disorder (i.e. no false negatives), and specificity to the probability that the test is negative in patients without the disorder (i.e. no false positives). Evaluation of renal function by measurement of BUN and serum creatinine concentrations is commonly performed in dogs and cats. However, both are, at best, only crude reflections of the glomerular filtration rate (GFR). They are not sensitive because neither value will rise above the normal range until there is loss of > 75% of renal function. Conversely, both values show lack of specificity in that they will rise under conditions that produce a decrease in GFR despite normal renal function (pre- or postrenal azotemia). BUN concentration is affected by numerous factors, including dietary protein intake, decreased renal blood flow, gastrointestinal hemorrhage. Serum creatinine is affected less by non-renal factors, but may be altered in animals with significantly decreased muscle mass.[1]

When BUN or serum creatinine are plotted against GFR, the curve is a rectangular hyperbola. Therefore, relatively large decreases in GFR will result in only minor increases in BUN or serum creatinine, often not exceeding the normal range. As GFR decreases, small changes will result in greater increases in BUN or serum creatinine.

Given the above issues, diagnosing subclinical renal disease or dysfunction is a challenge for veterinarians. In addition, alterations in renal function may be directly or indirectly important in many research studies, and changes in BUN and serum creatinine concentrations may not accurately reflect these changes.

GLOMERULAR FUNCTION

Urinary Clearance

Determining the GFR is the most sensitive and accurate indicator of renal function. There are several methods of estimating GFR using various filtration markers. An ideal filtration marker is freely filtered across the glomerulus, eliminated only by the kidneys with no systemic metabolism, not secreted or absorbed in the renal tubules, and not bound by protein in blood. Urinary clearance of a substance is defined as the rate at which the kidneys remove the substance from a given amount of plasma per unit of time. Inulin, a fructose polymer, is considered the "gold standard" as filtration marker because it is neither secreted nor reabsorbed.[1] However, availability may be limited, and commercial clinical assays are not readily available. Urinary clearance of endogenous and exogenous creatinine also yields accurate estimates of GFR in dogs.[2,3] However, there are practical concerns with the use of urinary clearance protocols. They are time consuming and require meticulous urine collection with urinary catheters or metabolic cages. Protocols that involve the use of urinary catheterization have the risk of causing an iatrogenic urinary tract infection.

Plasma Clearance

Measurement of GFR by plasma clearance of a filtration marker avoids some of the difficulties of urinary clearance techniques, and is more suitable in the clinical setting. Plasma clearance is defined as the volume of plasma that is cleared of a substance per unit of time. Single injection techniques are the most commonly used in clinical situations. In order for plasma clearance to accurately measure GFR, the substance must not only be excreted exclusively by glomerular filtration, but should achieve complete distribution in the body within a short period of time. If not, the process of distribution will contribute to the fall in plasma level and introduce error. The marker is injected IV, and its serum concentration is determined at various time points after injection. The GFR is determined by calculations using the plasma concentration of the marker at different time points.[4] Iohexol [5-7] and gadolinium [8] have characteristics of ideal filtration markers, similar to inulin. Others include radionuclide markers (^{125}I or ^{131}I-iothalamates, ^{51}Cr-ethylenediaminetetraacetic acid (EDTA) ,^{99qc}-diethylenetriaminepentaacetic acid (DTPA)[9] and creatinine.[10.11] The use of radionuclide markers in private practices may be limited by radiation safety regulations. Plasma clearance of creatinine appears to yield results that are a good estimate of GFR,[11] and is convenient in that creatinine levels are a standard test in diagnostic laboratories. However, creatinine is not available as a pharmaceutical. A solution must be prepared from chemical grade creatinine, raising concerns about sterility and quality control of the creatinine concentration.

Renal Scintigraphy

Renal scintigraphy can be used to measure GFR. It has the advantages of providing not only total GFR, but individual kidney GFR; it also does not require collection of urine. Disadvantages include the use of radioactive pharmaceuticals and the required licensing and regulation, expense of the gamma camera, and post-procedure isolation of the animal for a period of time. Tc-99m-DTPA is the most common radioisotope used for renal scintigraphy. It is injected intravenously with the animal positioned under a gamma camera. Multiple sequential images are taken for several minutes. Regions of interest are drawn around each kidney, and GFR calculated using the counts in the kidneys, background, and amount injected. Because there is a degree of subjectivity in the drawing of the region of interest, studies have shown within-observer and between-observer variations that can result in differences in GFR of up to 15% using manual drawing.[12]

Estimated Glomerular Filtration Rate

The lack of sensitivity of BUN and serum creatinine, and the practical issues with measuring GFR, are significant problems in human as well as veterinary medicine. Estimating equations (eGFR) have been developed and validated to avoid these problems. The Cockcroft-Gault equation, which estimates creatinine clearance (based on serum creatinine, age, sex, and body weight), was developed in the 1970s. However, the inclusion of weight in the equation as a measure of muscle mass can introduce significant error. A newer equation was developed from data collected during the Modification of Diet in Renal Disease study. This equation is corrected for human body surface area, and utilized serum creatinine concentration, age, sex, and race.[13] Further refinements of the MDRD equation continue to be developed.

Estimating equations are not available for animals. The large variation in body size and composition among dog breeds would make developing such an equation mathematically difficult. It may be more feasible for cats because of similarity in body size. One study found that GFR could be approximately estimated by using a graph of GFR vs. the reciprocal of the serum creatinine concentration in dogs.[14]

Cystatin C

Cystatin C is a 14-kD, nonglycosylated protein that is in the cystatin family of cysteine protease inhibitors. It is produced at a constant rate by all nucleated cells. It has properties of an ideal filtration marker, in that it is freely filtered through the glomerulus, and is not secreted by renal tubular cells. It is reabsorbed and metabolized completely by proximal tubular cells, so that little appears in the urine.[15,16] Because of this, urinary clearance of cystatin C cannot be measured. There is evidence in both humans and dogs that serum levels of cystatin C may provide at least as accurate an estimate of GFR as serum creatinine concentration, and may be more sensitive at detecting changes in GRF. [15-19] However, in dogs, some studies suggest that cystatin C may not be as sensitive as serum creatinine in cases with prerenal azotemia, or acute renal failure.[18] There is also conflicting evidence as to whether cystatin C levels are affected by muscle mass or body size.[18,19] These studies used different methodology for measuring serum cystatin C levels, Further studies using standardized assay techniques for both cystatin C and creatinine are needed to resolve these issues, and determine whether cystatin C is indeed a more sensitive and specific marker of renal function.

Proteinuria

Abnormal amounts of protein in the urine indicate loss of the normal glomerular filtration barrier. Low levels of proteinuria (microalbuminuria) may be seen in early stages of chronic kidney disease. Large amounts of protein are seen with diseases of the glomerulus. Proteinuria is discussed in detail in another presentation at this Forum.

TUBULAR FUNCTION

Urine Specific Gravity

Urine specific gravity is a simple test that yields significant information about renal tubular function, and overall renal function when performed in conjunction with BUN and serum creatinine. Unfortunately, it is frequently not performed because of the perceived difficulty or inconvenience in obtaining a urine sample. Although it has been said many times, it bears repeating and emphasizing that it is essential to obtain a urine specific gravity as well as BUN and creatinine in a sick animals prior to beginning fluid therapy.

Fractional Excretion of Substances

Fractional excretion (FE) refers to the percentage of filtered substance that appears in the urine. The classic method for determining FE involves urine collection over a 24 hours period as well as measurement of GFR. A simpler, "spot" method has been described using only single samples of blood and urine. Unfortunately, the spot method has not been shown to correlate with 24 hour excretion of various substances,[20-22] and use of this method may result in incorrect clinical conclusions.

Measurement of FE may be useful in evaluating suspected functional renal tubular disorders, such as Fanconi syndrome. They may also be used to help determine the cause of abnormal serum electrolyte or mineral levels, or to differentiate prerenal from renal azotemia.

REFERENCES

[1.] Braun J-P, et al . **Clinical Biochemistry of Domestic Animals** 6th ed. Elsevier 2008. 485. [2.] Finco DR,, et al. **Am J Vet Res** 1993:54:1575. [3.] Finco DR, et al. **Am J Vet Res** 1991:52:1029. [4.] Toutain PL, et al. **J Vet Pharmacol Therap** 2004:27:415. [5.] Finco DR, et al... **J Vet Intern Med** 2001:15:368 [6.] Brown SA, et al.. **Am J Vet Res** 1996:57:105 [7.] Bexfield NH, et al**. J Vet Intern Med** 2008:22(1):68. [8.] Nolan BG, et al. **Am J Vet Res** 2009:70:547. [9.] Heiene R, et al: **J Vet Intern Med** 1998:12:401 [10.] Labato MA, et al: **Res Vet Sci** 1991:50:253 [11.] Watson ADJ, et al: **J Vet Intern Med** 2002:16:22. [12.] Kampa N, et al. **Vet Radiol Ultrasound** 2006:47:212. [13.]Stevens LA, et al. **Med Clin N Am** 2005:89:457 [14.]Finco DR, et al. **J Vet Pharmacol Therap** 1995:18:418 [15.]Wehner A, et al:; **J Am Anim Hosp Assoc** 2008:44:131 [16.]Antognoni MT, et al: **Vet Res Commun** 2005:29(Suppl 2):265 [17]Roos JF, et al: **Clin Biochem** 2007 40:383. [18.]Almy FS, et al. **J Vet Intern Med** 2002:16:45. [19]Miyagawa Y, et al. **J Vet Med Sci** 2009:71:1169. [20.] Finco DR, et al. **Am J Vet Res** 1997:58:1184. [21.] Adams LG et al. **Am J Vet Res** 1991:52:718 [22.]Bartges JW, et al. **Am J Vet Res** 1994:55:472.

THE CHANGING PARADIGM OF PROTEINURIA

Gregory F. Grauer, DVM, MS, Diplomate, ACVIM (Small Animal Internal Medicine)
Manhattan, KS

Introduction

Persistent proteinuria with an inactive urine sediment is an established marker of chronic kidney disease (CKD).[1] There is also an association between renal proteinuria and progression of CKD in dogs and cats.[2-5] In addition, treatments that have reduced proteinuria have also been associated with slowed disease progression and/or improved survival in dogs and cats with CKD.[4-8] For these reasons screening for renal proteinuria and longitudinal monitoring of renal proteinuria have recently received renewed interest.

Proteinuria is a general term that includes any type of protein in the urine (e.g., albumin, globulins, Bence-Jones proteins). In both health and in disease, albumin is the predominate protein in urine in dogs and cats. Proteinuria/albuminuria of renal origin results from either a loss of selective filtration (glomerular proteinuria) or an impaired resorption of the filtered protein (tubular proteinuria).

The urine dipstick colorimetric test is the usual first-line screening test for the detection of proteinuria/albuminuria. False-positive reactions for protein are common with the dipstick test.[9] Many laboratories confirm positive reactions for protein on the dipstick test with the sulfosalicylic acid (SSA) turbidimetric test. Proteinuria detected by the dipstick and/or SSA screening tests that is thought to be of renal origin is often confirmed and quantitated using the urine protein/creatinine ratio (UP/C). All three of these tests are most sensitive to albumin. The recent development of species specific albumin ELISA technology that enables detection of low concentrations of canine and feline albuminuria has stimulated discussion about what level of proteinuria/albuminuria is normal and what levels may be associated with renal disease progression.

Screening Tests for Proteinuria

Proteinuria is routinely detected by semi-quantitative, screening methods, like the conventional dipstick colorimetric test (very common) and the sulfosalicylic acid (SSA) turbidimetric test (less common). The dipstick test is inexpensive and easy to use. This test primarily measures albumin, however both the sensitivity and specificity for albumin are relatively low with the dipstick methodology. False-negative results (decreased sensitivity) may occur in the setting of Bence Jones proteinuria, low concentrations of urine albumin, and/or dilute or acidic urine. The conventional dipstick test has a sensitivity level of approximately 30 mg/dl. False-positive results (decreased specificity) may be obtained if the urine is alkaline or highly concentrated, the urine sediment is active (pyuria, hematuria, and/or bacteriuria), or the dipstick is left in contact with the urine long enough to leach out the citrate buffer that is incorporated in the filter paper pad. False-positive results with the dipstick occur more frequently in cats compared with dogs but are common in both species.

The SSA test is performed by mixing equal quantities of urine supernatant and 3-5% SSA in a glass test tube and grading the turbidity that results from precipitation of protein on a 0 to 4+ scale. In addition to albumin, the SSA test can detect globulins and Bence Jones proteins. False-positive results may occur if the urine contains radiographic contrast agents, penicillin, cephalosporins, sulfisoxazole, or the urine preservative thymol. The protein content may also be overestimated with the SSA test if uncentrifuged, turbid urine is analyzed. False-negative results are thought to be less common in comparison with the conventional dipstick test due to the increased sensitivity of the SSA test for protein (approximately 5 mg/dl). Because of the relatively poor specificity of the conventional dipstick analysis, many reference laboratories will confirm a positive dipstick test result for proteinuria with the SSA test. Grading of both the color change on the dipstick test and the turbidity on the SSA test is subjective and therefore results can vary between individuals and laboratories.

Proteinuria detected by these semi-quantitative, screening methods has historically been interpreted in light of the urine specific gravity and urine sediment. For example, a positive dipstick reading of trace or 1+ proteinuria in hypersthenuric urine has often been attributed to urine concentration rather than abnormal proteinuria. In addition, a positive dipstick reading for protein in the presence of microscopic hematuria or pyuria was often attributed to urinary tract hemorrhage or inflammation. In both examples, the interpretation may not be correct. Given the limits of the conventional dipstick test sensitivity, any positive result for protein regardless of urine concentration may be abnormal (except in the case of false-positive results). Likewise, hematuria and pyuria have an inconsistent effect on urine albumin concentrations; not all dogs with microscopic hematuria and pyuria have albuminuria.[10]

Localization of Proteinuria

When proteinuria is detected by screening tests, it is important to try to identify its source (Table). Proteinuria may be caused by physiologic or pathologic conditions. Physiologic or benign proteinuria is usually transient and abates when the underlying cause is corrected. Strenuous exercise, seizures, fever, exposure to extreme heat or cold, and stress are examples of conditions that may cause physiologic proteinuria. The mechanism of physiologic proteinuria is not completely understood; however, transient renal vasoconstriction, ischemia, and congestion are thought to be involved.

Pathologic proteinuria may be caused by urinary or non-urinary abnormalities. Non-urinary disorders associated with proteinuria often involve the production of small-molecular-weight proteins (dysproteinemias) that are filtered by the glomeruli and subsequently overwhelm the reabsorptive capacity of the proximal tubule. An example of this "pre-renal" proteinuria is the production of immunoglobulin light chains (Bence Jones proteins) by neoplastic plasma cells. Genital tract inflammation (e.g., prostatitis or metritis) can also result in pathologic non-urinary proteinuria. Obtaining urine samples via cystocentesis reduces the potential for urine contamination with protein from the lower urinary tract.

Pathologic urinary proteinuria may be renal or non-renal in origin. Non-renal proteinuria most frequently occurs in association with lower urinary tract inflammation or hemorrhage (also referred to as post-renal proteinuria). Changes observed in the urine sediment are usually compatible with the underlying inflammation (e.g., pyuria, hematuria, bacteriuria, and increased numbers of transitional epithelial cells). On the other hand, renal proteinuria is often caused by increased glomerular filtration of plasma proteins associated with intraglomerular hypertension or the presence of immune complexes, structural damage, or vascular inflammation in the glomerular capillaries. Renal proteinuria may also be caused by decreased reabsorption of filtered plasma proteins due to tubulointerstitial disease. In some cases, tubulointerstitial proteinuria may be accompanied by normoglycemic glucosuria and increased excretion of electrolytes (e.g., acute tubular damage). Glomerular lesions usually result in higher magnitude proteinuria compared with proteinuria associated with tubulointerstitial lesions. Renal proteinuria caused by glomerular and tubular disease is most frequently accompanied by an inactive urine sediment, the exception being the presence of hyaline casts. In addition to glomerular and tubulointerstitial disease, renal proteinuria may be caused by inflammatory or infiltrative disorders of the kidney (e.g., neoplasia, and pyelonephritis) which are often accompanied by an active urine sediment.

Detection of Albuminuria

Albuminuria can be measured by point-of-care, semi-quantitative tests (e.g., the Heska E.R.D.® Urine test) and quantitative immunoassays at reference laboratories. Like proteinuria, albuminuria can be caused by pre- and post-renal disorders and therefore it is important to localize the source of albuminuria as discussed above. Microalbuminuria (MA) is defined as concentrations of albumin in the urine that are greater than normal but below the limit of detection using conventional dipstick urine protein screening methodology (i.e., $\leq$ 30 mg/dl). Urine albumin concentrations above 30 mg/dl are referred to as overt albuminuria and can often be detected using the urine protein/creatinine ratio (UP/C). Urine albumin concentrations can be adjusted for differences in urine concentration by dividing by urine creatinine concentrations. For example, a urine albumin/creatinine ratio > 0.03 is considered abnormal in people. Alternatively, urine can be diluted to a standard concentration, such as 1.010, prior to assay. In one study of dogs, normalizing urine albumin concentrations to a 1.010 specific gravity yielded similar results to the urine albumin/creatinine ratio.[11]

Comparison of Screening Tests, UP/C, and Quantitative ELISA for Detection of Canine and Feline Albuminuria

Compared to the ELISA, positive dipstick and SSA results ($\geq$ trace positive reaction) in canine urine samples have only moderate specificity and poor positive predictive value. These values improve when stronger positive dipstick and SSA results ($\geq$ 2+ positive) are compared to the ELISA with specificity and positive predictive values > 90%. In cat urine, both the dipstick and SSA test have poor specificity and positive predictive value regardless of the magnitude of the positive reaction. Urine protein/creatinine ratio results have high specificity for canine and feline albuminuria (> 99%), but relatively low sensitivity (< 30%).

Quantitation of Proteinuria

If the results of the screening tests suggest the presence of renal proteinuria/albuminuria, urine protein excretion should be quantified. This helps to evaluate the severity of renal lesions and to assess the response to treatment or the progression of disease. Methods used to quantitate proteinuria include the UP/C and immunoassays for albuminuria that are expressed as either urine albumin/creatinine ratios or in mg/dl in urine samples that have been diluted to a standard urine specific gravity (e.g., 1.010). Albumin $\geq$ 30 mg/dl in urine that has been diluted to a specific gravity of 1.010 will usually result in UP/C's above the normal range in cats and dogs. Urine that contains enough albumin to register > a medium reaction on the ERD test will also often have a UP/C above the normal range. The UP/C and urine albumin/creatinine ratios from spot urine samples have been shown to accurately reflect the quantity of protein/albumin excreted in the urine over a 24-hour period. Most studies have shown that normal urine protein excretion in dogs and cats is $\leq$ 10-30 mg/kg/24 hours and that normal UP/C's are $\leq$ 0.2. Urine protein/creatinine ratios are considered borderline for proteinuria if they fall between 0.2 and 0.4/0.5 in cats and dogs, respectively. Persistent proteinuria that results in UP/C's > that 0.4 (cats) and 0.5 (dogs), where pre- and post-renal proteinuria have been ruled out, are consistent with either glomerular or tubulointerstitial CKD. Urine protein/creatinine ratios > 2.0 are strongly suggestive of glomerular disease. It is possible that "normal values" for the UP/C will continue to be modified with additional research. For example, even the ultra-low level, single nephron proteinuria that can arise secondary to intraglomerular hypertension in hypertrophied nephrons in CKD, is abnormal in the face of what may be considered normal whole-body or whole-kidney proteinuria.

Monitoring Renal Proteinuria

Transient proteinuria/albuminuria is likely of little consequence and does not warrant treatment. On the other hand, persistent proteinuria/albuminuria (where lower urinary tract inflammation has been ruled out) indicates the presence of CKD. Persistent proteinuria/albuminuria can be defined as positive test results on $\geq$ 2 occasions, $\geq$ two weeks apart. Because persistent proteinuria/albuminuria can be constant or increase or decrease in magnitude over time, monitoring should utilize quantitative methods to determine disease trends and/or response to treatment. Changes in the magnitude of proteinuria should always be interpreted in light of the patient's serum creatinine concentration since proteinuria may decrease in progressive renal disease as the number of functional nephrons decrease. Decreasing proteinuria in the face of a stable serum creatinine suggests improving renal function, whereas decreasing proteinuria in the face of an increasing serum creatinine may suggest disease progression.

Implications of Proteinuria/Albuminuria

In addition to the classic complications of moderate to heavy proteinuria (hypoalbuminemia, edema, ascites, hypercholesterolemia, hypertension, and hypercoagulability), there is increasing evidence in dogs and cats that proteinuria is associated with glomerular and tubulointerstitial damage and progressive nephron loss. In cats with naturally occurring CKD, relatively mild proteinuria (UP/C's > 0.43[3] and $\geq$ 0.2[5]) appeared to be negative predictors of survival. Proteinuria has also been associated with decreased survival time in cats with systemic hypertension[4] and the development of azotemia in geriatric cats.[12] In dogs with naturally occurring CKD, the relative risk of uremic crises and mortality was approximately three time greater in dogs with UP/C's > 1.0 compared with dogs with UP/C's < 1.0.[2] In this study the risk of an adverse outcome was approximately 1.5 times greater for every 1 unit increase in UP/C and the decline in renal function was greater in dogs with higher UP/C's.[2]

Table: Localization of Proteinuria

Type of Proteinuria		Diagnosis
Physiologic/Benign Proteinuria		UP/C usually < 0.5, Compatible history
Examples include:	Change in exercise level	Intermittent/transient
	Fever	
	Seizures	
Pathologic Proteinuria		
Non-Urinary:		Variable UP/C
Examples include:	Congestive heart failure	History/PE/Echo
	Hemoglobinuria/Myoglobinuria	Urine remains red after centrifugation
	Dysproteinemia/Dysprtoetinuria	Serum/Urine electrophoresis
	Genital tract inflammation/hemorrhage	PE/Imaging/Urine sediment
Urinary - Non-Renal:		
Examples include:	Lower urinary tract inflammation (e.g., bacterial cystitis, cystoliths, polyps, neoplasia)	UP/C not indicated History/PE Urine sediment Imaging
Urinary- Renal:		
Examples include:	Renal parenchymal inflammation (e.g., pyelonephritis, renoliths, neoplasia)	Variable UP/C Urine sediment Imaging
	Tubular proteinuria	UP/C usually = 0.5-1.0
	Glomerular proteinuria	UP/C often $\geq$ 2.0 Inactive urine sediment with the possible exception of hyaline casts

REFERENCES

[1.] Lees GE, et al. **J Vet Intern Med** 2005; 19:377-385. [2.] Jacob F, et al. **J Am Vet Med Assoc** 2005; 226(3):393-400. [3.] Syme HM, **J Vet Intern Med** 2006; 20:528-535. [4.] Jepson RE, et al. **J Vet Intern Med** 2007; 21:402-409. [5.] King JN, et al. **J Vet Intern Med** 2007; 21:906-916. [6.] Grodecki KM, et al. **J Comp Pathol** 1997; 117:209-225. [7.] Grauer GF, et al. **J Vet Intern Med** 2000; 14:526-533. [8.] Mizutani H, et al. **J Vet Intern Med** 2006; 20:1074-1079. [9.] Grauer GF, et al. **J Vet Intern Med** 2004; 18:418-419 (abstract). [10.] Vaden SL, et al. **Vet Clin Pathol** 2004; 33:14-19. [11.] Lees GE, et al. **J Vet Intern Med** 2002; 16:353 (abstract). [12.] Jepson RE, et al. **J Vet Intern Med** 2009; 23:806-813.

GLOMERULAR FILTRATION RATE IN DOGS & CATS: WHERE ARE WE NOW?

Hervé P. Lefebvre, DVM, PhD, DECVPT
Toulouse, France

INTRODUCTION

"Glomerular filtration rate (GFR) is considered the single most useful and most sensitive test of renal function overall."[1] Indeed any decrease in GFR generally means that kidney disease is occurring or progressing. Assessment of GFR is therefore pivotal in evaluation of the severity and progression of renal diseases, especially chronic kidney diseases (CKD). In humans, GFR is used to stratify patients into five stages according to the CKD classification of the National Kidney Foundation Kidney Disease Outcomes Quality Initiative (NKF KDOQI ™). In small animal medicine, the staging proposed by the International Renal Interest Society (IRIS) is currently based on basal plasma creatinine, but the IRIS board considers that in the future, GFR measurement could be the major criterion for staging, as in humans.[2] Methods for GFR evaluation in dogs and cats were validated between the beginning of the 1980s and the early 2000s. Plasma clearance methods were developed as alternatives to urinary clearance methods and clearly offer new perspectives in clinical nephrology.[3] Some recent developments, potential issues and current challenges regarding GFR measurements using plasma clearance methods will be reviewed here.

BLOOD SAMPLING PROCEDURE AND STRATEGY

Plasma clearance methods require repeated blood samples over a period of several hours following intravenous administration of the marker. Optimally, the higher the number of blood samples and the longer the period of collection (ideally until the marker has been totally cleared by the kidneys), the more accurate the estimation of the area under the curve (AUC) and consequently the GFR (GFR=dose/AUC). In practice, GFR testing is nevertheless made easier if the number of blood samples and the period of time can be limited. A compromise between these two extreme situations therefore needs to be considered.

Repeated collection of 5-ml blood samples in miniature dogs or cats may be difficult and lead to excessive blood loss. The total volume of blood collected can be reduced if assays for markers requiring a very small plasma volume are used. For example, 0.2 mL of blood is sufficient for a plasma creatinine assay using an enzymatic method. Pediatric devices based on a capillary system have been proposed as an alternative to vacuum tubes for blood sampling and for plasma clearance methods.[4,5] Their potential advantages include absence of vein collapse, a limited blood volume sampled, and improved safety.

The timing of blood sampling should be aimed to minimize the proportion of AUC extrapolated. For this, the last sampling is especially critical, because it determines the percentage of extrapolated area, which is calculated from the slope of the elimination phase and the last measured concentration. A slight error in calculation of the slope may also induce an error in calculating clearance. Ideally, the extrapolated area should be < 20% of the total AUC. The lower the renal function, the more delayed the elimination of the marker. An additional related matter is that in clinical practice the GFR of the patient is unknown and therefore the optimum final blood sampling time is always uncertain. A pragmatic way, which is feasible when the assay can readily be performed in clinics (as with plasma creatinine) is to check the plasma concentration of the marker before discharging the animal so that a complementary blood sample can be obtained 1-2 h later if needed.

Limited sampling strategies cannot be empirically decided when testing the patient. For example, for exogenous plasma creatinine clearance, an optimal blood sampling strategy at 1 and 10 h post-administration provides a relatively accurate estimation of the AUC with a maximum error of -14% compared to the AUC determined from an 11-point kinetic profile. Inversely, blood sampling at inappropriate times can lead to errors exceeding 400%.[6] Correction formulas derived from those used in human medicine (e.g., the Brøchner-Mortensen formula) have also been proposed and applied in dogs and cats[7-11] to estimate the actual GFR value from the clearance estimated with the limited sampling strategy. Nevertheless a species-specific correction formula has only been established in dogs.[9]

REPEATED GFR MEASUREMENTS

One of the potential issues regarding repeated GFR measurements in the same patient is the within-individual between-day variability (i.e., the reproducibility of GFR measurement). If one dog is tested on two separate occasions, what is the minimal difference between the two consecutive measurements that can be interpreted as clinically relevant, i.e., higher than the difference due to the physiological and analytical variability? In previous studies using healthy or diseased animals, the between-day coefficients of variation were generally lower than 20% for most markers,[12-15] which can be considered reasonably low.

Repeated measurements in a given patient should be performed using the same plasma clearance method to avoid any misinterpretation due to methodological differences. Plasma clearance methods may differ according to the GFR value, as illustrated recently in healthy, hyperthyroid and renal-impaired cats.[16]

PHYSIOLOGICAL FACTORS OF VARIATION

Dietary factors which can affect renal function will not be considered here.

Age

In growing Beagle puppies, the mean GFR decreases by ~40% (from 4.1 to 2.5 mL/min/kg) between the age of 9 and 27 weeks.[17] The GFR was 87% higher in 2-month-old Beagle puppies than in 6–9 year-old Beagle dogs.[18] These results indicate that a progressive decline in GFR is normal in the growing pup and should not be interpreted as development of CKD. In kittens, the GFR is higher in kittens 9 to 19 weeks old than in younger or older kittens.[19]

In the adult dog, aging seems to have a limited effect as shown in longitudinal[20] and cross-sectional[10] studies. In the cat, plasma clearance of creatinine, but not iohexol, was about 25% lower in 9-12 year old cats than in 7-12 month old cats.[15] In another study in 51 cats, no correlation between plasma clearance of iohexol or creatinine, and age was evidenced.[11]

Therefore, a decline in renal function with aging seems limited in healthy dogs and cats. However, the potential clinical relevance of a GFR decrease in the aged animal needs further studies.

Body weight

An effect of body weight on GFR (plasma exogenous creatinine clearance) was reported in 2004 in a large healthy canine population (n=113).[21] In 118 adult healthy dogs, a significant negative linear relationship was also detected between body weight and GFR (plasma iohexol clearance).[10]

In 51 cats, a negative linear relationship was also evidenced between body weight and GFR (mL/min/kg, estimated by the plasma clearance of iohexol/creatinine). Nevertheless, the R^2 value was less than 20%.[11] In domestic shorthair cats, body weight also appears to slightly affect the plasma creatinine concentration: the mean±SD values for plasma creatinine concentration were 142±36 (1.6±0.4) and 161±22 μmol/L (1.8±0.2 mg/dL) for animals with body weights <4 kg and >5 kg, respectively.[22]

Breed

An effect of the breed on plasma creatinine concentration in dogs[23] and cats[24] was recently evidenced. Nevertheless, this does not mean necessarily that a breed-dependency of GFR exists as endogenous creatinine production and body hydration can also affect the plasma creatinine concentration. The Greyhound breed is a good illustration as increased plasma creatinine concentration in this breed appears not to be due to decreased GFR but to increased production of creatinine.[25-27]

The breed nevertheless could affect GFR, as shown recently.[28] GFR in German Shepherds (2.5±0.7 mL/min/kg) is lower than GFR in English Pointers and Setters (3.5±0.6 and 3.4±0.8 mL/min/kg, respectively). In this study, the age was similar in all 3 breeds but the body weight was higher in German shepherds. The effects of breed and body weight were therefore confounded and between-breed differences in GFR could be explained by differences in body weight.

INDEXATION OF GLOMERULAR FILTRATION RATE

In most studies, GFR is expressed in mL/min/kg. However, indexation of GFR values (mL/min) to the body weight does not allow for similar values of GFR (mL/min/kg) in dogs with different body weights (see above). Therefore, a "unique" cut-off value for clinical use in the overall canine population is not feasible as this value is dependent on the body weight of the patient.

A current challenge therefore is to define the most appropriate way to scale GFR. Body surface area (BSA) has been proposed as the reference for indexing physiological variables. However, the abandoning of such an indexation for GFR has been recommended[29] and the formulae used to estimate BSA in dogs are probably inaccurate.[30] Due to the limitations of correcting GFR for BSA, an indexation to ECFV has been proposed in animals and also humans because one of the major roles of the kidney is to regulate body fluid composition.[31,32] Nevertheless, standardization to ECFV did not produce substantial changes in the relationships between GFR estimates and body weight in adult dogs.[10] Moreover, when such an indexation was used, differences between puppies and adult dogs were still observed, but were inversed.[18] In a previous study in dogs, it was clearly shown that standardization to body weight, BSA or ECFV could produce quite different results and a biased clinical interpretation of the GFR value.[9]

There is thus a current need to establish an equation in the canine species to standardize GFR measurements in small animals. The only way is to use regression analysis on a large representative population of adult dogs (different breeds, body weights, age and sex). A simplified and alternative approach consists of stratifying the canine population into body weight categories and defining an "averaged" cut-off for the GFR estimates in each category. The main limitation of this approach is that the larger the range of body weights for a given category, the more inaccurate the corresponding cut-off values.

The issue in the cat is similar: it was shown recently that standardization to BSA resulted in the largest between- individual coefficients of variation (36%) compared to standardization to body weight (27%) or ECFV (24%).[11]

Indexation of GFR is also an issue in obese human patients, who are clearly different from lean individuals with similar body weights. The higher the body weight, the lower the GFR indexed to BSA (or the body weight in dogs). Other ways to index GFR have been tested unsuccessfully. An absolute, non-corrected GFR is currently recommended in obese patients.[33] No data regarding indexation of body weight in obese dogs or cats has been published.

REFERENCE INTERVALS

The definition of reference intervals for GFR in dogs and cats is a prerequisite for clinical applications. Such a definition is only possible in adult dogs (see above). It is generally admitted in small animal nephrology that the lower cut-off value is between 1.5 and 2.0 mL/min/kg. Based on the effect of physiological covariables and recent studies, it appears nevertheless that a unique cut-off value is not acceptable for the overall canine population and that stratification according to the body weight (but probably not the age) should be carried out.

From a previous set of data obtained in 113 dogs,[21] the animals were divided into 4 categories of body weight (Mini, Medium, Maxi and Giant; see above). The corresponding GFR values (mean±SD) were 3.7±0.5, 3.0±0.5, 2.5±0.4, and 2.4±0.6 mL/min/kg, respectively. It could therefore be considered that, if the hypothetical distribution of the GFR values was normal, the corresponding lower limits of the reference range would be 2.7, 2.0, 1.7 and 1.3 mL/min/kg. In another study including 118 adult healthy dogs, the reference ranges in different weight quartiles were 1.54-4.25, 1.29-3.50, 0.95-3.36, and 1.12-3.39 mL/min/kg, respectively, for body weight quartiles of 1.8-12.4, 13.2-25.5, 25.7-31.6, and 32.0-70.3 kg.[10] These results obtained with two different markers (creatinine and iohexol) demonstrate that according to the body weight, the lower cut-off value can be shifted to the right (small dogs) or to the left (Giant dogs). In other terms, choosing a value of 1.5 mL/min/kg to declare a dog as renal-impaired will be too low for small dogs (false negative results) and too high for giant dogs (false positive results). Breed may also affect the reference intervals. For example, tentative reference intervals for English Pointers, English Setters and German Shepherds are 2.3-5.1, 1.8-5.0 and 1.7-3.8 mL/min/kg.[28] However, establishment of breed-specific reference intervals in all the canine breeds seems unrealistic, except for the major breeds.

Reference ranges from a study performed in 51 cats have recently been proposed : [1.0-3.5 mL/min/kg] for plasma iohexol clearance and [1.3-3.8 mL/min/kg] for plasma creatinine clearance.[11] Interestingly, in terms of allometric relationships, the cut-off value for GFR in cats is lower than that in small dogs of similar body weight.

Although these preliminary results provide some estimates of reference intervals useful for veterinary nephrologists, further investigations in larger canine and feline populations are still needed.

NEW PERSPECTIVES: GFR MEASUREMENTS IN CARDIOVASCULAR AND ENDOCRINE DISEASES

GFR measurements may be interesting to assess renal function in small animal patients with non-primary renal diseases, as shown recently for endocrine and cardiovascular diseases.

Cardiovascular diseases

Recently, chronic kidney disease (CKD) was shown to be highly prevalent in humans with cardiovascular disease.[34] A very slight decrease in renal function may dramatically increase the cardiovascular risk in such patients.[34] Azotemia is a frequent finding in dogs with cardiac diseases.[35-38] In cardiac dogs, the degree of azotemia however was most often mild to moderate and not related to changes in cardiovascular variables.[37] Interestingly, the GFR in 24 dogs with chronic valvular disease was shown to be significantly lower in NYHA III-IV (1.7±0.7 mL/min/kg) than in NYHA I-II (3.1±0.8 mL/min/kg). Only 1/15 NYHA dogs in class I-II had a GFR below 2 mL/min/kg and only 2/9 NYHA class III-IV dogs had a GFR above 2 mL/min/kg.[37] Further investigations are needed to identify the cause of decreased GFR and its clinical relevance on the outcome of the cardiac disease.

Endocrine diseases

Interactions exist between endocrine systems and kidney function under both physiological and clinical conditions. Probably, one of the best examples in veterinary medicine is the effect of hyper- or hypothyroidism on renal function in small animals.[39]

Hyperthyroidism is the most common endocrine disorder in the aged cat and a prevalence of pre-existing CKD in hyperthyroid cats of between 14 and 40% has been reported.[39] Evaluation of kidney function in a hyperthyroid cat is pivotal as hyperthyroidism can mask and sometimes worsen CKD. CKD became clinically apparent after treatment of hyperthyroidism in about 17-39% of cats.[39] A challenge in such patients is to predict the risk of post-treatment azotemia. The pre-treatment GFR value was shown to be predictive of the development of post-treatment CKD.[40-42] An accurate evaluation of kidney function can be performed from 4 weeks after ^{131}I-treatment.[42,43]

GFR was shown to be decreased in experimentally-induced canine hypothyroidism whereas plasma creatinine concentrations remained unaltered.[44] In another study in 14 dogs with naturally occurring hypothyroidism, GFR was <2 mL/min/kg in all dogs, and increased significantly with levothyroxin treatment.[45] The potential effects of renal dysfunction and development of subclinical CKD on the clinical outcome of hypothyroid dogs require further investigation.

REFERENCES

[1]Finco DR, et al. **J Vet Intern Med** 1999;13:516. [2]Elliott J. 2007 **BSAVA Manual of Canine and Feline Nephrology and Urology**, 2nd ed; 159. [3]Ross L. **ACVIM Forum** 2010. [4]Reynolds BS, et al. **J Feline Med Surg** 2007;9;382. [5]Reynolds BS, et al. **J Am Anim Hosp Assoc** 2008;44:51. [6]Watson ADJ, et al. **J Vet Intern Med** 2002;16:22. [7]Brown SA, et al. **Am J Vet Res** 1996;57 :105. [8]Gleadhill A, et al. Res Vet Sci 1996;60:117. [9]Heiene R, et al. **J Vet Intern Med** 1999;13:587. [10]Bexfield NH, et al. **J Vet Intern Med** 2008;22:66. [11]Heiene R, et al. **Am J Vet Res** 2009;70:176. [12]Finco DR, et al. **J Vet Intern Med** 2001;15:368. [13]Finco DR. **Am J Vet Res** 2005;66:1046. [14]Miyamoto K. **Am J Vet Res** 2001;62:1332. [15]Van Hoek I, et al. **J Vet Intern Med** 2007;21:950. [16]Van Hoek IM, et al, **J Feline Med Surg** 2009;11:1028. [17]Lane IF, et al. **Am J Vet Res** 2000;61:577. [18]Laroute V, et al. **Res Vet Sci** 2005;79:161. [19]Hoskins JD, et al. **Am J Vet Res** 1991;52:1295. [20]Queau Y, et al. **J Vet Intern Med** 2007;21:598. [21]Lefebvre HP, et al. **J Vet Intern Med** 2004 ; 18:415. [22]Reynolds BS, et al. **Am J Vet Res** 2008;69:471–477. [23]Craig AJ, et al. **J Vet Intern Med** 2006 ;20:766. [24]Reynolds BS, et al. **J Vet Intern Med** 2009; [25]Drost WT, et al. **J Vet Intern Med** 2006;20:544. [26]Hill RC, et al. **J Vet Intern Med** 2007; :21:647. [27]Kukanich B, et al. **J Vet Pharmacol Ther** 2007;30:314. [28]Seguela J, et al. **J Vet Intern Med** 2009;23:764. [29]Turner ST, et al. **Am J Physiol** 1995;268:R978. [30]Price GS, et al. **J Vet Intern Med** 1998;12:267. [31]Gleadhill A, et al. **Res Vet Sci** 1995;59:118. [32]Peters AM, et al. **Eur J Nucl Med Mol Imaging** 2004;31:137. [33]Delanaye P, et al. **Nephrol Dial Transplant** 2005 ;20 :2024. [34]Ritz E, et al. **J Am Soc Nephrol** 2004;15:513. [35]Atkins CE, et al. **J Am Vet Med Assoc** 2002;221:654. [36]Boswood A, et al. **J Vet Cardiol** 2006;8:1. [37]Nicolle AP, et al. **J Vet Intern Med** 2007;21:943. [38]Gouni V, et al. **J Vet Cardiol** 2008;10:117. [39]Van Hoek I, et al. **Gen Comp Endocrinol** 2009;160:205. [40]Adams WG, et al. **Vet Radiol Ultrasound** 1997;38:231. [41]Becker TJ, et al. **J Am Anim Hosp Assoc** 2000;36:215. [42]Van Hoek I, et al. **Domest Anim Endocrinol** 2009;36:45. [43]Boag AK, et al. **Vet Rec** 2007; 161:711. [44]Panceira DL, et al. **J Vet Intern Med** 2009;23:1045. [45]Gommeren K, et al. **J Vet Intern Med** 2009;23:844.

A SIMPLIFIED METHOD FOR DETERMINATION OF GLOMERULAR FILTRATION RATE (GFR) IN CATS. N.C.Finch[1], H.Syme[1], J.Elliott[1], R.Heiene[2]. 1. Royal Veterinary College, London, UK. 2. Norwegian School of Veterinary Science, Oslo, Norway.

GFR is widely accepted as the best index of renal function. Plasma clearance (Cl) of iohexol is a validated method for GFR estimation in cats. Formulae which correct for the initial redistribution exponential can be applied to limited sampling strategies when measuring GFR. Such formulae have been developed for humans and dogs but no cat specific formula exists.

The measurement of plasma Cl of iohexol involved the intravenous administration of a bolus dose of iohexol (Omnipaque™):(647 mg/kg (300 mg iodine/kg), followed by the collection of blood samples via jugular venepuncture at the following time points:- 5, 15, 30, 60, 120, 180, 240 and 360 mins. The exact sampling time was recorded for each cat and used in the Cl calculations. Iohexol concentration was measured at two reference laboratories using a high performance liquid chromatography (HPLC) method. Results were provided as concentrations of both stereoisomers of iohexol, endo-iohexol (EnIox) and exo-iohexol (ExIox) and total iohexol (TIox) concentrations.

Pharmacokinetic analysis was performed using Phoenix WinNonlin6.0 software. Cl was calculated as dose/area under curve (AUC). AUC was determined by non-compartmental analysis. Cl was additionally calculated from a limited number of samples collected at 120, 180 and 240 mins using one-compartmental analysis. Cl values were standardised to bodyweight (kg). Agreement between the Cl of EnIox, ExIox and TIox was performed using Friedman's ANOVA and evaluating Bland-Altman plots. The change in the EnIox and ExIox ratio over time was explored using repeated measures ANOVA. A cat specific correction formula was developed by performing regression through the origin analysis of the 3 sample Cl data using a 2nd order polynomial model. Agreement between the corrected Cl and multisample Cl was performed using statistical methods as described above.

There was no significant difference between the Cl of EnIox, ExIox and TIox (P=0.423, N=20). The 95% limits of agreement and bias were small. There was no significant change in the EnIox and ExIox ratio over time (P=0.347, N=17). The derived correction formula (N=17) for a limited sampling technique was $Cl_{corrected} = 1.036(Cl_{uncorrected}) - 0.062(Cl_{uncorrected})^2$. Median(range) Cl for the multisample Cl and corrected Cl was 2.18(1.53-1.82) and 2.27(1.59-2.70) ml/min/kg respectively. The 95% limits of agreement were small (-0.26 – 0.26 ml/min/kg) and bias was negligible (0.0001 ml/min/kg).

The development of a cat specific correction formula will be useful in the future for the rapid and simple estimation of GFR in the cat. In addition this study shows that the Cl of the different stereoisomers of iohexol and TIox do not differ and can be used interchangeably in the correction formula.

EVALUATION OF D,L-METHIONINE AND ANTIMICROBIAL AGENTS FOR MEDICAL DISSOLUTION OF SPONTANEOUSLY OCCURRING INFECTION-INDUCED STRUVITE UROCYSTOLITHS IN DOGS J Bartges, T Moyers. The University of Tennessee College of Veterinary Medicine, Knoxville, TN.

Struvite urolith formation depends on increased urinary concentrations of magnesium, ammonium, and phosphate, and alkaluria. In dogs this is usually associated with a bacterial urinary tract infection (UTI) with urease-producing bacteria such as *Staphylococcus spp.* or *Proteus spp.* We hypothesized that utilization of a urinary acidifier (d,l-methionine, Methioform, Lloyd Inc.) and an appropriate antimicrobial agent without changing diet would be effective and safe in dissolving spontaneously-occurring, infection-induced struvite urocystoliths in dogs within 8 to 12 weeks.

Fourteen client-owned dogs with presumed infection-induced struvite urocystoliths were studied. Dogs were maintained on their regular diet. The UTI was treated with an appropriate antimicrobial based on culture and sensitivity results; antimicrobial therapy was continued during medical dissolution and for 2 weeks beyond radiographic evidence of urocystolith dissolution. A urinary acidifier[2] was administered during medical dissolution and for 2 weeks beyond radiographic evidence of urocystolith dissolution at an initial dosage of 100 mg/kg PO q12hr. Monitoring included plasma biochemical analysis, urinalysis, urine culture, venous blood gas analysis, and abdominal radiography performed every 4 weeks until radiographic evidence of urolith dissolution or until uroliths did not change over 2 successive re-evaluations at which time they were removed surgically.

Fourteen dogs met inclusion criteria and were included in the study; however, 3 dropped out due to inability of the owners to administer the treatment. Urocystoliths dissolved in 8 of 11 dogs with a median time of 2 months and range of 1 to 4 months. Average final dosage for dogs in which urocystoliths dissolved (n=8) was 97.3 ± 25.6 with a median of 104.2 and range of 56.8 to 125.0 mg/kg PO q12h In 3 dogs, urolith dissolution failure occurred at 3, 4, and 6 months; analysis of urocystoliths removed surgically revealed them to contain calcium oxalate. Dogs did not lose weight, experience metabolic acidosis, or have change in plasma biochemical parameters.

Administration of d,l-methionine at approximately 100 mg/kg PO q12h is safe and effective in dissolving presumed infection-induced struvite uroliths in dogs in combination with an appropriate anti-microbial agent without using a struvite dissolution diet. Successful dissolution occurs when uroliths decrease in size by at least 50% at the 1 month re-evaluation. If uroliths do not decrease in size by at least 50% at the 1 month re-evaluation, then consideration should be given to (1) lack of compliance, (2) inappropriate dosage, (3) difficulty in controlling the bacterial urinary tract infection, or (4) uroliths being composed of other minerals, most likely calcium oxalate, in addition to or instead of struvite.

CLASSIFICATION & DIAGNOSIS OF SEIZURE DISORDERS IN SMALL PATIENTS

Christopher L. Mariani, DVM, PhD, DACVIM (Neurology)
Raleigh, NC

DEFINITIONS & TERMINOLOGY

Seizures

Seizure – the clinical manifestation of excessive and/or hypersynchronous neuronal discharge within the brain; may be manifested as episodic impairment or loss of consciousness, abnormal motor phenomena, psychic or sensory disturbances, or autonomic signs such as salivation, vomiting, urination or defecation.[1,2] Synonyms – convulsion, ictus, fit. Seizures have been classified in the past according to the areas of the brain that they affect. Thus, they have been termed generalized or focal (partial) seizures, defined as follows:

Generalized – abnormal electrical discharges affecting the cerebral hemispheres bilaterally and typically causing symmetric signs. Several manifestations are possible, including generalized tonic-clonic ("Grand mal"), or exclusively tonic, clonic or atonic activity.[1,3]

Focal (Partial): abnormal electrical discharges of neurons in a focal area of the brain; may result in asymmetric motor or sensory signs such as twitching of the eyelids, lips, ears or one limb. Focal seizures may secondarily generalize to involve both hemispheres; that is, the electrical activity spreads across the cerebral cortex. This often gives rise to focal motor activity, which then evolves; this can often be recognized by owners.

Epilepsy

Epilepsy – a chronic neurologic condition characterized by recurrent seizures; has an intracranial origin.[1,2,4]

Idiopathic Epilepsy – Idiopathic epilepsy denotes "a syndrome unto itself"; that is, a well-recognized clinical condition of recurrent seizures without other neurologic signs and without any identifiable underlying cause.[4,5] It is considered to be an inherited or familial condition. Synonyms – primary epilepsy, genetic epilepsy.

Symptomatic Epilepsy – Epilepsy occurring secondary to an identifiable structural lesion within the brain (e.g., brain tumor, meningoencephalitis, cerebrovascular accident).[4,5] Synonyms – secondary epilepsy.

Probably Symptomatic Epilepsy – This condition is synonymous with acquired epilepsy and cryptogenic epilepsy. It implies that an acquired structural lesion within the brain is suspected to be causing the seizures, but cannot be detected with the available diagnostic tests (e.g., patients with a previous head trauma or meningoencephalitis and assumed scarring or fibrosis).[5] Synonyms – acquired epilepsy, cryptogenic epilepsy.

Reactive Seizures – Seizures that occur secondary to a systemic, extracranial disease process, which causes dysfunction in an otherwise normal brain (e.g., hypoglycemia, toxins).[3-5] Not considered to be epilepsy.

DIAGNOSIS

Seizures vs. Seizure Impostors

A number of other clinical conditions can mimic seizures, including syncope, acute vestibular dysfunction, tremors, narcolepsy/cataplexy, rapid eye movement (REM) behavior disorder, movement disorders (e.g., distemper myoclonus, episodic dyskinesia), and behavioral disorders.[6,7] The majority of patients presenting for seizures are normal in the veterinary hospital, and the clinician's first task is to decide whether the events in question are true seizures, or seizure "mimics". Once the clinician is reasonably certain that the events are true seizures, differential diagnoses may be considered. These can be broadly categorized into extracranial and intracranial etiologies.

Differential Diagnosis for Seizures

Extracranial Causes of Seizures

- Hypoglycemia – main causes are juvenile hypoglycemia, hyperinsulinemia (secondary to insulin overdose or an insulin-secreting tumor), and sepsis.
- Hypocalcemia – causes include eclampsia, hypoparathyroidism and ethylene glycol intoxication.
- Hyponatremia – results from loss of sodium-rich fluids (e.g. vomiting, diarrhea, diuretics), hypoadrenocorticism or increased water intake (psychogenic polydipsia).
- Hypernatremia – results from the loss of free water or sodium-poor fluids (diabetes insipidus, excessive panting, high temperatures), decreased water intake (primary adipsia or limited access) or rarely through ingestion of high levels of salt.
- Hepatic encephalopathy – results from altered filtration of gastrointestinal portal blood by the liver. Main causes are the presence of an anomalous portosystemic shunting vessel or severe parenchymal hepatic disease with secondary shunting.

- Hypertriglyceridemia – primarily a disease of miniature Schnauzers, which have an enzyme deficiency, allowing triglyceride accumulation in the blood.
- Nutritional disease – rare cause of seizures; historically thiamine deficiency has been implicated.
- Intoxication – a wide variety of toxins may lead to nervous system dysfunction and seizures. Examples include ethylene glycol, lead, heavy metals, metaldehyde, strychnine, and organophosphates.

Intracranial Causes of Seizures

- Degenerative disease – very rare causes of seizures; include lysosomal storage disorders and neuronal abiotrophies.
- Anomalous conditions – hydrocephalus is the most common and occurs primarily in toy and brachycephalic breeds. Less common conditions include epidermoid, dermoid, and arachnoid cysts, and lissencephaly.
- Neoplasia – common cause of seizures in older animals. Brain tumors may be primary (arise from brain itself) or secondary (metastatic or arise from adjacent structures [e.g., skull]). It is common to see seizures as the sole clinical sign of a brain tumor.
- Inflammatory disease – common cause of seizures in animals of any age. Encephalitis may be infectious (viral, fungal, protozoal, bacterial, rickettsial, or parasitic) or more frequently, non-infectious. Common presumed non-infectious causes in dogs include granulomatous meningoencephalitis (GME) and necrotizing meningoencephalitis (NME).
- Traumatic disease – seizures can occur immediately at the time of head trauma or as a late sequela after recovery.

Idiopathic epilepsy

Recent work in canine and human epilepsy have challenged traditionally held views on this disease in a number of ways; in fact, it is time we started considering this syndrome as a diverse group of genetic diseases, rather than a single isolated disease process. Seizures in animals with idiopathic epilepsy have traditionally been considered to be of the generalized tonic-clonic variety, although this is not always the case.[8-10] The seizures almost always begin between 1 and 5 years of age. One of the hallmarks of idiopathic epilepsy is its insidious onset; seizures initially occur weeks to months apart and gradually become more frequent. Many dogs progress to develop cluster seizures or status epilepticus. Although certain breeds of dogs are predisposed to the development of idiopathic epilepsy, this condition can probably occur in any dog. The inheritance pattern has traditionally been suggested to be complex and likely polygenic in nature, but some recent studies suggest alternate mechanisms of inheritance.[9,10] In the absence of a genetic test, diagnosis is based upon an appropriate signalment and description of seizure onset and character, and by ruling out other potential etiologic diagnoses with appropriate tests (typically cerebrospinal fluid [CSF] evaluation and brain imaging). Outside of the ictal and immediate post-ictal periods, the neurologic examination should be normal.

The presence of idiopathic epilepsy in cats is somewhat controversial, but most of this controversy is related to differences in terminology and misunderstanding of the epilepsy classification system. Cats frequently develop seizures that do not have an identifiable underlying cause.[11,12] According to the terminology used here, these cats would be diagnosed with probably symptomatic epilepsy in most cases, as there is little evidence suggesting the presence of heritable epilepsy in cats. However, some neurologists group all patients with seizures of unidentifiable etiology together as idiopathic (i.e., a presumed genetic origin is not required), which accounts for the confusion and apparent controversy.

Probably symptomatic epilepsy

Seizures may be focal, focal with secondary generalization, or generalized, and may occur at any age. All breeds of dog and cat can be affected. Although occasional focal neurologic deficits (e.g., reduced conscious proprioception on one side) are noted, the majority of these patients have a normal neurologic examination. The diagnosis is based on signalment, neurologic examination, and by ruling out other causes of seizures with appropriate diagnostic testing.

Causes of Idiopathic Epilepsy

In humans, a number of the known idiopathic epilepsies have been associated with mutations in ion channels or their receptors.[13] Dysfunction in these structures leads to abnormal movement of ions and associated potential differences across the membranes of excitable cells such as neurons. In addition to epilepsy, ion channel mutations are known to be responsible for a number of diseases that cause episodic nervous system or other organ dysfunction, such as congenital myotonia, episodic ataxia, and hyperkalemic periodic paralysis.[14] Such mutations are intriguing targets for further epilepsy research for a number of reasons; they generate electric currents that are paroxysmal in nature, they may involve a reorganization of channel subunits with development or maturation (potentially explaining the age-related onset of some epilepsies), and they are affected by many of the anticonvulsant medications in current clinical use. The next 10-20 years will likely result in the identification of the underlying etiology behind many canine idiopathic epilepsies, as mapping of the canine genome progresses and the tools available for comprehensive genome-wide screening improve.

Signalment and Commonly Associated Differential Diagnoses

Less than 1 year of age – intoxication, hypoglycemia, meningoencephalitis, hepatic encephalopathy, hydrocephalus, head trauma, lysosomal storage disease, other congenital disorders

From 1 to 5 years of age – idiopathic epilepsy, probably symptomatic epilepsy, head trauma, intoxication, meningoencephalitis, hydrocephalus, hepatic encephalopathy, cerebrovascular disease, cerebral neoplasia

Greater than 5 years of age – Cerebral neoplasia, probably symptomatic epilepsy, meningoencephalitis, cerebrovascular disease, hypoglycemia

Important History Questions to Ask – Character, onset, frequency and progression of seizures, evidence of asymmetry, evidence of prodrome or post-ictal period, neurologic status between seizures, vaccination status and travel history, previous illness and medications

Important Physical Examination Findings – Evidence of cardiovascular disease, evidence of hepatic disease, retinal (fundic) examination

Important Neurologic Examination Findings – Altered mentation, focal cranial nerve deficits, circling or turning, focal postural reaction or conscious proprioceptive deficits, visual deficits, cervical hyperesthesia

Diagnostic Testing and Plan

I find it useful to separate the available diagnostic tests into three different steps or stages as follows:

Step 1 – CBC, serum biochemistry, urinalysis
Step 2 – Thoracic and abdominal imaging, serum bile acids, toxin assays (cholinesterase, lead, ethylene glycol), Infectious disease titers, skull radiographs
Step 3 – CSF evaluation, computed tomography (CT) or magnetic resonance imaging (MRI), electroencephalography (EEG)

Step 1 should probably be performed on any patient with seizures. Step 2 consists of tests that are readily available to most veterinary practitioners and may be chosen in certain cases depending on the index of suspicion for a certain disease. Step 3 consists of specialized tests that specifically evaluate the nervous system, and are usually only available at specialty referral hospitals. The signalment, history and physical and neurologic examinations are critical when deciding which diagnostic tests are appropriate.

When to choose Step 3? Focal seizure activity, focal neurologic deficits, dogs less than 1 year or greater than 5 years, all cats, if owner desires

Utility of advanced diagnostic testing

Cerebrospinal fluid evaluation – analysis includes red and white blood cell counts, cytology and protein concentration. Pleocytosis is noted with most meningoencephalitis and occasionally other etiologies (e.g., neoplasia). Albuminocytologic dissociation (high protein, normal cell count) is a nonspecific change often seen with neoplastic disease, stroke, and degenerative conditions.

Magnetic resonance imaging – gold standard for imaging of the brain as it allows image capture in multiple planes, and manipulation of various acquisition parameters (sequences) allows highlighting of different tissues and pathology with varying water content or other characteristics. Specialized sequences include short T1 inversion recovery (STIR; removes fat signal, T2-weighting highlights pathology), fluid attenuated inversion recovery (FLAIR; T2-weighted, removes CSF signal but keeps edema and pathologic fluid signal), gradient echo/T2* (highlights hemorrhage), and diffusion-weighted imaging (detects differences in water diffusion such as in ischemic stroke).

Electroencephalography – captures and records electrical activity of cerebral cortex. May be able to capture paroxysmal discharges (e.g., spikes, sharp waves) which are highly suggestive of a seizure disorder; this may aid in the differentiation of seizures from "seizure mimics", localization of seizure activity to a specific area of the cerebrum, or in documenting ongoing seizure activity.

REFERENCES

1. Blume WT, et al. **Epilepsia** 2001;42:1212. 2. Engel J, Jr. **Epilepsia** 2006;47:1558. 3. Podell M. **Vet Clin North Am Small Anim Pract** 1996;26:779. 4. Thomas WB. **Vet Clin North Am Small Anim Pract** 40:161. 5. Engel J, Jr. **Epilepsia** 2001;42:796. 6. de Lahunta A, et al. **Compendium Contin Ed Pract Vet** 2006;28. 7. Podell M. **Vet Clin North Am Small Anim Pract** 2004;34:1435. 8. Berendt M, et al. **J Vet Intern Med** 1999;13:14. 9. Patterson EE, et al. **J Vet Intern Med** 2003;17:319. 10. Berendt M, et al. **J Small Anim Pract** 2009;50:655. 11. Barnes HL, et al. **J Am Vet Med Assoc** 2004;225:1723. 12. Schriefl S, et al. **J Am Vet Med Assoc** 2008;233:1591. 13. Steinlein OK. **Am J Med Genet** 2001;106:139. 14. Hirose S, et al. **Epilepsy Res** 2000;41:191.

MAINTENANCE THERAPY FOR THE ROUTINE & DIFFICULT TO CONTROL EPILEPTIC PATIENT

Christopher L. Mariani, DVM, PhD, DACVIM (Neurology)
Raleigh, NC

Deciding on a treatment plan for an animal with seizures depends on a number of factors, including the suspected etiologic cause of the seizures, the frequency and severity of the observed seizures, and the financial constraints or intentions of the owner.

ADDRESS THE UNDERLYING CAUSE

If an underlying cause of the seizures is known or suspected, it should be appropriately addressed, if possible. Thus, animals with hypoglycemia or electrolyte abnormalities may require no therapy other than correction of these deficiencies (or excesses). Likewise, patients with hepatic encephalopathy, hypertriglyceridemia or various intoxications may not require long-term anticonvulsant therapy if the primary disease is appropriately addressed, although they may benefit from shorter-term treatment with these drugs. In some cases, damage to the brain may lead to acquired (probably symptomatic) epilepsy, requiring long-term treatment.

Animals with intracranial diseases also benefit from addressing the underlying condition, although these patients are more likely to require maintenance anticonvulsant therapy. Thus, placement of a ventriculoperitoneal shunt for hydrocephalus, anti-inflammatory and/or antimicrobial medications for meningoencephalitis, surgery or radiation therapy for brain tumors, and other specific therapies address the underlying disease process and may reduce or eliminate the need for anticonvulsant therapy.

MAINTENANCE ANTICONVULSANT THERAPY

Maintenance anticonvulsant therapy is used as an adjunct in symptomatic epilepsy and is the cornerstone of therapy for patients with idiopathic or probably symptomatic epilepsy. The first question to address is: **When to start anticonvulsant therapy?** There are no hard and fast rules on this issue, and each patient much be approached individually. However, some guidelines apply. In general, maintenance therapy should be considered if:

- Seizures are more frequent than once every 6-8 weeks
- Seizures are obviously increasing in frequency
- Status epilepticus or cluster seizures occur
- Seizures last longer than 5 minutes
- Seizures are very severe or involve aggression towards the owner

The second question to address is: **Which anticonvulsant should I choose?** Historically in dogs, the two main initial options for therapy have been phenobarbital and (potassium) bromide. These medications are chosen because of their long history of use, apparent efficacy, ease of dosing, favorable pharmacokinetics and inexpensiveness. There is limited evidence to suggest that phenobarbital may be slightly more efficacious as a first line agent in the dog. Diazepam is not effective as a maintenance anticonvulsant in the dog due to a very short elimination half-life, and the development of tolerance within several weeks.[1] Some of the newer anticonvulsant medications can be effective as initial therapy, and the author uses zonisamide and levetiracetam with some frequency as first-line agents in dogs. Zonisamide is particularly attractive in this setting due to its low incidence of side effects and its relatively long half life, allowing twice daily administration.[2,3] However, published reports of efficacy in this setting are lacking in veterinary patients. Use of these newer drugs has been limited in the past by their expense when compared with traditional anticonvulsants, although generic versions of most of these newer generation drugs are now available at reduced costs. In the cat, phenobarbital (preferred) and diazepam are the historical maintenance drugs of choice. Bromide is an effective anticonvulsant in the cat, but is associated with a very high incidence of inflammatory lung disease, and is not recommended.[4,5] Diazepam should also be used with caution in cats, as it has been associated with idiosyncratic hepatic necrosis after oral administration.[6] Levetiracetam may be a reasonable choice in the cat if phenobarbital is not an option, although the drug must be administered three times daily.[7] There is limited information available on zonisamide in cats,[8] although side effects seem to occur more frequently in this species.[9]

Initial maintenance therapy for epileptic animals

- Phenobarbital – 2.5-3 mg/kg q 12 hours (Dogs or cats)
- Potassium bromide – 40-50 mg/kg/day q 24 hours or divided (q 12 hours) (Dogs)
- Zonisamide – 3-5 mg/kg q 12 hours (Dogs)
- Levetiracetam – 20 mg/kg q 8 hours (Dogs and Cats)
- Diazepam – 0.2-1.0 mg/kg q 12 hours (Cats)

Phenobarbital is available in generic tablets (15, 30, 60, 90, 100 mg) or suspension (3 and 4 mg/ml) formulations, as is diazepam (2, 5, 10 mg tabs; 1 and 5 mg/ml suspension). Zonisamide is available as 25, 50 and 100 mg capsules. Levetiracetam is available as 250, 500, 750 and 1000 mg tablets and a 100 mg/ml suspension. Bromide is typically compounded from the chemical grade salt, and complexed with potassium (KBr) or less frequently with sodium (NaBr). It should be noted that due to molecular weight differences between the cation, equal amounts of KBr and NaBr do not contain the same amounts of bromide, and therefore have different anticonvulsant potencies (250 mg KBr = 211 mg NaBr). KBr is available from a number of compounding pharmacies. Liquid formulations are preferred over capsules, as they facilitate dosage adjustments, and KBr is best administered with food to reduce gastrointestinal irritation. Dietary salt affects serum levels of bromide, and a constant salt level should be maintained in the diet.

Monitoring maintenance therapy

A complete blood count (CBC), serum biochemical evaluation and urinalysis should be performed before starting maintenance anticonvulsant therapy, both as part of the diagnostic evaluation (see previous talk) and as a baseline before starting therapy. In addition, the metabolism of these drugs varies between patients, and blood levels are essential to guide therapy. Steady state of a drug after regular oral dosing depends on its half-life in the body, and varies between medications and species. Monitoring times and desired blood levels are shown below.

Table 1. Monitoring Anticonvulsant Therapy

Medication	Time to Steady State	Desired Blood Levels
Phenobarbital	10-14 days	15-35 µg/ml
Bromide	3-4 months	1000-3000 µg/ml (or 1-3 mg/ml or 100-300 mg/dl)
Diazepam	5-10 days	Monitoring not typically performed
Zonisamide	3-5 days	10-40 µg/ml (extrapolated from humans)
Levetiracetam	1-2 days	5-45 µg/ml (extrapolated from humans)

These desired blood levels are a guide only, and must be interpreted in light of the resulting seizure frequency and clinical condition of the patient. When measuring therapeutic blood levels, as with any medications, a serum separator tube ("tiger top") should be avoided, as the separator device may bind the drug and artificially decrease the serum levels. After the establishment of acceptable therapeutic levels of the medication, it is generally recommended that blood levels along with a CBC, serum chemistry and urinalysis be monitored every 6-12 months or in the event of an acute change in seizure frequency or new onset of sedation, weakness or ataxia. Animals receiving phenobarbital may also benefit from pre-and post-prandial serum bile acid evaluation at these times to detect changes in hepatic function. Therapeutic monitoring is very important for phenobarbital and bromide, but has been utilized less frequently for the newer generation of drugs. As these newer medications are quite safe, they have generally been used to effect, and until recently, routine therapeutic monitoring for these drugs was not available.

Potential side effects of maintenance therapy

Side effects of phenobarbital include sedation, polyphagia, polyuria, polydipsia, weight gain, and in rare cases hepatic failure and blood dyscrasias. Although hepatic damage is a commonly cited concern with this medication, it is an unusual sequela, particularly if the serum levels are maintained below 35 µg/ml. It should be noted that increases in liver enzyme levels are expected with this medication due to enzyme induction, and this does not indicate hepatic failure. If this is a concern, serum bile acids should be evaluated. Side effects of bromide are similar for the most part, including sedation, polyphagia, polyuria, polydipsia, and weight gain but in rare cases also include pancreatitis. Vomiting related to the salt content can be minimized by administration with food. Diazepam may lead to sedation and polyphagia, and rare idiosyncratic reactions causing acute hepatic necrosis have been described in the cat.[6] Zonisamide and levetiracetam may both cause sedation, while the former may also cause vomiting and diarrhea.

DEALING WITH THE REFRACTORY EPILEPTIC

Monotherapy with one of the medications above controls an estimated 60-80% of epileptic dogs and the majority of cats. However, a number of animals will have their condition remain unchanged or worsen in the face of this therapy. In this situation, a number of additional steps may be considered:

- Ensure owner is administering drug correctly
- Ask about dietary changes, other medications or herbal preparations, and topical anti-parasite medications which may interfere with seizure control
- Reconsider diagnosis, pursue additional diagnostic testing
- Ensure optimal blood levels of maintenance drug
- Increase dosing frequency (Phenobarbital - from q12 h to q 8 h) if seizures occur at times corresponding to "trough" blood levels (base on therapeutic monitoring)
- Ensure female dogs have been spayed
- Add a second anticonvulsant drug

Regarding these points, many animals require blood levels of phenobarbital above 25 µg/ml for seizure control, although levels exceeding 35 µg/ml should be avoided. Although unusual, some animals receiving phenobarbital metabolize the drug very rapidly, and may benefit from dosing every 8 hours. Having the owner maintain a seizure diary is useful to document these cases, as seizures may occur during the expected "trough" period of drug metabolism and peak and trough serum levels may be beneficial in guiding therapy. Serum levels of bromide above 3000 μg/ml are tolerated in some dogs, especially when used as a monotherapy.

Adding a Second Anticonvulsant Drug

In the cat, diazepam may be added successfully to phenobarbital to control seizures. In the dog, a combination of phenobarbital and KBr (starting dose 20-30 mg/kg daily) is effective in controlling the majority of patients refractory to monotherapy with either drug alone. However, side effects are common with this protocol, and may be unacceptable to the owner. These include sedation, pelvic limb weakness and ataxia, polyphagia, polyuria and polydipsia. It should be noted that side effects may subside approximately 1-2 weeks after initiating the new drug, and so patience can pay off. Generally, a balance must be achieved between an acceptable seizure frequency and these side effects, although this may be impossible in some dogs. Generally, the best success is achieved by aiming for a serum bromide level between 2000-3000 mg/l and maintaining a lower phenobarbital level (e.g. 10-20 μg/ml).

If seizure control cannot be obtained with this combination of drugs, then other options exist. Many refractory dogs experience cluster seizures at varying time intervals, with relatively good control between cluster episodes. In this case, administration of rectal or nasal diazepam or other benzodiazepines (see next talk) may help to control the cluster events and avoid an emergency visit to the hospital. A third anticonvulsant medication may be added, and includes the following choices:

- Zonisamide (Zonegran) – 10 mg/kg q 12 hours (dogs only)
- Levetiracetam (Keppra) – 20 mg/kg q 8 hours
- Felbamate (Felbatol) – 15-20 mg/kg q 8 hours (dogs only)
- Gabapentin (Neurontin) – 10-30 mg/kg q 8 hours (dog) or q 8-12 hours (cat)
- Pregabalin (Lyrica) – 2 mg/kg q 8-12 hours, increasing 1 mg/kg/dose each week to a total of 3-4 mg/kg (dogs only)

These drugs have a variety of mechanisms of action, which appear to be different from phenobarbital and bromide, and patients may receive additional benefit from a multimodal antiseizure effect. Another advantage of the newer drugs is their improved side effect profile, as side effects are essentially limited to sedation (which tends to be less severe than that seen with either phenobarbital or bromide) and gastrointestinal side effects (vomiting, diarrhea) for some drugs. Felbamate is an exception, as there is concern with hepatic dysfunction, particularly when used in combination with phenobarbital. Elimination half-lives are relatively short, and drug steady state levels are reached relatively quickly with administration of a regular oral dose. The main disadvantages of these newer drugs are their expense (although most are now available generically, and costs are decreasing) and requirement for administration every 8-12 hours.

Of these newer generation drugs, the author generally prefers to use zonisamide or levetiracetam. Assays to measure blood levels of these drugs are available at a few select laboratories, but these drugs are often administered to effect. In some cases, success with the addition of an anticonvulsant drug may allow the eventual withdrawal of the initial medication, although this must be accomplished gradually and with caution. The author frequently uses these newer anticonvulsant medications (particularly zonisamide and levetiracetam) as the second drug choice (typically instead of bromide). Other potential interventions to consider in select situations include acupuncture and the administration of a hypoallergenic diet.

It is important to note that most of these drugs are evaluated in small, unblinded, open label trials in veterinary patients with inadequate responses to phenobarbital or combinations of phenobarbital and bromide. Patients are generally considered as responders if there is a 50% reduction in seizure activity after initiation of the drug. Under these conditions, 58% and 82% of dogs responded after initiation of zonisamide in 2 studies (mean 81% and median 70% reduction of seizures respectively),[10,11] 55% of dogs responded after addition of gapapentin,[12] and 64% responded after adding levetiracetam (overall 55% reduction of seizures).[13] In another study with addition of gabapentin to phenobarbital and bromide, there was a modest benefit noted characterized by an increase in the number of dogs that showed an increased interictal period.[14] However, caution must be urged in interpretation of these results for a number of reasons, including a placebo effect. A combined analysis of the only 3 available blinded, placebo-controlled trials evaluating interventions for canine epilepsy revealed a 26-46% reduction in seizures for animals receiving placebo and 29% of dogs had a 50% or greater decrease in seizure frequency ("responders").[15] Explanations for this placebo effect include a higher level of care or better adherence to medication administration as a result of taking part in a study, client or investigator bias, and regression to the mean. The latter term refers to the natural waxing and waning behavior of disease processes such as epilepsy. Owners tend to seek veterinary advice when such disorders are at their worst, and as a result, improvement over the short term is likely regardless of therapy. This may artificially increase the perceived benefit of the drug being evaluated. In addition, a "honeymoon effect" has been reported for several of these medications, where initial responders to the added medication have a subsequent increase in their seizure frequency.[11,13]

REFERENCES

[1.] Frey HH, et al. **Eur J Pharmacol** 1984;104:27. [2.] Orito K, et al. **J Vet Pharmacol Ther** 2008;31:259. [3.] Boothe DM, et al. **J Vet Pharmacol Ther** 2008;31:544. [4.] Boothe DM, et al. **J Am Vet Med Assoc** 2002;221:1131. [5.] Wagner SO. **J Vet Intern Med** 2001;15:562. [6.] Center SA, et al. **J Am Vet Med Assoc** 1996;209:618. [7.] Bailey KS, et al. **J Feline Med Surg** 2009;11:385. [8.] Thomas WB. **Vet Clin North Am Small Anim Pract** 2010;40:161. [9.] Hasegawa D, et al. **J Feline Med Surg** 2008;10:418. [10.] Dewey CW, et al. **J Am Anim Hosp Assoc** 2004;40:285. [11.] von Klopmann T, et al. **J Small Anim Pract** 2007;48:134. [12.] Platt SR, et al. **Vet Rec** 2006;159:881. [13.] Volk HA, et al. **Vet J** 2008;176:310. [14.] Govendir M, et al. **Aust Vet J** 2005;83:602. [15.] Munana KR, et al. **J Vet Intern Med** 2009;24:166.

TREATMENT OF CLUSTER SEIZURES & STATUS EPILEPTICUS

Christopher L. Mariani, DVM, PhD, DACVIM (Neurology)
Raleigh, NC

DEFINITIONS

- **Cluster seizures** – two or more seizures occurring within a 24-hour period.
- **Status epilepticus** – continuous seizure activity lasting longer than 5 minutes, or the occurrence of multiple seizures without recovery of baseline neurologic function between episodes.[1,2] Status epilepticus can be generalized or focal in nature, or in rare cases, can be nonconvulsive.[3]

PATHOPHYSIOLOGY AND GOALS OF TREATMENT

The vast majority of seizures are self-limiting events, with eventual spontaneous return to resting or baseline neurologic function. However, during status epilepticus, a variety of changes occur within cells and networks of cells that result in a situation where the seizure activity becomes self-sustaining. These changes may be independent of the initiating cause of the seizure, and involves a variety of molecular mechanisms. Repetitive seizures cause inhibitory $GABA_A$ receptors to move from the synaptic membrane to the cell interior, while excitatory N-methyl-D-aspartate (NMDA) receptors may be recruited to the cell surface.[1] Stores of inhibitory neurotransmitters may become depleted and increased expression of drug efflux transporters such as P-glycoprotein may occur.[4] After a period of time, these cellular alterations may lead to pharmacoresistance to first-line agents that would normally be effective in seizure termination at earlier phases, such as the benzodiazepines.

Generalized status epilepticus can lead to profound acidosis, hyperthermia, cardiac arrhythmias, hypoxia, neurogenic pulmonary edema, rhabdomyolysis, myoglobinuria, renal failure, cerebral edema, elevated intracranial pressure and neuronal necrosis, and therefore constitutes a medical emergency.[5] The goals of treatment are to stop the seizures, support systemic organ functions, and protect brain function. Finally, ongoing seizure activity/seizure control should be closely monitored. These goals are described in greater detail below.

1) Stop the Seizures

The most critical and pressing goal of therapy is to stop the seizures, by any means necessary. The initial drug chosen is usually a benzodiazepine (diazepam or midazolam), but depends on the suspected underlying cause.

- If hypoglycemia is suspected (juvenile toy breed dog, hunting dog or insulin overdose), administer 1-2 ml/kg of 50% dextrose intravenously (IV) diluted 1:1 in saline. Oral dextrose may be used in animals able to swallow when intravenous access is not readily achieved.
- In small or toy breed dogs that have recently whelped and are nursing puppies, the administration of calcium gluconate may be considered to address potential hypocalcemia.

Animals with known idiopathic or probably symptomatic (acquired/cryptogenic) epilepsy and those with unknown etiologies typically receive a benzodiazepine as the first-line drug.

- Diazepam (0.5 mg/kg) can be administered IV, intranasally or rectally to control seizures. The dose can be repeated twice, if necessary. Anticonvulsant action only lasts about 15-30 minutes, and therefore some form of longer acting therapy is required if the seizures stop. Midazolam can be substituted for diazepam in this scenario, and lorazepam (0.2 mg/kg) may also be considered. These drugs may also be administered IV or intranasally, but are not likely to be effective with rectal administration. In addition, unlike the others, midazolam can be successfully administered intramuscularly (IM).
- If the animal responds to a benzodiazepine bolus, phenobarbital may be considered for longer-term control. Naïve animals not previously receiving anticonvulsants can be loaded with 16-20 mg/kg divided into 4 doses and administered every 30-120 minutes (i.e., 4-5 mg/kg q 30-120 minutes). Epileptics already receiving phenobarbital may benefit from an additional "mini-loading dose" (5-10 mg/kg) depending on their serum levels of the drug. Phenobarbital should be continued at regular maintenance intervals (2-3 mg/kg IV, IM or PO q 12 hours or at regular dose) after this.
- Animals with severe cluster seizures or status epilepticus with some inter-ictal time (i.e., non-continuous) usually respond to a constant rate infusion (CRI) of diazepam (0.1-2.0 mg/kg/hour IV). The CRI can be started at the low end of the range (0.1-0.25 mg/kg) and gradually increased as necessary to control seizure activity. Once controlled, a seizure-free state is maintained for 12 hours, after which the infusion is gradually tapered (usually reduce dose by half every 4-6 hours) and stopped. The CRI can be administered with a syringe pump, if available, or by mixing with 0.9% saline in a small IV bag or Buretrol system. Diazepam is degraded by light and binds to plastic, and the syringe and tubing should be covered with brown plastic or aluminum foil, if possible. Midazolam can again be substituted in this scenario, and is less likely to cause thrombophlebitis.
- Animals with continuous, prolonged seizure activity or those refractory to benzodiazepines may receive pentobarbital (3-15 mg/kg IV to effect), if available. This drug induces general anesthesia, and is extremely effective in stopping the outward manifestation of the seizure. However, respiratory and cardiovascular function may be depressed, and these systems must be monitored very closely. Although intermittent bolus doses can be used, a CRI (2 mg/kg/hr adjusted to effect) may be more

effective. Similar to benzodiazepine CRIs, animals may be kept seizure free for approximately 12 hours, and then weaned from the drug. It can be difficult to distinguish recovery from pentobarbital anesthesia from overt seizure activity. However, paddling movements of the limbs typically indicate the former, while seizure activity is usually characterized by overt tonic or clonic muscle contractions. Electroencephalography, if available, can help to differentiate these two scenarios. This medication also reduces the metabolic requirements of the brain, and is considered to have neuroprotective effects.

- Propofol may be used as a substitute for pentobarbital if general anesthesia is required to control seizure activity. Due to its short duration of action, this drug must be given as a CRI (6 mg/kg initial bolus followed by 0.1-0.6 mg/kg/min). Substantial respiratory depression is frequent with this medication, and anesthesia must be closely monitored. In addition, propofol can have pro-convulsant effects in some patients. Some consider this to be the treatment of choice for patients in status epilepticus secondary to hepatic encephalopathy (typically after surgical repair of a portosystemic shunting vessel).
- If pentobarbital and propofol are not available, the use of an inhalant anesthetic (e.g., isoflurane or sevoflurane) to maintain general anesthesia should be considered as a last resort. Both require close monitoring of respiratory and cardiovascular parameters.
- Recently, a parenteral formulation of levetiracetam has been made available. Although its use in animals with status epilepticus or cluster seizures has been limited to date, it may prove useful in this role, based on reports in humans and preliminary experience in canine patients.[1,6] Pharmacokinetic studies in dogs suggest that a dose of 20-60 mg/kg IV results in blood concentrations within the range considered to be effective in humans (5-45 μg/ml) for greater than 8 hours.[7,8] Levetiracetam is approximately 100% bioavailable after IM administration and results in similar blood levels, although peak concentrations are not reached until about 40 minutes after the drug is given.[8]
- Reports of other medications for refractory status epilepticus are infrequent in veterinary medicine. There is a report of a dog with granulomatous meningoencephalitis and status epilepticus responding to intravenous ketamine infusion after failure to respond to diazepam and propofol.[9] This report follows several human case reports reporting similar efficacy for ketamine in the scenario of refractory status epilepticus, the rationale being blocking of NMDA receptors which may be responsible for the self-sustaining nature of this condition.[1,6,10] Additional therapies reported in refractory human cases include valproic acid, lidocaine, and topiramate.[6,11,12]

2) Support and Monitor Systemic Functions

As described above, status epilepticus can have profound effects on many body systems, and systemic functions must be closely monitored. These include:

- Mental status and level of consciousness
- Respiration, oxygen saturation and blood gases (if available)
- Cardiac rate and rhythm, blood pressure
- Body temperature
- Serum electrolytes, glucose, BUN and creatinine
- Fluid status and hydration
- Muscle damage and evidence of myoglobinuria (which may cause renal failure)

Intravenous fluid therapy is often indicated in order to maintain hydration, and may help prevent renal damage if myoglobinuria is a concern. As severe seizure activity may lead to non-cardiogenic pulmonary edema, thoracic radiographs, pulse oximetry, and blood gas analysis should be considered in animals with compromised respiration. Aspiration pneumonia is also a concern, particularly in large recumbent dogs. Oxygen therapy may be administered in some of these patients. Active cooling should be considered in animals that are severely hyperthermic. Basic supportive nursing care must be performed in recumbent and stuporous animals, including applying artificial tears/lubrication to the eyes, providing adequate bedding/padding, periodically changing body position, turning from side to side, and passive range of motion of the limbs.

3) Protect Brain Function

Prolonged, severe seizure activity can lead to cerebral edema, increases in intracranial pressure and neuronal necrosis. Select cases may benefit from oxygen therapy, mannitol (0.25-1.0 g/kg IV over 10-20 minutes) and furosemide (0.7 mg/kg IV, 15 minutes after mannitol) in order to address these effects. Compression of the jugular veins, coughing and sneezing all increase intracranial pressure, and should be avoided in animals where this is suspected to be increased. Therefore, jugular catheters, collection of blood from the jugular vein, neck bandages, nasogastric tubes, and nasal oxygen catheters should all be avoided. Intravenous lidocaine should be considered to reduce the coughing reflex if intubation is required. Elevation of the head approximately 30 degrees from the horizontal is a simple way to promote venous return from the brain and potentially reduce intracranial pressure. Pentobarbital administration, in addition to stopping seizure activity, also has the advantage of reducing cerebral metabolism, which can have neuroprotective effects in patients with status epilepticus.

4) Monitor ongoing seizure activity

Patients should be closely monitored to ensure the cessation of seizures and for the recurrence of seizure activity after initial therapy. This is typically done by visual observation and examination of animals for motor activity consistent with seizures.

Whenever possible, cessation of seizure activity should be confirmed electrophysiologically with the aid of electroencephalography (EEG). This may detect animals whose outward motor manifestations of the seizure activity have stopped, but who continue to have abnormal electrical brain activity, known as nonconvulsive status epilepticus (NSE). Although NSE has rarely been reported in veterinary patients,[9] this is likely a reflection of the infrequency with which veterinary clinicians perform EEG in this setting. The author has documented a number of canine patients with apparent NSE after prolonged convulsive status epilepticus or presenting with a primary complaint of altered mentation (Mariani, unpublished observations). An EEG is part of the routine diagnostic evaluation of human patients presenting with stupor or coma, and in the author's opinion, the same should be offered to veterinary patients wherever possible.

AT-HOME THERAPY FOR CLUSTER SEIZURES

Some owners can be taught to administer benzodiazepines in the home environment in order to reduce the number of seizures in dogs (or cats) prone to cluster seizure events. The goal is usually to prevent further seizures, reduce the number and severity of subsequent seizures, and avoid an emergency visit to the veterinary hospital. Diazepam has been used most often via the rectal route; a standard dose (0.5 mg/kg) can be administered, although in some animals on chronic phenobarbital therapy, a higher dose (1-2 mg/kg) may be required due to increased metabolism of the drug.[13,14] Drugs administered rectally in the dog undergo a substantial first-pass effect and hepatic metabolism as the majority of absorbed drug enters the portal circulation.[15] As a result, the bioavailability of diazepam after rectal administration is only about 2.7-7.4% at doses of 0.5 and 2.0 mg/kg respectively in dogs not receiving phenobarbital.[16] However, some anticonvulsant effect is achieved as the main metabolites of diazepam (desmethyldiazepam and oxazepam) posses 20-50% of the activity of the parent drug.[17-19] Lorazepam is unsuitable for rectal administration, as its primary metabolite (lorazepam glucuronide) does not have anticonvulsant activity.[15] Midazolam does have a metabolite with reported pharmacologic activity (1-hydroxymidazolam), although the contribution of this reported activity is controversial.[20]

Intranasal (IN) administration of benzodiazepines avoids several of the shortcomings of the rectal route. The IN route avoids substantial first pass metabolism, and drug is directly absorbed into the systemic circulation thorough the dense vascular plexus present in the nasal passages.[21] In addition, there is evidence for direct movement of drug through the cribriform plate and into the central nervous system.[22-24] Several studies suggest that the bioavailability of diazepam is much higher after IN administration than after rectal.[25,26] and this route has been used successfully by the author in several emergent clinical cases (0.5 mg/kg). Preliminary experience suggests that intranasal lorazepam (0.2 mg/kg) may also be useful as an alternative to rectal diazepam for at-home use by owners.[27] Intranasal or IM midazolam (0.5 mg/kg) is another option available for these scenarios.

REFERENCES

[1] Wasterlain CG, et al. **Epilepsia** 2008;49 Suppl 9:63. [2] Lowenstein DH, et al. **Epilepsia** 1999;40:120. [3] Engel J, Jr. **Epilepsia** 2006;47:1558. [4] Loscher W. **Epilepsia** 2007;48 Suppl 8:74. [5] Bassin S, et al. **Crit Care** 2002;6:137. [6] Abend NS, et al. **Pediatr Neurol** 2008;38:377. [7] Dewey CW, et al. **J Vet Emerg Crit Care** 2008;18:153. [8] Patterson EE, et al. **J Vet Pharmacol Ther** 2008;31:253. [9] Serrano S, et al. **J Vet Intern Med** 2006;20:194. [10] Mewasingh LD, et al. **Seizure** 2003;12:483. [11] Hattori H, et al. **Brain Dev** 2008;30:504. [12] Yildiz B, et al. **Pediatr Int** 2008;50:35. [13] Wagner SO, et al. **J Vet Pharmacol Ther** 1998;21:335. [14] Podell M. **J Vet Intern Med** 1995;9:68. [15] Podell M, et al. **J Vet Pharmacol Ther** 1998;21:158. [16] Papich MG, et al. **Am J Vet Res** 1995;56:1629. [17] Boothe DM. **Vet Clin North Am Small Anim Pract** 1998;28:411. [18] Frey HH, et al. **Pharmacology** 1982;25:154. [19] Randall LO, et al. **Curr Ther Res Clin Exp** 1965;7:590. [20] Johnson TN, et al. **Br J Anaesth** 2002;89:428. [21] Jones NS, et al. **Int J Clin Pract** 1997;51:308. [22] Hanson LR, et al. **J Neuroimmune Pharmacol** 2007;2:81. [23] Rapoport A, et al. **Headache** 2006;46 Suppl 4:S192. [24] Westin UE, et al. **Pharm Res** 2006;23:565. [25] Musulin SE, et al. **J Vet Pharmacol Ther** 2010 [in press]. [26] Platt SR, et al. **Am J Vet Res** 2000;61:651. [27] Mariani CL, et al. **J Vet Intern Med** 2003;17:402 [abstract].

NEW HORIZONS IN EPILEPSY DIAGNOSIS & THERAPY

Christopher L. Mariani, DVM, PhD, DACVIM (Neurology)
Raleigh, NC

DIAGNOSIS

Advances in neuroimaging

Advanced diagnostic imaging techniques have already revolutionized the practice of veterinary and human neurology and neurosurgery. Beginning with computed tomography (CT) and then magnetic resonance imaging (MRI), imaging techniques providing accurate spatial representation of the brain and other parts of the nervous system are now indispensable for quality neurology practice.[1] These techniques have allowed the identification of intracranial pathology of varying etiologies and guided therapeutic intervention for these same disease processes. As a result, therapies for symptomatic epilepsy have dramatically improved (e.g., brain tumors). However, imaging advances have also redefined the management of other forms of epilepsy and will continue to do so in the future.

A case in point involves children with cortical dysplasia, a group of disorders involving aberrant cortical neuron migration and development. Affected children frequently have severe epilepsy that is refractory to medical therapy. Prior to the age of advanced diagnostic imaging, an antemortem diagnosis would have been very difficult in most cases. However, current MRI technology allows the identification of small, dysplastic or malformed regions of cortex, which in some cases are amenable to surgical resection. Improvements in field strength of magnets have resulted in corresponding improvements in the resolution of lesion pathology. Similar advancements have been made in the detection of hippocampal sclerosis by MRI and the planning of subsequent temporal lobe resection.[2] Improvements related to conventional MR imaging that might further advance this field include improved coil design and the development of new sequences such as diffusion tensor imaging and arterial spin labeling.[3]

A variety of other novel neuroimaging techniques may contribute to the diagnosis of epilepsy lesions in the future, including functional MRI (fMRI), magnetic resonance spectroscopy (MRS), positron emission tomography (PET), magnetoencephalography (MEG), and single photon emission computed tomography (SPECT).[1]

THERAPY

New Anticonvulsant Drugs

In addition to investigating the utility of new anticonvulsant drugs in development for veterinary patients, veterinary practitioners can almost certainly do a better job of selecting currently available medications for patients with specific seizure syndromes. However, such improved selection awaits the unraveling of the mechanism of action of most veterinary epilepsies. Improved tools for gene discovery and mapping of the canine genome will likely result in the identification of a number of these mechanisms in canine patients.

The field of pharmacogenomics holds much promise for epilepsy patients as well as for medical patients in general.[4] Identifying individual variation in hepatic enzymes and other mechanisms of drug metabolism and distribution has the potential to open the door to individualized drug therapy. Such a strategy would potentially allow the identification of patients who might be, for example, "fast metabolizers" of a certain drug, before drug therapy is initiated. This might save large amounts of time and effort in finding effective therapies for individual patients, as well as substantially reducing patient morbidity and possibly even mortality. Efforts to identify pharmacogenetic differences in veterinary patients are underway.[5]

Between 1993 and 2007, a total of 11 new drugs were approved for seizure treatment in human patients in the United States. With this wealth of new options, many have predicted that the pipeline of new anticonvulsant medications would slow down or dry up due to saturation of the market. As yet, this has not occurred, and pharmaceutical companies are turning to new molecular strategies to identify effective anticonvulsant medications. In particular, altering the structure of known effective anticonvulsants to increase the affinity for the drug's receptor or to reduce toxicity may result in several new options for therapy. Drugs recently available or currently in active development include brivaracetam, fluorofelbamate, carisbamate, and lacosamide.

Surgical Therapy for Epilepsy

Surgical intervention is a well-established therapeutic option for certain subsets of human patients with refractory epilepsy. The most common procedure performed is temporal lobectomy,[6] although other surgeries have been utilized, including corpus callosotomy,[7] deep brain stimulation,[8,9] multiple subpial transections,[6] and focal resections of regions of cortical dysplasia. A large proportion of human patients with refractory epilepsy have seizures that originate in the temporal lobe, and mesial temporal sclerosis is the most common lesion accounting for temporal lobe pathology. For well-selected patients, temporal lobectomy can dramatically improve seizure control, in some cases even eliminating seizure altogether in patients previously refractory to multiple medical therapies.

Translation of this success in human epilepsy surgery to veterinary patients might be possible, but has substantial hurdles to surmount before this might be considered a viable option. First, and most importantly, the underlying mechanisms of disease in refractory canine and feline patients might be substantially different from human patients in many cases. Many refractory human patients experience complex partial seizures associated with temporal lobe pathology. Attempts to document similar pathology in veterinary patients has met with mixed success. Although lesions with some similarity have been shown in several series of cats in Europe,[10-12] such pathology is not consistently seen in dogs.[12] Secondly, most of the surgical techniques available in human patients, with the exception of corpus callosotomy, rely on precise mapping of the epileptic area of the cortex that is generating the seizures. Mapping typically begins with interictal EEG, followed by continuous video EEG monitoring to capture an ictal event, and finally invasive recording with dural and depth electrodes. Intraoperative electrophysiologic monitoring is also frequently employed.[13] Clearly if surgical options for non-symptomatic epilepsy are to be considered in the future for veterinary patients, dramatic improvements in electrophysiologic identification of areas of seizure initiation need to be made.

Despite these challenges, surgical intervention for canine epileptics has been reported. A technique for partial corpus callosotomy was developed in experimental dogs, and showed minimal perioperative morbidity.[14,15] However, long-term success of this intervention may be limited by post-operative scarring and eventual associated seizure recrudescence. The feasibility of deep brain stimulation has also been preliminarily investigated in dogs as a potential therapy for epilepsy.[3]

Epileptogenesis intervention

The process of epileptogenesis is a cascade of dynamic biological events that alters the balance between excitation and inhibition in neural networks. It involves multiple, progressive biochemical, anatomic, and physiologic changes that lead to recurrent seizures. Pharmacological strategies for the treatment of epilepsy have traditionally been aimed at suppressing the initiation or propagation of seizures, rather than the processes leading to epilepsy, and therefore, current drugs used for recurrent seizures are anti-ictal, as opposed to antiepileptic. There is no evidence to indicate that the administration of these drugs alters the long-term prognosis for seizure remission or provides antiepileptic effects against spontaneous unprovoked seizures.

The process of epileptogenesis likely involves a cascade of events at various levels of organization, from molecular (gene expression) to multicellular (altered neural networks) .. Specific mechanisms that may contribute include an imbalance of excitatory and inhibitory neurotransmission, alteration of neurotransmitter receptor expression and function, development of aberrant ion channels, formation of epileptic networks or circuits between different brain regions, and morphologic changes such as hippocampal sclerosis and axonal sprouting leading to aberrant neuronal synchronization. It has been known for some time that individuals with severe seizures have altered brain morphology through histological studies of post-mortem specimens and surgically excised tissue, and more recently through advanced imaging techniques such as MRI. Although traditional emphasis has been placed on neuronal loss, it is important to realize that many other abnormalities are found in the epileptic brain, including gliosis, alterations in axonal and dendritic plasticity, aberrant neurogenesis and molecular reorganization of neuronal and glial cell membranes and extracellular matrix.

REFERENCES

1. Shorvon SD. **Epilepsia** 2009;50 Suppl 3:39. 2. Urbach H. **Eur Radiol** 2005;15:494. 3. Dewey CW. An overview of deep brain stimulation (DBS) for treating seizure disorders. In: **American College of Veterinary Internal Medicine Forum**, San Antonio 2008. 4. Tate SK, et al. **Proc Natl Acad Sci** U S A 2005;102:5507. 5. Kennerly EM, et al. **Pharmacogenet Genomics** 2009. 6. Tellez-Zenteno JF, et al. **Brain** 2005;128:1188. 7. Jea A, et al. **Neurosurg Focus** 2008;25:E7. 8. Hamani C, et al. **Int J Neural Syst** 2009;19:213. 9. Lockman J, et al. **Neurol Clin** 2009;27:1031. 10. Brini E, et al. **J Feline Med Surg** 2004;6:377. 11. Fatzer R, et al. **J Vet Intern Med** 2000;14:100. 12. Schmied O, et al. **Vet Radiol Ultrasound** 2008;49:343. 13. Kuruvilla A, et al. **Seizure** 2003;12:577. 14. Bagley RS, et al. **Vet Surg** 1995;24:122. 15. Bagley RS, et al. **Vet Clin North Am Small Anim Pract** 1996;26:827.

DIAGNOSING LUNG DISEASE IN DOGS & CATS USING MECHANICS, BREATH & LUNG-SEEKING NANOPARTICLES

Kim A. Selting, DVM, MS, DACVIM & Carol Norris Reinero, DVM, PhD, DACVIM
Columbia, MO

OVERVIEW

Valuable physiochemical information can be determined from dogs and cats as they inhale and exhale. This lecture will focus on pulmonary mechanics and exhaled breath condensate to characterize and distinguish various non-neoplastic and neoplastic lung diseases. Different methods of evaluating mechanics and breath (including proteomic and metabolomic evaluation) will be discussed. Additionally, pilot data on lung-seeking nanoparticles will be shared.

PULMONARY MECHANICS

In humans, measurement of pulmonary function for obstructive and restrictive lung diseases is beneficial in diagnosis, stratification of disease severity, determination of response to therapy, and prognostication. Pulmonary function testing (PFT) is used most commonly for asthma and chronic obstructive pulmonary disease (COPD), but can also be used in the diagnosis of restrictive lung disorders and provide important details about the likelihood of respiratory complications prior to surgery. Simplistically, instruments are used which can measure airflow and volumes in and out of the lung. If airflow is impeded, this suggests an obstructive defect—eg, bronchoconstriction, edematous or inflamed airway walls with a narrow caliber, intralumenal airway secretions, a mass compressing large airways, etc. If air volumes are reduced, this suggests a restrictive lung disease such as infiltrative lung disease (infectious, non-infectious inflammatory or fibrotic, neoplastic), pleural cavity disease, and severe abdominal enlargement, etc.

Classically, in human medicine spirometry (measuring the amount of air entering and leaving the lungs) relies on a respiratory maneuver dependent on patient cooperation and effort as they are instructed to inhale and exhale forcibly through a mouthpiece. This type of pulmonary mechanics testing is not translatable to animals. Instead, efforts have been made to assess lung function using a variety of indirect and direct methods. Indirect methods include tidal breathing flow-volume loops using a tight fitting facemask, barometric whole body plethysmography (BWBP), and forced expiratory flow-volume curves using a thoracic compression technique[1-4]. Recently, there has been renewed interest in BWBP for airway diseases, although results must be interpreted cautiously as the index of airflow limitation measured by this technique called enhanced pause (Penh) has been criticized for not reflecting pulmonary resistance[5,6]. When evaluating airway diseases in which airway hyperreactivity is a key feature, compounds causing bronchoprovocation are used to constrict the airways (such as titrated doses of methacholine, histamine, or adenosine 5-monophosphate). Additionally, a positive response to a bronchodilator is an important means to distinguish asthma (which is episodic) from other disorders where airflow limitation is permanent (e.g., chronic bronchitis, chronic obstructive pulmonary disease).

Bronchoprovocation can be paired with indirect or direct measures of lung function. While indirect pulmonary mechanics measurements are appealing from the standpoint that they are minimally invasive, more accurate means to estimate airway resistance and responses to bronchoprovocants and bronchodilators rely on direct measurements. Direct measurements require anesthesia and range from minimally to highly invasive, with more invasive techniques including placement of an esophageal balloon or pleural catheter for measurement of transpulmonary pressures [7-9]. Equipment traditionally used for direct measurement of airflow limitation (differential pressure transducers and a pneumotachograph with integrative software) is impractical since it has no other purpose than to measure lung mechanics. All of these techniques require specialized training and expensive, dedicated equipment. Unfortunately none are routinely available for use in pets. However, recent pilot studies have been performed using ventilator-acquired pulmonary mechanics which have the potential to become clinically useful given that ventilators are commonplace in many specialty practices. In addition, mechanics measurements from ventilators are comparatively easier to acquire than traditional methods [10,11].

Examples of how ventilator acquired mechanics can be used include: 1. Demonstration of airway hyperreactivity in an experimental model of feline asthma in which cats are given a methacholine challenge. 2. Documentation of increased airway resistance after nebulization of N-acetylcysteine in cats and 3. Assessment of response to treatment in a dog with restrictive lung disease.

EXHALED BREATH CONDENSATE

Diagnosis of pulmonary parenchymal disorders has frequently relied on cytologic (fine-needle aspiration, lavage cytology) or histologic evaluation to discriminate inflammatory, fibrotic and neoplastic disorders. All of these techniques are invasive and associated with some risk to the patient. Non-invasive means to differentiate among lung diseases have clear appeal and demand. Analysis of exhaled breath condensate (EBC) represents one non-invasive technique to sample the health of the lung whereby compounds in the airways are released from the airway lining fluid and are exhaled in breath where they can be collected as the sample condenses in a cooled tube. EBC contains proteins, nucleic acids, and metabolic byproducts and can be analyzed using nuclear magnetic resonance (NMR) spectroscopy to characterize chemical fingerprints of metabolic processes. Additionally, proteomic profiling using surface-enhanced laser desorption/ionization time-of-flight (SELDI-TOF) may identify novel proteins present in the breath of diseased individuals.

Difference in metabolomics (profiling of metabolic pathway intermediates) in EBC have been seen in human patients with asthma and cystic fibrosis, as compared to normal. Pilot data using NMR from EBC in cats with experimentally induced asthma will be presented. Additionally, there is great interest in finding biomarkers for patients with lung cancer (primary or metastatic) which may be helpful for earlier detection and monitoring response to treatment. Dogs with spontaneously-occurring pulmonary neoplasia were recruited at the University of Missouri for EBC analysis. Results of NMR and SELDI-TOF evaluation will also be presented. The discovery of biomarkers of lung cancer in exhaled breath could allow earlier detection of metastasis thereby improving the efficacy of anticancer treatments, as well as provide a keyhole view of the pathophysiology of primary and metastatic lung cancer. By identifying biomarkers, we can also trace them to their origin and identify new therapeutic targets.

There are various ways to collect EBC, ranging from commercially available units such as the R-tube to homemade devices including chemical condensers.

LUNG SEEKING-NANOPARTICLES

Nanoparticles are smaller than usual particles of a given compound that are typically less than 30 microns in diameter and have potential for diagnostic and therapeutic medical applications. The advantage of nanoparticles is facilitated entry into cells. Nanoparticles, however must be stabilized to prevent agglomeration into larger particles. Historically, only toxic substances were used to accomplish this stabilization, but since the introduction of nanomedicine, stabilization has been attempted using substances that are natural and non-toxic. Of these compounds, some starches such as gum Arabic (a common food additive) seem to localize preferentially in the lungs. This may be a result of similarities to building blocks for surfactant, or other factors but the reason is not know. The nanoparticles can be tagged with a radiopharmceutical product for imaging and perhaps therapy.

FUTURE DIRECTIONS

These evaluation and treatment methods provide novel information (some in a non-invasive fashion) which are either complimentary to, or offer additional information, to currently available pulmonary diagnostics. The ultimate goal is to make each of the techniques adaptable to clinical practice, or at least to large referral practices. Further development is needed but preliminary results are promising.

REFERENCES

[1.] Bark H, Epstein A, Bar-Yishay E, et al. **Respir Physiol Neurobiol** 2007;155:49-54. [2.] Hoffman A, Dhupa N, Cimetti L. **Am J Vet Res** 1999;60:1487-1492. [3.] Amis T, Kurpershoek C. **Am J Vet Res** 1986;47:1002-1006. [4.] McKiernan B, Dye J, Rozanski E. **J Vet Intern Med** 1993;7:388-393. [5.] Bates J, Irvin C, Brusasco V, et al. **Am J Respir Cell Mol Biol** 2004;31:373-374. [6.] Kirschvink K. **Vet J** 2008;176:125-126. [7.] Dye J, McKiernan B, Rozanski E, et al.. **J Vet Intern Med** 1996;10:385-400. [8.] Norris Reinero C, Decile K, Berghaus R, et al. **Int Arch Allergy Immunol** 2004;135:117-131. [9.] Padrid P, Snook S, Finucane T. **Am J Respir Crit Care Med** 1995;151:184-193. [10.] Lee-Fowler T, Dodam J, Guntur V, et al. **27th Symposium of the Veterinary Comparative Respiratory Society** 2009. [11.] Rozanski E, Buckley G, Mazan M, et al. **J Vet Intern Med** 2009;23:727.

INHALATIONAL THERAPIES IN DOGS & CATS

Carol Norris Reinero, DVM, PhD, DACVIM, & Kim A. Selting, DVM, MS, DACVIM(O)
Columbia, MO

OVERVIEW

The airways can serve as a conduit for the delivery of many drugs when treating local or systemic diseases. This lecture will cover both targeted (lung selective) and diffuse treatments including bronchodilators, glucocorticoids, chemotherapy, and insulin in dogs and cats.

INTRODUCTION

While most drugs in veterinary medicine have traditionally been delivered by oral or injectable routes, inhalant delivery of aerosolized medication offers a number of benefits for local or systemic drug delivery. The lungs provide an enormous absorptive surface coupled with a permeable membrane, a low enzyme environment resulting in little drug degradation, avoidance of hepatic first-pass metabolism, and reproducible absorption kinetics. When the target of inhaled medications is the respiratory tract itself in the case of various infectious, inflammatory and neoplastic conditions, additional benefits include the potential for a high drug concentration directly at the site of disease with minimization of systemic toxicity, often at a fraction of the dose required if the same drug was administered through a systemic route. In humans, the most common indication for inhalant medications is for the treatment of airway diseases. While there are many studies in the medical literature regarding inhalational drug therapy in humans, there is comparatively little information in veterinary species.

Despite many advantages of inhalant drug delivery, there are inherent obstacles in using this route as well. Respiratory defenses are efficient at preventing particulates from reaching the lower airways. Only a small proportion of the administered medication reaches the lower airways, with a significant amount lost in the delivery device or deposited in the oropharynx. In translating this mode of therapy to veterinary species, voluntary "maneuvers" such as "breath hold" and "inhale" are not applicable (ie, dogs and cats cannot be trained to inhale maximally and hold their breath on command). Thus some of the devices used in humans cannot be used in small animals—e.g., dry powder inhalers. Adaptations of some devices facilitate their use in animals, and modified systems are now marketed for dogs and cats. Additionally, selecting systems that bypass certain obstacles (such as treating when intubated and coordinating drug delivery with inhalation) can significantly improve the effective delivery of drug to lower airways. Drug delivery by the aerosol route depends in part of respiratory depth and rate, tidal volume, and airflow rates, yet all of these may be negatively impacted by respiratory disease or by behavioral changes (e.g., shallow or slow breathing when a face mask is placed over the muzzle) . Additionally, not all drugs are suitable for aerosol delivery, and drugs themselves or preservatives contained in the drug preparation may cause airway irritation and possible bronchoconstriction (especially in cats) potentially worsening respiratory compromise.

TYPES OF DELIVERY SYSTEMS

There are two very basic types of aerosol delivery systems in common usage with applicability to dogs and cats: nebulizers and metered dose inhalers (MDI). In general, nebulizers deliver much smaller particles allowing deeper respiratory penetration and provide fluid along with drug. Metered dose inhaler devices deliver drug primarily to the larger airways.

Nebulizers utilize compressors to generate aerosols which can be inhaled without requiring positive pressure to propel droplets into the airways The basic nebulizer types include jet nebulizers and ultrasonic nebulizers. All nebulizers are not created equal and older jet and ultrasonic nebulizers are being replaced by newer vibrating mesh nebulizers. Vibrating mesh nebulizers are much more efficient at generating smaller droplets, so that less drug is needed to deliver the desired dose, which can help minimize the expense. These are also "gentler" with the drugs avoiding shear forces created by jet nebulizers. Additionally, novel delivery catheters are in development and allow certain advantages over standard nebulizers, such as the ability to pass the catheter through the biopsy channel of a bronchoscope to deliver therapy to an isolated lung lobe. While nebulized liquid can be administered to dogs and cats in a tent, closed aquarium-type container, or via face mask, it is highly likely that effective concentrations of drug will only be reached with a tight-fitting face mask or by insertion into a tracheotomy/ tracheostomy/endotracheal tube.

Metered dose inhalers are designed for at-home administration of aerosolized drugs. In humans with asthma and chronic bronchitis/COPD, they are the preferred routine route of delivery for glucocorticoid and bronchodilator medications. Particles delivered by MDI are larger than those created by nebulization, and thus do not penetrate as deeply into the respiratory tract. A traditional MDI consists of a mouthpiece and an actuator (holder) into which a canister of medication is inserted and actuated (depressed to release the medication). Humans are instructed to exhale maximally, insert the mouth piece, and simultaneously depress the canister and inhale as deeply as possible. After holding their breath for as long as possible, they exhale. It is expected that ~10% of each dose reaches the airways. In dogs and cats, spacer devices designed to fit the MDI have allowed their adaptation for use without the need for controlled respiratory maneuvers.

INDICATIONS FOR INHALANT DRUG DELIVERY

Inhalant drugs are most commonly used for treatment of local pulmonary disease. Examples of primary respiratory disease in which nebulizers or MDIs have been employed include infectious, inflammatory, and neoplastic conditions. Inhalant drugs can also be administered to treat systemic disease, taking advantage of the high surface absorptive area of the lung. Examples of drugs used include insulin, vaccines and genes (gene therapy).

Historically in small animal medicine, nebulizers were predominantly used for the treatment of respiratory infection. Nebulizers have long been used to provide airway humidification using sterile saline which is very helpful in case of pneumonia through decreasing viscosity of airway secretions to improving clearance from the lungs. Additionally, nebulizers may be used to administer antimicrobials. It is ideal to use antibiotics specifically made for inhalation (ie, lacking reactive additives/preservatives), however they may be cost prohibitive. Veterinarians have used drugs made for parenteral administration in nebulized solutions for the treatment of pneumonia or other respiratory infections, including *Bordetella bronchiseptica*. The most frequently used antibiotics for nebulization are the aminoglycosides. Few studies have documented efficacyor benefit over traditional routes of antibiotic administration. It is critically important that the nebulizer be kept meticulously clean as nebulization of bacteria colonizing the device would be devastating in a patient that already has a compromised lung.

In addition to administration of antimicrobials, there have been recent reports using nebulizers to administer bronchodilators [1,2], chemotherapeutics[3] and insulin[4] in dogs or cats. More studies need to be performed before this route can be considered a widespread viable alternative to traditional routes of administration.

Metered dose inhalers (MDI) have been advocated for the treatment of feline lower airway disease and occasionally for dogs with lower airway disease. The use of inhaled steroids can minimize systemic endocrine and immunologic effects of glucocorticoids [5,6]. This is particularly helpful with co-morbid conditions such as diabetes mellitus, congestive heart failure or immunosuppressive conditions. Metered dose inhalant delivery of steroids has been documented to reduce eosinophilic airway inflammation comparable to oral steroids in experimental asthma[7]. Using this model, it has been shown that increased drug amount delivered by MDI is not necessarily better; a low dose is equipotent to a high dose in the control of airway inflammation[8]. In comparison to inhaled bronchodilators which can be fast-acting, inhaled steroids do not work immediately , and therefore should not be used for emergent treatment of asthmatic cats (ie, status asthmaticus). One study in experimentally-induced asthma showed that bronchodilators delivered by MDI had limited efficacy in improvement of allergen-induced bronchoconstriction[9]. Importantly, in pet cats, struggle against a tight fitting face mask is likely when they are having difficulty breathing, thus preventing adequate delivery of an MDI (or perhaps even nebulized bronchodilators) into their lower airways. Therefore, until proven otherwise, injectable bronchodilators should be used during severe episodes of bronchoconstriction.

Inhaled chemotherapy above all other forms of inhaled drug treatment, capitalizes on the concept that that higher drug doses can be delivered to the lung and systemic effects minimized compared to intravenous administration. Clinical animal studies of inhaled chemotherapy exist and confirm efficacy and safety with some agents (doxorubicin, paclitaxel, and 5-fluorouracil) with approximately 25% of treated patients having an objective response. These drugs were delivered to dogs via mask or endotracheal tube. In human medicine, mask delivery is known to greatly limit the amount of aerosol that reaches the lung tissue. This also increases the environmental hazard and potential for human) exposure to cytotoxic drugs. Therefore, delivery by endotracheal tube, either aerosol or catheter, would be superior to mask delivery. Our group has recently completed two studies using targeted, intratracheal delivery of cisplatin with or without gemcitabine. Cisplatin-based chemotherapy protocols are the treatment of choice for primary lung cancer in people, and gemcitabine can work synergistically with cisplatin. Additionally, cisplatin is well suited to localized delivery. In normal dogs, we were able to show precise, localized delivery of drug, with no systemic adverse effects, and inflammation improved?? histologically in treated tissues only.

In addition to cytotoxic chemotherapy, the airways have been used to delivery immunotherapy. In dogs with spontaneously-occurring primary and metastatic lung cancer, delivery of liposome-encapsulated interleukin-2 (IL-2) resulted in some unexpected responses. Two of 4 dogs with metastatic pulmonary osteosarcoma had complete regression of their disease, including regression of extrapulmonary metastasis in one dog. Both were durable responses lasting more than a year. Pulmonary metastatic osteosarcoma is highly resistant to any form of therapy so this response was remarkable. Although this formulation of IL-2 is not commercially available, free IL-2 (Proleukin) can be purchased and administered by owners at home. This can be cost prohibitive but is very well tolerated.

Most dogs and cats with diabetes mellitus ultimately require lifelong injections of insulin. In human medicine, self-administration of injections for those with a fear of needles can be challenging, thus prompting studies on inhaled dry powder insulin as an alternative route of administration. Similarly, owner or pet compliance with insulin injections may not be optimal. Aerosol delivery of regular insulin to healthy cats could effectively decrease blood glucose concentrations, but the high dose required is unlikely to provide a clinically viable alternative for pets with diabetes mellitus[4].

FUTURE DIRECTIONS FOR AEROSOL DRUG DELIVERY

There are very few controlled scientific studies reporting on the efficacy of aerosol therapies in dogs and cats, in contrast to a plethora of information in textbooks, at continuing education meetings and empirical data. Due to the many questions still surrounding efficacy of drug delivery by aerosol, these drugs should probably be used as adjuncts in the treatment of animals with respiratory disease until scientific data supports their use . Perhaps the one exception is inhaled medications in cats with lower airway disease in which there have been several reports documenting both safety and efficacy for certain drugs; however, even for this condition, additional studies still need to be performed.

REFERENCES

[1.] Leemans J, Kirschvink N, Bernaerts F, et al.. **Vet J** 2009;180:236-245. [2.] Reinero C, Delgado C, Spinka C, et al. **Int Arch Allergy Immunol** 2009;150:43-50. [3.] Selting K, Waldrep J, Reinero C, et al. **Aerosol Med Pulm Drug Deliv** 2008;21:255-268. [4.] DeClue A, Leverenz E, Wiedmeyer C, et al**. J Feline Med Surg** 2008;10:519-522. [5.] Cohn L, DeClue A, Reinero C. **J Vet Intern Med** 2008;22:37-43. [6.] Reinero CR, Brownlee L, Decile KC, et al. **J Vet Intern Med** 2006;20:57-64. [7.] Reinero C, Decile K, Byerly J, et al. **Am J Vet Res** 2005;66:1121-1127. [8.] Cohn L, DeClue A, Cohen R, et al. **J Fel Med Surg** In press. [9.] Leemans J, Kirschvink K, Clercx C, et **alVet J** In press.

CHRONIC NASAL DISEASE IN DOGS: DIAGNOSIS & TREATMENT

Ned F. Kuehn, DVM, MS, DACVIM (SAIM)
Southfield, MI

IDIOPATHIC LYMPHOPLASMACYTIC (CHRONIC) RHINITIS

Idiopathic lymphoplasmacytic rhinitis (LPR) is a relatively common cause of chronic nasal disease in the dog. The definitive etiology of LPR remains undetermined; however, it is likely a stereotyped chronic inflammatory response to multiple precipitating factors. Inhaled aeroallergens and irritants likely play a primary role in development of this disease. Hypersensitivity to native commensal fungal organisms within the nose also may play a role in some patients. Dolichocephalic and mesaticephalic medium to large breed dogs and Dachshunds are typically affected. Chronic unilateral to bilateral mucoid to mucopurulent nasal discharge is often present, although some dogs may have mucohemorrhagic discharge or epistaxis. Obstruction to airflow through the nose may result from excessive mucous within nasal passages and turbinate mucosal edema. Lymphoplasmacytic inflammation may be present with other chronic nasal diseases (e.g., nasal neoplasia, fungal rhinitis or foreign body rhinitis), therefore it is imperative that these diseases be thoroughly excluded before a diagnosis of idiopathic LPR is entertained.

Nasal radiography is often not sufficient to differentiate chronic inflammatory rhinitis from neoplasia or fungal rhinitis because similar changes such as turbinate destruction and soft tissue opacification of the nasal passages and frontal sinus may be seen in all these diseases. Nasal CT or MRI is recommended because these imaging modalities greatly enhance the ability to differentiate inflammatory from neoplastic diseases.[1] Nasal CT findings in dogs with LPR are variable and may be completely unremarkable or only reveal diffuse edema of turbinate mucosa. Often mucous accumulation within air passages and sinuses is observed. Unilateral or bilateral mild to moderate turbinate destruction may also be seen and occasionally the turbinate destruction may be quite severe mimicking that seen with fungal rhinitis.[1,2] Destruction of the nasal septum, frontal sinuses or cribriform plate, or extension of disease into the nasopharynx or periorbital region are not expected with idiopathic LPR and should prompt investigation into the presence of fungal rhinitis or neoplastic disease.

Rhinoscopic abnormalities include unilateral or bilateral erythema or hyperemia and edema of the nasal mucosa with the presence of mucopus with air passages. Turbinate atrophy or loss is occasionally appreciated. Nasal samples for microbial culture are often not informative and not routinely recommended. Histologic changes include mild to severe lymphoplasmacytic inflammation with occasional infiltration of neutrophils or eosinophils. Turbinate remodeling or destruction may be absent or vary from mild to severe. There may be discordance between the right and left sides of the nasal cavity in the severity of histologic changes.

Treatment for idiopathic LPR is extremely frustrating with cure rarely achieved. Although this is not a life-threatening disease, owners of dogs so affected are often distraught by their pets nasal obstruction or the need to frequently clean up sneezed nasal discharge or hemorrhage throughout the house. Allergen avoidance is rarely helpful; however, avoidance of secondhand smoke can substantially reduce signs in some dogs. Controversy surrounds the use of systemic corticosteroid or immunosuppressive agents. I feel that these oral medications are seldom effective in controlling clinical signs. Topical glucocorticoid therapy with nasal steroid drops or aerosolized steroids administered using metered dose inhalers attached to a spacer and tightly fitting face mask have shown anecdotal promise in some dogs with LPR.[3] Antihistamine medications are rarely effective, but they occasionally slightly reduce the severity of nasal discharge. Long-term administration of antibiotics having immunomodulatory effects combined with nonsteroidal antiinflammatory agents can be helpful in some dogs. Doxycycline 3-5 mg/kg q12h, PO or azithromycin 5 mg/kg q24h, PO in combination with piroxicam 0.3 mg/kg q24h, PO is recommended. If distinct clinical improvement is observed within 2 weeks, daily piroxicam therapy is continued but the frequency of administration of doxycycline is reduced to once daily or azithromycin reduced to twice weekly. Therapy will likely be required for a minimum of 6 months if not indefinitely.

Investigation into the use of oral antifungal therapy in dogs' refractory to other therapeutic modalities may hold some promise in that hypersensitivity to ubiquitous fungi may play a role in this disease. Hypersensitivity to ubiquitous fungi is currently thought to play a role in some people with chronic rhinosinusitis. Nasal biopsies from dogs with LPR have been reported to display an elevated transcription of fungal genes as compared to dogs with nasal neoplasia using PCR techniques.[4] Preliminary experience with the administration of itraconazole 5 mg/kg q12h, PO for a minimum of 3-6 months has shown a dramatic beneficial improvement of clinical signs in some dogs.[5]

NASAL NEOPLASIA AND NASAL POLYPS

Nasal neoplasia is an important cause of chronic nasal disease in middle aged to older dolichocephalic and mesaticephalic dogs. In approximately one third of all dogs with chronic nasal disease, neoplasia is the underlying cause. Tumors of epithelial origin account for approximately two thirds of canine nasal neoplasms. The majorities of nasal tumors are malignant and primarily arise within the nasal cavity, although they occasionally may arise in the paranasal sinuses. Nasal tumors are locally invasive with local to widespread destruction of nasal turbinates seen initially, with invasion of septal, cribriform or facial bones later in the course of disease. Metastasis to regional lymph nodes or lung may occur, but this is rare and generally occurs in the very late stage of disease. Clinical signs are primarily related to obstruction of air flow through the nasal cavities, mucopurulent nasal discharge, epistaxis, sneezing, and reverse sneeze. Facial deformity or swelling, exophthalmia, or neurological signs may be seen as a result of tumor destruction of facial bones or cribriform plate. Facial pain and head shyness is rarely seen (unlike that with fungal rhinitis). In some patients, initial clinical signs may be very subtile with unexplained onset of snoring and occasional reverse sneeze reported.

Nasal radiographs are frequently limited in their ability to distinguish subtle neoplastic lesions as radiographic signs of early neoplasia overlap those seen with chronic and fungal rhinitis. Nasal CT or MRI are vastly superior imaging modalities for differentiating neoplastic from non-neoplastic disease, detection of bone destruction and extension of disease into surrounding structures. Retroflex nasopharyngoscopy is recommended to evaluate the nasopharyngeal region and identification of extension of tumor through the caudal nares. Anterior rhinoscopy in dogs with nasal neoplasia may reveal a mass lesion protruding within and occluding nasal air passages. Multiple biopsies of masses should be obtained to increase the likelihood of diagnosis as severe inflammation often surrounds nasal tumors. Frequently nasal tumors cannot be visualized due to hemorrhage or because their origin is in an inaccessible region of the nose. In these situations, nasal CT studies facilitate direction and location for blind biopsy of the affected region of the nose.

Radiation therapy is the treatment of choice for most nasal tumors. Thoracic radiographs are recommended when nasal neoplasia is identified prior to radiation therapy to rule out metastatic lung disease. To plan for radiation therapy, nasal CT or MRI is needed for staging and to delineate tumor boundaries. Surgery alone is ineffective with survival times similar to that observed in untreated dogs. Results with cryosurgery, either alone or in combination with radiation therapy, have been disappointing and not recommended. There is limited information on the response of nasal tumors to chemotherapy alone. Depending on the mode of radiation therapy available, approximate median survival times are between 16.5-23 months and approximate 1 year survival rates are between 54-60% in dogs with nasal neoplasia. Exenteration of the nasal cavity following accelerated radiotherapy significantly prolongs survival time in dogs with intranasal neoplasia over radiotherapy alone.[6]

Polyps within the nasal cavity are very rare in dogs. These are usually unilateral and rhinotomy is required for removal of the polypous tissue and surrounding conchae. Recurrence 1-2 years later is possible. To date, in all dogs I have seen with an initial diagnosis of a polyp, careful review of nasal CT has demonstrated localized turbinate destruction and following surgical resection of the polypous tissue the subsequent histologic diagnosis has been low-grade fibrosarcoma associated with moderate to severe chronic inflammation.

FUNGAL RHINITIS

Fungal rhinitis is a relatively common cause of chronic rhinitis in the dog within various geographic regions throughout North America. *Aspergillus fumigatus* is the most common cause of fungal rhinitis in dogs, but occasionally *Penicillium* spp and *Rhinosporidium seeberi* and very rarely *Cryptococcus neoformans* may cause disease. *Rhinosporidium seeberi* is associated with the growth of a granulomatous mass within the rostral nasal cavity. Cytology of tissue from these granulomatous masses is often diagnostic. Treatment for rhinosporidiosis is best accomplished by aggressive surgical resection of the granulomatous mass.

Nasal aspergillosis is most commonly seen in young to middle-aged dolichocephalic dogs. Affected dogs often present with copious unilateral or bilateral mucopurulent nasal discharge. Sneezing is common and may be accompanied by mild to severe epistaxis. Facial pain and depigmentation and ulceration of the nasal planum may be present. In contrast to nasal neoplasia, facial distortion is unusual in all but advanced cases of fungal rhinitis. Nasal CT or MRI studies along with rhinoscopic visualization of the nasal cavity are noteworthy for the presence of dramatic turbinate loss within the nasal cavity. Sinus involvement may be present. Invasion through the maxillary or palatine bones with extension into surrounding soft tissue structures is occasionally seen. A noninvasive form of nasal aspergillosis infrequently occurs in the dog and is characterized by compact masses of mycelia ("fungal balls") filling larger airspaces or frontal sinuses.

Diagnosis of nasal aspergillosis is confirmed by visualization of fungal plaques on nasal mucosa and demonstration of branching septate hyphae on cytologic or histologic samples from affected regions within the nose. There is high accuracy of cytology samples in the diagnosis of nasal aspergillosis or penicilliosis when collection is done under direct endoscopic visualization; whereas, there is poor value of samples collected by blind swabs or preparations from samples of nasal discharge.[7] Serologic tests positive for aspergillosis also support the diagnosis although negative results may occur even with extensive disease. Culturing nasal discharge material may be misleading because 30-40% of cultures from normal dogs and those with nasal neoplasia can yield *Aspergillus* or *Penicillium* species. Despite properly obtained samples there are some cases that fail to demonstrate fungal organisms. Repeated sampling or a trial of antifungal drugs may well be indicated in dogs with a high index of suspicion for fungal rhinitis.

The prognosis for treatment of nasal aspergillosis is fair to good, but relapses are possible necessitating re-treatment. Treatment of nasal aspergillosis has classically been approached with topical infusion of either clotrimazole or enilconazole, providing the cribriform plate is intact. Nasal CT or MRI studies are preferred over radiographs to evaluate integrity of the cribriform plate prior to local antifungal therapy. Topical therapy is more effective than orally administered antifungal agents. Clotrimazole (Lotrimin solution®, Schering Plough Corp.) is available over the counter as a 1% solution and enilconazole (Clinifarm-EC®, Sterwin Labs, Inc.) is provided as 13.8% concentrate which is diluted to a 1, 2, or 5% solution prior to instillation in the nasal cavity. Debridement and removal of diseased turbinate structures through the rostral nares prior to topical therapy will greatly improve response to treatment. Topical therapy with either drug alone is not effective in dogs in which the organism has invaded soft tissue structures adjacent to the nose. In these cases, topical therapy is recommended to be combined with systemic antifungal agents. Exploratory rhinotomy and turbinectomy prior to topical treatment or oral antifungal therapy is often detrimental and not recommended.

The topical application of enilconazole through surgically placed catheters into the frontal sinuses and nasal chambers has a success rate up to 90%. This procedure is quite distressing to most patients, and complications include premature removal of the catheters, subcutaneous emphysema, inappetence, and ptyalism. Because of these serious and frequent side-effects, topical therapy with clotrimazole is the treatment of choice. Clotrimazole is applied as a soak with the solution maintained in the nasal cavities for 1 hour with the patient under anesthesia. Side effects of clotrimazole therapy include severe pharyngitis and pharyngeal edema.

Recently another approach with excellent success rate, shorter treatment time, and low patient morbidity has been reported using a combination of clotrimazole irrigation and depot therapy.[8] Frontal sinus trephination is followed by a short, five-minute flushing of 1% topical clotrimazole solution followed by a 1% clotrimazole cream instilled as a depot agent into the frontal sinuses.

With this treatment protocol, 86% of dogs with nasal aspergillosis or penicilliosis established a cure from infection. Debridement of diseased nasal turbinates prior to treatment may result in a higher cure rate with this treatment procedure.

Oral antifungal agents have relatively poor efficacy against *Aspergillus* infection, but are recommended if the cribriform plate is penetrated. Oral antifungal agents are used in combination with topical agents if invasion of local bone and soft tissue structures is present. The newer azole derivatives have the best results, but side effects include anorexia, vomiting, lethargy, elevated BUN, skin ulcerations, fever and hepatotoxicity. Itraconazole (Sporanox®, Janssen) is recommended due to its low toxicity. Itraconazole 5 mg/kg q12h, PO given for 3-6 months may cure up to 60-70% of dogs with aspergillosis, although some studies have shown only marginal efficacy. Terbinafine (Lamisil®, Novartis) is also well tolerated and a dose of 5-10 mg/kg q12h, PO appears to have similar efficacy to itraconazole when given for 3-6 months. Fluconazole (Diflucan®, Pfizer) is an additional alternative with a cure rates up to 60% when given at a dose of 2.5-5 mg/kg q12h, PO for 3-6 moths. Voriconazole (Vfend®, Pfizer) is a new generation broad-spectrum antifungal agent that has both fungicidal and fungistatic activities *in vitro* against *Aspergillus* spp superior to that of fluconazole. Clinical experience with this drug is presently minimal, but a suggested dose for dogs is 5 mg/kg q12h. Multiple doses may result in increased metabolism of voriconazole subsequently reducing drug exposure; therefore, serum drug level monitoring may be advised for patients receiving voriconazole therapy.

NASOPHARYNGEAL STENOSIS

Nasopharyngeal stenosis is a rare disease that is characterized by scar tissue that forms a membrane or "webbing" above the soft palate and obstructs airflow throught the nasopharynx. Initially progressively worsening stertor is present with absence of nasal discharge. With severe stenosis, nasal discharge and extreme difficulty with nasal breathing is observed. Diagnosis is made using flexible endoscopy during evaluation of the nasopharyngeal region. Retroflex nasopharyngoscopy will often reveal circumferential stricturing or narrowing of a focal region of the nasopharynx. Early stenosis may be managed by stretching the affected region followed by corticosteroid therapy to reduce scar tissue and reformation of the stricture. Recurrence unfortunately is very common necessitating extensive surgical techniques to resect the stenotic region, or balloon dilation of the affected region. The placement of stents in the affected region following surgery or balloon dilation is often required to prevent re-stricture.

NASOPHARYNGEAL TURBINATES

Nasopharyngeal turbinates is a condition upon which the nasal turbinates protrude caudally through the caudal nares into the rostral nasopharyngeal region. Nasopharyngeal turbinates are the 4th most important complicating related component causing upper airway obstruction as part of the brachycephalic airway syndrome (elongation of the soft palate, stenotic nares, and everted laryngeal saccules are the 1st thru 3rd imported related components, respectively). The presence of nasopharyngeal turbinates results in variable degrees of upper airway obstruction with nasal breathing. Nasopharyngeal turbinates are seen with highest incidence in Pugs, and less frequently in French and English bulldogs, and Pekinese dog breeds. This condition is also seen occasionally in Himalayan and Persian cats. Retroflex nasopharyngoscopy should be performed in the evaluation of dogs with brachycephalic airway syndrome as this is a permanent and untreatable condition and may be a significant factor for constant airway obstruction in dogs so affected.

MISCELLANEOUS CASUSES OF CHRONIC RHINITIS IN DOGS

Parasitic rhinitis is uncommon. The treatment for nasal mites (*Pneumonyssus caninum*) is ivermectin 0.2 mg/kg, PO with the dose repeated in 2-3 weeks, or milbemycin 1 mg/kg PO once every 10 days for 3 treatments. The treatment for nasal nematodes (*Eucoleus [Capillaria] boehmi*) is not clearly defined although ivermectin 0.2 mg/kg, PO once has been reportedly effective. Allergic rhinitis is often mild, but if severe, antihistamines such as diphenhydramine, chlorpheniramine, or trimeprazine-prednisone (Temaril-P®, Pfizer Animal Health) may be prescribed to control symptomatology. In the rare situation of bacterial rhinitis due to *Bordetella bronchiseptica*, doxycycline 5-10 mg/kg q12h, PO for 2 weeks may be effective.

REFERENCES

[1]Lefebvre J, et. al. **J Sm Anim Pract** 2005;46:280. [2]Windsor R, et. al. **J Am Vet Med Assoc** 2004;244:1952. [3]Windsor R, et. al. **Clin Tech Sm Anim Pract** 2006;21:76. [4]Windsor RC, et. al. **J Vet Intern Med** 2006;20:250. [5]Kuehn NF. Prospective Long-term Pilot Study Using Oral Itraconazole Therapy for the Treatment of chronic Idiopathic (Lymphoplasmacytic) Rhinitis in Dogs. Clinical Research Abstract, **BSAVA Annual Congress** 2006. [6]Adams WM, et. al. **J Am Vet Med Assoc** 2005;227:936. [7]DeLorenzi D, et. al. **J Sm Anim Pract** 2006;47:316. [8]Sissener TR, et. al. **J Sm Anim Pract** 2006;47:312.

CHRONIC NASAL IN CATS: DIAGNOSIS & TREATMENT

Ned F. Kuehn, DVM, MS, DACVIM (SAIM)
Southfield, MI

IDIOPATHIC CHRONIC RHINOSINUSITIS

Chronic rhinosinusitis is an extremely significant disease, representing one of the two most common causes of sneezing and nasal discharge in cats.[1] Chronic rhinosinusitis may follow severe acute upper respiratory tract infection, particularly in kittens or adults cats exposed to an infected cat. Acute upper respiratory tract disease is presumed to be caused by infection with feline herpesvirus type 1, feline calicivirus, Chlamydia felis, or some combination of these agents. Although viral (especially feline herpesvirus type 1) or Chlamydia infection is often implicated as the initiating cause of acute rhinitis, the pathogenesis of idiopathic chronic rhinitis is unknown and it is uncertain whether acute rhinitis is related to the chronic syndrome. Even though cats with idiopathic chronic rhinosinusitis have similar clinical signs, the disease syndrome appears heterogenous among the population perhaps due to individual susceptibility to microbial pathogens, genetic characteristics of the inflammatory response, and environmental factors (e.g., stress) having a role in the development of clinical signs. The complex interrelationship of viral or bacterial organisms involved in the acute disease process combined with individual uniqueness of the inherent inflammatory response may be important factors in genesis of the chronic rhinitis. A thorough understanding of these characteristics would likely facilitate treatment recommendations for patients so affected.

Feline herpesvirus type 1 is estimated to account for the majority of cases of acute severe upper respiratory disease in cats, although calicivirus may be more prevalent in some populations. Feline herpesvirus type 1 is possibly an important pathogen for initiating chronic rhinitis. Experimental infection of germ-free cats with feline herpesvirus type 1 can cause severe upper respiratory disease in the absence of microbial flora. In the natural cat population, it is likely that some complex interrelationship between herpesvirus type 1 viral infection and the various bacterial organisms within the nose and host immunity play important factors influencing the severity and duration of upper respiratory disease. The role any one or combination of microbial agents plays in the pathogenesis of chronic rhinosinusitis is ultimately not understood.[2] Identification of bacterial or viral organisms via culture, antibody titers, or molecular techniques does not imply that the organism or organisms found are responsible for the clinical signs. In summary, a complex progression of microbial, physiological, anatomical, and immunological interactions are likely responsible for development of chronic rhinosinusitis.

Idiopathic chronic rhinosinusitis may be seen in cats of any age, although younger cats are most often afflicted. Affected cats with this disease syndrome typically have a recurrent history of chronic intermittent or progressive sneezing, stertor, and nasal discharge. The nasal discharge is usually copious, bilateral, and mucopurulent to purulent, although occasionally the discharge can be unilateral or intermittently contain blood. Systemic or ocular disease is generally not present. Older cats may develop anorexia due to loss of smell, which may exacerbate other underlying disease conditions (i.e., chronic renal failure, liver, or gastrointestinal disease). Physical examination findings are generally unremarkable other than abnormalities confined to the upper respiratory tract. The dental arcade (e.g., gingivitis, tooth root abscessation, oronasal fistula), hard palate (e.g., neoplasia, oronasal fistula), soft palate (e.g., nasopharyngeal mass or polyp), and regional lymph nodes (e.g., lymphadenopathy) are all structures that should be carefully evaluated. A fundic examination is recommended if Cryptococcosis is suspected.

Diagnostic studies should be undertaken once it is established that chronic refractory rhinosinusitis is present so that other treatable conditions may be excluded and to further characterize the disease present. A complete health profile with serology for feline leukemia virus and feline immunodeficiency virus is essential to assess systemic health. Titers for *Bartonella henselae* might be considered as this chronic disease may be a complicating immunosuppressive factor in some cats with chronic rhinosinusitis. General anesthesia is required for thorough oral examination and skull radiographs or advanced diagnostic imaging studies. The nasal passages, paranasal sinuses, and tympanic bulla are often involved in the disease process. The intra-oral radiographic view achieved using dental radiographic film often provides sufficient detail of both nasal cavities, although subtle changes in turbinate structures and distinction of soft tissue densities may be difficult to discern. Radiographs will generally show variable turbinate lysis and increased fluid density within the nasal cavities. Because the frontal sinuses are small in cats, frontal sinus projections can miss disease within these structures. In general, computed tomography provides better localization of lesions and determination of extent of disease within the nasal cavity, paranasal sinuses, and tympanic bulla than do radiographs.[3] Findings often seen with computed tomography include soft tissue attenuating densities within the nasal cavity, frontal and sphenopalatine sinuses and occasionally tympanic bulla; and, osteolysis of nasal and frontal bones, turbinate destruction, and asymmetry of the cribriform plate. Computed tomography is also superior for differentiation of inflammatory disease from neoplastic and fungal diseases.

Following completion of diagnostic imaging studies, the nasal cavity and nasopharyngeal region should be evaluated as completely as possible. The nasopharyngeal region and caudal nares can be evaluated with a flexible endoscope retroflexed around the caudal margin of the soft palate. Alternatively, the soft palate can be retracted cranially with a spay hook and a dental mirror is passed into the oropharynx to observe the nasopharyngeal region. Careful observation should be made for mass lesions projecting through the caudal nares or for masses or foreign bodes within the nasopharynx. Rostral rhinoscopy can be performed with either a small flexible endoscope or rigid arthroscopic equipment. Cats with idiopathic chronic rhinitis have mild to severe hyperemia of nasal mucosa, moderate to large amounts of mucoid to purulent discharge between turbinates, and varying severity of turbinate destruction (although in some cats the turbinates may be normal). The degree of observed abnormalities may vary between each side of the nasal cavity. However, visual appearance of the nasal cavity does not necessarily reflect on the presence or absence of underlying pathology. Histopathology is of critical importance and both sides of the nose should be biopsied. There may be considerable discordance between visualized abnormalities and those detected histologically.[4] Cats with relatively normal appearing nasal tissue may be found to have moderate to severe inflammation. Histopathology findings may include moderate to severe neutrophilic, lymphocytic, or pleiocellular inflammation with epithelial ulceration, turbinate destruction and remodeling, fibrosis, necrosis, and glandular hyperplasia. Histologic changes may be predominately unilateral rather than bilateral. Nasal biopsy samples or material cultivated from deep nasal aspiration should also be submitted for microbial culture and sensitivity. Aerobic, anaerobic, and Mycoplasma cultures should be requested. **Bacterial culture of nasal discharge is not recommended as secondary bacterial contaminates are typically isolated.** Following visualization of the nasal cavity and collection of biopsy materials, the nasal cavities should be suctioned and flushed with copious amounts of warm isotonic solution to remove secretions and provide temporary improvement in clinical signs. When flushing fluids within the nasal cavity, the tip of the nose should be pointed downward and the oropharynx packed with gauze to prevent aspiration of the lavage solution and nasal secretions into the trachea.

Cats with idiopathic chronic rhinosinusitis frequently prove to be very refractory to treatment. Broad spectrum antibiotics are often used for treatment of secondary bacterial colonization or infection within the nasal and paranasal sinus cavities. Recent work has disclosed that potentially pathogenic bacteria and a wider variety of bacterial species are isolated from cats with idiopathic chronic rhinitis than from cats without disease. Mycoplasma spp and anaerobic bacteria were only isolated from cats with idiopathic chronic rhinosinusitis. Antibiotic therapy should be determined from culture of nasal biopsy material or specimens collected from a deep nasal flush. Antibiotics recommended include doxycycline, clindamycin, amoxicillin-clavulanic acid, cefpodoxime, marbofloxacin, and azithromycin. Cats with idiopathic chronic rhinitis demonstrating an initial response to antibiotic therapy should have treatment continued for 6-8 weeks or longer. Antiviral therapy is not routinely recommended as substantiation regarding a primary role for feline herpesvirus type 1 in this disease syndrome has not been established. However, it is likely that feline herpesvirus type 1 may play a role in the disease of certain cats. Lysine (500 mg PO q12h) therapy may benefit some cats. Lysine replaces arginine in viral proteins rendering them nonfunctional and thereby reducing viral replication. For those cats demonstrating response to lysine, therapy may be continued indefinitely as it is safe to use in young or older cats. The role of immunosuppressant agents in management of idiopathic chronic rhinosinusitis is poorly understood. Glucocorticoids may exacerbate viral infection, cause recrudescence of viral shedding, or suppress immune response to bacterial infection. Piroxicam is a non-steroidal antiinflammatory agent that can ameliorate clinical signs in some cats with idiopathic chronic rhinosinusitis. Piroxicam is generally well tolerated at a dose of 0.3 mg/kg orally daily or every other day. Side effects may include anorexia, diarrhea, or vomiting. Piroxicam may also be combined with antibiotic therapy. Moisturization of nasal secretions promotes sneezing and evacuation of mucous from the nasal cavity. Some cats will tolerate the instillation of saline drops within the nasal cavity. Finally, there are no clinical studies with evidence to support the use of immune simulators (e.g., interferon) in the management of this disease.

NEOPLASIA

Nasal neoplasia and idiopathic chronic rhinosinusitis are the most common causes of chronic rhinitis in cats. Although there are exceptions, neoplasia is of greatest concern in cats > 8 years of age. Nasal lymphoma is most common with various carcinomas occasionally seen and other neoplasms encountered much less frequently. Radiation therapy is generally the treatment of choice.

LYMPHOPLASMACYTIC RHINITIS

Lymphoplasmacytic rhinitis is a relatively uncommon disease of unknown etiology in cats.[5] These cats do not generally have a history of a preceding acute episode of feline upper respiratory tract infection. The evolution of the disease is progressive with minimal signs of nasal disease early on. The disease is likely a hypersensitivity to any one or combination of environmental aeroallergens. Lymphoplasmacytic rhinitis should only be considered when all other causes for the chronic rhinitis are excluded. Lymphoplasmacytic inflammation may accompany other diseases, especially neoplasia. If the clinical findings are highly suggestive of nasal neoplasia, the diagnosis of lymphoplasmacytic rhinitis should be highly provisional and repeat biopsies or rhinotomy be strongly entertained. Nasal lymphoma should be considered in those cats with severe lymphoplasmacytic rhinitis. Immunohistochemical staining for B-cell and T-cell markers should be requested in cats with severe lymphoplasmacytic inflammation to rule out lymphoma.

FUNGAL RHINITIS

Fungal rhinitis is an uncommon cause of chronic rhinitis in cats, although in some geographic areas the disease is seen frequently. Nasal cryptococcosis is most common. Facial deformity of the rostral aspect of the nose is often striking in cats with cryptococcosis. Occasionally granulomatous tissue may be seen projecting through the external nares. The clinical signs of cryptococcosis are obstructive rhinitis with mucopurulent nasal discharge. Mucopurulent conjunctivitis may develop in some cats. Cytology can be highly diagnostic for nasal cryptococcosis. Fresh material from granulomatous lesions within the nose are gently smeared on a slide and stained with India ink. Cryptococcus spp. organisms are recognized as a thick, encapsulated round to oval yeast. Treatment may involve debridement of the granulomatous mass and treatment with itraconazole.

Sinonasal and sino-orbital[6] aspergillosis is comparatively rare in cats. Both *Aspergillus* spp and *Neosartorya* spp are fungal pathogens implicated in feline sinonasal fungal infection. The clinical signs are initially sneezing and copious unilateral or bilateral mucopurulent nasal discharge. Intermittent epistaxis is not uncommon. Sinonasal pain is often present. Anorexia and depression may develop as the disease progresses, especially with frontal sinus involvement or with bony lysis of the sinus, orbital wall, maxilla, or hard palate. Extensive turbinate destruction is present. Late in disease facial deformity or exophthalmia may be seen. The magnitude of bone lysis and determination of sinus, orbital, or maxillary bone involvement, and integrity of the cribriform plate is best demonstrated with computed tomography. The diagnosis of aspergillosis is made by direct observation of fungal plaques, identification of fungal elements in biopsy specimens or direct cytology obtained from affected tissue, or positive fungal culture of affected tissue. Treatment is challenging and antifungal therapies to consider include triazole drugs (itraconazole, posaconazole) either alone or in combination with terbinafine. Sinonasal infusion of clotrimazole is another option.

NASOPHARYNGEAL POLYPS

Nasopharyngeal polyps occur predominately in young cats or kittens. Nasopharyngeal polyps usually arise from the middle ear and grow down the eustachian tube to the nasopharyngeal region. Polyps occasionally may be visible in the external ear canal. Clinical signs are caused by direct obstruction of either the oropharynx or nasopharynx. Chronic rhinitis develops secondary to polyps within the nasopharyngeal region due to secondary bacterial overgrowth due to the lack of clearance of nasal secretions. Clinical signs observed in the early stage of disease include stertorous respiration (snuffling sound), gagging, and minimal sneezing or nasal discharge. As the polyps attain greater size, obstructive breathing and increased nasal discharge with sneezing are seen. These signs may be indistinguishable from those of idiopathic chronic rhinitis, especially in a young cat with a prior history of acute upper respiratory infection. For this reason, any cat with chronic rhinitis (especially those of young age) should be carefully evaluated for the presence of polyps.

The diagnosis of nasopharyngeal polyp is relatively straight forward. Oropharyngeal examination and palpation of the area above the soft palate for a mass or direct visualization of the nasopharyngeal region is diagnostic. Once the diagnosis of nasopharyngeal poly is made, skull radiographs (obtained with the cat under general anesthesia) or computed tomography of the bulla region should be performed. Careful evaluation of the osseous and tympanic bulla and petrous temporal bones is essential for evidence of middle ear inflammation or infections characterized by osseous bulla thickening, soft tissue densities within the tympanic cavity, or sclerosis of the petrous temporal bone. Radiographic evidence of middle ear inflammation or infection may not always be present, whereas computed tomography offers enhanced discrimination of middle ear involvement.

Nasopharyngeal polyps may be removed by traction avulsion, however recurrence is common. Bulla osteotomy is definitely indicated and should be performed in cats when radiographic or tomographic evidence of middle ear involvement is present. It has been advocated that bulla osteotomy should always be performed on the side of polyp origin, even without radiographic changes without the osseous bulla. For polyps with recurrence following traction avulsion, bulla osteotomy on the side of polyp origin should always be recommended to allow for removal of the origin of the polyp.

FOREIGN BODY RHINITIS

Nasal or nasopharyngeal foreign bodies are infrequent in cats and are usually due to blades of grass lodged within the nasal cavity. Seeds and grass awns are less frequently encountered. Clinical signs associated with foreign bodies within the nasopharyngeal region are often peracute with coughing, gagging, and hard swallowing response usually observed. With time, stertorous respiration, phonation changes, nasal discharge, and sneeze may develop with nasopharyngeal foreign bodies. Clinical signs associated with foreign bodies confined within the nasal passages may be acute or chronic and often consists solely of sneezing and nasal discharge. Cats having nasal foreign bodies not immediately expelled from the nose will develop progressively increasing nasal discharge, often attended by occasional hemorrhage. Severe granuloma tissue response may occur with long-term foreign bodies. Foreign bodies usually wind up within the nasal cavity by entry through the caudal nares. Foreign material is ingested and then either gagged or vomited, inadvertently transferring the foreign material into the nasopharynx and then through the caudal nares into the caudal nasal cavity.

Grass blades and similar material can usually be removed under direct examination of the nasopharynx or nasal cavities. Retroflex nasopharyngoscopy is diagnostic for nasopharyngeal foreign bodies. Alternatively, a spay hook may be used to retract the soft palate forward (with or without the aid of a dental mirror) to observed the nasopharyngeal region. Flushing the nose with copious amounts of warm saline may dislodge smaller particles (e.g., seeds) within the nasal cavity. The nose should be directed downward and the caudal oropharynx should be packed with gauze to prevent aspiration of saline and nasal secretions into the trachea. Rhinotomy may be required for longstanding cases with excessive granulation tissue or foreign bodies resistant to direct removal (e.g., grass awns).

NASOPHARYNGEAL STENOSIS

Nasopharyngeal stenosis is a rare complication of acute upper respiratory tract infections or following episodes of severe vomiting (e.g., aspiration rhinitis). Initially progressively worsening stertor is present with absence of nasal discharge. With severe stenosis, nasal discharge and extreme difficulty with nasal breathing is observed. Retroflex nasopharyngoscopy will often reveal circumferential stricturing or narrowing of a focal region of the nasopharynx. Early stenosis may be managed by stretching the affected region followed by corticosteroid therapy to reduce scar tissue and reformation of the stricture. Recurrence unfortunately is very common necessitating extensive surgical techniques to resect the stenotic region, or balloon dilation of the affected region. The placement of stents in the affected region following surgery or balloon dilation is often required to prevent re-stricture.

REFERENCES

[1]Henderson S, et. al. **J Fel Med Surg** 2004;6:245. [2] Johnson L, et. al. **J Am Vet Med Assoc** 2004;227:579. [3]Schoenborn W, et. al. **Vet Radiol Ultrasound** 2003;44:185. [4]Johnson L, et. al. **J Am Med Assoc** 2004;225:395. [5]Allen H, et. al. **J Am Anim Hosp Assoc** 1999;35:457. [6] Barrs VR, et.al. Sino-Orbital Aspergillosis: An Emerging Clinical Syndrome. In August JR (ed). **Consultations in Feline Internal Medicine, Volume 6**, Elsevier, St. Louis, MO 2010.

ASSESSMENT OF HORMONE ABNORMALITIES IN ADRENAL DISEASE

Patty Lathan, VMD, MS, DACVIM
Mississippi State, MS

Endocrine abnormalities in patients with adrenal disease are not limited to changes in adrenocortical hormones. As in people, alterations in both calcium metabolism and the hypothalamic-pituitary-thyroid axis have been documented in dogs with hypoadrenocorticism and hyperadrenocorticism.

Several recent articles have evaluated changes in calcium metabolism in dogs with adrenal disease. In 2005, Ramsey, et al, demonstrated that a population of patients with hyperadrenocorticism had increased parathyroid hormone levels and phosphorus concentrations, but similar calcium concentrations, relative to a control population.[1] In another report from the same institution, Tebb, et al, showed that PTH and phosphorus concentrations decreased significantly in Cushingoid dogs following treatment with trilostane, while calcium concentrations increased.[2] Thus, the authors hypothesized that PTH levels increased in response to increased phosphorus and/or decreased calcium concentrations. Additional studies are required to clarify the cause of adrenal secondary hyperparathyroidism in dogs.

In an effort to identify the cause of hypercalcemia associated with hypoadrenocorticism, Gow, et al, evaluated ionized calcium, PTH, PTHrp, and vitamin D metabolite concentrations in eight Addisonian dogs with hypercalcemia.[3] Total calcium concentrations were increased in all dogs, and ionized hypercalcemia was present in 5/7 dogs in which ionized calcium was measured. PTH levels were within reference range in 7/8 dogs; the PTH level was below reference range in one dog with ionized hypercalcemia. Vitamin D metabolite concentrations were variable. The authors concluded that hypercalcemia in some patients with Addison's disease cannot be "definitively explained" by elevations in PTH, PTHrp, or calcium metabolite concentrations.

In another recent study, Kenefick and Neiger attempted to identify alterations in thyroid hormone concentrations in dogs with hyperadrenocorticism.[4] Free thyroxine, total thyroxine, and thyroid stimulating hormone levels were evaluated in dogs with hyperadrenocorticism, prior to and following treatment with trilostane. Although TSH concentrations increased significantly following therapy, there was no change in tT4 concentration. Surprisingly, fT4 concentrations decreased. The authors concluded that suppression of TSH production may contribute to alterations in the hypothalamic-pituitary-thyroid axis that occur in the presence of glucocorticoids.

REFERENCES

[1]Ramsey IK, et al. **J Sm Anim Prac** 2005;46:531. [2]Tebb AJ, et al. **J Sm Anim Prac** 2005;46:537. [3]Gow AG, et al. **J Sm Anim Prac** 2009;50:426. [4]Kenefick SJ et al. **J Sm Anim Prac** 2008;49:139.

THE INCIDENTAL ADRENAL MASS: WHAT SHOULD YOU DO?

Audrey K. Cook, BVM&S, MRCVS, DACVIM, DECVIM-CA
College Station, TX

INTRODUCTION

As high quality diagnostic ultrasonography becomes a routine part of small animal medicine, practitioners now face a new dilemma: what should be done with the incidental adrenal mass? In human medicine, the term "incidentaloma" entered the lexicon about 20 years ago, and algorithms have been proposed for effective assessment and management of these lesions. In veterinary medicine, we do not have a consensus about the best approach for the incidental adrenal mass. This presentation will review the information in the veterinary literature, contrast this with data from the human field, and suggest a logical approach to the canine incidentaloma.

DEFINITON AND BACKGROUND

An incidental adrenal mass in one discovered during the evaluation of a patient with no symptoms or signs of adrenal disease. These patients may be clinically normal, or - more commonly – are undergoing diagnostic evaluation for non-adrenal disorders. In most cases, canine incidentalomas are found during abdominal ultrasonography, although some may be initially identified with more advanced imaging methods, such as computed tomography (CT) or magnetic resonance imaging (MRI).

In the human field, the term "mass" is generally reserved for lesions ≥1 cm in diameter.[1] Smaller lesions are often referred to as adrenal nodules. Similar criteria have not been established in veterinary medicine, and classifying these lesions based on size may be less appropriate due to substantial variations in patient size.

The incidence of incidentalomas in canines has not been reported. The incidence in human patients undergoing abdominal CT examination ranges from 0.35-5%, and is substantially impacted by patient age with likelihood of 0.2% in a patient 20-29 years compared to >7% in a patient over 70 years.[2,3,4]

DIFFERENTIAL DIAGNOSIS

There are two important but independent considerations regarding the incidental adrenal mass. First, is it benign (e.g., adenoma, granuloma) or malignant? Secondly, is it's functional status? Both questions must be answered to make an appropriate plan.

Assessing malignant potential

In human patients without extra-adrenal primary malignancy, 80% of incidentally discovered adrenal masses are benign adenomas.[5] It also seems that the likelihood of a mass being benign increases with age, meaning that lesions found in older human patients are less likely to be malignant when compared to those found in younger individuals. There is little data regarding this in canine patients, but one group of dogs with incidentalomas, necropsy findings were reportedly benign in 70%.[6]

Malignant adrenal masses may be primary tumors or metastatic lesions. Primary malignant adrenal tumors are routinely encountered in dogs. Although unilateral disease is more common, bilateral tumors may occur.[7] In addition, individuals with two distinct adrenal tumors have been reported (i.e., a cortical carcinoma in one adrenal and a pheochromocytoma in the other).[8] Although tissue biopsy is the only way to definitively determine malignancy, imaging may provide some useful evidence. Ultrasonographic findings suggestive of malignant behavior include invasion of adjacent structures and indistinct margins. In human patients, calcification, necrosis and hemorrhage are more commonly noted in malignant masses, but are not regarded as adequately specific for a definitive diagnosis. Tumor size appears to be a more useful guide in people, as most lesions less than 4 cm are ultimately found to be benign. Specific tissue findings on CT may also provide some indication of malignancy, based on the Hounsfield units, as benign lesions are usually rich in lipid. The rate of washout of IV contrast has also been identified as a useful indication of malignancy.[9,10] Although similar studies describing CT characteristics of canine adrenal masses are not presently available, CT imaging of the adrenal mass may provide some useful additional information in our patients and should probably be considered if the mass is large, the ultrasonographic appearance is questionable or rapid growth occurs.

Metastases to the adrenal glands are common in people, but have not been extensively reported in veterinary patients. However, in one study describing dogs with malignant neoplasia undergoing necropsy examination, metastatic lesions within the adrenal glands were noted in 21% of cases. Most of the primary tumors were carcinoma-type, and many were extra-abdominal in origin (pulmonary, mammary). Gross lesions were noted in 40% of affected dogs, and over half of those with adrenal metastases were affected bilaterally. Based on these findings, metastatic disease is a realistic differential for dogs with known malignant diseases or if masses are noted in both adrenal glands.

Prior to modern advances in imaging, guided aspiration of the mass was routinely performed in human medicine. This would be technically challenging in small dogs and carries the risk of hemorrhage and tumor seeding along the needle tract. In addition, cytological findings may be equivocal or complicated by necrosis or hemorrhage. However, it may be helpful to aspirate a large mass if information about malignant potential is needed to make a treatment decision or if metastatic or infectious disease is suspected.

Assessing functional status

An adrenal mass may arise from the medulla or the cortex. Medullary secretory tumors release catecholamines and are called pheochromocytomas. Cortical masses may release any of the steroid hormones or their various precursors. The most common product from a functional adrenal cortical tumor is cortisol, but aldosterone or one of the sex hormones may be produced instead.

Although dogs with functional adrenal tumors (both medullary and cortical) are expected to present with related clinical signs, this may not always be the case. In human beings, some reports suggest that up to 10% of patients with incidentalomas are found to have clinical manifestations of adrenal disease when evaluated more closely.[10] As these individuals are usually undergoing imaging because of unrelated symptoms and signs, indications of adrenal dysfunction may be overlooked initially or only identified after targeted questioning or examination.

Pheochromocytomas episodically release large amounts of catecholamines, resulting in intermittent hypertension or tachycardia. Symptoms reported in people include palpitations, pallor and positional hypotension: all of these manifestations will be missed in veterinary patients. As catecholamine release may be episodic, repeated measurements of blood pressure (BP) and heart rate (HR) may be necessary. In addition, repeated electrocardiograms (ECG) may help identify suggestive arrhythmias. The classic diagnostic test for pheochromocytoma in people is the measurement of 24 hour urinary catecholamines and their metabolites (metanephrine, normetanephrine, and vanillylmandelic acid).[1] However, these assays are not routinely performed by veterinary laboratories, as they are historically unreliable and normal ranges have not been established for dogs. In addition, the urine must be immediately acidified and refrigerated to prevent decay of the target metabolites, and secretion may be episodic. In the 1990s, interest arose in the human field in using plasma chromogranin A (CgA) concentrations to identify patients with pheochromocytoma. CgA is an acidic protein which is stored and released with various peptide hormones, including the catecholamines.[11] In people with familial pheochromocytoma, the CgA concentration may predict the size of the tumor, although renal failure may lead to false positives due to impaired clearance. Cross reactivity exists between canine and human CgA, and experimental reports suggest that reliable data may be obtained in dogs using human ELISA kits.[12] However, the author is not aware of any reports describing CgA concentrations in dogs with any from of neuroendocrine tumor. More recently, measurement of plasma levels of free metanephrines have been shown to be more sensitive than other biochemical markers in human patients, and avoids some of the limitations associated with the other test options.[13] Again, this diagnostic tool has not been evaluated in the dog. Canine pheochromocytomas are usually slow growing but may be vascular and are generally regarded as inherently malignant lesions. In addition to intermittent hypertension and tachycardia, sudden death has been associated with pheochromocytomas in dogs, either from acute cardiovascular issues or hemorrhage from the tumor. A patient with an incidentaloma and episodic hypertension or tachycardia should be regarded as a likely candidate for this diagnosis and surgical evaluation should be recommended. Pre-operative control of hypertension and pre-treatment with phenoxybenzamine may reduce intra-surgical complications and improve patient survival.[14]

Functional adrenal cortical tumors classically release cortisol and are expected to produce the signs of Cushing's syndrome / hyperadrenocorticism (HAC), e.g., polyuria, polydipsia, polyphagia, trunkal alopecia. In the early stages, owners may overlook some of these changes or attribute them to age, and careful questioning may be necessary to elicit this information. Careful attention to hair growth patterns, such as delayed regrowth after grooming or shaving, may be noted before true alopecia occurs. Routine laboratory evaluation may provide early evidence of cortisol excess, and this possibility should certainly be investigated further in dogs with supportive findings, such as dilute urine, proteinuria, hypercholesterolemia and increased alkaline phosphatase activity. In human medicine, patients are sometimes described as having 'subclinical' Cushing's; these patients do not report or manifest the classic picture but are shown to have autonomous secretion of cortisol. The term subclinical autonomous glucorticoid hypersecretion or preclinical Cushing's syndrome has been proposed for this scenario, which is the most frequent hormonal abnormality detected in people with incidentalomas.[1] It has been suggested that screening for adrenal cortical autonomy should be performed as part of the routine evaluation of all human patients with an incidental adrenal mass. This may not be a useful guideline for our patients unless there is clinical or laboratory evidence of adrenal dysfunction, or a concern about possible post-operative hypocortisolemia in a patient scheduled for adrenalectomy.

Although there is long-standing controversy about the merits of the various screening tests for HAC in dogs, the urine cortisol: creatinine ratio or the low dose dexamethasone suppression test (LDDST) are probably more sensitive than the adrenocorticotrophic hormone (ACTH) stimulation test in the identification of autonomous cortisol release from an incidentaloma. In fact, in earlier reports only about 60% of dogs with clinical signs of HAC and a confirmed function adrenal cortical tumor had an exaggerated response to exogenous ACTH.[15,16] This may be due to compromised expression of ACTH receptors by neoplastic cells or deranged intracellular synthetic and secretory pathways resulting in the release of precursor molecules rather than cortisol.

Failure to suppress cortisol production at 8 hours following administration of 0.01 mg/kg dexamethasone (i.e., the LDDST) supports autonomous release of cortisol. An elevated urine cortisol: creatinine ratio also supports excessive cortisol secretion but is less specific than the LDDST and should not be used as sole evidence of an functional adrenal cortical tumor. As the incidence of malignancy in dogs with confirmed functional cortical tumors is about 50%, documenting autonomous cortisol secretion increases the likelihood that the incidentaloma is malignant and should certainly prompt surgical evaluation.

In addition to cortisol, functional adrenal cortical tumors may release any of the corticosteroid precursors, androgens, estrogens, or aldosterone. In addition, a single adrenal cortical tumor may produce multiple hormones or precursors. In general, autonomous release of precursors to cortisol results in findings similar to those of classic HAC (i.e., changes in thirst or hair growth). Release of one of the sex hormones may result in very specific changes such as vulval enlargement in a spayed female dog, gynecomastia in a castrated male dog, or perianal adenomas in non-intact animals. An ACTH stimulation test with measurement of sex hormones and cortisol precursors should be performed if the clinical findings or history suggest adrenal dysfunction but the LDDST does not indicate autonomous release of cortisol. In fact, some dogs with increased precursor levels have suppressed cortisol production, and may need peri-operative glucocorticoid support until the contra-lateral gland returns to full function.[17]

Clinical signs secondary to aldosterone secretion may be subtle and include variable polydipsia, weight gain due to fluid retention, or retinal hemorrhage secondary to hypertension. Most veterinary reports of confirmed hyperaldosteronism have described hypernatremia and / or hypokalemia.[18,19,20] This is in contrast to the human field, in which the majority of patients with aldosterone-secreting tumors are diagnosed prior to the development of these classic electrolyte changes as part of a routine evaluation for essential hypertension.[21] If the serum electrolytes in a dog with an incidentaloma suggest hyperaldosteronism, serum aldosterone levels, both before and after the administration of exogenous ACTH may be measured. In human patients, definitive confirmation of hyperaldosteronism requires measurement of plasma renin activity or a fludrocortisone suppression test. Unfortunately, measurement of renin is not routinely performed in the dog and the reliability of the fludrocortisones suppression test has not been reported in this species. However, substantially elevated aldosterone levels in a euvolemic dog on a balanced diet would be highly suggestive of this disorder.

SUMMARY

Although there is limited information about the ultimate outcome for most dogs with an incidentaloma, it seems likely that the majority of incidental adrenal masses in this species are both benign and non-functional. However, each patient merits careful evaluation as early signs of endocrinopathy may be overlooked and appropriate tests should be considered if evidence of adrenal dysfunction is discovered. As endocrine activity is independent of malignant potential, advanced imaging or repeated evaluation may be necessary to identify malignant lesions and facilitate prompt intervention. A suggested algorithm for evaluating these patients is provided in Figure 1, but is intended as a general guideline only.

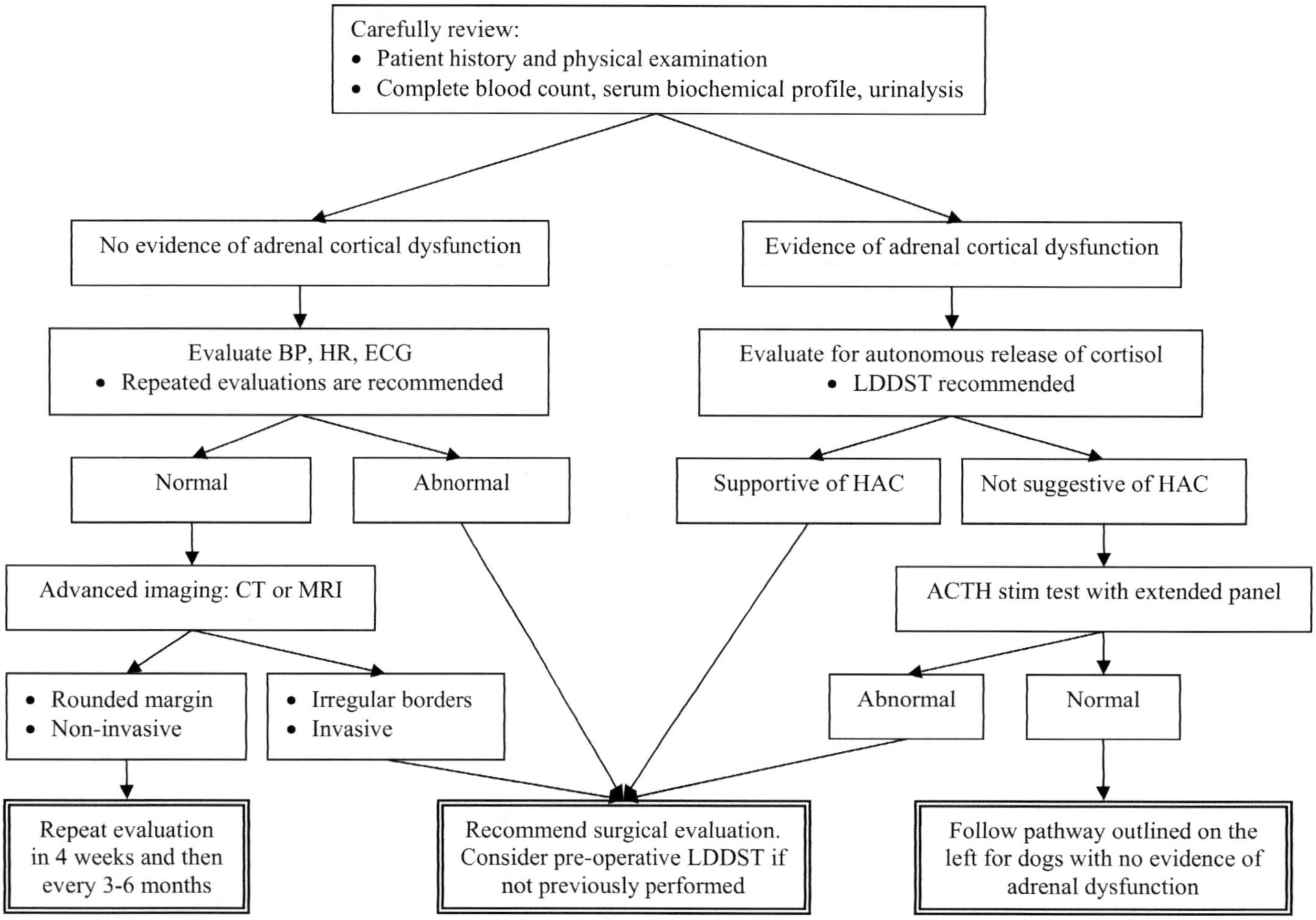

Figure 1. Suggested algorithm for a dog with unilateral incidental adrenal mass. See text for explanation of abbreviations

REFERENCES

[1]Young WF Jr. **New Eng J Med** 2007:356;601. [2]Higgins JC. **Clin Fam Pract** 2002;4:505. [3]Barzon L, et al. **J Urol** 2000;163:398. [4]Cook DM, et al. **Am J Med** 1996;101:88. [5]Ooi TC. **Can Med Assoc J** 1997;157:903. [6]Felman EC, et al. In: Feldman and Nelson. **Canine and Feline Endocrinology and Reproduction**. 3rd ed. St. Loius, Mo.: Elsevier, 2004:458. [7]Hoerauf A, et al. **J Am An Hosp Assoc** 1999;35:193. [8]Von Dehn BV, et al. **J Am Vet Med Assoc** 1995;207:322. [9]Mester U, et al. **J Nucl Med** 2006;47:32. [10]Mansmann G, et al. **Endo Rev** 2004;25:309. [11] Hsiao RJ, et al. **Am J Med** 1990;88:607. [12]Akiyoshi H, et al. **Am J Vet Res** 2005;66:1830 [13]Pacak K, et al. **Ann Int Med** 2001;134:315. [14]Herrara MA., et al**. J Vet Int Med** 2008;22:1333. [15]Peterson ME, et al. **J Am Vet Med Assoc** 1982;180:542. [16]Reusch CE, et al. **J Vet Int Med** 1991;5:3. [17]Syme HM, et al**, J Am Vet Med Assoc** 2001;219:1725. [18]Breitschwerdt EB, et al. **J Am Vet Med Assoc** 1985;187:841 [19]Johnson KD, et al. **J Vet Med Series A** 2006;53:467. [20]Behrend EN, et al. **J Am Vet Med Assoc** 2005;226:1662. [21]Fardella CE, et al. **J Clin Endo Met** 2000;85:1863.

RENAL HYPERTENSION: IS THE RENIN-ANGIOTENSIN-ALDOSTERONE SYSTEM RESPONSIBLE?

Harriet Syme BSc BVetMed PhD FHEA DACVIM DECVIM-ca MRCVS
London

Epidemiology of Renal Hypertension in Cats

Most cats that are diagnosed with hypertension (including those that have target organ damage, such as blindness) also have azotaemic chronic kidney disease (CKD).[1,2,3,4,5] It is generally assumed that in these cats hypertension has developed secondary to renal disease, rather than hypertension being the antecedent event. However, since in many cats the diagnoses of hypertension and renal failure are made concurrently, it is difficult to determine cause and effect.

Although azotaemic CKD has been diagnosed in the majority of the reported cases of hypertension in the cat (it is present in 70-75% in the series reported to date)[2,3,4,5] the converse is not true; many cats with azotaemic CKD have normal blood pressure.[6] It has also been observed that most cats presented with signs of systemic hypertension have relatively mild CKD. This observation is of interest because if the hypertension is directly caused by the renal failure the risk would be expected to increase as GFR progressively declines, and this does not seem to be the case. About 20% of cats diagnosed with systemic hypertension have idiopathic disease; they are non-azotaemic, non-hyperthyroid and do not have an alternative, obvious cause for their hypertension.[3,7] It is unclear whether in these cats hypertension could be associated with sub-clinical (i.e. non-azotaemic) CKD.

Epidemiology of Renal Hypertension in Humans

In humans, although hypertension is considered to be primary in 90-95% of cases, CKD is the most common cause of secondary hypertension. It is accepted that the kidneys can be both the 'villain and the victim' of hypertension. However, although the fact that malignant hypertension causes renal failure in humans has been accepted for decades, the importance of mild-to-moderate or 'benign' hypertension as a cause of renal failure is still debated.[8] In fact, there is now good evidence that hypertensive nephrosclerosis is over-diagnosed, especially in African Americans.[9] Although nephrologists generally agree that elevated systemic blood pressure exacerbates all forms of CKD in humans, speeding progression to end-stage renal disease (ESRD), the epidemiological evidence supporting mild to moderate essential hypertension as an initiator of kidney damage has always been weak. It may often be that the kidney injury generates the high blood pressure rather than the other way around.[10]

Guytonian Theory

Impairment of renal sodium excretion is assumed to be the central mechanism in the development of hypertension in animals with renal disease.[11] This is supported by a longitudinal study of a small cohort of human patients in which it was shown an increase in cardiac output preceded any changes in total peripheral resistance (TPR).[12] This change was accentuated in the subjects with the most severe renal failure. These findings appear to support Guyton's proposal that a persistent elevation in arterial pressure can only occur when there is a shift in the 'renal function curve' with disruption of the pressure-natriuresis relation.[13] The plasma volume of cats with CKD has not been systematically studied. In one very small study, plasma volume in four hypertensive cats with CKD (33.1 ml/kg) was not significantly different from that of the young, normal controls (29.3 ml/kg).[14] No normotensive cats with CKD were included for comparison in that study. Further characterisation of plasma volume in cats with CKD is required before conclusions can be drawn. However, blood volume is not the sole determinant of blood pressure in patients with renal disease. Of human patients with ESRD managed by dialysis, only about half are considered to be 'volume dependent' with hypertension that responds to ultrafiltration.[15] In the remainder other pathophysiological mechanisms, in particular activation of the renin-angiotensin system (RAS), are presumed to be of greater importance.

Role of the Renin-Angiotensin System

The RAS plays an integral role in maintaining normal BP, influencing both plasma volume and peripheral resistance. Human patients with ESRD that do not have a decrease in BP following ultrafiltration (and resultant contraction of circulating volume) have been termed 'renin-dependent' because many respond well to treatment with angiotensin converting enzyme inhibitors (ACE-inhibitors) or nephrectomy.[15] Correspondingly, components of the RAS are inappropriately increased (relative to plasma volume) in many patients with ESRD. This relation is not, however, always evident in patients with less severe renal failure, with no apparent increase in some studies. In spite of this, treatment with ACE-inhibitors may still produce beneficial haemodynamic effects. For example, in humans with diabetic nephropathy treatment with angiotensin II blockers produces a marked increase in renal blood flow, despite low plasma renin activity.[16] It is proposed that this effect is due to the inhibition of locally produced angiotensin II within the renal interstitium.

The role of the RAS in the development of hypertension in cats with CKD is unclear. It has been demonstrated in experimental studies using the partial nephrectomy model that cats respond to treatment with ACE-inhibitors with a drop in systemic BP as would be expected, although the change in BP that occurs is very small.[17] Favourable modulation of glomerular haemodynamics (decreased glomerular capillary pressure) was also demonstrated. These results support a role for RAS activation in the (relatively mild) hypertension that occurs in the remnant kidney model. However, whether these studies can be extrapolated to cats with naturally occurring disease is as yet unknown. The response of spontaneously hypertensive cats to treatment with ACE-inhibitors is generally poor, at least in terms of lowering systemic BP. In accordance with this clinical observation, plasma renin activity (PRA) has been normal or low in almost all of the renal hypertensive cats in which it has been measured.[18] In normal, healthy cats, PRA and aldosterone concentrations are reportedly very similar to values found in humans.[19]

Table 1: Response to RAS Inhibition in Feline Models of CKD

Drug & Response (mm Hg)	Δ	Reference
ACE-i		
SBP 115.3 ± 0.6 vs. 122.6 ± 0.3 (control, P=0.003)	7.3	Miller (1999)[20]
SBP 110.9 ± 1.0 vs. 127.2 ± 0.9 (PKD, P=0.003)	6.3	
SBP 136.5 ± 2.1 vs. 147.4 ± 3.1(P<0.05)	10.9	Brown (2001)[21]
SBP 129 ± 10 vs. 133 ± 11 (control, ns)	4	Mathur (2004)[22]
SBP 145 ± 9 vs. 153 ± 11 (remnant wrap, ns)	8	
Angiotensin Receptor Blockers (ARBs)		
Failed to block pressor response to AT I	7	Reynolds (2002)[23]
SBP102 ± 3 vs. 109 ± 4 (control, ns)	13	Mathur (2004)[22]
SBP140 ± 9 vs. 153 ± 11(remnant wrap, ns)		

Although the effects of RAS stimulation on hypertension and renal injury have been generally ascribed to the actions of angiotensin II, there is increasingly substantial evidence that aldosterone may also play a key role.[24] It has been shown that administration of exogenous aldosterone to rats results in hypertension and renal injury.[25] Recent studies have also demonstrated that many of the pathologic changes in the rodent remnant kidney model (hypertension, glomerulosclerosis and proteinuria for example) can be replicated by infusion of aldosterone, even when the actions of angiotensin II are pharmacologically inhibited.[26] Similarly, Quan and others demonstrated that in rodents that were partially nephrectomised, concurrent adrenalectomy ameliorated the resultant hypertension. [27] The results of these experiments suggest that aldosterone may be an important cause of hypertension in animals with CKD. Some, but not all, studies have shown increased plasma aldosterone concentrations in humans with renal failure.[28] Aldosterone concentrations were increased when compared with normal, healthy cats in one study of cats with renal hypertension suggesting that this hormone may be of key pathogenic significance.[29] A further study, not yet published in full, appeared to indicate that aldosterone concentrations were higher in hypertensive cats than in normotensive cats with similar degrees of renal dysfunction.[18]

Primary Hyperaldosteronism

Aldosterone (and other steroids with mineralocorticoid activity) exerts its main effects in the connecting tubule and collecting ducts where it increases the reabsorption of sodium and increases the excretion of potassium and hydrogen ions. Retention of sodium in turn results in plasma volume expansion, leading to an increase in cardiac output and hypertension. Primary aldosteronism is the term given to unregulated and excessive secretion of aldosterone by the adrenal cortex. Hyperaldosteronism, due to a functional adrenal tumour, has been recognised increasingly frequently in the cat. Interestingly, many of the cases have presented for signs related to hypokalaemia and not due to symptomatic hypertension.[30,31,32] Blood pressure measurements were not reported in all cases but were normal in at least some of these animals. The renal escape from the sodium-retaining effect of aldosterone is a well-recognised physiological phenomenon; the mechanisms known to contribute include increased renal interstitial hydrostatic pressure, endogenous release of renal autacoids, decreased sympathetic activity, and increased atrial natriuretic peptide levels.

Excessive aldosterone production as a cause of renal hypertension is classically an appropriate response to aberrant activation of the renin-angiotensin system (RAS), in which case it is referred to as secondary aldosteronism. Although plasma aldosterone concentration is reportedly slightly increased in cats with hypertension and renal failure, surprisingly plasma renin activity tends to be normal or even decreased in these animals suggesting that the disease may be primary.[18,29] The underlying cause for the hyperaldosteronism is unknown but this correlates with the clinical observation that hypertensive cats tend to have lower plasma potassium concentrations than normotensive cats with CKD.[6,33] Javadi and colleagues (2005)[34] reported that many hypertensive cats have micro-nodular hyperplasia of the adrenal cortex and suggest that this provides further evidence for primary hyperaldosteronism in this species. However, in a recent study we examined the adrenal glands of 67 cats (37 hypertensive and 30 normotensive) and found that adrenal hyperplasia was present in almost all of them (61/65).[35] Additionally two cats had adrenocortical adenomas. The degree of hyperplastic change did not differ between the hypertensive and normotensive cats.

REFERENCES

[1]Littman MP. (1994) **J Vet Intern Med** 8: 79-86. [2]Sansom J, et al. (1994) **J Small Anim Pract** 35: 604-611. [3]Maggio F, et al. (2000) **J Am Vet Med Assoc** 217: 695-702. [4]Elliott J, et al. (2001) **J Small Anim Pract** 42: 122-129. [5]Jepson RE, et al. (2007) **J Vet Intern Med** 21: 402-409. [6]Syme HM, et al. (2002) **J Am Vet Med Assoc** 220(12):1799-804. [7]Elliott J, et al. (2003) **J Vet Intern Med** 17: 754. [8]Luke RG.(1999)14: 2271-2278. [9]Freedman BI and Sedor JR (2008) **J Am Soc Nephrol** 19: 2047-2051. [10]Kestenbaum B, et al. (2008), **Ann Intern Med** 148: 501-508. [11].Navis G, et al. (2000). **Renal Disease**; 524-530. [12]Brod J, et al. (1983) **Clin Sci (Lond)** 64: 141-152. [13]Guyton AC. (1987) **Hypertension** 10: 1-6. [14]Hogan DF, et al. (1999) **J Vet Intern Med** 13: 249-249. [15]Misra M, et al.

(2000) 531-539. [16.]Price DA, et al. (1999) **J Am Soc Nephrol** 10: 2382-2391. [17.]Brown SA, et al. (2001) **Am J Vet Res** 62: 375-383. [18.]Syme HM, et al. (2002) **J Vet Intern Med** 16: 354-354. [19.]Javadi S, et al. (2004) **J Vet Intern Med** 18(5):625-31. [20.]Miller RH, et al. (1999) **Am J Vet Res** 60(12):1516-25. [21.]Brown SA, et al. (2001) **Am J Vet Res** 62(3):375-83. [22.]Mathur S, et al. (2004) **Am J Vet Res** 65(7):1006-13. [23.]Reynolds V, et al. (2002) **J Vet Intern Med** 16 (3): 341. [24.]Epstein M. (2001) **Am J Kidney Dis** 37: 677-688. [25.]Gomez-Sanchez EP, et al. (1996) **Steroids** 61: 184-188. [26.]Greene EL, et al. (1996) **J Clin Invest** 98: 1063-1068. [27.]Quan ZY, et al. (1992) **Kidney Int** 41: 326-333. [28.]Hene RJ, et al. (1982) **Kidney Int** 21: 98-101. [29.]Jensen J, et al. (1997) **Am J Vet Res** 58: 535-540. [30.]MacKay et al. (1999). **J Feline Med Surg** 1:117-122. [31.] Moore et al. (2000) **J Am Vet Med Assoc** 217: 213-215. [32.]Ash RA, et al. (2005) **J Feline Med Surg** 7(3):173-82. [33.]Sansom J, et al (2004) **Am J Vet Res** 65(2):245-52. [34.]Javadi S, et al. (2005) **Domest Anim Endocrinol** 28(1):85-104. [35.]Keele SJ et al. (2009) **J Vet Intern Med**23(6):1328.

PATHOGENESIS OF SECONDARY HYPERPARATHYROIDISM

Patricia A. Schenck, DVM, PhD
Lansing MI

RENAL SECONDARY HYPERPARATHYROIDISM

Secondary hyperparathyroidism occurs most commonly in association with chronic renal failure (CRF). Secondary hyperparathyroidism is characterized by elevated circulating parathyroid hormone (PTH) concentration with a serum ionized calcium (iCa) concentration either within the reference range, or below. While the serum iCa concentration is normal to low, serum total calcium (tCa) concentration is often elevated in CRF. The incidence of elevated serum tCa increases with the severity of azotemia. In 73 cats with CRF, serum tCa was increased in 8% with mild azotemia, 18% with moderate azotemia, and in 32% with severe azotemia.[1] However, serum iCa concentration does not show a strong association with the degree of azotemia.[2] Deleterious effects of hypercalcemia occur only if there are increases in serum iCa concentration, as iCa is the biologically active fraction of calcium.

Measurement of serum iCa concentration to assess calcium status in patients with CRF is crucial. Serum tCa should not be used, as serum tCa measurement incorrectly assessed iCa status in 36% of dogs[3] and 32% of cats[4] with CRF. In 490 dogs with CRF, based on serum tCa, 22% were hypercalcemic, 60% were normocalcemic, and 18% were hypocalcemic. When serum iCa was measured, 9% were hypercalcemic, 55% were normocalcemic, and 36% were hypocalcemic.[3] In 102 cats with CRF, serum tCa identified 19% as hypercalcemic, 71% as normocalcemic, and 10% as hypocalcemic. When serum iCa measurement was used, 30% were hypercalcemic, 60% were normocalcemic, and 10% were hypocalcemic.[4] The use of serum tCa "adjustment" formulas should also be abandoned. The use of the "adjusted tCa" value incorrectly assessed iCa status in approximately 53% of dogs and in 35% of cats with CRF. Serum tCa measurement or adjusted tCa measurement overestimates hypercalcemia and underestimates hypocalcemia in dogs; in cats with CRF, serum tCa measurement overestimates normocalcemia and underestimates hypercalcemia. Thus, iCa concentration must be directly measured to accurately assess calcium status in patients with CRF.

In 490 dogs with CRF, serum tCa was elevated in 22%, yet iCa concentration was elevated in only 9% of dogs.[3] This apparent discrepancy between tCa and iCa in dogs with CRF is due to an increase in the complexed calcium fraction. In CRF, organic anions such as citrates, phosphates, lactates, bicarbonates, and oxalates are capable of complexing with calcium. The complexed fraction of calcium accounted for 24% of serum tCa in dogs with CRF and elevated tCa, as compared to 11% in dogs with CRF and low serum tCa.[5]

EFFECTS OF PTH

PTH is a major uremic toxin. The toxic effects of PTH on tissues are most evident on bone (fibrous osteodystrophy, "rubber jaw"), but many tissues are adversely affected by high levels of circulating PTH. Initially, PTH increases as an adaptive mechanism for the control of calcium concentrations, but increased PTH levels become maladaptive when concentrations increase beyond a threshold of 3 to 5 times baseline. When PTH concentration reaches above 5 times baseline, severe bone disease occurs, and the threshold for injury to other tissues is reached. Some of the toxicity attributed to increased PTH may be directly due to a lack of adequate calcitriol, as there are calcitriol receptors in most tissues.

Central nervous system depression and detrimental effects on the peripheral nerves are common in CRF, caused in large part by excess PTH. Part of the anemia seen in CRF may also be caused by PTH excess. Abnormal lipid and carbohydrate metabolism can also be caused by excess PTH, and could contribute to the anorexia of uremia. Skeletal muscle myopathy causing weakness during uremia may be caused by both excess PTH, and by lack of calcitriol receptor activity within muscle cells. Some of the reduced physical activity of uremic dogs and cats may be attributed to bone changes of excess PTH. Obvious bone pain in animals is rare despite the presence of histologic bone lesions in chronic renal failure. Bone is the earliest tissue to exhibit histologic evidence of hyperparathyroidism perhaps due to its high concentration of PTH receptors. Increased renal cellular calcium content caused by increased PTH is toxic and has a role in the progression of chronic renal disease. Renal tubular cells have high concentrations of the PTH receptor and are among the body tissues affected earliest by excess PTH.

DEVELOPMENT OF RENAL SECONDARY HYPERPARATHYROIDISM: TRADITIONAL UNDERSTANDING

Parathyroid hormone (PTH) is the principal hormone involved in the minute-to-minute fine regulation of blood calcium concentration. It acts directly by influencing target cells primarily in the bone and kidney, and indirectly in the intestine to maintain circulating calcium at a sufficient level to ensure optimal functioning of cells. In general, the most important biological effects of PTH are to: (1) increase blood calcium concentration; (2) increase tubular reabsorption of calcium, resulting in diminished calcium loss into the urine; (3) increase bone resorption and numbers of osteoclasts on bone surfaces; and (4) in the kidney, to accelerate formation of calcitriol (1,25-dihydroxyvitamin D), the principal active vitamin D metabolite, through a trophic effect on 1α-hydroxylase enzyme in epithelial cell mitochondria in the proximal convoluted tubules.[6] Serum PTH, calcitriol, phosphorus and calcium concentrations are the principal regulators for renal calcitriol synthesis. The ability of PTH to enhance the renal reabsorption of calcium is of considerable importance in the maintenance of calcium homeostasis. The kidney serves an important role in the degradation and excretion of PTH, and it normally serves as the major source for clearance of carboxyl-terminal fragments of PTH by glomerular filtration.

Renal secondary hyperparathyroidism occurs when PTH synthesis and secretion becomes excessive as a result of kidney disease, and is the result of increased secretion of PTH by each parathyroid cell as well as increased number of cells due to parathyroid hyperplasia. A calcitriol deficit in uremic patients is the most important factor leading to the uncontrolled secretion of PTH. In renal disease, there are fewer healthy proximal tubule cells containing the mitochondrial 1α-hydroxylase enzyme system necessary to form calcitriol from precursor 25-hydroxyvitamin D. Nephron loss during CRF is estimated most commonly by the magnitude of increased serum creatinine. An association of increasing serum creatinine with diminished serum calcitriol in dogs has been shown. Decreased blood calcitriol lowers intestinal calcium absorption leading to hypocalcemia. As iCa concentration falls, the secretion of PTH is stimulated. In early CRF, a modified version of the calcitriol "trade-off" hypothesis emphasizes the role of deficits of calcitriol caused mostly by phosphorus inhibition of calcitriol synthesis. The increased PTH concentration can restore calcitriol and iCa in early stages of CRF when enough proximal tubular cells remain that are capable of calcitriol synthesis.

As glomerular filtration rate is further reduced in late chronic renal failure, greater increases of serum phosphorus occur, so that mass law interactions contribute to a decrease of iCa, stimulating further PTH production. A greater reduction in the activity of the 1α-hydroxylase responsible for calcitriol synthesis occurs as a consequence of the markedly increased serum phosphorus. In addition, the absolute loss of most of the proximal tubular cells makes adequate synthesis of calcitriol no longer possible. At this point, the markedly elevated PTH concentration that ensues is no longer able to restore calcitriol concentrations to normal.

TERTIARY HYPERPARATHYROIDISM

Tertiary hyperparathyroidism refers to the condition of a subset of patients with CRF that develop elevated iCa concentration and excessive PTH secretion that is not inhibited by high serum iCa concentration. It is likely that such patients had high PTH concentrations in association with normal or low serum iCa concentration (renal secondary hyperparathyroidism) earlier in the clinical course of CRF. Autonomous secretion of PTH from the parathyroid gland is unlikely, but the set-point for PTH secretion may be altered in CRF such that higher concentrations of iCa are necessary to inhibit PTH secretion.[7] Decreased serum calcitriol concentration, decreased number of calcitriol receptors in the parathyroid gland, and decreased calcitriol-vitamin D receptor interactions caused by uremic toxins may contribute to this increase in set-point. Decreased levels of the calcium receptor which both establishes the set-point and depends on calcitriol functionality for synthesis of its mRNA from parathyroid cell DNA may also contribute to the altered set-point. Dogs with elevated iCa concentration and marked hyperparathyroidism may have greater calcitriol receptor deficits in their parathyroid cells, which leads to poorly controlled PTH synthesis and parathyroid gland hyperplasia.[8] Deficient calcitriol functionality caused by vitamin D receptor deficits may also lead to calcium receptor deficits and the set-point elevations involved in tertiary hyperparathyroidism.

FIBROBLAST GROWTH FACTOR-23

Fibroblast growth factor (FGF)-23 is a recently discovered regulator of calcium and phosphorus metabolism, regulating phosphate homeostasis and calcitriol blood levels.[8] FGF-23 is a 251 amino acid protein that is synthesized and secreted primarily by osteoblasts. FGF-23 exerts its biological effects through activation of FGF receptors (FGF-R) as a Klotho/FGF-R complex. Klotho is a transmembrane glucuronidase capable of hydrolyzing steroid β-glucuronides. Klotho gene expression is only present in renal tubular cells, parathyroid cells and the choroid plexus. In the presence of Klotho, FGF-23 induces phosphorylation and activation of the extracellular signal-regulated kinase (ERK), which regulates an intracellular signal-transduction pathway.

FGF-23 is a major regulator of phosphorus, mediating phosphaturia. Phosphate reabsorption by the kidney is mediated through the Na/Pi IIa cotransporter on proximal tubular cells. FGF-23 reduces the abundance of the Na/Pi IIa cotransporter, decreasing the phosphate that is reabsorbed by the proximal tubules. If FGF-23 becomes inactivated, hyperphosphatemia occurs. FGF-23 may also inhibit gastrointestinal absorption of phosphate by reducing Na/Pi IIb cotransporter activity in the intestine. Humans with tumor-induced osteomalacia have an elevated circulating FGF-23 concentration with hypophosphatemia, severe muscle weakness, and bone pain. Tumor resection normalizes the FGF-23 concentration and phosphate concentration within hours. There are also genetic defects that lead to overstimulation of serum FGF-23 levels.

FGF-23 and PTH stimulate phosphaturia in a similar manner, but the relationship between FGF-23 and PTH regulation is still unclear. FGF-23 and Klotho are expressed in parathyroid glands, suggesting that PTH secretion may be regulated by FGF-23. There is evidence suggesting that FGF-23 decreases PTH mRNA transcription and PTH secretion in a dose-dependent manner. PTH may stimulate FGF-23 secretion by osteoblasts, as FGF-23 levels are increased in rodents with primary hyperparathyroidism.

FGF-23 decreases renal expression of 1α-hydroxylase, decreasing calcitriol synthesis, and increases the expression of 24-hydroxylase which controls calcitriol degradation. Less FGF-23 is required to decrease calcitriol synthesis, and decreases in calcitriol concentration occur prior to phosphaturia.[9]

The major physiologic function of FGF-23 is to regulate urinary phosphate excretion in accordance with dietary phosphate intake to maintain stable serum phosphorus levels. However, it is interesting that epidemiologic studies have shown no correlation between FGF-23 and serum phosphate levels in healthy patients. Clinical trials with phosphate loading have shown an increase in urinary phosphate excretion with either stable or slightly increased serum phosphorus concentrations.

FIBROBLAST GROWTH FACTOR-23 IN CRF

In CRF, FGF-23 concentrations increase gradually before an increase in serum phosphorus. In humans, there is a significant correlation between serum FGF-23 concentration and glomerular filtration rate. Since FGF-23 increases early in CRF, it may be a better indicator of early renal disease than other clinical parameters. It was originally thought that the increase in FGF-23 in CRF occurred as a result of decreased renal clearance. It is now thought that FGF-23 secretion is increased in CRF, and there is end-organ resistance to FGF-23 due to a deficiency of the Klotho cofactor. Klotho mRNA expression is decreased in kidneys of CRF patients. As a result of a downregulation of the Klotho/FGF-R complex in the parathyroid glands, an increase of circulating FGF-23 does not decrease PTH levels in CRF.[10]

In CRF, the increased secretion of FGF-23 may be a protective mechanism to stabilize serum phosphorus concentration with a decreasing number of nephrons. Thus FGF-23 increases urinary phosphate excretion, decreases gastrointestinal absorption, and decreases calcitriol synthesis through the inhibition of 1α-hydroxylase. These effects allow serum phosphorus to be maintained within a 'normal' range until CRF becomes more advanced.

The upregulation of FGF-23 in CRF results in early calcitriol deficiency which may initiate secondary hyperparathyroidism. Secondary hyperparathyroidism is typically the first metabolic complication noted in human patients with CRF, and occurs prior to hyperphosphatemia. In a small study of cats with CRF, there was no correlation between serum phosphorus concentration and PTH concentration; a number of cats with secondary hyperparathyroidism had normal serum phosphorus concentrations.[2]

Traditionally, secondary hyperparathyroidism has been thought to be the result of a loss of renal function with a direct reduction of 1α-hydroxylase, with a decrease in calcitriol production and oversecretion of PTH. However, with a reduction in tubular mass, both 1α-hydroxylase and erythropoietin secretion should be decreased, and evidence shows that erythropoietin secretion remains unaffected in earlier stages of CRF when decreased calcitriol is present. Through the actions of FGF-23, serum phosphorus concentration is maintained at the expense of calcitriol, and the decrease in calcitriol is greater than would be predicted from the numbers of nephrons lost. The early decrease in calcitriol stimulates PTH secretion and initiates secondary hyperparathyroidism.[8]

In several human studies, serum FGF-23 concentration was a better outcome predictor than serum phosphorus concentration in CRF. However, there have been no studies yet in dogs and cats to evaluate circulating FGF-23 concentrations in CRF.

At this time, there are no drugs available that directly lower FGF-23 concentrations. However, serum FGF-23 concentration can be indirectly lowered by the use of phosphate binders. From a few human studies, phosphorus binders appear to have beneficial effects beyond lowering of serum phosphorus concentration, but FGF-23 concentrations have not been evaluated in these studies.

REFERENCES

[1]Barber PJ et al. **J Small Anim Pract** 1998;39(3):108; [2]Davainis GM et al. **ESVIM/ESVE Meeting** 2001; [3]Schenck PA et al. **Am J Vet Res** 2005;66(8):1330; [4]Schenck PA and Chew DJ. **Can J Vet Res** 2010 (in press); [5]Schenck PA et al. **Am J Vet Res** 2003;64(9):1181; [6]Schenck PA et al. **Fluid Therapy in Small Animal Practice** 2005, 3rd ed: 122; [7]Goodman WG et al. **J Clin Endocrinol Metab** 1998;83(8):2765; [8]Seiler S et al. **Kidney International** 2009; 76(Suppl 114): 534 ; [9]Williams ME. Seminars in **Nephrology** 2009 ; 29(2) : 97; [10]Galitzer H. **Kidney International** 2009; December advanced online publication

DIFFICULT ESOPHAGEAL PROBLEMS: INFLAMMATION & STRICTURE

Michael Willard, DVM, DACVIM
Texas A&M University

ESOPHAGITIS

Esophagitis is much more common than many clinicians are aware. The difficulty partly arises from the fact that esophagitis can present with clinical signs that lead one to believe the dog is vomiting instead of regurgitating. Furthermore, mild esophagitis may only cause minor signs (mild regurgitation of mucus and phlegm) while severe esophagitis can cause so much pain that patients refuse to swallow water or even saliva. Because there can be so wide a range of clinical signs, it is easy to forget that esophagitis is a differential for a patient. It is critical to identify that esophagitis is present as delayed diagnosis can have serious clinical repercussions. Substantial inflammation of the esophageal mucosa causes muscular weakness by interrupting the reflex arcs within the esophagus and/or between the esophagus and the brain.[1] However, this weakness is not always reflected by finding megaesophagus. Most patients have very minor esophageal distention and yet can have major signs. Likewise, barium esophagrams can have relatively minor changes and not reflect the severity of the esophagitis. Esophagoscopy typically shows an edematous, reddened, bleeding esophageal mucosa, ± structure formation, making it the diagnostic method of choice to find esophagitis. However, in rare cases, there may be more subtle changes with thickening and discoloration (especially at the lower esophageal sphincter of cats).

Adding to this problem is the fact that there is such a wide range of causes of esophagitis. Severe esophagitis may be caused by anesthetic procedures in which animals are placed in dorsal recumbency and then have gastric acid pool in their esophagus for relatively long periods of time.[2,3] However, gastroesophageal reflux from any cause can be responsible.[4,5] Hiatal hernias occasionally are responsible for such reflux. Rare animals ingest caustic substances (e.g., lye), and some cats will lick caustic disinfectants off their fur. However, a surprisingly large number of animals are administered caustic substances by veterinarians. In particular, tetracyclines, NSAIDs, ciprofloxacin and clindamycin are recognized as having substantial potential to cause esophagitis. Pills and capsules are notorious for lodging in the esophagus of cats[6], and it is therefore not surprising that doxycycline is a recognized cause of esophageal stricture in cats. Esophagitis may also be secondary to any cause of protracted vomiting. In particular, parvovirus enteritis may cause such intense vomiting that esophagitis results. If a vomiting animal has the character of its vomitus change, which seems to suggest regurgitation, consider the possibility that esophagitis has occurred secondary to the persistent vomiting. Gastrinoma (a tumor which secretes gastrin and results in massive gastric acid secretion) also causes esophagitis because of the vast and unending amounts of acid the esophagus is exposed to as the dog continually vomits. Gastroesophageal reflux may be potentiated by, or even caused by, esophagitis (which may be caused by reflux in the first place). Thus, there may be a positive feedback loop, which can be hard to break (i.e., esophagitis causes more reflux which causes more esophagitis which causes more reflux which causes...). Rarely there can be spontaneous inflammation, as seen with eosinophilic esophagitis in dogs.[7] Brachycephalic dogs seem to have an increased incidence of gastroesophageal reflux, esophagitis and perhaps hiatal hernia.[8] Finally, esophageal foreign bodies typically cause varying degrees of esophagitis. The esophagus is far more susceptible to pressure necrosis from a foreign body than are the stomach or intestines.

You should seek to prevent further gastroesophageal reflux by keeping the stomach as empty as possible by using prokinetics such as metoclopramide or, preferably, cisapride. Studies in people show that cisapride is clearly more effective than metoclopramide. The only real advantage of metoclopramide is that it can be given by injection; a useful fact in animals that are regurgitating profusely. In addition, gastric acid secretion should be minimized and preferably abolished. H-2 receptor antagonists (e.g., cimetidine, ranitidine, famotidine) suppress gastric acid secretion, but they do not eliminate it. This is because they are competitive inhibitors. That means that there is constantly some degree of competition between the H-2 receptor antagonists and the stimuli for acid secretion. Omeprazole, lansoprazole and pantoprazole are non-competitive inhibitors of gastric acid secretion. Therefore, these drugs can be noticeably more effective and can cause near gastric anacidity.[9] You can try to achieve greater efficacy with the H-2 receptor antagonists by doubling or tripling their dose.

Sucralfate is of uncertain value in patients with esophagitis. Unless there is some gastric acid reflux into the esophagus (which you are desperately trying to stop in the first place), it is doubtful that the sucralfate is of much use. If you use it, it should be administered as a slurry.

A combination of omeprazole and cisapride seems to be the most effective medical treatment regime. Antibiotics are used to treat secondary infections, but nobody really knows if they do anything in this regard. Glucocorticoids have been thought to help retard fibrous connective tissue proliferation and cicatrix, but their effectiveness is uncertain (and they might predispose to infection). Placing a PEG tube seems to have some real advantages in patients with very severe disease. First, we will then know that the cisapride and omeprazole tablets will reach the stomach. Second, we will also know that the animal will receive its caloric and protein needs, and hopefully with less irritation to the esophagus than would have occurred otherwise.

If there is severe esophagitis, cicatrix may form and obstructions develop. Diagnosis of stricture is best accomplished by esophagoscopy IF the operator is familiar with such obstructions. It is surprisingly easy to pass a slender endoscope through a stricture and never recognize the stricture. It is also surprisingly easy to miss a partial obstruction due to a stricture with a barium esophagram. If you suspect a stricture and must use a barium esophagram to make the diagnosis, use barium mixed with solid food. Balloon-dilatation or bouginage is recommended if a stricture has occurred. Many animals need to have 2-6 dilatation procedures (all the while being treated for esophagitis), although some only need one procedure and some need more than 15. Do not try to resect the stricture unless you have had prior dilatation procedures fail.

Hiatal hernias may be more common than suspected. They can be difficult to diagnose, and one may need to put pressure on the abdomen during film exposure to try to push the stomach through the hernia and into the chest so that it can be diagnosed radiographically. Shar Pei's seem to have a relatively high incidence of hiatal hernias.

Gastrinoma is another cause of esophagitis. Dogs with this tumor usually vomit profusely and thus burn the esophagus by the large volumes of gastric acid that pass through it. Most affected dogs have vomiting, diarrhea, and weight loss. Documentation of increased serum gastrin concentration is necessary for diagnosis. Aggressive H-2 receptor antagonist therapy can be used, but omeprazole is preferred.

CICATRIX

Cicatrix (i.e., scarring) may occur after an episode of severe esophagitis from any cause (including foreign objects). It is particularly easy to miss this problem on a barium swallow if only liquid barium is used. If radiographs using liquid barium are nonrevealing, repeat the study using barium mixed with food, which is more likely to stop at a partial obstruction. Endoscopy is very good at finding these lesions; however, you must keep in mind the size of the patient as you evaluate the esophageal lumen. A partial stricture will be very obvious in a 10 lb dog or cat but may not be apparent in an 85 lb animal. Balloon-dilatation or bougienage is usually effective; it is also more likely to be successful than surgery and resection of the affected area. In general, surgical resection should be a last ditch resort and only used if esophageal ballooning[10] or bougienage[11] has failed despite repeated dilatations. However, you must use proper esophageal balloons because Foley catheters and endotracheal tubes with inflatable cuffs will often not allow you to dilate a dense or mature stricture. More difficult cases (i.e., those with extensive strictures or with concurrent severe esophagitis) may benefit from several newer techniques.

Endoscopic administration of intralesional steroids may help minimize reformation of the stricture.[12,13] Typically we inject1-2 ml of triamcinolone at the site of the stricture either before or after ballooning. Special needles are used to perform injections through the endoscope without damaging the biopsy/aspiration channel.

Another technique used in difficult strictures is to make 3-4 equidistant cuts into the stricture using an electrocautery device (i.e., either a snare or a knife) prior to ballooning. This helps the stricture to "break" open at multiple spots with the idea that there will be 3 or 4 smaller, less deep lacerations at the stricture site instead of one major, deep laceration that is more likely to restricture. However, you should not attempt to use electrocautery through an endoscope unless you have some training lest you cause too much trauma to the tissues or destroy your endoscopic equipment.

Another technique is to "paint" the site where the stricture was broken down with Mitomycin (NOT mithromycin C, there is a difference).[14] A 5 mg bottle is reconstituted and soaked up into a gauze sponge. Then this sponge is endoscopically placed on the site where the stricture was broken up for 5 min. Then it is flushed off with 60 ml of water.

Finally, for particularly difficult cases, stents may be placed in the esophagus.[15] Depending upon the type of stent, these must sometimes be sutured in place. Stents custom made for dogs may be obtained from the Infiniti corporation (http://www.infinitimedical.com/p_stents.html). The device that allows you to suture the stent in place by means of a flexible endoscope is made by Pare Surgical Inc (http://www.infinitimedical.com/p_stents.html).

The major point to remember is that if an animal starts to have problems days to weeks after anesthesia, consider strongly the possibility that an esophageal stricture has developed secondary to esophagitis.

REFERENCES

[1]Sellon RK, et al. Esophagitis and esophageal strictures. **Vet Clin N Am** 2003;33:945. [2]Wilson DV, et al. Postanesthetic esophageal dysfunction in 13 dogs. **J Am Anim Hosp Assoc** 2004;40:455. [3] Wilson D, et al. Effects of preanesthetic administration of morphine on gastroesophageal reflux and regurgitation during anesthesia in dogs. **Am J Vet Res** 2005;66:386. [4] Gualtieri M, et al. Reflux esophagitis in three cats associated with metaplastic columnar esophageal epithelium. **J Am Anim Hosp Assoc** 2006;42:65. [5] Han E, et al. Feline esophagitis secondary to gastroesophageal reflux disease: clinical signs and radiographic, endoscopic, and histopathologic findings. **J Am Anim Hosp Assoc** 2003;39:161. [6]Westfall DS, et al. Evaluation of esophageal transit of tablets and capsules in 30 cats. **J Vet Int Med** 2001;15:467. [7]Mazzei MJ, et al. Eosinophilic esophagitis in a dog. **J Am Vet Med Assoc** 2009;235:61. [8]Poncet C, et al. Prevalence of gastrointestinal tract lesions in 73 brachycephalic dogs with upper respiratory syndrome. **J Small Anim Pract** 2005;46:273. [9]Panti A, et al. The effect of omeprazole on oesophageal pH in dogs during anaesthesia. **J Small Anim Pract** 2009;50:540. [10]Leib MS, et al. Endoscopic balloon dilation of benign esophageal strictures in dogs and cats. **J Vet Int Med** 2001;15:547. [11]Bissett SA, et al. Risk factors and outcome of bougienage for treatment of benign esophageal strictures in dogs and cats: 28 cases (1995-2004). **J Am Vet Med Assoc** 2009;235:844. [12]Altintas E, et al. Intralesional steroid injection in benign esophageal strictures resistant to bougie dilation. **J Gastroenterol Hepatol** 2004;19:1388. [13]Fraune C, et al. Intralesional corticosteroid injection in addition to endoscopic balloon dilation in a dog with benign oesophageal strictures. **J Small Anim Pract** 2009;50:550. [14]Gillespie MG, et al. Role of mitomycin in upper digestive tract stricture. **Head and Neck** 2007;29: 12. [15]Fiorini A, et al. Self-expandable metal coil stents in the treatment of benign esophageal strictures refractory to conventional therapy: a case series. **Gastrointest Endoscop** 2000;52:259.

HARD TO RECOGNIZE / TREAT CAUSES OF REGURGITATION & VOMITING

Michael Willard, DVM, DACVIM
Texas A&M University

ACQUIRED ESOPHAGEAL OBSTRUCTION BESIDES CICATRIX

Esophageal foreign objects

Esophageal foreign objects usually consist of bones but may be rawhide treats, food, dental chew toys[1], toys, balls, rocks, wood, etc. They usually lodge at the thoracic inlet, base of the heart, or lower esophageal sphincter. A history of a patient that begins to regurgitate (as opposed to vomit) acutely is very suggestive of acquired esophageal obstruction due to a foreign object. These patients may continue to drink water, but they typically refuse solid food because the food bolus cannot pass by a partial esophageal obstruction and causes pain whenever it tries to. A casual, careless history that fails to raise the suspicion of regurgitation will typically lead the clinician to suspect an acute gastritis. However, the realization that the patient is regurgitating (as opposed to vomiting) should be a "red flag". Too often, a pet that has ingested a foreign object is treated conservatively while we wait and see if the supposed gastritis spontaneously resolves. This is problematic because foreign bodies can erode and perforate the esophagus much quicker than they would stomach or intestines.

Plain radiographs should be performed first. Bones are a common cause of obstruction, and plain films that are made with proper technique and then carefully evaluated are diagnostic in most cases. Remember that poultry bones are not as radiodense as the patient's bones, which means that excellent radiographic technique is required to see them. Foreign bodies in the esophagus can perfectly mimic pulmonary or mediastinal masses; you often cannot tell the difference with plain radiographs. If poor contrast in the region of the esophagus, pleural effusion or pneumothorax are seen, one must seriously consider esophageal perforation and mediastinitis. If plain films are not diagnostic, then contrast films can be performed. Barium provides better contrast, but iodide is safer if there is an unsuspected perforation. Esophageal perforation may occur at variable times after ingestion of a foreign object. Even a blunt object, if tightly lodged in the esophagus, can cause ischemia and perforation in 2-3 days. The prognosis for animals with esophageal perforation and severe mediastinitis is guarded to poor, depending upon their condition at the time of diagnosis.

Endoscopy is almost always the preferred method of removing foreign objects, but fluoroscopic[2] and surgical[3] techniques can be effective if the operator is well trained. Rigid endoscopes allow much more control of the foreign object and are preferred to flexible scopes for removal of these foreign objects. It is especially useful to be able to pull the object into a rigid endoscope and then withdraw it and the scope as a unit, thus protecting the esophagus. The main disadvantage of rigid endoscopes is that they are often not long enough in larger dogs. Finesse is required; brute force can easily lacerate/perforate the esophagus. If a large object or a bone cannot be easily dislodged, do not force it lest you perforate a previously intact esophagus; instead, you can use rigid equipment to "chew" it up and hopefully dislodge it. If that fails, passing a large Foley catheter behind the foreign object and inflating the balloon often helps; it distends the esophagus (thus freeing the foreign object) and then is used to pull the object out. If you cannot pull a foreign object out of the esophagus, you can try to push it into the stomach. However, do not push bones or other foreign objects into the stomach unless you are sure that it is smooth on the aborad side and will not further damage the mucosa. Finally, be careful if you insufflate the esophagus lest you rupture a weakened area in the mucosa and/or cause a fatal tension pneumothorax.

Fish hooks terrify many clinicians, but they can often be successfully removed endoscopically. Fish hooks have usually penetrated the mucosa (and sometimes the muscular tunics); you will often have to use rigid equipment to carefully force the tip of the hook back out of the mucosa. A small hole is left, but there are very seldom any complications. After removing the foreign object, retake plain chest radiographs to be sure that a pneumothorax (which would indicate a perforation) is not present. Antibiotics are indicated if there is substantial esophageal mucosal ulceration (and especially if you remove a fish hook which had been used with various baits that can harbor anaerobic bacteria). Depending upon the amount of damage, corticosteroids may be used to try to prevent cicatrix formation; however, it is not clear that they are effective. Rarely there may be severe hemorrhage.[4]

Tumors

Primary esophageal carcinomas occur rarely, although gastric carcinomas may spread into the lower esophagus. Most primary esophageal tumors are asymptomatic until they become very large. We have diagnosed a few primary esophageal tumors fortuitously when routine chest radiographs revealed a density in the diaphragmatic lung fields. Esophageal sarcomas are usually due to *Spirocerca lupi*.[5] Most esophageal tumors are secondary to mediastinal or thyroid tissues, and they cause esophageal obstruction by extramural pressure. Thyroid carcinomas may also invade the esophagus. The prognosis is usually poor. Leiomyomas of the lower esophageal sphincter are occasionally seen. Endoscopy is the best tool to find them. They are best seen when one looks at the lower esophageal sphincter from the stomach using a retroflexed view. It is worth looking for these tumors as they are potentially curable with timely surgery. In some cases, a small tumor at the site of the lower esophageal sphincter can cause gastroesophageal reflux with nausea and drooling.

Vascular ring anomalies

Congenital vascular ring anomalies are not always easy or straightforward. They may first be diagnosed when the patient is several years old. Sometimes they are only diagnosed if the patient ingests something that acts as a foreign body in the narrowed esophagus. In particular, it can be hard to distinguish vascular ring anomalies from an outpouching (not a true diverticulum) of the esophagus in the anterior mediastinal area due to weakness. If there is ever any doubt, endoscopy will confirm or deny the diagnosis and is much safer and easier than a thoracotomy.

HIATAL HERNIAS

Hiatal hernias may be more common than suspected. Shar Pei's seem to have a relatively high incidence of hiatal hernias.[6] They can be difficult to diagnose unless you know how to look for them. Sometimes seen on plain radiographs and simple barium contrast radiographs, the more occult cases sometimes need more aggressive diagnostics. Sometimes one must manually put pressure on the abdomen during film exposure to try to push the stomach through the hernia and into the chest so that it can be diagnosed radiographically. Endoscopic diagnosis is not always straightforward. You may need to put the endoscope into the stomach and retroflex it in order to see the abnormality. Even when found, the big question is whether the hiatal hernia is causing a problem or is an "innocent bystander". In particular, if you have an older dog or cat (i.e., > 1-2 years old) that just started having clinical signs, you should strongly consider that the hiatal hernia is a fortuitous finding that is not responsible for the clinical signs.

ESOPHAGEAL WEAKNESS

A common misconception is that all dogs with significant esophageal disease will have regurgitation and obvious changes on plain radiographs. It is critical to realize that while plain radiographs are often diagnostic, they miss a lot of cases. Barium contrast radiographs/fluoroscopy are the most definitive means of detecting this problem, but sometimes they can also miss the diagnosis. Many dogs have segmental esophageal weakness (especially in the cervical region) that is missed by radiographs of the thorax. It is important to include the entire esophagus including the cervical region when doing radiographs or fluoroscopy to look for esophageal dysfunction. Some animals have partial esophageal dysfunction that is sufficient to cause major clinical signs but is not seen on plain radiographs or even contrast radiographs unless fluoroscopy is used. Contrast studies on dogs with swallowing disorders need even more care.[7]

Acquired weakness

Acquired esophageal weakness is usually (but not always) easy to distinguish from obstruction radiographically, especially when a barium contrast radiograph is performed. However, the severity of the radiographic lesion (i.e., the degree of dilatation) does not always correlate well with the clinical severity. Acquired esophageal weakness is typically difficult to resolve because it is hard to find the underlying cause. Myopathy, neuropathy, myasthenia gravis, dermatomyositis, dysautonomia, esophagitis, Addison's disease, *Spirocerca lupi*[8], tick paralysis[9], central nervous system disease, or infiltrative non-obstructive esophageal tumors are possible causes. Generalized myopathies and neuropathies often affect the esophagus because it is composed of striated muscle in the dog. Signs of lower motor neuron disease in these patients are sometimes seen and can include loss of muscle mass, weakness, an inability to bark, or a change in the quality of the bark. Some clients report that their animal has laryngitis, which may seem likely because these pets typically have repeated respiratory infections due to aspiration pneumonia. Treatment of the myopathy or neuropathy should resolve the problem, but symptomatic therapy for the esophageal dilatation is indicated.

Generalized myasthenia gravis usually presents as weakness during exertion, which resolves after resting; however, generalized myasthenia can present in a variety of ways, including apparent lameness or permanent weakness. Electromyography and assay for circulating antibodies to acetylcholine receptors are the most definitive tests. Localized myasthenia in the dog is a syndrome in which the esophagus is the only muscle that is obviously weak. Up to 25-30% of dogs with acquired esophageal weakness have this syndrome. Third degree heart block may also be seen in some patients with megaesophagus due to myasthenia. This is diagnosed in dogs with esophageal weakness by detecting serum antibodies to acetylcholine receptors. The antibodies are relatively stable and require little special handling other than refrigeration. If myasthenia is strongly suspected but the titer is negative, it can be valuable to repeat the titer as they sometimes seroconvert later. You cannot perform an edrophonium response test to diagnose localized myasthenia. Myasthenia gravis will sometimes spontaneously resolve. Treatment for myasthenia gravis that does not spontaneously resolve may include anti-acetylcholinesterase drugs, corticosteroids and/or cytotoxic agents. Azathioprine and mycophenolate seem to be effective drugs for this purpose. In general, we try to avoid steroids as they seem to be associated with more problems. In really severe cases, we can place a percutaneous gastrostomy tube to support the patient and lessen aspiration while waiting for the drugs to have an effect. However, a gastrostomy tube will not prevent all aspiration as the dog is still swallowing saliva, which can be regurgitated and aspirated.

Hypoadrenocorticism may be responsible for causing esophageal weakness, even when the serum electrolytes are normal. This is especially true in standard sized, black poodles, but it can occur in any breed. Treatment for hypoadrenocorticism is steroids, which can make the esophagus start functioning again. However, if your diagnosis is wrong and you give steroids because you suspect the dog may have hypoadrenocorticism, all you are doing is making aspiration pneumonia and subsequent death that much more likely.

Idiopathic weakness

Idiopathic megaesophagus (i.e., either congenital megaesophagus or acquired megaesophagus for which a cause cannot be found) can only be treated with symptomatic therapy, which usually consists of feeding the animal 3-4 meals daily of gruel from an elevated platform and making the pet remain in the near vertical position from 5-10 minutes after eating. Near-vertical means just that. It is useless for the dog to just lift its head up while eating; it should be standing on its back legs. If necessary, use a portable ladder or put the dog in a large trash can to help it remain vertical during this time. This approach is a time-honored treatment, but it does not always work. Some animals with idiopathic esophageal weakness are controlled as well (or better) if they are fed free-choice dry food from an elevated platform. Some can even be fed from the floor. Free-choice feeding encourages the pet to eat small amounts of food throughout the day, thus avoiding intermittent large meals, which are more likely to be retained and further dilate the esophagus. If there is any esophageal motility remaining, the dry food may be easier for the esophagus to propel then gruel. It is difficult to predict which feeding regime will work best for a particular patient, and both of these feeding regimes may need to be tried. While most dogs with idiopathic megaesophagus die from aspiration, there are enough of them that respond well that it is very much worth trying. A reasonable percentage of dogs with idiopathic, congenital megaesophagus will spontaneously improve and have normal or near normal function.[10] You cannot predict response to therapy or spontaneous remission; all you can do is support the patient and see what happens.

Some individuals have tried using cisapride in selected patients with idiopathic esophageal weakness that does not respond well to nutritional modification. Theoretically, cisapride would not be expected to work in these animals because cisapride primarily works on smooth muscle and canine esophagus is striated muscle. Furthermore, cisapride is expected to tighten up the lower esophageal sphincter, thus making it harder for food to pass out of the esophagus and into the stomach. Perhaps cisapride helps patients when gastroesophageal reflux is part of the problem.

Some owners elect to have a permanent gastrostomy tube placed in the patient. This will not eliminate all regurgitation or aspiration, because the patient is still swallowing saliva, which will remain in the esophagus until it is regurgitated. However, gastrostomy tubes will help eliminate most of the regurgitation and can markedly prolong such a patient's quality, comfortable life.

Aspiration pneumonia is a major problem and cause of death in dogs and cats with esophageal weakness causing regurgitation. If alleviating the regurgitation by dietary therapy does not control the respiratory disease, then it must be controlled by antibiotics. A transtracheal wash with cytology and culture will help identify optimal antibiotics. Until culture results are known, use of broad-spectrum, bactericidal drugs (i.e., amikacin plus either cephalothin or amoxicillin; enrofloxacin plus amoxicillin or clindamycin) is recommended. In severe cases of aspiration pneumonia, one may have to bypass the esophagus with a gastrostomy tube to prevent further aspiration. These tubes can be placed with the aid of a flexible endoscope and be used for days to months.

CLINICAL APPROACH TO THE PATIENT WITH HEMATEMESIS OR GI BLEEDING:

There is often something in the history that is suggestive of the cause of the bleeding (e.g., use of NSAIDs, shock, etc). If that is the case, then it is often reasonable to begin appropriate therapy after requesting basic laboratory testing (e.g., CBC, serum chemistry panel) to determine the severity of the bleeding and if there are other diseases (e.g., hepatic disease, renal disease) present. Imaging (especially ultrasound) is typically appropriate but not necessarily imperative at this time. If the cause of the GI bleeding or hematemesis is not obvious, if the patient has not responded to 5-7 days of appropriate therapy, or if the bleeding is severe, then additional diagnostics are important and should be performed promptly. First, as stated above, it is wise to first eliminate coagulopathy with a platelet count and some measure of clotting factor adequacy. I typically request PT and PTT, but a mucosal bleeding time is a very useful screening test in these patients. Sometimes there is both a mucosal defect and a coagulopathy. In particular, if ehrlichiosis is possible, one must consider the possibility that the patient has what would normally be an insignificant mucosal defect but which is bleeding because of the effects of *Ehrlichia* spp. upon platelet numbers and their function. After coagulopathy has been eliminated, then imaging should be done if it has not already occurred, and ultrasound is especially important as it may reveal masses that can be aspirated percutaneously, thus avoiding the need for endoscopy/surgery. If these tests have not revealed the diagnosis, then gastroduodenoscopy is generally performed next. The specific reasons to do gastroduodenoscopy in a patient with GI bleeding are to:

a) determine if this is a case in which surgery can remove a defined number of ulcers (this is for cases that are bleeding and have not responded to medical management or cases that are bleeding so badly that one cannot wait on medical management). In these cases, it is important to be sure that bleeding is not due to widespread erosions that cannot be cured surgically. There is no relationship between the size of the mucosal defect and the amount of bleeding; patients with lots of small erosions often bleed as bad or worse than patients with ulcers. It is also important to determine the number and location of such ulcers, as they may be hard to find during a gastrotomy.

b) determine if there is a gastric tumor or some other infiltrate in a patient with GUE that is non-responsive to appropriate therapy.

c) determine the cause in a patient with GUE and no apparent cause on the history, physical examination, or routine blood work.

d) look for a cause of bleeding in a patient with GI blood loss of unknown cause.

It is important to note that endoscopy will not generally allow one to determine if an ulcer will or will not respond to medical management. In most cases, only by treating and observing the patient will you know.

REFERENCES

[1] Leib M, Esophageal foreign body obstruction caused by a dental chew treat in 31 dogs (2000-2006). **J Am Vet Med Assoc** 2008; 232:1021. [2]Moore AH. Removal of oesophageal foreign bodies in dogs: use of the fluoroscopic method and outcome. **J Small Anim Pract** 2001;42:227. [3]Sale C, et al. Results of transthoracic esophagotomy retrieval of esophageal foreign body obstructions in dogs: 14 cases (2000-2004). **J Am Anim Hosp Assoc** 2006;42:450. [4]Cohn LA, et al. Fatal hemothorax following management of an esophageal foreign body. **J Am Anim Hosp Assoc** 2003;39:251. [5]Ranen E, et al. Spirocerosis-associated esophageal sarcomas in dogs: a retrospective study of 17 cases (1997-2003). **Vet Parasitol** 2004;119:209. [6]Guiot L, et al. Hiatal hernia in the dog: a clinical report of four Chinese shar peis, **J Am Anim Hosp Assoc** 2008;44:335. [7] Bonadio CM, et al. Effects of body positioning on swallowing on esophageal transit in healthy dogs. **J Vet Int Med** 2009;23:801. [8]Mylonakis ME, et al. Clinical signs and clinicopathologic abnormalities in dogs with clinical Spirocerosis: 39 cases (1996-2004). **J Am Vet Med Assoc** 2006; 228:1063. [9]Campbell FE, et al. Megaesophagus in dogs with tick paralysis (Ixodes holocyclus). **Aust Vet Pract** 2001;31:75. [10]Bexfield N, et al. Esophageal dysmotility in young dogs. **J Vet Int Med** 2006;20:1314.

EPISODIC VOMITING & DIARRHEA

Michael Willard, DVM, DACVIM
Texas A&M University

BACTERIA & THE GI TRACT

Small intestinal bacterial overgrowth (SIBO) is a term that refers to a disease process that has been well documented in people. Affected people have > 10^6 CFU of bacteria per ml of fasting duodenal or upper small intestinal fluid. This large number of bacteria in an inappropriate place is attended with a constellation of clinical signs that may include weight loss, diarrhea, hypoalbuminemia, and various other changes. There are often structural problems (e.g., partial obstructions) or motility disorders that allow retention of ingesta, or there is decreased gastric acidity that allows excessive numbers of bacteria to enter the upper small intestine. This syndrome was extrapolated to dogs about 30 years ago, especially German Shepherd dogs. Since that time much has been written about this syndrome in dogs and cats, and multiple studies of the numbers of bacteria in normal and ill dogs and cats have been published. At this time, the accuracy of the term small intestinal bacterial overgrowth (SIBO) is dubious. In many dogs with bacterial disorders of the small intestines, it is probably an incorrect description. Clinically normal dogs can have > 10^8 CFU of bacteria in the upper small intestine, and this number is clearly in excess of what was initially accepted as being definitive for SIBO (i.e., > 10^6). Therefore, we will use the term "antibiotic-responsive enteropathy" (ARE) or "antibiotic-responsive diarrhea" (ARD) instead of SIBO.

Antibiotic-responsive enteropathy seems to be a relatively common problem in dogs. It can best be described as a syndrome in which there are substantial numbers of bacteria in the upper small intestines and the host responds to them in such a manner as to cause intestinal dysfunction. These bacteria are not usually obligate pathogens. Rather, they can be of any species, and *E. coli, Staph, Strep*, and *Corynebacterium* are particularly common aerobic/facultative anaerobic bacteria found in the upper small intestines, while *Clostridium* and *Bacterioides* are especially common anaerobic bacteria. These bacteria are probably commensals or they may represent contamination from ingested material, which are not eliminated by normal host defense mechanisms. The signs they produce, if any, seemingly depend upon at least two factors: a) which bacteria are present and b) how the host responds to them. The relationship of ARE to IBD is unclear, but it seems very likely that bacteria could be responsible for either initiating and/or perpetuating the intestinal inflammation we call IBD. The term "dysbiosis" has been suggested as the bridge between ARE and IBD – that is to say that having bacteria that are somewhat prone to cause problems (i.e., usually enterics such as *E. coli*) as opposed to having overt pathogens.

Antibiotic-responsive enteropathy is hard to definitively diagnose with laboratory tests.[1] Histopathology and cytology of the intestinal mucosa are extremely insensitive at detecting ARE. Serum cobalamin and folate concentrations have been used for diagnosis, and finding both a low serum cobalamin and an increased serum folate concentration had been considered to be relatively specific for ARE. However, we now know that serum cobalamin and folate concentrations are relatively insensitive and non-specific for detecting ARE. There are many dogs with chronic GI disease that respond to antibiotic administration but which have normal cobalamin and/or normal folate concentrations. It would seem that treatment for ARE is justified regardless of whether the serum cobalamin and folate concentrations are normal or abnormal, leading one to ask whether there is any benefit to measuring them to diagnose this disorder. Finding hypocobalaminemia or low serum folate levels is beneficial when looking for otherwise occult gastrointestinal disease. Supplementing cobalamin can clearly make cats feel better and cause their diarrhea to diminish. In fact, it is almost getting to the point where it is never wrong to give any sick cat cobalamin injections, regardless of blood values of the vitamin. While the value of supplementing cobalamin to cats is clear (in fact, it is almost never wrong to give any sick cat supplemental cobalamin), the clinical value of administering cobalamin to dogs with low serum cobalamin concentrations is very uncertain. Severe hypocobalaminemia has been suggested to be a poor prognostic sign in GI disease, exocrine pancreatic insufficiency[2] and perhaps other conditions (e.g., lymphoma)[3].

Culture of the small bowel was once considered the "gold standard", but this test is fraught with problems. It is technically hard to do it correctly. Samples must be obtained without contaminating them with oral secretions, and then they must be processed correctly in an expedient manner so as not to lose any anaerobic bacteria while not allowing the numbers of aerobes to increase. Many investigators have snap frozen fluid samples to culture them later, but such freezing appears to kill large numbers of bacteria, especially anaerobic bacteria. We now know that culture only detects about 30% of the bacteria in the gut; the other 70% cannot be cultured. This makes one seriously question the value of culture unless one is searching for a specific pathogen, and even then there are culture-less methods (e.g., PCR) that are probably much better.[4] Finally, as has already been said, just culturing bacteria from the small bowel does not allow one to make a diagnosis of a bacterial disease of the small intestines. There are rare patients that appear to have ARE and yet are resistant to treatment with commonly used antibiotics. Seemingly, these dogs may have one or two very resistant bacteria in their GI tracts, and culture may be required to determine what antibiotic will be effective. However, we have only seen this scenario twice, and we believe it to be very rare.

Because of the apparent difficulty in diagnosing ARE with lab tests, empirical antibiotic therapy is often chosen as a means to diagnosis instead of laboratory tests. The obvious drawbacks to this approach are a) clinical "response" of the patient to the administered antibiotics may be due to the antibiotics or may be due to something else, b) if the patient did not respond to the antibiotic, it may be that you used the wrong antibiotics, and c) even if the patient does have ARE, there may be yet another disease present (e.g., a tumor causing a partial intestinal obstruction) which predisposed the patient to the ARE.

Because any bacteria can be present in the upper small intestine, the species of bacteria in the upper small intestine may change from week to week, and we seldom know which bacteria we are treating, broad-spectrum antibiotics designed to lessen bacterial numbers seem to be indicated. You can never sterilize the GI tract. However, because clinical signs are due to a combination of large numbers plus an altered host response, simply lessening the numbers of bacteria often seems beneficial. Oral aminoglycosides were generally considered a poor choice to treat ARE because anaerobic bacteria (which have been suggested to be more of a problem) are resistant to aminoglycosides. However, this opinion is not clearly correct as there are occasional patients that clearly improve when given amikacin orally. Tetracycline is often effective; but giving tetracycline is inconvenient. Tetracycline must be administered alone (i.e., without any food) and yet be washed down with water to ensure that the capsule or tablet does not stick in the esophagus and cause esophagitis. Tylosin powder has also been useful and is revered by many clinicians.[5] Some clinicians like metronidazole. Metronidazole seems to have real benefit in many GI disorders, probably because it is so effective in eliminating many anaerobic bacteria. For patients that are extremely ill in which we need to know right now whether or not it will respond to antibiotics (i.e., that patient is so ill that you cannot take a chance of being 2-3 weeks from now and not having a response to therapy), I use a combination of enrofloxacin, cephalosporin and metronidazole. I did not say that I used this combination for long periods of time. I use this combination when I absolutely have to know whether or not I will have a clinical response within the next 2-3 weeks or take a chance on losing the patient.

Regardless of which drug is used, such a therapeutic trial should be performed for at least 2-3 weeks before a decision is made as to its efficacy. Remember, you must not only suppress the numbers of bacteria, but you must also allow the intestinal mucosa time to heal. Finally, it appears that concurrently feeding a high quality elimination diet can substantially enhance the efficacy of the antibiotic therapy.[6] Therefore, we now routinely use both in our therapeutic trials.

If the patient appears to respond to this therapeutic trial of elimination diet and antibiotics, then it appears best to continue everything unchanged for an additional 2-4 weeks to be sure that the patient responded to this therapy (as opposed to the patient having some fortuitous, transient response to who-knows-what). If the patient is still doing well at that time, then you either a) stop the antibiotics and see if they diet alone is sufficient to control signs or b) slowly wean the antibiotics to their lowest effective dose (e.g., once a day or even once every other day). It all depends upon how frequently the clinical signs occur. If the signs occur once every 2+ months, then it makes more sense to only treat when the patient is symptomatic. If the signs consistently recur within a few days of stopping the antibiotics, then you are probably stuck with treating almost constantly. This latter situation is one of the two times in veterinary medicine that I am aware of in which it is reasonable to look for the lowest effective dose of an antibiotic. Some patients only need antibiotic administration every 2 to 3 days in order to maintain control. In some cases, the patient will breakthrough and re-develop clinical signs after several weeks or months, and a different antibiotic must be used. If the decision is made to stop administering the antibiotics, then the owners should be warned that it is possible that the signs are likely to recur at some point. For ARE to occur, there is probably some defect in host defense mechanisms that allowed the commensal bacteria to cause the clinical signs, and this defect is unlikely to disappear. The question is how severe is the defect (i.e., is the dog likely to have problems continually or only once in a while)? You should warn the clients that they are likely to have to deal with this problem repeatedly, and you need to explain the difference between "cure" and "control".

It may be a good idea to routinely treat all dogs with chronic small intestinal disease for ARE, even if you have histologic evidence of IBD or other disease. I will treat for ARE almost every time I diagnose a dog with a malabsorptive disease since there is no test for ARE that is reliable in ruling this disorder out, including cobalamin and folate determinations.

PREBIOTICS & PROBIOTICS

Probiotic therapy is the administration of live bacteria, especially but not exclusively *Enterococcus*, *Lactobacillus* and *Bifidobacteria*. Upon administration, these bacteria do not usually colonize the bowel; rather, they apparently bind to toll-like receptors (TLRs) on the epithelial cells whereby they stimulate specific immune events in the GI tract. The TLRs are part of the innate immune system of the patient, and different TLRs bind to different parts of specific bacteria or their metabolic products. Herein lies the problem we currently face: different bacteria have different effects and we need to determine which bacteria (if any) are beneficial for which problem(s). Hopefully, we will find instances in which probiotics will be helpful, but at this time we are still exploring this issue. However, these is enough evidence to suggest that with further exploration we will hopefully find specific uses for these products.[7] In the US there is no FDA regulation of probiotics, which means that you do not know the quality of what you buy in the grocery store or health food store.[8] If you decide to attempt probiotic therapy, I recommend that you buy products made by major companies that make dog and cat food.

Prebiotics are dietary fibers that mammals cannot digest but which certain bacteria can metabolize. The idea is to give prebiotics that preferentially feed "good" bacteria that have a beneficial effect in the GI tract. In general, if the patient responds to prebiotic or probiotic therapy, it seemingly can take up to 4-5 weeks. However, in both cases, any beneficial effect disappears shortly after you stop administering the probiotic or prebiotic.

REFERENCES

[1]German AJ, et al. Comparison of direct and indirect tests for small intestinal bacterial overgrowth and antibiotic-responsive diarrhea in dogs. **J Vet Int Med** 2003;17:33. [2]Batchelor D, et al. Prognostic factors of canine exocrine pancreatic insufficiency: prolonged survival is likely if clinical remission is achieved. **J Vet Int Med** 2007;21:54. [3]Cook A, et al. Prevalence and prognostic impact of hypocobalaminemia in dogs with lymphoma. **J Am Vet Med Assoc** 2009;235:1437. [4]Suchodolski J, et al. Assessment of the qualitative variation in bacterial microflora among compartments of the intestinal tract of dogs by use of a molecular fingerprinting technique. **Am J Vet Res** 2005;66:1556. [5]Westermarck E, et al. Tylosin-responsive chronic diarrhea in dogs. **J Vet Int Med** 2005;19:177. [6]Westermarck E, et al. Effect of diet and tylosin on chronic diarrhea in beagles. **J Vet Int Med** 2005;19:822. [7]Kelley R, et al. Clinical benefits of probiotic canine-dervied *Bifidobacteium animalis* strain AHC7 in dogs with acute idiopathic diarrhea. **Vet Ther** 2009;10:121. [8]Weese JS. Microbiologic evaluation of commercial probiotics. **J Am Vet Med Assoc** 2002;220:794.

PROTEIN-LOSING ENTEROPATHIES

Michael Willard, DVM, DACVIM
Texas A&M University

HYPOALBUMINEMIA

When concerned with protein loss of any cause, one should measure serum albumin concentrations as opposed to the serum total protein concentration. Do not use human clinical pathology laboratories because their technology sometimes does not detect canine albumin; this means that they routinely report serum albumin concentrations of < 1.5 gm/dl in clinically normal dogs. If the patient has substantial hypoalbuminemia, the first step is to examine the skin for obvious lesions which can cause protein loss. Cutaneous lesions sufficient to cause such hypoalbuminemia are obvious; you should be able to just look at the patient and know if this is the problem or not. Next, hepatic function testing (e.g., resting and post-prandial serum bile acid concentrations) and a urinalysis are requested. If there is any doubt on the urinalysis, then a urine protein:creatinine ratio will quantify the magnitude of urinary protein loss. Severe hypoalbuminemia (i.e., < 2 gm/dl) in an animal with diarrhea suggests a protein-losing enteropathy (PLE); however, diarrhea (even when severe) in no way is sufficient to eliminate hepatic disease as the cause of the hypoalbuminemia. Furthermore, a very substantial number of dogs and cats with PLE do not have vomiting or diarrhea. Some only have ascites, and some are found fortuitously on routine blood work. This may be especially true of dogs with primary intestinal lymphangiectasia.

In general, once severe exudative cutaneous disease, protein-losing nephropathy, and hepatic insufficiency are eliminated, then PLE is a reasonable tentative diagnosis of exclusion in patients with a serum albumin < 2.0 gm/dl. Fecal concentrations of alpha-1 protease inhibitor can be used as a means of confirming PLE if there is confusion because of concurrent hepatic or renal disease. The major use for this test in clinical medicine seems to be the hypoalbuminemic patient in which you strongly suspect PLE (e.g., based upon it having severe diarrhea or having hypocholesterolemia), but which also has PLN and/or hepatic disease. However, there are several nuances about this test, especially collecting samples, that make it potentially difficult to interpret. We seldom need this test in clinical practice. Finally, contrary to what the textbooks say, PLE may be associated with a low, normal or increased serum globulin concentration.

Perhaps the most important point of this discussion is that while hypoalbuminemia has repeatedly been reported to be a poor prognostic sign in patients with chronic GI disease, there may be one or more subset(s) of patients that respond well to appropriate therapy.[1] Therefore, diagnosing PLE is not necessarily cause for despair. However, since many of these animals have severe alimentary tract disease that needs to be diagnosed promptly to maximize the chance for successful therapy, aggressive diagnostics are typically an appropriate recommendation. Although therapeutic trials can be chosen in place of classic diagnostic tests in many of the more common alimentary tract diseases (e.g., dietary allergy, dietary intolerance, antibiotic-responsive enteropathy, parasites), such an approach is generally ill-advised if the serum albumin concentration is less than 2.0 g/dl. This is true because it may be necessary to perform an antibiotic and/or dietary therapeutic trial for 3-6 weeks in order to ascertain if it is being effective, and a patient with severe PLE can become markedly worse in that time, especially if the serum albumin concentration is falling rapidly.

Any GI disease can cause PLE if it is severe enough. Many acute GI diseases cause PLE (e.g., parvoviral enteritis); however, these diseases typically are comparatively easier to treat than the chronic GI disease causing PLE. Therefore, the focus in this lecture is PLE in animals with chronic GI disease. The major causes of PLE in adult dogs tend to be intestinal lymphangiectasia, inflammatory bowel disease (IBD), alimentary tract lymphoma (LSA), and fungal infections (i.e., histoplasmosis and pythiosis). Other causes include alimentary tract ulceration/erosion, severe disease of intestinal crypts, antibiotic-responsive enteropathy, and parasites. The major causes of PLE in juvenile dogs tend to be parasites and chronic intussusception. Cats with PLE usually have IBD or alimentary tract lymphoma.

Many dogs with PLE have hypocholesterolemia. Pets with protein-losing nephropathies usually have hypercholesterolemia, while those with hepatic insufficiency often have hypocholesterolemia. Fecal examinations for parasites are appropriate. Although parasites are an uncommon cause of PLE in adult animals, pets in select environments (e.g., confined areas where patients can reinfect themselves) may incur substantial parasitic loads.

DIAGNOSTIC APPROACH TO PLE PATIENT

Once PLE has been diagnosed, intestinal biopsy is usually the ultimate means of establishing a diagnosis. Biopsy can be done via laparotomy, laparoscopy, or endoscopy. Feeding a small, fatty meal (use canned food, not dry, and add in cream or corn oil) the night before the procedure might (?) make it easier to diagnose lymphangiectasia. Flexible endoscopy, when done by someone who is trained in how to take diagnostic tissue samples and submit them is usually more than adequate to obtain diagnostic samples. However, if endoscopy will be used to biopsy the small intestines, it is preferable to first ultrasound the abdomen to make sure that there are no focal infiltrates that are out of reach of the endoscope, or which might be more easily diagnosed by ultrasound-guided fine needle aspiration. Furthermore, there are ultrasonographic changes (streaks in the submucosa) that can be diagnostic. Radiographs and barium series are seldom as sensitive as ultrasound. If flexible endoscopy will be done, one should biopsy both the duodenum and ileum. There have been numerous cases in which lymphangiectasia, IBD or LSA were obvious in the ileum but not in the duodenum. It is not necessary to enter the ileum with the endoscope to obtain a good tissue sample of the ileal mucosa.

Laparotomy and laparoscopy are good means of obtaining diagnostic samples, but it is surprisingly easy to procure non-diagnostic samples with these techniques (i.e., “full-thickness sample” is not synonymous with “diagnostic sample”). Endoscopy

does have the advantage of allowing one to visualize mucosal lesions that are "invisible" when looking at the serosa. In some cases, the
diagnosis can only be obtained by biopsy of these focal lesions. If full-thickness biopsies are obtained in severely hypoalbuminemic animals, then serosal patch grafting will minimize the risk of suture line leakage. A nonabsorbable or poorly absorbable suture (such as PDS) should also be used.

INTESTINAL LYMPHANGIECTASIA

Intestinal lymphangiectasia seems particularly common in Yorkshire terriers[2] and Soft-Coated Wheaten terriers, but may occur in any breed. Sometimes these dogs have distinct ultrasonographic findings: "streaks" in the mucosa that represent dilated lymphatics.[3,4] While histopathology is obviously the desired means of diagnosis, one can sometimes make a definitive diagnosis based upon grossly visible endoscopic findings (i.e., numerous, erratic, grossly engorged lacteals seen as large white blebs on the mucosa).[1] These lesions are "fragile" and apparently may be destroyed by biopsying them (both endoscopically and surgically) if the endoscopist or surgeon is not careful. It is important to note that lymphangiectasia can be a relatively localized disease in the intestines, being present in only the ileum or only the jejunum or only the duodenum; therefore, it is important to biopsy as much of the intestinal tract as possible. Furthermore, if one biopsies the intestines and cannot find a cause of PLE, sometimes lymphangiectasia can be tentatively diagnosed by elimination (i.e., by eliminating IBD, lymphoma, parasites, intussusception, fungal infections, etc).

Diagnosis by means of endoscopic biopsy is certainly possible if the endoscopist is trained in taking high quality tissue samples. However, recent work has demonstrated that poor quality mucosal biopsies (e.g., primarily villus tips or substantial "squash" artifact) makes is much more difficult or even impossible to find the lesions. If one is taking high quality tissue samples (i.e., total length of the villi plus subvillus mucosa down to the border of the mucosa and muscularis mucosa), it typically takes about 6-7 tissue samples to have 90-99% confidence in finding lymphangiectasia. However, it can take 5-7 times as many tissue samples to have the same assurance if you are obtaining poor quality tissue samples that primarily consist of villus tips.[5]

When doing endoscopy, it is important that ileal biopsies be taken in addition to the typical duodenal biopsies. We are finding that ileal biopsies often reveal lesions not found on duodenal biopsies. This is true for lymphangiectasia as well as lymphoma and other lesions. With basic training, an endoscopist should be able to obtain ileal biopsies endoscopically at least 85%+ of the time. Typically, ileal biopsies are often of higher quality than duodenal biopsies.

Therapy for intestinal lymphangiectasia revolves around an ultra-low fat diet, preferably with anti-inflammatory therapy designed to alleviate the lipogranuloma formation that commonly occurs within the intestinal wall and/or mesentery. Supplementation with medium chain triglyceride oil (MCT) used to be recommended because MCT oil supposedly bypasses intestinal lymphatics thus preventing further rupturing of the lacteals. Pancreatic enzymes were often added to the diet to ensure digestion of the medium chain triglyceride oil. MCT oil is seldom used anymore, probably because appropriate dietary therapy is usually more than sufficient. Feeding homemade diets that are highly digestible and ultra-low in fat (e.g., white turkey meat plus potato or rice) or feeding commercial diets is often very helpful in these patients. Commercial low fat diets can be used very successfully, but they need to have the lowest possible fat content. Such a diet can be so successful that it might occasionally be appropriate to use it as a therapeutic trial. Dogs with lymphangiectasia often show a marked increase in serum albumin concentration within 7-14 days of starting such a diet.

The importance of lipogranulomas in the intestinal wall and mesentery is uncertain.[6] However, it is hypothesized that some patients fail to respond to appropriate dietary therapy because of formation of very large or excessive numbers of lipogranulomas that so completely obstruct the intestinal lymphatics that even an ultra-low fat diet cannot prevent lacteal rupture. Therefore, once a diagnosis of lymphangiectasia is made (either by histology, grossly at endoscopy, or tentatively by response to an ultra-low fat diet), it seems to be appropriate to use anti-inflammatory therapy designed to prevent granuloma formation/enlargement. Prednisolone, azathioprine, and/or cyclosporin are commonly used for this purpose. I do not like prednisolone, simply because of all the side effects it has in these patients. Be aware that if you use cyclosporine, it is critical that you measure blood levels of the drug. Not only is there a major difference between patients in how much they absorb, but the bioavailability of the same product may change as the intestine heals.

If the serum albumin is very low (e.g., $\leq$ 1.3 gm/dl), one is often tempted to administer a plasma transfusion while waiting to see what effect the diet will have. However, it is exceedingly difficult to increase the serum albumin concentration by transfusing PLE patients with plasma because so much of the albumin is quickly lost out the gut. You would probably have to give at least two units of plasma to a 15 lb dog in order to raise the serum albumin from 1.0 gm/dl to 1.8 gm/dl, and sometimes you would have to give 3 or 4 units. If it is critical to raise the plasma oncotic pressure, then administering hetastarch may be preferred because it costs less than plasma and it stays in the intravascular compartment longer than albumin.

These patients may be at increased risk for hypomagnesemia, which may potentiate the problem of hypocalcemia.[7] At this time, we do not know how important it is to supplement magnesium to patients, but severe hypomagnesemia can be resolved by a constant rate infusion of magnesium sulfate.

Anecdotally, octreotide might help some patients that do not respond to more conventional therapy.[8,9]

CRYPT LESIONS

Lesions of the intestinal crypts have been recognized as being associated with PLE in dogs. We have identified two different lesions of the small intestinal crypts that can cause PLE. One type is characterized by crypts (usually duodenal) that are filled and somewhat distended with proteinaceous fluid and necrotic inflammatory cells.[10] While such dilated crypts can be found in many animals, including clinically normal dogs, finding large numbers of them in multiple tissue samples seems to be consistently associated with PLE. We do not know if this is a cause-and-effect relationship, or if the dilated crypts are simply a marker for some other process but are not causing the protein loss themselves. Several of these patients have responded to therapy with elemental diets, total parenteral nutrition, prednisolone, azathioprine, and/or metronidazole. We have seen this lesion associated with IBD as well as lymphangiectasia (especially in Yorkshire terriers).

A second type of crypt lesion, that appears to be less common than the first, if characterized by focal accumulations of mucus causing massive distention of the intestinal crypts.[11] This has been reported once before, and we have seen two such cases. The most important aspect of diagnosis seems to be the fact that the lesion may be very focal, almost appearing as ulcers when looking at the intestinal mucosa through an endoscope. Therapy similar to that used on animals with the other form of crypt lesion may be helpful. We have used cyclosporin, but do not know if it is helpful, or if the clinical response is due to the other drugs that the patient is receiving.

These lesions have not been commonly reported. Recent work has shown that these lesions are extremely easy to miss if poor quality endoscopic biopsies are performed. While 7-12 high quality tissue samples (i.e., full length of villi plus subvillus mucosa down to the level of the muscularis mucosa) will find these lesions 90-99% of the time, about 7 times as many tissue samples will be needed if poor quality samples primarily consisting of villus tips are submitted.[5]

ANTIBIOTIC-RESPONSIVE ENTEROPATHY (ARE)

We believe that we are starting to recognize ARE as a cause of PLE in dogs. We now have several patients that appeared to have marked increases in their serum albumin concentration associated with antibiotic therapy. However, because dietary change is often performed simultaneously with antibiotic therapy in these patients, cause-and-effect is not clearly established.

CHRONIC INTUSSUSCEPTION

Chronic intussusception is a relatively important and often missed cause of PLE in juvenile animals. The classic history is one of acute enteritis (e.g., parvoviral enteritis) that does not resolve as expected. The patient feels somewhat better, but continues to have diarrhea, and the serum albumin concentration gradually diminishes. It can be very hard to palpate an ileo-colic intussusception; abdominal ultrasound is clearly the preferred way to diagnose intussusception. Therapy is surgical.

PARASITES

Although uncommon, nematodes may cause PLE in adult animals if there are large numbers of them. Whipworms and hookworms in particular may occasionally be responsible for PLE in older dogs. Giardiasis has been reported to cause PLE in people.

REFERENCES

[1] Peterson PB, et al. Protein-losing Enteropathies. **Vet Clin N Am** 2003;33:1061. [2]Kimmel SE, et al. Hypomagnesemia and hypocalcemia associated with protein-losing enteropathy in Yorkshire terriers: five cases (1992-1998). **J Am Vet Med Assoc** 2000;217:703. [3] Sutherland-Smith J, et al. Ultrasonographic intestinal hyperechoic mucosal striations in dogs are associated with lacteal dilation. **Vet Radiol Ultra** 2007;48:51. [4]Kull PA, et al. Clinical, clinicopathologic, radiographic, and ultrasonographic characteristics of intestinal lymphangiectasia in dogs: 17 cases (1996-1998). **J Am Vet Med Assoc** 2001;219:197. [5]Willard MD, et al. Effect of sample quality on the sensitivity of endoscopic biopsy for detecting gastric and duodenal lesions in dogs and cats. **J Vet Int Med** 2008;22:1084. [6]Romanucci M, et al. Xanthogranulomatous inflammation of the small bowel in a dog. **Vet Path** 2008;45:207. [7]Bush WW, et al. Secondary hypoparathyroidism attributed to hypomagnesemia in a dog with protein-losing enteropathy. **J Am Vet Med Assoc** 2001;219:1732. [8]Markham KM, et al. Octreotide in the treatment of thoracic duct injuries. **Amer Surg** 2000;66:1165. [9]Filik L. et al. A case of intestinal lymphangiectasia successfully treated with slow-release octreotide. **Dig Liver Dis** 2004;36:687. [10]Willard MD, et al. Intestinal crypt lesions associated with protein-losing enteropathy in the dog. **J Vet Int Med** 2000;14:298. [11]Willard MD, et al. Protein-losing enteropathy associated with cystic mucoid changes in the intestinal crypts of two dogs. **J Am Anim Hosp Assoc** 2003;39:187.

ADVANCED COMPUTED TOMOGRAPHY

Tobias Schwarz, Dr. med. vet., DACVR, & Susanne Stieger-Vanegas, Dr. med. vet., PhD, DipECVDI
Edinburgh, Scotland, UK & Corvallis, OR

INTRODUCTION

Computed tomography (CT) has been a well established imaging modality for the small animal patient since the 1990's. In the new century the gaining popularity of small animal MRI has partially overshadowed the use of CT in particular for neurological patients. In recent years however the technological advances of CT, notably helical and multi-slice technology, have opened up an entire new dimension to small animal CT. Modern CT technology allows an extremely time efficient examination of every body part with a diagnostic value that is often superior or a match to contemporary MRI procedures. This presentation will demonstrate the highly versatile applicability and diagnostic value of CT for traditional and emerging CT indications for small animal patients.

TECHNICAL CONCEPTS AND CONSEQUENCES

Some technical details of image acquisition and reviewing are essential to understand CT potentials and limitations. Modern CT units offer very small *slice width* selections, which reduce volume averaging related image blurring. However, slice width is inversely proportional to *image noise*, which limits the use of sub-millimeter slice width images. This can be remedied to a point by increasing the *current* settings. The *pitch* in helical CT defines the speed of the table motion in relation to the selected slice width. Highly pitched image acquisitions result in reduced data sampling per patient volume and thereby introduce image blurring. In multi-slice CT all image data are generated by numerous very small detectors. The selected slice width reflects the number of detectors from which data are sampled collectively, called *binning*. This allows generation of thin- and thick-sliced images in parallel from the same data, without the need of rescanning the patient. Image reconstruction from raw data is performed via the Fourier transformation, a mathematical process that can be adapted using slightly different *reconstruction algorithms* (also called *filters* or *kernels*). Soft tissue structures and all post contrast studies require a medium frequency algorithm, whereas bone and mainly air-filled structures are best viewed with high frequency algorithm images. Multiple series with different algorithms can be reconstructed from the same raw data without the need of rescanning. Once the study is transferred and viewed on an external DICOM viewing station all of the above mentioned settings cannot be changed anymore. Proper reviewing of CT studies is an acquired skill and should be based on rational choices rather than aesthetic viewing pleasure. *Window width and level* on the Hounsfield scale are the CT equivalent of image contrast and brightness and are freely selectable on any DICOM viewing station. The window level should be roughly at level with the attenuation value of the structure of interest. The window width should be narrow (100 to 300 HU) for all soft tissues and all post-contrast scans and wide (500 to 1000 HU) for bone and air-filled structures. Thus there is a trinity of acquisition parameters and viewing principles that should be followed as a general rule: Obtain images of soft tissues with slightly thicker slices, medium frequency algorithm and view them in a narrow window setting. Obtain images of bone and air-filled structures with thin slices, high frequency algorithm and view them with a wide window setting. Any post contrast study requires settings for soft tissue to see contrast uptake.

HEAD

The most commonly performed head CT studies are for the *nose* and *ears*, where CT has been established as a sensitive, specific and time-efficient imaging modality.[1-3] To gain the highest diagnostic value from a nasal/ ear CT, a dual reconstruction pre contrast study with a thin slice and high frequency algorithm viewed with wide window settings for fine bone detail and a thick slice, medium frequency algorithm viewed with a narrow window setting for soft tissues is recommended, followed by a single reconstruction post contrast series that mirrors the soft tissue settings. It is advisable to scan the entire head and C1-C2, to include the mandibular and medial retropharyngeal lymph nodes. A follow-up thoracic metastasis check CT in case of suspected neoplasia will gain maximum diagnostic yield in the most time-efficient manner.

CT scanning of patients with suspected *brain* or *cranial nerve lesions* is an effective diagnostic tool. Pre- and post contrast soft tissue settings are essential, thin slice series can be added in case of suspected calvarial pathology, such as hyperostosis or lysis. Soft tissue or fluid-attenuating mass lesions and associated brain atrophy, compression or edema are well visible using CT and are sufficient for diagnostic purposes in most clinical settings. Falcial and cerebellar herniation can also be diagnosed on modern CT units on post contrast 1mm slice width images. Because of its high sensitivity for air, bone and hemorrhage, CT is the modality of choice for head trauma cases. Unless specific perfusion software is used, CT is insensitive for the diagnosis of non-hemorrhagic brain or spinal infarcts.

CT is an excellent diagnostic tool for *oral, dental, masticatory* and *periorbital lesions* with the additional advantage of offering CT assisted biopsy to sites that are otherwise inaccessible.[4,5]

SPINE

CT has superseded conventional myelography in the diagnosis and localization of neurologic deficits localized to the spine. In particular with multi-slice technology it is now possible to scan large spinal segments with a thin slice width over a short time period. CT is more sensitive than radiography for the detection of spinal fractures[6], lytic[7] and mineralized lesions.[8,9] Myelographic CT has shown superior sensitivity for lesion localization and lateralization than conventional myelography.[8,9] Multiplanar reconstructions

(MPRs) possible in any axis increase the level of certainty in a diagnosis. Optimized CT protocols using 1mm slice width images in axial acquisition mode have been published.[9] Survey CT is adequate to demonstrate the presence of mineralized extradural disk extrusions. Evaluation for the extent of spinal cord swelling or for evaluation of intramedullary or intradural-extramedullary lesions as well as compression from non-mineralized extradural lesions cranial to L5 requires a myelographic CT.[8] In the lumbosacral area, extradural fat provides sufficient image contrast for excellent CT diagnosis of any compressive lesion in the vertebral canal. While one can be confident from a negative myelographic CT that there is no surgical lesion, disease processes such as cord ischemia, FCE, myelomeningitis and degenerative cord changes are not detectable using CT.

THORAX

Computed tomography (CT) is a very valuable non-invasive imaging technique and is often used to further investigate the extent and origin of a lesion diagnosed on thoracic radiographs or on ultrasound. The fast technology advancement in CT especially in the area of multidetector scanning has also lead to a large increase in interest evaluating the heart and vascular structures of the thorax. Multi-detector scanners allow for very rapid scans; however to reduce respiratory motion while imaging the thorax and mediastinum a breath-hold technique should be applied. Performing a breath-hold technique reduces the misregistration between adjacent slices. Helical CT scanning has to be seen as a method to acquire volumetric datasets, which then can be used to create two-dimensional or three-dimensional images of varying thickness and in various planes anywhere within the scan volume. Often adequate images can be obtained without using contrast; however the use of contrast should be encouraged whenever possible to differentiate between vascular and non-vascular structures. In addition, the enhancement characteristics might help to differentiate between certain lesions.

Imaging the cardiac structures involves some additional important considerations, such as a high temporal resolution to minimize cardiac motion artifacts, a high spatial resolution to visualize small cardiac structures and fast volume coverage during breath-hold to minimize respiratory motion artifacts. In cardiac imaging during data acquisition, synchronization with the electrocardiogram is performed. To allow for oblique reconstruction of the image without loss of resolution isotropic voxels should be obtained. Overall the image quality in thoracic CT is very dependent on reducing artifacts related to cardiac and respiratory motion.[10]

Timing of the contrast application (nonionic contrast) in thoracic and especially cardiac imaging is very dependent on the area of interest and should be done with a test bolus or a bolus tracking technique. Patients with low cardiac output or when vascular abnormalities such as a shunt are present might have very variable arrival times of contrast in the area of interest, therefore it is crucially important to track the contrast in each individual patient. Usually two types of contrast timing methods are performed. One involves giving a small-volume test bolus of contrast to observe the arrival of contrast at the heart. The other involves scanning a specific area of interest over time and once the contrast density attenuation reaches a preset point, the scanner turns on and starts imaging; this technique is known as bolus tracking. These techniques and settings are dependent on scanner type, contrast type, contrast amount, delivery method, scan delay time and individual preferences.[10] In certain disease different contrast applications might be chosen, e.g. if an esophageal lesion is expected contrast might be applied intra-oral in addition to an intra-venous contrast application. CT has also been useful in evaluating lesions of the thoracic duct in canine patients with pleural effusion. To outline the thoracic duct a percutaneous injection of the intestinal lymph nodes with non-ionic iodinated contrast medium under ultrasound guidance is performed.[11]

Routine axial images of the thorax are suitable to evaluate several of the mediastinal structures such as the trachea, esophagus, aorta, cranial vena cava, which are oriented fairly perpendicular to the scanning plan and are therefore imaged in a cross section.[12,13]Axial images also allow evaluating all the mediastinal lymph nodes and lung parenchyma. Reconstructed images are useful in evaluating the aorta, great vessels, pulmonary vasculature and trachea in their longitudinal direction. Oblique reconstructed images of the heart are useful in evaluating cardiac abnormalities and the normal size of the cardiac structures.

The range of applications evaluating disease of the thoracic structures or pleural wall is endless. CT has been shown to be a great tool evaluating suspected airway or lung pathologies, where the extent of the disease or severity could not be determined on radiographs or in ultrasound.[14-18] CT-guided fine needle aspirates and tissue core-biopsies of intrathoracic lesions have shown to be fairly safe and useful diagnosing intrathoracal lesions.[19,20] CT has been useful in the diagnosis of a variety of cardiac disease such as pulmonary artery stenosis[21], arterio-venous fistula[22], vascular ring anomaly[23], pericardial mass[24] and pneumopericardium.[25] As verterinary cardiac CT is still in its infancy, multiple applications and the range of normal appearances of the heart in different species will likely be reported in the next few years.

ABDOMEN

Ultrasound is persistently one of the most common techniques evaluating tumors of the gastrointestinal tract or other abdominal organs, however due to the improvement of technology, CT is more commonly used to evaluate the extend and location of suspected or poorly defined abdominal masses. CT has proven to be very useful as a pre-surgical screening tool to evaluate the size and invasiveness of a mass or the presence and location of a shunt. CT is especially useful in evaluating lesions, which are not accessible with any other imaging technique.

In veterinary medicine most commonly intravenous iodinated contrast is used to evaluate the vascularity of abdominal abnormalities; however oral and/or rectal contrast can be used dependent on the indications of the CT scan. Especially in smaller sized large animals such as llamas, alpacas, ponies or miniature horses, where radiographic or ultrasonographic evaluation are limited, an abdominal CT post oral contrast administration might be used to evaluate the gastro-intestinal tract. Independent of the contrast application,[31] timing of the contrast is very important. Intraoral contrast might have to be given several hours before the abdominal CT is performed allowing the contrast to pass through the gastrointestinal tract. The delay time chosen for intravenous contrast agents is dependent on the area of interest and the suspected disease. Routinely at least one post contrast study with an approximately 60 sec delay should be included in a CT study of the abdomen. If a portal shunt is suspected timing can be performed in a similar fashion as described for the thoracic CT. After a pre-contrast helical CT study of the abdomen, either a small-volume bolus injection of contrast can be performed while scanning over a slice where the aorta, caudal vena cava and portal vein is present (usually between Th13-L2) and calculating the arrival time of contrast using a time versus attenuation curve. Another possibility is to place a region of interest with specified density attenuation in the portal vein or hepatic porta and once the pre-specified density is reached, the scanner turns on and starts imaging the specified region.[26]

Routine axial images of the abdomen provide exquisite anatomical detail and allow evaluation of the abdominal organs. Abdominal organs can be differentiated based on their anatomic location and their attenuation characteristics. Intravenous contrast application allows indentifying the caudal vena cava, aorta, portal vein and their respective branches.[27,28] Reconstructed images are useful in evaluating the length and size of an organ, the extent of a lesion, the exact location of a shunt or to follow the ureters from the kidney to the urinary bladder. Abdominal CT does not only provide anatomical information, but has the potential to be used for functional information such as the hepatic perfusion or glomerular filtration rate using a bolus injection of intravenous iohexol.[27,29]

Again the range of diagnostic benefits using abdominal CT is endless. Contrast enhanced CT (CECT) has shown potential differentiating benign from malignant liver or spleen masses.[30,31] It is a great technique determining the presence of portal shunts and to provide precise shunt location information for surgical planning.[27] CECT can be used to demonstrate the invasion of the local vasculature by tumors such as adrenal gland masses.[32] CT is an excellent imaging modality to image the pancreas. Acute pancreatitis and pancreatic abscesses and pseudocysts can be diagnosed and localized accurately. Using a dual-phase angiographic technique enhances the probability of detecting insulinomas, which are generally difficult to detect with any imaging technique.[33] Insulinomas demonstrated a strong arterial enhancement pattern using a dual-phase angiographic technique.[34,35] CECT is not only a great tool evaluating the kidneys for the presence of tumors,[36] but can also be used to evaluate the presence of ectopic ureters or the invasion of ureters by urinary bladder tumors. Ectopic ureters lack peristaltic features and can be visualized entering the urethra or vagina.[37,38]

REFERENCES

[1]Schwarz et al. **Vet Radiol Ultrasound** 2002;43:85. [2]Garosi et al. **Vet Radiol Ultrasound** 2003;44:137. [3]Reetz et al. **Vet Radiol Ultrasound** 2006;47:321. [4]Reiter, et al. **J Am Vet Med Assoc** 2007;231:924. [5]Soukup et al. 2009 **J Am Vet Med Assoc** 2009;235:710. [6]Kinns et al. **Vet Radiol Ultrasound** 2006;47:563. [7]Drost et al. **Vet Radiol Ultrasound** 1996;37:28. [8]Olby et al. **Vet Radiol Ultrasound** 2000;41:396. [9]Dennison et al. **2009 ACVR Proceedings, Vet Radiol Ultrasound** 2010 in press, [10]Drees et al. **Vet Radiol Ultrasound** 2009;50:74.[10]Naidich DP, et al. **Computed Tomography and Magnetic Resonance of the Thorax** 2007. [11]Johnson EG, et al. **Vet Surg** 2009; 38(3):361. [12]Matthew S, et al. **Vet Radiol Ultrasound** 2009; 50(6):622. [13]DeRycke LM, et al. **Am J Vet Res** 2005; 66(3):512. [14]Walker M, et al. **Vet Radiol Ultrasound** 1993; 34(2):93.[15]Bertolini G, et al. **Vet Radiol Ultrasound** 2009; 50(1): 80. [16]Seiler G, et al. **Vet Radiol Ultrasound** 2008; 49(6): 504. [17]Schultz RM, et al. **Vet Radiol Ultrasound** 2008; 49(3): 249. [18]Echandi RL, et al. **Vet Radiol Ultrasound** 2007; 48(3): 243. [19]Zekas L, et al. Vet Radiol Ultrasound 2005; 46(3): 200. [20]Tidwell A, et al. **Vet Radiol Ultrasound** 1994; 35(6):445. [21]MacGregor JM, et al. **Vet Radiol Ultrasound** 2006; 47(4):345. [22] Jenni SD, et al. **J Vet Intern Med** 2009; 23(1): 190. [23] Pownder S, et al. **J Vet Cardiol** 2008; 10(2):125. [24]Ben-Amotz R, et al. **J Small Anim Pract** 2007; 48(10): 596. [25]Leclerc A, et al. **J Am Vet Med Assoc** 2004; 224(5):710. [26]Frank P, et al. **Vet Radiol Ultrasound** 2003; 44(4):392. [27]Zwingenberger AL, et al. **Am J Vet Res** 2007; 69(9):970.[28]Caceres AV, et al. **Vet Radiol Ultrasound** 2008; 49(4):350. [29]O'Dell-Anderson KJ, et al. **Vet Radiol Ultrasound** 2006; 47(2):127. [30]Taniura T, et al. Hiroshima J Med Sci 2009; 58(1):17. [31]Fife WD, et al. **Vet Radiol Ultrasound** 2004; 45(4):297. [32]Schultz RM, et al. **Vet Radiol Ultrasound** 2009; 50(6):625. [33]Robben JH, et al. **J Vet Intern Med** 2005; 19:15. [34]Iseri T, et al. **Vet Radiol Ultrasound** 2007; 48(4):328. [35]Mai & Caceres **Vet Radiol Ultrasound** 2008;49(2):141. [36]Moe L, et al. **Vet Radiol Ultrasound** 1997; 38(5):335. [37]Rozear & Tidwell **Vet Radiol Ultrasound** 2003;44:155. [38]Samii V, et al. **J Vet Int Med** 2004;18:271.

NUTRITIONAL MYTHS & FALLACIES: TALKING TO AN OWNER

Sally C. Perea, DVM, MS, DACVN
Davis, CA

INTRODUCTION

The subject of pet food selection has always been a wide field with many opinions. Many pet owners feel that they can understand nutrition better than other fields in veterinary medicine.They may also feel that this is one aspect of their pet's health they can control. Because of this, many pet owners conduct internet research on nutrition information and often find information that is incorrect, misleading, or raises additional questions or creates confusion. This lecture will discuss some common misconceptions regarding pet food ingredients as well as areas of confusion that may arise when comparing pet food products.

PET FOOD INGREDIENTS

Meat, Poultry, Meat Meals, and By-products

It is important to help pet owners differentiate between truths and fallacies when educating them on what to feed their pets. When it comes to meat and poultry ingredients, there is a lot of confusion around the difference between meat, meat meal, and by-product ingredients. The Association of American Feed Control Officials (AAFCO) ingredient definitions clearly define what is included and excluded from these ingredients; however, when conducting online research on pet food ingredients one can find many additional definitions that create confusion and misunderstanding. The table below (Table 1) outlines the AAFCO definitions as well as misinformation that is commonly stated about these ingredients.

Table 1. Ingredient Definitions and Common Myths and Fallacies

Ingredient Name	AAFCO Definition[1]	Myths and Fallacies
Meat	"Clean flesh derived from slaughtered mammals and is limited to the part of the striate muscle which is skeletal or that which is found in the tongue, in the diaphragm, in the heart, or in the esophagus; with or without the accompanying and overlying fat and the portions of the skin, sinew, nerve and blood vessels which normally accompany the flesh." "The ingredients "meat" or "meat by-products" shall be qualified to designate the animal from which the meat or meat by-products are derived unless the meat or meat by-products are derived from cattle, swine, sheep, goats, or any combination thereof. For example, ingredients derived from horses shall be listed as "horsemeat" or "horse meat by-products".	Pet foods commonly contain horse meat, dog meat, and cat meat
Meat Meal	"The rendered product from mammal tissues, exclusive of any added blood, hair, hoof, horn, hide trimmings, manure, stomach and rumen contents except in amounts as many occur unavoidably in good processing practices. It shall not contain added extraneous materials not provided by this definition. The calcium level shall not exceed the actual level of phosphorus by more than 2.2 times."	Meat meals contain rendered whole animals, including hair, hoof, horn, hide trimmings, gastrointestinal contents, and even foreign materials.
Meat by-products	"The non-rendered, clean parts, other than meat derived from slaughtered mammals. It includes, but is not limited to, lungs, spleen, kidneys, brain, liver, blood, bone, partially defatted low temperature fatty tissue and stomachs and intestines freed from their contents. It does not include hair, horns, teeth and hoofs."	By-products include all of the unwanted waste parts, including gastrointestinal contents, horns, teeth, hoofs, diseased or cancerous animal parts, and foreign materials such as animal feed, fertilizer, industrial lubricants, soap, and rubber.
Poultry by-products	"Must consist of non-rendered clean parts of carcasses of slaughtered poultry such as heads, feet, viscera, free from fecal content and foreign matter except in such trace amounts as might occur unavoidably in good factory practice."	In addition to the above regarding meat by-products, poultry by-products include feathers.

Fats

In addition to meat and poultry ingredients, there are common misconceptions about oils and fats added to pet foods. For example, one internet site includes "fats" on their list of ingredients to avoid in pet foods, stating that "most fats added to pet food are indigestible."[2] While many such lists of ingredients to avoid are commonly compiled based on personal philosophies or preferences, this statement regarding fats is clearly a misunderstanding regarding the use of fats in pet foods and the nutritional benefits that they provide. Most fats added to pet food are highly digestible, commonly providing a digestibility of 90%.[3] Another example of the misperception of fat is seen in the statement, "… very similar to human foods, the higher the fat, the more unhealthy the diet."[4] While it is true that fats aid in palatability, fats are also required to meet essential fat and fatty acid requirements, and are important for the maintenance of normal health. Animal fats, such as chicken fat, as well as vegetable oils are commonly used to provide the essential fatty acid, linoleic acid. Fish oils are also commonly added to provide long-chain omega-3 fatty acids (EPA & DHA) that provide many potential health benefits.

Carbohydrates and Grains

Low carbohydrate and grain free pet foods are becoming more common and more popular among pet owners. While there are a variety of applications for these diets, it is important to remember that carbohydrates are not bad for all pets and that there have been no studies with evidence to suggest that feeding a food with a higher carbohydrate content will contribute to the development of conditions such as inflammatory bowel disease, urinary tract disease, diabetes mellitus or obesity. In fact, studies evaluating risk factors for obesity and diabetes mellitus in cats support the conclusion that total calories and physical activity play a more important role than carbohydrate intake.[5,6] Therefore, while low carbohydrate diets may be useful in the management of many conditions, this should not lead to the conclusion that carbohydrates contribute to the cause or onset of these conditions.

Another important consideration with pet foods in the "grain free" category is that not all grain free foods are necessarily lower in carbohydrate content. Many grain free products currently on the market replace grains with ingredients such as potato or peas. Potatoes and peas, as well as other fruits and vegetables, are primarily composed of carbohydrate calories. Therefore, if the dietary goal is to restrict or limit carbohydrate, one cannot rely on a grain-free claim alone, and must verify the carbohydrate content of the food.

COMMON CONFUSION WHEN COMPARIING PRODUCTS

Moisture and Percentages

One of the greatest areas of confusion in pet foods is when making nutrient comparisons in foods with different moisture levels. For example, dry foods commonly contain 10% or less moisture, canned foods contain 75-80% moisture; and fresh foods generally contain approximately 65% moisture. Because pet food guaranteed analyses are reported on an as fed basis, comparing pet food products with different moisture levels can be very confusing for pet owners. In addition, macronutrients (protein, fat, and carbohydrate) can also be expressed on a percent metabolizable energy (ME) basis, which is equivalent to the percent of calories provided by the given nutrient. Therefore, when recommendations are given to feed a specific percentage of protein, fat, or carbohydrate, it is important to clarify the units that are being discussed.

An example of how confusion can develop is seen with this example statement from one online discussion of pet foods, "In short, the two key variables in regards to this panel are the amount of protein and the amount of fat. It is important to have a solid amount of protein (15%) and a limited amount of fat (4%)." [4] Because the author of this statement refers to the "panel", it could be interpreted that they are referring to the as fed values of a pet food label. However, it is unclear if they are referring to a dry or canned product, as the example value given for protein (15%) would be very high for a canned food and very low for a dry food. Similarly, the value listed for fat (4%) is fairly low for a canned food, but much too low for a dry food.

UNDERSTANDING TERMS

Terms used in product materials or on labels, such as natural, organic, human grade, premium, gourmet, and holistic are variable regarding whether they have official definitions.

The AAFCO definition for natural is "a feed or ingredient derived solely from plant, animal or mined sources, either in its unprocessed state or having been subject to physical processing, heat processing, rendering, purification, extraction, hydrolysis, enzymolysis or fermentation, but not having been produced by or subject to a chemically synthetic process and not containing any additives or processing aids that are chemically synthetic except in amounts as might occur unavoidably in good manufacturing practices."[1] The term natural should only be used to describe products as a whole when all of the ingredients and components of ingredients meet this definition. An exception is made to allow the use of chemically synthesized vitamins, minerals, or other trace nutrients; however, a disclaimer must be present to inform the consumer that these supplemented nutrients are not natural.

The term "organic" refers to the conditions under which the ingredients used in products, and the product itself, were produced and must be consistent with regulations developed by the United States Department of Agriculture (USDA) National Organic Program. Current regulations covering organic products do not apply to pet foods, but the USDA is developing regulations for organic labeling of pet foods. The USDA Organic Pet Food Task Force issued a report in 2006 outlining some of their proposed regulatory changes.[7] The Task Force recommended that product composition requirements for organic pet food be similar to those for livestock, but that labeling categories be the same as for processed human food. These require a minimum of 70% organic ingredients for the use of the claim "made with organic", and at least 95% organic content for the use of the claim "organic". All products labeled as "made with organic ingredients" or "organic" must exclude ingredients that are genetically engineered, are produced using sludge or irradiation, or contain synthetic substances not on the National List of Allowed and Prohibited Substances as defined by the National Organic Standards Board (advisory to the federal Secretary of Agriculture). They also cannot contain sulfites, nitrates or nitrites, or include organic and non-organic forms of the same ingredients. In addition, products labeled "organic" cannot contain non-organic ingredients when organic sources are available. As no official rules have been adopted by AAFCO, there are some products that use the term "organic" more loosely than processed human food specifically with regard to the percent inclusion of organic ingredients.

Another term that is used is "human grade." Currently, there is not an official definition or regulation regarding the use of the term. According to AAFCO, this means that the term is not allowed.[1] However, there is no legal prohibition and the term continues to be widely applied to a variety of pet foods. In 2007, one pet food company was denied the right to sell their product in the state of Ohio due to the use of the term "human food grade."[8] The Ohio State Department of Agriculture argued that the term was misleading, and refused to grant a license to the company. The company followed with a lawsuit, which eventually sided in their favor due to the argument that the company had the constitutional right to make truthful statements about the quality of its pet food product on the label. So, despite the unclear and likely variable and inconsistent interpretation of the term, it enjoys popularity as pet owners look for foods they perceive have improved safety and quality, whether these qualities are true or not.

Other very commonly used descriptors include premium, super premium, and gourmet. These terms are primarily used as marketing tools, and have no official definitions. Products labeled as such can have widely variable nutritional profiles, micronutrient concentrations and ingredient types. They are not required to be of superior quality in any way or to contain different nutrient levels than other products.

REFERENCES

[1]Official Publication of the Association American Feed Control Officials. AAFCO, Oxford, IN, 2009. [2][WWW page]. http://forums.petlovers.com/articles/about/nutrition_basics, 2010. [3]Davenport DJ, et al. Gastrointestinal and Exocrine Disease. In: Hand ed. **Small Animal Clinical Nutrition, 4th Edition**. Missiouri, Walsworth Publishing, 2000: 729. [4][WWW page]. http://www.petfoodstory.com/pets/pet-food-nutrition-dos-and-donts.htm, 2010. [5]Backus RC, et al. **Brit J Nutr** 2007; 98: 641-650. [6]Slingerland LI, et al. **Vet J** 2009; 179: 247-253. [7]Interim Report of the National Organic Program, Organic Pet Food Task Force. [WWW page]. http://www.ams.usda.gov/nosb/meetings/OrgPetFood.pdf, 2006. [8]Petfood Industry. "Human food grade" to stay on petfood. [WWW page]. http://www.petfoodindustry.com/ViewContent.aspx?id=19032, 2010

MAKING PET FOODS AT HOME

Sally C. Perea, DVM, MS, DACVN
Davis, CA

INTRODUCTION

A 2007 survey of pet owners in the United States and Australia reported that non-commercial foods (table scraps, leftovers, or homemade foods) were fed as part of the main diet in 30.6% of dogs and 13.1% of cats, and approximately 7% of dogs received at least half their diet as home-prepared foods. [1] Since the time of this survey, interest in feeding home-prepared pet foods has been on the rise. Motivations to switch to a home prepared food vary, but may include increased control over ingredients selected, avoidance of artificial preservatives, a desire to pamper the pet, provide a more wholesome or nutritious diet, or to provide a medical benefit.[1] Following the widespread pet food recall in 2007 due to melamine contamination, and more recent recalls due to salmonella, safety is another motivating factor that may lead a pet owner to pursue a home prepared food. Finally, some pets may suffer from one or more medical conditions that cannot be adequately addressed with currently available veterinary therapeutic foods, and may therefore require a home prepared formulation. Whatever the motivating factor or requiring circumstances may be, it is important that veterinarians are able to recognize the potential concerns in feeding a home prepared food, and be able to provide guidance to clients that have elected this feeding style.

NUTRITIONAL ADEQUACY

One of the primary concerns with feeding home prepared foods is appropriate nutritional balance and adequacy. Of the pet owners feeding at least half of the diet from homemade foods in the survey mentioned above, only 30% reported use of a recipe designed for pets. Of these recipes 8 were from a veterinarian, 3 were obtained from the internet, and 5 were from other sources.[1] Although the nutritional adequacy of these diets were not evaluated in this study, it raises concerns on the potential misinformation available to the public regarding home prepared diets for dogs and cats. Even more concerning is the large portion of owners who fed home prepared foods that were not designed for pets. Another study evaluating the nutritional adequacy of home made pet diets from six published resources (49 maintenance and 36 growth diets) found that 86% of the diets had inadequate levels of various minerals, 62% had inadequate levels of various vitamins, and 55% had an inadequate protein or essential amino acid level.[2]

In addition to deficiencies in nutrient profiles, clinical nutrient deficiencies secondary to unbalanced home prepared foods have also been reported.[4,5,6] Inadequate calcium and improper calcium to phosphorus ratio is one of the most common problems reported in both dogs and cats fed unbalanced home prepared foods.[4,5] Calcium deficiency, with or without concurrent vitamin D deficiency, results in nutritional secondary hyperparathyroidism. Young growing animals will commonly develop long bone abnormalities, while adult animals have been reported to develop bone resorption of the mandible and maxilla, resulting in a rubber jaw syndrome. [4,5] As with most nutritional deficiencies, underlying calcium deficiency is not commonly recognized until calcium stores are significantly depleted. Because of this, pets on unbalanced diets may not display obvious clinical signs of underlying nutritional insufficiency, while functioning at suboptimal levels. Similarly, many underlying nutrient deficiencies, such as calcium deficiency, are not readily apparent on routine blood work because serum concentrations are tightly regulated even in the presence of severe dietary deficiency.

METHODS FOR EVALUATING NUTRITIONAL ADEQUACY

Chemical nutrient analysis is the ideal method to assess adequate nutrient composition of a home prepared food. Because of the cost associated with evaluating the full range of required nutrients (~$3,500), most home prepared diets are evaluated based on calculated nutritional analysis. The nutrient profiles of individual ingredients can be obtained from nutrient databases, such as the USDA Nutrient Database. The nutrient profiles of individual ingredients can vary widely based on the cooking method, and other factors such as the cut of meat, portion of the plant fed, and season in which the plant was grown. Because of this, a detailed diet history is the first and most important step in evaluating nutritional adequacy. Once a complete diet history is collected, the pet's daily diet can be evaluated with the use of formulation software or self developed spreadsheets.

While detailed evaluations should be conducted to carefully assess the nutritional adequacy and balance of a diet, there are simple items that can be evaluated to quickly assess general completeness. The diet should be evaluated for the inclusion of the ingredients required to provide all essential nutrients, including protein, essential amino acids, essential fatty acids, vitamins, and minerals. In general, the protein source should make up a minimum of 1/3 of diet by volume (higher amounts may be required for some vegetarian sources such as legumes). While dogs can generally perform well on vegetarian protein sources, providing the appropriate protein and amino acid levels with vegetarian ingredients is more challenging in cats. One reason for this is that vegetarian protein sources are generally limiting in essential sulfur-containing amino acids, such as methionine. Vegetarian protein sources also lack arachidonic acid, which is an essential dietary fatty acid for cats, but not for dogs. Arachidonic acid (20:4 n-6) is found primarily in animal fats, and is therefore difficult to provide in vegetarian diets. In addition to a source of arachidonic acid for cats, both dogs and cats require a source of linoleic acid (18:2 n-6). Linoleic acid is an essential fatty acid for dogs and cats and plays an important role in skin and coat health.[7] Animal fats often provide some linoleic acid, but most diets require an additional source to meet the pet's requirement. Vegetable oils such as corn oil, canola oil, safflower oil, and soybean oil can serve as dietary sources of linoleic acid. Of the readily available vegetable oils, corn oil has the highest concentration of linoleic acid. Although alpha-linolenic acid (18:3 n-3) has not previously been considered an essential fatty acid for dogs and cats, more recent studies suggest that providing alpha-linolenic acid in the diet provides some benefits for optimal health, especially in growing and reproducing animals.[7] Similarly docosahexaenoic acid (DHA) is now known to be essential for optimal brain and retinal development in puppies.[10,11] Although alpha-linolenic acid is commonly provided by terrestrial vegetable oil sources such as canola or flax seed oils, algal or fish oil sources must be used to supply long-chain omega-3 fatty acids such as eicosapentaenoic acid (EPA) and DHA in the diet.

After assessing for essential protein, amino acid, and fatty acid needs, an assessment should be made to identify a source of essential vitamins and minerals. Although many foods will contribute to the vitamin and mineral content of a diet, a supplemental form is generally required to ensure that recommended allowances are met. For those multivitamin and multimineral supplements that do require additional supplementation, calcium can be provided in a calcium only supplement (such as calcium carbonate or calcium citrate), or a calcium and phosphorus supplement (such as di- or tri-basic calcium phosphate or bone meal). Commonly, both a calcium only and a calcium and phosphorus supplement will be used in combination to create an appropriate balance of calcium to phosphorus. Bone meal is a readily available calcium and phosphorus supplement; however, it has fallen out of favor with many veterinary nutritionists due to concerns of lead contamination. In the United States, the only commercially available non-bone meal calcium and phosphorus supplement available is tribasic calcium phosphate (available under the brand name Posture®). When using this and other calcium supplements care should be taken to consider the impact of the vitamin D that is often added. This is especially true when using in combination with other vitamin and mineral supplements that already provide needed vitamin D levels.

In addition to calcium, other limiting nutrients in multivitamin/multimineral supplements may require additional supplementation. For example, vitamin B12 and zinc are common limiting nutrients when using One-A-Day® Maximum in home cooked diet formulations. Because simply adding higher quantities of the One-A-Day® can push other nutrients toward safe upper limits, an additional vitamin B12 or zinc supplement may be needed. Another nutrient that may require additional supplementation is choline. Choline serves as a component of choline phospholipids, and as a methyl donor for methylation reactions in the body. Other methyl donors, such as methionine, can serve as choline equivalents in the diet; however, because methionine can be a limiting amino acid, especially in reduced protein diets, additional supplementation with a choline supplement may be required. Finally, salts can be used in home cooked diets to provide additional sodium, chloride, potassium, and iodide. There are three basic salt supplements that are readily available for this use. The first is standard iodized salt, which provides sodium, chloride, and iodine. The second is salt substitute (such as Morton® Salt Substitute), which is a potassium chloride mixture, but provides no iodine. And the third is lite salt (such as Morton® Lite Salt™) that provides a 50:50 mixture of iodized salt and salt substitute.

INDICATIONS FOR HOME COOKED DIETS

When available, a complete and balanced commercial pet food is preferred to preparing a home cooked diet. However, the major advantages of home cooked diets are that the ingredients selected and nutrient levels of the diet can be tailored specifically to the patient's needs. This can become very important for patients that have multiple disease conditions that require multiple nutritional approaches that are not available in a single commercial food.

After exploring commercial options, if the choice is made to treat with a home cooked diet, the next step is to acquire an appropriate and nutritionally balanced home cooked diet formulation. An individual consultation with a board certified veterinary nutritionist will often provide the most guidance and individual tailoring of a diet to meet the patient's needs. Other veterinary software programs and published recipes are also options available to veterinarians looking for more standardized dietary formulations or recommendations. Formal consultations and the use of nutrition software are generally preferred to published recipes, as they provide more up-to-date nutritional strategies and can be formulated for the individual pet's specific caloric and nutrient requirements (Table 1). Most importantly one should create a diet that is complete and balanced for long-term feeding, and that has been evaluated by a qualified board certified veterinary nutritionist. As with all veterinary therapeutic diets, regular rechecks and monitoring of the patient is essential to ensure that the dietary therapy is meeting the patient's needs. Most veterinary nutritionists encourage feedback on how the home cooked diet formulation is working for the patient, and can provide reformulations and adjustments to the diet as needed.

Table 1. Nutrition Resources

Institutions with Clinical Nutrition Faculty and Residents Available for Consultation
University of California, Davis http://www.vetmed.ucdavis.edu/vmth/small_animal/nutrition/default.cfm 530-752-1387
University of Missouri http://www.vmth.missouri.edu 573-882-7821
University of Tennessee http://www.vet.utk.edu/clinical/sacs/nutrition.php 865-974-8387
Tufts Cummings School of Veterinary Medicine
http://www.tufts.edu/vet

508- 839-5395 x 84696
Home-Cooked Diet Formulations
Veterinary Nutritional Consultations http://www.petdiets.com Fax: 800-649-2043 Davis Veterinary Medical Consulting, Inc.* http://www.balanceit.com 530-756-3862 888-346-6362
Human Foods Nutrient Content Information
USDA Nutrient Database http://www.nal.usda.gov/fnic/foodcomp/search/

* Disclosure - DVM Consulting, Inc and Balance IT® are owned by the author

REFERENCES

[1]Laflamme, et al. **J Am Vet Med Assoc** 2008; 232(5):687-694. [2]Lauten SD, et al. **J Vet Intern Med** 2005; 19(3): 476-7. [3]Freeman LM, et al. **J Am Vet Med Assoc** 2001; 218(5): 705-9. [4]Tomsa K, et al. **J Small Anim Pract** 1999; 40(11):533-539. [5]de Fornel-Thibaud. **J Am Anim Hosp Assoc** 2007; 43: 52-60. [6]Niza MMRE, et al. **J Fel Med Surg** 2003; 3: 271-277. [7]Bauer JE. **J Am Vet Med Assoc** 2006; 229(5):680-684. [8]Cullinen K. **Med Health R I** 2006; 89(3):113. [9]Bauer JE. **J Am Vet Med Assoc** *2004*; 224(5): 669-675. [10]Heinemann KM, et al. **J Nutr** 2005; 135: 1960-1966. [11]Heinemann KM, et al. **J Am Vet Med Assoc** 2006; 228(5): 700-705.

THERAPEUTIC MISADVENTURES: CASES OF DRUG INTERACTIONS, DRUG TOXICITY, & OTHER DISASTERS: PART I

Lauren A. Trepanier, DVM, PhD, Dip. ACVIM, Dip. ACVCP
Madison, WI

INTRODUCTION

Most of the adverse drug reactions encountered in the clinic are dose-dependent. They may be an extension of the drug's intended pharmacologic effect (for example, hypotension from amlodipine), or unrelated to the drug's intended effect (e.g. retinotoxicity from fluoroquinolones). Dose dependent reactions are relatively predictable, and can often be prevented through therapeutic drug monitoring, if available. Dose dependent toxicity can also result from unanticipated adverse drug interactions.

In contrast, idiosyncratic drug toxicity reactions occur at therapeutic dosages, in a small proportion of the population. This type of toxicity does not occur in most patients even at high dosages (and is therefore not considered "dose-dependent" in the general population), but toxicity can increase with dosage among susceptible individuals. Idiosyncratic reactions are usually unrelated to the desired pharmacologic action of the drug (e.g. hepatotoxicity from diazepam), and may or may not involve an immunologic response (e.g. anti-drug antibodies in dogs with sulfonamide-induced blood dyscrasias). [1]

This workshop will present a series of cases that involved drug interactions, dose-dependent adverse reactions, or idiosyncratic drug toxicities, and will focus on the "lessons learned" from each patient.

BIOAVAILABILITY BLUES

Case One

A 1½ yo Mn Maltipoo presented for coughing and gagging of one week's duration. On physical exam, the dog was lethargic but responsive, 5% dehydrated, and mildly febrile at 102.8°. Increased bronchovesicular sounds were ausculted, with a respiratory rate of 50 breaths/min. Abdominal palpation was negative for pain or organomegaly. Chest radiographs showed a mild to moderate alveolar/interstitial pattern in the right middle lung lobe, consistent with aspiration pneumonia. Esophagram by fluoroscopy revealed proximal esophageal dysfunction. Resting cortisol was 4.2 µg/dl, ruling out hypoadrenocorticism. Serum for an acetylcholine receptor antibody test was submitted. Tracheal wash cytology showed suppurative inflammation and a positive culture for *E. coli*, sensitive to enrofloxacin. Following rehydration with IV fluids, the dog was discharged with enrofloxacin, along with omeprazole and sucralfate for possible reflux.

The fever, lethargy, and cough improved over the next 48 hours. The dog then became progressively more lethargic and weak, with an increased respiratory effort, and re-presented to the emergency service one week after discharge. The owners described attempts to eat, but trouble swallowing. At re-presentation, the dog was depressed, weak, and febrile at 103.9°. The dog was stabilized with oxygen and IV fluids. Repeat chest radiographs showed worsened pneumonia, and a repeat tracheal wash revealed an *E. coli,* now resistant to enrofloxacin. Meropenem (12 mg/kg SC q. 8 h) was instituted. The dog's pneumonia improved, and he began eating with less dysphagia, although he still appeared weak.

The dog's drug history was reviewed, and a likely drug interaction between enrofloxacin and sucralfate was noted. The owner stated that she had been giving the drugs two hours apart, as directed (but was giving sucralfate followed by enrofloxacin). Sucralfate was discontinued, and the pneumonia was successfully cleared with a two-week course of meropenem. The dog's dysphagia gradually improved without specific therapy.

<u>Case assessment</u>: Sucralfate and aluminum-containing antacids can form complexes with many other drugs in the GI tract, including fluoroquinolones, doxycycline, theophylline, and digoxin, and impair their absorption. [2] Sucralfate markedly decreases the absorption of fluoroquinolones, with only 4% bioavailability for ciprofloxacin when the drugs are given together in humans. [3] While this interaction can be lessened by giving the fluoroquinolone two hours *before* the sucralfate (bioavailability increases to 83%), the opposite regimen is not recommended (i.e. giving the sucralfate first, followed two hours later by the antibiotic) because of the persistence of sucralfate in the stomach. Furthermore, delayed gastric emptying may lead to altered fluoroquinolone absorption even when following the "other drugs two hours before sucralfate" rule. Finally, it is difficult for clients to comply with a two-hour drug separation regimen. I prefer not to send patients home on oral sucralfate when they are being given any other drugs besides antacids, because of the difficulty in avoiding an interaction. It is important to note that sucralfate delays, but does not decrease the extent of, the absorption of H_2 blockers in humans, so staggering the dosing of sucralfate with H_2 blockers is probably not necessary. [2] There are no studies available on bioavailability interactions between sucralfate and omeprazole.

In this dog, co-administration of sucralfate with enrofloxacin may have impaired enrofloxacin absorption, leading to subtherapeutic serum and tissue concentrations and proliferation of a resistant *E. coli* subpopulation. However, serum enrofloxacin concentrations were not measured, so this cannot be confirmed. Interestingly, this dog's acetylcholine receptor antibody test was positive, suggesting acquired myasthenia gravis as the cause of the weakness and proximal esophageal dysfunction. By the time the test results were returned, the dog's dysphagia had improved and his weakness was resolving, so long term treatment with an acetylcholinesterase inhibitor was not warranted. The dog's acetylcholine receptor antibody titer became negative over the next several months.

MALABSORPTION MALADY

Case Two

A 5 yo Fs Yorkie was evaluated for vomiting, inappetance, small bowel diarrhea, weight loss, and abdominal distension of 3 weeks progression. Physical exam revealed a 2 kg dog with a BCS of 3/9, a respiratory rate of 80 breaths/min, and palpable abdominal fluid. Blood work revealed panhypoproteinemia, hypocholesterolemia (66 mg/dl), and hypocalcemia (5.9 mg/dl, ionized calcium 0.83 mmol/L). Imaging revealed pleural effusion and ascites, both transudates on cytology. Protein losing enteropathy (lymphangiectasia) of Yorkshire terriers was suspected. Full thickness biopsies were not pursued because of the dog's severe hypoalbuminemia (<1.0 g/dl), pleural effusion, and frail body condition. After therapeutic thoracocentesis and stabilization in CCU on Hetastarch, treatment with immunosuppressive prednisolone suspension and oral calcium carbonate was instituted, with the understanding that lymphoma had not been ruled out.

On recheck one week after discharge, the dog's vomiting, inappetance, and small bowel diarrhea had resolved; however, the dog was progressively weaker. Worsened hypocalcemia (total 3.6 mg/dl, ionized 0.50 mmol/L) was documented, along with hypomagnesemia (0.7 mg/dl). The dog was hospitalized, and oral calcitriol (15 ng/kg BID), along with a CRI of calcium gluconate in Hetastarch, was instituted. Low dose aspirin and cyclosporine (modified) were also added because of low antithrombin levels (57%) and continued severe hypoalbuminemia (1.0 g/dl), respectively. After 4 days on IV calcium gluconate and oral calcitriol, the dog's total serum calcium was still low at 4.5 mg/dl, with a serum magnesium of 0.9 mg/dl, although weakness had improved. Treatment was changed to SC calcitriol at the same dosage (Calcijex® injectable formulation). In addition, despite some clinical improvement on oral prednisolone, the glucocorticoid regimen was changed to SC dexamethasone, at 1/7 of the prednisone dose to account for increased potency, also to guarantee absorption. Because SC cyclosporine was not an option, cyclosporine (modified) was continued orally with plans for monitoring of whole blood drug concentrations. Modified cyclosporine (found in Neoral and Atopica) was chosen because of its higher bioavailability compared to regular cyclosporine (e.g. Sandimmune) in dogs. [4]

Four days after changing to SC calcitriol, the dog's calcium had improved (total 5.9 mg/dl, ionized 1.03 mmol/L), with a serum magnesium of 1.0 mg/dl. The dog was clinically doing well, and was weaned from IV calcium gluconate and discharged on SC calcitriol, oral Tums, SC dexamethasone, oral cyclosporine (modified), and low dose aspirin. The dog was rechecked frequently for calcium, magnesium, albumin, and clinical status, and a good appetite and normal stools were reported at each visit. Two months after initiation of cyclosporine and SC calcitriol, the dog finally had a normal serum albumin (3.6 g/dl), ionized calcium (1.25 mmol/L), total magnesium (2.6 mg/dl), and cholesterol (177 mg/dl), but had continued moderate ascites. The dog is doing well 4 months after diagnosis on tapering dosages of dexamethasone and calcitriol, although the long-term prognosis is still guarded.

Case assessment: Low dose calcitriol has an elimination half-life of about 24 hours in dogs, [5] so steady state concentrations are reached in approximately 4 to 5 days, with clinical response usually seen sooner. Because of the inadequate response to oral calcitriol (which had been clinically effective in our hands in other Yorkies with lymphangiectasia), we suspected calcitriol malabsorption in this dog. This would make sense, since the etiology of the hypocalcemia and hypomagnesemia is thought to be intestinal vitamin D malabsorption, at least in part. [6] We did not supplement magnesium directly, because we have previously observed that magnesium in these dogs improves with calcitriol supplementation; however, magnesium supplementation may have been helpful. In retrospect, proactive parenteral calcitriol supplementation, started at the time of initiation of glucocorticoids, would have likely improved management. The dog's hypocalcemia worsened on prednisolone and oral calcium carbonate, despite improvement in stools and appetite. This may have been due in part to increased calciuresis from glucocorticoids. [7]

NOTHING BUT TROUBLE

Case Three

A 5 yo Mn Labrador mix was diagnosed with chronic hepatitis with bridging fibrosis on prior liver biopsy. Treatment with SAMe, ursodiol, colchicine, and a low-zinc diet brought some clinical improvement, but serum ALT activities continued to climb. Oral zinc was not tolerated. Prednisone led to unacceptable polyuria and polydipsia, so budesonide was substituted. However, dosages of budesonide (1 mg daily) that led to reductions in serum ALT activities were associated with progressive muscle atrophy. Cyclosporine at 5 mg/kg BID was added to allow dosage reductions in budesonide, to 1 mg every other day; this combination was associated with near normalization of ALT, although appetite was somewhat subdued. Appetite improved when the dosage of cyclosporine was reduced to 3 mg/kg BID, and the dog was finally doing well.

Several months after starting cyclosporine, the dog developed SC masses on the nasal planum. Excisional biopsy revealed pyogranulomatous inflammation with fungal hyphae. Fluconazole was started (5 mg/kg BID), and the dog was monitored for recurrence. A dosage reduction or discontinuation of cyclosporine was discussed, but the owners were reluctant to immediately change the regimen that had brought optimal clinical control of the dog's chronic hepatitis. Two weeks later, the dog developed inappetance. Serum ALT was 1295 IU/L, the highest ever. Since no changes had been made except for the addition of fluconazole, drug-induced hepatotoxicity was suspected. Fluconazole was discontinued; itraconazole was not substituted because it also carries a risk of hepatotoxicity.[8] Without antifungal prophylaxis, continued immunosuppression with cyclosporine could not be recommended given this dog's history, so cyclosporine was also discontinued. The dog's appetite normalized with these two changes, and serum ALT fell to previous moderate elevations (391 IU/L). The dog is now stable (6 years after diagnosis) on budesonide, SAMe, ursodiol, and colchicine, although muscle atrophy is considerable.

Case assessment: These owners were very dedicated and patient throughput various therapeutic misadventures. This dog developed dose-dependent side effects that were an extension of drug pharmacologic effects (PU/PD from prednisone, muscle atrophy from budesonide, secondary infections on cyclosporine), and dose-dependent effects that were independent of the drug's intended effects (inappetance from cyclosporine, and hepatotoxicity from fluconazole). Fluconazole reportedly has a lower risk of hepatotoxicity than itraconazole (at least in rodents), [9] which is why it was chosen in this dog with underlying liver disease. However, we have since observed several dogs (with Blastomycosis) that have developed reversible moderate to marked increases in serum ALT activities on fluconazole treatment. Monitoring of serum ALT appears to be important during fluconazole therapy in dogs.

REFERENCES

[1.] Lavergne SN, Trepanier LA. Anti-platelet antibodies in a natural animal model of sulphonamide-associated thrombocytopaenia. **Platelets** 2007;18:595-604. [2.] McCarthy DM. Sucralfate. **N Engl J Med** 1991;325:1017-1025. [3.] Van Slooten AD, Nix DE, Wilton JH, et al. Combined use of ciprofloxacin and sucralfate. **DICP** 1991;25:578-582. [4.] Fischer JR, Harkin KR, Freeman LC. Concurrent administration of water-soluble vitamin E can increase the oral bioavailability of cyclosporine a in healthy dogs. **Vet Ther** 2002;3:465-473. [5.] Rassnick KM, Muindi JR, Johnson CS, et al. In vitro and in vivo evaluation of combined calcitriol and cisplatin in dogs with spontaneously occurring tumors. **Cancer Chemother Pharmacol** 2008;62:881-891. [6.] Kimmel SE, Waddell LS, Michel KE. Hypomagnesemia and hypocalcemia associated with protein-losing enteropathy in Yorkshire terriers: five cases (1992-1998). J **Am Vet Med Assoc** 2000;217:703-706. [7.] Patschan D, Loddenkemper K, Buttgereit F. Molecular mechanisms of glucocorticoid-induced osteoporosis. **Bone** 2001;29:498-505. [8.] Legendre AM, Rohrbach BW, Toal RL, et al. Treatment of blastomycosis with itraconazole in 112 dogs. **J Vet Intern Med** 1996;10:365-371. [9.] Somchit N, Norshahida AR, Hasiah AH, et al. Hepatotoxicity induced by antifungal drugs itraconazole and fluconazole in rats: a comparative in vivo study. **Hum Exp Toxicol** 2004;23:519-525.

THERAPEUTIC MISADVENTURES: CASES OF DRUG INTERACTIONS, DRUG TOXICITY, & OTHER DISASTERS: PART II

Lauren A. Trepanier, DVM, PhD, Dip. ACVIM, Dip. ACVCP
Madison, WI

NOW YOU TELL ME

Case Four

A 7 yo Mn Persian cat was evaluated for two seizure episodes over the preceding weekend. The owners reported possible staring episodes over the previous few months. A male littermate had died after apparent status epilepticus the previous year. There were no know toxin exposures, and the cat was not on any medications. On physical exam, the cat was dull and post ictal (he had seizured again on the way to the hospital). The cat was otherwise hydrated, with a BCS of 5/9, pink mucous membranes, a normal CRT, normal rectal temperature, no murmur or gallop, and no evidence of abdominal masses. Retinal exam was normal. CBC revealed microcytosis (38.3 fl), a lymphocytosis, and occasional toxic neutrophils. The serum biochemical panel showed a low BUN (11 mg/dl), low normal albumin (2.8 g/dl), an increased ALT (365 IU/L), and increased SAP (276 IU/L). Total bilirubin was increased at 0.8 mg/dl, but the sample was slightly hemolyzed. Urinalysis revealed a few ammonium biurate crystals. Serum bile acids were markedly abnormal at 135 and 251 umol/L, as was blood ammonia (511 umol/L). The cat was treated with IV fluids, lactulose, and neomycin, followed by k/d diet once he improved neurologically. Clindamycin therapy was instituted while awaiting Toxoplasmosis titers.

Abdominal ultrasound revealed an enlarged and tortuous portal vein. Platelets, PT, and APTT were within reference ranges, and the cat underwent exploratory laparotomy. At surgery, the liver was noted to be discolored and irregular in texture, and the portal vein was indeed tortuous. A splenic portogram showed no extrahepatic shunting, and the abnormal vessel was not ligated. A liver biopsy was obtained, which revealed severe chronic lymphoplasmacytic and neutrophilic cholangiohepatitis with bridging fibrosis. Ursodiol and prednisone (5 mg twice daily) were added to the cat's treatment regimen.

On recheck one week later, the cat was doing well, and the ALT and bilirubin were normal, as was blood ammonia (36 umol/L). Toxoplasma IgM and IgG titers were returned negative. The cat was followed by the referring veterinarian, and the prednisone dosage was tapered over time to 5 mg once daily. On phone conversation one year later, the owner reported that the cat was doing very well, and that all medications had been discontinued. The owner mentioned that she and her husband had recently gotten a divorce, and that her ex had been giving the cat (and its deceased sibling) herbal supplements. Now that the cat was living with her, the supplementation had stopped.

<u>Case assessment</u>: This case was unusual in several respects: the late age of presentation of an initially suspected portosystemic shunt; the prior history of a sibling with similar clinical signs; the presence of acquired shunting at laparotomy in a cat; and the presence of hepatic encephalopathy in a cat with cholangiohepatitis, without jaundice. Our overall assessment was that this cat may have developed hepatotoxicity from one of the herbal supplements; however, the owner did not know which ones had been given, and repeat liver biopsy was not performed. Despite repeated drug history queries, the history in this case initially failed to reveal the administration of herbal supplements as a possible underlying cause of acute liver failure.

SULFA TROUBLE

Case Five

A 1 yo Fs Golden retriever was presented for stranguria and pollakiuria. Trimethoprim-sulfamethoxazole (15 mg/kg BID) was prescribed for 14 days. Twelve days after starting the drug, the dog became inappetant. Bloodwork revealed an ALT that was out of range for the analyzer, a bilirubin of 7.4 mg/dl, and 4,100 neutrophils. The dog was treated with IV fluids and was referred for evaluation two days later.

On presentation, the dog was depressed but responsive. Mucous membranes were jaundiced with oral ulcers present. Ocular exam was normal. The dog was afebrile with normal heart and respiratory rates. Abdominal palpation revealed cranial abdominal discomfort. There was edema along the ventral abdomen and the hocks. Both forelegs had been shaved and were bruised. CBC revealed 5,100 neutrophils, a normal HCT, and 40,000 platelets, with an increased MPV. The biochemical panel showed similar abnormalities from two days prior, with marked increases in ALT (4401 IU/L), SAP (3138 IU/L), and bilirubin (7.5 mg/dl), and decreases in albumin and BUN. The dog was treated with ½ strength saline with 2.5% dextrose plus KCl, famotidine, and doxycycline while awaiting Leptospira titers. Abdominal ultrasound revealed a subjectively enlarged liver with normal echogenicity, and no evidence of biliary obstruction, cholecystitis, or pancreatitis. Schirmer tear tests were normal.

Idiosyncratic sulfonamide toxicity was suspected, and trimethoprim-sulfamethoxazole was discontinued. The dog was treated with N-acetylcysteine (140 mg/kg IV, followed by 70 mg/kg IV q. 6 hours for 48 hours) as a glutathione precursor. The platelet count increased to 147,000 over 3 days, the oral lesions began to resolve, and the dog began to eat well. She was discharged on famotidine, SAMe, and 2 weeks of doxycycline. One week after admission, the platelet count was normal, the bilirubin was down to 4.8 mg/dl, and the ALT was 1269 IU/L.

<u>Case assessment</u>: This dog had a classic presentation for idiosyncratic sulfonamide toxicity: a delay in onset from drug initiation to clinical signs of typically 5 to 14 days, a transient neutropenia, transient or persistent thrombocytopenia, and evidence of both hepatocellular and cholestatic changes in liver enzymes. This dog was afebrile, as are about 50% of dogs with sulfonamide hypersensitivity. [1] In addition, the platelet count rose rapidly after drug discontinuation, without the need for glucocorticoids (although this dog was also treated with N-acetylcysteine). We have found anti-platelet antibodies in 90% of dogs with sulfonamide hypersensitivity and thrombocytopenia. [2] These antibodies may either target the platelet itself, or the platelet

only in the presence of drug; [2] we have yet to confirm whether a more rapid recovery after drug discontinuation is associated with the latter.

Sulfonamide hypersensitivity appears to be triggered by a reactive nitroso sulfonamide metabolite, which forms drug adducts that act as haptens. [3] We and others have shown that this metabolite can be reduced by antioxidants such as ascorbate, cysteine, and glutathione. [4,5] In addition, we have shown that sulfonamide-tissue adducts can be resolved *in vitro* in dog liver by exposure to either ascorbate or glutathione (Lavergne and Trepanier, unpublished observations). These findings provided the rationale for treatment of this dog with the glutathione precursor, N-acetylcysteine.

The incidence of sulfonamide hypersensitivity in dogs is not clear, but was reported to be 0.25% in one survey in the Netherlands. [6] Routine monitoring of blood-work is not recommended for idiosyncratic drug toxicities. Instead, careful clinical monitoring of dogs for any abnormal behavior by owners, followed by prompt communication, drug discontinuation, and clinical evaluation, is recommended. Because KCS from sulfonamides has a higher incidence (about 15%), [7] and appears to be dose- and duration-dependent, routine pre-treatment Schirmer tear testing, as well as repeat testing after 10 or more days of potentiated sulfonamides, is probably wise. Sulfonamide-induced KCS can be reversible if caught early.[7]

PHARMACOLOGY ROULETTE

Case Six

An 8 yo Mn Samoyed presented with a fever and cough, and was prescribed enrofloxacin and hydrocodone. One week later, signs had not resolved. A chest radiograph revealed a nodular interstitial pattern and tracheobronchial lymphadenopathy, and clindamycin was substituted. Two weeks later, a mucopurulent nasal discharge developed, and a course of doxycycline was added. The next day, the dog had two seizures, and phenobarbital was prescribed. The dog was referred for an MRI, which revealed multiple cystic masses in the brain. Prednisone was then added to the dog's regimen of clindamycin and phenobarbital. Five days later, a skin lesion was noted on the right muzzle, and Blastomyces organisms were found on skin cytology. The dog was treated with fluconazole 6 mg/kg BID, and prednisone, clindamycin and phenobarbital were continued.

Three weeks later, bloodwork revealed new liver abnormalities, to include an SAP of 4060 IU/L, ALT of 637 IU/L, and a bilirubin of 1.1 mg/dl. Phenobarbital was discontinued, and SAMe/milk thistle was instituted. One week later, the liver enzyme elevations had progressed and the bilirubin was 3.1 mg/dl. The dog became inappetant and was referred for evaluation.

On presentation, the dog was febrile and panting. There was scant mucupurulent nasal discharge and stertorous breathing. No heart murmur or arrhythmias were ausculted. Abdominal palpation was negative for pain, fluid, or organomegaly. There were multiple hairless erythematous skin lesions on the digits and perianal area, with linear ulcers on both sides of the tongue, and a focal ulcer on the hard palate. The dog appeared anxious but was cerebrally appropriate, with no localizing lesions on neurologic evaluation.

CBC revealed a stress leukogram with no left shift or toxic change. Platelet count and HCT were normal, with no schistocytes noted. The bilirubin had increased to 4.5 mg/dl, with a continued cholestatic pattern. The urine was positive for biluribin but negative for protein, with an inactive sediment. Abdominal ultrasound revealed a hyperechoic liver (consistent with recent phenobarbital and continued prednisone) with multiple hypoechoic nodules, a normal appearing gall bladder, and no evidence of dilation of the common bile duct. Chest radiographs showed no evidence of the previously identified pulmonary lesions (4 weeks into fluconazole therapy). Leptospirosis titers and cPLI were both submitted; all Lepto serovars were < 1:100 (drawn more than two weeks after hepatopathy noted), but cPLI returned later as > 1000 ug/L.

The owners were unable to re-hospitalize the dog or biopsy the liver after the cost of all of the preceding veterinary visits. The dog was drinking and hydrated. A drug-induced hepatopathy was suspected, or alternatively, infiltrative liver disease (although Blastomycosis does not typically affect the liver). An adverse reaction to fluconazole was considered possible, since liver enzymes had been normal at initial workup. Although phenobarbital hepatotoxicity could have a similar presentation, this would be unusual after only 3 weeks of treatment, and liver biochemistries had progressed despite phenobarbital discontinuation one week earlier.

Discontinuation of prednisone (which could impair clearance of Blastomyces) and clindamycin (which did not have a good rationale) was recommended. Fluconazole discontinuation was considered. However, given the brain lesions (which were presumed to be Blastomycosis, although CSF had never been evaluated), prior lung lesions and cough, and prior draining skin lesions, antifungal treatment was considered essential. Intravenous amphotericin B lipid complex was offered, but was not financially possible. In addition, brain concentrations of amphotericin B are poor.[8] Because hepatopathy from other azole antifungals has been shown to be dependent on plasma concentrations, [9] and fluconazole reaches high CNS concentrations,[9] we elected to continue fluconazole, but decrease the dosage by 50%, to 6 mg/kg once daily, and monitor closely.

The toe, perianal, and oral lesions were most consistent with a vasculitis, although biopsy was not allowed. These may have been either drug-induced, or due to fungal "die-off" in the first weeks of treatment for Blastomysosis. [10] Without the ability to further investigate these lesions, the plan was to carefully monitor their appearance daily on the streamlined drug regimen and a lower dosage of fluconazole.

The dog was followed at the referring vet and via fax and phone communication. The hyperbilirubinemia gradually resolved, as did the skin and oral lesions, over the next few weeks. The dog's appetite returned as the oral lesions resolved. On full recheck at the VMTH 3 months later, the dog's bilirubin was 0.1 mg/dl, the SAP was 138 IU/L, and the ALT was 88 IU/L. The dog was still being treated with fluconazole 6 mg/kg once daily, along with Denamarin and famotidine. Chest radiographs were normal, and no seizures had been noted. The dog continues to do well 1½ years after diagnosis, and is still on fluconazole. Subsequent MRI evaluations have shown almost complete resolution of previous brain lesions.

Case assessment: This case was notable for the prolonged survival of a dog with presumptive CNS Blastomycosis, for administration of multiple drugs which made evaluation of a new hepatopathy difficult, and for the resolution of an acute hepatopathy with only supportive care (Denamarin), discontinuation of clindamycin, and a dosage reduction of fluconazole.

Fluconazole has been associated with hepatopathy in both human patients and rodent models. [11,12] We have also observed hepatopathy in other dogs with treated with fluconazole (primarily those with Blastomycosis, but also Case Three in this series), although the pattern has typically been hepatocellular (fold increase in ALT > fold increase in SAP), not cholestatic. However, fluconazole can cause cholestatic hepatopathy in newborn infants, [13] and may have caused cholestasis in this dog.

Prednisone was thought to contribute to the increased SAP, but could not explain the hyperbilirubinemia. Clindamycin was a possible trigger for hepatotoxicity, but was considered least likely because it has not been reported to cause hepatopathy in dogs. However, a case of cholestatic hepatopathy associated with clindamycin was recently reported in a human patient, [14] and it is possible that clindamycin was a factor in this dog. There was certainly a risk in continuing fluconazole in this dog, given the rising bilirubin and skin lesions; however, we had few options at that time, and we had excellent cooperation and communication with the new referring vet and the owner in monitoring the dog in the first few weeks after discharge.

REFERENCES

[1.] Trepanier LA, Danhof R, Toll J, et al. Clinical findings in 40 dogs with hypersensitivity associated with administration of potentiated sulfonamides. **J Vet Intern Med** 2003;17:647-652. [2.] Lavergne SN, Trepanier LA. Anti-platelet antibodies in a natural animal model of sulphonamide-associated thrombocytopaenia. **Platelets** 2007;18:595-604. [3.] Trepanier LA. Idiosyncratic toxicity associated with potentiated sulfonamides in the dog. **J Vet Pharmacol Ther** 2004;27:129-138. [4.] Naisbitt D, Vilar F, Stalford A, et al. Plasma cysteine deficiency and decreased reduction of nitrososulfamethoxazole with HIV infection. **AIDS Research amd Human Retroviruses** 2000;16:1929-1938. [5.] Trepanier LA, Yoder AR, Bajad S, et al. Plasma ascorbate deficiency is associated with impaired reduction of sulfamethoxazole-nitroso in HIV infection. **J Acquir Immune Defic Syndr** 2004;36:1041-1050. [6.] Noli C, Koeman J, Willemse T. A retrospective evaluation of adverse reactions to trimethoprim- sulphonamide combinations in dogs and cats. **Veterinary Quarterly** 1995;17:123-128. [7.] Berger SL, Scagliotti RH, Lund EM. A quantitative study of the effects of Tribrissen on canine tear production. **J Am Anim Hosp Assoc** 1995;31:236-241. [8.] Bellmann R. Clinical pharmacokinetics of systemically administered antimycotics. **Curr Clin Pharmacol** 2007;2:37-58. [9.] Legendre AM, Rohrbach BW, Toal RL, et al. Treatment of blastomycosis with itraconazole in 112 dogs. **J Vet Intern Med** 1996;10:365-371. [10.] Nichols PR, Morris DO, Beale KM. A retrospective study of canine and feline cutaneous vasculitis. **Vet Dermatol** 2001;12:255-264. [11.] Chen SC, Sorrell TC. Antifungal agents. **Med J Aust** 2007;187:404-409. [12.] Somchit N, Norshahida AR, Hasiah AH, et al. Hepatotoxicity induced by antifungal drugs itraconazole and fluconazole in rats: a comparative in vivo study. **Hum Exp Toxicol** 2004;23:519-525. [13.] Aghai ZH, Mudduluru M, Nakhla TA, et al. Fluconazole prophylaxis in extremely low birth weight infants: association with cholestasis. **J Perinatol** 2006;26:550-555. [14.] Aygun C, Kocaman O, Gurbuz Y, et al. Clindamycin-induced acute cholestatic hepatitis. **World J Gastroenterol** 2007;13:5408-5410.

THE YEAR IN REVIEW: HEMATOLOGY:

Stephanie A. Smith DVM, MS, Diplomate ACVIM
Urbana, IL

1. Bianco D, Armstrong PJ, Washabau RJ. A prospective, randomized, double-blinded, placebo-controlled study of human intravenous immunoglobulin for the acute management of presumptive primary immune-mediated thrombocytopenia in dogs. **J Vet Intern Med**. 2009;23(5):1071-1078.

2. Helmond S, Polzin, D, Armstrong PJ, Finke M, Smith SA. Treatment of immune-mediated hemolytic anemia with individually adjusted heparin dosing in dogs. **J Vet Intern Med** 2010: In Press.

3. Horgan JE, Roberts BK, Schermerhorn T. Splenectomy as an adjunctive treatment for dogs with immune-mediated hemolytic anemia: ten cases (2003-2006). **J Vet Emerg Crit Care**. 2009;19(3):254-261.

4. Botsch V, Kuchenhoff H, Hartmann K, Hirschberger J. Retrospective study of 871 dogs with thrombocytopenia. *Vet Rec.* 2009;164(21):647-651

5. Brown HM, Berghaus RD, Latimer KS et al. Genetic variability of Cytauxzoon felis from 88 infected domestic cats in Arkansas and Georgia. **J Vet Diagn Invest**. 2009;21(1):59-63.

6. Dircks BH, Schuberth HJ, Mischke R. Underlying diseases and clinicopathologic variables of thrombocytopenic dogs with and without platelet-bound antibodies detected by use of a flow cytometric assay: 83 cases (2004-2006). **J Am Vet Med Assoc**. 2009;235(8):960-966.

7. Scott KC, Hansen BD, DeFrancesco TC. Coagulation effects of low molecular weight heparin compared with heparin in dogs considered to be at risk for clinically significant venous thrombosis. **J Vet Emerg Crit Care**. 2009;19(1):74-80.

8. Smart L, Jandrey KE, Kass PH, Wierenga JR, Tablin F. The effect of Hetastarch (670/0.75) in vivo on platelet closure time in the dog. **J Vet Emerg Crit Care**. 2009;19(5):444-449.

9. Tasca S, Carli E, Caldin M et al. Hematologic abnormalities and flow cytometric immunophenotyping results in dogs with hematopoietic neoplasia: 210 cases (2002-2006). *Vet Clin Pathol.* 2009;38(1):2-12.

10. Yeagley TJ, Reichard MV, Hempstead JE et al. Detection of Babesia gibsoni and the canine small Babesia 'Spanish isolate' in blood samples obtained from dogs confiscated from dogfighting operations. **J Am Vet Med Assoc**. 2009;235(5):535-539.

11. Rogers CL, Rozanski EA. Von Willebrand Factor Antigen Concentration in Dogs with Sepsis. **J Vet Intern Med** 2010; 24(1): 229-230.

CASE PRESENTATION: ATYPICAL IMMUNE-MEDIATED HEMOLYTIC ANEMIA IN A DOG

Andrew Mackin, BSc, BVMS, MVS, DVSc, FACVSc, DACVIM
Starkville, MS

CASE PRESENTATION

Relevant History

Katie was a six year old spayed female Australian Shepherd dog from rural western Alabama that presented to the Mississippi State University College of Veterinary Medicine (MSU-CVM) Animal Internal Medicine service with an eight day history of lethargy, anorexia, and fever. She originally presented to the referring veterinarian six days prior to referral with a three day history of lethargy and anorexia. Records from the referring veterinarian at the time of presentation reported that she was febrile, leukopenic, and anemic, with a packed cell volume (PCV) of 20%. The referring veterinarian started Katie on oral doxycycline (150 mg twice daily), gave several intramuscular injections of enrofloxacin and cefazolin, and referred her to MSU-CVM after her PCV decreased to 17% and she had not improved clinically.

Initial Investigation

Upon presentation as an emergency referral patient at MSU-CVM, Katie's physical examination revealed an elevated temperature (103.3° F), moderate depression, and slightly pale mucous membranes. Body weight (32 lbs) and body condition were normal. Initial testing consisted of a complete blood count, serum biochemistry, reticulocyte count, urinalysis, abdominal and thoracic radiographs, abdominal ultrasound, and a SNAP heartworm/*Ehrlichia* test.

Hematology

WBC/µL	4,700	(5,800-11,700)
Band	2,444	(<300)
Neutrophils	1,222	(3,000-7,100)
Lymphocytes	846	(1,100-5,100)
Monocytes	188	(<800)
Eosinophils	0	(<500)
Basophils	0	(0)
RBC, $\times 10^6$/uL	2.17	(5.6-8.5)
Hgb, g/dL	5.1	(13-19)
HCT, %	15.6	(38-59)
nRBC/100 WBC	2	(0)
MCV, fL	71.5	(63-73)
MCHC, g/dL	32.9	(33-37)
RDW, %	13.4	(13-16)
Platelets, $\times 10^3$/µL	19	(157-394)
Reticulocytes/µL	28,210	(<70,000)

Serum Chemistry

BUN, mg/dL	67	(9-27)
Creatinine, mg/dL	0.9	(0.7-1.2)
Glucose, mg/dL	67	(70-117)
ALT, IU/L	66	(21-67)
Chol, mg/dL	137	(123-363)
AP, IU/L	70	(14-72)
T. Prot, g/dL	6.4	(5.6-7.6)
Albumin, g/dL	3.6	(2.5-3.8)
Globulin, g/dL	2.8	(2.7-3.8)
Calcium, mg/dL	9.1	(9.1-12.0)
P, mg/dL	5.9	(3.3-6.0)
Sodium, mEq/L	148	(147-157)
Potassium, mEq/L	4.5	(4.4-5.3)
Chloride, mEq/L	113	(111-122)
Anion Gap, mEq/L	20.5	(15-25)
TCO_2, mEq/L	19	(18-24)

RBC morphology	Slight anisocytosis Slight poikilocytosis
Platelet estimate	Slightly decreased (estimate 128 $\times 10^3$ platelets/µL)

Urinalysis

Collection	cystocentesis
Color	amber
Urine SG	1.046
Prot	2+
Blood	-ve
Ketones	-ve
Bilirubin	moderate
pH	6.0
RBC/HPF	0
WBC/HPF	1-5
Casts/LPF	none
Crystals	none
Other	few lipid droplets

Abdominal and Thoracic Radiographs

Radiographs of the abdomen (three views) revealed no significant abnormalities. Thoracic radiographs (three views) were also within normal limits.

Abdominal Ultrasound

Abdominal ultrasound revealed that both the liver and spleen were moderately hypoechoic. The radiologist's rule-outs included extramedullary hematopoiesis, diffuse inflammation or diffuse neoplasia.

Heartworm/*Ehrlichia* Test

Heartworm/*Ehrlichia* SNAP test was negative.

FURTHER INVESTIGATION

Because of her marked anemia and depressed mental state, Katie was considered to be a candidate for transfusion with packed red blood cells. Shortly after admission, she was transfused with one unit of cross-match compatible universal donor packed cells. Prior to transfusion, samples were collected for a clinical pathologist review of a fresh blood smear, in-saline slide agglutination, and Coombs testing. Shortly after transfusion, Katie was sedated for bone marrow aspiration and core biopsy.

Clinical Pathologist Blood Smear Review

A blood smear from a six year old female Australian Shepherd is submitted for review. The red cell density appears moderately decreased consistent with a moderate anemia. Red cell morphology is within normal limits. No polychromasia or anisocytosis is observed indicating that the anemia is non-regenerative at this point. Leukocytes consist almost entirely of hyposegmented neutrophils with a mature chromatin pattern. Rare eosinophils have a hyposegmented nucleus as well. There is a lymphocytopenia. Platelets appear slightly decreased and occasional megaplatelets are observed.

Conclusion: *Moderate non-regenerative anemia; mild thrombocytopenia; Pelger-Huet anomaly.*

Slide Agglutination

Slide agglutination with saline dilution was performed using freshly collected pre-transfusion EDTA-anticoagulated fresh whole blood. Both observation for gross agglutination and microscopic examination for microagglutination were negative.

Coombs Test

Direct Coombs test was negative.

Bone Marrow Aspiration

Bone marrow aspirates were collected into EDTA anticoagulant under sedation from the iliac crest. Shortly after collection, marked gross agglutination of the anticoagulated marrow samples was noted. Gross agglutination failed to clear with saline dilution, and was confirmed as marked microagglutination of marrow red blood cells upon microscopic examination.

Cytology Report:
The bone marrow aspirates are of extremely high cellularity and good quality. The erythroid series is present in large numbers and all stages of maturity are represented in expected percentages. Occasional erythrophagocytosis of metarubricytes and rubricytes is observed. The myeloid series is present in lower numbers and all stages other than segmented neutrophils are present. The majority of the band neutrophils observed have a condensed mature chromatin pattern. The M:E ratio is decreased (0.3) most likely due to both erythroid hyperplasia and some degree of myeloid hypoplasia. Scattered megakaryocytes are observed. Bone marrow iron appears to be within normal limits. Increased numbers of plump spindle cells are observed and may indicate fibrosis. The red cells within the background exhibit microagglutination.

Diagnosis: *Erythroid hyperplasia; Pelger-Huet anomaly.*

Comments: *The erythroid hyperplasia observed may indicate that the anemia is about to regenerate. However, if no regeneration is observed in 1-2 days, a myeloproliferative disorder should be considered.*

Bone Marrow Core Biopsy

A bone marrow core biopsy was collected from the iliac crest immediately after collection of bone marrow aspirates for cytologic analysis.

Histopathology Report:
A bone marrow core biopsy is submitted. The bone marrow is markedly hypercellular with a cell to fat ratio of approximately 95:5. The marrow is populated by large numbers of erythroid series cells that include both blast type cells that cannot be identifies as to definitive lineage but also metarubricytes and numerous nucleated red blood cells. Only occasional deposits of hemosiderin are present. Numerous megakaryocytes are seen. Myeloid series cells are also identified.

Comments: *The erythroid series cells appear hyperplastic in the bone marrow core. The cellularity of the marrow is profound.*

Further Marrow Testing

Air-dried unstained bone marrow aspiration samples were submitted to Kansas State University for immunofluorescent staining for cell-associated immunoglobulins.

Result: *No bone marrow precursor cells showed specific/non-specific staining by anti-canine IgG. Isotype control antibody was also negative.*

DISCHARGE

Katie was bright and alert post-transfusion, with a PCV of 29%. Since she was stable, she was discharged without therapy, except for ongoing oral doxycycline, pending results of bone marrow histopathology and immunofluorescent staining. Short-term plans were to decide on therapy based on the results of marrow testing, to recheck a PCV and reticulocyte count in 3-4 days, and to repeat *Ehrlichia* serology in 2-4 weeks to detect potential seroconversion.

CLINICAL QUESTIONS STILL TO BE ADDRESSED

1. Based on final test results, what is/are the most likely diagnosis/diagnoses?

2. What treatment, if any, should be instituted?

Note:

Katie presented to the MSU-CVM in October of 2009, and only data from her first visit is presented in the preceding written case summary. Subsequent follow-up visits and response to treatment will be discussed during the case presentation at the ACVIM Forum.

CLINICAL PATHOLOGY OF RENAL PHYSIOLOGY

Mary B. Nabity, DVM, DACVP
College Station, TX

INTRODUCTION

The kidney influences the majority of analytes on a standard chemistry panel; however, this discussion will focus primarily on the clinical pathology indicators of renal disease. These renal biomarkers can be those detected in blood due to decreased removal by the kidney, and those present in the urine due to decreased function of the glomeruli, tubules, or both. However, no test is perfect, and a solid understanding of the tests for renal disease and function (and the non-renal factors that influence them) is key in accurate diagnosis of renal disease. The standard biomarkers of renal function used to diagnose kidney disease include serum creatinine and urea nitrogen, urine specific gravity, and the urine protein:creatinine ratio. However, do you always feel comfortable with what these tests are truly measuring, their interpretation, and the mechanisms causing an abnormal value? Do you remember to consider conditions other than renal disease that can affect these values? The goal of this session is to review the most commonly used indicators of renal function and those influencing factors that may be overlooked, underestimated, or unknown, leading to a misunderstanding or misinterpretation of these analytes. In addition, this session will introduce some new biomarkers and tests of renal disease that are currently undergoing investigation.

STANDARD TESTS OF RENAL FUNCTION

Serum creatinine and urea nitrogen (UN, BUN, SUN)

Serum creatinine and UN are used as endogenous markers of glomerular filtration rate (GFR). GFR is currently considered the gold standard of kidney function testing, and it represents the volume of plasma cleared of its marker substance by the kidney in a unit of time. At the glomerular level, GFR is determined primarily by the glomerular capillary hydrostatic pressure.[1] This stems largely from systemic arterial pressure, which explains why marked hypotension will decrease GFR, especially when local afferent and efferent arteriolar tone are unable to compensate for changes in systemic pressure. Similarly, increased tubular pressure due to a urinary tract obstruction will decrease GFR. GFR is also highly dependent on the glomerular capillary surface area and permeability, which is why lesions such as glomerulosclerosis and immune complex deposits decrease a single nephron's GFR, and non-functional nephrons decrease overall glomerular filtration.

The ideal molecule for measuring GFR does not exist, although nearly ideal solutes include inulin and iohexol. All current measures of GFR using nearly ideal solutes are relatively time-consuming and/or expensive and typically only performed at referral hospitals. That is why the less ideal (but pretty good) analyte, endogenous creatinine, is more commonly used as a rough "spot" estimation of GFR. Creatinine is produced at a relatively constant rate throughout the day, is not protein-bound, freely passes through the glomerular filtration barrier, does not undergo tubular reabsorption, and is minimally/insignificantly secreted by the tubules in veterinary patients. So why is it not an ideal indicator of GFR? Creatinine production is primarily determined by the muscle mass of the patient; therefore, it may overestimate renal function in cachectic, geriatric, and very young patients, and breed and weight must be considered in dogs.[2] Because of this, basing a normal vs. abnormal value on the reference interval will be inaccurate in many patients. For example, a value of 1.4 mg/dl is within most reference intervals used. However, this value may actually be lower than expected for a normal greyhound, whereas it may be significantly elevated in a thin, geriatric toy poodle. Even a creatinine of 1.0 mg/dl can be abnormally high in a puppy. Therefore, one of the most important concepts in creatinine measurement is that each animal serves as its best reference. While inter-individual variation can be quite large, intra-individual variation in a healthy adult animal is small over weeks to months, and even years.[3] Serial monitoring in a patient is therefore the best way to detect mild elevations in its creatinine value that may be associated with significant renal disease. Remember that with any serial monitoring, measurements must be performed by the same laboratory with the same instrument to minimize analytical variation.

In addition to muscle mass, several other factors influence creatinine levels. Tubular secretion of creatinine increases as blood levels rise, and this may overestimate GFR in renal failure patients.[4] Meat diets may increase serum creatinine up to 12 hrs after ingestion[3], and some sources report that severe rhabdomyolysis may also increase serum creatinine[4,5], but this is most likely because of myoglobinuric nephropathy. Creatinine should not be interpreted as an indicator of muscle damage or active breakdown. Never interpret increased creatinine by itself—always relate it to urine specific gravity before suspecting renal disease! Lastly, it is important to remember that serum creatinine is a rather insensitive indicator of renal disease, as >75% of kidney function must be lost in order to observe an increase, and it is particularly insensitive to marked decreases in GFR that occur early in the disease process.[3] Hence, small increases in creatinine at the low end of the reference interval are associated with large decreases in GFR.

UN is also considered an indicator of GFR; however, it is not reliable since its production is not constant, it undergoes tubular reabsorption, and it can be eliminated extrarenally. Decreased production of urea due to liver failure or a low protein diet may result in normal levels despite decreases in GFR. Alternatively, intestinal bleeding, high protein diets, and catabolic states can mildly increase UN despite a normal GFR. Therefore, it is primarily evaluated in conjunction with serum creatinine. The UN:creatinine ratio has been proposed by some as a way to help distinguish between different causes of azotemia, with a ratio >30 supporting prerenal azotemia.[5] This is because UN is highly reabsorbed by the tubules, so a decreased urine flow rate, as with hypovolemia and hypotension, can result in a higher increase in UN as compared to creatinine, which is not reabsorbed. However, there is a high degree of overlap in the UN:creatinine ratio with prerenal, renal, and postrenal causes of azotemia.[5] Therefore, strong conclusions based on a comparison of UN and creatinine should only be made when the UN is mildly to moderately increased but creatinine is clearly normal, or less commonly, when UN is decreased with a normal creatinine. In these cases, increased protein catabolism and severe liver disease, respectively, are most likely.

Urine specific gravity (USG)

USG as measured by refractive index is a reasonably good measure of urine osmolality, which reflects concentration of the urine. Most refractometers used in clinical practice have USG calibration scales based on the normal composition of human urine.[6] However, a refractometer with a calibration scale designed for veterinary patients is more accurate. These refractometers have one scale for dogs and large animals, and another scale for cats.[5] If the human or canine scale is used for cats, the USG will be overestimated. For example, a USG of 1.040 on the human scale corresponds to 1.034 on the feline scale.[6] Remember that marked proteinuria or glucosuria overestimates the USG (approximately 0.004 for every 1 g/dL) and can overestimate concentrating ability.[5] Specific gravity should never be determined using a reagent strip!

Clinical assessment of USG can be difficult, as it will be highly dependent on the hydration status of the patient and certain medications it might be receiving. For example, a normal dog that drinks more water than it needs to stay hydrated will have less concentrated urine, possibly even isosthenuria or hyposthenuria. Even a normally hydrated dog without renal disease can have a USG ranging from 1.015-1.045. Therefore, USG is of most help clinically in patients that are azotemic and/or polyuric, oliguric or anuric. An azotemic patient should have maximal antidiuretic hormone (ADH) release and response if the azotemia is solely due to dehydration. Therefore, in an obviously dehydrated patient that has not had prior fluid or drug therapy, the USG should be >1.035 in dogs and >1.040 in cats.[4] But what about those patients with creatinine at the upper end of the reference interval and USG between 1.015-1.030? This may indicate an earlier stage of renal failure, but thorough clinical evaluation and serial monitoring will be essential to determine whether renal failure is present. Situations in which a dehydrated patient could have a lower than expected USG without concurrent renal disease include the presence of medullary washout, diabetes insipidus, glucosuria, and hypoadrenocorticism.[5] If a patient is polyuric and not dehydrated but has a USG between 1.020-1.030, it indicates attempts by the kidney to conserve water. Whether or not early renal disease is present will require further monitoring. Remember that a USG of <1.007 indicates that the kidneys are able to dilute the ultrafiltrate, which argues against renal failure. In these cases, overhydration or diabetes insipidus (nephrogenic or neurogenic) is likely. However, it is important to keep in mind that diluting and concentrating ability of the kidney are not necessarily lost at the same time. If a patient that presents with azotemia and isosthenuria develops hyposthenuria with fluid administration, do not rule out renal failure if azotemia persists.

Urine protein:creatinine ratio (UPC)

UPC is defined as urine protein divided by urine creatinine, and it is an indicator of magnitude of proteinuria. It is becoming more commonplace in veterinary medicine as a marker of renal disease when found to be persistently elevated. It has also been shown to have prognostic significance in both dogs and cats, with a more rapid course to uremic crisis or death the higher the UPC.[7,8] Always remember, however, that to use UPC as an indicator of glomerular and/or tubular damage, prerenal and postrenal causes of proteinuria must be ruled out. The concept behind proteinuria as an indicator of renal damage is that the glomerular filtration barrier normally excludes most proteins the size of albumin and larger, and the proximal tubules efficiently reabsorb the smaller proteins that are allowed to pass through into the ultrafiltrate so that the amount of protein normally present in urine is quite small and typically results in a UPC much less than 0.5 in normal dogs and 0.4 in normal cats.[9] The glomerular filtration barrier is composed of an inner fenestrated endothelium lined by a basement membrane and surrounding epithelial cells called podocytes. Damage to any of these components, particularly the podocytes and the podocyte slit diaphragm, will result in increased passage of proteins into the ultrafiltrate, which can overwhelm the tubular resorptive capacity. Damaged tubules can cause proteinuria by reabsorbing less protein, although this in itself will result in a mild proteinuria as compared to glomerular damage. An increased UPC is currently one of the best non-invasive indicators for the presence of glomerular disease. In general, the higher the value the more severe or extensive the glomerular lesion. However, this is not always the case, and UPC may actually decrease with end-stage renal failure due to a decrease in the number of functioning nephrons. It is also generally expected that hypoalbuminemia will be present with an increased UPC, but whether or not this occurs in a given patient will depend on whether liver production of albumin (influenced by liver function and nutritional status) can replace that lost via the urine.

The urine protein measured for determining UPC represents the total protein present in the urine (albumin, globulins, Tamm-Horsfall protein, hemoglobin, myoglobin, and various low molecular weight proteins). Urine creatinine is used in this ratio in order to account for the marked variation in urine concentration that occurs in a given patient depending on hydration status. The justification for using creatinine as the normalizing factor in urine is based on the assumption that its production and elimination is in steady state (along with the other properties listed above that help make it a reasonable estimator of GFR). For this reason, the UPC is the most accurate estimate of the magnitude of proteinuria when the patient is relatively stable (i.e. renal function and muscle mass are not rapidly changing). When monitoring the UPC in a patient, however, it is useful to consider any changes in muscle mass from one timepoint to the next, as the UPC will be overestimated if the patient is muscle-wasted and therefore producing/excreting less creatinine as compared to when it was well-muscled. Also, as renal function declines, less creatinine will be excreted, resulting in higher serum creatinine. If steady state is reached after an initial decline (i.e., the difference between creatinine produced and that being excreted remains the same), then urine creatinine excretion can again be reliably used for serial comparisons. However, if renal function continues to decline, using creatinine to normalize urine concentration will overestimate the UPC to at least some degree.

Serial UPC measurements can be helpful when monitoring disease progression or drug therapy designed to reduce proteinuria, but when can one consider the value to be increasing or decreasing? The UPC is not a static number, even in dogs with an unchanging magnitude of proteinuria.[10] Also, results can be greatly affected by the methodology used. Even within a single laboratory using a single assay, results can vary if a different dilution factor is used to determine the urine protein concentration. Therefore, some degree of fluctuation is expected, and in one study, the UPC must increase or decrease by a minimum of 35% at high UPC values (near 12) and 80% at low UPC values (near 0.5) in order to indicate a significant change from the baseline value.[10]

Another common question is how urinary tract infection (UTI) and blood contamination affect the UPC? A recent study showed that in the majority of urine samples demonstrating pyuria, UPC was <0.5, and contamination with whole blood (1:400 dilution) that grossly discolored the urine did not result in a UPC > 0.5.[11] This argues that the minimal blood contamination from a cystocentesis will not affect UPC results. It also indicates that one should not disregard marked proteinuria when detected in a patient with a UTI. However, because the influence of inflammation can be quite variable, you must always rule out infection (or treat an infection and recheck the proteinuria) before attributing proteinuria to a renal cause, regardless of the sediment findings.

NEW TESTS OF RENAL FUNCTION

There is a well-recognized need for better non-invasive tests of renal disease, and this is an active area of research in dogs and cats. These areas of study include a search for a better serum marker of GFR as well as investigation of specific proteins in the urine in an attempt to detect early renal disease as well as to help localize renal damage.

Cystatin C

Cystatin C is a cysteine protease inhibitor that is produced at a constant level in all nucleated cells. It is freely filtered through the glomerulus since it is a low molecular weight protein, and it does not undergo tubular secretion. While it is almost completely reabsorbed by the renal tubules, it is broken down in the tubules so reabsorption doesn't increase plasma levels.[12] In people, it is not thought to be significantly affected by inflammation, muscle mass, or diet, but it may be increased by high thyroxine and cortisol levels as well as some malignancies.[13] In dogs, cystatin C may be influenced by age, weight, and dietary intake, although several conflicting studies exist.[14-16] It is stable in the blood at -80°C[15] and an automated assay has been validated in dogs[17], but its availability in veterinary medicine is still limited. The biological variation of cystatin C in dogs (inter- and intra-individual variation) is similar to creatinine.[18] In one study, cystatin C appeared to perform similarly to endogenous creatinine as a measure of GFR in dogs, although it did not seem to be a sensitive indicator of decreased GFR due to prerenal factors or in a dog with acute renal failure.[17] In other studies, cystatin C was a more sensitive indicator of early chronic kidney disease in dogs as compared to creatinine.[15,16] In the few canine studies so far, cystatin C does appear to be a reasonable alternative to creatinine in detecting decreased GFR due to renal disease. However, its value over creatinine and the influence of nonrenal factors on its level are still unknown in veterinary medicine. There are currently no published studies investigating cystatin C in cats.

Microalbuminuria (MA)

Since becoming commercially available several years ago, there has been a lot of confusion about the MA test regarding what it means and what to do if a positive result is obtained. MA is defined as a urine concentration of albumin between 1-30 mg/dL when USG is normalized to approximately 1.010. The concept behind MA is that even mild proteinuria undetectable by the urine dipstick or UPC level can indicate underlying renal (particularly glomerular) disease. Similar to the UPC, postrenal causes of proteinuria should be ruled out before MA can be attributed to renal disease, since urinary tract inflammation and hemorrhage can increase levels.

MA is rather nonspecific and can be due to renal disease secondary to glomerular or tubular damage as well as nonrenal diseases that result in endothelial damage or alter glomerular permeability (hypertension, metabolic disease, infectious and inflammatory diseases).[19] While persistent MA is considered an abnormal finding, it does not necessarily indicate progressive renal disease that will become clinically significant during the life of the animal. In general, MA is most useful in the following situations:

1. Patients with renal disease or diseases that predispose to renal disease, especially if the urine dipstick is negative for protein. In cats, the usefulness of microalbuminuria for detecting early renal disease as compared to dogs is unknown, since renal damage in cats is not as frequently associated with proteinuria.
2. Confirming significant proteinuria if the urine dipstick is Trace - 2+ or if the UPC is borderline (e.g. 0.3-0.5).
3. Monitoring treatment or progression of a disease process resulting in microalbuminuria (but not proteinuric enough to have an increased UPC).
4. Detecting occult disease in a clinically healthy animal, (esp. geriatric patients). Positive results should prompt evaluation for subclinical diseases, including tick-borne disease, dental disease, hypertension, etc.
5. Early detection of animals that might be affected with a hereditary renal disease.

What should be done about a positive result? If not already performed, a follow-up UPC is recommended to quantify the proteinuria. In general, MA correlates with the UPC, but one does not always predict the results of the other. In a study of 138 urine samples from dogs, 17% of the samples had a UPC 0.5-1, and the corresponding MA in these samples ranged from negative to very high.[20] In a study of urine from 122 cats, a similar finding was observed.[21] Finding microalbuminuria should prompt a search for some identifiable cause, and it justifies more careful monitoring as the animal ages. However, if no cause is found and microalbuminuria does not appear to be increasing, monitoring on a twice yearly or yearly basis is reasonable.

Tests for other specific urine proteins

Measurement of certain urinary proteins can aid in early detection of renal disease and localization of renal damage. Some of the proteins that have been evaluated in dogs with chronic kidney disease (CKD) include those that are released with tubular damage (N-acetyl-β-D-glucosaminidase (NAG), γ-glutamyltransferase), those that freely pass through the glomerulus but appear in the urine due to decreased tubular reabsorption (retinol binding protein (RBP), β2-microglobulin, vitamin D-binding protein, α1-microglobulin), and those that are increased due to glomerular damage (immunoglobulin (IgG), transferrin).[22] Tamm-Horsfall protein is normally produced by the distal tubule, and it has been shown to decrease in dogs with CKD.[23] All of these proteins are typically normalized to urine creatinine in order to account for varying degrees of urine concentration.

Of the above proteins, one of the most studied in veterinary medicine is NAG. NAG is a lysosomal enzyme present in proximal renal tubular cells, and it is released from the cells secondary to tubular damage. An increased urinary NAG:creatinine ratio has been reported in dogs with CKD as well as dogs with pyometra and acute tubular damage due to renotoxic agents.[24-27] Increases have also been found in cats with CKD and hyperthyroidism.[28-30] However, NAG did not appear to be a significant factor in predicting the development of azotemia in geriatric cats.[30]

RBP has also been evaluated in both cats and dogs. It is a 21 kDa protein that circulates in the blood both free and bound to protein. The free protein passes readily into the glomerular ultrafiltrate and is essentially completely reabsorbed by the proximal tubules in health. As tubular function decreases, reabsorption of this low molecular weight protein also decreases, resulting in increased urinary excretion. Smets, et al found the RBP:creatinine ratio to be increased in 10 dogs with CKD as compared to normal dogs, and it strongly correlated with serum creatinine and BUN.[24] Similarly, there was a strong correlation between RBP:creatinine ratio and the degree of azotemia in a colony of dogs with X-linked hereditary nephropathy, and in this study, urinary RBP was increased approximately 2 months before the development of azotemia (unpublished observations). Another study comparing dogs with renal disease with clinically healthy dogs found urinary increases in the high molecular weight proteins C-reactive protein and IgG, which supports glomerular damage, and RBP, which supports tubular damage.[31] While measurement of these proteins appears promising, further evaluation is needed to better establish their role in the early diagnosis and monitoring of renal disease.

Urine electrophoresis (SDS-PAGE, SDS-AGE)

Electrophoresis can be used to separate urine proteins on a gel based on their size (molecular weight). The theory behind its use with renal disease is that the size of proteins present in the urine can help localize where damage is occurring in the kidney. For example, high molecular weight proteins (>60-70 kDa) indicate glomerular damage, since these are normally blocked from entering the ultrafiltrate by the glomerular basement membrane, whereas low molecular weight proteins (<40-50 kDa) indicate tubular damage, since these are normally reabsorbed by the proximal renal tubules. In one study that compared urine SDS-AGE patterns with renal biopsies in 49 dogs with renal disease, sensitivity of the urine protein pattern for detecting glomerular and tubular damage was good, but specificity was relatively low.[32] We have also found good sensitivity of urine SDS-PAGE as compared to renal biopsies in dogs with renal disease; however, specificity is difficult to assess given the small number of cases without either glomerular or tubular damage (unpublished observations). Additional studies are needed to determine if urine SDS-PAGE provides useful information beyond that obtained from renal biopsies.

REFERENCES:

[1]Fried TA, et al **Arch Intern Med** 1983;143:787. [2]Concordet D, et al **Vet Clin Pathol** 2008;37(1):96. [3]Braun JP, et al **Vet Clin Pathol** 2003;32(4):162. [4]Latimer KS, et al **Clinical Pathology** 4th ed. 2003; 255. [5]Stockham SL, et al. **Fundamentals of Veterinary Clinical Pathology** 2nd ed. 2008; 437. [6]George JW. **Vet Clin Pathol** 2001;30(4):201. [7]Jacob F, et al. **J Vet Med Assoc** 2005;226(3):393. [8]Syme H, et al. **J Vet Intern Med** 2006;20(3):528. [9]Lees GE, et al. **J Vet Intern Med** 2005;19(3) :377. [10]Nabity MB, et al. **J Vet Intern Med** 2007;21(3):425. [11]Vaden SL, et al. **Vet Clin Pathol** 2004;33(1):14. [12]Grubb A. **Clin Nephrol** 1992;38:S20. [13]Westhuyzen J. **Ann Clin Lab Sci** 2006;36(4):387. [14]Braun JP, et al. **J Comp Clin Path** 2002;11:44. [15]Wehner A, et al. **J Am Anim Hosp Assoc** 2008;44:131. [16]Miyagawa Y, et al. **J Vet Med Sci** 2009;71(9):1169. [17]Almy FS, et al. **J Vet Intern Med** 2002;16:45. [18]Pagitz M, et al. **J Vet Intern Med** 2007;21:936. [19]Whittemore JC, et al. **J Am Vet Med Assoc** 2006;229(6):958. [20]Garner BC, et al. **Vet Clin Pathol** 2007;36(3):240. [21]Mardell EJ, et al. **J Feline Med Surg** 2006;8(4):269. [22]Raila J, et al. **J Vet Med** 2007;54(3):131. [23]Forterre S, et al. **J Vet Diagn Invest** 2004;16(4):271. [24]Smets PMY, et al. **J Vet Intern Med** 2010;24(1):65. [25]Sato R, et al. **J Vet Med Sci** 2002;64(4):361. [26]Heiene R, et al. **Res Vet Sci** 2001;70(2):129. [27]Grauer GF, et al. **J Vet Intern Med** 1995;9(5):324. [28]Lapointe C, et al. **J Vet Intern Med** 2008;22(5):1103. [29]Sato R, et al. **J Vet Med Sci** 2002;64(4):367. [30]Jepson RE, et al. **J Vet Intern Med** 2009;23(4):806. [31]Maddens BE, et al. **Vet Immunol Immunopathol** 2009:doi:10.1016/j.vetimm.2009.09.003. [32]Zini E, et al. **AJVR** 2004; 65:964.

CHALLENGING CASES: RENAL DISORDERS OF CATS & DOGS

Mary B. Nabity, DVM, DACVP
College Station, TX

INTRODUCTION

Renal disease is a common cause of morbidity and mortality in dogs and cats, and diagnosis and monitoring of renal disease depends on an accurate clinical history and clinical pathologic evaluation of blood and urine, with urine evaluation providing the most direct look into kidney function. The following cases are an attempt to critically evaluate common laboratory abnormalities observed with renal disease.

CASE 1

Signalment & History

Case 1 is a 7-year-old male castrated Boston Terrier that presented for polyuria and polydipsia that had progressed for several months. The patient had been diagnosed previously with allergic skin disease, but physical exam was otherwise unremarkable. Kidneys appeared normal on ultrasound examination.

Selected initial laboratory findings:	Result	Reference interval (RI)
Albumin (g/dl)	2.5	3-4.5
Calcium (mg/dl)	9.4	7.2-12.8
Phosphorus (mg/dl)	3.4	2.3-6.5
Urea nitrogen (UN)	16.3	8-30
Creatinine	0.8	0.5-1.4
Sodium (mEq/L)	151	141-156
Potassium (mEq/L)	4.0	3.8-5.5
Chloride (mEq/L)	116	109-124
Urine protein (mg/dl)	312	
Urine protein:creatinine (UPC)	4.4	<0.5
Urine specific gravity (USG)	1.030	

All other serum chemistry values and a complete blood count were within normal limits, except for a mild thrombocytosis. Other urinalysis findings included 4+ dipstick protein, rare hyaline casts, 0-5 WBC/hpf, and 10-20 RBC/hpf. Urine culture was negative.

Interpretation

Moderate proteinuria is present, indicated by the elevated UPC. The degree of hematuria is insufficient to explain the proteinuria. Therefore, the lack of significant sediment findings along with a negative urine culture supports localization of the proteinuria to the kidneys, and the degree of proteinuria indicates glomerular damage. The hematuria may be iatrogenic, secondary to cystocentesis, or it may be secondary to glomerular damage. The proteinuria is resulting in hypoalbuminemia, indicating that albumin loss exceeds nutritional support and production by the liver. The USG indicates that the renal tubules are able to concentrate the urine to some degree. Since UN and creatinine are well within the reference interval, this USG likely indicates adequate concentrating ability. Because proteinuria can interfere with the USG value by approximately 0.004 units per 1000 mg/dL protein, the USG is falsely elevated by 0.001 units, indicating that the true USG value is probably 1.029. After accounting for the effects of urine protein, conclusions regarding USG are not altered. The normal UN, creatinine, electrolytes and phosphorus support adequate glomerular filtration rate (GFR) and tubular regulation. Based on the above findings, benazapril was prescribed (2.5 mg/day). Monitoring was continued using a different laboratory (note the different reference intervals for this laboratory).

Number of weeks post initial exam	4 wks[a]	8 wks[b]	10 wks	12 wks[c]	16 wks[d]	24 wks[e]	30 wks	RI
Creatinine (mg/dl)	1.1	1.0	1.1	0.9	0.9	0.9	0.8	0.3-2
UN (mg/dl)	16	16	13	18	16	14	14	5-29
Albumin (g/dl)	1.9	1.6	2.0	2.0	1.5	1.9	1.9	2.4-3.6
Antithrombin (%NHP)	88.9	81.8	104	121	63.2	103	109	>114
Urine protein (mg/dl)	2345	1255	1408	1060	2320	954	1866	
Urine creatinine (mg/dl)	176.6	108.4	183.7	183.6	220.8	130.5	237.2	
UPC	13.3	11.6	7.7	5.8	10.5	7.3	7.9	
USG	1.036	ND	1.037	1.029	1.046	ND	1.052	
USG adjusted for proteinuria	1.027		1.031	1.025	1.037		1.045	
Urine dipstick protein	4+		4+		4+		4+	

[a]Benazapril increased to 5 mg/day + aspirin (5 mg/day); [b]Added amlodipine (2.5 mg/day) and omega 3 fatty acids
[c]Renal biopsy performed; doxycycline started at 100 mg/day for 2 wks; [d]Added azathioprine (25 mg every other day);
[e]Added cyclosporine (25 mg/day)

The above data supported adequate GFR throughout the monitoring period, and even when taking into account the urine protein interference on specific gravity, urine concentration is likely adequate. Urine protein loss roughly correlates with the degree of hypoalbuminemia and antithrombin activity. The UPC at 4 wks is markedly elevated compared to the initial evaluation and supports increased urine protein loss. While the UPC at 8 wks is not significantly lower than at 4 wks, the UPC at 10 and 12 wks is significantly lower, and the decreasing trend further supports a true decrease in the magnitude of proteinuria, likely treatment related. At 16 wks, there is a significant increase in UPC and decrease in albumin, indicating worsening proteinuria. The decreased UPC at 24 and 30 wks is likely a true decrease in magnitude of proteinuria compared to 16 wks, as albumin levels have improved slightly.

Note that urine creatinine roughly correlates with the USG, and the urine protein correlates with the dipstick results. It is important to consider absolute urine protein and creatinine values (not just the UPC) in order trust the results. For instance, if the urine dipstick indicates 4+ proteinuria but the corresponding UPC is 0.2, one of the results may be inaccurate. Few anecdotal instances of such a discrepancy have occurred when urine creatinine is >300-400mg/dl. It is possible that extremely high urine creatinine may falsely decrease urine protein results, so watch for discrepancies between any different tests performed for a particular analyte.

Outcome

The renal biopsy revealed mixed immune-mediated glomerulonephropathy (features of both membranous glomerulonephritis and membranoproliferative glomerulonephritis) with moderate lymphoplasmacytic interstitial nephritis and mild multifocal interstitial fibrosis. The patient was negative for Lyme disease, Rocky Mountain Spotted Fever, Bartonella vinsonii, Babesia canis, Bartonella henselae, Anaplasma, Ehrlichia canis, and heartworm disease. He is currently doing well and is continuing to be monitored closely.

CASE 2

Signalment & History

Case 2 is a 15-year-old male castrated DSH cat that presented for acute lethargy and twitching 2 hours after insulin administration for diabetes mellitus. He was diagnosed with diabetes 2 years ago, and has a 2-year history of anorexia and weight loss. He has exhibited polyuria and polydipsia for the past 6 months. Pertinent lab work includes:

	Result	RI
Glucose (mg/dl)	36	65-131
UN (mg/dl)	80	19-33
Creatinine (mg/dl)	2.5	0.8-1.8
Magnesium (mg/dl)	3.2	1.7-2.3
Phosphorus (mg/dl)	6.2	3.8-7.5
Plasma protein (g/dl)	8.2	6.1-7.7
Albumin (g/dl)	3.2	2.5-3.3
Globulin (g/dl)	4.9	2.3-3.8
Sodium (mmol/L)	149	144-155
Potassium (mmol/L)	2.1	3.5-5.1
Chloride (mmol/L)	110	113-123
CO_2 (mmol/L)	23	19-26
USG	1.030	
UPC	0.7	
Fructosamine	441	0-375

The urine dipstick revealed 2+ protein and 3+ blood. Glucose and ketones were negative. Sediment evaluation revealed 0-1 granular casts/lpf, 0-1 WBC/hpf, and 0-2 RBC/hpf. Urine culture was negative.

Interpretation

The marked hypoglycemia supports insulin overdose. There is a mild azotemia present along with inadequately concentrated urine. This is highly supportive of renal failure, although a prerenal component to the azotemia is likely, especially with albumin at the upper limit of the reference interval. Because this patient has undergone extensive weight loss and probably loss of muscle mass, the creatinine level may not be representative of the degree of renal function. Renal disease is also supported by the proteinuria and presence of granular casts. The proteinuria is localized to the kidney based on lack of evidence for prerenal and renal causes. The magnitude of proteinuria is at a level that could be due to either tubular disease or mild or focal glomerular disease. The granular casts indicate tubular damage. One consideration regarding the USG, however, is that despite insulin therapy this cat has likely had historic hyperglycemia, as supported by increased fructosamine. Therefore, osmotic diuresis may have been occurring previously, which could have resulted in medullary washout that is incompletely recovered at the time of urine sampling. The lack of glucose in the urine indicates that most of the urine present in the bladder at the time of sampling was produced after the hypoglycemic event. Therefore, USG does not need to be adjusted for glucose. The hypermagnesemia is secondary to decreased GFR. Hyperphosphatemia is also expected due to decreased GFR, and the lower than expected level is likely due to intracellular shifting secondary to insulin. Intracellular shifting secondary to insulin is also the most likely reason for the marked hypokalemia (due to activation of a Na^+-K^+-ATPase). However, chronic renal disease and osmotic diuresis (glucosuria) increase tubular flow and therefore potassium excretion, contributing to the hypokalemia. The hypochloremia indicates chloride loss in excess of sodium, possibly due to vomiting. Hyperglobulinemia supports immune stimulation.

Outcome

The patient recovered from the hypoglycemic episode, and the insulin dose was adjusted. Further serum chemistry monitoring indicated the presence of azotemia over several months, confirming the diagnosis of chronic renal disease.

PHYSIOLOGY OF ACID-BASE & ELECTROLYTE DISORDERS

John A. Christian, DVM, PhD
West Lafayette, IN

INTRODUCTION

Acid-base and electrolyte (AB-E) disorders seem to be approached with some trepidation by many; perhaps, because an organized approach is not obvious and key patterns for important abnormalities are not recognized. In this paper, a brief overview of individual electrolyte physiology will be integrated with acid-base profiling as it pertains to interpreting the routine clinical chemistry profile. A standardized approach to acid-base profiling will be emphasized.

Table 1 includes 6 acid base profiles as examples of patterns commonly seen in practice. It is recommended that these cases be interpreted before reading the text. After reading the text, reassess the results. Answers in brief are found at the end of the paper.

Table 1. Acid-Base Profiling – Case Examples

Parameter	Reference	Case #1	Case #2	Case #3	Case #4	Case #5	Case #6
TCO2	(13-24)*	9 L	10 L	26 H	10 L	31 H	8 L
Anion Gap	(9-18)*	14	25 H	22 H	37 H	11	13
Sodium	(138-148)*	145	143	141	148	136 L	143
Chloride	(105-117)*	127 H	113	97 L	106	95 L	125 H
Potassium	(3.5-5.0)*	4.8	5.1 H	4.0	5.2 H	3.1 L	3.3 L
Urine pH	(variable)	6.1	6.5	7.1	6.2	6.0	6.0

* (mmol/L)

For each case: (A) List acid base pattern(s), (B) List differential(s) or explain mechanism(s) for change, (C) List other findings, if any.

ACID-BASE PROFILING

The essential components for acid base profiling are listed in Table 2. Primary variables are those providing direct support for an acid-base disorder. Secondary variables are those that will predictably change as a consequence of an acid base disorder. Failure of these parameters to change or show an inappropriate change relative to the primary parameters often provides very important insights into the pathologic process.

Table 2. The Acid-Base Profile

Primary	Secondary
TCO2 Anion Gap Chloride (vs Sodium)	Potassium Urine pH

Primary Profile

TCO2: On routine chemistry profiles TCO2 is an estimate of bicarbonate. Bicarbonate and chloride are the two major anions in circulation and to maintain electroneutrality bicarbonate and chloride generally maintain an inverse relationship; an increase in one is associated with a decrease in the other. Increases in TCO2 reflect metabolic alkalosis which most commonly involves loss or sequestration of HCl rich gastric secretions [$NaCl + H_2CO_3 \rightarrow$ HCl (lost) + $NaHCO_3$ (retained)]. Metabolic alkalosis may also reflect compensation for respiratory acidosis, although blood gas analysis is needed to document this process. Decreases in TCO2 occur by 2 main metabolic mechanisms: Secretion (loss) or titration of bicarbonate. Losses of bicarbonate through secretion may occur in the GI track (e.g., diarrhea with bicarbonate rich secretions) or through the kidneys in certain types of renal disease or failure (e.g., renal tubular acidosis). Secretion acidosis is associated with chloride retention leading to increases in chloride relative to its usual association with sodium (see chloride below). This is sometimes called hyperchloridemic acidosis. Titrational metabolic acidosis involves the presence of organic acids that are titrated (buffered) by bicarbonate consumption. This process produces unmeasured anions that are recognized by an increase in the anion gap (calculated, see below). Careful evaluation of the TCO2 in light of the anion gap and chloride is essential to correctly categorize the type of metabolic acidosis and to aid in identifying mixed acid-base disorders (see below).

Anion Gap (AG): The AG is a calculated value based on the formula: $(Na^+ + K^+) - (TCO2^- + Cl^-)$, which is a subtraction of measured anions from measured cations. Under physiologic conditions this should always be a positive value. In simplified terms, the difference between the measured cations and the measured anions is the anion gap which mostly reflects the presence of unmeasured anions. In health the major unmeasured anion is albumin with smaller contributions from sulfates, phosphates, lactates, and possibly some globulins. In pathologic states there are a limited number of anions (endogenous or exogenous) that may be increased enough to elevate the AG. Five specific anions account for the vast majority of increased anion gaps. These include 3 endogenous anions: uremic acids (sulfates, phosphates, urates), ketoacids and lactic acid. Two exogenous anions of major importance include ethylene glycol poisoning (anionic by-products of metabolism), and salicylates (aspirin toxicity). The main usefulness of the anion gap then, is that increases are strongly associated with the presence of titratable acids (titrational metabolic acidosis) and that leads to a fairly narrow differential list to explain the change. Decreases in the anion gap are mostly due to decreases in the major unmeasured anion: albumin. Since albumin is a weak acid, this state has been described as hypoalbuminemic alkalosis. It is not considered a significant or treatable acid base disorder, other than what might be indicated based on the degree of hypoalbuminemia. However, the change may be reflected by a high normal or mildly increased TCO2 and, interestingly, chloride values may also trend upward (when compared to Na). Seeing TCO2 and Cl both tending toward elevations is highly unusual and the pattern should prompt evaluation of albumin status and possible hypoalbuminemic alkalosis. Negative values for the anion gap do not occur physiologically and most cases are explained by the presence of another halide (e.g., bromide) which is co-detected as chloride leading to anomalously high Cl readings.

Chloride (vs Na): Under most conditions sodium and chloride values tend to move in parallel. This reflects their relationship in maintaining electroneutrality. Whenever Cl values are not moving in concert with Na it generally indicates that the inverse relationship between Cl and bicarbonate is taking precedent over the Cl to Na relationship and strongly implies an acid base disturbance. Although often neglected in this respect, Cl can be a powerful tool in characterizing acid base disorders. A decrease in Cl vs Na implies a process is increasing bicarbonate (metabolic alkalosis). An increase in Cl Vs Na implies a process is decreasing bicarbonate (metabolic acidosis). Although not carefully validated experimentally, it appears that a good working rule of thumb is that Cl is typically within ±3 units of sodium – within their respective reference intervals. This is best evaluated by determining where Na & Cl are with respect to the middle of their reference intervals. To illustrate see Table 3:

Table 3. Is Chloride Following Sodium?

Analyte	Result	Reference Interval	Reference Mid-Point	Net Value
Na	148	(138-148)	143	Na is +5 from middle
Cl	106	(105-117)	111	Cl is -5 from middle

Interpretation: Cl is decreased 10 units vs Na. A decrease of more than 3 units indicates a process is increasing bicarbonate (i.e., metabolic alkalosis).

Secondary Profile

Potassium: Serum K levels may shift in response to acid base alterations. The degree of change is proportional to the change in blood pH. Since pH is not measured on the routine profile, acidemia cannot be conclusively identified. However, the magnitude of the process (acidosis or alkalosis) leading to a change in pH (acidemia or alkalemia) can be used as a rough indication of likely pH change. With acidemia excess hydrogen ions in circulation are moved intracellularly and to maintain electroneutrality K is moved out of cells leading to high normal or increased serum K. The opposite is true for metabolic alkalemia so that low normal or decreased K levels are anticipated. However, these shifts are considered to be relatively mild. Large swings in K should require that other differentials also be considered. A critical finding to recognize is hypokalemia in the face of acidosis. Although serum K is generally considered a poor reflection of total body stores, this pattern provides an exception. Since serum potassium should be increasing with acidosis, the presence of hypokalemia at this time strongly suggests that total body depletion of K had occurred and is a critical care finding.

Urine pH: Urine pH is affected by many variables including diet (carnivores have acidic urine, herbivores have alkaline urine), post-prandial alkaline tide, activity of urease producing bacteria, and alkalinization associated with prolonged storage and loss of CO2. However, in general, urine pH is also a crude indicator of acid-base status since it is a compensatory organ that excretes excess ions and retains depleted ions relative to acid base status. Therefore, urine pH should always be evaluated to see if it is shifting in an appropriate direction relative to acid-base status. Consistent inclusion of urine pH in the acid base assessment will occasionally reveal situations where urine pH is inappropriate for the blood acid-base status. Two conditions of importance include metabolic alkalosis with paradoxical aciduria and relatively rare renal tubular acidosis. The latter involves metabolic acidosis with either alkaline or inadequately acidified urine. See acid-base patterns below for more description of paradoxical aciduria.

ACID-BASE PATTERNS

Several specific acid base patterns can be recognized from the routine chemistry profile. It is important to be able to recognize these patterns because they point to general and sometimes specific etiologies/mechanisms or organs systems that may be affected in the disease process.

Metabolic Alkalosis

Pattern: The primary profile is characterized by an increase in TCO2, decrease in Cl vs Na, and normal anion gap. Depending on degree of change, the secondary profile may include a normal to low normal or mildly decreased K and a neutral to increased urine pH.

Significance: By far most commonly associated with gastric loss (gastric vomiting) or sequestration of HCl (e.g., obstruction).

Metabolic Acidosis, Secretional

Pattern: The primary profile is characterized by a decrease in TCO2, an increase in Cl vs Na, and a normal anion gap. Depending on the degree of change the secondary profile may include a normal to high normal or mildly increased K and a neutral to acidic urine pH.

Significance: Occurs when bicarbonate is lost (i.e., secreted) from the body as occurs with GI, especially pancreatic, secretions. These may be sequestered in cases of obstructions or lost in certain cases of diarrhea. Renal tubular loss is another important differential.

Metabolic Alkalosis, Titrational

Pattern: The primary profile is characterized by a decrease in TCO2, an increase in the anion gap and no change in Cl vs Na. Depending on the degree of change, the secondary profile may include a normal to high normal or mildly increased K and a neutral to acidic urine pH.

Significance: Indicates the presence of organic acids that are titrated with bicarbonate resulting in anionic salts, which directly reflect the increased anion gap. Only rarely will a significantly increased anion gap not be explained by one of the 5 following conditions: renal failure (uremic acids), ketoacidosis, lactic acidosis, ethylene glycol poisoning (antifreeze – glycolic acids), or salicylate (aspirin) toxicity. These should form an immediate differential list for an increased anion gap.

Mixed Metabolic Alkalosis and Metabolic Acidosis, Titrational

Pattern: The primary profile can be variable, depending on which process is stronger. With a pronounced alkalosis, the TCO2 may be increased and the Cl may be decreased vs Na. However, an increased anion gap would then be the only indication of the mixed disorder and the presence of titrational acidosis. With a pronounced titrational acidosis, the TCO2 would be decreased and the anion gap would be increased. In that case, the decrease in Cl vs Na would be the only indication of alkalosis. Several variations between these two extremes are possible. The secondary profile is difficult to predict since mixed disorders may lead to a balance in the compensations typical for K and urine pH. Blood pH is really necessary to predict an expected response.

Significance: See individual descriptions for metabolic alkalosis and titrational acidosis.

Metabolic Alkalosis with Paradoxical Aciduria

Pattern: The primary profile is characterized by an increase in TCO2, a decrease in Cl vs Na, and a normal anion gap. Depending on the degree of change, the secondary profile may include a normal to low normal or decreased K and an acidic urine pH. In many cases, however, K is mildly to moderately decreased, as that is often a component of the pathogenesis.

Significance: See alkalosis above for the increased TCO2. Paradoxical aciduria occurs when there is a need for Na conservation and yet there is a depletion of electrolytes needed to make the appropriate exchanges in the renal tubules (i.e., Cl, K). In the ascending loop where NaCl reabsorption is strong, if Cl is depleted (as occurs with gastric losses of HCl and ensuing alkalosis), HCO3 will be reabsorbed to maintain electroneutrality, resulting in more acidic urine and increasing alkalosis. If K is depleted in the distal tubules where aldosterone promotes Na-K exchanges, hydrogen ions will be swapped with the Na leading to more acidic urine and more alkaline blood. As long as the drive for Na absorption continues due to dehydration or sodium depletion or sequestration, this continues as a vicious cycle leading to greater alkalosis and maintaining electrolyte depletions. Restoring hydration and normonatremia are critical to resolving the cycle.

ANSWERS TO CASES IN TABLE 1

[NOTE: A = pattern(s), B = differentials/mechanisms, C = other findings]

#1: A.- Metabolic acidosis, secretional (↓ TCO2, ↑ Cl [+14] Vs Na) B.– Bicarb loss (diarrhea or renal loss?) C.– K, urine pH = OK

#2: A.- Metabolic acidosis, titrational (↓ TCO2, ↑ AG, normal Cl) B.- See ↑ AG, text C.- slight ↑K & aciduria appropriate

#3: A.- Mixed metabolic alkalosis (↑ TCO2, ↓ Cl [-12] Vs Na) + Metabolic acidosis, titrational (↑ AG) B.- Alkalosis: Upper GI loss or sequestration of HCl; Titrational acidosis (see text on AG) C.- Mid-range K and the urine pH are appropriate for a mixed disorder.

#4: A.- Mixed metabolic acidosis, titrational (↓ TCO2, ↑ AG) + mild metabolic alkalosis (↓ Cl [-10] Vs Na) B.- See #3. C.- Acidosis appears predominant based on TCO2, ∴ mild K ↑ and acidic urine are reasonable. Blood pH needed to confirm.

#5: A.- Metabolic alkalosis (↑ TCO2, ↓ Cl [-9] Vs Na) B.- See #1. C.- Acidic urine is paradoxical (metabolic alkalosis with paradoxical aciduria). Decreases in Na, K, & Cl = general electrolyte depletion which predisposes for this condition. (See text) Consider osmotic diuresis (e.g., diabetes mellitus) or a sequestration/third space process.

#6: A.- Metabolic acidosis, secretional (↓ TCO2, ↑ Cl [+14] Vs Na) B.- See #1. C.- ↓ K in face of acidosis strongly suggests total body depletion of K. Critical finding! (See text)

CHALLENGING CASES: ELECTROLYTES & ACID-BASE DISORDERS

John A. Christian, DVM, PhD
West Lafayette, IN

INTRODUCTION

Principles of integrated acid base and electrolyte interpretation as described in the previous session (#123) are carried forward to this session and applied to more comprehensive case profiles. Additionally, the impact of selected primary electrolyte disturbances on interpretation of the urinary profile will be illustrated.

Renal azotemia is defined as azotemia resulting from decreased glomerular filtration with inadequate urine concentration (SG: 1.030, dogs; 1.035, cats) that indicates decreased tubular function. In the vast majority of cases, renal azotemia is due to a loss of >75% of nephrons. Often the assumption is (correctly) that this is largely parenchymal disease which may or may not be to some degree reversible. Thus, the prognosis for return to renal function is initially guarded at best, especially if a specific etiology is not known. However, there are several conditions where primary electrolyte anomalies interfere with renal function without necessarily causing significant parenchymal damage. These conditions are often very reversible and if identified, can carry a favorable prognosis for return to normal renal function if treated at an early stage. These conditions often lead to a compromise in tubular concentrating function such that typical pre-renal mechanisms of azotemia will present with inadequate tubular concentrating function and therefore will appear as renal azotemia. A helpful term to describe this scenario, if identified, is 'secondary' renal azotemia because by definition the azotemia is 'renal' (inadequate concentration), but by mechanism the azotemia is typical of pre-renal conditions to which the tubules are temporarily prevented from responding. The pathophysiology of selected examples will be addressed as they apply to cases for discussion.

CASE #1

Background Information

The following data is from a 3.5 year old male dog with a limited history of weight loss and anorexia. The CBC was unremarkable and is not included. The urinary system profile including renal, electrolyte and acid base data are below.

Table 1. Chemistry and urinalysis data for Case #1

CHEMISTRY

Urea Nitrogen	76	H	7	-	32	mg/dl
Creatinine	3.1	H	0.5	-	1.5	mg/dl
Phosphorus	3.9		2.2	-	7.9	mg/dl
Calcium	15.7	H	9.7	-	12.3	mg/dl
Sodium	149	H	138	-	148	mmol/l
Potassium	4.3		3.5	-	5.0	mmol/l
Chloride	110		105	-	117	mmol/l
Total CO2	24		13	-	24	mmol/l
Anion Gap	20	H	9	-	18	mmol/l
Total Protein	6.3		4.8	-	6.9	g/dl
Albumin	3.1		2.3	-	3.9	g/dl
Globulin	3.2		1.7	-	3.8	g/dl

URINE

Specimen	Voided		pH	5.0
Color	Yellow		Protein	Trace
Turbidity	Hazy		Glucose	Neg.
Specific Gravity	1.018		Ketone	Neg.
WBC / HPF	3-4		Bilirubin	Neg.
RBC / HPF	neg		Occult Blood	Trace
Epith / HPF	1-2		Urobilinogen	0.2
Bacteria	1+			
Crystals	neg			
Fine gran casts/lpf	few			

Questions/Discussion

1. Classify the azotemia (explain your answer)
2. ID urine findings that are clearly abnormal in a voided sample. If there are equivocal findings, suggest follow up testing.
3. Interpret the acid-base profile. Identify abnormal patterns, if present. Explain the clinical significance of any findings.
4. Interpret any electrolyte changes that are not directly covered in the acid base section.

CASE #2
Data from a 2 year old male, neutered Bulldog with a primary history of vomiting

CASE #3
A 2 year old, FS, Shar-Pei is reported to have had a high BUN & creatinine, anorexia and vomiting for 2 weeks

CBC

Case #2:

Total Protein	8.7	H	6.0	-	8.0	g/dl
Hematocrit	61	H	37.0	-	55.0	%
Hemoglobin	20	H	12.0	-	18.0	g/dl
RBC	9.12	H	5.50	-	8.50	$x10^6 / \mu l$
MCV	67		60	-	75	fl
MCHC	32.7		32	-	36	g/dl
WBC	19.50	H	6.00	-	17.00	$x10^3 / \mu l$
Band	0.98	H	0	-	0.30	$x10^3 / \mu l$
Neutrophil	15.60	H	3.00	-	12.00	$x10^3 / \mu l$
Lymphocyte	0.98	L	1.00	-	5.00	$x10^3 / \mu l$
Monocyte	1.95	H	0.15	-	1.35	$x10^3 / \mu l$
Eosinophil	0.0	L	0.10	-	1.25	$x10^3 / \mu l$
Basophil	0.0		0	-	0.10	$x10^3 / \mu l$

Case #3:

Total Protein	6.6		6.0	-	8.0	g/dl
Hematocrit	41.2		37.0	-	55.0	%
Hemoglobin	15.6		12.0	-	18.0	g/dl
RBC	6.77		5.50	-	8.50	$x10^6 / \mu l$
MCV	60.8		60	-	75	fl
MCHC	37.9	H	32	-	36	g/dl
WBC	9.17		6.00	-	17.00	$x10^3 / \mu l$
Band	0.0		0	-	0.30	$x10^3 / \mu l$
Neutrophil	5.78		3.00	-	12.00	$x10^3 / \mu l$
Lymphocyte	2.02		1.00	-	5.00	$x10^3 / \mu l$
Monocyte	1.19		0.15	-	1.35	$x10^3 / \mu l$
Eosinophil	0.18		0.10	-	1.25	$x10^3 / \mu l$
Platelets	669		200	-	900	$x10^3 / \mu l$

CHEMISTRY

Case #2:

Glucose	125		67	-	132	mg/dl
Urea Nitrogen	109	H	7	-	32	mg/dl
Creatinine	2.3	H	0.5	-	1.5	mg/dl
Phosphorus	7.2		2.2	-	7.9	mg/dl
Calcium	10.6		9.7	-	12.3	mg/dl
Sodium	121	L	138	-	148	mmol/l
Potassium	4.3		3.5	-	5.0	mmol/l
Chloride	71	L	105	-	117	mmol/l
Total CO2	39	H	13	-	24	mmol/l
Anion Gap	15		9	-	18	mmol/l
Total Protein	7.7	H	4.8	-	6.9	g/dl
Albumin	4.3	H	2.3	-	3.9	g/dl
Globulin	3.4		1.7	-	3.8	g/dl

Case #3:

Glucose	100		67	-	132	mg/dl
Urea Nitrogen	50	H	7	-	32	mg/dl
Creatinine	2.9	H	0.5	-	1.5	mg/dl
Phosphorus	7.9		2.2	-	7.9	mg/dl
Calcium	17.8	H	9.7	-	12.3	mg/dl
Sodium	125	L	138	-	148	mmol/l
Potassium	7.6	H	3.5	-	5.0	mmol/l
Chloride	96	L	105	-	117	mmol/l
Total CO2	15		15	-	26	mmol/l
Anion Gap	22	H	9	-	18	mmol/l
Total Protein	7.5	H	4.8	-	6.9	g/dl
Albumin	4.5	H	2.3	-	3.9	g/dl
Globulin	3.0		1.7	-	3.8	g/dl

URINALYSIS

Case #2:

Specimen	Cysto
Color	Yellow
Turbidity	Clear
Specific Gravity	1.020
pH	6.0
Protein	Trace
Glucose	Neg
Ketone	Neg
Bilirubin	Neg
Occult Blood	Trace
WBC / HPF	1-2
RBC / HPF	1-2
Epith / HPF	1-2
Bacteria	Neg
Crystals	Mod. Amorph.
Casts	Neg

Case #3:

Specimen	Cysto
Color	Yellow
Turbidity	Hazy
Specific Gravity	1.015
pH	5.0
Protein	Trace
Glucose	Neg.
Ketone	Neg.
Bilirubin	Neg.
Occult Blood	Trace
Urobilinogen	0.2
WBC / HPF	0-2
RBC / HPF	0-2
Epith / HPF	0-2
Fine gran casts/lpf	few
Casts	Neg

Questions/Discussion: **(Cases 2 & 3)**

1. Interpret the CBC.
2. Classify the azotemia (explain your answer)
3. ID urine findings that are clearly abnormal. If there are equivocal findings, suggest follow up testing.
4. Interpret the acid-base profile. Identify abnormal patterns, if present. Explain the clinical significance of any findings.
5. Interpret any electrolyte changes that are not directly covered in the acid base section.

CASE 1 DISCUSSION

There is renal azotemia as evidenced by an increased BUN and Creatinine (azotemia, indicating decreased glomerular filtration) with an inadequate urine concentration (SG=1.018).

In any urine sample granular casts are clearly abnormal and indicate degeneration of renal tubular epithelial cells. Bacteria in a voided sample are equivocal and resubmission of a cystocentesis sample is recommended. Trace protein with a 1.018 SG is equivocal. If the trace proteinuria persists in a repeated cystocentesis without significant sediment findings, a UPC is indicated.

Acid-base parameters indicate a metabolic alkalosis based on a decrease in Cl vs Na (-7 units) and a high normal TCO2. Metabolic alkalosis usually develops from loss/sequestration of gastric HCl. Although not reported in the history, there is likely an upper GI disturbance associated with vomiting. There is also a titrational metabolic acidosis based on an increased Anion Gap (AG). The presence of a mixed disorder explains why the TCO2 is not actually elevated in the presence of alkalosis. A high AG indicates increased unmeasured anions, most likely due to the renal azotemia (uremic acids). (Other differentials include lactic acids, ketoacidosis, ethylene glycol and aspirin intoxication). The K and urine pH values are acceptable for a mixed acid base disorder.

The mild hypernatremia is most likely from hemoconcentration. Importantly, the marked hypercalcemia is not easily explained by the presence of renal azotemia, as most canine cases have normal to decreased Ca, and even when elevated it will not be this high. This indicates that the Hypercalcemia is a primary problem, which is probably contributing to the renal azotemia. Hypercalcemia interferes with ADH receptors leading to tubular concentrating dysfunction. This stage is a reversible change that progresses to metastatic mineralization that damages tubules (degeneration → casts) and eventually leads to interstitial fibrosis and primary renal azotemia. However, in the early stages, even with some degeneration, the lesions are likely reversible if the calcium is brought down. Differentials include humoral hypercalcemia of malignancy (HHM), hyperparathyroidism, Vitamin D toxicity and Addison's disease. This case had lymphoma with HHM. In summary, the azotemia should now be classified as 'secondary' renal azotemia because hypercalcemia is a relatively common and reversible cause of compromised tubular function.

CASE 2 DISCUSSION

The CBC reveals evidence in the erythron for relative polycythemia (dehydration). The leukon shows inflammation (neutrophilia with left shift, monocytosis), stress (lymphopenia, eosinopenia), and tissue demand for macrophages (monocytosis). Platelet data is missing.

Azotemia with inadequate urine concentration (1.020) indicates renal azotemia. However, evidence for dehydration indicates that there is also a pre-renal component. Urine parameters are near normal. Trace protein with 1.020 SG may warrant reevaluation to see if it is persistent. The dog appears quite ill and transient pre-renal (physiologic) proteinuria may be present. Urine pH must be evaluated with the acid base profile.

There is strong evidence for metabolic alkalosis (TCO2 = 39, Cl is decreased 10 units vs Na, anion gap is normal). The K is mid-reference range rather than showing a trend toward decreasing as expected, but there is no clinically significant association with that pattern. However, urine pH is acidic in the face of a strong alkalosis. Therefore, we have metabolic alkalosis with paradoxical aciduria. The strong alkalosis implicates upper GI loss or sequestration of HCl. Until Na concentrations and fluid volume are restored there will be continued efforts to retain Na by the kidneys. With Cl depleted there will be reabsorption of bicarbonate with Na to maintain electroneutrality, and the alkalosis and aciduria will be perpetuated.

The severe hyponatremia is in no way typical of primary renal azotemia. Although Na losses do occur with renal azotemia, there is also typically some water loss & dehydration that masks the loss of Na. Na this low implies a primary sodium disorder and GI losses or sequestration (3rd space) are strongly suspected. Addison's disease (renal loss) and possibly whipworm infestation should also be considered. Na values in the 120's are clearly associated with medullary washout and compromised tubular concentrating ability. With severe dehydration this Na is likely much lower than the measured 120. Therefore, this is a case of 'secondary' renal azotemia where restoration of Na levels should restore tubular function. This case was found to involve an upper GI foreign body obstruction. Fluid and electrolyte abnormalities were restored, the dog greatly stabilized and the foreign body was successfully removed with full recovery.

CASE 3 DISCUSSION

The CBC is pretty unremarkable, except perhaps for the lack of changes, including a stress response, in a seriously ill dog. Azotemia with inadequate urine concentration indicates renal azotemia. The main urine anomaly is granular casts indicating renal tubular degeneration. For trace protein see case #2. The primary acid base profile shows titrational metabolic acidosis based on high anion gap and is partially supported by the low normal TCO2. There are 5 main differentials for titrational acidosis (see above). Uremic acids and possibly lactic acids are good candidates in this case. Chloride, although decreased, is normal vs Na. In the secondary profile K is moderately elevated and exceeds values explained by acid-base shifts alone. Therefore, primary causes of hyperkalemia must be pursued. In this case, an oliguric renal azotemia could explain this change. But, in light of a severe/concurrent hyponatremia other differentials such as Addison's disease should be considered. The degree of hyponatremia is sufficient to cause medullary washout and secondary renal azotemia as in case 2. There is also a marked hypercalcemia that can lead to secondary renal azotemia (case 1) and could explain the casts. Primary differentials included humoral hypercalcemia of malignancy for the high calcium and Addison's disease for the hypercalcemia and most of the other changes. The final diagnosis was Addison's disease leading to secondary renal azotemia. The Ca along with all other abnormalities resolved rapidly upon treatment for Addison's and no neoplasia was found. Recognizing that severe hyponatremia or hypercalcemia cause secondary renal azotemia helps shift focus away from the kidney disease as the primary problem to pursue other primary diseases that secondarily affected renal function.

BONE MARROW COLLECTION, PREPARATION & ASSESSMENT IN CATS & DOGS

Robin W. Allison, DVM, PhD, Dipl. ACVP
Stillwater, OK

INTRODUCTION

Collection and preparation of diagnostic-quality bone marrow aspirates is an important skill for the veterinary practitioner. In most instances, slides prepared from bone marrow samples are sent to a veterinary clinical pathologist for evaluation and interpretation. As with other cytologic samples, sample preparation is of the utmost importance; inadequate samples or poorly prepared slides cannot be reliably interpreted by even the most skilled microscopist. The variety of normal (and potentially abnormal) cell types encountered in bone marrow cytology can be overwhelming for the novice; however, it is useful for the practitioner to at least recognize diagnostic versus non-diagnostic samples to avoid sending the latter for evaluation. Those with a real interest in cytology can take heart – bone marrow evaluation does become easier with practice! Although core biopsies may also be obtained for histologic evaluation, bone marrow aspirate cytology is preferred for evaluation of cellular morphology, and is the focus of these lectures. Core biopsy is most useful for assessing marrow cellularity, myelofibrosis, bone pathology, and focal lesions (inflammatory or neoplastic), and is best interpreted in conjunction with aspirate cytology.

INDICATIONS FOR BONE MARROW SAMPLING

The most frequent indication for obtaining a bone marrow aspirate is identification of persistent, unexplained abnormalities in peripheral blood. Additionally, the bone marrow may be evaluated when investigating potential malignancies (hematopoietic or not) and infectious diseases. Common indications include:

- Abnormal cells in peripheral blood* (early precursors, disorderly left shift, excessive nRBCs, abnormal morphology)
- Evaluation of iron stores (dogs)
- Fever of unknown origin
- Hypercalcemia*
- Hyperglobulinemia* (monoclonal or polyclonal)
- Neoplasia* – diagnosis, staging, monitoring
- Occult systemic infection
- Persistent, unexplained non-regenerative anemia, thrombocytopenia, or neutropenia*
- Persistent, unexplained pancytopenia*
- Persistent, unexplained leukocytosis, erythrocytosis, or thrombocytosis

*relatively common reason for bone marrow evaluation

CONTRAINDICATIONS

Complications related to bone marrow sampling are rare, provided appropriate attention to sterility and technique. Animals should be healthy enough to undergo sedation or anesthesia as required for restraint, although many animals require only local anesthesia. Animals with severe thrombocytopenia or coagulation disorders may be at risk for hemorrhage, but in reality this rarely occurs and can be managed with transfusion of fresh whole blood or appropriate components.

SAMPLE COLLECTION AND SLIDE PREPARATION

Commonly used sites for collection of bone marrow aspirates include the proximal humerus and trochanteric fossa of the femur in small dogs and cats, and the proximal humerus and iliac crest in large dogs. For core biopsies, the iliac crest is often used for large dogs and the wing of the ilium for small dogs and cats. Preparation of the site should include clipping hair and a surgical scrub; aseptic technique should be utilized during the procedure. The skin, subcutaneous tissue, and periosteum are infiltrated with a local anesthetic, and a stab incision is made through the skin. A variety of needles that incorporate a stylet are available for bone marrow aspiration; popular choices include 1 to 1.5 inch, 15-18 gauge Rosenthal or Illinois needles. The 18 gauge needles are used for small dogs and cats, and 15-16 gauge needles are used for dogs over 20 pounds. Some Jamshidi needles can be used for either aspiration or core biopsy. The needle is held firmly, making sure the stylet stays all the way in the needle so that the lumen is completely occluded, while the needle is advanced through the bone cortex with a back-and-forth twisting motion. Multiple resources are available describing and illustrating this procedure, including anatomic landmarks for needle placement.[1-4]

Many different methods have been described for preparing bone marrow aspirate slides. Some authors recommend using no anticoagulant and making slides immediately following aspiration. Bone marrow clots very rapidly, however, making this method difficult to master. In my experience, excellent quality bone marrow slides can be made by picking bone marrow spicules out of a petri dish containing anticoagulated bone marrow. The most convenient anticoagulant is ACD solution taken directly from a blood collection bag. A 2% solution of EDTA (sterile) can also be used, but is not available commercially. A small amount can be obtained by adding 0.5 mL sterile saline to a 7 mL EDTA tube (purple top blood tube), which contains 10.5 mg EDTA.

A 12-cc syringe containing 1-2 mLs of anticoagulant is used to aspirate the bone marrow sample once the needle is in place and the stylet has been removed. The syringe plunger must be pulled back firmly and briskly several times until marrow appears; the goal is to collect 1-2 mLs of sample. The syringe should be quickly disconnected from the needle and gently inverted several times to thoroughly mix the bone marrow sample with the anticoagulant, to ensure that it does not clot. (The stylet should be replaced in the bone marrow needle and left in place until the sample has been examined for spicules.) Once mixed, the syringe contents are expelled into a petri dish. Bone marrow spicules will appear as small irregular grey particles that tend to adhere to the bottom when the dish is tilted. There will also be clear, round, glistening fat globules present. A microhematocrit tube can be used to pick out one or two spicules, which are expelled onto a clean glass slide. If excess fluid is present, it can be quickly wicked away using absorbent paper. Another slide is placed on top of the sample, which will begin to spread, and then the slides are gently pulled apart horizontally. Excessive downward pressure must be avoided during this process to prevent cell lysis, which can result in non-diagnostic slides. Multiple slides can be prepared in this manner from an average bone marrow aspirate. If no spicules are visible in the dish, reaspiration may be attempted after advancing the needle (with stylet in place). If repeated attempts are unsuccessful, a bone marrow biopsy is recommended. Biopsy is more painful and requires sedation or anesthesia. Biopsies are obtained using larger bore, longer Jamshidi needles (11-13 gauge, 3-4 inches long). Once obtained, the core sample is pushed backwards out of the needle using a probe and placed in formalin. The core can also be gently rolled on a slide to provide a cytology sample prior to fixation.

At least one representative bone marrow aspirate slide should be stained and examined microscopically to ensure that an adequate sample has been obtained before sending unstained slides to a clinical pathologist. Standard Romanowsky stains that are used for in-clinic hematology (quick stains) are satisfactory, but bone marrow slides require a longer staining time than blood films because of the large number of nucleated cells present within the particles. Make sure slides are completely dry prior to staining, and that the stain is fresh. IMPORTANT TIP: Keep unstained slides away from formalin fumes (in a separate room from any formalin jars; do not ship unstained slides in the same container with formalin-fixed samples). Formalin fumes affect the staining properties of cytologic samples, and such exposure usually results in non-diagnostic slides. It is best to submit multiple slides to the laboratory (4-5 at a minimum, 8-12 is better). Additionally, a concurrent CBC and fresh blood film should be submitted for evaluation. Proper interpretation of the bone marrow slides is dependent upon current peripheral blood findings as well as a thorough history. The more information you can give the pathologist about your case the better job they can do for you. Include information about the duration of any peripheral blood abnormalities, other laboratory abnormalities, drugs/medications and any response to treatment, and significant physical exam findings. Pathologists in most laboratories are happy to speak with clinicians to discuss a challenging case.

CYTOLOGIC ASSESSMENT

Adequacy of the Sample

A properly prepared bone marrow cytology slide that contains cellular marrow particles is easily recognizable with the naked eye after staining – it will contain a central deeply basophilic area that appears granular (these are the marrow particles) surrounded by paler-staining blood . When observed microscopically at low power (10X objective), there will be very thick areas where cells are overlapping and hard to visualize, with thinner areas between the particles where cells are individualized and flattened; these are the areas where cell morphology will be evaluated. Make sure the majority of cells in those areas are intact; they should have distinct cytoplasmic borders and clearly defined nuclei, and there should be a minimum of streaming cellular debris from lysed cells.

Approach to Evaluation

Start by scanning the slide at low power (10X objective) to locate the marrow particles and assess their cellularity (proportion of cells versus fat). Because it dissolves in the stain fixative, fat appears as globular clear spaces between cells. Normal cellularity varies with the age of the animal; marrow particles from young animals may be 75% cellular, but most adult animals have 25-50% cellularity. Numerous marrow particles need to be evaluated on multiple slides because they can vary substantially from one area to another. Cellularity is reported as hypocellular (<25%), normocellular, or hypercellular (>75%). Relative numbers of megakaryocytes should also be assessed from low power. Estimation of megakaryocyte numbers can be difficult due to uneven distribution on the slide. A rough guideline is that a minimum of 3-5 megakaryocytes should be present in each particle from a normally cellular marrow. More than 10-20 per cellular low power field suggests hyperplasia. If megakaryocytes are absent or rare on multiple cellular slides, a core biopsy is recommended to confirm megakaryocytic hypoplasia. Iron stores can also be observed from low power in canine bone marrow; iron stores are not normally visible in feline bone marrow. Iron is usually seen as variably sized clumps of brown-black extracellular hemosiderin within particles, and may also be seen as blue-black material within macrophages. Several clumps of iron per particle are expected in normal adult dogs. An additional slide can be stained with Prussian blue to confirm presence or absence of iron, if needed.

While scanning, suitable areas should be identified for examination of nucleated cells at higher magnification. These areas are usually between the thick marrow particles, where there is minimal hemodilution and individual cells are nicely spread out so that their lineage can be identified. Because bone marrow is heterogeneous, it is important to evaluate multiple areas on more than one slide before coming to conclusions; it's not unusual to have a preponderance of cells of a single lineage in one area of the slide. Additionally, abnormal cell types are sometimes present in aggregates or clusters regionally on the slides; scanning for these abnormalities is important.

In addition to the estimates of cellularity, megakaryocyte numbers and maturation, and iron stores, a typical bone marrow cytology evaluation includes an estimation of the myeloid:erythroid ratio (M:E), evaluation of the maturation of both myeloid and erythroid series (including orderliness and completeness), examination of cellular morphology, notation of any other cell types that are present (normal or abnormal), and any other significant findings (microorganisms, prominent erythrophagocytosis, etc.). Most experienced cytologists do not do an actual differential cell count, although beginners may benefit from time spent doing a manual 500-1000 cell differential. With experience, accurate interpretations can be made by estimating relative myeloid and erythroid precursor numbers with careful attention to their maturation stages. It is sometimes helpful, however, to do an abbreviated differential consisting of myeloid cells, nucleated erythroid cells, lymphocytes, and plasma cells. Estimations of M:E ratios and evaluation of cellular morphology are performed using higher magnification, usually 40X and 50X-100X oil objectives.

Expected Normal Findings

Descriptions of cell types are inadequate without color images, which cannot be included here. However, there are several excellent references available to help with cell identification.[1,4,5] In general, cells of both myeloid and erythroid lineages become smaller with a lower nucleus:cytoplasm ratio as they mature. Normally, there are fewer of the most immature precursors present, with relative numbers increasing as the cells mature (**Fig. 1**). Blast cells are recognized by the presence of a nucleolus, and are normally fairly rare (<1-2% of cells).

Fig. 1. Normal pyramid of developing hematopoietic precursors in bone marrow. Relative cell numbers increase with maturity.

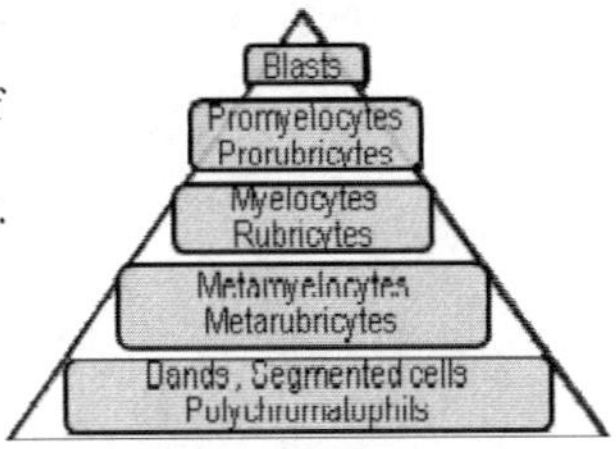

Erythroid precursors can often be identified by their deeply basophilic cytoplasm, round nuclei, and coarse nuclear chromatin relative to myeloid precursors. Recognizable precursors (from least to most mature) are rubriblasts, prorubricytes, rubricytes (basophilic, then polychromatophilic), metarubricytes, and polychromatophils. The last stage capable of division is the rubricyte. As hemoglobin accumulates in the maturing cells, cytoplasm becomes polychromatophilic and finally orthochromic in mature erythrocytes. The cytoplasm and nucleus should mature synchronously, with the nucleus becoming small and pyknotic (densely basophilic without visible chromatin structure) in polychromatophilic or orthochromic metarubricytes. Maturation is considered complete if polychromatophils (non-nucleated) are frequent. Polychromatophilic rubricytes and metarubricytes together are the most numerous nucleated erythroid precursors when maturation is orderly.

Myeloid precursors tend to be slightly larger and paler than erythroid precursors of similar stage. Most myeloid precursors are granulocytic, predominantly maturing to neutrophils. Nuclei are round to oval in most immature cells, becoming indented at the metamyelocyte stage before becoming band and finally segmented nuclei. Recognizable precursors (from least to most mature) are myeloblasts, promyelocytes, myelocytes, metamyelocytes, bands, and segmented granulocytes. The last stage capable of division is the myelocyte. Promyelocytes contain small magenta cytoplasmic primary granules. Specific granules (neutrophilic, eosinophilic, or basophilic) first appear at the myelocyte stage. Complete maturation is recognized by the presence of segmented neutrophils, which should also be the most numerous myeloid cells when maturation is orderly.

The M:E ratio is based on relative numbers of myeloid (which are predominately granulocytic) and nucleated erythroid precursors. In health this ratio can be somewhat variable, but is typically between 1:1 and 3:1. Because absolute numbers of myeloid and erythroid precursors are not generated, the M:E ratio MUST be interpreted with knowledge of both the marrow cellularity and concurrent CBC findings. For example, if the marrow is hypercellular and the animal has a normal peripheral neutrophil count, a decreased M:E ratio indicates erythroid hyperplasia. In a neutropenic animal, a normocellular marrow and decreased M:E ratio indicates myeloid hypoplasia. A hypercellular marrow with an increased M:E indicates myeloid hyperplasia, but if the animal is anemic then a concurrent erythroid hypoplasia exists. Examples of how to interpret M:E ratios will be given in the case presentations.

Megakaryocyte numbers are estimated at low power, and the relative maturation of megakaryocytes should also be noted. Megakaryoblasts are not often identified in normal bone marrow, as they appear similar to other blasts and are few in number. Promegakaryocytes are larger cells containing 2-4 nuclei and deeply basophilic cytoplasm. Still larger are basophilic megakaryocytes, which have a multilobulated nucleus with a moderate amount of basophilic cytoplasm. Mature megakaryocytes are very large cells with abundant granular cytoplasm (that resembles platelets) and multilobulated nuclei. Normally most megakaryocytes (>75%) observed are mature cells.

Other cells may be encountered in a normal bone marrow aspirate slide. Monocyte precursors are usually present in low numbers, but early precursors are often missed because they appear similar to early granulocyte precursors. Macrophages are normally present in low numbers. Small lymphocytes are usually observed in low numbers, typically representing <10% of cells; up to 20% have been reported in some healthy cats. Plasma cells are also normally present in low numbers (<2% of cells) but may be distributed unevenly. Stromal cells may be seen in particles that contain more fat; they are usually obscured in more cellular particles. Occasional osteoblasts or osteoclasts can be seen in marrow samples from young animals, but are rarely seen in adults. Mast cells are rare in normal bone marrow, but an occasional well-differentiated cell may be seen.

SUMMARY

Preparation of diagnostic-quality bone marrow aspirate slides is essential for an accurate interpretation, which must be made in conjunction with concurrent CBC findings and blood film evaluation. Bone marrow cytology is challenging, but clinicians with an interest in cytology will find their skills improve substantially with practice. Staining one representative slide and keeping it for comparison to the eventual pathologist report is one way to gain experience while ensuring the quality of slides being sent to a pathologist. If possible, practice cell identification on relatively normal bone marrow aspirates before trying to interpret abnormal samples. Invest in a reference with good photomicrographs!

REFERENCES

[1] Harvey JW. **Atlas of Veterinary Hematology**, 2001. [2] Grindem CB, et.al. **Vet Clin Small Anim** 2002;32:1313. [3] Grindem CB. **Vet Clin Small Anim** 1989;19:669. [3] http://www.vetmed.wsu.edu/resources/Techniques/bm.aspx, accessed 01/06/10. [4] Cowell RL, et.al. **Diagnostic Cytology and Hematology of the Dog and Cat**, 2008. [5] Thrall MA, et.al. **Veterinary Hematology and Clinical Chemistry**, 2004.

CHALLENGING CASES: BONE MARROW DISORDERS IN CATS & DOGS

Robin W. Allison, DVM, PhD, Dipl. ACVP
Stillwater, OK

INTRODUCTION

There are a limited number of potential diagnoses that can be made from a bone marrow aspirate slide. Often, a single bone marrow aspirate does not yield a definitive diagnosis, but may be consistent with several possible disorders that might be differentiated based on serial CBC and blood film evaluations, follow-up bone marrow aspirates, or response to treatment. Multiple good quality slides containing numerous marrow particles must be assessed before reaching conclusions, as bone marrow is heterogeneous and particles can vary considerably from slide to slide. Aspiration of an adequate sample may be difficult or impossible in some cases, in which case core biopsy should be pursued.

TERMINOLOGY

Terminology associated with bone marrow disorders can be quite confusing, and has not always been as specific in the veterinary literature as in the human literature. The following definitions are provided as a guideline, but there is considerable variation between authors.

- Dysmyelopoiesis/Myelodysplasia: A general term describing dysplastic conditions of hematopoietic cells resulting in abnormal morphology, which most often occur in myelodysplastic syndromes that may be congenital or develop secondary to another disease, drug or toxin.
- Dyserythropoiesis: Dysplastic changes in the erythroid line; includes nuclear fragmentation, megaloblasts, macrocytosis, nuclear:cytoplasmic maturation dysynchrony, siderotic inclusions, multiple nuclei or abnormal nuclear shapes.
- Dysgranulopoiesis: Dysplastic changes in the granulocytic line; includes giant neutrophils or other precursors, abnormal granulation, hypersegmentation, pseudo-Pelger-Huet cells, multinucleated precursors.
- Dysmegakaryopoiesis: Dysplastic changes in the megakaryocytic line; includes dwarf megakaryocytes with granulated cytoplasm and single nuclei, nucleus:cytoplasm maturation dysynchrony in larger cells, and abnormal nuclear morphology.
- Leukemia: Hematopoietic neoplasia in blood or bone marrow; may be acute or chronic.
- Lymphoproliferative Disorder: Usually refers to neoplastic proliferations, including acute lymphocytic leukemia (ALL), chronic lymphocytic leukemia (CLL), and multiple myeloma.
- Myelodysplastic Syndrome (MDS): Clonal disorders characterized by cytopenias and ineffective, dysplastic hematopoiesis; considered pre-leukemic conditions.
- Myeloproliferative Disorder: In veterinary literature, used to describe both chronic and acute myeloid leukemias, which include myeloid, erythroid, and megakaryocytic cell lines (versus chronic only in human literature). Some authors include myelodysplastic syndromes in this category.

DIAGNOSIS OF BONE MARROW DISORDERS

Erythroid Hyperplasia

Erythroid hyperplasia is recognized when the M:E ratio is decreased, particle cellularity is normal to increased, and the peripheral neutrophil count is not decreased. (If there is concurrent neutropenia, myeloid hypoplasia may be contributing to the decreased M:E ratio.) Usually maturation is orderly and complete (numerous polychromatophils), but there may be increased numbers of rubricytes and prorubricytes (left shift). Erythroid hyperplasia is a normal response to blood loss or erythrocyte destruction, and usually results in a regenerative anemia peripherally within a few days (effective erythroid hyperplasia). Erythroid hyperplasia may also be observed with primary or secondary erythrocytosis.

Ineffective erythroid hyperplasia occurs when there is a concurrent persistent nonregenerative anemia. Ineffective erythroid hyperplasia may be caused by nonregenerative IMHA, which results from targeting of erythroid precursors in the marrow. Typically there is erythroid hyperplasia and a left shift (sometimes with an apparent maturation arrest) and erythroid precursors may exhibit some dysplastic features. If reticulocytes or metarubricytes are targeted, these stages will be virtually absent from the marrow. Increased numbers of lymphocytes are common in cats and increased numbers of plasma cells are common in dogs with IMHA.[1] Myelofibrosis is frequent, and may cause poor quality aspirates.[2] Increased numbers of macrophages phagocytizing erythroid cells may be seen with immune-mediated disease.

Ineffective erythroid hyperplasia can occur with iron deficiency; iron stores are lacking, and metarubricytes and rubricytes may have mild dysplastic changes (scant, ragged cytoplasm). Ineffective erythroid hyperplasia with prominent erythrophagocytosis may occur in cats with disseminated histoplasmosis; a prolonged search may be required to identify the organisms in the bone marrow aspirate. Ineffective erythroid hyperplasia can also be part of a MDS or neoplastic process.

Myeloid Hyperplasia

Myeloid (granulocytic) hyperplasia is recognized when the M:E ratio is increased, particle cellularity is normal to increased, and the peripheral blood PCV is not decreased. (If there is concurrent anemia, erythroid hypoplasia may be contributing to the increased M:E ratio). Maturation is usually orderly and complete (many segmented neutrophils), but there may be increased numbers of less mature precursors (left shift) and depletion of the storage pool (fewer than expected segmented neutrophils). Myeloid hyperplasia is the normal response to inflammation, and usually results in an inflammatory leukogram within a few days (effective myeloid hyperplasia). Myeloid hyperplasia may also be observed with some paraneoplastic conditions, leukocyte adhesion deficiency, cyclic hematopoiesis in gray collies, early estrogen toxicity in dogs, and chronic myeloid leukemia (discussed later).

Ineffective myeloid hyperplasia occurs when there is a concurrent persistent peripheral neutropenia. Immune mediated neutropenia (IMN) results in myeloid hyperplasia, often with a left shift, and an apparent maturation arrest at the level of the cells being targeted (often segmented neutrophils). IMN is usually idiopathic, but may be drug-related in some cases. Ineffective myeloid hyperplasia may also occur with FeLV or FIV infection, and can be part of a MDS or neoplastic process.

Megakaryocytic Hyperplasia

Increased numbers of megakaryocytes are sometimes found in animals with secondary thrombocytosis, such as that associated with inflammation or iron deficiency, and are also expected with thrombocytopenia caused by platelet consumption or destruction. Such disorders include immune mediated thrombocytopenia, intravascular coagulation, vasculitis, and others. Megakaryocytic hyperplasia also occurs during recovery from a previous myelosuppressive episode. Sometimes increased numbers of less mature megakaryocytes are noted (left shift). Megakaryocytic hyperplasia can be part of a MDS or neoplastic process.

Generalized Marrow Hyperplasia

Generalized marrow hyperplasia is recognized when there are hypercellular marrow particles with a normal M:E ratio and increased numbers of megakaryocytes. This occurs whenever there is a regenerative anemia, inflammation, and reactive thrombocytosis, and in animals with combined IMHA-IMT and an inflammatory leukogram. However, bone marrow aspirates are rarely done on those animals.

Erythroid Hypoplasia

Erythroid hypoplasia is recognized when particle cellularity is normal to decreased, the M:E ratio is increased, and peripheral neutrophil numbers are not increased. (If there is anemia and neutrophilia, myeloid hyperplasia may be contributing to the increased M:E ratio). Erythroid hypoplasia is commonly seen with a nonregenerative anemia, and a specific cause is rarely identified on the bone marrow slide. Anemia of chronic/inflammatory disease (ACD) usually results in an increased M:E ratio due to both erythroid hypoplasia and myeloid hyperplasia (increased peripheral neutrophils); iron stores are often prominent. Bone marrow evaluations are not usually done for many causes of erythroid hypoplasia, such as ACD, endocrine disease, and renal disease. Erythroid hypoplasia may be observed in some cases of nonregenerative IMHA, depending upon the stage of precursors being targeted. (See Pure Red Cell Aplasia). Other causes of erythroid hypoplasia include chloramphenicol and FeLV; transient hypoplasia occurs in gray collies with cyclic hematopoiesis.

Pure Red Cell Aplasia

Pure red cell aplasia (PRCA) is a severe form of selective erythroid hypoplasia in which virtually all normal erythroid precursors are absent; particle cellularity is normal to decreased and a nonregenerative anemia is present. The M:E ratio is markedly increased, generally >75:1. Primary PRCA is likely due to immune-mediated destruction of early precursors; lymphocyte numbers are often increased in cats and plasma cell numbers are often increased in dogs.[1] Secondary PRCA can occur following treatment with recombinant human Epo. It has also been associated with FeLV infection, chloramphenicol administration in cats, parvovirus and rarely parvovirus vaccination.

Myeloid Hypoplasia

Myeloid (granulocytic) hypoplasia is recognized when particle cellularity is normal to decreased, the M:E ratio is decreased, and the PCV is not decreased. (If there is concurrent anemia, erythroid hyperplasia may be contributing to the decreased M:E ratio). Selective myeloid hypoplasia is an uncommon finding. Immune-mediated neutropenia may cause myeloid hypoplasia if early precursors are targeted. It may also be seen with some infections or drugs that more often affect multiple cell lines, such as parvovirus, FeLV, chemotherapeutic drugs, chloramphenicol, estrogen, griseofulvin, and phenylbutazone. Transient hypoplasia occurs in gray collies with cyclic hematopoiesis.

Megakaryocytic Hypoplasia

Selective megakaryocytic hypoplasia may occur secondary to a variety of drugs that more typically cause generalized bone marrow hypoplasia (chemotherapeutics, estrogens, griseofulvin, phenylbutazone, trimethoprim-sulfa, ribavirin, dapsone.) Immune-mediated destruction of early precursors is a rare cause.

Generalized Marrow Hypoplasia

Generalized marrow hypoplasia is recognized when multiple marrow particles are examined and consist predominately of fat (>75%) and stromal cells; erythroid, myeloid, and megakarytic precursors are markedly decreased to absent. When there is a concurrent peripheral pancytopenia, this is referred to as aplastic anemia. Typically, there are still lymphocytes and plasma cells present. Moderate numbers of mast cells have also been reported in aplastic bone marrow from dogs, which must not be mistaken for mast cell neoplasia.[3] Aplastic anemia has been associated with drugs or toxins including estrogen, chemotherapeutic agents, phenylbutazone, trimethoprim-sulfa, griseofulvin, phenobarbital, methimazole, and albendazole. Aplastic anemia also occurs with chronic ehrlichiosis, rarely with FeLV or feline parvovirus, and may be idiopathic. Marrow hypoplasia may also be associated with bone marrow necrosis from a variety of causes, myelofibrosis, and osteosclerosis; these conditions frequently require core biopsy for diagnosis.

Myelodysplastic Syndromes (MDS)

MDS are considered clonal disorders that eventually progress to leukemia. They are uncommon in dogs but are recognized more often in cats, especially those infected with FeLV. Hallmarks of MDS are ineffective hematopoiesis in the bone marrow with peripheral cytopenias, normal to increased blast percentages (but <30% of cells), and prominent dysplastic changes in one or more cell lines. Marrow particle cellularity is usually increased, but not always. Cats with FeLV-associated MDS usually have multilineage dysplasia with prominent erythroid maturation dysynchrony and megaloblasts. Basic categories of MDS were originally described for dogs and cats in a 1991 consensus paper,[4] but have been modified in various ways by different authors; at this point no clear consensus exists. Some of these categories include MDS-refractory cytopenias, MDS-excess blasts, MDS-erythroid predominance, and MDS-sideroblastic. The recommended cutoff for blast percentage may be decreased from,30% to <20% based on recent changes in the human classification system. [5] Regardless of the specific diagnosis, MDS must be differentiated from other conditions causing secondary myelodysplasia (such as drugs, inflammation, iron deficiency, neoplasia and immune mediated disorders) and congenital disorders (such as Poodle macrocytosis, congenital dyserythropoiesis of English Springer Spaniels, and cobalamin malabsorption in several breeds).

Neoplasia

Neoplasia in bone marrow may be of hematopoietic or non-hematopoietic origin. Hematopoietic neoplasia is broadly divided into lymphoproliferative and myeloproliferative categories. Histiocytic and mast cell neoplasia, although originating from bone-marrow derived cells, are generally not included in the myeloproliferative category. Non-hematopoietic tumors include sarcomas such as osteosarcoma or chondrosarcoma, and metastatic carcinomas.

Lymphoproliferative neoplasia includes acute and chronic lymphocytic leukemia, stage V lymphoma, and plasma cell myeloma. Lymphoid leukemias and plasma cell myeloma are usually recognized by the presence of large numbers of lymphocytes (>30%) or plasma cells (>15%), respectively, in the bone marrow. Diagnosis is straightforward when the marrow has been effaced by neoplastic cells or cell morphology is atypical, but can be problematic otherwise. Neoplastic lymphocytes may or may not contain obvious nucleoli; those that do can be confused with blasts of other lineages. With or without nucleoli, lymphocytes in ALL generally have more dispersed chromatin than normal small lymphocytes, and may exhibit other morphologic abnormalities. Neoplastic lymphocytes are often present in peripheral blood, but absolute lymphocyte numbers may not be increased. Lymphocytes in CLL have normal morphology and don't usually efface the marrow; numbers are often markedly increased in circulation. In addition, non-neoplastic diseases can result in peripheral lymphocytosis (e.g. ehrlichosis) and increased numbers of lymphocytes or plasma cells in bone marrow (e.g. ehrlichiosis, immune mediated disease). Marrow core biopsy or other advanced techniques (cytochemical staining, immunophenotyping, PCR for antigen receptor rearrangements) may be required for diagnosis. Importantly, large granular lymphocyte (LGL) leukemia, acute or chronic, generally originates in the spleen instead of bone marrow. LGL lymphoma, most often recognized in cats, appears to originate in the intestine and may metastasize to other organs and bone marrow.

Myeloproliferative neoplasia includes acute and chronic myeloid leukemia. Whereas peripheral blood leukocyte numbers are increased with chronic leukemias, this is not always the case with acute leukemias, which often present with peripheral cytopenias similar to MDS. These tumors are classified based on a 1991 consensus paper by the Animal Leukemia Study Group; classifications were adapted from the human FAB system.[4] The human classification system was revised by WHO in 2001. One significant change is that the blast percentage required for diagnosis of acute leukemias was dropped from >30% to >20%; some veterinary pathologists are now following those recommendations, although a consensus statement is currently lacking.[5] Acute leukemias defined for dogs and cats include: acute undifferentiated leukemia (AUL), acute myeloblastic leukemia without maturation (AML-M1), acute myeloblastic leukemia with maturation (AML-M2), acute promyelocytic leukemia (AML-M3), acute myelomonocytic leukemia (AML-M4), acute monocytic leukemia (AML-M5), acute erythroid leukemia (AML-M6), acute erythroid leukemia with erythroid predominance (AML-M6Er), and acute megakaryoblastic leukemia (AML-M7). There are specific guidelines available for diagnosis of each subtype of AML; morphologic appearance may not be sufficient to make these distinctions.[5] To date, AML-M3 has not been recognized in dogs or cats. Chronic myeloid leukemias include chronic myeloid leukemia (CML, usually neutrophilic, rarely eosinophilic or basophilic), chronic monocytic leukemia (CMoL), chronic myelomonocytic leukemia (CMMoL), polycythemia vera (primary erythrocytosis), and essential thrombocythemia. Blast counts are <30%, similar to MDS, but dysplasia is not a predominant feature and peripheral blood granulocyte/monocyte numbers are generally markedly increased; thus, CML is often difficult or impossible to distinguish from inflammation-induced myeloid hyperplasia in bone marrow. CML is often diagnosed by exclusion.

Histiocytic neoplasms in bone marrow include histiocytic sarcoma of either macrophage or dendritic cell origin. Generally other organs (lung, spleen, lymph nodes) are also involved. The macrophage variant is often markedly hemophagocytic.[6] These diseases must be differentiated from benign histiocytic proliferations in bone marrow such as occur with IMHA and hemophagocytic syndrome. Neoplastic histiocytic cells are rarely present in peripheral blood, however, primary dendritic cell leukemia involving bone marrow and blood has recently been described in a dog.[7]

Cautions

Importantly, a single bone marrow evaluation in an animal with peripheral cytopenias can be misleading. Differentials for a hypercellular, left-shifted marrow with maturation arrest include early recovery from previous insult, immune-mediated precursor destruction, MDS, and leukemia. Repeat bone marrow and/or peripheral blood evaluations are often necessary to distinguish them.

REFERENCES

[1] Weiss DJ. **J Comp Path** 2008;138:46. [2] Stokol T et.al. **JAVMA** 2000;216(9):1429. [3] Walker D et.al. **Vet Clin Pathol** 1997;26(3):106. [4] Jain NC et.al. **Vet Clin Pathol** 1991;20(3):63. [5] McManus PM. **Vet Clin Pathol** 2005;34(3):189. [6] Moore PF et al. **Vet Pathol** 2006;43:632. [7] Allison RW et al. **Vet Clin Pathol** 2008;37(2):190.

MYELODYSPLASTIC SYNDROMES IN CATS AND DOGS: IMPROVED PROGNOSTIC INDICATORS

Amy MacNeill, DVM, PhD, DACVP
Urbana, IL

INTRODUCTION

Myelodysplastic syndromes (MDS) are a group of disorders characterized by altered proliferation and differentiation of hematopoietic stem cells. Primary MDS and secondary MDS (due to treatment with drugs or exposure to radiation) are neoplastic diseases that must be distinguished from secondary dysmyelopoiesis, congenital dysmyelopoiesis, and other non-neoplastic conditions that lead to ineffective hematopoiesis. These non-neoplastic conditions are more common than MDS and are often immune mediated or secondary to administration of pharmacotherapeutics[1]. In human medicine, the minimal criteria needed to diagnosis MDS are the presence of at least 10% dysplastic cells with a specific myeloid lineage in the bone marrow, less than 20% blast cells in the bone marrow or peripheral blood, and a peripheral blood monocyte count under 100,000 cells per microliter[2]. Studies utilizing X-chromosome inactivation[3] and analysis of alterations in karyotype have established that the dysplastic cells in human MDS patients are a clonal population of cells. The two prognosis scoring systems used to predict disease progression in human cases of MDS rely heavily on specific abnormalities in the karyotype of the patient's neoplastic cells. A paucity of information is available regarding the abnormal karyotypes in cases of canine or feline MDS, so this type of scoring system currently cannot be applied to our small animal patients. However, the relatively recent inclusion of immunophenotyping as part of the diagnostic work-up for hematopoietic neoplasms of dogs and cats has greatly improved our ability to differentiate MDS from other hematopoietic neoplasms. As these techniques begin to be used routinely, our ability to predict disease progression in our patients will improve.

HISTORY OF THE DIAGNOSIS OF MDS[4]

In the early 1900's, MDS was recognized as a form of anemia that was unresponsive to treatment. The disease was first termed 'anemia pseudoaplastica'. This term was replaced in 1938 when a description of 100 cases of 'refractory anemia' was published. Discussion of MDS as a preleukemic condition was first started in 1949. In 1976, the French-American-British (FAB) classification system was proposed. The FAB classification scheme included discussion of acute leukemias and two subtypes of MDS. The establishment of the widely accepted FAB classification system allowed practitioners to compare the outcomes of human cases of MDS. Due to additional case reviews and recognition of several distinct biologic outcomes of myelodysplastic diseases, the FAB classification system was revised in 1982. This revision expanded the number of recognized forms of MDS from two subgroups to five subgroups. Use of the FAB classification system led to development of the International Prognostic Scoring System (IPSS) in 1997. This system is still considered an effective way to predict disease progression of MDS in humans. In 2001 the World Health Organization (WHO) reclassified the neoplastic hematopoietic diseases. One of the most important things that the WHO classification system changed was the requirement for a diagnosis of acute myeloid leukemia (AML) to include patients with •20% blasts in the bone marrow. This was a reduction from the recommendation of •30% blasts in the bone marrow for a diagnosis of AML in the FAB classification scheme. This change was based upon the recognition that disease progression in patients diagnosed with MDS who had 20-30% blasts in the bone marrow closely resembled disease progression in patients diagnosed with AML. In addition, these patients benefitted from the same chemotherapy protocols that are used with AML patients. The WHO classifications lead to the WHO-based prognostic scoring system (WPPS) in 2005. Most recently, the WHO-classification system was updated in 2008 to include additional chromosomal abnormalities that further differentiate disease subtypes and are strong prognostic indicators in human hematopoietic neoplasms.

A veterinary classification system for myeloid neoplasms was modeled after the FAB classification system in humans[5]. This classification system is still in use today and includes the classification of AML in patients with •30% blasts in the bone marrow, but not in patients with 20-30% blasts. Several veterinary clinical pathologists prefer to use WHO classification guidelines and diagnose AML in patients with •20% blasts in the bone marrow or peripheral blood. Neither of these cut-off values has been adequately linked to disease progression or prognosis in veterinary patients.

CLINICAL PRESENTATION OF PATIENTS WITH MDS[6]

Patients typically have peripheral cytopenias with normocellular or hypercellular marrow. Clinical signs are variable and are related to the specific cytopenia seen. Patients usually do not have organomegaly or lymphadenopathy. If MDS is diagnosed early in the disease course, animals may survive for months to years with supportive care. Cats diagnosed with MDS should be tested for feline leukemia virus infection.

DIAGNOSIS AND CLASSIFICATION OF MDS

In 2008, the WHO released an updated classification scheme for myeloid neoplasms and acute leukemias. This classification scheme includes many specific categories based upon chromosomal abnormalities found in humans with hematopoietic neoplasms. This limits its usefulness and application in veterinary patients. However, with the exception of the assessment of genetic mutations, the guidelines for using the WHO classification of myeloid neoplasms (Table 1)[7] can be followed to enhance diagnosis of these neoplasms in cats and dogs.

Table 1. WHO classification of myeloid neoplasm guidelines

Specimen requirements
Peripheral blood and bone marrow collected prior to therapy
Peripheral blood and bone marrow aspirate smears stained with Wright-Giemsa stain
Bone marrow biopsy
Bone marrow specimens for cytogenetic analysis and flow cytometry
Assessment of blasts
Determined in peripheral blood and bone marrow by visual inspection
Myeloblasts, monoblasts, promonocytes, and megakaryoblasts are counted as blasts
Flow cytometric assessment of CD34+ cells is not recommended as a substitute for visual inspection
If the aspirate is poorly cellular, immunohistochemical detection of CD34+ cells can be done on a bone marrow biopsy sample
Assessment of blast lineage
Multiparameter flow cytometry is preferred
Cytochemistry is not essential for all cases
Immunohistochemical stains applied to biopsy samples may be helpful
Assessment of genetic features
Complete cytogenetic analysis
FISH and RT-PCR are not essential for all cases
Mutational studies for *NPM1*, *CEBPA*, and *FLT3* mutations are recommended for cytogenetically normal acute myeloid leukemia
Mutational studies for *JAK2* mutations are recommended for *BCR-ABL1*-negative myeloproliferative neoplasms
Mutational studies for *KIT*, *NRAS*, *PTNP11*, and others should be performed as needed

To diagnose MDS, dysplastic changes in one or more of the hematopoietic cell lines (erythroid, myeloid, or megakaryocytic cells) must be identified in >10% of the cells in the affected lineage. In veterinary medicine, MDS is divided into non-erythroid dominant and erythroid dominant types based on the ratio of myeloid to erythroid precursors (M:E) in the bone marrow[6]. MDS is diagnosed in patients when the M:E>1 and there are <30% blast cells in the marrow. MDS with erythroid predominance (MDS-Er) has >50% erythroid precursors in the marrow, blast cells <30% of the non-erythroid cells, and blast cells plus pro-erythroblasts <30% of all nucleated cells. Additional classifications have been proposed including 'MDS-Eos' for patients with MDS and >10% eosinophils in the bone marrow. Many veterinary pathologists apply components of the 2001 WHO classification and further subtype MDS into MDS refractory cytopenia (MDS-RC) and MDS excess blasts (MDS-EB)[8]. MDS-RC has <50% erythroid precursors, a single-lineage cytopenia, dysplasia of the cytopenic lineage, and <5% blast cells. MDS-EB includes cases with multilineage dysplasia and 5-20% blasts. If the WHO classification is utilized, AML is diagnosed if blasts are •20% of the hematopoietic cells in the marrow. Note that secondary dysmyelopoiesis is common in dogs and should be ruled out before a diagnosis of MDS is made, as full recovery is often seen when the initiating cause of the dysmyelopoiesis (e.g. the drug or toxin) is removed.

The 2008 WHO classification lists nine subtypes of MDS and mentions one provisional subtype. Seven of these subtypes potentially could be applied to disease in cats and dogs. They are 1) refractory anemia (RA), 2) refractory neutropenia (RN), 3) refractory thrombocytopenia (RT), 4) refractory anemia with ring sideroblasts (RARS), 5) refractory cytopenia with multilineage dysplasia (RCMD), 6) refractory anemia with excess blasts (RAEB), and 7) myelodysplastic syndrome unclassifiable (MDS-U).

PATHOGENESIS OF MDS

Chromosomal abnormalities are found in many cases of human MDS and are thought to be the fundamental cause of the abnormalities in proliferation and differentiation. In some cases of MDS, these chromosomal changes are induced by carcinogens, radiation, or certain drugs. In most patients, no underlying cause for the genetic defect in the neoplastic cells can be documented. Subtle genetic alterations and haploinsufficiency of critical genes have been identified in some patients that were not initially identified as having chromosomal defects. There are several additional factors that appear to contribute to the pathogenesis of primary MDS[9]. Altered epigenetic control of gene expression seems to lead to development of MDS in some patients. It also has been noted that apoptosis is increased in dysplastic cells of MDS patients. This is in contrast to the neoplastic cells in AML patients, which are more resistant to apoptosis. Several molecules that promote apoptosis are present in higher levels in patients with MDS subtypes that have a low risk of progression to AML[10]. Abnormalities in the bone marrow microenvironment likely contribute greatly to development of MDS. Increased production of vascular endothelial growth factor (VEGF) and matrix metalloproteinase-9 have been reported[10]. Some MDS patients have decreased natural killer (NK) cell function which may contribute to reduced NK cell surveillance, allowing abnormal hematopoietic cells to accumulate. A subset of MDS patients may have CD8+ T cells that suppress hematopoiesis, contributing to the cytopenias that occur with the disease. Overexpression of transforming growth factor-beta (TGF) and stromal cell-derived factor-1 (SDF-1) may also suppress hematopoiesis in some patients. Increased production of reactive oxygen species has been linked to MDS and may contribute to pathogenesis by inducing mutations that lead to genomic instability.

PRONOSTIC INDICATORS

Two scoring systems are used in human cases of MDS to predict median survival and progression to AML. Both systems rely on detection of specific chromosomal defects so they cannot be used directly in cats or dogs. However, other components of these systems may be somewhat useful. For example, human MDS patients with <5% blasts are given a lower, more favorable score than patients with 5-10% blasts, and patients with 11-20% blasts are given a higher, less favorable score. Also, patients with multilineage cytopenias are assigned a higher score than patients with single lineage cytopenia. These distinctions could be easily determined in cats and dogs with MDS.

In addition to the established scoring systems, flow cytometry can assist in prognosis of MDS[12]. Recurrent phenotypic abnormalities have been recognized in human cases of MDS that correlate to prognosis. These abnormalities have not been incorporated into the WHO classification system due to the lack of a consensus statement on flow cytometric protocols. However, flow cytometry results will likely be used in the next update of the classification of myeloid neoplasms. This technique has a high potential for predicting outcome in canine and feline cases of MDS. Flow cytometric-based assays are rapidly developing and becoming more readily available to veterinarians for the diagnosis of hematopoietic neoplasms in dogs and cats. The findings from these diagnostic tests should be closely examined for correlation to disease progression in cats and dogs with MDS.

RESPONSE TO TREATMENT

The ability to predict response to treatment would be very useful in cats and dogs with MDS. In human patients, anemia associated with MDS is often treated with transfusions and/or erythropoietin (EPO). It is known that patients who require fewer transfusions or have serum EPO levels <500 U/L have a significantly higher rate of response to EPO[11]. Lenalidomide is recommended for MDS patients with the chromosome defect del(5q). This drug is antiangiogenic and anti-adhesive, it inhibits tumor necrosis factor alpha (TNF) production, induces growth arrest and apoptosis of malignant cells, modulates T cell function, and enhances cell lysis by NK cells[11]. However, severe neutropenia and thrombocytopenia are common side effects of lenalidomide treatment. Immunosuppressive therapies including administration of equine anti-thymocyte globulin or cyclosporine-A have been used in human MDS patients with some success. The administration of hypomethylating agents may be beneficial for patients with refractory anemia.

REFERENCES

[1.] Blue JT. **Toxicologic Pathology** 2003;31:44. [2.] Tefferi A and Vardiman JW. **N Engl J Med** 2009;361;19:1872. [3]Tefferi A, et al. **Blood** 1990;75:1770. [4.] Komrokji RS and Bennett JM. **Current Opinion In Hematology** 2007;14:98. [5] Jain NC, et al. **Vet Clin Path** 1991;20:63. [6.] McManus PM. **Vet Clin Path** 2005; 34:189. [7] Vardiman JW, et al. **Blood** 2009;114:937. [8.] Harvey JW. In Atlas of Veterinary Hematology. WB Saunders Co. 2001:175. [9.] Epling-Burnette PK and List AF. **Current Opinion in Hematology** 2009;16:70. [10.] Bernasconi P. **British Journal of Haematology** 2008;142:695. [11] Jadersten M and Hellstrom-Lindberg E. **Journal of Internal Medicine** 2008;265:307. [12]Valent P, et al. **Leukemia Research** 2007;31:727.

CLASSIFICATION OF ACUTE LEUKEMIAS

Amy MacNeill, DVM, PhD, DACVP
Urbana, IL

INTRODUCTION

Classification of acute lymphoid and myeloid leukemias in species of veterinary importance can be challenging. Examination of peripheral blood and bone marrow aspirates is essential in diagnosing hematopoietic diseases. The application of cytochemical stains to bone marrow aspirate smears has long been important in differentiating lymphoid from myeloid leukemia. The routine addition of immunohistochemical and immunocytochemical staining procedures has greatly enhanced the accuracy and specificity of the diagnosis of hematopoietic neoplasms. Flow cytometric phenotyping of cells in blood and bone marrow aspirates has recently been shown to be useful in the diagnosis of acute lymphoid and myeloid leukemias in many species including dogs[1&2], cats[3], horses[4], cattle[5&6], and an alpaca[7]. The utility of flow cytometry for classifying acute leukemias will likely expand as the antibodies used to detect the molecular markers of lymphoid and myeloid cells are validated in more species.

CLASSIFICATION OF HEMATOPOIETIC NEOPLASMS

In 1976, the French-American-British (FAB) classification system was proposed to standardize the diagnosis of acute leukemia in human patients. The FAB classification scheme included discussion of acute leukemias and two subtypes of myelodysplastic syndrome (MDS). The establishment of the widely accepted FAB classification system allowed practitioners to compare the outcomes of human cases of hematopoietic neoplasia. Due to additional case reviews and recognition of several distinct biologic outcomes of diseases, the FAB classification system was revised in 1982. The next reclassification of human hematopoietic neoplasms was published in 2001 by the World Health Organization (WHO). The WHO classification added disease categories based on cytogenetic analysis of hematopoietic neoplasms and included a category for diseases with aspects of both acute myeloid leukemia (AML) and MDS termed 'myelodysplastic/myeloproliferative neoplasms' (MDS/MPN).

The WHO classification of hematopoietic neoplasms was updated in 2008. This classification scheme combines cellular morphology, cytochemistry, immunophenotype, and genetics along with clinical features to define specific disease entities. Cell types classified as 'myeloid' include granulocytes, monocytes and macrophages, erythrocytes, megakaryocytes, and mast cells. Neoplastic transformation of precursor B lymphoblasts or T lymphoblasts is classified as acute lymphoid leukemia/lymphoma (ALL). The WHO classification has become more extensive as new information about disease pathogenesis and outcome is accumulated. The application of the WHO classification for non-human cases of acute leukemia is limited by the dramatic lack of cytogenetic studies in most species. That being said, studies in mice indicate that the genetic defects in leukemic cells of non-human species are likely to be similar to those that occur in humans that develop leukemia[8]. Eventually, cytogenetic evaluation of non-human cases of leukemia should enhance our ability to predict disease progression and response to therapy. At this time, classification based upon chromosomal abnormalities is not practical in veterinary medicine.

The veterinary classification system for myeloid neoplasms was modeled after the FAB classification system in humans[9]. This classification system is still in use today and includes the classification of AML in patients with •30% blasts in the bone marrow, but not in patients with 20-30% blasts. In 2001 the World Health Organization (WHO) reclassified the neoplastic hematopoietic diseases. One of the most important things that the WHO classification system changed was the requirement for a diagnosis of acute myeloid leukemia (AML) to include patients with •20% blasts in the bone marrow. This was a reduction from the recommendation of •30% blasts in the bone marrow for a diagnosis of AML in the FAB classification scheme. This change was based upon the recognition that disease progression in patients diagnosed with MDS who had 20-30% blasts in their bone marrow closely resembled disease progression in patients diagnosed with AML. In addition, these patients benefitted from the same chemotherapy protocols that are used with AML patients. Neither of these blast count cut-off values has been adequately linked to disease progression or prognosis in veterinary patients. Several veterinary clinical pathologists believe the WHO classification guidelines should be incorporated into the veterinary classification scheme and a diagnosis of AML should be applied to patients with •20% blasts in the bone marrow or peripheral blood. Regardless, the veterinary classification has proven useful in most species, as it is based on morphologic and cytochemical analysis of bone marrow samples and does not rely on either cytogenetics or immunophenotyping.

Note that for all of these classification schemes, blast cells include myeloblasts, monoblasts, promonocytes, and megakaryoblasts. Rubriblasts and precursor lymphoid cells are excluded from blast cell counts. In my opinion, an official revision of the veterinary classification of leukemia should be attempted as additional phenotypic studies are performed more routinely. At such a time, the utility of including patients with •20% blasts in the classification of AML should be thoroughly evaluated. It may be that the blast count cut-off used to diagnose AML in humans is appropriate to use in some non-human species, but not others. Table 1 compares the 2008 WHO classification of acute leukemias[10] to the classification typically used by veterinary pathologists[9&11].

Table 2. Comparison of the WHO classification and veterinary classification of acute leukemia

WHO Classification	Veterinary Classification	Description of the Veterinary Classification
AML with recurrent genetic abnormalities		
AML with myelodysplasia		
Therapy-related myeloid leukemia		
AML with minimal differentiation	AML M3	Atypical promyelocytes •30% of all nucleated cells
AML without maturation	AML M1	Myeloblasts •90% of all nucleated cells More mature granulocytes <10% of non-erythroid cells
AML with maturation	AML M2	Myeloblasts •30% of all nucleated cells More mature granulocytes >10% of non-erythroid cells
Acute myelomonocytic leukemia	AML M4	Myeloblasts and monoblasts •30% of all nucleated cells More mature granulocytes >20% of non-erythroid cells Promonocytes and monocytes >20% of non-erythroid cells
Acute monoblastic/monocytic leukemia	AML M5a	Monoblasts •30% of all nucleated cells Monocytes and promonocytes •80% of non-erythroid cells
	AML M5b	Monoblasts •30% of all nucleated cells Promonocytes and monocytes 30-79% of non-erythroid cells
Acute erythroid leukemia Erythroleukemia, erythroid/myeloid	AML M6	M:E<1 Myeloblasts and monoblasts •30% of non-erythroid cells
Pure erythroid leukemia	AML M6Er	M:E<1 Blast cells + Rubriblasts •30% of all nucleated cells
Acute megakaryoblastic leukemia	AML M7	Megakaryoblasts •30% of all nucleated cells
Acute basophilic leukemia		
Acute panmyelosis with myelofibrosis		
Myeloid sarcoma		
Myeloid proliferations related to Down syndrome		
Blastic plasmacytoid dendritic cell neoplasm		
Acute undifferentiated leukemia	AUL	Blasts •30% of all nucleated cells Blasts lack morphology and cytochemical staining that allow for designation of a myeloid or lymphoid cell origin
Mixed phenotype acute leukemia with specific genetic abnormalities		
B lymphoblastic leukemia/lymphoma	ALL – B cell	CD79a+ blasts •30% of all nucleated cells
T lymphoblastic leukemia/lymphoma	ALL – T cell	CD3+ blasts •30% of all nucleated cells

ADDITIONAL DIAGNOSTIC TESTS

Cytochemical Staining[12]

In most species, neutrophils are positive for peroxidase and Sudan black B stains. Neutrophils, basophils, and mast cells stain positive for chloroacetate esterase. Monocytes, lymphocytes, platelets, and megakaryocytes tend to be positive for -naphthyl-acetate esterases and -naphthyl-butyrate esterases, although this staining can be inhibited by NaF in monocytes. Platelets and megakaryocytes can also be stained for acetylcholinesterase and with Synacril black AN. Leukocyte alkaline phosphatase is less specific and may stain lymphoid and myeloid cells. Species-specific differences in cytochemical staining profiles will be discussed.

Immunophenotyping

Large numbers of immunophenotypic markers are applied to differentiate human hematopoietic neoplasms. The combination of several markers and recognition of subtle changes in expression of these molecules is used to classify neoplastic cells and predict prognosis in human patients[13&14]. Many of the markers are not validated for use in non-human species. The minimum recommended panel for diagnosis of subtypes of animal leukemia includes CD79a for B cells, CD3 for T cells, myeloperoxidase and CD11b for myeloid cells, CD41 for megakaryocytes, CD1c for dendritic cells, and CD34 for stem cells[9]. Additional markers improve the specificity of flow cytometric analysis and reduce the number of false positive results. Examples of these molecules include CD4 and CD8 for T cell subtyping, CD45 as a panleukocyte marker, CD14 on monocytes, CD21 on B lymphocytes, CD61 on platelets and megakaryocytes[1], and CD172a on myeloid cells[7]. There are species-specific differences in cell immunophenotypes, just as there are differences in cytochemical staining. Knowledge of these differences is critical for correct interpretation of flow cytometric analysis. Species-specific cell markers and other limitations of flow cytometric analysis will be reviewed.

REFERENCES

[1.] Tasca S, et al. **Vet Clin Path** 2009;38:2. [2.] Wilkerson MJ, et al. **Vet Immunol Immunopathol** 2005;106:179. [3.] Shelton GH and Ladiges WC. **Vet immunol Immunopathol** 1987;16:271. [4.] McClure JT, et al. **J Vet Intern Med** 2001;15:144. [5.] Letesson JJ, et al. **Vet Immunol Immunopathol** 1991;27:207. [6.] Aida Y, et al. **Am J Vet Res** 1992;53:1988. [7.] Steinberg JD, et al. **Vet Clin Path** 2008;37:289. [8.] Komeno Y, et al. **Journal of Cellular Physiology** 2009;219:529. [9.] Jain NC, et al. **Vet Clin Path** 1991;20:63. [10.] Vardiman JW, et al. **Blood** 2009;114:937. [11.] Raskin RE. **Vet Clinics of North America: Small Animal Practice** 1996;26:1023. [12.] McManus PM. **Vet Clin Path** 2005;34:189. [13.] Gluzman DF, et al. **Exp Oncol** 2008;30:102. [14.] Olsen RJ, et al. **Arch Pathol Lab Med** 2008;132:462.

TRANSFUSION REACTIONS: HOW TO RECOGNIZE, TREAT & AVOID THEM

Urs Giger, DACVIM, DECVIM-CA, DECVCP
Philadelphia, PA

Since the early 1980s, the use of blood product transfusions in the treatment of critically ill companion animals and in supporting dogs and cats undergoing surgery and minimally-invasive diagnostic and therapeutic procedures has increased tremendously. Blood components are, however, prepared from donor animals and are consequently a limited resource. This is despite the establishment of several commercial animal blood banks. Moreover, blood components are biologic products that possess the inherent ability to cause transfusion reactions. Fortunately, recent advances in modern blood banking techniques and transfusion medicine have markedly decreased the risks associated with transfusion therapy in companion animals. Animal blood banks, veterinary clinicians, nurses and technicians play an essential role in ensuring the safety and efficacy of transfusion therapy, especially the recognition of transfusion reactions. Guidelines for the selection of healthy blood donors, the preparation and storage of the blood products, blood compatibility testing, and the monitoring of transfusions have been established, and are constantly being modified and improved as our understanding of transfusion medicine evolves. This presentation will discuss the various transfusion reactions that may be observed in companion animals, as illustrated by case examples, and will provide insights into how to recognize, treat and preferably avoid such reactions.

Table 1. Transfusion Reactions in Companion Animals

Non-infectious Complications	**Infectious Complications**
Immune-mediated hemolytic reactions	Bacteremia and sepsis
DEA 1.1, 4, *dal*, other incompatibilities	Blood product or catheter related
AB, *mik*, other incompatibilities	Infectious disease transmission
Non-immune red cell hemolysis	Dogs: *Babesia spp.**
Freezing during storage, outdated blood	*Leishmania spp.**
Bacterial contamination	*Ehrlichia spp.*
Febrile non-hemolytic reactions (FNHTR, >1.5°F)	*Brucella canis*
Most common, self-limiting	*Anaplasma spp.*
(Hypothermia rare)	*Rickettsia rickettsii*
Allergic reactions	*Mycoplasma spp.*
Urticaria	*Hepatozoon spp.*
Anaphylaxis	*Dirofilaria immitis*
Nausea, vomiting	*Trypanosoma cruzi*
Related to feeding during transfusion	*Bartonella spp.*
Post-transfusion purpura (PTP)	
Transfusion-associated acute lung injury (TRALI)	Cats: FeLV, FIV, FCoV
Transfusion-associated circulatory overload (TACO)	*Mycoplasma spp.**
Cardiac & normovolemic patients	*Bartonella spp.*
Metabolic disturbances	*Cytozoon felis*
Hypocalcemia (liver failure, massive transf.)	*Ehrlichia spp.*
Hyperammonemia (stored red cell products)	*Anaplasma phagocytophilum*
Thromboemboli	*Toxoplasma spp.*
Air embolism	*documented transmission

There are many different causes of transfusion reactions, although acute hemolytic transfusion reactions resulting from blood type incompatibilities and the transmission of infectious agents are feared the most (Table). Transfusion reactions can be divided into two categories; 1. Infectious-disease complications and 2. Non-infectious serious hazards of transfusion (NISHOTs). Any blood component has the potential to induce a transfusion reaction, but red cell products, which invariably contain other cellular components, are most likely to do so.

There are over a dozen canine blood group systems and there persists controversy regarding the presence of clinically important naturally-occurring alloantibodies in dogs. Acute hemolytic transfusion reactions in dogs, however, have only been documented in previously transfused patients. In these cases, the reactions were demonstrated to be associated with either DEA 1.1 incompatibilities , incompatibilities involving the common red cell antigens DEA 4 and *dal*, or incompatibilities involving several as yet undefined red cell surface antigens. Sensitization becomes evident ≥4 days following a blood product transfusion. Pregnancy does not sensitize bitches and bitches that have previously whelped can safely be used as blood donors without additional compatibility testing.

In cats older than three months, strong naturally-occurring anti-A alloantibodies exist in type B cats, and weak naturally-occurring anti-B alloantibodies exist in type A cats. These alloantibodies can be responsible for serious acute hemolytic transfusion reactions. In cats, however, there appear to be other clinically important blood types, such as the *mik*-antigen. Currently, these are best detected by crossmatching or extended typing by reference laboratories.

Whilst many hemolytic transfusion reactions are the result of red cell alloimmune interactions, there are also other non-immunologic hemolytic transfusion reactions. Outdated or damaged (frozen, contaminated) red cell products may be hemolyzed and should never be administered. Interestingly, unlike humans, dogs and cats do not typically experience hemoglobin-induced nephropathies from intravascular hemolysis. In humans, transfusion–related acute lung injury (TRALI) is currently considered the most common cause of transfusion-related mortality. TRALI has been associated with the administration of plasma products that test positive for HLA and other leukocyte antibodies, which it is suggested might arise during pregnancy. The occurrence of TRALI in companion animals has been proposed, but remains unproven. Fortunately, many transfusion reactions are self-limiting, but some can be life-threatening particularly in severely compromised patients.

Most blood transfusion reactions occur acutely, that is during or within hours of the blood product administration. Delayed transfusion reactions may also be observed over days to weeks, as a result of the development of new alloantibodies or the transmission of infectious agents. In some cases, it can be difficult to determine whether apparent reactions or complications are a result of the patient's primary disease process, the various therapeutic measures being instituted, or a genuine transfusion reaction. In cases of a suspected transfusion reaction, any remaining blood product and fresh patient blood should be evaluated for compatibility and bacterial contamination (Gram stain, microbial culture, 16S rRNA PCR), as well as stored for any future analyses that may be indicated, and a transfusion reaction incident report compiled attempting to elucidate the underlying cause of the reaction.

Ideally, a transfusion monitoring record should be completed during any transfusion, which includes recording pre- and post-transfusion hematocrit, and serially obtained parameters such as temperature, heart rate, respiratory rate, pulse pressure and demeanor, thus facilitating the early recognition of transfusion reactions. Immediate cessation of the transfusion and supportive care are the only therapeutic measures for an acute transfusion reaction. Antihistamines and glucocorticoids are often administered to dogs and cats when an acute transfusion reaction is suspected, but there is insufficient evidence to support their efficacy. Limiting the risk of transfusion reactions, through the application of appropriate blood banking procedures and transfusion techniques, is therefore essential.

Most blood type incompatibilities and their associated transfusion reactions can now be readily avoided by DEA 1.1 or AB typing all canine or feline patients and donors. Crossmatching or extended typing is indicated under certain circumstances, particularly in patients receiving multiple transfusions As a result of the development of alloantibodies against blood types other than the ones routinely typed for, previously transfused animals must be crossmatched, even when transfused with blood from the same donor. In-clinic blood-typing and cross-matching assays are now readily available and simple to perform. Confirmation of these test results and extended typing are also now widely available through reference laboratories. The practice of testing patient and donor compatibility by administering a few milliliters of the donor's blood to the patient cannot be advocated as the transfusion of as little as 1 ml of incompatible blood to a patient has the potential to elicit a fatal transfusion reaction.

Febrile reactions are the most commonly observed complications, but are generally mild and transient unless related to hemolysis or sepsis. Febrile non-hemolytic transfusion reactions are likely induced by leukocyte and cytokines. The utilization of leukodepleted blood products, particularly when accomplished prior to storage, can reduce some of the hypersensitivity / allergenic and febrile reactions, but this filtration process is expensive.

Infectious disease transmission is now effectively limited by using only healthy donors and screening each donor routinely by antigen (PCR, blood smear) and serologic testing. Specific guidelines for infectious disease screening of blood donors were presented in the 2005 ACVIM consensus statement and comprehensive commercial blood donor infectious disease screening panels have recently been introduced that offer improved sensitivity and specificity.

It is evident that careful blood donor selection, stringent infectious disease screening, blood compatibility testing of all patients and donors, appropriate selection of the type of volume of blood product to be administered, and exceptional blood component processing and storage procedures, are essential measures in ensuring safe and efficacious transfusion medicine in our companion animals. Whilst the administration of blood products will never be completely without risk, remarkable progress has been made over the last two decades and promising new approaches are on the horizon.

ACKNOWLEDGEMENTS

The author would like to thank the many transfusion fellows and Penn Animal Blood Bank nurses, at the University of Pennsylvania, for their contributions. Part of the author's studies were supported by NIH 02512 and supplies from various companies related to blood banking and transfusion medicine.

REFERENCES

[1.] Alter HJ, Klein HG: The hazards of blood transfusion in historical perspective. **Blood** 11: 2617-2625, 2008. [2.] Auer L, Bell K: The AB blood group system of cats. **Anim Blood Groups Biochem Genet** 12: 287-297, 1981. [3.] Auer L, Bell K: Transfusion reactions in cats due to AB blood group incompatibility. **Res Vet Sci** 35: 145-152, 1983. [4.] Blais M-C, Berman L, Oakley DA, Giger U. Canine dal blood type: a red cell antigen lacking in some Dalmatians. **J Vet Internal Med** 21: 281-286, 2007. [5.] Callan MB, Jones LT, Giger U. Hemolytic transfusion reactions in a dog with an alloantibody to a common antigen. **J Vet Internal Med.** 9: 277-279, 1995. [6.] Gary AT, Richmond HL, Tasker S, Hackett TB, Lappin MR. Survival of Mycoplasma haemofelis and 'Candidatus Mycoplasma haemominutum' in blood of cats used for transfusions. **J Fel Med Surg** 8: 321-326, 2006. [7.] Giger U, Bucheler J: Transfusion of type-A and type-B blood to cats. **J Am Vet Med Assoc** 198: 411-418, 1991. [8.] Giger U, K Stieger, H Palos. Comparison of various canine blood-typing methods. **Am J Vet Res**. 66: 1386-1392, 2005.[9.] Giger U, Gelens CJ, Callan MB: An acute hemolytic transfusion reaction caused by dog erythrocyte antigen 1.1 incompatibility in a previously sensitized dog. **J Am Vet Med Assoc**. 206: 1358-1362, 2005. [10.] Giger U: Blood-typing and crossmatching. In: Bonagura JD, Twedt DC, eds. Kirk's Current Veterinary Therapy XIV. Saunders Elsevier Press; 260-265, 2009. [11.] Hackett TB, Jensen WA, Lehman TL, et al: Prevalence of DNA of Mycoplasma haemofelis, 'Candidatus Mycoplasma haemominutum', Anaplasma phagocytophilum, and species of Bartonella, Neorickettsia, and Ehrlichia in cats used as blood donors in the United States. **J Am Vet Med Assoc** 229: 700-705, 2006. [12.] Harrel KA, Kristensen AT: Canine transfusion reactions and their management. The Veterinary Clinics of North America Small Animal Practice: Canine and Feline

Transfusion Medicine 25: 1333-1364, 1995. [13.] Hohenhaus AE, Drusin LM, Garvey MS: Serratia marcescens contamination of feline whole blood in a hospital blood bank. **J Am Vet Med Assoc** 210: 794-798, 1997. [14.] Hohenhaus AE. Importance of blood groups and blood group antibodies in companion animals. **Transfus Med Rev**. 18(2): 117-126, 2004. [15.] Kessler RJ, Rankin S, Young S, O'Shea K, Calabrese M, Guldin A, Lipson N, Oakley DA, Giger U: Pseudomonas fluorescens contamination of a feline packed red blood cell unit and studies of canine units. **Vet Clin Pathol**. In press, 2009 [Epub] [16.] Kessler RJ, Reese J, Cheng D, Seth M, Hale AS, Giger U: DEA 1.1, 1.2, 3, 4, and 7, and Dal blood typing and crossmatching by gel column technique. **Vet Clin Path**. In press, 2010. [17.] Klaser DA, Reine NJ, Hohenhaus AE: Red blood cell transfusions in cats: 126 cases (1999). **J Am Vet Med Assoc** 226: 920-923, 2005. [18.] Melzer KJ, Wardrop KJ, Hale AS, et al: A hemolytic transfusion reaction due to DEA 4 alloantibodies in a dog. **J Vet Internal Med** 17: 931-933, 2003. [19.] Owens SD, Oakley DA, Marryott K, Giger U: Transmission of visceral Leishmaniasis through blood transfusions from infected English Foxhounds to anemic dogs. **J Am Vet Med Assoc** 219: 1076-1083, 2001. [20.] Stageman JR, Birkenheuer AH, Kruger JM, Breitschwerdt EB: Transfusion-associated Babesia gibsoni infection in a dog. **J Am Vet Med Assoc** 222: 959-963, 2003. [21.] Tocci LJ, Ewing PJ: Increasing patient safety in veterinary transfusion medicine: an overview of pretransfusion testing. **J Vet Emerg Crit Care** 19: 66-73, 2009. [22.] Wardrop KJ, Reine N, Birkenheuer A, et al: Canine and feline blood donor screening for infectious disease. **J Vet Intern Med** 19: 135-142, 2005.

INFLAMMATORY RESPONSES TO TRANSFUSION

Stephanie A. Smith DVM, MS, Diplomate ACVIM
Urbana, IL

INTRODUCTION

Although transfusion of erythrocytes is a potentially life-saving intervention in seriously ill patients, administration of red blood cell (RBC) transfusions is associated with some inherent risks for the recipient. These include the development of a variety of complications associated with antigenic incompatibility of the donor with recipient cells, transmission of infectious disease, and bacterial infection due to contamination of the RBC unit. These types of complications of transfusion therapy have been previously well-described in veterinary patients.

It must be recognized however, that RBC units also contain platelets and leukocytes, and that non-erythrocytic cells also have the potential to cause responses in the recipient. The type of reactions discussed here are a product not of transfusion of compatible donor erythrocytes, but rather a function of residual cells or plasma components within the RBC unit. Transfusion delivery to recipients results in exposure for the recipient to donor cells, chemokines, and cytokines produced by the cells within the stored units, and cell surface-associated and soluble antigens.

Transfusions in humans have been associated with increased mortality, as well as increased rates of multiple organ failure, cancer recurrence, and nosocomial infections.[1] Transfusion has also been determined to be the underlying cause of several specific clinical syndromes, including transfusion related acute lung injury (TRALI) and transfusion associated graft versus host disease (TA-GvHD). The mortality rates of dogs undergoing transfusion has been reported to range from 39-53% with most deaths attributed to the underlying disease process.[2,3] There are no published veterinary studies evaluating the impact of blood transfusion on morbidity and mortality in animals specifically as a consequence of transfusion.

Responses of the recipient immune-system to blood transfusion may occur as a function of several mechanisms: (1) immune recognition of donor leukocytes by the recipient via anti-leukocyte antibodies; (2) immune recognition of recipient tissues by donor lymphoctes; (3) immune destruction of incompatible donor platelets by recipient antibodies; (4) priming of recipient neutrophils by storage related changes in donor erythrocytes and leukocytes; (5) and passive transfer of donor biologic response modifiers (such as inflammatory cytokines).

CONTRIBUTORS TO THE PROCESS

Storage

Transfusion of aged vs fresh RBCs has been associated with the development of pneumonia, multiple organ failure and mortality.[4] A variety of studies have demonstrated that neutrophils from patients having received a stored transfusion are primed for superoxide and elastase release,[4] and have delayed apoptosis due to exposure to matrix metallinoproteases.[5] Donor cells are not required for priming to occur, as administration of supernatant plasma from stored RBCs produces a similar effect. Supernatant plasma from fresh RBCs does not result in recipient neutrophil priming; storage for > 2 weeks is generally necessary to cause the neutrophil priming effect.[4,6]

Neutrophils

The neutrophil has been implicated as the primary effector cell in the pathogenesis of inflammatory responses to transfusion. Neutrophils, when activated, produce toxic oxygen metabolites and convert oxygen to superoxide in a process called the respiratory burst. Certain molecules cause neutrophils to increase the rate and amount of production of superoxide in response to subsequent activation by a second stimulus. This process is called priming. Priming agents do not activate the respiratory burst, they simply augment it. Known priming agents include bacterial endotoxin, gamma interferon, and platelet activating factor (PAF).[7]

Platelet Activating Factor (PAF) and PAF-like lipids

RBC units, particularly stored RBC units, contain both PAF and PAF-like lipids. These lipids are produced by the action of intracellular and extracellular phospholipase A2, which hydrolyses cell membrane phospholipids into arachidonic acid and lysophospholipid. These lipids, which have a structure similar to that of PAF and can trigger PAF-like effects, are contained in the supernatant of stored RBC units. PAF-like lipids exist on membrane particles present in the supernatant as microvesicles.[7] As noted above, these molecules prime neutrophils for the respiratory burst, and also act as priming agents to increase IL-8 production by monocytes in response to a second stimulus.

Chemokines

Chemokines are small (8-10 kDa) secreted proteins that have roles in intracellular communication, cellular activation, and cell cycle regulation. Production of pro-inflammatory chemokines is upregulated in response to specific cellular signals. They have been grouped according to structural similarities and primarily classified based on occurrence of cysteine residues into CC chemokines and CXC chemokines. Receptors for these chemokines, found on target cell surfaces, are members of the G-protein coupled superfamily. Pro-inflammatory chemokines may accumulate in the supernatant of RBC units during storage due to either activation of contaminating leukocytes or to degranulation of contaminating platelets. Both the platelet-derived and leukocyte-derived chemokines accumulate progressively during storage, and prestorage leukoreduction prevents chemokine accumulation. CXC chemokines are implicated in the neutrophil activation that contributes to capillary leakage that is observed with TRALI.[8]

SPECIFIC CLUNICAL SYNDROMES IN HUMANS

Transfusion Related Acute Lung Injury (TRALI)

TRALI was first reported in the 1950's, but not recognized as a distinct clinical entity until the 1985. TRALI is a leading cause of transfusion-related death world-wide. TRALI occurs in 1/1333 to 1/5000 (per unit transfused), with a mortality rate of 5-35%.[9] All types of blood components have been implicated, but plasma containing components are most commonly associated with TRALI.[9,10] Interestingly, the use of male-predominant plasma in Great Britain has markedly decreased both numbers and severity of TRALI in that country.[9]

TRALI is a clinical diagnosis describing acute pulmonary insufficiency that occurs within 6 hours of receiving a transfusion. Patients present with tachypnea, cyanosis, dyspnea and hypoxemia (PaO_2/FiO_2 <300 mm Hg) with normal cardiac function. Radiographs indicated diffuse, fluffy infiltrates consistent with pulmonary edema, and post-mortem examination indicates widespread neutrophil infiltration in the lung, interstitial and alveolar edema, and other changes consistent with acute respiratory distress syndrome.[9]

The neutrophil is the primary effector cell in TRALI. The pathophysiology involves neutrophil-mediated endothelial cell cytotoxicity resulting in capillary leak. Several mechanisms have been proposed for TRALI and all are dependent on neutrophils. The currently favored model involves a two-event description where the initial event is a function of the clinical status of the recipient, and the second event is inherent to the blood-product and may be either donor-derived antileukocyte antibodies (against a recipient cognate antigen) or biologic response modifiers in the stored RBC units that cause priming of neutrophils.[9]

Transfusion Immunomodulation (TRIM)

The immunosuppressive effects of RBC transfusion were first recognized in the 1970's when it was reported that, in kidney transplant patients, improved allograft survival was associated with increased number of transfusions.[11] The beneficial effects of RBC transfusion on renal allograft survival have since been confirmed by animal data, observational clinical studies, and a large multicenter randomized controlled trial. Later studies evaluated the potential adverse impact of TRIM on recurrence of neoplasia. A large number of observational and retrospective reports have been published, with some indicating an adverse impact of transfusion on tumor-related outcome, while others report no effect. Randomized controlled-trails are limited due to ethical concerns, and the interpretation of such trials is complicated by regional variations in blood banking procedures resulting in different levels of leukocyte contamination of units (see buffy coat reduction vs leukoreduction below).

TRIM has also been implicated in post-operative infections. A variety of observational studies have indicated that patients receiving peri-operative transfusion have higher rates of bacterial infection, and that outcomes are poorer in those receiving transfusion. Interpretation of such studies is complicated by controversy over whether the need for transfusion is a marker for worse prognosis, or a cause of worse prognosis. Several randomized controlled studies have supported the interpretation that patients receiving leukocyte-containing transfusion(s) have increased risks of infection, multiple organ failure, or mortality.[11,12]

The TRIM effect has been ascribed to decreased helper to suppressor T-lymphocyte ratio, decreased natural killer cell function, defective antigen presentation, and a reduction in cell-mediated immunity.[11,12]

Transfusion-associated graft-versus host disease (TA-GvHD)

TA-GvHD is a rare complication of transfusion of blood components. It is caused by the attack of recipient tissues by blood-derived alloreactive donor lymphocytes. If the recipient's immune system is unable to destroy the donor lymphocytes in the transfused unit, donor cells will recognize the recipient's tissues as foreign, become activated, and release cytokines. These cytokines activate natural killer cells, macrophages, and other T- cells, resulting in the destruction of the targeted host (recipient) tissues.

Any blood component containing donor lymphocytes may cause TA-GvHD, but fresher components (<3 days) are more likely to contain viable lymphocytes. Previously frozen components are unlikely to contain a significant number of viable lymphocytes. TA-GvHD may occur when there is antigenic similarity between donor and recipient lymphocytes, allowing for the grafted lymphocytes to survive the host immunological response. Immunosuppression of the recipient may increase the likelihood of donor lymphocyte engraftment. Blood from relatives is associated with a higher risk due to antigenic similarity. Attack of the recipient's bone marrow by donor lymphocytes results in irreversible bone marrow failure associated with a grave prognosis due to lack of effective treatments. TA-GvHD can be prevented or reduced by removal or inactivation of contaminating lymphocytes. Lymphocytes can be killed through irradiation (widespread use) or use of pathogen-inactivation technologies, or can be removed by leukoreduction.

MODIFICATIONS OF RBC UNITS THAT REDUCE THESE EFFECTS

Pre-storage Removal of Leukocytes

Leukoreduction (LR) is standard of practice in human blood banking in Canada and much of Europe because it has been documented to reduce the incidence of transfusion reactions and inflammatory responses in recipients. Although not required in the US, LR is also widely employed in this country. LR is most commonly performed using in-line charged filters in the blood collection set. When the whole blood passes through the filter, the leukocytes (and platelets) stick to the filter and are removed from the whole blood. Blood banking standards require that a LR-unit contain < 5 leukocytes/µl.[13] Use of such filters has been validated for canine whole blood.[14]

Buffy coat reduction involves separation of the buffy coat portion of the unit after centrifugation. This approach has been common blood banking practice in Europe for much of the last 2 decades. The procedure results in removal of approximately 70-80% of the donor leukocytes.

Irradiation

Irradiation of blood components is in widespread use when blood is to be administered to specific at-risk populations (primarily the immunosuppressed). Due to regulatory issues and to the cost and size of the irradiating systems, irradiating is generally performed in the blood bank. Irradiation may shorten the shelf life of the component, and may cause minor damage to the erythrocytes resulting in excess potassium leakage.[15]

Pathogen-inactivation technologies

Strategies aimed at removing pathogens by targeting nucleic acids will also attack nucleated cells. Approved processes include solvent detergent treatment, methylene blue with exposure to visible light, and use of photochemicals (riboflavin or amotosalen) in conjunction with exposure of blood components to UV light.[15] Solvent detergent treatment of plasma in one country appears to have eliminated the occurrence of TRALI.[9]

Washing of Cells

Washing of RBC or other cellular components eliminates the mediators that are implicated in TRALI; ie those associated with the plasma fraction.[9] This approach is time-consuming and costly, however, and may delay transfusion for the critically ill. Decreasing the amount of residual plasma in RBC units may decrease, but not prevent the occurrence of TRALI.

INFLAMMATORY RESPONSES TO TRANSFUSION IN DOGS

Investigators at the University of Illinois recently completed a study which evaluated whether autologous RBC transfusion to normal dogs was associated with a subclinical inflammatory response. Random-source, clinically healthy medium and large breed dogs were divided into 2 groups. A whole blood unit was collected for processing with and without leukoreduction, followed by storage of the RBC unit for 21 days, at which time each dog received an autologous transfusion. Leukocyte counts, fibrinogen, and C-reactive protein were evaluated prior to collection of the unit, prior to administration of collected RBC units, and 1 day and 3 days following transfusion. There was a marked increase from baseline in WBC count, fibrinogen, and C-Reactive Protein in the dogs that received RBC units produced via standard methods (containing leukocytes and platelets), while there was no increase from baseline in any of the markers in the dogs that received leukoreduced RBC units (leukocytes and platelets removed). Antigen mediated effects could not have been responsible for the inflammatory response since all subjects received their own cells. This study suggested that transfusion of stored RBC units to dogs is associated with an inflammatory response, and that that response is a function of contaminating leukocytes and/or platelets in the RBC units.

REFERENCES

[1.]Blajchman MA. **J Trauma.** 2006;60(6 Suppl):S83. [2.]Callan MB, *et al.* **J Am Anim Hosp Assoc.** 1996;32(4):303. [3.]Kerl ME, *et al.* **J Am Vet Med Assoc.** 1993;202(9):1495. [4.]Biffl WL, *et al.* **J Trauma.** 2001;50(3):426. [5.]Frake PC, *et al.*. **Arch Surg.** 2006;141(4):396. [6.]Sparrow RL, *et al.* **Transfusion.** 2004;44(5):722. [7.]Cardo LJ, *et al.*. **Transfus Apher Sci.** 2008;38(2):117. [8.]Davenport RD. **Vox Sang.** 2009;96(3):183. [9.]Silliman CC, *et al.* **Blood Rev.** 2009;23(6):245. [10.]Silliman CC, *et al.* **Blood.** 2005;105(6):2266. [11.]Blajchman MA. **Hematology.** 2005;10 Suppl 1:208-14.:208. [12.]Vamvakas EC *et al.* **Blood Rev.** 2007;21(6):327. [13.]Dzik S *et al.* **Transfus Med Rev.** 2000;14(1):34. [14.]Brownlee L, *et al.*. **J Vet Intern Med.** 2000;14(4):412. [15.]Mintz PD, *et al.* **Bone Marrow Transplant.** 2009;44(4):205.

PLATELET TRANSFUSIONS: INDICATIONS & NEW PRODUCTS

Elizabeth B. Davidow, DVM DACVECC
Seattle, WA

Thrombocytopenia is seen commonly in emergency veterinary patients but platelet transfusions are used infrequently. There are limited publications in veterinary medicine about triggers for prophylaxis or therapeutic efficacy of platelet products in actively bleeding patients. Both the limited publications and infrequent use of these products are due largely to the complexity of processing and storing platelets for immediate use in all but the largest veterinary institutions. Recent advances in platelet storage may make platelet transfusion more readily accessible to veterinarians.

Platelet Physiology

Platelets were first noted to be a true blood component, rather than just slide debris, by Alfred Donne in 1842 and their role in coagulation was first proposed in the 1860s.[1] These cytoplasmic fragments originate from megakaryocytes located primarily in the bone marrow, but also in the lung, spleen, liver and kidney. Thousands of platelets are formed from branching and separation of megakaryocytes. This fragmentation process and the discoid structure of the platelets are driven by microtubules.

Platelets bind to exposed von Willebrand's factor and collagen in damaged endothelium. This binding is mediated by GP Ib-IX-V and GPVI complexes. Collagen binding causes intracellular signaling, leading to activation of the platelets with resultant platelet shape changes, changes in platelet receptor affinity and release of active molecules from the dense and alpha granules. Platelets aggregate to form the primary hemostatic plug through interactions of the activated GP11b-111a receptors with fibrinogen. Activated platelets also accelerate procoagulant activity and bind coagulation proteins to keep the reactions localized to the site of the injury.[2]

Average platelet survival in dogs is approximately 5-7 days while it is 8-10 days in people. In non disease states, platelets either become senescent and are removed by the spleen or are used in ongoing endothelial maintenance. Kinetic studies in people show that a set number of platelets are needed for endothelial maintenance resulting in a higher percentage that are needed in thrombocytopenic conditions. Thus, average platelet survival is correlated with the degree of thrombocytopenia.[3]

Causes of Thrombocytopenia and Thrombocytopathia

Thrombocytopenia can be caused by decreased production, accelerated removal, increased consumption, loss or dilution. The most common cause of marked thrombocytopenia in dogs is immune-mediated thrombocytopenia.[3,4] Other causes include infectious diseases such as Ehrlichiosis, bone marrow insults, neoplasia, disseminated intravascular coagulation (DIC), and blood loss.

Therapy with certain drugs can also result in thrombocytopenia. This can be a dose related toxicity or can be an idiosyncratic reaction. Azathioprine and chloramphenicol can lead to thrombocytopenia through a dose related affect on the bone marrow. In people, heparin can cause thrombocytopenia. This heparin-induced thrombocytopenia results from the development of antibodies against platelet factor 4 – heparin complexes. These antibodies lead to inappropriate platelet activation and consumption. It is most commonly seen in people who have received heparin previously. Drug therapy with potentiated sulfonamides can also result in an immune related thrombocytopenia in both dogs and people.[5]

In the inherited macrothrombocytopenia of Cavalier King Charles Spaniels, a genetic mutation in the beta-tubulin protein of the microtubules is known to be the cause.[6] Cavaliers affected with this condition can have platelet counts as low as 50,000/ul but often have no signs of bleeding. It is important to distinguish this condition from immune causes of platelet destruction as these animals do not require treatment.

In thrombocytopathic conditions, platelet numbers are normal but function is impaired. Acquired loss of platelet function is associated with medications and certain disease states. Non-steroidal anti-inflammatory drugs cause inhibition of the COX-1 enzyme, leading to decreased thromboxane generation and decreased platelet aggregation. Animals with both hepatic disease and uremia can have platelet function issues. The decreased function in hepatic disease thought to be due to changes in platelet phospholipids while uremia appears to impair platelet function by decreasing adhesion to the endothelium.[7]

Several hereditary thrombocytopathias have been identified in dogs and cats.[8] Glanzmann thrombasthenia is a genetic disease in which the GP 11b-111a receptor is decreased on missing. This disease has been documented in Great Pyrenees and Otterhounds. In Basset Hound thrombopathia, platelets do not aggregate normally due to a defect in calcium exchange. The gene has been identified and is autosomal recessive. Similar calcium exchange genetic defects have been found in 2 Eskimo Spitz and in Landseer-ECT. A selected defect of ADP content of granules was noted in a family of Cocker Spaniels. A family of German Shepherds was found to have a defect in which platelets did not express procoagulant activity. These dogs have signs more suggestive of a coagulopathy than a primary hemostatic defect but have normal coagulation screening tests and normal platelet function tests making diagnosis a challenge.

In one retrospective study at the Ontario Veterinary College in Guelph, O, Canada in 2001, 581 transfusions were given, but only 40 were for platelet support. All of the platelet support transfusions were in dogs, 50% for ITP and the rest for a combination of leukemias, aplastic anemias, DIC, splenic neoplasia, chemotherapy induced thrombocytopenia and thrombocytopenia associated with vitamin K antagonism toxicity.[3] In a recently completed multicenter trial looking at lyophilized platelets, 43 dogs received platelet transfusions. Of these, 36 had ITP or Evans syndrome, 3 had underlying neoplasia, 2 were in DIC, 1 had Ehrlichiosis, and 1 had Basset Thrombocytopathia.

Indications for Platelet transfusions and dosing

In humans, platelets are recommended for prophylaxis in any patient with a count less than 10,000/ul and in patients who need an invasive procedure with counts less then 50,000/ul. Platelets are recommended therapeutically in any actively bleeding patient with a count less than 20,000/ul. Platelets are also recommended in people with drug or hereditary impairment of platelet function who need an invasive procedure.[9, 10] However, ITP is considered a unique situation due to the rapid clearance of any administered platelets. In addition, platelets in ITP are often younger and hyperfunctional so that bleeding may not occur until counts are extremely low. Platelet transfusions are usually not recommended for prevention of hemorrhage in dogs with this disease.[10]

Interestingly, the risk of bleeding with thrombocytopenia is affected by the degree of anemia. A higher packed cell volume reduces the risk of bleeding through several mechanisms. RBCs scavenge nitric oxide which leads to increases in platelet activity. In addition, a higher hematocrit pushes platelets toward the endothelium and reduces sheer stress. Lastly, RBCs are known to contain mediators that increase platelet activity.[11] In cases of moderate thrombocytopenia and concurrent anemia, the risk of bleeding may be lessened with packed RBC transfusion alone. However, it has been recommended to increase the ratio of platelet and plasma transfusions to packed red blood cells in situations of massive blood loss. Two human retrospective studies of severe trauma have shown increased survival when these ratios were kept close to 1:1. [12, 13]

Lyophilized platelet products may also be used as hemostatic agents even in cases where platelet numbers and function are normal. In a recent study, 20 swine were subjected to a grade III liver injury. They were then given either a placebo or lyophilized platelets. Two hours later, the liver was repaired. There was 80% survival in the group that received the lyophilized platelets while only 20% survival in the group receiving placebo. One pig did have thrombi in other locations on necropsy indicating the need for more study before the products are used in clinical patients for this indication.[14]

The risk of bleeding from thrombocytopenia must be weighed against the risk of transfusion reaction especially when prophylactic platelet transfusions are considered. The reported rate of febrile reactions in people is 38% while 2% have a severe adverse reaction. In addition, alloimmunization does occur in both people and dogs which can lead to platelet transfusion refractoriness. This alloimmunization can occur with as little as 3 transfusions in dogs.[4]

Dosing recommendations have been made by calculating the expected rise in platelet count. The expected one hour platelet increment in dogs can be calculated as:

Total number platelets transfused x 0.51/blood volume, where blood volume is 85ml/kg.

The standard fresh platelet concentrate dose of 1 U per 10 kg was derived to aim for a 40,000/ul increase in platelet count.[3] Some investigators have looked at whether the interval between transfusions could be extended by using a higher dose initially. A meta-analysis in humans did show an increase in the interval between transfusions when a higher dose was used initially.[15]

Currently Available Products

Fresh whole blood is the product most veterinarians use when platelets are needed. A 500 ml unit of fresh whole blood obtained from a canine donor is estimated to contain 7 X 10^{10} platelets. A dose of 10ml/kg of whole blood would be expected to raise the platelet count about 10,000/ul..[3] The advantage of fresh whole blood is that no platelets are lost during separation. In addition, the platelets are less activated than in platelets obtained via centrifugation for concentrate.[16] Fresh whole blood at room temperature is considered safe for use for 4-8 hours. Refrigeration of human platelets, either in whole blood or in platelet concentrate, rapidly leads to platelet aggregation and activation . In addition, refrigerated platelet survival is half that of platelets held at room temperature. This effect is due to clustering of the vWF receptors in response to temperature leading to increased clearance by hepatic macrophages.[17] No studies have been done of canine refrigerated platelets to confirm if a similar response occurs.[4]

Fresh platelet concentrate has traditionally been made when initially processing whole blood. Whole blood is spun using a "soft" spin which separates the platelets into the plasma component. The plasma is expressed into a separate bag. This plasma is then known as platelet rich plasma (PRP). The PRP is then spun again to create a platelet concentrate (PC) and the plasma is removed and stored as fresh frozen plasma. One unit of PC is the amount made from one unit (500 ml) of whole blood but will contain a lower amount of platelets. A recent study showed a maximum in vivo platelet recovery of 80%.[18] The dose is normally calculated as 1 unit/ 10 kg. Fresh platelets must be stored in gas soluble bags at room temperature to remain active. They must also be constantly agitated and thus are kept on continuous rockers. Bacterial contamination is a concern at room temperature and storage is limited to 5 days.[10]

Fresh platelet concentrate can also be made through plateletpheresis. In this process, blood is taken from the donor, split into components using an apheresis machine, the desired component is retained, and the rest is returned to the donor. The process is more time consuming for the donor but many more platelets can be collected and sterility of the product is maximized. Dog donors have been trained to remain still for successful apheresis. An apheresis donation will produce a platelet concentrate containing 1-4 x 10^{11} platelets, 4-6 times the amount in a centrifuged platelet concentrate.[3] In addition, there is neglible WBC and RBC contamination.

Fresh platelets collected in this manner must also be kept constantly agitated at room temperature. Because of the higher sterility during collection, they can be stored for up to 7 days.[10]

Frozen platelet concentrate is made by stabilizing apheresed platelets with 6% DMSO or with 2% DMSO and Thrombosol. In original work done by Valeri, canine platelet recovery after freezing with 6% DMSO at -80 F was shown to be 70% with a ½ life of 2 days versus 3.5 days for fresh platelets.[19] The platelets were shown to be effective in halting active bleeding in thrombocytopenic dogs. A more recent study comparing 6% DMSO to 2%DMSO and Thrombosol showed only 49% and 44% platelet recovery respectively. Platelet half-life was confirmed to be about 2 days.[18] The frozen product must be thawed at room temperature.

Lyophilized platelets are currently in canine clinical trials through Animal Blood Resources International. Platelets are stabilized using a mild aldehyde crosslinking of membrane proteins and lipids. This process then allows for lyophilization and reconstitution with preservation of platelet structure and function. The lyophilized platelets can be stored for up to 24 months in the refrigerator, and are reconstituted with saline immediately prior to use. Research studies on these platelets have shown that they bind to collagen, von Willebrand factors and damaged endothelium. Receptors are activated normally and bind fibrinogen. Importantly, these platelets also retain the ability to increase procoagulant activity.[20] In a study of dogs on cardiac bypass, infusion of the lyophilized platelet product led to improvement in venous bleeding time that was most pronounced at 20-30 minutes after infusion.[21]

A prospective, multicenter trial was recently completed to study the use of a lyophilized platelet product (LYO) in dogs with thrombocytopenia. Dogs who met the entry criteria were randomized and received either lyophilized platelets or fresh platelet concentrate. Fresh platelet concentrate was sent to each hospital weekly to be on hand as needed for the study. Physical exam, hematologic, and coagulation status data was collected at baseline, one hour and 24 hours after transfusion. 34 dogs met the criteria for the study: 21 dogs were given LYO and 13 dogs were given fresh platelets. An additional 7 patients received LYO and 2 patients received fresh platelet concentrate either during or after the study as a compassionate use. . In total, 28 dogs had received lyophilized platelets at the time of this publication. Of the dogs receiving the fresh platelets, 13% were noted to have possible transfusion reactions within one hour of the transfusion – one with facial uticaria and one with panting. Of the dogs receiving lyophilized platelets, 7% had possible transfusion reactions – one with a 2 degree F rise in temperature but no other signs and one with tachycardia and pale gums noted one hour after the infusion that improved with further packed RBC administration. Although no significant difference in bleeding score or platelet count could be measured in dogs receiving lyophilized platelets, in at least 2 cases, active bleeding was observed to stop immediately after the transfusion of the LYO. The LYO was found to be easy to use and administer. An additional finding of the study was that keeping 2 units of aphresed fresh platelet concentrate on hand at all times at each of the 5 institutions was economically unfeasible on a long term basis.

REFERENCES

[1.] Boudreaux MK. ACVECC Post graduate Course 2010. [2.] Day M, Mackin A, Littlewood J. editors. Manual of Canine and Feline Haematology and Transfusion Medicine. Gloucester, UK: **BSAVA**, 2000. [3.] Abrams-Ogg ACG. **Vet Clin Sm Anim** 2003;33: 1401. [4.] Callan MB, Appleman EH. **JVECCS** 2009; 19: 401. [5.] Lavergne SN, Trepanier LA. **Platelets** 2007; 18: 595. [6.] Davis B, Toivio-Kinnucan M, Schuller S, Boudreaux MK. **JVIM** 2008; 22: 540. [7.] Brassard JA, et al. **J Lab Clin Med** 1994; 124:48. [8.] Boudreaux MK. **JAVMA** 2008; 233: 1251. [9.] Callow CR, et al. **Br J Haematol**. 2002; 118:677 [10.] Petrides M, Stack G, Cooling L, Maes LY. Practical Guide to Transfusion Medicine. 2nd ed. Bethesda, MD: **AABB press**, 2007. [11.] Valeri CR. **Transfusion** 2007; 47(suppl):206S [12.] Holcomb JB et al. **Ann Surg** 2008; 248: 447. [13.] Gunter OL et al. **J Trauma** 2008; 65: 527. [14.] Hawksworth JS. **J Thromb Haematol** 2009; 10: 1663. [15.] Cid J, Lozano M. **Transfusion** 2007; 47: 464. [16.] Gilmore G, et al. **Journal of Thrombosis and Haemostasis** 2005; Volume 3, Supplement 1: P0957 [17.] Hoffmeister, KM, et al. **Cell** 2003; 112: 87. [18.] Appleman EH, et al. **JVIM** 2009; 23:128. [19.] Valeri CR et al. **Cryobiology** 1986; 23:387. [20.] Bode AP and Fischer TH. **Artif Cells Blood Substit Immobil Biotechnol**. 2007; 35: 125. [21.] Bode AP et al. **Clin and App Thromb/Hemost** 2008; 14: 38.

FUTURE TRENDS IN TRANSFUSION MEDICINE

Anne S. Hale, DVM
Stockbridge, MI

INTRODUCTION

Prior to 1980, veterinary transfusion medicine was practiced rarely and with little or no predetermination. The early 80's saw the development of critical care specialists and commercial veterinary blood banks. Through the forethought and encouragement of early pioneers in veterinary transfusion medicine, NIH recognized the importance of funding educational efforts in this subspecialty. The Transfusion Medicine Academic Awards (TMAA) included veterinarians at University of Pennsylvania, Tufts University, Michigan State University, Washington State University and Kansas State University. These unique awards allowed for the development of an educational and research process that has led to our current state of understanding in veterinary transfusion medicine. Now that veterinary transfusion medicine has become a daily tool for practicing and academic veterinarians, the question should be asked where does the next decade lead in this emerging practice area.

THE DONOR DILEMMA

Veterinary transfusion medicine is inherently tied to the relationship between donors and recipients. As the need for blood products in daily practice has increased, the issue of donor recruitment has become more important. From the human experience we know that both volunteer and paid donors are required to allow for the blood product need in the United States to be met. Animal donors inherently bring concerns regarding their "right to choose", the definition of volunteer donor and compensation for owners when allowing their animal to donate. [1] A number of groups are currently debating the standard that veterinary transfusionists should utilize to verify the ethical use of animals as blood donors.[2] This debate should include not only veterinary professionals but animal welfare professionals to find healthy fair solutions for all donors participating in the supply side of veterinary medicine. Along with these discussions the next decade holds a number of expected enhancements to our current production and storage techniques that guard the precious donor resource and maximize our use of blood products.

EX VIVO PRODUCTION SYSTEMS

One method of solving the ethical and moral aspects of the donor dilemma is to engage in the production of erythrocytes and platelets through ex vivo expansion of cell lines. Recent progression of our understanding of stem cell (CD 34+) and late progenitor maturation processes present options for the production of cells through GMP production centers. Live animal donors would be required for initial stem cell harvest. This harvest can be obtained by cord blood, traditional bone marrow harvest and apheresis. Currently, Douay and others in France have described techniques capable of using sterile good manufacturing practices to culture red blood cells. These red blood cells have been evaluated and are capable of in vivo function similar to native adult erythrocytes.[3] The same process has been described for platelets and lymphocytes. These techniques will require large scale modification to provide economical, safe manufacture. Production systems of this nature would offer one way to solve the donor dilemma for veterinary transfusion medicine in almost every instance.

ARTIFICIAL BLOOD PRODUCTS

Blood products with the shortest storage half-life are the products that we have explored for artificial ways of production. Veterinary medicine has had the benefit of a FDA approved bovine-based free hemoglobin product. This product has been successfully used in multiple species as oxygen-carrying therapeutic. However, the company's attempt to fund a human product has led to a delay in production of the veterinary product. Long term availability of this product, Oxyglobin®, is currently unknown. Clinicians have used this product to mimic oxygen carrying capacity for 3-7 days. A number of other companies have developed other hemoglobin based products but to date none are licensed for use in animals.

Perflurocarbon based products may also be utilized as an oxygen therapeutic. Currently, Oxygent® and Oxycyte® are in phase II human clinical trials within the United States. These products have the advance of avoiding classical compatibility issues associated with red blood cells and demonstrate a stable, oxygen carrying substitute with a long term storage capacity. Many of these products are evaluated for FDA preclinical studies in the dog as a research model and show great promise in subhuman species.

Yet another way to produce an oxygen therapeutic like red blood cells is to use developing nanotechnology to create artificial erythrocyte. Single and double walled carbon and phospholipid nanotubes have been created to mimic cell wall structure. [4,5] This type of "nanocell" offers oxygen carrying capacity given the right set of internal additives and addresses the compatibility issue associated with red blood cells too. Eventually, a combination of free hemoglobin or perflurocarbon products and "nanocells" may provide a suitable artificial erythrocyte.

Artificial thrombocytes have met with early preclinical success. These microparticles have been demonstrated as successful platelet replacements in bench top research and rodents. Nanoplatelets may also be possible. However, due to the unique nature of the platelet membrane interaction with the endothelium, this type of manufacture is as of yet only theoretically proposed.

Finally, recombinant technologies allow for production of oxygen therapeutics, clotting factors and volume expanders are possible. Large scale production of human albumin by *E. coli* has been available for several years. NovoSeven®, a recombinant form of Factor VIIa, has been utilized in dogs to treat refractory Hemophilia A. As with other recombinant products based on human genetic sequence, extended use will likely lead to the development of inhibiting antibodies. Recombinant production does address several of the issues surrounding donor availability, first time compatibility and infectious disease control. This type of technology may ultimately lead to safe economical ways to produce genome specific recombinant product but currently these items are not species specific and relatively expensive for the veterinary patient.

GENE THERAPY

Many diseases currently requiring extensive transfusion support are the result of a genetic mutation or deletion. Early attempts to insert the correct copy of a gene to correct a defect or deficiency met with severe or fatal complications. However, recently a handful of successes have been described in the dog.

Hemophilia A and B

Margartitis et al reported the successful treatment of canine hemophilia (both types) through an adeno-associated viral vector inducing Factor VIIa continuous expression.[6] Previous uses of NovoSeven® have indicated that this type of continuous expression of Factor VIIa is a bypass for deficient Factor VIII or IX. This published report indicates that prothrombin time was shortened, TEG values were improved and no spontaneous bleeding was noted.

STEM CELL THERAPY

The dog has been a frequent research model for hematopoietic restoration post irradiation or chemotherapeutic depletion. Past attempts at managing clinical patients has been limited by the availability of compatible donors and transfusion support for the engraftment period. With the progression of transfusion services, available resource to manage the engraftment period now exists. Both autologous and allogenic transplants of stem cells have been utilized in the canine patient for management of myeloproliferative disease and lymphomas. Recent human studies suggest that in most incidences peripheral blood stem cells (PBSC) are just as effective as stem cells collected direct from bone marrow harvest. Apheresis technologies allow the collection of PBSC in domestic animals leaving only compatibility issues to stand in the way of successful allogenic PBSC transplantation/engraftment. As our understanding of Graft vs. Host disease increases, compatibility barriers will also disappear. Recently, the successful blockade of CD28 costimulatory pathway in the dog has led to tolerance to nonspecies red blood cells (sheep origin).[7] This breakthrough suggests that successful reconstitution with zero haplotype matching PBSC may someday be possible.

REGENERATIVE CELLULAR MEDICINE

Over the course of the last five years, a number of cellular "implants" have been used to stimulate natural healing properties and provide cellular matrices for restoration of tissue. These cellular "implants" are platelets (CD 61+) and stem cells (CD 34+). Because the earliest recognized resources for these cells were hematopoietic, regenerative medicine has been most closely associated with transfusion medicine. Currently the Center for Biologics Evaluation and Research (CBER) within the FDA is responsible for human transfusion products and regenerative cellular products. This close association makes it reasonable that veterinary blood banks will follow suit and participate in the emerging regenerative cellular medicine field in veterinary medicine.

As an example, many studies have utilized platelet rich plasma or platelet concentrates injected at the site to repair chronic ligamentous injury in the horse. A recent Google® search for Platelet rich plasma and therapy in the horse yielded 10, 300 sites. A large body of published studies exists to support the claims of repair.

Use of mesenchymal stem cells or PBSCs has also been used to repair injury. Several veterinary based companies are utilizing mesenchymal stem cells to treat osteoarthritis in the horse and dog. FDA Center of Veterinary Medicine oversight for study work in this area indicates its growing popularity in the veterinary community.

Recently, several Phase II clinical trials have been authorized by the FDA to evaluate stem cells and their ability to regenerate human cardiac muscle and spinal cord neurons. Early trial data suggests treatments of this nature are successful in improving fractional shortening in damaged hearts and reinitiating movement in paralyzed limbs.

Consistent harvest and storage of the cells will likely be the responsibility of the transfusion medicine center. Guidelines for collection, storage and reconstitution in subhuman species will need to be developed.

SHORT TERM FIXES FOR LONGTERM PROBLEMS

Veterinary transfusion medicine can look to the above as long term solutions for our growing needs. However, we must find short term "fixes" for these issues of availability, compatibility and safety within the next few years to keep the transfusion practice growing.

Availability

Veterinary blood banks and blood product providers are currently utilizing colony housed animals and "volunteer" donors to meet the product need. In order to maximize the resource, component therapies are expanding. Currently, one unit of whole blood can yield a unit of packed red blood cells and one unit of plasma. Apheresis technologies allow for a larger amount of cellular component or plasma to be collected safely from the donor. Commercial blood banks are working to provide fractionated plasma products further expanding the donor resource. Cellular therapies involving stem cells, platelets and leucocytes are also available on a limited basis.

Longer storage potential makes it possible for more veterinarians to use transfusion medicine as part of their daily practice. Newer storage options using unique nutrient solutions are allowing platelets and red blood cells to safely store for extended periods of time. Freeze drying and other long term frozen storage options are available now.[8] These techniques are even being explored for use with red blood cells extending the use period for our most critical transfusion resource.[9] Increasing the number of products provided per donation and changing storage stability are two methods of increasing availability.

Compatibility

The use of blood products in companion animals has greatly outpaced our ability to provide cage side compatibility decisions. Due to our lack of point of care determiners, compatibility testing was not provided to most transfusion recipients. A number of recent studies have indicated that our understanding of erythrocyte antigens and their relationship to compatibility are rudimentary at best. New generation technologies including gel tubes and erythrocyte membrane diffusion have been brought to the patient's side. This gives the veterinarian the opportunity to crossmatch and blood type both recipient and donor prior to transfusion.[10] Ultimately, it will be possible to determine the molecular blood group genotyping of individuals similar to routine practice in human medicine.[11] DNA microarrays can quickly determine status for a number of antigens at a very economical rate. Sequence is already available for feline erythrocyte antigens in the AB blood system.[12]

This information will allow the veterinary community to explore ways to maximize the efficient transfusion of red blood cells, platelets and plasma products. Collection of this data is important to help determine which erythrocyte antigens factor into the acute transfusion reaction and the delayed transfusion reaction. Outcomes to transfusion have rarely been reported before 1999. Once access to this data is collected, the veterinary transfusion community will have insight into economical ways in which to evaluate transfusion effectiveness and failure. Efficacy based on donor selection, production, storage, delivery and compatibility can provide the guidelines that veterinarians currently lack. Once an understanding has been gained, veterinary transfusion medicine can mature to the level of human transfusion medicine utilizing algorithms for first transfusion, multiple transfusions and complicating antibody scenarios.

To date, little has been done to evaluate the importance of lymphocyte antigens and their role in veterinary transfusion reactions. Molecular techniques have allowed us to determine lymphocyte antigens in most species. Exploration relating these antigens and disease resistance or susceptibility has just begun. Advancing compatibility through molecular and flow cytometric methods will allow clinicians to predict or associate disease processes like immune mediated hemolytic anemia by their lymphocyte antigen and erythrocyte antigen profile.[13]

Safety

Molecular assays in the form of DNA microassays are becoming more prevalent and economical. This type of infectious disease screening assists the blood banker in providing a safe product to the recipient. It also allows for more use of individually owned "volunteer" dogs as donors without compromising infectious disease surveillance.

Recently, several new pathogen inactivation systems have been approved in the European Union for platelets. Both products utilized the dog as their preclinical model suggesting that these products will safe and effective at lowering the bioburden in canine transfusion products and eliminating the need for exhaustive infectious disease testing on each blood product produced.[13]

REFERENCES

[1]DeLuca LA, et al. **J Appl Anim Welf Sci**. 2006; 9(2):129-41. [2]Hale AS, et al. **AAVBB Standards** 1st ed. 2006. www.aavbb.org. [3]Douay L, et al. **Methods Mol Biol**; 2009; 402:127-40. [4]Couzier T, et al. **Langmuir**. 2008 Nov 18; 24(22): 13173-81. [5]Ale I, et al. **Physics Letters A** 2003; 310(5-6):493-497. 6Maragritis P, et al. Blood 2009 Apr16:113(16):3649-50. [7]Graves SS et al. **Transplantation** 2009 Aug15; 88(3):317-22. [8]Appleman EH, et al. **J Vet Intern Med**. 2009 Jan-Feb; 23(1):138-45. [9]Fischer TH, et al. **Microsc Res Tech** 2004 Sep65 (1-2):62-71. [10]Giger U, et al. **Am J Vet Res** 2005Aug; 66(8):1386-92. [11]Veldhuisen B, et al. **Vox Sang** 2009: Oct; 97(3):198-206. [12]Millon LV, et al. **BMC Genetics** 2007:8(27). [13]Kennedy LJ, et al. **Tissue Antigens** 2006 Dec: 68(6):502-8. [13]Osselaer JC et al. **Transfusion** 2008; 48 (S1); 20A-21A

HIGHLIGHTS FROM DIGESTIVE DISEASES WEEK & VETERINARY CORRELATES - PANCREATITIS

Jörg M. Steiner, med.vet., Dr.med.vet., PhD, DACVIM, DECVIM-CA
College Station, TX

SESSION INTRODUCTION

The Digestive Diseases Week (DDW; www.ddw.org) is the premier scientific meeting in human gastroenterology worldwide. Every year close to 20,000 people attend the meeting and almost 6,000 research abstracts and hundreds of lectures and interactive sessions are being presented. Many developments that ultimately lead to paradigm shifts in the diagnosis and treatment of gastrointestinal diseases in human patients, and later on in veterinary patients with similar diseases, are originally presented during the DDW. Because of the dramatic impact of this meeting on both human and comparative gastroenterology, the Comparative Gastroenterology Society (CGS; www.cgs-online.org) has decided to highlight the DDW during this year's ACVIM Forum by reporting on new developments from this year's meeting in 4 different areas. The meeting was held from May 1st to May 5th 2010 in New Orleans. Because the proceedings for this session had to be written long before the DDW meeting, the authors have attempted to discuss various areas within the field they are reporting on where they anticipate developments that may ultimately have an impact on clinical veterinary medicine. Because of this, some areas may not be mentioned in this proceedings manuscript that will be discussed during the session and conversely, some areas discussed in this proceedings manuscript may not be discussed in as much detail during the session if the author feels that the DDW meeting has failed to provide significant developments in this specific area.

TOPIC INTRODUCTION

Pancreatitis remains of great significance in human gastroenterology and critical care, causing significant morbidity and mortality. Estimates for the incidence of human pancreatitis vary widely. The incidence of acute pancreatitis alone is estimated between 5 to 80 cases per 100,000 inhabitants.[1] Also, there are estimates that 20-40% of patients with acute pancreatitis have severe forms of pancreatitis and that 20-40% of those patients die of their disease process. While similar epidemiologic data are not available for dogs and cats, recent studies have clearly demonstrated that, upon necropsy, evidence of pancreatic inflammation is extremely common in both dogs and cats.[2-4] However, the degree of pancreatic inflammation that should be considered clinically relevant remains to be determined.

There are several areas in human pancreatology that have received special interest and that can be expected to be further discussed during this year's DDW meeting: classification of acute pancreatitis, autoimmune pancreatitis, early detection of severe forms of acute pancreatitis, fluid therapy in patients with acute pancreatitis, prophylactic antibiotic therapy in patients with acute pancreatitis, and alimentation in patients with acute pancreatitis. The following proceedings manuscript will focus on these 6 areas.

CLASSIFICATION OF ACUTE PANCREATITIS

Over the last few decades several international multidisciplinary symposia have been conducted to reach a universal classification system for pancreatitis in humans. The last such symposium was held in Atlanta in 1992 and has led to the so-called Atlanta classification system for pancreatitis.[5] This classification system defines acute and chronic pancreatitis as well as its complications. With a better understanding of the pathophysiology of pancreatitis and its complications, improvements in imaging modalities for the pancreas, and also with clinical use of the system, several areas of the Atlanta classification system were believed to require an update and the Acute Pancreatitis Classification Working Group was established. The working group published a suggested document for free review in April of 2008 (http://www.pancreasclub.com/resources/AtlantaClassification.pdf), which stimulated further discussion. Thus, the author would expect the working group to release and/or discuss a finalized document during the upcoming DDW meeting. While a finalized document has not been released, several important updates can be expected to be reported in this document. One important update is the recognition of pancreatitis as a two-phase disease. During the early phase, which usually lasts a week, the severity of pancreatitis is mostly related to organ failure. During the second phase, the disease either resolves or enters into a protracted phase that can last for weeks to months. This phase is associated with infectious complications and with changes in pancreatic and peripancreatic morphology. The consensus statement is also expected to suggest a classification system for acute pancreatitis based on the presence or absence of necrosis and/or infection. This determination is solely based on imaging with contrast-enhanced computed tomography (CECT), but the statement is expected to recognize that other imaging modalities may be used instead, such as magnetic resonance imaging, transabdominal ultrasound, or endoscopic ultrasound. Another important update expected is related to the definition for pancreatic and peripancreatic fluid collections. The term pancreatic abscess will likely be removed and the updated system will likely only recognize 3 types of pancreatic and peripancreatic fluid collections: acute peripancreatic fluid collections, pancreatic pseudocysts, and post-necrotic pancreatic/peripancreatic fluid collections. Each of these 3 types of fluid collections can then be further classified as either sterile or infected.

Veterinary Correlate

Unfortunately, a universal multidisciplinary classification system for pancreatitis is currently not available in veterinary medicine and different authors often use different classifications and definitions. The amendment of the Atlanta classification system will likely clarify some aspects of debate, but at the same time some issues will remain or newly arise when applying this human classification system to veterinary species. The recognition of acute pancreatitis as a two-phase disease might be of little relevance to dogs and cats. Most dogs and cats with acute pancreatitis are not maintained in an intensive care setting for many weeks or even months and thus the second phase is almost never experienced in dogs and cats, while in human patients this phase has great impact on the overall mortality of acute pancreatitis as an estimated 25-50% of all patients that die of severe acute pancreatitis are expected to die during this second phase. Also, the classification of acute pancreatitis based on the presence of necrosis is much more difficult in dogs and cats than it is in human patients. Three studies have evaluated the use of CECT for the diagnosis and classification of acute pancreatitis in dogs and cats and the results have been extremely disappointing for all three studies.[6-8] Currently, the only non-invasive diagnostic modality that provides some information about the presence of pancreatic necrosis in dogs and cats is transabdominal ultrasound, and neither the sensitivity nor the specificity of transabdominal ultrasound for the diagnosis of pancreatic necrosis have been systematically evaluated. The expected amendment of the classification for pancreatic and peripancreatic fluid collections would be expected to clarify these complications of pancreatitis in veterinary patients. There have been reports of pancreatic abscesses in dogs and cats, but not all of these patients reported in the literature would have fit the Atlanta definition for a pancreatic abscess in human pancreatitis patients.[9-11] Thus, abandonment of this term in veterinary medicine would also seem beneficial and may enhance our understanding of pancreatic and peripancreatic fluid collections in our patients.

AUTOIMMUNE PANCREATITIS

Over the last two decades a new form of pancreatitis, autoimmune pancreatitis, has been described in humans.[12] New findings about autoimmune pancreatitis are constantly being reported. Special areas of interest are the imaging characteristics of autoimmune pancreatitis, differentiation from other pancreatic and non-pancreatic conditions, diagnostic indicators for definitive diagnosis of the disease, and long-term follow up of patients.

Veterinary Correlate

Like many cases of chronic pancreatitis in dogs and cats, autoimmune pancreatitis in humans is characterized by lymphocytic-plasmacytic infiltration of the pancreas.[13,14] Human patients with autoimmune pancreatitis respond favourably to the administration of corticosteroids.[14] Recently, several clinicians have started to cautiously treat canine and feline patients with chronic pancreatitis with corticosteroids and have found this treatment strategy to be beneficial in a portion of these cases. However, further research is needed before this treatment strategy can be recommended for routine use.

EARLY DETECTION OF SEVERE FORMS OF ACUTE PANCREATITIS

Mortality of human patients with severe forms of acute pancreatitis is high during the initial phase of the disease. Thus, the need to quickly identify patients with severe forms of acute pancreatitis has long been recognized. Many severity scoring systems for early identification of severe forms of acute pancreatitis have been described. Some systems described have been extremely simple, such as a single measurement of the hematocrit, while other systems have been extremely complex and involve the collection of data regarding patient lifestyle, initial physical examination, initial biochemical evaluation, and advanced imaging modalities.

Veterinary Correlate

A few parameters and severity scoring systems have been tested for the prediction of disease severity in dogs and cats with pancreatitis. One report showed that serum calcium concentration was decreased in cats with severe forms of feline pancreatitis.[15] Another report used the number of organ systems that were compromised in dogs with acute pancreatitis to predict severity.[16] A recent report described a clinical severity index for use in dogs with pancreatitis.[17] Unfortunately, no parameter described thus far and no system evaluated have proved to be adequately sensitive and specific for identification of severe forms of acute pancreatitis in dogs or cats.

FLUID THERAPY IN PATIENTS WITH ACUTE PANCREATITIS

Fluid therapy has long been recognized to play a crucial role in the treatment of human patients with pancreatitis. Recent studies have focused on the level and timing of fluid therapy. These studies would suggest that fluid therapy has to be aggressive and initiated early on during hospitalization in order to have a maximum effect.

Veterinary Correlate

The need for fluid therapy has also long been recognized in dogs in cats. However, studies that assess the level and timing of fluid therapy are not available for dogs and cats.

PROPHYLACTIC ANTIBIOTIC THERAPY IN PATIENTS WITH ACUTE PANCREATITIS

The debate about the clinical efficacy of prophylactic antibiotic therapy in human patients with acute pancreatitis is old. Initial studies from the 70ties failed to show any benefit of prophylactic antibiotic administration, but a couple of studies in the late 90ties suggested a benefit in regards to occurrence of secondary infections and overall mortality. Infectious complications are estimated to be responsible for approximately 25-50% of all deaths associated with acute pancreatitis in humans, yet several recent meta-analysis studies have failed to show clear evidence of a beneficial effect of prophylactic antibiotic therapy.[1,18,19] Despite these findings many pancreatitis patients in human hospitals are still routinely treated with antibiotic prophylaxis. Thus, the topic remains very active and new findings can once again be expected at the upcoming DDW meeting.

Veterinary Correlate

There are no controlled studies about the benefit of antibiotic therapy in dogs and cats with pancreatitis. However, it is interesting to note that in sharp contrast to human pancreatitis patients, confirmed infectious complications are extremely rare in dogs or cats with pancreatitis. In fact, most dogs and cats that have been described in the literature as having a pancreatic abscess had sterile lesions. Taking into consideration the two phases described for pancreatitis in humans, it appears likely that most dogs and cats never reach the second phase, and thus would not reach a phase that would be associated with infectious complications. Therefore, it would seem rather unlikely that antibiotics should have any beneficial effect in dogs and cats with pancreatitis.

ALIMENTATION IN PATIENTS WITH ACUTE PANCREATITIS

Some key clinical signs in humans with pancreatitis include abdominal pain, nausea, and vomiting. Thus, traditionally, human pancreatitis patients have not been fed. At the same time, it has been recognized that pancreatitis is a highly catabolic disease and the need for alimentary support is now widely accepted. In the 80ties it was believed that total parenteral nutrition would be ideal for providing alimentation for patients with pancreatitis, but since then it has been demonstrated that enteral nutrition is preferable to total parenteral nutrition.[20] One recent meta-analysis study showed that enteral nutrition was associated with a significantly lower risk of infections, pancreatitis-related complications, single organ failure, multi-organ failure, and mortality.[20] A jejunostomy tube has been recognized to be associated with minimal exocrine pancreatic stimulation, but several studies have now shown that other routes of enteral alimentation, and more specifically nasogastric tubes, provide good alimentary support without causing major side effects.

Veterinary Correlate

Traditionally, dogs and cats with pancreatitis have been held off food, but over the last 10 years this practice has been questioned based on experiences in human patients with pancreatitis. Also, based on practices in humans, several veterinary clinicians have recommended alimentary support by total or partial parenteral nutrition. However, recently, a study in dogs with pancreatitis compared alimentation by total parenteral nutrition or esophagostomy tube in dogs. While there was no difference in mortality, disease severity decreased significantly more rapidly in the esophagostomy tube fed group.[21] These positive findings with enteral vs. total parenteral nutrition mirror recent findings in humans with pancreatitis.

REFERENCES

1. Wittau M, et al. **Hepatogastroenterology** 2008; 55:2233-223. 2. Steiner JM, et al. **Vet Ther** 2008; 9:263-273. 3. DeCock HEV, et al. **Vet Pathol** 2007; 44:39-49. 4. Watson PJ, et al. **J Small Anim Pract** 2007; 48:609-618. 5. Bradley EL. **Arch Surg** 1993; 128:586-590. 6. Gerhardt A, et al. **J Vet Intern Med** 2001; 15:329-333. 7. Forman MA, et al. **J Vet Int Med** 2004; 18:807-815. 8. Spillmann T, et al. **Proceedings of the 18th ACVIM Forum** 2000; 485-487. 9. Salisbury SK, et al. **J Am Vet Med Assoc** 1988; 193:1104-1108. 10. Simpson KW, et al. **J Sm Anim Pract** 1994; 35:93-99. 11. Stimson EL et al. **J Vet Int Med** 1998;9:202 (abstract). 12. Finkelberg DL, et al. **N Engl J Med** 2006; 355:2670-2676. 13. Newman SJ, et al. **J Vet Int Med** 2004; 18:488-493. 14. Church NI, et al. **Am J Gastroenterol** 2007; 102:2417-2425. 15. Kimmel SE, **J Am Vet Med Assoc** 2001; 219:1105-1109. 16. Ruaux CG, et al. **Aust Vet J** 1998; 76:804-808. 17. Mansfield CS, et al. **J Am Vet Med Assoc** 2008; 233:936-944. 18. Zhou YM, et al. **Hepatobiliary Pancreat Dis Int** 2005; 4:23-27. 19. Mazaki T, et al. **Br J Surg** 2006; 93:674-684. 20. Cao Y, et al. **Annals of Nutrition and Metabolism** 2008; 53:268-275. 21. Mansfield CS, et al. **Proceedings of the 19th ECVIM - CA Congress** 2009; 211.

HIGHLIGHTS FROM DIGESTIVE DISEASES WEEK & VETERINARY CORRELATES - ENTEROPATHOGENS

Jan S. Suchodolski, med.vet., Dr.med.vet., PhD
College Station, TX

SESSION INTRODUCTION

The Digestive Diseases Week (DDW; www.ddw.org) is the premier scientific meeting in human gastroenterology worldwide. Every year close to 20,000 people attend the meeting and almost 6,000 research abstracts and hundreds of lectures and interactive sessions are being presented. Many developments that ultimately lead to paradigm shifts in the diagnosis and treatment of gastrointestinal diseases in human patients, and later on in veterinary patients with similar diseases, are originally presented during the DDW. Because of the dramatic impact of this meeting on both human and comparative gastroenterology, the Comparative Gastroenterology Society (CGS; www.cgs-online.org) has decided to highlight the DDW during this year's ACVIM Forum by reporting on new developments from this year's meeting in 4 different areas. The meeting was held from May 1st to May 5th 2010 in New Orleans. Because the proceedings for this session had to be written long before the DDW meeting, the authors have attempted to discuss various areas within the field they are reporting on where they anticipate developments that may ultimately have an impact on clinical veterinary medicine. Because of this, some areas may not be mentioned in this proceedings manuscript that will be discussed during the session and conversely, some areas discussed in this proceedings manuscript may not be discussed in as much detail during the session if the author feels that the DDW meeting has failed to provide significant developments in this specific area.

ENTEROPATHOGENS

The most important pathogens causing diarrhea in humans are *Campylobacter* spp., nontyphoid *Salmonella,* toxin producing *E. coli* and *Shigella*, and rota-virus and noro-virus infections. In human medicine, bacterial enteropathogens lead to an estimated > 45,000 hospitalizations and > 1,500 deaths each year.[1] These infections are associated with billions of dollars in health care costs annually. Infections occur mostly in infants, as they typically require lower numbers of organisms compared to older individuals.[1] While the most common route of infection for some of these organisms are through food born transmission, animal to human transmission has been reported, making canine and feline enteropathogens of zoonotic interest.

The true prevalence of enteropathogens in dogs and cats has not been well examined. Furthermore, the exact role of pathogens in the development of GI disease is not fully understood. Some of the pathogens affecting humans are either rarely identified in dogs or cats (i.e., rota-virus, *Shigella, Helicobacter pylori*) or play a minor role (i.e., *Campylobacter jejuni, Salmonella*). The importance of other microorganisms (i.e., *Helicobacter* spp., Norwalk-like viruses, corona-virus, *Campylobacter* spp.) remains unclear. Some opportunistic pathogens (i.e., *Clostridium perfringens* and *E. coli*) are part of the normal intestinal ecosystem. Enteropathogens that have been associated with gastrointestinal signs in dogs and cats include *Campylobacter* spp., *Salmonella enterica* and enteropathogenic *E. coli.* However, these organisms have also been isolated from apparently healthy animals.

There is strong evidence that enteropathogens may also lead to the development of chronic gastrointestinal (i.e., IBD) and extra-intestinal disorders (autoimmune disorders), which can develop several years after acute gastrointestinal infection has passed. These are typically caused by the pathogen's effect on the host immune system. For example, *Campylobacter* spp. (but also other enteric or respiratory pathogens) have been associated with the development of Guillain-Barre Syndrome, a peripheral neuropathy. Approximately 25% of patients with this syndrome have had a recent *Campylobacter jejuni* infection. Furthermore, new epidemiological analyses have suggested that acute enteric infections (i.e., *Campylobacter* spp., *Salmonella enterica*) may be the inciting or an exacerbating factor for the development of IBD in humans.[2] Also, post infectious irritable bowel syndrome (IBS) is commonly caused by inflammatory enteropathogens in humans. Such data is currently lacking in dogs and cats.

New molecular studies have also demonstrated that the canine and feline intestinal microbiota is much more complex than previously believed, with the intestine harboring several hundreds or thousands of bacterial species, and a variety of different fungal organisms.[3,4] The intestinal microbiota plays a crucial role in the homeostasis of the immune system, and imbalances have been associated with gastrointestinal disease. Several recent studies have demonstrated a general dysbiosis in human, canine, and feline patients with IBD, implicating intestinal bacteria in the pathogenesis of chronic enteropathies, especially in genetically susceptible hosts.[5-7] Current theories for the development of chronic intestinal inflammation favor a combination of environmental factors, the intestinal microbiota, and a genetic susceptibility of the host (e.g., polymorphism in the NOD2/CARD15 gene of Crohn's disease patients). There is also new evidence that genetic susceptibility predisposes to infection with enteric organisms. Polymorphisms of specific genes of innate immunity (e.g., Interleukin-8 gene, lactoferrin gene, osteoprotegerin gene) have been associated with various diarrheal disorders (e.g., traveler's diarrhea in humans).[8] While small intestinal dysbiosis has been reported in dogs and cats with chronic enteropathies, underlying genetic susceptibilities are currently an area of intense research.

Intestinal dysbiosis in IBD

There is evidence that commensal intestinal microbiota play an integral part in the pathogenesis of IBD in humans, dogs and cats.[7,9,10] The microbiota is implicated as inflammation develops in areas with the highest bacterial counts and antibiotic therapy improves clinical signs. In rodents with genetic susceptibility, inflammation develops only in the presence of bacteria. It is suspected that intestinal inflammation causes a shift towards gram negative bacteria, perpetuating disease in genetically susceptible hosts. Intestinal inflammation may also lead to alterations in the immune system. For example, a subset of humans with Crohn's disease may be unable to effectively clear commensal or pathogenic bacteria. Most common microbial changes that are observed in intestinal inflammation are a decrease in the bacterial phyla *Firmicutes* and *Bacteroidetes*, and an increase in *Proteobacteria.* Furthermore, diseased individuals show a reduction in the diversity of *Clostridium* clusters XIVa and IV (i.e., *Lachnospiraceae, Ruminococcaceae,*

Faecalibacterium prausnitzii and *C. coccoides* subgroups), suggesting that these bacterial groups, important producers of short-chain fatty acids, may play an important role in maintenance of gastrointestinal health. Very similar changes (i.e., reduction in *Clostridium* clusters XIVa and IV) have been also observed in dogs and cats with IBD.[5,7,10] In Boxer dogs with histiocytic ulcerative colitis, an association with the presence of adherent and invasive *Escherichia coli* (AIEC) and inflammation was observed.[11] These AIEC isolates share similarities to AIEC isolates obtained from ileal tissues of humans with Crohn's disease.[11]

Protozoal infections

In humans, the prevalence of Giardiasis (*Giardia duodenalis)* is estimated at 2-6% in developed countries. Similar prevalence rates are observed in pet dogs (but up to 8% of healthy shelter animals can be positive). Of interest is that the observed genotypes differ between animals and humans. In human patients the most commonly observed *Giardia* genotypes are assemblage A-II and assemblage B. In contrast, in dogs the most commonly observed genotyopes are assemblage C and D. Assemblage F is the most common genotype observed in cats.[12] Assemblages C and D have never been isolated from humans. However, some of the human specific assemblages A (most commonly A-I) and B have been found in dogs and cats. Therefore, dogs and cats appear to be a rare source for human *Giardia* infection.[13] Giardiasis has also been associated with the development of post-infectious IBS syndrome in humans.

The most common species causing Cryptosporidiosis in humans is *Cryptosporidium hominis,* which causes watery diarrhea. Dogs most commonly harbor *C. canis* while cats harbor predominantly *C. felis.* Many dogs and cats are asymptomatic carriers, but infection has been associated with diarrhea. The prevalence rates for Cryptosporidiosis in cats ranges between 0-16% and in dogs between 0-7.4%. The zoonotic potential of canine and feline Cryptosporidiosis remains unclear. To date *C. canis* has not been identified in human patients, while *C. felis* has been identified in HIV+ patients. Severly immunocomprised people and children may be at risk for infection with *C. canis* and *C. felis.*[13]

Helicobacter pylori

Approximately 99% of humans are colonized by *Helicobacter pylori*, and only 1-4% of humans have large gastric *Helicobacter* – like organisms (HLO). In contrast, *Helicobacter* have been identified in up to 100% of dogs and cats, but animals harbor almost exclusively large HLO (most commonly *Helicobacter felis, H. bizzozeronii, H. salomonis, H. heilmanni, H. bilis, Flexispira rappini*). To date *H. pylori* has been identified only in a single colony of laboratory cats. Of importance is that HLO in dogs and cats may be located more intracellularly in the gastric epithelium, making eradication of HLOs more difficult and requiring more aggressive eradication protocols compared to humans.

***Campylobacter* spp**

In humans, *Campylobacter jejuni* is one of the most prevalent bacterial causes of acute diarrhea, with an estimated 2 million cases per year.[1] The most common route of infection is through contaminated food (often poultry, with contamination rates of up to 83%) and international travel. While the majority of humans suffer from mild self-limiting diarrhea, a few patients may develop chronic complications (Guillain-Barre Syndrome, reactive arthritis and iritis). While, there is concern in humans that *Campylobacter jejuni* may be transmitted through animal contact, there is little data evaluating the zoonotic potential of canine and feline *Campylobacter* spp. A recent study has shown that dogs may harbor some of the *Campylobacter jejuni* strains (MLST clonal complexes ST-45, ST-21, and ST-508) that are also commonly identified in humans.[14] While the exact mode of transmission of those strains remains unclear (e.g., common source such as contaminated water or a direct animal to human transmission), it is obvious that dogs and cats may constitute a zoonotic risk for human *Campylobacter* infection.

Veterinary Correlate

There is a high prevalence of *Campylobacter* spp. in canine and feline fecal samples, with observed prevalence rates as high as 42%, even in healthy animals. The true prevalence is difficult to establish because most published studies used different methods for detection of *Campylobacter* species (e.g., culture vs. PCR). However, *Campylobacter jejuni* appears to be much less common in dogs and cats compared to humans. In a recent study using a species specific PCR assay, *Campylobacter jejuni* was identified in only 3/187 dogs with diarrhea, in 0/77 healthy dogs, and in 0/171 cats. In contrast, approximately 50% of evaluated cats and 35% of dogs (regardless of health status) were positive for other *Campylobacter* spp., with *Campylobacter helveticus* more prevalent in cats, and *Campylobacter upsaliensis* more prevalent in dogs. *Campylobacter upsaliensis* strains isolated from dogs appear to differ from human isolates.

REFERENCES

[1]DuPont HL. **New England J Med** 2009; 361:1560-1569. [2]Gradel KO, et al. **Gastroenterol** 2009; 137:495-501. [3]Suchodolski JS, et al. **BMC Microbiology** 2009; 10:210. [4]Suchodolski JS, et al. **FEMS Microbiol Ecol** 2008; 66:567-578. [5]Xenoulis PG et al. **FEMS Microbiol Ecol** 2008; 66:579-589. [6]Packey CD et al. **Current Opinion Infect Dis** 2009; 22:292-301. [7]Janeczko S et al. **Vet Microbiol** 128(1-2), 178-193. 2008. [8]Mohamed JA et al. **J Infect Dis** 2009; 199:477-485. [9]Frank DN et al. **PNAS** 2007; 104:13780-13785. [10]Suchodolski JS et al. **Vet Microbiol** 2009; doi:10.1016/j.vetmic.2009.11.002. [11]Simpson KW et al. **Inf and Imm** 2006; 74:4778-4792. [12]Lebbad M et al. **Vet Parasitol** 2009; doi:10.1016/j.vetpar.2009.11.003. [13]Bowman DD et al. **Experim Parasitol** 2010; 124:121-127. [14]Parsons BN et al. **J Clin Microbiol** 2009; 47:3466-3471.

HIGHLIGHTS FROM DIGESTIVE DISEASES WEEK WITH VETERINARY CORRELATES: INFLAMMATORY BOWEL DISEASE

Dr. Karin Allenspach, Dipl ECVIM-CA PhD
London, UK

SESSION INTRODUCTION

The Digestive Diseases Week (DDW; www.ddw.org) is the premier scientific meeting in human gastroenterology worldwide. Every year close to 20,000 people attend the meeting and almost 6,000 research abstracts and hundreds of lectures and interactive sessions are being presented. Many developments that ultimately lead to paradigm shifts in the treatment of human patients with gastrointestinal diseases and later on in veterinary patients with similar diseases are originally presented during the DDW. Because of the dramatic impact of this meeting on both human and comparative gastroenterology the Comparative Gastroenterology Society (CGS; www.cgs-online.org) has decided to highlight the DDW by reporting on new developments from this year's meeting in 4 different areas. The meeting was held from May 1st to May 5th 2010 in New Orleans. Because the proceedings for this session had to be written long before the meeting, the authors have attempted to discuss various areas within the field they are reporting about where they anticipate developments that may ultimately have real impact in clinical veterinary medicine. Because of this, some areas may not be mentioned in this proceedings manuscript that will be discussed during the session and conversely, some areas discussed in this proceedings manuscript may not be discussed in detail during the session if the author feels that the DDW meeting has failed to provide significant developments in this field.

PATHOGENESIS OF CANINE IBD: THE INTERPLAY OF MUCOSAL IMMUNE SYSTEM WITH MICROBIOTA

The gut epithelium detects the presence of both commensal and pathogenic bacteria via Pattern Recognition Receptors (PRRs).[1] This group contains Toll-like receptors (TLRs) and Nucleotide Binding Oligomerisation Domain-like receptors (NODs).[2] These receptors recognize specific molecules termed Pathogen-Associated Molecular Patterns (PAMPs), which are conserved molecules found on bacteria and other infectious agents. Recognition of PAMPs by TLRs is a crucial part of the innate immune response to invading bacteria in the gut and initiates a complex intracellular signalling pathway culminating in the activation of the transcription factor NF-κB.[3] This results in the transcription and secretion of a variety of pro- and anti-inflammatory cytokines and chemokines from the cell bearing the TLR. If successful, immune cells activated by this response clear offending pathogens. Despite the fact that commensal and pathogenic bacteria share PAMPs, the immune system remains unresponsive towards commensal organisms present in the intestinal lumen in the normal case scenario.[4]

The inflammatory response which is normally only seen as a reaction towards pathogenic bacteria breaching the intestinal barrier is similar to the response seen in the mucosa of dogs affected with IBD. However, the response in IBD occurs in the absence of pathogens. It is believed that the innate immune system reacts to normal commensals in the intestinal lumen as if they were pathogens.

EVIDENCE OF INNATE IMMUNITY HYPER-RESPONSIVENESS IN CANINE IBD

TLRs have been shown to be upregulated in the intestine of human beings with Crohn's disease and ulcerative colitis. This may be either a consequence of the ongoing stimulation of TLRs by an altered microbiota or it may be a causal factor contributing to the pathogenesis of disease. Most human studies show that the mRNA and protein expression of TLR2 as well as TLR4 are increased in the intestine of people with active IBD.[5,6] In a recent clinical study at the Royal Veterinary College, London, we were able to show that dogs of any breed with clinically severe, active IBD express higher levels of TLR2 receptors in the duodenum compared to healthy dogs when measured by real-time PCR in whole endoscopic biopsies.[7] In addition, TLR2 expression was correlated with the clinical severity of IBD using the Canine Chronic Enteropathy Clinical Activity Index (CCECAI). However, TLR4 expression levels were similar to those in healthy canine intestine. Other studies have found that only a subgroup of dogs with IBD (the ones responding only to steroid administration) showed an increased expression of TLR2, TLR4 and TLR9 compared to healthy intestines when expression was measured by real-time PCR.[8] In further studies looking at German Shepherd dogs (GSD) with IBD, we have found that TLR4 expression was 60-fold higher in the duodenum, ileum and colon of dogs with IBD compared to samples from healthy dogs, however, TLR2 and TLR9 were similarly expressed as in healthy dogs.[9] These data show that it is important to look at similar phenotypes of dogs when choosing cases for such studies, as the results will vary depending on the severity of disease, the treatment response in the cases and the specific breed of dogs. In addition, care must be taken to compare studies using real-time PCR as the standardization method used depends on the reference genes, which need to be carefully chosen for each study. TLR2 has recently been shown to be overexpressed in the diseased intestine in mouse models of IBD.[10] TLR2 in this context is implicated in the homeostasis and repair of intestinal tissue after injury. It is therefore possible that the high expression of TLR2 found in dogs with IBD in the studies mentioned above could be a marker of intestinal inflammation and its physiological action is to downregulate ongoing inflammation.

TLR5 expression was seen to be consistently downregulated in the intestine of German Shepherd Dogs with IBD as compared to healthy dogs.[9] In mice and human beings, TLR5 is highly expressed in the healthy small intestine, with CD11c+ dendritic cells (DCs) in the lamina propria mucosae expressing most TLR5.[11] It is believed that this tolerogenic phenotype of DCs induces T regulatory cells and stimulates the production of anti-inflammatory cytokines such as IL-10 in response to flagellin.[12] In contrast, in intestinal inflammation characterized by the upregulation of Th1 and Th17 cytokines, CD11c- DCs express low levels of TLR5 but instead high levels of TLR4.[12] In this context, TLR4 is thought to be upregulated to compensate for the low TLR5 expression. It could be speculated that the differentially low expression of TLR5 and very high expression of TLR4 seen in the intestine in the GSD of our study indicates a similar compensatory role of TLR4, as gram-negative flagellated bacteria can also be recognized through binding of LPS by TLR4.

DYSBALANCE OF THE INTESTINAL MICROBIOTA IN CANINE IBD

Molecular studies on the intestinal microbiom in dogs of different breeds with IBD have found that members of the families *Enterobacteriaceae* and *Clostridiaceae* were enriched in the diseased intestine.[13] These bacteria are thought to contribute to the pathogenesis of disease in dogs as well as human beings with IBD.[14] However, in GSD with IBD, bacterial clones within the order *Lactobacilalles* were found in the duodenum significantly more frequently than in healthy dogs.[9] It appears that GSD with chronic enteropathies have a distinctly different microbiom from healthy dogs as well as from other breeds of dogs presenting with IBD, with over-representation of certain, traditionally labelled "beneficial" bacteria in the duodenum, specifically sequences of the order of *Lactobacilalles*.

GENETIC PREDISPOSITION IN GSD WITH IBD

Over the last decade, numerous genes have been found to be associated with an increased risk of development of IBD in human beings, many of them implicated in the innate immune response in the intestine.[15,16] Mutations in PRRs, such as NOD2, TLR4, IL-23 receptor and others have all been associated with IBD. In dogs, it has always been obvious to clinicians that IBD seems to have a genetic component. This is particularly evident in breeds like the Boxer, who is predisposed to histiocytic ulcerative colitis. Another example of a breed disposition is the German Shepherd Dog, which seems to be predisposed to antibiotic-responsive diarrhea. We recently performed a mutational analysis of the canine genes for TLR2, TLR4, TLR5 and NOD-2 in GSD with IBD. One of the three polymorphisms that we identified in the TLR5 gene (G22A) of GSD was evaluated further in a case-control study with over 50 cases and breed controls and was found to be significantly associated with IBD. In addition, 4 non-synonymous single nucleotide polymorphisms (SNPs) were identified in exon 4 of the canine NOD2 gene. The heterozygote genotype for all four NOD2 SNPs was found significantly more frequently in the IBD population ($p=0.04$, $OR=2.34$, $CI=1.03\text{-}5.28$) than in controls. These results were also mirrored in non-GSD breeds. The heterozygote genotype for all four SNPs was found significantly more frequently in a population of 96 dogs of different breeds with IBD compared to the non-GSD control population ($p=0.0009$, $OR=3.06$, $CI=1.55\text{-}6.05$).

WHERE TO GO FROM HERE?

We are still a long way from our goal of finding better ways of diagnosing and treating chronic enteropathies in dogs. The next step will involve investigations into the functional aspects of canine TLR and NOD2. In order to achieve this, current studies at the Royal Veterinary College are investigating signalling through canine TLR on a cellular level. These investigations will allow insights into the functional consequences of TLR polymorphisms observed in German Shepherd Dogs, and will help to elucidate the pathogenesis of antibiotic responsive diarrhoea and IBD in this breed.

If bacteria or bacterial products are identified which either stimulate the appropriate TLR response or inhibit an aberrant TLR-dependent NF-κB activation, then this knowledge can be used in prospective treatment studies in dogs with chronic enteropathies. Similar to the human system, it is possible that certain probiotic cocktails could be used as prophylactic treatment in dogs which are genetically predisposed to developing IBD, food allergy, or antibiotic responsive diarrhea. The identification of a specific bacterial product that reduces inflammation in the gut could serve as a treatment supplement or even as an adjuvant for potential vaccines against IBD in dogs.

REFERENCES

[1] Sandor F, Buc M. Toll-like receptors. III. Biological significance and impact for human medicine. **Folia Biol** (Praha) 2005;51:198-203. [2] Cario E. Bacterial interactions with cells of the intestinal mucosa: Toll-like receptors and NOD2. **Gut** 2005;54:1182-1193. [3] Magalhaes JG, Tattoli I, Girardin SE. The intestinal epithelial barrier: how to distinguish between the microbial flora and pathogens. **Semin Immunol** 2007;19:106-115. [4] Sirard JC, Bayardo M, Didierlaurent A. Pathogen-specific TLR signaling in mucosa: mutual contribution of microbial TLR agonists and virulence factors. **Eur J Immunol** 2006;36:260-263. [5] Cario E, Podolsky DK. Differential alteration in intestinal epithelial cell expression of toll-like receptor 3 (TLR3) and TLR4 in inflammatory bowel disease. **InfectImmun** 2000;68:7010-7017. [6] Szebeni B, Veres G, Dezsofi A, et al. Increased expression of Toll-like receptor (TLR) 2 and TLR4 in the colonic mucosa of children with inflammatory bowel disease. **Clin Exp Immunol** 2008;151:34-41. [7] McMahon LA, House AK, Catchpole B, et al. Expression of Toll-like receptor 2 in duodenal biopsies from dogs with inflammatory bowel disease is associated with severity of disease. **Vet Immunol Immunopathol** 2009. [8] Burgener IA, Konig A, Allenspach K, et al. Upregulation of toll-like receptors in chronic enteropathies in dogs. **J Vet Intern Med** 2008;22:553-560. [9] K Allenspach JS, FM McNeill, A House, A Hendricks, JM Steiner, D Werling. Characterization of the bacterial microflora and the innate immune response in German Shepherd Dogs with chronic enteropathies. **Abstract, ACVIM Forum** San Antonio, TX 2008. [10] Gibson DL, Ma C, Rosenberger CM, et al. Toll-like receptor 2 plays a critical role in maintaining mucosal integrity during Citrobacter rodentium-induced colitis. **Cell Microbiol** 2008;10:388-403. [11] Uematsu S, Fujimoto K, Jang MH, et al. Regulation of humoral and cellular gut immunity by lamina propria dendritic cells expressing Toll-like receptor 5. **Nat Immunol** 2008;9:769-776. [12] Uematsu S, Akira S. Immune responses of TLR5(+) lamina propria dendritic cells in enterobacterial infection. **J Gastroenterol** 2009;44:803-811. [13] Xenoulis PG, Palculict B, Allenspach K, et al. Molecular-phylogenetic characterization of microbial communities imbalances in the small intestine of dogs with inflammatory bowel disease. FEMS **Microbiol Ecol** 2008;66:579-589. [14] Darfeuille-Michaud A, Boudeau J, Bulois P, et al. High prevalence of adherent-invasive Escherichia coli associated with ileal mucosa in Crohn's disease. **Gastroenterology** 2004;127:412-421. [15] Dubinsky MC, Wang D, Picornell Y, et al. IL-23 receptor (IL-23R) gene protects against pediatric Crohn's disease. **Inflamm Bowel Dis** 2007;13:511-515. [16] Franchimont D, Vermeire S, El Housni H, et al. Deficient host-bacteria interactions in inflammatory bowel disease? The toll-like receptor (TLR)-4 Asp299gly polymorphism is associated with Crohn's disease and ulcerative colitis. **Gut** 2004;53:987-992.

HIGHLIGHTS FROM DIGESTIVE DISEASES WEEK WITH VETERINARY CORRELATES: MINIMALLY INVASIVE ENDOSCOPIC PROCEDURES

Keith Richter, DVM, DACVIM
San Diego, CA

SESSION INTRODUCTION

The Digestive Diseases Week (DDW; www.ddw.org) is the premier scientific meeting in human gastroenterology worldwide. Every year close to 20,000 people attend the meeting and almost 6,000 research abstracts and hundreds of lectures and interactive sessions are presented. Many developments that ultimately lead to paradigm shifts in the diagnosis and treatment of gastrointestinal diseases in human patients, and often eventually in veterinary patients with similar diseases, are originally presented during the DDW. Because of the dramatic impact of this meeting on both human and comparative gastroenterology, the Comparative Gastroenterology Society (CGS; www.cgs-online.org) has decided to highlight the DDW during this year's ACVIM Forum by reporting on new developments from this year's meeting in 4 different areas. The meeting was held from May 1st to May 5th 2010 in New Orleans. Because the proceedings for this session had to be written long before the DDW meeting, the authors have attempted to discuss various areas within the field they are reporting on where they anticipate developments that may ultimately have an impact on clinical veterinary medicine. Because of this, some areas may not be mentioned in this proceedings manuscript that will be discussed during the session and conversely, some areas discussed in this proceedings manuscript may not be discussed in as much detail during the session if the author feels that the DDW meeting has failed to provide significant developments in this specific area.

TOPIC INTRODUCTION

Minimally invasive endoscopic procedures have become commonplace in human medicine. These procedures include the use of both flexible and rigid endoscopes to accomplish in a less invasive manner than what could be performed in an open surgical procedure. Minimally invasive procedures have been developed for both diagnostic and therapeutic purposes. Advances in scoping equipment and instrumentation, and novel and innovative approaches have allowed this field to grow considerably in the last decade. This proceedings manuscript will describe recent topics from the 2009 DDW meeting. The author expects several more studies to be presented in the 2010 DDW meeting in these areas.

FLEXIBLE ENDOSCOPY

Many of the recent advances in endoscopy involve innovative methods to evaluate the small intestine beyond the duodenum. This has been termed small intestinal enteroscopy. Indications for this include the need to look for causes of occult or obscure gastrointestinal bleeding, occult intestinal neoplasia, to deploy intraluminal stents in the mid-small intestine, to facilitate endoscopic polypectomy in regions of the bowel that are beyond what could be examined with conventional endoscopy, and to evaluate the mucosal surface for a variety of intestinal diseases. In addition to the technique of using a long slender scope (termed push enteroscopy, analogous to what is commonly performed in small veterinary patients), other methods have been studied. One of these methods is double balloon enteroscopy. The technique uses two balloons, one attached to the tip of the endoscope and another attached to the tip of an overtube. The enteroscope and overtube are inserted through the mouth and passed in conventional fashion (that is, as with gastroduodenoscopy) into the small bowel. Following this, the endoscope is advanced a small distance in front of the overtube and the balloon at the end is inflated. Using the assistance of friction at the interface of the enteroscope and intestinal wall, the small bowel is accordioned back to the overtube. The overtube balloon is then deployed, and the enteroscope balloon is deflated. The enteroscope is advanced through the intestine by being held alternately by the balloon on the endoscope and the balloon on the overtube. The process is then continued until the entire small bowel is visualized. The double-balloon enteroscope can also be passed in retrograde fashion, through the colon and into the ileum to visualize the aboral end of the small bowel.

Another method of complete small bowel examination is with capsule endoscopy (also known as the "pill cam"). This technique involves swallowing of a wireless capsule with a battery powered light source and lens system. The images are transmitted to a receiver worn by the patient, and then assembled by computer to a viewable format. Although this technique is less technically demanding, it has the disadvantage of only being able to visualize the mucosa and not having the ability to obtain biopsy specimens. Capsule endoscopy is contraindicated in patients with swallowing disorders, known or suspected gastrointestinal obstruction, strictures, or fistulas.

A more recent method of small bowel examination is called spiral enteroscopy. In this technique, the scope is placed through an overtube with a smooth, compliant spiral located at the distal end of the overtube. By rotation of the overtube, the bowel is gathered and pleated in a rapid and controlled manner as the overtube is advanced. This also permits the enteroscope to be independently maneuvered while the overtube remains in place. This is especially useful for therapeutic procedures such as piecemeal polypectomy. The overtube holds the gathered small bowel in place while the enteroscope and specimen are completely removed through the overtube. The scope is then reinserted to its previous location to continue the procedure.

Several studies have recently described the pros and cons of these methods, and compared them in a variety of disease states, in particular for occult sources of bleeding and detection of neoplasia. The author expects follow-ups to these initial studies.

Veterinary Correlates

There are a variety of reasons to evaluate regions of the small intestine beyond the duodenum and proximal jejunum. Many of the indications described above in humans apply to animals as well. As clinicians, we occasionally encounter discordant biopsy results that do not match clinical suspicions. One possible explanation for this is the segmental nature of mucosal diseases seen in dogs and cats. Examples of this include intestinal lymphangiectasia in dogs and small intestinal lymphoma in dogs and cats. The ability to evaluate the intestine aboral to where conventional endoscopy is able to reach would be very helpful in some instances.

Double balloon enteroscopy was recently described in veterinary medicine. The technique was applied in two dogs of medium size, using both oral and retrograde approaches, and the authors were able to examine the surface of the mucosa of the entire small intestine of these patients.

Methods to prepare the bowel for capsule endoscopy may have applicability to preparing dogs and cats for colonoscopy. In the 2009 DDW meeting, a study was presented that described a protocol using a split dose of a polyethylene glycol-electrolyte solution plus simethicone for bowel cleansing prior to capsule endoscopy. Studies like this may suggest protocols that can be used in dogs and cats.

ENDOSCOPIC MANAGEMENT OF STRICTURES

Several studies have described newer approaches to the management of strictures. Newer stents are being studied to improve outcome in benign esophageal strictures. Two studies presented at the 2009 meeting described the use of biodegradable stents for this purpose. In addition, endoscopic methods of anchoring stents are being developed.

Veterinary Correlates

Recurrence of strictures after balloon dilatation is a frustrating problem in dogs and cats, and optimal methods to reduce the recurrence rate have not been defined. Several approaches have been anecdotally described in veterinary medicine, including elective pre-emptive follow up dilation attempts, intralesional triamcinolone injections, intralesional mitomycin-C injections, and intraluminal stent placement. The use of esophageal stents has recently been described in dogs with benign strictures, including the use of biodegradable stents. Studies are needed to define the role of stents, optimal stent material, and methods to anchor stents.

ADVANCES IN INTESTINAL BIOPSY

One of the recent advances in intestinal biopsy is the ability to procure full thickness biopsies endoscopically. This has been made possible by the use of devices to close full thickness defects deployed through the channel of an endoscope. These instruments are also being used for resection and anastomosis of the intestine. Several studies have examined wound healing characteristics, leakage rates, tensile strength, and complications using these methods.

Veterinary Correlates

There are several disease states in which full thickness intestinal biopsies would potentially be warranted. These include intestinal lymphoma in dogs and cats, feline infectious peritonitis (FIP), and intestinal lymphangiectasia in dogs. The diagnosis can potentially be missed in these disorders with analysis of endoscopically procured mucosal biopsies. For this reason, laparotomy is sometimes performed to obtain full thickness biopsies. Developing the techniques to obtain these biopsies in a noninvasive procedure using an endoscope in dogs and cats would be desirable.

LAPAROSCOPIC AND ENDOSCOPIC SURGERY

The use of laparoscopy has increased tremendously in human medicine during the last ten to fifteen years, including diagnostic and therapeutic procedures. The widespread acceptance of laparoscopy stems from the technological advances in equipment and the impressive results of this noninvasive modality compared with open surgery. Though some procedures performed in human beings will have little applicability in veterinary medicine, many of the procedures will gain acceptance due to their ease, effectiveness, and lack of morbidity. These include cholecystectomy, appendectomy, bile duct exploration, bowel resection, gastropexy, pyloromyotomy, fundoplication, and others.

More recently, Natural Orifice Transluminal Endoscopic Surgery (NOTES®) has been developed. These procedures involve the insertion of an endoscope into a natural orifice (such as the stomach, vagina, or colon), which is then used to perforate the wall to gain access to the peritoneal cavity. Procedures such as appendectomy, cholecystectomy, or hepatectomy can then be performed. This approach is thought to limit postoperative pain and morbidity.

Veterinary Correlates

Many of the same surgical procedures performed in humans could potentially be performed in dogs and cats. Controlled studies are necessary to define the role of all laparoscopic procedures in veterinary medicine. As clinicians and equipment manufacturers address technical limitations, many surgical procedures performed within a body cavity should be amenable to laparoscopic surgery.

REFERENCES

[1]Butala P, et al. **Gastroenterology** 2009; 136(5 Suppl): A-883. [2]Chung J, et al. **J Vasc Interv Radiol** 2010; 21(1): 152. [3]Fraune C, et al. **J Sm An Pract** 2009; 50(10): 550. [4]Horgan S, et al. **Gastroenterology** 2009; 136(5 Suppl): A-873. [5]Latorre R, et al. **Vet Rec** 2007; 161(17): 587. [6]Liu CD, et al. **Gastroenterology** 2009; 136(5 Suppl): A-914. [7]Podolsky ER, et al. **Gastroenterology** 2009; 136(5 Suppl): A-869. [8]Roberts KE, et al. **Gastroenterology** 2009; 136(5 Suppl): A-927. [9]Zehetner, J, et al. **Gastroenterology** 2009; 136(5 Suppl): A-913.

EPIDEMIOLOGY OF CRYPTOCOCCUS INFECTIONS

Dr. Jane Sykes, BVSc(Hons), PhD, DACVIM
Davis, CA

INTRODUCTION

Cryptococcosis is an important mycosis of humans and animals. It is the most common systemic mycosis of cats, which are approximately 10 times more susceptible than dogs. Cryptococcosis is most commonly caused by two encapsulated yeast species that are dimorphic, basidiomycetous fungi of the genus *Cryptococcus*, *Cryptococcus neoformans* and *Cryptococcus gattii. C. gattii* is considered an emerging pathogen of immunocompetent humans in the north-western United States, in contrast to *C. neoformans*, which is an opportunistic pathogen in humans.[1] Other *Cryptococcus* species have rarely been associated with disease in dogs and cats, including *Cryptococcus albidus* and *Cryptococcus laurentii*.[2,3] Infection is acquired following inhalation from the environment, and is not transmitted between animals. Occasionally, outbreaks of infection in humans and animals occur as a result of exposure to common environmental sources. Growing evidence suggests that the local epidemiology of cryptococcosis in dogs and cats may differ considerably between geographic locations worldwide.

LIFE CYCLE

The life cycle of *Cryptococcus* spp. involves vegetative growth as a haploid budding yeast, combined with the ability to transition to a filamentous form (*Filobasidiella* species), also known as a teleomorph. The teleomorphs of *C. neoformans* and *C. gattii* are *F. neoformans* and *F. bacillosporus*, respectively. Two mating types of *Cryptococcus* spp. exist, α and **a**, the α-type being vastly more prevalent in environmental and clinical samples. Under optimum conditions and in the presence of mating pheromones, the two mating types can fuse and adopt a dikaryotic filamentous state, which is known as the perfect state. This is followed by production of basidia, which are small club-shaped structures on which basidiospores form. Cells of the α mating type can also respond to dessication, nitrogen limitation, and pheromone signals and undergo asexual reproduction via haploid fruiting, which also involves filamentation and spore formation.[4] This may explain the relative high prevalence of the α mating type. When inhaled, basidiospores are thought to give rise to mammalian disease.[5] The mating of *Cryptococcus* spp. has never been observed in nature, but can be demonstrated under special laboratory conditions. Unlike other dimorphic fungi, the yeast phase of *Cryptococcus* is found under routine laboratory conditions and in mammalian host tissues.

MOLECULAR EPIDEMIOLOGY

Cryptococcus spp. were initially classified into 4 serotypes, A, B, C and D, based on antigenic differences in the cryptococcal polysaccharide capsule, and a hybrid serotype AD was also identified. *C. neoformans* contains serotypes A and D, and *C. gattii* contains serotypes B and C. With the introduction of a variety of genotyping methods, considerable genetic diversity has been identified within these serotypes. The results of PCR fingerprinting using microsatellite M13 or minisatellite $(GACA)_4$ primers, or amplified fragment length polymorphism (AFLP) analysis, have shown that the *Cryptococcus* species complex can be divided into 8 molecular types.[6] To some extent, the types correlate with *Cryptococcus* species and serotypes, and consist of VNI/AFLP1 and VNII/AFLP1A/1B (*C. neoformans* var. *grubii*, serotype A); VNIII/AFLP3 (AD hybrid); VNIV/AFLP2 (*C. neoformans* var. *neoformans*, serotype D), and VGI/AFLP4, VGII/AFLP6, VGIII/AFLP5 and VGIV/AFLP7 (*C. gattii*, serotypes B and C). Interspecies hybrid strains (AB and BD serotypes) have also been reported.[7] Differences in epidemiology, pathogenicity, clinical features and drug susceptibility have been associated with the species, variety, and molecular types.[6,8-10] *C. gattii* is recognized as a species distinct from *C. neoformans* based on differences in morphology, environmental niches, limited molecular identity, multiple gene geneologies, and inefficient cross-species mating with the production of sterile progeny.[11] *C. gattii* can be easily distinguished from *C. neoformans* in the microbiology laboratory using canavanine-glycine-bromthymol blue agar, although few veterinary diagnostic laboratories at the current time perform differentiation. Unfortunately, many publications and laboratory reports still refer to all *Cryptococcus* spp. isolates as *C. neoformans.* PCR fingerprinting is generally performed within specialized diagnostic or research laboratories. Other methods of genetic typing that have been applied to *Cryptococcus* spp. include restriction fragment length polymorphism (RFLP) of the *URA5* and *PLB1* genes, intergenic spacer (IGS) sequencing, multigene sequence analysis, multilocus sequence typing (MLST) and multilocus microsatellite typing (MLMT).[6] Because of the lack of standardization of these different methods, in 2007 the Cryptococcal Working Group I (Genotyping of *C. neoformans* and *C. gattii*) of the International Society for Human and Animal Mycology (ISHAM) was formed. The Working Group recommended a consensus multilocus typing scheme based on MLST using seven unlinked genetic loci for global strain genotyping.[6] This method was recommended because of its high discriminating power combined with a high degree of interlaboratory reproducibility.

Cryptococcus neoformans

C. neoformans has a worldwide distribution. The major environmental niche for *C. neoformans* is believed to be weathered bird (especially pigeon) guano, in which it mates abundantly.[12] It can also be found in decaying plant matter in hollows of certain trees.[13] *C. neoformans* passes through the gastrointestinal tract of pigeons, but systemic infection of pigeons is rare. Pigeons can also carry the fungus on their feathers, crops, or cloaca.[14] Pigeon guano provides an environment rich in many nitrogen-containing compounds including creatinine that favor cryptococcal growth. Cryptococci may remain viable for years in pigeon lofts, where accumulations of guano are protected from drying or sunlight. Macaws, swans, and parakeets have also been associated with *C. neoformans* carriage. The vast majority of human cryptococcal infections in human immunodeficiency virus-infected patients result from infection with *C. neoformans*, especially *C. neoformans* var. *grubii. C. neoformans* is the most common cryptococcal species isolated from humans, dogs and cats in southeastern Australia, accounting for more than 70% of infections.[15] The majority of dogs with cryptococcosis in California are also infected with *C. neoformans*, whereas infection of cats with this species appears to be rare.

Cryptococcus gattii

Originally thought to be a tropical or subtropical pathogen, *C. gattii* has also emerged as an important pathogen in temperate regions following outbreaks of disease in apparently immunocompetent individuals in British Columbia, Canada, and the Pacific Northwest of the United States.[14] Although to date it has been most commonly reported from Australia, New Zealand, Canada, and Central and South America, its presence has been reported in a growing list of other geographic regions including Austria, China, Congo, Germany, Greece, France, India, Italy, Japan, South Korea, the Netherlands, Spain, South Africa, the United Kingdom, and the United States. In Australia and North America, infections have been reported in dogs, cats, goats, horses, llamas, koalas, echidnas, marine mammal species, and exotic birds.[14] Studies to date have indicated that worldwide, genotypes VGI and VGII are prevalent, with VGIII and VGIV being less common. The outbreaks of disease in Canada and the Pacific Northwest have been associated with certain highly virulent molecular types of *Cryptococcus gattii*, primarily VGIIa, and to a lesser extent, VGIIb. In northern California, almost all cats with cryptococcosis have been infected with *C. gattii* genotype VGIII. In contrast, the prevalence of *C. gattii* infection is lower in cats from southeastern Australia, with most *C. gattii* isolates belonging to VGI. VGII predominates in southwestern Australia and the Northern Territory.[16,17] *C. gattii* VGI in Australia is associated with dead plant material in eucalyptus tree hollows. VGI is also found in countries where Australian eucalyptus trees have been exported. Nevertheless, the isolation of *C. gattii* from eucalyptus trees is rare outside Australia, despite extensive environmental sampling. Imported eucalyptus trees have not been associated with the environmental presence of *C. gattii* in Spain, central Africa, or Canada.[14] *C. gattii* has now been associated with over 50 tree species, the majority (77%) of which are angiosperms (hardwoods that are often deciduous). In British Columbia, Canada, *C. gattii* is consistently found on living tree bark and in air, and has also been found in freshwater and seawater.[18] Decayed wood hollows appear to be a specific ecologic niche for the organism. It has been suggested that *C. gattii* is an environmental pathogen that maintains its virulence attributes in the absence of a recognized host animal species.[14] Interestingly, approximately one-quarter of the cats with cryptococcosis seen at the author's teaching hospital are kept exclusively indoors. It is possible that organisms may be brought indoors on fomites such as shoes or soil used for indoor plants.

IMPLICATIONS FOR COMPANION ANIMAL AND HUMAN HEALTH

In humans, the lung is thought to be the primary site of infection with *Cryptococcus* spp. The nasal cavity has been suggested as the primary site of infection in cats and dogs, although some cats and dogs develop pulmonary or disseminated involvement without evidence of lesions within the nasal cavity. Differences in clinical presentation have been identified in humans infected with *C. gattii* versus those infected with *C. neoformans*, although host immune responses appear to play the predominant role in determining clinical presentation.[8-10] Patients infected with *C. neoformans* tend to be immunocompromised, in particular with acquired immunodeficiency syndrome (AIDS). Infections tend to be widely disseminated to a variety of organs, with a high prevalence of fungemia and funguria.[8,9] In contrast, patients infected with *C. gattii* are often immunocompetent. Fungemia and funguria are generally absent, and infections are frequently associated with mass lesions in the cerebrum and/or lungs (cryptococcomas). Patients infected with *C. gattii* are frequently treated for longer with antifungal drugs, antigen titers are higher, and morbidity, mortality, and severe neurologic sequelae are more frequent among patients infected with *C. gattii*.[9] Mouse models have shown that *C. gattii* may inhibit the migration of neutrophils to sites of infection, as well as the production of protective cytokines. As a result, it appears that *C. gattii* strains thrive in immunocompetent hosts by evading or suppressing immune responses that normally limit the progression of disease caused by *C. neoformans*.[19]

Evidence from cats and dogs from northern California infected with *C. neoformans* and *C. gattii* suggests that to some extent, similar patterns of infection may exist in companion animals. Overall, feline infections with *C. gattii* are associated with a better long-term prognosis, but infections are often difficult to cure with antifungal drugs, and high antigen titers tend to persist. The inflammatory response to infection is often absent or limited. Dogs tend to have widely disseminated disease, with a high prevalence of neurologic involvement and common involvement of sites such as the liver, myocardium, and kidneys, often with a marked inflammatory response to the organisms. Differences may also exist in clinical presentation between dogs infected with *C. gattii* and dogs infected with *C. neoformans*. Differentiation between *C. gattii* and *C. neoformans* by veterinary microbiology laboratories is encouraged for further understanding of the epidemiology, pathogenesis, and response to treatment of these organisms.

REFERENCES

[1]Dixit A, et. al. **Interdiscip Perspect Infect Dis** 2009;2009:840452. [2]Labrecque O, et. al. **J Vet Diagn Invest** 2005;17:598. [3]Kano R, et. al. **Med Mycol** 2008;46:75. [4]Hull CM, et. al. **Annu Rev Genet** 2002;36:557. [5]Velagapudi R, et. al. **Infect Immun** 2009;77:4345. [6]Meyer W, et. al. **Med Mycol** 2009;47:561. [7]Ngamskulrungroj P, et. al. **PloS One** 2009;4:e5862. [8]Speed B, et. al. **Clin Infect Dis** 1995;21:28. [9]Mitchell DH, et. al. **Clin Infect Dis** 1995;20:611. [10]Chen S, et. al. **Clin Infect Dis** 2000;31:499. [11]Kwon-Chung KJ, et. al. **FEMS Yeast Res** 2006;6:574. [12]Nielsen K, et. al. **Eukaryot Cell** 2007;6:949. [13]Lazera MS, et. al. **J Med Vet Mycol** 1996;23:127. [14]Springer DJ, et. al. **Emerg Infect Dis** 2010;16:14. [15]O'Brien CR, et. al. **Med Mycol** 2009;42:449. [16]Campbell LT, et. al. **Eukaryot Cell** 2005;4:1410. [17]McGill S, et. al. **Med Mycol** 2009;47:625. [18]Datta K, et. al. **Emerg Infect Dis** 2009;15:1185. [19]Cheng PY, et. al. **Infect Immun** 2009;77:4284.

OPPORTUNISTIC MYCOSES IN DOGS & CATS

Amy M. Grooters, DVM, DACVIM
Baton Rouge, LA

INTRODUCTION

Over the past three decades, opportunistic fungal pathogens have assumed an increasingly important role in human disease as immunocompromise associated with bone marrow transplantation, solid organ transplantation, and human immunodeficiency virus infection has become more prevalent. Because similar risk factors for immunocompromise have not been common in small animal patients, most of the mycoses encountered by veterinarians have continued to be those caused by the endemic fungal pathogens, such as *Blastomyces*, *Cryptococcus*, *Histoplasma*, and *Coccidioides*. However, in the past 5 years, the number of opportunistic fungal infections encountered by this author in veterinary patients receiving multi-agent immunosuppressive therapy (usually including cyclosporine) has increased significantly. In addition, several recent case reports have described opportunistic fungal infection in dogs receiving immunosuppressive therapy.[6,11,12] Because the majority of these patients have been undergoing treatment for immune-mediated or inflammatory medical disorders (such as granulomatous meningoencephalitis, IMHA, ITP, and immune-mediated polyarthropathy), it is important that internists who manage these diseases are familiar with the causes, clinical presentations, histologic findings, and treatment options associated with opportunistic mycoses.

CLINICAL PRESENTATION AND HISTOLOGIC CLASSIFICATION

The most common clinical presentation for opportunistic fungal infection in small animal patients is dermatologic. Infection most likely occurs via direct implantation of fungi from soil or other environmental sources. Lesions may be single or multifocal, and commonly include dermal or subcutaneous nodules, ulcerated lesions, and draining tracts. Dissemination of infection to non-cutaneous sites such as lungs, bone, or abdominal organs may occur, and is likely a sequela to cutaneous infection in most cases. Primary infection of non-cutaneous sites in the absence of skin lesions is uncommon, but has been described in patients with hyalohyphomycosis, and in a dog that developed zygomycotic pneumonia caused while receiving chemotherapy.

Because most opportunistic fungal infections will present as dermatologic disease, skin lesions that develop in small animal patients receiving immunosuppressive therapy should be biopsied for histologic evaluation without delay. Unlike the endemic mycoses (which can usually be identified in cytologic or histologic samples based on their characteristic morphologic features), opportunistic fungi require culture for definitive identification. However, even without culture, these fungi can be often placed into a general category based on their morphologic features in tissue (such as pigmentation, hyphal diameter, and frequency of septation), as well as the inflammatory response that they induce. These categories include ***phaeohyphomycosis*** (pigmented hyphal or yeast forms), ***hyalohyphomycosis*** (non-pigmented hyphal forms), ***eumycotic mycetoma*** (fibrosing granuloma with black or white tissue grains consisting of aggregates of pigmented or non-pigmented fungi, respectively), and ***zygomycosis*** (wide, infrequently septate, non-pigmented hyphae with non-parallel walls associated with pyogranulomatous and eosinophilic inflammation). Details regarding the histologic and cytologic characteristics of fungi that are most likely to cause cutaneous and deep fungal infections in dogs and cats are listed in Table 1. Although identification of a specific pathogen based on culture is ideal, classification of opportunistic mycoses based on the categories listed above is usually sufficient to allow the clinician to make a reasonable prediction about clinical course and prognosis, and to make good decisions about initiating therapy.

PHAEOHYPHOMYCOSIS

The term "phaeohyphomycosis" refers to cutaneous, subcutaneous, cerebral, or disseminated infections caused by pigmented (dematiaceous) fungi that contain melanin in their cell walls.[1,2,7,9] Infection usually results from traumatic implantation. Fungal genera that have been identified as agents of phaeohyphomycosis in veterinary patients include *Alternaria*, *Bipolaris*, *Cladophialophora*, *Curvularia*, *Exophiala*, *Fonsecaea*, *Moniliella*, *Phialophora*, *Ramichloridium*, and *Scolecobasidium*, among others. The most common clinical presentations in immunocompetent small animals are cutaneous distal extremity and nasal lesions in cats. The most common presentation in immunocompromised small animal patients appears to be multifocal cutaneous lesions in dogs being treated with multi-agent immunosuppressive therapy. In general, patients are presented with cutaneous nodules or a visible nasal mass. Infected tissues may appear grossly pigmented. Histologically, fungi that cause phaeohyphomycosis appear as dark-walled, irregularly septate hyphae or as yeast-like cells, solitary or in small groups or chains. The presence of melanin in the walls of lightly-pigmented hyphae can be confirmed by the examination of unstained sections, by lowering the microscope condenser during examination, or by utilization of a Fontana-Masson stain for melanin.

The lesions associated with phaeohyphomycosis tend to be locally invasive. Dissemination is not common in immuno-competent patients, but has been observed in patients receiving immunosuppressive therapy. Because pigmented fungi are often poorly responsive to medical therapy, aggressive surgical resection is the treatment of choice for phaeohyphomycosis when a lesion is solitary. The surgeon should attempt to obtain wide margins at the time of the initial surgery. Digit amputation is often indicated for lesions involving the distal phalanx. Phaeohyphomycosis has traditionally been difficult to treat medically, in part because melanin is a virulence factor. However, in this author's experience, cutaneous phaeohyphomycosis that occurs in dogs receiving immunosuppressive therapy can be resolved in many cases with itraconazole (10 mg/kg orally q24hr) administered for at least 6 months if the immunosuppressive therapy can be tapered quickly. Discontinuation of cyclosporine in these patients appears to be important for achieving a good outcome.

HYALOHYPHOMYCOSIS

The term "hyalohyphomycosis" refers to infections caused by fungi that are non-pigmented (transparent, hyaline) in tissue. Genera that have been described as agents of hyalohyphomycosis in veterinary patients include *Fusarium, Acremonium, Paecilomyces, Pseudallescheria*, *Sagenomella, Phialosimplex*, and *Scedosporium*, among others. By convention, infection caused by *Aspergillus* and *Penicillium* species are not included in the term hyalohyphomycosis because aspergillosis and penicilliosis can usually be identified as such based on their clinicopathologic features.

In general, hyalohyphomycosis is less frequently encountered in small animal patients than is phaeohyphomycosis. Infected animals are presented with lesions ranging from local disease confined to the skin, nasal mucosa, or cornea, to disseminated disease.[3,4,5,8,10] Traditionally, the disseminated and corneal forms have been most common. Therefore, animals that present with cutaneous lesions and no overt signs of systemic disease should still be evaluated for occult lesions in the chest and abdomen. However, in this author's experience, immunosuppressed patients that develop cutaneous hyalohyphomycosis may not have lesions at other sites, whereas immunocompetent patients that develop hyalohyphomycosis most often do have disseminated disease, or at least disease that is not confined to the skin.

Treatment of hyalohyphomycosis has traditionally been unrewarding because most patients have had disseminated disease. However, hyalohyphomycosis that develops while an animal is receiving immunosuppressive therapy seems to respond better to medical therapy (assuming that immunosuppressive therapy can be quickly tapered) than would have been expected based on previous experience with hyalohyphomycosis in immunocompetent patients, especially if the infection is confined to the skin. The prognosis should still be considered guarded, and aggressive antifungal therapy should be pursued. This author most often uses itraconazole (10 mg/kg orally q24hr) administered for at least 6 months. Other options include amphotericin B lipid complex or voriconazole.

MYCETOMA

The term "mycetoma" refers to localized mycotic or actinomycotic infections of the skin, subcutaneous tissue, muscle, or bone that are characterized by the presence of colonies or aggregates of organisms that form "grains" in tissue. Actinomycotic mycetomas are caused by bacteria such as *Actinomyces* sp and *Nocardia* sp, whereas eumycotic mycetomas are caused by fungi. Lesions result from traumatic implantation of soil organisms into tissue. The grains or granules associated with eumycotic mycetomas are characteristically pigmented (for black grain mycetoma, caused by dematiaceous fungi) or hyaline (for white grain mycetoma, caused by non-dematiaceous fungi), depending on the type of fungal pathogen involved.

Black grain mycetomas are most often caused by *Curvularia* species, and typically result in the development of chronic non-healing wounds and cutaneous nodules on the extremities. Lesions often develop weeks to months after a traumatic incident in the same area. Draining tracts are often present, and black grains may be observed in the exudate. White grain mycetomas, usually caused by *Pseudallescheria boydii* or *Acremonium* species, most often occur as body wall and/or intra-abdominal granulomas that develop subsequent to contamination following surgical wound contamination or dehiscence. Interestingly, the lesions associated with white grain mycetoma may not be evident until months or even a year or more after the surgical event. Affected dogs may be presented with a draining mass on the body wall, or may develop clinical signs of peritonitis. The treatment of choice for eumycotic mycetoma is aggressive surgical excision of infected tissues, including amputation if clinically indicated. Response to medical therapy is routinely poor. Dissemination of eumycotic mycetoma beyond local tissues is rare, but local extension of disease within the abdomen may be extensive.

ZYGOMYCOSIS

Pathogenic fungi in the class Zygomycetes that are well-documented causes of fungal infection in small animals include *Conidiobolus coronatus* and *Basidiobolus ranarum*. Both species are found in soil and decaying plant matter, and *Basidiobolus* species are also commonly isolated from insects and from the feces of amphibians and reptiles. Infection likely occurs by percutaneous inoculation of spores via minor trauma or insect bites, or by inhalation of spores. Basidiobolomycosis is a rare cause of ulcerative skin lesions in dogs. Conidiobolomycosis typically causes nasopharyngeal lesions with or without involvement of tissues of the face, retropharyngeal region, and retrobulbar space. Infected animals may present with signs of chronic nasal cavity disease, ulcerative dermatitis of the nasal planum, or ulcerative lesions of the hard palate. Fungal pneumonia caused by *Conidiobolus* spp. without associated nasopharyngeal disease has been described in a single dog that was receiving chemotherapy for disseminated sarcoma.[6] Interestingly, the fungal pneumonia resolved following treatment for 9 months with itraconazole despite the fact that chemotherapy was continued during this time.

The histologic features of zygomycosis are similar to those associated with pythiosis and lagenidiosis, but hyphal diameter tends to be significantly larger for *Basidiobolus* and *Conidiobolus* than for *Pythium insidiosum*. The treatment of choice for focal cutaneous lesions caused by zygomycosis is aggressive surgical resection followed with 2-3 months of itraconazole administered at 10 mg/kg q24hr. For non-resectable lesions, therapy with either itraconazole for at least 3-6 months is indicated. Other options would include amphotericin B lipid complex or voriconazole.

REFERENCES

[1]Abramo F, et al. **J Feline Med Surg** 2002;4:157. [2]Anor S, et al. **J Vet Intern Med** 2001;15:257. [3]Erne JB, et al. **J Am Vet Med Assoc** 2007;230:537. [4]Foley JE, et al. **J Vet Intern Med** 2002;16:238. [5]Grant DC, et al. **Med Mycol** 2009;47:221. [6]Hawkins EC, et al. **J Vet Intern Med**. 2006 20(6):1479. [7]Herraez P, et al. **Vet Pathol**2001;8:456. [8]Hugnet C, et al. **J Vet Diagn Invest** 2009;21:120. [9]Rajeev S, et al. **J Vet Diagn Invest** 2008;20:379. [10]Sigler L, et al. **Med Mycol** 2010 48(2):335. [11]Singh K, et al. **Vet Pathol** 2006;43:988. [12]Swift IM, et al. **Aust Vet J** 2006;84:431. [13]Watt PR, et al. **J Am Vet Med Assoc** 1995;207:67.

Table 1. Histologic and cytologic characteristics of small animal mycoses that cause cutaneous and deep tissue infections.

Disease	Causative Agents	Histologic & Cytologic Characteristics
Phaeohyphomycosis	*Alternaria, Bipolaris, Phialophora, Cladophialophora* (*Cladosporium*), *Curvularia, Exophiala, Fonsecaea, Moniliella, Ochroconis,Ramichloridium,* others	pyogranulomatous inflammation associated with pigmented, irregularly septate hyphae or yeast-like cells that may be solitary or may cluster in small groups or chains
Hyalohyphomycosis	*Acremonium, Fusarium, Geotrichum, Geosmithia, Paecilomyces, Phialosimplex, Pseudallescheria, Sagenomella, Scedosporium, Schizophyllum,* others	pyogranulomatous inflammation associated with hyphal elements that have non-pigmented (transparent, hyaline) walls
Mycetoma (black-grain)	*Curvularia*	pyogranulomatous inflammation associated with aggregates of pigmented fungal organisms (which appear grossly as pigmented tissue grains)
Mycetoma (white-grain)	*Pseudallescheria boydii, Acremonium*	pyogranulomatous inflammation associated with aggregates of non-pigmented fungal organisms (which appear grossly as non-pigmented tissue grains)
Pythiosis	*Pythium insidiosum*	pyogranulomatous and eosinophilic inflammation associated with broad (2-7 μm), infrequently septate hyphae
Lagenidiosis	*Lagenidium* spp	pyogranulomatous and eosinophilic inflammation associated with broad (4-25 μm), infrequently septate hyphae
Zygomycosis	*Conidiobolus* spp, *Basidiobolus ranarum*	pyogranulomatous and eosinophilic inflammation associated with broad (5-20 μm), infrequently septate hyphae with thick prominent eosinophilic sleeve
Sporotrichosis	*Sporothrix schenckii*	pyogranulomatous inflammation with round, oval, or cigar-shaped yeast forms, 5-9 μm long, extracellularly or within macrophages
Rhinosporidiosis	*Rhinosoporidium seeberi*	mixed inflammatory response associated with very large (300 μm) sporangia that contain many endospores; released endospores often visible in cytologic samples
Aspergillosis	*Aspergillus terreus, A. deflectus, A. flavipes, A. fumigatus*	suppurative to granulomatous inflammation associated with multiple, non-pigmented, 3-6 μm, septate hyphae with parallel walls and 45 degree angle branching
Candidiasis	*Candida albicans,* other *Candida* spp	suppurative inflammation with numerous 2-6 μm oval yeasts, pseudohyphae (chains of oval yeast cells), and true hyphae
Blastomycosis	*Blastomyces dermatiditis*	suppurative to pyogranulomatous inflammation with large (8-15 μm), spherical, thick-walled, broad-based budding yeasts
Cryptococcosis	*Cryptococcus neoformans*	granulomatous inflammation (may be minimal) with 3-7 μm pleomorphic, narrow-based budding yeasts surrounded by a variably thick (1-30 μm) polysaccharide capsule
Histoplasmosis	*Histoplasma capsulatum*	granulomatous inflammation with intracellular, 2-4 μm, round to oval yeast cells characterized by a basophilic center and clear halo
Coccidioidomycosis	*Coccidioides immitis*	pyogranulomatous inflammation with very large (20-200 μm), round, thick-walled spherules that at maturity contain many small (2-5 μm) endospores

UPDATE ON ANTIGEN TESTING FOR DEEP MYCOSES

Lawrence J Wheat, M.D.
Indianapolis, IN

INTRODUCTION

Several tests may be helpful for diagnosis of endemic mycoses. While a positive culture maybe the gold standard for diagnosis of endemic mycoses, cultures are infrequently performed in veterinary practice. Microscopy of cytologic or histopathology specimens using staining techniques appropriate for fungal organisms has been the basis for diagnosis of most cases of histoplasmosis and blastomycosis, and serology for coccidioidomycosis. More recently antigen detection has proven useful for diagnosis of blastomycosis [20]. The highest sensitivity for diagnosis by antigen detection is achieved by testing both urine and serum. Antigen also may be detected in respiratory samples collected by bronchoscopy or tracheal aspiration, providing a diagnosis in some cases with negative results in urine and serum. Antigen concentration declines during effective treatment, assisting in monitoring therapy.

BLASTOMYCOSIS

Endemic areas for blastomycosis include the Mississippi, Ohio, and Missouri River Valleys, the Eastern Seaboard, Southern Canada, and areas adjacent to the Great Lakes. Infection is acquired by the inhalation of mould spores found in the environment, and pneumonia is the most common clinical manifestation. Disseminated disease also is common, involving the skin, bone, urinary tract, eyes, or brain most often.

Microscopy

Examination of infected tissues or fluids has been the most common basis for diagnosis. The highest yield has been achieved by testing transthoracic fine needle aspiration specimens, trans-tracheal lavage fluid, lymph nodes and skin, positive in 69 to 97% of cases [3]. The pathology is characterized by a purulent or pyogranulomatous inflammatory process, and the yeast measures 8-12μm in diameter, with a thick wall and wide based bud.

Antigen Detection

Rapid diagnosis also may be established by detection of *B. dermatitidis* antigen in body fluids. Antigen was detected in the urine of 93% and in the serum of 87% of cases of blastomycosis in dogs, table 1 [20]. The highest sensitivity for diagnosis may be achieved by testing both urine and serum. Based on these findings about 10% of cases may be falsely negative: a negative result does not exclude the diagnosis. Cases have been identified in which the initial specimen was negative but a subsequent specimen was positive, supporting follow-up testing if the diagnosis remains under consideration or the illness progresses.

Cross-reactions occur with animals with histoplasmosis. In fact, the antigen detected in histoplasmosis and blastomycosis may be immunologically identical, supporting studies of the feasibility of the use of a single assay for diagnosis of both mycoses.

The antigen concentration declines with effective therapy earlier in the serum than urine [20], as summarized in table 2. Failure of antigen in serum to decline during the first month of treatment raises concern that itraconazole blood concentrations may be subtherapeutic (< 1 μg/mL) [23]. Some veterinarians monitor antigen levels for the purpose of deciding when to stop treatment.

Serology

Using agar gel immunodiffusion (AGID), serology has not been very useful for diagnosis of blastomycosis. In an evaluation of dogs with blastomycosis, comparing antigen detection to serology, only17% of cases were positive by AGID [20]. Others have reported seropositivity by AGID in 50% [3] to 83% of cases [1,18]. Using an investigational enzyme immunoassay [20], antibodies were detected in 76% of cases, supporting a potential role for serology in diagnosis of blastomycosis. Further studies are needed to define the role of serology for diagnosis of canine blastomycosis.

HISTOPLASMOSIS

Histoplasmosis is endemic in certain parts of North and South America, where it is common in both humans and animals. All mammals are susceptible to histoplasmosis, but cases have been reported most often in dogs, cats, and horses. Histoplasmosis is acquired by inhaling microconidia or hyphal fragments, and pneumonia is the most common manifestation. The other common manifestation in animals is progressive disseminated disease, usually accompanied with pneumonia.

Microscopy

In veterinary cases, the diagnosis has usually been based on demonstrating small yeasts measuring 2-4μm in diameter with narrow based budding. In humans the sensitivity of microscopy is about 35% in acute pulmonary histoplasmosis [22] and 67% in disseminated cases [11,12], largely based on results of respiratory specimens obtained by bronchoscopy [13].

Antigen detection

Although antigen detection is commonly used for diagnosis of histoplasmosis in animals, clinical studies have not been conducted. However, performance characteristics are expected to resemble those described in humans. Antigen detection is the most sensitive method for diagnosis of acute pulmonary [22] and disseminated histoplasmosis [2,11,12] in humans, positive in 83% and 99% of cases, respectively (table 1). The highest sensitivity is achieved by testing both urine and serum [21,22]. If only one specimen type is tested, up to 15% of cases may be missed. Antigen also may be detected in respiratory secretions obtained by bronchoscopy

[13], facilitating diagnosis in some cases with negative results in urine or serum. Detection of antigen in cerebrospinal fluid is useful in the diagnosis of central nervous system histoplasmosis [24].

Specificity is 99% excluding cases of blastomycosis or coccidioidomycosis, in which cross-reactivity is common. The antigen found in histoplasmosis cross-reacts almost completely with that found in blastomycosis [2,9], and to a lesser extent with that found in coccidioidomycosis [16]. Cross-reactions may be seen in less than 10% of immunocompromised patients with invasive aspergillosis [25], but not in candidiasis, cryptococcosis [26] or bacterial infections including tuberculosis.

Antigen concentration declines during effective treatment and increases with relapse, providing a tool for monitoring therapy [23]. Antigen concentration declines more rapidly in serum than in urine, providing an earlier indication of response to treatment [6]. Antigen concentration in serum should be negative and in urine below 4 ng/mL before stopping treatment [23]. Additional testing three months after stopping treatment should be considered to establish that the infection is not recurring.

Serology

In humans, serologic tests are positive in over 90% of cases of pulmonary histoplasmosis and about 60% of disseminated histoplasmosis. Serology, however, is usually negative during the first one to two months following acute infection [5] , and may be negative in immunocompromised patients with disseminated disease. The greatest role for serology is for diagnosis of subacute or chronic cases of pulmonary histoplasmosis [14]. Serology is not considered to be a reliable method for diagnosis of histoplasmosis in animals; but the evidence for that opinion is weak. Studies are needed to establish the accuracy of serology in animals.

COCCIDIOIDOMYCOSIS

Coccidioidomycosis is endemic in the semi-arid southwestern United States, northern Mexico and Central America. Infection occurs following inhalation of arthroconidia. Although often asymptomatic, the common clinical manifestations include pneumonia or disseminated infection, most often involving the skin, bone, or central nervous system. Microscopy and serology are most often used for diagnosis.

Microscopy

Examination of respiratory specimens or tissues may show characteristic spherules, which measure 20-150 µm in diameter, providing for a rapid diagnosis. However, cytology or histopathology may be falsely negative in up to one third of cases in humans [4,17]. In a review of dogs with coccidioidomycosis, histopathology was the basis for diagnosis in 25% of cases [15].

Serology

Serology using the AGID method is the basis for diagnosis in most cases. Serology may be falsely negative in the first month or two of acute infection, and positive results may persist for several years. Among cats with coccidioidomycosis, AGID was positive in 83%, and the complement fixation test was positive in 100%, at titers of 1:32 or greater in 89% [10]. Among dogs, the diagnosis was based upon positive AGID results in 75% of cases [15]. However, in a prospective serologic survey of asymptomatic dogs from an endemic area, 9% had elevated AGID titers. AGID titers were 1:16 or above in 10% of asymptomatic dogs and one third of the symptomatic cases [19]. Assuming a 10% prevalence of coccidioidomycosis among dogs undergoing serologic testing, a sensitivity of 95% and a specificity of 91%, the positive predictive value is 54%.

Antigen detection

Detection of antigen in the urine or serum has been reported in human cases [7,8]. The sensitivity was 71% and was highest if both serum and urine were tested [8]. In a study of dogs with coccidioidomycosis, conducted by Drs. A. Prahl (Tucson, AZ) and R. Greene (Phoenix, AZ), antigen was present in the urine or blood of 18.6% of cases (Table 1). The sensitivity was highest if both serum and urine were tested. Assuming a prevalence of 10%, sensitivity of 20% and a specificity of 98%, the positive predictive value is 71%. Reasons for the lower sensitivity for antigen detection in dogs with coccidioidomycosis as compared to blastomycosis most likely relates to the higher concentration of antigen in body fluids in dogs with these mycoses: over 1000 ng/mL in many cases of blastomycosis compared to less than 10 ng/mL in cases of coccidioidomycosis. Studies of dogs with coccidioidomycosis proven by microscopy or culture are planned to fully assess the role of antigen detection in coccidioidomycosis.

Table 1. Sensitivity and specificity of antigen detection for diagnosis of endemic mycoses.

	Humans			Dogs		
Parameter	Histo [2]	Blasto [9]	Cocci [7,8]	Histo[1]	Blasto [20]	Cocci[2]
Sensitivity						
Pulmonary	83%					
Disseminated	99%					
Pulmonary or disseminated		93%	71%		93%	19%
Specificity	99%	99%	99%		98%	99%

[1]Data not available as studies have not been conducted. [2]Unpublished data from a prospective study in conducted by Dr. Annalisa Prahl (Tucson, AZ) and Dr. Russell Greene (Phoenix, AZ).

Table 2. Clearance of antigen during itraconazole treatment for blastomycosis in dogs [20]

Months therapy	Serum			Urine		
	<10 units[1]	<4 units	Negative	<10 units	<4 units	Negative
1	77%	41%	0%	12%	8%	0%
3	100%	94%	90%	71%	32%	0%
6	100%	100%	100%	85%	77%	54%

[1]For specimens above 10 units at the start of therapy, antigen levels fell below the value indicated in the second row of table in the proportion of cases noted below, after 1, 3 and 6 months of treatment.

REFERENCES

[1]Arceneaux KA, et al. **J Am Vet Med Assoc** 1998; 213:658-64. [2]Connolly PA, et al. **Clin Vaccine Immunol** 2007; 14:1587-91. [3]Crews LJ, et al. **J Am Vet Med Assoc** 2008; 232:222-7. [4]Crum NF, et al. **Medicine** (Baltimore) 2004; 83:149-75. [5]Davies SF. **Sem Resp Infect** 1986; 1:9-15. [6]Drake KW and Adam RD. **Neurology** 2009; 73:1780-6. [7]Durkin M, et al. **Clin Infect Dis** 2008; 47:e69-e73. [8]Durkin M, et al. **Clin Vaccine Immunol** 2009; 16:1453-6. [9]Durkin M, et al. **J Clin Microbiol** 2004; 42:4873-5. [10]Greene RT and Troy GC. **J Vet Intern Med** 1995; 9:86-91. [11]Hage C, et al. **Clin Infect Dis** 2010; 50:122-3. [12]Hage CA, et al. **Clin Infect Dis** 2010; 50:85-92. [13]Hage CA, et al. **Chest** 2010; 137:623-628. [14]Hage CA, et al. **Semin Respir Crit Care Med** 2008; 29:151-65. [15]Johnson LR, et al. **J Am Vet Med Assoc** 2003; 222:461-6. [16]Kuberski T, et al. **Clin Infect Diseases** 2007; 44:e50-e54. [17]Sarosi GA, et al. **Semin Respir Infect** 2001; 16:238-41. [18]Seawell BW and Scalarone GM. **Mycoses** 1990; 33:375-81. [19]Shubitz LE, et al. **J Am Vet Med Assoc** 2005; 226:1846-50. [20]Spector D, et al. **J Vet Intern Med** 2008; 22:839-43. [21]Swartzentruber S, et al. **Clin Vaccine Immunol** 2009; 16:320-2. [22]Swartzentruber S, et al. **Clin Infect Dis** 2009; 49:1878-82. [23]Wheat LJ, et al. **Clin Infect Dis** 2007; 45:807-25. [24]Wheat LJ, et al. **Clin Infect Dis** 2005; 40:844-52. [25]Wheat LJ, et al. **Clin Vaccine Immunol** 2007; 14:638-40. [26]Zhuang D, et al. **Clin Vaccine Immunol** 2008; 15:392-3.

NOVEL ANTIFUNGAL DRUG THERAPIES FOR DEEP MYCOSES

Alfred M. Legendre DVM, MS ACVIM
Knoxville TN

With advances in organ transplantation, intense chemotherapy and human immunodeficiency virus infections, secondary fungal infections have become more common. This has fostered the development of new antifungal drugs. Some of the new agents are the lipid formulations of amphotericin B and the second generation of triazoles. The first generation of triazoles such as itraconazole and fluconazole remain a mainstay of antifungal therapy.

Amphotericin B (AMB) remains the "gold standard" drug for antifungal treatment especially in life threatening diseases. The lipid formulations have reduced the renal toxicity of amphotericin B but do not appear to have improved the effectiveness of treatment. The lipid formulations increase the cost of treatment by 10 fold or more. Preexisting renal insufficiency requires the use of the lipid formulations but a dog with normal renal function can be treated with standard amphotericin B which is not expensive. A dose of 0.5 mg/kg intravenously every 48 hours is recommended for dogs and a dose of 0.25 mg/kg every 48 hours for cats. A BUN should be checked before each dose is given because the renal toxicity varies greatly between individual dogs. The patient needs to be well hydrated before dosing but over hydration does not appear to be beneficial. Giving AMB over a period of 4 to 5 hours will reduce the likelihood of renal toxicity. If renal toxicity develops, a change to a lipid formulation can be used to reduce further renal damage. If the patient is stable, a change to an azole may be made. If lipid formulations are required, the dose of AMB should probably be increased to 2 mg/kg every 48 hours but there is little information in the veterinary literature about dosing the lipid formulations.

The mainstays of treatment for systemic fungal infections are fluconazole and itraconazole. Itraconazole is considered a superior drug for blastomycosis and histoplasmosis but fluconazole can be effective. The fluconazole will require a longer course of treatment to achieve a cure and few animals are likely to respond. Fluconazole is the preferred drug for cryptococcal infections and candidiasis especially when there is infection of the urinary tract. Fluconazole penetrates the eye and the central nervous system very well and is preferred when these sites are affected. Fluconazole is also considerably cheaper. Itraconazole is the preferred drug for most fungal infections. The use of the compounded itraconazole has become popular but presents serious problems. Itraconazole is a poorly absorbed drug even when brand name itraconazole is used. There may be 10 fold differences in serum concentrations in dogs receiving the same dose of drug. The capsules need to be given with food and the animals should not be receiving anti-acid medication. The solution of itraconazole is a cyclodextrin preparation that is better absorbed on an empty stomach. A recent example of the inconsistency of the compounded itraconazole was a dog with a poorly responsive saprophytic fungal infection. On compounded itraconazole the dog had a serum concentration of < 0.3 ug/ml which increased to 13.3 ug/ml when brand name itraconazole was used. If a favorable response to treatment with compounded itraconazole does not occurs, a serum drug concentration should be measured. Mira Vista Labs in Indianapolis, IN and the Fungus Testing Lab in San Antonio, TX can measure itraconazole concentration with a bioassay.

Terbinafine is a drug in a new class of antifungal agents that can be used in cryptococcal infections that have become resistant to the azoles. This has been valuable in the treatment of cats with central nervous system infections that were becoming symptomatic. A dose of 10 mg/kg per day can rectify the clinical signs though cats with CNS cryptococcus usually require treatment for life. Terbinafine treatment can be combined with itraconazole in the treatment of refractory fungal infections. Little data exist in the veterinary literature about the advantages of this combination.

Second generation triazoles such as voriconazole, which resembles fluconazole, and posaconazole, which is similar to itraconazole, have been developed for treating invasive aspergillosis. These agents are very expensive but have shown benefit in the treatment of systemic aspergillosis due to *Aspergillus terreus*. We have had good responses with posaconazole treatment of systemic aspergillosis without toxicity. Posaconazole appears to control the infection and resolve the clinical signs but does not eliminate the infection. Relapse of disease occurs within 4 to 6 weeks after termination of posaconazole treatment. The commercial cost of posaconazole for a German Shepherd is about $70.00 a day. It is likely that there is a defect in the immune response to aspergillosis in German Sheherds that does not allow them to clear the infection even when fungal growth is inhibited.

The new antifungal drugs offer new hope in the treatment of systemic fungal infections.

S-ADENOSYLMETHIONINE IN HEPATOCYTE GROWTH, APOPTOSIS & CANCER

Shelly C. Lu, M.D.
Division Los Angeles, CA

INTRODUCTION

S-adenosylmethionine (SAMe, also known as AdoMet and SAM) was discovered more than half a century ago as a methyl donor, but in recent years its role in hepatocyte growth, death, differentiation, injury, and neoplastic transformation has received increasing notice. In the US SAMe is widely used as a health supplement, but in Europe it is used therapeutically in cholestatic and alcoholic liver diseases. SAMe synthesis occurs in all mammalian cells in a reaction catalyzed by the essential enzyme methionine adenosyltransferase (MAT). SAMe is particularly important in the liver as most forms of chronic liver injury result in reduced SAMe biosynthesis due to decreased hepatic MAT activity. Chronic hepatic SAMe deficiency occurs in MAT1A knockout mice and leads to livers that are more susceptible to injury, development of steatohepatitis, and hepatocellular carcinoma (HCC). This session reviews hepatic SAMe metabolism, regulation of MAT genes and isoenzymes, consequences of chronic hepatic SAMe deficiency, and SAMe's effect on hepatocytes growth and apoptosis. New insights regarding how SAMe exerts its hepatoprotective action are provided and its potential role as a chemopreventive agent against HCC are being proposed.

HEPATIC SAMe METABOLISM

The liver plays a central role in the homeostasis of SAMe as the major site of its synthesis and degradation (**Fig 1**).[1] Up to half of the daily intake of methionine is converted to SAMe in the liver and up to 85% of all methylation reactions take place in the liver.[1] MAT is the enzyme responsible for formation of SAMe from methionine and ATP. SAMe is the link to three key metabolic pathways, polyamine synthesis, transmethylation, and transsulfuration. In polyamine synthesis, SAMe is decarboxylated and the remaining propylamino moiety is donated to putrescine to form spermidine and methylthioadenosine (MTA) and to spermidine to form spermine and a second molecule of MTA. In transmethylation, SAMe donates its methyl group to a large variety of acceptor molecules in reactions catalyzed by methyltransferase (MTs). S-adenosylhomocysteine (SAH) is generated as a product of transmethylation and is hydrolyzed to form homocysteine (Hcy) and adenosine through a reversible reaction catalyzed by SAH hydrolase. SAH is a potent competitive inhibitor of methylation reactions and prompt removal of adenosine and Hcy is required to prevent accumulation of SAH. Hcy can be remethylated to form methionine by two enzymes: methionine synthase (MS), which requires normal levels of folate and vitamin B_{12}; and betaine homocysteine methyltransferase (BHMT), which requires betaine, a metabolite of choline. Remethylation of homocysteine via MS requires 5-methyltetrahydrofolate (5-MTHF), which is derived from 5,10-methylenetetrahydrofolate (5,10-MTHF) in a reaction catalyzed by methylenetetrahydrofolate reductase (MTHFR). 5-MTHF is then converted to tetrahydrofolate (THF) as it donates its methyl group and THF is converted to 5,10-MTHF to complete the folate cycle. Hcy can also undergo the transsulfuration pathway to form cysteine (the rate-limiting precursor for glutathione (GSH)) via a two-step enzymatic process catalyzed by cystathionine β-synthase (CBS) and cystathionase, both requiring vitamin B_6. The transsulfuration pathway is particularly active in the liver, making SAMe an important precursor of GSH. All mammalian tissues express MAT and MS, whereas BHMT is limited to the liver and kidney. In the liver, SAMe inhibits MTHFR and MS and activates CBS. Thus, when SAMe is depleted, homocysteine is channeled to remethylation to regenerate SAMe; whereas when SAMe levels are high, homocysteine is channeled to the transsulfuration pathway.

Fig. 1 SAMe Metabolism

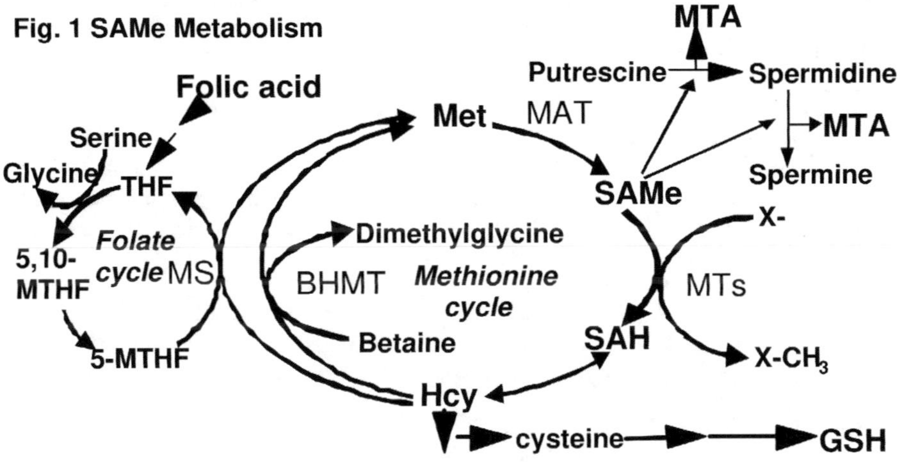

REGULATION OF MAT GENES & ISOENZYMES

In mammals, two different genes, MAT1A and MAT2A, encode for two homologous MAT catalytic subunits, α1 (forms either a dimer (MAT III), or a tetramer (MAT I)) and α2 (MAT II). A third gene, MAT2β, encodes for a regulatory subunit β that regulates MAT II. Normally differentiated liver expresses MAT1A while all extrahepatic tissues express MAT2A. MAT2A is induced in the liver during periods of rapid growth and dedifferentiation. In human HCC, MAT1A is replaced by MAT2A at the transcriptional level, which facilitates cancer cell growth.[2] MAT isoenzymes differ in kinetic parameters and regulatory properties with MAT II exhibiting lowest K_m for methionine but is susceptible to feedback inhibition by SAMe.[1] Thus, cells expressing MAT II have lower steady state SAMe levels as compared to cells expressing MAT I/III.[2] SAMe also exerts regulatory effects at the mRNA level of MAT genes; SAMe maintains MAT1A expression, whereas it inhibits the expression of MAT2A.[1]

CONSEQUENCES OF CHRONIC HEPATIC SAMe DEFICIENCY

It has long been recognized that patients with cirrhosis often have hypermethioninemia and delayed plasma clearance of methionine after intravenous injection.[3] Subsequent studies showed that this is due to a 50-60% decrease in the activity of MAT I/III,[4] which may contribute to decreased hepatic GSH levels in these patients (SAMe administration normalized GSH levels).[5] The decrease in MAT activity occurs by two mechanisms: pre-translational and post-translational. MAT1A expression is diminished in many end-stage cirrhotic patients independent of etiology.[6] MAT I/III (but not MAT II) can also be inactivated via covalent modification of a critical cysteine residue by nitric oxide and hydroxyl radical.[7] We found decreased MAT1A expression and reduced hepatic SAMe levels in patients hospitalized for alcoholic hepatitis.[8] Some of these patients have only liver fibrosis. Thus, decreased hepatic MAT1A expression and SAMe levels should not be considered as complications seen only with end-stage cirrhosis.

A model of chronic hepatic SAMe depletion is the MAT1A knockout (KO) mouse, which has increased susceptibility to steatosis and oxidative liver injury, spontaneous development of steatohepatitis and HCC.[9,10] We have identified thus far two mechanisms that contribute to the formation of HCC in this model. First is the expansion of liver progenitor cells, some of which have tumorigenic potential.[11] Using CD49f and CD133 as markers of liver stem cells we found a 4.5 to 5.5-fold increase in the proportion of CD49f+ and CD133+ cells within the CD45- (to remove hematopoietic contamination) nonparenchymal fraction as MAT1A KO mice aged from 6 to 18 months (no change in wild type controls). Clonal expansion of single CD133+CD49f+ cells revealed maintenance of bi-potency. After CD133+CD49f+ cells were injected into immune-deficient mice, 3 of 8 mice developed tumors of liver epithelial cells after 6-8 weeks. Thus, MAT1A KO mice have expansion of liver stem cells as they age. These cells express increased levels of several oncogenes and are tumorigenic in vivo. This is the first demonstration of adult liver stem cells possessing tumorigenic potential without the use of a carcinogen or manipulation of tumor suppressor or oncogene expression. It also demonstrates an important role of MAT1A deficiency in HCC pathogenesis. A second mechanism we have identified so far that contributes to HCC pathogenesis in the setting of chronic SAMe deficiency is related to genomic instability.[12] Genomic instability is known to contribute to HCC development and can occur when there is increased oxidative stress. The most common DNA lesions are apurinic/apyrimidinic (AP) sites. AP endonuclease 1 (APEX1) is a key enzyme that participates in the base excision repair of AP sites. APEX1 is induced by oxidative stress as a defense mechanism against genomic instability. In this work we asked the question whether there is genomic instability in the MAT1A KO livers and whether APEX1 is up-regulated. Indeed, we found increased genomic instability in the KO livers as early as one month of age. However, to our surprise, APEX1 was down-regulated, particularly at the protein level (by about 50%). We investigated the mechanism using primary hepatocytes from mice and humans and found that APEX1 protein is rapidly degraded during culture as MAT1A expression and SAMe levels fell. This can be prevented by SAMe treatment of the cells, which prevented the fall in SAMe level. Thus, a novel action of SAMe to prevent HCC is its ability to stabilize APEX1 protein, which is crucial in the defense against genomic instability. It also helps to explain why chronic hepatic SAMe deficiency predisposes to HCC.

SAMe'S EFFECTS ON HEPATOCYTE GROWTH & APOPTOSIS

In hepatocytes, SAMe levels are related to the differentiation status, being high in quiescent and low in proliferating hepatocytes.[1] Shortly after partial hepatectomy (PH) in rat livers, SAMe levels are dramatically reduced, coinciding with the onset of DNA synthesis and the induction of early response genes.[13] When this fall in SAMe after PH was prevented by exogenous SAMe administration, hepatocyte DNA synthesis was inhibited.[14] One of the molecular mechanisms of SAMe's growth inhibitory response is via inhibition of hepatocytes growth factor (HGF).[15] We discovered that SAMe can inhibit cell cycle progression through its ability to inhibit the HGF-mediated increase in AMP-activated protein kinase (AMPK) activity. HGF induces AMPK activation and stimulates the transport from nucleus to cytoplasm of HuR, a RNA-binding protein that increases the half-life of cell cycle genes such as cyclin A and cyclin D1; and SAMe blocked this effect. Thus, SAMe regulates cell cycle progression by blocking the ability of HGF to increase the activity of AMPK and nuclear to cytoplasmic translocation of HuR. MAT1A knockout mice have increased cell proliferation at baseline and increased cytosolic levels of HuR and cyclin D1.[15]

SAMe also modulates apoptotic response of hepatocytes but the effect depends on whether the hepatocyte is normal or malignant.[16] Interestingly, while SAMe is anti-apoptotic in normal hepatocytes, it is pro-apoptotic in hepatic cancer cells.[16,17] Using microarray, we found SAMe treatment induced Bcl-x expression. Bcl-x is alternatively spliced to produce two distinct mRNAs and proteins, Bcl-x_L and Bcl-x_S. Bcl-x_L is anti-apoptotic while Bcl-x_S is pro-apoptotic. We found that SAMe induced selectively Bcl-x_S in HepG2 cells by increasing alternative splicing. Furthermore, inhibitors of protein phosphatase 1 (PP1) blocked SAMe's ability to induce Bcl-x_S. PP1 modulates alternative splicing by dephosphorylating SR proteins. SAMe increased PP1 catalytic subunit (but not regulatory subunit) mRNA and protein levels and de-phosphorylated SR proteins. The effects of SAMe on Bcl-x_S, PP1 catalytic subunit expression and apoptosis were not seen in primary hepatocytes. Thus, SAMe and MTA can affect the cellular phosphorylation state and alternative splicing of genes, resulting in the induction of Bcl-x_S leading to apoptosis.[17]

Importantly, MTA, a key metabolite of SAMe in the polyamine pathway as well as a product of SAMe via non-enzymatic breakdown, also inhibited HGF's mitogenic activity and exerted the same differential apoptotic effects as SAMe.[1,16,17] SAMe is highly unstable with a half life of 12 hours at 37°C and pH 7.0 while MTA is stable and cell permeant.[18] At 37°C and pH 7.0, 1.3% of SAMe is converted to MTA per hour, or 31% after 24 hours.[18] In contrast to SAMe, MTA does not contribute to GSH synthesis, is not a methyl donor and inhibits methyltransferases.[19] These observations also suggest that many of the SAMe's pharmacologic effects are actually mediated via MTA, particularly if the effect is via an inhibition in methylation.

NEW INSIGHTS INTO THE HEPATOPROTECTIVE ACTIONS OF SAMe

Exogenous SAMe is well known to protect against acute and chronic liver injury in animal models.[20] In patients with liver disease, the strongest evidence is in those with acute cholestatic injury.[20] There are many ways that SAMe can exert its hepatoprotective action. Being a methyl donor, it can influence gene expression and improve membrane fluidity. SAMe was reported to inhibit the ability of lipopolysaccharide (LPS) to induce proinflammatory cytokines. SAMe is a precursor to GSH in the liver so it can enhance anti-oxidative defense capacity. We found that SAMe and MTA (at a lower dose) can both block LPS-induced tumor necrosis factor α (TNFα) expression in RAW (a murine macrophage cell line) and primary rat Kupffer cells at the transcriptional level without affecting nuclear factor κ B (NFκB) nuclear binding.[21] We examined whether the mechanism of SAMe and MTA's inhibitory effect on pro-inflammatory mediators might involve modulation of histone methylation, in particular that of histone 3 lysine 4 (H3K4), which is associated with actively transcribed genes because MTA has been reported to inhibit this. Using a chromatin immunoprecipitation (ChIP) assay, we found that LPS increased the binding of trimethylated H3K4 to the TNFα promoter and this was completely blocked by either SAMe or MTA pretreatment. Similar effects were observed with LPS-mediated induction of inducible nitric oxide synthase (iNOS). The effects of MTA in RAW cells were confirmed *in vivo* in LPS-treated mice. As described above, exogenous SAMe is unstable and converts spontaneously to MTA, which is stable and cell permeant. Treatment with SAMe doubled intracellular MTA and SAH levels. SAH also inhibited trimethylated H3K4 binding to TNFα and iNOS promoters. Taken together, the mechanism of pharmacologic SAMe's inhibitory effect on pro-inflammatory mediators is mainly mediated by MTA and SAH at the level of histone methylation.[21]

In addition to being a precursor of GSH, we recently found that SAMe can induce the expression of GSH synthetic enzymes. We found that the mRNA levels of glutamate-cysteine ligase (GCL), the rate-limiting enzyme in GSH synthesis, and GSH synthase (GS) are both markedly decreased during prolonged cholestasis induced by bile duct ligation (BDL) in mice. GCL is made up of a catalytic and a modifier subunit. Both subunits of GCL and GS are positively regulated by Nrf2 (a transcription factor) *trans*-activation of anti-oxidant response element (ARE) present in their promoters. During prolonged cholestasis, Nrf2 nuclear binding to ARE fell. SAMe and ursodeoxycholic acid (UDCA) administration can both protect against the fall in expression of GSH synthetic enzymes and GSH levels by preventing the fall in Nrf2 nuclear binding to ARE. However, when combined, they exerted additive effects and were more effective in the protection against necrosis, apoptosis, and fibrosis that occurred during BDL.[22] Importantly, ARE is found in many genes involved in anti-oxidant defense so that SAMe can enhance anti-oxidant capacity by this novel mechanism.

SAMe IN CHEMOPREVENTION & TREATMENT OF HCC

SAMe is known to chemoprevent against HCC in rodents fed different carcinogens by preventing carcinogen-induced hepatic SAMe depletion and hypomethylation.[1] In addition, SAMe is anti-apoptotic in normal hepatocytes but pro-apoptotic in hepatic cancer cells. This should make SAMe an attractive agent for both chemoprevention and treatment of HCC. We recently examined whether SAMe can be used to prevent establishment of HCC or as treatment of HCC in an orthotopic HCC model.[23] Due to the instability of SAMe, we had to deliver SAMe by continuous intravenous (IV) administration. SAMe at 150mg/kg/day IV raised hepatic SAMe levels over 10-fold after 24 hours. This regimen, started one day after injecting H4IIE cells into the parenchyma of the liver and continued for 10 days, was able to reduce tumor establishment and growth. However, if IV SAMe was started after HCC had already developed, it was ineffective in reducing tumor growth. Although plasma SAMe levels remained elevated, hepatic SAMe levels were minimally increased (30% higher). Chronic SAMe administration led to induction of hepatic methyltransferases, which prevented SAMe accumulation. Thus, SAMe administration was unable to maintain the high hepatic SAMe levels, which were likely needed to kill the liver cancer cells due to compensatory response it triggered to remove the excess SAMe. However, we uncovered that acute SAMe treatment of liver cancer cells was able to alter the expression of several genes involved in angiogenesis with the net effect of inhibiting angiogenesis. These changes were confirmed at the protein level and functionally in human umbilical vein endothelial cells. Thus, in addition to exerting a differential pro-apoptotic action on liver cancer cells, SAMe can inhibit angiogenesis and together these actions can prevent establishment of HCC. Although SAMe was ineffective as a treatment of already established HCC in this model, whether the same is true in human HCC remains to be examined because the major hepatic methyltransferase that is responsible for removal of SAMe, glycine N-methyltransferase, is often silenced in human HCC.

CONCLUSIONS

Taken together, SAMe is much more than just a methyl donor and when used therapeutically, it enhances anti-oxidant defense, inhibits the expression of inflammatory cytokine, inhibits growth and angiogenesis, and selectively kills cancer cells. While physiologic depletion leads to increased susceptibility to injury and malignant transformation, pharmacologic SAMe administration may exert effects that are actually due to MTA rather than SAMe. MTA inhibits hepatocytes growth and kills liver cancer cells at doses lower than SAMe. MTA is also more stable. However, SAMe is well established as a safe agent for chronic use in humans and MTA is not available for oral use and has not been tested in humans. Taken together, SAMe (and in the future possibly MTA) can offer additional benefits over UDCA in the treatment of chronic cholestatic liver diseases and both SAMe and MTA are great candidate chemopreventive agents for human HCC.

REFERENCES

1. Lu SC, et al. **J. Gastroenterol Hepatol** 2008;1:S73. 2. Cai J, et al. **Cancer Res**. 1998;58:1444. 3. Horowitz J, et al. **Gastroenterology** 1981;81:668. 4. Tsukamoto H, et al. **FASEB J.** 2001;15:1335. 5. Vendemiale G, et al. Scand. **J. Gastroenterol**. 1989;24:407. 6. Avila MA, et al. **J. Hepatol**. 2000;33:907. 7. Mato JM, et al. **FASEB J.** 2002;16:15. 8. Lee TD, et al. **Alcoholism: Clin. and Exp. Res.** 2004;28:173. 9. Lu SC, et al. **Proc. Nat. Acad. Sci.,** USA 2001;98:5560. 10. Martínez-Chantar ML, et al. **FASEB J** 2002;10.1096/fj.02-0078fje, published online June 7. 11. Rountree CB, et al.. **Hepatology** 2008;47:1288. 12. Tomasi ML, et al. **Gastroenterology** 2009;136:1025. 13. Huang ZZ, et al. **Am. J. Physiol.** 1998;38:G14. 14. Vázquez-Chantada M, et al. **Hepatology** 2009;49:608. 15. Martínez-Chantar ML, **Gastroenterology** 2006;131:223. 16. Ansorena E, et al. **Hepatology** 2002;35:274. 17. Yang HP, et al. **Hepatology** 2004; 40:221. 18. Chen H, et al.. **Gastroenterology** 2007;133:207. 19. Clarke SG. **The Enzymes** 2006;24:467. 20. Mato JM, et al. In: **Encyclopedia of Dietary Supplements** (1st edition; Coates PM, Blackman MR, Levine M, Moss J, White JD, [editors]). Marcel Dekker, New York, NY, 2005. http://www.dekker.com/sdek/abstract~db=enc~content=a713479736 21. Iglesias Ara A, et al. **Hepatology** 2008; 47:1655. 22. Yang HP, et al. **Hepatology** 2009;49:1982. 23. Lu SC, et al. **Hepatology** 2009;50:462.

CURRENT CONCEPTS RELATING TO THE CANINE BILIARY SYSTEM THROUGH THE EYES OF AN INTERNIST

SA Center, DVM, DACVIM, Professor, Cornell University
Ithaca, NY

This discussion is relevant to Small Animal Internists and is based on review of clinical case data, histopathology, bacterial cultures, and imaging findings on ~1,200 cases of canine liver disease, diagnosed at Cornell University over 2 decades, and also the current literature.

CANALICULAR CHOLESTASIS

The common histologic/cytologic observation denoting canalicular bile plugs has limited diagnostic and no prognostic significance. It occurs with a variety of disorders that may or may not directly involve hepatic or biliary structures (i.e., endotoxemia/sepsis, severe vacuolar hepatopathy, bile duct occlusion, and hemolytic jaundice).

BILE DUCTULE & INTRAHEPATIC BILE DUCTS

May be altered by inflammatory parenchymal or biliary pathology. Direct injury may cause duct destruction/involution (*ductopenia*). *Local or diffuse biliary obstruction* can lead to bile duct distention and serpiginous tortuosity, associated with epithelial dysplasia and circumferential peri-biliary fibrosis. *Hyperplasia of Epithelial Progenitor* "oval cells" (from terminal periductular cells) is a common response to hepatic inflammation/regeneration in diverse disorders. *Biliary epithelial hyperplasia* is not synonymous with cholangitis. Hepatic regeneration occurs by 3 mechanisms: 1) transdifferentiation of liver stem cells (LSCs) into hepatocytes or biliary epithelial cells, 2) oval cell proliferation, and 3) hepatocyte division. Because LSC and oval cells are stimulated by G-CSF extrameduallary hematopoiesis is observed with regenerative responses and oval cell hyperplasia. Hepatocytes divide at least 30-times with turnover (hepatocytes & biliary epithelium) in Zone 1 streaming toward Zone 3.

BACTERIAL INFECTION IN BILIARY DISEASE

Bacterial cultures (aerobic or anaerobic) reported in 190 canine samples (liver: n=166, bile: n=46; >80% US guided tru-cut) were "+"most commonly in patient with cholecystitis or biliary obstruction (20% [1/5] tissue, 62% [8/13] bile) vs "inflammatory" liver disease (6% [6/103] liver, 23% [7/30] bile). Cultures (aerobic, anaerobic) of liver/bile from > 425 dogs (Cornell, surgical/laparoscopic samples) were "+" in ~18%; 25% of "+" cultures were polymicrobial with primary GB or bile duct disorders and liver disease complicated by IBD predominating. Most "+"cultures involve *Enterobacteriaceae spp*. Since "+" results also may occur in health (if sub-cultured in enrichment media), results must be interpreted in light of histopathology, cytology, and patient status.

Bacterial Susceptibility-Hepatobiliary Infection

Enterobacterial isolates have a variable response to first generation amino-penicillins or cephalosporins, and >80% susceptibility to fluorinated quinolones, amikacin, or gentamicin; most enterococci are susceptible to penicillin or a fluorinated quinolone; most anaerobes to metronidazole. Thus, a combination of a potentiated penicillin, fluorinated quinolone, and metronidazole is recommended as standard therapy and has had a significant influence on patient survival (Cornell). Treatments should be modified in light of sensitivity results, patient response, or re-evaluation of bile/liver aspirate cultures. Antibiotics should be given within 1-hr before surgical interventions. GB bile and liver tissue should be cultured at surgery. Because bacteria are rarely visible on histopathology sections, cytology should be done to identify sepsis and bacterial morphology (gram staining). If infection remains suspected but unproven, FISH investigation of tissues for infectious agents is warranted.

Bile & Enteric Bacterial Translocation-Obstructive Jaundice

Because the blood supply of intrahepatic bile ducts is derived from systemic (hepatic arterial) and splanchnic (portal venous) circulations, hematogenously disseminated bacteria disperse in close proximity to the intrahepatic biliary tree. The normal hepato-biliary-enteric bacterial circulation allows translocated bacteria to unobtrusively egress in bile. A multitude of protective mechanisms thwart opportunistic infection and these are interrupted by extrahepatic bile duct occlusion (EHBDO) enabling infection and endotoxemia. Risk increases with severity and chronicity of jaundice. Impaired gut barrier function is multifactorial largely reflecting the absence of enteric bile: disruption of- 1) immunologic (e.g. reduced secretory IgA, altered intramucosal lymphocytes, mesenteric lymph nodes, & reticuloendothelial functions), 2) biological (normal intestinal flora provides colonization resistance), and 3) mechanical barrier (reduced epithelial & capillary endothelial cell integrity, loss of trophic effect on enterocytes, impaired enterocyte replication) functions.

EXTRAHEPATIC BILE DUCT OBSTRUCTION

The most common cause is pancreatitis (~50%) and primary GB disease (cholecystitis, GB mucocele [GM]). Other causes include cholelithiasis, mural/extramural ductal or GB neoplasia, trauma causing common bile duct (CBD) stenosis, and large duct choledochitis. Most dogs present for vomiting, inappetence, and jaundice; ~35% have cholecystitis and need cholecystectomy. Chronic canine EHBDO models biliary cirrhosis, causing peribiliary inflammation/fibrosis, hepatocellular/biliary epithelial necrosis, extensive architectural remodeling, portal hypertension, and acquired portosystemic shunting within 8 wks. Modeled EHBDO differs from clinical EHBDO where obstruction develops insidiously, may be incomplete, and is usually noted and treated before several weeks. EHBDO is affiliated with profound acute increases in total bilirubin, ALT, AST, ALP, and GGT that progressively rise during the first 2 wks; modest/marked increases in cholesterol also occur. Increased cholesterol can develop in all forms of cholestasis due to enhanced hepatic cholesterol synthesis and retention of bile lipids. While vitamin K_1 depletion may provoke bleeding tendencies secondary to fat malabsorption (acholic feces), this sequela is less common in dogs than cats. Of dogs with clinical EHBDO, ~18%

develop prolonged PT or APTT with APTT being slightly more sensitive. A thrombin clotting time rarely detects coagulopathy; fibrinogen is usually increased (~43%) reflecting inflammation. While coagulation tests often are unremarkable and preoperative bleeding uncommon, it is nevertheless prudent to supplement vitamin K_1 (0.5-1.5 mg/kg SC q 12 hrs x 3 doses) before invasive procedures. EHBDO, like any form of cholestasis, causes intrahepatic accumulation of potentially toxic bile acids, causing hepatocyte necrosis, apoptosis, and leading to biliary fibrosis and cirrhosis. In other forms of cholestasis, ursodeoxycholatic acid (UDCA) provides multiple beneficial effects (e.g., antiapoptotic, antifibrotic, cytoprotective, and anticholestatic). However, recent studies demonstrate detrimental effects of UDCA in EHBDO due to enhanced bile flow increasing cholangiolar biliary pressure leading to cell injury and "bile infarcts".

Biliary Decompression

EHBDO due to pancreatitis is often self-limiting (within 2 wks). When EHBDO does not resolve, temporary or permanent decompression of the biliary tree should be considered. While a high mortality rate (50%) has been reported for dogs with pancreatitis undergoing biliary diversion, results vary greatly among surgeons and hospitals. Pre-operative treatment with ticarcillin, enrofloxacin, and metronidazole, intravenous fluids, vitamin K_1, and judicious administration of fresh frozen plasma +/- abdominal lavage for bile peritonitis, has remarkably improved surgical survival. Pancreatitis complicates a decision to pursue surgical bypass, as surgery may exacerbate pancreatitis. Surgical biliary decompression may be complicated by peri- and post-operative hypotension due to endotoxemia and infection associated with the lack of enteric bile acids and hyperbilirubinemia.

Permanent Biliary-Enteric Anastomosis Decompression

Biliary decompression promptly resolves hyperbilirubinemia (10 days) with enzyme activities more gradually subsiding, unless cholangiohepatitis/cholangitis or choledochitis persist. Cholecystectomy should be done if necrotizing cholecystitis or cholecystoliths are identified, but must be carefully considered as this limits future cholecystoenterostomy. Long term survival with good life quality is possible with either cholecystojejunostomy or cholecysto-duodenostomy. The anastomotic stoma must be large enough to allow intestinal reflux to exit the GB (or ducts if choledochoenterostomy) and provide a margin of error for stoma contracture during healing (up to 50%). As in humans, stoma stricture and retrograde cholangitis are less likely when a stoma $\geq$ 2.5 cm in diameter at surgery as this reduces entrapment of enteric contents in the biliary tree (**sump syndrome**) causing secondary septic cholangitis. Mucosal apposition at the stoma is important to optimize aperture size.

Several methods for *temporary Biliary Decompression* in EHBDO have been explored. However, there is no large database guiding confident recommendations. In humans this remains a controversial topic where biliary stenting is used for short term alleviation of EHBDO but is associated with side effects that may complicate patient status (i.e., due to ascending infection, duct stenosis, stent occlusion,or the need for stent removal). Recent retrospectives conclude that preoperative biliary drainage in EHBDO has not improved case outcome, lending to a longstanding controversy. Reviews assessing the impact of preoperative endoscopic or radiologic drainage procedures describe high complication rates. Stent clogging or dislodgement predispose to cholangitis/sepsis, particularly in patients with local inflammatory disease that may invite bacterial translocation. Biofilms growing on implanted stents may augment retrograde infection.

Chronic Biliary Enteric Anastomosis

Hepatic injury cholecystoduodenostomy or -jejunostomy in adult dogs (n=6) confirmed ascending cholangitis and progressive peribiliary fibrosis (6 months) with suppurative cholangitis in 1 dog. Serum ALP increased within 1-2 wks of anastomosis chronically sustained. Radiographic contrast confirmed enteric reflux into the biliary tree. Since recurrent septic cholangitis/cholangiohepatitis can complicate permanent biliary-enteric decompression, owners should monitor for fever, inappetence, and vomiting, seeking veterinary assistance for cyclic illness and be enabled to provide chronic pulsatile antimicrobial treatment to control an ascending infection. Usually, illness is antibiotic responsive and transient. A CBC and serum biochemistry profile (liver enzymes) should be monitored quarterly and during acute illness.

DISORDERS OF THE GALLBLADDER

GB Mucocele (GM) is the most common GB disorder and is associated with breed predisposition (i.e., Shetland Sheepdogs, Miniature Schnauzers, Cocker Spaniels, Bichon Frise), median age ~10 yrs, and clinical illness of ~5 days with vomiting, inappetence, and vague abdominal pain predominating. Some dogs display cyclic inappetence, vomiting, and vague pain for months before diagnosis. Clinicopathologic indicators in symptomatic dogs include high liver enzymes (ALP [100%], GGT [86%], ALT [77%], and AST in [60%]), +/- hyperbilirubinemia, +/- hypercholesterolemia, and +/- hypertriglyceridemia. Aerobic bacteria may be isolated from bile or the GB wall. Risk factors include: hyperlipidemia/hypercholesterolemia (i.e., due to idiopathic, pancreatitis, nephrotic syndrome, endocrinopathies [typical/atypical hyperadrenocorticism, hypothyroidism] or feeding of a high fat diet), GB dysmotility, and cystic hyperplasia of the GB mucosa (may entrap mucin compromising mechanical GB cleansing). Tenacious mucin adherent to the GB mucosa corresponds to the hypoechoic "cortex" of an "kiwi" upon ultrasonography. Associated cholecystitis may reflect GB distention or reaction to retained biliary debris. A link between steroidogenic hormones and GB dysmotility is suspected. Dogs with risk factors may rapidly develop a mature GM on glucocorticoid treatment or feeding of a high fat diet or supplement. Vacuolar hepatopathy is common and requires diagnostic investigation. *Medical Treatment:* Asymptomatic dogs may benefit from UDCA (20-25 mg/kg/day, divided dose q12hr with food) and SAMe (Denosyl® 20-40 mg/kg/day, on an empty stomach, choleretic influence achieved with 40 mg/kg as GSH donor). Medical Rx requires sequential biochemical and US evaluations to monitor for GM progression (q6 wks). In addition to numerous beneficial effects of UDCA in liver disease, it may reduce lipid peroxidation and mucin secretagogue activity of GB bile (study of humans with cholelithiasis) and may improve GB motility. Improved clinical, clinico-pathologic, and US features indicate control of causal factors, but progression in any parameter recommends surgical intervention for cholecystectomy. *Surgical Treatment:* Cholecystectomy with the CBD flushed to remove obstructing debris is the treatment of choice. Broad spectrum antimicrobials should be initiated while cultures and biopsies are pending (see EHBDO). While cholecystectomy is an effective treatment, peri-operative mortality is high for symptomatic dogs with a ruptured GB complicated by sepsis (see EHBDO). *Elective Cholecystectomy* is recommended in the author's hospital for dogs with a firm US diagnosis of developing GM with GB dysmotility to avert a need for emergency surgical intervention. *Cholecystotomy* for removal of GB contents without cholecystectomy is NOT advised because GM has recurred in dogs so treated. Additionally, mural microscopic necrosis may be grossly inapparent causing post-operative GB rupture. After cholecystectomy, chronic choleretic therapy is sustained while risk factors are investigated and resolved or managed. Clinicopathologic abnormalities normalize after GB removal, except in dogs with cholangiohepatitis or unresolved endocrinopathies. Most dogs DO NOT require dietary protein restriction; feeding a protein restricted high fat diet can be detrimental if hyperlipidemia is an identified risk factor.

Dynamic US GB Study

GB dysmotility studies are conducted by measuring the GB volume before and sequentially (30, 45, 60, 90, and 120 min) after feeding a small meal (100 g) +/- 1.0 mg/kg erythromycin (motilin prokinetic). The normal minimal volume change ia 25%.

CHOLECYSTITIS

Cholecystitis is associated with vomiting, inappetence common signs, and fever indicating infection. Vague clinical signs could reflect episodic postprandial abdominal pain: panting, pacing, inability to get comfortable, and seeking a position of relief. Jaundice heralds focal transmural leakage of bile, overt bile peritonitis, or CBD obstruction. Some dogs are asymptomatic until acute clinical signs dictate emergency surgery. *Necrotizing cholecystitis* may accompany mature GM or cholelithiasis, reflecting GB ischemia secondary to luminal expansion. *Septic Cholecystitis* may develop secondary to EHBDO, IBD, pancreatitis, hyperadreno-corticism, diabetes mellitus, cholelithiasis, GM, cholecystoenterostomy, or biliary stenting. Single or polymicrobial infections with *Enterobacteriaceae* spp. +/- *Clostridium* or *Bacteroides* are most common. Bile and liver should be cultured for aerobic and anaerobic bacteria (combining cultures reduces expense) and cytology should be inspected for organisms. Broad spectrum antimicrobial treatment is advised (see EHBDO) but diminishes culture findings. Early diagnosis is based on US and clinical chemistry data. Pre-operative treatment with ticarcillin, enrofloxacin, and metronidazole, IV fluids, vitamin K_1, and judicious administration of fresh frozen plasma +/- abdominal lavage for bile peritonitis, has remarkably improved survival. *Emphysematous* cholecystitis is uncommon and reflects gas forming bacteria within the GB. This rare disorder is usually found in dogs that are predisposed to infection (e.g., diabetes, hyperadrenocorticism, cholelithiasis, IBD). In a few cases, emphysematous cholecystitis has resolved with medical intervention.

CONSEQUENCES OF CHOLESTECTOMY

GB removal has few side effects; episodic abdominal pain and diarrhea (fat malabsorption) being rare. Bile volume increases (reduced Na^+ resorption), the bile acid (BA) pool size reduces (~80% cats by 6-8 wks), and the enterohepatic BA circulation becomes continuous. A shift in BA pool composition reflects enteric bacterial dehydroxylation of trihydroxy primary to secondary BA (e.g. cholic acid → deoxycholate). CBD may appear larger.

CHOLELITHIASIS

Abdominal US increases the detection of cholelithiasis since most choleliths are clinically silent. Cholelithiasis is more common in middle-aged/older dogs and possibly small breeds; cholecystoliths predominate. Choleliths reflect lithogenic factors associated with 1) bile stasis, 2) bile supersaturation, 3) lith nidus formation (e.g. inflammation, neoplasia), and 4) lack of crystallization inhibitors (phosphatidylcholine).

Symptomatic Cholelithiasis

Signs include vomiting, anorexia, jaundice, fever, abdominal pain, and increased ALP activity. *"Biliary Colic"* pain is referred and evanescent with cholelith movement or impaction in the cystic duct, CBD, or at the sphincter of Oddi, causing intense but intermittent local and referred pain in humans. Discomfort is not exclusively postprandial, may persist for 15-30 minutes or up to several hours, and abates as a stone passes into the duodenum or back into a dilated CBD, or an impacted cystic duct stone returns to the GB lumen. Some dogs exhibit intermittent vague pain for years (similar to humans) before US diagnosis.

Cholelith Composition Reflects Etiopathogenesis

Calcium-carbonate liths are yellow and hard, reflect calcium supersaturation (infection, stagnation) associated with duct inflammation or saccular dilations (congenital malformations), and are radiodense. *Pigment Stones* (bilirubinate-calcium carbonate) are brown or black, insoluble, and reflect bilirubin deconjugation (brown, bacterial deconjugation in infection) or high bilirubin flux (hemolysis, other). Feeding a taurine and protein deficient diet to dogs rapidly provokes black calcium-bilirubinate GB choleliths. *Cholesterol-glycolipid liths* are tan and soft/gelatinous and predominate in GM. GB stasis (obstruction, hypomotility) increases GB mucin production (nucleation) and feeding a high fat diet may provoke GB cholesterol saturation, nucleation, & crystallization. Cholesterol aggregates in GB mucin-gel appears as US biliary sludge.

White Bile Syndrome

Large bile ducts or cystic duct obstruction can produce "white bile" (white=absent bilirubin pigment) when hepatic bile segregates from ductal/GB secretions. *Inspissated Bile* reflects bile stasis and dehydration and formation of inspissated dark green/black bile (calcium carbonate-bilirubinate concretions). *GB Mucocele:* reflects entrapment, retention, or local over-production of GB mucin & cholesterol. *Biliary Mucin:* Biliary hyperplasia & inflammation may increase ductular secretions (bicarbonate, mucin) with mucin histologically apparent. *Bile Acid Deconjugation:* While all biliary bile acids (BA) are conjugated, bacterial infection or low biliary pH can lead to deconjugation. Unconjugated BA are cytotoxic, alter vascular permeability, induce inflammation, and contribute to a "toxic BA syndrome", common to septic inflammation.

Treatment of Symptomatic Cholelithiasis

Requires surgical stone removal +/- cholecystectomy with medical treatment as described (cholecystectomy, EHBDO). Choleresis (UDCA, SAMe) facilitates clearance of biliary aggregates/microorganisms and may also attenuate inflammation. Whether administration of dehydrocholate (bicarbonate rich watery bile) has benefit has not been evaluated in dogs with spontaneous cholelithiasis. Liver biopsy determines propriety of immunomodulation for chronic inflammation but vitamin E (10 U/kg/day) can be safely used for its antioxidant and anti-inflammatory affects without biopsy.

CHOLERESIS

Enhanced flow of watery bile is a therapeutic goal in cholestasis except EHBDO before decompression. Bile formation is complex and dependent on multiple membrane transport systems. The influence of drugs on these systems is not well understood. Certain drugs modulate canalicular bile formation by inducing transport pumps (i.e., increased by glucocorticoids (and other steroid hormones) and structurally unrelated xenobiotics (e.g. phenobarbital, furosemide stimulates bile flow in dogs but can be detrimental if it induces dehydration). BA-dependent bile flow has a powerful direct linear influence at the canaliculus with non-micelle forming BA (e.g. dehydrocholate) bestowing the greatest effect. Ursodeoxycholic acid and dehydrocholic acid each induce choleresis and have been specifically studied in dogs. Oral UDCA (50mg/kg) increased bile flow by 70% in 1 hr and concentrations of phospholipid, cholesterol, BA, and bilirubin in canine bile. UDCA reduces biliary cholesterol and increases HCO_3^- in human bile, enhancing cholesterol but not pigment gallstones dissolution. Oral dehydrocholic acid (50 mg/kg) increased bile flow by 270% increasing biliary electrolytes & water in dogs. *Nor*UDCA, which is not yet available, has advantages compared to UDCA in the treatment of liver disease for mitigation of BA toxicity. *Nor*UDCA is relatively resistant to conjugation and undergoes cholehepatic shunting, which stimulates canalicular bile flow with direct stimulation of HCO_3^- rich hypercholeresis. This may exert a "flushing" action on injured biliary structures, removing and inactivating toxic compounds. UDCA-induced choleresis may disrupt cholangioles and aggravate liver injury in some conditions (e.g. EHBDO). *Nor*UDCA appears to induce BA metabolism and exportation via hepatic phase I and phase II detoxification systems and induction of hepatocellular BA transporters, which promotes bile acid detoxification to more water-soluble derivatives, facilitating their elimination in urine.

REFERENCES

[1.] Fickert P, et al: **Gastroenterology**,2006; 130(2):465 & **Gastroenterology** 2002;123(4)1238; [2.] Hofmann AF: **Hepatology** 2009;49(5):1403; [3.] Kline LW, et al: **Steroids**. 2008, 73(11):1174 & **Steroids** 2005;70(9):673; [4.] Martin RA, et al: **Vet Surg**, 1985;14:59; [5.] Mawdesley-Thomas et al. **Vet Rec** 1967;80(22):658. [6.] Nakagaki M, **Jap J Surg** 1982: 12(1)13; [7.] Ramstedt K, et al. **Am J Vet Res** 2008;69(5):647; [8.] Trauner M, et al: **Wien Med Wochenschr** 2008;158:542. [9.] Wagner K, et al. **J Vet Int Med**, 2007;21:417; [10.] Wang HH, et al. **Biochim Biophys Acta**. 2009; Oct 22, epub; [11.] Zollner G, et al.**Wien Med Wochenschr** 2006,156:380-385. [12.] Xu QW, et al. **Hepatology,** 1997:26(4):831.

CANINE BILIARY DISEASE: A SURGEON'S PERSPECTIVE

Philipp D. Mayhew BVM&S, MRCVS, DACVS
Davis, CA

Surgical diseases of the extra-hepatic biliary tract often present a very challenging decision making process to the small animal surgeon. The principal surgical syndromes in dogs include extra-hepatic biliary tract obstruction (EHBO), bile peritonitis, and gall bladder mucocele (GBM). These conditions can be present together or as separate syndromes and can result due to a plethora of underlying etiologies. In the past few years a significant expansion of the body of literature covering diseases of the extra-hepatic biliary tract in dogs has occurred. New developments in the pathophysiology of some of these conditions have been elucidated and several new techniques have been described that provide valuable additions to our armamentarium of procedures for management of these conditions.

GALL BLADDER MUCOCELES

GBM might currently represent the most common indication for surgical management of extra-hepatic biliary tract disease in dogs. The underlying lesion has been described as cystic mucosal hyperplasia. The cause of the condition remains largely unknown, but recent evidence has linked GBM to certain intercurrent diseases with a focus on endocrinopathies and other metabolic abnormalities. Certain genetic predispositions may also play a role as Shetland Sheepdogs were recently shown to be predisposed to gall bladder disease albeit not specifically to GBM formation.[1] Several previous studies have commented anecdotally on the relatively high incidence of endocrinopathies present in dogs with GBM.[1-3] In a study performed recently specifically to evaluate a possible association between GBM and hyperadrenocorticism (HAC) and hypothyroidism (HT) both conditions were found to have an association with GBM.[4] Dogs diagnosed with HT were 3 times more likely to have GBM than dogs without HT and the odds of having a mucocele in dogs with HAC was 29 times greater than in those without HAC. Overall, 21% of dogs with GBM had HAC, compared to 2% in the control group. In the case of HT, 14% of GBM dogs had the condition compared to 5% in the control group. However, the study was associated with some limitations and no causal relationship could be proven.[4]

Appropriate management of gall bladder mucoceles has seen significant debate and discussion in the literature recently. A recent report describes the successful medical management of two dogs with GBM that were followed ultrasonographically.[5] This interesting report opens the door for attempted medical management of dogs in which surgery is contraindicated due to co-morbidities or in cases where owners decline surgery for financial or other reasons. However, the authors of this report stated clearly that they believe GBM to be primarily a surgical disease. This belief is shared by most authors and is justified by the high levels of morbidity and mortality seen in cases that develop EHBO or bile peritonitis secondary to gall bladder rupture.[3,6]

The procedure of choice for GBM is generally accepted to be cholecystectomy, although one report includes a description of a significant percentage of dogs that underwent biliary rerouting procedures instead of cholecystectomy.[6] In the author's experience, most dogs do not have gall bladders that remain healthy enough to allow a viable cholecystoduodenostomy to be performed and progressive gall bladder wall necrosis can occur.[6] It is generally considered that the gall bladder wall is the source of mucus production and so the underlying cause appears to be removed when a cholecystectomy is performed. This view is backed by the observation that recurrence of mucocele formation has not been described post-operatively in dogs that have undergone cholecystectomy.

In dogs with GBM and concurrent evidence of partial or complete EHBO, such as ultrasonographic common bile duct distension or hyperbilirubinemia it is important to ensure that the common bile duct is free of congealed gall bladder mucus post-operatively. Concurrent EHBO has been documented in up to 30% of cases and based on laboratory data alone, functional obstruction may be present in an even higher percentage of cases.[3,6] Controversy exists as to the importance of ensuring common bile duct patency by flushing and/or catheterization of the duct. Some authors consider this very important and perform a duodenotomy for retrograde flushing in almost all cases.[6] Others consider that ongoing obstruction after cholecystectomy is rare in dogs and so flushing is not considered critical in all cases.[3] This author believes that in those dogs that have any evidence of hyperbilirubinemia pre-operatively or ultrasonographic evidence of cystic and/or common bile duct distension, ensuring bile duct patency intraoperatively is critical as we have observed patients where EHBO has persisted even after cholecystectomy in cases that were not flushed. Flushing can be performed in a normograde or retrograde manner or both.

Another controversial area in the management of GBM is how to treat incidentally discovered mucoceles or those not associated with EHBO or gall bladder rupture. Care must be taken to avoid confusing early mucocele formation with the presence of biliary sludge ultrasonographically, a common incidental finding in both dogs and cats. A careful history should also be taken in these cases as some apparently asymptomatic dogs will have mild to moderate clinical signs attributable to the presence of GBM, which can resolve after cholecystectomy.[7] No controlled studies exist to compare surgical and conservative management of incidentally discovered mucoceles and so each case must be considered on its own merit. In the author's opinion elective cholecystectomy is a reasonable procedure to perform in dogs with convincing evidence of mucocele formation in order to avoid the high levels or morbidity and mortality (22-40%) shown to occur in cases complicated by EHBO or bile peritonitis secondary to rupture, should these complications occur later on.[3,6]

Recently, the development of a laparoscopic cholecystectomy technique for dogs and its application to the treatment of gall bladder mucoceles has been described.[7] Indications for this technique include an uncomplicated gall bladder mucocele and symptomatic cholelithiasis/cholecystitis that is not associated with extra-hepatic biliary obstruction or gall bladder rupture. Until adjunctive techniques such as endoscopic common bile duct exploration or endoscopic retrograde cholangiography have been developed for use in clinical cases it is very important to use these strict case selection criteria to avoid complications, such as persistent EHBO post-operatively. This complication has not been described thus far, although few cases of laparoscopic cholecystectomy have been performed to date. Experience with minimally invasive surgery is required to perform this relatively complex laparoscopic intervention and certain pieces of specialized equipment are required to perform the procedure safely.[7]

EXTRA-HEPATIC BILIARY TRACT OBSTRUCTION

Extra-hepatic biliary tract obstruction is an uncommon diagnosis in dogs but when seen is often associated with significant systemic compromise that can present a considerable diagnostic and therapeutic challenge. The most common causes of EHBO in dogs include pancreatitis, neoplasia, gall bladder mucoceles, cholangitis, and cholelithiasis. Multisystemic organ dysfunction is often found in patient with EHBO. Multisystemic organ dysfunction in humans and experimental animal models have been attributed to the occurrence of systemic endotoxemia. It is hypothesized that the absence of bile salts in the intestinal lumen leads to bacterial overgrowth and endotoxin absorption. Impaired clearance of endotoxins due to impaired reticuloendothelial function is also thought to contribute to the development of systemic endotoxemia. Impaired myocardial contractility, hypotension, coagulopathies, gastrointestinal hemorrhage, renal dysfunction, and delayed wound healing have all been described as a result. Aggressive supportive therapy as well as early surgical intervention is advised in most patients as they rapidly become systemically compromised.

Traditional surgical options include biliary rerouting procedures such as cholecystoduodenostomy or cholecystojejunostomy. Rarely, sphincter-altering procedures may be used for very distal obstructions. Several reports have described an apparently time-sensitive development of often refractory intra-operative hypotension, especially in cats undergoing surgery for EHBO.[8,9] Some of the above mentioned endotoxin-mediated mechanisms are thought to be responsible. Also, experimental evidence demonstrating the refractory nature of this hypotension to vasopressive agents has been described. It appears therefore that every effort should be made to minimize surgical time in patients undergoing decompression of the extrahepatic biliary tract. Several recent procedures have been described that may decrease surgical time in these cases. The use of an endoscopic gastrointestinal anastomosis stapling device (EndoGIA) for creation of a cholecystoduodenostomy has recently been reported.[10] The authors fo this report suggested that surgical time might be shortened compared to traditionally hand-sutured cholecystoduodenostomy, which may in turn help minimize morbidity and mortality. Choledochal stenting has recently been described in both dogs and cats and has several potential advantages over biliary re-routing in patients with EHBO.[11,12] In addition to being a potentially more rapid technique, choledochal stenting also results in no long-term alteration in biliary tract anatomy. This may help to prevent complications associated with biliary rerouting such as dehiscence, stricture formation at the stoma, and ascending cholangitis/cholangiohepatitis. Choledochal stenting is potentially indicated in cases where functional obstruction exists, but a stent can be passed across the site of the obstruction and either removed after resolution of the pathology (e.g. pancreatitis) or to palliate lesions where a poor prognosis might be expected (e.g. neoplasia). In a report of 13 dogs treated with choledochal stents little morbidity was seen and in no cases did recurrence of EHBO occur in the post-operative period.[11] In cats 2/7 cases demonstrated re-obstruction of the biliary tract in the post-operative period and the use of stents in this species may be limited by the small size of the common bile duct.[12] It should be noted that no controlled studies exist comparing surgical times or morbidity and mortality rates for these different techniques and many other factors are likely to affect the outcome of these challenging cases. It seems reasonable to suggest, however, that the technique that achieves long-term drainage with minimal morbidity and that can be completed efficiently is likely to yield the best results.

The use of endoscopically-placed choledochal stents in humans is commonplace for a variety of conditions. This technique has now been described in a small cohort of normal dogs and in the future might hold promise for the treatment of certain types of EHBO.[13] This would allow decompression of the extrahepatic biliary tract to be achieved without the need for celiotomy. However, to date this technique has not been described in clinical cases and its use requires considerable expertise and the use of highly specialized equipment, such as a side-view duodenoscope and a sphincterotome.

Another option for the management of EHBO that does not result in permanent alteration of the biliary tract is the use of a cholecystostomy tube. Cholecystostomy tubes provide temporary diversion of bile from the gall bladder to the outside. When potentially reversible disease processes, such as pancreatitis, occur or when patients are judged to be poor candidates for prolonged anesthesia, establishing temporary biliary drainage may be of value. The use of pre-operative biliary drainage in high risk patients prior to definitive surgical intervention is used in humans, but remains controversial. In small animals, patients with EHBO that are systemically unstable, pre-operative drainage allows temporary drainage prior to definitive surgical intervention. Laparoscopic-assisted cholecystostomy tube placement has recently been shown to be technically feasible in dogs.[14] Placement appears to be more reliable with laparoscopic assistance than when the tube is placed with ultrasound guidance alone, although this maybe somewhat operator dependant.[14] The main complications for cholecystostomy tubes are premature obstruction and dislodgement. Early dislodgement with subsequent intraperitoneal bile leakage has also been reported.[14] Despite some suggestions that 5-10 days is sufficient for catheter tract maturation and leakage prevention, recent evidence suggests that maintenance of the catheter for 3-4 weeks may be more appropriate.[14]

BILE PERITONITIS

The most common underlying causes of bile peritonitis in dogs are trauma, necrotizing cholecystitis, and ruptured gall bladder mucoceles. The most reliable diagnostic test for bile peritonitis is evaluation of bilirubin concentration in a sample of the abdominal effusion and comparison to that in the serum. If the former is two or more times greater than the latter bile peritonitis can be diagnosed. This diagnosis is an immediate indication for surgical exploration as bile salts are a known cause of peritonitis and increase the local and systemic inflammatory response. It is therefore imperative to control the source of the leakage, copiously lavage the peritoneal cavity, and in most cases provide ongoing abdominal drainage. Two studies of clinical cases of bile peritonitis in dogs and cats have shown that mortality is significantly greater when septic bile is present compared to sterile bile.[2,15] It is therefore essential to culture the effusion and promptly administer appropriate antibiotics based on culture and sensitivity.

Leakage from the extra-hepatic biliary tree at a site other than the gall bladder occurs occasionally and can be very challenging to treat. Leakage from single hepatic ducts can be dealt with easily by simple ligation of both ends of the duct. Collateral circulation between liver lobes can be relied on to provide biliary drainage from the affected lobe. Defects in the common bile duct can be carefully sutured primarily with 4-0 to 6-0 monofilament absorbable suture material in a simple continuous or interrupted pattern. Frequently, necrotic tissue at the margins of the defect can make primary closure difficult. Traditionally it has been considered that morbidity is high in cases where choledochotomies or common bile duct tears have been sutured. However, a recent study has shown that post-operative dehiscence of a primary repair of common bile duct tears or choledochotomy incisions might not occur as often as might be expected.[16] In this study in 6/7 dogs that had either a choledochotomy incision primarily sutured or a common bile duct tear repaired dehiscence with subsequent leakage of bile did not occur and all animals in the study were discharged alive from the hospital. Augmentation of a common bile duct repair with a choledochal stent that spans the defect and exits through the major duodenal papilla may help with drainage and protect against stricture formation in these cases, although the benefit of this has not yet been proven in a controlled study in dogs.

REFERENCES

1. Aguirre AL, Center SA, Randolph JF, et al. Gallbladder disease in Shetland Sheepdogs: 38 cases (1995-2005) **J Am Vet Med Assoc** 231:79-88, 2007 2. Mehler SJ, Mayhew PD, Drobatz KJ, et al: Variables associated with outcome in dogs undergoing extrahepatic biliary surgery: 60 cases (1988-2002). **Vet Surg** 33:644-649, 2004 3. Pike FS, Berg J, King NW, et al. Gall bladder mucocele in dogs: 30 cases (2000-2002) **J Am Vet Med Assoc** 224:1615-1622, 2004 4. Mesich ML, Mayhew PD, Paek M et al: Gallbladder mucoceles and their association with endocrinopathies in dogs: a case control study**. J Small Anim Pract** 50:630-635, 2009 5. Walter R, Dunn ME, d'Anjou MA et al. Nonsurgical resolution of gallbladder mucocele in two dogs. **J Am Vet Med Assoc** 232:1688-1693, 2008 6. Worley DR, Hottinger HA, Lawrence HJ. Surgical management of gallbladder mucoceles in dogs: 22 cases (1999-2003). **J Am Vet Med Assoc** 225:1418-1422, 2004 7. Mayhew PD, Mehler SJ, Radhakrishnan A. Laparoscopic cholecystectomy for management of uncomplicated gall bladder mucocele in six dogs. **Vet Surg** 37:625-630, 2008 8. Mayhew PD, Holt DE, McLear RC, et al: Pathogenesis and outcome of extrahepatic biliary obstruction in cats. **J Small Anim Pract** 43:247-253, 2002 9. Buote NJ, Mitchell SL, Penninck D, et al: Cholecystoenterostomy for treatment of extrahepatic biliary tract obstruction in cats (22 cases (1994-2003). **J Am Vet Med** Assoc 228:1376-1382, 2006 10. Morrison S, Prostredny J, Roa D. Retrospective study of 28 cases of cholecystoduodenostomy performed using endoscopic gastrointestinal anastomosis stapling equipment. **J Am Anim Hosp Assoc** 44:10-18, 2008 11. Mayhew PD, Richardson RW, Mehler SJ, et al: Choledochal tube stenting for decompression of the extrahepatic portion of the biliary tract in dogs:13 cases (2002-2005). **J Am Vet Med Assoc** 228:1209-1214, 2006 12. Mayhew PD, Weisse C. Treatment of pancreatitis-associated extrahepatic biliary tract obstruction by choledochal st enting in seven cats. **J Small Anim Pract** 49:133-138, 2008 13. Berent A, Weisse C, Kochman M. Initial experience with endoscopic retrograde cholangiography and biliary stent placement in normal dogs. **Proceedings, 26th Annual ACVIM Forum**, San Antonio, TX 2008 14. Murphy SM, Rodriguez JD, McAnulty JF. Minimally invasive cholecystostomy in the dog: Evaluation of placement techniques and use in extrahepatic biliary obstruction. **Vet Surg** 36:675-683, 2007 15. Ludwig LL, McLoughlin MA, Graves TK et al: Surgical treatment of bile peritonitis in 24 dogs and 2 cats: a retrospective study (1987-1994). **Vet Surg** 33:644-649, 2004 16. Baker SG, Mayhew PD, Mehler SJ. Choledochotomy and primary repair of extrahepatic biliary duct tears in dogs and cats. **Proceedings, American College of Veterinary Surgeons Symposium**, Washington DC, October 2009

ADVANCES IN THE TREATMENT OF HYPERCALCEMIA – THE VETERINARY PERSPECTIVE

Dennis J. Chew, DVM & Patricia A. Schenck, DVM, PhD;
Columbus, OH & Lansing MI

SHOULD ALL PATIENTS WITH HYPERCALCEMIA BE TREATED?

Most patients are initially determined to have hypercalcemia based on evaluation of serum total calcium. Adverse effects of circulating calcium only occur when increased ionized calcium is the cause of the increased total serum calcium. There is considerable diagnostic discordance in the ability of serum total calcium to predict true ionized calcium status in both dogs and cats, so ionized calcium should be used as the gold standard to determine if the hypercalcemia is a damaging one or not.[1,2] The magnitude of the hypercalcemia, its rate of development, whether the serum calcium concentration is stable or increasing, and the modifying effects of other electrolyte and acid-base imbalances must be considered while developing a treatment plan. The clinical condition of the animal based on history and physical examination ultimately dictates how aggressive treatment should be. Concurrent presence of neurologic, cardiac, and/or renal dysfunction as well as calcium-containing uroliths increases the urgency for correction of hypercalcemia. Acidosis can magnify the effects of hypercalcemia at all concentrations of serum calcium by shifting more calcium to the ionized fraction. Soft tissue mineralization is potentiated when hyperphosphatemia exists at the same time as hypercalcemia.

Animals with rapid and progressive increases in serum calcium usually display serious clinical signs (vomiting, anorexia, dehydration, depression/coma, weakness, cardiac arrhythmia) that require aggressive therapy.[3] Supportive therapy is often necessary to decrease serum calcium concentration to a less toxic level while awaiting a diagnosis to be determined, for definitive treatment to permanently reduce serum calcium concentration, and for chronic management of hypercalcemia when the underlying cause cannot be removed. Treatments are designed to decrease the magnitude of hypercalcemia by increasing renal calcium excretion, inhibiting bone resorption, and decreasing calcium transport across the intestine, or some combination of these. Correction of extracellular volume contraction with subsequent dilution of circulating calcium is an initial goal of parenteral fluids, usually administered as 0.9% NaCl. Emergency rescue with sodium bicarbonate infusions can be used, rarely, to shift ionized calcium to the protein-bound fraction. Clinical signs often improve when ionized calcium decreases to some degree even when it does not return completely to normal.

Some animals with hypercalcemia have minimal to no observed clinical signs, or are normal following physical examination (e.g., primary hyperparathyroidism, idiopathic hypercalcemia of cats). Usually, the magnitude of hypercalcemia is minimal to moderate and stable in these instances, so urgent treatment to correct hypercalcemia is not warranted as there is time for further diagnostics and to observe the trend in serum calcium. If the hypercalcemia is known to be ionized, measures should be taken to restore normocalcemia even in the absence of obvious clinical signs, since clinical signs may develop and the magnitude of the hypercalcemia can become more severe. Even at stable levels of hypercalcemia, some patients will develop urolithiasis, renal mineralization, and chronic renal injury. Animals with weight loss, vomiting, constipation, or other vague GI signs along with chronic low-grade hypercalcemia often improve their clinical sings following restoration to normocalcemia. If present, polyuria and polydipsia usually greatly diminish following restoration of normocalcemia, if calcium-associated chronic renal injury has not been too great.

In studies by the authors,[1,2] about 20% of dogs and cats with chronic renal failure were found to have hypercalcemia based on serum total calcium. When ionized calcium was measured in the same animals, about 10% of dogs and 30% of cats were found to be hypercalcemic. This indicates that serum total calcium overestimates ionized hypercalcemia in dogs and underestimates ionized hypercalcemia in cats. The discrepancy between serum total and ionized calcium in dogs with chronic renal failure is due to an increase in the complexed fraction of calcium, possibly due to increased circulating organic anions capable of complexing calcium.[4] We suspect that the higher number of cats with ionized hypercalcemia during their chronic renal failure is in association with idiopathic hypercalcemia, a syndrome that occurs in cats but not in dogs. Tertiary hyperparathyroidism (elevated ionized calcium concentration and excessive PTH secretion that is not inhibited by high serum iCa concentration), aluminum-induced bone disease (elevated ionized calcium concentration and low PTH secretion), toxicity from calcium-containing intestinal phosphate binders, and excess effects of prescribed activated vitamin D compounds are considerations when ionized hypercalcemia is discovered in both dogs and cats with chronic renal failure (in addition to other differentials not associated with renal disease). Measurement of ionized calcium should be performed in all cases with chronic renal failure to determine if hypercalcemia based on serum total calcium is dangerous or not, and to determine if ionized hypercalcemia is a problem despite normocalcemia based on serum total calcium. Dogs and cats with chronic renal failure and increased ionized calcium are at increased risk for accelerated progression of their chronic kidney disease, presumably due to enhanced renal mineralization. As far as we know, those with increased total calcium and normal to low ionized calcium concentrations (high concentration of complexed or protein-bound calcium) are not at increased risk for further kidney damage, and do not need any specific therapy to lower circulating calcium.

STANDARD INITIAL TREATMENT – ACUTE STABILIZATION

The first step in treatment of severe hypercalcemia is with IV fluids to create volume expansion, dilution of calcium, and calciuresis. Fluid therapy alone will reduce the magnitude of pathological hypercalcemia somewhat, but is often not sufficient by itself to provide adequate amelioration of clinical signs or enough decline in circulating levels of calcium (hypoadrenocorticism and severe dehydration with hemoconcentration are exceptions). Usually the fluid of choice is 0.9% NaCl as it floods the renal tubules with sodium that competes for reabsorption of calcium resulting in calciuresis. Intravenous fluids increase renal perfusion, glomerular filtration rate, and calciuresis. Furosemide as a CRI or intermittent injection is a calciuretic diuretic that is often chosen next (thiazides should not be chosen as they promote renal calcium reabsorption). Furosemide CRI may be preferred since more urinary sodium and calcium as well as less urinary potassium were excreted compared to bolus intermittent injection in a small study of Greyhounds.[5] Care must be taken during furosemide therapy to avoid dehydration that can occur during intense diuresis. If the hypercalcemia is severe, salmon calcitonin can be added as treatment for the acute lowering of circulating calcium (within a few hours) due mostly to its effects that decrease osteoclastic bone resorption. The effects of calcitonin may only last a few days due to down-regulation of calcitonin receptors on the osteoclast, and some animals develop anorexia and vomiting as an adverse effect. Glucocorticosteroids can also lower circulating calcium levels by salutary effects on gut, bone, and kidney; they can also have dramatic effects to lower calcium levels in those with lymphoma, myeloma, or hematopoietic neoplasm from cytotoxicity, so their use should be avoided if possible until a definitive diagnosis has been made. Glucocorticosteroids also can exert beneficial effects in patients whose hypercalcemia is associated with hypervitaminosis-D or hypoadrenocorticism.

DIETARY CHANGE

A change in diet is not helpful for most cases with pathological hypercalcemia, since hypercalcemia most commonly develops from increased bone resorption along with some increased renal tubular reabsorption of calcium. A change to a diet low in calcium is helpful only if the mechanism responsible for the development of hypercalcemia features increased absorption of calcium across the gut, as happens in hypervitaminosis D. Renal diets are typically low in calcium and are somewhat alkalinizing, which can be helpful in those with hypervitaminosis D. All dairy products should be specifically avoided in those with hypervitaminosis D. In those with malignancy-associated hypercalcemia, gut absorption of calcium is usually already reduced due to effects of low PTH and calcitriol. Rarely some malignancies are associated with increased concentrations of calcitriol and hypercalcemia; in these instances, dietary restriction of calcium could be useful.

Attempts to reduce serum calcium concentration are sometimes successful following dietary modification in cats with idiopathic hypercalcemia; the mechanism for this beneficial effect has not been specifically studied. Mechanisms could include reduced intestinal absorption of calcium and reduced bone resorption of calcium. Intestinal absorption of calcium depends on the amount and bioavailability of calcium, dietary fiber type, the amount of fiber, and also other nutrients present in the diet. Feeding of increased dietary fiber has been reported to restore normocalcemia in some cats with IHC and calcium-oxalate urolithiasis,[6,7] but not in another study.[8] Intestinal absorption of calcium may be decreased with the feeding of high fiber diets by decreasing the gastrointestinal transit time. Most pet food manufacturers have taken this potential decrease in absorption of calcium into account, and have increased the quantity of dietary calcium in high fiber diets to enhance calcium absorption. Serum ionized calcium concentration and the degree of calciuria is increased in cats that eat acidifying and magnesium-restricted diets.[9] It is possible that in some cats hypercalcemia will resolve when fed a less acidifying diet. Renal diets are less acidifying and possibly alkalinizing so these may be chosen as therapy, but their degree of phosphorus restriction may enhance calcitriol synthesis that could blunt the salutary effects of the alkalinization.

BISPHOSPHONATES

Bisphosphonates are organic pyrophosphate analogues that inhibit bone resorption following deposition into areas of active bone turnover. Their presence in bone interferes with hydroxyapatite dissolution and exerts direct effects that reduce osteoclastic function. Induction of osteoclast apoptosis appears to be the main effect of bisphosphonates.[10] Bisphosphonates that contain a nitrogen side chain have enhanced antiresorptive activity, and are called aminobisphosphonates (pamidronate, zolendronate, ibandronate). Early generation bisphosphonates such as etidronate and clodronate are lacking in nitrogen side chains and so are less active. Intravenous administration of bisphosphonates is preferred over oral administration for treatment of severe hypercalcemia as some patients will be vomiting and cannot tolerate oral medication. Bisphosphonates are very poorly absorbed across the GI tract; usually less than 5% is absorbed and this can decrease to nearly zero with food in the stomach at the time of administration. High dose oral bisphosphonate treatment can achieve similar results as the IV route in some instances in people.[11] Administration of oral bisphosphonates can be considered for maintenance control of normocaclemia after IV doses or for initial treatment in those whose hypercalcemia is not severe. Following IV administration, there is a delay in the onset of action that lowers circulating calcium that may range from 1 to 5 days. The magnitude of effect in lowering circulating calcium concentrations is related to the initial level of hypercalcemia, the dose, route administered, and the specific bisphosphonate administered. The duration of initial beneficial effect following IV administration of bisphosphonates may last 1 to 4 weeks.

Acute renal injury/renal failure attributed to precipitation of insoluble calcium-bisphosphonate salts in the renal tubules was reported in the early days of IV bisphosphonate treatment in humans, often after multiple doses and some with pre-existing chronic kidney disease. This is very uncommon today due to efforts to ensure hydration before, during, and after bisphosphonate infusion. People with relatively severe azotemia can be safely given IV infusions of bisphosphonates when given slowly, and adequate hydration is maintained.[12] Acute intrinsic renal failure can occur in dogs given greater than or equal to a dose of 10 mg/kg, according to the package insert for pamidronate. Hypocalcemia can occur following treatment with bisphosphonates but this is not usually a clinical problem.

Intravenous pamidronate infusions at 1.0 to 2.0 mg/kg have been safely and effectively used to treat hypercalcemia in dogs and cats with malignancy, calcipotriene toxicity, granulomatous disease, and idiopathic hypercalcemia of cats. Beneficial effects were seen in these animals despite the presence of initial azotemia in over 40% of these cases.[13] It is important to deliver the pamidronate as an infusion over at least 2 hours during mild volume expansion after dehydration has been corrected. Intravenous pamidronate at 1.3 to 2.0 mg/kg was highly effective in lowering serum calcium in dogs with an experimental model of cholecalciferol intoxication,[14,15] and in dogs with clinical calcipotriene toxicity.[16] Though not specifically reported, we have used this dose range to treat dogs with primary hyperparathyroidism prior to surgical correction in an effort to reduce crisis-level hypercalcemia and to reduce post-operative "hungry-bone" syndrome.

There is limited information about the use of oral bisphosphonates in general, and none for control of hypercalcemia specifically reported in dogs or cats. A small number of cats with odontoclastic resorptive dental lesions was treated with oral alendronate at 9 mg/kg twice weekly orally for 27 weeks without development of adverse effects.[17] Once weekly oral alendronate reduced serum ionized calcium concentration in most cats with idiopathic hypercalcemia (IHC) in a pilot study conducted at The Ohio State University VMTH.[18] No side effects were documented in the 6 months of this study with an average weekly dose of 10 mg per cat. Care should be taken to ensure that tablet medication does not stick in the esophagus, as this is a known risk for erosive esophagitis in humans. Tap water PO following pilling is recommended to help lessen this possibility, as is "buttering" of the lips or paw top to encourage licking, salivation, swallowing, and increased transit of pills into the stomach. Oral bisphosphonate treatment can restore normocalcemia in some cats that fail to do so during diet and prednisolone treatment. Some cats have required up to 30 mg weekly to achieve normocalcemia. Oral bisphosphonate treatment has failed to achieve normocalcemia in a small number of cats with IHC. Several cats have been on alendronate treatment for years without known adverse effects. We have observed one cat with IHC to develop hypocalcemia and clinical signs. The salutary response during bisphosphonate treatment indicates that blunting of osteoclastic bone resorption can be effective to decrease serum ionized calcium but it does not prove that accelerated bone resorption is the underlying cause of IHC.

REFERENCES

[1]Schenck PA and Chew DJ. **Am J Vet Res** 2005;66(8):1330; [2]Schenck PA and Chew DJ. **Can J Vet Res** 2010 (in press); [3]Schenck PA et al. In:SP DiBartola, ed. **Fluid Therapy in Small Animal Practice** 2005, 3rd ed: 122; [4]Schenck PA and Chew DJ. **Am J Vet Res** 2003;64(9):1181; [5] Darcy B et al. **J Vet Intern Med** 2003;17:632–636; [6]Osborne CA. **Vet Clin North Am Small Anim Pract** 1996; 26(2):217-32; [7]McClain HM. J Am Anim Hosp Assoc 1999;35(4):297-301; [8]Midkiff AM. **J Vet Intern Med** 2000; 14(6):619-26; [9]Ching SV. **J Nutr** 1989;119(6): 902-915; [10]Body JJ et al. **Cancer Treat Rev** 1996;22:265-287; [11]Pecherstorfer M et al. **Treat Endocrinol** 2003; 2(4):273-292; [12]Machado CE et al. **Clin Nephrol** 1996; 45(3):155-9; [13]Hostutler RA et al. **J Vet Intern Med** 2005; 19:29-33; [14]Rumbeiha WK et al. **Am J Vet Res** 1998; 60:1092-1097; [15]Rumbeiha WK et al. **AVJV** 2000; 61:9-13; [16]Hare WR et al. **Vet Med** 2000; 95:770-778; [17]Harvey CE et al. 2004; [18]Hardy B unpublished observations.

IODINE TRANSPORT IN THE THYROID: ROLE IN DIAGNOSIS & TREATMENT OF NODULAR THYROID DISEASE & CANCER

Gregory A. Brent, MD
Los Angeles, CA

INTRODUCTION

Thyroid hormone is essential for normal development, especially brain and bone. It has important actions in the adult in essentially all tissues, but is especially important in metabolic regulation. Thyroid hormone is synthesized in the thyroid gland primarily as the prohormone, thyroxine (T4), and it is activated in tissues by the 5'-deiodinase enzymes to the active form, triiodothyronine (T3). Synthesis of thyroid hormone is regulated by thyroid stimulating hormone (TSH) secreted from the anterior pituitary under negative feedback regulation by T4 and T3. TSH is also regulated by Thyrotropin Releasing Hormone (TRH), secreted from the hypothalamus, which is also under negative feedback regulation from T4 and T3.

Iodine is required for the synthesis of thyroid hormone and must be derived from dietary intake. The sodium iodide symporter (NIS) is a membrane protein, expressed on the basal lateral portion of the thyroid follicular cell, and transports iodine into the thyroid (Figure 1). The iodine is transported at the apical portion of the cell into the colloid, where thyroid hormone is synthesized. The thyroid concentrates iodine 30-50-fold about the concentration in the circulation. NIS is expressed in other tissues, including salivary glands, choroid plexus, parietal cells of the stomach, and lactating breast. The primary regulator of NIS gene expression is TSH, which increases intracellular cAMP and promotes NIS gene expression as well as increased insertion into the membrane. NIS expression is increased in conditions of thyroid hormone excess, such as hyperthyroidism due to Graves' disease or in an autonomously functioning nodule, and has reduced activity in thyroid cancer. NIS is used in the diagnosis and treatment of a wide range of thyroid diseases.

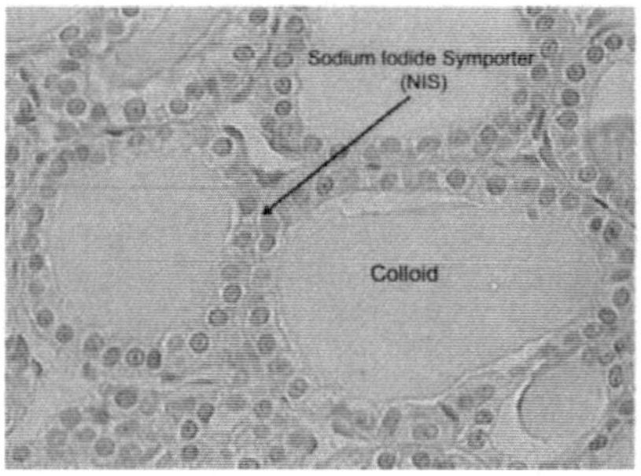

Figure 1-Thyroid Follicle
Thyroid follicle shown with layer of follicular cells and central colloid. The Sodium Iodide Symporter (NIS) is expressed at the basolateral surface (arrow) and transports iodine from the blood stream into the thyroid follicular cell. Iodine then goes through the apical portion of the cell into the colloid.

MECHANISM AND REGULATION OF IODINE TRANSPORT

NIS is a 13 transmembrane domain transporter that is a member of the sodium-dependent transporter family. Iodine is transported intracellularly, as 2 sodium ions (Na+) are pulled down the gradient from high extracellular to low intracellular concentration. The sodium potassium ATPase must be active, maintaining the sodium gradient, for iodine to be transported. Treatment with ouabain, which inhibits the sodium potassium ATPase, stops iodine transport. NIS is directly inhibited by perchlorate, which at one time was used to treat hyperthyroidism, and is an environmental contaminant, due mostly to use in manufacturing of rocket fuel. NIS is predominantly regulated by TSH, which binds to a G protein coupled receptor (GCPR), TSH Receptor (TSH-R) and increases intracellular cAMP (Table 1). There are several thyroid-specific transcription factors, PAX8 and TTF (thyroid transcription factors) 1 and 2, which also promote NIS gene expression. TSH-R can be activated by the usual ligand, TSH, but is also stimulated by TSH-R stimulating antibodies, as are found in Graves' disease. In autonomously functioning thyroid nodules, acquired mutations of the TSH-R gene confer constituitive activation. Activation of TSH-R by ligand, simulating antibody, or by constituitive activation, are all associated with an increase in NIS expression and functional activity.

Table 1-Activation of the Sodium Iodide Symporter (NIS)

Condition	Activator of TSH-Receptor	Action
Normal Regulation	TSH	Increase NIS gene expression and NIS protein to membrane
Graves' Disease	TSH-R stimulating immunoglobulin	Increase thyroid growth and thyroid hormone production
Toxic Nodule	Constituitive activating mutation of TSH-R gene	Increased thyroid growth and thyroid hormone production

NIS activity is also regulated by iodine itself, known as the *Wolff-Chaikoff Effect*. Administration of excess iodine in the iodine sufficient state reduces thyroidal iodine uptake within a few hours, likely due to down-regulation of NIS expression. This is used therapeutically as the most rapid way to reduce thyroid hormone release. There is an "escape" from this effect, however, when NIS activity and iodine uptake recovers, which usually occurs in the range of 5-7 days. In areas of iodine insufficiency, or in individuals with a multinodular goiter and regions with relative iodine insufficiency, administration of iodine can simulate thyroid hormone production and release, an effect known as *Jod-Basedow*. Iodine status, therefore, modulates the effect of iodine on the thyroid.

THYROID NODULES AND THYROID CANCER

Any thyroid enlargement is referred to as a goiter, and this can be the result of iodine deficiency, autoimmune response to thyroid autoantibodies, such as in Hashimoto's disease, or genetic susceptibility. A solitary nodule can be identified in an otherwise normal thyroid. With increasing sensitivity of imaging modalities, such as ultrasound, there are often multiple nodules found. The term multinodular goiter is usually used to describe this condition. The only medical therapy that is useful for solitary nodules or multinodular goiter is thyroxine supplementation to reduce TSH stimulation of growth. Exogenous thyroxine provides feedback to the anterior pituitary and reduces TSH. Thyroxine is primarily useful for nodules in the setting of iodine insufficiency or in the presence of thyroid autoantibodies (thyroid peroxidase, TPO), especially when the serum TSH is elevated. If the TSH is normal, there is usually no benefit of exogenous thyroxine therapy.

The majority of thyroid nodules (about 90%) are benign and cytology shows colloid and thyroid follicular cells. Ultrasound can be useful to evaluate nodules and usually shows a distinct area surrounded by a halo. Characteristics of the nodule, such as increased internal blood flow, irregularity of the borders, and coarse calcifications indicate an increased likelihood of malignancy, but none of these features are diagnostic. The definitive approach to a thyroid nodule is to perform a needle aspiration to evaluate cytology. About 10% of nodules are cancer, the primary types are papillary and follicular. Papillary thyroid cancer has very distinctive features on cytology. In contrast, follicular thyroid cancer is difficult to diagnose on the basis of cytology alone, as it can appear to be similar to benign nodules.

The distinctive features of thyroid cancer, in contrast to a benign nodule, are expansion of the nodule, spread through the thyroid capsule and extension into blood vessels. Papillary thyroid cancer typically spreads through the lymphatic system, to lymph nodes in the neck and then to the lungs. Follicular carcinoma expands outside of the thyroid capsule and then spreads hematogenously through the body. The primary treatment is radioiodine, although chemotherapeutic agents, especially inhibitors of angiogenesis factors and thymidine kinase, are useful in some circumstances.

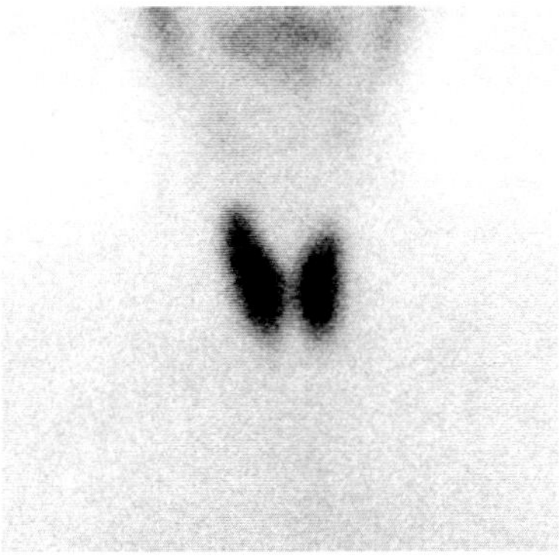

Figure 2-Normal Thyroid

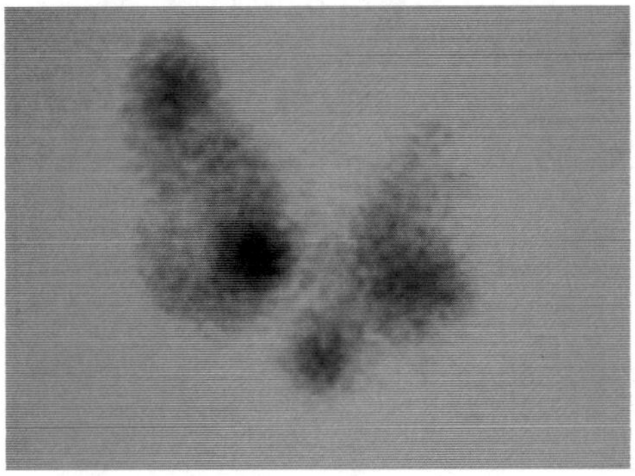

Figure 3-Multinodular Goiter

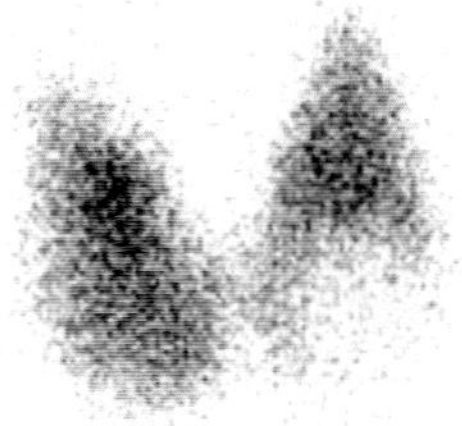

Figure 4-Hypofunctioning "Cold" Nodule

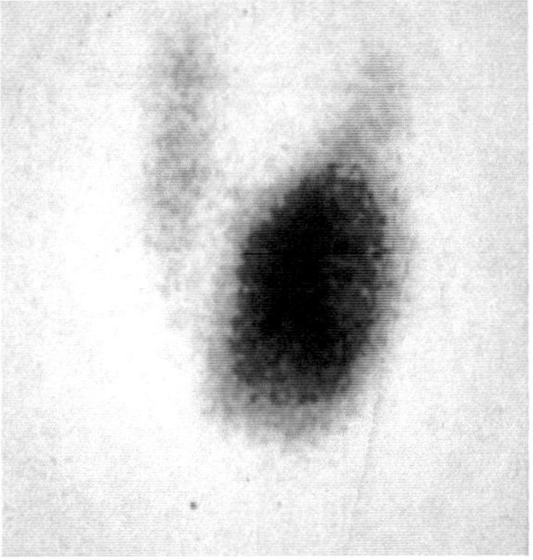

Figure 5-Hyperfunctioning "Hot" Nodule

IODINE TRANSPORT AND THE DIAGNOSIS AND TREATMENT OF THYROID DISEASE

Measurement of iodine transport and thyroid imaging is useful for the evaluation of hyperthyroidism and nodular goiter. Iodine imaging and uptake measurements are generally not useful in the evaluation of hypothyroidism, which is best assessed by measurement of thyroid function and thyroid autoantibodies.

The iodine isotope most often used for thyroid imaging is I^{123}, with a short half-life and low energy. Since the thyroid is very effective at concentrating iodine from the gastrointestinal tract, the isotope can be given orally, is absorbed and then taken up by the thyroid gland. The usual protocol is to administer the radioiodine and then return in 24 hours to measure the fraction of the administered dose taken up by the thyroid gland. The normal uptake is usually in the range of 15-30%, although if there is excessive exogenous iodine or thyroxine, the uptake is reduced. Conditions, such as Graves' disease, with high levels of NIS activity, are associated with increased uptake, typically in the 50-70% range. In addition to a measurement of uptake, an image of the thyroid gland is obtained. A normal thyroid image shows symmetric and uniform uptake in the two lobes (Figure 2). Multinodular goiter has an irregular pattern with areas of increased and decreased uptake (Figure 3). Most thyroid nodules have reduced uptake and are referred to as "cold" (Figure 4)." Most of these nodules are benign, but about 10% are thyroid cancer. There are also nodules that overproduce thyroid hormone, referred to as "hot" (Figure 5). These nodules are more intense in uptake on an imaging scan. Since excessive thyroid hormone is produced and suppresses TSH, the remainder of the gland is usually inactive or "cold."

Hyperthyroidism, either localized from a "hot" nodule or for Graves' disease, can be treated with radioiodine. For this purpose, a high energy long half-life isotope, I^{131} is used. The isotope is also given orally and then concentrated and retained in the thyroid gland. Destruction of the thyroid gland occurs over weeks to months, with maximum effect at 4-6 months.

Radioiodine can also be utilized to treat thyroid cancer spread after a thyroidectomy, although the intrinsic low activity of NIS must be overcome to augment activity and expression. The low activity of NIS in thyroid cancer can be increased after hyperstimulation with TSH. Thyroxine supplementation, required after a thyroidectomy, is discontinued. Exogenous TSH rises and when elevated for a sustained period of time, approximately 2-3 weeks, NIS can be stimulated sufficiently to use a therapeutic dose of radioiodine to image and treat metastatic tumor. Exogenous TSH (thyrogen) can also be administered, usually 2 doses given over several days. These maneuvers stimulate NIS expression sufficient for radioiodine treatment in most papillary thyroid cancer, but follicular thyroid cancer is less likely to concentrate radioiodine. A number of treatments have been used in an attempt to increase NIS expression in these cases, including retinoic acid, but the stimulation is usually modest.

REFERENCES

[1.] Bianco AC, Slavatore D, Gereben B, et al. Biochemistry, cellular and molecular biology, and physiological roles of the iodothyronine selenodeiodinases. **Endocr Rev** 2001; 23:38–39. [2.] Dohan O, De la Vieja A, Paroder V, et al. The sodium/iodide symporter (NIS): characterization, regulation, and medical significance. **Endocr Rev** 2003; 24:48-77. [3.] Kogai T, Ohashi E, Jacobs MS, Sajid-Crockett S, Kanamoto Y, Brent GA. Retinoic acid stimulation of the sodium/iodide symporter in MCF-7 breast cancer cells is mediated by the IGF1/PI3 kinase and p38 MAP kinase signaling pathways. J Clin Endocrinol Metab 2008; 93:1884-1892. [4.] Kogai T, Sajiid-Crockett S, Newmarch L, Liy Y-Y, Brent GA. Phosphoinositide-3-kinase (PI3K) inhibition induces Sodium/Iodide Symporter (NIS) expression in rat thyroid cells and human papillary thyroid cancer cells. **J Endocrinology** 2008;199:243-252. [5.] Kogai T, Taki K, Brent GA. Enhancement of sodium/iodide symporter expression in thyroid and breast cancer. **Endocr Relat Cancer**. 2006; 13:797-826. [6.] Brent GA. Graves' disease. N Engl J Med. 2008; 358:2594-2605. [7.] Kogai T, Brent GA. Regulation and enhancement of endogenous sodium iodide symporter (NIS) expression. In: Preedy VR, Burrow GN, Watson RR (eds). Comprehensive Handbook of Iodine: Nutritional, Biochemical, Pathological, and Therapeutic Aspects. San Diego: Academic Press/Elsevier Inc. 2009, 221-230.

THYROID NODULES, BUMPS, & LUMPS: WHEN IS MEDICAL THERAPY A USEFUL OPTION?

Mark E. Peterson DVM, DACVIM and Michael R. Broome DVM, DABVP
New York, NY

THYROID NODULES IN HUMAN PATIENTS

In humans, thyroid nodules are common, occurring in over half of individuals over 50 years of age (1). Over 95% of these thyroid nodules are benign, small to moderate in size, and are not associated with clinical or biochemical signs of hyperthyroidism. Some of these patients that present with a single thyroid nodule are later found to have many benign enlarged nodules within the thyroid (ie, multinodular goiter), some of which will develop signs of hyperthyroidism (toxic nodular goiter; Plumber's disease). Although the underlying cause of most human thyroid nodules is not known, a low-iodine diet, Hashimoto's thyroiditis, or prior radiation treatment can contribute to nodule development.

Historically, the most common medical treatment used in management of thyroid nodules in human patients is thyroid suppressive therapy, in which L-thyroxine is used to suppress pituitary TSH secretion and slow the growth of the thyroid nodule. In all human patients found to have a thyroid nodule, a major concern is the possibility of thyroid cancer, in which more aggressive therapy is required. The finding that thyroid replacement results in shrinkage of the nodule makes thyroid cancer less likely. Unfortunately, there is not clear evidence that treatment with L-thyroxine consistently shrinks thyroid nodules or even that shrinking small, benign nodules is necessary in most patients. Therefore, most authorities no longer recommend routine L-T4 suppressive treatment for thyroid nodules except in certain conditions (1).

THYROID NODULE/MASSES IN DOGS AND CATS

In most dogs and cats that present with a thyroid nodule or cervical mass, the pathogenesis and treatment are quite different from that of the human patient with a thyroid nodule (see Table 1). As discussed below, administration of T4 to suppress the goiter or nodule, as sometimes used in humans, is successfully used in some conditions (eg, some forms of congenital hypothyroidism). Most of the time, however, other treatments must be used, as dictated by the final diagnosis.

The most common situation for veterinarians to find a ventral cervical nodule would be in the middle to older cat (2). Some of these nodules are thyroid in origin; others are nonthyroidal in origin (eg, parathyroid or lymphatic nodules). Some are functional, some are non-functional or in the process of evolving toward overt hyperthyroidism (3). Indeed, the presence of an incidental mass in the cervical/thyroid area can sometimes be an enigma in companion animal practice. Surgical biopsy may not be justified if lesions are non-functional, which is often the case early in the clinical course when T4 determinations are in the reference interval.

Hyperthyroidism in Cats

Since hyperthyroidism was first reported 31 years ago, the prevalence of thyroidal nodules and the associated hyperthyroid state has been detected at an increasing frequency, with a prevalence now estimated to be as high as 2% of cats in general practice. Histopathology of affected thyroids usually reveals thyroid hyperplasia or benign thyroid adenoma (4); however, in a small percentage of cats (especially those with long-standing hyperthyroidism management with antithyroid drugs), thyroid adenocarcinoma is diagnosed (5). The time course of the progression of normal feline thyrocytes to hyperfunctioning adenomatous hyperplasia/adenoma and then to thyroid carcinoma is not known.

The presence of a palpable thyroid nodule is one of the cornerstones in securing a diagnosis of hyperthyroidism, but a functional hyperthyroid state must be confirmed on the basis of other characteristic clinical features, (ie, weight loss in the face of a good appetite). The diagnosis of hyperthyroidism can usually be confirmed by measurement of a single serum T4 concentration (5,6). In some cats with early hyperthyroidism or which have concurrent nonthyroidal disease, the T4 concentration may intermittently fluctuate into the normal range. If the T4 is high-normal or borderline, diagnostic options include repeating the T4 level or measuring a free T4 concentration (by equilibrium dialysis or equivalent assay).

Thyroid imaging (scanning) with pertechnetate (99m-Tc) is also a very useful diagnostic aids in the evaluation of hyperthyroidism in cats because it delineates functioning thyroid tissue (8.9). With pertechnetate thyroid scanning, a one-to-one ratio usually exists between the size and intensity of the salivary glands and the two thyroid lobes. In contrast, most cats with hyperthyroidism will have obvious enlargement of one or both thyroid lobes, together with an increased uptake of pertechnetate into the abnormal thyroid tissue, as compared to the salivary glands. Calculation of the percentage thyroidal 99m-Tc uptake can also be determined for additional diagnostic value. Another major usefulness of thyroid imaging, however, is in determining the extent of thyroid gland involvement and in detecting possible metastasis.

The treatment of choice for an individual hyperthyroid cat depends on several factors, including the age of the cat, presence of other major medical problems, availability of a skilled surgeon or I-131 facility, and owner's opinion and financial options (7). Of the 3 forms of treatment available, only surgery and radioactive iodine remove and destroy the adenomatous thyroid tissue, respectively, and thereby "cure" the hyperthyroid state. Use of an antithyroid drug (e.g., methimazole or carbimazole) will block thyroid hormone synthesis; however, since antithyroid drugs do not destroy adenomatous thyroid tissue, the nodule is not destroyed and will continue to grow in size. In some cats, benign thyroid adenoma or adenomatous hyperplasia appears to transform into thyroid carcinoma, especially after long-term treatment with antithyroid drugs (8).

Euthyroid Goiter in Cats

Experienced veterinarians can palpate ventral cervical nodules(s) in many middle-aged to older cats. It is not uncommon in 2010 to palpate a cervical nodule in a cat with no clinical or laboratory signs of hyperthyroidism (3). Possible differential diagnoses include early hyperthyroidism, thyroid cyst or cystadenoma (11,12), or non-functional thyroid adenoma or carcinoma (see Table 1). Nonfunctional thyroid carcinoma does occur but is rare in the cat. If an obvious cervical nodule is palpated in a cat with a normal T4 concentration, a fine needle aspirate should be considered to determine the tissue of origin. Thyroid imaging is again a very useful diagnostic aid in these cats and is considered by most to be the most sensitive means of diagnosis of early preclinical hyperthyroidism. In most of these cats, medical therapy is not indicated until clinical or biochemical hyperthyroidism develops.

Thyroid Cysts in Cats

Reports of cystic thyroid masses in cats are not uncommon, usually developing in cats with overt hyperthyroidism (11,12). Some of these masses may rapidly increase in size and rarely, signs related to local compression of adjacent structures (e.g., trachea) may develop in some cats. Diagnosis is based on palpation of the fluctuant mass, documentation of the cystic lesion with ultrasound, and aspirating the cystic fluid (which is typically serosanguinous). One can also determine T4 levels in the fluid to help confirm that the cystic lesion is of thyroid origin. Thyroid imaging can also be very helpful in determining if the cystic mass is of thyroid origin.

Treatment with either surgical excision or radioiodine has proven most successful in these cats. Long-term medical treatment of these cases is not recommended since the cystic lesions may increase greatly in size over time.

Congenital Hypothyroidism in Cats and Dogs

In dogs and cats, there are 2 major causes for congenital primary hypothyroidism: 1) thyroid dyshormonogenesis (defect in the biosynthesis of thyroid hormones); and 2) thyroid hypo- or aplasia (13). With thyroid dyshormonogenesis, the loss of T4's negative feedback effect on the pituitary leads to increased secretion of TSH, with resultant stimulation and hyperplasia of the thyroid gland. This can lead to an enlarged thyroid gland (goiter). In contrast, the thyroid would be small in animals with thyroid hypo- or aplasia.

Thyroid hormones are essential for normal post-natal development of the skeletal and nervous systems. Therefore, congenital hypothyroidism is characterized by disproportionate dwarfism and neurologic abnormalities. Obviously many signs observed in adult-onset hypothyroidism can also be present in affected puppies and kittens. On physical examination, hypothermia, bradycardia and palpable goiter may be present in those animals with thyroid dyshormonogenesis.

In young dogs and cats with suspected congenital hypothyroidism, thyroid scintigraphy can be a useful diagnostic aid since the thyroid image differs depending on the cause (9). With thyroid dysgenesis (hypo- or aplasia), no thyroid uptake is seen, whereas the thyroid lobes have a normal to increased size with normal to increase uptake in dogs and cats with thyroid dyshormonogenesis.

Treatment for congenital hypothyroidism consists of L-T4 supplementation; with therapy, the goiter will decrease in size as the circulating TSH concentration falls into the reference range (13). For congenital hypothyroidism, prognosis is rather guarded and will largely depend on the underlying etiology of the hypothyroidism and age at diagnosis.

Thyroiditis in Dogs

Autoimmune or lymphocytic thyroiditis is the main pathologic process causing hypothyroidism as an adult-onset condition in dogs. It is well known that at least 50% of primary hypothyroidism in dogs results from immune-mediated thyroiditis. Rarely, early in the course of the inflammatory process, palpable enlargement of the thyroid gland is possible in some of these dogs. As thyroiditis progresses, parenchyma is destroyed and replaced by fibrous connective tissue and the goiter resolves. Diagnosis of hypothyroidism and thyroiditis is based upon measurement of serum T4, cTSH, and thyroglobulin autoantibodies. Some of these dogs with thyroiditis will develop autoantibodies to T4, T3, or both (14). These autoantibodies produce spurious results when serum or plasma T4 or T3 are measured by radioimmunoassay, often resulting in an elevated apparent concentration of the thyroid hormones in affected dogs. Not uncommonly, these values will be markedly increased, causing the veterinarian to consider the possibility that the dog has hyperthyroidism due to thyroid neoplasia. The differentiation between hyperthyroidism and clinical hypothyroidism (due to thyroiditis) can usually be made on a clinical basis, as well as by measuring free T4 concentrations (which typically are low in this situation). If hypothyroidism is present in these dogs with lymphocytic thyroiditis, treatment with L-T4 supplementation is indicated.

Thyroid Tumors in Dogs

In dogs, thyroid tumors are common, representing approximately 1 to 4% of all canine neoplasms. As opposed to the relatively small, non-invasive thyroid tumors (i.e., adenomatous hyperplasia) associated with hyperthyroidism in cats, most clinically detected thyroid tumors in dogs are large, invasive carcinomas that are not hyperfunctional (i.e., do not produce hyperthyroidism). Most thyroid carcinomas in dogs arise from follicular cells, although up to a third of thyroid tumors may arise from parafollicular (C-cells), resulting in medullar C-cell carcinomas. Thyroid follicular carcinomas can be further subclassified as follicular, compact, compact-follicular (mixed), papillary, or undifferentiated, depending on the pattern of growth. Most canine thyroid tumors are mixed compact-follicular, whereas pure follicular or compact tumors or undifferentiated tumors are less common. Papillary carcinoma, a common tumor type in human patients, is very rare in dogs. Unfortunately, histological differentiation of these thyroid tumor types is generally of little help clinically when deciding on the best treatment for these carcinomas. In dogs with thyroid carcinoma, both local invasion of tumor into adjacent and distant metastasis are common. Approximately 50% of dogs with thyroid carcinoma have metastasis at time of diagnosis. During the natural course of disease, 65-90% of dogs with untreated disease will develop metastasis.

Many dogs with thyroid tumors are presented because the owner has noticed an enlargement of the neck. In >75% of dogs diagnosed in 1 survey, either the cervical swelling was the only reason for seeking veterinary care or the thyroid mass was detected by the veterinarian during an examination for another problem. Unlike the relatively small, freely movable thyroid tumors of the cat, most thyroid tumors in dogs are very large, easily palpable, and are well-embedded (fixed) into the surrounding soft tissues of the neck.

Thyroid imaging (scanning) is useful in the evaluation of dogs with thyroid masses because the procedure delineates differentiated thyroid tissue (9). 99m-pertechnetate (99m-Tc) thyroid scans do not provide direct information regarding thyroid function, but such imaging procedures do aid in demonstrating the location of abnormal thyroid tissue. Thyroid imaging can also be performed to help determine the extent of thyroid invasion or metastasis. Failure to identify distant metastatic sites with thyroid imaging, however, cannot prove that distant metastasis does not exist.

Treatment of thyroid neoplasia in dogs is dictated by the size of the primary tumor, extent of local tissue invasion, presence of detectable metastasis, presence of hyper- or hypothyroidism, and available options. Because most clinically detected thyroid tumors in the dog are malignant, treatment is rarely curative. Nevertheless, one should generally advise some form of treatment because palliative relief and increased lifespan can usually be achieved in most dogs with thyroid carcinoma. Surgery, chemotherapy, external beam radiation, and use of radioactive iodine therapy, alone or in combination, may be indicated depending on the individual dog. Medical control of the hyperthyroid state can be achieved by the daily administration of an antithyroid drug such as methimazole or carbimazole (5 to 15 mg/dog, twice, daily), but such medical treatment will not do anything to prevent tumor growth or metastasis.

TABLE 1. DIFFERENTIAL DIAGNOSIS FOR CERVICAL NODULES/MASS IN DOGS AND CATS

Category	Subcategory	Typical Clinical Syndrome	Species Affected
Neoplastic Disorders of Thyroid	Euthyroid goiter	None	Cat
	Thyroid cysts/cystadenoma	Euthyroid or hyperthyroidism	Cat
	Thyroid adenomatous hyperplasia	Hyperthyroidism	Cat
	Thyroid adenoma	Hyperthyroidism	Cat
	Thyroid malignancy/carcinoma	Usually hyperthyroidism	Cat
	Thyroid carcinoma	Usually nonfunctional	Dog
	Thyroid medullary carcinoma	Usually nonfunctional	Dog, Cat
Immune-Medicated Disorders of Thyroid	Thyroiditis (Hashimoto's)	Hypothyroidism	Dog, Cat?
Congenital Disorders of Thyroid	Congenital hypothyroidism	Hypothyroidism	Cat, Dog
	Juvenile hypothyroidism	Hypothyroidism	Cat, Dog
Endocrinopathies with Thyroid/Parathyroid Nodule	Acromegaly	Growth hormone excess; secondary thyroid enlargement	Cat, Dog
	Primary hyperparathyroidism	Hypercalcemia; parathyroid nodule palates like thyroid	Cat, Dog
Congenital Cysts in Cervical Region	Ultimobranchial cyst	None	Dogs, cats
	Thyroglossal cyst	None	Dogs, cats
	Branchial cyst	None	Dogs, cats
	Dermoid cyst	None	Dogs, cats
Nonthyroidal Neoplastic Disorders	Lipoma/Liposarcoma	None	Dogs, cats
	Lymphosarcoma	None	Dogs, cats
	Inflammatory conditions (e.g., abscess, granuloma)	None	Dogs, cats
	Salivary mucoceles	None	Dogs, cats
Non-thyroid Neoplastic Disorders	Regional soft tissue sarcomas, lymphoma, metastatic oral tumors	None	Dogs, cats

REFERENCES

[1]Gharib H, et al. **Eur J Endocrinol.** 2008;159 (5):493. [2]Paepe D et al. **J Feline Med Surg** 2008; 10 (6):558. [3]Ferguson DC, et al. **Consultations in Feline Internal Medicine**. 2006;206. [4]Gerber H, et al. **Vet Clin North Am** 1994; 24 (3): 541. [5]Hibbert A, et al. **J Feline Med Surg** 2009; 11(2):116. [6]Peterson ME, et al. **J Am Vet Med Assoc** 2001; 218: 529. [7]Mooney CT, et al. **BSAVA Manual of Canine and Feline Endocrinology** 2004; 95. [8]Broome M. **Clin Tech Small Anim Pract** 2006;21:10. [9]Taeymans O. et al. **J Vet Intern Med** 2007; 21 (4); 673. [10]Harvey AM, et al. **J Feline Med Surg** 2009; 11 (2):96. [11]Hofmeister E, et al. **J Am Vet Med Assoc** 2001; 219(2):190. [12]Phillips DE, et al. **J Am Anim Hosp Assoc** 2003; 39 (4):349. [13]Mooney CT, et al. **BSAVA Manual of Canine and Feline Endocrinology** 2004; 193. [14]Nachreiner RF, et al. **J Am Vet Med Assoc** 2002; 220:466. [15]Liptak JM. **Clin Tech Small Anim Pract** 2007; 22 (2): 75.

WHAT CONSTITUTES A DIAGNOSIS OF FELINE IDIOPATHIC CYSTITIS?

Jody Lulich, DVM, DACVIM, Carl Osborne, DVM, PhD, DACVIM John Kruger, DVM, PhD, DACVIM
Saint Paul, MN & East Lansing, MI

The lower urinary tract responds to disease with a limited number of clinical signs (i.e. hematuria, dysuria, pollakiuria, periuria and/or urinary obstruction) irrespective of the underlying cause(s). Because lower urinary tract signs may result from a variety of diverse disorders including: urinary tract infections (bacterial, parasitic, fungal), uroliths, urethral plugs, neoplasia, congenital or acquired anatomic abnormalities, trauma, or neurogenic causes, reaffirms the necessity of diagnostic procedures to identify, localize and redefine the underlying problem. However, for some young to middle aged cats, reasonable diagnostic evaluation failures to identify underlying cause(s). These cats are classified as having idiopathic lower urinary tract disease or idiopathic cystitis. Various presentations of idiopathic cystitis (acute self-limiting, frequently recurrent, chronic persistent, or obstructive) may be encountered (Table 1). Whether these presentations represent extremes in the spectrum of clinical manifestations associated with similar etiologic factors, or whether they represent entirely different mechanisms of disease is unknown.

DIAGNOSING IDIOPATHIC DISEASE

Owners of affected cats typically report the abrupt onset of various combinations of hematuria, dysuria, pollakiuria, periuria and/or urinary obstruction. Physical examination of cats with non-obstructive disease often reveals a small or minimally distended urinary bladder. The bladder wall may appear thickened and cats may vocalize or urinate following abdominal palpation. Because these findings are nonspecific, the diagnosis of idiopathic cystitis depends on excluding other known causes for lower urinary tract signs. To avoid perpetuating a potential misdiagnosis of idiopathic cystitis (table 2), we select tests such that "know diseases" are reasonable ruled out. We base our test selection on the cat's clinical presentation (table 1), the severity of disease, the number of cats in the house hold (e.g. multiple cat house holds are likely to have a behavioral component of inappropriate urination), and knowledge of previous invasive diagnostic and/or therapeutic interventions (e.g. urethral catheterization may be associated with urethral stricture months later). As of yet, there is no consistent specific diagnostic marker for feline idiopathic cystitis (table 3).

Table 1. Clinical Presentations of idiopathic lower urinary tract disease and suggested diagnostic tests to better confirm their idiopathic nature

	Non-obstructive, acute self limiting disease	Non-obstructive, frequently recurrent disease	Non-obstructive, chronic persistent disease	Obstructive disease
Relative prevalence	80 to 95%	2 to 15%	2 to 15%	15 to 25%
Probable confounding diseases necessary to rule out	Uroliths	Uroliths Inappropriate behavioral urination UTI	Uroliths Inappropriate behavioral urination UTI Anatomic abnormalities	Urethroliths Urethral Spasm (?)
Less likely confounding diseases to rule out	UTI Inappropriate behavioral urination	Foreign Body Blood Clot Neoplasia	Foreign Body Blood Clot Neoplasia	Strictures Blood clot Foreign body Spinal cord disease Neoplasia
Essential diagnostic considerations	Survey radiography Urinalysis	Survey radiography Urinalysis Urine culture Behavioral history	Survey radiography Urinalysis Urine culture Behavioral history Contrast urethrocystography	Survey radiography Urinalysis
Additional diagnostic considerations	Urine Culture Sonography	Contrast urethrocystography Cystoscopy	Cystoscopy Sonography	Urine culture Contrast urethrocystography Cystoscopy

- Urine culture should be performed prior to antibiotic administration
- Anatomic defects reported in the cat include Colourocystic fistula, ectopic uterine horn entering the urinary bladder, patent urachus, urachal diverticula, ectopic urethra, urethrorectal fistula.[1] A common acquired anatomic defect is urethral stricture formation.
- Foreign bodies recovered from the feline urinary tract have included radio-opaque (BB pellets, gun shot, etc.) and non radio-opaque (tooth picks, tip of artist paint brush, suture material, grass awns, etc.) objects.
- Spinal cord diseases altering normal feline micturition have included spinal column trauma and malformation causing spinal cord compression, and congenital and inherited myelodysplasia.

Table 2. COMMON DIAGNOSTIC MISTAKES THAT PERPETUATE A MISDIAGNOSIS OF IDIOPATHIC CYSTITIS

1. Assume That Persistent Hematuria Indicates Persistent Idiopathic Cystitis

Hematuria is a nonspecific laboratory abnormality that can result from pathologic processes affecting the lower urinary tract, upper urinary tract, genital tract and coagulation system. Hematuria can also be induced by trauma from bladder palpation, cystocentesis, or catheterization. To differentiate pathologic hematuria from iatrogenic hematuria, follow-up urinalyses should be collected by spontaneous micturition in non-absorbable litter. In addition, hematuria unaccompanied by concomitant lower urinary tract signs may originate from other locations.

2. Assume That Crystalluria Is The Cause Of Clinical Signs

Clinical evidence that microscopic crystals cause clinical signs has never been justified. Therefore, do not assume that microscopic crystals are the cause for lower urinary tract signs. Urinary crystals have been observed in urine of cats that are apparently healthy, in addition to those with: 1) uroliths, 2) urethral plugs, and 3) urinary tract disease without uroliths (e.g., feline idiopathic cystitis). Crystalluria is a marker of urine saturation and a risk factor for urolith and urethral plug formation.

3. Ignoring The Diagnostic Utility Of Survey Radiography

Uroliths are the second most common cause of lower urinary tract signs in cats. Ninety percent of stones are radio-opaque. Survey abdominal radiography is an essential test to provide a globular view of the abdomen documenting bladder size and location; urolith shape, size, and location; and spinal column abnormalities affecting coordinated micturition.

4. Utilizing Ultrasound As The Sole Method Of Diagnostic Imaging.

Although ultrasound is a very sensitive diagnostic tool to detect small uroliths, this diagnostic procedure will likely oerlook disease in the urethra.

5. Providing Antibiotics Prior To Culturing Urine

Urinalysis findings commonly overestimate and erroneously lead to a diagnosis bacterial infection of the urinary tract.[2] Urine cultured from appropriately collected and submitted samples remains the "Gold Standard" to diagnose urinary tract infections. Providing antibiotics to treat non-infectious disease with a high probability of spontaneous resolution (i.e. idiopathic cystitis) leads to the misperceptions that atypical bacteria are an underlying cause of clinical signs, urine cultures are unreliable, and future diagnostic tests during recurrent episodes of lower urinary tract signs are unwarranted (i.e. just provide more antibiotics).

Table 3. PARADIGMS IN THE DIAGNOSIS OF IDIOPATHIC CYSTITIS

1. Histopathologic Evaluation of the Urinary Bladder Wall

Exploratory celiotomy and cystotomy were at one time commonly employed to make the diagnosis of idiopathic cystitis. The constellation of mucosal erosions and ulceration, subcutaneous edema, hemorrhage, fibrosis, neovascularization, mastocytosis, mononuclear cell infiltration, and neutrophilic inflammation, were common light microscopic findings believed to be pathognomic for idiopathic cystitis. However, when bladder walls were similarly evaluated in cats with uroliths, the findings were identical.[3] Based on this information we do recognize the uselessness of bladder biopsy to make a diagnosis of idiopathic cystitis, and reserve this test to diagnosis bladder wall changes suspicious of benign or malignant neoplasia.

2. Cystoscopic evaluation of the urothelium.

At one time, submucosal petechial hemorrhages (i.e. glomerulations) identified cystoscopically were believed to be key in differentiating idiopathic cystitis from other lower urinary tract diseases. However, like histopathology, these lesions were soon determined to lack diagnostic specificity. The observation that these cystoscopic lesions persisted following clinical resolution of disease, further confounds the importance of glomerations as a disease marker.

3. Urodynamic Studies

Urodynamic studies have been rarely reported in cats with idiopathic cystitis. In one study of 12 female cats with idiopathic cystitis, bladder compliance was lower and maximal urethral closure pressure was higher when compared with a historical control group of healthy cats without idiopathic cystitis.[4] Unfortunately, this study did not compare urodynamic results to cats with other common lower urinary tract diseases. Until such information is proven useful, we do not recommend urodynamic studies to confirm a diagnosis of idiopathic cystitis.

4. Molecular Diagnostics

Limitations in standard diagnostic approaches has created intense interest in applying genomics and proteomics to better understand the etiopathogensis of disease. Genomics can provide information, linking the gene activity with disease. Proteomic technologies allow us to understand proteins and their modifications, elucidating properties of cellular behavior that may not be reflected in analysis of gene expression. These approaches are in their nascent stages. We look forward to the future when these techniques will improve our diagnosis and understanding of idiopathic cystitis.

REFERENCES

[1.] Lulich J, et.al. Urologic disorders in immature cats. **Vet Clin N Amer**. 1987;17:663. [2.] Swenson et.al. Wright-Staining of Urine Sediment: An Improved Method for Accurate Detection of Bacteriuria in Dogs **JAVMA** 2004;224:1282 [3.] Specht A , et.al. Litght microscopic features of chronic feline idiopathic cystitis. **JVIM** 2003 [4.] Westropp J, et.al. Urodynamic evaluations in cats with FIC **JVIM** 2009;23:689.

MULTIMODAL ENVIRONMENTAL ENRICHMENT THERAPY OF FIC

Dennis J. Chew, DVM and CAT Buffington, DVM, PhD
Columbus, Ohio

FIC PATHOPHYSIOLOGY OVERVIEW

Increased activation of the stress response system (SRS) appears to be an overarching feature in the pathophysiology of feline idiopathic/interstitial cystitis (FIC). In normal animals, stressors activate the outpouring of catecholamines and adrenal corticosteroids; corticosteroids then provide negative feedback to the release of ACTH and CRF and brainstem activation of catecholamines which turns off the response appropriately. Studies of cats with FIC have shown dysregulation of the SRS, in that release of catecholamines, is not constrained by outpouring of adrenal corticosteroids. Acute activation of the stress response system is appropriate, whereas chronic activation imposes additional "wear and tear" on the body.[1-3] The failure of adrenal corticosteroids to adequately restrain the stress response in FIC cats may help explain the waxing and waning nature of FIC, and how acute stressors create "flares" of LUTS and co-morbid disorders.[4]

Cats with FIC display a smaller increase in serum cortisol following ACTH stimulation compared to healthy cats, indicating less adrenocortical reserve, an effect that appears to be a consequence of small adrenal cortices[5], which appears to be hypoplasia in the absence of any lesions. The hypoplasia is likely a consequence of an in utero event following stress of the queen that suppressed ACTH-mediated development of the fetal adrenal gland at a critical time.

Unrestrained catecholamine release may contribute to neuro-inflammatory changes in various tissues, including the bladder. Excess catecholamines contribute to increased bladder permeability by effects on the urothelium allowing constituents of urine to gain access to the bladder wall with resulting activation of nociceptive afferent neurons; loss of contiguous epithelial cells also contributes to increased permeability. For example, a recent study found that mild to moderate stress for 8 days (12 hour fast followed by blood sampling, diet change, cage and housing location change) increased serum fluorescein (a biomarker for bladder permeability) concentration to a higher degree in cats with FIC than in normal cats. Fluorescein levels decreased following environmental enrichment for 3 weeks (move to a larger cage with a covered bed, 2 types of toys, larger litter pan, provision of canned in addition to dry food, increased interaction with humans, and music in the room) in both groups, but the levels remained higher in the cats with FIC. Catecholamine concentrations displayed the same pattern, suggesting that the increased catecholamines were related in some way to the increased bladder permeability to fluorescein.[6] These results provided the basis to study the effects of environmental enrichment on clinical signs in clinical cats with FIC.[7]

WHAT IS MULTIMODAL ENVIRONMENTAL MODIFICATION (MEMO) ?

Environmental enrichment refers to housing conditions that facilitate enhanced sensory, cognitive and motor stimulation.[8] We define MEMO as the implementation of changes in the cat's environment to attempt to reduce the likelihood of activating the SRS. It is not currently possible to identify the importance of any single factor implemented, so MEMO attempts to exploit as many enrichment opportunities into the environment as possible (and appropriate) in a step-wise manner. Plans for MEMO are developed after analysis of results of a detailed environmental questionnaire (available here: http://vet.osu.edu/2106.htm). Identification of stressors in general and those that act as specific triggers for a flare of clinical signs of recurrent FIC help identify specific MEMO opportunities.

A tailored treatment plan for each cat is created, since individual cats and environments vary widely. Some cats are extremely sensitive to what many owners and primary care veterinarians would consider trivial stressors, while other cats are able to adapt to severe stress without manifesting clinical signs. Decreased activity levels and increased hiding behaviors have been observed in cats living in impoverished (barren environments lacking in appropriate novelty and complexity) or unpredictable environments – we interpret this as part of the stress response. Some environments are chaotic, which also imposes an element of stress into a sensitive cat's life. Finding a balance between a barren and chaotic environment is a key MEMO objective. It is important to remember that cats are not pack animals like dogs and people; cats are not asocial (as are reptiles), but they have evolved as relatively independent creatures in hunting and general living arrangements.

The goal is to decrease the severity of clinical signs during an episode and to increase the interval between recurrent episodes. MEMO is a package of recommendations designed to reduce environmental or social stressors. These include general recommendations for all cats with FIC and then specific recommendations for some cats. It is important to remind owners that cats respond better to praise and treats than punishment. The objectives of MEMO are to empower the owner to understand how their efforts can contribute to reduction in LUT and other illness signs, provide guidelines the owner can use to manage the cat's environment, and reduce the cat's perception of threat from multiple potential environmental sources. These are achieved using effective, empathic communications skills and regular follow-up during the change process. Providing owners resources and recommendations is essential for the success of any MEMO program. Information for the owner can be provided verbally during the in-clinic consultation, but reading materials are recommended to provide reinforcement of the principals of good animal husbandry. We routinely recommend that owners read "From the Cat's Point of View" by Gwen Bohnenkamp and to visit the Indoor Cat Initiative website (www.indoorcat.org) as resources to consult during the change process. Also essential is a system of follow-up consultation and encouragement that is often best executed by animal technicians with special interest and training in this area (under the supervision of a veterinarian). After the initial in-clinic consultation has established a diagnosis of FIC, we recommend telephone follow-up approximately one week and one month later to determine how MEMO implementation is proceeding, and to suggest alternative or additional implementations. Monthly telephone rechecks are recommended for owners of particularly difficult FIC cases. Owners need to be made aware that a few months may be needed to implement the number of changes that may be necessary for their particular cat. It is important to not overwhelm the owner or cat with too many changes at one time. Usually one or at most two changes are attempted at the same time. It is our job to help the owner select which changes they feel the most comfortable making initially, and to adjust the plan as changes are found to work or not for this particular owner-cat pair. Appointments with the veterinarian or facilitating technician can be scheduled without the cat to facilitate goal planning and to decrease stress to the cat from an in-clinic visit.

Forty-six cats with recurrent FIC treated by MEMO were studied for 10 months. These cats went from a median of weekly LUTS to a median of 0 at the end of the study. No signs were referable to the lower urinary tract in 70 to 75% of these cats for the 10 months of this study ($p < 0.001$); improvement is some other organ systems was also noted. The MEMO recommendations most commonly followed were to read the prescribed client educational materials, increase the time spent with the cat, change to a canned diet, and add another litter box.[7] The multi-system benefits encountered following implementation of MEMO were consistent with a reduced output of the SRS in these cats. Mechanisms for the benefits in this study were not determined, but psychosocial stress is known to increase epithelial permeability[7] by decreasing the integrity of tight junctions between the cells; FIC cats have been shown to have increased uroepithelial permeability.[9]

WHEN and HOW SHOULD MEMO BE INSTITUTED ?

MEMO should be discussed and the first steps taken after the first episode of FIC. A cat that has been properly diagnosed with FIC will have the propensity to develop recurrent LUTS for all of its life depending on the magnitude of exposure to external stressors in the face of an altered SRS. We prefer to prevent future episodes of FIC with MEMO than to treat active flares. Treatment of the first episode or a flare during a recurrent episode includes provision of an analgesic and a tranquilizer/urethral antispasomodic for 5 to 7 days as our standard of care, though it is not known if this impacts the development of future episodes of FIC. We prescribe oral buprenorphine (injectable form aspirated into a syringe and given PO) at 10 to 20 micrograms per kg BID to QID depending on the severity of the clinical signs; we also provide oral acepromazine at 2.5 mg BID to TID. MEMO should be attempted BEFORE other drug therapy is prescribed since MEMO often works in cases, as described above, without the use of any drugs.

Change to a canned food diet is usually recommended as part of MEMO if this is not too stressful for the cat or the owners. Salutary effects of feeding canned foods may be due to elaboration of more dilute urine due to increased moisture – this could prove less noxious when gaining access to the highly permeable bladder wall of the FIC cat. Alternatively, hedonics or the mouth feel of moist food may change the neurobiology of the cat. A third mechanism that could explain this benefit is the ritual of feeding canned foods is substantially different than that of feeding dry foods and could favorably alter the owner's interaction with the cat. Adding additional water or broth to food from cans or pouches can provide additional water intake for some cats. Failure to ingest enough water therapeutically would be documented by a consistently very high urinary specific gravity (above 1.030 but often well above 1.050). MEMO is still often effective in our practice in cats in which diet change was not able to be implemented; diet change as the only environmental enrichment failed to prevent recurrence of clinical signs of FIC in 65% of the cases of a recent study.[10]

Litter box hygiene and management may be sub-optimal for some sensitive cats, a situation that provides a source of stress for some cats with FIC. We believe this to be the most frequently neglected concern for cats in general and in specific for those with FIC or other litter box issues. Optimal litter box management is often not discussed adequately with cat owners before problems with FIC develop. When FIC has been diagnosed, a detailed discussion about litterbox management is absolutely essential. An enhanced cleaning schedule should be prescribed, usually for scooping twice daily and to change the entire litter once weekly to ensure proper hygiene that will encourage cats to use the box. Some cats have an aversion to the litter that they associate with their painful urinary bladder and others have a preference for a particular litter substrate. It can be difficult at times to identify the best litter substrate for FIC cats – it requires sequential testing of the new candidate substrate compared to the old substrate in litter boxes next to each other to see if the cat has a preference or not. We recommend clumping unscented litter – clumping litter is more manageable for scooping and unscented litter is less aversive for some cats. Cats that enjoy a particular litter use it, stay a while, paw in the litter, and usually cover both their feces and urine deposits. We recommend the "rule of 1+1" for the number of litter boxes that should be available out of sight of each other for cats in the household, in that there should be one for each cat plus one more; this rule also applies for other resources such as food and water bowls. As or more important may be the location of the litterboxes – they should be available at several locations if there is a large living space. Litterboxes should be placed in accessible locations that allow for some privacy and away from noise and dogs. There are many nuances as to the suitability of the litterbox to any particular cat – such as size, depth, shape, hooding, and automatic cleaning devices.

Stress is likely increased for cats living in close confinement with small square footage for roaming. Available square footage can be increased for small living spaces by providing vertical places to climb and rest (perches to climb, view, and rest). Chaperoned visits to fenced in back yards or enclosed patios can also help with increased opportunity to explore novel spaces that may provide more interest to bored cats. Audio and video sensory stimulation (TV monitors and sound systems with music) may be useful to entertain and engage some cats while the owner is not present. Increased interaction between cats and their owners may increase the quality of life for those cats. This may be accomplished depending on the particular cat during grooming and petting, playing games with laser pointers, and simulating hunting of prey activities (feathered or tailed devices). Timing of these activities at dawn or dusk may be helpful since cats are generally more active at these times. Unrestricted outside activity is a controversial recommendation that we sometimes make – it may be the only change that will work to improve some cats with their FIC and use of the litterbox. Indoor restriction leads to increased length of life for cats, but not necessarily an increase in the quality of life for all cats, especially in they are living in un-enriched environments. The indoor environment for some house cats can be monotonous and predictable which can inappropriately activate the SRS.[11] Medically acquired illnesses are dose related to the amount of time that cats spend indoors, including hypertrophic cardiomyopathy, diabetes mellitus, obesity, hyperthyroidism, urolithiasis, odontoclastic resportive lesions, and

FIC.[12] Access to the outdoors seems to reduce stress in some cats, but stress can be increased for other sensitive cats, especially if the cat is released into an outdoor environment in which the density of cats nearby is high. The increased risk of death from infectious diseases, trauma, and toxins encountered during life outside must be balanced against the quality of life issues for some cats living indoors with medical diseases including FIC.

We always recommend MEMO before drug therapy is attempted. It is our impression that drug therapy, if necessary to control signs, works better in those in which MEMO has already been implemented. When MEMO has not been effective in reducing LUTS, we consider the use of tricyclic drugs such as amitriptyline or chlomipramine. When positive effects are seen, we taper the dose of these drugs gradually and stop them whenever possible after at least 3 months of use. The use of fluoexetine has proven useful in some cats with FIC. Specifics about this kind of drug therapy are beyond the scope of these notes. Drug therapy sometimes gives the owners an edge to implement more MEMO, which then lessens the need for long-term drug therapy. Most studies of FIC have found a high placebo-response rate. It may be possible to take advantage of this high placebo response by prescribing treats in the form of something like pill pockets. A positive interaction between the owner and the cat could occur during these "treatments" that could decrease stress. If there is no improvement after several weeks of providing the placebo treatment, a drug could be added to the treat vehicle.

REFERENCES

[1]McEwen BS. **Physiol Rev** 2007; 87(3): 873-904; [2]McEwen BS.. **Metabolism** 2008; 57 Suppl 2: S11-15; [3]McEwen BS. **Eur J Pharmacol** 2008;583(2-3): 174-185; [4]Buffington CA. **J Am Vet Med Assoc** 2006; 228(5): 722-725; [5]Westropp JL. **J Urol** 2003; 170(6 Pt 1): 2494-2497; [6]Westropp JL. **Am J Vet Res** 2006; 67(4): 731-736; [7]Buffington CA. **J Feline Med Surg** 2006; 8(4): 261-268; [8]Nithianantharajah AJ. **Nat. Rev.** Neurosci 2006;.7: 697–709; [9]Lavelle JP. Am **J Physiol Renal Physiol** 2000; 278(4): F540-553;[10]Gunn-More DA 2004; 6, 219-225; [11]Van Rooijen J. **Applied Animal Behaviour Science** 1991; 31, 283-286; [12]Buffington CA. **J Am Vet Med Assoc** 2006; 228(5): 722-725.

MEDICAL TREATMENTS OF FELINE IDIOPATHIC CYSTITIS

Joe Bartges, DVM, PhD, DACVIM, DACVN
Knoxville, TN

INTRODUCTION

Over the past decades, knowledge of specific causes and natural course of feline lower urinary tract diseases (LUTD) has increased, allowing for diagnostic and therapeutic effects to be directed towards identification and elimination of specific underlying causes. Feline idiopathic cystitis (FIC) is the most common cause of FLUTD in cats less than 10 years of age.[1,2] The biological behavior of FIC has not been evaluated by prospective studies of large populations of untreated cats; however, observations suggest that clinical signs of LUTD in many untreated cats subside within 5-7 days. Clinical signs may recur after variable periods of time and subside without therapy. Occasionally, cats with FIC have clinical signs that persist for weeks to months. It is unknown whether this chronic form represents an extreme in the spectrum of clinical manifestations associated with acute self-limiting disease caused by similar etiological factors or whether it represents a different mechanism of disease. Recurrent signs of LUTD in a cat with FIC may or may not be due to FIC as many causes of LUTD have similar clinical signs. Because of the variable biological behavior of FIC and due to the lack of identification of specific cause(s), many treatments have been attempted and recommended with variable degrees of "success". Evaluation of treatment in a placebo-controlled, randomized trial is essential in determining efficacy of the intervention. Some studies have demonstrated a "placebo" or "clinic" effect in up to 70% of patients; therefore, response to treatment in an uncontrolled situation does not imply cause-and-effect, only coincidence.

SUMMARY OF TREATMENTS EVALUATED

Date	Patient population	Proposed therapies	Assessment of therapy: Controlled study?	Efficacy*
1955[3]	Obstructed and non-obstructed males; +/- uroliths; +/- UTI	Antibiotics	No	ND
		Antispasmodics	No	ND
		Urinary acidifiers	No	ND
		Vitamin A	No	ND
		Testosterone	No	ND
		Hyaluronidase	No	ND
		Low magnesium and phosphorous diet	No	ND
1966[4]	Non-obstructed females; no uroliths; +/- UTI	Antibiotics	No	ND
		Urinary acidifiers	No	ND
1970[5]	Obstructed and non-obstructed males and females; no uroliths; +/- UTI	Antibiotics	No	ND
		Urinary acidifiers	No	ND
		Lactated Ringer's solution	No	ND
1972[6]	Non-obstructed males and females; no uroliths; no UTI	Antispasmodics	No	ND
		Urinary acidifiers	No	ND
		Sodium chloride	No	ND
		Moist food	No	ND
1975[7]	Non-obstructed males and females; no uroliths; no UTI	Urinary acidifiers	No	ND
		Lactated Ringer's solution	No	ND
		Topical phenol	No	ND
		Intravesicular Lactated Ringer's solution	No	ND
1977[8]	Non-obstructed males and females; no uroliths; +/- UTI	Antibiotics	No	ND
		Urinary acidifiers	No	ND
		Curecal-Feline#	No	ND
		Moist low ash diet	No	ND
1978[9]	Non-obstructed males and females; no uroliths; no UTI	Antibiotics	No	ND
		Urinary antiseptics	No	ND
		Urinary acidifiers	No	ND
		Sodium chloride	No	ND
		Megestrol acetate	No	ND
		Moist food	No	ND
1980[10]	Obstructed males; +/- uroliths; +/- UTI	Intravesicular copper coil	No	ND
1981[11]	Non-obstructed males and females; +/- uroliths; +/- UTI	Antibiotics	No	ND
		Urinary acidifiers	No	ND
		Sodium chloride	No	ND
		FCV/FHV-1 vaccine	No	ND
		Moist low ash diet	No	ND

1982[12]	Non-obstructed males and females; no uroliths; +/- UTI	Chloramphenicol (100mg PO q8hr x 5d)	Yes	NS
		Propantheline (7.5 mg PO once)	Yes	NS
		Lactated Ringer's soln (100ml SQ once)	Yes	NS
1990[13]	Non-obstructed males and females; no uroliths; no UTI	Antispasmodics	No	ND
		Urinary acidifiers	No	ND
		Sodium chloride	No	ND
		Diuretics	No	ND
		Corticosteroids	No	ND
		DMSO	No	ND
		Moist acidifying low-magnesium diet	No	ND
1995[14]	Non-obstructed males and females; no uroliths; no UTI	Antispasmodics	No	ND
		Potassium chloride	No	ND
		Moist food	No	ND
1995[15]	Non-obstructed males and females; IFLUTD**	Prednisolone (1.0 mg/kg PO q12hr x 10d)	Yes	NS
1995[16]	Non-obstructed males and females; IFLUTD**	Amitriptyline	No	ND
		Corticosteroids	No	ND
		Urohydrodistention	No	ND
1998[17]	Non-obstructed cats – gender not specified; IFLUTD**	Amitriptyline (10 mg PO q24hr x 12mo)	No	ND
1999[18]	Non-obstructed males and females; IFLUTD**	Moist and dry acidifying, low-Mg diet	No	ND
2003[19]	Non-obstructed males and females: IFLUTD**	Amitriptyline (5 mg PO q24hr x 7 d)	Yes	NS
2004[20]	Non-obstructed males and females, IFLUTD**	Feline facial pheromone spray	Yes	NS
2004[21]	Non-obstructed males and females, IFLUTD**	N-acetyl-glucosamine (125 mg PO q24hr x 6 mos)	Yes	NS
2006[22]	Non-obstructed males and females, IFLUTD**	Multimodal environmental modification	No	ND
2008[23]	Non-obstructed males and females: IFLUTD**	Pentosan polysulfate (3 mg/kg SQ, days 1, 2, 5 and 10)	Yes	NS
2009[24]	Non-obstructed males and females: IFLUTD**	Pentosan polysulfate (2, 8, or 16 mg/kg PO q12h x 6.5 mos)	Yes	NS

* Degree in reduction and/or severity of clinical signs in affected cats treated with a therapeutic agent compared to placebo treated or untreated controls. ND = not determined; NS = not significant; S = significant

#0.085% tetrasodium ethylenediamine tetraacetic acid (EDTA) and 0.06% sodium tripolyphosphate.

**A diagnosis of IFLUTD was based on exclusion of other known causes of hematuria and dysuria.

REFERENCES

[1]Buffington CA, et al.. **J Am Vet Med Assoc** 1997;210:46-50; [2]Kruger JM, et al. **J Am Vet Med Assoc** 1991;199:211-216; [3]Lamb WV. **Vet Med** 1955;50:26; [4]Holzworth J. In: **Kirk's Current Veterinary Therapy** 1966-1967;410-412; [5]Osbaldiston GW, et al. **Vet Med Small Anim Clinician** 1970;65:461-468; [6]Osborne CA, et al. **Canine and Feline Urology** 1972;361; [7]Taussig RA. **Fel Pract** 1979;5:52-54; [8]Engle GC. **Fel Pract** 1977;7:24-27; [9]Gaskell CJ, et al. **J Small Anim Pract** 1978;19:301-314; [10]Tucker HQ, et al. **Vet Med Small Anim Clinician** 1980;75:435-436; [11]Elcock L. **Feline Pract** 1981;11:6; [12]Barsanti JA, et al. **J Am Anim Hosp Assoc** 1982;18:387-390; [13]Ross LA. **Veterinary Medicine** 1990;85; [14]Ling GV. **Lower urinary tract diseases of dogs and cats**. 1995; [15]Osborne CA, et al. In: **Textbook of veterinary internal medicine. 5th ed.** 1999;1710-1746; [16]Buffington CA. **Urology** 2002;60:939; [17]Chew DJ, et al. **J Am Vet Med Assoc** 1998;213:1282-1286; [18]Markwell PJ, et al. **J Am Vet Med Assoc** 1999;214:361-365; [19]Kruger JM, et al. **J Am Vet Med Assoc** 2003;222:749-758; [20]Gunn-Moore DA, et al **J Feline Med Surg** 2004;6:133-138; [21]Gunn-Moore DA, et al **J Feline Med Surg** 2004;6:219-225; [22]Buffington CA, et al. **J Feline Med Surg** 2006;8:261-268; [23]Wallius BM, et al **J Feline Med Surg** 2009;11:409-412; [24]Chew DJ, et al. **J Vet Intern Med** 2009;23:690abstract.

STAGING PATIENTS WITH ACUTE KIDNEY INJURY: A NEW PARADIGN

Larry D. Cowgill, DVM, PhD, DACVIM
Davis & San Diego, CA

Precise definitions for acute kidney injury (AKI) have not been established in veterinary medicine. A myriad of definitions have been proposed in both human medicine and veterinary medicine based on sequential changes in glomerular filtration rate (GFR), azotemia, or urine production to provide objective criteria for the diagnosis. To emphasize the concept that AKI represents a continuum of renal injury, staging schemes recently have been proposed for human patients to stratify the extent and duration of renal injury and to predict outcomes. One staging system is categorized progressively into **R**isk, **I**njury, **F**ailure, **L**oss, and **E**SKD (RIFLE). A second slightly more restricted staging system emerged as an expert consensus from the Acute Kidney Injury Network (AKIN) in an attempt to reproducibly identify early AKI and classify its course and prognosis with regard appropriate diagnostic and therapeutic interventions. There is considerable overlap between both system, and criteria for each staging category are based ostensibly on insensitive markers of renal injury including abrupt changes in GFR, serum creatinine, urine output, and duration of signs. Preliminary validation of these staging concepts to categorize the extent of renal injury and to predict both morbidity and outcome has been demonstrated in human patients. Unfortunately, the criteria which define these staging schemes for humans rarely are applicable in animal patients in which the abruptness of the disease and the magnitude of changes in GFR, azotemia, or urine production are generally are unknown or quantitated.

Progressive (and often subtle) azotemia is the hallmark of AKI but is observed inconsistently in animals because the azotemia has not developed or is fully established at the time they are presented for evaluation or because serial observations have not been performed. Serum creatinine increases progressively with the severity of the renal injury or completeness of the outflow obstruction; however, reductions in lean body mass, decreased creatinine generation, and overhydration can lower the serum creatinine concentration and mask sequential changes. Despite its insensitivity and limitations as a biomarker of kidney injury, serum creatinine currently stands as the best, most time-tested, and most familiar marker of AKI. To date, no formal staging system has been established for animals with AKI to stratify or characterize the severity of the renal impairment as has been established recently for chronic kidney disease (CKD) in dogs and cats and AKI in humans. The IRIS staging system for CKD was developed as a consensus scheme to promote more uniform characterization and recognition of CKD in animals with the goal to better understand its pathophysiology and to facilitate its evaluation and rational management. Given the success and utility of the IRIS staging system for CKD, it seems tenable to adapt this schematic approach to classify and stratify the severity of AKI in dogs and cat. Unlike IRIS staging for CKD, the staging of AKI would not imply the kidney disease is stable or at steady-state. The "stage" represents a moment in the course of the disease and is predicted to change as the condition worsens or improves or transitions to CKD. Table 1 outlines a proposed staging scheme for AKI in dogs and cats based on serum creatinine, urine formation, and requirement for renal replacement therapy which is intended to facilitate the recognition, classification, functional stratification, and therapeutic decision making.

Stage I defines animals with historical, clinical, laboratory (biomarker, glucosuria, cylinduria, inflammatory sediment, microalbuminuria, etc.), or imaging evidence of acute kidney injury who are non azotemic and/or whose clinical presentation is readily fluid volume responsive. Stage I also include animals with progressive (hourly or daily) increases in serum creatinine within the non azotemic range. Presumptive guidelines could include a 0.3 mg/dl increase in serum creatinine within a 48 hour interval.

Stage II defines animals with documented acute kidney injury characterized by mild azotemia in addition to other historical, biochemical, or anatomic characteristics of AKI. This would include animals that have an increase from their baseline serum creatinine associated with pre-existing CKD.

Stages III, IV, and V define animals with documented AKI and progressively greater degrees of parenchymal damage and functional failure (uremia).

Each stage of AKI is further substaged on the basis of current urine production as oligoanuric (O) or nonoliguric (NO) and on the requirement for renal replacement therapy (RRT). The inclusion of substaging by urine production is based on the importance of the interrelationship of urine production to the pathological or functional contributions to the kidney injury and its influence on the clinical presentation, therapeutic options, and outcome of AKI. Substaging on the requirement for renal replacement therapy is established on the need to correct life-threatening iatrogenic or clinical consequences of AKI including severe azotemia, hyperkalemia, acid-base disorders, overhydration, oliguria or anuria, or to eliminate nephrotoxins. The requirement for renal replacement therapy could occur at any of the proposed stages. Substaging on the requirement for renal replacement therapy has similar clinical, therapeutic, and prognostic implications as for urine production to categorize the severity of the renal injury as well as its influence on outcome. As examples, a dog with anuric AKI secondary to leptospirosis with a presenting serum creatinine concentration 6.3 mg/dl who is managed with hemodialysis to correct life-threatening fluid overload, would be classified, AKI Stage IV, O, RRT. A puppy presenting for acute antifreeze exposure with ataxia, polyuria, acidosis, and a serum creatinine concentration of 1.8 mg/dl who is detoxified with hemodialysis would be classified, AKI Stage II, NO, RRT. In each example, the stage would be updated as the clinical and laboratory conditions change.

This scheme should be considered preliminary and remains subject to consensus revision and modification as appropriate biomarkers are validated to provide greater specificity and sensitivity to pathophysiologic phases of AKI. In the interim, as for human medicine, it has potential to better discriminate and categorize the clinical and therapeutic spectrum of acute kidney injury and promote earlier and more sensitive recognition of this syndrome.

Table 1. Proposed Staging System for Acute Kidney Injury in Dogs and Cats.

<table>
<tr><th>AKI Stage*</th><th>Serum Creatinine (mg/dl)</th><th>Clinical Description</th></tr>
<tr><td>Stage I</td><td>< 1.6</td><td>[a]Non azotemic AKI or volume-responsive AKI
[b]Historical, clinical, laboratory, or imaging evidence of renal injury
[c]Progressive non azotemic increase in serum creatinine; ≥0.3 mg/dl within 48 hours</td></tr>
<tr><td>Stage II</td><td>1.6 – 2.5</td><td>Mild AKI: Historical, clinical, laboratory, or imaging evidence of AKI and mild static or progressive azotemia</td></tr>
<tr><td>Stage III</td><td>2.6 – 5.0</td><td rowspan="3">Moderate to Severe AKI: Documented AKI and increasing severities of azotemia and functional renal failure</td></tr>
<tr><td>Stage IV</td><td>5.1 -10.0</td></tr>
<tr><td>Stage V</td><td>> 10.0</td></tr>
<tr><td colspan="3">*Each stage of AKI is further substaged on the basis of current urine production as oligoanuric (O) or non oliguric (NO) and on the requirement for renal replacement therapy (RRT); (see text)</td></tr>
</table>

REFERENCES

[1.] Bellomo R, et.al.; **Crit Care** 2004;8(4):R204. [2.] Kellum JA; **Crit Care Med** 2008;36(4 Suppl):S141. [3.] Kellum JA, et.al.; **Nephron Clin Pract**. 2008;109(4):182. [4.] Hoste EA, et.al.; **Contrib Nephrol** 2007;156:32 [5.] Mehta RL, et.al.; **Crit Care** 2007;11(2):R31 [6.] Molitoris BA, et.al.; **Nat Clin Pract Nephrol** 2007;3(8):439.

CONSERVATIVE MANAGEMENT OF ACUTE KIDNEY INJURY: PRESENT & FUTURE

David J Polzin, DVM, PhD, DACVIM
St. Paul, MN

OVERVIEW OF MANAGEMENT

In the clinical setting, acute kidney injury (AKI) is most often recognized by acute onset of azotemia/uremia or an acute change in urine volume. The timing of this recognition is unfortunate in that the principal opportunities to directly impact the magnitude and reversibility of the events affecting the kidney occur before the decline in renal function manifests clinically. Acute kidney injury is classically described as occurring in 4 stages: 1) initiation, 2) extension, 3) maintenance, and 4) recovery. Actual kidney injury occurs predominantly during the initiation and extension phases, whereas most patients are recognized during the maintenance phase. Thus, in patients presenting for acute uremia, the bulk of kidney injury has already occurred. Since the bulk of renal injury precedes the recognizable decline in GFR that leads to azotemia and uremia, an intensive search for markers of AKI that will identify renal injury as it begins. It is hoped that recognition of AKI during the initiation phase will allow application of therapies that may stop the injury, thus limiting the extent of renal damage. Until such markers and interventions become clinically available, conservative (i.e. non-dialytic management) of AKI is largely limited to minimizing the risk of additional renal injury, ameliorating signs of uremia, and keeping the patient alive long enough for any possible recovery of kidney function to occur.

At a minimum, treatment considerations in conservative management should include: 1) correction of fluid, electrolyte and acid-base deficits/derangements, 2) correction of gastrointestinal signs of uremia (H2 blockers, antiemetics, topical oral applications of 0.1-0.2% chlorhexidine solutions or gels), 3) provision of nutrition (microenteral nutrition, intravenous nutrition and/or oral nutrition, usually via a feeding tube initially), 4) enhancement of urine production in oliguric patients, 5) treatment of the primary renal disorder where possible (e.g. Leptospirosis), 6) prevention and treatment of infectious complications.

CORRECTING SHORT-TERM CAUSES OF PATIENT DEATH

Fluid Therapy

Volume depletion is a consistent finding in patients with uremic crisis. Some patients appear to be normally hydrated but have historical findings consistent with fluid losses. It is practical to assume these patients are subclinically dehydrated; carefully administer fluids to them at the rate of 1 to 3 percent of body weight on the premise that mild over hydration is less likely to be harmful than unrecognized volume depletion. Even mild volume depletion may promote renal injury in patients predisposed to acute kidney disease.

A substantial portion of volume depletion should be corrected over 2 to 6 hours unless the patient has known cardiac dysfunction, demonstrates intolerance to fluid administration, or becomes over hydrated (e.g. dyspnea, elevated neck veins, rales, S3 gallop rhythm, pulmonary edema). Administration of fluid at this rate will help to rapidly restore adequate renal perfusion. In addition, in patients with physiologic oliguria, urine volume will increase. Careful monitoring of the patient's response to fluid therapy is essential! If fluid overload occurs, the rate of fluid administration should be reduced or temporarily discontinued. Monitoring of central venous pressure may be necessary for patients with cardiac dysfunction or intolerance to fluid therapy.

A favorable clinical response and normalization of blood pressure usually indicate successful restoration of normal hydration. The goal of subsequent fluid therapy is to maintain fluid balance and prevent hypovolemia. Because even subclinically volume depletion may promote additional azotemia and renal damage, it may be desirable to maintain mild over hydration over the subsequent 6 to 12 hours in non-oliguric patients that have normal cardiopulmonary function. However, overt over hydration must be avoided in patients with persistent pathologic oliguria.

Patients may be predisposed to dehydration during maintenance and recovery phases of acute kidney disease because involuntary urine losses are often great. In order to prevent dehydration, the volume of parenteral fluids administered and oral fluids consumed should equal the sum of: 1) urine volume, 2) contemporary fluid losses (e.g. fluid lost via vomiting, diarrhea), and 3) insensible fluid losses (20 to 25 ml/kg/day). Because estimation of contemporary and insensible fluid losses may be inaccurate, serial determinations of body weight are commonly used to guide fluid therapy

Correcting Hyperkalemia

In absence of serious cardiotoxicity, mild to moderate hyperkalemia usually does not require specific treatment. However, hyperkalemia associated with serious cardiac arrhythmias is a potentially life-threatening condition requiring immediate therapy. Transient correction of hyperkalemic cardiotoxicity can be achieved by intravenous administration of calcium gluconate (10% solution; approximately 0.5 to 1.0 ml/kg body weight). The solution should be administered slowly over 10 to 20 minutes with electrocardiographic monitoring. The effect of such therapy usually lasts only 15 to 20 minutes. Other treatments that may be used in cardiotoxic emergencies are intravenous administration of glucose (with or without insulin) or sodium bicarbonate to attempt to redistribute potassium intracellularly. Typically, glucose is administered at a dose of 0.5 to 1.0 gm/kg body weight. Duration of effect is limited to several hours. Sodium bicarbonate may be administered at a dose of 0.5 to 1.0 meq/kg body weight, or at a dose calculated from the base deficit of the patient. In one recent study of the effect of this treatment on lowering potassium in humans with hyperkalemia due to acute kidney disease, bicarbonate therapy appeared to be of limited benefit.

Correcting Metabolic Acidosis

Because metabolic acidosis typically has little immediate adverse effect unless blood pH is below 7.20 (blood bicarbonate or total CO_2 concentrations less than 10 mEq/l), alkalinization therapy should be considered for patients with blood pH values less than 7.20. Immediate correction of metabolic acidosis is usually unnecessary when blood pH values exceed 7.20. Mild to moderate metabolic acidosis often improves following fluid therapy alone.

Metabolic acidosis is usually treated by administration of sodium bicarbonate, although potassium citrate is sometimes used when potassium and alkalizing effect are sought. The goal of alkalinization therapy is not to normalize acid-base balance, but rather to increase blood pH values to about 7.20, a level likely to prevent adverse cardiac effects of acidosis. Correction of blood pH to normal should be achieved gradually over several days.

DIURETICS - CONVERTING OLIGURIA TO NON-OLIGURIA

If oliguria persists despite correction of fluid deficits, therapy designed to increase urine volume is indicated because clinical management of non-oliguric patients is easier; it is not clear that this improves their prognosis appears. Use of diuretics or vasodilators in non-oliguric patients is generally discouraged because they are rarely needed and therefore unnecessarily expose the patient to risk of fluid and electrolyte depletion, additional renal injury, or adverse drug reactions. A therapeutic trial with diuretics is indicated for most patients with oliguric kidney disease. Furosemide and mannitol are the diuretics most commonly used. Patients that fail to respond to one of these diuretics may respond to the other or a combination of both.

Furosemide has been the most commonly used diuretic in canine and feline patients with oliguric kidney disease because it is relatively safe, readily available, and familiar to most veterinarians. Initially it should be administered intravenously at a dose of 2 mg/kg body weight. If no substantial diuresis develops within one hour after administration, the dose may be doubled (4 mg/kg). If this dose also fails to induce diuresis, the dose may be further increased to 6 mg/kg body weight. If diuresis still does not ensue, very large doses of furosemide, an alternative diuretic (e.g. mannitol), or the combination of furosemide and dopamine may be considered.

If furosemide successfully induces diuresis, it may be given by CRI or repeated at 8 hour intervals as needed to sustain diuresis and promote potassium excretion. However, the need for continued furosemide therapy must be considered in light of its potential adverse effects. It has been suggested that furosemide should not be used in patients with suspected gentamicin-induced acute kidney disease because it may promote aminoglycoside nephrotoxicity. Of primary concern is that diuretics do not induce volume depletion because it may promote additional hypoxic renal cell injury.

Mannitol has at least three theoretical advantages over furosemide: 1) it may enhance renal function by minimizing renal tubular cell swelling via its osmotic properties, 2) mannitol exerts its diuretic effects along the entire nephron and therefore may directly affect the proximal tubule, and 3) mannitol may expand the extracellular fluid volume. The major disadvantage of mannitol is the potential for vascular overload if oliguria persists. Therefore, mannitol should be avoided in over hydrated oliguric patients. Mannitol (20 or 25% solution) is administered intravenously over 5 to 10 minutes at a dose of 0.25 to 0.5 g/kg body weight. If substantial diuresis ensues, administration of mannitol can be repeated every 4 to 6 hours, or administered as a maintenance infusion (8 to 10% solution) during the initial 12 to 24 hours of treatment.

The use of dopamine for correcting oliguric/anuric acute kidney failure has been debated for decades. Studies of the effects of dopamine and furosemide on early uranyl nitrate-induced acute kidney disease in dogs suggest that combination therapy of dopamine with furosemide was more effective in inducing vasodilation and diuresis and preventing decline in GFR than either drug used alone. Descriptive studies in dogs have also suggested that dopamine may facilitate conversion from oliguria to non-oliguria. Preliminary work with fenoldopam, a selective DA1 agonist, has provided promising results in dogs and cats. The primary advantage of this drug is that it appears to increase renal medullary blood flow, an area of the kidney most likely to be exposed to hypoxia in ATN. In addition, cats do appear to have DA1 receptors in their kidneys. Reported dosage in cats has been a CRI of 0.5 mcg/kg/min. Further studies are needed to validate the safety and effectiveness of fenoldopam.

LIMITING CLINICAL SIGNS OF UREMIA

Cimetidine, zantac, or famotidine (H_2 blockers) are indicated to ameliorate uremic hemorrhagic gastritis because they blocks gastrin-stimulated gastric hyperacidity. Cimetidine (*Tagamet®*) is dosed at 5 mg/kg IM or slow IV, 1st dose; hereafter, 2.5 mg/kg q 12 h. Ranitidine (*Zantac®*) is given at 2.5 mg/kg q 12-24h. Famotidine (*Pepsid®*) is given at 0.5 to 1.0 mg/kg q 24h. Although almost uniformly used, there is precious little evidence supporting a role for H_2 blockers in uremia.

Maropitant citrate (Cerenia®) and/or serotonin 5-HT_3 receptor antagonists (e.g. ondansetron or dolesatron) are appropriate anti-emetics for patients with kidney disease. Suggested dosages are: Maropitant citrate (Cerenia® - 2 mg/kg daily orally or subcutaneously) Anzemet® (dolasetron – dosage = 0.3 to 0.6 mg/kg oral or slow IV once daily) or Zofran® (ondansetron – dosage = 0.6 to 1 mg/kg oral or IV every 12 hours). Metoclopramide has been shown to be substantially less antiemetic effectiveness in uremic humans than the serotonin 5-HT_3 receptor antagonists. Because it is a dopamine receptor antagonist, its role in patients with acute kidney injury has been questioned.

[1.] Cowgill, L. D. and T. Francey (2005). Acute Uremia. **Textbook of Veterinary Internal Medicine**. S. J. Ettinger and E. C. Feldman. Philadelphia, Elsevier Saunders. **2:** 1731-1751. [2.] Ross, L (2009): Acute Renal Failure, In, Bonagura JD and Twedt DC (eds): Kirk's **Current Veterinary Therapy XIV**. Philadelphia, Elsevier Saunders. pp. 879-882.

ACUTE KIDNEY INJURY: THE ROLE OF HEMODIALYSIS

Cathy Langston, DVM, DACVIM
New York, NY

INDICATIONS FOR DIALYSIS

In veterinary medicine, dialysis is used most commonly for acute kidney injury and less frequently for chronic kidney disease. Dialysis is indicated for patients with anuria or oliguria, life threatening fluid overload (i.e., pulmonary edema), or hyperkalemia. Dialysis is generally reserved until standard medical management has been attempted. Standard medical management includes adequate volume expansion and ensuring that the systemic blood pressure is adequate to perfuse the kidneys (> 80-100 mmHg systolic or > 60-80 mmHg mean arterial pressure). The use of diuretics is widespread in both human and veterinary medicine despite lack of conclusive evidence of a positive impact on outcome.(Bagshaw, et al. 2008) If loop diuretics, osmotic diuretics (i.e., mannitol), or other diuretics fail to induce adequate urine production within hours, early referral for dialysis is appropriate.

Uremic symptoms, progressive azotemia, or azotemia that does not improve with standard medical therapy are indications for dialysis, even if urine output is adequate or increased. In one human study, non-oliguric patients were referred for dialysis later during hospitalization compared to oliguric patients, and suffered a worse outcome.(Liangos, et al. 2005) Whether initiating dialysis early in the course of AKI improves outcome compared to later initiation is unresolved. Some studies have shown an advantage to early initiation, although others have not.(Gettings, et al. 1999; Bouman, et al. 2002; Liu, Himmelfarb et al. 2006) However, these studies are all confounded by differences in defining early vs. late, only one was randomized, and cohort size was generally small. In many situations, AKI is reversible, and because there are no studies that clearly predict which patients will need dialysis, it is possible that results of studies on early vs. late initiation will be biased towards early initiation because of inclusion of patients in the early group that did not need dialysis.(Waikar and Bonventre 2006) Despite the lack of an adequate study to answer this question, waiting until uremic symptoms are severe and patient condition has deteriorated may be disadvantageous, but the risk of early initiation in a patient that may not need dialysis must be weighed against the potential risks and cost of dialysis.

DIALYSIS DOSE

Prescribing the dialysis dose requires an ability to predict the adequacy of treatment. The most commonly used measure of clearance in veterinary hemodialysis is the urea reduction ratio (URR), which can be calculated by the formula: (pre-treatment BUN – post-treatment BUN)/pre-treatment BUN. The most commonly used measure of dialysis dose in humans is Kt/V, a dimensionless unit of measurement that incorporates the dialyzer clearance, time on dialysis, and volume of distribution.

When determining the dose of dialysis for acute kidney injury, the two major initial considerations are the rate of clearance and the intensity of treatment. When azotemia is severe, rapid clearance may induce dialysis disequilibrium syndrome. One method of limiting clearance during the initial few dialysis treatments is a short dialysis treatment time (i.e., 1.5- 2 hours), coupled with a moderately slow blood flow rate. An alternative approach would be sustained low efficiency dialysis (SLED) or CRRT. SLED is performed with an intermittent hemodialysis (IHD) machine, and involves a very slow blood flow rate for 6-24 hours. CRRT also uses a similar slow blood flow rate, markedly slower dialysate flow rate, and frequently relies on more convective clearance than most IHD prescriptions. Despite these differences, SLED and CRRT appear to have similar outcomes.(Berbece and Richardson 2006; Ghahramani, et al. 2008)

The hemodynamic stability of the patient is another factor that is generally included in the decision of dialysis modality. SLED and CRRT are generally preferred in the hemodynamically unstable patient. Rapid fluid removal in intermittent hemodialysis treatments can cause hypotension, which is thought to induce on-going renal damage.(Conger 1990) Slow or continuous fluid removal in SLED or CRRT intuitively is less likely to cause intermittent hypotension. Although clinical impressions support this, there is no evidence of superior hemodynamic stability in SLED or CRRT compared to IHD in randomized trials.(John, et al. 2001; Gasparovic, et al. 2003; Uchino 2008) Currently in veterinary medicine, the availability of therapy (IHD vs CRRT) is a major determining factor in which therapy is used, as few centers offer both.

Treatment intensity for acute kidney injury is an area of on-going research. In a recently published meta-analysis, modality (IHD vs CRRT) did not affect survival or renal recovery outcomes, although a subsequent study did find a lower rate of dialysis dependence with CRRT compared to IHD.(Bell, et al. 2007; Ghahramani, et al. 2008 Mehta, et al. 2001; Uchino, et al. 2007) However, intermittent IHD is less resource intensive. Daily 6-8 hour treatments (6-7 days a week) can be performed by day staff. Although some studies have shown a survival advantage with a higher dialysis dose, results have been inconsistent, and a recently published randomized trial showed no advantage to intensive therapy compared to less intensive therapy. (Ronco, et al. 2000; Bouman, et al. 2002; Schiffl, et al. 2002; Saudan, et al. 2006; Palvesky, et al. 2008; Tolwani, et al. 2008) During the recovery phase of acute kidney injury, a less intensive schedule (4-5 hour treatments three times a week, decreasing to twice weekly as function improves) may be adequate.

Guidelines for treatment intensity for veterinary patients have been established through clinical experience. After determining the desired percent reduction in urea (URR), the number of liters of blood to be processed to achieve the goal are based on weight of the patient. In early treatments, a slow rate of reduction decreases complications, so a longer treatment duration is selected, whereas in later treatments, a rapid blood flow rate can achieve the treatment goals faster (or achieve greater clearance in a set time interval). Once the desired amount of blood to be processed and the duration of treatment are selected, the average blood flow rate to achieve these goals can easily be calculated.

Limiting factors for blood flow rate are the catheter and the patient. Larger diameter catheters can provide much faster blood flow rates than smaller catheters. If the blood flow rate is less than desired due to catheter flow problems, the treatment time may need to be prolonged to achieve the desired clearance.

The desired volume of ultrafiltration depends on the patient's hydration status. Some patients may be 10 to 25% overhydrated, in the most severe cases. In severe cases, the excess fluid cannot generally be removed in a single short dialysis treatment because the rate of fluid removal should not exceed 20 ml/kg/hr, to avoid intradialytic hypotension. In those cases, SLED or CRRT may be advantageous. In human pediatric patients, fluid overload (>10%) is associated with a worse outcome.(Gillespie, et al. 2004; Goldstein, et al. 2005).

OUTCOME

Overall survival of dogs and cats treated with intermittent hemodialysis for AKI is 41-52%, but survival is dependent on etiology, with infectious and ischemic causes faring better than toxic causes, in general. (Langston, et al. 1997; Adin and Cowgill 2000; Francey and Cowgill 2002; Fischer, et al. 2004; Pantaleo, et al. 2004; Francey 2006) Of the non-surviving patients, about half of those die or are euthanatized due to extra-renal conditions (e.g., pancreatitis, respiratory complications). About a third of non-survivors are euthanized due to failure of recovery of renal function. On-going uremic signs, dialysis complications, and unknown causes account for the remaining patient deaths. Of the surviving patients, approximately half regain normal renal function (defined by normal serum creatinine concentration) and half have persistent chronic kidney disease. Although the majority of renal recovery has traditionally been thought to occur within four weeks of renal injury, we have seen patients with continued recovery over three to six months. Pet owners need to be prepared to continue treatment for at least 2 to 3 weeks before seeing signs of recovery (i.e., resumption of urine production in a previously anuric or oliguric patient, gradually decreasing accumulation of urea and creatinine between each dialysis treatment). Once the patient is stable (generally 1 to 2 weeks after starting dialysis), intermittent hemodialysis is suitable for outpatient therapy three to four days a week while waiting for renal recovery.

A clinical scoring system for outcome prediction of dogs with AKI receiving hemodialysis has been developed. The model uses data that is commonly available prior to instituting dialysis, although knowledge of the specific etiology, which is frequently not known initially, improves the accuracy of the outcome prediction. Although this scoring system has not yet been independently validated and should not be used in the decision-making process for an individual patient, use of scoring systems will hopefully help in determining appropriate candidates for dialysis.(Segev, et al. 2008)

REFERENCES

1. Adin C, et al. **J Am Vet Med Assoc** 2000;216(3):371-375, 2. Bagshaw S, et al. **Crit Care Med** 2008;36(4(Suppl)):S172-8, 3. Bell M, et al. (2007). **Intensive Care Med** 2007;33:773-780, 4. Berbece A, et al. **Kidney Int** 2006;70:963-968, 5. Bouman C, et al. **Crit Care Med** 2002;30(10):2205-11, 6. Conger J. **Seminars in Dialysis** 1990;3(3):146-148, 7. Fischer J, et al. **J Vet Intern Med** 2004;18(3):418, 8. Francey T. **Proc ARTS**, New York. 2006, pp. 230-233, 9. Francey T, et al. **J Vet Intern Med** 2002;16(3):352, 10. Gasparovic V, et al. **Renal Failure** 2003;25(5):855-862, 11. Gettings L, et al. **Intensive Care Med** 1999;25(8):805-13, 12. Ghahramani N, et al. **Nephrology** 2008;13(7):570-8, 13. Gillespie R, et al. **Pediatr Nephrol** 2004;19:1394-1399, 14. Goldstein S, et al. **Kidney Int** 2005;67(2):653-8, 15. John S, et al. **Nephrol Dial Transplant** 2001;16:320-327, 16. Langston C, et al. **J Vet Intern Med** 1997;11(6):348-355, 17. Liangos O, et al. **Nephron Clin Pract** 2005;99(2):c56-60, 18. Liu K, et al. **Clin J Am Soc Nephrol** 2006;1(5):915-9, 19. Mehta R, et al. **Kidney Int** 2001;60:1154-1163, 20. Palevsky P, et al. **N Engl J Med** 2008;359(1):7-20, 21. Pantaleo V, et al. **J Vet Intern Med** 2004;18(3):418, 22. Ronco C, et al. **Lancet** 2000;356(9223):26-30, 23. Saudan P, et al. **Kidney Int** 2006;70(7):1312-1317, 24. Schiffl H, et al. **N Engl J Med** 2002;346(5):305-10, 25. Segev G, et al. **J Vet Intern Med** 2008;22(2):301-308, 26. Tolwani A, et al. **J Am Soc Nephrol** 2008;19(6):1233-1238, 27. Uchino S. **Crit Care Med** 2008;36(Suppl):S238-S242, 28. Uchino S, et al. **Int J Artif Organs** 2007;30(4):281-292, 29. Waikar S, et al. **Clin J Am Soc Nephrol** 2006;1(5):903-4

CONTINUOUS RENAL REPLACEMENT THERAPY

Mark J. Acierno, MBA, DVM, DACVIM
Baton Rouge, LA

INTRODUCTION

Continuous Renal Replacement Therapy (CRRT) is a collection of four blood purification modalities that utilize both diffusion and convection to remove uremic toxins as well as correct electrolyte, fluid and acid-base imbalances associated with acute renal failure. It has also been employed in human medicine for the treatment of diuretic resistant heart failure. As the name implies, once a patient starts CRRT, the therapy continues until renal function recovers or the patient has been transitioned to intermittent dialysis.

Continuous Renal Replacement Therapy has several advantages over traditional intermittent dialysis. First, CRRT's continuous nature and ability to remove larger molecules more closely resembles normal kidney function.[1] Studies have demonstrated that CRRT provides better control of electrolyte and acid-base balance than intermittent dialysis.[1,2] Since intermittent dialysis uses diffusion to make large changes in relatively short periods of time, these systems require large amounts of sterile dialysate.[3] This dialysate must be produced "on-site" requiring the operation and maintenance of complex water purification equipment. CRRT's more steady continuous operation combined with its efficient use of convection and diffusion allows the use of commercially available pre-packaged fluids.

Blood purification

Historically, the basis of extracorporeal blood purification has been to divert patient's blood into a dialyzer where the blood is divided and passed though many straw-like semi-permeable membranes. These membranes are immersed in dialysate and substances can be made to diffuse in to or out of the blood by adjusting the composition of the dialysate. In CRRT, blood purification also takes place in a dialyzer; however, CRRT can utilize both diffusion and convection.[4,5] Convection also involves, dividing the patient's blood and passing it through many straw-like semi-permeable membranes; however, convection involves placing the patient's blood under pressure. This results in fluid (ultrafiltrate) being forced from the blood and across the semi-permeable membrane.[6,7] The ultrafiltrate, which is discarded, carries toxins, electrolytes and other molecules. The benefits of convection are that it can remove larger molecules from the blood than could be extracted by diffusion and it allows for more economical use of fluids.[5,7] Convection is technically challenging as fluids and electrolytes must be replaced with great accuracy or the patient can quickly become dehydrated, over hydrated, or suffer electrolyte imbalances. Fortunately, today's CRRT systems automate and monitor this process with great precision.

Modes of operation

CRRT combines the principals of diffusion and convection to provide four distinct modalities: Slow continuous ultrafiltration (SCUF), continuous veno-venous hemofiltration (CVVH), continuous veno-venous hemodialysis (CVVHD) and continuous veno venous hemodiafiltration (CVVHDF). These modes have been previously described.[8,9] Briefly, SCUF is a purely convective modality in which fluids removed from the patient are not returned. SCUF has been used to treat non-diuretic responsive fluid overload in people and may have clinical use in veterinary medicine.

Continuous veno-venous hemofiltration (CVVH) is a purely convective modality in which ultrafiltrate is replaced with a sterile, balanced, electrolyte solution. This solution, which is called "replacement fluid" can be added either before or after the dialyzer. Adding the fluid after the dialyzer is significantly more efficient in removing toxins; however, there is an increased risk of sludging and clot formation as the blood becomes hemoconcentrated. Since 20%- 25% of blood volume can be safely removed, we typically use a post filter configuration.

Continuous veno-venous hemodialysis (CVVHD) is a diffusive therapy, which closely resembles intermittent dialysis except that prepackaged fluids are used.

Continuous veno-venous hemodiafiltration (CVVHDF) combines the diffusive aspects of CVVHD with the convective properties of CVVH. Convection forces fluid out of the blood and across the semi-permeable membrane while the dialysate bathing the membranes results in the movement of small molecules by diffusion.

Studies have shown that the convective modalities such as CVVH and CVVHDF have a significant advantage in the removal of larger molecules from the blood; however, the role of these molecules in patients with acute renal injury is unclear. In addition, diffusive modalities (CVVHD) are just as effective at removing smaller molecules such as creatinine and urea. Nevertheless, we have had good successes treating patients with convection (CVVH).

Anticoagulation

Proper anticoagulation is essential to prevent clotting of the CRRT circuit. This is important as a clotted circuit can result in the loss of a significant amount of patient blood, requires replacement of the costly CRRT circuit, and represents time in which the patient's blood is not purified. There are two commonly used methods of anticoagulation. One involves systemic anticoagulation of the patient with heparin the other involves localized anticoagulation of the CRRT circuit with citrate. Both methods are associated with significant side effects. The chelation of calcium with citrate has been associated with significant hypocalcemia and metabolic alkalosis.[10-12] While systemic anticoagulation of the patient with heparin has a theoretical risk of hemorrhage, our personal experience has shown heparin to be safe, inexpensive and easy to monitor.

Indications for use

The most common indication for using CRRT is for patients suffering from acute kidney injury when there is a reasonable expectation that kidney function will recover in a relatively short period of time (days to weeks). We have used CRRT to treat animals with kidney injury secondary to Leptospirosis, melamine toxicity, heatstroke and acute tumor lysis syndrome. This has made managing anuric and oliguric patients significantly easier and provides time for renal function to return. CRRT can also be used in cases of acute kidney injury to stabilize patients so that they can either be transitioned to intermittent dialysis or transferred to a center that offers this modality. In human medicine, CRRT has also been used in the treatment diuretic resistant heart failure.

Patient Care

Providing properly trained, technically competent patient care twenty-four hours a day is one of the most difficult aspects of performing CRRT. Since an advanced knowledge of renal physiology and the mechanics of the CRRT unit are essential for treatment decisions, the pool of capable doctors and support staff is likely to be limited. In our hospital, a doctor and a technician or trained veterinary student is always by the patient's side. When patients require more than a few days of treatment, the staffing issues can become challenging.

Summary

CRRT is a new set of tools for providing support to animals with acute kidney injury. Its primary use is in cases where there is a reasonable expectation that the kidneys will return to function in the short term or for stabilization of patients that are to be transitioned to intermittent dialysis. Treatment of toxicosis and heart failure that is not responsive to diuretics may be other uses for CRRT. Although time, resource and personnel intensive, our experience has been that treating acute kidney injury patients with CRRT can be very rewarding.

REFERENCES

[1] Clark WR, et al. **J Am Soc Nephrol** 1994;4:1413-1420. [2] Bellomo R, et al. **Nephron** 1995;71:59-64. [3] Cowgill L, Maretzki C. In: Bonagura J, ed. **Kirk's Current Veterinary Therapy**. Philadelphia: W.B. Saunders, 1995;975-977. [4] Bellomo R, Ronco C. **Crit Care** 2000;4:339-345. [5] Clark WR, Ronco C. **Contrib Nephrol** 2004;144:264-277. [6] Golpher T. In: Bellomo R, BALDWIN I, Ronco C, et al., eds. **Atlas of Hemofiltration**. London: W.B. SAunders, 2002;15-18. [7] Ronco C, et al. **Nephrol Dial Transplant** 2001;16 Suppl 5:67-72. [8] Bellomo R, Ronco C. In: Bellomo R, Baldwin I, Ronco C, et al., eds. **Atlas of Hemofiltration**. London: W..B. Saunders, 2002;63-68. [9] Acierno MJ, Maeckelbergh V. **Compend Contin Educ Vet** 2008;30:264-272. [10] Landerville A, Seshadri R. **Journal of Veterinary Emergency and Critical Care** 2004;14:278-286. [11] Gibney N.11th Annual International Conference on Continuous Renal Replacement Therapies 2005;B30-B31. [12] Diehl S, Seshadri R. **J Vet Emerg Crit Care** 2008;18:370-382.

BEYOND THE CHEST FILM: ADVANCED CANINE & FELINE THORACIC IMAGING FOR THE SPECIALIST

Robert T. O'Brien, DVM, MS, DACVR
Urbana, IL

INTRODUCTION

Advanced imaging is becoming more and more readily available to general practices, either as a local referral practice or in-house in larger practices. Typically these procedures are performed only on the most complex cases and only under general anesthesia. Recent CT developments have allowed faster, safer and higher quality imaging in a clinical setting. This session will discuss nuclear medicine, MR and CT technologies for the diagnosis of respiratory diseases.

Nuclear Medicine Imaging

Nuclear medicine imaging has limited utility for the diagnosis of respiratory disease, with two important exceptions. The diagnosis of primary ciliary dyskinesia can be reached by placing a small drop (10 – 30 µl) of ^{99m}Tc-macroaggregated albumin (^{99m}TcMAA) or ^{99m}Tc-Sulfur Colloid into the caudal tracheal lumen under general anesthesia. Normal tracheal velocity in dogs in 20 normal dogs was 8.8 ± 3.3 mm/min. The diagnosis of pulmonary thromboembolic (PTE) disease has historically been made by demonstration of a perfusion deficit in a lobar artery on a nuclear angiogram. Imaging may be performed immediately following intravenous injection of 0.5 – 2.0 mCi of ^{99m}TcMAA and comparing the associated perfusion deficits with normal appearing radiographic or CT imaged lung fields. Unfortunately this diagnosis requires heavy sedation or anesthesia, often contraindicated in the clinical setting of a PTE.

Magnetic Resonance Imaging

MR is the mainstay of all neurological imaging. MR has exceptional soft tissue imaging capability and therefore been very valuable in the diagnosis of thoracic spine, body wall, mediastinal and other soft tissue body parts. Pulmonary nodules can occasionally be seen on the edge of thoracolumbar neurological MR studies. With the exception of angiography, MR has found little utility in pulmonary diseases. MR angiography shows great promise for advanced contrast and non-contrast angiography. This has been shown be useful in canine models for pulmonary embolic disease, although there are no clinical studies reported.

COMPUTED TOMOGRAPHY

The original computed tomographic systems were single slice sequential imaging machines. The gantry rotated about the patient in a few seconds then the patient couch moved forward a distance dictating the slice thickness and the gantry rotated again. These sequential systems were very slow. Often the imaging of a dog thorax required 60+ minutes for data acquisition and computer analysis for final display of the images.

All newer systems provide helical imaging capability. With helical (=spiral") imaging the patient moves continuously, without pause for computer analysis, and the entire batch of data is then analyzed after the scanning is completed. This provides faster imaging then sequential systems. However with single detector CT systems the scanning is still limited by rotation time (up to 1 second) times the thickness of each slice imaged (preferable <1mm) times the length of the body part scanned. So a 20cm long thorax might still require minutes to scan (i.e. 1 sec x 1 mm x 200mm = 200 seconds sec, = 3+ minutes). We can't stop the motion under these circumstances and patients are still anesthetized with single slice helical systems.

Newer CT system are multi-detector; 2, 4, 8, 16, 64, etc. With these systems we significantly decrease the time of imaging by 1) collecting many slices of data concurrently, and 2) faster rotation time (<0.5 seconds) resulting in dramatic shorter imaging times (0.5sec/slice x 1.0mm slice thickness x 200mm/16slices/rotation = 6 sec). Most of the imaging in dogs and cats can be performed in 5-15 seconds with a 16 slice helical CT protocol. Scanning is best performed with sub-millimeter slice thickness with overlap of the imaging sets. This will produce isotropic imaging.

Isotropic imaging means that the reconstructed images have the same image resolution as the original imaging plane. This makes MPR, 3-D, and virtual endoscopy CT imaging possible.

New CT technologies

Virtual CT imaging is possible for an array of hollow tubular organs. Because contemporary CT scanners offer isotropic, or near isotropic, resolution, display of images does not need to be restricted to the conventional axial images. Instead, it is possible for a software program to build a volume by 'stacking' the individual slices one on top of the other. The program may then display the volume in an alternative manner.

MPR Reconstruction

Multi-planar reconstruction (MPR) is the simplest method of reconstruction. A volume is built by stacking the axial slices. The software then cuts slices through the volume in a different plane (usually orthogonal). MPR is frequently used for examining the spine. Axial images through the spine will only show one vertebral body at a time and cannot reliably show the intervertebral discs. By reformatting the volume, it becomes much easier to visualize the position of one vertebral body in relation to the others.

Modern software allows reconstruction in non-orthogonal (oblique) planes so that the optimal plane can be chosen to display an anatomical structure. This may be particularly useful for visualizing the structure of the bronchi as these do not lie orthogonal to the direction of the scan.

For vascular imaging, curved-plane reconstruction can be performed. This allows bends in a vessel to be "straightened" so that the entire length can be visualized on one image, or a short series of images. Once a vessel has been "straightened" in this way, quantitative measurements of length and cross sectional area can be made, so that surgery or interventional treatment can be planned.

Surface Rendering

The operator chooses a threshold value of radiodensity. A threshold level is set, using edge detection image processing algorithms. From this, a three-dimensional model can be constructed and displayed on screen. Multiple models can be constructed from various different thresholds, allowing different colors to represent each anatomical component such as bone, muscle, and cartilage. However, the interior structure of each element is not visible in this mode of operation.

Volume Rendering

Surface rendering is limited in that it will only display surfaces which meet a threshold density, and will only display the surface that is closest to the imaginary viewer. In volume rendering, transparency and colors are used to allow a better representation of the volume to be shown in a single image—e.g. the bones of the pelvis could be displayed as semi-transparent, so that even at an oblique angle, one part of the image does not conceal another.

Where different structures have similar radiodensity, it can become impossible to separate them simply by adjusting volume-rendering parameters. The solution is called segmentation, a manual or automatic procedure that can remove the unwanted structures from the image.

Virtual CT Endoscopy

Virtual endoscopy of the upper airway, trachea, and central bronchi has been reported in a number of clinical settings. The images produced by the scanner must be processed into a "fly through" 3D image, using a cine program, which allows the user move through the bowel as if performing a normal endoscopic procedure. Anticipated improvements in CT resolution with the new multi-slice helical scanners should enhance the resolution of virtual endoscopy. The longitudinal resolution of multi-slice helical CT can be three times greater than that of single-slice helical CT. This will enable depiction of more distal airways.

CT IMAGING IN DOGS AND CATS

CT is the best imaging for the evaluation of many intra-thoracic diseases, including most pulmonary diseases. Many research studies and clinical reports have concluded that CT is superior to conventional radiography for the diagnosis of pulmonary thromboembolic diseases, detection of lung nodules and foreign bodies, characterization of bronchial disease, and characterization of diseases masked by current pleural or pulmonary disease. Case examples of these situations will be briefly presented.

SEDATED CT IMAGING

The need for anesthesia is a variable requirement now that CT scanning can be accomplished so quickly. Sedation that provides a minimally moving patient for a few minutes is now readily available for appropriate patient situations. All angiography for portosystemic shunt studies are performed on sedated patients in many practices. Upper and lower airway imaging in dogs, too large for a CT restraint device or that have behavioral characteristic negating a restraint device, can usually be imaged with sedation.

Nasal, laryngeal, tracheal, lung, intra-thoracic angiographic studies can all be easily and safely performed under sedation protocols. Examples of each of these studies will be presented. The most exciting diseases recently diagnoses with this type protocol includes brachycephalic syndrome, laryngeal collapse, bronchiectasis and bronchial collapse in additional to the more routine pulmonary studies. Virtual CT tracheoscopy and laryngoscopy imaging on awake (see below) or lightly sedated dogs results compare favorably with endoscopic studies, with significant reduced time, invasiveness and cost to the client.

AWAKE CT IMAGING

New technology has limited clinical utility if only applied in the abnormal physiology of an anesthetized or sedated patient. Cats and smaller dogs will not sit quietly on the CT couch. Even with very fast imaging, patients move, rendering the imaging nondiagnostic. A device is needed to minimize patient motion for this brief period. This device also needs to provide a nurturing environment for dyspneic, anemic, shocky or otherwise clinically compromised patients. The criteria for such a device includes the following:

Clinical

1. Oxygen
2. Catheter
3. Symmetrical
4. Transparent
5. Quick removal of patient
6. Secure closure
7. Non-heating
8. Low humidity
9. Portable
10. Cleanable

CT

11. No metal
12. Symmetrical
13. Restrictive
14. Low CT#
15. Rugged
16. Add Padding

Figure 1. Criteria for a CT imaging device.

Figure 2. Example device for awake CT imaging in cats and small dogs

One design (Figure 2) is a bivalved Plexiglas tube, with clear Plexiglas end plates and symmetrical end-plate ports for oxygen and i.v. catheters. The device limits a cat's movement without stressing sick patients. More than 100 cats and small dogs have been CT imaged using this device. Two normal cats were removed because of behavior problems. Both these cats were farm cats and otherwise healthy. No clinically sick cat has ever been demonstrably stressed by being in the device. Dogs have a more variable acceptance of the device. Pre-imaging acceptance testing decreases the likelihood of nondiagnostic imaging.

A clinical project has been completed comparing conventional radiography to CT imaging using a new device in emergency feline patients with clinical signs of respiratory disease; increased respiratory rate, noise or effort, cough, or open-mouthed breathing. This project included disease throughout the respiratory tract including nasal, pharyngeal, laryngeal, tracheal and intra-thoracic locations. CT was performed earlier in the work-up than radiographs. This was because the CT was deemed less stressful than radiography, effectively provided supplemental oxygen, and involved no handling of the patient. Radiographs are often quite stressful to patients during the handling and positioning process. No complications were experienced with the CT imaging. No cat has worsening of the clinical signs, needed to be removed from the device or developed any complications. The CT was performed in 6-12 seconds, on average. A small number of cases had repeated CT studies because of patient motion or because of the need for post-i.v. contrast CT imaging. Awake CT was judged superior to radiographs for the diagnosis of most upper airway, pulmonary, bronchial, pleural, mediastinal and body wall diseases. The diagnosis of heart size, and therefore heart failure, was equally difficult on CT imaging, unless augmented with i.v. contrast enhancement. With contrast enhancement, assessment of the left atrium luminal diameter and left ventricular wall thickness could be estimated.

Summary

Newer CT technologies offer the opportunity for faster, higher resolution imaging. These improvements will often result in a set of CT images that can be reconstructed into different planes, surface rendered 3-D or virtual endoscopic image sets. A device has been design and tested for awake imaging cats and smaller canine patients. Imaging case examples are provided demonstrating the utility of advanced awake CT imaging. The goal of all these new imaging technology is to provide an improved diagnosis minimizing client cost, patient morbidity and time. Initial studies offer hope that newer technologies and use of restraint devices may provide the window to improved CT imaging of dog and cat patients with respiratory diseases.

REFERENCES

1. Nemanic S, et. al. **J Vet Intern Med.** 2006 May-Jun;20(3):508-15. 2. Ogasawara N, et al. **J Magn Reson Imaging**. 2004 Jul;20(1):46-55. 3. Paoloni MC, et. al. **J Am Vet Med Assoc**. 2006 Jun 1;228(11):1718-22. 4. Schultz RM, et. al.**Vet Radiol Ultrasound**. 2008 May-Jun;49(3):249-55. 5. Schwarz T, et. al. **BSAVA** Manual of Canine and Feline Thoracic Imaging. 2008. 408pp. 6. Seiler G, et. al. **Vet Radiol Ultrasound**. 2008 Nov-Dec;49(6):504-8. 7. Suga K, et. al. **J Appl Physiol**. 2002 Jun;92(6):2439-51.

GETTING THE MOST FROM FELINE & CANINE RESPIRATORY CYTOLOGY

Anne Barger, DVM, MS, Diplomate ACVP
Urbana, IL

Respiratory cytology can be very useful in obtaining a diagnosis for an inflammatory or neoplastic disease. There are many techniques available for obtaining a specimen, depending on the area sampled. This proceeding will review cytology for different methods of obtaining a sample and different locations within the respiratory tract starting with the nasal cavity and finishing with the lung.

Normal cytology of the nasal cavity consists of several different cell types. The nasal vestibule is lined by stratified squamous epithelial cells which progresses in a transitional area to nonciliated cuboidal to columnar epithelium and eventually to a ciliated columnar epithelium. Within the submucosa of the caudal nasal cavity are lymphoid follicles, which consist of a mixed lymphoid population predominated by small lymphocytes with lesser numbers of large and intermediate sized lymphocytes and low numbers of plasma cells. Several types of glands, both serous and mucous are present and so cytology of the nasal cavity often contains thick mucinous material. Additionally, a mixed bacterial population is expected and these organisms commonly are extracellular, trapped in mucus or coating the cytoplasm of squamous epithelial cells.

There are several sample collection techniques utilized for cytology and these include; nasal swabs and flushes, fine needle aspirations of masses and imprints of biopsies. Nasal swabs generally sample the surface of the nostril and cytology often consists of mucus, bacteria and occasionally suppurative inflammation. Bacterial rhinitis is rarely a primary disease process but generally occur secondary to viral or fungal infection or presence of a soft tissue mass or foreign body. Therefore swabs often are not diagnostic for the underlying process. However if the patient has a disease of the superficial mucosa, such as can occur with cryptococcosis organisms may be identified with swabs. Nasal flushes are useful in sampling deeper in the nasal cavity but still only sample the surface epithelium and mucous. Generally samples obtained from patients with mucopurulent discharge consist of suppurative inflammation. Occasionally this technique is useful in diagnosing fungal disease such as cryptococcosis, aspergillosis or rhinosporidiosis. Fine needle aspirations of masses and imprints of biopsies can provide a higher diagnostic yield. If a mass is present, often the overlying epithelium is hyperplastic so the biopsy needs to incorporate some of the deeper tissue. Masses that deform the nose and induce lysis of bone often are good candidates for fine needle aspiration. Diagnosis of nasal carcinoma, lymphoma and osteosarcoma are all possible with cytology. Nasal carcinoma consists of clusters of neoplastic epithelial cells. These cells are round with a scant rim of basophilic cytoplasm which can make them difficult to differentiate from nasal lymphoma. Lymphocytes should not form cohesive clusters so this subtle feature can be helpful in differentiating these tumors. Additionally, immunocytochemistry with cytokeratin, CD3 and CD79 can also be helpful. If a neoplastic process is suspected, aspiration of the draining lymph node is useful. Nasal masses can be quite challenging to obtain a diagnostic aspirate or biopsy but if the tumor has already metastasized, diagnosis can be obtained in the lymph node. Diagnosis of some pathogens is enhanced by rhinoscopy and biopsy or cytology. For example, diagnosis of aspergillosis is much easier if the nasal plaques that are formed by the fungus are sampled directly. Often the fungal hyphae are embedded in thick mats of mucinous and necrotic cellular debris.

The larynx consists of cartilaginous plates lined by squamous and pseudostratified ciliated columnar epithelial cells. Cytology of the larynx often consists of mature squamous epithelial cells coated in horizontal chains of rod shaped bacteria (*Simonsiella sp.*). Generally, cytology of the larynx is performed when a mass is identified. These masses can be sampled by fine needle aspiration of brushing cytology. Tumors of the larynx are uncommon but include squamous cell carcinoma, lymphoma, oncocytoma, plasma cell tumor, chondrosarcoma and rhabdomyoma. Squamous cell carcinoma is an easily recognizable tumor with cytology. Clusters of epithelial cells are observed. These cells have moderate amounts of basophilic, occasionally keratinized cytoplasm with large round nuclei and prominent nucleoli. Perinuclear vacuolization is a common finding. Oncocytomas are most common in young dogs and often present as a well-circumscribed mass. Oncocytes are large foamy epithelial cells whose origin is uncertain. Most recent reports suggest the cells originate from seromucinous gland epithelial cells[1]. The behavior of this tumor is benign. Cytologically, rhabdomyomas appear very similar however rhabdomyomas are of skeletal muscle origin and immunostaining with vimentin and cytokeratin is useful to differentiate these two tumors. Chondromas and chondrosarcomas are also mesenchymal neoplasms. The cells are round to spindle shaped with basophilic cytoplasm with round nuclei. These cells often make abundant amounts of thick eosinophilic matrix which allows for diagnosis of this tumor.

The trachea and bronchi are lined by ciliated, nonciliated and mucus producing cells. The number of ciliated and mucous producing cells decrease deeper in the airway. Transtracheal washes (TTW) and bronchoalveolar lavage (BAL) can be used to evaluate the lower respiratory tract. The TTW is used to sample the larger airways including trachea and mainstem bronchi. TTWs from normal animals consist of ciliated columnar epithelial cells and few inflammatory cells, primarily macrophages and small lymphocytes arranged in a background of thick mucinous material. TTW samples can be obtained by inserting a catheter through an endotracheal tube or inserting a catheter directly between tracheal cartilaginous rings. The latter technique has a lower risk of oropharyngeal contamination. Cytological features of oropharyngeal contamination consist of mature squamous epithelial cells, often coated in a mixed bacterial population and often *Simonsiella sp.* bacteria. These organisms are specific for the oropharynx, as are the squamous epithelial cells. BALs are used to sample the smaller airways including bronchioles and alveoli. Alveoli consist of several cell types, type I and type II pneumocytes, endothelial cells, alveolar macrophages and interstitial fibroblasts. Type I pneumocytes are flattened squamous epithelial cells and type II pneumocytes are cuboidal epithelial cells. Cell counts are performed on BALs to determine the adequacy of the sample. BALs from healthy patients are predominated by vacuolated macrophages (alveolar macrophages) with low numbers of small lymphocytes and neutrophils. Up to 25% eosinophils are possible in healthy cats. The presence of high numbers of ciliated columnar epithelial cells suggests sampling of the larger airways such as in a TTW rather than a

BAL. Slides prepared for BAL and TTW should include direct smears and/or squash preparations of mucus clots and cytospin preparations to concentrate cell numbers. Concentrated direct smears of centrifuged samples are not as useful as cytospin preparations. Cells are more likely to lyse in manual preparations and these slides also underestimate the number of neutrophils[2].

Inflammatory lung disease results in multiple changes in cells from BAL and TTW. The epithelial cells appear more deeply basophilic and goblet cell hyperplasia is commonly observed. Increased numbers of inflammatory cells and in BALs, increased total cell counts and Churchman's spirals are common. With bacterial infections, aspiration pneumonia and fungal infections, increased numbers of neutrophils and lesser numbers of macrophages are identified. Bacterial infections are associated with degenerate neutrophils. Intracellular bacteria are commonly identified. In patients with aspiration pneumonia, a mixed bacterial population may be identified. Allergic and hypersensitivity responses in dogs and cats result in increased numbers of eosinophils and overall increases in cell counts. Chronic inflammatory responses may result in increased numbers of macrophages and lymphocytes in addition to neutrophils. Fungal pneumonia can result in suppurative to pyogranulomatous inflammation. Fungal organisms such as *Blastomyces dermatitidis* and *Histoplasma capsulatum* are easily identified on TTW and BAL samples. In people, cytology is considered a more reliable diagnostic tool than culture in diagnosis of *Blastomyces dermatitidis*[3]. *Pneumocystis carinii* is rarely diagnosed in dogs and has been reported in immunocompromised patients[4].

Neoplastic cells are occasionally observed in TTW and BAL samples but the tumors in these cases communicate with the airway. It is more common in these cases to see a significant inflammatory response and these cells may be more likely to exfoliate in the samples. Occasionally the neoplasm results in hemorrhage and erythrophagia, hemosiderin pigment and hematoidin crystals may be observed in the sample.

Lung aspirates are becoming more common, especially of mass lesions with ultrasound or CT guidance. In people, 84% of fine needle aspiration with CT guidance provided diagnostic material from lung aspirates[5]. A similar study was performed in paratracheal masses and 79% were diagnostic with a sensitivity of 97% and specificity of 100%[6]. Aspirates of healthy lung are of low cellularity. Low numbers of mesothelial cells, few macrophages and red blood cells may be seen, but many of the samples are acellular. Inflammatory and neoplastic diseases can be easily diagnosed in lung aspirates. Pulmonary carcinomas often present with a single lung mass. The samples are often highly cellular and consist of multiple clusters of neoplastic epithelial cells. These cells have moderate amounts of deeply basophilic cytoplasm and large round nuclei with stippled chromatin pattern and large prominent nucleoli. Marked anisocytosis and anisokaryosis are present. Often the cells contain discrete cytoplasmic vacuoles, sometimes a single large vacuole which displaces the nucleus, giving the cells a signet ring appearance. Less common lung tumors include histiocytic sarcoma, osteosarcoma and chondrosarcoma. These tumors do not exfoliate as well but can still result in a diagnostic specimen with fine needle aspiration. Sensitivity and specificity of diagnosis can be improved with cytochemical and immunocytochemical staining. Alkaline phosphatase staining of neoplastic cells is highly sensitive and specific for the diagnosis of osteosarcoma[7]. Staining with vimentin, CD18 and CD11b have been reported in the diagnosis of histiocytic sarcoma[8].

In conclusion, cytology of the respiratory tract is useful from nose to lung but has some strengths and weaknesses. Lung cytology of masses and disseminated disease can have excellent diagnostic yield. TTW and BALs can be useful for cytology but also in collecting samples for culture. Appropriate cytologic interpretation and appropriate slide preparation are essential to get the most out of the samples.

REFERENCES

[1]Doughty RW et al. **Vet Pathol** 2006;43:751-754. [2]Dehard S et al. **J Am Anim Hosp Assoc** 2008;44:285-94. [3]Patel AJ, et al. **Am J Surg Pathol** 2010;epub.[4]Watson PJ, et al.. **J Vet Intern Med** 2006;20;523-527. [5]Guimaraes MD, et al Clinics (Sao Paulo) 2009;64:1139-1144. [6]Barnett BP, et al. **Acta Cytol** 2009;53:672-678.[7]Barger A, et al. **Vet Pathol** 2005; 42:161-165. [8]Moore PF, et al. **Vet Pathol** 2006;43(5):632-645.

MODULATION OF THE ACUTE BRONCHOSPASM IN A MODEL OF FELINE ASTHMA. J. LEEMANS[1,2], N. KIRSCHVINK[1], P. GUSTIN[2]. 1. University of Namur, Faculty of Sciences, Namur, Belgium. 2. University of Liège, Faculty of Veterinary Medicine, Liège, Belgium.

Feline asthma is characterized by intermittent respiratory distress due to bronchoconstriction, non-specific bronchial hyperresponsiveness and airway remodeling at latter stages. In considering the pathogenesis of asthma, it is of major importance to control recurrent airway obstruction. In a preliminary *in situ* study, we demonstrated that different bronchodilators (BD) including β_2-adrenergic agonists (albuterol [AL], salmeterol [SLM]...), an anticholinergic (ipratropium bromide [IB]) and a methylxanthin derivative (theophylline) caused a dose-related relaxation of isolated feline bronchi. Our recent pre-clinical study conducted on healthy cats has highlighted the potential for BD delivered by metered-dose inhalers (MDI) to prevent muscarinic-induced bronchoconstriction. However, the clinical efficacy of inhaled BD in the management of acute asthmatic exacerbations remains to be explored. In continuity with our previous experiments, two randomized controlled crossover studies were designed to assess in Ascaris *suum* (AS)-sensitized cats the efficacy of (1) inhaled AL and IB in reversing allergen-induced bronchospasm and of (2) inhaled long-acting β_2–agonist SLM administered in a prophylactic fashion.

In the first study (n=5), four AS bronchial provocation challenges were carried out at 1 week intervals, followed by one of four treatment protocols delivered by MDI: SAL (100 μg), IB (20 μg), SAL/IB (100 μg/20 μg) or control (untreated). Responses to inhaled medications were assessed by barometric whole-body plethysmography and by computation of the area under the time–response curves of Penh from 0 to 60 or 120 min after drug administration (AUC_{0-60}, AUC_{0-120}). The times for Penh to half decrease ($T_{50\%}$) or to achieve nearly basal values ($T_{20\%}$) were also calculated. In the second study (n=5), two AS bronchial provocation tests were carried out whereby clinical score (Clin), respiratory rate (RR) and Penh were monitored prior to and repeatedly over the two hours following AS exposure. Cats were randomly assigned to receive either no treatment or SLM delivered by MDI (50 μg) 24, 12 and 1 hours before AS stimulation. The maximum AS-related changes in the aforementioned parameters (Max) were calculated as well as the area under the curve over the two first hours after allergen exposure (AUC_{2h}).

Inhalation of SAL, IB and SAL/IB via MDI failed to improve most Penh-derived parameters. No significant differences were found among the four study groups, with reference to the mean AUC_{0-60}, $T_{20\%}$ and $T_{50\%}$ values of Penh (P>0.05). Statistical comparisons between SLM-treated and untreated cats did not reach significance for MaxRR, MaxClin, MaxPenh, AUC_{2h}RR, AUC_{2h}Clin and AUC_{2h}Penh (P>0.05).

Under our experimental conditions, the inhaled BD (AL, IB, SLM) were of limited efficacy for the prevention and symptomatic treatment of allergic bronchospasm in cats sensitized to AS. This is in sharp contrast with the current and empirical-based recommendations that support inhaled BD in cats with naturally occurring asthma.

EVALUATION OF SOLUBLE-MESOTHELIN RELATED PEPTIDE AS A BIOMARKER FOR MALIGNANT MESOTHELIOMA IN DOGS. Sharp CR and Rozanski EA. Tufts Cummings School of Veterinary Medicine, North Grafton, MA.

Pleural effusion is a common presenting complaint in veterinary emergency rooms and can be caused by a wide variety of underlying diseases. Malignant effusions are associated with a poor prognosis but differentiation of malignant from benign disease is often difficult on the basis of pleural fluid cytology and routine imaging. Differentiating between normal, reactive and malignant mesothelial cells to refute or establish a diagnosis of malignant mesothelioma is particularly challenging. The objective of this study was to evaluate serum mesothelin related peptide (SMRP) as a biomarker for malignant mesothelioma in dogs. We hypothesized that the MesoMARK™ assay, designed to detect SMRPs in human serum, would cross react with canine SMRP and that dogs with malignant mesothelioma would have higher concentrations of SMRPs in both serum and pleural fluid, than dogs with pleural effusion in the absence of malignant mesothelioma. Additionally we hypothesized that dogs without pleural effusion would not have measurable serum concentrations of SMRPs.

Serum and pleural fluid was collected from dogs with confirmed mesothelioma (n=3) and pleural effusion of other etiologies (n=8). Additionally serum was collected from dogs without pleural effusion (n=5). Samples were stored at -20 to -80°C until analysis. A SMRP assay (MesoMARK™) was performed according to manufacturers instructions.

The group mean serum SMRP was significantly higher in dogs with mesothelioma compared to dogs with pleural effusion of other etiologies and normal dogs. There was no significant difference in SMRP concentration between dogs with pleural effusion of other etiologies and dogs without pleural effusion. Two of the three dogs with confirmed mesothelioma had extremely serum high concentrations of SMRP (202.9 and 246.5 nmol/L). One of these dogs, with pleural mesothelioma, had correspondingly high pleural fluid concentrations (182.1 nmol/L), while the other, with pericardial mesothelioma, did not. The other dog with confirmed mesothelioma had SMRP concentrations that overlapped with the non-mesothelioma groups. Six dogs in total did not have measurable SMRP concentrations; 4 in the non-mesothelioma pleural effusion group, and 2 in the non-pleural effusion group.

SMRPs were detected in canine serum and pleural fluid using the MesoMARK™ assay. In this small, pilot study, some dogs with malignant mesothelioma had elevated serum and pleural fluid SMRP concentrations, when compared to dogs with other causes of pleural effusion. SMRP is a potentially promising biomarker for malignant mesothelioma in dogs that warrants further investigation in a larger population of dogs with pleural effusion.

RESPIRATORY EVALUATION AND DISEASE IN NEONATAL CRIAS

Daniela Bedenice, Dr. med vet., Dipl. ACVIM, Dipl. ACVECC
North Grafton, MA

UPPER RESPIRATORY PATHOLOGY OF NEONATES

Choanal atresia is considered the most common congenital upper airway disease in South American Camelids. The choanae are the caudal openings of the nasal cavities into the nasopharynx.[1] Failure in the embryologic development of these openings may thus lead to unilateral or bilateral, membranous or osseous, and complete or partial choanal atresia.[2] Since camelids are primary nasal breathers rather than obligate nasal breathers, complete choanal atresia is not immediately fatal, but often presents with congenital respiratory distress.

Partial choanal atresia can be difficult to diagnose due to minimal or complete lack of clinical signs. In a study involving 72 animals (llamas, alpacas, and other related camelids), the reported incidence of choanal atresia was 10%. Of the affected animals, 64 (89%) had complete bilateral, 7 (10%) unilateral, and one had partial choanal atresia. However, it is possible that this study underestimated the true incidence of partial atresia, due to the absence of clinical signs in many patients. Choanal atresia is considered a heritable disorder with a complex mode of inheritance.[3]

Positive contrast rhinography, nasal catheter placement, endoscopy and computed tomography (CT) have all been used to diagnose choanal atresia. If surgical intervention is considered, CT could provide the spatial resolution necessary to differentiate the types of choanal atresia and estimate the length of the obstructive lesion.[1] Surgery may be beneficial in the membranous forms of choanal atresia, although the prognosis remains poor. In these situations, surgical access is achieved via a small incision at the nasoincisive notch, in the anesthetized and intubated cria. Membranous defects may thus be surgically opened using curved Metzenbaum scissors. The defect should then be dilated using an esophageal balloon catheter. Bougienage needs to be repeated several times under sedation over the following weeks.[4]

NEONATAL LOWER RESPIRATORY DISEASE

There is an apparent paucity of information regarding the clinical manifestation, etiology and outcome of neonatal cria pneumonia. Although pneumonia is a significant cause of morbidity and mortality in neonatal foals,[5,6] reports are limited to individual case reports in crias.[7] More specifically, pulmonary infection has been associated with gram negative sepsis (Salmonella typhimurium, Escherichia coli, Actinobacillus, and Klebsiella pneumoniae)[7,8] as well as opportunistic infection of the immune-compromised host. For example, BVDV is known to infect cells that are instrumental in the control of both the innate and acquired immune system. Secondary suppurative bronchopneumonia has thus been diagnosed in BVDV persistently infected crias.[9]

A single report has evaluated the significance of lower respiratory disease in critically ill neonatal camelids to date.[10] Lower airway disease [LAD] was identified in 61/182 (33.5%; 5 llamas, 56 alpacas) critically ill crias less that 4 weeks of age that were admitted to a referral center. A diagnosis of acute respiratory distress syndrome (ARDS) or acute lung injury (ALI) was based on a recent consensus statement[11] and could only be obtained in 6/61 (9.8%) crias with LAD. Of these, four crias survived to discharge. The most commonly identified concurrent diseases in crias with LAD included failure of passive transfer, systemic inflammatory response syndrome, clinical signs of prematurity, neonatal encephalopathy, confirmed sepsis and diarrhea. The latter study further identified specific indicators of LAD in hospitalized patients (manuscript in progress), as well as potential risks of mortality. A total of 142/182 (78%) critically ill crias were discharged alive from the hospital. Mortality was significantly higher in crias with LAD (39.3%) compared to animals without lower airway dysfunction (13.2%). Similar observations have been previously obtained in neonatal foals at the same cohort.[6] The latter study documented that 35% of neonatal foals with radiographic evidence of pulmonary disease at the time of admission did not survive to discharge, while the mortality of foals without respiratory illness was merely 14%.[12] The majority (54%) of deceased foals were euthanized due to futility of treatment. In contrast, 13/24 (54.2%) crias with LAD died naturally and 9/24 (37.5%) were euthanized. These data suggest that life-threatening respiratory illness may me more insidious in camelids compared to foals.

Open mouth breathing, severe depression, prematurity, recumbency and a diagnosis of systemic inflammatory response syndrome (SIRS) have been significantly associated with the development of lower respiratory disease in neonatal crias. It is important to remember that apparent signs of lung dysfunction may also be associated with non-respiratory conditions such as metabolic derangements (e.g. severe acidosis), pain, abdominal crisis, fever or high environmental temperatures and excitement.[13] Clinical signs of pulmonary diseases may thus be non-specific and difficult to differentiate by physical examination alone in camelids.

We postulate that infectious respiratory diseases in crias are usually part of a multiple-organ, systemic infection or sepsis, similar to observations in foals.[14] Based on a recent report of culture-positive, septic neonatal crias, the most common isolates may include Escherichia coli, Enterococcus spp., Listeria monocytogenes, and Citrobacter spp.[15] Development of respiratory disease in neonates is often related to abnormal perinatal respiratory development, abnormal parturition, aspiration and seeding of primary infectious agents or opportunistic pathogens.[14]

THE IMPACT OF IMMUNE FUNCTION

Pneumonia is commonly associated with a compromise in the immunologic protection of the neonate. Even in healthy newborns, reduced complement values, defective chemotaxis (directed migration of neutrophils or macrophages) and killing capacity of the neonatal neutrophil contribute to a relatively decreased defense against invading bacteria in other species. Furthermore, cellular immunity[16] and local pulmonary defenses may be reduced in neonatal animals. For example, an immature ciliary apparatus and the presence of fewer alveolar macrophages in neonates in comparison to adult horses lead to decreased bacterial clearance from the lungs.[17]

Failure of passive transfer of immunity (FPT) is a well known risk factor of neonatal infection, including respiratory disease.[18,19] Affected newborns are not only deprived of specific maternal antibody protection, but their neutrophil function is also seriously impaired.[20] Camelids have a thick layered epitheliochorial placenta, which prevents transplacental transfer of IgG. Pre-colostrum serum IgG levels are therefore low in camelids, with concentrations of 0.26 +/- 10.23 mg/ml. Maximum IgG levels are reached after 24 h of colostrum ingestion. The plateau concentration of 24.52 +/- 8.8 mg IgG/dl subsequently declines after 2-5 weeks. IgG concentrations above 10 mg/ml post partum indicate a successful passive transfer in camelids.[21] Although the neonatal camelid is immunocompetent at birth, it is immunologically naive and therefore dependent on the passively acquired humeral immunity.

Immune responses in llamas and alpacas are unique in that camelid IgG antibodies differ from all other known immunoglobulins. Camelids produce both conventional heterotetrameric IgG (containing paired heavy and light polypeptide chains) common to all vertebrates, as well as functional homodimeric heavy-chain antibodies that lack the light chains.[22] At present, three subclasses of camelid IgG have been identified (IgG_1: conventional antibodies; IgG_2; IgG_3), of which IgG_2 and IgG_3 do not have light chains.[23,24] IgG_2 and IgG_3 consist of dimers of short heavy chains that total approximately 75% of serum IgG. It has been suggested that these unconventional antibodies could represent an evolutionary advantage, being more efficient than conventional immunoglobulins to inhibit microbial enzymes, and thus exerting a more protective immune response against pathogens.[25] Whether these mechanisms enhance protection against neonatal respiratory infection, however, remains speculative.

PULMONARY FUNCTION AND GAS EXCHANGE

Arterial blood gas analysis is preferentially used to determine oxygenation and ventilation in neonatal crias. Technically, an arterial blood sample can most easily be obtained from the saphenous artery of the medial thigh. Normal arterial blood gas result have been documented in 14 adult, healthy male and non-pregnant female alpacas (mean weight: 62 ± 21kg, mean age: 4.5 yrs) as well as 12 healthy neonatal alpaca crias (2-7days old, mean 7.8 kg body weight):[13]

Normal arterial blood gas results in healthy neonatal (2-7 days) and adult alpacas

Variable	n	Mean	95% confidence interval
pH	14	7.47	7.46 – 7.48
Bicarbonate (mEq/L)	14	20.4	19.6 – 21.3
PaO_2 (adults)	14	104.6	100.7 – 108.4
Neonatal PaO_2	12	92	87.7 – 96.4
$PaCO_2$ (adults)	14	27.7	26.4 – 28.9
Neonatal $PaCO_2$	12	30.5	27.2 – 33.8

PaO_2 = partial arterial pressure of oxygen in mmHg, **$PaCO_2$** = partial arterial pressure of carbon dioxide in mmHg,

The effect of storage type (plastic vs. glass), temperature (iced at 4oC vs. 22oC room temperature) and blood gas tension (normoxemia, hypoxemia, hyperoxemia) was recently evaluated in arterial blood gases (ABG) of healthy alpacas, to facilitate ABG analysis under field conditions.[26] Statistically significant alterations in pH, PaCO2, lactate and bicarbonate were only associated with room temperature storage. However, mean changes in arterial blood gas parameters still remained below 3.40 mmHg for PaCO2, 0.5 mmol/L for lactate, 1.22 mEq/L for bicarbonate and 0.03 points for pH after two hour storage. In contrast, the PaO2 significantly decreased in both normoxic and hyperoxic blood samples, following storage in glass syringes at room temperature. Similar decreases in PaO2 were observed in hyperoxygenated blood stored at room temperature in plastic syringes. In contrast, PaO2 significantly increased after storage of hyperoxic blood samples in plastic syringes on ice. Overall, storage in glass syringes on ice resulted in the lowest variability of PaO2, with the highest mean PaO2 increase being 10.28 mmHg under these conditions. Based on these results, ABG from alpacas should be stored on ice in glass syringes, to reduce variability in PaO2 if analysis is delayed. In contrast to human reports, the PaO2 of cooled hyperoxygenated camelid blood increases following storage in plastic syringes.

Functional Residual Capacity (FRC) in neonatal animals

Non-invasive measurements of end-expiratory lung volume (or FRC) has been evaluated in both llamas and alpacas,[27,28] and may serve as an index of mechanical lung failure. Lung volume measurements thus assist in the diagnosis of lung fibrosis, airway inflammation, and inspiratory airway obstruction (e.g. laryngeal paralysis). FRC is defined as the volume of gas that remains in the respiratory system at passive end-expiration, and is determined by the balance of chest wall compliance and lung elastance. FRC is decreased in neonates that have a more compliant chest wall and in patients with restrictive lung diseases (e.g. pulmonary fibrosis, pneumonia) that will reduce final lung volume. In contrast, FRC may be increased in patients with airway obstructive diseases that result in gas trapping. The normal FRC in adult llamas was measured at 39.18 ml/kg (± 11.45).[28] In this study, FRC was positively correlated (p = .05) to body length (point of shoulder to ischial tuberosity) but not to weight, height, or circumference in adult llamas. Similarly, the mean FRC in adult alpacas was reported as 3.19 ± 0.53 L, which translated to 46.29 ± 7.50 ml/kg.[29] A positive correlation was observed between FRC and body weight in the study participants (*P = 0.003; r = 0.645)*. A mean FRC of 45.13 ml/kg (95% CI: 40.9 – 49.4) was similarly obtained in eleven clinically healthy alpaca crias (age: 1–18 days).[27]

REFERENCES

[1.] Nykamp SG, Dykes NL, Cook VL, et al. Computed tomographic appearance of choanal atresia in an alpaca cria. **Vet Radiol Ultrasound** 2003;44:534-536. [2.] Smith BB, Timm K, Huber M, et al. Choanal atresia in the alpaca **The Alpaca Registry Journal** 2000. [3.] Hathcock J, Pugh D, Cartee R, et al. Computed tomography of the llama head: technique and normal anatomy. **Vet Radiol Ultrasound** 1995;36:290-296. [4.] Whitehead CE. Management of neonatal llamas and alpacas. **Vet Clin North Am Food Anim Pract** 2009;25:353-366. [5.] Cohen ND. Causes of and farm management factors associated with disease and death in foals. **J Am Vet Med Assoc** 1994;204:1644-1651. [6.] Bedenice D, Heuwieser W, Solano M, et al. Risk factors and prognostic variables for survival of foals with radiographic evidence of pulmonary disease. **J Vet Intern Med** 2003;17:868-875. [7.] Adams R, Garry FB. Gram-negative bacterial infection in neonatal New World camelids: six cases (1985-1991). **J Am Vet Med Assoc** 1992;201:1419-1424. [8.] Anderson NV, Anderson DE, Leipold HW, et al. Septicemic salmonellosis in two llamas. **J Am Vet Med Assoc** 1995;206:75-76. [9.] Mattson DE, Baker RJ, Catania JE, et al. Persistent infection with bovine viral diarrhea virus in an alpaca. **J Am Vet Med Assoc** 2006;228:1762-1765. [10.] Bedenice D, Wesley J. Prognosis and outcome of critically ill neonatal crias. In: International Camelid Conference, Corvallis, OR 2009. [11.] Wilkins PA, Otto CM, Baumgardner JM, et al. Acute lung injury and acute respiratory distress syndromes in veterinary medicine: consensus definitions:The Dorothy Russell Havemeyer Working Group on ALI and ARDS in Veterinary Medicine. **J Vet Emergency Critical Care** 2007;17:333–339. [12.] Bedenice D, Paradis MR. ARDS and ALI in neonatal foals. In: **Dorothy Havemeyer Septicemia Workshop** 2009. [13.] Bedenice D. Approach to the critically ill camelid. **Vet Clin North Am Food Anim Pract** 2009;25:407-421. [14.] Paradis MR. Infectious diseases of the equine respiratory tract: from gestation to five months. **Veterinary Medicine** 1989;84:1174-1177. [15.] Dolente BA, Lindborg S, Palmer JE, et al. Culture-positive sepsis in neonatal camelids: 21 cases. **J Vet Intern Med** 2007;21:519-525. [16.] Prescott JF. Immunodeficiency and serious pneumonia in foals: the plot thickens. **Equine Vet J** 1993;25:88-89. [17.] Zink MC, Johnson JA. Cellular constituents of clinically normal foal bronchoalveolar lavage fluid during postnatal maturation. **Am J Vet Res** 1984;45:893-897. [18.] Bedenice D, Heuwieser W, Brawer R, et al. Clinical and prognostic significance of radiographic pattern, distribution, and severity of thoracic radiographic changes in neonatal foals. **J Vet Intern Med** 2003;17:876-886. [19.] McGuire TC, Pfeiffer NE, Weikel JM, et al. Failure of colostral immunoglobulin transfer in calves dying from infectious disease. **J Am Vet Med Assoc** 1976;169:713-718. [20.] Leblanc MM, Pritchard EL. Effects of bovine colostrum, foal serum immunoglobulin concentration and intravenous plasma transfusion on chemiluminescence response of foal neutrophils. **Anim Genet** 1988;19:435-445. [21.] Wernery U. Camelid immunoglobulins and their importance for the new-born -- a review. **J Vet Med B Infect Dis Vet Public Health** 2001;48:561-568. [22.] Muyldermans S, Lauwereys M. Unique single-domain antigen binding fragments derived from naturally occurring camel heavy-chain antibodies. **J Mol Recognit** 1999;12:131-140. [23.] Hamers-Casterman C, Atarhouch T, Muyldermans S, et al. Naturally occurring antibodies devoid of light chains. **Nature** 1993;363:446-448. [24.] Azwai SM, Carter SD, Woldehiwet Z. Immune responses of the camel (Camelus dromedarius) to contagious ecthyma (Orf) virus infection. **Vet Microbiol** 1995;47:119-131. [25.] Nguyen VK, Desmyter A, Muyldermans S. Functional heavychain antibodies in Camelidae. **Adv Immunol** 2001;79:261-296. [26.] Buckley GJ, Holmes KM, Bedenice D. The effect of different storage conditions and oxygen tensions on arterial blood gas analysis in healthy alpacas. In: Veterinary Comparative Veterinary Society Conference, Portsmouth, MA 2009. [27.] Paxson JA, Cunningham SM, Rush JE, et al. The association of lung function and plasma volume expansionin neonatalalpaca crias following plasma transfusion for failure of passive transfer. **J Vet Emergency Critical Care** 2008;18:601–607. [28.] Lascola KM, Hoffman AM, Mazan MR, et al. Respiratory mechanics in sedated and nonsedated adult llamas. **Am J Vet Res** 2007;68:676-684. [29.] Pacheco A, Bedenice D, Mazan MR, et al. Respiratory mechanics and bronchoalveolar lavage in healthy adult alpacas In: **VCRS**, Portsmouth, MA 2009.

RESPIRATORY EVALUATION & DISEASE IN ADULT CAMELIDS

Daniela Bedenice, Dr. med vet., DACVIM, Dipl. ACVECC
North Grafton, MA

A variety of primary lung diseases have been reported in llamas and alpacas in the peer-reviewed literature, although pulmonary dysfunction may be underestimated, due to subtle presenting signs and lack of routine functional analysis of the respiratory tractin this species. Diseases such as tuberculosis,[1] nocardiosis,[2] viral pneumonia (coronavirus,[3] adenovirus-associated pneumonia,[4] bovine herpesvirus type 1[5]), pulmonary neoplasia (adenocarcinoma,[6] adenosquamous carcinoma,[7] lymphosarcoma[8]), fungal pneumonia (pulmonary aspergillosis,[9,10] pulmonary histoplasmosis[11]), diaphragmatic paralysis,[12] rhodococcus equi-associated lymphadenitis,[13] endogenous lipid pneumonia[14] and cryptococcosis[15] have been identified to date.

The importance of contagious respiratory illness in camelids recently became apparent, when a novel acute respiratory syndrome [ARS] was recognized in California and the East coast of the United States in 2007.[3] Clinical signs of disease ranged from mild upper respiratory disease with influenza-like presentation to severe respiratory illness resulting in death. ARS was thus characterized by acute respiratory signs, high fever, and occasional sudden death, which was more common in pregnant alpacas. A novel coronavirus was subsequently recovered from lung tissue of a clinical case submitted to the California Animal Health and Food Safety Laboratory System (CAHFS). Furthermore, serologic results in 40 animals with a history of ARS and 167 controls, demonstrated a strong association between the presence of novel coronavirus antibodies, exposure to ARS and recent respiratory outbreaks, with an odds ratio of 121 (95% confidence interval: 36.54 and 402.84; $P < 0.0001$). Autopsy findings in affected animals identified severe pulmonary congestion and edema, often with marked pleural effusion. Histologically, there was severe pulmonary congestion and edema with a marked, diffuse, acute to subacute, interstitial to bronchointerstitial pneumonia. Salient features included free fibrin deposition within the lumen of terminal airways and alveoli, often with hyaline membrane formation. Although pathologic, histologic, and serologic results indicated a relationship between the novel coronavirus and the outbreak of ARS in alpacas, the Koch postulates to prove a true association between the virus and ARS still need to be fullfilled.[3]

THE RESPIRATORY TRACT OF NORMAL ADULT CAMELIDS

Camelids have 12 ribs,with lungs that most closely resemble those of horses. A cardiac notch separates the apical portion of the lung, but there are no lobes, except for a small accessory lobe of the right lung. The mediastinum is complete.[16] The resting respiratory rate of adult llamas or alpacas varies from 10-30 bpm, although normal lung sounds are often muted in heavily fleeced animals.

Standing **THORACIC RADIOGRAPHS** are generally limited to lateral views in adult camelids. A study of normal thoracic radiographs in llamas reported that pulmonary vessel visualization and definition is similar to that of the normal equine and better than in the bovine. The pulmonary vasculature was generally easily identified in the lung periphery and vessel edge definition was generally good. In only a few instances was peripheral pulmonary vascular visualization judged as poor. The degree of visualization of the caudal vena cava and descending aorta has been used as an indicator of the presence or absence of pulmonary infiltrates, which may mask vessel margin clarity and/or vessel visualization. The caudal vena cava was always seen in its entirety in the study of normal llamas. The dorsal margin was sharply defined in 15/16 subjects, while the ventral margin was only sharply defined in 7/16 llamas.[17]

The analysis of **PLEURAL FLUID** may be an important diagnostic tool in certain cases of respiratory illness as recently outlined in alpacas with ARS[3] and llamas with pulmonary neoplasia.[7] The preferred site of pleurocentesis is the 6th or 7th intercostal space, 10-15 cm dorsal to the ventrum of the sternum or 2-4 cm dorsal to the costochondral junction of the ribs in normal llamas.[18] Penetration can be made from either side and may utilize a 14-16 gauge needle, or cannula (2 inch length), inserted near the cranial border of the rib to avoid intercostal vessels. In normal llamas fluid should flow from the needle at an approximate depth of 2-3cm (1-1.5 inches). The pleural fluid of 17 clinically healthy adult llamas identified a normal total nucleated cell count of 200–1500 cells/ml with a mean of 576 +/- 361 cells/microliter. In these fluids, small lymphocyte, mononuclear cell, and neutrophil percentages ranged from 80%–100%, 0%–20%, and 0%– 10%, respectively. Pleural fluid total protein concentrations, determined by refractometry, ranged from < 2.5 to 3.5 g/dL. The pleural fluid glucose and lactate concentrations were very similar to the plasma concentrations in normal llamas, with a pleural fluid specific gravity of 1.0133 +/- 0.002, glucose concentration of 135.1 +/- 9.02 mg/dL and lactate of 2.95 +/- 1.34 mg/dL.[19,20]

The cytologic characteristics of tracheal fluid obtained by **TRANSTRACHEAL ASPIRATION** have also been described in 17 clinically healthy adult llamas.[19] The technique for transtracheal aspiration was similar to that used in horses. A 10 cm^2 area over the ventral mid-cervical region was clipped, aseptically prepared and locally infiltrated with Lidocaine HCl (2% solution). A stab incision was made through the skin with a number 15 scalpel blade to allow insertion of a 10-gauge, through-the-needle Winch catheter between the tracheal rings. Approximately 20 mlof sterile saline was injected through the catheter into the trachea, then immediately aspirated back into the syringe. Cytologic evaluation of TTA fluid revealed that the majority of cells were vacuolated macrophages (60-100%), with 0-40% neutrophils, and fewer lymphocytes (0-2 %), eosinophils (0-3%), and ciliated respiratory epithelial cells (0-10%).[19]

Cytological evaluation of the **BRONCHO-ALVEOLAR LAVAGE FLUID (BALF)** in healthy alpacas recently identified 58.52 +/-12.36% alveolar macrophages, 30.53 +/-13.78% lymphocytes, 10.95 +/-9.29% neutrophils, 0% mast cells and several ciliated epithelial cells. The percentage of alveolar macrophages and lymphocytes was therefore similar as previously reported in horses, while the mean percentage of neutrophils was higher in alpacas.[21] In this study all animals were anesthetized with a combined intramuscular injection of xylazine (0.5 mg/kg BW), butorphanol (0.05 mg/kg BW) and ketamine (5 mg/kg BW) and maintained in sternal recumbency with the head and neck extended vertically. A protective mouth gag was placed to allow visualization of the larynx with the aid of a laryngoscope. An enteral feeding tube (Mila International Inc, Erlanger, KY) containing a guide wire and measuring 250 cm in length and 6 mm in diameter was inserted into the animal's trachea and advanced until wedged. Following removal of the guide wire, 50 ml of sterile saline was instilled and immediately withdrawn for sampling. An additional 1-3 aliquots were instilled based on turbidity of the sample obtained. A lateral and a dorso-ventral thoracic radiograph were obtained at the end of the procedure to verify the location of the collection tube. The described BAL technique yielded samples with adequate cellularity and resulted in no complications in healthy alpacas, but it its use may be limited in clinical patients, as short-term, injectable anesthesia was required. [22]

HIGH ALTITUDE ADAPTATIONS

Llamas, and other South American Camelids, are native to the Andean altiplana region and show several evolutionary adaptations to environmental hypoxia. For example, the small elliptical red blood cells of camelids have a greater affinity for oxygen, a high hemoglobin concentration and high oxygen transfer conductance (facilitated by the small RBC size).[23] The oxygen dissociation curves of the blood of lamoids are thus shifted to the left, even in camelids adapted to low altitudes. Furthermore, llamas were identified to carry a higher blood oxygen content than, for example, sheep.[24] A high muscle myoglobin concentration, more efficient oxygen extraction at the tissue level[25] and high lactate dehydrogenase activity[26] further improve the camelid's resistance to hypoxemia. The overall interpretation of arterial blood gas results (see "RESPIRATORY EVALUATION AND DISEASE IN NEONATAL CRIAS" for reference ranges) follows the guidelines for other mammalian species. In the absence of alterations in the inspired oxygen fraction (FiO_2), arterial hypoxemia may result from four basic processes, including (1) hypoventilation, (2) diffusion impairment, (3) ventilation – perfusion inequality (VQ mismatch), and (4) shunt (intra- and extrapulmonary shunt).[27,28]

DIFFERENTIATING CAUSES OF DYSPNEA – EVALUATION OF RESPIRATORY DISTRESS AND DYSFUNCTION IN CAMELIDS

Respiratory distress is generally defined as outwardly evident, labored respiratory efforts or ventilation i.e. the clinically apparent inability to adequately ventilate and/or oxygenate. Primary inspiratory dyspnea may be related to alveolar hypoventilation (impaired air delivery to the lung) or inadequate gas exchange. Expiratory dyspnea may accompany inspiratory distress in any animal with severe impairment of gas exchange. However, primary expiratory dyspnea is usually associated with obstruction of the intrathoracic airways by mucous and bronchospasm (e.g. recurrent airway obstruction).

The evaluation of lung and airway disease in SAC may pose a particular challenge to veterinarians, as llamas and alpacas often merely display subtle clinical abnormalities even in the face of progressive or severe respiratory disease. Specific analysis of the breathing pattern in camelids has shown that asynchronous breathing is commonly observed during conditions of upper airway obstruction, phrenic nerve dysfunction, diaphragmatic or intercostal muscle paralysis and diaphragmatic fatigue due to chronic airway disease.[12,28]

Respiratory inductive plethysmography and pneumotachography (RIP)

The concurrent use of respiratory inductive plethysmography and pneumotachography (RIP) permits the noninvasive assessment of breathing pattern, flow limitation and ventilation as previously described.[29] This technique has been used to describe the breathing patterns of healthy alpacas,[22] llamas,[30] and a llama with diaphragmatic paralysis.[12] Evidence of lower airway obstruction may be identified by measured differences between plethysmographic and nasal flow over the first 25% of expiration (delta flow). The average delta flow recorded in healthy alpacas is near zero (0.18+/- 0.07 L/s)[22] as found in llamas (0.04 +/- 0.14 L/s).[30]

Forced Oscillatory Mechanics (FOM)

Novel non-invasive modalities such as Forced Oscillatory Mechanics (FOM) may assess the obstruction to airflow in non-sedated camelids. In short, the flow and pressure change of small air oscillations that are superimposed on the animal's spontaneous breathing via a face mask, are measured to calculate respiratory system resistance. This technique is easily tolerated in untrained camelids and may rapidly distinguish upper versus lower airway disease, without compromising the animal's breathing.

Functional Residual Capacity (FRC)

FRC is defined as the volume of gas that remains in the respiratory system at passive end-expiration, and is determined by the balance of chest wall compliance and lung elastance. FRC may thus serve as an index of mechanical lung failure and assist in the diagnosis of lung fibrosis, airway inflammation, and inspiratory airway obstruction (e.g. laryngeal paralysis). The normal FRC in adult llamas was non-invasively measured via helium dilution, with a mean FRC of 39.18 ml/kg (± 11.45).[30] In this study, FRC was positively correlated ($p = .05$) to body length (point of shoulder to ischial tuberosity) but not to weight, height, or circumference in adult llamas. Similarly, the mean FRC in alpacas was reported as 3.19 $\pm$ 0.53 L, which translated to 46.29 $\pm$ 7.50 ml/kg. A positive correlation was observed between FRC and body weight in the study participants ($P = 0.003$; $r = 0.645$).[22]

Non-invasive pulmonary function testing (PFT) has been validated in 10 non-sedated llamas to established normal parameters of respiratory mechanical function.[30] Measurements of pulmonary function included: functional residual capacity (FRC) via helium dilution, respiratory inductive plethysmography (RIP) to assess breathing pattern and flow limitations, esophageal-balloon pneumotachography and monofrequency forced oscillatory techniques (FOT: 1-7Hz) before and after sedation with 0.2 mg/kg xylazine IM. The following measurements of respiratory function were obtained in non-sedated llamas: FRC (5.60 +/-1.24 L), tidal volume (V_T = 1.03 +/- 0.3 L), dynamic compliance (C_{dyn}= 0.83 +/-0.4 L/cmH_2O), pulmonary resistance (R_L=1.42 +/-0.54 cmH_2O/L/s) and respiratory system resistance (R_{RS} in $cmH_2O/L/s$) at 1 Hz (2.4 +/-0.9), 2 Hz (2.3 +/-0.7) , 3 Hz (2.2 +/-0.6), 5 Hz (2.7 +/-0.7) and 7 Hz (2.5 +/-0.5) using FOT. Measurements of flow limitations via RIP were comparable to other species. However, xylazine sedation induced significant increases in R_L and maximal pleural pressure changes. A mean 127% increase in R_L and mean 116% increase in R_{RS} across 1-7 Hz, were observed following sedation. The magnitude of change in R_{RS} increased with decreasing impulse frequency, suggesting bronchoconstriction. Similar lung function tests were recently performed in non-sedated alpacas,[22] identifying the following results: FRC (3.19 +/-0.53 L), tidal volume (V_T =0.8 +/-0.13 L), respiratory system resistance (R_{RS} in $cmH_2O/L/s$) at 1Hz (2.70 +/-0.88), 2Hz (2.98 +/-0.70) , 3Hz (3.14 +/-0.77), 5Hz (3.45 +/-0.91) and 7 Hz (3.84 +/-0.93). The phase angle, as a measurement of thoracoabdominal asynchrony, was 19.59 +/-10.06° and the delta flow between nasal and plethysmographic flow measurements was 0.18 +/-0.07 L/s. The V_T, peak inspiratory and expiratory flows were significantly higher in cushed compared to standing animals. The authors of both studies concluded that PFT was well tolerated in untrained, awake camelids and may have extensive application in clinical practice. Due to alteration in respiratory function, testing should not be performed after xylazine sedation.

REFERENCES

1. Oevermann A, Pfyffer GE, Zanolari P, et al. Generalized tuberculosis in llamas (Lama glama) due to Mycobacterium microti. **J Clin Microbiol** 2004;42:1818-1821. 2. Chang CD, Boosinger TR, Dowling PM, et al. Nocardiosis in a llama. **J Vet Diagn Invest** 1993;5:631-634. 3. Crossley BM, Barr BC, Magdesian KG, et al. Identification of a novel coronavirus possibly associated with acute respiratory syndrome in alpacas (Vicugna pacos) in California, 2007. **J Vet Diagn Invest** 2010;22:94-97. 4. Galbreath EJ, Holland RE, Trapp AL, et al. Adenovirus-associated pneumonia and hepatitis in four llamas. **J Am Vet Med Assoc** 1994;204:424-426. 5. Williams JR, Evermann JF, Beede RF, et al. Association of bovine herpesvirus type 1 in a llama with bronchopneumonia. **J Vet Diagn Invest** 1991;3:258-260. 6. Ramos-Vara JA, Miller MA. Metastatic pulmonary adenocarcinoma in a llama (Lama glama). **J Vet Diagn Invest** 2002;14:328-331. 7. Ramos-Vara JA, Loiacono CM, Williams F, 3rd, et al. Pulmonary neoplasia in two llamas (Lama glama). **Vet Pathol** 2004;41:520-523. 8. Cebra CK, Garry FB, Powers BE, et al. Lym phosarcoma in 10 New World Camelids. **J Vet Intern Med** 1995;9:381-385. 9. Hughes K, Mueller K. Pathologic lesions of mycotic pneumonia in an alpaca following third compartment ulceration. **J Vet Diagn Invest** 2008;20:672-675. 10. Muntz FH. Oxalate-producing pulmonary aspergillosis in an alpaca. **Vet Pathol** 1999;36:631-632. 11. Woolums AR, DeNicola DB, Rhyan JC, et al. Pulmonary histoplasmosis in a llama. **J Vet Diagn Invest** 1995;7:567-569. 12. Bedenice D, Mazan MR, Kuehn H, et al. Diaphragmatic paralysis due to phrenic nerve degeneration in a llama. **J Vet Intern Med** 2002;16:603-606. 13. Hong CB, Donahue JM. Rhodococcus equi-associated necrotizing lymphadenitis in a llama. **J Comp Pathol** 1995;113:85-88. 14. Hamir AN, Andreasen CB, Pearson EG. Endogenous lipid pneumonia (alveolar histiocytosis) and hydrocephalus in an adult llama (Llama glama). **Vet Rec** 1997;141:474-475. 15. Bildfell RJ, Long P, Sonn R. Cryptococcosis in a llama (Lama glama). **J Vet Diagn Invest** 2002;14:337-339. 16. Viera RF, Sato Sato A, Nunez MQ. The lung and bronchial tree in alpacas. **Rev Fac Med Vet Univ Nac Mayor San Marcos (Lima)** 1968;22:54-60. 17. Mattoon JS, Gerros TC, Brimacombe M. Thoracic radiographic appearance in the normal llama. **Vet Radiol Ultrasound** 2001;42:28-37. 18. Fowler ME. Thoracocentesis. In: Fowler ME, ed. **Medicine and Surgery** of South American Camelids. Ames, IA: Iowa State University Press; 1989:44. 19. Gerros TC, Andreasen CB. Analysis of transtracheal aspirates and pleural fluid from clinically healthy llamas (Llama glama). **Vet Clin Pathol** 1999;28:29-32. 20. Tornquist SJ. Clinical pathology of llamas and alpacas. **Vet Clin North Am Food Anim Pract** 2009;25:311-322. 21. Couetil LL, Hoffman AM, Hodgson J, et al. Inflammatory airway disease of horses. **J Vet Intern Med** 2007;21:356-361. 22. Pacheco A, Bedenice D, Mazan MR, et al. Respiratory mechanics and bronchoalveolar lavage in healthy adult alpacas In: **VCRS**, Portsmouth, MA 2009. 23. Jurgens KD, Pietschmann M, Yamaguchi K, et al. Oxygen binding properties, capillary densities and heart weights in high altitude camelids. **J Comp Physiol B** 1988;158:469-477. 24. Moraga F, Monge C, Riquelme R, et al. Fetal and maternal blood oxygen affinity: a comparative study in llamas and sheep. **Comp Biochem Physiol A Physiol** 1996;115:111-115. 25. Benavides CE, Perez R, Espinoza M, et al. Cardiorespiratory functions in the fetal llama. **Respir Physiol** 1989;75:327-334. 26. Llanos AJ, Riquelme RA, Sanhueza EM, et al. The fetal llama versus the fetal sheep: different strategies to withstand hypoxia. **High Alt Med Biol** 2003;4:193-202. 27. West JB. **Pulmonary Pathophysiology. Baltimore: Williams & Wilkins**; 1982. 28. Bedenice D. Approach to the critically ill camelid. **Vet Clin North Am Food Anim Pract** 2009;25:407-421. 29. Miller C, Hoffman AM, Hunter J. Thoracoabdominal asynchrony failed to grade airway obstructions in foals. **J Appl Physiol** 2000;88:2081-2087. 30. Lascola KM, Hoffman AM, Mazan MR, et al. Respiratory mechanics in sedated and nonsedated adult llamas. **Am J Vet Res** 2007;68:676-684.

TRANSCRIPTOME ANALYSIS IN THE INVESTIGATION OF RUMINANT RESPIRATORY DISEASE

Tamara Gull, DVM, DACVIM, DACVPM, PhD
Stillwater, OK

INTRODUCTION

Global transcriptional analysis, or the investigation of types and numbers of cellular mRNA transcripts present in a sample, has become a widely-used investigative technique in research. There are several methods available with which transcriptomes may be probed, each with distinct requirements and advantages. Most studies using these methods have been done on human tissues and laboratory animal species, and some very intriguing results have been reported. Relatively little research has been published applying these techniques to ruminant respiratory disease, but a number of studies have used transcriptome analysis in the investigation of bovine reproduction, nutrition, meat quality, parasite resistance and Johne's disease.

MICROARRAY

The most commonly-used method of transcriptome analysis is the microarray. This technique uses small cDNA or oligonucleotide probes spotted onto a glass slide or built onto a matrix. The mRNA of the tissue or organism being investigated is transcribed into cDNA and labeled with a fluorescent dye. The labeled cDNA is hybridized with the probes on the slide and the intensity of fluorescence is measured. The intensity of fluorescence corresponds to the amount of target mRNA present. This technique works best with genomes that are fully sequenced, although even sequenced genomes suffer from limited annotation resulting in a number of array spots denoting genes or expressed sequence tags (ESTs) whose function is not known. Microarrays are known to be a less-sensitive tool than other gene expression analyses such as Northern blot or real-time quantitative polymerase chain reaction, but they allow a much larger number of genes or ESTs to be evaluated at a time. Most published microarray data has been applied to humans, laboratory animal species or human pathogens. Limited research has been published for the bovine, which will be discussed later. There are no published reports of microarray analysis in ovine, caprine or camelid respiratory disease.

MICROARRAY PLATFORMS AVAILABLE

Reported microarray platforms for investigators into the bovine genome include those from the W.M. Keck Center for Comparative and Functional Genomics at the University of Illinois (http://www.biotech.uiuc.edu/centers/Keck/) and the National Bovine Functional Genomics Consortium (NBFGC) and BOTL-series bovine total leukocyte arrays produced by the Center for Animal Functional Genomics at Michigan State University (http://www.cafg.msu.edu). Other commercially-available arrays are the Affymetrix Bovine GeneCHIP, the Agilent Bovine Gene Expression Microarray, and several bovine cDNA and oligonucelotide arrays from Ark-Genomics. Heterologous hybridization of bovine tissues onto non-bovine arrays has been reported, but hybridization efficiency is highly variable. No ruminant arrays other than bovine have been developed, although the Ark-Genomics bovine 20K cDNA array webpage (http://www.ark-genomics.org/) states that it is also suitable for use in sheep. In addition, ruminant immuno-inflammatory gene arrays have been developed at the Moredun Institute and used in sheep[1]. Additional noncommercial bovine arrays[2] have been applied in sheep. Pathogen arrays used in the investigation of bovine respiratory disease include those made by the Bacterial Microarray Group (BμG@s, http://www.bugs.sgul.ac.uk/), plus *Pasteurella multocida, Mycoplasma capricolum capricolum* and *Mycoplasma mycoides mycoides* (small colony) arrays, among others, produced by NimbleGen (http://www.nimblegen.com/).

SERIAL ANALYSIS OF GENE EXPRESSION (SAGE)

SAGE is a method of transcriptome analysis that is based on the identification of unique sequence tags from mRNA transcripts in a sample[3]. The individual tags (usually 9-14 bases long) are concatenated into long DNA molecules, and sequencing of the longer molecules reveals relative transcript numbers. SAGE does not require advance knowledge of the genome sequence, since expressed genes are sequenced after discovery; this allows the identification of new genes and does not limit the investigator to those sequences present on a microarray. However, the number of tags that require sequencing is dependent on the number of anticipated genes in the genome of the organism, as low-copy-number tags are less likely to be discovered.

MASSIVELY PARALLEL SIGNATURE SEQUENCING (MPSS)

MPSS is another method of transcriptome analysis that does not require knowledge of the genome sequence. This method uses microbead binding to capture cDNA sequences[4]. Repetitive cleavage of the bead-bound transcript by enzymes and identification of sequential 4-base sequences by fluorescently-labeled encoders yields unique 16-20 nucleotide sequences. This method may be more sensitive than SAGE at identifying unique or low-copy-number transcripts, as a MPSS library is typically 1 million signature tags as compared to a standard SAGE library of 50,000 tags. Both SAGE and MPSS permit specific counts of expressed sequences instead of the relative hybridization intensities of microarray, and thereby allow 'digital' output in place of the microarray 'analog' output.

GENERAL IMMUNITY STUDIES RELEVANT TO THE RESPIRATORY SYSTEM

Several studies have investigated aspects of the bovine immune response that may be relevant to respiratory system research. One laboratory has focused on the bovine $\gamma\delta$ T cell response to infection[5]. These investigators used SAGE to evaluate the response of $\gamma\delta$ T cells to infection in a series of experiments. They found that $\gamma\delta$ T cells respond to pathogen-associated molecular patterns (PAMPs) by upregulating a number of genes including MIP1α and 1β, IL-2Rα and CD69. These activated $\gamma\delta$ T cells were more able to proliferate in response to cytokine signals, even in the absence of antigen. Another microarray-based experiment[6] stimulated bovine γδ T cells with concanavalin A and human IL-2. This study found that both the Fas-mediated and mitochondrial apoptotic pathways were downregulated in stimulated cells.

Another study investigated the transcriptional response in bovine blood to orally-administered interferon-α[7]. The authors found upregulation in numerous transcripts, including genes associated with antigen processing and presentation, lymphocyte activation, leukocyte migration, immune modulation and apoptosis.

BOVINE TUBERCULOSIS (bTB)

Bovine Alveolar Macrophage Studies

A limited number of studies have been published examining gene expression of harvested bovine alveolar macrophages (BAMs) to mycobacterial strains. One study[8] compared gene expression of BAMs following infection with virulent or attenuated *Mycobacterium bovis* strains. They found that the virulent strain induced greater expression of chemokines, interleukin-8 and monocyte chemotactic protein-1, among other genes. Another laboratory infected BAMs with either *M. bovis* or *Mycobacterium tuberculosis*[9]. They found significantly higher expression of several chemokines including CCL3, CCL4, CCL5 and CXCL8 in *M. tuberculosis*-infected BAMs, while *M. bovis* infection induced higher expression of CCL23. They concluded that this differential expression could suggest how the *M. bovis*-infected BAM is able to avoid the host chemotactic response.

Bovine Peripheral Blood Mononuclear Cell Studies

The majority of recent studies investigating mycobacterial infection of bovine peripheral blood mononuclear cells (PBMCs) have been done by a group based out of University College, Dublin[10-13]. This group has performed a series of experiments looking at PBMCs harvested from a group of cattle naturally infected with *Mycobacterium bovis* and both without antigen stimulation and stimulated with mycobacterial purified protein derivative (PPD). This group found significant short-term (3-12 hours post-stimulation) downregulation in unstimulated infected PBMCs, with proinflammatory genes not found to be upregulated until 12 hours or later post-stimulation. The authors concluded that innate immune signaling may be repressed in chronically-infected cattle. When PBMCs from the same cattle were stimulated *in vitro*, rapid transient induction of proinflammatory genes was observed. The authors attributed this to removal of the PBMCs from an immunosuppressive environment (the infected bovine) and the presence of memory T cells. The authors have proposed a panel of four genes whose expression may permit differentiation of *M. bovis* infected and uninfected cattle.

Mycobacterial Studies

Several groups have investigated the mycobacterial pathogens using microarrays. An investigation into differential gene expression between *M. bovis* and *M. tuberculosis*, which have very similar genomes, found differences in cell-wall and secreted proteins, transcriptional regulators, lipid metabolism, PE/PPE family genes and toxin-antitoxin pairs[14]. Another study investigated highly-expressed mycobacterial genes as potential diagnostic antigens. This study identified the *M. tuberculosis* antigen Rv3615c as a diagnostic target in differentiating *M. bovis*-infected cattle from those vaccinated with BCG[15]. Investigation into the gene expression profile of the *M. bovis* BCG strain AN5[16] identified genes with decreased expression as compared to field strains of *M. bovis*. The authors concluded that the differential expression should not impact the use of the AN5 strain as a diagnostic reagent. Another study[17] examined a hypervirulent and an attenuated strain of *M. bovis* in bovine PBMCs, finding that the virulent strain expressed higher levels of genes involved in fatty acid metabolic processes, transcription factors and genes in the Mce protein family, which has been associated with virulence in *M. tuberculosis*. One other study[18] investigated the response of *M. tuberculosis* to bovine lung surfactant. The exposure of mycobacteria to surfactant induced increased expression of membrane-located lipases, suggesting that *M. tuberculosis* uses surfactant as a carbon source. Specific cell wall-component-cluster genes associated with interactions with alveolar macrophages and virulence were also upregulated; these genes are associated with protection of *M. tuberculosis* against host reactive nitrogen intermediates.

BOVINE VIRAL DIARRHEA VIRUS (BVDV)

In vivo studies

Analysis of white blood cells from persistently infected (pi) steers[19] compared to uninfected steers revealed 294 differentially-expressed genes, notably including 18 genes in the interferon pathway as well as genes in the cell cycle regulation and gene expression pathways. The authors described a robust innate antiviral response to the persistent infection. The same study investigated the immune response in blood from pi fetuses, transiently-infected (ti) fetuses and uninfected fetuses. They found that transiently-infected fetuses showed a much higher antiviral response than persistently-infected fetuses. The persistently-infected fetuses demonstrated expression levels comparable to pi steers.

Another study investigated the maternal transcriptional response in peripheral white blood cells to carriage of a pi, ti or uninfected fetus[20]. This study found significant expression differences in both innate and adaptive maternal immune responses for dams carrying both pi and ti fetuses, including MHCI and II antigen presentation, interferons α and β, T-cell receptor signaling, the B cell receptor pathway, NK cell signaling and leukocyte chemotaxis. Heifers carrying pi fetuses showed downregulation of the chemokine receptor CXCR4 and the T cell receptor signaling pathways, while ti-fetus-carrying heifers showed upregulation in the type I interferon pathway.

Cell culture studies

One microarray study on cultured bovine macrophages[21] using heterologous hybridization on a human array found a set of 26 genes differentiating cytopathic (cp) and noncytopathic (ncp) BVDV infection, which the authors suggested could be clinically useful. Another study[22] used SAGE in MDBK cells to discover >400 sequence tags differentially-expressed between infected and uninfected cells. Downregulated sequences coded for primarily host structural proteins and upregulated sequences coded for mostly protein translation and post-translational processing as well as protein transport genes. The same laboratory looked at ncp BVDV infection in a bovine B-lymphocyte cell line[23] and found a large number of untranslated proto-oncogene transcripts. In addition, this lab also demonstrated altered transcriptional responses in commercially-acquired bovine aortic endothelial cells found to be contaminated with ncp BVDV[24], leading to implications in labs using such commercial cell lines for research.

CONTAGIOUS BOVINE PLEUROPNEUMONIA (CBPP)

Preliminary investigations into bovine bronchoalveolar lavage gene expression[25] following infection with *Mycoplasma mycoides mycoides* (small colony) (MmmSC) has demonstrated a pattern of upregulation, primarily in Gene Ontology (GO) processes chemokine activity, negative regulation of cell proliferation and defense response. Upregulated genes also mapped to cellular (KEGG) pathways associated with cytokine-cytokine interaction, lectin-induced complement pathway and apoptosis. Particularly high levels of CCL2, IL-8, IL-1A and IL-1B were detected. Contrarily, genes in samples of bronchial epithelial tissue were mostly downregulated and mapped to GO processes associated with structural components and protein secretion and the KEGG pathways for oxidative phosphorylation and lectin-induced complement. Gene expression in whole blood samples was mildly increased and associated with GO processes G-protein signaling, cAMP second messenger and endonuclease activity. Top KEGG pathways active in blood samples included heparan sulfate biosynthesis and ethylbenzene degradation. Mechanistic genes, which may be key regulatory genes in a pathway, were identified for each cell type.

OTHER INFECTIOUS AGENTS

One paper investigated both host and viral gene expression in the lymph nodes of healthy versus malignant catarrhal fever-infected cattle[26]. This study found two regions of the viral genome were upregulated, one being a latency-associated nuclear antigen and the other having no predicted open reading frame. Infected host gene expression demonstrated low levels of IL-2 transcript and interleukin receptors as well as upregulation of many proinflammatory pathways including interferons, lymphocyte activation, cell proliferation and apoptosis.

REFERENCES

[1]McGuire K et al. **Vet Immunol Immunopathol** 2005; 105:259. [2]Tao W et al. **Vet Immunol Immunopathol** 2004; 101:1. [3]Velculescu VE et al. **Science** 1995; 270:484. [4]Reinartz J et al. **Brief Funct Genomic Proteomic** 2002; 1(1):95. [5]Jutila MA et al. **An Health Res Rev** 2008; 9(1):47. [6]Deng M et al. **Vet Immunol Immunopathol** 2005; 105:15. Namangala B et al. **J Interferon Cytokine Res** 2006; 26:675. [8]Wedlock DN et al. **Int Immunopharmacol** 2006; 6:957. [9]Widdison S et al. **Mol Immunol** 2008; 45:740. [10]Meade KG et al. **Vet Immunol Immunopathol** 2006; 113:73. [11]Meade KG et al. **BMC Genomics** 2007; 8:400. [12]Meade KG et al. **BMC Genomics** 2008; 9:447. [13]MacHugh et al. **Transboundary Emerg Dis** 2009; 56:204. [14]Golby P et al. **Microbiol** 2007; 153:3323. [15]Sidders B et al. **Inf Imm** 2008; 76(9):3932. [16]Garcia Pelayo MC et al. **Vet Microbiol** 2009; 133:272. [17]Blanco FC et al. **Microbes and Infection** 2009; 11:956. [18]Schwab U et al. **Microb Pathogenesis** 2009; 46:185. [19]Shoemaker ML et al. **J Interferon Cytokine Res** 2009; 29(1):23. [20]Smirnova NP et al. **Physiol Genomics** 2008; 36:129. [21]Werling D et al. **Vet Immunol Immunopathol** 2005; 108:157. [22]Neill JD et al. **Biologicals** 2003; 31:97. [23]Neill JD et al. **Virus Research** 2008; 135:326. [24]Neill JD et al. in Pirard M-H et al (eds) Animal Genomics for Animal Health. **Dev Biol (Basel)** 2008; 132:93. [25]Gull T et al. In process 2010. [26]Meier-Trummer CS et al. **PLoS ONE** 2009; 4(7):e6265.

RESPIRATORY MECHANICS AND BRONCHOALVEOLAR LAVAGE IN HEALTHY ADULT ALPACAS. Ana Pacheco, Daniela Bedenice, Melissa R Mazan, Andrew M Hoffman. Cummings School of Veterinary Medicine at Tufts University, North Grafton, MA.

The purpose of this study was to establish normal parameters of respiratory mechanical function and bronchoalveolar lavage (BAL) cytology in non-sedated, healthy adult alpacas.

Measurements of pulmonary function were performed in 16 healthy, untrained adult alpacas and included: functional residual capacity (FRC) via helium dilution; respiratory system resistance and reactance via forced oscillatory technique (FOT: 1-7Hz); and assessment of breathing pattern by use of respiratory inductive plethysmography (RIP). FOT and RIP results were obtained in both standing and cushed (sternally recumbent) animals. BAL was performed orotracheally under short-term anesthesia and analyzed following Toluidine blue and modified Wright staining.

The following mean +/- SD measurements of respiratory function were obtained in standing alpacas: FRC (3.19 +/-0.53L), tidal volume (V_T=0.8 +/-0.13L), respiratory system resistance (R_{RS} in $cmH_2O/L/s$) at 1Hz (2.70 +/-0.88), 2Hz (2.98 +/-0.70) , 3Hz (3.14 +/-0.77), 5Hz (3.45 +/-0.91) and 7 Hz (3.84 +/-0.93). The phase angle, as a measurement of thoracoabdominal asynchrony, was 19.59 +/-10.06° and the delta flow between nasal and plethysmographic flow measurements was 0.18 +/-0.07 L/s. The V_T, peak inspiratory and expiratory flows were significantly higher in cushed compared to standing animals. BAL cytology revealed 58.52 +/-12.36 % alveolar macrophages, 30.53 +/-13.78 % lymphocytes, 10.95 +/-9.29% neutrophils, 0% mast cells and several ciliated epithelial cells.

Pulmonary function testing was well tolerated in non-sedated cushed and standing alpacas. The technique developed for BAL was not associated with clinical complication and yielded adequate cellularity. BAL neutrophils were more abundant in alpacas than reported in horses.

EXPERIMENTAL INFECTION WITH PORCINE REPRODUCTIVE AND RESPIRATORY SYNDROME VIRUS (PRRSV) CHANGES PULMONARY FUNCTIONS IN PIGS. J Wagner, A Kneucker, M Glaser, P Reinhold. Friedrich-Loeffler-Institut, Jena, Germany.

PRRSV is one of the most pathogenic agents in respiratory diseases. The aim of this trial was i) to establish a respiratory infection model and ii) to evaluate the influence of PRRSV on pulmonary ventilation and gas exchange with non-invasive methods.

In this negative controlled study altogether 32 pigs were included and divided into four groups. Pigs of the challenged groups were experimentally infected (intranasally and intramuscularly) with PRRSV isolate VR-2332 (n = 8) or PRRSV isolate 205817 (n = 8), respectively, while pigs of the both control groups (n = 2 x 8) received 0.9% NaCl. For pulmonary function tests (PFT) the methods impulse oscilloresistometry, rebreathing (using carbon monoxide and helium) and volumetric capnography (Jaeger, Hoechberg, Germany) were applied via a tightly fitting face mask.

Compared to controls, pulmonary dysfunctions in pigs challenged with PRRSV were characterized by: i) alterations in pattern of breathing: significantly higher respiratory rates, significantly lower tidal volumes, increased minute ventilation ii) changes in respiratory mechanics: significantly elevated respiratory resistance (R) at frequency $\leq$5 Hz, this increase was more prominent during expiration (Rex) than during inspiration (Rin); significantly lower respiratory reactance (X) which became negative especially at frequencies below 10 Hz iii) changes in diffusion capacity: significantly reduced transfer factor for carbon monoxide iv) changes in CO_2-elimination: significantly reduced end-tidal carbon dioxide concentration and significantly reduced CO_2-volumes exhaled per breath.

Both challenged groups developed similar changes in pulmonary function but with different degree of severity, reflected in more worsening in breathing pattern and less impairment in respiratory mechanics in pigs challenged with PRRSV isolate 205817 compared to pigs challenged with PRRSV isolate VR-2332.

Acute infections with PRRSV were associated with obstructive and restrictive disorders as well as ventilatory asynchronisms leading to deteriorations in pulmonary gas exchange.

The work was funded by 'Akademie für Tiergesundheit e.V.' (Germany).

ANALYSIS OF CLINICAL PLATELET FUNCTION OF DOGS & CATS

Karl E. Jandrey, MAS, DVM, DACVECC
Davis, CA

INTRODUCTION

Platelets are anucleate, disc-shaped, cytoplasmic fragments that play an essential role in primary hemostasis, the initiation of a platelet plug, and the preservation of vascular integrity. Platelets are the smallest cellular component in the blood. Platelets circulate as small discs along the vessel edge. The exterior surface is highly reactive to external stimuli, ready to undergo a shape change to facilitate hemostasis on the damaged endothelial surface. Placed near the site of initial response to vascular injury, platelets react to stimuli, alter their shape, spread and adhere to the endothelium (and each other), secrete their granular contents (e.g. ADP, serotonin) to amplify the response, and interact with the hemostatic system to bind fibrinogen (to GP IIbIIIa) and solidify into a fibrin clot.

Alterations in platelet number as well as platelet function have profound implications for the maintenance of hemostasis and vascular health. Platelets are formed from megakaryocytes in the bone marrow under the influence of thrombopoietin (TPO), IL-3, IL-6, IL-11. TPO concentrations are constitutively expressed and vary with the platelet mass. Cytoplasmic extensions from the megakaryocyte (proplatelets) fragment into individual platelets that are released into the systemic circulation. The life-span of a platelet is approximately 7-10 days once released from the proplatelet. Thrombocytopenia is a relatively common finding in veterinary patients. However, platelet dysfunction is much less common. Many diseases contribute to defects in platelet function. In people, thrombocytopathia (reduced adhesion, altered aggregation, and poor clot retraction) is common secondary to neoplasia or dysproteinemias. In domestic species, the common causes of platelet hypofunction are renal or hepatic disease, zootoxins, and infectious or therapeutic agents. Decreased platelet function should be considered when superficial bleeding (petechiae, ecchymoses) occurs in the absence of profound thrombocytopenia. This review will provide information on the analyses available to the clinician for the diagnosis of decreased platelet function in dogs, cats, and horses.

HOW DO I KNOW MY PATIENT HAS A DEFECT IN PLATELET FUNCTION?

Two things are needed to suspect a patient to have thrombocytopathia. First, petechiae and ecchymoses are the most common compatible physical examination signs. Although less common, bleeding that mimics hemorrhage may be possible (epistaxis, hematochezia). In some patients, a familial history of bleeding or prolonged bleeding during the loss of deciduous dentition may be correlated to a platelet function defect. Often these patients are discovered to have a defect and may actually bleed profusely after venipuncture, surgery, (e.g. castration, ovariohysterectomy) or trauma. Secondly, secondary hemostatic testing is normal (i.e. prothrombin time [PT], activated partial thromboplastin time [aPTT or PTT]). These tests have no influence by platelets since they are best completed on platelet-free plasma. Some point-of-care coagulation analyzers use citrated whole blood and a weaker activator for stimulation of secondary coagulation. There is reportedly little impact on the PT/PTT with platelet count using this bench-top technology. The activated clotting time may be prolonged by a very low platelet count (usually less than 10,000/μL).

With normal tests of secondary coagulation, the next step in the diagnosis of thrombocytopathia is facilitated with a blood smear and manual platelet count.

Platelet Count

For all patients suspected of having a coagulation defect, a standard blood smear should be evaluated under a microscope for the estimation of a platelet count in the monolayer. A single platelet found on a 1000X field equates to ~15,000 platelets/μl in circulation. It is unlikely that spontaneous bleeding will occur until platelet counts drop below 20,000/μl. Platelet clumps will alter your interpretation of the estimated number of platelets. Many large platelet clumps will prevent an accurate manual platelet count; therefore, a general estimate will need to be determined (typically reported as less than adequate or adequate). If the platelet clumps are suspected to be due to poor venipuncture technique or an uncooperative patient, another sample obtained via an atraumatic venipuncture and blood smear may be evaluated. Blood samples obtained using a 3 syringe technique from indwelling catheters may not necessarily reflect the true platelet count of the patient and should be avoided if possible. Also, platelet numbers that are found to be low on automated, in-house blood cell counters should always be checked against a manual platelet count. Discordant results are often found because the automated cell counters are unable to count platelet clumps as individual cells, artificially lowering the platelet count.

Many tests of platelet function require a platelet count above 100,000/μl for accuracy. If the patient has greater than 100,000 platelets/μL and has signs of bleeding, the next test to consider is the buccal mucosal bleeding time (BMBT) to rule-in or out the most common thrombocytopathia (von Willebrand disease).

Buccal Mucosal Bleeding Time

When decreased platelet numbers do not account for bleeding, a buccal mucosal bleeding time (BMBT) is completed. This *in vivo* test is completed by the measurement of the time for a stable platelet plug to form from the time an incision is created on the upper lip of a dog or cat. An adaptation of the BMBT has been used on the skin on the forelimb in horses. The test uses a template device that creates 1 or 2 standardized superficial incisions (1mm deep by 5mm long). Normal BMBT results are less than 4 minutes. A prolonged result would be consistent with thrombopathia or von Willebrand disease. This test is highly operator dependent; variable results have been reported in animals and in humans. The most reliable results for the BMBT can be obtained by trained, experienced personnel. Internal reproducibility and consistency can be ensured through an increased frequency and attention to detail using the standard technique. BMBT measurements are poorly standardized, labor intensive, subjective, and have not been shown to correlate with clinical outcomes. In fact, this test has been all but abandoned in human medicine due to the lack of specificity and sensitivity to detect or predict clinical bleeding during surgery.

Alternative tests that have better correlation than the BMBT to platelet function with less concern for operator error have been developed and employed in human medicine. Some of these have been proven both experimentally and clinically in veterinary patients.

Platelet Function Analyzer-100®

The Platelet Function Analyzer-100® (PFA-100) is a bench-top instrument that evaluates platelet function in citrated whole blood. It has been shown to correlate with and predict surgical bleeding better than the BMBT. The system was introduced in the early 1990s and is easier, faster, and less expensive than other methods such as aggregometry and flow cytometry. The PFA-100 simulates primary hemostasis by aspirating citrate-anticoagulated whole blood under a high shear rate through a small aperture in a collagen membrane coated with platelet agonists (ADP or epinephrine). This design mimics the *in vivo* organization of the subendothelial matrix: the initial site of platelet deposition and aggregation. Closure time (CT) is the time it takes for a platelet plug to form and occlude flow. The CT is highly sensitive to qualitative and quantitative defects in platelet receptors that mediate adhesion (GPIb-V-IX) and aggregation (GPIIbIIIa).

The PFA-100 detects inherited, acquired, or induced platelet dysfunction. It has most commonly been used for analysis of coagulopathies (e.g. von Willebrand Disease) in dogs, horses, and humans as well as assessing aspirin therapy and affects of various fluids (saline, artificial colloids). Furthermore, the utility of the PFA-100 to assess closure time in response to various therapies is well-documented. Studies have shown that the PFA-100 can be used as an indicator of platelet function and dysfunction in many species (humans, dogs, pigs, horses, cats) that are not anemic (HCT <30%) or thrombocytopenic (<100,000 platelets/µl). In addition, the CT may be inaccurate when high hematocrits (>60%) or platelet counts (>500,000)/ µl) are present.

The cost of the machine (about the size of a small desktop printer) and disposable supplies may also limit its clinical use in private practice. Although the PFA-100 is a useful and accurate test, it is not the most sensitive for the detection and identification of the many discreet defects in platelet function due to the redundancy present in platelet membrane receptors invoked during activation for hemostasis.

Aggregometry

Turbidometric aggregometry was invented in the 1960s and is regarded as the gold standard for the diagnosis of primary hemostatic defects. It is able to detect many different aspects of platelet function and biochemistry through the use of agonists (ADP, thrombin, or collagen) at various concentrations. Platelet rich plasma from citrated whole blood is used in parallel chambers through which light is transmitted during the test. The opaque solution is stimulated to aggregate by way of the introduction of an agonist. As the platelets aggregate in response to the agonist, the sample becomes clearer as the platelets aggregate. The clarity of the sample is detected as an increase in the transmission of light across the sample. The difference between baseline and after agonist addition and aggregation is used to calculate the response. The specialized aggregometers graphically display the response as a curve that can be further analyzed to detect speed of aggregate formation. The major drawback is that it does not mimic *in vivo* physiologic conditions of adhesion, activation or aggregation to endothelial damage. Besides the limited availability to comparative hemostasis laboratories, it also requires large volumes of fresh blood (from patients with thrombocytopenia) and technical expertise. Aggregometry has been validated for use with equine, canine, and feline platelets as well as to assess a commercial frozen canine platelet concentrate.

VerifyNow®

To overcome some of the technical problems with aggregometry, alternatives such as the cartridge-based system of the VerifyNow® have been developed. This instrument measures platelet aggregation in whole citrated blood via changes in light transmission. The basis of this assay is that coated polystyrene microparticles will agglutinate in whole blood in direct proportion to the degree of platelet activation. Specific cartridges are employed to monitor for the effects on primary hemostasis by the following drugs: aspirin, P2Y12 inhibitors (clopidogrel), GP IIbIIIa antagonists (abciximab). These limitations and expense of the machine make it unlikely to be used in clinical veterinary medicine. This analyzer has received little use in research studies in domestic animals.

Impact®

The Impact-R™ is a cone and plate(let) analyzer (CPA) named after the shape and design of its disposable testing components. Impact™ is a mnemonic for Image analysis, Monitoring, Platelet, Adhesion, Cone and plate Technology. The R model is labeled and intended for research purposes. The Impact™ is currently undergoing FDA approval (expected in 2010) for clinical use in humans. The extent to which altered platelet function can be detected in companion animals has currently not been investigated. The Impact-R is a novel device for the assessment of platelet function under close to physiologic conditions. It has been validated for use in humans, pigs, mice and rats but not in dogs, horses, or cats. The Impact-R is used for the point-of-care measurement of whole citrated blood adhesion and aggregation of platelets under arterial flow conditions on a surface that mimics an extracellular matrix. Since platelets are the only cell to adhere to this matrix under these conditions, the elongated aggregates align in laminar flow lines. After simulation of blood flow conditions, the sample is stained and optically analyzed. The effects of variable hematocrit and platelet counts on results obtained by this analyzer have been investigated in people. Added benefits of this testing methodology that make it potentially useful for veterinary patients are: small sample volume, no blood processing, and fully automated and brief testing procedure. No studies were found that employed this technology in domestic animals.

Other viscoelastic analyzers

Developed more than 50 years ago, the thromboelastograph (TEG) measures the physical properties of clot formation. It is relatively rapid, point-of-care, and provides data on clot formation and lysis. Citrate anticoagulated whole blood is placed in the heated mobile cup of the analyzer. As the cup oscillates, a pin suspended in the blood reflects the tension of clotting blood to this cup. The derived measures include: R (reaction time in minutes), the time from initiation of coagluation to initial fibrin formation; K (clotting time in minutes), the time from the end of R until the amplitude of the tracing reaches 20 mm; α angle (in degrees), the angle tangent to the trace that represents the speed of clot formation; and MA (distance that is the widest point of the tracing in millimeters), that represents the final strength of the formed clot. Rotational thromboelastometry (ROTEM®) is an adaptation of the TEG that places the cup stationary and the pin oscillatory. The Sonoclot analyzer also measures changes in viscoelastic properties of a clot in whole blood upon activation. Because of the testing of whole blood, these viscoelastic analyzers are not specific for intrinsic platelet functional defects.

Flow Cytometry

Flow cytometry is a versatile tool to study many aspects of platelet function. It measures characteristics of a large number of individual cells. Basically, platelets can be labeled with a fluorescent antibody or ligand prior to single-file passage through the laser light beam within the flow cytometer. The intensity of the fluorescent light is proportional to the antigen density or specific characteristics of the cell being measured. Platelets can be analyzed in whole blood or platelet rich solutions. Processing techniques can induce activation of platelets, and technical expertise is essential for accurate and repeatable assays of platelet function. Species differences also exist in the response to preparation as well as activation by agonist. Some more common applications for platelet specific studies using flow cytometry applications include: the measurement of platelet activation, the diagnosis of specific platelet disorders, the monitoring of antiplatelet agents, and even the monitoring of thrombopoiesis.

Conclusions

Most platelet function tests available for humans have been used to diagnose or screen for platelet defects in veterinary patients. Most traditional and accurate tests pose technical challenges, are expensive, consume time, and require relatively large volumes of fresh blood. The lack of specialized hemostasis laboratories within close proximity to a veterinary practice has limited the widespread clinical application of all the technologies. Consultation with a specialist familiar with these tests is warranted when the challenge to diagnose platelet dysfunction present itself in a patient with hemorrhage despite adequate platelet numbers.

REFERENCES

[1.] Mammen EF, et al. **Semin Thromb Hemost** 1998;24(2):195-202. [2.] Callan MB, et al. **Am J Vet Res** 200162(5):652-658. [3.] Mischke R, et al. **Vet J** 2003;165(1):43-52. [4.] Wierenga JR, et al. **Am J Vet Res** 2007;68(6):605-609. [5.] Keidel A, et al. **Berl Munch Tierarztl Wochenschr** 1998;111(11-12):452-456. [6.] Jandrey KE, et al. **Vet Clin Path** 2008;37(4):385-388. [7.] Guillaumin J, et al. **Am J Vet Res** 2008; 69(12):1580-1586. [8.] Helenski CA, et al. **J Vet Int Med** 1987:1:24-28. [9.] Otto CM, et al. **J Am Vet Med Assoc** 2000;217:1500-1504. [10.] Donahue SM, et al. **J Vet Emerg Crit** Care 2005;15:9-16. [11.] Norris JW, et al. **J Vet Int Med** 2006;20:1450-1456. [12.] Welles EG, et al. **Vet Pathol** 1994;31:553-560. [13.] Guillaumin J, et al. **J Vet Emerg Crit** Care 2010 (in press).

COLLOID & COAGULPOATHIES: IS HYDROXYETHYL STARCH A CAUSE?

Karl E. Jandrey, MAS, DVM, DACVECC
Davis, CA

INTRODUCTION

Poor perfusion can easily be identified on physical examination. The physiologic response to shock is vasoconstriction which can be proved by the presence of the following parameters: delayed capillary refill time, pale mucous membranes, decreased mentation, tachycardia, poor pulse quality, and cool extremities compared to rectal temperature. A constellation of these findings upon triage examination signals the need for a restoration of effective circulating volume. This volume expansion for patients in shock can be accomplished through the judicious use of intravenous fluids. Short, large diameter catheters that can be placed quickly, most commonly in a peripheral vein, provide the access for this life-saving treatment. The large diameter facilitates faster infusion of shock fluid therapy. The fluid choice for volume resuscitation for patients in shock is often debated.

At their appropriate dosages, isotonic crystalloids and colloids are equally effective for the management of shock. Despite this some clinicians prefer to use artificial colloids instead of isotonic fluids. The volume expansion achieved at the end of the infusion depends on the volume administered. Due to differences in the behavior between isotonic crystalloids and colloids, the intravascular volume 30 minutes post-infusion depends upon the type of fluid used. Despite the many randomized, controlled clinical trials for resuscitation of humans with shock, there are no differences found in outcome between the fluids used. In subgroup analysis, isotonic fluid administration to trauma patients was associated with improved survival. A recent review stated that there was no reduction in the risk of death when colloids were provided for resuscitation from shock.

Synthetic colloid physiology and benefits

High molecular weight hydroxyethyl starch (HES) solutions are artificial colloid solutions used in the United States for plasma volume expansion and to increase colloid osmotic pressure in veterinary patients. These large starch polymers are suspended in isotonic 0.9% sodium chloride (Hetastarch® or Dextran-70®) or a modified Lactated Ringers solution (Hextend®). Commercially available HES solutions are classified by their mean molecular weight (MW) and degree of substitution (DS). The DS refers to the average number of hydroxyethyl groups per glucose unit within the branched-chain polymer. The most commonly used HES solution is HES 670/0.75 and therefore has an average MW of 670 kD and a DS of 0.75.

HES redistributes to the interstitial space at a much slower rate than isotonic crystalloids. As they increase the plasma colloid osmotic pressure, they facilitate interstitial fluid movement towards the intravascular space. As such, they can increase blood volume by a greater amount than what was administered. Thus effective restoration of circulating volume is possible with a smaller administered volume (compared to isotonic crystalloids). Silverstein et al showed that a 20ml/kg dose (over 5 minutes IV) of HES 670/0.75 experimentally increased blood volume to 25%, 36%, and 26% at the end of infusion, 30 minutes post-infusion, and 4 hour post-infusion, respectively.

Besides the effective expansion of circulating volume, the increase in colloid osmotic pressure is an important benefit in the management of hypoalbuminemia. The larger size of the molecules may also seal capillary pores in critically ill patients with capillary leak syndrome. This benefit is debated and is largely theoretical.

Adverse effects of synthetic colloids

Hemostatic defects are the most significant adverse effect of treatment with synthetic colloids. Hydroxyethyl starch solutions have been shown to compromise human platelet function *in vitro* and *in vivo* when measured using bench top (flow cytometry, platelet aggregometry) and bedside standard tests for platelet function (thromboelastography, Platelet Function Analyzer-100 [PFA]). Studies have shown that HES solutions with a higher MW (>200 kD) or a higher DS (>0.5) have a more pronounced effect on platelet function compared to other solutions with lower MW and DS. Clinically, some patient populations in human medicine have been shown to have higher rates of post-operative blood loss after HES administration. Anecdotally, veterinary surgeons suggest that there is an apparent bleeding diathesis in surgical patients after treatment with HES 670/0.75. To address this concern Wierenga *et al* completed an *in vitro* study to compare the effects of HES 670/0.75 and 0.9% sodium chloride solution on canine platelet function. The study used the PFA to demonstrate that HES 670/0.75, at a dilution of 1:3 with whole blood, significantly affected platelet function to a greater degree than hemodilution. Smart *et al* followed-up this study to answer the question about the *in vivo* effect of this HES on platelet function over time. The results of this study have shown that 20 mL/kg of HES 670/0.75 prolongs closure time on the PFA in dogs for at least 5 hours after administration. Even though the closure times were not back to baseline by 24 hours post treatment, the majority fell within the reference interval and were not different from controls. This effect may be due to a direct result of platelet dysfunction in addition to the effect of hemodilution.

Why do colloids have an effect on hemostasis?

In general patients that receive synthetic colloids have various critical illnesses from shock to sepsis to hypoalbuminemia. The concurrent diseases may actually have an impact on the patient's coagulation profile. Therefore the effect seen with or without HES solutions is variable.

The effects on coagulation secondary to HES solutions specifically are related to alterations in Factor VIII (FVIII) and von Willebrand Factor (vWF) concentration and function. FVIII is a glycoprotein that participates in the intrinsic pathway of coagulation. It is synthesized in the liver and secreted into the plasma where it circulates, in part, complexed with vWF. FVIII acts as a cofactor in blood coagulation accelerating the conversion of Factor X to Factor Xa (in the presence of Factor IXa, calcium, and phospholipids). FVIII levels increase dramatically in acute inflammatory diseases, much like fibrinogen. Decreased concentrations of FVIII have also been shown which accounts for the mildly prolonged coagulation times observed in humans after HES administration. Decreases in vWF and FVIII may occur due to binding with HES molecules. An accelerated elimination of the complex (HES:vWF:FVIII) is facilitated by the reticuloendothelial system whereas the complex without HES does not suffer the same fate.

vWF is a large multimeric glycoprotein which has 2 main functions: to mediate adhesion of platelets at the sites of vascular damage and to stabilize FVIII. vWF is released from the alpha granules of platelets and from Weibel-Palade bodies in the endothelium as a response to various stimuli. The most common clinical signs in dogs with von Willebrand Disease (vWD) are mucosal hemorrhages, bruising, and prolonged bleeding from surgical wounds, similar to other inherited or acquired defects in primary hemostasis. Concannon *et al* found decreased circulating levels of vWF and FVIII at various time points when dogs were treated with Dextran-70 (20ml/kg IV). This effect mimics Type 1 vWD. However, none of the tested dogs exhibited clinical signs of bleeding.

Platelet dysfunction in humans due to HES has been attributed to decreased platelet adhesion. A decreased agonist-induced expression and activation of platelet integrin $\alpha_{IIb}\beta_3$ induced by HES is another cause for altered hemostasis. Integrin $\alpha_{IIb}\beta_3$ on the surface of the platelet binds fibrinogen after a conformational change in this surface glycoprotein complex elicited by a rise in intracellular calcium upon activation/stimulation. The binding of fibrinogen is required for platelet aggregation and formation of a platelet plug. Patients with Glanzmann thrombasthenia have quantitiative and qualitative defects of integrin $\alpha_{IIb}\beta_3$ that impair the binding of fibrinogen (primarily) as well as vWF, fibronectin, and vitronectin (to a much lesser degree). It has also been shown that HES molecules coat the surface of the platelet. This obstructive impairment limits the binding of ligands to cell surface receptors. This can cause decreased platelet function independent of integrin $\alpha_{IIb}\beta_3$ blockade (see Figure 1).

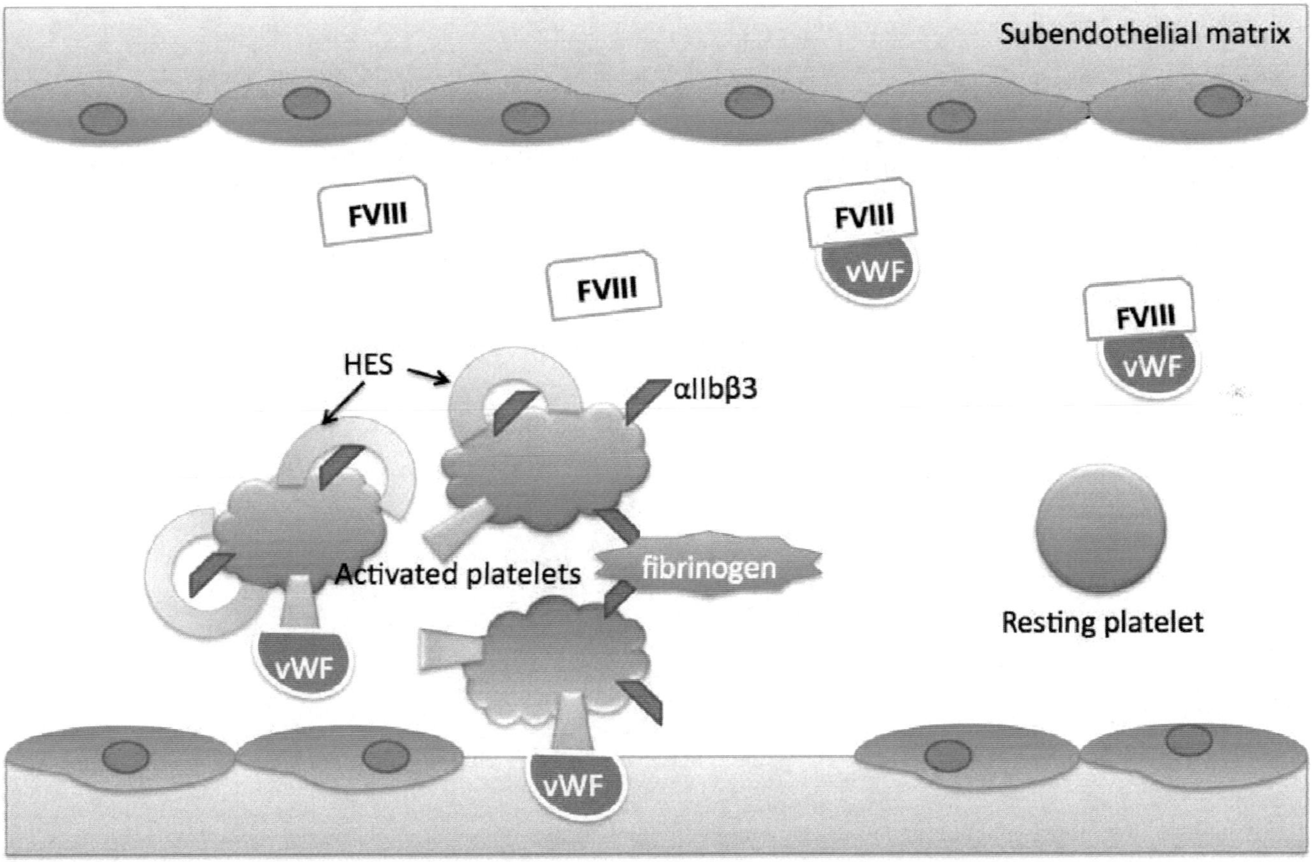

Figure 1: Platelet dysfunction due to hydroxyethyl starch (HES) solutions.
FVIII= Factor VIII, vWF= von Willebrand Factor, $\alpha_{IIb}\beta_3$= principle fibrinogen-binding integrin

The rate of HES degradation (based on the MW, DS and the position and hydroxyethylation on the carbon backbone) *in vivo* determines the degree and duration of platelet dysfunction. High DS and MW have slower clearance and will therefore affect hemostasis for a longer time. Recently (November 2008), a low molecular weight HES solution (Voluven®, 6% HES 130/0.4) obtained FDA approval in the United States. It is well established in Europe, where a wide array of HES solutions are commercially available, and in many papers in human medicine, that the low molecular weight HES solutions in general have minimal to no effect on hemostasis. The immediate and mid-range volume expansion effects are still present mainly due to the presence of a greater number of smaller osmotically active molecules. An occasional veterinary case report from Europe may reference Voluven® as part of therapy, but the reports do not focus on hemostatic issues.

Why would HES 130/0.4 have less impact on coagulation?

The maximum dosage of HES 130/0.4 in humans is 50 ml/kg/day. The lower molecular weight and less substituted HES solutions may result in lower plasma and tissue concentrations. This is one postulated reason to have less impact on hemostasis (particularly on FVIII/vWF complex and platelet aggregation) despite the larger dosage. In studies that compared HES 130/0.4 to HES 200/0.5 and/or gelatins, there was less blood loss and lower transfusion requirements in human patients undergoing orthopedic, cardiac, and abdominal surgery.

Remaining questions

An increased post-operative blood loss or increased transfusion requirement in human populations in some studies has been associated with HES treatment. However, surgical patients in one study were found to have no significant blood loss after treatment with HES. The hemostatic clinical impact (need for transfusions, length of stay, fiscal ramifications) of HES-induced platelet dysfunction in veterinary medicine is yet to be studied and published.

The use of HES solutions in patients already suffering from coagulation abnormalities should be approached with caution and knowledge. Colloids dilute out red blood cells as well as albumin. Isotonic crystalloids also do this, but the length of action and persistence proportionally last longer. There is no current literature in veterinary medicine to confirm or refute the use of HES as a therapeutic agent to pharmacologically modulate hemostasis.

Conclusion

The intent of this article is not to discourage the use of hydroxyethyl starch solutions. However, the clinician must have an awareness of the adverse clinical effects on hemostasis as well as the other beneficial effects (not discussed here) for colloid resuscitation and therapy. In summary, HES with high MW and DS decrease plasma coagulation FVIII and vWF as well as impair platelet adhesion to surface integrin $\alpha_{IIb}\beta_3$. Platelet interactions with vWF are also diminished. The HES with low MW and DS have little to no effect on hemostasis. The promise and utility of the newly-licensed HES solution in the USA remains to be documented in veterinary patients.

REFERENCES

[1.] Boldt J, et al. **Br J Anaesth** 2002; 89(5):722-728. [2.] Callan M, et al. **Am J Vet Res** 2001;62(5):652-658. [3.] Concannon KT, et al. **Am J Vet Res** 1992; 53(8): 1369-1375. [4.] Deusch E, et al. **Anesth Analg** 2004;99:665-668. [5.] Deusch E, et al. **Anesth Analg** 2003;97:680-683. [6.] Kozek-Langenecker SA. **Anesth** 2005;103(3):654-660. [7.] Smart L, et al. **J Vet Emerg Crit** Care 2009; 19(5):444-449. [8.] Treib J, et al. **Thromb Haemost** 1997;78:974-983. [9.] Wierenga JR, et al. **Am J Vet Res** 2007;68(6):605-609. [10.] Choi PT et al, **Crit Care Med** 1999;27:200-210. [11.] Schierhout G, et al. **Br Med J** 1998;316:961-964. [12.] Finfer S, et al. **N Engl J Med** 2004;350(22):2247-2256. [13.] Roberts I, et al. **Cochrane Database Syst Rev** 2004;18(4):CD000567. [14.] Silverstein et al. **J Vet Emerg Crit Care** 2005;15(3):185-192. [15.] Joubert KE, et al. **J Vet Emerg Crit Care** 2007;17(2):184-190.

HYPERCOAGULATBILITY IN CATS WITH HYPERTROPHIC CARDIOMYOPATHY

Karl E. Jandrey, MAS, DVM, DACVECC
Davis, CA

INTRODUCTION

Hypertrophic cardiomyopathy (HCM) is the most common type of heart disease in cats. The disease has been suggested to be heritable in some families of non-purebred cats, and is or is suspected to be hereditary in a number of cat breeds. Heritability has been proven in a colony of Maine Coon cats. Approximately half of these cats have a mutation in the cardiac myosin binding protein C gene.

Arterial thromboembolism (ATE) is a fatal complication and often the initial presenting sign of heart disease in cats with HCM. HCM sets up a constellation of abnormalities that predisposes these cats to thrombus formation. Virchow's triad describes these three major pathophysiologic mechanisms of hypercoagulability: blood flow stasis, endothelial dysfunction, and up-regulation of platelets and coagulation factors. While the pathology of HCM has been well studied with regard to changes in the myocardial organization, specific physiological changes that result in thromboembolism remain to be defined. Much has been published over the past several years to shed light on this subject. This article will review these findings from many research groups working to answer this complicated physiologic question.

Blood Stasis

Endothelial cell damage that results in endothelial activation, stasis of blood in the left atrium and aggregation of platelets and RBCs are all elements of Virchow's triad. The heart from a cat with HCM has severe left ventricular wall thickening. Papillary muscles may also become very thickened and prominent. The left ventricular chamber volume shrinks as the myocardial hypertrophy worsens. The left atrium subsequently enlarges. Schober *et al* demonstrated decreased left atrial appendage blood flow velocities in cats with cardiomyopathy compared to normal cats. Also associated with this finding was the likelihood of an increased left atrial size, a decreased left atrial function, a worse left ventricle dialstolic dysfunction and the presence of heart failure. This analysis found the left atrial appendage flow velocities to be a predictor of spontaneous echocardiographic contrast. Spontaneous echocardiographic contrast (SEC, aka "smoke) is generally attributed to blood stasis and has been shown to be the result of red blood cell clumping and the interaction of plasma proteins at low flow conditions. Although smoke was not correlated to hemostatic markers in this study, the potential for thromboembolism exists (there is an increased risk in humans) and may be higher in these cats.

Endothelial activation

Endothelin-1, a potent vasoconstrictor, is released from activated endothelium under conditions of flow-induced shear stress, hypoxia, and pharmacological agonists. Increased plasma endothelin-1 is found in human patients with heart disease and correlates with the degree of hemodynamic injury. Recently, Prosek *et al* have measured plasma endothelin-1 immunoreactivity in healthy cats and cats with varying types of cardiomyopathy. Their study demonstrated that the plasma endothelin-1 concentration from cats with various types of heart disease was more than double that of normal cats, strongly suggesting that endothelial activation is a definable component of feline cardiomyopathies. Endothelin-1, secreted by the endocardium, has been proposed to participate in paracrine signaling to the underlying myocardium. Endothelin-1 has also been implicated as a mediator of platelet activation and thus may play a role in platelet up-regulation in HCM.

Activated endothelium also induces the release of the pro-thrombotic molecules thrombin and vWF from Weibel-Palade bodies. The vWF released from activated endothelium has a high affinity for platelet GPIb-V-IX complex, further stimulating platelet activation. In addition to release of Weibel-Palade body contents, activated endothelial cells also translocate phosphotidyl serine (PS) to the outer leaflet of their plasma membrane (see below) and produce procoagulant endothelial cell microparticle.

Platelet upregulation and activation

Normally, the vascular endothelium expresses a large array of molecules that maintain an anti-thrombogenic potential through the maintenance of low resting arterial tone and inhibition of platelet activation. Nitric oxide (NO) secreted from endothelial cells causes relaxation of vascular smooth muscle and inhibits platelet adhesion, activation and aggregation. Endothelial cells secrete prostacyclin, which acts synergistically with NO to further inhibit platelet aggregation and adhesion. Normal endothelium also secretes an Ecto ADPase (CD39) that functions by degrading locally high concentrations of ADP, thus preventing ADP induced platelet activation.

During activation, endothelial cells acquire a procoagulant phenotype, expressing pro-thrombotic molecules on the cell surface, some of which are released into the plasma. This phenotype results in increased vasoconstriction and alterations of endothelial and smooth muscle organization of the endocardium and vasculature, conditions which have been shown to contribute to human left ventricular dysfunction and increased risk for thrombosis. Endothelial changes have also been shown to be associated with heart failure and pathologic cardiac remodeling. Endothelial activation is associated with significantly decreased expressions of NO and CD39.

During conditions of low flow and low shear, aggregated RBCs release ADP, further contributing to the activation of platelets and up-regulation of the coagulation system. Earlier studies have shown that platelets from cats with various types of heart disease have an increased sensitivity to some platelet agonists. Platelet hyper-aggregability has been documented in human patients with heart disease and vascular disease, where it has been shown to be predictive for increased risk of vascular events

Platelet activation is a multi-component system involving a complex series of events involving the interaction of platelets and endothelium, the end result of which is thrombus formation. The initial event in this process is the rolling and tethering of platelets to the endothelium through the binding of the GPIb-V-IX complex to subendothelial vonWillebrand factor (vWF). Our laboratory has previously identified this complex on the feline platelet surface. Binding of GPIb-V-IX leads to stable adhesion that initiates inside-out signaling which generates a rise in intracellular calcium and the translocation of PS to the outer leaflet of the platelet plasma membrane. PS flip is required for the binding of factors Va and Xa to the platelet surface and the subsequent formation of the procoagulant surface. The rise in intracellular calcium accompanies downstream signal transduction, resulting in the conformational change of integrin $\alpha_{IIb}\beta_3$ required for it to bind fibrinogen. As activation proceeds, platelets change shape and release alpha granule contents, such as platelet-endothelial cell adhesion molecule (PECAM)-1 and beta thromboglobulin into the extracellular milieu further contributing to the clot formation. Plasma levels of PECAM-1 have been shown to be consistently almost two-fold higher in human patients with heart failure than in controls. Activated platelets also release procoagulant microparticles (PMPs), that propagate the coagulation cascade by providing additional surfaces for the assembly of coagulation factors as well as interacting with platelets, leukocytes and activated endothelial cells. Production of PMPs has been implicated in the pathophysiology of vascular disorders in humans. While platelets are the main source of circulating procoagulant microparticles, leukocytes and endothelial cells can also contribute substantial numbers of microparticles in many pathophysiological situations.

Our lab investigated whether measurements from the Platelet Function Analyzer-100® (PFA) can identify hypercoagulability in cats with HCM when compared to healthy controls. This machine is a simple bench-top, point-of-care instrument to evaluate platelet function in citrated whole blood. The components structure mimics the *in vivo* arrangement of the site of platelet adhesion. These measures were unable to discern between healthy cats and cats with HCM despite being highly sensitive to defects in platelet quality. The machine in general seems more useful to identify hypocoagulable states.

Plasmatic markers of hypercoagulability

Brazzell and Borjesson studied antithrombin and D-dimer concentrations in cats with HCM and compared them to healthy controls and another population of critically ill cats. There was no significant difference between the groups. Cats with HCM tended to have higher antithrombin concentrations but this was not statistically significant. Bédard *et al* evaluated plasma markers of coagulation between a group of cats with HCM and a healthy control group. Their data showed that antithrombin was statistically lower in cats with HCM compared to controls. The measurements of thrombin-antithrombin complexes and D-dimer concentrations as well as prothrombin and partial thromboplastin times were not different between the groups. Although some individual cats in each group had measurements compatible with hypercoagulability, left atrial size and degree of HCM was not associated. Following this study, Stokol *et al* measured a spectrum of plasma markers of hemostasis in 3 groups of cats with HCM and compared them to healthy controls. The 3 groups consisted of: left atrial enlargement (LAE) alone, LAE + SEC, and LAE + ATE. Fibrinogen concentrations were statistically higher in cats with heart disease of any severity compared to healthy controls. Hypercoagulability was not associated with left atrial size or presence of heart failure. However, some individual cats (~50% with LAE+SEC and LAE+ATE) had criteria for hypercoagulability (high thrombin-antithrombin complexes, D-dimer and FVIII concentrations). Cats with LAE+ATE had higher vWF antigen concentrations likely due to the endothelial injury from the thromboembolism. This information taken together makes a discrete mechanism still unclear. However, these individual data highlight important aspects in the physiology of hypercoagulability in individual cats with HCM.

Conclusions

Despite our greater understanding of HCM, we are still largely unable to evaluate the potentially large pool of asymptomatic cats with the disease for potential thromboembolic risk. Simple non-invasive studies of blood platelets and soluble endothelial components secreted into the plasma, which can be obtained by venipuncture, will result in a better understanding of the role of platelets and endothelial cells in the development of thromboembolism secondary to HCM and possibly a better understanding of HCM itself. These studies are relevant to owners and veterinarians as they will help provide critical new insights into the development and assessment of new therapies to prevent thrombus formation and help provide longer and higher quality lives for cats with this disease. The development and use of these highly sensitive flow cytometric and ELISA assays may also be used as a prospective tool in the initial evaluation of any given population of cats. Veterinarians can easily use the information generated by these tests to follow platelet and endothelial function and develop a set of baseline values from which to evaluate potential prophylactic anti-thrombotic therapy, and the progression of disease. To define the sensitive and specific indicators of platelet and endothelium up-regulation in the pathogenesis of hypercoagulability associated with HCM may provide a mechanistic approach for improved diagnosis and therapeutic management.

REFERENCES

1. Ferasin L, et al. **J Feline Med Surg** 2003; 5(3):151-159. 2. Fujii Y, et al. **J Vet Med Sci** 2001;63(5):583-5. 3. Kraus MS, et al. **J Am Anim Hosp Assoc** 1999;35(4):293-6. 4. Kittleson MD, et al. **Circulation** 1999; 99:3172-3180. 5. Meurs KM, et al. **Human Mol Gen** 2005;14:3587-3593. 6. Liu SK, et al. **Am J Pathol** 1981;102:388-395. 7. Blann AD, et al. **Thromb Res** 2001;101(4):321-7. 8. Helenski CA, et al. **J Vet Intern Med** 1987:1:24-28. 9. Welles EG, et al. **Am J Vet Res** 1994;55:619-627. 10. Welles EG, et al. **Vet Pathol** 1994;31:553-560. 11. Jagroop A, et al. **Curr Vasc Pharm** 2003;3:393-399. 12. Trip MD, et al. **N Engl J Med** 1990;322:1549-1554. 13. Thaulow E,et al. **Circulation** 1991;84:936-938.. 14. Rush JE, et al. **J Am Vet Med Assoc** 2002;220:202-7. 15. Atkins CE, et al. **J Am Vet Med Assoc** 1992;201(4):613-8. 16. Laste NJ, et al. J Am Anim Hosp Assoc 1995;31(6):492-500 17. Baty CJ. **Vet Clin Small Anim** 1995;34:1227-1234. 18. Smith SA, et al. **J Vet Intern Med** 2003;17:73-83. 19. Chen J, et al. **Microcirculation** 2005;12:235-246. 20. Andrews RK, et al. **Thromb. Res.** 2004;114:447-453. 21. Tablin F, et al. **Am J Vet Res** 2002;62:195-201. 22. Jandrey KE, et al. **Vet Clin Pathol** 2008;37(4):385-388. 23. Prosek R, et al. **J Vet Intern Med** 2004;18:826-830. 24. Stokol T, et al. **J Vet Intern Med** 2008;22:546-552. 25. Bédard C, et al. **Vet Clin Pathol** 2007;36(2):167-172. 26. Schober KE, et al. **J Vet Intern Med** 2006;20:120-130. 27. Brazzell JL, et al. **Vet Clin Pathol** 2007;36(1):79-84.

THE THROMBOELASTOGRAPH IN CLINICAL PRACTICE

Karl E. Jandrey, MAS, DVM, DACVECC
Davis, CA

INTRODUCTION

In the last 10 years, thromboelastography has seen a rapid growth in application to study hemostasis in many disease states in the veterinary literature. This brief review of the technology and salient literature will focus on what is currently known in this field as it applies to veterinary internal medicine, emergency, and critical care. Although this is a valuable academic research tool to uncover questions of coagulation and hemostasis, it is not necessarily cost prohibitive. Due to its wide applicability to study all phases of the coagulation process (clot formation, strength, and fibrinolysis), thromboelastography may eventually be used in larger 24 hour, multi-specialty hospitals by internists, surgeons, and criticalists.

Technology

Thromboelastography (TEG) technology has been around for over 60 years (see Hartert 1948, currently distributed by Haemonetics Haemoscope Division., Niles, IL). After less frequent use due to the development of modern plasma coagulation testing (prothrombin and activate partial thromboplastin times, PT and PTT, respectively), TEG found a resurgence when rapid assessment was needed for complex medical and surgical diseases in humans (e.g. transplantation, bypass). Now it has wide application to aid in diagnosis, direction for transfusion support and decision-making, as well as a research tool to uncover complicated hemostatic defects in both hypo- and hypercoagulable states.

The principle behind TEG is based on the measurement of the physical viscoelastic characteristics of clots. Clot formation is monitored at 37°C in an osciallating plastic cylindrical cuvet (aka the cup) and a stationary suspended piston (aka the pin) that has a 1mm distance between each surface. The cup oscillates in either direction every 4.5s with a 1-s mid-cycle stationary period. The pin is suspended by a torsion wire. This transmits the torque perceived by the pin to the computer software. As clot formation ensues, fibrin fibrils physically link the pin to the cup. The rotation of the cup is transmitted from the clot and cup to the pin via the viscoelasticity of the clot. This data is displayed via the computer interface graphically in real time.

The difference between TEG® and rotational thromboelastometry (ROTEM®, Pentapharm GmbH, Munich, Germany) has been published (see Jackson). In brief, these technologies are both used for point-of-care coagulation analysis. Each company has a trademark on the name and technology. However, the ROTEM differs from the TEG by the fact that the pin is stationary and the cup rotates. The measures are similar but have different names. Since the TEG has a wider application in clinical and research use in the United States, the discussion herein will focus on the TEG.

Measurements

The following measurements are monitored by the TEG. Figure 1 shows the graphical output of this data from a healthy cat.

R: The reaction time is the period of time of latency (in minutes) from the time blood in placed in the TEG until the initial fibrin formation.

Reaction time generally reflects coagulation factor levels, but does not always correlate with PT and PTT.

K: The K-time is a measurement to a predetermined level (20mm) of clot strength (in minutes).

α: The alpha angle is a measure (in degrees) of the speed (kinetics) of fibrin build-up and crosslinkage, thus the speed of clot strengthening.

Both K and α impart similar information. Both are affected by the availability of: 1) fibrinogen, an important determinant in clot development, 2) Factor XIII (FXIII), which facilitates fibrin crosslinkage into a stable clot, and 3) platelets, to a much lesser extent than fibrinogen or FXIII. The α does not always correlate with fibrinogen concentrations.

MA: The maximum amplitude is a direct function of the maximum dynamic properties of fibrin and platelet bonding via GPIIbIIIa which represents the ultimate strength of the clot. MA (in mm) is measured at the peak of rigidity manifested by the clot at 45-90 minutes.

The MA is the assessment of a combination of the platelet count and function as well as fibrinogen activity. Another measure of clot firmness (**G**, in dynes/second) can be calculated from the MA using the equation $G=5000xMA/(100-MA)$.

LY30: This is a measure of clot stability. It measures lysis of the clot 30 minutes after the MA has been identified.

Whole blood is atraumatically collected, ideally with a Vacutainer® system, into 3.2% sodium citrate (1:9 v/v ratio). Earlier techniques did not use blood anticoagulation and needed to be completed patient-side. TEG should be completed within 2 hours of venipuncture. Sample handling to prevent rapid shifts in temperature, vibration or shock, and submission on ice should be avoided as these will all activate and alter platelet function. Multiple methods of activation of the blood for TEG analysis exist. Recalcification +/- tissue factor, as well as Factor XII have all been published as useful activators. There is no current standard between institutions or laboratories and each activator requires its own reference interval for each measurement.

Besides the assessment for hypocoagulable states, the TEG can also be used to uncover hypercoagulable states. Increased clot firmness (i.e. high MA value) is considered to predict post-operative thrombotic complications in human. Other applications include: monitoring of polycythemia, von Willebrand disease, individual factor deficiencies, antiplatelet therapies, hereditary platelet defects, hyperfibrinolysis, and hypo- or dysfibrinogenemias.

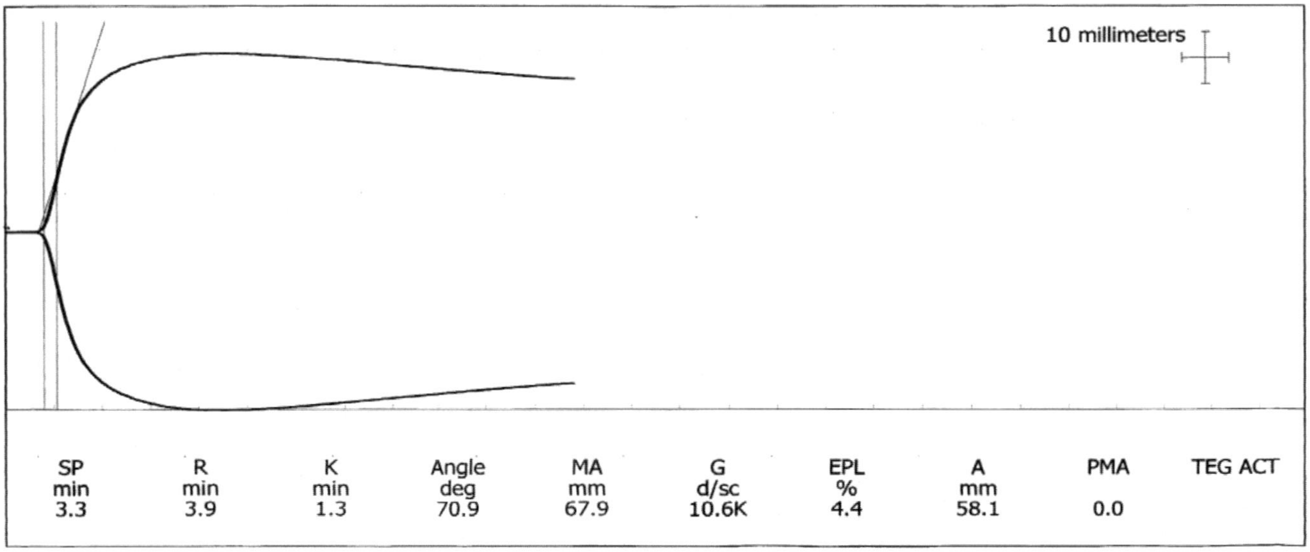

Figure 1: Normal feline thromboelastograph tracing printed from a Haemoscope TEG® 5000 Hemostasis Analyzer System using kaolin as an activator on citrated whole blood. Feline reference intervals for our laboratory are (see text for description): R= 1.5-4.4min, K= 1.0-2.8min, alpha angle= 59.2-79.8°, MA= 46.0-69.2mm.

Veterinary studies

Beginning with members of the Emergency and Critical Care Service at the University of Pennsylvania, TEG use has grown and has been employed since 2000 in many species and disease states. Otto *et al* described hypercoagulability in puppies with parvoviral enteritis. The increased MA measurements as well as decreased antithrombin concentrations were statistically different than the age-matched control group. Fibrinogen concentrations were also significantly higher than controls. The clinical assessment of a low antithrombin and a high fibrinogen are typical of a prothrombotic state. Four of nine puppies had clinical evidence of catheter-associated phlebitis or venous thrombosis. The use of the TEG was able to identify hypercoagulability but was unable to uncover the exact mechanism in this multi-factorial disease.

Brainard et al used the TEG and correlated its measurements with other measures of hemostasis (i.e. platelet aggregometry) in healthy dogs. It was found that MA and the alpha angle decreased when platelet aggregation decreased under the treatment with carprofen. Interestingly MA increased after treatment with deracoxib. This study proves that the coagulation system alters differently under the influence of various non-steroidal anti-inflammatory drugs (NSAIDs). Paired samples from a patient pre-and post-therapy may be indicated to identify alterations in hemostasis that may become clinical. The affect of NSAIDs on hemostasis is currently unknown in patients with illness.

Kristensen *et al* studied dogs with neoplasia using a newly-validated tissue factor-activated TEG protocol. This protocol detected both hypercoagulability (18/36, 50%) and hypocoagulability (6/36, 17%) in dogs with metastatic neoplasia. Of dogs with benign neoplasia, only 31% were hypercoagulable, none were hypocoagulable, and 69% were normal upon TEG analysis. Other parameters measured during this study included: PT, PTT, D-dimer, fibrinogen, and platelets. Dogs with malignant neoplasia had higher D-dimer concentrations and lower platelet counts. The complicated physiology of hemostasis may be viewed globally with an analysis like the TEG, but these data set the baseline for treatment studies to compare the effect of treatments.

The same group (see Wiinberg *et al*) used the tissue factor-activated TEG protocol to uncover disseminated intravascular coagulation (DIC) in dogs admitted to an intensive care unit with various diseases. The majority of dogs diagnosed with DIC using standard criteria were found to have hypercoagulability (22/50) more than hypocoagulability (11/50). However, the overall mortality of the hypocoagulable group was twice that of the hypercoagulable group. There was wide variation in TEG measurements overall. However, the study concluded that the TEG results can be clinically valuable to assess the hemostatic function in dogs with DIC.

Wiinberg *et al* showed in another study that the measures from the tissue factor-activated TEG correlated with clinical bleeding. In a prospective case-control study using G as their outcome measure, high positive (89%) and negative (98%) predicted value was correlated to dogs with clinical bleeding. These results were more favorable than traditional hemostatic measurements (i.e. PT/PTT). TEG is therefore a very useful screening test for dog suspected of having a bleeding disorder.

Abstract presentations at the International Veterinary Emergency and Critical Care Symposium, European Veterinary Emergency and Critical Care Symposium, and American College of Veterinary Internal Medicine Forum within the last year have also addressed the usefulness or application to the understanding of many disease states in many species. A study from Michigan State University proposed that the collection method of blood from healthy dogs would influence the TEG results. Blood was collected into syringes and transferred into anticoagulant sample tubes or directly into syringes containing anticoagulant. TEG analytical conditions were similar for both groups and flow cytometric analysis for P-selectin expression was also completed. Samples collected without anticoagulant in the syringe were found to be hypercoagulable. Neither sample technique activated platelets as indicated by P-selectin expression. This information should be used when assessment of TEG data is made. Another study from the same group assessed the influence of hematocrit on the TEG in healthy dogs. The major finding was that blood samples with a lower hematocrit were found to have a more hypercoagulable tracing. This must be accounted for when the interpretation of hypercoagulability may be due to both a disease state and anemia. Another abstract from the University of Georgia (UGA) presented for the evaluation of TEG in dogs with hyperadrenocorticism [Tufts University and the University of Pennsylvania (see Sinnott 200) completed a similar observational study in untreated dogs with immune-mediated hemolytic anemia]. Their primary outcome measure found that dogs with hyperadrenocorticism were not hypercoagulable compared to controls. Interestingly, they also found that a lower hematocrit inversely correlated with the MA. Another study from this group identified that hypocoagulable TEG results correlated with anti-Xa concentrations in horses treated with dalteparin. The same group (see Epstein 2009) investigated the use of a tissue factor-activated TEG in horses to find that this technique narrowed the individual variation in the measurements achieved by 4 independent operators. Another TEG analysis from The Ohio State University studied horses with severe colitis to find hypocoagulability consistent with prolongations in PTT. A Royal Veterinary College study in horses found significant differences in TEG measurements between blood collected into various anticoagulants and sample collection tubes. This data as well as collection method is important when comparing data from one laboratory to another. Our group used TEG to assess for hypercoagulability in cats with hypertrophic cardiomyopathy (HCM). Our data indicate that although individual cats may be hypercoagulable there is significant overlap in data between healthy cats and those with HCM. Another interesting *in vitro* study from UGA used anti-platelet drugs at various concentrations to determine if they had effects on healthy canine platelets.

These studies show the breadth of application and translational information that can be applied to veterinary patients both for clinical and research use. From this burst of recent information, we learn that the data obtained from the thromboelastograph can be used to monitor hemostasis in many species and in many conditions. Detailed methodological development is also important and raises the question for consistency of data for interpretation across laboratories. Currently there is no universally applied sample collection method, sample tube recommendation, or activator for all species. These parameters need to be considered when data assessment occurs for TEG-derived hemostatic analysis in the veterinary literature.

REFERENCES

[1.] Otto CM, et al. **J Am Vet Med Assoc** 2000; 217(10):1500-1504. [2.] Donahue SM et al. **J Vet Emerg Crit Care** 2005; 15(1): 9-16. [3.] Brainard BM et al. **Am J Vet Res** 2007; 68(3):251-257. [4.] Alwood AJ, et al. **J Vet Intern Med** 2008; 21(3):378-387. [5.] Kristensen AT, et al. **J Vet Intern Med** 2008; 22:140-147. [6.] Wiinberg B, et al. **J Vet Intern Med** 2008; 22:357-365. [7.] Wiinberg B, et al. **Vet J** 2009; 179:121-129. [8.] Jeger V, et al. **J Trauma** 2009; 66(4):1253-1257. [9.] Ganter MT, et al. **Anesth Analg** 2008; 106(5):1366-1375. [10.] Dai Y, et al. **Anesth Analg** 2009; 108(3):734-742. [11.] Chen A, et al. **Clin Lab Med** 2009; 29:391-407. [12.] Jackson GNB, et al **Anaesthesia** 2009; 64:212-215. [13.] Sinnott VB, et al. **J Vet Emerg Crit Care** 2009; 19(5):484-488. [14.] Epstein KL, et al. **J Vet Emerg Crit Care** 2009; 19(1):96-101

ADVANCES IN TOXICOLOGY: THE USE OF INTRA-LIPID THERAPY & HIGH-DEXTROSE INSULIN THERAPY

Justine A. Lee, DVM, DACVECC
Minneapolis, MN

INTRODUCTION

The use of intravenous lipid emulsion (ILE) or intravenous fat emulsion (IFE) has been well utilized in both human and veterinary medicine for decades as part of component therapy in the form of either total (TPN) or partial parenteral nutrition (PPN). More recently, the use of IFE has been utilized as an antidote for fat-soluble drug toxicity, particularly with local anesthesia overdoses in human medicine (i.e., bupivacaine),[1-4] where acute cardiopulmonary arrest has occurred. Numerous published case reports and experimental studies describe the beneficial use of IFE in the treatment of patients intoxicated with other drugs, including verapamil,[5] bupoprion, lamotrigine,[6] quetiapine, sertraline, calcium-channel blockers, and beta-blockers. Recently, IFE has been utilized in veterinary medicine also for cases of toxicity caused by administration or ingestion of lipid-soluble compounds including:

- Local anesthetics (i.e., bupivacaine, lidocaine)
- Clomipramine,[7] tricyclic antidepressants (TCA)
- Propranolol[8]
- Buproprion (i.e., Wellbutrin®, Zyban®) and others
- Muscle relaxants (i.e., baclofen, flexeril)
- Macrocyclic lactones (i.e., moxidectin, ivermectin)[9-11]

The exact mechanism of action by which IFE acts as an antidote and helps augments conventional resuscitation efforts is unknown. Currently, four potential hypotheses on how IFE is effective with fat-soluble toxicities include:

1. Providing myocytes energy substrates, thereby augmenting cardiac performance
2. Restoration of myocardial function by increasing intracellular calcium concentration
3. Acting as a "lipid sink" by sequestration of lipophilic compounds into the newly created lipid compartment within the intravascular space (a lipid or pharmacological sink). With this "lipid sink" hypotheses, compartmentalization of the drug into the lipid phase results in lesser free drug concentration available to tissues.
4. Increasing the overall fatty acid pool, which then overcomes bupivacaine inhibition of mitochondrial fatty acid metabolism.

Human and animal data

Extensive human literature exists on this topic, and the attendees are referred to the references listed below. As for veterinary medicine, there have not been any recent, prospective clinical trials evaluating the use of IFE; however, there is extensive animal literature on the use of IFE in *experimental* studies,[12,13] showing benefit for the following drugs: bupivacaine, propranolol, thiopental, verapamil, beta-blockers, clomipramine, and chlorpromazine. Recently, individual clinical case reports documenting the use of IFE in veterinary medicine been published (i.e., moxidectin);[10] however, blood levels for the toxic agent were not evaluated. Based on human literature and clinical response to therapy, the use of IFE has been recommended as a standard of care by both Pet Poison Helpline and the ASPCA Animal Poison Control for fat-soluble toxicities, particularly in severe, life-threatening overdoses with baclofen and ivermectin. There are anecdotal reports of the successful use of IFE with ivermectin, baclofen, and lidocaine toxicity in small animal veterinary medicine. Despite the paucity of data in veterinary medicine, clinical outcome with IFE has been successful.

Dosing

When using IFE therapy, it is important that veterinarians be aware of the different lipid emulsion formulations that are currently available in the market. Intravenous lipid emulsion is readily available at most human health care facilities and at veterinary emergency or specialty referral hospitals (where it may be used as part of PPN or TPN therapy). Currently, a commercially available antidote is available - LipidRescue 20%. More information can be found at: http://lipidrescue.squarespace.com. This product is specifically marketed as an antidote for toxicities in human medicine. Other IFE sources are readily available through veterinary suppliers (i.e., Baxter) as both a 10 and 20% solution. Currently, dose recommendations for IFE are extrapolated from human medicine and are listed below:

20% solution IFE dosing: Administer 1.5 ml/kg IV bloused over 1 minute, followed by 0.25 ml/kg/min for 30-60 minutes. The initial bolus could be repeated 1-2 times if no response to the initial bolus is obtained, with a goal to not exceed 8 ml/kg/day.

That said, this dose has been exceeded in clinical recommendations by Pet Poison Helpline without ill effect. A current retrospective case series evaluating the use of IFE by Pet Poison Helpline may hopefully clarify what therapeutic doses can be safely used.

Currently, when IFE is utilized for nutritional support, general guidelines are to not exceed 2 g/kg/day. However, some veterinary nutritionists utilize 20% IFE for sole PPN replacement by providing 100% of the resting energy requirement (RER) as lipid for short-term therapy (1-2 days). This would readily exceed the 2g/kg/day requirement. Therefore, based on the variability in dose of IFE used for PPN, the use of IFE in an acute setting for toxicity likely can exceed the current human recommendations. A comparison of doses for IFE and PPN are demonstrated below, and are designed to help indicate the variability in dosing:

General RER calculations:
Dog: 15 kcal/lb of dog; Cat: 20 kcal/lb of cat
20% IFE has 2 kcal/ml; 10% IFE has 1 kcal/ml; 20% solution = 200 mg/ml = 1g/5 mls

Using a 20% solution of IFE (osmolality 280 mOsm/L) to meet 100% of the RER, one can calculate the amount of lipid to administer:
Patient RER/2 = X mls of 20% IFE

Example:
11 lb cat
RER: 20 kcal X 11 lbs = 220 kcal/day
To meet 100% of the RER via 20% IFE (2 kcal/ml) = 110 ml/day
Lipid dose: **110 ml/day** = 4.6 ml/hour

Current human maximum recommendation of IFE: 8 ml/kg/day
11 lb cat = **40 ml/day**

Some nutritionalists recommend not to exceed: 2 g/kg/day
Cat: 2 g/kg/day X 5 kgs = 10 g/day = **50 ml/day**

Based on the variability in dosing, it is likely that IFE administration in the acute toxicology setting can warrant current increased dosing regimens to:

- Consider *additional* doses of 1.5 ml/kg IV over 30 min q. 4-6 hours X 24-36 hours until clinical signs resolve (based on clinical judgment) or
- Maintaining a CRI of 0.5 ml/kg/hour until clinical signs resolve.

Concurrent recommendations include the use of drug levels (serum or plasma) to evaluate the response to IFE therapy. This will aid in data collection for future retrospective study analysis. Time frames for drug levels include: time 0 (at the time of presentation to the health care facility), 30 minutes after administration of IFE, 1 hour post, 6, hours post, 12 hours post, and 24 hours post.

Side effects and potential complications of IFE administration

Deleterious effects from IFE are rare, and the benefits must out way the risks of use. Deleterious effects may be caused directly by the emulsion (egg-based, anaphylactoid reaction) or due to contamination of the infusate. Risks of microbial contamination and sepsis from IFE exist, as they do with the administration of TPN or PPN, but this rarely occurs when IFE is infused as a sole agent. Other known reported side effects seen in human medicine include: pulmonary toxicity, bacterial contamination, thrombophlebitis, fat embolism, hyperlipidemia, pancreatitis, immunosuppression, and hypersensitivity. Generally, IFE is well tolerated in the acute setting. In veterinary medicine, the side effects and known potential complications are likely similar, albeit more difficult to diagnose. Without the use of ventilation:perfusion scans, contrast-enhanced computerized tomography, etc., fat embolism is likely difficult to diagnose.

Patients should be carefully monitored during administration of IFE, particularly within the first 20 minutes, due to potential risks of acute, adverse "colloid reaction" or pyrogenic reactions. While rare (<1%), these reactions resemble anaphylactoid reaction and may include nausea, vomiting, fever, dyspnea and cyanosis. Another complicating factor with the use of IFE is "fat overload syndrome." This may be due to administration of large amounts of IFE, which then overwhelm the endogenous lipid clearance mechanisms. In human medicine, IFE administrate rates above 0.11g/kg/hr can be associated with fat complications. Clinical signs of "fat overload syndrome" in human medicine include hyperlipidemia, hepatomegaly, icterus, splenomegaly, fat embolism, thrombocytopenia, hemolysis, and prolonged clotting times. There is very little veterinary information on "fat overload syndrome." That said, severe lipemia has been seen in veterinary patients administered IFE. Severe hyperlipidemia may be easily detected on clinicopathologic testing (hematocrit tubes), and may affect total solid monitoring.

The potential use of heparin therapy (75-250 units/kg SQ q. 6) can be considered in cases where severe hyperlipidemia is present. However, as heparin infusion increases intralipid clearance, its effect on IFE therapy in the face of toxicity is unknown at this time. Partial thromboplastin time (PTT) should be carefully monitored prior to the use of heparin to ensure the patient is not already coagulopathic. If PTT exceeds 2-2.5X normal, the use of heparin should be either lowered (75 units/kg SQ q. 6-8) or discontinued. Currently, unless the patient is at risk for pancreatitis (i.e., Shetland sheepdog, miniature schnauzer, Yorkshire terrier, obesity, etc.), heparin therapy is not recommended with acute toxicosis and IFE treatment unless clinical signs or advanced diagnostics indicate otherwise, due to its unknown effect on the "lipid sink" hypothesis of IFE.

High-dose insulin (HDI) therapy

The use of HDI therapy (with regular insulin CRIs) has been recently advocated for the use of beta-blocker toxicity in both human and veterinary medicine to increase inotropy and treatment of hypoperfusion. Beta-blocker toxicity results in severe hypotension, bradycardia, cardiovascular collapse, shock, and death. Treatment traditionally includes fluid resuscitation, atropine, temporary pacemaker therapy, vasopressors, and inotropes.[14] Unfortunately, catecholamine vasopressors are often ineffective as they work on the same cell membrane receptors that are affected by beta-blockers.[14] Temporary cardiac pacemaker therapy and glucagon (which has inotropic properties) have previously been used in veterinary medicine,[15,16] but have been recently found to be less effective than the use of HDI therapy.[14] Glucagon was thought to reverse beta-blocker toxicity due to its ability to increase intracellular cAMP via a non-catecholamine receptor on the cell wall, thereby increasing myocardial cAMP.[14] Unfortunately, the effects of glucagon are variable, and fail as a single agent for the treatment of beta-blocker toxicity; currently, it is no longer considered the antidote for beta-blocker overdose. Other vasopressors have been evaluated for beta-blocker toxicity, including epinephrine and vasopressin. Holger et al evaluated the use of HDI at the dosing of 2 regular units/kg/hour, up to 10 units/kg/hour, compared to vasopressin (V) with epinephrine (E) (vasopressin 0.0028 units/kg/minute up to 0.014 units/kg/minute; epinephrine 10 mcg/kg/min up to 50 mcg/kg/minute) in a porcine model of beta-blocker toxicity.[14] In this study, glucose was concurrently administered in the HDI group when hypoglycemia < 60 mg/dL occurred.[14] Overall, all V/E pigs died with 90 minutes, while all HDI pigs survived. In the HDI group, cardiac output (CO) increased with the use of insulin therapy, while MAP, SBP, and systemic vascular resistance (SVR) all trended slightly downward. In the V/E group, CO dropped until death, while MAP, SBP, and SVR rose dramatically until 30-60 minutes, followed by an abrupt drop and then death. Overall, the use of HDI was superior to V/E in treating beta-blocker toxicity, and HDI was display marked inotropic properties when compared to the use of other vasopressors.[14] The attendees are referred to the references for further details about beta-blocker toxicity.

The exact mechanism of how HDI therapy works is unclear and currently unknown, and does not appear to be from a direct vasopressor effect. Rather, HDI's effectiveness may be due to insulin's ionotropic effects which increase perfusion by increasing cardiac output. While HDI does not increase SVR directly (thereby, no active increases in SBP or MAP are seen), clinical improvement in tissue perfusion is thought to occur with HDI. In addition, HDI helps promote aerobic metabolism in the myocardium,[14] and appears to promote the uptake and utilization of carbohydrates as an energy source. Insulin also prevents ischemic-reperfusion injury during shock states, and prevents apoptosis. Other hypotheses on how insulin works with beta-blocker toxicity is by enhancing glucose transport to the myocardium during the toxic state, improving intracellular calcium homeostasis by enhancing calcium channels, and stimulating catecholamine release.[14] HDI is also thought to enhance cardiac contractility and cardiac output.

Unfortunately, there is a paucity of veterinary literature evaluating the use of HDI in small animal medicine. Extrapolating from human medicine, currently, HDI treatment recommendations for toxicity cases include:

1) Central line placement for ease of patient care and blood glucose (BG) monitoring.
2) Check BG. Administer 0.5 ml/kg 50% dextrose IV if BG is <100 mg/dL for dog or < 200mg/dL for cat.
3) Administer regular insulin at 1 unit (U)/kg IV bolus, followed immediately with a CRI IV infusion of regular insulin at 2 units/kg/hr.
4) The regular insulin infusion can be increased 2 units/kg/hr every 10 minutes up to a maximum dose of 10 units/kg/hr.
5) Due to the HDI, carefully BG monitoring is imperative every 10 minutes while adjusting insulin dosing. Aggressive dextrose supplementation (typically > 5% dextrose CRI, IV) must be supplemented via the central line. Once insulin dosing has stabilized, the BG should continued to be monitored every 30-60 minutes.
6) Due to the intracellular shifting of potassium with insulin therapy, potassium levels should be frequently monitored every hour. Aggressive potassium supplementation should be utilized to maintain potassium within the low therapeutic range (3 mmol/L). Administer potassium chloride if $K^+ \leq 3.0$ mmol/L.
7) Once beta blocker toxicity has resolved, the regular insulin CRI should be slowly decreased by 1-2 units/kg/hr. Both potassium and dextrose supplementation will likely need to be continued for up to 24 hours, and careful monitoring should still be maintained. Likewise, frequent calcium levels should be monitored as the insulin CRI is decreased.

CONCLUSION

The use of newer modalities in veterinary toxicity may dramatically improve the overall outcome in toxicities that have been previously associated with a high morbidity, mortality, and cost (i.e., baclofen, ivermectin, etc.). The extra-label use of these new modalities should be judiciously used. Appropriate case selection is imperative.

REFERENCES

[1]Picard J, et. al. **Anaesthesia** 2009;64(2);122; [2]Rosenblatt MA, et. al. **Anesthesiology** 2006;105(1):217; [3]Spence AG. **Anesthesiology** 2007;107(3):516; [4]Weinberg G, et. al. **Reg Anesth Pain Med** 2003;28(3):198; [5]Young AC, et. al. **Resuscitation** 2009;5(80):591; [6]Sirianni AJ, et. al. **Ann Emerg Med** 2008;51(4):412; [7]Harvey M, et. al. **Ann Emerg Med** 2007;49(2):178; [8]Harvey MG, et. al. **J Med Toxicol** 2008;4(2):71; [9]Hopper H, et. al. **J Vet Int Med** 2002;16:89; [10]Crandell DE, et. al. **J Vet Emerg Crit Care** 2009;19(2):181; [11]Lallemand E, et. al. **J Vet Pharmacol Ther** 2007;30(5):375; [12]Van de Velde M, et. al. **Crit Care Med**. 1998;26(1):132; [13]Bania T, et. al. **Acad. Emerg. Med**. 2007;14:105; [14]Holger JS, et. al. **Clinical Toxicology** 2007;45(4):396; [15]Syring RS, et. al. **JVECC** 2008;18(1):75; [16]Costello MF, et. al. **JVECC** 2008;18(1):54; [17]Kerns W II, et. al. **Ann Emerg Med** 1997; 29(6):748.

COMPLICATIONS & CONTROVERSIES OF DECONTAMINATION: ACTIVATED CHARCOAL – TO USE OR NOT TO USE?

Justine A. Lee, DVM, DACVECC
Minneapolis, MN

ACTIVATED CHARCOAL (AC)

Activated charcoal (AC) has been used for over 100 years for the treatment of toxic ingestions, and continues to be the most common form of gastrointestinal decontamination for the poisoned patient.[1] Activated charcoal works by adsorbing poison in the gastrointestinal tract (GIT) and decreasing the extent of systemic absorption of the poison.[1] In order for AC to work, it must physically come into contact with the poison. To maximize AC absorption of the toxicant, the administration of AC should be implemented as soon as possible, as delayed administration reduces its effectiveness.[1]

Medicinal charcoal must meet BP, USP or similar standards for adsorption, microbial contaminants and purity.[1] Activated charcoal is created by controlled pyrolysis of petroleum, coconut shells, lignite (coal), peat, and wood, which then produces charcoal.[1] Once charcoal is created, it is activated by heating it at high temperatures (600-900°C) with steam, air, or carbon dioxide.[1] Finally, charcoal is washed with inorganic acids and dried. This activation process results in a small particle size, a large surface area (950 to 2,000 m^2/g), and a highly developed internal pore structure which contributes to adsorptive properties and extent of adsorption at equilibrium.[1] The carbon moieties (carbonyl, hydroxyl) of the adsorptive surface of AC contribute to the varying binding affinity.[1] For example, non-polar compounds bind to AC well, while heavy metals and alcohols (i.e., ethylene glycol) are not absorbed by AC; xylitol is also poorly bound to AC. In general, *in vitro* adsorption to AC reaches equilibrium in less than 30 minutes. While desorbption of toxicants to AC can occur, as binding to AC is a reversible process, this clinical effect is not known or well described.[1] The addition of a cathartic may be considered to help promote GIT transit time to help decrease the time for desorption to potentially occur. Activated charcoal may also contain preservatives, various types of cathartics, or even sodium bicarbonate or povidone, which may contribute to complications later discussed in this lecture.

The American Academy of Clinical Toxicology (AACT) and European Association of Poisons Centres and Clinical Toxicologists (EAPCCT) created a position paper in 1997 (revised again in 2004) which clearly stated that "*single-dose AC should not be administered routinely in the management of poisoned patients… [as]… there is no evidence that administration of AC improves clinical outcome.*"[1] Since then, the use of AC in human medicine has continued to steadily decline, according to the American Association of Poison Control Centers Toxic Exposure Surveillance System (from 7.7% to 5.9% in 2003). Despite human medicine's move away from AC, should veterinary medicine be following this similar declining trend in detoxification of poisoned patients? (Short answer: No).

HUMAN AND ANIMAL STUDIES

Numerous human and animal studies have evaluated AC administration and effect based on timing of ingestion. The position paper on single-dose AC evaluated the results of 122 studies, and compared 46 drugs, the absolute amount of charcoal (0.5-100 g) administered, and the time of administration (up to 360 minutes after ingestion).[1] Of these 122 cases, 84 (69%) had AC administered within ≤ 5 minutes.[1] In the studies where AC was administered within 30 minutes, there was a mean reduction in absorption of 51.7% (n=7); at 60 minutes, the mean reduction was 38.1% (n=16); at 120 minutes, the mean reduction was 34.5% (n=8); at 180 minutes, the mean result was 21.1% (n=3); at 240 minutes, the mean reduction was 29.3% (n=3), and at 360 minutes, the reduction was 14% (n=1).[1] The data from 48 comparisons involving 26 drugs, which all used at least 50 g of AC, showed a mean reduction at 30 minutes after dosing at 47.3% (n=3); the mean reduction at 60 minutes was 40.1% (n=12); at 120 minutes, 16.5% (n=3); at 180 minutes, 21.1% (n=3); and at 240 minutes, the mean reduction was 32.5% (n=2). Unfortunately, in these studies, certain factors could not be controlled (i.e., influence of food in the stomach, presence of a poison that may delay gastric emptying, etc.).[1] Based on these studies, the administration of AC in human medicine beyond one hour is generally not recommended; however, the beneficial potential beyond 1 hour cannot be excluded.

When it comes to animal studies, limited veterinary literature exists. Often, these studies are performed in mice and rats, and it is important to consider the differences in comparative anatomy, metabolism, GI motility, morphology, absorption rate and site, and elimination.[1] A recent prospective study evaluated the effect of AC alone vs. emesis and AC on carprofen absorption following experimental overdose in dogs, and found that AC alone is as effective.[a] Due to the paucity of prospective clinical research in veterinary medicine, further studies are warranted.

In veterinary medicine, it is almost impossible to administer AC within ≤ 5 minutes. Due to driving time (to the clinic), lapsed time since ingestion, time to triage, and the amount of time it takes to physically deliver AC (i.e., syringe feeding, orogastric tube, etc.), administration is often delayed to 30 minutes, up to an hour or more. As time since ingestion is often unknown (i.e., pet owner coming home from work to find their pet poisoned), decontamination is often a relatively benign course of action, provided the patient is not already symptomatic. As always, when administering *any* drug, it is important that benefits outweigh the risks, and that complications be prevented, when possible. In veterinary medicine, administration of AC with a cathartic as long as 6 hours out may still be beneficial with certain types of toxicosis, particularly if the product has delayed release (i.e., extended or sustained release) or undergoes enterohepatic recirculation (see multi-dose AC below). More importantly, the variable modalities of therapy [i.e., antidotes (such as digoxin-specific antibody fragments, 2-PAM, etc.), plasmaparesis, hemodialysis, mechanical ventilation, etc.], along with financial limitations of pet owners, warrant the aggressive use of AC in veterinary medicine, as this is often our last line of defense in adequate decontamination of a patient.

DOSING

Current recommended dosing for single dose AC is: 1-5 g of AC/kg with a cathartic (i.e., sorbitol) to promote GIT transit time.

Multi-dose Activated Charcoal

Certain situations or toxicities warrant multi-dose administration of AC. Drugs undergoing enterohepatic recirculation; drugs that diffuse from the systemic circulation back into the intestinal tract down the concentration gradient; or ingestion of sustained (SR), extended (XR), or long-acting release products will require multi-dose administration of AC; however, multi-dosing should *not* contain cathartics with additional doses, due to increased risks for dehydration and secondary hypernatremia via fluid losses from the GIT. Human studies have found that multi-dose AC significantly decreases the serum half-life of certain drugs, including antidepressants, theophylline, digitoxin, and phenobarbital. While veterinary studies are lacking, there is likely an added benefit from multi-dosing AC, provided the patient is well hydrated and monitored appropriately.

- Dose: 1-2 g of AC *without a cathartic* per kg of body weight, PO q 4-6 hours for 24 hours

AVAILABILITY

There are several types of AC commercially available (i.e., granules, capsules, tablets, aqueous suspension, powder), and the labeled directions should be followed appropriately for each specific type, as the dose is dependent on the product used and concentration. Many types already contain a cathartic within the AC (i.e., typically 70% sorbitol), and it is important to ideally have two types available (AC alone, AC with cathartic). Finally, some pet owners may have access to AC tablets and capsules; however, these are not generally recommended, as they are not as effective as AC liquid slurries or aqueous solutions. In human medicine, the aqueous suspension is significant more effective than tablets or capsules.

CONTRAINDICATIONS OF ACTIVATED CHARCOAL

Contraindications for AC include endoscopy (which would obscure visualization), abdominal surgery of the GIT, gastric or intestinal obstruction, gastrointestinal hemorrhage or perforation (due to pathology, caustic injury, etc.), recent surgery, late-stage presentation with clinical signs already present, dehydration, lack of borborgymi, ileus, hypernatremia, hypovolemic shock, compromised airway (risk for aspiration pneumonia), and ingestion of a caustic substance or hydrocarbon (due to increased risk for aspiration pneumonia). In patients that have an unprotected airway that are at risk for aspiration pneumonia (i.e., a depressed state of consciousness, excessive sedation, etc.), the use of AC is contraindicated without endotracheal intubation.

CATHARTICS

Cathartics are designed to increase the speed and transit time of the GIT, promoting fecal excretion of the toxin, but more importantly, decreasing the time allowed for toxin absorption through the GIT. Typical cathartics used in veterinary medicine comprise of osmotic cathartics: saccharide (i.e., sorbitol) vs. saline cathartics (i.e., sodium sulfate, magnesium citrate, or magnesium sulfate). Sorbitol is most commonly used, as it aids in the expulsion of the poison from the GIT, while also masking the grittiness and poor palatability of AC with its sweet taste. Side effects of sorbitol administration include vomiting, dehydration, secondary hypernatremia, abdominal cramping or pain, and possible hypotension. The contraindications for cathartics are similar to those that exist for AC. Another important consideration in veterinary medicine is that mineral oil is *no* longer recommended as a cathartic due to the high risks of secondary aspiration. In general, if AC does not contain a pre-existing cathartic (for the first dose administered), one can potentially add in sorbitol (70% solution) for a one dose dosing at: 1-2 ml/kg, PO, given within 60 minutes of toxin ingestion.

According to the position paper on cathartics endorsed by the AACT and EAPCCT, the use of cathartics in human medicine "has no role in the management of the poisoned patient and is not recommended as a method of gut decontamination."[2] If considered, the use of cathartics should be limited to a single dose to prevent known complications from multi-dosing of cathartics. There are no published clinical studies investigating the use of a cathartic (with or without AC) in reducing the bioavailability of drugs, or improving the clinical outcome of poisoned patients.[2] Finally, the position statement of cathartics warns about serious adverse effects from multi-dosing of cathartics, including dehydration and electrolyte abnormalities, including hypernatremia and hypermagnesemia (when using a sodium- or magnesium-containing cathartic). In addition, those patients with renal dysfunction or pre-existing dehydration may be at increased risk. In veterinary medicine, we still recommend a first dose dosing of AC with a cathartic; however, the use of multi-dose AC should *not* contain cathartics to prevent these clinically recognized, but rare, complications.

Unfortunately, there is limited literature evaluating the use of cathartics in veterinary medicine, and no studies prospectively evaluating its use in the clinical poisoned veterinary patient. Experimental animal studies do exist; in one study evaluating mice administered paraquat, the combined use of AC with magnesium citrate (30 minutes post) increased survival from 31% (controls) to 94%, with a significant $p<0.01$.[3] In a canine study evaluating sorbitol, mannitol, and AC administration in dogs given acetaminophen, the area under the curve (AUC) was 75% greater with cathartics and AC compared to charcoal alone ($p=0.07$), and the peak plasma concentration was 80.4% greater ($p=0.012$) after cathartics plus AC compared to AC alone.[4] When AC was given with sodium sulfate to rats treated with salicylate, pentobarbital, chlorpheniramine, and chloroquine, peak plasma concentrations were significantly reduced ($p<0.001$) compared to the control.[5] Finally, when sodium sulfate with superactivated AC was used in rats given a lethal dose of T-2 mycotoxin, survival and survival times increased ($p<0.01$).[6]

COMPLICATIONS

In both human and veterinary medicine, there are relatively rare reports of AC-related adverse events, especially when considering the widespread use of AC in both human and veterinary medicine.[1] Complications include respiratory complications (i.e., aspiration pneumonia, acute respiratory distress syndrome), corneal abrasions (if accidental contact to the eye occurs), intubation/endoscopy difficulty (due to the presence of AC in the oropharynx), constipation, vomiting, and fluid, electrolyte and acid-base abnormalities (including hypernatremia, hypokalemia, hypermagnesemia, and metabolic acidosis). In humans administered multi-dose AC, complications of ileus and small intestinal obstruction (from charcoal bezoars) requiring surgical intervention have been reported. When AC is used appropriately in a hydrated patient, the majority of adverse events are not directly related to AC; rather, it may be due primary to the cathartic affect, or from complications arising when aspiration of AC occurs or when AC is directly administered (iatrogenic) into the lung. Finally, additional preservatives or contents (i.e., povidone) in AC can result in severe respiratory problems. Rarely, anecdotal reports of hypernatremia have been clinically reported with AC administration. This is likely due to the sorbitol effect, which acts as a cathartic, resulting in excessive free water loss. Initially, the use of a one-time dose of a cathartic is recommended to help promote GIT motility, to prevent desorption from occurring between the toxin and the AC within the GIT. However, continued use of cathartics (with multi-dosing) can result in profound dehydration, fluid losses, and hypernatremia.

Burkitt et. al evaluated the use of a commercial AC, Actidose-Aqua, in a prospective, clinical trial in 6 dogs to determine if AC suspension containing propylene glycol (PG) and glycerol had any effects on serum osmolality, osmolal gap, and lactate.[7] Samples were also evaluated for acid-base status, and concentrations of sodium, potassium, BUN, and glucose. In this study, dogs (n=6) were given 4 g/kg of AC, and samples were taken before and 1, 4, 6, 8, 12, and 24 hours post-AC.[7] In this study, mean serum osmolality, osmolal gap, and lactate concentration were significantly increased after suspension administration compared to baseline, typically increasing at 1 hour after AC administration, peaking at 4 hours, and returning to baseline by 24 hours. Hypernatremia was not detected in any of the study dogs; however, mild hypokalemia was seen ($P<0.05$). Serum osmolality increased from 311 mOsm/kg at baseline to 353 mOsm/kg, osmolal gap increased from 5 to 52 mOsm/kg, and lactate concentration increased from 1.9 to 4.5 mmol/L after suspension administration ($P < .01$).[7] All 6 dogs drank frequently after AC administration, but water intake or urine output was not recorded. In addition, 3/6 dogs vomited within 1-3 hours of administration of AC, while 4/6 of dogs were lethargic.[7] While this study was important in evaluating complications from AC, this particular AC did not contain sorbitol or a cathartic (Actidose with Sorbitol). A prospective study evaluating the use of a cathartic with AC would be important to evaluate the prevalence of hypernatremia seen with AC and cathartic administration.

CLINICAL APPLICATION

Prior to administration of AC with or without a cathartic, the patient should be assessed for hydration status. Appropriate fluid supplementation (SQ, IV) should be used to prevent dehydration and hypernatremia. In addition, the concurrent use of parenteral (i.e., IV) anti-emetics should be considered, due to the high prevalence of vomiting from cathartic administration (or from the emetic previously used to decontaminate the patient): Cathartics should not be used in a dehydrated patient, due to the risks of voluminous fluid losses through the GIT and secondary hypernatremia. For patients receiving either multi-doses of AC with or without cathartics, serum sodium levels should be monitored, along with hydration status. Those patients that are dehydrated, vomiting, fasted, predisposed to hyperosmolality disorders (such as renal disease, poorly regulated diabetes mellitus, psychogenic polydyspia, diabetes insipitus, etc.)[7] may be more at risk for developing significant, clinical hypernatremia from the cathartic agent in AC.

CONCLUSIONS

In veterinary medicine, the primary treatment for toxicant exposure should be decontamination and detoxification of the patient to inhibit or minimize further toxicant absorption and to promote excretion or elimination of the toxicant from the body.[8] Knowledge of the underlying mechanism of action, the pharmacokinetics (including absorption, distribution, metabolism, and excretion), and the toxic dose of the toxicant are imperative in determining appropriate decontamination and therapy for the patient.

REFERENCES

[1]American Academy of Clinical Toxicology Position Paper: Single-Dose Activated Charcoal. **Clin Toxic** 2005;43:61; [2]American Academy of Clinical Toxicology Position Paper: Cathartics. **Clin Tox** 2004;42(3):243; [3]Gaudreault P, et. al. **Ann Emerg Med** 1985;14:123; [4]Van de Graaff WB et. al. **J Pharmacol Exp Ther** 1982;221:656; [5]Chin L, et. al. **Clin Toxicol** 1981;18:865; [6]Galey FD, et. al. **Toxicon** 1987;25:493; [7]Burkitt JM, et. al. **J Vet Intern Med** 2005;19(5):683; [8]Peterson ME. **Small Animal Toxicology**, 2nd Ed. St. Louis: Elsevier Saunders, 2006:127; [a]Schildt J, et. al. **IVECCS Proceedings**, unpublished data.

THE ROLE OF EDEMA IN SPINAL CORD INJURY

Yvette S. Nout, DVM MS PhD DACVIM DACVECC
San Francisco, CA

LESION DEVELOPMENT AFTER SPINAL CORD INJURY

Damage inflicted to the spinal cord occurs in 2 distinct phases. So-called "primary" injury refers specifically to the mechanical disruption of the spinal cord tissue including central nervous system (CNS) structures and the vasculature. This primary damage is followed by a period of "secondary" injury during which progression of tissue damage and cell death occurs that can last hours to weeks after initial insult [1-5]. Although treatment to reverse neurologic deficits resulting from initial primary injury does not currently exits, the period of secondary injury is considered to be the phase in which therapeutic measures could be taken to halt the process of ongoing cell loss. However, to date investigators have obtained limited results.

After SCI, the cord undergoes sequential pathological changes including hemorrhage, edema, axonal and neuronal necrosis, and demyelination, followed by cyst formation and infarction [4,6,7]. Many well-characterized mechanisms of cell damage have been shown to play a role during secondary damage, including 1) vascular compromise leading to reduced blood flow, loss of autoregulation, loss of microcirculation, vasospasm, thrombosis, edema, and hemorrhage; 2) electrolyte shifts, permeablility changes, loss of cellular membrane integrity, edema, and loss of energy metabolism; 3) biochemical changes including neurotransmitter accumulation, arachidonic acid release, free radical and prostaglandin production, and lipid peroxidation; and 4) apoptosis [3,5,8-11]. Free radical formation, excitotoxicity, endothelial dysfunction, activation of inflammatory mediators, thrombosis, vasospasm, and enzymatic alterations are some of the processes that occur during secondary injury that have the potential to increase local vascular resistance and reduce spinal cord blood flow [3,9,12]. Secondary expansion of the lesion cavity that occurs through necrotic and apoptotic pathways is associated with the invasion of immune cells and activation of the resident microglial population. Ischemia of the spinal cord appears to underlie many of the mechanisms of posttraumatic SCI and is the most common denominator resulting in neurological deficit after primary injury.

HYPERTONIC SALINE IN SHOCK AND TRAUMATIC BRAIN INJURY

Hypertonic saline (HS) has been widely investigated for resuscitation in shock [13-16] and for reducing intracranial pressure (ICP) to treat cerebral edema after traumatic brain injury (TBI) [15,17-19]. HS mobilizes free water from the intracellular into the extracellular space by osmotic force and reduction of peripheral vascular resistance. HS is a rapid volume expander and is commonly used to improve blood pressure and cardiac output during resuscitation from trauma with the goal of increasing delivery of oxygen to tissues. Hypertonic saline (7.5%) has been shown to be effective in expanding plasma volume and enhancing cardiac output in hemorrhagic shock [20,21]. Also, hypertonic saline has been shown to increase perfusion of the microvasculature, presumably through selective arteriolar vasodilation and be decreasing swelling of red blood cells and of the endothelium [22,23]. More recently, HS-dextran was shown to be effective in restoring splanchnic perfusion and microcirculation in a low cardiac output state (cardiac tamponade) [24] and it was shown to alleviate hepatic injury in hypovolemic rats [25]. HS-dextran also has antioxidant properties, which is suggested to be one of the reasons for improvement of the microcirculation seen in a shock model [26].

The deleterious effects of hypoxemia and hypotension on outcome after severe head injury have been clearly shown [27-30]. In fact, even a single episode of hypotension was associated with a doubling of mortality and increased morbidity. In cerebral edema, HS reduces brain swelling and thus ICP, by establishing an osmotic gradient between the intracellular and intravascular space. In addition, improved cerebral blood flow and increased delivery of oxygen cause a compensatory vasoconstriction and a reduction in cerebral blood volume, which further lowers ICP [18]. The result is improvement of cerebral perfusion pressure [15,18,19,31].

HYPERTONIC SALINE IN SPINAL CORD INJURY

In SCI, HS is not used specifically for its potential beneficial effects on the spinal cord directly. There have been a number of investigations into the use of HS in SCI [32-35], and although most studies report positive effects of HS, this has not led to widespread use of HS in human or experimental SCI. In 1994 Young et al showed that a single bolus administration of 7.5% HS (5 ml/kg) increased spinal cord blood flow and preserved spinal cord conduction during a 1 hour study period after SCI [35]. Follow up work showed that 7.5% HS (5 ml/kg bolus) resulted in a significant increase in spinal cord blood flow, but that this effect was very short-lived (5 minutes). All animals that received HS maintained somatosensory evoked potentials for the 4 hour study duration in contrast to untreated animals [34,36]. In a next study, animals were given 7.5% HS (5ml/kg bolus) 1 minute after injury and in the treated group, recovery of bladder and hind limb locomotor function was faster, higher locomotor outcome scores were measured, and attenuation of histopathological outcomes was seen [33,36]. In 2001 Legos et al showed that co-administration of 7.5% HS (5 ml/kg bolus) with methylprednisolone (MP) may enhance delivery of MP. Positive effects of the combinatorial treatment on survival and BBB score were recorded when compared to the control animals [32]. All these reports suggest HS would improve spinal cord blood flow much like it does in TBI. In addition, Spera et al (1998) showed that leukocyte adhesion after SCI was attenuated by HS. This antiinflammatory effect is similar to what has been shown in the liver after hemorrhagic shock [26]. In the liver HS improved microvascular perfusion and reduced endothelial cell swelling [37]. It has been suggested that HS may also reduce leukocyte swelling [38].

A recent study showed beneficial effects of HS on magnetic resonance imaging (MRI) indices of swelling, edema, and hemorrhage within the cord after experimental spinal cord injury in rats [39]. Immediately after SCI, MR images were obtained continuously for 8hrs. At 36 minutes after SCI, significant swelling of the spinal cord at the lesion center and extending rostrally and caudally was demonstrated by MRI. Also, at this time point, a hypointense core (hemorrhage) surrounded by a prominent ring of hyperintensity, characteristic of edema, was identified. Administration of HS resulted in increased serum [Na], attenuation of cord swelling, and decreased volume of hypointense core and edema at the last time points.

REFERENCES

[1] Allen AR. Surgery of experimental lesion of spinal cord equivalent to crush injury or fracture dislocation of spinal column. A preliminary report. **JAMA** 1911;57:878-880. [2] Allen AR. Remarks on the histopathological changes in the spinal cord due to impact. An experimental study. **J Nerv Ment Dis** 1914;41:141-147. [3] Tator CH, Fehlings MG. Review of the secondary injury theory of acute spinal cord trauma with emphasis on vascular mechanisms. **J Neurosurg** 1991;75(1):15-26. [4] Beattie MS, Li Q, Bresnahan JC. Cell death and plasticity after experimental spinal cord injury. **Prog Brain Res** 2000;128:9-21. [5] Beattie MS, Farooqui AA, Bresnahan JC. Review of current evidence for apoptosis after spinal cord injury. **J Neurotrauma** 2000;17(10):915-25. [6] Bresnahan JC, Beattie MS, Stokes BT, Conway KM. Three-dimensional computer-assisted analysis of graded contusion lesions in the spinal cord of the rat. **J Neurotrauma** 1991;8(2):91-101. [7] Noble LJ, Wrathall JR. Spinal cord contusion in the rat: morphometric analyses of alterations in the spinal cord. **Exp Neurol** 1985;88(1):135-49. [8] Amar AP, Levy ML. Pathogenesis and pharmacological strategies for mitigating secondary damage in acute spinal cord injury. **Neurosurgery** 1999;44(5):1027-39; discussion 1039-40. [9] Hall ED, Wolf DL. A pharmacological analysis of the pathophysiological mechanisms of posttraumatic spinal cord ischemia. **J Neurosurg** 1986;64(6):951-61. [10] Tator CH. Experimental and clinical studies of the pathophysiology and management of acute spinal cord injury. **J Spinal Cord Med** 1996;19(4):206-14. [11] Tator CH. Biology of neurological recovery and functional restoration after spinal cord injury. **Neurosurgery** 1998;42(4):696-707; discussion 707-8. [12] Tator CH, Koyanagi I. Vascular mechanisms in the pathophysiology of human spinal cord injury. **J Neurosurg** 1997;86(3):483-92. [13] Dubick MA, Bruttig SP, Wade CE. Issues of concern regarding the use of hypertonic/hyperoncotic fluid resuscitation of hemorrhagic hypotension. **Shock** 2006;25(4):321-8. [14] Moore FA, McKinley BA, Moore EE. The next generation in shock resuscitation. **Lancet** 2004;363(9425):1988-96. [15] Pinto FC, Capone-Neto A, Prist R, MR ES, Poli-de-Figueiredo LF. Volume replacement with lactated Ringer's or 3% hypertonic saline solution during combined experimental hemorrhagic shock and traumatic brain injury. **J Trauma** 2006;60(4):758-63; discussion 763-4. [16] Velasco IT, Pontieri V, Rocha e Silva M, Jr., Lopes OU. Hyperosmotic NaCl and severe hemorrhagic shock. **Am J Physiol** 1980;239(5):H664-73. [17] Baker AJ, Park E, Hare GM, Liu E, Sikich N, Mazer DC. Effects of resuscitation fluid on neurologic physiology after cerebral trauma and hemorrhage. **J Trauma** 2008;64(2):348-57. [18] Forsyth LL, Liu-DeRyke X, Parker D, Jr., Rhoney DH. Role of hypertonic saline for the management of intracranial hypertension after stroke and traumatic brain injury. **Pharmacotherapy** 2008;28(4):469-84. [19] Ziai WC, Toung TJ, Bhardwaj A. Hypertonic saline: first-line therapy for cerebral edema? **J Neurol Sci** 2007;261(1-2):157-66. [20] Moore EE. Hypertonic saline dextran for post-injury resuscitation: experimental background and clinical experience. Aust N Z **J Surg** 1991;61(10):732-6. [21] Muir WW, 3rd. Small volume resuscitation using hypertonic saline. **Cornell Vet** 1990;80(1):7-12. [22] Attuwaybi B, Kozar RA, Gates KS, Moore-Olufemi S, Sato N, Weisbrodt NW, Moore FA. Hypertonic saline prevents inflammation, injury, and impaired intestinal transit after gut ischemia/reperfusion by inducing heme oxygenase 1 enzyme. **J Trauma** 2004;56(4):749-58; discussion 758-9. [23] Mazzoni MC, Borgstrom P, Intaglietta M, Arfors KE. Lumenal narrowing and endothelial cell swelling in skeletal muscle capillaries during hemorrhagic shock. **Circ Shock** 1989;29(1):27-39. [24] Terajima K, Aneman A, Haljamae H. Haemodynamic effects of volume resuscitation by hypertonic saline-dextran (HSD) in porcine acute cardiac tamponade. **Acta Anaesthesiol Scand** 2004;48(1):46-54. [25] Ozguc H, Tokyay R, Kahveci N, Serdar Z, Gur ES. Hypertonic saline dextran alleviates hepatic injury in hypovolemic rats undergoing porta hepatis occlusion. **Shock** 2003;19(4):383-7. [26] Corso CO, Okamoto S, Ruttinger D, Messmer K. Hypertonic saline dextran attenuates leukocyte accumulation in the liver after hemorrhagic shock and resuscitation. **J Trauma** 1999;46(3):417-23. [27] Badjatia N, Carney N, Crocco TJ, Fallat ME, Hennes HM, Jagoda AS, Jernigan S, Letarte PB, Lerner EB, Moriarty TM and others. Guidelines for prehospital management of traumatic brain injury 2nd edition. **Prehosp Emerg Care** 2008;12 Suppl 1:S1-52. [28] Chesnut RM, Marshall LF, Klauber MR, Blunt BA, Baldwin N, Eisenberg HM, Jane JA, Marmarou A, Foulkes MA. The role of secondary brain injury in determining outcome from severe head injury. **J Trauma** 1993;34(2):216-22. [29] Fearnside MR, Cook RJ, McDougall P, McNeil RJ. The Westmead Head Injury Project outcome in severe head injury. A comparative analysis of pre-hospital, clinical and CT variables. **Br J Neurosurg** 1993;7(3):267-79. [30] Pigula FA, Wald SL, Shackford SR, Vane DW. The effect of hypotension and hypoxia on children with severe head injuries. **J Pediatr Surg** 1993;28(3):310-4; discussion 315-6. [31] Bratton SL, Chestnut RM, Ghajar J, McConnell Hammond FF, Harris OA, Hartl R, Manley GT, Nemecek A, Newell DW, Rosenthal G and others. Guidelines for the management of severe traumatic brain injury. II. Hyperosmolar therapy. **J Neurotrauma** 2007;24 Suppl 1:S14-20. [32] Legos JJ, Gritman KR, Tuma RF, Young WF. Coadministration of methylprednisolone with hypertonic saline solution improves overall neurological function and survival rates in a chronic model of spinal cord injury. **Neurosurgery** 2001;49(6):1427-33. [33] Sumas ME, Legos JJ, Nathan D, Lamperti AA, Tuma RF, Young WF. Tonicity of resuscitative fluids influences outcome after spinal cord injury. **Neurosurgery** 2001;48(1):167-72; discussion 172-3. [34] Spera PA, Vasthare US, Tuma RF, Young WF. The effects of hypertonic saline on spinal cord blood flow following compression injury. **Acta Neurochir (Wien)** 2000;142(7):811-7. [35] Young WF, Rosenwasser RH, Vasthare US, Tuma RF. Preservation of post-compression spinal cord function by infusion of hypertonic saline. **J Neurosurg Anesthesiol** 1994;6(2):122-7. [36] Tuma RF, Vasthare US, Arfors KE, Young WF. Hypertonic saline administration attenuates spinal cord injury. **J Trauma** 1997;42(5 Suppl):S54-60. [37] Corso CO, Okamoto S, Leiderer R, Messmer K. Resuscitation with hypertonic saline dextran reduces endothelial cell swelling and improves hepatic microvascular perfusion and function after hemorrhagic shock. **J Surg Res** 1998;80(2):210-20. [38] Spera PA, Arfors KE, Vasthare US, Tuma RF, Young WF. Effect of hypertonic saline on leukocyte activity after spinal cord injury. **Spine** 1998;23(22):2444-8; discussion 2448-9. [39] Nout YS, Mihai G, Tovar CA, Schmalbrock P, Bresnahan JC, Beattie MS. Hypertonic saline attenuates cord swelling and edema in experimental spinal cord injury: a study utilizing magnetic resonance imaging. **Crit Care Med** 2009;37(7):2160-6.

METHODS FOR NEURAL REPAIR AFTER SPINAL CORD INJURY

Yvette S. Nout, DVM MS PhD DACVIM DACVECC
San Francisco, CA

THERAPEUTIC CONSIDERATIONS

Based on the pathophysiology of events that occur after spinal cord injury (SCI) it is likely that single drug intervention would not be effective. Treatment to reverse primary injury does not currently exist and it is the assortment of pathophysiological processes that occur during the period of secondary injury that are considered the target for pharmaceutical intervention. In SCI, the period of secondary injury can be divided into three therapeutic windows [1]. The first 48 hours after acute spinal cord injury are dominated by the vascular and biochemical changes that occur within the spinal cord. The second period is a result of the effects of inflammatory cells that occurs within hours of injury and peaks around four days after injury. The third period occurs approximately one week after injury and is characterized by axonal regeneration and lesion repair. The goals of treatment are to stop the cascade of cellular events initiated by the traumatic insult, to protect spared neural tissue, and to promote regeneration. Surgical intervention is warranted when there is need for stabilization, fracture repair, or evidence of a compressive lesion, however, is not routine practice. The use of medical treatment to stabilize the patient should always be instituted before surgery is performed.

THERAPEUTIC STRATEGIES

Acute phase: Systemic hypotension is considered detrimental to recovery. Hypotension may exacerbate spinal cord hypoperfusion and ischemia, and maintaining systemic blood pressure has been shown to improve spinal cord perfusion. The current recommendation is to maintain euvolemic normotension, and because of sympathetic outflow disruption after cranial SCI, pressor therapy is commonly indicated in the treatment regimen [2,3]. Maintaining normal mean arterial blood pressure is also important to consider during stabilization of the acutely injured horse, and is particularly important when horses are placed under general anesthesia for various diagnostic/therapeutic procedures. To reduce edema, hyperosmolar substances such as hypertonic saline and/or mannitol are used with or without furosemide. Hypertonic saline attenuates cord swelling and severity of hemorrhage and edema detected by magnetic resonance imaging in the acute phase of unilateral cervical spinal cord injury in rats [4]. This study showed that at 36 minutes after SCI, significant swelling of the spinal cord at the lesion center and extending rostrally and caudally was demonstrated by MRI. A prominent ring of T2 weighted image hyperintensity, characteristic of edema, surrounded a hypointense core characteristic of hemorrhage. Administration of hypertonic saline resulted in increased serum [Na], attenuation of cord swelling, and decrease of volumes of hemorrhage and edema at the last time points during the study. In addition to this, treatment with antiinflammatory and free-radical scavenging drugs such as NSAID's, DMSO, and corticosteroids have a place in the treatment of CNS trauma. Many agents, that could target different aspects of the secondary injury mechanisms, have been investigated for use in SCI and have been reviewed [5,6]. Until now only methylprednisolone sodium succinate (MPSS) was shown to be efficacious in both animal models and humans [7-10], however, based on conflicting reports and small neurologic improvements, a number of reviews of the data have been published expressing concern with regards to the true beneficial effect of MPSS in acute SCI [11,12]. Controlling fever and seizure activity is important as well as maintaining appropriate antibiotic coverage and nutritional status.

For the subacute to chronic phase, currently investigated treatment strategies are those that promote regeneration, remyelination, and restoration of circuitry. These methods are based on for example: 1) altering the lesion milieu to create a less hostile environment in which axons can grow, (stem) cells can survive and differentiate, and cell function is restored (signaling, uptake, production); 2) providing exogenous substrate to enhance endogenous repair mechanisms (vector based delivery of growth factors, cell-based therapies); 3) rehabilitative strategies to stimulate spinal cord plasticity (interneuron activation and "re-learning" within the spinal cord). Currently the focus of much of the experimental spinal cord injury research is to develop, design, and test treatment strategies that are readily translatable for use in clinical situations. Many of these treatments are combinatorial strategies, in which multiple therapeutic elements are incorporated. An example is investigation of a delayed combined transplant strategy that included transplantation of glial restricted precursor (GRP) cells and administration of cAMP. We asked whether combining cAMP with a GRP cell transplant would improve functional recovery, similar to that observed after combining cAMP with Schwann cell transplants [13]. We used 51 adult, male rats. Groups: 1: age-matched control; 2: operated control; 3: GRP control; 4: cAMP control; 5: GRP cAMP. We used 12 rats in a short-term survival part of this study. Groups: 1: uninjured control; 2: operated control; 3: cAMP control. Our moderate contusion injury produced a very consistent lesion with the area of remaining tissue at the lesion center in the OP control group ($12.4 \pm 2.7\%$) in the same range as described previously for this level of injury ($9.9 \pm 4.8\%$). The presence of GRP cells resulted in a significant increase of tissue throughout the lesion region. The GRP cells survived, differentiated, and formed extensive transplants that were well integrated with host tissue in all animals that received a transplant. In animals that received exogenous cAMP, however, the area occupied by GRP cells at the lesion center and the volume of the GRP transplant throughout the lesion region were reduced. This study suggests that elevation of cAMP is not necessarily beneficial to survival of GRP cells. In conclusion, we showed no beneficial effects on functional outcome measurements of transplanting GRP cells and administering cAMP after SCI. Although elevation of cAMP has been reported to have beneficial effects in neural regeneration and may have beneficial effects on lineage differentiation of GRP cells, from our study it appears that cAMP are not always reduced after SCI and that administering cAMP may reduce survival or proliferation of GRP cells following transplantation into the injured spinal cord. Transplant strategies appear promising; however it seems appropriate to use combinatorial strategies in which cell transplants are combined with other therapies such as treatments that induce differentiation of progenitor cells into populations with desired functional characteristics, treatments that affect regeneration, and/or treatments that alter the spinal cord milieu.

Replication of experimental findings is considered an important step in moving towards finding a cure for SCI. Under NIH guidance, selected promising studies are chosen to be replicated by independent investigators at different institutions. Translation of treatments from bench to clinical situations can take many years, but may be faster when the treatment is already approved for another condition.

REFERENCES

[1.]Olby N. Current concepts in the management of acute spinal cord injury. **J Vet Intern Med** 1999;13(5):399-407. [2.]Ploumis A, Yadlapalli N, Fehlings MG, Kwon BK, Vaccaro AR. A systematic review of the evidence supporting a role for vasopressor support in acute SCI. **Spinal Cord** 2009. [3.]Casha S, Christie S. A systematic review of intensive cardiopulmonary management after spinal cord injury. **J Neurotrauma** 2009. [4.]Nout YS, Mihai G, Tovar CA, Schmalbrock P, Bresnahan JC, Beattie MS. Hypertonic saline attenuates cord swelling and edema in experimental spinal cord injury: a study utilizing magnetic resonance imaging. **Crit Care Med** 2009;37(7):2160-6. [5.]Kwon BK, Tetzlaff W, Grauer JN, Beiner J, Vaccaro AR. Pathophysiology and pharmacologic treatment of acute spinal cord injury. **Spine** J 2004;4(4):451-64.[6.]Baptiste DC, Fehlings MG. Pharmacological approaches to repair the injured spinal cord. **J Neurotrauma** 2006;23(3-4):318-34. [7.]Hall ED, Wolf DL, Braughler JM. Effects of a single large dose of methylprednisolone sodium succinate on experimental postt raumatic spinal cord ischemia. Dose-response and time-action analysis. **J Neurosurg** 1984;61(1):124-30. [8.]Behrmann DL, Bresnahan JC, Beattie MS. Modeling of acute spinal cord injury in the rat: neuroprotection and enhanced recovery with methylprednisolone, U-74006F and YM-14673. **Exp Neurol** 1994;126(1):61-75. [9.]Bracken MB, Shepard MJ, Holford TR, Leo-Summers L, Aldrich EF, Fazl M, Fehlings M, Herr DL, Hitchon PW, Marshall LF and others. Administration of methylprednisolone for 24 or 48 hours or tirilazad mesylate for 48 hours in the treatment of acute spinal cord injury. Results of the Third National Acute Spinal Cord Injury Randomized Controlled Trial. National Acute Spinal Cord Injury Study. **Jama** 1997;277(20):1597-604. [10.]Bracken MB, Shepard MJ, Collins WF, Jr., Holford TR, Baskin DS, Eisenberg HM, Flamm E, Leo-Summers L, Maroon JC, Marshall LF and others. Methylprednisolone or naloxone treatment after acute spinal cord injury: 1-year follow-up data. Results of the second National Acute Spinal Cord Injury Study. **J Neurosurg** 1992;76(1):23-31. [11.]Coleman WP, Benzel D, Cahill DW, Ducker T, Geisler F, Green B, Gropper MR, Goffin J, Madsen PW, 3rd, Maiman DJ and others. A critical appraisal of the reporting of the National Acute Spinal Cord Injury Studies (II and III) of methylprednisolone in acute spinal cord injury. **J Spinal Disord** 2000;13(3):185-99. [12.]Sayer FT, Kronvall E, Nilsson OG. Methylprednisolone treatment in acute spinal cord injury: the myth challenged through a structured analysis of published literature. **Spine J** 2006;6(3):335-43. [13.]Pearse DD, Pereira FC, Marcillo AE, Bates ML, Berrocal YA, Filbin MT, Bunge MB. cAMP and Schwann cells promote axonal growth and functional recovery after spinal cord injury. **Nat Med** 2004;10(6):610-6.

DIGITAL RADIOGRAPHY: READING THE NEW FILMS

Matthew Wright DVM, MS, DACVR
San Diego, CA

INTRODUCTION

Digital radiography is becoming very popular in veterinary medicine. Although not yet ubiquitous in the veterinary practice, internists and other specialists are routinely presented with digital radiographic studies. In many cases, digital radiographic images have a different appearance than traditional film radiography and this difference can lead to errors in interpretation or confusion on the part of the interpreter. In this lecture we will investigate, using case studies, ways to limit the confusion when interpreting digital radiographic studies.

APPROACHING A DIGITAL IMAGE

Interpreting digital radiographic images requires careful attention to detail when viewing studies. A proper monitor configuration, ambient lighting, an understanding of the controls used to manipulate images, and an understanding about the difference between DICOM images and exported JPEG images is essential for success when viewing digital radiographic images. Steps for success in configuring the reading environment include:

- Appropriate monitor selection
- Appropriate viewer selection and understanding of the basic functionality
- Importing outside studies to local PACS
- Avoiding interpreting exported JPEGS if at all possible
- Zooming images to fill the screen and retina.
- Selecting the appropriate ambient lighting

ARTIFACT VS ANATOMY

It is always difficult to determine if opacities present on digital radiographic images represent anatomy or, in many cases, digital artifact. Veterinarians must understand that image processing is a main determinant of image quality and what appears on the digital image may not represent real anatomy. Several case studies will be presented to address common issues regarding image processing that can be confused for anatomy. For example:

- Over processing vs. anatomy
- Understanding image exposure in digital radiography
- Understanding image latitude in digital radiography
- Introduction to common digital imaging artifacts that could be confused for anatomy

UPDATES ON INTRATHORACIC ULTRASOUND

Allison Zwingenberger, DVM, DACVR, DECVDI, MAS
Davis, CA

INTRODUCTION

Non-cardiac thoracic ultrasound has become an important part of clinical imaging practice. Although thoracic ultrasound is not normally the primary method of diagnosing thoracic lesions, it can be crucial in confirming or excluding a suspicious lesion seen on radiographs. In addition, the real-time, high-resolution properties of ultrasound allow for sampling of accessible lesions with fine precision. Lesions of the thoracic wall and axial skeleton, mediastinum, pleural space, diaphragm, and lungs can be visualized with ultrasound. The limiting factor is whether there is an air-filled structure between the point of contact and the lesion of interest. The imaging characteristics of thoracic disease, ultrasound imaging strategies, equipment, settings, and case examples will be discussed.

EQUIPMENT AND PREPARATION

Ultrasound machines are equipped with a variety of transducers for general use. The most applicable shapes to use in thoracic imaging are those with a small footprint. The spaces between ribs limit the imaging windows, and the smaller transducers are more maneuverable with a wider field of view. Lesions requiring higher frequency imaging may be investigated with a linear transducer if the contact point is large enough. Modern transducers with harmonic imaging capabilities reduce the artifact in the near field and allow for detailed examination of the body wall.

Lateral recumbency presents the operator with the largest surface area to work with when imaging cranial to the diaphragm. Animals should have their forelegs drawn well forward when accessing the cranial mediastinum and ribs. Sternal recumbency may be best for cranial lesions or small amounts of effusion, and dorsal recumbency may be used when investigating the diaphragm and pulmonary or vascular lesions that contact the diaphragmatic crura. Focal lesions may be imaged by clipping a limited area, judged by the projected access points as determined by thoracic radiographs. More extensive disease may require additional hair removal.

THORACIC WALL AND AXIAL SKELETON

Lesions involving the thoracic wall, especially the ribs, can be difficult to diagnose definitively on radiographs. The curvature of the thoracic wall reduces the chance that the radiographic beam will highlight an extrapleural sign or a focal rib lesion. Areas of suspicion can be investigated in more detail using thoracic ultrasound. The thoracic wall is made of skin, muscle, fat, and the parietal pleura. This is interspersed with ribs which form smooth, regular palisades with distal acoustic shadowing. The ribs and intercostal spaces should be followed from dorsal to ventral, critically appraising the hyperechoic cortical bone and the regularity of the soft tissues. Efforts should be made to localize the lesion on radiographs to target the area of interest for ultrasound.

Destructive and productive lesions of the ribs cause an irregular appearance to the cortical bone, periosteal reaction, and/or cortical lysis. Masses arising from the ribs often have a soft tissue component that extends into the surrounding tissues and may protrude into the pleural space. Examples of rib lesions include neoplasia, trauma, and fungal osteomyelitis (1-8).

Primary soft tissue tumors may also arise from the muscle and fat of the body wall. Lipomas are common masses that appear to have a marbled, regular, hyperechoic echotexture. They are generally well-circumscribed and have defined margins. Liposarcomas are not distinguishable from lipomas based on imaging characteristics. Fat necrosis may also cause a mass on the body wall. Necrotic fat is more heterogeneous in appearance, and may be surrounded by normal fat. Soft tissue sarcomas tend to be aggressive and large, and may involve skeletal structures. Although an estimate may be made in terms of extent and size, cross-sectional imaging is better suited to pre-surgical planning. More superficial tumors may also be investigated for size and extent of disease. Assessment of accessible local lymph nodes such as the axial and cranial mediastinal lymph nodes should be considered.

MEDIASTINUM

Lesions of the cranial mediastinum may be suspected based on radiographic findings. Some breeds store a large amount of fat in the cranial mediastinum, and ultrasound may be used to confirm or rule out a suspected mass. The normal cranial mediastinum can be imaged from the cranial rib spaces 1-5, or from the thoracic inlet. It may help to position the dog in sternal recumbency to minimize the volume in the peripheral lung lobes. The transducer should be positioned between the ribs above the sternum. Sagittal and transverse views can be used to obtain images. The vessels in the cranial mediastinum should be visible, surrounded by echogenic fat. Normal cranial mediastinal lymph nodes are not visible. The thymus may be seen as a granular structure in the ventral mediastinum in young animals.

Masses in the cranial mediastinum appear as hypoechoic or cystic structures that may displace surrounding structures (1-8). Common causes of cranial mediastinal masses include benign cysts, reactive lymph nodes, neoplastic lymph nodes, ectopic thyroid tissue, thymoma, metastatic thyroid carcinoma, neuroendocrine tumors, and others. A benign lesion may be granuloma, hemorrhage, or abscess secondary to migrating foreign body. Mediastinal masses are often associated with pleural effusion.

Two of the more common cranial mediastinal masses are enlarged lymph nodes and thymoma. Reactive and neoplastic lymph nodes are enlarged and hypoechoic. The number of lymph nodes in the cranial mediastinum varies. Thymomas tend to be lobular masses and may also be cystic with multiple, anechoic, fluid-filled compartments.

The caudal mediastinum is more difficult to evaluate due to the large volume of overlying lung. However, abnormalities involving the caudal esophagus or caudal vena cava may be seen with a transhepatic approach.

LUNG

The normal lung appears as a sharply delineated gas interface with deep reverberation echoes. This smooth surface should slide cranially and caudally with respiration against the stationary parietal pleura. Pneumothorax interrupts the contact of the two pleural surfaces, and the normal respiratory motion will not be seen (9). When alveoli are abnormal, the smooth interface becomes more irregular and ring-down artifacts are produced (10). Consolidation of lung results in a hypoechoic area with a relatively well-defined margin. Geographic or nodular areas may be seen.

Atelectasis results from recumbency or airway obstruction. Atelectic lung should be small in volume, and very regular in shape and echogenicity. In contrast, diseased lung, such as with pneumonia or lung lobe torsion, is enlarged with rounded margins (1-8). Atelectic lung may contain air bronchograms which appear as hyperechoic, branching structures. Lung masses often contain fluid bronchograms. Isolated pockets of air may result from focal necrosis or air trapping in lung lobe torsions (11).

Focal lung lesions can be imaged if the surrounding lung is atalectic, if they contact the visceral pleura, or if they are surrounded by pleural effusion. Masses a few centimeters from the pleural surface can be brought into contact by placing the dog in lateral recumbency with the affected side down, and imaging after turning them over, or from the ventral aspect if it is accessible. The transhepatic approach can be used for peripheral lesions in the caudal lung lobes.

PLEURAL SPACE

The pleural space is normally filled with a scant amount of fluid that is not ultrasonographically visible. Pleural effusions cause separation of the parietal and visceral pleura with anechoic or echogenic fluid. Transudates and chyle appear anechoic, and blood often appears echogenic. Any chronic effusion can result in fibrin strands that appear as thin, hyperechoic linear and globular structures. Fluid often collects around an area of localized pulmonary disease such as a lung lobe torsion. Effusions also increase the volume of thorax visualized with ultrasound, including the cardiac margins, the lung lobes, and the mediastinum. Thoracic ultrasound should be performed prior to thoracocentesis if the animal is stable.

Abdominal organs can intrude into the pleural space secondary to trauma. Diaphragmatic rupture can be diagnosed with ultrasound by lack of visualization of the hyperechoic, curved echo of the diaphragm, or by identification of abdominal organs cranial to the diaphragmatic border. Care must be taken not to mistake mirror image artifact for diaphragmatic rupture, as it can cause the appearance of abdominal organs in the thorax (12).

REFERENCES

[1.] Mattoon JS, et. al. **Small animal diagnostic ultrasound**. 2nd edition.2002;325. [2.] Moon MM. **Vet Clin Small Anim** 2009;39:733 [3.] Hecht S. **Atlas of small animal ultrasonography**. 2008;119. [4.] Reichle JK, et. al. **Vet Radiol Ultrasound** 2000;41:154. [5.] Stowater JL, et. al. **J Am Vet Med Assoc** 1989;195:514. [6.] Saunders HM, et. al.. Textbook of respiratory diseases in dogs and cats. 2004;72. [7.] Konde LJ, et. al. **Vet Radiol** 1991;32:178. [8.] Zekas LJ, et. al. **Vet Radiol Ultrasound** 2002;43:413. [9.] Lisciandro GR, et. al. **J Vet Emerg Crit Care** 2008;18:258. [10.] Louvet A, et. al. **Vet Radiol Ultrasound** 2008;49:374. [11.] D'Anjou MA, et. al. **Vet Radiol Ultrasound** 2005;46:478. [12.] Spattini G, et. al. **Vet Radiol Ultrasound** 2003;44:226

HOW TO GET THE LESION: US GUIDED THORACIC BIOPSY

Rachel E. Pollard, DVM, PhD, DACVR
Davis, CA

INTRODUCTION

Ultrasound-guided techniques facilitate aspiration of loculated fluid and biopsy of lesions of the lung, mediastinum, thoracic wall and pleural space provided that there is adequate contact between the lesion and the chest wall or pleural fluid is present[1,2]. This technique is especially useful when the lesion is close to vital structures. Recent thoracic radiographs should be compared with the ultrasound findings before interventional procedures, and spectral or color Doppler techniques are used to exclude masses of vascular origin and to avoid normal vascular structures. The size, type, and location of the lesion ultimately determine whether ultrasound guided biopsy is possible.

EQUIPMENT

Curvilinear or linear-array transducers can be used for ultrasound-guided biopsies. The near field is poorly seen with a sector scanner limiting its' usefulness to guide needle placement. However, when using a curvilinear or linear transducer, one must orient the transducer parallel to the ribs so as to prevent shadowing artifact that blocks visualization of the lesion.

In regards to needle selection, the first decision that must be made is whether to obtain a needle aspirate or a core biopsy. Aspiration procedures are intended to obtain material for culture or biochemical or cytologic analysis. A tissue-core biopsy is intended primarily to obtain material for histologic analysis, although samples can also be cultured, specially stained, or otherwise analyzed. In the thorax, aspirations are generally preferred with small solid masses (<2.5 cm diameter), pulmonary lesions, and lesions with high vascularity. Biopsy of larger masses (>2.5 cm) has been successfully done with spring-loaded biopsy devices although the risk of complications is greater than with aspiration. [3] Should a pulmonary mass be sampled, aspiration or core biopsy specimens should be taken from the non-aerated portions of the mass, avoiding blood vessels and the periphery of the lesion. It is important that needle travel not extend outside the mass during the biopsy. The decision to perform an aspiration or tissue-core biopsy and the type of needle selected are determined by the lesion to be sampled, the suspected abnormality, and the risk to the patient. Specific considerations based on lesion type are outlined in Table 1. At our institution, an aspirate is typically performed first regardless of lesion type and location while a biopsy is only performed if the aspirate is not diagnostic. It must be recognized that needle aspirate samples are smaller than those acquired from tissue-core biopsy and may not ultimately agree with the final histologic diagnosis. Aspiration of masses for cytologic analysis can usually be done with a 21- to 27-gauge, 1.0-1.5 inch injection needle attached to a 3 ml syringe. Tissue-core biopsies require 16- to 18-gauge needles for an adequate sample to be obtained. Spring-loaded biopsy devices provide far better samples than hand-deployed needles.

Table 1. Needle selection and additional considerations are outlined for different types of thoracic lesions.

Lesion Type	Needle selection	Additional considerations
Pleural fluid	21-25 gauge butterfly catheter	May require multiple punctures if fluid is pocketed into separate areas.
Pulmonary lesion	21-27 gauge needle on a 3 ml syringe	Aim for the center of pulmonary masses and a solid portion of consolidated lung to avoid pneumothorax.
Mediastinal mass	21-27 gauge needle on a 3 ml syringe; 16-18 gauge biopsy needle	Use color Doppler to identify and avoid the main vessels of the mediastinum.
Thoracic wall/rib mass	21-27 gauge needle on a 3 ml syringe	Look for a break in the cortex when sampling rib lesions. Sedation is necessary as puncture of the periosteum is painful.

PREPARATION OF THE PATIENT

The patient should be evaluated for hemostatic disorders before tissue-core biopsies when there is a significant risk of bleeding or when highly vascular lesions are being sampled. This assessment is made from the history, physical examination, laboratory work, and other diagnostic procedures. Screening tests such as platelet count, prothrombin time, partial thromboplastin time, and bleeding time may be indicated. A known coagulopathy should be corrected before the procedure. Plasma or whole blood transfusions may be helpful to allow a biopsy in certain cases. Routine needle aspirates usually require no special tests unless there is a high risk of bleeding.

The skin over the biopsy site should be clipped and prepared aseptically. In some cases, a needle aspirate can be performed with no chemical restraint. However, given the complexity of the thoracic anatomy, sedation is recommended in an anxious or heavily panting animal. Heavy sedation or general anesthesia is indicated for core biopsy of any thoracic lesion due to the risk of patient movement and the level of discomfort associated with the biopsy procedure.

If a thoracic lesion is difficult to see with ultrasound due to aerated lung in the near field, it is worthwhile to place the animal in lateral recumbency with the side of the lesion closest to the table. If allowed to recline for several minutes, the dependent lung will often become atelectatic, which may provide a window to the lesion. The animal is then quickly rolled into sternal or the opposite lateral recumbency so as to perform the biopsy from the side or non-dependant surface.

GENERAL TECHNIQUES

The shortest distance and safest path between the skin surface and the lesion or region to be sampled should be chosen. A separate needle should be used for each location if multiple biopsy samples are taken. Freehand puncture is accomplished by holding the transducer in one hand and inserting the biopsy needle with the other. [5] This technique is the most versatile and can be used for either aspiration or tissue-core biopsy acquisition. Needle visibility can be a significant issue when trying to assure accurate placement, particularly when sampling focal lesions. The needle is continuously visualized if it is kept within the plane of the beam. A larger bore needle is easier to visualize and can be used if there is no increase in complication risks. Better visibility may also be obtained by ensuring needle placement within the focal zone of the transducer. Wiggling the needle, moving the stylet in and out,[4] and rocking the transducer to better position the needle within the beam plane are additional techniques commonly used to assist with needle visualization (Figure 1). Practice is required to maintain the entire needle's length within the beam so that the tip can be accurately localized. A biopsy phantom made out of an agar-filled plastic intravenous fluid bag or barium enema bag may be useful to practice hand-eye coordination for freehand biopsy techniques. Olives or other suitable items for biopsy can be suspended within the agar phantom. The operator can practice keeping the needle within the plane of the beam and inserting needles at various angles with the freehand technique.

Aspiration of moderate to large amounts of fluid can be done with indirect or freehand guidance. A three-way valve and extension tubing can be used to remove large amounts of fluid. Over-the-needle plastic catheters or butterfly catheters can also be inserted to help decrease the risk of injury during drainage.

Solid lesions are aspirated by guiding the needle to the lesion under direct visualization. The needle tip is moved up and down in 1- to 2-cm excursions while maintaining several milliliters of negative suction. The needle can also be rotated during this process to help increase cytologic yield. [6] Suction should be released before removal of the needle to avoid sampling outside the lesion and to retain the contents primarily within the needle. The aspiration of blood should be avoided because it will dilute the sample. Immediate cytologic examination after each aspiration is preferred.

Tissue-core biopsies are performed by advancing the needle to the desired depth, while taking into account the additional distance that the needle might travel when deployed. This distance should always be known in advance so we recommend firing spring-loaded needles at least once prior to introducing it through the animals skin.

Figure 1. A needle has been introduced into the animal and the ultrasonographer is attempting to identify the location of the needle tip so as to direct it in to the lesion. A. The ultrasound beam is not aligned with either the needle or the lesion. The needle will appear as a hyperechoic 'dot' in the image. B. The beam is swept toward the lesion in an attempt to align with the needle and identify its' location. C. The lesion has been identified and is within the ultrasound beam but the needle still appears as a 'dot' on the image indicating that the needle tip is not likely to be in the lesion. D. The ultrasound beam is aligned with the needle and the needle appears as a hyperechoic linear structure. However, it is clear that the needle tip does not reside within the lesion. E. The needle has been backed out and redirected so that it is aligned with the ultrasound beam and will ultimately reach the lesion. (Image provided with permission by Mary Beth Whitcomb, DVM).

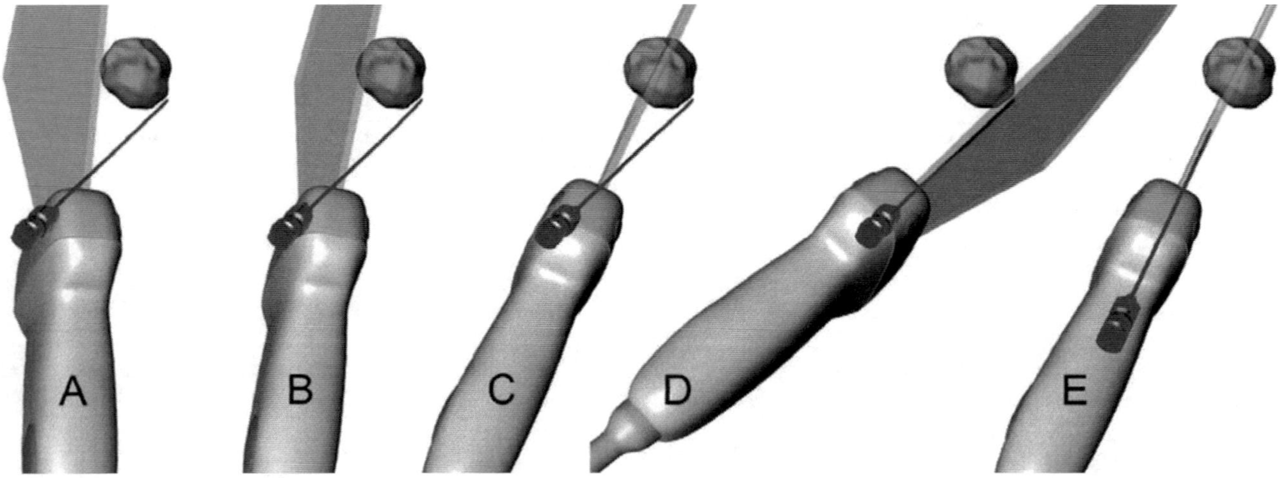

POTENTIAL COMPLICATIONS

The potential complications of ultrasound-guided biopsy vary with the experience of the operator, type of biopsy, size of the needle, and nature of the lesion. Complications are much less common than with blind techniques. Aspiration procedures should initially be performed with 21- to 27-gauge needles if there is a high potential for bleeding. This can be followed by a larger bore needle or tissue-core biopsy if required. Hemorrhage is usually minor and self-limited if there are no coagulation abnormalities and a small-gauge needle is used. If it is available, color Doppler imaging to help avoid vascular structures before biopsy will reduce complications. Pneumothorax is an additional potential complication when sampling thoracic lesions. At our institution, this is a rare occurrence as needle aspirates are more commonly performed than core biopsies. When pneumothorax occurs, it is typically minor and self-limiting. Seeding of tumor along the needle track has been infrequently reported in veterinary patients and appears to be associated with pulmonary adenocarcinoma in the cat. [7]

REFERENCES

[1] Stowater JL, Lamb CR. Ultrasonography of noncardiac thoracic diseases in small animals. **J Am Vet Med Assoc** 1989;195:514-520. [2] Wood EF, O'Brien RT, Young KM. Ultrasound-guided fine-needle aspiration of focal parenchymal lesions of the lung in dogs and cats. **J Vet Intern Med** 1998;12:338-342. [3] Parker SH, Hopper KD, Yakes WF, et al. Image-directed percutaneous biopsies with a biopsy gun. **Radiology** 1989;171:663-669. [4] Bisceglia M, Matalon TA, Silver B. The pump maneuver: an atraumatic adjunct to enhance US needle tip localization. **Radiology** 1990;176:867-868. [5] Papageorges M GP, Sande RD, Barbee DD. Ultrasound guided fine-needle aspiration: An inexpensive modification of the technique. **Vet Radiol** 1988;29:269-271. [6] Ferrucci JT, Jr., Wittenberg J, Mueller PR, et al. Diagnosis of abdominal malignancy by radiologic fine-needle aspiration biopsy. **AJR Am J Roentgenol** 1980;134:323-330. [7] Vignoli M, Rossi F, Chierici C, et al. Needle tract implantation after fine needle aspiration biopsy (FNAB) of transitional cell carcinoma of the urinary bladder and adenocarcinoma of the lung. **Schweiz Arch Tierheilkd** 2007;149:314-318.

CYTOLOGY OF LUNG & THORAX FINE NEEDLE ASPIRATIONS

Anne Barger, DVM, MS, Diplomate ACVP
Urbana, IL

Fine needle aspirations of the lung and thoracic masses can be challenging however with ultrasound and CT guidance, cytology of these tissues is becoming more common. Aspiration of solid masses and pleural fluid have a higher diagnostic yield than aspiration of diffuse lung disease; however, aspiration of diffuse or nodular lung disease can be beneficial in the diagnosis of *Blastomyces dermatitidis* and metastatic neoplasia.

Knowledge of the cytology of the healthy lung and thorax are important to appropriately interpret masses and diseased tissue. The pleural cavity is lined by mesothelium, which exfoliate readily when the lung is aspirated. These cells occur as uniform sheets of cells with distinct cell borders and uniform amounts of basophilic cytoplasm and round nuclei. These cells can be mistaken for sheets of epithelial tissue which may be misleading. The main stem bronchi are cartilaginous tubes lined by ciliated columnar epithelial cells. Smaller airways such as bronchioles are lined by simple columnar to cuboidal epithelial cells. Smooth muscle cells surround bronchi and bronchioles. The bronchioles end in the alveoli. These structures are composed of type I and type II pneumocytes, endothelial cells, alveolar macrophages and fibroblasts[1]. Type I pneumocytes are flattened squamous epithelial cells, and type II pneumocytes are cuboidal epithelial cells. Fine needle aspiration of a normal lung will be of low cellularity because alveoli will be the most likely structures aspirated, and those are primarily filled with air. However, low numbers of vacuolated macrophages, squamous and cuboidal epithelial cells and rare small spindle cells may also be identified.

Fine needle aspiration of lung can be performed when a mass, nodules or diffuse disease are present. Lung masses most commonly are tumors, either primary or metastatic. Primary lung tumors are predominantly carcinomas[2,3]. Bronchogenic adenocarcinomas, pulmonary carcinoma, adenosquamous carcinomas and squamous cell carcinomas have all been reported in the lung of dogs and cats. Less common primary tumors include chondrosarcoma, osteosarcoma, fibrosarcoma and histiocytic sarcoma[4,5]. Cytology of primary lung tumors and metastatic lung tumors will appear very similar and will be described together here. Carcinomas in the lung consist of clusters of neoplastic epithelial cells. These cells often show marked criteria of malignancy including anisocytosis, anisokaryosis, and prominent nucleoli. Depending on the tumor type, marked variation in the nuclear to cytoplasmic ratio may be seen, as well as signet ring shaped cells in adenocarcinomas. Pulmonary carcinomas exfoliate readily, and the samples are highly cellular which, is an unusual finding in healthy lung tissue. It is common to identify secondary inflammation associated with larger tumors, and it is important to determine that the level of atypia identified is enough to confirm the epithelial cells are neoplastic rather than dysplastic secondary to inflammation. Ultrasound or CT guidance is recommended to avoid excessively inflammatory or cavitated areas within the mass. Cavitation often occurs secondary to inflammation or necrosis. Cystic areas which may appear cavitated are seen commonly secretory tumors such as adenocarcinomas. The cytology from these areas will not be diagnostic for the neoplastic process.

Additional stains and techniques can be performed to further characterize the neoplastic cells. If an osteosarcoma is suspected, either primary or metastatic, alkaline phosphatase staining can be used with a sensitivity of 100% and specificity of 86%[6]. Immunocytochemical staining for transcription factor-1 can be used to differentiate primary lung carcinomas from metastatic tumors[7]. Other immunocytochemical stains such as vimentin, cytokeratin, CD18, CD3 and CD79, can be used to identify more anaplastic neoplasms.

Inflammatory disease within the lung may occur as a mass or diffuse disease. If the patient has diffuse disease, ultrasound guidance may not be as beneficial. Lung abscesses will consist of suppurative inflammation with degenerate neutrophils and intracellular bacteria will often be identified. Fungal pneumonia, such as *Blastomyces dermatitidis* or *Histoplasma capsulatum* is more likely to present as a pyogranulomatous inflammatory response. Fungal yeast organisms should be readily identified. *Blastomyces dermatitidis* organisms are generally observed extracellularly and are 8-14 microns with a double contoured wall, basophilic core and broad-based budding. *Histoplasma capsulatum* organisms are much smaller, 1-3 microns and are usually identified within the cytoplasm of macrophages. These organisms have a thin, nonstaining capsul with an eosinophilic, crescent shaped center. Eosinophilic inflammation can be associated with parasitic infestation or pulmonary eosinophilic infiltrate. Granulomatous inflammation is less common but can occur with atypical bacterial infections such as mycobacterium or primary hyphal forming fungus such as pythiosis or aspergillosis.

Mediastinal masses in dogs and cats include thymoma, thymic lymphoma, mediastinal lymphoma and cranial mediastinal carcinoma. Thymomas are more common in dogs than cats. The tumors are primarily found in the anterior mediastinum however they can extend from the neck to the posterior mediastinum. The neoplastic cell in this tumor is epithelial however, the epithelial cell is often difficult to find on cytologic specimens. Cytologically these tumors consist of a mixed lymphoid population (T cells) predominated by small lymphocytes with lesser numbers of large lymphocytes and varying numbers of mast cells. The mast cells are often overlooked on histopathology because the granules do not stain with H&E. Histopathologically, thymomas are classified as lymphocyte predominant, epithelial predominant or mixed. The epithelial cells are large, round cells with abundant clear cytoplasm. The majority of thymomas are benign; however up to one-third are reported to be locally invasive or have metastasized to draining lymph nodes. These are considered carcinomas.

The neoplastic cell in thymic lymphoma is the lymphocyte. Cytologically, thymic lymphoma appears similar to lymphoma in other locations. A uniform population of lymphocytes, large, intermediate or small is observed. The majority of lymphomas are large cell. These cells are larger than neutrophils, have a scant rim of basophilic cytoplasm and a large round nucleus with prominent nucleoli. Aspirates of thymic lymphoma are highly cellular and often of diagnostic quality unless the patient has been treated with glucocorticoids prior to sampling.

Mediastinal carcinomas have also been reported in dogs and cats. In cats, it has been a progression of thymoma to squamous cell carcinoma. In dogs, mediastinal carcinomas should include ectopic thyroid carcinomas[8]. Cytologically, these tumors will consist of clusters of neoplastic epithelial cells. These cells have a moderate amount of basophilic cytoplasm and round nuclei with stippled chromatin pattern and small but prominent nucleoli. The cells are fragile and often bare nuclei from ruptured cells are scattered throughout the sample. The majority of thyroid carcinomas in dogs are follicular; however, neuroendocrine medullary carcinomas have been reported. These tumors are very fragile and may consist of no intact cells but many bare nuclei loosely arranged on a sea of cytoplasm. Other neuroendocrine tumors that can occur in the thorax include heart based tumors such as chemodectoma and aortic and carotid body tumors. Occasionally these can be aspirated with ultrasound guidance, if they are large enough.

Pleural fluid is commonly aspirated and evaluated cytologically. Cell counts and protein are used to classify the effusions as transudate, modified transudate or exudate. Further classification is performed by evaluating the cytology or biochemistry of the effusion. Diagnosis of chylothorax is suspected with the gross evaluation of the fluid, further supported by increased small lymphocytes in the effusion and then confirmed by measuring triglyceride levels in the effusion and comparing it to serum levels. Cytologically chylothorax is predominated by small lymphocytes with lesser numbers of vacuolated macrophages, containing small discrete lipid vacuoles. The number of neutrophils will increase the longer the patient has the effusion or the more times the effusion is drained and reoccurs, so the cytology may be deceiving; therefore the triglyceride level is the best diagnostic test. Transudates generally are of low cellularity and cytologically varying numbers of vacuolated macrophages, neutrophils and small lymphocytes may be identified. Additionally, it is common to identify clusters of reactive mesothelial cells. These cells have abundant amounts of basophilic cytoplasm with round nuclei, binucleation or multinucleation are commonly observed and the cells have prominent nucleoli. Reactive mesothelial cells generally have an eosinophilic fringe at the periphery of the cytoplasm. It is important to differentiate these cells from neoplastic cells because reactive mesothelial cells can exhibit considerable criteria of malignancy. Modified transudates are not very specific and causes include cardiac disease, nonexfoliative neoplasia, lung lobe torsion and tissue inflammation. Generally these effusions have an elevated protein and cytologically are very similar to transudates. Exudates may be inflammatory or neoplastic in origin. Pyothorax is the most common inflammatory exudate and is predominated by degenerate neutrophils which often contain phagocytized bacteria. Extracellular bacteria may also be identified. Neoplastic effusions include metastatic carcinoma, lymphoma and mesothelioma. Lymphoma is often highly cellular and can be associated with chylothorax. The cells are large and round with prominent nucleoli and appear similar to lymphoma from solid tissues. In cats, it is common to observe small discrete vacuoles within the cytoplasm. Pyknotic and karyorrhectic cells are easily identified in these effusions. Carcinoma in effusions consists of large, irregularly shaped clusters with marked criteria of malignancy including nuclear molding, marked anisocytosis and anisokaryosis and very prominent, often very large nucleoli. Mammary carcinoma, adenocarcinomas of the lung and pancreatic adenocarcinoma are commonly effusive and the neoplastic cells exfoliate into the effusion. Metastatic carcinomas commonly contain a single large cytoplasmic vacuole which can displace and compress the nucleus giving the cells a signet ring appearance. Mesothelioma is also effusive and neoplastic cells fill the effusion. Cytologically these cells are very difficult to differentiate from carcinoma and immunocytochemical staining is necessary to differentiate these two tumors. Mesotheliomas are positive for cytokeratin and vimentin, which differentiates them from typical epithelial cells. Sarcomas do not commonly exfoliate into effusions. Tumors such as hemangiosarcoma are commonly associated with hemorrhagic effusions. Hemorrhagic effusions contain many red blood cells and macrophages containing hemosiderin pigment and/or phagocitized erythrocytes. Hematoidin crystals can also be observed. Samples with peripheral blood contamination rather than true hemorrhage often contain individual platelets or platelet clumps.

In conclusion, many disease processes can be definitively diagnosed with cytology and aspiration of difficult areas, such as lung and mediastinum, or aspiration of fluid from patients with minimal amounts of fluid can be greatly enhanced by imaging such as ultrasound or CT.

REFERENCES

[1]Maxie MG. Pathology of domestic animals 5th ed, 2007. **Elsevier Saunders, Edinburgh**. [2]Gordon IO, et al. **Arch Pathol Lab Med** 2009;133:1106-1115. [3]Wood EF, et al. **J Vet Int Med** 1998;12:338-342. [4]Dhumeaux MP, et al. **Can Vet J** 2009;50(7):755-578. [5]Moore PF, et al. **Vet Pathol** 2006;43(5):632-645. [6]Barger et al. **Vet Pathol** 2005;42:161-165.[7] Bettini G, et al. **Vet Comp Onc** 2009;7:28-37. [8]Liptak et al. **Vet Comp Onc** 2008;6:19-30.

CARDIAC EMBRYOGENESIS: NORMAL DEVELOPMENT

Harris Edward Durham Jr, CVT LATG, VTS (Cardiology)
Columbia, MO

INTRODUCTION

Any study of the heart needs to begin with an understanding of how the heart develops in the embryo. This knowledge is extremely valuable, especially when trying to evaluate a patient with congenital heart defects. If the heart does not develop properly or at least close-properly, the fetus will not survive to birth. This presentation will present an overview of a complex series of events that represents a field of study unto itself. During the formation of the heart *in-utero*, the orientation in the embryo changes several times which can add to the challenge of understanding the development. The terms cranial and caudal are used to indicate the orientation of the heart related to the embryo forebrain.

EARLY DEVELOPMENT

The development of the heart begins with the formation of the *cardiac tube.* The cardiac tube is situated in the pericardial cavity below the head of the embryo and begins as two tubes that fuse together creating a single lumen. It is anchored at the cranial end by the *ventricular aorta* and the caudal end by the *omphalomesenteric veins*. It is about this time that the heart begins to beat moving blood from the caudal to its cranial aspect. Since the cardiac tube is fixed at both ends, as the heart grows, it convolutes, folding over itself to form the convoluted *cardiac loop* at about 23 days. This rudimentary structure contains the *cardiac jelly.* The cardiac jelly is an acellular mass of glycoproteins and glyosaminoglycans that function as a basic valve in this early heart and will go on to become important in the development of the cardiac valves. Blood is circulated from the venous side at the caudal pole through the arterial pole at the cranial pole. During this stage of development the cardiac tube widens at the caudal aspect as the cranial begins to move caudal pushing the bottom to the side convoluting the structure into what eventually becomes the ventricles and begins the development of the aortic arches. As the cardiac loop grows, the venous portion gets pushed toward the cranial portion of the embryonic heart as the ventricular portion drapes down. The *sinus venosus* gets pushed cranial and has an atrial like structure in the cranial dorsal portion that collects blood returning to the heart, and a ventricular like portion at the caudoventral region with the bulbus cordis extending up to outflow tract which eventually becomes the *truncus arteriosus*. (Fig 1) From the ventral view, the two primitive atria can be seen projecting on either side of the truncus. The *atrioventricular sulcus* develops along the sagittal plane and the familiar structure of the heart begins appear, with multiple veins entering posteriorly allowing blood to drop caudally into the primitive ventricles to be ejected from the aortic arch at the cranial anterior. The narrowing between the primitive ventricle and atria created by bulboventriculer sulcus creates the *atrioventricular canal* (AV canal) with the cardiac jelly acting as a unidirectional valve.

SEPTATION

At about 27 days the heart begins to septate. There are three septa that develop simultaneously; the ventricular, the atrial and spiral septation of the truncus arteriosus into the aorta and the main pulmonary artery. At the same time the endocardial cushions develop from each side of the AV canal and in the center from the cardiac jelly. Ventricular septation begins at the apex with the muscular septum growing toward the atrioventricular junction. It will finally fuse with the central endocardial cushion (Fig 2) and leave on opening from the ventricles to the truncus arteriosus. An interventricular foramen persists until the complete septation of the truncus arteriosus. The beginning of the atrial septum also grows from the outer wall of the common septum toward the AV canal. This *septum primum,* nearly completely separates the two atria. The *ostium primum* is an opening at the distal end of this developing septum which allows blood to shunt from the right to left atrial sides. As it closes, a perforation (*ostium secundum)* develops higher up in the septum primum, creating the *fossa ovalis* allowing for the continued shunting of blood in the fetal heart. A second atrial septum then develops, the *septum secundum* overlapping the *foramen ovale* creating a functional valve over the foramen ovale allowing blood to only pass from the right atrium to the left in the fetal heart. (Fig 2) A spiraling septum develops in the truncus arteriosus that separates into the two great vessels exiting the heart, the pulmonary artery and the aorta. The spiral septum starts as two ridges of tissue that spiral down the long axis of the truncus arteriosus toward the ventricles. As the spiral septum grows, the truncus twists and divides into the two vessels, accounting for the crossing of pulmonary artery over the aorta in the adult heart. The interventricular foramen is ultimately obliterated by masses of tissue from the spiral septum, endocardial cushion tissue, and the muscular ventricular septum. The last phase of development is the formation of the cardiac valves. The AV valves are derived from folding of tissue in the ventricular walls and the endocardial cushions. The mesenchymal cells of the endocardial cushions transform into endocardial tissue forming the AV valves, which perforate along their cusps, then coalesce and form the chordae tendineae and papillary muscles. The semilunar valves are formed from masses of cardiac jelly in the aortic and pulmonary artery lumens seeded with mesenchymal cells covered by endocardial tissue.

FORMATION OF MAJOR VESSELS

Both the pulmonary artery and the aorta derive from the differentiating aortas of the cardiac loop in early cardiac development, and exit the cardiac loop from a single orifice which then divides to form a "V". This "V" then develops six pairs of aortic arches supplying different portions of the embryo. (Fig 3)) As the circulation system matures, the first, second and fifth pairs regress completely. The left fourth aortic arch forms the aorta and the sixth forms the pulmonary arteries. The right fourth aortic arch forms the base of the subclavian artery. The third pair develops into the internal carotid arteries. The distal portion of the left sixth aortic arch remains connected to the aorta to create the *ductus arteriosus.* Pulmonary veins grow from the left atrium to connect with the pulmonary venous drainage as the lungs are formed from the primitive foregut. The development of the lungs is completely independent of the heart.

FETAL CIRCULATION

The general arrangement of the adult circulatory system occurs as two circuits in series, that is two separate circulatory paths in which the end of one feeds into the beginning of the other. In the fetus, circulation occurs as two circuits in parallel, or two circulatory paths that cross over each other at strategic areas to incorporate placental blood flow and bypass the unused lungs. (Fig 4) As highly oxygenated blood enters the fetus from the placental artery, it passes to the liver where it mixes with the deoxygenated blood of the lower body. From there it travels to the right atrium and is shunted almost directly into the left atrium. There is very little mixing at this point with deoxygenated blood from the head and forelimbs due to the anatomic proximity of the posterior vena cava to the fossa ovalis which directs the blood almost directly into the left atrium. This blood is still highly oxygenated and goes to the left ventricle and out the aorta. At this point another mixing occurs with some highly oxygenated blood going to the head and some mixing with deoxygenated blood from the right ventricle via the pulmonary artery and ductus arteriosus. This mixed blood then goes to the lower body and to the placenta to be remixed into the maternal circulation. This mixed blood still carries enough oxygen to supply the lower extremities with all the nutrition they need. On the other end, the highly oxygenated blood from the left ventricle goes directly to the head and forelimbs to ensure a highly oxygenated blood supply to the developing brain. Deoxygenated blood from the head and forelimbs then return to the right atrium of the fetal heart out the pulmonary artery to mix as described earlier. The circulation remains like this until birth.

As mentioned earlier, during the last phases of the cardiac development, as the atrial septum develops, the fossa ovalis transforms into the foramen ovale. The foramen ovale develops as a one way valve through the intra-arterial septum that allows oxygenated blood to pass from the right atrium to the left, similar to the fossa ovalis, yet will close later to keep blood from shunting from the left atrium to the right in the adult.

It is at parturition and shortly thereafter that two systems in series finally develop. This process starts with the first breath, which drops pulmonary vascular resistance dramatically. Simultaneously the placental circulation is removed, which increases systemic vascular resistance and reverses the blood flow through the ductus arteriosus to move blood from the aorta to the pulmonary artery. The tissue of the ductus arteriosus is highly sensitive to oxygen. When exposed to the increased oxygen content of the arterial blood in the aorta, the musculature of the ductus arteriosus constricts and closes the communication. The net effect of these changes is to decrease the volume of blood flowing through the right side of the heart, and increase it in the left. This changes causes the pressure in the left atrium to increase which functionally closes the foramen ovale and finally separates the two circulations.

In the fetus the size of the individual ventricles is equal and their relationship stay the same for about ten days. Over the first few weeks of life, the systemic blood pressure raises and pulmonary pressures remain static, and this increased workload changes the size of the left ventricle. By about two weeks of age, the proportions of the two ventricles are that of the adult with the right ventricle being about one third as thick as the right. During the next four weeks cardiac growth converts from *hyperplasia* to *hypertrophy,* growth in cardiac mass by myocyte enlargement rather than division. From this stage on, essentially all increases in cardiac mass are related to hypertrophy.

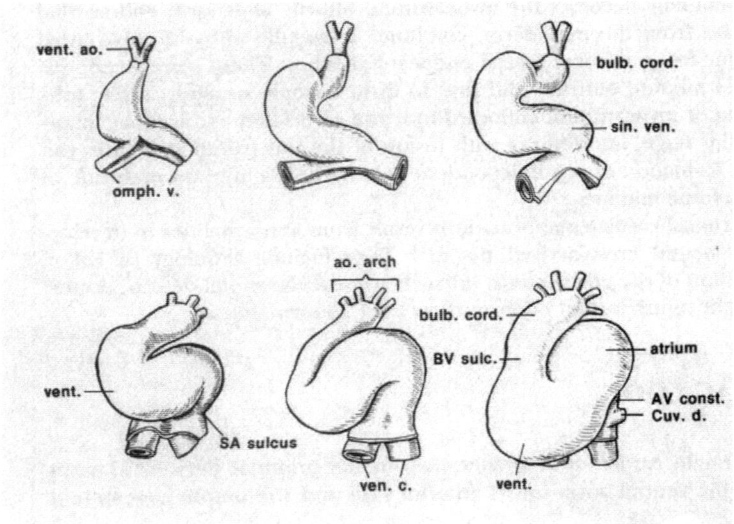

Figure 1

Ventral view of the cardiac tube developing into the convoluted cardiac loop.

ao.arch-aortic arches; AV const-atrioventricular constriction; bulb. cord.-bulbus cordis; BV sulcus-bulboventricular sulcus; Cuv. D-Cuvier's duct; omph. v.-omphalomesenteric veins; SA sulcus-sinoatrial sulcus; sin. ven.-sinus venosus; ven. c.-venae cavae;vent-ventricle; vent. ao.-ventricular aortic roots.

From *Textbook of Canine and Feline Cardiology*; Fox, Sisson, Moise 1999 Saunders

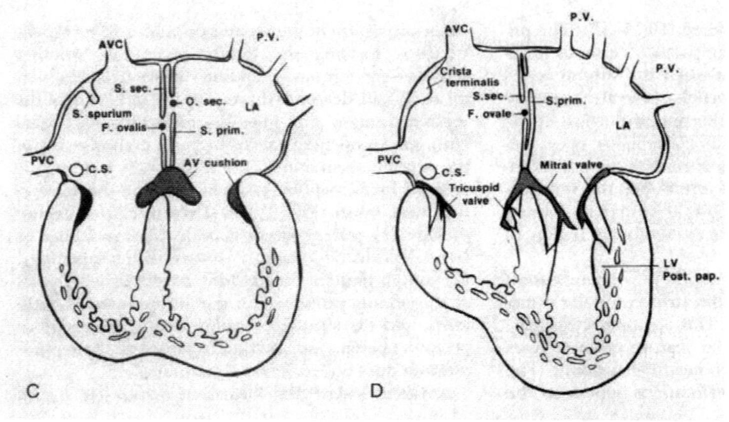

Figure 2

Development of the atrial septum showing fossa ovalis (diagram C) and foramen ovale (diagram D).
AVC- anterior vena cava; C.S.- coronary sinus; F. ovalis-fossa ovalis; F. ovale-foramen ovale; LA-left atrium; LV Post Pap-left ventricular posterior papillary muscle; O. sec.-ostium secundum; O. prim.-ostium primum; PV-pulmonary veins; PVC-posterior vena cava; RA-right atrium; S. sec.-septum secundum; S. prim.-septum primum; S. spurium-septum spurium
From *Textbook of Canine and Feline Cardiology*; Fox, Sisson, Moise 1999 Saunders

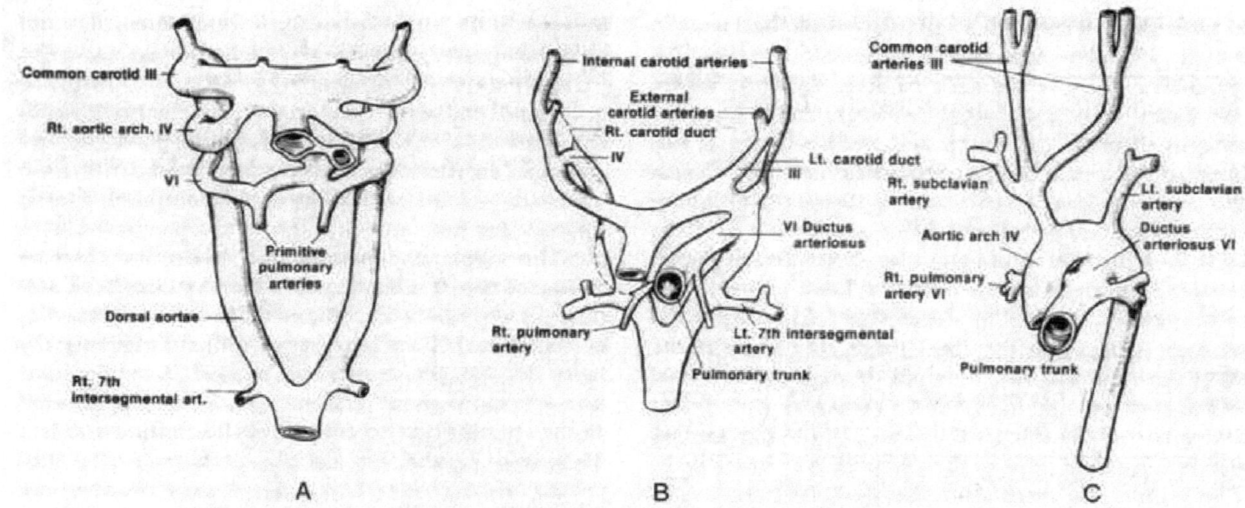

Figure 3

Development of the aortic arches in mammals
From *Textbook of Canine and Feline Cardiology*; Fox, Sisson, Moise 1999 Saunders

HEART FAILURE: STATE OF THE SCIENCE ON BEST TREATMENT PRACTICES

Romain Pariaut, DVM, DACVIM (Cardiology), DECVIM-CA (Cardiology)
Baton Rouge, LA

INTRODUCTION

Heart failure is a common complication of cardiovascular diseases. In dogs, it usually results from the progression of degenerative atrioventricular valve disease, dilated cardiomyopathy, and some congenital diseases such as patent ductus arteriosus, subaortic stenosis and pulmonic stenosis. In cats, heart failure is often the final pathway for hypertrophic and restrictive cardiomyopathy. Heart failure also affects nearly 5 million Americans. Accordingly, there is a large effort to identify new treatments for heart failure and define best treatment practices in human and veterinary cardiology. Those guidelines are now instantly available to medical professionals and the general public alike through the development of the Internet. However, it becomes increasingly difficult to assess the validity of such information; false or biased facts on treatments are commonly found amid data that was obtained from rigorous scientific methods and was submitted to peer-review prior to publication. Therefore, it becomes critical to evaluate the scientific evidence that supports our current recommendations for the treatment of heart failure.

PATHOPHYSIOLOGY

Heart failure is a progressive disorder characterized by a decline in the pumping capacity of the heart. In response, the body activates various compensatory mechanisms, which allows the animal to be free of clinical signs for an extensive period of time. However, the progression of the cardiac disease, chronic neurohormonal activation and fluid retention, may lead to a stage of decompensated heart failure with signs of pulmonary edema, ascites and pleural effusion.

Compensated heart failure

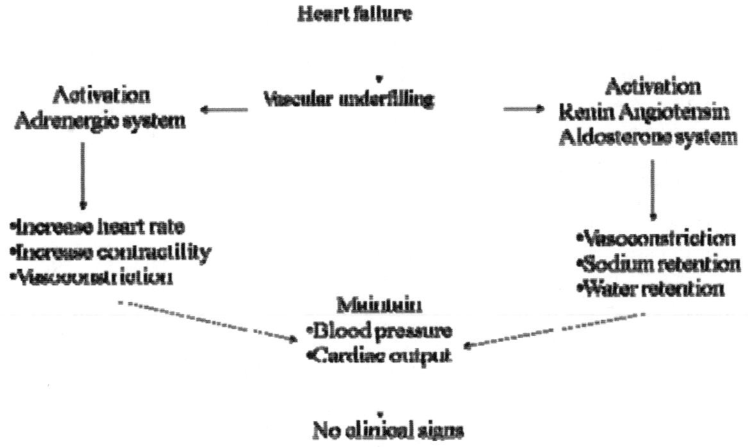

Figure 1. Pathophysiology of compensated heart failure

Vascular underfilling, which usually results from decreased cardiac output, is the main trigger for compensatory mechanisms through the activation of baro- and mechanoreceptors in the heart, the vasculature and the kidney. The initial compensatory mechanism is an activation of the adrenergic (also called sympathetic) system. This division of the autonomic nervous system is stimulated in response to a stress or a threat, the classic "Fight-or-flight-or-freeze response". It is responsible for systemic vasoconstriction, increase in heart rate and intensity of cardiac contraction, which together help maintain a normal cardiac output. The long-term response involves the stimulation of the renin angiotensin aldosterone (RAA) system. Angiotensin II is a potent vasoconstrictor and aldosterone causes the kidneys to reabsorb more water and sodium from the urine. Chronic activation of the adrenergic and RAA systems can be measured in animals with compensated congestive heart failure.[1]

Decompensated heart failure

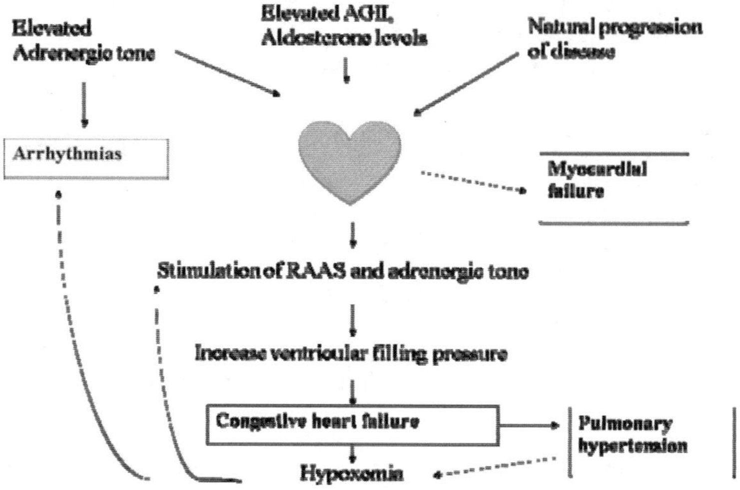

Figure 2. Pathophysiology of decompensated heart failure

Fluid retention (pulmonary congestion, peripheral edema) characterizes the decompensated stage of heart failure. Progression to the stage of congestive heart failure results from many factors:

- Natural progression of the disease. Most cardiac diseases in dogs and cats are inherited, which makes difficult to predict the rate of alteration of the cardiac function.
- Chronic activation of the adrenergic system promotes arrhythmias and has been shown to promote cardiac myocyte death.
- Chronic stimulation of the RAA system further impairs cardiac function by promoting cardiac muscle fibrosis.
- Fluid retention in the lungs (pulmonary edema), abdominal cavity (ascites) and organs, or pleural space (pleural effusion) interferes with breathing, and impedes on liver and kidney function.
- Hypoxemia caused by congestive heart failure further activates the adrenergic and RAA system, promotes arrhythmias.
- Pulmonary hypertension occurs in response to chronic pulmonary venous congestion.[2]

SOURCE OF INFORMATION FOR TREATMENT RECOMMENDATIONS

In veterinary medicine, treatment strategies are based from experimental or clinical studies in animals, but also frequently from data collected from the human medical literature. It is important to know the source that leads to treatment recommendations, as the level of scientific evidence to support them may be highly variable. The strongest evidence is provided by randomized placebo-controlled trials. Those are rare in veterinary medicine because of the cost and the difficulty to recruit enough subjects. Lower levels of evidence are provided by cases series, case reports or opinions of respected authorities, based on their clinical experience.[3]

TREATMENT STRATEGIES FOR HEART FAILURE

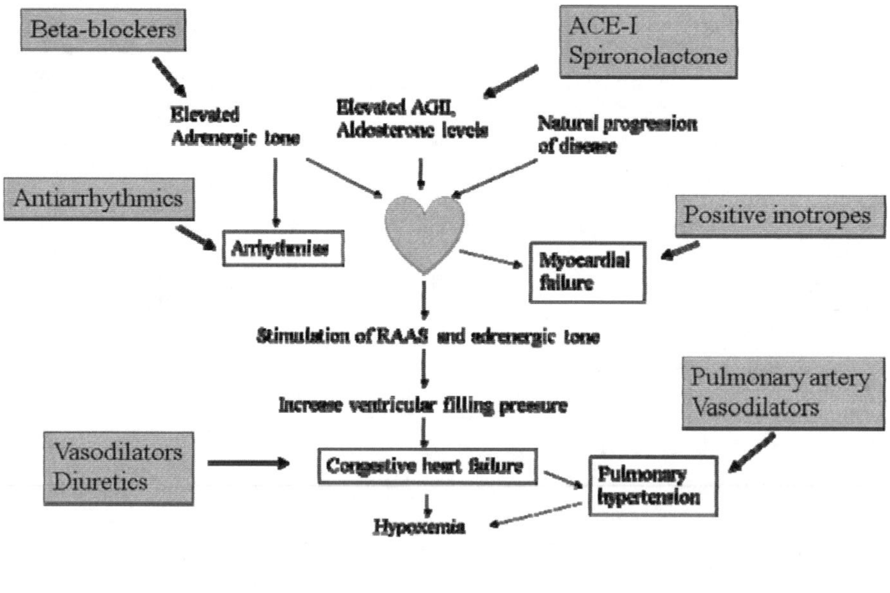

Figure 3. Treatment strategies for heart failure

TREATMENT STRATEGIES FOR ASYMPTOMATIC HEART FAILURE

Based on extensive clinical trials, the American Heart Association recommends the use of angiotensin converting enzyme (ACE) inhibitors in all patients with reduced cardiac contractility. There is also strong evidence that patients with a recent myocardial infarct should receive beta-blocker therapy. However, some data suggest that digoxin should not be used at this early stage of heart failure.[4]

Two placebo-controlled multicentre trials (SVEP and VetProof trials) have been conducted to assess the effects of Enalapril, an ACE inhibitor, in dogs with documented degenerative mitral valve disease but no signs of heart failure. Both trials failed to show a beneficial effect in delaying the onset of clinical signs. Therefore, the use of ACE-inhibitors cannot be recommended at an early stage of the disease in this group of dogs. The level of scientific evidence to support the use of ACE-inhibitors in dogs with asymptomatic dilated cardiomyopathy and cats with cardiomyopathy is much lower. A recent retrospective case series suggested that the use of Benazepril (ACE-inhibitor) may delay the progression of dilated cardiomyopathy in Doberman Pinshers.[5,6] As for feline cardiomyopathy, a prospective, blinded, placebo-controlled study in 26 Maine Coon cats and blinded, controlled study in 21 cats did not establish a role for ACE-inhibitors in the asymptomatic phase of the disease.[7,8]

There is currently no study to support the use of beta-blockers in dogs with compensated congestive heart failure. Beta-blockers have been used in cats with hypertrophic cardiomyopathy based on the limited information available in the human literature. There is no evidence of a beneficial effect.

TREATMENT STRATEGIES FOR CONGESTIVE HEART FAILURE

In humans, current guidelines recommend the use of ACE-inhibitors, beta-blockers and diuretics when patients experience symptoms of heart failure and their cardiac disease results from impaired cardiac contractility. There is also strong evidence that aldosterone antagonists (Spironolactone) improves outcome of patients at this stage of the disease. Similar treatments are recommended for heart failure with normal cardiac contractility with less evidence to support the guidelines.[9]

In dogs and cats with congestive heart failure, Furosemide is seen as the essential "life-saving" treatment.

Clinical trials have been conducted to evaluate the effects of ACE-inhibitors added to Furosemide in dogs with degenerative mitral valve disease. They convincingly demonstrated that this combination improves clinical signs and outcome in this group of patients. A recent placebo-controlled trial (Quest trial) has added to the information already available on the beneficial effects of the drug Pimobendan, an inodilator, for the treatment of congestive heart failure in dogs with degenerative valve disease. Dogs treated with Pimobendan and Furosemide reached the predefined endpoint later than dogs treated with Benazepril and Furosemide. There is currently not enough data to support the use of aldosterone antagonists and beta-blockers.

There are currently no studies that address the treatment of congestive heart failure in dogs with dilated cardiomyopathy and cats with hypertrophic cardiomyopathy. Conventional therapy includes Furosemide and ACE-inhibitors.

Recommendations for the emergency management of acute congestive heart failure is only based a few reports and expert opinions. They include: oxygen supplementation, repeated boluses or constant rate infusion of Furosemide, intravenous inotropes (Dobutamine) and vasodilators.

CONCLUSION

In a world where unlimited amount of information is directly available to everyone, it is important to be able to critically assess it and stay of the level of scientific evidence that supports our therapeutic recommendations for heart failure.

REFERENCES

[1] Sisson DD. **Vet Clin North Am Small Anim Pract** 2004;34(5):1105. [2] Oyama MA. **J Small Anim Pract** 2010;50(Suppl.1):3. [3] Hinchcliff KW. **J Vet Intern Med** 2010;24(1):8. [4] Goldberg LR, et al. **Circulation** 2006;113(24):2851. [5] O'Grady MR et al. **J Vet Intern Med** 2009;23(5):977;. [6] Haggstrom J et al. **J Small Anim Pract** 2010;50(Suppl.1):25. [7] Taillefer M, et al. **Can Vet J** 2006;47(5):437. [8] MacDonald KA, et al. **J Vet Intern Med** 2006;20(5):1093. [9] ACC/AHA 2005 Guidelines. **Circulation** 2005;112:1825.

ATRIAL FIBRILLATION: AN OVERVIEW OF THE MOST COMMON ARRHYTHMIA

Romain Pariaut, DVM, DACVIM (Cardiology), DECVIM-CA (Cardiology)
Baton Rouge, LA

INTRODUCTION

Atrial fibrillation (AF) is the most common arrhythmia encountered in veterinary and human medicine. In the United States, it affects 1% of the population and 10% of those older than 80 years old.[1] In animals, the disease is typically identified in middle to large breed dogs and horses. In recent years, the knowledge on AF has dramatically increased in human cardiology. We are also learning more on the pathophysiology and therapy of AF in animals.

MECHANISMS OF ATRIAL FIBRILLATION

Atrial fibrillation originates in the left and right atrium. It is characterized by a very rapid electrical activation of the left and right atrium, preventing them from contracting efficiently. In response to the impulses traveling through the atrioventricular node, the ventricles contract irregularly and usually rapidly.

In mammals, the pump function of the heart results from the sequential contraction of four cardiac chambers in response to an electrical impulse originating in the sinus node, which then travels within the atria to the atrioventricular node and finally within the ventricles. In AF, this regular rhythm is disrupted. In many cases the abnormal rhythm seems to originate in the region of the pulmonary veins, where some cells generate electrical impulses at a very rapid rate, which then propagate within the atrial tissue in the form of large waves of electrical current that simulate an "electrical storm". Some experimental recordings that display an image of the electrical impulses during atrial fibrillation reveal a large spiral wave, similar to the satellite picture of a hurricane.[2,3]

CAUSES OF ATRIAL FIBRILLATION

In dogs, AF is usually a complication of heart failure. Severe atrial dilation that is present in most dogs with advanced cardiac disease increases the risk of AF. Indeed, it appears that the larger the atrium, the higher the risk for AF. As a consequence, large breed dogs are more prone to AF than small breed dogs. Other factors, such as elevated sympathetic tone, inflammation or genetic predisposition are also involved in AF.

Atrial fibrillation may also occur in giant breed dogs that don't have evidence of congenital or acquired structural cardiac disease. This unusual form of AF is called "lone" AF. Once again, the size of the atrium is an important predisposing factor.[4]

Horses may also develop AF, which is oftentimes not associated with obvious cardiac disease. In this species, AF usually occurs at the end of a race. In many cases, it resolves by itself. Imbalance between elevated adrenergic tone and vagal tone seems to play a role in this form of AF.[5]

DIAGNOSIS

Most dogs diagnosed with AF are presented with signs of congestive heart failure. In some asymptomatic dogs with compensated heart failure, signs of congestion may be triggered by the onset of AF. Atrial fibrillation may be an incidental finding during routine physical examination in giant breed dogs with lone AF. It may also results in exercise intolerance or excessive panting during periods of faster heart rate. Atrial fibrillation is usually identified in horses evaluated for poor performance.

Auscultation

Atrial fibrillation is an irregularly irregular rhythm (Auscultation is often described as sounding like tennis shoes in a dryer) associated with femoral pulse deficits. In dogs with heart failure, AF is a rapid rhythm with an average heart rate that can reach 200 to 240 beats/minute. Most dogs with lone AF and horses have a normal heart. This is because their atrioventricular node is able to block most of the impulses initiated in the atria before they reach the ventricles. This filtering effect of the atrioventricular node (The only pathway for the electrical impulse between the atria and the ventricles) depends on the level of vagal tone. In dogs with heart failure, vagal tone is markedly decreased. As a result, more impulses are able to travel through the atrioventricular node and trigger ventricular contraction.

Electrocardiography

A persistent, rapid (>150 beats/min) and irregular rhythm is consistent with AF in dogs. P waves are absent, as they represent the organized propagation of the electrical impulse within the atria during normal sinus rhythm. Small F (fibrillation) waves may be seen as an undulation of the baseline. However, they are usually not visible as the filters selected on the ECG machine eliminate them. F waves are easier to identify in horses. The QRS complexes are usually normal in morphology, indicating that the electrical conduction is normal in the ventricles. The interval between the QRS complexes is always irregular.
With lone AF, the heart rate is usually normal.

MANAGEMENT

Treatment of AF is based on controlling the rapid heart rate that is responsible for low cardiac output and worsens heart failure. In people, AF markedly increases the risk of stroke, and most people with AF receive anticoagulants. This complication has not been reported in dogs and horses.

Treatment of persistent AF in dogs with heart failure

Conversion of AF into sinus rhythm is difficult because of the underlying cardiac changes. Therefore, treatment is based on a rate control strategy. Digoxin, Calcium channel blockers and Beta-blockers are typically used to slow the heart rate. Monitoring of the heart rate with ambulatory 24-hour Holter recordings is helpful to assess response to medications. It seems that an average of 100 to 130 beats/minute is consistent with adequate control of the arrhythmia.[6]

Treatment of lone AF in dogs

Most giant breed dogs with lone AF have no clinical signs and a normal average heart rate. Medical treatment is usually not required. In some dogs, digoxin is used to slightly decrease the average heart rate, or prevent excessive tachycardia during periods of exercise or excitement, when the adrenergic tone is high.

Treatment of acute and recent onset of AF in dogs

Acute onset of AF may occur in large breed dogs, typically after sedation or during anesthesia. Some of those episodes terminate by themselves. In other cases, electrical conversion with an external transthoracic defibrillation can be performed. Intravenous injection of lidocaine (2 mg/kg) has also been shown to terminate acute onset of AF triggered by elevated vagal tone.[7,8,9]

Treatment of AF in horses

Most horses don't have detectable structural cardiac disease and a normal resting heart rate during AF. Therefore, treatment of AF is not required. However, during exercise, horses with AF may have a very rapid heart rate, which results in poor cardiac output and poor performance. In horses, a rhythm control strategy (Conversion into sinus rhythm) is usually preferred. Oral administration of quinidine has been the standard treatment. However, it is associated with many side effects. Internal electrical cardioversion has been developed and provides an alternative to the medical treatment.[5]

CONCLUSION

Further research is needed to identify the best treatment strategies for veterinary patients with AF.

REFERENCES

[1]Kannel WB, et al. **Med Clin North Am** 2008;92:17. [2]Brundel BJJM, et al. **J Vet Cardiol** 2005;7:121. [3] Pariaut R, et al. **Am J Vet Res** 2008;69(11):1434. [4] Menaut P, et al. **J Vet Cardiol** 2005;7:75. [5]McGurrin et al. **J Vet Cardiol** 2005;7:109. [6] Gelzer AR, et al. **J Vet Intern Med** 2009;23(3):499. [7]Moise NS, et al. **J Vet Cardiol** 2005;7:143. [8] Pariaut R, et al. **J Vet Intern Med** 2008;22(6):1274. [9] Estrada AH, et al. **J Vet Cardiol** 2009;11(2):137.

ASSESSSMENT OF THE ANEMIC PATIENT

Kimberly Baldwin, LVT, VTS (ECC)
Spencer, NY

INTRODUCTION

Anemia is defined as the decrease in circulating red blood cells. This can occur due to decreased production, increased destruction (hemolysis) and through the loss (hemorrhage) of red blood cells. This lecture will review the factors impacting the loss of red blood cells, diagnostic tools to identify the primary disease, blood component considerations, and the technician's role in assessing and providing nursing care to the patient.

ANEMIC CRISIS

The patient in anemic crisis arrives in the veterinary clinic with a chief complaint of lethargy, progressive weakness, or collapse. The duration of clinical signs will vary with the underlying disease. A significant blood loss over hours will result in a hypovolemic state which could potentially lead to cardiovascular arrest. Quick assessment of heart rate, lung sounds, pulse quality, and mucous membrane color will provide useful information regarding the stability of the patient. Patients in an anemic crisis will present either extremely depressed or moribund with pale and/or icteric mucous membranes, tachypnea, tachycardia, with bounding pulses. Hepatomegaly, splenomegaly, a heart murmur and hematura may also be present.

STABLIZATION

Initial therapy should include airway and breathing support, fluid therapy for volume replacement, and identifying and controlling source of hemorrhage. A blood transfusion may be considered if adequate volume replacement is not able to provide sufficient tissue oxygen delivery and clinical signs persist. The largest bore peripheral catheter is placed in the vein for stabilization. Two peripheral catheters may be necessary based on the initial presentation of the animal. Avoiding the jugular vein for venous access or patient samples is warranted until further evaluation of the primary disease is determined. Ideally, the samples are drawn from the catheter before initiating therapy and should include PCV, TP, blood smear, CBC, chemistry, EDTA, Serum, and Citrated Plasma tubes for anticoagulation, blood typing, agglutination tests and titers. Fluid therapy should be initiated with a crystalloid solution at two to three times the fluid loss.

If massive blood loss has occurred a colloid solution should be considered to maintain intravascular oncotic pressure without administering large volumes of fluids. If intravenous fluid support does not provide optimal tissue oxygenation and improvement of clinical signs a blood transfusion should be considered. Hemoglobin is a critical component for oxygen transport. In human medicine hemoglobin concentration of less than 7g/dl is the primary transfusion trigger. Dogs and cats vary in their normal hemoglobin values making it difficult to develop a specific hemoglobin standard within veterinary medicine. A clinical assessment of respiratory rate, heart rate, and mentation may be better indicators then lab results. The benefits of a blood transfusion should be determined on an individual basis and outweighed against potential risks such as transfusion reactions, transmission of infectious disease and potential immunosuppression. A patient with long term chronic anemia is able to tolerate a very low PCV with little to no clinical signs. A patient with acute, moderate anemia may be severely compensated especially if hypoxaemia and hypoperfusion exist. Which component to choose should be based on PCV, and TP results. If only the packed cell volume (PCV) is low, a patient may only need packed red blood cells. If a patient has a reduction in both PCV, and TP with depressed clinical signs, packed cells, plasma or a whole blood transfusion may be required. Blood typing before providing a transfusion is mandatory for all feline species.

HISTORY AND PHYSICAL EXAM

After the initial assessment and stabilization has occurred, a more complete history and physical exam should be completed. The technician plays a key role in gathering the appropriate information needed to make a treatment plan for an animal with anemia.

Questions that allow the owner to recall pertinent information will lead to a better understanding and a quicker assessment of the patients needs. Specific questions to consider may be:

1. Signalment
2. Complete medical history
3. Vaccine history
4. Dietary indiscretions (onion, zinc)
5. Travel history
6. Environment in which the animal lives. Examples, indoors/outdoors, free roaming, leash walked, live in the city/on a farm
7. Exposure to any hazardous or household medications?
8. Is the animal on any medications, including, non prescriptions?
9. Has he received any form of component therapy before?

Staff members can create a checklist of questions to be handed to the client or asked when the patient arrives. This can be modified to the many disease processes we see within veterinary medicine.

The complete physical exam should include a thorough search for any clinical signs of anemia or hemorrhage and results should never be concluded based on a single parameter. Perfusion assessment is evaluated by examining mucous membrane color, capillary refill time, heart rate, and pulse rate, and strength. Bounding pulses suggest anemia. Prolonged capillary refill time may indicate poor perfusion or shock. Icteric may be due to hemolyis, hepatic or biliary disease. Petechiation may indicate a coagulopathy such as liver disease or DIC. Splenomegaly may be an indication of hemolyis, neoplasia, infectious disease or extramedullary hematopoiesis. Fever may indicate infectious disease

LABORATORY ASSESSMENT

PCV, TP

There are a number of in-house diagnostic test that can help identify the cause of the anemia. The first test that should be completed is packed cell volume (PCV) and total protein (TP). Dehydration and splenic contraction may falsely elevate the PCV where hemodilution may temporarily decrease PCV. If PCV and TP are both low, acute blood loss should be suspected. A low PCV with normal TP would be more consistent with hemolysis or decreased red blood cell production. The plasma should also be evaluated for hemolyis or icteric which could indicate intravascular or extravascular hemolysis respectively.

Blood Smear Evaluation

The blood smear can provide useful information regarding the type and cause of the anemia. When the patient is actively bleeding a blood smear can provide a quick assessment of platelet morphology and numbers. This can occur at the feathered edge of the blood smear. One platelet per oil immersion = 20,000 platelets. 11-25 platelets per high power field is normal. By checking the feathered edge first, you may see clumping. If the platelets are clumped an accurate count cannot be obtained. The presence of large platelets oftentimes indicates a regenerative process. RBC morphology should be evaluated next. RBC agglutination appears like grapelike clusters and strongly suggests AIMHA. Spherocytes are small, round erythrocytes with loss of central pallor. A large number may also indicate immune-mediated hemolysis. Heinz bodies appear as bulging red blood cells and can indicate damage to the red blood cell as indicated in toxins such as onion, or propylene glycol. Schistocytes are fragmented erythrocytes as a result of passage through narrow microvascular. Schistocytes typically represent microangiopathic causes of hemolysis such as caval syndrome, DIC, hemangiosacroma, and splenic torsion. Acanthocytes are erythrocytes with long spiny projections that are frequently seen with hepatic and splenic neoplasia. Polychromasia and anisocytosis may indicate the presence of a regenerative response. White blood cell numbers and proportions may indicate a left shift as a result of bone marrow stimulation. Parasites such as Babesia, and Mycoplasma should be screened when evaluated the blood smear.

Slide Agglutination Test

Agglutination is caused by the cross linking of antibodies to the red blood cell membrane. The slide agglutination test can distinguish from Rouloux formation and is a good diagnostic indicator of immune-mediated hemolysis. A drop of blood from a capillary tube or anticoagulated blood is placed on a slide with several drops of saline. Rouloux will be dispersed by the saline. Agglutination is demonstrated by obvious flecks within the blood. Weak agglutination does sometimes cause false negatives. Macroagglutination should also be evaluated in the EDTA. In severe cases it can be seen in the capillary tube.

Reticulocyte Count

Although this may not be an immediate diagnostic tool while assessing the anemic crises it is a critical test and the best indicator to determine bone marrow activity and regenerative response. The test is performed by using a few drops of blood which is mixed with .5% new methylene blue in physiological saline. The reticulocyte count during regenerative anemia will vary with the degree of anemia. The greater the stimulation on the marrow, the greater the reticulocytes. This is not the scenario for non-regenerative anemia therefore the degree of reticulocytes must be interpreted relative to the degree of anemia by calculation of the absolute reticulocyte count or a corrected reticulocyte count. Absolute is determined by (% of reticulocytes x RBC count). An absolute reticulocyte count of >60,000 μl is evidence of regeneration for the dog. The corrected reticulocyte count is calculated by using the following formula: corrected reticulocyte count% = (observed reticulocyte count) x (PCV of the patient/mean normal PCV for species). Average PCV for the dog is 45%, 37% for the cat. A corrected reticulocyte count above 1% reflects regenerative anemia.

REGENERATION OR NON-REGENERATION

When treating the anemic patient it is important to identify if the anemia is regenerative or non-regenerative. The reticulocyte count is the most useful diagnostic tool to determine this classification. Regeneration is most commonly seen when hemorrhage or hemolysis has occurred as seen in acute blood loss, AIHA, neoplasia, hereditary disorders, toxins, oxidative damage, and infectious disease. Non-regenerative anemia results from diminished production or ineffective erythropoieses or inadequate bone marrow production. Non-regenerative anemia is considered when the reticulocyte in the dog is < then 60,000μl –corresponding with the corrected reticulocyte of <1%. Caution should be taken when severe, acute blood loss or hemolysis is present keeping in mind the bone marrow response can sometimes take 2 – 3 days. Organ disorders (renal failure, endocrine disease), pure red cell aplasia, nutritional deficiencies (B12, iron), and bone marrow hypoplasia are non-regenerative diseases. Once a non-regenerative anemia is identified, a bone marrow aspirate is generally indicated.

CONCLUSION

Anemia is commonly seen in veterinary emergency and critical care. Patients may be brought to the clinic with a primary complaint of collapse or pale mucous membrane or the patient may have been hospitalized and acquires their anemia through a certain treatment or their underlying disease. Through a complete history and physical exam, monitoring and re-assessing ongoing parameters, and utilizing simple diagnostic tools the technician plays a critical role in the successful outcome of each patient we treat.

REFERENCES

[1.] **Veterinary Emergency and Critical Care Manual** – Karol Mathews [2.] **Manual of Canine and Feline Emergency and Critical Care**, Dr Leslie King, Dr Richard Hammond [3.] **Veterinary Emergency and Critical Care Medicine**, Murtaugh, Kaplan [4.] **Small Animal Critical Care Medicine** – Dr Silverstein, Dr Hopper [5.] Small Animal Emergency and Critical Care, **Hematologic Emergencies**, Ms. Oakley [6.] **Saunders Manual of Small Animal Practice**, Stephen Brichard and Robert Sherding [7.] **Veterinary Emergency Medicine Secrets**, Dr Wingfield [8.] **Critical Care Medicine** –Abstract, Vol 37, Issue 12, Pgs 3124-3157, Pub Dec 2009

PATHOGENESIS OF DIC: A ROCKY ROAD

Kim Zwerenz-Miks, MS, RVT, VTS (ECC SAIM)
Placerville, CA

Background

Dissemination intravascular coagulation (DIC) was first described in human medicine in a lecture by W.H. Seegers to the Cincinnati Academy of Medicine on April 18, 1950. DIC is also been referred to as consumptive coagulopathy, defibrination syndrome, or diffused intravascular clotting.

Disseminated intravascular coagulation (DIC) is a disorder characterized by massive systemic intravascular activation of coagulation, leading to widespread deposition of fibrin in the circulation which can compromise the blood supply to various organs, thus contributing to multiple organ failure. At the same time, the consumption of platelets and coagulation proteins resulting from the ongoing coagulation may induce severe bleeding. However, DIC is not a disease itself but is always secondary to an underlying disorder. In fact, a variety of clinical conditions may cause systemic activation of coagulation. Table 1 lists the diseases most frequently associated with DIC in dogs and cats. Bacterial infections, in particular septicemia, are the most common clinical conditions associated with DIC.

DISEASES AND CONDITIONS ASSOCIATED WITH DIC			
Inflammatory conditions	**Neoplasia**	**Infectious disorders**	**Miscellaneous**
Sepsis , Bacteria endocarditis, Leptospirosis, Canine infectious hepatitis (D) Babesiosis Dirofilariasis FIP	Hemangiosarcoma (D), Hemangioma, Metastatic thyroid carcinoma, Prostatic Adenocarcinoma, Lymphoma (C), Cholangiocarcinoma	Acute hepatic necrosis, Chronic active hepatitis, Pancreatitis, Hemorrhagic gastroenteritis, Suppurative dermatitis, Suppurative bronchopneumonia	Shock, Heatstroke, Venomous snakebite, Cirrhosis, aflatoxicosis, IMHA (D), GDV, CHF, Perioperative complications, Renal amyloidosis, Hepatic lipidosis(C), Pulmonary thromboembolism
Table 1			

Pathogenesis of DIC

DIC is essentially an exaggeration of the normal hemostatic mechanisms. Due to the abnormal overstimulation of the normal coagulation process, there is massive, rapid fibrin formation, resulting in widespread small thrombi in the microcirculation. The vast amount of clotting and the rapidity with which it occurs, platelets and clotting factors are quickly depleted, and the fibrinolytic process is stimulated to dissolve the fibrin clots. Fibrin degradation products (FDP's) are an end product of clot lysis and have an anticoagulant effect. See figure 1 for an overview.

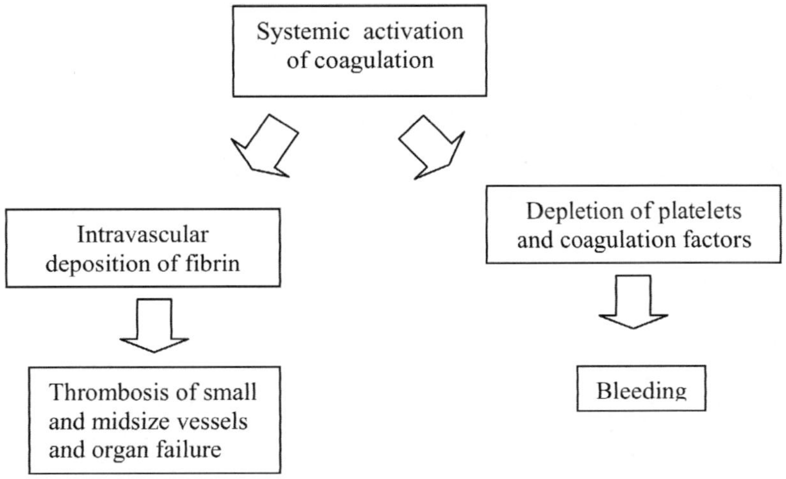

Fig. 1

<u>Some</u> of the mechanisms for triggering DIC include:

- Release of tissue factor or thrombo-plastic substances into the circulation
  - Often triggered by sepsis, severe tissue destruction (trauma, burns), pancreatitis, neoplasm (lymphoma, leukemia, and hemangiosarcoma).

- Injury to endothelial cells thus exposing subendothelial collagen
  - Endothelial injury can initiate DIC by release of tissue factor and promoting platelet aggregation to exposed collagen.
    - Causes include direct damage by trauma, vasculitis (FIP, lupus erythematosus), temperature extremes (burns, heat stroke), toxins, rickettsiae, viral infections, immunologic injury
- Release of Procoagulants from red blood cells (hemolysis)
  - Transfusions, extracorporeal circulation
- Release of procoagulants from other sources
  - Bacterial infections, gram-negative septicemia (E.coil & others), Viral infections, acute pancreatitis, fat embolism, hyperlipemia, acute hepatitis, direct damage by trauma, temperature extremes (burns, heat stroke), toxins
- Vascular stasis
  - Shock , massive pulmonary embolism

Once the coagulation cascade has been activated (throughout the microvascular) several events take place. Many of them happen simultaneously.

1. Primary and secondary hemostatic plugs are formed followed by formation of thrombi within the microcirculation of multiple organs such as brain, renal glomeruli and/or pulmonary capillaries leading to microinfarction and tissue hypoxia and eventual organ failure. During this period platelets are consumed in large quantities, leading to thrombocytopenia.
2. The fibrinolytic system is activated, resulting in clot lysis and the inactivation (or lysis) of clotting factors and impaired platelet function.
3. AT III and possibly proteins C and S are consumed in an attempt to halt intravascular coagulation, thus leading to "exhaustion" of the normal anticoagulants.
4. The formation of fibrin within the microcirculation leads to the development of hemolytic anemia as the RBCs are sheared by these fibrin strands (i.e., fragmented RBCs or schistocytes).

Clinical features

There are several clinical presentations in dogs with DIC; acute, subacute (subclinical) or chronic. Chronic DIC is found to be common in dogs with malignancy or possibly other chronic diseases. These patients tend to have diffused thrombosis and are described as having "compensated" DIC. This is demonstrated by an increase production of coagulation factors and platelets by the liver and bone marrow which results in the normalization of many of the coagulation parameters and the activated coagulation system is kept in check by the inhibitors. These patients are not actively bleeding.

In dogs with acute or subacute DIC will often demonstrate excessive hemorrhage, which is often clinically obvious and dramatic. The predominate clinical signs of bleeding are petechiae, ecchymosis, epistaxis, oozing from old injections sites or surgical sites, and gastrointestinal bleeding.

Evidence of thrombosis also may be seen, and symptoms are those associated with the system involved and vary with the extent of the vascular obstruction. For example, renal system involvement may cause hematuria, oliguria; central nervous system involvement, mentation changes (seizures, coma); integument involvement acrocyanosis; and so forth.

Hemorrhagic diathesis predominates commonly in acute DIC and thrombotic complications are the main feature of subacute or chronic DIC. Acute DIC may develop primarily (e.g. caused by sepsis), or secondary to decompensation of chronic DIC.

Acute DIC is more commonly observed in dogs and is rare in cats.

Diagnosis

The diagnosis of DIC is based on the presence of a predisposing condition (e.g., severe sepsis) combined with laboratory evidence of wide spread coagulation deficits. Diagnosis depends on a composite of coagulation studies revolving around the critically important fibrinogen-thrombin interaction. Typical laboratory findings include thrombocytopenia (or subsequentially decreasing platelet count), prolonged APTT (and possible ACT if factor deficiencies are severe enough) and PT, low fibrinogen, low antithrombin activity, and positive FDP and/or D-dimer results. Other laboratory findings might include: Regenerative hemolytic anemia with schistocytes and keratocyte formation, neutrophilia with left shift, neutropenia (rare), hemoglobinemia, and hypoalbuminemia.

The most sensitive test for positive diagnosis of DIC in dogs and horses are prolonged APTT, low antithrombin activity, and high FDP and/or D-dimer. Fibrinogen concentrations are often normal or even elevated as fibrinogen acts as an acute phase reactants with an increase in production occurring secondary to the underlying disease process. Cats with DIC may have prolonged aPTT and/or PT but may not have an increased FDP and/or D-dimer. They usually do not have low antithrombin activity or low fibrinogen concentrations.

Diagnosing DIC in patients with liver disease is difficult. Normally, the liver produces coagulation factors for the clotting process. With hepatic disease there is a decreased production of these factors, resulting in prolonged coagulation values. Also the liver's reticuloendothelial activities are impaired, resulting in decreased removal of activated clotting factors, fibrin, and other procoagulants all of which will cause abnormal values.

Treatment

Therapy for DIC is quite controversial and there is no consensus on the appropriate treatment regimes. It is agreed that the most important element of therapy is the identification and treatment of the underlying disease. Supportive therapy including intravenous fluids to maintain fluid volume and organ perfusion is vital. Besides correcting hypovolemia fluid therapy will also help prevent or alleviate vascular stasis, and help with the dilution of thrombin, FDP's and activators of fibrinolysis.

Opinions diverge when it comes to treating the coagulation defects. Currently, most authors recommend the administration of heparin (especially low molecular weight forms) in patients with thrombosis, although few controlled studies have been conducted to determine the benefit of heparin therapy. When heparin is used, it is given to inhibit clot formation. Heparin's antithrombin activity neutralizes free thrombin, thereby decreasing fibrin formation and the simulation of fibrinolysis. Aspirin is contra-indicated due to its effect on platelet function, as platelet function is inhibited in DIC by FDP's already and platelet hyperaggregability is not usually a cause for thrombosis in DIC.

When patients with DIC bleed, replacement of some or all blood constituents is indicated to replenish depleted coagulation factors and platelets. Plasma is preferred, but whole blood may be given of red cell infusion is necessary. The infusion of red cells carries a risk of hemolysis and thus exacerbation of DIC. Some authors have suggested that heparin injected into bags of fresh or fresh-frozen plasma (to activate antithrombin) prior to transfusion is beneficial; however this has not been substantiated by well-controlled studies. There have been recent studies in human medicine that suggest that the administrations of inhibitors, particularly protein C, are beneficial in treating DIC in sepsis. The product is not available for animals at this time.

Normalization of screening PT, APTT and FDP's usually denote successful therapy. The return of normal fibrinogen concentration is a reliable long-term indicator of heparin therapy.

Conclusion

Disseminated intravascular coagulation is a paradox. It's a disorder marked by excessive coagulability of the blood at the same time that the patient may be unable to form new blood clots. The patient may have symptoms of hypoxia and tissue damage secondary to the presence of unnecessary blood clots while at the same time, is bleeding from sites as clots are broken down. It is these very polar opposites that make DIC difficult to recognize and to manage.

Effective nursing care of the patient with DIC is enhanced by a through understanding of its pathophysiology and its clinical manifestations. Patient survival depends on early identification and treatment of DIC and its underlying cause. DIC does not follow a predictable pattern; signs and symptoms may vary depending on the severity of bleeding and extent of organ and tissue damage.

Early recognition is key, and nurses closely monitoring patients may first recognize early signs and symptoms, such as bruising, petechiae, mucosal bleeding, and oozing from venipuncture or IV sites. Compiling information, such as risk factors, clinical indicators, and laboratory results, may result in early detection of the onset of DIC. The earlier the syndrome is recognized, the more likely the underlying cause can be treated successfully.

REFERENCES

[1]Bick RL: Disseminated intravascular coagulation. Current concepts of etiology, pathophysiology, diagnosis, and treatment. **Hematol Oncol Clin N Am** 2003, 17:149-176. [2]Levi M, ten Cate H: Disseminated intravascular coagulation. **N Engl J Med** 1999, 341:586-592. [3]Levi M, de Jonge E, Van der Poll T, ten Cate H: Disseminated intravascular coagulation: state of the art. **Thromb Haemost** 1999, 82:695-705. [4]Levi M, de Jonge E, Meijers J: The diagnosis of disseminated intravascular coagulation. **Blood Rev** 2002, 16:217-223. [5]Mammen EF: Disseminated intravascular coagulation (DIC). **Clin Lab Sci** 2000, 13:239-245. [6]Levi M: Disseminated intravascular coagulation: what's new? **Crit Care Clin** 2005, 21:449-467. [7]Pathologic Basis of Disease, 6th ed: 1999. Cotran RS, Kumar V and Collins T (eds) S.B. Saunders Co. [8]Veterinary Laboratory Medicine, 3rd ed): 1994. Duncan JR, Prasse KW and Mahaffey EA (eds). Iowa State University Press [9]Carey MJ, Rogers GM: Disseminated intravascular coagulation: clinical and laboratory aspects. **Am J Hematol** 1998, 59:65-73.

DIAGNOSIS, TREATMENT, & LONG TERM MANAGEMENT OF CANINE IMHA

Colleen Ruderman, RVT, VTS (SAIM)
Atlanta, GA

INTRODUCTION

Immune mediated hemolytic anemia (IMHA), also called autoimmune hemolytic anemia (AIHA) is one of the most common causes of anemia in dogs. It consists of the immune system's removal of the patients own red blood cells through various pathways. The pathophysiology, diagnosis, and treatment can be complicated and the role of the veterinary technician is critical. Through understanding of the disease process, excellent nursing and laboratory skills, and client communication the technician plays an integral role in the survival of canine IMHA patients.

PATHOPHYSIOLOGY

IMHA occurs when the immune system attacks its own red blood cells causing anemia and associated complications. Tolerance is defined as the immune system's ability to recognize foreign particles from itself, and the lack of tolerance can lead to IMHA. Antierythrocyte antibodies will collect on the red blood cell (RBC) membrane and the immune system will target them for removal. This removal is typical done by the mononuclear phagocyte system (MPS). In this circumstance macrophages will remove the RBC causing extravascular hemolysis. Complement removal of RBC can also occur resulting in an intravascular hemolysis.

Primary and secondary classifications of IMHA exist. Primary IMHA is defined as idiopathic, or true immune disease with no known cause. Predisposition to other and concurrent immune disorders, such as immune mediated thrombocytopenia and lupus, can occur. Secondary IMHA occurs when hemolytic anemia has an identifiable or suspected inciting cause. Secondary IMHA causes most often include vaccinations and drug administration. Other causes included neoplasia and infectious (tick borne) diease. Some studies have shown up to 25% of canine IMHA patients being vaccinated within 30 days of presentation and a prevalence of IMHA in the spring/summer months adds to this theory, although there is no conclusive link[1,3,7]. Drug related IMHA occurs when the drug particles attach to the RBC inciting the immune response, or by increasing levels of antibody production inciting the RBC removal by complement. The most common medications that cause secondary IMHA are sulfonamides but can also occur with cephalosporins. Collection of a complete history including previous/current medications and date of any vaccination can help identify potential causes for secondary IMHA.

There are several complications associated with canine IMHA. Approximately 50% of canine IMHA patients will develop concurrent thrombocytopenia resulting in Evan's syndrome[3,7]. These cases are often complicated and require extensive monitoring due to the simultaneous loss of RBC through hemolysis, as well as the possibility of spontaneous bleeding that can occur with thrombocytopenia (Plt <50,000). In addition to Evan's syndrome the development of pulmonary thromboembolism (PTE) can occur in IMHA patients due to their hypercoagulable state. Disseminated intravascular coagulation (DIC) has also been a reported complication.

SIGNALMENT/CLINICAL SYMPTOMS

Typically young to middle aged dogs are affected and breed dispositions have been noted in the cocker spaniel, bichon frise, miniature pinscher, collie, Irish setter, and miniature schnauzer. No specific gender predisposition is noted however higher numbers of female dogs are seen with immune disorders. Clinical symptoms are those associated with anemia and include anorexia, icterus/pallor, lethargy/weakness/collapse, tachypnea, tachycardia, vomiting, and pigmenturia. Clinical symptoms can be mild to severe, and incidental diagnosis with routine lab work in the asymptomatic patient can occur.

LABORATORY FINDINGS

The complete blood count (CBC) will show a mild to severe anemia and may also show thrombocytopenia. Immune mediated thrombocytopenia (IMT) is a common complication. A reticulocyte count can often confirm a regenerative anemia; however in peracute cases, or immune disease against reticulocytes or RBC precursors, a non-regenerative anemia can also be seen. One third of IMHA cases are non-regenerative. Leukocytosis is also a common finding on presenting CBC and can be steroid induced or caused by inflammatory reactions and concomitant infections. It should be noted that autoagglutination is common in IMHA cases and can be seen at the time of blood collection, grossly within the EDTA vacutainer. Autoagglutination should be noted as several tests will be affected by its presence. Biochemistry panels can reveal hyperbilirubinemia as a result of RBC lysis and increased circulation of bilirubin. Elevated liver enzymes (ALT/ALKP) and azotemia can also be noted and can result from tissue hypoxia as well as dehydration. A urinalysis can reveal pigmenturia both from hemoglobin and bilirubin. When performing serum biochemistry panels it is important to note the level of hemolysis in the serum if present. This can be an indicator of intravascular hemolysis which is often associated with an increased severity of disease. Poor blood collection technique can produce hemolysis and may cause inaccuracies within the CBC as well. High-quality sample collection is imperative.

IMAGING

Abdominal radiographs and ultrasound can be used to rule out secondary causes for hemolytic anemia such as neoplasia and metallic foreign material which could represent zinc toxicity. Thoracic radiographs may reveal evidence of heartworm disease which can also cause hemolytic anemia, or a decreased vascular pattern associated with pulmonary thromboembolism may also be seen.

DIAGNOSTIC TESTING

The direct Coombs test is the diagnostic test for canine IMHA. This test detects antibodies that are attached to the RBC and is performed as a series of dilutions until agglutination occurs. False positive results are rare, but a negative result can occur as a result of low amounts of circulating antibodies and does not rule out IMHA. Autoagglutination will negate the need to submit a direct Coombs and it is important to identify. A slide agglutination test should be performed to rule out autoagglutination, and microscopic evaluation should follow to rule out rouleaux formation. It should be noted that a direct Coombs performed following a blood transfusion will not be accurate.

Blood smear evaluation can be a helpful diagnostic tool. Spherocytes are caused by the incomplete destruction of RBC during the elimination process. Spherocytosis occurs in 95% of IMHA patient and is easily seen a blood smear. Platelets and RBC parasites should also be considered when evaluating a blood smear in an IMHA patient. Evidence of regeneration should also be noted by identifying the presence of polychromasia, anisocytosis, and nucleated RBCs.

ADDITIONAL TESTING

There are several infectious causes for hemolytic anemia including Anaplasma, heartworm disease, Ehrlichia, and Rocky Mountain spotted fever. Infectious disease testing via tick borne disease panels, titers, and ELISA tests is performed to rule out infectious hemolytic anemia.

Complications of IMHA include PTE and DIC. Additional testing to help diagnose these potential complications includes blood gases, pulse oximetry, and coagulation testing (PT/PTT). If prolonged non-regenerative anemia and/or pancytopenia, bone marrow aspiration and/or biopsy may also be performed.

TREATMENT

In cases where diagnosis is early and hospitalization is not needed, treatment of canine IMHA consists of oral immune suppressive therapy. Prednisone, a glucocorticoid, is the mainstay of treatment and has the fastest onset of remission (3-7 days) when compared to other immune suppressive drugs. Prednisone will inhibit the destruction of RBCs by phagocytosis and complement, as well as reduce the amount of antierythrocyte antibody production. Side effects of prednisone include polyuria, polydypsia, polyphagia, panting, poor wound healing, muscle wasting, and increased susceptibility to infection. Single drug immune suppression may not be adequate for some patients and additional medications can be required for remission.

Other commonly used immune suppressive medications include cyclosporine and azathioprine. Cyclosporine suppresses lymphocyte and autoantibody proliferation and has properties that may inhibit the phagocytic action of macrophages. It is not myelosupressive and is often used in IMHA patients with non-regenerative anemia. Monitoring of therapeutic levels is typically recommended and performed at 48 hours followed by every 2-4 weeks until remission. Side effects include vomiting, diarrhea, anorexia, and can be minimized with twice daily dosing. It is important to note that several forms with varied bioavailability exist, and dosing requirements may vary based on the type of cyclosporine. Azathioprine reduces antibody production and decreases phagocytic action. It has a much slower onset of action (2-4 weeks) and side effects include gastrointestinal upset, myelosupression, hepatoxicity, cholestatic hepatopathy, and acute pancreatitis. Monitoring serum biochemistries for changes in liver values is recommended with the use of azathioprine.

There are several medications with limited clinical data that have been, or may be used for the treatment of IMHA. Leflunomide suppresses activated B and T cells and prevents antibody production has been used in the treatment for several immune mediated diseases. The side effects, which have limited documentation, may include decreased appetite, lethargy, and self limiting gastrointestinal upset. Danazol is an anabolic steroid that has been used to treat IMHA however due to expense and hepatoxicity it is not recommended for long term usage. Mycophenolate Mofetil and Mizoribine are medications that are currently used in transplant medicine and may be future treatments for canine IMHA.

For mild to severe IMHA hospitalization and intensive care is often needed and is continued until a stable PCV is achieved. General nursing and supportive care are essential for the survival of these patients and 24 hour care is recommended. Intravenous fluid therapy is needed in most cases and will correct dehydration as well as provide diuresis which has been shown to decrease hemoglobinemia which is associated with renal damage. Patients can often present with gastrointestinal upset and antiemetic medications may also be administered. Antibiotics are not always needed unless underlying infection is present, although doxycycline is often started while tick borne disease testing is pending. Monitoring patients for clinical signs of anemia (tachypnea, tachycardia, pallor, bounding pulses, and collapse) is needed to determine if a blood transfusion should be administered. A blood transfusion in the clinically anemic patient can prolong survival while immune suppressive medications are started.

Upwards of 70-90% of IMHA patients will require a blood transfusion[7], with a high number of patients requiring multiple transfusions. Blood typing and cross matching is recommended, especially in cases where multiple transfusions are needed, however the presence of autoagglutination will interfere with the ability to accurately perform these tests. Therefore, monitoring for transfusion reactions is imperative in these patients and use of only DEA 1.1 negative blood is recommended. Oxyglobin can be used if whole or packed RBC is not available as it will increase tissue profusion. In addition serial bloodwork (CBC, PCV/TS, serum biochemistry) is performed and tracked to monitor response to treatment, potential complications, and help identify relapse.

Intensive care will also provide a setting where other treatments may be initiated that may be of benefit in IMHA patients. Other therapies that can be given in a hospitalized patient include the administration of human immunoglobulin (IV IgG), which may help inhibit RBC phagocytosis. Rapid onset of action can decrease rapid hemolysis and induce a temporary remission while other immune suppression has yet to take effect. Some chemotherapy drugs have also been used in treatment of IMHA, such as cyclophosphamide (Cytoxan) and are used for their myelosupression although limited clinical data is available. In addition, due to a hypercoagulable state that occurs with canine IMHA, empiric anti-throbolic treatment with low dose aspirin, heparin, or low-molecular weight heparin (Enoxaparin, Dalteparin) may increase survival rates.

LONG TERM CARE

Once remission is achieved long term monitoring begins and treatment with immune suppressive medications is continued for a minimum of 3-6 months. Immune suppressive medications are slowly tapered to prevent relapse. Tapering schedules will vary and are dependent on the severity of disease, response to treatment, and the severity of side effects. Frequent CBC and serum biochemistry panels are used to monitor response to treatment and check for potential side effects if adjunct immune suppressive therapy is used. Tapering of medications will not typically begin until remission is achieved, and is then approximately every 2-4 weeks with continued monitoring of hematocrit and platelet levels if Evan's syndrome is present. Prednisone is often tapered at a rate of 25% every 2-4 weeks if remission is maintained. If relapse occurs it can be treated by increasing existing medication dosages and slower tapering is recommended when remission is achieved. Other long term considerations include occasional urinalysis to monitor for subclinical urinary tract infections which has been reported to occur with immune suppressive therapy.

CLIENT EDUCATION

Client education and compliance is extremely important in the treatment of immune disorders such as IMHA. The technician's role in client education can greatly affect a clients understanding of the disease and ability to comply. Potential complications, medications and their side effects, signs of relapse, and the length of treatment of IMHA should be discussed in detail. Avoidance of medications implicated in IMHA should be stressed and if vaccination is recommended to be discontinued, discussion of testing such as titers should be mentioned. Close communication with referring clinics and other caretakers involved with the patient is a necessity. Through the collection of detailed patient history, accuracy of laboratory results, extensive nursing care, and client education the veterinary technician can play a significant role in the survival of canine IMHA patients.

REFERENCES

[1.] Ettinger, S: **Textbook of Veterinary Internal Medicine**:2000; ed5:1793-1797, [2.]Cohan, M: **Vet Tech:**2005;26(9), [3.]Balch, A, Mackin A: **Compendium**: 2007; 29 (4), [4.]Hansen, K; **Vet Tech**; 2006; 27(7), [5.]Drellich, S, Bracker, K; **Compendium**: 2005; 27(7), [6.]Al-Ghazlat, S: **Compendium**: 2009, 31 (9), [7.]Sharp, C, Kerl, M: **Standards of Care**; 2008; 10(10), [8.]Hendrix, C, Sirois, M: **Laboratory Proceedures for Veterinary Technicians**; 2002, 279-282, [9.]Colville, T, Bassert, J; **Clinical Anatomy and Physiology for Veterinary Technicians**: 2008, 242-245, [10.]Wanamaker, B, Massey, K: **Applied Pharmacology**: 2009; 344-345

THE TECHNICIAN'S ROLE IN THE PREVENTION & CONTROL OF DISEASE IN THE SMALL ANIMAL HOSPITAL

Meri Hall, RVT, LATG, VTS (SAIM)
Athens, PA

Veterinary Technicians are an integral part of defense in the prevention of disease transmission. Isolation protocols, cleaning protocols and client education are all areas where the veterinary technician provides assistance to the veterinary team. Knowledge of the disease process is required to make informed day-to-day decisions. Veterinary technicians should be aware of the mode of transmission for individual diseases. The modes of transmission for viruses, bacteria and fungus are aerosolization, gastrointestinal (fecal/oral), vectors, fomites and direct contact. For hospitalized patients, aerosolization and fomites are of prime concern. Aerosolized particles can travel up to 20 feet and many microorganisms survive in the environment for hours to days. Good nursing care is imperative for prevention of nosocomial infections for all hospitalized patients. Postoperative patients have a lowered immune system and are very susceptible to bacterial infections. Animals that are unable to shift their weight, such as dogs with intravertebral disk disease should be turned to prevent decubital ulcers.

Human health care workers have adopted a universal protocol when treating patients. The premise is that all patients are potentially infectious and are treated as such until proven otherwise. This practice should be used in the veterinary setting as well. The use of some simple precautions will prevent the transmission of disease. Each time an animal is presented to the veterinary hospital, we should ensure the vaccine status of each patient. Obtaining good histories of animals with potential infectious disease can prevent the spread of disease to other animals.

If possible, dedicated examination rooms should be used for animals with immune compromised status, chemotherapy patients. Animals presented that are obviously ill with vomiting, diarrhea, coughing or upper respiratory symptoms should be placed into an examination room immediately to prevent animals in the waiting room to limit any contact with healthy animals. Use of effective hand washing, cleaning contact surfaces of examination tables, equipment and wearing gowns or jackets help to prevent the spread of disease. These patients should not be removed from the examination room (unless the animal is in distress) until it is determined that the cause of their symptoms is not due to infectious agents. Contact with minimal staff is suggested to prevent the potential for spread of disease to other patients. Staff who have interacted with these patients should limit their contact with any patients who could potentially become infected for the rest of their shift as much as possible.

Each hospital should have an isolation protocol and should be reviewed frequently to ensure that there are no "breaks" in the system. New staff should be thoroughly trained on the protocols before working with an infectious patient. It is recommended that the isolation protocol be posted outside and inside of the isolation area to aid the staff in adhering to the procedures.

Dedicated equipment, use of Personal Protective Equipment and disposable supplies should be used within the isolation area. No other equipment should be stored within this area. If equipment cannot be dedicated for isolation, it must be cleaned and disinfected prior to using it on another patient.

The following table is a sampling of some common microorganisms of importance to veterinary medicine.[, 3, 4, 5]

Microorganism	Mode of Transmission	Temperature	Time	Agent	Time
Feline rhinotracheitis virus	Aerosol	4°C / 39.2°F 25°C / 77°F 37°C / 98.6°F	154 days 33 days 3 hours	Heat 56°C/ 132.8°F 70% Alcohol Sodium hypochlorite 100% Iodine Phenol 0.5% Quat Formaldehyde Glutaraldehyde 0.78% Big	4-5 minutes 10 minutes 10 minutes 10 minutes 10 minutes 10 minutes 10 minutes 10 minutes 10 minutes
Canine Herpesvirus	Direct Contact Fomites	4°C / 39.2°F 37°C / 98.6°F	48 hours 48 hours	Heat-56°C / 132.8°F	4-5 minutes
Canine Distemper virus	Aerosol Body fluids Transplacental	-65°C/-85°F 4°C / 39.2°F 25°C / 77°F	7 years 7-8 weeks 3 hours	Heat- 60°C / 140°F Formaldehyde 0.2 % Quats	30 minutes 120 minutes 30 minutes
Canine Parvovirus	Fecal/oral	25°C / 77°F Very Stable in the environment	3 months or more	Heat-56°C / 132.8°F Sodium hypochlorite	15 minutes 15 minutes
Feline Panleukopenia virus	Fecal/oral	Very Stable in the environment	3 months or more	0.05% Phenol Sodium hypochlorite Formaldehyde Glutaraldehyde 0.78% Big	Rapid 10 minutes 10 minutes 10 minutes 10 minutes
Infectious Canine Hepatitis	Fecal/oral	4°C / 39.2°F 25°C / 77°F	2 months 14 days	Heat- 60°C / 140°F Ultraviolet	30 minutes 120 minutes

		37°C / 98.6°F	6 hours	Formaldehyde Sodium hypochlorite 100% Iodine Phenol	24 hours 10 minutes 10 minutes 10 minutes
Feline Calicivirus	Aerosol	7°C -20°C / 44.6°F - 68°F	10 days	Heat-50° C / 122° F Phenol (at 50%) Sodium hypochlorite Formaldehyde Glutaraldehyde	30 minutes 10 minutes 10 minutes 10 minutes 10 minutes
Canine and Feline Rotavirus	Fecal/oral	24°C / 75.2°F 37°C / 98.6°F	2 days 19 hours	Heat-60° C / 140° F 70% Alcohol Formaldehyde	30 minutes 25 minutes 56 minutes
Chlamydia	Aerosol Ocular discharge	0°C / 32°F 25°C / 77°F	24 hours 7 days	Heat- 60° C / 140° F 0.5% Quats	15 minutes 10 minutes
Mycoplasma	Aerosol	28° C / 82.4° F	21 days	Heat-55° C / 131° F 1% Phenol Formaldehyde	15 minutes <5 minutes <5 minutes
Gram + Staphylococci	Direct Contact Aerosol	4°C / 39.2°F	Several months	Heat-60° C / 140° F 1% Phenol	30 minutes 15 minutes
Brucella	Direct Contact	25°C / 77°F	3 months	Heat-60° C / 140° F 1% Phenol	15 minutes 15 minutes
Salmonella	Fecal/oral	25°C / 77°F	12-14 weeks	Heat-60° C / 140° F	20 minutes
Clostridium Spores		Very resistant in the environment		Heat-120° C /248 ° F 5% Phenol	10 minutes 10-12 hours
Leptospira	Direct Contact Urine	0-25° C / 32-77° F		Heat- 50° C / 122° F Iodine	10 minutes 10 minutes
Toxoplasma cysts	Fecal/oral	4°C / 39.2°F	68 days	Heat-56° C / 132.8° F Freeze	10-15 minutes
Isospora	Fecal/oral	25°C / 77°F	24 hours	Heat- 50° C / 122° F	4 hours
Canine Influenza H3N8	Aerosol Fomites	Surfaces Clothing Hands	48 hours 24 hours 12 hours	Sodium hypochlorite Quats	10 minutes 10 minutes
Canine Infectious Tracheobronchitis	Aerosol Fomites	28° C / 82.4° F	21 days	Heat-55° C / 131° F 1% Phenol Formaldehyde 0.78% Big	15 minutes <5 minutes <5 minutes 10 minutes
Influenza H1N1	Aerosol Fomites	Paper/cloth Surfaces Wet areas	8-12 hours 24-48 hours Up to 72 hours	Sodium hypochlorite Quats	10 minutes 10 minutes
Methacillin Resistant Staphylococcus Aureus	Direct Contact	4°C / 39.2°F	Several months	Heat-60° C / 140° F 1% Phenol Sodium hypochlorite Quats	30 minutes 15 minutes 15 minutes 15 minutes
Methacillin Resistant Staphylococcus Intermedius	Direct Contact	4°C / 39.2°F	Several months	Heat-60° C / 140° F 1% Phenol Sodium hypochlorite Quats	30 minutes 15 minutes 15 minutes 15 minutes

Sodium hypochlorite- household bleach Big- Biguanides Quats- Quaternary ammoniums Heat- Dry or Steam

There are some organisms that can survive for months in the environment. Removal of gross organic material is needed prior to sanitizing or disinfecting surfaces. Disinfection is the process of destroying all living tissue. Sanitization reduces the number of infective organisms to a level where they are considered safe from a public health standpoint.[1] Not every agent is effective against the microorganisms that cause disease. In addition, some agents have to be used at differing concentrations and contact times to be effective against certain microorganisms. Always check the product label for concentration, application method and contact time. Cleaning and disinfecting protocols should be reviewed on an annual basis or if there is an outbreak of disease in the veterinary hospital.

Floors, food bowls and bedding should be thoroughly cleaned and disinfected daily. Walls, cages and equipment should all be thoroughly cleaned and disinfected at least weekly. Examination tables should be sanitized between patients. Any examination room used for infectious patients should be thoroughly disinfected before using it again. This includes any surface that the owner has had contact with. This would include the reception areas and scales.

Any food bowls used for infectious patients should be disinfected in the isolation area and used for that patient during their hospitalization. Bedding should be washed separately and disinfected before being returned to genera use. Disposable shoe covers and footbaths with the appropriate agent should be used to prevent staff from carrying the microorganism throughout the hospital.

Disinfecting and sanitizing agents are divided into physical and chemical. Physical agents are heat- either dry or steam and light. All laundry in the veterinary hospital should be washed in hot water and the addition of the appropriate chemical agent should be added to the water, or alternatively the laundry can be soaked in the agent prior to washing.

Chemical agents are divided into eight main categories. There are several cleaning agents commercially available within each category. Some agents can be used for both sanitizing and disinfecting dependent upon the concentration or contact time.

Agent	Concentration	Presence of Organic Material
Alcohols	50-95% Ethanol, isopropyl	Ineffective
Aldehydes	Formaldehyde Glutaraldehyde	Some effect
Halogens	1:30 (sodium hypochlorite-bleach)	Ineffective
	Iodine	May be reduced
	Iodophor (betadine) 1-10% (1:9)	May be reduced
Oxidizing Agents	5-20% Hydrogen Peroxide Peracetic Acid	Some effect
Phenolics	Pine Sol (5% solution)	Effective
Quaternary ammonium	Roccal® D-256® Benzalkonium chloride	Not effective against non-enveloped viruses
Biguanides	Chlorhexidine (0.78%)	Ineffective
Ethylene Oxide		

Good client education is imperative to prevent the spread of disease. Educating owners is every veterinary team member's responsibility. Clients who have animals with infectious diseases should be given direction on disinfection and how to prevent the spread of disease to their other pets or future pets. If an animal has a zoonotic disease it is very important that those clients are given information on how to reduce their own risk of acquiring the disease. Special attention to clients whose animals are undergoing chemotherapy or are immune compromised is imperative. It is heartbreaking when animals become ill due to preventable circumstances.

Because each facility is unique, the staff of your facility must determine the protocol used. There are many resources available for the veterinary community to assist each facility in designing a specific protocol.

REFERENCES

[1.] Dvorak, Glenda, Disinfection 101.Available at: www.cfsph.iastate.edu/BRM/resources/Disinfectants/Disinfection101Feb2005.pdf. Assessed on 14 January, 2009. [2.] Greene, Craig E., **Infectious Diseases of the Dog and Cat**, W.B. Saunders, 1990. [3.] Crawford C. **Canine influenza: Frequently asked questions from veterinarians**. *University of Florida*; 2009. Available at: http://www.vetmed.ufl.edu/college/pr/documents/CanineinfluenzaFAQ.Veterinarians_000.pdf. Assessed on 14 January, 2009. [4.] www.enichols.com/downloads/cleaning_to_reduce_the_risk_of_h1n1_flu_virus.pdf. Assessed on 14 January, 2009. [5.] www.cdc.gov/ncidod/dhqp/ar_MRSA.html. Assessed on 14 January, 2009.

TRANSFUSION PRACTICE FOR HORSES

Sean Owens, DVM, DACVP
Davis, CA

INTRODUCTION

Blood transfusion is commonly performed in horses to correct anemia arising from acute blood loss secondary to trauma, surgical complications, ruptured uterine artery, guttural pouch mycosis, and neonatal isoerythrolysis.[1] The purpose of blood transfusion is to restore oxygen delivery to the tissues by increasing hemoglobin concentration and/or the restoration of circulating blood volume. Historically, whole blood has been used for the correction of anemia in equine patients. Until recently, little was published evaluating the viability of packed red blood cells (PRBCs) for equine transfusion. In 2008, Niinisto and colleagues found that equine erythrocytes (PRBCs) could be stored for at least 35 days before transfusion when stored in a saline-adenine-glucose-mannitol (SAG-M) solution.[2]

Little has been published on the post-transfusion survival of allogeneic or autologous RBCs in horses, or the effects of RBC storage solutions on post-transfusion RBC survival. In one study, the half-life of transfused autologous RBCs in foals, ranged from 7 to 18 days with a mean of 12 days.[3] Further, the lifespan of transfused allogeneic RBCs has been reported to be <4 days.[4,5]

Recently Owens and colleagues, using Biotin-NHS as a red cell label, found that the mean post-transfusion half-life of transfused autologous RBCs in adult Thoroughbred horses was 50, 45, 33 and 29 days for RBCs stored in CPDA-1 prior to transfusion, for 0, 1, 14 and 28 days respectively.[6] They concluded that autologous equine RBCs (whole blood) stored in blood bags containing CPDA-1 for up to 4 weeks are acceptable for transfusion. This work supports findings by Mudge[7] and colleagues suggesting that commercial CPDA-1 blood storage bags appear to be the optimal storage method for equine whole blood. Additionally, if a whole blood transfusion is being considered for a thrombocytopenic patient, standard blood collection bags are preferred as glass bottles containing ACD activate platelets by binding them to the glass surface.

In human medicine, the decision to transfuse is often based upon a "transfusion trigger." However, no such "trigger" has been validated in veterinary medicine for equine patients. In any event, transfusion therapy should be reserved for those patients with life threatening lack of oxygen delivery to tissues. Clinical signs consistent with poor oxygen delivery include tachycardia, tachypnea, pale mucus membranes, lethargy and weakness. As a guideline, transfusion therapy should be considered in unstable patients with a packed cell volume (PCV) < 12%, acute loss of 30-40% of circulating blood volume, or blood loss associated with collapse or ongoing hemorrhage.

The choice between transfusion of whole blood or PRBCs is influenced by the type of anemia (normovolemic vs. hypovolemic) and products available for transfusion. For instance, whole blood is preferred for animals with hemorrhagic disease, whereas for horses with anemia attributable to hemolytic disease or erythropoietic dysfunction, transfusion of PRBCs is often sufficient to improve tissue oxygenation while minimizing the likelihood of volume overload.[7]

PRETRANSFUSION TESTING

Horses have eight RBC groups or systems: A, C, D, K, P, Q, U, and T, of which the first seven systems are recognized by the International Society of Animal Blood Grouping Research.[8] One of the main functions of this society is to standardize the nomenclature of equine blood groups and to give international recognition to new specificities that have been determined by at least two of the participating laboratories. Blood groups are named with an upper case letter to denote the system and a lowercase letter to designate the factor, with factors representing specific antigenic sites within each blood group.

More than 30 different factors have been identified within these seven equine systems, and many more have been identified experimentally. Red cell antigens Ca, Aa, and Qa are clinically important for their role in transfusion reactions and neonatal isoerythrolysis. While there is no universal equine blood donor, the preferred donor is one that is negative for factors Aa, Qa, and Ca, in part to prevent inadvertent sensitization of brood mares against the two most common alloantigens involved in neonatal isoerythrolysis: Aa and Qa.[9]

Approximately 90% of horses have no naturally occurring RBC antibodies. However, of the 10% that do, Aa and Ca antibodies occur most often.[10] Equine red cell antigen alloantibodies are of two types, naturally occurring and acquired, and are either agglutinating, hemolytic, or both. Aa and Ca are both agglutinating and hemolytic antibodies, while Qa is a hemolytic anti-body. Horses that lack the Aa antigen frequently have anti-Aa serum antibodies, and horses that lack the Ca antigen typically have anti-Ca antibodies. One can therefore assume that any horse negative for factors Aa and Ca may have anti-Ca and anti-Aa serum antibodies. Anti-Aa antibodies are highly immunogenic and are capable of causing acute, severe reactions following incompatible transfusions.

Pre-transfusion testing is indicated to minimize the risk of incompatible red cell transfusions and to minimize immunizing transfusion recipients against red cell antigens they may lack. Clinically important antibodies against RBCs have been shown to develop in equine subjects following a single homologous blood transfusion.[11] Pre-transfusion testing consists of donor and recipient blood typing in conjunction with a blood crossmatch.

While most laboratories can perform a pre-transfusion crossmatch, current blood-typing methods are time-intensive, requiring 2.5–3.0 hours to complete, and need to be performed at 37°C, making this testing methodology impractical in an emergency setting. Given this difficulty, Owens and colleagues at University of California, Davis developed a rapid (15 minute) blood typing protocol, that is conducted at room temperature, for detecting the presence of Ca and Aa red cell surface antigens using equine-derived antisera.[12] This technique may be beneficial for pre-transfusion testing of equine patients in an emergency setting where complete equine blood typing services are not available.

At a minimum, a crossmatch should be performed prior to every red blood cell transfusion in equine patients. And, optimally, donor blood types should be known, transfusion recipients should be blood typed and a crossmatch performed prior to the administration of red cell products.

DONORS

With seven blood groups containing more than 30 different antigens (factors), more than 400,000 red cell phenotypes are possible in horses, which highlights the difficulty in selecting compatible donors for equine red cell transfusions.[13] Currently, the preferred blood donor is Aa-, Ca-, and Qa-negative. The health of the donor is vitally important. Vaccinations should be current for all donors and donors should be on a regular deworming schedule. Donors should have a negative Coggins test, as well as regularly scheduled care for their teeth, hooves, and sheaths. A complete blood count and serum biochemical profile should be performed annually on each donor. Screening for infectious diseases should based upon potential for exposure and the region of the country in which the donor is housed. Disease screening may include: Equine Infectious Anemia, EEE, WEE, WNV, EHV1, EHV4, Potomac Horse Fever, EVA, Babesia and VEE.

Blood donors should not have a history of whole blood transfusion, PRBC transfusion or previous pregnancies. When choosing a potential donor horse, thorough assessment of disease status (e.g., testing to detect EIA), size of the donor horse, and potential volume of blood to be harvested should be considered. It would be ideal to know the blood type of both the donor and recipient; at the least, the crossmatching compatibility status should be assessed to determine compatibility and minimize potential adverse reactions.

BLOOD COLLECTION

Four to eight liters of whole blood can be collected from a 450 Kg horse if the donor is healthy, has a hematocrit >40% and has not donated within the past 30 days. Blood should be collected into commercially available polyvinyl chloride blood bags containing CPDA-1. CPDA-1 has been shown to be the optimal solution for preserving equine red cell viability during storage.[14] Blood should be collected aseptically from a jugular vein, optimally through a large bore (10-14 gauge) catheter or needle.

ADMINISTRATION OF BLOOD AND PLASMA

Administer whole blood, PRBCs or plasma to the recipient using a standard blood administration set. Prior to administration, plasma should be thawed slowly using water that is no hotter than 102-104 degrees Fahrenheit. Blood products should never be thawed or warmed using a microwave oven. The blood administration set should be replaced after every 4 liters administered.

It is recommended to start the transfusion at an initial drip rate of 0.1 ml/kg body weight for the first 10 min using the standard blood delivery sets. Vital signs and temperature fluctuation should be monitored every 2 min. The recipient should be monitored continuously during the transfusion and the transfusion stopped if the recipient has any change in vital signs, paying close attention to any adverse reactions such as tachypnea, dyspnea, tachycardia, piloerection, urticaria, muscle fasciculations, restlessness, or sudden collapse. If temperature and vital signs do not change after the first 10 minutes, then the transfusion may be continued at a rate of 20 –30 ml/kg/hr. Following guidelines established by the American Association of Blood Banks, once a blood administration set is "spiked" into the blood bag, the transfusion of that product should be completed within 4 hours.

Should an adverse reaction develop, stop the transfusion, start IV isotonic crystalloid solutions, alert the patient's primary care clinician and begin the process of determining the mechanism underlying the adverse reaction.

TRANSFUSION REACTIONS

Anaphylactic reactions occur more commonly and are more severe when repeated transfusions from a single donor are administered. Potential complications of blood transfusion in horses and other species include transfusion reactions and transmission of infectious diseases. Transfusion reactions can vary in severity from mild urticarial reactions that require no intervention to acute hemolytic reactions to anaphylactic shock that requires emergency resuscitative procedures. Acute hemolytic transfusion reactions are rare. However, if suspected, the transfusion should discontinued, and fluid therapy instituted. The patient's serum should be checked for hemolysis and blood type and crossmatch results should be double-checked.

An anaphylactic reaction is very serious. In the event of an anaphylactic reaction, discontinue the transfusion immediately. Appropriate drug therapy and, in some instances, CPR may be required. Fortunately severe, life threatening transfusion reactions are infrequent. In any event, horse owners should be made aware of the potential risks associated with transfusion therapy prior to administration of blood products.

Despite the identification of erythrocyte surface antigens and crossmatching methods, there are still transfusion reactions in horses. In 2007, Hurcombe and colleagues reported adverse reactions in 7 of 44 transfusions (16%) in 31 cases in a retrospective study. Reactions ranged from urticarial reactions to anaphylactic shock. The authors concluded "the adverse effects of transfusion administration can be evident for all disease states and clinically important even when a compatible donor-recipient agglutination crossmatch status is determined."[15] The recognized risk of transmitting infectious disease in horses through transfusion is low, when compared with that reported for humans.

CONCLUSION

In summary, the following key points are important to remember with regard to equine blood transfusions: 1) The half-life of autologous equine red blood cell transfusions is greater than previously reported, 2) Equine whole blood and PRBCs may be stored for up to 35 days in CPDA-1, 3) Commercially available polyvinyl chloride blood bags containing CPDA-1 optimize storage of equine

RBCs, 4) A rapid equine blood typing protocol exists that may be used for emergent equine patients, 5) All equine patients should be, at a minimum, crossmatched prior to blood transfusion, 6) If a transfusion reaction is suspected, the transfusion should be stopped until the cause has been determined, and lastly, 7) In a perfect world all equine blood donors would be blood typed, all equine transfusion recipients would be blood typed, and a blood crossmatch would precede all blood transfusions.

REFERENCES

[1]Owens SD, et al. **Vet Clin Pathol** 2008;(37)1:49. [2]Niinisto K, et al. **The Veterinary Journal** 2008;(176):227. [3]Smith JE, et al. **JVIM** 1992;(6)3:183. [4]Kallfelz FA, et al. **Am J Vet Res** 1978;(39)4:617. [5]Gimlette TMD. **J Royal Soc Med** 1978;(71):576. [6]Owens SD, et al. **Am J Vet Res** 2010;In Press. [7]Mudge MC, et al. **Vet Surg** 2004(33):475. [8]The International Society for Animal Blood Group Research. Proceedings of the 20th International Conference on Animal Blood Groups and Biochemical Polymorphisms. **Anim Genet** 1987;18(suppl 1):1. [9]Bailey e. **Am J Vet Res** 1982;(43):1917. [10]Sellon D. **Schalm's Veterinary Hematology**. 5th ed. 2000:849. [11]Wong PL, et al. **Am J Vet Res** 1986;(47)2:2566. [12]Owens SD, et al. **Vet Clin Pathol** 2008;(37)1:49. [13]Slovis, NM et al. **Proc Assoc Equine Practnr** 2001;(47):266. [14]Mudge MC, et al. **Vet Surg** 2004(33):475. [15]Hurcombe SD, et al. **J Am Vet Med Assoc** 2007;(231)2:267.

ACUPUNCTURE & ANALGESIA FOR THE EQUINE PATIENT

Dr. Sarah le Jeune, DVM, DACVS, DECVS, CVA
Davis, CA

Acupuncture and Traditional Chinese Veterinary Medicine (TCVM) are being offered more and more commonly as an adjunctive clinical service to equine and small animal patients. Multiple veterinary institutions including the University of California, the University of Florida, Oregon State University, Virginia Tech, and Colorado State University, to name a few, are including acupuncture and TCVM in their clinical services and curriculum.

Acupuncture has been used to treat various medical conditions in humans and animals for over 3000 years in China[1]. Horses were among the first animals treated with acupuncture due to their importance in warfare and farming. Acupuncture and TCVM can easily be integrated into conventional diagnostic and treatment modalities to optimize clinical outcome.

Acupuncture and TCVM take into consideration the animal as a whole and not just as an isolated system; hence, they are often referred to as "holistic" medicine. The ancient Chinese theory is based on the premise that there are patterns of vital force called *Qi* throughout the body, which are related to the internal organs and musculoskeletal system. Pain or illness result when the flow of *Qi* is disrupted due to pathological factors, such as bacteria and viruses, trauma, stress, poor diet or other conditions. Acupuncture focuses on restoring the flow of *Qi* in the body by inserting very fine needles (acupuncture needles) into the skin at particular acupuncture points. Stimulation of these points resolves the blockage of *Qi*, relieving pain and restoring the body's ability to heal itself[1].

The ancient Chinese discovered 361 acupuncture points in humans and 173 in animals[1]. Modern research shows that acupoints are located in areas with a high density of free nerve endings, mast cells, small arterioles and lymphatic vessels[2]. Stimulation at acupoints induces the release of beta-endorphins, serotonin and other neurotransmitters[3,4,5]. Therefore, acupuncture's use for pain relief is well supported and elucidated by modern research studies. Acupuncture's effects on internal organs and on "balancing" the body as a whole is less well understood and more research is necessary to fully explain this ancient therapy.

The benefits of acupuncture have been widely recognized and integrated into Western veterinary and human medical practices. In November 1997, the National Institutes of Health (NIH) convened a panel of scientists to review the effectiveness of acupuncture. The result was the first formal endorsement of acupuncture by the NIH, stating: "There is sufficient evidence of acupuncture's value to expand its use into conventional medicine and to encourage further studies of its physiology and clinical value."[6]

Clinical trials indicate that acupuncture therapy can be effective as an adjunctive treatment in the following conditions[7-28]:
Musculoskeletal problems: Muscle soreness, back pain, neck pain, osteoarthritis, degenerative joint disease, obscure lameness, tendon/ligament problems, laminitis.
Neurological disorders: Seizure, laryngeal hemiplegia, facial and radial nerve paresis.
Gastrointestinal disorders: Diarrhea, impaction, chronic colic, gastric ulceration.
Other chronic conditions: Heaves (chronic obstructive pulmonary disease, recurrent airway obstructions, asthma), anhydrosis, uveitis, behavioral problems, Cushing's disease, hypo- or hyperthyroidism, infertility, renal failure, geriatric weakness, skin problems.
Performance enhancement and prevention of disease

Physiological effects induced by Acupuncture include pain relief, regulation of gastrointestinal motility, anti-inflammatory effects, immuno-regulation, endocrine and reproductive regulation, antipyretic effects and promotion of microcirculation[1,29].

Acupuncture methods include a diagnostic exam and treatment[1,29].
Diagnosis: An acupuncturist's diagnosis is based in part on conventional western methods: obtaining a thorough medical history, discussing the primary complaint and goals for improvement, performing a physical exam, blood work, radiographs, ultrasound etc. However, particular attention will be given to the tongue color and pulse quality of the animal and a diagnostic acupoint palpation exam will be performed which will identify particularly painful areas along acupuncture meridians translating to disease patterns.
Treatment: Modern acupuncture needles are 0.5 to 3 inches long, ultra-fine, and very flexible stainless steel. They are pre-sterilized, non-toxic and disposable. Depending on the condition being treated, 10 to 20 needles will be inserted and left in place for 5 to 30 minutes. Stimulation of acupoints by needles can be done by rotating the needles, or attaching electrodes to send a weak electrical current through the needles (electroacupuncture). Heating the needles with a burning moxa herb (moxabustion) may be indicated in certain conditions.

REFERENCES

1. Xie H: **Traditional Chinese veterinary medicine**, Beijing 1994, Beijing Agricultural University Press 2. Heine H: Anatomische Struktur der Akupunturpunkte, Dtsch Zschr Akup 31: 26-30, 1988 3. Pomeranz B: **Scientific basis of acupuncture**. In Stux G, Pomeranz B: Basics of Acupuncture, ed 4, Berlin 1998, Springer, pp 6-47 4. Chang HT: **Neurophysiological interpretation of acupuncture and analgesia**, Endeavor 4:92-96, 1980 5. Han JS, Terenius I: Neurochemical basis of acupuncture analgesia, **Ann Rev Pharmacol Toxicol** 22:193-220, 1982 6. NIH Consensus Statement Online 1997 November 3-5;15(5) http://odp.od.nih.gov/consensus/cons/107/107 statement.htm 7. Kim MS, Xie H. Use of electroacupuncture to treat laryngeal hemiplegia in horses. **Vet Rec**. 2009 Nov 14;165(20):602-4 8. Schofield WA. Use of acupuncture in equine reproduction. Theriogenology. 2008 Aug;70(3):430-4. **Epub** 2008 Jun 11 9. Habacher G, Pittler MH, Ernst E. Effectiveness of acupuncture in veterinary medicine: systematic review. **J Vet Intern Med**. 2006 May-Jun;20(3):480-8 10. Lankenau C. The effects of a single acupuncture treatment in horses with severe recurrent airway obstruction. **Equine Vet J.** 2006 Mar;38(2):109 11. Angeli AL, Joaquim JG, Takahira RK, Bulla C, Luna SP. Effect of acupuncture, electroacupuncture and Panax pseudoginseng on blood coagulation variables in horses. **Vet Rec**. 2005 Nov 19;157(21):662-4 12. Xie H, Colahan P, Ott EA. Evaluation of electroacupuncture treatment of horses with signs of chronic thoracolumbar pain. **J Am Vet Med Assoc**. 2005 Jul 15;227(2):281-6 13. Skarda RT, Muir WW 3rd. Comparison of electroacupuncture and butorphanol on respiratory and cardiovascular effects and rectal pain threshold after controlled rectal distention in mares. **Am J Vet Res.** 2003 Feb;64(2):137-44 14. Boldt E Jr. Use of complementary veterinary medicine in the geriatric horse. **Vet Clin North Am**

Equine Pract. 2002 Dec;18(3):631-6, ix [15.] Skarda RT, Tejwani GA, Muir WW 3rd. Cutaneous analgesia, hemodynamic and respiratory effects, and beta-endorphin concentration in spinal fluid and plasma of horses after acupuncture and electroacupuncture. **Am J Vet Res**. 2002 Oct;63(10):1435-42. Erratum in: **Am J Vet Res**. 2002 Nov;63(11):1584 [16.] Merritt AM, Xie H, Lester GD, Burrow JA, Lorenzo-Figueras M, Mahfoud Z. Evaluation of a method to experimentally induce colic in horses and the effects of acupuncture applied at the Guan-yuan-shu (similar to BL-21) acupoint. **Am J Vet Res**. 2002 Jul;63(7):1006-11 [17.] Fleming P. Nontraditional approaches to pain management. **Vet Clin North Am Equine Pract**. 2002 Apr;18(1):83-105, vii. [18.] Chan WW, Chen KY, Liu H, Wu LS, Lin JH. Acupuncture for general veterinary practice. **J Vet Med Sci.** 2001 Oct;63(10):1057-62 [19.] Nie GJ, Goodin AN, Braden TD, Wenzel JG. Luteal and clinical response following administration of dinoprost tromethamine or cloprostenol at standard intramuscular sites or at the lumbosacral acupuncture point in mares. **Am J Vet Res**. 2001 Aug;62(8):1285-9 [20.] Sumano Lopez H, Hoyas Sepulveda ML, Brumbaugh GW. Pharmacologic and alternative therapies for the horse with chronic laminitis. **Vet Clin North Am Equine Pract**. 1999 Aug;15(2):495-516, viii [21.] Dill SG. Acupuncture for gastrointestinal disorders. **Probl Vet Med**. 1992 Mar;4(1):144-54 [22.] Janssens LA. Trigger point therapy. **Probl Vet Med**. 1992 Mar;4(1):117-24 [23.] Steiss JE, White NA, Bowen JM. Electroacupuncture in the treatment of chronic lameness in horses and ponies: a controlled clinical trial. **Can J Vet Res**. 1989 Apr;53(2):239-43 [24.] White SS, Christie MP. Acupuncture used as an adjunct in the treatment of a horse with tetanus. **Aust Vet J.** 1985 Jan;62(1):25-6 [25.] Landholm JE, Mills LL. Use of acupuncture in treatment of laminitis in a horse. **Vet Med Small Anim Clin.** 1981 Mar;76(3):405-7 [26.] Feng KR. A method of electro-acupuncture treatment for equine intestinal impaction. **Am J Chin Med**. 1981 Summer;9(2):174-80 [27.] Hwang YC, Held JP. A preliminary report on an attempt to modify behavior of untamed ponies by acupuncture. **Am J Chin Med**. 1981 Spring;9(1):84-92 [28.] Gideon L. Acupuncture: clinical trials in the horse. **J Am Vet Med Assoc**. 1977 Jan 15;170(2):220-4 [29.] Schoen AM: **Veterinary Acupuncture: Ancient Art to Moderns Medicine**, ed 2, St. Louis 2001, Mosby Inc.

THERAPEUTIC OPTIONS IN ATRIAL FIBRILLATION

M. Kimberly J. McGurrin, BSc, DVM, DVSc, DACVIM
Guelph, ON, Canada

In the majority of cases, horses with atrial fibrillation (AF) have no apparent structural heart disease and no accompanying systemic illness. Establishment of sinus rhythm in these animals usually completely restores cardiovascular function and allows the animal to be returned to performance. Quinidine has been the primary option in the management of equine AF. Quinidine has low efficacy in chronic AF and has numerous side effects, additionally the agent is becoming more difficult to obtain. Over the last 5-7 years there has been increased interest into alternative or adjunctive options.

QUINIDINE

Quinidine is a Class Ia agent (primarily Na channel blocker). Oral and intravenous formulations are available. Quinidine sulphate is administered via nasogastric tube. Cannot be administered orally due to irritation. Quinidine gluconate is administered intravenously. It is important to monitor the ECG throughout treatment. Heart rate in excess of 80 bpm, widening of QRS beyond 125% of baseline, and abnormal complexes are all indicators to discontinue medication. Toxic effects are variable. Mild signs include nasal edema and mild depression. More severe signs include marked ataxia, hypotension, colic, diarrhea, seizures, sustained tachycardia, syncope and sudden death. Adverse effects are not necessarily dose dependent. Hypokalemia increases risk of torsades.

FLECAINIDE

Flecainide is a class Ic agent and has been investigated as a potential pharmacologic alternative treatment. It has a relatively higher efficacy and lower side effect profile in recent onset AF in humans. In horses, favourable response has been noted in short duration (15 minutes) experimentally induced AF following intravenous administration. However, clinically, the agent has not been reported as effective intravenously (none of 10 horses with naturally occurring AF responded and two had potentially dangerous dysrhythmias). There has been only one report of its successful use orally. This agent has proven pro-arrhythmic in clinical use and caution should be considered if this agent is nevertheless elected.

AMIODARONE

Amiodarone has a variety of actions but is generally considered as a class III agent (potassium channel blocker) and has been considered to be moderately effective in the management of human AF and has been noted to control rate even if it does not restore SR. It has been associated with worsening of pulmonary disease and causing pulmonary fibrosis. Neuropathies, bradycardia and hepatotoxicity have also been reported. Most of these concerns are heightened in humans because of chronic use. Oral bioavailability is low in the horse. There have been two reports on its (IV) administration in horses. Sinus rhythm was restored in 4/6 and 3/6, respectively. However, 3/6 horses had hindlimb weakness and 3/6 horses had transient increases in serum bilirubin levels in one study and 3/6 had diarrhea in the second. A treatment period of greater than 36 hours was associated with the adverse effects in both studies.

PROPAFENONE

Propafenone a class Ic agent has been reported to have high efficacy in recent onset AF in humans. A recent study investigating the intravenous use of propafenone in horses with AF of approximately 3 months duration (2 horses with natural AF, 4 with induced AF) reported no horses responded to treatment, despite levels within the low therapeutic range, and changes in the atrial fibrillation cycle length (indicating slowed conduction). The horses were subsequently treated successfully with quinidine.

DILTIAZEM

This class IV (calcium channel blocker) agent slows conduction velocity at the AV node. Diltiazem is not proposed as an antiarrhythmic to treat atrial fibrillation but as an adjunctive therapy for quinidine. Quinidine increases conduction through the AV node. The addition of diltiazem to a treatment regimen may aid in the control of the ventricular rate. This agent has been evaluated in healthy horses but has not yet been evaluated in horses with atrial fibrillation.

ELECTRICAL CARDIOVERSION

Presently, the only non-pharmacological option for the treatment of atrial fibrillation in the horse is electrical cardioversion. A successful shock depolarizes sufficient atrium to disrupt the fibrillating waves from perpetuating. Sinus rhythm is then restored. Placing electrodes so that sufficient atrial tissue is encompassed is therefore important, as well as generating sufficient energy so that the "targeted" myocardium is completely depolarized (influenced by electrode design as well as energy selection).

Successful transthoracic cardioversion in a horse has been reported. However, the horse was relatively small and concurrent antiarrhythmic pharmaceutical therapy was required. Energy levels of 2-4J/kg or higher are commonly used in humans and several authors based on efficacy data have recommended a directive towards this "high initial energy". The maximum energy level for commercial defibrillators is 360J. Extrapolation to horses would suggest that to achieve high success rates, multiple defibrillators would be needed (720J has been used in humans).

Transvenous electrical cardioversion has far lower energy requirements in humans with reported successful energies of less than 10J (0.14J/kg). These catheters are generally placed through a femoral vein approach under fluoroscopic guidance. The procedure is more technically challenging than transthoracic cardioversion and requires specialized catheters and is therefore considered only for refractory cases. Based on energy requirements this technique was considered for the horse and has since been adapted for use.

Cardiac catheterization is achieved in the standing horse. Introducer catheters (10g) are placed in the right jugular vein approximately 15 cm apart. Cardioversion catheters are then placed through the introducer catheters. The pulmonary artery (PA) catheter is placed first through the proximal introducer. It is advanced under intracardiac pressure and ultrasound guidance into the right heart and into the pulmonary artery (PA). Ultrasound guidance alone is used to guide placement into the left branch of the PA (approximately 125 cm in depth). The second catheter is then placed through the distal introducer and is advanced until its distal end is in the right ventricle (on pressure profile and on ultrasound). This catheter is then withdrawn until the ventricular waveform on the pressure profile is lost. This ensures that the electrode is positioned such that the electrode is in the right atrium (approximately 70 cm in depth). The catheters are sutured to the horse.

Once the cardioversion catheters are secured, the horse is induced under general anesthesia. The horse is positioned in left lateral recumbency. Lidocaine infusion is avoided, as it will increase cardioversion threshold. Thoracic radiography is performed to confirm electrode positioning. The PA catheter must not overly the cardiac silhouette, as this position would increase the risk of shock to the ventricle. Successful cardioversion has been achieved when the PA catheter either appears to curve behind the cardiac silhouette, or when it traverses slightly dorsad. The latter has been associated with lower energy levels.

Once the cardioversion catheters are confirmed as correctly positioned then cardioversion can be attempted. Vasopressive agents such as dobutamine should be withdrawn at least 5 minutes before shock delivery (these agents can be arrhythmogenic and could potentiate adverse response to shock application). The shocks should be synchronized with the QRS complex. Delivery of shock on the T wave has high risk of initiating ventricular fibrillation.

The author uses an initial energy level of 50J and a maximum energy of 300J, with 2 minute intervals between shocks to allow monitoring while allowing for a possible potentiating effect from the previous shock. At the Ontario Veterinary College, the author has treated 155 episodes of atrial fibrillation with this method; 153 responded to treatment. The mean energy level at conversion in these episodes was 156.8J (range 50-300J, 0.1-0.7 J/kg).

REFERENCES

Ohmura H, et al. **J Vet Med Sci.** 2000;62(7):711. Frye MA et al **J Am Vet Med Assoc.** 2002;220(7):1039. van Loon G et al **Equine Vet J.** 2004; 36(7):609. McGurrin MK et al. **J Vet Intern Med**. 2005;19(5):695. Schwarzwald CC et al;. **J Vet Intern Med**. 2005;19(5):703. McGurrin MK et al. **J Vet Cardiol.** 2005 Nov;7(2):109-19. Risberg AI et al. **J Vet Intern Med**. 2006 ;20(1):207. Schwarzwald CC et al. **J Vet Intern Med.** 2007;21(1):166. De Clercq D, et al. **Equine Vet J**. 2007;39(4):344. De Clercq D et al. **Vet J.** 2008;177(2):198-204. McGurrin MK, et al. **J Vet Intern Med** 2008;22(3):609. De Clercq D et al. **Am J Vet Res**. 2009;70(2):223.

TREATING LYMPHOMA IN CATS: NEW INSIGHTS

Laurel E. Williams, DVM, DACVIM (Oncology)
Raleigh, NC

INTRODUCTION

Lymphoma is the most common hematopoietic malignancy in the cat, with multicentric and gastrointestinal forms predominating. Given the systemic nature of most forms of lymphoma, combination chemotherapy is standard care for this cancer. Complete remission rates of approximately 30-65% and median remissions of approximately 5-9 months are reported with conventional chemotherapy, with some variation dependent on the form of lymphoma and the chemotherapy protocol used. Clinically relevant improvements in remission rate or duration will require a greater understanding of the various forms of lymphoma, treatment that is tailored for the individual, and consideration of multi-modality therapy.

ETIOLOGY

There is an association between FeLV infection and lymphoma. Approximately 25% of FeLV+ cats will develop lymphoma during their lifetime, and FeLV+ cats typically develop lymphoma at a younger age. FIV has also been implicated in the development of lymphoma, although its true contribution has been difficult to determine since FIV infection often occurs in conjunction with FeLV. The risk of developing lymphoma in FeLV+, FIV+ and FeLV+/FIV+ cats is approximately 60X, 5X and 80X greater, respectively, than in FeLV-/FIV- cats. The incidence of FeLV and FIV has declined over recent years and has led to a decrease in the incidence of FeLV-related lymphoma.

CLINICAL PRESENTATION

Lymphoma is a hetereogenous disease in cats and is most often described according to an anatomic scheme. Generalized peripheral lymphadenopathy, as is common in dogs, is rare in cats.

Anatomic site	Age	FeLV status	Presenting signs	Physical examination
Multicentric	Middle-older age	Usually (-)	Vary with organ(s) affected	Vary with organ(s) affected
Alimentary	Middle-older (ave 8 yrs) age	Usually (-)	Anorexia, weight loss, vomiting, diarrhea	Palpable abdominal mass
Mediastinal	Young (ave 2-3 yrs) age	Usually (+)	Tachypnea, dyspnea	Pleural effusion, non-compressible anterior mediastinum
Renal	Middle age	May be (+)	Anorexia, depression, PU/PD	Bilateral renomegaly
Spinal	Young-middle age	Usually (+)	Hindlimb parapesis	Hindlimb paraparesis
Nasal	Middle-older age	Usually (-)	Nasal discharge, epistaxis, facial deformity, sneezing	Upper airway congestion and discharge

BIOLOGIC BEHAVIOR AND DIAGNOSIS

Because lymphoma is a systemic disease, staging is important to determine the location and extent of involvement. Complete staging consists of a physical examination, complete blood count, serum biochemical profile, urinalysis, bone marrow aspiration, thoracic radiographs, and radiographs or ideally ultrasound examination of the abdominal cavity. Cytology can often provide a diagnosis, although histopathology and determination of tumor grade is also important. <u>High-grade, or lymphoblastic, lymphoma</u> is most common. This form of lymphoma is associated with rapid onset and progression. Affected cells are typically large and blastic in appearance and organomegaly and/or masses are often identified. <u>Low-grade, or lymphocytic, lymphoma</u> is most often reported in the gastrointestinal tract, although other organs may be affected. Unlike high-grade lymphoma, low-grade lymphoma is associated with a chronic insidious onset and disease course. Affected cells are well-differentiated and similar in appearance to small lymphocytes; histopathology is required for diagnosis. Nearly 50% of cats with low-grade lymphoma will have no obvious abnormalities on physical examination.

Ancillary diagnostics which may aid in diagnosis and/or yield additional prognostic information are increasingly available. <u>PARR (PCR for Antigen Receptor Rearrangement)</u> can be to determine whether lymphoid neoplasia is present, to determine the phenotype (B- versus T-cell), and to monitor for the presence of minimal residual disease in patients being treated for lymphoma. This test utilizes genomic DNA and PCR primers specific for B- and T-cell receptor gene segments in feline lymphocytes. The main advantage of PARR is its ability to differentiate a monoclonal (neoplastic) lymphoid proliferation from a polyclonal (reactive) lymphoid proliferation. In cats, sensitivity and specificity are better for the more common T-cell (89% and 80%, respectively) than B-cell neoplasias (60% and 70%, respectively). PARR can be run on a variety of tissues, including lymph node aspirates, bone marrow aspirates, cellular effusions, blood, cytology slides, formalin fixed tissues and fresh frozen tissues. It is always prudent to interpret PARR data in conjunction with other findings, including history, clinical signs, cytology, histopathology and flow cytometry (when available).

TREATMENT

Chemotherapy is considered standard therapy for the treatment of feline lymphoma, with the possible exception of nasal lymphoma, which may be treated with radiation therapy.

Single agent chemotherapy

Reports on single agent chemotherapy including prednisone, doxorubicin, and mitoxantrone describe response rates ranging from 9-40% with remission durations ranging from 3-12 months. While prednisone may be used as a single-agent, its use prior to the initiation of other chemotherapy use should be avoided since this may decrease response rate and duration to other agents.

Combination chemotherapy

Improved remission rates and duration can be achieved with combination chemotherapy. Various protocols have been reported with variations in scheduling, drug dosages, and dose intensity, although most utilize induction followed by maintenance chemotherapy. The most commonly used agents in these protocols include prednisone, L-asparaginase, vincristine, cyclophosphamide, and doxorubicin. Complete response rates range from approximately 30-65% with remission durations of approximately 5-9 months.

Evidence from human oncology has suggested that the use of discontinuous chemotherapy, rather than the conventional approach of using induction chemotherapy followed by maintenance chemotherapy, may provide comparable remission duration. Various reports in the veterinary literature described similar findings in dogs with lymphoma although reports in cats are lacking.

Radiation therapy

Definitive radiation therapy may be considered for nasal lymphoma, and disease free intervals of 6-69 months have been reported with this modality. It has been suggested, however, that radiation therapy, chemotherapy, or the combination of the two modalities may yield comparable survival times.

Since lymphoma is confined to the abdominal cavity in a large proportion of affected cats, it may be possible to administer targeted radiation therapy, enabling the administration of higher and potentially more effective cumulative doses than would be possible with half- or total-body irradiation while sparing non-affected tissues outside the abdomen. In a recent pilot study at North Carolina State University, 8 cats with high-grade lymphoblastic lymphoma confined to the abdominal cavity were treated with fractionated radiation therapy to a total dose of 15Gy. Treatment was well tolerated and yielded remission times comparable or superior to conventional chemotherapy protocols. Further prospective studies are underway to better determine the impact of this combined modality approach.

Oral chemotherapy

Given the chronic indolent course of low-grade lymphoma, oral chemotherapy consisting of chlorambucil and prednisone may be more appropriate than intensive injectable chemotherapy for this group of cats. Complete remission rates of 55-70% for a median of 20-30 months have been achieved with this approach.

REFERENCES

Spodnick GJ, et.al. **J Am Vet Med Assoc** 1992;200:373. Mooney SC, et.al. **J Am Vet Med Assoc** 1989;194:696. Ogilvie GK, et.al. **J Am Vet Med Assoc** 199;202:1839. Kristal et.al. **J Vet Intern Med** 2001;15:125. Cotter SM. **J Am Anim Hosp Assoc** 1983;19:166. Elmslie R, et. al. **Vet Radiol** 1991;32:277. Fondacaro JV, et.al. **Eur J Comp Gastroenterol** 1999;4:5-11. Kiselow MA, et.al. **J Am Vet Med Assoc** 2008;232:405.

ASPIRATING THE ANSWER: BONE MARROW ASPIRATE BASICS

Anjanette Price, BS, CVT, VTS (SAIM)
Springfield, OR

INTRODUCTION

Performing a bone marrow aspirate does not have to be just the responsibility of the veterinarian. Certified veterinary technicians can be trained to do this procedure and in turn make themselves more valuable team members.

INDICATIONS

Bone marrow aspirates can be done for various reasons. Some of those reasons include a nonregenerative anemia, any combination of peripheral cytopenias that may include a persistent or unexplained neutropenia, unexplained thrombocytopenia, or pancytopenia. Dysplastic changes, any persistent leukocytosis or severe thrombocytosis can also necessitate a bone marrow aspirate. Occasionally there may be immature hematopoietic cells in the peripheral circulation. A patient may also present with a lytic bone lesion, thus a cortical bone biopsy would be indicated in those cases.

Another reason bone marrow aspirates may be done is to stage neoplasia. Cytology on these cases will help the veterinarian and the pet owner know how aggressive the cancer is. Laboratory tests may reveal a patient is hypercalcemic or hypoproteinemic and a bone marrow aspirate may be indicated to rule out lymphoma, multiple myeloma, or a systemic infection.

CONTRAINDICATIONS AND COMPLICATIONS

There are few contraindications and complications when performing a bone marrow aspirate. The main contraindication is sedation, especially if the patient is already physically compromised. An uncommon complication is hemorrhage even if the patient is severely thrombocytopenic. Most patients require sedation for this procedure. Depending on the practices preference, general anesthesia or a propofol constant rate infusion may be used. A mild pre-medication may be administered to decrease the amount of propofol required. In patients that are at high risk of anesthetic complications, the procedure may be done with a local anesthetic.

SITES FOR COLLECTION

Several sites are used for bone marrow collection. The proximal humerus has its advantages because it is an easily accessible site in dogs and cats of all sizes. There is little overlying tissue and it provides a fairly large area for placement of the needle. The disadvantage is that it is close to the patient's head and if you are just using a local anesthetic there is a risk of being bitten if the patient perceives pain. The bone cortex is also thick and round, making it a little more likely that your needle may slip off the cortex. To localize the proximal humerus, palpate the greater tubercle. Insert the needle into the flat area of the craniolateral aspect of the humerus that is distal to the greater tubercle. Insert the needle perpendicular to the long axis of the humerus. Another potential location for performing an aspirate is at the iliac crest. It is easily accessible in thin dogs and large cats. In larger patients it may be more difficult to palpate because of larger amounts of overlying tissue. At this site, you need to localize the widest and most dorsal aspect of the wing of the ilium to insert your bone marrow needle.

The trochanteric fossa is a third possible bone marrow aspirate site that can be used on cats and small dogs. It is recommended to avoid using this location in obese animals. The trochanteric fossa is medial to the greater trochanter. Palpate the greater trochanter and insert the needle medial to it and into the trochanteric fossa. The needle should be parallel to the long axis of the femur.

PREPARATION AND PERFORMING THE PROCEDURE

There are three types of bone marrow aspirate needles: the Illinois sternal, the Rosenthal, and the Jamshidi. They come in various sizes and lengths, but are routinely a 16- or 18-gauge needle that is one to one and a half inches in length. Clean, glass microscope slides should be laid out in advance. A #11 scalpel blade should be available to make a stab incision at the location from which the aspirate will be taken. Preparation of the patient should be done with sterile technique. Clip the fur at the appropriate location and use surgical scrub and alcohol to cleanse the skin. Inject one to two milliliters of 0.5% marcaine along the periosteum then surgically scrub the location again. Donning surgical gloves, palpate the location in which you plan to insert your needle and make a stab incision over that location. Stabilize the collection site with your non-dominant hand and push the aspiration needle through the stab incision into the periosteum. Gently seat the needle on the bone and apply firm, clockwise-counterclockwise rotational pressure to advance the needle. Do not apply too much pressure or the needle may slide off the cortex, but apply enough pressure to move through the cortex or the needle may enter the subcutaneous tissues surrounding the site. If you slide off the cortex, reposition your needle and start again. Once you have successfully penetrated the cortex, your needle is in the marrow cavity and should be firmly seated.

Once the needle is seated, remove the stylet keeping it in the sterile field. Attach a 6-cc syringe to the needle and while holding the aspiration needle and syringe in one hand, make several firm, rapid and vigorous aspirates. When marrow first appears in the syringe, remove the syringe and immediately place a small amount on the slide. If no marrow appears, replace the stylet, advance, and reattempt the aspiration. If you reattempt the aspiration and once again no marrow appears, you may have advanced too far into the opposite cortex or this attempt may be a dry tap at which point a bone core biopsy may be indicated. After a successful bone marrow aspiration, remove the needle from the patient and apply direct pressure at the aspirate site.

SLIDE PREPARATION

Making diagnostic cytology slides is an art and takes practice. There are various techniques, but the slide-over-slide is one technique that will preserve your sample and decrease cells from rupturing when done appropriately. To do this, you place a drop of your sample in the middle of a slide. A second slide is placed over the sample in a perpendicular fashion. As this happens, the sample will begin to spread. Gently pull the second slide down the first slide smearing the sample. Do not put pressure on the slide or it will cause the cells to rupture. Bone marrow clots quickly so it is imperative that you do prepare the slide rapidly and efficiently. Laboratory submissions for the patient should include 10 to 15 slides with a concurrent complete blood count.

EVALUATING THE SAMPLE

Marrow samples are gritty with spicules that are white or grey. You should examine your sample under the microscope to ensure you have samples that are diagnostic. Bone marrow cells are immature cells. As the immature red blood cell matures it transforms from a proerythroblast to a polychromatic erythroblast to an erythrocyte and you should be able to visualize nucleated red blood cells microscopically. As granulocytes and agranulocytes mature, the myeloblast matures to a progranulocyte and then into basophils, eosinophils, and neutrophils while lymphoblasts mature to lymphocytes and monoblasts mature to monocytes. To evaluate white blood cells, you should be able to visualize band neutrophils under the microscope. The platelet maturation cycle begins as a megakaryoblast matures to a megakaryocyte and then they become platelets. Large megakaryocytes should be visualized microscopically to help you know you have obtained a good diagnostic sample.

CONCLUSION

Aspirate the answer and provide information that will enable the veterinarian to determine the best course of treatment for your patient. As a veterinary technician, successfully completing a bone marrow aspirate takes practice, but it is a rewarding accomplishment.

REFERENCES

[1.] Ettinger, et al. **Textbook of Veterinary Internal Medicine**, 6th ed. Elsevier Saunders, 2005; [2.] Rebar, A., **Hemogram Interpretation for Dogs and Cats**, Ralston Purina, 1998; 3. Cowell, et al. **Veterinary Clinical Pathology Secrets**, Elsevier Saunders, 2004.

FINE NEEDLE ASPIRATES OF LUMPS: HOW TO OBTAIN & EVALUATE CYTOLOGY SLIDES

Laura D. Garrett, DVM, DACVIM (Oncology)
Urbana, IL

INTRODUCTION

Cytology is a critical tool not only for the oncologist, but also for veterinarians in general practice. Cytology is easy, safe, fast, relatively inexpensive, and is often diagnostic. Cytology can be used to evaluate lymph nodes and masses (external or internal), and also fluids such as effusions and discharges. Results may change further diagnostic plans, treatment recommendations, or what a client chooses to do. For example, studies comparing evaluation of lymph nodes for metastatic disease via physical examination (PE) versus cytology showed 20-40% of patients felt to have normal or slightly enlarged lymph nodes on PE had metastatic disease found with cytology. Conversely, only 50-60% of patients with moderately to severely enlarged lymph nodes on PE had metastatic disease.[1,2] Thus, PE only of lymph nodes has poor sensitivity and specificity for metastatic evaluation. Cytology is of great benefit in these cases.

WHAT WARRANTS CYTOLOGY?

ANY LUMP OR BUMP SHOULD BE ASPIRATED!! Mast cell tumors, for example, can feel just like lipomas. We routinely "map" all of our patients. This procedure includes measuring and aspirating all masses. The masses are numbered on a body chart, and the fine needle aspiration results are recorded. Once a mapping has been accomplished, it stays in the animal's chart and can be referenced if an owner thinks a mass is growing or has identified a new mass. Thus, veterinarians unfamiliar with the case will not need to go through testing that has already been performed.

HOW TO OBTAIN A SAMPLE

Sample collection and preparation for cytology is simple and requires little equipment. Needles (25, 22, 20, or 19 gauge), 6 cc syringes, clean glass slides, and occasionally a scalpel blade for scraping are all that is needed. In vivo collection technique can be performed with a needle alone or with a syringe attached to the needle. Before obtaining the sample, make sure the slides are set out (at least 4-6) and the syringe is ready. Stabilize the mass in your non-dominant hand. For the needle alone technique, direct the needle into the mass; withdraw (yet stay within skin) and push back into the mass several times rapidly. Draw several cc air into the syringe, connect it to needle and expel needle contents onto a slide (make sure the bevel is down and the contents come out near the frosted end). For the needle plus syringe procedure, connect the needle and syringe, direct the needle into the mass, draw back plunger and release several times, with pressure released withdraw the needle from mass, disconnect the needle and draw several cc air into syringe, reconnect and expel needle contents onto the slide.

Slide preparation is key for good cytology results. The samples must be smeared out quickly to prevent clotting and need to be in a thin layer for best cellular evaluation. Even if only a small dot of sample is expelled onto the slide, smearing it is important. Cells that dry in a little round "puddle" of fluid maintain more of a 3-dimensional appearance, do not spread out their cytoplasm, and are difficult to assess. After the contents are expelled onto a slide, place a second slide onto the first cross-wise to the first slide) and pick up a SMALL amount of material. Pick up a third slide, gently place the second slide across the third slide to squash the sample flat and smear the sample by pulling the second slide away from the frosted edge of the third slide while maintaining slide contact with gentle pressure. The second slide needs to stay FLAT on the third slide (no edge to catch cells and drag them). Now, the second slide can be used again to pick up another small bit of sample from slide 1, and can be smeared onto a 4th slide, etc. When just a small amount of sample remains on slide 1, use slide 2 to smear it as well. Most aspirates will yield 3-6 good slides, allowing for in-house staining and evaluation of one slide while leaving several unstained for laboratory evaluation. Also, by making multiple smears, the samples are more thinly spread out which improves diagnostic ability. This is particularly helpful with very cellular aspirates.

Impression Smears

Cytology can also be of benefit even when a sample is being obtained and sent for histopathology. In some tumor types, particularly round cell tumors, cytology can provide a more precise diagnosis that histopathology. Thus, one may want to collect a cytologic sample in vitro, or after tissue has been removed from the patient. This can be done via impression smears or scrapes. Impression smears are best for round cell or epithelial tissues – tissues that exfoliate more easily. Without disturbing surgical margins, use a blade to cut a fresh section of tissue (maximum 1x1 cm). Gently blot the cut face on gauze or paper to rid it of excess moisture and blood. Press the blotted area onto the slide. Make several rows of impressions. If cellular yield is low with this approach, the scrape technique can be used. This may be best for mesenchymal tissues. Begin in the same manner as with the impression technique, but after blotting the sample, use the blade to scrape cells from the cut surface. Gently smear the collected cells from the blade across the slide in one direction.

STAINING AND EVALUATING

Preparation and staining of the slides is as follows. Air dry the slides. Standard hematologic stains are preferred for most purposes (Wright's, Wright's-Giemsa, Diff Quick, etc.) If submitting slides to an outside laboratory, submit them air-dried and unfixed, or fixed and stained. Do not submit them fixed and unstained or cell quality will diminish. Do not have cytology slides near formalin. Even the fumes can distort cytologic quality. Smears should be made and kept away from biopsy formalin jars.

Once the slide is stained it is ready for microscopic examination. Scan the slide on low power (10x) first to get an overall feel for the cellularity of the sample and then go to higher power to evaluate the cell types and individual cytologic features of the cells. The two most important questions as a practitioner to answer when looking at a slide are 1. Did I hit the intended tissue?

(salivary gland vs. lymph node, blood vs. bone marrow) and 2. Is it a good sample? (cells not all ruptured, enough cells for evaluation). After assessing these points, evaluate if the cells are from normal tissue. Note if inflammation (see cell types below) is present. If so, caution must be used in interpreting other cells on the slide. Inflammation can induce cells to look "malignant" when they are actually just reactive. Next ask if there are cells in abnormal places (non-lymphoid cells in lymph node, non-hepatic cells in liver). Finally, note the predominant cell type (as follows).

Inflammatory processes usually have a mixed population of cells including inflammatory and tissue cells. The inflammatory process is characterized by the cell types present. When neutrophils predominate (>85%), this is considered a suppurative response. The character of the neutrophils should be closely evaluated. If there is a bacterial infection, the neutrophils will often undergo karyolysis and the nucleus will appear pale and swollen. These cells are often called degenerative neutrophils. Also the cells should be closely examined for intracytoplasmic microorganisms. A pyogranulomatous response consists of a mixture of neutrophils, epitheliod macrophages and multinucleated giant cells. This type of inflammatory response is commonly observed with fungal infections or responses to foreign material. Granulomatous inflammation consists primarily of macrophages with multinucleated giant cells and lesser numbers of other inflammatory cells. This type of inflammatory response does not always exfoliate well and may be of low cellularity. Eosinophilic inflammation is more subtle. Cytologic specimens containing greater than 12% eosinophils are considered eosinophilic inflammatory lesions. Common causes include allergic or hypersensitivity response, parasitic infestation or paraneoplastic response.

SPECIFIC CELL CLASSIFICATIONS

Tissue cells are classified as epithelial, mesenchymal (connective tissue) or round cells.

Epithelial

Epithelial/glandular cells are found in the linings of tracts (respiratory, digestive/oral, urogenital), in skin and mucous membranes, and in glands (pancreas, thyroid, adrenal, prostate, mammary, etc.). Benign epithelial lesions are adenomas, while malignant versions are carcinomas. The cells are generally round and the sample is often highly cellular. The major defining feature of this cell population is clustering, or clumping, of cells. This feature can be observed on low power evaluation. Note that any very cellular aspirate can give the appearance that the cells of the sample are clustered together, even with mesenchymal masses. Assess less cellular regions to ascertain if in these areas cells are still bound together with other cells or if they do appear individually, as mesenchymal cells will. Round cells tend to be pressed together, as they exfoliate very well and thus provide cellular samples, but they do not tightly adhere to each other or swirl together to form 3-dimensional structures. Small spaces can often be seen between non-epithelial cells.

Neuroendocrine tissues such as thyroid and adrenal organs, while in the epithelial category, do have a few distinctive features. These cells tend to rupture very easily. Thus, samples from these tissues may have fields of naked nuclei in a smooth bluish background representative of the lost cytoplasm. Additionally, while most neuroendocrine tumors are malignant (thyroid carcinomas, insulinomas, etc), the cytologic appearance of these tumors is often very uniform with few features of malignancy. In these cases, knowing the site of origin of the mass and the fact that the tissue appears neuroendocrine is more important for predicting biologic behavior than are the malignant features of the cells.

Mesenchymal

Mesenchymal cells are found in connective tissues such as cartilage, bone, smooth and striated muscle, blood vessels, fibroblasts, etc. When benign, the masses are categorized as the cell of origin and then "oma" – e.g. fibroma. When malignant, the masses are termed sarcomas. Mesenchymal cells may not exfoliate as well as epithelial tissue; however, most mesenchymal masses will provide enough cells for cytologic evaluation, and some exfoliate very well. In benign lesions, it is not uncommon to see only 1-2 cells on a slide. The cells are commonly spindle shaped with wispy cytoplasmic borders. The nuclei are often oval in shape. Their cell shape, indistinct cytoplasmic borders, and the fact that they occur as single cells in the preparations characterize mesenchymal cells. At times mesenchymal masses have a round cell appearance, most notably with osteosarcoma. Malignant osteoblasts are notable for their "plasmacytoid" appearance in some cases.

Round cell

Round cells are called such based on their cytologic appearance: discrete, round cells with distinct cytoplasmic margins. This category includes lymphoid tissue, plasma cells, mast cells, histiocytomas, transmissible venereal tumors (TVT), and melanoma (which can have a round or mesenchymal appearance). Often the specific cell type can be cytologically determined in round cell tumors, unlike in epithelial or mesenchymal masses. The round cells will have round to bean-shaped nuclei, lacy to smudged chromatin and are generally small to medium sized cells. Individual features of the listed round cell types can help to provide a diagnosis. Lymphoblasts have a high nuclear:cytoplasmic (N:C) ratio, are as large or larger than neutrophils, have basophilic cytoplasm and open chromatin, and often have multiple nucleoli of varying sizes. Plasma cells have round, eccentrically located nuclei with ropy or clumped chromatin.

The cytoplasm is more abundant than in lymphocytes and has a dark blue/grey hue. A perinuclear clear zone is often apparent. The most notable feature of mast cells is the presence of pink/purple-staining cytoplasmic granules. The granules may be scant or so abundant that the cell nuclei are completely obscured. Often granules litter the background of the slide and may be mistaken for cocci. Mast cells are large round cells with a centrally located round nucleus ("fried egg" appearance). Mast cells secrete chemoattractant factors for eosinophils and occasionally fibroblasts. Some mast cell populations have granules which fail to take up Diff Quick stains;

these granules will often stain with other stains. If agranular cells with a "fried egg" appearance and eosinophils are noted on a mass aspirate, unstained slides should be sent to a laboratory that will use a Wright's stain to assess for granules. Histiocytomas are large cells with a moderate amount of blue-grey cytoplasm with clearing at the periphery of the cytoplasm. The cells can show extreme pleomorphism despite the tumor's benign behavior. Histiocytomas can be difficult to distinguish from some plasma cell tumors cytologically. The clinical appearance of the lesion is supportive of the diagnosis in these situations. TVTs also have a large amount of cytoplasm, usually grey-blue to blue in color, and can be quite pleomorphic. The defining feature of the TVT (other than the clinical presentation!) are unevenly distributed vacuoles in the cytoplasm. Melanoma can have a round cell appearance – assess for fine black/blue/green cytoplasmic granules. Amelanotic melanoma can be difficult to diagnose, yet even amelanotic masses will often have a few cells with pigment. Malignant melanomas are highly pleomorphic tumors with large cells that contain prominent and bizarre nucleoli.

FEATURES OF MALIGNANCY

Once the cell type has been characterized, the next step is to determine if the process is hyperplastic or neoplastic and if it is neoplastic, is it benign or malignant. It is important to evaluate for features of malignancy in epithelial or mesenchymal cells. The biologic behavior of the round cell tumors is based on cell type and site of the mass rather than specific cytologic features. Malignant criteria include: prominent and multiple nucleoli, anisocytosis, anisokaryosis, multiple nuclei, mitotic figures and nuclear molding. There are others, but these are the most common. When these criteria are present it is fairly easy to identify a malignant tumor; however, not all malignancies express these criteria. Therefore, cytology cannot always rule out a malignant process. Histopathology may be needed for diagnosis of malignancy. For example, soft tissue sarcomas are locally very aggressive yet may have a benign cytologic appearance. If ≥ 3 features of malignancy are present, then the lesion is most likely malignant. If malignant features are lacking, the lesion may still have a malignant behavior and a biopsy is indicated.

LOOK AT THE BACKGROUND

The final step in cytologic evaluation is examination of the material in the background. Proteinaceous material is eosinophilic and granular and is usually observed evenly distributed throughout the sample. Granules from ruptured mast cells or melanocytes must be distinguished from bacteria. Most cytologic samples contain some level of peripheral blood contamination. With the red blood cells come white blood cells. It is important to remember this when trying to distinguish inflammation from peripheral blood contamination. It may be necessary to do a CBC on the patient to determine the WBC in situations where increased numbers of WBC are noted on a cytology slide.

REFERENCES

1. Williams LE, et al. **J Am Vet Med Assoc**, 2003;222(9):1234-6. 2. Langenbach A, et al. **J Am Vet Med Assoc** 2001;218(9):1424-8.

PROPER ADMINISTRATION OF CHEMOTHERAPEUTIC AGENTS FOR EQUINE & SMALL ANIMALS

Dr. Tara Marmulak, Resident Pharmacist
Davis, CA

INTRODUCTION

Chemotherapy is being administered more and more commonly in veterinary clinics and, with this, comes exposure to hazardous medications. Technicians need to know what the risks are and how best to take precautions against contamination with these chemotherapeutic agents.

EXPOSURE FACTS

Employee exposure to chemotherapeutic agents can lead to cancer, reproductive toxicity, birth defects, and diminished health. Potential activities for exposure include drug preparation and administration, janitorial duties, waste handling, and kennel/barn husbandry activities. This exposure may occur via inhalation, skin contact, oral ingestion or accidental injection. Dermal contact and absorption appears to be the primary route of exposure[1]. Where the drug is not absorbed directly through the skin, the exposure may occur via hand to mouth contamination.

Many studies have been conducted looking at the exposure of hazardous drugs and their effects on healthcare workers. In 1979, the first study was released showing that pharmacists preparing and nurses administering chemotherapeutic agents were being exposed to these harmful agents[2]. Since then, many safety measures have been put in place, but even with these measures, exposure still occurs.

PERSONAL PROTECTIVE EQUIPMENT (PPE)

Anytime chemotherapeutic agents or patients treated with chemotherapy are handled, appropriate personal protective equipment (PPE) should be worn. PPE is not just protecting the wearer from the hazardous drug, it helps protect other employees working in the same facility. Using PPE, discarding or cleaning anything that might be contaminated and removing PPE when leaving the designated chemotherapy area, all reduce contamination of other areas of the facility and help protect employees. Recommended PPE while handling chemotherapy include:

- Gloves that are powder free, labeled and tested for use with chemotherapy drugs
- Gowns that are impermeable, closed-front, have cuffs, and intended for single use
- A NIOSH-approved respirator when there is risk of inhaling drug aerosols or if pregnant, breast-feeding or trying to conceive
- Face and eye protection when splashing is possible

It's important to always wear PPE even if it seems like a low risk of exposure activity. Studies have found that even the outside of new, unopened vials of chemotherapy straight from the manufacturer can be contaminated[3]. Since it's impossible to tell if something is contaminated with chemotherapy, always use precautions.

ESSENTIALS FOR CHEMOTHERAPY

Any clinic that uses chemotherapy should have in place protocols for personal protective equipment, drug administration, employee exposure, drug spills and extravasation of a chemotherapeutic agent. Chemotherapy preparation and administration should be in designated areas that have limited traffic, contain hazardous drug spill kits and specially marked hazardous waste containers. Because of the risks of exposure, there should be no food or drink and no cosmetic application within that area.

ERROR PREVENTION

The best means of preventing errors is to have trained, competent staff. Before ever preparing or administering chemotherapy, staff should be trained to know:

- Common chemotherapy drugs: mechanism of action, dosing range, administration
- Adverse effects and toxicities associated with chemotherapy including early identification, ongoing monitoring and principles of prevention and management
- Common chemotherapy protocols used within the practice
- Principles of safe handling of chemotherapy and risks of exposure with each handling procedure
- Chemotherapy preparation, storage and transportation

All chemotherapy should be provided on the basis of a documented and referenced protocol as this has been shown to reduce errors in prescribing. Each protocol should be of a standard format. It should include: the name of the protocol, tumor group and stage it is typically used for, usual number of cycles and total length of treatment, all chemotherapy to be used in the protocol (by generic name), route of administration of all drugs, all drug doses in appropriate units and method of calculation, drug vehicle and volume, rate and duration of administration, frequency of dosing and time intervals, laboratory tests required to monitor toxicities, dose modifications for each agent in the protocol, expected side effects and their management and supportive therapy that is required to be to be given with the protocol.

Orders for chemotherapy should be on pre-printed forms. Patients should be named by two identifiers, the weight, diagnosis, chemotherapeutic agent, and route of administration included. Orders should be clearly printed and no abbreviations used. Examples of dangerous abbreviations are included in the table below. Both the dose in mg/kg or mg/m^2, and the total dose should be written on the order to enable double checking.

Once the veterinarian writes the order, the person preparing the order must double check the calculations, and that the dose is appropriate for the diagnosis. Once prepared, the drug again needs to be double checked for accuracy. Is it the right drug? Is it the right dose? Is it the right patient? Is it the right route? Before administration begins, extravasation protocols should be reviewed and antidote drugs should be accessible.

In the drug preparation area, chemotherapeutics should be stored separately from other medications. Look-alike, sound-alike drugs should be separated and identified as higher risk drugs. Eg. VinCRISTINE - VinBLASTINE, CARBOplatin – CISplatin.

Table 1. Unsafe abbreviations and their safe alternatives

Abbreviation	Intended Meaning	Mistaken Meaning	Alternative
Trailing zero eg. 4.0 mg	4 mg	40 mg	Avoid using trailing zeroes
No leading zero eg. .9 mg	0.9 mg	9 mg	Always use a leading zero
U	Units	Mistaken as a zero	Write units
Ug or µg	Microgram	Milligram	Write out microgram
IU	International Units	IV	Write out units

Clinics should foster an environment that encourages a discussion of adverse events, errors and near misses. By discussing these events, the process of drug ordering, preparation and administration can be refined and safety improved in the future.

PREPARATION OF CHEMOTHERAPY

Ideally, all chemotherapy should be prepared in a biological safety cabinet (BSC) with negative air pressure and is vented to the outside to minimize employee exposure. However, the effectiveness of a BSC (class II) in preventing chemotherapy contamination of outside surfaces, is technique dependent and may not prevent all contamination. Closed-system transfer devices (example: PhaSeal, Tevadaptor, etc.) are another product that may be used to help reduce employee exposure. These devices can be used during chemotherapy preparation and administration to reduce environmental contamination.

For preparation and administration, luer-lock connections should be used. Prime i.v. sets before adding the hazardous drug to the bag. Once an i.v. set is attached to a bag containing chemotherapy, it should never be detached preventing the line from spilling any residual drug left in the line. It is recommended that after preparing the chemotherapy, the bag should be wiped down with a damp cloth or piece of gauze to remove any residue and prevent contamination of the receiving person's gloves. The readied product should then be placed in a sealable bag for transport to the patient being careful to prevent contaminating the outside of the bag.

DECONTAMINATION OF SURFACES

The area used for preparation and administration should be cleaned with water and a detergent following each patient. For the most part, chemotherapy is not deactivated by this process but is physically removed from the surfaces. Bleach, because of it's high pH, is often recommended for decontamination but it does not work with all classes of chemotherapeutics. Isopropyl alcohol is not effective. Surface Safe® is product that is available for decontaminating chemotherapy work surfaces. It works in a two step process using hypochlorite (bleach) and sodium thiosulfate. Unfortunately, there are currently no guidelines on the best method for decontaminating work areas, however, it is an important step in reducing the transfer of chemotherapy to other surfaces around the clinic.

POST CHEMOTHERAPY PRECAUTIONS

Precautions need to be in place even after administration of chemotherapy is complete. Active chemotherapeutic agents and their metabolites can be found in the urine, feces, vomit and sweat of a treated patient. The number of days that precautions need to be in place varies with each agent. A list of the chemotherapeutic agents used in the facility with the number of days requiring handling precautions should be posted. Policies should also be in place specifying what precautions should be taken.

Handling precautions should be used for all animals post chemotherapy. Studies have found in humans, that cyclophosphamide is partially excreted in the skin[4]. Once a patient is ready to go home, kennels should be treated by staff as being contaminated. The kennel should be washed with a detergent, and then rinsed while wearing PPE.

Urine and Vomit

Both urine and vomit should be treated as a hazardous drug spill for a time related to the specific drug being excreted. Gloves and a gown should be worn, the spill sopped up with a disposable towel, and the area cleaned with a detergent. Soiled bedding should be handled with gloves and washed with detergent. Owners should be told to bring their dog to an area that is away from high traffic areas to urinate. Another suggestion is to hose down the area where the dog urinates to dilute it. With horses, the stalls should be cleaned on the same schedule; however, workers should protect themselves by wearing disposable booties, gown and gloves. The dirty bedding and waste should be disposed of into a low traffic area. Equipment used to clean the stall should be hosed down and soaked with detergent.

Table 2. Hazardous drugs and their duration in urine and stool requiring protective handling –Extrapolated from human data[5,6].

Hazardous Drug	Duration of Detectable Drug Residue in Urine	Duration of Detectable Drug Residue in Stool
Asparginase (Elspar)	Not detectable	NA
Carboplatin (Paraplatin)	48 hours	NA
Chlorambucil (Leukeran)	48 hours	NA
Cisplatin (Platinol)	7 days	NA
Cyclophosphamide (Cytoxan)	72 hours	5 days after oral dose
Cytarabine (Cytosar)	24 hours	NA
Dacarbazine (DTIC)	6 hours	NA
Dactinomycin (Cosmegen)	5 days	NA
Doxorubicin (Adriamycin)	6 days	7 days
Fluorouracil (Adrucil)	48 hours	NA
Gemcitabine (Gemzar)	7 days	NA
Lomustine (CCNU)	24 hours	NA
Mechlorethamine hydrochloride (Mustargen)	48 hours	NA
Melphalan (Alkeran)	48 hours	7 days
Mitoxantrone hydrochloride (Novantrone)	6 days	7 days
Vinblastine sulfate (Velban)	4 days	7 days
Vincristine sulfate (Oncovin)	4 days	7 days
Vinorelbine (Navelbine)	4 days	7 days

Feces

Feces should also be handled with precaution because certain chemotherapeutic agents tend to be eliminated via the feces more than others. These agents include doxorubicin, cyclophosphamide, vinblastine, vincristine, mitoxantrone, melphalan, mechlorethamine and vinorelbine. On average, these medications are excreted in the feces from 5 to 7 days post administration in humans. The number of days of excretion in the dog and horse are not yet reported in the literature, but is thought to be similar. Contaminated waste should be bagged and disposed of as hazardous waste. Owners should be told to bag any "accidents" in the house and throw in the garbage or flush down the toilet while wearing gloves and wash hands afterwards. If outside, owners of treated pets should encourage their pet to use a low traffic area. The contaminated waste may then be picked up and disposed of in the trash. Alternatively, the area may be quarantined for a minimum of a week and disposed of at a later date when any chemotherapeutic activity is likely minimal.

Large Animal

Horses treated with systemic chemotherapy require additional measures. Owners should be instructed that they may continue to groom their animal as usual, but encouraged to wear gloves. Since chemotherapeutic agents such as doxorubicin are excreted in the sweat[7], owners should be counseled to use precautions and possibly avoid exercising the horse for the first few days post administration of doxorubicin.

CONCLUSION

Chemotherapeutic agents should be treated with caution. Anyone handling them should understand the risks involved in order to minimize their own exposure as well as protecting their coworkers. Although exposure cannot be eliminated altogether, it can be minimized through sound protocols and diligent handling.

REFERENCES

[1]Sessink PJ, et.al. **Arch Environ Health** 1994;49:165. [2]Falck K, et.al., **Lancet** 1979;313(8128):1250-1251. [3]Favier B, et.al. **J Oncol Pharm Pract** 2003;9(1):15-20. [4]Fransman W, et.al. **Ann Occup Hyg** 2004;48(3):237-244. [5]Cass Y, et.al., **Am J Health Syst Pharm** 1992;49:1957-1958. [6]**Facts & comparisons** 4.0 (CD). 58th ed. St. Louis: Facts and Comparisons, 2009. [7]Jacobi U, et.al. **Ann of Oncolgoy**. 2005;16(7):1210.

WHY AREN'T CATS VEGANS?

Liz Kelley, RVT
College Station, Tx

INTRODUCTION

We are often reminded of an exotic jungle creature when looking at a cat. They are miniature lions and tigers we have brought into our world who have captivated us with their grace and beauty. It is no wonder that they were once worshipped as gods. And it is no surprise that they are surpassing dogs as the most common house pet. Along with their exotic looks come unique qualities and needs that set them apart from other species. During this discussion we will focus on one of these unique needs, the nutritional and, more specifically, protein requirements that set them apart from dogs and other members of the family Carnivora. While dogs are omnivores, cats are strict carnivores and this makes for some dietary challenges for owners, particularly those who have chosen to live a vegan lifestyle.

DEFINITIONS

With today's ever growing focus on living a healthy life and conservation of the planet, vegetarianism has become a popular dietary choice. There are a few variations of vegetarianism, lacto-vegetarianism, lacto-ovo-vegetarianism, pesce-vegetarianism, pollo-vegetarianism and finally veganism. The first two incorporate dairy and egg products into their diets, while pesce- and pollo-vegetarians also will include fish or chicken. Vegans are more strict with their dietary choices choosing to not consume any meat or dairy products at all and some vegans may also avoid honey in their diet. In an effort to be more eco-friendly, many vegans will choose to avoid the use of leather, silk, wool, and down in their clothing. Living a vegan lifestyle and owning a cat present some very complicated challenges for both owner and cat. The biggest challenge, and probably the most difficult, is the cat's absolute nutritional and metabolic requirements as a strict carnivore. In order to maintain a complete or all-encompassing vegan lifestyle, these owners often decide to feed their cats a vegan diet. Because dogs are omnivorous, they can remain relatively healthy with a vegan diet. Cats on the other hand, cannot survive on plants, and, as strict carnivores, have some very stringent protein requirements. There are higher amino acid, fatty acid, and vitamin requirements that must be met. If cats are not fed a meat-based diet, these essential nutrients must be added into the diet.

ANATOMY AND PHYSIOLOGY

Beginning with anatomy, cats lack the molar and premolar that are responsible for grinding up food. They also lack amylase in their saliva. Amylase is an enzyme that is responsible for the breakdown of starches in the mouth during the chewing process as saliva is mixed with the food. A cat's stomach is also smaller and more simple in structure compared with that of a dog and when compared, by body weight, the gastrointestinal (GI) tract of a cat is roughly 4 times shorter than the GI tract of a dog. Also, cats have lower levels of disaccharidases to breakdown carbohydrates (CHO) and intestinal and pancreatic amylase is minimal. Cats can readily breakdown CHO for energy, but the minimal activity of amylase, the low levels of disaccharidases, and the shortened GI tract emphasize the carnivorous nature of cats. The increase of CHO in the diet can have negative effects on the GI flora which in turn affects protein digestibility. Cats also have a protein requirement that is 2-3 times greater than that of dogs, while kittens have a protein requirement that is 1.5 times higher than the protein requirement for puppies. Because a normal diet for cats is prey-based, they have evolved to use protein and fat preferentially as the primary energy source. Unlike omnivores, cats lack the ability to up- or down-regulate protein metabolism based on the amount of protein consumed in the diet, detrimentally affecting sick, injured, and healthy cats.

AMINO ACIDS

As part of the strict protein requirements, certain amino acids must be present in the cat's diet at levels that are much higher than those for dogs and other omnivores. These amino acids are taurine, arginine, methionine, cysteine, and tyrosine. Cats are unable to synthesize these amino acids and are not able to conserve them either.

Taurine

Taurine is essential for vision and the normal function of the cardiac, nervous, immune, and reproductive systems. Due their inability to store taurine and because of its obligate use in bile acid conjugation, taurine is required in large amounts by cats. It is only found in animal protein sources and is subject to reduced absorption in diets high in insoluble fiber. Deficiencies of taurine can take up to 3 years to manifest and may include blindness, dilated cardiomyopathy, and abortion.

Arginine

Arginine is an amino acid that is utilized in large amounts in the urea cycle. Like taurine, cats cannot synthesize arginine so they must obtain it from their diet. Also, like taurine, it is only found in animal protein sources. A deficiency in arginine results in hyperammonemia, and clinical signs can range from salivation and neurologic abnormalities to hyperesthesia, vomiting, tetany, coma, and death. Clinical signs of arginine deficiency can appear within hours after eating. This is an odd statement

Methionine and Cysteine

Methionine and cysteine are gluconeogenic amino acids in nature and are catabolized to pyruvate which is ultimately oxidized via the Cori cycle to energy. In dogs, these amino acids are primarily converted to taurine, homocysteine, and s-adenosylmethionine (SAMe) and its metabolites which act as antioxidants and free radical scavengers. For cats, methionine and cysteine are also used for the production of hair and felinine. A defiency of these two amino acids will result in crusting dermatoses, poor hair coats, and poor growth in kittens.

Tyrosine

Tyrosine is an amino acid that plays an important role in the synthesis and homeostasis of melanin. Melanin is a pigment present in the skin and black hair. Deficiencies of tyrosine will result in black hair turning reddish-brown and can be seen in older cats being fed a low protein diet due to renal insufficiency. It can be reversed by supplementing tyrosine back into the diet.

EFFECTS

Through the course of this discussion, the medical conditions associated with vegetarian and vegan diets have been listed, but the overall importance of a meat-based diet has not been discussed. Understanding how protein plays an important role in metabolism and energy expenditure is key to knowing how cats will be affected clinically by these types of diets. The role of protein in metabolism and energy expenditure has been extensively studied for the management of obesity. When cats are fed a high protein (meat-based) diet, the postprandial thermogenesis is increased which in turn increases protein turnover which then increases energy expenditure. What this means, simply speaking, is a high protein (meat-based) diet increases metabolism, even at rest. The higher protein levels also induce an increased sense of satiety for obese cats, which means they feel full longer and will lose weight by increasing fat loss while maintaining lean body mass.

Keeping in mind that cats are strict carnivores who cannot up- or down-regulate protein metabolism regardless of the amount of protein consumed in the daily diet and that cats will eat to satisfy their protein need, we can begin to discuss the effects of low protein diets. Remember, if the diet contains an appropriate amount of protein, there will be no negative effects to the cat. But, if the diet is low in protein, then the deficit will be made up by catabolizing the skeletal muscle. This puts a healthy cat at risk and gravely affects who that are ill or injured.

What does all of this information mean in a clinical setting? Combining the information of how vision, cardiac, neurologic, and immune systems are affected by insufficient protein levels and how protein affects metabolism and satiety, we can begin to formulate a clinical picture. What steps should be taken to ensure a complete health evaluation is done? When a "vegan" cat is presented for an overall health check, what, if any, clinical signs of nutritional deficiency will be present?

Starting with the first question, what steps should be taken to ensure a complete medical assessment? An in-depth medical history that includes diet history is the first important step. The diet history should include the trade name of the food, if it is commercially prepared or a complete list of ingredients, recipes, preparation, supplements, and flavoring or spices used in the diet. How often is the diet fed and how much is eaten at each meal are also important questions to ask. Baseline laboratory workshould be obtained including a complete blood count (CBC), chemistry profile, urinalysis, urine protein/creatinine ratio, and an amino acid profile should be obtained. Thoracic radiographs to evaluate heart size and a thorough ophthalmic exam should also be done.

The second question is more complicated. What, if any, nutritional abnormalities will the cat present with? The answers depend on a variety of factors. For instance, if the cat is relatively new to the vegan diet, there may be no abnormal physical findings and baseline laboratory tests, radiographs, and ophthalmic examination will be normal. However, if the animal has been fed a vegan diet over a long period of time, there may be many abnormal findings on physical examination and with the laboratory values. These abnormal findings could include, but are not limited to, weight loss, moderate to severe muscle mass loss, lethargy, dull mentation, poor hair coat, listlessness, ophthalmic abnormalities, and abnormal radiographic findings. All of the abnormal findings could be attributed to the protein deficiencie; however other diseases such as hyperthyroidism and chronic kidney disease must also be considered. The abnormal physical findings and laboratory results must be discussed in depth with the owner and this includes the dietary component of the findings.

The order of appearance of central retinal degeneration, dilated cardiomyopathy and other diseases associated with taurine and other amino acid deficiencies are difficult to determine. While the long term effects are well known and well studied, there is no time line correlation for their onset, especially given the many variables associated with how the diet is prepared.

HARD QUESTIONS AND THE CONCLUSION

One of the most common reasons people choose a vegan lifestyle, other than for religious reasons, pertains to the promotion of animal rights and the prevention of cruelty to animals raised for slaughter. It then becomes a question of ethics and morality. How does a person justify feeding a cat a meat-based diet when this goes against the owner's moral and ethical beliefs? Is it appropriate or fair to impose our personal beliefs on our pets? When asked another way, is it better to feed an inappropriate diet to a cat and save a farm animal from slaughter when the diet causes harm to the cat? How does a person balance the value of one animal over another? These are controversial and contentious questions but need to be addressed nonetheless. These issues cause great debate, even among the proponents for veganism. Because of these questions, many vegans choose to not own pets or cats in particular so that they can avoid the moral dilemma of providing an appropriate diet.

Based solely upon their nutritional requirements, cats simply cannot be vegans. It is a personal, human choice that should not be imposed upon an animal who has no human political agenda or human values. They are, after all, animals and have evolved to be hunters and to be meat-eaters. This is, as stated, a very difficult subject. The issue is encountered more and more often in veterinary clinics, and is a topic that requires education on the part of the veterinary community and on the part of the owner.

REFERENCES

[1]Bradshaw,J et.al. **Comp Biochem Physiol** 1996;114A(3):205, [2]Berschneider,H **Clin Tech Sm An Pract** 2002;17(1):1, [3]Michel, K, **Vet Clin Small An** 2006;369:1269, [4]Hoenig, M, et.al **Am J Phys Regul Integr Comp Phys** 2007;292:R227, [5]Markwell, P et.al. **Nutr Reser** 1995;15(1):53, [6]MacDonald, M et.al. **Ann Rev Nutr** 1984;4:521, [7]Laflamme, D **Vet Clin Small Anim** 2005;35:71[8]Laflamme, D et. al. Intern J Appl Res Vet Med 2005;3(2):62, [9]Morris, J et. al. **Am Soc Nutr Sci J Nutr** 2002;132:1648S, [10]Morris, **J Nutr Res Rev** 2002;15:153, [11]Zoran, D Text Vet Int Med 2010;7:652, [12]Zoran, D **JAVMA** 2002;221(11):1559, [13]Larson, A et al in **Consult Fel Med** 2010;6:127

PREBIOTICS & PROBIOTICS: WHAT CAN THEY DO FOR THE DOG & CAT

Ann Wortinger BIS, LVT, VTS (ECC, SAIM)
Belleville, MI

INTRODUCTION

It seems that everywhere we look these days we are seeing products that include either probiotics or prebiotics. This can be in the form of supplements, known as nutraceuticals, or in over the counter foods and therapeutic diets. According to the hype, these products can improve your immune system; improve your gastrointestinal function and any number of other things just short of emptying the litter box for you! But do you know what probiotics and prebiotics really are, and what they can actually do for our dogs and cats?

According to the currently adopted definition by the Food and Agriculture Organization (FAO) and the World Health Organization (WHO), probiotics are: "live microorganisms which when administered in adequate amounts confer a health benefit on the host" (Weese 2003) whereas prebiotics are "non-digestible food ingredients that selectively stimulate the growth and activities of specific bacteria in the gastrointestinal tract and exert beneficial effects on the host" (Pan 2009). The Food and Drug Administration (FDA) defines nutraceuticals as "nondrug substances produced in a purified or extracted form and administered orally to provide agents required for normal body structure and function with the intent of improving health and well-being" (Lerman 2007).

As we can see by all these definitions we are either talking about bacteria or substances that benefit bacteria in the intestinal tract. We all know that there are millions of bacteria that normally reside in the small and large intestines of normal, healthy animals. These bacteria help with digestion of food, maintain intestinal mucosal integrity, participate in metabolism, as well as local and systemic immune stimulation (Kelly 2006). When disease, parasites, stress or dietary changes affect these beneficial bacteria, imbalances can occur causing changes in all of the areas. One other common cause of changes within the gastrointestinal microflora is the use of broad-spectrum antibiotics to treat infections. When the gastrointestinal microflora is disturbed, potentially harmful and pathogenic bacteria can become more prolific and cause antibiotic-induced diarrhea (Kelly 2006).

The intestines

The intestines are seen as the largest immune system in the body, compromising ~70% of the total system. The mucosal barrier found in the intestines helps to block the entrance of most pathogenic bacteria into the body while allowing the entrance of permeable nutrients (Kelly 2006). Since most pathogens enter the body through the mouth and than the intestinal tract, these intestinal defenses must be working optimally to cope with the onslaught of foreign substances and pathogens to which it is constantly exposed (Kelly 2006).

This defense involves the coordination of three different systems within the intestines; the first is the resident intestinal microflora which provides an environment which favors the growth and functioning of beneficial bacteria, the second is the intestinal mucosa which provides a protective barrier against pathogenic bacteria and lastly the gut-associated lymphoid tissue (GALT) (Kelly 2006).

It would be difficult for us to affect the intestinal mucosa or the GALT, but we can help modulate the intestinal microflora to improve the environment the bacteria are living in allowing a positive impact on the dog or cat. There is substantial research being done looking at how enteral nutrition can be used to help improve not only the GALT but also the intestinal mucosa barrier. Hopefully, we will see the benefits of that research in the next few years. Currently we are using prebiotics and probiotics to help with this modulation.

Prebiotics

Typically, prebiotics are a type of carbohydrates called oligosaccharides though by definition non-carbohydrates can be classified as prebiotics. The ones used most frequently are classified as soluble fibers. These were first identified as a functional food in 1995 by Marcel Roberfroid. In 2007, Dr. Roberfroid clarified that only two classes of fructooligosaccharides fully meet the definition of a prebiotic. These would be oligofructose and inulin (Roberfroid 2007).

Oligofructose is a 2-8 chain fructose based saccharide molecule that undergoes fermentation fairly quickly in the colon providing nourishment to the bacteria in that area. Examples of sources for fructooligosaccharides (FOS) would be soybeans, oats, beets and tomatoes.

Inulin is a longer 9-64 chain fructose based saccharide that tends to be fermented more slowly benefiting bacteria farther down the colon. It can also be broken down into the FOSs by the intestinal bacteria to provide both FOS and inulin. Inulin can be found in Jerusalem artichoke, jicama and chicory root.

These two fibers are seen as minimally digestible because of their beta-bond base connecting the fructose molecules. Dogs and cats lack the enzyme in their intestines to break down the beta bond, instead they usually breakdown alpha saccharide bonds. The resident bacteria in these animals do have the ability to break this beta bond producing short chain fatty acids (SCFA). The SCFAs seen most commonly include acetate, proprionate and butyrate.

Plants do not have only one type of carbohydrate in them, so both FOS and inulin can be found in varying amounts in the same plant, while none of either may be found in others. Just adding fiber to the diet may or may not provide prebiotic effects, dependent on the types of fibers found in the product.

The SCFAs produced by the breaking of the beta bond through bacterial fermentation serve as an energy source to the colonocytes, act to lower the colonic pH, and stimulate sodium and water absorption (Pan 2009). One SCFA in particular, butyrate serves as the primary energy source for colonocytes, but may also directly enhance cell proliferation of normal cells, while suppressing proliferation of transformed cells. Because these are both fibers, there can be negative side effects to the addition to the diet. Both FOS and inulin have been shown to cause a decrease in fecal protein digestibility at higher levels (Pan 2009).

Studies have shown there are generally positive effects on gut microflora and host health as evaluated through gut integrity and bacterial colonization. Positive benefits have also been seen in animal performance as evaluated through digestion, body weight gain and feed efficiency when FOSs and fructose based inulins are added to the diet (Verdonk 2009).

Probiotics

Probiotics, rather than encouraging the growth of beneficial bacteria and suppressing the growth of pathogenic bacteria actually introduce these beneficial bacteria into the environment. To be able to function as a probiotic certain criteria must be met by the bacteria being used. They must be able to:

- survive the acid and bile found in the gastrointestinal tract
- adhere to the intestinal cells or transiently colonize various areas within the GI tract
- exclude or reduce pathogenic bacterial adherence
- produce acids, hydrogen peroxide and/or bacteriocins that antagonize the growth of pathogens
- coaggregate to help achieve normal balanced microflora population
- be safe, noninvasive, noncarcinogenic and nonpathogenic to the animal (Kelly 2006).

Initially, the gastrointestinal tract of a new born is sterile but is colonized within hours of birth. These bacteria find their individual niches within the intestinal tract and reach a state of equilibrium. Once this neonate "grace period" in over though, introduction of bacteria is significantly more difficult due to the gastric acidity and the introduction of bile acids to the chyme leaving the stomach. Though yogurt products contain beneficial bacteria, as do a number of other bacterial fermented products, these bacteria seldom survive the stomach and therefore do not provide any benefit to the intestinal bacteria (Weese 2003). The beneficial effects of probiotics depend on the number of live bacteria that make it to the intestines and are able to transiently colonize there (Kelly 2006).

The primary bacterial populations included in probiotics that are seen to benefit the cat and dog are the lactic acid bacteria, especially *Lactobacilli, Bifidobacteria* and *Entercocci*. These bacteria use fermentation to transform some sugars into organic acids, particularly lactic and acetic acids. These acids serve to lower the pH in the intestinal tract, and inhibit the growth of pathogenic bacteria (Kelly 2006).

Probiotics can benefit the intestinal microflora in a number of ways. They can increase fecal bacterial counts of good bacteria while decreasing the numbers of pathogenic bacteria. Some probiotics have been shown to minimize adherence to the intestinal epithelial cells and establishment of pathogenic bacterial populations. Compared to the large intestine the small intestines have poor microflora colonization and limited barrier protection against pathogens. Therefore, probiotics can exert an important beneficial effect in the small intestines (Kelly 2006).

Probiotics can also help to control diarrhea caused by bacterial overgrowth or parasitic infection. This uses the principals of competitive exclusion, competition for nutrients and binding sites and an increase in specific and nonspecific immune response (Kelly 2006).

Intestinal Bacteria

Beneficial	Pathogenic
Lactobacilli	Pseudomonas aeroginosa
Eubacteria	Proteus species
Bidifobacteria	Staphylococci
Enterococci	Clostridia
	Veillonellae

(Kelly 2006)

One of the biggest challenges for manufacturers of probiotic products is survivability of the bacteria. Most commercially available products suffer tremendous loss of activity during storage. After 5-6 months of storage, virtually no live organisms are present. Those bacteria that are present still have to survive the gastric pH and the duodenal bile acids before they can colonize the small intestines. To address these concerns, some manufacturers have developed microencapsulation processes to protect the bacteria while others provide a variety of bacteria to cover more bases.

CONCLUSION

Since prebiotics are used as both a source of nutrition for the colonocytes and as a fiber source for the animal, inclusion in the diet is fairly easy. The fiber source used should be listed on the ingredient panel, but may not state whether it is a source of FOS or inulin. Increasing your familiarity with the different fiber sources used, and their relative FOS and inulin contents can help when evaluating a diet. When is doubt, contact the manufacturer for further information.

Unlike the prebiotics, the use of probiotics is more of a transient process based on need rather than a long term process. Ideally, the body will supply the bacterial population to support the best intestinal health. But when this doesn't occur, or when challenges are expected that could affect intestinal health, probiotics can be easily added into the diet to help support the bacterial populations there. There have been no detrimental effects seen with long term prebiotic or probiotic use, unlike the use of antibiotics long term in animals.

When evaluating products to use for either prebiotics or probiotics ensure that there is research to support the claims given, that the product is proven to contain the stated levels of additives and that they have been proven to promote normal intestinal microflora. A review of probiotics in commercial dog and cat foods by Dr.Scott Weese, demonstrated that few products meet these guidelines (Weese 2003). Alternately, evaluation in a controlled clinical setting can provide firsthand knowledge of the effectiveness of the products used. By looking at these factors, you can ensure that you are providing the best product to support a happy, healthy intestinal tract with a hard working population of beneficial bacteria.

REFERENCES

[1.] Kelly M. The role of probiotics in GI tract health. **Nestle Purina PetCare Company**. 2006 [2.] Lerman A, Lockwood B: Nutraceuticals in veterinary medicine. **Pharmacy Journal** 2007 278:51 [3.] Pan X, Chen F, Wu T, Tang H, Zhao Z. Prebiotic oligosaccharides change the concentrations of short-chain fatty acids and the microbial population of mouse bowel. **Journal of Zhejiang University Science B**. 2009 10(4); 258-263 [4.] Roberfroid M. Prebiotics: The concept revisited. **Journal of Nutrition** 2007;137:830s [5.] Verdonk JMAJ, Shim SB, van Leeuwen P, Verstegan MWA. Application of inulin-type fructans in animal feed and pet food. **British Journal of Nutrition** 2005, 93 supplement 1 s125-s138 [6.] Weese, JS, Arroyo L. Bacteriological evaluation of dog and cat diets that claim to contain probiotics. **Canadian Veterinary Journal** vol. 44, March 2003

AVOIDING THE CONSEQUENCES OF MALNUTRITION: APPLICATIONS OF PARENTERAL & ENTERAL NUTRITION

Kara M. Burns, MS, MEd, LVT
Wamego, KS

Parenteral nutrition (PN) can be defined as the administration of nutrients by any means other than the gastrointestinal (GI) tract. Enteral nutrition (EN) is defined as providing nutrients through the use of some portion of the GI tract. The focus of this discussion will look at managing malnutrition through both means – PN and EN. PN is sometimes referred to as total or partial parenteral nutrition and these definitions are taken directly from the human literature. Total parenteral nutrition (TPN) implies that all energy and nutrient needs are met by the intravenous solution. These solutions are capable of maintaining a patient for months or years if necessary. Partial or peripheral parenteral nutrition (PPN) supply a portion of the energy and nutrient needs. These solutions are typically used for short durations (days to weeks). In veterinary medicine, the strict definition for TPN cannot be met because solutions that contain the total caloric energy needs do not meet the requirements for vitamins, macrominerals, and trace elements required for long-term feeding. As veterinary technicians it is important to realize the limitations of this treatment choice in regard to the critically ill patient.

The appropriate time to start nutritional support is determined by numerous factors. However, once the decision has been made to institute nutritional support, enteral feeding (utilizing the GI tract) is preferred to parenteral for several reasons. Enteral nutrition is more physiologic, prevents GI tract atrophy, reduces gut stasis, and limits bacterial overgrowth to list some of the underlying benefits. However, not every patient can tolerate enteral nutrition. Examples of patients who may be a good candidate for PN include patients with intractable vomiting, at increased risk for aspiration, poor anesthetic candidates, and those unable to tolerate the total caloric intake orally.

PARENTERAL NUTRITION

Administration of PN is achieved through a central or peripheral catheter. The utilization of a central venous catheter is preferred, especially when considering potential complications such as phlebitis and infection. When long term PN is indicated, central venous catheters comprised of polyurethane are recommended, as they do not have a predetermined removal time.

PN solutions are composed of carbohydrates, amino acids, lipid substrates, and other component. The carbohydrate is typically 50% dextrose. The use of a central intravenous line (e.g. jugular catheter) is necessary if the concentration of dextrose is greater than 10% in the PN solution. Dextrose provides a readily usable energy source for most patients. However, in critically ill patients insulin resistance may alter the body's ability to utilize this energy source. This is one of the reasons PN solutions rely on multiple energy sources to provide nutritional support. Crystalline amino acid solutions help maintain lean body tissue and positive nitrogen balance. They help the body maintain normal immune function, improve wound healing, and aid the function of many organs. These solutions are available in different concentrations from 3.5% to 15%. An 8.5% solution with or without electrolytes is the most common preparation utilized in veterinary medicine. These solutions lack taurine, but given the short duration of treatment, this deficiency is rarely a concern. Lipid emulsions provide the majority of the energy density of a PN solution. They supply essential fatty acids and are isotonic which helps decrease the osmolarity of the solution. These solutions are available in different concentrations from 10% to 30%. A 10% or 20% solution is the most common preparation utilized in veterinary medicine. Electrolytes, vitamins, and trace elements are other components that can be added to PN solutions. The most common electrolyte abnormalities that occur in patients receiving PN administration include hypokalemia and hypophosphatemia. These electrolytes can be supplemented in the PN solution. However, most patients receiving PN will also have a second line for crystalloid administration. Given the fact that these electrolyte abnormalities can change rapidly, it makes more sense to supplement them in the crystalloid solution rather than having to compound a new PN bag.

PN solutions can be compounded in-house, via a pharmacy, veterinary referral institution, or human health care company. Although the cost may be more to have the PN solution compounded at an outside facility, this may be a better way to obtain several days of treatment in a quick, aseptic manner. Typically all that is required by the outside facility is a nutritional calculation sheet and/or prescription from the veterinarian detailing the amounts of each substrate that needs to be added to the solution. If it is elected to compound the PN solution directly at your hospital, one of three methods may be utilized. The first and least desirable method uses a syringe to directly transfer each nutrient solution into a sterile, empty fluid bag. The concern with this technique is contamination of the system given the multiple transfers required to create the PN solution. The second option involves using a 3-in-1 sterile bag and gravity flow to create the PN solution. Each component of the PN solution can be connected to the bag and the nutrients are transferred directly by gravity flow. This method is faster and safer avoiding the potential for underlying contamination. However, transfer of exact quantities of nutrients is impossible and partially unused components are wasted. The final method involves utilizing a semi automated, closed-system, PN compounder. This method allows exact quantities of solution to be created and multiple bags can e made very quickly. The major downside of this method is the cost of the compounding machine. Once compounded, a PN solution can be stored for days or weeks at refrigerated temperatures.

Administration of PN solutions should be through a dedicated catheter placed in an aseptic manner. Whether the solution can be administered via a peripheral catheter or central line is directly related to the osmolarity of the solution. Simplistically, this decision is directly related to the dextrose concentration of the solution. As previously stated, a solution greater than 10% dextrose should be administered through a central line to avoid thrombophlebitis. Catheters composed of silicone, polyurethane, or tetrafluoroethylene are also recommended to avoid this complication. PN solutions are typically administered as a constant rate infusion over a 24-hour period. Once the solution is warmed to room temperature, it is recommended to utilize the entire solution in this time frame to prevent contamination and lipid particle destabilization. It is imperative that the healthcare team reevaluate the nutritional plan every day while administering PN. If the patient is not improving with this modality over several days and remains anorectic, options to institute enteral nutrition should be considered.

It is important for nursing care to include calculating the patient's caloric requirements. This is important because feeding more of any food than necessary has the potential for causing metabolic complications. A general rule to remember is most dogs and cats tolerate the food or solution that meets the resting energy requirement. Veterinary patients that are hospitalized have metabolic rates very near their RER. We can estimate the RER of hospitalized patients by using the following equations: < 2kg: RER = 70 x weight (kg)$^{0.75}$ Or > 2kg: RER = [30 x weight (kg)] + 70. Feeding patients at RER is a safe and rational approach as opposed to doubling basal requirements as this has been shown to result in overfeeding and potential complications including hyperglycemia, hyperammonemia, and hyperlipidemia. At minimum veterinary technicians should perform daily nutritional assessments to guide any needed adjustments to the nutritional plan of the hospitalized patient. At times a critically ill patient may require more frequent assessments.

Potential complications with PN administration can be classified into three main categories including mechanical, metabolic, and septic. Mechanical complications usually involve catheter related problems. Examples include occlusion, premature removal, line disconnection/ breakage, and/or thrombophlebitis. These problems can be avoided by the technician's strict adherence to aseptic technique and careful monitoring of patients. Metabolic complications are more likely to occur with PN solutions formulated to deliver total caloric requirements. The most common metabolic complication is hyperglycemia. Insulin therapy may be required to control this event. Other complications include hypertriglyceridemia, hyperammonemia, or electrolyte changes consistent with refeeding syndrome (e.g. hypokalemia, hypophosphatemia, hypomagnesemia). Reformulation of the PN solution is required if any of these problems occur. The most serious and potential life threatening complication is sepsis. Technicians need to use strict antiseptic technique when placing a catheter. Nursing management of catheters carrying hyperosmolar solutions containing amino acids requires special focus because the solution type is an excellent medium for colonization of bacteria. Antiseptic techniques and focused nursing care should be the same for all types of fluid administration, but is especially important in patients receiving PN. If signs of sepsis develop without an identifiable source, contamination of the solution and/or intravenous catheter should be suspected. A culture & sensitivity of both should be considered. It is due to this potential complication that many veterinarians recommend having the PN solution compounded at an outside facility.

ENTERAL NUTRITION

The gastrointestinal tract needs to be fed. The gut receives an overwhelming percentage of its nutrition from the chyme passing through it. In the small intestine, enterocytes utilize lumenal glutamine preferentially as their source of metabolic fuel. The colonocytes prefer butyrate, a short chain fatty acid formed by fermentation of lumenal carbohydrates. In the absence of these fuel sources, the gut epithelium slows growth and replication resulting in atrophy, necrosis and increased risk of bacterial translocation across the now abnormal gut barrier. A key to prevention of this potentially serious problem is providing nutrition support to the gut. One tenet of nutritional support is; if the gut works use it.

Force feeding sick, hospitalized animals is the best way to create food aversions. It is an inefficient feeding method resulting in more nutrients on the patient rather than in the patient. Typically force feeding will not provide a significant percent of needed calories and will lead to a false sense of accomplishment and continued inadequate intake.

Indwelling feeding tube is the method of choice if enteral assisted feeding is necessary for more than two days. After an indwelling tube had been placed, feeding is easier and less stressful. Nasoesophageal, esophagostomy, gastrostomy, and jejunostomy feeding tubes are the most commonly used. In animals undergoing laparotomy, placing gastrotomy or jejunostomy feeding tubes should be considered. The decision to use one tube over another is based on the anticipated duration of nutritional support (eg, days vs. months), the need to circumvent certain segments of the gastrointestinal tract (eg, oropharnyx, esophagitis, pancreatitis), healthcare team experience experience, and the patient's ability to withstand anesthesia (very critical animals may only tolerate placement of nasoesophageal feeding tubes).

Feeding Tube Options

- NE tubes are generally used for 3 – 7 days. Polyurethane tubes and silicon tubes may be placed in the caudal esophagus or stomach. An 8 Fr tube will pass through the nasal cavity of most dogs. A 5 Fr tube is more comfortable in cats. Anesthesia or tranquilization is not necessary (use topical ophthalmic anesthetics to numb the nasal cavity). These tubes can be used in patients considered anesthetic risks.
- Pharyngostomy and esophagostomy tubes vary from 8 to 16 Fr and may be placed in patients with disease or trauma to the nasal or oral cavity. These tubes can be used for long term in hospital or home feedings. Gastrostomy (mushroom-tipped, 16-22 Fr) can be placed either intraoperatively or percutaneously.
- Any tube that has been placed in the esophagus or stomach generally allows for bolus type meal feeding except in patients that vomit after each feeding. These patients will benefit from a slow continuous drip administered by a pump or gravity.
- Jejunostomy tubes (J-tubes, 5-8 Fr) are placed within the small intestine either surgically or endoscopically and are appropriate for cases where the stomach and proximal duodenum must be bypassed. Ideally food is administered at a slow, continuous drip delivered by a pump.

Nutritionists debate the exact formula for determining daily energy requirements in critically ill patients. A good starting point is to calculate the resting energy requirements for the patient's current weight. If ideal weight is used, underweight animals may be at greater risk for complications from overfeeding. Calorie intake may need to be adjusted to prevent weight loss or unintended weight gain. Food selection depends on tube size and location within the GI tract, the availability and cost of products and the experience of the clinician. Commercial foods available for enteral use in veterinary patients can be divided into two major types: 1) liquid or modular products and 2) blended pet foods. Nasal and jejunostomy tubes usually have a small diameter (<8 Fr.), which requires use of liquid foods. Orogastric, pharyngostomy esophagostomy and gastrostomy tubes have large diameters (>8 Fr.) and are suitable for blended pet foods.

Liquid foods are of two basic types: 1) elemental or monomeric and 2) polymeric. Foods said to be "elemental" are not truly elemental, but contain nutrients in small hydrolyzed absorbable forms and are best described as monomeric. The proteins are usually present as free amino acids, small dipeptides or tripeptides or larger hydrolyzed protein fractions. The fat source is often an oil of mixed (medium- and long-chain) fatty acids and the carbohydrate sources are mono-, di- and trisaccharides. There are several liquid foods on the human medical market that are positioned as monomeric or hydrolyzed diets and arc suitable when initially refeeding dogs and cats. These monomeric products are homogenized liquids that can be fed through any feeding tube including a J-tube. Monomeric foods are indicated in disease conditions such as inflammatory bowel disease, lymphangiectasia, refeeding parvoviral enteritis and pancreatitis cases and any other condition in which a patient's digestive capabilities are questionable.

Polymeric products contain mixtures of more complex nutrients. Protein is supplied in the form of large peptides (e.g., casein or whey). Carbohydrates are usually supplied as corn starch or syrup, and fats are provided by medium chain triglycerides (MCT) or vegetable oil. These foods require normal digestive processes and are appropriate for most veterinary clinical situations, especially when a small tube (<8 Fr.) has been placed and particular nutrient profiles are needed (e.g., low sodium, high protein, soluble fiber).

One of the leading liquid veterinary foods is a polymeric form that meets the current AAFCO nutrient allowances for adult dogs and cats. This product is a homogenized liquid containing I kcal/ml (4.2 kj/ml) and is usually accepted better than human liquid products containing MCT oil. This liquid food is the best option currently available in North America when small-diameter nasogastric and jejunostomy feeding tubes have been placed, or when continuous drip feedings are necessary. Several liquid milk replacer products are available; however, these products are not appropriate to feed to adult dogs and cats. They typically contain lactose, have high osmolarity, are lower in caloric density and do not meet AAFCO nutrient allowances for adult animals.

Module products are concentrated powdered or liquid forms of nutrients and are primarily supplemental. These products may be added to a liquid product to increase the concentration of a specific nutrient. There are protein, fat and carbohydrate modules (e.g., cascien powder, vegetable oil or corn syrup). For example, a protein modular product may be added to a human liquid product for an animal with high protein requirements. Soluble fiber can be added to these foods using psyllium husk fiber or pectin; however, these fibers may block the small side ports in 8-Fr. and smaller tubes.

Blended pet foods refer to commercial products nutritionally complete and balanced according to AAFCO allowances for dogs and cats. These products can easily be blended with a liquid to make a consistency that flows through a feeding tube. Some products have a blended texture, a high water content and very small particle size, whereas others are products that must be blenderized with water and may have to be strained to remove particulate matter.

The best recommendation when using the blended pet food method is to use a product that has been tested in feeding trials and is proven to be balanced and complete for dogs or cats. These products are more readily available, better tolerated and less expensive than the human liquid foods. These pet food products contain essential amino acids and essential micronutrients properly balanced to the caloric density of the food. Fewer medical complications (e.g., diarrhea) are likely to result. However, blended products are more likely to plug the feeding tube if the tube is not properly flushed after feeding. Patients may later consume the pet food orally, eliminating a diet change when the patient's appetite returns and the tube has been removed. These products are appropriate for patients in catabolic states that are using fat and protein substrates from body stores. When using small-diameter (<8 Fr.) feeding tubes, it will be necessary to dilute the pet food with water, which dilutes the caloric density. Blenderized moist veterinary therapeutic foods may have a place in assisted feeding of patients with specific disease conditions.

Calorie Content for some Prescription Diet® Blenderized foods

Feline Prescription Diet® k/d®	15.5 oz can + 1.5 cup water = 1 kcal/ml	18 Fr
Most all diets (k9 or feline) with ≥ 550 kcal / ~ 15 oz can will yield ~ 1 kcal/ml with above recipe		
Feline Prescription Diet® a/d®	2 cans + 50 ml water = 1 kcal/ml	10 Fr hard, 14 Fr easy
Feline Prescription Diet® l/d®	2 cans + 50 ml water = 1 kcal/ml	

The feeding schedule is often determined by the patient's ability to tolerate food and the logistics of feeding. Feeding an amount equal to the patients RER during the first 24 hours of food re-introduction, if physically tolerated, is recommended. Initially feeding one-third of the RER and then increasing the amount by one-third every 24 hours is a more cautious approach to initial feeding but isn't always necessary. Foods should be warmed to room temperature, but not higher than body temperature e before feeding.

Food boluses must be infused slowly (over approximately one minute per bolus) to allow gastric expansion. Daily food dosage should be divided into several meals according to the expected stomach capacity. Capacities for cats and dogs are 5 to 10 ml/kg body weight during initial food reintroduction. Maximum capacities as high as 45 to 90 ml/kg body weight have been measured in cats and dogs when fully re-alimented. Most often, meeting the patient's RER can be done in volumes far less than these maximums. Salivating, gulping, retching and even vomiting may occur when too much food has been infused or when the infusion rate is too fast.

Research in people has demonstrated that the stomach does not "shrink" during a prolonged fast, but rather the stretch receptors are more sensitive and stimulated by a smaller volume when refeeding occurs. Feeding should be stopped at the first sign of retching or salivating, the meal size reduced by 50% for 24 hours and then increased by 25% gradually. Foods provided via J-tubes must be infused slowly and often in either very small quantities or by a slow gravity drip or enteral pump with an hourly rate equal to RER/24 hours because the jejunum is volume sensitive.

Each meal must be followed by water flush to clear the feeding tube of food residue. When the patient is volume sensitive, it is important to know the minimum volume required to flush the tube. The patient's daily fluid requirement must also be met and additional tap water may be administered through the feeding tube to meet that requirement. Liquid oral medications may also be administered easily through feeding tubes. Plugged feeding tubes can be cleared by filling the tube with water or a nonalcoholic carbonated beverage and allowing time for the food plug to dissolve. In general, endport tubes are easier to maintain than sideport tubes because food tends to become trapped in the blind end of sideport tubes. All tubes except orogastric and nasoesophageal tubes require standard every-other-day bandage care.

In summary, PN and EN are viable nutritional choice for small animal patients. However, combined enteral and parenteral feeding is recommended when applicable to prevent intestinal hypertrophy and to facilitate healing by promoting intestinal growth. Proper patient selection is important to ensure an acceptable outcome. Proper nursing care and aseptic technique are also crucial to a positive patient outcome in both PN and EN. Nutritional requirements of hospitalized patients should be carefully monitored by the veterinary technician. Understanding some of the potential complications that can occur should enhance the effectiveness of PN and EN.

REFERENCES

[1.] Thomovsky E, et al. Parenteral nutrition: uses, indications, and compounding. **Compend Contin Educ Vet** 2007;29:76-8, 80-5. [2.] Thomovsky E, et al. Parenteral nutrition: formulation, monitoring, and complications. **Compend Contin Educ Vet** 2007;29:88-102. [3.] Chan DL , Freeman LM. Nutrition in critical illness. **Vet Clin North Am Small Anim Pract** 2006;36(6):1225-41. [4.] Proulx J. 2000. Nutrition in Critically Ill Animals. In **The Veterinary ICU Book**, 202-217. Jackson Hole, WY: Teton NewMedia. [5.] Saker K, Remillard RL. 2010. Critical Care Nutrition and Enteral-Assisted Feeding. In **Small Animal Clinical Nutrition 5th ed**, 439-476. Topeka, KS: Mark Morris Institue. [6.] Chandler ML,et al. Use of peripheral parenteral nutritional support in dogs and cats. **J Am Vet Med Assoc** 2000;216:669-73. [7.] Remillard RL, Saker K,. 2010. Parenteral-Assisted Feeding. In **Small Animal Clinical Nutrition 5th ed**, 477-498. Topeka, KS: Mark Morris Institue.

WHAT'S NEW IN THE TREATMENT OF CANCER IN SMALL ANIMALS

Laurel Williams, DVM, Dipl ACVIM (Oncology)
Raleigh, NC

INTRODUCTION

Evaluation and treatment of dogs and cats with cancer is increasingly common due to our expanding knowledge of tumor biology and treatment modalities and clients' willingness to pursue more advanced care for their pets. New developments in diagnosis and treatment are continuous. This lecture will focus on a few of these new developments

DIAGNOSTICS

The first step in making a diagnosis of neoplasia is aspiration cytology, although histopathology is needed for both definitive diagnosis and information on tumor grade. In addition to these commonly used tools, additional tests are available that can provide further information. In the case of round cell tumors, one such valuable tool is PARR (PCR for Antigen Receptor Rearrangement). This test can be used in both dogs and cats to determine whether lymphoid neoplasia present, to determine the phenotype (B- versus T-cell), and to monitor for the presence of minimal residual disease in patients being treated for lymphoma. This test utilizes genomic DNA and PCR primers specific for the V(D)J splice junctions of B- and T-cell receptor gene segments in canine and feline lymphocytes. The main advantage of PARR is its ability to differentiate a monoclonal (neoplastic) lymphoid proliferation from a polyclonal (reactive) lymphoid proliferation, although it is always prudent to interpret PARR data in conjunction with other findings, including history, clinical signs, cytology, flow cytometry and immunocytochemistry. In dogs, PARR has a reported sensitivity and specificity of approximately 90%. In cats, its sensitivity and specificity are slighty better for T-cell (89% and 80%, respectively) than B-cell neoplasia (60%, 70%, respectively). PARR can be run on a variety of tissues, including lymph node aspirates, bone marrow aspirates, cellular effusions, blood, cytology slides, formalin fixed tissues (25 micron sections) and fresh frozen tissues (25 micron sections at -80 degrees). Samples for PARR analysis may be submitted to the College of Veterinary Medicine, Clinical Immunology Laboratory, ATTN: Linda English, 4700 Hillsborough Street, Raleigh, NC 27606.

TREATMENT

Canine lymphoma

Chemotherapy is considered standard therapy for the treatment of canine lymphoma. However, the use of chemotherapy in combination with other treatment modalities is being investigated with the hope of improving remission rates and duration.

- Single agent chemotherapy

Early reports focused on the use of single agents including prednisone, L-asparaginase, cyclophosphamide, and doxorubicin. Response rates range from approximately 20-80% with remission durations of approximately 1-6 months.

- Combination chemotherapy

Attempts to improve remission rates and duration have been made through the use of combination chemotherapy. Numerous protocols have been reported with variations in scheduling, drug dosages, and dose intensity, although most have utilized induction chemotherapy followed by maintenance chemotherapy. The drugs most commonly used in these protocols include prednisone, L-asparaginase, vincristine, cyclophosphamide, and doxorubicin. Complete response rates range from approximately 65-90% with remission durations of approximately 6-11 months.

- Pulse chemotherapy

Evidence from human oncology has suggested that the use of discontinuous chemotherapy, rather than the traditional approach of using induction chemotherapy followed by maintenance chemotherapy, may provide comparable remission duration. Various reports in the veterinary literature have attempted to determine whether this is also true in veterinary oncology. Moore et al reported on 82 dogs treated with a 15-week discontinuous, or pulse, chemotherapy protocol consisting of prednisone, L-asparaginase, vincristine, cyclophosphamide, and doxorubicin. This group of dogs was compared to dogs previously treated with the same 15-week induction followed by maintenance chemotherapy. This study found no significant difference between the overall protocol remission duration between these two protocols. Similar findings were reported in a study by Garrett et al.

- Radiation therapy

Given the lack of appreciable differences among the various combination chemotherapy protocols, new approaches to the treatment of canine lymphoma are aimed at investigating adjuvant treatment modalities. Williams et al reported on 52 dogs treated with 11 weeks of induction chemotherapy consisting of prednisone, L-asparaginase, vincristine, cyclophosphamide, and doxorubicin followed by cranial and, three weeks later, caudal half-body radiation therapy. This protocol was well tolerated and the reported median overall remission was 486 days. Optimizations of combined modality therapy may yield improvements in maintaining long-term remission.

- Bone marrow transplantation

Given the apparent "ceiling" that has been reached with chemotherapy, another new treatment approach involves the use of bone marrow transplantation. Over the years, it has been shown that dogs can be given otherwise lethal doses of chemotherapy or radiation therapy and rescued via bone marrow transplantation, and this treatment has the potential to cure dogs of their disease. Based on this and under the direction of Dr. Steve Suter, North Carolina State University launched a bone marrow transplantation treatment unit in the fall of 2008. Dogs who have received combination chemotherapy and are in complete remission are eligible for transplantation. The transplantation protocol consists of an initial high-dose chemotherapy treatment to eradicate any remaining tumor cells, followed by daily administration of G-CSF to stimulate bone marrow stem cells. Peripheral blood mononuclear cells are then harvested via leukaphoresis, the dogs are treated with near lethal doses of radiation therapy and their harvested cells returned to them for engraftment and bone marrow recovery. Successful recovery and extended survival has been seen in many cases with new transplant patients being treated approximately every 2-3 weeks. Ongoing patient evaluation will be needed to determine the optimal role of this novel canine therapy.

Feline lymphoma

Chemotherapy is considered standard therapy for the treatment of feline lymphoma, with the possible exception of nasal lymphoma, which may be treated with radiation therapy.

- Single agent chemotherapy

Reports on single agent chemotherapy including prednisone, doxorubicin, and mitoxantrone describe response rates ranging from 9-40% with remission durations ranging from 3-12 months. While prednisone may be used as a single-agent, its use prior to the initiation of other chemotherapy use should be avoided since this may decrease response rate and duration to other agents.

- Combination chemotherapy

Improved remission rates and duration can be achieved with combination chemotherapy. Numerous protocols have been reported with variations in scheduling, drug dosages, and dose intensity, although most utilize induction followed by maintenance chemotherapy. The most commonly used agents in these protocols include prednisone, L-asparaginase, vincristine, cyclophosphamide, and doxorubicin. Complete response rates range from approximately 30-65% with remission durations of approximately 5-9 months.

- Radiation therapy

Building on what is known in humans and dogs, North Carolina State University recently completed a pilot study evaluating the use of targeted abdominal cavity radiation therapy for cats with high-grade lymphoblastic lymphoma confined to the abdominal cavity. As administered in this study, treatment was well tolerated and yielded remission times comparable or superior to conventional chemotherapy protocols. Further prospective studies are underway to better determine the impact of this combined modality approach.

Melanoma

Malignant melanoma is the most common oral tumor in dogs. Tumors are locally invasive with a moderate-high rate of metastasis to lymph nodes and lung. Conservative local resection is associated with poor tumor control and a median survival time of 3.5 months. Improved outcome and median survival of 9-10 months are achieved with more aggressive surgery, including maxillectomy or mandibulectomy, although both local recurrence and metastasis remain a problem. Radiation therapy yields similar outcomes, with overall response rates of 83-94% and median survival of 8 months reported using coarse fractionation. Given the rate of metastasis, chemotherapy should also be considered for patients with oral malignant melanoma. In one study, carboplatin chemotherapy produced an overall response rate of 28% when administered to dogs with malignant melanoma.

Rarely, spontaneous remissions are reported in melanoma patients, suggesting the immune system may play a role in controlling this tumor. Non-specific immune stimulants, such as IL-2 and IFN-α, have been shown to have limited activity against melanoma in human patients with a 12% overall response rate reported with IL-2. Antigen-specific therapies, such as vaccines, have also been evaluated for efficacy. In a trial reported in human patients, the combination of a melanoma peptide and IL-2 was associated with an overall response rate of 42%.

Anti-tumor vaccines exploit tumors cells frequent over-expression of normal tissue antigens. The immune system is conditioned to be naturally "tolerant" of these tumor cells, although vaccination may be used to break this tolerance. Tumor cells are more vulnerable to killing by vaccine-elicited immune effector cells than are their normal counterparts. Currently, Merial Pharmaceutical has a commercially available vaccine for melanoma which targets tyrosinase, an enzyme important in pigment production by melanocytes. In a clinical study, 9 dogs with melanoma in various locations were treated. All received four biweekly vaccinations, although the dosages varied (100 μg in 3 dogs, 500 μg in 3 dogs, and 1500 μg in 3 dogs). Two dogs with local control lived for 16 months. One dog with a pulmonary nodule had a complete remission, but later died of local recurrence. One dog had a lung nodule that did not progress for 4 years. No toxicity was noted in any dog.

LDM therapy

Angiogenesis is the process by which tumors induce new blood vessel formation and is essential for continued tumor growth and metastasis. In the absence of angiogenesis, tumors are restricted to sizes ranging from several microns up to 1-2 mm. The use of angiogenesis inhibitors has emerged as a clinically valid therapeutic target and complement to conventional chemotherapy in human oncology and may also have application in veterinary oncology. The advantages of using anti-angiogenic therapy include the rapid proliferation of tumor endothelial cells, the direct contact with the circulation and intravenous chemotherapy drugs, and the lack of genetic instability and mutation of endothelial cells leading to sustained responses to treatment. Many chemotherapy drugs can be cytotoxic to endothelial cells, although different schedules are required for optimal effectiveness. The use of LDM (low-dose metronomic) therapy is one such strategy. In one study, 9 dogs with stage II hemangiosarcoma were treated with LDM therapy (cyclophosphamide, etoposide, piroxicam). Median disease-free interval and survival (6) were comparable to that achieved with conventional doxorubicin chemotherapy.

CONCLUSION

Knowledge of conventional diagnostic and treatment approaches enables the practioner to care for the veterinary cancer patient. Familiarity with new developments allows one to stay current and progressive and provides a beneficial complement to conventional approaches. By seeking the balance between conventional and novel approaches, the goals of an extended good quality of life for the patient and a satisfied client can best be achieved.

REFERENCES

[1.] Burnett RC, et.al. **Vet Pathol** 2003;40:31. [2.] Werner JA, et.al. **Vet Pathol** 2005;42:596. Carter RF, et.al. **J Am Anim Hosp Assoc** 1987;23:587. [3.] Keller ET, et.al. **J Vet Intern Med** 1993;7:289. [4.] Moore AS, et.al. **J Vet Intern Med** 2001;15:348. [5.] Garrett LD, et.al. **J Vet Intern Med** 2002;16:704. [6.] Williams LE, et.al. **J Vet Intern Med** 2004;18:703. [7.] Mooney SC, et.al. **J Am Vet Med Assoc** 1989;194:696. [8.] Kristal O, et.al. **J Vet Intern Med** 2001;15:125. [9.] Cotter SM. **J Am Anim Hosp Assoc** 1983;19:166. [10.] Bostock DE. **Vet Pathol** 1979;16:32. [11.] Kosovsky JK, et.al. **Vet Surg** 1991;20:397. [12.] Wallace J, et.al. **Vet Surg** 1992;21:337. [13.] Bateman KE, et.al. **J Vet Intern Med** 1994;8:267. [14.] Blackwood L, et.al. **J Am Vet Med Assoc** 1996;209:98. [15.] Proulx DR, et.al. **Vet Radiol Ultrasound** 2003;44:352. [16.] Rassnick KM, et.al. **J Am Vet Med Assoc** 2001;218:1444. [17.] Bergman PJ, et.al. **Clin Cancer Res** 2003;9:1284. [18.] Lana S, et.al. **J Vet Intern Med** 2007;21:764. [19.] Moore PF, et.al. **Vet Immunol Immunopathol** 2005;106:167.

NEUROLOGICAL EMERGENCIES: EVALUATING HEAD TRAUMA IN SMALL ANIMAL PATIENTS

Theresa E. Pancotto DVM, MS
Blacksburg, VA

INTRODUCTION

Traumatic brain injury secondary to forces exerted on the cranial vault can be life threatening. The most commonly reported injuries in companion animals include motor vehicle accidents, trampling, falls, kicks, bites, and penetrating injuries[1,2,3,4]. Thorough initial evaluation will help provide medical staff with an accurate neurolocalization, estimate of prognosis, identification of life threatening concurrent injuries, and indications for appropriate therapy.

THE INJURED BRAIN

The major constituents inside the calvarium are the neural parenchyma, blood, and cerebrospinal fluid. In order to maintain cerebral homeostasis, an increase in any one component necessitates a decrease in another component as defined by the Monroe-Kellie Doctrine. Without compensatory changes, intracranial pressure (ICP) increases can lead to decreased oxygen and glucose delivery and neuronal damage[3,4]. Cerebral perfusion pressure (CPP), the net driving pressure for cerebral blood flow, is determined by the difference between mean arterial blood and intracranial pressure: CPP = MABP – ICP. Cerbral blood flow is determined by CPP and cerebrovascular resistance. In normal patients autoregulatory mechanisms function to adjust vascular tone to meet perfusion needs for MABP values between 50 – 150 mmHg. Injured brain is compromised and becomes more susceptible to ischemic injury during episodes of hypotension. Acute elevations in ICP may trigger a "Cushing's reflex". The initial drop in CPP causes massive sympathetic discharge to increase MABP via vasoconstriction and return CPP to normal. Baroreceptors in the aorta and carotid sinus respond to the elevated MABP by triggering reflex bradycardia. The presence of a Cushing's reflex indicates life-threatening increases in ICP and requires prompt treatment[3].

Primary Injury

Primary injury is the initial mechanical deformation of the calvarium and its contents. This includes concussion, contusion, laceration, hematoma, and fracture. Concussion is marked by brief loss of consciousness and has no histopathological change. Contusion is associated with hemorrhage and edema of brain tissue and clinical signs will vary with severity of the injury. Laceration is characterized by physical disruption to the neural parenchyma and may result in the formation of hematomas within the parenchyma or surrounding tissues. Clinical signs are often marked and may be focal[3].

Secondary Injury

Secondary injury is perpetuation of neuronal damage due to ischemia, excitotoxicity, free radical damage, and acidosis. Cerebral ischemia is worsened by systemic hypovolemia, systemic hypoxia, hypoventilation, or hyperventilation. The causes of exacerbated ischemia may be related to the initial injury affecting another body part, such as blood loss or pulmonary contusion, or from brain injury itself leading to ventilatory abnormalities. Concurrent spinal cord injury may also affect ventilation[5]. As neurons are damaged, excitatory neurotransmitters are released leading to further neuronal discharge, increased cellular activity, and depletion of energy stores; all causing further damage. Inflammatory mediators are also produced and can incite damage to the brain parenchyma and vasculature. Reperfusion leads to formation of free radicals that are particularly toxic to the nervous system because of its high lipid content. Accumulation of hydrogen ions and lactate from anaerobic metabolism leads to acidosis and cellular death[3,4].

PREREQUISITE VITALS

Patients must be evaluated globally to ensure systemic support of brain function. Extracranial injury can have a profound effect on outcome by altering blood pressure, gas exchange, and diaphragmatic function.

Cardiovascular Evaluation

Heart rate, heart rhythm, pulse quality, PCV/TS, mucus membrane color, and blood pressure should all be assessed on presentation[3,4].

Respiratory Evaluation

Airway patency, blood gas (PaO_2, $PaCO_2$), respiratory rate, respiratory character, mucus membrane color, and thoracic auscultation should all be assessed on presentation. If oxygenation status (PaO_2) is poor and does not improve with supplementation, pulmonary injury should be suspected[3,4].

Other

Blood glucose on admission has been shown to correlate with severity of head injury. Although persistent hyperglycemia (>24h) has been associated with a higher mortality rate in people, this relationship has not yet been proven in animals. Hyperglycemia has been demonstrated to worsen free radical production, excitatory amino acid release, cerebral edema, and vascular damage and iatrogenic hyperglycemia should be avoided[6].

A complete blood count and biochemical profile should be obtained as soon as possible to assess renal and hepatic function prior to initiating treatment, as medical therapies may compromise organ function in some animals[3]. Jugular venipuncture should be avoided.

NEUROLOGICAL EVALUATION AND THE MODIFIED GLASGOW COMA SCALE

The modified Glasgow coma scale[7] is a quantitative assessment of neurological function in small animals. It relies on evaluation in three primary areas: level of consciousness, brain stem reflexes, and motor activity/posture. Each category is scored 1-6 with 6 being normal in all categories and 1 being recumbent, non-responsive, and areflexic with dilated, unresponsive pupils and an absent oculocephalic reflex. The three scores are added and the total provides a 48 h survival prognosis[8]: 3-8 grave, 9-14 guarded, 15-18 good. Although the initial assessment is valuable, the trend over time is more indicative of final outcome.

In animals who maintain responsiveness and some degree of consciousness, a full neurological exam should be completed when they are systemically stable. The neurologic exam may yield results consistent with a focal intracranial injury. Such findings may support a diagnosis of an intra- or extra-axial hematoma. The patient should be monitored carefully and/or imaged as worsening signs may be alleviated with decompressive surgery.

ICP is not routinely measured in companion animal patients. Increased ICP is suspected based on clinical signs including the presence of a Cushing's reflex or signs referable to brain stem compression/herniation: bilateral, mydriatic, non-responsive pupils; loss of oculocephalic reflex; stupor.

Motor Activity	
Normal gait, normal spinal reflexes	6
Hemiparesis, tetraparesis, or decerebrate activity	5
Recumbent, intermittent extensor rigidity	4
Recumbent, intermittent extensor rigidity with opisthotonus	3
Recumbent, constant extensor rigidity with opisthotonus	2
Recumbent, hypotonia of muscles, depressed or absent spinal reflexes	1
Brain Stem Reflexes	
Normal pupillary light reflexes and oculocephalic reflexes	6
Slow pupillary light reflexes and normal to reduced oculocephalic reflexes	5
Bilateral unresponsive miosis with normal to reduced oculocephalic reflexes	4
Pinpoint pupils with reduced or absent oculocephalic reflexes	3
Unilateral, unresponsive mydriasis with reduced or absent oculocephalic reflexes	2
Bilateral, unresponsive mydriasis with reduced or absent oculocephalic reflexes	1
Level of Consciousness	
Occasional periods of alertness and responsive to environment	6
Depression or delirium, capable of responding to environment but response may be	5
Stupor, responsive to visual stimuli	4
Stupor, responsive to auditory stimuli	3
Stupor, responsive only to repeated noxious stimuli	2
Coma, unresponsive to repeated noxious stimuli	1

CONCURRENT INJURY

Concurrent injuries may be masked by initial stupor of the patient. Injuries to the caudal nasopharynx, cervical musculature, and cervical spine may be of particular importance as failure to address them could result in death from respiratory occlusion or failure (hypoventilation) or permanent non-ambulatory tetraparesis/plegia making the animal no longer an acceptable pet[5].

IMGAGING

Imaging is recommended for patients with localizing signs of brain dysfunction, moderate to severe deficits that are not responding to therapy, or worsening of neurological signs. Skull radiographs have low sensitivity but are helpful in some circumstances[3]. Although the overall incidence of concurrent cervical spinal injury has not been reported, depending on the type and location of injury, radiographic evaluation of the spine may be indicated. For example, young small breed dogs with higher incidences of atlanto-axial luxations should have concurrent spinal radiographs evaluated as any trauma could exacerbate a previously sub-clinical condition. Computed tomography (CT) is often the imaging modality of choice in traumatic brain injury. CT is a sensitive method of assessing fractures and detecting acute hemorrhage and brain edema[3]. Depending on the patient and equipment, multiple areas can be scanned in a single run without repositioning. In severely debilitated or rapidly worsening patients, CT can be performed without the use of general anesthesia. MRI provides more soft tissue detail than CT and in some instances is more sensitive for detecting subtle lesions.

FREQUENCY OF ASSESSMENTS

Serial monitoring is necessary to determine treatment and prognosis. The frequency of monitoring will depend somewhat on the patient's clinical signs since more critical patients may need continuous monitoring and neurological assessments every 30-60 min until clinical signs are improving.

REFERENCES

[1]Hopkins AL. **Vet Clin North Am Small Anim Pract** 1996;26:875-891. [2]Kolata RJ, et al. **J Am Vet Med Assoc** 1974;164:499-502. [3]Fletcher DJ **In: Bonagura JD, ed. Kirk's CVT XIV** 2009;33-37. [4]Syring, RS. **Vet Clin North Am Small Anim Pract** 2005;35:343-358. [5]Olby N, et al. **J Am Anim Hosp Assoc** 2002;38:321-326. [6]Syring RS, et al. **J Am Vet Med Assoc** 2001;218:1124-1129. [7]Shores A. **In: Kirk RW, ed. CVT X** 1983;847-854. [8]Platt SR, et al. **J Vet Int Med** 2001;15:581-584.

TOP TEN POTENTIAL DRUG INTERACTIONS IN DOGS & CATS

Lauren A. Trepanier, DVM, PhD, Dip. ACVIM, Dip. ACVCP
Madison, WI

In humans, the risk of adverse drug interactions multiplies as the number of administered drugs increases. Interactions can occur during IV drug administration, during oral absorption, at the target site, or during hepatic or renal elimination. Drug interactions may lead to loss of efficacy or increased toxicity. Although most of our knowledge of drug interactions comes from data in humans, many of these interactions are likely to occur in dogs and cats as well.

CIMETIDINE

Cimetidine is a major cytochrome P450 enzyme inhibitor, and decreases the clearance of many drugs, such as: chloramphenicol (dose-dependent leucopenia); lidocaine (GI and neurologic side effects); theophylline and aminophylline (GI upset, hyperactivity); warfarin, propranolol, diazepam, midazolam, and many others. Cimetidine may lead to toxicity of any of these drugs. Because of many potential cimetidine interactions, alternative H_2 blockers such as ranitidine, famotidine, or nizatidine (which are not P450 inhibitors at therapeutic concentrations), should be chosen over cimetidine for patients treated with multiple drugs. Ranitidine and nizatidine have the added advantage of prokinetic effects, which may counteract gastric atony in clinically ill patients.

SUCRALFATE

Aluminum-containing drugs such as sucralfate can form complexes with many other drugs in the GI tract, markedly decreasing drug absorption. This is a physicochemical interaction that is likely to occur in dogs and cats as it does in humans. Drugs that show decreased bioavailability with sucralfate include: tetracycline and doxycycline; theophylline, aminophylline, digoxin, and azithromycin. Sucralfate may decrease the efficacy of any of these drugs. Most notably, sucralfate markedly decreases the absorption of fluoroquinolones, with only 4% bioavailability for ciprofloxacin when the drugs are given together in humans. [1] While this interaction can be lessened by giving the fluoroquinolone two hours *before* the sucralfate (bioavailability increases to 83%), the opposite regimen is not recommended (i.e. giving the sucralfate first, followed two hours later by the antibiotic) because of the persistence of sucralfate in the stomach. As for H_2 blockers, sucralfate delays, but does not decrease the extent of, the absorption of H_2 blockers; therefore staggering of dosing is probably NOT required for sucralfate and H_2 blockers.

KETOCONAZOLE

Ketoconazole and itraconazole are best absorbed at acidic pH; therefore, do not combine these drugs with omeprazole, H_2 blockers, or other antacids. Interestingly, increased gastric pH does not affect the absorption of fluconazole. [2] Ketoconazole inhibits the cytochrome P450 enzyme, CYP3A, which has a wide substrate range and high potential for drug-drug interactions. Ketoconazole is also an inhibitor of p-glycoprotein, an important drug efflux transporter in the intestine, kidney, and biliary tree. Ketoconazole can therefore decrease the bioavailability and/or clearance of many drugs, such as digoxin, amitryptiline, midazolam, warfarin, and cyclosporine. The effects of ketoconazole to inhibit the clearance of cyclosporine can be exploited to allow lower doses of cyclosporine. Recommended dosages are cyclosporine, 5 mg/kg once to twice daily, depending on disease being treated, and ketoconazole, 10 mg/kg/day. Monitoring of ALT is recommended during ketoconazole therapy. Trough whole blood cyclosporine can be measured at steady state (by one week), just prior to the next dose. Target levels for immunosuppression in humans are 400-600 ng/ml, although lower concentrations may be associated with clinical responses in dogs and cats. Itraconazole, like ketoconazole, also inhibits the cytochrome P450 metabolism of the same drugs in humans.

FLUOROQUINOLONES

The oral absorption of fluoroquinolones is markedly impaired by any drugs that contain divalent or trivalent cations, such as aluminum hydroxide, aluminum carbonate, calcium carbonate, oral iron, and oral zinc, in addition to sucralfate. In humans and dogs, fluoroquinolones also inhibit the CYP1A2 metabolism of theophylline. This has lead to theophylline toxicity in humans. In dogs, enrofloxacin leads to higher plasma theophylline concentrations by about 30-50%.[3] The combination of a fluoroquinolone and theophylline could potentially lead to theophylline side effects in some dogs.

METOCLOPRAMIDE

As a dopaminergic (D2) antagonist and prokinetic agent, metoclopramide has several important drug interactions. Metoclopramide enhances the absorption of acetaminophen, aspirin, and alcohol overdoses via increased gastric emptying (shown in humans). Metoclopramide can theoretically lead to enhanced extrapyramidal side effects (tremor) in combination with phenothiazines (e.g. chlorpromazine, acepromazine), with selective serotonin reuptake inhibitors (e.g. fluoxetine), or with renal insufficiency. Interestingly, in humans metoclopramide reduces pain on injection of propofol, as well as the amount of propofol needed for anesthetic induction (by 20-25%), although the mechanisma are not clear. [4,5]Although metoclopramide is a dopamine antagonist, it has no effect on the use of dopamine for increasing urine output; this is mediated by D1 receptors.

CISAPRIDE

Like ketoconazole, cisapride is a substrate of CYP3A enzymes. Clarithromycin, erythromycin, and azole antifungals such as ketoconazole and itraconazole inhibit CYP3A, and can increase cisapride concentrations in humans. Such drug interactions led to ECG abnormalities (QT interval prolongation) and sometimes fatal cardiac arrhythmias in humans, which is why the drug was removed from the U.S. market. Cisapride can also cause similar ECG changes in dogs and cats at very high dosages. However, in one study in dogs, erythromycin did not alter cisapride pharmacodynamics.[3] More studies are needed in veterinary patients to determine the risk of clinically significant cisapride drug interactions.

FUROSEMIDE

Furosemide enhances the nephrotoxicity of amikacin and gentamicin; because of this, mannitol may be preferable to furosemide for treatment of acute renal failure caused by aminoglycosides. In combination with high dosages of furosemide, ACE inhibitors can cause hemodynamic changes that can lead to acute renal failure. Initial doses of ACE inhibitors should be conservative when furosemide is also instituted, and clinical status and renal function should be monitored over the first 1-2 weeks. Furosemide can lead to dehydration and pre-renal azotemia, which will decrease digoxin excretion. Furosemide also causes hypokalemia and hypomagnesemia, both of which exacerbate the cardiac toxicity of digoxin. Each of these interactions can lead to digoxin toxicity unless serum digoxin levels are monitored. Other furosemide-drug combinations can affect efficacy. Hypokalemia secondary to furosemide can blunt the antiarrhythmic effects of lidocaine. Serum potassium should be evaluated in patients with ventricular arrhythmias, and potassium supplementation should be considered if patients do not respond to lidocaine. Furosemide administration will also increase the renal loss of bromide, and lower serum bromide concentrations,[6] which may lead to seizure breakthrough.

OMEPRAZOLE

Omeprazole is an inhibitor of some cytochrome P450's in humans, and may inhibit the metabolism, and possibly increase the toxicity, of diazepam, midazolam, and warfarin. As an inhibitor of gastric acid secretion, omeprazole can also decrease the absorption of iron supplements, ketoconazole, and itraconazole (but not fluconazole, which does not require an acidic pH for absorption).[2]

PHENOBARBITAL

Phenobarbital is a major P450 enzyme inducer in humans and dogs. Phenobarbital speeds the metabolism of many drugs, including glucocorticoids, ketoconazole, clomipramine, lidocaine, etodolac, theophylline, digoxin, propranolol, and many others... Phenobarbital also induces mitotane clearance, and can lead to higher mitotane dosage requirements in dogs being treated for hyperadrenocorticism. Conversely, chloramphenicol is a major inhibitor of phenobarbital clearance, and can lead to sedation in dogs being treated with phenobarbital. In cats, however, phenobarbital causes minimal cytochrome P450 enzyme induction,[4,5] and therefore these P450-mediated drug interactions are unlikely in felines.

CLOMIPRAMINE

Clomipramine, which is a tricyclic antidepressant that inhibits norepinephrine reuptake, can potentially have pharmacologic interactions with monoamine oxidase inhibitors, which decrease the breakdown of norepinephrine and serotonin. For example, clomipramine, when used in combination with MAO inhibitors, can lead to "serotonin syndrome" (twitching, tremor, seizures) in humans.[7] Examples of veterinary MAO inhibitors include selegiline (L-deprenyl, in Anipryl®) and amitraz (Mitaban®, Preventic®), found in tick dips and collars. The potential for an interaction between clomipramine and these drugs has not been directly evaluated in dogs, but the Clomicalm® label recommends against clomipramine being given within 14 days of either L-deprenyl or amitraz. In addition to these pharmacologic interactions, the metabolism of clomipramine can be inhibited by ketoconazole or itraconazole, as well as by fluoxetine (Prozac), which can lead to increased clomipramine levels and cardiac conduction disturbances in humans.

Drug interactions in humans that may also affect dogs and cats

Drug	May increase the toxicity of:	May decrease the efficacy of:	Toxicity may be increased by:	Efficacy may be decreased by:
Cimetidine	Lidocaine, theophylline, diazepam, propranolol	Ketoconazole, itraconazole, iron supplements		
Sucralfate		Fluoroquinolones, tetracyclines, erythromycin, theophylline, digoxin		
Ketoconazole	Cyclosporine, warfarin, digoxin, amitriptyline, midazolam, cisapride			Antacids, H_2 blockers, omeprazole
Fluoro-quinolones	Theophylline			Sucralfate, iron, calcium, aluminum, magnesium
Metoclopramide	Ethanol, aspirin, or acetaminophen overdoses; propofol?	Probably *does not* counteract the renal effects of dopamine	Aceprozamine, fluoxetine (tremor)	
Furosemide	ACE inhibitors, digoxin, aminoglycosides	Bromide, lidocaine (via hypokalemia)	Aminoglycosides	NSAID's
Cisapride			Clarithromycin, erythromycin, ketonazole, itraconazole, fluoxetine	
Omeprazole	Diazepam, warfarin, digoxin	Ketoconazole, itraconazole, iron supplements		
Phenobarbital		Glucocorticoids, clomipramine, lidocaine, theophylline, digoxin, propranolol…	Chloramphenicol (increased sedation)	
Clomipramine	L-deprenyl, amitraz		Ketonazole, itraconazole, fluoxetine	

REFERENCES

[1] Van Slooten AD, Nix DE, Wilton JH, et al. Combined use of ciprofloxacin and sucralfate. **DICP** 1991;25:578-582. [2] Zimmermann T, Yeates RA, Riedel KD, et al. The influence of gastric pH on the pharmacokinetics of fluconazole: the effect of omeprazole. **Int J Clin Pharmacol Ther** 1994;32:491-496. [3] Intorre L, Mengozzi G, Maccheroni M, et al. Enrofloxacin-theophylline interaction: influence of enrofloxacin on theophylline steady-state pharmacokinetics in the beagle dog. **J Vet Pharmacol Ther** 1995;18:352-356. [4] Ganta R, Fee JP. Pain on injection of propofol: comparison of lignocaine with metoclopramide. **Br J Anaesth** 1992;69:316-317. [5] Page VJ, Chhipa JH. Metoclopramide reduces the induction dose of propofol. **Acta Anaesthesiol Scand** 1997;41:256-259. [6] Trepanier LA. Use of bromide as an anticonvulsant for dogs with epilepsy. **J Am Vet Med Assoc** 1995;207:163-166. [7] Gillman PK. Tricyclic antidepressant pharmacology and therapeutic drug interactions updated. **Br J Pharmacol** 2007;151:737-748.

NEUROLOGIC EXAMINATION & LOCALIZATION IN THE DOG: A CASE BASED APPROACH

Carissa D. Sparks, BS, RVT
West Lafayette, IN

NEUROLOGICAL EXAMINATION

The objective when performing a neurologic examination is to determine the presence or absence of neurologic disease and subsequently localize the lesion neuroanatomically. The functional divisions of the nervous system include the brain, spinal cord and neuromuscular system. Intracranial localization is segregated into the forebrain, brainstem, and/or cerebellum. Spinal cord localization is divided into C1-C5, C6-T2, T3-L3, and L4-S3. These divisions of the spinal cord require the clinical signs to be localized to the upper motor neuron (UMN) or the lower motor neuron (LMN) systems. Neuromuscular diseases are classified as neuropathies, myopathies, or junctionopathies based upon clinical signs and diagnostic testing. Multifocal disease should only be highly suspected if the deficits can not be explained by a single lesion.

Mentation

Observation of the patient's mentation, behavior and involuntary movements are essential for accurate localization. Mentation is classified as depressed, obtunded, stuporous, or comatose. Mental status is further classified as appropriate or inappropriate, based on the patient's environment and systemic health (e.g., pacing may be normal for a nervous animal at the vet clinic, but not for an animal at home). Behavior changes such as loss of learned behaviors, sudden aggression or disorientation are all important to document.

Posture and Gait

Posture and gait are often appreciated simultaneously. The trunk and limb position with respect to gravity are observed and the gait is evaluated from the rear, front and side views at both a walk and a trot. Abnormalities include, but are not limited to, leaning, falling, rolling, circling in tight or large circles, a wide or narrow based stance, or deviation of the spine. Gait abnormalities should be characterized as neurologic or non-neurologic. The most common neurologic gait abnormality is incoordination (ataxia). There are three types of ataxia: proprioceptive, vestibular, or cerebellar. In addition, patients may exhibit weakness (paresis) or paralysis (plegia). The prefixes mono-, para-, or hemi- are used to designate which limbs are involved in the paresis/plegia. It is important to note if a paretic patient is ambulatory or non-ambulatory and to assess whether non-ambulatory patients are paretic or plegic by moving them with support. If a patient is definitely plegic, then accurately differentiating the presence/absence of superficial/deep pain sensation is warranted.

Postural Reactions

Postural reactions evaluate conscious proprioception (the body's position in space). In addition to peripheral nerve pathways and spinal cord, postural reactions evaluate the patient's response and should not be confused with reflexes. Responses require cerebral involvement to process the stimulus; whereas reflexes do not require conscious involvement. There are several methods utilized to test conscious proprioception depending upon the species and size of the patient. Conscious proprioceptive placing ("knuckling"), hopping, hemistanding/hemiwalking, tactile placing, extensor postural thrust, and wheelbarrowing are the most commonly used conscious proprioceptive exams. These responses are recorded as normal, delayed, or absent. Proprioceptive deficits can be a sign of lesions in the peripheral nerve, spinal cord, brainstem, or contralateral forebrain.

Spinal Reflexes

Spinal reflexes encompass the myotatic reflexes, flexor withdrawal reflexes, cutaneous trunci reflex, and the perineal reflex. Each myotatic reflex examines a particular nerve and identifies a potential deficit within a specific spinal cord segment. Reflexes in the thoracic limbs include the biceps reflex (musculocutaneous nerve C6-C8), the triceps reflex (radial nerve C7-T1) and the extensor carpi radialis reflex (radial nerve C7-T1). Pelvic limb reflexes include the patellar reflex (femoral nerve L4-L6), the cranial tibial reflex (peroneal nerve L6-L-7) and the gastrocnemius reflex (tibial nerve L7-S1). *Myotatic reflexes* form monosynaptic pathways. The stimulus activates a stretch receptor sending an electrical transmission up the afferent nerve, where it synapses in the grey matter of the spinal cord. The impulse then travels back down the efferent nerve to the muscle resulting in a reflex.

The flexor withdrawal reflexes in the thoracic limbs involve all of the nerves in the brachial plexus (musculocutaneous, radial, ulnar, median, and axillary). The pelvic limb flexor withdrawal reflexes are innervated by the sciatic nerve. Flexor withdrawal reflexes form polysynaptic pathways. The stimulus activates afferent fibers to synapse with interneurons in the dorsal spinal cord. The flexor efferent neurons are activated resulting in flexion of the limb.

It is important to note that not all of the reflexes will elicit reliable results. The most reliable reflex in the thoracic limbs is the flexor withdrawal, and the most reliable in the pelvic limbs are the flexor withdrawal and the patellar (although reduced patellar reflexes can be observed in some geriatric patients or patients with stifle disease).

The cutaneous trunci reflex (a.k.a panniculus reflex) is also a polysynaptic pathway. This pathway requires superficial sensation to be intact. The stimulus sends an impulse through the afferent dermatomes to the spinal cord and cranially to spinal cord segments C8-T1. The efferent signal is sent through the brachial plexus, to the lateral thoracic nerve and back to the muscles bilaterally.

The perineal reflex examines the pudendal nerve (S1-S3) through pinching the perineal area and observing constriction of the anal sphincter.

It is important to note that reflexes are not responses and do not involve the cerebrum. The impulse does not include brain involvement for a conscious response; therefore it is a reflex. This is a very important fact to be made aware when testing a paraplegic patient's RESPONSE to a painful stimulus. *It is imperative not to mistake withdrawal of a limb for conscious sensory perception.*

Cranial Nerve Evaluation

There are twelve cranial nerves. There are several tests performed that will each assess one or more cranial nerves. For each of these tests, there will be an afferent (sensory/input) and efferent (motor/output) arm.

Olfaction (CN I) tests the ability to smell. Twitching of the nose is the efferent response to receiving an afferent stimulus of a pleasing aroma. It is important when testing CN I not to use a noxious odor because this will stimulate the trigeminal nerve and not the olfactory nerve. Assessing a unilateral deficit is not possible; therefore, this test is not commonly performed.

The menace response is a learned response and is typically not elicited in patients younger than 3 months of age. The afferent branch includes both the optic nerve (CN II) and the forebrain which perceives the menacing gesture. The efferent branch includes both the cerebellum and the facial nerve (CN VII) to coordinate closing of the eyelids.

Pupillary Light Reflex should be tested both directly and consensually. The afferent branch (CN II) senses the light and the efferent branch (oculocephalic nerve CN III) constricts the pupil. If the constriction is poor then consider a true lesion, iris atrophy, pharmacologic interference, or stress related sympathetic dilation. When evaluating anisocoria, it is important to both light and dark adapt to decipher which eye is abnormal.

There are two types of strabismus, resting and positional. Resting strabismus is continuous and is evaluated with the head in a neutral position to assess the efferent branch of CN III, trochlear (CN IV) and abducens (CN VI) nerves. A CN III lesion will result in a ventrolateral strabismus, a CN IV lesion will result in rotation of the globe, and a CN VI lesion will result in a medial strabismus. Positional strabismus only becomes apparent with a change in the head position and is due to a lesion in the vestibulocochlear nerve (CN VIII).

Nystagmus evaluates CN III, CN IV, CN VI and CN VIII. Normal physiologic nystagmus occurs during horizontal or vertical head movements. The fast phase should be in the direction of head movement. Abnormal physiologic nystagmus may occur with CN III, IV, VI and VIII lesions. Spontaneous nystagmus is the most common type of abnormal nystagmus, and occurs with peripheral (CN VIII) and central vestibular lesions. Spontaneous nystagmus is present while the head is stationary. Inducible or positional nystagmus is a variation of spontaneous nystagmus. If the brain has compensated for the CN VIII deficit and eradicated the spontaneous nystagmus, the patient can be turned on their back to induce nystagmus. This abrupt change in position briefly decompensates the patient and a spontaneous nystagmus may be appreciated. Nystagmus should be classified based on the direction of the fast phase and if it changes. This will help to differentiate between central or peripheral vestibular disease and identify the side of the lesion.

Facial sensation and the palpebral reflex test the trigeminal nerve (CN V) to sense touch, and CNVII to close the eyelids. The three branches of the trigeminal nerve are the ophthalmic, maxillary and mandibular nerves.

Nasal stimulation tests the ophthalmic branch of CN V and the forebrain. This test is a response to noxious stimulation of the medial nares and requires cerebral involvement. The motor branch of CN V is evaluated through palpation of the muscles of mastication and the evaluation of jaw tone.

The glossopharyngeal (CN IX) and vagus (CN X) nerves are evaluated while performing the gag reflex. Sensation of the pharynx causing contraction of the soft palate and swallowing are positive results.

The accessory nerve (CN XI) is perhaps the most challenging to evaluate. The trapezius muscles can be palpated for atrophy, but most often an EMG will be required to appreciate any abnormalities.

Tongue symmetry and movement are evaluated to assess the hypoglossal (CN XII) nerve. A lesion will show deviation of the tongue towards the lesion and atrophy of that side of the tongue.

Palpation/Sensory

Palpation of the axial and appendicular skeleton for symmetry, tone, strength, muscle mass, fractures, and hyperesthesia of the cervical, thoracic, lumbar, or lumbar-sacral spine will aid in localization and identify compression or inflammation of the aforementioned structures. Palpation for draining tracts, or fluid pockets is imperative to appreciate potential infections that enter the veins of the face, nasal cavity, orbit and upper teeth that drain directly into the cavernous sinus. Palpation of the cranium will help to identify open fontanelles of the cranium. Bladder size/function should also be noted. The order of functional loss for a compressive spinal cord lesion begins with conscious proprioception, progressing to motor and bladder dysfunction, and finally loss of superficial and deep pain perception. *Superficial pain* is evaluated by pinching the webbing between the toes. The patient should give a conscious response to the perception of the stimulus. If the patient does not respond to the superficial pain stimulus then the *deep pain* perception test can be performed by pinching across the bone of the digit. For upper motor neuron lesions, if a patient has voluntary motor function it is not necessary to test for superficial pain. If LMN disease is suspected then all of the digits should be evaluated individually. *A withdrawal reflex should NOT be confused for a response to painful stimuli.*

NEUROANATOMIC LOCALIZATION

Spinal Cord Localization

Lower motor neuron (LMN) and upper motor neuron (UMN) disease is classified based upon clinical signs. Both upper and lower motor neuron diseases will cause weakness, so other findings must be used for localization. Gait and posture should be observed to detect a proprioceptive ataxia, paresis or paralysis. *Lower motor neuron localization* occurs when the nerve impulse from the nerve cell body in the spinal cord (segments C6-T2 or L4-S3) cannot reach the effector muscle. As a result of this, there is a reduction or absence of the reflexes in the thoracic or pelvic limbs. The deficiency in the reflex arc can be a result of the afferent nerve, efferent nerve, or the cell bodies in the gray matter of the spinal cord. Lesions in these LMN structures are classified by hypo- or areflexia and hypo- or atonia and result in a decrease in motor function and rapid muscle atrophy.

Upper motor neuron localization occurs when the signal from cell bodies in the cerebrum and brain stem fail to connect with the lower motor neuron system. Lesions causing upper motor neuron signs to the thoracic and/or pelvic limbs can be due to brain or spinal cord segments C1-C5. Lesions causing UMN signs only in the pelvic limbs can be due to brain or spinal cord segments C1-C5, C6-T2, or T3-L3. Myotatic reflexes and muscle tone may be normal or increased due to loss of inhibition from upper motor neurons. Muscle atrophy occurs slowly due to disuse. Palpation of the spine for hyperesthesia and defining the cutaneous trunci cutoff are useful pieces of the exam in UMN cases.

Intracranial Localization

Intracranial localization is divided into three functional areas, forebrain, brainstem and cerebellum, based upon clinical signs. If an intracranial lesion is suspected, listing the abnormal components of the cranial nerve exam and comparing them to find the common component will help identify the localization.

Forebrain lesions often present with altered behavior, mentation or seizures. Posture is either normal, a head turn may be noted toward the lesion, or head pressing may occur. Gait is often normal, but compulsive walking in wide circles towards the lesion or pacing with lack of purpose may be apparent. Postural reactions may be absent contralateral to the lesion. Cranial nerve *reflexes* are usually normal but the *responses* may be abnormal. (Responses require forebrain involvement whereas reflexes do not). The menace response and response to noxious nasal stimulation may be deficient contralateral to the lesion. Spinal reflexes are intact. Cerebral disease can cause a referred hyperesthesia to the cervical spine. Micturation may be inappropriate due to the loss of learned behavior. In addition to the aforementioned abnormalities, lesions in the hypothalamus may also exhibit abnormal thirst, appetite and/or endocrine dysfunction.

Lesions in the *brainstem* are more likely to present with conscious disturbances if the reticular activating system is involved. Patients may be normal, or exhibit stupor/coma. Depending upon the severity of disease, a severe ataxia and tetraparesis or hemiparesis may be appreciated. Spinal reflexes are intact with brainstem disease. Postural reaction deficits are ipsilateral to the lesion. Cranial nerves III through XII may be affected and are ipsilateral to the lesion. Caudal brainstem lesions may cause central vestibular disease in which patients may present with a wide based stance, vestibular ataxia, ambulating in tight circles, falling, rolling, a head tilt, spontaneous nystagmus, and/or a positional strabismus. Although the neurological examination can be used to determine if the vestibular signs are a peripheral or central localization, central vestibular disease can mimic peripheral disease. For this reason, our examination may allow us to reliably determine that central disease is present, but rarely allows us to rule out central vestibular disease.

The *cerebellum* coordinates movement, but does not initiate movement. Lesions in the cerebellum will present with dysmetria, intention tremors and a cerebellar ataxia. One of the hallmarks of cerebellar disease is that these patients present *without* paresis. Postural reactions are normal, but awkward. Mentation is normal. Cranial nerve exam is normal with the exception of a potential absent menace response ipsilateral to the lesion.

Neuromuscular Localization

Neuromuscular disease should be differentiated from generalized systemic illness and spinal cord diseases. Some neuromuscular diseases may involve cranial nerves. *Neuropathies* affect either sensory or motor peripheral nerves, the latter of which produces LMN signs of hypo- or areflexia, hypo- or atonia. Proprioceptive deficits are variable. Rapid muscle atrophy is common. Cranial nerves may be affected. *Myopathies* typically present with normal reflexes and proprioception, but exhibit episodic or persistent weakness. Myopathies may also present with bilateral, symmetrical muscle atrophy or hypertrophy. Muscle pain (myalgia) may be present upon palpation. *Junctionopathies* present with normal sensation but weakness that is often exacerbated by exercise. Muscle atrophy is not appreciated, but muscle tone may be flaccid. Reflexes can be hypo- or areflexive depending upon the severity of the disease, but are often normal and fatigue with repetition. Neuropathies, myopathies, and junctionopathies are often difficult to discern from one another upon examination, but they are readily distinguished from intracranial and spinal cord disease. Diagnostic testing is often necessary to further characterize peripheral nervous system disorders.

CONCLUSIONS

The neurologic examination is complete when an accurate signalment and history have been obtained, and a thorough examination including gait, postural reactions, cranial nerve assessment and spinal reflexes have been performed. Clinical signs localize the lesion to the brain, spinal cord or neuromuscular system. The brain is further categorized into the cerebrum, brainstem or cerebellum. Spinal cord segments are divided into upper motor neuron (C1-C5, T3-L3) or lower motor neuron (C6-T2, L4-S3) segments based upon functionality and spinal reflex deficiency. The neuromuscular system is differentiated into neuropathies, myopathies, or junctionopathies. Results from the neurologic examination will guide toward a list of differential diagnoses, and determine what diagnostic tests to perform to obtain a definitive diagnosis.

REFERENCES

[1] de Lahunta et al, **Veterinary Neuroanatomy and Clinical Neurology 2009**, [2] Dewey CW, **A Practical Guide to Canine and Feline Neurology** 2008, [3] Oliver JE, et al. **Handbook of Veterinary Neurology** 1997, [4]Thomas WB,**Veterinary Clinics of North America: Small Animal Practice** 2000

INFLAMMATORY BRAIN & SPINAL CORD DISEASES

Christopher L. Mariani, DVM, PhD, DACVIM (Neurology)
Raleigh, NC

DEFINITIONS

- Meningitis: inflammation of the meninges (the pia, arachnoid and dura mater; protective membranes covering the brain and spinal cord).
- Encephalitis: inflammation of the brain.
- Myelitis: inflammation of the spinal cord.
- Meningoencephalitis: inflammation of the meninges and brain.
- Meningomyelitis: inflammation of the meninges and spinal cord.

INTRODUCTION

Meningoencephalitis and meningomyelitis are common problems in small animal medicine. They may affect any breed of dog or cat, and animals of multiple ages, although certain disease processes are more common in specific breeds and age ranges (see below). Small and toy breed dogs in particular are frequently afflicted with inflammatory central nervous system (CNS) disorders.

CLINICAL SIGNS

A variety of clinical signs may be seen with inflammatory CNS disease, depending on what area of the nervous system is affected. Involvement of the forebrain (cerebral cortex and thalamus) may result in seizures, altered mentation (stupor, coma, delirium, head pressing), vision abnormalities, and compulsive pacing or circling. Involvement of the brainstem (pons and medulla oblongata) may result in altered mentation, cranial nerve deficits (e.g., vestibular dysfunction [head tilt, nystagmus], facial nerve paralysis, swallowing difficulties, tongue paralysis), ataxia, and proprioceptive deficits. Spinal cord involvement leads to ataxia, paresis, and possibly segmental spinal reflex deficits. Most inflammatory CNS conditions result in pain due to involvement of the meninges. A hallmark of these conditions is the presence of multifocal neurologic signs; that is, involvement of multiple areas of the CNS (e.g., forebrain, brainstem and spinal cord).

The majority of small animals with meningoencephalitis or meningomyelitis have signs restricted to the nervous system. However, involvement of other organ systems is possible with some infectious agents (e.g., systemic fungal or bacterial infections, rickettsial disease). Contrary to what many believe, fever is quite uncommon in most meningoencephalitis patients.

DIAGNOSIS

The diagnosis of inflammatory CNS diseases in small animal patients involves consideration of the signalment, reported and observed clinical signs, and results of diagnostic testing. The most useful tests are advanced diagnostic imaging (particularly magnetic resonance imaging [MRI]), and cerebrospinal fluid (CSF) analysis. Although computed tomography (CT) is of some use in demonstrating brain lesions, MRI is the gold standard for brain imaging, and is much more sensitive in showing inflammatory lesions of the brain and spinal cord. Inflammatory lesions are typically hyperintense on T2-weighted, STIR, and FLAIR images, hypointense or isointense on T1-weighted images, and enhance to varying degrees after intravenous contrast administration (gadolinium compounds). Analysis of CSF in these patients typically shows elevated white blood cell numbers (pleocytosis) and protein. Cytology should be performed to determine the types of inflammatory cells present. Electrodiagnostic testing is less frequently used, but may provide useful functional information to complement the imaging and CSF results. Electroencephalography (EEG) and brainstem auditory evoked response (BAER) provide such information about the cerebral cortex and brainstem pathways respectively. Finally, testing for infectious disease agents may involve quantification of antibody titers or polymerase chain reaction (PCR)-based testing on blood and CSF samples.

THERAPY

The therapy of meningoencephalitis and meningomyelitis should be tailored to the specific etiology or suspected syndrome, while taking into account the severity of the clinical signs, comorbid disease processes, and financial constraints of the owner. Patients with identified infectious etiologies should be treated with appropriate antimicrobial medications, which may include antibiotics, antifungals, or antiprotozoals. Whenever possible, medications with good penetration of the blood-brain barrier should be selected for therapy. Such choices often include doxycycline, sulfonamide drugs, and fluconazole.

Corticosteroids are a mainstay of therapy for inflammatory CNS conditions, regardless of etiology. Although immunosuppression and long-term treatment with corticosteroids is usually contraindicated for infectious processes, short-term therapy can dramatically reduce the inflammatory response, and can be life-saving. In addition, corticosteroids form the cornerstone of therapy for the inflammatory conditions without an underlying identifiable etiology. Although many animals benefit from anti-inflammatory doses of these medications, immunosuppressive doses are often required for disease remission. The doses of corticosteroids are generally reduced over an extended period of time to the lowest dose that controls clinical signs.

Many clinicians add additional immunomodulatory or cytotoxic medications to the therapeutic regimen of patients with meningoencephalitis of unknown etiology (MUE). Such drugs include cytosine arabinoside (cytarabine, Ara-C), cyclosporine A, leflunomide, mycophenolate mofetil, lomustine (CCNU), and procarbazine. These medications are useful to provide additional immunosuppression through alternate mechanisms of action, and also to reduce the side effects associated with glucocorticoid therapy, which include polyuria, polydipsia, polyphagia, weight gain, panting, gastrointestinal ulceration, coagulopathies, pancreatitis, and steroid hepatopathy.

ETIOLOGIES AND SPECIFIC SYNDROMES

A variety of infectious agents may lead to nervous system dysfunction, including viruses, bacteria, fungal organisms, rickettsiae, protozoa, and parasites (see Table 1). Despite the number of potential infectious etiologies, these are rarely identified as the underlying cause of meningoencephalitis or meningomyelitis in small animals. Although cats are often presumptively diagnosed with an infectious etiology, dogs are much more likely to fit into one of the syndromes described below that have no identifiable infectious cause.

Table 1. Infectious Agents Causing Meningoencephalomyelitis in Small Animals

Category of Agent	Specific Organisms
Viral	Canine distemper virus, FIP, FeLV, Rabies, Pseudorabies, West Nile virus
Fungal	*Cryptococcus neoformans, Coccidiodes immitis, Blastomyces dermatidis, Aspergillus spp., Cladosporium spp.*
Rickettsial	*Ehrlichia canis, Rickettsia rickettsia, Anaplasma phagocytophilum*
Bacterial	*Streptococcus spp., Bartonella spp., Borrellia burgdorferi., Leptospira spp.*
Protozoal	*Neospora caninum, Toxoplasma gondii, Entamoeba histolytica*
Parasitic	*Cuterebra spp., Dirofiliaria immitis, Baylisascaris procyonis, Angiostrongylus spp., Prototheca spp.*

Steroid responsive meningitis-arteritis

As its name implies, this is a condition that involves the blood vessels of the meninges, leading to vasculitis and meningitis. It typically affects young (6-18 months), large breed dogs, particularly Boxers, Bernese Mountain dogs, Weimaraners, and Nova Scotia Duck Tolling Retrievers.[1,2] Beagles are also affected by this condition, which is also known as "Beagle Pain Syndrome". Affected animals usually present for moderate to severe neck pain, with a low head carriage and reluctance to move. True neurologic deficits are uncommon with this disorder, although chronically affected animals may develop ataxia or paresis if left untreated.[2] Diagnostic imaging is often unremarkable but may show meningeal enhancement.[2] Cerebrospinal fluid evaluation typically shows a profound neutrophilic pleocytosis (cell counts often exceed 1000 cells/μl). Dogs presenting in the acute stages usually respond completely to immunosuppression with corticosteroids, and additional immunosuppressive medications are rarely required.[1,3] However, more chronically affected animals may require additional immunosuppression and may have a more guarded prognosis.[2]

Generalized tremor (little white shaker) syndrome

This condition is commonly referred to as "little white shakers" or "shaker dog syndrome" due to its tendency to affect young toy breed dogs. However, it may affect dogs of any color or breed.[4] Encephalitis mainly involving the cerebellum is thought to result in the predominant clinical signs. Dogs show an obvious, constant tremor, which may worsen with movement. Additional neurologic signs are not usually found, although other cranial nerve deficits (e.g., vestibular signs), paresis, and visual deficits have been described.[5] This syndrome is often diagnosed based on clinical signs and response to therapy, and few descriptions of diagnostic imaging are available. Cerebrospinal fluid analysis usually shows a mild to moderate mononuclear pleocytosis, but may be normal.[4,5] Affected animals typically respond completely to corticosteroids and diazepam, which can usually be discontinued after a period of time.[4,5] However, recurrence of signs is possible.

Granulomatous meningoencephalitis (GME)

This syndrome is named after the profound granulomatous inflammation of the brain of affected patients, characterized by perivascular infiltrates of macrophages, lymphocytes and neutrophils.[6,7] The brainstem is usually the area most severely affected, and dogs typically present with signs of ataxia, paresis, cervical pain, and vestibular dysfunction. Other signs such as visual dysfunction, facial nerve deficits, dysphagia, pupillary abnormalities and seizures may also be seen.[6] Clinical signs may vary from relatively mild to very severe with rapid decompensation and death. Certain breeds appear to be predisposed to the development of GME, including the toy poodle and terrier breeds. However, any breed of dog may be affected, and dogs of all ages may be seen.

Lesions noted on MRI are hyperintense on T2-weighted and FLAIR images and enhance to varying degrees after contrast administration.[6] Lesions are often found in the brainstem, forebrain, and cervical spinal cord, while meningeal involvement is rarely noted on diagnostic imaging.[8] Evaluation of CSF typically shows a moderate to severe mononuclear pleocytosis, although neutrophils can occasionally dominate the cytology.[9] In rare cases the CSF is normal. The majority of patients will show improvement with corticosteroid therapy, although this improvement is often incomplete, and most animals require persistent long-term therapy. Relapses are frequent, particularly with attempts to reduce glucocorticoid doses. Many clinicians add additional immunosuppressive medications to improve control of clinical signs and reduce some of the side effects of the corticosteroids.[6,10-14] Radiation therapy may be beneficial for some patients with focal inflammatory lesions.[15] The long-term prognosis for these patients is guarded, as many relapse and eventually succumb to the disease.[15] However, responses to treatment are quite variable, and in the author's opinion, this syndrome likely represents several different disease processes with varying etiologies and outcomes.

Necrotizing meningoencephalitis (NME)

This condition was first reported in pug dogs, and is still often known as pug encephalitis.[16] However, other young, small and toy breed dogs are commonly affected with NME, including Maltese terriers, Chihuahuas, Shih Tzus, Boston terriers, Pekingese, Papillions, Pomeranians, and others.[6,17,18] The forebrain is primarily involved with infiltration of lymphocytes, plasma cells, and macrophages at the gray-white matter junction. Chronically affected animals develop necrosis of brain tissue with cavitation.[6] Seizures are a prominent clinical sign. Other signs include altered mentation, circling, visual dysfunction, proprioceptive deficits, cranial nerve deficits, and cervical pain. Brain imaging with MRI can show a variety of lesions including T2-weighted and FLAIR hyperintensities with variable contrast enhancement, loss of the distinction between gray and white matter, cavitated regions with CSF filling, and meningeal enhancement, predominantly in the forebrain.[6,19,20] The lesions are usually asymmetric, with evidence of mass effect and occasional brain herniation seen. Cerebrospinal fluid evaluation usually reveals a mild to moderate mononuclear pleocytosis, although cell counts are occasionally within normal limits.[16,18] The response to therapy is less clear with this syndrome, although some dogs appear to respond to therapies similar to those utilized for GME (see above). These include immunosuppressive corticosteroids as well as other immunosuppressive or cytotoxic therapies.[6] Prognosis is generally poor, although some dogs may survive for a year or longer before succumbing to the disease.

Necrotizing leukoencephalitis (NLE)

This syndrome is similar to NME, but predominantly affects the white matter of the forebrain and brainstem.[6] This is primarily a disease of Yorkshire terriers, although other breeds have been reported, including the French bulldog.[21] Affected dogs are typically young (6 months to 5 years of age). Clinical signs include seizures, altered mentation, visual deficits, circling, ataxia, and cranial nerve deficits. Magnetic resonance imaging usually reveals lesions within the white matter of the cerebrum and brainstem that are hyperintense on T2-weighted and FLAIR images, and hypointense on T1-weighted images with variable contrast enhancement.[22,23] Computed tomography may show areas of decreased density within the forebrain and brainstem.[24] CSF shows a mild mononuclear pleocytosis or may be normal. Therapy is similar to NME, as is the prognosis.

Eosinophilic meningoencephalitis and meningomyelitis

Inflammatory CNS disease with a substantial eosinophilic component has been reported with several infectious etiologies including fungal and protozoal meningocephalomyelitis. However, an idiopathic syndrome resulting in similar clinicopathologic findings but without an identifiable infectious etiology has also been described.[25-27] Young to middle-aged large breed dogs are predominantly affected and Golden retrievers and Rottweilers were overrepresented in some of these reports. Clinical signs may chiefly involve the brain, spinal cord or both regions. Peripheral eosinophilia was not found in the majority of cases described. Imaging of the CNS may show focal or multifocal lesions or may be normal.[27] Cerebrospinal shows a mild to marked pleocytosis with a substantial proportion of eosinophils (>20%). Most dogs respond dramatically and often completely to glucocorticoid therapy, which can often be discontinued after a period of time.[26,27] The long-term prognosis for the idiopathic condition is good in most cases.

REFERENCES

[1.] Behr S, et al. **J Am Anim Hosp Assoc** 2006;42:277. [2.] Tipold A, et al. **J Small Anim Pract** 2009; Epub before print, DOI: 10.1111/j.1748-5827.2009.00848.x. [3.] Cizinauskas S, et al. **J Small Anim Pract** 2000;41:295. [4.] Wagner SO, et al. **J Am Vet Med Assoc** 1997;211:731. [5.] Bagley RS, et al. **J Am Anim Hosp Assoc** 1993;29:141. [6.] Schatzberg SJ. **Vet Clin North Am Small Anim Pract** 2010;40:101. [7.] Braund KG. **J Am Vet Med Assoc** 1985;186:138. [8.] Cherubini GB, et al. **Vet Rec** 2006;159:110. [9.] Bailey CS, et al. **J Am Vet Med Assoc** 1986;188:418. [10.] Adamo FP, et al. **J Am Vet Med Assoc** 2004;225:1211. [11.] Gnirs K. **J Small Anim Pract** 2006;47:201. [12.] Zarfoss M, et al. **J Small Anim Pract** 2006;47:588. [13.] Coates JR, et al. **J Vet Intern Med** 2007;21:100. [14.] Pakozdy A, et al. **Vet Rec** 2009;164:89. [15.] Munana KR, et al. **J Am Vet Med Assoc** 1998;212:1902. [16.] Cordy DR, et al. **Vet Pathol** 1989;26:191. [17.] Higgins RJ, et al. **Vet Pathol** 2008;45:336. [18.] Stalis IH, et al. **Vet Pathol** 1995;32:230. [19.] Flegel T, et al. **Vet Radiol Ultrasound** 2008;49:419. [20.] Young BD, et al. **J Vet Intern Med** 2009;23:527. [21.] Timmann D, et al. **J Small Anim Pract** 2007;48:339. [22.] Lotti D, et al. **Vet Radiol Ultrasound** 1999;40:622. [23.] von Praun F, et al. **Vet Radiol Ultrasound** 2006;47:260. [24.] Ducote JM, et al. **Vet Radiol Ultrasound** 1999;40:617. [25.] Bennett PF, et al. **Aust Vet J** 1997;75:786. [26.] Smith-Maxie LL, et al. **J Vet Intern Med** 1989;3:167.[27.] Windsor RC, et al. **J Vet Intern Med** 2009;23:275.

ROLE OF THE TECHNICIAN IN EXAMINING & TREATING THE NEUROLOGIC HORSE

Yvette S. Nout, DVM MS PhD DACVIM DACVECC
San Francisco, CA

INTRODUCTION

Examination, assessment, and subsequent management of horses with neurologic disease may be difficult and challenging. The horse is a "fright and flight" animal and the combination of being debilitated with being in a strange environment surrounded by people, may cause anxiety or fear, both of which can contribute to dangerous situations. Furthermore, the horse's primary disease can lead to an abnormal mentation and subsequent abnormal and/or dangerous behavior. Safety of all involved should be considered when working with horses with neurological disease. Caution should also be taken for the risk of infectious and/or zoonotic diseases playing a role in the etiology of the disease, and appropriate measurements such as wearing protective clothing and minimizing human exposure, should be undertaken. Management of horses with neurological disease is typically a combination of supportive care and specific treatment aimed at the underlying or complicating disease processes [1,2]. Supportive care is mainly directed at protection of the animal, maintenance of adequate hydration and nutritional status, and physiotherapy. Care of down horses is time consuming and is more difficult if the horse is heavy.

NEUROLOGICAL DISEASES

Most common causes of neurological disease in the USA are cervical vertebral stenotic myelopathy (CVSM, Wobblers disease), equine protozoal myeloencephalitis (EPM), and trauma. Other diseases that are not uncommon include equine herpes virus myeloencephalopathy (EHM), botulism, equine motor neuron disease (EMND), equine degenerative myelopathy, vestibular disease, and encephalitides (such as caused by West Nile virus).

Through a series of diagnostic tests, the veterinarian will attempt to reach a final diagnosis and initiate treatment. Observation through a neurologic examination is usually the first diagnostic tool used in order to obtain information on how the horse responds to and moves within its environment. This most often requires the horse to be led by a person who is familiar with the neurological examination and can lead the horse through the various aspects of it. It is important that the horse has enough room to move and is not held too closely since this may obscure clinical signs. Ancillary tests may include radiographs, blood collection, cerebrospinal fluid collection, electroencephalogram (EEG), electromyogram (EMG), and myelographic examination. For each of these it is important that the persons working with the horse know what is expected and how to prepare for each of these tests. In addition, it is important to communicate information about the horse (zoönoses, danger) to other (veterinarians) working with the horse.

TREATMENT

Observation during regular time intervals is very important to look for changes in mentation and / or posture. Sometimes changes occur, such as seizure activity, or the horse is hanging its head with subsequent development of edema, or inability to stand up, or circling behavior, that are detrimental to the horse and / or to the treatment protocol (for example intravenous fluid therapy). Often it is recommended to handle these horses with a minimum of 2 people and if necessary protective clothing should be worn.

TREATMENT: PROTECTION

Prevention and care of further injury such as falling, fractures, muscle compartmental injury, peripheral nerve injury, and decubital ulcers are important aspects in the management of horse with neurologic disease. Protection may be accomplished by use of stalls with padded walls, adequate padding, and / or helmets or bandages. Damage to the eyes can occur directly by pressure to the eye or from bedding material. Furthermore, specific diseases may result in a horse's inability to blink (botulism, Horner's syndrome, facial nerve paralysis). The most common ophthalmic disorders seen in horses with neurologic disease are corneal ulcers and (fungal) keratitis. Eyes need to be examined daily and assessed and treated carefully. In addition, some horses need to be protected from themselves and this is most commonly done through administration of sedatives.

TREATMENT: FLUID AND DIET

Some neurological diseases can lead to dysfunction of swallowing or inability to reach the water bowl (recumbency). In these horses water intake is compromised and insufficient to meet their maintenance requirement. Maintenance requirements for down horses may be slightly lower (50 ml/kg/day) than what is considered normal. Fluid therapy can be provided intra-gastric by using a permanently placed small bore feeding tube or regular nasogastric intubation, or fluids can be administered intravenously. Depending on the dietary protocol that is used, electrolytes and glucose may be added to the fluids. Horses lose body condition rapidly when no nutrition is provided. Dramatic changes are often noticeable within 1 week. Adequate nutrition is important in horses that cannot eat and may be beneficial for wound healing [3] and maintaining muscular strength and adequate immunity. Diets and routes of administration can vary, depending on horse compliance, dietary tolerance, and budget. Dietary options can be divided in four alternatives [2]. Some horses are able to ingest a sufficient amount of roughage (± 2% of bodyweight) with or without grain while they are in sternal recumbency or standing with sling-assistance. The dietary intake and fecal output should be monitored closely, but in general no additional support is required. Many recumbent horses and horses that are dysphagic, however, are not able, or should not be allowed, to meet the required intake in this way. For these horses one can choose to use a soaked pellet gruel (30 Cal/kg/day), liquid diet (Osmolite®; Ross Products Division, Abbott Laboratories, Columbus, OH, USA or Critical Care Meals™; MD's Choice Inc., Louisville, TN, USA), or parenteral nutrition.

Down horses are prone to developing impactions of the cecum, large or small colon. Reduced fecal output may be a preliminary sign of impaction formation and the administration of laxatives such as magnesium or sodium sulfate, or polyethylene glycol containing solutions is then indicated. Some horses may develop ileus and this may have consequences for hydration and nutrition regulation. Urine and fecal retention may occur, for example in herpes myeloencephalitis and frequent emptying of rectum and bladder is then required. Alternatively an indwelling urinary catheter can be placed.

Adequate ventilation, maintaining a clean stall environment, and preventing aspiration of food, are important factors to reduce the chance of respiratory tract disease. A dysphagic horse should not be allowed to eat, but should be offered an alternative source of nutrition. Antimicrobial drug therapy should be initiated when respiratory disease is present and may be indicated in a dysphagic horse. Regular turning of the horse may minimize development of lung disease. In addition to decubital ulcers, the development of aspiration pneumonia is the most common complication seen in horses with botulism [4].

PHYSIOTHERAPY

Physiotherapy is an important aspect of supportive care and the rehabilitative process of injured horses. There is more and more evidence from experimental spinal cord injury research that spinal cord plasticity, repair, and regeneration do exist after injury [5]. From the rehabilitation point of view it is paramount to maintain a certain level of neurologic activity above and below the level of injury [6]. Physiotherapy can be provided in the recumbent animal by manipulating limbs depending on the horse's attitude or by assisting the horse to stand with a sling. Controlled exercise allows the unaffected parts of the nervous system to compensate for the affected parts by increasing strength and conscious proprioception. Exercise is especially helpful in improving weakness, ataxia, spasticity, and hypermetria. A sling is an important tool in the assessment and management of a recumbent horse. When a horse is unable to rise, but otherwise appears to have a normal mentation, i.e. unaltered behavior, it may be very helpful in the examination to assist the horse to stand with a sling. The presence of lameness, weakness, and ataxia as well as the number of affected limbs are much easier assessed when the horse is in an upright position. Furthermore, the horse may be able to stand freely and demonstrate clinical signs that may help forming a diagnosis. Further, using the sling to assist a horse to stand may provide valuable information for short-term management purposes. Deterioration or improvement of disease may be determined relatively easily by daily sling-assistance in order to provide short-term recommendations and prognosis. For the long-term management the sling can be useful, depending on the horse's primary disease and compliance. Horses may become used to using the sling as a "third wheel" and can be comfortably managed in the sling until they have regained sufficient control to ambulate freely. Periodic sessions of semi-standing with sling assistance may function as a form of physiotherapy and manipulation of limbs can occur during this time. Moreover, increasing the time a horse can be maintained in an upright position decreases the detrimental effects of continuous pressure on skin and muscle. The time a horse is up should also be used to clean and dry the horse and examine the horse's food and water intake. In addition, consistent, regular exercise (walking, trotting) is likely to be of benefit in horses with milder neurological deficits.

REFERENCES

[1.] McConnico RS, Clem MF, DeBowes RM. Supportive medical care of recumbent horses. **Compendium on Continuing Education for the Practicing Veterinarian** 1991;13(8):1287 - 1295. [2.] Nout YS, Reed SM. Management and treatment of the recumbent horse. **Equine Veterinary Education** 2005;7(6):416-432. [3.] Ferguson M, Cook A, Rimmasch H, Bender S, Voss A. Pressure ulcer management: the importance of nutrition. **Medsurg Nurs** 2000;9(4):163-75; quiz 176-7. [4.] Whitlock RH, Buckley C. Botulism. **Vet Clin North Am Equine Pract** 1997;13(1):107-28. [5.] Bareyre FM, Kerschensteiner M, Raineteau O, Mettenleiter TC, Weinmann O, Schwab ME. The injured spinal cord spontaneously forms a new intraspinal circuit in adult rats. **Nat Neurosci** 2004;7(3):269-77. [6] Sadowsky CL, McDonald JW. Activity-based restorative therapies: concepts and applications in spinal cord injury-related neurorehabilitation. **Dev Disabil Res Rev** 2009;15(2):112-6.

CURRENT APPROACHES TO SEIZURE MANAGEMENT

Karen R. Muñana, DVM, MS, DACVIM (Neurology)
Raleigh, NC

INTRODUCTION

Seizures are the most common neurological disorder seen in small animal practice. Because seizures are episodic and frequently not witnessed by the veterinary caregiver, a detailed and accurate history is extremely important in the initial assessment. Information that should be obtained from the client includes the age of onset, frequency and duration of seizures, description of the seizure, whether post-ictal signs are present and if the animal is normal between seizures. In addition, the animal's vaccination history, presence of any previous systemic illness or trauma, environmental history, and history of seizures in the family lineage should be ascertained. Historical and epidemiological information, as well as any abnormalities detected on examination, are useful in determining whether or not the episodes are seizures. Disorders that can be mistaken for seizures include syncope, transient vestibular attacks, narcolepsy/cataplexy and behavioral disorders. Cats in particular can display behavioral aberrations such as vocalizing, piloerection, attacking objects, startling and running that may not be immediately recognized as a seizure. The presence of other features characteristic of seizure disorders, such as facial twitching, salivation or secondary generalization often aid in the determination, and the owner should be queried as to whether any of these signs have been observed. If uncertainty persists as to whether or not an episode is a seizure based on a verbal description, it is often helpful to instruct the owner to attempt to obtain a video recording of an episode for review.

CAUSES OF SEIZURES

Epilepsy is a general term that refers to recurrent seizures of any cause. Primary or idiopathic epilepsy refers to seizures that occur for which an underlying cause is not found. Many breeds of dogs have a known predisposition to developing primary epilepsy, and evidence suggests that the disease may be heritable in many of these breeds. The onset of seizures is typically between 1 and 5 years of age. Cats have also been diagnosed with idiopathic epilepsy, although a genetic basis is less likely. Secondary or symptomatic epilepsy refers to seizures that are associated with a known disorder of the central nervous system, such as neoplasia, infection or trauma. Cryptogenic epilepsy refers to seizures that are suspected to be symptomatic but for which an etiology cannot be identified.

Seizure causes can be divided into extracranial and intracranial categories. Extracranial causes include metabolic or toxic disturbances that secondarily effect brain function. Metabolic causes that should be considered include hepatic disease, renal disease, electrolyte disturbances, hypoglycemia, hyperthyroidism, hyperlipoproteinemia, and polycythemia. Of the numerous toxins that can cause seizures, those that are most commonly incriminated include lead, organophosphates, strychnine, metaldehyde and ethylene glycol. Intracranial causes of seizures are the result of structural or functional abnormalities within the brain itself. This category include anomalous disorders such as hydrocephalus; neoplastic conditions, either primary brain tumors or metastatic disease; CNS infections, including viral, bacterial, fungal, protozoal, rickettsial and parasitic causes; inflammatory, non-infectious disease (meningoencephalitis of unknown etiology); cerebral trauma; and vascular causes. Initial diagnostic evaluation of an animal with seizures should be directed at identifying potential extracranial causes. If an extracranial cause is not identified, potential intracranial causes are pursued.

ANTIEPILEPTIC DRUG THERAPY

Antiepileptic medications are the mainstay of treatment for animals in which a primary disease process is not identified or cannot be treated definitively. The decision as to when to start treatment should depend on the frequency and severity of seizures, whether seizures occur in clusters, the severity of post-ictal signs, and the owner's expectations. As a general rule, the author recommends initiating treatment if seizures occur more frequently than once a month, or if the seizures occur in clusters. Client education is crucial to successful management of an epileptic animal. Before initiating therapy, the owner should be made aware that the goal of therapy is control of seizures, and that treatment is not curative. Consequently, therapy is often life-long, and the owner must appreciate the importance of administering medications on a regular schedule. Lastly, the client should be informed of potential side effects associated with the use of antiepileptic medications and understand the importance of regular evaluations to assess for any evidence of toxicity.

Treatment should be initiated with a single drug. Response to therapy is based on the seizure frequency once steady state serum levels have been achieved, the presence and severity of side effects, and serum drug concentrations (particularly for phenobarbital and potassium bromide). If seizures remain poorly controlled in an animal with serum drug concentrations within the low therapeutic range and no side effects from the medication, it is recommended that the dosage of the medication be increased. When drug concentrations reach the high end of the therapeutic range or unacceptable side effects develop, a second medication should be added to the treatment regime.

Phenobarbital

Phenobarbital remains one of the first line drugs to treat seizures in both dogs and cats. Initial oral dose is 2.0-2.5 mg/kg every 12 hours. Serum phenobarbital concentrations should be monitored 2-3 weeks after initiating therapy and after any dosage adjustment. Therapeutic serum concentrations have been established at 15-45 μg/ml. However, an optimum therapeutic levels of 23-30 μg/ml has been recommended in cats, in an attempt to maximize seizure control with the lowest potential for side effects.[1] Similarly, serum levels of 20-35 μg/ml are recommended in the dog, with the decrease in the high end of the range reflecting suggested changes to minimize the potential for hepatotoxicity.[2] Common side effects reported with phenobarbital administration include sedation, ataxia, polyuria, polydipsia, and polyphagia. Serious side effects include blood dyscrasias and liver toxicity, but these tend to occur infrequently and are reversible. Hepatotoxicity in dogs appears to be dose and time related, with most affected dogs maintaining serum phenobarbital concentrations of greater than 40 μl/ml.[2] It should be noted that it is common for dogs receiving phenobarbital to display increases in liver enzymes, and that this does not necessarily indicate the presence of clinically significant liver disease. However, dramatic increases or decreases in liver enzymes from one evaluation to the next, or increases in serum phenobarbital levels without a concomitant dosage increase, may suggest liver dysfunction and necessitate further evaluation. Animals receiving phenobarbital should be monitored at 6-12 month intervals with a general examination, hemogram, serum chemistry profile and serum phenobarbital concentrations.

Potassium Bromide

Potassium bromide was initially utilized in veterinary medicine as an add-on drug when therapy with phenobarbital had failed. However, it is now frequently recommended as a sole antiepileptic agent in dogs. Initial oral dose is 30 mg/kg once daily. The drug has a long half-life; steady state concentrations are not reached for approximately 3-4 months in dogs. When more immediate seizure control is needed, a loading dose of bromide can be administered to achieve a more rapid rise in serum concentration. A total oral loading dose of 400-600 mg/kg is divided into 4-5 doses and given with food over a period of 1-5 days. The therapeutic serum bromide concentration is 100-300 mg/dl (1.0 – 3.0 mg/ml). Levels between 100-200 mg/dl are often satisfactory when the drug is used in conjunction with phenobarbital, but values between 200-300 mg/dl are frequently required when potassium bromide is used as a sole antiepileptic agent. Bromide does not undergo liver metabolism and is excreted unchanged by the kidneys, where it competes with chloride for tubular reabsorption. Consequently, changes in chloride intake can alter serum bromide levels. Diets high in chloride result in a higher excretion rate of bromide and lower serum bromide concentrations, whereas a chloride-restricted diet will result in lower excretion rates of bromide and an increased serum bromide concentration. Because of this, it is important to keep treated animals on a stable diet. Common side effects of bromide in dogs include sedation, polyuria, polydipsia, polyphagia, and vomiting. Pancreatitis has also been reported as a potential side effect in animals being treated with bromide.[3]

Bromide has been used as an antiepileptic drug in cats, but there is data to suggest that seizures in cats might not respond to bromide as well as in dogs, and the incidence of adverse event is high.[4] Reported side effects include sedation, polydipsia, vomiting, weight gain and coughing. Coughing, which is due to airway disease, is characteristically acute in onset and may be associated with dyspnea. As such, this adverse event can be life threatening. Respiratory signs can resolve with discontinuation of the drug. Because of the potentially for serious adverse events, and the availability of newer antiepileptic drugs that are being used in veterinary medicine, the use of bromide in cats has declined.

Felbamate

Felbamate is a one of the newer antiepileptic medications approved for humans over the last 20 years. Although the drug appears to be efficacious, reports of aplastic anemia and hepatotoxicity have restricted its use in humans. Felbamate has been utilized in veterinary medicine as a second line drug for dogs when treatment with phenobarbital and potassium bromide has failed, and there is one report of its use as a sole agent to treat partial seizures in dogs.[5,6] Recommended oral dose is 15-60 mg/kg TID. The majority of the drug is excreted in the urine of dogs, but hepatic metabolism does occur leading to the possibility of drug reactions when phenobarbital and felbamate are administered concurrently. Side effects of felbamate in dogs include blood dyscrasias, sedation, nausea, vomiting, keratoconjunctivitis sicca and liver disease. It is recommended that complete blood counts and liver enzymes be monitored every 2-3 months during treatment to assess for adverse effects. Felbamate is used rather infrequently in dogs because of the potential for adverse events and drug interactions, and the expense of the medication. There are no reports on its use in cats.

Gabapentin

Gabapentin is another of the newer antiepileptic drugs that has been used in veterinary medicine. The drug is eliminated entirely by the kidneys in humans; however, gabapentin does undergo partial hepatic metabolism in dogs. The elimination half-life in dogs is 2-4 hours, which suggests that frequent administration is required to maintain steady state serum concentrations. There are two published reports on the use of gabapentin as an add-on antiepileptic drug in dogs.[7,8] The recommended oral dose is 10-20 mg/kg every 6-8 hours.[6,7] Primary side effects noted are sedation and ataxia. There are anecdotal reports of the use of gabapentin in epileptic cats, with a dosage of 5-10 mg/kg given orally every 8-12 hours most often utilized. Gabapentin is available in generic form, which has reduced the cost considerably since its introduction. The commercially available liquid formulation, which is often preferred when dosing cats and small dogs, contains 300 mg xylitol/ml. Because of the potential for xylitol toxicity,[9] a tablet or capsule form of gabapentin should be preferentially utilized in dogs. In the author's experience, gabapentin is the least effective of the newer antiepileptic drugs, particularly in dogs.

Zonisamide

Zonisamide is a sulfonamide derived antiepileptic drug that was approved by the FDA in 2000 as adjunctive treatment for seizures in humans. Zonisamide appears to have a favorable pharmacokinetic profile in dogs including rapid and complete absorption after oral administration and a long half-life (15 hours) that allows for twice daily administration. Although some hepatic metabolism occurs in the dog, 80% percent of the drug is excreted unchanged in the urine. The recommended oral dose in dogs is 5-10 mg/kg every 12 hours. The high end of the dose range is needed when used in combination with phenobarbital, because phenobarbital increases the metabolism of zonisamide. This is most likely a result of microsomal enzyme induction.[10] There are two recent studies on the use of zonisamide in epileptic dogs with refractory seizures.[11,12] Side effects described included sedation, ataxia and loss of appetite. There are anecdotal reports of the use of zonisamide in cats, with oral doses of 5 mg/kg every 12-24 hours most commonly utilized. A generic form of the drug is available, and the cost is now more comparable to that of other antiepileptic drugs used in veterinary medicine.

Levetiracetam

Levetiracetam is unique among the newer medications in that it has a different mechanism of action when compared to existing antiepileptic drugs. It also has a favorable pharmacokinetic profile, including rapid and complete absorption after oral administration, and minimal hepatic metabolism. However the half-life in both dogs and cats is short, necessitating frequent administration. There are reports in both dogs and cats on the use of levetiracetam as an add-on antiepileptic.[13,14] The recommended oral dose in both species is 20 mg/kg every 8 hours. Side effects include sedation in dogs, and lethargy and inappetance in cats. A generic form of the drug has recently become available, and the associated decrease in expense has enabled this drug to be a viable alternative for more pet owners. In addition, a parenteral form of levetiracetam is available, which can be used in the emergency treatment of seizures in dogs.

Pregabalin

Pregabalin is a newer generation drug in the same class as gabapentin, and is purported to be a more efficacious antiepileptic than its predecessor. There is a recent report on its use in dogs.[15] The recommended oral dose is 3-4 mg/kg every 8 hours; however, in order to minimize side effects, it is suggested that dogs initially be started on 2 mg/kg and the dose be increased by 1 mg/kg each week until the target dose is reached. Common side effects include sedation and ataxia. There are anecdotal reports of the use of pregabalin cats, with an oral dose of 1-2 mg/kg every 12 hours most commonly mentioned. Presently, the main disadvantage of this drug is expense.

ALTERNATIVE THERAPIES

Alternative, non-medical methods of seizure control have been explored for epileptic patients who have experienced poor control with antiepileptic drugs. The therapies that have proven the most successful in humans include vagal nerve stimulation and the ketogenic diet. Vagal nerve stimulation utilizes an implantable pacemaker device that delivers repetitive stimulation to the vagus nerve in the neck. This form of therapy can be utilized in dogs, but is prohibitively expensive for many owners.[16] Use of a strict high-fat, low-carbohydrate, low-protein ketogenic diet has been evaluated in dogs, but was not determined to have a beneficial effect on seizure control.[17] Other alternative forms of therapy include acupuncture and homeopathic remedies; to date, these have not been scientifically evaluated in epileptic dogs or cats.

HOME CARE FOR CLUSTER SEIZURES

Management efforts have also been directed at home treatment of cluster seizures to try to alleviate the need for repeated emergency veterinary care. Owners can be instructed to administer diazepam rectally to their dog, which has been shown to decrease the number of cluster seizure events over a 24-hour period.[18] The recommended diazepam dose is 1 mg/kg administered at the onset of seizures, and this can be given up to 3 times over a 24 hour period. Dogs receiving phenobarbital should receive a dosage of 2 mg/kg. Because diazepam is inactivated by light and adheres to plastic, it is best to dispense the drug in the original glass vial and instruct the owner to draw the required amount into a syringe when needed. A rubber catheter or teat cannula is then placed on the syringe for rectal administration. Suppository forms of the drug are available, but the pharmacokinetics of these products have not been studied in dogs to support their use. Additional doses of maintenance oral antiepileptic drugs, or pulse therapy with one of the newer drugs, have also been advocated to treat cluster seizures at home, but there have been no published reports documenting the results of such treatments.

CONCLUSIONS

Effective management of seizures depends on continued commitment from both the owner and the veterinary team. Animals on long-term antiepileptic medications require periodic evaluations to evaluate drug effectiveness and to monitor for the development of adverse effects. It is not uncommon for seizure frequency to increase over time, such that medication adjustments are required. However, with appropriate care, most dogs and cats can attain an acceptable level of seizure control and maintain a good quality of life.

REFERENCES

[1] Quesnel AD, et al. **J Am Vet Med Assoc** 1997; 210:72. [2] Dayrell-Hart B, et al. **J Am Vet Med Assoc** 1991; 199:1060. [3] Gaskill CL, et al. **Can Vet J** 2000; 41:555. [4] Boothe DM, et al. **J Am Vet Med Assoc** 2002; 221:1131. [5] Boothe DM. **Vet Clin NA Sm Anim Pract** 1998; 28:411. [6] Ruehlmann D, et al. **J Sm Anim Pract** 2001; 42:403. [7] Platt, SR, et al. **Vet Rec** 2006; 159:881. [8] Govendir M, et al. **Aus Vet J** 2005; 83:602. [9] Dunayer EK, et al. **J Am Vet Med Assoc** 2006; 229:1113. [10] Orito K, et al. **J Vet Pharmacol Therap** 2008; 31:259. [11] Dewey CW, et al. **J Am Anim Hosp Assoc** 2004; 40:285. [12] Von Klopmann T, et al. **J Sm Anim Pract** 2007; 48:134. [13] Volk HA, et al. **Vet J** 2008; 176:310. [14] Bailey KS, et al. **J Am Vet Med Assoc** 2008; 232:867. [15] Dewey CW, et al. **J Am Vet Med Assoc** 2009; 235:1442. [16] Munana KR, et al. **J Am Vet Med Assoc** 2002; 221:977. [17] Patterson EE, et al. **J Vet Intern Med** (abst.) 2005; 19:421. [18] Podell M. **J Vet Intern Med** 1995; 8:68.

INCREASING THE PERCEPTION OF YOUR VALUE WITHIN THE PRACTICE

Regina Fairman, LVT, VTS (ECC)
Auburn Hills, Michigan

INTRODUCTION

As a technician we are taught the skills that enable us to perform our jobs – phlebotomy, pharmacology, nursing, radiology, and client communication. What additional skills do you need in order to establish and retain your value? The current job market has changed drastically and is being described as competitive and sparse. Securing a new job has become an art, and requires skills that are not taught in technical school.

This will help you understand your job is the beginning of your career. Having a balanced career can be achieved by understanding and mastering the basic, but essential skills in the workplace, boosting or establishing new skills, and marketing your skills to your employer or potential employer.

THE ESSENTIALS

There are a few traits that every employee should understand, and it behooves that employee to have some of these qualities. The basics are honesty, dependability, professionalism, and adaptability. All of those traits are an extension of a larger trait – responsibility. As an employee, if you are responsible, all of the basics will follow.

As your career progresses, the basic skills should expand to include organization and planning, communication skills (listening, verbal, written) and basic computer literacy.

BOOSTING SKILLS

Today's job security comes from within ourselves. In order to boost your skills *you* must take the initiative to achieve. The first step is to earn your degree and become licensed/certified/registered. If you are already at that level, then establish which direction you want your career to go. To do this, you will need to examine your personal preferences and how strong your desire is to accomplish your goals. Establishing a career path will take a few weeks to a few months, will require research and should not be rushed. If your career path involves additional knowledge or skills, which it usually does, then attain the knowledge and the advanced training you will need. Advancement of skills can be done by attending professional seminars, and learning from the people that are more skilled than you. Advancing your career will require an assertive learning style.

THE UN-CLASSIFIEDS

As an employee, if you have already accomplished additional skills or certifications, what can you do if you want to up the ante? Here are some ideas of where to begin. These concepts or traits will help you transition into any position at any job. First, and foremost, establish a "can-do" mindset. If you aren't of this mindset, take the 10 day mental make-over. The mental make-over will lift the burden of negativity. If there are people around you that are negative, their negativity can be contagious and consuming; the mental make-over will move you above the negativity and past it. Become a mentor or role model to someone in your field. Being a mentor will help you establish your ability to teach and lead. It may also highlight the areas that you, as a mentor, may need to improve on even though you were not aware of it. Next, do not wait for your performance review to talk to your boss regarding your job duties. If there are no additional job duties at that time, your inquiry will let him/her know that you are interested when an opportunity does come along. When that additional job duty or position change is available you should appear enthusiastic and interested, thus you should always keep an updated copy of your resume available. You should update your resume every 6 months and also every instance that your job duties change. Position change or additional job duties should not be rushed. Always take calculate risks; apply for additional job responsibilities but not solely because it is available. Make sure the job or duties are something that you really want or are capable of doing. Uncalculated risks may set you up for failure, which is not the objective. Finally, learn the basics of conflict resolution, even if you don't utilize the skill in your current job/position.

FINAL THOUGHTS

Employability skills and personal values are the critical traits you need to succeed in the workplace. You can learn, cultivate, develop and maintain over your lifetime. To forward your career, document your skills, assess their value, and remember to market them to your employer.

REFERENCES

101 Ways to Recession Proof Your Career, Enelow, 2002; Promoting Yourself 52 Lessons for Getting to the Top and Staying There, Lancaster, 2002; Are You Paid What You're Worth, O'Malley, 1998; The Seasons of Your Career – How to Master the Cycles of Career Change, Sanborn and Ricci, 2003

WHAT'S SO "HOT" ABOUT RADIOACTIVE IODINE? : FELINE HYPERTHYROIDISM & TREATMENT OPTIONS

Kathi L. Smith, RVT, VTS (Oncology)
Portland, ME

INTRODUCTION

Feline hyperthyroidism (thyrotoxicosis) is the most common endocrinopathy in cats. It was first diagnosed in the late 1970s and has had an increased frequency in diagnosis due to: increased clinical awareness of the disease, increased improvement in diagnostic testing, increased feline population, the increased lifespan of pet cats and an increase in owner interest in seeking veterinary care. Thyrotoxicosis results from the excess secretion of the thyroid hormones. Thyroid hormones are the only iodinated organic compounds in the body. The only function for ingested iodine is for thyroid hormone synthesis. Ingested stable iodine is converted to iodide in the GI tract and is then absorbed into the circulation. Iodide is either concentrated or trapped by the active transport mechanism of the thyroid follicular cell. In the thyroid gland-iodide is oxidized to iodine- triiodothyronine (T_3) and thyroxine (T_4). These hormones are produced by multinodular adenomas (which appear as small, grape-like nodules that spread throughout the thyroid gland), adenomatous hyperplasia (large, solitary, non-encapsulated masses) or thyroid carcinomas (rarely seen in cats). Approximately 70-75% of cats have "benign" multinodular adenomas involving *both* thyroid lobes and roughly 30% of cats have functional adenomatous hyperplasia involving *one* thyroid lobe. Only 1-3% of cats with thyroid carcinoma have hyperthyroidism. The etiology for hyperthyroidism is not known at this time. Since hyperthyroidism has only been recognized as a disease entity since the late 1970s several possible explanations have emerged. Some studies suggest that the goitrogenic soy isoflavones, genistein, and daidzein which are widely found in commercially available canned and dry cat foods may have concentration levels that are high enough to cause negative effects. Environmental contaminates such as insecticides and flame retardants have also been implicated. Cat litter has also come into question. Although numerous studies are still ongoing, no one causative agent has been proven.

SIGNALMENT

Hyperthyroidism usually affects cats that are greater than 8 years of age with a reported age range of 4-22 years. There is a report out of the University of Illinois regarding an eight month old feline that developed hyperthyroidism. The mean age is 13 years. Less than 5% of cats are younger than 8 years of age. There is no gender or breed predilection although the Siamese and Himalayan appear to have a decreased risk for hyperthyroidism.

CLINICAL AND PHYSICAL EXAMINATION FINDINGS

The most common clinical and physical examination findings associated with hyperthyroidism include:

- Polyphagia/anorexia
- Polydipsia/polyuria
- Weight loss
- Tachycardia (>240 bpm)
- Hyperactive/difficult to examine
- Heart murmur
- Vomiting
- Diarrhea /bulky, foul smelling stool
- Aggressive behavior/easily stressed
- Palpable thyroid gland
- Skin changes: patchy alopecia, matted hair, minimal grooming
- Gallop cardiac rhythm
- Small kidneys
- Dehydrated/cachectic appearance
- Increased rectal temperature
- Premature cardiac beats
- Increased nail growth
- Depressed/weak
- Ventroflexion of the neck
- Thickened intestinal loops

DIAGNOSIS

Cats suspected of having hyperthyroidism should have a complete medical workup. This includes a complete blood count (CBC), serum biochemistries: minimally, albumin, alanine aminotransferase (ALT), alkaline phosphatase (ALK Phos), BUN, creatinine, glucose, total protein, electrolytes, total T_4, urinalysis, blood pressure, and lateral whole body radiograph.

The complete blood count is usually within normal limits or a mild increase in the packed cell volume (PCV) may be observed. Cats with decreased PCV's (20-30%) should be evaluated for other non-thyroid diseases.

Hepatic enzymes are usually (up to 90%) mildly to moderately elevated in cats with hyperthyroidism. Ultrasound guided aspirates/biopsies of the liver may be necessary to rule out primary hepatic disease.

Sodium and chloride levels are usually within normal limits but potassium levels may be decreased, so those patients with hypokalemia should be supplemented with potassium gluconate.

Approximately 20% of cats with hyperthyroidism have asymptomatic urinary tract infections so urinalysis should be included in all diagnostic workups for cats with suspected hyperthyroidism. Urine specific gravity is usually between 1.008-1.020 although there are many cats with specific gravities in the 1.040 and greater range.

Hyperthyroidism and renal insufficiency may be seen alone or in combination in geriatric cats. Hyperthyroidism does increase the glomerular filtration rate (GFR), and renal blood flow causing the kidney's to work overtime. When the thyroid is put into a normal state some degree of renal compromise can be "unmasked". This unmasking can result in increased BUN and creatinine. Hyperphosphatemia may also be noted.

Serum total T_4 is the diagnostic test of choice when ruling out potential hyperthyroidism. If the serum total T_4 is elevated and clinical signs match, then hyperthyroidism has been verified. If the serum total T_4 is within normal limits (or on the upper end of normal) and the patient has clinical signs suggestive of hyperthyroidism then a free T_4 by equilibrium dialysis is the next step. If the free T_4 is elevated then rechecking the total T_4 and free T_4 in a month-6 weeks is warranted. If they are both elevated then hyperthyroidism is a likely diagnosis. If the free T_4 is the only elevation then evaluation for non-thyroidal diseases should be considered as there is a population of cats that will have an elevated free T_4 without thyroid disease.

Systemic hypertension is common in cats with hyperthyroidism and it's not unusual to have blood pressures of 160-170 mmHg and greater. Cats with systemic hypertension should be evaluated for retinal hemorrhage and detachment.

Lateral whole body radiographs ("cat-o-grams") can be useful in ruling out cardiac enlargement (HCM vs. DCM), pleural effusion, pulmonary edema, kidney size, lymph node enlargement etc. If cardiomegaly is apparent then an echocardiogram to rule out primary cardiac disease is advised.

TREATMENTS

Medical Management

Medical management of hyperthyroidism consists of the use of antithyroid drugs. Antithyroid drugs block the formation of the thyroid hormones T_3 and T_4.The most common antithyroid drug is methimazole (Felimazole®). Other antithyroid drugs include propylthiouracil (PTU) and carbimazole (not available in the U.S.). These drugs only control the disease and must be taken daily to 3 times daily. Side effects associated with the use of methimazole drugs include vomiting, lethargy, anorexia, hematologic abnormalities (neutropenia, thrombocytopenia and rarely IMHA), hepatic enzyme abnormalities (↑ ALT, ALk Phos), facial excoriation, and myasthenia gravis have all been reported. Due to the potential negative side effects, CBC's and serum chemistries should be routinely monitored. The other problem with medical management of hyperthyroidism is owner compliance. Most patients require at least twice daily dosing. Many cats are refractory to oral treatments and owners become frustrated with trying to medicate their pets. Compounded methimazole (liquid and transdermal) may offer a solution but have issues of their own. Compounded pharmacies lack regulatory controls and the consistency/effectiveness of the product may differ from company to company and lot to lot. Dosing of methimazole is typically 2.5 PO q 12 hrs for 2 weeks then recheck CBC, serum chemistries and total T_4. Total T_4 levels should be in the normal to high normal range. Increasing dosages should be done slowly and cats should be rechecked (including bloodwork) every 2 weeks until the total T_4 is within the normal range. Very few patients require more than 5 mg PO q 12hrs.

Surgical Management

Thyroidectomy is an effective treatment for cats with diagnosed hyperthyroidism. However, surgery has several disadvantages. Hyperthyroid cats tend to be older/frailer and are potentially at increased risk for surgical and anesthetic problems. Since many of these cats have secondary cardiac disease, pre-operative treatment with antithyroid drugs and beta blockers is recommended. Potential adverse side effects include: damage to adjacent parathyroid glands (may cause a transient or permanent hypoparathyroidism and hypocalcemia which can be life –threatening resulting in increased cost and hospitalization), potential for regrowth of affected tissue, Horner's syndrome, laryngeal paralysis and the inability to remove ectopic thyroid tissue (usually in the anterior mediastinal area). Serum calcium levels should be run daily for 5 days post surgery. Hypocalcemia usually develop within 72 hours post thyroidectomy but clinical signs may not be evident for 7-10 days post surgery.

Radioactive iodine

The isotope radioiodine (^{131}I) is the treatment of choice for hyperthyroidism because of its high rate of success (95% cure rate with one treatment) and low rate of morbidity/mortality. ^{131} I is simple, safe, effective, requires no anesthesia, and the risk of hypoparathyroidism is eliminated. It is administered as a subcutaneous injection, an oral preparation or an IV injection. However, for safety purposes, the subcutaneous route is the preferred treatment. The isotope emits both β-particles and γ-rays. It is the beta particles that do much (80%) of the cellular damage. They also travel a maximum of 2 mm in tissue and their average path length is 400 µm so they are locally destructive (irradiate and destroys the hyperfunctioning tissue) but spare adjacent hypoplastic thyroid tissue, parathyroid glands and cervical structures. The majority of the radioiodine is concentrated in hyperplastic or neoplastic thyroid cells. Reduction in serum total T_4 is usually achieved within days to several weeks. The disadvantages of using radioactive iodine include: cost (the expense is "all up front"), possibility of a lengthy hospitalization (requirements for how long a patient must be hospitalized varies from state to state but the average is 3-10 days), regulatory compliance, proximity of an institution that provides the treatment, and the risk of post treatment-renal failure. A methimazole trial may be recommended for those patients that appear more at risk for post-treatment renal failure. Clients should be educated on all the potential health risks associated with hyperthyroidism and all the different treatment options in order to determine what the best therapy is for their pet.

REFERENCES

[1]Ward C, **Handbook of Small Animal Practice** 2008 5th ed.:459, [2] Mooney C, **Textbook of Veterinary Internal Medicine** 2005 6th ed.:1544, [3] Nelson R, **Small Animal Internal Medicine** 2009 4th ed.:745, [4]Cote E, **Clinical Veterinary Advisor Dogs & Cats** 2007: 552, [5] Feldman E, **Canine & Feline Endocrinology and Reproduction** 2004 3rd ed.:152

MAINTENANCE & TROUBLESHOOTING THE FLEXIBLE ENDOSCOPE

Regina Fairman, LVT, VTS (ECC)
Auburn Hills, MI

INTRODUCTION

As the cost of obtaining new equipment continues to rise and technology evolves, the need to maintain equipment in order to expand its lifespan becomes more focused on. It has been proven that proper care and handling saves money, especially when a potential problem is identified (and remedied) early, with 80% of all endoscope repairs and replacement being caused by improper care and handling.

Overview

The goal of this is to familiarize the support staff with the parts of the flexible endoscope, with its accessories, to establish a cleaning and maintenance schedule for your scope, and how to troubleshoot minor and major problems.

PARTS OF THE FLEXIBLE SCOPE AND ACCESSORIES

The body of the flexible endoscope is composed of three main parts, the light connector guide, the control section, and the bending section.

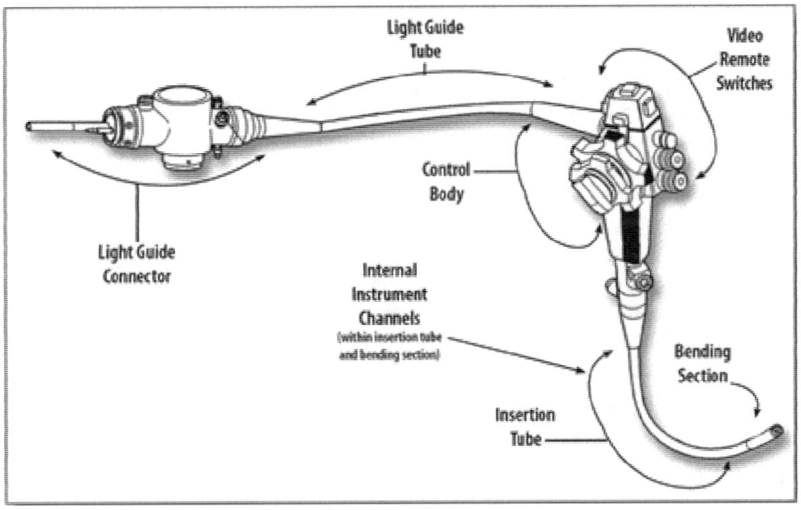

The Light Guide Connector (Video Connector)

This is the solid end of the endoscope that connects to the light source and video connector/processor. This is the part of the endoscope that is furthest from the patient. The video connector should not be exposed to water. The video processor has a cover that should be on whenever it is not attached to the tower. The light guide connector usually houses the parts that bring the light from the tower to the bending section and to the tip of the scope. The suction unit usually connects to the light guide connector via a side connection on the housing. The light guide connector and video connector have no moving parts and are composed of hard plastic and metal. This part of the scope is most frequently damaged from impact.

The Control Section

This part of the endoscope contains the position and function controls that help the doctors use the scope. The part that is held is ergonomically designed for ease of use. The control section has the angulation-control knobs (more commonly known as the wheels), the angulation lock, the air water valves, the suction valve, the remote switches (used for freezing images) and the biopsy valve (often referred to as the biopsy port).

The Bending Section

This is the section of the scope that is most fragile. The insertion tube is located between the control body and the bending section. The insertion tube is the long flexible tube inserted inside the body. The diameter and length varies with each flexible scope. The bending section is located distal to the insertion tube. The distal end of the bending section contains the video chip, fiber optics, optical glass lens system and the termination nozzle of the air/water/suction systems. Impact to the distal tip will potentially damage *all* of its components.

CLEANING AND MAINTENANCE

Establishing a care and maintenance protocol is vital for the care and longevity of the scope. Before adapting any standardized protocols, the user's manual should be referenced for care recommendations, along with any warranty information from the manufacturer. Attention should be given to whether or not the warranty is void in the event deviation from the recommended care occurs. The protocol that is established should be available in written format. The training of the support staff should be performed by the same person every time someone is trained. This is so the habits established are consistent, which avoids individual variation to the process. Next is a basic, step-by-step protocol.

Inspection

Inspection of the endoscope should be executed before and after each use. There are 3 components to the inspection process and should be performed in this order. *Visual* inspection is performed to the insertion tubing looking for peeling, cuts, blisters or adhesive separating. *Functionality* testing is next. The distal tip of the bending section should be articulated to ensure the scope can be used appropriately in imaging. Finally, *leak* testing should be done, which ensures fluid invasion has not occurred. Leak testing is done with a leak testing device. If any part of the inspection could be considered "failed", the endoscope should be considered not useable and should be sent for inspection and repair immediately.

Precleaning

After an appropriate inspection, a precleaning may be done. A precleaning involves wiping down the insertion tubing and control section to remove as much organic material as possible. A precleaning is arguably a non-necessary step if a mandatory, manual cleaning protocol is adhered to. If the scope has been used during the weekend, it should be reinspected and manually cleaned ASAP on Monday morning.

Manual Cleaning Process

The suggested steps for the manual cleaning process should be performed in this order post endoscopy. A leak test is only necessary if there has been trauma to the equipment during the scope. Immediately post-scope you should wipe down the exterior of the scope from the control section down with a cloth sponge to remove organic materials. Next, aspirate enzymatic detergent through the scope, the detergent must be diluted according to the label instructions in a sink or basin large enough to contain the scope without overcoiling. If the detergent is mixed too weak, it can actually not dissolve the organic materials, if it is mixed too strong, it can be damaging to the equipment. All detachable parts, removed for effective leak testing, should remain off during the cleaning process. With the scope immersed, brush the biopsy and suction channels, with channel cleaning brushes, until clean. Next flush the detergent through all internal channels using the appropriate manufacturer irrigation tubing. Rinse the scope with water thoroughly using the irrigation tubing. Finally, using the irrigation tubing, rinse/flush with 70% alcohol and purge with air. Hang to dry.

TROUBLESHOOTING THE SCOPE

Troubleshooting isn't a concept that is taught. Troubleshooting is accomplished when the user knows and is comfortable with care of the scope. In order to facilitate the process of troubleshooting the support staff should know that value of the scope. Also be very intimately familiar with the scope connections, know what and where your back-up equipment is and make sure it is functional, and very importantly, know who services your scope.

REFERENCES

Rutala WA et.al. **J Hosp Infect** 2004; 56: Suppl 2: S27; Nelson DB et, al. **Society for Healthcare Epidemiology of America** 2003; 27: 532; Muscarella LF. **Gastroenterology Nursing** 2004; 27: 271; Spach DH et, al. **Ann Intern Med** 1993; 118: 117; Association of PeriOperative Nurses, 2005; Martiny H et, al. **J Hosp Infect** 2004; 56 Suppl 2: S16

TOP 10 SMALL ANIMAL TOXINS: RECOGNITION, DIAGNOSIS, TREATMENT

Justine A. Lee, DVM, DACVECC
Minneapolis, MN

INTRODUCTION

Pet Poison Helpline, a 24/7 animal poison control located out of Minneapolis, MN, receives phone calls from both pet owners and veterinarians regarding toxicity cases from accidental or intentional misuse of OTC or prescription medications, common garden or outdoor toxins, and common household products. Below are the top 10 small animal toxins seen by Pet Poison Helpline. For further review on decontamination and specific treatment, attendees are referred to a veterinary toxicology book for more detailed review.

- In veterinary medicine, the primary treatment for toxicant exposure should be decontamination and detoxification of the patient.
- The goal of decontamination is to inhibit or minimize further toxicant absorption and to promote excretion or elimination of the toxicant from the body.
- Decontamination can only be performed within a narrow window of time for most substances; therefore, it is important to obtain a thorough history and time since exposure.
- Decontamination categories may include ocular, dermal, inhalation, injection, gastrointestinal (GI), forced diuresis, and surgical removal to prevent absorption or enhance elimination of the toxicant.

Selective serotonin re-uptake inhibitors (SSRIs)

Selective serotonin re-uptake inhibitors (SSRIs) are a class of medications that are commonly used in human medicine for depression. Common examples include drugs like fluoxetine (Prozac® in human beings; Reconcile™ in veterinary medicine), citalopram (Celexa®), escitalopram (Lexapro®), paroxetine (Paxil®), and sertraline (Zoloft®). These drugs are also used in veterinary medicine for a wide array of behavioral problems, including feline urine spraying, canine separation anxiety, lick granulomas, etc. In veterinary medicine, common clinical signs from SSRIs include sedation or central nervous system (CNS) stimulation, anorexia, and lethargy, even at therapeutic doses. SSRI drugs work by blocking the reuptake of serotonin in the pre-synapse, thereby increasing the levels of serotonin in the pre-synaptic membrane. Unfortunately, even small doses can cause increased levels of serotonin, resulting in "serotonin syndrome." Clinical signs of serotonin syndrome include: CNS stimulation, vomiting, tremoring, seizures, hyperthermia (secondary to tremoring and seizuring), diarrhea, abdominal pain, and mydriasis (dilated pupils). Treatment includes decontamination (ideally done at a veterinary hospital, due to the rapid onset of clinical signs), activated charcoal, hospitalization for sedation (i.e., with acepromazine or chlorpromazine), temperature regulation, IV fluid therapy, and supportive and symptomatic care.

Sleep aids

Sleep aids are often benzodiazepines or non-benzodiazepine hypnotics, and include drugs such as zolpidem (Ambien®) and eszopiclone (Lunesta®). These drugs work similarly to benzodiazepines (i.e., diazepam) as they potentiate GABA transmission, increasing frequency of chloride channel opening and resulting in inhibition of neuronal excitation. While these drugs result in sedation in humans, up to 40% of dogs ingesting toxic doses of sleep aids develop paradoxical CNS stimulation rather than expected depression. Clinical signs include CNS depression (i.e., depression, ataxia, weakness, paresis), CNS stimulation (i.e., hyperactivity, anxiety, agitation, panting, tremors), or even other signs like nausea, vomiting, diarrhea, and hyperthermia. Treatment includes decontamination, activated charcoal, and sedation (for CNS stimulation signs). However, in patients exhibiting CNS stimulation, benzodiazepines (i.e., intravenous diazepam) should *not* be used, as they may worsen the symptoms. Rather, the use of phenothiazines (i.e., acepromazine, chlorpromazine) or barbiturates (i.e., phenobarbital intravenously) should be used instead. In severe cases of respiratory or cardiac depression, the use of flumazenil, the reversal agent for benzodiazepines can be considered.

Amphetamines

Amphetamines, which are sympathomimetic agents, cause stimulation of the α and β-adrenergic receptors, and also stimulate release of serotonin and norepinephrine, which results in increased catecholamine stimulation in the synapse. Amphetamines come in legal and illegal forms. Legal forms include prescription medications for ADD/ADHD and weight loss, while illegal drugs include street drugs like methamphetamine or crystal meth. Toxicity results in CNS and cardiovascular stimulation,, including agitation, tremors, seizures, tachycardia, hypertension, secondary hyperthermia, mydriasis, vomiting, hypersalivation, and diarrhea. The oral lethal dose is low for amphetamines, and ranges from 10-23 mg/kg. Both clinical signs and treatment are similar to SSRI toxicity, and include IV fluids, cooling measures, sedation (i.e., with acepromazine or chlorpromazine), temperature regulation, blood pressure monitoring, and symptomatic and supportive care.

Topical toxins

One of the most dangerous topical toxins to be aware of is 5-fluorouracil (5-FU), which is a prescription anti-cancer medication used topically for humans with superficial basal cell carcinoma or actinic keratosis. Brand names include Efudex®, Carac®, Adrucil®, and Fluoroplex®. These drugs work by inhibiting DNA and RNA synthesis and production, and are rapidly absorbed from the gastrointestinal tract (GIT). Even tiny amounts can be very toxic, resulting in acute GIT and CNS symptoms. The prognosis with ingestion is grave in cats and guarded in dogs (with a reported survival in dogs of approximately 25%). Clinical signs can often be seen within 30 minutes up to 6 hours, and death has been reported in 7 hours. Clinical signs include acute nausea, vomiting, hemorrhagic diarrhea, abdominal pain, sloughing of the GIT, ataxia, severe and non-responsive seizures, and severe dose-dependent myelosuppression affecting all cell lines (i.e., pancytopenia resulting in leukopenia, thrombocytopenia, and anemia). A severe metabolic acidosis and signs of multi-organ failure can be seen. Typically, decontamination is not effective due to rapid onset of clinical signs. Treatment is supportive and includes anti-convulsant therapy, anti-emetic therapy, IV fluids to maintain perfusion to both the GIT and CNS, temperature regulation, antibiotic therapy (to help prevent sepsis from severe leucopenia), monitoring of baseline blood work to evaluate bone marrow and organ function (i.e., CBC, chemistry, venous blood gas), and symptomatic and supportive care. Death typically occurs due to secondary complications such as sepsis, intracranial hemorrhage (from severe thrombocytopenia), increased intracranial pressure (from severe seizures), etc. In general, a CBC should be performed every 3-4 days for at least 18 days, as it takes up to 3 weeks before all cell lines in the bone marrow return to normal.

Non-steroidal anti-inflammatory drugs (NSAIDS)

NSAIDs are competitive inhibitors of prostaglandin synthesis (cyclooxygenase or "COX") and result in decreased prostaglandin, which is important for normal homeostatic function (including maintaining renal blood flow, maintaining mucous production in the stomach, etc.). With NSAID toxicity, the GIT, kidneys, platelets, and CNS can be affected. Cats and certain breeds of dogs (i.e., German shepherds) seem to be more sensitive to NSAIDS, and must be treated aggressively. Common OTC human NSAIDS include drugs such as Advil®, Aleve®, certain types of Motrin®, etc. Common prescription veterinary NSAIDS include carprofen, deracoxib, etogesic, previcoxib, etc. Each NSAID has a different toxic dose. For example, ibuprofen results in GIT signs at doses as low as 16-50 mg/kg, while severe GI signs may be seen at 50-100 mg/kg. Renal compromise may be seen at doses of 100-250 mg/kg (resulting in potential acute renal failure), and fatalities have been reported at doses > 300 mg/kg. This differs tremendously from Aleve® (naproxen), where severe clinical signs can be seen at doses as low as 5 mg/kg. At experimental doses of 22 mg/kg orally once a day for 3 days, perforation of the intestines has been documented with naproxen. Each NSAID has a different half-life, route of excretion, and toxic dose, and an animal poison helpline should be contacted to identify what specific NSAID and toxic dose was ingested. Treatment includes decontamination, the use of activated charcoal (often multiple doses due to enterohepatic recirculation), GI protectants, aggressive IV fluid therapy (to help maintain renal blood flow), anti-emetic therapy, and symptomatic and supportive care. With high doses, anti-convulsants may also be necessary.

Xylitol

Xylitol is a natural sweetener found in small quantities in certain fruit. Xylitol has gained recent popularity because it is sugar-free, and is often found in diabetic snacks, foods, baked foods, and popular gums and candies. Unfortunately, there are variable amounts of xylitol in each product, and not all sources are disclosed by the company (how many grams of xylitol may be in each piece of gum). In general, > 0.1 g/kg of xylitol ingestion is considered toxic, and results in profound, sudden hypoglycemia from insulin stimulation. Higher doses are associated with acute hepatic necrosis, and aggressive decontamination and treatment should be initiated. With recent ingestion, decontamination should promptly be performed, provided the patient is not demonstrating any signs of hypoglycemia (which can be seen within 15 minutes of ingestion). Keep in mind that activated charcoal does *not* reliably bind to xylitol, and is not routinely recommended for this toxicity. Clinical signs include lethargy, weakness, vomiting, collapse, anorexia, etc. A blood glucose should be checked, and if hypoglycemic, a bolus of 1 ml/kg of 50% dextrose, diluted with an additional amount of 0.9% NaCl (in a 1:3 ratio) should be given IV over 1-2 minutes. Hypoglycemic patients should then be hospitalized for IV fluids [supplemented with dextrose (2.5 to 5% dextrose, CRI, IV)] and frequent blood glucose check should be performed every 1-4 hours.

Batteries

Battery ingestions occur quite frequently by dogs. This is often witnessed by the owner, or a chewed battery may be found by the owner. Often times, the pet owner may notice that the remote control is chewed on and the batteries are missing. When the casing for a battery is punctured, there is risk for alkaline or acidic material to leak out, resulting in severe ulceration to exposed tissues. The most common battery ingestion is of an alkaline dry cell battery (i.e., 9-volt, D, C, AA, AAA) or button/disc batteries. Alkaline dry cells (the majority of household batteries) contain potassium hydroxide or sodium hydroxide. When the compounds come in contact with tissue, liquefaction necrosis occurs causing deeply penetrating ulcers. In addition, newer types of "disc shaped" batteries can allow an electric current to pass to the tissues of the GIT as the battery is passed. This can cause a current-induced necrosis, resulting in tissue damage or even perforation of the oropharynx, esophagus, stomach or small intestine. Lithium button type batteries are the most dangerous, as one 3 volt battery can result in severe necrosis to the GIT or esophagus within 15-30 minutes of contact. Finally, certain batteries contain heavy metals (such as mercury, zinc, cobalt, lead, nickel or cadmium). Heavy metal toxicity can occur if the battery remains in the GIT for more than 2-3 days.

With any type of battery ingestion, the pet owner should seek veterinary attention immediately. A thorough oral exam and physical exam should be performed. Oral ulcerations may not be present for hours. The presence of black powdered material may be seen in the mouth, and occurs when dry cell batteries are punctured. The mouth should be thoroughly flushed and lavaged for 15-20 minutes with tepid tap water. A lateral abdominal radiograph (including the caudal esophagus in the chest) should be performed to evaluate the presence of the battery in the abdomen. Ideally, prompt removal should occur to prevent further corrosive injury. The use of endoscopy or surgery may be necessary. Emesis induction is not typically recommended, as corrosive injury may occur to the esophagus and oropharynx. Treatment includes removal of the battery, anti-ulcer medication (including H2 blockers and sucralfate) for 5-7 days, a bland or high-fiber diet, and analgesic therapy if necessary.

Fire starter logs

Fire starter logs typically do not pose a "toxicity" risk, but rather a foreign body obstruction (FBO) risk. Most types (i.e., Duraflame®) are made of compressed sawdust and wax, and do not break down in the stomach, resulting in a FBO. Rarer types of fire starter logs may contain heavy metals to provide a "color sparkle" to the fireplace. With recent ingestion, emesis induction should be performed to prevent FBO. If unknown ingestion or prolonged ingestion has occurred, abdominal radiographs should be performed to evaluate for the presence of gastric contents or FBO. If the material has passed out of the stomach, the use of a high-fiber diet, anti-emetic therapy, and careful monitoring (including clinical signs, radiographic evidence of obstruction, etc.) should be performed. With massive ingestions demonstrating evidence of FBO, surgical intervention may be necessary.

Hydrocarbons

Hydrocarbons consist of chemicals containing a hydrogen and carbon group as their main constituents. Examples include liquid fuels such as kerosene, engine oil, tiki-torch fuels, gasoline, diesel fuels, paint solvents, wood stains, wood strippers, liquid lighter fluids, asphalt/roofing tar, etc. These are often referred to as "petroleum distillates" based on their viscosity, carbon chain length, and lipid solubility. It is *contraindicated* to induce emesis with hydrocarbon toxicity, due to the risks of aspiration pneumonia. Due to the low viscosity of hydrocarbons, these compounds are more easily aspirated, resulting in respiratory injury and secondary infection. In general, hydrocarbons are GIT irritants, but can also be irritants to the respiratory system (if inhaled), eyes, and skin. Clinical signs include vomiting, nausea, tachypnea, and dermal or ophthalmic irritation. Typically, GIT irritation is self-limiting. Patients should be treated with anti-emetic therapy, possible subcutaneous fluid therapy (to assist in hydration), fasting (no food per os), and initiation onto a bland diet. Patients demonstrating any coughing, retching, or tachypnea post-ingestion should have chest radiographs performed to rule out aspiration pneumonia, of which treatment is supportive (i.e., oxygen therapy, IV fluids, antibiotic therapy, nebulization and coupage, etc.).

Bone or blood meal

Bone meal and blood meal are by-products from the meatpacking industry that are widely utilized as soil amendment products, fertilizer components, or as deer, rabbit and wildlife repellants. Bone or blood meal are "organic" compounds, and with the increased use of organic products in lawn and gardening, have increased exposure opportunities for animals. These are often considered low-level toxicities, but can result in FBO, severe pancreatitis, or GIT irritation with ingestion. A thorough history must be obtained from the pet owner, as these products are often mixed with more toxic agents (such as organophosphates found in rose fertilizers) which result in severe toxicity. Bone meal and blood meal are highly palatable to dogs and can result in unintentional, large ingestions. Tulip, daffodil and hyacinth bulbs may be dusted in bone meal when planted to fertilize and aid in repelling squirrels. The scent of bone meal may entice dogs to dig up newly planted bulbs and subsequently ingest both the potentially toxic bulb and bone meal. Large ingestions of bone meal can congeal into a solid ball in the stomach, resulting in a FBO. Decontamination is recommended with recent large ingestions or with dogs with a prior history of pancreatitis. Radiographs should be performed to determine if the material has passed out of the stomach, and to evaluate for the presence of gastric contents or FBO. With massive ingestions demonstrating evidence of FBO, surgical intervention may be necessary. In general, decontamination and symptomatic and supportive care are indicated.

CONCLUSION

Pet owners should be appropriately educated on how to pet-proof the house, and be trained on what common household products and kitchen items are poisonous. Pet owners should also be appropriately educated on crate training to help minimize toxin exposure. Once a pet is exposed to a toxicant, it is imperative to determine if emesis is appropriate and to understand when it may be contraindicated (i.e., symptomatic patient, delayed time since exposure, hydrocarbons, etc.). Knowledge of the underlying mechanism of action, the pharmacokinetics (including absorption, distribution, metabolism, and excretion), and the toxic dose of the toxicant are imperative in determining appropriate decontamination and therapy for the patient.

REFERENCES

1. Peterson ME, Talcott PA. **Small Animal Toxicology**. Elsevier, St. Louis, Missouri, 2006.

ADVANCED NURSING CARE OF THE EQUINE PATIENT TREATMENT METHODS & DIAGNOSTIC TOOLS

Kristin Miguel, RVT, VTS (LAIM)
Davis, CA

INTRODUCTION

Nursing care of the critically ill equine patient can be physically exhausting, dangerous and rewarding all at the same time. Due to the anatomical and physiological differences of the horse versus many other animal patients, they can provide nursing challenges even for the most experienced of technicians. Also, their size alone poses some risks to personnel even when they are absolutely healthy. Those that are reacting to pain (rolling, thrashing, kicking, etc.) or resenting treatment usually are unaware of their surroundings and the size difference between them and the personnel trying to care for them. This means that the technician needs to be very cognizant of safety for themselves, other personnel in the area, and to the patient as well.

Once an equine patient is triaged and then admitted into the intensive care nursing situation, a technician's role becomes vital. The technician becomes the spokesperson for their equine patient; monitoring vital signs, evaluating fecal and urine output, managing pain, addressing hydration status, and providing therapeutic treatments that are geared toward the patient's working diagnosis. Every patient that a technician has in their care needs to be under constant observation and basic values monitored so that even the most subtle of changes can be identified and reacted to in a timely and efficient manner.

TREATMENT METHODS

When receiving a patient into your care or taking over the care of a patient already hospitalized, it is critical to have an action plan in place. Thorough case transfer or discussion between the admitting veterinarian and the technician will help prepare for the technician's time with the patient. A brief history, signalment, and working diagnosis are the bare essentials for knowledge. All patients that come under the technicians care must then have a thorough physical exam performed, blood drawn for base line parameters and any prescribed medications and therapeutics started immediately. With a working diagnosis, the technician can also prepare themselves for worse case scenarios as their time with each patient progress, but be pleasantly surprised when the worst case doesn't occur. For example, if a patient is admitted as a post operative small intestine resection, the technician should be on the alert for continued pain, reflux, and potentially a long recovery. During the course of the technician's time with their equine patient, many duties and nursing care procedures will be performed: physical exams, blood analysis, catheter placement, fluid therapy, therapeutic administration, specialty procedures (naso-gastric intubation, oxygen delivery, sling management), feeding, bandaging and of course TLC. These are just to name a few.

Physical Exams

Physical exams consist of the following:

1. Heart rate asculted with a stethoscope: normal adult heart rate = 24-44 beats per minute; neonates = 40-60 beats per minute
2. Pulse quality
3. Respiratory rate: normal adult = 12-24 breaths per minute; neonates = 32-54 breaths per minute
4. Capillary Refill Time (CRT)
5. Mucous Membrane color
6. Gastrointestinal motility
7. Digital pulses
8. Rectal temperature: normal adult = 99.5 – 100.5 F; neonates = 100.5 – 101.5 F
9. Overall attitude and appearance

In addition, fecal and urine output needs to be measured. These vital signs should be checked frequently when caring for a critically ill patient and may become less frequent as the patient becomes more stable. Typically in our intensive care setting, thorough patient vitals and blood work are checked every 4 hours. The patient should be closely monitored in between these intervals. Many equine patients will not exhibit overt signs of pain or distress; the technician must be conscientious of the patient's subtle changes (dull eyes, raising a lip, muscle fasciculation, etc.) and act upon those changes in a timely manner. A flow sheet or other type of medical record should be used to document all information related to one patient.

Monitored Blood Values

Packed Cell Volume (PCV)
Total Plasma Protein (TP)
Electrolytes: sodium, potassium, ionized calcium, chloride
Blood gas: pH, CO2, O2, HCO3, acid base status
Glucose
Lactate

Other tests such as white blood count, joint or abdominal fluid analysis, Colloidal Oncotic Pressure (COP) and Osmolality may be requested once or as frequently as every 2 hours depending on the patient's condition and the need to assess these values. A technician should understand normal vs. abnormal blood values and should consult with the veterinarian when anything abnormal or outside the stated parameters occurs in the results obtained.

Once baseline values are established, the technician can begin the treatment plan. Most cases in the intensive setting will need intra venous access to provide fluids and medications. If a catheter has not been placed in the receiving/triage setting, then the technician will need to choose an appropriate vein (the jugular vein is the most accessible), pick the type of catheter that will be used: single lumen vs. multi port, short term vs. long term, length and gauge, and then follow aseptic procedures for placing an intra venous catheter.

Treatment Plan

Catheter care is very important in the overall care of the patient. The catheter and site will need to be monitored frequently to check for patency, heat, swelling, mechanical problems with the catheter, and any thing else that looks out of the ordinary. Any problems need to be addressed immediately, the catheter pulled and replaced with a new one if necessary. Complications at catheter sites (thrombophlebitis, infection, hematoma, etc.) can prolong the recovery of your patient. If the patient is not on fluids, but a catheter remains in place, appropriate flushing with heparinized saline at regular intervals should be considered a treatment, or catheters can be kept functional with a heparin lock (follow established practice protocols).

Fluid therapy is an integral part of most treatment plans whether the patient is an internal medicine case or other critical care case (colic, trauma, neurologic, etc.). The fluids chosen will help address the underlying disease process and resulting electrolyte imbalances, acid-base abnormalities, or protein losing incidents. The choice of fluids and additives will be derived from blood work results, physical exam values (dehydration status) and history of the patient. Once the veterinarian chooses a fluid administration plan, it is the technician's responsibility to see that the plan is followed and any adjustments needed addressed in a timely manner. Reactions to crystalloid fluid therapy do not often occur, but the technician needs to pay particular attention when administering blood products so that any transfusion reactions are immediately taken care of (stopping the infusion, alerting the veterinarian, possible epinephrine injection). Other fluid therapy considerations are the rate the fluids are being delivered and any additives added (knowledge of any potential complications from additives is essential).

Following the therapeutic treatment plan ordered by the veterinarian, the technician must be familiar with the particular therapeutics requested and the route with which to give them. Knowing antibiotics from anti-inflammatories, for example, is essential, and also what potential complication or reaction can arise from giving the requested therapeutic (anaphylaxis, neurologic signs, collapse, etc.). Basic knowledge of dosages, routes of administration and actions of the therapeutics will help the technician perform their nursing duties effectively and reduce the potential for harm or adverse reactions to their patients.

The technician may be asked to perform many other critical care procedures on their patient during the time they spend with them, and they should be proficient in performing these duties. For example, an equine patient that is admitted for colic that shows signs of pain may need to be tubed with a naso gastric tube to decompress their stomach or check for reflux. A neurologic patient may need to be treated in a sling and the technician will be responsible for comfort, care and maintenance of the patient. Many patients require bandaging either for support (lower limb support wraps) or to maintain the integrity of incisions (belly bands for abdominal surgery support). The technician will be responsible for proper application and monitoring of the bandage. Patient's that require oxygen therapy will need to be monitored closely for distress, occlusion of tubing, and that all delivery tubing remains intact and attached to patient.

Nutrition for the critically ill equine patient is another important part of the treatment plan. Ideally, horses should have free choice water and the feed that is offered to them should be kept as close to the same as their regular feedings at home. This is not often accomplished in the hospital setting due to the nature of the patient's disease or its surgical lesion. Certain conditions require that the patient receive nothing orally (food or water), while other conditions require that small amounts of soft, easily digestible feed is fed frequently (refeeding a post operative colic patient). No matter what the feeding regimen of the patient, the technician needs to observe what the patient eats or drinks, hunger or attitude towards the food, and any problems or inability to prehend or swallow. Some severe disease states may require the patient to be placed on intra venous feed (total parenteral nutrition, partial parenteral nutrition). In these instances, the technician will need to monitor any reactions or dramatic blood value changes; specifically glucose levels. Some patients may require tube feeding via naso gastric tube (milk to neonates, slurry for adult patients) which the technician will need to place and maintain. No matter what feeding and nutritional plan is outlined, it is very important for the technician to monitor what the patient takes in, any problems or reactions, and overall attitude of the patient has towards offered food and water.

DIAGNOSTIC TOOLS

Many critical care settings also use monitoring devices to aid in evaluation of effectiveness of treatment plans and methods that are employed on the case. In addition to using a stethoscope and visual observation of the patient, several monitors can give additional values that are useful in treating the patient.

1. Blood Pressure: systolic, diastolic and mean
2. ECG; cardiac function
3. Central Venous Pressure (CVP): circulatory regulation
4. Pulse Oximetry; oxygen saturation

The technician must have the ability to hook the patient up appropriately, trouble shoot the monitors and equipment when necessary for accurate readings, and understand the values that are obtained (normal vs. abnormal). The veterinarian must be notified immediately if any abnormal values are obtained.

SUMMARY

The technician's role in the nursing care of the critical equine patient is vital and demanding. The patients in their care are relying on them to provide the best care and make their healing process effective and as short in duration as possible. It is essential that the technician competently perform the prescribed treatment plan, document everything related to the patient's care and condition, and keep the lines of communication open between themselves and the attending veterinarian. They should be the consummate spokesman for their patient; any concern or change in condition should be addressed immediately. Last but not least, the technician should possess the skill to perform all of their duties with TLC (tender loving care!)

REFERENCES

Bassert and McCurnin **Clinical Textbook for Veterinary Technicians** 2010. Orsini and Divers **Manual of Equine Emergencies**; second edition; 2006. Saunders Comprehensive Veterinary Dictionary 1999. Reed et al **Equine Internal Medicine,** 2004; 273

DIAGNOSTIC & THERAPEUTIC APPROACH TO PLANT POISONINGS IN LARGE ANIMALS

Birgit Puschner, DVM, PhD, DABVT
Davis, CA

INTRODUCTION

Poisonous plants cause significant losses of large animals every year. There are several factors which contribute to a plant poisoning in a large animal. Foremost, there is the requirement that a sensitive species of animal ingest or otherwise be exposed to a toxic plant at an appropriate time. While there are many indications for plant poisonings, the most obvious cases involve sudden onset of disease in a group of animals. A diagnosis of poisoning is often supported when common feed or environmental conditions are found. A toxicosis is also suggested when the animal dies suddenly, since an acute poisoning is often the result of exposure to a toxic plant.[1] Undesirable effects may result from a single ingestion of a large amount of a poisonous plant, but some plants are so toxic that very small amounts may result in severe disease or death. Other plants cause chronic poisoning only after ingestion for weeks or months. The later situation may result in clinical signs long after exposure to the toxic plant material occured, and treatment may no longer be possible.

With few exceptions, large animals will not eat poisonous plants unless forced to by hunger. Thus, the single most important way to prevent poisoning is to use proper range and pasture management practices to provide ample forage and encourage consumption of non-toxic plants. Also, areas infested with poisonous plants should be avoided when trailing, holding, or unloading animals. Supplemental feed may protect animals if these conditions cannot be avoided, however there are instances (for example, herbicide applications) that may change palatability or increase toxicity in some plants. Additionally, if toxic weeds are embedded in pellets or cubes, or included in total mixed rations, animals may not be able to avoid ingestion.

Acutely toxic plants of specific concern to large animals include the following: cardiotoxic plants [oleander (Nerium oleander),[2] summer pheasant's eye (Adonis aestivalis),[3] foxglove (Digitalis purpurea), lily-of-the-valley (Convallaria majalis), dogbane (Apocynum spp.), some species of milkweed (Asclepias spp.), and grayanotoxin-containing plants (e.g. rhododendrons)], nitrate accumulating plants [pigweed (Amaranthus spp.), nightshades (Solanum spp.) and lamb's quarters (Chenopodium spp.)], cyanide containing plants [chokecherry (P. virginiana), California holly (Heteromeles arbutifolia), Juneberry (Amelanchior spp.), Sudangrass and Johnsongrass (Sorghum spp.), arrowgrass (Triglocin spp.), and eucalyptus (Eucalyptus spp.)], neurotoxic plants [tree tobacco (Nicotiana glauca), lupine (Lupine spp.), poison hemlock (Conium maculatum), and larkspur (Delphinium spp.)] and red maple (Acer rubrum).

Plants that result in chronic intoxications of large animals include: Pyrrolizidine alkaloid-containing plants [common groundsel (Senecio vulgaris), tansy ragwort (Senecio jacobaea), and fiddleneck (Amsinckia intermedia)], and locoweeds (Oxytropis and Astragalus spp.).

DIAGNOSTIC APPROACH

An accurate diagnostic approach is central to providing adequate treatment and preventing new cases. Correctly and rapidly diagnosing a plant poisoning is often extremely difficult, especially because there is no magic test for all possible plant toxins. In many cases of intoxications, initial clinical signs are non-specific (such as diarrhea or weakness) and post-mortem lesions may be absent. With a systematic approach to collecting all evidence, proper sampling techniques, and good communication between the clinician, technicians, client and laboratory, the usefulness of a toxicology investigation will be maximized. In addition, practitioners and technicians are encouraged to consult with a veterinary toxicologist who will help in the work-up of a poisoning case and provide advice on proper sample collection.

Acutely toxic plants

In many cases, the best way to support a diagnosis of an acute plant poisoning is to confirm the presence of a toxic plant in the animal's environment, to confirm that the plant has been ingested (by noting that the candidate plants have been chewed and/or finding plant fragments in gastrointestinal tract samples), and to correlate clinical findings, when possible, with those known to be associated with the suspect plant. If diagnostic tests are available, the diagnosis can be confirmed. This is especially important in insurance or legal investigations. Intoxication with an acutely toxic plant should be suspected if:

- there is a sudden onset of disease in a number of animals
- affected animals have a common feed source
- affected animals are kept under the same environmental conditions
- affected animals have recently been exposed to a new environment
- there has been a recent change in feed
- there are unusual weather conditions
- an uncommon clinical condition exists
- unexplained and/or sudden deaths have occurred

Even if a poisoning is suspected, the veterinarian has to remain a neutral observer and has to consider toxic as well as nontoxic etiologies. Thus, it is crucial to follow a systematic approach in order to reach an accurate diagnosis.[4,5] Important parts of this systemic approach include the following:

- obtain a complete history (including data on breed, age, sex, bodyweight, reproductive status, vaccine history, current medications, past medical history, boarding facilities, and presence of other animals and any abnormalities in them)
- obtain detailed information about recent feed or environmental changes and the timeline of events
- consult with a veterinary toxicology laboratory prior to the investigation to get expert advice on proper on-site investigation and sample collection (www.abvt.org or www.aavld.org). The veterinary toxicologist is uniquely qualified to discuss differential diagnoses, sampling techniques, and potential treatment options.
- assess the individual, affected animal as well as the herd or group of animals
- collect samples of suspect feed, water and other potential sources (do not pool samples[6])
- take pictures of suspect plant(s) and e-mail to a veterinary toxicologist or botanist for identification
- perform a complete physical examination and monitor rate of onset and progression of clinical signs
- perform a post-mortem examination or submit carcass to a veterinary diagnostic laboratory for evaluation
- collect specimens from affected live animals and if applicable, from dead animals (do not use formalin for samples submitted for toxicology testing; keep samples frozen or refrigerated; do not pool samples)

Table 1. Specimens Suitable for Toxicology Testing in a Suspect Acute Plant Intoxication

Toxin(s)	Toxic plant(s)	Suitable Specimens
Oleandrin	Oleander	Serum, urine, blood, gastrointestinal contents (lavage fluid, feces, stomach, rumen, cecal, colon contents), liver, cardiac muscle; Hay, pellets, total mixed ration, feed source
Grayanotoxins	Rhododendrons, azaleas, Japanese Pieris, mountain laurel, black laurel, fetter bush	Gastrointestinal contents (lavage fluid, feces, stomach, rumen, cecal, colon contents), urine; Hay, pellets, total mixed ration, feed source
Alkaloids (nicotine, lupine, coniine, taxine atropine, deltaline)	Tree tobacco, Lupine, Poison Hemlock, Yew, Jimsonweed, Larkspur	Gastrointestinal contents (lavage fluid, feces, stomach, rumen, cecal, colon contents), urine; Hay, pellets, total mixed ration, feed source
Nitrate/Nitrite	Pigweed, Lamb's quarters, nightshades, Sudangrass, Johnsongrass	Serum, ocular fluid (eyeball); Hay, pellets, total mixed ration, feed source
Cyanide	Choke cherry, California holly, Juneberry, Sudangrass, Johnsongrass	Whole blood (EDTA or heparin tube), gastrointestinal contents (lavage fluid, feces, stomach, rumen, cecal, colon contents), liver, skeletal muscle Hay, pellets, total mixed ration, feed source

Lesions associated with poisonings can be specific, nonspecific, or even absent, but in all cases of unexplained deaths, a complete postmortem examination should be performed, including thorough gross inspection of the entire body and all gastrointestinal contents. Before performing a necropsy, it is recommended to inquire about the legal status of the case so that any required documentation (chain-of-custody) and sampling (for insurance, legal, or other purposes) can be done. Samples collected during necropsy serve several important purposes, including histological, toxicological and microbiological examination. For histology, samples of major organs should be placed in fixative. For toxicology testing, separate samples have to be collected and frozen fresh, not fixed.[7] As with all samples for toxicology testing, they should not be pooled to avoid dilution and packed separately in whirlpak-type bags.[6] The use of rectal examination sleeves and gloves is discouraged, because sleeves and gloves may leak, and samples may wash together causing inadvertent pooling. In incidents of sudden deaths, careful inspection and sampling of GI contents is critical because unusual material or toxic plant parts may be identified. Generally, the most useful postmortem specimens for toxicology testing include gastrointestinal contents, liver, kidney, urine and brain. It is always best to collect as many samples as possible for toxicology testing at the time of necropsy, as it is much easier to discard unnecessary samples from the freezer than wishing additional samples were available for testing if histopathology suggests a toxic etiology. Many diagnosticians collect and hold toxicology samples until results of other tests (e.g., histopathology, bacteriology, and virology) are complete.

Chronically toxic plants

Chronic plant poisonings are extremely difficult to diagnose because the toxic plants may have been ingested over a period of weeks to months before clinical signs of illness develop and not at the time of presentation. Clinical signs of chronic plant intoxications are usually non-specific. Animals are often poor-doers, losing weight, showing poor reproductive performance, or showing a loss in production. In suspect chronic plant intoxications it is very important to obtain a complete history, including data on feed source, feed changes in the past few months, movement of animals to different environments, changes in the environment of the animals, and/or changes in supplementation. If available, it is most helpful to collect feed samples; especially feed that was offered weeks or months prior to the case presentation. Specimens from live, affected animals are unlikely to be useful for testing because toxin exposure was in the past. In chronic plant intoxications, most information can be gained from a postmortem examination by ruling out other causes for disease.

Animals with suspect pyrrolizidine alkaloid intoxication present a diagnostic challenge.[8] Stomach contents may help to identify plant matter in recent exposures but are usually not helpful in chronic disease. The diagnosis is generally based on history, clinical signs and necropsy (or liver biopsy) findings.

Animals with suspect locoweed intoxication are also difficult to diagnose. Horses are at greatest risk and in a recent exposure, a blood smear can provide useful diagnostic data.[9] The most accurate diagnostic approach is based on histopathological examination of tissues to characterize the pathognomonic lesions.

THERAPEUTIC APPROACH

Unfortunately, most plant poisonings have no specific treatments. Obviously, continued exposure to the suspect plant should be stopped immediately. In acute intoxications, routine decontamination procedures should be initiated. Administration of activated charcoal to adsorb plant toxins present in the gastrointestinal tract should be done as quickly as possible. Activated charcoal is administered as an aqueous slurry at a dosage range of 1-5 g/kg body weight (~1 g of AC per 5 ml of water). Multi-dose activated charcoal is beneficial for a number of plant intoxications, including oleander. Administration of a cathartic helps to hasten elimination of contents from the gastrointestinal tract. Commonly used cathartics include sodium sulfate (Glauber's salts), magnesium sulfate (Epsom salts), and sorbitol. Sodium or magnesium sulfate can be administered at 250-500 mg/kg body weight mixed in the AC slurry. Sorbitol (70%), also mixed in the AC slurry, can be administered at 3-ml/kg body weight. There is little need to administer a cathartic if significant diarrhea is already present. In addition, symptomatic and supportive care needs to be provided and should include administration of IV fluids and other medications, dependent on the actual toxin. For a few plant poisonings, specific antidotes may be indicated; the treatment of cyanide or nitrate intoxicated animals are examples.

The treatment of chronic plant intoxication is also non-specific. The length of exposure to a toxic plant is extremely important when evaluating such case. In most cases, decontamination with activated charcoal is not warranted. Clinical signs and pathological changes may be reversible until permanent damage has occurred. At the time, a horse or cow is diagnosed with pyrrolizidine alkaloid intoxication the liver may be too damaged and may not be able to regenerate. Therapeutic strategies include treatment for liver failure, but prognosis is guarded. Animals suffering from locoweed poisoning have a better chance to fully recover over a period of several weeks after the toxic plants are removed. However, horses with locoweed poisoning are at risk to developing clinical signs lifelong and have to be evaluated and handled carefully.

PREVENTION

Plant poisonings are best prevented by avoiding exposure to toxic plants in the first place. Hay and feed should be carefully inspected for contamination with potentially toxic weeds before the first feeding and owners should establish excellent working relationships with suppliers and growers. Awareness of poisonous plants growing in a certain geographical region and their associated clinical signs are instrumental in making a diagnosis, and initiating treatment. Many toxic plants are unique to a geographic location, or a specific time of year. It is often prudent to consult with a veterinary toxicologist at the earliest stage of a case, especially if one is available nearby.

SUMMARY

Plant intoxications can be devastating if toxic plants are grazed or unintentionally incorporated into hay or silage. The recognition of poisonous plants in hay or forage is essential to the prevention of plant poisonings in large animals. Veterinarians, diagnosticians, and technicians, along with producers and farm managers play important roles as guardians, and all can contribute important information to diagnose plant poisonings. Establishing an accurate diagnosis depends heavily on a systematic investigation. There are relatively few antidotes for plant toxins that can cause acute illness and death. However, early and appropriate decontamination and vigorous symptomatic and supportive care can result in recovery of acute intoxications. There are very limited treatment options for chronic plant poisonings and in many instances, prognosis is guarded. In order to prevent recurrences, it is critical to collect and evaluate all available data from a plant intoxication investigation and to generate a strategy for prevention.

REFERENCES

[1]Burrows GE et al. **Vet Clin North Am Food Anim Pract** 1989;5:263. [2]Galey FG et al. Toxicity and diagnosis of oleander (Nerium Oleander) poisoning in livestock, in Garland T, Barr AC (eds): Toxic plants and other natural toxicants. New York, **CAB International**, 1998, p 215. [3]Woods LW, et al. **Vet Pathol** 2004;41:215. [4]Galey FD: **Vet Clin North Am Food Anim Pract** 2000;16:409. [5]Galey FD: Diagnostic toxicology, in Plumlee KH (ed): Clinical Veterinary Toxicology (ed 1). Mosby, St. Louis, 2004, p 204. [6]Giorgi, M, et. al. **Vet Res Commun** 2004;28 Suppl 1:103. [7]Sato I, et. al. **J Toxicol Sci** 2006;31:191. [8]Hooper PT. Pyrrolizidine alkaloid poisoning – pathology with particular reference to differences in animal and plant species. In: Keeler RF,Van Kampen KR, and James LF, eds. Effects of poisonous plants in livestock. New York: **Academic Press**, 1978:161. [9]Stegelmeier BL et al. **J Nat Toxins** 1999;8:35.

DIAGNOSTIC SAMPLING FOR LARGE & SMALL ANIMALS – THE TECHNICIAN'S ROLE

Marika Pappagianis BS, RVT, VTS (LAIM)
Susan Cox, RVT, VTS (SAIM)

A routine responsibility for veterinary technicians is to collect or assist with the collection of biological samples for laboratory evaluation. Technicians also prepare samples for transport to laboratories and in many cases, perform the analysis on the sample.

It is important that the samples accurately reflect the condition of the animal so that the information obtained can be relied upon to help the veterinarian determine a diagnosis or accurately reflect the progress or deterioration of the patient and its response to therapy. If the samples have not been collected and handled properly, the results of analysis may not accurately indicate the real, current condition of the animal and may lead to misdiagnoses and inappropriate treatment.

Equine veterinary technicians are most frequently called upon to collect blood, urine and fecal samples. In the case of some biological samples, the technician's role is usually to assist while the veterinarian obtains the sample. The technician may be relied upon to perform some of these collections under the direction of the veterinarian. This is influenced by the experience and training of the technician, the level of comfort and trust established in the working relationship with the veterinarian and immediate needs.

The veterinary technician, whether assisting with a horse, dog, or cat, should be familiar with the procedures and be prepared to assist with the collection of biological samples. Depending on the species the technician is working with, the technician should be familiar with the collection of: Blood, Urine, Feces, Milk (colostrum), Semen, Abdominal fluid (abdominal tap, peritoneal tap, abdominal paracentesis, abdominocentesis, belly tap), Cerebrospinal fluid (cerebrospinal tap, CSF tap, Lumbosacral tap, LS tap, Atlanto occipital tap, AO tap, spinal tap), Pleural fluid (pleurocentesis, thoracentesis, thoracocentesis, pleural tap, chest tap), Synovial fluid (arthrocentesis, joint tap), Tracheal wash (transtracheal aspiration, trach wash, TTA), Bronchial wash (bronchoalveolar lavage, BAL), Bone marrow and Tissue biopsy.

Restraint of the animal is paramount for the safety of the technician, veterinarian and patient. The technician needs to be comfortable assessing each situation and must have a repertoire of assorted restraint methods that can be utilized as needed and must be able to adapt or modify the methods to suit the patients' temperament, medical condition, facilities and personnel available.

The least amount of restraint is desirable and a routine should be established to start with light restraint and modify to stronger methods as the situation requires. The technician needs to be observant of the response of the patient to the methods used and be prepared to back off or increase the intensity as the situation demands. The technicians must clearly communicate to the veterinarian if they see a need to change techniques so that the veterinarian can modify his or her actions for all involved to remain safe.

For adult horses, restraint methods most commonly utilized include use of a halter and lead rope, skin twitch (skin grab), twitch (nose twitch, lip twitch), shank (stud chain), lip chain, stocks and chemicals (tranquilizers, sedatives). The technician and veterinarian should make a point of being on the same side of the horse. If a troubling situation arises, the handler can direct the horse's head toward the handler thus turning the body and hind end of the horse away from the collector.

For foals, restraint options include holding in recumbence, securing foal between handler/collector and wall, placing an arm around the foal's chest and wrapping the other arm around the foal's rump, placing an arm around the foal's chest and holding firmly with the other hand at the base of the tail and chemical restraint.

For the restraint of canines and felines, safety and patient comfort are the main goals. A comfortable surface, such as a padded V-trough for such procedures as cystocentesis and ultrasonography will give the patient a feeling of security. Debilitated, chemically restrained or older patients require fleeces when placed in a lateral recumbent or dorso-ventral (DV) position for such procedures as paracentesis, venipuncture or urinary catheterization. Fractious animals may require **careful** placement of muzzles and/or chemical restraint.

Blood sample basics

The choice of blood collection tube or syringe containing an anticoagulant determines what analytes can be measured. Tests may require whole blood, serum or plasma. A variety of blood collection tubes are available. Some tubes contain no additives (clot tubes for obtaining serum), some contain preservatives which prevent a change in the specimen, and some contain anticoagulants that inhibit the formation of a clot. Prior to collection, the technician must know what type of sample is necessary for the desired lab tests and have the appropriate syringes and tubes prepared.

Choose the correct tubes/anticoagulant.

If samples must be immediately cooled, prepare a container of ice and water. Assess the condition of the veins, select a vein. Select an appropriate sized needle. (If an IV catheter is in place, withdraw a sufficient waste sample prior to sample collection). Insert needle in a manner to cause as little tissue damage as possible and aspirate smoothly. Aspirating blood too rapidly through a small gauge needle or causing tissue damage during collection can cause hemolysis. Components that are normally found in higher concentrations in red cells than in serum will falsely increase in the serum which can result in misleading lab results including but not limited to the following: increased potassium, increased protein, increased glucose, increased MCHC (mean cell hemoglobin concentration), decreased MCH (mean cell hemoglobin) and invalid coagulation profiles (1,2,5). Fill the collection tube or syringe adequately. A small amount of blood (less than 1 ml in a 5 ml tube or less than 0.5ml in a 2 ml tube) in a large tube containing anticoagulant can cause RBC's (red blood cells) to shrink which will falsely decrease the PCV (packed cell volume) and MCV (mean cell volume), falsely increase the protein and can lead to erroneous results of other blood constituents (1,2). Some labs may require tubes to be filled to a certain point on the tube's label, and past the vacuum point. It's good practice to always check with your reference lab when submitting a sample that is unfamiliar. If a variety of samples are needed, place the blood into tubes in a specific order to avoid contamination by additive from one tube to the next. Following is the suggested "order of draw":

1. Blood culture tubes (to maintain sterility)
2. Non additive tubes (serum tube, clot tube, red top tube, RTT)
3. Coagulation tubes (sodium citrate tube, blue top tube, btt)
4. Heparin tube (hep tube, green top tube, GTT)
5. Ethylenediaminetetraacetic acid K3 tube (EDTA tube, lavender top tube, LTT)
6. Oxalate / fluoride tubes (gray top tube)

(Adapted from Mosby's Diagnostic and Laboratory Test Reference, 3rd Edition)

Note – for coagulation profiles, collect a small amount of blood into a serum tube (even if no serum sample is required) as a waste sample prior to placing the blood into the sodium citrate tube. This is to clear the sample of tissue factors.

Gently invert the tube to mix the sample (rough mixing can lead to hemolysis). Apply pressure to the venipuncture site and observe for bleeding. If indicated, place a temporary bandage at the venipuncture site and remove ½-1 hour later. Label samples (if not labeled prior to collection). Process or analyze the sample promptly, submit to the lab and package for transport. Following sample collection the cells continue to be active, taking nutrients from the serum and producing waste products, so serum should be separated from the clot within 30 minutes of collection and analyzed, refrigerated or frozen (1,2). Prolonged contact of red cells with the serum by failing to analyze whole blood in a timely manner can result in the following: decrease in glucose levels (about 10% per hour), increases in phosphorus and potassium levels (2,5). Prolonged time in EDTA may change the appearance of neutrophils, resembling changes caused by a disease process thus leading to erroneous diagnoses. Don't freeze or overheat the samples (unless freezing serum after removing from clot) (1,5). Delays in analyzing blood gas samples without storing the samples in an ice water bath can change pH and gas results. Most biochemical parameters are stable in serum or plasma stored at 4 degrees C for at least 24 hours (2). If blood films are needed, they should be made within 15 minutes of collection to decrease in vitro morphologic changes in the blood cells (1). Do not refrigerate blood films as water condensation on the surface causes lysing of RBC's (1). Don't leave samples in bright light or in hot vehicles. Heat can destroy some chemicals and activate others and exposure to sunlight for an hour can decrease the bilirubin in a sample by up to 50% (5). Do not refrigerate clot tube before complete clot formation. The cold temperature inhibits good formation and retraction of clot. Do not centrifuge clot tube before complete clot formation, as a fibrin clot will form above the RBC's (1).

A vein is selected based on the condition of the vessel, the position and disposition of the animal, and/or the disease process. If repeated samples are anticipated, start with a more distal venipuncture site, with subsequent samples taken progressively more proximal on the vein or consider placement of an IV(1) or sampling catheter.

Blood Gas Samples from arteries

Heparinize a syringe (1 to 3 ml) by aspirating a small drop of heparin into the syringe (enough to fill the needle and appear in the hub), pull plunger back all the way, remove the needle and discard, place a new 25 g needle on syringe and expel all of the heparin from the syringe through the needle. This allows for the sharpest needle possible because the needle has not even punctured the top of a bottle of heparin. OR – use a commercially available blood gas syringe containing dry heparin. Clean the collection site with isopropyl alcohol. For severely compromised patients, clip hair, clean and disinfect the site with betadine solution and alcohol. Leave betadine solution on skin. Restrain patient. Feel for pulse (arteries pulsate, veins do not). Gently insert needle. When in artery, little aspiration is needed (very sick animals with poor blood pressure and poor perfusion will require aspiration). Collect a minimum of 1 ml if at all possible to ensure diagnostic values. Smaller amounts are OK when commercial syringes are used, as they are designed to eliminate dilutional errors. Remove needle from vessel. Tap out any air from syringe and place needle in rubber stopper of cap syringe with rubber top. Blood gas samples must remain anaerobic because exposure to air will modify the O_2 and CO_2 values. Hold firm pressure on the insertion site to prevent hematoma formation or, in small animals, place a moderately tight bandage at the venipucture site. Analyze sample ASAP or place in ice water bath.

Blood cultures

Are done to detect viable infectious agents (bacteremia, viremia, fungemia) in the blood.

To collect blood cultures: Shave the site. Perform a sterile prep (must prevent sample contamination with skin microflora). Collector wears sterile gloves. Use sterile needle and syringe. Assistant can open the syringe casing and drop the sterile syringe into the gloved hands of the collector. Assistant opens needle cap and places needle onto syringe and pulls the cap off. Needle is inserted into vein, blood is aspirated. Collector removes the used needle, replaces it with a new sterile needle and inserts needle into blood culture bottle. Label the bottle with patient identification, date, time and patient's temperature.

For small animals, blood cultures at the VMTH are collected at times 0, 10 minutes, 1 hour, and 70 minutes, preferably from 4 different sites. This protocol increases the probability of collection of bacteria. It is permissible to collect blood from an existing IV catheter if port is surgically prepped. As always, check with your clinician or reference lab for their recommended protocol.

Urine collection

Urine can be collected via free catch, cystocentesis (canine/feline) or bladder catheterization. For best culture results, a sample should be obtained via catheterization in the equine, and cystocentesis in the canine or feline. Many analytes can be measured on free catch urine samples. The sample is best collected mid stream to avoid the higher concentration of bacteria, mucous and cell debris that are flushed from the urethra and lower genital tract in the initial output (1,4,7). Note that fresh equine urine is slightly viscous due to the normal presence of mucous and cloudy due to the normal presence of calcium carbonate crystals (1, 2, 6).

Urine analysis (UA) should be performed on fresh urine within 20-30 minutes or the sample should be refrigerated to avoid changes in pH, cellular degeneration and bacterial growth (1, 2, 6). Bacteria present in the urine will multiply after collection. This produces urease which converts urea to ammonia and CO2 and causes the pH to increase (1). Urine cultures should be started within 2 hours of collection to reduce the likelihood of bacterial overgrowth by insignificant bacteria. Refrigeration of urine can change the specific gravity and interfere with the reagents on urine test strips. Therefore, refrigerated urine samples should be allowed to come to room temperature prior to analysis (1,5). Refrigeration tends to preserve most constituents but may alter pH and crystals and it is most desirable to analyze fresh urine. Refrigerate urine for up to 8 hours if necessary.

To prepare urine for a slide, centrifuge at a slow speed (1500 rpm for 5 minutes) to avoid disruption of cells and casts. Decant the supernatant from the sediment. Gently tap the tube to re-suspend the sediment and using a small pipette, a plain microhematocrit capillary tube or wooden stick, transfer a drop of suspended sediment onto a glass slide and cover with cover slip (1).

Fecal collection

Feces can be collected from the ground or directly from the rectum. Fresh feces yield more diagnostic information for parasite identification and culture results. Feces should be examined within a few hours, refrigerated or packed with ice packs for shipping to lab. Parasites can continue to develop in feces outside of the patient and may alter the results of analysis. When handling fecal material, the technician should always be aware of the zoonotic potential and should wear gloves, wash hands and clean counter surfaces well to avoid fecal/oral contamination. Collect from the ground using a gloved hand and place feces into a clean container.

Other diagnostic samples:

As listed earlier in this paper, many sampling procedures are usually performed by the veterinarian with the assistance of the technician. With length limitations of this discussion, we will highlight just a few comments that we believe the technician should be aware of and we refer the technicians to further reading in many excellent textbooks that will be listed at the end of this discussion.

Paracentesis (thoracocentesis/abdominocentesis) *

Samples are collected in both red top (clot) tubes and EDTA for possible culture and/or cytology. If performing a therapeutic paracentesis, a "mid-stream" sample is desired, as the first few drops may contain contaminants and if possible should not be included in the sample.

Arthrocentesis *

To minimize iatrogenic infection, sterile technique with surgical gloves and prep, such as chlorhexidine scrub is recommended. In small animals, the most distal sites (carpi, tarsi) should be attempted first.

Cerebrospinal fluid collection *

Alantooccipital (AO) and Lumbosacral (LS) taps are performed under general anesthesia in canine and feline patients. AO taps in horses are also performed under general anesthesia. For LS taps, attempt to keep horse standing as squarely as possible. These samples are collected in EDTA tubes.

Tracheal wash *

Tying the canine urinary catheter in a loose half hitch knot allows it to stay completely on the sterile field and makes it easier to control the distal tip, minimizing accidental contamination (1).

Bronchoalveolar lavage *

When preparing syringes, do NOT use bacteriostatic saline. If saline is warmed to body temperature, less bronchospasm results during infusion and thus more saline is retrieved. If sample can not be processed within 3-4 hours, refrigerate the sample or chill on ice. Refrigerated samples should be processed within 24 hours (4).

This paper has offered a very cursory overview of the role of veterinary technicians in obtaining diagnostic samples. Very detailed descriptions and discussions of procedures and related topics can be found in many excellent reference books.

REFERENCES

[1.] McCurnin, D., Bassert, J., editors: **Clinical Textbook for veterinary technicians**, ed 7, St. Louis, 2009, Saunders, chapters 16, 18, 20 [2.] Meyer, D., Harvey, J.: **Veterinary laboratory medicine interpretation and diagnosis**, ed 2, Philadelphia, 1998, Saunders [3.] Pagana, K., Pagana, T.: **Mosby's diagnostic and laboratory test reference**, ed 3, 1997, Mosby [4.] Schumacher,J., Moll, H.: **A manual of equine diagnostic procedures**, Jackson,2006,Teton NewMedia [5.] Pratt, P., editor: **Laboratory procedures for veterinary technicians**, ed 2, Goleta, 1992, American Veterinary Publications, chapters 3,5 [6.] Rose, R, Hodgson, D., editors : **Manual of equine practice**, ed 2, Philadelphia, 2000 Saunders, chapters 10, 17 [7.] Hanie, E.: **Large animal clinical procedures for veterinary technicians**, St. Louis, 2006, Mosby

EQUINE ZOONOTIC DISEASES & THE VETERINARY SETTING

K. Gary Magdesian, DVM DACVIM, DACVECC, DACVCP
Davis, CA

Zoonotic diseases have received renewed attention in human and veterinary medicine over the last several years as infectious disease control (IDC) has advanced; there is heightened awareness of methicillin resistant *Staphylococcus aureus* (MRSA), salmonellosis, and other infectious diseases that can act as nosocomial both and zoonotic pathogens. This review will focus on zoonotic diseases of horses.

GASTROINTESTINAL DISEASES

Salmonellosis

Gastrointestinal (GI) diseases are among the most severe and contagious zoonotic diseases. The priority for discussion is salmonellosis (*Salmonella enterica ssp enterica*, with many serovars or strain types such as Typhimurium or Agona). Horses and livestock species alike can shed *Salmonella* subclinically. The opposite end of the spectrum is highly fatal colitis and septicemia observed particularly in horses and neonatal animals, respectively. The group B and E salmonellae are among the most virulent in terms of causing disease, especially serovars Typhimurium, Anatum and Krefeld, however several groups of salmonellae can cause disease, including C1, C2, D, and E. *Salmonella enterica* serovar Abortusequi is the only host-adapted salmonellae in horses, but it has been eradicated from the United States. Clinical disease is a result of host (immunity, stress level, predisposing factors such as antibiotic use, hospitalization, long distance transport), agent (group, virulence level, numbers of organism), and environmental circumstances (ambient temperature, moisture, hygiene). Colonization with *Salmonella* can range from no disease to very severe colitis. Clinical signs of colitis include watery to bloody diarrhea with a systemic inflammatory response syndrome (SIRS). Horses with SIRS have tachycardia, tachypnea, fever or hypothermia, injected and toxic mucous membranes, cold extremities, lethargy, anorexia, and sometimes colic. Their lab work shows hemoconcentration, high blood lactate, acidosis, low total protein concentration, low white cell count, and clotting abnormalities. Chemistry panels can demonstrate organ dysfunction in the form of renal or liver disease. Foals with salmonellosis often develop septicemia and are at risk for septic joints, growth plates or bone. Without aggressive and early intensive care these infections have a high mortality. Complications include laminitis, thrombosed veins, weight loss and chronic diarrhea. Diagnosis is through fecal cultures or polymerase chain reaction (PCR). A series of 5 fecal cultures should be performed before a horse can be labeled as negative, and PCR should be performed on enrichment broth after an overnight incubation. Fortunately, horses do not become lifelong shedders, and shed for days to months (most commonly weeks) after infection.

Salmonella is very contagious to other high risk animals and humans. Sick, hospitalized horses, those that are stressed, and horses treated with antibiotics are at highest risk for infection. Immunocompromised humans, especially those with HIV, immunodeficiencies, cancer, or those treated with corticosteroids, are at risk, as well as infants, geriatrics, and pregnant women. Humans with salmonellosis develop GI signs (vomit, diarrhea, abdominal discomfort) and can develop septic shock. Because of these risks, horses with *Salmonella* should be strictly isolated, with a minimal number of personnel caring for them. They should be handled with barrier precautions (coveralls, impervious lab coats, booties or rubber boots, gloves, hair protection); hand washing and use of alcohol-based hand sanitizers should be emphasized. *Salmonella* is quite susceptible to most common disinfectants, including alcohol, chlorehexidine, betadine, bleach, phenolics, quarternary ammonium compounds, and higher peroxygen compounds. However, most of these are inactivated in organic matter, and disinfectants that are best suited to the farm setting are those that retain activity in the presence of organic debris, particularly phenolic compounds (e.g. TekTrol [Bio-Tek Industries, Atlanta, GA]; 1 Stroke Environ [Steris Corp, St. Louis, MO]), peroxygen compounds (e.g. Virkon S [DuPont, Wilmington, DE]), and some of the quarternary ammonium compounds. *Salmonella* is notoriously difficult to eradicate from the environment and requires diligent cleaning and disinfection. Serovar Newport (group C-2) is particularly hardy in the environment.

Cryptosporidiosis

Cryptosporidium primarily affects young animals, putting foals at greatest risk of disease. Horses and humans can shed the protozoa subclinically. Disease can range from mild diarrhea to severe enterocolitis. Immunocompromised people are at highest risk for infection, however even healthy people can develop illness with heavy exposure. GI signs predominate; foals with the disease develop very severe enteritis. Humans develop diarrhea, abdominal cramps, and fever. The disease can cause acute diarrhea with bloating, or chronic diarrhea that can last 2-4 weeks. Signs develop within 2-10 days after exposure. Diagnosis is through fecal analysis; because the organism is so small, it is easily missed on fecal flotation. To increase sensitivity of diagnosis, fecal samples should be submitted for acid fast stains, immunofluorescence (IFA), enzyme-linked immunosorbent assay (ELISA), or PCR. Treatment is usually supportive, with fluid, acid-base and electrolyte support. Withholding foals from nursing for 1 or more days, and providing parenteral nutrition instead is helpful. When allowed to nurse, provision of lactase enzyme (3000-6000 IU per foal every 3-6 h, PO) is helpful because affected foals can become transiently lactose intolerant. There are a few drugs available for severe cases that do not respond to standard therapy and for immunocompromised individuals. These include azithromycin, nitazoxanide, and paromomycin, although none have been studied specifically for *Cryptosporidium* infections in horses.

Cryptosporidiosis is highly resistant to common disinfectants. Therefore, scrubbing with detergents (for manual removal) is an important component to cleaning the environment. Afterwards, peroxide or peroxide compounds (e.g. Virkon) should be used as disinfectants.

Clostridial diseases

Clostridium difficile is another cause of diarrhea in foals and horses, as well as humans. It is unclear whether *C. difficile* can be transmitted between humans and horses, but it is certainly theoretically possible as a few of the strains have been found in both species. The single most important risk factor for *C. difficile* associated disease (CDAD) in both horses and humans is the use of antibiotics. However, neonatal foals and babies can develop *C. difficile* as a primary infection without specific risk factors other than group housing. Horses can become subclinically colonized with *C. difficile*, but they can also develop severe colitis and SIRS. Treatment varies with the clinical severity of disease. Mild cases can be treated by simply discontinuing the offending antimicrobial and ensuring that the horse eats and drinks adequately. More severe cases require IV fluid support, plasma, and metronidazole or other antibiotic to kill the clostridial organisms. Smectite (e.g. Biosponge [Platinum Performance, Buellton, CA]) has been shown to bind clostridial toxins in vitro, and it may be useful in the treatment of clostridial enterocolitis. A probiotic yeast, *Saccharomyces boulardii* might be beneficial for adult horses with CDAD.

Clostridial spores are very difficult to kill in the environment. Bleach (1:10 dilution with water) can reduce the number of spores. Higher peroxygen disinfectants, such as Virkon, are sporicidal. Adequate contact time is essential. Prior to bleaching, vigorous scrubbing with a detergent after removal of as much organic matter as possible is helpful. Horses should be isolated as above for salmonellosis.

Giardia

Giardia is a common protozoal cause of diarrhea in humans. It has not been well studied in horses. It is currently not believed to be a significant cause of diarrhea in horses. However, foals and horses with preexistent diarrhea can become colonized and shed *Giardia.* Horses shedding *Giardia* can therefore serve as sources of infection for humans. It is spread in the feces of shedders.

NEUROLOGICAL DISEASES

Rabies

Rabies can affect any mammal, but carnivores and bats are the primary species that spread it. Livestock can be infected but rarely transmit the virus. In fact, there are no documented cases of rabies in humans infected from horses. This fact should not lead to complacency, however, as it is certainly possible for a human to get rabies from a horse. The incubation period can be long (> 6 months), and that fact may make it difficult to definitively tie an infection to a horse exposure. Rabies is almost universally fatal.

Rabies infections have highly variable signs. Horses may be presented for colic, lameness, lethargy, fever, weakness, pain, behavioral change, or neurologic signs. The rule of thumb is that rabies is the great 'mimic'. The common feature about rabies cases is that they all progress, and most die within 10 days. The longest living case that I am aware of was at our hospital (16 days).

Vaccination for rabies is currently considered a core vaccine by the American Association of Equine Practitioners. It is a very effective means of prevention; however a vaccinated horse can still become infected with rabies. Other neurological diseases, such as equine protozoal myeloencephalitis or West Nile virus infection, cannot be differentiated from rabies based on clinical signs alone. Therefore, rabies must be on the differential diagnosis list of every acute (< 3 weeks' duration) neurologic case. Acute neurologic cases should be handled by a minimum number of staff, by only those who have been vaccinated for rabies themselves, and gloves should be worn at all times. Eye protection should be worn whenever contact with saliva (as during oral examinations) is possible. A list of people who have contacted the horse should be maintained. All such horses should be tested for rabies if they die or are euthanized. In addition, rabies is a good example of why infection control practices should be in place at all times, especially because rabid horses could have vague, colic, or lameness signs rather than typical neurological signs. Any bite wounds should be taken seriously and washed with soap and water for 15 minutes, then disinfected with alcohol or betadine.

There were 41 and 30 confirmed cases of rabies in horses in the US in 2007 and 2008, respectively (including Puerto Rico).[1]

SKIN AND SOFT TISSUE DISEASES

Methicillin Resistant *Staphylococcus aureus* (MRSA)

MRSA has received renewed attention in the last few years. It is a type of *Staph. aureus* with genetic resistance to the beta lactam family of antimicrobials, including penicillin, cephalosporins and methicillin. Many species of animals, including humans, horses and companion animals, can become colonized (no clinical signs) or infected (clinical disease) with MRSA. They can also pass the bacteria to and from one another. There are several different strains of MRSA that can affect both humans and horses. The most common one found in horses is USA 500, which is also sometimes found in humans, although it is a minor strain in terms of human disease. It appears to be well adapted to colonization of horses. Most healthy horses appear to clear this colonization without treatment, over a few weeks to 6 month period. It appears that horses are only transient carriers of MRSA.

When these carriers become ill or hospitalized they appear to be at risk for developing infections with MRSA.[2] Noncarriers can also develop MRSA infections, but carriers appear to be at particular risk. Clinical signs of infection in humans include dermatitis, abscesses, and more serious infections, including necrotizing pneumonia, sepsis, muscle and bone infections, as well as necrotizing fasciitis. In horses the signs include dermatitis and wound, incisional, joint or IV catheter site infections.[2] Just as with humans, horses can also develop deeper infections, including pneumonia, sepsis (especially foals), mastitis and bone/joint infections.

Superficial infections are treated topically with silver sulfadiazine, chlorhexidine, or mupirocin. Melaleuca oil (tea tree oil) has shown some promise for treatment of superficial infections. Deeper infections should be treated with systemic antimicrobials; however MIC susceptibility testing should always be performed due to highly variable and unpredictable susceptibilities. Chloramphenicol and amikacin are good first line antimicrobials for treatment of systemic infections, as most isolates are susceptible. Enrofloxacin also shows excellent efficacy in vitro, however the fluoroquinolone class of antimicrobials is not recommended for use in humans; the in vivo response is unpredictable and resistance can develop during treatment. The prognosis for horses with MRSA infections is good, with discharge reports of 79-84%.[3]

A somewhat alarming finding in the recent literature is that veterinary (veterinarians and technicians) and horse-related personnel have a relatively high subclinical colonization rate. Approximately 10-17 % of these people are colonized with MRSA, with the majority of isolates being USA-500, the same strain most often found in horses.[3] In evaluating risk factors for colonization, veterinarians that washed their hands between farms, and between infectious patients, were at significantly lower risk of being colonized. These findings demonstrate the importance of hand washing in preventing MRSA transmission. Alcohol based hand sanitizers are also effective; however any gross evidence of dirt or organic debris on hands necessitates washing with with soap and water.

Once a positive carrier or infected horse is identified, it should be isolated at the hospital or farm. Gloves, booties and barrier precautions (coveralls, lab coats) should be used. Feed and water containers should be separate. Wounds should be covered with bandages. MRSA is susceptible to all common disinfectants, but those with activity in the presence of organic matter should be used on farms where complete removal of dirt or bedding is virtually impossible. Most carrier horses will stop shedding MRSA in 3-4 weeks, but occasionally they will shed for up to 6 months. Testing for shedding should be commenced by culturing the nasal passages, as this is the most common site of carriage. Horses are considered negative when they have 2 consecutive negative cultures.

Caution should be used when handling MRSA horses because of documented zoonotic transmission.[4] In one report 3 technicians and students working with a septic foal developed skin lesions caused by MRSA. In another, a veterinary resident developed infection of a tattoo site after handling MRSA-positive horses. Immunocompromised humans should not handle MRSA positive horses.

SYSTEMIC DISEASES (SEPSIS AND ORGAN DISEASE)

Leptospirosis

Leptospirosis is a worldwide zoonosis. In horses it can cause acute renal failure and abortions. It can be passed to humans through contact with infected urine or placental/fetal fluids. Chronic infection in horses is associated with uveitis, although these horses do not pose a zoonotic risk because the infection is sequestered inside of the eye. Though zoonotic transmission from acutely septic horses is rare, handlers should be aware of it in order to avoid contact with spirochetes through infected body fluids. Clinical signs in humans include flu-like symptoms, meningitis, liver disease and renal failure. Doxycycline and penicillins are effective treatments.

OTHER ZOONOSES

The above discussed diseases are the most common and impacting zoonotic diseases that veterinary technicians should be aware of in the United States. Others include ringworm, anthrax, and brucellosis. Hendra virus is a concern in Australia. There are reports of *Rhodococcus equi* and *Streptococcus equi ssp equi* in immunocompromised humans, such as those with HIV.

REFERENCES

[1]Blanton JD, et al. **J Am Vet Med Assoc** 2009; 235(6):676. [2]Anderson ME, et al. **Equine Vet J** 2009;41(4):401. [3]Anderson ME, et al. **Vet Microbiol** 2008, 129(3-4):410. [4]Weese JS, et al. **Vet Microbiol** 2006; 114(1-2): 160.

PERIANESTHETIC ARRHYTHMIAS—WHAT IS IT & WHAT DO I DO NOW?

Katy W. Waddell, RVT, VTS (ECC, Anesthesia)
College Station, TX

Cardiac arrhythmias may be defined as an irregularity in the cardiac rhythm. They are typically classified according to origin, rate or regularity. Arrhythmias may occur from cardiac and/or extra cardiac reasons.

Bradycardia is commonly seen with the anesthetized patient. Commonly the cause is dose related to the anesthetic and/ or analgesic agents used. If the heart rate is low with a normal sinus rhythm and supporting blood pressure, be aware, be prepared to treat but be tolerant of a slow rate. If hypotension is noted in addition to a slow rate, an anticholinergic may be indicated to increase cardiac output by raising rate; this is particularly true with pediatric and juvenile patients.

Normal sinus rhythm is seen on the electrocardiogram (ECG) as a P wave associated with every QRS complex and a regular rhythm. In awake patients, the rate for canine patients will vary between 60 bpm to 180 bpm dependant on age, size and degree of body conditioning. The feline patient will have a rate that can vary between 140 bpm and 240 bpm depending on the anxiety level of the cat in the hospital setting.

Sinus bradycardia, a normal sinus rhythm with an abnormally slow rate, may be due to normal physiology i.e. – in natural sleep the heart rate may drop to the 30-40 bpm range, athletic dogs at rest may have a naturally low heart rate. An increase in vagal tone frequently results in a sinus bradycardia. In the anesthetized patient, bradycardia is defined as a heart rate less than 60 bpm for large/giant breeds and 60-70 bpm for the majority of the adult canine patient. It is important to remember that neonate and pediatric patients have a need for a higher heart rate as their blood pressure is rate dependent and will require intervention to maintain heart rate at a greater number of beats per minute to maintain blood pressures. The feline patient may be considered bradycardic with a heart rate of 100-120 bpm.

Potential causes for bradycardia:

1. High vagal tone- dose dependent, the use of Mu opioids such as oxymorphone, fentanyl and morphine and alpha 2 agonists will cause an increase in vagal tone thus reducing the heart rate.
2. Brachycephalic breeds- due to their anatomy.
3. Intubation- stimulation of the vagus nerve during sternal intubation.
4. GI disease/surgical manipulation-traction on the viscera.
5. Neurologic/spinal cord lesions
6. Ophthamological procedures-pressure or traction on the orbit.
7. Manipulation of vagus nerve during cervical/thoracic procedures.
8. Electrolyte imbalances i.e. hyperkelemia regardless of metabolic dysfunction.
9. Hypothermia
10. Excessive depth of anesthesia
11. Profound hypoxemia

First degree atrioventricular (AV) block occurs when there is a delay of conduction through the AV node which produces a prolongation of the PR interval of greater than 0.13 seconds in the canine and greater than 0.09 seconds in the feline.

Second degree atrioventricular block can be attributed to drug administration as well as inherent cardiac nodal disease. The atrial rate is faster than the ventricular rate. Second degree AV block if further divided into two types.

Mobitz type 1 (Wenckeback) is demonstrated by a progressive prolongation of the interval between the P wave and the QRS complex leading to occasional blocked QRS complexes.

Mobitz type II is seen as a regularly prolonged P to R interval with the QRS complex periodically dropping out. The block may respond to anticholinergic agents but a worsening of the block may be seen initially as the atria respond initially to the action of the anticholinergic.

In third degree AV block, there will be no association between the P waves and the QRS complexes. Frequently, this ventricular rate is based solely on escape or fusion beats and is lower than 40 beats per minute. This patient should be referred for a temporary or permanent pacemaker should anesthesia be indicated for diagnostics or a surgical procedure.

Tachyarrhythmia, bradyarrhythmia, also known as sick sinus syndrome, is more commonly seen in cocker spaniels, miniature schnauzers and dachshunds. The arrhythmias seen will include sinus bradycardia, sinus arrest, alternating brady-tachycardia and escape beats. Pacemaker therapy may also be indicated in these patients.

Treatments/interventions to manage peri-anesthetic bradyarrhythmias: Determine the origin of the arrhythmia, bearing in mind that there may be more than one factor affecting the rate.

1. Opioid administration- judicious use of anticholinergic agents should be designed to maintain the heart rate within normal parameters for the breed and age of the patient.
2. Brachycephalic breeds- a dose of an anticholinergic agent should be calculated and available to be administered during the post premedication and induction period should the heart rate drop to an unacceptable rate and is not supporting systemic blood pressure.

3. Patients with existing GI disease that are undergoing anesthesia for diagnostic or surgical procedures should also have an anticholinergic dose calculated and at hand to be administered should the need arise.

4. Neurologic patients are often found to be bradycardic. As long as the heart rate supports the systemic blood pressure, it is often not necessary to treat the heart rate with an anticholinergic agent.

5. Ophthalmology patients frequently are given additional agents intra-operatively i.e. neuromuscular depolarizing agents that can decrease heart rate. In addition to vagal stimulation provided by surgical manipulation this can be an unfortunate combination if not anticipated by the administration of an anticholinergic to maintain the heart rate within a normal parameter.

6. Manipulation of the vagus nerve during cervical or thoracic procedures will induce bradycardia that can be gauged by asking the surgeon to cease.
momentarily to judge the heart rate response, if it is inadequate, the administration of an anticholinergic would be advisable.

7. Electrolyte imbalances- these can be managed pre-operatively by having lab results available prior to induction of general anesthesia.

Hyperkalemia due to urinary blockage/ rupture of urinary bladder
Decreased elimination due to oliguric or anuric renal failure
Addison's disease that may not be stabilized
Potassium-sparing diuretics and ACE inhibitors
Excessive administration of supplements either oral or by IV fluid additive

Atrial standstill produces no P waves and little or no contractility of the atria is seen with echocardiography. This is non responsive to anticholinergic agents or exercise. This is cardiac in origin vs. extracardiac.

Tachyarrhymias may present as sinus tachycardia, atrial tachycardia, atrial fibrillation, ventricular tachycardia or supraventricular tachycardia. Tachycardias increases myocardial work and oxygen demand, as well as decreasing diastolic filling time. Approximately 70% of the perfusion of the myocardium occurs during diastole; therefore it is valuable to control excessive heart rates.

Potential causes of sinus tachycardia:

1. Administration of drugs
2. Patient awareness
3. Pain
4. Hypovolemia
5. Hypotension
6. Anemia
7. Hypoxemia
8. Hyperthermia
9. Myocardial disease
10. Hypercapnea

Ventricular arrhythmias are noted as premature contractions, tachycardia, and fibrillation. Again, these rhythms may be cardiac or extracardiac in origin.

Encountering arrhythmias in those patients with existing cardiac disease should not be unexpected depending on the nature of the disease. Those patients with conditions causing atrial enlargement are predisposed to the development of supraventricular arrhythmias such as supraventricular tachycardia (SVT), atrial premature contractions (APCs), atrial fibrillation, or atrial flutter while those with ventricular enlargement may be predisposed to ventricular premature complexes (VPCs), ventricular tachycardia, or ventricular fibrillation. Indeed the very act of providing anesthesia may predispose patients with normal myocardium function to show arrhythmias.

VPCs may be the result of poorly managed pain, electrolyte abnormalities, hypoxemia, as well as arise from underlying organic dysfunction. The first key is to recognize the arrhythmia, try to determine the cause and provide an intervention for correction.

Those patients exhibiting the occasional VPC, unifocal VPCs less than three in a row should not require intervention unless they are hemodynamically unstable. If the VPCs occur more often than approximately 10 per minute, occur as a "run" of more than three in a row, are multifocal or cause a decrease in systemic blood pressure, therapy should be instituted. A 2mg/kg bolus of lidocaine given IV should be the first line of treatment. Lidocaine is a class IB antiarrhythmic agent and works by inhibiting sodium channels. If the arrhythmia persists, up to four boluses, for a total dose of 8 mg/kg, may be delivered over a 15 minute time frame. If the bolus is given too rapidly, a fall in blood pressure may be seen. The onset of action is two minutes and has 10-20 minute duration. If the arrhythmia converts to a normal sinus rhythm, a lidocaine CRI may be initiated. The usual starting rate for arrhythmia control is calculated at 50 mcg/kg/minute and may be up titrated to 100 mcg/kg/minute. If a CRI is being administered, this may be considered as an adjunctive analgesic agent and it will have a minimum alveolar concentration (MAC) sparing effect thus allowing the anesthetist to decrease the vaporizer setting accordingly. Procainamide, a class IA antiarrhythmic agent, works by slowing conduction velocity and is used for the treatment of VPCs that are not controlled by lidocaine administration. Typically a dose calculated at 20 mg/kg is given IV over a 20 minute period or until the rhythm converts. If conversion is successful, a CRI may be delivered for further control. Long term treatment is usually achieved by using a beta blocking agent if indicated once the patient has recovered from anesthesia. Blood pressure monitoring is mandatory when administering Class I antiarrhythmic drugs as they may cause a precipitous drop in systemic pressure. If pain is the suspected cause of VPCs, a bolus of an opioid agent may be administered in an effort to control pain due to surgical manipulation.

Electrolytes should be evaluated prior to the administration of sedation or general anesthesia and correction initiated prior to the surgical or diagnostic procedure.

Hypoxemia may be evaluated by obtaining an arterial blood gas sample. If the patient is indeed hypoxic, attempt to correct by increasing inspired oxygen via flow by or nasal cannulas in the awake patient. The intubated patient may need to have an increase in tidal volume or ventilatory rate.

If there is a history of trauma, traumatic myocarditis may be suspected and managed by increasing inspired oxygen and medical management with lidocaine as an antiarrhythmic agent.

If the patient is manageable, it is always a good rule of thumb to pre-oxygenate ill patients by providing flow by oxygen for a 5 minute period prior to induction. A recent study done on healthy dogs showed that pre-oxygenation for 3 minutes prior to induction and capture of the airway improved/delayed hemoglobin oxygen desaturation from a time period of 60 seconds to 322 seconds. For the compromised patient, this can provide a greater degree of cardiovascular stability.

REFERENCES

[1.] G.L. Carroll editor, "**Small Animal Anesthesia and Analgesia**", Blackwell Publishing, 2008 [2.] L.P. Tilley, F. W. K. Smith, Jr., M.A Oyama, M.M. Sleeper, editors. "**Manual of Canine and Feline Cardiology**" 4th edition, W, .B. Saunders, Elsevier Publishers, 2008 [3.] G. Hanson, "**Designing an Anesthetic Protocol Workshop**", IVECCS technician proceedings, 2007 [4.] BSAVA "**Manual of Canine and Feline Aneasthesia and Analgesia**", second edition, editors Seymor, C. and Duke-Novakovski, T., 2007 [6.] Pascoe, Peter J., BSAVA "**Manual of Canine and Feline Aneasthesia and Analgesia**" second edition, Editors Seymour, C. and Duke-Novakovski, T., 2007 [7] Carroll and Martin, "**Veterinary Anesthesia and Analgesia**" third edition, Lumb and Jones, editors Tranquilli, W., Thurmon, J., and Grimm, K., Blackwell Publishing, 2007 [8.] Grubb, Tamara, "**Small Animal Anesthesia and Analgesia**", editor Carroll, G., Blackwell Publishing 2008

ANESTHETIC MANAGEMENT OF THE LEFT SIDED CARDIAC PATIENT

Katy W. Waddell, RVT, VTS (ECC, Anesthesia)
College Station, TX

Adjunctive agents used to support blood pressure

Anesthetizing a patient with cardiac disease requires a plan for the use of supportive measures to maintain adequate tissue perfusion. As in the case of left sided cardiac dysfunction patients, volume administration frequently is not an option to support blood pressure. In these cases, should a positive inotropic or pressor agent be indicated, the volume of the adjunctive agent required should be deducted from the volume of crystalloid administered to maintain a balanced hourly rate.

Cardiomyopathies

Cardiomyopathies, or pathologically affected heart muscle can occur in a variety of ways. These diseases are_functionally very different from one another, although ultimately the pathological effects on the body and treatment modalities are similar.

Dilated Cardiomyopathy

Dilated cardiomyopathy is most commonly seen in certain large breed dogs. This condition is characterized by elongation, weakening of the heart muscle, so that chamber size is increased, and contractility is compromised. Stroke volume is reduced and resulting in decreased blood pressure. Compensatory mechanisms for hypotension ensue. Chamber enlargement may alter the shape and function of the mitral valve, leading to valvular insufficiency. In addition, the chamber may be so enlarged as to affect cell communication and organized depolarization. Atrial fibrillation, premature ventricular contractions and ventricular tachycardia may develop. Whether heart failure is due to valvular insufficiency or cardiomyopathy, compensatory mechanisms in response to poor cardiac output are similar. The renin-angiotensin-aldesterone system is activated to release angiotensin II, which directly activates the sympathetic nervous system, increases secretion of antidiuretic hormome, stimulates the thirst center of the brain and releases alderserone, and directly causes vasoconstriction in an attempt to maintain systemic blood pressure. Treatment of heart failure from either cause is also similar, as it is aimed at minimizing this compensatory mechanism which becomes harmful in the chronic state. Medications to decrease heart rate and improve contractility, decrease blood volume by diuresis, and decrease vascular resistance against which the heart must pump are typically used

Digoxin is a positive inotrope which increase the concentration of calcium within the myocardial cells thus increasing the force of contractions and slows atrioventricular nodal conduction decreasing heart rate. A relatively new addition to management is pimobendan which elicits calcium sensitization of the myofilaments. Anti-arrhythmics such as procainamide, quinidine or mexiletine may be used if indicated. Calcium channel blockers and beta blockers may be used to help control supraventricular arrhythmias, but are contraindicated in the face of decreased cardiac contractility.

Diuretics such as furosemide are used for decreasing circulating blood volume. Angiotensin converting enzyme (ACE)inhibitors cause relaxation of the blood vessels, thereby decreasing systemic vascular resistance by blocking the conversion of angiotensin I to angiotensin II. Anesthetic protocols for patients with dilated cardiomyopathy should aim to avoid drugs that would increase heart rate or decrease contractility. Fluids must be delivered with caution to avoid volume overload. Ideally central venous pressure would be monitored to determine how well the heart is able to handle the circulating volume and fluids may be regulated accordingly.

Hypotension may be corrected with the use of constant rate infusion (CRI) opioid/benzodiazepines to reduce vaporizer setting thus avoiding the dose dependent peripheral vasodilation attributed to inhalant anesthetics. Positive inotropes or beta agonists such as dobutamine may be added if further support of adequate blood pressure is required. These drugs focus on improving contractility without increasing afterload. Dobutamine should be diluted and administered as a constant rate infusion, beginning at 2 mcg/kg/minute, and increased as needed. The use of dobutamine is not benign, as increased rate of administration may cause an increase in heart rate or arrhythmias may be noted as it acts as a positive inotrope. a At higher doses it may act as a vasocontrictor. Dopamine as a CRI may also be used to support blood pressure and is calculated to be delivered at a starting rate of 2 mcg/kg/minute. Ventricular arrhythmias may be exacerbated with the use of beta agonists; however, so close monitoring of the electrocardiogram is necessary.

Suggested protocols may include +/- glycopyrrolate at 0.011 mg/kg given subcutaneously, dependent on patient's heart rate
Hydromorphone or oxymorphone at 0.1 mg/kg or 0.05mg/kg given subcutaneously
Pre-medicants will be given 20-30 minutes prior to induction time
Should patients be of a brachycephalic breed where respiratory difficulties may be present, butorphanol at 0.2 mg/kg will be used in lieu of hydromorphone and the patient will be observed for any signs of distress.
Pre-oxygenation will occur prior to induction unless the patient becomes distressed by mask placement.
Induction agents:
diazepam /midazolam at 0.2 mg/kg given IV followed by:
etomidate at 1-2 mg/kg given IV to effect to facilitate placement of the endotracheal tube.
Anesthetic maintenance:
Inhalant gas anesthesia maintenance will be sevoflurane with oxygen.
Fentanyl at 0.8 mcg/kg/minute and midazolam at 8 mcg/kg/minute combined may be given IV via syringe pump as an anesthetic adjunct for a minimum alveolar concentration (MAC) sparing effect. This will allow the vaporizer setting to be reduced to between 0.25-1% if sevoflurane is the inhalant used.
Contraindicated: Ketamine, any alpha 2 agonist, and precaution should be taken with propofol administration.

Hypertrophic Cardiomyopathy

While hypertrophic cardiomyopathy (HCM) is not commonly seen in dogs, it is the most common heart disease in cats. Cardiac hypertrophy is common sequelae of hyperthyroidism, hypertension or is idiopathic, true HCM, but generally, its etiology is not considered genetic although some familial links are being elucidated. The disease is considered a disease of diastole and is characterized by inability of the heart to fill adequately with blood, thickening of the left ventricular wall and decreased chamber size. Relaxation of the heart muscle during diastole and decreased filling due to chamber size is affected thereby reducing cardiac output.

It is important to maintain heart rate within normal range in order to maintain cardiac output in the face of decreased stroke volume. Relaxation of the heart during diastole is important to promote optimal chamber filling, as well as perfusion to the heart muscle so that it may receive as much oxygen as possible to meet its increased demands. Anticholinergics should be avoided unless bradycardia occurs.

Mitral Valve Insufficiency

One of the most common cardiac disorders in dogs is mitral valve insufficiency. A normal mitral valve will close during ventricular contraction, thereby causing all blood to move forward to systemic circulation. A defective valve will allow some blood to be pumped back into the left atrium. This reduces forward flow, and thus cardiac output and systemic blood pressure. In addition, pressure in the left atrium will increase, making it more difficult for blood to enter it from the lungs. Resulting backpressure in the lungs may lead to pulmonary edema.

Compensatory mechanisms in response to lowered blood pressure will increase the total fluid volume in circulation and cause vasoconstriction. This, however, results in an increase in resistance against which the left ventricle must pump. Over time, the left ventricular muscle will may enlarge due to this "workout", but its efficiency continues to decline. Eventually, if left untreated, left-sided heart failure will result in life threatening pulmonary edema and inadequate tissue perfusion.

Treatment for significant mitral valve dysfunction consists of medications that cause relaxation of the blood vessels, such as ACE inhibitors, or other vasodilators making it easier for the left ventricle to pump blood forward. Diuretics (e.g. furosemide) are administered in order to reduce the volume load for the heart to pump, and remove fluid causing pulmonary edema. In addition, positive inotropes (e.g. digitalis) will increase contractility of the heart by increasing the concentration of calcium in heart muscle cells.

When anesthetizing a patient with mitral valve disease, one must be careful since the vasodilatory effects of many anesthetics may be exaggerated by the patient's existing medication. On the other hand, untreated mitral valve disease may benefit from the vasodilating effects of anesthetics. When pressors are needed, pure alpha agonists such as phenylephrine that cause vasoconstriction may exacerbate the back pressure and may result in pulmonary edema. Beta agonists such as dobutamine may be used to improve contractility. Fluids must be administered conservatively, to avoid volume overload and pulmonary edema. However, because these patients may have been on diuretics and also have the same issues a normal patient with regard to fluid needs, so there is a balance that must be achieved. Currently, the most effective way to monitor this balance is by measuring central venous pressure (which approximates the pressure of the blood entering the right atrium).

Maintaining heart rate within the normal range for these patients is important. Tachycardia should be avoided as it will increase the oxygen needs of the heart, so an anticholinergic is usually not recommended prophylactically as in a pre-med. However, low doses can be used if indicated by a bradycardia that is causing hypotension.

Opioids, benzodiazepines, and etomidate are considered good choices in an anesthesia protocol for these cases. Ketamine is usually avoided as it increases myocardial oxygen needs. Propofol should be avoided or used with extreme care due to its vasodilating and hypotensive effects. IV administration of an opioid/benzodiazepine combination immediately prior to induction with propofol can greatly reduce the amount of propofol required to capture the airway thus allowing the anesthetist to use this agent while avoiding the majority of the myocardial depressive effects.

Aortic stenosis

Aortic stenosis is usually a congenital abnormality. Stenoses may be subvalvular, valvular, or supravalular with subvalvular being the most common form in dogs. Subvalvular aortic stenosis (SAS) is caused by the development of extra fibrous tissue below the aortic valve in the development of the heart and can worsen during the first few months in life. The result is a partial obstruction of flow from the left ventricle through the aortic valve and out the aorta to systemic circulation. This obstruction causes increased left intraventricular pressure and may cause thickening of the left ventricular walls. Other sequelae include cardiac ischemia, arrhythmias, aortic or mitral regurgitation, and left-sided congestive failure.

Diagnosis is usually made at an early age due to a murmur caused by the increased turbulence of blood flow in the effected area, but the degree of disease may or may not correlate with the loudness of the murmur. Mildly affected dogs may lead fairly long lives, whereas severely affected animals develop CHF or may die suddenly of ventricular arrhythmias.

Surgical correction requiring open heart surgery and cardiac bypass is usually not an option and carries a poor prognosis. Medical management usually begins with the administration of beta blockers in order to limit myocardial oxygen consumption, prevent tachycardia and ventricular arrhythmias. However, as heart failure ensues, increased heart rate is necessary for maintaining cardiac output and beta blockers become counterproductive.

As the heart walls thicken and the muscle becomes less compliant cardiac output is dependent on adequate filling pressure. Thus, diuretics and venodilators must be used with caution. ACE inhibitors, calcium channel blockers, or other arteriolar dilators may also negatively affect the ability of the heart to produce adequate cardiac output due to worsening of the obstruction. Positive inotropes may also exacerbate outflow obstruction and ventricular arrhythmias.

- **Atropine/glycopyrrolate:**
- Will cause an increase in heart rate, contractility, cardiac output and myocardial oxygen consumption.

- Often there will be no change in blood pressure and a decrease in right atrial pressure
- **Phenothiazines –**
- Main cardiovascular effect = peripheral vasodilation
- Consequence – decreased blood pressure
- Some boxers – fainting/syncope seen due to vasovagal response (bradycardia and peripheral vasodilation)
- **Midazolam and diazepam:**
- Cause little or no myocardial depressant effects
- May see increase in heart rate due to excitation with inadequate use of adjunctive agent i.e. mu opioid
  - When combined with an opioid as a CRI can be utilitized to decrease MAC of inhalant agent.
- No analgesia provided
- **Alpha 2 agonist**
- Significant cardiovascular effects
- Vasoconstriction
- Bradycardia
- Decreased cardiac output
- **Thiopental**
- Reduction in blood pressure – peripheral vasodilation is the main action
- Commonly associated with ventricular arrhythmias – bigeminy rhythm not uncommon
- Contraindications:
- Patient with known cardiac disease – paricularly those with existing arrhythmias
- Trauma patients – think of traumatic myocarditis!
- Etomidate: no direct myocardial depression. Safe to use with cardiac, critical and septic patients.
- Cardiovascular stability may be better due to maintained barorecptor mediated responses.
- **Propofol**
- Does cause direct myocardial depression as well as decrease in systemic vascular resistance
- Decrease in contractility leads to increase in heart rate – will be transient – lasting several minutes
- Profound bradycardia has been noted (dose dependent)
- Use cautiously in patients with heart disease and hypovolemia
- **Mu opioids**
- Fentanyl
- Causes dose dependant bradycardia (increase in vagal tone)
- Bradycardia is responsive to anticholinergics – atropine/glycopyrrolate
- **Hydromorphone:**
- morphine-like agonist, primary activity at the mu receptors.
- Cardiovascular effects:
- bradycardia due to central vagal stimulation,
- alpha-adrenergic depression causing peripheral vasodilation, decreased peripheral resistance
- baroreceptor inhibition.
- **Morphine**
- No direct myocardial effect
- Dose dependent bradycardia – responsive to anticholinergics
- **Mixed agonist/antagonist agents**
- **Buprenorphine**: a partial mu agonist/antagonist. Slow onset of action, duration of 6-8 hours.
- Cardiovascular depression and respiratory depression not as profound as pure mu agonists.
- **Butorphanol: partial agonist/antagonist.** Similar to buprenorphine in cardiovascular/respiratory effects.
- Faster onset of action, shorter duration that buprenorphine.
- Recommended in multiple texts for premedication for cardiac patients due to minimal cardiac effects.
- **DISSOCIATIVE AGENTS**
- Will indirectly stimulate the cardiovascular system by increasing sympathetic tone
- cause an increase in heart rate, cardiac output, mean arterial pressure, pulmonary arterial pressure and central venous pressure.
- Increase in rate causes an increase in myocardial work and oxygen demand/consumption

SUGGESTED READING WOULD INCLUDE

1. Lumb and Jones' **Veterinary Anesthesia,** 3rd edition, Lippincott, Williams & Wilkins, 1996 2. **Small Animal Anesthesia and Analgesia**, G.L. Carroll, editor, Blackwell Publishing 2008 3. **Textbook of Medical Physiology**, A. C. Guyton editor, 8th edition, W. B. Saunders, 1991 4. **Manual of Canine and Feline Cardiology**, Tilley, Smith, Jr., Oyama, Sleeper, editors, 4th edition, W. B. Saunders, 2001 5. **Small Animal Clinical Pharmacology**, Maddison, Page, Church, editors, 2nd edition, W. B. Saunders, 2008 6. **Anesthesia workshop**, Hanson, Waddell, proceedings, IVECCS, 2007 7. **BSAVA** Manual of Canine and Feline Anaesthesia and Analgesia, second edition, C. Seymour and T. Duke-Novakovski , editors, BSAVA Publishing, 2007

TOCERANIB PHOSPHATE (PALLADIA™): A NEW TREATMENT OPTION FOR CANINE MCT

Chad M. Johannes, DVM, DACVIM (SAIM)
Brooklyn, NY

INTRODUCTION TO RECEPTOR TYROSINE KINASES

Research in molecular biology has greatly advanced our understanding of the complex cell signaling pathways present in normal cells which regulate cell growth, differentiation, survival, and death. One important group of cellular signaling proteins is the *tyrosine kinases*. While these proteins can be located on the cell surface, in the cytoplasm, or within the nucleus, the majority of the tyrosine kinases are found on the cell surface (called *receptor tyrosine kinases*). They are activated through the binding of growth factors, resulting in the generation of a series of events (downstream signaling) necessary for normal cell function. Several tyrosine kinases have also been shown to be important in the process of new blood vessel growth into tumors (*angiogenesis*).(*1*)

Alterations of tyrosine kinases have been found in many types of cancer in humans and animals. Dysregulation of these cell signaling pathways often leads to uncontrolled cell growth and subsequent tumor development, making them a promising target for therapeutic intervention.(*1*) Several inhibitors of these kinases (*tyrosine kinase inhibitors or TKIs*) are approved for the treatment of human cancers and many have demonstrated significant clinical efficacy.

TOCERANIB PHOSPHATE OVERVIEW

Toceranib phosphate is the first FDA-approved tyrosine kinase inhibitor developed specifically for canine cancer therapy. Toceranib is administered orally and indicated for the treatment of Patnaik grade II or III, recurrent, cutaneous mast cell tumor (MCT) with our without regional lymph node involvement in dogs.(*2*) Toceranib acts by inhibiting the tyrosine kinases *VEGFR-2* (vascular endothelial growth factor receptor-2) and *PDGFR-β* (platelet-derived growth factor receptor-β) which are important for blood vessel growth to tumors, as well as *KIT* (stem cell factor receptor) which is important for the development, survival, and proliferation of mast cells.(*3*) Mutations in KIT, resulting in continual activation of the receptor and subsequent cell signaling, have been identified in approximately 25% of grade II/III canine MCTs.(*4*)

In the toceranib pivotal clinical field study, MCTs in approximately 60% of the dogs disappeared, shrank, or stopped growing (*5*). Toceranib phosphate is administered at an initial dosage of 3.25 mg/kg (1.48 mg/lb) body weight, orally every other day. Dose reductions of 0.5 mg/kg (to a minimum dose of 2.2 mg/kg (1.0 mg/lb) every other day) and dose interruptions (cessation of toceranib for up to two weeks) may be utilized, if needed, to manage adverse reactions. Dose adjustments should be based on approximately weekly veterinary assessments for the first 6 weeks and approximately every 6 weeks, thereafter. The Client Information Sheet (found on the label) should be provided to clients with prescription.(*2*)

Adverse events (AEs) with toceranib phosphate can be serious, but most are mild or moderate and generally manageable, especially with early recognition. The most commonly reported AEs included diarrhea, anorexia, lethargy, vomiting, lameness, and weight loss (*5*). Active prevention of AEs is recommended via supportive medications as clinically indicated: H1 blocker (diphenhydramine), H2 blocker (famotidine), proton-pump inhibitor (omeprazole), antiemetic. Careful monitoring for gastrointestinal (GI) AEs is recommended. Serious, sometimes fatal, GI complications, including GI perforation, have occurred rarely in dogs treated with toceranib (*2,5*). If melena (dark tarry stools) or hematochezia (fresh blood in stool) is noted, toceranib should be stopped immediately and supportive care instituted (omeprazole, sucralfate, antibiotics, and/or fluids) as clinically indicated.

As toceranib can cause clinical signs similar to those seen with aggressive MCT (*5*), it is imperative to stop drug and re-evaluate the patient when signs are observed. AEs can often be managed with temporary discontinuation of drug and/or dose reductions. Dogs in poor health and/or with bulky MCT prior to starting toceranib are at a greater risk for developing AEs. Treatments to reduce tumor volume and improve patient's health status are recommended prior to initiating toceranib therapy (i.e., surgery, radiation therapy, chemotherapy, supportive medications). Referral to or consultation with an oncology specialist should be considered prior to initiating toceranib therapy. Owner education is important for successful use of toceranib and owners should not become complacent during toceranib therapy.

SUMMARY

Receptor tyrosine kinases play an important role in normal cellular processes and tumor angiogenesis. Dysregulation of some of these cellular signaling proteins has been associated with several forms of cancer, making them promising targets for therapeutic intervention. Toceranib phosphate is the first FDA-approved tyrosine kinase inhibitor developed specifically for canine cancer therapy and provides another treatment option for dogs with MCT.

REFERENCES

[1.] London C, Therapeutic Modalities for the Cancer Patient: Signal Transduction and Cancer. Withrow & MacEwen's **Small Animal Clinical Oncology**. 2007;242. [2.] **PALLADIA** (package insert). New York, NY; Pfizer Inc:2009. [3.] London CA, et al. **Clin Cancer Res.** 2003;9:2755. [4.] Webster JD, et al. **Neoplasia.** 2006;8:104. [5.] London CA, et al. **Clin Cancer Res**. 2009;15:3856.

PULSE-MEDIATED CHEMOTHERAPY OF SOLID TUMORS (ELECTROCHEMOTHERAPY): RECENT DEVELOPMENTS OF A NOVEL ANTICANCER THERAPY

Enrico P Spugnini, DVM, Ph. D., DACVIM (Oncology), DECVIM-CA (Oncology)
Rome, Italy

INTRODUCTION

Chemotherapy drugs are effective provided that they reach their site of action. The primary obstacle for chemotherapy agents is the cytoplasmic membrane, especially for lipophobic agents such as bleomycin. Electrochemotherapy (ECT) couples the administration of an anticancer agent (usually bleomycin or cisplatin), to the delivery of electric pulses having appropriate waveforms. The application of permeabilizing pulses leads to perturbation of the cell membrane, thus resulting in increased uptake of chemotherapeutic drugs, ultimately leading to cell apoptotic death.[1] Clinical trials evaluating humans treated with ECT for cutaneous neoplasms were first reported in the early 90s, and have been followed by many trials that confirm its potential for the treatment or palliation of dermatological tumors.[2-4] Recently, a standard operating procedure has been proposed.[5]

ELECTROCHEMOTHERAPY EQUIPMENTS

Different waveforms have been adopted by investigators: exponentially decaying, square, rectangular or biphasic. In general the number of pulses delivered is set in 8 single pulses applied per cm of tumor area at a voltage of 1300 V/cm (800 V/cm for intraoperative use), with a duration of 100 µs and a frequency of 1 Hz. Treatments are repeated until the whole tumor area is covered.

RESULTS IN VETERINARY MEDICINE USING SQUARE PULSES BASED ECT

The first clinical trial in veterinary patients involved the treatment of feline sarcomas recurring after radiation therapy using bleomycin and Interleukin 2 producing cells coupled with square pulses.[6] The authors did not report any clinical response but claimed prolonged control for cats undergoing ECT compared to untreated patients. More recently, CDDP or bleomycin based ECT has been reported to treat cutanous neoplasms in dogs and cats as well as sarcoids in horses.[7-9]

EVALUATION OF BIPHASIC PULSES BASED ECT IN PETS

Experimental design

Our group investigated in the past 8 years the feasibility of electrochemotherapy in companion animals affected by different neoplasms. We adopted trains of biphasic pulses delivered as a burst instead of single impulses. The electric waveform was generated by Chemopulse equipment. This device is built up by a toroidal core transformer generating a roughly rectangular pulse which is split in two halves that are sequentially driven to obtain a biphasic pulse. Pulses are not singularly produced but are created in bursts. The equipment allows choosing electric field intensity, number of pulses and pulse duration. The standard wave train used for this experiment was set to 8 pulses of 50 + 50 µs (biphasic pulse), and the electric field intensity was 1250 V/cm. The pulse repetition frequency is 1 Hz while the frequency of burst repetition is 1 kHz; resulting in a total burst duration of 7.1 ms. Bleomycin was locally injected at the tumor site and in the tumor's surrounding tissues. Five minutes after the injection, trains of biphasic pulses were administered through caliper electrodes as above described. Adherence of the caliper electrodes to the lesions was maximized using an electrophoresis gel. The treatment was repeated after 1 week.

RESULTS

In the first trial, ECT was used to directly attack neoplasms.[10] Overall response rate was 80%, with 40% long lasting remissions (in excess of 1 year). This phase I/II study led to the development of several electrodes specific for the different body districts.[11] On the basis of the preliminary studies, cohorts of dogs and cats affected by melanoma, soft tissue sarcoma mast cell tumor and nasal squamous cell carcinoma were enrolled in phase II studies.

Melanoma Ten patients with malignant melanoma (MM) of the oral cavity were enrolled in a phase II study. The overall response rate was 80% (median time to recurrence 6 months), with a 50% of patients in remission after 1 year.[12] Of interest, most of the long term responders experienced a vitiligo like discoloration at the treatment site, potentially suggestive of recruitment of the immune system after uncovering of deep antigens.

Soft tissue sarcoma A total of 72 cats affected by soft tissue sarcoma were assigned to three different groups: 1) surgery alone, 2) surgery coupled with intraoperative ECT, 3) surgery coupled with post operative ECT.[13] The median control times were respectively 4, 12 and 19 months. In this study were identified some prognostic factors such as previous treatment and tumor size. A cat with a recurring hemangiopericytoma was also referred at the time of the study, and is still in remission 2.5 years after the completion of the ECT treatment.[14] The only side effects were tumor dehiscence in 2 patients and focal inflammation in 2 patients. Systemic side effects were not reported among the enrolled patients.

In another study, adjuvant ECT was adopted to increase local control in a group of 22 dogs, obtaining a mean time to recurrence of 730 days, with 50% of the dogs still desease free at the time of writing.[15] In one dog with a large, unresectable sarcoma, ECT has been used to decrease tumor size and allow surgical excision.[16]

Mast cell tumor Adjuvant bleomycin-based ECT was used in 28 dogs with incompletely excised mast cell tumors. Overall response rate was 85%. The median time to recurrence was not reached. Estimated time to recurrence was 52.76±6.5 months.[17] One dog had wound dehiscence. Another study, enrolling 37 dogs, evaluated the efficacy of cisplatin for the treatment of incompletely excised canine mast cell tumors obtaining a mean disease-free interval of 1218 days with 78% of the patients experiencing local control at different times.[18]

Squamous cell carcinoma a small cohort of 9 cats with sun induced squamous cell carcinoma of the nasal planum were treated with intralesional bleomycin coupled with ECT with a rate of complete responses of 77%.[19]

CONCLUSION

Electrochemotherapy is a safe and efficacious approach to veterinary soft tissue and cutaneous neoplasms. Its low cost and ease of administration make it a valuable addition to the currently available oncological therapies. One advantage of this technique is the possibility of repeated treatments in case of local recurrence and the recruitment of the patients' immune system, as suggested by the low incidence of metastatic disease among our treated patients and by the tumor selection observed among some pets with recurring disease.

FUTURE DIRECTIONS

Ongoing studies are evaluating the efficacy of ECT when other drugs than bleomycin are used, in particular anthracyclines and their analogues (mitoxantrone) and platinum compounds such as CDDP and carboplatin. Studies are being performed to characterize the membrane alterations induced by trains of biphasic pulses in order to improve the efficacy of the technique.[20] Novel electrochemotherapy equipments adopting different electrical parameters are currently being developed. Finally, our group is assessing the potentials for electrogenetherapy using electric pulses to drive chemosensitizing plasmids within tumor cells.

REFERENCES

[1]Tounekti et al. **Cancer Res** 1993; 53: 5462. [2]Belehradek et a. **Cancer** 1993; 72: 3694. [3]Daskalov et al.**IEEE Eng Med Biol** 1999 ; 18 : 62. [4]Gehl et al. **Melanoma Res** 2000; 10: 585. [5]Mir et al. **Eur J Cancer** 2006 ; 4(S) : 14. [6]Mir et al. **Br J Cancer** 1997 ; 76 : 16. [7]Tozon et al. **Anticancer Res** 2001; 21 :2483. [8]Rols et al. **Bioelectrochem** 2002 ; 55 : 101. [9]Tozon et al. **Anticancer Res** 2005 ; 25 : 839. [10]Spugnini et al. **J Exp Clin Cancer Res** 2003 ; 22 : 571. [11]Spugnini et al. **J Exp Clin Cancer Res** 2005; 24 : 245. [12]Spugnini et al. **Melanoma Res** 2006; 16 : 23. [13]Spugnini et al. **Cancer Chemoter Pharmacol** Epub 2006; 2007; 59 : 375. [14]Spugnini et al. **Vet Rec** 2006; 159 : 598. [15]Spugnini et al. **In Vivo** 2007; 21: 819.[16]Spugnini et al. **Vet Rec** 2008; 162: 186. [17] Spugnini et al. **Anticancer Res** 2006; 26: 4585.[18] Spugnini et al. **Proceedings of the ACVIM Forum** 2009 :331.[19] Spugnini et al. **Vet J** 2009; 179: 117.[20] Spugnini et al. **Microsc Res Tech** 2007; 70: 1041.

CANCER CACHEXIA: PATHOPHYSIOLOGY & TREATMENT

Kara M. Burns, MS, MEd, LVT
Wamego, KS

Cancer - a disease which evokes a great deal of emotion. In dogs and cats, cancer is one of the most common causes of nonaccidental death.[1] Pet owners are acutely aware of cancer in humans, thus as healthcare team members, we need to remember to approach pets with cancer and their owners in a positive, compassionate, and knowledgeable fashion. Cancer cachexia is a complex neoplastic syndrome that adversely alters the patient's functional status. As a result of the cancer disease process, cancer cachexia manifests as weight loss, reduced food intake, and systemic inflammation. This session will take a deeper look at cancer cachexia and the role of nutrition in managing cancer cachexia.

Starvation vs. Cachexia

Cancer associated weight loss differs from that seen with simple starvation. In cachexia there is an equal loss of muscle and fat characterized by increased catabolism of skeletal muscle. During starvation, fat is mobilized first sparing muscle proteins, resting energy expenditure is decreased and glucose metabolism is reduced. In contrast patients with cancer cachexia have normal or elevated resting energy expenditures and glucose turnover. Adequate nutrition will halt and reverse the metabolic alterations that accompany simple starvation but will not completely reverse the metabolic disturbances associated with cancer cachexia.

Prevalence and Diagnosis

Cancer cachexia has been reported to affect 30-85% of human cancer patients. Cachexia is reported as the cause of death in approximately 20% of human patients with cancer.[2] Similar data are not available for veterinary patients. However, one study did document clinically relevant muscle wasting in 15% of dogs presented to a veterinary oncology service. In addition, 4% exhibited cachexia as defined by a body condition score of $\leq 3/9$ and 68% had documented weight loss of 5% to > 10%.[3] There are no specific criteria for diagnosis of cachexia in humans but the clinical features which characterize the syndrome include progressive involuntary weight loss with depletion of lean body mass, muscle wasting and weakness, edema, and declines in motor and mental function. There are no specific criteria for diagnosis of cancer cachexia in veterinary patients except weight loss. The diagnosis of cachexia in both human and veterinary patients is complicated by the fact that malnutrition occurs long before clinical signs are evident. Veterinary technicians should suspect cancer associated malnutrition in all tumor bearing animals even in the absence of documented weight loss.

Pathophysiology

Cancer associated malnutrition occurs as a consequence of an imbalance between the nutritional needs of the patient, the demands of the tumor and the availability of nutrients in the body. The competition for nutrients between the tumor and the host promotes a variety of metabolic disturbances including alterations in carbohydrate, lipid and protein metabolism. Cytokines play a key role as the main humoral and tumor derived factors involved in cancer cachexia and may be responsible for the majority of metabolic changes associated with cancer cachexia. Table 1 summarizes the effects of cytokines on nutrient metabolism in patients with cancer cachexia.

Table 1.

CHO	Increased resistance to insulin Increased glucose synthesis Increased Cori cycle activity (lactate recycling; net energy loss to tumor)
Protein	Increased protein breakdown (catabolism) Increased liver (acute phase proteins) and tumor protein synthesis
Lipid	Increased lipid mobilization Elevated levels of triglycerides

Therapeutic Strategies

Despite years of research in human medicine, practice guidelines for the prevention and treatment of cancer cachexia are lacking. Even less information is available for veterinary patients. The optimal therapy is to cure the underlying cancer; obviously this is not always possible. In veterinary medicine therapeutic strategies generally include management of anorexia, nutritional support / supplementation and omega-3 fatty acids.

Managing Anorexia / Hyporexia

Nutritional counseling should be a part of the pretreatment plan for all patients. Veterinary technicians should help owners understand the importance of measuring intake and should have a sequential plan for maintaining nutritional support throughout the treatment process. This may include simple strategies to increase consumption initially (see table) which may progress to assisted feeding if necessary. In a patient with a functional GI tract enteral feeding is preferred. Strategies for managing anorexia in dogs and cats have recently been thoroughly reviewed.[4] Table 2 is a brief review of those recommendations.

Table 2.

Create ambiance /improve service	Create a quiet comfortable feeding area away from disturbances.
Increase Palatability	
Increase moisture	Switch to canned food / add water to kibble. Caution: some cats may have texture preferences and actually prefer dry food
Increase fat	Higher fat diets are more energy dense so less food may be required to meet needs.
Increase protein	Dogs may select foods with higher protein level
Sweet and salty	Adding sweet flavor (No artificial sweeteners) as top dressing may increase palatability for dogs but not cats. Adding salt (if appropriate) may increase palatability for dogs but not cats.
Freshness, aroma, and food temperature	Provide 'fresh' canned / dry foods or home prepared foods. Warming foods to no greater than body temperature may also release aromas.
Rarity	Uncommon or rare food may entice some dogs and cats to eat. Best to offer foods that are uncommon but not completely novel.
Variety	If a variety of options are provided each should be carefully measured to ensure accurate accounting of food intake /preferences. This approach should not be used if food aversions are likely, so that the patient will not develop a learned aversion to all of the therapeutic options.
Drug administration	Never top dress food with medication or use the desired diet to administer medication. Avoid giving medications immediately before/ after a meal if possible.
Eliminate physical barriers	Be sure food bowls are accessible and account for any limitations in movement (raise bowls, remove E collars etc)
Appetite Stimulating drugs	No recommended as effects are unpredictable, intermittent and short lasting. If above approaches are ineffective assisted feeding is recommended.

FUNCTIONAL FOODS FOR PATIENTS WITH CANCER

Most dogs and cats do not have a requirement for soluble carbohydrates and fiber in their diet. Soluble carbohydrates may be poorly used by animals with cancer and can contribute to increased lactate production. Thus, soluble carbohydrates should comprise less than 25 percent of the food's dry matter content. Dietary protein should be highly digestible and exceed levels normally used for maintenance of adult animals because patients with cancer have altered protein metabolism and may suffer loss of lean muscle mass (cachexia). Protein levels should be 30 to 45 percent dry matter in foods for dogs with cancer and 40 to 50 percent dry matter in foods for cats with cancer. Some tumor cells have difficulty using lipids as a fuel source, while host tissues continue to oxidize lipids for energy. As mentioned previously, this finding led to the hypothesis that foods relatively high in fat may benefit animals with cancer compared with foods relatively high in carbohydrates. Pets in North America receive most of their nutrient intake from commercial dry pet foods. These foods are usually high in soluble carbohydrate (25 to 60 percent) and relatively low in fat (7 to 25 percent). These characteristics make most commercial dry foods inappropriate for nutritional management of animals with cancer. Omega-3 fatty acids, especially those found in certain types of fish and fish oil (eicosapentaenoic acid [EPA], docosahexaenoic acid [DHA]), are probably the most important nutraceuticals to consider for animals with cancer.

Alternative Feeding Strategies

It is recommended to feed a food (i.e., Prescription Diet® Canine n/d™) which contains high levels of dietary protein and fat to help meet nutritional demands of dogs undergoing chemotherapy or radiation treatment for cancer to increase survival time and improve quality of life. Studies indicate that approximately 10% of dogs may develop soft stools when initially placed on Prescription Diet® Canine n/d™ canned dog food. Of these dogs, about half will adjust to the product and develop normal stool consistency. The other half (about 5% of total cases) continue to have GI or stool consistency issues that pet owners find unacceptable. Several strategies are outlined below to help reduce potential discomfort for the pet and the owner.

GI Intolerance

1. Transition from the previous food to Prescription Diet® Canine n/d™ canned dog food over a minimum 10 to 14 day period. The longer transition period helps most patients adjust to the higher fat intake.

2. If unacceptable stools or discomfort persists despite a longer transition period, additional fiber can be added to the overall diet of the dog.

Additional fiber options include:

- Add psyllium husk fiber (Metamucil®); 1 teaspoon per 5 to 10 kg body weight with each meal.
- Replace approximately 10% of the total daily calories with Prescription Diet® *-Canine r/d® or Canine w/d® **dry** dog foods. Mix the dry food with canned Prescription Diet® Canine n/d™ at each meal.
- Add a high-fiber cereal to Prescription Diet® Canine n/d™ canned dog food by thoroughly mixing ½ to 1 cup of cereal per can of food (Post® 100% Bran™ cereal [28% fiber], Kellogg® All-Bran™ cereal [50% fiber], or General Mills® Fiber One™ cereal [43% fiber]).

Feline Cancer Feeding Strategies

The metabolic abnormalities associated with various forms of cancer in cats are poorly understood. However, the metabolic changes cats with cancer experience may be similar to those found in dogs with cancer and therefore similar feeding strategies should be employed.

Food Recommendations for Cats with Cancer

Cat Size lbs.	5	8	10	12	15
Cat Size kg	2.3	3.6	4.5	5.5	6.8
Recommended kcal/day	170	215	250	280	330

Prescription Diet® Canine/Feline a/d® can

Prot=45.7, Fat=28.7, NFE=16.5*

Arginine, mg/d	631	798	928	1,039	1,224
*Supplemental Arginine, mg***	*294*	*399*	*432*	*484*	*571*
Total n-3, mg/d	806	1,019	1,185	1,327	1,564

*nutrients expressed as % DMB

**amount of arginine to add to this product to provide equivalent amount delivered by Prescription Diet® Canine n/d™. L-arginine is available as 500 and 1000 mg tablets.

***number of fish oil capsules to add to this product to provide equivalent amount delivered by Prescription Diet® Canine/Feline a/d®; assumes fish oil capsules contain 1,000 mg/capsule with 300 mg EPA + DHA.

Note: Fish oil capsules should be given with each meal and/or broken open and mixed with food. Cost is approximately $0.05 to $0.10 per capsule.

Summary

It is imperative that veterinary technicians recognize and understand cachexia in cancer patients. It is also necessary for healthcare team members to employ nutritional strategies to help manage anorexia and the imbalance between the nutritional needs of the patient, the demands of the tumor and the availability of nutrients in the body. Feeding strategies should be aimed at alleviating the competition for nutrients between the tumor and the host which result in a variety of metabolic disturbances including alterations in carbohydrate, lipid and protein metabolism.

REFERENCES

[1.] Saker, KE, Selting, KA. Cancer. In: Hand MS, Thatcher CD, Remillard RL, Roudebush P, Novotny B. eds. **Small Animal Clinical Nutrition**, 5th ed. 2010:587-607. [2.] Hemming L, Maher D. Understanding cachexia and excessive weight loss in cancer. **Brit J Comm Nursing** . (2005) 10 (11); 492-495. [3.] Michel KE, Sorenmo K, Shofer FS. Evaluation of body condition and weight loss in dogs presented to a veterinary oncology service. **JVIM** (2004) 18; 692-695. [4] Delaney SJ. Management of anorexia in Dogs and Cats. **Vet Clin Sm An Pract** (2006) 36; 1243-1249.

MDR1 TESTING FOR PREVENTION OF CHEMOTHERAPEUTIC ADVERSE EFFECTS

Katrina L. Mealey DVM, PhD, DACVIM, DACVCP
Pullman, WA

What is MDR1?

MDR1 is a gene known as the multidrug resistance 1 gene. The new name for the MDR1 gene is ABCB1 (which stand for ATP-Binding Cassette subfamily1 gene). This gene codes for a protein called P-glycoprotein (P-gp). P-gp was first identified three decades ago in chemotherapy-resistant cancer cell lines where it was found to pump a variety of chemotherapeutic drugs out of the cancer cell[1]. Because these chemotherapeutic drugs were no longer able to kill tumor cells that expressed P-gp, the tumor cells were described as multidrug resistant (MDR). The way that P-gp is able to cause resistance to many anticancer drugs (rather than just a single class of drugs) is related to where it is expressed on the cell and how if functions. P-gp is expressed on the cell membrane and functions as a drug transport pump. If a drug that is recognized by P-gp diffuses across the cell membrane and enters the tumor cell, P-gp quickly and efficiently pumps that drug out of the cell. As you can imagine, tumor cells that express P-gp are difficult to kill with chemotherapy. Therefore, expression of P-gp by a tumor is often considered to be a poor prognostic indicator. When results from these studies first became known, many researchers believed that if P-gp's actions could be blocked, the war on cancer could be won. Areas of intense P-gp research in the early 1990's involved efforts to block P-gp function with pharmaceutical agents or monoclonal antibodies. By the mid 1990's, however, many other chemotherapy resistance proteins had been identified, some of which appeared to be more clinically important than P-gp. Furthermore, clinical trials of drugs that inhibited P-gp function failed to show significant improvement in survival times for cancer patients[2]. Consequently, interest in P-gp as a multidrug resistance pump waned.

At about the same time, researchers discovered that P-gp was expressed in non-neoplastic tissues where it appeared to play a role in transporting "xenobiotics" (natural and synthetic chemicals and drugs) out of the body or away from protected tissues such as the brain, placenta and others. More recent P-gp research involves determining P-gp's effects on drug absorption, distribution, and excretion[3]. It is this aspect of P-gp, particularly as it relates to cancer patients that will be discussed in the following sections.

P-glycoprotein expression and function in normal tissues

P-gp is normally expressed in various mammalian tissues including the lumenal border of intestinal epithelial cells, brain capillary endothelial cells, biliary canalicular cells, and renal proximal tubular epithelial cells. P-gp actively effluxes selected xenobiotics from within the cell back into the lumen of the intestine, brain capillary, bile canaliculus, or renal tubule, respectively[4,5]. P-gp is also expressed at high levels in the endothelial cells of the testis and placenta. Because of its location at these specialized tissues, P-gp is thought to function in a protective capacity for the organism by decreasing exposure to potentially toxic xenobiotics. P-gp has wide substrate specificity, but most substrates for P-gp are naturally-derived compounds including cyclosporines, many anticancer drugs, antiparasitics, and steroid hormones.

A major breakthrough in P-gp research, particularly with regard to its function in normal tissues, was the engineering of the mdr1a(-/-) knockout mouse. Initially, researchers were surprised, and somewhat disappointed, that these mice were healthy, fertile, and lived a normal life span. No abnormalities in anatomy were detected, nor were there any abnormalities in results of CBC's, serum biochemistry profiles, or other physiologic parameters. It was concluded that under normal laboratory conditions, P-gp was not essential for the basic physiologic functioning of an organism. Interest in the mdr1a(-/-) knockout mouse would have vanished had it not been for a serendipitous mite infestation. The infected mice were treated according to standard operating procedures for the facility by spraying with a solution of ivermectin. Within 24 hours, nearly all of the mdr1a(-/-) mice had died, whereas none of the wildtype or heterozygous mice showed any abnormal signs. Researchers ultimately determined that the blood brain barrier in mdr1a(-/-) mice was defective, allowing 100-fold greater concentrations of ivermectin to enter their brain as compared to wildtype mice[6]. Further investigation demonstrated that P-gp, as an integral component of the blood brain barrier, limits the brain accumulation of many other substrate drugs including vinblastine, dexamethasone, cyclosporin A, loperamide (Imodium®), and others.

In addition to increased brain penetration of P-gp substrates, mdr1a(-/-) mice also absorbed greater amounts of these drugs through their gastrointestinal tract. For example, oral bioavailability of the anticancer drug paclitaxel is 3-fold higher in mdr1a(-/-) mice than in wildtype mice. Similar results have been documented with other orally administered P-gp substrates, indicating that P-gp expressed on intestinal epithelium plays a major role in limiting absorption of many drugs. Studies in the author's laboratory have shown that P-gp is critical for excretion of drugs in the bile. Because several anticancer drugs (vincristine, doxorubicin and others) are normally eliminated in the bile, deficient P-gp function leads to accumulation of drug and increases toxicity.

P-glycoprotein deficiency in veterinary oncology patients

The author has identified a mutation in the canine MDR1 gene that is known to occur in many herding breed dogs[7]. From a clinical perspective, this MDR1 mutation causes complete loss of P-gp function in dogs that are homozygous for the mutation, and partial loss of P-gp function in dogs that are heterozygous for the MDR1 mutation. The percentage of other breeds affected are shown in Table 1.

Table 1. Dog breeds affected by the MDR1 mutation

Breed	Approximate Percent of Dogs Affected
Australian Shepherd	50 %
Collie	75 %
Longhaired Whippet	50 %
Shetland Sheepdog	10 %
Silken Wind	50 %
Mixed breeds of herding descent	35 %
Border Collies, English Shepherds, German Shepherds, Old English Sheepdogs	< 5%

P-gp-deficiency in veterinary patients has several important clinical implications for the cancer patient. P-gp-deficient patients are likely to experience toxicity at routine doses of chemotherapeutic agents that are P-gp substrates (i.e., doxorubicin, vincristine, vinblastine, etc.). For example, a collie with lymphoma developed myelosuppression and GI toxicity after treatments with vincristine or doxorubicin, even at lowered doses, but tolerated cyclophosphamide at the full dose. The patient was subsequently genotyped for the MDR1 mutation and was determined to be heterozygous (MDR1 mutant/normal). Historically, reports of severe doxorubicin sensitivity in collies and shelties have surfaced in online veterinary discussion forums. When questions of herding breed sensitivity to doxorubicin have been raised, it has previously been attributed to personality traits in the breeds. A prospective study was recently completed showing that dogs affected by the MDR1 mutation (either heterozygous or homozygous) were significantly more likely to develop neutropenia and thrombocytopenia after receiving vincristine than dogs that did not have the MDR1 mutation[8]. Thus, dogs with the MDR1 mutation should receive reduced dosages of vincristine and vinblastine in order to avoid adverse effects. In order to know which dogs have the MDR1 mutation, testing is recommended for all herding breed dogs and all mixed breed dogs (since a mixed breed dog may have a herding breed ancestor and thus, carry the MDR1 mutation). It is not possible to determine whether or not a dog is a herding breed cross simply by its appearance.

Drug Interactions

Importantly, virtually all of the alterations in absorption, distribution, and excretion of P-gp substrate drugs that have been documented in mdr1a(-/-) mice have been repeatable in wildtype mice when a highly effective P-gp inhibitor has been co-administered with the P-gp substrate. Thus, a P-gp-deficient state in mice (and in dogs) can result from either genetic alterations in the mdr1a gene, or it can be pharmacologically created. This can result in a serious drug interaction. If vincristine or doxorubicin is used to treat a dog with lymphoma, and a P-gp-inhibiting drug is also administered, delayed excretion of vincristine or doxorubicin may increase the risk of an adverse drug reaction (neutropenia, thrombocytopenia, anorexia, vomiting and/or diarrhea). Table 2 lists drugs that are either known or suspected to inhibit P-gp and should thus be avoided in dogs receiving P-gp substrate chemotherapeutic agents (i.e., vincristine, vinblastine, doxorubicin).

Table 2. Drugs known or suspected to inhibit P-gp

Ketoconazole (antifungal)
Cyclosporin (immunosuppressant)
Spinosad (flea preventive)
Tamoxifen (anti-estrogen)
Verapamil (calcium channel blocker)

Availability of MDR1 testing

Dogs can be easily tested for the MDR1 mutation. Either EDTA blood or a cheek swab can be submitted to the Washington State University Veterinary clinical Pharmacology Laboratory (WSU VCPL). A cheek swab testing kit can be ordered through the website [www.vetmed.wsu.edu/vcpl]; some veterinary hospitals order several testing kits to have on-hand. There is no charge for ordering a test kit. Alternatively, many veterinary hospitals simply print off the submission form from the website and submit EDTA blood (4 mL is required).

Conclusions

P-gp certainly contributes to multidrug resistance in a variety of tumors in veterinary patients. But it is involved in much more than multidrug resistance. Absorption, distribution, and elimination of many cancer drugs are affected by P-gp. Dogs with the MDR1 mutation are significantly more likely to develop neutropenia and thrombocytopenia after treatment with many chemotherapeutic drugs. In order to prevent serious adverse drug reactions, it is important that veterinary oncologists are aware of the consequences of P-gp-deficiency (the MDR1 mutation) in certain dog breeds, test for the MDR1 mutation, and modify treatment accordingly.

REFERENCES

1. Juliano RL and Ling V. **Biochim Biophys Acta**. 1976 455(1):152-62. 2. Leonard GD et al. **Oncologist** 2003 8:411-424. 3. Cascorbi I. **Pharmacol Ther** 2006;112:457-473. 4. Tran CDH et al. **J Pharm Sci** 2002;91:117-128. 5. Fellner S et al. **J Clin Invest**. 2002 Nov;110(9):1309-18. 6. Schinkel AH, et al. **Cell** 1994; 77:491-502. 7. Mealey KL et al. **Pharmacogenetics** 2001;11:727-733. 8. Mealey KL et al. **J Vet Int Med** 2008;22:996-1000.

CHANGE OF HEART: NUTRITION FOR THE CARDIAC PATIENT

Lisa M. Freeman, DVM, PhD, DACVN
North Grafton, MA

INTRODUCTION

The process of choosing an appropriate diet for an animal with cardiac disease involves examining the patient, the diet, and the owner's feeding practices and considering all the issues at hand. Generally, there is not a single "best" diet for all animals with cardiac disease. What might be the best diet for one animal may be contraindicated in another. Therefore, it is important to assess a number of different factors to determine which diet or diets might best suit a particular patient.

THE PATIENT

In general, the nutrients of concern in cardiac patients are calories, sodium, protein, potassium, and magnesium. However, patients with cardiac disease vary tremendously in terms of their clinical signs, laboratory parameters, and food preferences, and these will all affect diet selection. Some of the parameters that influence the appropriate diet choice for animals are listed below:

Clinical signs

Animals with congestive heart failure (CHF) will need more restriction of dietary sodium than will those with asymptomatic heart disease.

Physical examination findings

Some cats and dogs with cardiac disease will be overweight, while others will have weight and muscle loss (cardiac cachexia). In the case of an obese patient, a calorically restricted diet providing appropriate levels of other nutrients of concern should be selected while a calorically dense diet is required for patients with weight loss.

Concurrent diseases

Two thirds of dogs and more than half of cats with cardiac disease have concurrent diseases that might influence diet selection. For example, some dogs with CHF might also have chronic renal failure, and this might necessitate protein restriction (whereas patients with cardiac disease alone should not be protein restricted). Cats with hypertrophic cardiomyopathy (HCM) might also have concurrent struvite urolithiasis. These cats would need a diet that is sodium restricted but also one that has nutritional modification to reduce the risk of struvite urolith formation.

Laboratory results

Some dogs with CHF will be hypokalemic while others (especially those receiving angiotensin converting enzyme inhibitors) can be hyperkalemic. Diets vary in terms of the potassium content and this might also affect diet selection. Taurine deficiency is associated with feline dilated cardiomyopathy (DCM; although most current feline DCM cases are taurine independent) and some cases of canine DCM in certain breeds. If taurine deficiency is documented, taurine supplementation is warranted.

THE DIET

Based on these and other patient parameters, a diet or diets can be matched to the individual patient. For an animal with cardiac disease without CHF (ie, a dog with chronic valvular disease with no clinical signs or a asymptomatic cat with HCM), the author currently recommends only mild sodium restriction but counsels the owner to avoid diets high in sodium (<100 mg Na/100 kcal) and to avoid treats or table food that are high in sodium. Most owners are unaware of the sodium content of pet foods and human foods and need very specific instructions regarding which foods are appropriate. When CHF first arises, additional sodium restriction is recommended (<80 mg Na/100 kcal). Diets designed for animals with renal disease are not recommended for cardiac patients (unless there is concurrent severe renal failure) because of the protein restriction inherent in these diets. One must also be careful to avoid certain canine cardiac diets which are very protein restricted as this can contribute to muscle loss (cardiac cachexia). As CHF becomes more severe, more severe sodium restriction may allow lower dosages of diuretics to be used to control clinical signs. There currently are a number of commercial veterinary diets available that are specifically designed for dogs and cats with cardiac disease. The specific characteristics of these foods vary in sodium and chloride (as well as protein levels). Some also contain added taurine, carnitine, antioxidants, and omega-3 fatty acids. A cardiac diet may or may not be the optimal diet for an individual patient. For example, some diets may have the appropriate sodium levels but have characteristics which make them inappropriate for a given individual (eg, too low in protein for animals with cardiac disease, too high in calories for an obese patient, or too high in potassium for patients receiving an angiotensin converting enzyme inhibitor) so careful selection is important. However, diets designed for other diseases and even certain specific over-the-counter diets can be appropriate. Above all, the diet must be palatable enough that the animal will willingly eat it to aid in maintaining optimal weight.

In many cases, the nutrient modifications desired can be achieved through diet alone. In some instances, however, certain nutrients may need to be supplemented if they are either not in diet or not in the diet in high enough quantities to achieve the desired effect. For example, nutrients such as magnesium can become deficient in animals with CHF secondary to diuretic use. Since hypomagnesemia can cause refractory arrhythmias and muscle weakness, serial evaluation in an individual patient may be useful, especially in those patients with arrhythmias, and supplementation should be instituted when necessary. Other nutrients may act as pharmacologic agents, rather than just providing nutritional requirements. L-carnitine may help with energy metabolism in the heart. Coenzyme Q10 also is important for energy metabolism and is an antioxidant. Some veterinarians use coenzyme Q10 with apparent success in canine DCM. Nonetheless, controlled, prospective studies will be necessary to judge the true efficacy of this product. The omega-3 fatty acids, eicosapentaenoic acid (EPA) and docosahexaenoic acid (DHA), may be beneficial in animals with CHF to decrease production of inflammatory eicosanoids and cytokines, as well as reducing cardiac arrhythmias. While the optimal dose of omega-3 fatty acids for all animals with various cardiac diseases is not yet known, the author currently recommend a dosage of 40 mg/kg EPA and 25 mg/kg DHA in animals with cardiac disease. Fish oil supplements vary widely in the amount of EPA and DHA they contain, so it is important to know the exact amount in supplements recommended. Capsules that contain approximately 180 mg EPA and 120 mg DHA can be purchased over the counter at most human pharmacies and can be administered at a dose of 1 capsule per 10 pounds body weight. Fish oil should contain vitamin E as an antioxidant, but supplements which contain other nutrients are not recommended to avoid toxicities. Cod liver oil and flax oil should not be used. Also be aware that since dietary supplements are not regulated as stringently as drugs, quality control of all dietary supplements can be a big problem. Therefore, using resources such as Consumerlab.com to find specific supplement brands with good quality control is recommended.

FEEDING PRACTICES

Although the pet may have dietary preferences and dictate the types of food eaten to a certain degree, the owner also influences the pet's diet. Therefore, meeting the owner's expectations in terms of diet is important for animals with cardiac disease. Their pet's quality of life is of tremendous importance to owners of pets with cardiac disease so providing diets that are palatable and readily eaten is critical. Recent studies have also shown that, in dogs with cardiac disease, over 90% receive treats and table food (although only about one-third of cats with cardiac disease receive treats or table food). The method of administering medications also should be considered as many owners have difficulty administering medications without foods, and these foods often are high in sodium. Including this information in the overall diet plan is important to achieve success with nutritional modification.

The type of food preferred also varies between owners (and pets; eg, canned vs dry vs homemade) and must be taken into consideration. Finally, cost preferences should be considered as veterinary therapeutic diets may be out of the price range for long-term use by some owners. In these cases, lower priced alternatives should be offered.

IMPORTANT POINTS IN DIET SELECTION

1. As mentioned previously, there is usually not a single "best" diet for any patient. The author typically determines several diets that would be appropriate for an individual patient based on the patient, diet, and feeding practices. These diets are offered as choices for the owner and for the pet. Once they determine which diet the pet prefers, this diet can be used. Having specific dietary choices is particularly beneficial for more severely affected patients, in which a cyclical or selective loss of appetite is common.
2. For animals with CHF, don't change the diet until the patient is stabilized. Once the animal is home and stabilized on medications, a gradual change to a new diet can be made. Forced dietary changes when the animal is sick can induce food aversions.
3. Since most owners give treats (either pet treats or "people food"), be sure to specifically discuss treats with the owner. Most owners are unaware of treats that would be contraindicated (eg, high salt treats or table food). The author typically provides a list of foods that are appropriate and foods to avoid as treats (and to help administer medications) to assist the owner in wise selection.

MONITORING THE ANIMAL WITH HEART DISEASE

At recheck visits, a thorough physical examination should be done and renal function and electrolytes evaluated. Body weight should be recorded at each visit to assess for both weight loss from altered food intake. Assessment of muscle mass also is important because an animal can retain fat stores but lose lean body mass. Early detection of muscle loss is important to identify cardiac cachexia at an early stage at which intervention is more likely to be successful. A complete diet history should be obtained at each visit, including pet food, treats, table food, dietary supplements, and the method the owner uses to administer medications to ensure the entire diet is optimized.

REFERENCES

[1.] Freeman LM, Rush JE, Kehayias JJ, et al. Nutritional alterations and the effect of fish oil supplementation in dogs with heart failure. **J Vet Intern Med** 1998; 12: 440-448. [2.] Freeman LM, Rush JE, Cahalane AK, et al. Dietary patterns in dogs with cardiac disease. **J Am Vet Med Assoc** 2003; 223: 1301-1305. [3.] Freeman LM, Rush JE, Markwell PJ. Effects of dietary modification in dogs with early chronic valvular disease. **J Vet Intern Med** 2006; 20: 1116-1126. [4.] Freeman LM, Rush JE. Nutritional modulation of heart disease. In: Ettinger SJ, Feldman EC (eds). **Textbook of veterinary internal medicine, 7th ed.** St Louis: Saunders Elsevier, 2010: 691-696. [5.] Mallery KF, Freeman LM, et al. Factors contributing to the euthanasia decision in dogs with congestive heart failure. **J Am Vet Med Assoc** 1999;214:1201-1204. [6.] Slupe JL, Freeman LM, Rush JE. The relationship between body weight, body condition, and survival in dogs with heart failure. **J Vet Intern Med** 2008; 22: 561-565. [7.] Torin DS, Freeman LM, Rush JE. Dietary patterns of cats with cardiac disease. **J Am Vet Med Assoc** 2007; 230: 862-867

SYSTEMIC APPROACH TO THORACIC RADIOGRAPHS: WHY IS THIS DOG COUGHING?

Stacey B. Leach, DVM, Deborah M. Fine, DVM, MS, DACVIM (Cardiology)
Columbia, MO

INTRODUCTION

Coughing is a common presenting complaint in dogs and is seen with both respiratory and cardiovascular disease. Coughing is a reflex mechanism that occurs when cough receptors located throughout the respiratory tract (larynx, pharynx, and bronchi) are stimulated. Cardiac causes of cough include left-sided congestive heart failure with pulmonary edema and left mainstem bronchial compression secondary to left atrial enlargement. The most common cardiovascular diseases in the dog are chronic myxomatous degenerative valve disease (endocardiosis) and dilated cardiomyopathy. Non-cardiac causes of cough are much more numerous and include diseases of the upper and lower airways and pulmonary parenchyma. Upper airway diseases that can cause cough include collapsing trachea, infectious tracheobronchitis, tracheitis, foreign body, or laryngeal and tracheal masses. Lower airway causes include feline asthma as well as chronic or eosinophilic bronchitis. Cats rarely, if ever, cough secondary to heart failure. Pulmonary parenchymal diseases include infectious and non-infectious pneumonia, neoplasia, non-cardiogenic pulmonary edema, and various interstitial lung diseases such as pulmonary fibrosis. Additionally, the same breeds of dogs that are predisposed to developing myxomatous degenerative mitral valve disease are also predisposed to several respiratory disease including tracheal collapse and chronic bronchitis.

A thorough physical examination can often provide clues to the underlying disease process. Auscultatory findings of heart murmurs, gallops, or arrhythmias may suggest underlying cardiovascular disease; however, they do not rule out primary respiratory disease as the cause of cough. Wheezing may suggest lower airway involvement, while stridor may indicate upper airway involvement. Dogs with left-sided congestive heart failure typically have high sympathetic tone and are thus are almost invariably tachycardic on presentation. Thoracic radiography is an invaluable diagnostic test to help ascertain the cause of cough. Serial thoracic radiography is also useful for determining response to therapy or progression of disease.

THORACIC RADIOGRAPHIC INTERPRETATION

Technique and Positioning

Proper technique and positioning is vital to accurate interpretation. Inappropriate techniques can lead to underexposure (too light) or overexposure (too dark) which can lead to loss of detail or over-interpretation. Some degree of post-processing correction for exposure errors are available with digital radiography, however image quality may be compromised (loss of detail or grainy images).

Ideally, at least two orthogonal views should be obtained (i.e. a lateral projection and either a ventrodorsal (VD) or dorsoventral (DV) projection) at peak inspiration. Differences in the radiographic appearance of thoracic structures exists between the left versus right lateral projections and between the VD and DV projections, with one view being preferred over the other depending on the clinical situation.[1] For example, DV projections allow for evaluating the cardiac silhouette suspended in its normal orientation and allows for visualization of the caudal lobar vessels and detection of small amounts of pleural effusion. The VD projection allows for better visualization of the entire lung fields. The entire thorax from the inlet to the diaphragm should be included, centering on the cardiac silhouette. The dorsal arches of the ribs and costochondral junctions should overlap on lateral projections. The vertebral column and sternum should overlap on VD and DV projections. Breed conformation, body condition, positioning, phase of respiration, and hydration status should all be taken into account when evaluating thoracic radiographs.

Cardiac Silhouette

A systematic approach should be used when interpreting thoracic radiographs. The extrathoracic structures, cardiac size and shape, pulmonary vasculature, pulmonary parenchyma, and mediastinum should all be thoroughly evaluated. Thoracic radiographs can identify abnormalities in the size or shape of the cardiac silhouette and may indicate generalized cardiomegaly, specific chamber enlargement, or suggest pericardial effusion. It is important to compare the size of the cardiac silhouette to the size of the thorax. A thorax that is small due to breed conformation or excessive abdominal fat may appear to be enlarged, when in fact, the thorax is actually small. In lateral projections, the heart width typically spans 2.5 to 3.5 intercostal spaces and the height of the heart from apex to the carina should occupy no more than 2/3 of the thoracic cavity. On DV or VD projections, the width of the heart at its widest should be no more than the width of the hemi-thorax at the 9th rib. A more objective criteria for assessing cardiac enlargement utilizes the vertebral heart scale (VHS)[2]. Briefly, the VHS involves measuring the length of the short axis and long axis of the heart and superimposing these measurements over the vertebral column beginning at T4. The number of vertebral bodies occupied by the short and long axis measurements are summed to give a score. Normal dogs have a VHS of 8.5 to 10.6 although there are some breed variations. Normal cats have a VHS of 7.2 to 7.8. Often a combination of these methods must be used to fully interpret the cardiac size.

Specific chamber enlargement can also create abnormalities in the cardiac silhouette. On lateral views the cranial aspect of the cardiac silhouette is composed of the right atrium and ventricle, while the caudal aspect is composed of the left atrium and ventricle. The cranial dorsal region is composed of several overlapping structures including the right atrium, pulmonary artery, and aorta. The "cranial cardiac waist" is the division between these structures and the right ventricle. Loss of the cranial cardiac waist can be due to enlargement or dilatation of any of the cranial dorsal structures. The caudal dorsal region is defined by the left atrium.

The "caudal cardiac waist" is the division between the left atrium and ventricle, and is lost in conditions causing left atrial enlargement.

The clock face analogy is useful when evaluating the cardiac structures on DV and VD projections. Going clockwise, the aortic arch is located at the 11:00 to 1:00 o'clock position, the main pulmonary artery from 1:00 to 2:00, the left auricle from 2:00 to 3:00, the left ventricle from 3:00 to 6:00, the right ventricle from 6:00 to 9:00, and the right atrium from 9:00 to 11:00. The left atrium is superimposed over the center of the cardiac silhouette, just caudal to the carina. Specific chamber enlargement is often difficult to definitively determine with radiographs alone, and often echocardiographic examination is required. However, left atrial enlargement is usually easy to assess with thoracic radiography. On lateral projections, left atrial enlargement results in straightening of the caudal cardiac waist and a prominent dorsal bulge. On DV and VD projections, left atrial enlargement may cause outward displacement of the mainstem bronchi around the centrally located left atrium. In cats, left atrial enlargement is best appreciated in DV and VD projections where it may result in a widening of the heart base, commonly referred to as a "valentine-shaped" heart. It is important to note that lack of radiographic changes in the appearance of the cardiac silhouette does not rule out the possibility of heart disease.

Pulmonary Vasculature

Evaluation of the pulmonary vasculature is also essential for a diagnosis for congestive heart failure. On the lateral projections, the pulmonary vasculature is best visualized in the cranial lung fields. The pulmonary arteries are located dorsal to the bronchus, and the pulmonary veins are located ventral to the bronchus. On DV and VD projections, the pulmonary vasculature is best seen in the caudal lung fields with the pulmonary arteries located lateral to the bronchus and the pulmonary veins located medial to the bronchus. The pulmonary artery and veins should be equal in size, and on lateral projections be equal to the diameter of the 4th rib where they intersect. On DV and VD views, the pulmonary artery and vein should be equal to the diameter of the 9th rib where they intersect. Enlargement of the pulmonary arteries indicates pulmonary arterial hypertension while enlargement of the pulmonary veins is a hallmark feature of congestive heart failure. Symmetrical enlargement of the pulmonary arteries and veins in a young animal is consistent with a left to right cardiac shunt. This finding in an older animal may indicate left heart failure and concurrent pulmonary hypertension

Pulmonary Parenchyma and Airways

Evaluating the pulmonary parenchyma and airways for specific lung patterns is also vital in distinguishing between cardiac and noncardiac causes of cough. The basic abnormal radiographic lung patterns include interstitial, alveolar, and bronchial patterns. Interstitial lung patterns appear as an increase in the opacity of the lung fields with indistinct or hazy pulmonary vasculature. They can be structured (nodular in appearance) or unstructured and diffuse. This results from interstitial accumulation of fluid or cellular material. With further fluid accumulation, the interstitium becomes saturated and the fluid may spill into the airways, filling the alveoli. Alveolar patterns occur with pulmonary edema, hemorrhage, atelectasis, or from cellular infiltration of the alveoli. Alveolar lung patterns appear as border effacement of the pulmonary vasculature, heart and diaphragm, with the presence of air bronchograms (radiolucent bronchi surrounded by a soft tissue opacity indicative of either collapse of the alveoli or accumulation of fluid or cells in the alveoli).

Cardiogenic pulmonary edema is typically bilaterally symmetrical, most prominent in the caudodorsal and perihilar lung fields in dogs[3]. In cats, cardiogenic pulmonary edema is often located in the ventral lung fields, although a patchy, diffuse distribution can also be seen. Pulmonary edema secondary to non-cardiogenic causes can also have a similar caudodorsal and symmetrical distribution to cardiogenic pulmonary edema. Interstitial-to-alveolar patterns may also be seen with primary respiratory diseases such as pneumonia, pulmonary contusions, lung lobe torsion or atelectasis. The distribution of the alveolar pattern secondary to these types of disease processes tend to be asymmetrical.

Bronchial patterns appear either as an increased opacity of the bronchial walls (mineralization) or as thickened peribronchial infiltrates. This is sometimes described as "donuts" for end-on bronchi, and "tram lines" for long-axis views of bronchi. These findings are typical for inflammatory airway diseases such as chronic bronchitis or feline asthma.

CONCLUSIONS

Thoracic radiography in conjunction with a thorough physical examination is crucial in assessing the primary cause of a cough in veterinary patients. Radiographs can assist in determining if the cough is cardiac or respiratory in origin and can be useful in following progression and resolution of the disease process. Cases discussed in this session will provide examples of how thoracic radiographs help distinguish whether coughing is due to primary respiratory or cardiovascular disease and emphasize the common radiographic changes seen in these disease processes.

REFERENCES

[1] Brinkman, E.L. **J Am Anim Hosp Assoc**. 2006. 42: 440. [2] Buchanan, J.W. **Vet Clin N Am Small Anim Pract**. 2000. 30: 379. [3] Diana, A. **J Am Vet Med Assoc**. 2009. 235: 1058.

HOLTER MONITOR PLACEMENT & ANALYSIS

Anne Myers RVT, VTS (Cardiology)
Raleigh, NC

INTRODUCTION

A **Holter Monitor** is an ambulatory electrocardiogram (ECG) device worn at home by a patient which records and stores heart rate and rhythm for a 24 hour period or longer, and was first used on human patients about 50 years ago. Their use in veterinary medicine and by the NCSU-VTH Cardiology Service has become more common in the last 10 years, and the ability to place one on a small breed dog or a cat has become easier due to the development of smaller, lighter, digital monitors. They are extremely useful in providing much more information than a short ECG tracing done in hospital, especially if a patient is having an intermittent brady- or tachyarrhythmia, or a doctor is trying to differentiate between **seizures or syncope**, as these episodes can sometimes look similar to one another. Holter Monitors are becoming a standard diagnostic procedure for **cardiac arrhythmia screening**, **diagnosis, and for evaluating response to anti-arrhythmic drug** therapy, such as heart rate control for atrial fibrillation, or to monitor control of supraventricular or ventricular tachycardia. In a clinical setting, the veterinary technician with experience obtaining ECGs is a good candidate to perform Holter Monitor placement and "in house" analysis with an accompanying software program. The turnaround time for results is usually much quicker than sending the tracing to an outside vendor. There is plenty of information to share in order to make Holters an easy procedure for clients, their pets and the veterinary care team.

LOOP RECORDERS

There are other ambulatory ECG monitors occasionally used at NCSU that don't store a full 24 hour tracing, such as the **King of Hearts Express Cardiac Event Recorder** made by **Instromedix.** This small device with two chest leads continuously records a 5 minute ECG loop, and can be left on for over a week at a time. It is pre-set to save one or more loops after a button on the monitor is pressed. The data is then telephonically transmitted by the owner to a company that can print it out for a cardiologist on site to evaluate, or a software program is available for onsite analysis. An even longer term device is the **REVEAL PLUS 9526 Insertable Loop Recorder System** made by **Medtronic**. About the size of a pack of gum, it is surgically implanted under the skin on the left lateral chest, during heavy sedation or general anesthesia. It continuously records a several minute ECG loop, and when a hand-held device is activated by an owner, it can save 2 to 3 tracings. It can also be programmed to automatically save loops at a certain heart rate. To retrieve data, the patient is brought back in, and the tracing is printed and analyzed by the same type of computer used to check settings and events on a pacemaker battery. The device can be left in place for about a year.

BOXER ARVC

The middle-aged Boxer dog is the most common breed that Holter Monitors are used on at NCSU. Many veterinarians and technicians have heard of **Boxer Arrhythmogenic Right Ventricular Cardiomyopathy (ARVC).** ARVC occurs in humans also, and is the result of a fibro-fatty infiltrate of the right ventricular muscle. This disrupts normal electrical activity and conduction of impulses, and results in right-sided ventricular premature contractions (VPCs) that can lead to sustained runs of ventricular tachycardia, low cardiac output, syncope and even sudden death due to lethal arrhythmias. These complexes are early, wide, and positively deflected in Lead II. Boxers are placed on anti-arrhythmic drugs such as **Sotalol**, **Atenolol**, or **Mexiletine** and re-Holtered to evaluate response. Boxer ARVC is an inherited disease, and Dr. Kate Meurs at Washington State University has discovered a genetic mutation and offers testing of a blood sample or buccal swab to identify affected dogs. This test may not completely rule out the disease, as there may be more than one mutation that has not yet been identified, so a Holter Monitor may still be indicated if the test is negative.

PLACEMENT

Having equipment organized and in working order plus good communication skills are probably the two most important points for successful Holter Monitor placement. Treats available for the patient, if allowable, doesn't hurt either! First, take a few minutes to speak with the owner and interact with the patient. Explain the equipment, placement process and paperwork. Make sure the owner is aware their pet's hair will be clipped on both sides of the chest to get good contact for an accurate tracing. Have them watch the placement, as they will most likely be removing it at home the next day. With the **Lifecard CF** recorder, a **Spacelabs Healthcare** product, three electrode pads with leads attached are placed on the chest wall and connected to a small monitor that contains a battery and digital card to store the ECG data. **Remember to connect the electrode pads to the leads before placing them on the patient's chest.** It can cause patient discomfort when placing the pads on the chest wall first and then snapping the leads on. The operator is then able to view a 3 lead tracing **before** securing the pads with either 2" or 4" **Elasticon** tape wrapped around the chest. The start button is pressed to display the time the tracing began. The owner keeps an activity diary using the clock on the monitor and can trigger the system to mark a patient activated event by pressing a button. However, this isn't necessary to save any data, because the system automatically records and stores 24 to 48 hours regardless. Dogs typically wear a vest to protect the electrodes and leads and the monitor is placed in a small pocket at the top or inside the vest. Spacelabs offers a newer rechargeable Holter Monitor called the **Evo,** which does not require a battery or digital card.

PAPERWORK

Besides an activity diary, other paperwork associated with this outpatient test are an information sheet with contact names and numbers for trouble shooting problems, and directions on how to remove the monitor and leads (no scissors!).

REMOVAL

Most clients choose to forgo a second trip back to the clinic with their pet and will remove the monitor at home. To accommodate this process, they receive a bottle of adhesive tape remover at the time of placement and are given a pre-addressed **Fed Ex** box to return the monitor, activity diary and vest.

ANALYSIS

How long does it take to analyze a 24 hour Holter Monitor tracing with over 100,000 complexes? **Anywhere from 20 minutes to over 3 hours!** Begin by entering the patient information in the computer, and then download the ECG from the digital card. Using the **Spacelabs Impresario** software, the system will automatically analyze the entire tracing using standard presets. However, you should switch to the **interactive analysis program,** which allows you to change the presets and **"teach"** the system what to call a normal sinus beat, a supraventricular, or ventricular premature beat. Then it depends on many factors such as the quality of the tracing (**artifact** and patient activity such as running or playing can slow the system), and how arrhythmic the patient is. A patient with multi-focal VPCs can cause challenges in teaching the software to recognize all of the different forms, especially "**junctional**" premature beats that occur high in the right ventricle and closely resemble a sinus beat. Also, a patient with an extremely variable heart-rate or paroxysmal runs of an atrial or ventricular tachycardia can result in frequent changes to the "**blanking period,**" or the time in micro-seconds in which another "QRS" complex can be recognized after the one preceding it. To accurately count a rapid heart rate you would decrease the blanking period, and for slow heart rates increase it. The system also allows you to insert unrecognized beats. Other challenges include a large positive or negative "T" wave in some leads that cause double counting of complexes (try turning off that lead or increasing the T wave threshold), or low amplitude complexes which the system doesn't recognize (turn up the gain on the leads). Some basic advice to give when starting out is:

1. After downloading, try looking at a good portion of the tracing before changing the presets. You can page through about a minute a second, and get some idea of how arrhythmic the patient is.
2. Perform the scan with all 3 leads on. It will be label beats more accurately.
3. Try not to label complexes you are unsure of due to artifact. Just mark them as artifact, and concentrate on identifying beats on the clear parts of the tracing.
4. If you're having trouble teaching the software, ask a coworker for help, or call the helpline of the software company.
5. Try to avoid frequent changes to the presets. It can get confusing for the operator and the software.

CONCLUSION

Successful Holter Monitor placement and analysis starts with good equipment and software, a client and patient comfortable with the placement and removal process, and results in an accurately analyzed tracing by a veterinary technician or other professional for review by a veterinary cardiologist. More veterinary clients are asking for a higher standard of diagnostics and treatment for their pets with cardiac arrhythmias, and a Holter Monitor can provide this.

REFERENCES

[1.] Delmar Reynolds Impresario User's Guide copyright 2005 Hertford, UK & Irvine CA, USA [2.] Hanas S, et.al. **Jour Vet Card** 2009:11:17-22. [3.] Meurs KM. 14th Annual North Carolina Veterinary Conference, November 2009 Boxer Arrhythmogenic Right Ventricular Cardiomyopathy [4.] Ware WA. 2007 Cardiovascular Disease in Small Animal Medicine, Manson Publishing Ltd p.66, 142, 281.

FRACTIONAL SHORTENING: NOT ALL IT APPEARS TO BE

June A. Boon, MS
Fort Collins, CO

The heart's job is to pump enough blood to meet the metabolic needs of the organs. Cardiac systolic function is affected by preload, afterload and contractility. (Coordinated contraction seen with ischemic heart disease and distensibility–a factor in diastolic function--will not be discussed here.) As cardiac disease processes affect the heart, changes in preload, afterload and contractility affect the ability of the heart to do its job. Here we will look at how these parameters affect the ability of the heart to function properly, and how we can observe and evaluate these changes using echocardiography.

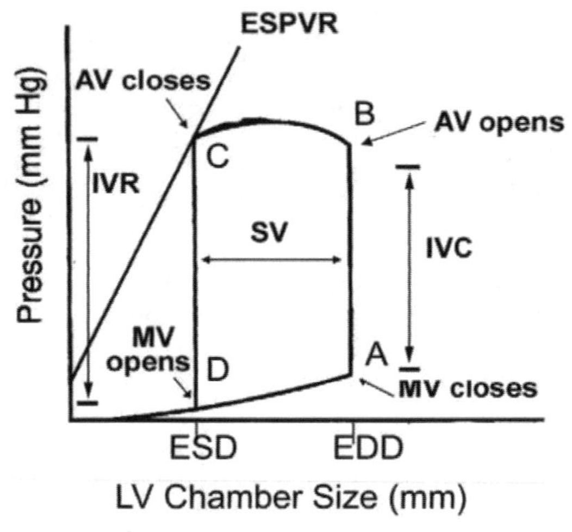

Figure 1: This figure shows the relationship between size, pressure, and function in the left ventricle. See text for details.

The cardiac cycle is comprised of the following phases: diastole (rapid ventricular filling phase and atrial contraction phase), iso-volumic contraction, ejection (contraction), and iso-volumic relaxation. These components can be shown on a pressure volume loop where dimension is used in place of volume (Figure 1). Point A represents the end diastolic dimension (EDD) or preload of the left ventricle (LV). The mitral valve (MV) closes at this point. The change in pressure between points A and B is isovolumic contraction (IVC) and represents the amount of pressure (afterload) needed within the LV chamber before the aortic valve can open at point B. The change in volume or dimension between points B and C corresponds to ventricular ejection. The aortic valve (AV) closes at point C. The pressure drop between points C and D corresponds to isovolumic relaxation, the pressure drop needed before LV pressure is low enough for the mitral valve to open and diastolic filling to start. Diastolic filling occurs between points D and A. Stroke volume (SV) is the change in volume between maximal volume at end diastole (EDD) and minimum volume at end systole (ESD). Fractional shortening (FS) is the change is dimension between end diastolic dimension and end systolic dimension and can be used in place of SV. The amount of work the heart undergoes is the area under the loop.

On an echocardiographic m-mode image of the left ventricle (Figure 2), point A on the pressure volume loop is found at the end diastolic dimension. Point B occurs where the left ventricular wall starts to move upward at the beginning of systole. End systole at the smallest left ventricular dimension is point C. Fractional shortening measured from m-mode images corresponds to the percent difference between end-diastolic dimension and end-systolic dimension. Hearts with normal myocardial function should change in dimension about 30 to 45 %.

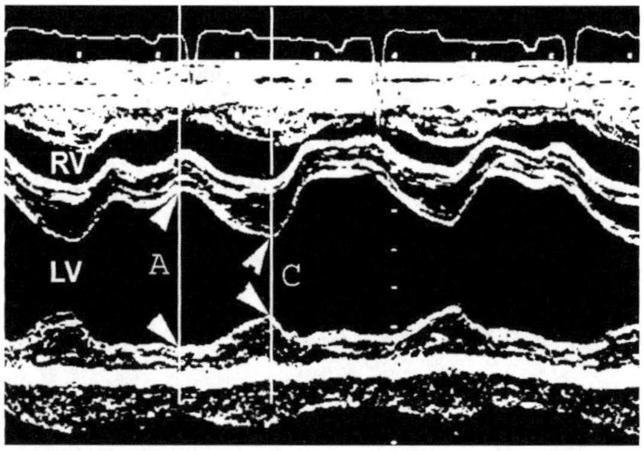

Figure 2: This m-mode of the left ventricle shows point A – the end diastolic dimension that corresponds to point A on the pressure volume loop. Point C, the end systolic dimension corresponds to point C on the pressure volume loop.

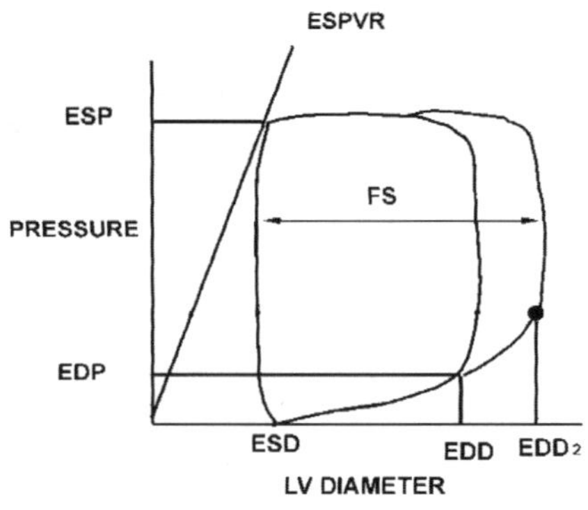

Figure 3. If the myocardium is able to contract normally (i.e. all physiologic components of the muscle work properly) then no matter how dilated the heart becomes (EDD_2), it will contract down to a normal systolic dimension (ESD).

Preload is the volume within the left ventricular chamber when it is maximally filled. An increase in preload puts more stretch on the myocardial fibers and their response is to contract more forcefully (Frank Starling's Law). Preload in the heart increases in response to leaks or shunts. As the volume of regurgitant flow or shunted blood compromises forward flow, the body responds by retaining sodium and water to increase the end-diastolic volume in an attempt to maintain normal forward cardiac output. Increases in preload are seen on the m-mode echocardiogram as an increased size at end-diastole (A – Figure 2).

When preload increases in the left ventricular chamber and there is increased stretch on the myocardial fibers the Frank Starling Law states that the ventricle will pump all of the blood within it to physiological limits. This means

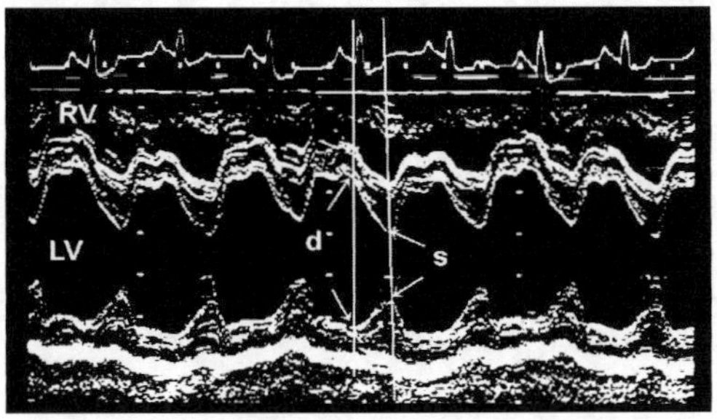

Figure 4: The left ventricular end-diastolic dimension (d) is increased in this heart secondary to mitral regurgitation. Myocardial function is normal since the systolic dimension (s) is normal and the fractional shortening is increased.

that if the myocardium is able to contract normally (i.e. all physiologic components of the muscle work properly) then no matter how dilated the heart becomes, it will contract down to a normal systolic dimension (Figure 3). There is a resultant change in end-diastolic volume, end diastolic pressure, stroke volume, and work, but no change in end systolic volume or end systolic pressure.

An M-mode image of the heart with increased preload (Figure 4) has an increased diastolic dimension and normal systolic dimension. The overall percent change in dimension–fractional shortening--is increased just as stroke volume is increased on the pressure volume loop. In any heart with increased left ventricular volume, if the myocardium is normal, fractional shortening should always be increased and systolic dimension should always be normal. If the systolic dimension is not normal–resulting in a fractional shortening that is not elevated–then there may be myocardial failure.

Decreased preload shifts the end-diastolic curve to the left. Diastolic pressure, diastolic volume, stroke volume and work are all decreased. End systolic pressure and volume remain unchanged (Figure 5). On m-mode images diastolic dimensions will be decreased, but if contractility is normal, the systolic dimension will be normal. The result is a fractional shortening that is decreased simply because there is not enough stretch on the myocardial fibers. Several things can cause volume to be decreased within the left ventricular chamber. Volume contraction occurs secondary to blood loss, dehydration, Addison's disease, pulmonary hypertension, and pulmonic stenosis.

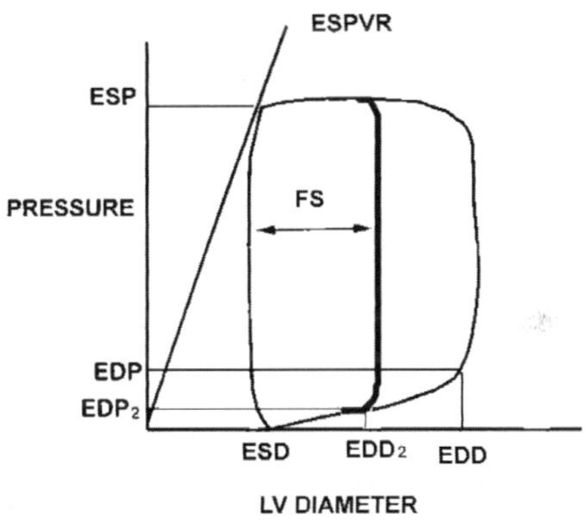

Figure 5. Decreased preload (EDD_2) results in a poorer fractional shortening. End systolic dimension (ESD) and contractility (ESPVR) remain unchanged.

Afterload is the pressure against which the cardiac muscle must contract. The degree of afterload is determined by wall stress. Systolic wall stress is calculated by systolic pressure times chamber radius at end systole divided by wall thickness at end systole. Therefore, afterload is increased when blood pressure is increased, when chamber volume is increased or when wall thickness is decreased.

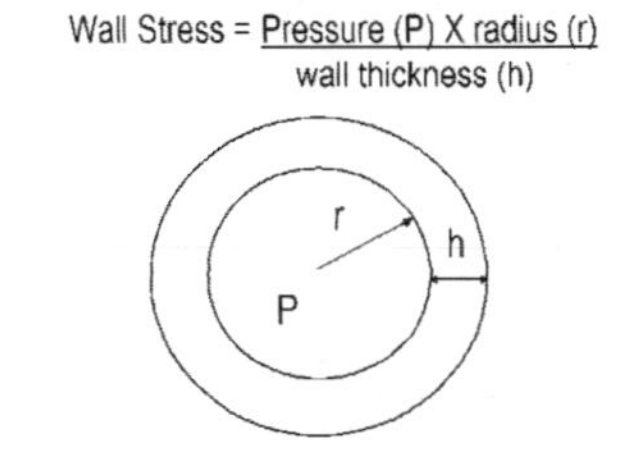

Figure 6: This diagram and equation represent LaPlace's Law.

Figure 6 illustrates a cross section of the left ventricular chamber. It shows the relationship of the factors affecting wall stress. Increased systolic pressure typically causes concentric hypertrophy of the myocardium. Sarcomeres are replicate in parallel, increasing wall thickness and normalizing wall stress. When sarcomeres have not had a chance to replicate, increased systolic pressure (blood pressure) prevents the heart from effectively shortening down to a normal systolic dimension.

When systolic pressure increases, the pressure volume curve shifts upward and right. The myocardium contracts until it reaches the volume that defines the pressure volume relationship (ESPVR) (Figure 7). There is no change in end diastolic volume, end diastolic pressure, work, or the end systolic volume relationship. There is an increase in end systolic pressure and a decrease in stroke volume.

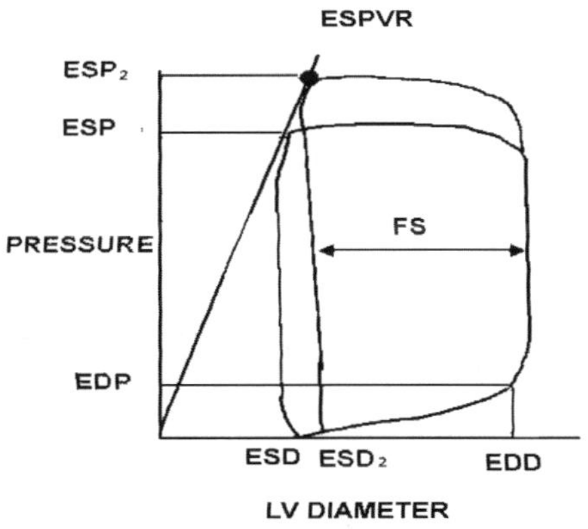

Figure 7. Increased afterload (ESP_2) increases systolic chamber size (ESD_2) and as a result function decreases because the muscle can only shorten to the degree that the ESPVR line indicates.

M-mode images of a left ventricle trying to contract against high afterload (high blood pressure) show decreased fractional shortening. This is not a reflection of poor contractility, but merely the effect of a relatively acute increase in blood pressure (Figure 8).

Decreases in left ventricular systolic pressure (afterload reduction) shift the systolic pressure curve downward and left. During systole, the systolic dimension decreases to meet the pressure volume relationship line. The change between the end diastolic volume and the end systolic volume is increased corresponding to fractional shortening.

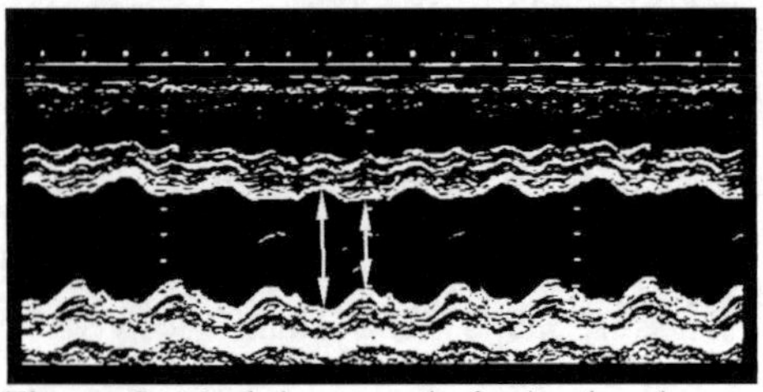

Figure 8: Both increased afterload and decreased contractility do not change the end diastolic dimension but do increase the systolic dimension, resulting in decreased percent change in dimension (fractional shortening).

When the contractility of cardiac muscle decreases before any remodeling of the heart occurs, (no dilation) the end diastolic volume will remain normal but the end systolic dimension will increase because the end systolic pressure volume relationship line shifts to the right (Figure 9). This reduces stroke volume and percent change in dimension. If inotropy increases, then the ESPVR shifts left, decreasing systolic dimension and increasing stroke volume (Figure 10). M-mode images and evaluation of poor contractility in hearts that have not remodeled (early myocardial failure) show normal diastolic dimensions and increased systolic dimensions with low fractional shortenings. These changes are similar to those seen with increased afterload.

In review, factors that reduce stroke volume and decrease fractional shortening are decreased preload, increased afterload, and decreased contractility. When looking at an m-mode with a subnormal fractional shortening, all three of these factors must be evaluated. Usually the most reliable parameter of left ventricular function is the systolic dimension. Regardless of the preload status, if myocardial contractility is normal, the systolic dimension will be normal.

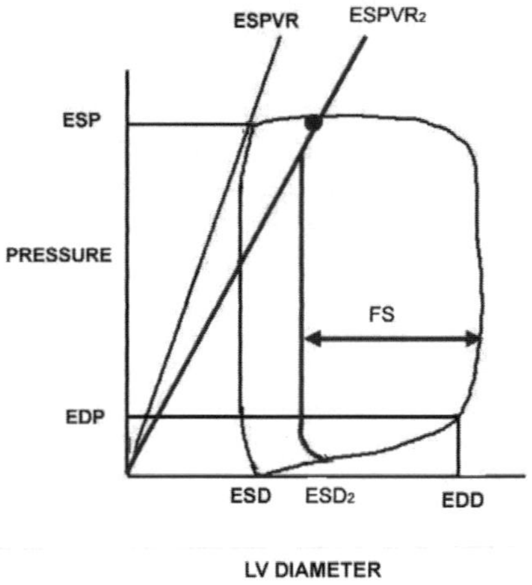

Figure 9. When contractility is impaired the ESPVR$_2$ (contractility) line shifts to the right. End systolic left ventricular dimension (ESD$_2$) will increase and fractional shortening decreases.

Left ventricular m-mode images provide us with several pieces of information regarding cardiac structure and function (Figure 2). Dimensions of chamber size and wall thickness are correlated to body size (kg or body surface area). Assessment of volume, wall and septal thicknesses should be made during diastole when the cardiac muscle is relaxed and pressure is at its lowest. Therefore, hypertrophy of the wall or septum is evident in the diastolic measurements, not systolic measurements. Assessment of function is made by looking at the systolic dimension. Increased systolic dimensions reflect decreased function not increased volume. Increases in systolic wall thickness may reflect increased function not necessarily hypertrophy.

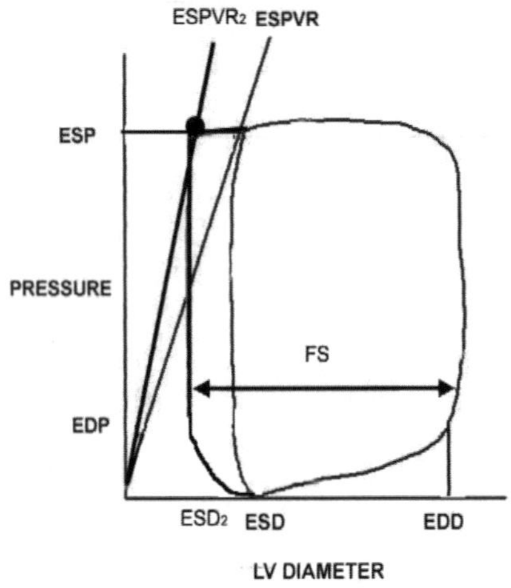

Figure 10. When contractility increases the ESPVR$_2$ (contractility) line shifts to the left. End systolic dimension (ESD$_2$) decreases and fractional shortening increases.

RECOMMENDED READING

1. Fox, P., D. Sisson, et al., Eds. (1999). **Textbook of Canine and Feline Cardiology: Principles and Clinical Practice**. Philadelphia, W B Saunders. 2. Kittleson, M. (1994). "Left ventricular function and failure - part I." **Comp Sm Anim** 16: 287-308. 3. Kittleson, M. (1994). "Left ventricular function and failure - part II." **Comp Sm Anim** 16: 1001-1017. 4. Lilly, L., Ed. (2007). **Pathophysiology of Heart Disease**. Baltimore, Lippencott Williams and Wilkins. 5. Mohrman, D. and L. Heller (2003). **Cardiovascular Physiology**. New York, Lange Medical Books / McGraw Hill.

ACVIM FORUM
As far as you want to go.

Research Abstract Program of the 2010 ACVIM Forum

Anaheim, California,
June 9 – 12, 2010
Index of Abstracts

ORAL PRESENTATIONS – Thursday, June 10

Time	#	Presenting Author	Abstract Title
SMALL ANIMAL – CARDIOLOGY**			
9:00 am	1	Sonja Fonfara	Cytokine and Matrix Metalloproteinase Expression in Blood Samples of Dogs With Congestive Heart Failure
9:15 am	2	Dennis Trafny	Cardiac Troponin-I Concentration is Elevated Pre and Post-Pacing in Dogs With Bradyarrhthmias: Is Myocarditis a Potential Etiology?
9:30 am	**3**	**Kristine Yee**	**Diagnostic Test Parameters in Cats With Heart Disease and their Correlation With NT-proANP, NT-proBNP and Troponin I Measurements**
9:45 am	4	Gemma Fraga Veloso	Expression of Urocortins in Canine Myocardium and Plasma Levels in Dogs With Cardiac Disease
BREAK			
10:30 am	5	Dan Ohad	Is the Cardio-Renal-Anemia Syndrome Prevalent in Dogs?
10:45 am	6	Andrea Lantis	Aldosterone Escape in Furosemide-Activated Circulating Renin-Angiotensin-Aldosterone System (RAAS) in Normal Dogs
11:00 am	7	Melanie Hezzell	Relationships Between Serum and Urine Aldosterone, Ventricular Remodelling and Outcome in Dogs With Mitral Valve Disease
11:15 am	8	Ian Jones	Flow Mediated Vasodilation in Canine Chronic Mitral Valve Disease
11:30 am	9	Inge Tarnow	Congestive Heart Failure in Dogs is Associated With Enhanced Platelet-Leukocyte Aggregates - A Marker for Platelet Activation
11:45 am	10	Inge Tarnow	Thromboelastography in Dogs With Asymptomatic Myxomatous Mitral Valve Disease
12:00 pm	11	Caroline Rasmussen	24-hour Electrocardiography in Clinical Healthy Cavalier King Charles Spaniels, Wire-Haired Dachshunds and Cairn Terriers
12:15 pm	12	Lisa Freeman	Development and Evaluation of a Quality of Life Questionnaire for Cats With Cardiac Disease

***Also see Cardiology abstracts 68 - 81 (Thursday, June 10, 2:15 pm – 6:00 pm)*

Time	#	Presenting Author	Abstract Title
SMALL ANIMAL – ONCOLOGY			
9:00 am	13	Luis Lembcke	Evaluation of Tyrosinase Expression in Canine and Equine Melanocytic Tumors
9:15 am	**14**	**Rebecca Brown**	**Expression of KIT in Canine Anal Sac Adenocarcinoma Using Tissue Immunohistochemistry**
9:30 am	**15**	**Courtney Mallett**	**Immunohistochemical Characterization of Feline Mast Cell Tumors**
9:45 am	**16**	**Shannon Parfitt**	**Radiosensitivity and Capacity for Radiation-Induced Sublethal Damage Repair of Canine Transitional Cell Carcinoma Cell Lines**
BREAK			
10:30 am	17	Jeffrey Phillips	Genetics of Osteosarcoma in the Scottish Deerhound
10:45 am	18	Masahiko Sato	Perfusion Method for Bone Marrow Cell Harvesting in Dogs
11:00 am	19	Sandra Axiak	Immunodysfunction in Dogs With Lymphoma

Boldface type indicates ACVIM Resident Research Award eligibility. Presentation times are subject to change.

SMALL ANIMAL – NEPHROLOGY/UROLOGY**

9:00 am	20	Anant Radhakrishnan	Urethral Stent Placement Using Digital Radiography: 10 Cases
9:15 am	**21**	**Amanda Blackburn**	**Urethral Stenting in Canine Patients With Urothelial Carcinoma: A Review of 41 Cases (2004-2008)**
9:30 am	22	Merja Leinonen	The Effect of Contrast Ultrasound on Kidneys in Cats
9:45 am	**23**	**Katie Baxter**	**Renal Disease in Cats Infected With Feline Immunodeficiency Virus**
BREAK			
10:30 am	24	Dominique Paepe	Screening of Ragdoll Cats for Kidney Disease: A Retrospective Evaluation
10:45 am	25	Shubhro Chakrabarti	Anemia Predicts Progression of Chronic Kidney Disease in Newly Diagnosed Azotemic Cats
11:00 am	**26**	**Emily Klosterman**	**A Case-Control Study of Nephrotic Syndrome in Dogs: 78 Cases**
11:15 am	27	Meaghan Whalen	Inhibition of P-Fimbriated *Escherichia coli* Adhesion in an Innovational Ex-vivo Model in Dogs Receiving a Bioactive Cranberry Tablet (Crananidin™)
11:30 am	28	Namphung Suemanotham	Immunohistochemical Expression of Cyclooxygenase (COX)-1 and -2 Enzymes in Feline Kidney Disease
11:45 am	29	Stephanie Lalor	Do Cats With Hypertension But No Ocular Lesions Have Increases in Plasma NT-ProBNP Concentration?
12:00 pm	30	Dominique Paepe	Comparison of Plasma Exo-Iohexol, Endo-Iohexol and Creatinine Clearances in Cats With Low, Normal and High Glomerular Filtration Rates
12:15 pm	31	Dominique Paepe	Limited Sampling Strategies to Estimate Glomerular Filtration Rate in Cats With Low, Normal and High Glomerular Filtration Rates

Also see Nephrology/Urology abstracts 105 - 109 (Thursday, June 10, 5:00 pm – 6:00 pm)

SMALL ANIMAL – HEMATOLOGY**

9:00 am	32	Mads Kjelgaard-Hansen	Increased Concentrations of Cytokines Related to Macrophage/Monocyte Activity and Recruitment are Risk Factors for Mortality in Primary Canine Immune-Mediated Hæmolytic Anemia
9:15 am	33	Robert Goggs	Serial Analysis of Coagulation Parameters in Dogs With Immune-Mediated Hemolytic Anemia (IMHA) Using Thromboelastography® (TEG)
9:30 am	**34**	**Alice Huang**	**Apparent Lack of Association Between Primary Immune-Mediated Thrombocytopenia and Recent Vaccination in Dogs**
9:45 am	**35**	**Sarah Flint**	**Thromboelastography in Dogs With Immune-Mediated Hemolytic Anemia Treated** With Prednisone, Azathioprine and Low-Dose Aspirin
BREAK			
10:30 am	36	Amir Kol	Serial Thrombelastographic Monitoring of Dogs with Multicentric Lymphoma
10:45 am	37	Alan Ralph	The Effect of Storage Temperature on Thrombelastrography Endpoints Using Citrated Canine Whole Blood
11:00 am	**38**	**John Thomason**	**Effects of Aspirin on Canine Platelet Cyclooxygenase Expression**
11:15 am	**39**	**John Thomason**	**Effects of Oral Cyclosporine on Platelet Activity in Normal Dogs**
11:30 am	40	Curtis Cathcart	The Effect of In Vivo Treatment With Acetylsalicylic Acid and Meloxicam on Platelet Aggregation in Cats
11:45 am	**41**	**Nicola Wilson**	**Clinical Investigation of a Point of-Care Hematology Analyzer in Dogs Receiving Chemotherapy**
12:00 pm	**42**	**Jason Reeder**	**Gross and Histologic Quality of Bone Marrow Core Specimens Obtained from the Same Site by a Combined Aspiration and Biopsy Technique in Dogs**
12:15 pm	43	Alice Defarges	Sternal Bone Marrow Aspiration in Clinically Normal Small Dogs

Also see Hematology abstracts 53 - 55 (Thursday, June 10, 11:45 am – 12:15 pm)

SMALL ANIMAL – ENDOCRINOLOGY

9:00 am	44	Tim Williams	Iatrogenic Hypothyroidism (IH) Contributes to the Development of Azotemia in Hyperthyroid Cats
9:15 am	45	Miguel Campos	Effect of Recombinant Human TSH on the Uptake of Radioactive Iodine (123I) by the Thyroid Gland in Healthy Beagles
9:30 am	46	Audrey Cook	Evaluation of Content of Compounded Trilostane Products
9:45 am	**47**	**Khristen Carlson**	**Optimization of a Test Protocol to Assess Aldosterone Secretory Capacity in Dogs**
10:00 am	47a	Stijn Niessen	Development of a Measure of Quality of Life for Cats with Diabetes Mellitus and their Owners: The DIAQol-pet ***(ECVIM Award Winner)***
BREAK			
10:30 am	48	Linda Martin	Comparison of Low-Dose and Standard-Dose ACTH Stimulation Tests in Critically Ill Dogs by Assessment of Serum Total and Free Cortisol Concentrations
10:45 am	**49**	**Elizabeth Lennon**	**Urine Sodium Concentrations in Hyponatremic Dogs With Hypoadrenocorticism**
11:00 am	**50**	**Siobhan O'Neill**	**Evaluation of Cytokines, Chemokines, and Hormones in Dogs With Diabetic Ketoacidosis and Uncomplicated Diabetes Mellitus**
11:15 am	51	Jacquie Rand	Glargine Administered Intramuscularly is Effective for Treatment of Feline Diabetic Ketoacidosis
11:30 am	**52**	**Sibylle Buob**	**An Intermittent Insulin Protocol Improves Metabolic Acidosis Faster Than a Continuous Rate Infusion of Regular Insulin in Feline Diabetic Ketoacidosis**

SMALL ANIMAL – HEMATOLOGY**

11:45 am	53	Charlotte Bjornvad	Obesity in Physically Inactive Domestic Shorthaired Cats is Associated With a Hypercoagulable State
12:00 pm	54	Rachel Rasmussen	Seasonality and Vector-Borne Disease Screening in Southern California Dogs With Immune Mediated Hemolytic Anemia
12:15 pm	55	Sara Zaldivar-Lopez	Chemistry Panel and Acid-Base Values in Retired Racing Greyhounds Using a Critical Care Analyzer

Also see Hematology abstracts 32 - 43 (Thursday, June 10, 9:00 am – 12:15 pm)

EQUINE**

9:00 am	56	Ashley Hill	Skin From Horses With Hereditary Equine Regional Dermal Asthenia (HERDA) Contains Collagen Crosslinking Patterns that Cause Reduced Tensile Strength
9:15 am	57	Robert Mealey	Use of an Equine Lentivirus Vector to Deliver Pathogen Genes to Dendritic Cells for Induction of Cytotoxic T Lymphocyte Responses in Horses
9:30 am	58	Dianne McFarlane	Neutrophil Function in Healthy Aged Horses and Horses With Pituitary Dysfunction
9:45 am	59	Dianne McFarlane	The Effect of Latitude of Residence, Breed and Pituitary Dysfunction on Seasonal Plasma ACTH and MSH Concentration
BREAK			
10:30 am	60	Teresa Burns	Characterization of Leptin and Adiponectin Gene Expression Patterns in Various Adipose Tissue Depots of Insulin-Resistant and Insulin-Sensitive Light Breed Horses
10:45 am	61	Kelly Chameroy	Effects of Metformin Hydrochloride on Glucose Dynamics During Transition to Grazing in Insulin-Resistant Horses
11:00 am	62	Andrew van Eps	Digital Hypothermia Inhibits Early Laminar Inflammatory Signaling in the Oligofructose Laminitis Model
11:15 am	63	Sophy Jesty	Continuous Cardiac Output Monitoring in Exercising Horses Using PulseCO®
11:30 am	64	Nicola Pusterla	Assessment of Congenital Neosporosis in Broodmares Originating from a Farm in California with High Prevalence of Antibodies to *Neospora hughesi*
11:45 am	65	NC Ringger	Comparison of 2 Biomarkers of Neuronal Injury in Neonatal Foals With Hypoxic-Ischemic Encephalopathy
12:00 pm	66	Kelsey Hart	Effects of Low-Dose Hydrocortisone Therapy on Measures of Innate Immune Function in Healthy Neonatal Foals
12:15 pm	**67**	**Johanna Elfenbein**	**Effects of Deferoxamine on Iron Elimination Following Blood Transfusion in Neonatal Foals**

Also see Equine abstracts 124 - 137 (Thursday, June 10, 2:15 pm – 6:00 pm)

SMALL ANIMAL – CARDIOLOGY**

2:15 pm	68	Serena Crosara	Use of Contrast Echocardiography for the Evaluation of Myocardial Perfusion and Pulmonary Transit Time in Normal Dogs. A Pilot Study
2:30 pm	69	David Prandi	A New Radiologic Measurement to Study Left Atrial Enlargement in Dogs With Mitral Insufficiency: Preliminary Study
2:45 pm	70	John MacGregor	Use of Pimobendan in 161 Cats (2006-2009)
3:00 pm	71	Sabine Riesen	Effects of Ivabradine on Invasive Indices of LV Function in Anesthetized Cats With Hypertrophic Cardiomyopathy
3:15 pm	72	Kathryn Meurs	A Splice Site Mutation in a Gene Encoding for a Mitochondrial Protein is Associated With the Development of Dilated Cardiomyopathy in the Doberman Pinscher
3:30 pm	**73**	**Mandi Schmidt**	**Combined Cutting Balloon and High Pressure Balloon Valvuloplasty for Dogs With Severe Subaortic Stenosis**
3:45 pm	**74**	**Joshua Stern**	**Familial Subvalvular Aortic Stenosis Appears to be an Autosomal Recessive Inherited Trait in Golden Retrievers**
BREAK			
4:30 pm	**75**	**Nonya Fiakpui**	**Pharmacokinetics of Fondaparinux in Healthy Cats**
4:45 pm	**76**	**Janet Olson**	**Minimally Invasive Per-Catheter Patent Ductus Arteriosus Occlusion in Small Dogs (< 3 kg): preliminary results**
5:00 pm	**77**	**Brian Maran**	**Identification of Beta-1 Adrenergic Receptor Polymorphisms in Cats**
5:15 pm	**78**	**Richard Cober**	**Effects of Ivabradine, A Selective I_f Channel Inhibitor, on Heart Rate in Healthy Cats**
5:30 pm	**79**	**Agnieszka Kent**	**Effects of Atenolol, Ivabradine and Pimobendan on Left Atrial Function: An Echocardiographic Study in Healthy Cats**
5:45 pm	**80**	**Justin Allen**	**NT-proBNP Levels in 30 Cats With Dynamic Right Ventricular Outflow Obstruction**
6:00 pm	81	James Andrade	Mitral Annuloplasty by External Plication in Dogs With Mixomatous Mitral Valve Degeneration

Also see Cardiology abstracts 1 - 12 (Thursday, June 10, 9:00 am – 12:15 pm)

SMALL ANIMAL – NEUROLOGY

2:15 pm	82	Cynthia Gaskill	Comparison of Phenobarbital and Potassium Bromide Monotherapies in the Treatment of Canine Epilepsy
2:30 pm	**83**	**Rossi House**	**Increased Blood Ammonia Concentration After Generalized Seizures in Dogs**
2:45 pm	84	Beverly Sturges	Safety, Performance and Efficacy of a Seizure Advisory System
3:00 pm	85	Paula Martin-Vaquero	Magnetic Resonance Imaging of the Canine Brain at 3.0 T and 7.0 T
3:15 pm	**86**	**Zachary Niman**	**Vascular Endothelial Growth Factor (VEGF) Levels in the Cerebrospinal Fluid of Dogs With Intracranial Disease**
3:30 pm	**87**	**Allison Haley**	**Investigation of Canine Primary Brain Tumors for the Presence of DNA from Herpes- and Polyomaviruses by Broadly Reactive Pan-Viral PCR (99 Cases)**
3:45 pm	**88**	**Mylene-Kim Leclerc**	**Interobserver Agreement and Accuracy in the Interpretation of Brain Magnetic Resonance Imaging in Dogs**
BREAK			
4:30 pm	**89**	**Lisa Lipitz**	**Central Nervous System Blastomycosis in 4 Dogs**
4:45 pm	**90**	**Lisa Gerlach**	**Plasma Cell Neoplasia of the Vertebral Column: Imaging, Clinicopathologic Findings, and Long-term Outcome With Treatment**
5:00 pm	91	Steven De Decker	Morphometric Dimensions of the Caudal Cervical Region in Doberman Pinschers With Clinical Signs of Disk Associated Wobbler Syndrome (DAWS), Clinically Normal Doberman Pinschers and Clinically Normal Foxhounds
5:15 pm	**92**	**Jason King**	**A Candidate Gene Approach to Myasthenia Gravis in Newfoundlands**

5:30 pm	**93**	**Laura Vasquez**	**Motor Unit Number Estimation Using the Incremental Method in Normal Dogs**
5:45 pm	**94**	**Ashley Bensfield**	**Lower Motor Neuron Dysfunction Consistent With a Peripheral Neuropathy in Young Bengal Cats**
6:00 pm	**95**	**Eva Furrow**	**Atypical Collapse in Exercising Labrador Retrievers Unaffected by the Dynamin-1 Mutation**

SMALL ANIMAL – RESPIRATORY

2:15 pm	96	Chick Weisse	Initial Short-Term Experience With Cisterna Chyli and Thoracic Duct Glue Embolization for Idiopathic Chylothorax in 2 Dogs and 1 Cat
2:30 pm	97	Emilie Krafft	Serum Endothelin-1 Concentration in Terrier Dogs With Idiopathic Pulmonary Fibrosis
2:45 pm	98	Claire Sharp	Endothelin-1 Concentrations in Bronchoalveolar Lavage Fluid from Healthy Control and Experimentally Asthmatic Cats
3:00 pm	99	Carol Reinero	Metabolomic Breath Analysis for Small Molecule Biomarkers in Airways of Experimentally Asthmatic Cats
3:15 pm	**100**	**Christine Cocayne**	**Subclinical Airway Inflammation Despite Oral Corticosteroid Therapy in Cats With Lower Airway Disease**
3:30 pm	101	Jérôme Leemans	Effects of Doxycycline on Bronchial Inflammation and Hyperresponsiveness and on Protease Activity in a Feline Model of Acute Asthma
3:45 pm	102	Jefferson Manens	Respiratory Functional Assessment Using Barometric Whole-Body Plethysmography in Healthy Growing and Aging Beagle Dogs
BREAK			
4:30 pm	103	Carol Reinero	Nebulized Acetylcysteine in Cats With Experimental Asthma
4:45 pm	104	Henna Heikkilä	Elevation of Procollagen Type III Amino Terminal Propeptide Concentration in Bronchoalveolar Lavage Fluid of Dogs With Idiopathic Pulmonary Fibrosis

SMALL ANIMAL – NEPHROLOGY/UROLOGY**

5:00 pm	105	Brice Reynolds	Effects of High Versus Normal Salt Diets on Renal Variables in Aged Cats: A 6-Month Blinded Randomized Prospective Study
5:15 pm	106	Doreen Houston	Efficacy of a Diet Designed With a Relative Supersaturation (RSS)
5:30 pm	107	Hasan Albasan	Evaluating Demographic Risk Factors for Urate Uroliths in Cats
5:45 pm	108	Ingrid van Hoek	Urinary pH is Not a Major Factor in Urinary CaOx Supersaturation of Healthy Cats
6:00 pm	109	Jennifer MacLeay	Struvite Urolith Dissolution in 5 Cats Fed a Maintenance Diet With 0.06% Magnesium, 35% Protein and a Targeted Urine pH of 6.2-6.4

Also see Nephrology/Urology abstracts 20 - 31 (Thursday, June 10, 9:00 am – 12:15 pm)

SMALL ANIMAL – INFECTIOUS DISEASE

2:15 pm	110	Katherine Stenske	Surveillance of Bacterial Contamination in Small Animal Hospital Environments in Kansas
2:30 pm	111	Elise Gingrich	Prevalence of Methicillin-resistant *Staphylococci* in a Northern Colorado Animal Shelter
2:45 pm	**112**	**Erika Meler**	**Hospital Prevalence and Risk Factors for Methicillin-Resistant *Staphylococcus* Infections in Small Animals at the Purdue University Veterinary Teaching Hospital (PUVTH), 2006-2009**
3:00 pm	113	Michael Lappin	Feline Herpesvirus-1 and *Mycoplasma* Spp. Infections in Cats With Acute Conjunctivitis in an Animal Shelter
3:15 pm	**114**	**Alyssa Mourning**	**Antigen Vs Antibody Testing for the Diagnosis of Blastomycosis Infection in Dogs**
3:30 pm	115	Daniel Foy	Correlation of Urine and Serum Blastomyces Antigen Levels With Clinical Remission in Dogs With Blastomycosis
3:45 pm	116	Mads Kjelgaard-Hansen	Monocyte Chemoattractant Protein-1 Concentration is a Marker of Mortality in Canine Babesiosis Caused by *Babesia rossi*
BREAK			
4:30 pm	117	Amelia Goddard	Mortality in Canine Babesiosis is Associated With a Consumptive Coagulopathy
4:45 pm	118	Michael Lappin	Effect of Vaccination on Results of Real Time PCR Assays for Adenovirus 2, Bordetella Bronchiseptica and Parainfluenza
5:00 pm	119	Paolo Silvestrini	Serum Cardiac Troponin I Concentration in Dogs With Leishmaniasis

SMALL ANIMAL – IMMUNOLOGY

5:15 pm	120	Claire Sharp	Inflammatory Biomarkers for Canine Sepsis
5:30 pm	121	Claire Sharp	Inflammatory Biomarkers for Sepsis in Cats
5:45 pm	**122**	**Todd Archer**	**Effects of Different Oral Doses of Cyclosporine on T-Lymphocyte Biomarkers of Immunosuppression in Normal Dogs**

SMALL ANIMAL – PHARMACOLOGY

6:00 pm	123	Lisa Singer	Pharmacokinetics of Leflunomide in Healthy Dogs

EQUINE**

2:15 pm	**124**	**Rosa Barsnick**	**Hormones of Energy Metabolism in Septic Foals: Associations between Insulin, Glucagon, Leptin, Adiponectin, Ghrelin and Growth Hormone**
2:30 pm	**125**	**Jamie Wearn**	**Effects of the Insulin Sensitizing Drug Pioglitazone on Indices of Insulin Homeostasis in Horses Following Endotoxin Administration**
2:45 pm	**126**	**Cody Alcott**	**Attenuation of Endotoxemia in Horses With a Ketamine Continuous Rate Infusion**
3:00 pm	**127**	**Linda Frellstedt**	**Endotoxin Tolerance Induced In Vitro in Equine Lymphocytes**
3:15 pm	**128**	**Lauren Wise**	**In Vitro Activity of Ponazuril Against *Theileria equi***
3:30 pm	**129**	**Joanie Palmero**	**Experimental Infection With *Bartonella henselae* and *Bartonella bovis* in Horses**
3:45 pm	**130**	**Robert Carmichael**	**Seroprevalence of *Anaplasma phagocytophilum*, *Borrelia burgdorferi* and *Ehrlichia* spp. in Horses in Oklahoma**

BREAK			
4:30 pm	**131**	**Katharyn Mitchell**	**Prevalence of *Salmonella* spp. in Healthy Foals, Crias and their Dams in the New York State Area**
4:45 pm	**132**	**Jennifer Sonis**	**Prevalence of Equine Herpesvirus 1 and Equine Herpesvirus 4 as Detected by Quantitative Polymerase Chain Reaction in Nasal Secretions of Febrile Horses**
5:00 pm	**133**	**Nora Nogradi**	**Field Efficacy Trial of An Intra-Rectally Administered Avirulent, Live Lawsonia Intracellularis Vaccine on Endemic Horse Farms in Central Kentucky**
5:15 pm	134	Pedro De Pedro	Development of Exogenous Lactate Clearance Curves in Normal Horses
5:30 pm	**135**	**Seika Hashimoto-Hill**	**Lactate Clearance in Horses With Acute Colitis**
5:45 pm	**136**	**Robert Carmichael**	**Pharmacokinetics of Ganciclovir After Single Intravenous Administration in the Adult Horse**
6:00 pm	137	Holly Bedford	Early Diagnosis and Treatment of Muscle Atrophy and Weakness Associated With Vitamin E Deficiency

Also see Equine abstracts 56 - 67 (Thursday, June 10, 9:00 am – 12:15 pm)

FOOD ANIMAL			
2:15 pm	**138**	**Charles Hinds**	**Serum Haptoglobin (HP)-Matrix Metalloproteinase 9 (MMP 9) Complex ELISA as a Marker of Septic Inflammation in Cattle: A Comparison to Hp or MMP 9 ELISA**
2:30 pm	**139**	**Manuel Chamorro**	**Experimental Evaluation of Transmission of Bovine Viral Diarrhea Virus (BVDV) by Horn Flies**
2:45 pm	**140**	**Katharyn Mitchell**	**Parelaphostrongylus tenuis Infection as a Cause of Meningiomyelitis in 5 Limousin Calves**
3:00 pm	**141**	**Davin Ringen**	**Utility of Pre-Partum Parameters for Predicting Intramammary Infection at Calving in First Calf Heifers on a Grazing Dairy**
3:15 pm	**142**	**Andrea Finnen**	**Comparison of Computed Tomography and Routine Radiography of the Tympanic Bullae in the Diagnosis of Otitis Media in the Calf**
3:30 pm	143	Jennifer Johnson-Neitman	Computed Tomography and Magnetic Resonance Imaging of the Normal Calf Brain
3:45 pm	144	Julie-Hélène Fairbrother	*Escherichia coli* is Associated With Persistent Mammary Infection in Canadian Dairy Cows
BREAK			
4:30 pm	145	Jérôme Leemans	Impact of Age and Breed on Incidence of Clinical Signs After Natural Infection With Bluetongue Virus Serotype 8 in Sheep
4:45 pm	146	Brandy Burgess	Methicillin Resistant *Staphylococcus* spp. in Commercial Pigs Used in Veterinary Student Training
5:00 pm	147	Elizabeth Dore	Risk Factors Associated With Transmission of *Mycobacterium avium* subsp. *paratuberculosis* (MAP) to Calves Within a Dairy Herd: A Systematic Review
5:15 pm	148	David Van Metre	Breeding Soundness Examination in Western Range Rams: 12,716 Cases (2000-2007)
5:30 pm	149	Jonathan Naylor	Inaccurate Rectal Temperature Measurement With Digital Thermometers: Causes and Solutions
5:45 pm	150	Timothy Potter	Integration of Pharmacokinetic and Pharmacodynamic Data for Florfenicol in Calves
6:00 pm	151	Sherry Cox	Pharmacokinetics of Intravenous and Intramuscular Tramadol in Llamas

ORAL PRESENTATIONS – FRIDAY, JUNE 11

SMALL ANIMAL – NUTRITION/METABOLISM			
8:00 am	152	Phill Toll	Effectiveness of Morphometric Measurements for Predicting Body Composition in Overweight and Obese Dogs
8:15 am	153	Angela Lusby	Effectiveness of BCS for Estimation of Ideal Body Weight and Energy Requirements in Overweight and Obese Dogs Compared to DXA
8:30 am	154	Charlotte Bjornvad	Obesity and Gender Affects Total Adiponection Concentration and Multimer Distribution in Domestic Cats
8:45 am	155	John Bauer	Lipid Metabolism and Satiety With Dietary Fiber and Carnitine in Obese Dogs
9:00 am	156	Nolan Frantz	Effect of Prescription Diet® Canine J/D™ on Clinical Measures, Cartilage Biomarkers, and Metabolomic Changes in Dogs With Osteoarthritis
9:15 am	**157**	**Isuru Gajanayake**	**Clinical Experience Using a Lipid-Free, Ready-Made Parenteral Nutrition Solution in 70 Cats and Dogs**
9:30 am	158	Stefanie Handl	Fecal Microbiota and Fecal Calprotectin and S100A12 Concentrations in Lean and Obese Dogs
SMALL ANIMAL – HEPATOLOGY			
8:00 am	**159**	**Dionne Ferguson**	**Cycad Intoxication in Dogs: Survival and Prognostic Indicators**
8:15 am	160	Angela Au	Hepatoprotective Effects of S-Adenosylmethionine and Silybin on Canine Hepatocytes
8:30 am	161	Hanna Dyggve	Evaluation of Promoter Alleles and Exon 2 of DLA-DQB1, -DQA1 and -DRB1 in Dobermans With and Without Hepatitis
8:45 am	**162**	**Allison Bradley**	**Feline Hyperbilirubinemia: A Retrospective Study of 180 Cases**
SMALL ANIMAL – GASTROENTEROLOGY			
9:00 am	163	Sally Bissett	Evaluation of a Human Virtual-Reality Endoscopy Trainer for Teaching Early Endoscopy Skills to Veterinarians
9:15 am	164	Anant Radhakrishnan	Comparison of Laparoscopic-Assisted Full Thickness Intestinal Biopsy versus Endoscopic Mucosal Biopsy in Cats: 11 Cases
9:30 am	165	Atsushi Tsukamoto	Ultrasonographic Evaluation of Vincristine-Induced Gastric Hypomotility and the Prokinetic Effect of Mosapride Citrate in Dogs
9:45 am	166	Stanley Marks	The Influence of Esomeprazole and Cisapride on Gastroesophageal Reflux During Anesthesia in Dogs
2:15 pm	167	Olivier Dossin	Plasma Citrulline Concentration is Decreased in Canine Parvovirus Enteritis
2:30 pm	168	Julien Dandrieux	Breed Predisopsitions for Severe Hypocobalaminemia and Relation to Folate Concentration in Dogs With Gastrointestinal Disease
2:45 pm	**169**	**Mary Tolbert**	**Efficacy of Oral Famotidine and Two Omeprazole Formulations for the Control of Intragastric pH in Dogs**

3:00 pm	**170**	**Alexandra Rosé**	**Causes, Usefulness of Clinical Investigations and Success of Antiemetic Therapy in Dogs Referred for Vomiting**
3:15 pm	**171**	**Fiona Tam**	**Safety and Palatability of Polyethylene Glycol 3350 as an Oral Laxative in Cats**
3:30 pm	**172**	**Lucie Goodwin**	**Evaluation of Hypercoagulability Using Thromboelastography (TEG) in Dogs With Protein Losing Enteropathy**
3:45 pm	173	Dottie Laflamme	Comparison of Two Canned Diets Designed for the Management of Feline Diarrhea
BREAK			
4:30 pm	174	Susanne Kilpinen	Determination of the Dosage Regimen of Tylosin in the Treatment of Canine Tylosin-Responsive Diarrhea
4:45 pm	175	Aarti Kathrani	CD11c Positive Dendritic Cells are Significantly Decreased in the Duodenum of Dogs With Inflammatory Bowel Disease
5:00 pm	176	Aarti Kathrani	Overdominant Single Nucleotide Polymorphisms in the Nucleotide Oligomerisation Domain Two (NOD2) Gene are Significantly Associated With Canine Inflammatory Bowel Disease
5:15 pm	177	Jan Suchodolski	Relationship of Mucosal Gene Expresion to Microbiota Composition in Dogs With Inflammatory Bowel Disease
5:30 pm	178	Nashwa Waly	Measurement of IL-12 (p40, p35), IL-23p19 and IFN-gamma; mRNA in Duodenal Biopsies of Cats With Inflammatory Bowel Disease and Healthy Controls using Quantitative Reverse Transcriptase Polymerase Chain Reaction (qRT-PCR)
5:45 pm	179	Melanie Craven	Mucosal Cytokine Profiling Reveals IL-6 Up-Regulation in Feline IBD and Alimentary Lymphoma
6:00 pm	180	Jevan Christie	Fecal Sensitivity as a Tool to Differentiate Between Non-Neoplastic and Neoplastic *Spirocerca Lupi* Nodules Using a Modified Centrifugal Flotation Method

POSTER PRESENTATIONS

On Display:	Thursday, June 10, 9:30 am - 4:30 pm; Friday, June 11, 9:30 am – 4:30 pm; Saturday, June 12, 9:30 am – 2:30 pm
Attended by Authors Eligible for ACVIM Resident Research Awards:	Thursday, June 10, 9:50 am – 10:30 am; Friday, June 11, 9:50 am – 10:30 am
Attended by ALL Authors – Wine & Cheese Reception:	Friday, June 11, 6:00 pm – 7:30 pm

#	Presenting Author	Abstract Title
SMALL ANIMAL – CARDIOLOGY		
181	Maria Helena Larsson	Time-Domain Signal-Averaged Electrocardiogram in Healthy German Shepherd and Boxer Dogs
182	Maria Helena Larsson	Time Domain High-Resolution Electrocardiography in Boxer Dogs With Arrhythmogenic Right Ventricular Cardiomyopathy and Dilated Cardiomyopathy
183	Aparecido Camacho	Heart Rate Variability in Boxer Dogs With Arrhythmogenic Right Ventricular Cardiomyopathy
184	Denise Schwartz	Six Minute Walk Test Standardization for Dachshund, Poodle and Labrador Retriever Dogs
185	Aparecido Camacho	Effects of Treadmill Training Over Autonomic and Hemodynamic Functions in Healthy Dogs
186	Masashi Mizuno	Effects of Running on the Renin-Angiotensin-Aldosterone System in Dog
187	Sara Granström	Occurrence of Hypertrophic Cardiomyopathy in a Large Cohort of British Shorthair Cats
188	Aparecido Camacho	Clinical Characterization of Hypertensive Hypertrophic Cardiomyopathy in Dogs With Chronic Kidney Disease (CKD)
189	Aparecido Camacho	Heart Rate Variability in Dogs With Mitral Endocardiosis or Natural Morbid Obesity
190	**Carley Saelinger**	**Comet-Tail Artifacts in Normal Dogs and Dogs With Cardiogenic Pulmonary Edema**
191	Takashi Ebisawa	Clinical Usefulness of Measuring Plasma Atrial Natriuretic Peptide Concentrations for Assessing the Severity in Dogs With Degenerative Mitral Valve Disease
192	Pierre Menaut	Circulating Natriuretic Peptides Concentrations in Hyperthyroid Cats
193	**Caryn Reynolds**	**Weekly Variability of Plasma NT-proBNP Measurements in Cats With and Without Heart Disease**
194	Aliya Magee	Use of Abciximab to Determine Platelet Reactivity in Healthy Cats
195	Carolina Carlos Sampedrano	Effects of High Versus Normal Salt Diets on Cardiovascular Variables in Healthy Aged Cats: A 6-Month Study
196	Takeshi Mizuno	Relationship Between Prognosis and Immune Response in Dogs After Mitral Annuloplasty
197	Shigeki Yamano	Endogenous Erythropoietin Levels and Iron Utilization in Dogs With Degenerative Mitral Disease
198	Yoko Fujii	Prevalence of Right to Left Shunt Due to Patent Foramen Ovale Concurrent with Pulmonary Stenosis in Dogs
199	Meg Sleeper	Dobutamine Stress Testing in Portuguese Water Dogs with Juvenile Dilated Cardiomyopathy
200	Sabine Riesen	Pharmacokinetics of Oral Ivabradine in Healthy Cats
201	Michael Katz	Thiamylal Anesthesia Reveals Predominant Role for the Central Mechanism of Respiratory Sinus Arrhythmia in the Dog
202	Lauren Calland	In-Hospital Electrocardiograph Versus 24-Hour Holter Monitor for Assessing Heart Rate in Dogs With Atrial Fibrillation
203	Ashley Saunders	Bradydysrhythmias Requiring Pacemaker Implantation in Chagas Positive Dogs
SMALL ANIMAL – ONCOLOGY		
204	Kensuke Nakamura	Contrast-Enhanced Ultrasonography With Sonazoid® for Characterization of Focal Splenic Lesions
205	Silvia Lucas	Evaluation of Oxidant/Antioxidant Total Status and Erythrocyte Antioxidant Defense in Cats With Lymphoma
206	**Elizabeth Lechner**	**Oxidative Stress in Dogs With Lymphoma Before and After Administration of Doxorubicin: A Pilot Study**

207	Thais Macedo	Erythrocyte Superoxide Dismutase (SOD) Concentration in Dogs With Multicentric Lymphoma
208	Archivaldo Reche Junior	Alimentary Lymphoma in Cats: Clinicopathological and Therapeutic Features of 32 Cases (2004-2009)
209	Hiroko Hiraoka	Identification of Methylguanine-DNA Methyltransferase in Relation to Lomustine Cytotoxicity in Canine Lymphoma Cell Lines
210	Clair Oliveira	Estrogen and Progesterone Receptors Expression in Canine Vaginal Leiomyoma
211	Krysta Deitz	Secreted Frizzled-Related Protein 2 Expression in Canine Thyroid Tumors
212	Jason Kidd	Determination of L-Asparaginase Antibody Titers in Dogs Being Treated for Lymphoma ***(VCS Award Winner)***

SMALL ANIMAL – NEUROLOGY

213	Karen Vernau	Dog Leukocyte Antigen Class II Association in Chihuahuas With Necrotizing Meningoencephalitis
214	Paula Martin-Vaquero	Magnetic Resonance Spectroscopy of the Canine Brain at 3.0 T and 7.0 T
215	Ryan Gallagher	Relationship of Magnetic Resonance Imaging to Surgical Findings in Dogs with Thoracolumbar Intervertebral Disc Disease
216	Fred Wininger	Magnetic Resonance Imaging of Plasma Cell Tumors in the Spinal Column of 10 Dogs
217	Steven De decker	Radiographic Vertebral Canal and Body Ratios in Doberman Pinschers With and Without Clinical Signs of Disk Associated Wobbler Syndrome (DAWS)
218	Jacqueline Johnson	Micromorphometry of the Canine Cervical Intervertebral Discs
219	**Stacy Dillard**	**In Vitro Biomechanical Comparison of Internal Fixation Techniques on the Canine Lumbar Spine**
220	Roberto Poma	Evaluation of Short-Term Electroencephalography in Dogs With Idiopathic Epilepsy
221	Roberto Poma	Long-Term Video-Electroencephalography in a Canine Epilepsy Monitoring Unit
222	Malte Schwartz	Risk Factors for Epileptic Seizures in Dogs With Intracranial Neoplasia
223	**Joseph Eagleson**	**Pharmacokinetics of a Novel Intranasal Midazolam Gel in Dogs**
224	**David Brewer**	**Zonisamide Therapy for Refractory Seizures in Five Cats With Presumptive Idiopathic Epilepsy**
225	**Michaela Cautela**	**Pharmacokinetics of Oral Pregabalin in Cats After Single Dose Administration**
226	Ryan Taggart	Evaluation of Cerebrospinal Fluid Lactate and Glucose Levels in Acute Canine Intervertebral Disc Disease
227	Cristian de la Fuente	Fibrinolytic Activity in Cerebrospinal Fluid (CSF) of Dogs With Different CNS Disorders
228	Christopher Mariani	Cytokine and Chemokine Levels in Serum and Cerebrospinal Fluid of Dogs With Inflammatory Central Nervous System Disease
229	Lindsay Boozer	Evaluation of Matrix Metalloproteinase-2 and -9 in the Cerebrospinal Fluid of Dogs With Intracranial Disease
230	Lindsay Boozer	Evaluation of Cytokine and Chemokine Levels in Canine Intracranial Neoplasia
231	**Katheryn Wolfe**	**Transcranial Magnetic Motor Evoked Potentials, Somatosensory Evoked Potentials, Spinal Evoked Potentials and Brainstem Auditory Evoked Potentials in Cavalier King Charles Spaniel Dogs With and Without Syringomyelia**
232	Simon Platt	Expression of Heat Shock Proteins, Ki-67 and Vascular Endothelial Growth Factor in Canine Intracranial Gliomas
233	Jill Narak	Distribution of Adeno-associated Virus Vector Serotypes 1, 9, and 10 in the Normal Feline Brain
234	Karen Vernau	Is Alaskan Husky Encephalopathy a Mitochondrial Disease?
235	Giosi Farace	Canine N-Terminal Prohormone Brain Natriuretic Peptide Concentration In Dogs Less Than One Year Old
236	Adam Schneider	Evaluation of a 14 and 16 Gauge Side-cutting Biopsy Needle and Four Different Aspiration Pressures Used for Obtaining Brain Tissue in Dogs
237	Nadia Shihab	Surgical Management of Hydrocephalus by Ventriculoperitoneal Shunt Placement in Dogs
238	Nozomi Takahashi	The Time Course of Plasma Phosphorylated Neurofilament NF-H Levels in Rodent and Canine Spinal Cord Injury Models
239	**Lauren Talarico**	**Acute Polyradiculoneuritis in Dogs: Utility of Clinical Parameters in Predicting Outcome**
240	Audrey Muguet-Chanoit	The Cutaneous Trunci Muscle Reflex in Normal Dogs
241	Audrey Muguet-Chanoit	The Cutaneous Trunci Muscle Reflex : A Predictor of Recovery in Dogs With Acute Thoracolumbar Myelopathies

SMALL ANIMAL – ENDOCRINOLOGY

242	Jo Ann Morrison	Comparison of Models for Predicting Renal Disease Following I-131 Therapy for Feline Hyperthyroidism
243	Jo Ann Morrison	Investigation of Prognostic Factors for the Development of Renal Disease Following I-131 Therapy in Feline Hyperthyroidism
244	Rebecca Hegstad-Davies	Breed-Specific Reference Intervals for Assessing Thyroid Function in Seven Dog Breeds
245	Hanna Johansson	A Novel Sandwich ELISA for the Measurement of Insulin in Feline Serum and Plasma
246	Christiane Weingart	Measurement of Beta-Hydroxybutyrate (HB) in Cats With Nonketotic Diabetes Mellitus, Diabetic Ketosis and Ketoacidosis Using a Portable Ketone Meter
247	Viviani De Marco	A Putative Gain-of-Function Mutation in CRHR1 Gene Causing ACTH-Dependent Hypercortisolism in a Poodle Dog
248	Viviani De Marco	Ultrasonographic Adrenal Glands Thickness Measurement in Dogs With Pituitary-Dependent Hyperadrenocorticism in Comparison With Normal Dogs Matched by Weight Body
249	Viviani De Marco	Prevalence of Metabolic Syndrome in Dogs With Pituitary-Dependent Hyperadrenocorticism

SMALL ANIMAL – GASTROENTEROLOGY

250	Ingrid Mõistus	Effect of Excess Dietary Protein and Carbohydrate on the Intestinal Microbiota in Healthy Beagles as Analyzed By Guanine-Plus-Cytosine-Based Profiling
251	Olivier Dossin	Effect of Omeprazole Treatment on Gastric and Duodenal Bacterial Populations in Healthy Dogs
252	Melanie Craven	*Wolinella succinogenes* and Not *Helicobacter* spp. are the Predominant Helicobacteraceae in the Canine Oral Cavity
253	Jan Suchodolski	Pyrosequencing Reveals a Diverse Fungal Microbiota in Canine and Feline Fecal Samples

254	Jan Suchodolski	Molecular Assay for the Detection of *Campylobacter* spp. in Canine and Feline Fecal Samples
255	Jose Garcia-Mazcorro	Predictors for Fecal Colonization of Probiotic Lactic Acid Bacteria in Healthy Cats and Dogs
256	Nora Berghoff	Intestinal Microbiota In Racing Alaskan Sled Dogs And Working Retrievers Before And After Exercise
257	Panagiotis Xenoulis	Intestinal *Tritrichomonas foetus* Infection in Cats: A Retrospective Study of 104 Cases
258	Michael Lappin	Assessment of Risk for Diarrhea in Young Dogs After *Giardia* Vaccine Administration
259	**Katherine Jones**	**Incidence of Bacteremia Following Gastric and Duodenal Endoscopic Biopsies Before and After Treatment with Omeprazole in Healthy Dogs**
260	Nora Berghoff	Fecal N-Methylhistamine Concentrations In Racing Alaskan Sled Dogs and Working Retrievers
261	Niels Grützner	Association Study in Cobalamin Deficient Chinese Shar Peis With an Increased Methylmalonic Acid Concentration
262	Romy Heilmann	A Method for the Quantification of Serum and Fecal Canine S100A12
263	Panagiotis Xenoulis	Serum Spec cPL® Concentrations in *Leishmania infantum*-Infected Dogs Treated With Meglumine Antimonate
264	Niels Grützner	Analytical Validation of a Radioimmunoassay for the Measurement of Canine Pancreatic Lipase Immunoreactivity in Chyme and Fecal Extracts From Dogs
265	Micah Bishop	Association Between Breed and Increased Serum Spec cPL® Concentrations in Dogs
266	Isabelle Ruhnke	Analytical Validation of a Commercially Available Immunoassay for Serum Gastrin Concentration in Cats
267	Nicole Luckschander	Flow Cytometrical Analysis of Canine Intestinal Intraepithelial Lymphocytes in Dogs With Chronic Enteropathies
268	Matthew Gaunt	Constipation Resolution With Administration of Polyethylene-Glycol Solution in Cats
269	Jacqueline Whittemore	Development of a Veterinary Flexible Endoscopic Training Device

SMALL ANIMAL – HEMATOLOGY

270	Katie Boes	Evaluation of Canine and Feline Hemograms Using the IDEXX ProCyte DX Analyzer Compared to the Abbott CellDyn 3700
271	Katie Boes	Comparison of Canine and Feline Manual Hematologic Determinations With Results from the IDEXX ProCyte DX Analyzer
272	Jennifer Owen	Evaluation of Canine and Feline Hemograms Using the IDEXX ProCyte DX Analyzer Compared to Manual Differential Counts and the ADVIA 120
273	Amelia Mellett	Prospective Evaluation of Clopidogrel, Ultralow-Dose Aspirin, or Both in Dogs With Primary Immune-Mediated Hemolytic Anemia
274	**Sarah Flint**	**Thromboelastography in Normal Dogs Treated With Prednisone, Aspirin and Prednisone and Aspirin Combined**
275	Curtis Cathcart	Treatment With Acetylsalicylic Acid or Meloxicam does not Affect Serotonin Secretion from Feline Platelets
276	Alice Defarges	Drill-Assisted 15-gauge Core Bone Marrow Biopsy in Clinically Normal Small Dogs
277	Maria Iazbik	Prevalence of Dog Erythrocyte Antigens in Retired Racing Greyhounds

SMALL ANIMAL – HEPATOLOGY

278	Meghan Respess	Portal Vein Thrombosis in Dogs
279	Barrak Pressler	Hepatic and Biliary Trace Element Concentrations in Healthy Dogs versus Dogs with Hepatopathies

SMALL ANIMAL – IMMUNOLOGY

280	Jonathan Fogle	CD4+CD25+ T Regulatory Cells from FIV Infected Cats Induce a Unique Anergic Phenotype in CD8+ Targets
281	Jacqueline Whittemore	Association Between Feline Antibody Responses to Crandell Rees Feline Kidney (CRFK) Cell Lysates, Alpha-Enolase, and Annexin A2 and Biochemical Abnormalities in 1,477 Privately-Owned Cats
282	Craig Webb	The Use of Flow Cytometry to Identify the Presence of Anti-Erythrocyte Antibodies in Anemic Cats

SMALL ANIMAL – INFECTIOUS DISEASE

283	**Jeanne Ficociello**	**Detection of *Bartonella henselae* IgM in Serum of Naturally Exposed Cats**
284	Sukullaya Assarasakorn	Prevalence of *Bartonella* Species, Haemoplasma Species, and *Rickettsia felis* DNA in Blood and Fleas of Cats in Bangkok, Thailand
285	Michael Lappin	Prevalence of *Bartonella* spp. DNA in Conjunctival Cells Collected from Shelter Cats With and Without Conjunctivitis
286	**Cristina Perez**	**Diversity of *Bartonella* Species in 62 Dogs Using the BAPGM Pre-Enrichment Culture Method**
287	**Christina Bradbury**	**Survival of *Bartonella henselae* in the Blood of Cats Used for Transfusion**
288	Tae-ho Kim	Recombinant *Bartonella* Adhesion a Stimulates Inflammatory Response and Angiogenesis in Human Umbilical Vein Endothelial Cells
289	Masato Sakuma	Molecular Survey of Possible Atovaquone-Resistant Variants of *Babesia gibsoni* in Dogs in Japan
290	JuiMing Lin	CD4+ and CD8+ T-Lymphocyte Measurements for Feline Immunodeficiency Virus (FIV) Positive Cats
291	Thomas O'Connor, Jr.	Microtiter Plate Format Elisa to Differentiate Sera from Feline Immunodeficiency Virus (FIV) Infected and Vaccinated Cats
292	Archivaldo Reche, Jr.	Occurrence of Chronic Kidney Disease in Cats Naturally Infected With Immunodeficiency Virus
293	Archivaldo Reche, Jr.	Adrenal Stimulation and CD4+ T Lymphocyte Subsets in Feline Immunodeficiency Virus Naturally Infected Cats
294	Mary Marcondes	Absence of Association Between *Leishmania* sp., FIV and FeLV Infections, in Cats From an Endemic Area for Visceral Leishmaniasis in Brazil

295	Mary Marcondes	Evaluation of the Skeletal Musculature of Dogs With Visceral Leishmaniasis by Means of Electromyography
296	Mary Marcondes	Histopathological Evaluation of the Skeletal Muscle of Dogs With Visceral Leishmaniasis
297	Alexandre Daniel	Flow-Cytometric Analysis of Peripheral Blood T Lymphocyte Subsets CD4+, CD8+ and CD4:CD8 Ratio in Maine Coon Cats With Chronic Gingivitis
298	**Kelly McCord**	**Methicillin-Resistant *Staphylococcus spp.* Contamination of Stethoscopes in a Small Animal Veterinary Teaching Hospital**
299	Joseph Blondeau	Killing of *Staphylococcus pseudintermedius* by Enrofloxacin and Its Active Metabolite Ciprofloxacin Using Clinically Relevant Drug Concentrations
300	Michael Lappin	Prevalence of Selected Zoonotic and Vector-Borne Agents in Dogs and Cats on the Pine Ridge Reservation
301	Rebecca Ruch-Gallie	Feline Herpesvirus-1 and Feline Calicivirus Molecular Diagnostic Assay Results in Adult Cats After Booster Vaccination With Modified Live Products
302	Jessica Midence	Assessing the Effect of Recent *Leptospira* Vaccination on Whole Blood Quantitative PCR Testing in Dogs
303	Alyssa Chandler	Assessing Renal Colonization in Dogs 15 Months after Receiving a Multi-Serovar Bacterin Based Vaccine, vs Unvaccinated Dogs, All Experimentally Infected With *Leptospira kirschneri* Serovar Grippotyphosa
304	Yasuyuki Endo	A Nation-Wide Molecular Epidemiological Survey of Feline Hemotropic Mycoplasma Infections in Japan
305	Jo Ann Morrison	Retrospective Investigation of Survival Factors in Canine Disseminated Histoplasmosis
306	Mellora Sharman	Distribution of Clotrimazole and Enilconazole in the Frontal Sinuses and Nasal Cavity of Dogs With Sinonasal Aspergillosis
307	Ramaswamy Chandrashekar	Canine Exposure to *Borrelia burgdorferi* and *Anaplasma phagocytophilum* in an Emergent Area: Use of a Multi-Target Elisa at Two Geographic Scales
308	Brett Stillman	Performance Comparison of Species-Specific Peptide-Based Assays With Immunofluorescence Assays for Detection of Canine Antibodies to *Anaplasma* and *Ehrlichia* spp.
309	Elise Gingrich	Common Enteric Pathogens in Cats Before and After Placement in an Animal Shelter
310	Andrea Wang	Prevalence of *Giardia* spp. and *Cryptosporidium* spp. in Veterinary Student Owned Dogs That Frequent Dog Parks
311	Shawn Bybee	Effect of *Enterococcus Faecium* SF68 Supplementation on Diarrhea in Cats Housed in a Northern Colorado Animal Shelter
312	Eran Dvir	C-Reactive Protein Concentration in Benign and Malignant Canine Spirocercosis

SMALL ANIMAL – NEPHROLOGY/UROLOGY

313	**Michael Zaid**	**Feline Ureteral Strictures: 10 Cases (2007-2009)**
314	**Jessica Clemans**	**Development of a High Performance Liquid Chromatography Serum Vitamin B6 Assay in Canines With Calcium Oxalate Urolithiasis**
315	Jessica Quimby	The Pharmacokinetics of Mirtazapine in Cats With Chronic Kidney Disease
316	Miki Nishida	Glomerular Filtration Rates Measured Using Pharmacokinetic Analysis of Blood Inulin in Healthy Cats
317	Miki Nishida	Glomerular Filtration Rates Measured Using Pharmacokinetic Analysis of Blood Inulin in Healthy Dogs
318	Bernard Schmidt	Longitudinal Follow Up of Cats With Reduced Renal Mass
319	Ryan Brock	Comparison of Urine Sodium Dodecyl Sulfate-Polyacrylamide Gel Electrophoresis (SDS-PAGE) With Renal Histologic Findings in Dogs with Renal Disease
320	Shiguang Yu	Age Associated Diminishing Urine Concentrating Ability in Apparently Healthy Cats
321	Adam O'Connell	Comparative Clinical Evaluation of BD Vacutainer™ Plus Urinalysis Preservative Tube and Plastic Tube for Urinalysis in Dogs and Cats
322	Michael Wood	Immunologic Testing of Dog Urine Requires Neutralization of Matrix Effects To Allow for Accurate Measurement of IL-6 Concentrations
323	Mary Nabity	Surface Enhanced Laser Desorption Ionization (SELDI) Analysis of Urine From Male Dogs in Early Stages of Progressive Tubulointerstitial Injury Due to X-linked Hereditary Nephropathy
324	Kayoko Harada	New Procedure for Implanting Perioneal Dialysis Catheters in Small Animals

SMALL ANIMAL – NUTRITION/METABOLISM

325	Nolan Frantz	Effect of a Test Food on Whole Blood Gene Expression in Cats With Appendicular Degenerative Joint Disease
326	Nolan Frantz	Effect of Prescription Diet® Canine J/D™ on Whole Blood Gene Expression in Dogs With Osteoarthritis
327	Dale Fritsch	An Open-Label, Prospective Study Evaluating the Response to Feeding a Veterinary Therapeutic Diet in Cats with Degenerative Joint Disease
328	Dale Fritsch	Improvement of Clinical Signs of Osteoarthritis in Cats by Dietary Intervention
329	Nolan Frantz	Effect of Prescription Diet® Canine B/D™on Inflammatory Biomarkers and Gene Expression Changes in Geriatric Dogs
330	Sharon Center	Lean Body Mass (LBM) Normalizes Daily Resting Energy Expenditure (REE) in Healthy Adult Cats (n=57) of Varied Body Conditions
331	Steve Zicker	Lipoic Acid Improves GSH:GSSG Ratio in Dogs Fed Lipoic Acid in Dry Dog Food
332	Lynda Melendez	Inducing Dietary Ketosis is Safe and Effective in Providing Weight Loss in Cats
333	Shiguang Yu	Body Weight is a Major Predictor of Body Composition in Grown Cats
334	Nolan Frantz	The Effect of Dietary Crude Protein and Lysine:Calorie Ratio on Body Composition and Kidney Measures in Healthy Adult and Geriatric Cats

335	Lynda Melendez	Maintenance of Lean Body Mass in Senior Cats Fed a Low Protein Therapeutic Renal Food
336	**Brianna Backlund**	**Effects of the Level of Dietary Protein on Renal Indices in Normal Adult Cats**
337	Dennis Jewell	Neutrophil Cell Killing Percent and Neutrophil-Related Gene Expression Decline With Age in Dogs
338	Dennis Jewell	Protein Concentrations Consistent With Controlled Protein Renal Foods Maintains Lean Body Mass in Senior Dogs
339	Caroline Horner	Effectivness of Subcutaneous Fluid Administration on the Reduction of Urolithiasis Risk as Measured by Relative Supersaturation (RSS) in Healthy Cats

SMALL ANIMAL – PHARMACOLOGY

340	Butch KuKanich	Pharmacokinetics of Hydrocodone and Its Metabolite Hydromorphone After Oral Hydrocodone Administration to Dogs
341	Jérôme Leemans	Characterization of In Vitro Relaxant Effects of β2-adrenoceptor Agonists on Isolated Cat Bronchi
342	Christophe Hugnet	Emodepside Sensitivity in Australian Shepherd is Associated With Deletion Mutation of the MDR1 Gene
343	Michael Lappin	Systemic Effects of a Commerical Preparation of Chondrotin Sulfate, Hyaluronic Acid and N-acetyl-D-glucosamine When Administered
344	Koya Fukunaga	Effect of St John's Wort on Pharmacokinetics of Cyclosporine and Cytochrome P450 Gene Expression in Dogs

SMALL ANIMAL – RESPIRATORY

345	Marta Planellas	Histological Properties of the Soft Palate Tissue: Comparison Between Human and Dogs With Upper Airway Obstruction
346	**Christine Venema**	**Histopathologic and Morphometric Evaluation of the Nasal Airways of Cats With Experimentally-Induced Asthma**

SMALL ANIMAL – OTHER

347	Astrid Bigio	Serum NT Pro-CNP Concentration Differentiates Sepsis From SIRS In Dogs
348	Ricardo Duarte	Prognostic Model for Critically Ill Dogs
349	Hanna Plickert	Evaluation of a Canine C-Reactive Protein (CRP) Point-of-Care Test
350	Elizabeth Spangler	Monitoring Hemostasis in Dogs Using Thromboelastography (TEG): Stability of TEG Values Measured in Healthy Dogs Over Three Consecutive Days

EQUINE

351	JoAnn Slack	Survey of Cardiac Arrhythmias and Myocardial Injury in Standardbred Racehorses
352	Adrienne Viljoen	Changes in Serum Cardiac Troponin I Concentration After Semen Collection in Stallions
353	Cristobal Navas de Solis	Severe Hemorrhage and Myocardial Injury in Horses
354	Kate Borer	Seasonal Influence on Insulin and Cortisol Results from Overnight Dexamethasone Suppression Tests (DST) in Normal and Previously Laminitic Ponies
355	Nicole du Toit	Determination of Serum Insulin and Insulin Resistance in Clinically Normal Donkeys and Donkeys With a History of Laminitis (Obese and Non-Obese)
356	Amy Schuver	Use of an Oral Sugar Test to Assess Insulin Sensitivity in Healthy and Insulin-Resistant Horses
357	Kelly Chameroy	Effects of Levothyroxine Sodium on Body Condition, Blood Measures of Metabolic Status, and Glucose Dynamics in Horses With Equine Metabolic Syndrome (EMS)
358	Kelly Chameroy	Validation of an Enzyme-Linked Immunosorbent Assay (ELISA) for Measurement of Active Glucagon-Like Peptide 1 (GLP-1) in Equine Plasma
359	Kelly Chameroy	Plasma Glucagon-Like Peptide 1 (GLP-1) Concentrations in Response to an Oral Sugar Test in Healthy and Insulin-Resistant Horses
360	Nichol Schultz	Assessment of Intramyofiber Lipid Content in Horses
361	Lais Costa	Variability of Nutrient Composition of Grass Hay Forage Fed to Horses in New England
362	Anne Wooldridge	Characterization of Adiponectin Protein Complexes in the Horse
363	Anne Wooldridge	High Molecular Weight Adiponectin is Reduced in Horses With Obesity and Inflammatory Disease
364	Teresa Burns	Characterization of Adipose Tissue Macrophage Infiltration in Insulin-Resistant and Insulin-Sensitive Light Breed Horses
365	Harold Schott, II	Urinalysis and Urine Culture Results in Aged Mares With and Without Pituitary Pars Intermedia Dysfunction
366	Ben Buchanan	Effects of Season on Adrenocorticotropin Hormone (ACTH) Concentrations in a Group of Horses Located in the State of Texas
367	Luciana Bergamasco	Electroencephalography Assessment in Adult, Geriatric and Geriatric Horses With Suspected Pituitary Pars Intermedia Dysfunction: A Preliminary Study
368	Alexandra Burton	The Molecular Epidemiology of Giardia Shedding in Horses and Alpacas in New York State
369	Alexandra Burton	Identification of the Cryptosporidium Horse Genotype from Foals in New York State
370	Jenifer Gold	Sequential Plasma Concentrations of Adrenocorticotropin and Cortisol in Term Septic Neonatal Foals Compared to Healthy Non-Septic Foals
371	Kelsey Hart	Daily Endogenous Cortisol Production and Hydrocortisone Pharamcokinetics in Healthy Neonatal Foals and Adult Horses
372	Melinda Himler	Thyroid Hormones in Healthy, Sick Non-Septic and Septic Foals
373	Nicola Pusterla	Comparison of Feces Versus Rectal Swabs for the Molecular Detection of *Lawsonia intracellularis* in Foals With Equine Proliferative Enteropathy
374	Nicola Pusterla	Oral Infection of Weanling Foals With an Equine Isolate of *Lawsonia intracellularis*, Agent of Equine Proliferative Enteropathy
375	Nicola Pusterla	Molecular Investigation of the Viral Kinetics of Equine Herpesvirus-1 in Blood and Nasal Secretions of Horses Following Corticosteroid-Induced Recrudescence of Latent Infection
376	Niamh Collins	Two-Dimensional and M-Mode Echocardiographic Findings in Healthy Thoroughbred Foals
377	Niamh Collins	Serum Triglyceride Concentrations at Admission in Critically Ill Foals Less Than Two Days of Age

378	Joan Carrick	The Incidence of Placentitis in the Dams of Foals Admitted to an Equine Neonatal Intensive Care Unit
379	Pamela Wilkins	The Effect of Intravenous Plasma Administration on Fibrinogen Concentration in Sick Neonatal Ffoals
380	Pamela Wilkins	Preliminary Results From a Prospective Multi-Center Study of the Association of Lactate Concentration With Survival in Sick Neonatal Foals
381	Pamela Wilkins	Arterial Blood Gas Parameters and Lactate Concentration in Normal Awake Standing Horses Receiving Intravenous Phenylephrine Infusions
382	Pamela Wilkins	Blood Gas Parameters in Normal Horses Receiving Intravenous Sodium Lactate Infusion
383	Lutz Goehring	EHV-1 Infection of Central Nervous System Endothelial Cell: Is PBMC Adhesion Necessary for Endothelial Cell Infection?
384	Allen Page	Use of Serial Serological and Fecal PCR Testing for Targeted Treatment of Weanling Foals on a Farm With Endemic *Lawsonia intracellularis*
385	Imogen Johns	The Normal Conjunctival Bacterial and Fungal Flora of Horses in the United Kingdom
386	Imogen Johns	Identification and Persistance of Antimicrobial Resistance in Fecal *Escherichia coli* in Horses Treated With Antimicrobials: Preliminary Findings
387	Steeve Giguere	Effects of a Comercially Available Immunostimulant on Cytokine Secretion and Prevalence of Pneumonia on a Farm With Endemic Infections Caused by *Rhodococcus equi*
388	Brandy Burgess	Nasal Shedding of Equine Herpesvirus-1 from Horses in an Outbreak of Equine Herpes Myeloencephalopathy in Western Canada
389	Gillian Perkins	RNA Interference Does Not Protect Against Experimental Equid Herpesvirus Type 1 (EHV-1) Infection
390	Veronique Lacombe	Classification of Epileptic Seizures in Horses: An Epidemiologic Study of 104 Cases (1988-2009)
391	Jérôme Leemans	Enzymatic and Oxidative Markers in Synovial Fluid and Chondrocyte Cultures
392	Teresa Burns	Role of Nitrosative Stress in the Pathophysiology of Experimentally-Induced Equine Laminitis: A Comparison of Two Models
393	Changaram Venugopal	Induction of Insulin Resistance in Equine Digital Vessel Rings
394	**Jamie Wearn**	**Pharmacokinetics of Pioglitazone in Horses After Multiple Dose Oral Administration**
395	Grace Forbes	Effects of Hyperimmune Equine Plasma on Clinical and Cellular Responses in a Low-Dose Endotoxemia Model in Horses
396	Margaret Wilson	Lidocaine Does Not Prevent Neutrophil Influx or Attenuate Lung Function in Heaves-Affected Horses
397	Sabrina Cummings	Effects of Cold Air Exercise on the Innate Immune Function of Equine Airway Leukocytes
398	Alicia Cattaneo	Endocrine Response to Resuscitation Solutions in Healthy Horses: 7.2% NaCl And 6% Hetastarch
399	Frank Andrews	Role of Cellular Sodium Transport in Nonglandular Equine Gastric Ulcer Disease
400	Britta Leise	Inflammatory Gene Expression in the Liver, Lung, and Kidney in the Carbohydrate Overload Model of Equine Laminitis
401	Nicholas Frank	Assessment of Two Experimental Models for Prolonged (48 Hours) Endotoxemia in Horses
402	Grace Forbes	Activin A in Plasma From Horses With Acute Gastrointestinal Disease: Biomarker Comparison With TNFa and Serum Amyloid A
403	Stephanie Smith	Results of Thromboelastometry Following Recalcification of Citrated Equine Whole Blood Are Strongly Influenced by Sample Storage
404	Lowella Heinecke	Inhibition of Prostaglandin E2 Production in Equine Chondrocytes by Pentosan Polysulfate and N-Acetylglucosamine Combinations
405	Alexandre Borges	Hepcidin Gene Expression and Plasma Iron Following Intramuscular Injection of Freund's Complete Adjuvant to Horses
406	Marian Little	Equine Serological Responses to Commercial Vaccines
407	Clare Ryan	Humoral Immune Responses to a Killed Adjuvanted Vaccine. Do Foals Respond Like Adult Horses?

ABSTRACT DISCLOSURE INDEX

Abstract	Presenter	Disclosure
3	Kristine Yee	Supported by a resident training grant from IDEXX Laboratories
45	Miguel Campos	Previously presented at the 19th Annual ECVIM Congress, Porto, Portugal, September 2009
46	Audrey K. Cook	Audrey Cook has done paid consulting work for Dechra, manufacturers of Vetoryl (trilostane). Susan Longhofer is an employee of Dechra. This study was funded by a generous unrestricted gift from Dechra to Dr. Audrey Cook.
47	Khristen Carlson	Previously presented at Auburn University, Phi Zeta Presentation
47a	Stijn Niessen	Previously presented at the 19th Annual ECVIM Congress, Porto, Portugal, September 2009
95	Eva Furrow	The EIC test was developed through financial support from the AKC Canine Health Foundation. A portion of the proceeds from the test will be returned to the AKC Canine Health Foundation to further its mission to improve the health of all dogs. Drs. Mickelson, Patterson and Taylor; and Ms. Minor, are owners of the pending patent for the EIC genetic test and a portion of the proceeds go toward patent royalties.
115	Daniel Foy	Dr. Wheat and Ms. Hackett are both employed by MiraVista Diagnostics where the Blastomyces urine and serum antigen level assay was conducted. MiraVista Diagnostics provided the funding to pay for these tests.
139	Manuel Chamorro	Previously presented at the Conference of Research Workers in Animal Disease. Chicago December 7, 2009
141	Davin Ringen	Previously presented at the Mastitis Research Workers Conference, 2009; National Mastitis Council (NMC) 2010
164	Anant Radhakrishnan	Previously presented at the 6th Annual Veterinary Endoscopy Society, Cancun, Mexico, March 2009
191	Takashi Ebisawa	Part of this study was supported by Monolis, Tokyo, Japan.
212	Jason Kidd	Previously presented at the VCS Annual Conference, Austin, Texas, October 2009
221	Roberto Poma	Previously presented at the American Epilepsy Society
255	Jose F. Garcia-Mazcorro	Partial funding for this study was provided by Nutramax Laboratories Inc.
265	Micah A. Bishop	Employye of the Gastrointestinal Laboratory where Spec cPL's are run.
270	Katie Mae Boes	Dr. Alan Rebar is a consultant for IDEXX Laboratories, Inc.
271	Katie Mae Boes	Dr. Alan Rebar is a consultant for IDEXX Laboratories, Inc.
272	Jennifer L. Owen	This research was funded by a grant from IDEXX.
280	Jonathan Fogle	Previously presented at the Keystone Symposia, Santa Fe, NM January, 2010
316	Miki Nishida	A part of this study was supported by FUJIYAKUHIN Co.Ltd., Saitama, Japan.
317	Miki Nishida	A part of this study was supported by FUJIYAKUHIN Co.Ltd., Saitama, Japan.
318	Bernard Hubert Schmidt	Bernard Schmidt is an employee of Bayer Animal Health GmbH which sponsored the studies from which the data reported were derived. Martin Murphy is an employee of Charles River Laboratories which was responsible for the performance of those studies.
327	Dale Allen Fritsch	This study was sponsored by Hill's Pet Nutrition, Inc. All authors are current or former employees of Hill's Pet Nutrition, Inc.
327	Dale Allen Fritsch	Previously presented at the 19th Annual ECVIM Congress, Porto, Portugal, September 2009
328	Dale Allen Fritsch	This study was sponsored by Hill's Pet Nutrition, Inc. All authors are or were previous employees of Hill's Pet Nutrition, Inc.
328	Dale Allen Fritsch	Previously presented at the 19th Annual ECVIM Congress, Porto, Portugal, September 2009
331	Steve Zicker	Authors are employees of Hill's Pet Nutrition, Inc.
350	Elizabeth Spangler	Previously presented at the Merck-Merial NIH National Veterinary Scholars Symposium, Raleigh, NC, August 2009
357	Kelly Chameroy	Partially funded by Lloyd, Inc., the manufacturer of ThyroL
387	Steeve Giguere	Consultant for Pfizer Animal Health on other unrelated products.
404	Lowella Fortuno Heinecke	Employee of Nutramax Laboratories
406	Dr. Marian G. Little	Two of the investigators are employees of Boehringer, who sponsored this research.

ABSTRACT #1

CYTOKINE AND MATRIX METALLOPROTEINASE EXPRESSION IN BLOOD SAMPLES OF DOGS WITH CONGESTIVE HEART FAILURE. Sonja Fonfara, Simon Tew, Peter Clegg, Joanna Dukes-McEwan. Faculty of Veterinary Science, University of Liverpool, Neston, UK.

Progression of acquired cardiac diseases differs between individual dogs and breeds. Myocardial extracellular matrix (ECM) remodelling is known to play an important role in the progression of cardiac diseases and pathogenesis of heart failure (CHF), with inflammation as an important co-contributor. However, the underlying pathophysiological mechanisms are still poorly understood.

Cytokine and ECM metalloproteinase-gene expression in blood samples of dogs presented in CHF ($n = 10$) and in dogs with no evidence of cardiac disease ($n = 10$) were analysed.

mRNA of the house-keeping gene GAPDH, interleukin (IL) 1, -2, -4, -8, -10, tumour necrosis factor α (TNF), interferon γ (IFN), transforming growths factor (TGF) β1 and β2 and the matrix metalloproteinases (MMP) 1, 2, 3, 9 and tissue inhibitors of metalloproteinase (TIMP) 1, 2, 3, 4 was analysed using quantitative RT-PCR.

Blood from dogs with CHF had increased levels of proinflammatory cytokines (IL-1, -2, -8, IFN), MMP1, -3 and TIMP3, whereas TNF, IL-10, TGFβ1, TIMP1 and TIMP2 were higher in dogs with no cardiac disease.

These results suggest an inflammatory state and evidence of myocardial remodelling in dogs with CHF, whereas dogs without cardiac disease revealed predominance of cardio-protective anti-inflammatory cytokines and TIMPs. Interestingly, TNF, known to be important in cardiac remodelling, was increased in normal dogs ($p = 0.019$). TIMP 3, a TNF inhibitor and regulator of resident fibroblasts, was significantly increased in dogs with CHF ($p = 0.049$), which suggests possible cardioprotective function.

ABSTRACT #2

CARDIAC TROPONIN-I CONCENTRATION IS ELEVATED PRE AND POST-PACING IN DOGS WITH BRADYARRHTHMIAS: IS MYOCARDITIS A POTENTIAL ETIOLOGY? DJ Trafny, MA Oyama, CA Reynolds, GE Singletary, C Wormser, GD Peddle. University of Pennsylvania, Philadelphia, PA.

Bradyarrhythmias requiring permanent pacemaker implantation usually affect geriatric dogs and are often due to high-grade to complete atrioventricular block (AVB) or sick sinus syndrome (SSS). Degenerative change has been previously postulated as the likely cause of such arrhythmias, however, myocarditis has been implicated as a cause of AVB in dogs. Plasma cardiac troponin I concentration [cTnI] is a sensitive and specific marker for myocardial injury. We sought to quantify [cTnI] in dogs with bradyarrhythmias before and after artificial pacing and to correlate [cTnI] with ECG diagnosis, HR, echo parameters, serology, and outcome. Medical records were reviewed between 2006 and 2009, and [cTnI] before and after pacemaker implantation was available in 14 dogs. Additional parameters, including heart rate, serology, presence of premature beats, echo measurements, and pacing modality were recorded. The ECG diagnosis included complete AVB ($n = 9$), SSS ($n = 3$), high grade 2nd degree AVB ($n = 1$), and atrial standstill ($n = 1$). Mean [cTnI] was elevated both pre- and post-pacing, and was significantly ($P = 0.037$) higher pre-pacing (1.99 ± 2.86 mg/dL) versus post-pacing (0.24 ± 0.29 mg/dL). [cTnI] decreased in 13 of 14 (93%) dogs post-pacing, however, in only 5 of 14 (36%) patients did the [cTnI] return to the normal reference range. Seven of 14 (50%) dogs underwent serologic testing for vector-borne diseases, including *Ehrlichia canis*, *Anaplasma phagocytophilum*, *Neospora caninum*, *Rickettsia rickettsii*, *Toxoplasma gondii*, *Borrelia burgdorferi*, and *Bartonella* spp. Two of 7 dogs (29%) yielded weak positive titers for *Borrelia burgdorferi*. Four of 5 dogs (80%) yielded strong serum titers for *Bartonella* spp. Of the 4 dogs with positive *Bartonella* titers, 3 had complete AVB and 1 had atrial standstill. Six dogs underwent recheck echocardiography at the time of their post-pacing cTnI sampling and of these, 4 of 6 (67%) were found to have progressive LV enlargement and myocardial failure as compared to their pre-pacing examination. The pre-pacing [cTnI] of these 4 dogs (0.36, 4.42, 3.59 and 10.66 ng/ml) were amongst the highest pre-pacing values obtained. Nine of 14 dogs (64%) had died or were euthanized at the time of abstract preparation (median time to death or euthanasia, 263 days; range, 47–1163 days). Baseline mean [cTnI] was higher in dogs that died of cardiac disease but this difference was not significant. [cTnI] is elevated in cases of symptomatic bradyarrhythmias and does not uniformly return to normal following pacemaker implantation. Myocarditis (especially from *Bartonella* spp.) may be an important cause of AVB and myocarditis in dogs. Additional prospective studies investigating *Bartonella* spp. and other causes of myocarditis, as well as the correlation of [cTnI] to post-pacing LV dysfunction and outcome are needed.

ABSTRACT #3

DIAGNOSTIC TEST PARAMETERS IN CATS WITH HEART DISEASE AND THEIR CORRELATION WITH NT-PROANP, NT-PROBNP AND TROPONIN I MEASUREMENTS. Stephen Ettinger, DVM*[1], Kristine Yee, DVM[1], Andrew Beardow, BVS[2], Giosi Farace, PhD[2], Justin Allen, DVM[1], Kirstie Barrett, DVM[1]. [1]California Animal Hospital Veterinary Specialty Group, Los Angeles, Ca. 90025. [2]Idexx Laboratories, Westbrook, ME 04092.

421 cats (727 samples) were presented to the California Animal Hospital Veterinary Specialty Group for evaluation of cardiac or other medical problems. All received complete cardiovascular examinations including measurement of cTnI, NT-proANP and NT-proBNP. No patient preselection occurred except for recruitment of normal cats of all ages for the study.

All patients received a complete physical examination; laboratory analyses included CBC, blood chemistry profile, total T4 level and urinalysis. The cardiac examination included thoracic radiography, VHS measurement, 2D-Doppler echocardiography and ECG. Resting blood pressures were measured when the cat was cooperative. Cardiac biomarkers were obtained from plasma samples taken at the same time as other cardiac studies were completed.

The aim of this study is to determine the correlation of the above markers with physical parameters such as echo-doppler. Non-parametric linear regression analysis of the biomarkers with specific echo derived measurements showed that NTproBNP has a stronger correlation with LA/Ao than Troponin I and NTproANP (0.53, $p < 0.0001$ vs. 0.43, $p < 0.0001$ and 0.42, $p < 0.0001$). IVS and LVPW have similar Spearmean's ρ values with NTproBNP and Troponin I. The latter is interesting since it is a marker of muscle damage rather stretch. NTproANP, another marker of stretch, has a much weaker correlation with these two measurements.

This would suggest that the disease processes in the cat involve muscle damage and ventricular stretch. It would also suggest that NT-proANP may not be as useful a measure of heart size change.

ABSTRACT #4

EXPRESSION OF UROCORTINS IN CANINE MYOCARDIUM AND PLASMA LEVELS IN DOGS WITH CARDIAC DISEASE. G Fraga Veloso[1], DG Ohad[2], AJ Francis[1], JM Vaughan[3], DG Brownstein[4], GJ Culshaw[1], WW Vale[3], AT French[1], PM Jamieson[1,4]. [1]Royal (Dick) School of Veterinary Studies, University of Edinburgh Hospital for Small Animals, Roslin, UK. [2]Department of Clinical Sciences, The Koret School of Veterinary Medicine, The Robert H. Smith Faculty of Agriculture, Food and Environment, The Hebrew University of Jerusalem, Rehovot, Israel. [3]The Clayton Foundation Laboratories for Peptide Biology, Salk Institute for Biological Studies, La Jolla, CA. [4]Centre for Cardiovascular Science, The Queen's Medical Research Institute, Edinburgh, UK.

Urocortin peptides are the endogenous ligands for the corticotropin-releasing factor type 2 receptor (CRFR2). They have potentially important roles in cardiovascular physiology in health and disease, and show promise as therapeutics for congestive cardiac failure. Differences in the expression of urocortins and CRFR2 receptor subtype between rodent and human heart have been observed; however no studies have been performed in the dog.

To characterise urocortin and CRFR2 subtype expression and distribution in canine heart, tissue from dogs without evidence of cardiovascular disease was analysed by RT-PCR and immunohistochemistry. mRNA expression of urocortin 1 and urocortin 3 and CRFR2α was present in all heart chambers, whereas urocortin 2 was found only in right atrium, and at barely detectable levels. CRFR2β was highly expressed in atrial tissue compared to other heart tissue. Immunohistochemistry demonstrated staining for urocortin 1 and urocortin 3, but not urocortin 2, in cardiomyocytes.

To assess the potential usefulness of circulating urocortins as a marker of heart disease, plasma samples from 45 dogs with cardiac disease and 15 controls were analysed by radioimmunoassay. Both urocortin 1 (controls 1.71 (1.08/7.92) pmol/L; cardiac disease 1.59 (0.42/7.33) pmol/L) and urocortin 3 (controls 0.48 (0.48/4.01); cardiac disease 0.48 (0.48/4.76) pmol/L) were measurable (Data expressed as median (minimum/maximum). However, the presence of cardiac disease did not significantly alter the levels of either peptide.

Therefore, while urocortins and CRFR2 are expressed in canine myocardium where they may play a role in the neurohumoral response to cardiac disease or failure, they do not appear to be a sensitive biomarker of cardiac disease in our canine patient population.

ABSTRACT #5

IS THE CARDIO-RENAL-ANEMIA SYNDROME PREVALENT IN DOGS?

D.G. Ohad, J. Berkowitz, T. Bdolah-Abram. The Koret School of Veterinary Medicine, The Hebrew University of Jerusalem, Rehovot, Israel.

Heart disease and chronic kidney disease often progress to become terminal conditions, even with optimal medical care. One factor potentially common to both conditions is anemia. A vicious cycle may exist between these three conditions, each triggering or exacerbating the other. We aimed at determining the prevalence of azotemia, hypokalemia, and anemia in canine heart disease, before and after the onset of pharmacotherapy.

Physical examination, echocardiography and blood work parameters were retrospectively collected from 223 consecutive dogs with heart disease, and compared to those from 248 dogs with no cardiac or renal dysfunction.

The mean creatinine levels in the study group were lower than those in the control group, while within normal range. The prevalence of azotemic dogs among heart patients was 24.1%. Mean hematocrit levels were higher in the study group compared to the control group, while within normal range as well. The prevalence of anemia among the cardiac study group was 22.8%, surprisingly lower than that in the control group. Six out of 156 cardiac patients (3.8%) were both anemic and azotemic. Cardiac disease severity and chronic cardiac therapy did not affect hematocrit, creatinine or potassium levels.

Impaired renal function and anemia is uncommon in canine heart disease. It remains important, however, to identify these conditions so that adjustment of an appropriate treatment regimen, as well as an appropriate prognosis is facilitated. Traditional medical therapy can likely be followed as long as these parameters remain stable and within normal range.

ABSTRACT #6

ALDOSTERONE ESCAPE IN FUROSEMIDE-ACTIVATED CIRCULATING RENIN-ANGIOTENSIN-ALDOSTERONE SYSTEM (RAAS) IN NORMAL DOGS.

AC Lantis[1], CE Atkins[2], TC DeFrancesco[2], BW Keene[2]. [1]Virginia-Maryland Regional College of Veterinary Medicine, Blacksburg, VA. [2]North Carolina State University College of Veterinary Medicine, Raleigh NC.

The purpose of this study was to evaluate the ability of the ACE-inhibitor, benazepril, to suppress circulating RAAS activation in normal dogs treated with furosemide. We hypothesized that benazepril would suppress plasma ACE activity and furosemide induced RAAS activation, as evaluated by urinary aldosterone:creatinine ratio.

Twelve healthy hound dogs were used. Group1 (control) received furosemide (2 mg/kg PO, BID) for 7 days. Group 2 received furosemide (2 mg/kg PO, BID) and benazepril (1 mg/kg every 24 hours) for 7 days. To measure the response of the RAAS to the medications, plasma ACE activity and urinary aldosterone:creatinine ratio (A:Cr) were obtained on days -1, -2, 1, 3 and 7.

Urinary aldosterone secretion, as indicated by the A:Cr, increased nearly 3-fold by furosemide monotherapy. This was maintained for the 7 days of the study. The addition of benazepril to group 2 dogs successfully suppressed plasma ACE activity but did not significantly reduce urinary aldosterone excretion.

We have shown that, while benazepril successfully blunted ACE activity, it did not reduce urinary aldosterone excretion. The absence of circulating RAAS suppression accompanied by ACE activity suppression defines aldosterone escape. The discrepancy between these data and the documented benefits of benazepril likely reflects the efficacy of this ACE inhibitor in suppressing tissue RAAS rather than circulating RAAS.

ABSTRACT #7

RELATIONSHIPS BETWEEN SERUM AND URINARY ALDOSTERONE, VENTRICULAR REMODELLING AND OUTCOME IN DOGS WITH MITRAL VALVE DISEASE.

MJ Hezzell, A Boswood, J Elliott. The Royal Veterinary College, London, UK.

Congestive heart failure develops in dogs with myxomatous mitral valve disease (MMVD) secondary to volume overload, in part mediated by the renin-angiotensin-aldosterone system. We hypothesised that aldosterone would increase with left ventricular size in volume overload and be negatively associated with survival.

Aldosterone concentrations (pg/ml) in serum and urine were measured in dogs with naturally-occurring MMVD and healthy controls using a commercially-available radioimmunoassay. Urine creatinine was measured by a commercial laboratory. Dogs with MMVD were examined twice, six months apart and categorised according to the percentage change in their normalised left ventricular diastolic diameter (%LVEDDN). Dogs with %LVEDDN $> 10\%$ were classified 'progressive', and the remainder 'non-progressive'. Results are reported as [Median (IQR)]. Comparisons of continuous variables were made between groups with Mann-Whitney and Kruskal-Wallis tests, proportions were compared with the Fishers' exact test and survival times (all-cause mortality) compared by the log rank test and Cox proportional hazard ratio (HR) [95%CI]. Correlations were assessed using Spearman's correlation coefficient.

Fifty-four dogs with MMVD (16 progressive, 38 non-progressive) and 20 control dogs were studied. There was no difference in age or weight between the progressive, non-progressive and control groups. There was no difference in baseline LVEDDN between progressive [1.56 (1.38–1.88)] and non-progressive groups [1.69 (1.49–1.79)] ($P = 0.256$).

At visit one there was no difference in serum aldosterone ($P = 0.220$) between progressive [59.39 (36.15–91.41)], non-progressive [44.38 (17.88–87.23)] and control [59.88 (29.79–136.21)] groups. There was no difference in urine aldosterone: creatinine ratio (UAC) ($P = 0.117$) between progressive [0.14(0.05–0.21)], non-progressive [0.07 (0.03–0.13)] and control [0.08 (0.06–0.14)] groups. There was no relationship between serum aldosterone ($P = 0.765$) and survival. The relationship between UAC and survival tended towards significance ($P = 0.054$).

At visit two serum aldosterone was higher in the progressive [87.69(31.68–192.14)] than the non-progressive group [40.70 (17.88–112.31)] ($P = 0.048$). UAC was higher in the progressive [0.17 (0.08–0.23)] than the non-progressive group [0.09 (0.05–0.15)] ($P = 0.014$). There was no difference in proportions of dogs receiving furosemide ($P > 0.999$) or ACE inhibitors ($P > 0.999$) between groups. UAC was negatively associated with survival ($P = 0.005$), HR = 0.33 [0.14–0.74]. The relationship between serum aldosterone and survival tended towards significance ($P = 0.062$). %LVEDDN $> 10\%$ was not associated with survival ($P = 0.214$). Serum aldosterone correlated with %LVEDDN ($P = 0.024$). UAC correlated with LVEDDN ($P = 0.006$) and %LVEDDN ($P = 0.007$).

In conclusion, serum aldosterone and UAC increase during ventricular remodelling. UAC appears to be negatively associated with survival.

ABSTRACT #8

FLOW MEDIATED VASODILATION IN CANINE CHRONIC MITRAL VALVE DISEASE. I.D. Jones, V. Luis Fuentes, A. Boswood, M.J. Hezzell, W. Moonamart, J. Elliott. Royal Veterinary College, University of London. U.K.

Diminished brachial artery flow-mediated dilation (FMD) is found in human cardiac diseases of differing etiologies, and is associated with increased cardiovascular mortality. The aim of this study was to compare FMD in normal dogs and in dogs with spontaneous chronic mitral valve disease (MVD).

11 normal and 19 client-owned dogs with MVD underwent FMD measurement. Longitudinal brachial artery images and blood flow velocities were acquired using 2D and spectral Doppler ultrasound, before and immediately following 5 minutes of arterial occlusion using a cuff inflated to > 200 mmHg around the antebrachium. The percentage change in diameter (FMD) was calculated using commercial software by comparing luminal diameter post-occlusion to baseline. Normalized left ventricular end-diastolic diameter (LVEDDN) was used as a surrogate marker of disease severity by dividing left ventricular end-diastolic diameter (LVEDD) by $(\text{bodyweight})^{0.294}$.

Data are reported as [median (range)], with significance set at $p < 0.05$. FMD measurement was possible in all normal and in 11 MVD dogs. Normal dogs were younger [4.8 (2.6–6.9) years] than MVD dogs [10.7 (4.9–16.2) years], $p < 0.001$, but there was no difference between groups in bodyweight, baseline diameter, or in increase in velocity time integral during reactive hyperemia. FMD was greater in normal dogs [7.7% (0.7–18.1%] versus MVD dogs [3.5% (0.5–7.2%)], $p = 0.05$. In MVD dogs, there was a negative correlation between FMD and LVEDDN ($r = -0.651$, $p = 0.03$).

FMD was reduced in dogs with MVD and is related to LVEDDN.

ABSTRACT #9

CONGESTIVE HEART FAILURE IN DOGS IS ASSOCIATED WITH ENHANCED PLATELET-LEUKOCYTE AGGREGATES - A MARKER FOR PLATELET ACTIVATION. I Tarnow[1], LH Olsen[2], SHS Andreasen[2], SG Moesgaard[2], CE Rasmussen[2], AT Kristensen[1], T Falk[2]. [1]Departments of Small Animal Clinical Sciences and [2]Animal and Veterinary Basic Sciences, Faculty of Life Science, University of Copenhagen, Denmark.

Chronic congestive heart failure (CHF) in humans is associated with abnormal hemostasis, and changes in hemostatic biomarkers carry a poor prognosis. CHF in dogs has been associated with plasma markers of hypercoagulability, however, platelet activation markers have not been investigated in dogs with clinical signs of heart disease. We hypothesized that platelet surface activation markers are higher in dogs with CHF compared to age-matched controls without clinical signs of heart failure.

Dogs with compensated congestive heart failure ($n = 21$; age 9.9 ± 2.8 years) caused by acquired heart disease and age-matched control dogs ($n = 19$; age 9.5 ± 2.0 years) without signs of heart failure were studied in parallel on same days. Platelet surface P-selectin (CD62P binding) expression, activation of platelet surface GPIIb-IIIa (CAP1 binding), monocyte-platelet (MPAs) and neutrophil-platelet aggregates (NPAs) defined as leukocytes positive for CD61 were measured by whole blood flow cytometry as markers of platelet activation. For every dog platelet activation was assessed in unstimulated samples, and in response to exogenously added adenosine diphosphate (ADP), thrombin and a synthetic thromboxane A_2 analogue + epinephrine. Complete blood count and biochemical screening were performed in all dogs.

Dogs with CHF had higher MPAs ($p = 0.014$) and NPAs ($p = 0.021$) compared to controls (2-way analysis of variance for repeated measurements).

Table [Abstract 9]

Agonist	Controls (%MPA)	CHF (%MPA)	Controls (%NPA)	CHF (%NPA)
None	10.4 ± 4.1	15.4 ± 13.0	8.8 ± 3.6	12.7 ± 6.8
ADP 20 μM	15.3 ± 10.2	24.4 ± 19.8	12.7 ± 5.7	20.5 ± 15.0
Thrombin 2 U/mL	80.3 ± 13.8	85.4 ± 10.9	70.1 ± 15.3	75.1 ± 16.8
Thromboxane A_2 + Epi	30.5 ± 25.6	57.8 ± 27.6	22.7 ± 21.8	46.9 ± 25.3

Platelet surface P-selectin expression was higher in dogs with CHF, however, this was not statistically significant. There was no difference in leukocyte or platelet count between groups.

In conclusion, congestive heart failure in dogs is associated with increased platelet-leukocyte aggregates probably reflecting increased circulating activated platelets and platelet hyper-reactivity. Prospective longitudinal clinical trials are needed to establish if anti-platelet drugs are beneficial in dogs with CHF.

ABSTRACT #10

THROMBOELASTOGRAPHY IN DOGS WITH ASYMPTOMATIC MYXOMATOUS MITRAL VALVE DISEASE. I Tarnow[1], LH Olsen[2], SG Moesgaard[2], AU Martinsen[1], AC Birkegaard[1], AT Kristensen[1] B Wiinberg[1]. [1]Departments of Small Animal Clinical Sciences and [2]Animal and Veterinary Basic Sciences, Faculty of Life Science, University of Copenhagen, Denmark.

Changes in platelet aggregation responses and von Willebrand factor multimer levels have been reported in dogs with asymptomatic heart disease. Alterations in hemostatic pathways may be involved in the progression of canine mitral valve disease by causing microthrombosis and vessel changes in the myocardium, and it is being debated whether dogs with heart disease could benefit from antithrombotic therapy. We hypothesized that a global hypercoagulability is present in dogs with asymptomatic mitral valve disease.

The study investigated Kaolin and Tissue Factor activated thromboelastographic (TEG) tracings in 3 groups of dogs: 11 Cavalier King Charles Spaniels (CKCS) with no or minimal mitral regurgitation (MR) (age 5.0 ± 1.8 years), 14 CKCS with severe MR (regurgitant jet size by echocardiography ≥ 50%) (age 7.1 ± 1.6 years), and 8 healthy Beagles with no or minimal MR (age 7.7 ± 2.3 years). Complete blood count (ADVIA 120) and biochemical screening were performed in all dogs. Differences between groups were analyzed by two-way repeated measures ANOVA and the effect of platelet count, hematocrit and leukocyte count on TEG values was analyzed.

Data revealed no significant differences between the 3 groups in any of the 4 analyzed TEG variables (R, K, Maximum Amplitude (MA) and angle) even when corrected for platelet count. Eleven of 25 CKCS had macrothrombocytopenia with platelet counts ranging from 45 to 96 × 10^9/L, and surprisingly there was an inverse correlation between platelet count and Tissue Factor activated MA ($r = -0.52$, $p = 0.008$) in this group of dogs. No correlation was found between hematocrit and TEG values, however, leukocyte count correlated with Tissue Factor activated values: MA ($r = 0.46$, $p = 0.007$), angle ($r = 0.46$, $p = 0.007$), and K ($r = -0.47$, $p = 0.006$).

These data do not support the hypothesis that a global hypercoagulability is present in CKCS with asymptomatic myxomatous mitral valve disease. Moreover, prospective longitudinal trials are needed to demonstrate a benefit of anti-thrombotic therapy in dogs with asymptomatic heart disease. The data supports previous findings that CKCS with macrothrombocytopenia have a normal coagulation response despite low platelet counts.

ABSTRACT #11

24-HOUR ELECTROCARDIOGRAPHY IN CLINICAL HEALTHY CAVALIER KING CHARLES SPANIELS, WIRE-HAIRED DACHSHUNDS AND CAIRN TERRIERS. CE Rasmussen[1], S Vesterholm[1], TP Ludvigsen[1], SG Moesgaard[1], HD Pedersen[2], J Häggström[3], LH Olsen[1]. [1]Department of Basic Animal and Veterinary Sciences, Faculty of Life Sciences, University of Copenhagen, Frederiksberg, Denmark. [2]Zealand Pharma, Glostrup, Denmark. [3]Department of Clinical Sciences, Swedish University of Agricultural Science, Uppsala, Sweden.

Studies regarding reference values in 24-hour electrocardiography (ECG) (Holter) in small dog breeds are sparse. Most ECG studies in dogs with myxomatous mitral valve disease (MMVD) are primarily based on short time ECG recorded at rest. Therefore, the aims of this study were to establish normal values for Holter recordings in three small dog breeds and to investigate if these variables were influenced by the risk these breeds have to develop MMVD.

This study included 54 privately owned dogs: 23 Cavalier King Charles Spaniels (CKCS) (at high risk to develop MMVD), 18 wire-haired Dachshunds (wD) (at moderate risk to develop MMVD) and 13 Cairn Terriers (CT) (at low risk to develop MMVD). All dogs were between two and nine years old and clinical healthy based on history, clinical examination, echocardiography, serum biochemistry and complete blood count. All dogs were Holter recorded for 24 hours using a CF-life recorder with electrodes placed in a two-lead precordial position. A blinded semiautomatic arrhythmia analysis with manual review and editing were performed on the Holter recordings using Pathfinder digital Holter analysis system (SPACELABS Healthcare, Washington, USA). Heart rate measured standardized at the clinical examination (HRex) and all Holter parameters obtained from the arrhythmia analysis were analyzed for influence by breed, age, gender, relevant heart rate (HR) parameters and bodyweight.

CKCS had a significantly higher HRex (mean 121 ± SD 12 bpm) compared to wD (94 ± 18 bpm) ($P > 0.0001$) and CT (102 ± 16 bpm) ($P = 0.0009$). The minimum HR (HRmin) measured during Holter recording was significantly higher in CKCS (51 ± 10 bpm) compared to wD (42 ± 7 bpm) ($P = 0.001$) and CT (45 ± 7 bpm) ($P = 0.05$). In addition, CKCS had a higher mean HR (HRmean) during Holter recording (83 ± 13 bpm) compared to wD (68 ± 8 bpm) ($P > 0.0001$) and CT (75 ± 10 bpm) ($P = 0.04$). Both HRmean ($P > 0.0001$) and HRmin ($P = 0.001$) were positively correlated with HRex. Second-degree atrioventricular-block (AV) and single, sporadic, uniform ventricular premature complexes (VPC) occurred in 63.0% and 40.7% of the dogs, respectively. The occurrence of second-degree AV-block and VPC showed no breed difference.

In conclusion, HR parameters were higher in CKCS compared to wD and CT. HRmean and HRmin were correlated with HRex. However, further studies are required to clarify the possible influence of high HR in the development of MMVD in CKCS. Second-degree AV-block and single, sporadic, uniformed VPC, were common in CKCS, wD and CT but not related to breed.

ABSTRACT #12

DEVELOPMENT AND EVALUATION OF A QUALITY OF LIFE QUESTIONNAIRE FOR CATS WITH CARDIAC DISEASE. LM Freeman, JE Rush, MA Oyama, SM Cunningham, KA MacDonald, DJ Hall, DJ Trafny. Tufts Cummings School of Veterinary Medicine, N. Grafton, MA; University of Pennsylvania School of Veterinary Medicine, Philadelphia, PA; VCA-Animal Care Center of Sonoma, Rohnert Park, CA.

Quality of life (QOL) is an important aspect of the management of animals with cardiac disease. Because owners often euthanize their animals with cardiac disease for quality of life issues, this parameter can influence survival time. Quality of life also can be a key outcome measure to assess the success of medical and surgical therapy of cardiac disease. Whereas many QOL assessment tools have been developed for use in people and dogs with cardiac disease, no validated QOL assessment tools are available for use in cats with cardiac disease. The purpose of this study was to develop and evaluate a questionnaire to assess QOL for cats with cardiac disease. Items related to clinical signs of cardiac disease in cats (eg, appetite, urination habits, interactions with others) were generated and a QOL questionnaire was developed. The questionnaire included 16 questions assessing owner perceptions of the cats' clinical signs using a 0–5 scale, as well as an overall quality of life item. To evaluate the QOL questionnaire, the instrument was self-administered by owners of cats with cardiac disease between June and November, 2009. An independent measure of disease severity [International Small Animal Cardiac Health Council (ISACHC) classification of heart failure] was recorded for each cat. The questionnaire was evaluated for internal consistency, construct validity, criterion-related validity, and test-retest reliability by Spearman rank correlation. The QOL questionnaire was completed by 75 owners of cats with cardiac disease and re-administered after an interval of 1 to 3 weeks to 10 owners. Mean age of the cats was 9.1 ± 4.9 years and the study included 51 male cats and 24 female cats. The most common underlying diseases were hypertrophic cardiomyopathy (n = 56), congenital diseases (n = 9), dilated or unclassified cardiomyopathy (n = 7), restrictive cardiomyopathy (n = 2), and arrhythmogenic right ventricular cardiomyopathy (n = 1). Cats were distributed between ISACHC class 1a (n = 3), 1b (n = 32), 2 (n = 27), 3a (n = 10), and 3b (n = 3). Internal consistency was tested by evaluating the relationship between individual questions and the total QOL score, which ranged from 0.06–0.57. Cronbach's alpha was 0.90, indicating high internal consistency. Correlation between overall quality of life and the total QOL score was 0.64 ($P < 0.001$). The ISACHC class was correlated with the total QOL score ($r = 0.43$; $P < 0.001$), suggesting moderate criterion validity. Among the 10 owners who completed the questionnaire twice to assess test-retest reliability, total scores on the two tests were highly correlated ($P = 0.94$; $P = 0.001$). We conclude that the QOL questionnaire for cats has good validity and reliability, and represents a promising new tool for assessment of quality of life in cats. Prospective evaluation of this QOL questionnaire in a larger population of cats with cardiac disease, including responsiveness to change, is ongoing.

ABSTRACT #13

EVALUATION OF TYROSINASE EXPRESSION IN CANINE AND EQUINE MELANOCYTIC TUMORS. J Phillips[1], C Noltenius[1], L Lembcke[1], D Grosenbaugh[2], T Leard[2]. [1]University OF Tennessee College OF Veterinary Medicine, Knoxville, TN. [2]Merial Pharmaceuticals, Athens, GA.

DNA vaccination against tumor specific proteins is a potential strategy for cancer therapy. In order to have a successful vaccine, one must identify tumor proteins that have a restricted expression pattern (i.e. tumor only) and can facilitate an appropriate immune response while not generating immune tolerance or significant auto-immunity. To this end, a vaccine has been developed for use in dogs with melanoma that has been shown to target the tumor-specific antigen, tyrosinase. Furthermore, pending the expression pattern of this tumor-specific antigen, this vaccine may also be considered in other appropriate species such as the horse. Unfortunately, the expression of this protein in canine (and equine) melanocytic tumors has not been described. In this report we describe the tumor-specific expression levels of tyrosinase in canine and equine melanocytic tumors. We also compare these levels to those found in normal cutaneous tissues from both species. Formalin-fixed, paraffin-embedded tumor tissues were collected from 50 dogs and horses previously diagnosed with a melanocytic tumor. RNA was isolated and quantified from these tissues using standard techniques. A TaqMan RT-PCR assay was then designed to amplify the exon 1/2 boundary. Relative expression levels ($\Delta\Delta$Ct) were determined and compared to both pigmented and non-pigmented cutaneous tissues. Results demonstrate high tyrosinase expression in both pigmented and non-pigmented melanocytic tumors with insignificant levels found in the normal skin. No significant difference in tyrosinase expression was found between the different canine or equine anatomic locations of melanocytic tumors. The specificity and sensitivity of tyrosinase expression was also compared to traditional methodology of immunohistochemistry with S100 and MelanA and found to be similar; correctly classifying all amelanotic melanomas and three undifferentiated sarcomas. In summary, this methodology is highly sensitive and specific for the detection of tyrosinase expression in both fresh and fixed tumor samples and has demonstrated the tumor-specificity of this antigen.

ABSTRACT #14

EXPRESSION OF KIT IN CANINE ANAL SAC ADENOCARCINOMA USING TISSUE IMMUNOHISTOCHEMISTRY RJ Brown, SJ Newman, DC Durtschi, AK LeBlanc. University of Tennessee College of Veterinary Medicine, Knoxville, TN.

Apocrine gland adenocarcinoma is an uncommon but highly invasive and metastatic malignancy, in which hypercalcemia of malignancy occurs in 25–50% of cases and causes life-threatening complications. Despite aggressive therapy, the prognosis for these patients remains poor, especially in cases with paraneoplastic hypercalcemia. Palladia™ (toceranib phosphate) is a receptor tyrosine kinase inhibitor that targets several members of the split kinase RTK family, including VEGFR-2 (Flk-1/KDR), PDGFR, and c-Kit. These membrane receptors are important for cell cycling, apoptosis, and angiogenesis which all play roles in carcinogenesis. Anecdotally, dogs with measurable anal sac adenocarcinoma have responded to treatment with Palladia, which was an unexpected finding in the investigational portion of Palladia's field trials in veterinary oncology. The objective of this study is to evaluate archived, paraffin-embedded canine anal sac adenocarcinomas for immunohistochemical detection of c-KIT.

The University of Tennessee Department of Pathobiology histopathology archives were searched for canine anal sac adenocarcinoma tissue samples that had been submitted between January of 2001 and September of 2009. A total of eighty two formalin-fixed paraffin-wax embedded tissue samples were found and included in the study. A single anatomic pathologist reviewed the samples to confirm the diagnosis. These samples were analyzed via immunohistochemistry for c-kit. Following staining for this receptor, samples were evaluated by the same anatomic pathologist for presence of c-kit and the staining pattern. Samples were assigned a grade based on the pattern of staining with: (0) < 50% of cells positive with weak labeling, (1) < 50% of cell positive with intense labeling, (2) > 50% of cells positive with weak labeling, (3) > 50% of cells positive with intense labeling.

Two of the eight two samples stained positive for c-Kit and were designated as having Grade 1 staining. These results suggest that c-Kit may not have a role in mediating the response of anal sac adenocarcinomas to Palladia. To further determine the role of Palladia in treating this disease, a similar immunohistochemical evaluation should be performed for VEGFR2 and PDGFR.

ABSTRACT #15

IMMUNOHISTOCHEMICAL CHARACTERIZATION OF FELINE MAST CELL TUMORS C.L. Mallett[1], N.C. Northrup[1], C.F. Saba[1], C.O. Rodriguez[2], K.M. Rassnick[3], T.L. Gieger[4], M.O. Childress[5], E.W. Howerth[1]. [1]University of Georgia College of Veterinary Medicine, Athens, GA. [2]University of California-Davis School of Veterinary Medicine, Davis, CA. [3]Cornell University College of Veterinary Medicine, Ithaca, NY. [4]Louisiana State University School of Veterinary Medicine, Baton Rouge, LA. [5]Purdue University School of Veterinary Medicine, West Lafayette, IN.

Feline mast cell tumors (MCTs) have variable clinical outcomes correlated with location. While immunohistochemical characterization of canine MCTs has been reported extensively, similar studies of feline MCTs of various locations are lacking. The purpose of this study was to describe expression of histamine, serotonin, and c-KIT in feline MCTs.

Sixty-one archived biopsy samples from client-owned cats with histologically confirmed MCTs in the skin, spleen, or gastrointestinal tract were included. Antibodies for histamine and serotonin were validated in feline tissues, and a c-KIT antibody previously validated in feline tissues was used. Immunohistochemical staining was used to determine histamine, serotonin, and c-KIT expression. For histamine and serotonin, immunoreactivity was recorded by cellular location and intensity. c-KIT staining patterns were determined based on a previously established system.

Seventeen primary splenic mast cell tumors, 15 primary gastrointestinal mast cell tumors, and 29 primary cutaneous tumors were stained. Three (17.6%) of 17 splenic tumors were positive for histamine, 0 were positive for serotonin, and 4 (23.5%) were positive for c-KIT. Eight (53.3%) of 15 gastrointestinal mast cell tumors were positive for histamine, 1 (6.67%) was positive for serotonin, and 1 (6.67%) was positive for c-KIT. Eight (27.6%) of 29 cutaneous mast cell tumors were positive for histamine, 0 were positive for serotonin, and 15 (51.7%) were positive for c-KIT.

In conclusion, expression of histamine, serotonin, and c-KIT was not consistent or predictable. Only 1 tumor showed staining for serotonin, suggesting that this is not a major vasoactive amine contained in feline MCT granules. Based on moderate histamine and c-KIT staining demonstrated, antihistamines and c-KIT inhibitors may play a role in the treatment of feline mast cell tumors.

ABSTRACT #16

RADIOSENSTIVITY AND CAPACITY FOR RADIATION-INDUCED SUBLETHAL DAMAGE REPAIR OF CANINE TRANSITIONAL CELL CARCINOMA CELL LINES. SL Parfitt[1], RJ Milner[1], ME Salute[1], JP Farese[1], NJ Bacon[1], FJ Bova[2], DA Rajon[2]. [1]Department of Small Animal Clinical Sciences, College of Veterinary Medicine, University of Florida, Gainesville, FL. [2]Department of Neurosurgery, McKnight Brain Institute, University of Florida, Gainesville, FL.

Further investigation into radiation therapy to treat canine transitional cell carcinoma (TCC) is warranted. Understanding the inherent radiosensitivity and repair capacity of TCC is necessary to optimize radiation protocols.

Three previously established TCC cell lines were evaluated (provided by Dr. Deborah Knapp, Purdue University). Using standard cell proliferation assays, we determined the surviving fraction at 2 Gy (SF_2), the α/β ratio, and the capacity for sublethal damage repair in response to radiation (SLDR) of the TCC cell lines. The cells were irradiated with a 6 MeV linear accelerator at doses of 0, 2,4,6,8, and 10 Gy.

The mean SF_2 for the cell lines was 0.67. The cell line survival fractions were compared using a two-way repeated measure analysis of variance. There was a statistical difference between radiation doses, but not between runs or cell lines. To estimate the α/β ratios for each cell line, the mean SFs for each cell line were regressed against dose by use of a standard linear quadratic model. The α and β values were derived from this equation when it was fitted to the dose-response curve. All 3 cell lines exhibited a moderate to low α/β ratio, with the mean being 4.0.

The relatively high SF_2 indicates that these cell lines are relatively radioresistant. The low α/β ratios suggest that these cell lines have a high repair capacity and would be sensitive to changes in fractionation schedules. These results indicate that radiation protocols utilizing higher doses with less fractionation may be more effective for TCC.

ABSTRACT #17

GENETICS OF OSTEOSARCOMA IN THE SCOTTISH DEERHOUND. J Phillips[1], L Lembcke[1], T Chamberlin[1]. [1]University of Tennessee College of Veterinary Medicine, Knoxville, TN.

Osteosarcoma is the most common malignant bone tumor in dogs and is characterized with both aggressive local behavior and high metastatic rates. A hereditary basis for canine osteosarcoma has long been suspect, given the disproportionate incidence in many breeds. The Scottish Deerhound is a breed of dog with a > 15% estimated incidence of osteosarcoma. In our previous work, we showed that a highly penetrant gene with dominant effect could explain most of the osteosarcoma phenotype within the Deerhound. Here we report the results of a whole genome linkage scan in a large multi-generation pedigree of Scottish Deerhounds. Clinical data was recorded for more than 1,200 Deerhounds and included spay/neuter status, height, weight, cancer history, and if applicable cause specific-mortality. Blood samples were collected from more than 500 of these dogs; 60 from this subset (40 affected) were selected for a whole genome linkage scan. Microsatellite markers (610 in total) were selected from CanFam 2.0 and included those described in the mapping set MSS-2/3 along with an additional 103 markers, resulting in an average spacing of 4.1 Mb. Multipoint linkage analysis using SimWalk2 was used to analyze genotyping results for evidence of linkage. SimWalk2 uses MCMC and simulated annealing algorithms to avoid limitations due to sample size or inbreeding. LOD scores (or location scores) of > = 3.0 were considered consistent with linkage. Using this approach we found strong evidence of linkage between the osteosarcoma phenotype and markers in our mapping set. Fine mapping and candidate gene analysis is currently underway.

ABSTRACT #18

PERFUSION METHOD FOR BONE MARROW CELL HARVESTING IN DOGS. M. Sato, Y. Goto-Koshino, Y. Fujino, K. Ohno, H. Tsujimoto. Department of Veterinary Internal Medicine, Graduate School of Agricultural and Life Sciences, The University of Tokyo, Tokyo, Japan.

Bone marrow transplantation is a promising strategy to compensate peripheral blood cytopenias after intensified cytoreductive treatments for malignancies. Many of its failures are due to insufficient blood cell recovery or graft-versus-host disease. To minimize peripheral blood interfusion, a major problem at harvesting bone marrow cells (BMCs), bone marrow perfusion method was recently developed in monkeys as a preclinical trial for humans. The present study was carried out to compare the perfusion method with conventional aspiration method with respect to cell composition and colony formation of the harvested BMCs in dogs.

BMCs were collected from humerus and femur of 7 Beagle dogs kept for experimental purposes. Flow cytometric analysis revealed that percentages of $CD3^+CD4^+$ and $CD3^+CD8^+$ lymphocytes indicative of the blood interfusion were 4.3 ± 2.7 and 4.2 ± 2.3, respectively, in BMCs obtained by perfusion method, significantly lower than those in BMCs obtained by aspiration method (15.3 ± 7.6 and 9.2 ± 3.6) (both $P < 0.001$). Percentage of $CD34^+$ cells was significantly higher by perfusion method (5.5 ± 1.9) than aspiration method (4.2 ± 1.8) ($P < 0.01$). Colony forming unit in culture (CFU-C) was significantly higher by perfusion method (63.7 ± 21.2) than aspiration method (21.9 ± 7.8) ($P < 0.01$).

The present study indicated that the perfusion method was more appropriate than aspiration method to harvest bone marrow progenitor cells for transplantation in dogs.

ABSTRACT #19

IMMUNODYSFUNCTION IN DOGS WITH LYMPHOMA.
BL Fowler, SM Axiak, AE DeClue. University of Missouri, College of Veterinary Medicine, Columbia, MO.

This study compares the inflammatory profile of normal dogs, dogs with lymphoma, and dogs with osteosarcoma to determine if dogs with an immune cancer (lymphoma) have a different inflammatory "foot print" then either dogs with non-immune cancer (osteosarcoma), or normal dogs. By characterizing inflammatory mediators we may also be able to increase dose intensity, and therefore median remission and survival times. We hope to provide a framework from which chemotherapy treatment can be based by using measurements of these mediators to predict which dogs may be at higher risk for sepsis. We hypothesize that dogs with lymphoma will have a blunted immune response compared to dogs with osteosarcoma or healthy dogs

Blood was collected from healthy controls (healthy; n = 3), osteosarcoma (OSA; n = 2), and dogs with lymphoma (LSA; n = 3), and Lipopolysaccharide (LPS), lipoteichoic acid (LTA), peptidoglycan (PG) and control (PBS) were used to stimulate production of TNF and IL-10 using a whole blood culture system. Data were reported using descriptive statistics.

Lipopolysaccharide, lipoteichoic acid and peptidoglycan induced TNF production from whole blood culture in all dogs. Whole blood from the LSA group produced less TNF in response to LPS, LTA and PG compared to the other groups (see figures).

These pilot data suggest that dogs with LSA have a blunted proinflammatory response to both gram positive and gram negative bacterial pathogen associated molecular patterns. This is initial evidence that dogs with LSA have immunodysfunction and that those differences may be exploited to select a specific patient population which might tolerate higher doses of chemotherapy with less chance of sepsis.

Figure [Abstract 19]

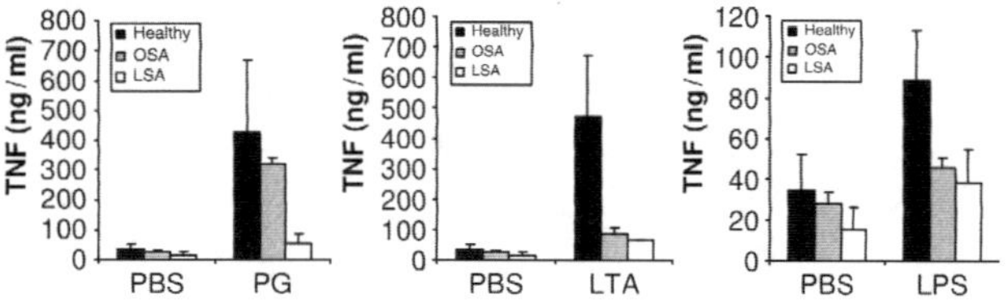

ABSTRACT #20

URETHRAL STENT PLACEMENT USING DIGITAL RADIOGRAPHY: 10 CASES.
Radhakrishnan, Anant. Bluegrass Veterinary Specialists, Lexington, KY.

Urethral stent placement for lower urinary tract obstruction has been recently described. Neoplasia of the urinary bladder, urethra, or prostate are the most common reasons for obstruction though granulomatous urethritis and urethral strictures have also been described. The development of an interventional technique was necessary due to morbidity associated with surgical intervention for treatment of tumors in the lower urinary tract while medical management does not alleviate symptoms associated with lower urinary tract tumors including obstruction. The technique previously described utilized fluoroscopic guidance for placement of urethral stents to restore luminal patency. However, digital radiography is more readily available in small animal practice and the ability to utilize it for urethral stent placement may allow for more widespread availability of this procedure.

10 client-owned patients (8 female, 2 male) presented for lower urinary tract disease. 8 patients (7 female, 1 male) were diagnosed with transitional cell carcinoma (TCC) of the bladder or urethra, 1 female patient was diagnosed with granulomatous urethritis, and 1 male patient was diagnosed with idiopathic detrusor-urethral dyssynergia. 3 patients (female) presented with urethral obstruction (2 TCC and 1 granulomatous urethritis). 6 TCC patients had significantly decreased luminal patency and therefore urethral stent placement was elected.

With cystoscopic guidance, a guidewire was passed into the urinary bladder. Guidewires and catheters were used to perform positive contrast cystourethrograms and sequential digital radiograph images were taken to determine the appropriate size and location for stent placement. All stents were placed transurethrally over the guidewire and sequential digital images were taken to monitor stent positioning and deployment.

Post-procedure urination was documented in 100%. All 10 patients were reported to have improvement in urine output post-procedure by their owners. 2 female patients with TCC developed urinary incontinence. Mild dysuria continued to be noted in 6/7 female TCC patients. Stent dislodgement into the urinary bladder occurred in one patient due to failure to maintain back tension on the stent. A second stent was placed in this patient without complication and the previous stent was removed via cystotomy. Seven patients are considered to have good-excellent outcome and three fair.

Digital radiography can be used for trans-urethral stent placement. The still imaging requires multiple images to be taken sequentially during the procedure to monitor movement and positioning of guidewires, catheters, and stents as well as to interpret the contrast cystourethrogram and determine luminal patency.

ABSTRACT #21

URETHRAL STENTING IN CANINE PATIENTS WITH UROTHELIAL CARCINOMA: A REVIEW OF 41 CASES (2004–2008).
A. Blackburn[1], A. Berent[2], C. Weisse[2], D.C. Brown[1]. [1]University of Pennsylvania, Philadelphia, PA. [2]Animal Medical Center, New York, NY.

Urethral stenting can be performed in a minimally invasive manner for immediate relief of malignant urethral obstructions in dogs. Traditionally, the most common cause for euthanasia in dogs with urethral/bladder/prostate tumors was local progression, resulting in severe stranguria/obstruction.

The purpose of this study was to review canine cases that received a urethral stent for obstructive urothelial carcinoma between 2004–2008. Post-stent incontinence rates were determined. Multiple factors were analyzed to determine effect on the incidence of post-stent incontinence: stent size, length, patient sex, tumor size/location, etc. Survival times after stent placement were determined.

Forty-one dogs (22 males, 19 females) were retrospectively evaluated. Incontinence rates, of any degree (mild or severe), were 47.6% and 75% in females and males, respectively, while only 28.6% of females and 20.8% of males were considered severely incontinent. The median survival time (MST) post-stent was 90 days. MST in dogs that did and did not receive non-steroidal anti-inflammatory drugs (NSAIDS) prior to stent placement were 181 and 54 days, respectively. MST for dogs that did and did not receive post-stent chemotherapy were 120 and 43 days, respectively. Dogs that had tumor extension into the bladder had a MST of 33 days, versus no bladder extension had a MST of 90 days.

In conclusion, post-stent incontinence rates were higher than previously reported, but a majority of these dogs were only mildly incontinent. Pre-stent use of NSAIDS and post-stent chemotherapy statistically increased survival, while extension of the tumor into the bladder significantly decreased survival.

ABSTRACT #22

THE EFFECT OF CONTRAST ULTRASOUND ON KIDNEYS IN CATS.
M Leinonen[1], M Raekallio[1], O Vainio[1], S Sankari[1], RT O'Brien[2]. [1]Department of Equine and Small Animal Medicine, Faculty of Veterinary Medicine, University of Helsinki, Finland. [2]Department of Veterinary Clinical Medicine, University of Illinois at Urbana-Champaign, IL.

The purpose of this study was to investigate the effects of contrast ultrasound (CEUS) on urine protein concentration, enzyme activities and hematuria in healthy cats.

The left kidney of eight, young (2–4 years) cats was imaged with CEUS (Esaote MyLab 70XVG) in a longitudinal plane using a 3–11 MHz linear transducer. The mechanical index was maintained at 0.04 and adjustable parameters were standardized. A 1.5 minute-long digital imaging series with a frame rate of 16–18 Hz was recorded after iv-bolus injections of contrast medium (0.1 ml Definity®). Plasma and urine samples were collected before CEUS and at 4 h and 48 h thereafter. Urine enzymatic activities, specific gravity, pH, protein, creatinine, albumin and sediment, and plasma urea and creatinine concentrations were evaluated. ANOVA with repeated measurements was used to compare the laboratory parameters at times 0, 4 h, 48 h. The significant changes were tested with post-hoc Student's T-test. Statistical significance was set at $P < 0.05$.

No changes were detected in serum samples, except a decrease in blood urea concentration at 48 h. A small rise in the N-acetyl-β-D-glucosaminidase (0.53 ± 0.35 to 1.43 ± 0.59 U/g) and gamma-glutamyl transferase (13.51 ± 3.28 to 18.19 ± 6.4 U/g) ratios was seen at 48 h, but not 4 h. This variation was less than the circadian variation (6.4 ± 2.7 U/g, 17.9 ± 13.5 U/g) reported in an earlier study in healthy cats.

These results suggest that CEUS can be safely used to assess kidney perfusion in cats without detrimental effects on kidneys.

ABSTRACT #23

RENAL DISEASE IN CATS INFECTED WITH FELINE IMMUNODEFICIENCY VIRUS. KJ Baxter[1], JK Levy[2], CH Edinboro[3], SL Vaden[4], MB Tompkins[4]. [1]University of Florida College of Veterinary Medicine, Gainesville, FL. [2]University of Florida Maddie's Shelter Medicine Program, Gainesville, FL. [3]Exponent, Inc. Health Sciences Group, Menlo Park, CA. [4]North Carolina State University College of Veterinary Medicine, Raleigh, NC.

Feline immunodeficiency virus (FIV) and human immunodefiency virus (HIV) are related lentiviruses that lead to immune dysregulation, opportunistic infections, inflammatory diseases, and neoplasia. Renal disease is diagnosed in up to 30% of HIV-infected humans and is the fourth most common cause of death associated with HIV. The aim of this study was to investigate the association between FIV infection and renal disease in cats.

Clinicopathological data were collected from FIV-infected (152) and FIV-noninfected (371) cats. Serum creatinine, urine specific gravity (USG), and urine protein:creatinine ratio (UPC) were recorded. In FIV-infected cats, CD4+ lymphocyte count and CD4+:CD8+ lymphocyte ratio were measured by flow cytometry. Renal azotemia was defined as a serum creatinine concentration > 1.8 mg/dL with a USG < 1.035. Proteinuria was defined as a protein:creatinine ratio > 0.4.

In univariate analysis, FIV-infection was associated with proteinuria (OR = 2.76, 95% CI 1.51–5.06, $P = 0.001$). Of those cats with a UPC > 1.0, 67% were FIV-infected. No significant association was detected between FIV-infection and renal azotemia ($P = 0.385$). There was a positive correlation between CD4+ lymphocyte count and UPC ($P = 0.01$), but not renal azotemia. No correlation was noted between CD4+:CD8+ lymphocyte ratio and renal disease.

These findings suggest that renal disease, characterized by proteinuria but not azotemia, is associated with FIV-infection. In HIV-infection, low CD4+ lymphocyte count is a risk factor for proteinuric renal disease; however, that does not appear to be true for FIV-infection.

ABSTRACT #24

SCREENING OF RAGDOLL CATS FOR KIDNEY DISEASE: A RETROSPECTIVE EVALUATION. D. Paepe, J.H. Saunders, V. Bavegems, G. Paes, L. Peelman, C. Makay, S. Daminet. Ghent University, Ghent, Belgium.

The Ragdoll is one of the most popular cat breeds worldwide. According to several breeder websites, Ragdolls are predisposed for chronic kidney disease (CKD), mainly due to polycystic kidney disease (PKD) and chronic interstitial nephritis (CIN). Although there is no evidence that Ragdolls are at increased risk for renal problems, screening of these cats is performed in several countries before breeding. In addition, information about the prevalence of PKD in Ragdoll cats is lacking. The aim of this study was to retrospectively evaluate the results of screening tests of Ragdolls at our institution to assess the prevalence of renal abnormalities in this breed.

All Ragdoll cats that were presented for PKD and/or CIN screening (abdominal ultrasonography, PKD genetic testing, renal values) between September 2001 and November 2009 were considered for inclusion in this retrospective study. Ragdolls were excluded if only the renal values were measured without undergoing an abdominal ultrasonography (US) to evaluate the kidneys and liver and/or genetic testing for PKD. PKD negative cats only evaluated by US needed to be at least 9 months to be included. The signalment, serum urea nitrogen and creatinine concentrations and US abnormalities of kidneys and liver were assessed. PKD was considered if at least one cystic structure was found in at least one kidney or the liver on US or if the cat was heterozygous for the PKD1 nonsense mutation on genetic testing. Cats with US changes compatible with CKD were considered at risk for CIN.

The study included 257 Ragdolls with a mean age of 774 days (range 90–3190). Abdominal US was performed on 249 cats: 7 cats were diagnosed with PKD, 22 cats had one or more abnormalities that could be compatible with CKD, 8 cats had other renal abnormalities with uncertain significance and 2 cats showed right unilateral renal agenesis or aplasia. None of the cats showed hepatic cysts. All 29 genetically tested cats were homozygous for the wild PKD1 allele, including 1 cat that was considered PKD positive on abdominal US. Renal values were measured in 142 cats: 68 had an increased serum creatinine concentration and 3 had both serum urea nitrogen and creatinine concentrations above the reference range. The mean serum creatinine concentration was 127.34 ± 30.67 μmol/L (reference range: 87–124 μmol/L).

Although this study is limited by its retrospective nature, the screening results for kidney disease of a large population of Ragdoll cats are evaluated for the first time. This study indicates that PKD occurs at a low prevalence in Ragdolls and that US findings possibly compatible with CIN can be detected in less than 10% of this Ragdoll population. Almost half of the cats had an increased serum creatinine concentration, therefore further examination will need to elucidate if this is due to a breed-specific feature or to decreased renal function.

ABSTRACT #25

ANEMIA PREDICTS PROGRESSION OF CHRONIC KIDNEY DISEASE IN NEWLY DIAGNOSED AZOTEMIC CATS. S Chakrabarti, H Syme, J Elliott. Royal Veterinary College, London, UK.

Chronic Kidney Disease (CKD) is common in geriatric cats but often appears stable for long periods of time. Several studies have evaluated prognostic markers in cats with CKD, but few have identified which ones precede disease progression. The aim of this study was to find a marker which would predict deterioration of renal function in cats newly diagnosed with CKD.

Feline CKD cases were recruited between 1992 and 2009 through geriatric cat clinics held at two first opinion practices. Diagnosis was based on concurrent findings of plasma creatinine > 2 mg/dl and urine specific gravity (USG) < 1.035, with persistence of azotemia for 2 weeks. All cases were first screened using objective criteria and then classified by two clinicians. Renal function was approximated using the reciprocal of plasma creatinine concentration (1/creatinine) and linear regression was used to assess changes over time. For cases to be considered progressive, a negative slope in 1/creatinine ($p < 0.1$) was required. For cases to be considered non-progressive, they were required to have no significant negative slope in 1/creatinine, at least 3 plasma creatinine values spread over 40 weeks follow-up, and no linear weight loss ($p < 0.05$). Graphs of 1/creatinine and bodyweight against time were plotted for cats meeting the above screening criteria. These were randomised and classified independently by the clinicians (HS and JE). Further cases were excluded where classifications disagreed or extreme fluctuations in 1/creatinine were observed. Cats which developed hyperthyroidism were excluded from the study and urine protein-to-creatinine ratios (UPCs) from urine samples with a positive urine culture outcome were not used in the analyses. Baseline variables at diagnosis were assessed for associations with progression using binary logistic regression. The odds ratio (OR), confidence interval (CI), or interquartile range (IQR) is provided in parentheses.

In total, 52 progressive and 42 non-progressive cases were identified out of 118 cases which passed the initial screening criteria. Lower packed cell volumes (PCVs) were associated with progression (OR: 0.916, 95%CI: 0.853–0.985, n = 90, p = 0.017). The median PCV for the progressive group was 32(IQR 29–36)% versus 36(IQR 34–41)% for the non-progressive group. The groups were matched for the degree of azotemia at diagnosis, with a median plasma creatinine concentration of 2.46(IQR 2.33–3.12) mg/dl for the progressive group and 2.51(IQR 2.26–2.92) mg/dl for the non progressive group. Plasma creatinine (n = 94, p = 0.544), urea (n = 94, p = 0.922), USG (n = 93, p = 0.216), phosphate (n = 93, p = 0.157), UPC (n = 45, p = 0.145) and potassium (n = 94, p = 0.544), did not predict CKD progression. The results of this study suggest that newly diagnosed cases of azotemic CKD which have low PCV values are more likely to progress. Thus, anemia may serve as a marker for cats which have more severe types of renal disease or may itself contribute to progression via hypoxic injury.

ABSTRACT #26

A CASE-CONTROL STUDY OF NEPHROTIC SYNDROME IN DOGS: 78 CASES. ES Klosterman[1], GE Moore[1], JF De Brito Galvao[2], SP DiBartola[2], RP Groman[3], JC Whittemore[4], SL Vaden[5], TL Harris[5], JK Byron[6], SR Dowling[6], DC Grant[7], GF Grauer[8], BM Pressler[1]. [1]Purdue Univ, W. Lafayette, IN. [2]OSU, Columbus, OH. [3]UPenn, Philadelphia, PA. [4]UTenn, Knoxville TN. [5]NCSU, Raleigh, NC. [6]U of Ill, Urbana, IL. [7]VMRCVM, Blacksburg, MD. [8]KSU, Manhattan, KS.

Nephrotic syndrome (NS) is an uncommon complication of glomerular disease defined as the concurrent presence of proteinuria, hypoalbuminemia, hyperlipidemia (i.e. hypercholesterolemia), and ascites or other third space fluid accumulation. Although NS is well-recognized in dogs, published information is limited to occasional mention of affected animals in larger case series of glomerular disease and sporadic case reports. The purpose of this study was to describe a larger population of dogs with NS, and determine whether or not the diagnosis of NS is associated with signalment or select clinicopathologic or histologic findings in patients with glomerular disease.

Medical record databases were searched at 8 participating institutions using the term 'nephrotic syndrome.' Records were reviewed by participants, with inclusion requiring the concurrent presence of all four criteria. Two non-NS dogs with glomerular disease were selected as controls for each NS dog, using the search terms 'protein-losing nephropathy,' 'glomerular disease,' and 'glomerulopathy'. Signalment, time until death, albumin, cholesterol, creatinine, BUN, sodium, UPC, and histopathologic diagnosis were recorded for each patient, and NS and control populations were compared using chi-squared or two-sample t-tests as appropriate.

78 dogs with NS and 156 control dogs were included in this interim analysis. Neither sex nor neuter status were associated with NS (p = 0.748). Mixed breed dogs and 11 pure breed varieties had > 5 animals within the 234 total dogs; however no breeds were over-represented in either group. Mean serum albumin in NS dogs was significantly lower than in control dogs (1.7 [95% CI 1.6–1.8] vs. 2.7 [95% CI 2.5–2.8] mg/dL, p < 0.001); mean cholesterol (373 [95% CI 357–390] vs. 313 [95% CI 289–338] mg/dL) and UPC (15.4 [95% CI 13.3–17.5] vs. 8.2 [95% CI 7.1–9.2]) in NS dogs were significantly greater than in control dogs (p ≤ 0.001). There were no significant differences in sodium, creatinine, or BUN between the two groups (p = 0.311, 0.854, and 0.919, respectively). Of the 38 NS dogs for which a histopathologic diagnosis was available, the most frequently identified diseases were membranous glomerulopathy (n = 10), amyloidosis (n = 8), and membranoproliferative glomerulonephritis (n = 8); however, no individual diagnosis was over-represented in NS dogs as compared to control dogs (p = 0.344). There was no difference in mean survival between groups (NS: 68 d [95% CI 18–117 d] vs. controls: 74 d [95% CI 35–113 d]).

These preliminary results suggest that neither signalment nor histopathologic diagnosis affect the likelihood of dogs with glomerular disease being concurrently diagnosed with NS, and that NS is not associated with a concurrent diagnosis of azotemia or survival time in dogs with glomerular disease. These findings are in contrast to those in people with NS, in whom specific diseases are more likely to result in NS, and the diagnosis of NS may be a negative prognostic indicator.

ABSTRACT #27

INHIBITION OF P-FIMBRIATED *ESCHERICHIA COLI* ADHESION IN AN INNOVATIONAL *EX-VIVO* MODEL IN DOGS RECEIVING A BIOACTIVE CRANBERRY TABLET (CRANANIDIN™). AB Howell[1], DW Griffin[2], MO Whalen[2]. [1]Rutgers University, Marucci Center for Blueberry Cranberry Research, Rutgers University, Chatsworth, NJ, USA. [2]Nutramax Laboratories, Inc. Veterinary Science Division, Edgewood, MD.

Use of cranberry juice decreased or eliminated recurrent urinary tract infections (UTIs) in women. In dogs, as with people *Escherichia coli* (*E. coli*) is the most common pathogen associated with UTIs. Cranberry proanthocyanidins (PACs) inhibit the attachment of the P- fimbriated *E. coli* (Pf *E. coli*) to uroepithelial cells. A supplement containing these PACs was assessed in six male dogs for anti-adhesion (AA) activity. Urine samples were tested in an *in vitro* assay for the ability of Pf *E.coli* to agglutinate human red blood cells. One Crananidin™ tablet containing 16 mg of bioactive PACs (mean dose 1.14 mg/kg bw) was administered daily for three weeks. Urine samples were collected on days 1, 3, 5, 7 and 21 at 0–3, 3–6, 6–12 and 21–24 hours post administration. AA assays were scored based on the following scale: 0 = no AA activity, 0.5 = 25%, 1 = 50%, 1.5 = 75% and 2 = 100% AA activity. Control urine demonstrated hemagglutination activity. AA activity appeared within 3 hours of Crananidin™ administration, peaked on Day 7 at 80 ± 26.5% and remained elevated on Day 21 at 73 ± 28.5% over the 24-hour period. No clinically significant clinical chemistry or hematology changes were observed. This innovational *ex-vivo* study shows that the bioactive PACs in Crananidin™ produce metabolites in the urine of dogs sufficient to reduce adhesion of Pf *E. coli*, and once daily usage may be beneficial in preventing *E. coli* associated UTIs in dogs.

ABSTRACT #28

IMMUNOHISTOCHEMICAL EXPRESSION OF CYCLOOXYGENASE (COX)-1 AND -2 ENZYMES IN FELINE CHRONIC KIDNEY DISEASE. N. Suemanotham, H. Syme, Y. Berhane, K. Smith, J. Elliott. The Royal Veterinary College, University of London, London, UK.

COX enzymes are involved in prostanoid synthesis, local mediators which modulate renal hemodynamics. Renal COX enzyme expression has not been described in cats, either with normal or diseased kidneys. The aim of this study was to test the hypothesis that dysregulation of renal COX enzymes is associated with progression in feline chronic kidney disease (CKD).

Formalin-fixed kidney sections were accessed from the RVC pathology archive for this study. Non-azotemic cats with normal renal histology that were euthanized for reasons unrelated to renal disease were defined as non-CKD group and were categorized as young (< 5 yrs) or old (> 9 yrs). Azotemic-CKD cats were diagnosed as having persistently elevated plasma creatinine (> 2 mg/dl) and urine specific gravity < 1.035 together with other CKD clinical signs. Azotemic cases were further classified. If plasma [creatinine] increased to > 5 mg/dl (IRIS stage 4) or by > 50% within 6 months of death, the case was categorized as progressive (CKD-P). Non-progressive cases (CKD-NP) had a follow-up of at least 24 wks and plasma [creatinine] changed by < 25%. COX immuno-staining was validated and assessed by numerical scoring (0 to 3). Ten or 20 representative renal structures (glomeruli (G), arterioles (aa), macula densa (MD), tubules (proximal (PT), distal (DT), Henle's Loop

(LoH), collecting ducts (CD)) were assessed per section and the grade point average intensity (GPAI) calculated and compared between groups using the Kuskal Wallis test.

In non-CKD cats, COX-1 immunostaining was seen in G, MD, DT and CD whereas COX-2 was seen in G, MD, PT and thick limb of the LoH. In general, CKD-P and CKD-NP showed significantly higher intensity COX-staining than young and old non-CKD cases ($p < 0.013$) but CKD-P did not differ from CKD-NP. This was true for COX-1 in the MD [CKD-NP 1.43(1.33–1.72); CKD-P 1.23(1.01–1.62) vs. young non-CKD 0.75(0.55–0.81); old non-CKD 0.82(0.8–1.13)], aa, PT and DT (data not shown). For COX-2 more intense staining was found in DT [CKD-NP 1.43(1.07–2.13); CKD-P 1.55(1.19–1.96) vs. young non-CKD 0.45(0.27–1.13); old non-CKD 0.86(0.42–1.09)], the thick limb of the LoH and CD (data not shown).

Up-regulation of COX isoenzymes in feline CKD supports their involvement in the compensatory mechanisms activated by nephron loss. No association was found between COX enzyme expression and progression of CKD. Further functional studies are required to understand the physiological importance of COX enzymes in the feline kidney and in the compensatory mechanisms that accompany loss of nephrons with age and CKD.

ABSTRACT #29

DO CATS WITH HYPERTENSION BUT NO OCULAR LESIONS HAVE INCREASES IN PLASMA NT-PROBNP CONCENTRATION?

SM Lalor, J Elliot, HM Syme. Royal Veterinary College, London, UK.

Plasma N-terminal pro-B-type natriuretic peptide (NT-proBNP) concentration has previously been found to be elevated in cats with hypertensive ocular disease, with NT-proBNP $\geq$ 202.7 pmol/L being both specific and sensitive for the detection of hypertension. As yet no studies have investigated plasma NT-proBNP concentration in hypertensive cats without ocular lesions. Since, in the absence of evidence of end organ damage, it can be difficult to clinically differentiate truly hypertensive cats from those with 'white-coat' hypertension, the aim of the present study was to investigate plasma NT-proBNP concentration as a diagnostic marker in these patients.

Sixty two cats were included in this study and were subdivided into four groups comprising; 12 normal cats, 14 normotensive (renal) azotemic cats, 13 hypertensive cats without ocular lesions and 23 cats with hypertensive ocular damage. Renal azotemia was defined as plasma creatinine > 2.0 mg/dl with a concurrent urine specific gravity of < 1.035. Systolic blood pressure (SBP) was measured indirectly using a Doppler flow detector and SBP > 170 mm Hg on multiple visits or in association with ocular lesions was taken as the definition of hypertension. Of the hypertensive cats only 3/13 (23.1%) without ocular lesions and 13/23 (56.5%) with hypertensive ocular lesions were azotemic. All cats were > 8 years of age. Plasma samples used in this analysis had been stored at −80 degrees since collection and were analysed using a NT-proBNP assay (VETSIGN Feline Cardio-SCREEN proBNP; Guildhay Ltd.). NT-proBNP concentrations in the normal and diseased groups were compared using non parametric statistical tests (Kruskal Wallis Test followed by post-hoc analysis using Mann Whitney U tests). Continuous data are presented as median [25th, 75th percentile].

NT-proBNP concentrations were significantly higher in cats with hypertensive ocular lesions (180.1 [7.0, 632.4] pmol/L), than in hypertensive cats (<7.0 [<7.0, 7.1] pmol/L; $P < 0.001$) and normal cats (< 7.0 [< 7.0, < 7.0] pmol/L; $P < 0.001$). All cats with NT-proBNP concentration > 202.7 pmol/L (n = 11) had hypertensive ocular lesions. Of the cats with hypertensive ocular lesions, 6/6 (100%) of those with SBP < 220 mm Hg. None of the hypertensive cats without ocular lesions had NT-proBNP > 202.7 pmol/L and only 2/13 of these cats had SBP > 220 mm Hg.

NT-proBNP concentrations are not significantly increased in hypertensive cats without ocular lesions. This may be because these cats are suffering from 'white-coat' hypertension or because they have less severe elevation in blood pressure or are less susceptible to end organ damage. Further studies using radiotelemetric implants to monitor blood pressure at home in hypertensive cats without ocular lesions are required to determine whether these cats have 'white-coat' hypertension.

ABSTRACT #30

COMPARISON OF PLASMA EXO-IOHEXOL, ENDO-IOHEXOL AND CREATININE CLEARANCES IN CATS WITH LOW, NORMAL AND HIGH GLOMERULAR FILTRATION RATES.

D. Paepe[1], I. vanHoek[1], H.P. Lefebvre[2], M. Campos[1], S. Croubels[1], S. Daminet[1]. [1]Ghent University, Ghent, Belgium. [2]National Veterinary School of Toulouse, Toulouse, France.

Determination of glomerular filtration rate (GFR) is the most reliable test to evaluate kidney function. Many plasma clearance markers can be used to estimate GFR, though recently iohexol and creatinine have been proposed as easily applicable clearance markers. A previous preliminary study suggested differences between clearance techniques to be related to the range of GFR estimate (van Hoek *et al.*, JFMS 2009;11:1028–1030). This study aimed to compare plasma creatinine, exo-iohexol and endo-iohexol clearance tests in a larger population of cats with low, normal and high GFRs.

Healthy cats and cats with hyperthyroidism (HT), diabetes mellitus (DM) and chronic kidney disease (CKD) that underwent a plasma exogenous creaitinine-iohexol clearance test (PEC-ICT) at the University of Ghent were considered for inclusion. The PEC-ICT was performed as described by van Hoek *et al.* (JVIM 2007;21:950–958). Cats were included only once if multiple PEC-ICTs were performed at different time points. The effect of health status on GFR was evaluated with a general linear model. A general linear model and Bland-Altman plots were used to compare GFR estimates according to the used clearance marker. All tests were performed with statistical software (Systat) and at the 5% significance level.

Forty-four cats were included and divided into 5 groups: 16 untreated HT cats, 6 DM cats, 10 CKD cats (6 were previously treated for HT, 1 cat had both CKD and DM), 6 young (7–12 months) healthy cats and 6 old (9–12 years) healthy cats. The mean GFR (expressed in mL/kg/min) was 2.33 ± 1.29 for exo-iohexol, 2.63 ± 1.39 for endo-iohexol and 2.65 ± 1.36 for creatinine clearance. The GFR estimates were significantly lower ($P < 0.001$) in the CKD group and higher ($P < 0.001$) in the HT group than in the other groups. A significant effect ($P < 0.001$) of the clearance marker on the GFR estimate was observed in healthy (young and old) and CKD cats, but not in HT and DM cats. The number of cats showing a difference in GFR estimate of > 0.5 mL/min/kg or > 25% was 17/44 and 14/44 between plasma creatinine and exo-iohexol clearance, 20/44 and 19/44 between plasma endo-iohexol and creatinine clearance and 25/44 and 19/44 between plasma endo- and exo-iohexol clearance, respectively. The Bland-Altman plots comparing endo-iohexol with the other 2 markers showed the largest discrepancies. Plasma creatinine clearance values were most often higher than those for exo-iohexol.

In conclusion, discrepancies between GFR testing methods may exist according to the clinical status of the animal. It is therefore recommended to use the same method during a patient's follow-up and separate cut-off values for each marker.

ABSTRACT #31

LIMITED SAMPLING STRATEGIES TO ESTIMATE GLOMERULAR FILTRATION RATE IN CATS WITH LOW, NORMAL AND HIGH GLOMERULAR FILTRATION RATES.

D. Paepe[1], H.P. Lefebvre[2], D. Concordet[2], I. vanHoek[1], P. Smets[1], S. Croubels[1], S. Daminet[1]. [1]Ghent University, Ghent, Belgium. [2]National Veterinary School of Toulouse, Toulouse, France.

Limited sampling strategies (LSS) to estimate feline glomerular filtration rate (GFR) are poorly documented and mainly assessed in healthy cats and/or using only one clearance marker. This study aimed to propose LSS in cats with normal, decreased and increased GFR for plasma exo-iohexol, endo-iohexol and creatinine clearance.

A combined plasma exogenous creatinine-iohexol clearance test was performed in 44 cats (16 hyperthyroid, 6 diabetic, 12 healthy, 10 with chronic kidney disease). The best LSS was determined by comparing 510 sampling time combinations to the entire 9 blood sampling times set for calculating AUC values by the trapezoidal rule. The maximum error was calculated for each best sampling time combination.

Optimal sampling times for 2-sample plasma clearance are 60 and 360 minutes for endo-iohexol and 60 and 480 minutes for creatinine and exo-iohexol, with a maximum error of −36.9%, −24.6% and 33.8%, respectively. Optimal sampling times for 3-sample plasma clearance are 30, 180 and 600 minutes for creatinine and 30, 120 and 480 minutes for exo- and endo-iohexol, with a maximum error of 11.4%, −18.2% and 18.2%, respectively.

Blood sampling timing is GFR marker dependent. The GFR estimate is more accurate for 3-sample compared to 2-sample plasma clearance for all markers. The error observed on GFR estimate was lower for LSS using creatinine as GFR marker.

ABSTRACT #32

INCREASED CONCENTRATIONS OF CYTOKINES RELATED TO MACROPHAGE/MONOCYTE ACTIVITY AND RECRUITMENT ARE RISK FACTORS FOR MORTALITY IN PRIMARY CANINE IMMUNE-MEDIATED HEMOLYTIC ANEMIA. M Kjelgaard-Hansen[1], R Goggs[2], B Wiinberg[1], DL Chan[2]. [1]Department of Small Animal Clinical Sciences, University of Copenhagen, Denmark. [2]Department of Veterinary Clinical Sciences, Royal Veterinary College, UK.

Canine immune-mediated hemolytic anemia (IMHA) is a pro-inflammatory disease with high mortality. However there is poor characterization of the inflammatory response in this disease and in particular, if the nature of the inflammatory process correlates with outcome. It was recently reported that acute phase proteins may be useful for monitoring disease activity, but the study did not provide prognostic information. Thus, the purpose of this study was to determine if excess pro- or anti-inflammatory cytokine activity related to specific cell populations and whether this was correlated with outcome in primary canine IMHA.

We prospectively enrolled twenty client-owned dogs with primary IMHA. Entry criteria included anemia (PCV $< 37\%$) in combination with autoagglutination, or a positive Coombs' test, or the presence of moderate to marked spherocytosis with no evidence of underlying disease. Six healthy dogs were included as controls. Blood samples were collected by jugular venipuncture at admission. Serum cytokine activity was assessed by a canine-specific multiplex assay (CCYTO-90K, Millipore, USA) for Interleukin-2 (IL-2), IL-4, IL-6, IL-7, IL-8, IL-10, IL-15, IL-18, Monocyte chemoattractant protein-1 (MCP-1), Granulocyte-macrophage colony stimulating factor (GM-CSF), Interferon-inducible protein-10 (IP-10), Interferon-gamma and Keratinocyte chemoattractant (KC). Thirty day mortality was recorded. The Mann-Whitney *U* test was used to compare cytokine level in controls vs. IMHA patients and survivors vs. non-survivors. Logistic regression was used to assess which cytokines were independently predictive of outcome in this population. $P < 0.05$ was considered significant.

Increased concentrations of several cytokines were observed in IMHA patients compared to controls; IL-2, IL-10, KC and MCP-1 representing a mixture of pro- and anti-inflammatory cytokines of various cellular origins. Thirty day mortality was 25% (5/20 dogs). Cytokines with prognostic significance were IL-15, IL-18, GM-CSF and MCP-1, which are all secreted by and strongly associated with macrophage/monocyte activity and recruitment. Logistic regression analyses suggest that increased IL-18 and MCP-1 concentrations are independently associated with mortality in this population.

A mixed cytokine response is present in IMHA patients compared to controls. Importantly, the two independent predictors of outcome in this population, IL-18 and MCP-1 are both secreted by activated macrophages and are involved in activation of macrophages and recruitment of monocytes. Our data suggest that excessive levels of mediators of macrophage activity and recruitment may serve as prognostic indicators for poor outcome in canine primary IMHA.

ABSTRACT #33

SERIAL ANALYSIS OF COAGULATION PARAMETERS IN DOGS WITH IMMUNE-MEDIATED HEMOLYTIC ANEMIA (IMHA) USING THROMBOELASTOGRAPHY® (TEG). R. Goggs[1], B. Wiinberg[2], M. Kjelgaard-Hansen[2], D.L. Chan[1]. [1]Department of Veterinary Clinical Sciences, Royal Veterinary College, UK. [2]Department of Small Animal Clinical Sciences, University of Copenhagen, Denmark.

Recent studies suggest TEG can identify hypercoagulability in dogs with IMHA. It has been proposed that relative hypocoagulability potentially due to a consumptive process might be a negative prognostic indicator in these dogs. Using TEG, this study aimed to prospectively characterise coagulation status in dogs with IMHA on immunosuppressive therapy over time and to identify if patients with relative hypocoagulability had a poorer prognosis.

We prospectively enrolled thirty client-owned dogs with primary IMHA; diagnosed on the basis of anemia (PCV $< 37\%$) in combination with autoagglutination, or a positive Coombs' test, or the presence of moderate to marked spherocytosis with no evidence of underlying disease. Patients with thrombocytopenia ($<$ 100,000 platelets/μl), patients pre-treated with aspirin, clopidogrel or heparin or patients pre-treated with glucocorticoids for > 3 days were excluded. Blood samples were collected by jugular venipuncture at admission and again 3 and 5 days later. Assays included prothrombin time (PT), activated partial thromboplastin time (aPTT), D-dimers, fibrinogen, antithrombin activity (AT), thrombin-antithrombin (TAT) complexes and recalcified, unactivated thromboelastography (TEG). Values for reaction time (R), clot formation time (K), clot formation angle (α) and maximum amplitude (MA) were recorded from TEG tracings. TEG tracings were compared with those obtained from 30 healthy control dogs.

At admission, canine IMHA patients had significantly shorter K times ($p < 0.001$), greater α angles ($p < 0.001$) and greater MA values ($p < 0.001$) than controls. Prior corticosteroid therapy did not significantly affect TEG values at admission. Between admission and day 5, R and K times and MA values reduced significantly ($p \leq 0.02$) and α angles increased significantly ($p = 0.003$). By day 5, all TEG variables were significantly different from controls ($p \leq 0.002$). No significant correlations were identified between R time and PT or aPTT. MA values were strongly correlated with fibrinogen concentrations ($R = 0.785$, $p < 0.001$) and moderately negatively correlated with PCV ($R = -0.376$, $p = 0.001$). 83.3% dogs and 80% dogs survived to discharge and to 30 days respectively. At admission MA values were significantly higher in survivors than in non-survivors at discharge and at 30 days ($p \leq 0.016$). Logistic regression analysis suggested each unit increase in MA was associated with an increased odds of 30-day survival of 1.13 (95% CI 1.02–1.25).

All dogs showed evidence of a hypercoagulable state. All TEG variables changed significantly between admission and day 5 but TEG tracings remained consistent with a hypercoagulable state 5 days after starting immunosuppressive therapy. Patients with higher MA values at admission were more likely to survive to 30 days post hospital admission. Our data supports the hypothesis that relative hypocoagulability may be a negative prognostic indicator in these patients.

ABSTRACT #34

APPARENT LACK OF ASSOCIATION BETWEEN PRIMARY IMMUNE-MEDIATED THROMBOCYTOPENIA AND RECENT VACCINATION IN DOGS. AA Huang, J Coe, GE Moore, JC Scott-Moncrieff. Purdue University School of Veterinary Medicine, West Lafayette, IN.

Vaccination is often cited as a potential cause of immune-mediated thrombocytopenia (ITP) in dogs. While this association has been documented in humans, particularly in children, this relationship has not been definitively established in dogs. This case-controlled study sought to document the presence of an association between recent vaccination and diagnosis of ITP in dogs and to evaluate whether vaccine type or number of vaccines administered were more likely to be associated with a diagnosis of ITP.

The medical records of 44 dogs diagnosed with primary ITP at the Purdue University Veterinary Teaching Hospital (PUVTH) over a 14 year period were reviewed and age-matched to a randomly selected control group of 88 dogs that were presented for reasons other than immune-mediated disease. Dogs with primary ITP were defined as cases with a platelet count of $< 40{,}000/\mu L$ and no discernible underlying cause of thrombocytopenia. The age, sex, and breed of patient and control groups were recorded. The time from vaccination to the date of presentation to the PUVTH, the type of vaccine given, and the number of vaccines given per visit for both groups were recorded and compared.

The median age at presentation for dogs with ITP was 7 years (range: 2 to 15 years). The majority of the ITP group was comprised of mixed breed dogs (39.4%); no pure breed was represented by more than 2 cases. The number of patients that were vaccinated within 42 days of presentation to the PUVTH did not differ between cases of primary immune-mediated thrombocytopenia and the control group. There was 9.3% of the immune-mediated thrombocytopenia cases vaccinated within 42 days of presentation compared to 16.5% of the control dogs ($p = 0.25$). The total volume of vaccine administered in the latest visit was not different between groups ($p = 0.41$).

These results do not support a temporal association between recent vaccination and the onset of primary immune-mediated thrombocytopenia.

ABSTRACT #35

THROMBOELASTOGRAPHY IN DOGS WITH IMMUNE-MEDIATED HEMOLYTIC ANEMIA TREATED WITH PREDNISONE, AZATHIOPRINE AND LOW-DOSE ASPIRIN.

S. Flint, A. Abrams-Ogg, S. Kruth, A. Bersenas, R. Wood. Ontario Veterinary College, University of Guelph, Guelph, Ontario, Canada.

Thromboembolic disease is a common and potentially fatal complication of immune-mediated hemolytic anemia (IMHA), and thromboelastography (TEG) has been proposed as a means of identifying hypercoagulability. The objective of this study was to characterize TEG in dogs with IMHA before and after treatment with prednisone, azathioprine and aspirin, and to evaluate its potential role in predicting survival.

Inclusion criteria for this prospective study were: informed client consent; moderate to severe anemia (hematocrit < 0.25 L/L), with evidence of spherocytosis and/or agglutination and/or positive Coombs test; no evidence of an underlying cause including neoplastic and infectious diseases. Exclusion criteria included pre-treatment for > 48 hours with corticosteroids, or treatment with heparin or clopidogrel. Blood was sampled for TEG on the day of admission to the hospital (day 0), and every 48 hours for the duration of hospitalization. TEG was performed on citrated whole blood without activator. The TEG parameters R, maximum amplitude (MA), G, coagulation index (CI) and Cl30 (% clot amplitude at 30 minutes) were evaluated. Complete blood count and serum biochemistry profile were performed on all dogs, and coagulation profile, activated clotting time (ACT) and D-dimer concentration were performed when there was sufficient sample. All dogs were treated with prednisone (2 mg/kg/day PO) or dexamethasone (0.25 mg/kg/day IV), azathioprine (2 mg/kg/day PO) and aspirin (0.5 mg/kg/day PO). Survival analysis was performed using Day 0 samples and a univariate logistical regression analysis.

Twenty dogs were enrolled in the study. The overall mortality rate was 50% (10/20). 16/20 (80%) were hypercoagulable based on a CI > 3. Of the dogs that died, only 2 had been discharged from hospital. TEG variables associated with mortality at 90 days, presented as odds ratio with 95% confidence intervals (OR[95%]), include decreased CI (0.48 [0.22–0.89], $p = 0.02$), decreased G-value (0.88[0.77–0.99], $p = 0.02$), and decreased MA (0.87[0.75–0.97], $p = 0.01$). Other factors associated with mortality at 90 days included prolonged prothrombin time (OR 4.06[1.12–33.78], $p = 0.02$), prolonged partial thromboplastin time (2.71[1.17–11.42], $p = 0.0035$)), degree of hyperbilirubinemia (1.02[1.00–1.05], $p = 0.03$), and prolonged ACT (1.09[1.02–1.2], $p = 0.01$). All dogs that appeared normocoagulable based on a CI within the reference interval died.

In conclusion, most dogs with IMHA appear hypercoagulable based on TEG, and those that are not hypercoagulable appear to have a poor prognosis. This may represent progression from a hypercoagulable to a hypocoagulable state due to disseminated intravascular coagulation, and resulting consumption of platelets and coagulation factors.

ABSTRACT #36

SERIAL THROMBELASTOGRAPHIC MONITORING OF DOGS WITH MULTICENTRIC LYMPHOMA.

A Kol, SL Marks, KA Skorupski, DL Borjesson. School of Veterinary Medicine, University of California, Davis, CA.

Venous thrombosis is a common finding in dogs with lymphoma. Routine coagulation assays are insensitive for the diagnosis of hypercoagulability. Thrombelastograph (TEG) is a point-of-care hemostatic analyzer that provides an overall view of coagulation and is sensitive for the diagnosis of hypercoagulability. We hypothesized that dogs with multicentric lymphoma have hypercoagulable TEG tracings and that alterations in TEG parameters would improve with the clinical state of remission. Our objective was to characterize TEG tracings in dogs with multicentric lymphoma from the time of diagnosis, throughout chemotherapy and at the point of clinical remission once chemotherapy was completed.

Sixteen client-owned dogs with naturally occurring, newly diagnosed multicentric lymphoma have been evaluated in this prospective cohort study. All dogs were clinically staged and treated with one of two doxorubicin-based protocols. Complete blood count and TEG were run at 4 times points during the course of the disease.

Fifty-six percent (9/16) of the dogs were hypercoagulable at initial presentation (G value > 8.1 dyn/cm^2) and 19% were hypocoagulable (G < 4.9 dyn/cm^2). After induction therapy, 50% (8/16) were hypercoagulable and none were hypocoagulable. Once weaned off corticosteroids, but still during chemotherapy, 83% (10/12) were hypercoagulable and none were hypocoagulable. After achieving clinical remission and completing their chemotherapy protocol, 71% (5/7) were hypercoagulable and none were hypocoagulable. Although most dogs tended towards hypercoagulability (increased G value) during chemotherapy treatment, there were no statistical differences in TEG parameters when analyzed serially between visits (repeated measures ANOVA, $p > 0.05$). At initial presentation, platelet count, hemoglobin and fibrinogen concentration correlated with G value. In summary, 75% of dogs with lymphoma present with a coagulopathy as detected by TEG with the majority of dogs becoming more hypercoagulable as treatment progressed. Clinical remission did not correlate with resolution of their hypercoagulable state.

ABSTRACT #37

THE EFFECT OF STORAGE TEMPERATURE ON THROMBELASTROGRAPHY ENDPOINTS USING CITRATED CANINE WHOLE BLOOD.

AG Ralph, BM Brainard, J Pittman, B Babski, A Koenig. University of Georgia, Athens, GA.

Storage temperature of citrated whole blood could alter thrombelastography (TEG) results through alterations in contact pathway activation. This study documents the effects of storage temperature of whole blood on TEG endpoints. We hypothesize that storage of blood at 37°C for 30 minutes prior to TEG testing will result in a shorter reaction time (R value), as compared to blood stored at room temperature or at room temperature with a 5 minute prewarming of sample to 37°C.

Blood was collected by jugular venipuncture from ten healthy dogs using a 19 gauge butterfly catheter. Collection tubes containing 3.2% sodium citrate were allowed to fill by vacuum, and randomly assigned to one of three groups prior to TEG testing: room temperature storage for 30 minutes, 37°C storage for 30 minutes, or room temperature storage for 25 minutes with 5 minutes of pre-warming to 37°C. All TEG samples were run at 37°C using 20 μL of calcium chloride added to 340 μL of citrated whole blood. Groups were analyzed using a one-way ANOVA or Kruskal Wallis one-way ANOVA.

There was no statistically significant difference between R values (mean = 5.75 +/− 1.705 minutes; $P = 0.362$), angles (mean = 42.323 +/− 9.738 minutes; $P = 0.710$), or maximal amplitudes (MA) (mean = 53.247 +/− 6.043 minutes; $P = 0.289$) for the three storage protocols. The temperature variables tested in this study resulted in similar endpoints on TEG testing, indicating that these temperatures all provide accurate results and are valid options.

ABSTRACT #38

EFFECTS OF ASPIRIN ON CANINE PLATELET CYCLOOXYGENASE EXPRESSION.

J. Thomason, A Mackin, J Stokes, K Mullins, L Pinchuk, R McLaughlin, S Pruett, C Langston, K Lunsford. Mississippi State University College of Veterinary Medicine, Starkville, MS.

Aspirin has been recommended to prevent pulmonary thromboembolism in canine patients with conditions such as immune-mediated hemolytic anemia. Aspirin resistance, however, is now well

Figure [Abstract 38]

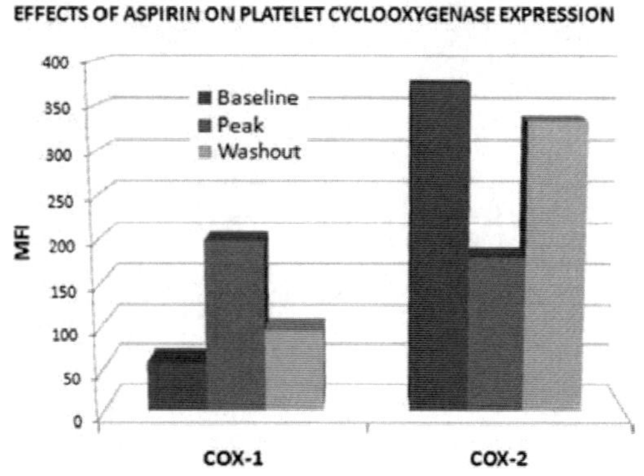

Figure [Abstract 39]

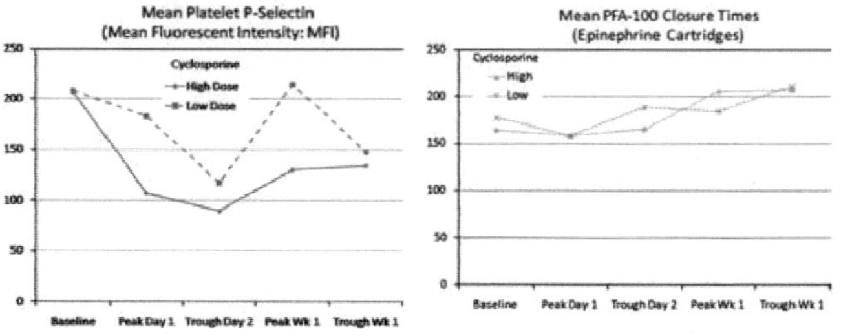

recognized in humans that develop thromboembolic complications despite appropriate therapy. Human aspirin resistance may potentially be associated with platelet cyclooxygenase-2 (COX-2), which provides an alternative pathway to platelet COX-1 for thromboxane A_2 synthesis. Human platelets were not believed to contain COX-2 until recently, when the intracellular presence of the enzyme was demonstrated via flow cytometry of permeabilized platelets. COX-2 has not as yet been demonstrated in canine platelets. Our study was designed to determine if canine platelets express COX-2, and to evaluate the effects of aspirin on COX enzyme expression.

Canine platelet COX-1 and COX-2 expression was quantified by flow cytometry. Platelets were permeabilized with a nonionic detergent to enable detection of intracellular COX-2. Platelets were evaluated in 8 healthy dogs before and after receiving aspirin for 10 days (anti-inflammatory dose, 10 mg/kg PO BID). COX-1 mean fluorescent intensity (MFI) increased in all dogs after aspirin administration, on average by 250% (range 63–476%). COX-2 MFI decreased in 6 dogs (average decrease 70%, range 56–80%), and increased in 2 dogs (increased by 8% & 116%). Washout values for COX-1 and COX-2 (2 weeks post-aspirin) approached baseline in most dogs. Platelet function was assessed by point-of-care analysis (Siemens PFA-100®) during aspirin administration, and was markedly impaired in all dogs (prolonged closure times, EPI cartridge).

Our study demonstrated that canine platelets express both COX-1 and COX-2, and that aspirin affects expression of both isoforms. Interestingly, aspirin caused increased COX-1 expression despite impairment of platelet function. Since our assay measured enzyme expression rather than function, increased COX-1 expression may be a response to aspirin-induced impairment of enzyme function. COX-2 expression tended to decrease in response to aspirin, but varied markedly between individual dogs.

Individual variations in platelet COX-2 expression in response to aspirin implicate variable responsiveness of the COX isoforms as a potential mechanism for aspirin resistance in dogs. Further investigation of the effects of aspirin at both high and low doses on COX expression and function (including thromboxane synthesis) is warranted in both normal dogs and clinic patients.

ABSTRACT #39

EFFECTS OF ORAL CYCLOSPORINE ON PLATELET ACTIVITY IN NORMAL DOGS. J Thomason, A Mackin, J Stokes, M Wallace, J Grady, L Pinchuk, S Pruett, C Langston, K Lunsford. Mississippi State University College of Veterinary Medicine, Starkville, MS.

Cyclosporine has become increasingly popular for treating canine immune-mediated diseases such as immune-mediated hemolytic anemia (IMHA). Cyclosporine has been shown to increase platelet procoagulant activity in humans, and may predispose to thromboembolic complications. Our study evaluated the effect of cyclosporine on platelet procoagulant activity in normal dogs.

Platelet reactivity was quantified by flow cytometric evaluation of platelet P-selectin, phosphatidylserine, COX-1, and COX-2. Platelet function was assessed via point-of-care analysis (Siemens PFA-100®), and clot formation was evaluated via viscoelastometry (Sienco SonoClot®). Platelets were evaluated before and after immunosuppressive doses of cyclosporine were given for 1 week (8 dogs; starting @ 10 mg/kg PO bid, dose adjusted to trough blood levels > 600 ng/mL); and again after 1 week of a cyclosporine dose approved for canine atopy (7 dogs; 5 mg/kg PO sid).

Mean expression of P-selectin decreased by 36% after a week of high dose cyclosporine, and by 29% at low dose. Mean phosphatidylserine expression decreased by 58% at high dose, and increased by 6% at low dose. Mean COX-1 decreased by 35% and 2% after a week at high and low dose, respectively, and mean COX-2 decreased by 31% at both doses, although on the 1st day of therapy at high doses there was a transient rise in COX-2 two hours after the 1st dose. Mean PFA-100® closure times using ADP cartridges increased by 61% at high dose, and decreased by 45% at low dose. Mean PFA-100® closure times using epinephrine cartridges increased by 26% and 21% at high and low doses, respectively. No significant change in SonoClot® results was seen.

Unlike human studies, our study suggests that cyclosporine does not tend to cause an increase in platelet reactivity during the first week of drug administration in dogs. Individual dog-to-dog variable responsiveness to cyclosporine may explain variations in drug pharmacodynamics, platelet thromboxane receptor function, thromboxane-mediated degranulation, and P-selectin expression. Further investigation of the effects of cyclosporine on platelet procoagulant activity with chronic administration is warranted both in normal dogs and in clinic patients.

ABSTRACT #40

THE EFFECT OF IN VIVO TREATMENT WITH ACETYLSALICYLIC ACID AND MELOXICAM ON PLATELET AGGREGATION IN CATS. C Cathcart, BM Brainard, SC Budsberg. University of Georgia College of Veterinary Medicine, Athens, GA.

This study documents the effect of acetylsalicylic acid (ASA; 5 mg/kg PO q48 h) and meloxicam (0.05 mg/kg PO q24 h) on feline platelet aggregation.

Each medication was given to 8 cats for 14 days in a randomized, placebo controlled, cross-over design. It was hypothesized that both NSAIDs would decrease platelet aggregation. Outcome measures included oral mucosal bleeding time (OMBT) and whole blood (impedance) aggregometry (WBA) using adenosine diphosphate (ADP; 10 µM) and collagen (5 µg/mL) as agonists. WBA was performed before drug administration on days 0, 7, 15, and 17. OMBT was measured on days 0 and 15. Blood was collected by jugular venipuncture using vacuum into tubes containing 3.2% sodium citrate. Groups were compared using repeated measures or paired tests appropriate for comparison and data distribution.

Compared to baseline, neither medication affected WBA at any time point. There was no difference between baseline WBA (ADP: 12.6 ± 5.7 ohm, Collagen: 22.9 ± 7.8 ohm) in any group ($P = 0.190$). Baseline OMBT did not differ between groups, and did not change from baseline in the placebo and meloxicam groups. OMBT, however, was significantly shorter on day 15 in the ASA group (from 73 ± 20 to 53 ± 7.5 sec, $P = 0.03$). Platelet counts were stable for the study duration.

At the doses studied, neither meloxicam nor ASA had an inhibitory effect on WBA or OMBT in cats. The use of ASA at the tested dose for platelet inhibition in cats should be reconsidered.

ABSTRACT #41

CLINICAL INVESTIGATION OF A POINT-OF-CARE HEMATOLOGY ANALYZER IN DOGS RECEIVING CHEMOTHERAPY. N Wilson, D Burney, M Endicott. Gulf Coast Veterinary Specialists, Houston, TX.

In-hospital diagnostic testing allows the practitioner to make rapid assessments and provide timely interventions. This is especially important in medical oncology where a CBC is performed prior to administration of each dosage of chemotherapy and patients are typically treated on an out-patient basis. Prior to administration of chemotherapy in dogs it is important to determine HCT, platelet count, total WBC count and neutrophil count in a rapid manner.

Impedance counter hematology instruments have traditionally been favored in oncology practices for their speed. Counts are performed based on cell size, thus the main draw back is poor accuracy. This is of particular concern in patients receiving chemotherapy since both the toxic nature of chemotherapy drugs on blood cells and the cancer itself can result in cytopenias, leukemias and other changes in blood cell morphology that may interfere with the accuracy of an impedance counter. A new hematology analyzer (IDEXX ProCyte Dx™) that uses laser flow cytometry and optical fluorescence in addition to impedance technology is under development as a point-of-care hematology analyzer. The CBC is performed on 30 µL of whole blood in less than 2 minutes. The purpose of this study was to determine the correlation of HCT, WBC, neutrophil count, and platelet count between the new IDEXX ProCyte Dx and the reference laboratory, in patients receiving chemotherapy.

Fresh EDTA anti-coagulated venous blood samples were collected from 80 dogs receiving chemotherapy at Gulf Coast Veterinary Specialists. Samples were run on the ProCyte Dx running developmental software within 2 hours and at the IDEXX Reference Laboratory (Sysmex XT-2000i V Automated Hematology Analyzer) in Dallas, TX within 12 to 24 hours of collection. All instruments were maintained according to manufacturer's specifications, including handling of test materials and performance of quality control procedures.

The regression scatter plot for the paired results showed a high degree of association between the two methods for all 4 parameters evaluated. Results were compared by Deming Regression and Bland-Altman plots using EP Evaluator v8. Correlation coefficients (R) and root mean square error (SEE) were: HCT ($R = 0.958$, $SEE = 1.680$), WBC count ($R = 0.995$, $SEE = 0.495$), neutrophil count ($R = 0.992$, $SEE = 0.565$), and platelet count ($R = 0.951$, $SEE = 41.8$).

The results suggest that the IDEXX ProCyte Dx is a reliable hematology analyzer that provides rapid and accurate results on a minimal volume of blood, even in patients undergoing chemotherapy or with leukemia. It had excellent correlation with the reference laboratory for HCT, WBC, and neutrophil counts and very good correlation for platelet counts.

ABSTRACT #42

GROSS AND HISTOLOGIC QUALITY OF BONE MARROW CORE SPECIMENS OBTAINED FROM THE SAME SITE BY A COMBINED ASPIRATION AND BIOPSY TECHNIQUE IN DOGS. JP Reeder, EC Hawkins, MC Cora, SL Marks, CB Grindem. North Carolina State University College of Veterinary Medicine, Raleigh, NC.

Evaluation of both a bone marrow aspirate and core is indicated for some dogs with hematologic disease. A combined technique for collecting both specimens from the same site can be performed by advancing a core biopsy needle through the cortex, aspirating, and then advancing the needle to obtain the core. Advantages include decreased procedural time, material costs and patient discomfort compared with independent collection. However, gross and histologic distortion of core specimens has been documented in people. The aim of this study was to determine if core biopsy preceeded by aspiration (combined technique) results in a core of decreased histologic quality compared with one obtained without pre-aspiration (direct technique).

Jamshidi needles were used to collect biopsies from 26 dogs immediately following euthanasia. Combined and direct techniques were performed on opposite humeri of the same dogs, with side and order randomized. Failed biopsy attempts were not repeated. Slides were prepared routinely, cut along the long axis of the core. Slides were evaluated in random order and blindly by a clinical pathologist (CBG) using a novel scoring system for predetermined criteria. The two slides from each dog were then paired (collection method masked) and overall histologic quality subjectively assessed to identify relative inferiority of either specimen. Comparisons involving numerical data were made by paired t-test for normally distributed data, or Wilcoxon signed rank test. Overall histologic quality was compared using Chi-square analysis (relatively inferior vs. not relatively inferior). A p value < 0.05 was considered significant.

Biopsies were unsuccessful in 4 of 26 (15%) dogs when using the combined technique and in 1 dog (4%) with the direct technique. Histologic comparisons were performed on 22 pairs of cores. Length of actual marrow obtained as measured on stained slides was significantly smaller using the combined technique (mean ± SD, 9.8 ± 5.6 mm) compared with the direct technique (13.7 ± 6.3 mm; $p = 0.011$). Specimens from the combined technique were of inferior overall quality compared with those from the direct technique ($p = 0.015$). Combined-technique cores were considered inferior in 14 dogs and not inferior in 8 dogs (similar in 2 dogs, superior in 6 dogs). Features that contributed to consideration of relative inferiority included intramedullary hemorrhage, crush or fragmentation of specimen, and zones of decreased or absent cellularity. No differences were identified between techniques for overall cellularity, megakaryocyte count, M:E ratio, or iron stores.

In conclusion, the combined technique of marrow collection yields core specimens with significantly shorter marrow length and poorer overall histologic quality compared with those obtained by the direct technique.

ABSTRACT #43

STERNAL BONE MARROW ASPIRATION IN CLINICALLY NORMAL SMALL DOGS. A. Defarges, A. Abrams-Ogg, D. Bienzle, R. Foster. Ontario Veterinary College, University of Guelph, Guelph, Ontario, Canada.

Sternal bone marrow aspiration in the dog is reported to be simple to perform, less painful than humeral or ilial aspiration, and to yield quality samples. It does not appear to have widespread clinical use. In this study the feasibility, safety, and sample quality of sternal aspiration in small dogs were investigated. Bone marrow aspirates were obtained by the same investigator from three sites in a randomized order from 17 clinically healthy beagles (weight range 6.5–11 kg), under general anesthesia following terminal abdominal surgery. For sternal aspirates, dogs were positioned in sternal recumbency and a 1-inch, 22–23 ga hypodermic needle attached to a 12cc syringe was inserted into the cranial manubrium, directed caudally along the long axis of the sternum. Aspirates for comparison were obtained from the greater tubercle of the right humerus and right iliac crest using an 18 ga Rosenthal needle attached to a 12cc syringe. Difficulty of site localization, needle insertion and advancement, and specimen aspiration, were scored from 1 (easiest) to 5 (hardest). Bone marrow smears were prepared by one investigator,

Table [Abstract 43]

Site	Attempts (number)	Time for last attempt (sec)	Difficulty (4–20)	Quality (1–5)	Cellularity (1–3)
Sternum	$1.7 \pm 0.3^{\#}$	50 ± 11	$6.2 \pm 0.5^{**}$	2.2 ± 0.3	2.3 ± 0.1
	(1–4)	(39–166)	(4–10)	(0–4)	(2–3)
Humerus	$1.0 \pm 0^{\#,\$}$	43 ± 6	$8.4 \pm 0.4^{**}$	2.8 ± 0.4	2.3 ± 0.1
	(1)	(20–94)	(4–10)	(0–5)	(2–3)
Ilium	$1.7 \pm 0.2^{\$}$	$122 \pm 49^{*}$	$12.0 \pm 0.5^{**}$	2.6 ± 0.4	2.3 ± 0.1
	(1–3)	(41–790)	(7–15)	(0–5)	(2–3)

$^{\#,\$}$Significantly different from each other;

*significantly different from both other sites;

**each site significantly different from both other sites (ANOVA, $p < 0.05$).

and these were subsequently reviewed by a pathologist blinded to aspiration site and dog. Sample quality was scored from 0 (no marrow particles) to 5 (excellent) according to the number of well-smeared marrow particles on the slide. Cellularity of the particles was scored from 1 (> 75% fat) to 3 (< 25% fat). Necropsies were performed to examine aspiration sites and for evidence of pneumothorax.

Results [mean ± SE (range)] were:

Megakaryocyte and lymphocyte/plasma cell proportions, myeloid:erythroid ratios, and iron stores were discordant between each site (Kappa), however still within reference intervals. At necropsy, sternal needle tracks could not be identified and there were no signs of pneumothorax.

In conclusion: 1) Sternal aspiration of bone marrow using a 22–23 ga hypodermic needle is easier to perform than humeral and iliac aspiration and is safe in small dogs. 2) Sternal aspiration provides samples of equivalent quality to humeral and iliac samples in clinically normal dogs.

ABSTRACT #44

IATROGENIC HYPOTHYROIDISM (IH) CONTRIBUTES TO THE DEVELOPMENT OF AZOTEMIA IN HYPERTHYROID CATS.

TL Williams, J Elliott, HM Syme. Royal Veterinary College, London, UK.

IH is reported to occur in some cats following treatment of hyperthyroidism (HTH). Hypothyroidism reduces GFR in other species, therefore IH could contribute to the development of azotemia in hyperthyroid cats with underlying mild chronic kidney disease. Our hypothesis was that the incidence of post treatment azotemia would be greater in cats with IH than euthyroid cats.

Records from two London-based first opinion clinics between 1999 and 2009 were reviewed to identify cats diagnosed with HTH without pre-treatment azotemia. Baseline age and plasma creatinine concentration was recorded. HTH was treated using anti-thyroid medication alone or in combination with thyroidectomy. Cats were monitored for the development of azotemia over a six month follow up period, and at the final visit the following plasma biochemical and physical examination data were recorded; total thyroxine (TT4) concentration, cholesterol concentration, creatinine concentration, alanine aminotransferase and alkaline phosphatase activities, packed cell volume (PCV), heart rate and body weight. Only cats with a TT4 < 40 nmol/l at the end of the follow up period were eligible for inclusion in the final analysis. Plasma TSH concentrations at the same time point were measured using the Immulite canine TSH assay. IH was defined as TT4 < 10 nmol/l in combination with TSH > 0.15 ng/ml. Cats of uncertain thyroid status (TT4 10–40 nmol/l and TSH > 0.15 ng/ml) were excluded from the analysis. The Fisher's Exact Test was used to compare the proportion of cats with azotemia between the IH and euthyroid groups. Continuous data are presented as median [25th, 75th percentile]. Clinicopathological parameters were compared between the IH and euthyroid groups using the Mann Whitney U test. Statistical significance was defined as $P < 0.05$.

Eighty-one cats were eligible for the study, of which 28 were classified with IH and 48 as euthyroid. Five cats were excluded as they were of uncertain thyroid status. No significant difference in age ($P = 0.063$) or plasma creatinine concentration ($P = 0.963$) at baseline was present between the two groups. Sixteen cats with IH (57%; 95% CI 39–75%) and fourteen euthyroid cats (29%; 95% CI 16–42%) developed azotemia. The proportion of cats with azotemia in the IH group was significantly greater than the euthyroid group ($P = 0.028$). Cats with IH had significantly lower plasma alkaline phosphatase activity (21 [14, 41] IU/L vs. 45 [30, 75] IU/L; $P < 0.001$), PCV (30 [27, 35]% vs. 36 [30, 40]%; $P = 0.004$) and heart rate (180 [160, 192] bpm vs. 189 [171, 201] bpm; $P = 0.034$) than euthyroid cats. Cats with IH also had significantly higher plasma creatinine concentrations (2.20 [1.86, 3.14] mg/dl vs. 1.85 [1.54, 2.18] mg/dl; $P = 0.01$). No other parameters were significantly different between the two groups.

The incidence of post treatment azotemia was significantly higher in cats with IH. Clinicopathological changes in cats with IH were consistent with those that occur in naturally occurring hypothyroidism in other species.

ABSTRACT #45

EFFECT OF RECOMBINANT HUMAN TSH ON THE UPTAKE OF RADIOACTIVE IODINE (^{123}I) BY THE THYROID GLAND IN HEALTHY BEAGLES.

M. Campos[1], K. Peremans[2], L. Duchateau[3], A. Dobbeleir[2], E. Vandermeulen[2], I. van Hoek[1], G. Paes[1], S. Daminet[1]. [1]Dept. of Med. and Clin. Biology of Small Animals, [2]Dept. of Vet. Med. Imaging and Small Animal Orthopaedics, [3]Dept. of Physiology and Biometrics, Ghent University, Belgium.

In human medicine recombinant human TSH (rhTSH) increases the thyroid radioactive iodine uptake (RAIU) allowing radioiodine (^{131}I) dose reduction and higher efficacy in the treatment of differentiated thyroid cancer and multinodular goiter. It can be expected that rhTSH has a similar effect in dogs. The goal of this study was to evaluate the effect of rhTSH, administered 24 h and 48 h before radioiodine (^{123}I), on the thyroid RAIU in healthy dogs.

7 euthyroid healthy beagles were randomly divided in 3 groups in a prospective, blinded, cross-over study. On week 1, one group received ^{123}I for a baseline RAIU, one group received 100 µg of rhTSH IV 24 h before ^{123}I and one group received 100 µg of rhTSH IV 48 h before ^{123}I. All dogs received 37 MBq of radioactive ^{123}I IV and thyroid RAIU was determined 8 h, 24 h and 48 h thereafter. The study was designed in such a manner that each dog received the 3 treatments and a wash-out period of 3 weeks was respected between them. Blood samples were taken for measurement of serum total thyroxine (TT4) at baseline, 6 h, 12 h, 24 h and 48 h after rhTSH administration.

rhTSH caused no significant change on thyroid RAIU. The overall mean thyroid RAIU significantly decreased during the study independent of the treatment. rhTSH significantly increased serum TT4 concentration, which peaked 6 h after rhTSH administration. No adverse effects of rhTSH were observed during the study.

Results suggest that 100 µg of rhTSH administered IV 24 h or 48 h before radioiodine have no influence on thyroid RAIU in healthy dogs. Further studies are needed to determine the best timing, dosage and route of rhTSH administration in healthy and thyroid carcinoma dogs.

Previously presented at the 19th Annual ECVIM Congress, Porto, Portugal, September 2009.

ABSTRACT #46

EVALUATION OF CONTENT OF COMPOUNDED TRILOSTANE PRODUCTS.

AK Cook[1], CD Nieuwoudt[1], SL Longhofer[2]. [1]Texas A&M College of Veterinary Medicine and Biomedical Sciences, College Station, TX. [2]Dechra Ltd, Shrewsbury, England.

Trilostane is licensed in the USA for dogs with hyperadrenocorticism. Although the brand product (Vetoryl®) is widely available, many independent pharmacies market compounded formulations.

Compounded trilostane was purchased from eight pharmacies. Twelve separate orders were placed with each over a six week period. Every order consisted of 30 capsules of 15 mg, 45 mg or 100 mg trilostane. Upon receipt, purchases were placed in unmarked vials and assigned a random identification code. For controls, a licensed pharmacist prepared capsules containing inert material (0 mg trilostane) or 15 mg of the brand product. These were placed in batches of 30 and labeled similarly to the compounded products. Four batches of Vetoryl 30 mg and Vetoryl 60 mg were also re-packaged and submitted as controls.

Ninety-six batches of compounded trilostane and 16 control batches were submitted for analysis; 2 or 3 capsules from each batch of 30 were randomly selected for determination of content. No trilostane was detected in the 0 mg controls; these were excluded from further analysis. The averaged percentage of label claim (%LC) for the each batch of compounded product ranged from 39–152.6%, with a mean of 97.0%. In all, 20.5% of compounded capsules contained < 85% or > 115% of LC. The averaged %LC for the active ingredient control batches ($n = 12$) was 96.1–99.6%, with a mean of 97.7%. The variance for the compounded product (234.65) was significantly different ($P < 0.0001$) compared to that of the control samples (1.27). In addition, there was substantial variation in averaged %LC from one batch to the next from particular pharmacies, e.g., a 45 mg capsule may contain either 17.5 mg or 56 mg of active ingredient.

These results indicate substantial unreliability in product content for compounded trilostane. In addition, substantial batch-to-batch variability was documented. Both factors may impact treatment outcome.

ABSTRACT #47

OPTIMIZATION OF A TEST PROTOCOL TO ASSESS ALDOSTERONE SECRETORY CAPACITY IN DOGS.

KJ Carlson, EN Behrend, LG Martin, RJ Kemppainen. Auburn University, College of Veterinary Medicine, Auburn, AL.

Current testing for serum aldosterone uses a 5 μg/kg dose of Cortrosyn with sampling before and at 60 minutes post. Preliminary studies showed that a 1 μg/kg dose of Cortrosyn yields peak aldosterone concentrations at approximately 20 minutes. Thus, with the current test protocol, the peak concentration of aldosterone secretion may not be accurately being measured, making the current test non-diagnostic. The purpose of our study was to determine the ACTH dose necessary to maximally stimulate aldosterone secretion as well as to determine the time of maximal aldosterone response in relation to the dose administered.

Ten healthy, privately owned dogs were used. The dogs were randomly divided into 3 groups. ACTH (Cortrosyn) was administered at 5, 10, and 20 μg/kg intravenously. Each group received a different dose of ACTH each day of the study with a washout period of at least 2 days in between injections. Three milliliters of blood were collected from the jugular vein before and 10, 20, 30, 40, 50, 60, and 90 minutes after Cortrosyn administration. Samples were placed in a red-top tube and were centrifuged after clotting; the serum was separated and stored at −80°C until analyzed by a previously validated radioimmunoassay (Coat-a-count aldosterone assay). Data were analyzed using a 2-way ANOVA with time and dose as the main effects. Since no dose effect was detected, the data were pooled and further analysis performed by repeated measures ANOVA on ranked data. Post hoc comparisons were done with the Tukey test. Significance was set at $p \leq 0.05$.

A significant increase in serum aldosterone concentration was found at all times in comparison to baseline. Serum aldosterone concentrations were consistently increased in each dog after administration of Cortrosyn. The concentrations at peak were approximately five fold that of baseline. The peak concentration for all doses was obtained at 30 minutes post-injection and was significantly higher than concentrations measured at 10, 60 and 90 minutes.

In conclusion, 5 μg/kg cosyntropin was the minimum dose in the study that stimulated maximal serum aldosterone concentration. Serum aldosterone concentrations were consistently increased in each dog. The time of peak response was at 30 minutes, which is different from the 60-min sampling time currently recommended and suggests current testing protocols may be suboptimal.

Previously presented at Auburn University, Phi Zeta Presentation.

ABSTRACT #47a

DEVELOPMENT OF A MEASURE OF QUALITY OF LIFE FOR CATS WITH DIABETES MELLITUS AND THEIR OWNERS: THE DIAQoL-pet.

S. Niessen[1,2], S. Powney[1], J. Guitian[1], A. Niessen[3], P. Pion[4], J. Shaw[2], D. Church[1]. [1]Veterinary Clinical Sciences, Royal Veterinary College, Hatfield, UK, [2]Diabetes Research Group, Medical School, Newcastle, UK, [3]Twan, Tilburg, Netherlands, [4]VIN, Davis, USA.

Defining treatment success in veterinary diabetes mellitus (DM) is generally limited to diminishing DM related clinical signs as well as demonstrating reductions in average blood glucose values. This study reports the design and application of a novel individualized, patient and owner-centered survey measuring perceived impact of DM on quality of life (QoL) of diabetic cats and their owners. Discussions and pilot-surveys were conducted with veterinarians, diabetic cat owners and human DM QoL survey designers, leading to the design of 29 specific DM-associated QoL questions (DIAQoL-pet). Each item was scored according to frequency with which it impacts upon lives of owners and their cat (range: all the time...never) and perceived importance (range: very important ... not at all important). The Item-Weighted-Impact-Score (IWIS) for each item was calculated by multiplying frequency and importance ratings; the Average-Weighted-Impact-Score (AWIS) by averaging all IWISs. Two overview questions measured overall QoL and diabetes-dependent-QoL. 221 owners of insulin-treated diabetic cats completed the survey. 87% of respondents originated from the UK and USA. The cats' average age was 12.4+/−3.1years (mean+/−standard deviation); body weight 5.7+/−1.9 kg; 64% were domestic short or long hair. Average insulin dose was 0.47+/−0.5 iu/kg and 82% received insulin 12 hourly. The 29-item DIAQoL-pet had high internal consistency reliability (Cronbach α=0.89) and could be summarized into an AWIS of −1.76+/−2.4. Areas reported as most negatively impacting on QoL (based on IWIS) included: 'boarding difficulties' (−4.67+/−5.3), 'owner wanting more control over DM' (−4.34+/−4.7), 'difficulties leaving cat with friends/family' (−4.21+/−4.7), 'worry about cat's DM' (−4.10+/−3.9), 'hypoglycemia worry' (−3.67+/−3.5), 'adapting social life' (−3.48+/−3.9), 'DM-related costs' (−3.04+/−3.8), 'adapting work life' (−3.03+/−3.7), 'pet's moods' (−2.87+/−2.9) and 'limiting activities like holidays' (−2.81+/−3.9). 51% considered their cat's present QoL 'as good as it could possibly be'; 49% felt their cat's life would be 'a little better' without DM. The DIAQoL-pet showed high internal consistency reliability and identified specific areas most negatively impacting on the QoL of a group of diabetic cats and their owners. This tool may prove useful for clinical use and whilst testing new diabetic treatment options.

Previously presented at the 19th Annual ECVIM Congress, Porto, Portugal, September 2009.

ABSTRACT #48

COMPARISON OF LOW-DOSE AND STANDARD-DOSE ACTH STIMULATION TESTS IN CRITICALLY ILL DOGS BY ASSESSMENT OF SERUM TOTAL AND FREE CORTISOL CONCENTRATIONS.

LG Martin,[1] EN Behrend,[1] MK Holowaychuk,[2] NL Graf,[3] HP Lee[1]. [1]Auburn University, College of Veterinary Medicine, Auburn, AL; [2]North Carolina State University, College of Veterinary Medicine, Raleigh, NC; [3]Washington State University, College of Veterinary Medicine, Pullman, WA.

Use of low-dose ACTH stimulation testing may be important in diagnosing mild degrees of adrenocortical insufficiency and a new syndrome, critical illness-related corticosteroid insufficiency (formerly known as relative adrenal insufficiency), which has been recognized in critically ill human and veterinary patients. Both 0.5 and 5 μg/kg doses of cosyntropin cause maximal adrenal stimulation with regard to serum cortisol concentrations in healthy dogs. The purpose of this study was to determine if low-dose ACTH stimulation testing was as accurate as standard-dose ACTH stimulation testing in determining adrenal function in critically ill dogs.

Two dose-response trials were performed in 35 critically ill dogs admitted to the ICU at Auburn University, North Carolina State University, or Washington State University which had diagnoses of sepsis, trauma, or GDV. The same dose-response trials were also performed in 10 clinically healthy dogs. Each dog was given 0.5 and 5 μg/kg of cosyntropin IV with a 3-hour wash out period between doses. Blood samples for determination of serum total and free cortisol concentrations were obtained before and at 60 minutes after each cosyntropin administration. Samples were centrifuged after clotting, and the serum was separated and stored at −20C until analysis. Data were analyzed using the Wilcoxon Signed Rank test, Mann-Whitney Rank Sum test, and Kruskal-Wallis One Way Analysis of Variance on Ranks as appropriate. Post-hoc analysis was performed using Dunn's Method; significance was set at the $p \leq 0.05$ level.

In the critically ill dogs, median post-ACTH serum total and free cortisol concentrations did not differ between doses. Every critically ill patient that was identified to have insufficient adrenal function (ACTH-stimulated serum total or free cortisol concentration below the reference range or no response to stimulation) by the standard-dose ACTH stimulation test was also identified by the low-dose test. Six additional dogs were identified with adrenal insufficiency by the low-dose ACTH stimulation test, which were not identified by the standard-dose test. Median post-ACTH serum free cortisol differed significantly for the 0.5 μg/kg dose when septic dogs were compared to trauma dogs. Median post-ACTH serum total and free cortisol concentrations differed significantly for both cosyntropin doses when critically ill dogs were compared to healthy dogs.

In conclusion, cosyntropin administered at a dose of 0.5 μg/kg IV appears to be at least as accurate as the administration of 5 μg/kg cosyntropin in determining adrenal function in critically ill dogs. In fact, the low-dose ACTH stimulation test may be more sensitive in detecting mild degrees of adrenal insufficiency in ill dogs.

ABSTRACT #49

URINE SODIUM CONCENTRATIONS IN HYPONATREMIC DOGS WITH HYPOADRENOCORTICISM.

EM Lennon, JB Hummel, SL Vaden. North Carolina State University, Raleigh, NC.

The purpose of this study was to characterize urine sodium concentrations in hyponatremic dogs with and without hypoadrenocorticism. In the hyponatremic state, the kidneys should conserve sodium; expected urine sodium concentrations are less than 25 mEq/L. However, in the case of aldosterone deficiency, the kidneys cannot conserve sodium; urine sodium concentrations are increased. Medical records of hyponatremic dogs that had urine sodium concentrations measured were reviewed. Dogs were excluded if diuretics, mineralocorticoids, or intravenous or subcutaneous fluids were administered prior to urine collection. Fifteen dogs were included; 7 were diagnosed with hypoadrenocorticism and 8 were diagnosed with hyponatremia of other etiology (control), including pyelonephritis, biliary mucocele, diabetic ketoacidosis, and heart disease. A Wilcoxon sum rank test was used to compare results between groups. Median urine sodium concentrations in hypoadrenocortical dogs was 125 mEq/L (interquartile range (IQR): 76.5–188.5) and in control dogs was 22 mEq/L (IQR: 7.5–58). Dogs with hypoadrenocorticism had significantly higher median urine sodium concentrations than control dogs ($P = 0.004$). Diagnoses of the two control dogs with urine sodium > 25 mEq/L included diabetic ketoacidosis with pyelonephritis, and biliary mucocele. These results suggest that urine sodium concentrations can be used to help to characterize the etiology of hyponatremia in dogs. Measurement of a single urine sodium concentration > 25 mEq/L in a hyponatremic dog is supportive of, but not diagnostic for, hypoadrenocorticism.

ABSTRACT #50

EVALUATION OF CYTOKINES, CHEMOKINES, AND HORMONES IN DOGS WITH DIABETIC KETOACIDOSIS AND UNCOMPLICATED DIABETES MELLITUS

SC O'Neill[1], KJ Drobatz[1], E Satyaraj[2], RR Pelker[2], RS Hess[1]. [1]Department of Clinical Sciences, Matthew J. Ryan Veterinary Hospital of The University of Pennsylvania, Philadelphia, PA and [2]Nestle Research Center, St. Louis, MO.

The pathophysiology of canine diabetic ketoacidosis (DKA) is incompletely understood. Dysregulation of inflammatory cytokines has been documented in humans with DKA or poorly regulated type I diabetes. The purpose of this study was to compare cytokine, chemokine and hormone concentrations in dogs with DKA before and after resolution of ketoacidosis, and to compare concentrations before treatment of DKA to those measured in dogs with uncomplicated diabetes mellitus (DM) and healthy dogs. GMCSF, IL-2, IL-4, IL-6, IL-7, IL-8, IL-10, IL-15, IL-18, IFNã, IP-10, TNFα, Monocyte Chemoattractant Protein-1 (MCP-1), Keratinocyte-derived chemokine (KC), GLP-1, glucagon, insulin, leptin, adiponectin, and resistin were assayed using the Milliplex MAP Canine kits (Millipore Corporation, Billerica, MA).

27 dogs were enrolled (9 DKA, 9 DM, and 9 healthy age-matched dogs). Inclusion criteria for the DKA group included a blood β-hydroxybutyrate (BOHB) concentration $\geq$ 2.0 mmol/L and a venous pH < 7.35. Dogs in the DM group had a blood BOHB concentration < 2.0 mmol/L and a venous pH ≥ 7.35. Data are presented as median (min-max) and all concentrations are in pg/ml.

Resistin concentrations were significantly higher in dogs with DKA before treatment compared to after resolution of ketoacidosis and compared to healthy controls (65,430.9 (13,096.9–158,563) vs. 45,905.9 (9,105.6–135,108.1), $p = 0.046$ and 18,424.5 (9,629.5–28,978.7), $p = 0.004$, respectively). IL-2 and IL-18 were significantly higher in dogs with DKA before treatment compared to after resolution of ketoacidosis (IL-2: 12.8 (2.4–274.7) vs. 5.6 (1.7–248.0), $p = 0.028$ IL-18: 32.1 (8.9–305.2) vs. 26.6 (4.4–155.5), $p = 0.025$). GMCSF was significantly lower in dogs with DKA before treatment compared to after resolution of ketoacidosis (67.2 (20.9–247.9) vs. 69.5 (20.9–225.8) $p = 0.046$). IL-8 was significantly higher in dogs with DKA (3,077.2 (1,089.6–18,059.7), $p = 0.001$) and DM (2,591.7 (1,405.3–3,583.0), $p = 0.009$) compared to healthy controls (1,064.7 (827.8–1,906.7)). MCP-1 was also significantly higher in dogs with DKA (264.3 (126.9–478.1), $p = 0.019$) and DM (284.7 (116.1–422.6), $p = 0.012$) compared to healthy controls (100.9 (64.9–393.6)). KC was significantly higher in DKA dogs (2,032.4 (1,173.4–4,163.8)) compared to DM (734.1 (416.7–2,579.0), $p = 0.019$) and healthy controls (634.7 (221.5–982.5), $p = 0.003$). Significant differences were not detected in concentrations of the other analytes that were measured.

In conclusion, KC and resistin were higher in DKA dogs compared to DM and healthy control dogs. Resolution of ketoacidosis resulted in significant decreases in resistin, IL-2, and IL-18. IL-8 and MCP-1 were higher in all diabetic dogs compared to healthy control dogs. GMCSF was higher at resolution of ketoacidosis. Dysregulation of inflammatory cytokines, chemokines, and hormone concentrations may be involved in the pathogenesis of DKA and DM, in dogs.

ABSTRACT #51

GLARGINE ADMINISTERED INTRAMUSCULARLY IS EFFECTIVE FOR TREATMENT OF FELINE DIABETIC KETOACIDOSIS.

RD Marshall[1,2], JS Rand[1], MN Gunew[2], VH Menrath[2]. [1]Centre for Companion Animal Health, Uni of Queensland, Australia, [2]The Cat Clinic, Brisbane, Australia.

Therapies recommended for treatment of DKA (diabetic ketoacidosis) in cats and the reported survival rates (78–82%) have not changed substantially in 20 years. There is no short-acting insulin registered for feline use and historically regular (crystalline) insulin is used for DKA. The glucose lowering effects of intravenous glargine and regular insulin in human patients are similar. The aim of this report is to describe outcome of administering glargine intramuscularly (IM) with or without subcutaneous (SC) glargine for initial stabilization of DKA in cats.

The study involved a retrospective case series presented to an exclusively feline veterinary clinic. Included cases had a diagnosis of DKA and were initially treated with IM glargine with or without SC glargine. Cats were diagnosed with DKA when the blood glucose concentration was greater than 16 mmol/L and signs of systemic illness, glycosuria (3–4+), ketosis (ketonuria and/or ketonemia (betahydroxybutyrate > 0.6 mmol/L)) and acidosis (serum bicarbonate < 12 mmol/L) were evident. Approximately half the cases (7/15) were primary access and half were referrals. The protocol for insulin therapy using glargine was adapted from published protocols using regular insulin IM for managing DKA in cats, but substituting glargine. Initial IM and SC doses of glargine were calculated on a per cat basis rather than per kg or on the degree of hyperglycemia. All cats were initially administered intramuscular glargine (1–2 U) and in 12/15 cats this was combined with SC glargine (1–3 U); in 3/15 cats SC glargine was begun later. Intermittent IM glargine was administered as required 2–22 hours later and SC glargine (1–2 U) every 12 hrs.

Cats were managed with subcutaneous glargine alone once appetite returned and dehydration had resolved. The outcome of using glargine IM in the initial stabilization of DKA cats was considered successful if ketosis resolved, appetite returned and the cat was discharged from hospital on SC glargine and survived for more than two weeks without requiring re-admission to hospital for management of DKA or related complications.

All 15 cats survived and were discharged from hospital (median 4 days; range 2–5 days) and one third (5/15) of cats surviving to discharge subsequently achieved remission (median time 20 days; range 15–29 days). Of the 12 cats treated with intermittent IM and SC glargine from the outset, half (6/12) were managed with SC glargine as their sole insulin therapy within 18 hrs.

This study demonstrates that glargine administered intramuscularly is an effective treatment for DKA in cats, and thus provides an alternative to regular insulin. It may also be more cost effective for owners because the same vial used for initial treatment of DKA can then be used for later management with subcutaneous glargine injections. A prospective randomized controlled trial is required to compare outcomes for intramuscular and intravenous administration of glargine and regular insulin in DKA cats.

ABSTRACT #52

AN INTERMITTENT INSULIN PROTOCOL IMPROVES METABOLIC ACIDOSIS FASTER THAN A CONTINUOUS RATE INFUSION OF REGULAR INSULIN IN FELINE DIABETIC KETOACIDOSIS. S Buob, OM Mahony, EA Rozanski, LM Freeman. Tufts Cummings School of Veterinary Medicine, North Grafton, MA.

Diabetic ketoacidosis (DKA) is a severe complication of diabetes mellitus with intensive treatment frequently required for a successful outcome. In people with DKA, studies have demonstrated the efficacy of subcutaneous (SQ) rather than intravenous administration of fast-acting insulin, thereby avoiding the costs of admission to the ICU and minimizing the length of hospital stay. A study in children with DKA has shown that the addition of subcutaneous insulin glargine to a continuous rate infusion (CRI) of regular insulin has led to a faster resolution of acidosis. The purpose of this study was to compare a simplfied regimen of intramuscular regular insulin and SQ insulin glargine to a CRI regular insulin protocol in cats with DKA.

Cats presenting with DKA defined as a metabolic acidosis accompaning hyperglycemia and ketonuria or ketonemia, were eligible for inclusion. Newly and previously diagnosed diabetics were included. Cats were randomized to receive insulin therapy with either a CRI of regular insulin adjusted via a sliding scale (CRI group) or a simplified protocol which included glargine insulin SQ twice daily and intramuscular regular insulin up to every 8 hours (Intermittent group). The primary endpoint was defined as time to resolution of metabolic acidosis, with secondary objectives including 1) Duration of hospitalization, 2) Nadir of hematocrit, 3) Hours until resolution of ketonemia. Baseline characteristics of the two groups were compared using independent t-tests while changes between the two groups over time were compared using ANOVA with repeated measures with a $p < 0.05$ considered significant.

Sixteen cats were enrolled ($n = 8$/group). There was no significant difference in age, weight, sex, pH at admission, ketonemia, or chemistry profile and complete blood count results between the two groups at baseline. Eleven of the sixteen cats enrolled in the study survived to discharge (CRI: $n = 6$; Intermittent: $n = 5$; $p = 0.29$). Cats in the intermittent group had a significantly faster time to resolution of metabolic acidosis [median = 16 hr (range, 6–28 hr)] compared to the CRI group [median = 38 hr (range, 11–96 hr); $p = 0.02$]. No significant differences were detected between the two groups for hospitalization time ($p = 0.10$), nadir of the hematocrit ($p = 0.65$), or resolution of ketonemia ($p = 0.50$).

These data support the use of an intermittent protocol with a long-acting insulin for cats with DKA. The intermittent protocol resulted in faster resolution of metabolic acidosis and was simple, likely requiring less technical time and supplies than the CRI protocol. These results support a more simplified but effective approach to treating cats with DKA that may be more cost-efficient for owners.

ABSTRACT #53

OBESITY IN PHYSICALLY INACTIVE DOMESTIC SHORT-HAIRED CATS IS ASSOCIATED WITH A HYPER-COAGULABLE STATE. CR Bjornvad, B Wiinberg, ST Traberg, AT Kristensen. Department of Small Animal Clinical Sciences, University of Copenhagen, Denmark.

Obesity predisposes to a prothrombotic state in humans, but whether obesity affect hemostasis in animals is unknown. The purpose of this study was to examine differences in hemostatic parameters including thromboelastography (TEG) between lean, overweight and obese client owned indoor confined physically inactive cats.

Following health screening by physical examination, urinalysis, hematology, coagulation and biochemical profiles, 73 cats (35 males and 38 females) were included in the study and informed consents were obtained. All cats were anaesthetized and body fat percentage (BF) was measured using Dual-Energy X-ray Absorbtiometry. A TEG analysis was performed with tissue factor (1:50,000) as activator. Data distribution was assessed using D'Agostino and Pearson omnibus normality test. The correlation between BF and individual TEG and traditional coagulation parameters (platelet count, PT, APTT, fibrinogen and D-dimer) was evaluated using Pearsons Rank Test. Significant difference between lean (BF $< 30\%$, $n = 12$), overweight ($30\% <$ BF $< 40\%$, $n = 25$) and obese (BF $> 40\%$, $n = 36$) cats were identified using ANOVA. Results are given as Median (range).

BF correlated positively with the rate of clot formation (α, $P < 0.03$) and absolute clot strength (MA, $P < 0.002$). Rate of initial fibrin formation (R) was fastest in obese cats (9.4 min (3.4–17) vs 11.5 min (3.9–20.5), pooled lean and overweight cats, $P < 0.05$). Significance was not achieved for other TEG parameters or traditional hemostasis parameters.

In conclusion, Obesity in cats is associated with a hypercoagulable state as measured by TEG and it is defined by both faster clot formation and increased clot strength. This may be linked to obesity-related low grade inflammation.

ABSTRACT #54

SEASONALITY AND VECTOR-BORNE DISEASE SCREENING IN SOUTHERN CALIFORNIA DOGS WITH IMMUNE MEDIATED HEMOLYTIC ANEMIA. R Rasmussen[1], P Slusser[2], E Chaplow[3], K Richter[4], S Hill[4], J Mott[5], B Werner[6], R Martin[7], D Kobayashi[8], D Feldman[9], L Kidd[1]. [1]Western University of Health Sciences College of Veterinary Medicine, Pomona, CA. [2]Animal Internal Medicine, San Diego, CA. [3]Animal Specialty Group, Los Angeles, California. [4]Veterinary Specialty Hospital, San Diego, CA. [5]Pasadena Veterinary Specialists, Pasadena, CA. [6]Pet Emergency Clinic, Thousand Oaks, California. [7]Brentwood Pet Clinic, Los Angeles, CA. [8]Advanced Veterinary Care Center, Lawndale, CA. [9]Los Angeles Veterinary Specialists, Los Angeles, CA.

IMHA is an important cause of morbidity and mortality in dogs. Genetic predisposition and triggers such as infection, neoplasia or drugs mediate the loss of immunological self-tolerance. IMHA is deemed "idiopathic" when no underlying cause is found. Idiopathic IMHA is seasonal in the Eastern US, suggesting undetected environmental triggers. It is not known if idiopathic IMHA is seasonal in Southern California.

Underlying disease testing often includes a serologic "tick panel" or a vector-borne disease PCR panel. However, causative agents may be excluded by such grouping. Furthermore the sensitivity of serology and PCR depend on the pathophysiology of the organism. For some organisms infection may be missed when testing is limited to one methodology.

We hypothesized that idiopathic IMHA is seasonal in Southern California, and that *Babesia, Bartonella, Mycoplasma* spp and *Anaplasma phagocytophilum* (organisms associated with IMHA) are not included vector-borne disease screening (VBDS) as often as *Rickettsia rickettsii, Borrelia burgdorferi* (organisms not associated with IMHA) and *Ehrlichia canis.*

To investigate the seasonality of idiopathic IMHA, cases were identified by searching patient databases from two large referral practices in San Diego and Los Angeles. Dogs presenting between June 2006 and June 2009 were included. Those with underlying disease were excluded. Increased incidence was predicted for May, June, July, September, October and November. To determine the frequency of VBDS, records of IMHA cases in Western University's student case log database for 7 affiliate teaching hospitals were examined. Anemia and spherocytosis, autoagglutination, or a positive Coomb's test were required for inclusion in both parts of the study.

25 cases of idiopathic IMHA occurred in the predicted months, while 11 cases occurred in other months (Chi squared test corrected for continuity, $p < 0.05$). 26 cases of IMHA were screened for vector borne disease. *E canis* ($n = 25$), *B burgdorferi* ($n = 25$) and *R rickettsii* ($n = 23$) were tested for nearly twice as often as other organisms. No patient was tested using both serology and PCR.

We conclude the incidence of idiopathic IMHA is seasonally variable in Southern California. Some causative organisms may be overlooked using currently available "panels" packaged by groups of organisms and methodology. The role of undetected tickborne illness in dogs with idiopathic IMHA deserves further investigation.

ABSTRACT #55

CHEMISTRY PANEL AND ACID-BASE VALUES IN RETIRED RACING GREYHOUNDS USING A CRITICAL CARE ANALYZER. S Zaldívar-López, N Westendorf-Stingle, MC Iazbik, E Cooper, CG Couto. The Ohio State University College of Veterinary Medicine, Columbus, OH.

Greyhounds have several clinicopathologic peculiarities that differ from other dog breeds. Since the adoption of retired racers is increasing every year, the knowledge of reference values in healthy Greyhounds is vital in daily clinical practice. Some biochemical values were reported decades ago in actively racing Greyhounds but, to our knowledge, values for full chemistry and acid/base panel have not been published in healthy retired racers using a critical care analyzer. The purpose of this study was to assess differences when compared to other breeds.

Samples were obtained from the jugular vein of healthy adult dogs (57 Greyhounds and 30 non Greyhounds) and placed into heparinized tubes. Chemistry full panel was performed with a Stat Profile® Critical Care Xpress Analyzer (Nova biomedical, Waltham, MA). Results obtained for each group were compared statistically using unpaired t-test or Mann-Whitney test, depending upon normality of the sample.

Greyhounds had lower potassium ($p = 0.0376$; median, 3.930 mmol/L), ionized calcium ($p < 0.0001$; mean, 4.616 mg/dl) and ionized magnesium ($p = 0.0027$; mean, 1.224 mg/dl) concentrations than the non-Greyhound group. In contrast, glucose ($p = 0.0005$; median, 109 mg/dl), creatinine ($p < 0.0001$; mean, 1.887 mg/dl), and standard bicarbonate ($p = 0.0111$; mean, 22.84 mEq/l) concentrations were higher in retired racers.

Some results of this study contrast with previous reports in racing Greyhounds.This study provides reference values for chemistry and acid/base panels in healthy adult Greyhounds, necessary to further detect abnormalities in patients.

ABSTRACT #56

SKIN FROM HORSES WITH HEREDITARY EQUINE REGIONAL DERMAL ASTHENIA (HERDA) CONTAINS COLLAGEN CROSSLINKING PATTERNS THAT CAUSE REDUCED TENSILE STRENGTH. A Hill[1], E Schwarz[2], E Cooke[1], A Rashmir-Raven[1], M Pasquali[3], CE Swiderski[1]. [1]Mississippi State University College of Veterinary Medicine, Mississippi State, MS; [2]ARUP Institute for Clinical and Experimental Pathology, Salt Lake City, UT; [3]University of Utah Departments of Pathology, Salt Lake City, UT.

Hereditary equine regional dermal asthenia (HERDA) is an autosomal recessive connective tissue disorder affecting horses of Quarter Horse lineage.[1] Although a mutation in the cyclophilin B gene has been linked to HERDA[2], its causal association with the disease is not yet documented. Clinical parallels between HERDA and the Ehlers-Danlos Syndromes (EDS) prompted our laboratory to investigate the intermolecular cross-links of collagen, specifically pyridinoline (PYD) and deoxypyridinoline (DPD), which give collagen its tensile strength. We have previously demonstrated that horses homozygous for the HERDA trait can be identified from birth by a significant elevation of urine DPD and DPD:PYD ratio[3] and have significant decreases in skin tensile strength which are not restricted to the dorsum.[4] These findings are similar to those of human EDS VIA.[5] The purpose of this study was to determine if the decrease in skin tensile strength of HERDA horses could be explained by a similar collagen cross linking pattern in the skin of HERDA horses as has been identified in their urine. The collagen cross linking pattern and amino acid (AA) composition of skin acid hydrolysates from 22 horses (6 homozygotes, 6 HERDA carriers, and 10 HERDA negative horses) was determined using reverse-phase HPLC and a Biochrom AA analyzer, respectively. Data subjected to one way ANOVA with a Bonferroni post hoc analysis indicate significantly lower concentrations of PYD and significantly higher DPD and DPD:PYD ratios in HERDA homozygotes versus negative or carrier horses. Hydroxylysine (HL) concentrations were significantly lower in affected versus negative and carrier horses. Total pyridinium cross links were not significantly different between groups. This alteration in the biochemical composition of HERDA skin, which is known to correlate to poor tensile strength in human EDS, is the first evidence to explain the clinical syndrome of skin fragility that is characteristic of HERDA. Furthermore, the significant decrease in HL (for the precursor of pyridinium crosslinks) in HERDA horses suggests that altered collagen lysine metabolism may be physiologically relevant to the clinical manifestation of the disease. (References on request)

ABSTRACT #57

USE OF AN EQUINE LENTIVIRUS VECTOR TO DELIVER PATHOGEN GENES TO DENDRITIC CELLS FOR INDUCTION OF CYTOTOXIC T LYMPHOCYTE RESPONSES IN HORSES. RH Mealey[1], RL Tallmadge[2], S Carpenter[3], SR Leib[1]. [1]College of Veterinary Medicine, Washington State University, Pullman, WA. [2]College of Veterinary Medicine, Cornell University, Ithaca, NY. [3]Department of Animal Science, Iowa State University, Ames, IA.

A key feature of effective immune responses against viruses and intracellular bacteria is the ability to destroy infected cells. Cytotoxic T lymphocytes (CTL) have this ability and are a critical component of the adaptive immune response. Unfortunately, eliciting protective CTL against several important intracellular pathogens by vaccination has been difficult in the horse. Efficient priming of naïve CTL requires that antigens be delivered to and processed intracellularly by dendritic cells, which are derived from bone marrow precursors and are the principle antigen presenting cells for initiating T cell responses. Equine infectious anemia virus (EIAV) is a lentivirus that causes persistent infection in horses, and CTL are critical in its control. Importantly, EIAV directly infects equine dendritic cells. Thus, a replication-defective vaccine vector based on EIAV could provide an effective means to elicit CTL in horses against a variety of intracellular pathogens. The purpose of this study was to determine if such a vector could elicit durable CTL responses in horses.

An EIAV Gag-expressing EIAV-based lentiviral vector was generated by co-transfection of 293 T cells with three plasmids: the EIAV viral core encoded by pEV53B; EIAV gag matrix and capsid by pSIN6.1p15/p26; and the $EIAV_{19}$ codon-optimized envelope encoded by pLGcoSU-TM. The vector was designated vSIN6.1-p15/p26 and was confirmed to be replication incompetent after initial infection of equine kidney (EK) cells by repetitive passage of supernatants. Dendritic cells were transduced by this vector as determined by RT-PCR and immunocytochemistry. Furthermore, transduced dendritic cells were capable of stimulating Gag-specific memory CTL from persistently EIAV infected horses, as determined by ^{51}Cr release CTL assays using EIAV-infected, Gag peptide-pulsed, and vSIN6.1-p15/p26-transduced EK targets.

Finally, 10^7 transducing units (TU) of vSIN6.1-p15/p26 were administered by a combination of ID and IM injections to each of two horses, and boosted three months later. Gag-specific CTL responses were elicited in both horses as early as 18 days after the first injection, and responses have persisted throughout five months of observation. Neither horse has developed clinical signs of EIA nor has viral RNA been detected in plasma using nested RT-PCR. Thus, robust and durable Gag-specific CTL responses were elicited by the vSIN6.1-p15/p26 vector, which was confirmed to be replication incompetent in vivo.

In conclusion, the EIAV-based lentiviral vector described here targets dendritic cells and safely elicits durable CTL responses in horses. This vector could prove to be an effective vaccine strategy for inducing protective CTL responses against a variety of important equine pathogens.

ABSTRACT #58

NEUTROPHIL FUNCTION IN HEALTHY AGED HORSES AND HORSES WITH PITUITARY DYSFUNCTION. Dianne McFarlane, Kristen McDaniel. Oklahoma St U, Stillwater, OK.

Equine pituitary pars intermedia dysfunction (PPID) is associated with an increased risk of bacterial infections, including sinusitis, pneumonia, abscesses and dermatophilus. Although little is known about the mechanism of immune dysfunction in horses with PPID, it is presumed to result from chronically increased plasma concentrations of immunosuppressive hormones, such as cortisol or α-melanocyte stimulating hormone (MSH). We hypothesized that neutrophil function is impaired in horses with PPID has a result of high plasma MSH concentration.

Three groups of horses were included, healthy young horses (mean age = 8.9 ± 4.0 yrs), healthy aged horses (mean age = 22.5 ± 4.5 yrs) and horses with PPID (mean age = 24.6 ± 5.7 yrs). Horses with PPID were selected based on clinical signs and confirmed by plasma α-MSH concentration. Heparinized blood (40 mls) was collected and neutrophils isolated by density centrifugation using Ficoll. Chemotaxis was assessed using IL8 and LTB4 as chemoattractants and calculating the percent migration of neutrophils through a 3 μ pore filter. Neutrophilic oxidative burst was determined by oxidation of dihydrorhodamine (DHR) and phagocytosis was measured by the uptake of Pansorbin PI (PanPI) using standard flow cytometry methodology. To determine if α-MSH suppressed neutrophilic oxidative burst, blood from 6 healthy horses was incubated with 1000 pmol/L MSH or an equal volume of PBS for 6 hours at 37°C, then processed for flow cytometry similar to other samples.

There was no difference in chemotaxis between the aged and PPID group (IL8: $P = 0.6$, LTB4: $P = 0.9$). Old horses had a greater chemotaxis than young horses (IL8: 77.1 ± 4.9 fluoroscent units (FU), n = 22 vs. 62.4 ± 5.0 FU, n = 14, $P = 0.05$; LTB4: 83.5 ± 6.6 FU, n = 22 vs. 63.8 ± 6.9 FU, n = 14, $P = 0.05$). There was no difference in phagocytosis among the three groups of horses and no affect of age on oxidative burst ($P = 0.6$). Horses with PPID had a decreased oxidative burst compared to healthy horses (mean burst: 242.2 ± 45.4 FU, n = 9 vs. 381.9 ± 39.5 FU, n = 26, $P = 0.06$). Whole blood incubated with MSH showed a significantly reduced oxidative burst compared to saline treated blood (mean reduction = 25.8%, $P < 0.05$). These data suggest high plasma α-MSH concentration may impair neutrophil oxidative burst and contribute to immunosuppression in horses with PPID.

ABSTRACT #59

THE EFFECT OF LATITUDE OF RESIDENCE, BREED AND PITUITARY DYSFUNCTION ON SEASONAL PLASMA ACTH AND MSH CONCENTRATION.

McFarlane D[1], Paradis MR[2], Zimmel D[3], Sykes B[4], Brorsen W[1], Sanchez A[2], Vainio K[4]. [1]Oklahoma State University, Stillwater OK. [2]Tufts University, N. Grafton, MA. [3]University of Florida, Gainesville, FL. [4]Helsinki, Finland.

Previous studies suggest season regulates equine pituitary pars intermedia (PI) function, affecting plasma ACTH and MSH concentration and confounding diagnostic testing for pituitary pars intermedia dysfunction (PPID). Studies to date have examined horses residing in a restricted geographic location, making it unclear if the seasonal effect is geographically biased. Preliminary data also suggests ponies have a greater increase in plasma PI hormone concentration in the fall compared to light horses. It is unknown if other "thrifty breeds" differ from light horses. It is also unclear whether seasonal hormone fluctation is abolished in horses with PPID. The goals of this study were to determine if month-to-month variation in plasma α-MSH and ACTH concentration in horses is influenced by geographic latitude of residence, breed or PPID.

Monthly plasma samples were collected over one year from healthy light breed horses residing in Florida, Massachusetts and Finland (n = 12/location). In addition, healthy Morgan horses (n = 13) and ponies (n = 9) were selected from MA and Fin, respectively. Horses with a plasma ACTH concentration > 50 pg/ml, MSH > 35 pmol/L, evidence of equine metabolic syndrome, or those pregnant or lactating were excluded. Horses with clinical signs of PPID were selected from MA and confirmed by either increased plasma ACTH or MSH concentration. Plasma ACTH and MSH concentration were determined by RIA. Cosinor analysis was used to determine the time of peak hormone concentration. Mean hormone concentration in fall and non-fall months were compared by ANOVA or t-test. Differences were considered significant if $P < 0.05$.

Horses in Finland had a significantly earlier fall peak of MSH (Aug 29) compared to horses in MA (Sept 10) or FL (Sept 15). ACTH rhythm was difficult to ascertain due to the relatively minimal seasonal fluctuation. Mean MSH concentrations were similar in all three locations, while ACTH concentration was greater in the fall in FL compared to MA ($P < 0.01$) and Fin ($P < 0.05$) and greater in MA than Fin ($P < 0.05$) in the non-fall months. Morgan horses had increased MSH and ACTH concentration compared with light horses from the same location during the fall months; there was no difference during the non-fall months. Hormone concentrations in ponies were not different than in horses during either season. In healthy horses, 8% of the non-fall samples had ACTH > 50 pg/ml and 1% had MSH > 35 pmol/L. In horses with PPID, 67% of non-fall samples had ACTH concentration > 50 pg/ml and 83% had MSH > 35 pmol/L. Both MSH and ACTH concentrations were significantly greater in the fall compared to spring in horses with PPID ($P < 0.01$). In summary, latitude of residence and breed may affect onset and/or amplitude of the seasonal increase in PI hormones. Horses with PPID continue to have seasonal variation in PI hormone concentrations.

ABSTRACT #60

CHARACTERIZATION OF LEPTIN AND ADIPONECTIN GENE EXPRESSION PATTERNS IN VARIOUS ADIPOSE TISSUE DEPOTS OF INSULIN-RESISTANT AND INSULIN-SENSITIVE LIGHT BREED HORSES.

Burns TA[1], Barsnick R[1], Toribio RE[1], Mudge MC[1], Geor RJ[2], McCutcheon LJ[2], Belknap JK[1]. [1]The Ohio State University, Columbus, OH. [2]Michigan State University, East Lansing, MI.

Obesity and insulin resistance have been associated with laminitis in equids. Adipose tissue (AT) produces many hormones (adipokines) that have important local and systemic effects on metabolism; two well-characterized adipokines are leptin and adiponectin. Leptin is an important regulator of satiety, and adiponectin can modulate inflammation and insulin sensitivity. In humans, expression of leptin and adiponectin in subcutaneous AT has been shown to decrease in reponse to chronic exposure to pro-inflammatory cytokines (PIC); however, acute inflammation increases AT leptin expression, suggesting a biphasic tissue leptin response to PIC. Recent work has shown depot-specific differences in PIC gene expression in equine AT, with nuchal ligament AT expression of IL-1 and IL-6 significantly higher than that of visceral and other subcutaneous AT depots. Depot-specific adipokine expression patterns (which may relate to PIC expression) have not been reported in the horse. This study evaluated gene expression of leptin and adiponectin in several AT depots of insulin-resistant (IR) and insulin-sensitive (IS) horses. Eleven mares receiving an identical diet (8 Quarter horses, 2 Thoroughbreds, and 1 Standardbred) and categorized as IR (insulin sensitivity [SI] = 0.58 ± 0.31 $L{\bullet}min^{-1}{\bullet}mU^{-1}$; n = 5) or IS (SI = 2.59 ± 1.21 $L{\bullet}min^{-1}{\bullet}mU^{-1}$; n = 6) based on results of an insulin-modified frequently-sampled intravenous glucose tolerance test were studied. Omental, retroperitoneal, and mesocolonic AT was collected via ventral midline celiotomy; nuchal ligament and tail head AT biopsies were collected via skin incision. Tissue samples were snap-frozen; plasma samples were collected concurrently. For each depot, total RNA was extracted and cDNA produced and analyzed via real-time quantitative PCR to quantify expression of leptin and adiponectin; plasma leptin concentrations were measured with a commercially-available radioimmunoassay (Millipore; Temecula, CA). Data were analyzed using one-way ANOVA for repeated measures or non-parametric equivalent (p = 0.05). The expression of leptin and adiponectin were significantly lower in nuchal ligament AT compared with other depots (p = 0.006 and p = 0.0001, respectively), with no observed effect of insulin sensitivity status. Plasma leptin concentration did not differ significantly between the IR and IS horses, nor did it differ between these horses (BCS = 7.2 +/− 0.7; scale of 1–9) and a lean historical cohort of horses (n = 13; BCS = 6.1 +/− 1.0) (p = 0.38). The finding of decreased leptin and adiponectin expression in nuchal ligament AT is in keeping with the previously reported PIC expression data from this depot and suggests a relatively stable (chronic) inflammatory state in the tissue. These data provide further support for the hypothesis that nuchal ligament AT displays unique biological behavior in the horse relative to other depots.

ABSTRACT #61

EFFECTS OF METFORMIN HYDROCHLORIDE ON GLUCOSE DYNAMICS DURING TRANSITION TO GRASS PADDOCKS IN INSULIN-RESISTANT HORSES. K Chameroy, N Frank, SB Elliott. Department of Large Animal Clinical Sciences, College of Veterinary Medicine, University of Tennessee, Knoxville, TN.

Metformin hydrochloride is used as a treatment for insulin resistance (IR) in equids. This drug may also be useful for controlling IR when insulin-resistant horses are reintroduced to pasture. We therefore tested the hypothesis that metformin alters glucose dynamics in insulin-resistant horses undergoing reintroduction to grazing. Six adult horses with chronic IR were transferred to grass paddocks after spending > 15 months confined in stalls with attached gravel turnout areas. Glucose dynamics were assessed using the combined glucose-insulin test (CGIT) before and 1, 2, 4, 6, 8, and 10 wk after horses were transferred to grass paddocks. After horses were housed in grass paddocks for 2 wk, they were treated with metformin (15 mg/kg q12h) for 2 wk, taken off treatment for 2 wk, and then treated with metformin at a high dosage (30 mg/kg q12h) for 2 wk. Horses were also reevaluated 2 wk later.

Time effects were significant ($P < 0.001$) for resting plasma glucose concentrations and area under the glucose curve (AUCg) values. Insulin results are pending. Mean glucose concentrations increased from 81.2 ± 3.4 to 99.4 ± 6.7 mg/dL over 1 wk and then decreased to 81.9 ± 4.0 mg/dL by 10 wk. No differences were detected between pre- and post-metformin mean values for resting glucose concentrations or AUCg at either dosage level. However, mean AUCg significantly decreased (18.6%) across the 10-week study period. Metformin did not affect glucose dynamics in this study, but AUCg decreased over time when insulin-resistant horses were transferred to grass paddocks.

ABSTRACT #62

DIGITAL HYPOTHERMIA INHIBITS EARLY LAMINAR INFLAMMATORY SIGNALING IN THE OLIGOFRUCTOSE LAMINITIS MODEL. A. Van Eps[1], B. Leise[2], M. Watts[2], C. Pollitt[1], J. Belknap[2]. [1]University of Queensland, St Lucia, Queensland, Australia; [2]Ohio State University, Columbus, OH.

Recent evidence suggests that laminar inflammatory events, similar to those previously reported in the black walnut extract model, also occur during the developmental phase of carbohydrate (CHO) induced laminitis. In the CHO model, early increases in laminar mRNA concentrations of the neutrophil chemokines CXCL8 and GRO are followed by increases in proinflammatory cytokines and monocyte chemokines. Digital hypothermia has been shown to reduce the severity of laminitis after induction with the CHO oligofructose (OF). Because hypothermia has been reported to reduce inflammatory injury in several diseases including traumatic brain injury in humans, we hypothesized that digital hypothermia inhibits inflammatory signaling during the development of laminitis induced with OF.

Eight horses were administered 10 g/kg OF by nasogastric tube and had one forelimb (CRYO) continuously cooled using ice and water. The other forelimb (NON-RX) was maintained at ambient temperature. Sections of dorsal laminae were harvested 24 h after OF administration for both snap freezing and paraffin embedding. mRNA was purified from 3 laminar sections of each forelimb, and cDNA produced for real time-quantitative PCR (RT-qPCR) analysis of mRNA concentrations of chemokines (IL-8, CXCL1), endothelial adhesion molecules (E-selectin, ICAM-1) proinflammatory cytokines (IL-1b, IL-6), COX-1, COX-2, and three housekeeping genes (used to generate a normalization factor). Data were analyzed using nonparametric tests.

When compared to control laminar samples (horses administered water instead of CHO during a previous study), CXCL1, IL-6, IL-8, and ICAM-1 were increased ($P < 0.05$) in the NON-RX laminae from the present study. Interestingly, digital hypothermia resulted in a marked decrease in laminar expression ($P < 0.05$) of the neutrophil chemokines (CXCL1 and CXCL-8 see graph), MCP-2, IL-1, and ICAM-1 when compared to the laminae from the untreated (NON-RX) limbs.

These data indicate that digital hypothermia effectively blocks early inflammatory events in the laminar tissue, including the expression of chemokines, proinflammatory cytokines and endothelial adhesion molecules. Inhibition of chemokines and adhesion molecules may result in decreased leukocyte emigration into the laminar tissue, an event likely to play an important role in laminar injury and failure in the septic horse.

Figure [Abstract 62]

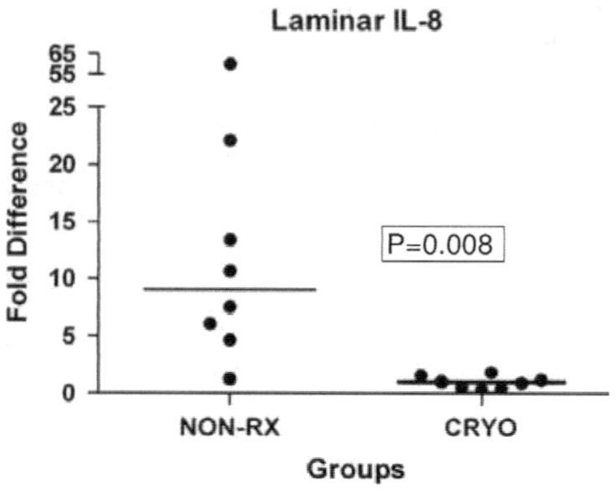

ABSTRACT #63

CONTINUOUS CARDIAC OUTPUT MONITORING IN EXERCISING HORSES USING PULSECO®. SA Jesty, DJ Fletcher, LM Mitchell, C Yaple, LV Soderholm. Cornell University College of Veterinary Medicine, Ithaca, NY.

The purpose of this study was to assess whether the relative changes in estimated cardiac output (CO) derived from uncalibrated pulse power analysis (PulseCO®) would predict changes in CO measured using an indicator-dilution method (LiDCO®), previously validated in exercising horses. Four adult healthy horses were studied. At baseline and at four stages of exercise, CO was measured using LiDCO® while continuous CO was estimated using PulseCO® applied to averaged pulse waveforms.

The ratio of the CO at each exercise level to CO at rest was compared between the LiDCO® and PulseCO® algorithms. Passing and Bablok regression analysis was used to evaluate the agreement between LiDCO® measured and PulseCO® estimated CO ratios at each exercise condition. Raw data and residuals were found to be normally distributed by the Kolmogorov-Smirnov test, meeting the requirements for the application of the regression analysis. A best fit slope and intercept, and 95% confidence intervals (CI) were determined by Passing and Bablok regression analysis (slope = 0.963, CI: .578–1.67; intercept = .360, CI: −1.68–1.69).

These data suggest that uncalibrated, signal averaged PulseCO® analysis accurately detects changes in CO in exercising in horses, but confidence intervals are wide. Further validation with larger numbers of horses is needed to verify this preliminary finding, and to determine if this minimally invasive technology will be useful for identifying subtle cardiac disease in poorly performing horses.

ABSTRACT #64

ASSESSMENT OF CONGENITAL NEOSPOROSIS IN BROODMARES ORIGINATING FROM A FARM IN CALIFORNIA WITH HIGH PREVALENCE OF ANTIBODIES TO *NEOSPORA HUGHESI*. N. Pusterla, P.A. Conrad, A.E. Packham, S. Mapes, C.J. Finno, I.A. Gardner, B.C. Barr, W.D. Wilson. School of Veterinary Medicine, University of California, Davis, CA.

The mode of transmission of *Neospora hughesi* has remained unknown; however, in comparison to bovine neosporosis caused by *N. caninum*, the parasite may potentially be transmitted transplacentally or orally via the ingestion of oocysts from a domestic or wild definitive host. The goal of this study was to investigate the possibility of congenital transmission of *N. hughesi* in a herd of horses with high prevalence of antibodies to this apicomplexan protozoal pathogen.

The study was performed over two years (2006 and 2007) and the population was composed of 60 adult Percheron broodmares living on a breeding farm in northern California. The study originated from the diagnosis of equine protozoal myeloencephalitis (EPM) due to *N. hughesi* in a 4-month old Percheron filly and from seroprevalence data showing that 15% of resident horses had antibodies to *N. hughesi* in 2005. Pre- and post-nursing serum of each foal, serum and colostrum of each periparturient mare and serum of each mare and foal pair collected at 3 months intervals were tested using an indirect fluorescent antibody test for *N. hughesi*. Further, whole blood, colostrum and placentae were tested for the presence of *N. hughesi* by real-time PCR.

The mares' seroprevalence at birth was 48% and 6.5% in 2006 and 2007, respectively. Colostral antibodies against *N. hughesi* were detected in 66% and 12% of the mares in the two study years. With the exception of three foals, all remaining foals were born seronegative to *N. hughesi* and passive transfer of colostral antibodies to *N. hughesi* was documented in 43% and 12% of the foals in 2006 and 2007, respectively. Three foals born from two different mares had pre-colostral antibodies titer ranging from 2,560 to 20,480. All three foals were born healthy and have remained healthy since birth. Two foals were born from the same dam that also gave birth to the weanling diagnosed with EPM in 2005. The third foal was born from a second mare with no previous foaling history at the farm. Seroconversion was documented in 11 foals and 6 mares over the two study years. All blood and colostrum samples tested PCR negative for *N. hughesi*. Only one placenta collected in 2007 from the mare with the two congenitally infected foals tested PCR positive for *N. hughesi*.

In conclusion, natural exposure to *N. hughesi* via the feces of as yet unidentified reservoir host was low in the study population; however, *N. hughesi* persisted in the horse population via transplacental infection.

ABSTRACT #65

COMPARISON OF 2 BIOMARKERS OF NEURONAL INJURY IN NEONATAL FOALS WITH HYPOXIC-ISCHEMIC ENCEPHALOPATHY. NC Ringger[1], S Giguère[2], PR Morresey[3], G Shaw[4]. [1]Los Angeles, CA; [2]College of Veterinary Medicine, University of Georgia, Athens, GA; [3]Rood & Riddle Equine Hospital, Lexington, KY; [4]McKnight Brain Institute, College of Medicine, University of Florida, Gainesville, FL.

Hypoxic ischemic encephalopathy (HIE) is a common disease affecting newborn foals but there are currently no ante-mortem diagnostic tests. Ubiquitin C-terminal hydrolase1 (UCH-L1) and axonal neuron-enriched neurofilament H (pNF-H) levels were measured to evaluate their efficacy as biomarkers of neuronal injury. Concentrations of UCH-L1 and pNF-H in plasma were measured by ELISA in 33 foals with a clinical diagnosis of HIE and 17 healthy foals. The performance of the assays for the diagnosis of HIE was assessed using receiver operating characteristic (ROC) curve analysis.

The overall diagnostic performance of UCH-L1 (AUC = 0.86; $P = 0.0001$) was significantly higher ($P = 0.001$) than that of pNF-H (0.52; $P = 0.497$) for the diagnosis of HIE. Median concentrations of UCH-L1 (6.57 ng/mL; 2.35 to 11.90 ng/mL) in foals with a clinical diagnosis of HIE were significantly ($P < 0.001$) higher than those of healthy controls (2.52 ng/mL; 1.4 to 4.01 ng/mL). Using a cut off of 4.01 ng/mL, the sensitivity and specificity of the UCH-L1 assay for the diagnosis of HIE were 70% and 94%, respectively. UCH-L1 has potential as a marker of brain injury in neonatal foals with CNS disorders.

ABSTRACT #66

EFFECTS OF LOW-DOSE HYDROCORTISONE THERAPY ON MEASURES OF INNATE IMMUNE FUNCTION IN NEONATAL FOALS. KA Hart, MH Barton, DJ Hurley. University of Georgia College of Veterinary Medicine, Athens, GA.

Low-dose hydrocortisone replacement therapy has been shown to improve outcome in people with Critical Illness-Related Corticosteroid Insufficiency. The objective of this study was to evaluate the effects of a proposed hydrocortisone replacement therapy regimen on *ex vivo* measures of immunity in healthy neonatal foals. A hydrocortisone dose of 1.3 mg/kg/day was divided and administered as six intravenous boluses to 11 healthy 2-day foals for 24 hours. This dose was halved for another 24 hours, halved again for 12 hours, and then discontinued. Peripheral blood leukocytes were collected from foals before treatment, at 48 hours, and 12 hours after cessation of therapy. A separate group of 12 age-matched foals received saline and served as the control group. Endotoxin-induced mononuclear cell gene expression of TNF-α, IL-1β, IL-6, IL-8, and IL-10 was measured by RTqPCR. Neutrophils were incubated with labeled, killed *S. aureus* or *E. coli* for assessment of phagocytic function, and with phorbol myristate acetate, zymosan, or endotoxin for measurement of reactive oxygen species (ROS) production. Data were compared between groups with Mann Whitney tests and ANOVA, with significance set at $P < 0.05$. No adverse effects were observed in any foals receiving hydrocortisone. Neutrophil phagocytosis and ROS production were similar in both groups. However, foals receiving hydrocortisone had significantly decreased endotoxin-induced production of TNF-α, IL-6, and IL-1β. These data suggest that this hydrocortisone replacement therapy regimen is both safe in foals and effective at ameliorating endotoxin-induced inflammatory cytokine expression without significantly altering innate immune responses needed to combat bacterial infection.

ABSTRACT #67

EFFECTS OF DEFEROXAMINE ON IRON ELIMINATION FOLLOWING BLOOD TRANSFUSION IN NEONATAL FOALS. JR Elfenbein[1], S Giguère[2], SK Meyer[1], LH Javsicas[3], DN Zimmel[1], LC Sanchez[1]. [1]University of Florida, Gainesville, FL. [2]University of Georgia, Athens, GA. [3]Saratoga Springs, NY.

Hepatic failure is one of the most common complications in foals requiring blood transfusions to treat neonatal isoerythrolysis. The strong association between development of liver failure and the volume of blood transfused as well as the nature of the histologic lesions suggest that iron toxicity is the likely cause of hepatic injury. The purpose of the study was to determine the effects of deferoxamine on iron elimination following blood transfusion in neonatal foals.

At 1–3 days of age, foals received either 3 L of washed packed mare's red blood cells (RBC) or 3 L of saline IV once. Foals were treated with deferoxamine (1 g) or saline (5 mL) subcutaneously twice daily for 14 days. Foals were randomly assigned to 1 of 3 groups: RBC/deferoxamine (deferoxamine; n = 5), RBC/saline (placebo; n = 5), or saline/saline (control; n = 3). Blood, urine, and liver biopsies were collected at selected time points for measurement of hematological, biochemical, and iron metabolism variables.

Hematocrit, RBC count, and hemoglobin were significantly higher in the 2 groups transfused with packed RBC compared to controls at all times. Biochemical variables were not significantly different between groups. Urine iron concentrations and fractional excretion of iron were significantly higher in foals treated with deferoxamine compared to controls and foals receiving a placebo. By day 14 post-transfusion, liver iron concentrations in foals treated with deferoxamine were significantly lower than that of foals receiving a placebo and similar to that of controls. This study demonstrates that deferoxamine enhances iron elimination and decreases hepatic iron concentration following blood transfusion in foals.

ABSTRACT #68

USE OF CONTRAST ECHOCARDIOGRAPHY FOR THE EVALUATION OF MYOCARDIAL PERFUSION AND PULMONARY TRANSIT TIME IN NORMAL DOGS. A PILOT STUDY. S Crosara[1], I Ljungvall[2], J Haggstrom[2], M Margiocco[3], M Borgarelli[3]. [1]University of Torino, Italy. [2]Swedish University of Agricultural Sciences, Uppsala, Sweden. [3]Kansas State University, KS.

Second generation echocardiographic contrast agents have been used in human medicine to assess myocardial perfusion (MP). More recently, contrast echocardiography has been proposed as a method to estimate pulmonary transit time (PTT) in dogs.

Six healthy dogs were enrolled in the study. Two boluses (0.03 ml/kg) and a CRI (0.035 ml/kg/min) of the contrast agent BR38® (Bracco Research SA, Switzerland) were sequentially administered to each dog. Ejection fraction (EF%), PTT and nPTT (PTT/mean RR) were measured after each bolus. EF% was also measured without contrast. The Real-time MP was performed in 4 regions of interest (ROIs) of the left ventricle: two basal and two mid-parietal. A qualitative assessment was also performed. A plasma sample was collected before and one hour after the study and cardiac Troponin I (cTnI) was analyzed by a high sensitivity assay (Beckman Coulter Accu TnI®). Intra- and inter-observer variability were calculated for the echocardiographic measurements.

There were no differences between the EF% values with and without contrast. Both PTT and nPTT values were similar to those previously reported. There was a good reproducibility for the real-time and the qualitative MP. The real-time MP had a coefficient of variation (CV%) less than 15% for the basal ROIs only, while it was higher for the mid-parietal ROIs. No differences were detected between the pre- and the post-contrast cTnI values. In conclusion, contrast echocardiography appears to be a feasible and safe technique for the evaluation of MP and PTT in healthy dogs.

ABSTRACT #69

A NEW RADIOLOGIC MEASUREMENT TO STUDY LEFT ATRIAL ENLARGEMENT IN DOGS WITH MITRAL INSUFFICIENCY: PRELIMINARY STUDY. X. Sànchez[1], D. Prandi[1], O. Domenech[2]. [1]Departamento de Medicina y Cirugía Animales, Universidad Autónoma de Barcelona, Bellaterra, Spain. [2]Clinica Veterinaria Gran Sasso, Milan, Italy.

Mitral insufficiency is the more prevalent heart disease in dogs and can induce left atrial enlargement. General heart enlargement can be monitored by radiographic objective measurement (VHS) but to date, no method to objectively estimate left atrial size on radiographs has been reported.

The aim of this prospective study was to evaluate the clinical usefulness of a new radiographic measurement, named Left Atrial Bisecting Line (LAB) to detect left atrial enlargement. LAB measurements were compared to left atrial to aortic root ratio (LA/Ao) and Vertebral Heart Size (VHS) in normal dogs and dogs with mitral insufficiency.

Physical exam, blood pressure, CBC, biochemistry panel, thoracic radiographs and echocardiographic exam were performed in 14 normal dogs (4.8–19 kg) and 28 dogs with mitral insufficiency (5.1–21.2 kg). LAB was obtained by bisecting a line from the 90 degree angle formed between the long and short cardiac axis lines used to calculate the VHS in right lateral thoracic view to the roof of the left atrium and measured in vertebral bodies length from the 4th thoracic vertebra to the nearest 0.1 vertebra (v). VHS was measured as described previously (Buchanan, 2000). LA/Ao ratio was obtained by two-dimensional ultrasonic mode from right parasternal short axis view at the level of the aortic valves in early diastole.

Results from all 42 dogs are summarized as follows:

ROC curves showed the values which provided greater sensibility and specificity for LAB, VHS and LA/Ao ratio, being respectively 1.95v, 11v and 1.75 (maximal reference values).

LAB not only showed good significant correlation with both LA/Ao and VHS but also appears to be a more specific radiographic measurement than VHS for evaluating left atrial size which can be clinically useful. LAB would bring to clinicians a simple and cost-effective tool to evaluate and monitor left atrial enlargement and help differentiating cardiac from respiratory origin of cough.

Table [Abstract 69]

	LAB	Ratio LA/Ao	VHS	Pearson Correlation	LAB	VHS	LA/Ao
Sensibility	96.4%	100%	82.1%	LAB r	1.00000	0.77614	0.75863
				p-value		< .0001	< .0001
Specificity	100%	100%	85.7%	VHS r	0.77614	1.00000	0.75959
				p-value	< .0001		< .0001
				LA/Ao r	0.75863	0.75959	1.00000
				p-value	< .0001	< .0001	

Further ongoing studies in a larger number of dogs are needed in order to validate these preliminary conclusions.

ABSTRACT #70

USE OF PIMOBENDAN IN 161 CATS (2006–2009). MacGregor JM,[1] Rush JE,[2] Laste N,[3] Malakoff R,[3] Cunningham SM,[2] Aronow N,[2] Hall DJ,[2] Williams J[3]. [1]Massachusetts Veterinary Referral Hospital, Woburn, MA. [2]Tufts University Cummings School of Veterinary Medicine, North Grafton, MA. [3]Angell Animal Medical Center, Jamaica Plain, MA.

Pimobendan is an inodilator that has been shown to be a useful treatment for dogs with congestive heart failure (CHF). There is limited data regarding the use of pimobendan in cats with heart disease. The purpose of this report is to describe use of pimobendan in a population of client-owned cats with naturally occurring heart disease. The medical records of three institutions were reviewed for cats receiving pimobendan for management of heart disease. Data were extracted from the medical records including disease process, presence of CHF, pimobendan dose, and certain echocardiographic data. The median age was 11 years (0.5–24.0 years) and cats were predominantly neutered males (n = 107) and domestic short or long haired cats (n = 129).

161 client-owned cats received pimobendan at a median dose of 0.23 mg/kg twice daily (range: 0.08 mg/kg BID - 0.42 mg/kg BID). Pimobendan use was discontinued in 5 cats. Discontinuation was due to recovery from the inciting cause in 2 cats, cost in 1 cat, difficulty administering medication in 1 cat and vocalization/agitation in 1 cat. Potential side effects were noted in 5 cats including unusual agitation in 2 cats, and anorexia, vomiting, and constipation in one cat each.

154 cats had active or historical CHF when pimobendan was first administered. In non-CHF cats, indication for pimobendan included reduced contractile function, post-surgical myocardial depression, and suspicion of imminent CHF. The inciting cause for CHF was hypertrophic cardiomyopathy in 62 cats, unclassified cardiomyopathy in 55 cats, and dilated cardiomyopathy in 27 cats. 110 cats (68%) had a fractional shortening < 40%, 107 (66%) had a left ventricular end systolic dimension > 1.0 cm and 8 (5%) cats had neither of these criteria but were judged to have systolic dysfunction based on subjective assessment of LV function on 2-D echo. Median left atrial M-mode diameter was 2.09 cm (0.98–3.43 cm).

148/161 cats were treated with furosemide either prior to or simultaneous to the initiation of pimobendan. 148 cats were treated with an angiotensin-converting enzyme inhibitor and 128 cats were treated with an anti-thrombotic agent concurrently with pimobendan.

Median survival of cats from first pimobendan administration was 102 days (0–870 days). Censored data included cats still alive (n = 47) and cats lost to follow up (n = 10). Fifteen cats died at home of unknown causes and 89 cats were euthanized. Contributing factors to euthanasia included signs of CHF (n = 33), aortic thromboembolism (FATE; n = 12), azotemia (n = 7), non-cardiac medical disorders (n = 7), anorexia (n = 5), neurologic signs (n = 4) and reasons not specified in the medical record (n = 21). Of 12 cats experiencing FATE during pimobendan administration, 5 had prior FATE. Eight cats with prior FATE experienced no recurrence while receiving pimobendan.

Pimobendan appears to be well tolerated in cats at a similar dose to the canine dose. Further study is needed to determine whether pimobendan is efficacious in managing feline heart disease.

ABSTRACT #71

EFFECTS OF IVABRADINE ON INVASIVE INDICES OF LV FUNCTION IN ANESTHETIZED CATS WITH HYPERTROPHIC CARDIOMYOPATHY. SC Riesen, KE Schober, DN Smith, CC Otoni, JD Bonagura. The Ohio State University, College of Veterinary Medicine, Columbus, OH.

Ischemia leading to fibrosis, diastolic dysfunction, congestive heart failure, and possibly sudden death is a pathophysiologic hallmark of HCM. Heart rate (HR), a major determinant of myocardial oxygen consumption, has become a novel treatment target in people with ischemic heart disease. The purpose of this study was to determine the acute effects of ivabradine, a selective HR-lowering drug, on LV function in cats with naturally-acquired HCM.

Six control cats and 6 cats with HCM were anesthetized, instrumented, and studied over a range of hemodynamic states induced by treatments with esmolol (200–400 μg/kg/min, IV), esmolol and dobutamine (5 μg/kg/min, IV), ivabradine (0.3 mg/kg, IV), and ivabradine and dobutamine. Cardiac output, +dP/dtmax, tau, and LV end-diastolic pressure were acquired and compared to baseline and control using 2-way repeated measures ANOVA.

Ivabradine reduced HR in control cats (mean±SD, −22 ± 10%, $P = 0.006$) and cats with HCM (−33 ± 4%, $P < 0.001$) without decreasing cardiac index ($P > 0.05$). The rate-pressure product decreased in cats with HCM after ivabradine (−31 ± 21%, $P = 0.009$). Ivabradine reduced +dP/dtmax only in cats with HCM (−29 ± 16%, $P = 0.01$). Tau did not change after ivabradine ($P > 0.05$) in both groups. However, ivabradine resulted in a significant increase in LVEDP (control +65 ± 40%, $P = 0.001$; HCM +54 ± 61%, $P = 0.006$).

Results suggest that ivabradine reduces HR and the rate-pressure product in a predictable manner, has minor negative inotropic effects, does not affect LV relaxation, increases LV end-diastolic pressure, and blunts the positive chronotropic effects of sympathetic stimulation in anesthetized cats with HCM. Studies in awake cats with HCM are needed to clinically validate our findings.

ABSTRACT #72

A SPLICE SITE MUTATION IN A GENE ENCODING FOR A MITOCHONDRIAL PROTEIN IS ASSOCIATED WITH THE DEVELOPMENT OF DILATED CARDIOMYOPATHY IN THE DOBERMAN PINSCHER. KM Meurs[1], S Lahmers[1], BW Keene[2], E Mauceli[3], G Acland[4], K Lindblad-Toh[3,5]. [1]Washington State University College of Veterinary Medicine, Pullman, WA. [2]North Carolina State College of Veterinary Medicine, Raleigh, NC. [3]Broad Institute of MIT and Harvard, Cambridge, MA. [4]Baker Institute, Cornell University, Ithaca, NY. [5]Uppsala University, Uppsala, Sweden.

Dilated cardiomyopathy (DCM) is a familial disease in the Doberman pinscher (DOBE) that often causes sudden death or heart failure. We have shown DOBE DCM to be inherited as an autosomal dominant trait. In humans, causative genes for familial dilated cardiomyopathy have been identified including cytoskeletal, sarcomeric and mitochondrial genes. We and others have excluded many of these genes as candidates for DOBE DCM. We hypothesized that a genome-wide SNP genotyping array might identify a genetic alteration associated with familial DOBE DCM.

DNA samples from 48 adult DOBEs with DCM (LVDD >4.8 cm and FS < 20%) and 48 apparently unaffected (≥ 10 years of age, normal physical examination) DOBEs were evaluated with a genome-wide SNP genotyping array. Association analysis suggested an area of statistical significance on canine chromosome 14). Fine-mapping of 13 additional SNPs within the region localized a signal to one strong candidate gene of potential cardiac importance. This gene encodes a mitochondrial protein that contributes to cardiac regulation of glucose metabolism.

Additional DNA sequencing of 121 DOBEs (64 affected, 57 unaffected) identified a 16 base pair deletion in a splice site region in affected dogs ($p = < 0.0001$). This deletion was also observed in 21/57 apparently unaffected DOBEs, but not in any of 100 unaffected dogs of 10 other breeds. Although the deletion was observed in some apparently unaffected dogs, their health status may reflect a later onset or variable penetrance of the disease. Affected dogs were either homozygous or heterozygous for the deletion. The deletion disrupts several splice site enhancer sites and leads to the presumed use of a cryptic splice site within the intron. Electron microscopy of myocardium from affected dogs demonstrated mitochondrial disruption. We conclude that a 16 base pair deletion in a splice site region of a gene encoding for a mitochondrial protein located on chromosome 14 is associated with the development of DCM in at least some DOBE families.

ABSTRACT #73

COMBINED CUTTING BALLOON AND HIGH PRESSURE BALLOON VALVULOPLASTY FOR DOGS WITH SEVERE SUBAORTIC STENOSIS. MK Schmidt, AH Estrada, HW Maisenbacher, RP Prošek, B Pogue, A Shih, JA Paolillo. University of Florida College of Veterinary Medicine, Gainesville, FL.

The aim of this study was to determine the feasibility and short-term effectiveness of cutting balloon valvuloplasty (CBV), followed by high pressure balloon valvuloplasty (HPBV) in dogs with severe subaortic stenosis (SAS). Eleven dogs < 30 months old with severe SAS (peak Doppler systolic left ventricular pressure gradient (PG) > 80 mmHg) underwent CBV and HPBV and were reexamined at one day (n = 11), one month (n = 11), and three months (n = 8) post-procedure. Under general anesthesia, cardiac catheterization was performed via the carotid artery. Left ventricular and aortic root pressures were obtained before and after valvuloplasty. An 8 mm cutting balloon catheter was inflated twice in the subaortic stenotic region in all dogs. A single high pressure balloon catheter was selected to be approximately 100% the diameter of the aortic annulus (mean balloon:aortic annulus ratio 0.96; range 0.86–0.99) and inflated in the subaortic stenotic region ≥ 2 times in all dogs. CBV combined with HPBV significantly reduced the intra-procedure peak-to-peak systolic PG in all dogs from 78.81 ± 46.45 (mean ± SD) to 36.10 ± 27.72 mmHg ($P < 0.0001$). Compared with baseline, mean peak Doppler systolic PG across the left ventricular outflow tract (162.42 mmHg; range 101.19–322.5 mmHg) was significantly decreased in all dogs examined at one day (93.2 mmHg; range 43.55–149.37 mmHg; $P < 0.0001$), one month (99.63 mmHg, range 69.0–157.33 mmHg; $P < 0.0001$), and 3 months (101.97 mmHg, range 73.0–154.9 mmHg; $P < 0.0001$). Clinical signs improved in the seven symptomatic dogs. CBV combined with HPBV is feasible with short-term effectiveness to significantly decrease the PG in dogs with severe SAS.

ABSTRACT #74

FAMILIAL SUBVALVULAR AORTIC STENOSIS APPEARS TO BE AN AUTOSOMAL RECESSIVE INHERITED TRAIT IN GOLDEN RETRIEVERS. JA Stern[1], KM Meurs[1], OL Nelson[1], SM Lahmers[1], LB Lehmkuhl[2]. [1]Washington State University, College of Veterinary Medicine, Pullman, WA. [2]Medvet Medical Center for Pets, Worthington, OH.

Subvalvular aortic stenosis is one of the most commonly reported canine congenital heart defects and is inherited in Newfoundland dogs and human beings. The golden retriever is over-represented in dogs with subvalvular aortic stenosis; however, the phenotype of this disease in golden retrievers and familial nature has not been described. We report the echocardiographic findings of subvalvular aortic stenosis and pedigree analysis of thirty-two golden retrievers.

Thirty-two adult golden retrievers diagnosed with subvalvular aortic stenosis (SAS) by Doppler echocardiography were retrospectively evaluated. All golden retrievers studied were previously evaluated by board certified ACVIM (cardiology) diplomates. Auscultation and echocardiography (2D, M-mode and Doppler) were performed. A subcostal continuous-wave Doppler aortic velocity ≥ 2.5 m/s and presence of a left basilar systolic ejection murmur was required for study inclusion. Three possible echocardiographic characteristics of SAS were recorded: evidence of aortic insufficiency (AI), a subvalvular ridge, or left ventricular hypertrophy (LVH). A disease status score was calculated by totaling the number of echocardiographic characteristics per subject. Descriptive, Spearman rank correlation and logistic regression statistics were performed.

Aortic velocities were as follows: range 2.5–6.8 m/s, median of 3.4 m/s and standard deviation 1.2 m/s. Presence of echocardiographic characteristics was distributed as follows: LVH 12/32, AI 20/32, and subvalvular ridge 20/32 dogs enrolled. Disease status score ranged from 0–3 with a median of 2. Spearman rank correlation revealed a statistically significant correlation between aortic velocity and disease status score ($r = 0.644$ $P < 0.0001$). For every 1 m/s increase in aortic velocity (> 2.5 m/s) a statistically significant odds ratio (OR) for demonstrating presence of subvalvular ridge,

LVH and AI was identified (OR = 2.95, P = 0.007; OR = 4.65, P = 0.0001; OR = 2.54, P = 0.01 respectively). The combination of LVH, AI and subvalvular ridge was reported in 10 of 11(90%) golden retrievers with aortic velocities $\geq$ 4.0 m/s.

Subvalvular aortic stenosis was observed in multiple generations of several families and appears familial. Evaluation of pedigree data has ruled out an autosomal dominant or X-linked dominant mode of inheritance as two unaffected parents produced affected offspring. X-linked recessive mode of inheritance is ruled out as an affected female produced an unaffected male offspring. An autosomal recessive pattern of inheritance seems most likely as the trait appears to skip generations and unaffected parents are observed to produce affected offspring.

Subvalvular aortic stenosis in the golden retriever is familial and best fits an autosomal recessive mode of inheritance. Characterization of SAS in golden retrievers suggests a similar phenotype to other reported breeds. Severity of stenosis determined by aortic velocity correlates well with the cumulative presence of echocardiographic characteristics (LVH, subvalvular ridge and AI).

ABSTRACT #75

PHARMACOKINETICS OF FONDAPARINUX IN HEALTHY CATS. NN Fiakpui[1], DF Hogan[1], T Whittem[2], HW Green III[1], EA Shipley[1], KA Sederquist[1]. [1]Purdue University, School of Veterinary Medicine, West Lafayette, IN; [2]University of Melbourne, Faculty of Veterinary Science, Werribee, Victoria, Australia.

Cardioembolic events are a devastating complication of myocardial disease in cats. Multiple antithrombotic agents have been used in cats but evidence of clinical efficacy with any agent is lacking while new antithrombotics are constantly being evaluated.

Arixtra® (fondaparinux) is a synthetic pentasaccharide and selective inhibitor of factor Xa. Unlike heparin or the low-molecular weight heparins, fondaparinux is a homogenous preparation of pentasaccharide units that has a long half-life and antithrombotic effects confined to Xa inhibition; rendering anti-Xa activity a reliable marker of drug effect.

An acute single dose trial was performed in four adult healthy cats where anti-Xa activity was measured at 0, 2, 4, 8, 12 and 24 hr after drug administration. Based on a pharmacokinetic model using these results two chronic (7 day) protocols were evaluated in 6 healthy adult cats. One thromboprophylactic protocol was of 0.06 mg/kg SC q 12 hr and the second was for thrombosis treatment using 0.2 mg/kg SC q 12 hr. For both protocols anti-Xa activity was measured at 2 and 12 hr after drug administration on days 1 and 7 and also at 12, 24 and 48 hr after the last dose of fondaparinux on day 7. Thromboelastography (TEG) and platelet aggregation studies were performed at the same time points.

Prophylactic and treatment protocols achieved median (range) peak anti-Xa activities of 0.59 mg/L (0.36–0.77) and 1.60 mg/L (1.52–1.76), respectively; approximating levels in humans. Median (range) trough anti-Xa activities [0.19 mg/L (0.00–0.37)] for the prophylaxis protocol also approximated levels in humans; the median (range) for the treatment protocol [0.16 mg/L (0.11–0.35)] was below levels in humans. While there was no significant effect on TEG parameters with the prophylactic protocol, there was a median 2.43 and 5.34 fold increase in the R and K, respectively ($p < 0.05$) along with a decrease of 56% and 28% in the α-angle and MA, respectively ($p < 0.05$) at peak on day 7 in the treatment protocol. There was a significant decrease in MA (9.8%, $p < 0.05$) 12 hr after drug discontinuation although anti-Xa activities were below the peak levels achieved in the prophylactic protocol. The changes in TEG parameters resolved by 24 hr after drug discontinuation. There was no change in platelet aggregation noted with either protocol.

A thromboprophylactic protocol of 0.06 mg/kg SC q 12 hr approximates anti-Xa activities seen in humans considered effectively treated with fondaparinux while a thrombosis treatment protocol of 0.2 mg/kg SC q 12 hr approximates only peak anti-Xa activities seen in humans. This study only provides a starting point for fondaparinux therapy in cats; clinical efficacy and final dosing protocol of fondaparinux for thromboprophylaxis or thrombosis treatment in cats must be determined through future large, multi-center clinical trials.

ABSTRACT #76

MINIMALLY INVASIVE PER-CATHETER PATENT DUCTUS ARTERIOSUS OCCLUSION IN SMALL DOGS ($\leq$ 3 KG): PRELIMINARY RESULTS. JLC Olson, AH Tobias, CD Stauthammer, KM Hohnadel. University of Minnesota College of Veterinary Medicine, St. Paul, MN.

Patent ductus arteriosus (PDA) occlusion with the Amplatz® Canine Duct Occluder (ACDO) is an extremely effective method for the treatment of PDA in dogs. However, the procedure requires the use of relatively large diameter delivery systems, which limits its application in small dogs. Consequently, we are investigating the use of a prototype low profile ACDO and an investigational 4F delivery catheter to occlude PDA in dogs weighing 1.5 to 3 kg.

Five dogs weighing 1.51 to 2.85 kg have been enrolled in the study to date. Femoral arterial access has been achieved in all cases using a cut-down technique, a 21 G micropuncture needle, and a 0.018 inch diameter wire guide. The delivery catheter is advanced over the wire guide and angiography identified type IIA PDA morphology in 4 dogs and type IIB in 1. Angiographic minimal ductal diameter ranged from 0.4 to 1.85 mm.

Low-profile ACDOs with waist diameters of 3 mm (n = 2) and 4 mm (n = 3) were deployed via the delivery catheter, and angiography demonstrated complete ductal occlusion in 4 dogs, and trivial residual flow in 1. These results were confirmed by echocardiography 1 day post-procedure. Complete PDA occlusion at 3 months was confirmed by echocardiography in the 1 dog in which this follow-up evaluation has been completed to date.

Further research and development is required before the low profile ACDO and investigational 4F delivery catheter can be made available for routine PDA occlusion in small dogs. However, our preliminary results suggest that minimally invasive per-catheter PDA occlusion with this device and delivery catheter is feasible and effective in dogs weighing 1.5 to 3 kg.

ABSTRACT #77

IDENTIFICATION OF BETA-1 ADRENERGIC RECEPTOR POLYMORPHISMS IN CATS. BA Maran, KM Meurs, S Lahmers, OL Nelson. Washington State University College of Veterinary Medicine, Pullman, WA.

Hypertrophic cardiomyopathy (HCM) is the most common cause of feline heart disease. Beta-adrenergic antagonists (beta-blockers) are one of the most frequently prescribed treatments for affected cats; however, there is significant variation in clinician preference regarding initiation of therapy and the anticipated response. In humans, two genetic polymorphisms within the Beta-1 adrenergic receptor (1-AR) gene have been shown to correspond to variable pharmacologic responses. Identification of specific 1-AR polymorphisms has allowed physicians to optimize beta-blocker therapy on a genomic basis. It is possible that cats also have polymorphisms that may impact individual response to therapy.

The hypothesis of this study is that polymorphisms are present in the feline 1-AR gene, which could result in an altered pharmacologic response to beta-blocker therapy. The objective of the study was to sequence the feline 1-AR gene in unrelated cats of several different breeds to evaluate for the presence of polymorphisms. The feline 1-AR gene was amplified with specifically designed and optimized primers using the feline and human 1-AR sequence, standard PCR techniques, and PCR-based genomic sequencing. DNA samples from 7 cats of 5 breeds (Domestic Shorthair, Maine Coon, Siberian, Ragdoll, and Bengal) were evaluated.

Two polymorphisms were identified within the 1-AR gene, each occurring in an apparently low frequency. One polymorphism was identified in four Siberians and two Maine Coons and consists of an A/C substitution; both heterozygous and homozygous animals were identified. The change did not alter the amino acid produced. A second polymorphism was identified in two Domestic Shorthair and four Ragdolls and consisted of an AA/CC substitution. Five animals were heterozygous and one animal was homozygous for this mutation. This alteration changed the amino acid produced from Proline to Glutamine at position 277 (Pro277Gln). Computer modeling (PolyPhen software) of this region predicts a damaging structural change to the 1-AR protein.

The clinical significance of this finding is unknown; however, the identification of1-AR polymorphisms in the cat could suggest variability in response to beta blocker therapy in the cat as it does in humans.

ivabradine reduced LAA inflow velocity (–21 cm/s, $p < 0.05$), but other effects on LA function were not significantly different from baseline. HR correlated weakly to LAA outflow velocity ($r^2 = 0.25$).

These data suggest that beta-blockade with atenolol could impair LA contractile function in cardiomyopathy; however, clinical studies are required to delineate this risk.

ABSTRACT #78

EFFECTS OF IVABRADINE, A SELECTIVE I_f CHANNEL INHIBITOR, ON HEART RATE IN HEALTHY CATS. RE. Cober, KE. Schober, CAT. Buffington, S. Riesen, JD. Bonagura. The Ohio State University, Columbus, OH.

One of the main determinants of ischemia, a pathophysiologic characteristic of hypertrophic cardiomyopathy (HCM), is heart rate (HR). Unwanted tachycardia may trigger decompensation in previously asymptomatic cats with HCM. The purpose of this study was to determine an effective oral dose of ivabradine, a pure negative chronotropic drug, in healthy cats.

In this single blinded, placebo controlled, randomized, fully-crossed study 8 healthy cats received placebo or one dose of ivabradine at 0.1 mg/kg, 0.3 mg/kg, or 0.5 mg/kg, PO. HR and blood pressure were monitored continuously for 24 hours via radiotelemetry after each treatment. Response to stress was studied twice by 15-min acoustic startle applied at baseline and 4 hours after drug. Statistical comparisons were made using a linear mixed model and 2-way repeated measures ANOVA.

Peak negative chronotropic effect was observed 3 hours after ivabradine administration. HR (min^{-1}) decreased significantly ($p < 0.05$) in a dose-dependent manner (mean±SD for placebo: 144 ± 20; ivabradine 0.1 mg/kg: 133 ± 22; ivabradine 0.3 mg/kg: 112 ± 20; ivabradine 0.5 mg/kg, 104 ± 11). HR (min^{-1}) was still reduced ($p < 0.05$) at 12 hours after ivabradine (0.3 mg/kg: 128 ± 18 and 0.5 mg/kg: 124 ± 16) compared to placebo (141 ± 21). HR response to acoustic startle was significantly ($p < 0.01$) blunted at all 3 doses of ivabradine. No effect of ivabradine on blood pressure was identified and no clinically discernable side effects were observed.

These findings indicate that ivabradine at 0.3 mg/kg and 0.5 mg/kg PO predictably lowers HR for at least 12 hours in healthy cats. Further studies in cats with HCM are needed.

ABSTRACT #79

EFFECTS OF ATENOLOL, IVABRADINE AND PIMOBENDAN ON LEFT ATRIAL FUNCTION: AN ECHOCARDIOGRAPHIC STUDY IN HEALTHY CATS. AM Kent, JD Bonagura, BA Scansen, S Riesen, KE Schober. College of Veterinary Medicine, The Ohio State University, Columbus, OH.

Arterial thromboembolism in cardiomyopathy is predisposed by left atrial (LA) enlargement and dysfunction. Drugs may further influence LA function via inotropic or chronotropic effects. We hypothesized that atenolol would depress, pimobendan enhance, and ivabradine minimally affect LA contractile function in cats.

Nine cats were studied prospectively using a double-blind, fully-crossed design. Cats were healthy based on physical examination, CBC, serum biochemistries, blood pressure, ECG, and 2D echocardiography. LA function was evaluated by echocardiography over four study periods: control, atenolol (1.8–2.5 mg/kg q12h), pimobendan (0.2–0.3 mg/kg q12h), and ivabradine (0.3–0.5 mg/kg q12h). Treatments were randomized and continued for 7 days prior to measurements. Peak LA appendage (LAA) inflow and outflow velocities; velocity and duration of the atrial reversal (AR) wave; and HR were measured. Treatments were compared by repeated measures ANOVA followed by Tukey's multiple-comparison test. Statistical significance was $p < 0.05$.

HR (min^{-1}) was 223.1 ± 27.3 at baseline (mean±SD) and significantly decreased following atenolol (146.2 ± 19.9) and ivabradine (125.3 ± 26.9). The inodilator pimobendan mildly increased LA function indices ($p > 0.05$). Atenolol significantly depressed LAA and AR velocities while prolonging AR duration relative to baseline, pimobendan, and ivabradine. Mean LAA outflow velocity was reduced by atenolol (–46 cm/s) compared to baseline. The I_f blocker

ABSTRACT #80

NT-proBNP LEVELS IN 30 CATS WITH DYNAMIC RIGHT VENTRICULAR OUTFLOW OBSTRUCTION. JW Allen, K Yee, MH Buckner, SJ Ettinger. California Animal Hospital Veterinary Specialty Group, Los Angeles, CA.

Dynamic right ventricular outflow obstruction (DRVO) is an echocardiographic finding that has been described as a physiologic cause of a systolic murmur in cats. NT-proBNP has been evaluated as a biomarker for the presence of cardiac disease in cats; levels are elevated in cats with hypertrophic cardiomyopathy, heart failure, and cats with renal insufficiency. As a murmur is a frequent indication for echocardiography, we sought to better characterize the relationship between DRVO and NT-proBNP levels, concurrent cardiac and urinary disorders, and clinical outcome.

Thirty client-owned cats with systolic murmurs were evaluated via echocardiography over a two year period. Doppler and 2-D echocardiographic studies were examined for evidence of DRVO and additional structural and electrocardiographic abnormalities according to previously identified criteria. Doppler BP, USG, serum BUN, Creatinine, TT4, and plasma NT-proBNP values were recorded. Medical history, concurrent diagnoses and communication with the owners or referring veterinarians were evaluated for the presence of non-cardiac disease. Follow up studies were available for 22 cats.

Cats with DRVO had a high prevalence of concurrent cardiovascular abnormalities (17/30, 57%), most commonly HCM (16/17, 94%). Eight cats died or were euthanized during the study period, with one cardiac cause of death. NT-proBNP was > 44 pmol/L in 26/30 cats (87%), all of which had either HCM, urinary tract disease, or both. The four cats with NT-proBNP < 44 pmol/L had isolated DRVO (no concurrent urinary or cardiac disease). 11/13 cats with DRVO and urinary tract disease (CKD, FLUTD, UTI) had NT-proBNP levels < 165 pmol/L. 7/8 cats with NT-proBNP > 754 pmol/L had severe cardiac disease, including the patient who died of cardiac causes.

NT-proBNP levels in cats with DRVO are associated with concurrent cardiac and/or urinary tract pathology. Isolated DRVO did not result in NT-proBNP elevation and was not predictive of future cardiac events in the cats in this report. In the absence of urinary tract disease, NT-proBNP may be helpful in diagnosing clinically significant cardiac disease in cats with DRVO.

ABSTRACT #81

MITRAL ANNULOPLASTY BY EXTERNAL PLICATION IN DOGS WITH MIXOMATOUS MITRAL VALVE DEGENERATION. JNBM de Andrade[1], CG Olivaes[1], AA Camacho[2]. [1]Universidade de Franca, Programa de Mestrado em Cirurgia e Anestesiologia Veterinária, Franca, SP, Brazil. [2]Universidade Estadual Paulista – Faculdade de Ciências Agrárias e Veterinárias – Departamento de Clínica e Cirurgia, Jaboticabal, SP, Brazil.

Mixomatous mitral valve degeneration (MMVD) is the most common cardiac disease in dogs. Once a dog is in congestive heart failure (CHF), survival time is variable and the long-term prognosis is guarded. Medical treatment is palliative only. The only definitive treatment is valve replacement or valve repair, which require cardiopulmonary bypass (CPB). The goal of this research is to describe and evaluate the mitral valve annuloplasty by external plication in dogs with CHF due to MMVD, performed without CPB.

Four ISACHC Class IIIa CHF dogs were submitted to the surgery, after the owner's consent was granted. After left thoracotomy and pericardiotomy, a buttressed double row running suture was inserted in the left ventricle myocardium parallel with the atrioventricular groove, avoiding injury to any coronary branch. Knots were tied when the left atrial thrill could not be palpable anymore. Every dog underwent physical examination, electrocardiography, echocardiography, and cardiac troponin I (cTnI) measurement before and after the procedure.

Two dogs had a decrease in the regurgitant jet after surgery (from severe to mild-to-moderate). One of them improved clinically during a 30-day post-surgery period of evaluation. Nevertheless the second dog died from acute renal failure two days after the surgical procedure. The cTnI level increased after the plication in both dogs and returned to basal values 30 days after surgery in the dog that survived. The two remaining dogs died trans-op because of myocardial rupture and uncontrolled hemorrhage.

The mitral annuloplasty by external plication can decrease the mitral regurgitant jet and improve clinical signs in dogs with MMVD and severe CHF. However this procedure has high mortality, at least in dogs with severe CHF. Further long-term studies in order to clarify if this surgery could avoid the progression of the disease in less-severe cases are suggested.

ABSTRACT #82

COMPARISON OF PHENOBARBITAL AND POTASSIUM BROMIDE MONOTHERAPIES IN THE TREATMENT OF CANINE EPILEPSY. CL Gaskill[1], WJ Kimber[2]. [1]University of Kentucky, Lexington, KY. [2]Toronto, ON.

The objective of this study was to compare safety and efficacy of phenobarbital (PB) and potassium bromide (KBr) monotherapies for treatment of canine idiopathic epilepsy. Specific aims were to determine which drug was associated with fewer adverse effects and which was more effective at controlling seizures over a one year period.

Sixty-two client-owned pet dogs with idiopathic epilepsy were enrolled in the study. Dogs were randomly assigned to either PB or KBr treatment groups (PB $n = 30$, KBr $n = 32$). Data collected at time of enrollment and at 1, 4 and 12 months after start of therapy included physical examination, detailed medical and seizure histories, serum biochemical and CBC analyses, and serum drug concentrations. Drug dosages were adjusted based on seizure control, adverse effects, and serum drug concentrations. Dogs with unacceptable seizure control or intolerable adverse effects not responsive to drug dosage alterations were removed from the study.

Analyses included comparisons between groups for signalment variables, biochemical and CBC values outside normal, proportions of individual adverse effects, proportions of dogs not completing the study for various reasons, seizure frequency and seizure severity. Associations between presence of adverse effects and risk factors including all signalment variables and drug dosage, serum drug concentration and other variables were evaluated using logistic regression models. Correlations between drug dosage and serum drug concentrations were determined. Poisson regressions and Hazard ratios were used to determine risk factors for increased seizure frequency and severity. Statistical significance was set at $P < 0.05$ throughout.

Results showed that PB monotherapy was associated with better seizure control and a lower incidence of adverse effects than was KBr monotherapy. Significantly more KBr-treated dogs were removed from the study due to poor seizure control or intolerable adverse effects (7 KBr-treated dogs vs. 1 PB-treated dog). Other significant differences between groups included overall higher seizure frequency and seizure severity scores with KBr treatment compared to PB, and higher incidence of marked lethargy with KBr. Prevalence of vomiting (23 KBr-treated dogs, 15 PB-treated dogs) and inappropriate defecation in the house (6 KBr-treated dogs, 1 PB-treated dog) were higher with KBr but P values were > 0.05. Other adverse effects noted in both groups included polyuria/polydipsia, polyphagia, anorexia, hyperactivity, aggression, ataxia, skin problems, pancreatitis, elevate amylase and lipase. Although differences were not significant between groups for these variables, in general prevalences were higher with KBr treatment.

In conclusions, study results indicate that PB is superior to KBr as a monotherapy for most cases of canine epilepsy, providing better seizure control and incurring fewer adverse effects over a one-year period of time.

ABSTRACT #83

INCREASED BLOOD AMMONIA CONCENTRATION AFTER GENERALIZED SEIZURES IN DOGS. RA House[1], TA Schubert[1], CA Meeks[2], RC Hill[1]. [1]University of Florida College of Veterinary Medicine, Gainesville, FL. [2]Florida Veterinary Specialists, Tampa, FL.

Transient post-ictal hyperammonemia has been reported in people, but has not been documented in the veterinary literature. This study was designed to determine whether blood ammonia concentrations are increased after generalized tonic-clonic seizures in dogs and the duration of the increase. We hypothesized that hyperammonemia is common after a generalized tonic-clonic seizure, but that the ammonia concentration returns to normal within one hour after cessation of seizure activity in dogs that do not have hepatic dysfunction.

Twelve client-owned dogs that had a generalized tonic-clonic seizure while at the University of Florida Veterinary Medical Center or Florida Veterinary Specialists were enrolled. Blood ammonia concentrations were measured immediately, twenty, and sixty minutes after the end of the seizure. Dogs with hepatic dysfunction were excluded.

Blood ammonia concentrations were higher ($p < 0.002$) immediately after the seizure than 20 or 60 minutes later; mean concentrations (+/− one standard deviation) were 32.3 ± 30.5 vs. 11.9 ± 10.8 and 9.4 ± 11.1 µmol/L, respectively. There was no evidence of a difference between concentrations measured 20 and 60 minutes post-ictally or between hospitals. Ammonia concentrations were higher than analyzer normal values in 4/12 (33%) dogs immediately after the seizure and none of the dogs 60 minutes later.

It appears that a transient hyperammonemia is common after a generalized seizure, but that the blood ammonia concentration returns to normal within one hour.

ABSTRACT #84

SAFETY, PERFORMANCE AND EFFICACY OF A SEIZURE ADVISORY SYSTEM IN 6 DOGS WITH EPILEPSY: A PILOT STUDY. B. Sturges[1], V. Ruedebush[2], D. York[1], P. Dickinson[1], K. Vernau[1], D. Sheffield[2]. [1]University of California – Davis, Davis, CA. [2]NeuroVista Corporation, Seattle, WA.

Epilepsy is a common neurological disorder affecting people and dogs. Chronic administration of anti-epileptic drugs (AEDs) is often associated with serious adverse effects. The NeuroVista Seizure Advisory System (SAS) is designed to provide real-time advisories indicating the likelihood of seizure occurrence. We hypothesized that the SAS could safely and effectively be used in dogs to provide advisories of high seizure susceptibility and the opportunity to intervene with the use of acute AEDs.

Sixteen monitoring electrodes were implanted subdurally over both cerebral cortices in 6 healthy research dogs diagnosed with epilepsy. Continuous, real-time intracranial EEG (iEEG) signals were received via electrodes by an implanted, rechargeable telemetry unit located in the thoracic body wall. iEEG data from this device was transmitted wirelessly to an external device placed in a vest and stored on replaceable data cards. Algorithms for detecting seizure activity were assessed in 3 of 6 dogs with recurrent seizures. The seizure detection algorithm detected all seizures with no false positive detections noted. Seizure prediction algorithms developed with human iEEG data are being assessed in the 3 dogs with recurrent seizures. 2 of 6 dogs have been seizure-free and one dog died from status epilepticus.

We conclude that implantable seizure advisory devices can be safely used in the dog to collect iEEG data needed to study seizure detection and susceptibility algorithms. Dogs may provide a useful large animal model of spontaneously occurring epilepsy allowing for translational human epilepsy research.

ABSTRACT #85

MAGNETIC RESONANCE IMAGING OF THE CANINE BRAIN AT 3.0T AND 7.0T. Paula Martin-Vaquero[1], Ronaldo C. DA Costa[1], Rita L. Echandi[1], Christina L. Tosti[2], Michael V. Knopp[2], Steffen Sammet[2]. [1]Department of Veterinary Clinical Sciences, College of Veterinary Medicine, The Ohio State University, Columbus, OH. [2]The Ohio State University Medical Center, Columbus, OH.

The purpose of this study was to compare the brain anatomy of clinically normal dogs using two magnetic resonance imaging

(MRI) units, a 3.0 T and a 7.0 T. We hypothesized that MRI at 7.0 T would allow superior resolution and provide superior image quality and anatomical detail.

Four healthy Beagle dogs with no history of neurologic disease were imaged with both scanners. Dogs underwent general anesthesia and were positioned in sternal recumbency for all imaging. A 3.0 T and a 7.0 T MRI units (Achieva, Philips Healthcare, Cleveland, OH) were used. Low-resolution turbo spin echo (TSE) T2-weighted images (T2-WI) were acquired in the transverse, dorsal and sagittal planes. High-resolution T2-WI and TSE proton density-weighted images (PD-WI) were obtained in the transverse and dorsal planes. A list of selected anatomical structures were identified and blindly assessed by 3 experienced reviewers for image contrast and definition. Overall image quality and the presence of artifacts were also evaluated. Quantitative assessment of regions of interest (ROIs) was also performed.

MRI at both 3.0 T and 7.0 T provided good spatial and contrast resolution of the canine brain anatomy. Although some structures were visualized with superior image contrast at 3.0 T MRI and other anatomical structures showed improved image contrast at 7.0 T, the majority of the neuroanatomical structures evaluated had comparable image contrast and definition at both 3.0 T and at 7.0 T. The overall image quality for low resolution T2-WI was comparable at 3.0 T and 7.0 T. High resolution T2-WI showed superior image quality at 3.0 T. Absent to mild artifacts were seen at 3.0 T and mild to moderate artifacts were noted at 7.0 T. ROIs analysis indicated slightly superior grey and white matter contrast at 7.0 T.

The results obtained in our study show that both 3.0 T and 7.0 T magnets offered high quality images and excellent definition of the canine cerebral cortex and selected anatomical structures. No remarkable differences were noted on the images obtained at 7.0 T. In our study, special importance was given to thalamocortical structures, which are the areas of the brain where seizure activity originates and spreads from, both in humans and dogs.

ABSTRACT #86

VASCULAR ENDOTHELIAL GROWTH FACTOR (VEGF) LEVELS IN THE CEREBROSPINAL FLUID OF DOGS WITH INTRACRANIAL DISEASE. ZE Niman[1], CL Mariani[1], L Boozer[1], J Campbell[1], S Dillard[1], PJ Early[1], R Gallagher[1], E MacKillop[2], KR Munana[1], NJ Olby[1]. [1]North Carolina State University College of Veterinary Medicine, Raleigh, NC. [2]Pittsburgh Veterinary Specialty & Emergency Center, Pittsburgh, PA.

Definitive diagnoses of intracranial diseases often remain elusive. Several markers have shown promise for systemic malignancies, but none have been fully evaluated in the cerebrospinal fluid (CSF) of veterinary patients in association with specific intracranial diseases. Current examination of CSF in canine veterinary practice is based on cellular examination and total protein measurement, though additional analyses may be useful. Vascular endothelial growth factor (VEGF) stimulates endothelial cell proliferation and has a pivotal role in tumor angiogenesis. VEGF was evaluated in the CSF of dogs with various intracranial diseases to probe its utility as a biomarker. CSF was prospectively banked in patients with intracranial disease during routine diagnostic evaluation or immediately post-mortem after euthanasia. VEGF concentrations were assayed using a commercially available ELISA kit, which has been validated for use in dogs. VEGF was readily quantifiable in canine CSF. Samples were subdivided into neoplastic (primary intracranial, $n = 28$), inflammatory ($n = 22$), and other neurologic disease ($n = 35$). There were significant differences in VEGF levels between all groups (mean values in pg/ml were 159.4, 50.9, and 2.2, respectively, $P < 0.0001$). Within the neoplastic group, there were significant differences in VEGF levels between specific malignancies ($P = 0.035$); choroid plexus (631.1) and glial cell (300.2) tumors demonstrated the highest average levels, followed by meningiomas (13.1), CNS lymphoma (3.0), and pituitary adenomas (2.12). Inflammatory brain diseases had variable VEGF levels (range 0–489.96), while other intracranial disease rarely had detectable VEGF in the CSF. These observations provide a potential basis for differentiating etiologies of intracranial masses.

ABSTRACT #87

INVESTIGATION OF CANINE PRIMARY BRAIN TUMORS FOR THE PRESENCE OF DNA FROM HERPES- AND POLYOMAVIRUSES BY BROADLY REACTIVE PAN-VIRAL PCR (99 CASES). Haley AC[1], Platt SR[1], Kent M[1], Miller, AD[2], Barber R[1], LI Q[1], Tong, S[3], Schatzberg SJ[1]. [1]University of Georgia College of Veterinary Medicine, Athens, GA. [2]Division of Comparative Pathology, Harvard Medical School, Southborough, MA. [3]Centers for Disease Control and Prevention, Division of Viral Diseases, Atlanta, GA.

Central nervous system (CNS) neoplasia has been reported in 1–3% of canine necropsies and occurs in approximately 14 in 100,000 dogs. Parallels exist between primary canine and human brain tumors including frequency, tumor types, classification, immunohistochemical markers, and tumor behavior. While it is generally accepted that both genetics and environmental factors contribute to CNS neoplasia, the etiologies for human and, particularly canine brain tumors, are largely unknown. Over the past decade, DNA and RNA viruses have been identified in various primary human brain tumor types. However, the causal relationship between CNS viral infection and primary brain tumors in humans has been a source of debate because of conflicting laboratory reports documenting the presence or absence of viruses in various brain tumor types.

Here, we hypothesize that viruses are *not* associated with primary canine brain tumors. Our objective was to apply broadly reactive, pan-herpes- and pan-polyomavirus consensus, degenerate oligonucleotide primers (CODEHOP PCR) to paraffin-embedded brain tumors to evaluate for viral DNA. An inherent limitation to specific PCR testing is that it may fail to identify atypical or novel viruses within a viral family, whereas CODEHOP PCR avoids a *priori* bias towards specific viral species. The utility of the CODEHOP PCR approach is that it would allow us to detect unsuspected or even novel herpes- and polyomaviruses in primary canine brain tumors.

DNA was extracted from 10 µm shavings from 99 paraffin embedded, formalin fixed archived primary canine brain tumors (meningiomas, gliomas, sarcomas, choroid plexus tumors, lymphoma, and ependymomas). PCR for the glyceraldehyde-3-phosphate dehydrogenase gene was completed for all 99 samples to confirm the integrity of extracted DNA. All samples were tested for the presence of herpes- and polyomaviruses using CODEHOP PCR methods. For each PCR reaction, rigorous positive and negative controls were included. PCR reactions were evaluated by 1.5% agarose gel electrophoresis, with amplicons assessed under ultraviolet light after the addition of ethidium bromide. All positive (porcine respiratory virus, JC virus) and negative controls (eg no template PCR, PCR of water extracted in parallel to brain tumor tissues) yielded expected results. None of the 99 samples were positive for herpes- or polyomavirus DNA.

The absence of herpes- and polyomavirus DNA in these 99 cases makes an association between these viral families and primary canine brain tumors unlikely.

ABSTRACT #88

INTEROBSERVER AGREEMENT AND ACCURACY IN THE INTERPRETATION OF BRAIN MAGNETIC RESONANCE IMAGING IN DOGS. Mylene-Kim Leclerc[1], Marc-Andre d'Anjou[1], Laurent Blond[1], Eric Norman Carmel[1], Ruth Dennis[2], Susan Kraft[3], Andrea Matthews[4], Joane Parent[1]. [1]Faculté de médecine vétérinaire, Université de Montréal, Saint-Hyacinthe, QC, Canada. [2]The Animal Health Trust, Newmarket, Suffolk, UK. [3]Veterinary Teaching Hospital, Colorado State University, Fort Collins, CO. [4]Veterinary Teaching Hospital, University of Tennessee, Knoxville, TN.

MRI is routinely used to investigate brain diseases in dogs, but the consistency and accuracy of radiology reports remain largely unknown.

Five board-certified veterinary radiologists with variable MRI experience interpreted MR (1.5 T) brain images of 44 dogs, acquired with transverse T2w, T2w-FLAIR and T1w-FLAIR; transverse, sagittal and dorsal T2w and T1w-FLAIR post-contrast. Imaging parameters that were scored included: lesion (present/absent); lesion characteristics (localization, mass effect, edema, hemorrhage, cavitation); contrast enhancement characteristics; and, most likely diagnosis (normal, tumoral, inflammatory, vascular, metabolic or toxic, or other). MRI diagnoses were determined initially without patient information, and then repeated providing history and signalment. For all cases and readers, MRI diagnoses were compared to final diagnoses established histopathologically (when available), or with other pertinent clinical data (CSF analysis, clinical response to treatment and/or MRI follow-up). MRI scores were compared between examiners with kappa statistics.

Interobserver agreement for presence of a lesion was substantial to excellent ($0.67 \leq \kappa < 0.86$). However, presence of edema was less consistently recognized ($0.30 \leq \kappa < 0.52$) or graded ($0.49 \leq \kappa < 0.78$). Presence and distribution of contrast enhancement was scored with moderate to substantial agreement ($0.40 \leq \kappa < 0.70$), whereas pattern and border definition were scored with more variability ($0.29 \leq \kappa < 0.70$). Interobserver agreement was moderate to substantial for establishing an imaging diagnosis without patient information ($0.56 \leq \kappa < 0.75$), and increased when patient information was provided ($0.65 \leq \kappa < 0.76$). Agreement between MRI and final diagnoses was substantial for all radiologists ($0.62 \leq \kappa < 0.73$).

In conclusion, while some imaging parameters such as edema are interpreted less consistently, radiologists are reasonably constant and accurate when providing diagnoses.

ABSTRACT #89

CENTRAL NERVOUS SYSTEM BLASTOMYCOSIS IN 4 DOGS. L Lipitz, H Rylander, LJ Forrest, D Foy. Department of Medical Sciences, University of Wisconsin-Madison, Madison, WI.

Blastomycosis is a systemic mycotic infection caused by *Blastomyces dermatitidis*. Central nervous system (CNS) infection is uncommon in dogs and the magnetic resonance imaging (MRI) characteristics of canine blastomycosis have not been previously reported. The purpose of this study is to retrospectively describe the clinical presentation, MRI findings, cerebrospinal fluid (CSF) analysis, and case outcome in four dogs with cytologically confirmed CNS blastomycosis.

The medical record database at the University of Wisconsin-Madison Veterinary Medical Teaching Hospital was searched for dogs with a diagnosis of blastomycosis that underwent MRI examination between 2000 and 2009. Data collected from the medical record included signalment, clinical history, physical and neurologic examination findings, results of thoracic radiography and CSF analysis, and results of MRI examination. Cases without cytologic diagnosis of blastomycosis were excluded.

Three dogs had multifocal intracranial granulomas and presented with seizure activity; one dog had a focal cervical spinal cord granuloma and presented with hemiparesis. Three dogs had signs of disease in other organ systems, including the lungs, eyes, and skin; one dog only had CNS involvement. Diagnosis of blastomycosis was made by cytologic examination of an intracranial mass, two pulmonary masses, and a skin lesion. On MRI examination, lesions were hypo- to isointense on T1-weighted (T1W) images and variably hyperintense on T2-weighted (T2W) images. On T1W images after intravenous gadodiamide administration, lesions most often strongly contrast enhanced; a small number of lesions ring-enhanced. Meningeal enhancement was present in all dogs with intracranial lesions and peri-lesional edema was seen in all cases on T2W images. Three dogs had CSF analysis and all had a neutrophilic pleocytosis and elevated total protein concentration. *Blastomyces* organisms were not detected in the CSF in any of the dogs. One dog was euthanized without treatment. Antifungal therapy and monitoring of response to treatment via serial MRI examination were initiated in three dogs, two of which survived. In one dog that survived, resolution of intracranial disease was seen on MRI examination after 6 months of therapy. In the other dog, the spinal granuloma reduced in size but persisted on MRI examination after 2 years of therapy despite resolution of clinical disease and normal CSF analysis.

Blastomycosis in the CNS is a rare disease with a guarded prognosis and antemortem diagnosis can be difficult. The ability to identify characteristics of blastomycosis using MRI examination and CSF analysis can provide clinicians with valuable diagnostic and prognostic information. Long-term therapy and monitoring is necessary and serial MRI examination is recommended since neurologic examination and CSF analysis may not identify persistent CNS infection.

ABSTRACT #90

PLASMA CELL NEOPLASIA OF THE VERTEBRAL COLUMN: IMAGING, CLINICOPATHOLOGIC FINDINGS, AND LONG-TERM OUTCOME WITH TREATMENT. L. Gerlach[1], B. Sturges[1], K. Vernau[1], P Dickinson[1], M. Knipe[1], W. Berry[2], R. LeCouteur[1]. [1]University of California – Davis, Davis, CA, [2]Southern California Referral Group, Irvine, CA.

Plasma cell neoplasms arise from malignant proliferation of B-lymphocytes and may affect the vertebral column as solitary plasmacytomas or multiple myelomas. The purpose of this study was to retrospectively evaluate treatment of these tumors with respect to long-term outcome. Our hypothesis was that surgical cytoreduction and stabilization combined with chemotherapy (melphalan/prednisone) provides significantly better outcome than no treatment or chemotherapy alone.

Twenty-one dogs diagnosed by histopathology with vertebral plasma cell neoplasia were studied. Dogs commonly presented for clinical signs of spinal pain, pelvic limb paresis, and lameness of months duration. On neurological exam, 5/21 dogs had spinal pain only while 11/21 dogs had a T3-L3 myelopathy. 18/21 dogs were ambulatory at the time of diagnosis. Spinal radiographs had bony changes consistent with vertebral neoplasia in 20/21 dogs, while 4/21 dogs had vertebral fractures. Extradural compression was evident in all dogs that had a myelogram (15/21). Images on 4 dogs diagnosed with MRI showed similar findings including enlarged vertebral venous sinuses, T2 hyperintensity within affected vertebrae, and T1 iso- to hyperintensity with mild contrast enhancement. Results of CSF analysis confirmed mild pleocytosis in 5/12 dogs and significant protein elevation in 10/11 dogs.

Seven dogs were treated with chemotherapy alone, 8 dogs with a combination of surgery and chemotherapy, and 6 received no treatment. Dogs treated with surgery and chemotherapy had longer survival times (median = 24 months) than dogs treated with chemotherapy alone (median = 5 months). Longer survival with surgery is likely due to the effects of cytoreduction and internal fixation on compromised vertebral structure.

ABSTRACT #91

MORPHOMETRIC DIMENSIONS OF THE CAUDAL CERVICAL REGION IN DOBERMAN PINSCHERS WITH CLINICAL SIGNS OF DISK ASSOCIATED WOBBLER SYNDROME (DAWS), CLINICALLY NORMAL DOBERMAN PINSCHERS, AND CLINICALLY NORMAL FOXHOUNDS. S De Decker, I Gielen, L Duchateau, H Van Bree, T Waelbers, V Bavegems, L Van Ham. Faculty of Veterinary Medicine, Ghent University, Merelbeke, Belgium.

The purpose of this study was to investigate if relative stenosis of the vertebral canal predisposes Doberman Pinschers to develop clinical signs of DAWS and to evaluate if morphometric MRI parameters could be identified to differentiate between clinically relevant and irrelevant spinal cord compressions.

Fifty-four dogs were prospectively studied: Doberman Pinschers with clinical signs of DAWS (n = 17), clinically normal Doberman Pinschers (n = 20), and clinically normal Foxhounds (n = 17). All dogs underwent MRI (0.2 T) of the caudal cervical spine. To evaluate relative vertebral canal stenosis, the following ratios were calculated at the cranial, middle, and caudal levels of C5, C6, and C7: canal occupying ratio of the spinal cord and cerebrospinal fluid (CSF) column, defined as the cross-sectional area (CSA) of respectively the spinal cord and CSF-column, divided by the CSA of the vertebral canal. To evaluate the degree of spinal cord compression, the following ratios were calculated at the site of most pronounced compression: compression ratio, defined as the smallest dorsoventral diameter of the spinal cord divided by the broadest transverse diameter; remaining spinal cord and CSF-column area, defined as the CSA of respectively the spinal cord and CSF-column at the compressed area divided by the CSA at the adjacent, non-compressed segment. To evaluate the amount of disk protrusion, the following ratios were calculated at the site of most pronounced compression: vertebral canal compromise ratio, defined as the protruded disk area divided by the CSA of the vertebral canal; dorsoventral vertebral canal compromise ratio, defined as the maximum dorsoventral diameter of the protruded intervertebral disk, divided by the dorsoventral diameter of the vertebral canal.

There was a global influence of assessed group of dog on the canal occupying ratio of the spinal cord at the cranial and caudal vertebral body level and for the canal occupying ratio of the CSF-column at the caudal vertebral body level. For each assessed canal occupying ratio, there was a significant difference between clinically affected Doberman Pinschers and clinically normal Foxhounds at the level of C7. The remaining spinal cord area was the only useful ratio to differentiate between dogs with clinically relevant and irrelevant spinal cord compression. ROC-statistics demonstrated that a remaining spinal cord area of 0.8 corresponded with a sensitivity of 0.9 and a specificity of 0.7 to discriminate between dogs with and without clinical signs of DAWS.

The results of this study suggest that relative stenosis of the caudal cervical vertebral canal predisposes Doberman Pinschers to the development of DAWS and that the CSA of the remaining spinal cord can be used to differentiate between clinically relevant and irrelevant spinal cord compressions.

ABSTRACT #92

A CANDIDATE GENE APPROACH TO MYASTHENIA GRAVIS IN NEWFOUNDLANDS. J King[1], K Vernau[2], N Safra[2], W Vernau[2], D Bannasch[2], B Sturges[2], NC Pedersen[2], GD Shelton[3]. [1]William Pritchard R. Veterinary Medical Teaching Hospital, Davis, CA. [2]University of California – Davis, Davis, CA. [3]University of California – San Diego, La Jolla, CA.

Acquired myasthenia gravis (MG) is an antibody-mediated autoimmune disease that occurs in dogs and people and is the most commonly diagnosed neuromuscular disease in dogs. Associations between genetic factors and MG are reported in people, but genetic studies have not been done in dogs. A familial predisposition for MG is reported in some dog breeds including the Great Dane and the Newfoundland, suggesting the possibility of a genetic association. In dogs, associations between the Dog Leukocyte Antigen (DLA) Class II genes and other autoimmune diseases are reported. The purpose of this study is to use a candidate gene approach to determine if there is a genetic association between specific DLA class II genotypes and MG in Newfoundland dogs.

DNA from 13 Newfoundland dogs (9 female, 4 male) with confirmed myasthenia gravis (positive acetylcholine receptor antibody titers) was collected. Dogs ranged in age from 11 months to 4 years. DNA from 38 healthy Newfoundland dogs greater than 7 years of age, with no clinical evidence of MG served as controls. The exon 2 regions of DLA-DRB1, DQA1, and DQB1 were sequenced on 8 of the affected and 9 of the healthy dogs. Genotype and haplotype assignment was made from this data, and affected and control animals were compared. Sequence data demonstrated a strong association with a specific DRB1 allele among affected dogs with a significant odds ratio of 8.6 ($p = 0.006$ CI: 1.7–42.6). Additional dogs in each group are being tested.

ABSTRACT #93

MOTOR UNIT NUMBER ESTIMATION USING THE INCREMENTAL METHOD IN NORMAL DOGS. L Vásquez[1], S Kanazono[1], JA Villamil[1], U Dhand[2], DP O'Brien[1], JR Coates[1]. [1]University of Missouri College of Veterinary Medicine, [2]School of Medicine, Columbia, MO.

Canine degenerative myelopathy (DM) is an adult onset fatal neurodegenerative disease. The initial upper motor neuron spastic paraparesis and general proprioceptive ataxia progress to a flaccid lower motor neuron tetraparesis. Recently, a missense mutation in the superoxide dismutase 1 gene was found to be a risk factor for DM suggesting that DM is homologous to human amyotrophic lateral sclerosis (ALS). Motor unit number estimation (MUNE) is an electrophysiologic technique for quantifying motor unit function and has proven useful in assessing disease progression in ALS patients. If the technique can reliably be established in normal dogs and applied to DM affected dogs, MUNE could be used as a biomarker for DM and an outcome measure for monitoring therapeutic trials. The objective of this study was to adapt the incremental method for MUNE in normal healthy dogs and establish a reference range.

A version of the McComas incremental stimulation technique was used to study MUNE in extensor digitorum brevis (EDB) muscles of nine neurologically normal dogs (age: 1–9 years; body weight 4.9–13.4 kg). Both pelvic limbs were evaluated. Monopolar stimulating needle electrodes were inserted caudal to the long digital extensor muscle tendon at the level of the tuber calcanei to stimulate the deep peroneal nerve. Direct evoked muscle potentials were recorded with the active surface electrode placed over the EDB motor point and the reference surface electrode over digit IV. The ground electrode was placed subcutaneously between cathode and recording electrode. Supramaximal compound muscle action potential (CMAP) negative peak area was first recorded. Then starting at the subthreshold level, stimulus intensity was slowly increased until the first all-or-none single motor unit potential (SMUP) was evoked. Successive small increments (0.026 mA, 50 µsec) were applied to elicit a total of 10 SMUP responses. The mean SMUP negative peak area was divided into the maximum CMAP negative peak area to yield the MUNE value. Multiple trials were performed on each limb to assess reproducibility.

EDB MUNE values were obtained from all dogs. The median CMAP area was 4.48 mV mS (range 2.00–9.32) with 25 and 75 percentile of 3.67 and 5.37, respectively. The median SMUP area was 0.09 mV mS (range 0.03–0.78) with a 25 and 75 percentile of 0.06 and 0.13, respectively. The MUNE median was 53.5 (range 8–154) with 25 and 75 percentile values of 30 and 71, respectively. There was no significant difference in MUNE between age groups older and younger than 7 years ($P = 0.301$), between right and left limbs ($P = 0.735$) or trials performed per limb ($P = 0.677$).

We show that EDB MUNE using the incremental method can be reliably recorded from normal dogs. These results provide potential to apply the described technique for longitudinal monitoring of lower motor neuron signs in DM affected dogs.

ABSTRACT #94

LOWER MOTOR NEURON DYSFUNCTION CONSISTENT WITH A PERIPHERAL NEUROPATHY IN YOUNG BENGAL CATS. AC Bensfield[1], J Evans[1], GD Shelton[2]. [1]Veterinary Neurological Center, Phoenix, AZ, [2]Comparative Neuromuscular Laboratory, University of California-San Diego, La Jolla, CA.

Between 2000–2008, muscle and/or peripheral nerve biopsies from 36 Bengal cats with neuromuscular weakness were submitted to the Comparative Neuromuscular Laboratory, University of California, San Diego. There is anecdotal information about juvenile onset polyneuropathy in Bengal cats, but more specific characterization with evidence based medicine is lacking. The aim of this retrospective study was to use this cohort of cats to characterize the clinicopathologic features and clinical course.

Medical records from all cases were evaluated. Parameters included history, physical and neurological examination, standard hematological and biochemical assays, testing for infectious diseases, electrodiagnostic evaluation, CSF analysis, and pathological changes in muscle or peripheral nerve biopsies. Clinical outcome was determined from either the medical record, through a survey questionnaire completed by the submitting veterinarian, or by phone consultation with the veterinarian or owners.

The mean age of onset was 10.8 mo (range 3–45, SD +/− 8.2) with a male predominance (23/36 cases). Progressive weakness and lower motor neuron dysfunction, most notably in the pelvic limbs, was most often reported. Of the 23 cases where infectious disease titers were reported, two had positive toxoplasma titers, two had mildly elevated coronavirus titers, and one was positive for feline leukemia virus on bone marrow IFA. Electromyography showed fibrillation potentials in 17 cats and positive sharp waves in 15 cats. Motor nerve conduction velocity or waveform was reported as decreased or abnormal in 12 of 21 cats (range 13–59 m/s). Histopathologic changes in muscle biopsies were consistent with denervation. A frequent finding was depletion of myelinated fibers in intramuscular nerve branches with variable mononuclear cell infiltration. Edema, onion-bulb formations and inappropriately thinly myelinated fibers were most common in peripheral nerve biopsies.

Information regarding treatment and recovery was available for 24 cats. Of these cats, 12 were treated with corticosteroids with or without additional medications or supplements, 7 received medications and supplements without corticosteroids, and 5 were not given any treatments. Complete recovery was reported in 12 cats and partial recovery in 9 cats. One or more relapses were reported in 8 cases. Although 4 cats were euthanized either at the time of diagnosis or during relapse, none of the cats died from the disease.

In conclusion, this study documents a peripheral neuropathy in young Bengal cats. The prognosis for recovery is good although relapses may occur. It is not clear if corticosteroids are of benefit as improvement and remission also occurred in untreated cats.

ABSTRACT #95

ATYPICAL COLLAPSE IN EXERCISING LABRADOR RETRIEVERS UNAFFECTED BY THE DYNAMIN-1 MUTATION. E. Furrow[1], K. Minor[1], S. Taylor[2], J.R. Mickelson[1], E.E. Patterson[1]. [1]University of Minnesota College of Veterinary Medicine, St. Paul, MN. [2]Western College of Veterinary Medicine, Saskatoon, SK, Canada.

Exercise-induced collapse (EIC) in Labrador Retrievers is a neurological disease manifested by limb weakness, ataxia, and collapse. Susceptibility to EIC is very strongly associated with homozygosity for a dynamin 1 gene (*DNM1*) mutation. However, a small population of collapsing Labradors are clear or carriers of the recessively inherited mutation.

The objective of this study was to compare characteristics of collapsing Labradors genetically susceptible to EIC based on *DNM1* genotype with those that are not. Our hypothesis was that dogs unaffected by the mutation have different clinical characteristics from those affected.

Collapse descriptions were obtained via surveys from 244 Labradors with reported collapse; 186 were homozygous for the *DNM1* mutation (genotype EE) and 48 were unaffected (NN or EN). All surveys were reviewed blinded to the *DNM1* genotype.

Several significant differences were found. The EE dogs were a fairly homogenous population exhibiting decreased muscle tone and strength originating in the pelvic limbs with collapse usually limited to these limbs. In contrast, NN and EN dogs were more diverse with a larger proportion exhibiting collapse in the thoracic limbs or all four limbs simultaneously and often resulting in paresis or ataxia without full collapse. Additionally, NN and EN dogs were more likely to have an abnormal mentation during collapse.

This study demonstrates that clinical features of collapse in Labradors that are clear or only carriers of the *DNM1* mutation differ from the classic collapse description in dogs homozygous for the mutation. The cause(s) of this "atypical" collapse has not been determined.

ABSTRACT #96

INITIAL SHORT-TERM EXPERIENCE WITH CISTERNA CHYLI AND THORACIC DUCT GLUE EMBOLIZATION FOR IDIOPATHIC CHYLOTHORAX IN 2 DOGS AND 1 CAT. C. Weisse[1], A Berent[1], J. Solomon[2], C. Cope[2], J. Rush[3], L. Rozanski[3]. [1]Animal Medical Center, NY, NY. [2]Hospital of the University of Pennsylvania, Philadelphia, PA. [3]Tufts University School of Veterinary Medicine, North Grafton, Mass.

Idiopathic chylothorax in small animals has an inconsistent response to medical and surgical management. Glue embolization of the thoracic duct through an abdominal lymphatic does not require entering the thorax, theoretically improves thoracic duct branch penetration, simultaneously embolizes the cisterna chyli to reduce potential for recanalization, and permits simultaneous lymphangiography to document complete occlusion.

The purpose of this study was to evaluate the resolution of idiopathic chylothorax and subsequent complications following glue embolization of the cisterna chyli and thoracic duct in dogs and cats. Requirements for inclusion included patients diagnosed with idiopathic chylothorax that had not received previous surgical therapy. Owner consent was required and follow-up information was collected through contact with the client or referring veterinarian.

Two dogs and 1 cat each received corn oil or heavy cream hourly for four hours pre-operatively. A ventral midline laparotomy was performed and an efferent lymphatic was catheterized. Lymphangiography was performed followed by D5W flush. A 1:1 or 2:1 combination of Lipiodol:cyanoacrylate was mixed and injected under fluoroscopic guidance. Lymphangiography was repeated to confirm complete occlusion of the thoracic duct; the procedure was repeated until complete occlusion confirmed.

Minor complications (3/3) included glue entering the venous system (2/3) and one catheter being glued in place. There were no major complications. There has been no evidence of significant recurrent chylothorax or chyloabdomen in these patients 21, 78, and 230 days post-operatively. Early experience suggests that glue embolization of the cisterna chyli and thoracic duct may be feasible in patients with idiopathic chylothorax.

ABSTRACT #97

SERUM ENDOTHELIN-1 CONCENTRATION IN TERRIER DOGS WITH IDIOPATHIC PULMONARY FIBROSIS. E. Krafft[(1)], HP. Heikkilä[(2)], P. Jespers[(3)], K. McEntee[(1,3)], D. Peeters[(1)], MM. Rajamäki[(2)], C. Clercx[(1)]. [(1)]Department of Clinical Sciences, Faculty of Veterinary Medicine, University of Liège, Belgium; [(2)]Department of Equine and Small Animal Medicine, Faculty of Veterinary Medicine, University of Helsinki, Finland; [(3)]Laboratory of Physiology, Faculty of Medicine, Free University of Brussels, Belgium.

Idiopathic pulmonary fibrosis (IPF) is an interstitial fibrotic pulmonary disease, mainly described in West Highland White Terrier (WHWT). The diagnosis is challenging and ultimately relies on lung histopathology. Identification of blood biomarkers specific for the disease would be very helpful. Endothelin-1 (ET1) has been shown to be elevated in the serum of humans with IPF. The aim of the present study was to compare serum ET1 level in healthy versus dogs with IPF.

Twelve dogs with IPF (eleven WHWT, one Scottish Terrier, mean age 12 years, range 8–15), thirteen privately-owned healthy WHWT from Finland (9 years, 3–14) and twelve healthy experimental beagle dogs from Belgium (7 years, 3–11) entered the study. IPF was diagnosed based on the presence of typical clinical findings, exclusion of other respiratory diseases, and confirmation by histopathology.

Serum ET1 concentration, determined by ELISA was significantly elevated in dogs with IPF as compared to both healthy beagle dogs and WHWT. Median ET1 was 2.32 pg/mL (interquartile range 2.05–3.38) in IPF dogs vs 1.56 (interquartile range 1.25–1.85) in WHWT ($p < 0.001$) and 1.28 (interquartile range 1.07–1.53) in beagle dogs ($p < 0.001$). No significant difference ($p = 0.066$) was detected between healthy beagle dogs and WHWT.

The present study shows that serum ET1 values are not significantly influenced by differences in breed, and suggest that ET1 might possibly be a blood marker of IPF in dogs. Comparison with serum ET1 concentration in dogs with other chronic pulmonary diseases is needed to assess the specificity of serum ET1 in the diagnosis of canine IPF.

ABSTRACT #98

ENDOTHELIN-1 CONCENTRATIONS IN BRONCHOALVEOLAR LAVAGE FLUID FROM HEALTHY CONTROL AND EXPERIMENTALLY ASTHMATIC CATS. CR Sharp,[a] T Lee-Fowler,[b] GJ Buckley,[a] EA Rozanski,[a] CR Reinero[b]. [a]Cummings School of Veterinary Medicine at Tufts University, North Grafton, MA. [b]University of Missouri, College of Veterinary Medicine, Columbia, MO.

Asthma is a common lower airway disorder in pet cats and there is a need to discover biomarkers which could be used to discriminate it from other airway disorders, monitor efficacy of treatment and provide prognostic information. Endothelin-1 (ET-1) is expressed primarily in the lung and has been implicated in the pathogenesis of inflammatory airway diseases through its pro-inflammatory, profibrotic and bronchoconstrictive properties. The objective of this study was to evaluate bronchoalveolar lavage fluid (BALF) of healthy cats and experimentally asthmatic cats for the presence of ET-1 to further our understanding of the pathogenesis of feline asthma and evaluate ET-1 as a potential biomarker. We hypothesized that experimentally asthmatic cats, but not control cats without airway inflammation, would have elevated concentrations BALF ET-1.

Bronchoalveolar lavage fluid (BALF) was collected from healthy control cats (n = 6) and cats with experimentally induced asthma (n = 5). The BALF total and differential cell count was evaluated for each cat. The concentration of immunoreactive ET-1 in BALF was determined using a capture ELISA (Biomedica Gruppe, Austria) according to manufacturer's instructions. The lower limit of detection (LD) is 0.625 fmol/mL. Two cats in the control group had BALF ET-1 concentrations below the LD of the assay; for statistical purposes the ET-1 concentration was reported as ET 0.625 fmol/mL. Data were analyzed using a Mann-Whitney U test with $p < 0.05$ considered significant. A Pearson product moment correlation was used to evaluate for correlation between BALF ET-1 concentration and BALF total cell numbers and eosinophil numbers.

The mean±SD BALF total cell numbers, eosinophil numbers and eosinophil percentages were significantly higher in asthmatic cats (2236 ± 80 cells/µl, 883 ± 559 cells/µl and 37 ± 12%) compared to control cats (476 ± 195 cells/µl, 22 ± 24 cells/µl and 4 ± 3%) (p = 0.004 for all). The BALF ET-1 concentration was also significantly higher in asthmatic cats (1.470±0.471 fmol/mL) compared to control cats (0.817± 0.168) (p = 0.009). In addition BALF ET-1 concentrations were positively correlated with both BALF total cell numbers ($R = 0.779$) and eosinophil numbers ($R = 0.793$); correlation was significant at the 0.01 level.

Cats with experimentally induced asthma had significantly increased BALF ET concentrations compared to control cats without evidence of airway inflammation. While this study provides promising pilot data regarding the potential utility of ET-1 as an asthma biomarker, future studies with larger numbers and cats with naturally developing asthma are needed. Additionally, since collection of BALF is invasive, analysis of ET in blood should be investigated.

ABSTRACT #99

METABOLOMIC BREATH ANALYSIS FOR SMALL MOLECULE BIOMARKERS IN AIRWAYS OF EXPERIMENTALLY ASTHMATIC CATS.

CR Reinero[a], VP Guntur[b], K Selting[a], C Tang[b]. [a]College of Veterinary Medicine, University of Missouri, Columbia, MO. [b]School of Medicine, University of Missouri, Columbia, MO.

For cats with asthma, clinically useful biomarkers to help with diagnosis, monitoring therapeutic efficacy and providing prognostic information are needed. Asthma in cats is believed to be allergic in etiology, orchestrated by T helper 2 cytokines which activate multiple inflammatory cascades and effector cells. This ultimately leads to airway inflammation, airway hyperreactivity and airway remodeling. Our laboratory has previously reported poor sensitivity of commercially available feline-specific cytokine ELISAs when used on bronchoalveolar lavage fluid in experimentally asthmatic cats and pet cats. We hypothesized that exhaled breath condensate (EBC) could be analyzed using nuclear magnetic resonance (NMR) spectroscopy for small molecules that would be manifest in higher quantities in experimentally asthmatic cats.

Cats were previously sensitized and challenged using Bermuda grass allergen (BGA). The asthmatic phenotype was confirmed by demonstrating eosinophilic airways inflammation using bronchoalveolar lavage fluid analysis (defined as eosinophils > 17%). To maintain the asthmatic phenotype, cats were administered an aerosol of BGA once weekly. On the morning of their BGA aerosol challenge, cats first had collection of EBC by placing them in an airtight plexiglass box fitted for an air compressor at one end and an outlet attached to a U shaped stainless steel tube at the other end. The stainless steel tube was submerged in an ice bath to collect exhaled condensate. Approximately 0.5 ml was collected from each cat over 30 minutes. Cats were then challenged with BGA by aerosol and 6 hours later had repeat collection of EBC. Samples were stored at -20°C until analysis. Immediately prior to ^{1}H NMR spectroscopy, EBC was sterile filtered (22 µm syringe filter), buffered to pH 7.0, and spiked with 1 µM of dimethyl-silapentane-sulfonate (DSS).

^{1}H NMR spectroscopy of EBC from BGA challenged cats resulted in changes among several small molecules when comparing pre-and post-allergen challenge samples. Formate and succinate showed a two-fold increase in signal post-allergen challenge. Lactate showed a 100-fold decrease in signal post-allergen challenge.

Our investigations suggest a difference in small molecule expression with allergen exposure in asthmatic cats. Results from this pilot study suggest that ^{1}H NMR spectroscopy of EBC can be used to identify differences in small molecule expression of cats after acute allergen challenge. Future studies are warranted to demonstrate differences between 1) non-asthmatic and asthmatic cats and 2) pre- and post-treatment in asthmatic cats. These assays and future studies will be used to identify small molecules as potential biomarkers of active inflammation and response to treatment.

ABSTRACT #100

SUBCLINICAL AIRWAY INFLAMMATION DESPITE ORAL CORTICOSTEROID THERAPY IN CATS WITH LOWER AIRWAY DISEASE.

CG Cocayne AE DeClue, CR Reinero. University of Missouri, College of Veterinary Medicine, Columbia, MO.

Recommendations for tapering of glucocorticoid therapy in cats with lower airway disease (eosinophilic or neutrophilic inflammation) have traditionally been based on resolution of clinical signs and radiographic abnormalities. In people, resolution of clinical signs is a poor predictor of airway inflammation. In cats, there are no reports on the sensitivity of clinical signs in predicting airway inflammation, in particular while receiving glucocorticoid treatment. Since assessment of the presence of clinical signs is often the driving force behind recommendations to alter therapy in cats with lower airway disease, it is essential to determine if they can reliably predict the presence of airway inflammation. We hypothesized that subclinical inflammation is present in cats with lower airway disease being treated with oral glucocorticoids.

Cats with clinical signs of lower airway disease (cough, wheeze, or expiratory respiratory distress) and a definitive diagnosis of lower airway disease diagnosed using bronchoalveolar lavage fluid (BALF) cytology (defined as eosinophils > 17% or neutrophils > 5%) and no evidence of pulmonary infection were enrolled. Cats were required to have complete resolution of clinical signs and BALF collected while asymptomatic and still receiving glucocorticoid therapy (minimum of 3 weeks in duration). Data were analyzed using a paired t-test with $p < 0.05$ considered significant and presented as mean±SE.

Nine cats were included in the study. The group percentage of BALF eosinophils and neutrophils prior to treatment were 25 ± 7% and 28 ± 9%, respectively. After the baseline BAL, variable treatments with antibiotics and bronchodilators were initiated by preference of the attending clinician and all cats had oral glucocorticoid therapy. Cats were treated with prednisolone (n = 8) or prednisone (n = 1) at a dose of 1.8 ± 0.2 mg/kg/day. There was no significant difference before and after treatment between the percent of BALF eosinophils (p = 0.616) or neutrophils (p = 0.808) despite complete resolution of clinical signs. After treatment, only 3/9 asymptomatic cats had complete resolution of airway inflammation based on BALF (eosinophils, 3 ± 2%; neutrophils, 4 ± 1%). The remaining 6/9 cats had persistent airway inflammation (eosinophils, 29 ± 11%; neutrophils, 43 ± 12%) despite resolution of their clinical signs.

Two-thirds of pet cats with lower airway disease treated with steroids had subclinical airway inflammation as determined by BALF cytology. Since persistent airway inflammation can lead to permanent structural and functional changes in the lungs, therapeutic decisions based solely on the resolution of clinical signs in cats receiving glucocorticoids should likely not be made. We conclude there is a need to reassess how we monitor efficacy of therapy in these cats. Future studies should focus on developing more comprehensive schemes of airway pathology including lung function testing and other, ideally less invasive tests of airway inflammation and remodeling.

ABSTRACT #101

EFFECTS OF DOXYCYCLINE ON BRONCHIAL INFLAMMATION AND HYPERRESPONSIVENESS AND ON PROTEASE ACTIVITY IN A FELINE MODEL OF ACUTE ASTHMA.

J. Leemans[1,2], N. Kirschvink[1], F. Bernaerts[2], C. Clercx[2], P. Gustin[2]. [1]University of Namur, Faculty of Sciences, Namur, Belgium. [2]University of Liège, Faculty of Veterinary Medicine, Liège, Belgium.

Matrix metalloproteinases (MMP) are zinc- and calcium-dependent endopeptidases involved in the pathogenesis of many disorders in which excessive degradation of extracellular matrix (ECM) occurs (e.g., periodontitis, asthma). In cats with experimentally induced asthma, acute exposure to allergen leads to significant increases in eosinophil counts and MMP-2 and -9 activities in the bronchoalveolar lavage fluid (BALF), suggesting that the enzymatic breakdown of the ECM might contribute to the allergen-induced bronchial inflammation. This study aimed at investigating the effects of doxycycline, a tetracycline antibiotic displaying MMP inhibitory activity, on allergen-induced airway inflammation and hyperresponsiveness in *Ascaris suum* (AS)-sensitized cats.

Eight AS-sensitized cats were enrolled in a randomized controlled three-way crossover study in which they underwent three AS-challenges at 1-month intervals. AS-stimulated cats were randomly assigned to receive either prednisolone (1 mg/kg PO q12h; gold standard), doxycycline hyclate (5 mg/kg PO q12h) or no drug. Each treatment was administered the day before (D_{-1}) and until 2 days following AS challenge (D_0, D_{+1}, D_{+2}), D_0 being the day of allergen exposure. Airway responsiveness (AR) towards inhaled carbachol was evaluated by means of barometric whole-body plethysmography prior to (D_{-5}) and 48 hours (D_{+2}) after AS challenge and was expressed as the provocative dose of carbachol increasing Penh to 300% of baseline values (C-$Penh_{300}$, in %). BALF and blood samples were collected 72 hours (D_{+3}) after AS stimulation. Treatment-related changes in AR, BALF cytology, total protein content and 8-iso-prostaglandin $F_2\alpha$ concentrations were assessed along with changes in plasma and BALF MMP-2 and -9 activities.

No significant differences ($P > 0.05$) in mean ΔC-$Penh_{300}$ ([C-$Penh_{300}$ D_{+2}] − [C-$Penh_{300}$ D_{-5}], in %) were found among treatment groups. Treatments did not show any significant differences among themselves, with reference to BALF total cell counts ($P > 0.05$). Mean BALF eosinophil % in cats treated with prednisolone (8.6 ± 1.7) were significantly lower, compared with values in control (40.1 ± 7.6, $P < 0.001$) and doxycycline (40.2 ± 6.7, $P < 0.001$) groups. Decreased BALF total protein content (mg/l) was found after steroid treatment (Untreated: 258.7 ± 37.7; Prednisolone: 120.8 ± 17.6, $P = 0.011$; doxycycline: 237.5 ± 31.5; $P > 0.05$). Results showed significantly lower plasma and BALF MMP-2 activities in prednisolone-treated cats than in other groups ($P < 0.05$). There were no significant differences among groups in BALF 8-iso-$PGF_2\alpha$ concentrations, nor in plasma and BALF MMP-9 activities ($P > 0.05$).

Oral doxycycline in the present experimental setting failed to decrease airway eosinophilic inflammation and bronchial hyperresponsiveness, suggesting that it should not be advocated as the sole therapy for feline allergic asthma.

ABSTRACT #102

RESPIRATORY FUNCTIONAL ASSESSMENT USING BAROMETRIC WHOLE-BODY PLETHYSMOGRAPHY IN HEALTHY GROWING AND AGING BEAGLE DOGS.

J Manens[1], M Bolognin[1], N Kirschvink[2], C Clercx[1]. [1]Department of Clinical Sciences, University of Liège, Belgium. [2]Animal Physiology, University of Namur, Belgium.

Barometric Whole-Body Plethysmography (BWBP) is a non-invasive pulmonary function test that has been validated in many species, including dogs. In cats, airway reactivity has been shown to decrease significantly in elderly animals. In dogs, however, effect of age on respiratory variables has not been reported. The objective of this study was to assess and compare BWBP respiratory function parameters (RFP) in beagle dogs at different ages. BWBP measurements were obtained in 3 different groups: (1) 10 young beagles evaluated twice, at 7 (1a) and at 16 months of age (1b); (2) 9 healthy adult beagles, aged 5 to 8 years (MA adults, Median = 7 y 6 m); (3) 8 healthy elderly beagles, aged 11 to 13 years (Old adults, Median = 12 y 6 m). Airway reactivity was also measured in all dogs using bronchial provocation test with nebulization of histamine at increasing concentrations (0.1 to 1.6%) until Penh (enhanced pause, parameter used as a bronchoconstriction index) exceeded 300% of baseline (H-Penh300).

Growth induced a significant increase in most parameters (see table), except for RR which was higher in group1a compared to group1b. Some parameters (TV/BW, Te, PIF/BW, PEF/BW) decreased significantly with aging. H-Penh300 was significantly higher in middle-aged adults compared to young adults.

Table [Abstract 102]

	Weight	RR	TV/BW	MV/BW	Te	Te/Ti	PIF/BW	PEF/BW	H-Penh300
7m(1a)	15.50±1.18	13.88±.2.03*	9,12±1.47*	125.79±28.22*	3.24±0.64*	3.24±0.64*	14.57±2.82*	7.45±2.06*	0.32±0.15
16m(1b)	15.40±1.713	11,19±1.74*	17,13±5.60*,°	188.75±53.45*	6.47±5.46*,°	4.99±2.66*,°	22.80±10.55*,°	13.13±5.17*,°	0.23±0.08°
MA adults(2)	14.06±1.62°	12.28±3.53	13,74±2.85	172.59±75.09	3.94±1.30	2.84±1.16	16.06±3.19	9.11±4.03	0.53±0.131°
Old adults(3)	16.78±1.53°	22,64±13.51	9,70±3.14°	183.58±54.97	2.57±1.91	2.00±1.05°	14.33±1.63°	6.45±1.17°	0.27±0.07

RR, respiratory rate (rpm); BW, body weight (kg); TV, tidal volume (ml); MV, minute volume (ml); Te, expiration time (s); Ti, inspiration time (s); PIF, peak inspiratory flow (ml/s); PEF, peak expiratory flow (ml/s)

**: Significant difference between groups 1a and 1b (paired t-Test or Wilcoxon Signed Rank Test).*

°: Significant difference between age groups (One Way Analysis of Variance Test or Kruskal-Wallis test)

This study shows that growing and aging processes have an influence on BWBP parameters. Pulmonary maturity, believed to occur around 1 year old in dogs, might explain the differences in growing dogs. Airway reactivity seems to be the lowest in middle-aged dogs. Modifications related to growing and aging are to take into account when analyzing BWBP RFP as part of a diagnostic investigation or treatment efficacy evaluation in dogs with pulmonary diseases.

ABSTRACT #103

NEBULIZED ACETYLCYSTEINE IN CATS WITH EXPERIMENTAL ASTHMA.

C Reinero[a], T Lee-Fowler[a], V Guntur[b], A DeClue[a], L Cohn[a], J Dodam[a]. [a]College of Veterinary Medicine and [b]College of Medicine, University of Missouri, Columbia MO.

N-acetylcysteine (NAC) is a mucolytic and antioxidant which is thought to cause bronchoconstriction in cats when delivered by aerosol. A recent abstract reported no adverse effects or alterations in enhanced pause (Penh) after delivery of 150 mg of NAC into a barometric whole body plethysmography chamber in 27 pet cats. However, delivery of NAC by nebulizing into a 38 L chamber likely resulted in minimal distribution to lower airways. Moreover, Penh is an indirect measure of airway resistance and may be insensitive to some changes in airway function. We hypothesized that in experimentally asthmatic cats aerosol delivery of NAC into an endotracheal tube with direct ventilator acquired mechanics would allow detection of airflow limitation.

Asthmatic cats (median %BALF eosinophils = 63%) were monitored during NAC administration (n = 4) and methacholine (MCh) challenges (n = 3) > 1 week apart. Cats were anesthetized with ketamine and propofol and mechanics measured by a ventilator set to neonatal mode (Engstrom Carestation, GE Healthcare, Fairfield, CT). Cats were intubated with 4 mm ID, 14 cm long endotracheal tubes and mechanically ventilated. Airway resistance (Raw) was continuously calculated by the ventilator. Sterile saline was delivered first for 30 seconds through an Aeroneb solo in-line nebulizer followed by 4 minutes of data collection (baseline). Delivery of NAC solution 10% for inhalation (American Regent, Inc, Shirley, NY) was followed by 4 minutes of data collection after each dose; the dosing scheme was 50 mg, 50 mg, 100 mg and 200 mg. Data were expressed as the % increase in baseline Raw. MCh challenge was performed by delivering doubling doses of MCh (from 0.0625–32 mg/ml) for 30 seconds followed by 4 minutes of data collection. MCh data were expressed as the concentration of MCh that increased Raw by 200% (EC200Raw).

The first 3 study cats had a % increase in baseline Raw after NAC of 123%, 159%, and 170%, and an EC200Raw of 2.7 mg/ml, 3.7 mg/ml and 12.8 mg/ml, respectively. The fourth cat had a % increase in baseline Raw after NAC of 158%. At least one adverse effect was noted in each of the first 3 cats: subjectively excessive tenacious oral secretions, increased lower airway secretions (crackles), and transient unilateral strabismus, respectively. The fourth cat died during anesthetic recovery immediately post-NAC challenge (necropsy pending).

Results of this pilot study are in contrast to the previous study which did not report changes in airflow limitation or other perceivable adverse reactions. It is unclear if the NAC caused or contributed to death in one asthmatic cat. We used a direct measure of airway resistance and used a higher dose of NAC (extrapolated from the human infant dosage of 200–400 mg) through the endotracheal tube. Extreme caution should be taken if using this drug in cats with airway disease, if at all.

ABSTRACT #104

ELEVATION OF PROCOLLAGEN TYPE III AMINO TERMINAL PROPEPTIDE CONCENTRATION IN BRONCHOALVEOLAR LAVAGE FLUID OF DOGS WITH IDIOPATHIC PULMONARY FIBROSIS. HP Heikkilä[1], E Krafft[2], P Jespers[3], K McEntee[2,3], MM Rajamäki[1], C Clercx[2]. [1]Department of Equine and Small Animal Medicine, Faculty of Veterinary Medicine, University of Helsinki, Finland. [2]Department of Clinical Sciences, Faculty of Veterinary Medicine, University of Liège, Belgium. [3]Laboratory of Physiology, Faculty of Medicine, Free University of Brussels, Belgium.

Idiopathic pulmonary fibrosis (IPF) is a chronic, progressive lung disease characterized by an abnormal deposition of collagen into alveolar walls. IPF affects mainly West Highland white terriers (WHWT). Procollagen type III amino terminal propeptide (PIIINP) is a marker of collagen type III synthesis and degradation. The aim of this study was to investigate the concentrations of PIIINP in bronchoalveolar lavage fluid (BALF) and serum of dogs with IPF and compare them with those of healthy controls.

Fourteen WHWTs and one Scottish terrier with IPF (mean age 12.4 years, range 8–15) and 13 healthy WHWTs (mean age 9.5 years, range 3–14) were included in the study. IPF diagnosis was based on clinical findings and exclusion of other respiratory diseases, and in ten dogs it was confirmed by histopathology. PIIINP concentrations were analyzed from BALF and serum by using a commercially available radioimmunoassay (^{Uniq}PIIINP RIA, Orion Diagnostica, Finland). Statistical significance was set at $P < 0.05$.

BALF PIIINP concentrations were significantly elevated in dogs with IPF (median 0.37 µg/L; interquartile range, IQ, 0.02–0.61 µg/L) compared with controls (median 0.02 µg/L; IQ 0.02–0.05 µg/L), $P = 0.033$.

Serum PIIINP concentrations did not differ between the groups. The median value in dogs with IPF was 6.03 µg/L (IQ 4.49–9.40 µg/L) vs. 7.48 µg/L (IQ 6.17–8.86 µg/L) in controls, $P = 0.573$. No correlation between serum and BALF PIIINP concentration was detected.

Our results suggest that elevated PIIINP concentrations in BALF, but not in serum, serve as a marker of collagen type III turnover in canine IPF.

ABSTRACT #105

EFFECTS OF HIGH VERSUS NORMAL SALT DIETS ON RENAL VARIABLES IN AGED CATS: A 6-MONTH BLINDED RANDOMIZED PROSPECTIVE STUDY. BS Reynolds[1], P Nguyen[2], V Chetboul[3], V Biourge[4], I Testault[5], H Dumon[2], C Carlos-Sampedrano[3], E Trehiou[5], J Abadie[6], HP Lefebvre[1]. [1]Clinical Sciences, National Veterinary School (NVS), Toulouse. [2]Nutrition/Endocrinology, NVS, Nantes. [3]Cardiology, NVS, Alfort. [4]Royal Canin, Research Center, Aimargues. [5]Atlantia, Nantes. [6]Pathology, NVS, Nantes, France.

Concerns have been expressed about adverse renal effects of high salt diets. The objective of this blinded study was to compare the effects of normal versus high salt diets on renal variables in aged cats.

Twenty healthy neutered cats (10.1 ± 2.4 y.) were included. Following baseline measurements, cats were randomly allocated in 2 groups according to their glomerular filtration rate (GFR), gender, age and body weight. One group was fed a high salt diet (1.30% sodium, 2.27% chloride as fed basis (4000 kcal ME/kg)) and the other one a control diet (0.35% sodium, 0.70% chloride). Body weight, urine specific gravity and protein/creatinine ratio, plasma urea, creatinine, electrolytes, calcium, phosphate, total proteins, systolic/diastolic arterial blood pressures and GFR were measured before, and again 3 and 6 months after diet implementation. Statistical analysis was performed using a general linear model.

None of the variables was affected by the dietary treatment.

In conclusion, over a six month period, a high salt diet has no adverse effects on renal function in healthy aged cats.

ABSTRACT #106

EFFICACY OF A DIET DESIGNED WITH A RELATIVE SUPER SATURATION < 1 TO DISSOLVE STRUVITE STONES IN THE FELINE BLADDER. D. Houston[1], H. Weese[1], M. Evason[1], I. van Hoek[2]. [1]Medi-Cal Royal Canin, Guelph, ON, [2]Royal Canin Research Center, Aimargues, France.

Previous studies have shown the efficacy of canned and dry magnesium-restricted, urine acidifying diets in dissolving struvite stones in cats; the mean times to dissolution in these studies were 26–36 days (canned food) and 34 days (dry food), respectively. Studies have shown that urine supersaturation is primarily responsible for the formation and dissolution of crystals within the urinary tract. A previous study showed the efficacy of a dry diet formulated to generate a urinary struvite relative supersaturation (RSS) < 1 (Royal Canin Urinary S/O), in dissolving feline struvite uroliths *in vitro*. The present study was undertaken to determine if a diet with efficacy for dissolving struvite stone *in vitro* also had efficacy of dissolving struvite stones *in vivo*.

Ten privately owned cats suspected of struvite urolithiasis based on radiographs taken by local practitioners were initially included. Cats were kept in their home environment and owners were instructed to feed Urinary S/O either wet (5 cases) or dry (5 cases) as the exclusive diet for the duration of the study. Investigational parameters at inclusion included a complete blood count, serum biochemistry, urinalysis, urine culture, and abdominal radiographs. Radiographs and urinalysis were repeated on a weekly basis and evaluated by two board certified internal medicine specialists to mark time of stone dissolution.

Eight spayed female and 2 neutered male cats with a mean age of 7.05 ± 3.34 years were included. Struvite stones dissolved in a median of 24 days (range 14–56 days) and of 19 days (range 10–39 days) in cats fed wet or dry food respectively.

Our preliminary results suggest that a diet designed to create urine undersaturated for struvite (RSS < 1) is effective both *in vivo* and *in vitro* in dissolution of struvite bladder stones. These results suggest that *in vitro* studies might be sufficient to predict dissolution times in vivo.

ABSTRACT #107

EVALUATING DEMOGRAPHIC RISK FACTORS FOR URATE UROLITHS IN CATS. H Albasan[1,2], CA Osborne[1], JP Lulich[1], LK Ulrich[1], LA Koehler[1], LL Swanson[1]. [1]University of Minnesota, College of Veterinary Medicine, Minnesota Urolith Center, St. Paul, MN. [2]Erciyes University, Faculty of Veterinary Medicine, Department of Internal Medicine, Kayseri, Turkey.

We hypothesize that several demographic factors are associated with increased risk for urate urolithiasis in cats.

Cases consisted of 5,072 cats with urate uroliths submitted to the Minnesota Urolith Center between January 1, 1981 and December 31, 20081. Controls consisted of 437,228 cats without urinary tract diseases identified in records of the Veterinary Medical Database (Urbana, IL) during the same period. Crude odds ratios (OR) and 95% confidence interval (CI) were calculated to test the hypothesis. Mixed breed, age group < 1 year, female, intact cats were the reference groups for statistical analysis.

Purebred cats had significantly higher risk of developing urate uroliths than cats of other breeds (OR > 2). Abyssinian, American Shorthair, Himalayan, and Manx cats had significantly lower risks of developing urate uroliths than did other breeds (OR < 0.5). Neutered cats were 12.2 times as likely to develop urate uroliths as were sexually intact cats (CI: 10.64 – 13.92). Mean ± SD age of cats with urate uroliths was 6.2 ± 4.3 (range: 0.3 to 16.8) years. Cats in all age groups had a significantly increased risk of developing urate uroliths. Cats ≥ 4 but < 7 years old had the highest risk of developing urate uroliths; they were 51.4 (CI: 40.59–62.40) times as likely to develop urate uroliths as were cats < 1 year old.

The prototypical cat with urate uroliths is a purebred neutered cat 4 to 7 years of old. This information may be helpful in predicting mineral composition of uroliths in vivo.

ABSTRACT #108

URINARY ph IS NOT A MAJOR FACTOR IN URINARY caox SUPERSATURATION OF HEALTHY CATS. I van Hoek[1], S Aladenise[1], H Dumon[2], C Tournier[1], F Garnier[1], C Venet[1], Y Soulard[1], S Vialle[1], V Biourge[1]. [1]Royal Canin Research Center, Aimargues, France. [2]National Veterinary School Nantes, France.

Acidifying diets for treatment of struvite uroliths are regarded as a risk factor for formation of calcium oxalate (CaOx) uroliths, by increasing urinary calcium concentration and excretion. This study investigated the relationship between urinary pH and CaOx relative supersaturation (CaOX-RSS) in healthy cats.

A dry expanded control diet served as basis for 12 diets inducing different urinary pH. Diets differed only by addition of various acidifying agents and their concentration. Diets were fed in random order for 2 weeks to 5 adult DSH cats (2 neutered males, 3 females [2 neutered]). Urinary CaOx-RSS was calculated using SUPERSAT™ software after measurement of pH, Ca, Mg, Na, K, NH_4^+, phosphate, citrate, sulfate, oxalate and uric acid in pooled urine of each cat from the last 5 days of each study period. Results were analyzed by repeated measurements ANOVA ($p < 0.05$) and expressed as mean ± SD (range).

Urinary pH was 6.1 ± 0.2 (5.7–6.9) and CaOx-RSS was 5.9 ± 2.0 (2.5–12.6). There was no significant effect of urinary pH on CaOx-RSS, urinary excretion of calcium, oxalate or citrate.

This study does not support a major role of urinary pH on CaOx-RSS. This suggests that is possible to formulate a diet to reduce both struvite and CaOx urine saturation.

ABSTRACT #109

STRUVITE UROLITH DISSOLUTION IN 5 CATS FED A MAINTENANCE DIET WITH 0.06% MAGNESIUM, 35% PROTEIN AND A TARGETED URINE PH OF 6.2–6.4.

J.M. MacLeay, I. Paetau-Robinson, D. Fritsch. Hills Pet Nutrition, Topeka, KS.

Although dissolution diets effectively eradicate struvite urocystoliths, we hypothesize that a maintenance food to prevent struvite crystalluria will also dissolve them. To test this hypothesis we fed 5 cats (3FS, 2MC) with struvite urocystoliths a dry kibble maintenance food formulated to provide 0.06% Mg (DM), 35% protein, and a calculated target urine pH (UpH) of 6.2–6.4. The food was fed exclusively in an amount to maintain body condition. The diagnosis of struvite urolithiasis was based on the radiographic density of uroliths. Weekly follow-up radiographs were performed until urolith resolution, enlargement or 8 weeks, whichever was sooner.

Initial mean USG = 1.047 (range 1.029–1.068) and mean UpH = 6.12 (range 5.82–6.30). Urine cultures were negative at screening in 4/5 cats. The fifth, was from urine obtained from the cat carrier. The cat was not treated since this was considered environmental contamination. A second cat became positive at the last visit after urolith dissolution and was treated with antibiotics. All cats had negative cultures when dismissed from the study. Infection with calculogenic bacteria was not observed.

Dissolution of the uroliths occurred in 4/5 cats. Radiographic resolution occurred in 2 wks ($n = 2$), 4 wks ($n = 1$) and 5 wks ($n = 1$). At resolution cats had a mean USG = 1.057 (range 1.043–1.078) and mean UpH = 6.44 (range 6.0–7.1). The fifth cat was taken to surgery at 6 weeks; the urolith was 100% ammonium urate.

This case series provides evidence that a maintenance dry kibble diet formulated with a restricted Mg content and moderate target UpH dissolves struvite urocystoliths. Further investigation is warranted to determine the implications of these findings in a larger population.

ABSTRACT #110

SURVEILLANCE OF BACTERIAL CONTAMINATION IN SMALL ANIMAL HOSPITAL ENVIRONMENTS IN KANSAS.

K Stenske, J Westbrook, K Lothrop, A Ghosh, L Zurek. Kansas State University, College of Veterinary Medicine, Manhattan, KS.

Contaminated veterinary equipment has the potential to spread infections if not cleaned properly between patients. We determined the prevalence of bacterial contamination of 4 surfaces from 10 small animal private practice hospitals in Kansas at 2 announced visits, 4 months apart. One swab each from a cage door, stethoscope, thermometer, and mouth gag, as well as a survey about cleaning protocols, were collected. Selective media for enterococci and enterics and one broad-spectrum medium were used. Species-specific PCR and *sodA* sequencing were used to identify enterococci. Antibiotic susceptibility was assessed using the disc diffusion method with Mueller-Hinton agar.

Enterococci were isolated from cage doors (7/10 hospitals), stethoscopes (7/10 hospitals), thermometers (4/10 hospitals), and mouth gags (1/10 hospitals), while enterics were rare. Enterococcal contamination was not consistent; only one hospital had substantial contamination at both visits. Enterococcal isolates selected for further analysis ($n = 192$) were mainly represented by *E. faecium* (40%), *E. hirae* (33%), and *E. faecalis* (23%). *Enterococcus faecium* were frequently resistant to enrofloxacin (79%), erythromycin (78%), ampicillin (77%), doxycycline (22%), nitrofurantoin (21%), and gentamicin (12%), while other enterococcal species were mostly susceptible to all nine antibiotics.

Seven of ten hospitals reported cleaning cage doors between patients, but only 2/10 allow 10-minute contact time with disinfectant. Fifty percent of veterinarians reported almost-never cleaning their stethoscopes. Contamination with multi-drug resistant enterococci is a concern and may increase risk of hospital-acquired infections. Cleaning cage doors, stethoscopes, and thermometers between patients and allowing 10-minute disinfectant contact time on surfaces is recommended to reduce bacterial contamination.

ABSTRACT #111

PREVALENCE OF METHICILLIN-RESISTANT STAPHYLOCOCCI IN A NORTHERN COLORADO ANIMAL SHELTER.

EN Gingrich, T Kurt, R Ruch-Gallie, MR Lappin. Colorado State University, Fort Collins, CO.

Methicillin-resistant *Staphylococcus aureus* (MRSA) and *S. pseudintermedius* (MRSP) are recognized as significant pathogens in veterinary medicine. To date there have been no studies examining MRSA or MRSP colonization rates in animal shelters in the United States. There is potential for animals to serve as a source of MRSA or MRSP re-infection in humans. The purpose of the study was to determine the prevalence of MRSA and MRSP colonization in cats and dogs housed in a northern Colorado animal shelter.

Nasal and perianal swabs were collected from 200 cats and 200 dogs in an open admission shelter. This included 100 cats and 100 dogs housed in respective stray wards and 100 cats and 100 dogs available for adoption. Samples from each animal were pooled and inoculated on selective media to isolate MRSA/MRSP. Antimicrobial sensitivity was confirmed via Kirby-Bauer disc diffusion. Overall MRSA/MRSP prevalence rates and prevalence rates by subgroup for MRSP were calculated.

MRSA (0.5%) or MRSP (3%) samples were isolated from dogs. The MRSA dog was housed in the stray ward. Of the MRSP, 33% (2/6) were obtained from dogs housed in the stray ward while 66.7% (4/6) were obtained from dogs on the adoption floor. The length of stay at the time of sampling for the MRSP dogs ranged from 1 to 8 days. The MRSA (0.5%) isolated from a cat was from a 3 month old kitten housed in the stray ward.

Results reveal a low prevalence of MRSA and are consistent with previous findings of 1–2% prevalence in dogs and cats in animal hospitals. The MRSP results were slightly higher than the 0–2% prevalence reported in other canine studies. The cat colonized with MRSA was housed with 4 MRSA negative littermates. These results suggest animals housed in a regularly cleaned shelter environment are not at greater risk of acquiring MRSA/MRSP and pose no more risk to adopters than animals in the general population.

ABSTRACT #112

HOSPITAL PREVALENCE AND RISK FACTORS FOR METHICILLIN-RESISTANT *STAPHYLOCOCCUS* INFECTIONS IN SMALL ANIMALS AT THE PURDUE UNIVERSITY VETERINARY TEACHING HOSPITAL (PUVTH), 2006–2009.

E. Meler[1], S. He[2], A. Dunham[1], L. Guptill[1]. [1]Department of Veterinary Clinical Sciences. [2]Department of Comparative Pathobiology, School of Veterinary Medicine Purdue University, West-Lafayette, IN.

Methicillin-resistant *Staphylococcus* (MRS) infections are an important problem in veterinary medicine. This unmatched case-control study tested the hypotheses that the prevalence of MRS infections of small animals presented to the PUVTH is increasing, and that breed, previous antibiotic treatment, previous surgery, hospital visit or ICU stay are risk factors for MRS infection.

The PUVTH microbiology results database was searched for *Staphylococcus* species infections between 1/1/2006 and 11/30/2009.

MRS was defined as an oxacillin-resistant *Staphylococcus*, polymerase chain reaction (PCR) positive for the *mec-A* gene. Data abstracted from medical records included case number, species, breed, age, gender, neuter status; *Staphylococcus* species; PCR result; site of sample collection (skin, lungs, urinary tract, wound/surgical site, other); date and season of MRS diagnosis; admitting department (dermatology, medicine, surgery); antibiotic use, surgical procedure, or veterinary visit $\leq$ 30 days prior to diagnosis; and ICU stay of $>$ 1 day prior to diagnosis. Hospital prevalence per year was determined by dividing the number of MRS infected animals by the total number of small animal visits to PUVTH per year. Only infections of dogs were considered for risk factor analysis. Controls were randomly selected from dogs with oxacillin-sensitive *Staphylococcus* infections (n = 64). Risk factors were evaluated by logistic regression using R (r-project.org) statistical software; $p < 0.05$ was considered significant. Abbreviations used: MRS *aureus* (MRSA), MRS *epidermidis* (MRSE), MRS *intermedius* (MRSI).

MRS were cultured from 59 clinically affected animals, including 55 dogs and 4 cats. Five dogs had MRS infection in more than 1 year and 4 had both MRSA and MRSE infections. There were 4 dogs and 1 cat with MRSA infections in 2006, 6 dogs in 2007, 14 dogs in 2008, and 18 dogs and 1 cat in 2009. There was 1 dog with MRSE infection in 2006, 2 dogs in 2007, 7 dogs in 2008, and 13 dogs and 2 cats in 2009. One dog had an MRSI infection (2007). Hospital prevalence increased significantly over time for MRSA ($p = 0.026$) and MRSE ($p = 0.002$). The dermatology department diagnosed more patients with MRSE than did other departments ($p = 0.002$); skin was the most common site of infection. Previous antibiotic treatment was associated with increased risk for MRSA infection ($p = 0.010$). Dogs undergoing surgery, visiting a veterinarian, or hospitalized in the ICU $\leq$ 30 days before MRS diagnosis were not at increased risk for MRS infection. MRS infection was not associated with season, age, gender, neuter status or breed.

To conclude, hospital prevalence of MRS infections of animals presented to PUVTH increased significantly from 2006 to 2009. MRSA infections were more common than other MRS, and MRSE infections were predominant in patients with dermatologic conditions. MRSI infection was uncommon. Finally, prior antibiotic use was a risk factor for MRSA infection.

ABSTRACT #113

FELINE HERPESVIRUS-1 AND *MYCOPLASMA* SPP. INFECTIONS IN CATS WITH ACUTE CONJUNCTIVITIS IN AN ANIMAL SHELTER.

D. Zirofsky[1], C.C. Powell[1], W. Rekers[2], M. Brewer[1], M.R. Lappin[1]. [1]Colorado State University, Fort Collins, CO. [2]Cat Adoption Team, Sherwood OR.

In previous studies, DNA of feline herpesvirus 1 (FHV-1) and *Mycoplasma* spp. were the most likely to be amplified by PCR assays from nasal discharges of cats with rhinitis and ocular discharges of cats with conjunctivitis. However, both organisms can colonize healthy cats and in the previous studies, only small numbers of samples from normal cats were included as controls. Thus, the significance of positive PCR test results for FHV-1 and *Mycoplasma* spp. in samples from clinically ill cats is unclear. The purpose of this study is to compare the prevalence rates for FHV-1 DNA and *Mycoplasma* spp. DNA in samples from cats with and without conjunctivitis.

Cats housed in two animal shelters in Oregon were used for the study. Cats with conjunctivitis with no obvious exogenous cause, with or without rhinitis were included in the study. For every two cats with conjunctivitis, samples were to be collected from a cat housed in the shelter without conjunctivitis. Conjunctival cells were collected by rubbing a sterile cotton swab gently in the conjunctival fornix of both eyes after application of topical 1% proparacaine to induce local analgesia. Samples were stored at −20C until shipped to Colorado State University where they were processed and assayed for total DNA as well as DNA of FHV-1 and *Mycoplasma* spp. using adaptations of previously published PCR assays. Differences in prevalence rates were compared by Fisher's exact test with significance defined as $p < 0.05$.

Overall, samples were collected from 69 cats with conjunctivitis and 30 cats without conjunctivitis. The results of the PCR assays are shown in the following table.

Table [Abstract 113]

Assay result	Conjunctivitis (n = 69)	Control (n = 30)	p value
FHV-1 alone	16 (23.2%)	8 (26.7%)	0.7917
Mycoplasma alone	21 (30.4%)	2 (6.7%)	0.0097
Both organisms	10 (14.5%)	1 (3.3%)	0.165
Neither organism	22 (31.9%)	19 (63.3%)	0.0043
Any FHV-1	26 (37.7%)	9 (30%)	0.5017
Any *Mycoplasma*	31 (44.9%)	3 (10%)	0.0006

Results of *Mycoplasma* spp. and FHV-1 PCR assay results do not always document a disease association. However, while both organisms can be amplified from cats with and without conjunctivitis, *Mycoplasma* spp. were amplified from a greater percentage of cats with conjunctivitis suggesting that this microbial group may be common primary pathogens in cats with this syndrome.

ABSTRACT #114

ANTIGEN VS ANTIBODY TESTING FOR THE DIAGNOSIS OF BLASTOMYCOSIS INFECTION IN DOGS.

A Mourning, EE Patterson, JK Paris, S Simmerson. University of Minnesota College of Veterinary Medicine, St. Paul, MN.

The objective of this ongoing study is to investigate the sensitivity and specificity of the urine and serum blastomycosis antigen tests (performed by MiraVista Diagnostics) for the diagnosis of blastomycosis in dogs and to compare the results with the established agar immunodiffusion (AGID) antibody test in clinically relevant populations of dogs with infiltrative respiratory disease.

In total, the study will involve the collection of serum and urine from 20 dogs definitively diagnosed with blastomycosis by cytology and/or histopathology and serum and urine samples from 20 dogs with a definitive diagnosis of other infiltrative pulmonary disease as non-fungal, respiratory disease controls. Dogs with histoplasmosis infection will not be included in this study. The testing laboratories are fully masked as to the disease status of each case.

To date the study has recruited 13 dogs with blastomycosis and 7 dogs with other non-fungal infiltrative pulmonary disease. The results for these 20 dogs produced a sensitivity of 100% and a specificity of 100% for both the urine and serum MiraVista antigen tests, compared with a sensitivity of 57% and a specificity of 100% for the AGID serum antibody test. It is anticipated that data for all 20 cases and all 20 respiratory controls will be compiled by mid 2010.

In conclusion, based on this initial data, the MiraVista antigen test appears to be a very sensitive and specific non-invasive means of detecting blastomycosis infection in dogs with respiratory disease.

ABSTRACT #115

CORRELATION OF URINE AND SERUM BLASTOMYCES ANTIGEN LEVELS WITH CLINICAL REMISSION IN DOGS WITH BLASTOMYCOSIS.

DS Foy[1], LA Trepanier[1], E Hackett[2], LJ Wheat[2]. [1]University of Wisconsin-Madison, Madison, WI. [2]MiraVista Diagnostics, Indianapolis, IN.

Commercially available urine and serum tests measuring Blastomyces antigen show high sensitivity and specificity in the diagnosis of blastomycosis in dogs. The purpose of this study was to monitor the levels of urine and serum Blastomyces antigen throughout antifungal therapy and evaluate their relationship with clinical remission.

Dogs were prospectively enrolled on the basis of: 1) cytologic or histopathologic confirmation of Blastomyces infection, 2) serum and urine samples obtained within 14 days of beginning antifungal therapy, and 3) adequate follow-up to the point of clinical remission. Dogs failing to survive the first month of treatment were censored from analyses. Dogs were evaluated monthly with a physical exam, fundic exam, chest radiographs (if abnormal at diagnosis), and additional specific imaging as indicated by clinical presentation. Dogs were considered to be in clinical remission when no active disease was identified on physical exam, and radiographs or other imaging were normalized, or static for at least one month. All dogs were treated with fluconazole or itraconazole until one month beyond clinical remission. Serum and urine samples were obtained at each visit for Blastomyces antigen levels via enzyme immunoassay; using laboratory reference ranges, Blastomyces antigen levels < 1.00 EIA units were considered negative.

Nineteen dogs (median age 5.0 years, range 1–11; 9 male [7 neutered] and 10 female [8 spayed]) were prospectively enrolled, 19/19 dogs were positive for Blastomyces urine antigen at the time of diagnosis (median 27.89 EIA units, range 1.98–46.10) while only 9/17 dogs were positive on the serum antigen test at diagnosis (median 1.16 EIA units, range 0.08–14.90). For dogs achieving negative urine antigen levels during treatment (n = 9), there was no significant difference in the time to negative urine antigen levels (median 117 days, range 32 - 230), compared to the time to clinical remission (median 130 days, range 73–215; $P = 0.60$). However, there was a significant difference in time to negative serum antigen levels (median 30 days, range 0–133) and clinical remission ($P = < 0.0001$). The median urine antigen levels at the time of clinical remission were 2.20 EIA units (range 0.31–10.04), and at treatment discontinuation (approximately one month later) were 1.17 EIA units (range 0.53–5.87). The sensitivity and specificity of Blastomyces urine antigen in predicting clinical remission were 41% and 92%, respectively.

These results indicate that a negative Blastomyces urine antigen level is associated with clinical remission with a high degree of specificity; work is underway to determine a cut-off with higher sensitivity. Serum Blastomyces antigen levels become negative quickly after the initiation of therapy and significantly prior to clinical remission, and should not be used to determine resolution of disease.

ABSTRACT #116

MONOCYTE CHEMOATTRACTANT PROTEIN-1 CONCENTRATION IS A MARKER OF MORTALITY IN CANINE BABESIOSIS CAUSED BY *BABESIA ROSSI*.

M Kjelgaard-Hansen[1], A Goddard[2], B Wiinberg[1], L Köster[2], JP Schoeman[2]. [1]Department of Small Animal Clinical Sciences, University of Copenhagen, Denmark. [2]Faculty of Veterinary Science, University of Pretoria, South Africa.

Babesia rossi infections in dogs are frequently complicated and patients have a high mortality rate within the first 24-hours after admission to the intensive care unit (ICU). Hypoglycemia and hypercortisolaemia at admission as well as persistent hyperlactataemia have been identified as strong prognosticators, indicating metabolic dysbalance to be important to outcome. Recently, the cytokine monocyte chemoattractant protein [MCP]-1 was identified as a strong predictor of mortality in human sepsis and has also been reported to increase in experimental canine sepsis. The aim of the study was to determine whether MCP-1 is a marker for mortality in complicated babesiosis caused by *B.rossi*.

This prospective, cross-sectional, observational study included 72 client-owned dogs diagnosed with canine babesiosis and admitted to the ICU of the Faculty of Veterinary Science, University of Pretoria. Diagnosis was based on morphological demonstration of the intra-erythrocytic trophozoites on blood films. Dogs with concurrent infections or with a history of treatment for an infection in the preceding 7 days were excluded. Infection with *B.rossi* was confirmed by PCR. Dogs infected with *B.vogeli*, or concurrent *Ehrlichia canis* or *Theileria* spp. infection, and dogs euthanized for reasons other than poor prognosis were excluded. Five healthy dogs were included as controls. Blood samples were collected at admission. Serum MCP-1 concentration was assessed by a canine-specific assay (CCYTO-90 K, Millipore, USA) and glucose by standard enzymatic methodology. Mortality during hospitalization was recorded. The Mann-Whitney *U* test was used to compare the cytokine level in controls vs. babesiosis patients and in survivors vs. non-survivors. Analysis of the area under receiver operating characteristic curve (ROC-AUC) was used to compare the prognostic abilities of MCP-1 and glucose. $P < 0.05$ was considered significant.

Mortality fraction was 19% (14/72), all within 24 h of admission. Significantly increased levels (median [range]) of MCP-1 was observed in babesiosis patients (431 pg/mL [99;20270 pg/mL]) compared to controls (118 pg/mL [85;634 pg/mL]) and in non-survivors compared to survivors (757 pg/mL [291;7951 pg/mL] and 421 pg/mL [99;20268 pg/mL], respectively). The prognostic abilities (ROC-AUC) of MCP-1 and glucose were similar (0.71 and 0.74, respectively, P = 0.65).

Increased concentrations of MCP-1 were observed in dogs infected with *B.rossi* compared to healthy controls and furthermore MCP-1 was seemingly a prognostic biomarker comparable to one of the strongest known predictors for outcome in this disease. Previous experimental studies have demonstrated MCP-1 as an important immunomodulator in sepsis and the results of this study indicate that dysbalances in immunoregulation play an important role in the outcome of babesia caused by *B.rossi*, in addition to the previously described metabolic dysbalances.

ABSTRACT #117

MORTALITY IN CANINE BABESIOSIS IS ASSOCIATED WITH A CONSUMPTIVE COAGULOPATHY.

A. Goddard[1], B. Wiinberg[2], J.P. Schoeman[1], A. Kristensen[2], M. Kjelgaard-Hansen[2]. [1]Department of Companion Animal Clinical Studies, University of Pretoria, South Africa. [2]Department of Small Animal Clinical Sciences, University of Copenhagen, Denmark.

The inflammatory response to infection can activate the coagulation system via complex interactions. If uncontrolled, this may lead to disseminated intravascular coagulation (DIC), which has been identified as a major risk factor for poor outcome in both human and canine medicine. Studies have indicated that patients naturally infected with *Babesia rossi* suffer from Systemic Inflammatory Response Syndrome (SIRS) with possible DIC, without clinical signs of bleeding. We hypothesized that a consumptive coagulopathy is present in dogs with *B. rossi* infection and that the extent of the coagulopathy is related to outcome.

This prospective, cross-sectional, observational study included 72 client-owned dogs diagnosed with canine babesiosis and admitted to the Intensive Care Unit (ICU) of the Faculty of Veterinary Science, University of Pretoria. Infection with *B.rossi* was confirmed by bloodsmear and PCR. Dogs infected with *B.vogeli*, or *Ehrlichia canis*, and dogs euthanized for reasons other than poor prognosis were excluded. Blood samples were collected at admission to hospital. Coagulation factor-, antithrombin- (AT), protein C activity and fibrinogen were assessed using an automated coagulometric analyzer (ACL top 500, Instrumentation Laboratory). D-Dimer was measured using an immunometric flow-through principle (D-Dimer Single test, Nycocard Reader II, Medinor A/S). Mortality during hospitalization was recorded. Levels between the non-survivors and survivors were compared using the Mann-Whitney U test. $P < 0.05$ was considered significant.

Mortality fraction was 19% (14/72). The individual median coagulation factor activity percentage was significantly lower in the non-survivors than the survivors: FII was 40.5% vs. 76.7% ($p < 0.01$); FV was 56.8% vs. 106.4% ($p < 0.001$); FVII was 46.9% vs. 73.4% ($p < 0.05$); FVIII was 60.3% vs. 109.7% ($p < 0.001$); FIX was 56.5% vs. 101.7% ($p < 0.001$); FX was 35.4% vs. 73.9% ($p < 0.001$); FXI was 43.4% vs. 74.9% ($p < 0.01$); FXII was 48.9% vs. 77.4% ($p < 0.01$). Median AT activity was not significantly different between the two groups. Median protein C activity was significantly lower in the non-survivors than the survivors; 44.0% vs. 74.0% ($p < 0.05$). Median fibrinogen level was not significantly different between the two groups; 3.7 g/L vs. 4.8 g/L ($p = 0.25$). The median D-Dimer value was significantly higher in the dogs that died than in the dogs that survived; 1.6 mg/L vs. 0.4 mg/L ($p < 0.01$).

This study demonstrated that mortality was associated with a consumptive coagulopathy in dogs infected with *B.rossi*. The presence of activated coagulation (decreased coagulation factor activity), inhibitor consumption (low protein C) and increased fibrinolytic activity (high D-dimers) in non-survivors of virulent canine babesiosis are indicative of DIC, possibly secondary to a hypercoagulable state caused by the inflammatory response.

ABSTRACT #118

EFFECT OF VACCINATION ON RESULTS OF REAL TIME PCR ASSAYS FOR ADENOVIRUS 2, *BORDETELLA BRONCHISEPTICA* AND PARAINFLUENZA.

S. Assarasakorn[1], V Scorza[1], S. Moroff[2], MR Lappin[1]. [1]Colorado State University, Fort Collins, CO. [2]Antech Diagnostics, Lake Success, NY.

Nucleic acid amplification techniques (PCR) are increasingly being used to document the presence of DNA or RNA of infectious agents in samples collected from dogs suspected to have infectious causes of respiratory disease. Dogs are often administered modified live vaccines (MLV) against some of the infectious agents including adenovirus 2, *Bordetella bronchiseptica*, and parainfluenza. The purpose of this study was to determine the effect of vaccination on results of the FastPanel™ PCR Canine Respiratory Disease Profile (Antech Diagnostics) for these agents.

Young adult, mixed sex, apparently healthy beagles retired from a breeding facility were housed in two separate groups of six. On Day 0, Group 1 was administered a MLV DA2PP and an inactivated coronavirus vaccine subcutaneously (SQ), an inactivated canine influenza virus vaccine SQ, and a MLV Parainfluenza and *B. bronchiseptica* vaccine intranasally (IN) and Group 2 was administered a MLV DA2PP vaccine SQ and an inactivated *B. bronchiseptica* vaccine SQ. On Day 14, all dogs were administered a canarypox vectored canine distemper virus, adenovirus 2 and parvovirus vaccine SQ and a MLV adenovirus 2, parainfluenza, and *B. bronchiseptica* IN. Swabs of the nasal and pharyngeal mucosa were collected prior to vaccination on Days 0 and 14 as well as on Days 3, 7, 10, 17, 21, 24 and 28 and were shipped on cold packs by overnight express for DNA/RNA extraction and assay in the FastPanel™.

Adequate DNA/RNA was present on all swabs. Nucleic acids of *B. bronchiseptica* (12 of 216 samples; 5.6%), parainfluenza (5 of 216 samples; 2.3%), or adenovirus 2 (6 of 216 samples; 2.8%) were amplified from some samples. DNA of *B. bronchiseptica* was amplified from two Group 1 dogs (Dogs 5 and 6) and five Group 2 dogs (Dogs 2–6) prior to vaccination. Over the remainder of the study, only three dogs were positive for *B. bronchiseptica* DNA on one sample date each (Group 1, Dog 2, day 17; Group 2, Dogs 2 and 4, day 17). Parainfluenza RNA was amplified only from Group 1 dogs (Dogs 1–3 and 5, day 3; Dog 2, day 7). Adenovirus 2 RNA was amplified from four dogs on one sample date (Group 1, dogs 4 and 5, day 17; Group 2; dogs 4 and 5, Day 17).

Nucleic acids of adenovirus 2, *B. bronchiseptica*, and parainfluenza can be amplified from adult beagles without respiratory signs of disease and so positive test results do not conclusively prove the agent was the cause of illness in dogs with clinical signs. However, the MLV vaccines studied here either did not induce positive test results or positive test results were short-lived suggesting that prior vaccination will have minimal effect on the positive predictive values of these assays in adult dogs. Further studies are indicated to define duration of positive test results in previously unvaccinated puppies.

ABSTRACT #119

SERUM CARDIAC TROPONIN I CONCENTRATION IN DOGS WITH LEISHMANIASIS. P. Silvestrini[1], M. Piviani[1], J. Alberola[1], A. Rodríguez-Cortés[1], M. Planellas[1], X. Roura[1], J.P. O'Brien[2], J. Pastor[1]. [1]Autonomous University of Barcelona, Spain; [2]University College of Dublin, Ireland.

Canine Leishmaniasis (CanL) is a severe systemic disease caused by *Leishmania infantum* and its prevalence in the Mediterranean basin is as high as 67%. Clinical signs are caused by the parasite itself, the circulating immune complexes and their deposition, leading to polyarthritis, uveitis, glomerulonephritis and vasculitis. It has been recently described an unusual presentation of CanL with a generalised vasculitis and myocarditis. Thus, CanL should be added to the differential diagnosis list of cardiomyopathy.

Cardiac troponin I (cTnI) has a high sensitivity and specificity for myocardial damage and it has never been investigated before in dogs suffering from CanL. CTnI concentration was measured in 160 dogs diagnosed of CanL by ELISA between January 2007 and December 2008. Sixty (37.5%) CanL positive dogs had cTnI values over the reference range. CTnI values showed statistical significant correlation with age ($r = 0.401$, n = 150), PCV ($r = -0.279$, n = 103), albumin ($r = -0.489$, n = 125), A/G ($r = -0.365$, n = 125), urea ($r = 0.195$, n = 131) and creatinine ($r = 0.245$, n = 130). Table shows number of animals (n) and median (Mdn) with the $P_{2.5}$–$P_{97.5}$ interval [P] for the studied parameters, grouped according to cTnI categorized as mildly, moderately or markedly increased (statistically signficant differences between groups are noted with different letters, $p < 0.05$).

Older, anemic, hypoalbuminemic animals and those with renal failure had higher cTnI values. High cTnI values were not associated with high *Leishmania* serology values. When cTnI values where correlated with age, creatinine, PCV and *Leishmania* serology in a multiple lineal regression model, the interaction of all four parameters only explains 7.5% of the variation in cTnI values, being the age the most important parameter affecting cTnI.

In conclusion our study shows that 37.5% of animals with CanL have an increase in cTnI values.

Anemia and renal failure affect cTnI, but age and albumin should also be considered. The reason for cTnI increase in animals with CanL might be related to the direct effect of *Leishmania* on myocytes and/or the presence of vasculitis associated to CanL, as previously described.

Table [Abstract 119]

	Age		PCV		Albumin		Urea		Creatinine		UPC		ELISA Leish	
CTnI Classification	n	Mdn[p]	n	Mdn [p]	n	Mdn [p]	n	Mdn[p]	n	Mdn[p]	n	Mdn[p]	n	Mdn[p]
Within reference values	96	5 [0.84–11]a	66	36.5 [21.7–52.3]a	84	2.82 [1.1–3.29]ac	86	32.35 [10.55–70.21]ac	86	0.94 [0.55–1.92]ab	34	0.45 [0.08–16.8]a	100	211.5 [43.53–324.1]
Midly (0.06 0.2)	34	7 [2 12]b	24	32 [18.8 55] b	27	2.07[0.6 3.46]bc	28	35.4 [16.1 376.2] c	28	1.01 [0.46 5.63] ab	13	3.73 [0.3 14.8]b	37	234 [43 322]
Moderate (0.2–1)	17	8 [4–13]b	10	29 [21–48]	11	1.39[0.74 2.67]b	13	144.2 [21.7–462.8]b	13	2.89 [0.47–7.87]ac	8	5.7 [0.7–11.3]b	20	246 [65–387]
Marked (> 1)	3	8 [8 13]b	3	37 [19.9 40]	3	1.73[1.72 2.04]bc	3	55 [29.1 91.8]	3	1.78 [1.16 2.9]bc	2	5.96 [2.85 9.06]	3	249 [45 300]

ABSTRACT #120

INFLAMMATORY BIOMARKERS FOR CANINE SEPSIS. CR Sharp[1,2], CH Chang[1], AE DeClue[1]. [1]University of Missouri, College of Veterinary Medicine, Columbia, MO. [2]Cummings School of Veterinary Medicine at Tufts University, North Grafton, MA.

Biomarkers are needed to aid in the diagnosis and prognostication of canine sepsis. We evaluated if TNF, IL-6 and IL-10 could be used as biomarkers for canine sepsis. Blood was collected from dogs with sepsis (n = 8), non-infectious systemic inflammatory response syndrome (SIRS) (n = 7) and healthy controls (n = 13) at hospital presentation.

Because plasma TNF, IL-6 and IL-10 were often undetectable, dogs were categorized as either having an inflammatory mediator response (i.e., inflammatory mediator detected in plasma) or not (i.e., inflammatory mediator not detected). Groups were compared using a Fisher exact test with $p < 0.05$ considered significant. Significantly more dogs in the sepsis and SIRS groups had TNF (sepsis, 8/8, $p < 0.001$; SIRS, 5/7, $p = 0.001$; control, 0/13) and IL-6 (sepsis, 4/8, $p = 0.047$; SIRS, 6/7, $p = 0.001$; control 1/13) response compared to the control group. There was no difference in the number of dogs with TNF or IL-6 response between sepsis and SIRS groups (TNF, $p = 0.21$; IL-6 $p = 0.28$) or IL-10 response among groups (sepsis 3/8; SIRS 2/7; control 9/13; $p > 0.15$). Mortality was 75% in the sepsis and 28.5% in the SIRS groups. Using Spearman rank order correlation, plasma TNF, IL-6 and IL-10 concentrations were not significantly correlated with mortality in the sepsis or SIRS groups.

TNF and IL-6 are detectable in plasma during infectious (i.e., sepsis) and non-infectious forms of SIRS in dogs while the presence of IL-10 in the plasma is variable during health and inflammation. Plasma TNF, IL-6 and IL-10 at hospital presentation do not differentiate sepsis from SIRS or predict mortality in dogs.

ABSTRACT #121

INFLAMMATORY BIOMARKERS FOR SEPSIS IN CATS. CR Sharp[1,2], C Delgado[1], A Honaker[1], CH Chang[1], AE DeClue[1]. [1]University of Missouri, College of Veterinary Medicine, Columbia, MO. [2]Cummings School of Veterinary Medicine at Tufts University, North Grafton, MA.

Sepsis is a serious problem in cats that is associated with morbidity and mortality rates of 22–79%. Understanding the inflammatory response to infection in cats may help in the diagnosis, treatment and prognostication of sepsis in this species. The purpose of this study was to compare the clinical and inflammatory parameters among cats with sepsis (n = 16), non-infectious systemic inflammatory response syndrome (SIRS) (n = 19) and healthy controls (n = 8) at presentation to the hospital.

Clinicopathologic abnormalities in the sepsis group included increased band neutrophil percentage, hyponatremia, hypochloremia, hypoalbuminemia, hypocalcemia, and hyperbilirubinemia. Comparing the sepsis and SIRS groups, the only significant differences in the CBC and plasma biochemical findings were the percentage of band neutrophils and plasma albumin concentration. Cats with sepsis had significantly greater plasma TNF activity than control and were more likely to have detectable concentrations of IL-6 than the control or SIRS group. Plasma IL-1β activity was not significantly different among groups and CXCL-8 was not detectable in the majority of cats. Mortality was greater for cats with sepsis (43.75%) than cats with non-infectious SIRS (26.3%). Plasma IL-1β, IL-6 and chloride concentrations were the only parameters evaluated that significantly correlated with mortality in the sepsis group.

This study demonstrated that cats are more likely to have band neutrophils and hypoalbuminemia than cats with SIRS and that sepsis is associated with TNF, IL-1β and IL-6 production in cats. Interleukin-1β and IL-6 may be promising prognostic biomarkers for feline sepsis; additional investigation is needed to confirm these findings.

ABSTRACT #122

EFFECTS OF DIFFERENT ORAL DOSES OF CYCLOSPORINE ON T-LYMPHOCYTE BIOMARKERS OF IMMUNOSUPPRESSION IN NORMAL DOGS.

T Archer, C Fellman, J Stokes, K Lunsford, L Pinchuk, S Pruett, C Langston, A Mackin. Mississippi State University College of Veterinary Medicine, Starkville, MS.

Cyclosporine is a potent immunosuppressive agent used to treat many canine diseases. Optimal dosing regimens remain unclear, primarily because standard methods that monitor effectiveness of immunosuppression have not been established. Our study was designed to evaluate a comprehensive panel of biomarkers of immunosuppression in 7 normal dogs at two different oral doses of cyclosporine. For our panel, we used flow cytometry to measure expression of cytokines (IL-2, IL-4, & INF-ã) and surface antigens (CD25 & CD95) on T-lymphocytes activated with PMA/ionomycin and concanavalin A, and quantitative real-time PCR to measure IL-2 and INF-α.

Dogs were first given a high dose of cyclosporine (starting dose 10 mg/kg bid), with doses adjusted to attain established immunosuppressive trough drug blood levels of at least 600 ng/mL (measured by HPLC). Biomarkers were evaluated on Day 8 of therapy, 8 hours after dosing. Flow cytometric expression of cytokines IL-2 & INF-ã was significantly reduced in all dogs, whereas results with cytokine IL-4 and surface markers CD25 & CD95 were more variable. After a washout period, dogs were given cyclosporine at the dose approved to treat atopy (5 mg/kg sid), and flow cytometric expression of IL-2 & INF-ã was again measured at steady state.

Mean T-cell expression of IL-2 & INF-ã measured by flow cytometry was significantly decreased below baseline values at both cyclosporine doses, although suppression of cytokine expression was much less marked at the lower dose of cyclosporine, and much more variable in individual dogs. No difference was detected between baseline and post-treatment IL-2 & INF-ã real-time PCR results 8 hours after drug administration (Day 8, high dose cyclosporine). However, a follow-up study in two of the dogs demonstrated suppression of IL-2 & INF-ã by real-time PCR at 4 hours post-dosing, demonstrating that, in contrast to cytokine expression as measured by flow cytometry, detectable effects of cyclosporine on T-lymphocyte cytokine mRNA as measured by PCR are more transient.

Our study confirms that cyclosporine suppresses selected biomarkers of immunosuppression in a dose dependent manner. Further investigation of the effects of cyclosporine on these biomarkers in normal dogs at different doses and blood levels of the drug, and in clinic patients, is warranted.

Figure [Abstract 122]

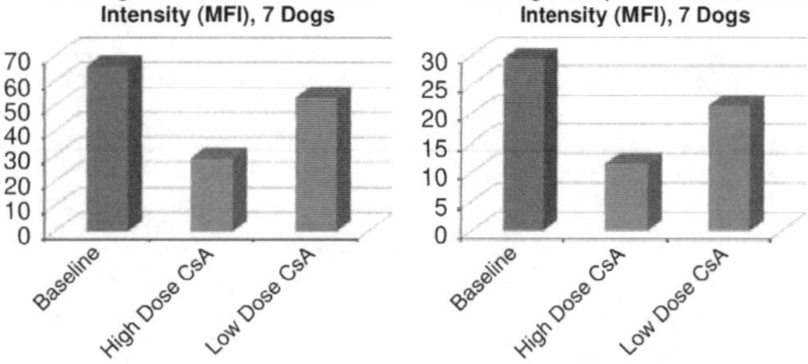

ABSTRACT #123

PHARMACOKINETICS OF LEFLUNOMIDE IN HEALTHY DOGS.

LM Singer[1], LA Cohn[1], CR Reinero[1], MG Papich[2]. [1]University of Missouri College of Veterinary Medicine, Columbia, MO. [2]North Carolina State University College of Veterinary Medicine, Raleigh, NC.

Dogs are susceptible to a wide variety of immune mediated diseases, many of which are life-threatening. Corticosteroids, usually the first drug of choice, cause important morbidity. Alternative and adjunctive immunosuppressive drug treatments have met anecdotal success. One such drug is leflunomide. It is converted to an active metabolite teriflunomide (A77 1726), which is responsible for its immunosuppressive and immunomodulatory effects. Currently there is no data available on the pharmacokinetics of this drug in dogs. The objective of this study was to determine the pharmacokinetics of leflunomide and its active metabolite teriflunomide in healthy dogs.

Four female dogs with normal CBC, chemistry and urinalysis were used. Heparinized blood samples were collected at hour 0,1,2,4,8,12, and day 1,2,4,6,8,10,12,14,16,17 after oral administration of 4 mg/kg leflunomide. The plasma was separated by centrifugation within 30 minutes of collection and frozen (−80°C) until analysis. Concentrations of leflunomide and its active metabolite, teriflunomide, were measured by HPLC analysis at NCSU. Concentrations of the parent drug, leflunomide, were undetectable in most samples. Pharmacokinetic analysis was subsequently performed on the metabolite teriflunomide. In these dogs, the mean terminal half life ($T_{1/2}$) was 21.33 (+/− 1.93) hrs and mean peak plasma concentration (C_{MAX}) was 5.01 (+/− 1.37) μg/mL. The mean residence time (MRT) of the metabolite was 26.8 hrs (+/− 10.8). These results were drastically different from what has been reported in people, in which the half-life of teriflunomide was 15 days, and the plasma concentration was 12 μg/mL three hours after a single loading dose of 100 mg per person (approximately 1.4 mg/kg). In dogs, when doses were simulated to steady-state (6 doses), the peak concentration attained after multiple doses was 7.3 μg/mL and the trough was 3.7 μg/mL, which is less than the therapeutic trough concentrations reported for people of 20 μg/mL.

A single dosage of 4 mg/kg of oral leflunomide in these dogs was well tolerated and resulted in a half-life that will permit 24 hour dosing intervals. The limited anecdotal information to date has indicated that dogs with immune mediated and inflammatory diseases have responded to an oral dosage of 4 mg/kg once daily, but these doses were not critically evaluated in controlled studies. In our study, we observed that plasma concentrations of the active metabolite obtained after 4 mg/kg of leflunomide reached a peak of only 5 μg/mL. Although a loading dose is used in people and has been suggested for dogs, loading is likely unnecessary in dogs because of the short half-life in this species. Steady-state concentrations in dogs, assuming pharmacokinetic values obtained from this study, will be achieved after once-daily dosing in approximately 4 days. Further controlled studies are needed to evaluate the therapeutic efficacy of this drug.

ABSTRACT #124

HORMONES OF ENERGY METABOLISM IN SEPTIC FOALS: ASSOCIATIONS BETWEEN INSULIN, GLUCAGON, LEPTIN, ADIPONECTIN, GHRELIN AND GROWTH HORMONE.

RJIM Barsnick[1], SDA Hurcombe[1], NM Slovis[2], KA Sprayberry[2], PA Smith[1], RE Toribio[1]. [1]The Ohio State University College of Veterinary Medicine, Columbus, OH. [2]Hagyard Equine Medical Institute, Lexington, KY.

Endocrine dysregulation of energy metabolism is well documented in critically ill human neonates, but limited information exists in critically ill neonatal foals. The purpose of this study was to investigate the association between hormones that regulate energy metabolism (insulin, glucagon, leptin, adiponectin, ghrelin and growth hormone) and their association with blood glucose and triglyceride concentrations in healthy and critically ill neonatal foals.

We hypothesized that concentrations of insulin and leptin will be positively correlated with serum glucose and negatively correlated with serum triglyceride concentrations. Glucagon, ghrelin, growth hormone (GH) and adiponectin will be positively correlated with triglycerides and negatively correlated with glucose concentrations. Further, ghrelin will be associated with GH.

Table [Abstract 124]. Correlation (*r*) between the hormones (all foals of the study population), $^{*}p < 0.05$

	Insulin		Glucagon		Leptin		Adiponectin		Ghrelin	
Variable	ρ	*P* value	ρ	*P* value	ρ	*P* value	ρ	*P* value	ρ	*P* value
Glucagon	−0.072	0.423	-	-	-	-	-	-	-	-
Leptin	0.02	0.828	0.164	0.067	-	-	-	-	-	-
Adiponectin	−0.098	0.277	0.031	0.727	−0.332	< 0.0001*	-	-	-	-
Ghrelin	−0.2	0.025*	0.453	< 0.0001*	−0.009	0.919	0.098	0.277	-	-
Growth Hormone	−0.198	0.027*	0.281	0.002*	−0.04	0.65	0.078	0.384	0.177	0.048*

We included 44 septic, 62 sick non-septic, and 19 healthy foals of < 7 days of age. Blood samples were collected on admission. Foals with a positive blood culture or a sepsis score ≥ 12 were considered septic. The Spearman rank order statistic was used to define correlations between variables.

Serum glucose concentrations were positively correlated with insulin and leptin (insulin only in septic foals), and negatively correlated with glucagon, adiponectin, ghrelin and growth hormone (growth hormone only in septic foals). Serum triglyceride concentrations were positively correlated with glucagon, ghrelin and growth hormone (glucagon and ghrelin only in septic foals). Glucose, triglycerides, insulin, ghrelin and glucagon were also associated with the sepsis score (data not shown). Correlations between the hormones are shown in table.

This study showed that energy endocrine dysregulation is frequent in septic foals as a result of increased caloric needs and systemic inflammation. We observed a complex hormonal response, as demonstrated by the association between hormones, glucose, triglycerides, and the sepsis score. Ghrelin and GH were associated with insulin and glucagon, supporting their role in energy regulation.

ABSTRACT #125

EFFECTS OF THE INSULIN SENSITIZING DRUG PIOGLITAZONE ON INDICES OF INSULIN HOMEOSTASIS IN HORSES FOLLOWING ENDOTOXIN ADMINISTRATION.

JG Wearn[1], JK Suagee[2], MV Crisman[1], MW Hulver[2], BA Corl[2], DR Hodgson[1], RJ Geor[2], LJ McCutcheon[1]. [1]Virginia-Maryland Regional College of Veterinary Medicine, Virginia Tech, Blacksburg, VA. [2]Virginia Tech, Blacksburg, VA.

Reducing insulin resistance may decrease the incidence of laminitis in horses with Equine Metabolic Syndrome. Pioglitazone, a thiazolidinedione class of anti-diabetic drug, modulates transcription of genes involved in glucose metabolism in humans. The ability of pioglitazone to influence insulin sensitivity and gene expression in equine tissues in an endotoxin-infusion model of induced insulin resistance was investigated.

Sixteen mature horses were randomly allocated to receive either 14 days of pioglitazone (1 mg/kg, orally, q24h, n = 8) or control (n = 8). On days 12 and 14 a frequently sampled intravenous glucose tolerance (FSIGT) test with minimum model analysis was performed and a skeletal muscle biopsy (middle gluteal) was harvested. Whole tissue RNA was extracted to determine, by qRT-PCR, the expression of insulin receptor and insulin responsive glucose transporter (GLUT4). On day 13, all horses received an endotoxin infusion (*Escherichia coli* 055:B5 lipopolysaccharide, 35 ng/kg, IV, over 30 min). Serial assessment of clinical and leukocyte parameters, indicies of insulin homeostasis and gene expression were compared prior to, and following, endotoxin infusion.

Heart rate (HR) and rectal temperature (RT) increased and white blood cell count (WBC) decreased following endotoxin infusion ($P < 0.001$), however maximal change from baseline did not differ between pioglitazone treated horses or their controls ($P = 0.3719$, $P = 0.6695$, $P = 0.3611$ HR, RT, WBC, respectively). Twelve days of pioglitazone did not increase insulin sensitivity ($P = 0.278$) nor protect against endotoxin induced insulin resistance ($P = 0.241$). Pioglitazone increased insulin receptor ($P = 0.018$), but not GLUT4 transcript levels in skeletal muscle. Endotoxin infusion reduced GLUT 4 transcript levels ($P = 0.059$), but had no effect on insulin receptor ($P = 0.396$) in both treated and control horses. Pioglitazone treatment had no effect on GLUT4 transcript levels ($P = 0.418$) following endotoxin infusion.

Although no physiologic changes in insulin sensitivity were identified following pioglitazone treatment, the finding of increased insulin receptor expression may warrant investigation of either the effects of a longer treatment period or in horses with obesity-associated insulin resistance.

ABSTRACT #126

ATTENUATION OF ENDOTOXEMIA IN HORSES WITH A KETAMINE CONTINUOUS RATE INFUSION.

CJ Alcott, DM Wong, BA Sponseller. Iowa State University College of Veterinary Medicine, Ames, IA.

Ketamine is a dissociative anesthetic routinely used in veterinary medicine. Ketamine has demonstrated positive cardiostimulatory effects, immunomodulation, decreased leukocyte-endothelial adhesion and analgesia in various species challenged by lipopolysacchride (LPS). A recent report in canids demonstrated significant blunting of tumor necrosis factor-α (TNF-α) production using subanesthetic doses of ketamine during periods of endotoxemia. *In-vivo* and *in-vitro* immunomodulation following endotoxemia has been demonstrated in mice through direct inhibition of nuclear factor kappa B and inhibition of LPS-induced TNF-α and interleukin-6 production in an equine macrophage cell line model, respectively. TNF-α is considered the earliest cytokine released during endotoxemia along with the induction of the vasoactive substance thromboxane (TXB_2). The aim of this prospective study was to determine if a subanesthetic continuous rate infusion (CRI) of ketamine could be safely administered and result in positive immunomodulation in horses with experimentally induced endotoxemia. Specifically, this pilot study addressed whether a subanesthetic dose of ketamine could attenuate both the LPS-induced production of TNF-α and TXB_2, and the induction of clinicopathologic findings of endotoxemia.

20 mares of various breeds were randomly assigned to two groups, control or ketamine treated. Each horse in the ketamine group received a subanesthetic CRI (1.5 mg/kg/h over 5 h) of racemic ketamine, following a loading dose, while the control group received an equal volume of 0.9% saline. Both groups were administered endotoxin (LPS, *Escherichia coli* 055:B5, 30 ng/kg over 30 min) 1 hour after initiation of the ketamine or saline CRI. Clinicopathologic parameters and sample collection were monitored at half-hour and hourly intervals relative to LPS administration.

All horses completed the study uneventfully. The most significant clinical findings were transient excitation in horses during the initial ketamine loading dose period. Respiratory rates were significantly elevated in the ketamine group at 1 and 6 hours post-LPS compared to controls. A neutrophilic leukocytosis was observed in both groups, with a significantly greater elevation in the ketamine group 5 hours post-LPS. No significant difference was observed between groups regarding plasma TNF-α production whereas a trend towards decreased TXB_2 levels was observed in the ketamine group. The targeted plasma steady state of ketamine (100 ng/ml) was attained.

This study indicates that a subanesthetic CRI of ketamine can be safely administered to endotoxemic horses. Although TNF- α levels were not lower in the ketamine treated group than controls, this study documented a trend in lower TXB_2 levels in ketamine treated horses. Results of this study may have been limited by endotoxin dose and ketamine treatment duration. Given these findings, further studies regarding the clinical utility of a subanesthetic ketamine CRI are warranted.

ABSTRACT #127

ENDOTOXIN TOLERANCE INDUCED IN VITRO IN EQUINE LYMPHOCYTES.

L Frellstedt, MO Furr. Marion DuPont Scott Equine Medical Center, Virginia-Maryland Regional College of Veterinary Medicine, Leesburg, VA.

Endotoxemia is responsible for the severe illness in horses associated with colic, diarrhea, pleuropneumonia, peritonitis and metritis. It is possible for the body to develop a state of 'tolerance', in which it becomes unresponsive to the endotoxin molecule. This has direct survival value, as laboratory rodents with induced endotoxin tolerance (ET) can survive massive doses of endotoxin which are fatal in normal mice. ET has been induced in horses in vivo; however, the cytokine responses associated with this have not been investigated. The purpose of this study was to describe a method for inducing ET in equine cells in vitro, and to describe the cytokine profile which is associated with the ET.

Blood was collected from 6 healthy horses and peripheral blood mononuclear cells were isolated. ET was induced by culturing cells with a variety of concentrations of endotoxin (0.1, 1.0 and 10 ng/ml), followed by a second endotoxin challenge (1.0, 10 and 100 ng/ml). The state of tolerance was identified by assaying cell culture media for TNF-α concentration (ELISA). ET was confirmed when the TNF-α production was reduced by greater than 75% compared to non-tolerized control cells. The relative mRNA expression of IL-10 and IL-12 was measured in tolerized and non-tolerized control cells by use of RT-PCR at the time of the second endotoxin challenge and 6, 12 and 24 hours after the second endotoxin challenge.

ET was induced in all cells exposed to the 2-step endotoxin challenge compared with non-tolerized cells. The most consistent results were achieved when cells were exposed to 1.0 ng/ml of endotoxin initially and then challenged with 10 ng/ml of endotoxin. The TNF-α production was reduced by greater than 85% in all cells exposed to this protocol compared with non-tolerized control cells.

For the second aim of this study ET was induced in equine cells by exposure to 1.0 ng/ml of endotoxin followed by challenge with 10 ng/ml of endotoxin. While not statistically signifi-cant, the relative mRNA expression of IL-10 in tolerized cells was increased two-fold when compared to non-tolerized control cells. The relative mRNA expression of IL-12 in tolerized cells was decreased significantly by fourteen-fold at 6 hours and five-fold at 12 hours after the second endotoxin challenge compared with non-tolerized control cells at the same time points.

This experiment demonstrated a reliable method for the *ex vivo* induction of ET in equine lymphocytes. A marked suppression of IL-12 production is associated with ET. The production of IL-10 does not appear to be related with the development of ET.

ABSTRACT #128

IN VITRO ACTIVITY OF PONAZURIL AGAINST *THEILERIA EQUI*.

LN Wise, M Ueti, MT Hines, LS Kappmeyer, DP Knowles. Department of Veterinary Microbiology and Pathology, Washington State University, Pullman, WA.

The equid hemoprotozoan parasite *Babesia equi* is endemic in most nations worldwide. Infected horses cause significant economic loss due to treatment, fatalities and restriction on trade with non-endemic nations. To date, no drug has been shown to completely eliminate infection and infected horses create a risk of transmission. Ponazuril, a metabolite of toltrazuril, has been used as an anti-protozoal agent to successfully treat equine *Sarcosystis neurona* and other domestic animal protozoa such as *Toxoplasma gondii*. In an attempt to establish the *in vitro* activity of this drug against *B. equi*, tissue cultures containing the Florida strain of *B. equi* were established. Ponazuril was diluted in dimethylsulfoxide (DMSO) and then added to the culture media to create 1% DMSO solutions with varying concentrations of ponazuril. Prior to treatment, all cultures were evaluated by light microscopy, combined and redistributed to yield separate culture wells with similar percent parasitemias. Treated media was created at 100 μg/mL, 250 μg/mL, 500 μg/mL, 750 μg/mL, and 1000 μg/mL concentrations. Control culture wells were also maintained with media containing only 1% DMSO. The treated media was substituted for the normal media during maintenance of the cultures once daily for 5 days. The cultures were treated in triplicate at each drug concentration. After the 5 day treatment period, routine maintenance of the cultures resumed. The cultures were evaluated via light microscopy of giemsa stained slides for percent parasitemia daily for the 5 treatment days and then daily for 30 days after discontinuation of the drug. Slides were prepared and labeled as to prevent evaluator bias. The control cultures continued to grow as expected. Degree of parasite inhibition was variable between drug concentrations. The 100 μg/mL inhibited parasite growth during the treatment period but the parasite was never completely eliminated. The 250 μg/mL, 500 μg/mL, 750 μg/mL, and 1000 μg/mL did completely eliminate detectable parasitemia during the treatment period. However, the 250 μg/mL culture growth recrudesced at 7 days post treatment in all cultures. No recrudescence of growth was noted in the 500 μg/mL or 750 μg/mL cultures during the 30 day period following treatment. The 1000 μg/mL percent parasitemia results were inconsistent. One of the 3 wells exhibited recrudescence of the parasite at 27 days after the cessation of treatment. This inconsistency was potentially related to the difficulty encountered with complete dissolution of this high concentration of drug within the culture media. However, the development of parasite resistance to the drug cannot be ruled out. Results indicate that ponazuril inhibited *B. equi* in cell culture with a minimum concentration of 500 μg/mL. Future research to evaluate this drug *in vivo* is warranted.

ABSTRACT #129

EXPERIMENTAL INFECTION WITH *BARTONELLA HENSELAE* AND *BARTONELLA BOVIS* IN HORSES.

J Palmero, N Pusterla, RW Kasten, S Mapes, NA Cherry, E Breitschwerdt, BB Chomel. School of Veterinary Medicine, University of California, Davis, CA.

Bartonella are emerging, potentially zoonotic pathogens, found in several wild and domestic animals, causing various clinical syndromes. *Bartonella* was not detected in horses until recently when serological and blood culture data suggested that horses could be naturally infected with *Bartonella sp*.

This study aimed to induce clinical signs, bacteremia and seroconversion in horses following experimental infection with *Bartonella henselae* or *Bartonella bovis*. Twelve healthy adult horses were divided into two groups of six. In Group One, an equine isolate of *B. henselae* was inoculated intradermally into four naïve horses, and a saline control was inoculated into two sentinel horses. In Group Two, this procedure was followed using *B. bovis*. Horses were examined daily for sixty days, and blood was collected regularly over 5 months for CBC, culture, PCR, and serology.

Beyond injection site abscess in two *B. henselae*-inoculated horses, *Bartonella*-associated clinical signs did not occur. One *B. henselae*-inoculated horse became blood culture and PCR positive. Three and one infected horse seroconverted to *B. henselae* and *B. bovis*, respectively. All sentinels remained PCR, culture, and serology negative.

In conclusion, experimental inoculation of *B. henselae* and *B. bovis* in horses did not induce clinical signs beyond injection site abscessation. The detection of *Bartonella sp.* in blood supports bacteremia and seroconversion. To be determined is if seroconversion was due to local replication of organisms or humoral immune response to inoculums. Additional research is warranted in healthy horses and in those displaying signs consistent with *Bartonella* infection to correlate positive tests with clinical signs.

ABSTRACT #130

SEROPREVALENCE OF *ANAPLASMA PHAGOCYTOPHILUM*, *BORRELIA BURGDORFERI*, AND *EHRLICHIA* SPP. IN HORSES IN OKLAHOMA.

RJ Carmichael, TC Holbrook, SE Little. Oklahoma State University, Stillwater, OK.

Both *Borrelia burgdorferi* and *Anaplasma phagocytophilum* have been associated with clinical disease in horses, and the frequency with which horses are exposed to these organisms is relevant information to regional veterinarians and horse owners. However, the seroprevalence of these agents in horses residing in the state of Oklahoma is currently unknown.

Serum samples from 120 horses in Oklahoma were randomly obtained from the Oklahoma Animal Disease Diagnostic Laboratory. Serologic testing was performed using IDEXX Laboratories Canine SNAP* 4Dx* which detects antibodies to *A. phagocytophilum*, *B. burgdorferi*, and *Ehrlichia canis*.

All horses were negative for *A. phagocytophilum*. One horse tested positive for *B. burgdorferi* exposure, and 12 horses (10 percent) had antibodies reactive to *E. canis*.

These findings indicate exposure of horses to *A. phagocytophilum* and *B. burgdorferi* is uncommon in Oklahoma. The relatively large percentage of horses with antibodies reactive to *E. canis* was unexpected. It is likely that these positive test results indicate cross-reactivity of the *E. canis* assay with *E. chaffeensis* or another *Ehrlichia* species; however, seroconversion of horses to these organisms has not previously been reported. Due to the significant exposure of horses in Oklahoma to a presently unidentified *Ehrlichia* species, further investigation with polymerase chain reaction and DNA sequencing is indicated to identify the organism(s) responsible.

ABSTRACT #131

PREVALENCE OF *SALMONELLA* spp. IN HEALTHY FOALS, CRIAS AND THEIR DAMS IN THE NEW YORK STATE AREA. KJ Mitchell, AJ Burton, DV Nydam, C Altier and GA Perkins. Cornell University, Ithaca, NY.

To determine the prevalence of *Salmonella* spp. shedding in healthy foals, crias and their dams in the New York State (NYS) area.

A cross-sectional study was conducted between January and August 2009. Single fecal samples were collected from mare-foal and cria-dam pairs on breeding farms in the NYS area. *Salmonella* cultures were performed within 24-h using standard enrichment techniques. For the apparent prevalence, the 95% confidence interval (CI) was calculated as the exact binomial.

A total of 349 samples were collected from 14 broodmare farms. The median number of animals per farm was 133 (range 32–523), with a median of 20 (range 10–76) animals sampled per property. Median age of the foals was 18 days (range 4–67) and mares was 10 years (range 4–31). *Salmonella* spp were not detected. The apparent prevalence was 0% (95% CI 0–1.1).

A total of 221 samples were collected from 14 alpaca breeding farms. The median number of animals per farm was 75 (range 45–330) with a median of 16 (range 10–18) animals sampled per property. Median age of the crias was 36 days (range 2–111), and dams was 5 years (range 2.5–19). *Salmonella enterica* serovar Typhimurium was isolated from a healthy 16 d cria. The estimated prevalence for crias was 1% (95% CI 0.02–5.0) and dams was 0% (95% CI 0–3.3).

The prevalence of *Salmonella* spp. in healthy foals, crias and their dams in the NYS area appears low, although serial fecal sampling may increase the likelihood of detection leading to a higher prevalence.

ABSTRACT #132

PREVALENCE OF EQUINE HERPESVIRUS 1 AND EQUINE HERPESVIRUS 4 AS DETECTED BY QUANTITATIVE POLYMERASE CHAIN REACTION IN NASAL SECRETIONS OF FEBRILE HORSES. JM Sonis, LS Goehring. Colorado State University College of Veterinary Medicine, Fort Collins, CO.

Infection of horses with Equine herpesvirus 1 (EHV-1) and Equine herpesvirus 4 (EHV-4) may cause a range of clinical diseases, including febrile respiratory disease, and typically results in establishment of latent infection. During periods of stress, such as transportation, re-housing, and hospitalization, latent virus can become reactivated, resulting in shedding of virus in respiratory tract secretions. Such reactivated virus may subsequently spread to susceptible in-contact animals.

The objective of this study was to determine the prevalence of shedding of EHV-1 and EHV-4 by quantitative polymerase chain reaction (qPCR) in nasal swab samples collected from horses that developed a rectal temperature of $\geq$ 101.3°F (38.5°C) while hospitalized in a referral hospital. We hypothesized that fevers are a result of EHV-1 or EHV-4 replication in the respiratory tract following either recrudescence of virus from latency or following horizontal virus transmission prior to or during admission into the hospital.

Over a nine-month period, nasal swab specimens of the ventral nasal meatus were collected from 56 febrile, hospitalized horses. The swabs were obtained within 12 hours of detection of a fever. Over the same time period and at least once a month, nasal swab samples were collected from 10 randomly chosen but afebrile horses within the hospital. After collection, swabs were immersed in virus transport medium and frozen at −20°C. For analysis, the samples were thawed, and routine DNA extraction was performed using a commercially available kit. A validated, qPCR assay was used to detect and quantify genomic DNA of EHV-1 and EHV-4.

Genomic DNA of EHV-4 was detected in the nasal swab specimen of one of 56 febrile horses (1.8%). EHV-1 DNA was not detected in any of the febrile horses. All samples collected from the afebrile horses were negative for both viruses. Considering the known high seroprevalence of EHV-1 and EHV-4 in the horse population, we anticipated finding a higher percentage of PCR positive nasal swab samples collected from febrile patients. Fevers detected were likely a result of the active disease processes for which the horses were hospitalized. It is also possible that inappropriate timing of sample collection relative to the onset of fever may have resulted in a low diagnostic sensitivity. Further studies are warranted to examine the frequency of Equine herpesvirus reactivation in horses in response to stressful conditions such as hospitalization.

ABSTRACT #133

FIELD EFFICACY TRIAL OF AN INTRA-RECTALLY ADMINISTERED AVIRULENT, LIVE *LAWSONIA INTRACELLULARIS* VACCINE ON ENDEMIC HORSE FARMS IN CENTRAL KENTUCKY. N Nogradi[1], NM Slovis[2], CJ Gebhart[3], KE Wolfsdorf[2], JL McCracken[2], CF Scoggin[4], B Toth[1], PH Kass[1], SM Mapes[1], ML Lundquist[2], N Pusterla[1]. [1]School of Veterinary Medicine, University of California, Davis, CA. [2]Hagyard Equine Medical Institute, Lexington, KY. [3]College of Veterinary Medicine, University of Minnesota, St. Paul, MN. [4]Claiborne Farm, Paris, KY.

Equine proliferative enteropathy (EPE) caused by *Lawsonia intracellularis* is an emerging disease with yet unaddressed preventative measures. We hypothesized that the intra-rectally administered avirulent live vaccine of *Lawsonia intracellularis* would prevent disease. Another question of interest was whether naturally infected foals would display lower total solid concentrations and lower average daily weight gains, as potentials markers of subclinical disease, when compared to vaccinated herd mates.

A total of 202 healthy weanling Thoroughbred foals were randomly assigned into two groups (vaccinated and non-vaccinated). Vaccinated foals received 30 mls of avirulent, live *Lawsonia intracellularis* vaccine intra-rectally twice, 30 days apart. Foals were monitored daily for clinical signs of EPE until yearling age, blood was collected monthly to determine concentration of total solids by refractometry and *L. intracellularis* specific antibodies by immunoperoxidase monolayer assay. Monthly weight measurements were performed and average daily weight gains were calculated on 116 horses.

There was no significant difference (P > 0.05) in the occurrence of the disease between the vaccinated (1 case) and the non-vaccinated group (3 cases). Eighty-two percent of vaccinated horses seroconverted following vaccine administration and 59% of non-vaccinated horses seroconverted to natural infection. Vaccinated foals maintained a significantly higher total solids concentrations (two-way repeated measures ANOVA; P = 0.023) throughout the study period when compared to naturally infected foals. No difference in average daily weight gain was found.

The small number of clinical cases precludes interpretation of vaccine efficacy. Vaccination may prevent subclinical disease, which is often characterized by lower concentration of total solids and decreased weight gain.

ABSTRACT #134

DEVELOPMENT OF EXOGENOUS LACTATE CLEARANCE CURVES IN NORMAL HORSES. P De Pedro[1], PA Wilkins[1], M McMichael[1], L Dirikolu[2], SC Clark-Price[1], KM Lascola[1]. [1]College of Veterinary Medicine, Departments of Veterinary Clinical Medicine, [2]Veterinary Biosciences, University of Illinois at Urbana Champaign.

Low clearance of exogenously administered lactate is predictive of non-survival in severely ill normo- or mildly hyperlactatemic human patients independent of other known risks. Hyperlactatemia, persistent hyperlactatemia and low change in lactate over time are predictive, albeit imperfectly, of non-survival in sick foals and horses. Development of an equine specific exogenous lactate clearance test should allow for both improved prognostic ability and exploration of causes of lactate dysmetabolism in horses.

500 mL of a solution containing 1 mmol/kg body weight Na-lactate in 0.9% saline was infused intravenously to 6 healthy horses over 15 min. Venous blood was collected in sodium citrate for lactate concentration [LAC] determination using the YSI Lactate Analyzer at 5, 10 and 15 min during infusion, at each minute for 15 min post-infusion and at 30, 45, 60 and 120 min following the start of infusion. [LAC] increased > 4-fold from baseline at the end of infusion (P = 0.005). A two compartmental model was used to specify the pharmacokinetic parameters of exogenous [LAC] including terminal elimination half-life (62.1 ± 19.9 min), central volume of distribution (279.3 ± 61.2 L), steady state volume of distribution (396.6 ± 110.2 L) and systemic clearance (4.49 ± 1.29 L min^{-1}). [LAC] remained ~60% above baseline (P = 0.008) at 60 min but was returned to baseline at 120 min.

Clearance of exogenous lactate appears slower in healthy horses when compared to published values for normo- or mildly hyperlactemic sick humans. Return to baseline is also delayed by comparison, underscoring the need for species specific test development.

ABSTRACT #135

LACTATE CLEARANCE IN HORSES WITH ACUTE COLITIS. S. Hashimoto-Hill, KG. Magdesian, PH. Kass. School of Veterinary Medicine, University of California, Davis, CA.

Blood lactate is a marker of tissue perfusion and inflammation in critically ill patients. In humans, lactate clearance has largely replaced single measurements in predicting mortality due to improved accuracy. The association between lactate clearance and outcome in acute colitis warrants further investigation. This retrospective study aimed to determine the prognostic value of lactate clearance in horses with acute colitis. Medical records for horses with colitis between 2003 and 2009 were reviewed. Early (admission to 4–8 h post) and late (admission to 24 h) lactate clearances were calculated. The sensitivity and specificity of different lactate clearance cutoffs were calculated. Two way ANOVA was used to evaluate differences in lactate between survivors and nonsurvivors and horses with and without systemic inflammatory response syndrome (SIRS). Mortality was compared between lactate clearers and nonclearers using Fisher's exact test and odds ratios.

96 cases were included. An early lactate clearance of ≥ 40% and a late clearance of ≥ 50% had the maximum sum of sensitivity and specificity. When these cutoffs were used, early and late lactate clearance were significantly associated with outcome (P = 0.02, OR: 3.35 [1.27–8.80] and P = 0.02, OR: 3.62 [1.26–10.41], respectively). Horses with SIRS had higher lactate concentrations over the first 24 h as compared to those without, however there were no significant differences in lactate concentrations between survivors and nonsurvivors. The association between lactate clearance and survival, and lactate concentrations over time with SIRS, suggest an importance in early identification and treatment of hyperlactatemia in colitis cases.

ABSTRACT #136

PHARMACOKINETICS OF GANCICLOVIR AFTER SINGLE INTRAVENOUS ADMINISTRATION IN THE ADULT HORSE. RJ Carmichael, C Whitfield, LK Maxwell. Oklahoma State University, Center for Veterinary Health Sciences, Stillwater, OK.

Equine herpes myeloencephalopathy (EHM) resulting from equine herpes virus (EHV) infection continues to be associated with high mortality in the horse. Therefore, appropriate prophylaxis and treatment is important to the equine industry. Previous pharmacologic research has focused on acyclovir and its prodrug, valacyclovir. However, the higher potency of ganciclovir makes it an attractive candidate for use in horses already infected with EHV. This study investigated the pharmacokinetics of ganciclovir after intravenous administration of 2.5 mg/kg of injectable ganciclovir as a slow bolus to five horses. Plasma ganciclovir concentrations were assessed using high-performance liquid chromatography and preliminary noncompartmental pharmacokinetic analysis was performed. Ganciclovir pharmacokinetic characteristics were similar to those of intravenous acyclovir, with a volume of distribution at steady state of 9.2 ± 2.5 L/kg and clearance of 4.0 ± 1.2 mL min^{-1} kg^{-1}. The pronounced multi-compartmental decay and extended elimination half-life of 24–58 hr also resembled those of acyclovir, and are expected to similarly produce accumulation of ganciclovir at steady state. Although up to one week of dosing may be required for steady state conditions to occur, a rapid onset of action would be required during an EHM outbreak. Nonparametric superposition predicted that a loading regimen of 2.5 mg/kg of intravenous ganciclovir every 8 hours for the first day, followed by 2.5 mg/kg every 12 hours, would maintain trough ganciclovir concentrations above the targeted IC_{50} of 0.1–0.4 µg/mL, within the first day of beginning therapy. Further studies are needed to examine the safety and efficacy of multiple doses of ganciclovir in horses.

ABSTRACT #137

EARLY DIAGNOSIS AND TREATMENT OF MUSCLE ATROPHY AND WEAKNESS ASSOCIATED WITH VITAMIN E DEFICIENCY. HE Bedford, SJ Valberg and AM Firshman. University of Minnesota, St Paul, MN.

Equine motor neuron disease (EMND) resulting from prolonged vitamin (vit.) E deficiency has a poor prognosis. Diagnostic criteria include vit. E < 2 µg/dl and neurogenic atrophy in the sacrocaudalis muscle. We hypothesized that an early stage of reversible myogenic atrophy may exist in horses with low serum vit. E prior to occult EMND. The first objective of this study was to retrospectively characterize clinical cases and muscle pathology of horses that had myofiber atrophy in the sacrocaudalis muscle. The second objective was to prospectively describe the clinical attributes and muscle pathology of 10 horses that presented to the UMN Equine Center with moderate weakness, muscle atrophy and low vit. E that successfully responded to treatment.

Objective 1: From records of the Neuromuscular Diagnostic Laboratory 36 horses were identified that had sacrocaudalis muscle fiber atrophy. Horses were of varying breeds, with 14 mares, 18 geldings and 4 stallions and a mean age of 12.5 yrs (range 1–30 yrs). Weight loss (13/3), muscle atrophy (15/36), weakness (17/36) and fasciculations (19/36) were reported. A clear diagnosis of EMND was made based on numerous angular atrophied fibers in 20 horses and 16 horses had a pattern of myofiber atrophy more typical of myogenic atrophy. Where measured, vit. E was below 2.5 µg/dl. Resectioning and re-evaluation of frozen sections of sacrocaudalis muscle from 9 horses with low vit. E revealed anguloid atrophy (9/9), angular atrophy (4/9) and moth eaten fibers in NADH stains indicative of oxidative stress (5/9). Thus, in 54% of cases with myogenic atrophy, moth eaten fibers could potentially have been early indicators of oxidative stress and EMND.

Objective 2: Horses of varying breeds, 6 geldings and 4 mares; mean age of 10.6 yrs (range 2 – 20 yrs) that were diagnosed with muscle weakness at the University of Minnesota Equine Center and had vit. E levels between 0.2–1.94 µg/dl were included. Clinical signs were subtle with generalized weight loss/muscle wasting (10/10), weakness (10/10) or fasciculations (3/10). Serum CK activities were elevated with a maximum of 5,805 U/L in 3/7 horses. Biopsy of gluteus medius and sacrocaudalis muscles obtained in 6/10 horses showed lesions in sacrocaudalis samples that included anguloid atrophy (6/6), rare angular atrophied fibers (3/6), macrophages (1/6) and moth eaten fibers in NADH stains (6/6). There was no clear evidence of neurogenic atrophy, but myopathic changes were evident. Horses were treated with 4,000 IU–15,000 IU vit. E/day and vit. E levels were normal (2–4 µg/dL) in horses where assessed (4/4) after 4 weeks and all horses had improved muscle mass and strength after 2–6 months.

Since muscle wasting and progressive weakness are insidious, clinical acumen, awareness of this early presentation of EMND and characterization of myofiber atrophy and moth eaten fibers in sacrocaudalis muscle biopsies are necessary to identify the disorder in its early stages and ameliorate clinical signs with vit. E therapy.

ABSTRACT #138

SERUM HAPTOGLOBIN (HP)-MATRIX METALLOPROTEINASE 9 (MMP 9) COMPLEX ELISA AS A MARKER OF SEPTIC INFLAMMATION IN CATTLE: A COMPARISON TO HP OR MMP 9 ELISA. Hinds CA,[1] G Bannikov,[1] PJ Rajala-Schultz,[2] F Silveira,[2] DM Rings[1], J Lakritz[1]. [1]The Ohio State University College of Veterinary Medicine, Dept. of Veterinary Clinical Sciences and [2]Preventive Medicine, Columbus, OH.

Serum Hp is an acute phase protein that has been identified in cattle with a variety of inflammatory conditions. Serum from healthy cattle contains very little haptoglobin. Matrix Metalloproteinase-9 (MMP 9) is a proteolytic enzyme released from neutrophil granules and other cellular sources including macrophages and has been reported to increase in the sera of sick animals. An Hp-MMP 9 protein complex was identified in neutrophil granules and the serum of cattle with acute bacterial sepsis (Bannikov, et al. 2007). The purpose of this study was to evaluate the utility of an Hp-MMP 9 complex ELISA in comparison to Hp or MMP 9 alone in the diagnosis of acute septic inflammation in cattle.

A total of 35 animals were sampled. Twenty two (22) were admitted to The Ohio State University VTH for evaluation of a variety of disorders and the remainder (13) were from OSU's Waterman Dairy. Animals were either bled for routine clinicopathologic work-up or as part of a herd infectious disease survey. Serum was harvested in routine fashion by centrifugation after clotting and frozen at −20C until analyzed. Animal age (3.5 ± 1.7 yr; range 0.4–8 yr), breed and sex (M = 3, F = 32) were recorded. Study animals were placed into 3 categories: 1) acute septic (12), 2) chronic disease (13), and 3) normal animals (10) based upon history, clinical examination, and the results of clinicopathologic evaluation as well as pathologic evaluation in those animals that died or were euthanized. Serum Hp was determined by commercial ELISA kits. Serum MMP 9 was determined using a monoclonal antibody (clone 10.1; MAb) to bovine MMP 9. Serum Hp-MMP 9 complexes were determined by ELISA using MAb 10.1 as a capture antibody and an HRP conjugated rabbit anti-bovine Hp antibody. Concentrations of analytes were arranged categorically and analyzed using Kruskil-Wallace One-way ANOVA on ranks.

Serum Hp concentrations in all study animals ranged from 0–945 ug/mL; MMP 9 ranged from 0.1–9.4 ug/mL; serum Hp-MMP 9 complexes range from 0–2.9 ug/mL. When grouped according to disease, serum Hp was higher in acute septic conditions (495 ± 268 ug/mL) compared to chronic (221 ± 198 ug/mL) and normal animals (4 ± 8.5 ug/mL); however, there was no statistical difference between acute and chronic animals. Serum MMP 9 concentrations were similar in acute (2.6 ± 2.4 ug/mL) and chronic disease (2.7 ± 1.5 ug/mL) and were significantly lower in normal animals (0.67 ± 0.9 ug/mL). Serum Hp-MMP 9 concentrations were significantly elevated in animals with acute septic conditions (1.4 ± 0.6 ug/mL) compared to animals with chronic disease (0.13 ± 0.23 ug/mL) and absent in normal animals (none detected). Serum Hp-MMP 9 concentrations were elevated in acute septic conditions; however, individual components of this complex were not reliable in defining acute septic disease in these study animals.

ABSTRACT #139

EXPERIMENTAL EVALUATION OF TRANSMISSION OF BOVINE VIRAL DIARRHEA VIRUS BY HORN FLIES. M. F. Chamorro, T. Passler, M. D. Givens, M. A. Edmondson, D. F. Wolfe, P. H. Walz. Departments of Clinical Sciences and Pathobiology, College of Veterinary Medicine, Auburn University, Auburn, AL.

The epidemiology of bovine viral diarrhea virus (BVDV), including its modes of transmission and maintenance has been well described. However, few studies have examined insects as indirect vectors of BVDV. The horn fly (*Haematobia irritans)* is the most prevalent insect affecting cattle in the southern US and this study sought to evaluate the potential for BVDV transmission by horn flies. Horn flies were collected from four persistently infected (PI) cattle and distributed into fly cage cylinders, with 25 flies per cage. An additional aliquot of 50 flies per PI animal was tested for BVDV by RT-PCR. Four principal and four control BVDV naïve calves were housed in individual isolations rooms. Two cages were attached to each animal. While cages on principal calves allowed flies to feed, a plastic barrier prevented feeding from control calves. After 48 hours, the cages were removed and fly homogenates from each calf were tested for BVDV by RT-PCR. Blood samples were collected from each calf on days 0, 3, 6–10, 14, 21, and 28 and tested by virus isolation and virus neutralization. Fly homogenates from three of four PI cattle were positive for BVDV by RT-PCR. Of the fly homogenates removed after 48 hours, three were positive from principal calves and four from control calves. Virus isolation and serology remained negative throughout the study. These results indicate that BVDV may be detected in horn flies that are feeding on PI animals, but they do not appear to be an important vector for BVDV.

Previously presented at the Conference of Research Workers in Animal Disease. Chicago, Dec. 7, 2009.

ABSTRACT #140

PARELAPHOSTRONGYLUS TENUIS INFECTION AS A CAUSE OF MENINGIOMYELITIS IN 5 LIMOUSIN CALVES. KJ Mitchell[1], T Stokol[1], J Peters-Kennedy[1], RW Gerhold[2] and TJ Divers[1]. [1]Cornell University, Ithaca, NY. [2]University of Georgia, Athens, GA.

Antemortem diagnosis of aberrant migration by Parelaphostrongylus tenuis (P. tenuis) has not been documented in cattle. This report describes an outbreak in limousin calves.

Calves from 2 beef farms presented for evaluation of progressive neurological signs. Neurological exams and lumbar cerebral spinal fluid (CSF) collection was performed. The 2 worst affected calves were euthanized and necropsied. Polymerase chain reaction (PCR) was performed on spinal cord.

The farms were 12 miles apart and had wetter pastures than normal. Large numbers of snails and slugs were observed around common feeding areas. A total of 5/29 calves between 3–5 month old were affected, with 2 heifers, 2 steers and 1 bull. Clinical signs were consistent with upper and lower motor neuron spinal cord disease and were present for > 1 m in 3 animals, with the remainder affected for < 10 d. CSF analysis revealed marked pleocytosis (mean 369 WBC/uL, range 113–1026 WBC/uL) and elevated protein (mean 81 mg/dL, range 66–101 mg/dL). Acute cases showed severe eosinophilic pleocytosis, chronic cases had a lymphocytic pleocytosis. Histopathology revealed lymphoplasmacytic and eosinophilic meningiomyelitis, with intramedullary lymphoid nodules containing nematodes. Histopathologic characteristics and PCR results, showing a faint band at 200 bp, were suggestive of P. tenuis. Surviving calves showed rapid improvement following treatment with anthelmentics and anti-inflammatories.

This case series identifies P. tenuis as a cause of weakness and ataxia in beef calves and highlights the need for vigilance against aberrant parasite migration in calves grazing wet, snail infested pastures. Duration of disease may affect CSF cytology and response to therapy.

ABSTRACT #141

UTILITY OF PRE-PARTUM PARAMETERS FOR PREDICTING INTRAMAMMARY INFECTION AT CALVING IN FIRST CALF HEIFERS ON A GRAZING DAIRY. DR Ringen, JR Middleton, TS Marshall. University of Missouri, College of Veterinary Medicine, Columbia, MO.

Primigravid dairy heifers are a source of entry for mastitis pathogens into the lactating herd. The average prevalence of intramammary infections (IMI) in heifers at calving ranges between 30–40%, with coagulase negative staphylococci as the predominant bacterial cause, followed by *S. aureus* or streptococci. Currently, there is no universally recommended procedure to predict IMIs in heifers at calving. Many producers are unwilling to collect prepartum secretions for fear of predisposing the mammary gland to an IMI. Some producers prophylactically treat all prepartum heifers to control IMI at calving. This study was designed to evaluate the utility of pre-partum samples (pre-partum mammary secretions and teat skin swabs) for predicting IMI in heifers at calving and to evaluate whether this sampling increases the likelihood of IMI post-calving.

Pre-partum samples were collected 2 weeks prior to expected calving date and quarter milk samples were collected once weekly for the 1st 3-weeks post-calving for bacterial culture. Heifers were systematically assigned to 1 of 3 study groups (n = 392 quarters each) as follows: 1) pre-partum secretions from all mammary quarters, 2) pre-partum secretions from two diagonal quarters, and 3) no pre-partum secretions collected. Skin swab samples were collected from all heifers. Bacteriology was performed according to NMC guidelines and isolates were categorized as coagulase negative staphylococci (CNS), coagulase positive staphylococci (CPS), streptococci, Gram-negatives, mixed infections (2 types) or contaminated samples ($\geq$ 3 types). Using phenotypic characterization of bacterial isolates the sensitivity (Se), specificity (Sp), positive (PPV) and negative predictive values (NPV) and accuracy were calculated to define the utility of pre-partum parameters for predicting IMI at calving. Mammary quarters sampled during the pre-partum period were compared with non-sampled quarters to determine if sampled quarters had a higher occurrence of post-calving IMI (Chi-square; $P < 0.05$).

Data from 262 mammary quarters were used to calculate Se, Sp, PPV, NPV and accuracy of pre-partum secretions for predicting an IMI at calving. The results were 0.81, 0.67, 0.36, 0.94, and 0.70, respectively. No CPS were isolated from teat skin, 92% (102/112) of mammary quarters that had a CNS on teat skin did not have a CNS IMI at calving, 3% (3/112) had an IMI, and 5% (6/112) were infected with a different bacterial pathogen. A greater proportion of mammary quarters (26%; 103/396) that were not sampled pre-partum had IMIs at calving than quarters that were sampled (19%; 61/327) ($P = 0.024$). Based on the calculated Se (0.81), pre-partum secretion culture is marginally useful for predicting IMIs in heifers at calving. However, given that 52% of quarters were culture negative prepartum and only 6% of these culture negative quarters acquired new infections after calving, prepartum sampling would be useful to producers for preferentially treating mammary quarters prepartum. Pre-partum sampling did not increase the likelihood of IMI at calving.

Previously presented at the Mastitis Research Workers Conference, 2009; National Mastitis Council (NMC) 2010.

ABSTRACT #142

COMPARISON OF COMPUTED TOMOGRAPHY AND ROUTINE RADIOGRAPHY OF THE TYMPANIC BULLAE IN THE DIAGNOSIS OF OTITIS MEDIA IN THE CALF.

A Finnen, L Blond, D Francoz, J Parent. University of Montreal Faculty of Veterinary Medicine, Saint Hyacinthe, Quebec Canada.

Otitis media appears to be a common disease of young calves however the true prevalence is unknown. Diagnosis is based upon clinical signs and radiography, which often can be difficult to interpret. The purpose of this study was to compare radiography and computed tomography in the diagnosis of otitis media in calves. Seventeen Holstein calves between 5 to 7 weeks of age were included. All calves following neurological examination were sedated with intravenous xylazine (0.5–1.0 mg/kg) for routine radiographs (3 views) and computed tomography of the tympanic bullae followed by necropsy. Based upon necropsy findings, 11 of 17 calves were affected with otitis media, 4 unilaterally and 7 bilaterally. Radiographic and computed tomographic changes associated with otitis media included increased soft tissue opacity within the bulla, thickening of the bulla wall, enlarged bulla, and osteolysis of the bulla wall and trabeculations. The most reliable radiographic change was lysis of the trabeculations within the tympanic bulla which was present in 92.3% of affected bullae. However, 50% of normal bullae were considered affected based upon radiography. On computed tomography images, increased soft tissue opacity within the bulla was present in all affected calves. Sensitivity of radiography and computed tomography was 72.2% and 94.4% and specificity was 50% and 100% respectively. The kappa value for agreement between the two imaging modalities was 0.29 indicating slight agreement. Results indicate that computed tomography is more specific and sensitive, faster to perform and easier to interpret than radiography in the diagnosis of otitis media in the calf.

ABSTRACT #143

COMPUTED TOMOGRAPHY AND MAGNETIC RESONANCE IMAGING OF THE NORMAL CALF BRAIN.

JL Johnson-Neitman, DS Biller, WC Cash, ML Jones, DE Anderson. Kansas State University College of Veterinary Medicine, Manhattan, KS.

The purpose of this study was to use computed tomography (CT) and magnetic resonance imaging (MRI) to provide descriptions of anatomic brain structures in normal calves.

Four normal Holstein calves, 23–28 days old, were included in the study. Calves had normal physical and neurological examinations, hematology, and biochemistry profiles. CT and MRI of the brain were performed after induction of general anesthesia. For MRI, pre-contrast T1-weighted, T2-weighted, T2-FLAIR, and T2*-weighted, and post-contrast T1-weighted and T2-FLAIR sequences were acquired. Following completion of imaging, cerebrospinal fluid was collected from the cerebellomedullary cistern and submitted for cytology and culture. All calves had cerebrospinal fluid within normal limits. Two calves were humanely euthanized while anesthetized, preserved, and the heads sectioned. The remaining two calves were recovered from anesthesia. Transverse sections were photographed and compared to the CT and MRI brain images.

Results conclude that the CT and MRI images correspond to normal gross anatomic structures. Computed tomographic images provide adequate evaluation of the ventricular system and excellent anatomic detail of bony structures, while MRI images provided greater parenchymal soft tissue anatomic detail of the brain when compared to CT.

It is concluded that CT and MRI are useful in the evaluation of the brain and surrounding structures when compared to conventional radiography and could be useful in the evaluation of intracranial disease in calves.

ABSTRACT #144

ESCHERICHIA COLI IS ASSOCIATED WITH PERSISTENT MAMMARY INFECTION IN CANADIAN DAIRY COWS.

JH Fairbrother, D Francoz, E Nadeau, S Messier. Faculté de Médecine Vétérinaire, Université de Montréal, Saint-Hyacinthe, Qc.

Mammary infections caused by *Escherichia coli* are generally considered to be of short duration and non contagious. However, in some cases, persistent infections seem to occur. Hence, it is important to characterize *E. coli* isolates from both single and recurrent excreting cows to elucidate bacterial differences which may be associated with clinical outcomes. The objectives of the present study were to confirm persistent infection of the udder in cows recurrently excreting *E. coli* from the same quarter, to identify possible spread of the organism between quarters and to examine antimicrobial resistance in these isolates. Antimicrobial resistance was also examined in isolates from transitory cases of clinical mastitis to compare results with persistent or contagious isolates.

E. coli isolates studied are part of a collection provided by the Canadian Bovine Mastitis Research Network (CBMRN). Milk samples were collected over a two year period, from the winter of 2007 to the summer of 2008. The National Cohort of Dairy Farms, part of the CBMRN, consisted of 91 herds in the first year and 75 in the second year. Cows of interest (n = 300) were selected and among these cows infected with *E. coli*, 22 cows repeatedly excreted the organism twice or three times from the same quarter. The first isolation was always associated with clinical mastitis. In addition, 16 cows were chosen because *E. coli* was isolated twice, but in different quarters. Persistent infection was confirmed by pulsed-field gel electrophoresis (PFGE). Susceptibility to ampicillin, cephalothin, ceftiofur, enrofloxacin, spectinomycin, tetracycline and trimethoprim-sulfamethoxazole (TMS) was tested by broth microdilution.

Of the 22 cows recurrently excreting *E. coli* from the same quarter, 18 were persistently infected with the same pulsotype. Proportion of persistently infected cows is 6% for this two year study. In two cows infected twice, but in different quarters, both isolates possessed the same pulsotype, suggesting contagion between quarters. Each persistently infected cow was matched with a singly infected cow for antimicrobial resistance comparison. In persistent strains the highest frequency of resistance was to ampicillin and tetracycline (both at 28%). In isolates from single cases, the highest frequency of resistance was to tetracycline (33%). None of the isolates were resistant to enrofloxacin. Excluding tetracycline, antimicrobial resistance was always higher, but not significant ($p > 0.1$), in persistent isolates: ampicillin (28% and 6%) cephalothin (22% and 0%), ceftiofur (11% and 0%), spectinomycin (6% and 0%) and TMS (11% and 6%). Isolates (4) associated with possible contagion did not show antimicrobial resistance.

In conclusion, most cows (82%) repeatedly excreting *E. coli* from the same quarter were persistently infected. However, a low potential of spread between quarters was detected (13%). Overall, antimicrobial resistance was low.

ABSTRACT #145

IMPACT OF AGE AND BREED ON INCIDENCE OF CLINICAL SIGNS AFTER NATURAL INFECTION WITH BLUETONGUE VIRUS SEROTYPE 8 IN SHEEP. N. Kirschvink[1], J. Leemans[1], B. Bolkerts[1], M. Raes[1], C. Baricalla[1], N. Mauguit[1], C. Saegerman[2]. [1]Animal Physiology, Veterinary Department, University of Namur, Belgium. [2]Epidemiology and Risk Analysis applied to Veterinary Sciences, Faculty of Veterinary Medicine, University of Liège, Belgium.

Serotype 8 of Bluetongue virus (BTV-8) emerged in 2006 in Northern Europe and infected ~90% of the Belgian sheep flocks in 2007. The aims of this study were to assess (1) the most frequent clinical signs presented by naturally-infected sheep (2) the impact of breed and age on the distribution of clinical scores (3) the relationship between the severity of clinical signs and the duration of antimicrobial and anti-inflammatory therapy.

The sheep flock (394 F, 49 M; composed by Texel, Ile de France and crossbreeds) of the University of Namur was thoroughly followed on a daily basis and clinical signs at detection of BTV-8 were recorded for each animal. An overall clinical score was calculated by summing up the scores for each relevant clinical sign (1 point/clinical score). Affected subjects were either treated with antibiotics alone (AB, long-acting tetracycline) or with antibiotics combined with steroidian (SAI, dexamethasone) or non-steroidian anti-inflammatory (NSAI, flunixin meglumine) drugs. The duration of treatment was estimated by the mean number of administrations per drug per treated animal.

Among 443 sheep, 203 (46%) were clinically affected by BTV-8 in summer 2007 (start week 29, end week 47; maximum of new clinical cases at week 31). The most frequent clinical signs were apathy (46%), salivation (45%), weight loss (31%), head edema (30%), followed by nasal discharge (16%), muzzle and lip lesions (12%), lameness (6%), submandibular edema (5%) and pulmonary signs (4%). Cyanosis of the tongue ("bluetongue") was observed in one ram only. Clinical scores were distributed as follows: score 1: 49%; score 2: 29%; score 3: 13%; score 4: 1%; score 5: 6%, score $\geq$ 6: 2%. Sheep younger than 3 years were more frequently and more severely affected than sheep older than 3 years ($P < 0.001$). Texel sheep (58%) and crossbreeds (44%) were more often affected than Ile de France (22%). The most severe clinical scores were recorded in crossbreeds (score 1: 42%; score 2: 25%; score $\geq$ 3: 33%) and Texel sheep (score 1: 46%; score 2: 31%; score $\geq$ 3: 23%), whereas Ile de France sheep were less severely affected (score 1: 65%; score 2: 19%; score $\geq$ 3: 16%). Finally, the most severe clinical scores were associated with longer duration of treatment with any of the administered drugs: [AB] score 1: 1.52, score 2: 1.84, score 3: 2.04; [NAIS] score 1: 1.41, score 2: 1.63, score 3: 1.65; [AIS] score 1: 1.11, score 2: 1.11, score 3: 1.00.

These field data recorded in a sheep flock after natural BTV-8 infection provide interesting information regarding incidence of the disease as well as type, frequency and severity of BTV-related clinical signs. These results also point out differences among breeds, Ile de France sheep being the less frequently and severely affected ones. Conversely, sheep younger than 3 years are more sensitive than others. Longer duration of treatment is needed in severely affected animals. Together, these findings might be helpful for early diagnosis and management of BTV-8 in sheep flocks.

ABSTRACT #146

METHICILLIN RESISTANT *STAPHYLOCOCCUS* SPP IN COMMERCIAL PIGS USED IN VETERINARY STUDENT TRAINING. BA Burgess[1], D Czerniak[1], M Jacobsson[2], D Bolte[1], D Smeak[2], JS Weese[3], PS Morley[1]. [1]Animal Population Health Institute, Colorado State University, Fort Collins, CO. [2]Department of Clinical Sciences, Colorado State University, Fort Collins, CO. [3]Department of Clinical Studies, Ontario Veterinary College, University of Guelph, Guelph, Ontario, Canada.

Research groups from many countries have recently identified methicillin resistant *Staphylococcus aureus* (MRSA) as a common colonizing agent in pigs, and that people with frequent contact (eg. veterinarians and producers) have increased risks for colonization and infection. MRSA has also been identified as one of the most common causes of outbreaks of nosocomial infections in veterinary hospitals. The Colorado State University College of Veterinary Medicine and Biomedical Sciences (CSU-CVMBS) utilizes commercial pigs in veterinary student training. These pigs are a potential source of environmental contamination as well as nosocomial and zoonotic infections. The objectives of this study were to estimate the frequency of MRSA colonization in young pigs used in teaching laboratories at the CSU-CVMBS and to evaluate associated environmental contamination.

Environmental samples were collected twice weekly from walls and hand contact surfaces of the holding facility (n = 28) and twice daily from floors and hand contact surfaces in the teaching laboratory (n = 140) using a commercially available electrostatic dust collection wipe (Swiffer, Procter and Gamble). Additionally, swabs were collected twice from each pig (n = 100) including nasal, perineal, and rectal samples. All samples were cultured for the presence of methicillin resistant *Staphylococcus* spp by a previously reported method.

Overall, the frequency of MRSA colonization in young pigs at the holding facility and teaching laboratory was 65.6% and 80%, respectively. The pigs arrived in two different groups (n = 50) one month apart. The frequency of colonization was markedly different between pig groups. There was environmental contamination as well as evidence of residual contamination between cleaning.

This study demonstrates a previously unrecognized risk associated with the use of young pigs in training and research in both veterinary and human medicine. Colonization of these pigs does not preclude their use rather it demonstrates the importance of biosecurity and hand hygiene measures to control the associated risks.

ABSTRACT #147

RISK FACTORS ASSOCIATED WITH TRANSMISSION OF *MYCOBACTERIUM AVIUM* SUBSP. *PARATUBERCULOSIS* (MAP) TO CALVES WITHIN A DAIRY HERD: A SYSTEMATIC REVIEW. E Doré[1], J Paré[2], S Buczinski[1], JP Roy[1], O Labrecque[3], G Côté[4], G Fecteau[1]. [1]Université de Montréal, Faculté de Médecine Vétérinaire, Saint-Hyacinthe, QC, Canada. [2]CFIA, Saint-Hyacinthe, QC, Canada. [3]MAPAQ, Saint-Hyacinthe, QC, Canada[4]INSA, St-Foy, QC, Canada.

Calves are considered more susceptible to MAP infection than adult cows. It is generally accepted that the young heifer group should be targeted to reduce the transmission of MAP in infected herds. Scientific evidence indicates that both colostrum and milk may contain MAP organism.

The objective of this study was to systematically review the scientific literature concerning risk factors associated with MAP transmission to calves. Relevant literature to the following questions was studied: 1) Does feeding pasteurized colostrum or milk replacer reduce the risk of transmission of MAP? 2) Does raising calves individually and separate from the adult reduce the risk of transmission of MAP? 3) Is there a relationship between the characteristics of the immediate neonatal calves' environment and risk of infection with MAP?

An electronic search of PubMed and CAB was conducted in September 2009. A first selection was based on the title of the manuscript. Papers published in non-peer-reviewed journals or not relevant to any of the three questions were eliminated. The abstracts of the remaining papers were read and retained if they appeared pertinent to any of the three questions. The remaining articles were reviewed by 3 authors and then assessed for pertinence..

The PubMed search yielded 399 references. Thirty-four were not in English, French or Spanish. Of the 365 papers remaining, 246 were discarded based on the title. Sixty-nine were eliminated after the abstract was read. The same selection was performed for CAB and yielded 7 additional articles. Fifty manuscripts were reviewed, of which 21 were considered relevant to the objective. The year of publication varied from 1992 to 2009 and originated from 10 different countries. The majority of the study designs was cross-sectional (n = 15). The number of farms in each study varied from 1 to 2,953. The definition of a positive herd was variable, but most often based on an ELISA results (n = 15). Thirteen papers addressed the question 1. Six concluded that feeding raw milk and/or colostrum represents a risk of transmission. Seventeen manuscripts addressed question 2. Nine of these concluded that raising calves individually and/or away from adult feces was protective. Finally, sixteen manuscripts pertained to question 3. Ten found that a better hygiene in the immediate environment of the calf was protective.

The results of the systematic review will be used to refine a questionnaire to estimate the risk of transmission in the neonatal period.

ABSTRACT #148

BREEDING SOUNDNESS EXAMINATION IN WESTERN U.S. RAMS: 12,716 CASES (2000–2007). DC Van Metre, J Fraser, PS Morley. Animal Population Health Institute, College of Veterinary Medicine and Biomedical Sciences, Colorado State University, Fort Collins, CO.

Breeding soundness evaluation of rams allows the veterinarian to assess the capacity of a ram to impregnate a defined number of ewes through natural service. The influence of ram fertility on pregnancy rate, lamb crop, and certain ranch economic indices is well established. Guidelines set by the Society for Theriogenology (SFT) are frequently applied to classify tested rams as satisfactory, questionable, or unsatisfactory potential breeders. The purpose of this retrospective study was to determine the reasons for unsatisfactory or questionable SFT classification in breeding soundness examination for 12,716 rams from over 100 flocks in CO, WY, and UT that were examined from 2000–2007.

Breeding soundness examinations consisted of a cursory physical examination, body condition score, palpation of the scrotal contents, scrotal circumference measurement, manual exteriorization of the penis, collection of semen by electroejaculation, microscopic evaluation of sperm motility and morphology, and, for selected rams, serologic testing for infection with *Brucella ovis* by ELISA. Rams determined to be seropositive for *B. ovis* were classified as unsatisfactory potential breeders.

Approximately 1 in 4 rams tested were classified as unsatisfactory or questionable potential breeders. The most common reasons for rams to be classified as such included (in decreasing order of frequency) abnormal sperm morphology, poor sperm motility, leukospermia, thin body condition, ulcerative posthitis, small scrotal circumference, abnormal palpation findings for one or both epididymides or testes, and seropositive status on *B. ovis* ELISA. Of all rams seropositive for *B. ovis*, nearly 50% had leukospermia and only 5% had palpable abnormalities of the epididymes or testes; importantly, nearly 30% of seropositive rams were otherwise normal on breeding soundness examination.

ABSTRACT #149

INACCURATE RECTAL TEMPERATURE MEASUREMENT USING DIGITAL THERMOMETERS: CAUSES AND SOLUTIONS. JM Naylor[1], R Streeter[1], P Torgerson[2]. [1]University of Ross School of Veterinary Medicine, St Kitts. [2]University of Zurich, Switzerland.

Taking the temperature has been an essential part of routine physical examination for over 100 years. Recently, some veterinarians have switched from the use of mercury (Cornell) thermometers to digital thermometers designed for use in people to measure rectal temperature. This abstract reports some of the common errors with digital thermometers designed for use in people and a method of obtaining reliable results.

Fifteen crossbred cows were used for student measurement of rectal temperature on 53 separate occasions over a 3 month period. Student's initially measured rectal temperature using techniques gained from previous experience, then retook the temperature following instruction to immerse the thermometer in the rectum to the beginning of the display window. Rectal temperature was also measured with a Cornell mercury thermometer completely inserted in the rectum. Student initial rectal temperature measurements were 38.46 +/− 0.52°C (mean +/− 1 SD, n = 53) which are approximately 0.5°C lower, and more variable, than those obtained with Cornell mercury thermometers 38.95 +/− 0.38°C. Values with digital thermometers inserted to the display window were 38.77 +/− 0.41°C, which are closer to the Cornell mercury thermometer method.

Separately, the accuracy of student digital thermometers was compared against traceable (reference) digital and mercury thermometers in a water bath. In general, most digital thermometers designed for human use were accurate but one showed a large error, possibly because it had been poorly maintained.

The affect of angle of insertion on rectal temperature measurement was determined to be of minor importance.

Accurate temperature measurements can only be obtained using digital thermometers if the thermometer is inserted to the window. The thermometer's accuracy should be checked periodically.

ABSTRACT #150

INTEGRATION OF PHARMACOKINETIC AND PHARMACODYNAMIC DATA FOR FLORFENICOL IN CALVES. TJ Potter, J Illambas, PK Sidhu, A Rassouli, A Rycroft and P Lees. Royal Veterinary College, London, UK

The objective of this project was to validate pharmacokinetic-pharmacodynamic (PK-PD) integration and modelling approaches to dose schedule design, by generating and correlating data on the PK and PD of five drugs (oxytetracycline, amoxicillin, florfenicol, marbofloxacin and tulathromycin) commonly used to treat calf pneumonia. This abstract describes the results for florfenicol.

Minimum inhibitory concentrations (MIC) for florfenicol were determined in Mueller Hinton Broth (MHB) and calf serum, using 5 overlapping sets of doubling dilutions of 6 clinical isolates of each of 2 organisms *M.haemolytica* and *P.multocida*. Time-kill relationships were established for florfenicol in broth and serum using multiples of MIC (0.25, 0.5, 1, 2 and 4 times). Pharmacokinetic data and *ex vivo* bacterial kill curves were established for three biological fluids; serum, exudate and transudate using a tissue cage model of inflammation. The PK study used 10 healthy female Aberdeen Angus Holstein Friesian Cross calves aged 3–5 months, with florfenicol (Nuflor) administered subcutaneously as a single dose at 40 mg/kg.

Mean MICs for *M.haemolytica* and *P.multocida* in MHB were 0.52 µg/ml and 0.39 µg/ml respectively and were not significantly different from those measured in serum (0.58 µg/ml and 0.48 µg/ml). *In vitro* time-kill studies in MHB indicated concentration dependent bactericidal activity of florfenicol with a 3-log reduction in colony forming units (cfu)/ml after 4 h incubation at 2 times MIC and a reduction to 33 cfu/ml following 8 h incubation at 4 times MIC. In the *ex vivo* study florfenicol exhibited bactericidal activity in serum, exudate and transudate with a 4-log reduction in cfu/ml after 4 h of incubation in samples collected at 12, 24 and 48 h post drug administration. Eradication occurred after 8 h incubation in these samples. Pharmacokinetic variables (mean ± SEM) were; C_{max} = 6.20 ± 1.56 µg/ml, T_{max} = 2.28 ± 0.13 h and AUC_{24} = 133.8 ± 6.6. h. µg/ml. Surrogate markers of antimicrobial activity are presented in Table 1. Targeted partial areas were used to express the ratio of average serum concentration relative to a targeted MIC for three consecutive 24 h periods after dose administration (Table).

The data indicates that, following a single subcutaneous dose, serum concentrations of florfenicol were above the MIC for both pathogens for greater than 72 h, and that at concentrations achieved under clinical conditions florfenicol exhibited bactericidal activity.

Table [Abstract 150]: PK-PD Integration and targeted partial areas for florfenicol in serum (mean values, n = 10)

	MIC (µg/ml)	Cmax: MIC	AUC/MIC (h)	Targeted Partial Areas ($C_{average}$: MIC) 0–24 h	24–48 h	48–72 h
M.haemolytica	0.58	9.64	230.7	6.33	2.97	1.53
P.multocida	0.48	11.65	278.8	7.65	3.58	1.85

ABSTRACT #151

PHARMACOKINETICS OF INTRAVENOUS AND INTRAMUSCULAR TRAMADOL IN LLAMAS. S. Cox, T. Martin-Jimenez, J Yarbrough, T. Doherty. University of Tennessee, College of Veterinary Medicine, Knoxville, TN.

The purpose of this study was to determine the pharmacokinetics of tramadol (2 mg/kg) and its metabolite M1 after intravenous and intramuscular administration to llamas. Tramadol is a centrally acting analgesic that possesses opioid agonist features that has been used to treat moderate to severe pain in humans. Camelids are a rapidly growing aspect of small ruminant agriculture in the US. The popularity and numbers of llamas being kept as pets, packing animals, and show animals has increased in recent years. With the increased numbers of animals there is an increased demand for veterinary service; however, there are no drugs approved by the FDA for use in camelids and pharmaceutical companies cannot economically justify seeking approval of drugs in this species. Six healthy adult male llamas were used to determine the metabolism of tramadol. Tramadol half-life, volume of distribution at steady state, clearance and AUC after intravenous administration were 4.43 ± 0.90 h, 4902.68 ± 1495.91 mL/kg and 1680.50 ± 141.05 mL/h/kg, 1193.65 ± 99.19 h ng/mL respectively. The bioavailability was 110 ± 21% and half-life 3.00 ± 0.15 h with intramuscular administration. M1 had a half-life of 10.40 ± 2.9 h and 7.71 ± 0.54 h following intravenous and intramuscular administration of tramadol, respectively. No adverse effects were observed. Administration of tramadol resulted in concentrations high enough to produce analgesia based on human data.

ABSTRACT #152

EFFECTIVENESS OF MORPHOMETRIC MEASUREMENTS FOR PREDICTING BODY COMPOSITION IN OVERWEIGHT AND OBESE DOGS. PW Toll,[1] I Paetau-Robinson[1], AL Lusby,[2] GA Henry,[2] CA Kirk[2]. [1]Hills Pet Nutrition, Topeka, KS. [2]University of Tennessee Department of Small Animal Clinical Sciences, Knoxville, TN.

In a prior study we demonstrated that current methods of estimating ideal body weight for weight loss feeding are inaccurate in dogs having ≥ 45% body fat. The purpose of this study was to develop simple and accurate methods of measuring body composition in clinical practice in overweight to morbidly obese dogs. Current morphometric measurement methods do not adequately predict body composition in obese dogs. Additionally, these methods tend to be difficult to use and have poor accuracy. This study evaluated two methods of predicting lean body mass (LBM) from animal morphology, radiographic measurement of skeletal size and external physical measurement of body size.

Thirty-six client-owned overweight or obese adult dogs (28% to 64% body fat) weighing 5 to 73.6 kg underwent dual energy x-ray absorptiometry (DXA) scanning to assess their body composition. Nine measurements were made from digital radiographs to assess skeletal size and 16 external physical measurements were made to assess body size. Body weight was also measured. Multiple regression analysis using DXA results as the dependent variable was used to develop regression equations for the prediction of LBM from the skeletal size and body size data. Several equations were developed that had high (> 0.9) correlation coefficients for both methods when the dogs were divided into 2 size groups (≤ 40 lbs and > 40 lbs). The best equations were applied to the data from the same group of dogs to evaluate how well values for individuals were predicted (within ± 10% of the DXA value). The best equations using skeletal size data resulted in a $r^2 = 0.99$ and a predictability (± 10%) ≥ 86% using ≤ 8 variables (including BW) for both size groups. The best equations using body size data resulted in a $r^2 = 0.99$ and a predictability (± 10%) = 100% using ≤ 8 variables (including BW and age) for both size groups.

Multiple regression analysis demonstrated that both methods could be used to predict LBM using a variety of regression equations. This approach shows great potential for the development of simple and accurate tools to be used in clinical practice. Further studies must be done to validate the equations in a population of dogs other than those used to develop the equations.

ABSTRACT #153

EFFECTIVENESS OF BCS FOR ESTIMATION OF IDEAL BODY WEIGHT AND ENERGY REQUIREMENTS IN OVERWEIGHT AND OBESE DOGS COMPARED TO DXA. AL Lusby,[1] CA Kirk,[1] PW Toll,[2] I Paetau-Robinson[2]. [1]University of Tennessee Department of Small Animal Clinical Sciences, Knoxville, TN. [2]Hills Pet Nutrition, Topeka, KS.

With an estimated 35–40% of pet dogs being overweight or obese, veterinarians must accurately assess energy needs to prescribe appropriate food doses. Body condition scoring (BCS) is the most popular and accessible method for estimating degrees of obesity in dogs, but these scales were designed to assess animals with body fat percentages below about 45%. Many obese dogs have body fat percentages of 50% and greater. This project compared the accuracy of using body fat percentages to the 5 and 9 point BCS systems for estimating ideal body weight and resting energy requirements (RER) in overweight dogs. Thirty-six healthy, client-owned dogs ranging from 5 to 73.6 kg underwent dual energy x-ray absorptiometry (DXA) scanning to assess their percentage of body fat (BF). BF percentage was then used to estimate ideal body weight, calculate RER, and classify each animal into the 5 and 9 point BCS system. Once a BCS was assigned, the median body fat percentage for each score was used to estimate ideal body weight and RER (5 point scale - 4 = 30%, 5 = 40%; 9 point scale - 6 = 30%, 7 = 35%, 8 = 40%, 9 = 45%). Based on DXA, body fat ranged from 28.3% to 63.7% with a mean of 45.9%. To assess the accuracy of BCS for moderately versus morbidly obese dogs, patients were divided into two groups: < 45% body fat (n = 15) and > 45% body fat (n = 21). Compared to DXA, estimations of ideal body weight were significantly higher using the 5 (23.0 vs. 19.2 kg) and 9 (21.1 vs. 19.2 kg) point BCS in dogs with BF > 45% ($p < 0.001$) but did not differ in dogs with < 45% BF ($p > 0.05$). DXA estimations of RER were also significantly lower than estimations using the 5 (709 vs. 616 Kcal/day) and 9 (665 vs. 616 Kcal/day) point scales in dogs with BF > 45% ($p < 0.001$), but did not differ in dogs with < 45% BF ($p > 0.05$). The results of this study demonstrate current BCS systems provide good estimates of ideal body weight and RER in dogs with less than 45% BF, but are inadequate for calculating RER and ideal body weight in morbidly obese dogs (BF > 45%). As a result, food dose calculations will be overestimated and this may affect weight loss efficacy. As canine obesity rates climb, we must develop new methods to assess our most obese patients and provide better weight management.

ABSTRACT #154

OBESITY AND GENDER AFFECTS TOTAL ADIPONECTIN CONCENTRATION AND MULTIMER DISTRIBUTION IN DOMESTIC CATS. CR Bjornvad[1], JS Rand[2], F Rose[2], HY Tan[2], PJ Armstrong[3], J Whitehead[2]. [1]Department of Small Animal Clinical Sciences, University of Copenhagen, Denmark. [2]The University of Queensland, Brisbane, Australia. [3]University of Minnesota, MN.

Adiponectin, an adipokine that attenuates insulin resistance, decreases with obesity in humans and cats. In particular, the more biologically active HMW (high molecular weight) isoform decreases to a greater degree in human subjects than the less active LMW (low molecular weight) isoform. We determined plasma concentrations of total, LMW and HMW adiponectin in client-owned, lean and obese adult, neutered, domestic shorthaired cats.

Following health screening by physical examination, urinalysis, hematology, coagulation and biochemical profiles, 73 cats (35 males and 38 females) were included in the study and informed consents were obtained. All cats were anaesthetized and body fat percentage (%BF) was measured using Dual-Energy X-ray Absorptiometry. Total adiponectin was determined using a commercial ELISA kit. LMW and HMW concentrations of adiponectin were determined by Western blotting after sucrose-gradient velocity centrifugation. Adiponectin multimer distribution (S_A = HMW/(HMW + LMW)) was calculated and linear regression used to assess the relationship between %BF, total adiponectin and multimer distribution. Differences between lean/normal weight (%BF < 35, n = 26) overweight

($35 < \%BF < 45$, n = 28) and obese cats (%BF > 45, n = 19) were evaluated by ANOVA. Results are given as mean ± SE.

Total adiponectin was lower in male than female cats (2.29 ± 0.62 vs 4.16 ± 0.60 ug/mL, $P < 0.04$) while there was no correlation with body composition. Increasing %BF correlated negatively with multimer distribution (S_A, obese, 0.55 ± 0.04; overweight 0.65 ± 0.03; and lean cats, 0.68 ± 0.03; $P < 0.02$).

In this study, obesity was associated with lowered HMW adiponectin while total adiponectin was not significantly decreased. Changes in adiponectin distribution parallel results obtained in obese humans. The lack of difference in total adiponectin between lean and obese cats was surprising and differs from a previous report. The observed gender difference correlates with observations in healthy human subjects but may not only be attributable to testosterone and estrogen levels because all study cats were neutered.

ABSTRACT #155

LIPID METABOLISM AND SATIETY WITH DIETARY FIBER AND CARNITINE IN OBESE DOGS. Y Mitsuhashi, JE Bauer. Companion Animal Nutrition Laboratory, and Intercollegiate Faculty of Nutrition, Texas A&M Univ., College Station, TX.

Obesity is a growing global problem in companion animals. Dietary fiber has been reported to possess physiological effects on satiety and hypolipidemia while carnitine acts to enhance fat utilization. Thus, this study was designed to evaluate the combined effects of dietary fiber and carnitine on 1.) short-term satiety and 2.) weight loss and lipid metabolism in obese Beagles.

Two diets were fed differing in fiber and carnitine content: Purina ONE® healthy weight formula (LFC) vs. Purina ONE® plus a fiber and carnitine containing canned dietary supplement (HFC, Diet Delite® mixer for canine weight control). Total dietary fiber and carnitine contents of the foods fed were 3.5 g/100 kcal and 0.5 mg.100 kcal (LFC) vs 4.5 g/100 kcal and 38.9 mg/100 kcal (HFC) respectively. Two studies were conducted. For the short-term satiety study, 14 female adult Beagles (body fat (BF) 37.5 ± 1.4% (SEM), body weight (BW) 13.4 ± 0.6 kg, were randomly assigned into 2 diet groups. Diets were fed at 8am and 3pm (7 h interval) or, during a second trail, at 8am and 11am (3 h interval) for 15 min each over a 2–3 day period. The amount of food offered was 1.2XMER at each feeding using a cross-over design with 3 day washout period. Blood samples were collected after the 8am feeding at 0, 45, and 120 min.Food intake and plasma Peptide YY (PYY) were measured as satiety markers. For the weight loss study, 7 obese Beagles were selected (BF 45.1 ± 1.6%, BW 15.2 ± 1.0 kg) and randomly assigned into 2 groups. The diets were fed once daily (ca. 60% of obese MER) for 42 days. Postprandial blood samples were collected at 60 min (days 1,28, and 42) for β-hydroxybutyrate analysis (a measure of fat oxidation) plus lipid, and lipoprotein analyses. Because the HFC diet contained high β-carotene due to its high pumpkin content, plasma retinoids were also measured. Food intake and BW were recorded daily and weekly, respectively. Repeated measures ANOVA was performed under $\alpha = 0.05$. Where there was significance, paired t-tests or one-wayAOV/Kruskal-Wallis tests with Tukey multiple comparisons were perfomed after checking normality by Shapiro-Wilks test.

In the satiety studies, daily energy intake during the 7 h interval study was not different beween diets. However, the HFC diet significantly decreased daily and 2nd feeding energy intake during the 3 h interval compared with the LFC diet ($P < 0.05$). Plasma PYY was not different between diets. In the weight loss study, all dogs lost BW; however, only the HFC diet significantly decreased both BW and BF at 42 days compared with day 1 ($P < 0.02$). Moreover, plasma postprandial β-hydoxybutyrate was signficiantly greater in the HFC than the LFC group at day 42. Plasma TG, total cholesterol (TC), lipoproteins (LPs), PYY, and retinoids did not change over time or with diet. In conclusion, the HFC diet provided satiety with lower energy (i.e. calorie) intake after a 3 h interval and increased β-oxidation thereby promoting more efficient weight and body fat loss without modifying TG, TC, LPs, and retinoid responses.

ABSTRACT #156

EFFECT OF PRESCRIPTION DIET® CANINE J/D™ ON CLINICAL MEASURES, CARTILAGE BIOMARKERS, AND METABOLOMIC CHANGES IN DOGS WITH OSTEOARTHRITIS. N.Z. Frantz, Kevin Hahn, Jen MacLeay, and Scott Mickelsen. Hills Pet Nutrition, Inc., Topeka, KS.

A study was conducted to evaluate the effect of Prescription Diet® canine j/d™ on clinical and biological measures when fed to dogs with osteoarthritis (OA). Thirty-one beagle dogs (initial weight, 13.5 ± 1.27 kg, age, 11.0 ± 2.23 years) with lameness and radiographic changes consistent with DJD in at least one joint were included in the study. All dogs were fed a control maintenance food for 28 days followed by canine j/d. Canine j/d contains increased levels of EPA and DHA, glucosamine, chondrotin sulfate, Vitamins C and E, and L-carnitine. Blood and urine samples were collected on the last day of the control food and after 14 days on canine j/d. Serum fatty acids, cartilage biomarkers, urine biomarkers, and plasma metabolomics were measured. Orthopaedic exams were performed every 14 days during the study and collars with activity watches were worn throughout the study. Dogs fed canine j/d had significantly higher serum eicosapentaenoic acid and docosahexanoic acid at day 14. After consuming canine j/d, arthritic dogs had improved orthopaedic scores at day 14, and reduced night activity compared to while on the control food. They also had lower mean serum concentrations of bone specific alkaline phosphatase, BAP:NTx ratio, CS846:KS ratio, C-reactive protein, and C1,2C at day 14 on canine j/d. As well, serum CPII, CPII:C2C ratio, IGF-1, keratan sulfate, urine LTB4, and urine PGE2 were elevated after consuming canine j/d. Arthritic dogs fed Canine j/d had lower plasma IL-2, IL-4, IL-6, IL-7, IL10, IL-15, IL-18, IFN-gamma, KC, and MCP-1. Metabolomic changes after consuming the canine j/d were consistent with decreased inflammation (kynurenine, histidine, 3-methyl-histidine, arachadonic acid). In summary, feeding dogs with arthritis Prescription Diet® canine j/d™ resulted in improved orthopaedic responses, reduced inflammation, and a reduction in cartilage and bone degradation biomarkers. Furthermore, the activity data may suggest improved comfort at night, similar to that observed with NSAIDs.

ABSTRACT #157

CLINICAL EXPERIENCE USING A LIPID-FREE, READY-MADE PARENTERAL NUTRITION SOLUTION IN 70 CATS AND DOGS. I Gajanayake, DL Chan. The Royal Veterinary College, Hatfield, Hertfordshire.

The aim of this retrospective study was to evaluate the clinical use of a lipid-free, ready-made parenteral nutrition (PN) solution in dogs and cats. Medical records of all dogs and cats that received this PN solution between November 2006 and September 2009 were reviewed and data pertaining to patient profile, time to initiation and duration of PN support, characterization of caloric and protein provision, complications, and outcome were recorded.

PN was provided to 61 dogs and 9 cats using a ready-made, lipid-free, amino acid (59 grams protein/L) and dextrose (100 g/L) solution. This PN solution also contains electrolytes, including potassium (20 mEq/L). The median time to initiation of PN support was 3 days (range 0–26 days) and this was administered via a peripheral catheter in most cases (n = 51). The median duration of PN administration was 60 hours (range 6–228 hours) in dogs and 36 hours (range 24–96 hours) in cats. The median level of protein provision was 6 gm/100 kcal (range 1–10 gm/100 kcal), while the median fraction of the resting energy requirements provided by the PN solution was 57% (range 9–100%). The median rate of glucose provided was 2.2 mg/kg/min (range 0.8–5.2 mg/kg/min).

Metabolic complications were seen in 36 patients, of which hyperglycemia (n = 19) and hyperkalemia (n = 15) were the most common. The median maximal glucose concentration post-initiation of PN in the patients that developed hyperglycemia was 135 mg/dL (range 120.7 – 477.5 mg/dL). The amount of glucose provided by the PN did not influence the development of hyperglycemia ($P = 0.4$). Azotemia developed following PN administration in 7 animals but the amount of protein provided did not influence the development of this complication ($P = 0.5$). Mechanical complications were seen in 27 patients, all of which involved administration via peripheral catheters. Suspected catheter infections were investigated via bacterial cultures of catheters and this resulted in a positive culture in 4 cases (all dogs).

Thirty-one (44%) patients were discharged from the hospital, whereas 17% died and 39% were euthanized. The median duration of hospitalization was 6.5 days (range 2–31 days). The occurrence of mechanical or septic complications did not affect the outcome of these patients, however patients without pre-existing hyperkalemia that developed this complication following PN administration were less likely to survive ($P = 0.015$). The presence of azotemia both prior to and following PN administration was associated with a poor outcome ($P = 0.016$ pre-PN and $P = 0.003$ post-PN). Pre-existing or the development of hyperglycemia was not associated with a poor outcome ($P = 1.0$). Concurrent enteral nutrition with PN was not found to influence outcome ($P = 0.73$).

This form of PN appears to be a viable alternative to compounded PN but may be associated with higher rates of complications compared with previous reports of PN in animals.

ABSTRACT #158

FECAL MICROBIOTA AND FECAL CALPROTECTIN AND S100A12 CONCENTRATIONS IN LEAN AND OBESE DOGS.

S Handl[1,2], RM Heilmann[2], AJ German[4], SL Holden[4], SE Dowd[3], JM Steiner[2], RW Grant[5], KS Swanson[5,6], JS Suchodolski[2]. [1]Institute of Nutrition, University of Veterinary Medicine Vienna, Austria. [2]Gastrointestinal Laboratory, Texas A&M University, College Station, TX. [3]Research & Testing Laboratory, Lubbock, TX. [4]School of Veterinary Sciences, University of Liverpool, UK. [5]Division of Nutritional Sciences, [6]Dept. of Animal Sciences, University of Illinois, Urbana, IL.

Recently, it has been shown that the intestinal microbiota is altered in obese mice and humans, and that obesity leads to a state of chronic subclinical intestinal inflammation. The aim of this study was to investigate differences in the fecal microbiota and in the fecal inflammatory markers calprotectin and S100A12 between lean and obese adult dogs.

Sixteen healthy lean adult pet dogs (median body condition score (BCS) 5/9, range: 4–5) were compared to 13 obese, but otherwise healthy, adult pet dogs (median BCS 8.5/9, range 7–9). The pet dogs were of various breeds and ages and were fed different commercial diets. Additionally, 9 healthy adult research Beagles were enrolled, five of which were fed *ad libitum* over 6 months to a mean BCS of 7.5/9 (range: 7.5–9) and 4 were fed the same diet restrictively and served as lean controls (mean BCS 4/9, range 4–5). Samples were taken at baseline (T0), after 3 months (T3), and after 6 months (T6). The fecal microbiota was analyzed by massive parallel 16S rRNA gene pyrosequencing. Fecal concentrations of calprotectin and S100A12 were measured using in-house radioimmunoassays. Data were analyzed using parametric or non-parametric tests when appropriate and contingency tables. Significance was set at $p < 0.05$.

Differences in the fecal microbiota between lean and obese dogs were mainly seen in the orders Clostridiales, Lactobacillales and Erysipelotrichales. In the pet dogs, obese animals had a higher proportion of the genus *Catenibacterium* ($p = 0.04$) and a lower proportion of *Streptococcus constellatus* ($p = 0.0005$) and *Clostridium* cluster II ($p = 0.01$). In the research dogs, proportions of Peptococcaceae ($p = 0.03$), *Clostridium bartlettii* ($p = 0.02$), and *C. sordellii* ($p = 0.02$) were higher, while proportions of Clostridiales Family XIII Incertae Sedis ($p = 0.02$) and *Anaerovorax* sp. ($p = 0.02$) were lower in the dogs fed *ad libitum* compared to the lean controls at T6. No significant differences ($p = 0.6$) were seen in fecal calprotectin concentrations between lean and obese pet dogs and all values were within the reference interval (2.9–137.5 μg/g). In the research dogs, fecal calprotectin concentrations were higher in the dogs fed *ad libitum* compared to the lean controls at T3 ($p = 0.03$) or T6 ($p = 0.02$). S100A12 was below the detection limit of 0.5 μg/g in all samples.

These results suggest alterations in the fecal microbiota of obese dogs. The most notable changes were observed within the orders Clostridiales and Erysipelotrichales (i.e. *Catenibacterium* sp.), which is similar to findings in obese mice and humans. Fecal calprotectin concentrations were significantly increased only in the research dogs fed *ad libitum*, possibly reflecting a state of inflammation of the intestinal tract during the early stage of obesity.

ABSTRACT #159

CYCAD INTOXICATION IN DOGS: SURVIVAL AND PROGNOSTIC INDICATORS.

D. Ferguson, M. Crowe, M. Acierno, F. Gaschen. Louisiana State University, Baton Rouge, LA.

Cycads (also called sago palms) are commonly encountered in various parts of the world. They are part of the native flora in tropical and subtropical regions. Additionally they have been extensively used in residential landscaping in the South of the USA. Cycads produce a variety of toxins that are poisonous to humans and animals. These toxins can be found in all parts of the plant, but seem to be present in highest concentration in the seeds and roots. After ingestion the main toxin cycasin is metabolized by the intestinal flora to its active compound, which is most likely responsible for GI and hepatic toxicity in dogs. The goal of this retrospective study of a cohort of affected dogs is to evaluate the outcome of cycad intoxication and identify possible prognostic factors.

Medical records of the LSU Veterinary Teaching Hospital were searched for cases of documented or suspected cycad ingestion presented over a 5 year period (2005–2009). All information including history, clinical signs, laboratory abnormalities, applied treatment modalities and survival was tabulated. Depending on the outcome, cases were assigned either to the surviving or the non-surviving group. Variables recorded at the time of presentation were compared between both groups using the Wilcoxon rank sum test or the Fisher's exact test. Additionally, all variables were evaluated as possible prognostic indicators in a univariate and multivariate Cox proportional hazard model.

A total of 45 cases were retrieved. Eating or chewing a cycad root or seed had been directly observed in 22 dogs, while there was strong evidence suggesting cycad intoxication in the remaining 23 cases. Nineteen (42.2%) dogs survived and 26 (57.8%) died or were euthanized as a direct consequence of cycad intoxication. The following significant differences were documented between survivors (Sv) and non-survivors (NSv) dogs ($p < 0.05$): median serum alkaline phosphatase was higher in NSv while median platelet count, serum glucose and serum albumin were lower in NSv, Additionally, coagulation times PT and PTT were more frequently prolonged in NSv than in Sv. Significant variables in the univariate model ($p < 0.05$) were platelet count, serum glucose and serum albumin, frequency of prolonged PT, and administration of fresh frozen plasma. However, dogs probably received plasma due to the severity of their condition. Time between ingestion and presentation, presence of neurological signs, or documentation of confirmed vs. suspect ingestion did not significantly affect the prognosis. Low platelet count and low blood glucose were negative prognostic factors in the multivariate analysis.

Cycad intoxication is associated with a much higher mortality rate than previously published. Several laboratory parameters may be helpful in determining the prognosis of intoxicated dogs on presentation. They likely reflect the severity of the liver function decrease, the presence of systemic inflammatory response syndrome, DIC, or a combination of these problems.

ABSTRACT #160

HEPATOPROTECTIVE EFFECTS OF S-ADENOSYLMETHIONINE AND SILYBIN ON CANINE HEPATOCYTES.

AY Au[1,2], JM Hasenwinkel[2], CG Frondoza[1,3,4]. [1]Nutramax Laboratories, Inc., Edgewood, MD. [2]Syracuse University, Syracuse, NY. [3]Johns Hopkins University, Baltimore, MD. [4]Mississippi State University, Mississippi State, MS.

Liver disease resulting from trauma or inflammation compromises the ability of the tissue to heal and regenerate. In dogs, chronic liver inflammation leads to life-threatening damage. The liver regenerative process relies on the critical participation of pro-inflammatory mediators which are responsible for priming hepatocytes to release growth factors necessary for regeneration. Hepatocytes, the main cellular constituent of liver, account for the regenerative capacity of liver. They also detoxify and metabolize noxious agents. We have developed a hepatocyte culture model that facilitates adhesion, proliferation, and maintenance of hepatocyte phenotype. We used this hepatocyte model to induce inflammation and oxidative stress with the cytokine interleukin-1 beta (IL-1). We then determined whether S-adenosylmethionine (SAMe) and silybin (SB), which are known to have anti-inflammatory and antioxidant properties, can attenuate this response. SAMe and SB are commonly used to support liver function and are found in commercially available Denamarin®. Both SAMe and SB have been reported to promote protein synthesis in hepatocytes, protect against oxidative stress by synthesizing intracellular glutathione (GSH) for detoxification, and suppress inflammation by inhibiting leukotriene production.

Primary canine hepatocytes (Celsis IVT; 5×10^5 cells/well) seeded in type I collagen-coated 6-well plates were incubated for 24 hrs with: (i) control media alone or (ii) combinations of clinically relevant concentrations of SAMe (30 or 2000 ng/ml) and SB (298 ng/ml). Cells were activated for 24 hrs with IL-110 ng/ml). Supernatants were analyzed for production of prostaglandin E2 (PGE_2) and the cytokine interleukin-8 (IL-8) by ELISA. Hepatocytes were lysed and analyzed for GSH content using a method modified from Prieto-Sagredo *et al.* (Photochemistry and Photobiology. 71:487–492. 2000). Immunostaining of cells plated on 8-well chambers (1×10^4 cells) for albumin and cytokeratin 8 (CK8) verified that they continued to express the liver phenotype. Statistical analysis was performed using one-way ANOVA and the Student-Neuman-Keuls post-hoc test with significance set at $p < 0.05$ using 2 to 3 animals.

Stimulation with IL-1 induced pro-inflammatory PGE_2 and IL-8 production while reducing GSH cellular content compared to non-activated controls. Pre-treatment with the combination of different concentrations of SAMe and SB significantly reduced PGE_2 and IL-8 production and maintained baseline GSH cellular content (all $p < 0.05$). The present study shows for the first time that pre-treatment with the combination of SAMe and SB inhibits induction of inflammation and oxidative stress. This observation supports their role in hepatoprotection. (Research supported by Nutramax Laboratories, Inc.)

ABSTRACT #161

EVALUATION OF PROMOTER ALLELES AND EXON 2 OF DLA-DQB1, -DQA1 AND -DRB1 IN DOBERMANS WITH AND WITHOUT HEPATITIS.

Dyggve H[1,3], Kennedy LJ[2], Meri S[3], Ahonen S[4], Speeti M[1]. [1]Department of Equine and Small Animal Medicine, University of Helsinki, Finland. [2]Centre for Integrated Genomic Research, University of Manchester, UK. [3]Department of Bacteriology and Immunology, Haartman Institute, University of Helsinki, Finland. [4]Department of Basic Veterinary Sciences, Department of Medical Genetics, Program in Molecular Medicine, Haartman Institute, University of Helsinki, Folkhälsan Institute of Genetics, Biomedicum Helsinki, Finland.

Doberman hepatitis (DH) is strongly associated with homozygous DLA class II haplotype DRB1*00601/DQA1*00401/DQB1*01303 and especially homozygous allele DLA-DRB1*00601. This DLA class II association clearly indicates a role for the immune system in the development of the disease. The DLA class II expression is primarily controlled at transcriptional level through the use of proximal promoters, several transcription factors and an essential co-activator, the MHC class II trans-activator (CIITA).

The aim of this study was to evaluate proximal promoters of DLA genes in Dobermans and the potential association of these promoters with DH. We included 35 DH cases and 15 controls in the study, all homozygous for DLA- DRB1*00601/DQA1*00401/DQB1*01303. The promoter regions of DLA-DRB, -DQA and –DQB genes were amplified using dog specific primers. Low levels of polymorphism were found within the DLA promoter regions, and no significant differences were noted between dogs with or without DH. However, when evaluating all the 50 homozygous DLA-DRB1*00601/DQA1*00401/DQB1*01303 Dobermans, a strong linkage disequilibrium was found between each promoter variant and exon 2 allele. In this breed, DLA-DRB1*;00601 and DLA-DQB1*01303 were clearly associated with 2 previously unrecognized promoter variants and DLA-DQA1*00401 was strongly associated with promoter allele DQAp*2.

The importance of hepatocytes presenting DLA class II antigens in connection with DH is unknown. However, an immune response is clearly maintained and modulated through the interaction between immune cells and cells of the target organ. Our results indicate that in Dobermans DLA alleles are in linkage disequilibrium with their promoters. However, in DH, the aberrant DLA class II expression by the affected hepatocytes is not due to variation in promoter sequences per se but indicates the involvement of additional factors.

Table [Abstract 162]

Diagnosis	n	Mean TB (mg/dL)	Survival (%)
Prehepatic	**8**	**4.18**	**75**
Hepatic	**70**	**3.54**	**47**
Neoplasia	15	2.45	47
Hepatitis	6	4.85	17
Hepatic lipidosis	24	4.30	29
Cholangiohepatitis	14	4.25	57
Unspecified	15	3.44	69
Posthepatic	**20**	**3.58**	**50**
Pancreatic disease	18	3.07	44
Biliary obstruction	4	5.90	75
Biliary neoplasia	2	0.45	100
Secondary	**122**	**1.21**	**65**
Sepsis /inflammation/fever	39	1.43	64
Presumed hepatic hypoxia	18	1.29	50
Feline infectious peritonitis	4	2.15	50
Miscellaneous	78	0.93	69

Note that total n is greater than 180, as some cats were diagnosed with multiple conditions. Data for cats that fit more than once within a larger category were only counted once for the purposes of the descriptive statistics.

ABSTRACT #162

FELINE HYPERBILIRUBINEMIA: A RETROSPECTIVE STUDY OF 180 CASES.

A Bradley, K McCord, J Quimby, D Twedt. Colorado State University College of Veterinary Medicine and Biomedical Sciences, Fort Collins, CO.

A retrospective study was conducted to characterize the diseases, clinicopathologic findings, and outcomes associated with hyperbilirubinemia in cats presented to Colorado State University Veterinary Teaching Hospital over a 39-month period.

Criteria for inclusion were a serum total bilirubin (TB, mg/dl) concentration > 0.2 on presentation and a complete medical record. History, physical exam findings, laboratory data, imaging, cytology, and histopathology were analyzed to identify the cause of hyperbilirubinemia.

The mean TB concentration for all 180 cats was 1.93 (range 0.3–24.9). Diagnoses were categorized as shown in the table. Non-survivors either died or were euthanized due to their disease. Patients that recovered or were lost to follow-up were counted as survivors. Overall survival rate was 62% (survivor TB = 1.64, non-survivor TB = 2.41). Survival rate was assessed for three TB levels: mild (TB < 0.5, 73% survival), moderate (TB = 0.5–1.9, 60% survival), and severe (TB ≥ 2.0, 45% survival).

Mildly elevated TB in cats with identifiable non-hepatic disease is likely clinically insignificant, whereas severely elevated TB was associated with primary causes of hyperbilirubinemia and lower survival rates.

ABSTRACT #163

EVALUATION OF A HUMAN VIRTUAL-REALITY ENDOSCOPY TRAINER FOR TEACHING EARLY ENDOSCOPY SKILLS TO VETERINARIANS.

SA Bissett, EC Hawkins, TL Hill, LA Degernes. North Carolina State University, Raleigh, NC.

Virtual-reality simulation is rapidly entering the mainstream of human medical training for teaching and competence assessment of technical skills, including endoscopy. Such trainers would be of great benefit in veterinary medicine through decreased laboratory animal use and flexible training time. However, only human virtual-reality endoscopy trainers (VRET) are available. The purpose of this study was to determine if a human VRET is an acceptable alternative to a traditional live-dog laboratory (LDL) for teaching early veterinary endoscopy skills based on feedback from trainees and comparisons of endoscopic performance between training methods.

Subjects were residents or interns who had performed < 10 endoscopic procedures. During a 4 week period, all subjects underwent: a 3-hour lecture; hands-on training with either a human VRET (AccuTouch® Endoscopy Simulator; n = 6) or a LDL (n = 6); 2 recorded esophagogastro-duodenoscopies (EGD) on live dogs for evaluation purposes; and lastly, hands-on training with the alternate method. Subjects were prospectively randomized for initial training method. Hands-on training by either method consisted of 2–3.5 hours of EGD, colonoscopy, and bronchoscopy. Subjects anonymously completed questionnaires with Likert scales prior to any hands-on training and post-training to rate aspects of their experiences. Prior endoscopy and computer game experience was also reported. Digital recordings of evaluation EGDs were subjectively assessed by 1 evaluator (SAB), blinded to training method. ANOVA was used to compare questionnaire responses and endoscopy performance, and statistical models controlled for order of training method (questionnaires) and order of evaluated EGDs (1st or 2nd), with $P \leq 0.05$ significant.

Of 12 subjects, 9 had no prior endoscopy experience and 4 had > 1 year of computer game experience. Subjects found the VRET to be less stressful than the LDL (P=0.02). Subjects initially VRET trained also found this method more useful for learning the different functions (knobs/dials) of the endoscope than subjects initially LDL trained (P=0.03). No significant differences were found for learning how to manipulate/"drive" the endoscope or perform an EGD procedure, or for rating the overall usefulness of the training method. All 12 subjects indicated the VRET was a useful and acceptable alternative to the LDL for early training. When performance was evaluated, subjects initially VRET trained had longer total procedure times than LDL subjects (P=0.02), and those with computer game experience had shorter procedure times (P=0.03). No significant differences were found for successful completion of the EGD, esophageal transit time, % time vision was maintained or the ability to stay centered in the esophageal lumen.

In conclusion, the human AccuTouch® VRET is an acceptable alternative to the LDL for early endoscopy training of veterinarians based on trainee feedback. Procedural time was shorter with LDL training, but was also shorter with prior computer game experience.

ABSTRACT #164

COMPARISON OF LAPAROSCOPIC-ASSISTED FULL THICKNESS INTESTINAL BIOPSY VERSUS ENDOSCOPIC MUCOSAL BIOPSY IN CATS: 11 CASES. Radhakrishnan, Anant. Bluegrass Veterinary Specialists, Lexington, KY.

Endoscopic mucosal biopsy is routinely performed in veterinary medicine. This minimally invasive procedure can be a very helpful diagnostic tool, however quality of samples can vary with operator experience. Furthermore, there is question regarding the reliability of mucosal samples for diagnosing intestinal disease in cats, particularly as it pertains to small cell gastrointestinal lymphoma. This study compares results obtained by gastroduodenoscopy with laparoscopic-assisted full thickness intestinal biopsy.

Eleven client-owned patients underwent gastroduodenoscopy followed by laparoscopy for evaluation of suspected gastrointestinal disease. Mucosal samples of the stomach and duodenum were obtained first. A two-port laparoscopic-assisted technique was used to obtain a full thickness jejunal biopsy. Mesenteric lymph node biopsy was obtained if lymphadenopathy was identified and samples of liver, spleen or other internal structures were obtained as clinically indicated.

Comparison of the full-thickness samples with endoscopic samples yielded comparable to identical results in 7 cases. In 2 cases, a more severe inflammatory description of a non-malignant process was obtained from the full-thickness samples. Two other patients were diagnosed with inflammatory bowel disease from the mucosal sample but with lymphoma by the full thickness, laparoscopically obtained sample. Of the 11 cases, four were diagnosed with lymphoma with two cases having identical results between the mucosal and full thickness samples.

Laparoscopic-assisted full thickness intestinal biopsy may be preferred over endoscopic biopsy for diagnosing intestinal disease in cats. Further study is needed to assess the consistency of endoscopic samples with full thickness samples for the diagnosis of intestinal lymphoma.

Previously presented at the 6th Annual Veterinary Endoscopy Society, Cancun, Mexico, March 2009.

ABSTRACT #165

ULTRASONOGRAPHIC EVALUATION OF VINCRISTINE-INDUCED GASTRIC HYPOMOTILITY AND THE PROKINETIC EFFECT OF MOSAPRIDE CITRATE IN DOGS. A Tsukamoto, K Ohno, T Tsukagoshi, K Nakashima, M Takahashi, Y Fujino, K Fukushima, H Tsujimoto. The University of Tokyo, Tokyo, Japan.

In humans, vincristine induces gastrointestinal motility disorders due to its neurotoxicity. The aim of this study was to evaluate the gastric motility after vincristine administration and the prokinetic effect of mosapride citrate in these dogs. We used ultrasonographic evaluation to assess daily changes of gastric motility in dogs. The validity of this method was confirmed by its correlation with gastric emptying assessed by the ^{13}C-octanoic acid breath test in advance. Five healthy dogs were administered vincristine (Oncovin) intravenously at a dose of 0.75 mg/m^2. Motility index (MI) of antral contraction was ultrasonographically evaluated before injection and daily over a 6-day period after injection. Gastrointestinal adverse events observed during the study were recorded. A significant decrease ($P < 0.05$) in MI was observed on day 3 (6.64 ± 0.38, mean ± SD) and day 4 (8.02 ± 0.72) compared with pre-injection (10.00 ± 0.62). Gastrointestinal adverse events were observed in 4 dogs (anorexia in 3 dogs, and vomiting, diarrhea, and hematochezia in 1 dog each). When 2 mg/kg of mosapride citrate (Pronamide) was orally administered to same dogs with vincristine and throughout the subsequent 5 days, MI improved significantly on day 3 (11.20 ± 0.52) and day 4 (11.11 ± 1.65). Furthermore, gastrointestinal adverse events occurred less frequently (vomiting and hematochezia in 1 dog each). In conclusion, administration of vincristine induces gastric hypomotility in dogs. Preventive administration of prokinetics such as mosapride citrate improves the hypomotility and may be effective for reducing the adverse gastrointestinal effects of vincristine.

ABSTRACT #166

THE INFLUENCE OF ESOMEPRAZOLE AND CISAPRIDE ON GASTROESOPHAGEAL REFLUX DURING ANESTHESIA IN DOGS. [1]AC Zacuto, [1]SL Marks, [2]J Osborn, [1]KL Douthitt, [1]KL Hollingshead, [1]K Hayashi, [3]PC Belafsky. [1]University of California, Davis, School of Veterinary Medicine, [3]University of California, Davis, School of Medicine, and [2]Sandhill Scientific, Inc., Highlands Ranch, CO.

Gastroesophageal reflux (GER) associated with general anesthesia is one of the most common causes of canine esophagitis and stricture. The ability of proton pump inhibitors and prokinetics to reduce intra-operative gastric and esophageal acid exposure is uncertain. The objectives of this study were to evaluate the effects of esomeprazole and cisapride on intraoperative gastric and esophageal pH, and on the frequency of acid and non-acid GER episodes.

Dogs undergoing elective orthopedic surgery procedures were randomized to receive either: 1) IV esomeprazole at 1 mg/kg 18 hours and 1-hour prior to induction; 2) IV esomeprazole and IV cisapride both at 1 mg/kg 18 hours and 1-hour prior to induction; or 3) saline placebo. Immediately following induction, a pH/impedance probe was endoscopically placed in the esophagus with the pH sensor 6 cm proximal to the gastroesophageal junction. The probe was attached to a recorder and data was downloaded onto a computer for interpretation and analysis.

Forty-five dogs were randomized. Mean (± SD) age of the cohort was 4.91 ± 3.13 yrs, and mean body weight was 26.74 ± 12.59 kg. The most common surgical procedures were extracapsular cruciate ligament stabilization (31%), tibial plateau leveling osteotomy (20%), and elbow arthroscopy (13%). There was no difference between age, gender, or body weight among groups ($p > 0.05$). The mean (± SD) gastric pH was 4.55 ± 1.98 for the placebo group, 7.39 ± 1.61 for the esomeprazole group, and 6.67 ± 1.30 for the esomeprazole plus cisapride group ($p < 0.05$). The mean (± SD) esophageal pH was 5.77 ± 1.54 for the placebo group, 6.41 ± 0.93 for the esomeprazole group, and 6.20 ± 0.84 for the esomeprazole plus cisapride group ($p > 0.05$). The mean $\pm$ SD number of reflux episodes was 0.71 ± 1.05 for the placebo group, 0.50 ± 0.94 for the esomeprazole group, and 0.29 ± 0.61 for the esomeprazole plus cisapride group ($p > 0.05$). The mean ($\pm$ SD) percentage of time that the esophageal pH was < 4 was $12.63 \pm 29.9\%$ for the placebo group, $1.65 \pm 6.17\%$ for the esomeprazole group, and $0.00 \pm 0.00\%$ for the esomeprazole plus cisapride group ($p > 0.05$).

Pre-operative intravenous esomeprazole and esomeprazole plus cisapride significantly increases intraoperative gastric pH. Trends suggest that pre-operative esomeprazole and esomeprazole plus cisapride may decrease intraoperative gastroesophageal reflux. The addition of cisapride to esomeprazole does not appear to offer additional benefit.

ABSTRACT #167

PLASMA CITRULLINE CONCENTRATION IS DECREASED IN CANINE PARVOVIRUS ENTERITIS.

O. Dossin[1], S.I. Rupassara[2], J.P. Schoeman[3], D. Williams[1] and P. Garlick[2]. [1]Department of Veterinary Clinical Medicine and [2]Department of Animal Sciences, University of Illinois, Urbana-Champaign, IL and [3]Department of Companion Animals Clinical Studies, University of Pretoria, South Africa.

Citrulline is an amino acid, synthesized almost exclusively by the enterocytes from luminally-absorbed glutamine. Several studies have shown that fasting plasma citrulline concentration is correlated to the functional length of the small intestine in human patients with short bowel syndrome, and to the severity of disease in patients with celiac disease or radiotherapy-induced gastrointestinal damage. To the best of our knowledge, plasma citrulline has never been studied in dogs with gastro-intestinal disease. Therefore the objective of this study was to document plasma citrulline concentration in canine parvovirus enteritis, a spontaneous canine model of acute and severe intestinal failure. Moreover, plasma citrulline concentration was also compared between dogs that died and dogs that survived.

The study enrolled 61 dogs with acute parvovirus enteritis and 14 age-matched, healthy control dogs. Parvovirus infection was confirmed by fecal electron-microscopy. The group of parvovirus infected dogs was subdivided into dogs that died ($n = 12$) and survivors ($n = 49$). An EDTA blood sample was collected from each dog at admission. The plasma was immediately separated by centrifugation at $+4°C$ and stored at $-80°C$ until measurement of citrulline. Citrulline was measured by liquid chromatography coupled to a tandem mass spectrometry (LC-MS/MS) after deproteinization with acetonitrile and derivatization with propyl chloroformate. Plasma citrulline concentration was compared between the three groups with a Kruskal-Wallis test and the *post-hoc* Dunn's test.

Plasma citrulline concentration was severely decreased in parvovirus-infected dogs (median 2.7 mol/L, range 0.5 to 28.9 mol/L) when compared to control dogs (median 53.8 mol/L range 11.4 to 96.1 mol/L, $p < 0.001$). The plasma citrulline concentration at admission was not statistically different ($p = 0.059$) between the dogs that died (median 2.1 mol/L; range 0.5–6.4) and survivors (median 2.8 mol/L; range 0.29–49.05). However, this statistical comparison was underpowered by a *post hoc* calculated power of 0.3.

This study is the first to report citrulline as a possible marker of enterocyte metabolic activity in dogs in a spontaneous model of acute intestinal failure. Further investigations of plasma citrulline as a possible marker of global enterocyte metabolic failure in other disease conditions should be considered. Such information may be of prognostic value in dogs with parvoviral and other enteropathies and may also help validate the value of therapeutic interventions such as enteral nutrition to provide substrate to the enterocytes in metabolic failure.

ABSTRACT #168

BREED PREDISPOSITIONS FOR SEVERE HYPOCOBALAMINEMIA AND RELATION TO FOLATE CONCENTRATION IN DOGS WITH GASTROINTESTINAL DISEASE.

JRS Dandrieux, P-JM Noble, LJ Halladay, AJ German. University of Liverpool, School of Veterinary Science, Neston, Wirral, UK.

Cobalamin is a hydrophilic vitamin involved in numerous metabolic processes. Marked hypocobalaminemia ($B12_{low}$) is observed with a variety of gastrointestinal disorders resulting in malfunction of multiple organ systems. This retrospective study was conducted to identify breed predispositions for $B12_{low}$ (< 100 ng/l) in dogs, with signs of gastrointestinaldisease, tested at the Comparative Gastroenterology Laboratory of the University of Liverpool, UK.

Entries from 1990 to 2002 were analyzed. Dogs with TLI ≤ 5.0 g/l and breeds with less than thirty different samples were excluded. Risk for $B12_{low}$ was determined by calculating odds ratio (OR) and 95% confidence intervals (CI) for each breed against the rest of the database. Significance was assessed using 2-sided Fisher's exact test ($P < 0.05$). Folate (< 3.5 g/l vs. ≥ 3.5 g/l) was then assessed in breeds with an increased risk of $B12_{low}$; proportions were compared using Fisher's exact test, with a modified Bonferroni correction for multiple comparisons.

A total of 13,069 records were reviewed and 9,997 dogs, 40 breeds, were considered for analysis after exclusion of EPI cases. $B12_{low}$ was present in 1,431 dogs (14.3%) with the following breeds identified to be at higher risk: Shar-Pei (SP; $n = 32$, OR: 18.2, CI: 7.9–47.0, $P < 0.0001$); Staffordshire Bull Terrier (SBT; $n = 123$, OR: 2.2, CI: 1.4–3.4, $P < 0.001$), German Shepherd Dog (GSD; $n = 2282$, OR: 1.5, CI: 1.3–1.7, $P < 0.0001$), and Mixed-breed dog (MIX: $n = 836$, OR: 1.2, 1.0–1.5, $P = 0.03$). In contrast, the following breeds were found to have a decreased $B12_{low}$ risk: Bullmastiff ($n = 45$, OR: 0.0, CI: 0.0–0.5, $P < 0.01$), Flat Coat Retriever ($n = 38$, OR: 0.2, CI: 0.0–1.0, $P = 0.04$), English Setter ($n = 59$, OR: 0.3, CI: 0.1–1.0, $P = 0.04$), Lurcher ($n = 62$, OR: 0.3, CI: 0.1–0.9, $P = 0.03$), Weimaraner ($n = 158$, OR: 0.4, CI: 0.2–0.7, $P < 0.001$), Old English Sheepdog ($n = 90$, OR: 0.4, CI: 0.2–1.0, $P = 0.03$), Boxer ($n = 487$, OR: 0.5, CI: 0.4–0.7, $P < 0.0001$), and Golden Retriever ($n = 684$, OR: 0.6, CI: 0.4–0.7, $P < 0.0001$). Concurrent low folate and $B12_{low}$ were observed in 0% of SPs, 2% of GSDs, 6% of MIX, and 4% of SBTs. Low folate without $B12_{low}$ was observed in 0% (SPs), 8% (GSDs), 12% (MIX), and 13% (SBT). SPs had significantly fewer cases with low folate concentrations than the other breeds predisposed to $B12_{low}$ ($P < 0.0001$ for both), whilst GSDs had significantly fewer low folate results than SBTs ($P < 0.0065$) and MIX ($P < 0.0001$).

This study shows that breed differences exist in the risk of developing $B12_{low}$. The SP is at highest risk, as previously demonstrated in North American dogs. The absence of concurrent low folate concentrations would be consistent with a carrier mechanism defect or disease process localized in the distal small intestine, rather than diffuse malabsorptive disease. Interestingly, other breeds not previously reported were also predisposed to $B12_{low}$ (SBT, GSD, and MIX). Further studies are warranted better to understand the underlying mechanisms for these different breed predilections.

ABSTRACT #169

EFFICACY OF ORAL FAMOTIDINE AND TWO OMEPRAZOLE FORMULATIONS FOR THE CONTROL OF INTRAGASTRIC pH IN DOGS.

K. Tolbert[1], S. Bissett[1], A. King[1], G. Davidson[1], M. Papich[1], L. Degernes[1], E. Peters[2]. [1]North Carolina State University, Raleigh, NC. [2]GlaxoSmithKline, Early Development Group, Biopharmaceutics Division, RTP, NC.

Omeprazole, a benzimidazole proton pump inhibitor, raises intragastric pH by binding to the parietal cell H^+,K^+-ATPase pump. In people, optimal gastroduodenal ulcer healing occurs at an intragastric pH of ≥ 3 and omeprazole has been shown to suppress gastric acid secretion more consistently than histamine-2 receptor antagonists. In dogs, effective clinical doses of anti-secretory drugs have not been established, but clinical experience suggests that the recommended oral omeprazole dose (0.7–1.0 mg/kg q24h) may not suppress acid secretion adequately. Additionally, only human or equine formulations of omeprazole are available for use in dogs and a comparison of their efficacy has not been conducted. The purpose of this study was to compare the gastric acid suppressing qualities of oral famotidine and two omeprazole formulations in dogs.

Six normal adult dogs were used in a randomized, crossover, open-label study. All dogs received omeprazole tablets (1.5–2.6 mg/kg q24h), omeprazole paste (Gastrogard® diluted in sesame oil at 40 mg/ml) (1.5–2.6 mg/kg q24h), famotidine tablets (1.0–1.3 mg/kg q12h), and placebo. Ambulatory pH monitoring (Bravo system®) was utilized to continuously record intragastric pH for four days (treatment days 4–7) followed by a 10-day washout period. Plasma omeprazole concentrations at 1 and 2 h post treatment (days 1, 7) were determined by HPLC with ultraviolet detection. Mean percentage of time the intragastric pH was ≥ 3 and 4 was compared among treatment groups using ANOVA. Tukey's Standardized Range test was used to determine which groups were different with $\alpha = 0.05$.

Mean percentages of time that intragastric pH was ≥ 3 and ≥ 4 were significantly ($p < 0.0001$) greater for dogs receiving omeprazole tablet and paste compared to famotidine and placebo. Famotidine and placebo were not significantly different from each other. Mean ± SD percentage of time the intragastric pH was ≥ 3 was 62.5 ± 6.1% for omeprazole tablet, 54.1 ± 8.3% for omeprazole paste, 21.5 ± 13.6% for famotidine, and 6.1 ± 16.8% for placebo. Mean ± SD plasma omeprazole concentrations on day 7 were similar in dogs receiving tablet (0.81 ± 0.67 mcg/ml) compared to paste (0.73 ± 0.63).

These results demonstrate that both omeprazole formulations provide superior acid suppression in dogs compared to famotidine or placebo. For once-daily administration, higher than commonly recommended oral doses of omeprazole is supported.

ABSTRACT #170

CAUSES, USEFULNESS OF CLINICAL INVESTIGATIONS AND SUCCESS OF ANITEMETIC THERAPY IN DOGS REFERRED FOR VOMITING. A Rosé, R Neiger. Small animal Clinic, Justus-Liebig Universität, Giessen, Germany.

The aims of this study were the following: find the most common diagnoses of dogs referred for vomiting, determine the usefulness of various diagnostic investigations and assess the effect of antiemetic therapy.

307 dogs referred for vomiting were analysed regarding signalment, history and clinical examination. Various diagnostic investigations were reviewed to assess their usefulness to reach a final diagnosis. The usefulness of a diagnostic investigation was categorized into the following four groups: enabled a diagnosis (ed); assisted a diagnosis (ad); no assistance, diagnosis reached by another procedure (dop); no assistance, no final diagnosis made (nand). The dogs were retrospectively assigned to one of the following diagnostic groups: gastrointestinal (stomach, small intestine, large intestine, nonspecific); abdominal (peritoneum, hepatobiliar, spleen, pancreas, renal, urogenital); systemic disease (metabolic, toxic, drug associated); neurologic; miscellaneous; no diagnosis. The following diagnostic investigations were looked at: haematology/biochemistry (CBC/BC), radiography (RX), ultrasound (US), urinalysis (UA), endocrine testing (ET), faecal analysis (FA), fine needle aspiration (FNA) and histopathology (Histo). Success of antiemetic therapy was grouped according to drug used and whether emesis stopped within 24 hours.

In 291 dogs (94.8%) a diagnosis was found and was classified into the following groups: gastrointestinal (43.1%), abdominal (26.2%), miscellaneous (19%), systemic (9.7%) and neurologic (4.8%). Usefulness of clinical investigations is shown in table 1. 65 dogs (21.1%) were treated with antiemetics (maropitant n = 50; metoclopramid n = 5). Therapy with maropitant and metoclopramid were successful within 24 hours in 24 dogs and 0 dogs, respectively. In 10 dogs therapeutic outcome was not clearly documented.

Over 50% of dogs referred for vomiting had extragastrointestinal diseases. Of the diagnostic investigations Histo, FNA and ET were the most useful. In 21.1% of dogs referred for vomiting antiemetic therapy was clinically indicated, but was successful only in half of dogs.

Table [Abstract 170] Number of dogs (%) in which clinical investigation was useful to find a diagnosis

Test	ed	ad	dop	nand
CBC/BC (301)	57 (18.9%)	57 (18.9%)	171 (56.8%)	16 (5.3%)
RX (273)	15 (5.5%)	26 (9.5%)	219 (80.2%)	13 (4.8%)
US (212)	20 (9.4%)	53 (25%)	130 (61.3%)	9 (4.2%)
UA (112)	11 (9.8%)	24 (21.4%)	71 (63.4%)	6 (5.4%)
ET (38)	8 (21.1%)	2 (5.3%)	24 (63.2%)	4 (10.5%)
FA (54)	10 (18.5%)	5 (9.3%)	39 (72.2%)	0 (0%)
FNA (66)	24 (36.4%)	10 (15.2%)	29 (43.9%)	3 (4.5%)
Histo (33)	26 (78.8%)	2 (6.1%)	5 (15.2%)	0 (0%)

ABSTRACT #171

SAFETY AND PALATABILITY OF POLYETHYLENE GLYCOL 3350 AS AN ORAL LAXATIVE IN CATS FM Tam, AP Carr, SL Myers. Western College of Veterinary College, Saskatoon, Saskatchewan.

Recurrent constipation is a common problem in cats. These cats often require long-term laxatives as part of their medical management to improve their quality of life. Despite the common use of these medications, there is very limited data in veterinary medicine regarding the efficacy, dosage and safety of different oral laxatives. Numerous studies have shown that polyethylene glycol oral preparation (PEG 3350) is one of the most effective and safe laxatives for treating chronic constipation in humans. Side effects in humans are rare, but include life-threatening hyponatremia. PEG is readily available in a powder form, which can be added to a cat's regular food. This aim of this study was to evaluate the safety and palatability of PEG for long term use in cats.

Six (6) healthy, research cats were included in this study. PEG3350 with electrolytes was administered by adding it to a commercial diet (canned and dry) and the dosage was titrated in each cat to achieve soft, but formed stool. A complete blood count and serum biochemistry panel with electrolytes were obtained prior to administration of PEG, as well as two and four weeks after PEG administration was initiated. Food intake, fecal scores (using a standardized fecal grading scale) and weights of the cats were monitored daily.

All cats completed the study with no significant side-effects. Sporadic vomiting was observed in one cat during the treatment. Soft to very soft stools were achieved in all animals. Food intake ($p = 0.86$) and body weight ($p = 0.92$) was not significantly influenced by PEG administration. Mild erythrocytosis was found in 1 of 6 cats, before and after the treatment ($11.3–12 \times 10^{12}$ /L, reference values: 6.89–10.8×10^{12}/L). No changes in serum sodium were found in any of the cats, however, 3 cats developed mild hyperkalemia (5.6–6 mmol/L, reference values: 3.9–5.5 mmol/L). The change in potassium was not statistically significant between any time points ($p = 0.11$ and $p = 0.06$). The cause was unknown, but may have been secondary to mild non-clinical dehydration. No clinical signs of hyperkalemia were observed in any of the cats.

This study showed that polyethylene glycol is a safe and palatable oral laxative in cats for long term use. Potential side effects may include hyperkalemia. Subclinical dehydration may also occur.

ABSTRACT #172

EVALUATION OF HYPERCOAGULABILITY USING THROMBOELASTOGRAPHY (TEG®) IN DOGS WITH PROTEIN LOSING ENTEROPATHY. L.V. Goodwin, R. Goggs, D.L. Chan, K. Allenspach. Department of Veterinary Clinical Sciences, Royal Veterinary College, University of London, UK.

Protein losing enteropathy (PLE) in dogs has anecdotally been associated with an increased risk of thromboembolic disease; however there is limited published evidence to support this association. The aim of this study was to determine whether dogs with PLE are hypercoagulable and if present, to characterize the nature of the hypercoagulable state using thromboelastography (TEG) in combination with conventional coagulation tests.

Fifteen client-owned dogs with PLE were prospectively enrolled. PLE was diagnosed on the basis of a history of gastrointestinal disease, panhypoproteinemia, exclusion of hepatic dysfunction and proteinuria and histopathological confirmation of a disease process associated with PLE. Dogs with PLE were scored using the canine chronic enteropathy clinical activity index (CCECAI). Assays included prothrombin time (PT), activated partial thromboplastin time (aPTT), D-dimers, fibrinogen, antithrombin (AT) activity and recalcified, unactivated TEG. Values for reaction time (R), clot formation time (K), clot formation angle (α) and maximum amplitude (MA) were recorded from TEG tracings. TEG tracings were compared with those obtained from 30 healthy control dogs (HC) of various breeds. A subset of 9 PLE patients were reassessed 4–24 days after initiation of immunosuppressive treatment for IBD. Wilcoxon Signed Rank tests were used to compare variables from PLE dogs with those from HC and to compare pre-and post treatment values in dogs with PLE. P values < 0.05 were considered significant.

When compared to HC, all dogs in the study were significantly hypercoagulable with decreased R (PLE: median 7.8, range 2.4–11.2;

HC: 14.1, range 9.1–20.3), decreased K (PLE: median 2.5, range 0.8–5.2, HC: 8.3, range 4.3–13.1), increased α (PLE: median 56.7, range 38.5–78.3, HC: 25.6, range 17–42.4), and increased MA (PLE: median 68.2, range 54.1–76.7, HC: 44.1, range 33.5–49) (all values $p < 0.001$). AT activity was borderline low in PLE dogs (median 65% (50–121%), reference range 65–145%), however, mean serum albumin concentration was severely decreased, mean 1.67 g/dl $\pm$ 0.5 (reference range 2.8–3.5 g/dl). Despite significant improvements in serum albumin (2.31 g/dl $\pm$ 3.5, $p = 0.01$) and CCECAI (pre-treatment: median 12 (6–17), post-treatment: 3 (0–8), $p = 0.003$), all 9 dogs re-examined remained hypercoagulable based on TEG.

Our data demonstrates that dogs with PLE are in a hypercoagulable state, which cannot be solely attributed to loss of AT but may be multifactorial in origin. Despite a good clinical response to treatment, dogs with PLE remained hypercoagulable and may therefore continue to be predisposed to thromboembolic complications even after clinical improvement.

ABSTRACT #173

COMPARISON OF TWO CANNED DIETS DESIGNED FOR THE MANAGEMENT OF FELINE DIARRHEA.

H. Xu, D. Laflamme, C. Cupp, Z. Ramadan, G. Long. Nestle Purina PetCare, St. Louis, MO.

Chronic diarrhea is a common problem in pet cats. In prior studies, more than half of cats with chronic diarrhea improved when fed an appropriate dry diet. The objective of this cross-over study was to evaluate the efficacy of a new canned diet relative to the leading veterinary-prescribed diet formulated for cats with chronic diarrhea.

Sixteen adult cats with chronic large and/or small bowel diarrhea were divided into two groups, and randomly assigned to test diet A[a] or B[b]. All cats were fed a common canned diet during baseline evaluations, then were fed their assigned test diet for 4 weeks. Following this, each cat was switched to the alternate test diet for an additional 4 weeks. No additional treatment for diarrhea was allowed during the study. Fecal scores (7 = very watery; 1 = extremely dry and firm) were analyzed daily during the last week on each diet. Serum biochemistry was assessed at baseline and at the end of each feeding period. Fresh fecal samples, collected during the final week of each period, were analyzed via PCR for Clostridia, Bifidobacteria, and Lactobacillus. Data were analyzed using repeated measures ANOVA and are presented as means +/− standard error.

Fifteen cats completed all phases of the study. Relative to baseline (5.3 +/− 0.37), fecal scores improved for both groups of cats, with significantly greater improvement seen in cats when fed diet B (4.4 +/− 0.37 v 3.6 +/− 0.37 for diets A and B, respectively; $p < 0.01$). Normal stools (fecal score 2 or 3) developed in 34.8% of cats fed diet B compared to 14.7% of cats fed diet A ($p < 0.05$; odds ratio 3.1). Fecal microflora did not differ by diet, although there was a significant diet $\times$ period interaction where *C. perfringens* was lower in cats fed diet B during the second period. Numerous serum values differed significantly after cats were fed diet A versus diet B: serum Ca, P, Mg, K, albumin, AST, cholesterol, total protein, globulin, T3 and urea nitrogen were lower in cats fed diet A, while serum glucose was greater in cats fed diet A.

In conclusion, in this study of cats with refractory diarrhea, a significant improvement in fecal quality was seen when cats were fed a therapeutic diet designed for cats with diarrhea. Improvement was significantly greater in cats fed diet B compared to diet A. Dietary changes should always be considered for cats with chronic diarrhea.

[a]Hill's Prescription Diet Feline i/d, canned, Colgate-Palmolive Co.

[b]Purina Veterinary Diet EN Gastroenteric Feline Formula, canned, Nestlé Purina Petcare

ABSTRACT #174

DETERMINATION OF THE DOSAGE REGIMEN OF TYLOSIN IN THE TREATMENT OF CANINE TYLOSIN-RESPONSIVE DIARRHEA.

S Kilpinen[1], T Spillmann[1], E Westermarck[1]. [1]Department of Equine and Small Animal Medicine, Faculty of Veterinary Medicine, University of Helsinki, Finland.

Tylosin-responsive diarrhea (TRD) refers to canine diarrhea that responds to tylosin in a few days, but reappears in many dogs a short time after discontinuing the treatment. The daily dosage recommendations for tylosin in canine enteropathies vary in textbooks from 30 to 80 mg/kg. A total of 17 client-owned dogs with diagnosed TRD responding to tylosin at a dose of 25/mg/kg q 24 h were enrolled in the study. The aim of our randomized, blinded prospective clinical trial was to assess if an even lower dose of tylosin is sufficient to stop the diarrhea. At the beginning of the trial all the dogs were taken off tylosin and had normal fecal consistency. A follow-up period of a maximum of two months commenced to determine whether diarrhea would reappear. The owners evaluated and marked down the fecal consistency of each stool according to previously published guidelines using a nine-point scale of one to five with half point intervals. No additional medical treatment or dietary changes during the trial occurred. During the two month follow-up 15/17 of the dogs developed diarrhea, reaching a fecal consistency of at least four on the scale. After onset of diarrhea, the dogs were randomly assigned to two different dosage groups receiving tylosin orally either 5 mg/kg ($n = 8$) or 15 mg/kg ($n = 7$) q 24h for seven days. The owners were blinded to the dosage. Treatment outcome was evaluated as the mean of fecal consistency scores assigned during the last three days of the treatment period. A responder was defined with a mean value of ≤ 3, a non responder of > 3. Of the dogs 8/8 responded to the 5 mg/kg dose, whereas 6/7 responded to the 15 mg/kg dose and one did not. It is noteworthy that 93% (14/15) of the dogs responding to tylosin at a dose of 25 mg/kg q 24h responded also to the 5 mg/kg or 15 mg/kg q 24h tylosin treatment. The study indicates that a suitable dose of tylosin for the treatment of canine TRD is thus less than the current recommendations in textbooks and could be as low as 5 mg/kg once a day.

ABSTRACT #175

CD11c POSITIVE DENDRITIC CELLS ARE SIGNIFICANTLY DECREASED IN THE DUODENUM OF DOGS WITH INFLAMMATORY BOWEL DISEASE.

A Kathrani[1], A House[1], D Werling[2], K Smith[2], O.A Garden[1], B Simbi[3], K Allenspach[1]. [1]Department of Veterinary Clinical Sciences, [2]Department of Pathology and Infectious Diseases and [3]Department of Veterinary Basic Sciences, Royal Veterinary College, University of London, UK

Dendritic cells are thought to play a vital role in the regulation of gut immunity and maintenance of tolerance to commensal microflora and dietary antigens. Disturbances in dendritic cell number and function may therefore be implicated in the breakdown of gut homeostasis seen in inflammatory bowel disease (IBD). In mice and humans, CD11c serves as a marker for dendritic cells, and cells expressing this marker have been shown to have similar morphological and functional characteristics in the canine system. The aim of this study was to quantify the expression of CD11c positive cells in the duodenum of dogs with and without IBD using immunofluorescence.

Endoscopic biopsies from the duodenum were obtained from 12 dogs diagnosed with IBD. Full thickness biopsies from the duodenum were taken from ten healthy beagle dogs upon euthanasia as controls. All biopsies were immediately frozen in liquid nitrogen and stored at −80°C until used. The fresh frozen sections were blocked with rabbit serum and treated with primary antibody (monoclonal anti-canine CD11c) followed by incubation with secondary antibody (FITC labeled rabbit anti-mouse IgG $F(ab)_2$ antibody). Cells positive for fluorescent staining were quantified in the duodenal sections. Counts were made in six areas measuring 400 μm by 300 μm (central villus tip, central mid-villus, left villus tip, left mid-villus, right villus tip and right central villus) in healthy and diseased sections and the total number of cells expressed in 720,000 μm^2. In addition, counts were also made in three additional areas (central crypt base, left crypt base and right crypt base) in the control sections so that a comparison could be made between the number of CD11c positive cells in the crypt and villus in dogs without IBD. Statistical comparison between areas in normal duodenal sections was made using Kruskal-Wallis test and a comparison between cases with and without IBD was performed using Mann-Whitney U test.

The number of CD11c positive cells in the duodenum of dogs with IBD was significantly reduced compared to controls ($p < 0.000$, median healthy controls: 61.5 cells/720,000 μm^2, IBD: 18 cells/720,000 μm^2). In the control dogs, CD11c positive cells were exclusively present in the lamina propria and there were significantly more CD11c positive cells in the crypt base compared to mid villus ($p = 0.001$) and villus tip ($p = 0.014$).

Our data suggests that an imbalance in intestinal dendritic cell populations may play a role in the initiation and/or the maintenance of chronic inflammation in canine IBD. The depletion of CD11c cells seen in the duodenum of dogs with IBD may reflect an increased migration of these cells to the local mesenteric lymph node to activate naïve T cells and this may be driving the continued T cell inflammation seen in this disease. Future studies will determine whether CD11c is useful as a marker for the diagnosis and monitoring of canine IBD.

ABSTRACT #176

OVERDOMINANT SINGLE NUCLEOTIDE POLYMORPHISMS IN THE NUCLEOTIDE OLIGOMERISATION DOMAIN TWO (NOD2) GENE ARE SIGNIFICANTLY ASSOCIATED WITH CANINE INFLAMMATORY BOWEL DISEASE.

A Kathrani[1], A House[1], D Werling[2], B Catchpole[2], A Murphy[3], A.J. German[4], H Lee[1], K Allenspach[1]. [1]Department of Veterinary Clinical Sciences, [2]Department of Pathology and Infectious Diseases and [3]Department of Veterinary Basic Sciences, Royal Veterinary College, University of London, UK, and [4]School of Veterinary Science, University of Liverpool, UK.

Nucleotide oligomerisation domain two (NOD2) is an important cytosolic pattern recognition receptor, recognizing muramyl dipeptide (MDP) from Gram-negative and Gram-positive bacteria. Although single nucleotide polymorphisms (SNPs) in several genes have been implicated to play a role in the etiology of Crohn's disease in humans, the most prominent are those in NOD2, with a relative risk of 44 and 38 for compound heterozygotes and homozygotes developing Crohn's disease respectively. Previous studies have shown up-regulation of NOD2 expression in a canine colonic epithelial cell line following challenge with various infectious stimuli, thus suggesting a potential role for this receptor in the pathogenesis of canine IBD. However, to date no studies have been performed to investigate the presence of SNPs in the NOD2 gene in dogs with IBD. The aim of this study was to investigate whether SNPs in the canine NOD2 gene are associated with IBD in German shepherd dogs (GSDs).

Four non-synonymous SNPs in exon 4 of the NOD2 gene (G1357A, C1578T, G1693C and A1885G), identified in a mutational analysis, were evaluated further in a case-control study using sequence based typing. Sequencing was performed using the ABI Prism Sequencing Kit (Applied Biosystems, UK) and analyzed using an ABI3100 automated sequencer (PE Applied Biosystems). Sequencing information from 48 GSDs with IBD from the UK were compared to a control group consisting of 55 GSDs from patients treated for non-inflammatory disease at the Royal Veterinary College, University of London (UK). In addition, genotyping of 192 non-GSD dogs were performed; 96 IBD cases and 96 non-breed matched control with no history of inflammatory disease.

All four SNPs were in complete linkage. In the GSD population, all four SNPs were found to be in Hardy-Weinberg equilibrium. When the GSD case population was compared to the GSD control group, the heterozygote genotype for all four SNPs was found to be significantly more frequent in the IBD population ($p = 0.04$, $OR = 2.34$, $CI = 1.03$–5.28). These results were also mirrored in the non-GSD breeds: The heterozygote genotype for all four SNPs was significantly more frequent in the IBD population compared to the non-GSD control population ($p = 0.0009$, $OR = 3.06$, $CI = 1.55$–6.05).

Our study suggests that the four SNPs in exon 4 of NOD2 when analyzed in an over-dominant model are significantly associated with IBD in GSDs as well as in other canine breeds with IBD. Our data may suggest a significant role of NOD2 in canine IBD. Further functional data would be needed to elucidate the exact role of NOD2 in canine IBD.

ABSTRACT #177

RELATIONSHIP OF MUCOSAL GENE EXPRESION TO MICROBIOTA COMPOSITION IN DOGS WITH INFLAMMATORY BOWEL DISEASE.

AE Jergens[1], D Nettleton[1], JS Suchodolski[2], M Wymore[1], V Wilke[3], S Dowd[4], JM Steiner[2], C Wang[1], MJ Wannemuehler[1]. [1]Iowa State University, Ames, IA; [2]GI Lab, Texas A&M University, College Station, TX; [3]University of Minnesota, St. Paul, MN; [4]Research and Testing Laboratory, Lubbock, TX.

Canine inflammatory bowel disease (IBD) is one form of chronic enteropathy thought to be caused by the interplay of genetic, immunologic, and enteric microbial factors. Although an important role for mucosal bacteria in the pathogenesis of IBD has been proposed, studies investigating the association between microbiota composition and host gene expression are lacking. Using unique molecular tools, we examined the relationship of differential gene expression (determined by Affymetrix DNA microarrays) to microbiota composition (determined by bacterial tag-encoded FLX 16S rRNA gene pyrosequencing) in intestinal tissues of healthy dogs and in dogs with IBD.

Samples of intestinal mucosa were collected by endoscopy or surgery from healthy dogs ($n = 6$) and dogs diagnosed with moderate-to-severe IBD ($n = 18$; CIBDAI ≥ 6). Total RNA was extracted and utilized for microarray analysis using the Affymetrix GeneChip® Canine Genome 2.0 Array. Pyrosequencing was performed on aliquots of cDNA from each dog.

Permutation testing yielded over 1730 probe sets differentially expressed (FDR $\leq 1\%$, expression change 2 fold or greater) between the IBD and control samples, suggesting that each canine group had a distinctive gene expression pattern. Analysis confirmed that a total of 1461 (84%) of these transcripts were down-regulated and included genes associated with protein synthesis, transcription/translation, and cell signaling. A lesser number of genes were up-regulated in IBD dogs, including genes encoding for calcium binding proteins, the matrix metallopeptidases, and immune function. Pyrosequencing yielded 152,000 16S rRNA gene sequencing tags. Intestinal inflammation in IBD dogs was accompanied by significant differences in composition of the intestinal microbiota. Specifically, the proportions of members of Bacteroidales and Clostridiales (i.e., genus *Faecalibacterium*, *Ruminococcus*, and *Dorea* within the Clostridium clusters IV and XIVa) were significantly reduced in the IBD dogs ($p < 0.01$). In contrast, proportions of Gamma-Proteobacteria (i.e., Moraxellaceae and Xanthomonadaceae) were significantly higher in the IBD dogs ($p = 0.012$).

In summary, transcriptional profiling documented differential gene expression that was associated with significant compositional differences of the intestinal microbiota in IBD dogs. Observed patterns of mucosal gene expression and altered microbial diversity in dogs are similar to those reported for humans with intestinal inflammation (i.e., IBD or infectious colitis). These data suggest that potential molecular targets for diagnostic monitoring and therapeutic intervention of disease prior to the onset chronic mucosal inflammation may exist in dogs with IBD.

ABSTRACT #178

MEASUREMENT OF IL-12 (P40, P35), IL-23P19 AND IFN-γ MRNA IN DUODENAL BIOPSIES OF CATS WITH IBD AND HEALTHY CONTROLS USING QUANTITATIVE REVERSE TRANSCRIPTASE POLYMERASE CHAIN REACTION (QRT-PCR).

N. E. WALY[1], V. Biourge[2], I. R. Peters[3], M.J. Day[3], C.R. Stokes[3], M. Bailey[3], T.J. Gruffydd-Jones[4]. [1]Department of Animal Medicine, Faculty of Veterinary Medicine, Assiut University, Egypt. [2]Royal Canin Research Centre, Aimargues, France. [3]Division of Veterinary Pathology Infection and Immunity, School of Clinical Veterinary Science University of Bristol, Langford, UK. [4]Feline Centre, Division of Companion Animal Sciences, University of Bristol, Langford, UK.

The aim of this study was to investigate the expression of genes encoding IL-12p35 and p40, IL-23p19 and IFN-γ in duodenal biopsies taken from cats with histological evidence of inflammatory bowel disease (IBD).

Cats referred to the Feline Centre, University of Bristol for chronic signs of gastrointestinal disturbance were examined and those fulfilling a set of inclusion criteria were recruited ($n = 13$).

Endoscopic duodenal biopsies were also collected from 21 healthy barrier-maintained cats which were used as controls. Formalin-fixed duodenal biopsies were processed for routine histopathology. Peroxidase immunohistochemistry was used to detect majorhisto-compatibility complex (MHC) class II expression. Quantitative RT-PCR was performed on snap-frozen duodenal biopsies to detect various cytokine gene expressions.

The majority of cats with IBD had moderate to severe colitis (histological score 2), even those cats without signs of large bowel disease. The density of cells expressing MHC class II within the lamina propria and the epithelial expression of this molecule by enterocytes were both slightly higher in cats with IBD than in healthy cats, but the difference was not significant ($p > 0.05$). IL-12p35 mRNA was significantly lower in the duodenal mucosa of cats with IBD compared with the mucosa of healthy cats ($p = 0.000$). In contrast, IL-23p19 mRNA expression was significantly higher in duodenal biopsies from IBD cats than healthy controls ($p = 0.001$). There was no significant difference in expression of IL-12p40 and IFN-γ mRNA ($p > 0.05$). The severity of histopathological infiltration suggestive of inflammation did not correlate with the levels of mRNA expression for any cytokine.

The results of this study suggest that IL-23 plays an important role in the pathogenesis of feline IBD.

ABSTRACT #179

MUCOSAL CYTOKINE PROFILING REVEALS IL-6 UP-REGULATION IN FELINE IBD AND ALIMENTARY LYMPHOMA. M Craven, RE Goldstein, RA Chiotti, A Greiter-Wilke, S McDonough, KW Simpson. College of Veterinary Medicine, Cornell University, Ithaca, NY.

Chronic enteropathies are widely thought to result from dysfunction of innate and/or adaptive mucosal responses to intestinal antigens in a susceptible individual. Lymphoproliferative disease is a well-recognised sequela to chronic intestinal inflammation in people, and though this has not been critically examined in dogs and cats, the increasing prevalence of intestinal T cell lymphoma in cats in particular could be preceded by chronic inflammation. The mucosal cytokine milieu is thought to play a pivotal role in neoplastic transformation in people, with IL6, TGFβ and TNFα implicated. There is growing evidence that the pro-inflammatory cytokine IL-6 in particular plays a central role in the intiation and maintenance of intestinal neoplasia in people. We hypothesize that intestinal cytokine responses differ in healthy cats versus those with IBD and lymphoma and that cytokine profiling might be a useful adjunct to disease diagnosis.

Endoscopic duodenal biopsies were obtained from 16 cats referred for evaluation of chronic gastrointestinal (GI) signs. Biopsies from 8 healthy DSH cats were included as controls. Tissue was collected into RNA later (cytokines), and formalin (histology and T cell receptor gamma, TCRG, gene analysis). Cytokine mRNA (IL-1α, IL-1β, IL-4, IL-6, IL-8, IL-10, IL-12p40, TGFβ, TNFα, IFNγ) was measured by probe-based real-time RT-PCR, normalised to GAPDH, and expressed as CT. TCRG gene clonality was evaluated by PCR amplification (CDR3) and PAGE on native and denatured PCR products, with samples classified as clonal (C), oligoclonal (OC), polyclonal (PC) and pseudoclonal (PS).

The histological diagnosis was lymphoma in 6 cases (4/6 confirmed as T cell by immunophenotyping), IBD in 7, and equivocal for IBD/lymphoma in 3. TCRG gene analysis in lymphoma cases was OC in 4/6, C 1/6, PC 1/6; in IBD cases was OC in 2/7, C 1/7, PC 3/7, PS 1/7; and in equivocal cases was OC in 2/3, and PC 1/3. Cytokine profiles for IL-1α, IL-1β, IL-4, IL-8, IL-10, IL-12p40, TGFβ, TNFα and IFNγ analyzed by histologic diagnosis were similar in lymphoma, IBD, and healthy cats. Amplification of IL-6 was significantly associated with the presence of GI disease ($p < 0.01$, Fisher's exact). IL-6 was quantifiable in only 2/8 healthy cats, (mean CT 13.15) versus 6/7 IBD (mean CT 3.34), 6/6 lymphoma (mean CT 2.83) and 3/3 equivocal cases (mean CT −0.32). There was no difference in IL-6 levels between IBD and lymphoma. When IBD and lymphoma were analyzed according to TCRG gene clonality there was no significant difference in IL-6 levels between OC/C and PC/PS. We conclude that mucosal IL-6 is a potentially useful adjunctive diagnostic marker for feline IBD and lymphoma. The IL-6 dominated intestinal cytokine milieu in feline enteropathies suggests that IL-6 mediated resistance to apoptosis may be associated with inflammation and neoplastic transformation, as has been recognized in people.

ABSTRACT #180

FECAL SENSITIVITY AS A TOOL TO DIFFERENTIATE BETWEEN NON-NEOPLASTIC AND NEOPLASTIC *SPIROCERCA LUPI* NODULES USING A MODIFIED CENTRIFUGAL FLOTATION METHOD. J. Christie, E. Dvir, L.L. van der Merwe. Companion Animal Clinical Studies, University of Pretoria, Pretoria, South Africa.

Spirocerca lupi (S.lupi) is a nematode with an indirect life-cycle with the dog as the predominant definitive host. Canine spirocercosis typically results in nodule formation in the caudal esophagus. It has been found that these nematodes shed eggs intermittently. Dogs may become re-infected, presenting with multiple nodules in various stages. Nodules can undergo neoplastic transformation into fibrosarcomas, osteosarcomas and sarcomas. It is reported that worms within these neoplastic nodules do not exist, are not viable or do not shed eggs. The goal of this study was to determine if fecal examination using a novel modified centrifugal flotation method was sensitive in differentiating neoplastic from non-neoplastic patients.

Feces were collected from 42 dogs which were diagnosed with spirocercosis using esophageal endoscopy at the Onderstepoort Veterinary Academic Hospital. A modified centrifugal flotation method was used as it has been shown to have the highest sensitivity and egg count. This method required centrifugation (1400 G) of the fecal solution (1 g feces mixed with 5 mls of sodium nitrate flotation solution, SG 1.22) for 10 minutes after which 0.1 ml of the supernatant was aspirated from the surface for microscopic examination. Nodules were classified as neoplastic based on histopathology ($n = 6$) and non-neoplastic based on regression in response to therapy determined by endoscopy or histopathology of the entire nodule ($n = 30$). Dogs with multiple non-neoplastic and neoplastic nodules were excluded ($n = 6$). A total egg count on the 0.1 ml supernatant was also performed on the first 27 cases (5 neoplastic and 22 non-neoplastic).

Fecal examination was positive in 67% of the non-neoplastic group which was not significantly different from the neoplastic group, where 33% were positive ($P = 0.13$, Chi Square Test). The overall sensitivity of detecting *S. lupi* eggs was 64%. The mean egg count was 1 ± 1.73 in the neoplastic group and 61 ± 95 in the non-neoplastic group ($p = 0.049$, Mann-Whitney Test).

The results show that dogs with neoplastically transformed *S. lupi* nodules do shed eggs but egg-shedding was higher in dogs with non-neoplastic nodules. This difference was not however significant. The number of eggs per case was significantly lower in the neoplastic group. This may indicate that neoplastic nodules may contain fewer worms or that these worms shed fewer eggs. The negative results in some of the non-neoplastic cases might be due to intermittent shedding and immature worms, factors which complicate such an analysis. The use of fecal examination as a tool to differentiate between dogs with neoplastic and non-neoplastic spirocercosis does not appear to be useful, although non-neoplastic cases tend to shed more eggs. The presence of multiple benign nodules concurrently with neoplastically transformed nodules in the 6 dogs (14%) which were excluded makes fecal examination unsuitable as a diagnostic tool in this group.

ABSTRACT #181

TIME-DOMAIN SIGNAL-AVERAGED ELECTROCARDIOGRAM IN HEALTHY GERMAN SHEPHERD AND BOXER DOGS. PPC Chamas[1], VMC Oliveira[1], FL Yamaki[1], MHMA Larsson[1]. [1]School of Veterinary Medicine and Animal Sciences – University of São Paulo, SP, Brazil.

Signal-averaged electrocardiogram (SAECG) is used for identification of ventricular late potentials (LP), low-amplitude electrical signals that occur in fibrous myocardium and are predictors of risk of sudden death. The objective of this study was to determine normal values of SAECG in healthy German Shepherd (GS; n = 21) and Boxer (B; n = 17) dogs. Animals were screened by physical examination, arterial pressure measurement, electrocardiography, echocardiography, thoracic radiography, laboratory tests, and 24-hour Holter monitoring. In face of normal results, dogs were submitted to SAECG performance without sedation. The parameters analyzed to detect late potentials were QRSd (width of filtered QRS complex), LAS < 40 μv (low amplitude signal in terminal portion of filtered QRS) and RMS 40 (root mean square of last 40 milliseconds of QRS complex). Statistical analyses was performed by T Student test ($P < 0,05$) to identify differences in each parameter between the two groups. Mean values of QRSd and LAS < 40 μv were not statistically different between the two breeds (GS = 77.67 ± 5.30 and 21.38 ± 2.94; B = 77.0 ± 5.17 and 20.06 ± 4.49), differently from the RMS 40, higher in German Shepherd dogs (GS = 385.8 ± 162.2; B = 287.4 ± 117.1). This difference could be explained by fibrofatty myocardial infiltration susceptibility in Boxer dogs that create fibrous areas of slow cardiac conduction and low amplitude signals appearance. Further studies of SAECG in dogs affected with arrhythmogenic and dilated cardiomyopathies are necessary to detect presence of LP and identify animals at risk for cardiac-related sudden death.

ABSTRACT #182

TIME DOMAIN HIGH-RESOLUTION ELECTROCARDIOGRAPHY IN BOXER DOGS WITH ARRHYTHMOGENIC RIGHT VENTRICULAR CARDIOMYOPATHY AND DILATED CARDIOMYOPATHY. PPC Chamas[1], FL Yamaki[1], MHMA Larsson[1]. [1]School of Veterinary Medicine and Animal Sciences – University of São Paulo, SP, Brazil.

Boxer dogs may have two forms of cardiomyopathy, the arrhythmogenic right ventricular cardiomyopathy (ARVC) and the dilated cardiomyopathy (DCM), both of them with frequent ventricular tachyarrythmias. Time domain high-resolution or signal-averaged electrocardiogram (SAECG) identifies ventricular late potentials (LP), low-amplitude electrical signals that occur in fibrous myocardium, markers of risk of sustained ventricular tachycardia and sudden death. The aim of this study was to evaluate the presence of late potentials on SAECG of Boxers with ARVC and DCM. Animals were screened by physical examination, arterial pressure measurement, electrocardiography, echocardiography, thoracic radiography, laboratory tests and 24-hour Holter monitoring. Depending on results of these evaluations, the animals were divided into three groups: normal (group 1; n = 22), ARVC (group 2; n = 20) and DCM (group 3; n = 14). All dogs were submitted to SAECG performance without sedation. Parameters analyzed to detect LP were QRSd (width of filtered QRS complex), LAS < 40 μv (low amplitude signal in terminal QRS) and RMS 40 (root mean square of last 40 milliseconds of QRS). The means and standard deviations of QRSd, LAS < 40μv and RMS 40 were, respectively, 76.09±5.75; 19.91±4.67; 343.0±171.4 (group 1), 77.75±7.46; 23.9±4.91; 245.6±232.3 (group 2) and 91.07±12.84; 29.79±10.13; 130.3±90.42 (group 3). Statistical analysis was performed by Dunnett's multiple comparison test that identified significant differences ($P < 0.05$) in all parameters between groups 1 and 3, and only in RMS 40 between groups 1 and 2. These results allow concluding that SAECG is a helpful test to detect LP in Boxers with DCM, but not in all dogs with ARVC.

ABSTRACT #183

HEART RATE VARIABILITY IN BOXER DOGS WITH ARRHYTHMOGENIC RIGHT VENTRICULAR CARDIOMYOPATHY. AA Camacho, E Zacché, FN Gava, EMG Ortiz, D Paulino Jr, JPE Pascon. Universidade Estadual Paulista – UNESP, Jaboticabal, São Paulo.

Arrhythmogenic right ventricular cardiomyopathy (ARVC) is a familial disease in Boxer dogs, first described by Harpster in the early 1980s. It is characterized by myocyte atrophy and fibrofatty infiltration in the ventricular myocardium, sometimes resulting in sudden death by ventricular arrhythmias. The involvement of the autonomic nervous system in the arrhythmogenesis is unclear. Thus, Boxer dogs affected by ARVC were submitted to analyses of heart rate variability in the time domain aiming to evaluate the influence of this disease on the autonomic nervous system of these animals.

Ten dogs were enrolled in the study. The ARVC group (n = 5) were 8 to 11 years old (mean age, 9.6 years), there were 1 male and 4 females and the range of VPCs per 24 hours was 1,006 to 8,346 VPCs/24 h. Five out of five dogs were asymptomatic and free from medical therapy. A second group (n = 5), comprised of 2 males and 3 females, 3 to 7 years old (mean age, 5.8 years), clinically normal non-Boxers dogs, with no VPCs per 24 hours, served as a control group. Chamber abnormality, valvular insufficiency or systolic dysfunction were not detected in any of this dogs. Twenty-four hours electrocardiography were obtained with a 7-lead, 3-channel electrode system and acquired with a digital recorder. Analyses of heart rate variability were performed in the domain of time. Six variables were assessed. Mean of the normal RRs, SDNN, SDANN, SDNN index, pNN50 e rMSSD. Dogs were kept in the same environment under the same conditions during the Holter registration. Data was submitted to a normality test and then analysis between the groups were performed by using unpaired Student's t-test.

In Boxer dogs affect by ARVC, mean RR (Affected, 538.4 ± 91.3; Control, 785.2 ± 62.6; P = 0.001), SDNN (Affected, 191.6 ± 98.3; Control, 316.8 ± 57.0; P = 0.03), SDNN index (Affected, 128.8 ± 76.7; Control, 266.4 ± 60.8; P = 0.01) and pNN50 (Affected, 23.3 ± 17.2; Control, 74.7 ± 4.5; P = 0.0001) were minor as compared to healthy control dogs. However, no differences were identified between SDANN (Affected, 146.4 ± 73.6; Control, 166.6 ± 25.0; P = 0.57) and rMSSD (Affected, 88.8 ± 35.0; Control, 121.8 ± 53.5; P > 0.28) of the groups.

Indexes based on the mean of individual RR intervals (SDNN, SDANN and SDNN index) represent the global variability and reflect both systems, sympathetic and parasympathetic. On the other hand, indexes based on the comparison between adjacent cycles (rMSSD and pNN50) reflect, predominantly, vagal tonus. Therefore, presented data suggest increase in the activity of the sympathetic nervous system and decrease in the activity of parasympathetic nervous system in dogs affected by ARVC, even in asymptomatic dogs. Such findings could help in the comprehension of arrhythmogenic mechanism of dogs affected by this disease.

ABSTRACT #184

SIX MINUTE WALK TEST STANDARDIZATION FOR DACHSHUND, POODLE AND LABRADOR RETRIEVER DOGS. AC Ferreira, PRR Melo, AM Mazini, VMC DE Oliveira, GT Goldfeder, DS Schwartz. FMVZ-USP (School of Veterinary Medicine, University of São Paulo) São Paulo, SP, Brazil.

Exercise intolerance is one of the most frequent complaints in dogs with heart disease, obesity and pulmonary disease, but it is difficult to quantify. The six minute walk test (6MWT) is used in humans to evaluate patient exercise capacity. It is an important component of quality of life assessment because it reflects the capacity to perform common daily activities, is easily performed, inexpensive, well tolerated and it has been adapted for dogs. The test is based on the maximum distance walked by the patient over six minutes. The purpose of this study was to standardize the 6MWT for healthy Dachshund, Poodle and Labrador Retriever dogs.

Twenty healthy dogs of each breed were enrolled based on complete clinical and cardiovascular evaluations. Morphometric variables were measured [as nose-to-tail length (L), thoracic (TC) and abdomen (AC) circumferences, rotula-to-heel distance (RH) among others]. The dogs were walked on a leash at their own pace, only guided by the conductor, for 6 minutes in a 30 meter hallway that was marked every 5 meters.

Labradors walked longer distances (604 ± 56 m; min = 520 m; max = 854 m; 95%CI 588–661; P < 0.001) than Poodles (495 ± 44 m; min = 415 m; max = 560 m; 95%CI 475–514) and Dachshunds (476 ± 42 m; min = 400 m; max = 556 m; 95%CI 459–498). In conclusion, regardless of anatomical differences between Poodles and Dachshunds, they walk similar distances. Based on a previously obtained equation (Distance = 55.3 + 8.3 TC + 0.9 L + 2.1 RH), predicted distances were overestimated for Labradors and Dachshunds, demonstrating that standardization is required for different breeds.

ABSTRACT #185

EFFECTS OF TREADMILL TRAINING OVER AUTONOMIC AND HEMODYNAMIC FUNCTIONS IN HEALTHY DOGS.
JPE Pascon, D Paulino-Júnior, E Zacché, FN Gava, EMG Ortiz, AA Camacho. College of Agricultural and Veterinary Sciences, São Paulo State University, Campus of Jaboticabal, Brazil.

Regular physical activity has been widely used in human cardiovascular therapy, promoting better autonomic control, heart function, life quality and decreasing sudden death risks. In dogs, however, there is not a standardized guideline to be used. This research has the goals of evaluating the effects of standardized treadmill training over the autonomic and hemodynamic functions of healthy dogs.

Six dogs (4 Beagles; 2 mixed breed), with mean weight of 13,1 Kg, were enrolled in this study. Twenty-four hour time domain heart rate variability (HRV), and echocardiography were analyzed before and after four weeks, five days a week, 40 minutes a day training. The intensity of training was individually determined by a maximal heart rate (MHR), observed in a maximal progressive effort test. Gradually, the intensity of training was increased in 50% of MHR, in the first week, to 60%, 70% and 80% of MHR in the second, third and fourth weeks, respectively. The paired *t* test was used to compare data before and after training. The improvement in functional capacity of the dogs was attested by a better performance on the physical test after training, achieving higher levels of intensity (9,0 km/h to 11,1 km/h; p = 0,0155) and decreasing the area under the curve of lactate (31,2 to 26,0; p < 0,0001). Increase in parasympathetic tone on HRV was verified by SDANN (155,5 ± 35,2 ms to 231,7 ± 46,7 ms; p = 0,0059), rMSSD (115,3 ± 50,4 ms to 181 ± 51,9 ms; p = 0,0118), amplitude of heart rate (188,5 ± 21,9 bpm to 200,3 ± 17,5 bpm; p = 0,0033), and by a higher percentage of respiratory sinusal arrhythmia on a maximal progressive effort test (> 50% until 2 km/h before training and 8 km/h after training). On the hemodynamic aspect, training lead to increase diastolic interventricular septum thickness (0,78 ± 0,12 cm to 0,91 ± 0,16 cm; p = 0,0294), decrease left atrial diameter (2,37 ± 0,22 cm to 2,15 ± 0,19 cm; p = 0,0369), left ventricular end-diastolic (4,47 ± 0,80 to 3,82 ± 0,62; p = 0,0068), and end-systolic (2,27 ± 0,39 to 1,88 ± 0,48; p = 0,0104) wall stress indexes, suggesting preload and afterload reduction. Improvement of diastolic function was confirmed by mitral E/A waves (1,42 ± 0,19 to 1,83 ± 0,46; p = 0,0467). No differences (p > 0,05) were detected on systolic function (EF%, SF%, left and right ejection time, left pre-ejection time, ejection index, and mean velocity of circumferential fibers shortening), left ventricular end-systolic and end-diastolic volume indexes, and on Tei index of myocardial performance. As observed in a human being, training results in better autonomic and hemodynamic control in healthy dogs. In a near future we expect that this therapeutic modality cold be helpful for cardiovascular improvement in the canine species.

ABSTRACT #186

EFFECTS OF RUNNING ON THE RENIN-ANGIOTENSIN-ALDOSTERONE SYSTEM IN DOGS.
M Mizuno, M Uechi, Y Inoue, S Kurihara, Y Kamiyama. Nihon University, Kanagawa, Japan.

Exercise and stress are important factors in the development of congestive heart failure. The present study evaluates the influence of exercise upon circulatory function and the renin-angiotensin-aldosterone system (RAAS) in healthy dogs. A placebo or benazepril hydrochloride was administered to four dogs and then heart rate and blood pressure were measured every 5 minutes for 30 minutes. Plasma renin activity, angiotensin-converting enzyme (ACE), angiotensin II (Ang II), aldosterone, adrenaline, noradrenaline and urinary aldosterone were measured in the dogs before and after running on a treadmill at 7 km/h for 10 minutes. Benazepril hydrochloride significantly (P < 0.05) decreased ACE (0.9 ± 1.0 U/l) and aldosterone (21.1 ± 16.1 pg/ml) compared with the placebo (4.0 ± 1.5 U/l, 42.9 ± 29.8 pg/ml). Plasma renin activity, Ang II, aldosterone and adrenaline levels increased during exercise. Heart rate and blood pressure significantly (P < 0.05) increased with both placebo and benazepril hydrochloride during exercise, and heart rate and blood pressure did not significantly differ between the two groups. These results suggest that the increase in heart rate and blood pressure during exercise is related to activation of the RAAS and sympathetic nervous system. Therefore, ACE inhibitors might modulate circulatory function and the RAAS during exercise.

ABSTRACT #187

OCCURRENCE OF HYPERTROPHIC CARDIOMYOPATHY IN A LARGE COHORT OF BRITISH SHORTHAIR CATS.
S. Granström[1], M. Nyberg Godiksen[2], M. Christiansen[2], J.L. Willesen[1], J. Koch[1]. [1]Department of Small Animal Clinical Sciences, University of Copenhagen, Denmark. [2]Department of Clinical Biochemistry and Immunology, Statens Serum Institut, Copenhagen, Denmark.

Familial hypertrophic cardiomyopathy (HCM) has previously been described in British Shorthair cats (BSH), but until now, no reports have been published on how prevalent the disease is within this breed. The aim of this study was to assess the occurrence of HCM in a large cohort of BSH and to evaluate the effect of gender, weight and age as potential risk factors to presence of the disease.

The study was conducted as a prospective study including all BSH presented at the Small Animal Hospital for HCM screening in the period of April 2006–August 2009. All cats were examined by the same two trained ultrasonographers using a Vivid 7 ultrasonographic system (GE Medical) with a 10 S phased array transducer (8–10 MHz). Measurements of the left ventricle were obtained by conventional 2D- and M-mode imaging of right parasternal long- and short axis views. Diagnosis of HCM was based on an overall assessment of echocardiographic findings, but cats were classified as to have a concentric hypertrophy if the interventricular septum (IVS) and/or left ventricular free wall (LVFW) measured > 5.5 mm in diastole. To rule out other causes of left ventricular concentric hypertrophy, a complete blood count, biochemical profile, thyroxin level and blood pressure were measured in affected cats. In the statistical analyses occurrence of HCM was expressed as a percentage and age, weight and echocardiographic variables as mean ± standard deviation (SD). A logistic regression analysis was used to test the effect of gender, weight and age on HCM as outcome and a p-value of < .05 was considered significant.

A total of 282 cats were examined, 189 (67.0%) females and 93 (33.0%) males. The average age of the cohort was 40 (± 29) months and the average weight was 4.5 (± 1.1) kg. Twenty-three cats (8.2%) were classified as HCM positive, 14 (4.9%) as equivocal and 242 (85.8%) as HCM negative. Three cats (1.1%) were diagnosed with other heart disease and excluded from further analysis. The average diastolic wall thickness of the IVS and LVFW in the HCM affected cats were 7.0 (± 1.2) mm and 7.1 (± 2.4) mm, respectively. In the HCM negative group the corresponding measurements were 3.9 (± 0.5) mm and 3.8 (± 0.5) mm. Male cats had a significantly higher occurrence of HCM (20.4%) compared with the females (2.1 %) corresponding to an odds ratio (OR) of 12.7 (95 % CI 4.2–38.6) for male gender (p < 0.001). No effect of weight and age on presence of HCM could be identified. Eighteen of the HCM positive cats had diffuse, symmetric hypertrophic changes of the entire left ventricle, whereas 5 had an asymmetric or regional hypertrophy of the left ventricular myocardium.

The conclusion of this study is that the BSH in our cohort had a high occurrence of HCM. Most affected cats presented with pronounced, diffuse hypertrophic changes affecting the IVS, LVFW and papillary muscles. As in many other breeds, male gender predisposed to development of the disease.

ABSTRACT #188

CLINICAL CHARACTERIZATION OF HYPERTENSIVE HYPERTROPHIC CARDIOMYOPATHY IN DOGS WITH CHRONIC KIDNEY DISEASE (CKD). D Paulino-Junior, JPE Pascon, E Zacché, FN Gava, EMG Ortiz, MG Sousa, AA Camacho. College of Agricultural and Veterinary Sciences, São Paulo State University, Campus of Jaboticabal, Brazil.

Chronic kidney disease (CKD) is characterized by progressive and irreversible lesions to the nephrons which result in renal failure and, often, in systemic arterial hypertension secondary to the chronic activation of compensatory mechanisms. Although the canine species is used as an experimental model for humans, information about cardiac function, compensatory enzymatic activation and biomarkers for myocardial lesion are scarce. The objective of this study was to clinically characterize the hypertensive hypertrophic cardiomyopathy in dogs with CKD. For this purpose, 30 dogs, males and females, with a mean age of 10 years and 16 kg of body weight, were distributed into two groups composed of healthy dogs (control – CG, n = 10) and dogs with CKD of glomerular origin (CKDG, n = 20). Both groups were evaluated using non-invasive blood pressure measurement, radiographs, electrocardiogram, echocardiogram, as well as measurement of cardiac markers as troponin I, creatine phosphokinase-MB (MB-CK) and enzymes as angiotensin converting enzyme (ACE), aldosterone and renin. The mean CG and CKDG values were compared by unpaired t-test. The highest values of systolic (CG: 138.40 ± 12.17; CKDG: 160.60 ± 24.25; p = 0.0113), mean (CG: 108.00 ± 17.29; CKDG: 126.20 ± 23.43; p = 0.0386) and diastolic (CG: 91.10 ± 19.32; CKDG: 112.20 ± 25.40; p = 0.0287) systemic arterial pressure and heart rate (bpm) (CG: 98.20 ± 27.02; CKDG: 125.10 ± 32.62; p = 0.0331) were seen in the CKDG. Left ventricular internal diameter in systole (cm/m^2) (CG: 3.70 ± 0.44; CKDG: 2.96 ± 0.92; p = 0.0240) and in diastole (cm/m^2) (CG: 5.48 ± 0.83; CKDG: 4.65 ± 1.18; p = 0.0576) were lower in the CKDG. A tendency of increase in the variables interventricular septum thickness in systole (cm/m^2) (CG: 1.72 ± 0.50; CKDG: 2.06 ± 0.42; p = 0.0632) and left ventricular mass (g/m^2) (CG: 126.80 ± 22.56; CKDG: 179.70 ± 87.22; p = 0.0723) was seen in the CKDG. The peak velocity of mitral A wave (m/s) (CG: 0.53 ± 0.10; CKDG: 0.41 ± 0.13; p = 0.0153) was lower in the CDKG, resulting in increased E/A wave ratio (CG: 1.19 ± 0.22; CKDG: 1.60 ± 0.27; p = 0.0003). Left ventricular end-systolic volume index (ml/m^2) (CG: 26.78 ± 10.48; CKDG: 17.86 ± 10.73; p = 0.0391) and mean velocity of circumferential fiber shortening (circ/s) (CG: 1.39 ± 0.40; CKDG: 1.75 ± 0.34; p = 0.0151) were respectively decreased and increased in the CKDG. Both ACE (U/L) (CG: 41.00 ± 6.25; CKDG: 81.00 ± 14.81; p = 0.0001) and renin (ng/ml/h) (CG: 0.14 ± 0.13; CKDG: 0.27 ± 0.13; p = 0.0180) were higher in the CKDG, but no differences were observed in Troponin I (ωg/L) (CG: 0.0080 ± 0.0042; CKDG: 0.1550 ± 0.3994; p = 0.2584) and MB-CD (U/L) (CG: 41.70 ± 6.30; CKDG: 48.70 ± 15.63; p = 0.1871) between groups. Therefore, dogs of the CKDG exhibited systemic arterial hypertension, reflex tachycardia, concentric hypertrophy of the left ventricle, significant increase in concentrations of renin and ACE, demonstrating the activation of the compensatory mechanisms, however, with no evident myocardial injury.

ABSTRACT #189

HEART RATE VARIABILITY IN DOGS WITH MITRAL ENDOCARDIOSIS OR NATURAL MORBID OBESITY. JPE Pascon, D Paulino-Júnior, E Zacché, FN Gava, EMG Ortiz, AA Camacho. College of Agricultural and Veterinary Sciences, São Paulo State University, Campus of Jaboticabal, Brazil.

Mitral endocardiosis (ME) and obesity are frequently observed in dogs and as in human patients, lower life time expectation and higher sudden death risks can be expected. Researches enrolling predisposed dog breed to develop ME or using experimental induction of mitral insufficiency indicated decrease in parasympathetic tonus. In the same way, obesity in dogs induced by supplementation of rich saturated fat diet seams to impair the autonomic control. Therefore, to verify in natural conditions we evaluated dogs with Ib functional class of endocardiosis or natural morbid obesity using time domain heart rate variability (HRV).

Thirty-two mature dogs (10 male; 22 female), with mean weight of 17,8 Kg, were enrolled in the study. The ME group (MEG – 5 male; 7 female) and obese group (OG – 1 male; 8 female) were compared to a control group (CG – 4 males; 7 females) composed by healthy dogs. Electrocardiograms were registered in the same day time (9 to 11am) and environment, using a three channel digital holter recorder. The variables mean NN, SDNN, SDANN, SDNNi, rMSSD and pNN50 were calculated from two hours of revised ECG. All 32 dogs were free from medical therapy, and other diseases associated with ME or obesity. Normality of data was attested by the Anderson-Darling test, and groups evaluated by ANOVA. MEG and OG were compared with CG by the Dunnett test with 95% probability. The mean values of mean NN (CG- 570 ± 51 ms/MEG - 580 ± 94 ms/OG - 650 ± 130 ms), SDNN (CG - 175,6 ± 46,8 ms/MEG - 169,5 ± 56,3 ms/OG – 192,3 ± 62,4 ms), SDANN (CG – 100,5 ± 33,9 ms/MEG – 93,8 ± 34,0 ms/OG – 100,0 ± 37,9 ms), SDNNi (CG - 139,9 ± 40,8 ms/MEG – 137,1 ± 40,1 ms/OG - 164,3 ± 58,1 ms), rMSSD (CG – 171,4 ± 51,3 ms/MEG – 170,5 ± 76,1 ms/OG – 222,1 ± 93,6 ms), and pNN50 (CG – 44,6 ± 13,7%/MEG – 51,5 ± 15,8/OG – 53,95 ± 13,5%) were not statistically different ($p > 0,05$) in comparison between the diseased dog groups (MEG or OG) and the healthy dog group (CG). Although small amounts of sudden death in initial stages of ME in dogs are described, possibly involving decrease in parasympathetic activity, our results allowed concluding that the Ib class of ME class had no influence on the autonomic nervous systems. Similarly, the natural morbid obesity in dogs had no effect on autonomic control, refusing the observations of induced obesity. These values should prove useful for future investigation about advanced phases of ME or obesity in this species.

ABSTRACT #190

COMET-TAIL ARTIFACTS IN NORMAL DOGS AND DOGS WITH CARDIOGENIC PULMONARY EDEMA. J. Pate, N. Rademacher, R. Pariaut, C. Saelinger, K. Strickland, L. Gaschen. Louisiana State University School of Veterinary Medicine, Baton Rouge, LA.

Pulmonary edema is the most common complication of congestive heart failure in dogs. Interstitial pulmonary edema, which usually precedes more severe alveolar edema, may not be radiographically detectable. In human medicine, transthoracic ultrasonography is used to detect narrow repetition artifacts, or comet-tail artifacts, that indicate the presence of edema in the lung.

The hypothesis of this study is that comet-tail artifacts can be identified using ultrasonography in dogs with cardiogenic pulmonary edema, and that these artifacts occur in greater numbers and are more widely distributed in dogs with congestive heart failure than in normal dogs.

Transthoracic ultrasonography was performed on a normal group of clinically healthy dogs and a heart failure group of dogs with radiographic and echocardiographic signs of congestive heart failure. Physical examination, two-view thoracic radiographs, two-dimensional and Doppler transthoracic echocardiogram and heartworm antigen test were performed on all dogs. During ultrasonography, dogs were gently restrained in a standing or seated position. The thorax was divided into 4 quadrants on each side, and video clips and still images of each quadrant were recorded. Distribution, characteristics and number of comet-tails were determined using Qlab software (Philips Healthcare, Bothell, WA). Statistical analysis was performed using SAS software version 9.1.3 (SAS Institute, Cary, NC).

31 dogs were recruited in the normal group and 6 dogs were included in the heart failure group. Transthoracic ultrasonography was easily and quickly performed on all dogs, with an average examination time of ten minutes or less. Comet-tails were detected in 32% of the normal dogs, with a mean of 2 tails per dog. No statistical association was found between the variables age, sex, side and distribution using a phi correlation coefficient.

In dogs with cardiogenic pulmonary edema, comet-tails occurred in greater numbers and were more widely distributed. In severe cases, comet-tail artifacts were so numerous that they became confluent and obscured the borders of the pleura. The presence of artifacts was consistent with the distribution of edema on radiographs.

Artifacts are rare and isolated in normal dogs, but numerous and widely distributed in dogs with congestive heart failure, as has been described in human patients. Further studies are needed to determine if the number and distribution of comet-tail artifacts in dogs with pulmonary edema correlates with the severity of clinical and radiographic signs.

Thoracic ultrasonography is a fast, easy, and effective method to characterize pulmonary edema in dogs with mitral valve disease. It may provide additional information in combination with traditional diagnostic tests for the assessment of dogs with heart failure, the most important being the detection of early signs of heart failure and monitoring of the response to treatment.

ABSTRACT #191

CLINICAL USEFULNESS OF MEASURING PLASMA ATRIAL NATRIURETIC PEPTIDE CONCENTRATIONS FOR ASSESSING THE SEVERITY IN DOGS WITH DEGENERATIVE MITRAL VALVE DISEASE. T Ebisawa, M Uechi, Y Ohta, S Yamano, T Mizuno, M Mizuno, K Harada, A Kasuya, T Sawada, M Endou, T Mizukoshi. Nihon University, Kanagawa, Japan.

The objective of this study was to assess the clinical usefulness of measuring plasma atrial natriuretic peptide (ANP) concentrations for assessing the severity of degenerative mitral valve disease (DMVD) in dogs. Eighty five dogs with MMVD and 16 healthy control dogs were prospectively examined from October 2009 to December 2009. ANP was measured using a chemiluminescent enzyme immunoassay (CLEIA). Results of the physical examination and echocardiogram were recorded for dogs with DMVD. ANP was significantly higher in dogs with DMVD (ISACHC Ia: 52.1 ± 45.8, Ib: 78.6 ± 55.5, II: 158.0 ± 132.7, IIIa: 306.7 ± 188.3) as compared with control dogs (12.9 ± 4.0). ANP rose significantly in the groups with progressive heart failure. Area under the receiver operating characteristic curve was 0.96 (95%CI, 0.93 to 0.99). An ANP concentration > 15.6 pg/mL could be used to discriminate dogs with DMVD from control dogs with a sensitivity of 95% and specificity of 81%. In dogs with DMVD, ANP was significantly correlated with heart rate, the left atrium/aorta ratio, and end-diastolic left ventricle internal diameters. The results indicate that ANP rose with an increase in the left heart load. The ANP may be useful for assessing disease severity in dogs with DMVD.

ABSTRACT #192

CIRCULATING NATRIURETIC PEPTIDES CONCENTRATIONS IN HYPERTHROID CATS. P. Menaut, D.J. Connolly, A. Klein, C. Pace, J. Elliott & H.M. Syme. Royal Veterinary College, London, UK.

The cardiac biomarkers N-terminal-pro-Brain Natriuretic Peptide (NT-pro-BNP) and N-terminal -pro-Atrial Natriuretic Peptide (NT-pro-ANP) have been shown to be of value in the diagnosis of heart disease and heart failure in cats. In humans, it is well known that natriuretic peptides are influenced by factors other than primary cardiovascular disease, including hyperthyroidism. The purpose of this study was to investigate natriuretic peptide concentrations in hyperthyroid cats before and after anti-thyroid treatment and to compare them to normal cats.

Hyperthyroid cats were recruited from two first opinion clinics in central London. All cats received a full physical examination, plasma biochemistry, total thyroxine (tT4), PCV and systolic blood pressure (SBP; Doppler technique) measurement. Hyperthyroid cats (tT4 > 55 nmol/l) were treated medically and/or surgically and cats achieving euthyroid status (tT4 < 40 nmol/l) were selected for inclusion in the study. A control population of old (> 9 years) normal euthyroid cats (n = 21) was also selected. NT-pro-ANP and NT-pro-BNP were assayed retrospectively on stored samples (VETSIGN Feline Cardio-SCREEN proBNP; Guilday Ltd. and proANP (1–98); Biomedica). The sample population was compared to normal euthyroid cats by using a Mann-Whitney test. Spearman's rho was determined to evaluate potential associations between heart rate, tT4, creatinine, SBP and natriuretic peptide concentrations. A Wilcoxon signed-rank test was used to compare circulating natriuretic peptides and other clinical parameters before and after treatment. Results are reported as median [25th, 75th percentile].

Eighty four cats were enrolled in the study. When compared to control euthyroid cats, hyperthyroid cats had significantly higher NT-pro-BNP (205.05 [69.7, 369.8] vs. 30 [12.2, 42.2] pmol/L, p < 0.0001) but not NT-pro-ANP concentrations. A significant association was found only between NT-pro-BNP concentration and tT4 (r = 0.29, p = 0.01) but not with any other parameter. A significant difference in NT-pro-BNP concentration (205.05 [69.7, 369.8] vs. 100.8 [58.5, 248.9] pmol/L, p < 0.0001), tT4 (152.7 [91.5, 173.3 vs. 20.6 [4.7, 19.4] nmol/L, p < 0.0001), heart rate (200 [190, 240] vs. 180 [160, 192] min^{-1}, p < 0.0001), creatinine (98.5 [80.05, 120.2] vs. 136.7 [110.5, 161] µmol/L, p < 0.0001) and PCV (35 [34, 40] vs. 33 [32, 38]%, p = 0.002) was found before and after treatment.

These findings demonstrate that hyperthyroid cats may have elevated NT-pro-BNP but not NT-pro-ANP concentrations compared to normal cats and that NT-pro-BNP but not NT-pro-ANP concentrations decrease following anti-thyroid treatment. Thyroid status should be taken into account when interpreting circulating NT-pro-BNP concentrations in cats.

ABSTRACT #193

WEEKLY VARIABILITY OF PLASMA NT-proBNP MEASUREMENTS IN CATS WITH AND WITHOUT HEART DISEASE. CA Reynolds[1], HB Kellihan[2], A Alivernini[1], GE Singletary[1], DJ Trafny[1], RL Stepien[2], MA Oyama[1]. [1]University of Pennsylvania School of Veterinary Medicine, Philadelphia, PA. [2]University of Wisconsin School of Veterinary Medicine, Madison, WI.

NT-proBNP plasma concentration is elevated in cats with heart disease and may be useful in detecting asymptomatic disease and in diagnosing congestive heart failure. Variability of NT-proBNP results could affect the diagnostic utility of the test by causing different diagnoses during different time points. We sought to evaluate the 7–10 day variability of plasma NT-proBNP in healthy cats and cats with cardiomyopathy and how this variability affected diagnosis of disease based on a previously proposed diagnostic threshold of ≥ 100 pmol/L.

Two populations of cats were recruited: healthy cats with no heart murmur, hypertension, or significant abnormalities on echocardiogram and cats with echocardiographic evidence of cardiomyopathy (CM). Plasma samples for NT-proBNP assay were obtained at baseline and repeated 7–10 days later. Cats were excluded if their clinical status or medications changed during the study period.

The study recruited 51 cats from 2 sites. The median NT-proBNP of the normal group (n = 38) at baseline (49 pmol/L; range, 49–98 pmol/L) was not statistically different from recheck (49 pmol/L [49–128]; p = 0.68). In 36 of 38 (95%) healthy cats, NT-proBNP at both sample points was < 99 pmol/L, and in 32 cats (89%) was < 50 pmol/L. The median NT-proBNP of the CM group (n = 13) at baseline (479 pmol/L [49–1211]) was not statistically different from recheck (421 pmol/L [49–965]; p = 0.20). In 10 of 13 (77%) cats NTpro-BNP was ≥ 100 pmol/L at both samples points. In 3 cats (23%) with cardiomyopathy, NT-proBNP measured < 99 pmol/L at both sample points. Diagnoses in these cats were mild focal left ventricular hypertrophy (n = 2) and excessive left ventricular moderator bands (n = 1); none had ever experienced clinical signs or required cardiac medication. The difference between the individual NT-proBNP values obtained at the two sampling points was < 10 pmol/L in 70% of cats, 11–100 pmol/L in 13%, and > 100 pmol/L in 16%. We conclude that 7–10 day variability of NT-proBNP measurements is low and unlikely to confound interpretation of NT-proBNP assay, however, cats with mild disease may demonstrate false-negative values and longitudinal studies in this population are needed.

ABSTRACT #194

USE OF ABCIXIMAB TO DETERMINE PLATELET REACTIVITY IN HEALTHY CATS. AN Magee, DF Hogan. Purdue University, School of Veterinary Medicine, West Lafayette, IN.

Cardiogenic embolism is a relatively common and devastating complication to cardiac disease in cats. While it is assumed that these cats have some sort of hypercoagulable state, none have been definitively identified although hyperactive platelets have been one suspected mechanism. The platelet glycoprotein IIb/IIIa (GP IIb/IIIa) receptor binding of fibrinogen is the hallmark of platelet aggregation regardless of agonist used. The GP IIb/IIIa antagonist abciximab (ReoPro®) has been shown to improve arterial flow in an *in vivo* feline model of arterial thrombosis, presumably due to an antiplatelet effect. Concentrations of 5 μg/ml abciximab *in vitro* achieve > 80% GPIIb/IIIa receptor blockade in humans and non-human primates. However, to date no study has been published which establishes the concentration of abciximab required to achieve > 80% GP IIb/IIIa receptor blockade in the cat.

The objective of this study was to determine the concentration of abciximab that would result in 50% inhibition of platelet aggregation in healthy cats *in vitro* and to use this response to identify cats with hyperactive platelets where there was reduced inhibition of platelet aggregation to abciximab.

Five healthy, male neutered, client-owned volunteer cats were used for the study. Blood was collected atraumatically from the jugular vein directly into 3.8% sodium citrate to a final ratio of 9 parts blood to 1 part citrate (v:v). Whole blood platelet aggregation was measured with 5 μM ADP and 1 μg/ml collagen as agonists. Maximal platelet aggregation (Ohms) was measured before the addition of abciximab (baseline) and after 10 minutes of incubation with 5 μg/ml, 20 μg/ml, 50 μg/ml and 100 μg/ml abciximab. The samples were also observed during incubation for spontaneous platelet aggregation due to the known possible outside-in signaling of abciximab.

For all abciximab concentrations, there were no significant changes noted in platelet aggregation compared to baseline. Spontaneous platelet aggregation was not seen at any concentration of abciximab.

Our results suggest that abciximab does not exhibit an antiplatelet effect in the cat even at concentrations 20 fold higher than those identified in humans. This is not unexpected given that abciximab is a monoclonal antibody directed against the human GP IIb/IIIa complex. However, our results are discordant with data from an *in vivo* feline model. This could suggest that abciximab may have antithrombotic effects outside of the inhibition of platelet-mediated effect on blood coagulation.

ABSTRACT #195

EFFECTS OF HIGH VERSUS NORMAL SALT DIETS ON CARDIOVASCULAR VARIABLES IN HEALTHY AGED CATS: A 6-MONTH STUDY.

V Chetboul[1], C Carlos Sampedrano[1], BS Reynolds[2], E Trehiou[1], P Nguyen[3], V Biourge[4], I Testault[5], H Dumon[3], J Abadie[6], HP Lefebvre[2]. [1]Cardiology, National Veterinary School (NVS), Alfort (France). [2]Clinical Sciences, NVS, Toulouse. [3]Nutrition/Endocrinology, NVS, Nantes. [4]Royal Canin, Research Center, Aimargues. [5]Atlantia, Nantes. [6]Pathology, NVS, Nantes.

Concerns have been expressed about adverse cardiovascular effects of high-salt diets. The objective of this blinded study was to compare the effects of normal- versus high-salt diets on systemic arterial blood pressure (ABP), ECG, and several echocardiographic and tissue Doppler imaging (TDI) variables in aged cats. Twenty healthy neutered cats (10.1 ± 2.4 years old) were included. Following baseline measurements, cats were randomly allocated in 2 groups according to their glomerular filtration rate, sex, age and body weight. One group was fed a high-salt diet (1.30% sodium, 2.27% chloride, 4000 kcal ME/kg) and the other one a control diet (0.35% sodium, 0.70% chloride). Systolic and diastolic ABP, heart rate, myocardial wall thicknesses, shortening fraction, left atrium/aorta ratio, and lastly, radial and longitudinal systolic and diastolic TDI variables assessed in the left ventricular free wall were measured before, and 3 and 6 months after diet implementation. Statistical analysis was performed using a general linear model.

All variables were unaffected by the diet.

In conclusion, over a 6-month period a high-salt diet has no adverse effects on ABP and cardiac function in aged healthy cats.

ABSTRACT #196

RELATIONSHIP BETWEEN PROGNOSIS AND IMMUNE RESPONSE IN DOGS AFTER MITRAL ANNULOPLASTY.

T Mizuno, M Uechi, T Ebisawa, M Mizuno, T Mizukoshi, K Harada. Nihon University, Fujisawa, Kanagawa, Japan.

We previously reported that dogs developed hypercytokinemia after mitral annuloplasty and were in a state of systemic inflammatory response syndrome (SIRS) that could have been alleviated by steroid administration before starting extracorporeal circulation (2008 and 2009 ACVIM FORUM). However, the complication rate of pulmonary edema is often high after mitral annuloplasty whereas the incidence of preoperatively elevated C-reactive protein (CRP) and white blood cell (WBC) counts is low. Such preoperative inflammation is thought to facilitate postoperative SIRS. The present study assesses the relationship between prognosis and preoperative inflammation in 20 dogs after mitral annuloplasty. Pre-surgical values for WBC, CRP, Interleukin (IL)-6 and IL-10 were obtained, and differences in preoperative clinical signs (ISACHC classification), cardiac murmur grade, LA/Ao, WBC, CRP, IL-6 and IL-10 between five dogs that died postoperatively (deceased group) and 15 that survived (surviving group) were determined. Mean age was significantly higher (12.3 ± 1.7 vs. 9.5 ± 1.8 y), and mean body weight was significantly lighter (4.3 ± 1.1 vs. 7.5 ± 2.7 kg) in the deceased, than in the surviving group. Neither WBC nor CRP significantly differed between the groups. Values for IL-6 and IL-10 were significantly higher in the deceased, than in the surviving group (IL6, 197.8 ± 198.1 vs. 4.3 ± 8.3; IL-10, 29.9 ± 47.3 vs. 2.9 ± 4.6; $p < 0.05$). These results suggest that besides age, increased IL-6 and IL-10 that indicate preoperative inflammation can help to predict postoperative prognosis.

ABSTRACT #197

ENDOGENOUS ERYTHROPOIETIN LEVELS AND IRON UTILIZATION IN DOGS WITH DEGENERATIVE MITRAL DISEASE.

S Yamano[1,2], M Uechi[1], K Tanaka[2], H Harada[1], Y Ukai[1], M Nishida[1], S Kohno[1]. [1]Nihon University, Kanagawa, Japan. [2)]Kanie animal hospital, Aichi, Japan.

We examined endogenous erythropoietin (EPO) and serum iron levels in 51 dogs with degenerative mitral valve disease (DMVD; Small Animal Cardiac Health Council (ISACHC) stages Ia, n = 19; Ib, n = 11; II, n = 14; IIIa, n = 7) and15 healthy dogs. Complete blood counts, serum creatinine, EPO, serum iron, total iron binding capacity (TIBC), unsaturated iron binding capacity (UIBC) and transferrin saturation (TSAT) were measured. Levels of EPO were significantly elevated in dogs with DMVD stages Ib, II and IIIa compared with healthy dogs (20.7 ± 8.6, 21.0 ± 9.2, and 21.5 ± 5.6 vs. 11.7 ± 2.0 mIU/mL; $P < 0.05$). Hematocrit did not differ between each stage. Red blood cell counts were significantly higher in dogs with DMVD stages Ia and Ib compared with healthy dogs (6.50 ± 0.82 and 6.52 ± 0.52 vs. 5.72 ± 0.67 × 106/μl; $P < 0.05$). Hemoglobin was significantly higher in dogs with DMVD IIIa than in healthy dogs (17.8 ± 5.71 vs. 13.9 ± 1.61 g/dl; $P < 0.05$). Serum iron concentrations were significantly lower in dogs with DMVD Ib than in healthy dogs (121 ± 29.4 vs.73.3 ± 42.6 μg/dl; $P < 0.05$) and TSAT was significantly lower in dogs with DMVD stages Ib and II than in healthy dogs (29.5 ± 8.11% and 30.6 ± 9.07% vs. 42.4 ± 6.80%; $P < 0.05$). These results suggest that mitral regurgitation modifies erythropoietin production.

ABSTRACT #198

PREVALENCE OF RIGHT TO LEFT SHUNT DUE TO PATENT FORAMEN OVALE CONCURRENT WITH PULMONARY STENOSIS IN DOGS.

Y Fujii, Y Nishimoto, T Aoki, H Takano. Azabu University, Kanagawa, Japan.

Right to left (R-L) shunt concurrent with pulmonary stenosis (PS) may cause an unexpected paradoxical embolism and worsening of cyanosis post balloon valvuloplasty due to augmented tricuspid and/or pulmonary regurgitation. The purpose of the study is to investigate the prevalence of R-L shunt due to patent foramen ovale (PFO) in dogs with PS. Clinical signs, echocardiography and CBC were evaluated in 33 dogs diagnosed with PS. Patent foramen ovale was diagnosed when intravenously injected microbubbles appeared at the left atrial level in PS dogs with intact atrial septum on echocardiogram (bubble-positive dogs). In addition, atrial septum was directly observed in 54 hearts from clinically healthy Beagle dogs who were euthanized due to other experimental purposes. Prevalence of R-L shunt due to PFO was 36.4% (12 out of 33 cases). In contrast, prevalence of PFO in healthy Beagle hearts was 0%. The pressure gradient across the pulmonary valve and relative thickness of the right ventricle (RVWd/LVPWd) were significantly increased in bubble-positive dogs compared with bubble-negative dogs. None of the dogs with mild or moderate PS (pressure gradient < 80 mmHg, n = 2) demonstrated R-L shunt. Clinical signs were frequently observed in bubble-positive dogs, although polycycemia was not a common abnormality. Tricuspid regurgitation was observed in 83% of bubble-positive dogs (48% in bubble-negative dogs). Transient worsening of cyanosis was observed in 2 out of 9 bubble-positive dogs immediately after balloon valvuloplasty. Diastolic hypertension due to thickened right ventricle and tricuspid regurgitation appeared to be responsible for the occurrence of R-L shunt due to PFO.

ABSTRACT #199

DOBUTAMINE STRESS TESTING IN PORTUGUESE WATER DOGS WITH JUVENILE DILATED CARDIOMYOPATHY. GE Singletary[1], MM Sleeper[1], LT Bish[2]. [1]School of Veterinary Medicine; [2]School of Medicine; University of Pennsylvania; Philadelphia, PA.

Juvenile dilated cardiomyopathy (JDCM) is an autosomal recessive inherited disease in the Portuguese water dog causing myocardial dysfunction that typically progresses to heart failure and death prior to four months of age. The autosomal recessive inheritance pattern suggests carrier dogs have 50% reductions in the mutant protein, however these dogs do not exhibit any cardiovascular abnormalities and lead normal life spans. Dobutamine stress testing (DST) can identify the extent of myocardial contractile reserve, and has been utilized to predict systolic function improvement with medical therapy in humans, dogs and mice with dilated cardiomyopathy. This study sought to investigate whether DST could identify subclinical myocardial dysfunction in carrier dogs with the genetic mutation for JDCM and to determine if affected dogs demonstrate reduced contractile reserve prior to development of clinical disease.

Thirteen age-matched (mean 8.7 weeks old) juvenile Portuguese water dogs from an existing colony were evaluated including 5 normals, 5 carriers and three affected dogs. Echocardiograms were performed at baseline and at incremental increases of dobutamine infusion ranging from 5–20 mcg/kg/min. Echocardiographic parameters assessed included measurements of left ventricular internal dimensions and wall thickness, flow velocities, as well as systolic and diastolic function parameters.

The only absolute variables which were significantly different between normal and carriers were pre-ejection period (PEP)/ejection time (ET) at 10 mcg/kg/min, and ET at 20 mcg/kg/min. Assessment of percentage change in echocardiographic variables at increasing dobutamine rates compared to baseline revealed that normal dogs had a significant response to dobutamine (fractional shortening (FS), heart rate) beginning at 10 mcg/kg/min, while the carrier and affected dogs did not have a significant FS response until 15 mcg/kg/min. Additionally, FS was lower for affected dogs compared to carrier and normal dogs at 10 mcg/kg/min, but not at other infusion rates.

Taken together, these findings suggest contractile reserve is maintained in dogs which carry and dogs which are affected with JDCM. However, compared to normal dogs, carriers and affected animals appear to require a higher dose of dobutamine to demonstrate a comparable increase in FS from baseline. To the authors' knowledge, this study is the first to present DST findings from normal puppies (all previous studies have reported on adult dogs). DST was capable of identifying affected, subclinical JDCM dogs, but not carriers. Differences between normal dogs and carriers were extremely subtle (and likely not clinically relevant), suggesting the causative mutation affects a protein, which if expressed at 50% normal levels, enables the affected animal to maintain normal cardiac function and a normal response to dobutamine. This finding will be important for researchers who continue to search for the causative mutation.

ABSTRACT #200

PHARMACOKINETICS OF ORAL IVABRADINE IN HEALTHY CATS. SC Riesen[1], KE Schober[1], KJ Lindsey[1], CA Carnes[2], W Ni[2], MA Phelps[2]. [1]The Ohio State University, College of Veterinary Medicine and [2]College of Pharmacy, Columbus, OH.

This study was designed to determine the pharmacokinetics of oral ivabradine (Procoralan), a negative chronotropic drug, in healthy cats after administration of a single dose and repeated doses.

Eight healthy female spayed experimental cats were used in this study. Ivabradine (0.3 mg/kg) was administered as a single dose on Day 1, followed by twice daily dosing on Days 2 through 5. Plasma samples were collected at the following time points on Day 1: predose, 0.25, 0.5, 1, 2, 4, 8, 12, and 24 hours after administration. Additional plasma samples were collected on Day 4 at pre-dose and 4, 8, 12, 24, and 36 hours after administration. Plasma was collected, frozen and analyzed by liquid chromatography-mass spectrometry-mass (LC/MS/MS). Both, the parent compound (S16257) and its major metabolite (S18982) concentration were measured. Individual plasma concentrations versus time from each cat were used in a noncompartmental analysis using the Phoenix WinNonlin 6.1 software.

Pharmacokinetic parameter estimations include C_{max} (60.60 ± 40.15 ng/mL for parent drug and 2.89 ± 1.62 ng/mL for metabolite), CL/F (1.30 ± 0.75 L/hr/kg for parent drug and 21.87 ± 14.03 L/hr/kg for metabolite), terminal half-life (6.42 ± 1.9 hr for parent drug and 3.29 ± 1.01 hr for metabolite), area under the plasma concentration-time curve from 4 to 24 hours (AUC_{4-24}) (183.29 ± 140.05 hr*ng/mL for parent drug and 13.88 ± 12.23 hr*ng/mL for metabolite). There was no significant difference for C_{max} and AUC_{4-24} on day one compared to day four for either parent drug or metabolite.

Single and repeated oral doses of ivabradine were clinically well tolerated and resulted in a plasma half-life suitable for 12 hour dosing intervals in healthy cats.

ABSTRACT #201

THIAMYLAL ANESTHESIA REVEALS PREDOMINANT ROLE FOR THE CENTRAL MECHANISM OF RESPIRATORY SINUS ARRHYTHMIA IN THE DOG. ML Katz[1], R Klucznik[2], GA Wyse[2], DT Lewis[1]. [1]University of Massachusetts, Amherst, MA: Department of Animal Science,[2]: Department of Biology.

Respiratory sinus arrhythmia (RSA) is marked in the dog. There are thought to be central mechanisms originating in the nucleus tractus solitarius and medulla (CNS), and peripheral mechanisms from the pulmonary stretch, intrathoracic pressure, and vascular pressure receptors, all of which affect vagal and sympathetic centers.a Based on clinical observations during the apneic period of thiamylal anesthesia in dogs that revealed sinus rhythm (SR) and then RSA without respiration, this study was performed to substantiate the derivation of RSA primarily from the CNS.

Fifty-one dogs of various breeds and both sexes were studied. Instantaneous heart rate, 2-lead EKG, and thoracic movement were recorded in all animals. Impedance pneumography detected any thoracic movement which might otherwise escape detection. All dogs were sedated with acepromazine (.55 mg/kg) and oxymorphone (.05 mg/kg), after which baseline recordings were taken. All dogs were then anesthetized by IV injection of thiamylal sodium (10 mg/kg over 15 seconds) to produce apnea, and recording was continued.

In all 51 dogs before thiamylal anesthesia, RSA was exhibited and consisted of an increase of at least 4 beats per minute associated with inhalation (4–50 bpm, 26.7 ± 40.3 [sd]). Large-chested dogs showed the greatest RSA. In general, amplitude of RSA was reduced in all dogs in plane 4 of stage III anesthesia. Cardiac rate returned to pre-thiamylal levels in all dogs after they resumed breathing.

Thiamylal produced tachycardia in all dogs within 21 ± 12 seconds, which reduced rapidly and then progressed to a SR or RSA. Cardiac arrhythmias occurred in 17 dogs with 5 exhibiting PVCs and 8 showing bigeminal rhythm. These arrhythmias often lasted longer than the apneic period.

Apnea occurred in 42 dogs within 29 ± 19 seconds after induction and lasted from 10 to 105 seconds, with an average length of 37 seconds. Fifteen of these dogs did not exhibit a return of RSA during apnea. Of the resulting 27 dogs who exhibited RSA during apnea, 10 at first had an SR which progressed to RSA in apnea before inspiration occurred. The apneic period was as long as 75 seconds, and RSA during apnea ranged from 15–50 seconds.

In summary, thiamylal anesthesia causes apnea by anesthetizing the respiration center, which allows tachycardia and, in some, arrhythmia. As anesthesia waned, tachycardia progressed to a normal SR, then to RSA, then to inspiration. Transition from SR to RSA during apnea and before inspiration strengthens evidence for the respiratory center as the origin of the RSA.

1: Berntson et al, *Psychophysiology*, 30 (1993).

ABSTRACT #202

IN-HOSPITAL ELECTROCARDIOGRAPH VERSUS 24-HOUR HOLTER MONITOR FOR ASSESSING HEART RATE IN DOGS WITH ATRIAL FIBRILLATION. LM Calland, HW Green, III, CM Van DE Wiele, K Sederquist. Purdue University School of Veterinary Medicine, West Lafayette, IN.

Atrial fibrillation (AF) is characterized by frequent disorganized atrial depolarizations typically associated with less productive atrial contractions, increased ventricular response rates and decreased cardiac output. In veterinary medicine treatment is most often aimed at decreasing heart rate in order to maximize stroke volume via pharmacologic intervention. Assessment of these therapeutic interventions to control heart rate is most often performed utilizing brief in-hospital electrocardiograms. However, twenty-four hour Holter monitor testing allows assessment of the dogs' heart rate in their normal environments. This study compares in-hospital electrocardiography (ECG) and at home 24-hour Holter monitor testing for assessment of heart rates in dogs with atrial fibrillation.

Sixteen dogs (9 males and 7 females) diagnosed with atrial fibrillation were assessed over 21 visits for inclusion in this study. Diagnostic and rhythm electrocardiograms were performed followed by an at home 24-hour Holter monitor test for each visit. Results were evaluated to determine minimum, maximum and average heart rates. Average heart rate for in-hospital ECG was calculated over 6 seconds.

Structural heart disease was noted in 93.75% of the participants with dilated cardiomyopathy (37.5%) and chronic valvular disease (43.8%) most commonly noted. Patent ductus arteriosus and tricuspid valve dysplasia was noted in one dog each and lone AF was diagnosed in one dog. Four mixed breed dogs, 3 Irish Wolfhounds, 3 Great Danes, and 2 Labrador Retrievers were included. There were one American cocker spaniel, Collie, Doberman pinscher, miniature Schnauzer, and Boston terrier each. The average weight of the dogs was 42.4 kg (6.7–77.3 kg). In-hospital ECG demonstrated atrial fibrillation as the lone rhythm diagnosis in 76% of the visits whereas additional significant arrhythmias were noted in 85.7% of Holter tests. Holter reported values were significantly lower than ECG calculated values for the mean minimum (69.5 ± 26.7 bpm vs 111.2 ± 43.1 bpm, p = 0.01), mean maximum (191.8 ± 28.3 bpm vs 224.24 ± 48.20 bpm, $p < 0.001$) and mean average (111.4 ± 32.5 bpm vs 166.33 ± 35.1 bpm, p = 0.001) heart rates respectively.

We conclude that Holter monitor testing demonstrates a lower heart rate for dogs with atrial fibrillation in their normal environment compared to in-hospital ECGs. This may have significant implications as it pertains to adjusting drug therapy in these dogs. Utilizing Holter monitor testing in conjunction with in-hospital ECG gives a more complete clinical assessment of dogs in atrial fibrillation.

ABSTRACT #203

BRADYDYSRHYTHMIAS REQUIRING PACEMAKER IMPLANTATION IN CHAGAS POSITIVE DOGS. MH Rector, AB Saunders, SG Gordon, A DeMaster, MW Miller. The Michael E. DeBakey Institute, College of Veterinary Medicine and Biomedical Sciences, Texas A&M University, College Station, TX.

Chagas disease (*Trypanosoma cruzi*) is a documented etiology of myocarditis in the southern United States, sometimes causing cardiac conduction abnormalities, arrhythmias and heart failure. This descriptive study reports clinical findings and outcome in Chagas positive (CP) dogs requiring pacemaker implantation. Of the 136 dogs requiring pacemaker implantation from January 2001 to December 2009, 76 had a Chagas titer performed. Of that subgroup, seven (9%) were CP, originating primarily from Southeast Texas. Six were male. Breeds included Pug, Australian Heeler, Brittany Spaniel, Standard Poodle, Miniature Schnauzer, Cavalier King Charles Spaniel and mixed. Median age was 3 years (range, 1–10). Dogs presented for syncope (n = 4), exercise intolerance (n = 2), bradycardia (n = 1). Heart failure was not diagnosed in any dog. Bradydysrhythmias in CP dogs included high grade 2nd or 3rd degree atrioventricular block (n = 6) and sinus bradycardia with 1st degree atrioventricular block (n = 1). Ventricular arrhythmias were documented in 4. Outcomes included euthanasia (n = 4) three of which were cardiac related, sudden death (n = 1), alive (n = 1), lost to follow-up 36 months post-pacemaker implantation (n = 1). Median survival was 12 months (range, 0–36). Dogs with sudden death or euthanasia related to cardiac disease were paced for 5, 12, 36 months. In one dog, the presence of multiform ventricular arrhythmias, troponin I of 17.8 ng/mL and difficulty maintaining capture with a temporary pacemaker led the owner to elect euthanasia prior to permanent pacemaker implantation. Bradydysrhythmias can result in clinical signs requiring pacemaker implantation in CP dogs. Progressive cardiac disease or sudden death are possible outcomes in this population.

ABSTRACT #204

CONTRAST-ENHANCED ULTRASONOGRAPHY WITH SONAZOID® FOR CHARACTERIZATION OF FOCAL SPLENIC LESIONS. K Nakamura, H Ohta, M Yamasaki, M Takiguchi. Department of Veterinary Clinical Sciences, Graduate School of Veterinary Medicine, Hokkaido University, Sapporo, Hokkaido, Japan.

An ultrasound contrast agent Sonazoid® allows for vascular and parenchymal imaging in the liver and spleen. Sonazoid®-enhanced ultrasonography improves the diagnostic accuracy to differentiate benign and malignant focal liver lesions in dogs. The utility of Sonazoid® for the differentiation of splenic lesions, however, has not been determined. Therefore, the purpose of this study was to clarify the utility of Sonazoid®-enhanced ultrasonography for the differentiation between benign and malignant focal splenic lesions. In a prospective study, Sonazoid®-enhanced ultrasonography was performed for 29 dogs with focal splenic lesions. Qualitative assessment of the enhancement pattern was performed in early vascular phase, late vascular phase, and parenchymal phase. Histology (n = 24) or cytology (n = 5) identified 13 benign and 16 malignant lesions. In the late vascular phase, hypoechoic pattern was significantly correlated with malignancy ($P = 0.001$) with a sensitivity of 81.3% [95% Confidence Interval (CI), 65.7–89.5%], a specificity of 84.6% [95% CI, 65.4–94.8%]. In the early vascular phase, hypoechoic pattern was significantly correlated with malignancy ($P = 0.02$) with a sensitivity of 37.5% [95% CI, 24.7–37.5%], and a specificity of 100% [95% CI, 84.2–100%]. On the other hand, there was no significant difference between malignant and benign lesions during the parenchymal phase. The detection of hypoechoic nodule in the early and late vascular phase with Sonazoid®-enhanced ultrasonography was strongly suspicious of canine malignant splenic lesion.

of oxidation approached those of the control population. This may be explained by increased oxidation caused by DOX or, alternatively, this trend may represent a movement toward normalcy with treatment.

ABSTRACT #205

EVALUATION OF OXIDANT/ANTIOXIDANT TOTAL STATUS AND ERYTHROCYTE ANTIOXIDANT DEFENSE IN CATS WITH LYMPHOMA. CF Pinto, SRR Lucas, TR Macedo, MLG Leal, CS Mori, R Ubukata. School of Veterinary Medicine and Animal Science - University of São Paulo, Brazil.

Oxidative stress plays a role in the pathogenesis of many systemic diseases, including tumors. Lymphomas comprise one of the most common groups of tumors in cats. The aim of this study was to evaluate the oxidant/antioxidant total status and erythrocyte antioxidant defense in cats with lymphoma. Blood samples were obtained from 17 cats with lymphoma, at the diagnosis and without previous treatment and 23 healthy cats (control group). Systemic oxidative status was measured by the analysis of malondyaldehyde (MDA) and total antioxidant status (TAS) was measured by the use of a commercial kit (Randox Laboratories Ltd). In red blood cells hemolysated, were determined reduced glutathione (GSH), glutathione peroxidase (GSHpx), glutathione reductase (GSHr) and superoxide dismutase (SOD), using commercial kits.(Randox Laboratories Ltd). Kolmogorov Smirnov followed by Mann Whitney or T student tests was used to compare the groups. Median of TAS was significantly lower ($p = 0.0039$) in cats with lymphoma (1.30 μmol/L) than in healthy cats (1.57 μmol/L). There was no difference in the MDA between the groups (mean 0.658 and 0.566 μM respectively). Mean of GSH was significantly lower ($p = 0.0337$) in cats with lymphoma (24.06 mg/dL) than in healthy cats (31.02 mg/dL). There were no difference in the enzymes GSHpx, GR and SOD.

In conclusion cats with lymphoma showed decrease in the total (TAS) and erythrocyte oxidative defense mechanisms, considering that GSH is one of the most important antioxidant. More studies are necessary to verify if this fact may contribute to prognosis and response to the treatment.

ABSTRACT #206

OXIDATIVE STRESS IN DOGS WITH LYMPHOMA BEFORE AND AFTER ADMINISTRATION OF DOXORUBICIN: A PILOT STUDY. ES Lechner[1], KL Cooke[1], KC Scott[1], KA Barabas[2], DM Lurie[3], RC Hill[1]. [1]University of Florida College of Veterinary Medicine, Gainesville, FL. [2]Southwest Florida Veterinary Specialists, Bonita Springs, FL. [3]Miami Veterinary Specialists, Miami, FL.

The relationship between lymphoma, chemotherapy, and oxidative stress in dogs is poorly understood. The effects of doxorubicin (DOX) are in part due to oxidative damage, and this drug is commonly used in protocols for the treatment of canine lymphoma. With the use of thiobarbituric acid-reducing substances (TBARS) as a measure of oxidation and Trolox equivalent antioxidant capacity (TEAC) as a measure of antioxidant capacity, the purpose of this study was to evaluate oxidative stress (TEAC/TBARS) in dogs with lymphoma before and after administration of DOX, compared with healthy controls. In this prospective, case-controlled study, TBARS and TEAC were measured in 14 dogs with lymphoma before and 5, 60, 120 minutes, and 2 weeks after treatment with DOX. Results were compared with those from 13 age, breed, and sex-matched healthy control dogs receiving equivalent volumes of 0.9% sodium chloride instead of DOX. Control dogs had higher mean TBARS (2.31 ± 0.67 μmol/L, range 1.40–4.42) compared to dogs with lymphoma (1.60 ± 0.40, range 0.90–2.71; $P = 0.0002$). TBARS values within the lymphoma group ($P = 0.0006$) increased over time. There was no difference in mean TEAC between or within treatment groups over time.

Healthy dogs had increased measures of oxidation compared to dogs with lymphoma. Following DOX administration, measures

ABSTRACT #207

ERYTHROCYTE SUPEROXIDE DISMUTASE (SOD) CONCENTRATION IN DOGS WITH MULTICENTRIC LYMPHOMA. TR. Macedo; SRR Lucas; R. Ubukata; CS. Mori; LCS Rodrigues; CF Pinto. School of Veterinary Medicine and Animal Science - University of São Paulo, Brazil.

Production of reactive oxygen species is the result of aerobic metabolism for energy production. Oxidative stress is an imbalance between oxidants and antioxidant defense mechanisms in favor of the oxidants. Proteins, lipids and DNA oxidative damage can contribute to aging, mutagenesis and carcinogenesis. SOD is an antioxidant enzyme that catalyzes the dismutation of superoxide to oxigen and hydrogen peroxide, protecting cells from injury induced by free radicals. Lymphomas comprise one of the most common groups of tumors in dogs. The aim of this study was to evaluate the erythrocyte superoxide dismutase (SOD) concentration in dogs with multicentric lymphoma (ML) when compared to healthy dogs. Blood samples were obtained from 24 dogs with ML, at the diagnosis and 20 healthy dogs (control group). Dogs with signs of concurrent disease or that were previously treated with prednisone or chemotherapy were excluded from the study. SOD was measured in red blood cells hemolysated by the use of a commercial kit (Ransod - Randox Laboratories Ltd). Kolmogorov Smirnov followed by Mann Whitney tests were used to compare the groups. Median was significantly lower ($p = 0{,}0168$) in dogs with multicentric lymphoma at the diagnosis (1.751,6 U/g Hb) than in healthy dogs (2.237,3 U/g Hb).

In conclusion dogs with multicentric lymphoma showed decrease in erythrocyte SOD concentration when compared to healthy dogs. Further studies are necessary to verifiy if this alteration may contribute to inadequate erythrocyte survival and the development of anemia in canine lymphoma.

ABSTRACT #208

ALIMENTARY LYMPHOMA IN CATS: CLINICOPATHOLOGICAL AND THERAPEUTIC FEATURES OF 32 CASES (2004–2009). A Reche Júnior[1], MCC Chalita[2], L Wang[2], CAG Júnior[2], CF Santos[3], MF Freitas[2]. [1]School of Veterinary Medicine – University of São Paulo, Brazil. [2]VetMasters – São Paulo – Brazil. [3]Hemovet – São Paulo – Brazil.

Alimentary lymphoma is the most common anatomic form of lymphoma in the cat. Our purpose is report a prospective analysis of feline alimentary lymphoma, presenting clinicopathological, therapeutical and prognostic data, from 32 cats with diagnosis of alimentary lymphoma seen at a private Cat Clinic in São Paulo – Brazil, from January 2004 to July 2009.

All the 32 cats were treated with a combination of L-asparaginase, prednisone and chlorambucil. Response to treatment was defined as either complete remission (CR), partial remission (PR) or no response (NR). The following variables were evaluated for their effect on response to the treatment and survival times: age, sex, breed, infection or not by feline immunodeficiency virus (FIV), clinical signs, histologic and immunophenotype classification of the lymphoma and chemotherapy toxicity. *Fisher's exact test* was used to determine the statistical influence of these factors on the response to the treatment. *Kaplan-Meier* survival statistics were used to determine overall survival times and *Cox regression* was used to determine the influences of these variables on the duration of survival time. A P value < 0.05 was considered significant.

Diagnosis of alimentary lymphoma was confirmed on the basis of microscopic examination of histologic ($n = 27/84.4\%$) or cytologic ($n = 5/15.6\%$) specimens. Sixteen (50%) were classified as low-grade, 11 (34.4%) as high-grade and 5 (15.6%) as intermediate-grade lymphoma. Immunophenotyping was performed on 10 tumors, 6 (60%) were classified as T-cell lymphoma and 4 (40%) as B-Cell lymphoma.

Serological testing for FIV and feline leukemia virus (FeLV) was performed in all 32 cats, six cats were FIV – positive and none of 32 cats were FeLV – positive. Four of 6 FIV-positive cats (66.7%) had high-grade lymphoma and 2 (33.3%) had low-grade lymphoma.

Historical findings included weight loss (n = 32/100%), inappetence (n = 19/59.4%), vomiting (n = 19/59.4%), diarrhea (n = 21/65.6%) and polyphagia (n = 5/15.7%).

Twenty-four cats (75%) went into CR with a median remission time of 655 ± 442 days (range 36–2010 days). Five cats (15.6%) achieved PR with a median remission time of 71 ± 43 days (range 35–127 days) and 3 cats (9.4%) had no response to treatment.

FIV-negative cats had significantly longer median survival time than FIV-positive cats (687 ± 446 vs 310 ± 240 days, $P < 0.05$). Also, cats with low-grade lymphoma had longer remission time compared to cats with high-grade lymphoma (684 ± 493 vs 171 ± 158 days, $P < 0.05$).

We can conclude that using a combination of L-asparaginase, prednisone and chlorambucil to treat cats with alimentary lymphoma was safe and resulted in a good outcome in most treated cats of the present study. Moreover, FIV-infection and histologic classification of lymphoma were prognostic indicators for these cats.

ABSTRACT #209
IDENTIFICATION OF METHYLGUANINE-DNA METHYLTRANSFERASE IN RELATION TO LOMUSTINE CYTOTOXICITY IN CANINE LYMPHOMA CELL LINES.
Koji Minami, Hiroko Hiraoka, Yuka Ogawa, Takako S. Miyama, Saori Umeki, Takuya Mizuno, Masaru Okuda. Yamaguchi University, Yamaguchi, Japan.

Lomustine is an alkylating agent in the nitrosourea subclass, and used for relapsed canine lymphoma. However, lack or reduction of its effectiveness has been a problem in the clinical use. In human medicine, methylguanine-DNA methyltransferase (MGMT) has been shown to be a cause of drug resistance against the nitrosoureas. The purpose of this study is to assess MGMT activity in canine lymphoma lines in relation to lomustine cytotoxicity.

Five canine T-cell lymphoma lines (CL-1, Ema, GL-1, Nody-1 and UL-1) and a human T-cell leukemia line (Jurkat), which has MGMT activity, were used. These lines were exposed to various concentrations of lomustine with or without O^6-Benzylguanine (BG), an inactivator of the MGMT, pretreatment and the IC_{50} values were estimated by a MTT assay. The IC_{50} values of Jurkat, CL-1 and Nody-1 against lomustine without BG pretreatment showed 45.0, > 50.0 and 36.5 M, respectively, but those were significantly decreased by BG pretreatment (18.5, 11.5 and 7.5 M, respectively). In contrast, the effect of BG pretreatment was not observed in Ema, GL-1 and UL-1. MGMT activity was assessed by a fluorometric oligonucleotide assay and was detected in Jurkat, CL-1 and Nody-1, but not in Ema, GL-1 and UL-1. Moreover, real-time PCR analysis revealed that the expression of *MGMT* in CL-1 and Nody-1 was more than 100 times higher than that in the other lines.

We conclude that MGMT is associated with the lomustine cytotoxicity in canine lymphoma lines. Our results suggest that lymphoma cells possessing MGMT activity may have low susceptibility against lomustine.

ABSTRACT #210
ESTROGEN AND PROGESTERONE RECEPTORS EXPRESSION IN CANINE VAGINAL LEIOMYOMA.
CM Oliveira. School of Veterinary Medicine – University of São Paulo, Brazil.

Leiomyomas are the most common benign tumors of vaginal smooth muscle. Leiomyoma is a round or oval tumor, well defined and encapsulated that occurs as solitary or multiple mass, sessile or pedunculate with intraluminal or extraluminal growth. Research consistently links hormone levels with increased rates of tumors. The estrogen and the progesterone are known promoters of tumor growth by increasing levels of growth factors with mitogenic activity. The cross-reactivity between antibody anti-human estrogen and progesterone receptors with canine estrogen and progesterone receptors makes possible to use humanś kit for the determination of these receptors in dog's tissue. Determining the status of estrogen and progesterone receptors in tumors may be an important tool for prognostic and therapeutic purposes.

The aim of this study was to evaluate the immunoexpression of estrogen (ER-α) and progesterone (PR) receptors in vaginal leiomyoma of dogs.

Twenty-two samples of histologically-confirmed vaginal leimyomas of bitches in anestrous and late diestrous were evaluated by immunohistochemistry. Progesterone serum levels were measured by radioimmunoassay in all 22 dogs to estimate the phase of the estrous cycle. The ER-α and progesterone PR receptors were evaluated by imminohistochemistry using the manufacturés protocol. Monoclonal anti- ER-α antibody 1D5 (M7047, Dako-Cytomation, Glostrup, Denmark) at a dilution of 1:50 was used to detect ER, and monoclonal anti-PgR antibody PgR 636 (M3569, Dako-Cytomation)) at a dilution of 1:50 was used to detect PR.

Of the 22 tumors, 15 (68,2%) did not express receptors for estrogen or progesterone and 7 (31.8%) expressed receptors, 1 expressed ER-α (4.54%) and 6 (27.27%) expressed PR. Thus, ER expression was lower than PR expression in vaginal leiomyoma of bitches in anestrus and late diestrus. Considering these results, one can conclude that using antiprogestagens, as a growth tumor inhibitor, in selected cases of vaginal leiomyoma can be promising.

ABSTRACT #211
SECRETED FRIZZLED-RELATED PROTEIN 2 EXPRESSION IN CANINE THYROID TUMORS.
K. Deitz[1], M. Formanek[1], K. Metivier[2], D. Moore[1], V. Wilke[2], M. Ackermann[1]. [1]Iowa State University College of Veterinary Medicine, Ames, IA. [2]University of Minnesota College of Veterinary Medicine, St. Paul, MN.

The purpose of this study was two-fold: 1) to determine differentially expressed genes in canine thyroid tumor samples when compared to normal thyroid tissue using microarray analysis and 2) to use immunohistochemistry (IHC) to confirm differential protein expression of one gene that was greater than 10-fold upregulated via microarray analysis.

For gene expression, dogs were prospectively enrolled with informed owner consent if they had a likely diagnosis of follicular thyroid carcinoma (FTC). The affected dogs then underwent a biopsy and a 1 cm^3 piece of tissue was saved in RNAlater for further analysis. Microarray analysis (5 FTC and 4 histologically normal thyroid glands) was performed using the commercially available Affymetrix GeneChip® Canine Genome 2.0 Array. Microarray data were imported into and analyzed using Genedata Expressionist Analyst®. A standard t-test was conducted to compare, for each probe set, the mean expression for the FTC group with that of the normal thyroid using a false discovery rate of $q < 0.05^2$. A probe set was considered to be differentially expressed if it had a fold change of 2 or greater, a q-value < 0.05 and had valid expression values for all samples. There were 610 differentially expressed probe sets corresponding to 208 unique genes with characterized proteins. Eight genes had a fold change greater than 10. One of these genes was secreted frizzled-related protein 2 (SFRP2), which was greater than 10 times up-regulated in all tumor samples versus normal thyroid, and was chosen for further analysis (IHC).

For IHC, twenty archived samples of thyroid neoplasia were chosen (including 5 samples from gene expression studies) along with 4 normal samples (also from gene expression studies) and analyzed with IHC using goat anti-SFRP2 antibody. All samples were given two separate scores for distribution and intensity of staining. Thyroid tumor samples had significantly higher distribution and intensity scores (p-value < 0.01) than the normal samples, which lacked immunoreactivity for SFRP2.

Results of this study indicate that SFRP2 is up-regulated in canine FTCs versus normal thyroid tissue at both the gene and protein level and may play a role in the pathogenesis of this tumor type.

ABSTRACT #212
DETERMINATION OF L-ASPARAGINASE ANTIBODY TITERS IN DOGS BEING TREATED FOR LYMPHOMA.
Kidd JA[1], Grover EN[1], Borgheiinck PR[1], Buntzman AS[1,2], and Hess PR[1]. [1]Dept. of Clinical Sciences, North Carolina State University College of Veterinary Medicine, Raleigh, NC. [2]Dept. of Microbiology and Immunology, University of North Carolina at Chapel Hill School of Medicine, Chapel Hill, NC.

L-asparaginase is a bacteria-derived enzyme that is commonly used in the treatment of canine lymphoma. The drug depletes extracellular asparagine, causing rapid apoptosis of malignant lymphocytes unable to synthesize this amino acid. Unfortunately, clinical resistance typically develops after repeated administration. Resistance to this foreign protein may be due in part to the development of antibodies, which can reduce enzyme activity by several mechanisms. We hypothesized that dogs exposed to L-asparaginase during treatment for lymphoma would form circulating antibodies against this chemotherapeutic agent. The objective of this project was to develop an ELISA to detect such antibodies and to analyze the changes in dogs' titers after one or more doses of L-asparaginase.

Plasma samples were collected at 0, 1, 2, 4, and 8 weeks after drug administration. Ninety-six-well plates were coated with L-asparaginase or control antigen (both from E.coli). Following overnight incubation with plasma, antibodies were detected using horseradish peroxidase-conjugated polyclonal anti-dog IgG and IgM antibodies. Titers were measured as the reciprocal of the highest dilution with an optical density (OD) greater than two standard deviations above the background OD observed in plasma from naïve dogs.

Robust L-asparaginase antibody responses were observed in a majority of canine lymphoma patients that had received multiple doses of the drug. Changes in IgG and IgM titers varied with dose number and with elapsed time after administration.

We conclude that circulating L-asparaginase antibodies are detectable via ELISA. In dogs with lymphoma treated with L-asparaginase, the presence of this humoral response may correlate with clinical resistance.

Previously presented at the VCS Annual Conference, Austin, Texas, October 2009.

ABSTRACT #213
DOG LEUKOCYTE ANTIGEN CLASS II ASSOCIATION IN CHIHUAHUAS WITH NECROTIZING MENINGOENCEPHALITIS.
KM Vernau, H Liu, RJ Higgins, PJ Dickinson, BK Sturges, W Vernau, RA LeCouteur, NC Pedersen. School of Veterinary Medicine, University of California Davis, Davis, CA.

Necrotizing meningoencephalitis (NME) in Pug Dogs is strongly associated by whole genome scanning and candidate gene sequencing with specific alleles and haplotypes in the DLA class II genes -DRB1, -DQA1 and -DQB1 (K. Greer et al., in press). NME also exists in Chihuahuas and we hypothesized that the same DLA class II genetic association occurs in both Chihuahas and Pug Dogs.

DNA from Chihuahuas and Pug Dogs with clinically and histologically confirmed NME was isolated from both fresh and archived paraffin embedded tissues. Healthy Chihuahuas and Pug Dogs with no clinical evidence of NME served as controls. Alleles of DLA-DQB1, -DQA1 and -DRB1 were determined by exon 2 sequencing. We reconfirmed the strong and specific DLA class II allele/haplotype association for NME in Pug Dogs. DLA class II genotyping of the Chihuahuas showed Chihuahuas to have more allelic diversity than Pug Dogs. There was no apparent association between specific DLA class II genotypes or haplotypes and NME in Chihuahuas. Therefore, NME in these two breeds either has no genetic associations in common, or if genetic associations exist, they are outside of the DLA class II genes or at other loci in the genome. The existence of a spectrum of immune encephalitides occurring predominantly in toy breeds, coupled with intense genetic selection for "toying", gives credence to the latter scenario.

ABSTRACT #214
MAGNETIC RESONANCE SPECTROSCOPY OF THE CANINE BRAIN AT 3.0 T AND 7.0 T.
Paula Martin-Vaquero[1], Ronaldo C. Da Costa[1], Rita L. Echandi[1], Christina L. Tosti[2], Michael V. Knopp[2], Steffen Sammet[2]. [1]Department of Veterinary Clinical Sciences, College of Veterinary Medicine, The Ohio State University, Columbus, OH, [2]The Ohio State University Medical Center, Columbus, OH.

Magnetic resonance spectroscopy (MRS) is a magnetic resonance imaging (MRI) based technique, which allows precise, non-invasive determination of the brain metabolism. The purpose of this study was to evaluate the feasibility of proton MRS (H^1 MRS) to study the brain of healthy dogs at 3.0 T and 7.0 T and to compare the results of the biochemical composition of the canine brain obtained at both magnetic field strengths.

Four healthy male beagle dogs (1.3–3 years, mean 2.3 years) were scanned in 3 T and 7 T human MR scanners (Philips Healthcare, Best, The Netherlands) under a single anesthetic procedure. The dogs were positioned in sternal recumbency for all data acquired. 3 T exams were conducted with an 8-channel receive-only phased-array coil operated in quadrature mode, and 7 T with a transmit-receive quadrature coil. A point resolved spectroscopy (PRESS) sequence that is suitable for common GABA editing schemes was used. Two PRESS localized single-voxel spectroscopy acquisitions with 1 cm^3 and 8 cm^3 volumes [samples = 1024; TE = 35 ms; TR = 2000 ms, NSA = 128] were acquired in the canine parietal cortex and cortical-ventricular area, respectively. The spectra were analyzed with the 3DiCSI software. Excitation water suppression was optimized and higher order pencil-beam shimming was used to improve B_0 homogeneity. Results were interpreted by a human physician and a physicist. Information regarding N-acetylaspartate (NAA), choline (Cho) and creatine (Cr) was recorded for both 3.0 T and 7.0 T.

The results obtained in our study show that 4 out of 4 dogs had excellent quality spectra for the small (1 cm^3) and large (8 cm^3) voxel at 3 T, while only 2 out of 4 dogs had high quality spectra at 7 T due to insufficient water suppression.

Our study found that MR spectroscopy in a clinical 3 T scanner using an 8-channel phased array extremity coil is an effective and robust method for the *in vivo* study of the canine brain, however, 7 T spectra was only successfully acquired in 2 out of 4 dogs. While 3 T appears to be a reliable option for canine spectroscopy, more advanced water suppression techniques would greatly improve the results of canine spectroscopy at 7 T.

ABSTRACT #215
RELATIONSHIP OF MAGNETIC RESONANCE IMAGING TO SURGICAL FINDINGS IN DOGS WITH THORACOLUMBAR INTERVERTEBRAL DISC DISEASE.
RM Gallagher[1], NJ Olby[1], DE Thrall[1], ID Robertson[1], W Lee[1], K Munana[1], P Early[1], CL Mariani[1], S Moore[2], Z Niman[1], S Dillard[1], J Campbell[1]. [1]North Carolina State University College of Veterinary Medicine, Raleigh, NC. [2]The Ohio State University, Columbus, OH.

The magnetic resonance imaging (MRI) findings in dogs with intervertebral disc disease have been previously described but have not been correlated to surgical findings. The objective of this study was to describe the MRI findings in dogs with surgically confirmed intervertebral disc disease (IVDD) and to compare imaging and surgical findings. In a prospective case series, MRI intensity of the spinal cord, and extradural compressive material were graded relative to normal spinal cord and the appearance and distribution of extradural material was described. At surgery the presence and location of fat, disc material and blood were noted. Disc material was described as mineralized or annular, consistency was noted and it was classified as acutely or chronically herniated.

Hemilaminectomies were performed on 49 disc spaces. Extradural compressive material was often heterogeneous in intensity and had a variable appearance at surgery. MRI appearance of blood was extremely variable making it difficult to differentiate from mineralized disc material prior to surgery. Assessed type of disc protrusion/extrusion (acute versus chronic) correlated well between MRI and surgical assessment. We conclude that it is difficult to distinguish between mineralized disc material and hemorrhage on MR images following acute disc herniations. However, chronicity of herniation can be predicted from MRI.

ABSTRACT #216

MAGNETIC RESONANCE IMAGING OF PLASMA CELL TUMORS IN THE SPINAL COLUMN OF 10 DOGS.

FA Wininger[1], SP Holmes[2], JR Coates[1], KA Selting[1]. [1]College of Veterinary Medicine, University of Missouri, Columbia, MO. [2]College of Veterinary Medicine, University of Georgia, Athens, GA.

We hypothesize that MR imaging of spinal plasma cell tumors (PCT) is more sensitive than survey radiography in the detection of solitary and multifocal lesions. A multicenter retrospective case series of 10 dogs with a cytologic or histologic diagnosis of PCT was used to identify the typical MR appearances of this tumor type. Tumoral signal characteristics were examined with a high-field strength MR using T1-weighted (T1W), T2-weighted (T2W), short tau inversion recovery (STIR), and T1W post-gadolinium sequences. Other variables evaluated were signalment, clinical signs, disease progression, serum globulin concentrations, radiographic findings, lesion location and presence of spinal cord compression. Radiologic interpretation was blinded with respect to MR imaging findings. Mean age was 12 years (range 9–14). Three of 10 dogs were Labrador retrievers. No sex predilection was noted. All cases presented for paraparesis and 8 dogs showed paraspinal hyperesthesia. Serum globulins in 8 dogs were outside the reference range (2.7–3.8 g/dl) with a mean concentration of 4.6 g/dl (SD +/− 1.16). Three dogs had focal and 5 dogs had multifocal evidence of axial skeletal osteolysis, while 2 had no identifiable lesions on radiography. All dogs had multiple identifiable spinal lesions on MR studies. All lesions identified on T2W were hyperintense relative to paraspinal musculature. Identified lesions on T1W were hypointense (4/10), isointense (3/10) and hyperintense (3/10). Degree of contrast enhancement in the affected vertebra was variable, with one showing no enhancement. STIR sequences were uniformly hyperintense and identified all lesions visible on radiography and other sequences. Importantly, in all 10 dogs, between two to twelve lesions were identified on STIR that was not seen on radiography. Eight dogs had lesions in the thoracic spine and 7 dogs in the lumbar spine, with none in the cervical spine. Sites deemed safe for biopsy were best identified using MR imaging. All dogs had 1–3 compressive spinal cord lesions not associated with fracture or subluxation, suggesting the presence of intraspinal expansile osteolysis. Compared to survey spinal radiography, MR imaging is a sensitive modality to detect vertebral sites of PCT. Lesion conspicuity is greatest on STIR sequences. PCT can be highly proliferative and expansile, resulting in invasion into the spinal canal.

ABSTRACT #217

RADIOGRAPHIC VERTEBRAL CANAL AND BODY RATIOS IN DOBERMAN PINSCHERS WITH AND WITHOUT CLINICAL SIGNS OF DISK ASSOCIATED WOBBLER SYNDROME (DAWS).

S De Decker, J Saunders, L Duchateau, P Pey, L Van Ham. Faculty of Veterinary Medicine, Ghent University, Merelbeke, Belgium.

Little is known about the etiology of DAWS. The relative size and shape of the vertebral canal have been proposed as potential risk factors for this disorder. The purpose of this study was to evaluate if selected vertebral ratios, representing vertebral canal stenosis, could be used to differentiate between Doberman Pinschers with and without DAWS.

Survey lateral radiographs of the cervical spine of 81 Doberman Pinschers with and 39 Doberman Pinschers without clinical signs of DAWS were retrospectively reviewed. Five measurements were made from the third (C3) to the seventh (C7) cervical vertebra: mid vertebral canal height (VCHm), vertebral body height (VBH), cranial vertebral canal height (VCHcr), caudal vertebral canal height (VCHcd), and vertebral body length (VBL). From these measurements, the following three radiographic ratios were calculated from C3 to C7: canal body ratio (CBR), defined as VCHm/VBH; longitudinal ratio (LR), defined as VCHm/VBL; caudal to cranial height ratio (CCHR), defined as VCHcd/VCHcr. The CBR and LR are suggested to represent vertebral canal stenosis. A high CCHR value indicates the presence of a funnel-shaped vertebral canal narrowed cranially in a lateral projection. All measurements were performed by the same person and a second person performed the same measurements on 65 different radiographs to evaluate interobserver agreement.

Mean global CBR and LR values for Doberman Pinschers with DAWS were significantly smaller when compared with the same values for Doberman Pinschers without DAWS. For CBR, this was also true for all assessed vertebrae seperately. The CCHR value for C7 was significantly larger in Doberman Pinschers with DAWS. The CBR was significantly smaller in male dogs. The values of the three vertebral ratios were significantly associated with the respective assessed vertebrae. Only the CCHR in Doberman Pinschers without clinical signs of DAWS was significantly influenced by age; each increase of age by one month resulted in an increase of CCHR value by 0.0018. There was a very good interobserver agreement.

In conclusion, the results of this study support the hypothesis that the size and shape of the vertebral canal are predisposing factors to develop clinical signs of DAWS. Moreover, our data suggest that Doberman Pinschers with a generalized relative stenosis of the vertebral canal combined with a funnel-shaped vertebral canal at the level of C7 have an increased risk to develop clinical signs of DAWS.

ABSTRACT #218

MICROMORPHOMETRY OF THE CANINE CERVICAL INTERVERTEBRAL DISCS.

J. Johnson, R. C. DA Costa, M. J. Allen. Department of Veterinary Clinical Sciences, The Ohio State University, Columbus, OH.

Pathology of the cervical discs may result in severe neck pain, thoracic limb lameness and paresis, ataxia, tetraparesis and tetraplegia. The pathogenesis of intervertebral disc degeneration is poorly understood, due in part to a lack of understanding of the histology and micromorphometry of normal cervical discs. The primary goal of this study was to undertake a detailed analysis of the micromorphometry of normal canine cervical discs, including gross structure and distribution of cellular components, and to identify differences between chondrodystrophic (CH) and nonchondrodystrophic (NCH) dogs.

All five cervical intervertebral discs were harvested from 4 chondrodystrophic and 4 nonchondrodystrophic dogs which were euthanized for reasons unrelated to this study. All dogs were under 2 years of age, skeletally-mature and free of radiographic abnormalities of the spine. The discs were decalcified in EDTA and two slides from each disc stained with H&E. Slides were scanned using an Aperio ScanScope and examined using ImageScope software. Direct measurements included: width of the dorsal and ventral annulus, width of the dorsal and ventral transitional zone, height and width of the nucleus, number and average width of the dorsal and ventral annular fibers, and nuclear and total disc area. Cells were counted in 1 mm^2 sections of outer and inner annulus fibrosus (AF), transitional zone (TZ) and nucleus pulposus (NP) and categorized according to total number, shape (round or elongate), and grouping (single, paired or clustered.) Statistical analysis was performed with SPSS and significance defined as $p < 0.05$.

The following micromorphomety measurements differed significantly between chondrodystrophic and nonchondrodystrophic dogs: ventral AF width:dorsal AF width (CH: 3.06 ± 1.10, NCH: 1.92 ± 0.96), number of ventral AF fibers (CH: 29.9 ± 6.1 vs NCH: 23.9 ± 44), dorsal AF fiber width (CH: 0.08 ± 0.03 vs NCH: 0.12 ± 0.04), and NP area:total disc area (CH: 17.24 ± 2.56 vs NCH: 25.36 ± 2.94). Cellularity decreased significantly from the outer to inner regions of the disc, while cell grouping (pairing or clustering) increased significantly. Cells in the outer regions were significantly more elongated, while those in the inner regions were significantly more rounded. No differences in any measurement attributable to disc level within the spine were noted.

Although the clinical significance of these differences in disc morphometry remains to be determined, the trends in cell distribution and cellular morphology are similar to those reported in humans. Cell clustering is a recognized feature of disc degeneration in humans, and future studies comparing normal with degenerative canine discs would be of interest. Improving our knowledge of normal cervical disc structure and organization will allow us to understand the mechanisms involved in the initiation and progression of intervertebral disc degeneration.

ABSTRACT #219

IN VITRO BIOMECHANICAL COMPARISON OF INTERNAL FIXATION TECHNIQUES ON THE CANINE LUMBAR SPINE.

S. Dillard, P. Early, S. Roe, J. Hicks and P. Mente. North Carolina State University College of Veterinary Medicine, Raleigh, NC.

The aim of this study was to determine the biomechanical characteristics of the canine lumbar spine in flexion and extension following three different internal fixation techniques. 1. Cortical screws and polymethylmethacrylate (PMMA), 2. SOP™ locking plate system with cortical screws placed bicortically and 3. SOP™ locking plate system with cortical screws placed unicortically. Fifteen canine cadaver lumbar spines from L2-L5 were collected from dogs ranging from 23–36 kg. A fracture was simulated in each lumbar spine at the L3-L4 vertebral space by performing bilateral facetectomies and discectomy. The spines were then stabilized with one of the three fixation techniques and subjected to incrementally increasing flexion and extension bending moments in a 4-point-bending hydraulic machine. Force and displacement were recorded during each loading cycle. Fixation techniques were compared on load to failure, defined as bending or breaking of the implant or to an overall displacement of greater than +/− 25 mm, which represented a disc space angle of 10 degrees. None of the spines fixed with the SOP plates with bicortically placed screws failed at the stabilized L3-L4 fracture site, but failed either due to fracture of the facets or an endplate of adjacent vertebrae. Four of the five specimens fixed with PMMA failed due to cracking of the PMMA and one failed due to fracture of adjacent facets. The PMMA group had less overall displacement of the entire L2-L5 spine as compared to the other groups; however, further analysis of the biomechanics at the L3-L4 disc space is warranted.

ABSTRACT #220

EVALUATION OF SHORT-TERM ELECTROENCEPHALOGRAPHY IN DOGS WITH IDIOPATHIC EPILEPSY.

S Arendse, R Poma, N MacLusky, B Hanna. Ontario Veterinary College, University of Guelph.

Idiopathic Epilepsy (IE) is the one of the most common neurological disease in dogs with a prevalence reported between 1 and 5%. Electroencephalography (EEG) is a useful tool in veterinary medicine for the determination and classification of seizures and for the evaluation of cerebral function in patients with intracranial disorders including IE. Based on human studies, first-time short-term EEG (less than 30 minutes) has a diagnostic yield to detect pathological activity between 22 and 25%. The objectives of this study were to evaluate the utility of a single outpatient short-term EEG recording (between 20 and 60 minutes) in detecting pathological activity in dogs diagnosed with IE and to determine whether the ability to detect pathological activity was influenced by the time between the last documented seizure and EEG recording.

Twenty-three dogs were included in the study. All dogs suffered from seizure activity (focal and/or generalized seizures) with initial onset between 7 months and 5 years of age. Neurological examinations, CBC, biochemistry profile and brain MRI were normal in all dogs. Based on these findings, a diagnosis of suspected IE was made. Upon presentation, each dog was sedated to effect with propofol (1–6 mg/kg IV) to allow the setting of a canine standardized EEG electrode montage. Using a routine video-synchronized EEG system, all subjects underwent a recording that lasted for a minimum of 20 minutes up to a maximum of 60 minutes. The period of time between the last documented seizure and the EEG recording was categorized as "acute" (up to 7 days) or "chronic" (greater than 7 days).

22% (5/23) of the recorded subjects presented pathological activity of which 80% (4/5) occurred within the first 20 minutes of recording, whereas the other 20% (1/5) occurred within a 60-minute recording period. Overall, 61% (14/23) of all dogs had EEG recorded within 7 days following the last documented seizure and 39% (9/23) of all dogs were recorded within the "chronic" period (after 7 days following the last documented seizure). Of the 5 dogs presenting pathological activity (5/23), 5/5 dogs (100%) were recorded within the "acute" phase (less than 7 days).

The results of this study suggests that short-term EEG recording is a useful diagnostic tool to detect pathological activity in dogs diagnosed with IE especially when the recording occurs within seven days from the last documented seizure. Sedation with propofol doesn't preclude the detection of pathological activity and it represents an ideal drug to use for initial sedation in canine EEG recording.

ABSTRACT #221

LONG-TERM VIDEO-ELECTROENCEPHALOGRAPHY IN A CANINE EPILEPSY MONITORING UNIT.

R Poma[1], JR Ives[2]. [1]Ontario Veterinary College, University of Guelph, Guelph, ON, Canada. [2]University of Western Ontario, London, ON, Canada.

Long-term video-synchronized EEG recording is a valid diagnostic tool to evaluate epileptic patients for seizure detection and for assessment of treatment during clinical and subclinical *Status Epilepticus* (SE). Canine epilepsy represents the most common neurological disease in veterinary medicine. Epileptic dogs are commonly affected by generalized and focal seizures. A canine epilepsy monitoring unit (canine EMU) was recently set up at the Ontario Veterinary College, University of Guelph to monitor epileptic dogs for seizure detection and long-term management of cases refractory to conventional antiepileptic drugs.

Two cases out of a broad series of recorded epileptic dogs will be presented to demonstrate practical applications of long-term video-synchronized EEG recording in canine epilepsy. Case 1: "Mojo" is an 8 m old MC Chihuahua that was referred to the VTH-OVC neurology service for continuous episodes of head and nose twitching. "Mojo" was admitted to the canine EMU for long-term video-EEG recording. A standard montage was used with 11 recording electrodes (subdermal wire electrodes). The electrodes were secured in place with a proper bandage to avoid movements and disconnection. The recording lasted several hours during which "Mojo" experienced multiple episodes of head and nose twitching that were electroencephalographically characterized by generalized spike-and-waves activity. A brain MRI was performed on the following day and no structural abnormalities was detected. Based on these findings, a diagnosis of absence-like juvenile myoclonic epilepsy was made and treatment with Phenobarbital was started.

Case 2: "Levi" is a 2 yo MC French Bulldog that was presented to the VTH-OVC neurology service in focal *Status Epilepticus* characterized by extreme confusion, foaming, compulsive circling and facial twitching. "Levi" was previously diagnosed with idiopathic generalized epilepsy and he was treated with Phenobarbital, Potassium bromide, and Levetiracetam. Upon presentation, "Levi" was admitted to the canine EMU for long-term video-EEG monitoring and management of focal SE. Constant polyphasic waves and spike-and-wave activity were recorded bilaterally and symmetrically. Treatment with diazepam was started along with repetitive intravenous boluses of Phenobarbital. Since no significant improvement was noted, Propofol infusion was initiated to suppress bursts of pathological cortical electrical activity. Drug dosages were adjusted in accordance to the video-EEG findings that were simultaneously recorded. Despite aggressive treatment with AEDs, pathological cortical electrical activity continued to subsist indicating a poor prognostic factor.

Long-term video-synchronized EEG recording performed in an appositely designed canine epilepsy monitoring unit can facilitate the investigation of different aspects of phenomenology of canine seizures including the determination of a seizure versus epileptic-like activity and the source of epileptic activity ("seizure focus"). In refractory epileptic patients or dogs presented with SE, long-term video EEG is essential to monitor subject's response to treatment with different AEDs and to establish a prognosis of successful outcome.

Previously presented at the American Epilepsy Society.

ABSTRACT #222

RISK FACTORS FOR EPILEPTIC SEIZURES IN DOGS WITH INTRACRANIAL NEOPLASIA. M Schwartz, CR Lamb, D Brodbelt, HA Volk. The Royal Veterinary College, London, United Kingdom.

Intracranial neoplasia is an important differential diagnosis for epileptic seizures in old dogs; however, not all intracranial neoplasias cause seizures. Most intracranial neoplasias do not act as primary seizure foci because they contain no neuronal cells; instead initiation of seizure activity is thought to depend on their effect on the adjacent neuronal tissue. The aim of this study was to identify morphologic features in magnetic resonance (MR) images that may represent risk factors for seizures.

Clinical records were searched for dogs that had full clinical history, histopathologically-confirmed intracranial neoplasia, and MR imaging of the brain. Occurrence of features in MR images in dogs with recent development of seizures was compared with those that did not have seizures by means of Fisher's exact and Mann Whitney test as appropriate. Level of significance was set at $P < 0.05$.

A total of 68 dogs were included in the study (42 with seizures; 26 without seizures). Neoplasias included meningiomas, gliomas, choroid plexus tumors, and metastatic or invasive tumors of extracranial origin. None of these tumor types was associated with a higher risk of seizure development. Dogs that had seizures (median age: 109 months, range: 9–156 months) were older than those without (median age: 89 months, range: 18–132 months) ($P = 0.03$). Dogs with neoplasias located within the frontal lobe were more likely to develop seizures ($P < 0.01$; Odds Ratio (OR) = 6.79, 95% Confidence Interval (CI) = 2.27–20.29), as were those whose neoplasias were associated with frontal lobe ($P = 0.02$; OR = 3.41, 95% CI = 1.22–9.52) or parietal lobe edema ($P < 0.01$; OR = 4.21, 95% CI = 1.49–11.92). Dogs with subfalcine and/or subtentorial herniation were more likely to develop seizures ($P = 0.02$; OR = 3.57, 95% CI = 1.28–9.96). A higher degree of enhancement after gadolinium administration was associated with seizures ($P < 0.01$).

Potential risk factors for seizures occurring in dogs secondary to intracranial neoplasia include tumor and/or peritumoral edema affecting the frontal lobe, increased supratentorial pressure, and marked enhancement after gadolinium administration. The cellular origin of an intracranial tumor appears not to be a factor.

ABSTRACT #223

PHARMACOKINETICS OF A NOVEL INTRANASAL MIDAZOLAM GEL IN DOGS. JS Eagleson,[1] SR Platt,[1] M Kent,[1] AC Freeman,[1] AC Haley,[1] P Nghiem,[1] AC Durden,[1] SJ Schatzberg,[1] B Zheng,[2] C White,[2] D Strong[2]. [1]University of Georgia, Colleges of Veterinary Medicine and [2]Pharmacy, Athens, GA.

The pharmacokinetics of a novel bioadhesive gel formulation of midazolam administered intranasally (IN) were compared to those of parenteral midazolam solution administered IN, intravenously (IV) and rectally.

Ten healthy beagles were randomly assigned to 4 treatment groups in a crossover design. Parenteral midazolam solution was administered IV to group 1, rectally to group 2, and IN to group 3 (0.2 mg/kg). A 0.2% hydroxypropyl methylcellulose midazolam gel formulation was administered IN to group 4 (0.2 mg/kg). Blood was collected from the jugular vein of each dog into tubes containing lithium heparin before and 3, 6, 9, 12, 15, 20, 30, 60, 120, 240, and 480 minutes following midazolam administration. The protocol was repeated 4 times with a 7-day wash out period between administrations. Each dog received all 4 forms of midazolam administration. Plasma concentration of midazolam was determined using high performance liquid chromatography.

Mean (+/− SD) peak plasma concentrations (C_{max}) of midazolam were 1.82 (+/− 0.278) g/mL (IV), 0.45 (+/− 0.09) g/L (IN gel), 0.21 (+/− 0.02) g/mL (IN solution) and 0.15g/mL (rectal). The C_{max} following IN gel administration was significantly higher than IN solution and rectal administration ($p < 0.0001$). Mean plasma concentrations of midazolam following IV administration were significantly higher at 3, 6, 9 and 12 minutes post-delivery when compared to all other methods of delivery ($p < 0.0001$). Mean plasma concentrations of midazolam following IN gel administration were significantly higher at 6, 9 and 12 minutes post-delivery when compared to IN solution and rectal administration ($p < 0.0001$). Time (+/− SD) to peak concentration (T_{max}) was 11.70 (+/− 2.63) minutes (IN gel), 17.50 (+/− 2.64) minutes (IN solution) and 39 (+/− 14.49) minutes (rectal). The T_{max} following rectal administration was significantly longer than both IN gel and IN solution administration ($p < 0.0001$). Mean bioavailability of midazolam was 70.4% (IN gel), 52.0% (IN solution) and 49.0% (rectal). Bioavailability following IN gel administration was significantly higher than IN solution and rectal administration ($p < 0.0001$). The half lives (+/− SD) were 121 (+/− 25.74) minutes (IV), 119.61 (+/− 22.83) minutes (IN gel), 118.66 (+/− 18.84) minutes (IN solution) and 115.83 (+/− 15.56) minutes (rectal).

Intranasal midazolam gel was superior to IN solution administration and rectal administration with respect to peak plasma concentration and bioavailability. Intranasal administration of this novel midazolam gel may be useful for treatment of seizures in dogs by owners or when intravenous access is not readily available.

ABSTRACT #224

ZONISAMIDE THERAPY FOR REFRACTORY SEIZURES IN FIVE CATS WITH PRESUMPTIVE IDIOPATHIC EPILEPSY. DM Brewer, S Cerda-Gonzalez, CW Dewey. College of Veterinary Medicine, Cornell University, Ithaca, NY.

Phenobarbital (PB) is currently the drug of choice for seizure management in cats due to its proven efficacy and acceptable side-effect profile. A subset of cats will not achieve acceptable seizure control with PB alone, necessitating the addition of a second anti-epileptic medication. Limited options are available in cats that are refractory to PB. The anti-epileptic drug zonisamide (ZNS) has been shown to be safe and well tolerated in cats. However, no clinical trials have been preformed to account for its efficacy.

A retrospective analysis was performed on 5 cats in which ZNS was used to treat refractory seizures. To qualify, cats needed to have clinical features consistent with idiopathic epilepsy, an accurate seizure log for at least 3 months prior to and after ZNS administration, a minimum average of 2 seizures/month and serum PB levels within the therapeutic range.

The mean ZNS dose was 11.54 mg/kg/day, range 6.14–17.0 mg/kg/day. The mean serum ZNS level was 19.44 mcg/ml, range 8.8–38.6 mcg/ml. The reduction in seizure frequency of all five cats was 65.5%. Three of five (60%) cats were classified as responders (> 50% reduction in seizure frequency) with a mean seizure reduction of 95.8%. Side effects consisted of mild lethargy in one cat and decreased appetite in one cat. No changes were noted in serum biochemistry or hemogram values within the 3 months of ZNS treatment.

Preliminary results indicate that ZNS is well tolerated and may be a useful therapy for refractory seizures in cats with presumptive idiopathic epilepsy.

ABSTRACT #225

PHARMACOKINETICS OF ORAL PREGABALIN IN CATS AFTER SINGLE DOSE ADMINISTRATION. MA Cautela[1], CW Dewey[1], WS Schwark[2], S Cerda-Gonzalez[1], BL Badgley[1]. [1]Cornell University College of Veterinary Medicine, Ithaca, NY. [2]Department of Molecular Medicine, College of Veterinary Medicine, Cornell University, Ithaca, NY.

Seizures are frequently encountered in cats, and there are limited treatment options for this species. Phenobarbital (PB) has long been considered the safest and most effective anticonvulsant for use in cats. Levetiracetam has recently been shown to be a safe and effective add-on to PB for presumed epileptic cats. There are also limited options for pain management in feline patients with suspected neuropathic pain (e.g., feline hyperesthesia syndrome). The purpose of this study was to investigate in healthy cats the pharmacokinetic profile of pregabalin (PG), a new medication (and structural analog of gabapentin), that has shown efficacy as both an anticonvulsant and pain medication in humans, and as an anticonvulsant in dogs.

Six healthy adult cats were included in the study. After physical and neurologic examination and minimum database (complete blood count, serum biochemistry, urinalysis), cats were administered one dose of oral PG (4 mg/kg), and serial blood samples were drawn via sampling catheter over a 36-hour period (t = 0,15,30,60,90,120,180, 240,360,480,720,1440, and 2160 min). Samples were analyzed via HPLC at a commercial laboratory. Resultant data were used to analyze pharmacokinetic parameters (e.g., Cmax, Cmin, Tmax, Tmin, elimination half-life, etc) using commercial software and the PG plasma vs concentration time curves.

The following parameters were evaluated (mean ± SD, range): area under the curve (AUC): 133.9 ± 71.5 (99.3–192.0 µg-h/ml), absorption half-life ($T_{1/2}$abs): 0.54 ± 0.44 (0.17–1.38 h), elimination half-life ($T_{1/2}$elim): 10.4 ± 2.6 (8.0–14.3 h), time above suspected minimal therapeutic concentration (2.8 µg/ml): 17.6 ± 6.2 (11.0–27.5 h), maximum serum concentration (C_{max}): 8.3 ± 1.6 (6.4–10.0 µg/ml), and time until C_{max} (T_{max}): 2.9 ± 1.2 (1.0–4.0 h). Predicted serum steady-state concentrations (maximum, minimum and average in µg/ml) were calculated for 12 hour (BID) dosing; CssMax: 16.2 ±5.8 (10.4–23.0), CssMin: 4.9 ± 4.1 (0.3–10.9), and CssAve: 9.5 ± 3.8 (5.9–15.5) and 24 hour (SID) dosing; CssMax: 10.8 ± 3.6 (7.0–16.1), CssMin: 2.5 ± 1.3 (1.4–4.5), and CssAve: 4.8 ± 1.9 (2.9–7.7) intervals.

PG at this dose had the resultant side effect of moderate sedation in 4/6 cats and achieved blood concentrations within the range considered efficacious for human seizure control (2.8–8.2 µg/ml). With $T_{1/2}$elim of 10.4 hours, BID dosing would likely be adequate for further clinical trials, which are warranted given the favorable pharmacokinetic profile of the drug. Based on our data, we theorize that a dose of 1–2 mg/kg BID of PG would be a reasonable starting dose in cats.

ABSTRACT #226

EVALUATION OF CEREBROSPINAL FLUID LACTATE AND GLUCOSE LEVELS IN ACUTE CANINE INTERVERTEBRAL DISC DISEASE.

R Taggart[1], M Maxwell[2]. [1]Purdue University College of Veterinary Medicine, West Lafayette, IN. [2]Mississippi State University College of Veterinary Medicine, Mississippi State, MS.

Elevations in cerebrospinal fluid (CSF) lactate and glucose have been demonstrated following experimentally induced spinal cord trauma. The purpose of this study was to determine whether CSF lactate and glucose levels are altered following acute intervertebral disc disease (IVDD) and to determine if those changes could be correlated with clinical signs and return to function.

A prospective study was designed utilizing three patient populations: normal dogs (n = 10), dogs with acute IVDD (n = 7), and dogs with neurologic signs unrelated to IVDD (n = 11). IVDD was confirmed with CT evaluation. Cerebrospinal fluid was collected via cisterna magna puncture in all dogs. Fluid lactate and glucose levels were analyzed within 30 minutes of collection using a common blood gas analyzer (i-Stat®) and a standard serum chemistry analyzer. All fluid samples were examined and those with cytologic evidence of infectious disease or demonstrating blood contamination (> 300 RBC/ul) were excluded. Groups were compared using ANOVA with $p < 0.05$ as significant. Glucose was ranked before ANOVA was performed.

The IVDD group included dogs with both thoracolumbar and cervical lesions with Grade 2–4 disease. The duration of clinical signs prior to presentation varied from less than 24 hours to two weeks. Dogs with IVDD received variable treatment with steroids prior to presentation. A variety of conditions including neoplasia, inflammatory disease, and idiopathic seizure disorder were represented in the group of dogs with neurologic signs without IVDD.

No significant differences in cerebrospinal fluid lactate or glucose levels were observed between the three groups.

There are several limitations to this study which may explain the lack of significant findings. Foremost, CSF was sampled via cisterna magna puncture only. The predominantly caudal flow of CSF may have limited the detection of biochemical alterations in samples collected cranial to the lesion in IVDD patients.

Table [Abstract 226]

Cerebrospinal Fluid Lactate and Glucose Levels

	IVDD	Other Neuro-logic	Normal
Lactate (mmol/L)	1.78 ± 0.36	1.61 ± 0.31	1.66 ± 0.22
Glucose (mg/dl)	82 ± 8	78 ± 22	72 ± 6

Data expressed as mean ± SD

ABSTRACT #227

FIBRINOLYTIC ACTIVITY IN CEREBROSPINAL FLUID (CSF) OF DOGS WITH DIFFERENT CNS DISORDERS.

C de la Fuente[1], L Monreal[2], S Añor[1]. [1]Neurology/Neurosurgery Service, [2]Equine Internal Medicine Service, Veterinary School, Autonomous University of Barcelona, Barcelona, Spain.

D-dimers are degradation products of cross-linked fibrin generated by the action of plasmin and, consequently, specific indicators of fibrinolysis. Plasma D-dimer concentration is a sensitive marker of systemic fibrinolytic activity and it is widely used for the diagnosis of thromboembolic disease in humans and animals. Inflammation and coagulation are interrelated processes that may affect each other. Inflammation leads to activation of coagulation, but coagulation also affects the inflammatory response. The aims of this study were: 1. To assess fibrinolytic activity in the CSF of canine patients affected by different neurological disorders by measuring CSF D-dimer concentrations, 2. To compare CSF D-dimer concentration between different etiological groups, and 3. To determine whether CSF D-dimer concentrations were related to plasma D-dimer concentrations. In addition, pre- and post-treatment CSF D-dimer concentrations were compared in dogs with steroid-responsive meningitis arteritis (SRMA) to determine if this could be a reliable indicator of response to treatment.

One hundred and forty CSF samples were collected from 133 dogs. Seven dogs were sampled two times (pre- and post-treatment): 5 SRMA, 1 suspected cerebrovascular accident (CVA), and 1 meningoencephalitis of unknown etiology (MUE). Citrated blood samples were obtained from all dogs at the time of CSF sampling. Following initial CSF analysis, CSF and plasma D-dimer concentrations were detected by a quantitative immunoturbidimetric latex agglutination assay. Dogs were grouped into eleven categories according to the disease process (vascular [mainly ischemic CVAs], idiopathic epilepsy, idiopathic non-epileptic disorder, traumatic CNS disorder, SRMA, MUE, extra-axial neoplasia [mainly meningiomas], intra-axial neoplasia [mainly gliomas], extra-dural compressive spinal cord disorder, degenerative spinal cord disease, and control group [non-neurologic diseases]), and median and range of CSF D-dimer values were calculated. Statistical differences in CSF D-dimer concentrations were found between different groups ($p < 0.05$), with the highest CSF D-dimer values in the SRMA and the extra-axial neoplasia groups. No relationship was found between CSF D-dimer and plasma D-dimer levels by lineal regression ($R^2 = -0.02$). Statistical differences were found between pre- and post-treatment CSF D-dimer concentrations in the SRMA group ($p = 0.016$). Results of this study indicate that CSF fibrinolytic activity in dogs can be measured using the D-dimer assay. Diseases such as SRMA and extra-axial neoplasms induce higher concentrations of CSF D-dimers, and CSF D-dimer levels do not seem affected by plasma D-dimer concentrations, which suggest the existence of intrathecal fibrinolytic activity in some inflammatory and neoplastic CNS disorders. CSF D-dimer levels can be a good indicator of response to treatment in cases of SRMA.

ABSTRACT #228

CYTOKINE AND CHEMOKINE LEVELS IN SERUM AND CEREBROSPINAL FLUID OF DOGS WITH INFLAMMATORY CENTRAL NERVOUS SYSTEM DISEASE. CL Mariani[1], LB Boozer[1], GB Cherubini[2], KM Vernau[3], J Campbell[1], S Dillard[1], PJ Early[1], R Gallagher[1], E MacKillop[4], KR Muñana[1], ZE Niman[1], NJ Olby[1], AO Inman[1], NA Monteiro-Riviere[1]. [1]North Carolina State University College of Veterinary Medicine, Raleigh, NC. [2]Dick White Referrals, Suffolk, United Kingdom. [3]University of California-Davis School of Veterinary Medicine, Davis, CA. [4]Pittsburgh Veterinary Specialty & Emergency Center, Pittsburgh, PA.

The underlying etiology and pathophysiology of encephalitis and myelitis in canine patients is usually enigmatic. Cytokines and chemokines are proteins that control the immune response through effects on leukocytes and other immune cells and evaluation of these molecules might provide insight into the mechanisms of inflammatory central nervous system (CNS) diseases. The purpose of this study was to examine cytokine and chemokine levels in peripheral blood and cerebrospinal fluid (CSF) of canine patients diagnosed with meningoencephalitis or meningomyelitis of unknown etiology (MUE). A canine-specific, multiplex assay (Milliplex MAP, Millipore) was used to analyze interleukin (IL)-2, IL-4, IL-6, IL-7, IL-8 (CXCL8), IL-10, IL-15, IL-18, interferon (IFN)-γ, tumor necrosis factor (TNF)-α, granulocyte macrophage-colony stimulating factor (GM-CSF), monocyte chemoattractant protein (MCP)-1 (CCL2), interferon-inducible protein (IP)-10 (CXCL10), and keratinocyte chemoattractant (KC or CXCL1) in these samples. Compared with control dogs, a number of patients with MUE showed elevations of IL-2, IL-7, IL-15, IL-18, GM-CSF, CXCL1, and CCL2 in serum and IL-2, IL-8, and CCL2 in CSF. These patterns suggest an environment favoring the proliferation and maintenance of lymphocyte populations (including memory cells) in the peripheral compartment (IL-2, -7, -15, CCL2), while promoting lymphocyte proliferation and survival within the CNS (IL-2, CCL2). IL-18 also supports lymphocyte proliferation and has been identified as an important cytokine contributing to CNS inflammation in other species. Serum GM-CSF and CSF CCL2 may help drive monocyte production in the periphery with subsequent recruitment into the CNS of dogs with MUE.

ABSTRACT #229

EVALUATION OF MATRIX METALLOPROTEINASE-2 AND -9 IN THE CEREBROSPINAL FLUID OF DOGS WITH INTRACRANIAL DISEASE. LB Boozer[1], CL Mariani[1], J Campbell[1], S Dillard[1], PJ Early[1], R Gallagher[1], E MacKillop[2], KR Muñana[1], ZE Niman[1], NJ Olby[1]. [1]North Carolina State University College of Veterinary Medicine, Raleigh, NC. [2]Pittsburgh Veterinary Specialty & Emergency Center, Pittsburgh, PA.

Matrix metalloproteinases (MMPs) are a family of enzymes responsible for maintenance and remodeling of the extracellular matrix throughout the body after cleavage from latent to active form. MMP-2 and -9 in particular have been implicated in the pathophysiology of a variety of disease processes, including tumor invasion and encephalitis. Latent MMP-2 is present in normal canine cerebrospinal fluid (CSF), but the presence of MMP-9 indicates pathology. This study evaluated levels of active and latent forms of MMP-2 and MMP-9 in the CSF of dogs with CNS disease. Samples of CSF from patients with intracranial meningiomas (n = 12), gliomas (n = 5), choroid plexus tumors (n = 3), lymphosarcoma (n = 3), and inflammatory CNS disease (n = 9) along with normal control animals (n = 6) were evaluated using gelatin zymography. All samples showed bands consistent with latent MMP-2. Active MMP-2 was not detected in any samples. Bands consistent with latent MMP-9 were seen in CSF from patients with meningioma (n = 1), glioma (n = 1), choroid plexus tumors (n = 3), lymphosarcoma (n = 3), and inflammatory CNS disease (n = 1). Bands matching active MMP-9 were seen with 1 meningioma, 1 glioma, 1 animal with inflammatory disease, 1 choroid plexus tumor, and 3 lymphosarcomas. Patients with detectable MMP-9 had higher nucleated cell counts in the CSF than those without ($P = 0.0009$). This data suggests that MMP-9 may play a role in the pathogenesis of a subset of intracranial tumors and encephalitis afflicting canine patients. Evaluation of MMPs in CSF may prove useful in the diagnosis, monitoring and prognosis of patients with intracranial disease.

ABSTRACT #230

EVALUATION OF CYTOKINE AND CHEMOKINE LEVELS IN CANINE INTRACRANIAL NEOPLASIA. LB Boozer, CL Mariani, J Campbell, S Dillard, PJ Early, R Gallagher, E MacKillop, KR Muñana, ZE Niman, NJ Olby, AO Inman, NA Monteiro-Riviere. North Carolina State University College of Veterinary Medicine, Raleigh, NC.

Tumor-immune system interactions are complex. Studies of brain tumors in humans and experimental animals have shown improved survival with immune cell infiltration; however, these tumors may create an immunosuppressive environment that impairs this response. Cytokines and chemokines mediate the immune response through effects on immune cells. The purpose of this study is to examine cytokine and chemokine levels in the CSF and serum of dogs with intracranial neoplasia. Canine-specific, multiplex cytokine/chemokine assays (Milliplex MAP, Millipore) were used to analyze transforming growth factor-beta (TGFb1, -b2, -b3), interleukin (IL)-2, IL-4, IL-6, IL-7, IL-8 (CXCL8), IL-10, IL-15, IL-18, interferon-gamma (IFN)-γ, tumor necrosis factor (TNF)-α, granulocyte macrophage-colony stimulating factor (GM-CSF), monocyte chemoattractant protein-1 (MCP-1/CCL2), and keratinocyte chemoattractant (KC/CXCL1) in samples from dogs with intracranial meningiomas (n = 23), choroid plexus tumors (CPT, n = 6), lymphosarcoma (LSA, n = 6), and gliomas (n = 13). Evidence of immunosuppression included CSF TGFb1 elevations in several dogs with gliomas or CPT, a trend towards increased CSF TGFb2 in dogs with tumors compared to control samples, and elevations in serum IL-4 and IL-10 in a few patients. Proinflammatory changes were mainly limited to elevations of serum IL-6 and IL-18 in some dogs. Cytokines associated with lymphocyte survival and proliferation (IL-2, IL-7, IL-15) were noted in the serum of a number of dogs with a variety of tumor types. There appeared to be consistent elevations of CCL2 and IL-8 in the serum and CSF of dogs with tumors when compared with controls, which may facilitate immune cell recruitment into the CNS.

ABSTRACT #231

TRANSCRANIAL MAGNETIC MOTOR EVOKED POTENTIALS, SOMATOSENSORY EVOKED POTENTIALS, SPINAL EVOKED POTENTIALS AND BRAINSTEM AUDITORY EVOKED POTENTIALS IN CAVALIER KING CHARLES SPANIEL DOGS WITH AND WITHOUT SYRINGOMYELIA. K Wolfe, S Nykamp, DG Allen, J Armstrong, R Poma. Ontario Veterinary College, University of Guelph, Guelph, ON, Canada.

Syringomyelia (SM) is a disorder characterized by fluid-filled cavitations of the spinal cord leading to parenchymal disruption and generation of severe neurological signs. The Cavalier King Charles Spaniel (CKCS) breed is overrepresented for this condition with an incidence rate of 50%. The aim of this study was to evaluate the validity of transcranial magnetic motor evoked potentials (TMMEP), somatosensory evoked potentials (SSEP), spinal evoked potentials (SEP) and brainstem auditory evoked responses (BAER) in the assessment of motor and sensory pathways throughout the central nervous system of CKCS dogs with and without SM diagnosed by MRI.

Fifty CKCS dogs were evaluated. Neurologic deficits were graded 0–3 (0 = normal, 1 = cervical or thoracolumbar pain on palpation, 2 = cervical or thoracolumbar pain on palpation plus evidence of neuropathic pain, 3 = ataxia or paresis). TMMEP, SSEP, SEP, and BAER were performed on all dogs under sedation with hydromorphone (0.05 mg/kg IV) and acepromazine (0.025 mg/kg IV). Onset latencies (msec) were measured for TMMEP, SSEP and SEP and I-V interpeak latencies were measured for BAER. Dogs were anesthetized for brain and spinal cord MRI (cervical, thoracolumbar, lumbar) including sagittal and transverse T2-weighted images.

Transverse T2-weighted image sequences were acquired on all regions of the spinal cord affected with SM to evaluate syrinx measurements (mm^2) at maximum size.

Based on neurologic score and MRI findings, all dogs were classified into group A (n = 16) (neurologic abnormalities and MRI-confirmed SM), B (n = 9) (neurologic abnormalities and no SM), C (n = 11) (normal neurologic examination and MRI-confirmed SM), and D (n = 14) (normal neurologic examination and no SM). There were no significant differences observed in the mean latencies of TMMEP, SSEP, SEP and BAER between groups A, B, C and D. However, a significant linear correlation was observed between the severity of neurologic dysfunction and size of the syrinx (p = 0.017; t-test procedure, SAS) with a larger syrinx being associated with more severe neurologic signs.

In conclusion, TMMEP, SSEP, SEP and BAER do not appear to be valuable tests in detecting functional abnormalities of the motor and sensory pathways throughout the central nervous system of CKCS dogs with and without neurological signs secondary to SM diagnosed by MRI.

ABSTRACT #232

EXPRESSION OF HEAT SHOCK PROTEINS, KI-67 AND VASCULAR ENDOTHELIAL GROWTH FACTOR IN CANINE INTRACRANIAL GLIOMAS. S.R. Platt, A.B. Yanke, A.C. Freeman, A.C. Haley, J. Eagleson, S.J. Schatzberg, M. Kent, E.W. Howerth. College of Veterinary Medicine, University of Georgia, Athens, GA.

Human brain tumors express heat shock proteins (HSP), which are associated with their degree of malignancy. The up-regulation of HSPs during tumor cell growth helps keep tumor proteins stable and therefore makes them a reasonable target for therapy. The aim of this study was to determine if canine intracranial astrocytomas and oligodendrogliomas express HSP 27, 72 and/or 90. An additional aim was to determine whether the expression of the HSPs was associated with Ki67 and/or Vascular Endothelial Growth Factor (VEGF) expression. Ki67 expression and VEGF up-regulation have been strong indicators of cell proliferation and angiogenesis during tumor growth, respectively.

Twenty-two formalin-fixed, paraffin-embedded canine brain tumors (8 astrocytomas, 2 oligoastrocytomas and 12 oligodendrogliomas) underwent immunohistochemical staining using anti-HSP 27, 72 or 90 antibodies. These tumor samples were also immunohistochemically stained for Ki67 and VEGF expression. Canine mammary carcinoma and squamous cell carcinoma tissues served as the control samples, as both have previously been shown to express HSPs. Four non-overlapping high power fields of each stained sample were selected and cell staining was analyzed using a semi-quantitative method. All analyses were performed using SAS V 9.2 (Cary, NC). Descriptive statistics of staining percentages were calculated for all tumors tested. Staining percentages were compared between tumor types and tumor type/grade subgroups by the Kruskal-Wallis test. All hypothesis tests were 2-sided and the significance level was $\alpha = 0.05$.

All three HSPs and VEGF were expressed in all tumor types. For all tumors pooled (n = 22), there was a positive correlation between HSP27 expression in tumor vessels and Ki67 expression in tumor tissue (r = 0.46, p = 0.0307). For astrocytomas (n = 8) there was a positive correlation between Ki67 and HSP 27 (r = 0.73, p = 0.0378). No such associations were found with HSP 72 or 90. No associations were found with any HSP variant when all oligodendrogliomas were grouped together. However, for grade 3 oligodendrogliomas (n = 7), there were strong positive correlations between Ki67 % and HSP72 % tumor (r = 0.84, r = 0.0169), and between Ki67 % and HSP72% (r = 0.795, p = 0.0325), and VEGF % and HSP90% (r = 0.79, p = 0.0362).

In conclusion, HSP expression was demonstrated in canine glial cell tumors. In the case of astrocytomas, the expression of HSP27 was associated with a more aggressive tumor, as noted by Ki67 expression. A similar association was found for aggressive oligodendrogliomas between HSP72 and Ki67. This study suggests that HSPs may have a role in the maintenance of aggressive canine gliomas and potentially represent a novel treatment target.

ABSTRACT #233

DISTRIBUTION OF ADENO-ASSOCIATED VIRUS VECTOR SEROTYPES 1, 9, AND 10 IN THE NORMAL FELINE BRAIN. Jill Narak[1], Nancy R. Cox[2], Carlos E. Sanchez[3], Miguel Sena-Esteves[4], and Douglas R. Martin[2]. [1]University of Tennessee, Knoxville, TN. [2]Auburn University, Auburn, AL. [3]Massachusetts General Hospital, Boston, MA. [4]University of Massachusetts, Worcester, MA.

This project evaluated the distribution of three serologically distinct adeno-associated virus (AAV) vectors within normal feline brains, to aid in determining the AAV serotype most useful for clinical trials in feline and human gangliosidosis gene therapy.

Using convection-enhanced delivery, nine healthy cats received an intrathalamic injection of 6×10^{11} genome equivalents of AAV genetically engineered to carry and express β-galactosidase complementary DNA (three cats received AAV serotype 1, three cats received AAV serotype 9, and three cats received AAV serotype 10). One month post-surgery, the cats were euthanized and necropsied. Brain tissue was harvested and frozen for qualitative analysis of β-galactosidase expression measured by fluorogenic enzyme assays and X-gal staining. Additionally, vector distribution was evaluated by quantitative polymerase chain reaction (qPCR) for AAV vector-specific sequences.

Qualitative analysis revealed that AAV serotype 10 produced the highest enzyme activity, but activity was detected over a narrow range compared to AAV serotype 1. AAV serotype 1 produced high enzyme activity over a wider distribution. Additionally, quantitative analysis via qPCR revealed that AAV serotype 1 achieved the greatest transduction. Although superior to a previously tested serotype (AAV serotype rh8), the AAV serotype 9 vector was not as effective as AAV serotypes 1 or 10. No evidence of vector toxicity was detected in any of the treated animals based on behavioral or histopathologic criteria. Based on the hypothesis that AAV serotypes that are efficient across species are most likely to be effective in humans, serotype comparisons were performed in mice and cats. In conjunction with data from mice, the results of this study led to selection of AAV serotype 1 for therapeutic experiments in gangliosidosis cats and for future human clinical trials. Information gained from this study also will be useful for serotype selection to utilize gene therapy to treat feline neurologic disease.

ABSTRACT #234

IS ALASKAN HUSKY ENCEPHALOPATHY A MITOCHONDRIAL DISEASE? KM Vernau[1], JA Runstadler[2], HJ Husan[2], RJ Higgins[1], C Ackerley[3], BK Sturges[1], PJ Dickinson[1], C Giulivi[1], GD Shelton[4], JM Cameron[3], BH Robinson[3]. [1]University of California – Davis, Davis, CA. [2]University of Alaska Fairbanks, Fairbanks, AK. [3]The Hospital for Sick Children, Toronto, ON, Canada. [4]University of California-San Diego, La Jolla, CA.

Alaskan Husky Encephalopathy (AHE) is an idiopathic, fatal central nervous system disease, reported in 13 dogs. AHE resembles Leigh syndrome in humans, caused by a number of abnormalities affecting mitochondrial energy generation. The goal of this study was to determine if AHE is caused by similar defects.

Six dogs with AHE (4 females, 2 males), 6–12 months of age, were examined. All dogs multifocal brain disease; bilaterally symmetrical thalamic, cerebral subcortical, midline cerebellar cortical, and medullary lesions were noted on MRI, and confirmed neuropathologically. Biochemically, no significant abnormalities were found in serum/CSF lactate and pyruvate levels, or in pyruvate dehydrogenase enzyme activity or 2-^{14}C pyruvate whole cell oxidation in fibroblasts, when compared with controls. Electron transport chain enzyme activities were within normal limits in muscle and liver (complexes I, II, I + III, II + III, IV and citrate synthase) but a reduction in complex IV and V activity was noted in the cerebral cortex and thalamus compared to controls.

Electron microscopy showed significant ultrastructure abnormalities in muscle and liver. Mitochondria were enlarged and pleomorphic (vacant mitochondria, disorganized cristae and concentric whorls). No genetic mutations were present in eleven candidate nuclear and mitochondrial genes known to cause Leigh syndrome in humans.

We conclude that there is no evidence to support a systemic defect within the TCA cycle, pyruvate dehydrogenase complex, or respiratory chain enzymes as the cause of AHE, but mitochondrial involvement is still indicated, suggesting the disease is unlike any known cause of Leigh Syndrome, and manifests only in certain tissues.

ABSTRACT #235

CANINE N-TERMINAL PROHORMONE BRAIN NATRIURETIC PEPTIDE CONCENTRATION IN DOGS LESS THAN ONE YEAR OLD. G. Farace[1], A. Beardow[1], S. Forney[2], K. Yee[2], J. Allen[2], SJ. Ettinger[2]. [1]IDEXX Laboratories, Inc., Westbrook, ME. [2]California Animal Hospital, Los Angeles, CA.

Human pediatric studies have shown that Brain Natriuretic Peptide (BNP) levels can be used to diagnose cardiac disease but that the cut off levels are different from those seen in adults. The aim of this study was to determine if N-Terminal Prohormone Brain Natriuretic Peptide (NTproBNP) concentrations in young dogs ($\leq$ 1 year old) differ from those of older dogs. The population recruited for the CAHVSG/IDEXX study was divided between those $\leq$ 1 year old and those $>$ 1 year old. The groups were then divided into normal dogs without cardiac disease and those with a congenital heart disease; dogs with acquired cardiac disease were not included in the analysis.

The 27 normal dogs $\leq$ 1 year old had a mean NTproBNP of 177 pmol/L (12 to 541 pmol/L) while the 133 normal dogs $>$ 1 year old had a mean of 372 pmol/L (0 to 2429 pmol/L). A Wilcoxon test shows that these two populations are significantly different ($p = 0.0012$). Comparing the two congenital groups the 58 dogs $\leq$ 1 year old had a mean of 867 pmol/L (0 to 5123 pmol/L) while the 67 dogs $>$ 1 year old with a congenital heart defect had a mean NTproBNP of 1251 pmol/L (0 to 8829 pmol/L). There is no significant difference between these two populations when tested with a Wilcoxon test ($p = 0.43$).

The difference in NTproBNP between dogs $\leq$ 1 and those $>$ 1 may suggest an age related difference in the normal population. However dogs with congenital heart disease appear to have similar NTproBNP concentrations regardless of age. This finding differs from the human studies.

ABSTRACT #236

EVALUATION OF A 14 AND 16 GAUGE SIDE-CUTTING BIOPSY NEEDLE AND FOUR DIFFERENT ASPIRATION PRESSURES USED FOR OBTAINING BRAIN TISSUE IN DOGS. AR Schneider, AV Chen, GJ Haldorson. Washington State University College of Veterinary Medicine, Pullman, WA.

Currently, stereotactic brain biopsies in dogs are performed with a 14 gauge side-cutting needle; however, this procedure is associated with a significant level of morbidity and mortality. In an attempt to reduce the risks involved with this diagnostic technique, the authors wanted to evaluate a larger gauge side-cutting needle. The purpose of our study was to evaluate the effectiveness of the 14 gauge and 16 gauge side-cutting biopsy needles at four different aspiration pressures for obtaining brain tissues for histopathology in dogs.

Twelve dog cadaver brains were used for sample collection. The cerebral cortex was randomly sampled using a combination of two different gauges of needles and four different aspiration pressures. These biopsies were evaluated grossly and with histopathology. The weight, surface area, architecture and degree of tissue artifact for each biopsy sample were evaluated.

A total of 96 biopsy samples were obtained (12 samples for each of the 8 combinations). Both the 14 and 16 gauge needles provided adequate tissue for histological interpretation (96–100%). There was a significant relationship between the amount of aspiration pressure and the amount of tissues obtained for both gauges ($p < 0.05$). However, the amount of aspiration pressure did not affect the degree of tissue artifact for both gauges ($P > 0.05$). Overall, we conclude the 16 gauge side-cutting biopsy needle is just as effective as the 14 gauge needle for obtaining brain tissues for histopathology in dogs. Based on Dunn's Multiple Comparison Test, the most effective aspiration pressure for both gauges was 2.0 ml.

ABSTRACT #237

SURGICAL MANAGEMENT OF HYDROCEPHALUS BY VENTRICULOPERITONEAL SHUNT PLACEMENT IN DOGS. N Shihab, E Davies, PJ Kenny, S Loderstedt, HA Volk. The Royal Veterinary College, London, United Kingdom.

Hydrocephalus is a common congenital or acquired neurologic condition in dogs and has been managed surgically for several decades. However, there is a lack of data and long term follow up on ventriculoperitoneal shunt (VPS) placement in dogs with hydrocephalus for which no active cause was identified. The objective of this study was to describe the clinical presentation and outcome of VPS treated dogs with hydrocephalus for which no active cause was determined.

The clinical records data base at a UK referral hospital was searched from June 2003–2009 to identify dogs diagnosed with hydrocephalus. Dogs were only included in the study if no active, ongoing cause was identified for the hydrocephalus, and who were managed with placement of a ventriculoperitoneal shunt. The clinical records of these cases were reviewed to determine the presenting and pre-operative clinical findings, initial post-operative outcome and re-examination findings. Additional follow-up information was obtained by telephone contact with the owners and referring veterinary surgeons.

Twelve dogs fulfilled the inclusion criteria. The median age at onset of signs was 10 months (2–111 months) and the median duration of clinical signs prior to diagnosis was 0.36 months (range 0.04–12 months). All dogs presented with signs of progressive forebrain dysfunction, seven with additional vestibular signs and three with multifocal spinal pain. Initial improvement was seen in all cases post surgery. The median time of follow up was 9.5 months (0.5–79 months). Three dogs were euthanised due to deterioration of their neurological signs after 2 weeks, 1 month and 5 months respectively. The only short-term complication reported after their 4 week re-examination was skull and cervical hyperaesthesia ($n = 4$). Other complications reported were undershunting due to shunt kinking ($n = 1$; 4 months post surgery) and development of seizures ($n = 1$; 6 months post surgery). Thus, sustained post-operative improvement was seen in nine dogs, only one of which required revision surgery.

In our study, sustained clinical improvement was seen in the majority of VPS treated cases of canine hydrocephalus for which no active underlying cause was identified. VPS placement should be considered for the management of hydrocephalus with no active underlying cause and progressive clinical signs.

ABSTRACT #238

THE TIME COURSE OF PLASMA PHOSPHORYLATED NEUROFILAMENT NF-H LEVELS IN RODENT AND CANINE SPINAL CORD INJURY MODELS. Nozomi Takahashi[1], Hiroaki Kamishina[1], Yui Kobatake[1], Jun Sasaki, Masaaki Katayama[1], Jun Yasuda[1], Reeko Sato[1], Hee-Myung Park[2], Byeong-Teck Kang[3], Gerry Shaw[4]. [1]Iwate University, Morioka, Iwate, Japan. [2]Konkuk University, Soul, South Korea. [3]Dongguk University, Goyang, South Korea. [4]University of Florida, Gainesville, FL.

Neurofilament (NF) is a cytoskeletal protein in axons known to be released into the blood after spinal cord injury (SCI). This characteristic makes NF-H a candidate as a biomarker for SCI. The objective of the study was to observe the time course of NF-H concentration in plasma after different severity of SCI as well as hindlimb locomotive functional recovery.

We created mild, moderate, and severe SCI in rats at T11 by MASCIS Impactor System. Canine SCI models were made at T13 using silicone balloon catheter compression methods ($n = 2$). For two weeks after SCI, ELISA assay was performed to quantify the concentrations of NF-H in plasma. Functional recovery of the hindlimbs was evaluated by use of BBB score in rats and the Olby score in dogs.

In rat SCI models, the concentration of NF-H had biphasic peaks at 16 hours and 3~4 days after SCI and declined thereafter in all severity. The overall NF-H concentration and the recovery of hindlimb locomotor function depended on the severity of the injury; the severe group being the worst and the mild group being the best. In canine SCI models, changes of post-injury NF-H levels showed different patterns in 2 dogs; NF-H levels increased sharply after injury with the peak on day-3 in first dog whereas the levels increased slowly with the peak on day-9 in the other dog. Significant motor recovery was not observed in either dog. These findings suggest that measurement of NF-H allows assessment of lesion severity and may aid in judging prognosis of SCI. NF-H is a unique biomarker of SCI and can be used to assess the efficacy of prospective neuroprotective agents in future SCI research.

ABSTRACT #239

ACUTE POLYRADICULONEURITIS IN DOGS: UTILITY OF CLINICAL PARAMETERS IN PREDICTING OUTCOME. LR Talarico[1], DJ Fletcher[1], CW Dewey[1], K Vernau[2]. [1]Cornell University College of Veterinary Medicine, Ithaca, NY. [2]University of California School of Veterinary Medicine, Davis, CA.

Acute polyradiculoneuritis (AP) is a sporadically occurring peripheral neuropathy of dogs of unknown etiology. A multicenter retrospective study is being undertaken to characterize the clinical disease course and outcome of dogs with AP and to identify prognostic indicators. A total of 86 cases of canine polyradiculoneuritis diagnosed between 1982 and 2009 were identified. Preliminary data analysis including 32 of these cases is reported here. Dogs that presented with an acute development of diffuse lower motor neuron signs and normal cranial nerves were included. Animals presenting for a relapse of a previous polyradiculoneuritis episode or clinical and/or diagnostic features consistent with myasthenia gravis, tick paralysis, botulism, or paraneoplastic neuropathies were excluded. Recorded data included: signalment, concurrent illnesses, vaccination history, raccoon encounters, number of days hospitalized, progression of clinical signs during hospitalization, and electrodiagnostic results. The severity of clinical signs at the time of presentation, rapidity of disease progression, secondary complications, survival to discharge, and clinical outcome were assigned numerical scores. Dogs that died before discharge had more rapid disease progression than those that survived (Wilcoxon rank-sum test, median progression score 1.75, 4.0 respectively, $p = 0.022$). There was no significant association between initial disease severity score and outcome. These results suggest that severity of disease at diagnosis is not prognostic in dogs with AP, but that dogs with rapid disease progression are less likely to survive.

ABSTRACT #240

THE CUTANEOUS TRUNCI MUSCLE REFLEX IN NORMAL DOGS. A. Muguet-Chanoit, K Babb, N.J. Olby. North Carolina State University, Raleigh, NC.

The cutaneous trunci muscle (CTM) is a thin but extensive sheet of skeletal muscle that covers most of the dorsal and lateral walls of the abdomen and thorax, and is innervated by the lateral thoracic nerve. Nociceptive stimulation of the skin elicits a CTM reflex contraction. This reflex is commonly used to localize spinal lesions, but there is little data available on the CTM reflex in normal dogs. The purpose of this study was to describe the sensory field, contraction strength and latency of the CTM reflex in normal dogs of different breeds and ages. The caudal border of the CTM reflex sensory field was evaluated in 100 dogs aged 2 to 10 years and 50 dogs aged 11 years or older. Hemostatic forceps were used to stimulate the skin two centimeters from the dorsal midline, starting at the level of the seventh lumbar (L7) dorsal spinous process and moving cranially until the CTM reflex was elicited. The border of the reflex was expressed as the corresponding vertebra identified by palpating the dorsal spinous processes. The lateral borders of the CTM reflex were evaluated in 50 dogs and the distance from the midline to the lateral border was expressed as a percentage of the hemi circumference of the trunk at each of four levels (the caudal border of the scapulae, the 13th thoracic and 3rd and 6th lumbar dorsal spinous processes). The mean and median values were calculated. Finally, the strength and latency of ipsilateral and contralateral contraction were described at the same four levels.

The CTM reflex was elicited in all dogs tested. Its caudal border was at L6 or 7 in 97% (97/100) of the 2 to 10 year old dogs and in 94% (48/50) of the old dogs. The caudal border was at L5 in three of the 2 to 10 year old dogs and at L1 and L3 in two old dogs. The lateral field of the CTM reflex occupied more than 50% of the hemi circumference of the trunk at each level tested, and was bilaterally symmetrical. The median percentage was 70% at the level of the scapulae, and it decreased caudally to 53% at the level of L6. The CTM contraction was more intense when the skin was pinched in the thoracic and cranial lumbar area than in the mid and caudal lumbar area in all dogs. There was no delay in response on the ipsilateral side, while the controlateral response was weaker than the ipsilateral response and occurred simultaneously or slightly later.

The CTM reflex is reliably elicited from L6/7 cranially in the majority of dogs although the caudal border of the reflex lay in the mid to cranial lumbar spine in two older dogs. The lateral borders of the CTM reflex extend further ventrally than previously believed; the dorsal cutaneous branches of the thoracic and lumbar spinal nerves provide the afferent arm of the CTM reflex and are reported to supply the dorsal third of the thorax and abdomen only. We conclude that either the sensory field of these cutaneous nerves is larger than believed, or that the lateral cutaneous nerves also contribute to the CTM reflex. This description of the normal CTM reflex provides baseline data for studies of the abnormal CTM reflex in dogs with thoracolumbar myelopathies.

ABSTRACT #241

THE CUTANEOUS TRUNCI MUSCLE REFLEX: A PREDICTOR OF RECOVERY IN DOGS WITH ACUTE THORACOLUMBAR MYELOPATHIES. A. Muguet-Chanoit, S. Moore, K. Babb, J.-H. Lim, R. Gallagher, S. Dillard, Z. Niman, J. Campbell, P. Early, C. Mariani, K. Munana, R. Longshore, C. Giovanella, P. Nghiem, C. Freeman, N.J. Olby. North Carolina State University, Raleigh, NC.

Acute intervertebral disc herniations occur most commonly between the eleventh thoracic and second lumbar vertebrae. Lesions in this location cause the caudal border of the cutaneous trunci reflex (CTMR) to advance cranially. We propose that changes in the caudal border of the CMTR in the immediate post operative period are predictive of outcome in dogs with acute, severe intervertebral disc herniations. The aim of this study was to evaluate changes in the CTMR in the post-operative period in dogs that have suffered a severe spinal cord injury due to an acute thoracolumbar disc herniation, and to compare these changes with the outcome.

The caudal border of the CTMR sensory field was studied in the immediate post-operative period in 42 dogs with thoracolumbar intervertebral disc herniations, paraplegia and loss of nociception at presentation. All participating dogs had cranial migration of the caudal border of the CTMR at presentation consistent with the localization of their spinal cord injury. The caudal border of the CTMR was assessed using hemostatic forceps to stimulate the skin bilaterally approximately two centimeters from the dorsal midline 24 hr after surgery, and at discharge. A complete neurological examination was repeated a minimum of three months after injury and the outcome of each dog was recorded as successful (voluntary motor present) or unsuccessful (euthanized or paralyzed). Changes in the CTMR caudal border (static, moved caudally or moved cranially) noted in the immediate post-operative period were then compared with the outcome using a Chi Square analysis.

Of the 42 dogs recruited, 30 dogs had improved three months after surgery, seven were euthanized because of ascending myelomalacia, and five remained paraplegic. Twenty-four of the dogs (57%) showed caudal movement of the caudal border of the CTMR post operatively. Twenty-two of these 24 dogs (92%) improved neurologically. Overall, the CTMR moved caudally in 73% (22/30) of the dogs that improved. The border of the CTMR moved cranially in 6 of the 7 dogs (86%) that developed ascending myelomalacia. There was a significant difference in outcome in dogs that had changes in the border of their CTMR (moving either cranially or caudally) and in dogs that had no change ($p < 0.0001$).

In the immediate post-operative period, the CTMR is an indicator of neurological recovery in dogs with severe acute intervertebral disc herniations. Caudal progression of the caudal border of the CTMR was associated with improvement in motor function in 92% of cases, whereas a cranial movement of the caudal border of this reflex was associated with the development of ascending myelomalacia in 86% of cases.

ABSTRACT #242

COMPARISON OF MODELS FOR PREDICTING RENAL DISEASE FOLLOWING I-131 THERAPY FOR FELINE HYPERTHYROIDISM. J Morrison[1], A Jergens[1], K Deitz[1], K Miles[1], R King[1], C Wang[1], M Winter[2], C Flatjord[1], A Hodnefield[1]. [1]Iowa State University, College of Veterinary Medicine, Ames, IA. [2]University of Florida, College of Veterinary Medicine, Gainesville, FL.

Hyperthyroidism is a common endocrine disorder in middle aged to older cats. Therapy with I-131 is considered to be curative for >

90% of affected cats. However, the hyperthyroid state may mask underlying renal insufficiency and rates of subsequent renal disease following I-131 therapy approach 30%. Initial prospective analysis of a population of hyperthyroid cats identified pre-treatment serum creatinine concentration and glomerular filtration rate (GFR, as determined by nuclear scintigraphy) to be significantly associated with the development of renal disease following I-131 therapy. A receiver operating characteristic (ROC) curve was developed for each of the following scores: pre-treatment creatinine concentration, pre-treatment GFR, and a linear combination score using weighted values of creatinine and GFR, i.e. (2.6953)(creatinine) – (0.9617)(GFR). The coefficients in the linear combination were determined by a logistic regression model with the two explanatory variables. Area under the curve (AUC) and sensitivity and specificity values were calculated for each model.

Forty-six hyperthyroid cats were treated with I-131 and 13/46 (28%) were determined to have renal disease following therapy. For cats that developed renal disease following I-131 therapy, mean pre-treatment creatinine concentration was 1.29 mg/dl (range: 0.6–2.2 mg/dl) and mean GFR was 2.01 ml/min/kg (range: 1.03–2.8 ml/min/kg). In cats that maintained normal renal function following therapy, mean pre-treatment creatinine concentration was 0.83 mg/dl (range: 0.2–1.7 mg/dl) and mean GFR was 2.91 ml/min/kg (range: 1.09–7.41 ml/min/kg).

Estimated AUC was highest for the combination model (0.853), followed by the creatinine concentration AUC (0.798) and the GFR AUC (0.769). The Youden index, defined as max (Se+Sp-1), was used to determine the optimal cutoff value for each of the 3 models. In the combination model, a combination score > 0.7818 (the cutoff value was chosen based on a cat with a pre-treatment creatinine concentration of 1.1 mg/dl and pre-treatment GFR value of 2.27 ml/min/kg) predicted the development of renal disease with 85% sensitivity and 85% specificity. For the creatinine only model, a pre-treatment creatinine concentration ≥ 1.4 mg/dl predicted the development of renal disease with 77% sensitivity and 76% specificity. In the GFR only model, a pre-treatment GFR value ≤ 2.18 mL/min/kg predicted post treatment renal disease with 85% sensitivity and 73% specificity.

The results of this analysis demonstrated the model based on both pre-treatment serum creatinine concentration and GFR afforded the best predictive power for the development of renal disease following I-131 treatment. This model was superior to models that evaluated the variables independently.

ABSTRACT #243

INVESTIGATION OF PROGNOSTIC FACTORS FOR THE DEVELOPMENT OF RENAL DISEASE FOLLOWING I-131 THERAPY IN FELINE HYPERTHYROIDISM. J Morrison[1], A Jergens[1], K Deitz[1], K Miles[1], R King[1], C Wang[1], M Winter[2], C Flatjord[1], A Hodnefield[1]. [1]Iowa State University, College of Veterinary Medicine, Ames, IA. [2]University of Florida, College of Veterinary Medicine, Gainesville, FL.

Hyperthyroidism is one of the most common endocrine disorders in middle aged to older cats. Renal disease may be masked by hyperthyroidism and subsequent I-131 therapy of hyperthyroid cats with subclinical renal dysfunction may precipitate progression to renal failure. The purpose of this study was to prospectively investigate potential prognostic factors for predicting the development of renal disease in spontaneous feline hyperthyroidism.

Fifty-two hyperthyroid cats were evaluated prior to I-131 therapy over a 48 month period. The diagnostic evaluation for all cats included: physical examination, CBC, serum biochemistry profile, urinalysis, total T4, thoracic radiographs, indirect blood pressure via Doppler manometry, single strand electrocardiography, thyroid scintigraphy with sodium pertechnetate (^{99m}TcO4-), and determination of GFR via ^{99m}Tc-Diethyltriaminepentacetic acid nuclear scintigraphy (^{99m}Tc-DTPA). Nuclear studies were performed in cats sedated with either midazolam/ketamine or butorphanol. Urine protein excretion was determined by either a urine protein:creatinine (U P:C) ratio if pathologic proteinuria was suspected on routine urinalysis, or urine was tested for the presence of microalbumin (E.R.D. – Health Screen, Heska) if proteinuria was not initially detected. Specific physical examination parameters recorded in all cats included age, weight, breed, gender, and heart rate. Repeat diagnostic testing occurred no sooner than 4 weeks post I-131 therapy whereby all cats minimally had a serum creatinine measurement determined. A diagnosis of renal disease was made on the basis of increased serum creatinine levels (> 1.6 mg/dl) coupled with urine specific gravity when available.

Forty-six cats completed I-131 therapy with 13/46 (28%) determined to have renal disease at the time of follow up. Pre-treatment serum creatinine levels were significantly ($p < 0.02$) associated with the development of renal disease as determined by logistic regression analysis. Mean pre-treatment creatinine concentration for all cats was 0.96 mg/dl (range: 0.2–2.2 mg/dl) and mean post-treatment creatinine was 1.65 mg/dl (range: 0.9–4.1 mg/dl). Pre-treatment GFR was less significantly ($p < 0.08$) associated with the development of renal disease. Mean GFR for all cats before treatment was 2.65 ml/min/kg (range: 1.03–7.41 ml/min/kg). The predicted chance for renal disease in a cat with a pre-treatment creatinine of 1.6 mg/dl and GFR of 2.25 ml/min/kg was 67%. No other patient factors were significantly associated with the development of renal disease post I-131 therapy.

The data of the present study confirms a 30% incidence of renal disease following radioactive iodine therapy of feline hyperthyroidism. Pre-treatment serum creatinine concentration was the variable most highly correlated with a negative outcome of renal disease.

ABSTRACT #244

BREED-SPECIFIC REFERENCE INTERVALS FOR ASSESSING THYROID FUNCTION IN SEVEN DOG BREEDS. RL Hegstad-Davies, SMF Torres, LC Sharkey, CA Muñoz-Zanzi, SC Gresch. University of Minnesota, College of Veterinary Medicine, St. Paul, MN.

We established breed specific reference intervals for serum thyroxine (T4), free T4 using equilibrium dialysis (FT4D), and thyrotropin (TSH) concentrations for golden retriever (GR), keeshond (K), Alaskan malamute (AM), Siberian husky (SH), samoyed (SAM), English setter (ES), and collie (C) breeds. Healthy, purebred, fasted dogs one to nine years of age were eligible for enrolment. Intact, neutered or spayed animals were eligible to participate. Female dogs that were in estrus, pregnant, pseudo-pregnant or lactating were excluded, as were dogs on medication known to affect thyroid function. A veterinarian reviewed the health history, performed a physical examination and supervised the interpretation of data from samples collected for a complete blood count, chemistry profile and urinalysis for all dogs prior to thyroid function testing.

Nine percent (57 of 621) of T4, and 7% of FT4D measured concentrations were less than the non breed-specific lower limit guideline for our laboratory (13 nmol/L TT4; 8 pmol/L FT4D). Four percent of all measured concentrations of TSH were above the conventional, non breed-specific upper limit (0.6 ng/ml) guideline.

By breed, percentage of samples containing less than 13 nmol/L T4 were: K:1.5%, AM:2.1%; collie:5.3%, SAM: 6%; SH:11%; GR: 12%; ES: 26%. For FT4D, the percentage of samples containing less than 8 pcmol/L for each breed was K: 0%; SAM: 0%; C: 4.2%, AM: 6%; SH: 7%; GR: 9%; ES: 25%. Less than 1% of samples had T4 or FT4D concentrations exceeding the non breed-specific reference interval limit guidelines (50 nmol/L T4; 40 pmol/L FT4D). For TSH, the percentages of samples exceeding the standard upper limit (normal < 0.6 ng/ml) were: AM/GR/ES: 0%; SH: 1%; K: 3%; SAM: 10%; collie: 16%.

For reference intervals established at the 95% confidence interval level, varibility of 2.5% at each limit is expected in a normal distribution. In this study, 9% of all purebred dogs had T4 values below the lower limit and in the English setter breed, this prevalence reached 25%. The increased prevalence of healthy purebred dogs with T4 and FT4D concentrations less than the conventional reference interval limit suggests the need for breed-specific reference intervals for evaluating thyroid function.

ABSTRACT #245

A NOVEL SANDWICH ELISA FOR THE MEASUREMENT OF INSULIN IN FELINE SERUM AND PLASMA. H Johansson, R Boman. Mercodia AB, Uppsala, Sweden.

Diabetes mellitus is one of the most common endocrine disorders in cats, a form that closely resembles human type 2 diabetes. Its incidence rate among cats appears to be increasing, probably due to an increase in obesity and a decrease in physical activity in the cat population. Obesity increases the risk for diabetes 3- to 5-fold. Diabetes can affect cats of all ages, but most cats are over six years of age when diagnosed. Diabetic cats may go into remission and studies have shown that different insulin therapy regimens may have an influence on this. Mechanisms underlying the development of insulin resistance in cats are currently unclear and further investigations are warranted.

The objective of this study was to validate an ELISA, optimized for quantitative measurements of feline insulin. The ELISA assay has recently been developed at Mercodia AB, Uppsala, Sweden.

The Feline Insulin ELISA is a simultaneous assay with a required sample volume of 10 µl. The incubation time is 120 minutes plus 30 minutes substrate incubation.

A validation of accuracy, sensitivity, specificity and precision was conducted for the ELISA. To validate the accuracy twelve feline samples were diluted with calibrator 0 and twelve additional feline samples were spiked with calibrator insulin. Capability of Detection was determined according to the methodology described in ISO11843. Cross reactivity of commercially available analogues was studied. The precision study included both repeatability and reproducibility. The repeatability study was performed with three feline samples, which were analyzed in four replicates at fourteen different occasions in one laboratory. The reproducibility or inter lab comparison was performed at four laboratories, two in Europe and two in North America. One control serum was measured in duplicate in a total of fifteen runs (1–5 runs in each laboratory).

The assay range is 5–700 ng/l and the Capability of Detection was determined to be 5 ng/l. Recovery upon dilution was 79–113% (mean 95%) and recovery upon addition was 93–122% (mean 107%). The cross reactivity of glargine was determined to 8.4%, detemir to 0.008%, caninsulin to 57.4%, aspart to 5.9%, lispro to 0.00000002% and glulisine to 0.0000007% No cross reactivity to c-peptide or proinsulin was found. The repeatability study showed an intra and inter assay variation of 4.0% and 7.2%, respectively. The reproducibility study showed an intra and inter assay variation of 3.9% and 9.4%, respectively. Statistical analysis was conducted using one-way ANOVA.

This novel feline insulin ELISA provides analytical performance, precision and reliability equal to methods currently used in human clinical research. This method enables the performance of additional mechanistic investigations, which may contribute to a more complete understanding of the development of feline diabetes.

ABSTRACT #246

MEASUREMENT OF BETA-HYDROXYBUTYRATE (HB) IN CATS WITH NONKETOTIC DIABETES MELLITUS, DIABETIC KETOSIS AND KETOACIDOSIS USING A PORTABLE KETONE METER. C Weingart[1], G Arndt[2], B Kohn[1]. [1]Small Animal Clinic. [2]Institute for Biometrics and Data Processing, Faculty of Veterinary Medicine, Freie Universität Berlin, Germany.

The diagnosis of diabetic ketoacidosis (DKA) is based on the detection of ketone bodies in plasma or urine.

The objective of this study was to evaluate a portable ketone meter (PKM) for the measurement of HB in peripheral and capillary blood in cats. Moreover, a cut-off value for the diagnosis of DKA should be established.

Cats presented between March and November 2009 were included in the study: 11 healthy cats, 23 cats (group 1) with nonketotic diabetes mellitus (n = 9) and diabetic ketosis (DK, n = 14) displaying hyperglycemia, glucosuria, and increased fructosamine concentrations without metabolic acidosis (in cats with DK plasma ketone bodies were present). Cats in group 2 (DKA, n = 15) had ketone bodies and were acidotic. HB concentrations were measured in whole blood by a PKM (Precision Xceed, Abbott GmbH, Germany) and were compared with a spectrophotometric method.

In the control cats the HB concentrations were 0 mmol/l with both methods. In cats of group 1 and 2 the median difference between both methods was −0.1 mmol/l (−5.5 to +1.9). The HB values measured by the reference method were higher in 78% of the cases if the PKM values were > 4 mmol/l. In cats of group 1 HB concentrations were 0 - 5.7 mmol/l (median 0.3), in cats of group 2 4.4 - 7.4 mmol/l (median 5.4).

The PKM is accurate for the quantitative measurement of HB concentrations < 4 mmol/l. A cut-off value of > 2.3 mmol/l revealed a 100% sensitivity and a 83% specificity to diagnose DKA.

ABSTRACT #247

A PUTATIVE GAIN-OF-FUNCTION MUTATION IN *CRHR1* GENE CAUSING ACTH-DEPENDENT HYPERCORTISOLISM IN A POODLE DOG. V. De Marco[1,2,3] LR Carvalho[1], AEC Billerbeck[1], PSL Oliveira[4] & BB Mendonca[1]. [1]Laboratório de Hormônios e Genética Molecular LIM/42, Faculdade de Medicina da Universidade de São Paulo, Brazil. [2]Universidade Guarulhos, SP, Brazil. [3]Pompéia Veterinary Hospital, São Paulo, Brazil. [4]Heart Institute (INCOR), Faculdade de Medicina da Universidade de São Paulo, Brazil.

Pituitary-dependent hypercortisolism (PDH) is a very common endocrinopathy in dogs, generally caused by an ACTH-secreting corticotroph adenoma although the underline pathogenesis is still unknown. There is a high incidence of PDH in Poodles and familial pituitary-dependent hypercortisolism have been identified suggesting a genetic involvement. An increased *CRHR1* expression was demonstrated in human and canine ACTH-secreting pituitary adenomas, despite the autonomous ACTH secretion and the low portal levels of CRH. In addition, prolonged exposure of human corticotroph adenoma cells to CRF does not result in receptor desensitization. The aim of the present study was to screen the *Crhr1* gene for mutations in genomic DNA from Poodles with PDH. Fifty Poodle dogs (33 female, median age of 8.71 years, range 1.5 to 14 years) with PDH were studied. The diagnosis was based on clinical signs (polyphagia, polyuria, polydipsia, abdomen enlargement, panting) and cortisol levels > 1,4 µg/dl after dexamethasone suppression test, plasmatic ACTH levels > 17 pg/ml, bilateral adrenal enlargement at abdominal ultrasound, hyperlipidemia, hyperphosphatasemia, low urinary density. Genomic DNA was isolated from peripheral blood, amplified by PCR using intronic primers to amplify 13 exons, and submitted to automatic sequence. The coding sequence was compared to the canine Boxer genomic DNA sequence available at NCBI site. We found the heterozygous allelic variation p.V97M (GTG > ATG) in exon 4 in one dog. To investigate if this variation is a mutation or a single nucleotide polymorphism, we studied fifty normal poodle dogs (32 female, median age of 9,38 years, range 6–16 years) and p.V97M was not found in one hundred normal alleles. The codon 97 is located in the extracellular aminoterminal domain of the CRHR1 and is extremely important for high affinity ligand binding. The molecular study of the quaternary structure of normal and mutated proteins showed a structural rearrangement of the mutated protein by changing the contact surface between the CRH and its receptor, Crhr1, resulting in a 17% higher binding energy comparing to the wild type. In conclusion, this study identified a Crhr1 gain-of-function mutation, probably responsible for ACTH-dependent hypercortisolism in a poodle dog of our cohort.

ABSTRACT #248

ULTRASONOGRAPHIC ADRENAL GLANDS THICKNESS MEASUREMENT IN DOGS WITH PITUITARY-DEPENDENT HYPERADRENOCORTICISM IN COMPARISON WITH NORMAL DOGS MATCHED BY WEIGHT BODY. V De Marco[1,2,3], RS Pereira[2], NK Kage[3], FA Santos[2], VK Ramos[2], S Buranello[3], AAL Jorge[1] & BB Mendonca[1]. [1]Laboratório de Hormônios e Genética Molecular LIM/42, Faculdade de Medicina da Universidade de São Paulo, Brazil. [2]Universidade Guarulhos, SP, Brazil. [3]Pompéia Veterinary Hospital, São Paulo, Brazil.

Pituitary-dependent hyperadrenocorticism (PDH) is a common canine endocrinopathy that results in bilateral adrenocortical hyperplasia and small breeds are at increased risk for developing PDH. Adrenal thickness (caudal polar width) evaluated by ultrasonographic is an important tool to determine adrenocortical hyperplasia and the current cut-off for PDH diagnosis is 0.75 cm.However, a markedly variability in adrenal thickness has been reported in normal dogs, from 0.2 to 0.7 cm, probably reflecting weight-associated variation. Our aim was to determine the size of the adrenal glands ultrasonographically in normal healthy dogs weighing $\leq$ 10 kg and compare these measurements in dogs with PDH with similar weight. Ultrasonographic adrenal gland evaluation was performed in 50 dogs with PDH (30 females and 20 males) from various breeds (30 Poodles, 7 Teckels, 6 Yorkshires terrier, 1 Maltese and 6 mongrel dogs) and in 109 healthy control dogs. The diagnosis of hyperadrenocorticism was based upon the characteristic signs and symptoms, such as polyphagia, polyuria, polydipsia, abdomen enlargement, panting, hyperlipidemia, hyperphosphatasemia, low urinary density and cortisol levels > 1,4 µg/dl after dexamethasone suppression test. PDH dogs had similar age in comparison with control ones (8.7 ± 2.6 vs. 7.7 ±3.4 y., $p = 0.07$), but they were heavier than controls (7.7 ± 1.4 vs. 5.9 ± 2.2 kg, $p < 0.001$). There was no difference in breed distribution between both groups. There was no difference between left and right adrenal thickness measurements, which was positively correlated with animal weight ($r = 0.41$, $p < 0.001$) and age ($r = 0.364$, $p < 0.001$) in control group but not in PDH group. All dogs with PDH had bilaterally symmetrical normal shape adrenal glands. Adrenal thickness measurement in PDH dogs (0.74 ± 0.12 cm) was significantly greater than adrenal thickness in control dogs (0.5 ± 0.11, $p < 0.001$). ROC curve analysis suggested that adrenal thickness measurement $\geq$ 0.59 cm has the highest accuracy for PDH diagnosis. Using this cutoff 45 (90%) of PDH dogs presented at least one enlarged adrenal, whereas 28/109 (25.7%) of control animals had adrenal thickness > 0.59 (sensitivity of 90%, specificity of 74%). Using the classical cut-off for adrenal thickness measurement of 0.75 cm, only 24 (48%) of PDH dogs of this cohort will be considered as having enlarged adrenals. In conclusion, our results indicate the need to correlate the thickness of the adrenal glands with the dog's body weight to improve the sensitivity of adrenomegaly detection in PDH dogs.

ABSTRACT #249

PREVALENCE OF METABOLIC SYNDROME IN DOGS WITH PITUITARY-DEPENDENT HYPERADRENOCORTICISM.

V. De Marco[1,2], VK Ramos[2], V. Oliveira[2], JS Silva[2], T Parraz[2], RS Pereira[2]. [1]Pompéia Veterinary Hospital, São Paulo, Brazil. [2]Universidade Guarulhos, SP, Brazil.

Metabolic syndrome is a cluster of metabolic abnormalities that increases the risk for type 2 diabetes and vascular disease. Pituitary-dependent hyperadrenocorticism (PDH) is one of the most common endocrinopathies in dogs. PDH, characterized by excessive levels of glucocorticoids, results in increased glucose and insulin levels which cause insulin-mediated effects on adipose tissue, ultimately promoting visceral adiposity, insulin resistance, dyslipidemia and hypertension. This group of abnormalities is observed in obese human patients with metabolic syndrome and at least three of these events must occur simultaneously in the same individual to confirm the metabolic syndrome diagnosis. The aim of this study was to evaluate the prevalence of metabolic syndrome in dogs with PDH. Seventy-seven dogs (54 females, mean age of 9.5 ± 2.6 years) from various breeds with untreated PDH were evaluated in the University Guarulhos Veterinary Hospital over a period of 4 years. The diagnosis of PDH was based on clinical signs (polyphagia, polyuria, polydipsia, abdomen enlargement, panting), cortisol levels > 1.4 µg/dl after dexamethasone suppression test, bilateral adrenal enlargement at abdominal ultrasound, hyperlipidemia, hyperphosphatasemia and/or low urinary density. The metabolic syndrome investigation was based on measurement of systolic blood pressure, fasting blood glucose paired with basal insulin levels, serum cholesterol and triglycerides. Blood pressure measurement was conducted in 62 dogs and fasting serum insulin was determined in 35 recent cases. Sixty per cent of the dogs ($n = 37/62$) had arterial hypertension with an average systolic pressure of 164 ± 24 mmHg. Elevated blood levels of cholesterol (mean ± SD, 382 ± 155 mg/dl) and triglycerides (181 ± 108 mg/dl) were observed in 74% ($n = 57/77$) and 75% ($n = 58/77$) of the cases, respectively. Absolute basal hyperinsulinemia (37 ± 27 µUI/ml) associated to normoglycemia (81 ± 11 mg/dl) was observed in 68.5% ($n = 24/35$) of the dogs. All animals with insulin resistance had dyslipidemia (increased blood levels of cholesterol or triglycerides). Hypertension and dyslipidemia was observed in 54.8% ($n = 34/62$) of the dogs in addition to visceral obesity. In conclusion, a significant number of PDH dogs of our cohort presented metabolic syndrome, confirming that hypercortisolism shares many characteristics of metabolic syndrome and should always be investigated in order to reduce morbidity and mortality of this endocrinopathy.

ABSTRACT #250

EFFECT OF EXCESS DIETARY PROTEIN AND CARBOHYDRATE ON THE INTESTINAL MICROBIOTA IN HEALTHY BEAGLES AS ANALYZED BY GUANINE-PLUS-CYTOSINE-BASED PROFILING.

I Mõistus[1], WM Vos[2,3], J Zentek[4], J Harmoinen[1], A Kettunen[5], T Rinttila[5], J Apajalahti[5], and T Spillmann[1]. [1]Department of Equine and Small Animal Medicine, University of Helsinki, Finland. [2]Department of Basic Veterinary Sciences, University of Helsinki, Finland. [3]Laboratory of Microbiology, Wageningen University, The Netherlands. [4]Institute of Animal Nutrition, Free University of Berlin, Germany. [5]Alimetrics Ltd., Espoo, Finland.

Considerable evidence suggests that food has a significant impact on the gastro-intestinal (GI) function and the microbial ecology of the canine GI tract. The aim of the study was to evaluate the influence of dietary carbohydrates and animal-derived proteins with high collagen concentrations on the canine colonic microbiota.

An experimental Greco-Latin square design with five healthy Beagles was utilized, feeding high-protein (HP, starch 54 g/kg, crude protein 609 g/kg), high-carbohydrate (HC, starch 438 g/kg, crude protein 194 g/kg), and balanced commercial (BC, starch 277 g/kg, crude protein 264 g/kg) diets for three weeks each. The BC diet was used for the baseline and washout phases, for 2 and 4 weeks, respectively.

The diets did not affect the body weight and were otherwise tolerated well, except during the HP diet all dogs developed diarrhea. Total fecal bacterial community was assessed by profiling the guanine-cytosine content (%G+C) by 72-hour cesium chloride (CsCl) equilibrium density gradient ultracentrifugation at 100 000×g. No major differences between the BC and HC diet groups were observed. Fecal samples from the HP diet had a lower abundance of microbes with %G+C between 33 and 41 when compared to the BC diet samples ($p = 0.003$). Moreover, the HP dietary samples contained a shoulder at the %G+C range between 46 and 50, which was completely lacking from the BC diet samples ($p = 0.02$). A low %G+C shoulder (range 25–29) was observed to be present with the HP diet and lacking with the BC diet ($p = 0.05$). The HP diet also resulted in a lower abundance of microbes with a %G+C between 33 and 40 as compared to the HC diet ($p < 0.01$). The shoulder peak present in HP diet samples at the %G+C range between 46 and 50 was completely absent in the HC samples ($p = 0.02$). In contrast, the HC diet favoured bacteria with %G+C higher than 60 ($p < 0.05$).

In conclusion, significant differences between the HP and HC diets were observed in the average %G+C profiles. The influence of the HP diet on canine fecal microbiota appears to be higher than the HC diet, probably due to the low precaecal digestibility of collagen. The phylogenetic position of the bacteria belonging to the most abundant %G+C peaks and their possible contribution to previously reported changes in fecal volatile fatty acid profiles and the development of diarrhea in the HP diet phase is currently under investigation by advanced sequence analysis of 16S rRNA amplicons.

ABSTRACT #251

EFFECT OF OMEPRAZOLE TREATMENT ON GASTRIC AND DUODENAL BACTERIAL POPULATIONS IN HEALTHY DOGS.

O. Dossin[1], K. Jones[1], S. Clark-Price[1], JM. Steiner[3], SE. Dowd[2], JS. Suchodolski[3]. [1]Department of Veterinary Clinical Medicine University of Illinois, Urbana, IL, [2]Medical and Testing Laboratories, Lubbock, TX, and [3]GI Lab, Texas A&M University, College Station, TX.

Several recent studies have shown that treatment with proton-pump inhibitors may increase the prevalence of infectious disease such as community acquired pneumonia or *Clostridium difficile* infection in humans. Antacids are also frequently used in dogs. However, their effect on the gastrointestinal microbiota has not been reported in this species. Therefore, the aim of this study was to document the effect of a two-week treatment course with omeprazole on bacterial populations of the stomach and proximal duodenum.

Eight healthy adult research dogs (4 males and 4 females) were enrolled into the study. The dogs received an average dosage of 1.1 mg/kg of omeprazole twice a day for 15 days. The pH of gastric fluid was measured and endoscopic gastric and duodenal biopsies were harvested 30 and 15 days before starting omeprazole, on the last day of treatment, and 15 days after the end of treatment. Biopsies were snap frozen, stored at −80°C, and subsequently processed for microbial identification by massive parallel 16S rRNA gene pyrosequencing. Indices of bacterial diversity and species richness were calculated. The proportions of bacterial groups were determined at each taxonomic level. Changes in the pH of the gastric fluid and changes in microbial diversity and composition were compared before and after treatment with Friedman and post-hoc Dunn's tests.

Gastric fluid pH was significantly increased ($p < 0.01$) at the end of the treatment period (median: 7.1; range: 4.2–8.1) when compared to pretreatment values (median: 1.7; range: 1.5–4.4). In the stomach, omeprazole treatment significantly increased the median proportions of the order Lactobacillales (from 0.05% to 0.8%) but these proportions returned to the baseline value 15 days after the end of the treatment. In the duodenum, omeprazole reduced the median proportion of the genus *Porphyromonas* (from 0.1% to 0.0%), and a rebound increase of the median proportions of *Helicobacter spp.* was observed 15 days after the end of the treatment period (from 0.0% to 10.6%). The median proportion of the genus *Comamonas* was also decreased 15 days after the end of the treatment when compared to the pretreatment value (from 0.0% to 0.1%). The indices of bacterial diversity and richness were significantly decreased two weeks after the end of treatment in the stomach but not in the duodenum

In conclusion, omeprazole treatment had a significant effect on the gastric fluid pH but overall did not induce major changes in the microbiome in either the stomach or the duodenum. However, the approach used here allows only the assessment of relative changes in the proportions of bacterial groups within the total population. Therefore, significant changes in the total number of bacteria in response to omeprazole treatment may not have been discovered.

ABSTRACT #252

WOLINELLA SUCCINOGENES AND NOT *HELICOBACTER* SPP. ARE THE PREDOMINANT HELICOBACTERACEAE IN THE CANINE ORAL CAVITY.

M Craven[1], C Recordati[2], V Gualdi[2], G Pengo[2], M Luini[2], E Scanziani[2], KW Simpson[1]. [1]College of Veterinary Medicine, Cornell University, Ithaca, NY. [2]Dipartimento di Patologia Animale, Universitá degli Studi di Milano, Italy.

The role of *Helicobacter* spp. (*H. felis* - HF, candidatus *H. heilmanni* - HH, and *H. bizzozeronii* -HB) in the pathogenesis of gastric disease in dogs and cats remains poorly understood, whereas in people, gastric colonization by *H. pylori* is causally associated with gastric mucosal inflammation, ulceration and neoplasia. The epidemiology and major mode of transmission of *Helicobacter* spp. is still unknown, despite > 20 years of research, but zoonosis is not considered an important aspect of *H. pylori* acquisition. Gastric infection with non-*H. pylori* species has, however, been associated with gastritis and MALT lymphoma in 0.25% to 6.0% of human *Helicobacter* infections, which suggests a zoonotic risk. This is supported by the recent PCR-based discovery of *Helicobacter* genus DNA in the saliva and dental plaque of > 70% of dogs (n = 27/38). The purpose of this study was to assess the potential risk for dog to human and dog to dog transmission of oral *Helicobacter* spp.

Saliva and dental plaque from 28 pet dogs and endoscopic gastric biopsies from a subset of 8/28 dogs (5 vomiting, 3 healthy) were evaluated for the presence of oral cavity *Helicobacter* DNA by PCR with *Helicobacter* genus primers C97 and C05 (yielding 1200 bp 16S rDNA amplicons). The 1200 bp PCR amplicons from the oral cavity and stomach of 8 dogs were cloned using the pGEM®-T Easy vector and introduced into *E. coli* DH5α. Plasmid DNA from up to 10 clones per sample was sequenced with M13F and M13R primers and sequence identity was determined by comparison to databases at the Ribosomal Database Project and NCBI. Phylogenetic analysis was performed using MEGA version 3.1.

Helicobacter DNA was detected in the oral cavity of 24/28 dogs. Sequence analysis revealed a predominance of *Wolinella succinogenes* in 8/8 dogs, representing 91.2% of oral cavity clones. Only 2/8 dogs were positive for oral *Helicobacter* DNA, with greatest sequence homology to HH and HF. Gastric biopsies were positive for *Helicobacter* DNA in 8/8 dogs, and sequence analysis was most homologous to HF in 50.8% of clones, HH in 34.9%, and HB in 14.3%. Phylogenetic analysis showed the HH sequences to be clearly distinct from HH Type 1, which is the predominant non-*pylori Helicobacter* spp. in human infections. Sequences from the stomach and saliva of one dog clustered very closely, suggesting similar strains of HF and HH at both sites. *Wolinella*-specific PCR of saliva in the other 20 dogs was positive in 16, and in a total of 24/28 dogs. In conclusion, *W. succinogenes* is the dominant *Helicobacteraceae* in the canine oral cavity, and these findings demonstrate lack of specificity of previously published *Helicobacter* genus-specific PCR primers. Our results suggest that contact with the oral cavity of dogs is unlikely to be an important source for the transmission of non-*H. pylori Helicobacter* infections in people.

ABSTRACT #253

PYROSEQUENCING REVEALS A DIVERSE FUNGAL MICROBIOTA IN CANINE AND FELINE FECAL SAMPLES.

JS Suchodolski[1], SE Dowd[2], JM Steiner[1]. [1]Gastrointestinal Laboratory, Texas A&M University, College Station, TX; [2]Research and Testing Laboratory, Lubbock, TX.

The gastrointestinal microbiota plays an integral role in gastrointestinal health and disease. The prevalence and phylogenetic description of fungal organisms and their role as part of the intestinal ecosystem have not yet been studied extensively in dogs and cats. The aim of this study was to characterize fungal organisms in canine and feline fecal samples using a culture-independent high-throughput sequencing technique.

Fecal samples were obtained from a total of 36 cats and 41 dogs of different ages and breeds and living in various home environments. Samples from each species were pooled, with each pool consisting of 6 samples. To evaluate inter-animal differences in the prevalence and phylogenetic composition, fecal samples from 12 healthy dogs were analyzed individually. Genomic DNA was extracted from all pooled and individual samples. Universal fungal primers were used to amplify a part of the fungal ribosomal DNA. The generated amplicons were identified by fungal tag-encoded FLX-titanium amplicon pyrosequencing (fTEFAP).

A total of 67,461 sequencing tags were analyzed across all samples. Obtained sequences were classified into 6 fungal phyla: Ascomycota (92.8% of all sequences), Basidiomycota (4.3%), Chytridiomycota (2.1%), Neocallimastigomycota (< 1%), Glomeromycota (< 1%), and Microsporidia (< 1%). Within the phylum Ascomycota, Dothideomycetes (35.7% of sequences) and Saccharomycetes (26.7%) were the predominant fungal classes. Further phylogenetic classification revealed 272 different fungal genera and 399 unique phylotypes across all samples. The most commonly observed fungal genera were *Catenulostroma* and *Candida* (each identified in 94.7% of dogs), followed by *Penicillium* (89.5%), *Aureobasidium* (84.2%), *Myrothecium* (78.9%), and *Bipolaris* (78.9%). Fungal sequences were identified in all 12 individually analyzed canine fecal samples (100%). Each of these samples contained multiple fungal phylotypes (median: 40, range: 10 to 92). Remarkable inter-animal differences were observed with each dog harboring a unique fungal profile. While most dogs harbored similar fungal genera, each animal had a very unique species distribution. The most commonly identified phylotypes were *Catenulostroma abietis* and an unclassified *Candida* sp. (each identified in 94.7% of dogs).

These results indicate the usefulness of culture-independent molecular techniques for the characterization of fungal phylotypes in the canine and feline intestine. These findings further suggest that fungal organisms are a part of the normal intestinal ecosystem in dogs and cats. In this study, each dog harbored a unique fungal profile on the species level. Further studies are warranted to determine the role of fungal organisms in gastrointestinal health and disease.

ABSTRACT #254

MOLECULAR ASSAY FOR THE DETECTION OF *CAMPYLOBACTER* SPP. IN CANINE AND FELINE FECAL SAMPLES.

JS Suchodolski, NM Gossett, KM Aicher, RM Heilmann, PG Xenoulis, JM Steiner. Gastrointestinal Laboratory, Texas A&M University, College Station, TX.

It has been reported that the canine and feline intestine harbors at least 14 different *Campylobacter* spp., with *C. jejuni, C. coli, C. ups-*

aliensis, and *C. helveticus* being the most common. To understand the role of the various *Campylobacter* spp. in the pathogenesis of intestinal disease, methods for their species identification are needed. The aim of this study was to develop a rapid and simple molecular assay for detection and differentiation of *C. jejuni, C. coli, C. upsaliensis*, and *C. helveticus*, and to evaluate their prevalence in fecal samples from dogs and cats with and without clinical signs of gastrointestinal disease.

Polymerase chain reaction (PCR) primers were designed to amplify conserved regions of the 16S rRNA gene shared by *C. jejuni, C. coli, C. upsaliensis*, and *C. helveticus* on DNA extracted from fecal samples. PCR amplicons were digested with a restriction enzyme (*DdeI*) and the resulting fragments were resolved by agarose gel electrophoresis. The four *Campylobacter* spp. were identified based on their distinct banding patterns. To validate the specificity of the assay, PCR amplicons were verified by sequencing. To test the reproducibility of the method, ten fecal samples were analyzed in four consecutive assays. To evaluate the prevalence of the specific *Campylobacter* spp., fecal samples were analyzed from healthy dogs ($n = 77$) and cats ($n = 33$), and dogs ($n = 187$) and cats ($n = 138$) with clinical signs of gastrointestinal disease. The proportions of dogs and cats with or without gastrointestinal signs that harbored *Campylobacter* spp. were analyzed using a χ^2-test (GraphPad Prism 5.0). Significance was set at $p < 0.05$.

Sequencing results revealed that the obtained banding patterns were specific for each individual *Campylobacter* species. Reproducible results were obtained in 100% of the repeated assays. *Campylobacter* spp. were significantly more prevalent in diseased dogs (53/187; 28.3%) compared to healthy dogs (11/77; 14.3%; $p = 0.015$). In the positive dogs, *C. upsaliensis* was the predominant species, identified in 10/11 (90.9%) healthy and 45/53 (84.9%) of diseased dogs. The prevalence of *Campylobacter* spp. was not significantly different between healthy cats (18/33; 54.5%) and diseased cats (57/138; 41.3%; $p = 0.169$). In the positive cats, *C. helveticus* was the predominant species, identified in 12/18 (66.7%) healthy and 39/57 (68.4%) diseased cats. *C. jejuni* was identified in only 3/53 (5.7%) diseased dogs and none of the healthy dogs or any of the cats. *C. coli* was not identified in any of the samples.

In conclusion, a rapid and reproducible molecular method for the detection and differentiation of *C. jejuni, C. coli, C. upsaliensis*, and *C. helveticus* was developed. The results suggest that *C. upsalienis* and *C. helveticus* are commonly present in fecal samples of dogs and cats with or without gastrointestinal signs. In contrast, *C. jejuni* has a low prevalence and was only identified in a small proportion (3/53) of dogs with gastrointestinal disease.

ABSTRACT #255

PREDICTORS FOR FECAL COLONIZATION OF PROBIOTIC LACTIC ACID BACTERIA IN HEALTHY CATS AND DOGS.

JF Garcia-Mazcorro[1], FM Speed[2], M Longnecker[2], DE Lane[2], K Krassavin[2], BJ Barney[2], JM Steiner[1], JS Suchodolski[1]. [1]Gastrointestinal Laboratory and [2]Department of Statistics. Texas A&M University, College Station, TX.

Lactic acid bacteria are commonly part of probiotic formulations as they are thought to have beneficial effects on gastrointestinal health. *In vivo* colonization properties of probiotics are generally not well understood and may depend on a number of variables. Here we investigated the role of age, body weight, as well as the concentrations of lactic acid bacteria at baseline as predictors for fecal colonization of bacterial species contained in a probiotic formulation (Proviable®, Nutramax Laboratories Inc.) in healthy cats and dogs.

Twelve cats (median age: 2.5 years, range: 1–7 years) and twelve dogs (median age: 3.5 years, range: 1–10 years) were enrolled. The median body weight was 4.9 kg (range: 3.8–5.9 kg) for cats and 26.4 kg (range: 5.0–36.8 kg) for dogs. A commercially available multi-species probiotic formulation (one capsule containing a total of 5×10^9 cfu *Bifidobacterium longum, Lactobacillus acidophilus, L. plantarum, L. rhamnosus, L. delbrueckii, Streptococcus salivarius*, and *Enterococcus faecium*) was administered daily for 21 days. *Lactobacillus, Bifidobacterium, Enterococcus*, and *Streptococcus* spp. were assessed by genus specific quantitative real-time PCR before (day 0), during (days 1, 8, and 17 in cats, and days 1, 5, and 17 in dogs), and after (days 26, 29, and 38 in cats, and days 23, 26, and 38 in dogs) probiotic administration. A repeated measures analysis using age, weight, and baseline populations of lactic acid bacteria as possible covariates was modeled by using an appropriate covariance structure using SAS version 9.2.

When compared with younger cats, older cats had significantly higher fecal concentrations of probiotic species during the administration period compared to the after administration period for all multiple comparisons. However, this effect was not associated with any particular probiotic bacterial group. Also, cats having fewer populations of lactic acid bacteria at baseline had significantly higher fecal concentrations of probiotic species during the administration period when compared to the after period for all multiple comparisons and for all the bacterial groups analyzed. Unlike in cats, in dogs age and concentrations of lactic acid bacteria at baseline did not influence fecal colonization with probiotic species. Moreover, body weight did not influence the fecal colonization with probiotic bacteria in either dogs or cats.

These results suggest that older cats are more likely to be colonized than younger cats by the lactic acid bacteria in the evaluated probiotic. Also, cats that had lower concentrations of lactic acid bacteria at baseline were more likely colonized by these probiotic species. Lastly, fecal colonization with Proviable® species was independent of individual weight (in both cats and dogs) or age (in dogs). Further studies are warranted to corroborate our findings using other probiotic formulations.

ABSTRACT #256

INTESTINAL MICROBIOTA IN RACING ALASKAN SLED DOGS AND WORKING RETRIEVERS BEFORE AND AFTER EXERCISE.

N Berghoff[1], SE Dowd[2], MD Willard[3], MS Davis[4], JM Steiner[1], JS Suchodolski[1]. [1]GI Laboratory, Texas A&M University, College Station, TX. [2]Research & Testing Laboratory, Lubbock, TX. [3]Department of Veterinary Small Animal Clinical Sciences, Texas A&M University, College Station, TX. [4]Center for Veterinary Health Sciences, Oklahoma State University, Stillwater, OK.

Several studies have reported a high prevalence of gastrointestinal (GI) disturbances in racing sled dogs during strenuous exercise, the cause of which is unknown. While a recent study has not shown an association between exercise and enteropathogens in sled dogs using bacterial culture, the global intestinal microbiome has not yet been studied. Thus, the aim of this study was to compare the intestinal microbiota in fecal samples obtained from racing sled dogs before and after a 300 mile race. To assess the effect of 2 different types of exercise, samples were also obtained from a group of working Retrievers before and after a field trial, during which dogs ran 4.4 miles.

Two fecal samples each were collected from 10 sled dogs and 9 Retrievers. Sample 1 was collected prior to, and sample 2 after exercise. For sled dogs, the sampling interval was 3 days, while the Retriever samples were obtained within 1 day. DNA was extracted from all samples, and analyzed by massive parallel 16S rRNA gene pyrosequencing. Diversity of the intestinal microbiota was evaluated by Simpson's diversity. Principal Component Analysis (PCA) was used to explore differences between pre- and post-exercise samples. Percentages of bacterial sequences in each sample were calculated and differences between pre- and post-samples were analyzed using a Wilcoxon signed rank test. Significance was set at $p < 0.05$.

The PCA indicated a difference between pre- and post-exercise samples in the sled dogs, where proportions of Lactobacillales were significantly more abundant in post-exercise samples in 8 of 10 dogs (median (range): 5.8% (1.9–68.1%) vs. 48.1% (14.4–85.7%), $p = 0.027$). An unclassified *Lactobacillus* sp. was found to be significantly more abundant in post-exercise samples in all 10 sled dogs ($p = 0.002$). Proportions of Clostridiales were significantly lower in 7 of 10 sled dogs post-exercise (median (range): 85.5% (20.4–95.6%) vs. 48.6% (11.9–84.2%), $p = 0.049$). In the Retrievers, Enterobacteriales were significantly more abundant post-exercise in all 9 dogs (median (range): 0.03% (0.00–0.15%) vs. 0.27% (0.02–0.81%), $p = 0.004$). Bacterial diversity was significantly higher in post-exercise samples in the sled dogs ($p = 0.026$), but not in the Retrievers.

These data suggest that strenuous exercise may lead to changes in the intestinal microbiota. A significant increase in Lactobacillales was seen in sled dogs only, which may be due to the duration of exercise and may be related to metabolic changes in the GI tract. It is possible that similar changes were not found in Retrievers because samples for this group were obtained within 1 day, which may be insufficient to show a change in the microbiota. Further studies are needed to clarify whether these changes observed contribute to clinical signs reported in the sled dogs or if they reflect an adaptation to physiological changes within the GI tract that occur during exercise.

ABSTRACT #257

INTESTINAL *TRITRICHOMONAS FOETUS* INFECTION IN CATS: A RETROSPECTIVE STUDY OF 104 CASES. PG Xenoulis, DJ Lopinski, LC Doyal, SA Read, JS Suchodolski, JM Steiner. Gastrointestinal Laboratory, Texas A&M University, College Station, TX.

Tritrichomonas foetus (*T. foetus*) is a flagellated protozoan parasite that has been recently recognized as an important cause of diarrhea in cats. It primarily colonizes the surface of the distal ileal and colonic mucosa, leading to chronic or intermittent diarrhea. Although many studies have confirmed the existence and pathogenicity of *T. foetus* in cats in many countries around the world, the history, clinical presentation, and response to treatment have not been described in a large number of naturally infected cats. The aim of this study was to collect and analyze clinical data from cats that were diagnosed with intestinal *T. foetus* infection.

The Gastrointestinal Laboratory's database was searched for the period 2006–2009, and cats that had tested positive for fecal *T. foetus* were identified. The diagnosis of *T. foetus* infection was based on detection of the microorganism's DNA in fecal samples, using a previously described PCR assay. The veterinarians who had submitted the fecal samples were contacted and asked to fill out a questionnaire for each infected cat. Information gathered involved signalment, medical history, clinical signs, diagnostics performed, concurrent conditions, and treatment.

Questionnaires for 104 cats living in North America were completed. A total of 24 breeds were represented and the median age of the cats was 12 months (range: 1–195 months). The median number of cats per household was 2 (range: 1–50), and 90.1% of the cats were exclusively indoor while 9.9% were both indoor and outdoor. Forty-six (44.2%) cats were females and 58 (55.8%) were males. The most common clinical sign was diarrhea (98.1%) with a median duration of 135 days (range: 1–2,880 days) and a fecal quality ranging from soft (40.9%) to watery (33.0%). Fifty-nine percent of cats had mucus in the feces, 45.5% had hematochezia, 42.9% had tenesmus, and 7% had melena. Forty-nine (59.0%) of 83 cats had diarrhea since adoption. Other clinical signs included anorexia (27.2%), depression (23.5%), weight loss or failure to gain weight (19.8%), vomiting (19.4%), fecal incontinence (14.9%), hyperthermia (13.8%), abdominal pain (8.9%), and increased appetite (2.9%). Fecal flotation had been performed in 85 cases, 3 (3.5%) of which were reported to be positive for coccidia. Of the 67 cats that were tested for *Giardia* spp., 15 (22.4%) were positive. A total of 49 cats had completed treatment with ronidazole, 29 of which (59.2%) showed clinical response to treatment. Sixteen (32.7%) cats had either partial or no improvement, or a relapse shortly after discontinuation of treatment; 10 (62.5%) of these cats had been treated with dosages of ronidazole of 30 mg/kg, q24h, for 2 weeks (n = 4) or lower (n = 5), while one cat had been treated for only 1 week. In 4 (8.1%) cats, the treatment was discontinued due to neurologic signs (n = 3) and/or anorexia (n = 3).

In this study, diarrhea was the predominant clinical sign of natural intestinal *T. foetus* infection in cats. Other clinical signs, such as anorexia, depression, and weight loss, were also commonly present. Most treatment failures were associated with the use of lower dosages of ronidazole.

ABSTRACT #258

ASSESSMENT OF RISK FOR DIARRHEA IN YOUNG DOGS AFTER *GIARDIA* VACCINE ADMINISTRATION. E Lund[1], M Yang[1], K Faunt[1], P Shearer[1], M Lappin[2]. [1]Banfield, The Pet Hospital, Portland, OR. [2]Colorado State University, Fort Collins, CO.

An inactivated *Giardia* vaccine has been used but little is known concerning the impact of this vaccine on the occurrence of diarrhea in naturally exposed puppies.

A retrospective cohort study was performed using medical record data from puppies seen at Banfield, The Pet Hospital. Puppies (n = 6,000) < 6 months of age without gastrointestinal disease seen in the North Central, North East and South Central regions between January 1st and July 31st, 2008 were studied. All puppies were administered at least two doses of a distemper, parvovirus and adenovirus 2 containing vaccine (DA2P); two doses of *Giardia* vaccine were administered to 3,000 of the puppies. A Cox proportional hazards model was used to conduct the analysis and determine likelihood (hazard ratio) for diarrhea and days to diarrhea by *Giardia* vaccine status. Variables included in the model were *Giardia* vaccination status, age (at last DA2P vaccine), breed, gender, region, wellness plan status at entry into study, number of encounters after vaccination, concurrent diagnoses/examination findings. In addition, the numbers of puppies that were positive for *Giardia* cysts or *Giardia* antigen were compared between groups during the study period.

Puppies administered *Giardia* vaccine were more likely to have diarrhea between 2 months after and within 12 months of vaccination (hazard ratio = 1.2, 95% confidence interval = 1.1, 1.5) when compared to puppies without *Giardia* vaccination. Average days to diagnosis of diarrhea were 132 days for *Giardia* vaccinates (n = 464) vs. 120 days for non-vaccinates (n = 303). There were no differences in *Giardia* cyst or *Giardia* antigen detection rates between groups.

There is insufficient evidence to suggest that the canine *Giardia* vaccine is efficacious in preventing diarrhea or infection with *Giardia* spp. Analysis of a larger dataset, with a greater number of positive tests, may provide additional evidence.

ABSTRACT #259

INCIDENCE OF BACTEREMIA FOLLOWING GASTRIC AND DUODENAL ENDOSCOPIC BIOPSIES BEFORE AND AFTER TREATMENT WITH OMEPRAZOLE IN HEALTHY DOGS. K. Jones[1], O. Dossin[1], S. Clark-Price[1], C. Maddox[2]. [1]Department of Veterinary Clinical Medicine and [2]Pathobiology Department, College of Veterinary Medicine, University of Illinois, Urbana, IL.

The incidence of endoscopy induced bacteremia has been extensively evaluated in human medicine. However, to the best of our knowledge, there is no information available in the veterinary literature about the incidence of bacteremia in dogs undergoing routine endoscopic procedures. Anti-acid medications are frequently used in canine medicine. Studies have suggested that proton pump inhibitor treatment might induce duodenal bacterial overgrowth with oral or fecal type bacteria in humans. We hypothesized that omeprazole treatment would increase the risk of bacterial translocation in dogs. The goal of this study was to evaluate the incidence of detectable bacteremia utilizing conventional blood culture techniques following gastroduodenoscopy with multiple biopsies before and after treatment with omeprazole

Endoscopy with gastric and duodenal biopsies were performed on 8 healthy research dogs (4 males, 4 females) 30 and 15 days prior to treatment with omeprazole (1.1 mg/kg twice daily), on day 15 of the treatment period and 14 days after the end of treatment. Fifteen endoscopic biopsies were performed in both the stomach along the greater curvature and in the descending duodenum caudal to the papilla. Blood samples were collected using aseptic technique from each dog at the time of anesthesia induction, immediately following the end of the endoscopic procedure and 24 hours following endoscopy. The samples were immediately inoculated into blood culture bottles and monitored for both aerobic and anaerobic growth.

A total of 96 aerobic and 96 anaerobic blood cultures were evaluated. There were 2 positive aerobic cultures and no growth was detected in the anaerobic cultures. *Lactobacillus sp* was cultured from a single sample prior to endoscopy at the end of the treatment period with omeprazole and *Staphylococcus intermedius* was cultured from a 24 hour post endoscopy sample 2 weeks after the end of the treatment period. Both dogs with positive culture did not show any clinical signs.

The overall incidence of bacteremia as detected by positive blood culture following routine gastroduodenoscopy with biopsy in healthy research dogs is low and likely of little clinical significance. Treatment with omeprazole for 2 weeks did not appear to have a significant effect on incidence of bacteremia after endoscopic biopsies. It is concluded that routine prophylactic antibiotic treatment in dogs undergoing gastrointestinal endoscopy is not necessary.

ABSTRACT #260

FECAL N-METHYLHISTAMINE CONCENTRATIONS IN RACING ALASKAN SLED DOGS AND WORKING RETRIEVERS. N Berghoff[1], MD Willard[2], MS Davis[3], JS Suchodolski[1], JM Steiner[1]. [1]GI Laboratory, Texas A&M University, College Station, TX. [2]Department of Veterinary Small Animal Clinical Sciences, Texas A&M University, College Station, TX. [3]Center for Veterinary Health Sciences, Oklahoma State University, Stillwater, OK.

Vigorous exercise, such as running a marathon, has been associated with gastrointestinal disturbances in people. Similarly, previous studies have revealed a high prevalence of gastrointestinal abnormalities in racing sled dogs during strenuous exercise. Affected dogs frequently have diarrhea, and up to 61% of dogs develop gastric mucosal erosions and/or ulcers. Increased intestinal permeability has also been demonstrated. The cause of these changes is currently unknown. N-methylhistamine is a stable metabolite of histamine and has previously been shown to be increased in fecal samples from dogs with gastrointestinal disease. The purpose of this study was to determine if fecal concentrations of N-methylhistamine are increased in sled dogs after a 300 mile race. A group of working Retrievers was also investigated before and after a 4.4 mile field trial to determine if the intensity of exercise would influence the outcome.

Two fecal samples each were collected from 10 sled dogs and 10 working Retrievers. The sled dogs had undergone exercise training for several months, but were rested for 2 days prior to the race. In both groups of dogs, sample 1 was collected prior to exercise and sample 2 immediately thereafter. Samples were stored frozen until analysis. N-methylhistamine concentrations were measured using a stable isotope dilution gas chromatography-mass spectrometry method (control range: 8.5–175.5 ng/g). Pre- and post-exercise N-methylhistamine concentrations were compared within groups using a Wilcoxon signed rank test, and between groups using a Mann-Whitney U test. Significance was set at $p < 0.05$.

Median fecal N-methylhistamine concentrations in sled dogs were significantly higher after racing (pre-race: 176.8 ng/g feces; range: 83.3–711.1 ng/g; post-race: 2,405.6 ng/g feces; range: 28.3–5,628.9 ng/g; $p = 0.020$). In contrast, there was no significant difference between the median pre- and post-exercise N-methylhistamine concentrations in the Retrievers (pre-exercise: 82.1 ng/g; range: 5.7–304.0 ng/g; post-exercise: 109.5 ng/g; range: 0.0–561.7 ng/g; $p = 0.432$). Median fecal N-methylhistamine concentrations were significantly higher in sled dogs vs. Retrievers both before (176.8 vs. 82.1 ng/g; $p = 0.023$) and after exercise (2,405.6 vs. 109.5 ng/g; $p = 0.012$).

These results indicate that strenuous exercise may increase fecal N-methylhistamine concentrations in sled dogs. This increase might be caused by histamine release due to mucosal mast cell degranulation in the gastrointestinal tract. However, it is not clear whether this may contribute to the clinical signs reported in previous studies. It should be noted that breed or population differences between sled dogs and Retrievers could not be excluded and further studies are needed to clarify a potential impact. The lack of a significant change in N-methylhistamine in the Retrievers suggests that mild exercise does not induce gastrointestinal histamine release.

ABSTRACT #261

ASSOCIATION STUDY IN COBALAMIN DEFICIENT CHINESE SHAR PEIS WITH AN INCREASED METHYLMALONIC ACID CONCENTRATION. N Grützner[1], ND Halbert[2], N Berghoff[1], MA Bishop[1], JS Suchodolski[1], JM Steiner[1]. [1]Gastrointestinal Laboratory and [2]Department of Veterinary Pathobiology, Texas A&M University, College Station, TX.

Using the canine minimal screening set 2 (MSS-2) and the Affymetrix v2 Platinum canine single nucleotide polymorphism (cSNP) array, serum cobalamin deficiency in the Chinese Shar Pei (Shar Pei) has recently been linked to a genomic region on canine chromosome 13. Based on human and veterinary studies, an increased serum methylmalonic acid (MMA) concentration has been suggested to reflect cobalamin deficiency at the cellular level. Therefore, the aim of this study was to re-evaluate previous MSS-2 and cSNP array results from Shar Peis with decreased serum cobalamin and increased serum MMA concentrations, and Shar Peis with serum cobalamin and MMA concentrations within the reference intervals.

Whole blood and serum samples were analyzed from unrelated Shar Peis. Serum MMA concentrations (reference interval: 415–1,193 nmol/L) and serum cobalamin concentrations (reference interval: 252–908 ng/L) were measured, and genomic DNA was extracted from whole blood. A total of 313 microsatellite markers from the MSS-2 and 49,633 markers of the Affymetrix v2 Platinum cSNP array were evaluated. For statistical analysis of the MSS-2 results, a Fisher's exact test was used with significance set at $p < 1 \times 10^{-4}$. An association analysis toolset (Plink v1.07) for cSNP array analysis and a Bonferroni correction for multiple statistical comparisons were used to evaluate the significance of any potential association. Significance was set at a p-value of $< 1.0 \times 10^{-6}$.

Sixty-one Shar Peis were included in this study, with 20 (32.8%) dogs having an undetectable serum cobalamin concentration (< 150 ng/L) and an increased serum MMA concentration (> 1,193 nmol/L). The MSS-2 was analyzed for all 61 Shar Peis. Data from the cSNP array were available for 34 of these dogs, with ten dogs (29.4%) having an undetectable serum cobalamin concentration and an increased serum MMA concentration. Allele 283 of the MSS-2 marker FH3619, located on chromosome 13 (22,877,770 Mb), was found significantly more frequently in Shar Peis with an undetectable cobalamin concentration and an increased MMA concentration (29 of 40 alleles, 72.5%) than in Shar Peis with normal serum cobalamin and MMA concentrations (14 of 82 alleles, 17.1%; $p = 3.2 \times 10^{-9}$). Analysis of the cSNP array revealed two significant markers (21,600,902 Mb; $p = 4.1 \times 10^{-7}$ and 22,937,592 Mb; $p = 4.5 \times 10^{-7}$, respectively) located in close proximity to FH3619 on chromosome 13.

In conclusion, the analysis of the MSS-2 and cSNP array revealed one region on chromosome 13 that is significantly associated with undetectable serum cobalamin and increased serum MMA concentrations. The region identified in this study is approximately 5 Mb downstream of a previously identified region associated with this disorder.

ABSTRACT #262

A METHOD FOR THE QUANTIFICATION OF SERUM AND FECAL CANINE S100A12. RM Heilmann, DJ Lanerie, JS Suchodolski, JM Steiner. Gastrointestinal Laboratory, Texas A&M University, College Station, TX.

S100A12 is a Ca^{2+}-binding protein that has been proposed to play a central role in the innate and acquired immune response. S100A12 has been reported to be increased in serum and/or plasma in human patients with various inflammatory disorders and has also been suggested to be a sensitive and specific marker for inflammatory bowel disease (IBD) in humans. An immunoassay for the measurement of S100A12 is currently available for use in humans, but antibodies against the human protein do not cross-react with canine S100A12 (cS100A12). Given the current lack of both sensitive and specific markers for canine patients with chronic inflammatory diseases, the aim of this study was to develop and analytically validate a radioimmunoassay (RIA) for the quantification of cS100A12 in serum and fecal specimens from dogs.

Antiserum against purified cS100A12 was raised in rabbits. A tracer was produced by ^{125}I-labeling of cS100A12. A competitive double antibody RIA was developed and analytically validated by determining assay working range, analytic specificity, dilutional parallelism, spiking recovery, and intra- and inter-assay variability. Control intervals for serum and fecal cS100A12 were established from 29 and 24 healthy dogs, respectively, using the central 95th percentile.

The working range of the assay for serum samples in a 1:20 dilution and for fecal extracts assayed in a 1:800 dilution was 118,654 µg/L and 0.5346.2 µg/g, respectively. There was no cross-reactivity with the cS100A8/A9 protein (calprotectin) complex, the closest structural analogue, over a wide range of the assay. Observed to expected ratios (O/E) for serial dilutions ranged from 97.2–146.8% (mean ± standard deviation [SD]: 122.5 ± 14.9%) for 6 different serum samples at dilutions of 1:12, 1:24, 1:48, and 1:96, and from 72.5–133.3% (mean ± SD: 103.9 ± 14.3%) for 4 different fecal extracts serially diluted from 1:100 to 1:800, four different fecal extracts diluted from 1:200 to 1:1,600, and 2 different fecal extracts diluted from 1:400 to 1:6,400. O/E for spiking recovery ranged from 87.8130.4% (mean ± SD: 100.6 ± 6.5%) for 7 different serum samples and 8 different spiking concentrations, and from 84.8–143.8% (mean ± SD: 103.7 ± 10.6%) for 7 different fecal extracts and 8 different spiking concentrations. Intra-assay coefficients of variation (CV) for 7 different serum samples were 1.5, 8.1, 6.3, 4.5, 6.6, 5.7, and 6.2%, and 2.3, 3.8, 5.5, 3.8, 4.2, 6.1, and 7.4% for 7 different fecal samples. Inter-assay CV for 7 different serum samples were 5.6, 5.9, 7.8, 5.8, 5.1, 3.3, and 6.5%, and were 5.3, 5.9, 5.7, 4.0, 2.9, 7.5, and 8.7% for 7 different fecal extracts. The control intervals for serum and fecal cS100A12 were established as 32145 μg/L and 0.51.5 μg/g, respectively.

We conclude that the RIA for the measurement of canine S100A12 described here is sensitive, linear, accurate, precise, and reproducible, and will facilitate further studies to evaluate the clinical usefulness of the quantification of S100A12 in serum and fecal samples from canine patients with inflammatory conditions.

ABSTRACT #263

SERUM SPEC cPL® CONCENTRATIONS IN *LEISHMANIA INFANTUM*-INFECTED DOGS TREATED WITH MEGLUMINE ANTIMONATE. PG Xenoulis[1], MN Saridomichelakis[2], MK Chatzis[2], JS Suchodolski[1], JM Steiner[1]. [1]Gastrointestinal Laboratory, Texas A&M University, College Station, TX. [2]Clinic of Medicine, School of Veterinary Medicine, University of Thessaly, Karditsa, Greece.

Canine visceral leishmaniosis caused by *Leishmania infantum* (*L. infantum*) is very common in Mediterranean countries and Latin America, and is considered an emerging disease in the United States. One of the most commonly used drugs for the treatment of canine leishmaniosis is meglumine antimonate (MA), which belongs to the group of pentavalent antimonials. These drugs have been associated with pancreatitis in humans infected with *Leishmania* spp., and it has been speculated that pancreatitis might also occur as a side effect of MA in dogs. The aim of this study was to measure serum Spec cPL® concentrations in dogs with leishmaniosis during treatment with MA, and to compare them with those of dogs with leishmaniosis not treated with this drug.

A total of 36 non-uremic dogs with leishmaniosis, living in Greece, were prospectively enrolled into the study. The diagnosis of leishmaniosis was based on a compatible clinical presentation and positive serology testing (SNAP® Leishmania Test or IFAT) in all 36 dogs, with (n = 34) or without (n = 2) microscopic detection of the parasite in lymph-node aspirates. Of the 36 dogs, 25 (Group A) were treated with a combination of MA (Glucantime®; 100 mg/kg, SC, q24h; 1 dog was treated with 75 mg/kg, SC, q12h) and allopurinol (Zylapour®; 10 mg/kg, PO, q12h) for 28 days, while 11 dogs (Group B) were treated with allopurinol alone (10 mg/kg, PO, q12h) for 28 days. Clinical signs were recorded each day by the owner. Blood samples were collected at timepoint 0 (before treatment) and at 14 and 28 days after the initiation of treatment. Serum Spec cPL concentrations were measured in all samples and a concentration of ≥ 400 μg/L was considered suggestive of pancreatitis. Data were analyzed using Mann-Whitney and Friedman's tests.

A total of 23 dogs (14 in Group A and 9 in Group B) completed the study. Another 7 dogs (6 in Group A and 1 in Group B) had only one follow-up sample. None of the 36 dogs had a Spec cPL concentration ≥ 400 μg/L or clinical signs suggestive of pancreatitis at timepoint 0. In Group A, none of the 19 dogs re-evaluated on day 14, and none of the 15 dogs re-evaluated on day 28 had a Spec cPL concentration ≥ 400 μg/L. Similarly, in Group B, none of the 10 dogs re-evaluated on day 14, and none of the 9 dogs re-evaluated on day 28 had a Spec cPL concentration ≥ 400 μg/L. There was no significant difference ($p = 0.344$) of the median Spec cPL concentration among the 3 time points for the 14 dogs in Group A that completed the study. Median Spec cPL concentrations for each time point were not significantly different between the two treatment groups. No clinical signs suggestive of pancreatitis were observed in any of the dogs treated with MA.

In the present study, MA treatment in dogs with leishmaniosis did not result in clinical or biochemical (serum Spec cPL) evidence of pancreatitis. Larger scale studies are needed to determine whether pancreatitis may occur as a rare complication in dogs treated with MA.

ABSTRACT #264

ANALYTICAL VALIDATION OF A RADIOIMMUNOASSAY FOR THE MEASUREMENT OF CANINE PANCREATIC LIPASE IMMUNOREACTIVITY IN CHYME AND FECAL EXTRACTS FROM DOGS. N Grützner[1], I Mõistus[2], JS Suchodolski[1], T Spillmann[2], JM Steiner[1]. [1]Gastrointestinal Laboratory, Texas A&M University, College Station, TX. [2]Faculty of Veterinary Medicine, University of Helsinki, Helsinki, Finland.

The exocrine pancreas synthesizes enzymes (e.g., pancreatic lipase) and zymogens of digestive enzymes (e.g., trypsinogen), stores them in zymogen granules, and releases them into the pancreatic duct system in response to gastrointestinal hormones. To measure canine pancreatic lipase immunoreactivity (cPLI) in small intestinal chyme or fecal extracts in response to varying dietary components such as protein, fat, and carbohydrates, an assay with a high analytical sensitivity is needed. The aim of this study was to analytically validate a previously described radioimmunoassay (RIA) for the measurement of cPLI in chyme and fecal extracts from dogs.

Chyme and fecal specimens were collected from healthy Beagle dogs with permanent jejunal nipple valve fistulas living in a research colony. Both sample types were extracted according to the protocol for the canine fecal elastase-1 assay (ScheBo® Biotech). A previously described competitive double antibody RIA was used for the measurement of cPLI concentrations in canine chyme and fecal extracts, and was analytically validated by determination of assay sensitivity, dilutional parallelism, spiking recovery, and intra- and inter-assay variability.

The analytical sensitivity of the RIA was 11.9 μg/L with a calibration range between 6.3 and 400 μg/L. Chyme and fecal extracts were both analyzed in a 1:2 dilution. Observed to expected (O/E) ratios for serial dilutions ranged from 70.4% to 108.3% with a mean ± standard deviation (SD) of 88.6% ± 10.3% for three different chyme extracts at dilutions of 1:1, 1:2, 1:4, and 1:8, and from 80.9% to 115.3% with a mean ±SD of 93.8% ± 12.3% for three different chyme extracts at dilutions of 1:2, 1:4, 1:8, and 1:16. O/E ratios for serial dilutions ranged from 91.1% to 119.9% with a mean ± SD of 105.6% ± 9.3% for six different fecal extracts at dilutions of 1:2, 1:4, 1:8, and 1:16. O/E ratios for spiking recovery ranged from 85.3% to 120.4% with a mean ±SD of 102.5% ± 9.6% for six different chyme extracts and from 78.9% to 167.1% with a mean ±SD of 106.2% ± 19.3% for six different fecal extracts in a 1:2 dilution. Intra-assay coefficients of variation (CV) for six different chyme extracts and six different fecal extracts were 7.6%, 4.2%, 3.0%, 6.1%, 4.6%, and 2.6%, and 8.6%, 6.6%, 2.8%, 3.1%, 3.8%, and 3.2%, respectively. Inter-assay CVs for six different chyme extracts and six different fecal extracts were 8.9%, 9.8%, 7.1%, 8.1%, 7.7%, and 8.6%, and 8.1%, 7.6%, 6.0%, 4.7%, 8.5%, and 6.8%, respectively.

In conclusion, the RIA for the measurement of cPLI evaluated here is sensitive, linear, accurate, precise, and reproducible for use with chyme or fecal extracts from dogs. These data will facilitate the measurement of chyme and fecal cPLI concentrations in response to various dietary components.

ABSTRACT #265

ASSOCIATION BETWEEN BREED AND INCREASED SERUM SPEC cPL® CONCENTRATIONS IN DOGS. MA Bishop, PG Xenoulis, JS Suchodolski, JM Steiner. Gastrointestinal Laboratory, Texas A&M University, College Station, TX.

Serum Spec cPL® concentration is considered to be a sensitive and specific marker for pancreatitis in the dog. The aim of this study was to retrospectively evaluate a large population of dogs that had their serum Spec cPL concentration measured and determine whether a breed predilection for having an increased Spec cPL result could be identified.

Serum Spec cPL concentration results from dogs whose serum was submitted to the Gastrointestinal Laboratory at Texas A&M University over a 3 year period (2006–2009) were evaluated retrospectively. Dogs that had a serum Spec cPL concentration $\geq$ 400 µg/L (considered suggestive for pancreatitis) or a Spec cPL concentration within the reference interval (RI; $\leq$ 200 µg/L) were included, provided that information on the breed, age, and sex was available. Samples with a Spec cPL concentration in the questionable range (between 201–399 µg/L) and breeds with < 15 individuals were also excluded. A total of 16,604 dogs (representing 96 breeds) fulfilled the inclusion criteria. Of these, 3,482 (21.2%) had a Spec cPL concentration $\geq$ 400 µg/L. A Mann-Whitney *U* test was used to compare ages between dogs with increased and normal Spec cPL concentrations. A two-sample test of proportion was used to compare proportions of dogs that were male or female within each group. A Fisher's exact test was used to compare the proportion of dogs (within each breed) with increased Spec cPL concentrations compared to the total study population (excluding the breed of interest). Odds ratios and their 95% CI were calculated. The *P* values were corrected for multiple testing using Benjamini and Hochberg's False Discovery Rate. In dogs where the odds ratio was > 1, a logistic regression model was fitted to adjust for the differences in age and sex. Models were tested for goodness-of-fit using the Hosmer-Lemeshow χ^2 test. Additionally, models were tested for multicolinearity, overdispersion, and interaction between variables.

Dogs with Spec cPL concentrations $\geq$ 400 µg/L were significantly older (median age: 9 years) than dogs with a Spec cPL in the RI (median age: 6 years; $P < 0.0001$). There were no differences in the proportion of males vs. females in the dogs with a Spec cPL in the RI ($P = 0.1216$), however there were significantly more females than males in the increased Spec cPL group ($P < 0.0001$). After adjusting for age and sex using logistic regression, 11 breeds had increased odds ratios for having an increased Spec cPL concentration. The top five breeds were Miniature Pinschers (OR: 3.64), Boxers (3.49), Keeshonds (3.24), Pomeranians (2.53), and Eskimos (2.02).

These results demonstrate that some breeds have a predilection for having an increased Spec cPL. Many breeds with an increased odds ratio in this study are similar to those reported in previous studies, evaluating breed prevalence for pancreatitis in dogs. Interestingly, three of the top five breeds encompass the group of breeds known as the German Spitzs. Further studies are warranted in dogs with confirmed pancreatitis.

ABSTRACT #266

ANALYTICAL VALIDATION OF A COMMERCIALLY AVAILABLE IMMUNOASSAY FOR SERUM GASTRIN CONCENTRATION IN CATS. I Ruhnke, JS Suchodolski, VM Schwierk, JM Steiner. Gastrointestinal Laboratory, Texas A&M University, College Station, TX.

While gastrinomas are rarely reported in cats, they should be considered as a possible rule-out in patients with unexplained chronic vomiting, hematemesis, anorexia, weight loss, and/or gastric ulceration. Due to the non-specific clinical signs and the lack of specific tests, diagnosis of gastrinomas in cats remains challenging. Measurement of serum or plasma gastrin is routinely used for the diagnosis of this disease in humans and dogs. The aim of this study was to analytically validate a commercially available immunoassay intended for use in humans for the measurement of gastrin in feline serum samples, and to determine a reference interval for this assay in cats.

For the analytical validation of the assay, serum gastrin concentrations were measured using a commercially available immunoassay (IMMULITE® 2000 Gastrin immunoassay, Siemens Healthcare Diagnostics Ltd., Deerfield, IL) in leftover feline serum samples submitted to the Gastrointestinal Laboratory. Samples with a range of serum gastrin concentrations were selected for validation. The minimum detection limit of the assay was reported by the manufacturer to be 10 pg/dL. Validation of the assay included determination of dilutional parallelism, spiking recovery, and intra- and inter-assay variability. To determine the reference interval, fresh serum samples were collected from 20 healthy pet cats. The reference interval was established using the central 95th percentile in this group of cats after food had been withheld for 12 hours.

Observed to expected ratios for dilutional parallelism ranged between 76.2% and 84.9%, with a mean $\pm$ SD of 77.9 $\pm$4.9% in three samples (concentrations: 34.9 pg/dL, 49.7 pg/dL, and 63.1 pg/dL) at 1:2 and 1:4 dilutions. Observed to expected ratios for spiking recovery ranged from 98.2% to 112.3% with a mean $\pm$ SD of 104.9 $\pm$ 5.6% for four samples (concentrations: 13.3 pg/dL, 31.2 pg/dL, 33.0 pg/dL, and 68.8 pg/dL). The coefficients of variation (CV) for intra-assay variability were: 5.9% (13.2 pg/dL mean sample value), 9.8% (19.3 pg/dL), 5.7% (38.8 pg/dL), and 3.3% (54.0 pg/dL). The CV for inter-assay variability were: 5.7% (13.2 pg/dL mean sample value), 6.2% (19.2 pg/dL), 4.1% (39.7 pg/dL), and 4.1% (55.2 pg/dL). The reference interval for feline serum gastrin concentration was established as < 10–39.5 pg/dL.

These data demonstrate that the gastrin immunoassay described here is linear, accurate, precise, and reproducible for use in cats. Further studies regarding the clinical utility of this assay for the diagnosis of feline gastrinomas are warranted.

ABSTRACT #267

FLOW CYTOMETRICAL ANALYSIS OF CANINE INTESTINAL INTRAEPITHELIAL LYMPHOCYTES IN DOGS WITH CHRONIC ENTEROPATHIES. N Luckschander[1], IA Burgener[2], S Schneider[3], N Corazza[3], T Brunner[3]. [1]Department for Companion Animals and Horses, Clinic for Internal Medicine, University of Veterinary Medicine, Vienna, Austria; [2]Department of Clinical Veterinary Medicine, Vetsuisse Faculty, University of Berne, Switzerland; [3]Division of Experimental Pathology, Institute of Pathology, University of Berne, Switzerland

Although several lines of evidence point towards an important role of uncontrolled activation of the intestinal intraepithelial lymphocytes (IEL) in canine chronic enteropathies (CE), little is currently known about intestinal lymphocyte populations in diseased dogs. Thus, the aims of this project were 1) the investigation of intestinal endoscopical biopsies for the isolation of IEL and flow cytometry techniques, 2) the comparison of IEL composition between healthy dogs and dogs suffering from chronic enteropathies (CE), 3) the monitoring of changes in the IEL subpopulations during the disease course.

For the phenotypical characterization mainly multi-colour staining flow cytometry techniques were used. Full thickness biopsies were taken from healthy adult Swiss Beagle dogs (n = 7) and compared to endoscopical biopsies from the same dog. These results (healthy controls) were compared to endoscopical biopsies of dogs suffering from CE (n = 11), which were enrolled into a clinical prospective study. These dogs were grouped according to the success of therapy in a "food responsive group" (FRD n = 7) and an inflammatory bowel disease group (IBD n = 4).Flow cytometry analyses of the endoscopical biopsies were performed before (n = 11) and 6–10 weeks after therapy (n = 10).

No significant differences in the lymphocyte subpopulations between lymphocytes isolated from full thickness biopsies versus endoscopical biopsies could be observed. No significant differences could be shown comparing T cell populations from healthy control dogs with dogs suffering from chronic entropathies. However, a significant decrease ($p < 0.05$) in the percentage of CD4+ TCRαβ + IEL in dogs suffering from IBD, compared to healthy control dogs, was found. Furthermore, there was also a significant difference between CD4+ TCRαβ + IEL from FRD versus IBD dogs. There was no significant change in the IEL subpopulation in chronic enteropathy dogs, FRD and IBD dogs before and after therapy.

In summary, it could be shown that endoscopical biopsies from healthy and diseased dogs can be used for flow cytometrical characterization of intestinal intraepithelial lymphocytes. These preliminary data indicate that flow cytometry techniques may not only provide a diagnostic tool in the characterization of the subtypes of canine chronic enteropathies but could give also further insights into the pathomechanisms of canine IBD.

ABSTRACT #268

CONSTIPATION RESOLUTION WITH ADMINISTRATION OF POLYETHYLENE-GLYCOL SOLUTION IN CATS. AP Carr, MC Gaunt. Western College of Veterinary Medicine, Saskatoon, Saskatchewan.

Constipation and opstipation is commonly seen in cats. In some instances this may be the result of megacolon. Other potential causes include spinal cord disorders, pelvic anatomical abnormalities, foreign body, neoplasia, strictures and dehydration to name a few. Recurrent constipation may eventually contribute to the development of megacolon. The impacted cat can become systemically ill with signs such as anorexia, vomiting, abdominal pain, and straining.

Initial therapy for constipation/opstipation is directed at removing the hardened feces and providing supportive care if needed. A variety of therapies can be used including rectal suppositories, enemas, oral laxatives, and prokinetic agents. In some instances manual disimpaction is required. In humans the use of PEG 3350 has been described to be highly successful in the treatment of severe constipation or fecal impaction both in adults and children. Side effects are rare, though occasionally these solutions can result in the syndrome of inappropriate antidiuretic hormone release (SIADH). This retrospective study describes the use of PEG 3350 (Colyte®) infusions as the predominant therapy resulting in the resolution of constipation in 9 cats. PEG 3350 is commonly used in cats for colonoscopy preparation where dosages of 50 to 125 ml/kg have been recommended.

All 9 cats were admitted to the University of Saskatchewan Veterinary Teaching Hospital and received intravenous fluid therapy as well as various supportive care measures. If cats were on laxatives, these were continued at previously used dosages. All cats had clinical signs consistent with clinical constipation/impaction. Seven of the cats had multiple previous episodes of constipation. Diagnosis was established by an appropriate history, physical examination findings and radiographs (6 of 9). One cat received a 5 ml/kg warm water enema at presentation with no fecal production 12 hours prior to administering PEG 3350. One cat received a single warm water enema concurrent to PEG 3350 administration (20 ml/kg). PEG 3350 (Colyte®) was administered via naso-esophageal tube at rates between 6 and 10 ml/kg/hr. The median total dose given was 80 ml/kg (range 40–156 ml/kg). Median time to significant defecation was 8 hours (range 5 to 24). All cats had either radiographs or palpation to confirm that all firm feces had been passed. Side effects were limited to vomiting in one cat, however the cat had been vomiting prior to presentation. Hyponatremia was not noted in 5 cats where blood work after administration was obtained.

Use of PEG 3350 solution (Colyte®) appears to be an effective alternative or adjunctive therapy for the treatment of constipated/impacted cats.

ABSTRACT #269

DEVELOPMENT OF A VETERINARY FLEXIBLE ENDOSCOPIC TRAINING DEVICE. KC Kottkamp, JC Whittemore. University of Tennessee, Knoxville TN.

Minimally-invasive endoscopic procedures are increasingly performed by small animal practitioners because of continuing education opportunities, affordability of endoscopic equipment, and high public demand. Successful operation of both rigid and flexible endoscopes requires technical expertise only attainable through practice. Incorporation of endoscopic simulators into veterinary medical training has been limited by cost. The purpose of this study was to develop an affordable veterinary flexible and rigid endoscopic training device (FRED).

A commercial stuffed animal was acquired. Removable upper and lower gastrointestinal tracts were constructed from rubber tubing and plicated balloons. Modifications to the modules were made based upon achievements and problems identified during scoping trials to optimize scoping difficulty and realism.

Normal canine gastrointestinal anatomy and basic haptic feedback were achieved. Standard endoscopic views, including retroflexed examination of the cardia, evaluation of the lesser incisura and entry into the pyloric outflow tract, were realistic and attainable. Other structures of acceptable accuracy, excluding coloration, were the lower esophageal sphincter, duodenal papilla, and ileocecocolic area. The dexterity and manipulative skills required for performance of endoscopic procedures on live small animal patients were necessary for successful navigation of the facsimile gastrointestinal tract.

The FRED prototype provided a realistic simulation of gastrointestinal endoscopy. Modules were easily removed for cleaning. Further studies will be necessary to determine whether use of the simulator is associated with improved procedural time, decreased complications and improved outcome in live animal patients. Additional training modules are being developed to expand use of the FRED for other minimally invasive procedures.

Table [Abstract 270]

Test	Canine Result Intervals[1]	Canine R	Canine SEE	Canine Mean Bias[1]	Feline Result Intervals[1]	Feline R	Feline SEE	Feline Mean Bias[1]
RBC	1.38–9.26	0.997	0.123	0.041	2.20–11.28	0.993	0.235	−0.05
HGB	3.4–22.3	0.997	0.27	0.09	3.64–16.00	0.997	0.226	−0.07
HCT[2]	7.60–60.70	0.992	1.341	−1.261	10.80–45.80	0.970	2.048	0.722
MCV	57.3–101.9	0.966	1.72	−3.91	32.1–63.4	0.880	2.36	1.53
MCH	18.6–33.6	0.919	0.67	−0.12	10.5–20.4	0.947	0.43	−0.04
MCHC	24.9–36.9	0.568	2.00	1.89	28.4–37.2	0.575	1.83	−1.17
WBC	0.50–83.60	0.997	1.038	−0.046	2.30–40.40	0.985	1.101	0.17
PLT	3–827	0.968	36.3	−40.6	23–812	0.832	57.2	−67.8

[1]Units for each test result: RBC, × 10^6/μl; HGB, g/dl; HCT, %; MCV, fl; MCH, pg; MCHC, g/dl; WBC, × 10^3/μl; PLT, × 10^3/μl.
[2]Sample number (n): Dogs, 122; Cats, 72.

ABSTRACT #270

EVALUATION OF CANINE AND FELINE HEMOGRAMS USING THE IDEXX PROCYTE DX ANALYZER COMPARED TO THE ABBOTT CELLDYN 3700. KM Boes, NK Rosen, JA Christian, AH Rebar. Purdue University, School of Veterinary Medicine, West Lafayette, IN.

The ProCyte Dx is an in-house hematology analyzer that utilizes impedance technology and laser flow cytometry with optical fluorescence capabilities, and provides a complete hemogram within two minutes. The purpose of this study was to compare the IDEXX ProCyte DX with the Abbott CellDyn 3700 for routine hemogram parameters (RBC, HGB, HCT, MCV, MCH, MCHC, WBC and PLT) in healthy or ill dogs and cats. Within one hour of sample collection, EDTA-anticoagulated whole blood from 164 dogs and 143 cats was analyzed using the ProCyte DX (developmental software version 1.0) and CellDyn 3700 (software version 1.3). ProCyte DX and CellDyn 3700 results, including abnormal, flagged results, were compared by Deming Regression and Bland-Altman plots using EP Evaluator v8 software; test result intervals, correlation coefficients (R), root mean square errors (SEE) and mean biases are presented in the table below. There are strong linear relationships between most ProCyte DX and CellDyn 3700 results. Parameters with correlation coefficients < 0.9 may be explained by either differences in methodology (MCV and MCHC) or specific technical issues with feline platelet clumping. These data indicate that results from the IDEXX ProCyte DX correlate strongly with a reference laboratory analyzer, and that the ProCyte DX is capable of providing a rapid and accurate hemogram for the canine and feline samples.

ABSTRACT #271

COMPARISON OF CANINE AND FELINE MANUAL HEMATOLOGIC DETERMINATIONS WITH RESULTS FROM THE IDEXX PROCYTE DX ANALYZER. KM Boes, NK Rosen, JA Christian, AH Rebar. Purdue University, School of Veterinary Medicine, West Lafayette, IN.

The ProCyte Dx is an in-house hematology analyzer that utilizes impedance technology and laser flow cytometry with optical fluorescence capabilities, and provides a complete hemogram within two minutes. The purpose of this study was to compare IDEXX ProCyte DX results with manually determined PCV, reticulocyte and leukocyte differential values. Within one hour of sample collection, EDTA-anticoagulated blood from 164 dogs and 143 cats was analyzed using the ProCyte DX (developmental software version 1.0). From these, manual PCV values were determined for 122 canine and 72 feline samples. Reticulocyte counts were performed on 146 canine and 135 feline samples using new methylene blue-stained blood films and the Miller disc method. Experienced technicians performed 400-cell leukocyte differential counts on all samples. These percent differential counts were applied to the total leukocyte counts obtained from a validated hematology analyzer (Abbott CellDyn 3700, software version 1.3) to determine absolute differential counts. ProCyte DX and manual results, including abnormal, flagged results, were compared by Deming Regression and Bland-Altman plots using EP Evaluator v8 software; test result intervals, correlation coefficients (R), root mean square errors (SEE) and mean biases are presented in the table below. There are strong correlative relationships between the ProCyte DX

Table [Abstract 271]

	Canine				Feline			
Test	Result Intervals[1]	R	SEE	Mean Bias[1]	Result Intervals[1]	R	SEE	Mean Bias[1]
PCV	6.00–60.00	0.990	1.495	-.571	11.00–46.00	0.966	2.237	0.794
%RETIC	0.00–11.40	0.953	0.731	0.58	0.00–21.79	0.994	0.19	0.16
#RETIC	0.00–324.12	0.929	28.86	30.65	0.00–479.38	0.971	11.73	11.33
#NEUT	0.04–68.93	0.967	2.39	−1.05	0.13–32.10	0.981	0.97	−0.63
#LYM	0.06–65.12	0.897	3.13	0.70	0.03–21.76	0.944	1.01	0.60
#MONO	0.05–13.41	0.450	6.00	0.45	0.01–4.03	0.768	0.41	0.09
#EOS	0.00–14.03	0.985	0.23	0.01	0.00–3.75	0.963	0.21	0.07
#BASO	0.00–2.07	0.727	0.13	0.01	0.00–0.43	0.731	0.07	0.05

[1]Units for each test result: PCV and %RETIC, %; #RETIC, #NEUT, #LYM, #MONO, #EOS and #BASO, $\times 10^3/\mu l$.

and manual methods. Parameters with correlation coefficients < 0.9 may be partially explained by the limitations of manual methods. These data indicate that results from the IDEXX ProCyte DX correlate strongly with most common manual hematology methods, and that rapid and accurate PCV, reticulocyte and leukocyte differential values are possible for canine and feline samples.

ABSTRACT #272

EVALUATION OF CANINE AND FELINE HEMOGRAMS USING THE IDEXX PROCYTE DX ANALYZER COMPARED TO MANUAL DIFFERENTIAL COUNTS AND THE ADVIA 120. JL Owen, HL Wamsley, BL Conrad, JW Harvey. University of Florida College of Veterinary Medicine, Gainesville, FL.

The purpose of this study was to compare the IDEXX ProCyte DX with the Bayer ADVIA 120 and manual leukocyte differential counts in dogs and cats that were clinically normal or ill. EDTA-anticoagulated venous blood samples were submitted for routine hematology analysis; 148 dog and 58 cat samples were analyzed within 1-hour of blood collection using the ProCyte DX (developmental software version 1.1) and ADVIA 120 (software version 3.1.8.0-MS). A 200-cell manual differential leukocyte count was performed independently by 2 experienced veterinary hematology technicians on Wright-Giemsa-stained blood films. Results were compared by Deming Regression using EP Evaluator v8; correlation coefficients (R) were recorded.

There was a strong linear relationship between the ProCyte DX and the comparison methods over a broad range of hematology values that included both clinically normal and ill animals. Correlation coefficients (R) between the ADVIA 120 and the ProCyte DX for dogs and cats, respectively, were 0.997 and 0.996 for erythrocyte counts, 0.993 and 0.988 for calculated Hct, 0.993 and 0.994 for Hb, 0.936 and 0.896 for MCV, 0.829 and 0.670 for MCHC, 0.962 and 0.823 for platelet counts, and 0.996 and 0.994 for total leukocyte counts. Correlations between manual microcentrifuged Hct and the Hct calculated by the ProCyte Dx for dogs and cats, respectively, were 0.990 and 0.983. Correlations between manual absolute leukocyte differential counts and the ProCyte Dx for dogs and cats, respectively, were 0.964 and 0.987 for neutrophils, 0.656 and 0.700 for lymphocytes, 0.666 and 0.820 for monocytes, 0.980 and 0.974 for eosinophils, and 0.428 and 0.752 for basophils.

These data indicate that canine and feline complete blood cell counts obtained using the in-house analyzer, IDEXX ProCyte DX, correlate strongly with hematology results obtained from a reference laboratory and that the ProCyte DX is capable of providing a rapid and accurate five-part leukocyte differential count.

ABSTRACT #273

PROSPECTIVE EVALUATION OF CLOPIDOGREL, ULTRA-LOW-DOSE ASPIRIN, OR BOTH IN DOGS WITH PRIMARY IMMUNE-MEDIATED HEMOLYTIC ANEMIA. AM Mellett, AC Fenimore, RK Nakamura, D Bianco. Red Bank Veterinary Hospital, Tinton Falls, NJ.

The authors hypothesized that the use of clopidogrel (Plavix®), alone or in combination with ultralow-dose aspirin (ULDA), will be safe and improve survival compared to ULDA alone in dogs diagnosed with primary immune-mediated hemolytic anemia (pIMHA). A prospective, randomized pilot study with three treatment groups was conducted in client-owned dogs with newly diagnosed pIMHA and treated with prednisone and azathioprine at standard therapeutic doses. Dogs with severe thrombocytopenia (<40,000 platelets/μL), inability to tolerate oral medications, or concurrently receiving other anticoagulants or immunosuppressive agents were excluded. The primary endpoint was 90-day survival. Secondary endpoints evaluated were survival at discharge, frequency of hemorrhage and gastrointestinal signs.

After informed owner consent was obtained, 24 dogs with pIMHA were randomly assigned to 3 treatment groups (clopidogrel, ULDA, or both) with a 1:1 allocation ratio. A clopidogrel single loading dose of 10 mg/kg PO was initiated on day 1, followed by a maintenance dose of 2 mg/kg PO q24 hrs for 89 days. Dogs allocated to the ULDA-group received a dose of 0.5 mg/kg PO q24 hrs for 90 days. Other immunosuppressive agents were not used during the first 14 days of therapy. All dogs were monitored for any evidence of hemorrhage or gastrointestinal signs over the 90-day study period.

One dog in each treatment group did not survive to discharge. At day 90, 6 out of 8 dogs in the clopidogrel group, 7 out of 8 dogs in the clopidogrel/ULDA group, and 6 out of 8 dogs in the ULDA group were alive. Survival analysis revealed no significant difference between the three treatment groups at discharge and 90 days. No evidence of hemorrhage was noted in any dog. A self-limiting diarrhea was observed in some dogs following the clopidogrel loading dose, but no significant difference in the frequency of gastrointestinal signs was identified between the three treatment groups.

Results of this study suggest that there is no difference in survival to discharge or at 90 days in dogs with pIMHA treated with clopidogrel, ultralow-dose aspirin, or both. Limitations of this pilot study include small sample size, lack of blinding, lack of placebo-controls, difficulty of monitoring effectiveness of therapy, and long-term follow-up data (>90 days). Larger prospective, multi-center, double-blinded, placebo-control studies are needed to evaluate efficacy and safety of clopidogrel in dogs with pIMHA.

ABSTRACT #274

THROMBOELASTOGRAPHY IN NORMAL DOGS TREATED WITH PREDNISONE, ASPIRIN, AND PREDNISONE AND ASPIRIN COMBINED. S. Flint, A. Abrams-Ogg, S. Kruth, A. Bersenas, R. Wood. Ontario Veterinary College, University of Guelph, Guelph, Ontario, Canada.

The objective of this study was to evaluate the effects of prednisone, aspirin and the combination of the two drugs on thromboelastography (TEG) in normal dogs.

Sixteen healthy purpose-bred hounds were randomized to four treatment groups: prednisone (2 mg/kg PO, q24h), aspirin (0.5 mg/kg PO, q24h), prednisone and aspirin together at the aforementioned doses, and control. All dogs were treated for 7 days, and blood was collected for TEG on days 0, 2, 4 and 6. The TEG parameters R, K, Angle, maximum amplitude (MA), G, and Cl60 (% clot amplitude at 60 minutes) were evaluated. Data are reported as mean ± SD. Analysis was by ANOVA for repeated measures.

Aspirin alone did not alter any TEG parameters. For treatment effect, G (dyn/cm^2) increased in the prednisone group (7315 ± 540, p = 0.025), and prednisone/aspirin group (7709 ± 528, p = 0.0067) compared to controls (5421 ± 566); MA (mm) also increased in the prednisone group (58.8 mm ± 2.1, p = 0.03), and prednisone/aspirin group (60.2 mm ± 2.0, p = 0.01) compared to controls (52.0 mm ± 2.0). For treatment-by-time effect, Cl60(%) increased in the prednisone group on day 6 (99.8 ± 0.3), compared to day 6 controls (90.7 ± 6.9, p < 0.0001) and baseline (93.3 ± 5.7, p < 0.0001). Cl60 also increased in the prednisone/aspirin group on day 6 (96.9 ± 4.3) compared to baseline (93.4 ± 4.5, p = 0.007) and day 6 controls (90.7 ± 6.9, p < 0.003).

In conclusion, prednisone treatment at 2 mg/kg/day for 7 days, with and without aspirin, increased clot strength and decreased clot lysis in normal dogs.

ABSTRACT #275

TREATMENT WITH ACETYLSALICYLIC ACID OR MELOXICAM DOES NOT AFFECT SEROTONIN SECRETION FROM FELINE PLATELETS. BM Brainard, C Cathcart, SC Budsberg. University of Georgia, Athens, GA.

Serotonin release from dense granules of activated platelets is implicated in vasoconstriction and clot growth associated with aortic thromboembolic disease (ATE) in cats. This study was designed to evaluate the effect of treatment with acetylsalicylic acid (ASA; 5 mg/kg PO q48 h) and meloxicam (0.05 mg/kg PO q24 h) on post-aggregatory serotonin production in feline whole blood.

Each medication was given to 8 cats for 14 days in a randomized, placebo controlled, cross-over design. It was hypothesized that NSAID therapy would decrease platelet serotonin secretion following whole blood platelet aggregation (WBA) using adenosine diphosphate (ADP; 10 µM) or collagen (5 µg/mL) as agonists. WBA was performed before drug administration on days 0 and 7, and also on day 15. Blood was collected by jugular venipuncture using vacuum into tubes containing 3.2% sodium citrate. Serotonin was measured in plasma collected after WBA with each agonist using a commercially-available ELISA assay (IBL Inc.). Statistical analyses were performed using repeated measures or one-way ANOVA.

Compared to baseline, neither medication affected platelet serotonin secretion, regardless of WBA agonist. Baseline serotonin was not different between groups (mean $\pm$ SD (ADP): 822.1 $\pm$ 456.8 ng/mL, mean $\pm$ SD (collagen): 1511.8 $\pm$ 552.5 ng/mL), consistent with published values. Platelet count remained stable for the duration of the study.

At the tested doses, neither meloxicam nor ASA decreased post-aggregatory serotonin release in feline whole blood. Further studies evaluating isolated platelets may be indicated.

ABSTRACT #276

DRILL-ASSISTED 15-GAUGE CORE BONE MARROW BIOPSY IN CLINICALLY NORMAL SMALL DOGS. A. Abrams-Ogg, A. Defarges, D. Bienzle, R Foster. Ontario Veterinary College, University of Guelph, Guelph, Ontario, Canada.

The diameter of 11–13 ga needles makes bone marrow core biopsy difficult in some small dogs. In this study, biopsies of the left humeral head and iliac crest were taken under anesthesia using a modified 1-inch, 15 ga needle attached to a drill (EZ-IO® Intraosseous Infusion System, Vidacare) from 17 healthy beagles weighing 6.5–11 kg. These were compared to biopsies taken from a different site on the left humeral head using a 2-inch, 13 ga Jamshidi needle. Biopsies were performed in randomized order by one investigator. Collection was repeated until a sample $\geq$ 5 mm long was obtained. Difficulty of site localization, needle insertion, and needle advancement were scored from 1 (easiest) to 5 (hardest). Biopsy sections were reviewed by a pathologist blinded to biopsy site and dog. Biopsy quality was scored from 1 (0 intratrabecular spaces free of artifact) to 5 ($\geq$ 5 spaces). Biopsy sites were examined at necropsy.

All biopsies captured hematopoietic tissue. Results for the procedures [mean $\pm$ SE (range)] were:

All intact spaces had normal hematopoiesis. Quality significantly correlated (Spearman) with cell density (rho = 0.66), cellularity (rho = 0.73), and megakaryocyte counts (rho = 0.69). These three parameters in 13 ga biopsies were underestimated by 75%, 69%, and 60% in 15 ga humeral, and by 77%, 83%, and 54% in ilial biopsies, respectively, and their medians were significantly different (Friedman). Myeloid:erythroid ratio was not significantly correlated between 13 ga and 15 ga biopsies (rho = 0.34–0.41). Lymphocytes/plasma cells, iron stores, and cell maturation and morphology were not different between sites. At necropsy, humeral heads had grossly moderate-to-marked hemorrhage, while ilial crests had no or minimal hemorrhage. The 13 ga biopsy sites were easily identified, while 15 ga biopsy sites were difficult to identify.

In conclusion: 1) Drill-assisted 15 ga humeral biopsy is easier to perform than manual 13 ga humeral biopsy. 2) The risk for a lower quality biopsy is greater with a 15 ga needle, especially from the iliac crest. 3) A 15 ga biopsy may quantitatively underestimate hematopoiesis, which may be due to limited sample quality, but is qualitatively representative in normal dogs.

Table [Abstract 276]

Site, needle	Attempts (number)	Time for last attempt (min)	Difficulty (3–15)	Quality (1–5)	Length on slide (mm)
Humerus, 13 ga	1.4 $\pm$ 0.2 (1–3)	1.4 $\pm$ 0.1* (0.8–2.8)	6.6 $\pm$ 0.4 (4–10)	3.9 $\pm$ 0.3** (1–5)	13.2 $\pm$ 1.3** (7–23)
Humerus, 15 ga	1.6 $\pm$ 0.3 (1–5)	1.6 $\pm$ 0.1 (1.0–3.5)	4.2 $\pm$ 0.3** (3–7)	1.9 $\pm$ 0.3 (1–5)	9.1 $\pm$ 1.0 (3–16)
Ilium, 15 ga	2.5 $\pm$ 0.4 (1–6)	2.7 $\pm$ 0.1* (1.1–11.7)	7.0 $\pm$ 0.4 (5–10)	1.4 $\pm$ 0.1 (1–3)	7.3 $\pm$ 0.5 (1–11)

*Significantly different;

**significantly different from the other two sites (ANOVA, $p < 0.05$).

ABSTRACT #277

PREVALENCE OF DOG ERYTHROCYTE ANTIGENS IN RETIRED RACING GREYHOUNDS. Iazbik MC, O'Donnell M, Marin L, Zaldivar S, Hudson D, Couto CG. The Ohio State University, College of Veterinary Medicine, Columbus, OH.

The current blood groups in the dog are designated as dog erythrocyte antigen (DEA) 1.1, 1.2, 3, 4, 5, and 7. Due to the immunogenic characteristics of the DEA 1.1 and 1.2, we evaluated their frequency in 206 Greyhounds and 66 non-Greyhound dogs. The ages ranged between 1 and 7 years (mean 4.0 years for Greyhounds and 3.2 years for non-Greyhounds), and there were 103 females (75 Greyhounds and 28 non-Greyhounds) and 169 males (131 Greyhounds and 38 non-Greyhounds). One hundred and thirteen Greyhounds and 24 non-Greyhounds were evaluated for the presence of naturally occurring antibodies against all the known DEAs.

Prevalence of DEA groups were compared between Greyhounds and non-Greyhounds using contingency tables and Fisher's exact test. Significance was set at $p < 0.05$.

The prevalence of 1.1 antigen was significantly lower in the Greyhounds than in the non-Greyhounds ($p < 0.0001$), as was the prevalence of 1.1/1.2 antigen ($p < 0.0001$). There were no significant differences between Greyhounds and non-Greyhounds for the frequency of 1.2, 3, 4, 5, or 7 antigens. As expected, there were significantly more "universal donors" in the Greyhound than in the non-Greyhound group ($p < 0.0001$). Serum from 137 dogs (113 Greyhounds and 24 non Greyhounds) was evaluated for the presence of naturally occurring antibodies and none of them were positive.

ABSTRACT #278

PORTAL VEIN THROMBOSIS IN DOGS: 29 CASES. MJ Respess[1], TE O'Toole[1], CL Rogers[2], CRL Webster[1]. [1]Tufts Cummings School of Veterinary Medicine, North Grafton, MA. [2]Cornell University College of Veterinary Medicine, Ithaca, NY.

Canine portal vein thrombosis (PVT) has been infrequently reported. Medical records of 29 cases with PVT were evaluated. PVT was documented by ultrasound (24) or direct examination at surgery (3) or necropsy (2). There was no breed or sex predilection. Median age was 7.5 years (1 month-13 years). Pre-existing conditions included hepatic (6), immune (4), endocrine (2), renal (1), and inflammatory disease (1), protein-losing enteropathy (2), neoplasia (1), and sepsis (1). Seven dogs had previous surgery or procedures including splenectomy (3), intestinal surgery (1), coil occlusion for intrahepatic portosystemic shunt (1), liver lobectomy (1), and laparoscopic liver biopsy (1). Eleven dogs had a history of prior corticosteroid administration. New diagnoses at time of PVT were neoplasia (5) and ITP (1).

Presenting complaints included vomiting (13), inappetence (10), diarrhea (10), and acute collapse (9). Ten dogs presented in shock. Presenting signs included acute abdominal pain (8), abdominal distension (8), elevated body temperature (4), and petechiae (3). Twenty-seven dogs had CBC and serum chemistry analysis. Thrombocytopenia was present in 16 dogs (range, 2,000–187,000/µL), 10 of which had concurrent anemia. Liver enzymes were elevated in 20 dogs and serum bilirubin was elevated in 12. Twenty dogs had ascites, total protein of which was <2 mg/dL in 12, and 3.7 mg/dL in 1.

Treatment included anticoagulation (16), surgery (2) and thrombolysis (1). Seventeen dogs survived to discharge, 10 were euthanized due to grave prognosis and 2 died.

In summary, PVT may result from hypercoagulability associated with certain disease states or medications.

ABSTRACT #279

HEPATIC AND BILIARY TRACE ELEMENT CONCENTRATIONS IN HEALTHY DOGS VERSUS DOGS WITH HEPATOPATHIES. BM Pressler[1], GE Moore[1], RP Favier[2]. [1]Purdue University, West Lafayette, IN. [2]Utrecht University, Utrecht, The Netherlands.

Storage and loading of trace elements onto transport proteins is oftentimes performed by the liver, and when absorbed in excess of dietary requirements, the route of excretion of a number of these elements is via bile. Primary copper (Cu) hepatopathy has been described in a number of breeds, and dogs with liver disease of other etiologies are noted oftentimes to have excess hepatic Cu concentrations. In addition, other trace elements have been implicated in canine diseases (iron [Fe] and aluminum [Al]) or have been suggested as therapies for various disease (zinc [Zn], vanadium [V], and chromium [Cr]). The purpose of this study was therefore to determine liver and bile concentrations for select trace elements in healthy dogs and dogs with various hepatopathies, and to determine whether concentrations are correlated with one another or between liver and bile.

Paired liver and bile samples were collected from 37 overtly healthy hound-type adult dogs (group 1A), 5 healthy, adult COMMD$^{+/-}$ Beagle-Bedlington terrier crossbreed dogs (group 1B), and 18 client-owned dogs with a variety of non-primary Cu hepatopathies (group 2). 4 adult COMMD$^{-/-}$ Beagle-Bedlington terrier dogs previously confirmed to have primary Cu hepatopathy (group 3) had bile collected for analysis, but liver was only available for copper assay. All dogs in group 1A, 1B, and 3 were being fed commercial dry adult maintenance dog foods, whereas group 2 dogs were fed a variety of diets based on owner and referring veterinarian preference. Group 3 dogs had been receiving penicillamine (10 mg/kg PO q12h) which was discontinued 6 days prior to obtaining liver and bile samples. Liver and bile Al, cobalt (Co), Cr, Cu, Fe, lead (Pb), V, and Zn concentrations were assayed by a commercial laboratory, and differences analyzed by two-sample t-test with significance designated at $p < 0.05$; clinical significance was designated at $|rho| > 0.5$.

The only significant correlation noted within liver from group 1A dogs was between Cu and Zn ($p < 0.001$; rho = 0.7). However, multiple positive correlations were noted within bile, including Al with Co, Fe, Pb, V, and Zn; Co with Cu, Fe, Pb, V, and Zn; Cu with V and Zn; Fe with Zn, Pb with V and Zn, and V with Zn; the only negative correlations noted were Cr with Pb and V. Trace elements with significant, positive correlations between liver and bile in 1A dogs were Al, Co, and Cr ($p < 0.001$ and rho 0.6 for all). Liver Co was higher in 1B dogs than 1A, and 1A + 1B dogs (i.e. healthy dogs) had significantly higher liver Co than 2A dogs and significantly higher bile Co than group 2 + 3 dogs (i.e. hepatopathy dogs). Liver Cu in group 3 dogs was significantly higher and bile Cu was significantly lower than any other group. There were no significant differences between groups in liver or bile Al, Cr, Fe, Pb, V, or Zn concentrations.

These results suggest that several trace elements may influence concentration of other elements, and that bile concentration of Al, Co, or Cr can potentially be used to predict liver concentrations. COMMD$^{+/-}$ Beagle-Bedlington terriers, although asymptomatic, may have an uncharacterized defect that results in liver cobalt accumulation.

ABSTRACT #280

$CD4^+CD25^+$ T REGULATORY CELLS FROM FIV INFECTED CATS INDUCE A UNIQUE ANERGIC PROFILE IN $CD8^+$ TARGETS. Jonathan E. Fogle, Mary B. Tompkins, Wayne A. Tompkins. North Carolina State University, College of Veterinary Medicine, Raleigh, NC.

Early impaired $CD8^+$ and $CD4^+CD25^-$ T cell immune responses, reminiscent of anergy, correlate with activation of $CD4^+CD25^+$ T regulatory cells (Treg) in AIDS lentiviral infections. Using the FIV model, we reported previously that $CD4^+CD25^+$ T regulatory cells in both the acute phase and long-term, asymptomatic phase of infection are constitutively activated and suppress $CD4^+CD25^-$ and $CD8^+$ T cell immune responses. T cell anergy is defined by G_1 cell cycle arrest and failure to proliferate in response to appropriate stimulation. Because Treg cells are known to induce G_1 cell cycle arrest in T cell targets; we asked whether Treg cells altered the expression of cyclins, cyclin-dependent kinases and their inhibitors that regulate progression from G_1 to S phase. To address this question, lymphocytes were obtained from SPF negative control cats or cats chronically infected with $NCSU_1$ FIV and sorted by FACS into $CD4^+CD25^+$ and $CD8^+$ populations. Following either 12 or 24 hour co-culture with $CD4^+CD25^+$ cells, $CD8^+$ target cells were examined by western blot for changes in cyclins D_3, E and A, retinoblastoma protein (Rb), as well as the cyclin dependent kinase inhibitors (CDKIs) $p15^{ink4b}$, $p21^{cip1}$, and $p27^{kip1}$. Following co-culture with $CD4^+CD25^+$ Treg cells, we have observed up-regulation of $p21^{cip1}$ and cyclin E with down-regulation of cyclin D_3, and cyclin A in $CD8^+$ cells from FIV^+ cats; while little change was noted in $p15^{ink4b}$ and $p27^{kip1}$ expression. As expected, $CD8^+$ targets from FIV-negative control cats were relatively quiescent with little up-regulation of $p21^{cip1}$ and cyclin E as was observed in the FIV^+ cats. Hyper-phosphorylation of Rb is consistent with irreversible commitment to S phase; and further investigation showed there was a lack of Rb phosphorylation in $CD8^+$ target cells consistent with late G_1 cell cycle arrest. We are currently performing FACS cell cycle analysis in $CD8^+$ targets, following $CD4^+CD25^+$ co-culture, to confirm G_1 cell cycle arrest. Collectively these data suggest that $CD4^+CD25^+$ Treg cells are activated in FIV^+ cats in vivo and induce anergy of $CD8^+$ T cell targets by disruption of normal G_1 to S regulatory mechanisms.

Previously presented at the Keystone Symposia, Santa Fe, N, M January 2010.

ABSTRACT #281

ASSOCIATION BETWEEN FELINE ANTIBODY RESPONSES TO CRANDELL REES FELINE KIDNEY (CRFK) CELL LYSATES, ALPHA-ENOLASE, AND ANNEXIN A2 AND BIOCHEMICAL ABNORMALITIES IN 1,477 PRIVATELY-OWNED CATS. JC Whittemore[1], JR Hawley[2], SV Radecki,[2] MR Lappin[2]. [1]University of Tennessee, Knoxville TN. [2]Colorado State University, Fort Collins, CO.

By Western blot immunoassay, we previously identified three immunodominant CRFK antigens (α-enolase, annexin A2, and macrophage capping protein/cap G) that are widely distributed throughout tissues and highly conserved between species. We have shown that SQ administration of FVRCP vaccines containing CRFK proteins induces CRFK, α-enolase and annexin A2 antibody levels detectable by ELISA. The purpose of this study was to assess for associations between CRFK, α-enolase and annexin A2 antibodies by ELISA and biochemical abnormalities in privately-owned cats.

Residual sera submitted for biochemical analysis for 1,477 privately-owned cats were acquired. Indirect ELISAs for antibodies to CRFK cell lysate, α-enolase and annexin A2 were performed. Albumin, bilirubin, creatinine, calcium, chloride, cholesterol, globulin, glucose, phosphate, potassium, sodium, and total protein concentrations as well as activities of creatine kinase (CK), alkaline phosphatase (ALP), alanine aminotransferase (ALT), and aspartate transferase (AST) were compared to CRFK, α-enolase and annexin A2 %ELISA values by multiple logistic regression.

Significant negative associations were identified between CRFK antibodies and glucose, creatinine, and ALP activity; α-enolase antibodies and glucose, albumin and ALP activity; and annexin A2 antibodies and glucose and albumin. A significant positive association was found between CRFK antibodies and bilirubin.

Anti-CRFK, α-enolase, and annexin A2 antibodies are correlated with some biochemical abnormalities. The associations between bilirubin and creatinine concentrations and CRFK, but not α-enolase or annexin A2, antibodies suggests antibodies against additional CRFK proteins may be clinically relevant. Further studies evaluating the significance of these results are necessary.

ABSTRACT #282

THE USE OF FLOW CYTOMETRY TO IDENTIFY THE PRESENCE OF ANTI-ERYTHROCYTE ANTIBODIES IN ANEMIC CATS. ML Hart, S Dow, CB Webb. Department of Clinical Sciences, Colorado State University, Fort Collins, CO.

The purpose of this study was to determine if a flow cytometry protocol could identify feline anti-erythrocyte antibodies attached to red blood cells (RBC) of cats with a presumptive diagnosis of immune-mediated hemolytic anemia (IMHA).

EDTA-preserved whole blood from 3 anemic cats with a presumptive diagnosis of IMHA was processed concurrently with samples from non-anemic cats. Ten microliters of whole blood was washed 3 times and the RBCs incubated in fluorescence activated cell sorting buffer alone or with goat anti-feline IgM or IgG FITC-conjugated antibodies (Serotec). The cells were washed and analyzed using a Cyan flow cytometer. Untreated cells were used to set the threshold for non-specific fluorescence.

Three cats (2, 5, and 16 years of age) with a regenerative anemia and a presumptive diagnosis of IMHA were positive for anti-erythrocyte antibodies using the described protocol. The packed-cell volume of the anemic cats ranged from 6% to 17%. One cat was FeLV+, tested Coombs positive, and showed 2 + saline agglutination. The second cat also showed marked RBC saline agglutination and icteric serum. A third cat was both anemic and thrombocytopenic. All anemic cats were negative by PCR for M. haemofelis and M. hemominutum. All 3 anemic cats had RBCs that were positive by flow cytometry for IgM (mean 8.0% +/− 3.3 std dev) and 2 that were positive for IgG (mean 3.4% +/− 1.8) class anti-erythrocyte antibodies. The non-anemic cats' RBCs stained a mean of 0.5% +/− 0.4 with anti-feline IgM and 0.3% +/− 0.3 with anti-feline IgG.

The conclusion from this study is that flow cytometry can be used to detect feline anti-erythrocyte antibodies in clinical cases of anemia in cats. This should be an important tool in future studies of the prevalence and pathophysiology behind immune-mediated hemolytic anemia in the feline species.

ABSTRACT #283

DETECTION OF *BARTONELLA HENSELAE* IGM IN SERUM OF NATURALLY EXPOSED CATS. J Ficociello, C Bradbury, A Morris, MR Lappin. Department of Clinical Sciences, Colorado State University, College of Veterinary Medicine and Biomedical Sciences, Fort Collins, CO.

Bartonella henselae has been associated with illness in cats and people. However, results of blood culture, PCR on blood, or *B. henselae* IgG in serum have not correlated to the presence or absence of clinical disease. In addition, results of *B. henselae* IgG in serum is poorly predictive of bacteremia. We recently validated a *Bartonella* spp. IgM ELISA using serum collected sequentially from cats experimentally infected by exposure to *B. henselae* infected *Ctenocephalides felis*. The purpose of this study was to determine whether *B. henselae* IgM results can be used to predict clinical bartonellosis or bacteremia.

Paired samples from cats with and without fever (n = 184), paired samples from cats with and without stomatitis (n = 108), and samples from healthy shelter cats (n = 51) that had been used in previous studies of *Bartonella* spp. IgG antibodies and *Bartonella* spp. PCR had been stored at − 80 C. The samples were thawed and assessed in the *Bartonella* spp. IgM assay. The distribution of the paired sample results was evaluated by logistic regression to determine odds ratios and 95% confidence intervals. Wilcoxon's rank sum test was used to determine whether *Bartonella* spp. IgM ELISA titer magnitude was associated with presence of fever or stomatitis. Significance was defined as $p < 0.05$. Results were used to calculate PPV and NPV for presence of disease and for presence of bacteremia as defined as a positive PCR assay result.

Results from the analyses revealed no statistical association between *Bartonella* spp. IgM (positive or negative) or IgM titer magnitude and the presence of fever or stomatitis. The PPV and NPV of *Bartonella* spp. IgM and disease were approximately 40% in all analyses. The PPV and NPV of *Bartonella* spp. IgM and bacteremia based on PCR results were approximately 15% and 85%, respectively.

Bartonella spp. IgM antibodies do not correlate with the presence of fever or stomatitis and are not accurate for the prediction of the *Bartonella* spp. bacteremia.

ABSTRACT #284

PREVALENCE OF *BARTONELLA* SPECIES, HAEMOPLASMA SPECIES, AND *RICKETTSIA FELIS* DNA IN BLOOD AND FLEAS OF CATS IN BANGKOK, THAILAND. S Assarasakorn, JK Veir, MR Lappin. Department of Clinical Sciences, Colorado State University, Fort Collins, CO.

While *Ctenocephalides felis* infestations are common in Thailand, little is known about the prevalence of flea-associated pathogens. The purpose of this study is to determine the prevalence of *Bartonella* spp., hemoplasmas, and *Rickettsia felis* in client-owned cats and their fleas in Bangkok, Thailand.

Blood and fleas were collected from cats that were randomly selected during June and August 2009 from a referral veterinary hospital and 2 private clinics in Bangkok. The samples were stored at −20°C until being shipped to Colorado State University for analysis. Total DNA was extracted from blood and pooled fleas, and was assessed in conventional PCR assays that amplify the DNA of these particular pathogens.

A total of 153 blood samples and 50 flea pools were collected from cats. *Ctenocephalides felis* was identified in 99.1% of all fleas. The overall prevalence rates for *Bartonella* spp. in blood and flea samples were 16.3% and 32.0%, respectively. *Bartonella henselae* and *Bartonella clarridgeiae* were the common *Bartonella* spp. amplified. The overall prevalence rates for haemoplasmas in blood and flea samples were 22.9% and 34.0%, respectively. '*Candidatus* M haemominutum' and *M. haemofelis* were the haemoplasmas amplified. *Rickettsia felis* DNA was not amplified from any blood or flea sample. Combinations of *Bartonella* and haemoplasma DNA were amplified from some cats.

The results demonstrate that *Bartonella* spp. and haemoplasma spp. infections are common in client-owned cats and cat fleas in Bangkok and suggest that further study of associations of these infections with clinical syndromes is indicated in the area.

ABSTRACT #285

PREVALENCE OF *BARTONELLA* SPP. DNA IN CONJUNCTIVAL CELLS COLLECTED FROM SHELTER CATS WITH AND WITHOUT CONJUNCTIVITIS. CL McInnis[1], L Clark[1], M Brewer[1], CC Powell[1], W Rekers[2], P Dingman[3], MR Lappin[1]. [1]Colorado State University, Fort Collins, CO. [2]Cat Adoption Team, Sherwood, OR. [3]University of Florida, Gainesville, FL.

Conjunctivitis is a common, painful condition of cats that has the potential to induce pathologic corneal changes and consequently, in severe cases, compromise vision. *Bartonella* spp. have been implicated as a cause of a variety of feline ocular abnormalities including conjunctivitis. However, studies definitively confirming the association between feline conjunctivitis and *Bartonella* are lacking. The primary purpose of the study was to determine the prevalence of *Bartonella* spp. DNA in conjunctival cells collected from cats with and without conjunctivitis.

Participating shelter veterinarians were asked to collect blood (1 ml in EDTA; 2 ml for serum separation) and conjunctival swabs from cats with clinical evidence of conjunctivitis prior to administration of antibiotics. For every two cats with conjunctivitis, samples were requested from a healthy cat in the same shelter without conjunctivitis. Samples were stored at −20C until shipped to CSU for assessment in a conventional PCR assay that amplifies DNA of common *Bartonella* spp. and an ELISA for *Bartonella* spp. IgG antibodies. The DNA prevalence rates and seroprevalence rates were compared between groups using Fisher's exact test.

Of the 70 cats with conjunctivitis, 0 had *Bartonella* DNA amplified from conjunctival swabs, but 10 of the 41 cats tested (24.4%) had *Bartonella* DNA amplified from blood and 23 of the 41 cats tested (56.1%) had *Bartonella* spp. IgG in serum. Of the 43 control cats, 0 had *Bartonella* DNA amplified from conjunctival swabs, but 10 of the 27 cats tested (37.0%) had *Bartonella* DNA amplified from blood and 22 of the 27 cats tested (81.5%) had *Bartonella* spp. IgG in serum. There were no significant differences in the prevalence of *Bartonella* spp. DNA amplified from conjunctival swab samples or from whole blood samples ($p = 1.0$, $p = 0.115$, respectively). Cats without conjunctivitis were significantly more likely to have *Barton-*

ella spp. IgG detected in their serum than cats with conjunctivitis (p = 0.02).

The results of this study indicate that while *Bartonella* spp. infection was common based on the blood results, there was no association between *Bartonella* DNA in blood and conjunctivitis and *Bartonella* DNA was not amplified from conjunctival cells. These results fail to confirm an association between *Bartonella* spp. and feline conjunctivitis and suggest that *Bartonella* may not be a cause of this syndrome.

ABSTRACT #286

DIVERSITY OF *BARTONELLA* SPECIES IN 62 DOGS USING THE BAPGM PRE-ENRICHMENT CULTURE METHOD.
C. Pérez[1], PPVP Diniz[2], RG Maggi[1], EB Breitschwerdt[1]. [1]Intracellular Pathogens Research Laboratory, Center for Comparative Medicine and Translational Research, College of Veterinary Medicine, North Carolina State University (NCSU), Raleigh, NC. [2]Western University of Health Sciences, College of Veterinary Medicine, Pomona, CA.

The diagnosis of bartonellosis in dogs can be extremely challenging and frequently requires the use of specialized culture techniques, in combination with PCR. The objectives of this study were to: (1) describe the cumulative diagnostic experience using the BAPGM (*Bartonella* alpha Proteobacteria growth medium) pre-enrichment culture and PCR platform; (2) to describe an expanding spectrum of *Bartonella* species identified in diagnostic samples from sick dogs; and (3) to characterize the relationship between infection and *Bartonella* seroreactivity.

Between 2003 and 2009, 924 samples from 663 dogs were submitted to the NCSU-CVM-VBDDL for testing using the BAPGM platform, of which 62 (9.3%) dogs were infected with one or more *Bartonella* sp., which included *B. henselae* (*Bh*, 46.8%), *B. vinsonii* subsp. *berkhoffii* (*Bvb*, 25.8%), *B. koehlerae* (*Bk*, 4.8%), *B. volans-like* (*Bvl*, 3.2%) and *B. bovis* (*Bb*, 1.6%). Co-infection with more than one *Bartonella* sp. was documented in 11 dogs (17.7%). Serology results were available for 48.3% of dogs infected with *Bh* (42.9% were *Bh* seroreactive and 21.4% were seroreactive to another tick-borne pathogen [TBP, includes *Anaplasma* spp., *Babesia canis, Borrelia burgdorferi, Ehrlichia canis*, and *Rickettsia rickettsii*]), 25% of the dogs with *Bvb* (25% were *Bvb* seroreactive, 50% were seroreactive to another TBP), and 22% of the dogs that were co-infected with *Bh* and *Bvb* (50 % seroreactive to *Bh* or *Bvb* and 50% were seroreactive to another TBP).

The results of this study, in conjunction with recent seroprevalence studies, indicate that: (1) dogs are more frequently exposed to and infected with *B. henselae*, as compared to *Bvb* or other *Bartonella sp;* (2) co-infection with more than one *Bartonella* sp. is not uncommon in sick dogs; (3) and although infrequent, dogs can be infected with *Bk*, *Bb*, or a novel *Bartonella* sp. (*Bvl*). Based upon serology, 25–50% of infected dogs were concurrently seroreactive to a known tick-borne pathogen and for reasons that remain unclear, 50% of infected dogs did not have detectable *Bartonella* antibodies.

ABSTRACT #287

SURVIVAL OF *BARTONELLA HENSELAE* IN THE BLOOD OF CATS USED FOR TRANSFUSION.
CA Bradbury[1], M Green[1], M Brewer[1], E Breitschwerdt[2], M Lappin[1]. [1]Colorado State University College of Veterinary Medicine and Biomedical Sciences, Fort Collins, CO. [2]North Carolina State University College of Veterinary Medicine, Raleigh, NC.

Bartonella henselae can be a significant pathogen of cats, is considered a significant zoonotic disease, and can be transmitted by IV administration of infected blood. Blood used for transfusions is often stored in citrate-phosphate-dextrose-adenine (CPDA-1) solution for up to one month and it is currently unknown if *B. henselae* remains viable in blood stored in this solution. The purpose of this study was to determine whether *B. henselae* survives after storage in blood preserved in CPDA-1.

Two specific pathogen free cats were inoculated intravenously with *B. henselae*. When infection was confirmed via a *Bartonella* spp. PCR assay, blood was collected from each cat and placed into commercially available sterile, closed-collection systems that contained CPDA-1 and stored at 4°C. On days 0, 3, 7, 10, 14, 17, 21, 24, and 28, 2 ml samples were aseptically removed from each bag. The 2 ml were divided into 0.5 ml aliquots and each was plated separately onto four plates per time point. The plates were incubated and checked daily for 28 days. When colonies characteristic of *B. henselae* were identified, they were confirmed as *B. henselae* by PCR.

Bartonella henselae was cultured from all samples on all plates from each time point. Due to the large numbers of colonies isolated, it could not be determined whether the organism replicated within the solution.

The results of this experiment confirm that *B. henselae* remains viable when stored for up to 28 days in CPDA-1 solution and suggests that cats used as blood donors should be screened for this organism.

ABSTRACT #288

RECOMBINANT *BARTONELLA* ADHESION A STIMULATES INFLAMMATORY RESPONSE AND ANGIOGENESIS IN HUMAN UMBILICAL VEIN ENDOTHELIAL CELLS.
Tae Ho Kim[1], Dong Ha Bhang[2], Sung Jin Ko[1], Hee Jeong Youn[1], Joon Seok Chae[1], Hee-Woo Lee[2], Hwa Young Youn[1]*, Suh-young Sohn[1] and Ye-rin Coh[1]. [1]Department of Veterinary Internal Medicine, College of Veterinary Medicine, Seoul National University, Seoul, Republic of Korea. [2]KRF Priority Zoonotic Disease Research Institute, College of Veterinary Medicine, Seoul National University, Seoul, Republic of Korea.

Bartonella henselae is a zoonotic pathogen causing cat scratch disease and bacillary angiomatosis. We investigated the role of BadA as a pathogenic factor on inflammatory response and angiogenesis in HUVECs to help understanding the bacterial pathogenesis. We prepared the recombinant BadA protein using the pET-28a Expression System and evaluated the expression of pro-inflammatory related adhesion molecules and cytokines, and VEGF in HUVECs incubated for 6, 12, 24 and 48 h after treatment of BadA (200 ng/ml) using comparative RT-PCR and ELISA analysis. We also investigated the proliferation of HUVECS and quantitative adherence of U937 monocytes to HUVECs after BadA treatment (200 ug/ml). BadA exhibited very minimal endotoxic activity in HUVECs incubated for 24 and 48 h at a concentration of 100–1000 ng/ml. BadA also showed the enhanced ability on proliferation of HUVECs and on adherence of U937 monocytes to HUVECs. The expression of E-selectin, ICAM-1, VCAM-1, IL-8 and MCP-1 mRNA were significantly increased in HUVECs from 6 to 48 h after BadA treatment ($p < 0.05$). The expression levels of IL-6, IL-8 in HUVECs culture supernatant was significantly higher at 24 h and 48 h after treatment of BadA than control ($p < 0.05$), which is in accordance with the results of RT-PCR analysis. Interestingly, the expression of VEGF mRNA and protein was significantly upregulated at 24 h in HUVECs after BadA treatment ($p < 0.001$). Our data show that BadA stimulates inflammatory response in HUVECs as well as production of VEGF, implicating BadA plays an important role in *Bartonella* pathogenesis including cat scratch disease and bacillary angiomatosis.

ABSTRACT #289

MOLECULAR SURVEY OF POSSIBLE ATOVAQUONE-RESISTANT VARIANTS OF *BABESIA GIBSONI* IN DOGS IN JAPAN.
M Sakuma[1], K Fukuda[2], K Takayama[2], Y Kobayashi[2], A Setoguchi-Kinoshita[1], Y Endo[1]. [1]Kagoshima University, Kagoshima, Japan; [2]Adtec Co., Ltd., Oita, Japan.

Combination therapy using atovaquone (ATV) and azithromycin (AZM) is effective for the treatment of acute onset *Babesia gibsoni* infection. However, the incidence of relapse was high, with some clinical cases resistant to further treatment with ATV and AZM.

Potentially ATV-resistant *B. gibsoni* variants that possessed mutations in a putative ATV-binding site of cytochrome b were predominant in these ATV-resistant cases. In this study, we conducted a survey of the *cytochrome b* (*cytb*) gene of *B. gibsoni* isolated from clinical cases to determine the prevalence of potential ATV-resistance variants.

Ninety-two blood samples were collected from dogs with clinically confirmed *B. gibsoni* infection and no treatment history with anti-protozoal drugs. The nucleotide sequence of *cytb* from the dominant *B. gibsoni* strain in each case was determined by direct sequencing. Many single nucleotide substitutions were detected but 12 non-synonymous substitutions were identified in *cytb*: M43V (1.1%), M43I (1.1%), A108T (2.2%), M121I (3.3%), I136V (1.1%), A216T (1.1%), V220I (6.5%), I226V (12.0%), A276V (1.1%), A290V (1.1%), I303V (8.7%), and P310S (16.3%). The principal ATV-resistance substitution, M121I, was detected in three cases and other substitutions possibly related to ATV-resistance were also detected.

This survey determined that potentially ATV-resistant *B. gibsoni* strains are present in dogs in Japan. Although the frequency of these resistant strains was low, the emergence of drug-resistant *B. gibsoni* strains is likely to complicate therapy. Therefore, the establishment of screening methods to identify such strains and the development of new therapeutic strategies are required to limit the spread of babesiosis.

ABSTRACT #290

CD4$^+$ AND CD8$^+$ T-LYMPHOCYTE MEASUREMENTS FOR FELINE IMMUNODEFICIENCY VIRUS (FIV) POSITIVE CATS.

J Lin[1], M Goyette[1], M Beall[1], G Kociba[1], D DeNicola[1], A Litster[2], C Deka[1]. [1]IDEXX Laboratories, Westbrook, ME. [2]School of Veterinary Medicine, Purdue University, West Lafayette, IN.

FIV infection can cause declining CD4 + T-lymphocyte counts and an immunodeficiency syndrome in infected cats. The aim of this study was to develop a rapid and accurate flow cytometry method to characterize the key lymphocyte subtypes involved in the infection and immune response in FIV-positive cats.

Blood samples were obtained from 47 FIV-positive shelter cats, 10 FIV vaccinated cats and 203 FIV-negative cats. IDEXX Feline Combo test kits were used to determine FIV status. Fresh whole blood was incubated with CD4, CD5, CD8, CD21 and CD61 antibody conjugates. CD61 (platelet marker) was used to exclude platelet interference, common in feline samples due to large platelets and/or platelet aggregates. The CD8 antibody used in this method was purchased from AbD Serotec (CD8Alpha-Beta, clone vpg9). Erythrocytes were then lysed using PharmLyse (BD) and the mixture was centrifuged. The resultant supernatant was removed and resuspended with 1×PBS/1%FBS before further centrifugation and resuspension. The resuspended cells were run on a flow cytometer (Accuri C6). CD4 and CD8 bivariate plots were used to determine percentages of CD4+ T cells and CD8 + T cells.

Precision of the method was evaluated using 10 FIV-negative samples comprising same-day same-sample 3-run precision tests and same-sample 3-day precision tests. The coefficient of variation (CV) for FIV-negative CD4/CD8 same-day same-sample 3-run precision tests averaged 3.15% and CV for same-sample 3-day precision tests was 9.3%.

CD4/CD8 ratios for FIV-positive shelter cats measured in this study were significantly lower ($n = 47; 0.89 \pm 0.52$ mean ± SD) than for FIV-negative cats ($n = 203; 1.75 \pm 1.2$ mean ± SD; $p < 0.0001$) and reflected previously published values for FIV-infected cats. The CD8+ T-cell population showed two subsets, 'CD8 low intensity' and 'CD8 high intensity'. The 'CD8 low intensity' cells appeared in markedly higher numbers in FIV-positive cats compared to FIV-negative cats.

The cellular analysis method reported here may be useful in staging FIV disease progression by determining T-lymphocyte subset characteristics in serially collected blood samples from infected cats. Combining CD4/CD8 ratio and the percent of 'CD8 low intensity' T-cells on a single bivariate plot may offer added benefits in such analysis.

ABSTRACT #291

MICROTITER PLATE FORMAT ELISA TO DIFFERENTIATE SERA FROM FELINE IMMUNODEFICIENCY VIRUS (FIV) INFECTED AND VACCINATED CATS.

TP O'Connor[1], J Saucier[1], P Andersen[1], MR Lappin[2] and R Chandrashekar[1]. [1]IDEXX Laboratories, Inc., Westbrook, ME, [2]Veterinary Medical Center, Colorado State University, Fort Collins, CO.

Feline immunodeficiency virus (FIV) is a feline-specific retrovirus that can produce chronic immunodeficiency disorders in cats. Infected cats have a world-wide distribution; infection rates in North America have been shown to vary from 2.8% for cats routinely tested at veterinary clinics and shelters to 7.4% for symptomatic and/or at risk cats from veterinary clinics. A formalin-inactivated whole virus vaccine (Fel-O-Vax®, Fort Dodge Animal Health, Fort Dodge, IA) was introduced in 2002 and produces an antibody response that cannot be distinguished from the response produced in naturally-infected cats using serological assays. Investigations into assays which differentiate infected from vaccinated cats are ongoing and pose a significant challenge.

We have evaluated a competition-format microtiter plate ELISA using monoclonal antibodies specific for different epitopes of FIV antigen. The HRP-labeled monoclonal antibodies bound to microtiter wells coated with FIV antigen were competed by sera from naturally infected cats but less so by sera from vaccinated cats, particularly if outside of a 15 week period immediately following vaccination. The capacity of this assay to correctly identify FIV infection in a field population was determined by testing samples from a non-vaccinated shelter cat population and comparing results to the licensed FIV antibody ELISA (PetChek® FIV Ab Assay, IDEXX Laboratories, Westbrook, ME). Each of the 84 positive and 42 negative samples was correctly identified by the competition assay. Temporal samples from known non-infected, vaccinated cats were produced using 3 groups of cats (5–6 cats per group) each vaccinated with a different vaccine lot administered 3 times over a 6 week period. Serum from each of the cats was reactive in the FIV Western blot assay and FIV Ab ELISA (PetChek® FIV Ab, IDEXX Laboratories, Westbrook, ME) starting 2–4 weeks following administration of the initial dose of vaccine. In the competition assay, sera from vaccinated cats were reactive for a period of 15 weeks or less following vaccination and were then non-reactive. Five cats that received the full vaccine series (3 doses) in 2003 and again in 2009 were non-reactive in the competition assay prior to and 9 weeks following administration of the second vaccine series.

These results show that following vaccination, the antibody response to the FIV epitopes defined by these monoclonal antibodies was transient and was limited to a period of less than 16 weeks. While this preliminary data is encouraging, further development efforts are needed to generate a robust and reliable diagnostic assay.

ABSTRACT #292

OCCURRENCE OF CHRONIC KIDNEY DISEASE IN CATS NATURALLY INFECTED WITH IMMUNODEFICIENCY VIRUS.

A Ávila, A Reche Júnior, MM Kogika, AGT Daniel, A Merlo, CAG Junior, BM Teixeira, MK Hagiwara. School of Veterinary Medicine – University of São Paulo, Brazil.

Cats naturally infected with the feline immunodeficiency virus (FIV) develop a syndrome that share common characteristics with the human immunodeficiency virus (HIV) infection. For this reason, felines are considered a promising model for the study of HIV infection. HIV associated nephropathy is a common and concerning complication in human beings, resulting in progressive renal insufficiency. Likewise, clinic-pathological findings in FIV - naturally infected cats suggest a renal involvement.

This study was undertaken to evaluate the occurrence of chronic kidney disease (CKD) in cats infected with FIV and to verify a possible association between both diseases. To achieve these goals, 44 cats (27 females and 17 males) were prospectively monitored every 4 months for 18 months or until death. Of these cats, 20 were FIV-infected (10 females and 10 females) and 24 (17 females and 7 males) were FIV-uninfected. None of them were infected with feline leukemia virus. Every 4 months all cats were submitted to physical examination and serum creatinine and urinary protein:creatinine ratio (UP/C) measurements. Urinary protein:creatinine ratio was performed as described by Bradford, 1976. The association between the categorized variables was evaluated with the *Pearson's chi square* test or the *Fisher exact test*. A level of significance of 5% ($P < 0.05$) was adopted in the tests

The occurrence of CKD in FIV-infected cats was higher than in FIV-uninfected cats (45% vs 25%, $P < 0.05$). Proteinuria in at least one moment of evaluation was observed in 60% of FIV-infected cats and in 26.1% of non-infected cats ($P < 0.05$). Considering persistent proteinuria as the observation of urinary protein excretion in at least 3 consecutive moments, FIV-infected cats exhibited occurrence of 30.8% compared to 6.7% in the non-infected cats ($P < 0.05$). Besides that proteinuria was significantly associated with death among FIV-infected cats with CKD ($P < 0.05$). It was observed an association between death and CKD only in the cats infected with FIV ($P < 0.05$).

In conclusion, this study demonstrates a high occurrence of CKD in FIV-infected cats and proteinuria is associated with increased risk of death in these cats. The association between death and CKD in FIV-infected cats suggests FIV may contribute for the worsening of CKD, resulting in a quicker organic dysfunction and marked reduction of survival.

ABSTRACT #293

ADRENAL STIMULATION AND CD4+ T LYMPHOCYTE SUBSETS IN FELINE IMMUNODEFICIENCY VIRUS NATURALLY INFECTED CATS. RA Oliveira, A Reche Júnior, AGT Daniel, D Ramos, JM Monteiro, P Viau, CAG Júnior, JR Kfoury Júnior, CA Oliveira. School of Veterinary Medicine – University of São Paulo, Brazil.

Abnormalities of serum basal cortisol levels in human immunodeficiency virus (HIV)-infected patients have been documented in the early as well as late stages of infection, ranging from subclinical alterations to frank adrenal insufficiency. The etiologies of the hypothalamic-pituitary-adrenal (HPA) axis derangement in these patients have not yet been established. The feline immunodeficiency virus (FIV)-infected cat is a model for HIV infection due to similarities in biology and clinical disease, but so far nothing has been done to evaluate the HPA axis of FIV-infected cats.

This study was undertaken to assess the adrenal function of FIV-infected cats using ACTH stimulation test and correlate the results with blood CD4+ cell counts.

To achieve these goals, 13 (7 males and 6 females) FIV-naturally-infected cats and 14 (7 males and 7 females) FIV-uninfected and healthy cats were used. All the 27 cats belong to the same rescued colony. All the 13 FIV-infected cats were asymptomatic at the moment of ACTH stimulation tests were performed.

Serum cortisol levels before and 60 minutes after an exogenous ACTH stimulation test were measured by radioimmunoassay in FIV-infected cats as in FIV-uninfected cats. The ACTH stimulation test was performed at the same day and period of the day in all 27 cats. The blood CD4+ cell counts were determined by flow cytometry in all cats.Unpaired *t* test was used to compare the cortisol and CD4+ counts between groups. A P value < 0.05 was considered significant. Results are presented as means±standard deviation (SD) where appropriate.

The basal serum cortisol levels were not statistically different between FIV-infected or uninfected cats (1.650 ± 1.391 vs 1.914 ± 1.603, $P = 0.653$).

The serum cortisol levels 60 minutes after the ACTH injection were not different between FIV-infected or uninfected cats (12.391 ± 3.600 vs 11.978 ± 3.111, $P = 0.752$).

The serum cortisol levels increased significantly after the ACTH stimulation in the FIV-infected group ($1.914 \pm 1{,}603$ vs 12.391 ± 3.600, $P < 0.0001$) as well as in the FIV-uninfected cats (1.650 ± 1.391 vs 11.978 ± 3.111, $P < 0.0001$).

The blood CD4+ cell counts were significantly lower in FIV-infected cats than in FIV-uninfected cats (809.43 ± 665.24 vs 1469.60 ± 823.07 $P < 0.05$). Although no correlation was observed between serum cortisol levels and blood CD4+ counts in FIV infected cats as well as in FIV-uninfected cats ($P < 0.05$).

We can conclude that FIV-infected cats, differently than HIV-infected patients, may not have adrenal dysfunction.

ABSTRACT #294

ABSENCE OF ASSOCIATION BETWEEN *LEISHMANIA* SP, FIV AND FELV INFECTIONS, IN CATS FROM AN ENDEMIC AREA FOR VISCERAL LEISHMANIASIS IN BRAZIL. LSV Sobrinho[1], CN Rossi[1], MD Laurenti[2], VMF Lima[1], TAC Costa[1], JP Vides[1], AAD Gomes[1], ET Braga[1], M Marcondes[1]. [1]São Paulo São Paulo State University, Araçatuba, São Paulo, Brazil.

Sporadic reports of myositis can be found among clinical alterations caused by visceral leishmaniasis in dogs. The goal of this study was to determinate electromyographic alterations in dogs naturally infected with *Leishmania chagasi*, with no locomotor alterations. The electromyography was made in four muscles (triceps brachii, extensor carpi radialis, biceps femoris and gastrocnemius) of 25 adult dogs from the Zoonosis Control Center (ZZC) from Araçatuba, São Paulo, Brazil. The insertional activity was reduced in 19 dogs (76%). During electrical activity at rest, spontaneous fibrillation potentials, positive waves and repetitive complex discharges were identified in 13 animals (52). During the evaluation of the voluntary electrical activity, 18 dogs (72%) had an altered recruitment in at least one of the studied muscles, evidenced by the increase in the recruitment speed and the presence of small polyphasic potentials of short duration. The electromyographic results encountered in this study allowed the identification of 13 animals (52%) with chronic polymyositis; two (8%) at acute stage; four (16%) whose muscles had both acute and chronic inflammation; three (12%) whose electromyography was normal and three (12%) with isolated electromyographic alterations in the tested muscles, thus not allowing the determination of a polymyositis diagnosis. These alterations led to the conclusion that dogs with visceral leishmaniasis, even with no muscular alterations at physical examination, may show an impairment of the muscular system that is compatible with chronic or acute polymyositis, confirmed by electromyography.

There are divergences in the literature concerning the susceptibility of domestic cats to *Leishmania* sp. infection and its possible association with infection by feline leukenia virus (FeLV) and feline immunodeficiency virus (FIV). The purpose of the present study was to determine if there was an association between Leishmania, FIV and FELV infections in 211 adult cats from Araçatuba, São Paulo, Brazil, an endemic area for visceral leishmaniasis. Leishmaniasis diagnosis was based on serology by ELISA (cut-off: 0.365), IFAT (cut-off: 1:40) and direct parasitological examination by needle aspiration biopsy of lymph nodes and bone marrow. Serology for FIV and FeLV was determined by the SNAP Combo FeLV Ag/FIV Antibody Test (IDEXX Laboratories). From the 211 cats, 16 (7.58%) had amastigote forms of *Leishmania* sp either in lymph nodes or in bone marrow without the presence of antibodies against *Leishmania chagasi*; two (0.94%) were both serologically and parasitologically positives and 12 (5.68%) were only serologically positives, totalizing 30 (14.21%) infected cats. From the 30 positive animals, in four (13.33%) were detected antibodies against FIV, and in one (3.33%) we detected the FeLV without antibodies against FIV. Twenty five (83.33%) cats were infected by *Leishmania* sp without FIV or FeLV infections, and eight (26.66%) cats had antibodies against FIV without evidence of *Leishmania* sp infection. These results suggest the absence of association between *Leishmania* sp., FIV and FeLV infection ($p > 0.005$).

Originally presented at the 2009 ACVIM Forum & Canadian Veterinary Medical Association Convention, Montreal, Canada.

ABSTRACT #295

EVALUATION OF THE SKELETAL MUSCULATURE OF DOGS WITH VISCERAL LEISHMANIASIS BY MEANS OF ELECTROMYOGRAPHY. GC Ferraro, MH Bueno, DE Camargo, CB Simões, CM Vieira, DC Costa, AAD Gomes, LSV Sobrinho, M Marcondes. São Paulo State University, Araçatuba, São Paulo, Brazil.

Sporadic reports of myositis can be found among clinical alterations caused by visceral leishmaniasis in dogs. The goal of this study was to determinate electromyographic alterations in dogs naturally infected with *Leishmania chagasi*, with no locomotor alterations. The electromyography was made in four muscles (triceps brachii, extensor carpi radialis, biceps femoris and gastrocnemius) of 25 adult dogs from the Zoonosis Control Center (ZZC) from Araçatuba, São Paulo, Brazil. The insertional activity was reduced in 19 dogs (76%). During electrical activity at rest, spontaneous fibrillation potentials, positive waves and repetitive complex discharges were identified in 13 animals (52). During the evaluation of the voluntary electrical activity, 18 dogs (72%) had an altered recruitment in at least one of the studied muscles, evidenced by the increase in the recruitment speed and the presence of small polyphasic potentials of short duration. The electromyographic results encountered in this study allowed the identification of 13 animals (52%) with chronic polymyositis; two (8%) at acute stage; four (16%) whose muscles had both acute and chronic inflammation; three (12%) whose electromyography was normal and three (12%) with isolated electromyographic alterations in the tested muscles, thus not allowing the determination of a polymyositis diagnosis. These alterations led to the conclusion that dogs with visceral leishmaniasis, even with no muscular alterations at physical examination, may show an impairment of the muscular system that is compatible with chronic or acute polymyositis, confirmed by electromyography.

Originally presented at the 2009 ACVIM Forum & Canadian Veterinary Medical Association Convention, Montreal, Canada.

ABSTRACT #296

HISTOPATHOLOGICAL EVALUATION OF THE SKELETAL MUSCLE OF DOGS WITH VISCERAL LEISHMANIASIS. GC Ferraro, GF Machado, MH Bueno, DE Camargo, CB Simões, CM Vieira, DC Costa, AAD Gomes, LSV Sobrinho, M Marcondes. São Paulo State University, Araçatuba, São Paulo, Brazil.

Canine visceral leishmaniasis is a disease that can cause muscular atrophy. The ethiopathology of muscular alterations remains uncertain, but is often attributed to the catabolic nature of the disease. Aiming to identify the nature of muscular lesions, biopsies of the triceps brachii, extensor carpi radialis, biceps femoris and gastrocnemius muscles were made from 25 dogs naturally infected with *Leishmania chagasi*. The biopsies were performed under general anesthesia, with the protocol accepted by the Ethical Commission from the Veterinary Medicine Course from the São Paulo State University. The samples were evaluated by the hematoxylin-eosin stain. Only one animal (4%) had all muscular samples within normality. In two dogs amastigote forms of the parasite were identified in muscular fibers. From 25 animals with visceral leishmaniasis, 20 (80%) showed degeneration and necrosis of muscular fibers; twelve (48%) developed a mononuclear inflammatory infiltrate, suggesting a myositis of an immunological nature; 15 (60%) developed a reducing in the diameter of the muscular fibers; 14 (56%) showed perimysial and endomysial fibrosis and 10 (40%) displayed adipose tissue in the endomysial and perimysial region, suggesting that the inflammatory myopathy was in a chronic stage. The histopathological alterations identified in the dogs of this study therefore confirm the occurrence of polymyositis in this disease and discard the possibility of a myoatrophy solely due to the catabolic nature of the disease.

Originally presented at the 2009 ACVIM Forum & Canadian Veterinary Medical Association Convention, Montreal, Canada.

ABSTRACT #297

FLOW-CYTOMETRIC ANALYSIS OF PERIPHERAL BLOOD T LYMPHOCYTE SUBSETS CD4+, CD8+ AND CD4:CD8 RATIO IN MAINE COON CATS WITH CHRONIC GINGIVITIS. AGT Daniel, A Reche Júnior, CAG Júnior, A Pellegrino, RV Bosch, JM Monteiro, JR Kfoury Júnior. School of Veterinary Medicine – University of São Paulo, Brazil.

Chronic inflammatory diseases of feline oral cavity are extremely common in daily clinical practice. Several factors may contribute in the development of gingivitis in cats: diet, oral conformation, infectious diseases and specific breed characteristics (juvenile gingivitis of Maine Coons). The etiology and pathogenesis of gingivitis in Maine Coons are still to be determined. Dysfunctions in the cellular immune responses have been reported in human patients with various forms of periodontal disease. This study was undertaken to assess the possibility that an abnormality in the systemic T lymphocyte subsets is involved in the etiology of chronic gingivitis of Maine Coon cats.

Peripheral blood samples were obtained from 43 client-owned Maine Coon cats, 20 with chronic gingivitis (scores 3 to 4 – WATERS et al., 1993), and 23 without gingivitis. All 43 cats were negative for FIV and FeLV. CD4+ and CD8+ T lymphocytes counts were determined by flow cytometry. Mann-Whitney test was used to statistic analyze. A P value < 0.05 was considered significant.

Maine Coons with gingivitis had lower levels of CD4+ T lymphocytes than Maine Coons without gingivitis ($1{,}132.6 \pm 905.7$ vs $1{,}891.2 \pm 1{,}240.9$, $P < 0.05$), but no significant difference was observed in CD8+ count ($1{,}405.2 \pm 1{,}509.2$ vs 859.96 ± 480.92, $P > 0.05$). The CD4:CD8 ratio were also lower in Maine Coons with gingivitis (1.3 ± 1.2 vs 2.29 ± 1.01, $P < 0.05$).

The results showed that there is a correlation between the cellular immunity (based on lower CD4+ count and lower CD4:CD8 ratio) and gingivitis in Maine Coon cats. However, it is impossible to know if these findings are the cause or consequence of chronic gingivitis.

Studies of CD4:CD8 ratios in the peripheral blood of human patients with periodontal disease have yield disparate results. While there are reports that the CD4:CD8 ratio is decreased in patients with juvenile periodontitis, in others the CD4:CD8 ratio does not appear to be consistently altered. Thus, the utility of immunological measures as diagnostic aids for periodontal disease in general has not been established.

Important points are highlighted, as a possible familiar or breed predisposition to low cellular immunity and consequent oral disease; or other factors that can influence the gingivitis in this breed (catteries environment, respiratory viral agents) and these may cause the onset of oral disease and subsequent decrease on cellular immunity. Further studies are necessary to investigate the pathogenesis of chronic gingivitis in Maine Coon cats and correlated factors.

ABSTRACT #298

METHICILLIN-RESISTANT *STAPHYLOCOCCUS* SPP. CONTAMINATION OF STETHOSCOPES IN A SMALL ANIMAL VETERINARY TEACHING HOSPITAL. K McCord, K Stanton, D Bolte, D Hyatt, K Lunn. Department of Clinical Sciences and the Veterinary Diagnostic Laboratory, Colorado State University Veterinary Medical Center, Fort Collins, CO.

Methicillin-resistant *Staphylococcus aureus* (MRSA) and *S. pseudintermedius* (MRSP) are emerging pathogens in companion animals. MRSA is associated with nosocomial infections, and the organism can be transmitted by fomites. Human studies have shown staphylococcal contamination rates of stethoscopes of 66% to 100%. We hypothesized that at least 60% of stethoscopes at the CSU-VTH are contaminated with > 20 colony-forming units (CFU) of *Staphylococcus spp.* per square centimeter, with a small percentage of these being MRSA and MRSP. The purpose of this study was to quantify CFU of staphylococci present on the diaphragms of 100 stethoscopes and determine the percentage of organisms that are MRSA or MRSP.

Blood agar was used as the culture substrate. An MRSA enrichment broth-soaked swab was rubbed across the surface area of the stethoscope diaphragm and cultured in broth.

22% of all stethoscopes cultured positive for *S. aureus* (7%) or *S. pseudintermedius* (16%). One stethoscope produced both species. MRSA was cultured from 2%, and MRSP was cultured from 5%, of stethoscopes. Staphylococci of any species were found on 93% of stethoscopes. 6% were contaminated with > 20 CFU of any Staphylococcus species. Only 3% were contaminated with > 20 CFU of *S. aureus* and *S. pseudintermedius*. One stethoscope each was contaminated with > 20 CFU of MRSA and MRSP.

Staphylococci were present on most stethoscopes, but only a small percentage were contaminated with methicillin-resistant organisms, as hypothesized. These results indicate that stethoscopes are potential fomites in the spread of nosocomial infections, and regular disinfection of this instrument should be employed.

ABSTRACT #299

KILLING OF *STAPHYLOCOCCUS PSEUDINTERMEDIUS* BY ENROFLOXACIN AND ITS ACTIVE METABOLITE CIPROFLOXACIN USING CLINICALLY RELEVANT DRUG CONCENTRATIONS. J.M. Blondeau, S.D. Borsos, B.J. Blondeau. Royal University Hospital and University of Saskatchewan, Saskatoon, SK, Canada.

Enrofloxacin (ENR) is has low minimum inhibitory concentration (MIC) values against susceptible clinical isolates *S. pseudintermedius* (SI). In dogs and cats, ENR is metabolized to ciprofloxacin (CPR). Both ENR and CPR circulate in treated animals (in varying percentages of total drug) but the contribution each active agent contributes to killing of Si remains largely unknown. We determined the rate and extent of killing of Si by ENR and CPR alone and in combination based on clinically relevant drug concentrations.

For MIC testing, 10^5 cfu/ml of SI strains were exposed to doubling drug concentrations of ENR or CPR in Mueller-Hinton (MH) broth and following incubation under ambient conditions, the lowest concentration preventing growth was the MIC. For kill experiments, 10^5 cfu/ml were exposed to either 2.1 or 4.1 ug/ml (maximum serum drug concentrations based on dosages) of total drug (ENR + CPR in varying ratios respectively 50:50, 60:40, 70:30) and the reduction in viable cells ($\log_{10}$ and % kill) were recorded at 5,10,15, 20,25,30,60,120 and 180 minutes after drug exposure. All experiments were conducted in triplicate and the results averaged.

For Si isolates, ENR and CPR MIC values ranged from 0.063 to 0.125 ug/ml - being lowest for ENR. When 10^5 cfu/ml were exposed to 2.1 ug/ml of ENR or CPR, growth to a 0.19 $\log_{10}$ reduction (growth – 36% kill) was seen after 15 minutes of drug exposure – less killing was seen with CPR. For ENR/CPR drug ratios (2.1 ug/ml total drug) of 50:50, 60:40 and 70:30, a 0.09–0.44 log10 reduction (9–64% kill) was seen following 15 minutes of drug exposure. When 10^5 cfu/ml of SI was exposed to 4.1 ug/ml of ENR or CPR, growth to a 0.26 $\log_{10}$ reduction (growth –44% kill) was seen following 15 minutes of drug exposure – less killing was seen for CPR. For ENR/CPR drug ratios (4.1 ug/ml total drug) of 50:50, 60:40, 70:30, a 0.05 – 0.34 $\log_{10}$ reduction (12–54% kill) was seen following 15 minutes of drug exposure; 25–64% kill by 30 minutes. Following 180 minutes of drug exposure, 88- > 99% of bacteria were killed by ENR, CPR and all ENR/CPR ratios investigated.

ENR MIC values were low against the SI strains tested – 0.063 ug/ml. ENR killed SI strains more efficiently than CPR. ENR is metabolized to CPR and both active compounds simultaneously circulate in treated animals. Investigating the killing of SI with ENR/CPR in combination is useful. ENR and CPR were not antagonistic and may be additive. Enhanced killing may be related to differences in intrinsic activity of the compounds against SI strains and both drugs in combination enhance the potency of the commercial formulation. Killing was equivalent to or greater than each compound alone when ENR and CPR were together in the kill assays. Rapid killing may serve to reduce the likelihood for resistance selection and bacterial eradication has been linked to clinical recovery.

ABSTRACT #300

PREVALENCE OF SELECTED ZOONOTIC AND VECTOR-BORNE AGENTS IN DOGS AND CATS ON THE PINE RIDGE RESERVATION. AV Scorza, MR Lappin. Department of Clinical Sciences, Colorado State University, Ft.Collins, CO.

The objective of this study was to determine the prevalence of intestinal parasites and vector borne agents of dogs and cats in the Pine Ridge Reservation, South Dakota.

Fecal samples were examined for parasites by microscopic examination after fecal flotation and for *Giardia* and *Cryptosporidium* by a commercial immunofluorescence assay (IFA). PCR was performed on *Giardia* and *Cryptosporidium* IFA positive samples and the isolates were genotyped using the glutamate dehydrogenase and by the heat shock protein-70 genes, respectively. Cat sera were tested for antibodies against *Bartonella* spp., *Toxoplasma gondii*, and FIV antibody, as well as antigens of FeLV and *Dirofilaria immitis*. Dog sera were tested for antibodies against *T. gondii*, *Borrelia burgdorferi*, *Ehrlichia canis*, and *Anaplasma phagocytophilum* as well as *D. immitis* antigen. Total DNA extracted from blood was assessed in PCR assays for DNA of *Rickettsia* spp., *Bartonella* spp., *Ehrlichia* spp., *Anaplasma* spp., haemoplasmas, and *Babesia* spp. (dogs only).

In the 84 dog fecal samples, *Giardia* spp was the most common parasite (32% by IFA; 11.9% by fecal float), followed by *Taenia* spp. (17.8%), *Toxascaris leonina* (9.5%), *Toxocara canis* (7.1%), *Cryptosporidium* spp. (7.1%), hookworms (3.5%), and *Cystoisospora canis* (2.3%). All nine feline fecal samples were negative. Nine *Giardia* spp. isolates were sequenced and were dog genotypes (5 were assemblage C and 4 were assemblage D). Four of the six *Cryptosporidium* spp. could be sequenced and were *C. canis* . Antibodies against *T. gondii* were detected in 15% of the 82 dog sera but all sera were negative for other antibodies as well as *D. immitis* antigen. Of the 32 cat sera, 37.5% had antibodies against *Bartonella* spp. and 6% had antibodies against *T. gondii* . FeLV antigen was detected in 10% of the cats but all were negative for *D. immitis* antigen and FIV antibodies. None of the 92 dog blood samples or 39 cat blood samples was PCR positive for DNA of the target agents.

Enteric parasites in dogs and *Bartonella* spp. antibodies in cats were common on this reservation. As many of the agents have zoonotic significance, this information should be used to develop a preventive medicine plan could be implemented for the dogs and cats on this, and similar reservations.

ABSTRACT #301

FELINE HERPESVIRUS-1 AND FELINE CALICIVIRUS MOLECULAR DIAGNOSTIC ASSAY RESULTS IN ADULT CATS AFTER BOOSTER VACCINATION WITH MODIFIED LIVE PRODUCTS. R Ruch-Gallie, JK Veir, MR Lappin. Colorado State University, Fort Collins, CO.

Molecular diagnostic assays (PCR) are now commonly being offered by veterinary diagnostic laboratories. Cats with respiratory diseases thought to be from feline herpesvirus-1 (FHV-1) and feline calicivirus (FCV) are being tested, but current molecular assays do not differentiate vaccine strains from natural strains. The purpose of this study is to determine the impact vaccination has on FHV-1 and FCV molecular diagnostic assay results from nasal and pharyngeal samples collected from healthy cats after inoculation.

Nasal and pharyngeal samples were collected on days -7, -3, 0, 3, 7, 10, 14, 17, 21 and 28 from 12, young adult, previously vaccinated, apparently healthy, mixed sex cats housed in two groups in separate research wards On day 0, six cats were administered a modified live FVRCP vaccine intranasally (Feline UltraNasal®, HESKA Corporation) and six cats were administered a modified live FVRCP vaccine SQ (Purevax Feline 3RCP; Merial). Total RNA and DNA were extracted from the sample using a commercially available kit. Presence of FHV-1 DNA, FCV RNA, and total RNA/DNA on each sample was determined using previously published assays.

FHV-1 DNA was amplified from 2 of 237 adequate samples (0.8%); one cat in the SQ group was positive on one sample (pharynx, day 0) prior to vaccination and one cat in the IN group was positive on one sample (pharynx, day 3) after vaccination. FCV RNA was amplified from 10 of 237 adequate samples (4.2%); one cat in each group was positive on one sample (pharynx, day 0) prior to vaccination. After vaccination, three cats in the SQ group were positive for FCV RNA on one sample (Cat 1- nasal day 21; Cat 2-pharynx day 10; Cat 3-pharynx day 3) and four cats in the IN group were positive for FCV RNA on multiple samples (Cat 1-pharynx days 14 and 17; Cat 2-nasal and pharynx day 14 and pharynx day 21; Cat 4-pharynx day 7 and 14; Cat 6-nasal day 28 and pharynx day 21).

FHV-1 DNA and FCV RNA can be amplified from samples collected from apparently healthy, vaccinated cats and which could represent current subclinical infection or shedding of vaccine strains. These findings should be considered when attempting to interpret results of molecular diagnostic assay results from clinically ill cats.

ABSTRACT #302

ASSESSING THE EFFECT OF RECENT *LEPTOSPIRA* VACCINATION ON WHOLE BLOOD QUANTITATIVE PCR TESTING IN DOGS. JN Midence, AM Chandler, and RE Goldstein. College of Veterinary Medicine, Cornell University, Ithaca, NY.

The inability of traditional serologic testing, such as microscopic agglutination testing (MAT), to provide reliable data in acute cases of human and canine leptospirosis has led to the development of novel testing modalities. Currently, the most commonly used novel technique is quantitative real-time PCR (qPCR). This modality, performed on whole blood, has been used successfully in the diagnosis of human and animal disease and is currently offered commercially by a number of veterinary laboratories. The common usage of available bacterin based canine *Leptospira* vaccines presents a unique challenge for the use of whole blood qPCR as a diagnostic tool, since recent vaccination might induce an artifactual false positive result if the targeted DNA fragment is present within the vaccine. This is especially problematic in the weeks after vaccination when high vaccinal MAT titers already complicate the diagnosis of true infection. The aim of this study was to assess whether the 2 currently available (in the US) 4 serovar vaccines induce a positive qPCR reaction in healthy recently vaccinated dogs. Following owner consent – 20 student and staff owned dogs at Cornell University, College of Veterinary Medicine were enrolled in the study. Each dog was randomly assigned to 1 of 2 vaccine groups: A purified sub-unit vaccine (Fort Dodge Animal Health®) or a microfiltered bacterin vaccine (Pfizer Animal Health®). Both vaccines included *Leptospira interrogans* serovars Pomona, Canicola and Icterohaemorrhagiae and *Leptospira kirschneri* serovar Grippotyphosa. A very sensitive (has been shown to be able to identify 1–2 *leptospiral* bacterial copies per sample) 16S qPCR assay previously validated for use in canine blood was utilized. Whole blood for qPCR and serum for MAT testing was collected from each dog prior to vaccination (day 0), as well as 3 and 7 days after vaccination and weekly thereafter for 8 weeks. Negative (canine blood, mastermix, water) and positive controls (cultured organisms diluted in canine blood and a standard curve) were run in triplicate on each plate as were the experimental samples. Both Leptospira vaccines produced a positive qPCR reaction when assayed undiluted or diluted 1:100 in canine blood. Vaccines not including Leptospira antigens, run as controls, were negative. All blood samples drawn at all time points for all vaccinated dogs were negative for the presence of Leptospira DNA. These results suggest that recent vaccination with the commonly used canine vaccines in the United States does not interfere with the use of qPCR for the identification of acute Leptospira infection in dogs.

ABSTRACT #303

ASSESSING RENAL COLONIZATION IN DOGS 15 MONTHS AFTER RECEIVING A MULTI-SEROVAR BACTERIN BASED VACCINE, VS UNVACCINATED DOGS, ALL EXPERIMENTALLY INFECTED WITH *LEPTOSPIRA KIRSCHNERI* SEROVAR GRIPPOTYPHOSA. AM Chandler, and RE. Goldstein. College of Veterinary Medicine, Cornell University, Ithaca, NY.

The ability of vaccines to prevent renal colonization and urinary shedding for at least 1 year after vaccination is an important objective when vaccinating dogs against leptospirosis. Little is known about the degree of success of many vaccines in achieving this goal. The aim of this study was to assess the effect of vaccination on renal colonization in experimentally infected dogs more than a full year following vaccination. Forty young dogs were divided into 2 groups: Vaccinates and controls. Vaccinates received 2 doses, 3 weeks apart of a multi-valent inactivated bacterin vaccine against 4 serovars of *Leptospira* spp. including *Leptospira kirschneri* serovar Grippotyphosa. Controls were unvaccinated. The groups were age and sex matched, and the dogs were allowed to co-mingle during the study. An intraperitoneal and subconjunctival challenge with *Leptospira kirschneri* serovar Grippotyphosa was given approximately 15 months after the second vaccine. The dogs were sacrificed on day 35 following challenge. Blood was tested by microscopic agglutination test (MAT), a whole cell *Leptospira* ELISA and Western blot and blood and urine were cultured for *Leptospira* spp. on multiple days. Renal histopathology was evaluated and scored in a blinded fashion from all dogs. Real-time quantitative PCR (qPCR) was performed on DNA purified from 2 separate urine samples as well as from frozen renal tissue from all dogs in a blinded fashion using a16S qPCR assay previously validated for use in canine blood (L16). All dogs were determined to be serologically negative by MAT (for 4 serovars), Western blot and ELISA at the time of challenge. Vaccination significantly reduced the scores of qPCR (urine and renal tissue), renal histopathology, and blood, urine, and kidney reisolation ($p < 0.001$). Renal qPCR was positive in 10/21 control cases, and 2/19 vaccinates. Positive urine reisolation or qPCR results were not observed in any vaccinates, while 16/21 controls were positive on reisolation and 11/21 controls were positive on qPCR. This study demonstrates that vaccination can protect dogs from leptospiral colonization of the kidneys and urinary shedding even when the challenge occurs more than a year after vaccination and at the time of negative serologic tests including MAT.

ABSTRACT #304

A NATION-WIDE MOLECULAR EPIDEMIOLOGICAL SURVEY OF FELINE HEMOTROPIC MYCOPLASMA INFECTIONS IN JAPAN. M Tanahara[1], S Miyamoto[1], T Nishio[1], Y Yoshii[1], M Sakuma[1], Y Sakata[2], A Setoguchi-Kinoshita[1] and Y Endo[1]. [1]Kagoshima University, Kagoshima, Japan. [2]Merial Japan Co., Ltd., Tokyo, Japan.

Hemoplasma (hemotropic mycoplasma) causes hemolytic anemia in infected domestic cats. There are three species of Mycoplasma known to cause hemoplasmosis, *Mycoplasma haemofelis* (Mhf), Candidatus *M. haemominutum* (CMhm) and Candidatus *M. turicensis* (CMt). Mhf is known to be more pathogenic than the other two species. Most infected cats become asymptomatic carriers; however, immunosuppressed cats are likely to experience acute onset of disease. Previously, an epidemiological study concerning hemoplasma infections in cats was carried out in Japan. The number of cats included in this study was small and clinical information was limited. In the present study, we conducted a nation-wide molecular epidemiological survey of feline hemoplasmosis to investigate its infection rate, clinical characteristics, and the identity of hemoplasma species.

Blood samples were collected from cats admitted to 47 hospitals located in each prefecture from March–October 2008. Upon admission, the signalment and clinical background of these cats were recorded. Total DNA was extracted from blood samples, and PCR analysis to distinguish the three causative agents of hemoplasma was performed. The final number of cats included in this study was 1,770; 465 (26.3%) were infected with one or more species of *Mycoplasma*. The infection rate for each species was 5.1%, 21.0% and 6.7% for Mhf, CMhm and CMt, respectively. Co-infection was observed in 98 cases (5.5%) with Mhf+CMhm evident in 28 cases (1.6%), Mhf+CMt in six cases (0.3%) and dual CMhm+CMt infection apparent in 50 cases (2.8%). All three species were present in 14 cases (0.8%). The infection rate was generally lower than that reported in the previous Japanese study. This might be due to the difference of examined cat population. Furthermore, the infection rate obtained in this study was twice as high as those reported in European countries.

Several risk factors of hemoplasma infection were also detected in examined cats. Male and adult cats with a history of bite wounds possessed a significantly higher risk of hemoplasma infection. Additionally, feline immunodeficiency virus and feline leukemia virus infections were coincidently observed with hemoplasma infection. These results seemed to suggest that the major infection route of hemoplasma may be direct transmission.

This study clarified that hemoplasma is still widespread in Japan. The majority of cats do not exhibit anemia related to hemoplasma infection; however, the cats exposed to high risk factors should be examined for hemoplasma infection to avoid acute onset of the disease in future.

ABSTRACT #305

RETROSPECTIVE INVESTIGATION OF SURVIVAL FACTORS IN CANINE DISSEMINATED HISTOPLASMOSIS. J Morrison, J Clemans, C Wang. Iowa State University Veterinary Medical Center, Ames, IA.

Histoplasma capsulatum is the causative agent of an endemic mycosis that may result in disseminated, life threatening disease. Young dogs appear to have an increased susceptibility to infection and clinical signs of systemic disease are commonly due to enteric or respiratory involvement. The purpose of this study was to retrospectively analyze confirmed cases of canine disseminated histoplasmosis to identify potential factors correlated to survival.

Twenty seven cases of disseminated histoplasmosis were identified over a 13 year period (1996–2009) and 25 were included in the analysis. Infections were defined as disseminated on the basis of multiple organ involvement identified on necropsy, or clinical evidence of infection beyond the respiratory tract. Historical information collected included the use of immunomodulatory (eg. corticosteroid) therapy and the length of time of clinical signs prior to diagnosis. Upon presentation, all cases had the following recorded: age, weight, gender, rectal temperature, and the presence or absence of respiratory signs. Clinical data, at initial presentation, consisted of: HCT/PCV, WBC and platelet counts, serum glucose, total calcium, total protein, hepatic enzyme (ALT, ALP) and total bilirubin levels. Single versus multiple agent antifungal therapy was recorded, as was concurrent anti-inflammatory therapy with prednisone. Survival times were calculated from the time of diagnosis to death. Censoring time was calculated from diagnosis to last known follow up for dogs still alive.

Twenty one of the 25 dogs (84%) were female and 12 dogs (48%) died or were euthanized as a result of disseminated histoplasmosis. Three dogs did not receive any antifungal therapy (1 died before therapy could be administered and 2 were euthanized at the time of diagnosis due to advanced disease and poor prognosis). Median survival time was 760 days (range 1 to $> 2{,}555$ days). Cox proportional hazard models were used to analyze the relationship between survival hazard and the multiple explanatory variables. The model with the best goodness of fit contained five explanatory variables that were significantly associated with survival hazard: length of time of clinical signs ($p < 0.01$) [mean 79 days (range 1–365 days)]; rectal temperature ($p < 0.04$) [mean 102.5°F (range 98.8–105.3°F)]; WBC count ($p < 0.04$) [mean 17.8×10^3/ul (range 4.8–40.1×10^3/ul)]; total calcium ($p < 0.01$) [mean 9.36 mg/dl (range 7.1–10.8 mg/dl)]; body weight ($p < 0.01$) [mean 15.9 kg (range 1.8–31.8 kg)]. Rectal temperature and WBC count were positively associated with survival hazard. Length of time of clinical signs, total calcium, and body weight were negatively correlated with survival hazard.

This study identified 5 variables significantly associated with survival hazard in cases of disseminated canine histoplasmosis. This data is also being used to formulate a scoring system that will be applied to cases to further assist in determining prognosis.

ABSTRACT #306

DISTRIBUTION OF CLOTRIMAZOLE AND ENILCONAZOLE IN THE FRONTAL SINUSES AND NASAL CAVITY OF DOGS WITH SINONASAL ASPERGILLOSIS. M J Sharman[1], Z Lenard[2], CS Mansfield[1]. [1]Department of Veterinary Clinical Sciences, Murdoch University, Perth, WA, Australia. [2]Veterinary Imaging Centre, Perth Veterinary Specialists, Perth, WA, Australia.

Multiple topical treatments are often required for clinical cure of sinonasal aspergillosis. Inadequate distribution and retention of antifungal agents may influence outcome. The objective of this study was to evaluate distribution and retention of two commonly used antifungal agents in the nasal cavity and frontal sinuses of dogs with confirmed sinonasal aspergillosis using computed tomography (CT).

Nine dogs were alternately treated with clotrimazole (1% in polyethylene glycol) or enilconazole (10 mg/ml; Imaverol®, Austrichter). Frontal sinuses were trephined, debrided and flushed with saline. Each agent was mixed with iohexol to a 5% solution and administered via the frontal sinuses and nasal cavity with dogs in sternal recumbency. Five minutes after completion of infusion CT of the nasal cavity was performed. Distribution and retention was scored from 1–4 at the canine tooth, premolar four, cribriform plate and frontal sinus on both sides, to a maximum of 32.

Distribution and retention was excellent in the rostral nasal cavity in 8/9 dogs, but variable in the caudal nasal cavity. Retention within the frontal sinuses was poor in 7/9 cases. No obvious difference in retention between the two agents was observed, despite the difference in viscosity. Moderate to marked leakage into pharyngeal packing was observed in three dogs but did not consistently result in a lower overall score.

Distribution and retention of liquid antifungal agents within the frontal sinuses is poor in clinical cases treated with trephination. This may influence outcome unless the sinuses are periodically refilled.

ABSTRACT #307

CANINE EXPOSURE TO *BORRELIA BURGDORFERI* AND *ANAPLASMA PHAGOCYTOPHILUM* IN AN EMERGENT AREA: USE OF A MULTI-TARGET ELISA AT TWO GEOGRAPHIC SCALES. P Rand[1], E Lacombe[1], B Cahill[1], S Elias[1], and R Chandrashekar[2]. [1]Maine Medical Center Research Institute, South Portland, ME. [2]IDEXX Laboratories, Westbrook, ME.

Two canine serosurveys were conducted using a multi-target ELISA to detect canine antibodies to the agents of Lyme disease (*Borrelia burgdorferi* [Bb]) and anaplasmosis (*Anaplasma phagocytophilum* [Ap]). The first was to determine the statewide distribution of the transmission of these agents, and the second to estimate the degree of transmission in an island community considering the need for vector control. Since first recorded in Maine in the mid 1980s, these ticks have become established and abundant in south coastal areas, have invaded inland to the higher elevations, and have recently been found in northern Maine. Annually reported human cases of Lyme disease have increased rapidly within the last 10 years to nearly 1000 in 2008. We first detected Ap in Maine deer ticks in 1995. Currently ~20 human cases of human anaplasmosis are being reported annually. Canine Lyme disease, already prevalent in the last decade, has become a major veterinary issue in the southern half of the state, and horses have become infected as well. Canine anaplasmosis is now increasingly diagnosed in south coastal clinics.

In 2007, during the heartworm testing season, we conducted a statewide canine serosurvey using an in-clinic ELISA which detects *Dirofilaria immitis* antigen and antibodies to Bb, Ap, and *Ehrlichia canis* (SNAP® 4Dx® test, IDEXX Laboratories, Westbrook, ME). Seventy-eight veterinary clinics participated, providing results on 1087 dogs. All counties were represented. Canine Ap positivity varied sharply geographically. It was highest in south coastal towns ($\leq$76.5%), ~5–8% along the midcoast, and zero in interior and northern Maine. In Maine's southernmost county, where Lyme disease is hyperendemic, 36.0% of unvaccinated dogs were Bb positive, while ~4% were positive in the state's 3 northern counties (including dogs in the northernmost town). Lyme vaccination varied widely in 2007; dogs with a history of vaccination were half as likely to be seropositive for Bb.

In 2009, we repeated this protocol both on a Lyme endemic island and in nearby mainland towns. Six veterinary clinics provided 4Dx data on 65 island dogs and 573 off-island dogs from 30 mid-coastal towns. Island and mainland dogs were 3.1% and 3.4% positive for Ap, respectively, far less than in south coastal towns previously. In the 25% of both island and mainland dogs that were unvaccinated for Bb, Bb seropositivity was 25.0% and 18.9%, respectively. Thus, Bb transmission frequency on the island is approaching ~70% of the level seen earlier in Maine's most hyperendemic county.

The use of SNAP® 4Dx® test, a multi-target test has demonstrated a sharp geographic gradient in the transmission of Ap in an emergent area, a confirmation of the extent of Bb transmission in Maine, and a quantification of transmission risk to inform a concerned community.

ABSTRACT #308

PERFORMANCE COMPARISON OF SPECIES-SPECIFIC PEPTIDE-BASED ASSAYS WITH IMMUNOFLUORESCENCE ASSAYS FOR DETECTION OF CANINE ANTIBODIES TO *ANAPLASMA* AND *EHRLICHIA* SPP. B.A. Stillman[1], M.J. Beall[1], P. Shields[1], B. Hegarty[2], E. Breitschwerdt[2], R. Chandrashekar[1]. [1]IDEXX Laboratories, Inc, Westbrook, ME. [2]College of Veterinary Medicine, North Carolina State University, Raleigh, NC.

The immunofluorescence assay (IFA) is a well-accepted method used to test human and animal serum samples for evidence of vector-borne infection. While IFAs usually possess very good

Table [Abstract 308]

Experimental Infection	IFA			ELISA				
	Aph	Eca	Ech	Aph	Apl	Eca	Ech	Eew
A phagocytophilum (Aph)	**Pos** (6/6)	**Pos** (2/6)	**Pos** (1/6)	**Pos** (6/6)				
A platys(Apl)	**Pos** (4/6)	**Pos** (2/6)	**Pos** (4/6)		**Pos** (6/6)			
E canis (Eca)	**Pos** (2/6)	**Pos** (6/6)	**Pos** (5/6)			**Pos** (6/6)		
E.chaffeensis (Ech)		**Pos** (2/2)	**Pos** (2/2)				**Pos** (2/2)	
E.ewingii (Eew)		**Pos** (1/3)	**Pos** (3/3)					**Pos** (3/3)

sensitivity, they have been shown to have reduced specificity due to antibodies which cross-react with common proteins among closely related organisms. The goal of this study was to compare the IFA with ELISA-based assays that use peptides derived from species-specific proteins or protein sub-domains found in the following organisms of interest: *Anaplasma phagocytophilum* (*Aph*), *Anaplasma platys* (*Apl*), *Ehrlichia canis* (*Eca*), *Ehrlichia chaffeensis* (*Ech*), and *Ehrlichia ewingii* (*Eew*). We compared the serological results from these five assays with those from three commercially available IFAs using samples derived from dogs experimentally infected with each of the five pathogens (23 dogs tested in total). At least three time points were tested; day 0, day 14–30 pi, day 60–90 pi. All samples from day 0 of the respective infection study tested negative on the peptide-based assays, but two were seropositive for *E. canis* by IFA. The following table summarizes by assay the number of dogs testing positive for antibody at a minimum of one time point post-infection.

None of the peptide-based assays demonstrated evidence of cross-reactive antibodies as a result of other infections. Significant cross-species reactions ($\geq$2 dogs with IFA titers > 1:64) were identified in *A. platys*-infected dogs with *Aph* IFA, *E. chaffeensis*-infected dogs with *Eca* IFA, and *E. canis, A. platys, and E. ewingii*-infected dogs with the *Ech* IFA. While cross-reactivity within a genus is to be expected, reactivity across genera typically occurred at a lower endpoint dilution. Our data also suggests these peptide assays possess comparable sensitivity to IFA. The assays derived from species-specific analytes may offer an alternative to IFAs with comparable sensitivity, improved specificity and greater ease-of-use.

ABSTRACT #309

COMMON ENTERIC PATHOGENS IN CATS BEFORE AND AFTER PLACEMENT IN AN ANIMAL SHELTER.

EN Gingrich[1], AV Scorza[1], CM Leutenegger[2], M Ferrero[2], MR Lappin[1]. [1]Colorado State University, Fort Collins, CO. [2]IDEXX Laboratories, West Sacramento, CA.

Diarrhea is a common problem in shelter cats and enteric pathogens can contribute significantly to this problem. Stress and other factors of the shelter environment may contribute to infection and shedding rates of enteric pathogens. The purpose of this study was to determine the prevalence of common enteric pathogens in cats housed in a shelter in north central Colorado.

Fecal samples were collected from 46 healthy cats within 24 hours of entrance into the shelter and again seven days after admission. At Colorado State University, feces were assayed by centrifugal flotation and microscopic examination for parasites, IFA for *Giardia* cysts and *Cryptosporidium* oocysts, ELISA for *Clostridium difficile* toxin A/B (CD toxin), ELISA for *C. perfringens* enterotoxin (CP toxin), and culture for *Salmonella* spp. and *Clostridium* spp. An aliquot of each sample was sent to a collaborating laboratory (IDEXX) for PCR assays to amplify DNA or RNA of feline coronaviruses (Corona RNA), feline panleukopenia virus (Panleuk DNA), *Tritrichomonas foetus*, *Giardia*, *Toxoplasma gondii*, *C. perfringens* enterotoxin A (CP DNA), *Cryptosporidium* spp., and *Salmonella* spp.

Table [Abstract 309] Fecal assay results in cats housed in an open admission animal shelter

	Isospora	*Giardia*	CD toxin	CP toxin	CP DNA	*Salmonella* culture	Corona RNA	Panleuk DNA
Entry	3	1	1	2	7	1	14	4
Day 7	1	1	0	2	3	1	14	13

The results of all assays are listed in Table. Only the panleukopenia PCR assay was positive in samples from more cats on Day 7 than on entry into the shelter.

The results are shown as number of positive cats. The cat with *Giardia* positive results was positive by all three assays on entry and only by PCR on day 7.

Feline panleukopenia virus was the only pathogen with increasing numbers of positive results after admission to the shelter which is likely to indicate acquisition of the virus after entrance to the shelter, either through infection or vaccination. The stress of housing or potential for pathogen exposure did not appear to affect the enteric pathogen shedding rates in these cats.

ABSTRACT #310

PREVALENCE OF *GIARDIA* SPP. AND *CRYPTOSPORIDIUM* SPP. IN VETERINARY STUDENT OWNED DOGS THAT FREQUENT DOG PARKS.

A Wang, R Ruch-Gallie, P Lin, AV Scorza, MR Lappin. Department of Clinical Sciences, Colorado State University, Fort Collins, CO.

In a previous study of feces from dogs in Colorado, 16.2% were positive for agents with zoonotic potential; *Giardia* spp. and *Cryptosporidium* spp. were the most prevalent and are both infectious immediately when passed in feces. Dog parks are very popular in urban areas, but to our knowledge, no studies have attempted to correlate visits to dog parks and risk of colonization by these enteric parasites. The purpose of this study is to determine whether dog park visitation is associated with an increased prevalence of *Giardia* spp. and *Cryptosporidium* spp. in dogs owned by veterinary students at Colorado State University.

Feces were submitted by the veterinary student owners with a completed survey form detailing dog park attendance rates, fecal character scores, and other clinical information. Feces were examined for parasite eggs, cysts, and oocysts by microscopic examination and for *Giardia* spp. and *Cryptosporidium* spp. by a commercial IFA. Positive samples were genotyped using the glutamate dehydrogenase, b-giardin and triose phosphate isomerase genes for *Giardia* spp. and the heat shock protein-70 gene for *Cryptosporidium* spp.. Prevalence rates in dogs that did and did not attend dog parks were compared between by Fisher's exact test with significance defined as $p < 0.05$.

Feces from 52 veterinary student dogs that attend dog parks and 42 veterinary student dogs that did not attend dog parks were analyzed. In the dogs that attended dog parks, five (9.6%) were *Giardia* spp. positive. Two of these dogs were coinfected with *Cryptosporidium* spp. and one had diarrhea. *Giardia* spp. and *Cryptosporidium* spp. were not detected in veterinary student dogs that did not attend dog parks. Sequence analysis was available from four *Giardia* spp. isolates (two assemblage D, one assemblage C, one assemblage C and D mixed infection) and one *Cryptosporidium* spp. isolate (*C. canis*). The prevalence rates between the two populations were not statistically different (Fisher's exact test; $p = 0.062$). The overall *Giardia* spp. prevalence rate (5.3%) in veterinary student dogs was not significantly different from a previous study of healthy, client-owned dogs in the same city (24 of 220 dogs; 11.4%; $p = 0.1405$).

While all *Giardia* spp. positive veterinary student dogs visited dog parks, the results were not significantly different from dogs that did not visit dog parks. The failure to show statistical differences could merely reflect the small sample size. All *Giardia* spp. and *Cryptosporidium* spp. were the host adapted genotypes suggesting the infections originated from other dogs and that the dogs were unlikely to be zoonotic health risks.

ABSTRACT #311

EFFECT OF *ENTEROCOCCUS FAECIUM* SF68 SUPPLEMENTATION ON DIARRHEA IN CATS HOUSED IN A NORTHERN COLORADO ANIMAL SHELTER. SN Bybee, V Scorza, MR Lappin. Department of Clinical Sciences, Colorado State University, Fort Collins, CO.

Numerous studies have evaluated the effects of probiotics in veterinary medicine with conflicting results. In previous studies with *Enterococcus faecium* strain SF68 (FortiFlora, Nestle Purina PetCare), we showed that supplemented healthy cats had increased %CD4+ lymphocytes over time and supplemented cats with chronic FHV-1 infection had less conjunctivitis over time than cats administered a placebo. The purpose of this study was to determine if administration of SF68 lessens diarrhea in cats housed in a shelter.

Stray or feral cats housed in two separate rooms were all fed a standardized diet and were administered SF68 or a placebo mixed with their food daily for eight weeks. A standardized fecal score system was applied to all samples daily by a person masked to the treatment groups and the percentages of cats with diarrhea of ≥ 2 days duration were calculated. A generalized linear mixed model using a bionomial distribution with treatment being a fixed effect and the room being a random effect was used to assess for statistical differences between treatment groups. Significance was defined as $p < 0.05$.

The percentage of cats with diarrhea ≥ 2 days was 7.7% for the probiotic group and 20.7% for the placebo group. This result was significantly different ($p = 0.0297$).

These results suggest that administration of SF68 to cats housed in shelters may lessen the numbers of days with diarrhea. As this was a short term study, this effect was likely from probiotic influences on intestinal flora rather than systemic immune enhancing effects.

ABSTRACT #312

C-REACTIVE PROTEIN CONCENTRATION IN BENIGN AND MALIGNANT CANINE SPIROCERCOSIS. V Mukorera[1], E Dvir[1], LL Van der Merwe[1], A Goddard[1]. [1]Faculty of Veterinary Science, University of Pretoria, Pretoria, South Africa.

C-reactive protein (CRP) is one of the major positive acute phase proteins in the dog. It is produced in response to infection and inflammation, but is non specific as it is part of the innate host response. In humans, CRP has shown potential as a diagnostic and prognostic marker in a variety of neoplasms namely Hodgkin's lymphoma amongst others. In dogs, CRP is elevated in multicentric lymphoma and is useful in predicting complete remission status. *Spirocerca lupi* (*S. lupi*) is a nematode of canidae which commonly infests the esophagus, forming a nodule which can undergo neoplastic transformation into a sarcoma. Both benign and malignant nodules show histopathological evidence of inflammation. Currently the determination of malignant transformation is based on histopathology samples obtained via endoscopy, an invasive and expensive procedure. Samples obtained are often falsely negative due to the necrotic periphery of the tumor. The purposes of this study were to determine if serum CRP concentration is (1) elevated in spirocercosis, (2) could be used to differentiate benign from neoplastic cases and (3) could be used to monitor response to treatment in dogs with benign spirocercosis.

Medical records of dogs diagnosed with spirocercosis, at the Onderstepoort Veterinary Academic Hospital, University of Pretoria, were reviewed for the period 1998–2009. Twenty eight confirmed benign cases and 14 confirmed malignant cases were selected. Six benign cases had follow up serum samples collected at least 2 weeks after treatment was commenced. Twenty one dogs presented for routine ovariohysterectomy were used as controls. Serum CRP concentrations were determined as a batch on serum stored at −80°C, using an automated human CRP Turbidometric Immunoassay, previously validated for use in dogs. Statistical analysis was performed using the Student's *t*-test.

The mean CRP concentration of the benign cases was 60.41±48.02 mg/l (range 0–190.6) and that of the malignant cases was 76.52±44.79 mg/l (range 3.1–152–6), both values were significantly higher ($p<0.001$) than the control group (13.39±17.94 mg/l). There was no statistically significant difference in the CRP concentrations between malignant and benign cases ($p=0.3$). The mean CRP concentration for the six follow up sera was 1.73±2.8 mg/l (range 0–6.6) compared to 86.96±66.6 mg/l (range 0–157.3) before treatment. The convalescent sera had statistically significant lower concentrations of CRP compared to the concentrations at diagnosis ($p=0.01$).

Serum CRP concentration is elevated in both the benign and malignant forms of *S. lupi* cases confirming the intense inflammation observed on histopathology but it can not be used to differentiate between benign and malignant cases. CRP concentrations could potentially be used to monitor response to treatment in benign spirocercosis cases in the absence of other causes of inflammation.

ABSTRACT #313

FELINE URETERAL STRICTURES: 10 CASES (2007–2009). MS Zaid[1]; AC Berent[2]; C Weisse[2]; A Caceres[1]. [1]University of Pennsylvania School of Veterinary Medicine, Philadelphia, PA. [2]The Animal Medical Center, New York, NY.

Feline ureteral obstructions have emerged as a common problem with increasing prevalence in clinical practice over the past few decades. Ureteral strictures are rarely reported as a cause of these obstructions. The purpose of this study is to describe a series of feline patients with ureteral strictures, including historical and diagnostic findings, treatment modalities, and clinical outcomes. Data from the medical records was retrospectively reviewed for patients with suspected ureteral strictures based on ultrasonagraphy, ureteropyelography, surgical exploration, or histopathology. Ten cats were diagnosed with a ureteral stricture based on consistent findings from at least two of the above criteria.

At presentation, median age was 12 years, median creatinine concentration was 3.7 mg/dl, and median renal pelvis size was 10 mm. 6/10 cats had ultrasonagraphic hyperechoic periureteral tissue at the stricture site. Four cats had a circumcaval ureter. Six cats underwent ureteral stenting, two had surgery alone, and two had no intervention. 7/8 cats had reductions in creatinine preceding discharge and six had persistently improved creatinine and renal pelvic parameters at their most recent re-examination. All patients survived to discharge with a median hospitalization of 10.5 days. Median survival time was > 294 days (14–858 days) with 6/10 cats still alive.

Ureteral strictures may occur in cats secondary to ureteral surgery, a circumcaval ureter, impacted ureterolithiasis, or for unknown causes. With appropriate and timely intervention, the prognosis for long-term survival is good. In addition to ureteral reimplantation or ureteronephrectomy, ureteral stenting or subcutaneous ureteral bypass may be considered as therapeutic options.

ABSTRACT #314

DEVELOPMENT OF A HIGH PERFORMANCE LIQUID CHROMATOGRAPHY SERUM VITAMIN B6 ASSAY IN CANINES WITH CALCIUM OXALATE UROLITHIASIS. J Clemans, P Imerman, J Phillips, K Deitz, A Jergens. Iowa State University College of Veterinary Medicine, Ames, IA.

Calcium oxalate urolithiasis now represent the second most common canine urolith in North America. As in people, these stones have a high rate of recurrence with a suspect genetic predisposition. The risk factors for these stones are poorly understood and current medical management does not preclude recurrence. Vitamin B6 deficiency is a risk factor in humans but this has not been evaluated in the canine population. The goal of the study was to develop an assay to detect Pyridoxyl-5-Phosphate (PLP) (active form of vitamin B6) in canine serum and assess dogs with and without calcium oxalate stones. A preliminary study of vitamin levels in predisposed versus non-predisposed breeds was also performed.

A High Performance Liquid Chromatography (HPLC) assay was developed using pooled canine serum from the ISU Veterinary Medical Center clinical pathology laboratory. Serum from dogs with stones and without stones were light protected and stored at −80°F within 12 hours of collection. Serum concentration of PLP was obtained using a standard curve with known concentrations of research grade PLP in water for each run. Recovery was also measured for each run using a known concentration of PLP in serum. Runs with recoveries lower that 80% were not allowed. For initial assessment for differences in serum concentration, dogs were separated into CaOx group (3 dogs) and non-CaOx group (4 dogs).

Vitamin B6 concentrations in canine serum were unstable at 2° and −15°C. Storage at these temperatures caused a rapid loss of 20% in two days and 50% at 14 days respectively. When stored at −80°C for 14 days, approximately 100% remained. Freezing and thawing the samples resulted in minimal recovery loss after three cycles. The pilot study revealed that the dogs with calcium oxalate urolithiasis were significantly lower (p value < 0.05) in vitamin B6 concentrations than those without stones, with mean concentrations of 29 ppb (SD = 18) and 123 ppb (SD = 70) respectively. Preliminary results comparing serum concentrations of dogs of a predisposed breed (mean levels 55 ppb) to dogs not of predisposed breeds support a lower vitamin B6 level in predisposed breeds.

An HPLC assay was successfully utilized to reliably measure serum vitamin B6 levels in canine serum and a deficiency relative to dogs without urinary stones was present in calcium oxalate forming dogs. A difference in vitamin concentrations was supported in an evaluation of a breed predisposed to these stones and dogs not predisposed. Vitamin B6 deficiency may prove to be a risk factor for calcium oxalate urolithiasis and further evaluation is currently underway.

ABSTRACT #315

THE PHARMACOKINETICS OF MIRTAZAPINE IN CATS WITH CHRONIC KIDNEY DISEASE.

JM Quimby, DL Gustafson, KF Lunn. Department of Clinical Sciences, Colorado State University, Fort Collins, CO.

This study was designed to determine the pharmacokinetics of mirtazapine, an appetite stimulant, in cats with chronic kidney disease (CKD) and age-matched controls (AMC) after administration of a single oral 1.88 mg dose.

Two cats with stable CKD from each IRIS Stage (II, III and IV) were utilized. Six age-matched cats with normal CBC, chemistry and urinalysis served as controls. Blood samples were collected prior to and 0.5, 1, 1.5, 2, 4, 8, 24, and 48 hours after oral administration of 1.88 mg of mirtazapine. Mirtazapine concentrations were measured by liquid chromatography coupled with tandem mass spectrometry. Non-compartmental pharmacokinetic modeling was performed.

Increased vocalization and affection were the only side effects noted. Mean half-life was 16.4 +/− 5.1 hours (CKD) and 12.3 +/− 1.8 hours (AMC). Mean peak plasma concentration was 110.6 +/− 30.8 ng/ml (CKD) and 79.6 +/− 21.7 ng/ml (AMC). Mean area under the curve (AUC) was 770.7 +/− 225.5 ng/ml•hr (CKD) and 555.5 +/− 175.4 ng/ml•hr (AMC). Mean clearance was 0.61 +/− 0.1 L/hr/kg (CKD) and 0.79 +/− 0.16 L/hr/kg (AMC). Groups were compared with the Mann-Whitney test and a statistically significant difference in AUC and clearance was found ($p = 0.03$ for both). The Spearman rank test identified a significant negative correlation between creatinine and clearance ($p = 0.002$).

CKD delays the clearance of mirtazapine. A single low dose of mirtazapine was well tolerated and resulted in a half-life that is compatible with 48 hour dosing intervals in cats with CKD.

ABSTRACT #316

GLOMERULAR FILTRATION RATES MEASURED USING PHARMACOKINETIC ANALYSIS OF BLOOD INULIN IN HEALTHY CATS.

M Nishida, M Uechi, K Harada, S Yamano, Y Ukai, S Kohno. Nihon University, Kanagawa, Japan.

The present study evaluates methods of measuring inulin clearance as a marker of glomerular filtration rate (GFR) in healthy cats without collecting urine. We evaluated inulin plasma clearance (C_{PLASMA}) and inulin infusion clearance (C_{INF}) in eight healthy cats and compared GFR values that were obtained from simultaneous measurements of standard inulin clearance (C_{STD}) and C_{INF}. We collected blood and urine three times every 30 minutes from 45 min of starting a continuous inulin (INULEAD$^{®}_{INJ}$) infusion. C_{STD} was calculated from inulin levels in urine and the serum. [C_{STD} = urine inulin (mg/ml) × urine volume (ml/min) /serum inulin (mg/ml)/body weight (kg)]. C_{INF} was calculated from serum inulin level and infusion dose [C_{INF} = infusion dose (mg/min)/serum inulin (mg/ml)/body weight (kg)]. Inulin was administered to the cats as a bolus and AUC was determined from the transition of the inulin blood level. C_{PLASMA} calculated from the inulin dosage was divided by the AUC. Values for C_{STD} ($n = 8$), C_{INF} ($n = 7$) and C_{PLASMA} ($n = 7$) were 3.01 ± 0.57, 2.98 ± 0.52 and 2.83 ± 0.60 ml/min/kg, respectively. We found a correlation between C_{STD} and C_{INF}, but C_{PLASMA} did not significantly correlate with either C_{STD} or C_{INF}. These results indicated that C_{INF} is clinically useful for measuring GFR and urine sampling is not required in healthy cats.

ABSTRACT #317

GLOMERULAR FILTRATION RATES MEASURED USING PHARMACOKINETIC ANALYSIS OF BLOOD INULIN IN HEALTHY DOGS.

M Nishida, M Uechi, K Harada, S Yamano, Y Ukai, S Kohno. Nihon University, Kanagawa, Japan.

The present study evaluates methods of measuring inulin clearance as a marker of glomerular filtration rate (GFR) in healthy dogs without collecting urine. We evaluated inulin plasma clearance (C_{PLASMA}) and inulin infusion clearance (C_{INF}) in 18 healthy dogs and compared GFR values that were obtained from simultaneous measurements of standard inulin clearance (C_{STD}) and C_{INF}. We collected blood and urine three times every 30 minutes from 45 min of starting a continuous inulin (INULEAD$^{®}_{INJ}$) infusion. C_{STD} was calculated from inulin levels in urine and the serum. [C_{STD} = urine inulin (mg/ml) × urine volume (ml/min) /serum inulin (mg/ml)/body weight (kg)]. C_{INF} was calculated from serum inulin level and infusion dose [C_{INF} = infusion dose (mg/min)/serum inulin (mg/ml)/body weight (kg)]. Inulin was administered to the dogs as a bolus and AUC was determined from the transition of the inulin blood level. C_{PLASMA} calculated from the inulin dosage was divided by the AUC. Values for C_{STD}, C_{INF} and C_{PLASMA} were 4.19 ± 0.77, 3.95 ± 0.48 and 4.13 ± 0.66 ml/min/kg, respectively. We found a correlation on C_{STD} vs. C_{INF} ($r = 0.57$, $P < 0.05$), C_{STD} vs. C_{PLASMA} ($r = 0.55$, $P < 0.05$), and C_{INF} vs. C_{PLASMA} ($r = 0.57$, $P < 0.05$). These results indicated that C_{INF} and C_{PLASMA} may be clinically useful for measuring GFR and urine sampling is not required in healthy dogs.

ABSTRACT #318

LONGITUDINAL FOLLOW UP OF CATS WITH REDUCED RENAL MASS.

BH Schmidt[1], FJ van der Staay[2], M Murphy[3]. [1]Bayer Animal Health GmbH, Monheim, Germany. [2]University of Utrecht, Utrecht, The Netherlands. [3]Charles River Laboratories Preclinical Services Ltd, Ballina, Ireland.

Surgical renal ablation is a well established technique for modeling feline chronic kidney disease (CKD). The aim of the present retrospective study was to follow up a group of animals subjected to such renal mass reduction over a period of about 4.5 years, with a special focus on the development of azotaemia, uremia, hyperphosphataemia, anemia, serum accumulation of the nephrotoxin indoxyl sulfate, clinical symptoms of CKD, and survival.

Towards the end of 2004, fourteen healthy young adult cats (7 males and 7 females) underwent surgical two-stage 11/12 renal ablation. Body weight was determined and blood samples were collected in the fasting state before the start of surgery as well as repeatedly between February 2005 and June 2009. All assessments were either clinical check-ups or were performed to obtain baseline values for subsequent testing of specific interventions. The minimum wash-out period between such studies was 10 months. Between successive studies, the animals were group-housed by gender. They were fed commercially available canned moist maintenance diets with mean analyzed values of 8.9% protein and 0.36% phosphorus in the original substance. Drinking water was always available ad libitum. Changes in body weight and clinical chemistry endpoints were statistically analyzed by a within subjects ANOVA with observation time points as levels of the repeated measures factor. Because time points were not equally spaced, the levels were weighted with the number of weeks elapsed since the start of surgical renal ablation. Changes were considered as statistically reliable when the associated probability did not exceed 5% ($p < 0.05$).

At the first time point of post-surgical assessment, all cats showed a slight reduction of body weight compared to pre-surgical values and elevated serum levels of urea and creatinine. Five animals were lost to follow up due to an acute uremic crisis necessitating euthanasia. In the remaining survivors, clinical symptoms of CKD were not observed until the last observation. Body weights stabilized and even increased over time. Serum creatinine and urea values slightly decreased from post-surgical values and remained mildly increased without further exacerbation during the observation period. Serum inorganic phosphate and total calcium levels decreased over time but remained well within the reference range. White and red blood cell counts were normal at all observation time points with a slight but significant decrease of erythrocyte counts over time. Blood hemoglobin levels and packed cell volume also steadily decreased to achieve levels in the low reference range at the end of the 4.5 year observation period. The majority of animals showed mild accumulation of indoxyl sulfate in blood without significant changes over time.

The results confirm that cats subjected to 11/12 renal ablation may well compensate their renal function over more than four years without clinically relevant signs of CKD progression.

ABSTRACT #319
COMPARISON OF URINE SODIUM DODECYL SULFATE-POLYACRYLAMIDE GEL ELECTROPHORESIS (SDS-PAGE) WITH RENAL HISTOLOGIC FINDINGS IN DOGS WITH RENAL DISEASE. MB Nabity,[1] R Brock,[1] JS Brown,[1] R Cianciolo,[2] C Hunter,[1] GE Lees[1]. [1]Texas A&M University, College Station, TX. [2]North Carolina State University, Raleigh, NC.

Previous studies have shown that molecular weight (MW) of urine proteins separated by SDS-PAGE correlates well with renal biopsy findings, considering proteins ≥ 60–70 kilodaltons (kDa) indicative of glomerular damage and proteins < 40–50 kDa indicative of tubular damage. However, there is a lack of consistency in these studies regarding the method of urine normalization. The objectives of this study were to: 1) compare urine SDS-PAGE results using urine normalized to either specific gravity (SG) or protein concentration to determine which normalization method corresponds best with renal biopsy; 2) compare visual versus software analysis of the gels.

Urine samples submitted to the Texas Veterinary Renal Pathology Service in conjunction with renal biopsies from 25 client-owned dogs with renal disease were analyzed. Each sample was normalized to both SG (diluted to 1.0025) and protein concentration (0.3 ug/ul) and run on 4–12% non-reducing SDS-PAGE gels. Gels were stained and scanned followed by analysis using BioNumerics. Protein bands were assigned a MW based on standards, and density of each band as a percent of total density of all bands was determined. For software analysis, the sum of band densities within 3 categories was determined: low MW bands (LMW: 0–40 kDa), intermediate MW bands (IMW: 40–70 kDa), and high MW bands (HMW: 70–300 kDa). Glomerular damage was identified by ≥ 20% density of HMW bands or ≥ 60% density of IMW bands. Tubular damage was identified by ≥ 15% density of LMW bands or < 60% density of IMW bands. Gels were also analyzed visually, based on the number and intensity of bands within each MW category. Corresponding renal biopsies were assigned a numerical severity score of 0–3 for glomerular damage and 0–4 for tubular damage. Sensitivity (Sn) and specificity (Sp) of the urine SDS-PAGE analysis for determining glomerular and/or tubular damage were determined.

Based on histologic evaluation, 21 (84%) of the dogs had evidence of both glomerular and tubular damage, 2 (8%) had glomerular damage only, and 2 (8%) had tubular damage only.

In conclusion, urine SDS-PAGE has high sensitivity and moderate to high specificity for detecting tubular and glomerular damage as compared to renal biopsies using urine normalized to either SG or protein concentration. Visual assessment of the gels is comparable to software analysis for glomerular damage but is more sensitive for tubular damage.

Table [Abstract 319]

	Glomerular damage					Tubular damage			
	Visual analysis		Software analysis			Visual analysis		Software analysis	
	SG	Protein	SG	Protein		SG	Protein	SG	Protein
Sn	91%	96%	91%	91%	Sn	87%	96%	74%	57%
Sp	100%	50%	50%	100%	Sp	50%	50%	50%	100%

ABSTRACT #320
AGE ASSOCIATED DIMINISHING URINE CONCENTRATING ABILITY IN APPARENTLY HEALTHY CATS. S Yu. Hill's Pet Nutrition, Inc., Topeka, KS.

Feline chronic kidney disease is a slowly progressive disease with a high prevalence in older cats. Oftentimes, it is not diagnosed until cats become azotemic. The objective of the present study was to investigate age associated changes in clinical markers associated with kidney function in apparently healthy cats.

A cross sectional study was conducted with 96 apparently healthy American domestic shorthair cats, 29 males and 67 females. Cats used in the study were either castrated or spayed except two intact female cats. The age of cats ranged from 1.7 to 15.9 years. Healthy cats were defined as with normal physical examination by a licensed veterinarian, normal serum chemistry profile, normal complete blood count, and normal serum thyroxine level within three months prior to the study. All cats were immunized against feline panleukopenia, calici and herpes viruses, and rabies. Blood and urine samples were collected after the cats had been fed with a nutritionally complete and balanced dry cat food for at least 3 weeks. Cats had free access to water during the study. Food was withdrawn overnight before the sample collection. Serum creatinine, urea nitrogen and phosphorus (Hitachi 912 Automatic analyzer) as well as urine specific gravity (refractometer), creatinine (Hitachi 912 Automatic analyzer), and microalbumin (E.R.D.-HealthScreen) were measured.

Both urine specific gravity and urine creatinine were inversely associated with age in cats. Urine specific gravity had a significant drop after cats were 9.8 years old (95% CI: 7.1, 12.5). Urine creatinine was positively associated with urine specific gravity ($r = 0.79$, $p < 0.01$). Age had little effect on serum creatinine, urea nitrogen, and phosphorus. Urine microalbumin was not affected by age either.

These data show that urine concentrating ability diminishes with age as indicated by decreasing urine specific gravity. The diminishing urine concentrating ability escalated after about 10 years of age in cats, suggesting that urine specific gravity might be an early marker for kidney disease in apparently healthy cats.

ABSTRACT #321
COMPARATIVE CLINICAL EVALUATION OF BD VACUTAINER™ PLUS URINALYSIS PRESERVATIVE TUBE AND PLASTIC TUBE FOR URINALYSIS IN DOGS AND CATS. E. Jillings, A. O'Connell, S. Forsyth, N. Cogger & K.E. Hill. Institute of Veterinary, Animal and Biomedical Sciences, Massey University, Palmerston North, NZ.

Urinalysis is an important part of veterinary clinical practice. However the storage of urine samples can influence the presence of crystal formation in the urine sediment. With the advent of centralization of many of the human medical laboratories, urinary preservative tubes have become popular. To date there has been no validation of these tubes in dogs and cats. The aim of this study was to validate the use of the BD Vacutainer™ plus urinalysis preservative tube (UAP), as compared to the standard plastic tube using urine from dogs and cats.

Fresh morning urine samples were obtained from 6 dogs and 5 cats. Urine was obtained by free catch sample for 1 cat and 5 dogs and by antipubic cystocentesis for 1 dog and 4 cats. At least 15 mls of urine was obtained from each animal. Within 15 minutes of urine collection, the urine taken from each dog was transferred into three UAP tubes and three plastic tubes, and the urine collected from each cat was transferred into one UAP tube and one plastic tube. Between 7–8 mls was added to the UAP tubes and at least 2 mls (range 2–20 mls) into the plastic tubes.

A fresh urine sample was analysed within 60 minutes of collection from both the plastic and UAP tubes. Within 30 minutes of collection, the remaining aliquots were stored at room temperature for the UAP urine preservative tubes and at 4°C for the plastic tubes. Samples were further analysed at 6 hours and 24 hours after collection.

Urinalysis were performed using an automated dipstick analyser (Clinitek 50 by Bayer, using Multistix 10 SG urinary dipsticks by Siemens). Urine was then centrifuged for 5 minutes using the Heraeus multifuge 1 s and Heraeus Megafuge 1.0. Visual micropsy was performed using a Nikon microscope. Urine specific gravity (SG) was performed by refractometer (Reichert Temperature compensated hand held refractometer), and pH was measured using the Metler Toledo seven easy pH meter.

Welch two sample t-test and Wilcoxon Rank sum were used to determine significant differences between time and analysis. Statistics programme R was used.

There was no significant difference in biochemistry, SG, or pH between the two urine tubes at any time. At 24 hours the UAP tube samples had significantly less crystalline debris than the refrigerated plastic tubes (p-value = 0.005328).

The BD Vacutainer^{TM} plus urinalysis preservative tube can be used for urine preservation in cats and dogs without affecting the biochemical parameters. The BD Vacutainer^{TM} plus urinalysis preservative tube is better than the refrigerated plastic tube for preventing in vivo crystalline debris formation.

ABSTRACT #322

IMMUNOLOGIC TESTING OF DOG URINE REQUIRES NEUTRALIZATION OF MATRIX EFFECTS TO ALLOW FOR ACCURATE MEASUREMENT OF IL-6 CONCENTRATIONS. M. Wood, E. Breitschwerdt, S. Vaden, S. Nordone. North Carolina State University College of Veterinary Medicine, Raleigh, NC.

The measurement of cytokine concentrations within body fluids provides a means to recognize subclinical and/or unresolved infectious or inflammatory states in patients. In the urinary tract such information could be useful in identifying asymptomatic bacteriuria, recurrent infections, and non-infectious inflammatory conditions. Complicating the development of reliable testing is the inherent biological nature of urine. Urine can vary widely in composition, as indicated by the range of pH and urine specific gravity (USG) observed in healthy patients. Other contributing variables include urinary protein and carbohydrate moieties that create a matrix capable of hindering immunologic testing modalities, such as ELISA. The purpose of this study was to optimize a sandwich ELISA capable of accurately measuring IL-6 within the urine of dogs. IL-6 is one of the primary inflammatory cytokines produced in the urothelium following exposure to bacterial virulence factors. Standard dilutions of a capture polyclonal goat anti-canine IL-6 antibody and a detection biotinylated goat anti-canine IL-6 antibody were utilized to optimize the chemiluminescent quantification of canine recombinant IL-6 within a dynamic range of 4 ng/ml to 15 pg/ml. Urine obtained from purpose bred beagles and/or clinical samples spiked with 4 ng/ml of IL-6 failed to replicate the standard curve with cytokine quantitative recoveries as low as 30% in neat samples and < 70% after 1:64 dilutions. To counteract nonspecific binding of IL-6 to matrix components, reagent diluents composed of bovine serum albumin combined with goat serum, proprietary calcium chelators, and/or proprietary non-mammalian proteins were tested. Diluents containing calcium chelation combined with non-mammalian proteins achieved the best results, improving quantitative recovery of IL-6 in the spiked samples to > 75% for all dilutions. The increased recovery came at the cost of reducing the range of IL-6 detection to 4 ng/ml-62.5 pg/ml. The effects of pH and USG variability on the ELISA were also assessed. Identical urine samples with an adjusted pH range of 5.5 to 9.5 revealed no appreciable effect on urine IL-6 measurements with quantitative recoveries all within +/− 15% of each other. USG was more variable as spiked undiluted samples with USG < 1.015 had significantly (p < 0.01) lower quantitative recoveries compared to samples with USG between 1.030 and 1.040. There was no significant difference when comparing undiluted samples with USG of 1.015–1.040, however the quantitative recoveries did vary from 55%–85%. The use of matrix neutralizing agents was able to improve a previously unacceptable ELISA allowing for the measurements of IL-6 within the urine of dogs. The testing is imperfect as urine variables such as USG can alter IL-6 recovery and hence may underestimate IL-6 concentrations by as much as 45% for samples with USG < 1.015. For many tests this would be diagnostically unacceptable however given that IL-6 is not found within the urine of healthy individuals, absolute values are less critical than relative amounts of cytokine present.

ABSTRACT #323

SURFACE ENHANCED LASER DESORPTION IONIZATION (SELDI) ANALYSIS OF URINE FROM MALE DOGS IN EARLY STAGES OF PROGRESSIVE TUBULOINTERSTITIAL INJURY DUE TO X-LINKED HEREDITARY NEPHROPATHY. MB Nabity, GE Lees, JM Steiner, JS Suchodolski. Texas A&M University, College Station, TX.

There is a need for non-invasive methods to reliably identify tubulointerstitial disease in its early stages in dogs with chronic kidney disease (CKD). Proteomic techniques are powerful tools that can aid in the discovery of new biomarkers of renal injury. The objective of this study was to identify predictive biomarkers of early tubulointerstitial injury in urine of male dogs with X-linked hereditary nephropathy (XLHN) using SELDI, which is a proteomic technique that combines chromatography with mass spectrometry to detect proteins and protein fragments.

Urine samples from male dogs (n = 6) with XLHN were analyzed. Dogs with XLHN have a gene defect that causes the type IV collagen in their glomerular basement membranes to be abnormal. Affected males develop glomerular lesions and progressive tubulointerstitial injury that lead to end-stage CKD during adolescence. Samples were compared at two timepoints: 1) at the onset of a urine protein:creatinine (UPC) of persistently > 2; and 2) at the onset of a serum creatinine ≥ 1.2 mg/dl (mildly azotemic). Albumin was removed from samples by affinity chromatography and samples were concentrated with ultrafiltration. Samples were normalized to protein concentration and applied to cationic (CM10), metal-binding (IMAC30), and hydrophobic (H50) chips with a SPA or CHCA matrix applied. Optimization was performed for protein concentration, buffer ratio, matrix type and concentration, and laser intensity. Statistical significance was determined using a paired Wilcoxon signed rank test. Principal component analysis (PCA) was also performed.

Peak reproducibility was acceptable for the IMAC30 and CM10 chips using the SPA matrix and was marginally acceptable for the H50 chip using the CHCA matrix. The greatest number of total peaks was obtained with the IMAC30 and CM10 chips (123 and 101, respectively). Among all chips, there were 53 unique, well-defined, statistically significant peaks and/or peak regions. Of these, 33 peaks had higher intensity (implying greater abundance) in the samples obtained during the later timepoint (mass range: 2–157 kDa) while 20 peaks had higher intensity in the samples obtained during the earlier timepoint (mass range: 3–22 kDa). PCA analysis showed a clear difference between the two timepoints. Peaks and/or peak regions at approximately 4.4, 5.0–5.8, 8.5, 9.7, 10.4–10.9, 11.5, 13.0–13.7, 20.6, 25, 51, and 100 kDa accurately distinguished between the two timepoints and were identified by more than one chip type.

In summary, the CM10 and IMAC chips using a SPA matrix produced the best results for these modified urine samples. Regarding the statistically significant peaks, more abundant protein was generally present as renal disease progressed, especially for those peaks > 20 kDa. Peaks at 11.5 and ~21 kDa are similar to those identified as markers of renal disease in a previous veterinary study, while the remaining peaks may serve as candidates for novel markers of early renal disease.

ABSTRACT #324

NEW PROCEDURE FOR IMPLANTING PERIONEAL DIALYSIS CATHETERS IN SMALL ANIMALS. K Harada, M Uechi, S Yamano, Y Ukai, M Nishida, S Kohno. Nihon University, Kanagawa, Japan

Renal failure is treated by peritoneal dialysis, but the catheters often cause difficulties. We designed a new method of implanting disk-type PD catheters and compared it with the conventional straight catheters. Four healthy dogs were anaesthetized using isoflurane, placed in the dorsal position and disinfected. A catheter was inserted through a small incision made 3~5 cm above the umbilicus. The tip of the cath-

Figure [Abstract 324]

eter was positioned at the bladder side for the conventional procedure, and the catheter disk was positioned between the liver and the diaphragm during the new procedure. Peritoneal dialysis was performed using 500 ml of dialysate and the amount drained and time required for drainage were measured. We collected > 100% of the infused dialysate within 30 minutes using the new method, whereas the catheter became obstructed using the conventional procedure which rendered dialysate collection impossible. These results indicate that our new procedure is effective for performing peritoneal dialysis in small animals.

ABSTRACT #325
EFFECT OF A TEST FOOD ON WHOLE BLOOD GENE EXPRESSION IN CATS WITH APPENDICULAR DEGENERATIVE JOINT DISEASE. N.Z. Frantz, Kevin Hahn, Jen MacLeay, and S. AL-Murrani. Hills Pet Nutrition, Inc., Topeka, KS.

This study was conducted to evaluate the effect of a test food on whole blood gene expression profiles when fed to cats with appendicular degenerative joint disease (DJD). Thirty-one domestic short hair cats (initial weight, 4.2 ± 1.27 kg, age, 12.1 ± 2.98 years) with lameness and radiographic changes consistent with DJD were included in the study. All cats were fed a control food for 28 days followed by a test food. The test food contained increased levels of DHA and EPA, methionine, manganese, Vitamins C and E, and L-carnitine. Whole blood samples were collected on the last day of the control food and after 14 days on the test food. After consuming the test food for 14 days, cats with DJD had decreased expression of the 6 genes (INDO, ANXA1, MMP8, CD79a, ODC1, and STXBP2) previously identified to be up-regulated in DJD versus healthy geriatric cats. These genes are associated with cartilage degradation, B cell function, polyamine synthesis, and protein trafficking. In summary, feeding the test food to cats with DJD resulted in reversal of the gene expression patterns previously observed in the blood of cats with DJD versus healthy geriatric cats after 14 days. Furthermore, this study helps increase the understanding of how the test food may aid in improving mobility in cats with DJD at the molecular level.

ABSTRACT #326
EFFECT OF PRESCRIPTION DIET® CANINE J/D™ ON WHOLE BLOOD GENE EXPRESSION IN DOGS WITH OSTEOARTHRITIS. N.Z. Frantz, Kevin Hahn, Jen MacLeay, and S. AL-Murrani. Hills Pet Nutrition, Inc., Topeka, KS.

A study was conducted to evaluate the effect of Prescription Diet® canine j/d™ on selected arthritis related genes in whole blood when fed to dogs with osteoarthritis (OA). Thirty-one beagle dogs (initial weight, 13.5 ± 1.27 kg, age, 11.0 ± 2.23 years) with lameness and radiographic changes consistent with OA in at least one joint were included in the study. All dogs were fed a control maintenance food for 28 days followed by Prescription Diet® canine j/d™. Canine j/d contains increased levels of EPA and DHA, glucosamine, chondrotin sulfate, Vitamins C and E, and L-carnitine. Whole blood samples were collected on the last day of the control food and after 14 days on canine j/d. Improved orthopedic exam scores were noted in these dogs after 14 days on canine j/d. After consuming canine j/d for 14 days, OA dogs had increased expression of 22 genes (IL-6, ADAMTS-4, IFNG, HAS2, BGN, SOX-9, ADAMTS-5, MMP3, ACP5, IL1A, TNC, HAS3, COMP, IGF-1, GHR, Xaa-Pro Peptidase, RANKL, SMAD7, PGE2, TLR9, PLOD1, and SCL2A9) that were previously demonstrated to be down-regulated in OA versus healthy geriatric dogs and decreased expression of ANXA1 that previously was shown to be up-regulated. In summary, feeding canine j/d to dogs with osteoarthritis resulted in reversal of the gene expression patterns previously observed in the blood of arthritic versus healthy geriatric dogs after 14 days. Furthermore, this study helps increase the understanding of how Prescription Diet® canine j/d™ improves mobility in arthritic dogs at the molecular level.

ABSTRACT #327
AN OPEN-LABEL, PROSPECTIVE STUDY EVALUATING THE RESPONSE TO FEEDING A VETERINARY THERAPEUTIC DIET IN CATS WITH DEGENERATIVE JOINT DISEASE. A. Sparkes[1], J. Debraekeleer[2], D. Fritsch[2], K. Hahn[2]. [1]Animal Health Trust, Suffolk, UK. [2]Hill's Pet Nutrition, Inc., Topeka, KS.

The aim of this study was to determine whether a therapeutic diet with increased levels of manganese, methionine, n-3 fatty acids, and reduced n-6:n-3 fatty acid ratio was effective in the management of cats with historical, clinical and radiographic signs of appendicular degenerative joint disease (DJD).

47 adult cats were recruited prospectively from 21 veterinary clinics in Europe for an open-label clinical trial. The cats underwent a veterinary examination that included scoring the severity of overall arthritic changes on a scale of 1–5 for the presence of lameness, joint swelling, reduction in joint movement and pain on joint manipulation. Owners were also asked to score (on a scale of 1–5) seven aspects of their cat's mobility including the cat's willingness and ability to jump, presence of stiffness or pain, the amount of time spent playing and interacting, grooming behavior and overall activity levels. Cats were transitioned onto the therapeutic diet over 7 days and then fed the diet exclusively for 30 days after which time the clinical evaluation and questionnaires were repeated. Following feeding of the therapeutic diet, there were significant improvements seen in all of the parameters evaluated. The most marked and frequent changes seen were an improved ability to jump, reduced stiffness, increased activity, reduced lameness and reduced pain on joint manipulation. The results of this study suggest that feeding the therapeutic diet may have a significant impact on the morbidity associated with feline DJD and that comparison to a scientific control diet is warranted to confirm these observations.

Previously presented at the 19th Annual ECVIM Congress, Porto, Portugal, September 2009.

ABSTRACT #328
IMPROVEMENT OF CLINICAL SIGNS OF OSTEOARTHRITIS IN CATS BY DIETARY INTERVENTION. D. Fritsch[1], T. Allen[2], A. Sparkes[3], C. Marion[1], K. Hahn[1]. [1]Hill's Pet Nutrition, Inc., Topeka, KS. [2]Dechra Pharmaceuticals PLC, Overland Park, KS. [3]Animal Health Trust, Suffolk, UK.

The objective of this study was to determine the safety and efficacy of an experimental food with increased levels of manganese, methionine, n-3 fatty acids as an aid in the management of clinical signs associated with feline osteoarthritis. The food was tested in a randomized, double-blind, controlled clinical trial involving 172 client owned adult cats with clinical signs and radiographic changes consistent with osteoarthritis. Cats were assigned to receive the experimental food or a control food for 12 weeks. The cats were evaluated at 0, 4, 8, and 12 weeks of feeding. Cat owners evaluated 15 signs associated with mobility and behavioral changes, and veterinarians conducted orthopedic evaluations to assess overall arthritic condition, lameness, joint swelling, range of motion, and pain on palpation of each affected joint.

Based on the veterinarians' evaluation of the cats presenting with moderate or severe arthritis, a significantly higher proportion fed the experimental food (61%) showed improved arthritic conditions within four weeks compared to the control food (37%). Cats presenting with mild arthritis signs at the start of the study responded less favorably to the dietary therapies when compared to the moderate or severe arthritic cats based on veterinary orthopedic evaluations and pet owner assessment of overall arthritic condition.

Given that cats presenting with moderate or severe arthritis responded positively and more rapidly to the experimental food than to the control food within 4 weeks, the role of n-3 fatty acids, methionine and manganese in the management of cats with arthritic signs requires further investigation.

Previously presented at the 19th Annual ECVIM Congress, Porto, Portugal, September 2009.

ABSTRACT #329

EFFECT OF PRESCRIPTION DIET® CANINE B/D™ ON INFLAMMATORY BIOMARKERS AND GENE EXPRESSION CHANGES IN GERIATRIC DOGS N.Z. Frantz, K.A. Hahn, and S. AL-Murrani. Hills Pet Nutrition, Inc., Topeka, KS.

A study was conducted to evaluate the effect of Prescription Diet® canine b/d™ on inflammatory biomarkers and gene expression in healthy geriatric dogs. Twenty-nine geriatric beagle dogs (initial weight, 13.51 ± 1.66 kg, age, 10.7 ± 2.33 years) were included in the study. All dogs were fed a control maintenance food for 28 days followed by Prescription Diet® canine b/d™. Canine b/d contains increased levels of omega 3 fatty acids, lipoic acid, antioxidants from a fruit and vegetable blend, Vitamins C and E, and L-carnitine. Serum and whole blood samples were collected on the last day of the control food (day 0) and after 14 days on canine b/d. Inflammatory and hormone biomarkers as well as gene expression changes were measured. RNA was extracted according to the procedures provided in the PAXgene Blood RNA Kit Handbook (Qiagen, Valencia, CA). RNA was hybridized to an Affymetrix GeneChip Canine-2 Genome Arrays and normalized using Robust Multi-Array Average. Transcripts having a $P < 0.05$ (following a false discovery rate adjustment value of 0.2) and a fold change range of at least 1.25 were considered different. After consuming canine b/d, geriatric dogs fed Canine b/d had lower IL-4, IL-6, IL-10, KC, and total cytokines. No differences in hormone biomarkers were detected. Geriatric dogs fed Canine b/d also had 1123 genes that were up or down regulated compared to day 0. Of these, genes associated with amyloid beta plaque formation were down-regulated while genes associated with clearance of beta amyloid plaque were up-regulated. In addition, genes associated with neurotransmitter signaling and cell adhesion were significantly up-regulated when geriatric dogs were fed Canine b/d compared to the day 0 on the control. This study provides gene expression evidence that support the benefits observed in geriatric dogs for memory and behavior associated responses previously shown with Prescription Diet® canine b/d™.

ABSTRACT #330

LEAN BODY MASS (LBM) NORMALIZES DAILY RESTING ENERGY EXPENDITURE (REE) IN HEALTHY ADULT CATS (n = 57) OF VARIED BODY CONDITIONS. SA Center[1], KL Warner[1], JF Randolph[1], JJ Wakshlag[1], GD Sunvold[2]. [1]College of Veterinary Medicine, Cornell University, Ithaca, NY; [2]Proctor & Gamble, Pet Care Research & Development, Lewisburg, OH.

Non-sedated healthy adult cats (5.1+/−1.7 kg) withheld from food for 12-hrs were used to 1)validate an open-flow indirect calorimetry (IC) system determining REE, 2) validate a multi-frequency bioelectrical impedance analysis (MF-BIA) device estimating total body water (TBW) for LBM calculation, and 3) normalize REE against LBM.

Using 5-point body condition scoring, 7 under-conditioned (≤ 2.5/5), 29 normal-conditioned (3.0–3.5/5), and 21 over-conditioned (≥ 4.0/5) cats in stable body weight were studied. Cats had continuous IC measurements recorded over 45-min after acclimation to a plexiglass chamber; cage flow regulated with mass flow controllers was confirmed with nitrogen dilution. IC was calibrated using span gases and alcohol combustion. Lowest values (least-observed-metabolism) were analyzed. IC intra- and inter-assay comparisons were made using CV%, ANOVA, and Bland-Altman analyses (bias, 95% limits of agreement). Daily energy expenditure (DEE; n = 12) was estimated using doubly-labelled water. Criterion references (deuterium, sodium bromide) validated MF-BIA to determine scaling factors for resistivity settings (iterative model fitting). Optimal body posture, electrode positions, and body-segment path were determined. Criterion reference and MF-BIA volume comparisons were made using paired T-tests, Bland-Altman analyses, and Pearson's correlation test. Intra- and inter-assay MF-BIA volumes were compared using ANOVA for repeated measures and CV%. $P \leq 0.05$ was considered significant.

Mean REE (176.3 +/− 52.8 kcal) had 30% inter-individual CV% with intra- and inter-assay CV% ranging from 1.5–14%; differences were not significant. REE represented 61.3 +/− 14.3% DEE. Sternal posture with electrode positioning and body-segment path in neck-tail configuration was optimal. Inter-assay CV% ranged from 6.6–10.1% for TBW, 2.8–6.9% for extracellular water, 8.7–12.1% for intracellular water, and 6.6–10.1% for LBM. MF-BIA LBM significantly correlated with the criterion reference ($P < 0.0001$). Over- and under-conditioned cats had significantly lower and higher REE (kcal)/kg compared to normal-conditioned cats, respectively ($P \leq 0.001$). However, differences disappeared when REE was normalized with LBM. For normal-, under-, and over-conditioned cats, REE (kcal)/kg was 36.6 +/− 7.7, 50.0 +/− 9.7, and 28.5 +/− 6.3, whereas REE (kcal)/LBM (kg) was 54.7 +/− 18.7, 54.8 +/− 12.9, and 54.7 +/− 18.8, respectively.

IC and MF-BIA systems validated herein reasonably estimate REE and LBM in non-sedated cats and suggest that REE should be normalized using LBM rather than allometric scaling formulas.

ABSTRACT #331

LIPOIC ACID IMPROVES GSH:GSSG RATIO IN DOGS FED LIPOIC ACID IN DRY DOG FOOD. SC Zicker, DE Jewell D Joshi, S Davidson. Pet Nutrition Center, Hill's Pet Nutrition, Topeka, KS.

Lipoic acid is a coenzyme of several enzyme complexes involved in energy metabolism. It also functions as a biologic antioxidant attributable to its thiolane ring structure which undergoes oxidation-reduction chemical reactions. Previous work has shown lipoic acid to act as an antioxidant in dogs but it is unknown what the lowest effective supplementation is for antioxidant benefits. As such the antioxidant effects of graded levels of lipoic acid inclusion in dry dog food were evaluated in 50 dogs. Racemic alpha-lipoic acid was added to a control food (C) at levels of approximately 50 ppm, 100 ppm, 150 ppm and 300 ppm on a dry matter basis for administration to the five different groups (n = 10 per group). All other dietary constituents were held constant in the formulation. All 50 dogs (M = 25, F = 25) were fed the control food for 2 months prior to allocation into groups. Serum chemistry, complete blood count, vitamin E, GSH:GSSG ratio, malondialdehdyde, 8-hydroxydeoxyguanosine, CD4:CD8 ratio, body weights and fecal scores were evaluated monthly for 3 months. Data were analyzed with a repeated measures procedure with day 0 and concentration of vitamin E in serum as covariates. The concentration of vitamin E in serum was used as a covariate because lipoic acid has been shown to spare vitamin E. Adjusted means of the GSH:GSSG ratio for the different groups were as follows; B = 13.1 ± 2.2, 0 = 16.8 ± 2.5, 100 ppm = 16.3 ± 1.8, 150 ppm = 19.3 ± 1.7, 300 = 16.9 ± 1.8. Supplementation with lipoic acid resulted in a significant increase ($P < 0.05$) of GSH:GSSG ratios in foods for adult dogs at all inclusions administered above control. There were no significant differences between groups in which lipoic acid was added for the previous GSH:GSSG ratios, however, the highest adjusted mean occurred at the 150 ppm inclusion level. No significant improvements were noted in the other outcomes measured and, no adverse changes were observed in either clinical chemistry or complete blood count analyses at any time points. In conclusion, addition of racemic lipoic acid to dry dog food resulted in a significant antioxidant benefit with no adverse effects noted at doses lower than 150 ppm as previously reported. Specific biologic benefits other than antioxidant activity remain to be determined.

ABSTRACT #332

INDUCING DIETARY KETOSIS IS SAFE AND EFFECTIVE IN PROVIDING WEIGHT LOSS IN CATS. Melendez L, Schoenherr WD, Hahn KA. Hill's Pet Nutrition, Inc. Topeka, KS.

A prior study showed that a high protein, low carbohydrate, moderate fiber food induced ketosis in obese cats during weight loss. The objective of this study was to determine the effect dietary-induced ketosis on body composition during weight loss in obese cats.

16 obese cats (body condition score > 4; 5 point scale) were randomized to be fed a dry or wet formulation of a high protein, low carbohydrate, moderate fiber food for 24 weeks. Food intake was restricted for the duration of the experiment to cause weight loss. Daily food amount was based on ideal body weight (20% body fat) energy requirements (kcal offered per day = 0.7 × (70 × initial body weight $(kg)^{75}$)). The amount of food offered daily to each animal was calculated by dividing the amount of calories to be offered by the caloric density of the food (kcal/kg). Use of this equation was estimated to allow animals to lose body weight at a rate of 0.5 to 1.5% of their initial body weight per week. Initial body composition was determined by DEXA. Ideal body weight was estimated by calculating fat-free body mass from the DEXA analysis and adding 20% fat to this total. Both groups were compared for rate of weight loss, body compositional changes as measured by DEXA, and clinical chemistry changes every 4 weeks. The study concluded when a cat attained ideal body weight or 24 weeks.

Obese cats in both food groups had increased serum levels of β-hydroxybutyrate (BHBA) by week 8. Cats fed the wet food maintained higher levels of BHBA throughout the study. Cats fed the dry food had BHBA levels return to baseline by week 12. No other changes in clinical chemistry were observed.

The rate of weight loss did not differ between the two formulations (wet, −7.3 g/d; dry, −6.9 g/d). Most of the body weight change was related to change in body fat tissue (dry, −7.6 g/d; wet, −6.5 g/d). Cats fed the wet formulation lost very little lean tissue (−0.8 g/d; or 4% compared to baseline) during the 24 week study. Cats fed the dry formulation gained a small amount of lean tissue 0.8 g/d) during the study.

These data indicate that a low-fat, high-fiber, high-protein diet induces ketosis, provides effective weight loss, and maintains lean body mass.

ABSTRACT #333

BODY WEIGHT IS A MAJOR PREDICTOR OF BODY COMPOSITION IN GROWN CATS. S Yu. Hill's Pet Nutrition, Inc., Topeka, KS.

The objective of the present study was to investigate the effects of age, gender, and body weight on the body composition in grown cats.

A cross sectional study was conducted with 95 healthy grown American Shorthair cats, 29 male and 66 female. All cats used in the study were either castrated or spayed except two intact female cats. The age of cats ranged from 1.7 to 15.9 years. Healthy cats were defined as with normal physical examination by a licensed veterinarian, normal serum chemistry profile, normal complete blood count, and normal serum thyroxine level within three months prior to the study. All cats were immunized against feline panleukopenia, calici and herpes viruses, and rabies. Cats were housed with 10 - 12 other cats in spacious rooms with natural light that varies with seasonal changes. Cats experienced behavioral enrichment through interactions with each other, by daily interaction and play time with caretakers, large windows and sun porches to watch the natural landscape and access to toys. Cats were fed with a nutritionally complete and balanced dry cat food for at least 3 weeks before body composition measurement. Food was continuously available throughout the day until their daily caloric requirements were met. The body composition was measured with a dual x-ray absorptionmetry (Hologic; Hologic, Inc, MA).

Body weight significantly affected ($p < 0.05$) body composition of grown cats. Age and gender had no effect ($p > 0.05$) on the body composition. Bone mineral content ($r = 0.93$, $p < 0.01$), bone mineral density ($r = 0.67$, $p < 0.01$), total lean body mass ($r = 0.86$, $p < 0.01$), total fat body mass ($r = 0.87$, $p < 0.01$), and percent fat body mass ($r = 0.64$, $p < 0.01$) were positively while percent lean body mass ($r = -0.63$, $r < 0.01$) was negatively associated with body weight in grown cats.

These data show that body weight is a major predictor of the body composition of healthy grown cats. Age and gender have little effect on the body composition of neutered grown cats.

ABSTRACT #334

THE EFFECT OF DIETARY CRUDE PROTEIN AND LYSINE:CALORIE RATIO ON BODY COMPOSITION AND KIDNEY MEASURES IN HEALTHY ADULT AND GERIATRIC CATS. N.Z. Frantz and R.M. Yamka. Hills Pet Nutrition, Inc., Topeka, KS.

A study was conducted to evaluate the affect of dietary crude protein and lysine:calorie ratio in healthy adult and geriatric cats on body composition and kidney health. The study utilized 25 healthy adult (5.6 ± 0.81 years of age (range 5–7)) and 25 geriatric American short hair cats (11.1 ± 2.1 years of age (range 8–15), initially 4.64 ± 1.18 kg.Cats were blocked by age, gender, urine specific gravity, and body fat and assigned to one of five experimental maintenance foods (34% crude protein (CP), 34% CP with ideal amino acid balance, 34% CP with 5.5 lysine:calorie ratio, 44% CP, or 54% CP) for the 180 day feeding period. Chemistry screens, glomerular filtration rate, serum bone biomarkers, serum and urine kidney biomarkers, DEXA scans, and microalbuminuria were evaluated at day 0, 30, 90, and 180. No differences in kidney function as measured by glomerular filtration rate, microalbuminuria, urine specific gravity, serum creatinine, urine creatinine, or urinary biomarkers of kidney function were detected. Body weight and fat changes over time were unaffected by dietary treatment. Most interestingly, cats fed the 44% and 54% CP treatments had decreased bone mineral density ($P < 0.01$) at day 180, bone specific alkaline phosphatase (BAP) ($P < 0.05$) at day 90, and BAP:NTX (N-terminal telopeptide of type I collagen) ratio ($P < 0.05$) at day 90 indicating greater bone turnover compared to the 34% CP 5.5 lysine:calorie ratio food. This study suggests that feeding foods containing crude protein levels of 34–54% do not affect kidney health, lean muscle mass, or fat mass in healthy adult and geriatric cats. However, feeding high CP foods (44% and 54%) appears to have negative consequences on bone mineral density through increased bone turnover.

ABSTRACT #335

MAINTENANCE OF LEAN BODY MASS IN SENIOR CATS FED A LOW PROTEIN THERAPEUTIC RENAL FOOD. R.M. Yamka and L. Melendez. Hill's Pet Nutrition, Inc., Topeka, KS.

The objective of this study was to determine the effect of feeding a low protein therapeutic renal food on lean body mass in senior cats. Hill's Prescription Diet® Feline k/d® dry (28.2% crude protein, 22.6% crude fat 0.52% phosphorus and 0.24% sodium on dry matter basis (DMB)) was fed for maintenance (average intake 46.2 ± 2.6 grams) to twelve senior cats (average age = 10.05 ± 2.20 years (range 7 to 13 years); average weight = 5.49 ± 1.03 kg; average creatinine = 1.23 ± 0.07 mg/dL) for a period of four months. Cats were group housed and had free access to water. Additionally, cats were offered enrichment toys, received routine grooming, and had daily opportunities for socialization with other cats and people. Dual-energy x-ray absorptiometry (DXA; DXA-QDR-4500, Hologic, Inc., Waltham, MA) scans were performed at months 0, 2 and 4 to assess the effect of the food on body weight, total fat mass, lean body mass and bone mineral content during the study. Cats consuming this food maintained total body weight, total fat mass, lean body mass and bone mineral content throughout the study. In summary, senior cats fed a low protein (28.2% crude protein on DMB) therapeutic food for maintenance were able to maintain body weight and lean body mass.

Table [Abstract 336]

Diet	BUN*	Albumin*	ALT*	Creatinine**	Phosphorus***
Purina NF	17.57 (± 0.28)	3.10 (± 0.03)	49.82 (± 5.50)	1.59 (± 0.03)	4.85 (± 0.06)
Purina DM	24.39 (± 0.27)	3.30 (± 0.03)	88.35 (± 5.52)	1.47 (± 0.03)	4.66 (± 0.06)

*$p < 0.0001$; **$p = 0.0002$; ***$p = 0.0011$.

ABSTRACT #336
EFFECTS OF THE LEVEL OF DIETARY PROTEIN ON RENAL INDICES IN NORMAL ADULT CATS. B Backlund, DL Zoran, JE Bauer, MB Nabity, B Norby. Texas A&M University, College Station, TX.

Established reference intervals for biochemical analytes in normal adult cats are likely based on values from cats consuming diets with a low to moderate level of protein and may not accurately reflect values in cats consuming a diet high in protein. If cats consuming high protein diets have higher serum renal values, they may be inappropriately switched to a renal diet early in life based on suspicion for renal disease. The objective of this study was to determine whether healthy cats eating diets high in protein have higher serum renal values without a corresponding change in renal function.

Twenty-four ~3-year-old adult neutered female cats in a research colony were evaluated. Twenty-three cats were assessed to have a normal body condition score of 5/9 and to be clinically normal by physical exam, complete blood count, serum biochemistry profile, and urinalysis. They were FeLV/FIV negative. One cat was excluded due to renal insufficiency. A single cross-over design feeding study was used, with the cats acting as their own controls. The two diets compared were Purina DM (54.24% protein) and Purina NF (28.67% protein). The diets were selected based on their protein contents and made similar with respect to carbohydrates and total fat content by adding 4.7 g vegetable oil/100 g diet to the Purina NF diet. All diets were complete and balanced to meet the known dietary needs of young adult cats. Cats were fed one diet for the first 4 weeks and the second diet for the final 4 weeks, with a 2-week washout period in between these two periods. Hills adult maintenance was used as the washout diet. Blood urea nitrogen (BUN), creatinine, phosphorus, albumin, and alanine aminotransferase (ALT) were measured at biweekly intervals. The effect of the two diets on these analytes was assessed using a linear mixed model framework including cat identification nested within sequence as a random effect. Diets, treatment periods, a diet and treatment interaction term, and baseline measurements after the run-in and wash-out periods were included in the models. Normality of the residuals was assessed visually. Results are presented as least squared means ± standard errors.

All values measured were statistically significant between the two diets but remained within their respective reference interval.

In summary, dietary protein concentration appears to have a statistically significant effect on several serum biochemical analytes, and in particular BUN and albumin concentration are both increased by a diet with high protein content. However a longer feeding period may be needed to demonstrate clinical significance.

ABSTRACT #337
NEUTROPHIL CELL KILLING PERCENT AND NEUTROPHIL-RELATED GENE EXPRESSION DECLINE WITH AGE IN DOGS. Dennis E. Jewell[a], Rachel M. Chinn[b], William J. Vorachek[c], M. Elena M. Gorman[b], Jean A. Hall[b]. [a]Pet Nutrition Center, Hill's Pet Nutrition, Inc., Topeka, KS. [b]Department of Biomedical Sciences, College of Veterinary Medicine, and [c]Department of Environmental and Molecular Toxicology, College of Science, Oregon State University, Corvallis, OR.

This study evaluated the effects of age on cell killing ability of neutrophils and on the expression levels of specific mRNA neutrophil migration or killing function genes. Three groups of Beagle dogs were evaluated in a cross sectional study design: puppies less than 1 year of age (n = 15), adults 1 to 7 years of age (n = 37), and senior adults over 8 years of age (n = 25). The study protocol was reviewed and approved by the Institutional Animal Care and Use Committee, Hill's Pet Nutrition, Inc., Topeka, KS, USA. Neutrophil phagocytosis and killing of *Lactococcus lactis* was assessed, and the relative abundance of mRNAs for myeloperoxidase (MPO), L-selectin, interleukin-8 receptor (IL-8R), and interleukin-1β-converting enzyme (ICE) was determined. Results showed a significant effect of age on percent bacterial killing by neutrophils with phagocytic ability at 4 and 10 years of age reduced by 25% and 39% (compared to 1 year of age), respectively. In addition, younger dogs had significantly higher levels of mRNA for IL-8R, L-selectin, and ICE. These results suggest that older dogs have depressed innate immune responses compared with younger dogs, which may contribute to increased morbidity and mortality with aging. Additional studies are needed to determine how dietary modifications can enhance neutrophil functions in adult and senior Beagle dogs.

ABSTRACT #338
PROTEIN CONCENTRATIONS CONSISTENT WITH CONTROLLED PROTEIN RENAL FOODS MAINTAINS LEAN BODY MASS IN SENIOR DOGS. Dennis E. Jewell, Lynda D. Melendez. Pet Nutrition Center, Hill's Pet Nutrition, Inc., Topeka, KS.

Two studies were used to evaluate the effect of dietary protein varying from 15% to over 30% dry matter (DM) protein on the senior dog's ability to maintain lean body mass (LBM). All dogs were exercised daily, and were provided with regular opportunities for socialization and environmental enrichment. Dog enrichment included toys in their enclosures, paired housing for a portion of each day, and trips to an indoor and/or outdoor group socialization area. The study protocol was reviewed and approved by the Institutional Animal Care and Use Committee, Hill's Pet Nutrition, Inc., Topeka, KS, USA. In the first study, 70 beagle dogs (average age 13.8 years) were fed one of five different foods with protein concentrations, with or without added amino acids, varying from 15.4 to 30.7% dry matter (DM) protein. In all treatments, over the six month period there was a numeric increase in lean body mass. There was no linear or quadratic effect of protein or amino acid supplementation on LBM. Blood urea nitrogen (BUN) increased linearly with dietary protein concentration, and normal albumin, hematocrit and hemoglogin were maintained. In the second study a food designed for nutritional support of animals with renal disease (Prescription Diet® k/d®) with 14.8% (DM) protein was fed for four months to 34 healthy adult dogs (average age 13.0 years). This resulted in an increase in LBM ($P < 0.05$), a decrease ($P < 0.05$) in creatinine and BUN, and no change in albumin. In conclusion these data show that controlling dietary protein concentration to levels (15% DM) of highly digestible protein, used in some foods designed for aid in the management of renal disease is sufficient to meet the protein requirement of the senior dog.

ABSTRACT #339
EFFECTIVENESS OF SUBCUTANEOUS FLUID ADMINISTRATION ON THE REDUCTION OF UROLITHIASIS RISK AS MEASURED BY RELATIVE SUPERSATURATION (RSS) IN HEALTHY CATS. CG Horner, CA Kirk, G Galyon, JW Bartges. University of Tennessee, Department of Small Animal Clinical Sciences, Knoxville, TN.

The quantity of increased fluid intake needed to produce a significant risk reduction for urolith formation in cats has not been defined. The purpose of this study was to determine the effect of graduated increases in fluid intake on urinary relative supersaturation (RSS) for calculogenic substances in healthy cats.

Six healthy adult cats fed a commercial dry maintenance food were used. Cats were housed separately in metabolic cages and allowed free access to food and water. Water intake was measured daily for 3 weeks (wk). During wk 2 and 3, 0.9% saline was administered daily to each cat at 1 × or 2 × their individual basal water intake determined in wk 1. All fluids were administered subcutaneously (SQ) in 2 divided doses using a standard drip administration set and a 20 g needle. A 24 h urine collection was performed at the end of each 7 day period for analysis of metabolic analytes used to calculate RSS for struvite (RSS struvite), calcium oxalate monohydrate (RSS caox mono), and RSS for calcium oxalate dihydrate (RSS caox di) using EQUIL 1.51b. Statistical analysis was performed using ANOVA and Kruskal-Wallis tests.

Table [Abstract 339]

Parameter	1st collection	2nd collection	3rd collection
Volume (ml/kg/d)	11.30 +/− 3.83 a	17.48 +/− 2.08 a	27.12 +/− 7.70 b
PH	6.35 +/− 0.15 a	6.32 +/− 0.34 a	6.06 +/− 0.22 a
RSS struvite	0.066 +/− 0.040 a	0.192 +/− 0.232 a	0.019 +/− 0.018 b
RSS caox mono	6.91 +/− 3.92 a	3.14 +/− 2.16 ab	2.29 +/− 1.74 b
RSS caox di	2.85 +/− 1.55 a	1.43 +/− 0.755 ab	1.01 +/− 0.73 b

Mean water intake (mls/kg/day) for all cats was 21, 28.6, and 48.6 in wk 1, 2, and 3, respectively. Water intake was significantly increased only between wk 1 and 3. See table for Urine Volumes, pH and RSS. Subscripts indicate significant differences between weeks.

In conclusion, increased fluid intake increased urine volume in all cats. Increasing fluid intake to twice basal intake was required to significantly increase urine volume and decrease RSS.

ABSTRACT #340
PHARMACOKINETICS OF HYDROCODONE AND ITS METABOLITE HYDROMORPHONE AFTER ORAL HYDROCODONE ADMINISTRATION TO DOGS.

B KuKanich and J Paul. Kansas State University, College of Veterinary Medicine, Manhattan, KS.

The purpose of the study was to examine the plasma concentrations and pharmacokinetics of hydrocodone and its active metabolite hydromorphone after oral hydrocodone administration to healthy dogs.

The study was approved by the Institutional Animal Care and Use Committee at Kansas State University. Six healthy Greyhounds dogs, 3 male and 3 female, were administered commercially available tablets containing hydrocodone bitartrate 10 mg and acetaminophen 325 mg to the nearest half-tablet. The dose of hydrocodone bitartrate was 0.5 mg/kg PO. Blood was collected from jugular catheters and plasma was separated and analyzed for hydrocodone and hydromorphone concentrations by liquid chromatography and mass spectrometry. The limit of quantification was 1 ng/mL for both analytes.

All dogs tolerated oral hydrocodone well. None of the dogs vomited or became notably sedated. The mean maximum plasma concentration (CMAX) of hydrocodone was 11.7 ng/mL occurring at 0.7 hours with a terminal half-life of 1.6 hours. The mean CMAX of hydromorphone was 5.2 ng/mL occurring at 1.4 hours with a terminal half-life of 3.1 hours. Hydromorphone concentrations were above 1.9 ng/mL in all 6 dogs from at least 30 minutes through 8 hours post hydrocodone administration.

In conclusion, oral hydrocodone was well tolerated in dogs at a dose of 0.5 mg/kg hydrocodone bitartrate. Plasma concentrations of hydromorphone were detected through 8 hours. Further studies assessing the analgesic efficacy of oral hydrocodone are needed.

ABSTRACT #341
CHARACTERIZATION OF *IN VITRO* RELAXANT EFFECTS OF β_2-ADRENOCEPTOR AGONISTS ON ISOLATED CAT BRONCHI.

J. Leemans[1,2], N. Kirschvink[1], P. Gustin[2]. [1]University of Namur, Faculty of Sciences, Namur, Belgium. [2]University of Liège, Faculty of Veterinary Medicine, Liège, Belgium.

Although inhaled β_2-adrenoceptor agonists are currently recommended for the preventive and curative treatment of acute bronchospasms in cats with naturally occurring asthma, their efficacy on cat airways remains uncertain and poorly documented. As a preliminary approach to this issue, our study aimed at comparing on isolated feline bronchi the potency, efficacy and intrinsic activity of β_2-mimetics commonly used in human medicine.

Feline bronchi were obtained from euthanized client-owned cats free from respiratory tract disease at clinical investigation and necropsy. Bronchial rings were contracted with acetylcholine 1.2×10^{-4} mol/l to induce 80% of the acetylcholine maximal contraction. Dose-response curves to isoproterenol (ISO), salbutamol (SAL), fenoterol (FEN), formoterol (FOR) and salmeterol (SLM) were established by cumulative additions of β_2-adrenoceptor agonists. The remaining active tension was abolished by theophylline 3×10^{-3} mol/l producing a maximal relaxation. The potency (concentration of drug causing 50% of maximal relaxation induced by theophylline 3×10^{-3} mol/l [EC_{50}]) and efficacy (maximal effect [E_{max}], % of the action of theophylline 3×10^{-3} mol/l) of each compound were determined. The intrinsic activity (IA) was expressed as the ratio between E_{max} of the studied drug and E_{max} of ISO (reference compound).

All β_2-mimetics relaxed feline bronchi in a concentration-dependent manner with the following rank order of potency: FOR > FEN ≥ ISO > SAL. SLM potency was not determined since its relaxant effects failed to reach 50% of the relaxation induced by theophylline 3×10^{-3} mol/l. SAL and SLM displayed less maximal relaxant effect than the other β_2-agonists. Significant differences in E_{max} were found between ISO and SAL or SLM. Accordingly, the IA of both compounds was lower compared with that of FOR and FEN. E_{max} and -log EC_{50} values are reported in the following table as means ± SEM († significantly different from ISO):

β_2-mimetics were effective at reducing muscarinic-induced tone in isolated feline bronchi. ISO and FEN are the most efficacious β_2-agonists, FOR the most potent one. FEN and FOR act as full agonists of β_2-adrenoceptors, SAL and SLM as partial agonists. These encouraging results deserve further investigations in cats with experimentally induced or naturally occurring asthma.

Table [Abstract 341]

	n	$-\log EC_{50}$	E_{max}	IA
ISO	6	7.78 ± 0.21	88.63 ± 3.96	1
SAL	6	6.80 ± 0.13	73.16 ± 4.18†	0.83
FEN	4	7.84 ± 0.16	89.11 ± 3.04	1.01
FOR	7	9.06 ± 0.20	82.55 ± 5.52	0.93
SLM	6	–	48.26 ± 6.45†	0.54

ABSTRACT #342
EMODEPSIDE SENSITIVITY IN AUSTRALIAN SHEPHERD IS ASSOCIATED WITH DELETION MUTATION OF THE MDR1 GENE.

C. Hugnet[1], X. Pineau[2], F. Buronfosse-Roque[2], S. Queffélec[2]. [1]Clinique vétérinaire des Lavandes, La Bégude de Mazenc, [2]Veterinary Pharmacovigilance Center of Lyon, Ecole Nationale Vétérinaire de Lyon, Marcy L'Etoile, France.

Profender modified-release tablets (Bayer HealthCare) have been authorized and commercialized in Europe since 2008: this product contains two active substances, emodepside and praziquantel. Emodepside is a member of a new class of cyclic desipeptide compounds, acting at the neuromuscular junction by stimulating presynaptic receptors belonging to the secretin receptor family, resulting in paralysis and death of the parasites. The approved indication is "For dogs suffering from, or at risk from, mixed parasitic infections caused by roundworms and tapeworms...". The applicant identified that emodepside is a P-gp substrate drug, and tolerance studies in dogs showed that its margin of safety is lower in Collies than in common breeds.

The multidrug resistance gene (MDR1/ABCB1) produces P-glycoprotein (P-gp) that belongs to the ATP Binding Cassette protein superfamily. P-gp is involved in the active cellular efflux of a large number of drugs and transports many structurally and pharmacologically unrelated hydrophobic compounds. In the central nervous system (CNS), P-gp is found in the capillary endothelial cells that form the blood brain barrier. So, the role of P-gp is particulary important for the protection against accumulation in the CNS of P-gp substrate drugs and their subsequent neurotoxicity. P-gp is also disposed on the biliary canalicular membrane of hepatocytes and on the apical side of enterocytes. A 4-bp deletion mutation of MDR1/ABCB1 gene was described in ivermectin-sensitive Collies and many other dog breeds (e.g Australian Shepherds, White German Shepherds). This mutation generates a premature stop codon, preventing synthesis of the complete and active protein product.

Profender was administered as recommended (correct dose and 12-hours fasting period) to 5 client-owned Australian Shepherd pups (2 males, 3 females) of the same litter (4-month-old, 9 to 10 kg b.w). Of these 5 pups, only one male dog developed adverse reaction 3 hours later after ingestion of the product: vomiting, mydriasis, ataxia, muscle tremors and convulsions. Symptomatic therapy was sufficient to obtain resolution of symptoms over the 10 hours hospitalization period. This dog was tested homozygous for the MDR1/ABCB1 mutation. The other dogs of this litter were heterozygous (n = 3) or homozygous for the wild allele (n = 1). Similar observations concerning 18 dogs of these specific breeds (Australian Shepherds, Collies, White German Shepherds) were reported in France with this product. All the tested dogs were homozygous for mutant allele. Emodepside (associated with praziquantel) was prescribed at dosages that cause no adverse effects in normal or heterozygous dogs.

In purebred dogs of these breeds, veterinarians should be aware that even at licensed dose, adverse reactions may occur with emodepside-based products. It is advisable to test these dogs for the MDR1/ABCB1 mutation, before prescribing Profender modified-release tablets.

ABSTRACT #343

SYSTEMIC EFFECTS OF A COMMERICAL PREPARATION OF CHONDROTIN SULFATE, HYALURONIC ACID AND N-ACETYL-D-GLUCOSAMINE WHEN ADMINISTERED PARENTERALLY TO HEALTHY CATS. MR Lappin, JK Veir, CB Webb. Colorado State University, Fort Collins, CO.

Glycosaminoglycans are administered to cats for the treatment of a variety of inflammatory conditions including idiopathic cystitis and osteoarthritis. However, little information is available concerning effects of various products on lymphocyte profiles, cytokine production, or anti-oxidant capabilities of treated cats. The purpose of this study was to determine whether parenteral administration of an aggregans of chondrotin sulfate, hyaluronic acid, and N-acetyl-D glucosamine has systemic effects in healthy cats.

Polyglycan® (Arthrodynamic, Lexington, KY) was administered to eight healthy, young adult, mixed sex cats at 0.1 ml/lb intramuscularly (IM) on days 0, 4, 8, 12, and 16. On days 0, 14, 28, and 56, blood was collected and the reduced-to-oxidized glutathione ratio (GSH:GSSG) was determined as a measure of oxidative stress and the anti-oxidant capacity of plasma (Trolox units) were determined using previously published assays. %CD4 + lymphocyte counts, %CD8+ lymphocyte counts, CD4 +/CD8 + ratio, and CD44 lymphocyte counts were determined using flow cytometry. Th1 and Th2 cytokine expression by PBMC cells was measured by real time PCR assays (IL1, IL2, IL4, IL6, IL10, IL12, TNF alpha, IFN alpha). On days 0 and 56, complete blood cell counts, serum biochemical panels and urinalyses were performed. The data were analyzed by repeated measures ANOVA with time as the only fixed effect. Where the effect of time was significant ($P < 0.05$), comparisons were made between day 0 and day 14, 28 and 56.

Other than evidence of mild discomfort from the IM injections, no other clinical side-effects were noted. There were no significant changes in CBC, biochemical, or urinalysis findings over time. Significant decreases over time were detected for GSH:GSSG ratio (day 28), %CD4+ lymphocytes (days 14 and 56) and CD44 lymphocytes (days 14, 28, and 56). Significant increases over time were detected for IFN-alpha (days 28 and 56) and TNF alpha (days 28 and 56). Anti-oxidant capacity for all 8 cats was increased by Day 14 (Range 1.75% to 148.0%; median percentage increase = 23.2%) but the mean results was not statistically significant when compared to Day 0 ($P = 0.0514$). IL10 concentrations increased over time for 6 of 8 cats but the change in mean results over time was not statistically significant ($p = 0.2868$).

The results indicate that IM administration of this aggregans of glycosaminoglycans induces systemic effects in cats. Some of the findings (decreased oxidative stress, increased anti-oxidant capability, and increased IL10 concentrations) could be of potential benefit to cats with inflammatory diseases and suggest that controlled studies of clinically affected cats are indicated.

ABSTRACT #344

EFFECT OF ST JOHN'S WORT ON PHARMACOKINETICS OF CYCLOSPORINE AND CYTOCHROME P450 GENE EXPRESSION IN DOGS. K. Fukunaga[1], H. Sahara[2], K. Orito[1]. [1]Department of Physiology II and [2]Department of Biology, School of Veterinary Medicine, Azabu University, Kanagawa, Japan.

St John's Wort (SJW) has been used for the treatment of nuisance barking of dogs. In humans, SJW alters the pharmacokinetics and/or clinical response of a variety of clinically important drugs by upregulation of CYP metabolism. The effect of SJW on drugs metabolism is, however, unclear in canine medicine. Therefore, we investigated the effect of multiple administration of SJW on the pharmacokinetics of cyclosporine (CsA), which has been used for atopic dermatitis was examined. The effect of multiple SJW on the cytochrome P450 (CYP) mRNA expression was also examined.

After SJW (300 mg PO q24h) was administered for 4 weeks, a single cyclosporine (5 mg/kg PO) was administered in 4 Beagle dogs (SJW dogs). As a control, cyclosporine was administered without multiple SJW administration in other 4 Beagle dogs (control dogs). The whole blood concentration of CsA were determined using a high performance liquid chromatography. The liver and the intestine were taken from these dogs and their CYP mRNA expression were measured by the quantitive real time RT-PCR.

The maximum whole blood concentration and the area under the curve of CsA in SJW dogs were lower than those in control dogs. The hepatic CYP2C21 and 3A12 expression and the intestinal CYP2B11 and 3A12 expression in SJW dogs were higher than those in control dogs.

The present results suggest that SJW may decrease the whole blood concentration of CsA by upregulation of hepatic and intestinal CYP metabolism in dogs.

ABSTRACT #345

HISTOLOGICAL PROPERTIES OF THE SOFT PALATE TISSUE: COMPARISON BETWEEN HUMAN AND DOGS WITH UPPER AIRWAY OBSTRUCTION. M Planellas[1], R Cuenca[1], R Velarde[1], P Huguet[2], J Pastor[1], S Lavin[1], J Lorente[3]. [1]Faculty of Veterinary Medicine, Universitat Autònoma de Barcelona (Spain); [2]Anatomopathology Department and [3]Otorrinolaringology Department, Hospital General Universitari de la Vall d'Hebron (Spain).

English and French bulldogs are characterized by abnormal anatomy of upper airways (UA) that leads to an obstructive syndrome. These breeds have been proposed as an animal model for obstructive sleep apnea syndrome (OSAS) in humans. The purpose of this study is to compare qualitative histopathological changes and muscle fiber type distribution of the soft palate in brachycephalic dogs with obstructive UA disease with those changes present in human patients with OSAS. All samples were obtained under permission of owners and human authorization and have been accepted by the human and veterinary ethical commission.

Eleven human patients with OSAS undergoing uvulopalatopharyngoplasty and 21 bulldogs undergoing soft palate resection were included in the study. Control tissues were collected from four human patients and seven dogs with no related disorders at the time of necropsy. Samples were stained with hematoxylin-eosin and Masson trichrome. Immunostaining with antibodies directed against myosin heavy chain Type II isoform was performed using standard techniques in a subset of the samples. Histological comparative findings were: 1) mucosal edema was present in both human and canine pathologic samples; (2) when severe, edema and increase endomysial connective tissue was seen within muscle fascicles; (3) fiber group atrophy was present in half of human and canine pathologic samples and absent in controls; (4) a predominance of muscle fiber type II versus type I was observed in the soft palate of canine and human patients and controls; (5) quantitative differences of fiber types was not seen between canine and human patients and controls; (6) fat infiltration within the muscle was present in similar percentage in human pathologic and control samples but was absent in canine samples.

As previously suggested by others authors, UA obstructive events have the potential to cause muscle and nerve injury due to mechanical trauma induced by pharyngeal pressure swings, vibration, and muscle contractions. The main findings of our study support this scenario. The presence of fiber group atrophy, suggesting neurogenic lesion, edema and fibrosis observed exclusively in human and canine pathologic soft palate samples probably appears in response to chronic hyperstimulation and vibration. Those lesions could contribute to UA collapsibility. On the other hand, several authors described an increased proportion of fast-twitch type II fibers in UA dilator muscles of patients with OSAS, due to overuse. Our findings indicate that there is a similar proportion of fiber type II in the soft palate between patients and controls, suggesting that soft palate presents a different muscular activity in comparison with other UA dilator muscles. Our findings manifest that there are important similarities between human patients and bulldogs with an UA obstructive syndrome, confirming this breed as a good animal model for human OSAS.

ABSTRACT #346

HISTOPATHOLOGIC AND MORPHOMETRIC EVALUATION OF THE NASAL AIRWAYS OF CATS WITH EXPERIMENTALLY-INDUCED ASTHMA. SA Carey[1], CM Venema[1], KJ Williams[1], LJ Gershwin[2], CR Reinero[3]. [1]Michigan State University College of Veterinary Medicine, East Lansing, MI. [2]University of California at Davis School of Veterinary Medicine, Davis, CA. [3]University of Missouri College of Veterinary Medicine, Columbia, MO.

Rhinitis and asthma often co-exist in people, and there is growing experimental evidence to suggest that they are two points along a continuum of airways disease. The potential relationship between rhinitis and asthma in cats has not been investigated. We have developed an experimental model of feline asthma, using the aeroallergen Bermuda Grass Allergen (BGA), which histologically replicates the spontaneous condition. We hypothesized that aeroallergen sensitization and challenge evokes morphologic changes in the nasal airways of cats (airway eosinophilia, mucous cell hyperplasia) that mimic those observed in the bronchial airways of feline asthmatics.

Five mixed breed cats were sensitized and challenged via aerosolization with BGA to create an asthmatic phenotype. Four cats were similarly treated with phosphate buffered saline to serve as controls. Nasal and pulmonary tissue samples were collected at necropsy. The nasal cavities of all cats were processed for light microscopy and morphometric analyses. Inflammatory cell influx in the nasal mucosa was quantified using numeric cell densities. The volume density of intraepithelial mucosubstances (e.g., goblet cells) was quantified using image analysis.

All control and asthmatic cats exhibited mild rhinitis, consisting primarily of neutrophils with rare eosinophils, in the non-ciliated transitional epithelium (NTE) and respiratory epithelium (RE) lining the septum and turbinates of the rostral nasal cavity. There was no significant difference in the magnitude (numeric cell density) of rhinitis observed along the mid-septum between control and asthmatic cats. The volume density of intraepithelial mucosubstances in the dorsal meatus was not significantly different between control ($2.858 \pm 1.038\,nl/mm^2$) and asthmatic ($3.863 \pm 0.720\,nl/mm^2$) cats. There were no significant nasal morphologic or inflammatory differences observed between control and asthmatic cats. These preliminary results suggest that the nasal airways of BGA-sensitized cats are not significantly altered during the development of aeroallergen-induced feline asthma.

ABSTRACT #347

SERUM NT PRO-CNP CONCENTRATION DIFFERENTIATES SEPSIS FROM SIRS IN DOGS. A Bigio, AE DeClue, K Osterbur, C Sharp. University of Missouri, College of Veterinary Medicine.

Rapidly differentiating sepsis from non-infectious causes of systemic inflammatory response syndrome (SIRS) in dogs is challenging. C-type natriuretic peptide (CNP) is expressed by the vascular endothelium in response to stimuli including lipopolysaccharide and is significantly increased in humans with sepsis. We investigated if the N-terminal fragment of CNP (NT-pCNP), a more stable compound with greater potential for clinical utility, could serve as a biomarker for sepsis in dogs.

Dogs that presented to the MU VMTH were enrolled in the sepsis group ($n = 11$) if they fulfilled $\geq 2/4$ SIRS criteria (temperature ≤ 100 or ≥ 103 F; heart rate ≥ 160 bpm; respiratory rate ≥ 40 bpm; WBC $\geq 12{,}000$, $\leq 4{,}000$ cells/uL or $\geq 10\%$ bands) and had confirmed infection. Dogs in the SIRS group ($n = 17$) fulfilled $\geq 2/4$ SIRS criteria without evidence of infection. Healthy dogs served as controls ($n = 49$). Serum was obtained at presentation and frozen at -80°C. The NT-pCNP assay was performed by a commercial laboratory (Veterinary Diagnostics Institute). Data were compared using an ANOVA on ranks and post-hoc Dunn's MCP with $p < 0.05$ considered significant.

Dogs in the sepsis group had significantly greater serum NT-pCNP concentrations (median, Q1, Q3; 31.2, 10.5, 34.6 pmol/l; $p < 0.001$) than dogs in the SIRS group (4.7, 2.9, 8.1 pmol/l or healthy controls (6.4, 4.8, 8.8 pmol/l). Using a cut off point of 10.2 pmol/l, serum NT-pCNP had a good sensitivity (82%) and specificity (94%) for differentiating sepsis from non-infectious SIRS in dogs.

Serum NT-pCNP appears to be a useful diagnostic test to differentiate sepsis from non-infectious SIRS in dogs.

ABSTRACT #348

DEVELOPMENT OF A MULTIVARIABLE PROGNOSTIC MODEL FOR CRITICALLY ILL DOGS: PRELIMINARY RESULTS. FG Ponce[1], RC Jorge[1], ETS Martins[2], TB Gimenes[1], MCAD Valle[1], R Duarte[1]. [1]Hospital Veterinário Pompeia, São Paulo, SP, Brazil. [2]Universidade Vila Velha, Vila Velha, ES, Brazil.

The objective of this study was to determine the probability of death with different combinations of common point-of-care tests in a heterogeneous population of critically ill dogs and construct a prediction model from multiple variables.

Data were collected from 84 dogs admitted to a referral intensive care unit. Dogs were of various breeds. Forty-four dogs were females and 41 were males with ages varying from 2-months to 13-years old (mean age: 6 years). The most common diagnostic categories were sepsis ($n = 20$); metabolic diseases ($n = 15$), multiple trauma ($n = 12$); neurologic diseases ($n = 10$); and oncologic emergencies ($n = 7$). Thirty-one dogs died or were euthanized during hospitalization (mortality: 37%).

Venous blood pH, partial pressure of CO_2 (PCO_2), electrolytes (sodium, potassium and chloride), blood urea nitrogen (BUN), blood glucose and lactate were measured on admission, using point-of-care instruments (iStat, Abbott Point of Care, East Windsor, NJ and Accutrend Plus, Roche, Mannheim, Germany). The values of plasma bicarbonate, base excess and anion gap were calculated automatically. All variables were included as candidate predictors for inclusion in the model.

A multivariable logistic regression model was developed to predict the risk of death. We selected the predictors in the model using the backward procedure. Initially, all the candidate predictors selected were included. A sequence of hypothesis tests was applied to determine whether a given predictor should be removed from the model. The candidate predictors were selected into the model if $P < 0.05$ and removed from the model if $P > 0.10$.

The main prognostic predictors of death were BUN ($P = 0.09$), pH ($P = 0.06$) and PCO_2 ($P = 0.01$). Risk score from a logistic regression model to predict the risk of death (the adjusted model) can be described as: (risk score) $= 38.4430 + 0.0134 \times \text{BUN} - 4.9311 \times \text{pH} - 0.0930 \times PCO_2$.

Point-of-care testing is frequently requested as part of the evaluation of critical care patients. These tests are frequently used to improve decision making and to guide therapy. These preliminary findings suggest that blood pH and BUN are the most important predictors of death in a heterogeneous population of critically ill dogs. Future steps include validation of the model in a larger sample and applying the model in samples from other facilities.

ABSTRACT #349

EVALUATION OF A CANINE C-REACTIVE PROTEIN (CRP) POINT-OF-CARE TEST. H Plickert[1], B Kohn[1], L Brunnberg[1], R Einspanier[2]. [1]Small Animal Clinic, [2]Institute of Veterinary Biochemistry, Faculty of Veterinary Medicine, Freie Universität Berlin, Germany.

The objective of this study was to evaluate a CRP point-of-care test (Lateralflow Immunoassay, TECOmedical AG, Switzerland) for qualitative CRP measurements in healthy and diseased dogs.

Blood samples of 69 client-owned dogs were available for testing: 16 healthy dogs and 53 dogs suffering from different infectious, inflammatory or neoplastic diseases. CRP was measured in heparinized whole blood samples and serum with the point-of-care test; an ELISA (Tridelta Development Ltd, Kildare, UK) validated for canine samples was used as a reference method. CRP reference values for healthy dogs range from 0.8–5.4 µg/ml; the cut-off for the point-of-care test is 5 µg/ml, visible by a color change.

The serum CRP values of all dogs measured with the ELISA ranged from 0.3 to ≥ 350 μg/ml (median 39.2) (table). The sensitivity and specificity of the point-of-care test compared to the reference method using whole blood was 96% and 100%, respectively, and 100% and 94%, respectively, using serum samples.

In summary the point-of-care test was capable of discriminating between normal and increased CRP concentrations with a high sensitivity and specificity and may therefore be introduced as an inflammation screening test.

Table [Abstract 349]: Results of qualitative CRP measurements.

	CRP test		ELISA		Sens.	Spec.
	pos	neg	pos	neg	(%)	(%)
Healthy n = 16						
Blood	2	14			100	100
Serum	4	12	2 0.4–10.2 μg/ml median 1.55 μg/ml	14	100	86
Sick n = 53						
Blood	48	5			98	100
Serum	49	4	49 0.3– ≥ 350 μg/ml median 67.5 μg/ml	4	100	100

ABSTRACT #350

MONITORING HEMOSTASIS IN DOGS USING THROMBOELASTOGRAPHY (TEG): STABILITY OF TEG VALUES MEASURED IN HEALTHY DOGS OVER THREE CONSECUTIVE DAYS.

Elizabeth Spangler, Emily Galliers. College of Veterinary Medicine, Auburn University, AL.

Thromboelastography (TEG) is a method for evaluation of coagulation that detects viscoelastic changes in whole blood during polymerization of fibrin, and can be used to identify both hypocoagulable and hypercoagulable states in dogs. Application of TEG for serial monitoring of clinical patients assumes that changes in TEG values reflect a change in patient status. The goal of this project was to evaluate the validity of this assumption by examining whether or not TEG produces consistent results when used to monitor hemostasis in clinically healthy dogs over a period of three consecutive days. Additionally, we evaluated the impact of elapsed time after venipuncture (15 minutes vs. 1 hour) on TEG results and also compared results obtained with and without use of kaolin to activate coagulation for TEG analysis. Blood was collected from 10 healthy dogs on three consecutive days. Complete blood cell counts were performed daily; Hct% and WBC counts were within the reference interval in all samples. Thrombocytopenia was reported in four samples, and evaluation of a blood smear revealed platelet clumping in each case. The platelet count estimated from the smears was adequate. A coagulation profile, including measurement of the prothrombin time (PT), activated partial thromboplastin time (aPTT), fibrinogen concentration and antithrombin activity (AT), was completed on the first day for each dog and all values were within reference intervals.

TEG was performed using citrated whole blood with addition of calcium to initiate clot formation. Parameters that were evaluated included reaction time (R), clotting time (k), rate of clot formation (angle), and maximum clot strength (MA). Results were analyzed using one-way ANOVA for repeated measures and a t-test or Mann-Whitney test for group comparisons. No statistically significant differences were found for TEG parameters of individual dogs measured on three consecutive days, and all values fell within established reference intervals. Comparison of TEG performed at 15 minutes vs. 1 hour post-venipuncture, and comparison of TEG performed with and without use of kaolin also showed no significant differences.

Thromboelastography is a reliable method for monitoring hemostasis in dogs over a period of three days and clinically significant changes in the measured TEG values are likely due to changes in patient status. Storage of citrated whole blood at room temperature for up to one hour does not result in a significant change in TEG results for healthy dogs, and use of an activator such as kaolin may not be needed. Further experiments should be performed to assess the impact of time after venipuncture and use of a coagulation activator on TEG in patients with abnormal hemostasis.

Previously presented at the Merck-Merial NIH National Veterinary Scholars Symposium, Raleigh, NC, August 2009.

ABSTRACT #351

SURVEY OF CARDIAC ARRHYTHMIAS AND MYOCARDIAL INJURY IN STANDARDBRED RACEHORSES.

J Slack, L Soma, R Boston, VB Reef. University of Pennsylvania, New Bolton Center, Department of Clinical Studies, Kennett Square, PA.

Cardiac arrhythmias are a recognized but poorly characterized problem in Standardbred racehorses. This study was designed to examine the prevalence, frequency and risk factors for cardiac arrhythmias and to look for evidence of myocardial injury evidenced by elevated cTnI in racing Standardbreds.

Electrocardiographic recordings (ECGs) were obtained over a 12 week period from Standardbred horses racing at a single racetrack. ECGs were obtained on admission to the racetrack and within 20 minutes of completing the race using a clipless digital recording device that obtains a non-standard single lead ECG. Blood was obtained by random selection for venous blood gas/electrolyte (n = 727 race entries) and cTnI determination (n = 754 race entries) prior to racing. Logistic regression analysis was used to determine whether age, gender, gait, furosemide administration, ambient temperature, race classification, lifetime mark, venous blood gas parameters, or electrolyte concentration was associated with tachyarrhythmias.

A total of 8658 ECG recordings were obtained from 463 trotters and 1,398 pacers entered in 4335 racestarts. Six horses developed atrial fibrillation post race (prevalence = 0.32%, frequency = 0.14%), one horse had atrial tachycardia prior to racing (prevalence = 0.05%, frequency = 0.02%), and four horses had significant ventricular arrhythmias either pre or post race (leading to disqualification and/or suspension; prevalence = 0.21%, frequency = 0.09%). Only 3 horses had prerace cTnI values above the reference range. There was no association between the presence of tachyarrhythmias and any of the parameters evaluated. Lack of association may be due to the small number of horses with arrhythmias.

ABSTRACT #352

CHANGES IN SERUM CARDIAC TROPONIN I CONCENTRATION AFTER SEMEN COLLECTION IN STALLIONS.

A Viljoen[1], MN Saulez[1], A Carstens[2], M Schulman[3], PN Thompson[4].
[1]Section of Equine Medicine, [2]Section of Diagnostic Imaging, Department of Companion Animal Clinical Studies, [3]Section of Reproduction, [4]Section of Epidemiology, Department of Production Animal Studies, Faculty of Veterinary Science, University of Pretoria, South Africa.

Increased concentration of cardiac troponin I [cTnI] is associated with piroplasmosis, ventricular tachycardia and short-term high intensity exercise in horses. Anecdotal reports exist of acute death in stallions during and immediately after breeding and this may be due to myocardial injury. We hypothesized that [cTnI] would increase following semen collection.

Fourteen clinically-healthy, phantom-trained stallions were studied. Echocardiography was performed prior to semen collection and heart rate recorded prior to, during and after semen collection using cardiac telemetry. Fractional shortening (FS) was determined within 5 min following semen collection. [cTnI] was measured using a chemiluminescent immunoassay (Access AccuTI®) pre-semen collection and at 4, 6, 12 and 24 hrs post-semen collection. After log-transformation of data, changes in [cTnI] post- *vs* pre-semen collection were analyzed using a paired *t*-test. Linear regression analysis was used to determine the effect of predictors (heart rate, age, weight and FS) on change in [cTnI]. A P-value < 0.05 was considered significant.

[cTnI] was available for eleven stallions (mean ± SD age 11 ± 7 yrs and weight 510 ± 86 kg). Compared with pre-semen collection

[cTnI] (0.023 ± 0.045 ng/ml), mean [cTnI] was significantly increased at 4 (0.041 ± 0.062 ng/ml; P = 0.018) and 6 hrs (0.054 ± 0.070 ng/ml; P = 0.034), but not at 12 (0.031 ± 0.045 ng/ml; P = 0.286) and 24 hrs (0.028 ± 0.047 ng/ml; P = 0.581) post-semen collection in eight stallions. No significant association was found between the measured predictors and increased [cTnI] at 4 and 6 hrs post-semen collection.

Data suggest that [cTnI] may increase after semen collection. This may not be of clinical significance, but should be considered in stallions with suspected cardiac disease.

ABSTRACT #353

SEVERE HEMORRHAGE AND MYOCARDIAL INJURY IN HORSES.

Cristobal Navas de Solis, Barbara Dallap Schaer, JoAnn Slack. New Bolton Center, University of Pennsylvania, Kennett Square, PA.

Severe hemorrhage causes a decrease in circulating volume and insufficient oxygen delivery to the heart. The clinically relevant effects of this are poorly described in the horse. This study was designed to examine the occurrence of myocardial injury and arrhythmias in horses with either severe or mild controlled hemorrhage and determine if increased cTnI concentrations are associated with arrhythmias or outcome.

Eight adult horses with severe hemorrhage were studied prospectively. Serial cTnI concentrations and continuous electrocardiograms were obtained on admission and during the first 48–96 hours of hospitalization. Four blood donors (mild controlled hemorrhage) were studied in parallel. Rank sum measures for nonparametric data were used to determine associations between cTnI concentration, arrhythmias and outcome. Cardiac troponin I concentration was stratified into a scoring system.

All horses with severe hemorrhage had increased cTnI concentrations (median 0.85 ng/ml, range 0.11–29.90 ng/ml; reference range < 0.07 ng/ml); Six developed significant arrhythmias (multifocal ventricular ectopy and or supraventricular premature depolarizations) that warranted therapy or monitoring and three were euthanized. All blood donors had normal cTnI concentrations and electrocardiograms. There was an association between severe hemorrhage and increased cTnI concentration ($p < 0.001$), and between severe hemorrhage and presence of arrhythmias ($p = 0.019$). There was no association between survival and maximal cTnI, cTnI category or presence of arrhythmias ($p \geq 0.055$).

All horses with severe hemorrhage had myocardial injury and clinically significant arrhythmias were identified frequently. Horses with severe hemorrhage may benefit from cardiac monitoring. A larger population is needed to quantify risks or suggest specific monitoring and therapeutic interventions.

ABSTRACT #354

SEASONAL INFLUENCE ON INSULIN AND CORTISOL RESULTS FROM OVERNIGHT DEXAMETHASONE SUPPRESSION TESTS (DST) IN NORMAL AND PREVIOUSLY LAMINITIC PONIES.

KE Borer[1], NJ Menzies-Gow[1], Y Berhane[1], SR Bailey[2], PA Harris[3], J Elliott[1]. [1]Royal Veterinary College, London, UK; [2]University of Melbourne, Australia; [3]WALTHAM Centre for Pet Nutrition, Leics., UK.

DSTs are used to screen ponies for PPID and abnormalities in control of insulin secretion and may aid identification of ponies susceptible to laminitis. Cortisol responses are affected by season, being higher in summer/autumn. The aim of this study was to measure the cortisol and insulin responses to a standard DST in non-laminitis prone (NP) and previously laminitic ponies (PLP) at four different times of year in the UK.

Standard 19 hour overnight DSTs were performed on the same group of 6 NP and 6 PLP (although not laminitic at the time of testing) in spring (April), summer (July), autumn (October) and winter (December) in southern England, whilst the ponies were maintained out at pasture. None of the ponies were obese or showed any clinical signs of PPID. Serum cortisol analysis was by chemiluminescent assay at a commercial laboratory (Beaufort Cottage Laboratory, UK). Insulin was analysed by radioimmunoassay (Coat-A-Count, Siemens) previously validated in the authors' laboratory. Results were analysed using 1-way and 2-way ANOVAs and are presented as mean±SD. Significance was set at $p < 0.05$.

No suppression of [cortisol] occurred in PLP in autumn. In summer, mean post-injection [cortisol] was above the reference range (< 25 nmol/l) in both groups. Individual ponies (both NP and PLP) did not suppress [cortisol] normally in winter, however all individuals had normal cortisol suppression in spring. There were no significant differences in basal [insulin] between NP and PLP at any season. In NP, serum [insulin] post-dexamethasone was significantly higher in spring and summer compared to winter. Serum [insulin] post-dexamethasone in PLP in spring and summer (966 ± 488 µIU/ml and 844 ± 820 µIU/ml respectively) was significantly higher than PLP in winter (26 ± 16 µIU/ml) although not significantly higher than PLP in autumn (211 ± 143 µIU/ml). Serum [insulin] post-dexamethasone in PLP in spring and summer was significantly ($p < 0.001$) higher than NP at all seasons. The increase in serum [insulin] in response to dexamethasone was significantly greater in PLP vs. NP in spring (726 ± 422 µIU/ml vs. 147 ± 183 µIU/ml respectively; $p < 0.01$) and summer (678 ± 651 µIU/ml vs. 184 ± 193 µIU/ml respectively; $p < 0.05$) although the fold change in [insulin] was not significantly different between groups at any point. Serum [insulin] in winter in both groups was low and responded minimally to dexamethasone injection (NP pre:10 ± 5 µIU/ml, post:28 ± 16 µIU/ml; PLP pre:13 ± 9 µIU/ml, post:26 ± 16 µIU/ml).

Abnormal cortisol responses were seen in both NP and PLP in summer, autumn & winter. All ponies showed normal cortisol suppression in spring. PLP displayed exaggerated serum insulin responses to dexamethasone in spring and summer, which were significantly different from the same ponies in winter and from NP at all times of year. In winter there was a minimal change in [insulin] post-dexamethasone in both groups. In autumn and winter, the insulin responses to dexamethasone could not distinguish between NP and PLP.

Previously presented at the AAEP Laminitis Research Workshop, November 2009.

ABSTRACT #355

DETERMINATION OF SERUM INSULIN AND INSULIN RESISTANCE IN CLINICALLY NORMAL DONKEYS AND DONKEYS WITH A HISTORY OF LAMINITIS (OBESE AND NON-OBESE).

N du Toit, AF Trawford. The Donkey Sanctuary, Sidmouth, UK.

Insulin resistance has been implicated as a predisposing factor in horses that suffer from recurrent laminitis. A similar metabolic syndrome in donkeys with recurrent laminitis is proposed. Three hundred and ninety four new donkey arrivals at The Donkey Sanctuary, UK were evaluated over a 12 month period (October 2008–November 2009). Age, sex, body condition score (BCS out of 5), whether they were obese (BCS ≥ 3.5), history of previous laminitis and the occurrence of laminitic bouts within one month of sampling were recorded. Glycaemic status was assessed by means of resting glucose and insulin analysis and the derivation of surrogate (proxy) values. General linear models were performed to determine significant associations. There were 114 (28%) females and 280 (72%) geldings; 75 (19%) had a history of previous laminitis, 43 (11%) had a laminitic episode within one month of sampling and 122 (31%) were classed as obese. There was a statistically significant difference in mean serum insulin(µIU/ml), RISQI (reciprocal of the square root of insulin; $[mU/L]^{-0.5}$) and MIRG (modified insulin-to-glucose; $[mU_{ins}]^2/10.L.mg_{gluc}$ [Insulin < 50 µIU/ml]) values between donkeys classified as obese(20.86) and non-obese(7.31; $P < 0.001$), previously laminitic (23.83) and not (8.52; $P < 0.001$), and post-sampling laminitic (22.84) and not (10.08; $P < 0.001$). There was also a significant difference in insulin and MIRG values between male (9.15 and 4.28) and female donkeys (17.2 and 5.13; $P \leq 0.014$) with higher values in females, due to more female donkeys being obese and having a history of laminitis ($P < 0.001$). This study has determined mean insulin values (4.95) and associated insulin sensitivity proxy measurements (RISQ 0.6; MIRG 3.79) in non-obese donkeys with no history of previous laminitis in the UK, and has also shown that there is an association between insulin resistance/metabolic disease and laminitis observed in donkeys.

ABSTRACT #356

USE OF AN ORAL SUGAR TEST TO ASSESS INSULIN SENSITIVITY IN HEALTHY AND INSULIN-RESISTANT HORSES. Schuver A, N Frank, K Chameroy, SB Elliott. Department of Large Animal Clinical Sciences, College of Veterinary Medicine, University of Tennessee, Knoxville, TN.

Insulin resistance (IR) is associated with obesity and laminitis in equids, so diagnostic tests are required to identify affected animals. An intravenous glucose tolerance test (IVGTT) can be used to diagnose IR, but oral tests are easier to perform and may be more applicable to field conditions. Dextrose powder can be used for this purpose, but requires mixing into solution and administration via nasogastric tube. We therefore selected Karo Original Light Corn Syrup (10g sugar/30 mL syrup) for this study because of its availability and palatability. We tested the hypothesis that an oral sugar test (OST) can be used to diagnose IR in horses. Ten horses with IR and 8 healthy horses were included in the study and each animal underwent an IVGTT (50 mg/kg dextrose) and OST (50 mg/kg sugar; 15 mL per 100 kg) under fasting conditions on consecutive days in random order. Dextrose was administered intravenously as a 50% solution and syrup was administered via dose syringe. Blood samples were collected at 15 min intervals for 60 min during the IVGTT and every 30 min for 120 min during the OST.

Mean area under the curve values from the IVGTT and OST were 27% and 28% higher respectively for glucose, and 10-fold and 6-fold higher respectively for insulin when insulin-resistant horses were compared with healthy control animals. Insulin-resistant horses were identified with the OST, so this test may be used by collecting a single blood sample 60 to 90 minutes after syrup is administered by the owner.

ABSTRACT #357

EFFECTS OF LEVOTHYROXINE SODIUM ON BODY CONDITION, BLOOD MEASURES OF METABOLIC STATUS, AND GLUCOSE DYNAMICS IN HORSES WITH EQUINE METABOLIC SYNDROME (EMS). K Chameroy, N Frank, SB Elliott. Department of Large Animal Clinical Sciences, College of Veterinary Medicine, University of Tennessee, Knoxville, TN.

Obesity, regional adiposity, insulin resistance (IR), and laminitis are components of Equine Metabolic Syndrome (EMS). Previous work has shown that levothyroxine sodium induces weight loss and improves insulin sensitivity in healthy horses kept on a weight reduction diet. The purpose of this study was therefore to assess the efficacy of this treatment in horses with EMS kept on a maintenance diet. It was hypothesized that levothyroxine sodium would significantly alter physical characteristics, glucose dynamics, and markers of metabolic status in horses with EMS. Ten horses with obesity, IR, and a history of laminitis were evaluated over a 9-month period. Seven horses received levothyroxine sodium (Thyro L) orally at a dosage of 48 mg/day (6 months) and then 72 mg/day (3 months) once daily in 0.25 kg oats, while 3 other horses received oats alone. Body weights, morphometric measurements, and frequently-sampled intravenous glucose tolerance tests were performed at 0, 3, 6, and 9 months.

Analysis of variance analysis revealed significant time × treatment effects for within-horse change in body weight ($P = 0.049$) and mean neck circumference ($P = 0.027$), but no significant change in BCS ($P = 0.350$). Mean body weight decreased by 30.8 ± 14.0 kg over 9 months in treated horses and by 6.5 ± 24.6 kg in the three control animals. Treatment did not affect area under the glucose curve values (time × treatment; $P = 0.511$). Insulin results are pending. Levothyroxine induces weight loss and reduces neck circumference in horses with EMS, even when animals are kept on a weight maintenance diet.

ABSTRACT #358

VALIDATION OF AN ENZYME-LINKED IMMUNOSORBENT ASSAY (ELISA) FOR MEASUREMENT OF ACTIVE GLUCAGON-LIKE PEPTIDE 1 (GLP-1) IN EQUINE PLASMA. K Chameroy, N Frank, A Schuver. Department of Large Animal Clinical Sciences, College of Veterinary Medicine, University of Tennessee, Knoxville, TN.

Glucagon-Like Peptide 1 (GLP-1) is an incretin hormone secreted by the small intestine that plays a role in satiety and stimulates insulin release from pancreatic β-cells. The objective of this study was to validate a commercially available (Millipore Corporation) human active GLP-1 enzyme-linked immunosorbent assay (ELISA) for use with equine plasma. Resting blood samples were collected from one insulin-resistant horse and one healthy horse. Plasma was collected with EDTA and treated with dipeptidyl-peptidase IV inhibitor. Dilution, linearity, and lowest limit of detection were assessed using sample buffer included in the kit and plasma from the insulin-resistant horse. For spike and recovery, the high standard was diluted in equine plasma from one healthy horse. Intra-assay variability was calculated from quadruplicate measures of GLP-1 in blood collected at 3 time points during oral sugar tests in 3 healthy horses. Inter-assay variability was assessed from concentrations of quality control solutions provided in the kit on 3 plates.

Dilution of equine plasma resulted in a linear plot of GLP-1 concentrations ($r^2 = 0.995$). The lowest limit of detection for the assay was 1.7 pmol/L because this was the value obtained for the sample buffer. Overall recovery was 98.3% for active GLP-1 in equine plasma. Inter-assay coefficient of variation (CV%) values of 2.3% and 2.6% were obtained for two the quality control solutions. Mean ± SD intra-assay CV% was 9.5 ± 5.1%. Results indicate that GLP-1 can be measured in equine plasma using the ELISA evaluated in this study.

ABSTRACT #359

PLASMA GLUCAGON-LIKE PEPTIDE 1 (GLP-1) CONCENTRATIONS IN RESPONSE TO AN ORAL SUGAR TEST IN HEALTHY AND INSULIN-RESISTANT HORSES. K Chameroy, N Frank, A Schuver. Department of Large Animal Clinical Sciences, College of Veterinary Medicine, University of Tennessee, Knoxville, TN.

Glucagon-Like Peptide 1 (GLP-1) is an incretin hormone secreted by the small intestine in response to feeding. Studies in other species indicate that obesity and insulin resistance are associated with alterations in GLP-1 concentrations after feeding. An enzyme-linked immunosorbent assay for human active GLP-1 has recently been validated for use with equine plasma in our laboratory. We therefore tested the hypothesis that healthy and insulin-resistant horses differ in their GLP-1 response to an oral sugar test (OST). Eleven healthy horses and six insulin-resistant horses underwent an OST after overnight fasting. After baseline blood samples were collected, 50 mg sugar per kg body weight was administered as Karo Original Light Corn Syrup (10 g sugar/30 mL syrup) via dose syringe. Blood samples were then collected at 15, 30, 45, 60, 75, 90, 105, and 120 min. Active GLP-1 was measured in EDTA plasma treated with dipeptidyl-peptidase IV inhibitor.

Mean GLP-1 concentrations significantly increased ($P < 0.001$) during the OST, but individual responses varied widely. Median (range) maximum percent increase in GLP-1 concentration after syrup administration was 67% (-15% to 188%) in healthy horses and 148% (35% to 286%) in insulin-resistant horses. Metabolic status did not affect GLP-1 concentrations ($P = 0.623$), percent change ($P = 0.167$), fold change ($P = 0.161$), or area under the curve in response to sugar administration. Healthy and insulin-resistant horses did not differ with respect to GLP-1 responses to Karo syrup in this study, but specific nutrients should be tested in the future.

ABSTRACT #360

ASSESSMENT OF INTRAMYOFIBER LIPID CONTENT IN HORSES. NE Schultz, SJ Valberg, M Lucio. University of Minnesota, College of Veterinary Medicine, St. Paul, MN.

Lipid storage myopathies (LSMs) are common causes of rhabdomyolysis and weakness in humans and are characterized by accumulation of lipid droplets predominantly in high oxidative muscle fibers. The normal range of lipid storage in muscle biopsies from horses is not well characterized and excessive myofiber lipid storage has only been reported in a small number of horses. The first objective of this study was to document the normal range of myofiber lipid storage and the second objective was to screen horses with myopathies of unknown origin for potential LSMs.

Objective 1: Frozen sections of gluteal biopsies from 65 Quarter (QH) /Paint Horses with no history of neuromuscular disease and 7 thoroughbreds on a high fat diet were stained with Oil Red O for lipid and scored (0 to 3) for 1) number of fibers containing lipid droplets, 2) density of lipid droplets within the fibers and 3) size of the lipid droplets. The sum of the biopsy score (0–9) was calculated for each horse. Ninety four % of control horses (61/65) had a lipid score < 6.0 and 6.1% (4/65) had a maximum score of 6.0. Thoroughbred scores ranged from 3.5–5.5 with a mean of 4.7. Results suggested a score of 6.0 or greater was compatible with high myofiber lipid storage.

Objective 2: From Neuromuscular Diagnostic Laboratory records at the University of Minnesota horses with exertional and non-exertional rhabdomyolysis were identified which were negative for glycogen synthase 1 and ryanodine receptor mutations. In horses with exertional rhabdomyolysis, 2/18 (11%) thoroughbreds, 5/63 (7.9%) QHs with normal PAS stains and 6/80 (7.5%) QHs diagnosed with type 2 PSSM had a score of 6.0 or greater. The mean score for thoroughbreds was 2.5 ± 0.6, for QH 1.1 ± 0.3 and for QH with type 2 PSSM 1.0 ± 0.2. In Quarter Horse foals with a history of non-exertional rhabdomyolysis (NER) 13/57 (22.8%) had a score of 6.0 or greater [2.6 ± 0.4]. Among horses of various breeds experiencing NER, 4/17 (24%) had a score of 6.0 or greater [3.8 ± 0.7] and of these 3 had a history and distribution of myonecrosis compatible with a diagnosis of atypical myopathy. Urine organic acid and plasma acylcarnitine profiles obtained from two Midwestern horses suspected to have atypical myopathy revealed a pattern of marked elevations in plasma acylcarnitines and urine organic acids consistent with multiple acyl-CoA dehydrogenase deficiency (MADD), previously reported in European cases of atypical myopathy.

Results indicate that in addition to an acquired LSM caused by MADD, potential genetic causes of LSM may also exist in horses. Oil Red O stains on fresh frozen muscle biopsy sections, urine organic acid and plasma acylcarnitine profiles during an episode of rhabdomyolysis and measurement of plasma and muscle carnitine concentrations may become valuable tools to screen for defects in lipid metabolism.

ABSTRACT #361

VARIABILITY OF NUTRIENT COMPOSITION OF GRASS HAY FORAGE FED TO HORSES IN NEW ENGLAND.

L.R.R. Costa, P. Hermida, M.R. Paradis. Department of Clinical Sciences, Cummings School of Veterinary Medicine, Tufts University, North Grafton, MA.

Horses have limited ability to utilize poor-quality forages because fermentation by microflora occurs in the hindgut. It is often assumed that grass hay fed to horses provides 0.9 Mcal/lb and contains 10 to 12% protein on a DM basis. Other variables are important in hay quality and adequacy. This study reports a comparison of nutrient composition of 50 samples of grass hay consumed by horses in New England.

The hay samples were collected during farm visits to 36 premises between July through November 2009. The premises consisted of 16 training/boarding facilities, 20 small farms/residences. The number of horses per premise ranged from 1 to 50 (median 6). All hay samples were analyzed by Equi-Analytical Laboratories, Ithaca, NY. The results are reported as dry matter basis for each component.

Only 52% (26/50) of the hay sampled had acid detergent fiber (ADF) ≤ 35%, and considered of good quality. Whereas, 6% of the hay sampled had ADF > 45% (poor quality). The digestible energy (DE) ranged from 0.84 to 1.12 Mcal/lb, mean 0.96. The crude protein ranged from 4.6 to 18.4%; 20/50 had CP < 10% and 22/50 had CP > 12%. Nonfiber carbohydrates (NFC) ranged from 11.6 to 26.5% (mean 17.95%). Water soluble carbohydrates (WSC) ranged from 7.3 to 23.4% (mean 11.36%).

The results of our report demonstrate considerable variability in the nutritional composition of the grass hay, highlighting the importance of hay analysis in order to provide adequate nutrition to horses.

ABSTRACT #362

CHARACTERIZATION OF ADIPONECTIN PROTEIN COMPLEXES IN THE HORSE.

AA Wooldridge[1,3], R Applegate[1,3], Q Zhong[1,3], EP Plaisance[2,3], JS Taintor[1], RL Judd[2,3]. [1]Departments of Clinical Sciences and [2]Anatomy, Physiology and Pharmacology; Boshell Diabetes and [3]Metabolic Disease Research Program, Auburn University College of Veterinary Medicine Auburn, AL.

Adipokines are essential regulators of insulin and glucose. Adiponectin is the only adipokine reduced with obesity and is an insulin sensitizer. In species studied, adiponectin circulates as high molecular weight (HMW, 360–540 kd), low molecular weight (LMW), and trimeric/dimeric/monomeric complexes. HMW adiponectin is the metabolically active complex and is an indicator of insulin sensitivity. Only total adiponectin has been measured in the horse. The purpose of this study was to identify and characterize adiponectin protein complexes in the normal horse.

Sera were collected from lean (body condition score 4–6), healthy geldings (n = 19) and mares (n = 15), aged 2–15 years. Western blot analysis under denaturing and/or reducing conditions demonstrated complexes ≤ 180 kd. An enzyme linked immunosorbant assay (ELISA) was validated and used to quantify HMW complexes. HMW values were analyzed for normality and compared between mares and geldings using a 2-tailed unpaired t-test with $p < 0.05$.

In equine serum, adiponectin was present as a multimer consisting of LMW (180 kd) and trimeric (90 kd) complexes under denaturing conditions and dimeric (60 kd) and monomeric (30 kd) complexes under denaturing and reducing conditions. Mean HMW concentration was 8044 ± 1508 ng/mL in mares and 6597 ± 778 ng/mL in geldings. The mean for all was 7342 ± 744 ng/mL. There was no difference in HMW adiponectin levels between mares and geldings.

Adiponectin multimers are present in normal horses, with HMW adiponectin in high concentrations. A greater understanding of the role of adiponectin in equine metabolism will provide insight into the pathophysiology of diseases associated with metabolic derangements.

ABSTRACT #363

HIGH MOLECULAR WEIGHT ADIPONECTIN IS REDUCED IN HORSES WITH OBESITY AND INFLAMMATORY DISEASE.

AA Wooldridge[1,3]; DR Taylor[1]; Q Zhong[1,3], R Applegate[1,3]; EP Plaisance[2,3]; RL Judd[2,3]. [1]Departments of Clinical Sciences and [2]Anatomy, Physiology and Pharmacology; [3]Boshell Diabetes and Metabolic Disease Research Program, Auburn University College of Veterinary Medicine Auburn, AL.

Adipokines have essential roles in metabolic regulation and in the inflammatory response. Adiponectin is an adipocyte-derived hormone with insulin sensitizing, anti-inflammatory, and vascular protective properties. The metabolically active form, high molecular weight (HMW), is low in humans with obesity, insulin resistance, and inflammation. We hypothesized that HMW adiponectin levels would be reduced in horses with obesity or inflammatory conditions.

Serum HMW adiponectin was measured (using a previously validated ELISA) in normal horses and compared to horses with clinical signs and/or laboratory findings consistent with inflammatory disease or obesity. All horses were between 2 and 15 years old. Inflammatory disease was scored using an increasing severity scale of 1–3. Natural log transformed data were compared between groups (normal, obese, all inflammatory, and scored groups (1, 2, 3)) using one-way analysis of variance (ANOVA) and Dunnett's multiple comparison test, $p < 0.05$.

HMW adiponectin concentrations were as follows: normal (n = 37): 7342 ± 744 ng/mL; obese (n = 15): 3595 ± 1108 ng/mL; score1 (n = 29): 5409 ± 706 ng/mL; score2 (n = 46): 4334 ± 423 ng/mL; score3 (n = 9): 6859 ± 1299 ng/mL. Horses with inflammatory disease (all scores) had lower HMW adiponectin than normal horses. Obese horses and horses with mild (score1) and moderate (score2) inflammation had significantly lower HMW adiponectin than normal horses.

HMW adiponectin is reduced in horses with obesity and mild to moderate inflammatory disease. Because of its effects on the immune response and vascular function, adiponectin may have a role in the pathophysiology of complications of obesity and inflammatory disease.

ABSTRACT #364

CHARACTERIZATION OF ADIPOSE TISSUE MACROPHAGE INFILTRATION IN INSULIN-RESISTANT AND INSULIN-SENSITIVE LIGHT BREED HORSES. Burns TA[1], Geor RJ[2], Mudge MC[1], McCutcheon LJ[2], Belknap JK[1]. [1]The Ohio State University, Columbus, OH. [2]Michigan State University, East Lansing, MI.

Insulin resistance has been associated with predisposition to laminitis in horses and ponies and frequently accompanies nutritional obesity in this species. In humans and rodents, macrophage infiltration of adipose tissue (AT) correlates with adiposity and degree of tissue and systemic insulin resistance. These cells adopt an inflammatory phenotype, elaborate pro-inflammatory cytokines, and perpetuate adipose tissue IR in obesity. Recent reports have established depot-specific pro-inflammatory cytokine expression profiles in equine adipose tissue, but the relationship of this expression to macrophage infiltration is unknown. This study quantified the macrophage content of several AT depots of insulin-resistant (IR) and insulin-sensitive (IS) horses. Eleven mares (8 Quarter horses, 2 Thoroughbreds, and 1 Standardbred), categorized as IR (insulin sensitivity [SI] = $0.58 \pm 0.31\,L \cdot min^{-1} \cdot mU^{-1}$; n = 5) or IS (SI = $2.59 \pm 1.21\,L \cdot min^{-1} \cdot mU^{-1}$; n = 6) based on results of an insulin-modified frequently-sampled intravenous glucose tolerance test were studied. Omental, retroperitoneal, and mesocolonic AT was collected via ventral midline celiotomy; nuchal ligament and tail head AT biopsies were collected via skin incision. Tissue samples were snap-frozen and fixed in formalin. For each depot, total RNA was extracted and cDNA analyzed via real-time quantitative PCR to quantify expression of monocyte chemoattractant protein-2 (MCP-2), and tissue samples were immunohistochemically stained for CD163 (macrophage surface antigen). The number of CD163 (+) cells in n = 10 40× fields per slide was recorded for each depot. Data were analyzed using one-way ANOVA for repeated measures or non-parametric equivalent ($p = 0.05$). The expression of MCP-2 was significantly higher in omental AT compared with other depots ($p = 0.0003$), and omental AT of IR horses showed greater expression than that of IS horses ($p < 0.0001$). The total number of CD163 (+) cells was not significantly different between depots ($p = 0.06$), and no effect of insulin sensitivity was observed. CD163 (+) cells were further differentiated based on their location within the tissue (interstitial vs. perivascular). There was no difference in the number of interstitial cells between depots ($p = 0.48$); however, omental adipose tissue contained a significantly greater number of perivascular cells than other depots ($p = 0.04$). These data suggest a role for MCP-2 as a macrophage chemoattractant in equine AT; omental AT may have a more dynamic macrophage population than other depots. Importantly, CD163 (+) macrophages may not be the most numerous or important population in equine AT, as increased depot-specific expression of pro-inflammatory cytokines (nuchal ligament) is not accompanied by an increased number of CD163 (+) cells.

ABSTRACT #365

URINALYSIS AND URINE CULTURE RESULTS IN AGED MARES WITH AND WITHOUT PITUITARY PARS INTERMEDIA DYSFUNCTION. H Schott II. Michigan State University College of Veterinary Medicine, East Lansing, MI.

Polyuria and polydipsia (PU-PD) is one of the clinical signs reported with pituitary pars intermedia dysfunction (PPID). The cause(s) of PU-PD are not well understood but may include osmotic diuresis consequent to glucosuria, a decrease in antidiuretic hormone release (acquired central diabetes insipidus), or direct effects of cortisol on glomerular filtration rate. The aims of this study were: 1. to determine prevalence of PU-PD based on a owner reported assessment score (n = 25); 2. to measure urine concentration and perform reagent strip urinalysis (n = 20); and 3. to examine for bacteriuria (n = 22) in a group of 25 aged mares (17–24 years) being screened for PPID by clinical examination and an overnight dexamethasone suppression test (DST). Blood was collected by jugular venipuncture for serum biochemical analysis and urine was collected by bladder catheterization (without sedation) shortly after hospital admission.

None of the 25 mares had azotemia. Of 19 mares that had DST results supportive of PPID, PU-PD scores were 0 (normal, n = 6), 1 (mild, n = 3), 2 (moderate, n = 9), and severe (n = 1) for a total of 11/19 horses (prevalence 58%) with PU-PD as compared to scores of 0 (n = 5) and 2 (n = 1) in the six mares that had DST results that were not supportive of PPID. Measures of urine concentration tended to be lower in PPID-affected mares but differences were not significant. Not surprisingly, urine specific gravity values were highly correlated ($r = 0.64$, $p < 0.01$) with urine osmolality values.

All urine samples were alkaline (pH 7.5–8.5) and pigmenturia was not detected in any sample. Glucosuria was detected in two of the 19 PPID-affected mares (curiously with serum glucose concentrations of 95 and 102 mg/dl) and none of the non-PPID mares. Two of the 19 PPID-affected mares had bacteruria (50–100,000 CFU/ml group D non-enterococcus Streptococcus spp. in one mare and < 100 CFU/ml Streptococcus equi subspecies equi and 100–500 CFU/ml Corynebacterium spp. in another mare). One of the six non-PPID mares also had bacteriuria (10–50,000 CFU/ml Enterococcus spp.). Although not assessed in a quantitative manner, it is noteworthy that more than 50% of owners of PPID-affected horses reported mild to severe PU-PD despite most horses having measures of urine concentration that would be considered within reference ranges.

Table [Abstract 365]

Group	Specific gravity mean (range)	Osmolality mean (range)
PPID, n = 15	1.033 (1.014–1.044)	1016 (490–1311)
non-PID, n = 5	1.038 (1.034–1.043)	1178 (1006–1434)

ABSTRACT #366

EFFECTS OF SEASON ON ADRENOCORTICOTROPIN HORMONE (ACTH) CONCENTRATIONS IN A GROUP OF HORSES LOCATED IN THE STATE OF TEXAS. B Buchanan,[1] N Frank[2], SB Elliott[2]. [1]Brazos Valley Equine Hospital, Navasota, TX. [2]Department of Large Animal Clinical Sciences, University of Tennessee, Knoxville, TN.

We have previously reported a seasonal increase in adrenocorticotropin hormone (ACTH) concentrations in horses located in Tennessee, with higher concentrations detected in September and October. However, it was not known whether these findings were specific to this geographical location. We therefore studied a group of horses located in Texas to test the hypothesis that the same seasonal alterations in ACTH concentrations occur in two different geographical locations. Fifteen horses (9 mares, 4 geldings, and 2 stallions) with a median (range) age of 15 (4–38) years were included in the study. Appaloosa (n = 1), Arabian (7), Quarter horse (5), and Tennessee Walking Horse (1) breeds were represented and one horse was a crossbred. Horses were sampled under fed (hay or grass) conditions approximately every month across a 12-month period (September 2007 to August 2008). Hormones and metabolites were measured in stored serum and plasma samples at the end of the study. Data were analyzed by repeated measures ANOVA.

One horse suffered from diabetes mellitus, so its data were excluded from statistical analyses. Plasma ACTH concentrations were significantly affected by season (time effect; $P = 0.027$), with a significantly higher mean ± SD concentration (184.5 ± 313.3 pg/mL) detected in September, compared with eight other months. Three horses had ACTH concentrations > 35 pg/mL in the spring/summer, indicating pituitary pars intermedia dysfunction. There was no effect of season on serum insulin concentrations. We conclude that horses in another geographical location exhibit higher plasma ACTH concentrations during the late summer and autumn.

ABSTRACT #367

ELECTROENCEPHALOGRAPHY ASSESSMENT IN ADULT, GERIATRIC AND GERIATRIC HORSES WITH SUSPECTED PITUITARY PARS INTERMEDIA DYSFUNCTION: A PRELIMINARY STUDY. L Bergamasco, L Beard, D Rankin, BM DeBey. Kansas State University College of Veterinary Medicine, Manhattan, KS.

The purpose of this study was to establish normal electroencephalography (EEG) parameters in horses. The specific aims were 1) validate the use of the spectral analysis EEG in adult normal anesthetized horses 2) compare EEG data to aged normal horses and 3) evaluate EEG of horses with suspected pituitary pars intermedia dysfunction (PPID) with comparison to clinical, clinicopathologic, and histopathologic findings. This protocol was approved by the Institutional Animal Care and Use Committee at Kansas State University (KSU).

Horses donated to the KSU for problems other than intracranial diseases constituted the three study groups (adult = 5; geriatric = 5; PPID = 2). Complete physical, neurologic and blood examinations were performed. EEG was performed under general anesthesia. A 12 channel montage was used to record EEG for 20 minutes. The horses then were euthanized and immediately after cerebrospinal fluid were collected and submitted for cytologic examination. Postmortem examinations of the central nervous system were performed.

EEG visual inspection and spectral analysis were performed. Statistical analyses included normality tests, ordinary ANOVA and multiple comparison tests. All data were expressed as means ± standard error of the means. Statistical significance was designated a priori as $P < 0.05$.

Significant differences between the three groups for the all four EEG frequency bands and the total power were reported.

These results may influence health care recommendations for horses as they age. Differences identified may have important implications for the distinction between benign age-associated changes and age-related changes that predispose to disease.

ABSTRACT #368

THE MOLECULAR EPIDEMIOLOGY OF GIARDIA SHEDDING IN HORSES AND ALPACAS IN NEW YORK STATE.
AJ Burton[1], DV Nydam[1], K Mitchell[1], TK Dearen[2], DD Bowman[1], L Xiao[2]. [1]College of Veterinary Medicine, Cornell University, Ithaca, NY. [2]Division of Parasitic Diseases, CDC, Atlanta, GA.

Giardia duodenalis (syn. *G. intestinalis*, *G. lamblia*) is a zoonotic flagellate parasite infecting mammals and causing gastroenteritis and diarrhea. To date, little is known about the prevalence, genotypes and zoonotic potential of *Giardia* affecting horses and alpacas in North America.

A cross-sectional prevalence study was conducted in the New York State area, USA between February 25th and August 24th 2009. Fecal samples were collected from 349, 1–10 week old foals and their dams on 14 different broodmare farms and from 220, 1–16 week old crias and their dams on 15 different alpaca breeding farms. All fecal samples were screened for *Giardia* using a direct immunofluorescence assay (DFA). DNA extraction and PCR analysis of the triose phosphate isomerase (*tpi*) and glutamate dehydrogenase (*gdh*) genes were performed on all DFA positive samples. PCR positive samples were subtyped by DNA sequencing of the *tpi* and *gdh* genes. A Wilcoxon Rank Sum test was used for data with continuous outcomes and Fisher's Exact Test used for comparison of dichotomous variables. Alpha was set at 0.05.

By DFA, 27/175 (15%) of the foal, 22/174 (12.6%) of the mare, 18/110 (16.3%) of the cria and 7/110 (6.4%) of the adult alpaca fecal samples were positive for *Giardia*. There was no significant age difference between *Giardia* positive foals (median = 18, range = 6–46 days) and negative foals (median = 18, range = 4–65 days), p = 0.99, nor *Giardia* positive crias (median = 26, range = 15–70 days) and negative crias (median = 38, range = 2–111 days), p = 0.2. However, *Giardia* positive adult alpacas were significantly younger (median = 3, range = 3–4 years) than negative (median = 6 years, range = 2.5–11 years) adult alpacas p = 0.02. There was no significant age difference between *Giardia* positive and negative mares. There was no significant difference in fecal consistency (score 1–3) between *Giardia* positive and negative foals or crias (p = 0.2 and p = 0.3 respectively). PCR analysis plus DNA sequencing of the *tpi* and *gdh* genes the identified assemblages A and B in the equine *Giardia* positive samples and assemblages A and E in the alpaca *Giardia* positive samples.

Both assemblages A and B are infectious to humans and therefore equine and alpaca *Giardia* infections should be considered zoonotic. As *Giardia* shedding did not appear to be associated with diarrhea in foals or crias, the zoonotic potential of apparently healthy animals should not be discounted.

ABSTRACT #369

IDENTIFICATION OF THE CRYPTOSPORIDIUM HORSE GENOTYPE FROM FOALS IN NEW YORK STATE.
AJ Burton[1], DV Nydam[1], K Mitchell[1], TK Dearen[2], DD Bowman[1], L Xiao[2]. [1]College of Veterinary Medicine, Cornell University, Ithaca, NY. [2]Division of Parasitic Diseases, CDC, Atlanta, GA.

Cryptosporidium spp. are zoonotic Apicomplexan protozoan parasites that may infect mammals, birds, reptiles, amphibians and fish. To date, little is known about the prevalence, genotypes and zoonotic potential of *Cryptosporidium* spp. affecting horses, especially in North America.

A prospective study was conducted in New York, USA between February 25th and May 1st 2009. Fecal samples were collected from 175 1–10 week old foals and their dams (174) on 14 different broodmare farms. All fecal samples were screened for *Cryptosporidium* spp. using a direct immunofluorescence assay (DFA). DNA extraction and PCR-RFLP analysis of the small-subunit (SSU) rRNA gene were performed on all the foal samples. PCR positive samples were subtyped by DNA sequencing of the 60-kDa glycoprotein (gp60) gene. A Wilcoxon Rank Sum test was used to compare the sets of continuous data and a Fishers Exact Test used to compare dichotomous variables. Alpha was set at 0.05.

On DFA, 13/175 (7.4%) foal samples and 3/174 (1.7%) mare samples were designated positive for *Cryptosporidium* spp., whereas on SSU rRNA-based PCR, 9/175 (5.1%) foal samples were positive. *Cryptosporidium* PCR-positive foals were significantly older (13–40 days, median age of 28 days) compared with negative foals (4–67 days, median 18 days, p = 0.02). The number of foals with diarrhea or soft feces was not significantly different between positive and negative foals (p = 0.09). PCR-RFLP analysis of the SSU rRNA gene and DNA sequencing of the gp60 gene identified the parasite as subtype VIaA14G2 of the horse genotype.

This is the first report of a group of foals affected with the *Cryptosporidium* horse genotype, which has recently been detected in humans. As other contemporary molecular studies have identified *C. parvum* in foals, it seems that equine cryptosporidiosis should be considered a zoonosis.

ABSTRACT #370

SEQUENTIAL PLASMA CONCENTRATIONS OF ADRENOCORTICOTROPIN AND CORTISOL IN TERM SEPTIC NEONATAL FOALS COMPARED TO HEALTHY NON-SEPTIC FOALS.
JR Gold[1], ND Cohen[1], TH Welsh Jr[2]. [1]Texas A&M University, Department of Large Animal Clinical Sciences, College Station, TX. [2]Texas A&M University, Department of Veterinary Integrative Biosciences, College Station, TX.

Critical illness-related corticosteroid insufficiency (CIRCI) occurs frequently in critically ill human patients, and is associated with increased mortality in patients with sepsis and septic shock. CIRCI is caused by dysfunction of the hypothalamic-pituitary-adrenal axis (HPA). Data regarding the prevalence of HPA axis dysfunction in critically ill neonatal foals is increasing: however, the clinical significance and the method for best assessing HPA dysfunction remain unclear. The purpose of this study was to compare temporal changes in secretion of HPA axis hormones (adrenocorticotropin [ACTH] and cortisol) between hospitalized, term, septic neonatal foals, and term, healthy, neonatal foals at a referral hospital.

Blood samples were obtained from term septic foals (n = 15) < 7 days of age admitted to Texas A&M University, Large Animal Teaching hospital at admission (T = 0), 24(T = 1), 48(T = 2), 72 (T = 3) and 96 (T = 4) hours, plasma was separated by centrifugation, and then stored at -80C until assayed. Foals were defined septic based upon having a positive blood culture, a sepsis score of ≥ 11 or both. Blood was sampled similarly in healthy term foals < 7 days of age obtained from Texas A&M Horse Center (n = 4) or from term foals (n = 11) admitted to the teaching hospital for reasons other than sepsis. Endogenous basal ACTH and cortisol concentrations were measured using validated immunoassays.

Data were analyzed using linear mixed-effects models to account for the repeated measures among foals. Significance was set at $P < 0.05$. There were no significant temporal changes in ACTH concentration over time and the temporal pattern did not differ significantly between septic and healthy foals. However adjusting for effects of time, ACTH concentrations were significantly greater among septic foals than healthy foals ($P = 0.031$) The estimated magnitude of increase in ACTH was 1.5 units (95% confidence interval, 1.2 to 1.94 units) in septic foals relative to healthy foals. Significant effects of sepsis and time on cortisol concentrations were noted, but the temporal changes did not vary between septic and healthy foals. At each time-point, cortisol concentrations were significantly ($P < 0.05$) greater among septic foals than healthy foals. The estimated magnitude of increase in cortisol was 19.2 units (95% confidence interval, 10.2 to 24.3 units) in septic foals relative to healthy foals. Accounting for effects of sepsis, cortisol concentrations were significantly ($P < 0.05$) lower at T = 1, 2, 3, 4 than at admission.

These findings demonstrate that ACTH and cortisol concentrations are elevated in septic foals compared to healthy foals. The elevated hormone concentrations are expected in critically ill foals. However, cortisol concentrations at admission were significantly greater than at other times in all foals. How this temporal change affects the HPA axis is unclear, but this change did not impact survival in this study population.

ABSTRACT #371

DAILY ENDOGENOUS CORTISOL PRODUCTION AND HYDROCORTISONE PHARMACOKINETICS IN HEALTHY NEONATAL FOALS AND ADULT HORSES. KA Hart,[1] L Dirikolu,[2] DC Ferguson,[2] MH Barton[1]. [1]University of Georgia, College of Veterinary Medicine, Athens, GA. [2]University of Illinois, College of Veterinary Medicine, Urbana, IL.

Hydrocortisone replacement therapy improves outcome in septic people with critical illness-related cortisol insufficiency (CIRCI), and may also benefit septic foals with CIRCI. However, calculation of an appropriate hydrocortisone replacement dose requires understanding of both endogenous and exogenous cortisol metabolism, which are poorly understood in foals. This study was designed to compare daily endogenous cortisol production and the pharmacokinetics of an intravenous bolus dose of hydrocortisone between neonatal foals and adult horses. Two milliliters of blood were collected every 15–20 minutes for 24 hours from 4 healthy 2-day-old foals and 4 healthy adult horses for determination of serum cortisol concentrations. Then, 1 mg/kg of hydrocortisone was given intravenously to the same 4 foals and to 4 additional healthy horses 12 hours after suppression of endogenous cortisol production with dexamethasone (0.08 mg/kg IV). Cortisol and hydrocortisone concentrations were determined using a previously validated chemiluminescent immunoassay. Data were compared between groups with Mann Whitney tests, with significance set at $P < 0.05$. Mean 24-hour cortisol concentration was significantly lower in foals ($2.0 \pm 0.4\,\mu g/dl$) than horses ($2.6 \pm 0.6\,\mu g/dl$), but daily cortisol production rate was significantly greater in foals (0.671 ± 0.032 mg/kg/day) than horses (0.215 ± 0.040 mg/kg/day). Foals had significantly greater hydrocortisone volume of distribution and clearance and significantly lower peak hydrocortisone plasma concentrations ($Vd_{ss} = 1.92 \pm 1.11$ L/kg; $Cl_s = 1.39 \pm 0.11$ L/kg/hr; peak = $85.5 \pm 11.8\,\mu g/dL$) than horses ($Vd_{ss} = 0.58 \pm 0.15$ L/kg; $Cl_s = 0.35 \pm 0.07$ L/kg/hr; peak = $846.7 \pm 378.0\,\mu g/dL$). These considerable differences in cortisol metabolism between horses and foals may contribute to the development of CIRCI in septic neonatal foals, and should be considered during hydrocortisone replacement therapy in foals.

ABSTRACT #372

THYROID HORMONES IN HEALTHY, SICK NON-SEPTIC AND SEPTIC FOALS. M Himler[1], A Cattaneo[1], RJIM Barsnick[1], SDA Hurcombe[1], R Rathgeber[2], RE Toribio[1]. [1]The Ohio State University College of Veterinary Medicine, Columbus, OH. [2]Hagyard Equine Medical Institute, Lexington, KY.

Sepsis is the number one cause of foal mortality. We have shown that endocrine disorders are frequent in septic foals and often associated with mortality. Thyroid hormones (TH) are important in growth and development (neural, musculoskeletal), maturation (lung, gastrointestinal, cardiovascular), and metabolism (energy, thermogenesis). Decreased TH concentrations in newborn foals are associated with musculoskeletal abnormalities, incoordination, weakness, respiratory distress, hypothermia, failure of transfer of passive immunity, and likely sepsis.

Since THs are important for several developmental and metabolic processes, the goal of this study was to measure serum total triiodothyronine (tT3), free triidothyronine (fT3), total thyroxine (tT4) and free thyroxine (tT4) concentrations in septic, sick non-septic and healthy foals, and to determine their association with severity of sepsis and survival. We hypothesized that THs will be lower in septic foals (euthyroid sick syndrome) compared to sick non-septic and healthy foals, and associated with severity of sepsis and mortality. Blood samples were collected on admission from 53 septic, 100 sick non-septic and 26 healthy foals of < 7 days of age. Foals were considered septic if they had a positive blood culture or a sepsis score ≥ 12.

Serum tT3, fT3, tT4, and fT4 were lower in septic foals compared to healthy foals ($P < 0.01$), while tT3 and fT3 were lower in septic foals than in sick non-septic foals ($P < 0.05$). Serum fT3, tT3, and fT4 concentrations were lower in non-septic compared to healthy foals. All THs were negatively correlated with the sepsis score. THs were also associated with white blood cell count, IgG, albumin, glucose, and triglyceride concentrations. Within septic foals, serum fT3, tT3 and fT4 were lower in foals that did not survive.

These findings indicate that euthyroid sick syndrome (non-thyroidal illness syndrome) is frequent in septic foals, associated with severity of sepsis and mortality, as well as with the metabolic needs (energy status) of these foals. We believe that low TH concentrations in septic foals are a consequence of the severity of the disease and likely a reflection of system inflammation.

ABSTRACT #373

COMPARISON OF FECES VERSUS RECTAL SWABS FOR THE MOLECULAR DETECTION OF *LAWSONIA INTRACELLULARIS* IN FOALS WITH EQUINE PROLIIFERATIVE ENTEROPATHY. N. Pusterla[1], S. Mapes[1], C. Johnson[1], N. Slovis[2], A. Page[2], C. Gebhart[3]. [1]School of Veterinary Medicine, University of California, Davis, CA. [2]Hagyard Equine Medical Institute, Lexington, KY. [3]University of Minnesota, College of Veterinary Medicine, St. Paul, MN.

The diagnosis of equine proliferative enteropathy (EPE) is based on the presence of specific clinical, clinicopathological and sonographic abnormalities coupled with the serological detection of antibodies to *Lawsonia intracellularis* and/or the molecular detection of *L. intracellularis* in feces. Rectal swabs are generally considered the sample of choice for the PCR detection of *L. intracellularis* in pigs due to the ease of collection. The purpose of this study was to compare the molecular detection rate of *L. intracellularis* between feces and rectal swabs collected from foals with suspected EPE. Case selection was based on the presence of EPE-associated clinical signs, hypoproteinemia/hypoalbuminemia and positive titer (≥ 60) against *L. intracellularis* by immunoperoxidase monolayer assay (IPMA).

Forty-two fecal samples and rayon-tipped rectal swabs collected from EPE-suspected cases were processed for DNA purification using an automated nucleic acid extraction system. The purified DNA was then analyzed by real-time PCR for the presence of the *aspartate ammonia lyase* (*AAL*) gene of *L. intracellularis*. Absolute quantitation was calculated using a standard curve for *L. intracellularis* and expressed as copy numbers of *AAL* gene of *L. intracellularis* per microliter of purified DNA. Furthermore, a real-time PCR assay targeting a universal sequence of the bacterial *16S rRNA* gene was used as quality control and as an indicator of fecal inhibition.

DNA was successfully extracted from all feces and rectal swabs based on positive PCR signals for the universal bacterial *16S rRNA* gene. The overall PCR accuracy for the detection of *L. intracellularis* was 90% with 38 foals testing PCR positive in feces (33 samples), rectal swabs (32) or both (27). Six foals tested PCR positive only in feces, while 5 tested positive only in rectal swabs. Feces yielded a significantly higher *AAL* gene copy number of *L. intracellularis* (mean±SD = 908 ± 479 *L. intracellularis AAL* gene copies/μl DNA) than rectal swabs (50 ± 27 *L. intracellularis AAL* gene copies/μl DNA; Wilcoxon signed-rank Test, $P < 0.0001$). Feces and rectal swabs tested PCR negative from 4 EPE suspected foals.

In conclusion, the results showed that feces yielded similar PCR positive results with a higher *L. intracellularis AAL* gene load than rectal swabs. By analyzing dual samples, the PCR accuracy for *L. intracellularis* increased from 76 and 79% for rectal swabs and feces, respectively, to 90%. Rectal swabs should be considered as an alternative sample type for EPE-suspected patients with decreased or no fecal output. Further, the diagnosis of EPE should be based on integrating clinical, clinicopathological, sonographic and laboratory results and not solely on PCR results.

ABSTRACT #374

ORAL INFECTION OF WEANLING FOALS WITH AN EQUINE ISOLATE OF *LAWSONIA INTRACELLULARIS*, AGENT OF EQUINE PROLIFERATIVE ENTEROPATHY.

N. Pusterla[1], S. Wattanaphansak[2], S. Mapes[1], J. Collier[1], J. Hill[1], M. DiFrancesco[1], C. Gebhart[2]. [1]School of Veterinary Medicine, University of California, Davis, CA. [2]University of Minnesota, College of Veterinary Medicine, St. Paul, MN.

The purpose of the study reported here was to establish an animal model in order to study sequential clinical, laboratory and ultrasonographic parameters associated with experimental equine proliferative enteropathy (EPE). Further, a question of interest was whether experimentally infected foals would shed enough *Lawsonia intracellularis* in the environment to potentially infect sentinel herdmates.

Recently weaned foals were assigned to either the challenge (n = 3), the sentinel (n = 3) or the control (n = 2) group. Foals were experimentally infected via intragastric inoculation of 3×10^{10} *L. intracellularis* organisms grown in cell culture. Each experimentally infected foal was housed with a sentinel foal in order to assess feco-oral transmission. All foals were monitored daily for the development of clinical abnormalities and were weighed once weekly for the duration of the study (90 days). Abdominal ultrasound examination was performed weekly. Feces were collected every other day for 60 days, then weekly for an additional 30 days for the quantitative molecular detection of *L. intracellularis*. Blood was collected weekly for hematological, biochemical and serological analysis.

Only challenged foals developed transient clinical signs of proliferative enteropathy consisting of anorexia, depression, fever, cowpie feces and peripheral edema. Two challenged foals developed transient hypoalbuminemia. Fecal shedding of *L. intracellularis* was first detected in the challenged foals between days 12 and 18 post-inoculation and lasted for 7 to 21 days. Sonographic abnormalities were observed only in one challenged foal and consisted of mild increase in abdominal free fluid and moderately thickened small intestinal wall. Seroconversion was documented in all challenged foals and in one sentinel foal. The average daily weight gain of 4 infected foals (3 challenged and one sentinel foal) was significantly lower (Mann Whitney Test, $P = 0.0286$) than the average daily weight gain of 4 non-infected foals (2 control and 2 sentinel foals).

In conclusion, clinical EPE of variable degrees was induced in all foals infected with *L. intracellularis*. The clinical signs, abnormal biochemical and sonogarphic findings were transient and self-limiting. The study results emphasize the need to institute infectious disease control measures in order to prevent spread of *L. intracellularis* from infected to susceptible animal. Further, subclinical disease may be associated with decreased daily weight gain in infected weanling foals.

ABSTRACT #375

MOLECULAR INVESTIGATION OF THE VIRAL KINETICS OF EQUINE HERPESVIRUS-1 IN BLOOD AND NASAL SECRETIONS OF HORSES FOLLOWING CORTICOSTEROID-INDUCED RECRUDESCENCE OF LATENT INFECTION.

N. Pusterla[1], S.B. Hussey[2], S.M. Mapes[1], C. Johnson[1], J. Collier[1], J. Hill[1], D.P. Lunn[2], W.D. Wilson[1]. [1]School of Veterinary Medicine, University of California, Davis, CA. [2]College of Veterinary Medicine and Biomedical Sciences, Colorado State University, Fort Collins, CO.

Latency and reactivation are thought to be critical features of the epidemiology of equine herpesvirus-1 (EHV-1) infections. Recrudescence of latent EHV-1 with subsequent viral shedding via nasal secretions is a potential source of infection for naïve horses and has been implicated in outbreaks occurring in closed populations. The objective of this study was to examine the viral kinetics of EHV-1 following corticosteroid-mediated reactivation of experimentally induced latency in adult horses. A question of further interest was whether reactivated EHV-1 from a latently infected horse could potentially infect sentinel herdmates.

Four horses previously infected with EHV-1 four months prior to study commencement received dexamethasone (100 mg/horse IV) on 5 consecutive days. Four naïve horses served as sentinels and had direct nose-to-nose contact with the latently infected horses during the entire study period (60 days). All horses were monitored daily for the development of clinical abnormalities such as depression, anorexia, fever, nasal discharge, cough and enlarged submandibular lymph nodes. Whole blood and nasal secretions were collected daily for the molecular detection and cell culture of EHV-1. Serum was collected weekly for the detection of antibodies against EHV-1.

All latently infected horses showed transient molecular detection of EHV-1 in blood and nasal secretions following administration of dexamethasone, but only one horse developed EHV-1 associated signs. Molecular evidence of lytic infection was detected only in blood and nasal secretions from one latently infected horse, while viral cultures remained negative for all horses in this group. Three out of 4 latently infected horses displayed an increase in antibody levels against EHV-1. None of sentinel horses developed EHV-1 associated signs, viremia, viral shedding or seroconversion.

In conclusion, EHV-1 was successfully reactivated following corticosteroid administration in latently infected horses. However, transmission of reactivated virus to sentinel horses was unsuccessful. Failure to effectively transmit EHV-1 to susceptible sentinel horses may have resulted from the low level and short period of viral shedding in latently infected horses.

ABSTRACT #376

TWO-DIMENSIONAL AND M-MODE ECHOCARDIOGRAPHIC FINDINGS IN HEALTHY THOROUGHBRED FOALS.

N.M. Collins[1], C.M. Marr[2], L. Palmer[2]. [1]Scone Equine Hospital, Scone, New South Wales, Australia. [2]Rossdales Equine Hospital, Newmarket, United Kingdom.

Echocardiography is a non-invasive diagnostic technique that can provide valuable information in foals such as the identification of congenital cardiac abnormalities and infectious conditions such as endocarditis or pericarditis. It also provides a considerable amount of functional and morphological information in critically ill foals. Sequential examinations can be used to assess pathological alteration of cardiac dimensions and function in response to the presence of cardiac disease. Reference values for many of the conventional echocardiographic variables of adult horses have not been established in healthy Thoroughbred foals. In order to detect changes in the structure and function of the heart produced by disease, it is important to determine accurately the effect of normal growth and development on echocardiographic variables. The aims of this study were (1) to obtain measurements for a series of routinely used echocardiographic variables in foals at increasing age intervals to establish reference values for the breed, and (2) to establish the correlation between age and the measured echocardiographic variables in healthy Thoroughbred foals.

Nineteen healthy Thoroughbred foals (11 male, 8 female; 61.7 ± 6.5 kg at day 7) were examined on days 7, 14, 28, 56, 84 and 112 of life. The foal's body weight and auscultatory findings were recorded at each examination. Two-dimensional (2-D) and M-mode echocardiographic examinations were performed by one of two operators with the foal standing and unsedated. Summary statistics (including the mean, standard deviation, 95% confidence interval for the means) for the 12 echocardiographic variables at each of the increasing age intervals were determined. Simple linear regression analysis was used to evaluate the relationship between the echocardiographic variables and age.

No murmurs greater than grade 3 of 6 were recognised in any of the foals during the examinations. Early systolic murmurs (grades 1–3/6), with point of maximum intensity over the left heart base, were frequently identified (54.7% of total examinations). Reference values were determined for 12 routinely used 2-D and M-mode echocardiographic variables in the 19 healthy Thoroughbred foals at increasing age intervals. Although there was a trend for each of the echocardiographic variables to increase with increasing age ($P < 0.001$);

only the 2-D measurement of aortic diameter ($r^2 = 0.89$; $p < 0.001$) and the 2-D measurement of left atrial diameter ($r^2 = 0.80$; $p < 0.001$) showed a strong linear relationship with age during the study period. The data from this study provides reference echocardiographic variables that may assist in the interpretation of the results of echocardiographic examinations of Thoroughbred foals of similar ages with suspected congenital or acquired cardiac disease.

ABSTRACT #377

SERUM TRIGLYCERIDE CONCENTRATIONS AT ADMISSION IN CRITICALLY ILL FOALS LESS THAN TWO DAYS OF AGE. N.M. Collins[1], J.E. Axon[1], J.B. Carrick[1], J. Kiehne[1], D. Feary[2]. [1]Scone Equine Hospital, Scone, New South Wales, Australia. [2]University of Sydney Veterinary Hospital, New South Wales, Australia.

Disturbances of lipid metabolism that result in the accumulation of triglycerides in blood are suspected to occur in several disease processes of the critically ill neonatal foal (e.g. sepsis, prematurity and perinatal asphyxia syndrome). Hypertriglyceridemia has been reported in foals receiving lipid-containing total parenteral nutrition solutions and was recently associated with non-survival (Meyers et al. 2009). The aims of this prospective study were to compare serum TG concentrations in critically ill foals at admission to an intensive care unit (ICU) with age-matched healthy foals and to investigate the association of serum TG concentrations with historical information, clinical and diagnostic findings, severity of illness and survival.

Serum TG concentrations were determined in 69 critically ill Thoroughbred foals less than 2 days of age admitted to the Scone Equine Hospital ICU and in 52 age-matched healthy Thoroughbred foals during August to November 2009. Serum TG concentrations were measured using a Vitros 350, Ortho Clinical Diagnostics.

Serum TG concentrations were significantly higher in sick foals admitted to the ICU (median 34 mg/dl, interquartile range (IRQ) = 21 to 89, n = 69) than in the healthy foals (median 22 mg/dl, IQR = 15 to 32, n = 52) ($p < 0.001$). The admission serum TG concentration of hospitalized foals with a gestational age of less than 320 days (median = 70 mg/dl, IQR = 44 to 224, n = 10) was higher than the serum TG concentration of hospitalized foals with a gestational age greater than 320 days (median = 32 mg/dl, IQR = 20 to 71, n = 55) ($P = 0.0084$). The foal's gender, age at admission, length of hospital stay, history of dystocia or placentitis was not correlated with the foal's serum TG concentration at admission.

The admission serum TG concentration was higher for hospitalized foals that had cold extremities, poor peripheral pulses, an indirect mean arterial pressure < 70 mmHg, systemic inflammatory response syndrome (SIRS), a sepsis score > 11, and required cardiovascular support (vasopressors and/or inotropes) than ICU foals that did not have these findings ($p < 0.05$). Foals with a serum creatinine concentration greater than 2.27 mg/dl had a higher admission serum TG concentration (median = 50 mg/dl, IQR = 25 to 101, n = 45) than foals with serum creatinine concentration less than 2.27 mg/dl (median = 27 mg/dl, IQR = 16 to 38, n = 24) ($p = 0.0089$). Foals that died were more likely to have a high TG concentration than foals that were discharged from the hospital ($p < 0.0007$, odds ratio 1.012). Logistic regression analysis showed the measurement of serum TG concentration at admission to be a good parameter to assess survival, the presence of SIRS and the need for cardiovascular support. All these parameters had areas under the receiver operating characteristic curves of greater than 0.75. In conclusion, measurement of serum TG concentrations at admission to the ICU was shown to have diagnostic value in critically ill foals in the first two days of life.

ABSTRACT #378

THE INCIDENCE OF PLACENTITIS IN THE DAMS OF FOALS ADMITTED TO AN EQUINE NEONATAL INTENSIVE CARE UNIT. J.B. Carrick, J.E. Axon, N. Collins. Scone Equine Hospital, Scone, New South Wales, Australia.

Placentitis is a critical determinant of neonatal foal health; however there are no reports of the incidence of placentitis associated with admission to a neonatal intensive care unit. Eighty six foals admitted during 2009 to the Scone Equine Hospital Intensive Care Unit (ICU) less than 72 hours old were included in the analysis. All placentas were examined and recorded with the foal admission, treatment and survival data in a spreadsheet which was then transferred to a statistics program for analysis (JMP 7, SAS Institute Inc, Cary NC, USA). Contingency analysis, Fishers Exact Test, Wilcoxon/Kruskal-Wallis Tests or Logistic Regression was used as appropriate.

Abnormal placentas were recorded for 17 of the 86 foals (19.8%) admitted to the ICU. Foals admitted to the ICU that were delivered from mares with a normal placenta were more likely to be older at admission to the ICU than foals from mares with placentitis ($p = 0.001$, unit OR 1.3). All foals from mares with placentits were admitted to the ICU before 9 hrs of age. The probability that a foal had a gestational age of < 320 days was greater if the mare had placentitis than if the mare had a normal placenta ($p = 0.028$, RR 3.82, 95% CI 1.3–11.7). The probability that the delivery was complicated by dystocia was greater if the mare had placentitis compared to mares with normal placentas ($p = 0.0015$, RR 2.29; 95% CI 1.5–3.5) There was no association between the presence of placentitis and gender, the length of hospital stay or the assessment of the foal as dysmature.

Foals from mares with placentitis were more likely to have perinatal asphyxia syndrome ($p < 0.001$, RR 1.9, 95% CI 1.5–2.3) than foals from mares with a normal placenta. There was no association of placentitis with positive blood culture, pneumonia, fractured ribs, renal failure, infectious gastrointestinal disease, umbilical infection, uroperitoneum, infectious orthopedic disease, or noninfectious orthopedic disease. Foals from mares with a normal placenta were more likely to have an initial IgG of > 800 mg/dL than foals from mares with placentitis ($p = 0.0002$, RR 5.3, 95% CI 1.4–19.7). Foals from mares with placentitis were more likely to have a low blood glucose ($p = 0.0003$, unit OR 1.6) and less likely to have a low serum creatinine concentration ($p = 0.0003$, unit OR 0.99) than foals from mares with normal placentas. The area under the receiver operator characteristic curves for glucose and creatinine were 0.79 and 0.84 respectively. Foals from mares with placentitis were more likely to have a sepsis score ≥ 11, require parenteral nutrition, require inotropes or vasopressors, be recumbent at admission, have poor peripheral pulses and cold extremities than foals from mares with a normal placenta ($p < 0.05$). However, foals from mares with placentitis had a similar probability of survival as foals from mares with normal placentas.

These data indicate that placentitis increases the probability of prematurity, dystocia, perinatal asphyxia syndrome and failure of transfer of passive immunity. Foals from mares with placentitis are more likely to present with signs of shock and more likely to require intensive supportive therapy but have the same probability of survival as foals from mares with a normal placenta.

ABSTRACT #379

THE EFFECT OF INTRAVENOUS PLASMA ADMINISTRATION ON FIBRINOGEN CONCENTRATION IN SICK NEONATAL FOALS. AR Hollis, PA Wilkins, B Tennent-Brown, JE Palmer, RC Boston. School of Veterinary Medicine, Department of Clinical Studies; University of Pennsylvania.

Fibrinogen concentration [FIB] is measured in sick neonatal foals as an indication of inflammation, disease severity and response to treatment. These foals are generally administered plasma in order to treat failure of passive transfer, as treatment for sepsis, and to provide additional colloidal oncotic support. Transfused plasma contains fibrinogen ($[FIB]_{Plasma}$) from the donor animal. This study was undertaken to determine if $[FIB]_{Plasma}$ affects post-transfusion $[FIB]_{Foal}$ in sick neonatal foals.

[FIB]Plasma of comercially prepared fresh frozen plasma was determined using citrated plasma provided by the manufacturer for each specific plasma lot. [FIB]Foal was determined in 31 foals receiving plasma transfusions at admission and 2 and 6 hours following tranfusion completion. Two foals with extreme values were removed from analysis as outliers. Not all data points were available for all foals. Kruskal-Wallis and regression analysis were used to assess differences between sample periods and associations, significance $P \leq 0.05$.

Median $[FIB]_{Plasma}$ was 245 mg/dl (Range 152–378) and was not outside the normal range for any lot. $[FIB]_{Foal}$: Admission 277 mg/

dl (164–674); 2-Hr 322 mg/dl (220–621); 6-Hr 323 mg/dl (202–506) and were not significantly different. ~30%(8/27) and 48%(12/25) of foals had decreased $[FIB]_{Foal}$ from admission at 2-Hr and 6-Hr respectively and had larger admission $[FIB]_{Foal}$ ($P = 0.026$ and $P = 0.031$). Neither $[FIB]_{Foal}$ nor change in $[FIB]_{Foal}$ between sampling times were associated with administered $[FIB]_{Plasma}$. $[FIB]_{Foal}$ in sick foals does not appear to be impacted by transfusion of commercial plasma with normal fibrinogen concentration except for possible dilution.

ABSTRACT #380

PRELIMINARY RESULTS FROM A PROSPECTIVE MULTI-CENTER STUDY OF THE ASSOCIATION OF LACTATE CONCENTRATION WITH SURVIVAL IN SICK NEONATAL FOALS.

A Borchers[1], PA Wilkins[1], PM Marsh[2], JA Axon[3], J Read[4], C Castagnetti[5], L Pantaleon[6], C Clark[7], L Qura'n[8], R Belgrave[9], C Schwartwald[10], M Levy[11], D Bedenice[12], M Saulez[13], RC Boston[1]. [1]University of Pennsylvania; [2]Hagyard Equine Medical Institute; [3]Scone Veterinary Hospital; [4]Equine Medical Center of Ocala; [5]University of Bologna; [6]Peterson and Smith Equine Hospital; [7]University of Illinois; [8]Mid Atlantic Equine Medical Center; [9]University of Zurich; [10]Purdue University; [11]Tufts Cummings School of Veterinary Medicine; [12]University of Pretoria.

The concentration of lactate [LAC] in the blood of sick equine neonates at admission ($[LAC]_{ADMIT}$) and other time points is associated with survival in several retrospective studies, with larger [LAC] associated with non-survival. We hypothesized that this association would remain when examined prospectively.

Thirteen university and private equine referral hospitals on 3 continents enrolled 643 foals over the 2008 foaling season of both hemispheres. Signalment, historical, clinical and clincopathologic data were entered into a standardized spreadsheet by a single veterinarian at each institution. Spreadsheets were unified into a single data set for analysis. Not all data were available for all foals. Data were analyzed using Kruskal-Wallis and regression analysis, $P \leq 0.05$.

There were more colts ($N = 354$) than fillies ($N = 277$) enrolled; there was no difference in $[LAC]_{ADMIT}$ by sex. Neither gestational age nor age at admission were significantly different between survivors and non-survivors. Median $[LAC]_{ADMIT}$ was significantly larger in non-survivors ($N = 120$, 5.79 mmol/L, $P < 0.001$) compared to survivors ($N = 466$, 3.55 mmol/L). The difference in [LAC] between groups remained statistically significant through 120 hrs post-admission. $[LAC]_{ADMIT}$ was significantly greater in foals born following premature placental separation ($P < 0.001$) and those with a history of dystocia or Caesarian section delivery ($P < 0.001$). Mean arterial pressure had a small ($R^2 = 19.1\%$) but significant ($P < 0.001$) association with $[LAC]_{ADMIT}$. $[LAC]_{ADMIT}$ was not significantly different in blood culture positive *vs* negative foals.

This large prospective study supports the findings of several smaller retrospective studies regarding the utility of initial and repeated [LAC] measurement in the management of sick foals.

ABSTRACT #381

ARTERIAL BLOOD GAS PARAMETERS AND LACTATE CONCENTRATION IN NORMAL AWAKE STANDING HORSES RECEIVING INTRAVENOUS PHENYLEPHRINE INFUSIONS.

PA Wilkins[1], M McMichael[1], SA Smith[2], SC Clark-Price[1], KM Lascola[1], P DePedro[1]. [1]College of Veterinary Medicine Department of Veterinary Clinical Medicine, [2]College of Medicine Department of Biochemistry, University of Illinois at Urbana Champaign.

Phenylephrine infusion (PEI) is used clinically in horses with colic and suspected left dorsal displacement of the colon to reduce splenic size. Blood gas analysis (BG) and lactate concentration [LAC] are used in the evaluation and monitoring of horses with colic. The aim of the study was to determine the effect of PEI on BG parameters and [LAC] in adult healthy horses.

Arterial blood samples from 6 normal horses were collected into heparinized arterial BG syringes for BG and [LAC] analysis via a catheter in the transverse facial artery prior to PEI at $6\,\mu g\,kg^{-1}\,min^{-1}$ phenylephrine for 15 min. BG and [LAC] were repeated 0, 30 and 120 min after PEI completion. Statistical comparisons were by repeated measures ANOVA $p \leq 0.05$.

pH, PaO_2, [LAC] and Na^+ were unchanged. $PaCO_2$, HCO_3 and BE_{ecf} decreased with PEI but returned to pre-infusion baseline by 120 min. Glucose, Ca^{++} and Mg^{++} increased with PEI but returned to pre-infusion baseline at 120 min. K^+ was decreased at 30 and 120 min post-PEI. Cl^- was increased at 0 min and returned to baseline at 30 and 120 min.

PEI does not change arterial pH, PaO_2, [LAC] or Na^+. Mild hyperglycemia appears to be associated with PEI. Other blood gas parameters may be altered with PEI and should be considered when evaluating horses receiving PEI.

ABSTRACT #382

BLOOD GAS PARAMETERS IN NORMAL HORSES RECEIVING INTRAVENOUS SODIUM LACTATE INFUSION.

PA Wilkins, P DE Pedro, M McMichael, KM Lascola, SC Clark-Price. College of Veterinary Medicine Department of Veterinary Clinical Medicine, University of Illinois at Urbana Champaign.

Clearance of exogenous sodium lactate can be used as a prognostic indicator in sick humans. Sick horses frequently have blood gas (BG) and electrolyte abnormalities. The aim of the study was to determine the effect of sodium lactate infusion (NLI) on BG and electrolyte parameters, should this test prove useful in horses.

Venous blood samples from 6 normal horses were collected into heparinized syringes for BG analysis via a catheter in the right jugular vein prior to NLI at 1 mmol/kg over 15 minutes. BG was repeated 0, 30 and 120 min after NLI completion. Statistical comparisons were by Friedman analysis $p \leq 0.05$.

pH, PCO_2, PO_2, Cl^- and glucose were unchanged. HCO_3^-, BE_{ecf} and Na^+ increased with NLI and did not return to pre-infusion baseline by 120 min. K^+, Ca^{++} and Mg^{++} decreased with NLI and also failed to return to baseline by 120 min. Lactate increased approximately 4-fold and returned to baseline at 120 min. Calculated strong ion difference (SID) increased and failed to return to baseline. Total protein (A_{TOT}), albumin and phosphorus were not significantly changed despite appearing decreased by inspection.

NLI does not change pH, PCO_2, PO_2, Cl^- or glucose. The physiochemical acid-base approach identifies PCO_2, weak acid buffers (A_{TOT}) and SID –which includes Na^+ and lactate- as independent determinants of pH. It is possible the infusion of lactate and Na^+ was balanced by smaller changes in other acid-base determinants. Hypokalemia, hypocalcemia and hypomagnesemia may potentially be worsened with NLI and should be monitored when considering NLI.

ABSTRACT #383

IN VITRO EHV-1 INFECTION OF CENTRAL NERVOUS SYSTEM ENDOTHELIAL CELLS: IS PBMC ADHESION NECESSARY FOR ENDOTHELIAL CELL INFECTION?

LS Goehring. Colorado State University College of Veterinary Medicine, Fort Collins, CO.

There is general consensus that EHV-1 viremia is a pre-requisite for Equine Herpesvirus-associated Myelopathy (EHM) in horses. We hypothesized that the rate of central nervous system (CNS) endothelial cell (EC) infection is key to the development of EHM. Further, CNS-EC infection occurs after cell-cell contact between virus-infected peripheral blood mononuclear cells (PBMC) and the CNS-EC.

. . . To study potential cell-to-cell infections we used monolayers of CNS-EC and EHV-1 (Ohio 03)–infected PBMC from 9 horses. For each horse, experiments were conducted in quadruplicate, and duplicate wells were co-cultured either with or without EHV-1 neutralizing antibodies (VN antibodies). EC-monolayers and *in vitro* infected PBMC in culture wells were placed on a laboratory rocker in an incubator (5% CO2; 37°C) for 4 hours (phase 1), allowing PBMC to roll over the EC monolayer. Then, the monolayer was thoroughly washed to remove all (non-adhered) PBMC; then incubated for another 72 hours with all wells now containing VN antibodies (phase 2). Viral cytopathic effect (CPE) was quantified by plaque count. The amount of added VN antibodies was pre-determined in an experiment where EHV-1 –infected PBMC were placed in the medium of CNS-EC monolayers but without contact, and no CPE was observed. The rationale being that all 'free virus' will be blocked by VN antibodies. The data was evaluated by regression analysis for fit and variance. A p-value of < 0.05 was used to determine significant differences in plaque counts observed after co-incubation with or without VN antibodies during initial 4 hour incubation.

... Plaques developed in all wells. However, a plaque reduction between 10 and 90% was observed when VN antibodies were co-cultured during phase 1 of the experiment *($p < 0.05$)*. While other effects may have been responsible for this reduction, these findings may also suggest that individual PBMC - EC combinations exhibited stronger or weaker adhesions. A plaque reduction of 10% in the VN antibody – containing wells may therefore be an indicator of 'efficient' or intensified cell-cell contact or adhesion, and a 90% plaque reduction may be suggestive of 'poor' adhesion.

Future studies will focus on co-cultivating PBMC and CNS-EC with cytokines or adhesion-blocking antibodies and in the presence of VN antibodies to determine the role of adhesion molecules in CNS-EC infection with EHV-1.

ABSTRACT #384
USE OF SERIAL SEROLOGICAL AND FECAL PCR TESTING FOR TARGETED TREATMENT OF WEANLING FOALS ON A FARM WITH ENDEMIC *LAWSONIA INTRACELLULARIS*.

A. Page[1], N. Slovis[1], C. Gebhart[2], K. Wolfsdorf[1], S. Mapes[3], N. Pusterla[3]. [1]Hagyard Equine Medical Institute, Lexington, KY. [2]University of Minnesota, College of Veterinary Medicine, St. Paul, MN. [3]School of Veterinary Medicine, University of California, Davis, CA.

Lawsonia intracellularis (*LI*), the causative agent of equine proliferative enteropathy (EPE), is an emerging disease with an increasing economic importance in the equine industry, which has placed an emphasis on early detection and prompt treatment of mildly affected horses. The purpose of this study was to utilize multiple diagnostic modalities, along with regular monitoring, to recognize *LI*-exposed weanlings on an endemic farm and target these horses for additional testing to identify and treat subclinically affected horses to prevent development of clinical EPE.

Monthly serum samples for the immunoperoxidase monolayer assay (IPMA) were and bimonthly feces for rtPCR for the aspartate ammonia lyase gene were collected from a total of 53 weanlings on an *LI*-endemic farm from August 2008 to January 2009. An IPMA titer of ≥ 60 was regarded as a positive result. Any weanling with positive IPMA or PCR tests, or clinical signs of EPE, had biochemistry blood work submitted. A total of 32/53 (60.3%) horses had positive IPMA titers and 19/32 of these horses were treated for EPE. Of these 32 horses, only 5 showed outward manifestations of EPE; 2/5 horses were hospitalized for supportive care and 1/2 hospitalized horses were euthanized. The number of horses that were hospitalized during this period was less than the year before when no laboratory diagnostics were regularly used.

In conclusion, IPMA testing for *L. intracellularis* appears to be a useful tool when combined with careful monitoring, as well as targeted biochemical monitoring, for detecting subclinical and mild cases of EPE.

ABSTRACT #385
THE NORMAL CONJUNCTIVAL BACTERIAL AND FUNGAL FLORA OF HORSES IN THE UNITED KINGDOM.

I Johns[1], N Menzies-Gow[1], K Baxter[1], H Booler[2] and C Hicks[1]. [1]Royal Veterinary College, North Mymms, [2]Astrazeneca, Pharmaceutical Division, Macclesfield, United Kingdom.

Ulcerative keratitis is a frequent clinical condition affecting horses throughout the world. Although trauma is typically believed to be the initiating cause, secondary bacterial and fungal infections can result in serious sequelae such as corneal malacia. The organisms which initially infect a corneal defect are believed to originate from the normal conjunctival flora. These organisms can vary according to geographic location and season. Knowledge of these isolates, and their antimicrobial sensitivity patterns, can aid practitioners in their choice of initial empirical treatment.

The aim of this study was to describe the normal conjunctival flora in horses in the southern United Kingdom at three different time points throughout the year. Conjunctival swabs were obtained from 60 clinically normal horses in May (n = 35), July (n = 15) and November (n = 10). Routine bacterial and fungal culture was performed and sensitivity to 6 antimicrobials (chloramphenicol, ciprofloxacin, fusidic acid, gentamicin, neomycin, oxytetracycline) determined for bacterial isolates using the Kirby-Bauer disc diffusion method.

Bacteria were isolated from 58% of horses, and fungi from 15%. The majority were gram negative. (53%) The most frequently isolated bacterial species were *Acinobacter* spp (35% of isolates), *Staphylococcus* spp (26%) and *Micrococcus* spp (15%). Fungal isolates were *Aspergillus* spp, *Mucor* spp and *Absidia* spp. The presence of bacteria or fungi was not affected by season. Compared to similar studies in other countries, gram negative isolates were more commonly isolated, and fungi less commonly. Based on the current study, appropriate antimicrobials to prevent bacterial infection of corneal ulcers include topical gentamicin or tetracyclines.

ABSTRACT #386
IDENTIFICATION AND PERSISTENCE OF ANTIMICROBIAL RESISTANCE IN FECAL *ESCHERICHIA COLI* IN HORSES TREATED WITH ANTIBIOTICS: PRELIMINARY FINDINGS.

Johns I, Good L, Verheyen K, Rycroft A. Royal Veterinary College, North Mymms, United Kingdom.

The development of resistance by bacteria to antimicrobials is a major threat to the health and welfare of humans and animals. Antimicrobial use is considered to be the key driver in the development of antimicrobial resistance (AMR). The development of AMR can be evaluated by identifying resistance patterns of commensal bacteria, which can potentially act as pathogens and can also provide a vast reservoir of mobile resistance genes.

This study aims to document the development and persistence of AMR in commensal fecal *E coli* isolates from horses treated with antibiotics. Fecal samples were obtained from 16 client-owned hospitalized horses treated with systemic antibiotics on the 1st (TI) and 5th (T2) days of treatment. A third sample was obtained from 9 horses 2 weeks after treatment. (T3) Antibiotic sensitivity testing using the Kirby-Bauer disc diffusion method was performed on three *E coli* colonies from each sample. The percentage of isolates resistant to each of 8 antibiotic drug classes was compared at each time point. Isolates resistant to cefpodoxime were evaluated for Extended Spectrum β-lactamase (ESBL) production using a combination disc method.

A substantial increase in resistance to all classes except for carbapenems was identified between T1 and T2. (Table) By T3, the percentage of resistant isolates had decreased for all classes, but had not returned to pre-treatment levels for penicillins, tetracyclines, fluoroquinolones and sulfonamides. Multiple drug resistance was identified in 71% of horses at T2 and 44% of horses at T3. *E coli* from all horses treated with cephalosporins showed cephalosporin resistance by T2. All but one of these showed ESBL production.

Commensal fecal *E coli* rapidly display antimicrobial resistance. Resistant isolates may persist for more than 2 weeks after treatment.

Table [Abstract 386]

Antimicrobial class	Resistant isolates % T1 (n = 48)	Resistant isolates % T2 (n = 45)	Resistant isolates % T3 (n = 27)
Penicillins	0	63	19
Aminoglycosides	0	53	0
Tetracyclines	5	68	30
Cephalosporins	0	63	0
Fluoroquinolones	0	16	15
Sulfonamides	29	74	56
Carbapenems	0	0	0
Chloramphenicol	0	63	0

ABSTRACT #387

EFFECTS OF A COMMERCIALLY AVAILABLE IMMUNOSTIMULANT ON CYTOKINE SECRETION AND PREVALENCE OF PNEUMONIA ON A FARM WITH ENDEMIC INFECTIONS CAUSED BY *RHODOCOCCUS EQUI.* S Giguère[1], RP Franklin[2], ND Cohen[3], AE Kalyuzhny[4]. [1]University of Georgia, Athens, GA. [2]Weatherford Equine Medical Center, Weatherford, TX. [3]Texas A&M University, College Station, TX. [4]R&D Systems, Minneapolis, MN.

Previous studies have shown that treatment of neonatal foals with inactivated parapox ovis virus (IPOV) enhances phagocytic cell activity against *ex vivo* infection with *Rhodococcus equi*. The objectives of the present study were to determine if administration of IPOV can decrease the incidence of pneumonia and increase the frequency of IFN-gamma and IL-4 secreting cells.

Fifty-nine foals were randomly assigned to 2 treatment groups (IPOV or placebo) prior to birth. At 24–48 h of age, foals received 2 mL of either IPOV or a placebo by IM injection. Injections were repeated 24 h and 8 days later. The frequency of IFN-gamma and IL-4 secreting cells was measured using a validated ELISPOT assay on blood mononuclear cells collected when the foals were between 2 and 14 days of age. Foals were monitored daily for clinical signs of pneumonia and biweekly for lung lesions by ultrasonography.

The proportion of foals that developed clinical or ultrasonographic evidence of pneumonia was not significantly different between IPOV (16 of 28) and placebo (14 of 31). IFN-gamma and IL-4 secreting cells were detected in only 21 and 15 foals, respectively. There was no significant effect of treatment with IPOV on the frequency of IFN-gamma and IL-4 secreting cells. There was no significant association between IFN-gamma or IL-4 secretion early in life and subsequent development of pneumonia. The odds of detecting IFN-gamma (5.1; 95% CI: 1.5–15) and IL-4 (3.5; 95% CI: 1.1–12) were significantly higher in foals $>$ 7 days than in younger foals.

ABSTRACT #388

NASAL SHEDDING OF EQUINE HERPESVIRUS-1 FROM HORSES IN AN OUTBREAK OF EQUINE HERPES MYELOENCEPHALOPATHY IN WESTERN CANADA. BA Burgess[1,2], N Tokateloff[1], K Poirier[1], S Manning[1], K Lohmann[1], DP Lunn[2], SB Hussey[2], PS Morley[2]. [1]Department of Large Animal Clinical Sciences, Western College of Veterinary Medicine, University of Saskatchewan, Saskatoon, Saskatchewan, Canada. [2]Animal Population Health Institute, Colorado State University, Fort Collins, CO.

In horses, Equine herpesvirus–1 (EHV-1) typically causes mild respiratory disease, but is also associated with late term abortion, neonatal foal death, and neurologic disease or Equine Herpes Myeloencephalopathy (EHM). Recently, EHM has been described as a "potentially emerging disease" by the USDA. The absence of information regarding shedding in horses with naturally occurring disease, and the knowledge that latent carriage complicates management of this disease, makes development of objective quarantine recommendations for those managing outbreaks of this disease very difficult. In fact, there is little consensus for quarantine recommendations or for the release of an imposed quarantine. The American Association of Equine Practitioners currently recommends a 28-day quarantine since the last suspected infection. The objectives of this report were to describe an outbreak of EHM in western Canada during the spring of 2008 and to evaluate the duration of nasal shedding of EHV-1 in horses affected with EHM during this outbreak.

This outbreak involved horses from three different facilities (a boarding facility, a private farm, and a large animal referral hospital) linked by a common exposure, a riding clinic. In total, 111 horses were naturally exposed to EHV-1 with 21 developing clinical signs of EHM. Nasal swabs were collected daily from 16 of the 21 horses demonstrating clinical signs of EHM. Samples were evaluated via a previously reported quantitative polymerase chain reaction technique specific for the gene sequence of glycoprotein B of EHV-1. Of the horses tested, the last day EHV-1 nasal shedding was detected in any horse was Disease Day 9 with 6 of 13 shedding until Disease Day 6. EHV-1 was detected in nasal secretions of 43.8% of horses tested on Disease Days 0 to 3. There was high variability with significant nasal shedding and severity of neurologic signs.

Based on the findings of this study, in the absence of laboratory testing of animals, we recommend biosecurity measures be utilized when managing EHM cases a minimum of 9 days beyond the onset of clinical signs. EHV-1 nasal shedding does not appear to be highly correlated with the severity of clinical signs. In addition, this report illustrates the potential for EHV-1 to spread between farms and within a large animal referral hospital reinforcing the need for biosecurity measures to be a normal part of daily operations at equine boarding and medical facilities.

ABSTRACT #389

RNA INTERFERENCE DOES NOT PROTECT AGAINST EXPERIMENTAL EQUID HERPESVIRUS TYPE 1 (EHV-1) INFECTION. Gillian A. Perkins,[a] Nicola Pusterla,[b] Hollis Erb,[a] Nikolaus Osterrieder[a,d]. [a]Cornell University, Ithaca, NY, [b]University of California Davis, CA, [c]Institut für Virologie, Freie Universität Berlin, Berlin, Germany.

Small interfering (si) RNAs targeted against essential genes of EHV-1 glycoprotein B (sigB3) and origin binding protein (siOri2) reduce EHV-1 replication *in vitro* and in mice. An EHV-1 challenge infection showed a decrease in severity of neurological symptoms in horses treated with sigB3/ori2 12-h before and 12-h after EHV-1 infection but no difference in the nasal viral load or viremia. We tested whether metaphylactic treatment with EHV-1-specific siRNAs at a higher dose and frequency would reduce viral load and clinical symptoms in EHV-1-infected horses.

Horses were infected intranasally (IN) with 1×10^7 PFU of EHV-1 virus strain Ab4 (day 0). Seven horses were given 5-nmols each of sigB3 and siOri2 IN 24-h pre- and 12, 24, 36 and 48-h post-EHV-1 infection. The other 6 horses were given 10-nmols of an irrelevant siRNA against firefly luciferase (siLuc). Horses were monitored daily and blood and nasal swabs taken on days -1, 1 to 10, 12, 14 and 21 post-infection (p.i.) for virus isolation and qPCR. Wilcoxon's rank-sum tests were performed.

All horses developed serous nasal discharge and a biphasic fever p.i. Three sigB3/Ori2 and 2 siLuc horses had neurologic signs beginning day 6 p.i. There was no significant difference between groups in clinical signs, nasal viral shedding or viremia.

The sigB3/siOri2 regimen did not protect horses against EHV-1 experimental infection.

ABSTRACT #390

CLASSIFICATION OF EPILEPTIC SEIZURES IN HORSES: AN EPIDEMIOLOGIC STUDY OF 104 CASES (1988-2009). S Mosseri, SM Reed, VA Lacombe. The Ohio State University, Columbus, OH.

Although many studies have been performed to classify epileptic seizures (ES) in humans and small animals, no similar epidemiologic study is available in horses. Therefore, the objectives of this study was to classify ES in horses (all ages except neonates) presented for seizure disorders at The OSU Equine Veterinary Teaching Hospital between 1988 and 2009, following small animal classification and the current guidelines of the International League Against Epilepsy.

ES were classified by type and etiology, based on history, clinical observations, diagnostic investigations (e.g., electroencephalograms, CSF and computed tomography imaging of the head) and postmortem examinations (when available) in 104 horses.

Regarding their type, ES were defined as primary generalized in 22% of horses, secondary generalized in 24% of cases, simple and complex partial in 43% of cases, and unclassified in 11% of cases. Regarding their etiology, symptomatic ES, with underlying structural brain lesions (including brain tumor, abscess, traumatic or infectious meningoencephalitis), were identified in 41 horses and suspected in 48 horses. In contrast, reactive ES were only reported in 2 cases. Epilepsy (i.e., recurrence of $\geq$ 2 ES) was identified in 72 cases, and further classified as symptomatic (24 cases), cryptogenic (i.e., probably symptomatic, 42 cases), idiopathic (i.e., suspected genetic predisposition, 1 foal) or unknown in 5 cases.

This study is the first attempt to classify ES and epilepsy in a referral equine population. As in humans, such classifications are dynamic concepts but could help clinicians in the establishment of a diagnosis, a therapeutic plan or prognosis.

ABSTRACT #391

ENZYMATIC AND OXIDATIVE MARKERS IN SYNOVIAL FLUID AND CHONDROCYTE CULTURES. M. Daix[1], J. Leemans[1], S. Schulsse[1], L. Wiggers[1], B. DE Moffarts[2], N. Kirschvink[1]. [1]Veterinary Department, University of Namur, Namur, Belgium; [2]Pavesco Twydil, Basel, Switzerland.

Articular enzymatic and oxidative stresses are involved in the pathogenesis of osteoarthritis (OA) and can be assessed by different markers, e.g. matrix metalloproteases 2 and 9 (MMP-2 and MMP-9) and F2-Isoprostane. Nutraceuticals have been used for prevention and treatment of OA, and there are *in vitro* evidences of their positive effect. The objective of this placebo-controlled study was to evaluate in clinically healthy ponies whether a 6-week oral supplementation, with a glucosamine-chondroitine sulfate-harpagophytum procumbens complex (G-CS-HP), would affect MMP-2 and MMP-9 activity in synovial fluid (SF) as well as MMP-2 activity and F2-Isoprostane concentration before and after interleukine-1β (IL-1β) stimulation in supernatant of chondrocyte cultures (CC).

Sixteen young (2 to 4 years) and clinically healthy Belgian ponies were enrolled in the study before being euthanized for anatomy teaching purposes. Eight ponies received a daily G-CS-HP oral supplement and eight ponies received a placebo. SF and chondrocytes were collected from fetlocks within 2 hours after euthanasia. MMP-2 and MMP-9 activity was determined in SF and CC supernatants by zymography, and F2-Isoprostane was determined by ELISA. Data were submitted to analysis of variance. A p value < 0.05 was considered statistically significant.

The G-CS-HP supplement group showed significantly reduced MMP-9 activity in SF and significantly increased MMP-2 activity. In CC supernatant, MMP-2 activity was also significantly higher before and after IL-1β stimulation in the treatment group than in the control group. F2-Isoprostane concentrations were significantly lower before and after IL-1β stimulation in CC supernatants from supplemented ponies.

These data suggest that the oral supplementation with G-CS-HP tested decreased MMP-9 and increased MMP-2 activity in SF. *Ex vivo* results on CC support these findings, since CC supernatant has an increased MMP-2 activity. They also suggest that the inflammatory response of chondrocytes in supplemented animals may be reduced, as shown by a decreased oxidative response after IL-1β stimulation.

ABSTRACT #392

ROLE OF NITROSATIVE STRESS IN THE PATHOPHYSIOLOGY OF EXPERIMENTALLY-INDUCED EQUINE LAMINITIS: A COMPARISON OF TWO MODELS. Burns TA,[1] Nuovo GJ,[2] Watts MR,[1] Belknap JK[1]. [1]College of Veterinary Medicine, The Ohio State University, Columbus, OH. [2]College of Medicine, The Ohio State University, Columbus, OH.

Oxidative stress, occurring secondary to elaboration of reactive oxygen and nitrogen species by activated leukocytes, has been shown to play an important role in initiation and perpetuation of multiple organ dysfunction and failure in the setting of sepsis in multiple species. In the septic horse, the laminae are the most frequent tissue undergoing injury, and evidence of oxidative stress has been reported in an experimental black walnut extract (BWE) laminitis model. Laminitis induced with soluble carbohydrate (CHO) may be more similar to clinical disease than the BWE model due to the fact that CHO-induced animals more commonly proceed to laminar failure. However, the role of oxidative stress in the CHO model has not been reported. Banked laminar tissue collected at various time points from animals subjected to BWE and CHO protocols were used in this study. State-of-the-art immunohistochemistry (IHC) for 3-nitrotyrosine (3-NT; marker of nitrosative stress) was optimized using the Ventana Ultraview Universal™ system with Fast Red as the chromogen on paraffin-embedded laminar sections from CHO horses (control, $n = 8$; developmental, $n = 6$; Obel grade 1 [OG1], $n = 6$) in order to identify and quantify the cell types positive for 3-NT. BWE laminar sections were stained as a positive control (control, developmental, and late time points; $n = 1$). Very few 3-NT-positive cells were visualized in any of the samples from the CHO model compared to the large number of cells observed in the BWE model; within the CHO samples, significantly fewer positive cells were noted in OG1 samples than controls ($p = 0.04$). Positive cells in the CHO model were predominantly interstitial leukocytes and endothelial cells and were limited to the laminar dermis (in contrast to the BWE model, where prominent laminar epithelial signal is present by OG1 lameness). To quantitatively assess laminar 3-NT concentrations, a commercially-available competitive chemiluminescent plate ELISA (Millipore; Temecula, CA) was used to assess total protein extracts of pulverized BWE and CHO laminar samples. No significant difference in 3-NT concentration was observed between control ($n = 8$), developmental ($n = 6$), and OG1 ($n = 6$) samples from CHO horses. The concentration of 3-NT in BWE samples was significantly higher at the 1.5 h time point ($n = 5$) than at the 0 h (control, $n = 5$), 3 h ($n = 4$), or OG1 ($n = 5$) time points ($p = 0.028$). In keeping with previous reports, these data do not support the hypothesis that nitrosative oxidative stress plays a prominent pathophysiologic role in CHO-overload laminitis and only support a transient nitrosative stress in the BWE model of laminitis. Thus, nitrosative stress (and, possibly general oxidative stress) does not appear to be the central event leading to laminar injury and failure in the CHO model of laminitis.

ABSTRACT #393

INDUCTION OF INSULIN RESISTANCE IN EQUINE DIGITAL VESSEL RINGS. CS Venugopal, RE Beadle, EP Holmes, SC Eades. Louisiana State University School of Veterinary Medicine, Baton Rouge, LA.

The purpose of the study is to evaluate an *in vitro* model of insulin resistance which is a cause of laminitis in horses. Digital palmar vessel segments were collected from clinically healthy horses immediately after euthanasia. Four arterial and four venous rings (3 mm wide) were prepared from each horse. Each ring was mounted in a tissue bath containing Tyrode's solution at 37°C by fixing one side of the ring to the floor of the bath and the other end to a force transducer interfaced with a polygraph. One ring of each vessel type was used as a control. Each of the remaining 3 rings of each vessel type was incubated for 30 min with insulin (to induce resistance), Wortmannin (PI3 kinase blocker), and PD 098059 (MAP kinase blocker), respectively. The rings were given 2 g initial tension and allowed to equilibrate for 45 min. The bath solution was gently replaced at every 15 min, drugs were added and the ring tension was readjusted to 2 g except at the final wash. After the incubation period, the rings were contracted with phenylephrine. When the contractile response reached a plateau, a single dose of insulin was added to the bath and the response of each ring was monitored for thirty min. Insulin relaxed the control rings and those treated with PD 098059 but contracted the rings pretreated with insulin and Wortmannin. The present study demonstrated successful induction of insulin resistance in arterial rings. However, the responses of venous rings were inconsistent and unreliable. The study suggests that the induction of insulin resistance in equine digital arterial rings is reliable and gives a good *in vitro* model for studying insulin resistance in laminitis. The study also suggests that MAP kinase may be involved in the pathology of vessels in insulin resistance.

ABSTRACT #394

PHARMACOKINETICS OF PIOGLITAZONE IN HORSES AFTER MULTIPLE DOSE ORAL ADMINISTRATION. JG Wearn[1], MV Crisman[1], JL Davis[2], RJ Geor[3], DR Hodgson[1], JK Suagee[3], M Ashraf-Khorassani[3], LJ McCutcheon[1]. [1]Virginia-Maryland Regional College of Veterinary Medicine, Virginia Tech, Blacksburg, VA. [2]Department of Clinical Sciences, College of Veterinary Medicine, North Carolina State University, Raleigh, NC. [3]Virginia Tech, Blacksburg, VA.

Pioglitazone is a thiazolidinedione class of anti-diabetic agent with proven efficacy in increasing insulin sensitivity in humans with non-insulin dependent diabetes mellitus, a syndrome of insulin resistance sharing some similarities with equine metabolic syndrome. The purpose of this study was to determine the pharmacokinetics of pioglitazone in adult horses following multiple dose oral administration.

Pioglitazone hydrochloride (1 mg/kg) was administered *per os* for 11 doses at 24-hour intervals in an inert carrier (unflavored gelatin), mixed immediately prior to administration and plasma samples were collected. Initially, a pilot study was performed using one horse; and thereafter the drug was administered to six horses. Samples were analyzed by liquid chromatography with tandem dual mass spectrometry. The maximum plasma concentration (mean ±SD) was 509.1 ± 413.5 and 448.1 ± 303.5 ng/mL achieved at 1.88 ± 1.39 and 2.83 ± 1.81 h following oral administration of the first and eleventh dose, respectively. Apparent half life ($t_{1/2}$) was 9.94 ± 4.57 and 9.63 ± 5.33 h of the first and eleventh dose respectively. There was no significant difference noted between pharmacokinetic parameters calculated on the first and eleventh doses. No adverse effects were noted in any of the horses.

This study showed that in normal horses pioglitazone at a dose of 1 mg/kg administered orally at 24 h dosing intervals results in plasma drug concentrations and total drug exposure approximating, but slightly below, those considered therapeutic in human medicine (maximal plasma concentration 617 ± 129 ng/mL).

ABSTRACT #395

EFFECTS OF HYPERIMMUNE EQUINE PLASMA ON CLINICAL AND CELLULAR RESPONSES IN A LOW-DOSE ENDOTOXEMIA MODEL IN HORSES.

G Forbes, SR Bailey, S Church, CJ Savage. University of Melbourne, Victoria, Australia.

Endotoxaemia is a major cause of equine morbidity and mortality. Endotoxin (LPS) plays a pivotal role in many pathophysiological states, inducing a cytokine mediated hyper-inflammatory cascade after activating mononuclear leukocytes. Hyperimmune plasma products from horses immunised against a mutant J5 strain of E. coli are currently used in the treatment of endotoxaemia. The purpose of this study was to determine the effects of commercially available hyperimmune equine plasma on the clinical and leukocyte responses in a low dose LPS challenge. Secondly to determine whether hyperimmune plasma decreases TNFα production and/or bio-availability of TNF, which may be reduced by soluble TNFα receptors (sTNFαR) in the plasma.

Six adult standardbred horses were used in a randomised cross over design. Purified LPS (E. Coli 055:B5; 30 ng/kg) was administered intravenously after pre-treatment with hyperimmune equine plasma (Equiplas; Plasvacc Pty Ltd; 2 L) or equivalent volume of saline. Each horse underwent both treatments (in random order), with a washout period of at least 14 days between each experiment. Clinical signs (including rectal temperature, heart rate and respiratory rate) were monitored at frequent intervals (15 min for 2 hours then every 30 min to 6 hours). Jugular blood samples were taken at the same time points. Leukocyte and platelet counts were performed using a Coulter Counter (model Z1; Coulter Electronics Inc.). Total TNFα was measured in heparinised plasma by ELISA assay (Endogen) using equine recombinant TNF for the standard curve. Bioavailable (unbound) TNFα was measured by a cell survival bioassay (L293 mouse fibroblasts).

LPS caused a significant increase in rectal temperature, peaking at 39 ± 0.2°C at 210 min; other clinical signs were mild. Hyperimmune plasma did not significantly affect rectal temperature, heart rate or respiratory rate. Blood leukocyte counts decreased from 12.4 ± 1.6 × 10^9/L at time 0 to a nadir of 4.1 ± 0.8 at 105 min, in controls, before recovering back to baseline values by 6 hours. Blood leukocyte counts were not significantly affected by hyperimmune equine plasma, reaching a nadir of 5.4 ± 1.1 × 10^9/L and recovering to 13.1 ± 0.2 × 10^9/L at 6 hours. Plasma TNFα concentrations measured by ELISA peaked at 881.9 ± 58.9 pg/ml in saline treated horses after 90 min, and hyperimmune equine plasma administration did not significantly affect the peak TNFα concentration (812.4 ± 126.1 pg/ml). However, the bioavailable (unbound) TNFα concentration measured by bioassay was significantly reduced by hyperimmune plasma pretreatment (1044.4 pg/ml ± 199.9 pg/ml compared with 1373.9 ± 107.6 pg/ml in saline treated controls; p = 0.05).

These data suggested that although the hyperimmune equine plasma did not significantly reduce the signs of endotoxemia in this experimental model, the bioavailable fraction of TNFα did appear to be moderately reduced. This effect is consistent with the presence of substances such as sTNFR in the hyperimmune plasma. Further work is required to confirm the presence and identity of these immune modulating factors.

ABSTRACT #396

LIDOCAINE DOES NOT PREVENT NEUTROPHIL INFLUX OR ATTENUATE LUNG FUNCTION IN HEAVES-AFFECTED HORSES.

ME Wilson, C Berney, A Behan, NE Robinson. Equine Pulmonary Lab, Large Animal Clinical Sciences, Michigan State University, East Lansing, MI.

In models of acute lung injury lidocaine has beneficial anti-inflammatory actions including reduced neutrophilic inflammation. Further, it reduces inflammation and improves lung function in asthmatics. The present study tested the hypothesis that lidocaine would attenuate pulmonary inflammation and function changes in heaves-affected horses.

In a randomized cross-over design, 6 heaves-susceptible horses in remission received intravenous lactated ringers solution or lidocaine hydrochloride 2% with at least 4 weeks pasture between treatments. Baseline clinical score (CS, grade 0–8) and pleural pressure change (ΔPpl_{max}) were measured, and bronchoalveolar lavage fluid (BALF) cytology and CBC were assessed before treatment (during remission). Treatments were delivered as continuous infusions 4 hours before and for 68 hours during exposure to hay/straw challenge, after which measurements were repeated. Serum lidocaine concentrations were monitored daily. Data analysis was by 2-way repeated measures ANOVA ($p < 0.05$).

At baseline, there were no significant differences between treatments in CS, ΔPpl_{max} and BALF cytology. Lidocaine concentration was consistently > 1000 ng/ml, and, as reported previously in humans, was associated with a significant reduction in peripheral eosinophils. Following challenge CS and ΔPpl_{max} increased significantly from 2.4 to 4.5 and 4.5 to 12.7 cmH_2O, respectively, but there was no difference between treatments. Total cells in BALF were significantly increased after exposure, but contrary to our hypothesis the cell numbers were significantly greater after lidocaine. Neutrophil percentage in BALF increased but there were no differences between treatments.

We conclude lidocaine has no anti-inflammatory effects in heaves.

ABSTRACT #397

EFFECTS OF COLD AIR EXERCISE ON THE INNATE IMMUNE FUNCTION OF EQUINE AIRWAY LEUKOCYTES.

Sabrina L Cummings[1], Michael S. Davis[1], Katherine K. Williamson[1], Virginia Buechner-Maxwell[2]. [1]Oklahoma State University Center for Veterinary Health Sciences, Stillwater, OK. [2]Virginia-Maryland Regional College of Veterinary Medicine, Blacksburg, VA.

Breathing cold air during strenuous exercise has been shown in multiple species to stimulate airway inflammation, producing a syndrome known as "ski asthma". Paradoxically, strenuous exercise is also known to increase susceptibility to respiratory viral infections. Studies in our laboratory on the cellular and biochemical mechanisms of ski asthma have described a distinct pattern of cytokine expression, most notably upregulation of Th2 cytokines that might translate to suppression of cell-mediated immunity (CMI), thus increasing the susceptibility of the respiratory tract to pathogens that are normally cleared through CMI. The purpose of this study was to test the hypothesis that horses exercised while breathing cold air causes functional impairment of airway innate immune function.

Horses (n = 11) were examined in a randomized block crossover design after exercising under controlled environmental conditions: breathing room temperature air and breathing subfreezing air. Exercise challenges consisted of 5 min of walking (3.5–4 m/s), 5 min of trotting (4.5–6.0 m/s), and 2 min of cantering (7.8–9.0 m/s), titrated to the individual horse to achieve peak heart rate of 160 bpm. Airway leukocytes were harvested 5 and 48 hours post exercise challenge via bronchoaveolar lavage. Oxidative burst capacity of mononuclear cells and neutrophils was measured by flow cytometry, using phorbol-myristate acetate (PMA) to stimulate oxidative burst (or omitting the PMA for measurement of basal oxidative burst), and dichlorofluorescin diacetate to convert the oxidative burst to a fluorescent signal. Phagocytic capacity was measured in the same cell populations via flow cytometry. All cells were incubated (stimulated samples at 37°C and unstimulated samples at 4°C) with 1 micron FITC beads added to the samples.

At 5 hr after exercise while breathing cold air, basal neutrophil oxidative burst was increased ($p < 0.0001$) and stimulated phagocytic capacity of neutrophils was decreased ($p < 0.025$). No significant effect of inspired air temperature was observed in any leukocyte function assay 48 hours post exercise.

The increase in neutrophil oxidative burst activity following cold air exercise may be an important mechanism contributing to damage of the airway lining, leading to pulmonary mechanical dysfunction. Increased expression of Th2 cytokines after cold air exercise (previously demonstrated in this model), may have some functional relationship with the downregulation of phagocytic capacity of neutrophils. Leukocyte function assays more specific to viral pathogen resistance may be needed to determine whether there is increased susceptibility to viral pathogens as a result of cold air exercise.

ABSTRACT #398

ENDOCRINE RESPONSE TO RESUSCITATION SOLUTIONS IN HEALTHY HORSES: 7.2% NACL AND 6% HETASTARCH.

AA Cattaneo[1], J Rossetto[1], MS Himler[1], SDA Hurcombe[1], CW Kohn[1], HC Schott II[2], RE Toribio[1]. [1]The Ohio State University College of Veterinary Medicine, Columbus, OH. [2]Michigan State University College of Veterinary Medicine, East Lansing, MI.

Horses that are critically ill often require urgent fluid therapy to correct hypotension, hypovolemia, and volume depletion. Hypertonic sodium solution (7.2% NaCl; HSS) and 6% hetastarch (HET) are two fluids solutions used to assist in these conditions. HSS increases blood volume and pressure and HET increases plasma oncotic pressure to reduce capillary leakage/fluid loss. Although the principles of fluid resuscitation relate to volume expansion, from a physiological standpoint it is likely that the hypothalamic-pituitary-adrenal axis (HPAA), which modulates vasomotor tone, blood pressure and inflammation is involved. Upon HPAA stimulation, the hypothalamus secretes vasopressin (AVP) and corticotropin releasing hormone (CRH), the pituitary releases adrenocorticotropin (ACTH), which in turn stimulates the adrenal gland cortex to release cortisol and aldosterone.

Based on our understanding of the equine HPAA, we hypothesized that the administration of HSS or HET will induce an endocrine response to retain water and sodium (volume expansion), which will be associated with increased arterial and central venous pressure.

Ten healthy horses were given HSS (4 ml/kg), HET (10 ml/kg), or 5% dextrose (10 ml/kg) in a cross-over design. Arterial and venous catheters were placed for solution administration, blood collection and blood pressure measurements. Blood samples were collected at −10, 0, 1, 3, 5, 10, 30, 60, 90, 120, 150, 180, 240, 270, 300, 330, 360 minutes, 12 and 24 hours. AVP, ACTH, aldosterone, cortisol, hematocrit (HCT), central venous pressure (CVP), and colloid oncotic pressure (COP) were determined. Urine samples were collected every 30 minutes.

HSS induced a rapid AVP response that was followed by a mild increase in ACTH and cortisol, while aldosterone concentrations decreased. CVP increased while HCT and COP decreased from HSS. HET increased AVP, ACTH, cortisol and aldosterone concentrations. While CVP increased and HCT decreased with HET (as HSS), COP increased with HET and remained high for 24 h. Dextrose decreased COP and HCT, increased CVP but had no effect on the hormones. Urine output increased and specific gravity decreased with all three solutions.

We conclude that the acute response to resuscitation solutions, in addition to rapid volume expansion, is also the result of changes in hormones that modulate vascular tone and fluid volume. Future studies will address the HPAA endocrine response in critically ill hypotensive horses.

ABSTRACT #399

ROLE OF CELLULAR SODIUM TRANSPORT IN NONGLANDULAR EQUINE GASTRIC ULCER DISEASE.

FM Andrews[1], A Peretich[2], RE Reese[2], L Abbott[2], M Dhar[2]. [1]Equine Health Studies Program, Louisiana State University School of Veterinary Medicine, Baton Rouge, LA. [2]University of Tennessee College of Veterinary Medicine (UTCVM), Knoxville, TN.

Gastric ulcers in the nonglandular stomach of horses are common and are likely caused by the presence of stomach acids. Recently, volatile fatty acids (VFAs), byproducts of the fermentation of carbohydrate by resident stomach bacteria, have been shown to decrease sodium transport across nonglandular mucosa, which leads to cell swelling and eventual ulceration. However, the role of the cellular sodium transport, specifically sodium-potassium adenosine triphosphatase (NAKA), has not been elucidated. The purpose of this study was to determine the role of NAKA and sodium transport in the pathogenesis of nonglandular gastric ulcer disease.

Samples of nonglandular gastric mucosa were collected from horses ($n = 10$) donated to the UTCVM for diseases not related to the gastrointestinal tract. Nonglandular (NG) tissue bioelectric properties were measured in an Ussing Chamber system while exposed to normal Ringers solution (NRS, control), a VFA mixture containing VFA concentrations found 2 hours after feeding a sweet feed diet (0.5 Kg/Kg BW) (VFA Mixture-AF) and a VFA mixture containing high acetic (40 mM) (VFA Mixture-HAc), in NRS, at pH 1.5, 4, or 7. After NRS or VFA exposure, tissues were removed from the Ussings Chamber system and divided into half. Half of the tissue was examined under light microscopy for swell swelling and half the tissue was immediately frozen in liquid nitrogen for determination of mRNA-expression the β1 subunit of NAKA gene using real-time PCR.

NG mucosa exposed to NRS (pH 4.0 and 1.5) showed immediate significant decrease in PD, followed by a decrease in Isc and R. NG mucosa, when exposed to the VFA mixture-AF, showed a decrease in PD and Isc, similar to tissue exposed to NRS at the same pH. However, NG mucosa exposed to VFA mixture-HAc showed an immediate decrease in PD and Isc. Nonglandular cell swelling was noted in tissues exposed to the VFA mixture-AF. When the tissues were all grouped together there was no difference in relative expression of mRNA for NAKA, however, when horses were divided by age group, for horses < 5 years of age ($n = 5$) relative expression of mRNA for NAKA was significantly decreased in tissues exposed to the VFA mixtures, whereas, for horses > 12 years of age ($n = 5$) relative expression of mRNA for NAKA was significantly increased compared to controls. Results suggest that VFAs present in gastric juice from excessive grain feeding, at pH ≤ 4.0, result in changes in barrier function and that this barrier function may be related to alternations in NAKA and may be age-dependent. This may account for the increased prevalence of nonglandular ulcer disease in younger horses.

ABSTRACT #400

INFLAMMATORY GENE EXPRESSION IN THE LIVER, LUNG, AND KIDNEY IN THE CARBOHYDRATE OVERLOAD MODEL OF EQUINE LAMINITIS.

BS Leise[1], M Watts[1], PJ Johnson[2], SJ Black[3], JK Belknap[1]. [1]The Ohio State University College of Veterinary Medicine, Columbus, OH. [2]University of Missouri College of Veterinary Medicine, Columbia, MO. [3]University of Massachusetts, Amherst, MA.

In the human with endotoxemia/sepsis, a systemic inflammatory response syndrome (SIRS) occurs, leading to injury and dysfunction/failure of several organs including most commonly the liver, lungs, kidney, and intestinal tract. Sepsis in the adult horse, the most common cause being gastrointestinal diseases such as enterocolitis or colon torsions, most commonly leads to injury of the digital laminae in the form of laminitis. Renal dysfunction/failure can occur in some cases (especially with enterocolitis) and mild liver dysfunction may be noted, but it is much less common to have pulmonary dysfunction/failure. A similar SIRS has been reported in the black walnut extract (BWE) model of laminitis evidence by leucopenia, leukocyte migration into the tissues and increased inflammatory gene expression in the lung and liver. Interestingly, the carbohydrate overload (CHO) laminitis model, a model that more closely reflects the severity of laminar injury observed in clinical sepsis-related laminitis cases, does not have the same SIRS-like events (i.e. no leucopenia) as the BWE model. Therefore, the purpose of this study was to determine inflam-

Table [Abstract 400]. Fold-changes in mRNA concentration compared to control samples for same organ (P < 0.05), NS = not significant

	IL-1β	TNFα	IL-6	IL-8	IL-10	COX1	COX2
Liver DEV	↑ 4.1	↑ 8.6	NS	↑ 8.6	↑ 2.3	NS	NS
Lung DEV	NS	NS	NS	NS	NS	NS	NS
Kidney DEV	NS	NS	NS	NS	NS	NS	NS
Liver OG1	↑ 4.2	↑ 7.6	NS	↑ 9.6	↑ 2.1	NS	↑ 5.7
Lung OG1	↑ 3.2	NS	NS	NS	↑ 2.6	NS	↑ 2.6
Kidney OG1	↑ 1.8	NS	NS	NS	NS	NS	NS

matory signaling events in hepatic, pulmonary, and renal tissue in the CHO laminitis model. Archived snap frozen samples were used from eighteen horses administered either water (CON group, n = 6, euthanized 24 h after water administration) or 17.6 g of starch/kg body weight via nasogastric tube. Horses administered starch were euthanized after a two degree increase in rectal temperature (DEV group, n = 6) or at the onset of Obel grade 1 lameness (OG1 group, n = 6). Real time-quantitative PCR for IL-1β, IL-6, IL-8, IL-10, TNFα, COX-1 and COX-2 was performed (see Table 1). Interestingly, although the kidney is the most likely to be injured of the three organs in equine sepsis, the liver underwent the greatest inflammatory signaling, most likely reflecting function of this organ in processing toxins from the gastrointestinal tract. Furthermore, increases in inflammatory gene expression of the organs were less than previously reported for the primary "target tissue" in the septic horse, the lamina.

ABSTRACT #401

ASSESSMENT OF TWO EXPERIMENTAL MODELS FOR PROLONGED (48 HOURS) ENDOTOXEMIA IN HORSES.

L Tadros, N Frank. Department of Large Animal Clinical Sciences, College of Veterinary Medicine, University of Tennessee, Knoxville, TN.

Lipopolysaccharide (LPS) is detected in the blood of horses after experimental alimentary carbohydrate overload and may contribute to the development of laminitis by inducing systemic inflammation. Models for endotoxemia have focused upon single bolus infusions of LPS in the past, but clinical endotoxemia has a greater magnitude and duration. The objective of this preliminary study was therefore to assess two models of prolonged (48 h) endotoxemia.

Twelve healthy adult horses were randomly assigned to receive a continuous rate infusion (CRI) of LPS (80 μg total over 48 h; n = 4), a saline CRI (80 mL/h for 48 h; n = 4), or 8 boluses of LPS (10 μg q6h for 48 h; n = 4). Physical parameters were recorded every 4 h and intravenous glucose tolerance tests were performed at –24, 8, 24, 32, and 48 h.

A treatment × time effect ($P < 0.001$) was detected for rectal temperature, with LPS groups differing significantly from the saline group. Mean rectal temperature was significantly higher than baseline from 4 to 12 h and again at 20 h, but did not increase significantly again after this point. With the exception of one horse, all animals treated with LPS exhibited tachycardia. Laminitis was not detected. A significant time effect was detected for area under the glucose curve, but groups did not differ significantly. No differences were detected between the two models and most horses appeared to be more tolerant of LPS after 24 h. These models increased the duration of systemic inflammation, but did not match the severity of naturally occurring disease.

ABSTRACT #402

ACTIVIN A IN PLASMA FROM HORSES WITH ACUTE GASTROINTESTINAL DISEASE: BIOMARKER COMPARISON WITH TNFα AND SERUM AMYLOID A.

G Forbes, S Bailey, L Fennell, S Church, CJ Savage. University of Melbourne, Victoria, Australia.

Acute gastrointestinal disease in the horse may be associated with a number of different processes including strangulations, impactions and inflammatory lesions. Endotoxemia may be a concurrent factor in a number of these conditions, affecting morbidity and mortality. Available medical treatments for systemic inflammation associated with endotoxemia may not ameliorate all of the inflammatory pathways, and other therapies are being investigated to more effectively target leukocyte activation. The evaluation of biomarkers of inflammation and clinical disease will help in understanding and assessing these conditions and the effectiveness of novel therapeutic agents. The aim of this study was to measure Activin A, a member of the TGFβ family of proteins which has been validated as a plasma marker of endotoxemia in other species, in acute gastrointestinal disease in horses.

Thirty horses of various breeds presenting to the University of Melbourne Equine Hospital with signs of acute gastrointestinal disease were examined in this study. Conditions were grouped into strangulating (large or small intestinal strangulations), non-strangulating (including impactions, displacements, intussusceptions) or inflammatory (colotyphlitis, peritonitis) categories, based on standard diagnostic criteria. Ten normal horses were used as controls. Jugular venous blood samples were obtained on admission, and serum and plasma stored at −80°C. Samples were assayed for TNFα (Equine TNFα ELISA kit; Endogen), serum amyloid A (SAA ELISA Kit; Tridelta) and activin A (Quantikine Activin A ELISA kit; R&D Systems).

Horses in the inflammatory, strangulating and non-strangulating categories had significantly elevated serum SAA levels compared with controls, with greatest concentrations recorded in the inflammatory group (median 709, range 276–1480 μg/ml vs normals 0–40 μg/ml). A similar trend was observed in the TNFα concentrations, although this was not statistically significant. Plasma activin A was significantly greater in the inflammatory gastrointestinal disease category (median 55.2, range 7.8–264.7 pg/ml) than the non-strangulating category (median 24.7, range 0.2–31.0 pg/ml) or the normal horses (median 5.9, range 0.4–23.2 pg/ml). There was a significant correlation between activin A and TNFα concentrations in individual horses (p = 0.003), but not between activin A and SAA concentrations (p = 0.32).

Although these markers of inflammation could not be used alone to distinguish between different causes of gastrointestinal disease, these results suggest that they might be useful in monitoring the severity of inflammation and/or endotoxemia in such cases. Activin A has not previously been measured in the horse, and plasma concentrations of activin A were significantly increased in gastrointestinal disease, where this protein may play a role in inflammation and endotoxemia, as in other species. This protein may be a useful biomarker for evaluating novel therapies for endotoxemia.

ABSTRACT #403

RESULTS OF THROMBOELASTOMETRY FOLLOWING RECALCIFICATION OF CITRATED EQUINE WHOLE BLOOD ARE STRONGLY INFLUENCED BY SAMPLE STORAGE.

TM Rossi[1], SA Smith[2], PA Wilkins[1], MA McMichael[1]. [1]College of Veterinary Medicine or [2]College of Medicine, University of Illinois, Champaign-Urbana, IL.

Previous reports of thromboelastography or thromboelastometry (TE) have employed recalcification to assess coagulation in citrated equine whole blood. The objective of this study was to determine whether equine citrated whole blood undergoes changes prior to initiation of TE via recalcification which impact the results obtained with TE. Whole blood was collected by atraumatic venipuncture from 11 normal horses into siliconized glass tubes containing 3.2% citrate (9:1 blood/citrate) and held at 37°C for variable times prior to initiation of TE. TE parameters evaluated were clot time (CT), clot formation time (CFT), alpha angle (α), and maximum clot firmness (MCF).

Mean (SD) results were:

Sample holding had a profound impact on results, with increasing coagulability developing over the 30 min time period.

Equine citrated blood undergoes marked *ex vivo* contact activation after collection. When recalcification alone is used to initiate clotting, coagulation is entirely dependent on the FXIIa production that occurs in the sample *ex vivo* after collection. Since contact ac-

Table [Abstract 403]

Holding Time (Min)	CT (s)	CFT (s)	α (°)	MCF (mm)
1	1324 (222)^{b,c,d}	431 (136)^{b,c,d}	37 (12)^{b,c,d}	47 (6)^{b,c}
10	789 (123)^{a,c,d}	265 (86)^{a,c,d}	50 (7)^{a,c,d}	51 (8)^{a}
20	581 (80)^{a,b,d}	178 (44)^{a,b,d}	58 (5)^{a,b,d}	53 (6)^{a}
30	449 (120)^{a,b}	154 (52)^{a,b}	61 (7)^{a,b}	51 (7)

$p < 0.05$ as compared to a: 1 min, b: 10 min, c: 20 min, d: 30 min

tivation is known to be irrelevant for *in vivo* hemostasis, recalcification of citrated samples in the absence of a clotting trigger is not a suitable representation of *in vivo* hemostatic capability.

ABSTRACT #404

INHIBITION OF PROSTAGLANDIN E2 PRODUCTION IN EQUINE CHONDROCYTES BY PENTOSAN POLYSULFATE AND N-ACETYLGLUCOSAMINE COMBINATIONS.

Heinecke, LF[1], Grzanna, MW[1], AU, AY[1,2], Ownby, SL[1], Rashmir, A[3], Frondoza, CG[1,3,4]. [1]Nutramax Laboratories, Inc., Edgewood, MD. [2]Syracuse University, Syracuse, NY. [3]Mississippi State University, Mississippi State, MS. [4]Johns Hopkins University, Baltimore, MD.

Lameness attributed to osteoarthritis (OA) is a major cause of reduced performance and economic burden in the equine industry. Pathological changes in osteoarthritic joints include excess production of prostaglandin E2 (PGE_2). This inflammatory mediator up-regulates metalloproteinases (MMPs) that degrade cartilage as well as suppresses aggrecan synthesis and total proteoglycan accumulation. It also contributes to the development of pain in OA. The compounds pentosan polysulfate (PPS) and glucosamine in the form of glucosamine sulfate, glucosamine chloride or N-acetyl glucosamine (NG) have been used for the management of equine OA. These compounds have been documented to promote chondrocyte anabolic activity and are considered disease modifying. However, their anti-inflammatory activity is less understood. To address this question, we tested whether these compounds can inhibit the inflammatory response using an equine chondrocyte culture model. We also determined whether the PPS-NG mixture when combined with an avocado soybean unsaponifiables (ASU), glucosamine (GLU), and chondroitin sulfate (CS) mixture would exert a more potent anti-inflammatory effect. ASU, GLU, and CS individually have been shown to provide chondroprotective effects *in vivo* and *in vitro*.

Equine chondrocytes (5×10^5 cells/well) were incubated for 24 hrs with: control media alone, mixtures of clinically relevant concentrations of ASU + GLU + CS, PPS + NG, or the combination of ASU + GLU + CS and PPS + NG. Cells were activated for 24 hrs with IL-1 10 ng/ml). Supernatants were analyzed for PGE_2 production by ELISA. Immunostaining of chondrocytes for type II collagen and aggrecan verified that they continue to express the cartilage phenotype. Statistical analysis was performed using one-way ANOVA and the Tukey post-hoc test with significance set at $p < 0.05$.

Activated chondrocytes pre-treated with different concentrations of PPS + NG significantly reduced PGE_2 synthesis ($p < 0.001$). Similarly, the ASU + GLU + CS mixture elicited a significant decrease in PGE_2 production ($p < 0.001$). Combining the PPS + NG mixture with the ASU + GLU + CS significantly potentiated the inhibitory effect on cytokine-induced PGE_2 production ($p < 0.001$). Our finding reinforces earlier reports that these compounds exhibit both anti-inflammatory and anabolic activity. Superior anti-inflammatory activity may be achieved with the combination of these compounds while minimizing adverse side effects such as life-threatening gastrointestinal complications. The responsiveness of chondrocytes to this combination supports their potential utility for the management of joint inflammation in horses and other species. (Research supported by Nutramax Laboratories, Inc.)

ABSTRACT #405

HEPCIDIN GENE EXPRESSION AND PLASMA IRON FOLLOWING INTRAMUSCULAR INJECTION OF FREUND'S COMPLETE ADJUVANT TO HORSES.

JP Oliveira Filho[1], PR Badial[1], PHJ Cunha[1], DQ Cagnini[1], JR Silva[1], JP Araújo Jr[2], TJ Divers[3], NJ Winand[3], AS Borges[1]. [1]School of Veterinary Medicine and Animal Science UNESP – Univ Estadual Paulista, Botucatu, SP-Brazil. [2]Institute of Biosciences UNESP – Univ Estadual Paulista, Botucatu, SP. [3]Cornell University, College of Veterinary Medicine, Ithaca, NY.

Hepcidin has been found to be the key regulator of iron metabolism. Hepcidin up-regulation causes hypoferremia during systemic inflammatory disorders in some species. Recent work has shown that equine hepcidin is predominantly expressed in the horse's liver. In this study, hepcidin gene expression was determined in liver and bone marrow of healthy horses following intramuscular administration of Freund's complete adjuvant (FCA).

Inflammation was induced in 4 normal horses by two intramuscular injections of 5.0 ml of FCA into the neck muscle at 0 and 12 hours. Clinical evaluations as well as fibrinogen and plasma iron were measured at multiple time points between 0 and 96 hours after the first injection. Liver and bone marrow biopsies were taken immediately prior to (control) and 12 and 24 hours after the first FCA injection. Expression of the hepcidin gene in liver and bone marrow was quantified by real-time PCR.

Mild swelling was observed at the injection sites. Plasma iron was significantly decreased from the pre-inoculation level at 8 h. The lowest plasma iron was observed at 36 h (30% initial value). Fibrinogen did not increase until 36 h and peaked at 48 h. Liver hepcidin messenger RNA relative expression was up-regulated (two fold) at 12 h post first FCA injection and remained increased at 24 h. Bone marrow hepcidin relative expression was not influenced by FCA injection.

It was concluded that FCA administration caused rapid onset of hypoferremia and this was likely the result of up-regulated hepatic hepcidin gene expression.

ABSTRACT #406

EQUINE SEROLOGICAL RESPONSES TO COMMERCIAL VACCINES.

MG Little[1], HGG Townsend[2], JP Little[3], RM Stenbom[1], SB Hussey[4], DP Lunn[4]. [1]Boehringer Ingelheim Vetmedica, St. Joseph, MS, [2]Western College of Veterinary Medicine, Saskatoon, SK, [3]Clarkrange Veterinary Services, TN, [4]Colorado State University, Fort Collins, CO.

Equine vaccination is an important method of prophylaxis against several important infectious diseases. Serological responses to vaccination can predict efficacy and be used to compare product performance. The study described here was performed to determine whether contemporary equine vaccines in common use show significant performance differences between products as determined by serological responses.

The study was conducted in 31 6–9 mth old horses housed on the same farm, with no history of vaccination or disease. Horses were assigned to three vaccination groups (each group was vaccinated with products from a single manufacturer; n = 8 in each group) or a control group (n = 7). Vaccination antigens included Eastern & Western Equine Encephalitis (EEE & WEE), tetanus, influenza virus and EHV-1 & 4 in each group. Products evaluated were sourced from Boehringer Ingelheim Vetmedica, Fort Dodge Animal Health, and Intervet Animal Health. Horses were vaccinated on two occasions, 3 weeks apart, and serum was collected at the times of vaccination, and approximately monthly thereafter until 3 months after the second vaccination. Serological responses were determined by ELISA (tetanus and influenza), serum neutralization titer (EHV-1&4), and plaque reduction assay titer (Eastern & Western Encephalitis). Two horses were lost from the study due to unrelated events. For statistical analysis, data values for each animal were summed over the entire time course of the experiment to deal with repeated measures. The summed values for each horse were not normally distributed and so were ranked, and a one-way ANOVA performed on the ranked data. Means were compared using Tukey's test. Statistical significance was reported when p values ≤ 0.05.

The control group showed no evidence of sero-conversion to any antigen, indicating no natural exposure to any antigen studied. Significant differences between vaccination groups and controls were identified for each antigen, except EHV-4, to which no vaccination response was detected. However, responses were frequently significantly different between vaccination groups, and in the case of tetanus, EHV-4, EEE & WEE, one or more vaccination groups were not different from controls.

This study demonstrates that on the basis of serological response to vaccination, there are significant differences amongst commercially available products in North America, and some products did not produce serological evidence of an immune response to vaccination. These results indicate that during the process of vaccine selection, differences in product efficacy predicted by serological response should be considered.

ABSTRACT #407

HUMORAL IMMUNE RESPONSES TO A KILLED ADJUVANTED VACCINE. DO FOALS RESPOND LIKE ADULT HORSES? C. Ryan[1], S. Giguère[2]. [1]University of Florida, Gainesville, FL. [2]University of Georgia, Athens, GA.

The purpose of this study was to compare serum IgM and IgG subclass responses of foals to that of adult horses following administration of a killed adjuvanted vaccine. Eleven 3-day-old foals, fifteen 3-month-old foals, and six adult horses were vaccinated intramuscularly twice at 3 week intervals with a killed adjuvanted vaccine containing antigens from type II bovine viral diarrhea virus, infectious bovine rhinotracheitis virus, parainfluenza-3 virus, and bovine respiratory syncytial virus. A cattle vaccine was chosen to avoid any effects of maternal antibody on immune responses. Serum samples were obtained immediately prior to vaccination and 3 weeks after each vaccination. Antigen-specific serum IgGa, IgGb, IgG(T), and IgM concentrations were measured using a capture ELISA.

No adverse reactions to the vaccine were observed. Both groups of foals and adult horses responded to the vaccine with a significant increase in IgGa and IgG(T) concentrations. In contrast, only adult horses and 3-month old foals mounted significant IgGb and IgM responses. Concentrations of IgGb were significantly higher in adult horses than in both groups of foals. Concentrations of IgM and IgG(T) were significantly different between all groups with the highest concentrations in adult horses followed by 3-month-old foals and finally 3-day old foals. Concentrations of IgGa were not significantly different between groups. The results of this study indicate that the humoral immune responses to this killed adjuvanted vaccine are modest in newborn foals. Although humoral immune responses improve with age, 3-month-old foals do not respond with the same magnitude as adult horses.